ABSTRACTS OF PAPERS

Part 2

213th ACS National Meeting
0-8412-3500-7

American Chemical Society

San Francisco, CA

April 13-17, 1997

American Chemical Society
DIVISION OF INDUSTRIAL & ENGINEERING CHEMISTRY

213th ACS National Meeting

San Francisco, CA
April 13-17, 1997

R.C. Gatrone, Program Chairperson

SUNDAY MORNING

- **Ligand Design for Ion Separations**
R.A. Sachleben, Presiding   Papers 1-6

- **Biotechnological Products Derived from Crustacean Exoskeletons**
S.K. Burgess, Presiding   Papers 7-11

SUNDAY AFTERNOON

- **Ligand Design for Ion Separations**
R.E. Barrans Jr., Presiding   Papers 12-17

- **Industrial Chemistry Award Honoring R.D. Sydansk**
Cr(III)-Carboxylate-Complex Crosslingked Oilfield Polymer Gels
D.L. Gibbons, T.P. Lockhart, Presiding   Papers 18-23

SUNDAY EVENING

- **Posters**
R.A. Bartsch, Presiding   Papers 24-32

- **Posters**
D.J. Pruett, J.G. Pruett, Presidining   Papers 33-35

- **Posters**
L. Nutildnez, A.H. Bond, Presiding   Papers 36-55

MONDAY MORNING

- **Ligand Design for Ion Separations**
R.E. Barrans Jr., Presiding   Papers 56-61

E.V. Murphee Award Honoring A.W.Westerberg
Process Systems Engineering
L.T. Biegler, Presiding   Papers 62-67

MONDAY AFTERNOON

- **Ligand Design for Ion Separations**
R.A. Sachleben, Presiding   Papers 68-73

- **Stabilization, Formulation, and Delivery of Proteins and Nucleic Acids**
M. Bruner, T.W. Randolph, Presiding   Papers 74-78

MONDAY EVENING

- **Sci-Mix**
L. Nunez, A.H. Bond, Presiding   Papers 36-55

TUESDAY MORNING

- **Ligand Design for Metal Ion Separations**
R.A. Sachleben, Presiding   Papers 79-84

- **Current Trends in Applied Chemistry**
Industrial/Academic Interface in Separation Science
R.D. Rogers, Presiding   Papers 85-89

- **E.B. Barnes Award**
P.M. Norling, Presiding   Papers 90-95

TUESDAY AFTERNOON

- **Recognition with Imprinted Polymers**
R.A. Bartsch, Presiding   Papers 96-101

- **Current Trends in Applied Chemistry**
Industrial/Academic Interface in Separation Science
R. D. Rogers, Presiding   Papers 102-106

- **Separation Science and Technology Award Honoring C. Judson King, III**
R.W. Rousseau, Presiding   Papers 107-110

WEDNESDAY MORNING

- **Recognition with Imprinted Polymers**
M. Maeda, Presiding   Papers 111-114

- **Ground and Surface Water Remediation**
  D.J. Pruett, Presiding   Papers 115-121

- **Separation Science and Technology Award Honoring C.Judson King, III**
  M.R. Etzel, Presiding   Papers 122-127

WEDNESDAY AFTERNOON

- **Recognition with Imprinted Polymers**
  R.H. Fish, Presiding   Papers 128-132

- **Ground and Surface Water Remediation**
  J.G. Pruett, Presiding   Papers 133-139

- **Separation Science and Technology Award Honoring C.Judson King, III**
  D.D. Frey, Presiding   Papers 140-144

THURSDAY MORNING

- **Recognition with Imprinted Polymers**
  M. Maeda, Presiding   Papers 145-148

- **Separation Science and Technology Award Honoring C.J.King, III**
  D.H. Mohr, Presiding   Papers 149-153

THURSDAY AFTERNOON

- **Recognition with Imprinted Polymers**
  R.A. Bartsch, Presiding   Papers 154-157

- **Separations in the Biological Industries**
  S. Behrens, M. Burns, Presiding   Papers 158-164

DIVISION OF INDUSTRIAL & ENGINEERING CHEMISTRY

001.
STRUCTURAL CRITERIA FOR THE DESIGN OF METAL-SELECTIVE LIGANDS. Benjamin P. Hay, Pacific Northwest National Laboratory, Richland, Washington 99352.

Over the past several years, a novel application of molecular mechanics calculations has yielded relationships between crown ether structures and their binding affinity for the alkali and alkaline earth cations. These relationships have been used successfully to improve the performance of a metal-ion selective sequestering agent in an industrial separation process. This talk will present recent developments and applications of molecular mechanics in metal-selective ligand design.

002. COMPUTER-BASED PREDICTION OF FORMATION CONSTANTS AS A TOOL IN LIGAND DESIGN. Robert D. Hancock, IBC, American Fork, UT 84003

A computer program LOGKEST is presented that predicts formation constants of complexes of 75 metal ions in aqueous solution with a wide variety of ligands. LOGKEST makes predictions in three stages. 1) A dual-basicity equation is used to predict formation constants for unidentate ligands. 2) Chelate effect equations predict formation constants for complexes of multidentate ligands containing only five-membered chelate rings from $logK_1$ values obtained in step 1, and 3) changes in complex stability induced by changes in ligand architecture, such as change of chelate ring size, addition of C-alkyl substituents, addition of groups bearing neutral oxygen donors, or presence of macrocyclic rings, are accounted for. LOGKEST can predict formation constants for many presently unknown ligands. A limiting factor can be prediction of effects in step 3. Prediction of novel chemistry by the program, and use of MO and MM calculations in step 3 is discussed.

003. THEORETICAL STUDY OF COPPER/THIACROWN INTERACTIONS. S. E. Hill and D. F. Feller, Environmental Molecular Sciences Laboratory, Pacific Northwest National Laboratory, Richland, Washington 99352

The binding behavior of macrocycles containing sulfur donors to transition metal cations is markedly different from that of polycyclic ethers complexing with alkali and alkaline earth cations. The hard cation-donor interaction tends to favor primarily electrostatic bonding. The influences of d-orbitals and electron-electron correlation, however, promote strongly hybridized bonding between the soft cation and donors. We will present results of quantum mechanical gas-phase models of copper cations complexing with sulfur-containing analogs of crown ethers.

004. LIGAND DESIGN FOR ION SEPARATIONS: CRYSTAL STRUCTURES OF SUBSTITUTED DICYCLOHEXYL-18-CROWN-6 ISOMERS. Robin D. Rogers, Department of Chemistry, The University of Alabama, Tuscaloosa, AL 35486; Mark L. Dietz and E. Philip Horwitz, Chemistry Division, Argonne National Laboratory, Argonne, IL, 60439; Andrew H. Bond, Eichrom Industries, Inc., 8205 S. Cass Ave., Suite 107, Darien, IL 60561.

Dicyclohexyl-18-crown-6 and its substituted derivatives have shown great promise for selective separations of such ions as Sr^{2+} and Pb^{2+}. There have been few, if any, reports, however, which detail the isomeric composition of the crown ether used in a particular investigation. As part of a detailed study on the effects of isomerism and substituents on the selectivity and efficacy of DCH18C6 extractants, we have investigated the crystal structures of several isomers of this crown ether and two isomers of its di-*tert*-butyl derivative. The structures of DCH18C6 (Isomer C), two polymorphs of DCH18C6 (B), DCH18C6•2CH$_3$CN (B), (^{t}Bu)$_2$DCH18C6 (A), (^{t}Bu)$_2$DCH18C6•2CH$_3$CN (B), [(NH$_4$)(DCH18C6)]Cl•H$_2$O (A), and [Pb(OH$_2$)$_2$(DCH18C6)][ClO$_4$]$_2$ (B) will be discussed in this presentation. These structures will be compared with several other DCH18C6 complexes of Sr^{2+} and Pb^{2+} reported in the literature.

005.
THE EFFECT OF IMMOBILIZED LIGAND GEOMETRY ON METAL ION AFFINITIES
Spiro D. Alexandratos, Latiff A. Hussain, and Subramanian Natesan.
University of Tennessee, Department of Chemistry, Knoxville, TN 37996.

The synthesis of a new series of ion exchange resins will be presented. The principal difference among the resins is the geometry of the ligands, all of which have a common functional group. In one example, it is seen that the metal ion complexing ability of alpha-keto-phosphonic acids is very different from the beta- and gamma- homologues.

006. **METAL ION SEPARATIONS USING HYDROPHOBIC ANIONS: ASPECTS OF LIGAND DESIGN.** Rebecca M. Chamberlin,[1] Kent D. Abney,[1] Jennifer F. Clark,[2] and Steven H. Strauss,[2] (1) Nuclear and Radiochemistry Group, MS J514, Los Alamos National Laboratory, Los Alamos, NM 87545; (2) Department of Chemistry, Colorado State University, Fort Collins, CO 80523.

Metal ion extraction using hydrophobic anions has been investigated by several researchers for remediation of Cs-137 and Sr-90 in nuclear waste. The rich derivative chemistry of the cobalt *bis*-dicarbollide anion makes it amenable to systematic studies of the relative importance of anion structure, solvent, and synergists on the extraction selectivity and efficiency. Halogenation or alkylation of cobalt dicarbollide strongly influences the anion's solubility and stability but has little effect on extraction properties. Alkali metal selectivity depends primarily on solvent, while alkaline earth selectivity is driven by the concentration and molecular weight of polyethylene glycol synergists. Additional aspects of ligand design, including a simple extraction and recovery cycle based on redox-active metal centers, will be discussed.

007. EXTRACTION OF CAROTENOPROTEINS FROM CRUSTACEAN WASTES, N. F. Haard, Institute of Marine Resources, University of California, Davis, CA 95616

The red, green and blue hue of marine invertebrates is due to the presence of carotenoprotein complexes. The principal pigment of these complexes is the xanthophyll called astaxanthin. An enzymic method was developed to isolated carotenoproteins from the epithelial tissue of crustacean exoskeleton. Species examined include Arctic shrimp (*Pandulus borealis*), snow crab (*Chionocetes opilio*) and American lobster. An excellent yield (>90%) of intact carotenprotein is obtained by digestion of shrimp waste with pure bovine trypsin. Low temperature digestion of exoskeletons with pure Atlantic cod trypsin gives better carotenoprotein yield because of the high molecular activity of cod trypsin. However, the presence of minor impurities of chymotrypsin and other proteinases in trypsin results in rapid degradation and poor yield of carotenoproteins. Astaxanthin of intact carotenoproteins is less susceptible to auto-oxidation and attack by digestive dioxygenase activity than the free pigment. Feeding trials with rainbow trout showed that dietary carotenoproteins are much better assimilated in skin and muscle tissues than free astaxanthin.

008. ENZYMATIC FUNCTIONALIZATION OF CHITOSAN. G. Kumar, L. Shao, G.F. Payne, Department of Chemical and Biochemical Engineering and Center for Agricultural Biotechnology, University of Maryland Baltimore County, 5200 Westland Blvd. Baltimore Maryland, 21227

Chitin, a polymer of N-acetylglucosamine, is a major structural polymer of crustacean shells. De-acetylation of chitin yields chitosan which has interesting physical, chemical and biological properties. In our studies, we are examining how the properties of chitosan can be altered by enzymatically modifying the primary amine group of chitosan. Specifically, we are using the enzyme tyrosinase to graft various functional groups onto chitosan. Results on the grafting of phenols, polyphenols, and peptides will be reported.

009. TEXTILE MATERIALS FROM CHITOSAN. S. M. Hudson, Fiber and Polymer Science Program, Box 8301, NC State University, Raleigh, NC 27605

Chitosan is well known to have excellent fiber and film forming properties. An alternative method of forming fibrous objects is the shear precipitation of polymers from solution. Chitosan is ideally suited for the process, which yields short cotton-like staple fibers. These chitosan fibers are highly oriented and crystalline and are easily processed into nonwoven structures and papers. These products are described, along with modified forms of the chitosan, such as obtained by graft copolymeriz4tion of vinyl monomers to the fiber surfaces. These chitosan products are expected to have a number of interesting application in the field of medical textiles.

010. CALCIFICATION INHIBITORY PROTEINS IN CALLINECTES SAPIDUS PREMOLT CUTICLE TISSUE: ISOLATION, CHARACTERIZATION AND IMPLICATIONS FOR COMMERCIAL USE Sybil K. Burgess, Yen Xu, Melissa Goodman and David Straight, Department of Chemistry, University of North Carolina at Wilmington, 601 S. College Road Wilmington, NC 28403.

The premolt cuticle of the Atlantic Blue Crab, Callinectes sapidus, contains proteins which delay the onset of calcium carbonate precipitation. These proteins have been isolated and characterized by our laboratory. Data will be presented on the protein purification protocol, the chemical and physical characteristics of these proteins and their putative use as industrial mineralization inhibitors.

011. THE IMPACT OF CRUSTACEANS ON THE GLOBAL CARBON AND CALCIUM CYCLES. J. V. Schloss, Department of Medicinal Chemistry, University of Kansas, Lawrence, Kansas 66045 and Sybil K. Burgess, Department of Chemistry, University of North Carolina at Wilmington, Wilmington, NC 28403

A global perspective of carbon and calcium cycles in the biosphere and the role played by crustaceans and other marine invertebrates will be discussed. Contrast will be drawn between irreversible [greenhouse depleting] and reversible effects on the deposition of carbon dioxide from the biosphere. Mechanisms for maintenance, the effects of pollutants and their interaction with carbon deposition will be examined.

012. TIGHT-BINDING LIGANDS FOR NEW SEPARATIONS TECHNOLOGIES. Daryle H. Busch, Anne K. McCasland, and Stephen J. Archibald, Chemistry Department, University of Kansas, Lawrence, KS 66045

Ligands that bind metal ions very strongly offer special advantages for separations where high dilution or large competing equilibrium constants limit the utility of weaker ligands. However, such tight-binding ligands are rarely used in conventional separations because of their relatively slow rates of metal ion binding and dissociation. Chemical processes that convert tight-binding ligands to more weakly binding structures accelerate the rates of dissociation while interconversions of the opposite kind may accelerate the rates of binding. Examples will be presented and the various rate and equilibrium relationships will be explored.

013. PROBLEMS OF LIGAND DESIGN FOR BIOLOGICAL APPLICATIONS. Mark M. Jones, Department of Chemistry, Vanderbilt University, Nashville, Tennessee 37235.

The central problem in the design of ligands for biological applications is that of satisfying both the coordination chemistry restraints and the structural features needed to obtain the desired biological properties. The coordination chemistry restraints have been examined in greatest detail and both empirical and theoretical approaches are available. The concurrent satisfaction of the biological requirements involve manipulation of the molecular features governing metabolism, tissue, organ and organelle targeting, and toxicity. The development of chelating agents for the mobilization of cadmium from its internal deposits in the kidney and liver illustrate the restraints introduced by biological systems on basic chelating structures needed to remove cadmium from intracellular cadmium-binding macromolecules. Important unsolved problems in this field will be discussed including the desiderata for more effective agents for the treatment of intoxication with lead, mercury and other toxic heavy metals.

014. ENHANCED METALS SEPARATION USING FUNCTIONAL GROUPS WITH NITROGEN DONOR ATOMS. Arup K. SenGupta, Yuewei Zhu and Arthur Kney, 13 East Packer Avenue, Fritz Engineering Laboratory, Lehigh University, Bethlehem, PA 18105

Regardless of their exact chemical compositions, all polymeric chelating exchangers use oxygen, nitrogen, phosphorus, and sulfur as principal donor atoms (Lewis bases) in their functional groups to attain enhanced heavy-metal-ion affinity. In all such metals separation processes, two type of interactions, namely, Coulombic and Lewis acid-base, are commonly present. Our studies show that by designing chelating exchangers with only nitrogen donor atoms, Coulombic interaction can be completely suppressed resulting in an increased heavy-metal-ion affinity over alkali- and alkaline-earth-metal cations. These nitrogen donor atoms are present primarily as pyridyl groups which are border-line soft Lewis bases. Consequently, high affinity toward border-line soft metals, such as Cu(II), Ni(II), and Pb(II), is observed over hard cations like Fe(III), Al(III), and Ca(II). These chelating exchangers with nitrogen donor atoms are also capable of removing dissolved heavy metals in the presence of strong ligands through formation of ternary complexes.

015. CROWN ETHERS AND CRYPTANDS DERIVED FROM FUNCTIONALIZED PENTACYLO-[5.4.0.0^{2,6}.0^{3,10}.0^{5,9}]UNDECANE-4,11-DIONES. A NEW CLASS OF IONOPHORES FOR SELECTIVE ION COMPLEXATION. **Alan P. Marchand**, Kaipenchery A. Kumar, Artie S. McKim, Sulejman Alihodzic, and Hyun-Soon Chong, Department of Chemistry, University of North Texas, Denton, TX 76203-0068.

Details concerning the syntheses and metal ion complexation properties of systems **1-3**, below will be presented and discussed.

1a (n =1)
1b (n = 2)
1c (n = 3)

2a: X = Y = O
2b: X = O, Y = S

3

016. *PARTIAL CONE* CALIX[4]ARENE-CROWN-6 ETHERS.*
Goutam Das, Richard A. Sachleben, Tanneguy Descazeaud, Jeffrey C. Bryan, and Bruce A. Moyer, Chemical & Analytical Sciences Division, Oak Ridge National Laboratory, Oak Ridge, TN 37831-6119

Partial cone calix[4]arene crown ethers represent a new category of lariat ethers. In general, calix[4]arene crown ethers have been previously shown to be efficient extractants for metal cations. Appropriately substituted *cone* and *1,3-alt* conformers of calix crowns exhibit selectivities for Na^+ and Cs^+, respectively. The *partial cone* conformers have not been as thoroughly studied. We have developed an efficient method of synthesis of *partial cone* calix[4]arene-crown ethers and have synthesized a variety of *partial cone* calix[4]arene crowns. Solution and solid-state structures of representative examples have been determined by ^1H NMR spectroscopy and X-ray crystallography. Relative affinities of these *partial cone* calix[4]arene crowns for alkali metal ions have been investigated by solvent extraction techniques.

*This research was sponsored by the Division of Chemical Sciences, Office of Basic Energy Sciences, U.S. Department of Energy, under contract number DE-AC05-96OR22464 with Oak Ridge National Laboratory, managed by Lockheed Martin Energy Research Corp.

017.
METAL-TRIGGERED CYCLOREVERSION OF DIANTHRACENES John Desper* and Shusen Sun, Department of Chemistry, Vanderbilt University, Box 1822-B, Nashville TN 37235.

We have prepared dianthracene chromophores that contain metal-binding groups at the 2- and 3- positions (using the anthracene numbering scheme.) These molecules were designed so that the rate of spontaneous cycloreversion to anthracene could be modulated by binding of a metal ion. The design of these molecules is similar to known metal-chelating anthracene molecules by Bouas-Laurent and Desvergne. By incorporating groups into the peripheral positions, instead of the central 9- and 10- carbons, difficulties caused by unfavorable monomer-dimer equilibrium ratios and susceptibility to oxidation to the quinone are avoided. The design of the molecule provides a well-defined coordination sphere for metal ions, such that the rate of cycloreversion should be influenced by the size of the chelated metal ion. We will show that the rate of cycloreversion is influenced by the presence of metal ions in the medium.

018.
POLYMER GELS CROSSLINKED WITH CHROMIUM(III) CARBOXYLATE COMPLEXES. Robert D. Sydansk, Petroleum Technology Center, Marathon Oil Co., Littleton, Colorado 80122

An overview will be presented of the chemistry and development of oilfield polymer gels that are chemically crosslinked via ligand exchange using Cr(III) carboxylate complexes. This effective and inexpensive aqueous gel technology has rapidly evolved to become a widely applied and effective polymer gel technology for use in improving oil recovery. These gels have been used to: 1) produce more oil, 2) reduce production of undesired large volumes of water and gas and, consequently, operating costs, and 3) help mitigate environmental concerns. The chemistry of this chemically robust and versatile polymer gel technology will be reviewed. The presentation will include a review of the chemistry employed to achieve a broad range of gel onset times over a wide range of temperatures. Also, the chemistry involved in producing a broad range of gel properties will be presented. A brief discussion will be given on the low toxicity and environmental aspects of these gels.

019. SOME REACTION PARAMETERS IMPORTANT TO FORMATION AND STABILITY OF CHROMIUM(III)-CROSSLINKED POLYMER GELS. Robert H. Lane, Northstar Technologies International, Anchorage, Alaska 99515

Rates of formation of polymer gels formed by crosslinking partially hydrolyzed polyacrylamides with chromium(III) are critically dependant on temperature and both the degree of hydrolysis of the polymer and availability of reactive chromium(III) species. Likewise, stability of the crosslinked product is dependant on temperature and on the presence, nature and amounts of ligating species that can potentially compete with polymer for binding chromium(III). This paper will review some important relevant aspects of small molecule chromium(III) substitution chemistry, and relate them to behavior of chromium(III)-crosslinked polymer gels in oil-producing formations.

020. **CHEMICAL CHARACTERIZATION OF CHROMIUM(III) ACETATE / POLYACRYLAMIDE GELS.** J. E. Tackett, Marathon Oil Company, 7400 S. Broadway, Littleton, CO 80122

A variety of spectrometric techniques were used to increase our understanding and control of chromium(III) acetate / polyacrylamide gel systems. FTIR allowed measurement of carboxylate content in polyacrylamide down to 0.2 mole%. FTIR, ^{2}HNMR, and FAB mass spectrometry provided insight into the chemical structure of chromium(III) acetate complexes and their hydrolysis products in aqueous solution. HNMR and visible spectroscopy were used to monitor competing reactions during the gelation process. All of this information contributed to the development of polymer gel systems whose kinetics and physical properties can be adjusted to meet the unique demands of a variety of oil field applications.

021. GELATION OF POLYMERS WITH CR(III)-CARBOXYLATES FOR OIL RECOVERY APPLICATIONS, A. Moradi-Araghi and K. B. Fox, Phillips Petroleum Company, Bartlesville, OK 74004

The use of Cr(III)-ligand complexes has been widely explored in recent years in production of gels for conformance control in oil recovery applications. Extensive studies in this and other laboratories show that solutions of partially hydrolyzed polyacrylamides, xanthan gums and carboxymethylcellulose (CMC) polymers at low concentrations can be crosslinked with Cr(III)-ligand complexes such as chromic acetate, chromic propionate and chromic malonate under various conditions to produce a range of versatile gels suitable for many oilfield applications. This paper will describe parameters which have pronounced effects on the rate and quality of the produced gels. Such parameters include polymer type, molecular weight as well as anionic character of polymer, polymer and Cr(III) concentrations, salinity and pH of the solution as well as the temperature. At a given polymer and Cr(III) concentration, when other parameters are kept constant, the rate of gelation is reduced when the ligand in Cr(III)-ligand complex is varied from acetate, to propionate, malonate or citrate respectively.

022.
LIGAND-BASED STRATEGIES FOR DELAYING GELATION AT HIGH TEMPERATURE.
T. P. Lockhart, Eniricerche SpA, 20097 San Donato Milanese, Italy

Cr(III) acetate/polyacrylamide gels have proven themselves an effective tool for reducing excess water production from petroleum wells. The short gelation delays of Cr(III) acetate gelants at higher temperature, however, limit their applications in the deep reservoirs that have been extensively developed over the past decade. Our laboratory has explored the use of carboxylate ligands with greater Cr(III)-complexing power than acetate in order to reduce the rate of polymer crosslinking. This talk will examine the chemistry underlying the ligand-based approach to increasing gelation delays. Examples showing the importance of achieving the proper balance between the stabilities of the Cr(III)-ligand complex and the Cr(III) polymer crosslink complex will be given. In this regard, the special reaction chemistry displayed by malonate at high temperature will be described. Finally, it will be shown that the kinetic and thermodynamic stabilization of Cr(III) by complexing ligands also greatly improves Cr(III) propagation in porous media.

023. UPTAKE AND GELATION REACTIONS of Cr(III) OLIGOMERS WITH POLYACRYLAMIDE. C.L. Dona, Terracon Environmental, Inc., 7810 NW 100th St., Kansas City, MO 64153, D.W. Green, and G.P. Willhite, Department of Chemical and Petroleum Engineering, University of Kansas, Lawrence, KS 66045-2223

Reaction rates of monomeric, dimeric and trimeric Cr(III) with partially hydrolyzed polyacrylamide(2.3% carboxyl groups) were measured. Oligomers (dimeric and trimeric chromium) were prepared in relatively pure form using liquid chromatography. The oligomers were allowed to react with polymer, quenched and then separated using equilibrium dialysis to determine the amount of chromium uptaken by the polymer. In the pH range of 4-5, Cr(III) concentrations in the range of 8-30 ppm and polymer concentrations in the range of 4000-20000 ppm, reactions rates increased with increasing oligomer size, concentration and pH. In the pH 4 monomer uptake runs, an uptake model was developed based on assumptions of elementary reactions and low oligomerization rates. This model did not fit the dimer or trimer data. Gelation rates inferred from rheological measurements were found to follow closely the Cr(III) reaction rates.

024. A LEAD ION-IMPRINTED POLYMER AND ITS APPLICATION IN A LEAD ION-SELECTIVE ELECTRODE. Xiangfei Zeng and George M. Murray, Department of Chemistry and Biochemistry, University of Maryland, Baltimore County, Baltimore, MD 21228-5398.

Lead ion-imprinted polymers that exhibit enhanced selectivity for Pb(II) ion have been synthesized. A lead ion-selective electrode based on the synthesized polymer has been prepared. The electrode exhibits a linear response to lead ion with a slope of 33 mV per decade which is close to the theoretical Nernstian slope. Selectivity coefficients of the electrode were determined by different methods and effects of the pH of the test solution and composition of the membrane were studied.

025. A NOVEL BIFUNCTIONAL ORGANOPHOSPHORUS MONOMER FOR METAL ION-IMPRINTED POLYMERS BY SURFACE TEMPLATE POLYMERIZATION. Masahiro Yoshida, Kazuya Uezu, Masahiro Goto and Fumiyuki Nakashio, Department of Chemical Science and Technology, Kyushu University, Hakozaki, Fukuoka 812-81, Japan.

Surface template polymerization is a powerful technique for the ready preparation of polymeric materials containing recognition sites of predetermined specificity at the surface of the polymers. This technique has the advantage of effective recognition for aqueous soluble guests, such as metal ions. In the technique, the interfacial activity of functional monomers is a dominant factor for successful preparation of highly selective imprinted polymers. It is also important to fix recognition sites rigidly and to make strong interactions between functional monomers and imprint molecules. From this point of view, we have designed novel bifunctional organophosphorus monomers as a recognition monomer, and prepared highly selective Zn(II)-imprinted polymers by surface template polymerization with a W/O emulsion. Adsorption behavior of zinc and other metal ions onto the polymers was examined, and the template effect was characterized.

026. ENHANCING THE SELECTIVITY OF MOLECULARLY IMPRINTED POLYMERS. Yohei Yokobayashi, Scott McNiven and Isao Karube, Research Center for Advanced Science and Technology, The University of Tokyo, 4-6-1 Komaba, Meguro-ku, Tokyo 153.

One of the main problems encountered in non-covalent molecular imprinting is that the molecularly imprinted polymer (MIPs) posses binding cavities with a wide range of affinities for the template molecule. This has been termed the "polyclonality" of MIPs and is responsible for the exaggerated tailing of the peaks (especially of the template molecule) due to a diminishing number of sites with increasing binding strengths. We present the first successful attempt to reduce the polyclonality of MIPs. By judicious methylation of the polymer in the presence of the original template molecule, the selectivity of the polymer was enhanced by more than 10% over the untreated polymer. Moreover, we demonstrate that the presence of the template molecule during the methylation actually protects the specific sites for alkylation.

027. SOLVENT DEPENDENT SPECIFIC DRIVING FORCES IN THE MOLECULAR RECOGNITION IN IMPRINTED POLYMERS, Börje Sellergren and Christian Dauwe, Department of Inorganic and Analytical Chemistry, Johannes Gutenberg University Mainz, J.-J.-Becherweg 24, 55099 Mainz, Germany

Triazine herbicides were used as model templates for a basic study of the molecular recognition process in imprinted polymers. These templates have the benefit of being structurally similar in the hydrogen bonding part while possessing widely different basicities and hydrophobicities. This facilitates a systematic evaluation of the influence of the hydrophobic effect and the strength of hydrogen bonding between monomer and template, on the molecular recognition properties of the materials. If these effects contribute specifically to the observed recognition a change in solvent composition in the rebinding step is expected to change the driving force in this process. In aqueous-poor solvent systems we found that the selectivity for the template increased with its Broensted basicity whereas in aqueous-rich systems, affinity and selectivity increased with template hydrophobicity. An analogy is found in the specific hydrophobic effect often present in the binding of small molecules to biological macromolecules.

028. MOLECULAR IMPRINTED MEMBRANES HAVING A CROSS-LINKED GEL LAYER PREPARED BY PHOTOGRAFT POLYMERIZATION. Takaomi Kobayashi, Hong Ying Wang and Nobuyuki Fujii, Department of Chemistry, Nagaoka University of Technology, Kamitomioka, Nagaoka 940-21, Japan

Photograft polymerization was applied as a novel approach to the introduction of imprint sites of a template, theophylline, into a cross-linked polymeric gel formed on a membrane support. The asymmetric membrane of polyacrylonitrile with covalently bound dithiocarbamate photosensitive groups was prepared by a phase inversion method. Photo-irradiation to the membrane surface in the presence of the template methylenebisacrylamide, and acrylic acid results in the formation of a gel layer containing the template. After removal of the template from the membrane, theophylline or caffeine was filtered through the imprinted membrane. Evidence is presented that theophylline is taken into the imprinted sites of the gel network. However, the amounts of caffeine taken into the membrane are lower than that of theophylline. The results demonstrate that the gel network on the membrane surface records the shape of the template molecule during photograft polymerization.

029. A METAL ION-TEMPLATED POLYMERIC SENSOR FOR LEAD. Anton Bzhelyanskiy, Amanda Jenkins, and G.M. Murray, Department of Chemistry, University of Maryland, Baltimore County, Baltimore MD 21250 and O. M. Uy, The John Hopkins Applied Physics Research Lab, Laurel MD 20723

A novel device is being designed to determine lead in aqueous solutions for application to environmental and biological samples. Lead-templated polymers have been developed as selective sequestering agents and have been fabricated into membranes and used in ion-selective electrodes. Fluorescent lead complexing agents are now being used to develop a metal ion-templated polymer to serve as a sensing element for an optical device for the determination of lead ions. The lead complex shifted ligand fluorescence will be excited by a compact laser and monitored via an optical fiber by either a filtered photometer or a monochromator. Sensitivity based on fluorometric detection coupled with the chemical selectivity of a templated polymeric film will result in a large dynamic range of measurement and low limit of detection.

030. SEPARATION OF RARE EARTH METALS BY IMPRINTED POLYMERS UTILIZING SURFACE TEMPLATE POLYMERIZATION. Tsuyoshi Hara, Masahiro Yoshida, Kazuya Uezu, Masahiro Goto and Fumiyuiki Nakashio, Department of Chemical Science and Technology, Kyushu University, Hakozaki, Fukuoka 812-81, Japan..

An efficient separation process for rare earth metals is needed to provide novel advanced materials for various electronic, optical, and magnetic devices. Solvent extraction is well known to be an effective method for separation of the metals on an industrial scale. However, this process requires a large number of stages in a series of mixer-settlers to obtain high-purity products because these elements behave almost identically. We have developed surface-templated polymers for the separation of rare earth metals. The imprinted polymers were prepared by surface template polymerization with dioleylphosphoric acid as a functional monomer. The separation of rare earth metal ions with the polymers was conducted, and the template effect was characterized by comparison with that in ordinary solvent extraction for the metal ions.

031. DEVELOPMENT OF STEROID SENSORS USING MOLECULARLY IMPRINTED POLYMERS. Soo-Hwan Cheong, Scott McNiven, Kazuyoshi Yano and Isao Karube, Research Center for Advanced Science and Technology, The University of Tokyo, 4-6-1 Komaba, Meguro-ku, Tokyo 153.

The molecular imprinting technique was used to synthesize polymers having a high affinity for testosterone. These polymers thus function as artificial receptors for the drug. Various polymerization conditions were examined in order to determine their influence on the selectivity of the receptors. Using a suite of similar steroids with differing shapes and functionalities, we were able to identify features of the molecules which affect their affinity for the polymer matrix. The efficacy of covalent- and non-covalent imprinting methods was also compared. Our results illustrate a number of basic guidelines for the synthesis of molecularly imprinted polymers. Our optimized testosterone receptor bound testosterone more than 4 times more effectively than a non-imprinted polymer and afforded separation factors of between 3.3 and 9.6 for a number of structurally similar steroids.

032. BINDING ASSAYS FOR DRUGS AND HERBICIDES USING MOLECULARLY IMPRINTED POLYMER PARTICLES AS RECOGNITION ELEMENTS IN DIFFERENT ASSAY FORMATS; K. Haupt and K. Mosbach, Department of Pure & Applied Biochemistry, Lund University, Chemical Center, POB 124, S-22100 Lund, SWEDEN

Molecular imprinting technology has been successfully applied to create recognition sites for certain target molecules in synthetic polymers. The resulting molecularly imprinted materials have been used as stationary phases for chromatographic separation of molecules with closely related structures (e.g., optical isomers), as recognition elements in sensors, in catalytic applications or, instead of antibodies, in immunoassay-type analyses (see Mosbach, K. and Ramstrom, O. 1996, The emerging technique of molecular imprinting and its future impact on biotechnology, Bio/Technology 14, 163). The latter application appears especially promising as synthetic polymers can be imprinted with compounds against which it is difficult or impossible to raise natural antibodies, e.g., small or non-immunogenic molecules, immunosuppressive drugs, etc. This presentation will include recent developments of assay systems for drugs and herbicides for medical and environmental analyses and their use in water-based and organic solvents. In addition to the "conventional" radioligand binding assay, new assay formats that avoid the use of radiolabelled compounds will be described.

033.
TITANOSILICATES AND THEIR STRUCTURAL ANALOGS FOR THE ENVIRONMENTAL REMEDIATION OF CESIUM AND STRONTIUM. Elizabeth A. Behrens, Damodara Poojary, Abraham Clearfield, Department of Chemistry, Texas A&M University, College Station, TX, 77843.

Recently, several Hanford storage tanks leaked radioactive waste into the surrounding environment, thereby spawning concern in the scientific community for proper and economical disposal options. One way to achieve this goal is by using highly selective inorganic ion exchangers. A particular area of interest in our research group is a titanosilicate material whose framework contains alternating metal ions that create a 3-dimensional tunnel structure with charge-neutralizing cations in the face centers. The potassium and sodium titanosilicates show high selectivities for strontium and cesium in the presence of ppm-levels of alkali and alkaline earth metals. Altering selectivities towards different cations may be achieved by 1). changing the charge-balancing cation or 2). synthetically altering the framework composition with smaller or larger metal ions. New and existing synthetic variations of this material will be presented along with preliminary ion exchange data for groundwater remediation and some nuclear waste applications.

034.
CELLS ON FIBERS FOR POLLUTION CONTROL. Robert Clyde, Clyde Engineering, New Orleans LA 70174

Thousands of gallons of radioactive metals are in tanks which are leaking. From Hanford, for example, they are leaking into the Columbia River which goes to Portland. Many people use resins to capture the metals, but resins are not effective at low concentrations of uranium. The bacterium Pseudomonas aeruginosa can capture uranium in minutes and this is covered in Clyde-Whipple patent 4,530,763. Other bacteria capture cesium, strontium and selenium. Lead is over the limit in many large cities and this can be removed in seconds. White rot fungus degrades many pollutants and it grows on old cardboard boxes. When buried in soil, air is entrapped for growth of the fungus. Patent 5,256,570 tells how to get oxygen into liquid quickly. H2S can be degraded by the same method. Cells also grow on foam and when a stainless screen is put around it, it is made heavier so it can be fluidized.

035.
NEW SORBENTS AND SORPTION TECHNOLOGIES FOR LIQUID RADIOACTIVE WASTES MANAGEMENT D.V. Marinin V.A. Avramenko, V.Yu. Glushchenko, V.I. Sergienko, V.A. Vasilevskiy, V.V. Zheleznov, Laboratory of Sorption Processes, Institute of Chemistry FEDRAS, Vladivostok 690022, Russia

Advantages of sorption technologies for liquid radioactive wastes (LRW) management were shown. Modified sorbents highly selective to the most dangerous long-living radionuclides (first of all, Cs-137 and Sr-90) have been developed. The sorbents developed allow to clean LRW of various types (including high-salted LRW and those containing colloid particles) down to accepted radionuclides concentrations within one stage. Comparison between selective sorbents and ion-exchange resins efficiency for large-scale LRW management was given. Parameters and results of work of respective sorption filter unit are presented.

036. PRACTICAL POLLUTION PREVENTION: AN EXCITING NEW SUBDIVI-SION OF I&EC. S.S. Seelig, Seelig and Associates, Inc., 964 Old Meadow Court, Carol Stream, Illinois 60188-3345 and G. Kohl, Dow Corning Corporation, Mail #128, 3901 S. Saginaw Road, Midland, Michigan 48686

Practical Pollution Prevention (P3), a relatively new subdivision of the ACS Division of Industrial and Engineering Chemistry (I&EC) sponsored its first symposium during the 1995 Fall ACS meeting in Chicago. The intent of P3 is to spearhead programming in the applied areas of pollution prevention, waste/resource management, green technology, full-cost accounting and life cycle screens. The organizational committee is made up of members from industry, government and academia. The I&EC Division has supported the P3 subdivision, supplying both money and an advisory committee of former chairs of the division. Special symposia for 1997 include a joint effort with the USEPA on New Chemical Science / Engineering Technology for Pollution, Emerging Technologies in Hazardous Waste Management, and Industrial Cleaning Without Chlorofluorocarbons at the Fifth Chemical Congress meeting of North America. Future plans include sponsoring a P3 Award for this technical area and the establishment of a topical newsletter. All new members are welcome.

037. JOIN THE SEPARATION SCIENCE AND TECHNOLOGY SUBDIVISION OF THE ACS I&EC DIVISION! J.F. Scamehorn. Institute for Applied Surfactant Research, University of Oklahoma, 100 E. Boyd, Rm. T335, Norman, Oklahoma, 73019.

Separations represent the key to most industrial chemical processes and lie at the heart of analytical chemistry. Active areas in separation science include new chromatographic techniques, improved ion exchange resins, membrane separations, separating transuranic elements, separating proteins, and developing new ligands for metal ions. Improving selectivity of separation methods and separating extremely dilute components from streams are two major themes of current interest. The SST subdivision promotes activities in all areas of separations relevant to chemists and to chemical engineers and membership improves your productivity in laboratory or large scale industrial separations. The subdivision sponsors technical sessions at ACS National and other meetings, coordinates activities with AICHE, and publishes a biannual newsletter.

038.
CORROSION STUDY OF CANDIDATE APT ENGINEERING MATERIALS. **Doris K. Ford,** Los Alamos National Laboratory Los Alamos, NM 87545

The anticipated needs for tritium have called for an analysis of proton accelerator technologies (APT) as an alternative to reactor production. To produce tritium with an accelerator, a neutron spallation target is bombarded with high energy protons (800-1200MeV). To minimize its temperature and to provide a neutron moderator, the spallation target is immersed in a cooling water loop. Initially the water in the loop is distilled and deionized, but with irradiation radiolysis products are formed. The most stable species is hydrogen peroxide ($H2O2$) which is a long lived radiolysis product and is formed from the combination of two OH radicals. In this poster we describe our current progress to measure the corrosion rates of candidate APT materials in the laboratory to simulate the effects of radiological environments (such as effects of pH, hydrogen peroxide, temperature, metal-chlorides, and flowrates). Future efforts for control and monitoring of corrosion rates of these materials at the Los Alamos LANSCE facility will also be presented.

039. A CLOSED LOOP SYSTEM FOR THE CONVERSION OF URANIUM TURNINGS TO URANYL OXY-HYDROXIDE, Trudi M. Forman, Nancy N. Sauer, Wayne H. Smith, Gregory Ogden; Chemical Sciences and Technology, Los Alamos National Laboratory, Los Alamos, NM 87545

The machine shops at Los Alamos National Laboratory generate up to 5 Kgs of uranium turnings daily. Presently, the turnings are packed in diesel fuel in 55 gallon drums and shipped off site for treatment and disposal. In response to a request for an in-situ generator treatment plan, a three-part closed loop system has been designed to dissolve the turnings and leave them in a non-reactive form for either storage or disposal. The system uses electrochemically generated sodium hypochlorite to dissolve the turnings, converting them to uranyl oxy-hydroxide precipitate. The precipitate is continually centrifuged to separate the liquids from solids. The supernant, spent hypochlorite, feeds into the electrochemical cell, the hypochlorite is regenerated and pumped back into the dissolution reactor. This closed loop system accomplishes both conversion of the uranium turnings to an acceptable form and minimizes the treatment wastestream.

040. THE REMOVAL OF PRECIOUS METALS BY CONDUCTIVE POLYMER FILTRATION, M. E. Cournoyer, P. C. Stark, S. M. Trujillo, R. E. Aguino, Chemical Science and Technology Division, Los Alamos National Laboratory, Los, Alamos, NM 87545

The growing demand for platinum-group metals (PGM) for several applications within the DOE complex and in industry, the need for modern and clean processes and the increasing volume of low grade material for secondary PGM recovery have a direct impact on the industrial practice of recovering and refining of precious metals. With precipitation-dissolution methods being the most common method of recovery, there is a tremendous need for advanced metal ion recovery and waste minimization techniques. Here at Los Alamos there is an integrated program in ligand-design and separation's chemistry for recovery of actinide and toxic metals from variety of process streams. In the present investigation, a novel hollow fiber membrane (CP1) is characterized and its selectivity for PGM reported. In addition, a continuous single step unit operation is proposed for the removal, concentration and recovery of platinum from catalytic and electroplating process streams is proposed.

041.
HYDROTHERMAL PROCESSING OF ACTINIDE CONTAMINATED ORGANIC WASTES. **Laura A. Worl,** Steven J. Buelow, Loan A. Le, Dennis D. Padilla, James H. Roberts, D. Kirk Veirs, MS-E510, Los Alamos National Laboratory, Los Alamos, NM 87545

Hydrothermal oxidation is an innovative process for the destruction of organic wastes, that occurs above the critical temperature and pressure of water. The process provides high destruction and removal efficiencies for a wide variety of organic and hazardous substances. For aqueous /

organic mixtures, organic materials, and pure organic liquids hydrothermal processing removes most of the organic and nitrate components (>99.999%) and facilitates the collection and separation of the actinides. We have designed, built and tested a hydrothermal processing unit for the removal of the organic and hazardous substances from actinide contaminated liquids and solids. Here we present results for the organic residues generated at the Los Alamos National Laboratory Plutonium Facility.

042. REACTION OF CARBON DISULFIDE AND O-PHENYLENE DIAMINE CATALYZED BY QUATERNARY AMMONIUM SALT IN THE PRESENCE OF POTASSIUM HYDROXIDE, Maw-Ling Wang, Biing-Lang Liu, Department of Chemical Engineering, National Tsing Hua University, Hsinchu, Taiwan 30043 ROC

The synthesis of 2-mercaptobenzimidazole (MBI) from the reaction of carbon disulfide and o-phenylene diamine catalyzed by tetrabutyl-ammonium bromide in the presence of KOH was carried out in a homogeneous-phase solution using protic and aprotic solvent. Water in a small amount was added to dissolve KOH. The role of KOH in the reaction was examined by conducting several independent experiments, which include a timed reaction of o-phenylene diamine in the presence or absence of tetrabutyl-ammonium bromide catalyst in various sequential orders. A homogeneous-phase reaction mechanism was proposed which supports the kinetic reaction model. The reaction was situated in protic and aprotic solvent and are described by zeroth-order and pseudo-first order rate law, respectively. The effect of the amount of KOH on the conversion of o-phenylene diamine was also studied. The reaction is greatly accelerated both in the presence of tetrabutylammonium bromide catalyst and potassium hydroxide. Satisfactory explanations were provided for the enhanced catalytic reaction of carbon disulfide and o-phenylene diamine by catalyst compared with KOH. Other reaction conditions also affected the conversion.

043. RHEOLOGICAL MODELS OF SAVANNAH RIVER SITE HLW RADIOACTIVE SLUDGES. Bao C. Ha, SRTC, Westinghouse Savannah River Co., Aiken, SC 29808

Knowledge of the rheology of radioactive sludge slurries at the Savannah River Site (SRS) is necessary in order to ensure that they can be retrieved from waste tanks and processed for final disposal. Rheological properties of Tank 42 and Tank 51 radioactive sludges at various solids concentrations were measured remotely in the Shielded Cells at the Savannah River Technology Center (SRTC) using a modified Haake Rotovisco viscometer. The experimental results indicated that the radioactive sludge slurries could be easily adjusted to meet yield stress and viscosity limits of the Defense Waste Processing Facility (DWPF) design basis. A pseudo-Bingham plastic fluid model has been used successfully to approximate the viscosity and yield stress behavior of the radioactive sludge slurries. Theoretical models, such as the Bingham model parameters correlating the reduced yield stress and consistency of solids in liquid suspension as a function of the volume of insoluble solids, will also be further discussed.

044. RECOVERY OF POLYELECTROLYTE AS USED IN POLYELECTROLYTE-ENHANCE ULTRAFILTRATION (PEUF) FOR THE REMOVAL OF MULTIVALENT IONS FROM AQUEOUS STREAMS, Ahmadali Tabatabai, John F. Scamehorn, Sherril D. Christian, Institute for Applied Surfactant Research, The University of Oklahoma, Norman, OK 73019

The effectiveness of a novel membrane-based separation process using polyelectrolytes for the removal of multivalent ions from aqueous streams is shown. In polyelectrolyte-enhanced ultrafiltration (PEUF), the contaminant is removed from an aqueous feed stream and concentrated in a low-volume stream. The PEUF process is effective in the removal of multivalent ions with rejections up to 99.7%.

The methods of recovery of the polyelectrolytes used in the PEUF process and the importance of the polyelectrolyte recovery for economic feasibility of the process is emphasized in this study.

045. AN AB INITIO STUDY OF THE SELECTIVE BINDING OF ALKALI AND ALKALINE EARTH METAL CATIONS FOR CROWN ETHERS. D.F. Feller, E.D. Glendening, M.A. Thompson, S.E. Hill, Environmental Molecular Sciences Laboratory, Pacific Northwest National Laboratory, Richland, Washington 99352

Crown ethers are a class of macrocyclic polyether that are known to preferentially bind certain cations in potentially complex mixtures of other cations. They are being considered as candidates for separating radioactive cesium (137Cs) and strontium (90Sr) from the high level waste tanks on the Hanford Nuclear Reservation. We present the results of RHF and Mp2 calculations on a variety of crown ethers, including 12-crown-4, 18-crown-6 and Clark Still's 18-crown-6 derivative. Gas phase calculations show that 18-crown-6 most strongly binds lithium, but our results indicate that potassium selectivity is recovered with the inclusion of even a small number of waters of hydration in the calculations. Several polarized Gaussian basis sets were used in an effort to derive convergence estimates for the binding enthalpies. The only available experimental values, obtained from FT-ICR experiments by A.R. Katritzky, et al, Rapid Comm. Mass Spect.,6,25(1992), are over 30 kcal/mol smaller than our best estimate for K+/18-crown-6.

046. METALS SEPARATION USING SOLVENT EXTRACTANTS ON MAGNETIC MICROPARTICLES, L. Nuñez, Argonne National Laboratory, 9700 S. Cass Ave., Argonne, IL 60439 and M. Pourfarzaneh, CORTEX-BIOCHEM, Inc., 1933 Davis St., STE 321, San Leandro, CA 94577.

The magnetically assisted chemical separation program was initially funded by DOE EM-50 to develop processes for the efficient separation of radionuclides and other hazardous metals. This process has stimulated the partnership between industry and ANL for many applications related to hazardous metal problems in industry. In-tank or near-tank hazardous metals separation using magnetic particles that have selective coating is a new approach to the problems of metal removal and recycling [of industrial (e.g., mining, printing circuit board, plating)] corrosive waste streams. This concept of coated magnetic particles promises simple, compact processing at very low costs and employs mature chemical separations technologies to remove and recover hazardous metals from aqueous solutions. The selective chemical extractants are attached to inexpensive magnetic carrier particles. Surfaces of small particles composed of rare earths or ferromagnetic materials are treated to retain chemical extractants (e.g., TBP, CMPO, quaternary amines, carboxylic acid). After selective partitioning of contaminants to the surface layer, magnets are used to collect the loaded particles from the tank. The particles can be regenerated by stripping the contaminants and the selective metals can be recovered and recycled from the strip solution. This process and its related equipment are simple enough to be used for recovery/recycling and waste minimization activities at many industrial sites. Both the development of the process for hazardous and radioactive waste and the transfer of the technology will be discussed. This work supported by the U.S. Department of Energy, Office of Environmental Management, under contract W-31-109-ENG-38.

047. **Polymer Modification of Resorcinol-Formaldehyde (R-F) Ion-Exchange Resin.** T.L. Hubler[1], J.A. Franz[1], J.J. Yang[2], and J.M. Shreeve[2]. Pacific Northwest National Laboratory, MS P8-38, P.O. Box 999, Richland, WA 99352; [2]University of Idaho, Department of Chemistry, Moscow, ID 83844.

Resorcinol-Formaldehyde (R-F) resin is a candidate regenerable ion-exchange resin for removal of radioactive cesium from highly alkaline waste tank supernates at both the Hanford and Savannah River sites. Previous investigations into the structure/function relationships of R-F resin have shown that the R-F resin undergoes facile oxidation to produce *para*-quinones, with loss of ion-exchange sites, hence lowered performance of the resin for cesium ion-exchange. In this report, we give the results of our studies into polymer resins prepared using 4-methylresorcinol and 4-fluororesorcinol. The reaction of 4-methylresorcinol with formaldehyde formed oligomeric structures, while a mixture of 4-fluororesorcinol, phenol, and formaldehyde produced a non-soluble resin in aqueous/alkaline conditions. The 4-fluororesorcinol resin underwent significant nucleophilic displacement of the fluorine substituent to give oxidized resins with lower ion-exchange performance.

Pacific Northwest Laboratory is operated for the U.S. Department of Energy by Battelle Memorial Institute under Contract DE-AC06-76RLO 1830.

048.

EVALUATION OF THE PERFORMANCE OF TWO NEW EQUATIONS OF STATE IN CONJUNCTION WITH A SIX-COMPONENT SOUR NATURAL GAS MIXTURE S. M. A Wong, H. Mahgerefteh, and I. Economou, Department of Chemical and Biochemical Engineering, University College London, London, United Kingdom, WC1E 7JE

In the past fifty years, several equations of state have been proposed for predicting equilibrium data for multi-component hydrocarbon systems. The most popular equations include Soave Redlich-Kwong (SRK) and Peng-Robinson (PR) cubic equations of state (CEOS). It is well-known that these are inadequate for properly predicting liquid densities. Recently, two new equations, a three-parameter CEOS proposed by Twu-Coon-Cunningham and a non-cubic EOS based on the Simplified Perturbed Hard-Chain Theory have been developed. They have been shown to provide good predictions in conjunction with pure component hydrocarbons. In this paper, the performance of both equations will be evaluated against SRK and PR CEOS by comparing predictions of densities and equilibrium constants with experimental data for a six-component sour natural gas mixture at sub-ambient temperatures in the pressure range 27 - 80 bar.

049. SOLVENT EFFECTS ON STEREOCHEMISTRY, L. Rhodes, M. Conner, R.C. Gatrone, Department of Chemistry, Wilkes University, Wilkes-Barre, PA 18766

The results of an investigation into the stereochemical effect of the chosen solvent for the Grignard reaction will be reported.

050. ATTEMPTS TO PREPARE OXASEMIBULLVALENE, T. Gibbs and R.C. Gatrone, Department of Chemistry, Wilkes University, Wilkes-Barre, PA 18766

Our approach to the synthesis of Oxasemibullvalene (1) will be reported

1

051.
STERIC STRAIN IN THE NUCLEOPHILIC ADDITION REACTION TO CARBONYLS, K. Wilk and R.C. Gatrone, Department of Chemistry, Wilkes University, Wilkes-Barre, PA 18766

The effect of steric strain during the reaction of hydroxylamine with ketophosphonates

052.
THE ACIDITY OF STRONG ACID POLYMERIC CATALYSTS MEASURED USING SOLID STATE 13C NMR. **Eric G. Lundquist,** Kebede Beshah, Research Laboratories, Rohm and Haas Company, Spring House, PA 19477

Using methods originally developed by Haw (J. Amer. Chem. Soc. 1994, 116, 1962-1972) and Farcasiu (J. Amer. Chem Soc. 1993, 115, 10901-10908), the acidity of Strong Acid Amberlyst® polymeric catalysts has been determined by measuring the chemical shifts of probe molecules (acetone, mesityl oxide, acetyl chloride, isobutylene) adsorbed on the catalyst in the solid state 13C NMR. Unlike previous indicator methods, this solid state 13C NMR technique allows for the determination of acidity values in a nonaqueous medium. The impact of water on the measured acidity of Amberlyst® catalysts and on the activity in alkylation reactions will be discussed.

053. **4,4'-DIPYRIDYL AS A SYNTHON FOR THE DESIGN AND SYNTHESIS OF POROUS SOLIDS.** **Richard P. Swatloski**, Scott T. Griffin, Lillian M. Rogers, and Robin D. Rogers, Department of Chemistry, The University of Alabama, Tuscaloosa, AL 35487

Our group is attempting to use the tenets of crystal engineering to design and synthesize porous solids for specific separations. We are currently synthesizing complexes of 4,4'-dipyridyl and 4,4'-dipyridyl-like ligands capable of holding metal ions in rigid two dimensional structures, and then locking these two dimensional structures into three dimensional porous solids. This presentation will focus on our crystallographic results as we attempt to define the structural features necessary to prepare rigid channels of specific sizes.

054. **A DIRECT COMPARISON OF METAL ION PARTITIONING IN LIQUID/LIQUID PEG-BASED ABS AND RETENTION ON ABEC™ RESIN. HOW GOOD IS THE ANALOGY?** **Scott T. Griffin** and Robin D. Rogers, Department of Chemistry, The University of Alabama, Tuscaloosa, AL 35487

We have been studying metal ion partitioning in ABS formed when polyethylene glycol (PEG)-2000 is salted out by certain water structuring inorganic anions (e.g., OH^-, SO_4^{2-}, CO_3^{2-}, PO_4^{3-}) to form an upper PEG-rich phase and a lower salt-rich phase. We have extended this study to include an $\underline{A}$queous $\underline{B}$iphasic $\underline{E}$xtraction $\underline{C}$hromatographic resin (ABEC™) containing covalently bound monomethylated PEG. In order to further characterize the relationship between the liquid/liquid and resin separations, we are conducting detailed comparative studies between these systems. In this presentation, we will discuss our current results.

055. **PARTITIONING OF DYES AND METAL/DYE COMPLEXES IN AQUEOUS BIPHASIC SYSTEMS.** **Jennifer A. Nicol**, Heather D. Willauer, Scott T. Griffin, and Robin D. Rogers, Department of Chemistry, The University of Alabama, Tuscaloosa, AL 35487

As part of our overall research effort to develop new separation technologies for aqueous biphasic systems to obviate the need for volatile organic solvents, we have begun to investigate the removal of water soluble dyes and their metal ion complexes from textile wastes. In this presentation we will focus on our investigation of the partitioning behavior of acid red #4, orange G_1, and plasmocorinth B between the polymer-rich and the salt-rich phases of aqueous biphasic systems. Our preliminary results adopting this separation to ABEC™ resins will also be presented. This work is supported by the Division of Chemical Sciences, Office of Basic Energy Sciences, Office of Energy Research, U.S. Department of Energy.

056. **ANION BINDING BY MACROCYCLIC LIGANDS AND THEIR METAL COMPLEXES,** A. E. Martell, Q. Lu and David A. Nation, Department of Chemistry, Texas A&M University, College Station, Texas 77843-3255.

The protonated forms of the macrocyclic dinucleating ligand OBISDIEN, and its analogs BFBD, BPBD, and MXBD bind anions through hydrogen bonds and coulombic forces. Their mononuclear and dinuclear metal complexes bind anions through coordinate metal bonds, hydrogen bonds and coulombic forces. To be effectively bound anions must have the appropriate size, shape and donor groups to be recognized and fit in the space provided by the macrocyclic ligands. This paper will describe the selective binding of oxalate, phosphate, pyrophosphate and triphosphate anions by these macrocyclic ligands, and the analogous complexes formed with the nucleotides AMP, ADP, and ATP.

057.
STEROID-BASED RECEPTORS FOR INORGANIC ANIONS. Anthony P. Davis, Shay Broderick, John G. Gilmer, Justin J. Perry, and Robert P. Williams, Department of Chemistry, Trinity College, Dublin 2, Ireland

The steroid nucleus is large, rigid, readily available, and capable of a wide variety of substitution patterns. It is thus well-suited for exploitation as an organising framework in supramolecular chemistry. We have been exploring the possibilities of cholic acid, which is one of the least expensive steroidal starting materials and is furnished with an especially useful pattern of functionality. This presentation will focus on the design, synthesis and investigation of macrocyclic and podand-type anionophores derived from cholic acid. Operating through preorganised arrays of H-bond donor centres, these electroneutral and lipophilic molecules have achieved remarkable affinities for inorganic anions. Their suitability for "phase transfer" applications may lead to their exploitation in sensors, catalyst systems and biochemical investigations.

058. Cyclic and Acyclic Polypyrroles. Novel Receptors for Anionic Substrates. Jonathan L. Sessler, Andrei Andrievsky, Vladimír Král, Petra Sansom, Kevin Shreder, Philip A. Gale, Pierfrancesco Morosini, and Brent L. Iverson, Department of Chemistry and Biochemistry, University of Texas, Austin, TX 78712.

Expanded porphyrins, such as sapphyrin (e.g. **1**) and turcasarins (e.g. **2**), when protonated, act as anion binding agents *par excellence*, chelating a range of negatively charged species including DNA. This chemistry will be reviewed in this lecture. Also presented will be other novel approaches to anion chelation, especially those based on the use of neutral calix[4]pyrroles (e.g. **3**) and open-chain "bipods" (e.g. **4**).

059. NEW LIGAND SYSTEMS FOR ANION COMPLEXATION
Jerry L. Atwood,[a] K. Travis Holman,[a] and Jonathan W. Steed[b]
[a]Department of Chemistry, University of Missouri-Columbia, Columbia, MO 65211;
[b]Department of Chemistry, King's College London, Strand, London, UK WC2R 2LS

A new class of host molecules for the supramolecular complexation of anionic guest species has been developed. In particular, air- and water-stable organometallic hosts based upon calixarenes and the related macrocycle cyclotriveratrylene (CTV) have been synthesized and characterized. X-ray crystallographic results clearly demonstrate the inclusion of anionic guest species such as BF_4^-, I^-, $CF_3SO_3^-$, ReO_4^-, etc. within the ostensibly electron-rich bowl-shaped cavities of both types of host as a result of cooperative effects arising from the presence of two or more metal centers arranged around a common binding pocket. The new compounds show the ability to discriminate between anions on a size and shape selective basis. Solution complexation studies have revealed binding constants as large as 6,000. New generation hosts with multi-site binding capabilities are expected to exhibit tight binding and high selectivity. Applications of this chemistry to sensor devices and for environmental waste remediation are under investigation.

060. SUBSTITUENT EFFECTS IN THE COMPLEXATION AND EXTRACTION OF SODIUM SALTS OF MONOVALENT ANIONS BY BIBRACHIAL DIBENZO-14-CROWN-4 LARIAT ETHERS.*
Richard A. Sachleben, John H. Burns, Tamara J. Haverlock, Peter V. Bonnesen, Jeffrey C. Bryan, and Bruce A. Moyer; Chemical and Analytical Sciences Division, Oak Ridge National Laboratory, Bldg. 4500S, MS-6119, P.O. Box 2008, Oak Ridge, TN 37831-6119

The structure of coordinating sidearms in the family of bibrachial dibenzo-14-crown-4 (DB14C4) ethers plays an important role in controlling the interaction between a coordinated sodium cation and the counteranion. Investigations of DB14C4 and derivatives bearing functional groups at the 6,13 positions and/or *tert*-octyl substituents on the benzo groups have been undertaken.

X-ray crystal structure studies revealed examples of inner- and outer-sphere cation-anion interactions. In the case of *syn*-6,13-DB14C4-diol, hydrogen bonding interactions between the alcohol functionality and the anion are important. In solution studies, the extractability of the sodium salts of seven monovalent anions by lipophilic DB14C4 derivatives in 1,2-dichloroethane and 2-octanone is strongly influenced by sidearm structure, both in terms of the overall extraction efficiency and the anion selectivity. The anion selectivity in particular could be rationalized in terms of the interactions observed in the crystal structures.

*This research was sponsored by the Division of Chemical Sciences, Office of Basic Energy Sciences, U. S. Department of Energy, under contract number DE-AC05-96OR22464 with Oak Ridge National Laboratory, managed by Lockheed Martin Energy Research Corp.

061. MOLECULAR RECOGNITION OF ANIONIC AND NEUTRAL SUBSTRATES BY SILICA BOUND CALIX[4]PYRROLE. Jonathan L. Sessler, John W. Genge, Philip A. Gale. Department of Chemistry and Biochemistry, University of Texas, Austin, Texas, 78712.

Recently, it has been shown that the tetrapyrrolic macrocycles, calix[4]pyrroles, are capable of binding anions and neutral substrates in solution with varying degrees of affinity. Here, we report that calix[4]pyrrole derivatized silica gel, **1**, when used as an HPLC solid support, effected the separation of various anionic species, nucleotides, N-protected amino acids, neutral aromatic substrates and various oligonucleotides.

1

062. CONTRIBUTIONS OF ARTHUR WESTERBERG TO THE ADVANCEMENT OF PROCESS SIMULATION AND DESIGN, J. D. Seader, Department of Chemical Engineering, University of Utah, Salt Lake City, Utah 84112

Computer-aided process simulation and design began almost 40 years ago, soon after the introduction of large-scale digital computers and FORTRAN, a compiled programming language. For almost 30 years, Arthur Westerberg has been a major contributor to its theoretical development. His research, which is chronicled here in the context of the historical development of the subject, has addressed all of the major problems involved in developing fast and reliable computer-aided methods, including nonlinear algebraic and differential equation solving, flowsheeting architecture, modeling environment, and expert systems.

063. A SYSTEMATIC MODELING AND FORMULATION FRAMEWORK FOR PROCESS SYNTHESIS, Ignacio E. Grossmann and Hector Yeomans, Department of Chemical Engineering, Carnegie Mellon University, Pittsburgh, Pennsylvania 15213

Two major steps in the mathematical programming aproach to process synthesis are the development of a process superstructure and the formulation of a discrete/continuous optimization problem. It is shown in this paper that a systematic framework for these two steps can be developed using state-task-equipment networks for the formulations. It is shown that existing optimization synthesis models can be systematically derived with this approach.

064. ART WESTERBERG: UNDER PRESSURE. M. F. Doherty, Department of Chemical Engineering, University of Massachusetts, Amherst, Massachusetts 01003

Synthesis of nonideal distillation systems plays a major role in industrial processes. This talk highlights recent advances in these synthesis procedures, particularly for azeotropic and reactive distillation. Also, featured is the pioneering work of Andrecovich and Westerberg in developing the representations and insights to these synthesis strategies. Several design variable interactions are considered, especially the role of pressure in the synthesis of heat integrated distillation columns.

065. PROCESS SYNTHESIS IN INDUSTRY. Lionel O'Young, Mitsubishi Chemical America, Inc., San Jose, California 95134

Process integration and the synthesis of heat recovery networks has become a staple in the process industries with a long track record of commercial successes. This talk reviews the application of process synthesis methodologies at Mitsubishi. Parallels will also be drawn to the pioneering work of Art Westerberg in the areas of heat and process integration.

066. EXPERIMENTAL AND NUMERICAL STUDY OF E-BEAM EVAPORATION OF TITANIUM. M. A. McClelland, T. C. Meier, D. B. Braun, L. V. Berzins, K. W. Westerberg, T. M. Anklam, Lawrence Livermore National Laboratory, Livermore, California 94550

The e-beam evaporation of titanium alloys is an important step in the formation of metal matrix composites for aircraft components. In a typical process, an e-beam is used to heat and evaporate alloy from the top of a rod which is pushed from below through a water-cooled cylindrical crucible. A pool of liquid metal at the top of the rod extends to the cold crucible wall where it forms an interface with complex thermal - mechanical characteristics. The liquid metal in the pool circulates strongly as a result of buoyancy and capillary forces. A finite element model was developed for this evaporation system which includes the effects of fluid flow and energy transport in the pool and conduction in the solid. The deformation of the liquid-vapor and solid-liquid interfaces are tracked using a mesh which deforms along spines parallel to the rod axis. A comparison is made between the finite element simulations and the measurements.

067. THE CREATION OF COMPUTER-BASED ENVIRONMENTS TO SUPPORT DESIGN, A. W. Westerberg, Department of Chemical Engineering, Carnegie Mellon University, Pittsburgh, Pennsylvania 15213

We examine our work on two computer-based design support environments: ASCEND and n-dim. ASCEND supports the creation, debugging and solving of complex equation-based models. It features three major components: an object-oriented modeling language, an interactive user interface and a suite of solvers. The n-dim system supports collaboration among members of geographically distributed design teams by managing their information and tools in a common environment. We will discuss the goals for these systems, how we created them and how we continue to test and improve them. We discuss several of the technical issues we continue to encounter.

068.
THE TRANSITION FROM NATURE TO ACTINIDE EXTRACTION AGENTS. A BIOMIMETIC APPROACH TO THE DEVELOPMENT OF NEW ACTINIDE (IV) EXTRACTION TECHNOLOGIES
Kenneth N. Raymond, Thomas R. Mohs, Vadim V. Romanovski, Alan C. Veeck, David J. White, Jide Xu, Darleane C. Hoffman, Department of Chemistry, University of California Berkeley, Berkeley, CA 94720; and Lawrence Livermore National Laboratory, Livermore, CA 94550

Using siderophores (microbial iron chelating agents) as models, and the many chemical similarities between Fe(III) and Pu(IV), highly selective actinide (IV) sequestering agents have been prepared. These have included the functional groups found in siderophores (catecholate, hydroxamate, hydroxypyridonate). The selectivity of such agents is being used to develop new separation processes for metal cations. Inparticular, the treatment of nuclear waste at DOE's Hanford, WA site presents many challenging separation problems. One of these is the separation of parts-per-million concentrations of the generally long-lived actinide isotopes from larger volumes of nonradioactive chemical waste. A series of solid-supported chelating resins and liquid/liquid extractants have been synthesized and characterized. The solid supports are based on derivatives of Merrifield's peptide resin, and both types of extractants employ the hydroxypyridinone functionality. The background of this chemistry will be reviewed and a progress report presented.

069.

THE SOLVENT EXTRACTION OF AMERICIUM(III) BY 2,6-BIS[(DIPHENYLPHOSPHINO)-METHYL]PYRIDINE N,P,P' TRIOXIDE FROM NITRIC ACID AND HYDROCHLORIC ACID SOLUTIONS. **Evelyn M. Bond** (1), Udo Engelhardt(1), Timothy P. Deere(1), Brian M. Rapko(1), John R. FitzPatrick(2), and Robert T. Paine(1), (1)University of New Mexico, Albuquerque, NM, 87131; (2)Los Alamos National Laboratory, CST-7, Los Alamos, New Mexico 87545

The liquid/liquid extractions of Am(III) from nitric acid and hydrochloric acid solutions with chloroform solutions of 2,6-bis[(diphenylphosphino)methyl]pyridine N,P,P' trioxide will be described. Americuim(III) extracts well from high concentration nitric acid solutions (D>3000 at 6M nitric acid) and can be back extracted from the organic phase at 0.01M Nitric Acid. Americium(III) exhibits modest extraction from hydrochloric acid solutions (D=2.2 at 5M hydrochloric acid) and can be back extracted from the organic phase at 0.1M hydrochloric acid. The ligand dependency data suggest that two ligand molecules are coordinated to americium in the nitric acid system and three ligand molecules are coordinated to the americium in the hydrochloric acid system.

070.

SYNTHESIS AND EVALUATION OF WATER-SOLUBLE CHELATING POLYMERS FOR THE SELECTIVE REMOVAL OF ACTINIDE METAL IONS. **Thomas W. Robison (1),** Peter C Stark (1), Barbara F. Smith (1), Rowena R. Gibson (1), Hollie K. Jacobs (2), and Amudhu S. Gopalan (2), 1. CST-12, J-569, Los Alamos National Laboratory, Los Alamos, NM 87545, 2. Department of Chemistry and Biochemistry, New Mexico State University, Las Cruces, NM 88003.

A major goal of our research program is to develop polymer supported ion specific ligand systems for the removal of actinides and other hazardous metals from wastewaters. The advantage of water-soluble polymers in metal ion separation processes is that the homogeneity of the system allows for more rapid exchange kinetics than ion exchange or chelating resins. A number of water-soluble chelating polymers have been synthesized by the functionalization of commercially available polyamine precursors with various ligand moieties such as hydroxamates. The ability of these polymers to complex with metal ions to give soluble complexes which can be separated and concentrated by ultrafiltration under different pH conditions have been examined.

071.

EXTRACTION OF F-ELEMENTS BY NOVEL PHOSPHINE OXIDE/PYRIDINE N-OXIDE LIGANDS
Brian M. Rapko Pacific Northwest National Laboratory, Box 999, Battelle Blvd., Richland, WA 99352

Complexation of the tridentate neutral ligand 2,6-Bis(Diphenylphosphine oxide)methyl pyridine N-oxide (1) has been shown previously to displace anionic ligands such as nitrate in the solid state. This remarkable feature prompted an investigation into the extraction chemistry of f-elements with this ligand. Extraction of chloroform solutions of 1 with Eu and Am as a function of ligand concentration, metal concentration and nitric acid concentration have been performed. The results of these and other extraction measurements will be presented and discussed, together with available extraction data concerning this ligand with other actinides as time permits.

072.

DICARBOLLIDE COMPLEXES OF THE ACTINIDES. **Kent D. Abney (1)**, Francis M. de Rege (1), Daniel Rabinovich (2), Rebecca M. Chamberlin (1), Brian L. Scott (1), and Wayne H. Smith (1). (1) Los Alamos National Laboratory, Los Alamos, NM 87545; (2) Department of Chemistry, University of North Carolina at Charlotte, Charlotte, NC 28223.

The development of alternative low temperature processes to prepare transuranium element storage forms has the potential to dramatically alter the current DOE nuclear materials management strategies. To this end, we report here the ligand design, synthesis, and characterization of actinide dicarbollide complexes and their relevence to advanced nuclear material storage forms. We have to date prepared several uranium and thorium complexes in the (IV) oxidation state and several uranium (III) complexes. Both mono and di-substituted actinide dicarbollide complexes have been prepared. Metathesis and transmetalation reactions of these complexes have been investigated as well as their potential for making ceramic precursors.

073.

DEVELOPMENT OF ANION-EXCHANGE RESINS FOR SEPARATIONS OF ACTINIDES. **M.E. Barr**, G.D. Jarvinen, S.F.Marsh, R.A. Bartsch, Nuclear Materials Technology Division, Los Alamos National Laboratory, Los Alamos, NM 87545, and the Department of Chemistry and Biochemistry, Texas Tech University, Lubbock, TX 79409

The evolution of these new ion-exchange resins hinges upon an understanding of the chemistry of actinide metal complexes as they exist in solution and when interacting with the cationic sites. Molecular modeling of electrosatic interactions between the metal complex and the cationic substrate is used to help visualize uptake mechanisms and,eventually, to predict optimized coordination sites. Thus, these new maerials integrate the fields of ion-specific chelation and ion-exchange technology. Synthesized resins contain cationic sites designed to facilitate the uptake of plutonium nitrato complexes from high-nitratesolutions,and they exhibit distribution coefficients that are up to 10 times higher than those observed for commercial resins. Additionally, some new resins show surprisingly large Kd's for anionic complexes of Am(III) and U(VI).

074.

FORMULATION AND BIOPHYSICAL CHARACTERIZATION OF RECOMBINANT RAT AND MOUSE GROWTH HORMONES. Gautam Sanyal, Dorothy Marquis-Omer, Sarah K.Watson, Christine P. Chan, Leonard J. Rubenstein, Srinivasa Prahalada and David B.Volkin. Merck Research Laboratories, West Point, PA 19486

There is intense interest in growth hormone secretagogue as a potential therapeutic agent for various clinical conditions. For comparative purposes, it is desirable to have available growth hormone proteins from different species that could be used as animal models. Recombinant rat and mouse growth hormone (r-rGH and r-mGH) genes were cloned, expressed, and shown to produce proteins similar in primary structure to their pituitary derived counterparts. Both r-rGH and r-mGH contain 190 amino acids, and they differ in sequence by only three uncharged amino acids. However, their physical behaviors in solution, with regard to aggregation, differ significantly. Circular dichroism (CD) and FTIR spectroscopy suggest that the secondary structure contents of r-rGH and r-mGH are very similar to that reported for r-hGH, although the extent of sequence homology between the human and the rodent growth hormones is approximately only 65%. Intramolecular disulfide bonding pattern of r-rGH also appears similar to that of r-hGH. The effects of varying pH, temperature and formulation conditions on protein structure and intermolecular association were examined by (CD), FTIR, fluorescence and dynamic light scattering techniques. Based on these studies, stabilizing buffer and pH conditions were identified, that minimized aggregation, maintained the properly folded structure and *in vivo* bioactivity. Frozen (-70°C) formulations in these solutions manifest excellent long-term stability, as judged by analytical as well as biological assays. Lyophilized formulations were also developed in which the protein maintained proper structure and biological function, although they did not offer any additional advantage over the frozen liquid formulations for further *in vivo* studies.

075.

THE MULTIMERIZATION OF ANP IN THE FREEZE-DRIED STATE, Shiaw-Lin (Billy) Wu, DeMei Leung, Losha Tretyakov, Jie Hu, Andrew Guzzetta, Y. John Wang, Scios Nova, Mountain View, CA.

Auriculin® anaritide (ANP) is a peptide with 25 amino acid residues and one intra-chain disulfide bond. The freeze dried products of ANP in different stability protocols (varied with lyophilization cycles, storage temperatures, and excipients) were examined by SEC, SDS-PAGE, and LC-MS to determine the amount and size of the multimer formations at different times of storage. Under harsher storage conditions and longer storage times, we observed an increase in multimer formation. We observed two kinds of multimers, one reducible and the other non-reducible, as shown by SDS-PAGE and SEC. The mass of the reducible multimers is consistent with disulfide-linked multimers. The non-reducible multimers were mainly dimers, as shown by SDS-PAGE and SEC. MS showed that the mass of the non-reducible dimer was different from the disulfide-link dimer. Klibanov and Langer et al. proposed that the multimerization was initiated by β-elimination to form a free thiolate ion (HS-) and dehydroalanine. We postulate the HS- then catalyzes ANP to form the disulfide linked multimers, and the dehydroalanine forms a non-disulfide linked dimer through reaction with amino groups or the side chain of tyrosine. These disulfide linked multimers were formed first through dimer, then trimer, tetramer, and so on to the large multimers. The multimerization is minimized by the removal of moisture, the addition of bulking agents such as mannitol, and buffers such as acetate, which minimize the HS- formation. The detailed mechanism and the pathway of the multimerization will be discussed.

076.
SECONDARY STRUCTURAL PROTECTION OF FREEZE-DRIED LYSOZYME IS LOST AT HIGH INITIAL TREHALOSE CONCENTRATIONS. S. D. Allison1 B.S. Chang2, T.W. Randolph3, and J.F. Carpenter1. 1School of Pharmacy, University of Colorado, Denver, CO, 2Amgen Inc. Thousand Oaks, CA, and 3Dept. Chem. Engineering, University of Colorado, Boulder.

We used FTIR spectroscopy to monitor secondary structure and H-bond density of lyophilized lysozyme. At sub monolayer water contents, H-bonding with increasing sugar content. H-bonding between sucrose and lysozyme was greater for all mass ratios than the extent of H-bonding between trehalose and lysozyme. Secondary structure improved with increasing trehalose or sucrose up to 5:1 sugar:protein. As the trehalose to lysozyme mass ratio increased beyond 7.5:1, native secondary structure decreased. Lysozyme freeze-dried from 1.2 M trehalose was more damaged than lysozyme freeze-dried without excipients. Trehalose did not crystallize in dry samples at either 5:1 or 20:1 trehalose. Glass transition temperatures Tg' of freeze-concentrated lysozyme-trehalose solutions measured by saturation transfer EPR spectroscopy were higher than Tg' of the same solutions measured by DSC. We conclude that amorphous phase separation of trehalose causes loss of native protein structure.

077. DEVELOPMENT OF A STORAGE STABLE LIQUID FORMULATION FOR A RECOMBINANT HEMOGLOBIN. Bruce A. Kerwin, Douglas G. Looker, Edward Hess, Patricia Revilla-Sharp, Somatogen Inc., Boulder, CO. 80301 and Michael J. Akers, Lilly Research Laboratories, Indianapolis, IN 46285.

The recombinant human hemoglobin, rHb1.1, is a trimeric protein composed of two β-globins and two genetically fused α-globins (di-α globin). These subunits, together with four heme prosthetic groups form a hemoglobin molecule which has improved oxygen delivery properties when compared to donated red blood cells. Storage of hemoglobin in the oxygenated form (oxyHb) may be associated with protein aggregation, cystine formation between Hb's and oxidation of the heme iron from the ferrous (Fe^{2+}) to the ferric (Fe^{3+}, met-Hb) state. Oxidation leads to heme release, hemichrome formation and precipitation. These sequelae can be circumvented by storage under deoxygenated conditions (deoxyHb). Purified rHb1.1 contains low levels of met-rHb1.1 which cannot bind and release oxygen and is less stable than the ferrous form. To minimize the met-rHb1.1 content in the formulation, sodium ascorbate was added. Under deoxygenated conditions, the ascorbate ion reduces the ferric Fe centers to the ferrous state. It was found that thorough deoxygenation of the rHb1.1 solution prior to addition of ascorbate was required. In the presence of low levels of residual oxygen, an ascorbate/oxygen mediated reaction results in modification of the beta globin primary structure. It was also determined that formulating the deoxyHb at pH 8 induced aggregation, eventually leading to precipitation. In contrast, at pH 7 the protein remained stable for greater than 5 months of storage at 4°C. Together, these data led us to a formulation in which the met-rHb1.1 levels remained at <2% for up to 1 year at 4°C with no detectable precipitation of the protein.

078.
DEVELOPPMENT OF A SMALL SCALE MODEL FOR STUDYING STABILITY OF LARGE VOLUME PROTEIN SOLUTIONS DURING FREEZING. Alexandros Fotopoulos Peter F. Levy, Process Development, Biogen Inc., 14 Cambridge Center, Cambridge, MA 02142

During the manufacture of biological products, process intermediates and bulk materials containing proteins are subjected to large scale freeze-thaw cycles for either storage or handling purposes. The effect of the freezing process on the product quality and product recovery is usually assessed during the development phase using small quantities of protein solutions without considering scale-up factors of the freezing process. In this presentation, two case studies will be discussed, in which heat removal properties and heat transfer information for production freezers were used to develop a scale-down model for quick and effective assessment of the large scale freeze-thaw cycles on protein stability. In this model, thermal profiles are reproduced at a small scale using a controlled rate freezer. This technique allows the investigation of precipitate and soluble aggregate formation of proteins during freezing. The controlled rate freezer was used to examine the effects of freezing rate, final temperature, and defrost cycle characteristics on the product stability. Comparable results between the bench scale experiments and larger scale freeze-thaw experiments were obtained. Data were predictive of results obtained during full scale manufacturing.

079. **Selective Sensors for Metal Ions Based on Discriminating Ion-Macrocycle Interactions: The 5-Chloro-8-hydroxyquinoline-Substituted Azacrown Ethers.** J.S. Bradshaw, P.B. Savage, X.X. Zhang, A.V. Bordunov and R.M. Izatt, Department of Chemistry and Biochemistry, Brigham Young University, Provo, UT 84602

Diaza-18-crown-6 (DA18C6) containing two 5-chloro-8-hydroxyquinoline (CHQ) side arms attached through CHQ positions 7 (Ligand **1**) and 2 (Ligand **2**) were prepared in good yields from DA18C6. Attachment of CHQ to DA18C6 increases the metal ion selectivity of the macrocycle, and provides for a measurable response to ion binding. For example, ligand **1** exhibits selectivity for Ni(II) over Cu(II) while DA18C6 is selective for Cu(II) over Ni(II) in methanol. Ligand **2** displays strong complexes with K^+ and Ba^{2+} but not Mg^{2+}. Ligand **1** fluoresces strongly in the presence of Zn^{2+} but not with Na^+ or K^+ allowing the measurement of Zn^{2+} concentrations in dilute solutions. The ligand **1**-Mg^{2+} complex has a U.V. absorption band that is 6 times greater than that of Mg^{2+}-CHQ allowing the measurement of Mg^{2+} concentrations in very dilute solutions. The synthesis and ion selectivities of these and other sensor molecules will be discussed.

080. METAL ION SEPARATIONS BY PROTON-IONIZABLE LARIAT ETHERS. Richard A. Bartsch, Sheryl N. Ivy, Vincent J. Huber, Jianping Lu, and Wladyslaw Walkowiak, Department of Chemistry and Biochemistry, Texas Tech University, Lubbock, TX 79409-1061.

Attachment of functional side arms which contain acidic groups to crown ethers gives proton-ionizable lariat ethers. Presence of the acidic group markedly enhances the efficiency of metal ion extraction since an electroneutral complex may be formed without concomitant transfer of an aqueous phase anion. When oriented over the polyether cavity, the side arm may also increase the selectivity of metal ion extraction by preorganization of the binding site. Results for competitive solvent extractions of alkali metal cations by lariat ether carboxylic acids and phosphonic acid monoalkyl esters will be presented. In addition, a new class of proton-ionizable lariat ethers will be introduced. In N-(R')sulfonyl lariat ether carboxamides, the acidity of the proton-ionizable group may be varied by changing the R' group. Certain N-(R')sulfonyl lariat ether carboxyamides are found to exhibit extremely high selectivities in competitive solvent extraction of alkali metal cations.

081.
ACID TREATMENT OF HYDROXYALKYLBENZO CROWN ETHERS. TOWARD IMPROVED EXTRACTANTS FOR CESIUM FROM ACIDIC NITRATE MEDIA. **Richard E. Barrans Jr. (1),** Mark L. Dietz (1), Mark P. Jensen (1), Susan Rhoads (1), David J. Rausch (2), E. Philip Horwitz (1) 1. Chemistry Division, Argonne National Laboratory, 9700 South Cass Ave., Argonne, IL 60439 2. Department of Chemistry, Benedictine University, 5700 College Rd., Lisle, IL 60532.

Aqueous acid treatment of a hydroxyalkyl-substituted dibenzo-18-crown-6 dramatically enhances its ability to extract cesium from 4M nitric acid into tributyl phosphate. This transformation was characterized by MALDI-TOF mass spectrometry. The product's structure was studied by independent syntheses of isomers and analogues, and their metal extraction properties and stabilities were evaluated. Work supported by the U. S. Department of Energy, Office of Basic Energy Sciences, Division of Chemical Sciences, under contract W-31-109-Eng-38.

082. CHEMICAL MODIFICATION OF CELL WALL COMPOSITION OF ALGAE AND APPLICATIONS FOR THE SEPARATION OF HEAVY METALS, Young Je Yoo, Yong Hwan Kim, Jae Yeon Park, and Kwang Myung Cho, Department of Chemical Engineering, Seoul National University, Seol 151, Korea

Novel biosorption technology for the separation of heavy metals was developed through chemical modification of the cell wall composition of algae. Marine brown algae, Undaria pinnatifidia, showed the sorption capacity of 450 mg lead/g of biomass. To elevate the sorption capacity of biomass, various functional groups such as xanthate, phosphate and oxime were introduced to the cell wall. The sorption capacity reached about 1000 mg lead/g biomass. Furthermore, this modified biomass could adsorb heavy metals without the interference of alkaline metal ions such as calcium and magnesium. Instrument analysis using EXD and XRD confirmed that the ions were adsorbed on the surface of the cell wall. Adsorbed lead could be recovered using EDTA with high efficiency and this biosorbent could be used repeatedly and economically. Kinetic analysis of the biosorption phenomena and applications of immobilized biosorbent for heavy metal separation in a packed-bed reactor will be discussed.

083. **Complexation of Metal Ions by Azacrown Ethers Bearing an 8-Hydroxyquinoline Sidearm.** Xian Xin Zhang, Jerald S. Bradshaw, Andrei V. Bordunov, and Reed M. Izatt, Department of Chemistry and Biochemistry, Brigham Young University, Provo, UT 84602

Thermodynamic quantities (log K, ΔH, and $T\Delta S$) for the interactions of six azacrown ethers containing 8-hydroxyquinoline (CHQ) sidearms with Na^+, K^+, Ba^{2+}, and Cu^{2+} were determined by a calorimetric titration in methanol at 25 °C. The results show that these ligands form stable complexes with the cations studied. Azapyridino-18-crown-6 (AP18C6) (**1**) and aza-18-crown-6 (A18C6) (**3**) which have CHQ attached through its position 7 (next to the OH group) show high selectivity for Cu^{2+} (log K values 8.12 and 9.44, respectively) over Na^+, K^+, and Ba^{2+} by more than four orders of magnitude. On the other hand, these same ligands (**2** and **4**) which have CHQ attached through its position 2 (next to the quinoline nitrogen group) form more stable complexes with Na^+, K^+, and Ba^{2+} but less stable complexes with Cu^{2+} than do ligands **1** and **3**. All ligands interact more strongly with K^+ than with Na^+. The K^+/Na^+ selectivity by ligand **4** is equal to 1.5 log K units. All complexation reactions display negative enthalpy changes. In most cases the entropy changes are also negative, indicating that formation of the complexes is enthalpy driven.

084.

THE DESIGN OF MONOVALENT CATION SELECTIVE LIGANDS. **R.C. Gatrone**, R. Kutz, J. Stevens, and M. Shahda, Department of Chemistry, Wilkes University, Wilkes-Barre, PA 18766

A new class of ligands which have selectivity for potassium over calcium and strontium have been prepared. Computer assisted modelling was used to design the molecular parameters necessary.

085. SYNTHESIS OF A NOVEL ION EXCHANGE RESIN WITH DIPHOSPHONIC ACID LIGANDS

Spiro D. Alexandratos[+], E. Philip Horwitz[‡], Andrzej W. Trochimczuk[+], Ralph C. Gatrone[+], Darrell W. Crick[+], and Renato Chiarizia[‡]. [+]University of Tennessee, Department of Chemistry, Knoxville, TN 37996. [‡]Argonne National Laboratory, Chemistry Division, 9700 S. Cass Avenue, Argonne, IL 60439.

The polymerization of the tetraalkyl ester of vinylidene diphosphonic acid has led to the preparation of a novel ion exchange resin which displays, among other properties, the ability to complex actinide ions from highly acidic solutions. As a 1,1-disubstituted monomer, tetraethyl vinylidene diphosphonate does not polymerize easily. The development of the suspension polymerization technique which was later scaled up at Eichrom Industries will be detailed.

086. EICHROM'S DIPHONIX® RESIN: PRODUCTION-SCALE APPLICATIONS IN RADIOACTIVE WASTE TREATMENT AND IRON CONTROL IN COPPER ELECTROWINNING. Michael J. Gula,[1] Frank Chang,[1] David B. Dreisinger,[2] and E. Philip Horwitz.[3] [1]Eichrom Industries, Inc., 8205 S. Cass Ave., Darien, IL 60561; [2]Department of Metals and Materials Engineering, The University of British Columbia, Vancouver, B.C., Canada V6T 1Z4; [3]Chemistry Division, Argonne National Laboratory, Argonne, IL 60439

Eichrom's Diphonix® resin has been phased through synthetic scale-up, pilot testing, and production installation in radioactive waste treatment and hydrometallurgical applications. The geminal diphosphonic acid groups of Diphonix resin allow selective retention by cation-exchange and/or chelation. The resin is effective at low pH where sulfonic and carboxylic acid resins are ineffective.

Diphonix resin has been used in nuclear facilities to reduce actinide concentrations in radioactive waste effluents and to reduce waste volumes. The high retention of iron(III) by Diphonix resin in acidic sulfate media has led to an installation capable of removing one ton of iron per day from a copper electrowinning stream. This iron control process diminishes cobalt losses in the electrowinning circuit and significantly reduces operating costs. The authors will discuss the development of these Diphonix resin applications.

087. APPLICATIONS OF ORGANIC NAME REACTIONS IN INDUSTRY.
O. S. Fruchey, Hoechst Celanese, Corpus Christi, Texas 78469

Hoechst Celanese's entrance into over the counter Pharmaceuticals started with acetaminophen and ibuprofen. The unique application of several organic name reactions lead to the development of new processes to produce these over the counter analgesics with exceptional atom economies. Commercialization of these processes at Hoechst Celanese's Bishop plant has made this small Texas town the analgesic capital of the world. The talk will show how industry takes reactions originally developed at universities and applies them to the production of chemicals.

088. MAGNETIC PARTICLE SEPARATION PROCESS FOR HAZARDOUS AND RADIONUCLIDE ELEMENTS, L. Nuñez, Argonne National Laboratory, 9700 S. Cass Ave., Argonne, IL 60439 and M. Pourfarzaneh, CORTEX-BIOCHEM, Inc., 1933 Davis St., STE 321, San Leandro, CA 94577.

The magnetically assisted chemical separation program was initially funded by DOE EM-50 to develop processes for the efficient separation of radionuclides and other hazardous metals. This process has stimulated the partnership between industry and ANL for many applications related to hazardous metal problems in industry. In-tank or near-tank hazardous metals separation using magnetic particles that have selective coating is a new approach to the problems of metal removal and recycling [of industrial (e.g., mining, printing circuit board, plating)] corrosive waste streams. This concept of coated magnetic particles promises simple, compact processing at very low costs and employs mature chemical separations technologies to remove and recover hazardous metals from aqueous solutions. The selective chemical extractants are attached to inexpensive magnetic carrier particles. Surfaces of small particles composed of rare earths or ferromagnetic materials are treated to retain chemical extractants (e.g., TBP, CMPO, quaternary amines, carboxylic acid). After selective partitioning of contaminants to the surface layer, magnets are used to collect the loaded particles from the tank. The particles can be regenerated by stripping the contaminants and the selective metals can be recovered and recycled from the strip solution. This process and its related equipment are simple enough to be used for recovery/recycling and waste minimization activities at many industrial sites. Both the development of the process for hazardous and radioactive waste and the transfer of the technology will be discussed. This work supported by the U.S. Department of Energy, Office of Environmental Management, under contract W-31-109-ENG-38.

089. APPLICATIONS OF MAGNETIC PARTICLES IN INDUSTRY, M. Pourfarzaneh, Cortex-Biochem, Inc., 1933 Davis Street, STE 321, San Leandro, CA 94577

Magnetic particles applications to solve current industrial needs will be discussed. Furthermore, magnetic particle processes for hazardous and radioactive waste problems will be addressed.

090. DISCONTINUOUS INNOVATION: HOW IT REALLY HAPPENS. P. M. Norling, R. J. Statz, DuPont, Experimental Station, P. O. Box 80328, Wilmington, DE 19880-0328

Many books and articles have been written about the "best practices" in making major innovations happen and how organizations introduce substantially new products. The "actual practices" may be quite different. We examine the introduction of "Surlyn" ionomer resins and the efforts required to overcome several deaths of the product. Here was a case where financial estimates said this was not the product to pursue; here was the case where there were several important lessons: (1) we saw unique properties and technology in search of markets -- the gradual switch from technology driven research to market driven research, (2) the properties that DuPont felt were important to the marketplace were quite different from the properties that the customer came to value. This learning was important and difficult, and (3) it took much dedication, commitment, patience, and courage on the part of many unique individuals for "Surlyn" to survive and become the commercial success that it is today.

091. THE EXTERNAL PERSPECTIVE IN TECHNOLOGY MANAGEMENT. D. L. Ransley, 3125 Withers Avenue, Lafayette, CA 94549

Today's technology managers are too busy sawing to sharpen the saw, to paraphrase Covey. It is my thesis that the external perspective does not get appropriate attention in that scenario. I tend to think in terms of a hierarchy of tiers, as follows: Tier I-The external customer - identifying unarticulated needs, Tier II - External technology watching - ongoing, long-term awareness of factors that can impact the business positively or negatively, Tier III - External collaborations - ranging from contributions to universities to joint ventures, Tier IV - Personal networking - as a planned process.

This presentation provides some high level thoughts about the success factors for each of these tiers. The conclusion is reached is that good person-to-person interactions, trust and respect are critical and are common to each tier. These are attributes exhibited by our honoree.

092. LINKING BUSINESS AND TECHNOLOGY STRATEGIES FOR GROWTH. David A. Duke, P. O. Box 3507, Park City, Utah 84060

The management of Science and Technology for business growth requires a tight linkage of strategic directions. Every business needs a strategy which includes answers to such key questions as, "What business are we in? What is changing? What do we want our position to be? and How do we get there?" The technical community needs to define and develop core capabilities that enable the business success.

The development of common business/technology strategies, use of portfolio management, innovation plans, technology roadmaps, and other processes will be discussed. A number of examples from Corning Incorporated's experience will be used. These include fields ranging from Opto Electronics to technical ceramics and cover R/D&E product and process development, and manufacturing.

093. RESEARCH - SWING FOR THE FENCES. T. C. MacAvoy, Darden School of Business, University of Virginia, Charlottesville, VA 22906

Most successful technoloyg-based companies were founded on breakthrough ideas or inventions. In their maturity, many such companies chose to follow less risky research strategies. Those companies that continue to be leaders, however, pursue a research portfolio rich in high risk/high gain "home run" projects.

In this context, research means "exploratory" research, aimed at finding new phenomena or at solving very difficult but important problems. Such research projects have highly uncertain outcomes. When considered from the viewpoint of option theory, however, they can be seen as having much greater long range potential than less risky projects whose outcome is more predictable. When such high risk/ high gain projects are part of a diversified portfolio of such projects, indeed the entire portfolio is of greater potential value than a supposedly "safe" portfolio. To assure that a "home run" research portfolio is truly successful, it is valuable to follow five conditions.

094. CPP, A ROUTE TO NEW BUSINESS DEVELOPMENT, T. T. Wojcik, Hoechst Celanese Corp., P. O. Box 32414, Charlotte, NC 28232

Customer-Pull Partnership (CPP) is a new approach for finding business opportunities by combining the complementary competencies of two different companies; one with strengths in chemical technology and materials science and the other with strength in product development, marketing, and new application development. The premise of CPP is that companies need both good idea sources and a disciplined system for developing the ideas into viable business concepts and ultimately new businesses. Successful management of the partnership increases both the quality of the ideas and it keeps both partners focused on the milestones and strategic imperatives of the objective. This presentation will provide insights about how to set up a viable CPP and how to manage all of the important components such as: defining the fields of interests, selecting a lead project, developing and specifying the dollar value of a portfolio of concepts, and managing the cross-company teams.

095. GLOBAL RESEARCH AND GROWTH. Douglas, Frank L. Global Research, Hoechst Atkiengesellschaft, Building K 607, HNR Research, D-65926, Frankfurt, Germany.

ABSTRACT NOT AVAILABLE
AT THIS TIME

096. MOLECULAR IMPRINTING IN POLYMERS - NEW OPPORTUNITIES IN SEPARATION AND CATALYSIS. G. Wulff, Institut für Organische Chemie and Makromoleculare Chemie, Heinrich Heine Universität Düsseldorf, 40225 Düsseldorf, Germany

A crosslinked polymer is prepared in the presence of interacting monomers around a molecule that acts as a template. After removal of the template, an imprint of specific shape which contains functional groups capable of chemical interactions remains in the polymer. A survey of new developments in our institute regarding this procedure is presented. We have recently developed new types of binding-site interactions which are noncovalent, stoichiometric (due to high binding constants), and do not show the disadvantages of other types of noncovalent binding. The polymeric carrier has also been varied considerably. New soluble systems are of particular interest in this respect. Furthermore, new catalytic systems have been designed which show high esterolytic activity and Michaelis-Menten Kinetics.

097. IMPORTANT CONSIDERATIONS IN THE „DESIGN" OF RECEPTOR SITES USING NONCOVALENT IMPRINTING, Börje Sellergren, Department of Inorganic and Analytical Chemistry, Johannes Gutenberg University Mainz, J.-J. Becherweg 24, D-55099 Mainz, Germany

Receptor sites capable of distinguishing between molecules having minor structural differences can be prepared by noncovalent imprinting of templates in network polymers. Initially, these polymers are prepared considering functional group complementarity between a functional monomer and the template.

The performance of these „first generation" materials is often unacceptable and a careful optimization of the variables in the imprinting process is therefore needed in order to reach the desired level of affinity and selectivity for the target compound. Depending on the separation requirements, the imprinted materials are further associated with problems, i.e. non-linear adsorption isotherms, slow mass transfer kinetics, low sample load capacities, that may need to be solved. In the above context, this talk will address options available at each stage in the process of preparing and evaluating an imprinted polymer. Recent examples of how affinity may be enhanced by applying pressure during polymerization and how the hydrophobic effect may contribute, either specifically or nonspecifically, to the observed binding will be given.

098. MOLECULAR RECOGNITION OVER FOOTPRINT CAVITIES. Kensaku Morihara, Department of Chemistry, Faculty of Science, Nara Women's University, Nara 630, Japan.

Molecularly imprinted sites on a silica gel (alumina) surface, "footprints", are complementary cavities with Lewis acid sites at their bottoms. These sites exhibit catalytic behavior with tailored specificities, which distinguish footprints from the imprinted adsorption sites in vinyl polymers. Our studies of enantioselective catalysis over chirally imprinted cavities reveal that there are two enantioselective mechanisms. The major mechanism allows a chiral molecule to bind in two postures, which produces low enantioselectivities. Another mechanism is observed for binding over the exclusive cavities that accept only one binding posture of a chiral molecule, which brings highly enantioselective catalysis. Our investigation of these binding postures provides useful information on how to design templates for cavities to provide fine molecular recognition.

099. DESIGN OF IMMOBILIZED METAL SITES IN POROUS ORGANIC HOSTS: EVIDENCE FOR SITE ISOLATION. A. S. Borovik,* John F. Krebs, Anjal Sharma. Department of Chemistry, The University of Kansas, Lawrence, KS 66045-0046.

Copolymerization of molecular assemblies into organic hosts is an effective way to fabricate new materials having desirable properties. The assembly of the molecular species prior to polymerization is advantageous because of the greater control of the structure and amount of species incorporated into the polymer and the possibility of regulating their microenvironments. We are using copolymerization techniques to design network polymers that can stabilize metal-small molecule adducts that can not be isolated at room temperature in solution. As part of the stabilizing process, the polymer matrix needs to minimize the intermolecular interactions between immobilized complexes in order to prevent undesirable chemistry. In this talk we discuss our findings on the monodispersal of metal complexes in network polymers and their ability to stabilize carbon monoxide, dioxygen and nitric oxide adducts.

100. SELECTIVE ADSORPTION OF METAL IONS TO SURFACE-TEMPLATED RESINS PREPARED BY EMULSION POLYMERIZATION USING A FUNCTIONAL SURFACTANT. Yoshifumi Koide, Department of Applied Chemistry, Kumamoto University, Kumamoto 860, Japan.

Monomer-type surfactants which can function as a ligand, 10-(p-vinylphenyl)decanoic acid (**Rac**) and 2-(p-vinylbenzylamino)alkanoic acid (R_nNAc), have been used as emulsifiers for the preparation of surface-templated resins. The surfactants adsorb at the oil-water interface and emulsify divinylbenzene-styrene monomers (DVB-St) (γ_{cmc} was 3 mN m^{-1} for $R_{18}NAc$ of the highest emulsion) and the resulting resins were evaluated. The emulsion polymerization using a $K_2S_2O_8$ initiator (70°C) or by irradiation with γ-rays gave fine particles which were 200-300 nm in diameter. The metal-imprinted resins prepared with **Rac** were 1.8 times more effective than the unimprinted resins, and Zn^{2+}-imprinted resins showed highly effective adsorption of Zn^{2+}. Such surface-template effects were also seen for metal-imprinted resins prepared with R_nNAc, but the effect was sensitive to the alkyl chain length. The $R_{18}NAc$ resin was the most effective. The Cu^{2+}/Zn^{2+} ratio in competitive sorption was 3.7 for the Cu^{2+}-imprinted resins prepared with **Rac** and 4.2 by $R_{18}NAc$. The metal-imprinted resins can be used repeatedly.

101. METAL ION-IMPRINTED POLYMERS PREPARED BY SURFACE TEMPLATE POLYMERIZATION WITH W/O EMULSIONS. Kazuya Uezu, Masahiro Goto and Fumijuki Nakashio. Department of Chemical Science and Technology, Kyushu University, Hakozake, Fukuoka 812-81, Japan

Metal ion-imprinted polymers may be prepared by surface template polymerization, which is emulsion polymerization utilizing a functional monomer, an emulsion stabilizer, a polymer matrix-forming comonomer, and a print molecule. Such surface-templated polymers offer a high adsorption rate for metal ions as well as high selectivity. For this technique, the interfacial activity of the functional monomer is vital for the selection of a suitable metal ion-binding amphiphile. It is also an important factor for the firm attachment of the functional monomer onto the polymer matrix. We have succeeded in preparing highly selective ion-imprinted polymers by two approaches: 1) design of functional monomers, and 2) post-irradiation with γ-rays to make the polymer matrix more rigid. Adsorption behavior of several metals by the newly synthesized polymers has been investigated. The imprinting effect on the adsorption ability and selectivity will be discussed.

102. **Crown Ethers: The Search for Selective Cation Ligating Agents.** Reed M. Izatt and Jerald S. Bradshaw, Department of Chemistry and Biochemistry, Brigham Young University, Provo, UT 84602

Using a combined synthesis and reaction thermodynamics approach, the authors have searched for crown ether ligands that selectively complex specific metal ions. Many interesting examples will be reported. For example, 1,10-Dithia-18-crown-6 has a high selectively for Hg^{2+} over Ag^{2+} while changing one sulfur atom for a pyridone nitrogen reverses the selectivity to Ag^+ over Hg^{2+}. In another case, 1,10-Diaza-18-crown-6 (DA18C6) with two 5-chloro-8-hydroxyquinoline (CHQ) groups attached through their positions 7, has a high affinity for Mg^{2+}, Cu^{2+}, and Ni^{2+} while DA18C6 with two CHQs attached through their positions 2 has high affinities for K^+ and Ba^{2+} but does not complex with Mg^{2+}. Attaching appropriate ligands to solid supports allows the removal, separation and recovery of specific metal ions. Examples of the use of supported ligand systems for separations of commercial interest will be presented and discussed.

103. **From Academic Lab to Commercialization: The IBC Experience.** Steven R. Izatt, Kryzysztof E. Krakowiak, Ronald L. Bruening, IBC Advanced Technologies, Inc., American Fork, Utah, 84003

IBC Advanced Technologies, Inc. has successfully commercialized a broad range of products based on Molecular Recognition Technology (MRT). A unique commercialization model based on a series of customer-focused activities has been employed. IBC's experience of combining research from academic labs with the skills of business partners and key end-users to develop high value-added products will be discussed.

104. ABEC™ RESINS: FROM AQUEOUS BIPHASIC NOVELTIES TO SELECTIVE AQUEOUS BIPHASIC EXTRACTION CHROMATOGRAPHIC RESINS FOR METAL IONS. **Robin D. Rogers** and Scott G. Griffin, Department of Chemistry, The University of Alabama, Tuscaloosa, AL 35487; E. Philip Horwitz, Chemistry Division, Argonne National Laboratory, Argonne, IL, 60439; Andrew H. Bond, Michael J. Gula, and Frank Chang, Eichrom Industries, Inc., 8205 S. Cass Ave., Suite 107, Darien, IL 60561.

We have recently been allowed patents on the use of polyethylene glycol-grafted resins for chaotropic metal ion separations based on the concepts of aqueous biphasic extraction. The concept and realization of the technology was developed by collaborative research between university and national laboratories. Commercialization of the technology was accomplished by licensing the technology to a specialty separations company itself started based on the foundation of technology transfer. This tripartite interaction was facilitated by the joint interest and membership of all involved parties in the Industrial & Engineering Chemistry Division.

This presentation will highlight the research which led to the adaptation of aqueous biphasic partitioning to ABEC™ resins for metal ion separations. The role of the I&EC Division and its Separation Science & Technology Subdivision in facilitating such technology transfer will also be discussed.

105. EICHROM'S ABEC™ RESINS: ALKALINE RADIOACTIVE WASTE TREATMENT, RADIOPHARMACEUTICAL, AND POTENTIAL HYDROMETALLURGICAL APPLICATIONS. Andrew H. Bond,[1] Michael J. Gula,[1] Frank Chang,[1] Robin D. Rogers,[2] and E. Philip Horwitz.[3] [1]Eichrom Industries, Inc., 8205 S. Cass Ave., Darien, IL 60561; [2]Department of Chemistry, The University of Alabama, Tuscaloosa, AL 35487; [3]Chemistry Division, Argonne National Laboratory, Argonne, IL 60439

Eichrom's ABEC™ resins selectively extract certain anions from high ionic strength acidic, neutral, or strongly alkaline media, and solute stripping can be accomplished by eluting with water. ABEC resins are stable to pH extremes and radiolysis and operate in high ionic strength and/or alkaline solutions where anion-exchange is often ineffective.

Potential applications of the ABEC materials include heavy metal and ReO_4^- separations in hydrometallurgy and purification of perrhenate, iodide, and iodate in radiopharmaceutical production. Separation of $^{99m}TcO_4^-$ from its $^{99}MoO_4^{2-}$ parent and stripping with water or physiological saline solution have been demonstrated for radiopharmaceutical applications. Removal of $^{99}TcO_4^-$ and $^{129}I^-$ from alkaline tank wastes has also been successfully demonstrated. The authors will discuss the scale-up studies, process-scale testing, and market development of this new extraction material.

106. DEVELOPMENT OF A HIGHLY EFFICIENT TECHNOLOGY FOR SEPARATION OF ORGANIC FROM AQUEOUS SYSTEMS USING FINE MESH SCREENS. Misak A Dzhragatspanyan, William Greene, SpinTek Membrane Systems Inc., Huntington Beach, California

A novel approach to coalescence and separation of organic compounds from aqueous solutions has been conceived using finely woven mesh screens. Benchscale testing has shown fine mesh screens as valid media for coalescing fluids with both organic-in-aqueous or aqueous-in-organic liquids. Separation efficiency was determined for oil-in-water and water-in-oil systems for more than 95% rejection. Flow rates were 2000-3000 gfd (gallons/ft^2/day) using a 375x2400 woven mesh screen. Expanded research includes testing done in two ways: a) in a horizontal matrix tank vs. a vertical matrix tower, and b) in an upflow matrix tower vs. a downflow matrix tower. The mechanism of the coalescing process at the surface of the screen is discussed in detail as well as the ability to reverse a dead end flow.

107.

RECOVERY OF POLAR ORGANICS FROM AQUEOUS SOLUTION. C. Judson King, Department of Chemical Engineering, University of California, Berkeley, CA 94720

Hydrophilic organic solutes with volatilities less than, or comparable to, that of water are particularly difficult to remove from aqueous solutions. The problem is made more difficult if the solute has a low tendency to solidify. Substances in this category include carboxylic acids, multi-OH compounds such as glycols and glycerol, and some nitrogen-bearing compounds. Methods for recovering such substances from water are evaluated and classified according to range of applicability. Complexation and reactivity facilitate the separation. In the case of carboxylic acids complexation with amine-bearing extractants or solid sorbents is effective, and substantial insight into equilibria for such separations has been gained recently. Selective complexation is difficult for multi-OH compounds, because of the similarity of hydrogen-bonding characteristics between -OH groups and water. Viable processing approaches should incorporate effective means of regeneration, for which several different methods are worthy of consideration.

108. R&D DRYING NEEDS IN INDUSTRIAL PRACTICE
Larry R. Genskow and Srinivas Achanta, The Proctor and Gamble Company, 4550 Este Avenue, Cincinnati, Ohio 45232

The R&D needs and opportunities in industrial drying are explored, primarily from the standpoint of consumer products. Opportunities in drying will be divided into several key areas -- Innovative Drying Opportunities, Dryer Design and Scale-up Methods, Tools for Product Analysis, Understanding of Secondary Transformation, and Education.

Drying of consumer products is somewhat unique and challenging in that the final product attributes are considerably influenced by the drying process itself. During drying, other important secondary transformations also occur which influence product quality such as shrinkage, biological degradation, loss of flavor and aroma. Maintaining high product quality standards is of the utmost importance and several gaps in the technology need to be addressed. Innovation, design and scale-up, and product analysis will be discussed in this context as well as special education needs.

109. CHEMICAL COMPLEXATION: NEW ADSORBENTS AND NEW APPLICATIONS: R.T. Yang, N.D. Hutson, J. Padin, N. Chen, J.H. Lindner, E. S. Kikkinides, and R. Foldes, Department of Chemical Engineering, University of Michigan, Ann Arbor, MI 48109

Based on principles of chemical complexation, new solid adsorbents are made for the following gas separations: Olefin-paraffin separation, acetylene separation, and air separation. Molecular orbital calculations are performed to obtain an understanding of the adsorption bonds and to guide design of the sorbents. For olefin-paraffin separation, sorbents with highly dispersed Cu^+ or Ag^+ provide high selectivities and capacities for olefins. Similarly, surface Fe^{2+}, Co^{2+} and Ni^{2+} ions bond acetylene selectively over olefins and paraffins. These two classes of sorbents are based on π-complexation. For air separation, we are making two different types of sorbents, N_2-selective or O_2-selective sorbents. Ag^+-containing zeolites are prepared for N_2 complexation. Immobilized (and stable) cyanocobaltates on various substrates are synthesized for O_2 complexation. Adsorption-desorption reversibility, favorable rates, and stability are important considerations and have been achieved in these sorbents.

110. REMOVAL AND RECOVERY OF H_2S FROM HOT COAL GAS USING $CaCO_3$. L.A. Fenouil, M.W. Brooks and S. Lynn, Department of Chemical Engineering, University of California, Berkeley

Millimeter-sized particles of limestone can be used to sorb H_2S from coal gas at temperatures of the order of 1000 °C and pressures of the order of 30 atm. The minimum in the H_2S concentration, which is of the order of 100 to 200 ppm, comes at the calcination temperature of the limestone. However, the kinetics of the sorption is much faster, and limestone utilization is much more complete, at temperatures a few degrees above calcination. In the aqueous regeneration process that has been developed, CaS is dissolved by H_2S solvated with an alkanol amine. $CaCO_3$ is precipitated as 10 μm crystals by mixing the CaS solution with amine-solvated CO_2 and H_2S of 90+ % purity is recovered in a stripper.

111. THE EMERGING TECHNIQUE OF MOLECULAR IMPRINTING AND ITS FUTURE IMPACT ON BIOTECHNOLOGY. K. Mosbach, Pure and Applied Biochemistry, Chemical Center, Lund University, P.O. Box 124, S-221 00 Lund, Sweden.

The technique of molecular imprinting allows the formation of specific recognition and catalytic sites in macromolecules by the use of templates. Molecularly imprinted polymers have been employed in an increasing number of applications where molecular binding events are of interest. These include: (i) the use of molecularly imprinted polymers as tailor-made separation materials; (ii) antibody and receptor binding sites mimics in recognition and assay systems; (iii) enzyme mimics for catalytic applications; and (iv) recognition elements in bio-sensors. The stability and low cost of molecularly imprinted polymers make them advantageous for use in analysis as well as in industrial-scale production and application.

112. TFMAA: A NOVEL FUNCTIONAL MONOMER FOR NON-COVALENT MOLECULAR IMPRINTING. T. Takeuchi and J. Matsui, Faculty of Information Sciences, Hiroshima City University, Hiroshima 731-31, Japan.

Regarding the molecular imprinting principle, the selection of the functional monomer is of great importance to realize molecularly imprinted polymers with high affinity and selectivity. In non-covalent molecular imprinting, methacrylic acid (MAA) has been demonstrated as the most useful functional monomer. Recently, we have introduced a new functional monomer, 2-(trifluoromethyl)acrylic acid (TFMAA), on the rationale that a more acidic functional monomer could be preferable for imprinting basic template molecules. Agrochemicals and drugs with high basicity were examined as test template molecules, and the use of TFMAA resulted in imprinted polymers with high affinity and selectivity. It was found that TFMAA shows "taste" in template molecules that is different from MAA's. Other research activities on molecular imprinting performed in our laboratory will be also mentioned.

113. APPLICATIONS OF MOLECULARLY IMPRINTED POLYMERS TO BIOACTIVE MOLECULES. Scott McNiven, Soo-Hwan Cheong, Raphael Levi, Yohei Yokobayashi, Kazuyoshi Yano and Isao Karube, Research Center for Advanced Science and Technology, The University of Tokyo, 4-6-1 Komaba, Meguro-ku, Tokyo 153, Japan.

We have recently developed an HPLC based sensor for the antibiotic chloramphenicol whereby a dye conjugate is displaced from a molecularly imprinted polymer (MIP) by the substrate. The sensor is immune to interference from chloramphenicol derivatives and is only one third as responsive to thiamphenicol, an almost identical analogue. The dynamic range of the system easily spans the therapeutic range of the drug and further applications of this system will be discussed. Other work concerns the development of an artificial steroid receptor which is selective for testosterone. We have thoroughly investigated the effect of various polymerization conditions on the selectivity of the MIP and will discuss the implications. Furthermore, we report a novel post-polymerization treatment by which the selectivity of this polymer can be further enhanced.

114. DEVELOPMENT OF ARTIFICIAL HOST POLYMERS WHICH MIMICK BIOCHEMICAL INTERACTIONS. Kazuyoshi Yano and Isao Karube, Research Center for Advanced Science and Technology, University of Tokyo, 4-6-1 Komaba, Meguro-ku, Tokyo 153, Japan.

Separations using molecularly imprinted polymers (MIPs) which mimic biological interactions are reported. Multiple hydrogen bonding interactions between nucleotide bases play a crucial role in stabilizing the structure of double-stranded DNA. A novel functional monomer designed to form multiple hydrogen bonds with barbiturate templates has been synthesized and used for molecular imprinting. Liquid chromatographic analysis shows that a cyclobarbital-imprinted polymer possesses a distinct preference for the template molecule. Another example which mimicks the interactions between peptide bonds in α-helical or β-sheet domains of proteins is introduced. A MIP which utilizes a L-valine derivative as a functional monomer is shown to exhibit stereoselectivity for a number of dipeptides.

115.
REMOVAL OF OXYANIONS FROM AQUEOUS SYSTEMS USING POLYMER FILTRATION (WATER-SOLUBLE METAL-BINDING POLYMERS WITH ULTRAFILTRATION).
Yvonne C. Rogers Eric C. Santos, Thomas W. Robison, Rowena R. Gibson, and Barbara F. Smith, CST-12, G740, Los Alamos National Laboratory, Los Alamos, NM 87545

The group V-A and VI-A elements exist in the environment as oxyanions. They enter the environment from both agricultural and industrial sources. The Kestersen reservoir in California has been the recipient of selenium-rich agricultural runoff, making the reservoir hazardous to wildlife. Arsenic and antimony are used in pesticides and as hardeners in alloys, bearings, and in storage batteries. All these metals are emitted during smelting operations. We have been able to remove As(V)/Se(V) from the 100 ppm level to very low levels using water-soluble metal-binding polymers with ultrafiltration. Arsenic contaminated drinking water from Europe was treated and the As level was successfully lowered to better than drinking water standards (<0.1 ppm). We will report on the development and testing of new water-soluble polymers for binding oxyanions.

116.
POLYMER-ASSISTED ULTRAFILTRATION FOR REMOVAL AND CONCENTRATION OF BORIC ACID. **Bryan M. Smith,** Christopher N. Bowman and Paul Todd, Department of Chemical Engineering, University of Colorado, Boulder CO 80309-0424

Polymer-assisted ultrafiltration is an extraction technique whereby boric acid may be separated and concentrated from aqueous systems. The boron-contaminated feed is mixed with a concentrated solution of a polymer containing chelating groups for borate. The borate is complexed by the ligands on the polymer and retained by an ultrafiltration membrane having a suitable molecular weight cut off. Two polymers have been designed, synthesized, characterized, and tested. These polymers have been regenerated, reused, and appear to suffer no degradation of performance through several adsorption/desorption cycles. We have evaluated this separation technique relative to the competing technique for boron removal, adsorption onto a commercial resin. The economics and potential commercial applications were also assessed.

117. LOW-RANK COAL AS A NOVEL MATERIAL FOR REMEDIATING HEAVY METAL AND RADIONUCLIDE WASTE. J.H. Kuhr and J.D. Robertson, Department of Chemistry and Center for Applied Energy Research, University of Kentucky, Lexington, Kentucky, 40506-0055.

The use of low-rank coal as an inexpensive ion exchange material for waste remediation is being investigated. A process utilizing the natural cation exchange capacity, mineral content, and calorific content of the low-rank coal could be used to produce a non-leachable vitrified material suitable for disposal and to provide energy for the remediation process. The kinetic and thermodynamic properties, and effect of pH and charge density on the ion exchange process of low-rank coals has been investigated. The exchange process was found to follow first-order kinetics and to proceed rapidly; equilibrium was reached within 10-20 minutes. The Langmuir adsorption isotherm was found to describe the exchange of several metals and radionuclides. Results on the stability of the vitrified material and the application of this system toward "real-world" samples will also be presented.

118.
APPLICATION OF ELEMENT AND SPECIES SPECIFIC X-RAY ABSORPTION SPECTROSCOPY TO RADIONUCLIDE CHARACTERIZATION P. G. ALLEN D. K. Shuh, J. J. Bucher, and N. M. Edelstein. Chemical Sciences Division, Lawrence Berkeley National Laboratory, Berkeley, CA 94720.

X-ray absorption spectroscopy (XAS) is an element-specific, noninvasive structural technique which can determine the oxidation state and local molecular structure for almost any atom. Because the information derived from XAS is frequently unobtainable by other methods, XAS has become a valuable analytical tool for studying the speciation of radioactive contaminants. In this talk, we discuss investigations on the mechanism of Tc immobilization in cement wasteforms which have shown that blast furnace slag and other sulfide-containing additives effectively reduce Tc(VII) to Tc(IV), a less soluble form. In other studies, the structures of the uranyl ion, and Np(IV) have been studied in HCl solutions and in the presence of Dowex anion exchange resin. Probing the solution species directly can provide valuable information on the molecular structures, speciation distributions, and formation mechanisms in these systems.

119.

A NEW HYBRID INORGANIC SORBENT FOR HEAVY METALS REMOVAL. **Arup K. SenGupta** Yi-min Gao and Arthur Kney, Environmental Engineering Program, Fritz Engineering Laboratory, Lehigh Univerisity, Bethlehem PA 18015

Iron oxyhydroxides, commonly known as ferrihydrites, are known to remove trace amounts of dissolved heavy metals at alkaline pH. However, at pH < 6.0, sorption capacities of ferrihydrites diminish drastically due to competition from hydrogen ions. A new Hybrid Inorganic SORBent

(HISORB) has been synthesized and characterized and its crystalline phases have been identified. HISORB removes dissolved metals in the pH range 3-11 and strongly prefers heavy metals (Zn, Cu, Pb) over competing Ca and Na ions. Its heavy-metal affinity is due to iron oxides and other solid phases. HISORB is regenerable and relatively inexpensive. It possesses most of the desirable properties of polymeric chelating exchangers commonly used for selective removal of heavy metals. HISORB appears to be a commercially attractive heavy-metal-selective sorbent. Sorption mechanisms and regeneration properties of HISORB will be presented.

120. USE OF POLYELECTROLYTE-ENHANCED ULTRAFILTRATION TO REMOVE CHROMATE FROM POLLUTED GROUND WATER. S. Sriratana, J.F. Scamehorn, S. Chavadej, C. Saiwan, K.J. Haller, S.D. Christian and E.E. Tucker. Institute for Applied Surfactant Research, University of Oklahoma, 100 E. Boyd, Rm. T335, Norman, Oklahoma, 73019.

Polyelectrolyte-enhanced ultrafiltration (PEUF) is a process which can be used to remove multivalent ions from water. In PEUF, a polyelectrolyte of opposite charge to the target ion is added to the water to bind the ion to be removed. The solution is then treated using ultrafiltration with membrane pore sizes small enough to reject the polymer and bound ion. In this study, chromate (CrO_4^{2-}) is removed from water using poly(diallyldimethyl ammonium chloride) with an average molecular weight of 240 K as the polyelectrolyte. In the absence of other added electrolytes, chromate rejections of up to 99.8% were observed. The presence of added NaCl reduces the chromate rejection substantially. A study of the flux of the system yielded a gel concentration of 0.55 M cationic polyelectrolyte. This high gel concentration and high rejection means that the ultrafiltration can produce a concentrated, low-volume waste stream, and a purified stream containing chromate at low concentration.

121.
SYNTHETIC INORGANIC ION EXCHANGERS FOR SELECTIVE Cs^+ AND Sr^{2+} IONS REMOVAL FROM CONTAMINATED GROUNDWATER. Anatoly Bortun, Lyudmila Bortun and Abraham Clearfield, Texas A&M University, College Station, TX 77843, USA

A comparative investigation of the ion exchange performance towards Cs^+ and Sr^{2+} ions of more than 20 novel synthetic inorganic adsorbents prepared with the use of a "soft chemistry" approach was carried out. The exchangers studied were amorphous and crystalline (with layered and framework structures) titanium, zirconium and tin phosphates, alkali metal titanium, zirconium and niobium silicates, mixed titanium-manganese oxides, sodium micas, etc. The best performance for Cs^+ removal from groundwater simulant (in mg/L): Ca 100; Mg 10; Na 15; Cs 4.8-5.5, Sr 4.6-4.8, pH=6-9, was exhibited by sodium micas ($K_d > 4 \cdot 10^5$ ml/g; IEC 150 mg Cs/g), hydrous sodium zirconium trisilicate ($K_d > 4 \cdot 10^5$ ml/g; IEC 150 mg Cs/g), layered δ-SnP (K_d $2 \cdot 10^5$ ml/g; IEC 190 mg Cs/g) and two layered and framework sodium titanium silicates (K_d $1-3 \cdot 10^5$ ml/g, IEC 45-60 mg Cs/g). No exchangers with selectivity for Sr^{2+} close to that for Cs^+ ions were found. The best adsorbents tested: sodium niobium silicate, semicrystalline sodium titanium silicates and hydrous titanium manganese dioxide, show a strong dependency on the pH and are able to purify 1,000-5,000 volumes of Sr^{2+} contaminated groundwater (K_d 10,000-50,000 ml/g).

122.
THE EVOLUTION OF CHROMATOGRAPHY. **Edwin N. Lightfoot,** Department of Chemical Engineering, University of Wisconsin, Madison, WI 53706

The evolution of chromatographic and allied separations is followed from their first appearance in the laboratory of Michael Tswett to the industrial scale operations of the modern biotechnology industry. The purpose is to identify effective means of equipment design, and chromatography proves an attractive model, because it is both a compact subject and well documented. In retrospect much early chemical engineering effort, largely devoted to detailed analysis of idealized models, had little impact. This was primarily because the engineers, with very few exceptions, had little contact with chromatographic practice and were not even aware of the real problems being faced in the field. Recent work, which is beginning to have impact, is based on realistic approximations and close association with equipment manufacturers and users. Simple but reliable design procedures and a wide variety of innovative solutions are now appearing, and these will be discussed in terms of underlying thermodynamic and transport principles.

123. TAILORED SOLVENTS FOR MINIMIZING POLLUTION IN REACTION AND SEPARATION PROCESSES. Linda K. Molnar, Julie Sherman, Bain Chin*, Stephen L. Buchwald*, T.Alan Hatton, Depts. of Chemical Engineering and *Chemistry, Massachusetts Institute of Technology, Cambridge, MA 02139

Solvents are necessary as reaction and separation media in pharmaceutical and fine chemical syntheses but can lead to unwanted environmental contamination through volatile emissions and discharge with aqueous waste streams. A new approach to ameliorate these problems is suggested based on tailoring existing solvents to minimize their volatility and water solubility, either by attaching a hydrophobic tail to the solvent, or attaching the solvent to a polymer backbone. The concepts have been investigated using a class of model replacement solvents based on tetrahydorfuran (THF). The solvents have been shown to have significant advantages over conventional solvents used in published procedures for a reaction sequence important in the synthesis of an HIV-1 protease inhibitor.

124. MEMBRANE APPLICATIONS FOR AQUEOUS PHASE SEPARATIONS. Norman N. Li, NL Chemical Technology, Inc., Mt. Prospect, IL

Recent developments of industrial applications of membranes for aqueous phase separations are reviewed. These developments include sugar separation, glycerol separation, and ground water purification where humic acid separation is a major problem.

The membrane materials as well as process designs will be described. Future trends of developments will be discussed.

125. SPECIFIC INTERACTIONS FOR SOLVENT-BASED SEPARATIONS
Prof. Charles A. Eckert and Barry L. West, School of Chemical Engineering and Specialty Separations Center, Georgia Institute of Technology, Atlanta, Georgia 30332-0100

Specific intermolecular interactions between the solute and a solvent or cosolvent play an important role in designing solvent-based separation processes, such as for example liquid extraction, extractive distillation, crystallization, and supercritical extraction. Such interactions can frequently be deduced not only from thermodynamic measurements, but also from spectroscopic data of various types. This work discusses the measurement, correlation, and modeling of specific interactions, such a hydrogen bonds or charge-transfer interactions, and show how these can be used to enhance distribution coefficients in separation processes.

126. PUSHING THE FRONTIER OF SEPARATION SCIENCE. D. W. Savage, Corporate Research, Exxon Research and Engineering Company, Annandale, NJ 08801

A retrospective look at the sweep of developments over the past 30 years in separation science and technology, and the changing business considerations and technology needs driving them will be presented. Selected developments from the petroleum refining/petrochemicals industry will be described. These include (I) hindered amines for selective acid gas separations, (ii) polymer membranes for aromatics separation from fuels, (iii) adsorbents for benzene removal from gasoline, and (iv) separations for environmental restoration and pollution prevention. Future challenges, including the closer integration of thinking about separation precess performance/cost and environmental considerations will be discussed.

I&EC

127. CRYSTALLIZATION FOR SEPARATION AND PURIFICATION. Ronald W. Rousseau, School of Chemical Engineering, Georgia Institute of Technology, Atlanta, Georgia 30332-0100

Attention will be given in this presentation to the use of crystallization for the purification of specialty chemicals, which are often characterized as high-value-added compounds and most often produced through batch operations. When the specialty chemicals are used for pharmaceutical purposes, purity requirements are necessarily stringent and there is often limited ability to modify existing processes to effect product improvements. Relationships between crystal purity and process protocol will be explored and, where possible, supported by a mechanistic framework. The present work will address a new separation technique involving the control of nucleation to produce protein crystals of enhanced purity.

128. IMPRINTING WITH METAL COMPLEXES: SELECTIVE ADSORBENTS AND SENSORS FOR AQUEOUS MEDIA. Frances H. Arnold, Division of Chemistry and Chemical Engineering 210-41, California Institute of Technology, Pasadena, CA 91125.

Ligand exchange on metal complexes provides a powerful and versatile route to recognition in imprinted polymers that is particularly useful in aqueous media where weaker electrostatic interactions are ineffective. Molecular imprinting is used either to assemble metal complexes to complement metal coordinating sites on the target molecule or to create specific binding sites through shape recognition. Here I will discuss recently developed chiral ligand-exchange adsorbents and polymers for recognition and sensing of sugars in complex biological media. Chiral adsorbents for underivatized amino acids are prepared by coating silica particles with a polymer imprinted with an *achiral* Cu(II)-chelating monomer. Racemic phenylalanine is resolved on the resulting materials (α ~1.6), demonstrating that the molecular imprinting process is capable of imparting the chiral selectivity. To prepare polymers for measuring glucose concentrations in biological samples, we used the fact that glucose chelation to various metal ion complexes results in the net release of protons at alkaline pH. Imprinted polymers rebind glucose and release protons in proportion to the glucose concentration. This approach, which combines metal coordination/chelation for sugar binding, ligand exchange for signal transduction, and molecular imprinting to enhance selectivity, could be used to prepare materials for selective sensing of a variety of target carbohydrates.

129. MOLECULARLY IMPRINTED MATERIALS - THEIR USE IN SEPARATIONS, IMMUNOASSAY-TYPE ANALYSES AND SYNTHESES. O. Ramström and L. Ye, Pure and Applied Biochemistry, Chemical Center, Lund University, P.O. Box 124, S-221 00 Lund, Sweden

In recent years, the development of new, attractive applications of imprinted materials have enabled an advance in molecular imprinting technology. New methodologies have been implemented on the basic technology and increasing efforts are being made to widen the horizon. In addition to futher developments in the separation sciences, imprinted materials have increasingly been used as recognition elements in immunoassay-type analyses, in biosensor-like devices and in catalytic applications. This presentation will point to some recent developments in the use of imprinted materials obtained by self-assembly using only non-covalent inntreactions. In particular, the use of such materials in aqueous media, in radio-immunoassays and as auxiliary agents in enzymatic syntheses will be discussed.

1. Mosbach, K. and Ramström, O. 1996. The emerging technique of molecular imprinting and its future impact on biotechnology, Bio/Technology 14, 163.

2. Ramström, O., Ye, L. and Mosbach, K. 1996. Artificial antibodies to corticosteriods prepared by molecular imprinting, Chem. Biol. 3, 471.

130. IMPRINTING OF PROTEINS ON POLYMER-MODIFIED DNA FOR AFFINITY SEPARATIONS. Daisuke Umeno, Masashi Kawasaki and Mizuo Maeda, Department of Chemical Science and Technology, Faculty of Engineering, Kyushu University, Fukuoka 812-81, Japan

Double helical DNA can provide binding selectivity. Some successful separation techniques have been developed in which double helical DNA is used as an affinity ligand for DNA-binding molecules. We have reported the use of thermo-responsive DNA conjugates for the separation of DNA-binding proteins. As a strategy to improve this system, bioprinting has now been applied to the photochemical modification of plasmid pBR322 DNA (4,363 bp) with poly(N-isopropylacrylamide) (poly NIPAAm) which has a Lower Critical Solution Temperature (LCST) of 31 °C. The resulting DNA-poly NIPAAm conjugate was found to precipitate above the LCST and was easily collected by centrifugation. On the other hand, accessibility of DNA-binding molecules to the conjugate decreased gradually with the degree of modification. However, we have found that the presence of a restriction endonuclease (EcoRI) during the polymer modification produced accessibility of the enzyme to the conjugate, in comparison with the conjugate prepared in the absence of the EcoRI. Thus application of an "imprinting" technique to the poly NIPAAm modification of DNA has produced a novel separation material with enhanced selectivity for the DNA-binding proteins.

131. IMPRINTING OF NUCLEOSIDES AND NUCLEOTIDES IN POLYACRYLAMIDE GEL. M. Akashi, K. Ito and T. Serizawa, Department of Applied Chemistry and Chemical Engineering, Kagoshima University, Kogoshima 890, Japan.

Specific cavities which can incorporate nucleosides or nucleotides were prepared in a crosslinked polyacrylamide hydrogel by free radical copolymerization of acrylamide, the nucleobase vinyl monomer, methacrylamide-benzeneboronic acid (MABB), and N,N'-methylenebisacrylamide in the presence of a nucleoside or nucleotide as the imprinting molecule. After washing out the imprinted molecules, their incorporation into the hydrogel was evaluated by measuring the concentration of imprinted molecules in the supernatant solution. The results suggest that when both 9-vinyladenine and MABB were used as the affinity ligands and adenosine or thymidine was imprinted, hydrogels were formed with specific cavities incorporating the imprinted molecules. The application of this system in capillary gel electrophoresis will be discussed.

132. PHOSPHATE AND PHOSPHONATE RECEPTORS IN SILICATE MATERIALS. C. E. Daitch[†], D. J. Rush*, K. J. Shea*, and D. Y. Sasaki[†]
[†]Sandia National Labs, Albuquerque, NM 87185, *UC Irvine, Chem. Dept., Irvine, CA 92717

Molecular imprinted silicate materials for phosphates and phosphonates have been prepared via the sol-gel method. Guanidine functionality, with its unique combination of electrostatic and hydrogen bonding ability, were used as selective binding points in imprinted receptor cavities. We prepared the monomer 1-trimethoxysilylpropyl-3-guanidinium chloride and incorporated it into TEOS xerogels with a phosphonate template molecule. Spectroscopic analyses, including solid state ^{29}Si NMR, found that the guanidine monomer was quantitatively incorporated into the gel.
Binding affinities of various phosphate and phosphonate substrates were determined for templated and non-templated materials using HPLC analysis and a modified Bartlett assay. Molecular memory was observed and these results will be discussed along with comparative results of imprinted xerogels functionalized with amine and trimethylammonium groups.

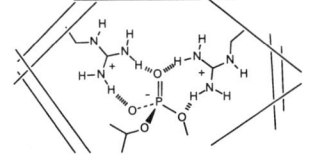

This work was performed at Sandia National Laboratories, supported by the US Department of Energy under Contract No. DE-AC04-94AL85000.

133.

OVERVIEW OF THE U.S. DEPARTMENT OF ENERGY'S EFFICIENT SEPARATIONS AND PROCESSING CROSSCUTTING PROGRAM AND THE TECHNOLOGIES FOR REMEDIATION OF GROUND AND SURFACE WATER. Kurt D. Gerdes, U.S. Department of Energy, Germantown MD, 20874, William L. Kuhn, Pacific Northwest National Laboratory, Richland WA 99352, Ian Tasker, Waste Policy Institute (WPI), Gaithersburg MD 20878, and James F. McGlynn, Science Applications International Corporation (SAIC), Gaithersburg MD 20878

Within the U.S. Department of Energy's (DOE) Office of Science and Technology is the Efficient Separations and Processing Crosscutting Program (ESP-CP). This a needs driven technology development program for chemical and physical separations and separations-related processing to address DOE's environmental restoration and waste management problems. The program is comprised of: Short-lived Radionuclides, Long-lived Radionuclides, Heavy Metals, and Waste Processing and Treatment. This paper will present an overview of the ESP-CP, past successes, and the technologies being developed to address problems throughout the DOE complex.

134.

TECHNETIUM SEPARATION FROM AQUEOUS SOLUTIONS USING POLYMER FILTRATION.
Norman C. Schroeder, Jason R. Ball, Thomas W. Robison, Rowena R. Gibson, and Barbara F. Smith, CST-11, J-514, Los Alamos National Laboratory, Los Alamos, NM 87545.

Water-soluble, metal-binding, polymers that possess functional groups with high selectivity for technetium have been developed for ground and waste waters remediation. When combined with ultrafiltration, a new homogeneous all aqueous-based technology for metals removal/recovery, called Polymer Filtration, becomes available. Technetium distribution coefficients experiments were obtained with the polymers from simple solutions, high nitrate simulants, and DSSF simulant. We have completed a preliminary proof-of-principal evaluation of Polymer Filtration technology for removal of technetium-99 from Paducah Gaseous Diffusion Plant contaminated groundwater simulant.

135.

TECHNETIUM OXIDATION STATE ADJUSTMENT FOR HANFORD WASTE PROCESSING.
Norman C. Schroeder, Glenn D. Whitener, CST-11, J-514, Los Alamos National Laboratory, Los Alamos, NM 87545, Professor Kenneth R. Ashley, and Ahn P. Truong, Chemistry Dept., East Texas State University, Commerce, TX 75429.

Technetium-99 is a major fission product resulting from the fission of plutonium or uranium. An estimated 2000 Kg of technetium is being stored in the Hanford waste tanks. As TcO_4^-, technetium is a very mobile species in the environment. This characteristic, along with its long half-life, 213,000 years, causes technetium to be a major contributor to the long-term risk associated with storage of low level waste forms. One option for mitigating this hazard is to separate technetium and vitrify it with high level waste components. Separation technologies rely on TcO_4^- being present in the waste stream. We have demonstrated that as much as 70% of the technetium in the Hanford supernates is not TcO_4^-. We will discuss this information and the feed adjustment chemistry we are developing to set technetium to TcO_4^- for processing.

136. RESINS FOR SELECTIVE SORPTION OF TECHNETIUM FROM GROUNDWATER. Gilbert M. Brown,[1] Peter V. Bonnesen,[1] Derek J. Presley,[1] Laurie M. Bates,[1] Bruce A. Moyer,[1] Spiro D. Alexandratos,[3] Vijay Patel,[3] Baohua Gu,[2] and Liyuan Liang,[2] Chemical and Analytical Sciences Division[1] and Environmental Sciences Division,[2] Oak Ridge National Laboratory, Oak Ridge, TN 37831, and Department of Chemistry, University of Tennessee,[3] Knoxville, TN 37996

A new class of anion exchange resins with improved selectivity and sorptive capacity for the pertechnetate anion (TcO_4^-) as compared to commercially available resins has been prepared and evaluated, both in batch-equilibrium and flow-through column testing conditions using a groundwater test solution. The resins have been designed and optimized for removal of pertechnetate from contaminated groundwater containing nanomolar concentrations of pertechnetate and sub-millimolar quantities of inorganic anions like nitrate, sulfate, and chloride. Equilibrium distribution coefficients for the sorption of pertechnetate to several classes of resin as a function of both time and electrolyte concentration will be discussed. The selectivity and exchange kinetics as a function of the chemical and physical structure of the resins will be addressed.

This research was sponsored by the Efficient Separations and Processing Crosscutting Program, Office of Technology Development, Office of Environmental Management, U. S. Department of Energy, under contract number DE-AC05-96OR22464 with Oak Ridge National Laboratory, managed by Lockheed Martin Energy Research Corp.

137.

IN-SITU MINERALIZATION OF ACTINIDES WITH PHYTIC ACID. **Kenneth L. Nash,** Mark P. Jensen, Lester R. Morss, Evan H. Appelman, Mark A. Schmidt, Sarah Friedrich, Argonne National Laboratory, Chemistry Division, Argonne IL 60439-4831

A new approach to the remediation of actinide contamination is described. A hydrolytically unstable organophosphorus compound, phytic acid, is introduced into the contaminated environment. In the short term (up to several hundred years), phytate acts as a cation exchanger to absorb mobile actinide ions from ground waters. Ultimately, phytate decomposes to release phosphate and promote the formation of insoluble phosphate mineral phases, considered an ideal medium to immobilize actinides, as it forms compounds with the lowest solubility of any candidate mineral species. This overview will discuss the rate of hydrolysis of phytic acid, the formation of lanthanide/actinide phosphate mineral forms, the cation exchange behavior of insoluble phytate, and results from laboratory demonstration of the application to soils from the Fernald site.

138.
COMPARISION OF URANIUM REMOVAL FROM GROUNDWATER BY SORBENTS. **Becky Sams**, Lockwood Greene Technologies/Jacobs EM Team, 125 Broadway Ave., Oak Ridge, TN, 37830; K. Thomas Klasson, David Watson, Oak Ridge National Laboratory; Jon Blount, Jacobs EM Team, Oak Ridge, TN, 37830

Several sorbents have been tested for the capability of uranium removal from two very chemically different groundwaters. Sorbents evaluated in the study include granular activated carbon, peat moss, ion exchange resin (all commercially available) as well as innovative products not commercially available. Screening experiments on all of the sorbents identified the most promising sorbents, which have been carried forward for isotherm and column studies. For the most promising sorbents, studies showed that uranium could be removed to below analytical detection limits. The effect of competing ions is also discussed.

139.
TREATABILITY STUDIES FOR THE REMOVAL OF CESIUM AND STRONTIUM FROM INEL TEST AREA NORTH CONTAMINATED GROUNDWATER Terry A. Todd, Ken N. Brewer, Troy G. Garn, and Rich D. Tillotson, Lockheed Martin Idaho Technologies Co., Idaho Falls, ID 83415

Treatability studies to determine the efficacy of ion exchange sorbents in columns and 3M web cartridges for the removal of cesium and strontium from actual contaminated groundwater at the INEL Test Area North were performed. Sorbents tested included PELLX-137, potassium cobalt hexacyanoferrate, and crystalline silicotitanates (IONSIV-911) for Cs removal and Ionac C-250, sodium titanate, and IONSIV-911 for strontium removal. High capacities and removal efficiencies for all cesium sorbents were observed. Capacities and removal efficiencies for the strontium sorbents varied between sorbents tested. The groundwater contained millimolar concentrations of calcium and magnesium, which compete for exchange sites with strontium in some of the sorbents tested.

140. ABSORPTION OF CARBON DIOXIDE INTO AQUEOUS MIXTURES OF METHYLDIETHANOLAMINE AND DIETHANOLAMINE. Edward B. Rinker, Sami S. Ashour and Orville C. Sandall, Department of Chemical Engineering, University of California, Santa Barbara, CA 93106-5080

A penetration theory diffusion and reaction model was developed and used to interpret CO_2 absorption data into aqueous blends of a tertiary and a secondary amine. The model considers nine independent reversible chemical reactions; however, the two most significant reactions for CO_2 are the carbamate formation reaction with the secondary amine (diethanolamine) and the catalysis of the hydrolysis reaction by the tertiary amine (methyldiethanolamine). Most of the pertinent physicochemical properties needed for the model such as CO_2 diffusivity, CO_2 solubility and the kinetic rate constants were measured in this work. Absorption rate measurements in a laminar-liquid jet absorber and in a stirred-cell absorber were compared to model predictions.

141.
MERCURY ABSORPTION IN AQUEOUS OXIDANTS CATALYZED BY MERCURIC CHLORIDE. **Gary T. Rochelle,** Lingbing Zhao, Department of Chemical Engineering, The University of Texas at Austin, Austin, TX 78712-1062

The absorption of elemental Hg vapor into aqueous solution containing $HgCl_2$ was measured in a stirred cell contactor at 25°C and 55°C. In 0.8 M HNO_3, the reaction is first order in Hg and Hg^{++}, respectively. The overall second order rate constant is given by:
$$k_2 = 2.90e9 \exp(-1765/T)$$
In 0.8 M HNO_3 with the addition of H_2O_2, the reaction is first order in Hg, H_2O_2 and Hg^{++}, respectively. The overall third order rate constant is given by:
$$k_3 = 2.13e23 \exp(-10110/T)$$
The addition of Fe^{2+} or Fe^{3+} has no immediate effect on mercury removal.
In 0.8 M HNO_3 with the addition of $K_2Cr_2O_7$, the reaction is first order in Hg, $Cr_2O_7=$ and Hg^{++}, respectively. The overall third order rate constant is $4.3e8$ $M^{-2}s^{-1}$ at 25°C.

142.
MASS TRANSFER AND SORPTION KINETICS LIMITATIONS IN ADSORPTIVE MEMBRANE BIOSEPARATIONS. **Mark R. Etzel** University of Wisconsin, 1605 Linden Drive, Madison, WI 53706

Adsorptive membranes are a new separation technology designed to bypass the fundamental limitations of columns packed with gel beads. However, entirely new limitations are encountered with membranes. These limitations can be minimized by developing a basis for proper design, operation, and application. In this presentation, the precise influence of key design and operating parameters on the performance of adsorptive membrane systems will be discussed. Focus will be placed on immunoaffinity and ion exchange membranes. For immunoaffinity membranes, the sorption kinetics between the free solute and the immobilized ligand, not the pressure drop, limit the flow rate through thin membranes. For ion exchange membranes, the time for solute mass transfer through the boundary layer to the pore wall, not the sorption kinetics or pressure drop, limits the flow rate. These limitations occur during both the breakthrough and elution curves.

143.
MEMBRANE REACTORS FOR DEHYDROGENATION REACTIONS: CONCEPTUAL DESIGN ISSUES. **Donald H. Mohr** Chevron Research and Technology, 100 Chevron Way, Richmond, CA 94802

A number of industrially important dehydrogenation reactions are carried out in processes where equilibrium limited reactions are followed by distillation and recycle of the unconverted feed. Membrane reactors hold the promise of increasing conversion by combining the reaction and separation in one unit operation. Conceptual process designs for accomplishing the separation in the reactor are compared to the cost of traditional separation and recycle. The costs of a membrane reactor process include creating a driving force to pull hydrogen across the membrane, and recovering the hydrogen for its chemical or fuel value. Combustion of the hydrogen is a promising method of providing a driving force, provided that the heat of combustion is utilized in the reactor design. Preliminary experimental tests of this concept will be presented along with a performance analysis based on equilibrium and mass balance relationships.

144.
PERFORMANCE OF PACKED BED ADSORBERS: SOME PRACTICAL CONSIDERATIONS. **Curtis L. Munson** Chevron Research & Technology, 100 Chevron Way, Richmond, CA 94802-0627

Packed bed adsorbers remain a workhorse of separations technology, especially in areas of application such as removal of VOCs from air. Some practical considerations in the design of full-scale packed bed adsorbers will be considered in this presentation. Non-ideal effects such as imperfect working capacity, adsorbate reactivity, and ill-behaved adsorption isotherms will be considered with regard to bed performance.

145. **MOLECULAR IMPRINTING. THE *DE NOVO* SYNTHESIS OF MACROMOLECULAR BINDING AND CATALYTIC SITES.** Kenneth J. Shea, Department of Chemistry, University of California, Irvine, CA 92697-2025

Molecular imprinting is a technique for the creation of macromolecular binding sites for small organic molecules. The process utilizes a template or imprinting molecule to organize polymerizable monomers prior to their copolymerization with crosslinking monomers. The resulting polymers are highly crosslinked network materials. Following removal of the imprinting molecule, the solid polymers are found to selectively rebind the imprinting molecule. The application of this technique for the creation of small molecule receptors and designed catalysts will be discussed.

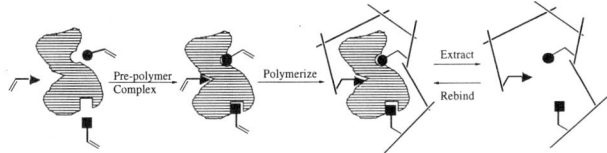

146. **TEMPLATED POLYMERS FOR THE SELECTIVE SEQUESTERING AND SENSING OF METAL IONS.** Xiangfei Zeng and G. M. Murray, Department of Chemistry, University of Maryland Baltimore County, Baltimore, MD 21250.

Polymeric ionophores have been prepared by the copolymerization of styrenic monomers with metal ion complexes possessing polymerizable ligands. The polymers were ground to a fine mesh and the accessible metal ions were removed by acid washing to leave cavities templated for specific metal ions. The polymers were tested and characterized for metal ion sequestering. Polymers were sieved and various uniform particle sizes were incorporated in polyvinyl chloride (PVC) membranes and used as ion-selective electrodes. Electrodes prepared from untemplated polymers, unpolymerized complex and free ligand were also constructed and tested. The templated polymers were superior to other the forms in the electrodes. Fluorescent metal ion complexing agents are now being used to develop metal ion-templated polymers to serve as sensing elements for optical devices. Sensitivity based on fluorometric detection coupled with the chemical selectivity of a templated polymeric film results in a large dynamic range of measurement and low limit of detection.

147. **METAL ION TEMPLATED POLYMERS. SYNTHESIS AND STRUCTURE OF N-(4-VINYLBENZYL)-1,4,7-TRIAZACYCLONONANE-METAL COMPLEXES, POLYMERIZATION OF THE METAL-MONOMER COMPLEXES WITH DIVINYLBENZENE, AND METAL ION SELECTIVITY STUDIES OF THE DEMETALATED RESINS.** Richard H. Fish, Lawrence Berkeley National Laboratory, University of California, Berkeley, CA 94720

The selective removal of toxic metal ions from aqueous waste sites represents a new direction in environmental inorganic chemistry. We will introduce a relatively new approach in separation science: the metal ion templated polymer process entails the synthesis, metal ion complexation, and polymerization of monomer-metal complexes for selective removal of metal ions from aqueous solution. While the concept has been reported, there is a general lack of knowledge concerning the structures of the monomer-metal complexes, the polymerization chemistry, and parameters concerning selectivity; i.e., template metal ion versus stability constants of the metal-ligand complexes in the polymer matrix. We will discuss our initial results concerning the synthesis of mono- and tris-N-(4-vinylbenzyl)-1,4,7-triazacyclononane ligands, their Zn^{2+} complexes, and the bulk polymerization of these Zn^{2+} templated monomers with a crosslinking agent, divinylbenzene. Selectivity studies with the demetalated bulk polymers for Cu^{2+}, Zn^{2+}, Ni^{2+}, Co^{2+}, Mn^{2+}, and Fe^{3+} ions will be presented. These studies were funded by the EM/OST/OTS/ESP program of the DOE.

148. **A PHYSIOCHEMICAL STUDY ON THE ORIGIN OF THE IMPRINTING EFFECT.** T. Miyajima, K. Sohma and S. Ishiguro, Department of Chemistry, Faculty of Science, Kyushu University, Fukuoka 812-81, Japan and M. Ando, N. Nakamura, M. Maeda and M. Takagi, Department of Chemical Science and Technology, Faculty of Engineering, Kyushu University, Fukuoka 812-81, Japan.

A Cu^{2+}-ion selective resin was synthesized by a template polymerization technique using oleic acid as a host surfactant and divinylbenzene as a matrix-forming monomer. Acid-dissociation and complexation equilibria of both Cu^{2+}-ion templated and non-templated resins were studied potentiometrically by the use of a glass electrode and a Cu^{2+} ion-selective electrode. These equilibria are highly sensitive to the added salt concentration levels together with the degree of dissociation of the resin which arises from the polyelectrolyte nature of the carboxylated resins. By correction for electrostatic non-ideality, the microscopic information on the complexation of the resins has been extracted and the intrinsic imprinting effect has been evaluated. The origin of the imprinting effect will be discussed by comparison of the complexation behavior of these two-dimensional polyelectrolytes with a linear (one-dimensional) polyelectrolyte.

149. CHROMATOFOCUSING OF PROTEINS. Douglas D. Frey, John Strong, and Ronald Bates, Department of Chemical and Biochemical Engineering, University of Maryland Baltimore County, Baltimore, MD 21250

Gradient elution chromatography is widely used for protein purification at the analytical, preparative, and process scales. This presentation discusses a particular form of ion-exchange gradient chromatography, termed chromatofocusing, which involves the use of an internally generated, retained, pH gradient formed using multicomponent mixtures of adsorbed buffering species. Existing chromatofocusing techniques and their limitations are described along with improved chromatofocusing methods which have been developed through the use of experimental and theoretical investigations of the phenomena which underlie the separation mechanism in these systems.

150.
SELECTIVE MEMBRANES FOR THE SEPARATION OF DIFFICULT MIXTURES: INVESTIGATION OF COMPLEX TRANSPORT MECHANISMS
Timothy A. Barbari, Richard L. Ward, and Denise L. Farmer
Department of Chemical Engineering, The Johns Hopkins University, Baltimore, MD 21218

Polymeric solution-diffusion membranes have been developed for a variety of gas and liquid separations, such as O_2 removal from N_2, CO_2 recovery from CH_4, the dehydration of organics, and VOC recovery from air. In each case, the components to be separated differ significantly in one or more physical properties, such as diffusivity, polarity, or condensability. More difficult separations, such as ethylene from ethane or acetic acid from water, require a polymer that can selectively transport one of the components across the membrane or a polymer matrix that can dictate the extent of solute-solute interactions during transport. To determine the polymer characteristics necessary to achieve these separations, an experimental technique is required that can elucidate transport mechanisms at a molecular level. FTIR-ATR spectroscopy will be discussed as a means of investigating these complex mechanisms in polymeric membranes.

151.
REVERSIBLE COMPLEXATION OF BIOTIN LABELED CELLS AND OLIGONUCLEOTIDES USING IMMOBILIZED SILVER IONS. **Antonio A. García** Jaime Ramírez-Vick, Sarah Johnson, Sanjay Agarwal, Arizona State University, Tempe, AZ, 85287-6006

Non-radioactive labeling of biological molecules has become quite popular during the past 16 years mostly due to advances in molecular biology and more specifically due to the development of biotin-streptavidin methods and the increasing sophistication of immunological methods. These non-radioactive labeling techniques also provided opportunities for purifying complex, protein or biological fluids for clinical diagnostics and commercial production. We have developed a substitute for streptavidin using a simple metallo-organic molecule which: (1) is less expensive; (2) is stable at room temperature; (3) is inherently free from bacterial contamination due to the use of silver ions; and (4) can be used to reversibly bind biotin conjugates using very mild, biocompatible conditions. Our latest results for selective binding of biotin labeled oligonucleotides and T cells as well as elution using NaCl will be discussed.

152.
THE LONG-TERM DISSOLUTION/DESORPTION CHARACTERISTICS OF RESIDUALLY TRAPPED PETROLEUM HYDROCARBON CONTAMINANTS IN SOIL. **William G. Rixey** Sanjay Garg, Pravin Murkute, University of Houston, Department of Civil and Environmental Engineering, Houston, TX 77204-4791

It has been observed in both the laboratory and the field that contaminant concentrations emanating from a waste reach an apparent asymptotic concentration with time as leaching progresses. This has been attributed to the relatively slow release of a fraction of the contaminant in the waste. The results of experiments designed to characterize the entire leaching history (from close to the solubility to as low as the MCL) on organic compounds (e.g., BTX and naphthalene) from soils containing a residually, trapped hydrocarbon phase with an emphasis on the rate-limited portion of the leaching curve will be presented. The relative contributions of rate-limited dissolution from a residually trapped hydrocarbon phase and rate-limited desorption from soil on long-term leaching behavior will be assessed.

153.
EFFECT OF CARBON ADSORBENT SURFACE PROPERTIES ON THE UPTAKE AND ACETONE REGENERATION OF PHENOL. **James E. Kilduff, Department of Environmental and Energy Engineering, Rensselaer Polytechnic Institute, Troy, NY 12180,** C. Judson King, Department of Chemical Engineering, University of California, Berkeley, CA

The uptake of phenol by carbonaceous adsorbents and subsequent regeneration by acetone leaching was investigated as a function of carbon source material, pH and adsorbent surface properties. At pH 9, under conditions which promote oxidative coupling reactions, 20 to 80% of the total uptake was irreversible (not acetone extractable). Oxidation of carbon surfaces with concentrated HNO3 increased surface acidity as measured by Boehm titration and reduced phenol uptake, but increased the reversible uptake. Reversibility was correlated with total acidity, reaching an upper plateau at about 1 meq/m2. Uptake of water vapor by surface treated carbons correlated with surface acidity, suggesting that selectivity for water may contribute to the suppression of oxidative coupling reactions.

154. MOLECULARLY IMPRINTED POLYMERIC MEMBRANES FOR OPTICAL RESOLUTION. Masakazu Yoshikawa, Department of Polymer Science and Engineering, Kyoto Institute of Technology, Matsugasaki, Kyoto 606, Japan

The development of membranes which exhibit optical resolution is an important technological objective. Except for their optical activity, enantiomeric compounds have the same physicochemical properties. Therefore a chiral microenvironment must be introduced into the molecular recognition sites in the membrane. In our study, oligopeptide derivatives, which are expected to produce chiral microenvironments are adopted as the recognition sites for optical resolution. A molecular imprinting technique is used to obtain polymeric membranes for enantioselective separations. In this technique, the "molecular memory" of the template is introduced into the membrane at the same time as the polymeric membrane is being formed from solution. We have found that membranes which contain an oligopeptide residue from a L-amino acid and imprinted with a L-amino acid derivative, recognizes the L-isomer in preference to the corresponding D-isomer, and vice versa. By use of electrodialysis, the amino acid optical isomer which is preferentially adsorbed in the membrane is selectively permeated.

155. MOLECULARLY IMPRINTED MEMBRANES BY A PHASE INVERSION METHOD. T. Kobayashi and N. Fujii, Department of Chemistry, Nagaoka University of Technology, Kamitomioka, Nagaoka 940-21, Japan.

We will summarize recent results for the phase inversion method of preparing molecularly imprinted membranes. The technique is applied to membrane preparation from acrylonitrile-acrylic acid copolymers, which contain both acrylic acid (AA) residues as the functional site and acrylonitrile residues as the membrane formation site. A copolymer cast solution of dimethyl sulfoxide containing the template is coagulated in a poor solvent, water. It will be shown that the imprinting properties of the copolymer membrane depend on the template concentration in the cast solution, the AA content in the copolymer and the coagulation temperature. Evidence for interaction between the template and AA segments will be shown by characterization of the imprinted membranes.

156. DEVELOPMENT OF UNIFORMLY SIZED, MOLECULARLY IMPRINTED STATIONARY PHASES FOR HPLC. Ken Hosoya and Nobuo Tanaka, Department of Polymer Science, Kyoto Institute of Technology, Kyoto 606, Japan.

Molecularly imprinted polymers have been utilized for specific molecular recognition in HPLC and many useful stationary phases have been developed. Although these stationary phases show quite high separation factors between the template molecule and its enantiomer or isomers, the reported column efficiencies are not very high. This is because the imprinted polymers were prepared using a bulk polymerization technique and the resultant polymer block must be ground and sieved to be used as a HPLC packing material. We have developed uniformly sized, molecularly imprinted polymer particles to avoid these time-consuming procedures and to achieve higher column efficiencies. A two-step swelling and polymerization technique affords uniformly sized, polymer particles. When an appropriate template is involved in the system, molecularly imprinted polymer particles with excellent size uniformity can be obtained. These stationary phases show comparable separation factors with much high column efficiency so the resolution is much better in comparison with stationary phases prepared by bulk polymerization technique.

157. AN IN-SITU APPROACH TO MOLECULARLY IMPRINTED "TAILOR-MADE" SEPARATION MEDIA. J. Matsui and T. Takeuchi, , Faculty of Information Sciences, Hiroshima City University, Hiroshima 731-31, Japan.

Molecularly imprinted polymers have been recognized as stable and easy-to-handle substitutes for antibodies and receptors in affinity-type separations, such as liquid chromatography. However, since the polymers have been obtained as blocks in most cases, several tedious and time-consuming procedures, such as grinding, sieving, sizing and column packing, are required to use the polymers as chromatographic stationary phases. This presentation describes an in-situ approach to prepare molecularly imprinted polymer rods for immediate use as chromatographic separation media. The polymer preparation was carried out inside liquid-chromatography columns in the presence of a template molecules to obtain rod-type imprinted polymers. The resultant polymer rods have been used for chiral or selective separation of drugs. This new approach for preparation can be expected to make molecular imprinting an easier and more versatile technique for preparing "tailor-made" affinity media for liquid chromatography.

158.
PROTEIN PURIFICATION FROM COMPLEX BIOLOGICAL MIXTURES USING LOW MOLECULAR WEIGHT DISPLACERS IN ION EXCHANGE SYSTEMS K.A. Barnthouse, A. Shukla, S.M. Cramer, Isermann Department of Chemical Engineering, Rensselaer Polytechnic Institute, Troy, NY 12180

The purification of proteins by displacement chromatography offers high selectivity, simultaneous purification and concentration, high throughputs, and straightforward scalability. The discovery that low molecular weight displacers can be successfully employed for protein purification in ion exchange systems has generated significant interest in this technology. In this paper, results are presented on the purification of proteins from complex biological mixtures using novel low molecular weight displacers. Methods development protocols based on the steric mass action model are presented for complex displacement separations. Case studies are given on the use of selective displacement chromatography as an alternative to traditional initial step elution processes. Results are also presented on the use of high resolution displacement chromatography for the resolution of extremely difficult mixtures such as variants and glycoforms. Methods development protocols for complex displacement separations are also presented. These case studies with biological mixtures obtained from industrial collaborators demonstrated the efficacy of displacement technology for difficult bioseparation problems and show how the productivity in ion exchange systems can be significantly improved when moving from a gradient to a displacement operation.

159.
SCREENING OF ANION EXCHANGE GELS FOR THE PURIFICATION OF AN INSULIN ANALOGUE. Arne Staby, Novo Nordisk A/S, Hagedornsvej 1, 2820 Gentofte, Denmark.

For the anion exchange purification of an insulin analogue, a number of chromatographic gels have been tested in semi-preparative scale. The gels studied are designed for protein purification at medium to high pressures in semi-preparative or preparative scale and have different base matrix and ligand chemistry. The particle size of the various gels were nominally in the range of 15-35μm. The examination employing several buffers, eluting salt systems, pH values, and gradients were otherwise performed under identical conditions of load, temperature, organic solvent content, and flow. The gels have mainly been on the market for less than three years or are not commercially availiable yet.

160.
A MULTI-STAGE CHIRAL EXTRACTION MODEL. Jürgen Koska, Charles A. Haynes, Biotechnology Laboratory and Department of Chemical Engineering, 237 Wesbrook Bldg., 6174 Univ. Blvd., University of British Columbia, Vancouver, BC, V6T 1Z3, Canada

A shift towards enantiomerically pure therapeutics is driven by new FDA regulations. Pharmaceutical industries are therefore in need of new large scale chiral separation systems, preferably continuous. We are studying a 2-phase chiral extraction system [Takeuchi et al., 1984] which is based on selective complexation of an enantiomer to a water insoluble ligand and extraction into an organic phase. A multi-chemical equilibrium model has been developed for this 2-phase extraction system to predict enantiomer partitioning. The model was then extended to investigate the extraction performance of multi-stage chiral 2-phase extraction devices. Model predictions were compared with data from a multi-stage chiral extraction system utilizing hollow-fibre membranes. The data show that the model is a powerful tool for the design and prediction of multistage chiral 2-phase extraction system performance.

161.
POLYMER DIFFUSION WITHIN CHROMATOGRAPHIC MATERIALS: A PULSED-FIELD GRADIENT NMR STUDY. **David H. Reeder**, Peter W. Carr, Michael C. Flickinger, Alon V. McCormick, Univeristy of Minnesota, 421 Washington Ave. SE, Minneapolis, MN, 55455

We report efforts to ascertain reasonable models for describing diffusion of large solutes within the pores of chromatographic materials. Pulsed-field gradient NMR has been used to measure intra-particle diffusion of non-adsorbing polymers of controlled molecular weight within materials with a range of pore structures. Current theories are found to be insufficient. Hydrodynamic interactions between the solute and the pore surface can be qualitatively described by existing theories, but there are unaccounted for dependencies on chain flexibility and pore structure. Hydrodynamic interactions between contained solutes are unchanged from those in free solution. Most significantly, we quantify how the pore structure of the material can be used to optimize the diffusivity. Theory and simulation for diffusion within random and ordered colloidal assemblies will be presented to examine the role of material structure in chromatographic performance.

162. PROCESS CONSIDERATIONS IN SCALE-UP OF BIOACTIVE PROTEIN PURIFICATION FROM BOVINE CEREBRAL CORTEX TISSUES. Natraj Ram, Matthew B. Herwig, Amir Ahmadi, Antonio R. Moreira, Department of Chemical and Biochemical Engineering, University of Maryland Baltimore County, 1000 Hilltop Circle, Baltimore, MD 21250

A scaled-up process has been developed to purify a novel protein extremely effective in synchronization of mammalian cell growth cycles. While development of the process is a challenge in itself, other factors such as the reproducibility, the processing time, removal of growth enhancers and removal of proteases have been identified as being crucial in determining the effectiveness of the protein in mammalian cell cultures. These key factors have been accounted for in this scaled-up process and currently a process that can produce cell-culture grade protein has been developed and established.

163. OPTIMIZATION OF HYDROPHOBIC INTERACTION CHROMATOGRAPHY FOR PROCESSING OF RECOMBINANT PROUROKINASE. Donald A. Ross[1], Alexander H. Chu[1], Casey Bunker[2], Abbott Laboratories[1], North Chicago, IL 60064, Center for Biotechnology[2], Northwestern University, Evanston, IL 60201.

Retention in Hydrophobic Interaction Chromatography (HIC) is generally controlled by varying salt concentration during protein purification. However, many other variables may contribute to the retention mechanism of proteins in HIC such as pH, temperature, and type of resin ligands. The effects of salt and protein concentrations, as well as flow rate and load amount, on the performance of the HIC used for the purification of recombinant prourokinase(r-ProUK) were investigated. It was observed that although higher salt concentration gives higher mass recovery, the product purity decreases with increasing salt concentration. One particular area of interest is the increasing mass recovery observed with increasing r-ProUK concentration, while the presence of other impurities in the r-ProUK load does not negatively impact the mass recovery.

164.
IMPACT OF PARTICLE SEGREGATION ON CHROMATOGRAPHIC BREAKTHROUGH BEHAVIOR, K. Seibert[1] and M.A. Burns[2], [1]Merck & Co., Inc., P.O.Box 2000, Rahway, NJ 07065, [2]Department of Chemical Engineering, University of Michigan, 3022 H. H. Dow Building, Ann Arbor, MI 48109

The ability to directly process fermentation broths in fluidized beds with their inherent low pressure drop is extremely advantageous. Unfortunately, fluidized beds typically show a great deal of liquid and solids mixing compared to fixed bed adsorbers. Stabilization of the beds through use of a particle size distribution reduces this mixing and improves the performance of the adsorber. Simulation results will be presented that predict the location of particles in fluidized beds based on a Monte Carlo type method. Breakthrough predicitons from a mathematical model that incorporate these spatial positions will be compared to experimental results.

American Chemical Society
DIVISION OF INORGANIC CHEMISTRY

213th ACS National Meeting

San Francisco, CA
April 13-17, 1997

T.E. Bitterwolf, Program Chairperson

SUNDAY MORNING

- **Frontiers in Organometallic Chemistry I**
  T.E. Bitterwolf, Presiding — Papers 1-7

- **Bioinorganic Spectroscopy Tutorial II**
  E.I. Solomon, Presiding — Papers 8-11

- **Photochemistry**
  D.S. Williams, Presiding — Papers 12-24

- **Group 13 Chemistry**
  D.A. Atwood, Presiding — Papers 25-36

- **Metallocenes**
  K.M. Kane, Presiding — Papers 37-48

- **General Solid State I**
  A.B. Bocarsly, Presiding — Papers 49-59

SUNDAY AFTERNOON

- **Frontiers in Organometallic Chemistry I**
  R.B. King, Presiding — Papers 60-66

- **Bioinorganic Spectroscopy Tutorial II**
  K.O. Hodgson, Presiding — Papers 67-71

- **Transition Metals in Novel Environments**
  D.C. Crans, Presiding — Papers 72-84

- **General Organometallic Chemistry II**
  L.J.S. Vosejpka, Presiding — Papers 85-97

- **Bimetallic & Cluster Compounds**
  S. Harris, Presiding — Papers 98-110

- **General Solid State I**
  M.L. Lynch, Presiding — Papers 111-123

SUNDAY EVENING

- **Poster Session** Papers 124-321

MONDAY MORNING

- **Novel Ligands**
  D.R. Click, Presiding Papers 322-326

- **Transition Metal-Main Group Multiple Bonds**
  C. Newton, Presiding Papers 327-331

- **Metal Complexes to Macromolecules**
  R.M. Roat, Presiding Papers 332-336

- **Phosphorus Chemistry**
  C.M. Barnes, Presiding Papers 337-341

- **Award Session II** Papers 342-344

MONDAY AFTERNOON

- **Symposium Honoring Bercaw / Green I**
  M.E. Thompson, Presiding Papers 345-355

- **Bioinorganic Spectroscopy I**
  B. Hedman, Presiding Papers 356-361

- **Materials Science I**
  Molecular Precursors and New Approaches to Materials Synthesis
  D.A. Keszler, T.E. Mallouk, Presiding Papers 362-367

- **General Main Group Chemistry I**
  G.H. Purser, Presiding Papers 368-376

- **Bioinorganic Model Compounds**
  M.M. Millar, Presiding Papers 377-388

- **General Solid State I**
  J.T. Spencer, Presiding Papers 389-397

MONDAY EVENING

- **Sci-Mix** Papers 126,127,131, 132,137,140,149,150,154,163, 167,170,174,180,187,191,194,212,219,228,229,232, 236,245,250,252,264,267,280,283,302,304,313,398-421

TUESDAY MORNING

- **Award Session II** Papers 422-425

TUESDAY AFTERNOON

- **Symposium Honoring Bercaw / Green II**
  G. Parkin, Presiding Papers 426-436

- **Bioinorganic Spectroscopy Symposium I**
  J.E. Penner-Hahn, Presiding
  Papers 437-442

- **Materials Science Symposium I**
  Building Blocks and Hierarchical Materials
  T.E. Mallouk, A. Clearfield, Presiding
  Papers 443-448

- **Organometallic Photochemistry & General Transition Metal**
  R.M. Chamberlin, Presiding
  Papers 449-461

- **Joint Inorganic / Analytical Symposium Honoring Marjam Behar**
  M.Y. Darensbourg, T.M. Loehr, Presiding
  Papers 462-468

- **General Transition Metal Chemistry**
  A.D. Brown, Presiding
  Papers 469-480

TUESDAY EVENING

- **Poster Session**
  Papers 398-400, 481-510, 401-411, 511-564, 412-421, 565-628

WEDNESDAY MORNING

- **Symposium Honoring Bercaw / Green II**
  P.J. Shapiro, Presiding
  Papers 629-640

- **Bioinorganic Spectroscopy Symposium I**
  P.J. Stephens, Presiding
  Papers 641-645

- **Materials Science Symposium I**
  Functional Materials I
  D.A. Keszler, M.M. Lerner, Presiding
  Papers 646-652

- **Symposium Honoring the Late Sir Geoffroy Wilkinson I**
  A.R. Barron, M.L.H. Green, Presiding
  Papers 653-661

- **Coordination Chemistry, Metal Chelates and Metal Dioxygen Complexes I**
  G. McLendon, Presiding
  Papers 662-669

- **Mechanism and Catalysis**
  G. Mann, Presiding
  Papers 670-682

WEDNESDAY AFTERNOON

- **Transition Metal Chemistry**
  C.M. Hockensmith, Presiding
  Papers 683-695

- **Bioinorganic Spectroscopy Symposium I**
  K. Wieghardt, Presiding
  Papers 696-700

- **Materials Science Symposium I**
  Functional Materials II
  D.R. Talham, M.J. Sailor, Presiding
  Papers 701-706

- Symposium Honoring the Late Sir Geoffroy Wilkinson I   Papers 707-715

- Coordination Chemistry, Metal Chelates and Metal Dioxygen Complexes I
R.Bogucki, Presiding   Papers 716-725

- General Organometallic Chemistry II   Papers 726-736

THURSDAY MORNING

- Organometallic Chemical Vapor Deposition
A.W. Apblett, Presiding   Papers 737-749

- Bioinorganic Spectroscopy Symposium I
R.L. Musselman, Presiding   Papers 750-762

- Materials Science Symposium I
Bio-Inspired and Uninspired Materials: From Nanoscale to Macroscale
M.S. Whittingham, M.E. Davis, Presiding   Papers 763-768

- Catalysis
D.S. Glueck, Presiding   Papers 769-780

- Coordination Chemistry, Metal Chelates and Metal Dioxygen Complexes I
E.H. Abbott, Presiding   Papers 781-792

- Bioinorganic Chemistry - Porphryns
K. Skidmore, Presiding   Papers 793-803

THURSDAY AFTERNOON

- General Main Group III
D.B. Grotjahn, Presiding   Papers 804-815

- General Bioinorganic Chemistry
M.A. Lopez, Presiding   Papers 816-827

- General Solid State I
S.W. Keller, Presiding   Papers 828-833

- Lanthanide / Actinide
K.M. Schaab, Presiding   Papers 834-845

- Solid State Synthesis
Z.J. Zhang, Presiding   Papers 846-856

- Chemistry of Heavy Metals
E.R. Birnbaum, Presiding   Papers 857-865

DIVISION OF INORGANIC CHEMISTRY Inc.

001. NEW METALLACYCLOBUTANE COMPLEXES OF Nb AND Ta

Kazushi Mashima,† Michitaka Kaidzu,† and Akira Nakamura,†‡ Department of Chemistry, Graduate School of Engineering Science,† and Department of Macromolecular Science, Graduate School of Science,‡ Osaka University, Toyonaka, Osaka, Japan 560

Based on the unique reactivity of the 14 electronic metal fragment, Cp*(1,3-diene)M : M = Nb, Ta; di- and tri-substituted metallacyclobutane complexes were prepared by the thermal reaction of di-alkyls, Cp(diene)MR$_2$, with olefins. Structural and NMR data are obtained for the niobacyclobutanes prepared from norbornene and acenaphthylene. The observed chemical properties are correlated with the ROMP catalysis of these complexes.

002. FORMATION AND REACTIVITY OF THE METAL-CENTERED RADICAL CATION Ru(PPh$_2$(CH$_2$)$_2$PPh$_2$)(CO)$_3$$^{+\cdot}$ Steven J. Skoog, John P. Campbell, and Wayne L. Gladfelter*, Department of Chemistry, University of Minnesota, Minneapolis, MN 55414

During the catalytic carbonylation of nitroaromatics, the radical cation, [Ru(dppe)(CO)$_3$]$^{+\cdot}$, is one of the possible intermediates in the initial oxidation of Ru(dppe)(CO)$_3$ (**1**). The zero valent metal complex **1** can be oxidized electrochemically or chemically with Cp$_2$Fe(PF$_6$) to form the highly reactive radical cation **1**$^{+\cdot}$, which rapidly dimerizes to form [Ru$_2$(dppe)$_2$(CO)$_6$]$^{2+}$ (**2**). A CV of **2** in CH$_2$Cl$_2$ gave a two electron reduction wave with a return oxidation wave corresponding to **1**. The dimer was isolated and characterized by single crystal X-ray diffraction, and shown to have an extremely long unsupported Ru-Ru bond (3.0413(6) Å) with one phosphine lying along, and another lying orthogonal to the M-M axis. Steric forces dictate the structure and the ligands assume a staggered geometry. The dimer is fluxional ($\Delta H^\ddagger$ = 11 kcal/mol; $\Delta S^\ddagger$ = -4 eu; by ^{31}P{^{1}H} lineshape analysis) and the exchange is proposed to take place through a bridging mechanism based on ^{13}C{^{1}H} DNMR. The dimer undergoes disproportionation thermally and photochemically in CH$_3$CN to form **1** and [*trans*-Ru(dppe)(CH$_3$CN)$_2$(CO)$_2$]$^{2+}$, as well as halogen atom abstraction reactions forming *fac*-[Ru(dppe)(CO)$_3$Cl]$^+$ via M-M bond homolysis.

003. THE CHEMICAL REACTIVITY OF PHOSPHORUS-BRIDGING CARBONYL COMPOUNDS: A NOVEL APPROACH TO HOMOGENEOUS CATALYSIS? R. B. King, Department of Chemistry, University of Georgia, Athens, Georgia 30602

The use of a metal carbonyl "matrix" allows the stabilization of phosphorus-bridging carbonyl groups in air-stable readily synthesized iron carbonyl derivatives of the general type (R$_2$NP)$_2$COFe$_2$(CO)$_6$ (R$_2$N = *bulky* dialkylamino group particularly iPr$_2$N). Such compounds undergo the following types of reactions: (1) Attack of nucleophiles such as organolithium compounds, alkylidenephosphoranes, etc., on the phosphorus-bridging carbonyl group under mild reaction conditions to give alcohols [(R$_2$NP)$_2$C(R)OH]Fe$_2$(CO)$_6$ or unusual rearrangement products thereof; (2) Extrusion of the phosphorus-bridging carbonyl group at ~100°C to give a (R$_2$NP)$_2$Fe$_2$(CO)$_6$ unit with reactive sites on the phosphorus atoms which can be trapped by a variety of reagents. For example, alcohols trap (R$_2$NP)$_2$Fe$_2$(CO)$_6$ as (R$_2$NPH)(R'OPNR$_2$)Fe$_2$(CO)$_6$ whereas the (R$_2$NP)$_2$Fe$_2$(CO)$_6$ unit can also add to the C≡N of acetonitrile or benzonitrile; the C=O of aldehydes, ketones, or even maleic anhydride; and the C=C of acrylonitrile. The phosphorus-bridging carbonyl extrusion chemistry of (R$_2$NP)$_2$COFe$_2$(CO)$_6$ suggests the possibility of designing novel catalyst systems in which reactive vacant sites are generated at phosphorus atoms rather than transition metal atoms.

INOR

004. THE CHEMISTRY OF PLATINUM GROUP METALS WITH FUNCTIONALIZED YLIDES
U.Belluco, R.A.Michelin, M.Mozzon, G.Facchin, R.Bertani, L.Zanotto, Istituto Chimica Industriale, Facoltà Ingegneria and Centro Chimica Metallorganica,C.N.R., Via Marzolo 9, Padova (Italy) 35131

Functionalized ylides represent useful synthons for a wide variety of organic and organometallic systems. Here we describe the reactivity of some phosphorus and arsenicum carbonyl stabilized ylides towards a series of palladium and platinum complexes bearing ancillary ligands with different steric hindrance, in order to determine the factors that influence the coordination mode of the ylides : through the oxygen atom or the more nucleophilic methyne carbon. The whole results indicate that a high steric hindrance around the metal appears to address the coordination mode through the oxygen atom. Furthermore, the chemistry of unprecedented ylide-isocyanide systems with transition metal substrates is presented. Finally, the organometallic chemistry of ketenylidenetriphenylphosphorane is described with the first complete crystallographic determination of a η^1-ketenyl-Pt-derivative.

005. NEW MULTIAMIDO LIGANDS FOR EARLY TRANSITION METAL CHEMISTRY Richard R. Schrock, Scott W. Seidel, Jesse C. Lee, jr, Lan-Chang Liang, George E. Greco, Myra B. O'Donoghue, Steven M. Reid, Robert Baumann, Timothy H. Warren, Department of Chemistry, Massachusetts Institute of Technology, Cambridge, Massachusetts 02139

We have been developing new multidentate amido ligands for early transition metal chemistry, e.g. trianionic triamidoamine ligands, $[(RNCH_2CH_2)_3N]^{3-}$ (R = $SiMe_3$ or C_6F_5). New ligands include triamido ligands of the type $[(RNCH_2CH_2)_2NR']^{3-}$ and diamido/donor ligands such as $[(RNCH_2CH_2)_2NR']^{2-}$, $[(RN\text{-}o\text{-}C_6H_4)_2NR']^{2-}$, and $[(RN\text{-}o\text{-}C_6H_4)_2O]^{2-}$. Ligands where R or R' is not a silyl group have proven to be of greater potential in view of their robust character. The metals involved in this chemistry include Ti, Zr, Hf, Ta, Mo, and W. Recent syntheses and applications will be discussed.

006. PHOTOLYSIS OF $CpRe(CO)_3$ AND $CpM(CO)_4$, WHERE M = Nb AND Ta, IN NUJOL MATRICES AND SUPERCRITICAL ETHYLENE. AN UNEXPECTED ETHYLENE POLYMERIZATION CATALYST. T. E. Bitterwolf[1], J. T. Bays[1], S. Gallagher[1], J. C. Linehan[2], C. R. Yonker[2], (1) Department of Chemistry, University of Idaho, Moscow, ID 83844, (2) Battelle Pacific Northwest Laboratory, P. O. Box 999, Richland, WA 99352.

Nujol matrix photochemical studies of $CpRe(CO)_3$ establish the primary photochemical event to be loss of carbon monoxide to form an electron deficient species. $CpM(CO)_4$, where M = Nb or Ta, give both $CpM(CO)_3$ and $CpM(CO)_2$ species upon photolysis in the Nujol matrix. Upon photolysis of these compounds in supercritical ethylene, IR bands associated with $CpM(CO)_{n-1}$(ethylene) are observed as well as bands associated with $CpM(CO)_{n-2}$(ethylene)$_2$. White precipitate, formed under photochemical conditions, has been shown to be polyethylene.

007.
The Heck Reaction: Mechanistic Investigations

Wolfgang A. Herrmann, Claus-Peter Reisinger
Technische Universität München
Lichtenbergstr. 4
D-85748 Garching b. München (Germany)
fax: Int.+8928913088

Palladium-catalyzed carbon-carbon coupling reactions are versatile procedures in organic synthesis with growing interest for fine chemical syntheses. Heck reactions are the most important class of these reactions, but the understanding of single steps in the catalytic cycle is surprisingly poor beyond the standard text book mechanism. This failure became obvious in the context of kinetic investigations when we developed two new catalyst systems, i.e. phospha-palladacycles and imidazol-2-ylidene palladium complexes. These catalysts exhibit the highest activities reported in the literature.

008. STRUCTURE OF ENZYME INTERMEDIATES BY ENDOR SPECTROSCOPY. Brian M. Hoffman, Department of Chemistry and BMBCB, Northwestern University, Evanston, IL 60208

Electron-nuclear double resonance (ENDOR) spectroscopy can provide NMR spectra of *all* the nuclei that comprise a paramagnetic center. When CW and pulsed ENDOR studies of frozen-solution samples are carried out so as to generate 2-D, field-frequency datasets, and these are analyzed with theoretical procedures we have developed for determining hyperfine and nuclear-quadrupole tensors, it is possible to determine the composition, connectivity, coordination geometry, and even metrical parameters of metalloenzyme active sites, including reactive intermediates as well as ground states. This procedure will be explained and illustrated through a discussion of Intermediate-X of ribonucleotide reductase and/or allylbenzene-inactivated chloroperoxidase (CPO). The former is a high-valence intermediate formed by the reaction of dioxygen with a carboxylate-bridged diferrous center. The latter is a product of the reaction of allylbenzene with hydrogen-peroxide-oxidized CPO, where the substrate has formed a covalent bond to the heme.

009.
COMBINING MÖSSBAUER, EPR AND MAGNETIC SUSCEPTIBILITY STUDIES
Eckard Münck and Karl E. Kauffmann, Department of Chemistry, Carnegie Mellon University, Pittsburgh, PA 15213

^{57}Fe Mössbauer spectroscopy is a powerful tool to study the electronic structure of exchange-coupled systems, in particular when used in conjunction with techniques such as electron paramagnetic resonance (EPR) and magnetometry. For Kramers systems the correlation between EPR and Mössbauer spectroscopy has been exploited for many years. More recently, with the advent of integer-spin EPR, it has become fruitful to combine the two techniques for non-Kramers systems as well. The study of synthetic complexes often involves polycrystalline samples. These materials often yield well resolved Mössbauer spectra, reflecting a homogeneous environment; however, intermolecular exchange and alignment of crystallites at low temperatures by an applied magnetic field pose particular problems which can be addressed and solved by combining Mössbauer studies with magnetization studies. The lecture will focus on the methodology of combining these three techniques.

010. SINGLE- AND MULTIDIMENSIONAL NMR STUDIES OF PARAMAGNETIC METAL COMPLEXES. F. Ann Walker, Department of Chemistry, University of Arizona, Tucson, AZ 85721

The proton (and ^{13}C and ^{15}N) NMR signals of paramagnetic metal complexes (and metalloproteins) can often, but not always, be detected by suitable modification of the standard experiments that are a part of the software of any modern NMR spectrometer. These spectra potentially hold a vast amount of important information concerning the electronic and geometrical structure of the complex, because the NMR spectra of paramagnetic complexes are exquisitely sensitive to metal-proton distance and to the molecular orbital into which the unpaired electron is delocalized. Successfully obtaining 1D spectra is usually not difficult, unless the relaxation times T_1 and T_2 are considerably shorter than 1 ms, because the short relaxation times of the nuclei of the ligand allow rapid recycle times. Thus, even though lines may be broad, good S/N can be obtained in a relatively short time. To successfully obtain 2D spectra of paramagnetic complexes, however, requires careful optimization of the experimental parameters, in line with the relaxation times T_1 and T_2 of the signals of interest. For heteronuclear 2D experiments such as HMQC and HMBC, experiments that include gradient pulses greatly improve S/N, making it possible to detect weak correlations.

011. METALLOPROTEIN CRYSTALLOGRAPHY. Hans C. Freeman, School of Chemistry, University of Sydney, N.S.W. 2006, Australia.

The steps in the crystal structure analysis of a biological macromolecule are (i) the discovery of conditions that lead to the formation of suitable crystals, (ii) measurements of the intensities of X-ray beams reflected ('diffracted') by the crystals, (iii) the determination of the phases of the X-ray reflections, leading directly to the calculation of an electron-density map, (iv) the interpretation of the electron-density map, resulting in a molecular model, and (v) the refinement of the molecular model to make it as accurate and precise as the available data allow. A final step, whose importance has only recently received the attention that it deserves, is the assessment of the accuracy and precision of the results. Recent years have witnessed spectacular advances in all these procedures. Systematic protocols are replacing trial-and-error in crystallisation experiments. Diffraction measurements can be made more rapidly and with smaller crystal specimens by using the intense, wavelength-tunable X-ray beams from synchrotron radiation sources, and by using new types of detector. Low-temperature methods prolong the life of crystal specimens in the X-ray beam, and improve the quality of the diffraction data. There is an increasing use of multiple-wavelength anomalous dispersion and molecular replacement methods for phase determination. Improved refinement strategies capable of producing more accurate structures have been developed. There has been progress in the detection of outright errors, and estimates of the precision of crystallographic results are becoming more objective and quantitative. A survey of these developments will be presented.

012.

LONG-LIVED, LUMINESCENT PHOTOEXCITED STATES OF D0 GROUP 5 IMIDO COMPOUNDS. **Darryl S. Williams,** Andrey V. Korolev, Department of Chemistry, Wayne State University, Detroit, MI, 48202

Building on recent observations that d0 group 5 imido compounds of the form M(NR)Cl3(dme) (M = Nb, Ta) exhibit intense, long-lived luminescence in solution at room temperature, we have shown that an energy gap law correlation exists in this series of compounds. As the imido alkyl substituent is changed, absorption and emission energies change, while rates of radiative and non-radiative decay remain invariant and change as well, respectively. The medium frequency acceptor mode responsible for this correlation appears to be the (mainly) M-N stretch (which is coupled to alkyl vibrations). Physical evidence for this correlation and insights into the electronic structure of these compounds through electronic spectroscopy will be discussed. Results indicating facile photoinduced electron transfer reactions using these chromophores will be presented.

013.

ELECTRON EXCHANGE COUPLING AND THE PHOTOPHYSICS OF TRANSITION METAL COMPLEXES. **James K. McCusker,** Department of Chemistry, University of California at Berkeley, Berkeley, California, 94720-1460

Two research areas in our group which involve the study of the effects of electron exchange coupling on the photophysical and photochemical properties of transition metal complexes will be discussed. The first focuses on metal-ligand exchange interactions in chromium-semiquinone/

catechol complexes as a means of detailing the effects of exchange coupling in electronic excited states on radiative and non-radiative decay dynamics. The second area looks at electron transfer and involves the synthesis and spectroscopy of exchange-coupled metal clusters as components of electron donor-acceptor assemblies. Our interest in these systems is to examine the possibility of whether exchange interactions -- more specifically, their modulation -- can play a functional role in mediating biological electron transfer in polynuclear transition metal active sites.

014.

SPECTROSCOPIC AND ELECTROCHEMICAL PROPERTIES OF MIXED-METAL SUPRAMOLECULAR COMPLEXES WITH APPLICATIONS AS MOLECULAR DEVICES FOR PHOTOINITIATED ELECTRON COLLECTION.
Karen J. Brewer, Sumner W. Jones, Elizabeth R. Bullock, and Jeff A. Clark, Department of Chemistry, Virginia Polytechnic Institute and State University, Blacksburg, VA 24061-0212. kbrewer@chemserver.chem.vt.edu

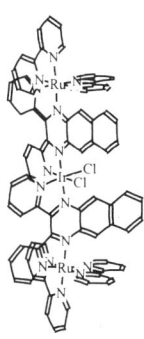

Recent work in our group has focused on the design, construction and analysis of supramolecular complexes which are applicable to the development of molecular devices for photoinitiated electron collection. The design of such systems is based on being able to dictate the light absorbing and redox properties of various component of our molecular device within a supramolecular architecture. This presentation will discuss the changes we observe in the spectroscopic and electrochemical behavior of our monometallic "building blocks" as they are incorporated into a polymetallic framework. In addition, the talk will cover how the application of mixed-metal, mixed-ligand systems allows us to design and construct supramolecular complexes with predictable and desirable redox properties. This work is supported by NSF CHE-9313642 and CHE-9632713.

015.

PHOTOCHEMICAL AND DNA BINDING STUDIES OF PLATINUM-RUTHENIUM BIMETALLIC COMPOUNDS Matthew Milkevitch, Eric Brauns, Pat Boyer, Hannah Storrie, Brenda W. Shirley and Karen J. Brewer, Departments of Chemistry and Biology, Virginia Polytechnic Institute and State University, Blacksburg, VA 24061.

In recent history, studies have shown that ruthenium, rhodium, nickel and cobalt compounds incorporating planar aromatic ligands with extended π systems intercalate DNA. In addition, rhodium, nickel and cobalt complexes of this type are capable of photoinitiated cleavage. Recently in our laboratory, we have synthesized four compounds of the general formula $[(bpy)_2M(BL)PtCl_2]Cl_2$, (where M= Ru^{II} and Os^{II}, and BL= 2,3 bis-(2-pyridyl)quinoxaline and 2,3 bis-(2-pyridyl)benzoquinoxaline). The rationale behind the development of these compounds is to produce a DNA binding agent capable of multiple modes of binding while also being capable of photoinitiated cleavage. Gel electrophoresis studies indicate that these compounds avidly bind DNA. Information on synthesis, characterization, DNA binding and photochemical studies follows in this presentation. This work is supported by NSF CHE-93 13642 and CHE-9632713.

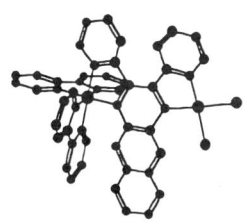

016. VALENCE TAUTOMERIC COBALT O-QUINONE COMPLEXES AS DUAL-MODE MOLECULAR DEVICES. Daniel Ruiz and David N. Hendrickson, Department of Chemistry and Biochemistry, University of California at San Diego, La Jolla, CA 92093-0358.

Bistable metal complexes with two redox-active ligands that can interconvert between two stable states in response to external stimuli have been proposed as potential building blocks to obtain molecular switching devices. Cobalt o-quinone complexes are excellent candidates since they have two valence tautomeric isomers $[Co^{III}(N-N)(SQ)(Cat)]$ and $[Co^{II}(N-N)(SQ)_2]$ (where N-N is a bidentate nitrogen ligand and SQ^- and Cat^{2-} are, respectively, the semiquinonate and catecholate forms of 3,5-di-tert-butyl-1,2-benzoquinone), with different net magnetic moments and optical properties. It has been shown that both tautomers can interconvert reversibly in response to external perturbations such as temperature or pressure and photochemical excitations. Here we propose a new dual-mode switching device by combining the valence tautomeric equilibrium with the reversible redox behavior of both isomers that allow us to control the critical temperature (T_c), the chromic properties as well as the intramolecular electron transfer phenomena.

017. **MAGNETIC FIELD EFFECTS ON THE PHOTOLYSIS OF ALKYL-COBALT COMPLEXES.** Ettaya N. Natarajan and Charles B. Grissom, Department of Chemistry, University of Utah, Salt Lake City, Utah 84112.

An external magnetic field can alter the rate of thermochemical, photochemical, and biochemical reactions with radical pair (RP) intermediates. The biologically important cofactors adenosylcobalamin (AdoCbl) and methylcobalamin undergo photolysis of the Co-C bond to give a RP, but only AdoCbl exhibits magnetic field dependent recombination. The β-oxygen of the ribofuranose ring appears to be necessary for fast recombination in the geminate cage. In alkylcobalamins, the magnetic field dependence observed in the range 0-0.7 T is entirely accounted for by changes in the efficiency of HFI. Magnetic field effect (MFE) studies on model complexes such as alkylcobaloximes and alkylCosalen complexes may help in understanding the mechanism of MFEs on Co-C bond homolysis. In alkylcobaloximes, only benzylcobaloxime exhibits a MFE and not ethylcobaloxime. This may be due to β-hydrogen abstraction in the ethyl radical. Both benzylcobaloxime and ethylCosalen complexes exhibit CIDNP upon photolysis and identifies the precursor excited state for the Co-C bond homolysis as the singlet state.

018. **Photoinitiated Sulfur Atom Transfer to a Quadruply Bonded Molybdenum-Tungsten Complex** Sara A. Helvoigt, Daniel S. Engebretson, Daniel G. Nocera, Michigan State University East Lansing, MI 48824

We have utilized quadruply bonded metal-metal complexes with the general formula, $M_2Cl_4P_4$, (M= Mo, W; P = PMe_2Ph) to initiate a multielectron atom transfer reaction. Two photon spectroscopy establishes that excitation to a low energy metal-localized excited state initiates electron transfer from one metal to the other, resulting in a mixed valence excited state, M^{II}-$M^{II} \rightarrow M^{III}$-$M^{I}$. This mixed valence excited state is predisposed to two electron reductions at M^I. A significant result to date is that excitation with visible light (λ = 375- 495 nm) prompts sulfur atom transfer from $SPPh_3$ to $MoWCl_4(PMe_2Ph)_4$ to give $W(S)Cl_4(PMe_2Ph)$; the homobimetallic complexes $M_2Cl_4(PMe_2Ph)_4$ (M = Mo, W) are photochemically inert. This result illustrates a notable example of multielectron photochemistry from a discrete excited state. The photophysics of heteronuclear bimetallic complexes as compared to homonuclear bimetallic complexes will also be discussed.

019. **STEP-SCAN FT-IR ABSORPTION DIFFERENCE TIME-RESOLVED SPECTROSCOPY STUDIES THE EXCITED STATE ELECTRONIC STRUCTURES OF d^6 TRANSITION METAL POLYPYRIDINE COMPLEXES.** G. D. Smith and R. A. Palmer, Department of Chemistry, Duke University, Durham, NC 27708. P. Chen, K. M. Omberg, D. A. Kavaliunas, J. A. Treadway and T. J. Meyer, Department of Chemistry, UNC-Chapel Hill, Chapel Hill, NC 27599-3290.

Step-Scan FT-IR Absorption Difference Time-Resolved Spectroscopy (S^2FT-IR ΔA TRS) has been used to study the excited state electronic structure of several d^6 transition metal polypyridine complexes, including a series of ester-substituted bipyridine complexes. Insight on the distribution of electron density in the excited state is obtained by comparing the transient ΔA spectra of the MLCT excited state in the region of the ν(CO) of the metal-bound carbonyl or ester group to those of the one-electron oxidized and reduced forms of the complexes. ΔA timeslices were collected with 10 ns time resolution.

020.
A COMPREHENSIVE STUDY OF THE ZWITTERIONIC EXCITED STATE OF THE DELTA MANIFOLD OF QUADRUPLY BONDED COMPLEXES. **Daniel S. Engebretson**, George E. Leroi and Daniel G. Nocera, Department of Chemistry and the LASER Laboratory, Michigan State University, East Lansing, MI 48824

A comparative study of the one- and two-photon spectroscopy of quadruply bonded metal dimers of the type $M_2X_4L_4$ (M_2 = Mo_2, W_2, MoW; X = Cl, Br, I; L = PMe_3 or $AsMe_3$) has been performed. Two-photon fluorescence excitation allows access to the δ^{*2} state, formally forbidden by one-photon selection rules. The valence bond (VB) description of bonding suggests this doubly excited δ^{*2} state to be highly zwitterionic, which we have verified by experimentally determining one- and two- electron energies. This data, in conjunction with data previously determined for the $\delta\delta^*$ state, allows us to fully elucidate the energies of the states in the δ/δ^* manifold. The systematic effect of ligand substitution on these energies is analyzed.

021. PHOTOLUMINESCENCE OF GOLD(I) COMPLEXES. Zerihun Assefa, John P. Fackler, Jr., Jennifer M. Forward, and Tiffany A. Grant, Department of Chemistry and Laboratory for Molecular Structure and Bonding, Texas A&M University, College Station, TX 77843-3255

In recent years the photochemistry and photophysics of Au(I) complexes has come to be of interest. A large variety of complexes have been synthesized that exhibit luminescent properties both in the solid state and in solution. Typically the focus has been on trying to understand what types of complexes will exhibit luminescence and why it occurs. The sulfonated phosphine ligand, TPPTS (trisodium tris(*m*-sulfonatophenyl)phosphine), has been utilized to form a three-coordinated Au(I) in $Na_8[Au(TPPTS)_3]$ which exhibits luminescence in aqueous solution. The complex dissociates in solutions with low dielectric constants and the luminescence is lost. The synthesis, characterization and luminescence studies of this and related complexes will be discussed.

022. EVIDENCE FOR EXTENSIVE EXCITED-STATE METAL-METAL INTERACTIONS IN PLATINUM(II) α,α'-DIIMINE COMPLEXES. K. R. Kendrick and G. A. Crosby, Department of Chemistry, Washington State University, Pullman, WA 99164-4630

Spectroscopic measurements of the excited-state properties of $Pt(bpy)(CN)_2$ and the red form of $Pt(bpy)Cl_2$ are yielding new information on the lowest emitting levels of these complexes in the solid state. Studies of the temperature dependences of the emission lifetimes indicate that the emitting manifolds reflect an electronic symmetry higher than the molecular geometry. The zero-field splittings are similar to those found in several bimetallic platinum(II) complexes that have been attributed to a $^3(d\sigma^* \leftarrow p\sigma)$ configuration localized between adjacent metal ions. These data support an assignment of the broad symmetric emission bands observed from Pt(α,α'-diimine)(1-(ethoxy-carbonyl)-1-cyanoethylene-2,2'-dithiolato) complexes as $^3(d\sigma^* \leftarrow p\sigma)$ also. The evidence for these assignments will be summarized.

023.
A CHARGE-SEPARATED STATE BASED ON COPPER.
Ruthkosky, M.; Kelly, C. A.; Meyer, G. J., Department of Chemistry, Johns Hopkins University, Baltimore, MD, 21218.

A series of $Cu(N-N)(PPh_3)_2(PF_6)$ compounds, where N-N is bipyridine or phenanthroline, have been synthesized and characterized. These compounds exhibit metal-to-ligand-charge-transfer transitions (d-π^*) near 370nm and photoluminesce at room temperature in dichloromethane. The excited-state lifetimes of the bipyridine compounds last for ~100ns, while the phenanthroline excited-states last twice as long (~200ns). These lifetimes decay biexponentially, suggesting emission from two different excited-states. The shorter-lived component for the $Cu(phen)(PPh_3)_2PF_6$ compound has been deteremined to be ~20ps. Substitution at the 2 and 9 positions of the phenanthroline ligand increase the copper complex excited-state lifetime to ~10μs and a single-exponential decay suggests emission from one state. A donor-acceptor compound based on a cuprous triphenylphosphine bipyridine compound covalently attached to a methyl viologen moiety has also been prepared and characterized. Excited-state absorption spectroscopy has been used to quantify the lifetime of the charge-separated state.

024.
Dynamic Quenching of Porous Silicon Excited States. Minh C. Ko and <u>Gerald J. Meyer</u> Department of Chemistry, Johns Hopkins University, Baltimore, MD 21218

The excited states of porous silicon materials are dynamically quenched by donors and acceptors in fluid solution at room temperature. The quenching processes have been explored by steady state and time resolved photoluminescence and absorption spectroscopies. The kinetics are complex but, are well described by the Kohlrausch-Williams-Watts function. Inverse Laplace transformation techniques have been utilized to recover the underlying distribution of excited state relaxation rates and provide some insights into the dynamic quenching mechanism(s).

025. FOUR- AND FIVE-COORDINATE ANIONIC ALUMINUM. <u>David. A. Atwood</u>, and Drew Rutherford, Center for Main Group Chemistry, Department of Chemistry, North Dakota State University, Fargo, North Dakota 58105.

Primary amido alanes $[R_2AlNHR']_2$, can be deprotonated and metallated by R"Li to yield the unique derivatives, $[R_2AlNLi(thf)_nR']_2$ where n = 1-3 depending on the steric requirements of R and R'. By increasing the steric bulk of R and R' the potentially π-bonded species, $[R_2Al=NR]^-$ $[Li(thf)_n]^+$ might be accessible. As the size of these groups are increased the reaction with R"Li does, indeed, yield anionic species. However, the aluminum atom is also alkylated and the resulting complexes are of the form, $[R_2R"AlNHR']^-$ $[Li(thf)_4]^+$. A systematic study of these 4-coordinate complexes, and a discussion of their impact on the search for 5-coordinate aluminum imides (incorporating a tetradentate (N_2O_2) salen ligand), will be presented.

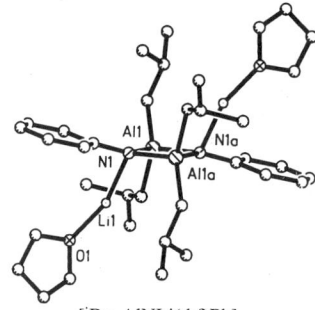

$[^iBu_2AlNLi(thf)Ph]_2$

026. COMPUTATIONAL INVESTIGATION OF THE EXISTENCE OF BP DOUBLE BONDS IN X_2BPY_2 COMPOUNDS, <u>Thomas M. Gilbert</u> and Steven M. Bachrach, Department of Chemistry, Northern Illinois University, DeKalb, IL 60115

Several examples of X_2BPY_2 compounds have been synthesized, and a few of these exhibit relatively short BP distances and planar P atoms. These latter have thus been characterized as containing B-P double bonds. However, *ab initio* calculations by Schaefer and Allen on the model compound H_2BPH_2 indicated that its molecular ground state structure exhibits a fairly long BP distance and a pyramidal P atom, inconsistent with such a characterization. A transition state structure 4.8 kcal above the ground state displayed BP double bonding.
We have greatly expanded on this work by computational examination (MP2/MP4 methods, 6-31G* basis sets) of more complex X_2BPY_2 compounds containing donor and acceptor groups (OMe, F, NO_2, Me, BH_2, allyl), in homoleptic and mixed stoichiometries. Several molecules mimic the Schaefer/Allen results; however, some intriguing differences are observed which, taken in total, tend to suggest that BP double bonds do not exist.

027. SYNTHESIS AND CHARACTERIZATION OF THE INDIUM (III) COMPLEX OF A NOVEL AMINOTHIOL (N_3S_2) LIGAND. <u>Jiafang He</u>[†], Sunita Saluja[†], Robin Rogers[§], Lynn C. Francesconi[*], [†]Hunter College of the City University of New York, New York City, New York 10021, [§] University of Alabama, Tuscaloosa, Alabama 35487.

A neutral indium complex of β,β'-[Bis(2-mercapto-2-methylpropyl)bis(2-aminoethyl)amine] has been synthesized at room termperature with isothiocyanate as an ancillary ligand. The complex has been characterized in the solid state by X-ray crystallography in addition to other techniques. NMR studies suggest the presence of two species in DMSO solution. The structural characterization and the chemistry of the new complex will be discussed. Derivatives of the ligand will also be presented.

028.
Insertion Reactions of GeGl$_2$ in transition metal-chlorine and hydrogen bonds. **A.C. Filippou**, J.G. Winter, G. Kociok-Kohn

NO ABSTRACT AVAILABLE

029. **Cyclogallanes as a Function of Ligand Steric Loading**

R. Chad Crittendon, Xiao-Wang Li, Hank Su, and Gregory H. Robinson
Department of Chemistry, The University of Georgia, Athens, GA 30602-2556.

The landscape of Group 13 organometallic chemistry has been considerably enhanced with the recent discovery of cyclogallanes—cyclic gallium metal triangles containing Ga–Ga bonds—by this laboratory. In addition, these interesting substances display traditional aromatic character—a property we have referred to as *metalloaromaticity*. Thus far, two examples of cyclogallanes have been prepared and characterized: $M_2[(Mes_2C_6H_3)Ga]_3$ (M= Na, K; Mes = 2,4,6-Me$_3$C$_6$H$_2$). In an effort to further assess the steric loading capabilities of cyclogallanes and to further examine this interesting phenomena a number of other substituted aryl ligands have been examined. Synthesis, properties, and characterization of new organometallic compounds will be described.

030. **Monovalent and Multiply Bonded Gallium Complexes Supported by the Tris(3,5-di-t-butylpyrazolyl)hydroborato ligand, [Tp$^{Bu^t2}$]. Syntheses and Structures of [Tp$^{Bu^t2}$]Ga and [Tp$^{Bu^t2}$]GaE (E = S, Se, Te).** Matthew C. Kuchta and Gerard Parkin*, Department of Chemistry, Columbia University, New York, New York 10027.

[Tp$^{Bu^t2}$]Ga has been synthesized by metathesis of [Tp$^{Bu^t2}$]Na with "GaI". [Tp$^{Bu^t2}$]Ga is the first discrete (*i.e.* not mixed valent) monovalent gallium compound to be structurally characterized in the solid state. [Tp$^{Bu^t2}$]Ga undergoes oxidative addition reactions with the elemental chalcogens, sulfur, selenium and tellurium, to give the terminal chalcogenido complexes [Tp$^{Bu^t2}$]GaE (E = S, Se, Te). [Tp$^{Bu^t2}$]GaE (E = S, Se, Te) are the first simple examples of valence multiple bonding to gallium. The structures of [Tp$^{Bu^t2}$]GaE (E = Se, Te) have been determined by X-ray diffraction.

031.
SYNTHESIS AND REACTIVITY OF THE NOVEL ALUMINOXANE [Mes*AlO]$_4$. R. J. Wehmschulte, P.P. Power, Department of Chemistry, University of California, Davis, CA 95616

The reactions of [Mes*AlH$_2$]$_2$ with oxygen donors like DMSO, Ph$_2$SO and water yield the novel aluminoxane [Mes*AlO]$_4$. (1). The structure of 1 features a unique planar, eight-membered Al-O ring and is currently the only aluminoxane [RAlO]$_n$ with three coordinate aluminum and two coordinate oxygen atoms. Reactions of 1 with Lewis acids such as AlMe$_3$ and AlEt$_3$ give crystalline compounds, whose structures may be of relevance to the still unknown structure of methylaluminoxane.

032. SYNTHESIS AND CHARACTERIZATION OF POTENTIAL PRECURSOR COMPOUNDS CONTAINING THE GALLIUM-ANTIMONY BOND. Richard L. Wells* and Edward E. Foos, Department of Chemistry, Duke University, Durham, NC 27708-0346; Louise M. Liable-Sands and Arnold L. Rheingold, Department of Chemistry, University of Delaware, Newark, DE 19716; Peter S. White, Department of Chemistry, University of North Carolina at Chapel Hill, Chapel Hill, NC 27514.

In an attempt to prepare potential single-source precursors to GaSb, as well as to study the formation and structural environment of the gallium-antimony bond, several new compounds have been prepared. The 1:1 reaction of t-Bu$_3$Ga with Sb(SiMe$_3$)$_3$ resulted in the formation of the Lewis acid-base adduct t-Bu$_3$Ga•Sb(SiMe$_3$)$_3$, while 1:1 reaction of Mis$_2$GaCl (where Mis = (trimethylsilyl)methyl-) with Sb(SiMe$_3$)$_3$ yielded the dehalosilylation product [Mis$_2$GaSb(SiMe$_3$)$_2$]$_2$. The synthesis and characterization of these compounds will be presented, along with the results of the series of reactions involving t-Bu$_x$GaCl$_{(3-x)}$ and Mis$_x$GaCl$_{(3-x)}$ (x = 1-3) with Sb(SiMe$_3$)$_3$.

033. SYNTHESIS OF DISUBSTITUTED DERIVATIVES OF THE a^2 ISOMER OF THE [B$_{20}$H$_{18}$]$^{4-}$ ANION. R. A. Watson-Clark, C. B. Knobler, M. F. Hawthorne, University of California at Los Angeles, Los Angeles, California 90095-1569

During our search for compounds suitable for use in boron neutron capture therapy, we have synthesized several new derivatives of the [a^2-B$_{20}$H18]$^{4-}$ anion. The apical-apical (a^2) isomer of the polyhedral borane anion [a^2-B$_{20}$H18]$^{4-}$ reacts with oxalyl chloride in dichloromethane to produce a bis-carbonyl anion, [a^2-B$_{20}$H$_{17}$(CO)$_2$]$^-$. The two carbonyl substituents are located on the equatorial boron belts farthest removed from the B-B bond linking the two cages. The a^2 isomer of the carbonyl anion easily rearranges to the equatorial-equatorial (e^2) isomer under basic conditions. The [a^2-B$_{20}$H$_{17}$(CO)$_2$]$^-$ anion reacts with sodium azide to produce a bis-isocyanate species, [a^2-B$_{20}$H$_{17}$(NCO)$_2$]$^{3-}$, and with water to form a bis-carboxylic acid species, [a^2-B$_{20}$H$_{17}$(COOH)$_2$]$^{3-}$. The reactions, ^{11}B NMR spectra, and structures of these and related species will be discussed.

034. PREPARATION AND STRUCTURE OF 5-(9-BORA[3.3.1]BICYCLONONYL)-(5,BBN-μ-HYDRIDO)-TRIDECAHYDRODECABORANE: CONVERSION TO 5-BBN-B$_{10}$H$_{13}$ AND 1,8-BIS-(DIMETHYL-AMMONIUM) NAPTHALENE[η^4-6,5,10,9-BBN-B$_{10}$H$_{12}$]. A. N. Bridges; J. Liu; R. G. Kultyshev; D. F. Gaines; S. G. Shore, Department of Chemistry, The Ohio State University, Columbus, OH 43210

Na[B$_{10}$H$_{13}$] reacts with 9-Br-BBN (BBN = 9- bora[3.3.1]bicyclononane) in CH$_2$Cl$_2$ to produce (5-BBN)-(5,BBN-μ-hydrido)-B$_{10}$H$_{13}$ (1) in high yield. Solid state studies show that the BBN unit in 1 rejects a symmetrical bridging interaction on an edge of the decaborane cage between the B(5) and B(6) atoms. Instead, it favors an off-center interaction localized near the B(5) vertex in tandem with an agostic interaction of the B(5) terminal hydrogen to the BBN boron atom. NMR studies suggest this structure is maintained in solution. Upon reaction of **1** with Et_2O, the BBN unit moves into a terminal position at the B(5) vertex to give 5-BBN-$B_{10}H_{13}$ (**2**). Reaction of **1** with the Proton Sponge, 1,8-bis-(dimethylamino)napthalene, in CH_2Cl_2 gives the unusual species PSH[η^4-6,5,10,9-BBN-$B_{10}H_{12}$] (**3**) (PSH = protonated Proton Sponge). Solid state studies of **3** confirm migration of the BBN unit to a symmetrical position over the decaborane face in preparation for insertion into the cluster. To our knowledge the novel interaction of the BBN fragment with the decaborane cluster in **1**, and the semi-insertion interaction observed in **3**, are the first examples of such interactions observed between monoboron moieties and boron hydride systems.

035. LIQUID CRYSTALS CONTAINING BORON CLUSTER. Zbynek Janousek, Andrew G. Douglass, Kerim A. Bairamov, Ryan M. Nunley, Piotr Kaszynski, *Organic Materials Research Group*, Department of Chemistry, Vanderbilt University, Nashville, Tennessee 37235.

We have been developing a new class of liquid crystalline materials based on boron *closo*-clusters. Theoretical models suggest that these 3-dimensional, highly polarizable σ-aromatic moieties can be used to induce coaxial dipole moments as large as 16 D and to engineer mesogens for NLO applications.[1] Mounting experimental evidence supports these theoretical predictions and indicates that calamitic materials incorporating these boron clusters are nematogenic.

Here we will present the preparation and properties of liquid crystals containing 10- and 12-vertex *closo*-boranes which include highly polar mesogenic derivatives of monocarboranes. We will also discuss the significance of conformational preference in stabilization of liquid crystalline phases.

1. Kaszynski, P.; Huang, J.; Jenkins, G. S.; Bairamov, K. A.; Lipiak, D. *Mol. Cryst. Liq. Cryst.* **1995**, *260*, 315.

036. MOLECULAR BUILDING BLOCKS FOR THE SYNTHESIS OF PHOSPHATE AND ARSENATE MATERIALS OF ALUMINUM AND GALLIUM
Mark R. Mason,* R. Mark Matthews, Alisa M. Perkins, James D. Fisher, Mark S. Mashuta, and John F. Richardson, Department of Chemistry, University of Louisville, Louisville, KY 40292, and Ashwani Vij, Department of Chemistry, University of Idaho, Moscow, ID 83844.

We have initiated the synthesis of molecular building blocks suitable for the construction of new phosphate and arsenate materials. Our approach has been to react an alkylaluminum or alkylgallium reagent with a phosphonic or an arsonic acid. Reactions with phoshonic acids yield *cis-trans* mixtures of the cyclic products [$R_2M(\mu_2$-$O_2P(OH)R'_2)]_2$ (**1**) (M = Al, Ga) which are converted to [$RM(\mu_3$-$O_3PR')]_4$ (**2**) via thermolysis. Compounds **2** possess a cubic $M_4P_4O_{12}$ core analogous to the D4R building units common to several $AlPO_4$ and $GaPO_4$ materials. Reaction of phenylarsonic acid with $^tBu_3Al$ similarly yields the cubic compound [$^tBuAl(\mu_3$-$O_3AsPh)]_4$ (**3**), whereas reaction of phenylarsonic acid with $^tBu_3Ga$ yields the bicyclic compound [$^tBu_2Ga(\mu_2$-$O_2AsPh)]_2(\mu_2$-$O)$ (**4**). All compounds have been characterized by NMR spectroscopy, mass spectrometry, and elemental analysis. The structures of six cubic compounds **2**, the structures of two of the cyclic precursors **1**, and the structure of **4** have been verified by X-ray crystallography. Details of the synthesis, characterization, and solution dynamics of these new compounds will be presented.

037.
SYNTHESES AND CHARACTERIZATION OF η^5-SILACYLCOPENTADIENYL AND η^5-GERMACYLCOPENTADIENYL TRANSITION METAL COMPLEXES.
Jeffrey M. Dysard, T. Don Tilley*, Department of Chemistry,University of California at Berkeley, Berkeley, CA 94720

Our group has long had an interest in the stabilization of reactive silicon fragments through coordination to transition metals. This strategy has been successfully employed in our labs to produce the first examples of η^5-sila- and η^5-

germacyclopentadienyl compounds. Structural as well as spectroscopic evidence supports the aromatic nature of these heterole rings. We have recently developed routes to the first early transition metal sila- and germa-aromatic complexes, including the first structurally characterized η^5-silacyclopentadienyl transition metal complex, $[(C_5Me_5)(C_4Me_4SiSiMe_3)ZrCl_2]$. Synthetic routes to these types of compounds will be described, as well as bonding, structure, and reactivity studies of the new complexes.

038. **SYNTHESIS AND REACTIVITY OF CATIONIC GROUP 4 METAL COMPLEXES WITH AMIDO-FUNCTIONALIZED CYCLOPENTADIENYL LIGANDS**
P.J. Sinnema, K. Liekelema, A. Meetsma, B. Hessen, J.H. Teuben, Department of Organic and Molecular Inorganic Chemistry, University of Groningen, Nijenborgh 4, 9747 AG Groningen, the Netherlands.

Convenient synthetic routes have been found to prepare a family of alkylamido-functionalized cyclopentadienyl Group 4 metal hydrocarbyl complexes of general formula $[C_5H_4(CH_2)_nNR']MR_2$ (M = Ti, Zr; R = Me, Ph, CH_2CMe_3, CH_2Ph; R' = Me, i-Pr, t-Bu). Thermolysis of the Ti-phenyl and Ti-neopentyl derivatives in the presence of PMe_3 yields stable titanium benzyne and alkylidene complexes respectively. Reaction of the benzyl derivatives with the Lewis-acidic borane $B(C_6F_5)_3$ produces thermally stabile cationic benzyl complexes $\{[C_5H_4(CH_2)_nNR']M(CH_2Ph)\}[PhCH_2B(C_6F_5)_3]$. In this presentation the solution dynamics and cation-anion interactions of these compounds will be discussed, together with their activity in the catalytic polymerization of olefins.

039. **GROUP 4 METALLOCENE POLYOXOANIONS.** Robert H. Beer[1] and Emil Radkov[2], Departments of Chemistry, [1]Fordham University, Bronx, NY 10458 and [2]Columbia University, New York, NY 10027

Group 4 metallocene polyoxoanions were prepared by reacting $Cp_2M(OTf)_2$ (M = Ti, Zr, Hf) with $PW_{11}O_{39}NbO^{4-}$ to form $Cp_2M(PW_{11}O_{39}NbO)_2^{6-}$. An X-ray crystallographic study and a solution ^{17}O labelled nmr study of the zirconocene derivative supports its formulation as two polyoxoanion fragments bridged by a metallocene moiety. $Cp_2M(PW_{11}O_{39}NbO)_2^{6-}$ reacts with simple ions, e. g. Cl$^-$, Br$^-$, CN$^-$, $CH_3CO_2^-$, to restore $PW_{11}O_{39}NbO^{4-}$. The reaction of chlorotrialkylsilanes with $Cp_2M(PW_{11}O_{39}NbO)_2^{6-}$ leads to $PW_{11}O_{40}Nb(OSiR_3)^{3-}$ (R = Me, iPr). $Cp_2Zr(PW_{11}O_{39}NbO)_2^{6-}$ reacts with aluminum pthalocyanine chloride, PcAlCl, to form the oxo bridged conjugate PcAl-O-NbPW$_{11}O_{39}^{3-}$.

040.
COORDINATION OF TROPIDINYL ANION TO ZIRCONIUM(IV) AND STUDY OF THE POLYMERIZATION OF ALKENES BY THE RESULTING COMPLEXES. Gino G. Lavoie, Robert G. Bergman, Department of Chemistry, University of California, Berkeley, California, 94620-1460

A derivative of the natural product tropine was used to investigate a novel coordination environment about transition metals. In zirconium(IV) complexes of the tropidinyl ligand, the monoanionic bicyclic ligand is bound to the metal center via a two-electron nitrogen dative bond and a four-electron allyl moiety, as observed in the X-ray structure. The activity of these resulting zirconium complexes towards the polymerization of alkenes was explored.

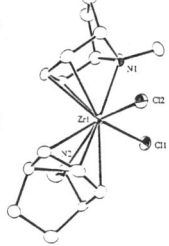

041. REDUCTION OF NITROGEN OXIDES TO DINITROGEN BY WELL-DEFINED ZIRCONIUM(II) COMPLEXES
Kristopher P. McNeill, Robert G. Bergman, and Richard A. Andersen
Department of Chemistry, University of California, Berkeley, California, 94720

Exposure of solutions of Cp_2ZrL_2 (Cp= cyclopentadienyl, L = trialkylphosphine) to NO gas results in production of $[Cp_2ZrO]_n$ and N_2. The zirconium oxide may be trapped by Me_3SiCl and Cp_2ZrMe_2 to give $Cp_2Zr(OSiMe_3)(Cl)$ and $(Cp_2ZrMe)_2O$, respectively. Similar results are obtained when N_2O is used as the oxidant. These reactions occur very quickly even at -100 °C with half-lives under 1 s. Mechanistic details will be presented. Development of a catalytic system will be discussed.

042. SYNTHESIS AND REACTIVITY OF A MONOMERIC TITANOCENE SULFIDO COMPLEX
Zachary K. Sweeney, Jennifer L. Polse, Richard A. Andersen, and Robert G. Bergman, Department of Chemistry, University of California, Berkeley, California, 94720-9989.

Reaction of $Cp*_2Ti(C_2H_4)$ with H_2S or elemental sulfur in the presence of pyridine forms the sulfido complex **1**. This complex functions as a selective nucleophile in reactions with allyl chlorides and α,β-unsaturated aldehydes. Treatment of **1** with H_2 results in the reversible formation of $Cp*_2Ti(H)SH$ (**2**). Evidence for the intermediacy of a dihydrogen complex in the hydrogen exchange reactions of **2** will be discussed.

043. EVIDENCE FOR A PYRAMIDAL d⁰ YTTRIUM METALLOCENE. Charles P. Casey,* Maureen A. Fagan, Susan L. Hallenbeck, J. Monty Wright, Department of Chemistry, University of Wisconsin, Madison, WI 53706.

The d⁰ yttrium pentenyl chelate **1**, a model for the proposed but unobserved intermediates in Ziegler-Natta alkene polymerization, has been synthesized and studied. A fluxional process that exchanges the diastereotopic Cp* ligands was observed by variable temperature NMR spectroscopy. Cp* exchange is proposed to occur via rapid and reversible alkene dissociation followed by rate limiting inversion of the pyramidal d⁰ yttrium center.

044. NOVEL C_2-SYMMETRIC *ansa*-CALCICENES via REDUCTIVE COUPLING OF FULVENES.
K. M. Kane, P. J. Shapiro, University of Idaho, Moscow, ID 83844, and A. L. Rheingold, University of Delaware, Newark, DE 19716

The reductive coupling of dimethylfulvene with activated calcium to give the *ansa*-calcicene $Me_4C_2(C_5H_4)_2Ca$ has been shown to yield considerable amounts of the impurity $Ca(Cp-iPr)_2$. Further examination of this reaction suggests that abstraction of hydrogen by the nascent anion radical accounts for the impurity. We have developed a high yield route to novel C_2-symmetric *ansa*-calcicenes via the reductive coupling of (6-aryl)fulvenes. The major product of phenylfulvene reacted with calcium in THF is *trans*-1,2-$(PhH)_2C_2(C_5H_4)_2Ca(THF)_2$, the structure of which has been confirmed by X-ray crystallography. The synthesis of C_2-symmetric calcicenes from other fulvenes will be presented.

INOR

045. C$_2$-SYMMETRIC *ansa*-CHROMOCENES: SYNTHESIS AND REACTIVITY
K. M. Kane and P. J. Shapiro, University of Idaho, Moscow, ID 83844.

Our attempts to prepare new *ansa*-chromocenes from the dilithium and disodium salts of X–(CpH)$_2$ type ligands resulted in poor yields of the *ansa*-complexes. We have developed a high yield route to C$_2$-symmetric *ansa*-chromocenes from novel *ansa*-calcicenes. For example, CrCl$_2$ reacts with *trans*-1,2-(PhH)$_2$C$_2$(C$_5$H$_4$)$_2$Ca(THF)$_2$, in THF (in the presence of C≡N-*t*Bu) to give *trans*-1,2-(PhH)$_2$C$_2$(C$_5$H$_4$)$_2$Cr(C≡N-*t*Bu) in yields greater than 80%. The synthesis and reactivity of these new Cr(II) and Cr(III) *ansa*-complexes will be presented.

046. INTRAMOLECULAR ALLYL REARRANGEMENTS IN C$_S$- AND C$_2$- SYMMETRIC ANSA-METALLOCENES.
Michael B. Abrams, Jeffrey C. Yoder, Cyrille Loeber, Michael Day, and John E. Bercaw.
Arnold and Mabel Beckman Laboratories of Chemical Synthesis, California Institute of Technology, Pasadena, California 91125

A series of novel dimethylsilylene-bridged, *bis*-(cyclopentadienyl) based ligand arrays have been developed for preparation of *meso*- and *racemo*-complexes of scandium and yttrium. The (η^3-C$_3$H$_5$) derivatives have been synthesized, and variable-temperature proton NMR spectroscopy indicates dynamic behavior associated with the allyl moieties. Two processes have been found to operate concurrently in these systems: (1) rotation of the η^3-allyl about the metal-allyl bond, and (2) generation of an unstable η^1-allyl complex, rotation about the resultant allyl C-C single bond, and regeneration of the η^3-species. The use of rigid ancillary ligand arrays has permitted, for the first time, quantitative information to be obtained on the relative and absolute rates of these processes. The effects of solvent, ligand array, and transition metal on the rates will be discussed.

047. FUNCTIONALIZED MACROCYCLES VIA ZIRCONOCENE COUPLING.
Jonathan R. Nitschke and T. Don Tilley, Dept. of Chemistry, UC Berkeley, Berkeley, CA 94720

The coupling of diynes into macrocyclic species via the Negishi reagent, "Cp$_2$Zr" generated from Cp$_2$ZrCl$_2$ and ^{n}BuLi, has recently been reported by our group. More recently, we have determined that this zirconocene-coupling chemistry can be extended to the synthesis of functionalized macrocycles containing bipyridyl units. The synthesis of these species will be discussed, along with their assembly into macromolecular architectures via co-coordination of metal ions to the bipyridyl units. Alternative strategies for introducing functional groups into these macrocyclic species will also be presented.

048. THE SYNTHESIS OF CAGED MOLECULES BY ZIRCONOCENE COUPLING OF TRI(ALKYNYLARYL)SILANES . Feng-Quan Liu, T. Don Tilley*, Department of Chemistry, University of California at Berkeley and Lawrance Berkeley Laboratory, Berkeley, CA 94720

The coupling of two alkynes by zirconocene produces a metallacyclic compound. Recently, we have shown that the zirconocene coupling of diynes produces either polymers or macrocycles. We have now shown that this carbon-carbon bond-forming reaction can be utilized to construct cage-like molecules. More specifically, the zirconocene coupling of tri(alkynylaryl)silanes results in cage compounds. Upon treatment with acid the resulting zirconacyclopentadiene complexes are hydrolyzed to give the corresponding organic cage compounds and zirconocene dichloride. An attempt to control the size of the cage through modification of the length of the triyne spacer will be discussed. The characterization, thermostabilities and reactivities of the compounds will also be described.

049. **LAYERED DOUBLE HYDROXIDES : PRECURSORS OF NEW HYBRID MATERIALS**

V. Prevot, C. Forano, J.P. Besse
Laboratoire Physico-Chimie des Matériaux, CNRS 444
Université Blaise-Pascal,63177 Aubière Cedex, FRANCE

During the last few years, Layered Double Hydroxides, expressed as $[M^{II}_{1-x}M^{III}_x(OH)_2]^{x+}[A^{m-}_{x/m} \cdot nH_2O]^{x-}$ where A^{m-} is an exchangeable anion, have received considerable attention owing to their widespread use as anion exchangers, sorbents and catalysts. These lamellar host structures, with their high anionic exchange capacity (around 250 meq/100g) can take up a great amount of organic compounds such as fatty or dicarboxylic acids salts, alkyl or arene sulfonates and phosphonates by different ways of synthesis: coprecipitation, reconstruction or exchange reactions. With inorganic anions, like silicate, chromate and sulfate, we have recently evidenced a grafting of the guest anions onto the gallery surface by condensation with OH groups of the layers. With the aim to compare the reactivity of organic guest molecules in order to immobilize them on LDH mineral support, we have prepared a series of organic anions (X) containing $Zn_3Al(OH)_8X \cdot nH_2O$ and $Zn_2Cr(OH)_6X \cdot nH_2O$ LDH and studied the evolution of their chemical and structural properties under different thermal treatements. In the case of the tartrate anion, two LDH phases with a perpendicular and a parallel orientation of the anions in the interlayer domains are obtained. The last one is strongly favoured by the hydroxyl groups on the hydrocarbon skeleton. A grafting of the organic anions on the layer seems to occur at higher calcination temperature. This study has been extended to LDH's intercalated with different hydroxyl carboxylate anions. On the basis of the IR spectroscopy, x-ray diffraction, thermal analysis and ^{13}C NMR, the structure and the grafting of this new hybrid materials will be discussed.

050.
LOW-TEMPERATURE, FLUX-GROWN, NA-DOPED LAMNO3: MAGNETIC AND TRANSPORT PROPERTIES. **Newell R. Washburn** and Angelica M. Stacy, Department of Chemistry, University of California, Berkeley, CA 94720

Crystalline powders of Na-doped LaMnO3 have been precipitated from molten sodium hydroxide at 450°C. Elemental analysis by wavelength dispersive spectroscopy indicates that Na substitues for Mn giving LaMn1-xNaxO3+d, with x = 0.2. The pseudocubic lattice parameter of 3.916 Å is larger than that of the undoped material, consistent with substitution at the Mn-site. Analysis of magnetic susceptibility gives a ferromagnetic Curie-Weiss temperature of 130 K, which is to be compared with the onset of weak ferromagnetism near 300 K. Low-temperature magnetization measurements suggest a model of canted antiferromagnetic clusters. Transport data will also be presented and compared with magnetic data.

051.
A NEW POLYTYPE OF ALKALI-RHODATES: $K_{0.5}RhO_2$, $Rb_{0.2}RhO_2$, AND Cs_xRhO_2. **Barbara A. Reisner**, Anne Dolbecq, and Angelica M. Stacy, Department of Chemistry, University of California, Berkeley, CA 94720.

The synthesis and structure of a series of alkali-rhodates, $K_{0.5}RhO_2$, $Rb_{0.2}RhO_2$, and Cs_xRhO_2, are reported. The rhodates were precipitated from fluxes of their respective alkali-hydroxides at temperatures between 500-700 °C. The materials are layered and consist of alkali-metal atoms in trigonal-prismatic coordination between sheets of edge-sharing rhodium oxide octahedra. Because of alkali non-stochiometry, these are mixed-valent materials and contain rhodium in a high formal oxidation state. Powder X-ray diffraction data were used to refine the structure of $K_{0.5}RhO_2$ using Rietveld profile analysis. The rubidium and cesium analogs are highly reactive towards water. Depending on synthetic conditions, $K_{0.5}RhO_2$ preferentially forms its hydrated or anhydrous form. Variation in structural properties with hydration will be discussed.

052.
STRUCTURES AND PROPERTIES OF ALKALINE-EARTH METAL ALUMINUM BORATES. **Douglas A. Keszler** and Ki-Seog Chang, Oregon State University, Department of Chemistry and Center for Advanced Materials Research, Gilbert Hall 153, Corvallis, OR 97331-4003

The materials MAl2(BO3)2O (M = Ca, Sr, Ba) comprise a unique family of structures characterized by layers of planar BO3 triangles that are bridged by MO6 groups and Al2O7 units containing a linear Al-O-Al interaction. The BO3 groups in the Sr and Ba derivatives are oriented to produce substantial second-order nonlinear optical susceptibilities. These features as well as UV- and VUV-excited luminescence, ionic conductivity, and thermal-expansion properties will be discussed.

053. SYNTHESIS AND PROPERTIES OF CROWN ETHER PILLARED AND FUNCTIONALIZED LAYERED ZIRCONIUM PHOSPHONATE. B. Zhang and A. Clearfield*, Department of Chemistry, Texas A&M University, College Station, Texas 77843-3255.

A Series of crown ether functionalized unbridged and bridged layered zirconium phosphonates have been synthesized by direct reaction. The crown ether precursors used are 1-aza-15-crown-5 and 1,4,10,13-tetraoxa-7,16-diazacyclooctadecane. The phosphonic acids, prepared by a Mannich type reaction, were further reacted with the zirconium hexafluorozirconate ion in the presence of phosphoric acid. The studies suggest that when the ratio of H_3PO_4 to phosphonic acid is below 1, a zirconium layered compound is formed in which phosphonates bond with zirconium similar to that found for α-$Zr(HPO_4)_2 \cdot H_2O$ while phosphates bridge zirconiums in a mode like PO_4 in γ-$Zr(PO_4)(H_2PO_4) \cdot 2H_2O$. Futher addition of phosphoric acid results in an α-type layered compound. Ion exchange behavior of these compounds will be presented.

054. STRUCTURE AND PROPERTIES OF $Yb_{14}MnSb_{11}$ and $Yb_{14}MnBi_{11}$. **Julia Y. Chan**[§], David J. Webb[‡], Susan M. Kauzlarich[§]
[§]Department of Chemistry, University of California, Davis, CA 95616
[‡]Department of Physics, University of California, Davis, CA 95616

Rare earth transition metal complexes $Yb_{14}MnPn_{11}$ (Pn = Sb, Bi) have been prepared stoichiometrically from the elements in high yield in a welded tantalum tube, sealed in a fused silica ampoule, at 1000-1200 °C. These compounds are isostructural with the Zintl compound $Ca_{14}AlSb_{11}$ and crystallize in the the tetragonal space group $I4_1/acd$ (Z = 8). Single-crystal X-ray data were refined for $Yb_{14}MnSb_{11}$ (a = 16.615(2) Å, c = 21.935(4) Å, V = 6022(3) Å^3, R_1 = 3.58 %, wR_2 = 8.5 %), $Yb_{14}MnBi_{11}$ (a = 16.989(3) Å, c = 22.248(3) Å, V = 6420(3) Å^3, R_1 = 6.25 %, wR_2 = 16.3 %). The structures of these compounds are compared with $Ca_{14}MnPn_{11}$. Magnetization measurements performed on powder samples have shown that $Yb_{14}MnSb_{11}$ and $Yb_{14}MnBi_{11}$ order ferromagnetically at 60 K. The effective moments are as expected for $Yb_{14}MnSb_{11}$ and $Yb_{14}MnBi_{11}$ with μ_{eff} = 4.9(1) μ_B, consistent with a Mn^{3+} ion. $Yb_{14}MnSb_{11}$ and $Yb_{14}MnBi_{11}$ show a μ_{sat} = 6 μ_B at 5 K, higher than expected. Synthesis, structure, electronic, and single crystal magnetic data will be presented and discussed.

055. SYNTHETHIC OPTIMIZATION AND CHARACTERIZATION OF $Na_2Ti_2Pn_2O$ COMPOUNDS (Pn=As, Sb). **E. A. Axtell, III** and Susan M. Kauzlarich, Department of Chemistry, University of California, Davis, CA, 95616

$Na_2Ti_2Sb_2O$ and $Na_2Ti_2As_2O$ are members of a small class of compounds known as pnictide oxides, since both O^{2-} and Pn^{3-} (Pn=As, Sb) anions are present, with no Pn-O bonding observed. Although the structures and a synthesis of these materials have been reported,[1] in our hands, production of a single phase material has proven to be challenging. The optimal conditions for their synthesis will be presented. These remarkable compounds possess Ti in the 3+ oxidation state (d^1 electronic configuration) in the presence of oxygen. The Ti atoms sit on a square lattice. Ti is bound to two oxygen atoms and four pnictogen atoms. Exchange could occur both through Ti-O-Ti linkages and from weak Ti-Ti bonding (2.93Å). For this reason, magnetic measurements have been undertaken for the two compounds. There is low temperature peak in the T dependendent magnetism of the As analogue and rising susceptibility as RT is approached. There is a rise the suceptibility of the Sb analogue at 100K. The field dependent data for both compounds is nonlinear. These compounds are of interest as possible "ladder" compounds and may be of relavance to d^1 superconductivity.
  1. A. Adam, H-U. Schüster, *Z. Anorg. Allg. Chem.* **1990**, *584*, 150.

056. TRANSFORMATIONS OF MOLECULAR PRECURSORS TO ONE-, TWO-, AND THREE-DIMENSIONAL PHOSPHATE MATERIALS. Claus G. Lugmair, T. Don Tilley, Department of Chemistry, University of California, Berkeley, Berkeley, California 94720-1460, and the Chemical Sciences Division, Lawrence Berkeley Laboratory, 1 Cyclotron Road, Berkeley, California 94720

The preparation of multicomponent metal oxides from molecular precursors has several advantages including better control over stoichiometry and homogeneity. Another advantage is the ability to modify the structure of the final solid state material by coordination of an appropriate structure-modifying agent to the precursor. Our effort in this area has included the preparation of a zinc cluster containing the $O_2P(O^tBu)_2$ ligand as a precursor to $Zn_xP_yO_z$ materials. This precursor can be transformed under mild conditions to a polymer or a layered material without loss of the organic groups. At low temperatures (134 °C) the organic groups thermally eliminate isobutylene and water to form $Zn_xP_yO_z$ materials. The structure of the resulting inorganic material can be modified by the use of templating agents. Attempts to prepare ordered phosphate materials containing Al, Zr, and Ti will also be described.

057. HIGHLY CHARGED DENDRITIC MACROMOLECULES AND THEIR USE AS NANOSCOPIC BUILDING BLOCKS. Joshua W. Kriesel, T. Don Tilley, Department of Chemistry, University of California, Berkeley, Berkeley, California 94720-1460, and the Chemical Sciences Division, Lawrence Berkeley Laboratory, 1 Cyclotron Road, Berkeley, California 94720

There has been significant recent interest in the use of dendrimers as nanoscopic building blocks. Dendrimers are ideal building blocks for molecular scaffolds due to their high degree of structural regularity and there readily modifiable surface functionality. We have synthesized highly charged dendrimers based on a carbosilane backbone and cationic ruthenium atoms at the periphery. These cationic ruthenium dendrimers have been used as building blocks in superlattice structures by linkages to organic polyanions. The synthesis and characterization of these highly charged dendrimers and supperlattices derived therefrom, will be discussed. Additionally, possible applications of these supperlattices as host structures will be described.

058. **Phosphine-Functionalized Tetrathiafulvalenes as Precursors to Redox-Active Inorganic-Organic Extended Materials.** Kim R. Dunbar,[a]* Marc Fourmigué,[b] Calvin E. Uzelmeier,[a] and Giulio Grandinetti.[a] [a]Department of Chemistry, Michigan State University, East Lansing, MI 48824 and [b]Institut de Materiaux (IMN, UMR CNRS 110), Université Nantes, F-44072 Nantes Cedex 03, France.
Incorporation of transition metal ions into a conjugated polymer backbone to form coordination polymers is the subject of considerable interest in view of the possible electrical and magnetic properties of such chains. Thus, it has been reasoned that the introduction of metal ions into a conjugated chain could lead to partially-filled bands and possibly metallic conductivity. In this vein, suitably designed multidentate ligands bearing the extra redox functionality of a tetrathiafulvalene core are particularly attractive. These include both the coordinating ability of the selected TTF's substituents and the capability of the TTF core to be readily oxidized to the cation radical. By studying the reactivity of a new class of phosphines stabilized by TTF, our groups have designed a program to bridge the areas of organic conductors and transition metal phosphine chemistry. The new ligands under investigation include monodentate $\{P(TTF)_x(Ph)_{3-x}$ (x = 0-2)$\}$, bidentate $\{ortho$-$TTF(Me)_2(PPh_2)_2\}$ and tetradentate $\{TTF(PPh_2)_4\}$ phosphines. Substitution reactions with dinuclear metal-metal bonded complexes with ortho-$TTF(Me)_2(PPh_2)_2$ (referred to as P2) and $TTF(PPh_2)_4$, lead to molecules with interesting redox and structural properties. Compounds have been prepared from $[Re_2Cl_8]^{2-}$, $[Rh_2(CH_3CN)_{10}]^{4+}$ and $[M(CH_3CN)_n]^{2+}$ (n = 6, M = Fe, Co, Ni, Pd,; n = 4, M = Pt). X-ray structures of key molecular and oligomeric compounds reveal interesting solid-state interactions that may be exploited in the context of charge-transfer π-stacking as for example in the Rh(I) species $[Rh(P2)_2]^+$ which crystallizes in one dimensional chains with close intermolecular contacts (3.5 Å) between TTF substituents.

059. ENCAPSULATION OF TETRATHIAFULVALENE IN ZEOLITES AND MESOPOROUS HOSTS. Kathleen Hoekstra and Thomas Bein,* Department of Chemistry, Purdue University, West Lafayette, IN 47907, USA.

We are interested in the design and properties of nanoscale electronic conductors. Previously, we have synthesized conducting organic polymers within zeolitic frameworks and within MCM-41 channels. Charge transfer salts of tetrathiafulvalene (TTF) crystallize into segregated stacks that exhibit metallic conductivity. We now investigate the assembly of TTF salts within three-dimensional inorganic hosts. Thermal analysis, nitrogen sorption, and optical spectroscopies (IR, VIS-NIR) demonstrate encapsulation of the pi donor within zeolite Y and siliceous mesoporous MCM-41. Infrared spectra exhibit a broadened and shifted C-H stretch region for TTF after encapsulation, and a band characteristic of dimerized TTF cation radicals after exposure to iodine. A sharp singlet observed by ESR indicates the presence of radical species even prior to iodine treatment, and is tentatively associated with the acidity of the host surface. The influence of the nature of the host on the encapsulated TTF products will be discussed.

060. *s*-INDACENYL AS A BRIDGING LIGAND IN HOMO AND HETERO-BIMETALLIC COMPLEXES OF Rh(I) AND Cr(0).

Metal atoms that are held in close proximity exibit unique properties and many ligands have been tested for their ability to coordinate two or more metals in an effort to search for new chemical or physical properties. In previous studies [1] we described a series of heterobimetallic indenyl complexes of Cr(0) and Rh(I) and we found that the reactivity of both metal centers is greatly increased in comparison with that exhibited by the monometallic complexes.

As a part of an effort to compare the structure and bonding of indenyl complexes to the rigid tricyclic *s*-indacenyl, we shall present results concerning the synthesis and characterization of (*s*-indacenyl)-bis(RhL$_2$) and Cr(CO)$_3$- (5-hydro-*s*-indacenyl)-RhL$_2$ complexes (L$_2$ = COD; L = CO).

Attemps to prepare the trimetallic complexes Cr(CO)$_3$-(*s*-indacenyl)-bis(RhL$_2$) will be also discussed.

[1] C. Bonifaci, G. Carta, A. Ceccon, A. Gambaro, S. Santi, A. Venzo, *Organometallics* 1996, 15, 1680; C. Bonifaci, A. Ceccon, A. Gambaro, P. Ganis, L. Mantovani, S. Santi, A. Venzo, *J. Organomet. Chem.* 1994, 475, 267.

061. HYDROGEN BONDING IN ORGANOMETALLIC COMPLEXES
Ben P. Patel and Robert H. Crabtree, Yale Chemistry Dept., 225 Prospect St., New Haven CT 06520-8107.

The hydrogen bonding groups -NH$_2$ and -N=CHR have been incorporated into the 2-position of benzoquinoline. The resulting ligands Hbq-NH$_2$ and Hbq-NCHR (=LH) cyclometalate readily to give [Ir(L)(H)(H$_2$O)(PPh$_3$)$_2$]$^+$, where the H-bonding group is *cis* to the labile aqua ligand. When a variety of ligands, such as F$^-$, H$^-$, or NCR, replace the H$_2$O, their reactivity is strongly affected by H-bonding interactions from the -NH$_2$ or -N=CHR groups.

062. SELECTIVE HETEROATOM ASSISTED LITHIATION AND DENDRITIC MOLECULES, Gerard van Koten, Johann T.B.H. Jastrzebski, Department of Metal-Mediated Synthesis, Utrecht University, 3584 CH Utrecht, The Netherlands.

Heteroatom assisted lithiation of arenes with alkyllithiums is an important reaction in our synthetic program directed to the synthesis of metallodendrimers. The selective ortho- or α-lithiation of 1,3-(Me$_2$NCH(R))$_2$C$_6$H$_4$ (I) will be reported. Reaction of I (R=H) with BuLi gives dimeric 2,6-(Me$_2$NCH$_2$)$_2$C$_6$H$_3$Li whereas I (R=Me) gives novel [2,6-(Me$_2$NCHMe)$_2$C$_6$H$_3$Li]$_2$[BuLi]$_2$ which constraints a unique CLi$_2$C'$_2$LiC ladder structure. The latter finding is related to the larger steric contaigns of the monoanion of I (R=Me). These results have been important for the design and synthesis of larger branched molecules that carry at their surfaces reactive C-Li funtionalities that can be used for the synthesis of catalytically useful metallodendrimers by transmetallation reaction procedures. The unique structural features of these aggregated multi-aryllithium compounds will be discussed.

063. From 2H-Phospholes to BIPNOR, a New Efficient Biphosphine for Asymmetric Catalysis.

F. MATHEY
Laboratoire "Hétéroéléments et Coordination " URA 1499 CNRS, DCPH, Ecole Polytechnique, F-91128 Palaiseau Cedex, France.

For many years now, we have studied the 1H-/2H-phosphole equilibrium and its synthetic applications. Upon reaction with alkynes, the 2H-phospholes yield the corresponding 1-phosphanorbornadienes. As ligands of rhodium(I), these phosphines show some potential in catalytic hydrogenation and hydroformylation of alkenes. Starting from 3,3',4,4'-tetramethyl-1,1'-biphospholyl and tolan, we have similarly obtained the corresponding 2,2'-bis-(1-phosphanorbornadienyl) (BIPNOR) with two chiral, non racemisable, phosphorus at the bridgeheads. The pure enantiomers of BIPNOR appear to be efficient ligands in asymmetric hydrogenation of C=C and C=O double bonds and in other related reactions.

064. ORGANOMETALLIC COMPOUNDS AS SINGLE-SOURCE PRECURSORS TO NANO-COMPOSITE MATERIALS: AN OVERVIEW. Joseph P. Carpenter, C. M. Lukehart, Stephen B. Milne, and Frank E. Jones, III, Department of Chemistry, Vanderbilt University, Nashville, Tennessee 37235; S. R. Stock, School of Materials Science and Engineering, Georgia Institute of Technology, Atlanta, Georgia 30332; James E. Wittig, Department of Applied and Engineering Sciences, Vanderbilt University, Nashville, Tennessee 37235; Bobby D. Jones and Robert Glosser, Department of Physics, University of Texas at Dallas, Richardson, Texas 75083; Donald O. Henderson and R. Mu, Physics Department, Fisk University, Nashville, Tennessee 37208; and Jane G. Zhu, Solid-State Division, Oak Ridge National Laboratory, Oak Ridge, Tennessee 37831.

Molecularly doped silica xerogels are prepared via sol-gel procedures by the addition of either main-group or transition metal organometallic compounds containing bifunctional ligands to conventional sol-gel recipes. These bifunctional ligands contain distal alkoxysilyl groups, so that the dopant molecules become covalently incorporated into the silica xerogel matrix as it is being formed. Subsequent thermal treatment under reducing or oxidizing-then-reducing conditions leads to the decomposition of molecular precursor and formation of a nanoparticulate material highly dispersed throughout the xerogel matrix. The formation of nanocomposites containing nanoclusters of metals, semiconductors, or binary alloys will be presented along with characterization data. Factors permitting the formation of nanoparticulates with stoichiometric precision will be discussed within this "molecules to nanocomposite" synthetic strategy.

065. SYNTHESIS AND REACTIVITY OF PALLADIUM CLUSTER COMPOUNDS. D. M. P Mingos, Department of Chemistry, Imperial College of Science, Technology and Medicine, South Kensington, London SW7 2AY, UK.

[$Pd_2(dba)_3$] has proved to be a very useful reagent for the synthesis of a wide range of cluster compounds containing tertiary phosphines and π-acid ligands such as CO, SO_2 and CNR. In the presence of controlled quantities of tertiary phosphines [$Pd_2(dba)_3$] is also effective at activating C-X bonds and this has led to a number of novel metal-metal bonded compounds. Furthermore [$Pd_2(dba)_3$]/P^tBu_3 reacts with $CFCl_3$ to give the unusual μ^3 methylidyne complex [$Pd_4(\mu^3$-CF)Cl_3 ($P^tBu_3)_4$]. The cluster reacts with H_2 under mild conditions to form CFH_3. This system therefore stoichiometrically converts $CFCl_3$ into CFH_3 and serves as an interesting model for heterogeneous palladium catalysts which are used for the conversion of CFC's into HFC's.

066. THE THIRD ROUTE TO THE FORMATION OF Si-O-Si- GROUP AND SILOXANE STRUCTURES. SILOXANES THROUGH SILANONES. M.G. Voronkov. Irkutsk Institute of Organic Chemistry, Siberian Branch RAS, 1 Favorsky Street, Irkutsk, 664033, Russia

In the organic and inorganic chemistry of silicon there is a prevailing concept of two most important pathways to the formation of disiloxane group Si-O-Si, a basic unit of all the siloxane structures. These routes are: intermolecular reactions of homo- and heterofunctional condensation of the corresponding Si-functional monomers and oligomers: SiOR + ROSi≡ → ≡SiOSi≡ + ROR; R = H, alkyl, acyl. SiMe₃, NO₂, etc. (1). SiOR + XSi≡ → ≡SiOSi≡ + RX ; X = halogen, alkoxyl, acyl. hydrocarbonyl, etc.(2). The author suggests a new concept on the formation of Si-O-Si group, according to which in the presence of two geminal reactive substituents OR and X at the silicon atom the conversions analogous to reactions 1 and 2, can proceed intermolecularly involving the generation of the corresponding silanone: >Si(OR)X → >Si=O + RX (3). This concept is supported theoretically and by abundant experimental data.

067. RECENT ADVANCES IN ELECTRONIC ABSORPTION SPECTROSCOPY. F. Tuczek, Institut für Anorganische Chemie und Analytische Chemie, Johannes Gutenberg Universität Mainz, D-55099 Mainz, Germany

Electronic coupling within multimetallic sites in metalloproteins is in general described phenomenologically in terms of a spin-Hamiltonian. However, exchange interactions are found in the manifold of excited states as well and are in general much larger than in the ground state. In binuclear copper systems, optical absorption spectroscopy has been used to directly evaluate these effects by the detection of characteristic shifts and splittings of bridging ligand-to-metal charge-transfer bands. A valence-bond configuration interaction (VBCI) model is described which parametrizes these interactions in terms of transfer integrals between metal d and ligand frontier orbitals. Since the charge-transfer excited-state electronic interactions determine the properties of the ground state, optical absorption spectroscopy can be used to evaluate ground-state spin-Hamiltonian parameters and in particular obtain information about superexchange pathways.

068. RECENT ADVANCES IN MAGNETIC CIRCULAR DICHROISM SPECTROSCOPY.
Elizabeth G. Pavel and Edward I. Solomon.
Department of Chemistry, Stanford University, Stanford, CA 94305-5080

Magnetic circular dichroism (MCD) spectroscopy has proven to be a useful tool for investigating the electronic and magnetic properties of metal centers in bioinorganic systems. A brief background of this important technique will be presented, followed by a discussion of the information available from the observed excited-state transitions. Ground-state electronic structure information is obtained through probing the temperature- and magnetic field-dependence of the MCD intensity at saturating limits, or saturation magnetization behavior. This variable-temperature, variable-field (VTVH) MCD data can be analyzed to extract ground-state sublevel splittings and g-values, as well as band polarizations. More recently, the methodology to analyze VTVH MCD data of non-Kramers $S = 2$ systems has been developed for both negative zero-field-splitting ($M_S = \pm 2$ ground state) and positive zero-field-splitting ($M_S = 0$ ground state), including the effects of temperature-independent $\mathcal{B}$-terms, z-polarization, and low-lying excited states. MCD and VTVH MCD have proven particularly valuable for investigating high-spin ($S = 2$) non-heme Fe^{2+} sites, which are EPR-silent and have very weak absorption extinction coefficients. Application of this methodology to non-heme enzymes will be presented to demonstrate how this technique provides active site geometric and electronic structure information which can be used to probe oxygen and substrate reactivity, provide insight into catalytic mechanism, and/or explore structure/function relationships.

069. RECENT ADVANCES IN RESONANCE RAMAN SPECTROSCOPY. Thomas M. Loehr, Department of Chemistry, Biochemistry, and Molecular Biology, Oregon Graduate Institute of Science & Technology, Portland, Oregon 97291-1000

Recent advances in Raman spectrometry that have had significant impact on studies of metalloenzymes include the CCD detector, the freeze-trapping of reaction intermediates, and time-resolved spectroscopic methods. The revolutionary charge-coupled device (CCD) photodetector is robust and offers low noise, high sensitivity, fast response, and a wide spectral window with up to 1100 pixels. The study of reaction intermediates has benefited by the adaptation of stopped-flow methodology with freeze trapping in cold liquid isopentane. Time-resolved spectroscopy can be used to follow the kinetics of the appearance and disappearance of chemical species. Examples will be given of the application of these methods in the study of metalloenzymes, with particular emphasis on the use of isotopic labeling for the study of oxygenated intermediates.

070. X-RAY SPECTROSCOPY S. P. Cramer, H. X. Wang, C. Bryant, X. Wang, C. Ralston, U. Bergman, M. Grush, G. Meigs, Department of Applied Science, University of California, Davis, CA 95616 and Lawrence Berkeley National Laboratory, Berkeley, CA 94720

EXAFS has been a valuable tool for bioinorganic structure analysis for two decades. Dramatic improvements in x-ray optics, detectors, and sources now permit many other types of x-ray experiments. Transition metal L-edge spectroscopy in the 500 - 1000 eV region reveals rich information about electronic structure. X-Ray magnetic circular dichroism (XMCD) offers powerful sum rules that can be used to determine spin and orbital angular momentum. XMCD also permits selectivity for the paramagnetic components of a metalloprotein. High resolution x-ray fluorescence spectroscopy is sensitive to oxidation state and spin state, and it complements the information revealed in absorption spectra. It also allows spin-polarized EXAFS (SPEXAFS) on dilute samples. This talk will illustrate the application of these techniques to bioinorganic samples. Specific systems include blue Cu and Fe-S proteins, Ni hydrogenases and CO dehydrogenases, nitrogenase, and the Mn in photosystem II.

071. ELECTRONIC STRUCTURE CALCULATIONS: DENSITY FUNCTIONAL CALCULATIONS OF SPIN POLARIZATION, SPIN AND CHARGE TRANSFER, AND SOLVENT/ENVIRONMENTAL EFFECTS IN TRANSITION METAL COMPLEXES. L. Noodleman, J. Li, C.L. Fisher, M.R. Nelson, D. Bashford, D.A. Case[a], D.N. Hendrickson, D.M. Adams[b], [a]Department of Molecular Biology, The Scripps Research Institute, La Jolla, CA 92037 and [b]Department of Chemistry and Biochemistry, University of California at San Diego, La Jolla, CA 92093-0358.

The basic principles of spin polarized density functional calculations for transition metal complexes will be summarized. The original of spin polarization splitting will be analyzed, as will related phenomena including electron relaxation, and spin and charge transfer. These concepts will be illustrated by examples, including transition metal-semiquinone complexes, the Mn containing active site of $Mn^{3+,2+}$, $Fe^{3+,2+}$ ions in aqueous solution, and iron-sulfur clusters. Calculations of optical spectra of the transition metal quinone complexes, and redox potentials of the Mn and Fe complexes will be presented. We will also discuss the self-consistent-reaction-field (SCRF) methodology for coupling a quantum cluster to a "classical" environment which can include solvent/protein or "extended ligands". The limitations of a "classical environment" will be analyzed. Applications to 2Fe2S proteins will also be introduced.

072. **VANADIUM(V) COMPLEXES OF POLYDENTATE AMINOALCOHOLS: FINE-TUNING COMPLEX PROPERTIES.** Debbie C. Crans* and Iman Boukhobza, Department of Chemistry, Colorado State University, Fort Collins, CO 80523-1872.

The structure, stability and lability of the mononuclear vanadium(V) complexes were examined upon modification of the aminoalcohol ligand. We chose to examine a series of 13 complexes that contain a five-coordinate vanadium atom and are well-behaved and characterized. The factors expected to affect these properties include electron donating effects of amine, sterics and solvation. The ligands were chosen so that the five-coordinate complex geometry was maintained and this was indeed the case for all but one complex. The effects of the stability of electron donating properties of the amine, partially reflected by the pK_a value of the protonated amine, were observed. Sterics were found to effect the stability of the complex and able to both stabilize and destabilize the complex depending on the nature of the interaction. Solvent interaction, on the other hand, proved to be important when the hydrofobicity of the ligand increased. Selected complexes were examined in further detail with respect to their lability; surprisingly little difference was found in the dynamic processes examined in this work. The stability of these complexes, on the other hand, was sensitive to electron donation of the amine, sterics and solvation. General rules for designing vanadium(V) complexes of this type are summarized.

073. **Syntheses and Reactivity of a Novel (μ-oxo)diiron(III) Complex.** G. Musie, M. Y. Darensbourg*, J. H. Reibenspies. Depart. of Chem., Texas A&M University, College Station, Tx 77843

Synthetic designs for the (μ-oxo)diiron(III) complexes of N,N'-(bis-2-methyl-2-mercaptopropane)-1,5-diazacyclooctane show very different rates as indicated below. Metal vs. ligand-based oxidations, redox properties and preliminary results of mechanistic studies relating the two syntheses of the first well-characterized (μ-oxo)diiron(III) complex in N_2S_2 ligation will be discussed.

074. **SYNTHESIS, CHARACTERIZATION AND REACTIVITY OF MOLYBDENUM DINITROGEN COMPLEXES CONTAINING A SILYLATED TRIAMIDOAMINE LIGAND.** Myra B. O'Donoghue, Richard R. Schrock, Department of Chemistry, Massachusetts Institute of Technology, Cambridge, MA 02139

The reaction between [N$_3$N]MoCl ([N$_3$N]$^{3-}$ = [(Me$_3$SiNCH$_2$CH$_2$)$_3$N]$^{3-}$) and excess Mg powder in THF under dinitrogen (1 atm) yields the diazenido species ([N$_3$N]Mo-N=N)$_2$Mg(THF)$_2$ (1). (1) can be oxidized to give [N$_3$N]Mo(N$_2$) (2), the neutral terminal dinitrogen complex. Upon heating (2) converts to the bridging dinitrogen complex, [N$_3$N]Mo-N=N-Mo[N$_3$N] (3). (1) reacts with various transition metal halides to give heterobimetallic bridging dinitrogen complexes and/or (2). Among the complexes characterized by X-ray crystallography are examples of Fe/Mo and V/Mo dinitrogen complexes.

075. ORGANOMETALLIC COMPLEXES OF MOLYBDENUM AND TUNGSTEN CONTAINING TRIAMIDOAMINE LIGANDS. S. W. Seidel, R. R. Schrock, and W. M. Davis. Department of Chemistry 6-331, Massachusetts Institute of Technology, Cambridge, MA 02139

Alkylation of tungsten(IV) and molybdenum(IV) chloride complexes containing triamidoamine ancillary ligands yields high spin alkyl complexes, [N$_3$N]MR ([N$_3$N]$^{3-}$ = [(R'NCH$_2$CH$_2$)$_3$N]$^{3-}$). The tungsten complexes rapidly undergo an α,α-double dehydrogenation reaction which yields the corresponding tungsten(VI) alkylidynes with loss of molecular hydrogen. The molybdenum alkyls are much more stable with respect to loss of dihydrogen, with alkylidyne formation only occurring at elevated temperatures and only with alkyls with relatively weak C-H$_\alpha$ bonds, such as neopentyl. Mechanistic studies using alkyls with only one alpha proton, such as cyclopentyl, have been performed in an attempt to understand the details of how this remarkable reaction works. Additional studies of the magnetic behavior of these complexes have been carried out to determine why the molybdenum complexes lose dihydrogen so much more slowly than the tungsten complexes.

076. LINEAR TRACTABLE COBALT(III) β-DIKETONATE POLYMERS. Ronald D. Archer, Emmanuelle Darmon, M. Gabriele Menges,* and Ven O. Ochaya, Department of Chemistry, University of Massachusetts, Amherst, MA 01003-4510.

Three series of linear tractable cobalt(III) β-diketonate condensation polymers have been prepared using two different synthetic strategies to avoid three dimensional intractability. With reference to the figure, when X= CHO and Y = H or X = H and Y = F, condensation with SCl$_2$ and S$_2$Cl$_2$ only occurs with the other two 2,4-pentanedionato (acac$^-$) ligands and Z = S and S-S, respectively. Characterization by NMR end-group analysis, gel permeation chromatography, and viscosity measurements indicate number-average molecular masses in excess of 10^4. Alternatively, the cobalt complex where X = H and Y = H, but the other two ligands have 3-CHO substituents, condenses with diamines, such as 1,4- or 1,3-diaminobenzene, to form low molecular weight polymers, where Z = -CH=NRN=CH- and R = C$_6$H$_4$ or an alkyl group. The polymers have some solubility in polar organic solvents.
*Univ. Mainz (Germany) exchange student

077.
CARBON-CARBON BOND-FORMING REACTIONS WITH GROUP 4 TROPOCORONAND COMPLEXES. **Michael J. Scott** and Stephen J. Lippard, Department of Chemistry, Massachusetts Institute of Technology, Cambridge, Massachusetts 02139

Utilizing the N$_4$-macrocyclic tropocoronand ligand system, a variety of organometallic group 4 metal complexes have been prepared and structurally characterized. The ability to change readily the size of the binding cavity without altering its electronic properties has afforded complexes with identical coordination environments but divergent stereochemistries at the metal center. The reactivity of these complexes has been probed with a range of substrates including carbon monoxide and isocyanides. In addition, unprecedented coupling reactions of substrates including isonitriles, isocyanates, and ketones with [Hf(TC-3,5)(η^2-OC(CH$_2$Ph)$_2$)], the product from the reaction of carbon monoxide with [Hf(TC-3,5)(CH$_2$Ph)$_2$], will be described.

H$_2$(TC-n,m)

078. **LINKAGE ISOMERISM IN Pd(II) COMPLEXES OF MACROCYCLIC LIGANDS.** Savitri Chandrasekhar[*a], Steve R. Rettig[b] and Alexander McAuley[*c]
[a]Chemistry, SWGC, Memorial University of Newfoundland, Corner Brook, NF Canada A2H 6P9.
[b]Department of Chemistry, University of British Columbia, Vancouver, BC, Canada V6T 1Z1.
[c]Department of Chemistry, University of Victoria, Victoria, BC, Canada V8P 3V6.

The synthesis, solution and solid state structure of Pd(II) *bis* complex of 1,5-dithia-8-azacyclodecane,[10]-aneS$_2$N are reported. Two spatial linkage isomers for the Pd(II) *bis* complex of [10]-ane S$_2$N have been obtained. The red isomer consists of four S atoms, two from each macrocyclic ligand to form a square planar Pd(II) complex with *syn* endodentate N atoms. Crystals of the purple isomer, [Pd([10]-aneS$_2$N)$_2$](PF$_6$)$_2$ are orthorhombic, space group, *Pccn*, a = 12.060(4), b = 19.295 (4), c = 11.913(2) Å, Z = 4. The structure was solved by Patterson and Fourier syntheses and was refined by full-matrix least-squares procedures to R = 0.033 and R_w = 0.031 for 2976 reflections with $I \geq 3\sigma(I)$. The purple Pd(II) isomer is essentially square planar with axial *syn* endodentate S atoms. Solution NMR suggests strong interaction from the axial S atoms with the central Pd(II) ion. Cyclic voltammetry and EPR spectroscopy have been used to study the redox reactivity of the Pd(II) complexes.

079. **EARLY TRANSITION-METAL MIXED ARSENIC-SULFUR COMPLEXES FROM *Cyclo*-(CH$_3$AsS)$_{3,4}$.** Omar M. Kekia and Arnold L. Rheingold, Department of Chemistry and Biochemistry, University of Delaware, Newark, DE 19716

While the transition-metal chemistry of the group 15 and 16 homocycles, e.g., *cyclo*-(RAs)$_n$ and *cyclo*-S$_8$ has been studied in detail, little attention has been given to the transition-metal derivatives of the mixed group 15/16 heterocycles, e.g., *cyclo*(RAsO)$_n$, *cyclo*-(RAsS)$_n$ and *cyclo*-(RSbS)$_n$. Our present work with *cyclo*-(CH$_3$AsS)$_{3,4}$ is an extension of our previous work with *cyclo*-(RAs)$_n$ and *cyclo*-(RAsO)$_n$. Using *cyclo*-(CH$_3$AsS)$_{3,4}$ as a source of mixed arsenic-sulfur ligands we have prepared and characterized a number of early transition-metal complexes, some of which have analogues in the pure arsenic and sulfur chemistry, but several others are novel complexes including single and double cubane analogue clusters. The product of the reaction of Cr(CO)$_6$ **1** shown here is a ring expansion product, one of the many diversified complexes isolated from this heterocyclic ligand system.

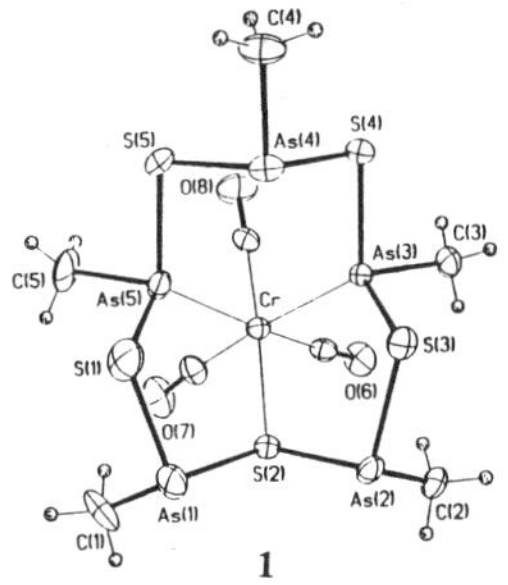

080. **CYCLOTRIVERATRYLENE DERIVATIVES AS SCAFFOLDS FOR MULTIMETAL CONSTELATIONS.** Daniel Stasko Dr. D. Scott Bohle, University of Wyoming, Laramie, Wy, 82071

The rigid, bowl shaped veratrole trimer, Cyclotriveretrylene (CTV), has only recently been exploited as a ligand for the formation of inorganic complexes, either through the functionalization of the CTV unit with chelating pendant groups or use of CTV in the formation of arene complexes. Extension of CTV's complexing ability to the inherent catecholate moiety, derived from cleavage of pendant methoxy groups, has yet to be explored. Here we report an examination of the CTV's versitility throught the syntesis of several CTV derivatives, CTV(OH)3(X)3, where X=OH, NH2, and SH, and the subsequent formation a number of Platinum constelation-type complexes of the form CTV(OX Pt L)3, where L=bidentate phosphine ligands, as well as the analoguos cathecolate, amidophenolate, and monothiochatecholate 'monomer' derivatives.

INOR

081. STEREOSPECIFICITY AND SELECTIVITY IN THE ASSEMBLY OF BINUCLEAR METAL COMPLEXES.
Eric Enemark and T. Daniel P. Stack, Department of Chemistry, Stanford University, Stanford, CA 94305-5080.

Our research in self-assembled metal complexes of a systematically-varied family of bis-catecholamide ligands provides a unique opportunity to assess the importance of simple steric interactions in the process of self-assembly. The specific form of metal complex is dependent on the chirality of the ligand backbone. Two different resolved chiral ligands generate stereospecific helices. While one of these ligands demonstrates self-recognition from a racemic mixture, the other displays no preference. The important factors governing such self-selectivity are addressed.

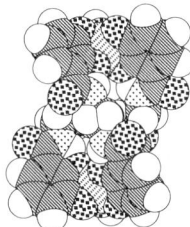

082.
CIRCULAR HELICATES. **Bernold Hasenknopf** Jean-Marie Lehn, Universite Louis Pasteur, Institut Le Bel, 4 rue Blaise Pascal, 67000 Strasbourg, France

The self-assembly of tris-bipyridines with iron(II) salts yields oligonuclear supramolecular architectures with the structural features of circular double helices. By variation of the counter-anion and the bridging ligand, tetra-, penta-, and hexanuclear complexes have been obtained. Their self-assembly proceeds after the initial formation of a trinuclear triple helicate.

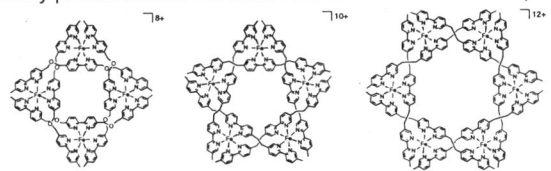

083.
COLUMNAR MESOPHASES BASED UPON NON-DISCOID SHAPED METAL DIKETONATE COMPLEXES. **S. T. Trzaska**, H. Zheng, T. M. Swager*, Department of Chemistry and Laboratory for Research on the Structure of Matter, University of Pennsylvania, Philadelphia, PA 19104-6323 and Massachusetts Institute of Technology, Cambridge, MA 02139-4307

Several new non-discoid shaped metal diketonate complexes exhibit columnar mesophases (Φ) commonly seen in planar disc shaped molecules. Rhodium and iridium β-diketonate compunds (I) align in an antiparallel arrangement to give a Φ_h phase. Octahedral cobalt, chromium, manganese, and iron complexes with low aspect ratios have also been found to display novel Φ_h phases. The synthesis and liquid crystalline properties of these and related compounds will be discussed.

084.
SYNTHESIS AND STABILITIES OF MONONUCLEAR AND DINUCLEAR IRON(II) AND IRON(III) COMPLEXES INCORPORATING A NEW MACROCYCLIC LIGAND.
Zheng Wang, Arthur E. Martell*, Ramunas J. Motekaitis and Joseph Reibenspies, Department of Chemistry, Texas A&M University, College Station, Texas 77843-3255

The reaction of the non-template [2+2] condensation between 2,6-diformylpyridine and 2,6-bis(aminomethyl)-p-cresol leads to a new dinucleating 24-membered macrocyclic ligand L. The crystal structure of L•4HCl•6CH$_3$OH was determined by x-ray crystallographic methods. The ligand maintains dimeric intergrity in both iron(II/II) and iron(III/III) states, while facilitating the formation of bridging μ-phenolate diiron cores. Potentiometric studies indicate that a variety of mononuclear and dinuclear iron(II) and iron(III) complexes form through pH 2 to 11 in aqueous solution. The protonation constants and stability constants for all the complexes identified will be discussed. The mechanism of the formation of both dinuclear iron(II) and iron(III) complexes are proposed.

085. NUCLEOPHILIC ATTACK OF SULFUR YLIDES AT η^2-ACYL COMPLEXES OF ZIRCONIUM.
Curtis White and L. J. Smith Vosejpka, Department of Chemistry, Alma College, Alma, MI 48801.

There have been a limited number of examples of nucleophilic attack on transition metal η^2-acyl complexes, however none of these reactions involve direct attack at the acyl carbon. It appears that the barriers to successful nucleophilic attack of a carbanion at the acyl carbon of an η^2-acyl complex are 1) the possibility of deprotonation and 2) the stability of the intermediate adduct. The reaction of the η^2-acyl complex, **1**, with sulfur ylides was studied. The phenyl substituted zirconium η^2-acyl represents an excellent target for nucleophilic attack as it contains no enolizable alpha protons.

The results of this investigation and their relation to the Brook rearrangement of silyl aryl ketones will be presented

086. TUNING REDOX POTENTIALS IN METAL COMPLEXES. Huang, J.[a]; Wright, S.[a]; Yap, G. P. A.[b]; Rheingold, A. L.[b]; Walters, M. A[a]. [a]*Department of Chemistry, New York University, New York, NY 10003;* [b]*Department of Chemistry, University of Delaware, Newark, DE*
Redox variation among the iron-sulfur and other metallo redox proteins is thought to be strongly influenced by hydrogen bonding and/or dipole interactions with the redox site. Much recent data on hydrogen bonding and dipole effects has been derived from redox-active Mo thiolate model complexes. Evidence will be presented that demonstrates the importance of polar effects in the modulation of redox potentials in redox active molybdenum thiolate complexes. The compounds to be discussed are Mo[BH(Me$_2$pz)3](NO)(SRS), where R = -(CH$_2$)$_6$-, -(CH$_2$)$_5$-, -(CH$_2$)$_2$CON(CH$_3$)(CH$_2$)$_2$-, -(CH$_2$)$_2$CONH(CH$_2$)$_2$-, -(CH$_2$)$_2$CONH(CH$_2$)$_3$-, and -(CH$_2$)$_3$CONH(CH$_2$)$_3$-; listed in the order of increasing redox potentials in the range -1.013 to -0.785 V relative to SCE.

087. SPECTROSCOPIC DETECTION OF A HYDROGEN FLUORIDE COMPLEX OF IRIDIUM. Ben P. Patel Robert H. Crabtree, Yale University, 225 Prospect St., New Haven, CT, 06520-8107

The incorporation of a non-ligating hydrogen bonding functionality into the ligand sphere of $[IrH(bq-NH_2)(PPh_3)_2(FH)][BF_4]$ (bq-NH$_2$ = 2-amino-7,8-benzoquinoline) leads, through an intramolecular Ir-F-H...NH$_2$ hydrogen bond, to the stabilization of a hydrogen fluoride ligand at 183 K. The formulation Ir-F-H...NH$_2$ was defined by ^{1}H, ^{31}P and ^{19}F NMR spectroscopy, and supported by a $^1J_{HF}$ coupling constant of 440 ± 5 Hz.

088. NEW BORATABENZENE ZIRCONIUM COMPLEXES. Arthur J. Ashe, III, Saleem Al-Ahmad and Jeff. W. Kampf, Department of Chemistry, University of Michigan, Ann Arbor, MI 48109-1055.

When activated by methylaluminoxane bis(boratabenzene) zirconium dichlorides (**1**) can serve as good olefin polymerization catalysts with activity comparable to Cp$_2$ZrCl$_2$. Thus it is of interest to prepare analogous bridged boratabenzene complexes (**2**). We have prepared several *ansa* boratabenzene complexes with one and two atom bridges (E).

089. SOLVENT AND LIGAND EFFECTS ON THE REACTIVITY OF ODD ELECTRON ORGANOMETALLIC RADICALS STUDIED BY PHOTO MODULATED VOLTAMMETRY (PMV). Denis C. Barbini and Wayne E. Jones, Jr., Dept. of Chemistry, Binghamton University, Binghamton, New York 13902-6016.

The coordination of a 17-electron radical by 2-electron donating solvents or ligands results in the formation of a 17/19 electron radical equilibrium. A study of the effects of solvent modification, reactivity, and ligand substitution on the reduction potential within an extended family of organometallic radicals of the type $M(CO)_3(C_5R)\cdot$ (where M = Mo, W and R = H$_5$, H$_4$CH$_3$, (CH$_3$)$_5$, (CH(CH$_3$)$_2$)$_3$H$_2$) by PMV leads to insight into this equilibrium. Two reduction potentials are observed in the frequency dependent analysis of $W(CO)_3(C_5H_5)\cdot$ with 5 mM PPh$_3$ in DMF solution. The reduction of the 17-electron species occurs at 0.04 V and the reduction of the 19 electron species at ~ -0.9 V vs. SCE. Additional information on the equilibrium kinetics can be gained from the PMV phase shift. For the PPh$_3$ system, the results can be correlated to the results of time-resolved spectroscopy to confirm our method.

090.
STERIC VS. ELECTRONIC EFFECTS IN THE SPIN-STATE EQUILIBRIA OF CYCLOPENTADIENYL AND INDENYL COMPLEXES OF CHROMIUM AND MANGANESE. **Jason S. Overby** and Timothy P. Hanusa, Department of Chemistry, Vanderbilt University, Nashville, TN 37235

Previous work in our laboratory has determined that the spin state of various substituted manganocenes can be influenced by the amount of steric bulk present in the complex, e.g., $[Mn(iPr)_4C_5H]$ displays a persistent high-spin state, whereas $[Mn(iPr)_3C_5H_2]$ undergoes a spin-crossover above room temperature. As an extension of that work, we have prepared other manganocenes and chromocenes with variously substituted cyclopentadienyl and indenyl ligands, and have isolated and characterized the first high-spin complex of Cr(II) containing solely π-bound ligands. The synthetic, magnetic and structural features of these compounds will be discussed.

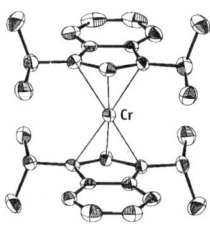

091. MAIN CHAIN ORGANOMETALLIC POLYMERS: POLY(ARYLENE-1,1'-FERROCENYLENE)S. M. David Curtis and Glen Southward, Department of Chemistry, University of Michigan, Ann Arbor, MI 48109-1055.

Conjugated polymers with metals in the main chain are of interest for a variety of device applications, e.g. thin film transistors, optical switches, etc. due to their interesting properties, including optical non-linearity, conductivity, and magnetic properties. We will report on the synthesis and characterization of a series of polymers and oligomers of the type **1**, poly(1,4-arylene-1,1'-ferrocenylene), and **2**, poly(2,5-thienylene-1,1'-ferrocenylene). Characterization will include conductivity, cyclic voltammetry, optical and magnetic behavior of neutral, and partially or fully oxidized polymer or oligomer.

092. TRANSITION METAL-CATALYZED RING-OPENING POLYMERIZATION OF SILICON-BRIDGED [1]FERROCENOPHANES. **John B. Sheridan**, Karen Temple, Alan J. Lough and Ian Manners. Rutgers University, 73, Warren St, Newark, NJ 07102 and University of Toronto, 80, St George St, Toronto, Ontario, M5S 3H6, Canada.

Monitoring of the $Pt(1,5-cod)_2$ (**1**) catalyzed ring-opening polymerization (ROP) of silicon-bridged ferrocenophanes by ^1H NMR spectroscopy revealed the initial formation of a [2]platinasila-ferrocenophane (**2**). This species can be prepared from **1** and $Fe(\eta-C_5H_4)_2SiMe_2$ (**3**), was fully characterized and acts a pre-catalyst for the ROP of **3** and other silacyclobutanes. A key step in the mechanism is believed to involve dissociation of the 1,5-cod ligand.

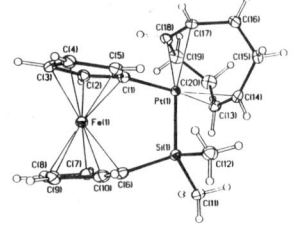

093. **ELECTROCHEMICAL AND STRUCTURAL STUDIES OF OLIGO(FERROCENYLSILANES) AS MODELS FOR POLY(FERROCENYLSILANE).** Ron Rulkens[a], Alan J. Lough[a], Ian Manners[a]*, Sherri R. Lovelace[b], Casey Grant[b], William E. Geiger[b]*
[a]Dept. of Chemistry, University of Toronto, Toronto, Ontario M5S 1A1
[b]Dept. of Chemistry, University of Vermont, Burlington, Vermont 05405-0125

Linear oligo(ferrocenylsilanes) were synthesized by anionic ring-opening oligomerization of the silicon-bridged [1]ferrocenophane fcSiMe$_2$ (fc = Fe(η-C$_5$H$_4$)). The pure, individual oligomers with between two and nine ferrocene units were isolated by chromatography. The structure of the linear pentamer represents a valuable model for the high polymer. Electrochemical studies indicate that "odd" and "even" oligo(ferrocenylsilanes) show different redox behavior, both series approaching that of the high polymer. spectroelectrochemical experiments show an intervalence electron transfer absorption for mixed-valent oligo(ferrocenylsilanes).

094. SYNTHESIS OF FERROCENYLENE-TETRAFLUOROPHENYLENE CO-OLIGOMERS.

Michael J. Lane and Paul A. Deck*, Department of Chemistry, Virginia Polytechnic Institute and State University, Blacksburg, Virginia 24061-0212.

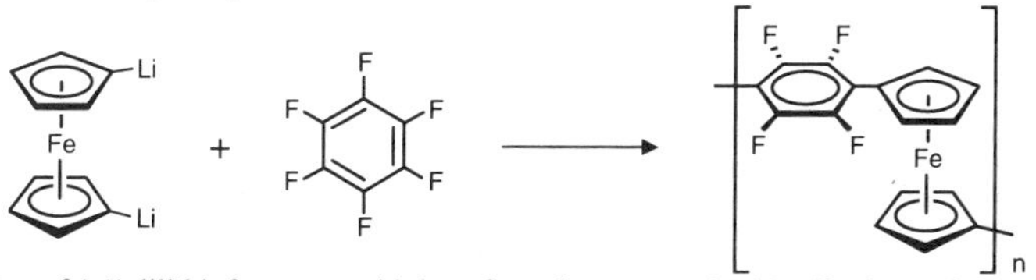

The reaction of 1,1'-dilithioferrocene with hexafluorobenzene at 25 °C affords a red-orange solid after aqueous workup. Characterization is vitiated by poor solubility and volatility, but analytical data are most consistent with a mixture of linear oligomers having the repeat unit shown above. This paper summarizes our progress toward the isolation and characterization of this and related materials.

095. BORON-SILICON EXCHANGE REACTIONS OF TRIMETHYLSILYL METALLOCENES.

Paul A. Deck,* Travis S. Fisher, and J. Sloan Downey. Department of Chemistry, Virginia Polytechnic Institute and State University, Blacksburg, Virginia 24061-0212.

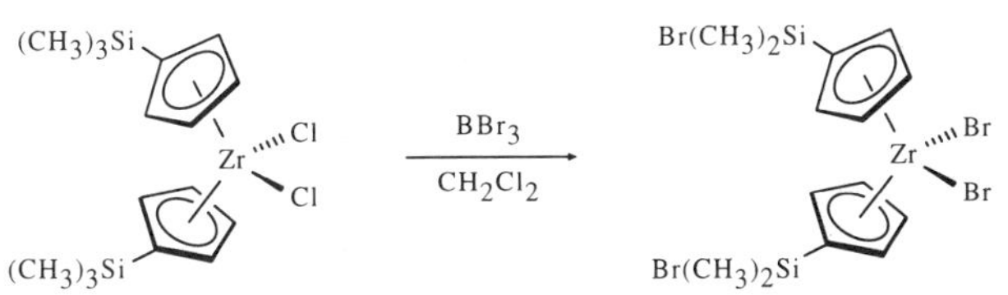

When treated with boron tribromide, trimethylsilyl-substituted group 4 metallocenes undergo selective bromo-demethylation of each trimethylsilyl substituent to afford the corresponding bromodimethylsilyl-substituted metallocene dibromides as shown above. In contrast, trimethylsilyl-substituted ferrocenes readily cleave the trimethylsilyl substituents to afford dihaloborylferrocenes as expected.

096. SYNTHESIS AND CHARACTERIZATION OF CONJUGATED METALLO-COPOLYMERS FOR ELECTRON TRANSFER APPLICATIONS. Biwang Jiang, Ross Niver, S. Yang, Wayne E. Jones Jr.*, Department of Chemistry, Institute for Material Research and Integrated Electronics and Engineering Center State University of New York at Binghamton, Binghamton, NY 13902-6016

A chlorine precursor copolymer was synthesized from the monomers of 2,5-bis(chloromethyl)-1,4-dihexyloxybenzene and [5,15-bis(mesityl)-10,20-bis(4-chloromethylphenyl)porphyrin]zinc(II) in different ratios. The precursor copolymer is soluble in common organic solvents and was characterized by NMR, IR, UV-vis, TGA and Elemental analysis. High molecular weights (Mn=770,000) were measured by GPC. The arylene vinylene copolymer film was formed through thermo-conversion. Doping of undrawn films of the conjugated copolymers with iodine vapor led to conductivities as high as 2 S cm^{-1}. The conjugated porphyrin copolymer also demonstrate substantial photoactivity including energy transfer, electron transfer and photoconductivity. X-ray data demonstrate long range order within thin films of these materials and recent results suggest their possible use in LED applications.

097. IMIDO/SULFUR EXCHANGE REACTIONS AND THE FORMATION AND REACTIVITY OF ZIRCONOCENE DIAZAMETALLOCYCLOBUTANES. Rebecca L. Zuckerman, Robert G. Bergman, Department of Chemistry, University of California, Berkeley, California, 94720.

Treatment of imido zirconocene complex Cp$_2$Zr(THF)(=N^{t}Bu) with CS$_2$ generated ^{t}BuNCS and (Cp$_2$ZrS)$_2$ in high yield. Reaction of Cp$_2$Zr(THF)(=N^{t}Bu) with R'NCS, R' = ^{t}Bu, Ph, Me, afforded (Cp$_2$ZrS)$_2$ and the corresponding carbodiimides R'NCN^{t}Bu.

Zirconocene imido complexes Cp$_2$Zr(THF)(=NR), R = ^{t}Bu, 2,6-Me$_2$-C$_6$H$_3$, and 2,6-^{i}Pr$_2$-C$_6$H$_3$, have been found to react with various symmetrical carbodiimides to give zirconocene diazametallocyclobutanes in a [2 + 2] cycloaddition reaction. The steric requirements for the [2 + 2] cycloaddition reaction will be discussed. Further reactivity of the diazametallocyclobutane complexes and studies of the mechanism of these reactions will be presented.

098. CARBENE MIGRATION IN A CLUSTER COMPLEX: OBSERVATION OF AN INTERMEDIATE CONTAINING A μ$_3$-CH$_2$ LIGAND INVESTGIGATED BY TWO-DIMENSIONAL (^{1}H, ^{13}C) and (^{1}H, ^{103}RH) NMR SPECTROSCOPY. F. Holger Försterling*, Craig E. Barnes§ and Christian Griesinger*
§Department of Chemistry, University of Tennessee, Knoxville TN 37996 and *Institut für Organische Chemie, Universität Frankfurt, D-60439 Frankfurt, Germany.

The carbene ligand in the cluster complex Cp*Rh(CpCo)$_2$(μ-CO)$_2$(μ-CH$_2$) is mobile between the two Rh-Co edges as shown by NOESY data. These data also reveal the presence of an intermediate in this process. {^{1}H,^{13}C} NMR experiments suggest that this species contains a μ$_3$-bridging rather than a terminal carbene ligand. Additional IR and multinuclear NMR experiments used to eludicate the structure of this intermediate will be discussed.

099.
HYDROGEN ATOM ABSTRACTION FROM ORGANOMETALLIC HYDROGEN SOURCES: FIRST KINETICS OF H ATOM ABSTRACTION BY BENZYL RADICAL FROM "Mo-S-H". Jerome C. Birbaum, James A. Franz and John C. Linehan, Pacific Northwest National Laboratory, P. O. Box 999, Richland, WA 99352.

Hydrogen transfers are very important in catalytic and stoichiometric organometallic reaction schemes, yet few kinetic studies of the transfers from a metal hydride to an organic radical have been performed. To our knowledge there are no kinetic studies of hydrogen atom transfers from atoms attached to metals such as sulfur or oxygen. We present the first such study with the extremely fast hydrogen atom transfer from $(CpMoSSH)_2$ to benzyl radical. The rate constant for this reaction was found to be greater than 10^6 $M^{-1}s^{-1}$ at room temperature.

100. BONDING AND REACTIVITY IN TRANSITION METAL SULFUR "BUTTERFLY" CLUSTERS. Lora Bergert and Suzanne Harris, Department of Chemistry, University of Wyoming, Laramie, WY 82071-3838

Curtis and coworkers have demonstrated that the "butterfly" cluster $(Cp')_2Mo_2Co_2S_3(CO)_4$ ($Cp' = CH_3C_5H_4$) has the ability to remove sulfur from thiophene.[1] This results in the formation of the "cubane" cluster $(Cp')_2Mo_2Co_2S_4(CO)_2$. While no intermediate has been isolated in this reaction, reactions with other nucleophiles lead to intermediate clusters of the form $[Mo_2Co_2S_3(CO)_4(Cp')_2(X)]$ (X = CNMe, PMe$_3$). We have carried out Fenske-Hall molecular orbital calculations on the butterfly cluster, the cubane cluster, and several intermediate clusters. Our results show that while the Co centers in the butterfly cluster are very electron rich, the LUMO is Co centered and provides a pathway for coordination of thiophene, ultimately leading to desulfurization of the thiophene. The orbital structure of the resulting cubane suggests that Co is oxidized in the transformation from butterfly to cubane. The different reactivity of the methyl isocyanide and phoshine intermediates (ligand substitution in the butterfly versus reversion to the parent butterfly) can be related to the ability of the ligands to remove electron density from the Co center.

[1] Curtis, M. D.; Riaz, U.; Curnow, O. J.; Kampf, J. W.; Rheingold, A. L.; Haggerty, B. S. *Organometallics* **1995**, *14*, 5337-5343.

101. PHOTOELECTRON SPECTROSCOPY OF DIMETAL TETRAFORMAMIDINATES. Dennis L. Lichtenberger and Matthew A. Lynn, Department of Chemistry, University of Arizona, Tucson, AZ, 85721; Malcolm H. Chisholm, Department of Chemistry, Indiana University, Bloomington, IN 47405.

The recent preparation of a variety of dimetal tetraformamidinate complexes by the Cotton group has added to the number of known complexes that contain a metal-metal multiple bond surrounded by four *bis* chelating ligands in a paddlewheel fashion (i.e., D_{4h} symmetry). We present the HeI and HeII photoelectron spectra of $M_2(form)_4$ (M = Mo, W, Ru; form = any of a series of substituted formamidines) to establish the metal- and ligand-based ionizations of these systems. Although the δ ionization from the quadruply bonded species is well resolved and is found in the 5-6 eV range, depending on the substitution of the formamidinate ligand, the metal-based σ and π ionizations lie among several that are ligand-based. The various formamidinate substitutions do not aid in resolving the metal-based ionizations from those that are ligand-based. Similar spectra are observed for the diruthenium tetraformamidinate complexes, with the lowest energy ionizations corresponding to the removal of an electron from the δ* and π* metal-based molecular orbitals.

102. **The Electronic Structure of d^4-Metal-Metal-η^1-Acetylides.** Dennis L. Lichtenberger, Teresa S. Wu and Andrew B. Uplinger, University of Arizona, Tucson, AZ 85721.

Gas-phase ultraviolet photoelectron spectroscopy was used to investigate the electronic structure of d^4 metal-metal-η^1-acetylides of the general formula $M_2(P)_4(C≡C-R)_4$ [M = Mo, P = PMe_3, R = H, Me, $SiMe_3$]. The metal-metal π and δ orbitals are able to have filled/filled interactions with the acetylide π orbitals, but can also by symmetry donate electron density into the empty acetylide π* orbital. All of the valence ionizations of the acetylide complex are found to be substantially destabilized from those of the parent chloride complex. The energy difference between the metal-metal δ and metal-metal π ionizations was found to decrease in going from the parent chloride complex to the trimethylsilylacetylide complex, which can be interpreted in terms of the influence of some backbonding into the acetylide π* orbitals.

103. TITLE Presenting Author Secondary Author, Address, City, State, Zip

Abstract: RIGID ROD COMPLEXES WITH PARAMAGNETIC MANGANESE END GROUPS.
Vasily V. Krivykh and Heinz Berke, Institute of Inorganic Chemistry, University of Zürich, Winterthurerstr. 190, CH-8057 Zürich, Switzerland

There is growing interest in rigid rod complexes due to their expected unusual magnetic and electronic properties. The rigid rod, which is defined as an inflexible, electronically conducting polyatomic unit, may function as a terminal or spacer group in di- and polynuclear complexes. In our approach, we sought to incorporate paramagnetic, redox-active and therefore polarizable centers into these species. The following series of manganese derivatives

$L_nMn-X]^{0,+}$ $L_nMn(-X)_2]^{0,+}$ $L_nMn-X-MnL_n]^{0,2+}$ $[L_nMn(-X-MnL_n)_m]^{0,x+}$

X = C≡C–R, C≡C, C≡C–C≡C, C≡C–⟨◯⟩–C≡C ; L_n = bis(dimethylphosphino)ethane (dmpe) and/or cp, $CH_3C_5H_4$, I

were prepared starting from cp_2Mn or its methyl derivative. In facile exchange reactions one or two cp rings can be replaced by one or two dmpe ligands and by the acetylenic moiety. The resulting mono-, binuclear and polymer Mn(II) complexes undergo reversible oxidation to Mn(III) species. In a unique way cpMn(dmpe)C≡CR complexes can dimerize and thus form a new redox-active series of dinuclear cpMn(dmpe)C=C(R)-C(R)=CMn(dmpe)cp compounds. The structural, electronic and magnetic properties of these complexes will be discussed.

104. **Formation and Characterization of C_2-Bridged Heterodinuclear Complexes.** Xinhong Gu, and Michael B. Sponsler, Department of Chemistry and W. M. Keck Center for Molecular Electronics, Syracuse University, Syracuse, NY 13244-4100.

Attempts to generate -CH=CH- bridged dinuclear complexes through the addition of metal hydrides to an iron acetylide uniformly failed and instead provided high yields of the -C≡C- bridged complexes. The reactions of Cp*(dppe)Fe-C≡CH with Cp_2ClZrH, DIBAL, and catecholborane each produced hydrogen and complexes **1, 2**, and **3**, respectively. The roles of acetylide acidity and steric interactions in these reactions will be examined through comparison with related reactions, and the reactivity of the new complexes will also be presented.

105.
AN INTERESTING NiPt-ACETYLENE COMPLEX, B. Richter, J. A. Davies, Department of Chemistry, University of Toledo, Toledo, Ohio 43606

The reaction of [NiPt(μ-CO)(μ-dppm)$_2$Cl$_2$] with excess MeOC≡CMe in toluene yields the 1,2-dimetallaolefin, [NiPt(μ-MeOC=CMe)(μ-dppm)$_2$Cl$_2$], **1**. Determination of regiochemistry by X-ray crystallography proved to be impossible as the heavy atom positions were best modelled with occupancies of 50 % Ni / 50 % Pt. The ^{1}H NMR spectrum of **1** in CD$_2$Cl$_2$ allowed measurement of the coupling constant between ^{195}Pt and protons of the acetylenic methyl group. The magnitude of $^3J_{Pt,H}$ strongly suggests formation of the regioisomer where the -OMe substituent of the acetylene is oriented to the Ni side and the -Me substituent is oriented to the Pt side.

Oligomerisation of MeOC≡CMe was found to be promoted by **1** with formation of two isomeric benzenes, 1,3,5- and 1,3,4-C$_6$(CH$_3$)(OCH$_3$)$_3$, and substituted cyclooctatetraenes.

106. CHARACTERIZATION OF CARBON DIOXIDE REACTION PRODUCTS OF A TITANIUM DINUCLEAR COMPLEX. Jinfeng Ni,[a] Yuhua Qiu,[a] Tamara M. Cox,[a] Cynthia A. Jones,[a] Chala Berry,[a] Laura Melon,[a] and Simon Bott.[b] [a]Department of Chemistry and Physics, Texas Woman's University, Denton, Texas 76204. [b]Department of Chemistry, University of North Texas, Denton, Texas 76203

Reaction of [(η^5-C$_5$H$_5$)Ti(μ-H)]$_2$(μ-η^5:η^5-C$_{10}$H$_8$) with CO$_2$ (3 atm) leads to the formation of the bis-formato complex [(η^5-C$_5$H$_5$)Ti(μ-η^1(O)CHO)]$_2$(μ-η^5:η^5-C$_{10}$H$_8$) (**1**) which has been characterized by X-ray crystallography. Complex **1** contains monodentate bridging formate ligands. The difference between vibrational frequencies υ_{as}(CO$_2$) and υ_s(CO$_2$) is 455 cm^{-1}, the largest ever observed for a monodentate formate ligand. The synthesis, structure, and IR spectra of **1** will be discussed. The reaction of [(η^5-C$_5$H$_5$)Ti(μ-H)]$_2$(μ-η^5:η^5-C$_{10}$H$_8$) with CO$_2$ at low pressure will also be discussed.

107. USE OF THE *E*- AND *C*-BASED DUAL PARAMETER SUBSTITUENT MODEL TO IDENTIFY A SMOOTH MECHANISTIC CHANGE IN THE REACTION OF ORGANIC DISULFIDES WITH A COBALT-CENTERED METAL RADICAL
Mark A. Aubart and Robert G. Bergman, *Department of Chemistry, University of California, Berkeley, CA 94720*

The monomeric heterobimetallic radical Cp$_2$Ta(μ-CH$_2$)CoCp (**1**) reacts with diaryl disulfides (ArSSAr) at room temperature to afford the corresponding diamagnetic monothiolate complexes, Cp$_2$Ta(μ-CH$_2$)Co(SAr)Cp (**2**). Kinetic and solvent effect studies suggest that the reaction is an associative bimolecular process, whose rate is only modestly enhanced in polar solvents. A Hammett σ/ρ substituent constant analysis yields a nonlinear correlation. Application of the *E*- and *C*-based dual parameter substituent model to the relative rates provided a more informative analysis, and suggested that the mechanism changes smoothly as the electronic character of the disulfide changes. The reaction of cobaltocene, CoCp$_2$, with disulfides in different solvents and with varying substituents was also examined for comparison with the reaction of **1** with disulfides. The results of these studies and how they relate to the intimate mechanism of disulfide activation by paramagnetic organometallic compounds will be presented.

Cp″Ta⇌Co-Cp (Cp below) + 0.5 ArSSAr ⟶ Cp″Ta⇌Co(Cp)(SAr)
1 **2**

108.
BIMETALLIC COMPLEXES OF 1,3,5,7-TETRA-*TERT*-BUTYL-*S*-INDACENE. Dermot O'Hare, Douglas R. Cary, Catherine G. Webster, Mark J. Drewitt, Stephen Barlow, Inorganic Chemistry Laboratory, South Parks Road, Oxford OX1 3QR, UK

Several mono- and bimetallic complexes of the *anti*-aromatic ligand 1,3,5,7-tetra-*tert*-butyl-*s*-indacene (Ic') have been prepared. The structures and electronic properties of these complexes will be discussed, particularly the ability of coordinated metal centers to interact through the conjugated ligand. Cyclic voltammetry has shown that there is strong metal-metal interaction in both $[(OC)_3Fe]_2Ic'$ and $(CpCo)_2Ic'$. We have examined the chemistry of this ligand with chromium, manganese and nickel, as well as oxidized and reduced forms of some of these compounds. Magnetic studies of bimetallic, particularly mixed-valence, species will be presented. In addition, the possibility of extending these complexes to coordination polymers and/or charge-transfer salts exhibiting interesting magnetic and electronic properties will be discussed.

109. ELECTRON TRANSFER ON THE VIBRATIONAL TIME SCALE IN PYRAZINE BRIDGED DIMERS OF TRI-RUTHENIUM CLUSTERS Clifford P. Kubiak, Tasuku Ito, and John Washington, Department of Chemistry, Purdue University, West Lafayette, Indiana, 47907

The mixed valence states of the pyrazine bridged hexaruthenium clusters $Ru_3(\mu_3$-$O)(CH_3CO_2)_6(CO)(L)(\mu$-$C_4H_4N_2)Ru_3(\mu_3$-$O)(\mu$-$CH_3CO_2)_6(CO)(L)$ (L = 4-dimethylamino-pyridine (**1**), pyridine (**2**), 4-cyanopyridine (**3**)) undergo intramolecular electron transfer on the vibrational time scale. Two discrete and reversible single electron reductions are evident by cyclic voltammetry in the redox chemistry of **1-3**. The splitting of the reduction waves, ΔE, depends on the electronic coupling H_{AB} between the triruthenium centers, and varies from 220 mV for **3** to 440 mV for **1**. The magnitude of the electronic coupling H_{AB} is found to strongly influence the IR spectra of the mixed valence states of **1-3** in the $\nu(CO)$ region. In the case of relatively weak electronic coupling (**3**), two $\nu(CO)$ bands are clearly resolved. In the case of strong electronic coupling (**1**), these bands broaden to a single $\nu(CO)$ absorption band. These data allow the rates of electron transfer for the mixed valence states of **1** and **2** to be estimated as $6 \pm 1 \times 10^{11}$ s^{-1} and $2 \pm 1 \times 10^{11}$ s^{-1}, respectively. Electron transfer in the mixed valence state of **3** is slower than the vibrational time scale

110. NMR AND OTHER TECHNIQUES FOR UNDERSTANDING CENTERED ZIRCONIUM HALIDE CLUSTERS, INSIDE AND OUT. Jerry D. Harris and Timothy Hughbanks*, Department of Chemistry, Texas A&M University, College Station, TX 77843.

Solution chemistry of the centered octahedral zirconium halide clusters, $[(Zr_6Z)Cl_{12}L_6]^{m-}$, Z = H, Be, B, C, N, Mn, Fe, Co, has continued to develop over the past few years. One of the problems that has plagued the area is product characterization; unless the products were crystalline they have remained uncharacterized, aside from elemental analysis. In an effort to find a routine method for analyzing the products in solution, NMR of the interstitial element has been investigated. NMR offers the advantages of one singlet per cluster species and each cluster species has a different chemical shift. In addition, different isomers of the same composition have different chemical shifts. Both solid state and solution NMR have been investigated for the different interstitials. This work has established some benchmark chemical shift ranges for the different interstitials that will prove useful for future work in this area.

111.
MOLECULAR LEVEL ENGINEERING OF SOLID STATE OXIDE POROUS MEMBRANES: APPLICATIONS TO SEPARATIONS AND DEGRADATION TECHNOLOGIES G.F. Strouse, B.I. Swanson, Los Alamos National Laboratory, CST-1, MS G755, Los Alamos, NM

Controlled preparation of molecular level engineered porous oxide membranes is critical for the development of novel devices. Such membranes find applications in catalytic and separation technologies. We present recent data on the surface modification of ITO and TiO_2 through the use of a patterned, self-assembly approach. The latter approach is shown to afford control over surface density and lateral distribution. The oxide substrates are deposited onto porous alumina membranes by CVD deposition. We present SEM and AFM images of a porous oxide membrane that has been surface modified by self-assembly. By modifying the surface with molecular recognition components and photo-active materials, we demonstrate the applicability of selective photo-oxidation at a molecularly engineered material. Probing the molecular array by surface enhanced Raman provides information about the binding and surface reactivity.

112.
COLOUMB DRIVEN PHASE TRANSITIONS IN A SINGLE CRYSTAL QUASI-1-D ELECTRONIC MATERIAL. B.I. Swanson, G.F. Strouse, Los Alamos National Laboratory, CST-1, MS G755, Los Alamos, NM

The MX materials $[Pt(en)_2I_2][Pt(en)_2](ClO_4)_4$, where en is per-deuterated ethylenediamine, represents a new structural type for the MX family. The material, which crystallizes in the C2/m monoclinic space group, forms sheets of ordered and disordered 1-D chains. The material has two observable phase transitions at 160K and 120K, which results in a a 3-D ordered material in an acentric C2 space group at 4K. The phase transitions are driven by changes in the coloumbic and hydrogen bonding interactions in the disordered sheet, resulting in re-organization of the ordered sheet. By comparison of the crystallographic, vibrational, and acoustic data, the phase transitions can be structural interpreted as arising from a discommensurate to commensurate phase transition at 120K, and an ordering transition resulting in the loss of the mirror plane at 160K. A theoretical model supporting the coloumbic model for the phase transitions is proposed.

113.
SYNTHESIS AND CHARACTERIZATION OF NEW QUASI-ONE-DIMENSIONAL CHAIN MATERIALS OF THE TYPE $[PT(EN)_2][PT(CN)_4X_2]$ J.A. Brozik, B.I. Swanson, and B.L. Scott, Chemical Science and Technology Division, Los Alamos National Laboratory, Los Alamos, NM, 87545

A new series of neutral, mixed-valent, quasi-one-dimensional chain materials with the general formula $[Pt(en)_2][Pt(CN)_4X_2]$ (where en = ethylenediamine and X = Cl, Br, I) have been synthesized. The chemical structures have been determined by X-ray crystallography. Raman, IR, UV-Vis absorption, and luminescence measurements have been carried out on these compounds in the solid state. These results will be discussed in terms of the intervalence charge transfer bands and the charge density wave (CDW) strengths of these extended low dimensional materials.

114.
IRON SANDWICH POLYOXOANIONS. SYNTHESIS, CHARACTERIZATION, X-RAY CRYSTAL STRUCTURES AND CATALYTIC OXIDATION STUDIES

Xuan Zhang, Dean C. Duncan, Qin Chen and Craig L. Hill, Department of Chemistry, Emory University, Atlanta, GA 30322.

Tetranuclear ferric sandwich compounds with the formula, $[Fe^{III}_4(H_2O)_2(PW_9O_{34})_2]^{6-}$ and $[Fe^{III}_4(H_2O)_2(P_2W_{15}O_{56})_2]^{12-}$ have been prepared via rational syntheses from the corresponding Keggin and Wells-Dawson lacunary species. The complexes have been characterized by IR, ^{31}P NMR, elemental analysis and single crystal diffraction analysis. These complexes have also been tested as oxidation catalysts for the reaction of hydrocarbons with hydrogen peroxide.

115.
SYNTHESES AND CHARACTERIZATIONS OF CHIRAL TETRAHEDRAL COBALT PHOSPHATES WITH ZEOLITE ABW AND RELATED FRAMEWORKS. **P. Y. Feng,** X. H. Bu, S. H. Tolbert, and G. D. Stucky, Department of Chemistry, University of California, Santa Barbara, California 93106

The hydrothermal syntheses, X-ray crystal structures and magnetic properties of a family of 3-D chiral framework cobalt phosphates are presented. Two different types of framework structures are described: One has the same topology as the well-known zeolite ABW framework and represents the only pure cobalt phosphate framework with a known zeolite topology; The other exhibits a framework connectivity which can be considered as a hybrid of tridymite and ABW frameworks. The transition from one framework type to the other is affected by the size of extra framework cations. These new chiral materials were obtained by systematic chemical variation, accurate control of solution pH values, and use of non-aqueous solvents. The enantiomorphic purity and twinning of these salts are correlated with the cation type and synthesis conditions.

116.
MULTICOMPONENT SELF-ASSEMBLY: THE CONSTRUCTION OF RIGID-RACK MULTIMETALLIC PSEUDOROTAXANES. **Hanadi Sleiman,** Paul N.W. Baxter, Jean-Marie Lehn, Karri Airola and Kari Rissanen, Institut LeBel, Universite Louis Pasteur, Laboratoire de Chimie Supramoleculaire, 4, Rue Blaise Pascal, 67000 Strasbourg, France; Department of Chemistry, American University of Beirut, Beirut, Lebanon; Department of Chemistry, University of Jyvaskyla, Jyvaskyla, Finland

The spontaneous self assembly of a new class of rigid rack-type multimetallic complexes comprising a linear array of metal ions is reported. Thus, macrocyclic phenanthroline ligands efficiently thread onto rigid-rod linear polytopic ligands as directed by copper(I) ions to yield multimetallic [3]- and [4]-pseudorotaxanes.This construction process is one of a few examples of inorganic self-assembly processes where two different ligand types are brought together as directed by metal ions to form the final architecture.The multicomponent self-assembly process as well as spectroscopic and structural properties of these pseudorotaxanes will be described.

117.
MODIFYING THE PORE SURFACES OF MCM-41 AND MESOPOROUS KANEMITE. **Blake J. Aronson,** Andreas Stein, Department of Chemistry, University of Minnesota, 139 Smith Hall, 207 Pleasant St. SE, Minneapolis, MN 55455

The use of long-chain surfactants as templating agents enables a relatively mild synthesis of high surface area (800-1500 sq. m/g) mesoporous (2.0-25 nm) silicates with ready control of the pore size. Through proper modification of the pore surfaces, these materials have a wide range of potential applications, including catalysis and electronics. Titanium dioxide, a large band-gap semiconductor and versatile photocatalyst, has been grafted onto the pore surface of MCM-41 and a mesoporous form of kanemite by reacting titanium tetrachloride with the as-synthesized mesostructured silicate. The grafting of managenese oxide and titanium sulfide, which are layered systems of interest for lithium batteries, has also been explored. The formation of mono- and multilayers has been analyzed by powder XRD, nitrogen adsorption, solid state NMR, Raman, and UV/Visible spectroscopies.

118. TURNING DOWN THE HEAT: THE MOLTEN SALT SYNTHESIS OF LAYERED PEROVSKITE SOLID ACIDS. Margret J. Geselbracht, Leo S. Macdonald, and Russell J. Scarola, Department of Chemistry, Reed College, Portland, OR 97202.

Layered perovskites have captured much attention recently due to their potential use as solid acid catalysts. These compounds are typically prepared by solid state reaction of the constituent carbonates and oxides at very high temperatures. In this work, the synthesis of the layered perovskite, $KCa_2Nb_3O_{10}$, was investigated in molten salt fluxes. This material, typically synthesized at 1200°C, was prepared with equivalent purity in excess molten KCl at 900°C. The molten salt synthesis was further used in attempts to dope the niobium site of the layered perovskite with tantalum. Strong evidence will be discussed for a previously unreported series of layered perovskites, $KCa_2Nb_{3-x}Ta_xO_{10}$ ($0 \leq x \leq 0.6$). After ion-exchange of the potassium ions for protons in 6M HNO_3, X-ray powder diffraction revealed that the tantalum-doped compounds intercalated several basic amines including pyridine (pK_a 5.25). Thus early evidence suggests that the acidities of the tantalum-doped analogs are at least comparable to that of the niobium parent compound.

119.
OPTICAL NONLINEARITIES OF LATTICE INCLUSION COMPOUNDS. **Songping D. Huang,*** R. Tom Xiong, and Lyzmarie Andino, Department of Chemistry, University of Puerto Rico, San Juan, PR 00931

This paper reports the design, construction and characterization of molecule-based, self-assembling lattice inclusion compounds exhibiting second-order optical nonlinearities. Through inclusion into the noncentrosymmetric 2D or 3D crystal lattice formed by divalent metal ions such as Zn(II), Cu(II), Cd(II) and Ni(II), and 4,4'-bipydine, organic chromophores with large hyperpolarizabilities can be aligned to maximize the nonlinear response of such supramolecular assemblies. Representative examples of the new nonlinear optical materials synthesized at this laboratory will be presented and discussed in terms of their X-ray structures, thermal and orientational stability of the host-guest materials.

INOR

120.
TWO CRYSTALLINE FORMS OF LOW-SPIN [Fe(TMP)(5-MeHIm)$_2$]ClO$_4$. RELATIVE PARALLEL AND PERPENDICULAR AXIAL LIGAND ORIENTATIONS. **W. Robert Scheidt**,[¶] Orde Q. Munro,[¶] Judith A. Serth-Guzzo,[¶] Ilona Turowska-Tyrk,[¶] K. Mohanrao,[¶] F. Ann Walker,[§] and Peter G. Debrunner.[†]
[¶]Department of Chemistry and Biochemistry, University of Notre Dame, Notre Dame, Indiana, 46556. [§]Department of Chemistry, University of Arizona, Tucson, Arizona, 85721. [†]Department of Physics, University of Illinois, Urbana, Illinois, 61801.

Two crystalline forms of [Fe(TMP)(5-MeHIm)$_2$]ClO$_4$ (TMP = *meso*-tetramesitylporphyrin) with different relative axial ligand orientations have been synthesized and characterized using X-ray crystallography, Mössbauer, and EPR spectroscopy. The form with a relative perpendicular axial ligand orientation manifests a more ruffled porphyrin core conformation, shorter Fe–N$_{porph}$ and Fe–N$_{ax}$ bond lengths, and a more symmetric electronic structure than the form with a relative parallel axial ligand orientation. Molecular mechanics calculations indicate that the form with a relative perpendicular axial ligand orientation is the more stable on steric grounds. The calculations also show that a rhombic distortion of the coordination geometry of the metal should be observed when the axial ligands adopt a relative parallel axial ligand orientation and eclipse a pair of *trans* Fe–N$_{porph}$ bonds.

121.
SOLID-STATE CHEMISTRY OF PALMITIC ACID - SODIUM PALMITATE PHASE COMPOUNDS. M.L. Lynch , Procter & Gamble Company, Sharon Woods Technical Center, 11520 Reed Hartman Highway, Cincinnati, Ohio, 45241

A number of different palmitic acid-sodium palmitate phase compounds of the general chemical formula- Na$_x$H$_y$P$_{x+y}$, have been isolated and characterized by X-ray diffraction, calorimetry, FTIR, and solid-state 13C NMR. The data demonstrate an interesting hydrogen bonding arrangement between acid and carboxylate head groups in crystals. This is evident from the existence of two distinct carbonyl carbons. They differ from the carbonyl carbons in both pure palmitic acid and sodium palmitate crystals; although one carbonyl carbon is still carboxylic acid-like while the other is carboxylate ion-like in nature. These crystals demonstrate a rich array of thermal transitions upon heating- including polymorphic and peritectic transitions. XRD and calorimetric measurements suggest that thermal disproportionation of these crystals results in the formation of other palmitic acid-sodium palmitate phase compounds.

122. DIELECTRIC AND FT-IR REFLECTANCE SPECTROSCOPIC ANALYSIS FOR A SERIES OF LITHIUM-FLUORO-BORATE GLASSES. A. Burns, Department of Chemistry, Kent State University-Stark Campus, Canton, Ohio, 44720

Impedance and FT-IR reflectance spectroscopic data for a series of lithium-fluoro-borate glasses will be presented. Conductivities and activation energies were determined from the temperature-dependent dielectric data. The temperature dependence of the impedance data has been further analyzed according to two well known dielectric models: the Cole-Cole and Havriliak-Negami. Results of these analyses will be presented, along with an analysis of which model is more appropriate. An interpretation of the fitting parametes as it relates to the structure of these glasses, which were examined by FT-IR reflectance spectroscopy, will also be presented.

123.
SURFACE MODIFICATION OF POROUS SILICON BY QUINONOID COMPOUNDS. **THERESA F. HARPER,** ANSELM P. HALL, AND MICHAEL J. SAILOR*, DEPARTMENT OF CHEMISTRY AND BIOCHEMISTRY, UNIVERSITY OF CALIFORNIA, SAN DIEGO, LA JOLLA, CALIFORNIA 92093-0358

The surface of luminescent porous silicon can be derivitized by reaction with quinone compounds. The mechanism of derivitization presumably occurs by transfer of a hydride ion from the porous silicon surface to the quinone, followed by attact at silicon by the resulting hydroquinone anion. The resulting modified surface is characterized by infrared spectroscopy. The effect of quinone surface modification of the visible photoluminescence spectrum of porous silicon will be presented. In addition, the effect surface modification has on the ability of porous silicon to sense other molecular adsorbates will be discussed.

124.
Improved Synthesis and Chemistry of Group V (Nb and Ta) Substituted Cyclopentadienyl Tetracarbonyl Derivatives. Skip Gallagher and Thomas E. Bitterwolf, Department of Chemistry, University of Idaho, Moscow, ID 83844.

Reaction of $(C_5H_4R)MCl_4$ under a pressure of carbon monoxide produced the corresponding $(C_5H_4R)M(CO)_4$ in high yields (where R is either electron donating or withdrawing). This multi-gram route produces abundant material for further chemistry. Substitution chemistry using traditional ligands (e.g. PPh_3) appears to afford additional stability to these carbonyl complexes.

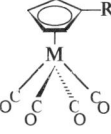

125. REDUCTION REACTIONS OF NEW COPPER(I) COMPLEXES WITH A NAPHTHYL-APPENDED NS_2-CYCLODECANE LIGAND

William S. Striejewske and Rebecca R. Conry*
Department of Chemistry/216, University of Nevada, Reno, Nevada 89557

New copper(I) complexes have been prepared which incorporate a novel multidentate ligand, N-(2-(1-naphthyl)ethyl-1-aza-4,8-dithiacyclodecane (**L**, shown at right), an NS_2-heterocycle with a pendant naphthalene group. The synthesis and characterization of the new ligand and its copper(I) complexes will be discussed, including the X-ray crystal structure of $[LCu(CH_3CN)]PF_6$.

In addition, results from chemical and electrochemical reduction experiments of the copper(I) complexes will be presented. The nature of the reduced product is being probed, for instance to distinguish such possibilities as (**L**)Cu(0) or (**L**$^-$)Cu(I) as well as to investigate its reactivity toward carbon dioxide.

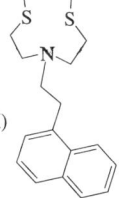

125A COPPER(I) HALIDE, HYDRIDE AND RELATED COMPLEXES WITH THE TRIDENTATE LIGAND BIS[2-(2-PYRIDYL)ETHYL]PHENYLPHOSPHINE

Rebecca R. Conry* and Yang Liu
Department of Chemistry/216
University of Nevada, Reno, Nevada 89557

New copper(I) complexes have been prepared which incorporate the tridentate ligand, bis[2-(2-pyridyl)ethyl]phenylphosphine (NPN, shown at right). The synthesis and characterization of copper(I) complexes with the empirical formulas [(NPN)CuX] and [(NPN)CuL]PF_6, where X is an anionic ligand such as bromide and L is a neutral ligand such as acetonitrile will be discussed, including the X-ray crystal structure of [(NPN)Cu(PPh$_3$)]PF_6.

NPN

126. KINETICS AND MECHANISM OF THE THERMAL SCISSION OF ETHYLENE FROM CpNb(CO)$_3$(η^2-C$_2$H$_4$) IN SUPERCRITICAL ETHYLENE. J. Timothy Bays and Thomas E. Bitterwolf, Department of Chemistry, University of Idaho, Moscow, ID 83844-2343. John C. Linehan and Clement R. Yonker, Pacific Northwest National Laboratories, PO Box 999 Richland, WA 99352.

Ethylene dissociation from CpNb(CO)$_3$(η^2-C$_2$H$_4$) was observed *in situ* by FT-IR to follow first order kinetics in supercritical fluid (SCF) ethylene solvent over the temperature range from 30°C to 80°C. The rate expression for the this reaction is log(k/s^{-1}) = (12.5 ± 1) - (24.6 ±1)/θ, θ = 2..303 RT kcal/mol. We present these kinetics along with reaction schemes for the thermal formation of CpNb(CO)$_4$ in the presence of CO and the photolytic formation of CpNb(CO)$_3$(η^2-C$_2$H$_4$) from CpNb(CO)$_4$ in SCF ethylene.

127. PHOTOCHEMICAL C-H/Si-H ACTIVATION AND PHOTOSUBSTITUTION REACTIVITY OF CpRh(CO)$_2$ (Cp = η^5-C$_5$H$_5$) IN FLUID SOLUTION. Nicholas Dunwoody and Alistair J. Lees. Department of Chemisry, State University of New York at Binghamton, New York 13902-6016

The photochemistry of CpRh(CO)$_2$ (Cp = η^5-C$_5$H$_5$) involving ligand substitution and intermolecular C-H and Si-H bond activation processess has been investigated in decalin at 293K. Absolute quantum efficiencies (ϕ_{cr}) have been determined following 458-nm laser excitation. In the absence of entering ligand the *trans*-Cp$_2$Rh$_2$(CO)$_3$ complex is identified as the major photochemical reaction product; this species is formed with low quantum efficiency. When excess triethylsilane (Et$_3$SiH) is present in solution the CpRh(CO)$_2$ complex is converted cleanly on irradiation to the CpRh(CO)(SiEt$_3$)H photoproduct. The ϕ_{cr} values exhibit saturation-type kinetics with [Et$_3$SiH] in the range 0.001-0.3 M. Kinetic analysis of the ϕ_{cr} data illustrates that the photoproduced solvated CpRh(CO) intermediate is scavenged competitively by Et$_3$SiH and CpRh(CO) under these solution conditions. When triphenylarsine (AsPh$_3$) and triphenylphosphine (PPh$_3$) ligands (L) are present in the hydrocarbon solution and the monosubstituted CpRh(CO)L photoproducts are formed cleanly and completely. The ϕ_{cr} values are dependent on [L] in the range 0.05-0.3M, revealing that L reacts via a competitive scavanging process for the photoproduced intermediate. The mechanistic differences in the ligand substitution and C-H/Si-H activation reactivities of the photoproduced intermediates are discussed.

128. RATE LAWS FOR PHOTOCHEMICAL REACTIONS OF METAL COMPLEXES WITH CHLORINATED SOLVENTS. Patrick E. Hoggard, Department of Chemistry, Santa Clara University, Santa Clara, California, 95053.

Photochemical reactions of metal complexes with (and in) chlorinated solvents have been found sometimes to be initiated through light absorption by the metal complex and sometimes through light absorption by the solvent. In the latter case, radicals produced through carbon-chlorine bond homolysis attack the metal complex, often leading to products that cannot be distinguished from those expected for metal-initiated photoreactions. If the reaction is metal-initiated, the rate of the reaction will depend on the fraction of light absorbed by the metal complex, f_R (R is for reactant). If, however, the reaction is solvent-initiated, the dependence will be on the fraction of light absorbed by the solvent, f_S. A typical photochemical rate law for a metal complex reacting with the solvent takes the form

$$-\frac{d[R]}{dt} = a(I_0 f_X)^m [R]^n$$

where X = R or S, m = ½ or 1, and n = 0, ½, or 1. We have found reactions that follow several different rate laws

129. DENSITY FUNCTIONAL CALCULATIONS OF DINUCLEAR AND TETRANUCLEAR COMPLEXES CONTAINING METAL-METAL QUADRUPLE BONDS. Craig D. Hamilton and Bruce E. Bursten, Department of Chemistry, The Ohio State University, Columbus, Ohio 43210

Density functional calculations have been applied to dinuclear complexes of the type $M_2(O_2CR)_4$, where M = Mo or W. The results of these calculations are in good agreement with the experimental electronic absorption spectra. Chisholm and coworkers have reported the electronic spectra of related complexes where two dimers have been covalently linked by a bridging oxalate ligand to form the "dimer of dimers", $[M_2(O_2C\text{-}t\text{-}Bu)_3]_2(\mu\text{-}O_2CCO_2)$. Our calculations on the "dimer of dimers" model system, $[M_2(O_2CH)_3]_2(\mu\text{-}O_2CCO_2)$ (M = Mo or W), indicate that some of the bands in the absorption spectrum are due to vibrational structure.

130. PHOTOCHEMICAL AND PHOSPHINE REACTIVITY STUDIES OF DINUCLEAR ORGANOIRON COMPLEXES. Marsi E. Shapiro and Bruce E. Bursten, Department of Chemistry, The Ohio State University, Columbus, Ohio 43210

The low temperature matrix photochemistry of $Cp_2Fe_2(CO)_2(\mu\text{-}CO)_2$ (**1**, Cp = $\eta^5\text{-}C_5H_5$), $Cp^*_2Fe_2(CO)_2(\mu\text{-}CO)_2$ (**2**, Cp^* = $\eta^5\text{-}C_5Me_5$), and $CpCp^*Fe_2(CO)_2(\mu\text{-}CO)_2$ (**3**) was studied. The double-CO-loss product of **1**, $Cp_2Fe_2(CO)_2$ (**4**) previously has been reported. Compounds **2** and **3** also generate double-CO-loss products, $Cp^*_2Fe_2(CO)_2$ (**5**) and $CpCp^*Fe_2(CO)_2$ (**6**), respectively, in low temperature matrices of neat 3-methylpentane (3MP). **4**, **5**, and **6** are highly unsaturated compounds that formally have unsupported Fe-Fe triple bonds. The CO-stretching frequencies for **6** are halfway between those of **4** and **5**, which is empirical evidence for their proposed structure. Subsequent irradiation of **5** in frozen 3MP generates $Cp^*_2Fe_2(\mu\text{-}CO)_2$ (**7**), a bridged double-CO-loss product. The exploratory solution chemistry of **1**, **2**, and **3** with $P(n\text{-}Bu)_3$ will also be presented.

131.
LONG LIVED CHARGE SEPARATION IN INTRAZEOLITIC RU TRIS-BPY/VIOLOGEN SYSTEMS. **Marcello Vitale,** Norma B. Castagnola, Nancy J. Ortins and Prabir K. Dutta, Dept. of Chemistry, The Ohio State University, Columbus, OH 43210

The back-electron transfer in photogenerated viologen radical-Ru(III) tris-bpy couples in zeolite Y was followed by time-resolved diffuse reflectance spectroscopy as a function of viologen concentration for a variety of viologens. At low viologen concentrations (1 every 5 supercages), lifetimes of a few hundred nanoseconds were observed. At a loading level of 1.2 viologens per supercage, the charge separated species appeared to persist for a much longer time. After 100 microseconds from photoexcitation, 40% to 20% of the initial amount of viologen radicals was still detected, along with a comparable depletion signal for Ru(II). Competition between back-electron transfer and zeolite-modified viologen self-exchange is proposed as the mechanism for the observed long-lived charge separation. This result opens new possibilities for solar energy conversion cycles leading to hydrogen production from water.

132. PHOTOELECTRON STUDY OF THE FE-FE INTERACTION IN *BIS*(FULVALENE) DIIRON; [FULVALENE = (η^5-C_5H_4 : η^5-C_5H_4)] Dennis L. Lichtenberger and Huajun Fan, Department of Chemistry, University of Arizona, Tucson, Arizona 85721.

The *bis*(fulvalene)diiron complex has interesting possibilities for both through-space and through-ligand metal-metal interactions. Gas phase UV-PES has been used to experimentally determine the electronic structure interactions in this complex. The PES of the fulvalene complex is compared to that of ferrocene to observe the effects of dimerization with the fulvalene ligand. The ionizations resulting from the d_{z^2} orbitals on each metal each occur in a single sharp ionization band. The high resolution of the experiment helped to reveal that there is little Fe-Fe splitting. The delocalization of the fulvalene ligand electronic structure across the two fulvalenes is evidenced by one ligand ionization which is totally separated from the other ionizations, because this *bis*-fulvalene combination is the incorrect symmetry to mix with any metal combinations.

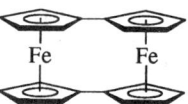

133. THE BINDING OF THE $Ru_3(CO)_9$ FRAGMENT TO BENZENE, FULLERENES, AND GRAPHITE: THE INTERACTION OF A METAL-CONTAINING MOIETY WITH AROMATIC C_6-CONTAINING SURFACES. Dennis L. Lichtenberger and Matthew A. Lynn, Department of Chemistry, University of Arizona, Tucson, AZ, 85721.

Shapley and Hsu have recently prepared $Ru_3(CO)_9(\eta^2,\eta^2,\eta^2-C_{60})$, the first complex observed to possess a metal-containing moiety bound in a hexahapto fashion to a fullerene. Via the Fenske-Hall molecular orbital method, we have examined the electronic structure of this complex, $Ru_3(CO)_9(\eta^2,\eta^2,\eta^2-C_6H_6)$, and $Ru_3(CO)_9(\eta^2,\eta^2,\eta^2-C_{70})$ to explain the shorter metal-carbon distances observed in the fullerene-bound systems. Although the modes of bonding are similar among the three complexes (i.e., Dewar-Chatt-Duncanson-type interactions between similar valence fragment orbitals), it is necessary to view the fullerenes as possessing a band-like electronic structure with a greater number of frontier orbitals that are available for binding. We have also examined the surface interactions of the $Ru_3(CO)_9$ moiety with graphite for comparison with the benzene- and fullerene-ligated systems and present experimental and computational results.

134. THE ELECTRONIC STRUCTURE OF d^6 METAL-ACETYLIDES. Dennis L. Lichtenberger,[†] Sharon K. Renshaw,[†] Andrew B. Uplinger,[†] R. Morris Bullock,[‡] and Mark G. Humphrey.[‡‡] [†]University of Arizona, Tucson, AZ 85721; [‡]Brookhaven National Laboratory, Upton, New York, 11973; [‡‡]Australian National University, Canberra, ACT 0200, Australia.

Gas-phase ultraviolet photoelectron spectroscopy has been used to investigate the electronic structure and bonding interactions of d^6 piano-stool metal-acetylides of the general formulas $CpML_2C\equiv C$-R [M = Ru, L = PMe_3, R = H, Me, ^tBu, C_6H_5] and $CpML_2C\equiv C$-p-C_6H_4-NO_2 [M = Fe, Ru, L = CO; M = Ru, L = PMe_3]. Previous studies of analogous $CpFe(CO)_2C\equiv C$-R complexes found that the filled-filled interaction between the metal d electrons and the acetylide π bond electrons dominates the shift of the first valence ionizations, and that backbonding of the metal d electrons into the acetylide π* orbitals is very small. It is found here that the change to the second row transition metal and the substitution of phosphines for the carbonyls makes the metal more electron rich, but does not change the basic description of the metal interaction with the acetylide.

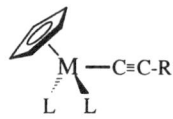

135. PHOTOELECTRON SPECTROSCOPY OF CHIRAL THIOLATE COMPLEXES OF THE TYPE CpM(NO)(L)SMe (M= Re or Mn, L= P'Pr$_3$, PPh$_3$, or PMe$_3$). Nicolai Burzlaff[1], Dennis L. Lichtenberger[2], Wolfdieter A. Schenk[1] and Kristie R. Winfield,[2] [1]Institut für Anorganische Chemie, Universität Würzburg, Am Hubland, D. 97074 Würzburg, [2]Department of Chemistry, University of Arizona, Tucson, AZ 85721.

The valence electronic structures of d^6 chiral thiolate compounds of the general type CpM(NO)(L)SMe (M= Re or Mn, L= P'Pr$_3$, PPh$_3$, or PMe$_3$) and the isoelectronic complex CpFe(CO)$_2$SMe are investigated via gas phase photoelectron spectroscopy (PES). The assignments of the ionization bands are aided by analysis of He I and He II cross section trends and by modeling with Fenske-Hall molecular orbital calculations. A previous study of the isoelectronic complexes of the type CpFe(CO)$_2$SR (Inorg. Chem. 1988, 27, 191), where R= C$_6$H$_4$-p-Z; Z= OMe, H, Cl, CF$_3$, NO$_2$, uses calculations and PES to show that the HOMO of these species is an antibonding dπ-pπ interaction between the filled metal d orbital and the sulfur 3p lone pair. Presented here is an analogous study of the Re and Mn nitrosyl, phosphine complexes.

136. PHOTOELECTRON SPECTROSCOPY OF METHYL-SUBSTITUTED FERROCENES. Gary P. Darsey, Dennis L. Lichtenberger and Julia K. Moberg. Department of Chemistry, University of Arizona, Tucson, AZ 85721.

High resolution gas phase He(I) and He(II) photoelectron spectra various methyl substituted ferrocenes have been collected. Detailed vibration progressions are evident. in the lowest energy ionizations that were not resolved in previous studies. An analysis of the electronic effects of methyl substitution on the cyclopentadienyl (Cp) ring shows that the methyl group does not act as a stronger σ donor than hydrogen as commonly assumed. The filled-filled π interaction between the methyl and Cp ring effectively destabilizes the π orbitals of the Cp ring, resulting in a more covalent interaction between the ring π orbitals and the metal d orbitals. This information, when combined with gas phase XPS measurements, indicates that the negative charge on the Cp carbon atoms is reduced when a hydrogen is replaced by a methyl, due to the increased donation to the metal.

137. QUENCHING ELECTRONICALLY EXCITED Ru^{II} by Co^{III} - TOWARD PHOTOREDOX THERAPY. Ofer Blum, Rose M. Carlos, Thomas J. Meade, Jay R. Winkler and Harry B. Gray; Beckman Institute, California Institute of Technology, Pasadena, CA 91125.

[Co^{III}(acacen)L$_2$]$^+$ (acacen = bis-acetylacetonate ethylene-diimine, L = NH$_3$, 2-Me-imidazole) is a proven drug against ocular herpes. We are exploring now the possibility of using these Co^{III} complexes as photochemical enzyme inhibitors. Using a light source, we hope to generate the more reactive d^7 [Co^{II}(acacen)] in the presence of an enzyme target. After binding, the Co^{II} should oxidize to form a kinetically inert Co^{III}-target complex. A simple inorganic photochemistry system was devised to examine the viability of this proposal. Coordinating solvents were employed as enzyme simulators, Ru^{II} complexes as sensitizers, and several Co^{III} complexes as the potential pro-drugs. While this system did not work as we had hoped (it is a rare example of pure diffusional energy transfer quenching of Ru^{II*} by Co^{III}), it did provide guidelines for the development of photoredox therapy based on similar coordination compounds.

138. **NEW BINUCLEAR RE(I)/CU(II) COMPLEX AS A POSSIBLE LUMINESCENT SENSOR.** Marianne E. Riklin, Dat Tran, Peter C. Ford*, Department of Chemistry, University of California, Santa Barbara, CA 93106.

Recent work in our laboratory has shown that tetrahedrally distorted Cu(II) complexes can be reduced by nitric oxide to form the corresponding cuprous complexes. This is the case for [Cu(2,9-dimethyl-1,10-phenanthroline)$_2$]$^{2+}$ [*Inorg. Chem.* 1996, *35*, 2411-12] or for [Cu(bis[2-(3,5-dimethyl-1-pyrazolyl)ethyl]amine)]$^{2+}$ (**1**). With the intention of developing a sensitive probe for NO, we have designed and synthesized the heterometallic complex **2**, formed by a [Re(CO)$_3$Br(1,10-phenanthroline)] luminescent fragment bound to the reactive Cu(II) complex **1**. We expect that the presence of NO may enhance the emission of the Re lumophore, due to less efficient quenching by the diamagnetic Cu(I) metal center compared to that of the paramagnetic Cu(II) center.

139. **DIFFUSE REFLECTANCE SPECTROSCOPY UNDER A REACTIVE GAS: REACTION OF NO WITH A FERRI PORPHYRIN SOL-GEL MATERIAL.** Malcolm A. De Leo, Leo Van Wuellen and Peter C. Ford Department of Chemistry, University of California at Santa Barbara, Santa Barbara, CA 93106.

Solid materials can be characterized with the aid of UV/vis diffuse reflectance spectroscopy. This technique affords excellent sensitivity while requiring small samples. An original diffuse reflectance cell (V. I. Srdanov, to be published) modified to allow spectroscopy under a reactive gas atmosphere has been utilized. This permits observation of changes occurring during the course of the reaction with a reactive gas. The conversion of an iron(III) tetraphenylsulfonatoporphyrin in a tetraorthosilicate glass (Fe(III)(TPPS)·TEOS) to its nitrosyl derivative (Fe(III)(TPPS)(NO)·TEOS) by reaction with nitric oxide was used to check the cell's performance. This reddish brown glassy solid (λ_{max} 398 nm, Soret) turns emerald green (421 nm) under a nitric oxide atmosphere.

140. INVESTIGATIONS OF NITRIC OXIDE BINDING TO WATER SOLUBLE FERRIPORPHYRINS. L. E. Laverman and P. C. Ford, Department of Chemistry, University of California at Santa Barbara, Santa Barbara, California 93106

Recent interest in the physiological roles of nitric oxide has prompted the investigation into the thermodynamics and dynamics of nitric oxide binding for two heme model compounds Fe^{III}(TPPS)(H$_2$O)$_2{}^{3-}$ (TPPS=tetrakis(p-sulfonatophenyl)porphine) and Fe^{III}(TMPS)(H$_2$O)$_2{}^{4-}$ (TMPS=tetrakis(sulfonatomesityl)porphine). Nitric oxide binding kinetics have been studied by nanosecond laser flash photolysis over the temperature range 298 to 313 K. The equilibrium constant for each compound has been determined both by laser flash photolysis and by spectrophotometric titration. Results of preliminary experiments show that ΔH for ligand exchange is -2.4 kJ/mol and -18.9 kJ/mol for Fe^{III}(TPPS) and Fe^{III}(TMPS) respectively and that the second order association rate constant, k_{NO}, is a factor of five times faster for Fe^{III}(TMPS). Rate constants and equilibrium constants will be presented and discussed.

$$PFe^{III}(H_2O)_2 + NO \underset{k_d}{\overset{k_{NO}}{\rightleftharpoons}} PFe^{II}(H_2O)(NO^+) + H_2O$$

141. TIME RESOLVED INFRARED SPECTROSCOPIC STUDIES OF INTERMEDIATES OF MIGRATORY INSERTION REACTIONS OF COBALT CARBONYLS. Julienne G. Rabor, Brian Lee, Bill Boese, Peter C. Ford*, Department of Chemistry, University of California Santa Barbara, CA 93106

Time Resolved InfraRed (TRIR) spectroscopy has been used extensively as a tool in determining the structure and kinetic behavior of transient species. In this work, TRIR and other photochemical techniques are used to study the photolysis of the acyl cobalt complex $CH_3C(O)Co(CO)_3(PPh_3)$. The reactions studied are of interest because they are analogous to crucial steps in the cobalt carbonyl catalysis of olefin hydroformylation. UV photolysis of $CH_3C(O)Co(CO)_3(PPh_3)$ in various organic solvents results in the photoejection of CO and the formation of the corresponding methyl complex $CH_3Co(CO)_3(PPh_3)$. Transient IR spectra of the intermediate species formed in this reaction have been obtained in different solvents. Kinetic studies have revealed that there is no competitive trapping by CO to reform the starting material, indicating that the unimolecular methyl migration pathway is faster than the bimolecular reaction with CO. Reaction mechanisms and intermediate structures proposed to interpret TRIR results will be presented.

$$CH_3C(O)Co(CO)_3L \underset{+CO}{\overset{h\nu, -CO}{\rightleftarrows}} CH_3C(O)Co(CO)_2L \longrightarrow CH_3Co(CO)_3L$$

142. RUTHENIUM PORPHYRIN ASSISTED DISPROPORTIONATION OF NITRIC OXIDE: MECHANISTIC INVESTIGATIONS. Ivan M. Lorkovic, Katrina M. Miranda, Peter C. Ford*, Department of Chemistry, University of California, Santa Barbara, CA 93106

(Octaethylporphyrin)Ru(CO)(HOEt) and (tetraphenylporphine)Ru(CO)(HOEt) react with nitric oxide in the solid phase and in organic solvents to give Ru(P)(NO)(OR) (R=H, Et), **1**, and Ru(P)(NO)(ONO), **2**. Although **1** and **2** represent oxidation of one and two equivalents of NO, respectively, we find that **1** does not react with NO to give **2**. We describe experiments, including GC N_2O detection, to establish the stoichiometry and mechanism of this multistep transformation, formally written as $4NO \rightarrow NO^+ + ONO^- + N_2O$.

143.
EXCHANGE COUPLING INTERACTIONS IN METAL-QUINOID COMPLEXES: SYNTHESIS AND GROUND STATE PROPERTIES OF METAL-QUINONE DYADS. **Daniel E. Wheeler** and James K. McCusker*, Department of Chemistry,University of California at Berkeley, Berkeley, California, 94720-1460

The influence of magnetic exchange interactions on the magnetic and optical properties of metal-quinone complexes is being explored. The present work deals with metal-quinone dyads of the form $[M(Q)L]n+$, where Q is a reduced form of an orthoquinone and L is a tetradentate ligand such as tris-(2-aminoethyl)amine (TREN) and M is chromium(III), cobalt(III) or zinc(II). The synthesis and ground state structural, spectroscopic and magnetic properties of several complexes of this general type will be described.

144.
PROBING THE EFFECTS OF DELOCALIZATION ON THE EXCITED STATE EVOLUTION OF TRANSITION METAL COMPLEXES. **Niels H. Damrauer,** Thomas R. Boussie, and James K. McCusker*, Department of Chemistry University of California at Berkeley, Berkeley, CA, 94720-1460

For transition metal complexes exhibiting photo-induced metal-to-ligand charge transfer states, the degree of delocalization of the acceptor ligand plays an important role in determining excited state properties and dynamics. This work addresses the excited state evolution of a series of metal complexes based on the [Ru(bpy)$_3$]$^{2+}$ moiety where substituents at the 4,4' positions of the bypyridine ligands have been used to "tune" the degree of excited state delocalization. The synthesis of these ligands and metal complexes in this series is described. Both static and nanosecond time-resolved spectroscopies are used to determine how delocalization effects the long-lived excited states in these systems. Spectroscopy on an ultrafast time scale is then used to address how delocalization manifests itself in the initial evolution of the Franck-Condon state.

145.
EXCITED-STATE DYNAMICS OF ASYMMETRIC RUTHENIUM POLYPYRIDYL COMPLEXES. **Aimee E. Curtright** and James K. McCusker*, Department of Chemistry,University of California at Berkeley, Berkeley, California, 94720-1460

Ruthenium polypyridyl complexes of the general form [Ru(bpy')(bpy")$_2$]$^{2+}$, where bpy' and bpy" are substituted bipyridines, are being studied to assess the impact of electronic structure variations on ultrafast photo-induced excited state dynamics. These compounds allow for systematic variation of the excited-state electronic structure, since changes in ligand substituents (e.g. CH$_3$ to CF$_3$) result in changes in the relative vertical displacement of the corresponding potential energy surfaces. Selective excitation of a specific ligand and subsequent energy and electron transfer processes may then be studied. New synthetic strategies for preparing compounds containing both electron donating and withdrawing ligands will be described. In addition the steady-state and time-resolved spectroscopic properties of several complexes of this type will be presented.

146.
THE OPTICAL SPECTROSCOPY OF U3-OXO METAL CLUSTERS: A MODEL FOR GAIN QUENCHING IN SOLID-STATE LASER MATERIALS. **Eric A. Juban** and James K. McCusker*, Department of Chemistry,University of California at Berkeley, Berkeley, California, 94720-1460

Concentration-dependent gain quenching in metal-ion-doped solid-state laser materials is a well-known phenomenon. Our research proposes that the formation of metal clusters, and the accompanying changes in electronic structure due to electron exchange effects, are at least partly responsible for the observed decrease in lasing efficiency. To examine this possibility, we are studying compounds of the form [M$_3$O(O$_2$CR)$_6$L$_3$]$^+$, where M=Cr(III), Ti(III), etc. The identities of the metal centers, as well as the R and L ligands, are systematically varied. Both static and time-resolved spectroscopies are then used to determine the effects of cluster nuclearity and electron exchange on the radiative and nonradiative decay rates important for achieving population inversion in laser materials. Results from several systems of this general type will be described.

147. CHEMISTRY OF THE HEXABORANE(10) ANALOGUE $(PPh_3)_2(CO)OsB_5H_9$: FORMATION AND CHARACTERIZATION OF LEWIS BASE ADDUCTS AND A NEW DEGRADATION REACTION FOR METALLABORANE CLUSTERS. Lawrence Barton, Jonathan Bould, Hong Fang, Kevin Hupp and Nigam P. Rath. Department of Chemistry, University of Missouri-St. Louis, St. Louis, MO 63121, USA.

The metallahexaborane $[(PPh_3)_2(CO)OsB_5H_9]$ (I) reacts with PPh_3 in refluxing CH_2Cl_2 to afford $[(PPh_3)_2(CO)OsB_5H_9](PPh_3)$ (II) in 57% yield. The reaction is reversible and the original metallaborane is recovered on thermolysis at 50°C. NMR spectra for II are very similar to those for $[(PPh_3)_2(CO)OsB_4H_8]$ (III) and suggest a structure with an *exo*-$BH_2\cdot PPh_3$ moiety bonded to a basal B atom. This formulation is supported by high resolution mass spectrometry and elemental analysis. The species II tends to degrade to III with concomitant release of $BH_3\cdot PPh_3$. A crystal structure determination of the PPh_2Me analogue of II confirms the structure as an osmapentaborane skeleton with a pendant $BH_2\cdot PR_3$ group. The use of other phosphines gives similar products. The position of the equilibrium and the tendency to degrade to III is dependent on the base strength. The results suggest a possible mechanism for the formation of 1-$[Fe(CO)_3]B_4H_8$ and 4-$[Fe(CO)_3]B_5H_9$ from $Fe(CO)_5$ and B_5H_9. They also suggest that this mode of reaction represents a new general degradation process for cluster molecules.

148. CONTROLLED SYNTHESIS OF CHROMIUM AND MOLYBDENUM METALLABORANE CLUSTERS. PREPARATION AND CHARACTERIZATION OF $(Cp*Cr)_2B_5H_9$, $(Cp*Mo)_2B_5H_9$ AND $(Cp*MoCl)_2B_2H_5(B_2H_5)$
Simon Aldridge, Thomas P. Fehlner* and Maoyu Shang
Department of Chemistry and Biochemistry, University of Notre Dame, Indiana, 46556, USA.

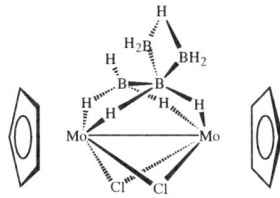

$(Cp*MoCl)_2B_2H_5(B_2H_5)$, 4

Reaction between the unsaturated cluster $(Cp*Cr)_2B_4H_8$, **1**, and $BHCl_2\cdot SMe_2$ at 55°C allows systematic cluster expansion giving $(Cp*Cr)_2B_5H_9$, **2**, by addition of a BH fragment. Control of cluster nuclearity is also achieved in the corresponding molybdenum system, reduction of $[Cp*MoCl_2]_2$ with $LiEt_3BH$, followed by reaction with $BH_3\cdot thf$ yielding $(Cp*Mo)_2B_5H_9$, **3**. Direct reaction of $[Cp*MoCl_2]_2$ with $BH_3\cdot thf$ also gives rise to the Mo(II) cluster, **3**, together with the Mo(III) species $(Cp*MoCl)_2B_2H_5(B_2H_5)$, **4**, both in good yield. The structure of **4** as determined by X-ray diffraction reveals a highly unusual structure in which two Cp*Mo units are bridged by two chlorides and a single $B_4H_{10}^{2-}$ ligand.

149. SYNTHESES AND STRUCTURES OF CLUSTER CARBOXYLATES GENERATED FROM REACTION OF GROUP(IV) ALKOXIDES WITH HETERONUCLEAR CLUSTER ACIDS, $C_5H_5(CO)_2M(CO)_6Co_2(\mu_3\text{-}CCO_2H)$ (M=Mo, W). X. Lei, H. Shimomura, M. Shang and T. P. Fehlner, Department of Chemistry and Biochemistry, University of Notre Dame, Notre Dame, IN 46556

In our continuing studies on the effects of transition metal cluster substituents on metal carboxylate coordination chemistry, two new heteronuclear trimetal cluster acids, $C_5H_5(CO)_2M(CO)_6Co_2(\mu_3\text{-}CCO_2H)$ (abbreviated as GCO_2H; M = Mo, W) were prepared. The reaction of GCO_2H with group(IV) alkoxides $M'(OR)_4$ (M' = Ti, Zr; R = CH_2CH_3, $CH(CH_3)_2$) have been examined. Five new compounds with formulae $[Ti_4(\mu_3\text{-}O)_4(OR)_4(\mu\text{-}GCO_2)_4]$ or $[Zr_2(\mu\text{-}OH)_2(\mu\text{-}GCO_2)_2(GCO_2)_4]$ have been structurally characterized.

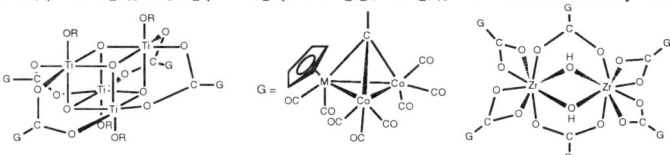

150. CATALYTIC DEOXYGENATION OF EPOXIDES WITH (Cp*ReO)$_2$(μ-O)$_2$ AND CATALYST DEACTIVATION. <u>Kevin P. Gable</u> and Fedor Zhuravlev, Department of Chemistry, Oregon State University, Corvallis, OR 97331-4003.

Attempts to achieve catalytic epoxide deoxygenation with Cp*ReO$_3$/PPh$_3$ (generating *in situ* (Cp*ReO)$_2$(μ-O)$_2$) gave modest turnover numbers for alkenes with electron-withdrawing substituents. These catalytic reactions failed to give notable turnover numbers for alkenes with electron rich substituents. The deactivation of the catalytic system was traced to conproportionation of (Cp*ReO)$_2$(μ-O)$_2$ with Cp*ReO$_3$ to form (Cp*Re)$_3$(μ-O)$_6$$^+$(ReO$_4$$^-$)$_2$ at the elevated temperatures needed for deoxygenation.

151. d^1 RHENIUM(VI) DIOXO HYDROTRIS(1-PYRAZOLYL)BORATE COMPLEXES. <u>Darin D. DuMez</u>, and James M. Mayer, Department of Chemistry Box 351700, University of Washington, Seattle, WA 98195-1700

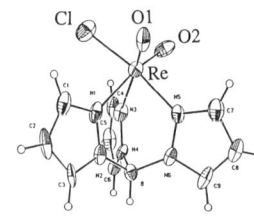

A number of d^1 rhenium cis-dioxo compounds, (HBpz$_3$)ReO$_2$(X) (HBpz$_3$ = hydrotris(1-pyrazolyl)borate; X = Cl, Br,I) have been synthesized from (HBpz$_3$)ReO(X)(OTf) and pyridine-N-oxide in good yield. (HBpz$_3$)ReO$_2$(Cl) was shown to have an E$_{1/2;\ Re(VI)/Re(VII)}$ of 0.93 V vs. Cp$_2$Fe in acetonitrile. Alkoxide complexes can be made by the addition of MOTf (M = Ag, Li), followed by LiOR (R = Me, Et) resulting in the relatively stable (HBpz$_3$)ReO$_2$(OR). The general nature and reactivity of these rhenium cis-dioxo compounds, especially pertaining to the better understanding of the reactive (HBpz$_3$)ReO$_2$(X)$^+$ species, will also be presented.

152. RHENIUM(V) OXO -HALO, -ALKYL, AND -ARYL COMPLEXES BEARING A HYDROTRIS-(3,5-DIMETHYL-1-PYRAZOLYL)BORATO LIGAND, <u>Yoshihiro Matano</u>, Todd Northcutt, and James M. Mayer, Department of Chemistry, University of Washington, Seattle, WA 98195-1700

Rhenium(V) oxo bis(halide) compounds (Tp*Re(O)(Cl)(X); Tp* = hydrotris(3,5-dimethyl-1-pyrazolyl)borato and X = F, Cl, and I) have been cleanly synthesized from Tp*ReO$_3$. Reaction of Tp*Re(O)(Cl)$_2$ with diethylzinc yields Tp*Re(O)(Et)(Cl), while the reaction with phenylmagnesium chloride affords Tp*Re(O)(Ph)(Cl). Metathesis reactions of the halides with AgOTf or Me$_3$SiOTf result in the formation of the corresponding triflates Tp*Re(O)(X)(OTf), Tp*Re(O)(Et)(OTf), and Tp*Re(O)(Ph)(OTf). Reactions of these rhenium(V) oxo -halo, -alkyl, and -aryl complexes will be presented and compared with the reactivity of analogous (hydrotris(1-pyrazolyl)borato)rhenium(V) oxo complexes.

153. **RHENIUM IMIDO ALKYL AND ARYL COMPLEXES**
W. Stephen McNeil, Darin D. DuMez, and James M. Mayer
Department of Chemistry, University of Washington, Seattle, WA, 98195-1700

Complexes of the general formulae TpRe(N-*p*-tol)X$_2$, TpRe(N-*p*-tol)(R)(X), and TpRe(N-*p*-tol)R$_2$ (Tp = hydrotris(pyrazolyl)borate; R = Et, Ph, OEt; X = Cl, I, O$_3$SCF$_3$) have been prepared via metathesis reactions from the dihalo precursors, a representative example of these complexes being shown at the right. The general properties and reactivity of these compounds reflects the more electron-rich nature of the imido species as compared to their known oxo derivatives. Oxygen-atom transfer reactions to these imido complexes will be presented and also compared to those of the oxo analogues.

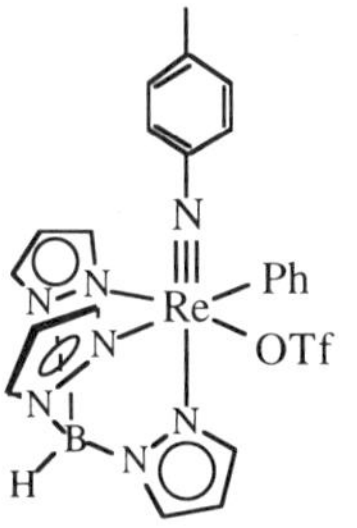

154. **REACTIVITY OF A Cu^{III} COORDINATION COMPOUND WITH WEAK C-H BONDS.** Teresa J. Blubaugh, James M. Mayer, Department of Chemistry, University of Washington, Seattle,WA. 98195-1700

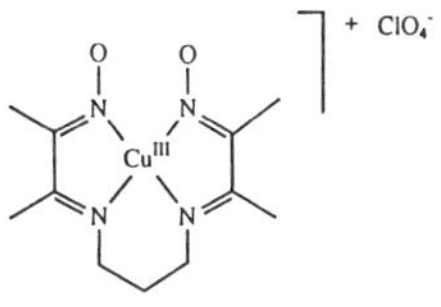

[Cu^{III}Pre]ClO$_4$ shown to the left, has a strongly oxidizing metal center and a basic site. On this basis it is proposed that [Cu^{III}Pre]ClO$_4$ can abstract hydrogen atoms from organic substrates by proton-coupled electron transfer. This coordination compound reacts with dihydroanthrance (DHA) to form anthracene in 40% yield. Kinetic data reveal that the reaction is first order in [Cu^{III}Pre]ClO$_4$ and first order in DHA with no evidence of an induction period. Although [Cu^{III}Pre]ClO$_4$ is a d^8 complex with no unpaired electrons it is predicted to be able to abstract hydrogen atoms from weak C-H bonds based on its ability to form a strong bond with hydrogen. The bond strength is calculated to be 83 kcal/mol based on its pH dependent reduction potential.

155. **ACTIVATION OF O-H BONDS BY A COPPER(III)-PEPTIDE COMPLEX.** Andrea M. Collier and James M. Mayer, Department of Chemistry, University of Washington, Seattle, WA 98195-1700.

The copper-peptide complex, Cu^{III}(H$_{-2}$Aib$_3$)(**1**), where Aib denotes the amino acid α-amino isobutyric acid, abstracts hydrogen atoms from substrates with weak O-H bonds. The chemistry of this complex with a series of substituted phenols has been studied. The kinetics, products, and mechanistic implications of these reactions will be discussed. No reaction of **1** is observed with C-H bonds of comparable strength.

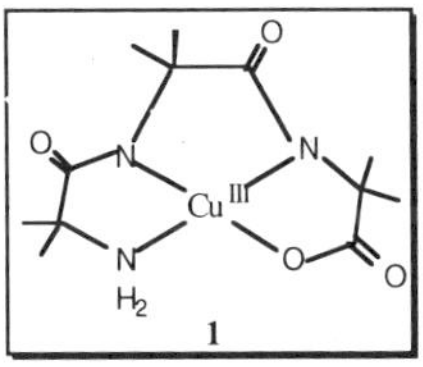

156.
THE EFFECT OF LIGAND STRUCTURE ON THE POLYMERIZATION BEHAVIOR OF UNBRIDGED METALLOCENES Jennifer L. Maciejewski Petoff, Michael D. Bruce, Robert M. Waymouth, Md. Athar Masood, Department of Chemistry, Stanford University, Stanford, CA, 94305

Bis(2-arylindenyl)zirconium dichlorides activated by methylaluminoxane produce elastomeric polypropylene with a stereoblock microstructure. In an effort to experimentally explore the importance of the 2-arylindene ligand in producing these materials, a series of catalysts was investigated which had significant electronic and steric deviations from the parent ligand architecture. In particular, we sought to examine the influence of aryl stacking on the polymerization behavior of these complexes. We discovered that the bis(2-arylindenyl) framework is essential for the production of elastomeric polypropylene by unbridged metallocenes. Slight changes in the fundamental catalyst structure upset the delicate balance between ligand rotation and monomer insertion resulting in the generation of amorphous polymers.

157. CLEAVAGE OF N-N AZOARENE BONDS AND DEOXYGENATION OF NITROGEN OXIDES BY A Ta(III) METALLOCENE.
Mark A. Aubart and Robert G. Bergman, *Department of Chemistry, University of California, Berkeley, CA 94720*

The Ta(III) metallocene $Cp_2Ta(L)R$ (**1**; R = Me, 2-thienyl; L = PMe_3, ethylene) cleaves N-N bonds in azoarenes and N-O bonds in nitric oxide and nitrous oxide. The reaction of **1** with azoarenes leads to imido complexes $Cp_2Ta(NAr)R$ (**2**) or to π-complexes $Cp_2Ta(\eta^2$-azoarene)R depending on the reaction conditions and substituents on the azoarene. The reaction of **1** with NO or N_2O leads to mixtures of monomeric $Cp_2Ta(O)R$ (**3**) and $(Cp_2TaR)_2O$ depending on the stoichiometry. Nitrous oxide (in the case of the reaction of **1** with NO) and dinitrogen have been identified as products of the latter reaction by gas chromatography and infrared spectroscopy. Mechanistic studies of these reactions have shed light on the nature of N-N and N-O bond activation.

$$Cp_2Ta\overset{O}{\underset{R}{\diagdown}} \quad \underset{-L, N_2O}{\overset{2\ NO}{\longleftarrow}} \quad Cp_2Ta\overset{L}{\underset{R}{\diagdown}} \quad \overset{0.5\ N_2Ar_2}{\underset{-L}{\longrightarrow}} \quad Cp_2Ta\overset{NAr}{\underset{R}{\diagdown}}$$

$$\mathbf{3} \qquad\qquad\qquad \mathbf{1} \qquad\qquad\qquad \mathbf{2}$$

158. STRUCTURAL ASPECTS OF BENZYL, CYCLOPROPYL AND TRIMETHYLSILYLMETHYL TITANOCENE DERIVATIVES
Nicos A. Petasis,* and Yong-Han Hu, *Department of Chemistry and Loker Hydrocarbon Research Institute, University of Southern California, Los Angeles, CA 90089-1661*

The thermolysis of dialkyl titanocenes is an important method for the in situ generation of titanocene alkylidenes, which react readily with carbonyls, alkynes and other functional groups. In an effort to understand the factors controlling the relative reactivity the various alkyl titanocene substituents, we have carried out an X-ray crystal structure study of several of these derivatives. The structures of these compounds, including the mixed cyclopropyl trimethylsilylmethyl titanocene (**1**) will be discussed.

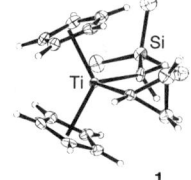

1

159. REACTIONS OF [Pt$_2$(μ-CO)(CO)$_2$(μ-dppm)$_2$] WITH PRIMARY SILANES AND DISILANES. April Hennis and Serge Schreiner, Department of Chemistry, Randolph-Macon College, Ashland, VA, 23005.

Reaction of [Pt$_2$(μ-CO)(CO)$_2$(μ-dppm)$_2$] (1) with the primary silanes phenylsilane, methylsilane, ethylsilane, n-hexylsilane, and n-octylsilane in a 1:1 ratio yields complexes formulated as [Pt$_2$(μ-SiRH)(CO)$_2$(μ-dppm)$_2$], R=Ph, Me, Et, n-Hex, n-Oct. These products have been characterized by multinuclear NMR techniques and infrared spectroscopy. A reaction mechanism for the formation of the products has been proposed. Low temperature NMR studies show the formation of an unsymmetrical intermediate containing an agostic H-Pt-Si interaction. (1) appears to be unable to activate the silanes hexamethyldisilane, 1,1,2,2-tetrachloro-1,2-dimethylsilane, and 1,2-diphenyltetramethyldisilane. Reaction of (1) with the disilanes 1,3-disilabutane and disilylbenzene yield bimetallic silicon-containing complexes, which will be discussed.

160. LIGAND MEDIATED C-F, C-C, AND C-H BOND ACTIVATION IN COBALTOCENE COMPLEXES Brian K. Bennett and Thomas G. Richmond*, Department of Chemistry, University of Utah, Salt Lake City, Utah 84112.

The cyclopentadienyl ligand is the site of C-F activation of C$_6$F$_6$ by cobaltocene to afford the known complex (exo-C$_6$F$_5$-η^4-C$_5$H$_5$)Co(η^5-C$_5$H$_5$). Interestingly, cleavage of the phenyl C-C bond occurs at 60 °C to give C$_6$F$_6$H and cobaltocenium cation. Analogously, (η^5-C$_5$Me$_5$)$_2$Co reacts with C$_6$F$_6$ at -50 °C and undergoes C-C cleavage at room temperature. Protons and Lewis acids also induce competitive C-C and C-H bond scission under very mild conditions.

161. REDOX-PROMOTED REACTIONS OF SOME COBALTACYCLOBUTENE COMPLEXES Jitendra Malik, Bella S. Fong, Lee V. Gray, Joseph M. O'Connor,* and Bernadette T. Donovan-Merkert,* Department of Chemistry, The University of North Carolina - Charlotte, Charlotte, North Carolina 28223-0001 and Department of Chemistry and Biochemistry, University of California, San Diego, La Jolla, California 92093-0358.

Chemical or electrochemical oxidation of **1** results in rapid formation of a furan product, whereas oxidation of **2** or **3** affords radical cations that are relatively stable on the cyclic voltammetry time scale. Details of these and other redox-promoted reactions of **1-3** will be presented.

2 R = C$_6$H$_5$
3 R = p-MeC$_6$H$_4$

INOR

162. **ACTIVATION AND INSERTION OF CE_2 (E = O,S) BY DIVALENT AND TRIVALENT ORGANOSAMARIUM COMPOUNDS.** Christopher A. Seibel, Joseph W. Ziller, Robert J. Doedens, and William J. Evans, Department of Chemistry, University of California, Irvine, CA 92697-2025.

The activation and insertion chemistry of CE_2 (E = O,S) with divalent and trivalent organosamarium compounds has been investigated. The divalent compounds $(C_5Me_5)_2Sm(THF)_x$ (x = 0,2) reductively couple CO_2 to oxalate, $C_2O_4^{2-}$, while COS reductively disproportionates to COS_2^{2-} and CO. The trivalent $[(C_5Me_5)_2Sm]_2N_2Ph_2$ inserts only one molecule of CO_2 producing the asymmetrical complex $(C_5Me_5)_2Sm[\mu-\eta^2:\eta-PhNN(CO_2)Ph]Sm(THF)(C_5Me_5)_2$. The trivalent organosamarium allyl compounds $Cp^*_2Sm(\eta^3-CH_2CH=CHR)$ (R= H,Me,Et) insert CO_2 to produce $(C_5Me_5)_2Sm(O_2C-CH_2CH=CHR)$ complexes. In coordinating solvents, these compounds are monometallic with $Sm-\eta^2-O_2CR$ bonding. In non-coordinating solvents, these compounds are bimetallic with $Sm_2(\mu-\eta^2-O_2CR)$ connectivity. These structural types can be readily interconverted. CS_2 and COS also form insertion products with organosamarium allyl compounds, but the resulting compounds are monometallic in both coordinating and non-coordinating solvents.

163. **THE UTILITY OF ALKOXIDE LIGANDS IN ASSEMBLING POLYZIRCONIUM HOMO- AND HETEROMETALLIC COMPLEXES.** Mohammad A. Ansari, Joseph W. Ziller, and William J. Evans, Department of Chemistry, University of California, Irvine, CA 92697-2025.

Zirconium is a desirable metallic component in a variety of materials which have useful practical properties. Attempts to manufacture defect-free zirconium-containing materials using molecular zirconium chemistry as a route alternative to the usual thermal/mechanical techniques are hampered by the fact that there are few polymetallic and polyheterometallic zirconium precursors in the literature. Using alkoxide ligands we have synthesized and structurally characterized a series of new polymetallic zirconium complexes including trimetallic complexes $[(RO)_2Zr]_3(\mu-OR)_3(\mu_3-Cl)(\mu_3-O)$, (R = CH_2CMe_3) and $[(R'O)_2Zr]_3(\mu-OR')_2(\mu_3-OH)(\mu_3-OR')(\mu_3-O)$, (R' = CMe_3) and the mixed-metal decametallic complex, $\{Na_2[(EtO)Zr]_3(\mu-OEt)_7(\mu_3-OEt)_2(\mu_5-O)\}_2$. Synthesis and characterization of these and some heteropolymetallic complexes involving Zr/Nd and Zr/Y combinations will be presented.

164. **SYNTHESIS AND TWO ELECTRON OXIDATIVE ADDITION REACTIVITY OF SAMARIUM(II) COMPLEXES INVOLVING THE CYCLOOCTATETRAENIDE LIGAND.** Robert D. Clark, Mohammad A. Ansari, Joseph W. Ziller, and William J. Evans, Department of Chemistry, University of California, Irvine, CA 92697-2025.

Sm(II) chemistry is currently focused on two main types of compounds, the iodide complex, $SmI_2(THF)_5$, and the C_5Me_5 complexes, $(C_5Me_5)_2Sm(THF)_x$ (x = 0-2). We report here efforts to extend the ligand set for Sm(II) to the cyclooctatetraenide ligand. $[(C_5Me_5)Sm(\mu-I)(THF)_2]_2$ reacts with two equivalents of $K_2(C_8H_8)$ to form the new Sm(II) complex $K[(C_5Me_5)Sm(C_8H_8)](THF)_n$ (1). $[(C_5Me_4R)Sm(\mu-I)(THF)_2]_2$ reacts with one equivalent of $K_2(C_8H_8)$ in toluene to form the bimetallic species $[(C_5Me_4R)Sm(THF)]_2(\mu-C_8H_8)$ [R = Me (2), Et (3)]. X-Ray crystallographic analysis of the diglyme adduct of 2, $[(C_5Me_5)Sm(diglyme)]_2(\mu-C_8H_8)\cdot 2THF$ (4), crystallized from THF, revealed that a planar cyclooctatetraenide ligand is sandwiched between two $(C_5Me_5)Sm(diglyme)$ units. The bimetallic 2 undergoes two electron oxidative addition reactions, as might be expected for a bimetallic Sm(II) species, but with cyclooctatetraene and C_5Me_5Cl, both oxidative addends add to a single Sm center. 2 and 3 can be desolvated to yield $[(C_5Me_4R)Sm]_2(\mu-C_8H_8)$ (R = Me (5), Et (6)). Hence, 5 and 6 are related to 2 and 3 as $(C_5Me_5)_2Sm$ is related to $(C_5Me_5)_2Sm(THF)_2$.

165.
SYNTHESIS AND STRUCTURAL STUDIES OF BULKY INDENYL COMPLEXES OF THE LATE TRANSITION METALS. **Jason S. Overby** and Timothy P. Hanusa, Department of Chemistry, Vanderbilt University, Nashville, TN 37235

Although transition metal complexes containing the indenyl ligand ($C_9H_7^-$) have been known for almost as long as ferrocene, little work has been done with sterically bulky indenes and their metal compounds. We have employed various isopropyl- and trimethylsilyl-substituted indenyl ligands in the preparation of iron, cobalt and nickel complexes. The bulkiness of these ligands confer interesting solution and solid-state properties on their complexes. Synthetic, magnetic and conformational features of these compounds (e.g., the structure of $[C_9(iPr)H_6]_2Co$ at right) will be discussed.

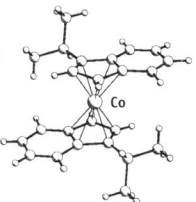

166.
A HIGHLY ISOPROPYLATED NICKELOCENE: SYNTHESIS, CHARACTERIZATION AND STABILIZATION OF REACTIVE ANIONS. **Jason S. Overby** and Timothy P. Hanusa, Department of Chemistry, Vanderbilt University, Nashville, TN 37235

Recent work in our laboratory has focused on the use of the sterically bulky and highly electron-donating ligands tri- and tetraisopropylcyclopentadiene. Many main-group and transition metal complexes containing these ligands have been prepared, including hexaisopropylnickelocene, $(Cp^{3i})_2Ni$. Despite the presence of the six isopropyl groups, the structure of $(Cp^{3i})_2Ni$ displays little distortion when compared to unsubstituted Cp_2Ni. The bulky rings serve to shield the metal center from attack, however, and we have used this "encapsulation" effect to form nickelocenium salts containing reactive anions such as fluoride and trifluoromethanesulfonate (triflate). General synthetic, magnetic and structural features of these compounds will be discussed.

167. FERROCENYL-BASED S_3 FACE-CAPPING LIGANDS Peter Schebler,[†] Charles G. Riordan,*,[†] Louise Liable-Sands[§] and Arnold L. Rheingold[§] Departments of Chemistry, [†]Kansas State University, Manhattan, KS 66506-3701 and [§]University of Delaware, Newark, Delaware 19716

The poly(methylthiomethyl)borate ligands, exemplified by **1**, developed in these laboratories have been extended to include derivatives with ferrocenyl substituents on boron, **2**. This report will detail the synthesis, structure and spectroscopic properties of metal complexes derived from this new ligand class. We are keenly interested in developing complexes that may exhibit charge-transfer properties, bulk magnetic behavior or that may provide building blocks for supramolecular arrays.

168. REGIOSELECTIVE HYDROGENATION OF TRANS-CINNAMALDEHYDE IN AQUEOUS BIPHASIC SYSTEM, CATALYZED BY RUTHENIUM AND OSMIUM COMPLEXES WITH meta-SULFOPHENYLDIPHENYLPHOSPHINE (TPPMS). Francisco López-Linares,* Marcias Gonzáles and Daniel E. Páez. INTEVEP S.A. Research and Thecnological Support Center of Petróleos de Venezuela, Apdo. 76343, Caracas 1070A, Venezuela.

The regioselective hydrogenation of trans-cinnamaldehyde to cinnamyl alcohol was catalyzed by ruthenium and osmium complexes having the water soluble ligand TPPMS with TPPMS = $(C_6H_5)_2P(m-C_6H_4SO_3Na)$. Complexes of the type $MCl_2(CH_3CN)_4$ (M=Ru and Os) stabilized with TPPMS, can act as catalyst precursor in the hydrogenation of transcinnamaldehyde under mild reaction conditions in an aqueous biphasic system. The selectivity of these precursors have an influence with the temperature, hydrogen pressure and ionic strength. With the Ru complex, the Substrate/Catalyst ratio (S/C) has a remarkable effect on the selectivity to cinnamyl alcohol. At S/C=25 the yield of the unsaturated alcohol is 90% (6 h). A comparison between complexes with TPPMS and TPPTS, showed that the tensoactivity properties observed for the TPPMS has a tremendous favorable effect in the yield of cinnamyl alcohol when it is compared to the analog trisulfonated phosphine TPPTS that behaves more like electrolite under similiar conditions.

169. **Intermetal Interactions Across Non-aromatic Bridges: Effects of Electron Density, Ligand Bulkiness, and Bridge Structure.** <u>Xinhong Gu</u>, Anne Marie Spuches, Beth A. Etzenhouser, and Michael B. Sponsler, Department of Chemistry and W. M. Keck Center for Molecular Electronics, Syracuse University, Syracuse, NY 13244-4100.

The diiron complex [Cp*Fe(dppe)]$_2$(μ-CH=CHCH=CH, Cp* = η^5-C$_5$Me$_5$) has been prepared and oxidized to its radical cationic and dicationic forms. The mixed-valence cation, like the Cp (C$_5$H$_5$) analog, is a delocalized ion. However, the bridge-mediated interaction is significantly increased in the Cp* complex, as evidenced by cyclic voltammetry. Comparisons among these and other mixed-valence complexes from our group and from the literature will be presented with an evaluation of several effects on intermetal interaction: electron density at the metal centers, steric size of ligands, and structure of the bridge.

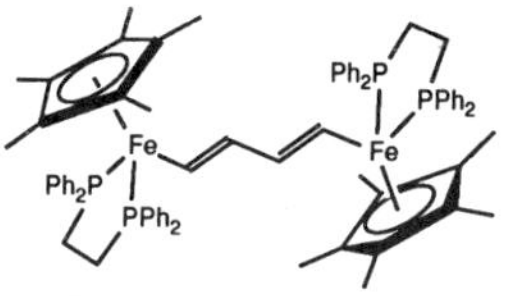

170. CUPRATE CATALYZED ADDITION OF ORGANOLITHIUM COMPOUNDS TO HIGHLY SUBSTITUTED EPOXIDES. <u>David Bartley</u>, Laura J. Smith Vosejpka, Department of Chemistry, Alma College, Alma, MI 48801 and Cynthia L. Rand, Pharmaceutical Process Group, The Dow Chemical Company, Midland, MI 48640.

During the design of a synthesis of steroid **1**, Rand and co-workers developed a ring opening reaction of a tetrasubstituted epoxide that utilizes a cuprate catalyzed transfer of a trimethylsilylpropargyl (TMSP) ligand.

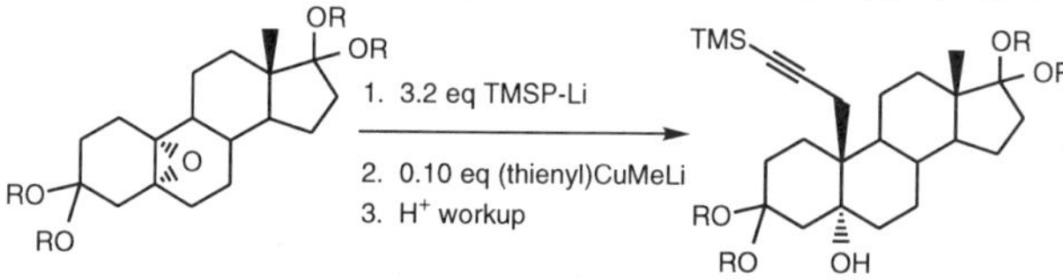

The key features of this reaction are the *in situ* generation of the mixed, higher order cuprate species, L$_1$L$_2$CuRLi$_2$ (L$_1$ = thienyl, L$_2$ = methyl), and the high yield of the ring opening product with very little elimination product observed. The results of our investigation of the scope of this reaction as applied to a series of highly substituted epoxides will be presented.

171. SYNTHESIS OF LABELED ACYL AND ALKYLIRON COMPLEXES FOR PHOTODECOMPOSITION STUDIES. <u>Todd D. Clarke</u> and Laura J. Smith Vosejpka, Department of Chemistry, Alma College, Alma, MI, 48801.

We have been interested in the mechanism of the photoreactions of ω-haloalkyiron complexes to form alkenes as shown below.

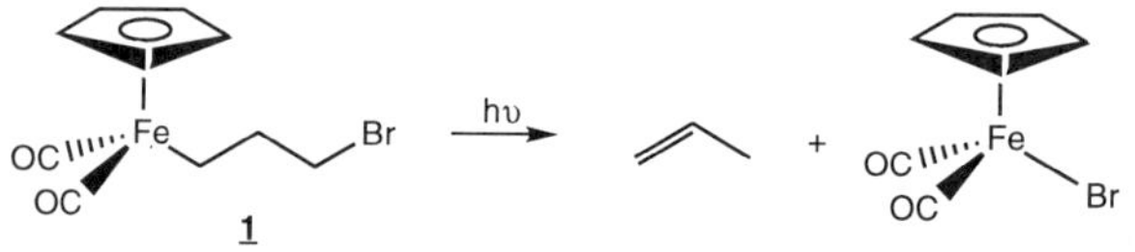

Earlier results have suggested that the photodecomposition does not follow a pathway similar to that taken by unsubstituted alkyliron complexes. We will present the results of our current attempts to synthesize deuterium labeled analogs of **1** and also the behavior of these labeled complexes under photolysis conditions.

172. A RAMAN STUDY OF THE METAL-METAL STRETCHING VIBRATIONS IN TRIPLY-BONDED $M_2(OR)_6$ AND M_2R_6 (M = Mo, W) COMPOUNDS, John C. Littrell, Chad E. Talley, Michael A. Vance and Richard F. Dallinger, Department of Chemistry, Wabash College, Crawfordsville, IN 47933; Thomas M. Gilbert, Department of Chemistry, Northern Illinois University, DeKalb, IL 60115.

The elucidation of the metal-metal stretching frequencies in triply-bonded complexes of the type $L_3M \equiv ML_3$ (M = Mo,W) has been a long-unresolved issue. Chisholm and Cotton first reported the Raman spectra for the complexes $M_2[N(CH_3)_2]_6$ (M = Mo,W) in 1976, but a clean assignment of the metal-metal stretching frequency was not possible due to coupling between the metal-metal stretch and low frequency ligand vibrations. We have acquired Raman spectra for 16 pairs (M = Mo, W) of alkoxy complexes of the type $M_2(OR)_6$ and 5 pairs of alkyl complexes of the type M_2R_6. Many pairs of compounds with the same ligand set show very similar Raman spectra, except for one band which dramatically changes in frequency between the Mo and W compounds. We have assigned this band, which occurs at ~365 cm^{-1} in the dimolybdenum compounds and ~285 cm^{-1} in the ditungsten compounds, as the metal-metal stretching frequency $\nu_{M \equiv M}$. Several of our compounds, however, did not have clearly observable metal-metal stretching frequencies. We will present our Raman data for the various M_2L_6 complexes and discuss the relationship between the observation of a clean $\nu_{M \equiv M}$ and the structural features of the molecules.

173. CHIRAL PHOSPHINOFERROCENES, S.K. McAlister; P.F. Brandt, Department of Chemistry and Physics, Western Carolina University, Cullowhee, NC 28723.

Chiral phosphines are very important species in chemistry for their ability to interact with other chiral molecules. Two important examples of chiral phosphines can be cited in the fields of catalysis and polymers (tacticity). For this reason, it is of interest to develop high yielding syntheses of chiral phosphines. The following results lend themselves to important early studies on this topic. We have found that an *in-situ* reaction of $PhP(Cl)NEt_2$ with dilithioferrocene yields a mixture of **Ia** and **IIa** in a 1:2 ratio upon work-up. Oxidation of **Ia** and **IIa** with S_8 allowed isolation and characterization of **Ib** and **IIb**. Single crystal X-ray diffraction of **Ib** showed a racemic mixture while **IIb** showed the d,l pair of isomers. $^{31}P\{^1H\}$ NMR spectra of **IIb** revealed two peaks in a ratio of 90:10 presumed to be d,l:meso isomers.

Ia, E = lone pair
Ib, E = S

IIa, E = lone pair
IIb, E = S

174. FERROCENYL REDOX SWITCHES FOR MODULATING REACTIONS OF $(ARENE)Cr(CO)_3$ COMPLEXES. L.K.Yeung, S. J. Schofer, Y.K. Chung and D. A. Sweigart, Departments of Chemistry, Brown University, Providence, RI 02912 and Seoul National University, Seoul 151-742, Korea

It has been shown that the oxidation of compound (1) where R = R' = H, occurs first at the ferrocenyl moiety. Never the less, there is sufficient activation of the neighboring 18-electron chromium center so that rapid ligand substitution and addition occur with nucleophiles such as $P(OEt)_3$. In this sense, the ferrocenyl group acts as a redox switch. The degree of activation at the chromium was postulated to be related in a simple and quantitative manner to the potential difference between the iron and the chromium centers. The present paper describes the electrochemistry and reactivity of several versions of (1), with R= H, R' = Me and R= Me, R'= H. These modifications allow the potential difference between the ferrocenyl switch and the chromium to be tuned and provide a test for the proposed activation model. X-ray structures of a variety of complexes of type (1) will be presented and discussed.

175. SYNTHESIS OF (η5-C$_5$(CH$_3$)$_5$)(η5-C$_5$H$_5$)Zr(R)OR* (R = CH$_3$; R* = (CH$_3$)$_2$CPh, CH$_3$CHPh) COMPLEXES. **Walden Lee**, Eva L. Huang and Wayne Tikkanen. Department of Chemistry and Biochemistry, California State University, Los Angeles, CA 90032

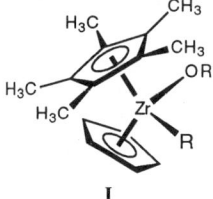

I

Cationic bis[cyclopentadiene]zirconium(IV) complexes have been shown to induce asymmetric Diels-Alder reactions and stereospecific olefin polymerizations. A new series of cationic zirconocene complexes that contain one pentamethylcyclopentadiene(Cp*) ligand and one unsubstituted cyclopentadiene (Cp) ligand are currently under development. A route towards these compounds are pseudotetrahedral Cp*CpZr(R)OR*,**1**, where R=CH$_3$ and R*=(CH$_3$)$_2$CPh, CH$_3$CHPh. Addition of two equivalents of CH$_3$MgBr with Cp*CpZrCl$_2$ yields the dimethyl analogue followed by the addition of one equivalent of R'OH affords chiral **1**. The synthesis and characterization of **1** will be discussed. Future work will involve conversion of **1** to the corresponding cation and evaluation as catalysts for stereospecific 4+2 addition and polymerization.

176.
ANSA-CHROMOCENE DERIVATIVES Frank Schaper and Hans-Herbert Brintzinger, University of Konstanz, Postfach 5560 M738, D-78434 Konstanz, Germany

The synthesis and crystal structure of a ring-bridged and permethylated chromocene, Me$_2$Si(C$_5$Me$_4$)$_2$Cr, is described. The crystal structure of this paramagnetic compound is surprisingly similar to that of unbridged decamethylchromocene. By reaction with CO or isonitrile, the corresponding diamagnetic carbonyl and isonitrile complexes Me$_2$Si(C$_5$Me$_4$)$_2$Cr(CO)$_x$ (x=1,2), Me$_2$Si(C$_5$Me$_4$)$_2$Cr(CN-xylyl) were synthesized and structurally characterized by X-ray diffraction. The kinetics of the interconversion reactions of the carbonyl complexes were studied by use of high-pressure IR and NMR techniques. These interconversions include hydrogen migration and surprisingly fast ring flip reactions.

177.
THE CHEMISTRY OF NEW BIDENTATE P,N-PALLADIUM(II) ALKYL COMPLEXES Peter H.P. Brinkmann, Gerrit A. Luinstra, University of Konstanz, Postfach 5560 M738, D-78434 Konstanz, Germany

The synthesis of new bidentate P,N-Palladium(II) alkyl complexes is described. Reaction with CO gives the Pd-acyl complexes. The order of the various CO insertion rates was obtained from competition experiments. We found that aryl phosphines insert faster CO than alkyl derivatives and that with amines faster rates were obtained than with pyridines. Rigid systems are also advantagous for high insertion rates. The subsequent olefin insertion into the Pd-acyl bond is also discussed.

178.
SYNTHESIS OF NEW ZIRCONOCENE HYDRIDES WITH AROMATIC AND ALIPHATIC SULPHONATE LIGANDS Nicola S. Hüsgen, Gerrit A. Luinstra, University of Konstanz, Postfach 5560 M738, D-78434 Konstanz, Germany

The synthesis of zirconocene sulphonate hydrides Cp$_2$ZrH(OSO$_2$R) (R= n-hexyl, n-butyl, i-propyl) and (MeCp)$_2$ZrH(OSO$_2$R) (R = Ph, p-tolyl, p-chlorphenyl, m-CF$_3$-phenyl, 3,4-dichlorophenyl) will be described. These readily obtainable zirconocene hydrides are soluble in THF and very reactive towards alkenes and alkynes. They represent an attractive alternative to Schwartz´s hydrozirconation reagent Cp$_2$Zr(H)Cl. The steric bulk of the aliphatic substituent R in Cp$_2$ZrH(OSO$_2$R) determines the regioselectivity of hydrozirconation. The influence of the various electron-withdrawing aromatic sulphonate ligands on the metal-center in (MeCp)$_2$ZrBH$_4$(OSO$_2$R) and (MeCp)$_2$ZrH(OSO$_2$R) was studied by NMR- techniques and compared to the analogous Cp$_2$Zr-sulphonate-complexes.

179. ORGANOMETALLIC COMPLEXES OF BENZOPORPHYRINS. E. D. Sternberg and D. Dolphin, Department of Chemistry, University of British Columbia, Vancouver, B.C., Canada

Iron carbonyl complexes of 1,3 and 1,4-diene systems were first extensively investigated by Birch in the 1960's. We have noticed in the literature a dearth of π-organometallic complexes which are directly in conjugation with the porphyrin aromatic system. Only recently has Senge et al. (Angew. Chem. Int. Ed. 1996, 35, 1923) described a direct π-complex to the pyrrole of metallated porphyrin. We have generated metal carbonyl complexes of unusual chlorins and porphyrins (Morgan et al. Chem. Commun. 1984, 1047) which result in structure of type I. These are rather more intricate species than here to fore described. We have begun to investigate their structure and chemistries.

M_1 = Zn, M_2 = Fe(CO)$_3$
PMe = methypropionate

180. SYNTHESIS AND REACTIVITY OF IRIDIUM FISCHER CARBENE COMPLEXES.
Hans F. Luecke and Robert G. Bergman. Department of Chemistry, University of California, Berkeley, California, 94720-9989

A series of cationic carbene complexes [Cp*(PMe$_3$)(H)Ir=C(OMe)(R)]$^+$OTf$^-$ (R=Ph, ^{t}Bu, H) has been synthesized by the treatment of the hydrido acyl complexes Cp*(PMe$_3$)IR(H)(C(O)(R)) **2** with MeOTf in CH$_2$Cl$_2$. These complexes can be deprotonated using lithium 2,2,6,6-tetramethylpiperidide to form the neutral carbene complexes Cp*(PMe$_3$)Ir=C(OMe)(R). The reactivity of these carbene complexes with a variety of nucleophilic and electrophilic substrates will be presented, including attempts to utilize the neutral complexes as precursors to carbyne complexes.

181. η^2-N$_2$-TITANIUM DIAZOALKANE COMPLEXES: IMIDO VS. CARBENE REACTIVITY AND MECHANISTIC STUDIES OF N$_2$ LOSS
Anne Warren Kaplan, Richard A. Andersen, Robert G. Bergman. Department of Chemistry, University of California, Berkeley, California, 94720-9989

Our group recently reported the synthesis of the η^2-N$_2$ diazoalkane complex Cp*$_2$TiN$_2$C(H)SiMe$_3$ which loses N$_2$ to form a transient carbene complex. A series of diazoalkane complexes of varied stability have been prepared and the mechanism of N$_2$ loss has been investigated. Studies of these unusually reactive diazoalkane complexes have revealed two major reactivity manifolds: as a carbene complex precursor (Path 1) or as an imidotitanium complex (Path 2).

182. **INVESTIGATION OF COMPLEXES CONTAINING THE STRONG π-DONOR Cp*W(NO)(PPh$_3$) FRAGMENT.** David J. Burkey, and Peter Legzdins, Department of Chemistry, The University of British Columbia, 2036 Main Mall, Vancouver, Canada, V6T 1Z1

The synthesis of several low-valent tungsten complexes, Cp*W(NO)(PPh$_3$)(L) (L = neutral two-electron donor) will be presented. Complexes of this type can be generated by the ligand-induced reductive elimination of arene in Cp*W(NO)(H)[η^2-PPh$_2$C$_6$H$_4$] (see below), or by substitution of one PPh$_3$ ligand in Cp*W(NO)(PPh$_3$)$_2$. Characterization of these complexes reveals that the Cp*W(NO)(PPh$_3$) fragment is an exceptionally strong π-donor, capable of binding weakly π-acidic ligands such as ketones, esters and nitriles in an η^2-fashion. Mechanistic details of the reactions to produce the Cp*W(NO)(PPh$_3$)(L) complexes will also be presented.

L = CNCMe$_3$ CN(c-C$_6$H$_{11}$)
Me$_2$C=O   (CH$_2$)$_4$C=O
Me(OEt)C=O

183. **ELECTROPHILIC ATTACK AT THE NO LIGAND IN NEUTRAL AND ANIONIC GROUP 6 NITROSYL COMPLEXES.** Peter Legzdins, Steven F. Sayers and W. Brett Sharp, Department of Chemistry, The University of British Columbia, 2036 Main Mall, Vancouver, B.C., Canada V6T 1Z1

Study of the reactivity of nitrosyl ligands with electrophiles has previously encompassed only complexes in which the NO ligand is bridging or bent. Electrophilic attack at linear, terminal nitrosyl ligands in Group 6 half-sandwich complexes is the topic of this presentation. For example, the neutral bis(alkyl) species Cp*W(NO)(CH$_2$SiMe$_3$)$_3$ **1** reacts with the hard electrophile H$^+$ to form the isolable nitrosyl adduct **2**. Further reaction of **2** in acetonitrile results in the protonolysis of an alkyl ligand to form the solvated cation **3**.

184. A THEORETICAL ANALYSIS OF A SERIES OF POLYENE-BRIDGED DIFERROCENE COMPOUNDS, {((η^5-C$_5$H$_5$)Fe(η^5-C$_5$H$_4$))$_2$μ-(CH)$_x$}^{y} (x = 2, 4, 6, ..., y = 0; x = 1, 3, 5, ..., y = +1). HollyAnn Harris; Department of Chemistry, Creighton University, Omaha, NE 68178-0104

Recent experimental work[1] on polyene-bridged ferrocene systems suggests that electronic coupling between the two metal centers can be greatly enhanced by the presence of odd-alternant polyene bridges as opposed to the more common, symmetrical, even-alternant polyene bridge. We have performed MO calculations on several bridged ferrocene systems using the Fenske-Hall MO method in order to compare the electronic structures of the even-bridged systems with the odd-bridged systems. The results of these calculations will be presented.

1. Tolbert, L.M.; Zhao, X.; Ding, Y.; Bottomley, L.A. *J. Amer. Chem. Soc.*, **1995**, *117*, 12891-12892.

185. **A MECHANISTIC PROBE INTO THE NATURE OF THE INTRAMOLECULAR [4+2] DIELS-ALDER CYCLOADDITION OF PHOSPHOLES TO COORDINATED UNSATURATED PHOSPHINES, PHOSPHOLES AND ARSINES**
Kalyani Maitra, Vincent J. Catalano, and John H. Nelson*, Department of Chemistry/216, University of Nevada-Reno, Reno, Nevada 89557-0020

Previously our research group has reported the syntheses of conformationally rigid asymmetric bidentate and tridentate ligands by metal promoted [4+2] Diels-Alder cycloaddition reactions of coordinated phospholes with a variety of dieneophiles. In the present research, we have synthesized bidentate ligands coordinated to molybdenum via the stereoselective [4+2] Diels-Alder addition reaction of 1-phenyl-3,4-dimethylphosphole (DMPP) with unsaturated tertiary phosphine, phosphole and arsine complexes e.g. Ph_2E-$trans(CH_2CH=CHCH_3)Mo(CO)_5$, where E = P, As. The stereochemistry of the products provides insight into the mechanistic pathway for such [4+2] intramolecular cycloaddition reactions i.e., concerted or stepwise. Interestingly, we have found that the phosphole (DMPP), undergoes a [1,5] phenyl migration at 145 °C prior to or concomitant with the [4+2] intramolecular cycloaddition reaction with the alkene moiety of the dieneophile to give a new class of conformationally rigid bidentate ligands.

186. **DIASTEREOSELECTIVITY OF CHLORIDE SUBSTITUTION REACTIONS OF (S_{Ru}, R_C)-$[(\eta^6$-BENZENE)Ru(N,N-DIMETHYL-α-PHENYLETHYLAMINE)Cl]$; KINETIC STUDIES AIMED AT UNDERSTANDING THE MECHANISM OF THESE SUBSTITUTION REACTIONS.** Heather D. Hansen, Kalyani Maitra, Vincent J. Catalano, and John H. Nelson*, Department of Chemistry/216, University of Nevada - Reno, Reno, NV 89557-0020

The diastereoselectivity of chloride substitution reactions at the chiral Ru center in (S_{Ru}, R_C)-$[(\eta^6$-benzene)Ru(N,N-dimethyl-(α-phenylethyl)amine)Cl] has been previously investigated in our lab with halides and N- and P-donor ligands. These reactions proceed predominantly with retention of configuration at the ruthenium center with stereoselectivity (%de) increasing with increasing bulk of the incoming ligand. A dissociative mechanism has been proposed. Recent studies have involved substitution of Cl$^-$ with other anionic ligands (SCN$^-$, N$_3^-$, 5-methyltetrazolate, and 5-benzyltetrazolate); these reactions also proceed with predominant retention of configuration at Ru, with 60 to 97 % de. Current research is aimed at measuring overall rates of these substitution reactions to support (or disprove) a dissociative mechanism. Also, the tetrazolate complexes will be prepared by [1,3] dipolar cycloaddition reactions of the azide complex [(η^6-benzene)Ru(N,N-dimethyl-α-phenylethylamine)N$_3$] with the appropriate nitrile to assess whether the products are formed under kinetic or thermodynamic control.

187. **RUTHENIUM SILYL FRAGMENTATION: A MASS SPECTROMETRY STUDY OF A SERIES OF RUTHENIUM SILYL COMPLEXES.** Frederick R. Lemke and Elaine Saulinskas, Department of Chemistry, Ohio University, Athens, Ohio 45701-2979.

The formation of metal–silicon multiple bonds generally involves the removal of substituents from the silicon in metal silyl complexes containing metal–silicon single bonds. Mass spectrometry has been used to predict the feasibility of the removal of these substituents. Along these lines, we have been studying the fragmentation pattern of a series of ruthenium silyl complexes, $Cp(PMe_3)_2Ru$–SiR_3 (SiR_3 = $SiCl_3$, $SiHCl_2$, SiH_2Cl, SiH_3, $SiMeCl_2$, $SiMeHCl$, $SiMeH_2$, $SiMe_2Cl$, $SiMe_2H$, $SiCl_2Cp^*$, SiH_2Cp^*), by mass spectrometry. Three fragmentation pathways are observed: silylene formation due to loss of a substituent, silylene (SiR'$_2$) elimination, and phosphine loss, and are very dependent on the substituents on silicon. For example, $Cp(PMe_3)_2RuSiCl_2Cp^{*+}$ readily loses Cp^{*-} to form the silylene cation $Cp(PMe_3)_2Ru=SiCl_2^+$. $Cp(PMe_3)_2RuSiH_3^+$ prefers silylene (SiH$_2$) lose to give $Cp(PMe_3)_2RuH^+$ while $Cp(PMe_3)_2RuSiHCl_2^+$ readily loses PMe$_3$ to give $Cp(PMe_3)RuSiHCl_2^+$. Details describing fragmentation preferences and fragmentation characteristics will be described.

188. **IMIDAZOLIDINYL CARBENES WITH PENDENT FERROCENYL SUBSTITUENTS.** B. Bildstein, M. Malaun, H. Kopacka, and K. Wurst; Inorganic Chemistry Department, University of Innsbruck, A-6020 Innsbruck, Austria, Europe.

Stable carbenes of the Wanzlick/Arduengo-type are currently under active investigation. In this contribution we report synthetic approaches to metallocenyl substituted imidazol-2-ylidenes which are of interest as novel redox-active ligands. Preparative, structural, and electrochemical properties of the imidazolium precursors and of transition metal carbene complexes will be presented.

189. PHOSPHORUS-PHOSPHORUS COUPLING CONSTANTS IN BISPHOSPHINE TRICARBONYL IRON COMPLEXES, Fe(CO)$_3$LL'. Richard L. Keiter,[a] John William Benson,[a] Ellen A. Keiter,[a] Travis A. Harris,[a] Matthew W. Hayner,[a] Carol A. Boecker,[a] Douglas E. Brandt,[a] Arnold L. Rheingold,[b] and Glenn P.A. Yap,[b] a. Department of Chemistry, Eastern Illinois University, Charleston, IL 61920; b. Department of Chemistry, University of Delaware, Newark, DE 19716.

Mixed-ligand complexes, *trans*-Fe(CO)$_3$LL' [L = PPh$_3$, L' = PPh$_2$Me, PPhMe$_2$, PMe$_3$, PPh$_2$Et, PEt$_3$, PPh$_2$H; L = PMe$_3$, L' = PEt$_3$, PPh$_2$Et, PCy$_3$; L = PEt$_3$, L' = PPh$_2$Me; L = PPh$_2$H, L' = PPh$_2$CH=CH$_2$, PPh$_2$Et] have been obtained from the step-wise reaction of phosphines with Fe(CO)$_3$(BDA) (BDA = benzylideneacetone) or Fe(CO)$_3$(AsPh$_3$)$_2$ and from the reaction of phosphine with Fe(CO)$_4$PPh$_3$ in the presence of base. A strong negative correlation exists between $^2J_{PP}$ coupling constant values and the sum of the phosphine pK_a values. By application of quantitative analysis of ligand effects (QALE), it has been shown that $^2J_{PP}$ correlates strongly with both χ and E_{ar} but not with θ. Although a near perfect fit is obtained from the three-parameter equation, a statistical analysis suggests that for this small data set it has no predictive advantages over the one-parameter pK_a model. An X-ray structure of solid state *trans*-Fe(CO)$_3$(PEt$_3$)(PPh$_3$) shows equal Fe-PEt$_3$ and Fe-PPh$_3$ bond distances implying that bond strength equalization may occur when two rather different phosphines occupy trans coordination sites.

190. REACTIVITY STUDIES OF A BIS-ALKYLIDENE COMPLEX (RCH$_2$)$_4$Ta$_2$Cl$_2$(=CHR)$_2$ (R = SiMe$_3$). Liting Li, Xiaozhan Liu, Ziling (Ben) Xue, Department of Chemistry, University of Tennessee, Knoxville, TN 37996-1600

The title compound (**1**) is found to decompose to an alkylidene-alkylidyne complex (RCH$_2$)$_3$Ta$_2$Cl$_2$(=CHR)(≡CR) (**2**) through α-hydrogen elimination. This reaction follows first-order kinetics. The addition of LiSiR$_3$ to **1** leads to the formation of a bridging bis-alkylidyne complex (RCH$_2$)$_4$Ta(μ-CR)$_2$ (**3**) and the elimination of HSiR$_3$. (Me$_3$P)$_2$(Cl)Ta(μ-CR)$_2$Ta(Cl)(CH$_2$R)$_2$ (**4**), a stable product of the reaction of **1** with PMe$_3$, is found to react with 1 equiv of LiCH$_2$R to produce (Me$_3$P)$_2$(RCH$_2$)Ta(μ-CR)$_2$Ta(Cl)(CH$_2$R)$_2$ (**5**). Subsequent Cl$^-$ substitution by another equiv of CH$_2$R$^-$ gives **3**.

191. SOL-GEL SENSOR MATERIALS FOR CONCENTRATED STRONG ACIDS Leonardo R. Allain*, Ziling (Ben) Xue* and Michael J. Roberts**, Department of *Chemistry and Department of **Electrical Engineering, University of Tennessee, Knoxville, TN 37996

An optical sensor for high acidity (1 - 11 M [H$^+$]) measurements has been developed. The sensor is based on the spectrophotometrical response of an organic dye such as bromocresol purple entrapped inside the pores of a sol-gel silica thin film. The preparation of the sol-gel sensor films by acid- and base-catalyzed methods will be presented. These sol-gel films were coupled to a flow-cell system and a fiber optic spectrometer to allow *in-situ* measurements. They show short response time (sec) with low levels of hysteresis. The results of reproducibility and reversibility tests will also be discussed.

192. UNEXPECTED EASY AND SIMPLE FORMATION OF A RUTHENIUM-METHYLIDENE COMPLEX Montserrat Oliván, Kenneth G. Caulton, Department of Chemistry, Indiana University, Bloomington, IN 47405-4001

Methylidene complexes can be synthesized by different methods, mainly *via* α-hydrogen abstraction, methylidene transfer from phosphoranes or diazomethane and metathesis of ethylene. We have found that RuH$_2$(H$_2$)$_2$L$_2$ (L = PCy$_3$) reacts with an excess of CH$_2$Cl$_2$ to form RuCl$_2$(=CH$_2$)L$_2$. Using CD$_2$Cl$_2$, only RuCl$_2$(=CD$_2$)L$_2$ is formed. Additional studies toward the identification of the mechanism of this apparently simple reaction and further reactivity of the product will be presented.

193. OXIDATIVE ADDITION OF "Ru(CO)L$_2$" TO Si-C(sp) BONDS D. Huang, K. Folting, J. C. Bollinger, K. G. Caulton Department of Chemistry and Molecular Structure Center, Indiana University, Bloomington, IN 47405-4001

Deprotonation of RuH(OTf)(CO)L$_2$ with LiN(SiMe$_3$)$_2$·OEt$_2$ generates a "ruthenate", LiRu(OTf)(CO)L$_2$, which is a masked form of 14-electron species, "Ru(CO)L$_2$". In the presence of Me$_3$SiCCSiMe$_3$, oxidative addition of "Ru(CO)L$_2$" to the Si-C bond occurs to form Ru(SiMe$_3$)(CCSiMe$_3$)(CO)L$_2$. The reactivity of Ru(SiMe$_3$)(CCSiMe$_3$)(CO)L$_2$ will also be discussed.

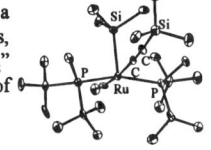

194. MECHANISTIC ASPECTS OF ALKYL REARRANGEMENTS IN SCANDOCENE COMPLEXES: MODELS FOR CHAIN EPIMERIZATION IN OLEFIN POLYMERIZATION. Paul J. Chirik and John E. Bercaw. Arnold and Mabel Beckman Laboratories of Chemical Synthesis, California Institute of Technology, Pasadena, California 91125

The isotopically-labeled, propyl scandocenes, DpSc(PMe$_3$)(^{13}CH$_2$CH$_2$CH$_3$) (I) and Cp*$_2$Sc(^{13}CH$_2$CH$_2$CH$_3$) (II) {Dp = Me$_2$Si[η^5-C$_5$H$_3$(CMe$_3$)]$_2$; Cp* =η^5-C$_5$Me$_5$} undergo rearrangement to form, DpSc(PMe$_3$)(CH$_2$CH$_2$^{13}CH$_3$) and Cp*$_2$Sc(CH$_2$CH$_2$^{13}CH$_3$), respectively. For (I), the isomerization process was found to be competitive with olefin insertion, yielding 2-methyl-1-pentene, depending on monomer concentration. Addition of unlabeled propylene to either (I) or (II) results in rapid crossover between free olefin and the alkyl fragment. Model studies employing internal olefins have also added insight into the mechanistic aspects of the rearrangement process. The relevance to stereo- and regioerror formation in metallocene-based olefin polymerization will be presented.

195.
MECHANISTIC INVESTIGATIONS ON SYNDIOSPECIFIC POLYMERIZATION OF α-OLEFINS CATALYZED BY DOUBLY BRIDGED ZIRCONOCENES.
Dario Veghini and John E. Bercaw
Arnold and Mabel Beckman Laboratories of Chemical Synthesis,
California Institute of Technology, Pasadena, California 91125

Group IV metallocenes/methylalumoxane systems catalyze the polymerization of α-olefin, in certain cases with extremely high stereospecificity and activity. The main interactions operating in the C-C bond forming step during the olefin insertion are reasonably well understood. Less clear are the mechanisms by which this stereocontrol fails, thus the key contacts which govern the formation of the different stereoerrors. Syndiospecific C$_S$-symmetric doubly bridged metallocene catalysts have recently been prepared in these laboratories. The reaction of methylalumoxane activated catalysts with deuterium labeled propene have been investigated to establish the kinetic isotope effect on the polypropene microstructure and the deuterium distribution in the polymer produced. The implications on the mechanisms of stereoerror formation will be discussed.

196. AMINO-FLUORENIDE DERIVATIVES OF GROUP IV METALLOCENES.
Stephen A. Miller and John E. Bercaw.
Arnold and Mabel Beckman Laboratories of Chemical Synthesis, California Institute of Technology, Pasadena, CA 91125

New Group IV compounds containing a 9-dialkylaminofluorenide ligand have been prepared. Their properties and reactivity have been investigated. For example, the metallocene (9-Me$_2$N-fluorenyl)$_2$ZrCl$_2$, in the presence of methylaluminumoxane (MAO), is an active catalyst for the polymerization of propylene to polypropylene. The syntheses of 9-dialkylfluorene compounds are described, as are the syntheses of group IV compounds via the corresponding 9-dialkylaminofluorenide lithium salts.

197. DENSITY FUNCTIONAL STUDIES OF ZIEGLER NATTA CATALYSIS
Christopher G. Brandow and John E. Bercaw.
Arnold and Mabel Beckman Laboratories of Chemical Synthesis, California Institute of Technology, Pasadena, CA 91125

Various Group IV and Group III ansa metallocenes were modeled using density functional methods and their catalytic properties evaluated. The systems studied include a tradtional cationic group IV metallocene, Group III neutral metallocene, a zwitterionic metallocene with an anionic boron linking the two cyclopentadienyl rings, and a new doubly linked cationic Group IV metallocene. These studies extend previous electronic structure calculations by studying propylene insertion into an ethyl bond on a substituted metallocene in order to more accurately assess the energetics of stereospecific propylene polymerization.

198. REACTIVITY OF A WATER-SOLUBLE PHOSPHONATE FUNCTIONALIZED PHOSPHINE LIGAND WITH $Co_2(CO)_8$ Michael C. Miggins and D. Andrew Knight*, Department of Chemistry, The George Washington University, 725 21st Street, N.W. Washington, District of Columbia 20052.

p-Triphenylphosphine monophosphonate (p-TPPMP) is highly water-soluble due to the presence of the anionic phosphonate moiety. The reactivity of p-TPPMP with $Co_2(CO)_8$ has been studied with the intent of preparing novel catalytically active water-soluble transition-metal complexes. The resulting complexes react with $ZnCl_2$ to yield organometallic polymers.

$$Co_2(CO)_8 \xrightarrow[L]{\text{Toluene, } H_2O} [Co(CO)_3(L)_2]^+[Co(CO)_4]^- + Co(CO)_6(L)_2$$

$$L = Ph_2P\text{-}\langle C_6H_4 \rangle\text{-}PO_3Na_2$$

199. PREPARATION of $[(TACN)Rh(H)(R)(L)]^+$. UNUSUALLY STABLE ALKYL HYDRIDE COMPLEXES of RHODIUM. Renjie Zhou, Chunming Wang, and Thomas C. Flood.* *Department of Chemistry, University of Southern California, Los Angeles, California 90089-0744.*

(Tacn)RhR$_3$ (tacn = 1,4,7-triazacyclononane; R = Me, Et, Ph, Vi) have been prepared by alkylation of (tacn)RhCl$_3$ with excess RLi followed by treatment with MeOH. Standard techniques convert the Me and Et derivatives to [(tacn)Rh(D)R(L)](BAr$^F_4$) (L = PMe$_3$; BAr$^F_4$ = $^-$B[C$_6$H$_3$-3,5-(CF$_3$)$_2$]$_4$). These react with C$_6$D$_6$ at 80 °C to form [(tacn)Rh(D)(C$_6$D$_5$)(L)](BAr$^F_4$) in high yields. The half-lives of these cleanly first-order reactions at 80 °C are 1.2 hr (R = Et) and 7.3 hr (R = Me). However, heating at 56 °C scrambles the deuterium into the alkyl group without detectable alkane loss. This difference in rate between C-H oxidative addition and alkane loss from the proposed alkane-complexed intermediate is unprecedented and offers promise for investigation of the dynamical properties of alkane complexes.

200. REACTIONS OF (Cn)Rh(CH=CH$_2$)$_3$ AND -RhMe(CH=CH$_2$)$_2$ WITH PROTIC ACIDS. THE RELATIVE MIGRATORY APTITUDE OF METHYL AND VINYL TO AN (Ethylidene)Rh ALKYLIDENE CARBON. Hongshi Zhen, Chunming Wang, and Thomas C. Flood.* *Department of Chemistry, University of Southern California, Los Angeles, California 90089-0744*

Synthesis and characterization of CnRh(CH=CH$_2$)$_3$ and CnRhMe(CH=CH$_2$)$_2$ (Cn ≡ 1,4,7-trimethyl-1,4,7-triazacyclononane) are reported. Treatment of these compounds with HOTf (triflic acid), [H(OEt$_2$)$_2$][BAr$^F_4$] (BAr$^F_4$ = $^-$B[C$_6$H$_3$-3,5-(CF$_3$)$_2$]$_4$), or HCl in organic solvents as well as with methanol as solvent and acid yields [CnRh(R)(η^3-CH$_2$CHCH-Me)]$^+$ (R = Me, CH=CH$_2$). No intermediates can be observed by NMR at low temperature. Use of CD$_3$OD as acid/solvent leads to formation of [CnRh(CH=CH$_2$)(η^3-CH$_2$CHCH-CH$_2$D)] X (X = CD$_3$O$^-$ assumed) as the only isomeric product, suggesting [CnRh(=CH-CH$_2$D)-(CH=CH$_2$)$_2$]$^+$ as a probable intermediate. In proposed intermediate [CnRh(=CH-CH$_3$)Me-(CH=CH$_2$)]$^+$ a competition between methyl and vinyl for migration to the carbene carbon is established that is dominated by vinyl migration.

R = Me, CH=CH$_2$

HX
CH$_2$Cl$_2$
-80 to 22 °C

201. CATIONIC INDENYL COMPLEXES OF NICKEL. Rainer Vollmerhaus, Isabelle Dubuc, Francine Bélanger-Gariépy, and Davit Zargarian, Department of Chemistry, University of Montréal, Montréal, Québec, Canada H3C 3J7.

The cationic complexes [(Ind')Ni(PPh$_3$)L]$^+$ (Ind'= indenyl, 1-Me-indenyl; L= MeCN, PPh$_3$, PMe$_3$) can be prepared from the precursor neutral complexes (Ind')Ni(PR$_3$)Cl (R= Ph, Me) via Cl$^-$ abstraction in the presence of L. The coordinated MeCN in the cationic complexes can be replaced by stronger ligands such as phosphines and pyridine, but not by weaker ligands such as simple olefins. Generation of the cations in non-coordinating solvents and in the presence of reactive olefins (e.g., styrene and norbornylene) generates the olefin complexes which catalyze the polymerization of the olefin in a relatively rapid reaction. These and other reactivities will be discussed in terms of the electrophilicty of the Ni centre and the variable hapticity of the indenyl ligands.

202. GOLD(I) MEDIATED THIOLATE/DISULFIDE EXCHANGE REACTIONS. M.A. DiLorenzo, S. Ganesh, A. E. Bruce, M. R. M. Bruce, Department of Chemistry, University of Maine, Orono, Maine 04469

Reactions of gold(I) complexes with disulfide are important because they have potential biochemical relevance, as well as significance for surface applications of gold. This poster will discuss results of thiolate-disulfide exchange reactions of mono and dinuclear gold(I) phosphine thiolate complexes with organic disulfides. The reaction of Au$_2$(dppm)(SC$_6$H$_4$CH$_3$)$_2$ and (SC$_6$H$_4$Cl)$_2$ has been monitored by ^{1}H NMR and preliminary kinetic measurements indicate that the reaction is first order in disulfide and mixed order in gold complex. Conductivity measurements of Au$_2$(dppm)(SC$_6$H$_4$CH$_3$)$_2$ suggest it forms a charged species in solution. A cationic, thiolate-bridged gold(I) species has been postulated as the form reactive to disulfide. The possible role of phosphine exchange in facilitating the reaction will also be discussed.

203. REACTION OF A LOW VALENT METAL COMPLEX WITH A CHELATING ORGANIC AZIDE. P. S. Runnels, A. E. Bruce, and M. R. M. Bruce, Department of Chemistry, University of Maine, Orono, Maine 04469.

We have been investigating the reaction of iron(0) tricarbonyl benzylideneacetone and 8-azidoquinoline with the goal of isolating and characterizing a low valent iron-nitrene complex. In the course of kinetic and mechanistic investigations, we have found indirect evidence for an iron-hydride intermediate and a nitrene intermediate. The reaction is first order in iron and zero order in azide. When enough water is available, the major product of the reaction is 8-aminoquinoline. If the reaction is initially totally dry and then water is added, there is another mechanistic path leading to a mixture of 8-aminoquinoline and another quinoline-type product.

204. ARYLETHENYL COBALTICINIUM AND RHODICINIUM SALTS- A NEW CLASS OF NON-LINEAR OPTICAL MATERIALS. John E. Sheats, David Latini, Kenneth T. Micai, Albert Lang, Andrew Jones, Christy Geraci, and David S. Sheats, Chemistry Dept., Rider University, Lawrenceville, NJ 08648

Knoevenagle condensation of mono and dimethyl cobalticinium and rhodicinium salts Ia,b (M=Co$^+$, Rh$^+$) with aromatic aldehydes leads to highly conjugated, brightly-colored dyes which show a high degree of optical non-linearity (IIa-e). Compound Ib also forms a substantial yield of the bridged species IIIa,b which arises from a Knoevenagle reaction of one CH_3 group followed by an intramolecular Michael addition to close the bridge.

$[(C_5H_4X)_2]M^+PF_6^-$ $\xrightarrow{ArCHO}$ $[(C_5H_4X)_2]M^+PF_6^-$

Ia X=CH_3, H
b X=CH_3, CH_3

IIa X= -CH=CH-p-$C_6H_4NMe_2$, H   d X= -CH=CH-p-$C_6H_4NMe_2$
b X= -CH=CH-p-$C_6H_4OCH_3$, H   e X= -CH=CH-p-$C_6H_4OCH_3$
c X= -CH=CH-4-$C_{10}H_6NMe_2$, H   f X= -CH=CH-4-$C_{10}H_6NMe_2$

$[C_5H_4M^+C_5H_4CH_2CHArCH_2]PF_6^-$ IIIa Ar = p-$C_6H_4NMe_2$
b Ar = p-$C_6H_4OCH_3$

205. SYNTHESIS OF MONOSUBSTITUTED COBALTICINIUM AND RHODICINIUM SALTS VIA NUCLEOPHILIC ADDITION FOLLOWED BY HYDRIDE ELIMINATION. John E. Sheats and Lawrence J. Kennedy, Department of Chemistry, Rider University, Lawrenceville, NJ 08648

Synthesis of monosubstituted cobalticinium salts uncontaminated by disubstituted derivatives has long proved difficult because the cobalticinium nucleus is resistant to any kind of electrophilic substitution. We have obtained monosubstituted derivatives by nucleophilic addition of alkyl or aryl lithium reagents followed by thermal isomerization and hydride abstraction. The procedure and the applicability of this method to other types of cobalticinium derivatives will be discussed.

R = CH_3, n-C_4H_9, C_6H_5

206. A STRUCTURAL AND THEORETICAL STUDY OF THIOLATO DERIVATIVES OF NICKEL INDENYL COMPLEXES. Talin Kadkhodazadeh, Francine Bélanger-Gariépy, Christian Reber, and Davit Zargarian, Department of Chemistry, University of Montréal, Montréal, Québec, Canada H3C 3J7.

The complexes monomeric complexes (Ind')Ni(PPh$_3$)(SR) (Ind'= indenyl and 1-Me-indenyl; R= Ph, Et, Me) and the dimers [(Ind')Ni(PPh$_3$)(μ-SPh)]$_2$ have been prepared and characterized spectroscopically. The solid state structure of the monomer (1-Me-Ind)Ni(PPh$_3$)(SPh) has been determined and shows a "slipped" indenyl ligand. The Ni-S bond distance is virtually equivalent to that found in the analogous Cp complex. This poster will outline the structural and spectroscopic data we have obtained for these compounds, as well as the results of some EHMO calculations.

207. CATIONIC CHROMIUM(VI) AND MOLYBDENUM(VI) BIS(IMIDO) COMPLEXES OF 1,4,7-TRIAZACYCLONONANE AND ITS 1,4,7-TRIMETHYL DERIVATIVE: CRYSTAL STRUCTURE AND ELECTROCHEMISTRY. **Michael C.W. Chan**, F.W. Lee, K.K. Cheung and C.M. Che, Department of Chemistry, The University of Hong Kong, Pokfulam Road, Hong Kong

The first Group 6 d(0) imido complexes containing 1,4,7-triazacyclononane (tacn) and 1,4,7-trimethyl-1,4,7-triazacyclononane (Me3tacn) have been prepared. The molecular structures of [M(tacn)(NtBu)2Cl]ClO4 (M=Cr, Mo) are described. Electrochemical studies by cyclic voltammetry reveal a reversible oxidation couple which is proposed to be imido-centered.

208. SYNTHESIS AND REACTIVITY OF BISSULFONAMIDE COMPLEXES OF TITANIUM. P. J. Walsh*, P. Gantzel, S. Pritchett, D. Ho San Diego State University, San Diego, CA 92182

Asymmetric bissulfonamide based transition metal and main group catalysts have been used as chiral auxiliaries in several catalytic asymmetric processes including addition of dialkylzinc reagents to aldehydes, cyclopropanation of allylic alcohols and the Diels-Alder reaction. However, little is known about the structure and reactivity of these complexes. We wish to report the synthesis, structure and reactivity of bissulfonamido complexes of titanium. In this system, the bissulfonamido ligand binds in a tetradentate fashion to the titanium center providing a rigid C_2 symmetric framework. The relevance of these complexes to asymmetric catalysis will be discussed.

209. GROUP 13/14 AMIDINE DERIVATIVES, REACTIONS AND APPLICATIONS

Damian R. Aris, James Barker[†], Paul R. Phillips, Bronwyn Syed and Malcolm G. H. Wallbridge.* Department of Chemistry, University of Warwick, Coventry, CV4 7AL, U. K. [†]Assocd. Octel Co. Ltd., South Wirral L65 4HF, U. K.

There is considerable current interest in metal-amidine bonded systems, since they function both as catalysts and as precursors to metal nitrides. The reactivity of the amidine [RHNC(R)NR] is dependent upon the substituent R group, with R varying as H, alkyl, CF_3, and aryl, e. g.

$$AlMe_3 + 2[H_2NC(CF_3)NH] \rightarrow [Me_2AlN(H)C(CF_3)NC(CF_3)NH] + CH_4 + NH_3$$

The reaction products from other amidines with a variety of Group 13/14 systems will be presented, together with comments on the reasons for the different types of reactions observed. The thermal properties, and decomposition of some of the products will also be described.

210. MANIPULATION OF NONRADIATIVE PATHWAYS IN RUTHENIUM POLYPYRIDINE EXCITED STATES. Joseph A. Treadway, Thomas J. Meyer, Department of Chemistry, University of North Carolina, Chapel Hill, NC 27599-3290.

Recent advances in the general preparation of Ru(II) complexes incorporating three different bipyridyl units have allowed us to probe in a subtle way the additive effects of ligand structural and electronic characteristics on properties such as absorbance, emission, photo-reactivity, and excited state lifetime. We have applied observations about electronic asymmetry to the preparation of a class of photo-inert Ru(II) bis(pyridine) complexes, and observations of rigidity and delocalization effects in the acceptor ligand to control light absorption and excited state lifetime. This has led to the preparation of chromophores which absorb broadly throughout the UV and visible, emit in the near-IR, exhibit uncharacteristically long excited state lifetimes, and are photochemically stable.

211. CATALYTIC SURFACES FROM MOLECULAR ASSEMBLIES BASED ON RUTHENIUM POLYPYRIDYL COMPLEXES. John A. Moss and Thomas J. Meyer, Department of Chemistry, University of North Carolina, Chapel Hill, NC 27599

Catalytic oxidation chemistry based upon mono- and di-oxo polypyridyl complexes of the types $cis\text{-}[(bpy)_2Ru(=O)_2]^{2+}$ and $cis\text{-}[(bpy)_2Ru(py)(=O)]^{2+}$ (bpy = 2,2'-bipyridine) has been developed in solution. We have extended this chemistry to carbon, tin-doped indium oxide, and titanium dioxide surfaces modified with adsorbed and electropolymerized ruthenium-oxo containing thin films.

Incorporation of the ruthenium-oxo moiety into molecular assemblies such as dimers and soluble polymers allows the fabrication of new thin films where the nature of the molecular assembly and its surface attachment help dictate the properties of the resulting catalytic films. Electrochemical and spectroelectrochemical methods have been utilized to investigate the properties and electrocatalytic reactivity of the films.

212.
NICKEL AND PALLADIUM CATALYST SYSTEMS FOR THE ADDITION POLYMERIZATION OF CYCLOPENTENE. Jerald Feldman, Stephan J. McLain, Mark F. Teasley, Lin Wang, Lynda K. Johnson, Maurice Brookhart, DuPont Company, Central Research and Development, Wilmington, DE 19880-0328 and Department of Chemistry, University of North Carolina at Chapel Hill, Chapel Hill, North Carolina 27599-3290

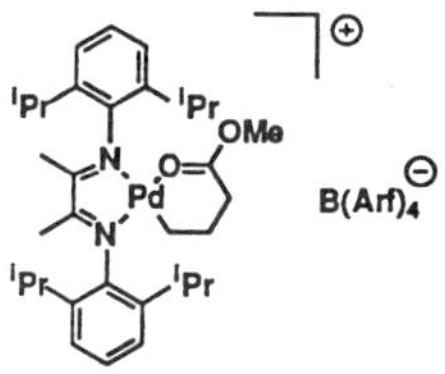

Ni and Pd complexes containing α-diimine ligands are catalysts for the addition polymerization of cyclopentene to give a high melting, melt-processible polymer. Both well-defined, single component catalysts such as that shown at left and easily accessible, multicomponent catalysts such as $Pd(OAc)_2/\alpha$-diimine/$AlEt_3/B(C_6F_5)_3$ are effective. The effect of ligand sterics/electronics, choice of metal, and choice of anion/cocatalyst on catalyst activity, polymer molecular weight, and polymer melting point will be discussed.

213. Co(III)-CYCLEN CATALYZED HYDROLYSIS OF P-NITROPHENYL PHOSPHATE. J. Rawlings, Auburn U at Montgomery, Montgomery AL 36117 , and A. C. Hengge, Enzyme Institute, U of Wisconsin, Madison, WI 53705

The Co(III) complex of cyclen (1,4,7,10-tetraazacyclododecane) efficiently promotes the hydrolysis of p-nitrophenylphosphate (pNPP) at neutral pH. The species involved in this reaction have been characterized by ^{31}P NMR and analytical data. At pH 5, the initial complex between Co(III)-cyclen and pNPP is a relatively stable dimer which contains monodentate pNPP. At pH 7, a coordinated hydroxide has formed, which attacks pNPP, releasing nitrophenol and forming a dimeric Co(III)-cyclen-phosphate species. The inorganic phosphate in this species bridges the two cobalt and has either 3 bonds (^{31}P NMR at 31 ppm) or 4 bonds (^{31}P NMR at 38.1 ppm) to cobalt. The mechanism of the Co(III)-cyclen reaction with the diester ligand ethyl p-nitrophenylphosphate has also been investigated. (Supported by NIH grant GM 18938)

214.
CATALYSIS OF ELECTRON-TRANSFER REACTIONS OF SIMPLE RUTHENIUM AMMIME COMPLEXES VIA NON-COVALENT INTERACTIONS WITH DISSOLVED ELECTROLYTES ACTING AS BRIDGES. Vivian Luo, Jeff C. Curtis , University of San Francisco, San Francisco, CA 94117.

Stopped-flow kinetic studies of pseudo-self exchange reactions such as the one shown below

$(NH_3)_5Ru(II)py^{2+} + (NH_3)_5Ru(III)3Fpy^{3+} \xrightarrow{k_{e.t.}} (NH_3)_5Ru(III)py^{3+} + (NH_3)_5Ru(II)3Fpy^{2+}$

indicate that certain electrolytes in solution, such as the conjugated dicarboxylic anion muconate, can have a rate enhancing effect on the reaction *far in excess* of what can be explained on the basis of simple ionic strength effects alone. We postulate catalysis of the electron-transfer reaction *via* non-covalently associated bridged species in solution such as the one depicted below,

$[Ru(II)py (NH_3)_5 \cdots OOC-CH=CH-CH=CH-COO \cdots (NH_3)_5Ru(III)3Fpy]^{3+}$

215. ELECTROCHEMICAL INVESTIGATIONS OF SOLVENT EFFECTS ON HYDROGEN-BONDED ADDUCT FORMATION BETWEEN CYANO- AND ETHYLENEDIAME-COMPLEXES OF RUTHENIUM IN SOLUTION. Gene Arbatin, Jie Yang, Dhehinie Seneviratne, Jeff C. Curtis, University of San Francisco, San Francisco, CA 94117.

Hydrogen-bonded adducts between ruthenium complexes such as the ones shown below are known to form in solvents of low electron donicity,

2 $(bpy)_2Ru(II)(CN)_2$ + $(en)_2Ru(III)(bpy)^{3+}$ ----> $[((bpy)_2Ru(II)(CN)_2)_2(en)_2Ru(III)(bpy)]^{3+}$

and they exhibit I.T. spectra in the near infra-red. The interaction between the parent complexes can also be observed in the form of an electrochemical wave-repulsion effect. The magnitude of this interaction responds to the electron donicity of the solvent and essentially drops off to zero in strong-donor solvents where hydrogen bonding between the solvent and the en hydrogens competes with adduct formation.

216. DRIVING FORCE DEPENDENCE ON ELECTROCATALYTIC OXIDATION OF ALCOHOLS BY $Ru(O)(trpy)(R_2dppi)^{2+}$

Vincent J. Catalano[†*], Ryan A. Heck[†], Michael G. Hill[‡] and Chad Immoos[‡]

[†]Department of Chemistry, University of Nevada, Reno, NV 89557
[‡]Department of Chemistry, Occidental College, Los Angeles, CA 90041

The stoichiometric and electrocatalytic oxidation of alcohols by Ru(IV)=O species is well known. It has been postulated that driving force and catalytic site preorganization could be factors determining the rate of catalysis. Previous work has been inconclusive as to the effect of driving force. This work seeks to address this through a series of Ru(IV)=O complexes electochemically generated from $Ru(trpy)(R_2dppi)H_2O$ (where R may be H, Me, or Cl, trpy is 2,2',:6,6':2" terpyridine, and dppi is 3,6-di-2-pyridylpyridazine). The substituted ligands have methyl (Me_2dppi) or chloro (Cl_2dppi) groups in the 6 and 6' positions of the pyridyl rings. The substituted and unsubstituted ligands provide a series of complexes that differ only in the redox potential of the ruthenium center. The rate of catalysis can then be studied solely as a function of driving force of the catalyst.

217. ELECTRONIC AND STERIC EFFECTS OF CHIRAL BIDENTATE DIPHOSPHINES IN ASYMMETRIC HYDROFORMYLATION AND HYDROGENATION REACTIONS. Cheng Chao PAI, Albert S. C. CHAN*. Department of Applied Biology and Chemical Technology. The Hong Kong Polytechnic University, Hong Kong.

Four new chiral diphosphine ligands containing electron-rich or electron-deficient groups were synthesized from (2S, 4S)-pentandiol.

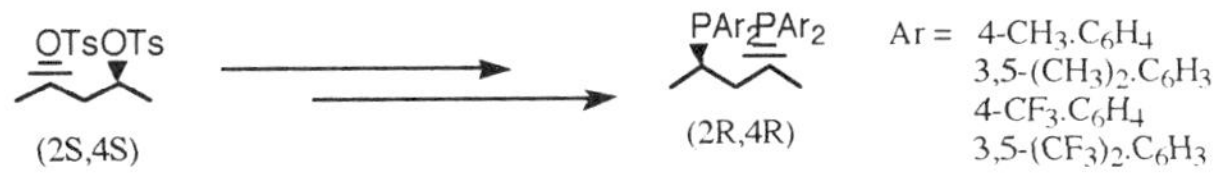

OTsOTs (2S,4S) ⟶ PAr_2PAr_2 (2R,4R)   Ar = $4-CH_3-C_6H_4$, $3,5-(CH_3)_2-C_6H_3$, $4-CF_3-C_6H_4$, $3,5-(CF_3)_2-C_6H_3$

The influence of the steric and electronic effects on the rhodium-catalyzed asymmetric hydroformylation of styrene and on the asymmetric hydrogenation of methyl (Z)-2- acetamidocinnamate will be presented.

218. MECHANISTIC STUDY OF THE Ru(S-BINAP)-CATALYZED HYDROGENATION OF 2-(6-METHOXY-2-NAPHTHYL)ACRYLIC ACID. Chih-Chiang Chen[1], Ting-Ti Huang[1], Albert S. C. Chan*[1] W. T. Wong[2,1] [1]Department of Applied Biology and Chemical Technology, The Hong Kong Polytechnic University, Hong Kong; [2]Department of Chemistry, The University of Hong Kong, Hong Kong

The molecular structure of Ru(S-BINAP)(acac)(MNAA)(MeOH) [where MNAA = 2-(6-methoxy-2-naphthyl)acrylic acid] was determined by single crystal x-ray diffraction. This complex was found to give faster rate of reaction than other Ru(BINAP) type catalyst precursors such as

Ru(BINAP)(OAc)2 and Ru(BINAP)(acac)2 in the hydrogenation of 2-(6-methoxy-2-naphthyl)-acrylic acid. Details of the synthesis, characterization, and application of this catalyst will be presented.

219. SOLUBLE CHIRAL POLYMER-BOUND BINAP LIGAND AND ITS APPLICATION IN ASYMMETRIC HYDROGENATION. Qinghua Fan, C. H. Yeung, Yiching Li and Albert S. C. Chan*, Department of Applied Biology and Chemical Technology, The Hong Kong Polytechnic University, Hong Kong

Recyclable homogeneous hydrogenation catalysts were prepared by using chiral polyester-bound phosphine ligands. The phosphine groups were bound to the polyester support by the copolymerization of (2S,4S)-pentanediol, 5,5'-diamino-BINAP and terephthaloyl chloride. The polyester-bound BINAP ligand thus produced is readily soluble in toluene or a toluene-methanol mixture (v/v>2/3) but is insoluble in methanol. The polymer-bound BINAP-Ru(II) complexes, being soluble in a toluene-methanol mixture (v/v=3/2), catalyzed the hydrogenation of 2-(6-methoxy-2-naphthyl)acrylic acid to give naproxen with better catalytic activity and enantioselectivity than those for the 5,5'-diamino-BINAP-Ru(II) catalyst. These polymer-bound catalysts were easily separated in methanol and the recycled catalyst was found to maintain the same reactivity and enantioselectivity as the freshly prepared catalyst.

220. ASYMMETRIC ADDITION OF DIETHYLZINC TO ALDEHYDES CATALYZED BY TITANIUM-BINAPHTHOL COMPLEXES AND DERIVATIVES. Fu-Yao Zhang, C.W. Yip, Albert S.C. Chan*, Department of Applied Biology and Chemical Technology, The Hong Kong Polytechnic University, Hong Kong

Recently, the enantioselective addition of diethylzinc to aldehydes has attracted much attention. Among the different types of catalysts used, chiral titanium alkoxide complexes (e.g. Ti-TADDOL complexes) have been found to be highly effective in this reaction. In contrast, titanium-binaphthol complexes, which can be conveniently prepared from commercially available starting materials, have not been reported to be useful in this reaction. Herein, we will report our preliminary results of the enantioselective addition of diethylzinc to a variety of aldehydes catalyzed by titanium-binaphthol complexes and derivatives. Moderate to excellent enantioselectivity (60-99%) and high conversion (100%) were obtained. The effects of reaction conditions on the enantioselectivity of the reaction will be presented. A possible mechanism will also be proposed.

221. **Catalytic Dehydrogenation of Linear Alkanes and Ethylbenzene by a Dihydrido Iridium P-C-P Pincer Complex.** Chrystel Hagen[a], Mukta Gupta[a], William C. Kaska[b], and Craig M. Jensen[a], Departments of Chemistry, [a]University of Hawaii, Honolulu, HI 96822 and [b]University of California, Santa Barbara, CA 93106.

The P-C-P pincer complex, $IrH_2\{C_6H_3\text{-}2,6\text{-}(CH_2PBu^t_2)_2\}$ has been found to be a highly active and robust catalyst for the transfer dehydrogenation of linear alkanes. The dehydrogenation of n-octane to octenes in the presence of the hydrogen acceptor, t-butylethylene has been carried out at temperatures as low as 150°C. We have also found that the functionalized substrate, ethylbenzene can be catalytically dehydrogenated to styrene. The details of our studies of these catalytic reactions will be presented.

222.
RELATIVE RATES FOR CARBON-SULFUR BOND-FORMING REDUCTIVE ELIMINATION WITH SP, SP^2 AND SP^3-HYBRIDZED HYDROCARBYL GROUPS. INSIGHTS INTO THE TRANSITION STATE IN THE REDUCTIVE ELIMINATION OF SULFIDES. Grace Mann and John F. Hartwig*, Department of Chemistry, Yale University, P.O. Box 208107, New Haven, CT 06520-8107

We report the synthesis, characterization and reactivity of palladium thiolato complexes of the type, [(DPPE)Pd(R)(SR')], with different hybridized ligands, R, that undergo reductive elimination to form dialkyl sulfides, alkyl aryl sulfides, alkyl vinyl sulfides and alkyl alkynyl sulfides. Reductive elimination to form alkyl sulfides is fastest from compounds with sp^2-hybridized alkenyl and aryl ligands, slower for those with sp-hybridized alkynyl ligands, and slowest for those with sp^3-hybridized alkyl ligands. The ability of the organic product to coordinate to the metal complex seems to dictate reaction rates. Electronic effects for the formation of diaryl sulfides showed that reductive elimination is fastest for electron deficient aryl groups, R, and electron rich aryl thiolates, SR'. Effects of phosphine concentration, solvent, and chelating phosphines were studied with [(DPPE)Pd(S-t-Bu)(Me)] for comparison with the analogous thiolato aryl system. No thiyl radical was detected. Kinetic studies have eliminated a mechanism involving association of PPh$_3$ and argue against a mechanism involving partial dissociation of the chelated phosphine.

223.
MECHANISM OF PLATINUM-CATALYZED HYDROPHOSPHINATION OF ACRYLONITRILE Igor V. Kourkine, David S. Glueck, Department of Chemistry, 6128 Burke Lab, Dartmouth College, Hanover NH 03755.

The complex Pt(dppe)(stilbene) is a catalyst precursor for addition of the P-H bond of PHR$_1$R$_2$ across the C=C double bond of acrylonitrile (hydrophosphination) to afford the functionalized phosphines PR$_1$R$_2$(CH$_2$CH$_2$CN) [**1**, R$_1$ = H, R$_2$ = Mes*; **2**, R$_1$ = R$_2$ = Mes; Mes* = 2,4,6-(t-Bu)$_3$C$_6$H$_2$, Mes = 2,4,6-Me$_3$C$_6$H$_2$]. Studies of the reaction mechanism, which involves oxidative addition of the P-H bond to the metal center, selective olefin insertion into the Pt-P and not the Pt-H bond, and reductive elimination of the product, will be discussed.

224.
INVERSION AT PHOSPHORUS IN PLATINUM PHOSPHIDO COMPLEXES David S. Glueck,[a] Denyce K. Wicht,[a] Louise Liable-Sands,[b] Arnold L. Rheingold.[b] (a) Department of Chemistry, 6128 Burke Lab, Dartmouth College, Hanover NH 03755. (b) Department of Chemistry, University of Delaware, Newark DE 19716.

In order to investigate the barrier to phosphorus inversion in asymmetric phosphido ligands, platinum complexes with enantiomerically pure chiral diphosphine ligands (P~P*) were prepared. The complexes Pt(P~P*)(R)(PR$_1$R$_2$) exist as a mixture of diastereomers which can interconvert by inversion at phosphorus. The cationic precursors [Pt(R-Tol-Binap)(Me)(PHMeMes*)][BF$_4$] (R-Tol Binap = (R)-(+)-2,2'-Bis(di-p-tolylphosphino)-1,1'-binaphthyl, Mes* = 2,4,6-(t-Bu)$_3$C$_6$H$_2$, **1**) and [Pt(S,S-Chiraphos)(Me)(PHMeMes*)][BF$_4$] (S,S-Chiraphos = (2S,3S)-(-)-Bis(diphenylphosphino)butane, **2**) were prepared. Treatment of **1** and **2** with an excess of base yields the corresponding phosphido complexes Pt(R-Tol-Binap)(Me)(PMeMes*) (**3**) and Pt(S,S-Chiraphos)(Me)(PMeMes*) (**4**). The characterization, reactivity, and phosphorus inversion in these complexes will be described.

$$[\text{Pt}]\begin{array}{c}\text{PH(Me)(Mes*)} \\ \text{Me}\end{array} \Big]\text{BF}_4 \xrightarrow{\text{base}} [\text{Pt}]\begin{array}{c}\text{P(Me)(Mes*)} \\ \text{Me}\end{array}$$

1,2   **3,4**

1,3 [Pt] = Pt(R-Tol-Binap)
2,4 [Pt] = Pt(S,S-Chiraphos)

225. Specify Equipment Required for Presentation Other than 2" x 2" (35 MM) slide or overhead (transparency) projector (Subject to approval from the Program Chair)

PHOSPHAAZAALLENE DIMERIZATION AND PHOSPHAALLENE ISOMERIZATION: CATALYSIS BY ZEROVALENT PALLADIUM AND PLATINUM COMPLEXES Marie-Anne David,[a] John B. Alexander,[a] David S. Glueck,[a] Glenn P. A. Yap,[b] Louise M. Liable-Sands,[b] Arnold L. Rheingold.[b] (a) Department of Chemistry, 6128 Burke Lab, Dartmouth College, Hanover NH 03755. (b) Department of Chemistry, University of Delaware, Newark DE 19716.

Dimerization of the phosphaazaallene Mes*PCNPh [Mes* = 2,4,6-(t-Bu)$_3$C$_6$H$_2$] and isomerization of the phosphaallene Mes*PCCPh$_2$ is catalyzed by structurally similar Pd and Pt(0) phosphine complexes, respectively. The dependence of this reactivity on the metal and ligands suggests that phosphacumulene coordination and phosphine dissociation are key steps in the catalyses.

226.
P-H AND P-C BOND ACTIVATION OF DIMESITYLPHOSPHINE BY A Pt(0) COMPLEX Igor V. Kourkine, Matthew Sargent, David S. Glueck, Department of Chemistry, 6128 Burke Lab, Dartmouth College, Hanover NH 03755.

The reaction of Pt(dppe)(stilbene) with PHMes$_2$ (Mes = 2,4,6-Me$_3$C$_6$H$_2$) initially affords the phosphido-hydride complex Pt(dppe)PMes$_2$(H) (**1**), which slowly decomposes in solution to the phosphido-aryl complex Pt(dppe)(PHMes)(Mes) (**2**). Independent synthesis of **2** and the reaction mechanism will be discussed.

PALLADIUM-MEDIATED P-C BOND FORMATION Michael A. Zhuravel, Ryan D. Sweeder, David S. Glueck Department of Chemistry, 6128 Burke Lab, Dartmouth College, Hanover NH 03755.

227. The complexes [Pd(DPPE)(R)(PPh$_2$H)]BF$_4$ (R = CH$_3$, C$_6$H$_5$, CF$_3$, C$_6$H$_4$CH$_3$) were synthesized and characterized by NMR. Addition of LiN(SiMe$_3$)$_2$, followed by DPPE, gave phosphines of general formula PPh$_2$R and Pd(DPPE)$_2$, presumably by reductive elimination from Pd(DPPE)(R)(PPh$_2$).

228.
CP*(2PHIND)ZRCL2 AS THE CATALYST PRECURSOR FOR THE PRODUCTION OF ELASTOMERIC POLYPROPYLENE. Raisa L. Kravchenko, Robert M. Waymouth, Department of Chemistry, Stanford University, Stanford, CA, 94305

Unbridged mixed-ring complex (pentamethylcyclopentadienyl)(2-phenylindenyl)zirconium dichloride produces elastomeric stereoblock polypropylene when activated by methylaluminoxane. The microstructure of the resulting polymer is sensitive to polymerization temperature and propylene concentration. The mechanism of the stereoblock formation will be discussed in relation to other homogeneous metallocene systems known to produce elastomeric polypropylene.

229. PRESSURE DEPENDENCE OF ASYMMETRIC CATALYSIS IN SUPERCRITICAL FLUOROFORM. Philip G. Jessop,[a] Jiangliang Xiao,[b] Takao Ikariya,[b] Ryoji Noyori[b,c] [a]Department of Chemistry, University of California, Davis, CA 95616, [b]ERATO Molecular Catalysis Project, Research Development Corporation of Japan, 1247 Yachigusa, Yakusa-cho, Toyota 470-03, Japan. [c]Permanent address: Department of Chemistry, Nagoya University, Chikusa, Nagoya 464-01, Japan.

Pressure has very little effect on the enantioselectivity of asymmetric catalysis in liquid solvents except at extremely high pressures because activation volumes of reactions in non-compressible solvents are too small. However, supercritical fluids (SCF) with their high compressibilities offer the possibility of achieving pressure-dependent enantioselective reactions at low or moderate pressures. This concept has been applied previously to uncatalyzed Diels-Alder reactions. In our laboratories we have investigated the use of transition metal-containing chiral homogeneous catalysts in supercritical fluoroform. The results to be described include the pressure dependence of the enantioselectivity of the alkylation of benzaldehyde by $ZnEt_2$/DAIB and the hydrogenation of unsaturated carboxylic acids.

230. *ANSA*-METALLOCENE-MEDIATED OLEFIN HYDROBORATION.
Scott J. Stoudt and Milton R. Smith, III,* Department of Chemistry, Michigan State University, East Lansing, MI 48824-1322

The *ansa*-metallocene complex, *rac*-(EBTHI)Ti(CH$_2$)$_4$ (**1**), was prepared from *rac*-(EBTHI)TiCl$_2$ and 1,4-dilithiobutane. The complex, which is in equilibrium with **2**, efficiently catalyzes the hydroboration of olefin substrates. The chemo- and regioselectivities of this process will be highlighted, as well as the significance of olefin structure and borane "tuning" for effective catalysis by this early transition-metal system.

231. METATHESIS REACTIONS OF B-B AND IR-H BONDS.
Carl N. Iverson and Milton R. Smith, III,* Department of Chemistry, Michigan State University, East Lansing, MI 48824

HBCat (Cat = $O_2C_6H_4^{2-}$) and CatB-BCat are useful synthons in hydroboration and diboration processes. B-X activation by transition metal complexes are fundamental steps in stoichiometric and catalytic reactions involving borane reagents. Recently, we have explored the reactivity of both HBCat and X_2B-BX_2 with the Ir(III) dihydride, Cp*Ir(PMe$_3$)H$_2$. We have found Cp*Ir(PMe$_3$)H$_2$ sequentially reacts with CatB-Bcat to generate Cp*Ir(PMe$_3$)(BCat)(H) and Cp*Ir(PMe$_3$)(BCat)$_2$. Details of these reactions and comparisons to other systems will be presented.

232. MECHANISTIC STUDIES OF α-OLEFIN POLYMERIZATION USING Pd(II)- AND Ni(II)-BASED DIIMINE CATALYSTS. Daniel J. Tempel, Lynda K. Johnson, Christopher M. Killian, and M. Brookhart, Department of Chemistry, University of North Carolina, Chapel Hill, NC 27599-3290

Recently our group has developed cationic Pd(II) and Ni(II) catalysts that polymerize ethylene, α-olefins, and internal olefins. α-Olefins can insert into a metal-alkyl bond followed by isomerization via "chain-straightening" to give polymers with 1,ω-enchained monomer units:

$$[M]^+\!-\!\underset{P}{\overset{(CH_2)_{\overline{n}}CH_3}{\overset{\omega}{|}}}\xrightarrow{\text{2,1-insertion}}\xrightarrow[\text{migration}]{\text{chain}}[M]^+\!-\!\overset{\omega}{C}H_2-(CH_2)_{n+1}\!-\!\overset{1}{C}H_2\!-\!P\xrightarrow{\text{insertion}}\text{1, }\omega\text{-enchainment}$$

This process leads to "chain-straightened" poly(α-olefins) with polyethylene-like segments that contain fewer branches than are expected from simple 1,2 monomer enchainment. The degree of branching is highly dependent upon the mode of α-olefin insertion. We report here studies employing both ^{13}C-labelled α-olefins and ^{13}C-labelled [Pd]-CH$_3^+$ precursors which address issues concerning these processes. Specifically, low temperature NMR spectroscopic studies will be described which reveal rates of migratory insertion in the alkyl olefin complexes, the nature of the metal alkyl cation formed after a single insertion, as well as its dynamic behavior (chain-running) and the variation of 1,2- versus 2,1-insertion as a function of ligand structure.

233. ALKANE OXIDATION PROMOTED BY THERMAL DECOMPOSITION OF Co(III) PEROXY COMPLEXES. **John M. Rowland**, Cattien V. Nguyen, and Pradip K. Mascharak*, Department of Chemistry and Biochemistry, University of California, Santa Cruz, California 95064

The mechanisms of decomposition of metal peroxides have been studied extensively for their use in catalytic oxidations of hydrocarbons, as well as in biological oxidation processes. We have studied the mechanism of oxidation of hydrocarbons promoted by thermal decomposition of Schiff base (salen) complexes containing the Co(III)-OO^{t}Bu unit (salen = N,N'-ethylenebis(salicylideneiminato)). In these complexes, changes in both the ligand environment and the axial base bring about a change from cleavage of the Co-O bond to homolysis of the O-O bond in the Co-OO^{t}Bu unit. Thus, with different complexes, formation of either a *tert*-butylperoxyl radical or a *tert*-butoxyl radical is observed. As a result, different distributions of products are obtained (detected by GC analysis) from the same alkane substrate when different cobalt complexes are employed. Results of this study suggest a strong correlation between specific bond cleavage and the ligand environments.

234. ALKANE OXIDATION WITH STRUCTURALLY CHARACTERIZED COBALT (III)-ALKYLPEROXIDE COMPLEXES: ROLE OF LIGANDS IN THERMAL DECOMPOSITION PATHWAYS. **Ferman A. Chavez**, Cattien V. Nguyen, Marilyn M. Olmstead, and Pradip K. Mascharak*, Department of Chemistry and Biochemistry, University of California, Santa Cruz, California 95064

Two [(L)Co(III)-*tert*-butylperoxide] complexes namely, [Co^{III}(Py$_3$P)(OO^{t}Bu)] .2CH$_2$Cl$_2$ (**1**) (Py$_3$P^{2-} = N,N'-bis(2-(2-pyridyl)ethyl)-pyridine-2,6-dicarboxamido) and [Co^{III}(PyPz$_2$P)(OO^{t}Bu)] .2CH$_2$Cl$_2$ (**2**) (PyPz$_2$P^{2-} = N,N'-bis(2-(1-pyrazolyl)ethyl)-pyridine-2,6-dicarboxamido) have been synthesized to study the role of the ligand in the decomposition of the Co-O-O-^{t}Bu moiety. Thermal decomposition of **1** and **2** in presence of alkanes results in stoichiometric oxidation of the C-H bonds. When heated in presence of excess TBHP, catalytic oxidation of alkanes is observed. Careful scrutiny of the oxidized products under aerobic and anaerobic conditions reveals that in case of **1**, homolysis of the O-O bond of the Co-O-O-^{t}Bu moiety occurs exclusively while with **2**, homolysis of the Co-O bond is the predominant reaction. The change in the locale of homolysis is attributed to the difference in the donor strengths of the pyrazole Ns vs. pyridine Ns in the ligand frames.

235. REDUCTIVE ELIMINATION OF ETHANE FROM (dppe)PtMe$_4$. **D. M. Crumpton** and K. I. Goldberg, Department of Chemistry, Box 351700, University of Washington, Seattle, WA 98195-1700

Thermolysis of (dppe)Pt(CH$_3$)$_4$ (**1**) in solution and in the solid state leads to reductive elimination of ethane and formation of (dppe)Pt(CH$_3$)$_2$. The thermal behavior of **1** in the solid state was monitored by DSC and TGA. Kinetic studies in solution have been carried out in several solvents. Temperatures greater than 120°C are required to observe reasonable rates of reaction. The higher temperatures required for elimination of ethane from **1** relative to other bisphosphine Pt(IV) complexes is attributed to the lack of a ligand which can easily undergo dissociation. Results from kinetic studies and other experiments to explore the mechanism of C-C elimination from **1** will be discussed.

236. MECHANISTIC STUDIES OF MIGRATORY INSERTION REACTIONS INVOLVED IN THE COPOLYMERIZATION OF ETHYLENE AND CO BY BIDENTATE PHOSPHINE Pd(II) COMPLEXES. M. S. Brookhart, John. S. Ledford and Peter S. White, Dept. of Chemistry, UNC-CH, Chapel Hill, NC 27599.

Variation in the alkyl backbone length as well as steric bulk of the phosphine substituents of bidentate ligands of cationic Pd(II) ethylene/CO copolymerization catalysts (below) has produced dramatic differences in a series of rates of migratory insertion reactions. The free energies of activation for migratory insertion reactions of $[(P-P)Pd(Me)(CO)]^+[BAr'_4]^-$, $[(P-P)Pd(Me)(\eta^2-C_2H_4)]^+[BAr'_4]^-$, and $[(P-P)Pd(CH_2CH_3)(\eta^2-C_2H_4)]^+[BAr'_4]^-$ have been determined for a variety of bidentate phosphine ligands. A significant number of these reactive intermediates have been isolated, including the novel β-ethyl agostic complex $[(DIPPP)Pd(CH_2CH_2\text{-}\mu\text{-}H)]^+[BAr'_4]^-$

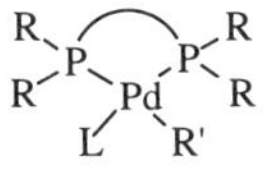

R = Phenyl, Methyl, Isopropyl
R' = Alkyl, Acyl
L = C_2H_4, CO

237. ENFORCING GEOMETRIES ON METAL COMPLEXES THROUGH THE USE OF LIGANDS WITH BIPHENYL BACKBONES. Mitchell R. Malachowski, Ash Kasto and Mark Adams, Department of Chemistry, University of San Diego, San Diego, CA 92110

The biphenyl backbone consists of two aromatic rings that are not co-planar. As a result, when donor atoms are present on the rings, they will take up positions that are not in the same plane. In this work, two series of biphenyl-based ligands which incorporate nitrogen and sulfur or nitrogen only donors were prepared. Copper, zinc, and iron complexes of these N_2S_2 or N_6 ligands were synthesized and characterized. The biphenyl rings impart on the metal complexes geometrical constraints which force them into specific geometries. For the N_2S_2 systems, the goemetries are tetrahedral-like, while for the dinuclear N_6 complexes, the biphenyl ring keeps the two metal centers separated at relatively large distances. Comparisons to the copper-containing proteins hemocyanin and tyrosinase along with the blue copper proteins will be made.

238. SYNTHESIS AND REACTIONS OF ORGANOMETALLIC TUNGSTEN COMPLEXES CONTAINING WATER-SOLUBLE CHELATING NITROGEN LIGANDS. Danielle R. Prud'homme, Andrea M. Pellerito, Amy J. Rogowski, and Brian P. Buffin, Department of Chemistry, University of Detroit Mercy, Detroit, MI 48221

The chemistry of water-soluble organometallic compounds has received increasing interest of late because of their potential use in environmentally-benign biphasic catalysis. Previous work has focused primarily on organometallic compounds that contain hydrophilic phosphine ligands. We have developed a novel class of water-soluble nitrogen-based ligands that contain ionic sulfonate or phosphonate groups with properties that differ from their phosphorous counterparts. The synthesis, characterization, and reaction chemistry with W(0) carbonyl complexes will be presented.

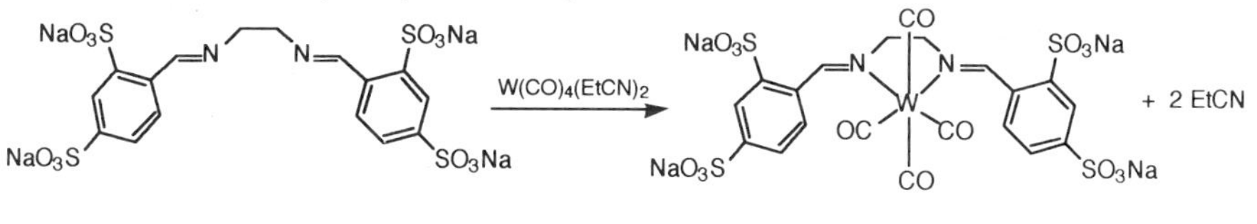

239. SCHIFF-BASE EXPANDED PORPHYRINS.
Sylvie Meyer, Jonathan L. Sessler, Steven J. Weghorn, Tarak D. Mody. Department of Chemistry and Biochemistry, University of Texas at Austin, Austin, TX 78712.

The synthesis of the diformylbipyrrole-derived macrocycles 1 (a-d) has been reported previously and was suggested to be mediated *via* an anion template effect. We have synthesized another type of Schiff-base expanded porphyrin, the diformylterpyrrole-derived macrocycle 2. Compound 2 is large enough that one could contemplate using it as ditopic receptor and/or as a receptor for large neutral molecules. Compound 3 has also been synthesized and may show interesting ditopic coordination properties. The properties of these systems are currently under investigation and the results of these studies will be presented.

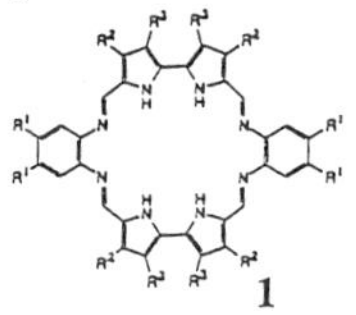

a. R^1 = H, R^2 = Et, R^3 = Me
b. R^1 = Me, R^2 = Et, R^3 = Me
c. R^1 = OMe, R^2 = Et, R^3 = Me
d. R^1 = OMe, R^2 = n-Pr, R^3 = H

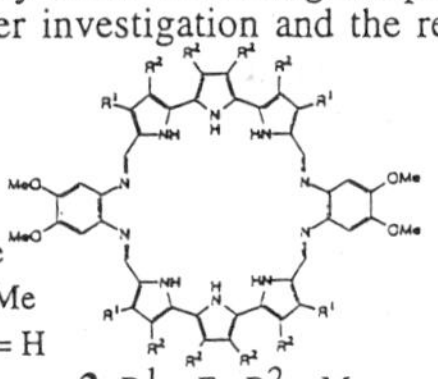

1

2 R^1 = Et, R^2 = Me

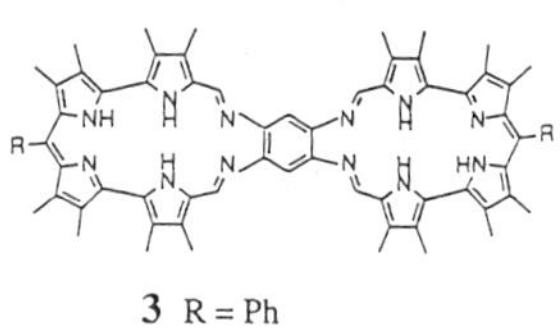

3 R = Ph

240. **SELF-ASSOCIATION OF SYNTHETIC DINUCLEOTIDE GC AND GG**
Jonathan L. Sessler and Ruizheng Wang
Department of Chemistry and Biochemistry,
University of Texas at Austin, Austin, TX 78712

It is well known that single-stranded DNA self-associates to make complementary double-stranded helical structures through Watson-Crick base pairing. It also has been found that single-stranded DNA containing short guanine-rich motifs will self-associate to form guanine quadruplexes by Hoogsteen pairing. In order to mimic some features of these two processes, two systems containing GC and GG have been synthesized. The synthesis and studies of self-association behaviors of these systems will be presented.

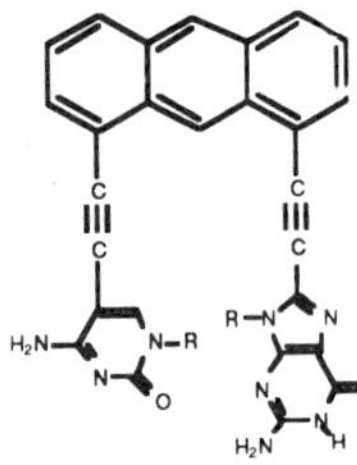

241. LANTHANIDE(III)-TEXAPHYRIN HYDROLYSIS OF RNA. **C.B. Black**, Darren Magda and Jonathan L. Sessler Department of Chemistry and Biochemistry, The University of Texas at Austin, Austin, Texas 78712; Pharmacyclics Inc., 995 East Arques, Sunnyvale, California 94086.

The hydrolysis of RNA under physiological conditions has long been a desired goal in the treatment of genetically based diseases. It has recently been reported that a series of lanthanide(III)-substituted texaphyrins (1) (Ln-Tx) are kinetically stable and, when modified with oligodeoxynucleotide, can act hydrolytically to site-specifically cleave ssRNA (Magda, et al., *J. Am. Chem. Soc.*, **1994**, *116*, 7439). In an effort to further improve the efficacy of this system, kinetic and thermodynamic measurements were carried out using the Gd(III), Eu(III), Dy(III) and Lu(III) texaphyrin derivatives with ssRNA. Through titration and kinetic experiments, evidence for Ln-Tx binding through a templated-stacking mode as well as a binding affinity trend (Dy>Gd>Eu>Lu) was observed. Kinetic experiments using ssRNA show a similar trend. A reactivity model is proposed where the dominant factor in reactivity is the coordination capability of the central lanthanide cation.

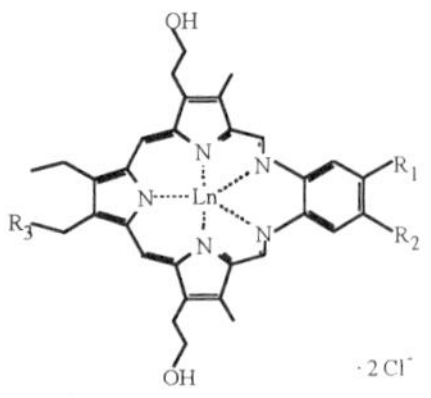

242. STEREOSPECIFICITY AND SELF-RECOGNITION IN THE ASSEMBLY OF BINUCLEAR COPPER COMPLEXES OF SEMI-RIGID BIS-BIDENDATE LIGANDS.
Eric Enemark, Md. Athar Masood, and T. Daniel P. Stack, Department of Chemistry, Stanford University, Stanford, CA 94305-5080.

The conformationally restricted binucleating N_4 ligand, *trans* 1,2-diimino-bis(6-bromo-2-pyridylmethyl)cyclohexane self-assembles with Cu(I) to form a 2:2 distorted helical complex in a system that displays both stereoselectivity and self-recognition based on chirality. A racemic mixture of ligands self-selectively sorts to form a pair of homochiral, stereospecific species in solution and the solid state. The design features responsible for such assembling behavior will be discussed.

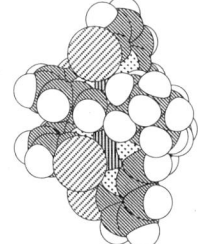

243. METAL AND NONMETAL COMPLEXES OF CALIX[5]ARENES, J. Li, M. Fan, H. Zhang, and M. Lattman, Department of Chemistry, Southern Methodist University, Dallas, TX 75275-0314.

Complexes of calix[4]arenes are well-known due to the ready availability of the starting material. Much less is known about the complexing ability of calix[5]arenes. The larger cavity of the latter should lead to a greater variety of complex types. We herein report the reactions of para-*tert*-butylcalix[5]arene (R = *tert*-butyl) with halides of arsenic, antimony, tin, niobium, tantalum. Complexes are formed in which three and five of the oxygens of the calixarene are bound to the central atom. The syntheses, spectral properties, and structures of these derivatives will be discussed.

244. SILICON-CONTAINING CALIX[4]ARENES: BARRIERS TO INTERCONVERSION OF "CONE" AND "PARTIAL CONE" ISOMERS, M. Fan, G. Ford, H. Zhang, and M. Lattman, Department of Chemistry, Southern Methodist University, Dallas, TX 75275-0314.

We have recently obtained evidence that two isomers of the silyl-calix[4]arene compound illustrated (R = *tert*-butyl and H) exist: the cone, with the hydroxyl group pointing in the same direction as the three oxygens bound to silicon, and the partial cone, in which the hydroxyl is flipped in the opposite direction. We have studied the interconversion of these isomers in both solution and the solid-state. For the following solution equilibrium (at 150 °C in 1,2-dichlorobenzene),

$$\text{partial cone} \underset{k_{-1}}{\overset{k_1}{\rightleftharpoons}} \text{cone}$$

$K = 0.59$ with approximate first-order rate constants $k_1 = 0.025$ m^{-1} and $k_{-1} = 0.042$ m^{-1} (R = *tert*-butyl). Conversion of the partial cone to the cone (R = H) is complete in the solid-state at 230 °C in 3 h. No solid-state transformation is observed for R = *tert*-butyl. Energy barriers to this equilibrium and mechanisms will be discussed.

245.
ZIRCONIUM COMPLEXES OF CHELATING ALKOXIDES. **P. Shao,** D.J. Berg. Department of Chemistry, University of Victoria, P.O.Box 3065, Victoria, B.C., V8W 3V6.

In this contribution we report synthesis of a series of chelating amino-dialkoxides (I and II) and cyclopentadienyl alkoxides (III). Zirconium cationic alkyls with ligand I show some interesting reaction chemistry including activity as 1-hexene polymerization catalysts. Fluorinated groups (2,4-bis(trifluoromethyl)phenyl) greatly reduce the alkoxide's electron donating ability. Thus, zirconium complexes with ligands II and III show remarkable Lewis acidity and function as high activity catalysts for vinyl ether polymerizations.

246.
TEACHING AN OLD LIGAND NEW TRICKS: SUBSTITUTION AT A METHYL CARBON SITE OF ACETYL-ACETONATE Deirdre I. Arnold, F. Albert Cotton, John H. Matonic, Carlos A. Murillo Laboratory for Molecular Structure and Bonding and Department of Chemistry, Texas A&M University College Station, TX 77843 Mail Stop 3255

Metal acetylacetonates have been studied extensively for over a century. Numerous electrophilic substitution reactions involving oxygen-bound metal acetylacetonates have been reported in which the substitution occurs at the methine carbon site of the chelating diketonate. In our ongoing work with transition metal formamidine complexes we have found that a melt reaction between vanadyl acetylacetonate and diphenylformamidine produces oxobis[(6-phenylamino)hexane-2,4-dionato-$\underline{O},\underline{O}$']vanadium(IV), VO(Phad)$_2$, which contains a new diketonate ligand. This ligand arises from the substitution of two methyl hydrogen atoms from a methyl group of a chelating acetylacetonate ligand by a formimidoyl fragment cleaved from a formamidine. We believe that the formamidine cleavage takes place after initial binding of the ligand at the sixth coordination site of the vanadium atom. Our ongoing investigations into the mechanism of this highly unusual substitution reaction are presented.

247. CHIRAL CIS α QUADRIDENTATE COMPLEXES FOR LEWIS ACID CATALYSIS.
Scott A. Savage, Daniel R. Derringer, Christina Ng, Jaydeep J. S. Lamba, Laura Johnston, Cassandra L. Fraser* Department of Chemistry, University of Virginia, Charlottesville, VA 22901

With the appropriate choice of chelate ring sizes and substituents, it is possible to control both the topology and absolute stereochemistry of metal complexes of quadridentate ligands. Despite the fact that these principles of inorganic stereochemistry have been known for some time, octahedral metal complexes with C_2 symmetric cis α topologies have been little exploited in asymmetric catalysis. A series of chiral quadridentate ligands have been prepared by sequential alkylation of chiral diamines first with pyridyl or quinoline-based reagents, followed by non-chelating groups, alkyl and aryl, possessing a range of steric and electronic properties. These ligands were reacted with metal chloride salts (M(II) = Mn, Co, Ni, Cu, Zn; M(III) = Fe) to generate a series of complexes. Syntheses as well as structural and physical characterization of this series of compounds will be described. Preliminary investigations of their ability to serve as Lewis acid catalysts will also be presented.

248. INFLUENCE OF SULFUR CONTAINING DONOR GROUPS ON THE COORDINATION GEOMETRY OF PHOSPHORUS AND SILICON COMPOUNDS. Robert R. Holmes, A. Chandrasekaran, David J. Sherlock, Roberta O. Day, Joan A. Deiters, and Natalya V. Timosheva. Department of Chemistry, Box 34510, University of Massachusetts, Amherst, MA 01003-4510.

Previous work on a series of cyclic oxyphosphoranes has shown that oxygen atoms of sulfonyl groups and sulfur atoms are capable of acting as donor ligands with the result that an increase in the coordination geometry with displacements from trigonal bipyramidal to octahedral take place. These trajectories provide a useful model for the course of nucleophilic substitution reactions. In current work, we have obtained the first series of cyclic tri- and tetracoordinated phosphorus compounds that experience an increase in coordination geometry to pseudo trigonal bipyramids and trigonal bipyramids, respectively, as a result of sulfur donor action. These are compared with related cyclic oxysilane compounds that form five-coordinated structures as a result of similar interactions. Also a comparison is made with tri- and tetracoordinated phosphorus compounds that undergo an increase in coordination via P–N interactions. With the knowledge that phosphorus in coordination states from three to five is capable of a change in geometry in the presence of donor groups containing S, O, and N (previously known) suggests that this occurrence may be important in designing mechanistic criteria for nucleophilic displacement reactions. In particular, nearby residues at active sites of phosphoryl tranfer enzymes having these donor atoms may participate in altering activated state geometries as well as that of substrates.

249. POTENTIOMETRIC AND NMR STUDIES OF THE COMPLEXATION OF AZACROWNETHER-PHOSPHONATES

Martina Peters and Thomas A. Kaden, *Department of Chemistry, University of Basel, Spitalstrasse 51, CH-4056 Basel (Switzerland)*

The pK_H-values and complexation constants with M = Pb(II), Cd(II) and Zn(II) of **1-3** were determined by potentiometric titrations. The values indicate that besides mononuclear species MLH_x with different protonation degree, also binuclear complexes M_2LH_x are formed. Furthermore to study their structures in solution ^1H-, ^{13}C- and ^{31}P-NMR measurements were run as a function of pH.

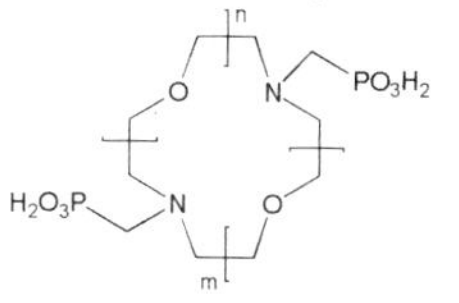

1 m = n = 1
2 m = 1, n = 2
3 m = n = 2

250. SYNTHESIS OF TRISUBSTITUTED 1,4,7,10,13-PENTAAZACYCLOPENTADECANE DERIVATIVES USING EASILY REMOVABLE CARBAMATE PROTECTIVE GROUPS

M. Russell and A. D. Sherry, Department of Chemistry, University of Texas at Dallas, Richardson, Texas 75083-0688.

1,4,7,10,13-Pentaazacyclopentadecane **1** was reacted with benzylchloroformate (BZOC-Cl) to give the bisprotected derivative **2** that proved to be useful in the synthesis of various trisubstituted "pentaaza" derivatives. Protonation and metal-ligand stability constants were determined.

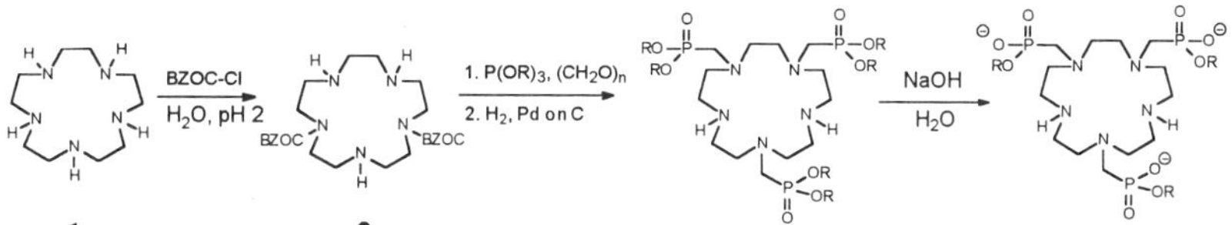

251. SYNTHESIS OF MONO- AND TRISUBSTITUTED CYCLEN DERIVATIVES USING EASILY REMOVABLE CARBAMATE PROTECTIVE GROUPS

Z. Kovacs and A. D. Sherry, Department of Chemistry, University of Texas at Dallas, Richardson, Texas 75083-0688.

1,4,7,10-Tetraazacyclododecane **1** reacted with 2-(*tert*-butoxycarbonyloxyimino)-2-phenylacetonitrile (BOC-ON) to give selectively the trisprotected derivative **2** that proved to be very useful in the synthesis of various mono- and trisubstituted cyclen derivatives.

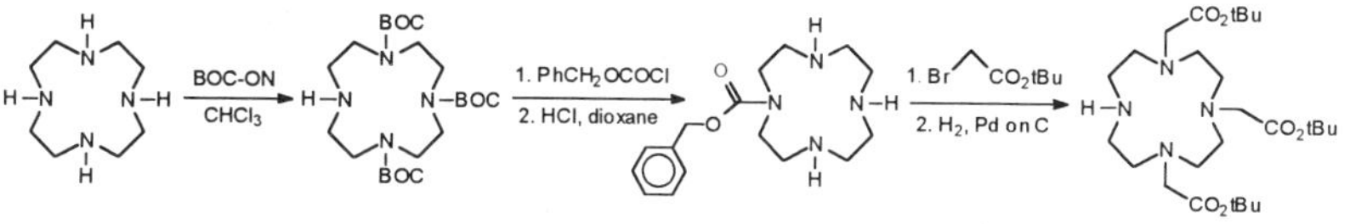

252. **SYNTHESIS AND COORDINATION CHEMISTRY OF NEW CHELATING PHOSPHINO-KETONE and PHOSPHINO-IMINE LIGANDS.** Brian T. Rasley, Robert E. Bachman* and Robert J. Kulawiec*, Department of Chemistry, Georgetown University, Washington DC 20057-1227.

A number of novel bidentate keto-phosphine and imino-phosphine ligands have been synthesized from a common phosphino-nitrile precursor. Synthetic strategies toward these ligands and their coordination chemistry with Ru(II), Pd(II), and Au(I) will be discussed, along with structural and spectroscopic trends. Representative X-ray crystal structures of new ligands and complexes will be presented.

X = O, NR'

253. THE SYNTHESIS AND CHARACTERIZATION OF $W_2(\mu\text{-}H)_2Cl_4(\mu\text{-}dppm)_2$. Thomas. E. Concolino and Judith L. Eglin, Department of Chemistry, Mississippi State University, Mississippi State, MS 39762. Edward J. Valente, Department of Chemistry, Mississippi College, Clinton, MS 39058. Jeffrey D. Zubkowski, Department of Chemistry, Jackson State University, Jackson, MS 39217.

The synthesis of $W_2(\mu\text{-}H)_2Cl_4(\mu\text{-}dppm)_2$, marks the first time that a bulk scale preparation of a bridging dihydride W(III-III) species has been reported. The compound retains the characteristic edge-sharing bioctahedral (ESBO) geometry with a W-W bond distance of 2.3918(7) Å. Several ESBO complexes have been prepared but only two with dihydride bridges. The complex has been characterized with $^{31}P\{^1H\}$, ^1H NMR and UV-vis spectroscopic techniques. X-ray crystallographic studies yield structural information for the complex which crystallizes in the monoclinic space group C2/c with a = 17.493(5) Å, b = 17.052(4) Å, c = 16.223(4) Å, β = 93.86(2)°, V = 4828(2) Å^3 and Z = 4 with R1 = 0.0448 and wR2 = 0.0823.

254. **NEW MONO- AND BI-NUCLEAR TITANIUM IMIDO COMPLEXES SUPPORTED BY ARYLOXIDE LIGANDS: FINE CONTROL BY ORTHO-SUBSTITUENTS**

Philip E Collier and Philip Mountford*
Department of Chemistry, University of Nottingham, Nottingham, U.K. NG7 2RD.
Email: Philip.Mountford@Nottingham.ac.uk

Reaction of the titanium imido complexes [Ti(NR)Cl$_2$(py)$_3$] (R = Bu^t, 2,6-C$_6$H$_3$Me$_2$ or 2,6-C$_6$H$_3$Pr$^i_2$) with two equivalents of lithium aryloxide Li[O-2,6-C$_6$H$_3$R'$_2$] (R' = Me, Pr^i or Bu^t) affords the monomeric four- or five-coordinate, or dimeric four-coordinate complexes [Ti(NR)(O-2,6-C$_6$H$_3$R'$_2$)$_2$(py)$_n$]$_m$ (m = 1, n = 1 or 2; m = 2, n = 0) depending on the identity of R and R'. The X-ray structures of [Ti$_2$(μ-NBu^t)$_2$(O-2,6-C$_6$H$_3$Me$_2$)$_4$] and [Ti(NBu^t)(O-2,6-C$_6$H$_3$Pr$^i_2$)$_2$(py)$_2$] are discussed. Extended Hückel molecular orbital calculations for models of [Ti$_2$(μ-NBu^t)$_2$(O-2,6-C$_6$H$_3$Me$_2$)$_4$] show that the distortion of the μ-imido Bu^t substituents from co-planarity with the Ti$_2$N$_2$ core in this and related species may be attributed to a second-order Jahn-Teller distortion.

255. EARLY TRANSITION METAL IMIDO AND OXO COMPLEXES SUPPORTED BY N, N'-BIS(TRIMETHYLSILYL)BENZAMIDINATE LIGANDS

Peter J Stewart, Alexander J Blake and Philip Mountford*
Department of Chemistry, University of Nottingham, Nottingham, U.K. NG7 2RD.
Email: Philip.Mountford@Nottingham.ac.uk

The mononuclear N, N'-bis(trimethylsilyl)benzamidinate-supported titanium imido complexes [Ti(NR){PhC(NSiMe$_3$)$_2$}Cl(py)$_2$] [R = Bu^t, 2,6-C$_6$H$_3$Me$_2$ (X-ray structure) or 2,6-C$_6$H$_3$Pr^i$_2$] exist in temperature-dependent equilibrium with their mono(pyridine) homologues. The relative contributions of enthalpy and entropy effects to the position of these dissociative equilibria have been determined and a number of organometallic and inorganic derivatives of these complexes are reported. The reaction of two equivalents of Li[PhC(NSiMe$_3$)$_2$] with [Nb(E)Cl$_3$(L)$_2$] (E = NBu^t, L = py; E = O, L = THF) to give the fluxional derivatives [Nb(E){PhC(NSiMe$_3$)$_2$}$_2$Cl] (E = NBu^t or O, X-ray structure) is also discussed. Addition of one equivalent of Li[PhC(NSiMe$_3$)$_2$] to [Nb(NBu^t)Cl$_3$(py)$_2$] gives a mixture of [Nb(NBu^t){PhC(NSiMe$_3$)$_2$}$_2$Cl] and [Nb(NBu^t){PhC(NSiMe$_3$)$_2$}Cl$_2$(py)] (X-ray structure).

256. TRANSITION METAL COMPLEXES OF DIAMINE MESOCYCLES. Thomas N. Sorrell, Hongping Yuan, and Oliver Fazio, Department of Chemistry, The University of North Carolina, Chapel Hill, NC 27599-3290 USA

Di- and trivalent manganese, iron, and cobalt complexes have been prepared with pentadentate ligands having a framework based on the 6-hydroxy-1,4-diazacycloheptane or 7-hydroxy-1,5-diazacyclooctane ring systems. Presented are the structures of these complexes as well as their reactions with other ligands.

257. The Structure of Na[Fe(H$_2$O)(EDTA)]·2H$_2$O Jeffrey D. Zubkowski[1], Edward J. Valente[2], Crystal Lewis[3] and Dale L. Perry[3] [1]Department of Chemistry, Jackson State University Jackson, MS 39217, [2]Department of Chemistry, Mississippi College, Clinton, MS 39058, [3]Lawrence Berkeley National Laboratory,University of California, Berkeley, CA 94720

Details are reported for the synthesis and x-ray single crystal structure of Na[Fe(H$_2$O)(EDTA)]· 2 H$_2$O. The complex compound consists of yellow, cube-like monoclinic crystals (in space group Cc) with cell constants of a = 8.898(2), b = 11.928(4), and c = 15.045(4). The D$_{calc}$ is 1.779 mg/m^3, Z = 4, R$_1$= 0.0373. The structure is compared to others containing iron-oxygen and iron-nitrogen linkages, and the hydrogen bonding for the complex is discussed.

258. Synthesis and Characterization of Cu$_2$(O$_2$CCH$_2$CH$_2$COCH$_3$)$_4$(OC$_5$H$_4$NH)$_2$ and Cu$_2$(O$_2$CCH$_2$CH$_2$COCH$_3$)$_4$(NC$_5$H$_4$CN)$_2$. D'Angelo Washington, and Jeffrey D. Zubkowski, Department of Chemistry, Jackson State University, Jackson, MS 39217-0510 and Edward Valente The Department of Chemistry, Mississippi College, Clinton, MS 39058.

We have become interested in γ- and δ-oxocarboxylic acids as ligands in the copper systems. The parent oxoacids may exist in solution as an equilibrium between open and cyclic lactol (pseudoacid) forms depending on their structures. The reactions of copper levulinate and two para substituted pyridines have been performed. The products of the reaction of Cu$_2$(LEV)$_4$, {LEV=O$_2$CCH$_2$CH$_2$COCH$_3$}, with 4-hydroxypyridine and 4-cyanopyrdine produced Cu$_2$(LEV)$_4$(OC$_5$H$_4$NH)$_2$ and Cu$_2$(LEV)$_4$(NC$_5$H$_4$CN)$_2$ respectively. This paper will present the details of the synthesis and characterization of two compounds. Both compounds have been studied using IR, magnetic sucseptibility and single crystal x-ray diffraction.

259. **COMPLEXATION BETWEEN Ni(II) AND THIOCYANATE IN WATER AND AQUEOUS METHANOL.** N. Aldahhan, C.M. Richter, C. Walkup and H. B. Silber, Chemistry Department, San Jose State University, San Jose CA 95192-0101.

The addition of thiocyanate to Ni(II) solutions causes a color change from green to blue-green, attributed to the formation of three complexes, $NiSCN^+$, $Ni(SCN)_2$ and $Ni(SCN)_3^-$. The Beer-Lambert relation at three wavelengths (396, 659 and 721nm) was used to measure the extinction coefficients of free Ni(II) and free SCN^- as a function of temperature. Absorbance data in the presence of thiocyanate results in deviations from Beer's Law and these absorbance data as a function of solvent and temperature were converted to complexation constants using the Gaizer computer program. The variation of the complexation constants as a function of temperature provides enthalpy and entropy data. These results are compared to those in the analogous Co(II) system, where a geometry change accompanies the addition of the fourth thiocyanate ligand. This research is supported by ACS PRF Type B, NIH Minority Access to Research Careers (MARC), and NIH-Minority Biomedical Research Support (MBRS) grants.

260. **EU(III) - ALANINE COMPLEXATION IN AQUEOUS METHANOL MIXTURES.** Yen Nguyen and Herbert B. Silber, Chemistry Department, San Jose State University, San Jose CA 95192-0101.

The interactions between Eu(III) and alanine in methanol:water mixtures are being investigated as a function of solvent and temperature. Equilibrium constants at each temperature in 20% methanol-80% water mixtures are higher than the equilibrium constants in 100% water. Only one to one complexation has been observed. The equilibrium constants are being obtained at higher methanol fractions. The thermodynamic data for the 20% methanol aqueous system are similar to those of the aqueous system. These results will be compared to those found for Eu(III) with glycine under similar conditions. This research is supported by ACS PRF Type B and NIH-Minority Biomedical Research Support (MBRS) grants.

261. **PREPARATION, PURIFICATION AND CHARACTERIZATION OF DINUCLEAR RUTHENIUM(II) COMPLEXES SEPARATED BY SPACERS OF VARIABLE DISTANCE.** Youxiang Wang, Greg Y. Zheng, Willie J. Perez, Connie L. Huber and D. Paul Rillema, Department of Chemistry, Wichita State University, Wichita, KS 67260.

Four polypyridyl ligands were synthesized based on the reaction between 4,5-diazafluorenone and diamino organic compounds with glacial acetic acid as the catalyst. Dimeric "$Ru(bpy)_2^{2+}$" complexes containing the four ligands were then synthesized and purified by a special technique to remove luminescing impurities. The resulting complexes displayed emission in a glass at 77K, but none at room temperature in solution. In all cases, ligand based reductions were found at potentials < 1V vs. SSCE in acetonitrile. The $Ru^{3+/2+}$ couples fell in the range of 1.32 to 1.41 V vs. SSCE and were two one-electron processes separated by 60 mV, or less. The low-lying optical transition associated with metal-to-ligand charge transfer occurred near 450nm, or at slightly higher energy, in contrast to "$Ru(bpy)_2^{2+}$" complexes containing other heterocyclic ligands where the optical transition shifted to lower energy.

262. **INVESTIGATION OF THE SPECTROSCOPIC AND ELECTROCHEMICAL PROPERTIES OF OSMIUM (II) COMPLEXES WITH RIGID C_n-BRIDGED POLYPHOSPHINES** Jeffrey V. Ortega, Dengfeng Xu, Steven R. Woodcock, Bo Hong* Department of Chemistry, University of California, Irvine, CA 92697

The construction of artificial supramolecular organizations from a discrete number of molecular components has been an area of ongoing research interest. Through the linkage of various prefabricated molecular components with light-related properties, structurally organized and functionally integrated photochemical molecular systems can be obtained. In this study, polyphosphines with rigid C_n-bridges (either cumulenic or acetylenic, Figure 1) will be incorporated in the construction of Os-based supramolecular systems for the following purposes: (a) to obtain long-lived ^{3}MLCT excited states; (b) to study the photoinduced electron/energy transfer across rigid C_n-bridges with various chain lengths; (c) to investigate the redox chemistry and electronic coupling across the rigid C_n-bridged polyphosphine spacers.

Figure 1

263. SYNTHESIS AND CHARACTERIZATION OF{[(bpy)$_2$Ru(dpp)]$_2$RhCl$_2$}(PF$_6$)$_5$: A DEVICE FOR PHOTOINITIATED ELECTRON COLLECTION
Elizabeth R. Bullock, Sharon M. Molnar, and Karen J. Brewer. Department of Chemistry. Virginia Polytechnic Institute and State University, Blacksburg VA, 24061-0212.

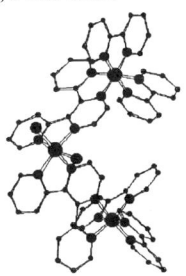

A series of trimetallics has been devised to function as devices for photoinitiated electron collection. These incorporate light absorbing chromophores (LA) as the end units on the molecule, with an electron collecting metal center (EC), LA-EC-LA. In this study, the functioning of the system {[(bpy)$_2$Ru(dpp)]$_2$RhCl$_2$}(PF$_6$)$_5$ will be investigated (dpp= 2,3 bis(2-pyridyl)pyrazine). The characterization of this compound by bulk electrolysis and spectroelectrochemistry will be discussed.
This work is supported by NSF CHE-9313642 and CHE-9632713.

264. SELF-ASSEMBLY OF TETRAMERIC PORPHYRIN AGGREGATES VIA COORDINATED TRANSITION METAL BISPHOSPHINE AUXILIARY. Jun Fan, Bogdan Olenyuk and Peter J. Stang*, Chemistry Departmnet, University of Utah, Salt Lake City, UT 84112
Preparation and characterization of multicomponent porphyrin aggregates via self-assembly with Pd (II) and Pt (II) bisphosphine auxiliaries will be described. Use of zinc complex of the porphyrin under similiar conditions results in the formation of heterobimetallic Zn(II)-Pd(II) or Zn(II)-Pt(II) porphyrin tetramers. Possible use of these assemblies as new models of energy transfer will be discussed.

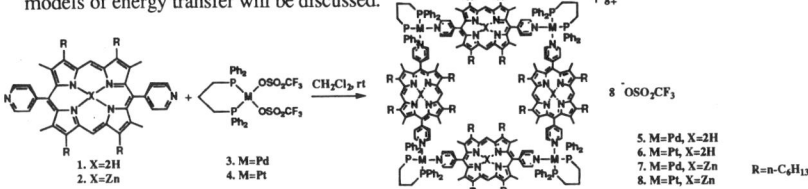

1. X=2H
2. X=Zn
3. M=Pd
4. M=Pt
5. M=Pd, X=2H
6. M=Pt, X=2H
7. M=Pd, X=Zn
8. M=Pt, X=Zn
R=n-C$_6$H$_{13}$

265. BIMETALLIC PLATINUM AND PALLADIUM COMPLEXES: BUILDING BLOCKS FOR SUPRAMOLECULAR MACROCYCLES. Christopher J. Kuehl, Joseph Manna, and Peter J. Stang, Department of Chemistry, University of Utah, Salt Lake City, Utah 84112.

We are currently interested in the preparation of *trans*-bimetallic linkages with variable geometric, steric and solublity properties for the design of potentially useful supramolecular structures. Such linkages were synthesized by double oxidative-addition of Pt(PEt$_3$)$_4$ into dihaloaryl organic compounds posessing a specific geometry amenable to the desired shape. This approach is utilized in the preparation of more hindered, self-assembling structures such as triangles, squares, and rhomboids.

266. METALMEDIATED CYCLOTRIMERIZATION OF PHOSPHAALKYNES. S. Leininger, Department of Chemistry, Univeristy of Utah, Salt Lake City, UT 84112; M. Regitz, Department of Chemistry, University of Kaiserslautern, 67663 Kaiseslautern, Germany

Since the pioneering work of Reppe in the late 40's the cyclo-oligomerization of acetylenes to furnish benzene derivatives is well known. In contrast no cyclotrimers of the isoelectronic phosphaalkyne have been known. Using cyclooctatetraene metal complexes of group IV we are able to synthesize and characterize the first triphospha benzene and triphospha Dewar benzene derivatives. Depending on the reaction conditions we are also able to isolate tetraphosphabarellen complexes as well as phosphaalkyne trimer complexes and the first known triphosphacyclobutadiene complex. In all cases reaction of the metal complexes with mild halogenating reagents furnishes the free triphospha benzenes and tetraphosphabarellenes as the only products. Due to the polarized P/C-double bonds in this systems the triphospha benzene derivatives are highly reactive and undergo a variety of cycloaddition reactions.

267. MODULAR SELF-ASSEMBLY, CHARACTERIZATION, AND HOST-GUEST CHEMISTRY OF NANOSCALE ORGANOMETALLIC ARCHITECTURES. Joseph Manna, Christopher J. Kuehl, and Peter J. Stang, Department of Chemistry, University of Utah, Salt Lake City, Utah 84112 and David C. Muddiman and Richard D. Smith, Environmental Molecular Sciences Laboratory, Pacific Northwest National Laboratory, Richland, Washington 99352.

The supramolecular synthesis and chemistry of organic macrocycles has been the focus of considerable study for over thirty years. In contrast, the chemistry of analogous inorganic and organometallic macrocycles is in it infancy; little is know about the stability, spectroscopic and physical properties, and chemistry of these species. We will report on the design of several unique supramolecular macrocycles and the characterization of these species by a range of spectroscopic techniques, including electrospray-ionization Fourier transform ion cyclotron resonance spectrometry. Preliminary data concerning the host-guest chemistry of these macrocycles will also be presented.

268. PHOSPHA-FERROCENES AS LIGANDS: SYNTHESES, STRUCTURES, COORDINATION CHEMISTRY, AND ELECTROCHEMISTRY. C. Müller,[a] R. Bartsch,[b] A. Fischer,[b] and P.G. Jones[b]
[a] Department of Chemistry, The University of Utah, Salt Lake City, Utah 84112
[b] Institut für Anorg. und Analyt. Chemie der Technischen Universität, Braunschweig, Germany

Syntheses and structures of penta- and hexaphosphorus analogues of ferrocene have been described recently. Unlike their simple ferrocene analogues, these complexes have further ligating potential towards other transition metal centers by virtue of the availability of the ring phosphorus lone-pair electrons that are not involved in the η^5-coordination. We now describe syntheses, structures and the electrochemistry of coordination compounds of the triphospha-ferrocene [Fe(η^5-C$_5$Me$_5$)(η^5-C$_2{}^t$Bu$_2$P$_3$)]. In the ruthenium complex [Fe(η^5-C$_5$Me$_5$)(η^5-C$_2{}^t$Bu$_2$P$_3$)Ru$_3$(CO)$_9$] two adjacent phosphorus atoms of the η^5-C$_2{}^t$Bu$_2$P$_3$ ring are interlinked by a ruthenium carbonyl cluster in which all three ruthenium atoms interact with the phosphorus atoms. The tetrametallic nickel complex [Fe(η^5-C$_5$Me$_5$)(η^5-C$_2{}^t$Bu$_2$P$_3$)Ni(CO)$_2$]$_2$ represents the first example of intermolecular interlinkage of two phospha-ferrocene systems by two metal centers.

269. CHIRAL NANOSCOPIC FRAMEWORKS VIA SELF-ASSEMBLY.
Bogdan Olenyuk and Peter J. Stang.* Department of Chemistry, University of Utah, Salt Lake City, UT 84112.

A new family of chiral conformationally rigid nanoscale-sized assemblies with nearly 90° turns is prepared by utilizing coordination of a variety of polydentate heterocyclic ligands to optically active chelated Pt(II) and Pd(II) bisphosphanes as well as to chiral bimetallic organoplatinum complexes. The resulting species represent structurally well-defined aggregates of a unique shape, containing large, molecule-sized cavities. The specific design outlines, preparation, characterization as well as potential applications of these highly symmetrical yet chiral macromolecules will be addressed.

270. DESIGN AND SYNTHESIS OF SELF-ASSEMBLING CATIONIC LATE TRANSITION METAL MACROCYCLIC HEXAMERS. Neal E. Persky and Peter J. Stang*, Department of Chemistry, University of Utah, Salt Lake City, Utah 84112.

Several years ago we reported the self-assembly of cationic tetranuclear Pt(II) and Pd(II) macrocycles. Recently we reported the synthesis of two novel organometallic nanodimensional macrocycles using *trans*-platinum organometallic linear linkages. By making use of these new *trans* linkers, organic based pyridine corners with the appropriate angle, and our assembly methodology we have extended our research to include the self-assembly of hexameric platinum macrocycles. Several examples of these macrocycles using different organic corners to connect the organometallic linkers will be presented along with other examples of hexameric macrocycles with metals at the corners in conjunction with organic linkers as well as hexamers with alternating organometallic and organic corners.

271. TRANSPORT PROPERTIES AND FAB MASS SPECTROMETRY OF TRANSITION METAL HOST-GUEST MOLECULAR SQUARE COMPLEXES. Jeffery A. Whiteford, Kuanchiang Chen, Elliot M. Rachlin, and Peter J. Stang*, Department of Chemistry, University of Utah, Salt Lake City, Utah 84112

Several types of molecular squares have been synthesized to date: 1) homonuclear macrocycles comprised of transition metals at the corners; 2) heteronuclear mixed-neutral charged transition metal species; 3) hybrid iodonium-transition metal squares; 4) silicon macrocycles; and 5) nanoscale organometallic macrocycles. Presumably, cationic molecular squares should be useful for complexation with electron-rich guests. We will discuss the utility of cationic molecular squares as liquid phase transfer vehicles and their guest binding characteristics. Since these molecular squares are in the molecular weight range of of ~3000-9000 amu, and possess considerable void space in the solid state, they are often difficult to characterize. The development of a FAB mass spectrometry method for the observation of these host-guest complexes will be presented.

272. PALLADIUM(II) AND PLATINUM(II) COMPLEXES OF A PENDANT-ARMED MACROCYCLE.
Savitri Chandrasekhar[*,a] and David Miller[b]
[a]Chemistry, SWGC, Memorial University of Newfoundland, Corner Brook, NF Canada A2H 6P9.
[b]Department of Chemistry, Memorial University of Newfoundland, St. John's NF Canada A1C 5S7.

Pd(II) and Pt(II) complexes of the pendant-armed macrocyclic ligand 1,4,8,11-tetrakis(2-pyridylmethyl)-1,4,8,11-tetraazacyclotetradecane have been synthesized and characterized by elemental analysis, electrospray mass spectroscopy, ^{1}H and ^{13}C NMR spectra. In all previously reported cases, either binuclear complexes are obtained with the metal ions outside the macrocyclic ring or only three of the nitrogen atoms on the cyclam ring are coordinated. In the present case the Pd(II) and Pt(II) ions form mononuclear complexes and are coordinated to the four nitrogen atoms of the cyclam ring forming a square planar structure and the nitrogen atoms from the pyridyl moiety remain uncoordinated. Chemical shift values have been assigned based on the analysis of the ^{1}H-^{13}C HETCORR spectra. Cyclic voltammetry and EPR spectroscopy have been used to study the redox reactivity of the complexes wherein the flexible pendant arms may provide axial coordination.

273. ELECTRON TRANSFER BEHAVIOR AND SOLID STATE STRUCTURES OF THE HELICAL COBALT COMPLEXES OF THE OPEN-CHAIN TETRAPYRROLE LIGAND, OCTAETHYLBILINDIONE. Saeed Attar, Alan L. Balch, Andrew Ozarowski, Pamela Van Calcar, and Krzysztof Winkler, Department of Chemistry, University of California, Davis, California 95616

While octaethylbilindione (H$_3$OEB), a biliverdin analog, undergoes an irreversible two-electron oxidation, its cobalt complex undergoes a series of reversible, one-electron transfer processes with [Co(OEB)]^n (n = +1, 0, -1, -2) observed electrochemically. The most highly oxidized member of this series has been isolated as its triiodide salt, [Co^{II}(OEBOx)]I$_3$ • 0.5CH$_2$Cl$_2$, and characterized by X-ray diffraction. Furthermore, addition of pyridine to this complex, which shows a typical axial EPR spectrum with well-resolved cobalt hyperfine splitting, converts the complex into one with a Co^{III}/ligand radical electronic structure as shown by the EPR spectrum with g = 1.98, 2.02.

274. OLD AND NEW BUILDING BLOCKS FOR MOLECULAR MAGNETISM RESEARCH
Roger D. Sommer, Brenda J. Conklin and <u>Gordon T. Yee</u>
Department of Chemistry and Biochemistry, University of Colorado, Boulder CO 80309

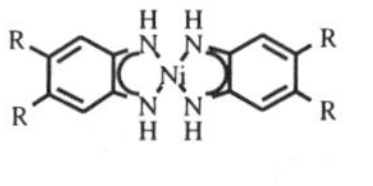

We have found that diiminenickel (II) complexes of the type shown to the right are viable one-electron donors for charge-transfer ferromagnetic salts. In particular, when R = CH_3 and the acceptor is TCNE, the resulting product possesses two unpaired spins per formula unit and orders ferromagnetically below 10 K. Previously, these complexes were investigated as donors for electrically conductive materials with limited success. Varying the identity of the peripheral R group determines the donor strength. Whereas for conductivity, partial charge transfer is desirable, for magnetism, integral charge transfer is required. Acceptors we have examined include TCNE, TCNQ and DDQ. Our progress in synthesizing analogs of TCNE of the types shown below will also be discussed.

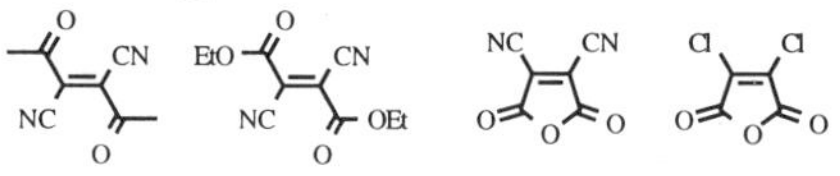

275. THE SYNTHESIS, CHARACTERISATION AND ELECTRONIC STRUCTURE OF o-BENZOQUINONEDIIMINE(DICHLORO-BIS-TRIPHENYLPHOSPHINE)RUTHENIUM(II).
<u>Diego Venegas-Yazigi</u>, Juan Costamagna, Ramon Latorre and A. B. P. Lever, Depts. of Chemistry, York University, Toronto, Ontario, Canada, and the Universidade de Santiago de Chile.

o-Benzoquinonediimine is a non-innocent ligand, known to form with ruthenium(II), complexes in which there is substantial $d\pi$-π* mixing [1]. Redox series can be synthesized where the formal oxidation state of the ligand varies from -2 to zero. Previously we have probed how the electronic structures of the ruthenium complexes varies with the net oxidation level of the species. The question then arises- what will happen to the electronic description of these species if the ruthenium centre is made very much more electron rich favoring thereby much greater $d\pi$-π* mixing. The bis(bipyridine) co-ligands of the previous study [1] are replaced here by $(PPh_3P)_2Cl_2$ making the ruthenium about 0.7 V richer. The consequences of this enhanced π-donicity are discussed here. [1] Masui, H.; Lever, A. B. P.; Dodsworth, E. S *Inorg. Chem.* **1993**, *32*, 258; Masui, H.; Auburn, P R. Lever, A. B. P. *Inorg. Chem.*, **1991**, *30*, 2402.

276. PREPARATION AND REACTIVITY OF SOME NIOBIUM(IV)TRISAMIDO PSEUDOHALIDE COMPLEXES. M. G. Fickes and C. C. Cummins*, Department of Chemistry, Massachusetts Institute of Technology, Cambridge, MA 02139

New Niobium tris(amido) complexes containing N-*tert*-butylanilido ligands have been prepared by reaction with the paramagnetic $Nb(Cl)(NRAr)_3$. Reduction of the isocyanato complex, $Nb(NCO(NRAr)_3$, leads to complexes with an uncommon η^2 binding mode at the metal center. Preparation, characterization and reactivity of $Nb(NCO(NRAr)_3$ and other pseudohalide complexes will be discussed.

277. CONTINUOUS REDOX AND SPECTROSCOPIC TUNING OF DINUCLEAR COMPOUNDS: CHLOROTETRAKIS(μ-N,N'-DIARYLFORMAMIDINATO)DIRUTHENIUM(II, III); Chun Lin, Tong Ren,* Edward J. Valente, and Jeffrey D. Zubkowski; Departments of Chemistry, Florida Institute of Technology, Melbourne, FL 32901, Mississippi College, Clinton, MS 39058; and Jackson State University, Jackson, MS 39217

Synthesis and characterization are reported for a series of novel Ru_2^{5+} compounds supported by diarylormamidinate (ArNC(H)NAr-), where the aryl is either XC_6H_4-, with X as o-OMe (**1**), p-OMe(**2**), m-OMe(**3**), p-Cl(**4**), m-Cl(**5**), m-CF$_3$(**6**), p-CF$_3$(**7**), or 3,4-Cl$_2$C$_6$H$_3^-$ (**8**), or 3,5-Cl$_2$C$_6$H$_3$- (**9**). The redox potential ($E_{1/2}(Ru_2^{6+}/Ru_2^{5+})$), cathodic peak potential (E_{pc}) associated with the Ru_2^{5+}/Ru_2^{4+} couple, and energy (E_{op}) of one of the visible absorptions for the series correlate with the Hammett constant of aryl-substituents according to the following equations: $\Delta E_{1/2}(Ru_2^{6+}/Ru_2^{5+}) = 88.9 \cdot (8\sigma_x)$ mV, $\Delta E_{pc}(Ru_2^{5+}/Ru_2^{4+}) = 69.3 \cdot (8\sigma_x)$ mV, and $\Delta E_{op} = 269 \cdot (8\sigma_x)$ cm^{-1}. The electrochemical behavior of the current series is consistent with an ECE mechanism, and the chemical step (transfer of Cl$^-$) is regulated by the polar substituent X. Structural and magnetic properties of the series will also be discussed.

278. Spectroscopic and Kinetic Studies of $[L(NH_3)_5Ru]^{3+}$ (L = purine nucleosides, nucleotides)
M.J. Clarke, V.M. Bailey, K.J. LaChance-Galang, P.E. Doan
Merkert Chemistry, Boston College, Chestnut Hill, MA 02167,
Department of Chemistry, Northwestern University, Evanston, IL

The ^{1}H NMR spectra of complexes involving the paramagnetic metal center, $[(NH_3)_5Ru^{III}]$, coordinated at ring nitrogens have been examined with pyridine, purine, nucleoside and nucleotide ligands along with the ^{31}P NMR of the nucleotide complexes and EPR of representative complexes. Variations in the spectra have been investigated as a function of the coordination site and pH. Pseudocontact and contact shifts have been calculated for various protons and an attempt made to correlate sugar conformations in coordinated 5'GMP, 5'IMP, Guo and Ino with paramagnetically induced shifts. The autoxidation of $[L(NH_3)_5Ru^{III}]$, where L = Guo, dGuo and 1MeGuo, to the corresponding 8-oxo complexes under atmospheric oxygen is first order in the complex and [OH$^-$]. For L = Guo, k = 6.6 x 10^{-5} M^{-1} s^{-1}, $\Delta H^* = $ 58 kJ/mol and $\Delta S^* = $ -124 J/mol °K.

279.
SYNTHESIS, CHARACTERIZATION AND OXIDATIVE POLYMERIZATION OF NI(II) AND PD(II) PYRROLES. **Timothy W. Hanks,** Mathew Mathis, Wayne Harsha, Department of Chemistry, Furman University, Greenville, SC, 29613.

We have recently found that transition metal complexes of pyrrole, (Mx-pyrrolyl), are able to undergo oxidative polymerization to form conducting polymers with properties similar to that of the parent complex polypyrrole. Here, we discuss recent advances where Mx is either a Ni(II) or a Pd(II) complex containing a bi- or tridentate, planar aromatic ligand. The structures of these new complexes will be described and compared to quantum mechanical models. Oxidation of the complexes to polymers can be achieved with chemical oxidants. The structure of the resulting materials will be discussed in terms of spectroscopic data and calculations.

280.
DESIGN OF CONDUCTING REDOX POLYMERS: POLYTHIOPHENE-RU(BIPY)$_3^{n+}$ HYBRID MATERIALS **S. Sherry Zhu** and Timothy M. Swager* Department of Chemistry, Massachusetts Institute of Technology, Cambridge, MA 02139 and Department. of Chemistry, University of Pennsylvania

The thiophene substituted Ru(bipy)$_3^{2+}$ complexes, **I** and **II**, were synthesized and shown to undergo polymerization with electrochemical oxidation. Electrochemical investigations of the resultant polythiophene-Ru(bipy)$_3^{n+}$ hybrid materials reveal relatively high redox conductivities due to rapid self-exchange between the Ru centers.

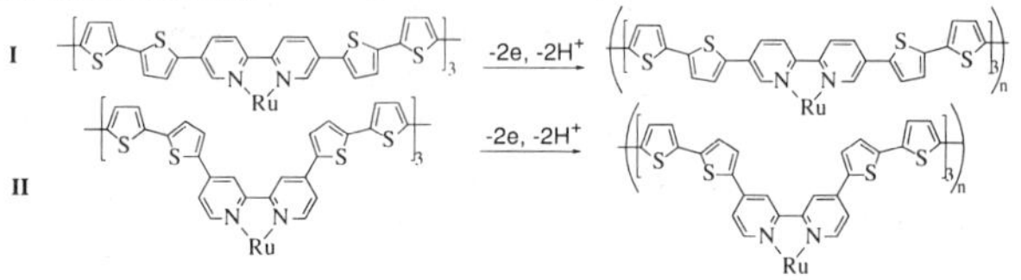

281.
NEW CONDUCTING POLYMER SENSORS: POLYTHIOPHENE TRANSITION METAL HYBRIDS. **R. P. Kingsborough** and T. M. Swager*, Department of Chemistry, Massachusetts Institute of Technology, Cambridge, MA 02139

Schiff base complexes (1) and phthalocyanines containing electrochemically polymerizable thiophene substituents have been synthesized and studied electrochemically. The thiophene substituents and diamine bridges have been varied to tune the oxidation potential of the polymer. Perturbing the electronics of the metal center by binding small molecules is currently being investigated as a means to making chemoresistive sensors.

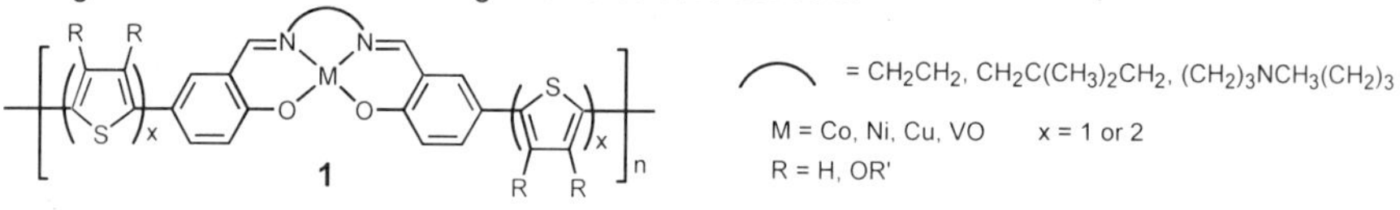

⁀ = CH_2CH_2, $CH_2C(CH_3)_2CH_2$, $(CH_2)_3NCH_3(CH_3)_3$
M = Co, Ni, Cu, VO   x = 1 or 2
R = H, OR'

282. ELECTRON TRANSFER REACTIONS OF PORPHYRIN-PARAPHENYLENE-QUINONE SYSTEMS: PROBING THE ROLE OF MEDIUM PROPERTIES IN MODULATING DONOR-ACCEPTOR ELECTRONIC COUPLING
Rebecca M. Allen, Matthew A. Kellett, and Michael J. Therien*
Department of Chemistry, University of Pennsylvania Philadelphia, PA 19104-6323

Porphyrin-spacer-quinone systems are synthesized employing paraphenylene spacers in which the electronic properties are varied. The effect of changing the tunneling medium redox properties on electronic coupling is measured using photoinduced charge separation and thermal charge recombination reactions. The synthesis, optical, electrochemical, and excited-state properties of these systems will be discussed along with their photoinduced electron transfer kinetics, measured by time-resolved transient absorption spectroscopy.

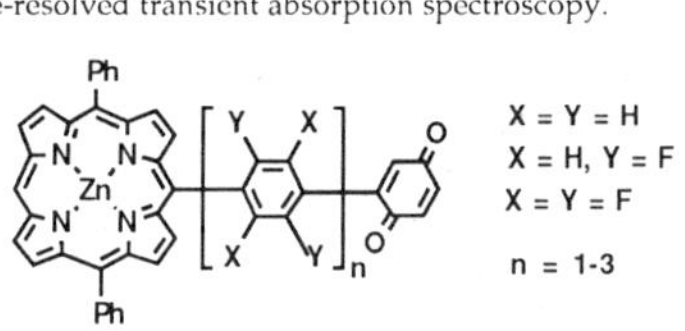

X = Y = H
X = H, Y = F
X = Y = F
n = 1-3

283.
SYNTHESIS AND CHARACTERIZATION OF COFACIAL PORPHYRIN-QUINONE SYSTEMS TO INVESTIGATE ELECTRONIC COUPLING AND PHOTOINDUCED ELECTRON TRANSFER THROUGH STACKED AROMATIC ARRAYS

Peter M. Iovine, Matthew A. Kellett, Michael J. Therien*, *Department of Chemistry, University of Pennsylvania, Philadelphia, Pennsylvania, USA, 19104-6323*

Metal-Mediated cross-coupling has enabled the sythesis of conformationally rigid porphyrin-based D-A systems for the study of fast electron transfer between cofacial aromatic rings.

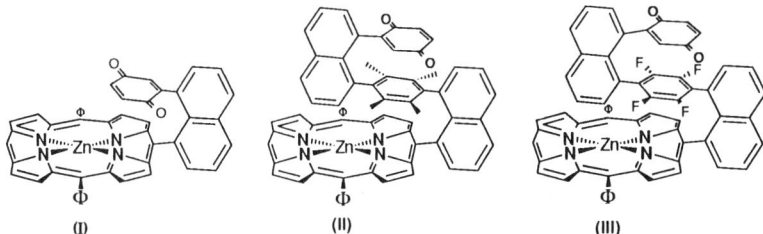

(I)   (II)   (III)

Such model systems, in which prophyrin and quinone are separated only by the sum of their van der Waals radii (I) provides the prototype for a series of molecules designed to probe the electonic coupling modulated through stacked pi interactions (II),(III).

284. SYNTHESIS, STRUCTURE AND REACTIVITY OF BENZAMIDINATE STABILISED VANADIUM(III) COMPLEXES
E.A.C. Brussee, A. Meetsma, J.H. Teuben, Department of Organic and Molecular Inorganic Chemistry, University of Groningen, Nijenborgh 4, 9747 AG Groningen, the Netherlands.

The N,N'-bis(trimethylsilyl)benzaminate ligand is a useful stabilizing ligand for early transition-metal centers, as has previously been shown for Group 3 and 4 metals and recently also for vanadium. The synthesis and reactivity of tris-, bis- and mono-benzamidinate complexes of vanadium(III) is described. For example, the benzonitrile adduct $(Me_3SiNCPhNSiMe_3)_2VCl \cdot PhCN$ reacts with Li-benzamidinate to give the complex $(Me_3SiNCPhNSiMe_3)_2V[Me_3SiNCPhNC(Ph)=NSiMe_3]$ in which the benzonitrile molecule is inserted into the benzamidinate N-Si bond. Reactions of benzamidinate vanadium chlorides with alkylating agents will also be discussed.

285. CHEMISTRY OF VANADIUM(V) IMIDO ALKYL COMPLEXES WITH AN AMIDO-FUNCTIONALISED CYCLOPENTADIENYL ANCILLARY LIGAND
P.T. Witte, A. Meetsma, B. Hessen, Department of Organic and Molecular Inorganic Chemistry, University of Groningen, Nijenborgh 4, 9747 AG Groningen, the Netherlands.

The (*iso*propylamido)ethyl-functionalised cyclopentadienyl ligand is readily introduced on vanadium(V) by the reaction of the neutral ligand with the V(V) imido complex $(Me_2N)_2V(NtBu)Cl$, in which dimethylamine is eliminated. The chloride complex $(C_5H_4C_2H_4NiPr)V(NtBu)Cl$ can be alkylated to produce the alkyl derivatives $(C_5H_4C_2H_4NiPr)V(NtBu)R$ (R = Me, CH_2CMe_3, CH_2SiMe_3), of which the methyl derivative has been characterized by single crystal X-ray diffraction. The methyl complex reacts with the Lewis-acidic borane $B(C_6F_5)_3$ in benzene to give the contact ion pair $(C_5H_4C_2H_4NiPr)V(NtBu)[(\mu-Me)B(C_6F_5)_3]$. In more polar solvents, the reactivity of the V=N-bond in the 16-electron cationic vanadium imido species $[(C_5H_4C_2H_4NiPr)V(NtBu)]^+$ can be studied.

286. ELECTROCATALYTIC PROPERTIES OF POLY-Co-TETRAAMINOPHTHALOCYANINE MODIFIED ELECTRODES. Min Huang, Peter W. Faguy, Richard P. Baldwin, Jo nathan Shaw, Jorge Pavez, Jianping Wang, Francois Bailly, and Robert M. Buchanan, Department of Chemistry, University of Louisville, KY 40292.

Chemically modified electrodes have been used extensively as sensors, in electrochromic devices, and as electrocatalysts. Modifiers such as phthalocyanine complexes are particularly interesting as electrode mediators because their redox properties are easily tuned, and they display resistance to chemical degradation by strong acids and bases. We have found that tetraamino cobalt pthalocyanine (CoTAPc) oxidatively electropolymerizes on glassy carbon electrodes forming electroactive films of poly-CoTAPc of controllable thicknesses. In addition, these poly-CoTAPc modified electrodes have been shown by rotating-disk voltammetry to electrocatalytically reduce O_2 to H_2O_2 and H_2O, via parallel two-electron and four-electron pathways. Details concerning the electropolymerization process and the structure of the resulting electroactive polymers will be presented along with a complete evaluation of the O_2 reduction process.

287. SYNTHESIS, PROPERTIES AND CATALYTIC BEHAVIOR OF A NEW UNSYMMETRICAL DINUCLEAR ZINC(II) COMPLEX. Robert M. Buchanan;[a] Nicholas Daubresse,[a,b] Mark S. Mashuta,[a] John F. Richardson,[a] Alain Rabion,[b] and Jean-L. Seris[b]
[a]Department of Chemistry, University of Louisville, KY 40292 and [b]CSN, Elf Aquitaine, Inc.

A new unsymmetrical μ-alkoxo and μ-acetato bridged dinuclear zinc(II) complex has been prepared as a model of the active site of alkaline phosphatase proteins. The compound is prepared by reacting zinc(II) acetate with bis(N,N-2-pyridinylmethyl)aminoethan-2-ol. The crystal structure of the complex shows that the two zinc atoms are in dissimilar coordination environments (Zn(1) is five-coordinate, N_3O_2, and Zn(2) is six-coordinate, N_3O_3). Room temperature ^{1}H NMR studies show that the μ-alkoxo and terminal alcohol ligands are rapidly exchanging in CD_3CN, and the dynamic exchange process has been thoroughly studied by variable temperature NMR methods. The dinuclear zinc(II) complex catalyzes the hydrolysis of activated phosphodiesters in non-aqueous solutions. Both mono and diphosphoester cleavage products have been detected. Details of the crystal structure, VT ^{1}H NMR and the hydrolysis of activated phosphodiesters will be presented.

288. 2-(2-PYRIDYL)-4-CARBOXYQUINOLINE: A FUNCTIONALIZED DIIMINE LIGAND FOR RUTHENIUM(II)
R. J. Morgan[1]*, R. J. Donovan, Yakov Bass[1] Oma Morgan[2] and A. D. Baker[2]
[1]Department of Chemistry, Long Island University, Brooklyn N.Y, 11201.
[2]Department of Chemistry and Biochemistry, Queens College CUNY 11367.

The title compound I is available in a single convenient step via a Pfitzinger reaction, and forms luminescent stable chelates to ruthenium (II). The presence of the carboxylate group allows the preparation of a myriad of interesting derivatives, and represents a point of attachment, allowing its complexes to participate in supermolecular applications. Several different ligands derived form I have been prepared, including optically active diimine compounds containing amino acid residues. Presented will be the preparation of I, its derivatives, and its optically pure bis-bipyridine ruthenium (II) complex ions.

289. COPPER(II) COMPLEXES OF POLYAMIDE MACROCYCLIC LIGANDS, M. A. Sanchez, C. L. and G. H. Rawji, Department of Chemistry, Southwestern University, Georgetown, TX 78626.

This paper describes the synthesis of several polyamide macrocyclic ligands and their copper(II) complexes, and their characterization by a combination of the following techniques: IR, UV-Vis, MS, NMR, ESR, elemental analysis and magnetic measurements. The acidity constants of the ligands were determined by potentiometric titrations. Molecular modelling studies conducted on the ligand and metal complexes provided useful structural information on the variation in cavity size of the ligand and on the geometric environment of the metal in the complex. These results will be discussed in comparison to those obtained from the above methods.

290. CIS EFFECTS IN COBALT CORRINOIDS: THE CRYSTAL STRUCTURES OF 10-CHLORO-AQUACOBALAMIN PERCHLORATE, 10-CHLORO-CYANOCOBALAMIN AND 10-CHLORO-METHYLCOBALAMIN. <u>Kenneth L. Brown</u>, Shifa Cheng, and Xiang Zou, Department of Chemistry, Ohio University, Athens, OH 45701, Jeffrey D. Zubkowski, Department of Chemistry, Jackson State University, Jackson, MS 39217, Edward J. Valente, Department of Chemistry, Mississippi College, Clinton, MS 39058, Leanne Knapton and Helder Marques, Department of Chemistry, University of the Witwatersrand, Johannesburg, South Africa

The crystal structures of the title compounds show an unusual disorder in the C pyrrole ring consistent with a two-site occupancy in which the major conformation has the C46 and C47 methyls in the usual orientation while the minor conformation has both methyls pseudo-equatorial. NMR measurements suggest this disorder persists in solution as a C ring flip. In 10-Cl-CH$_3$Cbl, the axial bond lengths are quite similar to those in CH$_3$Cbl, but the equatorial Co-N bond lengths are on average 0.05 Å shorter. In 10-Cl-CNCbl, the Co-C bond is 0.10(2) Å longer and the cyanide C-N bond is 0.06(2) Å shorter than in CNCbl, suggesting weaker cyanide coordination. In 10-Cl-H$_2$OCbl$^+$, the Co-O bond length is the same as in H$_2$OCbl$^+$, and the coordinated water is hydrogen bonded to the c side chain carbonyl oxygen as in H$_2$OCbl$^+$. NMR observations suggest that this hydrogen bond persists in solution. The picture which emerges is one of competitive electron donation to the corrin by the axial ligand, X, and the Cl substituent, with resonance electron donation of the latter being more important than inductive electron withdrawal.

291. QUANTITATIVE ASSESSMENT OF C SIDE CHAIN THERMAL MOTIONS ON THE ENTROPY ACTIVATION FOR Co-C BOND HOMOLYSIS IN NEOPENTYLCOBALAMIN. <u>Shifa Cheng</u>, Kenneth L. Brown*, Chemistry Department, Ohio University, Athens, OH 45701.

NpCbl-8-butanamide permits a quantitative assessment of the influence of thermal motions of the c-side chain on the entropy activation for carbon-cobalt bond thermal homolysis in neopentylcobalamin (NpCbl). The kinetics of the thermolysis were studied in aerobic aqueous solution at temperatures between 15 and 45 °C. After correction of the observed first-order rate constants for the presence of the essentially unreactive base-off species using an established NMR method, an Eyring plot yields the activation parameters $\Delta H^{\ddagger}_{on}$ = 26.7 ± 0.1 kcal mol^{-1} and $\Delta S^{\ddagger}_{on}$ = 13.2 ± 0.2 cal mol^{-1} K^{-1}. While the enthalpy of activation is slightly reduced (6%) from that of NpCbl, the entropy of activation is reduced by 6.1 ± 0.6 cal mol^{-1} K^{-1}, or 32 ± 3 %. The c side chain thus contributes about one-third of the total entropic activation of NpCbl for carbon-cobalt bond homolysis.

292. MULTINUCLEAR MAGNETIC RESONANCE STUDY OF ZINC(II) AND CADMIUM(II) COMPLEXES WITH ISOTHIOCYANATE. <u>Anthony Fratiello</u>, Vicki Kubo-Anderson, Eddie Curtis, Jr., Tai Mao, Richard Perrigan, Victor San Lucas, and Annie Wong, Dept. of Chemistry & Biochemistry, Cal State Univ., Los Angeles, CA 90032
A multinuclear magnetic resonance study of zinc(II) and cadmium(II) complexes with isothiocyanate ion is underway. At -120°C, ligand exchange is slow enough to permit the observation of separate NMR signals for coordinated and free ligand, and, in the Cd(II) solutions, this metal-ion in each complex. This observation permits the development of detailed structures of the complexes, including the number of species formed; the competitive coordinating ability of different ligands, specifically water and NCS$^-$ in these systems; the ligand binding site; the concentration dependence of the complexes; and spin coupling effects. The complexes of both metal-ions are being studied by ^{1}H, ^{13}C, and ^{15}N NMR spectroscopy. In the ^{1}H spectra, the intensity of the coordinated water signal decreases dramatically as NCS$^-$ is added, indicating the complexing process in each system involves more than a simple 1:1 ligand replacement. In Zn^{2+}-NCS$^-$ solutions, the ^{13}C and ^{15}N spectra reveal signals for several species, possibly arising from a combination of hexa- and tetracoordinated complexes. In the Cd^{2+}-NCS$^-$ solution spectra, the ^{13}C and ^{15}N signals for the complexes are three line patterns, corresponding to a doublet (^{111}Cd and ^{113}Cd J-coupling) and a dominant central peak (bonding to inactive Cd isotopes). The ^{113}Cd NMR spectra correlate well with the ^{13}C and ^{15}N NMR data, although the J-coupling patterns could not be resolved. At anion to metal-ion mole ratios exceeding 4:1, only the signals for Zn(NCS)$_4^{2-}$ and Cd(NCS)$_4^{2-}$ are observed. At this point, it appears that the Zn(II) and Cd(II) coordination mechanisms are similar, that is, the tetracoordinated species is favored in both cases.

INOR

293. MULTINUCLEAR MAGNETIC RESONANCE STUDY OF COMPARATIVE LANTHANIDE COMPLEXING. Anthony Fratiello, Vicki Kubo-Anderson, Deborah Lee, Richard Perrigan, Amy Sharp, Stephanie Stoll, and Kenneth Wong, Department of Chemistry & Biochemistry, California State University, Los Angeles, CA 90032

A study of the trivalent lanthanide complexes with nitrate ion in aqueous mixtures is being continued using nitrogen-15 (^{15}N) NMR. At low temperature, ligand exchange is slow enough to permit the detection of resonance signals for bound nitrate and this anion in bulk medium. This observation has been possible with ^{15}N-enriched nitrate solutions of paramagnetic Ce(III)-Eu(III) and Er(III)-Yb(III), as well as diamagnetic La(III) and Lu(III). For the paramagnetic cations, the ^{15}N NMR chemical shifts induced by complexing vary from 10 to 1,600 ppm, reflecting the magnitude of their magnetic moments. The ^{15}N NMR spectra of Eu^{3+}-NO_3^- solutions in water-acetone-Freon reveal four coordinated nitrate signals arising from the mono- and dinitrato complexes and two higher species. Signal areas and a comparison to the Ce(III)-Sm(III) results indicate these higher complexes may be the tetra- with either the penta- or hexanitrato. Preliminary ^{15}N NMR results for Tm^{3+}-NO_3^- and Yb^{3+}-NO_3^- solutions are comparable in that signals for four complexes are observed. However, the chemical shifts, due to the strong paramagnetism of these ions, are extremely large and widely separated. For the Yb(III) solutions, signals for two additional complexes, closely spaced near the bulk anion peak, are observed and appear to arise from a different type of Yb^{3+}-NO_3^- interaction. A previous study of the Er(III) system indicated only two complexes are formed over the entire NO_3^- concentration range. These results may indicate a different type of coordination behavior between the lighter and heavier lanthanides.

294.

INTER- AND INTRAMOLECULAR HYDROGEN AND PHOSPHINE LIGAND EXCHANGE: DYNAMIC BEHAVIOR STUDIES OF PLATINUM COMPLEXES WITH 2-D NMR SPECTROSCOPY METHODS. **Shi LOU**, Ursula Simonis, Jane V. Z. Krevor; Department of Chemistry and Biochemistry, San Francisco State University, 1600 Holloway Avenue, San Francisco, CA 94132

The hydrogen and phosphine ligand exchange of several platinum phosphine hydride complexes {[$Pt_2Cl_2(\mu$-H)(μ-dppm)$_2$]PF_6 (1), $Pt_2Cl_2(\mu$-dppm)$_2$(2), [$Pt_2H_2(\mu$-H)(μ-dppm)$_2$]Cl(3), [$Pt_2H_2(\mu$-Cl)(μ-dppm)$_2$]Cl(4), [$Pt_2HL(\mu$-dppm.)$_2$]Cl (L= dppm(5), PPh_3(6)), and trans-PtHCl(PPh_3)$_2$]} have been investigated. The ^{1}H and ^{31}P NOESY experiment, a 2-D NMR spectroscopy method, was utilized for studying the dynamic ligand exchange of these platinum complexes. Cross correlations between satellite peaks and central peak allowed the direct observation of inter platinum ligand exchange. The mechanism of ligand exchange for each of the systems is presented.

295.

PLATINUM(II) COMPLEXES OF CROWN THIOETHER AND MIXED OXA-THIA LIGANDS: ANALYSIS BY NUCLEAR MAGNETIC RESONANCE *David F. Galas*, Gregory J. Grant, Department of Chemistry, The University of Tennessee at Chattanooga, Chattanooga, Tennessee 37403.

We have prepared a series of homoleptic platinum(II) complexes which contain crown thioether ligands such as 1,4,7-trithiacyclononane (9S3) and mixed oxa-thia crowns such as 1,4-dithia-7-oxacyclonane (9S2O). Non-homoleptic Pt-thioether complexes involving halide ligands have also been prepared. This paper describes the use of ^{195}Pt and ^{13}C NMR as an analytical tool for the characterization of these Pt(II) complexes. There is a strong correlation between the ligand environment around the platinum center of the complexes and the observed ^{195}Pt chemical shift.

296.

COBALT(III) COMPLEXES OF CROWN HEXATHIOETHERS *Gregory J. Grant*, Department of Chemistry, The University of Tennessee at Chattanooga, Chattanooga, Tennessee 37403.

We have prepared a series of homoleptic, octahedral Co(III) complexes which contain crown hexathioether ligands such as 1,4,7,11,14,17-hexathiacycloeicosane (20S6). The complexes have been characterized by electronic spectroscopy, cyclic voltammetry, and both carbon-13 and cobalt-59 NMR. There is a correlation between the ligand field strength of the complexes and the observed ^{59}Co chemical shift. Our NMR data also show that a single stereoisomer is formed exclusively. A single crystal x-ray structure of the [Co(20S6)](BF_4)$_3$ supports the NMR data in that the stereoisomer obtained is the one in which the two six-membered chelate rings are found in a *cis* arrangement around the octahedral Co(III) center.

INOR

297. TRANSITION METAL COMPLEXES OF *EXO*-CALIXARENES. Hongping Yuan and Thomas N. Sorrell, Department of Chemistry, The University of North Carolina, Chapel Hill, NC 27599-3290 USA

New *exo*-calix[4]arenes were prepared from bisphenol A in three steps. Covalently-linked derivatives were generated by the treating the *exo*-calix[4]arene with bromochloromethane or ethylene ditosylate. NMR spectra taken over a wide temperature range show that the unlinked or partially-linked *exo*-calixarenes are conformationally mobile. The fully linked substances are conformationally locked. A study of metal ion binding by these new *exo*-calixarenes will be reported.

298. MOLECULAR AND ELECTRONIC STRUCTURES OF LOW-SYMMETRY NITRIDO COMPLEXES OF MANGANESE AND CHROMIUM. **Christopher J. Chang,** William B. Connick, Donald W. Low, and Harry B. Gray, Department of Chemistry and the Beckman Institute, California Institute of Technology, Pasadena, California 91125

The single-crystal-polarized absorption and circular dichroism spectra of nitridomanganese(V) and analogous nitridochromium(V) complexes have been investigated. As revealed by crystal structure analyses, these molecules have a distorted square-pyramidal geometry. Their electronic structure is readily described using a ligand field model for axially-compressed systems.

299. **PREPARATION AND CHARACTERIZATION OF BIMETALLIC COMPLEXES WITH A BIS(DIKETONE) PHENYLENE LIGAND**
Grace Monis, Peter I. Djurovich, Herbert B. Silber, Department of Chemistry, San Jose State University, San Jose, CA 95192-0101.

The *p*-bis(diketone)phenylene ligand (**I**) can potentially bind two metal centers and also undergo a two electron reduction to form a conjugated quinoidal bridge.

This presentation will describe the synthesis and characterization of bimetallic complexes of (**I**) with either ML_n^{m+} = Ru(bipy)$_2^{2+}$ or Rh(ppy)$_2^+$. Comparison of the UV-vis and electrochemical properties of the bimetallic complexes with their mononuclear diPhacac derivatives will be made to assess the degree to which the phenylene bridge influences the electronic interaction between the metal centers. This work is supported by ACS PRF and NIH MBRS funds.

300. **SYNTHESIS AND EQUILIBRIUM STUDIES OF μ-DIHALO-DIMETALLIC COMPLEXES OF RHODIUM AND IRIDIUM**
Peter F. Huefner, Peter I. Djurovich, Herbert B. Silber, Kenneth I. Hardcastle[a], Ricardo Duque[a] and Mandana Visi[a], Department of Chemistry, San Jose State University, San Jose, CA 95192-0101; [a]Department of Chemistry, California State University-Northridge, Northridge, CA 91330.

Chlorobis[(6-isopropyl-8-quinolyl)dimethylsilyl]rhodium complexes exist as either 5- or 6-coordinate species in both solution and the solid state. The 5-coordinate Rh(*i*Prdmsiqn)$_2$Cl complex adopts a distorted trigonal bipyramidal geometry with a compressed Si–Rh–Si angle of 86°. The Rh(*i*Prdmsiqn)$_2$Cl complex reacts with [Rh(COD)Cl]$_2$ to form a heteroleptic dimer complex, Rh(*i*Prdmsiqn)$_2$(μ-Cl)$_2$Rh(COD) that exists in equilibrium with the two starting materials in CHCl$_3$ or CH$_2$Cl$_2$ solution. Similar equilibrium mixtures are formed with Rh(*i*Prdmsiqn)$_2$Cl and [Ir(COD)Cl]$_2$ or [Ru(p-cymene)Cl$_2$]$_2$. The 5-coordinate geometry and monomer-dimer equilibrium both stem from the strong Si–Rh σ-bonding present in the complex. Additional data from x-ray crystallographic and equilibrium studies with dibromo-bridged metal dimers will also be presented.

301. FOUR-COORDINATE LEWIS ACIDIC ZINC, Amy L. Singer and David A. Atwood*, Center for Main Group Chemistry, Department of Chemistry, North Dakota State University, Fargo, North Dakota 58105

Our interest in main group Lewis acids has prompted us to investigate zinc complexes of tetradentate -N_2O_2 Schiff base ligands. These complexes can coordinate to a large variety of substrates such as alcohols, carbonyls, and nitrogen based donors. The interactions of these complexes with various Lewis bases will be described. Several have been structurally characterized. Salophen(^{t}Bu)Zn•py (shown) consists of a five-coordordinate distorted square pyramidal geometry with the pyridine occupying the axial position. The relevance of these complexes to Lewis acid catalysis will also be described.

302. TOWARD *NIDO*-CARBORANYLMETHYLATION. Cynthia L. Beno, C.B. Knobler, M.F. Hawthorne, Department of Chemistry, UCLA, Los Angeles, California 90095.

Derivatives of $[nido\text{-}7\text{-}(CH_2X)\text{-}7,8\text{-}C_2B_9H_{11}]^-$ have been developed in our laboratory which apparently have the ability to provide electrophilic reaction centers in nucleophilic substitution reactions. The addition of nucleophiles to the boron atoms of the open face of the *nido*-carborane cage accompanied by formation of an *exo*-C-methyl group has also been observed. These results, taken together, suggest that the dicarbollide ion forms an energetic fulvene-like intermediate capable of reaction with nucleophiles at the *exo*-methylene carbon center. We anticipate that such reagents will be powerful tools in synthetic chemistry, enabling the addition of a *nido*-carborane cage to appropriate nucleophiles. The syntheses of the novel "*nido*-carboranylmethylation" reagents will be presented along with the products of their reactions with various nucleophiles, characterized by mass spectrometry, proton and boron NMR spectroscopy, and single crystal X-ray analysis.

303.
DEUTERATION AND CLOSURE OF DECABORANE (14). J. A. Dopke, A. N. Bridges, A. G. Turoso, D. F. Gaines, Department of Chemistry, University of Wisconsin at Madison, Madison, Wisconsin 53706.

Electrophilic substitution chemistry has a rich history in its application to boron hydride clusters. Deuterium transfer from deuterated aromatic solvents has been previously shown to occur in the presence of $AlCl_3$ to generate 1-DB_5H_8 regiospecifically. We have recently verified that $B_{10}H_{14}$, decaborane(14), is deuterated by C_6D_6 and C_7D_8, first in the 2,4 positions, then somewhat more slowly in the 1,3 positions at room temperature in the presence of $AlCl_3$. The deuterated decaborane species has been used as a mechanistic probe to investigate our postulated "fold-over" mechanism for conversion of *nido*-$B_{10}H_{12}^{2-}$ to *closo*-$B_{10}H_{10}^{2-}$, a process that occurs in good yield only in the presence of dimethoxyethane (glyme). Closure of alkylated decaborane(14) derivatives to *closo*-1-$RB_{10}H_9^{2-}$ will also be discussed.

304. Monofunctionalization of Cubeoctameric Silsesquioxanes: How Do You Effect Monofunctionalization of an Octafunctional Molecule?

Darvong Soulivong and Frank J. Feher
Department of Chemistry, University of California, Irvine, California 92697

Over the past several years, cubeoctameric silsesquioxane frameworks (e.g., **1**) have attracted attention as precursors to a wide variety of interesting molecules. The synthetic manipulation of highly functionalized silsesquioxane frameworks presents a number of interesting challenges, especially if it is desirable to effect the modification of only one functional group. This paper will examine the general problems associated with the monofunctionalization of an octafunctional silsesquioxane framework, and it will describe our efforts to devise general protocols for effecting such transformations.

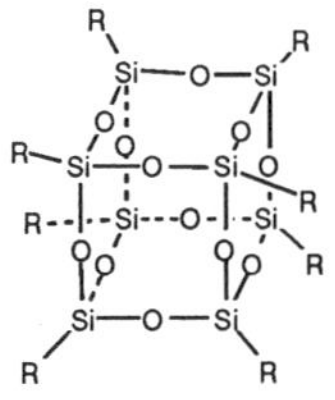

305.

ORGANOMETALLIC DERIVATIVES OF THE ALKALINE-EARTH METALS WITH NON-CYCLIC LIGANDS. Melanie J. Harvey, Jason S. Overby and Timothy P. Hanusa, Department of Chemistry, Vanderbilt University, Nashville, TN 37235

The recent revival of interest in organometallic compounds of the heavy alkaline-earth (Ae) metals calcium, strontium, and barium has centered on cyclopentadienyl compounds, including metallocenes Cp′$_2$Ae and mono(cyclopentadienyl) compounds Cp′AeX. New types of non-cyclic analogues are also accessible that extend the range of compounds and reactivity that can be obtained from the heavy Group 2 elements. The reaction of alkali metal salts of substituted allyl and pentadienyl anions with the alkaline-earth diiodides yields thermally stable alkaline-earth bis(allyl) and bis(pentadienyl) complexes. The synthesis, structures and reactions of these and related compounds will be discussed.

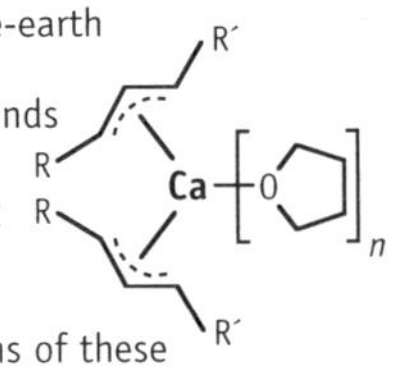

306.

MACROCYCLIC STABILISATION OF P- AND F- BLOCK CATIONIC SPECIES Marcus P. Spry Gerald R. Willey, University of Warwick, Coventry, UK, CV4 7AL

For both p- and f- block metal chlorides, sequential halide abstraction using SbCl5 in acetonitrile is a convenient route to the formation of cationic species.

These can be stabilised in situ by macrocyclic ligands such as crown ether, crown thioether, cryptands, aza, and 'mixed' ONS donor systems. Recent X-ray structure determinations include [SbCl2(Me3-9N3)][SbCl6], [Er(12-Crown-4)2(H2O)][SbCl6]3, [Er(12-Crown-4)2(MeCN)][SbCl6]3, and [(12-Crown-4)2(H3O)][SbCl6]. The coordination profile and general reactivity of these cationic complexes is under investigation.

307. NEW CLASSES OF MONOMERIC AND DIMERIC TIN (II) ALKOXIDES AND AMIDES Rimo Xi and Lawrence R. Sita, Department of Chemistry, The University of Chicago, 5735 South Ellis Avenue, Chicago, Illinois, 60637

A variety of new classes of fully characterized monomeric and dimeric tin (II) alkoxides and amides have been synthesized by both carbon dioxide and isocyanate metathesis and by redistribution processes. These include the monoamide complex **I** and the heteroleptic bisalkoxide **II**, both of which retain structural integrity in solution. The kinetics of the metathesis reactions and the relative thermodynamic stabilities of the new tin (II) complexes will be presented along with spectroscopic data pertaining to the characterization of these new compounds.

308. PHOSPHAZANE REDOX CHEMISTRY: NEW TETRAPHOSPHAZANES. R. M. Hands, S. A. Katz, R. C. Haltiwanger, A. D. Norman, Department of Chemistry and Biochemistry, University of Colorado, Boulder, CO 80309.

Dichlorophosphine ($RPCl_2$; R = Me, Ph), reactions with excess $1,2\text{-}(NH_2)_2C_6H_4$ in toluene at 95 °C yield, instead of expected phosphazane condensation oligomers/polymers, products of oxidation/reduction reactions. The $1,2\text{-}(NH_2)_2C_6H_4$ reaction with $PhPCl_2$ yields the new **I** and **II**, along with $(PhP)_{4,5}$, whereas reaction with $MePCl_2$ yields only $(MeP)_{4,5}$. **I** and **II** are novel $\lambda^3\text{-}\lambda^5\text{-}\lambda^3$ cis,trans- and cis,cis- tetraphosphazanes, readily crystallized as disulfide derivatives. Mechanisms of reaction involving possible participation of intermediate diimine $1,2\text{-}(NH)_2C_6H_4$ and tetraphosphazane $[C_6H_4(NH)_2]_2PPh$, and applications to subsequent new syntheses will be described.

309.
ALKYLPHOSPHONATE DERIVATIVES OF TRANSITION, ALKALI AND ALKALINE EARTH METALS AT HIGH EXTERNAL PRESSURES. **R.D. Markwell**, I.S. Butler, L.M. Dickinson, W. Gao and L. Reven, Chemistry Department, Mcgill University, 801 Sherbrooke Street West, Montreal, Quebec, Canada, H3A 2K6

Alkyl phosphonate complexes form lamellar polymeric structures showing phase changes of interest in energy storage applications. They are also of interest as simple models of the behaviour of alkyl phosphonate layers on surfaces. The length of the alkyl chain has a singular effect on the properties of these materials. Formation of crystalline, amorphous and intermediate meso phases and the thermal interconversion of these phases are predicated on the alkyl chain. Under the influence of applied external pressures, these phase changes are also observed. Interestingly, in some examples, two different materials co-exist over a pressure range and can reversibly interconvert, implying the presence of two different crystalline polymorphs spanning a significant pressure regime. The results for a series of chain lengths are presented.

310.
CYCLIC INTERMEDIATES ISOLATED FROM THE SYNTHESIS OF ALKYLENE-BRIDGED POLYSILSESQUIOXANES Joseph P. Carpenter, Douglas A. Loy, Kenneth J. Shea, John Greaves, Peter K. Dorhout, Sandia National Laboratories, Albuquerque, NM, 87185-1407, Department of Chemistry, University of California, Irvine, CA 92717-2025, Department of Chemistry, Colorado State University, Fort Collins, CO 80523

The hydrolysis and condensation of α,ω-bis(triethoxysilyl)alkanes with acid catalysts normally leads to polymeric gels within a few hours. However, when the alkylene-bridge is short (ethyl, propyl, or butyl) gelation times of days or even months have been observed. Investigations into the initial condensation reactions revealed that highly condensed bicyclic and tricyclic dimers are major products when the alkylene bridge is short. Some of these dimeric compounds have been isolated as crystalline solids. These intermediates have been characterized by ^{1}H and ^{13}C NMR and x-ray crystallography. This work was supported by the United States Department of Energy under Contract DE-AC04-94AL85000.

311.
MOLECULAR MECHANICS (MM3) CALCULATIONS ON BENZOCROWN ETHER COMPLEXES OF THE ALKALI AND ALKALINE EARTH CATIONS. **Linrong R. Yang** Benjamin P. Hay, Pacific Northwest National Laboratory, WA 99352

Abstract: The new metal-ligand feature of MM3 has been extended to benzocrown ether complexes of alkali and alkaline earth cations. Over 50 complexes were compared with the crystal structures retrieved from Cambridge Crystal Database. The results agree with experimental data. The averages of absolute deviations between experimental and calculated structural features are: metal-oxygen bond length, 0.03 Å; Metal-oxygen-carbon angles, 4.1°; and Metal-oxygen-carbon-carbon angles: 5.1°. Development of structure-function relationships is in progress.

312.
THE DINITRAMIDE, O_2N-N-NO_2^- AND NITROCYANAMIDE, O_2N-N-CN$^-$ ANIONS, Mark A. Petrie, Karl O. Christe, William W. Wilson, Harvey H. Michels, Jeffrey C. Bottaro, Richard Gilardi, Robert Bau, Hughes STX, Phillips Laboratory, Edwards AFB, CA 93535, United Technologies Research Center, East Hartford, CT 06108, SRI International, Menlo Park, CA 94025, Naval Research Laboratory, DC 20375, Loker Hydrocarbon Research Institute, University of Southern California, Los Angeles, CA 90089

The dinitramide, O_2N-N-NO_2^- and nitrocyanamide, O_2N-N-CN$^-$ anions are highly energetic species of great interest for halogen free, high performance, solid propellants. The IR and Raman bands of these species have been measured and assigned with the help of ab initio calculations at the HF/6-31G* and MP2/6-31+G* levels of theory. Crystal structures and vibrational spectra suggest a strong counterion and solvation dependence on the geometry of these anions. The reactive chemistry of these anions will also be presented.

313. **FORMATION OF $[B_9H_{11}Cl_3]^-$ BY INSERTION OF THE ELECTRON PRECISE ANION $[BCl_3H]^-$ INTO THE ELECTRON DEFICIENT BORANE B_8H_{10}.** Ewan J. M. Hamilton, Jianping Liu, Edward A. Meyers, and <u>Sheldon G. Shore</u>, Department of Chemistry, The Ohio State University, Columbus, OH 43210

The reaction between $[Me_4N][B_9H_{14}]$ and BCl_3 produces $B_{10}H_{14}$, H_2, and $[Me_4N][BCl_3H]$. Another major species also forms, accounting for approximately 45% of the total boron in the system. This was previously proposed to possess a polymeric structure, $1/n\{B_8H_{10}\}_n$, supported by the reaction stoichiometry and its lack of volatility. Polymerization of small boranes has long been suggested as a possible route to alleviation of electron deficiency. However, it has now been determined that polymerization of "B_8H_{10}" does not occur in this instance. Instead, insertion of $[BCl_3H]^-$ into the B_8H_{10} moiety results primarily in the $[6,7,8-B_9H_{11}Cl_3]^-$ anion, which has been structurally characterized as the $[Me_4N]^+$ salt. The anion is analogous to $[B_9H_{14}]^-$. An interesting feature is the migration of the chlorine atoms on the B_9 cage. This clearly requires cleavage and reformation of at least 2 B-Cl bonds by a mechanism which has not yet been elucidated.

314. **SYNTHESIS AND CHARACTERIZATION OF AMINO ACID CARBORANES** Ye Wu and <u>William Quintana</u>, Department of Chemistry and Biochemistry, New Mexico State University, Las Cruces, NM 88003-8001.

We have continued to explore synthetic pathways leading to the preparation of carboranes which contain amino acids units within them. Using well established synthetic protocols we have been able to prepare *ortho*-substituted carboranes which include glycine, alanine and phenylalanine units coupled via urea linkages. In addition, we are in the process of developing a synthetic route for the preparation of the compound presented below.

$$Cl^- \ ^+H_3NCH_2C\text{---}C\text{---}CH_2COOH$$
$$B_{10}H_{10}$$

Experimental details, structural and spectroscopic data for this family of compounds will be presented and discussed.

315. **A CONVENIENT METHOD FOR HETERODISUBSTITUTION OF *p*-CARBORANES.** <u>Piotr Kaszynski</u> and Zbynek Janousek, *Organic Materials Research Group*, Department of Chemistry, Vanderbilt University, Nashville, Tennessee 37235.

In our pursuit of liquid crystals based on boron clusters[1] we have developed a convenient method for heterodisubstitution of *p*-carboranes. The differentiation of carborane carbon atoms is achieved by monosilylation with triphenylsilyl chloride. The disilylated product is easily removed and converted to the starting carborane which is recovered along with the unreacted material. The monosilylated carborane and its C-alkylated derivatives are easy to purify solids containing phenyl chromophores.
Here we report the preparation and synthetic application of triphenylsilyl-*p*-carborane.

1. Kaszynski, P.; Huang, J.; Jenkins, G. S.; Bairamov, K. A.; Lipiak, D. *Mol. Cryst. Liq. Cryst.* **1995**, *260*, 315.

316. PHOTOPHYSICAL PROPERTIES OF ARYLATED BORON CLUSTERS. Andrew G. Douglass, Ryan M. Nunley, Erik Brady, Piotr Kaszynski, *Organic Materials Research Group*, Department of Chemistry, Vanderbilt University, Nashville, Tennessee 37235.

We have demonstrated that 1,10-dipyridinedecaborane(10) and its derivatives exhibit strong charge transfer bands at about 320 nm that were assigned to cage-to-ring excitations.[1] Similar CT bands found in aryl derivatives of 10- and 12-vertex *p*-carboranes are conversely attributed to ring-to-cage electronic transitions. Neutral unsymmetrical derivatives of decaborate $B_{10}H_{10}^{-2}$ undergo a change in dipole moment upon excitation, which results in a modest solvatochromic effect.

Here we present absorption and emission spectra and solvatochromism of some aryl derivatives of *closo*-clusters. The photophysical behavior of the compounds is supported by results of *ab initio* calculations.

1. Kaszynski, P.; Huang, J.; Jenkins, G. S.; Bairamov, K. A.; Lipiak, D. *Mol. Cryst. Liq. Cryst.* **1995**, *260*, 315.

317.
FORMATION OF Os^{II}-µ-N_2-$Os^{II,III}$ BRIDGING DINITROGEN DIMERS *via* REDUCTION OF OSMIUM (VI) NITRIDES.

Kostas D. Demadis and Thomas J. Meyer
Department of Chemistry, Venable and Kenan Laboratories CB# 3290, University of North Carolina at Chapel Hill, Chapel Hill, NC 27599-3290.

Transition metal dimers containing bridging dinitrogen ligands occupy an important and well established theme in the inorganic literature. We recently discovered that chemical or electrochemical reduction of the well-defined complex *trans*-[Os^{VI}(N)(tpy)Cl_2](PF$_6$) (tpy=terpyridine) affords the *trans,trans*-[Cl_2(tpy)Os^{II}-µ-N_2-Os^{II}(tpy)Cl_2]^o dimer in excellent yields. This compound undergoes reversible one-electron oxidation to give the mixed-valence *trans*-[Cl_2(tpy)Os^{II}-N≡N-Os^{III}(tpy)Cl_2](X) (X=PF$_6^-$, BF$_4^-$, I$^-$) dimer. The identity of the latter was unequivocally established by X-ray crystallography (Figure). Details regarding the chemical and electrochemical behavior of the above compounds as well as spectroscopic characterization data (including infrared, near infrared and UV-vis) will be presented.

318.
THE ELECTRONIC EFFECTS OF HALIDE METATHESIS IN GROUP 5 IMIDO COMPOUNDS TA(NR)X3L2. **Darryl S. Williams**, Ramez A. Elgammal, M. J. Heeg, Department of Chemistry, Wayne State University, Detroit, MI, 48202

Building on recent observations that d0 group 5 imido compounds of the form M(NR)Cl3(dme) (M = Nb, Ta) exhibit intense, long-lived luminescence in solution at room temperature, we examine the effects of changing the coordination sphere on electronic structure. There are significant and somewhat unpredictable electronic changes as alkyl, aryl, or alkoxide ligands are exchanged for the halide(s). Compounds with varying alkyl and alkoxide ligands have been prepared and changes in their excited state properties and structural data compared to the trihalides will be used to infer changes in electronic structure using absorption, emission, emission quantum yield and emission lifetime measurements. New photochemistry and the usefulness of chemical intuition in predicting these electronic changes will be discussed.

319. REACTIONS OF THIONITRITES AND ALKYL NITRITES WITH METALLOPORPHYRINS OF THE GROUP 8 METALS. George B. Richter-Addo, Geun-Bae Yi, and Li Chen, Department of Chemistry and Biochemistry, University of Oklahoma, Norman, OK 73019

Thionitrites (RSNO, S-nitroso) and alkyl nitrites (RONO, O-nitroso) possess vasodilator action. Although RSNO and RONO compounds may decompose in protic solvents to release NO, there is some evidence that their vasodilator action may not involve such prior dissociation of NO. Evidence is accumulating that RSNO compounds may interact directly with soluble guanylate cyclase. We have investigated the reactions of various RSNO and RONO compounds with metalloporphyrins of the group 8 metals, and the results of our chemical and spectroscopic studies will be presented and discussed.

320. NITROSAMINE COMPLEXES OF OSMIUM PORPHYRINS. Li Chen, and George B. Richter-Addo, Department of Chemistry and Biochemistry, University of Oklahoma, Norman, OK 73019

We have previously demonstrated nitrosamine binding to iron and ruthenium porphyrins. Our current research involves the use of osmium porphyrins for discrete σ-O diethyl nitrosamine binding to the metal center.

321. FORMAL ADDITION OF S- AND O-NITROSO COMPOUNDS TO METALLOPORPHYRINS. Geun-Bae Yi, Li Chen and George B. Richter-Addo, Department of Chemistry and Biochemistry, University of Oklahoma, Norman, OK 73019

Biologically important S- and O- nitroso compounds undergo unprecedented formal trans-additions to the metal center in ruthenium porphyrin complexes to give nitrosyl thiolates and alkoxides, respectively. The products have been isolated and have been fully characterized.

322. **NON-CYCLOPENTADIENYL LIGANDS FOR OLEFIN POLYMERIZATION CATALYSTS.** Damon R. Click, Brian L. Scott, and John G Watkin. CST-18, Los Alamos National Laboratory, Los Alamos, NM 87544, USA.

The use of metallocene-type catalysts for olefin polymerization has been investigated extensively over the last two decades. We have been examining alternative non-cyclopentadienyl ligand sets which may be utilized together with Group IV or lanthanide metal centers to form active polymerization catalysts. Our investigations have focused upon the synthesis of a series of "tied-back" bis-amide ligands bearing electron withdrawing substituents, in order to exert control over the electrophilicity of the metal center. The synthesis of the new ligands, formation of metal complexes containing these ligands, and use of the metal complexes as catalysts for polymerization of alpha-olefins will be described.

323. **BITE ANGLES MAKE THE CATALYST.** Piet W.N.M van Leeuwen, Paul C.J. Kamer and Mirko Kranenburg, J.H. van 't Hoff Research Institute, University of Amsterdam, Nieuwe Achtergracht 166, 1018 WV, Amsterdam, the Netherlands.

We have found that valence angles above 100°, enforced in a bidentate ligand, have an enormous effect on selectivity and reactivity of a range of catalysts by stabilizing tetrahedral or trigonal geometries. The trigonal bipyramidal complex $RhH(CO)_2(P-P)$ is a highly active and selective catalyst for the hydroformylation of alkenes. Square planar intermediates of divalent nickel and palladium catalysts can be destabized with respect to zerovalent tetrahedral intermediates. The newly developed ligands yield fast nickel catalysts for the hydrocyanation reaction. The reaction turned out to be extremely sensitive to the precise value of the bite angle; the same was found for palladium catalysed reactions such as cross-coupling and allylic substitution. Fast and highly selective catalysts have been obtained.

324. **SYNTHESIS OF WATER-SOLUBLE PHOSPHONATE-FUNCTIONALIZED PHOSPHINE LIGANDS** Terence L. Schull[a], Kate R. Richardson[a], James C. Fettinger[b], and D. Andrew Knight*[a], [a]Department of Chemistry, The George Washington University, 725 21st Street, N.W. Washington, DC 20052. [b]Department of Chemistry and Biochemistry, University of Maryland, College Park, MD 20742.

Anionic phosphonate groups can impart a high degree of water solubility to phosphine ligands, which bind exclusively to palladium(II), platinum(II), and rhodium(I) through the phosphine. Water-soluble phosphonate-functionalized triarylphosphines **1a-c** have been prepared, and their physical properties and application as ligands for transition-metal mediated catalysis have been studied. Recent work has focused on the synthesis of phosphonate-functionalized phosphines such as **2**, which have potential as ligands in micellar catalysis.

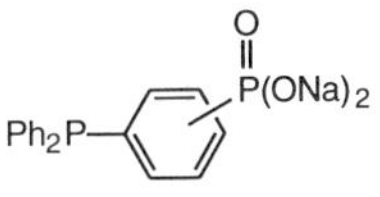

Ph_2P—⟨⟩—$P(O)(ONa)_2$ a = *para*
                              b = *meta*
                              c = *ortho*

1a-c

Ph_2P—(CH$_2$)$_n$—PO_3Na_2

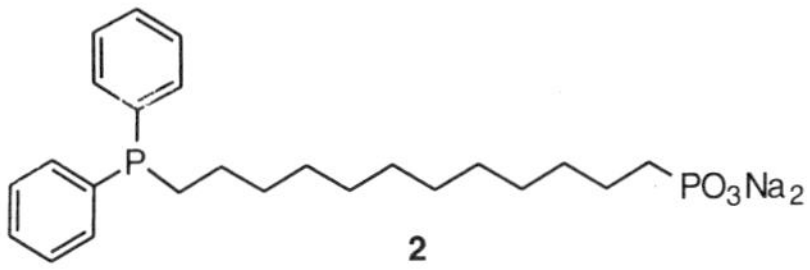

2

325. Asymmetric Catalysis Based on New Chiral Phosphine Ligands.
<u>Xumu Zhang</u>*, Ping Cao, Zhaogen Chen, Qiongzhong Jiang, Yutong Jiang, James Longmire, Michael Terry, Dengming Xiao, and Guoxin Zhu
Department of Chemistry, The Pennsylvania State University, University Park, PA 16802

Asymmetric phosphine ligands have played a significant role in the development of novel transition metal catalyzed asymmetric reactions. Over thousand chiral phosphine ligands have been made. High selectivities have been observed in some asymmetric reactions. However, there are many reactions where current ligandsare not very efficient in terms of activity and selectivity. We have developed several families of new chiral phosphines for asymmetric catalysis. These includes chiral monophosphines, chiral tridentate ligands and novel chiral bidentate phosphines with rigid backbones. Excellent activitities and enantioselectivities (up to 99 % ee) have been observed in several asymmetric reactions such as hydride transfer reaction, hydrogenation and allylic alkylation.

326. ELECTRON WITHDRAWING SUBSTITUENTS ON EQUATORIAL AND APICAL PHOSPHINES HAVE OPPOSITE EFFECTS ON THE REGIOSELECTIVITY OF RHODIUM CATALYZED HYDROFORMYLATION. Charles P. Casey*, <u>Evelyn J. Lin Paulsen</u>, Bernd R. Proft, Eckart W. Beuttenmüller, Lori M. Petrovich, Gregory T. Whiteker. Department of Chemistry, University of Wisconsin-Madison, Madison, WI 53706.

A series of diphosphines containing electron-withdrawing aryl groups have been synthesized and studied in rhodium-catalyzed hydroformylation of 1-hexene. It has been found that in diequatorial chelating diphosphines, such as BISBI, the effect of electron-withdrawing groups is to increase n:i ratio while in apical-equatorial chelating diphosphines, such as DIPHOS, electron-withdrawing groups decrease the n:i ratio. In the case of T-BDCP, which exists as a mixture of diequatorial and apical-equatorial isomers, electron-withdrawing groups moderately increase the n:i ratio.

327. RING-OPENING REACTIONS OF [(η-C$_5$Me$_5$)V(N)Cl]$_2$ WITH LEWIS ACIDS. <u>Claire Newton</u>, Arnis Aistars, Joseph W. Ziller, and Nancy M. Doherty,* *Department of Chemistry, 516 Physical Sciences 1, University of California, Irvine, CA 92697-2025*

Reactions of [(η-C$_5$Me$_5$)V(N)Cl]$_2$ and (η-C$_5$Me$_5$)V(NSiMe$_3$)Cl$_2$ with Lewis acids will be presented. The V$_2$N$_2$ core of the vanadium(V) nitrido dimer has proven very favorable and robust. Substitution of the chloride ligands occurs readily maintaining the cyclic structure, the dimer is unreactive towards Lewis bases, and thermolysis of the silylimido compound produces the dimer in high yield. We observe a facile, clean reaction of [(η-C$_5$Me$_5$)V(N)Cl]$_2$ with a variety of Lewis acids—BCl$_3$, AlCl$_3$, SbCl$_5$, NbCl$_5$, TaCl$_5$—resulting in abstraction of a nitrido ligand and ring opening of the V$_2$N$_2$ core to afford a cationic divanadium(V) mononitrido complex. We will present the details of this chemistry, including the characterization of [(η-C$_5$Me$_5$)$_2$V$_2$(N)Cl$_4$][AlCl$_4$] by a combination of NMR and IR spectroscopy, X-ray crystallography, and reaction studies, and related reactions of (η-C$_5$Me$_5$)V(NSiMe$_3$)Cl$_2$.

328. SILYLAMIDO COMPLEXES OF EARLY TRANSITION METALS. Tomislav I. Gountchev and T. Don Tilley, Dept. of Chemistry, UC Berkeley, Berkeley, CA 94720

The chemistry of d^0 early transition metal complexes with non-cyclopentadienyl ancillary ligands has recently attracted much attention. This talk will present the synthesis, structural characterization, and reactivity studies of some complexes of chelating C_2 - symmetric silylamido ligands. The relatively high reactivity of the N–Si bond leads to a number of interesting transformations involving the ligand. A Ta complex containing an unusually highly bent imido ligand has been isolated and characterized, and the reactivity of the Ta=N double bond investigated. This compound undergoes a sequence of addition / elimination reactions with Si–H bonds in silanes, which have been the subject of kinetic and mechanistic studies.

329. REACTIVITY OF TITANIUM DINITROGEN COMPLEXES. PREPARATION OF COMPOUNDS CONTAINING Ti=E, $Ti_2(\mu-E)_2$, AND $Ti(\eta^2-E_2)$ FUNCTIONALITIES. John R. Hagadorn and John Arnold, Department of Chemistry, University of California, Berkeley, California, 94720

Chemical reduction of Ti(IV) amidinates generates a highly reactive Ti(II) species that can be trapped with N_2 giving the bridging-dinitrogen complex, $(L_2Ti)_2(\mu-N_2)$ (L = PhC(NSiMe$_3$)$_2$). This dark blue compound reacts readily with planar Lewis bases to form 6-coordinate adducts without displacing the N_2 ligand. Reaction with oxygen and sulfur sources reveals a wide range of chemistry which allows for the isolation of novel complexes containing Ti=E, $Ti_2(\mu-E)_2$, and $Ti(\eta^2-E_2)$ (E = O, S) functionalities.

330. ELECTROPHILIC TUNGSTEN (II) METHYLENE CARBENE COMPLEXES: ADDUCT FORMATION, METHYLENE TRANSFER, AND CATALYSIS OF AZIRIDINE FORMATION FROM IMINES AND ETHYL DIAZOACETATE. T. Brent Gunnoe, Luis Casarrubios, Matthew L. Surgan, P. S. White, and Joseph L. Templeton*, Department of Chemistry, University of North Carolina, Chapel Hill, NC 27599-3290.

Treatment of the methyl complex Tp'(CO)(PhC$_2$Me)W-Me (**1**) with trityl cation yields the persistent methylene carbene complex [Tp'(CO)(PhC$_2$Me)W=CH$_2$][PF$_6$] (**2**) (Tp'=hydridotris-(3,5-dimethylpyrazolylborate)). The carbene (**2**) displays electrophilic behavior as evidenced by addition of nucleophiles and methylene transfer to electron rich olefins. Upon binding to the carbene (**2**), aryl imines are activated towards nucleophilic attack by ethyl diazoacetate (EDA). The net result of addition of an excess of an aryl imine followed by an excess of EDA to a solution of **2** is catalytic aziridine formation. Attempts at resolution of the methyl complex (**1**) into single enantiomers and subsequent stereoselective reactions will be discussed.

331. **CARBENE COMPLEXES WITH THE BENZIMIDAZOL-2-YLIDENE LIGAND**
F. Ekkehardt Hahn and Volker Langenhahn, Institut für Anorganische und Analytische Chemie, Freie Universität Berlin, Fabeckstraße 34-36, D-14195 Berlin, Germany and Wolf Peter Fehlhammer, Deutsches Museum, Museumsinsel 1, D-80538 München, Germany

Carbene Complexes with ligands based on imidazol have been studied intensively. We have investigated the preparation of benzimidazol-2-ylidene complexes. 2-Azidophenyl isocyanide reacts with $[Cr(CO)_5THF]$ to yield the isocyanide complex **1**, which upon reaction with triphenylphosphine and subsequent hydrolysis yields the carbene complex **2** The utility of benzimidazol-2-ylidene complexes of type **2** for the template synthesis of multidentate carbene ligands will be discussed.

332. **PLATINUM(II) CATALYSIS OF PLATINUM(IV) COMPLEXES IN REACTIONS WITH NUCLEOBASES** Rosette M. Roat, Department of Chemistry, Washington College, 300 Washington Ave., Chestertown, MD 21620

Reactions of tetrachloroamineplatinum(IV) complexes with nucleobases are speeded up dramatically by additions of small amounts of the analogous dichlorodiamineplatinum(II) complex. This behavior may be explained by the chloride bridging mechanism first elucidated for platinum(II) and (IV) chloroammine complexes by Basolo and co-workers in the 1950s. Since known platinum(II) and (IV) antitumor drugs target nucleobases of DNA to effect cell death in malignant cells, the reactions discussed may shed light on platinum antitumor compound reaction mechanisms. The spectroscopic methods of choice to study these systems are ^1H, ^{195}Pt, and ^{31}P NMR.

333. **PHOTOINDUCED ELECTRON TRANSFER IN ETHIDIUM-MODIFIED DUPLEXES.**
Shana O. Kelley, R. Erik Holmlin, Eric D.A. Stemp, and Jacqueline K. Barton, Division of Chemistry and Chemical Engineering, California Institute of Technology, Pasadena, CA 91125

The ability of the DNA double helix to mediate long-range electron transfer reactions has been systematically studied as a function of distance in a series of small DNA duplexes covalently modified with two intercalators, Et' and Rh(phi)$_2$bpy'$^{3+}$. In these systems, fluorescence quenching attributed to electron transfer occurs at distances up to 30 Å. The quenching observed in these systems is contingent on the presence of the highly ordered DNA π-stack, while insensitive to solution conditions. Temperature-dependent quenching yields illustrate that the π-stack of the double helix electronically couples the acceptor and donor brought together by the complementary strands to which they are tethered. In duplexes containing a highly disruptive single base pair mismatch, large decreases in the quenching efficiency are observed. In fact, observed quenching yields are more sensitive to these stacking perturbations than to donor-acceptor separations; the distance dependence of this reaction is shallow. We propose that electron transfer requires a fully stacked duplex and that the variation in quenching with increasing donor-acceptor separation reflects the greater probability of destacking with increasing duplex length.

334. DESIGN, SYNTHESIS AND CONJUGATION OF LARGE METAL CLUSTER COMPLEXES AS IMMUNOPROBE LABELS. Frederic R. Furuya, James F. Hainfeld, Carol M. R. Halsey and <u>Richard D. Powell</u>, Nanoprobes, Incorporated, 25 East Loop Road, Suite 124, Stony Brook, NY 11790-3350

The large "Nanogold" gold cluster complex has a core of gold atoms 1.4 nm in diameter, and may be visualized in the transmission electron microscope (TEM). Previously, it was covalently conjugated to a variety of antibodies and other targeted biological probes, and used to localize specific biomolecules within cells and tissue specimens for high-resolution electron microscope observation. We now describe the preparation of synthetically modified forms of this cluster in which chemically functionalized tris (aryl) phosphine ligands are used to enable multiple sequential cross-linking reactions, to attach different fluorophores, and to control cluster physical and chemical properties. Spectroscopic and fluorescence data will be presented for the fluorescently conjugated gold clusters. These cluster labels have been cross-linked to antibody Fab' fragments and to streptavidin, and the resulting probes used to detect antigens in blots and localize cell components microscopically.

335. CHARGE TRANSFER THROUGH THE DNA BASE STACK: CHEMISTRY AT A DISTANCE. **R. E. Holmlin**, P. J. Dandliker, D. B. Hall, and J. K. Barton, Division of Chemistry and Chemical Engineering, California Institute of Technology, Pasadena, CA 91125

Double helical DNA consists of a linear array of electronically coupled, π-stacked aromatic heterocycles contained within a polyanionic sugar-phosphate backbone. Does the DNA base stack facilitate efficient electron transfer over long-range? Photoinduced electron transfer between metallointercalators bound to DNA is remarkably fast and appears to be insensitive to distance. In order to test this sensitivity to distance, we developed a multistep, heterogeneous coupling method in which a metallointercalator is tethered to an oligonucleotide attached to a solid support. Rhodium-modified duplexes permit the investigation of long-range guanine oxidation and thymine dimer repair in which DNA not only mediates electron transfer over as much as 34 Å, but serves as reactant as well. By extending this approach, we now prepare oligonucleotides with an electron donor (Os) and an electron acceptor (Rh) attached to the same strand.

336. SYNTHESIS AND CHARACTERIZATION OF ^{99}Tc P246: A TECHNETIUM PEPTIDE. Yongyong Zheng[†], Judit Bartis[†], Michael Blumenstein[†], Catherine Costello[¥], Mark DeRosch[§], John Lister-James[§], <u>Lynn C. Francesconi</u>[†*]. [†]Department of Chemistry, Hunter College of the City University of New York, [¥]Boston University Medical School, Mass Spectrometry Resource, [§]Diatide, Inc., Londonderry, New Hampshire

The peptide, P246, has the sequence shown, where Apc is S-(3-aminopropyl)cysteine, an arginine surrogate and ACM is the acetamidomethyl protecting group. P246 is a monomer of peptide P280; the ^{99m}Tc labeled P280, in clinical trials, provided excellent images of deep vein thrombosis. We have prepared the ^{99}TcO[P246] complex by reacting [^{99}TcOCl$_4$](N(C$_4$H$_9$)$_4$) with the P246 peptide in DMF and water. Purification was accomplished by preparative reverse phase HPLC. Two dimensional NMR spectroscopic data, Infrared data and Mass Spectrometry data are consistent with the [^{99}Tc^V=O]$^{3+}$ unit binding to the three amide nitrogen donor atoms (gly 11, gly 12, cys 13) and the thiolate sulfur (cys 13) at the amide terminus and also show that the ACM groups are retained in the technetium complex. The COSY and TOCSY spectra of the ^{99}TcO[P246] complex show only 10 NH-CH cross peaks for amino acids 1-10. The amide nitrogen atoms of amino acids, gly 11, gly 12, and cys 13 deprotonate on complexation with the [Tc^V=O]$^{3+}$ group.

337. CYCLIC AMINE ADDUCTS OF *META*-DITHIOPHOSPHATE ESTERS AND THEIR POTENTIAL AS CHIRAL BUILDING BLOCKS. Christopher M. Barnes and D. Scott Bohle*, Department of Chemistry, University of Wyoming.

The pyridine adducts of the *meta*-dithiophosphoryl halides have long been known as precursors to a variety of cyclic and acyclic esters and amides of dithiophosphoric acid. We report the preparation of the related homochiral 3-dialkyl-1,3,2-oxazaphospholidine disulfides **1**, $S_2P(OCH(Ph)CHNR_2)$ (R = methyl, benzyl, both (4S,5R) and (4S,5S) diastereomers). These compounds react with alcohols and sulfoxides to yield dithiophosphate diesters and diastereomers of P-chiral OSP(OCH(Ph)CHNR$_2$, **2**, respectively. We evaluate the potential for enantioselective synthesis of chiral dithiophosphate diesters $HS_2P(OR^*)_2$ and cyclic chiral thiophosphate esters $HSOP^*(OR^1)(OR^2)$ from these novel phosphorus reagents.

338. ZIRCONOCENE PHOSPHIDO- AND PHOSPHINIDENE COMPLEXES.
E. Urnezius, J. D. Protasiewicz, Department of Chemistry, Case Western Reserve University, Cleveland, Ohio 44106-7078

Deprotonation of the new hindered primary phosphine 2, 6-dimesitylphenylphosphine (DmpPH$_2$) leads to Li(H)PDmp. Reaction of Li(H)PDmp with [Cp$_2$ZrCl$_2$] leads to [Cp$_2$ZrCl(HPDmp)] (**1**) and [Cp$_2$Zr(HPDmp)$_2$] (**2**) depending on specific reaction conditions. Reaction of Li(H)PDmp with [Cp$_2$ZrMeCl] in the presence of PMe$_3$ affords the phosphinidene complex [Cp$_2$Zr(=PDmp)(PMe$_3$)] (**3**). Compounds **1-3** have been structurally characterized and their chemistry will be described.

339. STERICALLY ENCUMBERED META-TERPHENYL DIPHOSPHENES.
S. Shah and J. D. Protasiewicz, Department of Chemistry, Case Western Reserve University, Cleveland, Ohio 44106-7078

The hindered dichlorophosphines DmpPCl$_2$ and DmtPCl$_2$ (Dmp = 2, 6-mesitylphenyl, Dmt = 2, 6-mesityl-4-methylphenyl) react with magnesium to afford coupling of the phosphinidenes and form DmpP=PDmp (**1**) and DmtP=PDmt (**2**), respectively. These orange materials undergo facile one electron reduction to stable purple radical anions [ArPPAr]$^-$, which have been characterized by EPR spectroscopy. The reduction reactions are reversible in THF as ascertained by electrochemical experiments. Direct reductive cleavage of the P=P bonds in **1** and **2** has been achieved.

340.
A SHORT P(III)-P(V)-P(III) TRIPHOSPHORUS CHAIN IN A NOVEL ENCAPSULATED HEXACOORDINATED PHOSPHORUS (V) CATION.
Ronald G. Cavell, Hongyan Luo and Robert McDonald
Dept. of Chemistry, University of Alberta, Edmonton, Alberta, Canada, T6G 2G2

Reaction of *bis*(o-(trimethylsilyloxy)phenyl)phenylphosphine, (PhP{OT}$_2$), with PCl$_5$ gave the disubstituted cationic triphosphorus chloride salt (eqn). The crystal structure revealed two facially bound P{OO}$^-$ tridentate ligands defining an elongated octahedral coordination at the P(V) center. Four phenolate oxygen atoms form the equatorial plane (with slightly long bonds (1.762 Å) but normal angles of 89.5(2)°). The two P(III) donor atoms are trans axially bound with the shortest yet recorded P(III)-P(V) bonds (2.2023 (10) Å). This compound simultaneously represents a rare example of strong mixed valence triphosphorus binding and also a hexacoordinate cationic phosphorus center.

Acknowledgement: We thank NSERC of Canada for financial support.

341. **SYNTHESIS AND CHARACTERIZATION OF NEW TRIFLUOROVINYL ETHER DERIVATIVES OF PHOSPHORUS AND SILICON.** Robert H. Neilson, Junmin Ji, Sridevi Narayan-Sarathy, James Oxley. Department of Chemistry, Texas Christian University, Fort Worth, TX 76129; and Dennis W. Smith, Jr., Central Research and Development, The Dow Chemical Company, Freeport, TX 77541.

A wide variety of phosphorus and silicon halides react smoothly with either the Grignard or organolithium derivative of *p*-bromophenyltrifluorovinyl ether to afford new mono- and/or disubstituted products.

M = MgBr, Li

n = 1; E = -SiMe$_2$Cl, -SiMe$_2$CH=CH$_2$, -P(R)N(SiMe$_3$)$_2$, etc.

n = 2; E = SiMe$_2$, SiMe$_2$(CH$_2$)$_2$SiMe$_2$, -P(R)- [R = Ph, *t*-Bu], etc.

The preparative details of these reactions as well as the structural characterization (by multi-nuclear NMR spectroscopy) of several representative products will be presented.

342.
FUNCTIONAL ANALOGUES OF THE OXYGEN BINDING/ACTIVATING HEME PROTEINS. James P. Collman, Department of Chemistry, Stanford University, Stanford, California, 94305-5080

Metal porphyrins which are capped with macrocyclic superstructures have been prepared. A series of iron(II) derivatives have been studied in the context of equilibrium binding of CO and O_2 (M values). The relative CO vs O_2 affinities are dramatically affected by the inner cavity superstructure. Another series having iron(II) or Co(II) in the porphyrin and copper(I) in the macrocycle have been studied as functional analogues of the O_2 binding site in cytochrome *c* oxidase. One compound catalyzes the catalytic 4e$^-$ reduction of O_2 at physiological pH (7.3).

343. METALLOCENE CATALYSTS FOR ZIEGLER-NATTA POLYMERIZATION
John E. Bercaw.
Arnold and Mabel Beckman Laboratories of Chemical Synthesis, California Institute of Technology, Pasadena, CA 91125

The evolution of catalysts for stereoselective polymerizations of α-olefins will be traced. Early single component catalysts consisting of permethylscandocene alkyls permitted studies of chain propagation and transfer by β H elimination, but they are capable of polymerizing only ethylene. Second generation *ansa*-scandocenes produce dimers from α-olefins, and proved well-suited for investigations of the bonding interactions in the transition structure for chain propagation. Third generation scandium alkyls bearing linked cyclopentadienyl and amide ligands produce low molecular weight, but atactic poly-α-olefins. More recently, chiral yttrocene catalysts have been prepared that yield very highly isotactic polypropylene. Enantiomerically pure C_2-symmetric yttrocenes have been used to establish the diastereomeric transition structures for propylene enchainment. The principles garnered from these studies have been used to develop new types of doubly-linked, *ansa*-zirconocene catalysts for the production of highly syndiotactic polypropylene.

344. DEVELOPMENTS IN THE SYNTHETIC CHEMISTRY OF TRANSITION METALS, Malcolm L.H. Green, Inorganic Chemistry Laboratory, South Parks Road, Oxford, OX1 3QR, UK

The following topics will be discussed.

(i) The changes of reactivities consequent upon the introduction of a single atom carbon bridge in the chemistry of bent-metallocenes.

(ii) New and unexpected chemistry of the strong Lewis acid $B(C_6F_5)_3$ with transition metal-alkyl, hydrido and oxo compounds.

(iii) The synthesis of nanotube-transition metal systems, including metal containing single wall nanotubes.

345. APPLICATIONS OF HYDROGEN BONDING IN METAL COMPLEXES
Robert H. Crabtree, Yale Chemistry Dept., 225 Prospect St., New Haven CT 06520-8107.

Hydrogen bonds of a new type, Re-H...H-N, have been discovered in which protonic and hydridic hydrogens attract one another. The bond energy (4-6 kcal/mol) is substantial and the H...H distance is short (1.8Å). Using ligands with H-bond functionality allows N-H...π H-bonds to be seen for some nitrile complexes. The H-bonding also accelerates nitrile hydrolysis. H-bonding also allows isolation of an HF complex of type Ir-F-H...N, where J(H,F) = 440 Hz by NMR.

346.
ZWITTERIONIC ZIRCONOCENES. Warren E. Piers, Yimin Sun, Rupert E. v H. Spence, Masood Parvez, Steven J. Rettig and Glenn P. A. Yap. Department of Chemistry, University of Calgary, 2500 University Dr. NW, Calgary, Alberta T2N 1N4.

The reactions of the highly electrophilic bis-(pentafluorophenyl)borane and tris-(pentafluorophenyl)borane with two classes of zirconocene compounds are described. Phosphine stabilized olefin complexes react with these two boranes to yield highly reactive zwitterionic compounds with a borate counterion attached to the end of the alkyl ligand of a cationic zirconium center. Reaction of these boranes with group 4 metal members of the long known "tuck-in" family of metallocenes result in a different group of zwitterionic metallocenes which incorporate a borate counterion attached to one pentamethylcyclopentadienyl ligand. The structures, reactivity and olefin chemistry of several of these zwitterionic species will be discussed.

347. DIRECT CATALYTIC FUNCTIONALIZATIONS OF METHANE AND ETHANE BY BREAKING CARBON-CARBON AND CARBON-HYDROGEN BONDS. Ayusman Sen, Department of Chemistry, The Pennsylvania State University, University Park, Pennsylvania 16802, U. S. A.

Methane and ethane are the least reactive and most abundant members of the hydrocarbon family with known reserves approaching that of petroleum. Thus, the selective oxidative functionalization of these alkanes to more useful chemical products is of great practical interest. This presentation reports several highly catalytic systems for the low temperature hydroxylation and hydroxycarbonylation of C-H and C-C bonds of these alkanes. Using our catalytic systems it is possible to selectively hydroxylate and/or hydroxycarbonylate the alkanes in the presence of water, carbon monoxide, and dioxygen. Whether hydroxylation or hydroxycarbonylation predominates depends on the nature of the solvent. Most significantly under certain reaction conditions, C-C bond activation competes with C-H bond activation. The scope and the mechanism of these reactions will be discussed.

348. THE EFFECT OF ANCILLARY LIGANDS ON THE PREFERENCE FOR BRIDGING VERSUS TERMINAL NITRIDO LIGANDS IN EARLY TRANSITION-METAL COMPOUNDS. Nancy M. Doherty, *Department of Chemistry, 516 Physical Sciences 1, University of California, Irvine, CA 92697-2025*

Vanadium(V) nitrido compounds with varying ancillary ligands have been prepared. Fluoroalkoxides favor terminal nitrido compounds, aryloxides produce labile nitrido-bridged dimers, and cyclopentadienides yield stable cyclic nitrido dimers. The patterns in structure, bonding, and reactivity will be discussed with emphasis on the effect of competing π-donor ligands on metal-nitrido bonding. Extension of this model to the preparation of nitrido-bridged compounds of other group 5 and group 6 metals will be described. The influence of the reagent used to deliver the nitrido ligand—azides versus amines versus amides, silylated versus protonated—on the course of the reactions will also be discussed.

349. THE EFFECT OF CHELATE RING SIZE ON THE HETEROLYTIC CLEAVAGE OF DIHYDROGEN BY PALLADIUM TRIPHOSPHINE COMPLEXES. Calvin J. Curtis, Daniel L. DuBois, Alex Miedaner and Sheryl A. Wander, National Renewable Energy Laboratory, 1617 Cole Blvd., Golden, CO 80401

Pd(ttpE)(CH$_3$CN)$^{2+}$ (**1**) reacts with dihydrogen in DMF or acetone to give an equilibrium mixture of the hydride **2** and H$^+$. The analogous complex with two-carbon bridges connecting the phosphorus atoms shows no reaction under identical conditions. Kinetic studies show that the reaction is first-order in both Pd complex and H$_2$, and inhibited by ligands that bind more tightly than CH$_3$CN, suggesting that H$_2$ reacts directly with a three-coordinate Pd intermediate. The activation of H$_2$ by **1** can be attributed to both steric and electronic effects produced by the larger chelate bite in **1** compared to the analog with two-carbon linkages in the ligand backbone. Similar results using Ni(II) complexes with tetradentate phosphine ligands will also be presented.

$$[\text{Ph P}(PEt_2)_2\text{Pd-NCCH}_3]^{2+} + H_2 \rightleftharpoons [\text{Ph P}(PEt_2)_2\text{Pd-H}]^+ + H^+ + CH_3CN$$

          1   **2**

350. INORGANIC/ORGANIC THIN FILMS AND THEIR USE IN PHOTOVOLTAIC DEVICES. <u>Mark E. Thompson,</u> Elena Dana, Jonathan Snover, Bryan Koene, Ousama Abbas, Department of Chemistry, University of Southern California, Los Angeles, CA 90089

Stable photo-induced charge separation has been observed in layered zirconium viologen-phosphonate compounds, *i.e.* $M_2(O_3PCH_2CH_2(bipyridinium)CH_2CH_2PO_3)X_6$, (M = Zr, Hf, Sn; X = Cl, Br, I). In order to utilize this charge separated state we have investigated the growth of these and related materials as thin films. Films were prepared with multilayers of donor substituted and acceptor substituted zirconium phosphonates on a variety of substrates (*e.g.* Au, Si, ITO coated glass, *etc.*). When irradiated these multilayer films produce a photocurrent with a maximum current density of 5 µA/cm^2. The direction of current flow is controlled by the order the films are deposited in. The quantum efficiency for absorbed light is estimated to be ~ 5%. We will discuss our most recent results aimed at increasing the quality of the thin films, the photovoltaic conversion efficiency and shifting the active wavelengths into the visible part of the spectrum.

351.

NOX UPTAKE AND DECOMPOSITION ON RAPID HEATING BY HETEROPOLYTUNGSTATES, A MECHANISTIC STUDY. **Andrew M. Herring,** Robert L. McCormick, Sukritthira Boonerung, and Michael S. Graboski, Colorado Institute for Fuels and High Altitude Engine Research, Colorado School of Mines, Golden, CO, 80401.

For diesel fueled and other lean-burn engines there is currently no practical strategy for elimination of NOx from the exhaust. While commercial strategies do exist for stationary sources, they are expensive and inefficient. Recently Phosphotungstic acid has been shown to have activity for NOx absorption and decomposition in the presence of oxygen (Moffat, J. B. et al. J.Catal. 1995, 152, 179. Yang, R.T. et al. J.Catal. 1995, 157, 76). We have examined the chemistry of NO absorption in this material and extended these studies to other heteropolytungstic acids. We find that heterotungstic acids based on the Keggin structure will aborb as many NO molecules as they have acidic protons as counterions. The mechanism of NOx binding and reactivity based on our extensive infra-red, solid state NMR and x-ray studies will be discussed.

352.

THE CONTROL OF POLYMER PROPERTIES THROUGH RATIONAL CATALYST DESIGN
<u>Eric J. Moore,</u> Andreas B. Ernst. Amoco Chemical Company, 150 Warrenville Rd., Naperville, IL 60566

Non-bridged bis-(2-arylindenyl) metallocenes have been reported to afford elastomeric polypropylene when activated by methylaluminoxane. To explain the elastic properties of this material, it has been suggested that the metallocene oscillates between isospecfic and aspecific states on the time scale of chain propagation, producing a stereoblock polymer in which isotactic and atactic polypropylene blocks are linked together in the same polymer chain. By proper choice of ligand substituents and process conditions, a variety of polymer microstructures can be achieved. These microstructures, in turn, control the properties of the resulting polymers, which range from highly elastic to plastomeric. A model which incorporates aryl-stacking interactions of the ligands and ligand substituents explains many of the experimental results obtained to date. This model will be discussed in light of experimental observations and the use of molecular modeling to predict polymer microstructure and, hence, polymer properties will be described.

353. **HYDROCARBON OXIDATION BY MANGANESE μ-OXO DIMERS.** Kun Wang and James M. Mayer, Department of Chemistry, Box 351700, University of Washington, Seattle, WA 98195-1700

[L$_2$Mn(μ-O)$_2$MnL$_2$](PF$_6$)$_3$ (**1**, L = 1,10-phenanthroline) reacts with dihydroanthracene (DHA) in acetonitrile solution to give anthracene and small amounts of anthrone and anthraquinone. Oxidation of a mixture of DHA and d_{12}-DHA reveals an isotope effect of 4.2 at 55 °C. Reaction of **1** with fluorene forms bifluorenyl and 9-fluorenone, and in the presence of CBrCl$_3$ 9-bromofluorene is also produced. These data indicate the presence of radical intermediates, formed by hydrogen atom abstraction from the weak C–H bonds of the substrates. Kinetic data for the reaction of DHA with **1** have been modeled with a mechanism in which **1** abstracts H• from DHA to form 9-hydroanthracenyl radical (HA•) and [L$_2$Mn(O)(OH)MnL$_2$]$^{3+}$ (**2**), which can also abstract H• from DHA to form [L$_2$Mn(μ-OH)$_2$MnL$_2$]$^{3+}$ (**3**). HA• is rapidly oxidized by **1** and **2** to give anthracene. The ability of **1** and **2** to abstract hydrogen atoms is as predicted based on the O–H bond strengths in **2** and **3**. The possible relevance of these results to heterogeneous oxidations and metalloenzyme reactions will be discussed.

354. STYRENE INSERTION INTO THE ZR-H BOND OF ZIRCONOCENE HYDRIDE SULPHONATES Gerrit A. Luinstra, and Nicola S. Hüsgen, University of Konstanz, Postfach 5560 M738, D-78434 Konstanz, Germany

Reaction of Cp$_2$ZrH(OSO$_2$-Ph-X) (X = 4-Me, H, 4-Cl, 3-CF$_3$, 3,4-Cl$_2$) and styrene gives a kinetically controlled mixture of Cp$_2$ZrCH$_2$CH$_2$Ph- and Cp$_2$ZrCHMePh(OSO$_2$-Ph-X). Details of the transition state of the insertion of styrene into the Zr-H bond were obtained from a kinetic study of the equilibration of the kinetic to the thermodynamic mixture.

An explanation for the disparate selectivities for 1,2- vs. 2,1-insertion to and rates of β-hydrogen elimination from both zirconocene alkyl isomers of the various aryl sulphonate derivatives will be offered.

355. New Ligand Designs for the Early Transition Elements, Michael D. Fryzuk*, Jason B. Love and Steven J. Rettig, Department of Chemistry, University of British Columbia, 2036 Main Mall, Vancouver, B. C., CANADA, V6T 1Z1

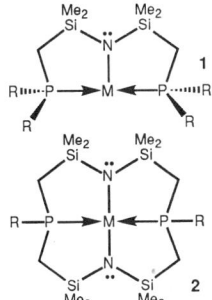

In an effort to change the reactivity patterns of certain transition metal complexes we have been examining the effect of combining different donor types into chelating and macrocyclic assemblies. The tridentate amido-diphosphine ligand **1** has been shown to bind to both the early and the late transition elements. In particular, we have been interested in the ability of group 4 complexes that incorporate this ligand to activate and functionalize dinitrogen. To prevent phosphine dissociation, we have designed and synthesized the macrocyclic assembly **2** that contains two amido donors and two phosphine ligands in the 12-membered P$_2$N$_2$ ring. This P$_2$N$_2$ donor set coordinates to Zr(IV) to generate the mononuclear complex ZrCl$_2$[P$_2$N$_2$] ([P$_2$N$_2$] = [PhP(CH$_2$SiMe$_2$NSiMe$_2$-CH$_2$)$_2$PPh]) which can be reduced under N$_2$ by potassium graphite to produce the dinuclear dinitrogen complex, {[P$_2$N$_2$]Zr}$_2$(μ-η^2-N$_2$). This complex is the first ever dinitrogen derivative that undergoes a reaction with H$_2$ and certain silanes without release of the N$_2$ moiety. An intermediate with a bridging side-on dihydrogen unit has also been isolated.

356. SPECTROSCOPIC INVESTIGATIONS OF PROTEIN-FOLDING DYNAMICS. F. A. Tezcan, J. R. Telford, D. W. Low, G. A. Mines, J. R. Winkler, and H. B. Gray, Beckman Institute, California Institute of Technology, Pasadena, CA 91125

Understanding how the secondary and tertiary structures of proteins are formed from nonnative conformations is a continuing challenge for theory and experiment. A new method for initiating protein folding by electron-transfer (ET) chemistry has been developed that permits measurements in the nanosecond-to-millisecond time range. Thermodynamic analyses indicate that many redox-active proteins will be amenable to this approach. We have used laser-flash initiated ET reactions and time-resolved spectroscopy to probe the kinetics of folding a high-potential heme protein, ferrocytochrome c, in the presence of guanidine hydrochloride (GuHCl). Multiple kinetic phases have been observed in the protein from horse heart. Transient spectra suggest that heme-axial-ligand substitution can occur in less than 10 μs; subsequent dynamics on the 100-μs and 10-ms time scales involve subtle changes in heme spectra that are consistent with protein folding.

357. SPECTROSCOPIC STUDIES OF [3Fe-4S] ←→ [4Fe-4S] INTERCONVERSION IN *AZOTOBACTER VINELANDII* FERREDOXIN I. Philip J. Stephens, Department of Chemistry, University of Southern California, Los Angeles, CA 90089-0482.

Electronic Absorption, Circular Dichroism and Magnetic Circular Dichroism, Proton NMR and Mossbauer Spectroscopies have been used to illuminate [3Fe-4S] ←→ [4Fe-4S] cluster interconversion in the electron transfer protein *Azotobacter vinelandii* Ferredoxin I and its mutants.

358. RESONANCE RAMAN SPECTROSCOPY OF BLUE COPPER AND IRON-SULFUR PROTEINS. R. S. Czernuszewicz, Department of Chemistry, University of Houston, Houston, TX 77204

Proteins containing iron bound to inorganic sulfide and/or cysteine residues (Fe-S proteins) or copper bound to a single cysteine and two histidine residues (blue Cu proteins) are the most prevalent metalloprotein electron transfer agents in nature. Resonance Raman (RR) spectroscopy, a technique that selectively enhances vibrations of the M–S(Cys) chromophore, holds great promise for characterizing iron- and copper-cysteinate sites. Several structural motifs have been established by X-ray crystallography for iron-sulfur proteins, and the Fe-S RR patterns have been found unique and distinctive for the different structural types. However, the reported RR spectra of Fe-S proteins in the same coordination state show marked differences among the 250–450 cm^{-1} Fe–S(S^{2-}) and Fe–S(Cys) stretching vibrations, which undoubtedly are due to protein-imposed distortions of the Fe-S clusters. The differences in the Cu–S(Cys) interactions are clearly evident in the RR spectra of blue Cu proteins as well. Moreover, the RR spectra of these relatively simple chromophores are unusually complicated, showing three or more prominent bands of variable frequencies and intensities around 400 cm^{-1}. Inasmuch as the physical mechanism for electron transfer in proteins is believed to be a vibration-induced electron tunneling, we have undertaken a systematic RR study of a series metal-cysteinate sites aimed at elucidating the exact origin of RR bands being monitored. A review of recent results will be presented.

359. **BIMETALLIC ACTIVATION IN CO DEHYDROGENASE; EVIDENCE FROM METAL-LIGAND RESONANCE RAMAN BANDS.** D. Qiu,[a] M. Kumar,[b] S. W. Ragsdale,[b] and T. G. Spiro,[a] [a]Department of Chemistry, Princeton University, Princeton, NJ 08544, [b]Department of Biochemistry, Beadle Center, University of Nebraska, Lincoln, NE 68588-0664.

Resonance Raman spectra of CO dehydrogenase contain bands associated with the Ni-FeS clusters, and also bands arising from exogenous ligands, including CO, CN and coenzyme-A. These bands have been identified via isotopic substitution. The use of metal isotopes permits identification of the site of ligand binding. CO binds to Fe, not Ni, en route to incorporation into the acetyl group, while the attacking methyl group binds to Ni. Cyanide, an inhibitor of CO oxidation, binds simultaneously to Fe and Ni at Center C, in a manner that suggests that CO again binds to Fe, and is attacked by Ni-bound OH$^-$ during oxidation. Thus, both oxidative and reductive processes appear to proceed by similar bimetallic mechanisms. It is proposed that CO binds to Fe in both Center A and Center C and is alternatively attacked by an Ni-bound nucleophile [oxidation] or electrophile [reduction].

360. EXAFS STUDIES OF THE Cu_A CENTERS OF HEME-COPPER OXIDASES: A UNIQUE METAL-METAL BONDED COPPER CLUSTER. N. J. Blackburn[*], J. A. Fee[†], S. de Vries[‡], Y. Lu[§], C. Dennison[¶], and G. W. Canters[¶]. [*]Oregon Graduate Institute of Science and Technology, [†]University of California, San Diego, [‡]Delft University of Technology, [§]University of Illinois, Urbana-Champaign, [¶]Leiden Institute of Chemistry.

The Cu_A centers of cytochrome c oxidases are unique examples of a new type of binuclear copper cluster. X-Ray crystallography of enzymes from beef heart, *Paracoccus*, and the engineered cyoA fragment of the quinol oxidase of *E. coli* have provided a structural description of the site. The coppers are bridged by two cysteine ligands and have an extremely short Cu–Cu distance of ~2.4 Å. X-Ray absorption spectroscopy, which had previously predicted the short Cu–Cu distance, has been used to further refine the structural details of the site in both the oxidized and reduced forms. Both the metrical parameters and the temperature dependence of the Debye-Waller factors exhibit subtle differences between the oxidized and reduced proteins which suggest that the short distance may be the result, in part, of a weak metal–metal bond. These studies have been extended to include Cu_A derivatives of the blue proteins azurin and amicyanin produced by "loop-directed mutagenesis" in which the Cu_A-binding sequence has been introduced into the blue copper proteins.

361. THE ELECTRONIC STRUCTURE OF Cu_A: A NOVEL MIXED-VALENCE DINUCLEAR COPPER ELECTRON-TRANSFER CENTRE

A.J. Thomson, J.A. Farrar, P. Lappalainen, P.M.H. Kroneck, M. Saraste, W.G. Zumft. Centre for Metalloprotein, Spectroscopy and Biology, School of Chemical Sciences, University of East Anglia, Norwich, UK.

Cu_A, an electron transfer centre present in cytochrome c oxidase, COX, and nitrous oxide reductase, N_2OR, is a dimeric copper complex with two cysteine thiolates bridging the metal ions and two terminal histidine residues. The centre cycles between the mixed-valence state [Cu(I),Cu(II)] and the reduced state, [Cu(I),Cu(I)]. The EPR, optical absorption, low temperature magnetic circular dichroism and CD spectra of three proteins containing the mixed-valence state of Cu_A, one in the enzyme N_2OR and in COX and the other, referred to as Cu_A,* in a site engineered into a soluble domain of subunit II of the quinol oxidase in *Escherichia coli* show that Cu_A in COX and N_2OR have identical electronic structures with complete valence delocalisation whereas Cu_A* shows partial trapping of the valences. Cu_A is a highly covalent planar rhomb which provides an effective path, with low reorganisation energy, for electron transfer from cytochrome c to the cytochromes of COX.

362. METAL-ORGANIC CHEMICAL VAPOR DEPOSITION ROUTES TO METAL OXIDE FILMS FOR SUPERCONDUCTING ELECTRONICS AND OTHER APPLICATIONS. Tobin J. Marks, Department of Chemistry, the Materials Research Center, and the Science and Technology Center for Superconductivity, Northwestern University, Evanston, IL 60208-3113.

If volatile molecular precursors and suitable gas-phase deposition chemistry can be devised, metal-organic chemical vapor deposition (MOCVD) offers an attractive approach to the growth of high-T_c superconducting (HTS) thin films and those of other oxides for useful devices. Potential advantages of this film growth method *vis-à-vis* physical deposition techniques include greatly simplified apparatus, high deposition rates, amenability to coating complex shapes, tailorable film uniformity, adaptability to highly oxidizing deposition conditions, and the possibility of synthesizing metastable phases. This lecture describes the design and synthesis of volatile coordination compounds for the formation of thin films of HTS and related materials by MOCVD. Combined with the appropriate methodologies for deposition and subsequent processing, highly oriented, phase-pure thin films of the $YBa_2Cu_3O_x$, $(BiO)_2Sr_2Ca_{n-1}Cu_nO_{2n+2}$, and $(TlO)_mBa_2Ca_{n-1}O_{2n+2}$ families of superconductors can be readily prepared. Equally important to HTS device fabrication is the ability to grow low dielectric, HTS lattice- and thermal expansion-matched insulators to serve as substrates, seed layers, buffers, and interlayers. The MOCVD growth of perovskite $YAlO_3$, $PrGaO_3$, Sr_2AlTaO_6, and $LaSrGaO_4$ layers as well as multilayers with HTS materials is discussed. Similar strategies can be employed for the growth of ferroelectric and photonic oxide thin films.

363. A MOLECULAR PRECURSOR ROUTE TO OXIDE NETWORKS AND NANOCOMPOSITE MATERIALS. AN ALTERNATIVE TO THE SOL-GEL METHOD. T. Don Tilley, Claus G. Lugmair, Kai Su, and Karl W. Terry, Department of Chemistry, University of California, Berkeley, Berkeley, California 94720-1460, and the Chemical Sciences Division, Lawrence Berkeley Laboratory, 1 Cyclotron Road, Berkeley, California 94720

Advanced solid-state materials with useful properties increasingly involve intricate 3-dimensional networks, characterized by complex stoichiometries (e.g., in ceramic superconductors such as $HgBa_2Ca_2Cu_3O_{8+\delta}$) and/or metastable architectures (e.g., in zeolites). New generations of materials will undoubtedly result from chemically directed, low-temperature synthetic routes. Our approach involves use of synthesis, coordination chemistry, and condensation reactions for construction of novel 3-dimensional networks. Primary targets have been oxide-based materials, which are derived from tailored, oxygen-rich precursor molecules. Initial directions have been based on metal complexes of the siloxide ligand $OSi(O^tBu)_3$ and the phosphate ligand $O_2P(O^tBu)_2$, which eliminate isobutylene and water cleanly at remarkably low temperatures (100-200°C) to form $M_xSi_yO_z$ or $M_xP_yO_z$ materials. Attempts are made to take advantage of the chemistry of network formation and the homogeneity of the resulting systems to control the course of phase transformations at higher temperatures.

364. *ORDERED ARRAYS OF TRANSITION METALS WITH POLYNITRILE DONORS: STRUCTURAL, ELECTRONIC AND MAGNETIC PROPERTIES*. Kim R. Dunbar, Gary M. Finniss, Robert V. Heintz and Hanhua Zhao, Department of Chemistry, Michigan State University, East Lansing, Michigan 48824

Chain, layered and three-dimensional assemblies of transition metals bridged by redox-active organonitrile molecules are being synthesized from soluble molecular species. Central to these studies is the use of acetonitrile metal complexes that undergo substitution reactions with organic molecules containing :N≡C- or :C≡N- functionalities. Among the classes of macromolecular compounds being investigated are covalent donor-acceptor compounds with transition metals sigma-bonded to polynitrile acceptors such as TCNQ and related molecules. As opposed to charge-transfer salts that involve the stacking of planar radicals, direct attachment of the organic and metal components is favored to yield a co-assembly of metal and organic radicals. Characterization tools include X-ray techniques as well as infrared spectroscopy, magnetic susceptibility, microscopy and conductivity (SEM, TEM). The effects of various media (*e.g.*, alcohols, nitriles, water) on the properties of the products will be discussed in terms of the different structures formed in these solvents.

"Innocent" nitriles such as MeCN are also being used to design new ionic materials with transition metals. We have discovered that cationic Rh atoms form an unprecedented solvated "wire" with only MeCN ligands. Prospects for generalizing the chemistry of this 1-D system are excellent, as there is considerable potential for "fine-tuning" by varying the R group on the nitrile and the identity of the anion, thereby altering properties such as conductivity and magnetic interactions.

365. USING DIFFUSION LENGTH AS A SYNTHETIC TOOL TO CONTROL STRUCTURE AND COMPOSITION OF CRYSTALLINE MATERIALS. Myungkeun Noh, Marc Hornbostel, Christopher D. Johnson, David C. Johnson, Materials Science Institute & Department of Chemistry, University of Oregon, Eugene, Oregon 97403

Slow solid state diffusion rates have historically been viewed by solid state chemists as a problem which needs to be overcome. By using elementally modulated reactants in which composition can be controlled on an Ångstrom lengthscale, slow solid state diffusion rates become a synthetic advantage. Controlled crystallization of elementally modulated reactants was used in the rational synthesis of a series of kinetically stable, crystalline skutterudites. Nucleation of the skutterudite structure from an intimately mixed amorphous intermediate occurs with a large exotherm on annealing at temperatures above 200 °C regardless of the ternary metal. All of the new skutterudite compounds were found to decompose exothermically on higher temperature annealing showing that they are thermodynamically unstable with respect to disproportionation. We will also discuss the synthesis of a series of crystalline superlattice compounds containing an integral number of intergrown transition metal-dichalcogenide layers. These compounds $[TaSe_2]_m[TiSe_2]_n$ have been prepared through controlled crystallization of Ti/Se/Ta/Se elementally modulated reactants. The structure of the initial modulated elemental reactant controls the number of unit cells (m and n) of each component compound in the repeating unit. Annealing at elevated temperatures results in the interdiffusion of the superlattice structure leading to the formation of the random solid solution. This rational synthesis approach of this class of superlattice compounds permits the tailoring of physical properties as a function of compositional layer thicknesses and native properties of the parent compounds.

366. LESSONS FROM THE IMMUNE SYSTEM: FROM CATALYSIS TO MATERIALS. P.G. Schultz, Department of Chemistry, University of California, Berkeley, California 94720

Living organisms are remarkable in their ability to generate complex structures with functions ranging from gene regulation and the immune response to photosynthesis and catalysis. The machinery of the cell when combined with the tools and principles of chemistry, can be used to create molecules and assemblies of molecules with properties not yet found in nature. A number of such examples of the synergistic use of biology and chemistry will be discussed ranging from the generation of selective catalysts and drug discovery to materials discovery and the synthesis of nanostructures.

367. SYNTHESIS AND CHARACTERIZATION OF GROUP IV-IV ALLOY HETERO-STRUCTURES USING MOLECULAR CHEMISTRY. J. Kouvetakis, Department of Chemistry, Arizona State University, Tempe Arizona 85287

Growth of new heteroepitaxial binary and ternary alloy semiconductors and ordered phases based on group IV (C,Si,Ge) via molecular chemistry and ultrahigh vacuum CVD will be presented. Examples include random alloys in the $Si_{1-x-y}Ge_xC_y$, $Ge_{1-x}C_x$ and $Si_{1-x}C_x$ systems, as well as ordered structures in the Si-C and Ge-C systems. Metastable SiGeC alloys offer the prospect of strain compensation in the pseudomorphic SiGe and provide the possibility of changes in the bandgap to values greater than that of Si. We demonstrate the use of a C-H free carbon source,$C(SiH_3)_4$, to grow epitaxial alloys containing 4-6 at.% C, the highest C content incorporated in crystalline SiGeC. Non stoichiometric Ge-C combinations are also candidates for bandgap engineering as well as lattice matching with Si. UHV-CVD reactions of $(GeH_3)_{4-x}CH_x$ with GeH_4 produce crystalline $Ge_{1-x}C_x$ with up to 7 at.% C concentration. The creation of very unusual trihalogermyl methanes, $(GeX_2Br)_4C$, and the corresponding hydride analogs used in Ge-C growth will be described.

367A SELF-ASSEMBLY IN THE SYNTHESIS OF POROUS SILICATES. Mark E. Davis, Department of Chemical Engineering, California Institute of Technology, Pasadena CA 91125

Porous materials have found widespread use as adsorbents and catalysts. The objective in synthesizing new porous materials is to precisely control the atomic-level structure of these materials to provide appropriate macroscale properties. This talk begins with a brief overview of the importance of porous materials. The controlled synthesis of zeolites (microporous materials) is then described with emphasis on the assembly process and how it can be manipulated. The synthesis of zeolites involves structure-directing organic molecules that assist in the organization of the inorganic components. Extension of this assembly process to the use of molecular aggregates as structure-directing agents allows the formation of mesoporous materials. Finally, the outlook for synthesizing materials by design through controlling the self-assembly process is discussed.

368.
THE PRODUCTION OF NITRATE DURING THE DECOMPOSITION OF AQUEOUS TRICHLORAMINE. **Gordon H. Purser** Cristin E. Moran, Todd D. Jaeger, Department of Chemistry, The University of Tulsa, Tulsa, Oklahoma 74104

The use of trichlorotriazinetrione, the active ingredient in many swimming pool chlorinating tablets, produces the undesirable byproduct, trichloramine. The base-catalyzed decomposition of trichloramine is known to produce hypochlorite ion, chloride ion and nitrogen gas. The decomposition products of the uncatalyzed hydrolysis reaction have not been reported previously, and are the subject of this presentation. In contrast to the stoichiometry of the base catalyzed hydrolysis, a significant amount of nitrate ion is produced in the uncatalyzed hydrolysis reaction . This is significant as it suggests that the uncatalyzed decomposition may occur via multiple pathways. Further evidence of multiple pathways is the observation that the addition of nitrate ion affects the rate of the hydrolysis reaction. The latest model of uncatalyzed trichloramine decomposition will be presented.

369.
LOW OXIDATION STATE COMPLEXES OF GROUP 3 ELEMENTS. THE FIRST ISOLATED EXAMPLES OF SCANDIUM IN FORMAL OXIDATION STATES OF (I) AND (II).
Polly L. Arnold, F.Geoffrey N. Cloke,* Peter B. Hitchcock, John F. Nixon.*
School of Chemistry, University of Sussex, Brighton, E. Sussex, BN1 9QJ, UK.

Cocondensation of scandium vapour with 2,2-dimethylpropylidynephosphine, [Me$_3$CC≡P], affords two highly unusual compounds:

- An organometallic complex of Sc(I), the triple decker sandwich [(η^5-^{t}Bu$_2$C$_2$P$_3$)Sc(μ-η^6:η^6-^{t}Bu$_3$C$_3$P$_3$)Sc(η^5-^{t}Bu$_2$C$_2$P$_3$)].
- An organometallic complex of Sc(II), the phosphorus-substituted scandocene [Sc(η^5-^{t}Bu$_3$C$_3$P$_2$)$_2$].

The former, the first example of a tripledecker complex of scandium, exhibits a remarkably low valence electron-number of 22. A single crystal X-ray diffraction study reveals, for the first time, structural information for a triphosphabenzene moiety ligated in an η^6 fashion to a metal.

The latter, generated in the same experiment, is the second observation of scandium(II) in our laboratory, isolation of the product has proved possible in this instance. Magnetism and EPR studies are used to investigate the system.

Related cocondensation reactions with elements of group 3 and the lanthanides have also resulted in the isolation of some equally novel compounds.
[Arnold, P. L.; Cloke, F. G. N.;* Hitchcock, P. B.; Nixon, J. F.* *J. Am. Chem. Soc.* **1996**, *118*, 7630.]
[Cloke, F. G. N. *Chem. Soc. Rev.* **1993**, 17.]

370.
STERIC VS. ELECTRONIC EFFECTS IN THE STRUCTURES OF GROUP 14 ORGANOMETALLICS: HOW RELEVANT IS THE LONE PAIR? David J. Burkey, Jason S. Overby, **Timothy P. Hanusa**, Department of Chemistry, Vanderbilt University, Nashville, TN 37235 and John C. Huffman, Molecular Structure Center, Department of Chemistry, Indiana University, Bloomington, Indiana 47405

A "stereochemically active" lone pair of metal valence electrons has been thought to play a critical role in the reactions and structures of Group 14 organometallic complexes. It thus came as a surprise when the structure of the stannocene [Sn(C$_5$(iPr)$_4$H)$_2$] (*Organometallics* **1995**, *14*, 11) was found to match that of the bent [Ca(C$_5$(iPr)$_4$H)$_2$], which has no metal valence electrons. Furthermore, Lawless recently reported the structures of two Group 14 metallocenes with *parallel* rings, [(Sn,Pb)(C$_5$Me$_4$(SiMe$_2$Bu^t))$_2$] (see *Organometallics* **1996**, *15*, 3905). We have been reevaluating the chemical and structural relevance of the lone pair in Group 14 complexes, and will discuss the structure of [Pb(C$_5$(iPr)$_3$H$_2$)$_2$] (which has parallel rings), a reinvestigation of the originally reported structure of [PbCp$_2$], and the characterization of bis(indenyl) complexes of Pb(II).

371. STRUCTURAL AND SOLUTION CHEMISTRY OF THE PRIMARY NITRENE SOURCES [(TOSYLIMINO)IODO]ARENES ArINSO$_2$Ar'.
M. A. Boucher, D. Macikenas, ^{1}J. D. Protasiewicz, Department of Chemistry, Case Western Reserve University, Cleveland, Ohio 44106-7078.

[(Tosylimino)iodo]benzene (PhINTs) is an extremely important primary nitrene source for many transition metal catalyzed olefin aziridation reactions. We are studying the solid state structures of this and related nitrene precursors ArINSO$_2$Ar'. Single crystal X-ray diffraction have revealed extensive aggregation in the solid state for Ar= 2-Me-C$_6$H$_4$, Ar '= 4-Me-C$_6$H$_4$ (**1**); Ar= 3-Me-C$_6$H$_4$ Ar'= 4-Me-C$_6$H$_4$ (**2**); Ar= 3-Me-C$_6$H$_4$, Ar'= C$_6$H$_5$ (**3**); Ar= 3-Me-C$_6$H$_4$ Ar'= 4-NO$_2$-C$_6$H$_4$ (**4**); and Ar= 4-Me-C$_6$H$_4$, Ar'= 4-Me-C$_6$H$_4$ (**5**). The structures of 1-5 will be contrasted to those reported for Ar= Ph, Ar'= 4-Me-C$_6$H$_4$ (PhINTs) and Ar=2,4,6-Me$_3$-C$_6$H$_2$, Ar'=4-Me-C$_6$H$_4$ (MesINTs). The solution phase properties and reactivity of these species will also be presented.

372.
HYDROGEN BONDING IN ANION RECOGNITION: A READILY AVAILABLE NEUTRAL ACYCLIC HALIDE RECEPTOR. **Konstantinos Kavallieratos,** Susan R. de Gala, David J. Austin and Robert H. Crabtree*, Department of Chemistry, Yale University, New Haven, CT 06511-8118.

Isophthalic diamides, m-$C_6H_4(CONHAr)_2$ (Ar = Ph, **1**; p-n-BuC_6H_4, **2**), readily synthesized on a large scale, have been shown to bind halide ions with an association constant K_a (for **2**) of 6.1×10^4 M^{-1} (Cl$^-$), 7.1×10^3 M^{-1} (Br$^-$) and 5.0×10^2 M^{-1} (I$^-$) determined by ^{1}H-NMR spectroscopy. The structure of the complex [PPh$_4$]$_2$[**1**.Br][Br].CH$_2$Cl$_2$, was determined by X-ray crystallography and shows a 1:1 complexation of bromide ion exclusively via rarely structurally characterized N-H...Br hydrogen bonds and a [PPh$_4$]Br of crystallization. The acyclic diamide adopts an unusual *syn-syn* non planar binding conformation. The presence of hydrogen bonding in the 1:1 complex was confirmed in solution by FT-IR and NMR spectroscopy. The anion binding properties of this and other simple artificial receptors are expected to allow applications in a wide variety of situations.

373. ANION SELECTIVE RECOGNITION AND SENSING BY NOVEL TRANSITION METAL RECEPTOR SYSTEMS. P. D. Beer, Inorganic Chemistry Laboratory, University of Oxford, South Parks Road, Oxford, OX1 3QR.

The synthesis and design of positively charged or neutral electron deficient receptor molecules designed to non-covalently bind anionic guest species is an area of ever increasing research activity. Anions play numerous fundamental roles in biological and chemical processes and some are of environmental concern. We have initiated a research programme aimed at the design and construction of innovative spectral and electrochemical sensory reagents for anions based on novel transition metal organometallic and coordination receptor systems. The talk will discuss our latest results in this field, focusing on new acyclic and macrocyclic systems which selectively recognise and sense dihydrogenphosphate, chloride and carboxylate anions.

374. A pH-REGULATED CHEMICAL OSCILLATOR: THE HOMOGENEOUS SYSTEM OF HYDROGEN PEROXIDE-SULFITE-CARBONATE-SULFURIC ACID IN A CSTR. Glen A. Frerichs, Department of Chemistry, Westminster College, Fulton, MO 65251, and Richard C. Thompson, Department of Chemistry, University of Missouri-Columbia, Columbia, MO 65211

Periodic oscillations in both pH and the potential of a platinum electrode have been observed in the H_2O_2-Na_2SO_3-Na_2CO_3-H_2SO_4 system using a continuous-flow stirred tank reactor. Aperiodic (possibly chaotic) behavior was also found under certain conditions, especially in the bifurcation region. The system was studied over a wide range of conditions. Phase diagrams delineating the behavior of the system under various conditions will be presented. A proposed model for the system will be discussed. Acceleration of the oscillation frequency by addition of carbonic anhydrase confirms the importance of the CO_2 hydration and H_2CO_3 dehydration reactions in the mechanism.

375.
CHARACTERIZATION OF Cu_6 AND CENTERED Cu_8 DIALKYLDITHIOPHOSPHATE CLUSTERS, AND A COMPARISON TO THE CENTERED Zn_4 DIALKYLDITHIOPHOSPHATE CLUSTER. **R. T. Stubbs,** J. P. Fackler, Jr., Department of Chemistry, Texas A&M University, College Station, TX 77843-3255.

It has been observed that $Zn_4S(dtp)_6$, $Cu_8S(dtp)_6$, $[Cu_8Br(dtp)_6]^+$, and $Cu_6(dtp)_6$ (dtp = dialkyldithiophosphate) are closely related structurally, and are analogous to the basic beryllium acetate structure. The stoichiometry of the metals is very different, but the ligand frameworks are virtually identical. The sulfur atoms for all of these complexes are arranged in an almost regular icosahedron around a cubic core. For the three structure types characterized, the metal atoms are in a tetrahedral, cubic, or antiprismatic arrangement. The tetranuclear zinc and octanuclear copper(I) clusters need a central anion to balance charge, and in the case of the centered halides a salt is formed. The Cu_6 cage is neutral, and the copper atom positions are slightly disordered. This disorder exists because two of the eight cubic sites are vacant. We are reinvestigating the centered Zn dtp cluster for evidence of similar disorder, since four of the eight cubic sites are unoccupied. The tetranuclear and hexanuclear complexes with their vacant cubic sites may be used as templates to isolate mixed metal and/or mixed valence cubic clusters.

376. RELAXOMETRIC DETERMINATION OF THE PROTOTROPIC EXCHANGE RATE OF THE COORDINATED WATER MOLECULE IN PARAMAGNETIC COMPLEXES. M. Botta S. Aime, A. Barge, D. Parker, A.S. De Sousa, Department of Chemistry I.F.M., University of Torino, 10125 Torino, Italy

The elucidation of the mechanisms operating the nuclear magnetic relaxation enhancement of the solvent water protons is of primary importance in the design of novel contrast agents for Magnetic Resonance Imaging. Analysis of the magnetic field and temperature dependence of the 1H and 17O relaxation data has shown that the tripositively charged Gd(III) tetraamide macrocyclic complexes considered in this work display exceedingly slow exchange rates of the coordinated water molecule. This has allowed the assessment of the contribution due to the prototropic exchange to the overall solvent protons relaxation rate by a NMR relaxometric method. Evidence has been gained of the occurrence of tight ion-pairs which affect considerably the exchange of the coordinated water.

377. THE DESIGN AND SYNTHESIS OF POLYDENTATE METAL-THIOLATE COMPLEXES WITH FUNCTIONAL SUBSTITUENTS IN THE SECONDARY COORDINATION SPHERE. Michelle Millar and Dao Hinh Nguyen, Department of Chemistry, State University of New York at Stony Brook, Stony Brook, New York 11790-3400.

The secondary coordination environment of metal ions in proteins can significantly influence the reactivity properties of the metal centers in metalloproteins. We seek to mimic this phenomenon by incorporating substituents *ortho* to the thiol group in the polythiolate ligand, [P(*o*-$C_6H_4(SH))_3$]. We report the acquisition and characterization of a series of Ni complexes in which the substituent adjacent to the thiol group is modified. As an example, a compound where R is the amide group [-$CONHCH(CH_3)_2$] is shown. Other examples where R = an amine, alcohol, carboxylate or ester substituent will be presented. In particular, the impact of **hydrogen bonding interactions** of the substituent with the sulfur atom will be discussed.

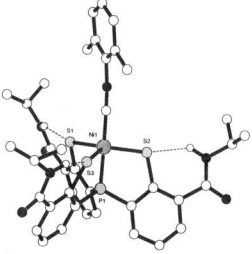

378. IRON COMPLEXES OF *TRIS* THIOLATE AMINE AND PHOSPHINE TRIPOD LIGANDS. Hua-Fen Hsu, Duncan Quarless, Jr., and Stephen A. Koch, Department of Chemistry, State University of New York at Stony Brook, Stony Brook, NY 11794-3400

A series of complexes of iron with *tris* thiolate tripod ligands have been synthesized and structurally and spectroscopically characterized. These complexes are potential models for iron-sulfur coordination modes in metalloproteins. With the phosphine thiolate ligand, *tris*(3-phenyl-2-thiophenyl)phosphine (PS3*), monomeric trigonal bipyramidal and octahedral compounds are obtained with Fe(II), Fe(III) and Fe(IV) and ligand including CN$^-$, CO, CNR, NO, SR, N_2H_4. Parallel studies with the amine trithiolate tripod ligand, $N(CH_2$-*o*-$C_6H_4S)_3$, will also be reported.

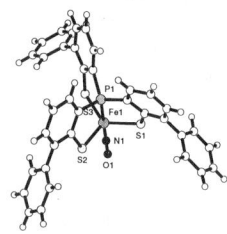

[Fe(PS3*)(NO)]

379. METAL COMPLEXES WITH TETRADENTATE CHELATING PHENOLATE LIGANDS. Jungwon Hwang and Stephen A. Koch, Department of Chemistry, State University of New York at Stony Brook, Stony Brook, NY 11794-3400

The coordination chemistry of various metals such as Fe(III), Ga(III), In(III), V(V), Cu(II) and other transition metals with a new polyphenolate tripod ligand, *tris*-(2-hydroxybenzyl)amine, has been investigated. These phenolate complexes synthesized have either trigonal bipyramidal or octahedral coordination center with ancillary ligands including N-Methylimidazole, DMF, and 1,10-Phenanthroline. The structural and spectroscopic properties of these complexes will be discussed to serve as models for the metal-tyrosinate coordination in a number of metalloenzymes.

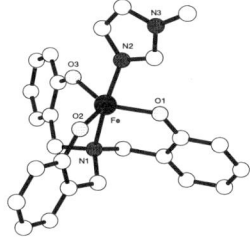

Fe(III)(NO3)(N-Me-Imid)

380. METAL THIOLATE MONOMERS AS MODELS FOR THE ACTIVE SITE OF HYDROGENASE. R. Harmony Voorhies, Michelle Millar and Susan T. Beatty, Department of Chemistry, State University of New York at Stony Brook, Stony Brook, New York 11794-3400.

We are exploring the coordination chemistry of polydentate thiolate ligands to prepare Ni(II), Ni(III) and Ni(IV) as possible models for the active site in hydrogenase enzymes. Interactions between the metal centers and thiolates in metalloenzymes are mimicked through a series of structurally and spectroscopically characterized complexes. In the Ni(II) compound, M(PS2H)$_2$, the SH groups are hydrogen bonded to the lone electron pair in the d_{z^2} orbital on Ni. Deprotonation of the two SH groups gives a Ni(II) compound formulated as [Ni(II)(PS2)]$^{2-}$ (see Figure), which has two anionic thiolate groups positioned outside the primary coordination sphere of the Ni center. Correlations between the chemistry of the various Ni(II), Ni(III) and Ni(IV) species will be made.

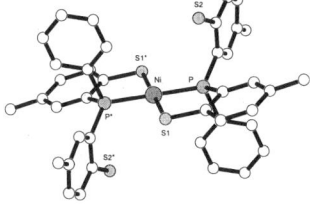

381. STRUCTURE AND BIOLOGICAL SIGNIFICANCE OF THE FIRST WELL-CHARACTERIZED VANADIUM(V) BROMO-COORDINATION COMPLEX. Petra N. Turowski, Louise-S. W. Githiora, Erin J. Andersson, Jerry P. Jasinski, Yu Li, *Departments of Chemistry, Smith College, Northampton, MA 01063 and Keene State College, Keene, NH 03431*

Vanadium(V)-bromide interactions are of importance because they presumably play a role in the mechanism of vanadium bromoperoxidase, an enzyme found in algae, fungi, and lichen. The only V(V)-bromo compound known before was the unstable oxohalide species VOBr$_3$. We prepared bromooxobis(N-phenylbenzohydroxamato)vanadium(V) and isolated dark purple monoclinic prisms whose structure was determined by X-ray crystallography. The structure of its isomorphous chloro-analogue was also determined. Both compounds crystallize as distortional isomers, where the halo atom takes the place of the oxo ligand and *vice versa* in a significant proportion of the molecules. The spectroscopic characterization and reactivity of these potential haloperoxidase model complexes will be discussed.

382. MODELS FOR THE WOC NEAR PHOTOSYSTEM II: REACTIVITY OF TETRANUCLEAR Mn COMPLEXES WITH BIOLOGICALLY RELEVANT MOLECULES. Aromí G.; Aubin, S.; Folting, K.; Hendrickson, D. N.; Christou, G. Department of Chemistry and Molecular Structure Center, Indiana University, Bloomington, IN, 47405, and Department of Chemistry, University of California at San Diego, La Jolla, CA, 92093-0358.

Photosynthetic oxidation of water is catalyzed by an oxo-manganese tetramer located near Photosystem II (PSII). With no crystal structure at hand, extensive research effort has been made to elucidate the structure and mechanism of action of the Water Oxidation Complex (WOC) with other techniques. We have developed a series of complexes with a highly distorted cubane $[Mn_4O_3X]^{6+}$ core, and O-based peripheral ligation. The $[Mn_4O_3]$ partial cubane is one of the structural units consistent with EXAFS data on the WOC. The site-specific reactivity of these molecules at the μ_3-X position has been studied, and monitored by ^{1}H- and ^{2}H-NMR spectroscopy. We have investigated reactions with biologically-relevant ions (AcO$^-$, F$^-$, Cl$^-$, Br$^-$, NO$_3^-$) and small molecules (ROH, H$_2$O) using X-ray crystallographic and various physicochemical techniques to characterize the resulting new clusters. These results, including the fixation and deprotonation of H$_2$O by a manganese tetramer, will be described.

383. SYNTHESIS AND STRUCTURAL CHARACTERIZATION OF Cu(I) AND Fe(II) CARBOXYLATE-BRIDGED DIMETALLICS AND THEIR REACTIVITY TOWARD DIOXYGEN. **Daniel D. LeCloux** and Stephen J. Lippard, Department of Chemistry, Massachusetts Institute of Technology, Cambridge, MA 02139

The preparation of multinuclear complexes of transition metals in low oxidation states for the reductive activation of dioxygen is of fundamental and practical importance in bioinorganic chemistry. Many metalloenzymes, including soluble and particulate methane monooxygenase (sMMO and pMMO) and dopamine β-monoxygenase, employ this strategy to carry out oxygen atom transfer reactions. As part of our efforts to assemble structural and functional mimics of this class of metalloenzymes, we have synthesized a series of dicopper(I) and diiron(II) complexes with the *m*-xylylene-diamine bis(Kemp's triacid imide) (H$_2$XDK) ligand and its propyl and benzyl analogs. Exposure of some of these complexes to dioxygen at low temperature has afforded stable intermediates. Characterization of these intermediates and their reactivity toward substrates will be presented.

384. **C-H BOND ACTIVATION BY A FERRIC METHOXIDE COMPLEX: A MODEL FOR THE RATE DETERMINING STEP OF LIPOXYGENASE.** Robert T. Jonas and T. Daniel P. Stack, Department of Chemistry, Stanford University, Stanford, CA 94305-5080.

Iron complexes involved in C-H hydrogen atom (H-atom) abstraction reactions in biological systems are generally thought to proceed through high-valent iron-oxo or ferric-peroxy species. In contrast, the non-heme iron lipoxygenase enzymes (LOs) are proposed to abstract H-atoms from 1,4-diene substrates by a ferric-hydroxide species in the rate determining step. The ferric species is subsequently reduced by the H-atom to a ferrous-water complex. Using a new non-charged, pentadentate pyridine ligand (**PY5**), the relevant ferric complex [(**PY5**)Fe^{III}(OMe)]$^{2+}$ (Figure) has been synthesized and characterized. The model complex is reduced by 1,4-dienes to the ferrous-methanol complex, with the observed reactivity most consistent with a H-atom abstraction mechanism.

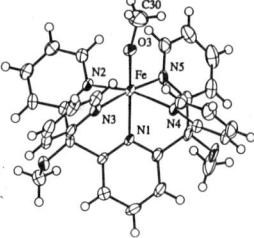

385. FUNCTIONAL MODELING OF GALACTOSE OXIDASE
Yadong Wang and T. Daniel P. Stack
Department of Chemistry, Stanford University, Stanford, CA 94305-5080

Galactose oxidase (GOase) is a mononuclear copper enzyme which catalyzes the oxidation of primary alcohols to aldehydes with subsequent reduction of dioxygen to H_2O_2. The active state of GOase is EPR silent because it contains a modified tyrosyl radical strongly antiferromagnetically coupled to a Cu(II) center. Several structural attributes of the known GOase metal coordination site have been incorporated into a family of mononuclear Cu(II) complexes. Systematic variation of the ligands has allowed the necessary structural features for reactivity to be identified. Of these complexes, only a specific subset can be oxidized to moderately stable, EPR silent species; each of these complexes is capable of oxidizing one equivalent of an alcohol to an aldehyde. A single complex (Figure) is found to be catalytically active with either a strong chemical oxidant or dioxygen. This reactivity is similar to that proposed for GOase.

386. AZIDE AND NITROSYL BOUND NITRILE HYDRATASE MODELS.
J. J. Ellison, S. C. Shoner, and J. A. Kovacs
Department of Chemistry, University of Washington
Seattle, WA 98195-1700

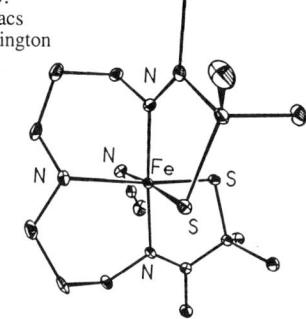

Nitrile Hydratases are proposed to contain six-coordinate low-spin Fe^{3+} centers ligated by two cysteinates and three histidines, and can bind a variety of substrates. In order to model the reactivity of this system, a coordinatively unsaturated Fe(III) complex with thiolate and nitrogen (imine and amine) ligation has been synthesized. This five-coordinate complex binds NO and reversibly binds azide (structure at right). Comparisons between model complexes and nitrile hydratase enzymes (epr, IR, and UV-vis data) will be presented.

387. COORDINATION CHEMISTRY AND SOLUTION THERMODYNAMICS OF COPPER COMPLEXES WITH TRIPODAL TETRAMINE LIGANDS.
F. E. Hahn and A. Dittler-Klingemann Institut für Anorganische und Analytische Chemie, Freie Universität Berlin, Fabeckstraße 34-36, D-14195 Berlin, Germany

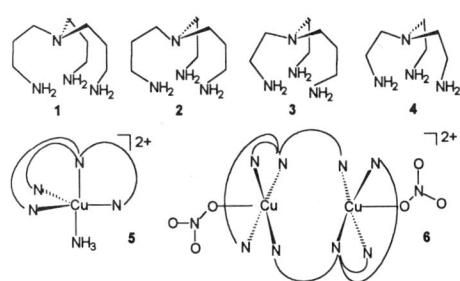

We have studied the coordination chemistry of the symmetric und unsymmetric tripodal tetramine ligands 1-4 with Cu(II),[1] Cu(I) and Ni(II). Ligands 1-3 form mononuclear complexes 5 at high pH but yield dinuclear complexes 6 at neutral pH. The stability constants of the Cu complexes as well as the reasons for the difference in nuclearity depending on the pH will be discussed based on potentiometric and UV-VIS spectroscopic data.

F. E. Hahn and A. Dittler-Klingemann, *Inorg. Chem.* **1996**, 35, 1996.

388. NOVEL IRON COMPLEXES CONTAINING A MULTIDENTATE SULFUR LIGAND: A STEP TOWARD NEW ACTIVE SITE MODELS FOR NITROGENASE
T. Adrian George and Jodi D. Niemoth-Anderson, Chemistry Department, University of Nebraska-Lincoln, Lincoln, Nebraska 68588-0304.

One of the intriguing features revealed by the crystal structure of the MoFe protein of nitrogenase by Kim and Rees is the pseudo-trigonal planar geometry of most of the iron atoms in the FeMo cofactor. In an effort to provide a structural model of the FeS_3 geometry, we have employed a tripodal tris-thiolate ligand, $P[(C_6H_3\text{-}3\text{-}Me_3Si\text{-}2\text{-}S^-)_3]$, PS_3. Reaction of the trilithio salt with $FeCl_2$ led to a series of complexes during extended workup. One of the products has been crystallographically characterized as a stable iron (IV) complex, $[FeCl(PS_3)]$. The crystal structure shows the iron in a trigonal bipyramidal arrangement nearly coplanar with the three equatorial sulfur atoms; each molecule has C_{3v} symmetry. Electrochemistry of $[FeCl(PS_3)]$ in 0.1M TBA-PF_6 in CH_2Cl_2 reveals a reversible one-electron reduction at -0.27V and an irreversible reduction at -1.39V relative to the ferrocenium/ferrocene couple (0.00V). These are believed to be due to the Fe^{+4}/Fe^{+3} and Fe^{+3}/Fe^{+2} couples, respectively. A dimer, $[Fe(PS_3)]_2$, has also been crystallographically characterized revealing the PS_3 ligand in a unique bonding arrangement. These results and others will be discussed, including other compounds with this ligand.

389.
MOLECULES TO CHAINS TO NLO MATERIALS: TITANYL IODATE. **Marc D. Schrier** and Galen D. Stucky*, Department of Chemistry, University of California, Berkeley, CA 94720, *Department of Chemistry, University of California, Santa Barbara, CA 93106.

Nonlinear optical materials have enjoyed an ever increased role in the opto-electronics industry. In our studies on the extended titanyl chain and its impact on second-order hyperpolarizability, we wished to use an asymmetric molecular anion to control the (-Ti=O-Ti=O-) chain geometry. Under mild hydrothermal conditions we have prepared a family of titanium iodates building stepwise from molecules to extended structures. Among them we report the first all-trans titanyl solid state material that crystallizes in an acentric spacegroup and exhibits an exceptionally large second order nonlinear optical response. We believe that the all trans titanyl chains give rise to a large net delocalized bond polarization and are largely responsible for the contribution to the second-order nonlinear susceptibility. Results of powder second harmonic generation measurements with a Nd:YAG laser and the optical window will be reported.

390.
CONTROLLED DOPING IN ALKALI AND ALKALINE EARTH METAL DOPED LANTHANUM COPPER OXIDE SUPERCONDUCTING MATERIALS PREPARED FROM MOLTEN ALKALI METAL HYDROXIDE FLUXES. **Marc D. Schrier** and Angelica M. Stacy, Department of Chemistry, University of California, Berkeley, CA 94720

A variety of alkali and alkaline earth metal doped lanthanum copper oxide superconductors have been reported in the literature. The preparation of homogeneous and crystalline electronic materials with stringent doping control is essential prior to further study and application. Here we report the enhanced doping control observed with this system when crystallized from molten alkali metal hydroxide fluxes. This reproducible control is achieved through control of the oxidizing potential of the flux with a variety of cation and anion additives as well as rigorous control of the temperature and gaseous environment over the flux. The doping control achieved is monitored by powder X-ray diffraction and microprobe elemental analysis. Magnetism trends within and between these systems will also be explored.

391. ORGANIC FUNCTIONALIZED MESOPOROUS MATERIALS. M. H. Lim, A. Stein, Department of Chemistry, University of Minnesota, Minneapolis, MN 55455-0431

We present a synthesis of mesoporous silicate with covalently bound vinyl groups. The co-condensation of vinyltriethoxysilane with tetraethoxysilane in an aqueous solution of cetyltrimethylammonium bromide results in a vinyl-functionalized hexagonal mesophase (v-MCM-41). By using this v-MCM-41 as a starting material, we will show its reactivities in bromination and other organic reactions, and estimate the reaction rates. In our previous study, the bromination of the vinyl groups occurred at an extremely slow rate implying that most vinyl group are present within the mesopore channels. However, since many factors, such as diffusion, steric hindrance, induced dipole interactions etc., can be the reason for the rate deceleration compared to olefins even with bulky protecting groups, it is important to determine which factor contributes the major effect. These will be discussed.

392. PULSED LASER DEPOSITION OF MESOPOROUS MOLECULAR SIEVES FOR THIN FILM APPLICATIONS. M. E. Gimon-Kinsel, K. J. Balkus, Jr., Department of Chemistry, The University of Texas at Dallas, PO Box 830688, Richarson, Texas 75083-0688

Mesoporous molecular sieves possess a uniform arrangement of channels (pore diameter 20 - 100 Angstrom). Extensive research is focused on discovering new mesoporous phases, improving current syntheses, and characterizing the physical and chemical properties of mesoporous molecular sieves in order to put them to practical use. Thin films made from mesoporous molecular sieves may be useful from the standpoint of sensor fabrication. This presentation will focus on our efforts to generate thin films from transition metal oxide mesoporous molecular sieves. Included in the discussion will be a careful evaluation of the optimized conditions for thin film growth and the preliminary results for a capacitive-type gas sensor made using the mesoporous molecular sieve thin film as the dielectric phase.

393. DETERMINATION OF HYDROGEN DENSITY OF STATES IN AMORPHOUS SILICON FILMS USING FRACTIONAL EVOLUTION EXPERIMENTS, A.J. Franz and J.L. Gland, Dept. of Chemical Engineering, University of Michigan, Ann Arbor, W.B. Jackson, Xerox Palo Alto Research Center, Palo Alto, CA

Hydrogen plays an important role in the electronic behavior, structure and stability of amorphous silicon films. Therefore, determination of the density of hydrogen binding energies (hydrogen density of states-DOS) and correlation of the H DOS with the electronic film properties are important research goals. We have developed a novel method for determination of hydrogen DOS in silicon films, based on fractional evolution experiments. Fractional evolution experiments are performed by subjecting a silicon film to a series of linear alternating heating and cooling ramps, while monitoring the hydrogen evolution rate. The fractional evolution data can be analyzed using two alternative procedures to yield complementary results. Using a rigorous, mean-field evolution model, we demonstrate the applicability of the two approaches to obtaining the hydrogen DOS in silicon films. We further validate both methods by analyzing experimental fractional evolution data for an amorphous silicon carbide film. Both types of analysis yield the same density of states for the a-Si:C:H film, and the energy difference between maxima in the density of states corresponds closely to the difference between bonding energies of hydrogen with silicon and carbon found in the literature.

394. INVESTIGATIONS INTO THE CHEMICAL MECHANISMS INVOLVED IN THE PYROLYTIC FORMATION OF METAL BORIDE THIN FILMS BY CHEMICAL VAPOR DEPOSITION. John A. Glass, Jr., Shreyas S. Kher, and James T. Spencer,* Department of Chemistry and the W. M. Keck Center for Molecular Electronics, Center for Science and Technology, Syracuse University, Syracuse, New York 13244-4100.

The formation of metal boride thin films by Chemical Vapor Deposition (CVD) is an area of current technological and scientific interest. These materials have attracted attention primarily due to their breadth of unique physical properties and structural diversity. Recently, we have explored the use of a number of boron-containing compounds to prepare a large variety of both pure metal and metal boride thin film materials. This CVD chemistry appears to constitute a highly efficient method for the formation of polycrystalline transition metal and lanthanide metal boride thin films.
The fundamentally important chemical processes and reactions in the CVD formation metal boride

films from boranes has not, however, been previously well investigated. In our recent work, which will be presented here, we have explored several of the important mechanistic details of the CVD of metal borides from gas phase boranes and metal halide precursors. Critical reactions to the deposition have been found to occur both in the gas phase and on the substrate surface. The studies which have provided valuable insights into these complex chemical reactions will be presented here.

395. **THE PHOTOCHEMISTRY OF LARGER BORANE AND METALLABORANE CLUSTER COMPLEXES.** Ralf Littger, Jesse Taylor, and James T. Spencer,* Department of Chemistry and the W. M. Keck Center for Molecular Electronics, Center for Science and Technology, Syracuse University, Syracuse, New York 13244-4100.

The photochemistry of organometallic species has been intensively investigated and has been found to provide efficient methods for the preparation of otherwise inaccessible complexes. The photochemistry of boranes and metallaboranes has until recently, however, not been explored in detail. Our previous work in this field has centered on the synthetic and structural aspects of photochemical processes of relatively small metallaborane complexes, such as [Fe(η^5-C_5H_5)$(CO)_2B_5H_8$]. Recently, we have begun to explore the photochemical reactions of larger metallaborane clusters, specifically those involving the B_9 and B_{18} clusters, in an attempt to relate the photochemistry of these larger cages with the known photoreactions of the related smaller metallaboranes and other organometallic cluster species.

In one aspect of the work to be presented, we have prepared the bis-B_9 cluster complexes of the group 12 metals (Zn, Cd, and Hg). These new complexes have been characterized crystallographically and have been found to be photoreactive to give new cluster species. Also, the Co(CO)$_3$ substituted conjuncto B_{18}-cluster has been prepared and characterized crystallographically. Details of the syntheses, structures, and photoreactions of these and related metallaborane clusters will be presented in detail.

396. **MOCVD-DERIVED INSULATING METAL OXIDES FOR THIN-FILM SUPERCONDUCTING APPLICATIONS.** J. A. Belot, T. J. Marks, J. Lei, R. P. H. Chang, J. L. Schindler, C. R. Kannewurf, The Science and Technology Center for Superconductivity, Northwestern University, Evanston, IL 60208.

Thin-film high temperature superconducting devices based on $YBa_2Cu_3O_{7-\delta}$ are approaching the "realization stage" in development. Critical to this emerging technology is the ability to deposit oxide insulators as buffer or intermediate layers in multilayer structures. The choice of oxide dielectric must satisfy stringent requirements concerning lattice and thermal mismatch, chemical compatibility, crystalline orientation, and surface roughness. Recently, our efforts have focused on c-axis oriented thin film insulators of tetragonal quaternary perovskites having the general formula LnSrGaO$_4$ (Ln = La or Pr). These materials have proven to be excellent buffer layers for high-quality $YBa_2Cu_3O_{7-\delta}$ thin films, and these results will be communicated.

LnSrGaO$_4$ =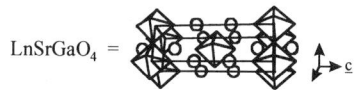

397. **NMR INVESTIGATION OF LITHIUM-INTERCALATED CARBON MATERIALS FOR BATTERY APPLICATIONS.** S.E. Hayes, A.K. Cheetham, H. Eckert, Materials Research Laboratory, University of California, Santa Barbara, CA 93106, and W.R. Even, Sandia National Laboratories, Livermore, CA 94551.

Because of their chemical stability and relative inertness, lithium-graphite intercalation compounds have garnered considerable interest as a substitute for metallic lithium or lithium alloy-based anodes in rechargeable solid electrolyte batteries. Recently, significantly higher intercalation levels have been reported for disordered and turbostratic carbon materials. We are investigating the effect of the microstructure of these carbon materials on the intercalation process and the local environments and spatial organization of the lithium ions. Our study focuses on the relationship between microstructure and performance of these materials, as addressed by solid state nuclear magnetic resonance (NMR). The aim of this work is to characterize the intercalated materials, to distinguish sites where lithium binds or associates preferentially, and to determine surface structures which may lead to material failure. We have used ^{7}Li magic angle spinning NMR to identify lithium sites within the carbon framework and correlate them with performance factors, and double resonance techniques to monitor dipolar interactions between the Li and the framework species (C, N, H).

398. PULSED EPR STUDIES OF MN(III)MN(IV) MODEL COMPOUNDS. David W. Randall,[1] Bradley E. Sturgeon,[1] Michael K. Chan,[2] William H. Armstrong,[3] Andrew K. Gelasco,[4] Vincent L. Pecoraro,[4] R. David Britt[1]; [1]Department of Chemistry, University of California, Davis, CA 95616, [2]Department of Chemistry, University of California, Berkeley, CA 94720, [3]Department of Chemistry, Boston College, Chestnut Hill, MA 02167, [4]Department of Chemistry, University of Michigan, Ann Arbor, MI 48109

Advanced pulsed EPR techniques reveal information about the electronic structure of trapped valence Mn(III) Mn(IV) model compounds for metalloproteins such as Manganese Catalase and Photosystem II, which both contain multinuclear Mn complexes. ESEEM and ESE-ENDOR of liganding nitrogen atoms indicate two distinct nitrogen hyperfine couplings. The dipolar hyperfine interaction of protons on the ligand with a dinuclear exchange coupled metal cluster can be used to elucidate the relative orientation of the protons observed with ^{1}H ESE-ENDOR. Finally, ^{55}Mn ESE-ENDOR on oxo and alkoxo bridged dinuclear Mn complexes gives insight into the effects of these bridging units on the electronic structure. The model compound results are compared to spectra from the tetranuclear Mn cluster of the oxygen evolving complex in PS II.

399. NITRIC OXIDE ADDUCTS OF THE REDUCED NON-HEME IRON CENTER IN SOLUBLE METHANE MONOOXYGENASE. **David E. Coufal** and Stephen J. Lippard, Department of Chemistry, Massachusetts Institute of Technology, Cambridge, MA 02139

The multi-component soluble methane monooxygenase (sMMO) enzyme system catalyzes the oxidation of methane to methanol utilizing a non-heme diiron center located at the active site of the sMMO hydroxylase. A key step in this reaction is the binding of dioxygen to the reduced diiron center. Nitric oxide is also capable of reacting with the reduced diiron center to form species that model aspects of the dioxygen-binding chemistry. The resulting sMMO-nitrosyl complex has been studied by using EPR, ENDOR, Mössbauer and EXAFS spectroscopy. This work was supported by NIGMS and carried out in collaboration with the laboratories of K.O. Hodgson, B.M. Hoffman and B.H. Huynh.

400. SYNTHESIS OF A MODEL FOR THE MOLYBDENUM-IRON INTERACTION IN SULFITE OXIDASE EMPLOYING SULFUR COORDINATION AT THE MOLYBDENUM CENTER. **Mark Turner** and John H. Enemark, Department of Chemistry, University of Arizona, Tucson, Arizona 85721.

An unsymmetrically substituted tetra-mesityl porphyrin has been synthesized in which one mesityl group has been replaced by a 3,4-difluorobenzene ring. Alkylthiodehalogenation chemistry introduces a bis-thioether moiety at the periphery of this porphyrin. The dealkylation chemistry to facilitate the coordination to an oxo-Mo(V) fragment and insertion of metals into the porphyrin core will be discussed. The sulfur ligation at the molybdenum center more closely mimics sulfite oxidase than previous related bimetallic models that utilized a pendant catecholate group to coordinate the molybdenum center.

401. LONG CHAIN ACID BINDING TO CYTOCHROME C. M.D. Johnson and Aaron M. Rowland, Department of Chemistry and Biochemistry, New Mexico State University, Las Cruces, New Mexico 88003

Long chain fatty acids bind to both the reduced and oxidized forms of cytochrome c and produce significant changes in the visible spectrum. We have found that this binding affects the protein's ligand exchange and electron transfer properties. In particular, its interaction with cytochrome c oxidase is modified. Dramatic differences between the interaction of the protein with the monomer and micellar forms of the fatty acids have also been observed. This poster will present the binding kinetics, ligand exchange kinetics of cytochrome c with SDS and oleic acid.

402. DNA CLEAVAGE BY IRON(III) COMPLEXES OF THE HYDROLYSIS PRODUCTS OF THE ANTIOXIDANT CARDIOPROTECTIVE AGENT, ICRF-187, AND ITS DESMETHYL DERIVATIVE, ICRF-154. Thomas J. Magliery[†], Lizabeth K. Vitellaro, Ndeye Khady Diop, and Rosemary A. Marusak, Department of Chemistry, Kenyon College, Gambier, OH 43022.
[†]Current address: Department of Chemistry, University of California, Berkeley, CA 94720.

ICRF (Imperial Cancer Research Fund) compounds are well known for their anti-cancer properties and their amelioration of anthracycline-induced cardiotoxicity. Intracellular hydrolysis of the parent diimide compound to the open diacid diamide form followed by chelation of anthracycline-bound iron(III) is the proposed mechanism by which ICRF drugs exert their antioxidant effects. Thomas et al. (*Biochem. Pharm.* **1993**, *45*, 1967), however, have shown that Fe-ICRF compounds, like FeEDTA, produce hydroxyl radicals via the Fenton reaction. This work reports on DNA cleavage by Fe-ICRF complexes in the presence of H_2O_2 and ascorbate. DNA cleavage is dependent on metal chelate concentration and chelate structure; cleavage differences may be explained by differences in the solution chemistry of these complexes. A solid state structure and the solution chemistry of Fe-ICRF complexes will be presented. The implication of these results on therapeutic limits for ICRF drugs will be discussed.

403. CONFORMATIONAL EFFECTS OF PERIPHERAL SUBSTITUENTS AND AXIAL LIGANDS IN STERICALLY CROWDED PORPHYRINS. Daniel J. Nurco, Kevin M. Smith, Department of Chemistry, University of California, Davis, California 95616 and Jack Fajer, Department of Applied Science, Brookhaven National Laboratory, Upton, New York 11973

Crystallographic results for peripherally substituted porphyrins illustrate effects of steric crowding, axial ligation and π-π interactions on the structures of metalloporphyrins that begin to resemble the architectures found for porphyrinic prosthetic groups and chromophores *in vivo*. 1) Metalloporphyrins with multiple peripheral substituents can adopt planar, saddled, ruffled or "wavy" conformations that persist upon axial ligation of the metal. 2) In porphyrins with substituents at all peripheral positions, the substituents form pockets that force the axial ligands to align in specific orientations relative to each other and to the axes defined by the porphyrin nitrogens. 3) The enforced alignment of the axial ligands caused by the substituent pocket prevents aromatic ligands from rotating around the coordination axis with intriguing consequences: the axial ligands can be forced to tip off axis because of steric crowding by adjacent molecules in the crystal lattice. 4) In porphyrins with meso substituents only, the axial ligands are free to rotate but π-π interactions and cofacial stacking between adjacent molecules in the crystal also induce off-axis ligand tilts. (Work supported by Dept. of Energy and National Science Foundation.)

404. INVESTIGATION OF RNA HYDROLYTIC CLEAVAGE BY Ce^{3+} AND La^{3+}. George C. Yeh and James K. Bashkin*. Department of Chemistry, Washington University, Campus Box 1134, One Brookings Drive, St. Louis, MO 63130-4899.

In studies of RNA hydrolysis, one major problem is the choice of an appropriate substrate. Long RNA molecules easily degrade and give complicated mixtures of hydrolysis products. RNA dinucleotides thus are commonly used in such studies, but these lack the polymeric and polyanionic character of many biologically important RNA molecules. We have been using the embedded RNA (embRNA) assay (*J. Am. Chem. Soc.*, **1996**, *118*, 6822) as part of our studies of the mechanisms of RNA hydrolytic cleavage by metal ions and complexes. This assay uses chimeric substrates that contain a single RNA nucleotide in a DNA oligomer (see also McLaughlin et al., *Nucleic Acids Res.*, **1995**, *23*, 4753). We report here our use of the embRNA assay to investigate hydrolytic cleavage of RNA by aqueous Ce^{3+} and La^{3+}. Dependence of the reaction rate on pH and $[Ln^{3+}]$ will be discussed and comparisons made to RNA cleavage studies that use RNA dinucleotides as substrates.

405. ENGINEERING NEW TEXAPHYRINS FOR X-RAY TREATMENT OF CANCER. Nicolai A. Tvermoes, William E. Allen, Tarak Mody, and Jonathan L. Sessler. *Department of Chemistry and Biochemistry, The University of Texas at Austin, Austin, Texas 78712-1167, and Pharmacyclics Inc., Sunnyvale, California 94086-4521.*

The gadolinium texaphyrin [Gd-T2BET]$^{2+}$, **1**, has shown the capability to increase the cytotoxic effects of radiation; human clinical trials are currently underway to evaluate the efficacy of [Gd-T2BET]$^{2+}$ as a radiation sensitizer for X-ray therapy (XRT). The sensitization may arise from [Gd-T2BET]$^{2+}$ acting as an "electron sponge," thereby maximizing the amount of hydroxyl radicals available to perform cell damage. To test this hypothesis the reduction potential of the parent [Gd-T2BET]$^{2+}$ was systematically altered by changing the electronic properties of the peripheral groups of the texaphyrin (e.g., **2** and **3**).

406. MAPPING AN ANTIBODY METAL-BINDING SITE BY NUCLEAR DECAY. **Rakesh Mogul,** Douglas P. Greiner, and Claude F. Meares, Chemistry Department, University of California, Davis, CA 95616

Metal complexes can be used to study protein structure by chemically modifying accessible sites. (S)-[p-nitrobenzyl]-EDTA(In-111) was bound specifically to antibody CHA255 and allowed to undergo radiolysis. Indium-111, which decays by electron capture, produces reproducible, site-specific fragmentation of both the heavy and light chains. The products were analyzed by denaturing polyacrylamide gel electrophoresis, followed by blotting and immunostaining with antibodies that recognize either the C-terminus of the heavy chain or the C-terminus of the light chain. The crystal structure of the complex shows several residues from the antibody heavy and light chains within van der Waals distance of the chelate (R.A. Love et al., Biochemistry 32:10950-10959 [1993]). The heavy chain yields a single ≈37 kD C-terminal fragment, suggesting efficient cleavage at the histidine residue coordinated directly to the metal (His99H). The light chain yields at least four distinct C-terminal fragments, consistent with peptide loops located farther from the metal. Accurate assignment of the cleavage sites and correlation with the crystal structure will provide calibration of the (evidently very short) range over which nuclear decay acts to efficiently cleave the polypeptide backbone.

407. EFFECTS OF REMOTE PROTONATION IN NON-HEME METALLOENZYMES
M. Tyler Caudle, Jeff W. Kampf, Paul G. Rasmussen, and Vincent L. Pecoraro. Dept. of Chemistry, University of Michigan, Ann Arbor, MI 48109-1055.

Our recent work on the structure and reactivity of biologically relevant manganese coordination compounds has led to a unique series of mononuclear and dinuclear Mn(III) and Mn(IV) complexes containing the 4,5-dicarboxyimidazole (H$_3$DCBI) unit. In the dinuclear [Mn^{III}(1,3-bis(3,5-di-*t*-butylsalicylimino)propane)]$_2$H(DCBI) complex (right), the imidazole group bridges the two metal ions in a mode analogous to the bridging histidine in Cu/Zn superoxide dismutase. Dissociation of the proton hydrogen bonded between the two carboxylate groups lowers the redox potential and modulates the reactivity of the site. The effect of protonation on the EPR and magnetic parameters of this center in multiple oxidation states is being explored to address whether the interconversion between the g=2 multiline and g=4.1 EPR signals in the S$_2$ state of the oxygen-evolving manganese cluster in Photosystem II could arise from protonation chemistry at a site remote from but coupled to the metal cluster.

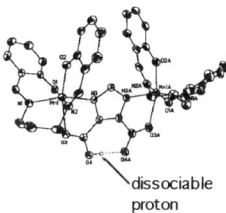
dissociable proton

408.
RECENT ADVANCES IN TETRAAZAPORPHYRIN SYNTHESIS. **J. P. Fitzgerald**, C. M. Holloway, Department of Chemistry, U. S. Naval Academy, Annapolis, MD 21401

In recent years general syntheses of planar tetraazaporphyrins bearing alkyl and aryl substituents, which confer solubility in organic solvents, have appeared in the literature. In order for tetraazaporphyrin chemistry to develop further, there is a need for more complex, second generation derivatives. Examples of second generation tetraazaporphyrins might include 1) unsymmetrical analogs, 2) those bearing bulky substituents, 3) perhalogenated derivatives or 4) water-soluble complexes. This poster will present recent work from our lab in this area, primarily on the synthesis of perfluorosubstituted tetraazaporphyrins.

409.
INVESTIGATIONS OF THE PATTERN OF UNPAIRED ELECTRON SPIN DENSITY DISTRIBUTION IN PARAMAGNETIC MODEL HEMES BY TWO-DIMENSIONAL NMR SPECTROSCOPY. **Joanne Marquez** and **Ursula Simonis**, Department of Chemistry and Biochemistry, San Francisco State University, San Francisco, CA 94132

We have synthesized several low-spin *bis*-imidazole and *bis*-cyanide ligated mono-*ortho* dialkylcarboxamidotetraphenylporphyrinatoiron(III) complexes to study the overlapping effects of unsymmetrical substitution and hindered axial ligand rotation on the unpaired electron density distribution in the porphyrin macrocycle as revealed by the pattern of the pyrrole-H resonances. In both sets of model hemes, the electronic effect of the substituent results in up to 8 distinct resonances for the pyrrole protons. However, compared to the imidazole complexes, the resonance spread is significantly decreased in the cyanide derivatives, suggesting that an axial ligand plane has a major effect on the pyrrole resonance pattern. To unambiguously assign the pyrrole-H resonances of the *bis*-cyanide ligated compounds for determining the electron density distribution in the porphyrin macrocycle, these compounds were investigated by two-dimensional NMR spectroscopy experiments including ROESY, CLEAN TOCSY, and gradient COSY.

410.
EARLY EVENTS IN CYTOCHROME C FOLDING. **Jason R. Telford**, F. Akif Tezcan, Jay R. Winkler, Harry B. Gray; Arthur Amos Noyes Laboratory, California Institute of Technology, Pasadena, CA 91125

Under certain denaturing conditions (e.g., 3.5 M guanidine HCl, pH 7.0, 298K) the reduced form of cytochrome c (cyt c) is stable and fully folded, whereas the oxidized form is fully unfolded. Thus, reduction of oxidized denatured cyt c under these conditions will trigger protein folding. Several photochemical methods have been employed to reduce ferric cyt c on timescales ranging from microseconds to milliseconds. The kinetics of reduction and protein folding were monitored by transient absorption spectroscopy. Our results suggest that the earliest events following reduction are ligand exchange reactions involving the heme iron.

411.
^{13}C NMR RELAXATION STUDIES OF HCO_3^- IN SOLUTION
D. Ohliger, R.E. Forster, & K. Wroblewski
Department of Biochemistry and Biophysics, University of Pennsylvania

Carbon-13 spin-lattice relaxation times, T_1, have been measured in aqueous $H^{13}CO_3^-$ at pH 9 with the inversion recovery technique under different conditions. This allowed separation of various relaxation mechanisms such as dipole-dipole, spin rotation, and chemical shift anisotropy from the total spin-lattice relaxation rate.

T_1 data were acquired in water at temperatures between 280 and 325°K, at frequencies of 21.7 and 125 MHz, and these were found to be 22 +/- 2 sec. We expect that this T_1 rate is weakly affected by changing conditions, but these times may be within the error of our measurement. The relaxation times observed in ^2H_2O ranged from 22 sec. to 78 sec. at different temperatures. The relaxation mechanism appears to be dominated by intermolecular dipolar relaxation of the carbon spin by solvent water molecules.

Relaxation data obtained are relevant to chemical exchange studies of the reversible hydration-dehydration of $^{13}CO_2$ and $H^{13}CO_3^-$ as catalyzed by the zinc enzyme Carbonic Anhydrase V.

412. EXTRACTION AND RECOVERY OF NUCLEAR FISSION PRODUCTS USING HYDROPHOBIC, REDOX-ACTIVE ORGANOMETALLIC COMPLEXES. Jennifer F. Clark,[1,2] Rebecca M Chamberlin,[1] and Steven H. Strauss,[2] (1) Nuclear and Radiochemistry Group, MS J514, Los Alamos National Laboratory, Los Alamos, NM 87545; (2) Department of Chemistry, Colorado State University, Fort Collins, CO 80523.

We are investigating a new strategy for liquid-liquid extraction and recovery of Cs-137 and Sr-90 from aqueous nuclear waste and contaminated groundwater. The extractants are hydrophobic, redox-active derivatives of $Ni(C_2B_9H_{11})_2$ and $FeCp(C_2B_9H_{11})$. They are activated for extraction by reduction to the corresponding monoanions. The cesium and strontium cations are recovered from the organic phase by re-oxidation (deactivation) of the anionic extractants back to the neutral species, which also allows for recovery and reuse (i. e., recycling) of the extractants. The reduction potentials of the parent compounds are 0.18 V and -0.08V respectively (acetonitrile, SCE). Therefore, the derivatives can be easily reduced and re-oxidized by many common substances. The redox potentials, relative solubilities in organic solvents and aqueous solution, and D values for extraction and recovery, which vary as a function of substituent groups on the ligands, will be discussed.

413. EFFECT OF LIGAND CONFORMATION UPON THE ELECTRONIC STRUCTURE OF $[Fe_4S_4]^{1+/2+}$. Michael J. Knapp, and David N. Hendrickson. Department of Chemistry and Biochemistry, University of California at San Diego, La Jolla, CA 92093-0358.

The electronic structure of $[Fe_4S_4]^{1+/2+}$ is of much current interest. By varying the bulkiness of the arythiolate ligand in $[Fe_4S_4(SAryl)_4]^{2-/3-}$, we have investigated the effects of ligand conformation upon the electronic structure of these clusters. EPR, magnetic susceptibility, and electronic absorption spectroscopy have ben employed to characterize the ground and excited states. The results are discussed in light of current models for the $[Fe_4S_4]^{1+/2+}$ cluster.

414. INVESTIGATIONS OF TRINUCLEAR RHENIUM (III) HALIDES BY ELECTROSPRAY IONIZATION AND LASER DESORPTION IONIZATION MASS SPECTROMETRY. Nancy Carter Dopke, Paul M. Treichel, Department of Chemistry, University of Wisconsin, Madison, WI, 53706

As part of our synthetic work on trinuclear rhenium (III) clusters, mass spectrometry was utilized as a characterization tool that gives detailed information concerning the composition of the cluster. Electrospray ionization (ES) and laser desorption ionization (LD) were used to obtain the negative-ion mass spectra of trinuclear rhenium (III) halide salts. In the ES spectrum of $Cs_3[Re_3Cl_{12}]$ the most abundant peak is $Re_3Cl_{10}^-$. This peak is also observed in the LD spectrum of this species, however, the most abundant peak is $Re_2Cl_7^-$. In LD spectra higher mass aggregates such as $Re_5Cl_{12}^-$ and $(Re_3Cl_9)_4^-$ are also seen. Fragmentation involving chloride loss is seen using both techniques. In crude samples, peaks corresponding to oxygen-containing rhenium (III) species are also observed. We will discuss the usefulness of the techniques and implications of the results obtained for the trinuclear systems.

415. METAL ALKOXIDE DOPANTS FOR A PYRIDINE BASED SOLUTION PROCESS TO METAL OXIDES Timothy J. Boyle*, Catherine D. Buchheit, Gregorio J. Moore, Randall T. Cygan, and Husam N. Al-Shareef Sandia National Laboratories, Advanced Materials Laboratory 1001 University Blvd. SE, Albuquerque, NM 87106≠.

Recently, we have developed a simple and rapid process for synthesizing precursor solutions for lead containing ceramic materials, termed the "Basic Route to PZT" (or BRP) process. The BRP process offers several advantages over standard methods, including: rapid solution synthesis (< 10 minutes), use of commercially available materials, film production under ambient conditions, ease of dopant dissolution at high concentrations (< 1 min.), and no heating requirements during solution synthesis. The synthesis and characterization of a series of Nb, La and Sn alkoxide dopants which were added to BRP solution will be reported along with the effect they have on film morphology and ferroelectric properties. Due to the "chemical simplicity" of the BRP solution, computer modeling has been initiated.

416. **LIGHT-HARVESTING DENDRITIC ASSEMBLIES: SYNTHESIS, PHOTOPHYSICAL STUDY, AND SURFACE ATTACHMENT**
Kimberly D. Edwards, Travis P.S. Thoms, Mai Zhou, Bo Hong*
Department of Chemistry, University of California, Irvine, CA **92697**

We are investigating the systematic synthetic approach, photophysical study, and surface attachment of dendritic assemblies bearing transition metal-based peripheral photosensitizers and central luminophores. Our present goal is the synthesis of poly(phenyl benzyl ether) dendrimers with terpyridyl or isonitrile linkages to the peripheral photosensitizers and phosphine or ether linkages to the central luminophore. Various monodendron wedges have been synthesized and bound to photosensitizers. Also, we are studying the ground state and excited state properties of dendritic assemblies using electronic absorption, steady-state and time-resolved emission spectroscopy. Furthermore, we are exploring the covalent attachment of dendritic assemblies to various surfaces. In this report, osmium-based units have been attached to quartz surfaces and studied by electronic absorption and X-ray photoelectron spectroscopy.

417. ELECTROCHEMISTRY OF INTERCALATORS BOUND TO A DNA-MODIFIED ELECTRODE: ELECTRON TRANSFER THROUGH THE π-STACK OF DNA.
Nicole M. Jackson,[†] Michael G. Hill,[†] **Shana O. Kelley,**[‡] Jacqueline K. Barton.[‡] [†]Occidental College, Department of Chemistry, Los Angeles, CA 90041. [‡]California Institute of Technology, Department of Chemistry and Chemical Engineering, Pasadena, CA 91125.

The electrochemistry of DNA-bound species has been investigated using double-stranded DNA oligonucleotides derivatized with a hexanethiol linker. These modified duplexes have been covalently adsorbed onto gold electrodes, and the resultant surfaces characterized by a variety of electrochemical methods, as well as ellipsometry and a ^{32}P-radioactive tagging assay. Binding constants for several redox-active organic and inorganic intercalators to the DNA-modified surface have been determined by chronocoulometry. The voltammetry of these intercalated species, both non-covalently associated to the electrodes, and covalently attached to specific sites on the surface-bound duplexes, suggests that the π-stack of DNA is extremely efficient at mediating charge migration over long molecular distances. These results are in complete accord with previous photophysical studies which point to DNA as an effective medium for propagating long-range electronic interactions.

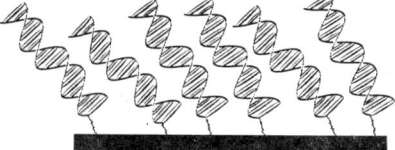

418. MOLECULAR RECOGNITION IN SUPPORTED MONOLAYERS: INCLUSION OF SURFACE-CONFINED FERROCENE-BASED GUESTS BY CYCLODEXTRIN HOSTS. **W. E. Cleland, Jr.**, Sukanta Bhattacharyya, R. C. Sabapathy, and C.L. Hussey, Department of Chemistry, University of Mississippi, University, Mississippi 38677.

Self-assembled monolayers formed by chemisorption of organosulfur compounds on gold surfaces represents a unique method for fabricating chemically modified electrode surfaces. As part of our interest in the design and preparation of supported monolayers with molecular recognition sites, a few ferrocenylalkyl disulfides were synthesized and self-assembled on vapor deposited Au(111) electrode surfaces. Host-guest interactions were observed between these ferrocene-based surface-confined guests and different cyclodextrin hosts. These interactions are manifested primarily by a decrease in the voltammetric current for the ferrocenyl redox center. The preparation, electrochemical behavior and infrared spectroscopy of these host-guest systems will be presented.

Fc = Ferrocenyl, = cyclodextrin

419. MAGNETIC CHARACTERIZATION OF SINGLE-MOLECULE MAGNETS: Mn_{12} AND Mn_4 MOLECULES. <u>Sheila M.J. Aubin</u>, Hilary J. Eppley, George Christou and David N. Hendrickson. Department of Chemistry and Biochemistry, University of California, San Diego, La Jolla, CA 92093-0358 and the Department of Chemistry and Molecular Structure Center, Indiana University, Bloomington, IN 47405-4001.

Considerable effort is being directed towards the preparation of single-molecule magnets which could potentially be used in storing high density magnetic data. High-field EPR and magnetic properties are presented for a series of Mn_{12} and Mn_4 molecules that display novel magnetic properties. These complexes display magnetic hysteresis loops and frequency dependent out-of-phase AC magnetic susceptibility (χ''_M) peaks, properties that are commonly observed in single domain magnets as a result of long range ordering of spins. Unlike single domain magnets, these Mn_{12} molecules behave as single-molecule magnets due to a high-spin ground state and large single-ion magnetic anisotropy. Mn_{12} molecules can display one or two out-of-phase AC magnetic susceptibility peaks. The origin of these peaks will be discussed. Steps are observed in the magnetic hysteresis loops at non-incremental values of field. The nature of these steps may be due to field-tuned quantum tunneling.

420.

SELF-REDUCING COMPLEXES FOR CHEMICAL VAPOR DEPOSITION OF COPPER. Michael Stewart, Andrew Maverick. Department of Chemistry, Louisiana State University, Baton Rouge, Louisiana 70803.

We have been studying Chemical Vapor Deposition (CVD) of copper metal films using copper(II) precursors. Recently we have used precursors that contain reducing agents bound as ligands. For example, alcohol adducts of bis(1,1,1,5,5,5-hexafluoropentane-1,3-dionato) copper(II) have been used to perform CVD in our laboratory. We now report that $Cu(hfac)_2 \cdot 2N_2H_4$, originally prepared by Bublik et al. in 1984, can also be used as a copper CVD precursor. The complex is transported by an inert carier gas (N_2) and deposition occurs at 120°C. Related work has been to use adducts of ligands similar in reducing character to hydrazine. Complexes of hydroxylamine and N-methylhydroxylamine have been made and are being tested for self-reducing CVD ability.

421.
ELECTROCHEMICAL AND PHOTOCHEMICAL STUDIES OF Ru(II) POLYPYRIDYL POLYMER FILMS. **Patricia L. Richardson**, Ana R. Guadalupe, and Jorge L. Colón*, Department of Chemistry, University of Puerto Rico, P.O. Box 23346, San Juan, PR 00931

We are studying compounds that can play the role of photosensitizers and quenchers like in the photosynthetic apparatus for solar energy conversion. Our strategy consists of immobilizing the appropriate components in a polymer matrix with the ultimate goal of building solid state and regenerable photoelectrochemical cells. We are preparing films of copolymers of Ru(II) and Zn(II) polypyridyl complexes using two different strategies. In the first case, polymeric films were prepared directly on $SnO_2(In)$ surfaces by electropolymerization while in the second case they were prepared by radical polymerization using a TEMPO adduct to control the polymer molecular weight and site distribution. This presentation includes the polymer syntheses and characterization both in solution and on the electrode surface, their photophysical and photochemical properties and preliminary quenching studies.

422. COMPOUNDS THAT CONTRACT ON HEATING. Arthur W. Sleight, Department of Chemistry, Oregon State University, Corvallis, OR 97331-4003.

Four mechanisms for negative thermal expansion in oxides will be discussed. Mechanism 1 is based on polyhedra becoming more regular with increasing temperature, causing *average* metal–oxygen bond lengths to decrease. Mechanism 2 is based on polyhedra which share edges in sheets or share faces in chains. Normal thermal expansion of certain bonds actually pulls these sheets or chains closer together with increasing temperature. Mechanism 3 is found in network structures where interstitial cations change position with changing temperature. Mechanism 4 is based on the transverse thermal motion of oxygen in metal–oxygen–metal linkages pulling the metal atoms closer together with increasing temperature. Examples of the four mechanisms are (1) $PbTiO_3$, (2) $Mg_2Al_2Si_5O_{18}$, (3) $LiAlSiO_4$, and (4) ZrW_2O_8. Mechanisms 1, 2, and 3 are only known to occur in anisotropic materials. Mechanism 4 can occur in cubic or amorphous materials leading to isotropic negative thermal expansion. Usually this mechanism only occurs at temperatures well below or well above room temperature. Thermal expansion in cubic ZrW_2O_8 and HfW_2O_8 is unique in that it is strongly negative over a very broad temperature range, 0.3 to 1050 K.

423. ELECTRIDES: FROM 1D HEISENBERG CHAINS TO 2D PSEUDO-METALS. James L. Dye, Department of Chemistry, Michigan State University, East Lansing, Michigan 48824.

Electrides are ionic salts in which the cations are complexed by cryptands or crown ethers and the "anions" are electrons. The crystal structures of five electrides are known and are similar to alkalides, except that the anion sites are "empty". Theory and experiment strongly support a model in which the "excess" electrons are trapped in these cavities and interact through connecting channels. Measurements of optical, alkali metal NMR, and EPR spectra, magnetic susceptibilities, and conductivities have provided much data that can be correlated with the structures. Three electrides have essentially 1D chains of cavities connected by channels through which the electrons communicate. These compounds yield susceptibilities that are well described by a 1D Heisenberg model, and there is EPR evidence of spin-wave character. The electride $K^+(cryptand[2.2.2])e^-$ has a 2D array of cavities and channels. It appears that hole-doping may be responsible for its near-metallic conductivity. A mixed-crown-ether electride has a complex 3D collection of cavities and channels dominated by a six-cavity ring. This talk will focus on the unusual and varied electronic and magnetic properties of electrides. (Supported in part by NSF Grant DMR 94-02016 and the MSU Center for Fundamental Materials Research.)

424. A STUDY OF HYDROBORATES: METAL COMPLEXES TO BIOLOGICALLY ACTIVE BORANE ANALOGS. K.W. Morse, Western Washington University, Bellingham, WA 98225.

Hydroborates and their derivatives are well known for their reducing and catalytic behavior, and, more recently, certain hydroborate derivatives have shown unusual biological activity. Modification of the hydroborate presents the potential for changing such behavior and activity (as was observed with the change in reducing behavior of tetrahydroborate compared to cyanotrihydroborate.) Our studies of metal complexes of tetrahydroborate and (alkylcarbonyl)trihydroborate afforded the first evidence of a singly bridged M-H-B bond, a rare occurrence at the time of its discovery. In addition to the first accurate characterization of an unsupported metal-hydrogen-boron bridge bond by neutron and X-ray diffraction studies, the dependence of the linkage on ligand strength, steric factors and complexing metal will be discussed. Temperature dependent studies in selected complexes verified the slowing of the fluxional process and the existence of bridging and non-bridging hydrogens at low temperatures.

Iron, cobalt, and chromium complexes of trimethylamineboranecarboxylate (($CH_3)_3NBH_2CO_2^-$) resulted in monodentate, bidentate-chelating, or bidentate-bridging compounds, all bonded through oxygen. These complexes were observed to be cytotoxic in murine and human tumor cell lines. The potential of biological activity in borane derivatives led us to develop a variety of (aminomethyl)phosphonate-borane derivatives with the goal of producing a boron analog of glyphosate. The results of these studies will be presented.

425. ELEMENTS OF CURRICULUM REFORM: PUTTING SOLIDS IN THE FOUNDATION. Arthur B. Ellis, Department of Chemistry, University of Wisconsin-Madison, Madison, WI 53706

The Ad Hoc Committee for Solid-State Instructional Materials has prepared a collection of materials for incorporating solids into the chemistry curriculum. These products illustrate connections of chemistry to other disciplines and methods for integrating advanced materials and high-tech devices into classes and laboratory experiments. Student familiarity with and interest in modern devices and materials, ranging from LEDs to memory metal, provides an engaging link for communicating fundamental chemical concepts. Demonstrations of some of these products and their offshoots will be presented, and their implications for curriculum reform discussed.

426. Sterically-Demanding *Tris*(imidazolyl)phosphine and Alkoxy-*bis*(pyrazolyl)hydroborato Zinc Complexes as Synthetic Analogues of Carbonic Anhydrase and Thermolysin. Prasenjit Ghosh, Conor Dowling, Clare Kimblin, Cetywa Powell, and <u>Gerard Parkin</u>*, Department of Chemistry, Columbia University, New York, New York 10027.

The sterically-demanding *tris*(imidazolyl)phosphine ligand [Pim$^{Pr^i,Bu^t}$] has been used to synthesize the zinc hydroxide complex {[Pim$^{Pr^i,Bu^t}$]ZnOH}+, which is the first structurally characterized four-coordinate monomeric zinc hydroxide complex supported by imidazole functionalities. As such, {[Pim$^{Pr^i,Bu^t}$]ZnOH}+ is an excellent structural model for the active site of carbonic anhydrase. Structural models for the NNO donor array provided by peptide backbone of thermolysin, namely [η^3-(R$_2$CHO)Bp$^{Bu^t,Pr^i}$]ZnX (X = Me, I), have been constructed by insertion of ketones and aldehydes (R$_2$CO) into one of the B-H bonds of [Bp$^{Bu^t,Pr^i}$]ZnX.

427. **SONOCHEMICAL SYNTHESIS OF INORGANIC MATERIALS.**
K. S. Suslick, M. M. Fang, M. Mdleleni, T. Hyeon, and J. Ries School of Chemical Sciences, University of Illinois at Urbana-Champaign, Urbana, IL 61801.

High intensity ultrasound has found new applications in the synthesis of unusual inorganic materials, due in large part to the extreme conditions of cavitation: ~5000 K, ~1000 atm, with heating and cooling rates that exceed 10^9 K/s. The sonolysis of volatile organometallic precursors during cavitation produces clusters a few nm in diameter, and we have utilized this for the sonochemical syntheses of nanophase metals, alloys, carbides, sulfides, supported heterogeneous catalysts, and nano-colloids. Such nanostructured solids are active heterogeneous catalysts for various reactions. Most recently, we have discovered a new synthesis of nanometer-sized transition metal colloids of iron, cobalt, and iron-cobalt alloy colloids. Sonication of iron pentacarbonyl, tricarbonylnitrosylcobalt, or appropriate mixtures of these two precursors in polyvinylpyrrolidone or oleic acid solutions in n-hexanol generated black colloidal solutions. The particle size ranges from 3 to 7 nm as determined by TEM. Diffuse ring patterns in the electron microdiffraction showed that these metal particles are amorphous as initially prepared. Elemental analysis by TEM-EDX revealed that the Fe-Co alloys are homogeneous in atomic composition. Magnetic studies show that the colloids are superparamagnetic and function as useful ferrofluids.

428.
OLEFIN POLYMERISATION CATALYSIS USING ORGANOMETALLIC COMPLEXES IN LAYERED AND CHANNEL INORGANIC HOSTS. **D. O'Hare,** J. Tudor, Inorganic Chemistry Laboratory, South Parks Road, Oxford, OX1 3QR

Layered and mesoporous silicates can be intercalated with homogeneous olefin polymerisation catalyst precursors such as rac-ethylenebis(indenyl)zirconium dichloride. These new organometallic-inorganic composites in the presence of methylaluminoxane (MAO) are active ethylene and propene polymerisation catalysts. In some cases, high molecular weight isotactic polypropylene with a unique spherulite morphology can be produced.

429. SYNTHESIS AND COMPLEXATION OF CYCLOPENTADIENES WITH POTENTIALLY LIGATING SUBSTITUENTS. Terence A. Nile, Department of Chemistry, University of North Carolina at Greensboro, Greensboro, North Carolina 27412

This presentation will review our recent results in the area of cyclopentadienes substituted with potentially ligating functional groups. We have synthesized cyclopentadienes with sustituents such as pyridyl, thienyl, bipyridyl, dimethylamino, etc. which have either one or two carbon or benzyl spacers between the cyclopentadiene and the functional group. In addition the steric bulk of the cyclopentadiene has been increased by the use of trimethylsilyl ring substituents or by methyl substituents on the functional groups. We, along with our collaborators, have demonstrated both inter- and intramolecular coordination of these functional groups in a variety of cyclopentadienyl metal complexes. The synthesis and properties of a selection of ligands and complexes will be presented.

430. CHROMIUM OXO ALKYLS - ORGANOMETALLIC CHEMISTRY IN HIGH OXIDATION STATES. Alexandra Hess, Rainer Papp, Louise Liable-Sands, Arnold L. Rheingold and Klaus H. Theopold, Department of Chemistry and Biochemistry, University of Delaware, Newark, DE 19716.

We are exploring the reactions of chromium alkyls in the formal oxidation states II and III with oxygen atom transfer reagents like O_2, Me_3NO, PhIO, and N_2O. Oxo alkyls of chromium in all of the remaining higher oxidation states (i.e. IV, V, and VI) have thus been prepared and characterized. Changes of reactivity with increasing oxidation state are of interest in connection with the uses of chromium in catalysis (olefin polymerization and oxidation). For example, Tp$^{t\text{-Bu,Me}}$Cr-Ph reacts with PhIO to yield Tp$^{t\text{-Bu,Me}}$Cr(O)-Ph, while its reaction with O_2 proceeds via an observable intermediate and ultimately produces Tp$^{t\text{-Bu,Me}}$Cr(O)-OPh. The structures and dynamical features of these molecules will be described. Another set of reactions concerns the oxidation (by O-transfer) of chromacyclobutanes. Thus metallacycles of the type Cp*(py)Cr(CH$_2$)$_2$X(CH$_3$)$_2$ (X = C, Si) react with Me$_3$NO to yield a variety of novel organometallic structures combining oxo-ligands and metallacycles. The reactivity of these molecules will be discussed.

431. TRIAZACYCLOHEXANE COMPLEXES OF TRANSITION METALS FOR OLEFIN POLYMERIZATION. R. D. Köhn, M. Haufe, G. Seifert, G. Kociok-Köhn, Inst. f. Anorg. u. Analyt. Chemie, Technische Universität Berlin, D - 10623 Berlin, Germany

In the search for ancillary ligands enforcing unusual coordination geometries in complexes with potential catalytic activity, we investigated the coordination chemistry of N-substituted 1,3,5-triazacyclohexanes (R$_3$TAC). X-ray crystallography on several complexes has shown that this ligand coordinates facially to a variety of transition metals with N-M-N bite angles of around 60° and severely misdirected nitrogen lone pairs deviating from the direct N-M bond by 25-45°. [(R$_3$TAC)CrCl$_3$] reacts with {MeAlO}$_x$ (MAO) to solutions that polymerize ethylene. As a model for the active species we attempted to prepare complexes [(R$_3$TAC)MR]$^{n+}$. As a first structural model, [(R$_3$TAC)ZnEt]$^+$ was prepared from ZnEt$_2$ and R$_3$TAC/PhNMe$_2$H$^+$. Several complexes (R$_3$TAC)CrR'$_3$ were prepared. However, we were unable to isolate the de- -ed cationic Cr(III) complexes by protonolysis. Similar Cr(II) complexes were also investigated.

432. HEXAGONALLY PACKED MESOPOROUS MATERIALS. D. M. Antonelli, Department of Chemistry, University of Sussex, Falmer, Brighton, UK BN1-9QJ and J. Y. Ying, Department of Chemical Engineering, Massachusetts Institute of Technology, Cambridge, MA 02139

Since the discovery of the mesoporous silicate molecular sieve MCM-41 in 1992 the formation of organic/inorganic mesostructures from new materials has been a subject of great interest. This silicate material consists of a hexagonally-packed array of inorganic tubules ranging from 20-100Å in diameter and is formed by templating of inorganic oxides onto a self-assembled liquid crystal mesostructure. Mesoporous materials are not only of tremendous interest as catalysts or catalytic supports in processes involving large molecules which are unable to diffuse effectively through the smaller microporous channels of zeolites, but also to nanoelectronic device production, and gas sorption processes. This talk will focus on the synthesis of transition metal oxide analogues of MCM-41, termed M-TMS1 (M = Nb, Ta, Ti), which represent the first transition metal oxide based zeolites. The compositional extension of the mesostructures to the transition metal oxide-based M-TMSI family provides further flexibility in tailoring of oxidation states, coordination chemistry and electronic properties.

433. CATIONIC ALUMINUM ALKYL COMPLEXES INCORPORATING AMIDINATE LIGANDS. R. F. Jordan and M. P. Coles, Department of Chemistry, University of Iowa, Iowa City, Iowa 52242.

A novel strategy for the design of cationic aluminum alkyl species has been developed based on compounds containing amidinate ligands. Complexes of the general formula {RC(NR)$_2$}AlR'$_2$ have been synthesised and converted to [{RC(NR)$_2$}AlR']$^+$ and [{RC(NR)$_2$}AlR'(L)]$^+$ cations, employing methods previously developed for group 4 transition metal complexes.

434. MECHANISM OF β-H TRANSFER IN d⁰ DIALKYL COMPLEXES: SYNTHESIS OF W(VI) OLEFIN COMPLEXES. Steve Wang, Carlos G. Ortiz, Khalil A. Abboud, and James M. Boncella Department of Chemistry, and Center for Catalysis, University of Florida, Gainesville, Fl 32611-7200

A series of electronically and coordinatively unsaturated W(VI) dialkyl complexes, **1**, that contain β-H atoms have been synthesized and are remarkably stable to β-H elimination. In the presence of trimethyl phosphine, **1**, are converted into the olefin complexes, **2** and alkane, via a β-H transfer process. Mechanistic studies on this transformation indicate that the transfer occurs via a β-H *abstraction* which nvolves direct transfer of the β-H from one alkyl group to the other without formation of a metal hydride intermediate. The reaction chemistry of the alanogous Mo complexes will also be discussed.

R = Et, *i*-Bu, CH$_2$CH$_2$Ph

435. **CYCLOPENTADIENYL, INDENYL AND BIS(CYCLOPENTADIENYL) TITANIUM IMIDO COMPOUNDS**

Simon C Dunn, David A Robson and Philip Mountford
Department of Chemistry, University of Nottingham, Nottingham, U.K. NG7 2RD.
Email: Philip.Mountford@Nottingham.ac.uk

The versatile synthons [Ti(NBu^t)Cl$_2$(4-NC$_5$H$_4$R)$_n$] (R = H, n = 2 or 3; R = Bu^t, n = 2) are entry points to half-sandwich η^5-cyclopentadienyl derivatives [Ti(η^5-C$_5$R'$_4$R")(NBu^t)Cl(py)], the η^5-1,2,3-trimethylindenyl species [Ti(η^5-C$_9$H$_4$Me$_3$)(NBu^t)Cl(4-NC$_5$H$_4$Bu^t)] and the bis(η^5-cyclopentadienyl) derivative [Ti(η^5-C$_5$H$_5$)$_2$(NBu^t)(py)]. Treatment of [Ti(η^5-C$_5$Me$_4$R)(NBu^t)Cl(py)] (R = Me or Et) with Na[C$_5$H$_5$] gives the mixed-ring derivatives [Ti(η^5-C$_5$H$_5$)(η^5-C$_5$Me$_4$R)(NBu^t)Cl(py)]. Addition of Li[C$_9$H$_7$] to [Ti(η^5-C$_5$H$_5$)(NBu^t)Cl(py)] gives the η^5-cyclopentadienyl, η^3-indenyl analogue [Ti(η^5-C$_5$H$_5$)(η^3-C$_9$H$_7$)(NBu^t)(py)]. The fluxional behavior of the mono- and bis-(cyclopentadienyl) derivatives is reported. The reaction of [Ti(η^5-C$_5$H$_5$)$_2$(NBu^t)(py)] with H$_2$S or [Ti(η^5-C$_5$H$_5$)$_2$(SH)$_2$] to form the μ-sulfido species [Ti(η^5-C$_5$H$_5$)$_2$(μ-S)]$_2$ is also discussed.

436.

CHEMISTRY OF THE DIVALENT SAMARIUM SUPPORTED BY TRIS-PYRAZOLYLBORATE LIGANDS, M. R. J. Elsegood, Dept of Chemistry, The University, Newcastle upon Tyne, NE1 7RU, U.K. A. C. Hillier, S.Y. Liu, **Andrea Sella,** Dept of Chemistry, University College London, 20 Gordon Street, London WC1H 0AJ, U.K.

The pyrazolylborates are proving a useful alternative to cyclopentadienyls. At the large lanthanide centres they are useful in controlling the coordination sphere giving lower coordination numbers and unusual geometries. In this paper we compare the reactivity of Sm(Tp*)$_2$ (Tp* = hydro-tris(dimethylpyrazolyl)borate) with that of decamethylsamarocene, considering its reactions with dichalcogenides, halocarbons, and metal-metal single bonds. Reaction of Sm(Tp*)$_2$ with ArEEAr give good yields of complexes (Tp*)$_2$SmEAr the structure of which depends on the size of the chalcogen- the thiolate is molecular [(Tp*)$_2$SmSPh] while the tellurolate crystallizes with a salt-like structure [(Tp*)$_2$Sm][Te$_3$Ph$_3$]. Analogous results are obtained in reactions of Sm(Tp*)$_2$ with halogens or halocarbons. Reaction of Sm(Tp*)$_2$ with metal-metal bonded transition metal carbonyls gives a range of structures depending on the system. [MCp(CO)$_3$]$_2$ (M= Cr, Mo, W) are cleaved to give the isocarbonyl bridged complexes [(Tp*)$_2$Sm-OC-MCp(CO)$_2$]. Co$_2$(CO)$_8$ gives the simple salt [(Tp*)$_2$Sm][Co(CO)$_4$]. With Mn$_2$(CO)$_{10}$, the unusual salt {[(Tp*)$_2$Sm]$_2$(μ-O$_2$CH)]$^+$ [Mn(CO)$_5$]$^-$ is isolated. A mechanism for the formation of this product will be discussed.

INOR

437. RECENT DEVELOPMENTS IN THE CHEMISTRY OF CUBOIDAL AND CUBANE-TYPE IRON-SULFUR CLUSTERS. R. H. Holm, C. A. Crawford, Jiesheng Huang, S. Mukherjee, J. W. Raebiger, and B. Segal, Department of Chemistry, Harvard University, Cambridge, Massachusetts 02138

Results pertinent to the title clusters will be presented. These include synthesis and properties of heterometal cubane-type clusters MFe_3S_4 derived from the Fe_3S_4 cuboidal clusters, $(OC)_3MFe_3S_4$ clusters with emphasis on redox chemistry, and covalently linked Fe_4S_4, Fe_4S_5, and MFe_3S_4 clusters. Structural and other properties will be considered in relation to the metal clusters in nitrogenase.

438. DOUBLE EXCHANGE AND VIBRONIC TRAPPING IN METAL CLUSTERS. J. J. Girerd, G. Blondin, Laboratoire de Chimie Inorganique, URA CNRS 420, Université Paris-Sud, 91405 Orsay, France

The concept of Double Exchange (DE) was introduced by solid-state physicists C. Zener, P.W. Anderson, H. Hasegawa and P.G. de Gennes in their study of $La_{1-x}Ca_xMn(III)_{1-x}Mn(IV)_xO_3$ manganites. DE generally applies to mixed-valence systems and basically corresponds to parallel alignment of core spins under the influence of the spin of an itinerant extra electron. We will give an orbital description of this effect. Vibronic trapping tends to inhibit electron transfer and thus DE. Use of the trivial equality $(S+0.5)^2 = S^2 + S + 0.25 = S(S + 1) + 0.25$ allows one to make the connection between second-order DE effects and Exchange. Examples will be given pertaining to Fe-S protein field. Remembering that DE was introduced for Mn(III)Mn(IV) manganites, the possibility of DE in a Mn(III)Mn(IV) pair of the Mn-O cluster of the photosynthetic Oxygen Evolving Center will be considered.

439. PULSED EPR STUDIES OF THE MANGANESE CLUSTER OF THE OEC AND ITS INTERACTION WITH TYROSINE RADICAL $Y_Z^\bullet$
R. David Britt, Department of Chemistry, University of California, Davis, CA 95616

^{55}Mn Electron Spin Echo (ESE) ENDOR experiments indicate that the manganese cluster of the photosystem II oxygen-evolving complex is tetranuclear. ESE-ENDOR and ESE envelope modulation (ESEEM) demonstrate that the redox active tyrosine $Y_Z^\bullet$ is sufficiently close to the Mn cluster for a significant magnetic interaction to exist between the tyrosine radical and the paramagnetic Mn cluster. This close proximity has led to models in which the tyrosine radical abstracts protons or hydrogen atoms from substrate waters bound to the Mn cluster [Gilchrist et al., (1995) *PNAS* 92:9545-9549; Hoganson et al., (1995) *Photosyn. Res.* 46:177-184]. Results of further pulsed EPR experiments designed to test these models will be presented.

440. **PROPERTIES OF VALANCE DELOCALIZED $[Fe_2S_2]^+$ CLUSTERS IN MUTANT FORMS OF *C. PASTEURIANUM* 2Fe FERREDOXIN**
Evert C. Duin,[†] Brian R. Crouse,[†] Jacques Meyer,[‡] and Michael K. Johnson[†], [†]Department of Chemistry and Center for Metalloenzyme Studies, University of Georgia, Athens, GA 30602, [‡]CEA, Département de Biologie Moléculaire et Structurale, CENG, 38054 Grenoble Cedex 9, France

The $[Fe_2S_2]^+$ clusters in the Cys56Ser and Cys60Ser mutant forms of *Clostridium pasteurianum* 2Fe ferredoxin exist in two distinct forms with $S = 1/2$ and 9/2 ground states. The properties of both forms have been investigated by the combination of EPR, variable-temperature MCD and resonance Raman spectroscopies. The results indicate that the $S = 1/2$ and 9/2 species correspond to valence localized and valence delocalized forms, respectively, and afford unique insight into the properties of a valence delocalized $[Fe_2S_2]^+$ fragment. The origin of valence delocalization in these mutants and the implications for the electronic and magnetic properties of higher nuclearity Fe-S clusters will be discussed.

441. **NMR OF FeS PROTEINS.** I. Bertini, Department of Chemistry, University of Florence, 50121 Florence, Italy

^{1}H NMR can be successfully performed on $Fe_2S_2^+$, $Fe_3S_4^+$, $Fe_4S_4^{1+/2+/3+}$ containing proteins. Analysis of the data based on an exchange coupling model provides a detailed pictures of the valence distribution within the polymetallic center and within the protein frame. A model for the folding/unfolding of Fe-S proteins is also proposed.

442.
THE ELECTRONIC STRUCTURE OF HETEROTRINUCLEAR THIOPHENOLATE BRIDGED COMPLEXES: LOCALIZED VS DELOCALIZED MODELS K. Wieghardt, T. Glaser, E. Bill, T. Weyhermüller, V. Schünemann, and A. X. Trautwein, Max-Planck-Institut für Strahlenchemie, D-45470 Mülheim, and Institut für Physik, Medizinische Universität Lübeck, D-23528 Lübeck, Germany

We have synthesized linear trinuclear complexes of the general formula [LFeMFeL]n+ where M=Fe (n=2,3,4), Co (n=2,3), Cr (n=1,2,3), and Ni (n=2,3,4). The metals are all octahedrally coordinated and bridged by six thiophenolate ligands. The complexes were characterized by X-ray crystallography, UV-Vis, IR, electrochemistry, MS, and NMR. In order to elucidate the electronic structures we have measured the temperature-dependent susceptibilities, Mössbauer, and EPR spectra. From temperature-dependent Mössbauer spectra varying degrees of electron delocalization for the mixed valent species were established. An understanding of their electronic structures is provided within the frame of 'double exchange' mechanism.

443.
QUANTUM DOT STRUCTURES USING SEMICONDUCTOR NANOCRYSTALLITES. **Moungi G. Bawendi** Department of Chemistry, The Massachusetts Institute of Technology, Cambridge, MA 02139

Nanometer size particles of semiconductors can be fabricated with narrow size distributions and with surfaces passivated with a wide variety of organic groups. These particles have electronic properties which are discrete and tunable due to quantum confinement effects, hence the label "quantum dots." In this talk we discuss the CdSe system. We begin with a brief description of the synthesis and characterization of non-interacting dots. We then turn to structures of dots, including close packed colloidal crystals of dots, glassy thin films, and dots combined with a bulk semiconductor or with semiconducting or insulating polymers in thin film systems. We address issues of energy and charge transfer in these heterostructures.

444. VISUALIZATION OF INTERCALATION PROCESSES AT THE ATOMIC LEVEL. Michael J. McKelvy* and William S. Glaunsinger,°* *Center for Solid State Science and °Department of Chemistry and Biochemistry, Arizona State University, Tempe, Arizona 85287-1704

The mechanisms of intercalation processes have been observed for the first time at resolutions down to the atomic scale using dynamic high resolution transmission electron microscopy. In this work, the transition metal disulfides TiS_2 and TaS_2 have been reacted with Ag, Hg and NH_3 to form unique insertion compounds. The following reaction processes have been elucidated using these model materials:

intercalation: $(4y'/3 + y'')NH_3 + 2H\text{-}TaS_2 \rightarrow (NH_4^+)_{y'}(NH_3)_{y''}TaS_2^{y'-} + (y'/6) N_2$
deintercalation: $Hg_{1.24}TiS_2 \rightarrow TiS_2 + 1.24\ Hg$
structural: $Ag_{0.17}TiS_2$ (stage 1) $\rightarrow Ag_{0.17}TiS_2$ (stage 2)

This research has demonstrated that intercalation can be a complex process that can be dramatically influenced by defects.

445. NEW HORIZONS IN GRAPHITE CHEMISTRY - PILLARED STRUCTURES AND NANOCOMPOSITES. Sherry Zhang, Zhengwei Zhang, and Michael M. Lerner, Department of Chemistry, Oregon State University, Corvallis, OR 97331

The large family of known graphite intercalation compounds (GICs) includes more members than for most layered hosts, and graphite is unique in incorporating a broad range of both cations and anions. Nevertheless, because the intercalants have be primarily small and highly-symmetric ions, these GICs cannot be employed in many applications that benefit from nanoporous structures or functional intercalants. Recent reports have described the first indications of pillared graphite structures and polymeric intercalants, compounds that have well-known analogs in layered metal oxides and sulfides. This talk will focus on these recent developments in graphite chemistry, and provide the details of simple methods for preparing pillared graphites in quantity.

446. NANOPOROUS MOLECULAR SANDWICHES: PILLARED TWO-DIMENSIONAL HYDROGEN-BONDED NETWORKS WITH ADJUSTABLE POROSITY. V. A. Russell, C. C. Evans, W. Li, and M. D. Ward*. Dept. of Chemical Engineering and Materials Science, University of Minnesota, 421 Washington Ave. SE, University of Minnesota, Minneapolis, MN 55455

Crystal engineering of molecular materials commonly is frustrated by the absence of reliable structural paradigms that enable the systematic design of crystal lattices with predictable structure and desirable function. This can be attributed, at least partially, to the absence of robust supramolecular motifs that serve as synthons for the assembly of crystal lattices. We describe here a novel class of molecular crystals based on 2-D hydrogen-bonded networks, comprising guanidinium ions and the sulfonate groups of alkane- or arenedisulfonate ions, which are robust because of their ability to flex about (guanidinium)N-H...O(sulfonate) hydrogen-bonding hinges. The disulfonate ions act as pillars to connect opposing hydrogen-bonded sheets, forming nanoporous galleries with one-dimensional channels defined by the molecular pillars. The flexibility of the hydrogen-bonding network allows the galleries to adapt to changes in the steric requirements of guest molecules that fill the nanoscale voids. The robustness of the 2-D networks effectively reduces crystal engineering to the last remaining dimension, enabling rational and systematic design of nanoporous materials whose gallery heights can be adjusted predictably by the choice of the disulfonate pillar. This design strategy may lead to new host-guest solid state materials for molecular separations, chemical reactions performed in molecularly confined environments, and optoelectronic applications.

447.
CYANOGEL SOL-GEL PROCESSING AS A NOVEL APPROACH TO NEW MATERIALS.
Andrew B. Bocarsly, Marija Heibel and Stefanie Sharp, Frick Laboratory, Princeton University, Princeton, NJ, 08544-1009

Traditionally, sol-gel processing has been limited to a select group of oxide materials due to the lack of known inorganic hydrogel systems which might be utilized as ceramic precursors. Recently, we reported that a new broad class of hydrogels can be synthesized based on the reaction of tetrachlorometalate with a wide variety of transition metal cyanometalate complexes. We term these materials cyanogels, based on the use of a bridging cyanide ligand as a structural motif within the gel. Xerogels generated from these materials offer a new entrance into the processing of gel systems to generate ceramics and metal alloys containing a variety of transition metals. To date, systems based on Pd-Co and Sn-Fe have been explored and provide a basis for the application of cyanogel chemistry to materials synthesis..

448. DECORATING THE CHANNELS OF POROUS COORDINATION SOLIDS.
O. M. Yaghi, Hailian Li, C. E. Davis and T. L. Groy, Department of Chemistry and Biochemistry, Arizona State University, Main Campus, PO Box 871604, Tempe, AZ 85287-1604

Cationic, anionic and neutral porous networks constructed from transition metal ions and highly symmetric organic molecules have been produced in crystalline form. The pores are occupied with molecular guests that can be removed, thus allowing the inclusion of other species into the channels. Networks based on 4,4'-bipyridine and 1,3,5-benzenetricarboxylate have been used to demonstrate the utility of coordination chemistry in tailoring the channels for the selective binding of aromatics, alcohols, and anions. We have found that such processes occur with high selectivity due to the presence of coordinatively unsaturated metal ions and weak intermolecular forces within the pores. This presentation will show that 1-D, 2-D and 3-D networks can be prepared as porous solids capable of reversible binding of guests and selective inclusion that is not only based on shape and size of incoming guests but also on their electronic affinity to the channels. The synthesis, structure and inclusion properties of these solids including strategies for decorating the channels with organic π-systems and coordinately unsaturated metal centers will be presented.

449. PSEUDO-TRIGONAL PRISMATIC VS. PSEUDO-OCTAHEDRAL STRUCTURES FOR THE HALF-SANDWICH DERIVATIVES Cp*MH$_3$(dppe)
(M = Mo, W). Brett Pleune, Rinaldo Poli, and James Fettinger, University of Maryland, College Park, MD 20742

The syntheses of compounds Cp*MH$_3$(dppe) (M = Mo, **1**; W, **2**) are described. The X-ray structure of **1** shows an unprecedented pseudo-trigonal prismatic geometry. Protonation with HBF$_4$·Et$_2$O in Et$_2$O at room temperature affords [Cp*MH$_4$(dppe)]$^+$ salts (M = Mo, **3**; W, **4**), which are classical tetrahydrido species with distorted pseudo-pentagonal bipyramidal structures. Decomposition of **3** in MeCN at room temperature selectively affords [Cp*MoH$_2$(MeCN)(dppe)]$^+$, **5**, which is fluxional with inequivalent H and P nuclei, consistent with a pseudo-trigonal prismatic structure analogous to that of **1**. Further protonation of **5** in MeCN or direct protonation of **1** affords two isomers of complex [Cp*MoH(MeCN)$_2$(dppe)]$^{2+}$, **6** and **7**. The X-ray structure of **6** shows a highly distorted pseudo-octahedral geometry with relative *trans* MeCN ligands, while the geometry of **7** has been elucidated by solution NMR studies. Hydrogen scrambling and/or isomerization processes for the various compounds are rationalized via facile pseudo-Bailar twists interconnecting pseudo-trigonal prismatic and pseudo-octahedral geometries.

450. **RESOLUTION OF SOLVENT-CAGED AND DIFFUSIVE REACTIONS OF (η^5-C$_5$H$_4$COCH$_3$)Mn(CO)$_2$.** Tien Jie Jiao and <u>Ted Burkey</u>, Department of Chemistry, The University of Memphis, Memphis, Tennessee, 38152-6060.

We are studying the photolysis of (η^5-C$_5$H$_4$COCH$_3$)Mn(CO)$_3$ as a model to understand potential photochromic compounds. Flash photolysis studies of (η^5-C$_5$H$_4$COCH$_3$)Mn(CO)$_3$ with tetrahydrothiophene (THT), indicate that THT reacts in two kinetically distinct pathways. The lifetime of the first pathway is shorter than the instrument response time (< 100 ns), while the lifetime of the second pathway is dependent on ligand concentration (2.4 x 10^6 M^{-1} s^{-1}). This kinetic behavior is distinct from that observed for other metal carbonyls. The enthalpy of CO dissociation is 31 kcal/mol and THT addition is -22 kcal/mol. The magnitude of these values are 16 and 12 kcal/mol, respectively, less than analogous reactions for (η^5-C$_5$H$_5$)Mn(CO)$_3$ suggesting that Mn-L bond enthalpies are very sensitive to cyclopentadienyl substituent effects These results will be compared to those for other complexes and ligands.

451. **ORGANOMETALLIC NOBLE GAS COMPOUNDS IN SOLUTION AT ROOM TEMPERATURE**
<u>Michael W. George</u>, Xue-zhong Sun, David C. Grills and Martyn Poliakoff
Department of Chemistry, University of Nottingham, University Park, Nottingham NG7 2RD, UK
e-mail: mike.george@nottingham.ac.uk

We have combined of time-resolved infrared spectroscopy and supercritical fluids to characterise, for the first time, organometallic noble gas complexes in solution at room temperature. Photolysis of metal carbonyl complexes in supercritical noble gas solution results in ejection of CO and coordination of the rare gas, eg W(CO)$_6$ → W(CO)$_5$(Xe). The noble gas complexes have comparable stability to organometallic alkane complexes. We found that CpRe(CO)$_2$(Xe) is less reactive toward CO than all of the reported organometallic alkane complexes except CpRe(CO)$_2$(n-heptane)

452. **SONOLUMINESCENCE FROM METAL CARBONYLS.**
<u>W.B. McNamara III</u>, Y.T. Didenko, K.S. Suslick School of Chemical Sciences, University of Illinois at Urbana-Champaign, Urbana, IL 61801

Acoustic cavitation results in extraordinary transient conditions inside the collapsing bubble. In addition to interesting chemical effects (sonochemistry), cavitation also produces light emission. Such sonoluminescence from cavitating clouds of bubbles ("multibubble sonoluminescence", MBSL) in room temperature liquids closely resembles flame emission, and this has been used to measure the effective temperature and pressure of species formed during the cavitation process. Effective emission temperatures have been obtained for MBSL from excited state metal atom emission (from sonolysis of several metal carbonyls in silicon oil). Effective pressures can be obtained from line shifts of the metal atom emission. The effective transient conditions formed during cavitation of bubble clouds are 5000 K and ≈ 1000 atmospheres.

453. **GAS-PHASE PHOTOELECTRON SPECTROSCOPY OF CpNiNO AND Cp*NiNO: A "RESOLUTION" TO THE CONTROVERSY.** Dennis L. Lichtenberger and John R. Pollard, University of Arizona, Tucson, AZ 85721.

The utilization of different source energies in photoelectron spectroscopy is an effective method for determining the orbital character of transition metal complexes. Problems can arise when the resolution loss that accompanies higher energy ionization sources also effects the apparent relative intensities of overlapping ionization bands. Recent publications suggest different assignments for the gas-phase photoelectron spectrum of CpNiNO (Inorg. Chem. 35, 2504 and 2515). In order to eliminate resolution factors, we collected very high resolution HeI and HeII data (13 meV Ar $^2P_{3/2}$) on CpNiNO. In addition, Cp*NiNO was synthesized and HeI and HeII spectra were collected to observe how the more electron-donating $C_5(CH_3)_5$ ring would affect the ionization bands. These results show that experimental resolution plays an extremely important role in the interpretation of the CpNiNO spectra.

$R = H, CH_3$

454. **RATES AND MECHANISMS OF COBALT DICARBOLLIDE EXTRACTANT "DEACTIVATION" BY NITRIC ACID AND SODIUM HYDROXIDE.** Rebecca M. Chamberlin, Anthony B. Pinkerton, Paul K. Hurlburt, and Kent D. Abney. Chemical Science and Technology Division, Los Alamos National Laboratory, Los Alamos, NM 87545.

Cation exchange of cesium and strontium for nuclear waste remediation is mediated by cobalt dicarbollide and its derivatives. It has long been known that the parent anion, $[Co(C_2B_9H_{11})_2]^-$, loses most of its extraction capacity after prolonged contact with nitric acid, while boron-halogenated derivatives are indefinitely stable. Our studies indicate that this deactivation process is autocatalytic in nitric acid concentrations above 3 M, promoted by addition of nitrite, and inhibited by addition of azide or hydrazine. Reaction of $[Co(C_2B_9H_{11})_2]^-$ with nitric acid yields a mixture of anionic and zwitterionic boron-substituted products. Reaction rates of halogenated and alkylated cobalt dicarbollide with sodium hydroxide were also measured and found to correlate with the electron-withdrawing character of the substituents. The implications of these results for the use of cobalt dicarbollide in nuclear waste treatment will be addressed.

455. CONDENSATION POLYMERIZATION OF COBALT DICARBOLLIDE. S.A. Fino; G.J. Balaich, Department of Chemistry, Chemistry Research Center, United States Air Force Academy CO 80840-6230. R.M. Chamberlin; K.D. Abney, Chemical Science and Technology Division, Mail Stop J514, Los Alamos National Laboratory, Los Alamos, NM 87545.

Polymers containing the cobalt dicarbollide anion, $[Co(C_2B_9H_{10})_2]$, in the main chain have potential application as cation exchange materials for the selective removal of $^{137}Cs^+$ and $^{90}Sr^{+2}$ from nuclear wastes. Our research is directed at the synthesis of difunctionalized monomers suitable for condensation polymerization using the dilithio complex $[Co(LiC_2B_9H_{10})_2]$, as a starting material. Reaction of $[Co(LiC_2B_9H_{10})_2]$ with CO_2/H^+ in the thf gives the dicarboxylic acid complex, $[Co(C_2B_9H_{10})_2(COOH)_2]$. Two methods for the conversion of the dicarboxylic acid to a polyamide derivative, $[Li]_n[OC(C_2B_9H_{10})Co(C_2B_9H_{10})CONH(CH_2)_6NH]_n$ have been investigated. The spectroscopic properties of the monomers as well as the characterization of the polyamide by light scattering photometry and GPC methods will be discussed. Reaction of the dialcohol monomer, $[Co(C_2B_9H_{10})_2(C_2H_4OH)_2]$, formed by reaction of $[Co(LiC_2B_9H_{10})_2]$ with CH_2OCH_2/H^+ in thf, with isocyanates provides entry into polyurethane chemistry and will also be discussed.

456.
AQUEOUS CHEMISTRY OF THE FIVE-COORDINATE ZIRCONOCENE DICHLORIDE COMPLEX $PhP(CH_2CH_2C_5H_4)_2ZrCl_2$. O. J. Curnow, J. R. Butchard, S. J. Smail, Department of Chemistry, University of Canterbury, Private Bag 4800, Christchurch, New Zealand

Phosphine ligands and cyclopentadienyl ligands are two of the most important classes of ligands in organometallic chemistry. The bis(cyclopentadienyl)-phosphine ligand $[PhP(CH_2CH_2C_5H_4)_2]^{2-}$ provides the first example of a ligand containing two cyclopentadienyl substituents and one phosphine group that is able to chelate in a tridentate fashion. The preparation and characterisation of the novel five-coordinate zirconocene dichloride complex $PhP(CH_2CH_2C_5H_4)_2ZrCl_2$, in which the P atom is also coordinated to the Zr atom and tethered to both cyclopentadienyl rings, will be presented along with its aqueous chemistry which has been investigated using x-ray crystallography, ^{31}P-NMR spectroscopy and Electrospray MS.

457. COMPARISON OF THE PHOTOELECTRON SPECTRA OF LMo(X)(3,4-TOLUENEDITHIOL) (L=HYDROTRIS(3,5-DIMETHYLPYRAZOLYL)BORATE) (X=O, NO, S): THE REINTRODUCTION OF NeI AS A SUITABLE IONIZATION SOURCE, Barry L. Westcott and John H. Enemark, Department of Chemistry, University of Arizona, Tucson, AZ 85721

Fixed ionization sources for photoelectron spectroscopy (PES) have traditionally consisted of HeI and HeII radiation; however, these sources cannot distinguish between Mo and S ionizations because their relative photoionization cross-sections do not differ appreciably between the two energies. A NeI ionization source should be useful for distinguishing between ionizations from Mo and S, as the photoionization cross-section for S increases fourfold at the NeI energy as compared to the HeI, whereas the Mo cross-section remains relatively constant for these two ionization energies. Only limited use has been made of NeI radiation in the past because of the problems presented by the spin orbit coupled line 0.12 eV below the main line. We report here the HeI and NeI spectra of three structurally similar molybdenum dithiolene complexes. The results show that the ligand X and the formal oxidation state of the metal have little effect on the position of the first ionization potential, suggesting that the Mo-S interaction is the dominant factor. We propose that the S donors create an "electronic buffer" for the Mo center of these complexes, and that a similar effect occurs in molybdenum containing enzymes.

458. PHOTOCHEMISTRY OF ROUSSIN'S SALTS, A STRATEGY FOR NITRIC OXIDE DELIVERY. James Bourassa and Peter C. Ford*, Department of Chemistry, University of California, Santa Barbara, CA 93106.

Roussin's black salt (RBS), $Fe_4S_3(NO)_7^-$, and Roussin's red salt (RRS), $Fe_2S_2(NO)_4^{2-}$, have attracted interest due to their possible biological relevance. The quantitative photochemistry and flash photolysis of the compounds have been studied. RBS photodecomposes ($\phi = 1.1 \times 10^{-3}$ in aerated H_2O, λ_{irr} = 313 - 546 nm) to ferrous ions, sulfide and NO, which later form insoluble ferric precipitates, elemental sulfur and nitrite. RRS cleanly photoconverts to RBS, sulfide and NO ($\phi = 0.39$ in aerated H_2O, λ_{irr} = 313 - 546 nm). In deaerated CH_3CN, RRS photochemically forms the RBS dianion, $Fe_4S_3(NO)_7^{2-}$, which is oxidized by NO to form the monoanion and N_2O. The amount of nitric oxide photoreleased by both compounds has been determined with NO sensitive electrodes. Proposed mechanisms as well as photochemical NO delivery will be discussed.

459.
PHOTOCHEMICAL REPROCESSING OF URANIUM. **T. Mark McCleskey,** Erin E. Fahrenhrog-Hallman, Trudi M. Foreman, Tracey Franckom, Carol J. Burns, and Nancy S. Sauer, Chemical Science and Technology Division, Los Alamos National Laboratory, MS-J514, Los Alamos, NM, 87545.

We have studied the photochemical reduction of uranyl to uranium dioxide for the potential application of reprocessing uranium. Reductive quenching of the uranyl excited-state occurs in the presence of either oxalate or formate to give uranium dioxide as the product. Despite the same observed product, we have found that the quenching mechanisms are different for oxalate and formate. The solid uranium dioxide product has been characterized by IR, powder XRD and SEM. The solid shows unique features in terms of crystallinity and particle size, which make it ideally suited for further conversion steps.

460. NITROGEN, OXYGEN AND CHLORINE ATOM TRANSFER REACTIONS INVOLVING EARLY TRANSITION METAL PORPHYRINS. Xiaotai Wang and L. Keith Woo*, Department of Chemistry, Iowa State University, Ames, Iowa 50011-3111

Metalloporphyrin complexes have been very useful in studying inner sphere electron transfer processes. An atom transfer reaction between the V(III)-porphyrin complex (TTP)VCl (TTP = 5, 10, 15, 20-tetra-p-tolylporphyrin dianion) and (OEP)Mn≡N (OEP = octaethylporphyrin dianion) led to the observation, by ^{1}H NMR, of the first vanadium porphyrin nitride complex (TTP)V≡N. New divalent metalloporphyrins (TTP)M(THF)$_2$ (M =V, Ti) have allowed us to study oxygen and chlorine atom transfer reactions with organic sulfoxides, epoxides and chlorides. Some representative reactions are shown in equations (1) through (3). The scope of these single and two-electron redox processes as well as the mechanisms will be discussed.

$$(TTP)Ti(THF)_2 + Me_2S=O \longrightarrow (TTP)Ti=O + Me_2S + 2THF \quad (1)$$

$$(TTP)Ti(THF)_2 + ClCH_2CH_2Cl \longrightarrow (TTP)TiCl + CH_2=CH_2 + 2THF \quad (2)$$

$$(TTP)Ti(THF)_2 + \text{PhCH(O)CH}_2 \longrightarrow (TTP)Ti=O + \text{PhCH=CH}_2 + 2THF \quad (3)$$

461.
TIME-RESOLVED INFRARED SPECTROSCOPY OF MLCT EXCITED STATES. **K. M. Omberg** J. R. Schoonover, R. B. Dyer, and T. J. Meyer, Department of Chemistry, University of North Carolina-Chapel Hill, Chapel Hill, NC 27599-3290 and Bioscience and Biotechnology Group, Division of Chemical Science and Technology, MS J586, Los Alamos National Laboratory, Los Alamos, NM 87545

Time-resolved, step-scan Fourier transform infrared spectroscopy (TRIR) has been applied to the MLCT excited states of polypyridyl complexes of Ru^{II}, Os^{II}, and Re^{I}. TRIR is complementary to time resolved resonance Raman spectroscopy, but has certain unique advantages; both chromophoric and non-chromophoric ligands can be observed as can non-Raman-active asymmetric modes. Application of TRIR to [Ru(bpy)$_3$]$^{2+*}$ and [Re(4,4'-(X)$_2$bpy)(CO)$_3$(4-Etpy)]$^{+*}$ gives insight into excited-state electronic structure at the metal and at the peripheral and acceptor ligands.

462. SIDEROPHORE-MEDIATED IRON TRANSPORT IN MICROBES: FORM AND FUNCTION. Kenneth N. Raymond and Zhiguo Hou, Department of Chemistry, University of California, Berkeley, CA 94720-1460

Most aerobic and anaerobic bacteria and fungi synthesize and excrete siderophores for solubilization and transport of iron. The very stable Fe^{3+} siderophore complex is taken up via membrane receptors that are generally stereoselective, primarily recognizing the metal ion coordination center. Since iron is required by all except a few microorganisms, and often is **the** limiting nutrient in microbial growth, understanding siderophore-mediated iron transport is a significant problem in both bioinorganic chemistry and medicine. In recent years it has become increasingly clear that iron plays a major role in several disease states, notably bacterial infections.

This paper will review siderophore-mediated iron transport and will then focus on recently characterized siderophores, their structure and function. In particular, alcaligin is a 20-membered macrocyclic dihydroxamate siderophore that is a potent tetradentate iron chelator. Although first isolated from an aquatic algae it was subsequently found to be the siderophore of two pathogenic bacteria, *Bordetella pertussis* and *B. Bronchiseptica*, pathogens of the upper respiratory tract of humans and domestic mammals, respectively. The unique preorganization of the ferric alcaligin structure has a dramatic effect on its stability constants. The solution thermodynamics and other properties of this and related compounds will be described in detail.

463.

FERRITIN mRNA AND PROTEIN: The many dimensions of bioinorganic chemistry. Elizabeth C. Theil, Dept. Biochemistry, North Carolina State University, Raleigh, NC 27695-7622.

Ferritin mRNA is recruited for protein synthesis in animals when cytoplasmic levels of iron are high. (Ferritin mRNA is transcribed from DNA and used immediately for protein synthesis in plants when cytoplasmic levels of iron are high.) The regulatory mRNA structure in the ferritin mRNA of animals, the IRE is a folded helix loop. Transition metal complexes (TMCs) of redox active metals (Cu, Ru, Rh, Fe) have been very useful in detecting/probing the effect of mutation on subdomains of RNA within the IRE. Different sizes and shapes of the metal ligands recognize features of the IRE three-dimensional structure and metal dependent free radicals cleave the RNA to report the site of the interaction and can also show the outline of protein binding sites.

Ferritin protein is the only protein known which directs a phase transition of a metal ion in solution to a solid phase. The purpose is to concentrate iron as nanoparticles inside the soluble protein because the solubility of free Fe(III) is $\sim 10^{-11}$ times below the concentration needed by aerobic cells. Hydrated Fe(II) ions are oxidized and transported through the protein to nucleation sites on the protein interior where the mineral phase forms by hydrolysis of hydrated Fe(III). The metal ions are apparently guided by binding sites on the protein itself which can control the rate of oxidation hydrolysis. Tyrosine ligands (R-phenyl-OH) can be involved in fast rates specific to H-type (fast) ferritin subunits while all ferritin subunits have clusters of glutamate ligands (R-COOH) necessary for nucleation. The nuclearity and level of hydration of Fe(III) as it passes through the protein and the exact path are currently being studied. (part support NIH-DK020251).

Reference
Waldo, G.S. and Theil, E.C. (1996) Ferritin and Iron Biomineralization. IN: COMPREHENSIVE SUPRAMOLECULAR CHEMISTRY, Vol. 5, (K.S. Suslick, vol. ed.), Pergamon Press, Oxford, U.K., pp. 65-89.

464.

REAGENTS FOR THE DELIVERY OF THERAPEUTIC METALS TO CANCER SITES. **Claude F. Meares** Chemistry Department, University of California, Davis, CA 95616

Monoclonal antibodies and engineered protein fragments that target radiotherapy to human cancer are now available. A number of different bifunctional chelating agents have been developed to tag monoclonal antibodies with radiometals for in vivo diagnosis and therapy. Because of their nuclear properties, metallic radioisotopes such as In-111, Y-90, and Cu-67 are particularly useful for labeling tumor-targeting molecules. Challenges still to be overcome include physiological barriers such as transport into tumors, bone marrow irradiation, and immunogenicity. But the chemistry of radionuclide conjugation is now well developed, there are cures in model systems, and there are promising clinical results. The methodology of radiolabeling has advanced from the original conjugation of bifunctional chelating agents to proteins followed by addition of the metal, to pre-labeled antibodies in which a carrier-free radiolabeled reagent is conjugated to the protein, and to pre-targeted antibodies in which a radiolabeled metal chelate is administered to the patient after the antibody has been allowed to localize in the tumor. Recent results will be described.

465.

METALLOBIOCHEMISTRY AT THE ENDOTHELIAL CELL SURFACE: ANGIOTENSIN CONVERTING ENZYME. James F. Riordan, Ctr. Biochem. Biophys. Sci. Med. Harvard Medical School, Boston, MA 02115

The apical surface of the endothelial cell, which faces the lumen of the vascular system and is in constant contact with circulating blood, is a thriving locus of essential metallobiochemistry. Not only does it figure prominently in the cellular uptake of metals destined to be incorporated into metalloproteins but it is also the attachment site of a variety of metalloenzymes, most notably angiotensin converting enzyme (ACE), endothelin converting enzyme, and various aminopeptidases, as well as membrane-protein-solubilizing proteases. ACE is unusual in that it contains two catalytically active zinc sites, both of which act in blood pressure regulation and are the targets of antihypertensive ACE inhibitors. The structure and function of endothelial ACE and its relationship to testicular ACE will be discussed in the context of other zinc enzyme active sites.

466. REVERSIBLE NO-CARRYING HEME PROTEINS FROM THE SALIVA OF BLOODSUCKING INSECTS. **F. Ann Walker**, José M.C. Ribeiro, Jesus G. Valenzuela, Donald E. Champagne, John R. Anderson, Andrzej Weichsal, Celia Balfour and William R. Montfort, Departments of Chemistry, Entomology and Biochemistry, University of Arizona, Tucson, AZ 85721

We have discovered that insects from two very different families have cherry-red saliva glands which contain abundant nitrosylheme proteins. In each case, the heme is in the Fe(III) form, and the NO-loaded protein is EPR silent. NO dissociates reversibly from these ferriheme centers upon dilution or increase in pH from that of the saliva (~ 5.5) to that of the blood of the victim (~7.4), allowing the insects to release NO into the blood stream of the victim to cause blood vessel dilation and to prevent platelet aggregation. Recent cloning and expression of the major nitrophorin from *Rhodnius prolixus* (the "kissing bug") has allowed us to obtain large amounts of this protein for spectroscopic and X-ray crystallographic studies. We have shown that the ^{1}H NOESY spectra of the native and recombinant proteins are identical, and the rates of NO release from the two are within experimental error. The bedbug, *Cimex lectularius*, also has a NO-releasing protein, but its molecular weight, protein sequence, and probable heme proximal ligand are different from those of the *Rhodnius* protein, suggesting that these two insects arrived at the use of NO heme proteins to aid in obtaining a sufficient blood meal by convergent evolution.

467. BIOINORGANIC CHEMISTRY IN PASADENA DURING THE MARJAM BEHAR YEARS. Sunney I. Chan, Department of Chemistry, California Institute of Technology, Pasadena, CA 91125

Through her interest in my work on cytochrome c oxidase the past fifteen years, Dr. Marjam Behar has been an inspiration to me and my students to continue to work toward an improved understanding of the structure and function of this interesting and important enzyme. In this lecture, I will highlight our current knowledge of how this molecular machine works, specifically focusing on those aspects of the problem that relate to the proton pumping function of the enzyme. New results on other terminal oxidases that bear on this issue will also be presented. These results will be discussed in light of the recently reported X-ray structures of the protein.

468. REFLECTIONS ON A CAREER IN RESEARCH TEACHING AND SCIENCE ADMINISTkATION. M.G. Behar, Division of Research Grants, National Institutes of Health, Bethesda, Maryland 20892-7852

From 1962-79 I was Director of the Core Facility for Analytical Chemistry in the Center for Research in Anesthesia at the University of Penn. I was the only chemist in a group of physicians investigating the effects of anesthesia on cerebral blood flow (CBF). Human volunteers were used, who were given Kripton 85 to inhale, before and during anesthesia. Arterial and yugular venous blood samples were taken at set intervals. By measuring the concentration of Kr 85, by scintillation counting, in these samples, CBF could be calculated. My role was in the development of micro analyticasl techniques to measure the state of oxygenation and the metabolic status of the brain during these studies. Measurements included oxygen and carbon dioxide contents, and partial pressures ,glucose, lactate, pyruvate,hemoglobin, electrolytes (bromide, fluoride) calcium, adenosine, inosine, xanthine, hypoxanthine, kallikrein, bradykinin, histamine, morphine, barbiturates and other anesthetics concentrations. I was also responsible for training and advising anesthesia residents, fellows, and medical students, on bioanalytical chemistry techniques needed for their research projects, as well as training and supervising research laboratory technicians. In 1980 I came to NIH as a Health Scientist Administrator, the Executive Secretary of the Bioanalytical and Metallobiochemistry (BMT) Study Section a position I held until 1984, when I moved to Special Review Section to review a variety of applications in chemistry including R01, SBIR (small business innovative research), S10's (shared instrumentation), P41 (research resources), P01 (program project). Duties include setting up the review panels consisting of distinguished scientists in the appropriate discipline, running the SSS meetings, giving the members proper orientation on the scientific merit review process, and writing the summary statements. Both my careers, the first one in research and teaching and the last one in science administration have been equally challenging and rewarding. My experiences at Penn have helped me to gain appreciation for the quality of the peer review process an its impact on the advancement of science.

469. SYNTHETIC AND SPECTROSCOPIC APPROACHES TO THE COPPER/QUINONE COUPLED CENTER IN AMINE OXIDASES.
Jochen Rall and Wolfgang Kaim, Institut für Anorganische Chemie der Universität, Pfaffenwaldring 55, D-70550 Stuttgart, Germany

The ambivalent behaviour of copper/quinone complexes in terms of a possible valence tautomerism has received wider attention because a Cu^I/semiquinone intermediate was reported to occur in the O_2-related part of the catalytic cycle of amine oxidase enzymes (Dooley et al., Nature 349 (1991) 262).
Varying the ancillary ligand(s) L in synthetic copper/o-quinone complexes $(L)Cu^{n+}(Q^{n-})$ has made it possible to tune the spin distribution in these compounds as studied by EPR spectroscopy. With nitrogen donor co-ligands L, the $Cu^{II}(Q^{2-})$ form was observed in all instances. In contrast, monodentate or chelating thioethers favour the $Cu^I(Q^{\bullet-})$ form as evident from small $^{63,65}Cu$ coupling, semiquinone hyperfine splitting and from g factors which are closer to the free electron value. Employing a bidentate ligand with imidazole-N and thioether-S donor atoms, we could observe a temperature dependent valence tautomer equilibrium, as observed previously only for the native enzyme.

470. NEW DIRHODIUM(II) CARBOXYLATE COMPLEXES WITH O,N-CHELATING AND N, N, N-TRIDENTATE LIGANDS. Angelica D. Brown, John C. Bollinger, John C. Huffman, George Christou, Department of Chemistry, Indiana University, Bloomington, Indiana 47405

Dirhodium tetracarboxylate compounds, $Rh_2(O_2CR)L_2$ (R=alkyl), have been found effective against several tumors in mice. These complexes are capable of binding to DNA, RNA and proteins in cancerous cells. The detailed mechanism of activity remains unknown. In order to increase our understanding of the reactions of dinuclear Rh complexes with the kinds of donor ligands found in nucleotides, we have been investigating the use of simple chelate groups. In this study, dirhodium carboxylate complexes with O, N-chelating ligands (6-methylpicolinate, 2-pyridylacetate, 2-(hydroxyethyl)pyridine, and 2-(hydroxymethyl)pyridine) and tridentate ligands have been synthesized and characterized. In several of the complexes the bidentate ligands are coordinated to the dirhodium core in an axial-equatorial fashion, and in the presence of coordinating solvents the ligands migrate to the equatorial sites to make the thermodynamically stable form, and the solvent molecules occupy the axial positions. The rate and mechanism for this transformation were investigated by UV-vis and ^{1}H-NMR spectroscopies.

471. REACTIONS OF NICKEL(II) WITH 8-HYDROXYQUINOLINE SALTS: STRUCTURAL AND MAGNETIC PROPERTIES OF NOVEL CUBANE AND TRINUCLEAR NICKEL AGGREGATES. Kevin S. Kolack, David N. Hendrickson[‡], Ziming Sun[‡], John C. Huffman[†], Kirsten Folting[†], William E. Streib[†], and George Christou* Department of Chemistry, Indiana University and [†]the Molecular Structure Center, Bloomington, IN 47405, and [‡]the Department of Chemistry, University of California at San Diego, La Jolla, CA 92093

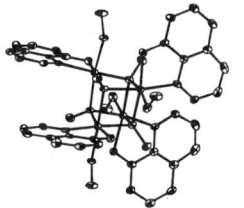

Reaction of nickel(II) salts with the chelating ligand 8-hydroxyquinolinate under mild conditions results in novel complexes in which the ligand adopts some unusual binding modes. Subtle differences in reaction conditions produce $Ni_4(OR)_4$ cubane complexes (one example at right) or trinuclear Ni_2Na complexes. The magnetic behavior of the complexes is evaluated in light of the structural data.

472. BINARY AND MIXED LIGAND Cu(II) COMPLEXES CONTAINING PHOSPHONOFORMIC AND PHOSPHONO ACETIC ACIDS. Y. Lakshmi Kumari and Srinivas Matta Mohan, Department of Chemistry University College for Women, Osmania University, Hyderabad 500 195, INDIA.

The formation of binary and mixed ligand complexes (MLA) containing Cu(II), phosphono formic (PFA, L) or phosphonoacetic acid (PAA, L) and a second ligand (A) viz., α-alanine, serine, methionine, phenylalanine, tryptophan, tyrosine, lysine, histidine, aspartic acid, 2,2'-bipyridyl, 1,10-phenanthroline, 5-nitro 1,10-phenanthroline, 2,9-dimethyl 1,10-phenanthroline, ethylenediamine, oxalic acid, pyrocatechol, malonic acid or 5-sulfosalicylic acid was investigated by a potentiometric method at 37.0°C and μ=0.15 M (KNO_3). Formation constants were quantitatively evaluated using the computer programme SCOGS. The parameters $\Delta \log K$ and $\Delta \log \beta_{MLA}$ were used to compare the relative stability of the mixed ligand and the corresponding binary complexes (MA). The mutual influence of the two ligands L and A on the stability of the mixed ligand complexes was assessed. pH-metal species profiles for various binary and mixed complexes were generated from the equilibrium data using the computer programme BEST. The antiviral activity of the ligands PFA and PAA was correlated with their ability to form stable metal complexes.

473. EPR AND NMR STUDIES OF 2,4-PENTANEDIONE BIS([4]N-METHYLTHIO-SEMICARBAZANATO) LIGAND AND ITS COPPER COMPLEX. Abbas Pezeshk[a], Marcus C. Durrant[b], and Raymond L. Richards[b]. [a]Department of Chemistry, Moorhead State University, Moorhead, Minnesota 56563, and [b]Nitrogen Fixation Laboratory, John Innes Center, Norwich Research Park, Colney Lane, Norwich, NR4 7UH, UK.

In this work we report the synthesis, structure, EPR, and NMR properties of 2,4-pentanedione bis([4]N-methylthiosemicarbazanato) ligand, and its copper(II) complexe. Reaction of 2,4-pentanedione with semicarbazide readily cyclized leading to the formation of pyrazoline. Following the reaction of pyrazoline with copper, the title complex was isolated as a result of a ring-opening process of the ligand. The ligand and its copper complex were characterized by multinuclear NMR and EPR, respectively. The EPR spectrum of the complex exhibits a typical copper hyperfine pattern with approximately axial symmetry. The EPR parameters obtained are a_{iso} = 0.0042 cm^{-1}, g_{ave} = 2.067, $g_{\parallel}$ = 2.133, $g_{\perp}$ = 2.034, $A_{\parallel}^{Cu}$ = 0.0186 cm^{-1}, $A_{\perp}^{Cu}$ = 0.0038 cm^{-1}, $A_{\parallel}^{N}$ = 0.0012 cm^{-1}, $A_{\perp}^{N}$ = 0.0013 cm^{-1}.

474. DIMERIC BIS[BIS(BIPYRIDINE)RUTHENIUM(II)] 1,2,4,5-TETRAIMINO 3,6-DIKETOCYCLOHEXANE REDOX SERIES; A ZINDO SEMI-EMPIRICAL MOLECULAR ORBITAL ANALYSIS

By Hitoshi Masui, Scott S. Fielder and A. B. P. Lever., Department of Chemistry, York University, North york (Ontario), Canada, M3J 1P3

The title complex is a binuclear complex in which two bis(bipyridine)ruthenium units are linked via a flat 1,2,4,5-tetraimino-3,6-diketocyclohexane bridge, isolated as the 4+ cation. This binuclear species undergoes ligand-based reductions, chemically or electrochemically, giving the 3+ and 2+ charged species. The X-ray structure, electrochemistry, NMR, ESR , UV/VIS, resonance Raman, and IR spectra were analyzed where possible, giving an electronic model of the binuclear species and its redox products. This presentation will emphasize the electronic spectra and molecular orbital description of these species as eludidated by the ZINDO method.

475. REACTIVITY OF DINUCLEAR MOLYBDENUM(II) CARBOXYLATE SYSTEMS WITH FACIALLY COORDINATING NITROGEN CHELATES. Elizabeth F. Day, John C. Huffman, Kirsten Folting, and George Christou, Department of Chemistry and the Molecular Structure Center, Indiana University, Bloomington, IN 47405

Tridentate ligands have been used to model proteins and other biological ligands, as well as provide a pronounced chelate effect. This research focuses on the ability of these ligands to coordinate in two equatorial and one axial site of a dinuclear metal carboxylate compound. Some dinuclear carboxylate compounds have shown anticancer activity, but the method of action is still unknown. By restricting the binding of additional ligands to the metal complex by use of a tridentate ligand, we strive to model the possible interactions of DNA bases in the remaining equatorial sites. In this study, 1,4,7-triazacyclononane and the trispyrazolylborate anion have been used to enforce a cis arrangement of incoming ligands or a chelating binding mode for bidentate ligands. The resulting compounds, despite a short M-M separation and steric congestion, have well-behaved equatorial sites and are conducive to nucleobase coordination. The structure determination and spectroscopic characterization of biscarboxylate compounds with a tridentate ligand and the remaining equatorial sites occupied by 2,2'-bipyridine allow for interesting comparisons of these compounds. The results of this study will be reported.

476. OXOVANADIUM COMPLEXES: SOME NOVEL FEATURES. C.P. Rao and A. Sreedhara, Department of Chemistry, Indian Institute of Technology Bombay, Powai, Mumbai 400 076, India.

Oxovanadium chemistry is rather a well developed subject due to its multidimensional growth. However, in view of the role of vanadium in biological systems, chemistry and biomimetic aspects of oxovanadium complexes have become important and taken a renaissance. Recently we have synthesized and characterized, including single crystal X-ray, a number of complexes of Schiff's base, their reduced forms and some mixed ligand ones with mononuclear mono- and dioxovanadium(V) and dinuclear monooxovanadium(V) centers having geometries of distorted tbp, sp and Oh types. Several interesting correlations have been drawn based on the structural and spectral data. This has provided a lead point towards the suggestion of the type of metal center present in the enzyme *bromoperoxidase*. Weak interactions such as offset $\pi...\pi$, C-H$...\pi$ and hydrogen bonding, lead to the formation of supramolecular arrangement in the crystal lattices of some of these compounds. During the process, metal recognition and selective extraction of VO^{3+} has been achieved by acyclic Schiff base containing alkoxo groups and its -CN- reduced counterpart. The molecular sensor property of the Schiff's base ligand is demonstrated for vanadium through conductivity studies. The catalytic bromination activity of these complexes towards xylene cyanol and phenol red are demonstrated using UV-Vis and ^{51}V nmr studies. The relevance of oxovanadium complexes as model molecules for biological processes will also be discussed.

477. SUPRAMOLECULAR STRUCTURE OF TRINUCLEAR METAL COMPLEXES OF A HALF UNIT LIGAND. Mohamed M. Aly and Najat I. Al-Shatti, Chemistry Department, Project SC 0069, Faculty of Science, University of Kuwait, KUWAIT.

The half unit ligand (HL) was prepared from the (1 + 1) condensation of isonitrosoacetylacetone with p-phenylenediamine in ethanol. The 2:1 molar ratio reaction of (HL) with the divalent metal acetate (M = Ni, Co or Cu) produced complexes L_2M with the amino group of the ligand available for coordination with similar or dissimilar metal ions as in (I). Spin-spin interaction is operative between the metal ions in (I)

M_a, M_b or M_c = Ni (II), Co(II) or Cu(II)

for M_a = Ni ; 2 H_2O

478. MIXED OXIMATO-ACETATO COMPLEXES OF DIVALENT NICKEL AND COPPER. Mohamed M. Aly and Najat I. Al-Shatti, Chemistry Department, Project SC 0069, Faculty of Science, University of Kuwait, KUWAIT.

High-spin Ni(II) and normal magnetic moment Cu(II) complexes of type (I) are characterized. The acetato complexes (R = -CH_3) was prepared from the template reaction of isonitrosoacetylacetone with o-phenylenediamine in the presence of the metal acetate (1 : 1 : 1 molar ratio) in ethanol. The acetato complex had reacted with the haloacetic acid (RCO_2H; R = -CH_2Cl, -$CHCl_2$ or -CCl_3, 1:1 molar ratio in chloroform) to produce the acetato replacement products. The coordination modes of the ligands are established.

M = Ni(II) or Cu(II)
R = -CH_3, -CH_2Cl, -$CHCl_2$ or -CCl_3

479. USE OF VARIOUS BETA-DIKETONATE COMPLEXES WITH IRON FOR HYDRODESULFURIZATION, S. J. Eng and A. E. Martell, Department of Chemistry, Texas A&M University, College Station, Texas 7784-3255, USA

Hydrodesulfurization (HDS), the removal of hydrogen sulfide from natural gas, by oxidation of H_2S to S by air is represented by the overall reaction:
$$H_2S + \tfrac{1}{2} O_2 \rightarrow \tfrac{1}{8} S_8 + H_2O$$
The formation of the unwanted soluble byproducts (such as SO_x and $S_2O_4^{2-}$), which are not as conveniently separated from the reaction mixture as is elemental sulfur, can be avoided through separation of the overall reaction into two steps with an Fe(III), Fe(II) chelate as a redox catalyst, as follows :
$$H_2S + 2\,Fe^{III}L_n \rightarrow 2\,Fe^{II}L_n^- + \tfrac{1}{8} S_8 + 2 H^+$$
$$2\,Fe^{II}L_n^- + \tfrac{1}{2} O_2 + 2 H^+ \rightarrow H_2O + 2\,Fe^{III}L_n$$
The redox catalyst is easily recovered and recirculated, and the sulfur removed in the form of elemental sulfur. Various β-diketonate complexes of iron have been proposed as potential redox catalysts in HDS. This presentation will discuss the study of the stability constants and the redox potentials of iron β-diketonate complexes, along with their ability to function as redox catalysts in HDS.

480. **BEYOND THE AURIDE ION: TRIANGULATED GOLD NETWORKS AND ETHANE-LIKE STRUCTURAL UNITS IN BINARY AND TERNARY ALKALI METAL GOLD INTERMETALLICS.** R. B. King, Department of Chemistry, University of Georgia, Athens, Georgia 30602

The range of known structurally characterized alkali metal gold intermetallics now extends far beyond the simple "auride" derivatives CsAu and RbAu containing the discrete auride ion, Au^-, isoelectronic with Hg^0. The intermetallics M_2Au_3 (M = K, Rb), Rb_3Au_7, and MAu_5 (M = K, Rb) contain infinite triangulated anionic gold networks which can be constructed from linear sp-hybridized gold atoms linked to each other through multicenter bonding within Au_4 tetrahedra (for Rb_3Au_7), Au_4 butterflies (for M_2Au_3), or Au_5 trigonal bipyramids (for MAu_5). The infinite anionic gold networks in other intermetallics provide clear examples of four-coordinate gold in the alkaline earth aurides $AeAu_2$ (Ae = Ca, Sr) and NaAuGe as well as six-coordinate gold in $NaAu_2$ and $NaAuIn_2$. Many ternary gold intermetallics (E = Ge, In) also contain ethane-like E_2Au_6 units with unbridged E–E bonds (e.g., $K_4Au_7Ge_2$, NaAuGe, and $NaAuIn_2$) or diborane-like E_2Au_6 units with E–E bonds bridged by two gold atoms (e.g., $Na_2Au_6In_5$). More complicated gold structures are found in other Na-Au-In ternary intermetallics including $Na_2Au_6In_5$ and $Na_8Au_{11}In_6$.

481.

OUTER SPHERE REACTION OF DIOXYGEN WITH A DIIRON(II) COMPLEX: FORMATION AND PROPERTIES OF A MIXED-VALENCE OXO BRIDGED COMPLEX Sonha C. Payne, Karl S. Hagen*, Department of Chemistry, Emory University, Atlanta GA, 30322

Model complexes of the diiron(II) forms of non-heme proteins and enzymes interact with dioxygen to form diiron(III) oxo bridged complexes. The unstable transient green intermediates observed during reactions of [Fe2(OH)(RCOO)2(Me3tacn)2]+ (1 R = CF3 or CH3) can be stabilized using the bulky bridging carboxylate, R = Ph3. This intermediate is the first mixed-valence Fe(II)Fe(III) oxo bridged complex to be crystallized. It's magnetic properties and Mössbauer, paramagnetic NMR and Resonance Raman spectral properties will be presented and compared to the diiron(II) hydroxy bridged and diiron(III) oxo bridged complexes. The reactivity of these binuclear complexes towards dioxygen and other oxygen sources will be compared to the related mononuclear complexes, [Fe(Ph3COO)(Me3tacn)]+ and [Fe(H2O)(Ph3COO)(Me3tacn)]+.

482. **MODEL COMPLEXES AND HYDROLYSIS REACTIONS FOR UNDERSTANDING HEMOZOIN FORMATION.** D. S. Bohle. S. K. Madsen, Department of Chemistry, University of Wyoming, 82071.

During the intraerythrocytic trophozoite stage of malaria, a microcrystalline polymer is formed, called malaria pigment or hemozoin. The synthetic analogue is β-hematin which has been characterized by IR, UV-Vis, ESR, Mössbauer, magnetic susceptibility and MCD. X-ray synchrotron radiation has been used to show malaria pigment and β-hematin (P-1) are identical. These physical methods have been used to confirm that the iron is high spin (S=5/2) and coordinated to one of the propionic acid side chains of another porphyrin, while the other is involved in hydrogen bonding.

Interactions of hemozoin and other porphyrinic intermolecular interactions such as π-π, metal-ligand bonds or hydrogen bonding (with the exception of hemin) in hemes are unknown. A preliminary study was done where a series of model complexes have been used to examine the possible mechanisms of formation of malarial pigment and their hydrolysis reactions have been studied to examine the stability of the iron-propionate linkage. Iron protoporphyrin IX dimethylester complexes have been synthesized and their reaction with water was monitored for the rate of formation of μ-oxo dimer. Rate constants were calculated using pseudo-first order kinetics for the decarboxylation reaction.

Although there are many mechanistic models proposed in the literature about dimerization, there are only speculations about the structure of aggregates. Exploitation of these physical properties could lead to a possible mechanism of heme polymer formation. These studies should be a better basis for a model of malaria pigment formation. However, during the intraerythrocytic state of malaria, the conditions under which an insoluble polymer is formed and precipitates, whether facilitated by histidine rich proteins or enzymatically, are still unknown.

483. MULTIPLE-EDGE XAS STUDIES OF IRON-COPPER BRIDGED MOLECULAR ASSEMBLIES RELEVANT TO HEME-COPPER OXIDASES. **Hua Holly Zhang,**[†] Adriano Filipponi,[§] Andrea Di Cicco,[‡] Michael J. Scott,[‖] R. H. Holm,[‖] Britt Hedman[†] and Keith O. Hodgson[†]
[†]Department of Chemistry and Stanford Synchrotron Radiation Laboratory, Stanford University, Stanford, CA 94305
[§]European Synchrotron Radiation Facility, B.P.220, F-38043 Grenoble, France
[‡] Dipartimento di Matematica e Fisica, Università degli Studi di Camerino, Camerino (MC), Italy
[‖] Department of Chemistry, Harvard University, Cambridge, MA 02138

A series of heme-based molecular complexes relevant to heme-copper oxidases have been studied. These molecular complexes, containing a bridge unit [Fe^{III}-X-Cu^{II}] with X=O^{2-}, OH^- and CN^-, have similar structural fragments around the metal centers but differ significantly in the bridge structure, especially the bridge angle. This allows a comparative study on the angle-dependence of multiple-scattering (MS) effect. The cyanide complex contains a unique four-body Fe-C-N-Cu bridge which offers an opportunity to study long-range MS interactions (~5Å) between the Fe and Cu centers. EXAFS data of these complexes at both Fe and Cu K edge have been analyzed using the newly-expanded GNXAS and a simultaneous multiple-edge algorithm. Results illustrating the impact of the bridge angle on MS effects and the determination of strong long-range MS interactions in the four-body case will be presented, along with the discussion of the application to heme-copper oxidases.

484. STABLE SIX-COORDINATE MONONUCLEAR MONOOXO-MOLYBDENUM(VI) CENTERS AS MODELS OF DMSO REDUCTASE. Partha Basu, Julie Graff and John H. Enemark, Department of Chemistry, University of Arizona, Tucson, AZ 85721.

Pterin-containing molybdoenzymes take part in important environmental processes such as global sulfur, nitrogen and arsenic cycles by catalyzing an oxo-transfer reaction at the molybdenum center (see below). In the process the molybdenum center shuttles between a oxosulfido or dioxo-Mo(VI) center and a monooxo-Mo(IV) center. However, in enzymes such as DMSO reductase, it is believed to shuttle between a monooxo-Mo(VI) and desoxo-Mo(IV) center. To date there has been no report of an oxygen atom transfer reaction from a $[MoO]^{4+}$ center. We have initiated a detailed exploration of such chemistry. Herein we discuss the syntheses and characterization of stable six-coordinate mononuclear monooxo-Mo(VI) complexes and their reactivity with oxygen abstracting agents.

$$X + H_2O \rightleftharpoons XO + 2H^+ + 2e$$

485. UNDERSTANDING FE K-EDGE XAS PRE-EDGE FEATURES: A MULTIPLET ANALYSIS OF THE 1s→3d FINAL STATES. Tami E. Westre, Pierre Kennepohl, Jane G. DeWitt, Britt Hedman, Keith O. Hodgson, and Edward I. Solomon, Department of Chemistry and Stanford Synchrotron Radiation Laboratory, Stanford University, Stanford, CA 94305, USA.

X-ray absorption Fe K-edge data on high-spin and low-spin ferrous and ferric model complexes with varying geometries were measured and analyzed to establish a detailed understanding of Fe 1s→3d pre-edge features and their sensitivity to the electronic and geometric structure of iron sites. Intensity distributions and energy splittings of the pre-edge features vary with spin and oxidation state. A methodology for the interpretation of observed energy splitting and intensity distribution of the 1s→3d pre-edge features is developed for high-spin ferrous and ferric complexes in octahedral, tetrahedral and square pyramidal environments and for low-spin ferrous and ferric complexes in octahedral environments. In each case, the allowable many-electron excited states are determined using ligand field theory. The calculated energies of the excited states are compared to experimental splitting and the distribution of intensity in the pre-edge is discussed. Contributions from both quadrupole and dipole (from 3d–4p mixing) intensity mechanisms are determined for all cases. The amount of 4p mixing into the Fe 3d manifold is determined experimentally and compared to theoretical estimates from density functional calculations. This methodology is presented as an important aid in the interpretation of the Fe K-edge XAS of iron model complexes and non-heme iron enzymes.

486. CHLORIDE DEPLETION BLOCKS NH_2OH REACTIVITY AND AFFECTS THE STRUCTURE OF Mn IN THE S_1 STATE OF PHOTOSYSTEM II.
Pamela S. DeMarois, Rui Mei, Pamela J. Riggs-Gelasco, Chun-Hui Xu, James E. Penner-Hahn and Charles F. Yocum, Department of Biology, Department of Chemistry and Biophysics Research Division, University of Michigan, Ann Arbor, MI 48109

Photosystem II preparations depleted of extrinsic 23 and 17 kDa proteins were exposed to NH_2OH in the absence of Cl^- at pH values ranging from 5.5 to 9.0. Increased pH decreases NH_2OH inhibition of O_2 evolution; the minimum activity inhibition occurs at pH 7.5. A similar block of NH_2OH inhibition is detected at pH 6 as well after repeated washings in Cl^- free buffer. We interpret this as evidence that under Cl^- depleted conditions NH_2OH is unable to effect rapid reduction of PSII Mn. This behavior is freely reversible. Addition of either Cl^- or Br^- to samples at pH 7.5 restores activity inhibition and the appearance of Mn^{2+}. Other anions (F^-, SO_4^{2-}) are ineffective at promoting NH_2OH reduction. We have further characterized the consequences of Cl^- depletion using X-ray absorption spectroscopy. Samples depleted of Cl^- either by extensive washing (pH 6.0) or by exposure to pH 7.5 were examined to determine whether any changes in the structure of the tetranuclear Mn cluster of PSII could be detected. Both treatments cause significant alteration only in the 3.3 Å peak. This feature has been attributed either to a Mn-Mn interaction or to a Mn-(Mn+Ca) interaction in samples with Cl^- present. The increased disorder in this interaction suggests a disruption of the Mn environment by Cl^- depletion.

487.

THE EFFECT OF ORTHO-SUBSTITUENTS ON THE REDUCTION POTENTIALS AND COMPLEX STABILITIES OF UNSYMMETRICALLY-SUBSTITUTED TETRAPHENYL-PORPHYRINATOIRON(III) COMPLEXES. **Xiaodong Ding**, F. Ann Walker, Department of Chemistry, University of Arizona, Tucson, AZ 85721

Three meso-ortho-phenyl substituted porphyrins, 5,10,15-tris(4-methoxyphenyl)-20-(2,6-dichlorophenyl)porphyrin, 5,10,15-tris(4-methoxyphenyl)-20-(2-fluorophenyl)porphyrin and 5,10,15-tris(4-methoxyphenyl)-20-(2-trifluoromethylphenyl)porphyrin, were synthesized and purified. Cyclic voltammetry was used to measure the redox potentials of these models in dimethylformamide with TBAP (0.1 M) as electrolyte. The overall equilibrium constants, $\log\beta_2^{III}$ for iron(III) and $\log\beta_2^{II}$ for iron(II), have been calculated based on the redox potentials of the iron(III/II) couple as a function of N-methylimidazole concentration. The values of $\log\beta_2^{III}$ for the (o-F)-, (o-CF$_3$)- and (2,6-Cl$_2$)(p-OCH$_3$)3TPPFe(III) complexes are 6.65, 6.20 and 5.69. The values of $\log\beta_2^{II}$ are 5.92, 5.46, and 5.18. In both cased the order is F > CF$_3$ > 2Cl. These results do not follow either the expected (based on p-substituents) electronic effect or the steric effect; CF$_3$ has the strongest electron-withdrawing ability and F the weakest, while CF$_3$ is the largest and F the smallest ortho-substituent.

488.
OXYGENATION REACTIONS OF NITROSYL MYOGLOBIN. **Ernst V. Arnold,** Scott D. Bohle, University of Wyoming, Laramie, Wyoming, 82071

In our efforts to model the down regulation of soluble guanylyl cyclase by nitric oxide, NO, the oxygenation kinetics of nitrosyl myoglobin, MbNO, have been studied. Although exhibiting isobestic behavior, the kinetic data has the best Marquardt fit with an A→B→C model after singular value decomposition. The observed rates at 30° C and pH = 7.0 are; $k_1 = 4.63\times10^{-4}$ s^{-1} ± 4.3×10^{-6} and $k_2 = 7.85\times10^{-4}$ s^{-1} ± 1.5×10^{-5}. The proposed mechanism is attack of oxygen at the nitrogen of NO forming an iron bound peroxynitrite intermediate which then dissociates to ultimately yield metmyoglobin and nitrate. This possibly represents a new *in vivo* pathway for the formation of peroxynitrite.

489. MODELS FOR THE REACTION INTERMEDIATES OF METHANE MONOOXYGENASE AND RIBONUCLEOTIDE REDUCTASE. Yanhong Dong,[§] Yan Zang,[§] Elizabeth C. Wilkinson,[§] Karl Kauffmann,[‡] Eckard Münck,[‡] Lawrence Que, Jr.[§], Department of Chemistry and Center for Metals in Biocatalysis,[§] University of Minnesota, 207 Pleasant St. SE, Minneapolis, MN 55455 and [‡] Carnegie-Mellon University, 4400 Fifth Ave., Pittsburgh, PA 15213

In our efforts to model the oxygen activation chemistry of methane monooxygenase (MMO) and ribonucleotide reductase (RNR), two high-valent bis(μ-oxo)Fe^{III}Fe^{IV} intermediates were discovered in the reaction of H_2O_2 with [Fe$_2$(O)(OH)(L)$_2$](ClO$_4$)$_3$ complexes (L = tris((5-R$_3$-2-pyridyl)methyl)amine (**5-R$_3$-TPA**) and N-((6-methyl-2-pyridyl)methyl)-N,N-bis(2-pyridylmethyl)amine (**6-Me-TPA**)) (Que and Dong, *Acc. Chem. Res.* **1996**, *29*, 190-196). These intermediates not only resemble the spectroscopic properties of the high-valent intermediates of MMO and RNR R2 but also carry out oxidation reactions analogous to those attributed to the high-valent intermediates. These observations have led us to propose that the key oxidizing species in MMO and RNR may be a bis(μ-oxo)diiron complex. Further studies have uncovered that the first intermediate in the reaction of H_2O_2 and the (μ-oxo)diiron(III) complex is a (μ-oxo)(μ-1,2-peroxo)diiron(III) complex, which then decays to the bis(μ-oxo)Fe^{III}Fe^{IV} species probably via an Fe^{IV}Fe^{IV} state. Evidence for these transformations will be presented.

490. **SPECTROSCOPIC INVESTIGATION OF THE INTERACTION OF DNA WITH THE FERROUS SITES OF BLEOMYCIN AND BLEOMYCIN DERIVATIVES**
Kelly E. Loeb,[a] Jeffrey M. Zaleski,[a] Cynthia Hess,[b] Sidney M. Hecht,[b] and Edward I. Solomon[a],*

[a] *Department of Chemistry, Stanford University, Stanford, California 94305-5080*
[b] *Department of Chemistry, University of Virginia, Charlottesville, Virginia 22901*

Mononuclear non-heme iron active sites exist in a variety of systems which perform important biological functions requiring dioxygen. To provide insight into the structural requirements for reactivity of the anticancer drug Bleomycin (BLM), the active site geometric and electronic structures of this glycopeptide, as well as natural and synthetic derivatives, have been investigated by optical absorption and magnetic circular dichroism spectroscopies. The observed spectral changes in the d→d ligand field transitions and the Fe^{2+}→pyrimidine charge transfer envelope are correlated with variations to the ligand framework which govern the ability (or inability) for the perturbed complexes to form activated BLM, the reactive intermediate that may be responsible for DNA degradation. The interaction of the ferrous site of BLM with DNA has been explored using calf thymus DNA and self-complementary oligomers designed with specific binding and cleavage sites. Interaction with DNA causes large spectral differences associated with the direct involvement at the metal site. Taken together, these studies aid in understanding the steps in the catalytic mechanism on a molecular level with particular emphasis on the effect on the iron site of BLM due to the interaction of DNA.

491.
EXCHANGE COUPLING WITH TRIVALENT TIN- AND GERMANIUM-IONS IN LINEAR TRINUCLEAR COMPLEXES WITH CORE STRUCTURES FE(III) M(III) FE(III) (M = GE, SN)
T. Glaser, E. Bill, T. Weyhermüller, and K. Wieghardt, Max-Planck-Institut für Strahlenchemie, D-45470 Mülheim, Germany

We have synthesized linear trinuclear complexes of the general formula [LFeMFeL]n+ where M could be a transition metal ion or a group 14 ion. The metals are all octahedrally coordinated and bridged by six thiophenolate ligands. From the study with M = transition metal ion we have elucidated a high degree of electron delocalization in the trinuclear core. In this contribution we will show that this behaviour also exists with M = Sn and Ge but not with M = Pb. In order to elucidate the electronic structures we have measured the temperature-dependent susceptibilities, 57-Fe-Mössbauer spectra with and without applied field,119-Sn-Mössbauer spectra, and EPR spectra. In the paramagnetic complex [LFeSnFeL]3+ an S =1/2 EPR signal is detected with 117- and 119-Sn hyperfine satellites known from organic R3Sn radicals. This indicates an exchange coupled Fe(III) Sn(III) Fe(III) system.

492. SPECTROSCOPIC AND MECHANISTIC STUDIES OF A SOYBEAN LIPOXYGENASE IRON COORDINATION MUTANT, N694H. Ted R. Holman*, J. Zhou[#], E. Sigal, E.I. Solomon[#], Dept. of Chemistry, U.C. Santa Cruz, CA 95064*, Stanford U., CA 94305[#]

Lipoxygenases are widely distributed throughout the plant and animal kingdoms and have been implicated in plant germination, human asthma and arthritis, to name a few. They contain an active site, iron atom which is essential for activity and catalyze the incorporation of dioxygen into 1,4-*cis,cis*-pentadiene containing fatty acids. There are two classes of proposed mechanisms for LO catalysis which have fundamental differences in the role of the iron. Our objective is to correlate structural and functional data of active site mutants to better understand the enzymatic mechanism.

In this poster we shall present spectroscopic and enzymatic data of an iron ligation mutant in soybean lipoxygenase (N694H) in order to better understand the role of the iron in the mechanism of dioxygenation. We have probed this mutant with EPR and MCD and determined that the His_{694} is ligated to the iron to form a 6-coordinate octahedral geometry. Preliminary crystallographic data also suggests the His is an iron ligand. In conjunction, we have also determined that the enzymatic activity has been reduced to 2% that of the WT enzyme. A unified theory which explains the structure/function relationship of the iron atom in the lipoxygenase mechanism will be discussed.

493. PROTON NMR SPECTROSCOPY AS A PROBE OF DICOPPER(II) CENTERS.
Richard C. Holz, Leila Ustynyuk, Ernestene Lee, Guanjing Chen. Department of Chemistry and Biochemistry, Utah State University, Logan, Utah 84322-0300.

Enzymes containing dinuclear copper centers play important roles in nature such as the oxidation of organic molecules coupled to the reduction of dioxygen, reduction of nitrogen oxides, dioxygen transport, and hydrolysis chemistry. Consequently, the characterization of their structure and function is a problem of outstanding importance. A fundamental and, as yet, largely unexplored issue is the determination of the structural and magnetic properties of dinuclear copper centers in biological systems using ^1H NMR spectroscopy. ^1H NMR is a natural technique to probe these systems because only protons proximate to the paramagnetic center are affected. We have recorded both one- and two-dimensional ^1H NMR spectra on a series of magnetically diverse µ-phenoxo and µ-alkoxo dicopper(II) complexes and assignments of all of the observed signals have been obtained. Moreover, magnetically-coupled dicopper(II) centers have a singlet and a triplet state that differ in energy by the exchange constant (J). The population distribution between these two levels is dependent upon the temperature as well as the magnitude of J. Therefore, the temperature dependence of observed hyperfine shifted ^1H NMR signals have been correlated with the magnitude of the exchange coupling between the two Cu(II) ions. The magnetic and structural properties of the hyperactive Cu(II)-substituted aminopeptidase from *Aeromonas proteolytica* have been deduced from ^1H NMR studies and will be discussed.

494.

DELOCALIZED FE(II/III)2-COMPOUNDS WITH S=9/2
Carsten Krebs, Eckhard Bill, Thomas Weyhermüller, Phalguni Chaudhuri, and Karl Wieghardt
Max-Planck-Institut für Strahlenchemie, D-45470 Mülheim, Germany

The magnetic properties of mixed-valent transition metal compounds are determined by the competition of electron transfer and magnetic superexchange. In order to get a deeper understanding of these phenomena we have synthesized two compounds containing a facesharing bioctahedral Fe(II/III)2 unit of the general formula [M(III)-Fe(II/III)2-M(III)] wih M=B, Fe. Both compounds have extensively been studied by different techniques (EPR, 57-Fe-Mössbauer, temperature- and field-dependent magnetization measurements) which reveal that the excess electron is fully delocalized over both iron ions. This favors the parallel alignment of the spins (S=9/2) within the Fe(II/III)2 unit due to a double exchange mechanism. Thus [B(III)-Fe(II/III)-B(III)] has a S=9/2 ground state. In [Fe(III)-Fe(II/III)2-Fe(III)] the terminal ferric ions are coupled antiferromagnetically to yield a S=1/2 ground state.

495. ACTIVATION OF DIAZENE COORDINATED TO FE(II)–SULFUR COMPLEXES. Nicolai Lehnert, Felix Tuczek, Institut für Anorganische Chemie und Analytische Chemie, Johannes Gutenberg-Universität Mainz, D–55099 Mainz, Germany.
Andreas Hennige, Dieter Sellmann, Institut für Anorganische Chemie der Universität Erlangen–Nürnberg, D–91058 Erlangen, Germany.

The electronic and vibrational properties of two diazene–bridged iron(II) complexes with sulfur ligand sphere are described. Using SCF-Xα-SW calculations, the MO–scheme of two corresponding model complexes has been calculated. Importantly, the diazene LUMO also is the LUMO of the complexes and is located between the fully occupied t_{2g} and empty e_g orbitals of iron. The consequences of this bonding description with respect to the further reduction of diazene both in the model systems and on the iron–molybdenum cofactor of the enzyme nitrogenase are considered. The intense absorption at about 600 nm of both complexes causes strong enhancement of several diazene vibrations in the Raman experiment. The resonance Raman spectra obtained with different isotopes and the results of a normal coordinate analysis of the central Fe-N_2H_2-Fe unit are presented.

496. INORGANIC HOST-GUEST CHEMISTRY IN POLYNUCLEAR CARBOXYLATO-BRIDGED-IRON(II) CLUSTERS. Sanjay K. Mandal, and Lawrence Que, Jr.*, Department of Chemistry and Center for Metals in Biocatalysis, 207 Pleasant St. SE., University of Minnesota, Minneapolis, MN 55455

In the course of exploring the iron(II) coordination chemistry of the BPG ligand, we have synthesized a trinuclear carboxylate-bridged-iron(II) complex, [Fe$_3$(O$_2$CCH$_3$)$_3$(BPG)$_3$] (1), (H-BPG = bis(2-pyridyl) glycine) that can act as a host for a fourth metal ion, forming a number of novel tetranuclear complexes, [MFe$_3$(O$_2$CCH$_3$)$_3$(BPG)$_3$](ClO$_4$)$_n$ (M = Fe, n = 2, 2; M = Fe, n = 3, 3; M = Ga, n = 3, 4). The host-guest chemistry can be monitored by ^{1}H NMR spectroscopy. Both the host and the host-guest complexes exhibit a number of well-resolved isotropically shifted resonances, and their spectral features are very distinct. Compounds 1 and 2 have been structurally characterized by X-ray diffractometry. Compound 1 contains an unprecedented triangular structural motif that is held together with carboxylate bridges. Compound 2 is a tetrairon aggregate which consists of three crystallographically identical Fe(O$_2$CCH$_3$)(BPG) units coordinated to a central Fe^{II} ion. EPR spectroscopic properties of these complexes will also be discussed.

497.

1-D AND 2-D PARAMAGNETIC NMR OF NON-PLANAR LOW-SPIN IRON(III) PORPHYRIN COMPLEXES. **Hiroshi Ogura,** F. Ann Walker, Department of Chemistry, University of Arizona, Tucson, AZ 85721; Craig J. Medforth, Kevin M. Smith, Department of Chemistry, University of California, Davis, CA 95616

Hemes are surprisingly flexible, and a significantly distorted conformation is accessible by the forces exerted by surrounding proteins. Iron(III) octaethyltetraphenylporphyrin ([Fe(III)OETPP]$^+$) is a saddled macrocycle that could serve as a model for distorted hemes. Two low-spin [Fe(III)OETPP]$^+$ complexes, one with N-methylimidazole (N-MeIm) as the axial ligand and the other with 2-methylimidazole (2-MeIm), have been made, and their NMR spectra measured as a function of temperature. A spectral assignment for the N-MeIm complex has been made through 2-D techniques. The 1-D spectra of this complex show a significant deviation from Curie behavior, and also evidences of chemical exchange even at very low temperature. The 1-D spectra of the 2-MeIm complex indicates that the ligand coordinates readily at low temperature in spite of the steric hindrance by the methyl group next to the coordinating nitrogen atom.

498. KINETIC AND THERMODYNAMIC INVESTIGATIONS OF SULFITE OXIDASE USING LASER FLASH PHOTOLYSIS. A. Pacheco, J. T. Hazzard, J. H. Enemark and G. Tollin, Department of Chemistry, University of Arizona, Tucson, AZ 85721.

Laser flash photo produced 5-deazariboflavin semiquinone (dRFH) is a very useful reagent for the rapid in-situ reduction of proteins. We use this method to generate the 1-electron reduced form of sulfite oxidase (SO), which undergoes subsequent internal electron transfer (IET) between the Fe and Mo centers. We previously reported the qualitative behavior of the IET net rate constant under various conditions; however, until recently our efforts to quantitatively analyze this process were stymied by the close overlap between the SO reduction and IET steps, which makes meaningful least-squares fitting of the data very difficult. We have solved this problem by simultaneously fitting the data collected under a wide variety of conditions, subject to several restrictions. In this way we can obtain the rate constants for IET between Fe and Mo in both directions, and consequently the equilibrium constant for the process.

499. AN EXAFS INVESTIGATION OF THE ZINC BINDING SITE IN COBALAMIN-INDEPENDENT AND COBALAMIN-DEPENDENT METHIONINE SYNTHASE Katrina Peariso[‡], Julio C. Gonzalez[§], Celia W. Goulding[§], Rowena G. Matthews[§], and James E. Penner-Hahn[‡], Department of Chemistry, Department of Biological Chemistry and the Biophysics Research Division, University of Michigan, Ann Arbor, MI 48109

Cobalamin-independent (MetE) and cobalamin-dependent (MetH) methionine synthases catalyze the transfer of a methyl group from methyltetrahydrofolate to homocysteine. Although these enzymes show no sequence homology, both have been shown to contain zinc. Zinc extended x-ray absorption fine structure (EXAFS) spectra were measured on both MetE and the N-terminal 71 kDa region of MetH before and after the addition of homocystiene. In both native forms, the zinc site appears to be four coordinate with mixed sulfur and nitrogen/oxygen ligation ($Zn(N/O)_2S_2$ for MetE and $Zn(N/O)S_3$ for the MetH region). The addition of homocysteine to both enzymes results in displacement of one of the nitrogen/oxygen ligands by the homocysteine sulfur. These results directly implicate zinc in the binding of homocsteine for both MetE and MetH enzymes.

500. **SPECTRAL STUDIES OF COPPER DIOXYGEN COMPLEXES**
Elna Pidcock, Edward I. Solomon, Honorio V.Obias, Kenneth D. Karlin
Department of Chemistry, Stanford University, Stanford, CA 94305-5080

Copper proteins have been shown to perform a variety of oxidation reactions involving dioxygen, and there is considerable interest in the development of synthetic analogues that mimic aspects of the biological reactions. Protein active sites often utilise novel coordination chemistry and have interesting physical and spectroscopic properties, for example the magnetically coupled binuclear copper centers in hemocyanin, an oxygen carrier, are known to bind peroxide in an unusual μ-η^2-η^2 "side-on" geometry. Tyrosinase, a mono-oxygenase contains a similar Cu_2O_2 moiety. Resonance Raman couples electronic absorption and vibrational spectroscopies and is a sensitive probe of the copper oxygen interaction (the "side-on" bound peroxide has a characteristic low O-O stretch at approx. 740cm^{-1}). It is the nature of the copper-oxygen bonding that is critically important to the understanding of the protein functionality. Currently we are using resonance Raman to study a number of binuclear copper-peroxide complexes which exhibit spectral features perturbed from those observed in oxyhemocyanin and oxytyrosinase. We are especially concerned with how the electronic and vibrational features relate to the geometry of the Cu_2O_2 core and the state of the O-O peroxide bond.

501.
MAGNETO-STRUCTURAL CORRELATION IN PIVALATO BRIDGED MIXED-METAL COMPLEXES Eva Rentschler, Karl Wieghardt, Grigore A. Timko, Nicolai V. Gerbeleu, Max-Planck Institut für Strahlenchemie, D-45470 Mülheim, Germany and Institute of Chemistry of the Academy of Science of Moldavia.

In order to get further insight into the structural conditions which determine the magnetic interactions between metal ions, magnetic properties of transition metal complexes, in which the metal atoms are bridged by oxo-, hydroxo- as well as carboxylato groups are the subject of many studies. Since it is unambigous that the coupling constants in heteronuclear complexes are not only influenced by the spin states of the metal atoms involved, but are even more strongly influenced by their charges, we have started to investigate a series of isostructural homo- and heterovalent heteronuclear complexes, in which the carboxylato groups are pivalato ions.

502. 2-D NMR INVESTIGATIONS OF THE ROTATION OF AXIAL LIGANDS IN SIX-COORDINATE LOW-SPIN Fe(III) AND Co(III) TETRAPHENYLPORPHYRINATES.
Tatjana Kh. Shokhireva, Nikolai Shokhirev, Konstantin Momot, Jayapal Reddy Polam, F. Ann Walker Department of Chemistry, University of Arizona, Tucson, AZ 85721,

Phase sensitive NOESY/EXSY experiments have been utilized to measure the rates of axial ligand rotation for [TMPFe(2-MeImH)$_2$]H$^+$ and [TMPCo(2-MeImH)$_2$]$^+$ and several related complexes as a function of temperature. The derivations of the expressions for EXSY cross peak volumes as a function of mixing time, longitudinal relaxation time, and chemical exchange rate constant have been extended to the case of four-site chemical exchange having a single rate constant. The rate constants were calculated using a computer fitting program developed in this laboratory. The temperature dependence of the rate constants was used to calculate the transition state enthalpy and entropy for these complexes. The values of ΔH are very similar for all Fe(III) complexes, and ΔS values are close to zero. Nevertheless, the combined differences in these activation parameters led to rate constants for ligand rotation at 25° C ranging from 1.1×10^5 to 1×10^4 sec^{-1}. For the [TMPCoL$_2$]$^+$ complexes the values of ΔH are very similar, but the values of ΔS are rather negative, which lead to rate constants at 25° C of 14 and 4 sec^{-1}, respectively.

503. MÖSSBAUER SPECTROSCOPIC AND KINETIC CHARACTERIZATION OF FERRIC CLUSTERS FORMED IN H-CHAIN FERRITIN MINERALIZATION. P. Tavares,[1] A. S. Pereira,[1] S. G. Lloyd,[1] D. Danger,[2] D. E. Edmondson,[3] E. C. Theil,[2] and Boi Hanh Huynh,[1] Departments of [1]Physics and [3]Biochemistry, Emory University, Atlanta, GA 30322 and [2]Department of Biochemistry, North Carolina State University. Raleigh, NC 27695

Ferritin is an iron storage protein comprised of 24 subunits organized into a hollow protein shell. It directs the reversible transition of Fe^{2+} ions in solution into Fe^{3+} ions in the ferrihydrite core within the protein cavity. Ferritin subunits can be categorized into two classes (H and L) according to the rates of Fe mineralization (fast and slow, respectively) associated with the subunits; Ferritin proteins with different ratio of H and L subunits coincide with tissue differences in iron physiology. Applying the rapid freeze-quench technique and using small amounts of iron (36 Fe/ferritin molecule), we were able to monitor the mineralization process of H chain ferritin at early stages by Mössbauer spectroscopy. The rate of Fe^{2+} oxidation was found to parallel the rate of formation of ferric-tyrosine species observed previously (Waldo and Theil (1993) *Biochem.* **32**, 13262). On oxidation of Fe^{2+}, four distinct ferric clusters (three dimers and one trimer) are formed at comparable rates. Taken together, these results suggest that multiple processes occur in H chain ferritin mineralization and that tyrosine residues are involved in most, if not all, of the initial processes.

INOR

504. SINGLE CRYSTAL EPR STUDIES OF LMo(O)(1,2-BENZENEDITHIOL) (L=HYDROTRIS(3,5-DIMETHYLPYRAZOLYL)BORATE): IMPLICATIONS FOR THE MOLYBDENUM CENTER OF SULFITE OXIDASE. Barry L. Westcott, Ish Dhawan, Arnold Raitsimring and John H. Enemark, Department of Chemistry, University of Arizona, Tucson, AZ 85721.

The first single crystal EPR study of an oxo-Mo(V) complex with a single diothiolene ligand is reported. The structurally similar nitrosyl complex, LMo(NO)(3,4-toluenedithiol) was used as a diamagnetic host lattice. In this study, we have determined the magnitude and direction of the components of the principal g and A tensors. The largest g has a value of 2.004 and lies in the Mo-S_2 plane, nearly perpendicular to the Mo-O vector. This large g value is not observed in the analogous alkoxide complexes, and can be attributed directly to contributions from sulfur spin orbit coupling. These findings suggest a significant amount of S character in the metal singly occupied molecular orbital (somo), and give some insight into Mo-S interactions, including those observed in molybdenum-containing enzymes.

505. CIRCULAR DICHOISM AND MAGNETIC CIRCULAR DICHOISM STUDIES OF BINUCLEAR NON-HEME IRON ENZYMES
Yi-Shan Yang and Edward I. Solomon.
Department of Chemistry, Stanford University, Stanford, CA 94305-5080

Circular dichorism (CD) and magnetic circular dichroism (MCD) spectroscopies have been used in order to probe the geometric and electronic structures of binuclear non-heme iron active sites. Ligand field transitions of both ferric and ferrous sites can be observed in the NIR region, and they can be interpreted in terms of ligand field theory. A theoretical protocol has also been applied to variable temperature variable field (VTVH) MCD data to probe the zero field splitting (ZFS) and g_{eff} values of the ground state and the excited sublevel energies. These data are further interpreted in terms of a spin Hamiltonian which includes ZFS at the iron center, combined with the exchange coupling (J) between the irons. Results have been obtained from the applications of these protocols to both the binuclear ferrous site of Δ^9 desaturase and the mixed-valence site of uteroferrin, and mechanistic implications are discussed.

506. EPR AND ENDOR STUDIES OF NATIVE AND METAL-SUBSTITUTED FERREDOXINS ISOLATED FROM THE HYPERTHERMOPHILIC ARCHAEON *Pyrococcus furiosus*.
J. Telser, Roosevelt University, Chicago, IL 60605; H. Huang, P. Brereton, M.W.W. Adams, Dept. of Biochemistry, U. of Georgia, Athens, GA 30602; B. Koehler, Z.H. Zhou, M.K. Johnson, Dept. of Chemistry, U. of Georgia, Athens, GA 30602; and B.M. Hoffman, Dept. of Chemistry, Northwestern U., Evanston, IL 60208.

A novel 4Fe ferredoxin has recently been isolated from the hyperthermophilic archaeon *Pyrococcus furiosus*. In *Pf*-Fd, one Fe lacks cysteinyl coordination. This unique Fe can be easily removed to yield protein containing a $[Fe_3S_4]^{0/+}$ cluster (*Pf*-Fd-3Fe-ox). Under reducing conditions, this cluster binds exogenous metal ions allowing preparation of a series of isotopologs of *Pf*-Fd-4Fe-red (S= ½): $[^{57}FeFe_3S_4]^+$, $[Fe^{57}Fe_3S_4]^+$, and $[^{57}Fe_4S_4]^+$ as well as of *Pf*-Fd-3Fe-ox (S= ½): $[^{57}Fe_3S_4]^+$.

We have used EPR and electron nuclear double resonance (ENDOR) spectroscopy to investigate these isotopologs in both the native state and the cyanide-bound form for *Pf*-Fd-4Fe-red, which contains a single CN⁻ bound to the unique Fe. The ^{57}Fe hyperfine coupling constants so obtained suggest that one of the $Fe^{2.5+}$ pair binds cyanide, which then becomes Fe^{2+}.

^{1,2}H ENDOR allowed investigation of constitutive and solvent-exchangeable hydrogens in *Pf*-Fd. These results are compared to those for related protein and model systems, and show typical cysteinyl coordination, but relatively little solvent accessibility to the $[Fe_4S_4]$ cluster of this hyperthermophile. *Pf*-Fd globally enriched in ^{15}N allowed ^{15}N ENDOR identification of hydrogen-bonding from peptide N-H to cysteinyl and/or sulfide S.

507.
CORRELATIONS BETWEEN MAGNETIC EXCHANGE COUPLING AND THE REDOX PROPERTIES OF OXO-BIS-CARBOXYLATO-BRIDGED METAL COMPLEXES. **Brandon T. Weldon** and James K. McCusker*, Department of Chemistry, University of California at Berkeley, Berkeley, California, 94720-1460

The electron transfer characteristics of molecules are intimately related to their electronic structure. Electron exchange coupling is one key factor in the electronic structure of polynuclear metal clusters. We are interested in the possible correlation between magnetic exchange coupling and the electron transfer characteristics of systems containing a metal cluster as the electron transfer medium. The present work focuses on the effect of exchange interactions on the redox properties of molecules of the form [Fe(III)M(u-X)(u-O2CMe)2L2]n+, where M = Fe(III), Zn(II), Co(III) and X = O(2-) or MeO(-). The synthesis, magnetic, and electrochemical properties of this series of molecules will be described.

508.
An ESEEM Study of the Substrate Binding Site of Photosystem II

Dee Ann Force, Gary A. Lorigan, David W. Randall, Keri Clemens, R. David Britt

Department of Chemistry, University of California, Davis, CA 95616-0935

We are using Electron Spin Echo Envelope Modulation (ESEEM) spectroscopy to probe the water-binding pocket of the Oxygen Evolving Complex (OEC) of Photosystem II. A series of small isotopically-labelled alcohols which vary in length and bulkiness have been added to Photosystem II membranes. These alcohols are being studied under a variety of conditions, including a range of alcohol concentration and membrane treatment steps, in order to obtain a dimensional estimate of the binding pocket.

509.
S=1/2 AND 9/2 DIIRON COMPLEXES FORMED FROM IRON(II) COMPLEXES Karl S. Hagen*, Alain Diebold, Sonha C. Payne, Alex Kitaygorodskiy, and Abdelaziz Elbouadilli, Department of Chemistry, Emory University, Atlanta GA, 30322

Variations in the bridging groups in diiron(II) model complexes of the diiron(II) forms of non-heme proteins results in very different reactivity of these complexes towards dioxygen and other oxidants. Using tacn or Me3tacn and mono- or dicarboxylates prepared by condensing Kemp's triacid with various diamines form complexes that react very differently with dioxygen. These differences can be ascribed to supramolecular differences in structures of these complexes. We shall describe the conditions that afford intermediates with either S = 1/2 or 9/2 ground states on the reaction path to diiron(III) oxo bridged complexes and relate these to mixed-valence species detected in ribonucleotide reductase.

510. COMPUTATIONAL ANALYSIS OF PTERIN DERIVATIVES WITH APPLICATIONS TO METALLOENZYMES. W. A. Wehbi, N. V. Shokhirev, B. Fischer, P. Basu and J. H. Enemark, Department of Chemistry, University of Arizona, Tucson, Arizona 85721 and Lehrstuhl für Analytische Chemie, Ruhruniversität Bochum, Bochum, Germany.

The pterin-containing molybdenum cofactor is involved in redox reactions in the biological metabolism of carbon, nitrogen, and sulfur. This complex contains a molybdenum atom coordinated to a tricyclic pterin ring system via a dithiolene moiety. We have investigated the electronic properties of several redox states of pterins, such as the dihydropterins and tetrahydropterin derivatives by semiempirical and *ab initio* methods in an effort to rationalize chemical properties during redox reactions. Preliminary results indicate that the tetrahydropterin derivatives have similar electronic structures and that coordination to molybdenum has a marked effect on their frontier orbitals.

511.
MECHANISTIC STUDIES OF TOPA QUINONE BIOGENESIS. **Christy E. Ruggiero** David M. Dooley, Montana State University, Dept. Of Chemistry and Biochemistry, Bozeman, MT, 59717

Copper-containing amine oxidases are among a growing class of enzymes that contain a post-translationally modified amino acid as an active-site redox cofactor. The cofactor in amine oxidases, topa quinone (TPQ), is formed from a tyrosine residue in a highly specific, self-processing reaction that requires both copper and dioxygen. We have shown the copper binds into the active site of the unprocessed (tyrosine-containing) protein and that this bound copper is required for TPQ formation. These and other results of recent kinetic and spectroscopic studies on the mechanism of TPQ formation will be presented.

512. ZEOLITE "WAX IMPRESSIONS": SYNTHESIS OF ORGANIC MOLECULAR SIEVES USING INORGANIC TEMPLATES. Stacy A. Johnson, Elaine S. Brigham, Patricia J. Ollivier, Gina L. Bowden, and Thomas E. Mallouk*, Department of Chemistry, The Pennsylvania State University, University Park PA 16802

Organic molecular sieves have been prepared by a novel templating method. Aluminosilicate zeolites with 12-ring channels were impregnated with phenol, which was converted to an intrazeolitic novolac resin by reaction with gaseous formaldehyde. The resin was cured under inert atmosphere and the zeolite template etched away using aqueous HF, to leave the "wax impression" structure. These organic molecular sieves have very high surface area and a narrow distibution of pores in the 6 Å range. The mesopore (15 - 30 Å) volume depends on the identity of zeolite template. Electronically conducting microporous carbons can be prepared by pyrolyzing the polymers prior to removal of the zeolite template. X-ray diffraction patterns suggest an open structure composed of high aspect ratio graphene ribbons.

513.
SITE-SPECIFIC METALLATION OF NUCLEIC ACIDS VIA INCORPORATION OF CHELATE-FUNCTIONALIZED NUCLEOSIDES. **Thomas W. Welch**, Jeffrey J. Rack and Thomas J. Meade, Division of Biology and The Beckman Institute, California Institute of Technology, Pasadena, CA 91125.

Electron-transfer rates comparable to those encountered in modified proteins have been observed between ruthenium-based donors and acceptors coordinated to 2'-aminouridine nucleosides at the 5' ends of a double-stranded octanucleotide. While coordination of [Ru(bpy)2(imidazole)]2+ to the 2'-amino- functionality proceeds with high yield, the instability of [Ru(NH3)4(py)(H2O)]2+ at high pH dictates the use of buffer conditions in which the 2' amino-group is protonated, thus severely limiting the yield. To exploit the chelate effect for enhanced yield and stability, we have prepared N2'-(2-pyridylmethyl)- and N2'-(2-imidazolylmethyl)-2'-aminouridine derivatives. Characterization of the modified nucleosides, model complexes with [(bpy)2Ru]2+ and [(NH3)4Ru]3+ fragments, and metallated oligonucleotides will be presented.

514. A DETERMINATION OF THE REDUCTION POTENTIAL OF THE N- AND C-BOUND IRON CENTERS IN HUMAN TRANSFERRIN AT LOW pH. D. Kraiter*, P. Aisen#, and A. L. Crumbliss* *Department of Chemistry, Duke University, Durham, NC 27708 and # Department of Physiology and Biophysics, Albert Einstein College of Medicine, New York, NY 10461.

Human transferrin is a bilobal protein, with two iron(III/II) redox active sites (C and N terminal lobe) with almost identical coordination environments. We present a spectroelectrochemical determination of the formal redox potentials of diferric (Fe_2Tf), and Fe(C)Tf and Fe(N)Tf monoferric, transferrins in an anaerobic optically transparent thin layer cell, using methylviologen as an electron transfer mediator at 25 ºC, 0.5 M KCl and pH = 5.8. Values of -331, -307, and -292 mV (*vs* NHE) were found for (Fe_2Tf), Fe(C)Tf, and Fe(N)Tf, respectively. These results indicate that the redox centers are only slightly different and do no interact with each other. Some evidence of Fe scrambling between the N and C sites was obtained in the form of Nernst slopes < 1. Our results are in agreement with previously published results and support recent findings of kinetically active or significant anion binding sites. These results support the observation that background electrolyte ions (Cl^-) play an important role in the activity of the redox centers.

515. CHARACTERIZATION OF CATALYTIC ACTIVITY OF IRON PORPHYRINS IN THE LARGE-PORE ALUMINO SILICATE MCM-41. **Rachel Narehood Austin** LeRae Graham, John Pham, Martin Piech, Volker Schünemann, A.X. Trautwein, Ivonne Rietjens, M. Boersma, Cees Veeger, Dominique Mandon, Raymond Weiss, Department of Chemistry, Bates College, 5 Andrews Rd. Lewiston ME 04240; Institut für Physik, Med. Universität Lübeck, 23562 Lübeck, Germany; Department of Biochemistry, Agricultural University, 6703 HA Wageningen, The Netherlands;Institut Le Bel, Université Louis Pasteur, 67070 Strasbourg, France

Larger-pore alumino silicates may serve as suprastructrures in which to embed fully synthesized metalloporphyrins to study both the influence of the suprastructure on heme intermediates and catalytic activity. Results from attempts to use MCM-41 as a host for a series of metalloporphyrins, including ICP-AES data that indicate percent Fe in the zeolite after impregnation of the metallated compounds and results from assays to determine the efficiency and selectivity of these heterogeneous systems as catalyst will be presented.

516. BIOMOLECULAR CLEAVAGE IN AN LpNA^{2+}/FeEDTA$^-$ MODEL SYSTEM. Elizabeth M. Boon and Rosemary A. Marusak; Department of Chemistry; Kenyon College; Gambier, OH, 43022.

In the presence of an oxidant and a reductant, FeEDTA$^-$ complexes are known to rapidly catalyze protein cleavage. This scission has been reported to take place by oxidative or hydrolytic-type mechanisms; sometimes both are observed in the same system. Three mechanisms for iron-mediated protein cleavage have been proposed, but the driving force for one over the others is still speculative. In an attempt to further investigate the chemistry in these scission processes, we have expanded investigations initiated by Marusak and Meares (unpublished results) in an LpNA^{2+}/FeEDTA$^-$ model system. Studies varying ionic strength, ascorbic acid concentration, metal ion, chelate structure, substrate, and solvent have helped determine how reaction environment affects product formation. Products have been isolated by HPLC and identified by mass spectroscopic analysis. The data will be interpreted in light of the three proposed mechanisms for protein cleavage.

517.
ORGANOTINS AS POTENTIAL LARVACIDES AND/OR INSECTICIDES TO CONTROL MALARIA. Charnita Whitmyer*, George Eng* and Barbara Sina'. *Department of Chemistry, University of the District of Columbia, Washington, DC 20008 and 'Department of Entomology, University of Maryland, College Park, MD 20742.

The larvicidal and insecticidal activities of several triphenyl- tributyl-, and tricyclohexyltins were tested on the *Anopheles Stephensi* adult female mosquito and their larvae. The *Anopheles Stephensi* is the primary vector of the parasite causing human malaria. Development of a more effective larvacide/insecticide would be of tremendous value in the efforts to reduce the spread of this highly debilitating and often deadly disease. Preliminary larvae studies using triphenyltin chloride indicated that the LD_{50} value for this compound is 10 ppm. Twenty-four hour insecticidal studies were performed using triphenyl-,tributyl-and tricyclohexyltin chlorides, tricylcohexyltin acetate as well as bis-triphenyltin oxide on the adult female mosquitoes. The results from the study indicated that the toxicity of the organotins was a function of the contact time between the toxicant and mosquitoes, rather than on the organic substituent on the tin atom.

518. FORMATION AND REACTIVITY OF FLUOROALKYL COBALT PORPHYRIN COMPLEXES   Yaping Ni and Bradford B. Wayland   *Department of Chemistry, University of Pennsylvania, Philadelphia, Pennsylvania 19104-6323*

A series of fluoroalkyl cobalt porphyrin complexes ((Por)Co-Rf) are prepared through reactions of fluoroalkenes with porphyrinatocobalt(II) ((Por)Co$^{II\cdot}$) and alkyl radicals (·C(CH$_3$)(CN)R) formed from dialkylazo precursors. The reactions go through the additions of a transient cobalt(III) hydride ((Por)Co-H) with fluoroalkenes, which generally occur with high regioselectivity corresponding to formation of the most stable fluoroorganic radicals. A single crystal X-ray structure of (OEP)Co-CFClCF$_2$H and the reactivity of porphyrin cobalt(III) fluoroalkyl complexes((Por)Co-Rf), in comparison to the corresponding porphyrin cobalt alkyl complexes((Por)Co-R), will be presented.

519.   RECONSTITUTION OF THE PORPHYRIN ISOMER PORPHYCENE INTO THE "HEME POCKET" OF MYOGLOBIN. Christopher J. Fowler, Tetsuo Takimuri, and Jonathan L. Sessler. Department of Chemistry and Biochemistry, University of Texas at Austin, Austin, Texas, 78712

Fe(III) Etioporphycene (1) has been reconstituted into horse heart apomyoglobin in the monomeric form. The absorption spectrum of purified rMb(1) has both the protein peak (280 nm) and the Fe(III) porphycene peaks (385 nm, 560 nm, 617 nm) in the ratio A_{385}/A_{280} equal to 2.6. The binding site of Fe(III) porphycene was determined by the competition reaction for apoMb against the native prosthetic group (Fe(III) protoporphyrin). These results show that the binding site of 1 is in fact the "heme pocket" of myoglobin. The structure of the porphycene reconstituted myoglobin is measured with paramagnetic ^{1}H NMR. Other physiochemical properties investigated are the autoxidation rate and oxygenation properties using Co. As the porphycene 1 has solubility limitations, porphycene 2 is being synthesised to correct for these problems.

520. SYNTHESIS AND CHARACTERIZATION OF A NOVEL EXPANDED PORPHYRIN. Julian M. Davis, Jonathan L. Sessler, Vincent Lynch. Department of Chemistry and Biochemistry, The University of Texas at Austin, Austin, Texas 78712-1167.

Expanded porphyrin macrocycles are important because they possess characteristics such as anion binding that their porphyrin cousins do not. The novel expanded porphyrin isomer **1** and its mono-oxa analog **2** were recently synthesized by a 1 + 1 condensation of a bis-alpha free terpyrrole and a diformyl dipyrromethane. Studies using **1** and known macrocycles sapphyrin and smaragdyrin are currently underway to examine the effects of pyrrole connectivity on metallation and anion binding properties.

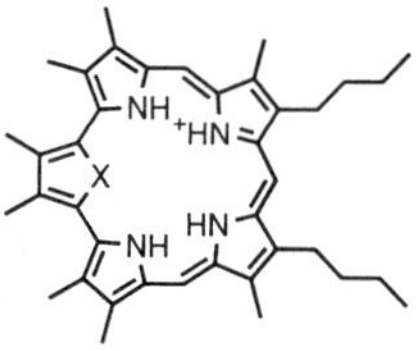

1 X=NH

2 X=O

521. SYNTHESIS OF PYRROLE SUBSTITUTED ANSA-FERROCENES. M. Scherer, J. L. Sessler*, A. Gebauer, V. Lynch; Department of Chemistry and Biochemistry, The University of Texas at Austin, Austin, Texas 78712

Pyrroles and ferrocenes are versatile building blocks used for the synthesis of numerous ligands and receptors. Here, we wish to present the synthesis of systems incorporating both of these moieties. The new key starting compound, **1**, is used to make more complex, ferrocene-containing systems (*e.g.*, **2**). Following hydrogenation of **2**, a strapped ansa-metallocene **3** is prepared by an easy amide bond-formation reaction. It possesses an unusual trans symmetry as revealed by X-ray crystallography.

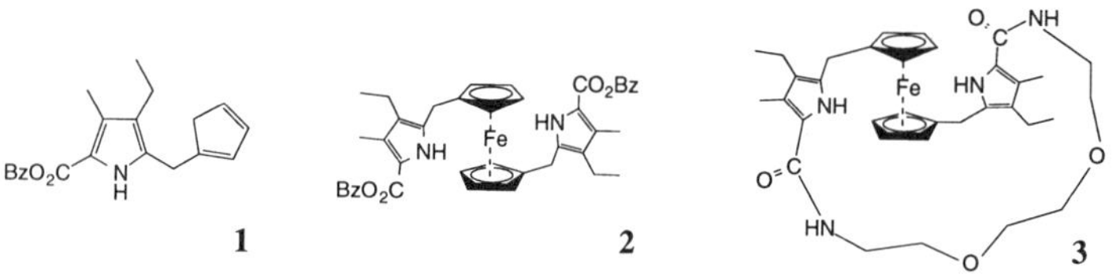

522. β-SUBSTITUTED CALIX[4]PYRROLES: NEW CHEMISTRY AT THE *C*-RIM.
Philip A. Gale, William E. Allen, John Genge, Nicolai Tvermoes, Vincent Lynch and Jonathan L. Sessler. Department of Chemistry and Biochemistry, University of Texas at Austin, Austin, Texas 78712-1167.

Over the last year, calix[4]pyrroles have been shown by us to be selective anion binding agents. We are currently interested in adding extra functionality to the calixpyrrole skeleton. Two strategies have been pursued in the synthesis of β-substituted calix[4]pyrroles: firstly, synthesis from 3,4-disubstituted pyrroles and ketones and secondly, direct modification of the *C*-rim of a pre-synthesized calix[4]pyrrole. Thus, new calix[4]pyrroles (**1 - 4**) have been produced. The anion binding strengths of **1, 2** and **3** are reduced relative to their parent calix[4]pyrrole macrocycles, whereas compound **4** shows a higher affinity for anions than the β-free analogue.

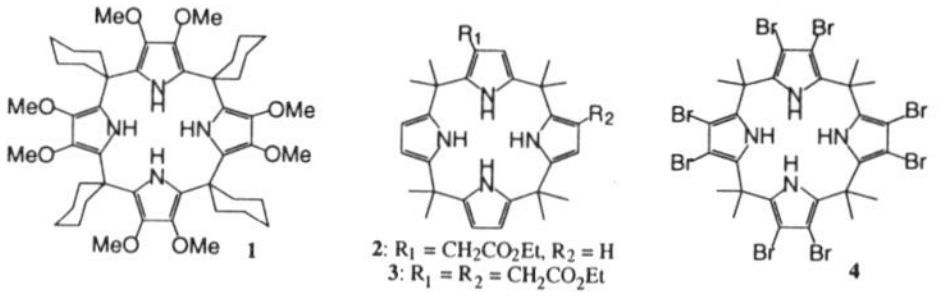

2: $R_1 = CH_2CO_2Et, R_2 = H$
3: $R_1 = R_2 = CH_2CO_2Et$

523. NOVEL COMPLEXES OF OSMIUM AS MOLECULAR PROBES FOR DNA. **J. A. Yao**, R. E. Holmlin, and J. K. Barton, Division of Chemistry and Chemical Engineering, California Institute of Technology, Pasadena, CA 91125

Bound to DNA, Os(phen)$_2$dppz^{2+} derivatives exhibit emission maxima $\geq$ 740 nm and excited-state lifetimes in the 10 ns regime. To explore these molecules as a new class of luminescent DNA probes we incorporated simple modifications in the ancillary phen or bpy ligands as well as the intercalating dppz ligand of Os(L^1)$_2$(L^2)$^{2+}$ and Os(L^1)(L^2)(L^3)$^{2+}$ complexes to selectively tune elecrochemical, spectroscopic and DNA-binding properties. Absorption titrations are consistent with intercalative binding to DNA with high affinity ($K_B \sim 10^6$ M^{-1}) for the family of compounds. By steady-state and time-resolved emission spectroscopy, the emission yield and excited-state lifetime of the complexes depend sensitively on the ligand architecture. Studies with calf thymus DNA, poly d(AT) and poly d(GC) reveal that these properties also depend on the DNA sequence. Cyclic voltammetry indicates that the Os(3+/2+) couple is affected by derivatization of the ancillary ligands but not the intercalating ligand. The phenazine reduction is shifted toward the negative with incorporation of methyl substituents at the 7,8 or 6 positions of the dppz ligand.

Δ-Os(L^1)(L^2)(L^3)$^{2+}$

---

524. SYNTHESES, STRUCTURES AND CYTOTOXICITY OF PLATINUM (II) DINUCLEAR COMPLEXES. Yukifumi Dohta†, C. Scott Browning, Petri Rekonen, Masato Kodaka, Ken-ichi Okamoto†, and Hiroaki Okuno, Biomolecules Department, National Institute of Bioscience and Human-Technology, 1-1 Higashi, Tsukuba, Ibaraki 305, Japan. † University of Tsukuba, Tsukuba, Ibaraki 305, Japan.

We synthesized platinum (II) dinuclear complexes with 1-alkyluracil (R=Me, Et, n-Bu, Bzl, CH$_2$Naph), 1-alkylthymine (R=Me, Et, n-Pr), succinimide, 3,3-dimethylglutarimide and 3-ethyl-3-methylglutarimide, to search novel platinum complexes with high cytotoxicity. Here, the properties of the complexes (hydrophobicity and electrophilicity), which are thought to dominate the cytotoxicity, were varied. The Pt(II) complexes with 1-alkyluracil ligands were synthesized via the coupling of cis-[Pt(NH$_3$)$_2$(DMF)$_2$]$^{2+}$ and cis-[Pt(1-RUra)$_2$(NH$_3$)$_2$]. The other complexes were prepared by the reaction between cis-[Pt(NH$_3$)$_2$(H$_2$O)$_2$]$^{2+}$ and the corresponding ligands. Some of the complexes gave good crystals and were subjected to X-ray structural analysis. The complexes with head-to-head form showed high cytotoxicity, while those with head-to-tail form were inactive.

---

525. FACTORS CONTROLLING CYTOTOXICITY OF CIS-DIAMMINEPLATINUM(II) DINUCLEAR COMPLEXES. Yukifumi Dohta†, Petri Rekonen, Tomoko Okada, Masato Kodaka, and Hiroaki Okuno, Biomolecules Department, National Institute of Bioscience and Human-Technology, 1-1 Higashi, Tsukuba, Ibaraki 305, Japan.
† University of Tsukuba, Tsukuba, Ibaraki 305, Japan.

We investigated the effect of hydrophobicity and exchange rate of leaving ligands of cis-diammineplatinum(II) dinuclear complexes upon the cytotoxicity. As the leaving ligands, 1-alkyluracils, 1-alkylthymines and cyclic imides were chosen. Hydrophobicity of the dinuclear complexes were evaluated by calculated logP of the ligands and by the capacity factor of the complexes obtained by HPLC. Exchange rates of leaving ligands were measured in saline. Cytotoxicity was examined against S-180 and L1210 cell lines. The cytotoxicity of the dinuclear complexes with head-to-head arrangement seems to depend on the combination of the hydrophobicity and the exchange rate. It seems that some fraction of the head-to-head complexes affects cancer cells in their intact form, while the other fraction is transformed to CDDP or to the complexes with serum proteins.

526. **IMIDO, HYDRAZIDO, AND AMIDO GROUP 4 METALLOPORPHYRINS.** Joseph L. Thorman and L. Keith Woo*, Dept. of Chemistry, Iowa State University, Ames, IA 50011-3111.

We recently reported the synthesis and characterization of (TTP)Ti=NR (TTP = tetratolylporphyrinato; R = phenyl, tolyl, t-butyl, trimethylsilyl, cyclohexyl). New metalloporphyrin imido complexes for Zr and Hf have been prepared. In addition, hydrazido metalloporphyrin complexes of Ti and Zr and bis amido metalloporphyrin complexes of Hf have been synthesized. The characterization and reactivity of these complexes will be discussed.

$$(TTP)M=NNR_2 \xleftarrow{H_2N_2R_2} (TTP)MCl_2 \xrightarrow{LiNHR^*} (TTP)M=NR^*$$
M = Ti, Zr; R = CH_3, C_6H_5 M = Ti, Zr, Hf; $R^* = C_6H_2$-2,4,6-t-Bu_3

527. DNA-MEDIATED ELECTRON TRANSFER RATES THROUGH SINGLE VS. DOUBLE STRANDED OLIGONUCLEOTIDES. **Elizabeth S. Krider,** Thomas W. Welch, Jeffrey J. Rack, and T. J. Meade, The Beckman Institute, California Institute of Technology, Pasadena, CA 91125.

Electronic coupling interactions are significant to understanding electron transfer mechanisms through modified DNA. In order to determine the role of base sequence and π-stacking on electron transfer rates, we have prepared an oligonucleotide which contains two sites for covalent attachment of donor and acceptor complexes. Located at the strand's termini, these attachment sites consist of nucleosides functionalized at the 2' position of the ribose ring. The attachment of the ruthenium complexes, [Ru(NH3)4(pyridine)]2+ and [Ru(bpy)2(imidazole)]2+, does not affect the ability of the oligonucleotide to hybridize with its complementary strand. ET rates through this novel oligonucleotide both in the presence and absence of the complementary strand are reported.

528. BIOMIMETIC AND BIOINTERACTION STUDIES OF OXOVANADIUM(IV & V) COMPLEXES. A. Sreedhara and C.P. Rao, Department of Chemistry, Indian Institute of Technology Bombay, Powai, Mumbai 400 076, India.

Use of small molecules as structural and functional models for biological systems is not a new concept anymore. This presentation will deal with the role of oxo-complexes of V(IV) and V(V) in the context of their ability to show substrate bromination, RNase inhibition, DNA cleavage, protein synthesis inhibition, lipid peroxidation and cytotoxicity. A few vanadyl complexes have shown potential to inhibit RNase activity by acting as transition state analogue and at the same time not inhibiting the DNase. Experiments with plasmid DNA indicated cleavage of the original form which is facilitated in the presence of H_2O_2 due to the involvement of oxygen-based radicals in the process. The bromination activity of oxovanadium(V) complexes in $HClO_4$ and H_2O_2 demonstrated the catalytic behaviour and the formation of VO_2^+ followed by oxo-peroxo vanadium species in the reactions. Xylene cyanol and phenol red are used as substrates for bromination and the studies were followed by UV-Vis absorption and ^{51}V-nmr spectra. Some oxovanadium(IV) complexes have shown protein synthesis inhibition in rabbit reticulocytes. Some of these compounds have also shown lipid peroxidation and cytotoxicity as studied with rat hepatocytes.

529. KINETICS AND MECHANISM OF Fe(III) COMPLEXATION BY LIPOPHILIC 3-HYDROXY-2-METHYL-1(γ-STEAROAMIDOPROPYL)-4-PYRIDINONE (HMSP). Yanlong Shi,* Edward M. Eyring*, Rudi van Eldik,‡ Gang Liu,∥ Fred W. Bruenger∥ and Scott C. Miller∥ *: Department of Chemistry, University of Utah, Salt Lake City, Utah 84112; ‡ : Institute for Inorganic Chemistry, University of Erlangen-Nürnberg, 91058 Erlangen, Germany; ∥: Radiobiology Division, School of Medicine, University of Utah, Salt Lake City, Utah 84112

The kinetics of Fe(III) complexation by lipophilic 3-hydroxy-2-methyl-1(γ-stearoamidopropyl)-4-pyridinone (HMSP) were studied when [Fe(III)] >> [HMSP] in MeOH/H_2O mixed solvent and [Fe(III)] << [HMSP] in MeOH, respectively. When Fe(III) is in excess, the observed rates depend on $[Fe(III)]_{tot}{}^2$ and on the reciprocal of [H^+] and decrease with increasing pressure. $\Delta V^\ddagger$ values are around +8.0 cm^3 mol^{-1}. A mechanism consisting of the complexations of the hydrolyzed monomer $Fe(H_2O)_5OH^{2+}$, and dimer species, $Fe_2(H_2O)_7(\mu\text{-}OH)_2OH^{3+}$ by HMSP is proposed. This mechanism is supported by the solvent effect and the work of other researchers. When HMSP is in excess, 1:3 $Fe(HMSP)_3$ is formed and three kinetic steps on different time-scales are observed. An "intermolecular chelate ring-closure" mechanism is proposed, differing from the "intramolecular chelate ring-closure" complexation previously reported for the formation of ferrioxamine B.

530. METAL BINDING TO A PEPTIDE MODEL OF THE COPPER-BINDING REGION OF LYSYL OXIDASE.
Faina Ryvkin and Frederick T. Greenaway. Department of Chemistry, Clark University, Worcester, MA 01610.

Lysyl oxidase is a 29kDa Cu(II)-containing quinoprotein responsible for catalyzing the oxidation of the ε-amino group of lysine side chains in collagen and elastin, ultimately leading to the formation of cross-links in these important structural proteins. Although the location of the lysine tyrosylquinone cofactor has been recently reported, the copper-binding ligands have not been identified. We report the results of a study of metal binding to a 34-amino acid peptide model of lysyl oxidase designed to help identify the copper-binding region. We will present the results of molecular modelling, CD, and 2-D NMR studies of the peptide confirming the existence of a stable conformation of the peptide in solution, and of NMR, EPR, CD, and UV-visible spectroscopic characterization of the copper(II) and cobalt(II) binding sites.

531.
PROTECTION OF METAL-CARBOXYLATE BOND FROM DISSOCIATION BY INTRAMOLECULAR NH···O HYDROGEN BONDS. **Y. Yamada,** N. Ueyama, T. Okamura and A. Nakamura, Department of Macromolecular Science, Graduate School of Science, Osaka University, Toyonaka, Osaka 560 Japan

Metal-carboxylate binding sites are often seen in metalloproteins. In many cases, these carboxylate proteins have NH···O hydrogen bonds from main chain amide proton to the coordinating carboxylate oxygens. To elucidate the role of the NH···O hydrogen bonds, we synthesized and determined the structures of the anionic tetrakis(carboxylate) mononuclear complexes, $[M^{II}\{OCO\text{-}2,6\text{-}(t\text{-}BuCONH)_2C_6H_3\}_4]^{2-}$ (M = Ca, Zn, Mn), which have intramolecular NH···O hydrogen bonds. The ligand exchange reaction between $(NEt_4)_2[Zn^{II}\{OCO\text{-}2,6\text{-}(t\text{-}BuCONH)_2C_6H_3\}_4]$ and large excess of [2,4,6-$Me_3C_6H_2COOH$] in $CDCl_3$ at 30 °C was found to be absent. On the contrary, the reaction between $(NEt_4)_2[Zn^{II}\{OCO\text{-}2,4,6\text{-}Me_3C_6H_2\}_4]$ and 4 equimolar 2,6-(t-$BuCONH)_2C_6H_3COOH$ gave a quantitative yield of $(NEt_4)_2[Zn^{II}\{OCO\text{-}2,6\text{-}(t\text{-}BuCONH)_2C_6H_3\}_4]$. The results indicate that the intramolecular NH···O hydrogen bonds protect the M-O bond from dissociation.

532.
SYNTHESES AND PROPERTIES OF CYTOCHROME P-450 MODEL COMPLEXES WITH HELICAL CYSTEINE-CONTAINING PEPTIDE. **T. Ueno,** N. Nishikawa, S. Moriyama, N. Ueyama, and A. Nakamura, Department of Macromolacular Science, Graduate School of Science, Osaka University, Toyonaka, Osaka 560, Japan

Cytochrome P-450 model peptide complexes having helical structure, [Fe^{III}(OEP)(Ac-LCXAFLLLLLALFL-OMe)] {OEP = octhaethylporphyrin, X = Leu (**1**), Pro (**2**)} were synthesized and characterized. The helical structure of the peptides was confirmed by the NMR spectra of the cysteine-containing-peptides in THF-d_8. The ligation of cysteine at the axial position in P-450 model complexes was established using MCD, UV-visible, ^{1}H NMR and ESI-MS. Redox potentials of Fe(III)/Fe(II) of **1** and **2** are –0.54 and –0.55 V (vs. SCE in CH_2Cl_2), respectively. The values are higher than those of [Fe^{III}(OEP) (tetrapeptide)] {tetrapeptide = Z-CLGL-OMe (–0.59 V) and Z-CPLA-OMe (–0.63 V)}. These results indicate that change of the secondary structure of the peptide ligand contributes to the regulation of redox reactivity during the catalytic oxidation

533.
EFFECT OF MUTATIONS IN THE PROBABLE I-HELIX OF CYTOCHROME P4504A1 ON HEME-SITE STRUCTURE AND PRODUCT FORMATION. **E.A. Dierks,** S.C. Davis, and P.R. Ortiz de Montellano, School of Pharmacy, Department of Pharmaceutical Chemistry, University of California, San Francisco, CA 94143-0446

The cytochrome P4504A enzymes catalyze the omega-hydroxylation of fatty acids of varying chain lengths, while other P450 enzymes catalyze the more thermodynamically favorable (omega-1)-hydroxylations. Little is known about the structure of the 4A enzymes, although model studies have suggested the heme site must be quite hindered in order to preferentially hydroxylate fatty acids at the omega-position. Sequence alignment of the rat P4504A1 with the heme domain of the bacterial P450102 led to identification of two possible residues in the I-helix that could restrict access to the heme. Mutations at these positions to the residues present in P450102 yielded three P4504A1 mutants, E325A, D328E, and E325A/D328E. The effects of these mutations on product formation/specificity, heme-site structure, and substrate specificity were studied.

534.
SYNTHESIS AND CHARACTERIZATION OF PORPHYRINS BEARING POLAR SUBSTITUENTS AND THEIR METAL COMPLEXES. **Elizabeth Louie and Ursula Simonis,** Department of Chemistry and Biochemistry, San Francisco State University, San Francisco, CA 94132

The synthesis of water-soluble porphyrins and metalloporphyrins is of great interest due to their potential applications as pharmaceutical drugs. Porphyrins are used as photosensitizers in photodynamic therapy, and they have shown to be inhibitors of HIV replication. Porphyrins have also been used to localize and treat tumors. Recently, it has been found that paramagnetic metalloporphyrins can be used as MRI contrast agents. To better understand the role of water-soluble porphyrins and metalloporphyrins as pharmaceutical agents, we have synthesized unsymmetrically substituted mono para-carboxylato-, sulfonato-, or phosphonatotetraphenylporphyrins and their corresponding metal complexes. These porphyrins and their metal derivatives were characterized by UV/vis and NMR spectroscopy to determine how the substituents affect the electron density distribution in the porphyrin macrocycles of these compounds.

535.
NMR SPECTROSCOPIC STUDIES OF MODEL HEMES EMBEDDED IN MICELLES. **Sandra Ried and Ursula Simonis,** Department of Chemistry and Biochemistry, San Francisco State University, San Francisco, CA 94132

To better understand biological processes involving heme proteins, such as electron transfer reactions and mechanisms by which hemes are transported across membranes, porphyrins and hemes have been studied in model membranes. Although model hemes are predicted to bind to lipid bilayers and micellar interfaces with a preferential orientation, their precise location inside micelles or lipid bilayers is essentially unknown. To investigate porphyrin-micellar interactions, we were able to embed 5-(p-carboxy)phenyl-10,15,20-triphenylporphyrin into tetradecyltrimethylammonium bromide (TTAB), cetyltrimethylammonium bromide (CTAB) and Triton-X micelles. In contrast, the porphyrin did not incorporate into sodium dodecylsulfate (SDS) micelles. Low- and high-spin Fe(III) complexes behaved similarly. UV/vis and NMR spectroscopy were used to determine the preferred orientation of the porphyrin in the micelles, which will be discussed.

536.
GEOMETRY OF BINDING OF RUO(TPY)(BPY)$^{2+}$ TO DOUBLE STRANDED DNA. **Brian T. Farrer** H Holden Thorp, University of North Carolina, NC, 27514

RuO(tpy)(bpy)$^{2+}$ binds DNA in both the major and minor grooves. From the major groove it oxidizes guanine residues and causes strand scission upon piperidine treatment. From the minor grove, it cleaves the DNA via 1'-hydride abstraction. In order to modify the selectivity between sugar and base oxidation, substituents were placed in the 4,4'-positions on the bipyridine ligand and in the 4'-position of the terpyridine ligand in the hopes that the binding to the minor and major grooves would be effected differently. Electronic effects of the substiuents were studied using mononucleotides and single stranded DNA. The binding of RuO(X-tpy)(Y$_2$-bpy)$^{2+}$ to double stranded DNA was studied using gel electrophoresis.

537. CHEMOENZYMATIC SYNTHESIS OF HEME PROTEIN MODELS Donald W. Low, Grace Yang, Jay R. Winkler and Harry B. Gray, Arthur Amos Noyes Laboratory and the Beckman Institute, California Institute of Technology, Pasadena, California 91125, USA

A reverse proteolysis strategy is used to covalently modify the C terminus of microperoxidase-9 (MP9), the heme nonapeptide obtained from tryptic hydrolysis of horse-heart cytochrome c. We have found that trypsin will efficiently catalyze the addition of a wide variety of amino acid amides to MP9 in mixed aqueous-organic solvent mixtures, resulting in novel "mutant" microperoxidase-10 fragments. We have prepared and isolated histidine (H23MP10), methionine (M23MP10), tyrosine (Y23MP10) and tryptophan (W23MP10) mutants. Coupling yields of up to 50% were observed, and reactions proceeded cleanly with negligible side products. The H23MP10 exhibits a pH dependent spin state equilibrium, with a pK ~ 4. This change has been monitored by optical and resonance Raman spectroscopies. Both are consistent with the acid form having a his-aquo coordinated high-spin heme, and the basic form having a bis-his coordinated low-spin center. M23MP10 and Y23MP10 were not found to exhibit similar spin state changes that would indicate thioether or phenolate coordination. Tryptophan fluorescence can be readily detected from W23MP10, and may prove to be a useful probe of peptide tail dynamics in MP fragments.

538.
BINDING STUDIES OF CHROMIUM(III) COMPLEX TO FERRITIN. **Carmen M. Barnes** *Elizabeth C. Theil, Kenneth N. Raymond, Department of Chemistry, University of California Berkeley, Berkeley, CA 94720, *Department of Biochemistry, NCSU, Raleigh, NC, 27695-7622

Ferritin, the universal iron storage protein, is assembled from 24 subunits, producing channels with 3-fold and 4-fold symmetry. Mammalian ferritins are formed from mixtures of two subunit types, H and L. The H-chains, in contrast to L-chains, can catalyze the rapid oxidation of Fe(II) to Fe(III). In order to investigate the routes of iron incorporation, [Cr(TREN)(H2O)(OH)]2+ has been used as inhibitor of Fe(II) transport. Given the size and kinetic inertness of the Cr(III) complex, it cannot penetrate the protein shell. Based on spectroscopic evidence, the chromium compound inhibits Fe(II) uptake in Horse spleen Ferritin (HoSF) and recombinant Bullfrog H Ferritin (HF). The interaction between the Cr(III) complex with HF, HoSF and Bullfrog L-Ferritin (LF) was examined by denaturing gel electrophoresis. As evidenced by the formation of cross-linked subunits, binding of the Cr(III) complex occurs specifically in HF to an intersubunit site.

539.
SYNTHESIS, STRUCTURE AND PROPERTIES OF NICKEL CATECHOLATES. **Christian Brückner,** Dana L. Caulder and Kenneth N. Raymond, Department of Chemistry, University of California at Berkeley, Berkeley, CA 94720-1460.

Catechols, as exemplified by the paramagnetic nature of the iron(III) triscatecholate complex, have been regarded to be weak field ligands. In this light it seems surprising that a 2,3-dihydroxybenzoic acid amide derivative reportedly forms a diamagnetic square planar complex with nickel(II). There are only few reports on nickel(II) catecholates and most of these - even partially contradicting - reports date back over 50 years. Because of this, and the general interest in catchol as a ligand, we have prepared nickel(II) complexes of catechol and catechol derivatives such as 2,3-hydroxyterephtalamides. We report on the studies of their physical, electrochemical and chemical properties and the X-ray crystal structure of the nickel(II) biscatecholate as its dipotassium salt.

540.
COORDINATION NUMBER INCOMMENSURATE CLUSTER FORMATION. **Dana L. Caulder,** Ryan E. Powers and Kenneth N. Raymond, Department of Chemistry, University of California, Berkeley, CA, 94720

While there is a great deal of interest in the synthesis of supramolecular assemblies, the examples to date have largely been fortuitous accidents. This paper will describe a model for the symmetry-driven assembly of clusters, and thereby provide a rational approach, or recipe, for the synthesis of supramolecular clusters. On the basis of a symmetry-driven process that relies on incommensurate lock-and-key interactions, M2L3 helices, M4L6 tetrahedra, M4L4 tetrahedra and higher symmetry clusters can be pre-designed and synthesized. These supramolecular clusters will be discussed in terms of their structural, kinetic and thermodynamic characteristics.

541.
FE-TRENSAM: A SALICYLATO ANALOGUE OF THE SIDEROPHORE ENTEROBACTIN. **Seth M. Cohen** , Michel Meyer, Kenneth R. Raymond, Department of Chemistry, U.C. Berkeley, Berkeley, CA 94720

Enterobactin is a tris-catecholate siderophore synthesized by bacteria in order to incorporate vital ferric iron from their environment to support metabolic processes. Enterobactin complexes ferric iron in a chiral, delta conformation complex with unrivaled stability. Previous work has suggested that at lower pH the enterobactin catechols become sequentially mono-protonated and the structure changes to a salicylato mode of binding. This poster reports the synthesis and spectra of a novel ligand that froms a tris-salicylato complex with aluminum and ferric ion, with the latter structurally characterized by single crystal x-ray diffraction. The ligand, tris(2-hydroxybenzoyl aminoethyl)amine (TRENSAM), forms a psuedo-three-fold symmetric complex in P2(1)/c. The ferric ion is bound by the phenolic and amide oxygens in three, six-membered chelate rings. The spectra and structural properties of this ligand and its metal complexes will be presented, as well as implications for iron release mechanisms of enterobactin.

542.
HEXADENTADTE HYDROXYPYRIDONATE GADOLINIUM COMPLEXES WITH IMPROVED WATER SOLUBILITY. **Sharad Hajela,** Christopher J. Sunderland, Jide Xu, and Kenneth N. Raymond, Department of Chemistry, University of California, Berkeley, CA 94720.

We have recently reported that multidentate 3-hydroxy-2(1H)pyridinone (3,2-HOPO) complexes of gadolinium have promising potential as a new class of agents for MRI contrast applications. The hexadentate TREN-Me-3,2-HOPO complex with gadolinium displays excellent stability, selectivity, and water proton relaxivity. Current studies are aimed at altering the solubility properties of the tris-HOPO ligand system by the use of modified TREN derivatives. In particular, we have recently succeded in synthesizing homochiral TREN derivatives bearing hydroxymethyl substituents on the 2-positions of the TREN arms. As expected, further elaboration of this new hydrophilic scaffolding affords a tris-HOPO gadolinium complex with improved water solubility compared to the parent TREN-linked species. Synthetic details and physical properties of the new compounds will be presented.

543.
STABILITITY AND SELECTIVITY OF TRIS-BIDENTATE 3,2-HYDROXYPYRIDONATES WITH GADOLINIUM. **Christopher J. Sunderland,** Kenneth N. Raymond, University of California at Berkeley, Berkeley, California, 94720.

The TREN-Me-3,2-HOPO complex of Gd(III) has recently shown promise as the start of a new class of neutral MRI contrast reagents. The low solubility of this complex in water, however, requires that more water soluble derivatives be developed.

Molecular modeling (MM2) studies have been undertaken which indicate that substantial freedom of the ligand scaffold structure should be available when designing water solubilizing capping amines that are readily synthetically accessible. The stability of some TREN-Me-3,2-HOPO analogs have been studied to ascertain if these modeling results are realized in vitro. Spectrophotometric titrations for the Gd - TPT-Me-3,2-HOPO system, for example, have been performed (ß110 = 21.4, ß111= 24.4). This indicates that the ligand has essentially the same stability with Gadolinium as the ligand TREN-Me-3,2-HOPO and so should provide another valid parent structure for the synthesis of functionalized (water solubilized) HOPO containing ligands.

544.
TRIPODAL 3-HYDROXY-2(1H)-PYRIDINONES AS IRON SEQUESTERING AGENTS

Jide Xu, K.N. Raymond, University of California, Berkeley, Berkeley, CA 94720

A family of tripodal N-methyl-3-hydroxy-2(1H)-pyridinone ligands has been developed as metal ion sequestering agents. Synthetic, structural, thermodynamic and electrochemical studies of the ferric complexes formed with these ligands will be presented.

TREN-Me-3,2-HOPO   TR223-Me-3,2-HOPO   TR332-Me-3,2-HOPO   TRPN-Me-3,2-HOPO

545. PROBING VANADIUM(V)-CARBOHYDRATE COMPLEXES. Debbie C. Crans* and Feibo Xin, Department of Chemistry, Colorado State University, Fort Collins, CO 80523-1872.

The complex formation between vanadate and a series of pentoses, including ribose, arabinoase, lyxose and xylose, as well as fructose in aqueous solution, was examined at pH 8. The studies were carried out using ^{1}H and ^{51}V NMR spectroscopy. Entertaining the possibility that the complexes formed in pentose solution are all of the 2:2 type of complex formed between vanadate and diols, we were able to quantify the formation constants of pentoses knowing merely the contributions of furanose and pyranose of free pentose in aqueous solution. The formation constants for the vanadium complex with the pure furanose and pyranoses were obtained from furanosides and pyranosides, and with these values as endpoints we were able to fit pentoses containing variable concentrations of furanose, pyranose and open chain in free ligand. Our predictions were verified using ^{1}H NMR spectroscopy as well as literature reports. We are even able predict the formation constant for vanadium complexes and/or isomer fractions for other pentoses and ketoses using this analysis.

546.
YEAST CYTOCHROME C FOLDING TRIGGERED BY PHOTOINDUCED ELECTRON TRANSFER. **F. Akif Tezcan,** Jason R. Telford, Jay R. Winkler, Harry B. Gray; Arthur Amos Noyes Laboratory, California Institute of Technology, Pasadena, CA 91125

S. cerevisiae cytochrome c (cyt c) mutants with single or double point mutations have been examined. Guanidine hydrochloride (Gdn.HCl)-induced static denaturation experiments show that reduced folded cyt c is more stable than the oxidized folded protein as predicted by electrochemical data. Thus, there is a range of Gdn.HCl concentrations where electron injection into the oxidized unfolded protein initiates folding. The point mutations affect the overall stability of cyt c towards denaturation. The folding kinetics of these mutants were followed across a range of Gdn.HCl concentrations using transient absorption spectroscopy in the millisecond to second time regime. A linear relationship between the folding rate and free energy is observed.

547. A COMPARATIVE STUDY OF LOW-TEMPERATURE AND TIME RESOLVED FTIR DIFFERENCE SPECTRA OF CYTOCHROME OXIDASES
James A. Bailey, Sandra L. Mecklenburg, Gina M. MacDonald, Andromachi Katsonouri, Robert B. Gennis, R. Brian Dyer, and William H. Woodruff, Chemical Sciences and Technology Division (CST-4), Los Alamos National Laboratory, Los Alamos NM 87545; School of Chemical Sciences, University of Illinois, Champaigne-Urbana IL.

Difference FTIR spectroscopy was used to study protein changes associated with carbon monoxide (CO) photodissociation in two heme-copper oxidases, Cytochrome c Oxidase (CcO) isolated from bovine heart muscle and Cytochrome bo (Cbo) isolated from Escherichia coli. Low temperature infrared data were obtained before and after photodissociation of CO from the heme iron in CcO. The resulting light-minus-dark difference spectra contain changes associated with the transfer of the CO ligand from iron to copper, and any accompanying protein and cofactor changes associated with this exogenous ligand transfer. Time-resolved (TR) difference data were obtained at room temperature after photolysis of CO from the heme iron. This data reveals protein and heme changes occurring on the timescales of 5 us to 20 ms. TR difference FTIR data obtained on the two species of heme-copper oxidases is discussed.

548. SYNTHESIS OF Pt(IV)-TETRAPHENYLPORPHYRINS VIA DIRECT OXIDATIVE ADDITION. RaeAnne E. Falvo, Van T. Nguyen, D. Michele Smith, Larry M. Mink. Department of Chemistry, California State University, San Bernardino, California 92407. Robert K. Boggess, Heather Washburn. Department of Chemistry and Physics, Radford University, Radford, VA 24142-6949.

Platinum(IV) porphyrins of the type $[Pt^{IV}(para\text{-}X)_4TPP]L_2$, (TPP = tetraphenylporphyrin; X = H, OCH_3; L = Cl, Br), have been synthesized by direct halogenation. The Pt(IV) porphyrins are synthesized from their corresponding $Pt^{II}(para\text{-}X)_4TPP$ counterparts by direct oxidative addition of $Cl_{2(gas)}$ and $Br_{2(l)}$. The brominated complexes, $[Pt^{IV}(para\text{-}X)4TPP]Br_2$, exhibit slight bathochromic shifts in the U.V. Soret bands ($\approx$ 4 nm), and increased sheilding of all ^1H NMR resonances ($\approx$ 0.02 ppm) with respect to their chlorine analogues. Voltammetric results of the bromide complexes will be compared to previously reported data on the chlorine complexes.

549. REACTIONS OF ZIRCONIUM(IV) WITH N-ALKYLPHENOTHIAZINES: SYNTHESIS OF COMPLEXES. N. M. Made Gowda, Q. T. Ahmed, and Y. B. Kim, Department of Chemistry, Western Illinois University, 1 University Circle, Macomb, Illinois 61455

Reactions of Zirconium(IV) with N-alkylphenothiazines (NAPTZs) have been investigated in aqueous medium. Some NAPTZs find extensive applications in the field of medicine as psychotherapeutic and antiemetic drugs. The NAPTZ ligands used include: promethazine hydrochloride (PM.HCl), chlorpromazine hydrochloride (CP.HCl), ethopropazine hydrochloride (EP.HCl), trifluoperazine dihydrochloride (TF.2HCl), and thioridazine hydrochloride (TR.HCl). Based on the complexation reactions of the ligands with $ZrCl_4$ and $Zr(SO_4)_2$ in the presence or absence of hydrochloric or sulfuric acid, several complexes have been prepared and characterized. The new zirconium(IV) complexes are less soluble in common solvents such as water, MeOH, and EtOH and more soluble in DMF and DMSO. They are nonhygroscopic and stable solids at room temperature. We will propose tentative structures for the new products based on their analytical, conductance, magnetic, and spectroscopic data.

550. EFFECTS OF MUTATIONS IN PLASTOCYANIN ON THE KINETICS ON THE GATED ELECTRON TRANSFER WITH ZINC CYTOCHROME C. Maja M. Ivković-Jensen,[a] Simon Yeung,[b] Örjan Hansson,[b] and Nenad M. Kostić,[a] [a]Department of Chemistry, Iowa State University, Ames, Iowa 50011, and [b]Department of Biochemistry and Biophysics, Lundberg Laboratory, Göteborg University and Chalmers University of Technology, S-41390 Göteborg, Sweden

Kinetics of photoinduced electron-transfer reaction between zinc cytochrome c and mutants of plastocyanin is studied in the temperature range 273.3-302.9 K at pH 7.0. Because the electron transfer is faster than the rearrangement of the diprotein complex, this rearrangement is the rate-limiting step, the one actually observed in kinetic experiments. The mutations of plastocyanin are designed to explore the observed configurational fluctuation. The mutations are within or near the acidic patch (Asp42Asn, Glu43Asn, Glu59Lys/Glu60Gln), within the hydrophobic patch (Leu12Asn, Leu12Glu), and in between the two patches (Gln88Glu). The enhancement of the rate with Gln88Glu implies that the zinc cytochrome c molecule moves in between the two patches of plastocyanin to improve the intrinsic electron-transfer reactivity. The activation parameters are obtained for the electron-transfer reaction, gated by the rearrangement. The change in enthalpy of activation is mainly responsible for the change in the reaction rate.

INOR

551. TOLERANCE OF BRINE SHRIMP (*ARTEMIA SALINA*) TO SEVERAL ORGANOTIN COMPOUNDS. Hai Nguyen, Nwaka Ogwuru, George Eng. Department of Chemistry, University of the District of Columbia, N.W. Washington, D.C. 20008.

The brine shrimp (*Artemia salina*) assay has been used to determine the toxicity of various compounds and plant crude extracts in aquatic environments. This assay was used to assess the toxicity of several organotin compounds, including organotin derivatives of sucrose phthalic esters. The phthalic ester derivatives should constitute a more soluble biocide, and they are expected to be taken up more readily into tissues and organs. Toxicity evaluations were done by determining the effective dose for 50% mortality (LC_{50} or ED_{50}) after 24 hour exposure time to the nauplii and adult stages of the brine shrimp. Analysis of the data was done by probit method of Miller and Tainter and the Reed-Muench method. Preliminary studies have shown that the mean 24 h LC_{50} for triphenyltin chloride using the first nauplii stage (48-72 hours after hatching) is 1.2 ppm. The mean 24 h LC_{50} using the second nauplii stage is 0.52ppm. The toxicity of these compounds is expected to correlate to the exposure time, solubility, uptake into tissues, and developmental stages of the brine shrimp.

552. ENTEROBACTIN-BASED HYBRID LIGANDS FOR Fe(III). R.J.A. Ramirez, E.L. Bolaños, J.Y. Nagasawa, T. Lau, Z.Sun, and C.G. Gutiérrez, Dept. of Chemistry & Biochemistry, California State University, Los Angeles, California 90032-8202.

The siderophore enterobactin (**1**), produced by *E. coli* and other gram negative bacteria to solubilize environmental Fe(III) and transport it into the bacterial cell, exhibits the strongest binding for ferric ion of all natural substances (log K_{ML}= 49). The tri-L-serine lactone backbone appears to contribute strongly to the ligand's efficacy by preorganizing the free ligand, and allowing for relatively strain-free binding of the ferric ion in the complex. We have found that methyl *N*-trityl-L-serinate is cyclooligomerized in refluxing xylene by the action of 2,2-dibutyl-1,3,2-dioxastannolane as template to mixtures of macrocyclic lactones through thermodynamically controlled processes, with the triolide as the major lactone product (85%) after 24 h. The tri-*N*-trityltriserine lactone platform has been converted into enterobactin (**1**). We also report here the first synthesis of new hybrid synthetic ligands **2-5** which incorporate the enterobactin triserine lactone nucleus and binding units other than *N*-2,3-dihydroxybenzoyl.

553. MOLECULAR DYNAMICS OF DURENE CYCLOPHANE HEMES. Andrew Strelzoff, and M.A. Lopez, Department of Chemistry & Biochemistry, California State University, Long Beach, CA 90840-3903

Estimating the contributions of polar and steric effects towards the binding of ligands to heme models systems[1] and hemeproteins[2] has been done using the AMBER software package. We have carried out molecular dynamics calculation of a series of durene cyclophane heme model systems[3] with the aim of understanding the motion of the durene strap. We find that the motion of the strap correlates with the vibrations of the porphyrin plane. These porphyrin plane vibrations oscillate between two saddle shape geometries and is dependent on the size of the strap.

(1) Lopez, M. A.; Kollman, P. A. *J. Am. Chem. Soc.* **1989**, *111*, 6212-6222. (2) Lopez, M. A.; Kollman, P. A. *Protein Science* **1993**, *2*, 1975-1986. (3) David, S.; James, B. R.; Dolphin, D.; Traylor, T. G.; Lopez, M. A. *J. Am. Chem. Soc.* **1994**, *116*, 6-14.

554. TRIMETHYLAMMONIUM PORPHYRINS. Jose Ramirez, and M.A. Lopez, Department of Chemistry & Biochemistry, California State University, Long Beach, CA 90840-3903

The discrimination between dioxygen and CO binding to hemeproteins and heme model systems is affected by the polarity of the binding site.[1,2] We are extending this to include a cation in the local binding site. Our work is toward the synthesis of a porphyrin model system having a trimethylammonium ion at the binding site. We have two approaches to this end: 1) Building a tetra-(2-trimethylammoniumphenyl)porphyrin, and 2) building a trimethylammoniumphenanthrene porphyrin. Our work so far has been in the convertion of 2-nitroaniline to orthonitrotrimethylanilium and are now working with the porphyrins. We will report on our progress to date.

(1) David, S.; James, B. R.; Dolphin, D.; Traylor, T. G.; Lopez, M. A. *J. Am. Chem. Soc.* **1994**, *116*, 6-14.
(2) Traylor, T. G.; Koga, N.; Deardurff, L. A. *J. Am. Chem. Soc.* **1985**, *107*, 6504-6510.

555. DISCRIMINATION IN BINDING BETWEEN CO AND DMSO TOWARD HEMES IN DMSO SOLVENT. John Escobar, Cynthia Ybarra, and M.A. Lopez, Department of Chemistry & Biochemistry, California State University, Long Beach, CA 90840-3903

We have used kinetic,[1] thermodynamic, and computational methods to study the binding of CO toward ferrous hemes in dmso/imidazole mixtures. We find by kinetic and semi-empirical methods that 1,2-dimethylimidazole discriminates against the binding of dmso and favors the binding of CO. We will present all of our methods used to study the binding between iron(II) protoporphyrins and dmso, CO, and imidazoles.

1. Lopez, M.A. Ybarra, C.D., Hyatt, S.D., *Inorganica Chim. Acta,* (1995), *231*, 121-131.

556. HAMMETT TREATMENT OF SUBSTITUENT EFFECTS ON CO BINDING TO FERROUS HEMES. Martha de la Rosa, and M.A. Lopez, Department of Chemistry & Biochemistry, California State University, Long Beach, CA 90840-3903

We have measured CO dissociation rates and CO binding constants to a series of tetra para substituted ferrous tetraphenyl porphyrins and have found that increasing electron donation to the iron leads to and increase in CO dissociation. This is consistent with the recent report of El-Kasmi[1] who reported on the effect of electron donation toward both CO and dioxygen dissociation rates. Preliminary findings show that thermodynamic binding of CO decreases with increasing electron donation toward the iron atom. We present our findings to date.

(1)El-Kasmi, D.; Tetreau, C.; Lavalette, D.; Momenteau, M. *J. Am. Chem. Soc.* **1995**, *117*, 6041-6047.

557. AROMATIC VERSUS AQUEOUS SOLVENT EFFECTS ON LIGAND BINDING TO IRON PORPHYRINS. Nancy Gardner and M.A. Lopez, Department of Chemistry & Biochemistry, California State University, Long Beach, CA 90840-3903

Hemeproteins bind ligands in aqueous solvent, however, the majority of ligand-binding studies to heme model systems have been conducted in toluene or benzene,[1] or in aqueous detergent[2] As such, the effect of the apoprotein on the ligand binding to hemeproteins in water is not known. We are in the process of measuring binding CO constants to both hydrophilic and hydrophobic ferrous hemes to determine the effect of solvent. We have used displacement equilibria to measure the CO binding constants and will report our results to date.

(1) David, S.; James, B. R.; Dolphin, D.; Traylor, T. G.; Lopez, M. A., *J. Am. Chem. Soc.* **1994**, *116*, 6-14.
(2) Traylor, T. G., Accounts Article, *Accts. Chem. Res.* **1981**, *14*, 102.

558. **Probing The Electronic Effects of Redox Chemistry in Ni Complexes with S-Donor Ligands by Ni and S K-edge X-ray Absorption Spectroscopy.**

Suranjan B. Choudhury, Gerard Davidson and Michael J. Maroney
Department of Chemistry, University of Massachusetts, Amherst, MA 01003-4510.

X-ray Absorption Spectroscopy (XAS) has played a central role in obtaining information about the structure of Ni sites in hydrogenases and other redox proteins containing Ni. In the case of hydrogenases, the Ni K-edge energy is not sensitive to the redox state of the enzyme, and EXAFS analysis indicates the presence of S-donor ligands with Ni-S distances of ~2.22 Å that are also insensitive to the redox chemistry.[1] To understand the role of Ni and S in the redox processes, we have examined a series of 15 Ni complexes with S-donor ligands including thiolates, thioethers, thiocarboxylates, and sulfinates by a combination of Ni and S K-edge XAS spectroscopy. These spectra provide a measure of changes in the electron density on S or Ni, and in the Ni-S distance upon changes in the redox state of the complex covering formal oxidation states of Ni from II to IV.

(1) Z. Gu, et al. *J. Am. Chem. Soc.* **1996**, *118*, 11155 - 11165.

559. **Copper Complexes With Adenosine Receptor Agonists**, Chester M. Mikulski, Holly DiGiosaffatte and Wei Li, Department of Chemistry & Physics, Beaver College, Glenside, PA 19038

In an attempt to generate new potent adenosine receptor agonists, we have recently conducted synthetic and characterization studies of 2'-deoxyadenosine (2'DADO) and 5'-deoxyadenosine (5'DADO) complexes with 3d metal perchlorates and halides. These studies have now been extended to include the isolation of novel complexes of N^6-cyclopentyladenosine (CPA), N^6-cyclohexyladenosine (CHA), N^6-phenyladenosine (P-ado) and 2-phenylaminoadenosine (Pha-ado) with copper (II) perchlorate. Products were synthesized by refluxing the hydrated metal perchlorate with the adenosine derivatives in mixtures of 1:1 triethyl orthoformate (teof)-ethanol. Spectral, magnetic and conductance studies indicate that CPA, CHA and P-ado ligands all form octahedral complexes of the $[CuL_2(OClO_3)_2(H_2O)_2]$ type. Evidence supports binding of L through the N(7) imidazole nitrogen. The Pha-ado complex with Cu^{+2} is a 1:1 adduct with bonding sites occupied by either aquo or ethanol ligands.

560. **SELECTIVITY OF VANADIUM BROMOPEROXIDASE WITH INDOLE DERIVATIVES.** Anne Baldwin, Matt Simpson, Michelle Trester, Gretchen E. M. Winter, Richard Tschirret-Guth, Alison Butler, Department of Chemistry, University of California, Santa Barbara, CA, 93106-9510

Vanadium Bromoperoxidase (V-BrPO), isolated from the marine brown alga *Ascophyllum nodosum*, catalyzes the peroxidative halogenation of organic substrates. The reaction proceeds through a bromonium intermediate, either free HOBr or an enzyme bound "Br+" species. Our previous work on competitive bromination kinetics at pH 6.5 showed that methylindole derivatives bind to V-BrPO, blocking the release of HOBr or other diffusable bromine species. Current work on the pH dependence of the reaction of methylindole derivatives with V-BrPO shows a decrease in binding at higher pH. We have also investigated the regioselectivity of the reaction of V-BrPO with indole derivatives.

561. PEROXIDATIVE HALOGENATION CATALYZED BY TRANSITION-METAL-ION-GRAFTED MESOPOROUS MCM-48 AND MCM-41 MATERIALS. Håkan Carlsson, Mark Morey, Jerrylaine V. Walker, Galen D. Stucky, Alison Butler, Department of Chemistry, University of California, Santa Barbara, CA 93106-9510

The catalytic activities of mesoporous materials, which have large surface areas (1000-1400 m^2/g) and variable pore diameters (20-100 Å) are attracting much attention. Such activities include hydroxylation of benzene, epoxidation of cyclic alkenes, and the oxidation of di-*tert* - butylphenol. Catalytic halogenation reactions are presently carried out with haloperoxidase enzymes. For instance, vanadium bromoperoxidase (V-BrPO) catalyzes the bromination of a variety of organic substrates under mild conditions using hydrogen peroxide as an oxidant of bromide. We have found that, like V-BrPO, titanium(IV)-grafted mesoporous silicate materials (e.g., Ti/MCM-48 and Ti/MCM-41) efficiently catalyze peroxidative halogenation reactions in aqueous solution at neutral pH or in organic solvents. In addition other transition metal incorporated mesoporous materials are active. Characterization, comparative rate data and mechanistic considerations will be presented.

562. **STRUCTURAL PARAMETERS OF METAL COMPLEXES OF NOVEL HEXADENTATE LIGANDS BASED ON THE CIS-1,3,5-TRIAMINOCYCLOHEXANE FRAMEWORK.** N. Ye,[#] R.P. Planalp,[#*] R.D. Rogers,[†] and M.W. Brechbiel,[§] Departments of Chemistry, University of New Hampshire,[#] Durham, NH 03824 and Northern Illinois University,[†] DeKalb, IL 60115, and the Chemistry Section, Radiation Oncology Branch, N.I.H.,[§] Bethesda, MD 20892.

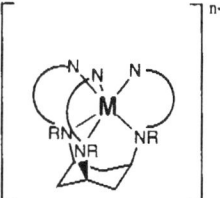

Complexes of the ligand system pictured, in which the pendant arms on nitrogen may include –CH_2-pyridyl and –CH_2–C(H)(R)–NH_2, have been prepared for M = Ni^{II}, Cu^{II}, Zn^{II}, Cd^{II}, Ga^{III}, and In^{III}. Structural distortions of the ligand geometry as a function of metal radius will be described through X-ray crystallographic data.

WITHDRAWN

563.
SMALL PEPTIDE INTERACTIONS WITH METALS.
Roger G. Harrison, Matthew J. Hinton, Brent C. Moore, Department of Chemistry and Biochemistry, Brigham Young Univeristy, Provo, Utah 84602

Protein amino acid residues such as histidine, tyrosine, glutamic acid, and cysteine bind metal ions and dictate metal reduction potentials and metal reactivity in metalloproteins. In metalloenzymes, active site pockets are formed around a metal center by the protein and selective reactivity is then carried out by the enzyme. Amino acids arranged in small peptide units also coordinate metals and promote certain metal electronic characteristics. We have designed and synthesized small peptides with histidine for metal binding and observed complex formation upon peptide addition to metal ions. Metal promotion of peptide helix formation was investigated and will be presented. Sites of metal-peptide interaction and strength of metal-peptide binding will also be presented.

$$M(n+) + peptide \longrightarrow M(n+)\text{-peptide}$$

564.
CO/IMINE INSERTION INTO PALLADIUM METHYL BONDS: A POTENTIAL ROUTE TO POLYPEPTIDES. Rania D. Dghaym Bruce A. Arndtsen; Department of Chemistry, McGill University, 801 Sherbrooke St. W., Montreal, Que. H3A 2K6

Palladium catalyzed co-polymerizations are being investigated as a potential synthetic route to peptides and their derivatives. The alternating insertion of imine and CO into the palladium carbon bond of L2Pd(Me)X (L2 = bidentate ligand) is expected to yield a polyamide backbone, in analogy to olefin/CO co-polymerization systems. Initial studies have shown that the addition of imine to L2Pd(Me)X results in imine coordination via through nitrogen. While these complexes are inert towards insertion, addition of CO and mild heating results in the clean formation of chelated palladium bound amides L2Pd[C(H)TolN(R)COMe]+X-. This represents the novel product of insertion of imine into the palladium-acyl bond. The unique thermodynamics, intermediates, and insertion regiochemistry of this reaction will be discussed, as well as the potential extension of this system into an imine/CO co-polymerization process.

565. SYNTHETIC PATHWAYS FOR THE PREPARATION OF HIGH SURFACE AREA MESOPOROUS HYDROTALCITE-LIKE COMPOUNDS.

François MALHERBE, Claude FORANO and Jean-Pierre BESSE

Laboratoire Physico-Chimie des Matériaux, CNRS URA 444,
Université Blaise Pascal, 63177 Aubière Cédex, France

In this work we have focused on the surface and porosity properties of [Mg-Al-CO$_3$] and [Mg-Al-Cl] obtained by the coprecipitation method. Different synthetic routes are used as well as post-synthesis treatments in order to measure how far can textural modifications be induced in these materials. For example, by performing the direct synthesis of [Mg-Al-CO$_3$] in mixtures composed of water and an organic solvent, or by regenerating calcined samples in similar mixtures, the surface areas, pore volumes and pore sizes are shown to be greatly affected by some solvents : synthesis in ethylene glycol causes an increase of 81 % in the surface area while in glycerol it results in the induction of microporosity (30 %) and reconstruction in glycerol give a high surface area mesoporous material. On the other hand we have also investigated the influence of these organic solvents on the textural properties during anion exchange reactions. Through analysis of their BET nitrogen adsorption/desorption isotherms, unpillared [Mg-Al-CO$_3$] proved to be mostly mesoporous while [Mg-Al] hydrotalcite pillared with hexacyanoferrate (III) exhibited high microporosity, with this caracteristic being greatly influenced when the exchange reactions are carried out in organic media. SEM images provided evidence of the microtextural modifications undergone by the products throughout these different treatments.

566.
KINETIC AND MECHANISTIC STUDIES OF THE REDUCTION OF URANIUM(VI) BIS(IMIDO) COMPLEXES. **R. Chris Schnabel** and Carol J. Burns. Chemical Science and Technology, Los Alamos National Laboratory, Los Alamos, NM 87545.

We have been exploring the synthesis, structure and reactivity of high-valent organo-f-element complexes supported by multiply-bonded functional groups such as organo(imido) ligands. Depending on the nature of the R group in the uranium(VI) bis(imido) complex $Cp^*_2U(NR)_2$, differing rates of hydrogenation to produce the uranium(IV) bis(amido) complex $Cp^*_2U(NHR)_2$ are observed. Silanes also reduce the uranium(VI) bis(imido) complexes. The addition of phenylsilane to $Cp^*_2U(NR)_2$ yields the uranium(IV) amido complex $Cp^*_2U(NHR)(NRSiH_2Ph)$. We have undertaken kinetic studies varying the R group on the organo(imido) ligand to elucidate the nature of the uranium imido bond.

567. NEW OCTANUCLEAR Ln CLUSTERS CONTAINING HYDROXO AND POLYSELENO LIGANDS. Christopher G. Pernin, James A. Ibers, Department of Chemistry, Northwestern University, Evanston, Illinois 60208-3113

The lanthanoid clusters $Ln_8(DMF)_{13}(\mu_4\text{-}O)(\mu_3\text{-}OH)_{12}(Se_3)\text{-}(Se_4)_2(Se_5)_2$ (DMF=N,N-dimethylformamide; Ln=Y,Eu,Gd,Yb) have been prepared by the reaction of the appropriate $LnCl_3$ salt dissolved in tetrahydrofuran with a solution of K_2Se_n in DMF. The structure as determined by X-ray crystallography consists of a bicapped octahedron of Ln(III) atoms with each face capped by a hydroxo group. A $\mu_4\text{-}O$ atom, three types of polyselenide chains, and thirteen DMF molecules complete the coordination spheres of the lanthanides. The $Ln_8(DMF)_{13}(\mu_4\text{-}O)(\mu_3\text{-}OH)_{12}(Se_3)(Se_4)_2(Se_5)_2$ cluster (shown to the right without the DMF's) is the only product isolated under a variety of reaction conditions.

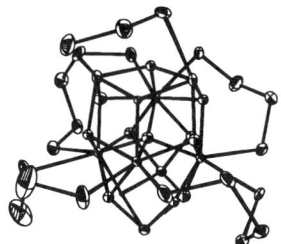

568. STRUCTURE AND STABILITY OF URANIUM(VI) COMPOUNDS FORMED IN CARBONATE CONTAINING SOLUTION. S.M. Kitten, M.P. Neu, and W. Runde*, Chemical Science and Technology Division, Los Alamos National Laboratory, Los Alamos, NM 87545.
Structural and thermodynamic properties of actinide(VI) compounds in aquatic systems are of particular importance to understand and predict their geochemical behavior. We investigated the system $U(VI)/H_2O/NaCl/Na_2CO_3$ by a multi-method approach to determine the structure of U(VI) solution species and solid phases. Monomeric and polymeric carbonato complexes, $UO_2(CO_3)_n^{2-2n}$ and $(UO_2)_3(CO_3)_6^{6-}$, and chloro complexes, $UO_2Cl_n^{2-n}$, have been found to be the major soluble complexes in carbonate containing solution, and their stability has been investigated as a function of carbonate, chloride, and other electrolyte concentrations. We have applied spectroscopic techniques, including Raman, UV-Vis, laser-induced fluorescence, and EXAFS, and powder and single crystal X-ray diffraction to investigate the structure of U(VI) chloride and carbonate complexes and to determine significant species' formation constants. Single crystals of $(K\text{-}18\text{-}Crown\text{-}6)UO_2Cl_4$ and $(18\text{-}Crown\text{-}6)UO_2Cl_2(H_2O)_2$ have been obtained from aqueous solution and characterized. Three major types of U(VI) carbonates structures have been found: monomeric $UO_2(CO_3)_n^{2-2n}$ solution species with $n = 1 - 3$, planar D_{3h} trimetallic anion $(UO_2)_3(CO_3)_6^{6-}$ formed at $U(VI) : CO_3^{2-}$ concentration ratios of 1 : 3 in solution and solid state, and polymeric $UO_2(CO_3)_n$ layers found in the solid state of rutherfordine, UO_2CO_3. $Na_4AnO_2(CO_3)_3 \cdot xH_2O(s)$ has been found to form at high carbonate concentrations and its structure will be presented.

569.
OPTICAL STUDIES OF THE STOICHIOMETRY AND THERMODYNAMICS OF PLUTONIUM (IV) NITRATE COMPLEXES IN HIGH IONIC STRENGTH ACIDIC SOLUTIONS. J. M. Berg, R. B. Vaughn, M. R. Cisneros, D. K. Veirs and C. A. Smith., Nuclear Materials Technology Division, Los Alamos National Laboratory, Los Alamos, New Mexico, 87545

We will show that a rigorous analysis of optical absorption spectra of actinide f-f transitions can quantify unusually complicated, multi-component complexation equilibria, using our analysis of the aqueous plutonium(IV) nitrate system as an example. The aqueous solution chemistry of plutonium in the +4 oxidation state includes the simultaneous formation of many different complexes of the type $Pu(NO_3)_n(H_2O)_x$. In a series of acidic Pu(IV) solutions ranging from 2 to 22 molal ionic strength and 0 to 22 molal nitrate concentration, we have identified and quantified at least five Pu(IV) species and have determined upper limits for the maximum abundances of two other possible complexes. We will discuss the strong ionic strength dependence formation constants for the Pu(IV) nitrate complexes and the effects of several different electrolyte cations.

570. SYNTHESIS AND REACTIVITY OF NEW ORGANOMETALLIC COMPOUNDS OF THE LANTHANIDES
A. Mandel and J. Magull, Department for Inorganic Chemistry, University of Karlsruhe, D-76128 Karlsruhe

New benzyl complexes of the Rare Earths can be prepared by using either chelating donor ligands or substituted cyclopentadienyl ligands. From the reaction of YBr_3 with ^{n}BuLi and TMEDA in toluene, $[(tmed)Y(Bz)_2(\mu-Br)_2Li(tmed)]$ **1** can be isolated. With Cp*, $[(Cp^*)_2Y(Bz)(thf)]$ **2**, $[(Cp^*)Ln(Bz)_2(thf)]$ (Ln = Gd **3**, Sm **4**) and $\{[(Cp^*)_2Sm(Bz)_2][K(thf)_2]\}_\infty$ **5** can be synthesized. The compounds $[(Cp^*)Ln(Bz)_2(thf)]$ show a γ-agostic interaction between one Benzyl ligand and the lanthanide centre and a crystallographic disordered Cp* ring. In the complex $[(C_5{}^tBuMe_4)Gd(Bz)_2(thf)]$ **6**, the substituted Cp*-Ligand is not disordered.

Reactions of the obtained benzyl complexes with small molecules were carried out to examine insertion in the Ln–C bond. The reaction of the complex $[(Cp^*)_2Yb(Bz)(thf)]$ with CO leads to $[K(thf)_6][(Cp^*)_6Yb_6(O)_6Br]$ **8** and $[(Cp^*)K(thf)]_\infty$. With N_2O, insertion of one molecule N_2O in the Ln–C bond is observed. The dimeric product $[(Cp^*)_2Sm(\mu-C_6H_5CH_2NNO)]_2$ **7** contains the benzyldiazotate ligand $(C_6H_5CH_2NNO)^-$.

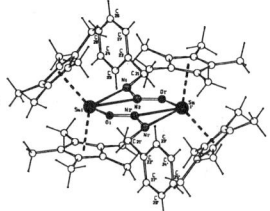

Figure 1: Structure of **7**

571. THE SYNTHESIS AND CHARACTERIZATION OF LANTHANIDE DIOLATES. Sandra M. Hick and Eric J. Voss.* Department of Chemistry, Southern Illinois University at Edwardsville, Edwardsville, IL 62026-1652.

Many metal/sugar complexes have been studied in aqueous solution, but a large portion of this field remains unexplored. Our research group is interested in the study of carbohydrate complexes of the lanthanide metals. Since diols can be thought of as simple models of sugars, this research has focused on making lanthanide diolate complexes in nonaqueous solution. Synthetic routes employed include reactions of diols with tris[bis-(trimethylsilyl)amido]lanthanide complexes and diol/alkoxide exchange reactions. These precursors are easily made and are readily soluble in organic solvents. Preliminary studies include interactions of the sterically bulky diols pinacol and 1-phenyl-1,2-ethanediol with tris[bis(trimethylsilyl)amido]ytterbium(III). The synthesis, isolation, and characterization of these complexes by multinuclear NMR will be presented. This research is supported by grants from the Southern Illinois University at Edwardsville Summer Research Fellowship, Funded University Research, and Research Grants for Graduate Students funds.

572. **LANTHANIDE TRIOLATES: SIMPLE MODELS OF METAL/SUGAR COMPLEXES** John M. Hardimon and Eric J. Voss.* Department of Chemistry, Southern Illinois University at Edwardsville, Edwardsville, IL 62026-1652.

Carbohydrates are very important molecules to study because of their role in energy storage in living species, for cell recognition, and as a renewable source of raw materials for organic syntheses. Our recent research has focused on the synthesis, isolation, and characterization of lanthanide triolates, simple models of metal/sugar complexes. Glycerol and 1,2,4-butanetriol were reacted with tris[bis(trimethylsilyl)amido]dysprosium(III) in attempts to make compounds containing triolate ligands. The bis(trimethylsilyl)amido synthetic precursor was chosen because it has been previously characterized and is readily soluble in several organic solvents. Preliminary results, including the preparation and characterization of precursors and the triol substitution reactions will be reported. This research is supported by grants from the Southern Illinois University at Edwardsville Summer Research Fellowship, Funded University Research, and Research Grants for Graduate Students funds.

573. **LANTHANIDE ALDITOLATES: SIMPLE METAL/SUGAR COMPOUNDS DERIVED FROM LANTHANIDE ALKOXIDES.** Joseph R. Hardimon and Eric J. Voss.* Department of Chemistry, Southern Illinois University at Edwardsville, Edwardsville, IL 62026-1652.

Our research group is interested in the synthesis and characterization of organometallic carbohydrate/metal complexes. The alditols, straight chain sugars with only hydroxyl functionality, can be viewed as the simplest of carbohydrates. The strategy employed to make these new compounds involves the preparation of a series of tris[bis(trimethylsilyl)amido]lanthanide precursors, reactions with sterically bulky alcohols to make metal alkoxides, and ligand exchange reactions between the alkoxide groups and alditols. The preparation, purification and characterization of the lanthanide alkoxide complexes will be presented, as well as preliminary results from alditol exchange reactions. This research is supported by grants from the Southern Illinois University at Edwardsville Summer Research Fellowship, Funded University Research, and Research Grants for Graduate Students funds.

574. PLUTONIUM SOLUBILITY AND SPECIATION UNDER HYDROTHERMAL WASTE TREATMENT CONDITIONS. S.D. Reilly, M.P. Neu*, W.H. Runde, Chemical Science and Technology Division, Los Alamos National Laboratory, Los Alamos, NM 87545.

One of the most complex problems concerning environmental management and restoration of plutonium production sites is the treatment and disposition of mixed and transuranic wastes. Hydrothermal oxidation, which completely oxidizes a wide variety of organic and other hazardous materials, is rapidly developing as a promising technology for the treatment and volume reduction of actinide-containing waste. Information on the speciation and solubility of plutonium under process conditions will facilitate the development of separation techniques for remaining radionuclides. Such a strongly oxidizing environment will generate plutonium(VI); and upon organic destruction, hydrothermal reactor solutions will likely be concentrated in carbonate. We are investigating the solubility and speciation of the plutonium(VI) carbonate system under expected process effluent conditions. Based upon solution thermodynamic studies (potentiometric and spectrophotometric), we determined that the monocarbonato species, $PuO_2CO_{3(aq)}$, has the largest relevant stability range. We prepared the corresponding solid phase, $PuO_2CO_{3(s)}$, and characterized the pale tan solid using powder XRD, EXAFS, and diffuse reflectance. We are measuring the solubility of PuO_2CO_3 in NaCl solution as a function of pH and ionic strength. Thus far we have determined the solubility product of PuO_2CO_3 to be, log K_{sp} = -13.0 (0.1 m NaCl) and log K_{sp} = -12.4 (2.1 m NaCl). Results from solution thermodynamic and structural studies will be presented.

575. MECHANISTIC STUDIES ON THE STEPWISE HYDROGENATION OF QUINOLINES COMPLEXED TO TRIOSMIUM CLUSTERS. Doug Kolwaite, Edward Rosenberg, K.I. Hardcastle, Joana Ciurash, Rauf Farraz, Ricardo Duque. Chemistry Departments of The University of Montana, Missoula, MT 59812 and California State University, Northridge, CA 91330.

The stepwise reduction of the carbocyclic ring in the electron deficient quinoline complexes, $Os_3(CO)_9(\mu_3-\eta^2-C_9H_5RN)(\mu-H)(R=H, 6-CH_3)$ have been studied using deuterium labeling experiments. Initial reduction takes across the C(5)-C(6) bond with H^-/H^+. Experiments with D^-/H^+ for R=H, 6-CH_3 reveal the formation of equal amounts of the *cis*- and *trans*- isomers. Further reduction with a second mole H^-/H^+ results in formation of $Os_3(CO)_9(\mu_3-\eta^2-C_9H_8RN)(\mu-H)_2$ where the carbocyclic ring is fully reduced. Labeling experiments show that H^- attack is at C(7) of the quinoline ring. Reduction of the related electron deficient complex $Os_3(CO)_9(\mu_3-\eta^2C_9H_8N)(\mu-H)$, where the heterocyclic ring is partially reduced, results in initial attack by H^- at C(2) and not at C(5). Further reduction of related quinoline complexes leads to C-N bond cleavage. The results of labeling experiments using D_2 will be presented.

576. ACTIVATION OF QUINOLINES TOWARDS NUCLEOPHILIC ADDITION/SUBSTITUTION BY TRIOSMIUM CLUSTERS. Brian Bergman, Edward Rosenberg, K.I. Hardcastle, Joana Ciurash, Mandana Visi. Chemistry Departments of The University of Montana, Missoula, MT 59812 and California State University, Northridge, CA 91330.

The reaction of the electron deficient clusters $Os_3(CO)_9(\mu_3-\eta^2C_9H_4NRR')(\mu-H)(R=R'=H; R=H, R'=4-Me; R=6-CH_3, R'=H; R=H, R'=4-Me; R=6-Me, R'=H, R'=4-OMe, R=H; R=6-OMe, R'=H)$ with the carbon based nucleophiles $Nu^-(Nu^-=(CH_3)_2CC\equiv N^-, CH_2C(O)-O^tBu, CH(SCH_2CH_2S-), Ph^-, Bz^-, n-Bu)$ leads to nucleophilic addition across the C(5)-C(6) bond after quenching with acid. In the case of the 6-substituted derivatives, stereospecific addition is realized with only the *cis*-addition product being obtained. When the 5-position is substituted with a halogen, nucleophilic substitution or nucleophilic addition across the C(3)-C(4) bond is observed depending on the halogen and the steric bulk of the nucleophile. Applications of this new synthetic methodology to ring forming reactions will be presented.

577. SYNTHESIS, STRUCTURE AND REACTIVITY OF TRIMETALLIC CLUSTERS OF 5,6-BENZOQUINOLINE. Ryan Smith, Edward Rosenberg, K.I. Hardcastle, Joana Ciurash. Departments of Chemistry, The University of Montana, Missoula, MT 59812 and California State University, Northridge, CA 91330.

The reaction of 5,6-Benzoquinoline with $Os_3(CO)_{10}(CH_3CN)_2$ yields the 46e$^-$ cluster $Os_3(CO)_9(\mu_3-\eta^2-C_{13}H_8N)(\mu-H)$(1) in good yield after decarbonylation. Reaction of 1 with $H^-/H+$ gives $Os_3(CO)_9(\mu_3-\eta^2C_{13}H_9N)(\mu-H)_2$(2). Reaction of 1 with D^-/H^+ reveals that initial attack of the nucleophile is at C(9). Reaction of 1 with bulky nucleophiles leads to nucleophilic addition across the C(4)-C(3) bond of the heterocyclic ring. Decarbonylation of the phosphine derivative $Os_3(CO)_9(\mu-\eta^2C_{13}H_8N)(\mu-H)PPh_3$ leads to the formation of an electron precise σ-π-vinyl complex $Os_3(CO)_8(\mu_3-\eta^2C_{13}H_8N)(\mu-H)PPh_3$(3) instead of the usual 46e$^-$ species. The reactions of 1 and 3 with a range of nucleophiles will be presented.

578. KINETICS AND MECHANISM OF HYDROGEN TRANSFER IN METAL CLUSTER COMPLEXES. Kirk Hash, Richard J. Field, Edward Rosenberg. Department of Chemistry, The University of Montana, Missoula, MT 59802.

Kinetic studies on the reaction of $Os_3(CO)_{10}(\mu-X)_2$ (X=H or D) and CF_3CN which yields $Os_3(CO)_{10}$ ($\mu-\eta-F_3C(X)=N)(\mu-X)$(1 or 1-d) and $Os_3(CO)_{10}(\mu-\eta^2-CF_3C=N(X)(\mu-X)$(2 or 2-d) have been completed. The reaction is first order in cluster and CF_3CN, with the rate determining step being formation of the adduct $Os_3(CO)_{10}(CF_3CN)(\mu-X)(X)$. However, the ratio of 1:2 shows a significant temperature dependence with

2 being favored at lower temperatures. A significant kinetic deuterium isotope effect is noted for **2-d** but not for **1-d**. This evidence, taken together with an analysis of the activation parameters obtained from the variable temperature kinetics (-20 to +55°C) suggests that there is a significant tunneling component in the formation of **2** but not **1**.

579. REACTIONS OF $[PW_{11}O_{39}NbO]^{4-}$ WITH ELECTROPHILIC SULFONIC ACID DERIVATIVES. Robert H. Beer[1] and <u>Emil Radkov</u>[2], Departments of Chemistry, [1]Fordham University, Bronx, NY 10458 and [2]Columbia University, New York, NY 10027

The reactions of the mixed-metal Keggin ion $PW_{11}O_{39}NbO^{4-}$ with Group 14 sulfonic acid derivatives Me_3EOTf (E = Si, Sn, Ge, Pb; Tf = O_2SCF_3) form $PW_{11}O_{39}Nb(OEMe_3)^{3-}$. The Sn species was found to be the most hydrolytically stable of the $PW_{11}O_{39}Nb(OEMe_3)^{3-}$ series. Treatment of $PW_{11}O_{39}Nb(OEMe_3)^{3-}$ with tetra-n-butylammonium halide removes the Me_3E^+ group quantitatively. Other sulfonic acid derivatives, Tf_2O, Ms_2O (Ms = $(MeO)_2SO$), Bu_2BOTf and ROTf (R = Me, Et) deoxygenated $PW_{11}O_{39}NbO^{4-}$ to produce the oxo-bridged species $[PW_{11}O_{39}NbO]_2O^{6-}$. Alcoholysis of this compound yielded $PW_{11}O_{39}Nb(OR)^{3-}$ complexes (R = Me, Et, iPr, tBu, Bn). Crystallographic and solution ^1H, multinuclear, and ^{17}O labelling nmr studies indicate that derivatization of $PW_{11}O_{39}NbO^{4-}$ occurs exclusively at the terminal NbO oxygen atom.

$$[[PW_{11}O_{39}NbO]_2O]^{6-} \xleftarrow{Tf_2O,\ Ms_2O,\ ROTf,\ or\ Bu_2BOTf} [PW_{11}O_{39}NbO]^{4-} \xrightarrow[E=Si,Sn,Ge,Pb]{Me_3EOTf} [PW_{11}O_{39}NbOEMe_3]^{3-}$$

580.
SYNTHESIS AND CHARACTERIZATION OF ORGANOMETTALIC CHALCOGENIDE CLUSTER COMPOUNDS. **C. Paul Lai,** Ivonne L. Rivera, and Songping D. Huang*, Department of Chemistry, University of Puerto Rico, San Juan, PR 00931

Despite many successful applications of the hydro(solvo)thermal synthesis in solid state and materials chemistry, the concept of carrying out chemical reactions in the sealed ampoule or autoclave at pressures and temperatures higher than ambient values has remained foreign to most organometallic chemists. We have initiated a study of exploring the organometallic chalcogenide cluster chemistry under the hydro(solvo)thermal conditions. In this presentation, representative examples of tansition metal organometallic sulfide, selenide and telluride clusters will be given. The scope, applicability and generality of the hydro(solvo)thermal synthesis to other organometallic cluster compounds will be discussed.

581.
SYNTHESIS, X-RAY STRUCTURE, SPECTROSCOPIC AND LUMINESCENT PROPERTIES OF THE PENTANUCLEAR EUROPIUM COMPLEX EU5(OH)5(DBM)10 (DBM=DIBENZOYLMETHIDE). **R. Tom Xiong,** Songping D. Huang*, Department of Chemistry, University of Puerto Rico, San Juan, PR 00931

As part of our studies of probing the $^5D_o \rightarrow {}^7F_o$ emmission band of Eu(III) under the multiple ligand field environments, the title compex was synsesized and structurally characterized. The compound crystallizes in the space group P21/c with a=22.223(5), b=17.427(1), c=35.708(5) Å, V=13,614(4) Å^3, Z=4. In this report, the spectroscopic characterization including ESCA, Eu-151 Mössbauer as well as the low-temperature luminescent studies will be presented.

582. SYNTHESIS AND CHARACTERIZATION OF MACROCYCLIC HEXANUCLEAR METAL CLUSTER. Myoung Soo, Lah; Byunghoon, Kwak; Hakjune, Rhee; Soonheum, Park Department of Chemistry, College of Science, Hanyang University 1271 Sa-1-dong, Ansan, Kyunggi-do, 425-791 Korea

The macrocyclic hexanuclear metal cluster, [Mn(III)6(FSHZ)6(MeOH)6], was prepared using a potential trianionic pentadentate ligand N-formyl salicylhydrazide(FSHZ). The disc shaped macrocyclic metal cluster with a pseudo C3i symmetry has the dimension of 1.8 nm in diameter and of 0.8 nm in thickness. The complex has a large cavity in the center of the cluster. The pentadentate ligand is not only bridging the metal ions using a hydrazide N-N group but also enforcing the stereochemistry of the metal ion as a propeller configuration due to the meridianal coordination of the ligand to the metal ion. Details of the structural and spectroscopic properties of the cluster will be discussed.

583. MAGNETIC EXCHANGE INTERACTIONS BETWEEN SINGLE UNPAIRED-ELECTRON CENTERS IN OCTODECA-, PENTADECA- AND NONANUCLEAR COMPLEXES. Ziming Sun, Sheila M. Aubin, David N. Hendrickson, Dept. of Chemistry, University of California at San Diego, La Jolla, CA 92093-0358; Gail B. Karet, John C. Bollinger, George Christou, Dept. of Chemistry, Indiana University, Bloomington, IN 47405-4001; Xiao-Ming Chen, Yu-Luan Wu, Yan-Sheng Yang, Dept. of Chemistry, Zhongshan University, Guangzhou 510275, PRC.

The interest in the magnetism of polynuclear metal complexes spans from industrial applications as magnetic materials to understanding magnetic and electronic structures. The magnetic properties of a series of octodecanuclear heterometallic complexes, $[M_6Cu_{12}(OH)_{24}(H_2O)_{18}(pyb)_{12}(ClO_4)]$-$(ClO_4)_{17} \cdot nH_2O$ (M = Y, Ce, Nd, Gd, Dy; pyb = pyridine betaine), and two polyvanadium complexes, $(Et_4N)_5[V_{15}O_{36}]$ and $(Et_4N)_3[V_9O_{19}(OAc)_5]$, have been studied. Both types of complexes possess d^1 or d^9 single unpaired-electron metal centers, where the magnetic exchange interactions are supported by μ-hydroxo and μ-oxo ions, respectively. With well-shielded $4f$ electrons on lanthanide ions, only antiferromagnetic interactions are observed between the copper ions in the M_6Cu_{12} complexes. The two polynuclear V complexes have valence-trapped V^{IV} and V^V ions as well as electronically delocalized μ-oxo-bridged $(V^{IV.5})_2$ pairs, as reflected in their magnetochemical properties.

584. THE ELECTROCHEMICAL AND SPECTROELECTROCHEMICAL INVESTIGATION OF Nb(V) IN ACIDIC MEDIA AND THE FORMATION OF TRINUCLEAR CLUSTERS Nb3(3.67+). Caroline M. Hoadley, and Vladimir Katovic, Department of Chemistry, Wright State University, Dayton, OH, 43435.

The electrochemical properties of Nb(V) in acidic media were investigated using cyclic voltammetry, spectroelectrochemistry and constant potential coulometry. In concentrated H_2SO_4, Nb(V) is reduced by an ECC mechanism to give Nb3(3.67+) trimer. The reaction pathway involves the reduction of Nb(V) to Nb(III) species which reacts with the starting materials to form the Nb3(3.67+) trimer, $Nb_3O_2(SO_4)_6(H_2O)_3^{5-}$. The electrochemical reduction of Nb(V) in CH_3SO_3H and HCl indicates that the reduction pathway is similar to the one observed in H_2SO_4 and leads to the formation of Nb3(3.67+) metal clusters. The electrochemical data of Nb(V) in HCl, CH_3SO_3H, CF_3SO_3H, oxalic and tartaric acid will be presented.

585. SYNTHESIS AND CHARACTERIZATION OF TECHNETIUM AND RHENIUM POLYOXOTUNGSTATES BY ^{31}P AND ^{183}W NMR SPECTROSCOPY. A. Venturelli, and L. C. Francesconi, Department of Chemistry, Hunter College of the City University of New York, New York, NY, 10021

In our investigation of potential matrices for the immobilization of Technetium-99, an isotope in radioactive waste, we are studying the incorporation of ^{99}Tc in polyoxotungstate frameworks. For comparison, we are also studying the chemistry of the third row congener, rhenium. We have prepared the technetium complex of $[\alpha\text{-}2\text{-}P_2W_{17}O_{61}]^{10-}$ (defect Wells-Dawson structure) by reaction of NaTcO(eg)$_2$ (where eg = ethylene glycol) with $K_{10}[\alpha\text{-}2\text{-}P_2W_{17}O_{61}]$ at pH 2.8. The multinuclear NMR data and IR spectroscopic data are consistent with the $[TcO(\alpha\text{-}2\text{-}P_2W_{17}O_{61})]^{7-}$ formulation. The analogous $[ReO(\alpha\text{-}2\text{-}P_2W_{17}O_{61})]^{7-}$ complex was synthesized by the reaction of K_2ReCl_6 with $K_{10}[\alpha\text{-}2\text{-}P_2W_{17}O_{61}]$ in an aqueous medium. The technetium and rhenium complexes of the Keggin derivative $[\alpha\text{-}PW_{11}O_{39}]^{7-}$ have also been synthesized and characterized by ^{31}P and ^{183}W NMR spectroscopy.

586. HYDROTHERMAL SYNTHESIS AND CHARACTERIZATION OF A NOVEL LAYERED ZIRCONIUM PHOSPHATE $Zr_2O_3(HPO_4) \cdot nH_2O$. Lyudmila Bortun, Anatoly Bortun and Abraham Clearfield, Texas A&M University, College Station, TX 77843, USA

Several novel crystalline layered zirconium phosphates with composition $Zr_2O_3(MPO_4) \cdot nH_2O$ (M=Na, K) were prepared by the partial hydrolytic decomposition of amorphous or crystalline α-zirconium phosphates in alkaline media under mild hydrothermal conditions. Two hydrogen forms of the novel zirconium phosphate (ψ-ZrP) - $Zr_2O_3(HPO_4) \cdot 1.5H_2O$ and $Zr_2O_3(HPO_4) \cdot 0.5H_2O$ were prepared by acid treatment of the salt forms. All the synthesized materials were characterized by elemental analysis, TGA, X-ray, IR and ^{31}P MAS NMR spectroscopy. It was found that ψ-ZrP has a layered structure and that it exhibits extremely high hydrolytic stability in alkaline media; the phosphorus release in 1M NaOH being lower than 0.11-0.13 mmol PO$_4$/g in comparison with the complete hydrolytic decomposition of α-ZrP or γ-ZrP. The ion exchange behavior of the novel ψ-ZrP towards alkali, alkaline-earth, and some di- and trivalent metal cations in different model solutions was studied over a pH range 2-14.

587. SYNTHESIS AND CHARACTERIZATION OF TWO NOVEL FIBROUS TITANIUM PHOSPHATES OF FORMULA $Ti_2O(PO_4)_2 \cdot 2H_2O$. S. Khainakov, M.A. Villa-Garcia, J.R. Garcia, J. Rodriguez, Universidad de Oviedo, Oviedo 33071, Spain, A. Bortun, D. Poojary, L. Bortun, A. Clearfield, Texas A&M University, College Station, TX 77843, USA

Two novel fibrous titanium phosphates with framework structure of formula $Ti_2O(PO_4)_2 \cdot 2H_2O$ have been synthesized under mild hydrothermal conditions. New compounds were characterized by elemental analysis, TGA, X-ray, IR and ^{31}P MAS NMR spectroscopy. The crystal structure of one of them solved by powder diffraction method. Their porous structure was studied by N_2 adsorption. It was found that fibrous titanium phosphates are thermally stable materials possessing ion exchange properties. Their adsorption behavior towards alkali, alkaline-earth, and some di- and tri-valent metal cations was studied over a pH range 2-14 and an explanation of the experimental facts is proposed.

588. TEMPERATURE-DEPENDENT RAMAN STUDY ON LiTaO$_3$ CRYSTALS. Zhaohui Wang, Department of Physics, Harbin Normal University, Harbin, 150080, P.R. China

In this paper, the result of temperature-dependent Raman study on the vibrational modes in LiTaO$_3$ single crystals are presented. In the ferroelectric phase, all the 13-Raman active modes were observed. It is found that the two low-frequency A1 modes show significant changes during the temperature increasing process. "Modes softening" and "frequency-repulsion" effects were observed and discussed.

589. **AN X-RAY ABSORPTION SPECTROSCOPIC STUDY OF THE TEMPERATURE AND PRESSURE DEPENDENCE OF THE ELECTRONIC SPIN-STATES IN SEVERAL IRON(II) AND COBALT(II) TRISPYRAZOLYLBORATE COMPLEXES.** C. Hannay, M. J. Hubin-Franskin, and F. Grandjean, University of Liège, Belgium, V. Briois, J. P. Itié, and A. Polian, LURE, France, Gary J. Long, University of Missouri-Rolla, and S. Trofimenko, Du Pont Polymers.

An Fe and Co K-edge x-ray absorption study at 295 K and above and 77 K of $Fe[HB(pz)_3]_2$, $Fe[HB(3,5-(CH_3)_2pz)_3]_2$, $Fe[HB(3,4,5-(CH_3)_3pz)_3]_2$, and their analogous cobalt(II) complexes, where pz is 1-pyrazolyl, reveals that the first complex is low-spin below ca. 400 K and high-spin above this temperature, the second is, as expected, low-spin at 77 K and high-spin at 295 K, and the third complex, and all of the cobalt complexes, remain high-spin at all temperatures. A high-pressure room-temperature study of the first complex shows that it remains low-spin up to 85 kbar, whereas the second complex shows a spin-crossover between zero and 30 kbar from the high-spin to the low-spin state. The energies of the metal 4p virtual orbitals are found to be very sensitive to pressure and to the metal electronic spin-state. An EXAFS analysis of the cobalt complexes indicates both that they are all structurally very similar, with the expected high-spin cobalt to nitrogen bond distances, and that they remain high-spin upon cooling from 295 to 77 K. The XANES spectra of the cobalt complexes, obtained between ambient pressure and 95 kbar, indicate that they also undergo a pressure induced reversible spin-state transition at high pressure.

590. **AN X-RAY ABSORPTION SPECTRAL STUDY OF THE $Ce_2Fe_{17-x}M_x$ SOLID SOLUTIONS, WHERE M IS Al OR Si.** D. Vandormael and F. Grandjean, University of Liège, Belgium, V. Briois, LURE, France, D. P. Middleton, Philips Research Laboratories, the Netherlands, K. H. J. Buschow, University of Amsterdam, the Netherlands, and Gary J. Long, University of Missouri-Rolla.

The XANES spectra at the Ce L_{III}-edge of the $Ce_2Fe_{17-x}Al_x$ solid solutions show a two peak structure characteristic of the $4f^1$ and $4f^0$ configurations of Ce, a structure which indicates that Ce is in a mixed valent state in these compounds. All the XANES spectra have been consistently and excellently fit with one sigmoidal function and two Gaussian broadened Lorentzian functions. The Ce spectroscopic valence obtained from the relative areas of the two peaks decreases from 3.64 to 3.43 between x = 0 and 9, and correlates linearly with the Ce site volume. This correlation confirms[1] that the Ce valence is strongly dependent upon steric effects. In contrast, the Ce valence obtained from the XANES spectrum of $Ce_2Fe_{14}Si_3$ is not determined by steric effects and indicates, in agreement with other measurements[2,3] and calculations,[4] that Si covalently bonds with its near neighbors. EXAFS measurements at the Fe K-edge of the $Ce_2Fe_{17-x}Al_x$ solid solutions indicate, in agreement with the neutron diffraction results,[5] that the average distance between Fe and its first and second neighbor shells increases with x.
[1]T. W. Çapehart et al., Appl. Phys. Lett. **65**, 3151 (1994); [2]D. P. Middleton et al. J. Appl. Phys. **78**, 5568 (1995); [3]D. Vandormael et al., J. Appl. Phys. accepted for publication; [4]M. Z. Huang and W. Y. Ching, J. Appl. Phys. **79**, 5545 (1996); [5]S. R. Mishra et al., J. Appl. Phys. **76**, 5383 (1994).

591.
STEREOISOMERICALLY CONTROLLED INORGANIC ARCHITECTURES: SYNTHESIS OF EXTENDED ENANTIO- AND DIASTEREOMERICALLY PURE RUTHENIUM MOLECULAR WIRES AND DISKS FROM ENANTIOPURE BUILDING BLOCKS. Kenneth Wärnmark, Organic Chemistry 1, Department of Chemistry, Lund University, P.O. Box 124, S-221 00 Lund, Sweden and Olga Heyke, Jim A. Thomas, Paul N.W. Baxter and Jean-Marie Lehn*, Laboratoire de Chimie Supramoléculaire, Institut Le Bel, Université Louis Pasteur, 4 rue Blaise Pascal, F-67000 Strasbourg, France

Diastereoisomerically pure compounds are important in the construction of polymetallic molecular structures because diastereoisomers have different spatial orientation which may lead to differences in their physical properties such as rates of energy- or electron transfer. We are constructing enantio- and diastereomerically pure rigid π-conjugated rods and disks using the appealing enantiomerically pure building-block approach. These molecules constitute potential molecular wires and antenna systems. The presence of a π-deficient aromatic central ligand should facilitate the excited state electron transfer from the metals to the extended ligand. Work is also directed towards metal doped polypyridine chains, where the physical properties of the chain, such as conductivity, will be altered upon metal ion coordination.

592.

Formation of Silica-, Tungstate- and Molybdate- Surfactant Mesophases Studied by in situ Energy Dispersive X-Ray Powder Diffraction. Dermot O'Hare, Stephen O'Brien, Robin J. Francis, Stephen Price and Andrew M. Fogg, Inorganic Chemistry Laboratory, South Parks Road, Oxford OX1 3QR, UK

The formation of silica-surfactant mesophases that are precursors to the mesoporous materials MCM-41 and FSM-16 were followed in real time using a synchrotron X-ray facility developed by our research group at the Daresbury SRS, UK. The facility has been designed to record powder diffraction data of the phases as they form in situ, from solutions, gels or via intercalation processes. The technique is a direct method of following the formation processes and can therefore provide a means of elucidating the mechanisms involved. A wide range of temperatures and hydrothermal conditions are possible. The preparation of Keggin cluster tungsten (VI) oxide and molybdenum (VI) oxide mesophases using quaternary ammonium salts, altered pH and ammonium metatungstate or molybdate respectively produced lamellar and hexagonal phases that could be observed using this technique. A review of the study will be presented.

593.

IN-SITU ENERGY DISPERSIVE DIFFRACTION STUDIES ON THE HYDROTHERMAL SYNTHESIS OF MICROPOROUS MATERIALS. Dermot O'Hare, Robin J. Francis, Stephen J. Price, Stephen O'Brien, Andrew M. Fogg, Inorganic Chemistry Laboratory, South Parks Road, Oxford OX1 3QR, UK

An experimental apparatus has been developed which enables hydrothermal reaction to be studied in-situ using time resolved energy dispersive X-ray diffraction at temperatures up to 250°C and autogenous pressures up to 20 bar. We have used this apparatus to study the synthesis of a number of classes of microporous materials. Results of studies of the syntheses of microporous gallophosphates, metal sulfides, and aluminophosphates will be presented. These studies have demonstrated the existence of several previously unidentified crystalline intermediates, and data on the kinetics of formation of these phases, in addition to the final phases, has been obtained. The results of these studies and a discussion of the possible mechanisms of these reactions will be presented.

594. PREPARATION OF SURFACE MODIFIED INORGANIC LAYERED COMPOUNDS.
H. Tagaya, S. Ogata, H. Morioka, M. Karasu, J. Kadokawa and K. Chiba, *Department of Materials Science and Engineering, Yamagata University, 4-3-16 Jonan, Yonezawa, Yamagata, 992 Japan.*

Surface modified inorganic layer compounds were prepared by the reaction of water treated zinc / aluminium layered double hydroxide (Zn/Al LDH) and amorphous metal hydroxides with organic compounds such as organic oxychlorides. Their layer structures were similar to those of the LDHs except the products of amorphous metal hydroxides with aromatic oxychlorides such as benzoyl chloride. Interlayer spacings of the reaction products were 7.7 ~ 26.7 Å depending on the size and number of function groups of organic compounds and the amounts of incorporated organic compounds. It was considered that, in the reaction of metal hydroxide with organic compounds, obtained nano composites were crystallized by self assembly.

595. **Photoisomerization of Sulfonated Spiropyran in Silica Matrix.**
T.Nagaoka, H.Tagaya, T.Hori, J.Kadokawa, M.Karasu and K.Chiba,
Department of Materials Science and Engineering, Yamagata University, 4-3-16 Jonan, Yonezawa, Yamagata 992, Japan.

Photoisomerization reactions of spiropyran incorporated within silica matrix by sol-gel route were studied. No photochromic properties were observed when spiropyran alone was incorporated within silica matrix. The presence of organic anions such as 4-ethylbenzene sulfonic acid (EBS), dodecylbenzene sulfonic acid (DBS) were essential for stable photoisomerization. The photochemical properties such as reversibility and thermal stability of the matrix which included spiropyran were correlated with the structure and quantity of organic anions. In the case of DBS/SP-SO_3^- (mole ratio=50), the photochemical properties of the thin film were better than those of other films.

596. RADICAL ANION SCHIFF BASE SALTS: SOLUTION AND SOLID STATE CHARACTERIZATION.

Andrew S. Ichimura, Qingshan Xie, Lawrence P. Szajek, James E. Jackson, John L. McCracken and James L. Dye
Department of Chemistry, Michigan State University, East Lansing MI 48824.

Radical anion salts in the condensed phase present possibilities for materials with interesting magnetic and electric properties. The reaction of alkali metals with hexa-imine macrobicyclic cryptands with aromatic linking groups can be controlled stochiometrically to produce mono-, di-, and trianion salts. For comparison, a single strand of the macrobicycle, methyl-(3-methyliminomethyl-benzylidene)-amine (I), was reduced in solution to produce the mono- and dianion. Solution EPR and ENDOR studies of the monoanion in different solvents with and without cryptand[2.2.2] reveal complex behavior which can be attributed to dimer formation. Frozen solutions at cryogenic temperatures display recognizable triplet signals with a temperature dependence characteristic of a singlet ground state. Magnetic susceptibility measurements on adducts of I with potassium are in progress. Optical spectra of I in various solvents and related Schiff base cryptands will also be presented. Current efforts are focused on crystal growth and characterization by susceptibility, X-ray and EPR.

597. STRUCTURE AND PROPERTIES OF A NEW CESIDE: LITHIUM CRYPTAND [2.1.1] CESIDE

Rui H. Huang, Michael J. Wagner, Andrew S. Ichimura, Qingshan Xie, John L. McCracken and James L. Dye
Department of Chemistry and Fundamental Material Research, Michigan State University, East Lansing, MI 48824.

Cesides are of interest because estimates of the polarizability and molar volume suggest that they may be on the metallic side of the insulator-metal transition.[1] The structure of the new ceside features the huge cesium anion forming zigzag chains along one crystal axis with a Cs^--Cs^- distance of only 6.0Å, which is 1Å shorter than the diameter of the anion. The crystals grow as long needles along the Cs^--Cs^- chain direction and have a low melting point. The properties of the new compound, such as conductivity, EPR, NMR and magnetic susceptibility will be presented.
1. Herzfeld, K. F. Phys. Rev. 1927, 29, 701-705.

598. CHARACTERIZATION OF THE VAPOCHROMIC BEHAVIOR OF COMPLEX DOUBLE SALTS [Pt(ARYLISOCYANIDE)$_4$][Pt(CN)$_4$], (WHERE ARYLISOCYANIDE = p-CN-C$_6$H$_4$-C$_n$H$_{2n+1}$; n = 6-14) USING X-RAY POWDER DIFFRACTION, IR/NIR AND UV-VIS SPECTROSCOPY. Carolyn E. Anderson, C. L. Exstrom, J. R. Sowa Jr., C. A. Daws, D. E. Janzen, M. K. Pomije, and Kent R. Mann.* *Department of Chemistry, University of Minnesota, Minneapolis, MN 55455.*

We are studying platinum complex double salts for use as environmental sensors of Volatile Organic Compounds (VOCs). The salts exhibit rapid reversible spectral shifts in the UV-VIS and NIR when exposed to VOC vapors. X-ray powder diffraction studies of these materials show a reversible change in the lattice constants of the tetragonal unit cell. The combination of X-ray diffraction and spectroscopy provides valuable insights into the mechanism of vapochromic behavior.

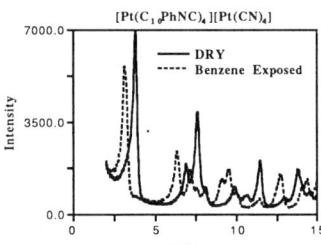

599. PROBING THE CHEMICAL TOPOGRAPHY OF CLAY–CONDUCTING POLYMER ASSEMBLIES: REAL–TIME STUDIES OF THE POLYMERIZATION OF ORGANICS ON HECTORITE SURFACES USING SCANNING FORCE MICROSCOPY AND PHASE CONTRAST IMAGING. Michael E. Hagerman, Jennifer L. Attuso, Ben P. Reynolds, and Michael P. Eastman, Department of Chemistry; Timothy L. Porter, Department of Physics, Northern Arizona University, Flagstaff, Arizona 86011

Investigations of the chemical topography of polymer-clay composites are important in the development of nanochemical materials including chemical sensors and inorganic-organic conductive polymer assemblies. Cu(II)-exchanged hectorite thin films exhibit remarkable properties as catalytic hosts for the spontaneous polymerization of a variety of organic guests including benzene, aniline, thiophene, and pyrrole. The polymerization of these guest species occurs both on the surface and in the intergallery regions of the clay. We have shown that Cu(II) cations play a pivotal role in promoting the oxidation of organic guest monomers and the subsequent polymerization reactions. Real–time scanning force microscopy (SFM) studies employing novel phase contrast imaging of these polymerization reactions provide crucial information on the coalescence and growth of guest moieties on clay surfaces. These in–situ SFM studies offer unprecedented views of the polymerization process and may prove essential in the development of nanochemical sensing devices based on inorganic–organic composite materials.

600. LOW TEMPERATURE, IN-SITU CHEMICAL VAPOR DEPOSITION OF TANTALUM AND ITS NITRIDES FROM TANTALUM(V) BROMIDE, Xiaomeng Chen, Gregory Peterson, Harry Frisch, and Alain E. Kaloyeros, New York State Center for Advanced Thin Film Technology and Department of Physics, The University at Albany-SUNY, Albany, New York, 12222, Barry Arkles, Gelest Inc., Tullytown, Pennsylvania, 19007

Emerging microelectronics technologies demand new diffusion barriers between the different metallization levels to prevent undesirable inter-layer interaction and diffusion. In the case of copper based interconnect schemes, Ta and its nitrides are considered among the most suitable adhesion promoters / diffusion barriers. Our work, accordingly, has focused on the development of low-temperature (<450^{0}C), in-situ, sequential CVD process for the growth of ultrathin (<250Å) films of Ta and its nitrides for such applications. The process employs simple inorganic Ta sources, such as tantalum(V) bromide, to grow Ta or TaN by smoothly and reversibly changing the type of reactant used. This paper presents progress to-date in the identification and optimization of plasma-assisted CVD (PACVD) and thermal CVD processes for the growth, respectively, of Ta and TaN. In particular, microchemical and microstructural analyses showed pure β-phase Ta and TaN with good conformality in sub-quarter-micron device structures.

601. MAGNETIC AND MECHANICAL ALIGNMENT PROPERTIES OF SOME TELLURIDES Jian H. Zhang, Department of Chemistry, Xavier University of LA, New Orleans, LA 70125; Feng Chen, Department of Chemistry, Rider University, Lawrenceville, NJ 08648.

The magnetic and mechanical alignment properties have been determined on four ternary tellurides to study the relationship between the alignment property and the crystal structure. Both $NbFeTe_2$ and $NbNiTe_2$ have a simple orthorhombic structure with their magnetic alignment direction [010] perpendicular to the mechanical alignment direction [001]. While $NbCoTe_2$ and $TaCoTe_2$ exhibit monoclinic structure, and can be aligned along [100] under either magnetic or mechanical force. A quantitative evaluation on their magnetic and mechanical properties using polefigure analysis will be presented. In addition, the magnetic properties on the samples both parallel and perpendicular to the alignment direction will be discussed in correlation with the X-ray diffraction data.

We acknowledge the support from the Research Corporation.

602. GOLD ELECTRODE MODIFICATION WITH SURFACE-CONFINED MONOLAYERS OF FERROCENE-BASED 'ACCEPTOR-σ-DONOR' SYSTEMS. W.E. Cleland, Jr., Sukanta Bhattacharyya, Z. He, and C.L. Hussey, Department of Chemistry, University of Mississippi, University, Mississippi 38677.

The modification of Au(111) electrodes with aminoalkyl disulfides produces self-assembled monolayers with a reactive amino tail-group. These chemisorbed monolayers are then utilized to produce surface-bound electron acceptor-sigma bridge-electron donor systems through either a single-step or a two-step surface reaction. 1,4 - Benzo- and napthoquinones are used as electron acceptors, and ferrocene is used as electron donor. The preparation, electrochemical behavior and infrared spectroscopy of these surface-confined monolayers will be presented.

603. NOVEL METAL-RICH PHOSPHIDES: SIMILARITIES BETWEEN CO, NI AND P. H. Kleinke, H. F. Franzen, Iowa State University and Ames Laboratory, Ames, IA 50011

A new class of metal-rich phosphides is accessible by arc-melting of cold-pressed mixtures of MP, M and M' and subsequent annealing at temperatures above 1400°C (M = early-transition-metal; M' = late-transition-metal). A common feature is that the M atoms form a 3D metal sublattice with phosphorus and M' atoms in trigonal prismatic voids. Besides strong M–M, M–M' and M–P bonding interactions, short M'–P bonds can occur between neighboring M_6P and M_6M' prisms, prohibiting mixed M'/P occupancies. A partial substitution of P by Co was observed on a P position without Co-P contacts in the structure of Hf_5CoP_3, whereas an analogous substitution did not take place in case of the isostructural Hf_5NiP_3. On the other hand, exchanging of each second Ni atom of HfNi by P led to the synthesis of $HfNi_{0.5}P_{0.5} \equiv Hf_2NiP$. Thus, the similar behavior of M' and P in these compounds provides a tool to tailor solid state materials.

604. **Chemisorption of Organosulfur Molecules onto Au(111): New Insights into the Chemisorption Process.**

Frank J. Feher, Mark H. Dishner and John C. Hemminger
Department of Chemistry, University of California, Irvine, California 92697

The popular mechanism for SAM formation by thiols on Au(111) invokes deprotonation of the thiol (RSH) to form a thiolate on the surface (RS-Au). This process creates gold vacancy islands, or "pits", in the surface. We have examined the details of this process by performing STM studies with a variety of organosulfur species, including thiophene and methanethiol. Adsorption of thiophene on Au(111) produces a monolayer which exhibits uniformly distributed vacancy islands (i.e., "pits"). The monolayer quickly anneals to produce a SAM comprised of large domains adopting the unit cell $(2\sqrt{19} \times \sqrt{3})R30°$. Adsorption of MeSH produces monolayers with vacancy islands which are first created within the hcp regions of the $(22 \times \sqrt{3})$ gold reconstruction. Ostwald ripening and further annealing produces well-ordered domains that have the same surface structures observed for other RSH-based SAMs on Au(111). Our results demonstrate for the first time that the structure of a fully-formed thiol monolayer on Au(111) is influenced by the original reconstruction of the underlying gold surface, and they provide compelling support for the suggestion that the gold vacancy islands observed in Au(111) supported monolayers are not the result of chemical etching.

605. OXIDATION-REDUCTION OF FE,CE:LITHIUM NIOBATE. R. Uhrin, Deltronic Crystal Industries, Inc., 60 Harding Avenue, Dover, NJ 07801

Practical applications of lithium niobate and other crystals for volume holographic data storage applications will require sensitivity not only in the blue spectral region to maximize storage density but also at longer wavelengths in the visible and near-IR for other applications as well. Laser diodes already developed at longer wavelengths makes $Rh:BaTiO_3$ and other crystals promising candidates for compact and economical devices. Co-doping is known to shift to longer wavelengths the response of singly-doped crystals and $Fe,Ce:LiNbO_3$ is a likely choice. The optimum optical response for a given photorefractive sample is a function of the +4/+3 and +3/+2 valence state ratios for the active ions. Results of REDOX experiments performed to optimize the peak absorption for $Fe,Ce:LiNbO_3$ are presented.

606. **STUDY AND CHARACTERIZATION OF LOW TEMPERATURE TUNGSTEN OXYNITRIDE PHASES STABLIZED BY DOPING WITH Fe, Co, AND Ni**

Julia Truszkowski, Feng Chen, Department of Chemistry
Rider University, Lawrenceville, NJ 08648

We recently isolated new low temperature tungsten oxynitride phases which were believed to be stabilized by doping with Fe, Co, and Ni. Since most tungsten containing binary and ternary transition metal nitrides/oxynitrides can only be formed and stabilized at temperature ranging 600 °C- 900 °C, we will report our study on the relationships between the phase stability and the incorporation of the transition metal content. In addition the study of the relationship between the nitrogen content and reaction conditions will be also reported.

607.
MOLECULAR COMPLEX FORMATION: 2-CHLORO-4-NITROANILINE AND POLY(ETHYLENE OXIDE). **C.C. Evans,** M.D. Ward, Department of Chemical Engineering and Materials Science, University of Minnesota, Minneapolis, Minnesota 55455

One approach to the design of materials for nonlinear optics involves polymer films which contain polar chromophores. Noncentrosymmetry, necessary for second harmonic generation, is achieved by electric field poling of the chromophores above the Tg of the matrix. However, once the film is cooled below the Tg and the field is removed, the net alignment decays over hours. This reorientation of the chromophores may be inhibited if they interact strongly with the polymer. Here we report the formation of a complex between poly(ethylene oxide) (PEO) and 2-chloro-4-nitroaniline, a polar chromophore. Infrared spectra of films and differential scanning calorimetric data for bulk samples show that a 3:1 complex forms between ethylene oxide repeat units and 2-chloro-4-nitroaniline. Preliminary spectroscopic data suggests that bonds along the polymer backbone have a much greater degree of trans configuration in the complex than in pure PEO.

608. SYNTHESIS AND IMMOBILIZATION OF FUNCTIONALIZED LONG CHAIN MOLECULES ON MICA SURFACE, S.S.Siltchenko and D.H.Busch, Chemistry Department, University of Kansas, Lawrence, KS, 66045.

The chemical immobilization of long chain molecules on the mica surface can lead to materials with important applications in separation and sensor science. Using a silanization / hydrosilation pathway, two kinds of functionalized silanes [N-monoethyl-21-docoseneamine-dimethylethoxysilane (AMES) and N-monoethyl-21-docosenoamine-triethoxysilane (AES)] were deposited on molecularly smooth muscovite mica surfaces. In both cases, the functional groups were secondary amines. An atomic force microscope (AFM) study showed that the AES molecules were bound and oriented perpendicular to the surface. Both monolayer islands (clusters) and individual molecules were observed on the surface.

$(C_2H_5O)_3Si(CH_2)_{22}NHC_2H_5$ - (AES); $(C_2H_5O)(CH_3)_2Si(CH_2)_{22}NHC_2H_5$ - (AMES)

609. MOLECULAR STAPLING--PREPARATION OF A [3]-ROTAXANE BY OXIDATIVE DIMERIZATION OF 2-(9-ANTHRACENYLMETHYLAMINO)-ETHANETHIOL THREADED THROUGH A CROWN ETHER. A.G. Kolchinski, N.W. Alcock, R.A. Roesner and D.H. Busch

The spontaneous threading of monomeric substituted ammonium salts into large crown-ether (A.G. Kolchinski, D.H. Busch, N.W. Alcock. *J.Chem.Soc. Chem. Comm.* **1995**, 1289; P.R. Ashton, P.T. Glink, J.F. Stoddart, P.A. Tasker, A.J.P. White and D.J. Williams, *Chem.Eur.J* **1996**, *2*, 729, and references therein) has been exploited to staple together pairs of macrocycles, producing [3]-rotaxanes. In search of effective amine/crown pairs, some twenty crown-ethers with large rings (21-members and greater) were tested with several primary and secondary ammonium salts under a variety of experimental conditions. This led to the identification of a crown-amine couple that is highly effective for [3]-rotaxane preparation, 2-(9-anthracenylmethylamino)-ethanethiol and dibenzo-24-crown-8. The iodine oxidation of this threaded precursor provided the [3]-rotaxane in high yield, which has been fully characterized. The two macrocycles have saddle conformations and are meshed together, their boundary resembling the seam of a tennis ball.

610. MOLECULAR MODELING (CHARMm) STUDIES OF LIGAND BINDING WITHIN THE CAVITY OF UNBRIDGED CYCLIDENES. E.V. Rybak-Akimova, K. Kuczera, D.H. Busch, University of Kansas, Department of Chemistry, Lawrence, KS 66045

Lacunar cyclidene ligands are known to protect the 6th coordination site at the metal ion from solvent or axial ligand binding, thus providing a vacant coordination site for small molecule (O_2, CO) binding. Even in the absence of the bridge that produces the lacuna, 16-membered cyclidene macrocycles preferentially exist in a saddle-shaped conformation. The possible protective role of the cavity in unbridged cyclidenes was evaluated using the CHARMm molecular modeling package. Atomic charges were obtained from ZINDO calculations (Mulliken population analysis), force field parameters were altered to fit *ab initio* HF/6-31G* frequences of model molecules. The results demonstrate an increase in the strain energy of the macrocycle upon binding the imidazole derivatives within the cavity, which is partially compensated by a favorable electrostatic term.

611.
PREPARATION OF WATER-SOLUBLE TUNGSTATE PRECURSORS. **Larry E. Reinhardt** and Allen W. Apblett., Department of Chemistry, Tulane University, New Orleans, LA, 70118

Metal tungstates serve in a variety of important materials and catalytic applications. For example, ZrW_2O_8 is a ceramic that undergoes a negative thermal expansion when heated and is of considerable interest to the manufacturing sector. However, the preparation of metal tungstates can be difficult and/or energy intensive. The most direct route, precipitation from a water-soluble metal salt with an aqueous solution of the tungstate anion, is hampered by the lack of overlap of the pH ranges in which the metal cations and the tungstate anion are stable. As well, preparation of tungstates via solid state methods requires very high temperatures and long reaction times. It is further complicated by the fact that volatilization of WO_3 is pronounced at such high temperatures. We have developed a synthetic process that utilizes aqueous chemistry for the preparation of water-soluble precursors for the metal tungstates that consist of tungstate esters of metal gluconate salts.

612.
SYNTHESIS AND PROPERTIES OF YTTRIUM METALORGANIC POLYMERS. **Galina K. Todorova** and Allen W. Apblett, Department of Chemistry, Tulane University, New Orleans, LA, 70118

Metaloxane polymers are a class of materials that have significant industrial importance in electronic, optical, mechanical, biomedical and preceramic applications. We have developed a novel synthetic route to yttrium-containing metalorganic polymers using yttrium oxychloride as a precursor which contains a preformed yttrium-oxide polymeric backbone. YOCl is readily converted to yttrium oxyacetate by reaction with neat tributyltin acetate. Reaction of the acetate polymer with various other acids such as gluconic acid, 4-hydroxybenzenesulfonic acid, di-(2-ethylhexyl) phosphoric acid, and 2-[2-(2-methoxy)ethoxy]ethoxyacetic acid in appropriate solvents results in replacement of the acetate with a new pendant ligand. In this manner, the physical and chemical properties of the polymers may be readily varied so that materials that are either moldable or hard, or are soluble in water or organic solvents may be prepared. In

613.
SOLUTION GROWTH OF CDS THIN FILMS. **Allen W. Apblett** Marc L. Breen, Daniel K. Schwartz and John T. Woodward, Department of Chemistry, Tulane University, New Orleans, LA, 70118

The solution growth technique is an attractive, inexpensive, water-based method for the preparation of CdS thin films for application in photovoltaic cells. The method involves the slow generation of sulfide ions via the controlled hydrolysis of thiourea or thioacetamide in the presence of a soluble cadmium salts and a chelating ligand. Typically two types of films are formed, a primary compact mirror-like film and a secondary layer of larger particles which is loosely attached to the the first layer. The mechanism of growth is believed to either involve attachment of colloidal particles to the surface or an ion-by-ion deposition process. We have used atomic force microscopy to prove conclusively that the latter mechanism is the correct one. Furthermore, we have used novel, strongly binding chelating agents to prepare dense CdS thin films with superior optical transmittance.

614.
APPLICATION OF LIQUID METAL CARBOXYLATES TO THE SYNTHESIS OF YBCO. **Allen W. Apblett** and Edwin H. Walker, Department of Chemistry, Tulane University, New Orleans, LA, 70118

Salts of 2-[2-(2-methoxy)ethoxy]ethoxyacetic acid have the remarkable property of being liquids at room temperature. Furthermore, they are excellent solvents for other metal salts so they may be readily applied to the preparation of multicomponent oxides such as the YBCO superconductor. Such liquid precursors may be used for the synthesis of thin films via metallo-organic deposition and can also potentially be applied to the preparation of fibers or wires. However, the low ceramic yield causes significant problems with the latter application. An alternative approach is to use a mixture of superconducting powders and the liquid precursor. In this way, a plastic material is obtained that has a high ceramic yield and may readily be formed into wires and other shapes.

615.
ELECTROCHEMICAL SYNTHESIS OF (BA,K)BIO3 WITH THE KSBO3 STRUCTURE. **D.Y. Jung, P.D. Han, S.R. Wilson*, D.A. Payne** Department of Materials Science and Engineering, Seitz Materials Research Laboratory, Beckman Institute, and School of Chemical Sciences*, University of Illinois at Urbana-Champaign, IL 61801

Electrochemical crystal growth at low temperatures is an approach for the synthesis of materials which would be thermodynamically and kinetically unstable at higher temperatures. Single crystals of $(Ba_{0.05}K_{0.95})BiO_3$ with an unusual oxidation state of Bi(V) were obtained electrochemically from molten KOH at 255°C. The working potential for the synthesis was 0.7V when a Bi-reference electrode was used. $(Ba_{0.05}K_{0.95})BiO_3$ was determined in space group Im-3(No. 204) with Z=12. X-ray diffraction data at room temperature were refined to a = 10.0204(2)Å, V = 1006.1(1)Å3 and R = 2.4% for 225 unique reflections. The Bi(V) content was determined by thermogravimetric analysis.

616.
MODULATION OF LIGHT-INDUCED BIVECTORIAL ELECTRON FLOW IN A HETEROSUPRAMOLECULAR TRIAD Geoffrey D. Will S.Nagaraja Rao, Donald Fitzmaurice, University College Dublin, Dept. of Chemisrty, Belfield, Dublin 4, Ireland

We will describe a heterosupramolecular system in which the direction of electron flow can be modulated. This system consists of a TiO2 nanocrystallite to which has been chemisorbed a ruthenium complex covalently linked to a viologen. Photoexcitation of the ruthenium yields an electron which can reduce either the viologen or the TiO2 nanocrystallite. By modulation of the bulk properties of the semiconductor nanocrystallites either potentiostatically and photostatically the direction of electron flow and the acceptor that is reduced may be determined. Specifically under open circuit conditions or absense of band gap irradiation the excited state electron on the ruthenium reduces available conduction band states in the TiO2 nanocrystallite. On the other hand, at sufficiently negative applied potentials or under bandgap irradiation electron injection into the TiO2 nanocrystallite by the excited state ruthenium is energetically unfavourable and the viologen is reduced. Implications of the above for the modulation of light-induced electron flow in molecular devices is discussed.

617.
VISIBLE LIGHT-INDUCED TRANSMEMBRANE ELECTRON TRANSFER Robert W. Hoyle Donald Fitzmaurice, University College Dublin, Dept. of Chemistry, Belfield, Dublin 4, Ireland

We will describe the preparation and characterization of mechanically robust transparent nanoporous-nanocrystalline unsupported TiO_2 (anatase) membranes. The modification of a membrane will also be described. Briefly, a ruthenium based visible-light sensitser is adsorbed in the pores of the membrane. An ordered monolayer of an appropriately modified viologen is then deposited on one face of the sensitised membrane using Langmuir-Blodgett techniques. It will be shown that visible light excitaton of the Ru-TiO2-Viologen assembly results in charge-injection from the electronically excited sensitiser into the conduction band of the semiconductor membrane and subsequent reduction of the electron acceptor. Possible significance of these findings for development of a practical water splitting device are considered.

618. LOW-TEMPERATURE SOLUTION-PHASE SYNTHESIS OF NANOCRYSTALLINE INDIUM SULFIDE PHASES. Jennifer A. Hollingsworth and William E. Buhro, Department of Chemistry, Washington University, St. Louis, MO 63130

We have conducted solution-phase reactions between t-Bu_3In and H_2S at 164°C to 203°C and have obtained in high yield an indium sulfide phase that is, to the best of our knowledge, previously unknown. This phase (~25 nm) was evident in the XRD powder pattern only if the protic "catalyst", thiophenol, was added to the reaction mixture prior to reflux; amorphous material was otherwise obtained. In addition to the unknown phase, nanocrystalline orthorhombic InS (30-60 nm) has been produced when 10 mole% In° was added to the reaction mixture -- regardless of the presence of a catalyst. The unknown phase transforms above 400°C to orthorhombic InS (major component) and monoclinic In_6S_7. TEM images show rods, tapered fibers and platelets which exhibit identical, hexagonal spot patterns (EDS: 50:50 +/- 4), dense regions with complex spot patterns indicative of multiple crystallites, and some amorphous equiaxed particles. SAD suggests the presence of superstructure.

619. NEW ORANGE AND RED ORGANIC LIGHT-EMITTING DEVICES (OLEDS) USING ALUMINUM TRIS(5-HYDROXYQUINOXALINE)   Yujian You[a]; Andrei Shoustikov[a]; Mark E. Thompson[a,*]; Stephen R. Forrest[b] and Paul E. Burrows[b].   [a] Department of Chemistry, University of Southern California, Los Angeles, CA 90089. [b] Center for Photonic and Optoelectronic Materials, Princeton University, NJ 08544

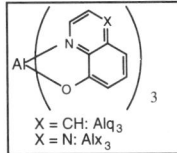
X = CH: Alq_3
X = N: Alx_3

Vacuum-deposited, single-heterojunction OLEDs have been made. The electron transporting layer consists of aluminum tris(8-hydroxyquinoline) (Alq_3) or aluminum tris(5-hydroxyquinoxaline) (Alx_3) (host) doped with a red fluorescent dye or C_{60} (guest). Current-voltage characteristics, UV-visible, photoluminescence and electroluminescence spectra have been measured and will be discussed. The devices emit orange or red light upon applying low voltage (< 15 V). Correlations observed in the energy transfer efficiency between the host and guest, and the degree of overlap between spectra of the emission of the host and the absorption of the guest, suggest Förster energy transfer is dominant. To the best of our knowledge, this is the first report of Alx_3 employed as an efficient energy transfer host in OLEDs.

620. SYNTHESIS AND PHOTOPHYSICAL PROPERTIES OF NOVEL MATERIALS FOR PHOTOCHEMICAL ENERGY CONVERSION. Ousama Abbas, Bryan E. Koene, and Mark E. Thompson,* Department of Chemistry, University of Southern California, Los Angeles, California 90089-0744

Stable photo-induced charge separation has been observed in layered zirconium phosphonate thin films. Films were prepared with multilayers of donor substituted and acceptor substituted zirconium phosphonates on a varietyof substrates (e.g. Au, Si, ITO coated glass, etc.). When irradiated these multilayer films produce a significant photocurrent. In order to improve the degree of charge separation in these thin film materials we have prepared a novel 4-amino stibazole complex and examined it by UV-vis and fluorescence spectroscopy. The solvatochromic properties of the complex were studied, and show that this complex has a high degree of intramolecular charge transfer. The pyridyl amine was prepared as an asymmetric bisphosphoante derivative and used for the growth of metal bisphosphonate and related thin films. We will discuss the growth of these multilayer thin films, as well as thier use in photochemical energy conversion.

621. MULTILAYERED ZIRCONIUM PHOSPHONATES AS CATALYSTS FOR SMALL MOLECULE REDUCTION/OXIDATION. Shannon C. Rice, Mark E. Thompson, Department of Chemistry, University of Southern California, Los Angeles, CA 90089

Our work deals with zirconium-(viologen phosphonate-phosphate) multilayer structures containing intrapore platinum and palladium particles. These porous materials can be made by hydrothermal synthesis as microcrystalline solids or grown as films on silica or derivatized polymer microspheres. Previously these materials have been shown to catalyze the production of hydrogen peroxide from H_2 and O_2. In this poster we will discuss the synthesis and characterization of the supported materials and our current efforts focusing on selective reduction and oxidation of small molecules using such catalysts. Specifically, hydrogenation of aromatic groups in the presence of esters and other potentially reducible groups will be considered.

622. SYNTHESES AND CHARACTERIZATIONS OF THREE POLYMORPHIC SODIUM COBALT PHOSPHATES. **X. H. Bu,** P. Y. Feng, S. H. Tolbert and G. D. Stucky, Department of Chemistry, University of California, Santa Barbara, California 93106

The syntheses, X-ray crystal structures, magnetic properties and phase transitions of three polymorphic sodium cobalt phosphates are presented. Unlike three commonly occuring polymorphs of $SiO2$ or $AlPO4$ which are all tetrahedral frameworks, the divalent cobalt cation in three cobalt phosphate polymorphs adopts tetrahedral, trigonal bipyramidal and octahedral configurations, respectively. The thermal analysis shows that the red trigonal bipyramidal phase transforms into nearly colorless octahedral phase, which transforms into the blue tetrahedral phase at even higher temperature. The framework topology of these phases is related to that of zeolites. The magnetic properties of three polymorphs are consistent with the structural features such as the presence of Co-O-Co clusters and chains.

623. **NMR STRATEGIES FOR THE STRUCTURAL INVESTIGATION OF LITHIUM-INTERCALATED GLASSY CARBONS.** **Sophia E. Hayes,** A.K. Cheetham, H. Eckert, Materials Research Laboratory, University of California, Santa Barbara, CA 93106, and W.R. Even, Sandia National Laboratories, Livermore, CA 94551.

Recently, disordered and turbostratic carbon materials have become a focus of interest for rechargeable lithium batteries, because they appear to allow for significantly higher intercalation levels. Previous studies have shown that emulsion-derived copolymers of poly-methacrylonitrile/divinylbenzene are stable, highly crosslinked systems and can serve as suitable polymeric precursors for the disordered carbon materials used as electrodes. We are investigating the effect of the microstructure of these carbons on the intercalation process, the role of matrix atoms other than carbon, and the local environments and spatial organization of the lithium ions. Solid state nuclear magnetic resonance (NMR) is an ideal structural tool for addressing these questions, being an element-selective, inherently quantitative method, sensitive to local environments. To this end, we have used several magnetic resonance techniques to study structural features in the lithium intercalates of these microporous glassy carbons, as well as the non-intercalated species, including cross polarization and magic angle spinning NMR of ^{13}C, ^{15}N, ^{1}H, and ^{7}Li nuclei.

624. STRUCTURAL CHARACTERIZATION AND MAGNETIC PROPERTIES OF $Ba_2MnZn_2As_2O_2$. **Michael D. Price,** Dianna M. Young, Stephanie L. Brock, Enos A. Axtell, III, Susan M. Kauzlarich, Department of Chemistry, University of California, Davis, Ca, 95616

A new compound, $Ba_2MnZn_2As_2O_2$ has been synthesized. Neutron diffraction data indicates a tetragonal cell (*I4/mmm*.)with lattice parameters of $a = 4.24279(5)$, $b = 4.24279(5)$, $c = 19.5094$ (3) Å , $wR_p = 9.32\%$, $R_p = 6.71\%$. This compound crystallizes in the layered $Sr_2Mn_2As_2O_2$ structure type . The structure is distinguished by two layers, one consisting of a transition metal in a square planar coordination to oxygen and the other consists of an anti PbO - like layer in which a transition metal is tetrahedrally coordinated to arsenic. In the transition metal arsenic layer the arsenic displays square pyramidal coordination to zinc. The two distinguishing layers are separated by Ba^{2+} ions that serve to balance the charge of the MnO_2^{2-} layers and the $Zn_2As_2^{2-}$ layers. The neutron diffraction refinement is consistent with Mn in the square planar site. The structural model will be present along with temperature dependent magnetic data.

625. **PYROLYTIC DECOMPOSITION STUDIES OF ORGANOPHOSPHORUS SOURCE COMPOUNDS FOR THE MOMBE FORMATION OF INDIUM PHOSPHIDE THIN FILMS.** John A. Glass, Yexin Tan and James T. Spencer,* Department of Chemistry and the W. M. Keck Center for Molecular Electronics, Center for Science and Technology, Syracuse University, Syracuse, New York 13244-4100.

The decomposition pathways for twenty organophosphorus source compounds were studied through the application of statistical, thermodynamic and quantum mechanical (semiempirical MNDO) methods to understand the chemical behavior of these phosphorus compounds under CVD and MOMBE conditions. In addition, the thermal decomposition of four organophosphorus InP CVD source molecules was examined by *in situ* mass spectrometry; dichloro-*t*-butyl phosphine (**1**), cyclohexylphosphine (**2**), phenylphosphine (**3**), and diphenylvinylphosphine (**4**). The β-H elimination pathway appears to be the dominant decomposition pathway for compounds **1** and **2** at the onset of decomposition, while at higher temperatures the homolytic bond fission pathway effectively competes with the β-H elimination pathway. For compound **3**, the predominant decomposition pathway at lower temperatures was α-elimination while at higher temperatures, the homolytic bond fission pathway was again found to effectively compete with this pathway. The results of these investigations will be presented in detail.

626. **SYNTHETIC AND MECHANISTIC ASPECTS OF THE CHEMICAL VAPOR DEPOSITION OF METAL BORIDE THIN FILMS FROM BORANE PRECURSORS.** Shreyas S. Kher, Yexin Tan, John A. Glass, Jr., and James T. Spencer,* Department of Chemistry and the W. M. Keck Center for Molecular Electronics, Center for Science and Technology, Syracuse University, Syracuse, New York 13244-4100.

In recent years, the formation of metal boride thin films by CVD has received increasing research attention. Interest in the preparation, theoretical modeling and solid state characteristics of metal borides is particularly due to their wide structural diversity and potential applications to many materials problems, from extremely refractory coatings to advanced electronic and magnetic materials. Metal-organic chemical vapor deposition (MOCVD) is clearly one of the best methods for the formation of these materials since it circumvents many of the problems associated with other technologies. We have used a number of boron-containing CVD precursors, especially the boranes, to prepare many pure metal and metal boride thin film materials. In this work, the formation of polycrystalline transition metal and lanthanide metal boride thin films from borane precursors has been explored. These films have been extensively characterized by SEM, AES, EDXA, XRD, AA, TEM and other techniques. In addition, we have explored several of the most important mechanistic details of the CVD of metal borides from gas phase boranes and metal halides. The results of these investigations will be presented in detail.

627. **PHOTOCHEMICAL REACTIONS OF BORANE AND METALLABORANE CLUSTER COMPLEXES.** Ralf Littger, Jesse Taylor, George Rudd, and James T. Spencer,* Department of Chemistry and the W. M. Keck Center for Molecular Electronics, Center for Science and Technology, Syracuse University, Syracuse, New York 13244-4100.

The photochemical decarbonylation and rearrangement reactions of several small σ-metallated pentaborane(9)-based metallaborane clusters have been investigated previously. These reactions have been found to produce new borane species in good yield. Our recent work has been aimed at extending these initial explorations to other borane cage species. As an example to be presented here, among the smaller boron hydride compounds, the *arachno*-octahydrotriborate(1-) anion, $B_3H_8^-$, shows exceptional oxidative and hydrolytic stabilities. Indeed, one synthetic route to this borane is the reaction of pentaborane(9) and $NaOCH_3$ in methanolic solutions. Several photoinduced reactions of the octahydrotriborate(1-) anion have been observed from the irradiation of solutions of $(Bu_4N)B_3H_8$ with a variety of organic species, including alcohols, halohydrocarbons, ketones and aldehydes. Details of the photoinduced synthetic and mechanistic aspects of the octahydrotriborate(1-) anion and also the photochemistry of several larger B_9 and B_{18} metallaborane cluster will be presented here.

628.
EFFECTS OF THE ALKYL CHAIN OF THE CARBOXYLIC ACID IN THE PREPARATION OF SILVER CERMETS VIA AN AMMONIUM SOAP ROUTE. **David A. Robinson** Chatham College, Woodland Road-Box 598, Pittsburgh, PA 15232

Silver/Ceramic microcomposites have been prepared by an ammonium soap route using 2-ethylhexanoic acid. Previous work on metal/ceramic microcomposites with copper, nickel and cobalt via the same route revealed that the alkyl chain of the carboxylic acid influences the coating of the metal onto the ceramic. In this work, we are examining the influences of carbon chain length and the branching of the alkyl chain of the carboxylic acid on the preparation of silver/ceramic microcomposites. SEM, optical micrographs and X-ray data will be presented.

629.
BORYL–CYCLOPENTADIENYL–TITANIUM AND –ZIRCONIUM COMPLEXES.
P. J. Shapiro, D. S. Stelck, J. T. Golden, and S. A. Larkin, Department of Chemistry, University of Idaho, Moscow, ID, 83844-2343

Boryl groups have been appended directly to the cyclopentadienyl rings of half-sandwich and bent-sandwich complexes of titanium and zirconium in order to promote intramolecular participation of these Lewis-acidic moities in the chemistry of the transition metal. A versatile synthetic approach to these complexes has been found which has allowed the incorporation of a variety of boryl substituents into these cyclopentadienyl–group IV complexes as pendant groups, as linkages between the rings of bimetallic structures, and as interannular bridges between the rings of *ansa*-metallocenes.

630. SONOCHEMICAL SYNTHESIS OF INORGANIC MATERIALS.
K. S. Suslick, M. Fang, M. Mdleleni, T. Hyeon, and J. Ries School of Chemical Sciences, University of Illinois at Urbana-Champaign, Urbana, IL 61801

High intensity ultrasound has found new applications in the synthesis of unusual inorganic materials, due in large part to the extreme conditions of cavitation: ~5000 K, ~1000 atm, with heating and cooling rates that exceed 10^9 K/s. The sonolysis of volatile organometallic precursors during cavitation produces clusters a few nm in diameter, and we have utilized this for the sonochemical syntheses of nanophase metals, alloys, carbides, sulfides, supported heterogeneous catalysts, and nano-colloids. Several nanostructured solids are active heterogeneous catalysts for various reactions. Most recently, we have discovered a new synthesis of nanometer-sized transition metal colloids of iron, cobalt, and iron-cobalt alloy colloids. Sonication of iron pentacarbonyl, tricarbonylnitrosylcobalt, or appropriate mixtures of these two precursors in polyvinylpyrrolidone or oleic acid solutions in n-hexanol generated black colloidal solutions. The particle size ranges from 3 to 7 nm as determined by TEM. Diffuse ring patterns in the electron microdiffraction showed that these metal particles are amorphous as initially prepared. Elemental analysis by TEM-EDX revealed that the Fe-Co alloys are homogeneous in atomic composition. Magnetic studies show that the colloids are superparamagnetic and function as useful ferrofluids.

631. WATER-SOLUBLE, NON-IONIC, ALKYLDIPHOSPHINES: SOME COORDINATION CHEMISTRY WITH IRON (II)
David K. Lyon and Warren K. Miller Bend Research, Inc. 64550 Research Rd., Bend, OR 97701-8599

The recent synthesis of water-soluble bis[di(hydroxyalkyl)phosphino]ethane ligands has given us the opportunity to develop their transition-metal chemistry.[1-3] Recently, we have explored the chemistry of these ligands with iron(II) salts. Some of this chemistry parallels related non-aqueous chemistry--e.g., formation of *trans*-Fe(diphosphine)$_2$Cl$_2$ compounds from FeCl$_2$ and two equivalents of the water-soluble diphosphines. However, novel iron-diphosphine chemistry that has been observed includes formation of water-stable hydrido-iron-bis(diphosphine) compounds. Details of the synthesis, characterization, and reactivity of the iron(II)hydroxyalkyldiphosphine compounds will be presented.

1. Nieckarz, et al. *Inorg. Chem.* **1996**, 35, 1721.
2. Baxley, et al. *J. Mol. Catal.* **1996**, in press.
3. Baxley, et al. *Inorg. Chem.* **1996**, 35, in press.

632. TRANSITION-METAL SILYLENE COMPLEXES: SYNTHESIS AND REACTIVITY. T. Don Tilley, Paulus Wanandi, Greg Mitchell, and Steven K. Grumbine, Department of Chemistry, University of California, Berkeley, Berkeley, California 94720-1460

For decades, transition-metal silylene complexes have proven to be synthetically elusive. Thus, many attempts to prepare $L_nM=SiR_2$ species have proven unsuccessful, apparently due to the strong Lewis acidity of the silylene silicon or the inherent instability of M=Si double bonds, which makes isolation of stable complexes difficult. This situation has seemed all the more frustrating by the fact that analogous carbene, germylene, stannylene, and plumbylene complexes are well known. The lack of isolable silylene complexes which can be isolated and scrutinized has also hindered attempts to define the role that these species may have in observed (or planned) metal-mediated transformations. We have addressed the synthetic problem described above with attempts to design and prepare stable silylene complexes. Our strategy involves use of electron-rich metal fragments such as $Cp^*(PMe_3)_2Ru^+$ (Cp^* = η^5-C_5Me_5) and $(PR_3)_2(H)Pt^+$ to stabilize an inherently electron-poor silylene ligand. In this way, we have recently isolated the first examples of transition-metal silylene complexes. Reactivity patterns for these new chemical species will be discussed, and routes to osmium silylene complexes will be described.

633. **REACTIVITY OF THE CATIONIC TANTALUM IMIDO COMPLEX $[Cp^*_2Ta(=NCMe_3)][B(C_6F_5)_4]$**

Robert E. Blake, Jr., David M. Antonelli, Larry M. Henling and John E. Bercaw*; The Arnold and Mabel Beckman Laboratories of Chemical Synthesis, California Institute of Technology, Pasadena, CA 91125 Kenneth I. Hardcastle; California State University, Northridge, Northridge, CA 91330

The reactivity of the title complex with a variety of small molecule substrates has been explored. Observed products are the result of formal [2+2] additions across the Ta=N bond or C-H bond activation reactions.

634. BIS-BORATABENZENE COMPLEXES CONTAINING GROUP IV METALS. Guillermo C. Bazan, Caroline Kowal, George Rodriguez, Jonathan Rogers, Richard Barnhardt Department of Chemistry, University of Rochester, Rochester NY 14627-0216.

Group 4 complexes of general composition $[C_5H_5B-R]_2MCl_2$ (R = NMe_2, OCH_2CH_3, Ph; M = Ti, Zr, Hf) have been prepared and structurally characterized. The substituent on boron influences the orbital overlap between the metal and the boratabenzene ligand. There is π overlap between strong donors, such as in $[C_5H_5B-NMe_2]_2ZrCl_2$ and the boron atom. Boron is therefore isolated from the metal and the rest of the boratabenzene ligand binds in a pentadienyl-like fashion. Poor donors, such as Ph, force boron to interact with zirconium, removing electron density from the metal. Modulating the electron density at the metal via the boron atom influences greatly the polymerization activity of catalysts formed by activation with methylaluminoxane. The rates of β-hydrogen elimination increase with decreasing boron-substitutent overlap. For $[C_5H_5B-Ph]_2ZrCl_2$/MAO mixtures, α-olefins are created under one atmosphere of ethylene. The rate of monomer consumption in these reactions is comparable to polymerization reactions mediated using standard metallocene systems, *i.e.*, Cp_2ZrCl_2/MAO.

635. SYNTHESIS AND REACTIVITY OF ANSA-MONOCYCLOPENTADIENYLAMIDO GROUP 4 COMPLEXES. L. Kloppenburg and J. L. Petersen, Department of Chemistry, West Virginia University, Morgantown, WV 26506-6045.

An increase in the Lewis acidity and a reduction of the steric congestion at the electrophilic metal center in 16-electron group 4 metallocene complexes can be accomplished by the replacement of the two Cp rings with a bifunctional cyclopentadienyl ligand containing an appended amido functionality. Although the spotlight on group 4 metal complexes of this type remains focused on their industrial development as a new class of Ziegler-Natta olefin polymerization catalysts, recent work in our laboratory has been directed toward understanding how the availability of an additional vacant orbital at the metal center influences the chemical behavior of these electrophilic complexes. The results of our reactivity studies of $[(C_5Me_4)SiMe_2(N\text{-}t\text{-}Bu)]ZrMe_2$ with CO_2, isocyanides, and primary amines will be described and illustrate three distinctly different reaction pathways for this 14-electron complex.

636. NITRENE GENERATION FROM THE REACTIONS OF BULKY AZIDES WITH Ni COMPLEXES. Beatrice Lin and Gregory L. Hillhouse,* Department of Chemistry, The University of Chicago, Chicago, IL 60637.

Aryl azides (RN_3, R = Ph, p-Tol) react cleanly with (bpy)$NiEt_2$ (**1**) to give the corresponding Ni(II) amides (bpy)Ni(NREt)(Et)$_2$ (**2**) in ~95% isolated yield. Surprisingly, the sterically bulkier mesityl azide ($MesN_3$) reacts with **1** to give only traces of products derived from analogous Ni amides (i.e., 2% MesN(Et)H) but significant amounts of products derived from mesityl nitrene, including azomesitylene (MesN=NMes, 30%) and mesitylamine ($MesNH_2$, 62%).

637. CONNECTING CHEMISTRY WITH CHEMICALS: USING LEAD(II) IODIDE TO UNIFY GENERAL CHEMISTRY AND INORGANIC CHEMISTRY. David K. Erwin, Department of Chemistry, Rose-Hulman Institute of Technology, Terre Haute, Indiana 47803

Undergraduate chemistry courses provide concepts and principles that are often compartmentalized and not unified into a complete picture of chemistry. The synthesis, characterization, and study of lead(II) iodide can be applied to many of the principles encountered in both general and inorganic chemistry. Topics including solid and solution reactions, precipitation, equilibrium, thermodynamics, electrochemistry, atomic theory considerations, and spectroscopy can be tied together with lead(II) iodide. Approaches and their outcomes will be discussed.

638. REVERSIBLE TIN-TO-PALLADIUM TRANSMETALATION: EVIDENCE FOR A NUCLEOPHILICALLY ASSISTED PATHWAY. W. D. Cotter, L. Barbour, K. McNamara, R. Hechter, C. M. Algozo, Carr Laboratory, Department of Chemistry, Mount Holyoke College, South Hadley, MA 01075.

The arylbis(phosphino)triflato palladium complex **1** (Venanzi, Rimml, 1983) undergoes stoichiometric transmetalation with 2-tributylstannylfuran at room temperature in dry acetone. Treatment of the palladium furyl product with 2-tributylstannyl triflate regenerates 2-tributylstannylfuran; the transmetalation event is reversible. Other stannane substrates (e.g., vinyltributyltin) do not undergo stoichiometric transmetalation, instead reversibly forming an adduct with the cationic palladium center. Complex formation is strongly enhanced in the presence of water at low concentrations. **1** catalyzes the hydrolysis of unsaturated and heteroaromatic stannanes in the presence of stoichiometric concentrations of water. Phenyltributyltin is unreactive with **1**, even under extreme conditions. The transfer mechanism suggested by these observations is one involving two reversible events: pre-coordination of an unsaturated substrate to the transition metal center, followed by nucleophilic attack (by external nucleophile or solvent) at tin.

639. SYNTHESIS AND CHARACTERIZATION OF NEW IMIDO COMPLEXES OF MOLYBDENUM. Antonio Pastor, Francisco Montilla, Agustín Galindo and Ernesto Carmona. Department of Inorganic Chemistry, University of Sevilla. 41071-Sevilla, Spain.

This talk will describe the synthesis and chemical properties of a number of new d^0, d^1, d^2 molybdenum imido derivatives containing the 2,4,6-trimethylphenylimido ligand (Nmes). Several compounds have been characterized by X-Ray crystallography and their solid-state structures will be discuss. The bis(imido) $MoCl_2(Nmes)_2(dme)$ has been used as starting material for the preparation of new d^0 bis(imido) and oxo-imido complexes. Compound $Mo(Nmes)Cl_3(dme)$, obtained by reaction of $MoCl_2(Nmes)_2(dme)$ with $MoCl_4(THF)_2$, is a suitable source for the synthesis of d^1 mono(imido) complexes. New d^2 imido derivatives have been prepared using $MoCl_2(Nmes)(PMe_3)_3$ as starting material.

640. SYNTHESIS, CHARACTERIZATION, AND REACTIVITY OF [((i-Pr)$_2$P(CH$_2$)$_3$P(i-Pr)$_2$)(PCy$_3$)Pd-H][OR] Pedro J. Perez and Emilio E. Bunel Central Research & Development, E.I. du Pont de Nemours and Co. Wilmington, DE 19880, USA

Transition metal hydrides play a crucial role in a variety of catalytic and stoichiometric reactions. Despite the extensive information available on the characterization of metal hydrides, very little is known about their role in catalytic transformations. Searching for unusual metal hydrides we found that Pd(0) phosphine complexes such as (DIPPP)PdPCy$_3$ (DIPPP = (i-Pr)$_2$P(CH$_2$)$_3$P(i-Pr)$_2$) react with alcohols and phenols to give cationic square planar palladium hydrides of structure [(DIPPP)(PCy$_3$)Pd-H]$^+$. For all the alcohols and phenols studied we found that the protonation reaction is an equilibrium, and that K$_{eq}$ depends only on the pKa of the alcohol or phenol. For a variety of (DIPPP)PdL complexes we found that the basicity of the metal is determined by the nature of the ligand L. Reaction of [(DIPPP)(PCy$_3$)Pd-H]$^+$ (1) with CO gives [(DIPPP)Pd(μ-H)(μ-CO)Pd(DIPPD)]$^+$ (2) and reaction with ethylene gives (DIPPP)Pd(CH$_2$=CH$_2$) (3). (DIPPP)PdPCy$_3$ catalyzes, under mild conditions, the reaction of ethylene, carbon monoxide and phenol to give phenoxypropionate. In-situ high pressure NMR experiments show that under catalytic conditions only complexes 2 and 3 are detected in addition to (DIPPP)Pd(CO)$_2$. The mechanism of these transformation along with the structures of several intermediates will be discussed in detail.

641. SPECTROSCOPY OF THE DIIRON CENTER IN sMMO. Ann M. Valentine, David E. Coufal, Sonja Komar-Panicucci, Stephen J. Lippard, Department of Chemistry, MIT, Cambridge, MA, 02139

Soluble methane monooxygenase converts methane and dioxygen to methanol and water at a non-heme diiron center coordinated by two histidine and four glutamate residues from the protein as well as water and hydroxide ion. Neither the resting, (μ-hydroxo)(μ-aqua)(μ-carboxylato)diiron(III), or reduced, bis(μ-carboxylato)diiron(II), forms of the protein have optical spectroscopic features which can be used to characterize the protein. Intermediates in the reaction cycle, however, have been identified by continuous and discontinuous kinetics methods employing optical, Mössbauer, and resonance Raman spectroscopic methods. The characterization of these transients has been greatly facilitated by the synthesis of model complexes having similar spectroscopic features. Although the mixed valent, $Fe^{II}Fe^{III}$ form of the protein is not active in the catalytic cycle, its EPR and ENDOR spectral features make it a valuable species for study. Recently we have used these methods in collaboration with the laboratories of Brian Hoffman and Roman Davydov to provide evidence for the binding of acetate and methanol to the dinuclear iron center, to interrogate structural changes at this center upon binding of the coupling protein B, and to examine products of the reactions of small molecules with the reduced, diiron(II) protein.

642.

STRUCTURE AND FUNCTION OF THE Cu CLUSTERS IN pMMO. Sunney I. Chan, Noyes Laboratory 127-72, California Institute of Technology, Pasadena, CA 91125

The particulate methane monooxygenase (pMMO) from Methylococcus capsulatus (Bath) has been purified to homogeneity. It is a copper protein containing several trinuclear copper clusters. Several lines of evidence, including data from electron paramagnetic resonance spectroscopy and magnetic susceptibility measurements, have been used to show that these copper ions are arranged into groups of three's, with the ions within each cluster positioned in sufficiently close juxtaposition that the motions of the spins are correlated in their oxidized states. Based on a detailed study of the turnover chemistry, we have concluded that the enzyme consists of 2 catalytic (C) clusters and 3 electron transfer (E) clusters. For optimal activity, the copper ions within the C-clusters must be fully reduced; and those within the E-clusters at least partially so. Unlike the soluble methane monooxygenase, which is a non-heme iron protein, the copper methane monooxygenase exhibits unusual hydroxylation chemistry with various alkanes as substrates. These experiments have revealed unprecedented regioselectivity and stereoselectivity in the hydroxylation chemistry mediated by the pMMO. Finally, in an attempt to define the nature of the copper clusters further, we have examined the chemistry of the enzyme with CO, NO, as well as with acetylene. Results have been obtained in various redox states of the enzyme and during enzyme turnover in the catalytic cycle.

643. X-RAY ABSORPTION SPECTROSCOPY OF MN ENZYMES. Eileen Y. Yu[1,2], Pamela J. Riggs-Gelasco[2], Timothy L. Stemmler[2], Charles F. Yocum[3], and James E. Penner-Hahn[2], [1]Biophysics Research Division, [2]Department of Chemistry, and [3]Department of Biology, The University of Michigan, Ann Arbor, MI 48109-1055.

Manganese redox enzymes are crucial in a variety of biological systems including the photosynthetic oxidation of H_2O to O_2, the disproportionation of hydrogen peroxide, and the disproportionation of superoxide. The Mn sites in several of these systems have been characterized using x-ray absorption spectroscopy. The OEC shows ca. two 2.7 Å Mn-Mn interactions characteristic of di-μ-oxo bridged Mn dimers. These binuclear-like sites can be selectively reduced by the appropriate choice of reductant, leading to formation of two spectroscopically distinct reduced species. A similar di-μ-oxo bridged binuclear site is found in superoxidized Mn catalase, although with nearest neighbor ligation that is clearly distinct from that in the OEC. The Oec and reduced Mn catalase contain EXAFS detectable Mn-Mn interactions at 3.3-3.4 Å; similar features are *not* found in other binuclear Mn enzymes. The similarities and differences between these binuclear sites will be discussed. Recently, we have found that treatment of the OEC with fluoride and turnover of Mn catalase in the presence of fluoride both lead to the reversible formation of species in which the Mn site has been reduced, as judged by the XANES energy. Possible interpretations of these observations will be discussed.

644. EPR AND MCD STUDIES OF OXOMOLYBDENUM CENTERS IN SULFITE OXIDASE AND RELATED MODEL COMPOUNDS. John H. Enemark, Department of Chemistry, University of Arizona, Tucson, Arizona 85721

Oxomolybdenum enzymes play key roles in sulfur, nitrogen and carbon metabolism. The active site of each of these enzymes contains at least one novel pterin moiety that coordinates to the molybdenum through a cis-enedithiolate group. Sulfite oxidase (SO) is thought to contain one such ligand per molybdenum. The Mo(V) state of SO gives distinctive EPR spectra that are sensitive to pH and anions. Electron spin echo envelope modulation (ESEEM) spectra in phosphate buffer provide unequivocal evidence for phosphate coordination to molybdenum. Magnetic circular dichroism (MCD) studies of model compounds containing the $[Mo^VO]^{3+}$ core and chelating dithiolates provide evidence for low energy ($\sim$10,000 cm^{-1}) πS to Mo d_{xy} charge transfer bands. A single crystal EPR investigation of an oxo-Mo(V)-monodithiolene complex has enabled the g components to be related to molecular geometry. The spectral properties of these models are consistent with the large g-values for the Mo centers of enzymes.

645. **X-RAY ABSORPTION SPECTROSCOPY STUDIES OF NITROGENASE.**
Keith O. Hodgson and Britt Hedman Department of Chemistry and Stanford Synchrotron Radiation Laboratory, Stanford University, Stanford, CA 94305

An understanding of the electronic/geometric structure and reactivity of the active site in the nitrogenase enzyme in various redox and reactive states remains a topic of current interest. X-ray absorption edge experiments can directly probe the electronic environment of the FeMo-cofactor as well as the P-clusters. Analysis of EXAFS data utilizing a multiple scattering approach based on theoretical parameters (called GNXAS) provides detailed metrical information about longer-range interactions in the polynuclear FeMo clusters. Studies including substrate/inhibitor interactions, single crystal XAS and in-situ electrochemical experiments will be described in this presentation. The work to be described is supported by NSF and the XAS and SAXS data were measured at the Stanford Synchrotron Radiation Laboratory which is supported by the U.S. Department of Energy and the NIH.

646. COMPARATIVE CRYSTAL CHEMISTRY AND SUBSOLIDUS PHASE RELATIONS IN SYSTEMS OF INTEREST FOR MICROWAVE DIELECTRIC CERAMICS; T.A. Vanderah, Ceramics Division, National Institute of Standards and Technology, Gaithersburg, MD 20899

Chemical systems of interest as dielectric ceramics for microwave communications applications include complex oxides featuring (Ti,M)-O polyhedral matrices with larger A-cations such as Ba^{2+}, Sr^{2+}, or Ca^{2+}. The required properties of these oxides include high dielectric constants, high Q factors (i.e. low dielectric losses), and zero or low temperature coefficient and frequency dependence of the dielectric properties. The results of recent experimental studies of the $BaO-Fe_2O_3-TiO_2$ and $AO-TiO_2-M_2O_5$ (A=Sr,Ba; M=Nb,Ta) ternary systems will be presented. Trends in compound formation and crystal chemistry among these systems will be described.

647.
CHEMICAL REACTIONS AT SILICON SURFACES. **M. J. Sailor**, Eric J. Lee, Theresa F. Harper, Theodore W. Bitner, James S. Ha, University of California, San Diego, Department of Chemistry and Biochemistry, La Jolla, California, 92093-0358

Luminescent porous silicon has attracted attention for potential use in electroluminescent, photovoltaic, and chemical sensor devices. We have been developing chemical reactions to functionalize the silicon surface in order to investigate the relationship between the surface and the emissive properties of the quantum nanocrystallites within the porous silicon matrix. Reactions that involve chemical and electrochemical oxidation of silicon-silicon bonds and abstraction of surface hydride species will be presented. The chemical reactions allow the introduction of chemical specificity to sensors based on the phenomenon of photoluminescence quenching.

648. CONTROL OF THE BULK MAGNETIC PROPERTIES OF MANGANOPORPHRYIN-BASED MAGNETS, *Joel S. Miller*, Department of Chemistry, University of Utah, Salt Lake City, UT 84112 and Arthur J. Epstein, Department of Physics and Department of Chemistry, The Ohio State University, Columbus, OH 43210-1106.

A new class of coordination polymer-based magnets comprised of alternating [Mn(III)(porphryrin)]$^+$ and [A]$^{\bullet-}$ (A = TCNE, TCNQ, C$_4$(CN)$_6$, chloranil) have been prepared and structurally and magnetically characterized. The [A]$^{\bullet-}$ is *trans*-μ_2-bound to two adjacent Mn(III) sites. Adjacent sites exhibit antiferromagnetic coupling between the S = 1/2 [A]$^{\bullet-}$ quantum spin and the S = 2 Mn(III) classical spin and magnetic order occurs below ca. 22 K. Use of octaethylporphyrin leads to a antiferromagnet whereas use of tetraphenylporphryin and substituted tetraphenylporphryin leads to magnets with interchain ferromagnetic coupling and ferrimagnetic behavior. The differing magnetic behavior is attributed to enhanced ferromagnetic dipolar interactions for the latter class of magnets.[1] The magneto-structure correlation as well as recent developments in this class of materials will be discussed.

[1] C. M. Wynn, M. Gîrtu, W. B. Brinckerhoff, K-I. Sugiura, J. S. Miller, A. J. Epstein, submitted.

*Supported in part by: US DOE-DMS Grant Nos. DE-FG03-93ER45504 and DE-FG02-86BR45271 and NSF Grant No. CHE9320478.

649. A NEW FAMILY OF COMPOUNDS SHOWING GIANT MAGNETORESISTIVE EFFECTS. Susan M. Kauzlarich, Department of Chemistry, University of California, Davis, CA, 95616

Our extensive study of the unusual magnetic and electronic properties of the Zintl compounds, $A_{14}MnPn_{11}$ (A = Ca, Sr, Ba; Pn = As, Sb, Bi) has been extended with the synthesis of new compounds, $Eu_{14}MnPn_{11}$ (Pn = Sb, Bi). These compounds are isostructural with $Ca_{14}AlSb_{11}$ (space group: I4l/acd). Both show magnetic ordering; the Sb compound is apparently ferromagnetic (T_C = 92 K) whereas the ordering for the Bi compound (T = 36 K) cannot be simply classified as either ferro- or antiferromagnetic. The temperature dependent magnetic, electronic and magnetoresistive properties will be presented and discussed in light of structural and electronic trends in this series and other compounds that exhibit giant magnetoresistive effects.

650. THE EFFECTS OF SURFACE CHEMISTRY ON DEGRADATION OF PHOSPHORS FOR FIELD EMISSION FLAT PANEL DISPLAYS*. Paul H. Holloway, Department of Materials Science and Engineering, University of Florida, Gainesville, FL 32611-6400

While a number of technologies are being pursued to compete with liquid crystals in flat panel displays, one of the most promising are field emission displays (FEDs). An introduction to this technology will be given to illustrate the reasons for this interest. Light is emitted by cathodoluminescence from these displays, and one of the biggest challenges is to produce phosphors which will operate efficiently for long times in the FEDs. A primary concern has been the lifetime and degradation, and this talk will illustrate a previously unknown degradation mechanism involving electron beam-stimulated surface chemical reactions. Specifically, it will be shown that electron beam dissociation of residual gases (H_2, H_2O, CO and CO_2) can result in deposition and/or removal of C, and reactions with ZnS-based phosphors which lead to removal of S from the surface and replacement of ZnS with ZnO. The resultant decrease in CL brightness is unacceptable in FEDs. Methods to reduce or negate this degradation will be discussed.

*-This work supported by DARPA Grant MDA972-93l-0030 through the Phosphor Center of Technology.

651.
RATIONAL SYNTHESIS OF SILICON NANOWIRES. **Alfredo M. Morales** and Charles M. Lieber, Department of Chemistry and Chemical Biology, Harvard University, Cambridge, MA 02138

The synthesis of one dimensional structures is an ongoing challenge for solid state chemists. The synthesis of one dimensional silicon is especially important considering the preponderance of Si in electronic devices and expected novel properties of nanostructures. Herein we report the first synthesis of bulk Si nanowires with diameters between 2 and 30 nm and lengths exceeding 1 micron. The silicon nanowires have been prepared via laser ablation of metal containing silicon targets. Systematic studies of the products obtained as a function of different experimental parameters show that nanowires grow via a laser assisted vapor-liquid-solid mechanism in which metal silicide nanoparticles act as the critical catalysts. The implications of this growth mechanism with respect to the synthesis of other one dimensional materials will be discussed. In addition, electrical and optical measurements of these nanowires will be reported.

652. **LANGMUIR-BLODGETT ANALOGS OF LAYERED TRANSITION METAL PHOSPHONATES: MAGNETIC MONOLAYERS AND "DUAL-NETWORK" ASSEMBLIES.** Gail E. Fanucci, Garrett E. Granroth, Mark W. Meisel, Melissa A. Petruska, Candace T. Seip and Daniel R. Talham, Department of Chemistry and Department of Physics, University of Florida, Gainesville, Florida 32611.

The Langmuir-Blodgett (LB) method is perhaps the earliest technique to afford molecular level control over the dimensions of supramolecular assemblies. However, physical properties requiring long-range structural order, such as superconductivity or magnetic order, have been elusive in LB films. We have developed a method for introducing long-range structural order into LB films that uses the inorganic lattice energy of known solid-state layered structures. LB films of a series of divalent metal octadecylphosphonates will be described. Octadecylphosphonate films with Mn^{2+}, Mg^{2+}, Co^{2+} and Cd^{2+} form with stoichiometry $M(O_3PC_{17}H_{37}) \cdot H_2O$ and have the same extended lattice structure as the known $M(O_3PR) \cdot H_2O$ solids. The Ca^{2+} film forms as $Ca(HO_3PC_{18}H_{37})_2$. The $Mn(O_3PC_{17}H_{37}) \cdot H_2O$ film under-goes a magnetic ordering transition at 13.5 K resulting in a "weak ferromagnet". Attempts to prepare "dual network" assemblies, where both the organic and inorganic components add function to the LB films, will also be presented.

653.
THE LIFE AND TIMES OF GEOFFREY WILKINSON F. A. Cotton Texas A&M University, Department of Chemistry, College Station, TX 77843-3255

From the day in 1951 when he delivered his first lecture in Chem 150 at Harvard University until a telephone conversation some ten days before his death, I had the opportunity to work with and know G.W. as few if any others have. He was a remarkable man in many respects, and there are lessons to be learned - as well as simple pleasure to be had - by examining not just the things he accomplished, but how he went about his work. I will try to convey a little of the flavor and not just the cold facts regarding his distinguished career.

654. **RHENIUM AND TECHNETIUM ORGANOHYDRAZIDE CHEMISTRY: TRUTH AND FICTION** Alan Davison,* Melissa Hirsch-Kuchma, Terrence Nicholson, and Alun G. Jones,* *Department of Chemistry, Massachusetts Institute of Technology, Cambridge, MA 02139 and Department of Radiology, Harvard Medical School, Boston, MA 02115.*

The chemistry of organohydrazide ligands with transition metals has been evolving steadily since the 1960s when catalytic nitrogen fixation was the principle concern for investigators in this area. Since then, these ligands have been shown to have useful properties for a number of chemical studies. Their structural and electronic versatility make them extremely useful probes for examining new ligand systems. They can adopt a number of bonding modes to accommodate various co-ligand requirements. Like the molecule hydrazine, they are powerful reducing agents, actively pumping electrons into high oxidation state precursors, being oxidized in the process. Conversely, the diazonium salts can effectively serve as wells for electrons from low oxidation state metal precursors and are subsequently reduced in the process. Unfortunately, the established literature frequently contains misnomers which tend to confuse oxidation state assignments, greatly complicating any discussions pertaining to this area of chemistry. We intend to discuss some of the chemistry of the group 7 organohydrazide complexes.

655. **CHEMISTRY AND MICROSTRUCTURE OF AN ORGANCERAMIC COMPOSITE**
O. O. Popoola, P. G. Desai, W. M. Kriven and J. Francis Young,* *Dept. Mater. Sci. Engr., Univ. Illinois, Urbana IL.*

A new class of organic-inorganic composites are formed by reactions between calcium aluminate and poly(vinyl alcohol) at near ambient temperatures. Processing under high shear creates a composite in which residual calcium aluminate particles become the close-packed filler phase in a binding matrix, which is an interpenetrating biphase composite. One phase is a cross-linked PVA and the other is an "interphase", which is itself a nanocomposite of calcium aluminate hydrate embedded in an amorphous matrix of aluminum hydroxide and cross-linked PVA. The chemistry involved in forming this unique microstructure will be discussed.

656. **RECENT ADVANCES IN TRANSITION METAL MOLECULAR CATALYSTS.** J. A. Osborn*. *Laboratoire de Chimie des Metaux de Transition et de Catalyse, Universite Louis Pasteur, Strasbourg, Cedex, France*

The Metal-Oxo functional group is of great importance for its role in catalytic processes as diverse as that found in the heterogeneous catalytic oxidation occurring on metal oxide surfaces or in biological oxidation mechanisms such as those proposed in P450 cytochromes.
We will trace the research project from its origins in the studies on the reactivity of certain Mo(VI) dioxo complexes, which were designed to model on a level the active species supposedly involved in the heterogeneously catalyzed oxidation and ammonoxidation of propene (SOHIO process). These studies resulted in the discovery of new molybdenum catalyzed reactions such as the rearrangement of allyl alcohols and oxidation of organic substrates. These observations then led to the development of other high valent oxo complexes or Re(VII), Ru(VII) and Os(VIII) as catalysts, and, in particular, for certain oxidation reactions using molecular oxygen as the oxidizing source.

657. **DEUTERIATION USING ANALOGS OF WILKINSON'S CATALYST** Fred Jardine *Department of Mathematical and Environmental Sciences, University of East London, Romford Road, London E15 4LZ, England*

Because it does not undergo isotopic exchange with other components of the catalytic system, Wilkinson's catalyst is used for cis-deuteriation of alkenes. More searching investigations have revealed that some scrambled deuteriation (i.e., incorporation of deuterium at other sites in the product) and polydeuteriation occur. These by-reactions show that the two atoms of deuterium are not added simultaneously to the alkene bond but that a relatively longlived alkyl intermediate is formed. The reactions of this intermediate give rise to the various products, the lower the stability of the intermediate alkyl, the greater the participation of the often unwanted by-reactions. The instability of the intermediate alkyl can arise from both steric and electronic causes. By replacing the chloro or triphenylphosphine ligands of the original catalyst with suitable analogs, it is possible to control the participation of undesirable side-reactions in the deuteriation of alkenes. Additionally it is possible to increase the proportion of dideuteriated product from alk-1-enes by taking steps to encourage the formation of a 1-alkyl intermediate.

658. EXTRACTION OF HEAVY METALS WITH CALIX[4]ARENES.
D. Max Roundhill, Department of Chemistry, Texas Tech University, Lubbock, TX, 79409-1061.
The extraction of heavy metals from soils and waters in an important environmental problem. We will present data showing that chemically modified calix[4]arenes can be selective extractants for heavy metal ions. The metal ions the will be discussed are Hg^{2+}, Cd^{2+}, Pb^{2+}, Hg_2^{2+}, $MeHg^+$, Au^{3+}, Ag^+, Pt^{2+}, and Pd^{2+}, and their selectivities with different calix[4]arenes presented.

659. **DIRECTED SYNTHESIS OF DIPHOSPHINE-BRIDGED HETEROBIMETALLIC COMPLEXES.** Joel T. Mague* and Z. Lin, *Department of Chemistry, Tulane University, New Orleans, LA 70118*

The use of the "metalloligands" CpMCl[η-1-MeN(PF$_2$)$_2$]$_2$ (M = Fe, Ru), *fac*-ReBr(CO)$_3$(η-1-L$_2$) [L$_2$ = MeN(PF$_2$)$_2$, Me$_2$PCH$_2$PMe$_2$] and CpNi(CN)(η-1-Ph$_2$PCH$_2$PPh$_2$) for the directed synthesis of heterobimetallic complexes with, for example, Ru, Rh, Ir and Pt is described. Typical products obtained include the heterobimetallic complexes: CpFe[μ-MeN(PF$_2$)$_2$]$_2$MCl$_2$(PPhMe$_2$) (M = Rh, Ir), *fac*-Re(CO)$_3$[μ-MeN(PF$_2$)$_2$]$_2$PtBr(PPh$_3$), [*fac*-Re(CO)$_3$(μ-Me$_2$PCH$_2$PMe$_2$)$_2$Pt(PPh$_3$)]Br and CpNi(μ-Ph$_2$PCH$_2$PPh$_2$)Pt(CN)(PPh$_3$). The structures and spectroscopic data for the complexes will be discussed.

660. **DOUBLE C-H BOND ACTIVATION OF ALKANES BY AN ALKYLIDENE NITROSYL COMPLEX OF TUNGSTEN.** Peter Legzdins and Elizabeth Tran, Department of Chemistry, The University of British Columbia, 2036 Main Mall, Vancouver, B.C., Canada V6T 1Z1.

The activation of carbon-hydrogen bonds by coordinatively unsaturated metal complexes continues to be an area of considerable research activity. This presentation will deal with the utility of Cp*W(NO)(=CHCMe$_3$) for effecting such activation. The alkylidene complex may be generated thermally from the 16e dialkyl precursor, Cp*W(NO)(CH$_2$CMe$_3$)$_2$, and can be trapped with PMe$_3$ as the 18e adduct. More interestingly, it effects the double activation of alkane C-H bonds as illustrated below for cyclohexane. The scope of this reactivity and pertinent mechanistic details of these transformations will also be presented.

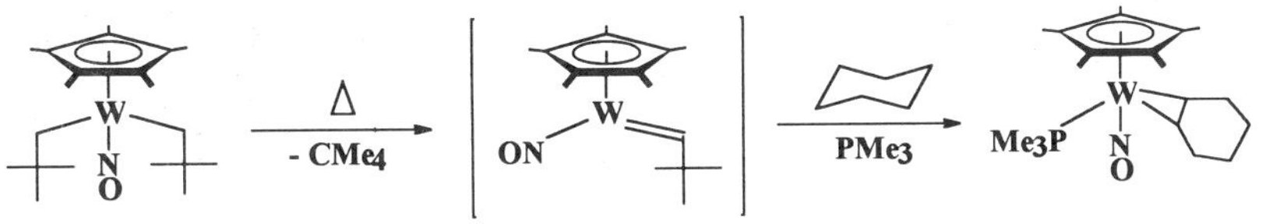

661. **SPIN EQUILIBRIA IN MANGANOCENES; NEW EXAMPLES TO AUGMENT THE CLASSICAL STUDIES OF GEOFFREY WILKINSON** Chadwick D. Sofield and Richard A. Anderson,* *Chemistry Department and Chemical Sciences Division Lawrence Berkeley Laboratory, The University of California, Berkeley, CA 94720*

Solid state magnetic studies on some new substituted manganocenes show that, in some cases, an equilibrium exists between high and low spin electronic states. The origin of this phenomenon will be described.

662. CHEMISTRY AT TEXAS A&M UNIVERSITY. E. A. Schweikert, Department of Chemistry, Texas A&M University, College Station, Texas 77843

A phenomenal intellectual building project was engineered by Arthur E. Martell in the 1970's at Texas A&M. A brief account of the exceptional development of a chemistry program will be presented. The lessons to be drawn are relevant to the challenges facing chemistry today in academia.

663. TEMPLATED SYNTHESIS OF N_2S_2 MACROCYCLES M. Y. Darensbourg*, G. Musie, D.C. Goodman, D.K. Mills, I. Font, M. Pala, and J. H. Reibenspies. Dept. of Chemistry, Texas A&M University, College Station, TX 77843

The nickel-mediated template effect has been used to prepare a series of N_2S_2 and N_2S_2Y macrocycles as well as open chain penta- and hexadentate ligands from the precursor dithiolates **Ni-1** and **Ni-1***. Characterization of the series by x-ray crystallography, electrochemistry (stabilization of $Ni^{II/I}$ and $Ni^{II/III}$ couples), and metal exchange properties will be presented. The ligand dependent-electrocatalytic production of H_2 in the presence of $N_2S_2Ni(I)$ will be discussed.

Ni-1:   R = H;
Ni-1*: R = Me

664. **IRON-SPECIFIC SEQUESTERING AGENTS: NATURAL TO SUPRANATURAL.** Kenneth N. Raymond and Jide Xu, Dept. of Chemistry, University of California, Berkeley, CA 94720

The chemistry of microbial iron sequestering agents has largely been developed in the last twenty years. Siderophores, which incorporate hydroxamate, catecholate and occasionally other groups (e.g., aminocarboxylate, thiohydroxamate or hydroxypyridonate) form extraordinarily stable complexes of high spin Fe^3. This affinity and selectivity of the siderophores make good targets for therapeutic iron removal agents. Desferrioxamine B (Desferal®) has been marketed by Ciba-Geigy as the only generally accepted therapy for human iron overload for over 30 years. The search for other chelating agents has especially focused on orally active compounds, since lack of oral activity is one of DFO's problems.

Very strong hydroxypyridonate iron complex agents have been developed. Their characterization and solution thermodynamics will be reviewed. The structure and stability of some new complexes will be presented and the relationship of stability and structure, including previous siderophore and synthetic analog structures, will be presented.

Kenneth N. Raymond and Jide Xu in *The Development of Iron Chelators for Clinical Use*, R. J. Bergeron and G. M. Brittenham, Eds., CRC Press, Inc.: Boca Raton, FL, 1994, pp 307-327.
Kenneth N. Raymond, Jide Xu, US Patent Application Serial No. 08/285,640

665. SOLUTION EQUILIBRIA LEADING TO THE FORMATION OF METAL-METAL BONDS IN PARTIALLY OXIDIZED BISOXALATOPLATINATE(II) SYSTEMS.
B. J. Keller[a], S. O. Dunham[a], E. S. Peterson[b] and E. H. Abbott*[a]
a. Montana State University, Bozeman, MT 59717 b. Idaho National Engineering Laboratory, Idaho Falls, ID 83415

The chemistry of solutions containing partially oxidized bisoxalatoplatinate(II) anions, $[Pt(ox)_2]^{2-}$, has been investigated. As previously reported, chains of the bisoxalatoplatinate fragment are formed and the platinum ions are linked to one another by metal-metal bonds. We have identified four polynuclear species and have assigned formulas to them. In these complexes the average oxidation numbers for the platinum ions are Pt^{3+}, $Pt^{2.67+}$, $Pt^{2.5+}$ and $Pt^{2.4+}$, respectively. A reaction scheme for the polymerization is proposed. It entails the formation of a dinuclear Pt^{3+} complex and subsequent addition of bisoxalatoplatinate(II), $Pt(ox)_2^{2-}$, ions to the Pt^{3+} species, sequentially forming the higher oligomers.

666. POLARIZED XAFS MEASUREMENTS ON ORIENTED SINGLE CRYSTALS OF CARBONMONOXYMYOGLOBIN. **Hans C. Freeman**, Paul J. Ellis and Dashuang Shi, School of Chemistry, University of Sydney, N.S.W. 2006, Australia.

Polarized X-ray absorption spectra have been recorded for single crystals of sperm-whale carbonmonoxymyoglobin (MbCO) in six orientations. Both the absorption-edge features and the XAFS are strongly orientation dependent. In contrast with X-ray absorption measurements on MbCO in solution, where the contributions of the porphyrin atoms are dominant, the symmetry properties of the crystals enable the XAFS contributions of the axial ligand atoms to be resolved. The geometry of the heme–Fe–C–O site has been determined by fitting a 3-dimensional model simultaneously to the six sets of XAFS data. The value of the Fe–C–O angle will be compared with a series of values from published X-ray and neutron crystal structure analyses.

Such structural studies of metalloproteins rest on foundations to which A. E. Martell made seminal contributions. In the 1950s and 1960s, when the characterization of metalloproteins was still in its infancy, Martell and collaborators undertook a series of studies on the equilibria of metal-peptide complexes in solution. Many of the metal-peptide species which they identified later became the subjects of crystal structure analyses at the University of Sydney and elsewhere. This aspect of Martell's early work still underpins much of our understanding of metal-protein interactions.

666A MY ASSOCIATION WITH ARTHUR MARTELL. Kazuo Nakamoto, Department of Chemistry Marquette University, Milwaukee, Wisconsin 53201-1881 USA

I worked with Arthur Martell at Clark University (1957-1961) and Illinois Institute of Technology (1961-1967). I would like to reflect on my association with Arthur during these years.

667. CHELATE REAGENTS FOR BIOLOGICAL APPLICATIONS. K. Ueno, Dojindo Laboratories, Tabaru 2025-5, Mashiki-machi, Kumamoto 861-22, Japan

Since the first report of EDTA in mid-1930s, chelate reagents have found widespread use in our life because of their unique property to bind metal ions, and their use continues to be central to industrial applications of analytical reagents. Recent applications, however, focus more on biological use as "bifunctional" chelate compounds, which are highlighted by intracellular fluorescent metal probes that have made enormous contribution in understanding cellular functions, Meares reagents that have large potentials in radioimmunoimaging as well as radioimmunotherapy, and nickel chelates with affinity for histamine which are most popular technique to purify recombinant proteins in molecular biology. In an effort to increase their applicability, we have been trying to develop and commercialize those compounds; this talk will discuss some of their recent applications along with brief historical background.

668. COMMERCIAL APPLICATIONS OF IRON COORDINATION CHEMISTRY. S. A. Bedell, The Dow Chemical Company, Freeport, Texas 77541

Iron complexes are valued in commercial applications because of their versatility, low cost and ease of disposal. The development of three processes will be reviewed. The first involves the use of low valent nitrosyl complexes for the catalytic dimerization of butadiene. The other two processes involve the use of aminopolycarboxylic acid complexes of iron for purification of gas streams. One involves the formation of Fe(II) nitrosyl complexes for NO removal. The last process discussed utilizes ferric chelates for oxidative removal of hydrogen sulfide.

669. METAL CHELATES FOR MEDICAL IMAGING AND THERAPY. Michael J. Welch, Division of Radiological Sciences, Washington University Medical Center, St. Louis, MO 63110.

Metal chelates have important applications in nuclear medicine, magnetic resonance imaging and radiation therapy. The authors laboratory has collaborated with Professor Martell for 15 years on the development and assessment of new metal chelates. Chelates for indium and gallium have been designed for nuclear medicine imaging, copper chelates for imaging and therapy and gadolinium chelates as contrast enhancing agents for magnetic resonance imaging. A series of chelates with different stability constants and different lipophilicities have been synthesized to evaluate the structure activity relationships of these compounds when injected into animals. A series of gallium and copper chelates will be presented as examples of how stability lipophilicity and charge alter the biodistribution and metabolic disposition of metal chelates in vivo. Chelates have been developed that can be attached covalently to proteins and peptides. Chelates for indium and copper have been evaluated in human trials.

670. HIGHLY EFFECTIVE RHODIUM PHOSPHINITE CATALYSTS FOR HOMOGENEOUS ASYMMETRIC HYDROGENATION REACTIONS. Albert S. C. Chan[a], Wen-hao Hu[a], Chak-po Lau[a], Yao-zhong Jiang[b], Ai-qiao Mi[b], Jian Sun[b] and Ming Yan[b]. [a]Department of Applied Biology and Chemical Technology, The Hong Kong Polytechnic University, Hong Kong and [b]Chengdu Institute of Organic Chemistry, China.

In the pursuit of effective asymmetric hydrogenation catalysts, rhodium and ruthenium complexes containing chiral phosphine ligands have been found to give the best results. In contrast, most of the catalysts containing chiral phosphinite ligands reported in the literature are relatively ineffective, giving lower optical yields for the desired products. In this presentation we will report a new class of chiral phosphinite ligands which form extremely enantioselective catalysts for the asymmetric hydrogenation of enamides. At ambient temperature and under one atmosphere of H_2, most prochiral enamides are smoothly hydrogenated to give the desired product in high e.e.'s. The best e.e. for the hydrogenation products has been found to be over 99.9%. Details of the synthesis, characterization, and catalytic properties of the new ligands and catalysts will be presented.

671. MECHANISTIC ASPECTS OF THE CO-LABILIZING ABILITY OF THE GLYCINATE LIGANDS IN ZERO VALENT CHROMIUM AND TUNGSTEN DERIVATIVES.
J. D. Draper, D. J. Darensbourg, Department of Chemistry, Texas A&M University
College Station, TX. 77843

Several amino acid derivatives of chromium(0) and tungsten (0) have been synthesized. The complexes have been characterized in solution by ^{13}C NMR and infrared spectroscopies, and in the solid state by X-ray crystallography. The geometry of the metal anion is, in each case, that of a distorted octahedron consisting of four carbonyl ligands and a puckered five-membered chelate ring, bound through the nitrogen atom and one of its oxygen atoms. Investigations of the lability of the carbonyl ligands have been carried out. The lability is proposed to be due to base assisted removal of a proton from the amine ligand leading to a substitutionally labile amide transient species. The isotope effect (k_H/k_D) and the activation parameters for the intermolecular exchange of CO in [W(CO)$_4$(O$_2$CCH$_2$NH$_2$)][Et$_4$N] were determined and found to be consistent with the proposed mechanism. Other ligands with mixed oxygen and nitrogen donor atoms, such as pyrimidine and SALEN derivatives, are under investigation.

672.
MECHANISTIC STUDIES ON THE OXIDATION OF ALCOHOLS AND AMINES BY A RUTHENIUM-OXO COMPLEX. **Jeffrey J. Rack,** Jay R. Winkler, and Harry B. Gray Beckman Institute, and the California Institute of Technology, Pasadena, CA 91125.

We are working on catalysts that could be incorporated into the anodes of direct oxidation methanol fuel cells. In order to achieve reasonable thermodynamic efficiency in the fuel cell a homogeneous oxidation catalyst must operate at or near 0 V vs. SCE, which poses a major challenge. In the present study, a ruthenium(IV) oxo complex, [Ru(bpy)2(O)(H2O)]2+, has been used to oxidize methanol to formaldehyde operating at 400 mV. The oxidation of additional substrates including ethanol, isopropanol and other simple amines has been investigated. The kinetics and mechanisms of these reactions have been examined and an unusual isotope effect (kH/kD) of 12 has been observed using the ROD isotopomer in place of the ROH alcohol substrate. These results and others will be discussed with regard to the design of better alcohol oxidation catalysts.

673. **SOLUTION THERMOCHEMISTRY OF TANTALUM ALKYL AND ALKYLIDENE COMPLEXES. THE FIRST EXPERIMENTAL ANALYSIS OF METAL-CARBON DOUBLE BOND ENERGETICS** Lubin Luo, Liting Li, and Tobin J. Marks*, Northwestern University, Department of Chemistry, Evanston, IL 60208-3113

The measurement of heat for the elimination reaction: $Ta(CH_2SiMe_3)_5 \rightarrow$ $(Me_3SiCH_2)_3Ta=CHSiMe_3 + SiMe_4$ (1) along with the reaction enthalpies obtained through the titration calorimetry of the reactions: $Ta(CH_2SiMe_3)_5 + 5\ I_2 \rightarrow Ta\ I_5 + 5\ ICH_2SiMe_3$ (2) and $Ta(CH_2SiMe_3)_5 + 1\ I_2 \rightarrow (Me_3SiCH_2)_4TaI + ICH_2SiMe_3$ (3) allows extraction of D(Ta=C) in $(Me_3SiCH_2)_3Ta=CHSiMe_3$. D(Ta=C) = 128(2) kcal/mol; D_1(Ta-C) = 45 (1) kcal/mol which represents the first D(Ta-C) in $Ta(CH_2SiMe_3)_5$ complex, obtained from reaction 3; $\bar{D}$(Ta-C) = 67.2 (6) kcal/mol, obtained from reaction 2. The driving force for the formation of the alkylidene is thus quantified. D(Ta=C) is compared with reported theoretical results for other alkylidene complexes, and is used to analyze the thermochemistry of several reaction classes including olefin metathesis, C-H bond activation, and other alkylidene forming processes.

674. **Osmium and Ruthenium Catalyzed Cyclotrimerization of Acetylene: An Unexpected Result.** G.J.Irvine, W.R.Roper* and L.J. Wright. Department of Chemistry, The University of Auckland, Private Bag 92019, Auckland, NEW ZEALAND.

Transition-metal catalyzed oligomerization of acetylene and substituted alkynes is one of the oldest and best studied reactions of organometallic chemistry. Recent studies which center on the reactions of osmium and ruthenium hydride complexes, $MHCl(CO)(PPh_3)_2$ (M = Os, Ru) with excess acetylene led to the formation of a complex which possesses a novel 5-methylene-2-cyclopenten-1-yl ligand (1). This moiety results from a metal mediated cyclotrimerization of three molecules of acetylene with the incorporation of the hydride ligand. Details of the sythesis, characterization, reactivity and structure of $OsCl(\eta^5-C_6H_7)(CO)(PPh_3)_2$ will be presented along with a likely mechanism of formation.

675. CATALYTIC POLYMERISATION OF ALKENES USING NOVEL GROUP 4 AMIDO COMPLEXES. Suzanne Tinkler*, David J. Duncalf and Andrew McCamley[†]. Department of Chemistry, University of Warwick, Coventry, CV4 7AL, U.K.

Group 4 metallocene compounds are of current interest as alkene polymerisation catalysts. We are developing novel group 4 amido complexes as viable alternatives to these systems. The bidentate amide ligands [N'HCH$_2$CH$_2$N'H] and [N'H(CH$_2$CH(CH$_2$)$_4$CHCH$_2$)N'H] (where N' =NSiMe$_3$) are prepared by action of SiMe$_3$Cl on the lithiated amine. The ansa- metallocene analogue [1] is prepared via a metathesis reaction, described below.

$TiCl_4(thf)_2\ +\ 2\ [N'Li(CH_2)_2N'Li] \longrightarrow \begin{bmatrix} N' & N' \\ & Ti & \\ N' & N' \end{bmatrix} \xrightarrow{TiCl_4(thf)_2} 2 \begin{bmatrix} N' & Cl \\ & Ti & \\ N' & Cl \end{bmatrix}$
[1]

When reacted with an excess of methylaluminoxane (MAO), [1] is found to be an active catalyst for the polymerisation of ethene (Tinkler, et. al., J. Chem. Soc., Chem. Commun., **1996**, in press). Current experiments are being directed towards the incorporation of chirality into the amide ligand, and the subsequent asymmetric synthesis of stereoregular polymers and chiral organic compounds. [†]Deceased

676. A HIGHLY EFFICIENT, ONE-STEP SYNTHESIS OF CONSTRAINED GEOMETRY POLYMERIZATION CATALYSTS VIA ALKANE ELIMINATION, AND ISOLATION OF THE CORRESPONDING CATIONIC COMPLEXES. You-Xian Chen, Peng-Fei Fu, Charlotte L. Stern, and Tobin J. Marks, Department of Chemistry, Northwestern University, Evanston, Illinois 60208-3113

A highly efficient synthesis of constrained geometry (CGC) and tetramethyl Cp phenolate (TCP) group 4 catalysts has been developed. Characteristics of this approach include an one-step synthesis (to the final dialkyl complex), high isolated yield (90%), a convenient synthetic temperature (60°C), and simple work-up process (extration and filtration). Isolation, characterization, and solution dynamics studies of the cationic CGC and TCP benzyl cationic complexes activated by $B(C_6F_5)_3$ and $Ph_3CB(C_6F_5)_4$ reveal anion η^n-coordination and ring-flipping processes. These complexes are highly active for α-olefin polymerization and copolymerization.

677. Catalytic Dehydrogenation of Cycloalkanes to Arenes and Tetrahydrofuran to Furan by a Dihydrido Iridium P-C-P Pincer Complex. Mukta Gupta[a], William C. Kaska[b], and Craig M. Jensen[a], Departments of Chemistry, [a]University of Hawaii, Honolulu, HI 96822 and [b]University of California, Santa Barbara, CA 93106

The P-C-P pincer complex, $IrH_2\{C_6H_3\text{-}2,6\text{-}(CH_2PBu^t_2)_2\}$ (1) catalyzes the transfer dehydrogenation of cycloalkanes to cycloalkenes and arenes in the presence of the hydrogen acceptor, t-butylethylene. The dehydrogenation of cyclohexane produces cyclohexene and benzene. Turnover numbers of 44 and 54 respectively are obtained after 1 h at 150 °C and 86 and 77 after 0.5 h at 200 °C. This catalytic system also dehydrogenates methylcyclohexane to methylcyclohexenes and toluene; decalin to octahydronaphthalenes, tetrahydronaphthalene, and naphthalene; and tetrahydrofuran to dihydrofurans and furan but lower levels of activity are observed. The details of our studies of these catalytic aromatization reactions as well as the synthesis and characterization of 1 will be presented.

678. Reactivity of the Alkyl Hydride Complex, $IrH(CH_2CH_2CMe_3)\{C_6H_3\text{-}2,6\text{-}(CH_2PBu^t_2)_2\}$ with Alkanes. Do W. Lee, Chrystel Hagen, and Craig M. Jensen, Department of Chemistry, University of Hawaii, Honolulu, HI 96822.

The alkyl hydride complex $IrH(CH_2CH_2CMe_3)\{C_6H_3\text{-}2,6\text{-}(CH_2PBu^t_2)_2\}$ (1) has been prepared by the reaction of t-butylethylene with $IrH_2\{C_6H_3\text{-}2,6\text{-}(CH_2PBu^t_2)_2\}$ (2) at 25°C in cyclohexane. The dihydrido P-C-P pincer complex, 2 has been found to be a highly active catalyst for the transfer dehydrogenation of alkanes to alkenes and cycloalkanes to arenes in the presence of t-butylethylene. Thus we can infer that formation of 1 is the first step of these catalytic reactions. The isolation of 1 has allowed us to directly study the subsequent, key alkane activation step of the catalytic sequences. The reactivity of 1 with alkanes and the implications of these reactions on the mechanism of alkane dehydrogenation will be discussed.

679. Pd(II)-CATALYZED ADDITION POLYMERIZATIONS OF 3,3-DIALKYLCYCLOPROPENES. Stephen Rush, Annette Reinmuth and Wilhelm Risse, University College Dublin, Department of Chemistry, Belfield, Dublin 4, Ireland

(η^3-Allyl)Pd-complexes bearing bulky (-)-sparteine and C2-symmetric bisoxazoline ligands catalyze the olefin addition polymerization of highly strained cyclopropene derivatives. This reaction contrasts with ring-opening olefin metathesis (ROMP) in that the cyclic structure of the monomer is retained. Polymers composed of triangular repeating units with molecular weights M_n(GPC) above 10000 are obtained. The steric bulk of the chelating ligand and the size of the monomer substituents influence the polymerization behavior. ^{13}C-NMR analysis reveal a partially stereoregular polymer microstructure with 1,2-cis-linked repeating units and a moderate excess of meso-diads.

INOR

680. (η^3-ALLYL)Pd(II) AND Pd(II) NITRILE CATALYSTS FOR THE ADDITION POLYMERIZATION OF NORBORNENE DERIVATIVES WITH FUNCTIONAL GROUPS. Joice P. Mathew, Annette Reinmuth, Noeleen Swords and <u>Wilhelm Risse</u>, University College Dublin, Department of Chemistry, Belfield, Dublin 4, Ireland

Cycloaliphatic polyolefins with functional groups are obtained by Pd(II)-catalyzed homo- and copolymerizations of norbornene derivatives. Pd(II)-nitrile catalysts $[Pd(RCN)_4][BF_4]_2$ (with R = CH_3 and C_2H_5) quite selectively polymerize the exo-isomers of 80/20 endo/exo-mixtures of alkanoic acid esters of 2-(hydroxymethyl)bicyclo[2.2.1]hept-5-ene. In situ prepared (η^3-allyl)palladium complexes with associated tetrafluoroborate and hexafluoroantimonate ions are substantially more active and lead to high monomer conversions in addition polymerizations of norbornene derivatives containing a high amount of endo-isomers. Addition polymers containing 0.2 carboxylic acid groups per repeating unit are obtained from the copolymerization of equimolar mixtures of norbornene and endo-bicyclo[2.2.1]hept-5-ene-2-carboxylic acid.

681.
REDUCTIVE ELIMINATION OF AMINES AND ETHERS FROM PALLADIUM ALKOXIDES. CATALYTIC FORMATION OF ETHERS FROM DPPF-LIGATED Ni AND Pd SYSTEMS. COMMENTS ON THE REDUCTIVE ELIMINATION OF ALCOHOL FROM Ir(III). <u>Grace Mann</u> and John F. Hartwig*, Department of Chemistry, Yale University, P.O. Box 208107, New Haven, CT 06520-8107

The palladium alkoxide complex, $[(DPPF)Pd(p-C_6H_4-t-Bu)(O-t-Bu)]$ (**1**), was prepared to test its ability to undergo reductive elimination of ethers and N-H activation to produce amido intermediates in palladium-catalyzed amination of aryl halides. Complex **1** reacted with amines to produce $[(DPPF)Pd(p-C_6H_4-t-Bu)(NRR')]$ which underwent reductive elimination of arylamines. Complex **1** resisted reductive elimination of ether, but analogs with electron withdrawing groups in the p-position of the Pd-bound aryl group underwent rapid C-O bond forming reductive elimination of ethers. The reductive elimination of aryl ether was directly observed for $[(DPPF)Pd(p-C_6H_4-C(O)H)(O-t-Bu)]$ which was isolated free of NaO-t-Bu. A combination of $Pd(dba)_2$ or $Pd(PPh_3)_4$ and DPPF catalyzed the addition of Na-O-t-Bu or NaOTBDMS to electron poor aryl chlorides or bromides to form aryl ethers in 58-69% isolated yields. While alkoxides containing β-hydrogens were inefficient in forming alkylaryl ethers with Pd(0), the use of $Ni(COD)_2$, DPPF, and $NaOCH_3$ formed methoxy aryl ethers in 76% yield. In contrast to previous reports of the P(p-tol)$_3$-analog, the Ir(III) hydroxo complex, $[Ir(CO)(PPh_3)_2(OH)(Me)(I)]$, was stable at room temperature and did not undergo reductive elimination of MeOH, but rather reacted with CD_3I to produce CD_3OH.

682. **THE THERMODYNAMICS OF ADDITION TO Ir(I) COMPLEXES. C-H BOND ADDITIONS AND A REMARKABLY STRONG Ir-C BOND.** <u>Glen P. Rosini</u>[a], Alan S. Goldman[a], Chunbang Li[b], and Steven Nolan[b], Departments of Chemistry, [a]Rutgers Univ., New Brunswick, NJ 08903, and [b]Univ. of New Orleans, New Orleans, LA 70148

The thermodynamics of addition of various substrates (CO, H_2, H-C_6H_5, H-Si(OEt)$_3$, H-C≡C-R) to complexes containing the Ir(P^{i}Pr$_3$)$_2$Cl fragment have been determined, either calorimetrically or by equilibrium studies. Determination of the absolute Ir-N_2 BDE of Ir(P^{i}Pr$_3$)$_2$(N$_2$)Cl, in conjunction with measurements on a series of substitution reactions, allows the calculation of numerous absolute reaction enthalpies. It has been found that benzene addition to Ir(P^{i}Pr$_3$)$_2$Cl is exothermic by 22 kcal/mol, while addition to the d^8-square planar Ir(P^{i}Pr$_3$)$_2$(CO)Cl is endothermic by 5 kcal/mol. The latter value represents the first direct measurement of the energetics of C-H addition to a Vaska-type complex. The BDE of an Ir-C≡C(CO$_2$Me) bond was determined to be 123 kcal/mol.

683. **TRANSITION METAL-SALICYLATE COORDINATION COMPLEXES AND SYSTEMIC ACQUIRED RESISTENCE IN PLANTS**, Vlad Gorengaut and <u>Christa M. Hockensmith</u>, Department of Chemistry, University of Illinois at Springfield, Springfield, IL 62794

Salicylic acid and its derivatives are well known as naturally-occurring constituents of plants. Salicylates help to induce flowering in certain plants and are associated with a phenomenon called Systemic Acquired Resistance (S.A.R.), a defense mechanism in which resistance to attack by a second pathogen occurs after exposure to an initial pathogen (viruses, bacteria, fungi, *etc.*). It is known that after infection, the amount of salicylic acid in plant tissue increases rapidly. Plant tissue levels of the acid and its derivatives may be determined by fluorescence spectroscopy when the salicylates are complexed with metal cations to form coordination complexes. Detection of these coordination complexes and their characterization by UV-vis, NMR, and IR as well as other spectroscopic methods will be presented.

684. **Coordination Compounds With Sterically Strained or Unstrained Oxyligands as Precursors to Strategic Materials** Ricardo Gonzalez, and <u>Christa M. Hockensmith</u>, Department of Chemistry, University of Illinois at Springfield, Springfield, IL 62794

Resorcinates (and resorcinate-like ligands) in which the donor atoms are in the *ortho-para* positions on the ring system, and catecholates (and catecholate-like ligands) in which the donor atoms are in the *ortho-meta* positions illustrate the difference between a strained-ring coordination compound and an unstrained-ring metal-ligand complex. The unstrained ring system produces a monomeric precursor without bridging or polynuclear metal centers. However, the coordination compound formed from the metal and strained ring system does not. Characterization of these coordination compounds by IR, UV-vis, mass spectrometry, NMR, solid-state NMR, X-ray diffraction, and thermal analysis will be presented in addition to analysis of the usefulness of the strained *versus* unstrained oxyligand precursors.

685. RECENT DEVELOPMENTS IN RUTHENIUM TRISPYRAZOLYLBORATE CHEMISTRY. <u>K. Kirchner</u>, Institute of Inorganic Chemistry, Technical University of Vienna, Getreidemarkt 9, A-1060-Wien, AUSTRIA

Ruthenium trispyrazolborate (Tp) complexes have been known for over 20 years. However, the few systematic studies hitherto done on the complexes of the RuTp fragment do not allow definite conclusions to be drawn as to the factors that determine their formation, stability and reactivity patterns. In this talk I will discuss the reactivity of new RuTp complexes based on experiments and MO calculations, and their application to catalytically operating processes. Emphasis will be given to vinylidene complexes and their role as catalyst precursors in C-C and C-O coupling reactions.

686. RHENIUM(III) AND RHENIUM(V) COMPLEXES WHICH CONTAIN THE $(C_6F_5NCH_2CH_2)_3N$ LIGAND: SYNTHESIS, CHARACTERIZATION, AND REACTIVITY OF NEUTRAL AND CATIONIC SPECIES INCLUDING UNUSUAL PHOSPHINE HYDRIDES. <u>S. M. Reid</u>, R. R. Schrock, W. M. Davis, Department of Chemistry, Massachusetts Institute of Technology, Cambridge, Massachusetts 02139

The smooth reduction of the trigonal bipyramidal complex $\{[N_3N]ReBr\}Br$ by Mg in the presence of a variety of donor ligands gave robust rhenium (III) amide complexes, $[N_3N]Re(L)$, one of which gives rise to H/D exchange ($[N_3N] = (C_6F_5NCH_2CH_2)_3N^{3-}$). Ligands L may be phosphines, pyridine, tetrahydrothiophene, acetonitrile, N_2, H_2, and CO, olefins, and silanes.

Several of these complexes were protonated to give Re(V) species of the general form $\{[N_3N]Re(L)H)\}BAr_4^F$. Spectroscopic and X-ray crystallographic studies reveal that the unusual bonding of the hydride is governed by the nature of L.

687.
NICKEL(0), COPPER(I), AND ZINC(II) COMPLEXES WITH 2,9-DIMETHYL-1,10-PHENANTHROLINE - AN ISOELECTRONIC ISOSTRUCTURAL SERIES. **Alexander J. Pallenberg** Thomas M. Marschner, and David M. Barnhart, ProCyte Corporation, 12040 115th Ave. NE, Suite 210, Kirkland, WA 98034

Structures and physical properties of dmphen (2,9-dimethyl-1,10-phenanthroline) complexes with d^{10} metals Zn(II), Ni(0), and Ag(I) were determined. These, together with the more well known Cu(I) compounds, constitute a rare example of a Ni(0), Cu(I), Zn(II) isoelectronic, isostructural series. The first homoleptic zinc dmphen compounds, $[ZnNO_3(dmphen)_2]NO_3$ and $Zn(dmphen)_2(BF_4)_2$, were prepared and analyzed by X-ray crystallography. The nitrate is five-coordinate, with one nitrate ion coordinated. The tetrafluoroborate, of the same stoichiometry, exhibits the pseudotetrahedral $Zn(dmphen)_2^{2+}$ cation, which is isoelectronic and isostructural with the more well known $Cu(dmphen)_2^+$ cation. $Ni(dmphen)_2$ was prepared for the first time. NMR experiments show that this compound is diamagnetic, with chemically equivalent dmphen ligands. Finally, $Ag(dmphen)_2BF_4$ was prepared, which extends the isostructural series described above to a second row transition metal. The structure of this compound was determined by X-ray crystallography.

688.
ORGANOHYDRAZIDE CHEMISTRY OF GROUP 7 METALS. **Terrence Nicholson**, Melissa Hirsch-Kuchma, Alan Davison and Alun G. Jones. Department of Chemistry, Massachusetts Institute of Technology, Cambridge, MA 02139 and Department of Radiology, Harvard Medical School, Boston, MA 02115.

The organohydrazines are proving to be very useful ligands for the reduction of pertechnetate and perrhenate. These precursors, in oxidation state VII, are effectively reduced by four electrons to give the Tc(III) and Re(III) complexes $[MCl_3(HN=NC_5H_4N)(N=NC_5H_4NH)]$, {where M = Tc, Re}; which contain two of the organohydrazide ligands. Concomitant to the metal's reduction is a two electron oxidation of each of these two ligands, which coordinate in an oxidized form. This four electron transfer reflects the strongly reducing nature of the organohydrazide precursor, a feature shared with the parent molecule hydrazine. The two forms of the coordinated organohydrazide ligands are a neutral, bidentate organodiazene ligand which is chelated through the pyridine nitrogen atom and a neutral, but zwitterionic, pyridiniumdiazenido ligand. We intend to elaborate on this chemistry as well as discuss the substitution chemistry of these robust synthetic precursors with phosphines.

689.
RHENIUM AND TECHNETIUM ORGANOHYDRAZIDE COMPLEXES FORMED FROM THE REACTION OF HYDRAZINOPYRIMIDINE. **Melissa Hirsch-Kuchma,** Terrence Nicholson, Alan Davison, and Alun G. Jones; Department of Chemistry, Massachusetts Institute of Technology, Cambridge, MA, 02139 and Department of Radiology, Harvard Medical School, Boston, MA 02115.

We have been exploring the chemistry of potentially chelating organohydrazines with rhenium and technetium metal centers. The organohydrazide complexes are formed by a two electron reduction of the metal by the organohydrazine 2-hydrazino-4(trifluoromethyl)pyrimidine. In addition to the redox chemistry, these diamagnetic, d^4 octahedral complexes display some interesting electronic characteristics. The organohydrazide ligand on these complexes can coordinate in both monodentate and bidentate fashions and in protonated and unprotonated forms. Depending on the degree of protonation, the organohydrazide acts as a neutral or monoanionic ligand.

690.
ACTIVATION OF MOLECULAR OXYGEN BY MNII COMPLEX. ISOLATION OF A STABLE TRIGONAL BIPYRAMIDAL MNIII-(OH) COMPLEX. **Zahida Shirin,**¶ Victor G. Young, Jr.,‡ and A. S. Borovik,*¶. ¶ Department of Chemistry, University of Kansas, Lawrence, KS 66046, ‡ Department of Chemistry, University of Minnesota, Minneapolis, MN 55455

The activation of molecular oxygen by manganese ions is an important process in biological and catalytic processes. We have isolated and characterized a stable monomeric Mn(III) complex with a terminally bonded hydroxo ligand whose oxygen atom is derived from molecular oxygen. A mechanism is proposed that includes the formation of a dinuclear peroxo-bridged Mn(III) intermediate. Homolytic cleavage of the O-O bond generates a metal-oxo intermediate which is competant to abstract hydrogen atoms from the solvent. The observed reactivity at the Mn(II) ion is controlled, in part, by the three-fold symmetric cavity formed by a new tris-N-amidate tripodal ligand, tris(N-cyclopentylcarbamoylmethyl)amine. We will discuss in this talk the reactivity of Mn(II) ion with oxygen and characterization of intermediates.

691.
EVIDENCE FOR THE TEMPLATE EFFECT AND SITE ISOLATION WITHIN METALLO-NETWORK POLYMERS: THE USE OF COBALT(III) SCHIFF BASE COMPLEXES AS CO-MONOMERS. **John F. Krebs,** A. S. Borovik* Department of Chemistry, University of Kansas, Lawrence, KS, 66045

Copolymerization of molecular assemblies into porous organic hosts is an effective way to fabricate new materials having desirable properties. The assembly of the molecular species prior to copolymerization is advantageous because of the greater control of the structure and amount of species incorporated into the polymer and the possibility of regulating their microenviroments. Cobalt Schiff base complexes are being used, through formation of the Co(III)-superoxide species, to probe the effect of ligands on the architecture of the metal sites within the porous organic host. Evidence for site isolation and the relation of oxygen binding capacity of the polymers with respect to ligand orientation will be presented.

692. IRON(III) SALICYLALDIMINATO PRECURSORS: THE NEUTRAL SALICYLALDIMINE COMPLEX Fe(salHNR)Cl$_3$. Robert H. Beer[1] and Jennifer L. Kisko[2], Departments of Chemistry, [1]Fordham University, Bronx, NY 10458 and [2]Columbia University, New York, NY 10027

The reaction of FeCl$_3$ with N-alkylsalicylaldimine (salHNR) forms Fe(salHNR)Cl$_3$ in which salHNR acts atypically as a monodentate donor. A crystallographic study reveals the ligand as a phenolate oxygen bound to a tetrahedral iron center with a protonated non-coordinating imine. When Et$_3$N is added to Fe(salNR)Cl$_3$ or included in its synthesis, the salicylaldiminato complex Fe(salNR)$_2$Cl is formed. The zwitterionic form of salHNR is, therefore, likely to be an important coordination mode in intermediates formed during the reaction between FeCl$_3$ and Schiff base ligands.

693.
SYNTHESIS, CHARACTERIZATION AND REACTIVITY OF EARLY TRANSITION METAL NEO-PENTOXIDES. Timothy J. Boyle[*], Todd M. Alam, Eric R. Mechenbeir, Brian Scott, and Joseph W. Ziller Sandia National Laboratories, Advanced Materials Laboratory 1001 University Blvd. SE, Albuquerque, NM 87106[≠].

Titanium neo-pentoxide was isolated by the alcoholysis exchange between $Ti(OCHMe_2)_4$ and ONep. The molecule, $[Ti(ONep)_4]_2$, was characterized using X-ray analysis and solution ^{17}O, and $^{47,49}Ti$ NMR spectroscopy. This dinuclear complex is the smallest $Ti(OR)_4$ isolated wherein each metal center is 5-coordinated. The molecule is highly soluble in standard solvents and volatile. The reactivity of this compound has been undertaken to compare with the ubiquitous $Ti(O\text{-}i\text{-}Pr)_4$. The various compounds isolated and further NMR studies will be reported. Analogous routes to other $M(ONep)_n$ will be reported as well.

[≠] This work was supported by the U.S.D of E. under contract DE.-AC04-94AL85000.

694. ELECTROCHEMICAL AND MAGNETIC PROPERTIES OF THE NOVEL MOLYBDENUM(III) COMPLEX, $[Et_4N][HB(pz)_3MoBr_3]$. S. E. Lincoln, H. Voorhies, K. Donohoue, L. Kovacevic, University of Portland, Portland, Oregon, 97203

Examples of Mo(III) complexes remain rather uncommon in the literature due to their instability towards oxidation. Earlier work of the author reported that coordination of the trispyrazolylborato ligand can stabilize Molybdenum(III), allowing the formation of unusual air-stable complexes. We now report the characterization of electrochemical and magnetic properties of the novel complex, $[NEt_4][HB(pz)_3MoBr_3]$ **1**, synthesized in our laboratory. Magnetic susceptibility measurements yield a magnetic moment of $3.81\mu_B$. The magnetic susceptibility and electronic spectrum of **1** is consistent with an $S = 3/2$ ground state. Cyclic voltammograms of **1** in CH_3CN give a reversible oxidation at 0.76 V [Mo(III)/Mo(IV)] and a non-reversible oxidation at 1.78 V [Mo(IV)/Mo(V)] vs Ag/AgCl. Square wave analysis confirms the [Mo(III)/Mo(IV)] potential at 0.76 V. The high positive Mo(III)/Mo(IV) potential of 0.76 V explains the robust nature of $[NEt_4][HB(pz)_3MoBr_3]$, which resists air oxidation both as a solid and in solution. Further studies are in progress to synthesize Mo(III) complexes containing substituted $HB(pz)_3$ ligands and to compare their electrochemical properties.

695. METAL COMPLEXES OF A NEW BIS(SALICYLIDENE) BICYCLIC DIAMINE LIGAND. Mark P. Sweet, Kathleen M. Barkigia,[+] Ronnie C. Mease and Suresh C. Srivastava. Brookhaven National Laboratory, Medical Department, Building 801,[+] Department of Applied Science, Building 815, P.O. Box 5000, Upton, NY 11973-5000.

The rigidity of cis-endo-2,3-diaminobicyclo[2.2.2]oct-5-ene (**1**) locks the nitrogens into spatial positions that are favorable to metal binding and should result in thermodynamically and kinetically more stable metal ligand complexes. Condensation of **1** and salicylaldehyde affords the N,N'-bis(salicyclidene)-cis-endo-2,3-diamino-bicyclo[2.2.2]oct-5-ene (**2**) in high yield. X-ray diffraction of **2** shows that the nitrogens are cis and endo and that the phenolic hydrogens are hydrogen bonded to the nitrogens. The ligand **2** was found to form neutral complexes with divalent metals such as cobalt, nickel, and copper. X-ray crystal studies of the nickel complex (which contains one $CHCl_3$ per Ni) showed a square planar arrangement of the coordinating groups around the metal. The nickel and the salicyclidene moieties roughly lie in the plane defined by carbons 2 and 3 and the two nitrogens. This work was supported by both the OHER and the Division of Chemical Science, U.S. DOE under Contract # DE-AC02-76CH00016.

696. SPECTROSCOPIC SIGNATURES OF THE HIGH-VALENT Fe_2O_2 DIAMOND CORE. Lawrence Que, Jr., Elizabeth C. Wilkinson, Lijin Shu, Yanhong Dong, and Yan Zang, Department of Chemistry, University of Minnesota, 207 Pleasant St. SE., Minneapolis MN 55455

A $Fe_2(\mu-O)_2$ diamond core structure has been proposed as the common high-valent intermediate in the redox cycles of nonheme diiron enzymes such as methane monooxygenase, ribonucleotide reductase, and fatty acid desaturase (Que and Dong, *Accts. Chem. Res.* **1996**, *29*, 190-196). This hypothesis is based on synthetic precedents such as $[Fe_2(\mu-O)_2L_2]^{3+}$, where L is a tetradentate tripodal pryridine ligand. Such diamond cores can be recognized by two spectroscopic signatures: a Raman vibration at 600-700 cm^{-1} arising from a breathing mode of the diamond core and an Fe-Fe distance of less than 3 Å as observed by EXAFS. Analysis of the spectroscopic properties of the synthetic precedents and their implications to the structures of high-valent intermediates in the enzyme cycles will be discussed.

697. PROTON CONTROL IN THE REDUCTION OF O_2 BY CYTOCHROME OXIDASE. Gerald T. Babcock, Michelle Pressler, and Denis Proshlyakov, The LASER Laboratory and Department of Chemistry, Michigan State University, East Lansing, Michigan 48824

Cytochrome oxidase is the terminal protein in respiratory electron transfer and activates O_2 for reduction to water. Recent work has shown that, with this enzyme, rate limitation in the activation, reduction, and protonation of dioxygen in the Fe/Cu binuclear active site of the enzyme is conferred by the proton-transfer reactions.[1] Physiologically, this situation insures efficient coupling of the oxidase proton pump to the driving O_2 reduction chemistry. A manifestation of this kinetic scheme is that intermediate species in the redox reaction build up to detectable levels during the activation process and can be detected by a number of time-resolved spectroscopic techniques. We have used time-resolved resonance Raman spectroscopy in combination with Gibson-Greenwood flow flash to take advantage of this situation and to determine the structure of the various intermediates that occur and their temporal behavior. Several intermediates have been assigned, both by us and others, and consensus has emerged on these species.[1,2] At least one intermediate, which has an oxygen-isotope sensitive vibration at 356 cm^{-1}, remains enigmatic, and recent work on this species will be summarized.

1. Ferguson-Miller, S. and Babcock, G. T., *Chem. Revs.*, in press.
2. Kitagawa, T. and Oguri, T., *Prog. Inorg. Chem.*, in press.

698. INTERMEDIATES IN NON-HEME IRON DIOXYGENASE CATALYSIS. John D. Lipscomb, Allen M. Orville, Richard W. Frazee, and Douglas H. Ohlendorf, Department of Biochemistry, Medical School, University of Minnesota, Minneapolis, Minnesota 55455.

Protocatechuate 3,4 dioxygenase (3,4-PCD) sequentially binds its substrates 3,4-$(OH)_2$-benzoate (PCA) and O_2 before catalyzing oxidative aromatic ring fissure. Crystal structures of 3,4-PCD complexes with 9 aromatic inhibitors, 2 substrates, and CN$^-$ have been solved and reveal likely intermediates in the progress of substrate binding to the active site Fe^{3+}in accord with previous EPR, Mössbauer, EXAFS, resonance Raman, and CD/MCD studies. The trigonal bipyramidal Fe^{3+}site in resting 3,4-PCD has Tyr and His ligands in axial positions and His, Tyr, OH$^-$ in the equatorial positions. It is hypothesized that PCA first binds to the active site, but not the iron. Then, PCA moves closer to the Fe^{3+}so that the C4-O$^-$ binds to the Fe^{3+} and the C3-OH binds in a H-bonding pocket. This results in a shift to octahedral Fe^{3+}coordination with OH_2 and the PCA C4-O$^-$ in the equatorial plane. Subsequently, PCA slides "upward" so that: i) the C3-O$^-$ becomes the equatorial ligand, ii) the C4-O$^-$ displaces the axial Tyr to form an Fe-PCA chelate, and iii) simultaneous steric displacement of the OH_2 and 90° rotation of the displaced Tyr opens a space for O_2 next to PCA. It is postulated that electrophilic O_2 attack on the dianionic PCA leads to oxygen insertion and ring opening. As a test, a mutant 3,4-PCD was constructed in which the axial Tyr ligand was replaced by His, which cannot bind well in the alternate binding position of the Tyr. 1000-fold slower turnover was observed. Stopped flow kinetics showed that the decreased rate was due to much slower substrate binding and product release steps. Spectroscopic evidence for each of the proposed intermediates in the reaction cycle was obtained.

699. MÖSSBAUER CHARACTERIZATION OF DINUCLEAR IRON-CLUSTER INTERMEDIATES IN BIOLOGICAL OXYGEN ACTIVATION REACTIONS. Boi Hanh Huynh, Department of Physics, Emory University, Atlanta, GA 30322

Rapid freeze-quench method was used to trap reaction intermediates in the reaction of O_2 with the dinuclear iron clusters in the soluble methane monooxygenase from *Methylococcus capsulatus* (Bath) and in the R2 subunit of the ribonucleotide reductase from *Escherichia coli*. Mössbauer and EPR spectroscopies were used to characterize these intermediates. For methane monooxygenase, the initial intermediate formed on reaction of O_2 with the diferrous cluster is a diferric peroxo complex which undergoes further structural and electron reorganization to form a high-valent, formally diferryl, complex termed Q. In the case of R2, reaction of O_2 with the diferrous center in the presence of reducing agents generated a mixed valent, formally ferric-ferryl, complex termed X, which is capable of oxidizing the proximal Y122 to its radical form with formation of the resting diferric cluster. Based on these observations, a unified scheme for the activation of O_2 by diiron cluster-containing proteins is proposed. Site-selective mutants of R2 were engineered and produced to test the proposed mechanism. Spectroscopic properties of the diiron intermediates and preliminary data on the mutant studies will be presented. (This work involves research teams from the laboratories of Drs. J. Martin Bollinger, Jr., Dale E. Edmondson, Stephen J. Lippard, JoAnne Stubbe and the presenting author.)

700. **OXYGEN INTERMEDIATES IN COPPER ENZYMES: GEOMETRIC AND ELECTRONIC STRUCTURE/FUNCTION CORRELATIONS**
Edward I. Solomon, Department of Chemistry, Stanford University, Stanford, CA 94305-5080

Laccase, Ascorbate Oxidase and Ceruloplasmin couple four 1 electron oxidations of substrate to the four electron reduction of dioxygen to water. These enzymes contain a trinuclear copper cluster site (comprised of a coupled binuclear Type III center and a Type II center) and additional blue copper centers. Spectroscopic studies on the Type III center demonstrate that it is fundamentally different from the coupled binuclear copper site in hemocyanin and tyrosinase and that the trinuclear cluster is the minimum structural unit required for dioxygen reduction. This reaction generates a peroxy-trinuclear copper intermediate that has been extensively spectroscopically studied (by absorption, EPR, SQUID magnetic susceptibility, variable-temperature MCD, isotope ratio mass spec., and XAS) and found to have a different geometric and electronic structure from the peroxide intermediate formed in the reversible binding and activation of dioxygen by hemocyanin and tyrosinase. The peroxy-trinuclear cluster intermediate is further reduced to generate a second intermediate which had been thought to be a three electron reduced oxygen radical, but is in fact the four electron reduced product which decays to the resting enzyme. These studies have defined geometric and electronic structure/function differences for activation versus reduction of peroxide and generated a detailed molecular mechanism for the four electron reduction of dioxygen to water by the multicopper oxidases.

701. MIXED VALENT MANGANESE PEROVSKITES : RELATIONS BETWEEN STRUCTURE AND COLOSSAL MAGNETO RESISTANCE PROPERTIES. B. Raveau, A. Maignan, C. Martin, V. Caignaert, M. Hervieu and Ch. Simon, Laboratoire CRISMAT, ISMRA-Université de Caen, 6 boulevard du Maréchal Juin 14050 CAEN Cédex - France.

The perovskite manganites $Ln_{1-x}A_xMnO_3$ (A = Ca, Sr, Ba) have been extensively studied these last three years for their colossal magnetoresistance (CMR) properties, showing that resistance ratios R_0/R_H, can reach values of several orders of magnitude. We discuss in a first step two factors which play a prominent role in the magnetic and transport properties of these materials : the hole career density which corresponds to the mixed valence Mn^{3+}/Mn^{4+} and the average size of the interpolated cation (A, Ln). The neutron diffraction study of these compounds versus temperature, coupled with high resolution electron microscopy sheds light on several features that are of capital importance for the understanding of the properties of these materials : variation of the Jahn Teller effect at the transition, existence of local charge ordering at room temperature, evidence of local charge ordering at low temperature for several of these compounds such as the "x = 1/2" phases, but also for $Nd_{0.7}Ca_{0.3}MnO_3$. In a last step, the doping of the Mn sites with cations of various valence is studied. The substitution of divalent and trivalent cations for manganese in the oxide $Pr_{0.5}Sr_{0.5}MnO_3$ leads to an extension of the antiferromagnetic state at the expense of the ferromagnetic state, whereas the doping with tetravalent element suppresses the AFM state, leading to a FMM-PMI transition. The doping of the insulating phase $Pr_{0.60}Ca_{0.40}MnO_3$ is spectacular : a PMI to FMM transition is induced in this phase, in the absence of a magnetic field by doping the Mn site with a trivalent cation such as Fe, Al, Ga, and by a divalent cation such as Mg. The appearance of such an effect for a phase characterized by a small size A cation (Pr, Ca) is compared to the field induced metal-insulation transition observed in several manganites.

702. MAGNETIC AND PHOTONIC PHASE-CONTROL IN PEROVSKITE MANGANITES. Y.Tokura, Department of Applied Physics, University of Tokyo, Tokyo 113

Strong coupling among spin, charge, orbital, and lattice degrees of freedom produces a variety of electronic and magnetic phases in perovskite manganites. One such example is a novel competition between the ferromagnetic metallic and charge-ordered insulating states. An external magnetic field can cause the collapse of charge-ordering, namely the 1:1 real-space ordering of Mn^{3+} and Mn^{4+}, into the ferromagnetic metallic state, producing a gigantic change in resistance of more than several orders of magnitude. In addition to chemical and magnetic control of the electronic phases, we report here on our latest attempts of exotic phase-control, for example by means of current injection, X-ray irradiation, and laser pulse irradiation.

This work was done in collaboration with Y.Tomioka, H.Kuwahara, A.Asamitsu, K.Miyano, and B.Keimer, and was supported in part by NEDO and the Ministry of Education.

703. ELECTRICAL PROPERTIES OF $NiS_{2-x}Se_x$ SINGLE CRYSTALS: FROM MOTT INSULATOR TO PARAMAGNETIC METAL. X. Yao* and J. M. Honig. Department of Chemistry, Purdue University, West Lafayette, IN 47907-1393. T. Hogan and C. Kannewurf. Department of EESC, Northwestern University, Evanston, IL 60208. J. Spałek. Institute of Physics, Jagiellonian University, Ulica Reymonta 4, 30-059 Kraków, Poland.

Electrical resistivity (ρ) and Seebeck coefficient measurements (α) are reported for $NiS_{2-x}Se_x$ single crystals in the range $0 \le x \le 0.71$. There is a general trend toward increasing metallicity with increasing x. In the range $0.38 \le x \le 0.51$ a pronounced rise of ρ with temperature (T) is observed where the antiferromagnetic insulating phase changes over to the paramagnetic insulating phase. The analysis of α vs. T curves suggests that in the low temperature insulating or metallic state both holes and electrons participate in charge transport. It is emphasized that the many changes in electrical characteristics occur without significant alterations in the pyrite crystal structure, and that physical properties are greatly altered by adjustment of the anion sublattice while the cation sublattice remains intact. The results concerning electrical transport are interpreted qualitatively on the basis of a Hubbard splitting of the subbands at the transition to the paramagnetic insulating (semiconducting) state.

* Present Address: Beckman Institute, University of Illinois, 405 N. Mathews Avenue, Urbana, IL 61801

704. NONSTOICHIOMETRY AND DOPING OF ZINC OXIDE. Arthur W. Sleight and Ruiping Wang, Department of Chemistry, Oregon State University, Corvallis, OR 97331-4003.

When zinc oxide is heated under highly reducing conditions, it becomes red and develops significant electrical conductivity. The usual explanation is that zinc interstitials have formed in red zinc oxide with the composition $Zn_{1.0003}O$. The conduction is then explained assuming that the Zn interstitials are singly or doubly ionized. Based on our studies, we conclude that the Zn interstitials are not ionized. Instead, the 4s electrons associated with a Zn interstitial are trapped in an electron pair bond between a zinc atom at a normal lattice site and an interstitial, i.e., a $(Zn_2)^{2+}$ cluster forms. The red color can be due to excitation from this deep trap to the conduction band. The rise in conductivity when ZnO is reduced is due to impurities such as Si and Al. These impurities are compensated by extra oxygen unless it is removed by reduction. Purposely doping of zinc oxide with F, B, Al, In, or Ge can produce highly conducting materials. We have obtained a room temperature conductivity of 300 S/cm in ZnO:Ga powders. This is 1000 times higher than previously reported for a doped zinc oxide powder.

705. **FERROMAGNETISM AND SPIN GLASS BEHAVIOR IN ISOMORPHOUS TETRAAZAPORPHYRINS. Gordon T. Yee**, Brenda J. Conklin and Scott P. Sellers, Department of Chemistry and Biochemistry, University of Colorado, Boulder, CO, 80309

The magnetic properties of a family of square planar tetraazaporphyrin complexes have been determined by ac and dc SQUID magnetometry. Most interesting are those of iron octaethyltetraazaporphyrin which is a canted ferromagnet with Tc = 5 K and its manganese analog which shows complex spin-glass-like behavior. Powder diffraction has been used to show that these two phases are isostructural. Also, an additional phase exists for each of thse compounds which shows no cooperative magnetic properties. These results are compared to those of the corresponding phthalocyanine complexes.

706. HETEROSUPRAMOLECULAR CHEMISTRY : NON-COVALENT SELF-ASSEMBLY AND SELF-ORGANISATION OF MOLECULAR SCALE DEVICES CONTAINING MOLECULAR AND NANOCRYSTALLITE COMPONENTS. L. Cusack, A. Gorelov, X. Marguerettaz, N. Rao, R. Rizza, J. Wenger and D. Fitzmaurice, Department of Chemistry, University College Dublin, Belfield Dublin 4, Ireland.

Recent years have seen the development of powerful techniques and methodologies for the self-assembly and self-organisation of supermolecules and for the preparation of metal and semiconductor nanocrystallites. Together, these developments allow us explore the possibility of self-assembling and self-organising complex structures containing both molecules and nanocrystallites, referred to as heterosupermolecules. This talk will describe the self-assembly of a heterosupramolecular electron donor-acceptor complex in solution from a surface modified semiconductor nanocrystallite that can recognise and selectively bind, by complementary hydrogen bonding, an appropriately modified molecular acceptor in solution. Band-gap excitation of the semiconductor nanocrystallite results in electron transfer to the acceptor in the heterosupermolecule. Also described will be the extension of these studies to permit the self-assembly in solution of two semiconductor nanocrystallites, again by complementary hydrogen bonding, and their subsequent self-organisation to form an ordered nanocrystallite array, or superlattice. Some possibly general advantages of heterosupramolecular chemistry will be discussed.

707. **A NEW APPROACH TO COMPOSITES OF SEMICONDUCTOR NANOPARTICLES WITH POLYMERS.** David J. Cole-Hamilton,[a]* Steven Haggata,[a] Nigel L. Pickett,[a] Douglas F. Foster,[a] and John R. Fryer,[b] [a]*School of Chemistry, University of St. Andrews, St. Andrews, Scotland, and* [b]*Chemistry of Glasgow, Glasgow, Scotland.*

MOVPE growth of Zn/CdSe using Me_2M and H_2Se is plagued by problems of prereaction which leads to deposition of the semi-conductor all over the reactor. Addition of pyridine to the gas phase can control this to some extent. It will be shown that the action of the pyridine involves controlling particle growth in the gas phase leading to the production of nanoparticle material. This understanding has led to the design a new solution based process to the production of nanoparticles (quantum dots) embodied in a polymer matrix, in which the size distribution is relatively narrow and the average size can be controlled. It has been necessary to design new polymers by the catalytic modification of polybutadienes. These polymers act to control the particle growth and encapsulate the final particles. Because of low-lying acceptor orbitals in the pyridine based polymers, we have developed soluble polymeric analogs of phosphine oxides since free phosphine oxides are known to be excellent for stabilizing particle surfaces and for providing an insulating sheath that allows light emission. Preliminary light absorption and emission studies of these particles will be discussed.

708. **NEW ROUTES TO NITRIDES OF GALLIUM AND TITANIUM** C. J. Carmault, A. H. Cowley, G. Hair and R. A. Jones,* *Department of Chemistry and Biochemistry, The University of Texas at Austin, Austin, TX 78712*

Several new molecules which provide new routes to the solid state materials GaN and TiN will be described. A general feature is the use of the azide (N_3^-) ligand which can eliminate dinitrogen and acts as the source of nitrogen in the final material. Examples of these molecules include $Ga(N_3)_3L_3$ (L = pyridine, PR_3, AsR_3 etc.), $[Ti(NMe_2)_2(N_3)_2]_n$ and $Ti(NMe_2)_2(N_3)_2(py)_2$. Use of the compounds for the deposition of thin films of GaN and TiN will be described.

709. **MACROCYCLIC MULTIDENTATE PHOSPHINES: SYNTHESIS AND COMPLEXES.** Peter G. Edwards*, James S. Fleming, Florent Ingold and Sudantha Liyanage. *Department of Chemistry, University of Wales, Cardiff, P. O. Box 912, Cardiff CF1 3TB UK.*

Selective oxidation of the Cr/Mo(0) 12-aneP$_3$R$_3$ template complexes allows the stereospecific and high yield liberation of the free 12-aneP$_3$R$_3$ macrocycles, thus constituting the first steroselective synthesis of triphosphorus macrocycles. The functions (R) are readily varied allowing the selective incorporation of alkyl groups as well as pendant functions including ethers, thioethers, amines and phosphines. Structural data indicate the new macrocycles to be flexible and capable of facially capping co-ordination polyhedra of metals of widely varying radii. Simple complexation reactions with a range of metal halide and organometallic precursors give rise to a series of new complexes. The synthesis, structural and other properties of the macrocyles and complexes will be discussed.

710. **NEW FIVE-, SIX-, AND SEVEN-COORDINATE ORGANOZIRCONIUM COMPLEXES.** Gregory S. Girolami and Melissa J. Nelsen, School of Chemical Sciences, University of Illinois at Urbana-Champaign, Urbana, IL 61801

In 1989, we reported the synthesis and structure of [Li(tmed)]$_2$[ZrMe$_6$], the first example of a trigonal prismatic compound with six unidentate, electronically innocent ligands. In order to further explore the geometries of six-coordinate d^0 complexes of the transition metals, we have synthesized the new phenyl complex [Li(Et$_2$O)]$_2$[ZrPh$_6$]. Despite the presence of orbitals on the phenyl ligands that are of the right symmetry to serve as π-donors, the trigonal prismatic geometry of the [ZrPh$_6^{2-}$] anion clearly shows that the phenyl groups in this molecule are not acting as π-donors. Attempts to prepare related pentafluorophenyl complexes led instead to the isolation of [Li(12-crown-4)$_2$]$_2$[ZrCl$_2$(C$_6$F$_5$)$_4$] and [Li(tmed)$_2$][Li(tmed)]$_2$-[ZrF$_2$(C$_6$F$_5$)$_5$]; the first adopts an octahedral structure owing to the π-donor character of the chloride ligands, while the latter adopts a regular pentagonal bipyramidal structure. Addition of LiCH$_2$SiMe$_3$ to the phosphine complexes ZrCl$_4$(PP) and HfCl$_4$(PP), where PP is 1,2-bis(dicyclohexylphosphino)ethane (dcype) or 1,2-bis(diisopropylphosphino)ethane (dippe), affords an unusual five-coordinate zirconium alkyl: [Li(dcype)$_2$]-[Zr(CH$_2$SiMe$_3$)$_5$]. The structure of the anion is best described as a distorted square pyramid; the cation consists of a lithium atom surrounded by a distorted tetrahedral array of two diphosphine ligands.

711. **SYNTHESIS, CHARACTERIZATION AND CATALYTIC BEHAVIOR OF IRON/ZINC SPECIES INVOLVED IN GIF-TYPE OXIDATION OF HYDROCARBONS.** Bharat Singh,[a] Jeffrey R. Long,[b] Georgia C. Papaefthymiou,[c] and Pericles Stavropoulos[a]. *[a]Department of Chemistry, Boston University, Boston, MA 02215; [b]Department of Chemistry, Harvard University, Cambridge, MA 02138; [c]Francis Bitter Magnet Laboratory, MIT, Cambridge, MA 02139.*

This study explores structural/functional similarities as well as discrepancies between iron species mediating oxidation of hydrocarbons in Gif-type systems and analogous iron sites in hydroxylase components of biological systems. This exploration reveals that dinuclear iron units are key components of both oxidized and reduced iron-containing sites in Gif solutions. These sites consistently interconvert upon chemical redox cycles and may thus sustain catalytic turnovers. Catalytic oxidation of hydrocarbons, under GIF^{IV} conditions (Zn/O$_2$), affords product profiles and kinetic isotope effects that will be contrasted to those obtained with biological and artificial hydroxylases in light of the continuing debate with respect to the active oxidant(s) involved in Gif-type systems.

712. **CHEMISTRY AND CATALYSIS OF Mo(II) and Mo(III) COMPOUNDS CONTAINING ONLY METAL-CARBON BONDS.** Rinaldo Poli* and Li-Sheng Wang, *Department of Chemistry and Biochemistry, University of Maryland, College Park, MD 20742.*

Reductive allylation of CpMoCl$_2$(η^4-C$_4$H$_6$) affords CpMo(η^3-C$_3$H$_5$)(η^4-C$_4$H$_6$), **1**, the simplest possible compound containing a 3-, a 4-, and a 5-electron π ligand on the same transition metal. Compound **1** has been observed in three distinct isomeric forms with different relative configurations of the allyl and diene ligands, each of which has been isolated and studied independently. One-electron oxidation of these affords four different observed isomers of [**1**]$^+$. Isomerization processes involving partial ligand dissociation are faster for [**1**]$^+$ than for **1**. The catalytic activity of **1** and [**1**]$^+$ for butadiene dimerization and polymerization will be shown, as well as their reactivity with HX and with neutral ligands.

713. **DESIGN OF LATENT LEWIS ACID COMPOUNDS OF ALUMINUM: INTER- VERSUS INTRA- MOLECULAR STABILIZATION** Julie A. Francis,[a] C. Niamh McMahon,[a] Simon G. Bott,[b] and Andrew R. Barron[a]* *(a) Department of Chemistry, Rice University, Houston, Texas 77005 (b) Department of Chemistry, University of North Texas, Denton, TX 76203.*

The role of alumoxane co-catalysts in the Ziegler-Natta type polymerization of olefins is derived from the cage's *latent Lewis acidity*. If new catalyst and co-catalyst systems are to be designed, then the molecules must show high latent Lewis acidity. We will report on a catalyst design based on latent Lewis acidity. Intramolecularly stabilized heterocyclic compounds of the general formula, $R_2Al[X'(CH_2)_nXR'_n]$ (I) where R, R' = alkyl, X' = CH_2, O; X = N, P, As, O, S. The intra- versus inter-molecular stabilization is controlled by the choice of substituents (X, E, R, and R') and the ring strain (as controlled by n). The catalytic activity of these compounds will be investigated.

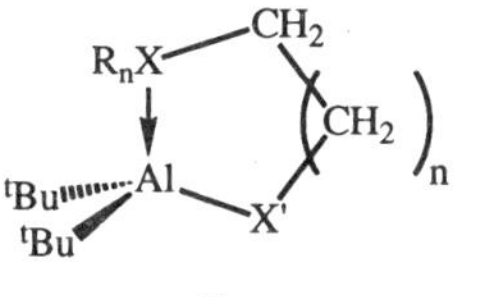

714.

HOMOLEPTIC TRANSITION METAL TELLUROLATES AND SELENOLATES. John Arnold, Christopher P. Gerlach, and Stephen P. Wuller. Department of Chemistry, University of California, Berkeley, CA 94720-1460

The synthesis, structure and reactivity of a range of complexes of general formula $M(ER)_n$, where M = a transition metal, E = Se and Te, and R = alkyl and silyl, will be discussed and the X-ray structures of a range of d^0, d^1, and d^2 derivatives will be presented. A view of the X-ray crystal structure of $V[TeSi(SiMe_3)_3]_4$ is shown here.

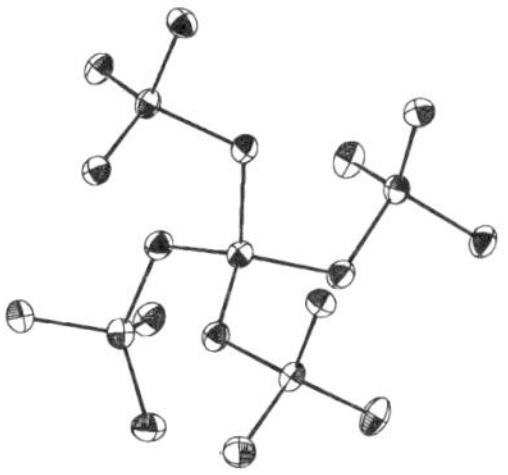

715. **HIGH OXIDATION STATE IMIDO COMPOUNDS OF MANGANESE** Andreas Danopoulos and Geoffrey Wilkinson, *Department of Chemistry, Imperial College of Science, Technology and Medicine, South Kensington, London, SW7 2AY, UK.*

The synthesis of $Mn(NBu)_3Cl$ (**1**) the first non-oxo compound of Mn(VII), opened an interesting new area of imido chemistry of manganese in oxidation states V-VII. Chloride substitution in compound **1** gave phenoxides, thiophenoxides, amides and aryls of Mn(VII). Reduction under specific conditions afforded $Mn(NBu)_4Li_2(dme)_2$, the imido analogue of manganate dianion, or $Mn_2(NBu)_6$, a homoleptic Mn(VI) dimer. Interaction of compound **1** with BuN(H)Li gave $[Mn(N)(NBu)_3]Li_2$ and alkylations with R_2Zn gave dimeric alkyls of Mn(V). Methods leading to imido bridged linear oligomers have been found. Structural data of some of the new compounds as well as mechanistic considerations will be discussed.

716. METAL DEPENDENT FOLDING AND ASSEMBLY OF NATURAL AND DESIGNED PROTEINS. George McLendon, Department of Chemistry, Princeton University, Princeton, NJ 08544

Thermodynamic data are reported for metal binding by zinc dependent nucleic acid proteins, as well as for metal templated designed proteins. Biological and chemical principles of metal recognition will be discussed.

717. CONFORMATIONALLY-CONSTRAINED OCTADENTATE LIGANDS BASED UPON ALL CIS-TETRASUBSTITUTED CYCLOPENTANES. Takenori Tomohiro[a], Pamela Orchard[ab], Horst Puschmann[a], and Stephen Cooper[ac], [a]Inorganic Chemistry Laboratory, University of Oxford, Oxford OX1 3QR, England. [b]Amersham International plc, White Lion Road, Amersham HP7 9LL, England. [c]Mallinckrodt Inc., 675 McDonnell Blvd., Hazelwood, MO 63134.

Medical applications of coordination complexes typically require exquisitely high stability constants to preclude deposition of metal ions in the body. Few such ligands exist for metals with coordination numbers higher than six. We will present our work on the all cis-cyclopentane-1,2,3,4-tetracarboxylic acid framework as a conformationally-constrained platform for assembly of pre-organized octadenate ligands for such medically important metal ions as Gd(III) and In(III).

718. PYRIDOXAL AND PYRIDOXAMINE CATALYSIS IN SYNTHETIC BILAYER MEMBRANES: METAL ION EFFECTS. Yukito Murakami, Department of Chemical Science and Technology, Kyushu University, Fukuoka 812-81, Japan

Model studies on the vitamin B_6-dependent enzymes have a long history extended over a half century, originating from pioneering work by Snell and his co-workers, and metal ions coordinated to the intermediate Schiff base derived from an amino acid are capable of promoting the vitamin B_6-dependent reactions non-enzymatically in the absence of an apoprotein as originally demonstrated by Martell et al. We constituted catalytic bilayer membranes as artificial vitamin B_6-dependent enzymes in combinations of peptide lipids, hydrophobic vitamin B_6 derivatives, and metal ions. Such a catalytic bilayer membrane constructed with a peptide lipid bearing an L-lysine residue, a hydrophobic pyridoxal derivative quaternized at the pyridyl nitrogen, and copper(II) ions exhibited turnover behavior for the transamination between L-phenylalanine and pyruvic acid with a high enantiomeric excess. A similar catalytic membrane behaved as an artificial tryptophan synthase which produces tryptophan derivatives by β-replacement of serine with indoles. Formation of β-phenylserine from glycine and benzaldehyde (an aldolase-type reaction) was promoted by a similar catalytic membrane using zinc(II) ions in place of copper(II). Both metal-ion catalysis and a specific microenvironment effect provided by bilayer membranes are quite important to simulate reactions catalyzed by the vitamin B_6-dependent enzymes.

719. EQUILIBRIUM STUDIES OF REACTIONS OF 4-CHLORO-1,2-DIAMINO-BENZENE, 4-METHYL-1,2-DIAMINOBENZENE, 4-METHOXY-1,2-DIAMINOBENZENE AND 4,5-DIMETHYL-1,2-DIAMINOBENZENE WITH DINUCLEAR COBALT(II)-1,4,7,13,16,19-HEXAAZA-10,22-DIOXACYCLOTETRACOSANE. Neiva D. Rosso,[1] Ricardo Nunes,[2] Bruno Szpoganicz[2] and Arthur E. Martell[3]* [1]Department of Chemistry, UPF, RS, Brazil. [2]Department of Chemistry, UFSC Florianópolis, SC 88040-900, Brazil. [3]Department of Chemistry, Texas A&M University, College Station, Texas 77843-3255.

Potentiometric equilibrium studies and UV-vis spectrophotometric measurements were carried out to determine the degree of binding of 4-chloro-1,2-diaminobenzene (ClDB), 4-methyl-1,2-diaminobenzene (MeDB), 4-methoxy-1,2-diaminobenzene (MeODB) and 4,5-dimethyl-1,2-diaminobenzene (diMeDB) with dicobalt(II)-OBISDIEN complexes in the absence and presence of dioxygen. The log of oxygenation constants are: 2.37, 3.49, 3.51 and 3.71 for ClDB, MeDB, MeODB and diMeDB, respectively and they increase with the protonation constants of these substrates.

720. COMPLEXATION EQUILIBRIA IN THE TERNARY SYSTEM CONTAINING ALUMINUM(III), CITRIC ACID, AND TRANSFERRIN. W. R. Harris, Zhepeng Wang, and Yahia Hamada, Department of Chemistry, University of Missouri-St. Louis, St. Louis, MO 63121.

The speciation of aluminum in serum involves a competition between the iron transport protein transferrin and low molecular weight chelating agents such as citric acid. Binding constants have been measured for aluminum-transferrin and aluminum-citrate individually. We have also conducted direct competition studies between transferrin and citrate to calculate effective Al-citrate binding constants at physiological pH and μM total Al. The $[Al_3(H_{-1}cta)_3(OH)]^{4-}$ trimer, which dominates at higher Al concentrations, is not detected in the competition studies. Both the direct competition experiments and speciation calculations indicate that citrate should bind only about 2% of serum Al under normal physiological conditions. Thus Al-citrate is not the primary low molecular weight Al complex in serum.

721. IRON BASED LIQUID REDOX SULFUR RECOVERY PROCESSES: A REVIEW. D. McManus, Wheelabrator Clean Air Systems, Inc., 2 W. Main St., Plainfield, IL 60544

The industrial use of iron as a regenerable oxidant for the conversion of hydrogen sulfide to sulfur is traced from inception as solid phase iron oxide processes to modern, homogenous redox catalysts employing amino polycarboxylate chelated iron solutions.
Current processes contact a broad range of hydrogen sulfide bearing gas streams with a dilute, mildly alkaline iron chelate is regenerated to the active iron (III) state by aeration in the same or a separate vessel, depending on the need to prevent admixture of air with the desulfurized process gas. Process flow diagrams and both beneficial and detrimental process chemical reactions are described.
The scope of process applications is reviewed and advantages and limitations relative to alternate technologies are discussed. Future expectations based on current research and market projections are presented.

722. KINETICS AND MECHANISM OF OXYGEN ATOM TRANSFER REACTION IN THE FORMATION OF $Ru^V=O$ (edta)$^-$ COMPLEX : A REACTIVITY SCALE FOR THE OXIDANTS. M. M. Taqui Khan, Department of Chemistry, Osmania University, Hyderabad, A.P., INDIA 500007

The oxidation of [Ru^{III}(EDTA)(H_2O)]$^-$ $\underline{1}$ with single oxygen atom donors (viz. H_2O_2, PhIO, KHSO, NaOCl, PyN-Oxide) was studied spectrophotometrically by following the development of characteristic peak of the $Ru^V=O$ (EDTA)$^-$ cxo-complex $\underline{2}$. The activation parameters were calculated in terms of a mechanism involving an intramolecular oxygen atom transfer from oxidant to complex $\underline{1}$. A reactivity scale has been set for the oxidants ClO^-, PyO, H_2O_2, $KHSO_5$, and C_6H_5IO on the basis of $\Delta G^{\ddagger}$ values.

723. BIOINORGANIC CHEMISTRY AND THE FIRST-YEAR STUDENT. J.A. Marcello, Department of Chemistry, Sheldon Jackson College, Sitka, Alaska 99835

One of the challenges in teaching chemistry to beginning students is introducing these students to current research topics and concepts. How does one talk about state of the art chemistry to students who are learning who are learning the basics? The interaction of inorganic elements and compounds with biological systems (and their model systems) defines bioinorganic chemistry in its broadest sense. Since bioinorganic chemistry is related to the inner workings of life, it affords a wonderful opportunity to demonstrate the complex workings of the chemical world to the everyday life of all students. Thus bioinorganic chemistry is a wonderful mechanism for the teaching of chemistry to the beginning student.

724. POLYNUCLEAR MIXED-VALENCE COMPLEXES. D.A. Rockcliffe* and P.M. Henry** *Kentucky State University, Frankfort, KY 40601. **Department of Chemistry, Loyola University of Chicago, Chicago, IL 60626.

Ruthenium(III) and iron(III) complexes of 1,1,1-trifluoro-4-ferrocenyl-2,4-butanedionato 1-phenyl-3-ferrocenyl-1,3-propanedionato and 1-ferrocenyl-1,3-butanedionato were prepared and examined electrochemically and spectroscopically. The ruthenium complexes showed reversible M(III)/M(II) and Fc/Fc^+ couples with potentials which were generally greater than those for the corresponding iron complexes. The near infrared-visible spectra of the ruthenium(III) complexes exhibited intervalence transfer bands. The range of values for the calculated delocalization parameters indicates that these mixed-valence complexes are weakly coupled, valence isolated class II compounds of the Robin and Day classification scheme. None of the related iron(III) complexes displayed intervalence transfer bands in the near infrared-visible region. This presentation will describe the preparation, electrochemical, mixed valence and solvent dependence characteristics of the mixed-valence complexes. The implication of the data on the delocalized nature of the complexes will be discussed.

725. ION RECOGNITION AND SEPARATIONS. Reed M. Izatt and Jerald S. Bradshaw. Department of Chemistry and Biochemistry, Brigham Young University, Provo, UT 84602 and IBC Advanced Technologies, Inc., P.O. Box 98, American Fork, UT 84003.

The design of ion selectivity into separations systems is of continuing interest. The need for hosts with high ion selectivity is increasing as more stringent environmental requirements are set in place and as greater emphasis is placed on simplifying element recovery and purification processes. Recent investigations at BYU aimed at the design and synthesis of host macrocycles capable of highly selective interactions with specific metal ions will be discussed. Results at IBC involving the commercial use of a solid phase extraction technique to selectively separate ions from industrial streams will be presented.

726.

SYNTHESIS, STRUCTURE AND FUNCTIONALIZATION OF ANIONIC MOLYBDENUM TRIS-AMIDO COMPLEXES SUPPORTING THE LINEAR DIATOMICS CO, CN, AND DINITROGEN. **Jonas C. Peters,** Aaron L. Odom and Christopher C. Cummins, Department of Chemistry, Massachusetts Institute of Technology, Cambridge, MA 02139

Mo(NRAr)$_3$ complexes (where R = *tert*-butyl or adamantyl and Ar = 3,5-dimethylphenyl) serve as excellent substrates for binding and activating small diatomic molecules such as dinitrogen, cyanide and carbon monoxide. Anionic complexes of the general form [X$_3$Mo(N$_2$)]$^-$, [X$_3$Mo(CN)]$^-$ and [X$_3$Mo(CO)]$^-$ can be prepared in convenient and high yielding syntheses. The complexes (AdArN)$_3$Mo(N$_2$)Na(THF)$_3$, (AdArN)$_3$Mo(CN)Li(THF)$_3$ and [(^{t}BuArN)$_3$Mo(CO)Na(Et$_2$O)]$_2$ have been structurally characterized. Each provides an intriguing starting point for reactivity studies aimed at functionalization of the terminal heteroatom, redox chemistry, and cleavage of the bound diatomic fragments. New strategies aimed at cleaving the C-N and C-O bond in [X$_3$Mo(CN)]$^-$ and [X$_3$Mo(CO)]$^-$ will be discussed in detail.

727. **GAS/SOLID REACTIVITY OF MOLECULAR SOLIDS.** Montserrat Oliván, Alexei Marchenko, Joseph Coalter and <u>Kenneth G. Caulton</u>, Department of Chemistry, Indiana University, Bloomington, IN 47405

The molecules $Ru(CO)_2L_2$, $RuHCl(CO)L_2$ and $Ru(H)_2Cl_2L_2$ (L = bulky phosphine) represent unsaturated species in systematically varied oxidation states. The reactivity of these crystalline solids with gaseous reagents (H_2, HCl, CO, O_2, Cl_2, etc.) has been studied, with the goal of using solid-state immobilization to detect <u>primary</u> reaction products, as well as products whose formation constant is low in solution at 25 °C. This work also reveals the mechanism of propagation of the "reaction front" in the solid, and has long-range implications for the development of new, broad-spectrum chemical sensors.

728. **OPERATIONALLY 14-ELECTRON RU(II) COMPLEXES** <u>D. Huang</u>[a], J. C. Huffman[a], W. E. Streib[a], O. Eisenstein[b], K. G. Caulton[a] a. Department of Chemistry and Molecular Structure Center, Indiana University, Bloomington, IN 47405-4001. b. Laboratoire de Structure et Dynamique des Systèmes Moléculaire et Solides, UMR5636, Université de Montpellier 2, 34095 Montpellier Cedex 5, France

Salt metathesis reaction between $RuR(OTf)(CO)L_2$ (R = H, Ph, OTf = O_3SCF_3, L = P^tBu_2Me) and $NaBAr'_4$ (Ar'_4 = $3,5\text{-}C_6H_3(CF_3)_2$) yields $[RuR(CO)L_2][BAr'_4]$. The X-ray structure of $[RuPh(CO)L_2][BAr'_4]$ reveals a cation with two agostic interactions between the metal center and the ^tBu C-H bonds. Nonetheless, these complexes react like 14-electron Ru(II) species by facile dissociation of the agostic ligands. The comparison of experimental structural data with computational results of model complexes $[RuR(CO)(PH_3)_2]^+$ will also be presented.

729. **NEW ENTRY TO HYDRIDO/HALO/VINYLIDENES FROM RUTHENIUM POLYHYDRIDES** <u>Montserrat Oliván</u>,[a] Odile Eisenstein,[b] Kenneth G. Caulton,[a] a) Department of Chemistry, Indiana University, Bloomington, IN 47405-4001, USA. b) Laboratoire de Structure et Dynamique des Systèmes Moléculaires et Solides, UMR 5636, Université de Montpellier 2, 34095 Montpellier Cedex 5, France

The reaction of RuH_3XL_2 (X = Cl, I; L = P^tBu_2Me) with terminal alkynes, $RC\equiv CH$, in a molar ratio 1:2 produces the vinylidene complexes $RuHX(CCHR)L_2$ (R = Ph, $SiMe_3$) and $RCH=CH_2$. Computational studies (*ab initio* DFT(B3LYP) methods) on the model $RuHCl(CCH_2)(PH_3)_2$ show the unusual Y-shape of these molecules. Possible mechanism of formation, catalytic activity, preliminary reaction studies as well as related osmium chemistry will be discussed.

INOR

730. NEW DISILANYL PLATINUM HYDRIDES: ATTEMPTS TO STABILIZE SILYL-SILYLENE INTERMEDIATES VIA INTERNAL BASE COORDINATION. R. Landtiser, J. T. Mague, M. J. Fink, Department of Chemistry, Tulane University, New Orleans, LA 70118

The rearrangement of disilanyl platinum hydrides to bis(silyl) platinum compounds presents a fundamental mechanistic question in silametallic chemistry. Evidence will be presented which supports the intermediacy of silyl-silylene intermediates. A number of new disilanyl platinum hydrides bearing heteroatoms have been synthesized in order to trap these putative intermediates via internal base coordination. The results of these trapping attempts will be discussed.

R=Ph
R'=Ph or Me
X = alkoxy, o-(α-dimethylamino)tolyl

731. The Synthesis of Novel Metallacyclynes and Their Metal Complexes

Daming Zhang, David B. McConville, Claire A. Tessier*, Wiley J. Youngs*
Chemistry Department, University of Akron, Akron, OH 44325-3601

Metallaalkynes have drawn great attentions because of their potential electrical properties and nonlinear optical properties. Among them, metallacyclynes provide the possibilities for polynuclear metal complexes which are important for catalysis. A tetrakynyl platinum complex (N^{n}Bu$_4$)$_2$[Pt(OBET)$_2$] has been synthesized lately in our group. The dianion has a planar structure and would appear to form one dimension electrical conductors. This complex contains two regions for binding additional metals: the two pockets and the two tweezers. These two regions provide different complexation selectivity toward metal species according to their size and coordination geometry preferences.

732. SYNTHESIS AND REACTIVITY OF AN IRIDAPYRYLIUM COMPLEX. Jonathan M.B. Blanchard and John R. Bleeke, Department of Chemistry, Washington University, St. Louis, MO 63130.

A novel "metallapyrylium" complex, [CH=C(Me)–CH=C(Me)–O=Ir(PEt$_3$)$_3$]$^+$BF$_4$$^-$ (1), has been synthesized by oxidation of the metallaoxacyclohexadiene precursor, CH=C(Me)–CH=C(Me)–O–Ir-(PEt$_3$)$_3$(H), with Ag$^+$BF$_4$$^-$. Compound **1** reacts with a variety of neutral and anionic substrates, including acetone, nitrosobenzene, chloride, and trimethylphosphine. The products of these reactions have been fully characterized by NMR spectroscopy and X-ray crystallography.

733.
SYNTHESIS AND REACTIVITY OF TRIGONAL BIPYRAMIDAL RHENIUM ARYL COMPLEXES.
Christopher A. Morse, Alan Davison, and Alun G. Jones; Department of Chemistry, Massachusetts Institute of Technology, Cambridge, MA, 02139 and Department of Radiology, Harvard Medical School, Boston, MA 02115.

The majority of rhenium complexes with η^1-aryl ligands either contain only one aryl group or have several aryl groups with oxo or imido functionalities present. Previously, there were only two known complexes of the form Re(aryl)$_3$L$_2$, where L = PEt$_2$Ph, and no work had been done concerning their reactivity. This research investigates the reactivity and syntheses of compounds of the type Re(aryl)$_3$L$_2$ and other polyaryl rhenium species. A variety of rhenium(III) starting materials have been used to synthesize the title complexes, which include the trigonal bipyramidal Re(THT)$_2$(p-tolyl)$_3$ (THT = tetrahydrothiophene).

734. **DIRECT ATTACK OF R⁻ AT AN OSMIUM NITRIDO TO FORM A d^4 IMIDO.** Thomas J. Crevier and James M. Mayer, Department of Chemistry, University of Washington, Box 351700, Seattle, WA, 98195-1700.

The reaction of TpOs(N)Cl$_2$ (Tp=hydrotris(1-pyrazolyl)borate) with traditional arylation reagents (PhMgCl, PhLi, ZnPh$_2$) leads to the formation of [TpOs(NPh)Ph$_2$]⁻, a rare example of a d^4 imido complex. Quenching with D$_2$O produces TpOs(NDPh)Ph$_2$; with MeOTf, TpOs(NMePh)Ph$_2$ is formed. Mechanistic studies rule out the participation of free phenyl radicals in the formation of the imido and implicate direct attack of R⁻ at a nitrido ligand which is unprecedented. The electrophilicity of TpOs(N)Cl$_2$ is also demonstrated by its reaction with a variety of other nucleophiles such as PR$_3$ and S$_8$. Synthesis of TpOs(N)Ph$_2$ can be achieved through other routes, including arylation of TpOs(N)Cl$_2$ by phenylcuprate at low temperature.

735. AMIDO DERIVATIVES OF NICKEL INDENYL COMPLEXES. Isabelle Dubuc, Francine Bélanger-Gariépy, and <u>Davit Zargarian,</u> Department of Chemistry, University of Montréal, Montréal, Québec, Canada H3C 3J7.

The complex (1-Me-Ind)Ni(PPh$_3$)Cl has been used as a precursor to prepare the corresponding amido (NR$_2$) derivatives (1-Me-Ind)Ni(PPh$_3$)(NR$_2$). The electronic nature of the R groups plays a determining role in these syntheses. In general, strongly electron withdrawing groups are required to stabilize the Ni-N bond, presumably because of the destabilizing N→Ni π-donation. The solid state structure of the phthalimido derivative has shown a significantly "slipped" indenyl ligand and a relatively short Ni-N bond. The kinetic robustness of this bond is also evident from its unreactive character. The presentation will discuss these and other structural and reactivity features of these complexes.

736. SYNTHESES AND PROPERTIES OF ZEROVALENT TITANIUM CARBONYLS CONTAINING ALKOXY AND ARYLOXY LIGANDS, [Ti(CO)$_4$(μ-OR)]$_2$²⁻. Paul J. Fischer, Pong Yuen, Victor G. Young, Jr., and John E. Ellis, Chemistry Department, University of Minnesota, Minneapolis, MN 55455

Reactions of [Ti(CO)$_6$]²⁻ [1] with alcohols and phenols provide formally unsaturated titanium complexes of the general formula [Ti(CO)$_4$(μ-OR)]$_2$²⁻, the first known Ti(O) alkoxides and aryloxides. Spectroscopic and chemical properties of the methoxy bridged dimer will be discussed in detail and compared to corresponding properties of the saturated complex [Ti(CO)$_4$C$_5$H$_5$]⁻.[2] Structural characterization of the phenoxy bridged dimer, [Ti(CO)$_4$(μ-OPh)]$_2$²⁻, reveals that each titanium has trigonal prismatic geometry and that the eclipsed Ti(CO)$_4$ units closely resemble those present in [Ti(CO)$_4$C$_5$H$_5$]⁻. All of the data we have presently suggest that the bridging RO ligands in this unprecedented class of metal carbonyl dimers function as 5 electron donors.

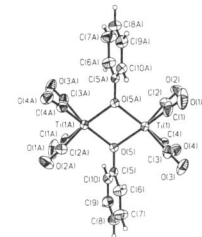

[1] J. E. Ellis and K. M. Chi, *J. Am. Chem. Soc.* (1990) *112*, 6022.
[2] J. E. Ellis, S. R. Frerichs, and B. K. Stein, *Organometallics* (1993) *12*, 1048.

737. **NANOSTRUCTURED MATERIALS: SONOCHEMICAL SYNTHESIS, CHARACTERIZATION, AND CATALYTIC STUDIES.** Millan M. Mdleleni, Taeghwan Hyeon, and Kenneth S. Suslick*, School of Chemical Sciences, University of Illinois at Urbana-Champaign, Urbana, IL 61801

Ultrasonic irradiation of slurries of Mo(CO)$_6$ in alkane solvents leads to the formation of molybdenum carbide (Mo$_2$C). When these irradiations are performed in the presence of Co$_2$(CO)$_8$ cobalt cemented molybdenum carbides (CoMoC) are obtained. Cemented carbides of various Co to Mo_2C ratios were prepared by adjusting the $Mo(CO)_6$ to $Co_2(CO)_8$ ratios in the initial slurry prior to irradiation. Molybdenum sulfide (MoS_2) was also prepared by sonicating $Mo(CO)_6$ in the presence of sulfur. The XRD, EDX, XPS, SEM, TEM, BET, and elemental analysis were used to characterize these compounds. These high surface area, high porosity materials catalyze several industrially significant processes. The synthesis, characterization, and catalytic properties in cyclohexane dehydrogenation and thiophene hydrodesulfurization of these compounds will be discussed.

738. ELECTROCHEMICALLY GENERATED FULLERENE FILMS. David A. Costa, Alan L. Balch, Krzysztof Winkler, W. Ronald Fawcett, Department of Chemistry, University of California, Davis 95616

A great deal of research has been applied to the formation of polymeric fullerene materials. Our efforts have been focused on the electrochemical formation of fullerene based polymers. The electrochemical reduction of fullerene epoxide produces stable, adherent, electrochemically active films on the electrode surface. This talk will present recent results which demonstrate that polymeric fullerene films can be generated through the direct electroreduction of C_{60} in the presence of trace amounts of O_2. Films formed in this manner have similar redox properties to those formed from $C_{60}O$ although a greater degree of electrode deactivation is observed Electron transfer within the films is shown to be mediated by the incorporation of ferrocinium ions into the film structure. The structures of films formed from $C_{60}O$ and $C_{60} + O_2$ are compared by IR, UV-VIS and electron microscopy.

739. SYNTHETIC ADVANCES IN VOLATILE METAL-ORGANIC SOURCES FOR MOCVD-MEDIATED OXIDE FILM GROWTH. J. A. Belot[1], C. R. Reedy[1], C. L. Stern[1], T. J. Marks[1]*, G. Yap[2], A. L. Rheingold[2], Departments of Chemistry, Northwestern University[1], Evanston, IL, 60208 and the University of Delaware[2], Newark, DE, 19716.

The metal-organic chemical vapor deposition (MOCVD) of metal-oxide thin films is a rapidly developing field with applications ranging from high temperature superconductors to ferroelectric materials. A critical issue in the continued implementation and success of MOCVD is the availability of suitable metal sources displaying attributes such as high, stable vapor pressure, appropriate reactivity, longevity with respect to repeated thermal cycling, and low cost. Presented will be our recent synthetic advances in volatile metal sources, particularly those involving group II and group IV complexes containing fluorinated and non-fluorinated β-diketonate and Schiff-base ligands.

740. SYNTHESIS OF TRANSITION AND MAIN GROUP METAL NITRIDES USING A MICROWAVE GENERATED NITROGEN PLASMA. Hans-Conrad zur Loye, Joel D. Houmes, Department of Chemistry and Biochemistry, The University of South Carolina, Columbia, SC 29208.

TiN, AlN and GaN were synthesized by microwave generated plasma nitridation of the oxide precursors. It was shown that some binary nitrides that cannot be synthesized by ammonolysis reactions of the respective oxides for either kinetic or thermodynamic reasons, can be prepared using a hydrogen/nitrogen plasma. This techniques extends the range of oxides that can be used as precursors for nitride synthesis. These and other compounds synthesized by microwave generated nitrogen plasmas will be discussed.

741. PROBING THE CVD CHEMISTRY OF Pt(PF$_3$)$_4$ AND OTHER TRIFLUOROPHOSPHINE PRECIOUS METAL COMPLEXES Cynthia L. Hammill and Ronald J. Clark, Department of Chemistry, Florida State University, Tallahassee, Florida 32306

The chemical vapor deposition of precious metals to form thin films is of considerable importance to electronics. Several organometallic precursor compounds have been developed for the CVD of platinum and other precious metals. In most cases, elevated source temperatures are required, but compounds such as Pt(PF$_3$)$_4$ and HRh(PF$_3$)$_4$ are far easier compounds to handle. They are liquids at room temperature with vapor pressures approaching 40 torr. Therefore, one does not have to contend with the problem of slow decomposition. Data will be presented concerning the purity of the platinum deposits. Phosphorus contamination is often a problem, but the chemistry that produces the impurity is unclear. Bulk deposition studies are being done to supplement the use of thin-film surface techniques. Studies thus far show the product from bulk deposition to be pure platinum based upon powder x-ray diffraction. In addition, the molar quantity of PF$_3$ gas expected from the reaction has been found. However, preliminary investigation using other methods indicates phosphorus is present. These methods are being pursued further.

742. THE GAS PHASE PHOTOCHEMICAL REACTIONS OF Fe(PF$_3$)$_x$(CO)$_{5-x}$ WITH VARIOUS OLEFINS, Melanie M. Bower and Ronald J. Clark, Department of Chemistry, Florida State University, Tallahassee, Florida 32306

In most respects, carbon monoxide and phosphorus trifluoride are quite similar as ligands in mixed CO-PF$_3$ metal complexes. The iron species Fe(PF$_3$)$_x$(CO)$_{5-x}$ are typical of the compounds of concern. The similarity among the compounds extends to the physical properties, structure, and to spectroscopic properties such as IR and NMR. However, preliminary studies show that there is a noticeable difference in the way individual iron compounds interact photochemically with the two ligands C$_2$H$_4$ and C$_2$F$_4$. For instance, the compound of the formula olefinFe(PF$_3$)$_3$(CO) is formed easily and in good yield with C$_2$H$_4$ but the formation of the analogous tetrafluroethlyene analog is not readily demonstrated. All current data points to the CO as being the primary ligand to be lost during photolysis. In addition, there appears to be a considerable difference in the rate at which the two olefins ligands add to the coordinately unsaturated species in the formation of the stable compound. Our further observations concerning this chemistry will be presented.

743.
DETECTION OF HALOGENATED HYDROCARBONS BY TIN OXIDE THIN FILM SENSORS PREPARED BY PHYSICAL VAPOR DEPOSITION-THERMAL OXIDATION. S.H. Park, B.R. Shaw, K.E. Creasy, and S.L. Suib, Department of Chemistry, University of Connecticut, Storrs, Connecticut 06269-4060

Tin oxide thin film sensors prepared by physical vapor deposition (PVD)-thermal oxidation and sol-gel method are tested to detect halogenated hydrocarbons. Responses of tin oxide films are unique to several chlorinated methanes (CH$_2$Cl$_2$, CHCl$_3$, and CCl$_4$) in O$_2$. Sensing behavior to 1% CCl$_4$ at different temperature (100, 200, 300, and 400 °C) is studied. The best operating temperature is 300 °C for maximum sensitivity. The films are stable at operating conditions (300 °C, O$_2$) and give reproducible responses to chlorinated methanes for more than two weeks. Film resistance changes are related to the concentrations of gases. Response times (to reach 90% of maximum intensity) with 1000 ppm chlorinated methanes at 300 °C are similar and less than 40 sec. Surface accessible oxygens appear to play an important role in the sensing mechanism. Film response to these chlorinated methanes is chemically selective (independent). Test results of other halogenated hydrocarbons with different MnO$_x$ and SnO$_x$ films and their preparation and characterization will also be presented.

744.
PREPARATION OF METAL OXIDE MATERIALS USING LIQUID METAL CARBOXYLATE PRECURSORS. **Allen W. Apblett** and Edwin H. Walker, Department of Chemistry, Tulane University, New Orleans, LA, 70118

We have recently discovered a novel class of metal carboxylates which are liquids at room temperature. These metal salts bear polyether organic residues and their physical properties make them highly conducive to the preparation of ceramic films and fibers. Furthermore, the liquid salts are excellent solvents for other metal salts such as nitrates. The resultant solutions are readily converted upon pyrolysis to multi-metalic oxide phases at fairly low temperatures due to the high homogeneity of the cation distribution in the liquid. The preparation of a variety of metal oxide ceramics such as spinels and related materials in this manner will be reported.

745. SYNTHESIS OF TUNGSTEN NITRIDES AND CARBIDE NITRIDES. Zhihong Zhang and Robert E. McCarley, Ames Laboratory U.S.D.O.E., Iowa State University, Ames, IA 50011.

Efforts to prepare the unknown, metastable tungsten(VI) nitride WN_2 have proceeded with various reactions of $WNCl_3$ conducted under mild conditions. Reactions of $WNCl_3$ with Zn_3N_2 in sealed tubes at 400°C frequently failed because of explosions. Under careful management of composition of the reaction mixture and temperature programming, however, the reactions were successfully completed to provide amorphous W_3N_5. Annealing of the latter at 800°C in a sealed silica tube resulted in decomposition into a mixture of W and hexagonal WN, as well as growth by chemical transport of a few crystals of the new nitride carbide WN_2C, based upon single crystal x-ray structure analysis. Reaction of $WNCl_3$ with Ca_3N_2 in acetonitrile at 80°C results in a precursor which can be converted at 500°C into the new metastable cubic-WN with the rock salt structure. The explosive azido precursor $WN(N_3)_3 \cdot xCH_3CN$ can be obtained by reaction of $WNCl_3$ with NaN_3 in acetonitrile at 20°C, and successfully converted to nitrido products *in situ*. Some reactions of $WN(N_3)_3 \cdot xCH_3CN$ will be discussed.

746. SYNTHESIS, STRUCTURE, AND MOLECULAR ORBITAL STUDIES ON EARLY TRANSITION METAL COMPLEXES BEARING PYRAZOLATO, TRIAZOLATO, AND TETRAZOLATO LIGANDS. Charles H. Winter, Ilia A. Guzei, Anwar G. Baboul, and H. Bernhard Schlegel, Department of Chemistry, Wayne State University, Detroit, Michigan 48202.

As part of our program relating to the preparation of molecular precursors to metal nitride films, we have been exploring the synthesis of early transition metal complexes that possess only nitrogen ligands in the coordination sphere. 5-Membered nitrogen heterocycles, such as pyrazoles, triazoles, and tetrazoles, are particularly attractive ligands due to their ease of preparation and anticipated low-energy decomposition routes. We will present the synthesis and structural characterization of titanium, zirconium, and tantalum complexes bearing η^2-pyrazolato ligands. η^2-Pyrazolato ligand coordination is extremely rare in d-block transition metals. Molecular orbital calculations have been conducted on several model pyrazolato complexes and the results will be overviewed. Finally, synthetic and molecular orbital studies on early transition metal triazolato and tetrazolato complexes will be discussed.

747. SYNTHESIS OF LANTHANIDE COMPLEXES WITH NITROGEN COORDINATION SPHERES. Charles H. Winter and Dirk Pfeiffer, Department of Chemistry, Wayne State University, Detroit, Michigan 48202.

We are interested in the development of volatile lanthanide complexes for use in the fabrication of doped semiconductor films by chemical vapor deposition and related techniques. To avoid the introduction of undesired dopants into the films, we have restricted synthetic studies to complexes bearing nitrogen coordination spheres and with ligands containing only nitrogen, carbon, and hydrogen. We will describe synthetic approaches to lanthanide complexes containing several nitrogen-rich donors, including pyrazolato and triazenido ligands. Crystal structures of representative complexes will be overviewed. Finally, trends in volatility of these new complexes will be presented.

748. ORGANIC DISULFIDES AS DISULFIDE ION SOURCES: POSSIBLE IMPLICATIONS FOR THE CHEMICAL VAPOR DEPOSITION OF METAL SULFIDE FILMS. Charles H. Winter, Peggy J. McKarns, and T. Suren Lewkebandara, Department of Chemistry, Wayne State University, Detroit, Michigan 48202.

There is considerable interest in the development of selective routes to thin films of binary inorganic materials. In sulfur materials, the sulfide ion (S^{2-}) is can be replaced by the disulfide ion (S_2^{2-}) to afford sulfur-rich phases. While sulfide materials are generally obtained under the high temperatures used in most film depositions, phases containing the disulfide ion have been reported when organic disulfides are used as the sulfur sources. We will describe the reactions of titanium tetrachloride, niobium pentachloride, and tantalum pentachloride with organic disulfides. In reactions with tert-butyl disulfide, complexes are obtained that contain S_2^{2-} and tBuSS$^-$ ligands, which result from loss of tert-butyl groups from tert-butyl disulfide. The characterization and crystal structures of these complexes will be presented. Additionally, the fate of the tert-butyl groups will be overviewed. Finally, we will interpret our results in light of film depositions that yield sulfur-rich phases.

749. SYNTHESIS, CHARACTERIZATION, AND REACTIVITY OF 1,4-DI-TERT-BUTYL-2,5-BIS(CHLOROMAGNESIO)BENZENE. Charles H. Winter and Cathrine E. Reck, Department of Chemistry, Wayne State University, Detroit, Michigan 48202.

Despite the abundance of aromatic Grignard reagents containing one chloromagnesio group, there are very few well characterized aromatic compounds that contain two or more magnesium substituents. We will describe a synthetic approach designed to afford 1,4-di-tert-butyl-2,5-bis(chloromagnesio)benzene, which entails selective dimercuration of 1,4-di-tert-butylbenzene followed by mercury-magnesium exchange using methylmagnesium chloride. Isolated "1,4-di-tert-butyl-2,5-bis(chloromagnesio)benzene" exists as a mixture of oligomers formed by shifting of the Schlenk equilibrium. Results of NMR studies on these oligomers will be presented. The reactivity of "1,4-di-tert-butyl-2,5-bis(chloro-magnesio)benzene" will also be overviewed. Finally, attempts to prepare tetramagnesiated and hexamagnesiated benzene derivatives will be described.

750.
Li^+ TRANSPORT PATHWAYS IN SH-SY5Y NEUROBLASTOMA CELLS BY NMR SPECTROSCOPY. J. Nikolakopoulos[*], C. Zachariah, and D. Mota de Freitas, Department of Chemistry, Loyola University Chicago, Chicago, IL 60626.

We have developed and utilized a ^7Li nuclear magnetic resonance method to identify the pathway(s) by which Li^+ is transported into and out of perfused neuroblastoma cells embedded in agarose gel threads. This work has an impact on the knowledge available on the bioinorganic chemistry of Li^+ and its use in the treatment of manic depression. Lithium influx and efflux pathways have been characterized in red blood cells. However, these pathways may not apply to neuronal tissue. In neuroblastoma cells, chemical equilibrium and ionic strength are maintained by ion channels, a pathway which is not present in red blood cells. Rate constants for both Li^+ efflux and influx were obtained in the absence and presence of ion transport activators or inhibitors. The use of veratridine led to altered rates of Li^+ transport, thus identifying Na^+ channels in SH-SY5Y cells as the primary pathway mediating Li^+ transport.

751. RESONANCE LIGHT SCATTERING FOR THE STUDY OF MOLECULAR ASSEMBLIES. R. F. Pasternack, P. J. Collings, J. C. DePaula and E. J. Gibbs, Department of Chemistry, Swarthmore College, Swarthmore, PA 19081

Light scattering experiments typically involve measurements away from absorption bands but, for species that aggregate, large enhancements can be observed at wavelengths characteristic of the aggregated species. The RLS effect is observed as increased scattering intensity at or very near the *wavelength of absorption of an aggregated molecular species*. The effect can be enhanced by several orders of magnitude when strong electronic coupling exists among the chromophores. In addition to its sensitivity, this technique allows for selective observation of aggregates even in multicomponent systems where a large fraction of monomers or other aggregates exist. The physical principles underlying RLS will be discussed, and the advantages and limitations of the technique reviewed.

752. SINGLE-CRYSTAL POLARIZED SPECULAR REFLECTANCE SPECTROSCOPY OF INTENSE ELECTRONIC TRANSITIONS. Ronald L. Musselman, Department of Chemistry, Franklin and Marshall College, Lancaster, PA 17604, and Edward I. Solomon, Department of Chemistry, Stanford University, Stanford, CA 94305.

Knowledge of the polarizations of electronic transitions is helpful in making assignments. Polarized specular reflectance spectroscopy is a valuable tool for studying allowed charge-transfer transitions (with ε greater than 1000 $M^{-1}cm^{-1}$) in metalloprotein model complexes where crystals cannot be ground thin enough to obtain an absorption spectrum. Absorbance data may be retrieved from the reflectance data through a Kramers-Kronig transformation, $\theta(\omega_j) = \frac{(2\omega_j)}{\pi} \int_0^\infty \frac{\ln r - \ln r_j}{\omega_j^2 - \omega^2} d\omega$ where θ is the phase change upon reflection, r is the reflectivity, and ω is circular frequency. Epsilon is a function of θ and r. We have used this technique for determining the polarizations of transitions in several complexes, including copper chloride dimers and μ-peroxo cobalt dimers as models for hemocyanin, several μ-oxo iron dimers as models for hemerythrin, and are currently studying μ-oxo manganese dimers and copper μ $\eta_2-\eta_2$ peroxo dimers.

753. **Magnetic and Moessbauer-Investigations on Dimeric and Tetrameric Valence-Delocalized Fe(II)-Fe(III)-Complexes.** C. Saal, W. Haase, J. Ensling, P. Gütlich, S.K. Dutta, S. Mohanta, K. Nag, E. Duin, M.K. Johnson,

The magnetic susceptibility measurements, NIR-absorption and Moessbauer-spectroscopy was used to investigate the valence-delocalization and localization of three closely related dimeric and one tetrameric-Fe(II)-Fe(III)-complexes. The latter consists of two dimeric subunits. The Moessbauer-results give a strong hint to valence delocalization in two of the complexes - according to class III in the Robin-Day classification scheme. The two remaining compounds are valence trapped at low temperatures but delocalization seems to become more important at higher temperatures. The magnetic structure of these compounds was investigated by susceptibility measurements and analyzed using isotropic- and double-exchange. These results are in good agreement with those obtained from Moessbauer spectroscopy.

754. LIGAND ELECTROCHEMICAL PARAMETERS AND BIO-INORGANIC CHEMISTRY.
A. B. P. Lever, Department of Chemistry, York Univerrsity, North York (Toronto), Ontario, Canada, M3J 1P3

Ligand electrochemical parameter theory [1] has proven itself to be a very useful procedure both as a predictive strategy and in extracting additional information from electrochemical data. While initially applied to metal localized redox sites, it has recently been shown applicable to ligand reduction sites [2] and indeed probably has general redox utility. It is also of value in analyzing electronic spectra, specifically charge transfer spectra, through electrochemical/optical synergic relationships. The parameters are intrinsic to the ligand [3] and hence in principle should be applicable to problems in bio-inorganic chemistry.
We explore here potential applications, which are not restricted to electrochemistry and charge transfer spectroscopy but may also impact other techniques such as Mossbauer spectroscopy and EXAFS and also bio-inorganic site reactivity etc. [1]. Lever, Inorg. Chem., 1990, 29, 1271. [2]. Dodsworth.; Vlcek, ; Lever, . Inorg. Chem. 1994, 33, 104 [3]. Fielder.; Osborne,; Lever,.; Pietro,. J. Am. Chem. Soc. 1995, 117, 6990.

755.
SPECTROSCOPIC STUDIES OF HIGH-VALENT M=O SITES IN BIOLOGY. **Martin L. Kirk**, The University of New Mexico, Albuquerque, NM, 87131-1096.

High-valent M=O active sites are found as key catalytic intermediates in heme catalases and peroxidases, and in pterin containing molybdenum enzymes. The function of these M=O sites ranges from their involvement in coupled proton-electron transfer to formal oxo-transfer reactions. We present recent ground state and excited state spectroscopic data for model oxoferryl porphyrin π-cation radical (cpd I), oxoferryl porphyrin (cpd II), and oxomolybdenum complexes and discuss the implications of their unique electronic structure with regard to the role of M=O units in enzymatic catalysis.

756. LIGAND K-EDGE X-RAY ABSORPTION SPECTROSCOPY APPLIED TO BIOINORGANIC SYSTEMS
Kendra Rose Williams[1], Edward I. Solomon[1], Britt Hedman[2], Keith O. Hodgson[1,2]
[1] Department of Chemistry, Stanford University, Stanford, CA 94305
[2] Stanford Synchrotron Radiation Laboratory, Stanford University, Stanford, CA 94309

The technique of ligand K-edge x-ray absorption spectroscopy is being developed as a spectral probe of ligand-metal bonding, which is analogous to superhyperfine coupling in EPR, but is also applicable to EPR inactive sites and to ligands with zero nuclear spin. Previously, it was established that open-shell metals, such as Cu, bound to Cl give a pre-edge feature in the Cl K-edge spectrum which corresponds to a transition from the Cl 1s to the half occupied $d_{x^2-y^2}$ derived HOMO (*J. Am. Chem. Soc.* **1990**, *112*, 1643). Due to the localized nature of the Cl 1s orbital, this transition derives its intensity from the Cl 3p character mixed into the $d_{x^2-y^2}$ orbital due to bonding. Thus, the intensity of the pre-edge feature quantitates this covalent mixing. This methodology has been extended to metals with more than one d-hole and also to metals with S ligation. With the ligand K-edge methodology established, proteins such as the blue-Cu site in plastocyanin, and the Cu_A site of cytochrome *c* oxidase have been investigated. Also, the S K-edge methodology is now being applied to Fe-S systems including rubredoxin and Fe_nS_n clusters.

757. "EPR STUDIES OF AN ARTIFICIAL LINEAR MnIIIMnIV 3 CLUSTER WITH A S = 1/2 GROUND STATE. RELEVANCE TO THE OEC", G. Blondin, R. Davydov, C. Philouze M.-F. Charlot, S. Styring, B. Akemark, J.-J. Girerd, A. Boussac, A. W. Rutherford

NO ABSTRACT AVAILABLE

758. ELECTRONIC STRUCTURE OF Cu_A-TYPE CENTERS IN CYTOCHROME c OXIDASE, BLUE-COPPER MUTANTS, AND RELATED MODEL COMPLEXES.
Daniel R. Gamelin,[a] Kendra Rose Williams,[a] Louis B. LaCroix,[a] Robert P. Houser,[b] William B. Tolman,[b] Ton C. Mulder,[c] Simon de Vries,[c] Marjorie Ang,[d] Yi Lu,[d] Britt Hedman,[a] Keith O. Hodgson,[a] Edward I. Solomon[a]

[a]*Department of Chemistry, Stanford University, and/or Stanford Synchrotron Radiation Laboratory, Stanford, CA 94305,* [b]*Department of Chemistry, University of Minnesota, Minneapolis, MN 55455,* [c]*Department of Microbiology and Enzymology, Technical University Delft, 2628 BC Delft, The Netherlands,* [d]*Department of Chemistry, University of Illinois, Urbana, IL 61801*

Delocalized mixed-valence $[Cu_2(SR)_2]^+$ dimers are involved in biological electron-transfer processes contributing to mitochondrial and bacterial respiration (cytochrome c oxidases (CcO)) and denitrification (nitrous oxide reductases (N_2OR)). Using a combination of absorption, magnetic circular dichroism, resonance Raman, and x-ray absorption spectroscopies, electronic-structural descriptions have been developed for a series of Cu_A-type centers obtained from CcO as well as engineered into the blue-copper protein azurin. Spectroscopic results also allow description of the electronic properties of two structurally-characterized $[Cu_2(SR)_2]^+$ and $[Cu_2(SR)_2]^{2+}$ synthetic model complexes. These results are used to define the Cu_A orbital ground state and allow a detailed description of the factors contributing to valence delocalization and electron transfer in these systems.

759. SENSITIZED NEAR INFRARED EMISSION FROM YTTERBIUM(III) VIA ELECTRON TRANSFER (ET). NEW LANTHANIDE ION PROBES OF LONG-RANGE ET IN PROTEINS, W. DeW. Horrocks, Jr., Department of Chemistry, The Pennsylvania State University, University Park, Pennsylvania, 16802

Irradiation of the tryptophan residue in codfish parvalbumin (290 nm) to which two Yb^{3+} ions are bound results in emission from the $^2F_{5/2}$ state of the ion (977 nm). This phenomenon is attributable to an electron transfer from the singlet excited state of tryptophan, Trp*, to a Yb^{3+} ion producing Yb^{2+} and $Trp^{\cdot}$. The back electron transfer restores the system, but leaves some Yb^{3+} in its excited, emissive state. Both Eu^{3+} and Yb^{3+} (the first and second most easily reduced lanthanides) quench Trp fluorescence and measurement of Trp fluorescence lifetimes provides electron transfer rate constants. A straightforward method of determining the nuclear rearrangement energy, λ, and the electronic-coupling strength, H_{AB}, of Marcus theory from these measurements will be presented. (Supported by the National Institutes of Health.)

760. IDENTIFICATION OF PUTATIVE PEROXIDE INTERMEDIATES OF PEROXIDASES BY ELECTRONIC STRUCTURE AND SPECTRA CALCULATIONS, Gilda Loew and Danni Harris
Molecular Research Institute, Palo Alto, California 94304

In the enzymatic cycle of this family of oxidative metabolizing heme proteins, hydrogen peroxide is required to transform the ferric resting state to the catalyltically active, Compound I, species. An iron heme peroxide complex has been proposed as a key transient intermediate in this reaction and electronic spectra observed prior to Compound I formation in both wild type and R38L horse radish peroxidase–C (HRP-C) have been attributed to it. There are however significant qualitative differences in these observed spectra. Specifically, there is an extra hyper-Soret band in the 300-450 nm region of the wtHRP-C spectra, while the R38L mutant exhibits a normal Soret band, not very different from the resting state. In the absence of any additional information, it is not possible from these spectra alone to identify the species that gives rise to them or to understand these differences. To further identify the origin of these spectra and their differences, the INDO/RHF/CI quantum chemical method has been used to calculate the electronic structure and spectra of two possible peroxide-perroxidase intermediates, one with HOOH and the other with OOH as the heme iron ligand. Comparison of the calculated spectra with the observed spectra, led to the identification of the transient species responsible for each of these spectra and could account for the observed differences between them. The "hyper-porphyrin" spectrum observed in the wtHRP-C experiments was found to originate from the OOH- form of this transient intermediate in a low spin ground state. The normal Soret observed in the R38Lmutant experiment originated from the neutral peroxide form in a high spin ground state. These results allow for the first time a one to one correspondence to be made between the observed spectra and the species that could be causing them. They also provide further evidence for transient peroxide intermediates in the pathway from the resting state to Compound I of peroxidases.

761. NMR STUDIES OF PROTON DONATION IN REDUCED SUPEROXIDE DISMUTASE: TYR 34 CAN EXPLAIN THE PK OF 8.5. D. L. Sorkin, D. K. Duong and A.-F. Miller, Department of Chemistry, The Johns Hopkins University, 3400 N. Charles St., Baltimore, MD 21218

We have exploited NMR spectroscopy to observe the paramagnetically shifted resonances of the active site residues of reduced Fe-superoxide dismutase (Fe-SOD), and obtained the first spectroscopic evidence for the pK near 9 that has been predicted for over ten years (Bull & Fee, 1985, J. A. C. S. 107, 3295). We find that the pK has a value of 8.5 at 30 °C and does not affect the active site conformation. However, by measuring saturation transfer from water we show that the pK is associated with a greater than 100-fold increase in protection against solvent exchange for one of the active site protons. By contrast, mutant Fe-SOD in which Tyr 34 is replaced by Phe lacks both the chemical shift and exchange rate signatures of the pK. Thus, the conserved Tyr 34 can account for the long-predicted pK of reduced Fe-SOD, and may be one of the proton donors to substrate. Supported by ACS-PRF 28379-G4 and N.S.F. MCB-9418181.

762. SPECTROSCOPIC CHARACTERIZATION OF DISCRETE Cu^{III} OXIDE COMPLEXES: FULL REDUCTION OF DIOXYGEN BY SIMPLE DIAMINE CU(I) COMPLEXES. A. P. Cole; P. Mukherjee; V. Mahadevan; Z. Hou; D. E. Root; M. J. Hensen; Solomon, E. I. and T. Daniel P. Stack, Department of Chemistry, Stanford University, Stanford, CA 94305-5080

Copper/dioxygen coordination chemistry is remarkably diverse as are the associated spectroscopic features of the complexes. Much interest in characterizing copper/oxygen intermediate species results from the biologicial use of copper to bind or activate dioxygen. The reaction of dioxygen with simple diamine Cu(I) complexes in aprotic solvents (-80° C) yields an intriguing class of discrete, thermally sensitive Cu^{III} oxide complexes. Both 2:1 and 3:1 Cu:O_2 complexes are formed dependent on the nature of the diamine ligand. Each type of complex appears to display full reduction of dioxygen resulting in bond scission; the 2:1 and 3:1 complexes are best formally described as a diamagnetic $[Cu^{III}{}_2O_2]^{2+}$ core and a paramagnetic, localized valence $[Cu(III)Cu(II)_2O_2]^{3+}$ core, respectively. Spectroscopic characterization of these complexes will be presented and related to the structures.

763. IMPRINTING STRUCTURAL INFORMATION FROM ORGANIC MOLECULES ONTO INORGANIC LATTICES. R. Haushalter, NEC Research Institute, 4 Independence Way, Princeton, NJ 08540, and J. Zubieta, Department of Chemistry, Syracuse University, Syracuse, NY 13244

Unlike the rational point-to-point connection of molecules in solution that has led to the successful synthesis of organic molecules, polymers and biological materials, the preparation of 3-D inorganic solids is more obdurate because of the insolubility and lack of reactivity at low temperatures of the starting materials, coupled with fact that the solid grows simultaneously in all three dimensions. To overcome these difficulties, we have employed polar organic molecules - which can be prepared by the rational methods of organic synthesis - coupled with the use of low temperature (<200 °C) hydrothermal synthetic methods that allow the use of organics and solubilization of inorganic starting materials, to synthesize a wide variety of novel organic-inorganic hybrid materials. These generalized concepts have been utilized to prepare such diverse materials as organically templated transition metal oxides, phosphates, borates and halides, the structures of which will be discussed in terms of hydrogen bonding and hydrophobic-hydrophilic interactions between the organic templates and the framework atoms.

764. SELF-ASSEMBLY IN THE SYNTHESIS OF POROUS SILICATES. Mark E. Davis, Department of Chemical Engineering, California Institute of Technology, Pasadena CA 91125

Porous materials have found widespread use as adsorbents and catalysts. The objective in synthesizing new porous materials is to precisely control the atomic-level structure of these materials to provide appropriate macroscale properties. This talk begins with a brief overview of the importance of porous materials. The controlled synthesis of zeolites (microporous materials) is then described with emphasis on the assembly process and how it can be manipulated. The synthesis of zeolites involves structure-directing organic molecules that assist in the organization of the inorganic components. Extension of this assembly process to the use of molecular aggregates as structure-directing agents allows the formation of mesoporous materials. Finally, the outlook for synthesizing materials by design through controlling the self-assembly process is discussed.

765.

HYDROTHERMAL SYNTHESIS OF NEW VANADIUM OXIDES. **M. Stanley Whittingham,** Thomas G. Chirayil, Fan Zhang, and Peter Y. Zavalij, Chemistry Department and Materials Research Center, SUNY at Binghamton, Binghamton, NY 13902-6016

A wide range of vanadium oxides with new structures have been synthesized using hydrothermal methods; the tetramethyl ammonium ion (TMA) was used as a structure directing agent. The compounds formed depend critically on the pH of the reaction medium; with vanadium pentoxide, TMA and LiOH six different compounds, including the layer structures $TMAV_4O_{10}$, $TMAV_3O_7$, and $Li_xVO_2 \cdot H_2O$, are formed from pH 10 to pH 2.5 when acetic acid is used for pH control. At the lowest pHs highly crystalline fibrous vanadium oxides are formed. A range of double layer V_2O_5 structures are formed when iron, zinc and copper are also present in the reaction medium. These new layered vanadium oxides readily and reversibly undergo redox reactions with lithium and other ions and molecules, either chemically or in electrochemical cells; e.g. giving layered VO_2, $(C_8H_{17}NH_2)_nVO_2$ and $TMAV_3O_8$. They also undergo ion-exchange reactions. Supported by NSF-DMR.

766. THE NANOGAS 1 EXPERIMENT: DOES MICROGRAVITY INFLUENCE SELF-ASSEMBLY? Homayoun Ahari, Carol Bowes, Neil Coombs, Ömer Dag, Tong Jiang, Alan Lough, <u>Geoffrey A Ozin</u>, Srebri Petrov, Igor Sokolov, Atul Verma, Gregory Vovk and David Young, Materials Chemistry Research Group, Chemistry Department, 80 St. George St., University of Toronto, Toronto, Ontario, Canada, M5S 3H6.

In this lecture I will describe the results for the templated synthesis and crystal growth of nanoporous tin(IV) sulfides in µG and 1G environments.[1] The µG experiments were conducted on-board the June 1996 Endeavor STS 77 NASA Space Shuttle flight. The main objective of the mission was to evaluate the building-block assembly and crystal growth steps that are under the influence of a near zero gravity situation. There is compelling evidence that the materials harvested in space display a marked improvement in their bulk and surface crystallinity, optical quality, crystal habit, smoothness of the crystal faces, and registry of the porous layers. [1] Materials Science in Microgravity, Canadian Space Agency Newsletter, 1996.

767. SHAPE AND PHASE TRANSFORMATIONS IN THE NANO- TO MACRO-SCALE CONTINUUM. <u>G. D. Stucky</u>, C. C. Landry, S. H. Tolbert, Q. Huo, A. Firouzi, M. Janicke, and B. F. Chmelka, Materials, Chemistry, and Chemical Engineering Departments, University of California, Santa Barbara CA 93106; S. Schacht and F. Schüth, Institut für Anorganische Chemie, Universität Frankfurt, Frankfurt, Germany

Non-equilibrium modification of interfacial interactions in multicomponent silicate-surfactant mesophases has been used to promote and control transformations between different morphologies and macroscale topologies. These interactions include framework charge density, structure frustration due to the uneven response of organic versus inorganic domains, co-solvent effects, and changes introduced by silicate polymerization kinetics under hydrothermal conditions. Time-resolved *in situ* X-ray diffraction on silicate-surfactant mixtures has been used to monitor mesostructural changes during hydrothermal polymerization of silicate species in the presence of alcohol co-solvents. These studies reveal the existence of a new intermediate phase in a hexagonal (MCM-41) —> cubic (MCM-48) epitaxial mesophase transition. Diffraction data and theory strongly suggest that this new phase originates from fluctuations along the 6-fold axis of the hexagonal phase which ultimately transforms into the 3-fold body diagonal of the Ia3d unit cell of MCM-48. The probable transformation mechanism, along with the structure and properties of this new phase, will be described. More recent work on the nature of phase transformations involving bicontinuous mesophases and the use of these transformations and defect disclinations in materials synthesis will also be presented.

768. DESIGN STRATEGIES IN MINERALIZED BIOLOGICAL MATERIALS. J. Aizenberg, S. Albeck, G. Falini, Y. Levi, S. Weiner, L. Addadi, Department of Structural Biology, Weizmann Institute of Science, Rehovot 76100, Israel.

Organisms have been producing mineralized skeletons for hundreds of millions years. They have thus evolved many different strategies for improving the materials used, at almost all hierarchical levels from Ångstroms to millimeters. Key components of biological materials are the macromolecules, which are intimately involved in controlling nucleation, growth, and texture of the single crystals as well as in shaping them and adapting the mechanical properties of the mineral phase to function. In bivalve mollusk shells, acidic glycoproteins have been recently shown to be responsible for the stabilization of a given polymorph of calcium carbonate, calcite or aragonite. In some calcareous sponge spicules and ascidian spicules, different glycoproteins are responsible for stabilizing an amorphous calcium carbonate phase, which is very unstable under non-biogenic conditions. Much can still be learned from studying the principles of structure-function relations of biological materials. Some of this information may also provide new ideas and strategies for improved design of synthetic materials.

769.
ALKENE INSERTION AND REDUCTIVE ELIMINATION REACTIONS OF PLATINUM AND PALLADIUM PHOSPHIDO COMPLEXES David S. Glueck,[a] Denyce K. Wicht,[a] J. Mulei Nthenge,[a] Belinda M. Lew,[a] Louise Liable-Sands,[b] Glenn P. A. Yap,[b] Arnold L. Rheingold.[b] (a) Department of Chemistry, 6128 Burke Lab, Dartmouth College, Hanover NH 03755. (b) Department of Chemistry, University of Delaware, Newark DE 19716.

The metal-mediated formation and cleavage of phosphorus-carbon bonds has been proposed as important steps in catalytic phosphination and hydrophosphination reactions, but has rarely been observed directly. In addition, in many of the industrially useful processes in which the active catalyst contains ancillary phosphine ligands, P-C cleavage has been cited as a cause of catalyst deactivation. Metal phosphido-alkyl complexes are often invoked as intermediates in these processes. We recently reported the insertion of an olefin exclusively into the Pt-P bond in a series of phosphido-alkyl complexes as well as P-C reductive elimination in a related Pd species. The mechanistic aspects of these reactions will be discussed as well as model studies of their application in the asymmetric catalytic synthesis of chiral phosphines.

[M] = Pt(dppe), Pd(dppe)

770. PALLADIUM-CATALYZED DECARBONYLATION OF A PHOSPHAKETENE: DIPHOSPHAUREYLENE INTERMEDIATES IN DIPHOSPHENE FORMATION Marie-Anne David,[a] Denyce K. Wicht,[a] Michael A. Walters,[a] David S. Glueck,[a] Glenn P. A. Yap,[b] Louise M. Liable-Sands,[b] Arnold L. Rheingold.[b] (a) Department of Chemistry, Dartmouth College, Hanover NH 03755. (b) Department of Chemistry, University of Delaware, Newark DE 19716.

Zerovalent palladium complexes catalyze decarbonylation of the phosphaketene Mes*PCO [Mes* = 2,4,6-(t-Bu)$_3$C$_6$H$_2$] to form the diphosphene Mes*P=PMes*. P-P bond formation is proposed to occur via reductive elimination/CO extrusion from diphosphaureylene intermediates, in a reaction whose rate is dramatically affected by the ancillary ligands.

771. FACILE THERMAL ACTIVATION OF SMALL MOLECULES BY A TUNGSTEN VINYL COMPLEX. Sean A. Lumb, Jeff D. Debad, and Peter Legzdins, Department of Chemistry, The University of British Columbia, 2036 Main Mall, Vancouver, Canada, V6T 1Z1

Details of the novel thermal activation chemistry displayed by Cp*W(NO)(CH$_2$SiMe$_3$)(CPh=CH$_2$) **1** in the presence of a variety of small molecules will be presented. For example, the double C-H activation of Et$_2$O yields the functionalized metallacyclopentane complex **2**. In contrast, C-H activation is not observed during the thermolysis of **1** in neat MeCN, the reaction yielding the diazametallabicyclic complex **3**. A mechanism for the two modes of reactivity will be substantiated by the results of kinetic and mechanistic studies.

772. CATALYTIC HYDROACYLATION BY COBALT(I) OLEFIN COMPLEXES Maurice Brookhart* and Christian P. Lenges, Department of Chemistry, University of North Carolina at Chapel Hill, Chapel Hill, North Carolina 27599-3290

The combination of C-H activation with C-C bond formation remains an important objective in homogeneous catalysis. The formation of ketones from aldehydes and olefins illustrates this reactivity and has been discussed in detail for the intramolecular formation of cyclopentanones using Rh-based systems described by Bosnich et al. (*JACS*, 1994, *116*, 1821).

We report here the catalytic *inter*- and *intra*molecular hydroacylation using labile Co(I) bis-olefin complexes of the type [C$_5$Me$_5$Co(C$_2$H$_3$R)$_2$] which convert aromatic and aliphatic aldehydes, regioselectively, to ketones under ambient conditions. A detailed mechanistic investigation of this reaction will be presented which includes an identification of the resting state of catalysis.

773. **NEW ENTRIES TO AND NEW REACTIONS OF FLUOROCARBON LIGANDS** D. Huang, P. R. Koren, E. R. Davidson, K. G. Caulton Department of Chemistry, Indiana University, Bloomington, IN 47405-4001

Reaction of RuHF(CO)L$_2$ (L = P^{t}Bu$_2$Me) with Me$_3$SiCF$_3$ yields RuHF(CF$_2$)(CO)L$_2$. The formation of this product is rationalized by computational study. Rearrangement of RuHF(CF$_2$)(CO)L$_2$ occurs at 25 °C in THF to give RuF(CF$_2$H)(CO)L$_2$. Evidence is offered that the mechanism for this migration is pre-equilibrium dissociation of L. Reaction of saturated RuHF(CF$_2$)(CO)L$_2$ with CO at 25 °C is remarkably fast to give RuH(CF$_3$)(CO)$_2$L$_2$. This is attributed to the fast equilibrium (see drawing) of RuHF(CF$_2$)(CO)L$_2$, and RuH(CF$_3$)(CO)L$_2$ supported by an ^{19}F spin saturation transfer experiment.

774. **The Significance of Borane Tuning in Titanium Catalyzed Hydroborations**
Douglas H. Motry, Scott J. Stoudt, Aimee G. Brazil and Milton R. Smith, III, *Department of Chemistry, Michigan State University, East Lansing, MI 48824*

Hydroborations mediated by early transition metal complexes can provide product selectivities that complement those observed in uncatalyzed hydroboration processes. A common problem that confronts early transition metal systems is incompatibility between the metal and borane reagent. For example, selectivities that are observed in stoichiometric transformations are not always realized when reactions are run under catalytic conditions. We have found that the substituents that ligate boron are critical to the catalytic viability in early metal systems. Recent results concerning metal catalyzed hydroborations will be presented.

775. THE MECHANISM OF INSERTION OF ETHYLENE INTO AN Ir-OH BOND. CATALYTIC TANDEM ACTIVATION OF THE REACTANTS BY TWO [η^5-Cp*(Ph)IrPMe$_3$]$^+$ COMPLEX FRAGMENTS. JOACHIM C. M. RITTER AND ROBERT G. BERGMAN DEPARTMENT OF CHEMISTRY, UNIVERSITY OF CALIFORNIA, BERKELEY CA 94720.

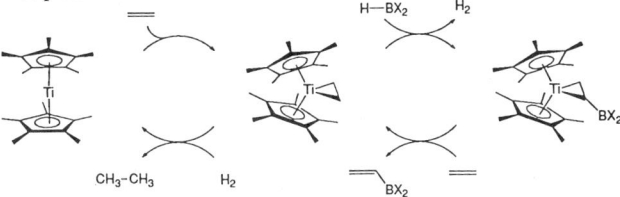

The Wacker process is used to make about 4 million tons of aldehydes from alkenes each year. The critical step in this transformation involves the insertion of an alkene into a Pd-OH bond. The reaction of the iridium hydroxide Ph[Ir]OH (**1**) ([Ir] = Cp*IrPMe$_3$) with ethylene at 25 °C to give Ph[Ir]CH$_2$CH$_2$OH (**2**) can serve as a homogeneous model for the insertion reaction. Mechanistic studies provide evidence that this process is catalyzed by cation **6**. In this reaction, both ethylene and OH$^-$ are brought together by two individual Ph[Ir]$^+$ fragments. In a second catalytic cycle, Ph[Ir]$^+$ cations catalyze the turnover of Ph[Ir]CH$_2$CH$_2$OH (**2**) and Ph[Ir]OH (**1**) to give Ph[Ir]CH$_2$CHO (**3**), Ph[Ir]H (**10**) and H$_2$O. This transformation involves the Meerwein-Ponndorf-Verley type reaction of an alkoxide intermediate with Ph[Ir]$^+$ cations, a new mechanism for β-hydride elimination of coordinatively saturated alkoxides.

776. COMPETITIVE REDUCTIVE ELIMINATION REACTIONS FROM PT(IV) CENTERS TO FORM CARBON-CARBON AND CARBON-OXYGEN BONDS. K.I. Goldberg, B.S. Williams, and A. W. Holland, Department of Chemistry, Box 351700, University of Washington, Seattle, WA 98195-1700

Pt(IV) complexes dppePtMe$_3$X (X = OAc, OAr) have been synthesized and characterized. An X-ray crystal structure of the acetate derivative has been obtained. Solution thermolyses of these compounds lead to competitive carbon-carbon and carbon-oxygen reductive eliminations to form ethane and methyl acetate or methyl aryl ethers. The relative product distributions are highly dependent on solvent polarity. Kinetic and mechanistic studies of these carbon-carbon and carbon-oxygen reductive elimination reactions will be presented.

777.
USING CARBON-HYDROGEN BOND ACTIVATION FOR THE SYNTHESIS OF TERTIARY-ALKYL IRIDIUM COMPLEXES. **Peter J. Alaimo**, Bruce A. Arndtsen, Robert G. Bergman, University of California, Berkeley, CA, 94720-9989

The iridium(III) complex (1) reacts quickly and cleanly with a variety of aldehydes to give products that have been characterized as (2). This reaction works well for certain tertiary aldehydes, giving the first two structurally characterized tertiary-alkyl iridium complexes. Synthetic, spectroscopic, and mechanistic details concerning this general transformation will be presented.

778. C,O– REDUCTIVE ELIMINATIONS FROM Ni(II) COMPLEXES. R. Han and G. L. Hillhouse,* Department of Chemistry, The University of Chicago, Chicago, IL 60637.

The oxanickelacycles [(PMe$_3$)NiOCH$_2$CH$_2$CMe$_2$C$_6$H$_4$]$_2$ (1), (bpy)NiOCH$_2$CH$_2$CMe$_2$C$_6$H$_4$ (2) and (dmpe)NiOCH$_2$CH$_2$CMe$_2$C$_6$H$_4$]$_2$ (3) react at ambient temperature with O$_2$ to undergo oxidatively-induced O,C-reductive elimination to give 4,4-dimethylchroman (4) in ~40 % isolated yield. In the absence of external oxidants, at 100 °C **1**, **2**, and **3** undergo β-H elimination followed by C,H-reductive elimination to give 3-methyl-3-phenylbutyraldehyde (5) as the major organic product. The generation of formates from **1** and paraformaldehyde will also be described.

779. **FACTORS AFFECTING THE STABILITY OF PLATINUM(II) σ-ADDUCTS.**
Shannon S. Stahl, Jay A. Labinger and John E. Bercaw.
Arnold and Mabel Beckman Laboratories of Chemical Synthesis,
California Institute of Technology, Pasadena, CA 91125

Several complexes of the type trans-$(PCy_3)_2Pt(H)X$ and [trans-$(PCy_3)_2Pt(H)L$]$^+$ BArF$^-$ (X = halides, H, CH_3, Ph, SiH_3, CN; L = CO, PMe_3, 4-picoline; BArF = B(3,5-$C_6H_3(CF_3)_2$)$_4$) have been treated with a proton source, HX' (X'$^-$ = Cl$^-$, OTf$^-$, BArF$^-$), at low temperature. The site of protonation varies depending on the nature of the ligand *trans* to the hydride as well as the counterion, X'$^-$; both dihydrogen adducts and platinum(IV) dihydride species have been identified. These results provide insight into the factors which promote oxidative cleavage of platinum(II)-coordinated σ-bonds such as those in dihydrogen or alkanes. σ-adducts have been proposed as intermediates in platinum(II)-catalyzed alkane oxidation.

780. **C-H ACTIVATION AT CATIONIC PLATINUM(II) CENTERS.**
Matthew W. Holtcamp, Jay A. Labinger, and John E. Bercaw.
Arnold and Mabel Beckman Laboratories of Chemical Synthesis, California Institute of Technology, Pasadena, CA 91125

The addition of one equivalent of [H(solvent)$_n$][BAr$_f$] {BAr$_f$ = B(3,5-$C_6H_3(CF_3)_2$)$_4$} to (tmeda)Pt(CH_3)$_2$ {tmeda = tetramethylethylene diamine} in pentafluoropyridine results in the formation of [(tmeda)Pt(CH_3)(NC_5F_5)][BAr$_f$], **1**. At 85 °C **1** reacts with $^{13}CH_4$ forming [(tmeda)Pt($^{13}CH_3$)(NC_5F_5)][BAr$_f$]. Reactions of **1** in pentafluoropyridine were also carried out with benzene yielding [(tmeda)Pt(C_6H_5)(NC_5F_5)][BAr$_f$]. The reactions of **1** with methane-d_4, cyclohexane-d_{12}, toluene-d_8, or benzene-d_6 in pentafluoropyridine yield isotopomeric mixtures of CH_nD_{4-n}, implicating alkane σ-complex intermediates. The addition of one equivalent of [H($Et_2O)_2$][BAr$_f$] to (tmeda)Pt(CH_3)$_2$ in diethyl ether yields [(tmeda)Pt(CH_3)(OEt_2)][BAr$_f$], which at room temperature liberates methane with formation of [(tmeda)Pt(=C(CH_3)(OCH_2CH_3)(H)][BAr$_f$].

781. **Phosphodiester hydrolysis catalyzed by metals and Functional Groups.**
Thomas C. Bruice, Department of Chemistry, University of California, Santa Barbara, CA 93106

Studies carried out in the last few years which deal with the general-acid/general-base and Lewis acid/metal bound hydroxide ion hydrolysis of phosphodiesters will be described.

782.

The Role of X-ray Crystallography in Solution Coordination Chemistry, Abraham Clearfield, Department of Chemistry, Texas A&M University, College Station, TX 77843.

The need for crystal structure solutions became apparent in brief discussions I had with Professor Martell concerning several aspects of his work in solution chemistry. Initial work was carried out by Arthur's students Jim Timmons and Eric Neiderhoffer. These results proved to be enlightening and several will be discussed in the context of the solution chemistry. One cannot be sure of deducing the exact species in solution since changes may occur on dissolution. However, with the advent of powerful synchrotron sources it may soon be possible to determine the structures of the dissolved species and have a more complete picture of the chemistry of the system.

783. **THE Ga(III) AND In(III) COMPLEXES OF TRIS(2-MERCAPTOBENZYL)AMINE**
R. J. Motekaitis, A. E. Martell, Department of Chemistry, Texas A&M University College Station, Texas 77843-3255, USA, and M. J. Welch, The Edward Mallinckrodt Institute of Radiology, Washington University, School of Medicine, St. Louis Missouri 63110, USA

Tris(2-mercaptobenzyl)amine NS3 (H_3L) was synthesized by S. A. Koch, et al. (JACS, 1995, *117*, 8468). The ligand was studied by potentiometric titration in 70% ethanol/water (v/v) at 25.00 (5) °C and μ = 0.100 M (KCl). The successive log protonation constants found are 7.88 (3), 6.90 (2), 6.21 (1) and 2.53 (1). The Ga(III) and In(III) complexes were also studied in the same medium, and at low pH two species were detected with each metal ion: ML and MHL^+. The log normal formation constants ($M^{3+} + L^{3-}$ = ML) found are: 20.5 (1) for Ga^{3+} and 21.2 (1) for In^{3+}. The chelate log protonation constants (ML + H^+ = MHL^+) are 2.0 (1) for Ga^{3+} and 1.8 (1) for In^{3+}. The equilibrium constants will be discussed in terms of the probable coordinate bonding in the complexes formed and will be compared to the analogous NO3 ligand. The biodistributions of the Ga(III) and In(III) chelates will be described.

784. CHELATION THERAPY FOR TREATING IRON OVERLOAD. C. A. Tyson, Toxicology and Metabolism Laboratory, SRI International, Menlo Park, California 94025.

The clinical usefulness of chelation therapy in the management of iron-loading anemias has been well demonstrated since the reintroduction of desferrioxamine (DFO) some 25 years ago. Despite problems with patient compliance because of the need for repeated, lengthy intravenous or subcutaneous infusions of DFO and with its high costs, this iron chelator remains the only agent in widespread clinical use. The National Institute of Diabetes and Digestive and Kidney Diseases (NIDDK) has sponsored a number of research initiatives to develop orally effective, less expensive substitutes. Several hundred candidate compounds have been screened over the last two decades using animal models *in vitro* and *in vivo* or combinations of both. These studies have led to the identification of several interesting compounds of possible clinical usefulness. The most promising appear to be polyanionic amines, aryl hydrazones, and 3-hydroxypyrid-4-ones. This presentation will review the current status of the preclinical development and clinical evaluation of these compounds. Supported in part by NIH Contract No. N01-DK-3-2201.

785.
STEREOSELECTIVE BINDING OF OPTICALLY ACTIVE AMINO ACIDS BY THE CU(II) AND Ni(II) CHELATES OF N-CARBOXYMETHYL-L-HISTIDINE.

M.H. Kim and R. Nakon*, Department of Chemistry, West Virginia University, Morgantown, WV 26505.

Formation constants for the binding of D-and L-amino acids to the Ni(II) and Cu(II) complexes of N-carboxymethyl-L-histidine (N-Cm-L-ht) indicate that the Ni(II) chelate preferentially binds D-amino acids while the Cu(II) one favors the L-isomers. Models indicate the "reverse" stereoselectivity can be explained in the different conformations that are adopted by N-Cm-L-ht when bound to Ni(II) and Cu(II). Visible spectra indicate [Ni(N-Cm-L-ht) (AmAc)$^-$] is octahedral while the Cu(II) analog is five coordinate with all three nitrogen donors in the square plane of Cu(II). The order of stereoselectivity is val>phe>leu~thr>ala. The [(Cu(N-Cm-L-ht)AmAc$^-$)] chelates undergo a remarkable reaction in basic solution, dechelation of the aminoacidate ligand upon imidazole proton ionization (eq 1).

OH^- + [Cu (N-Cm-L-ht)(D-AmAc)$^-$]$^-$ ⇌ [Cu(H_{-1} N-Cm-L-ht) H_2O)$^-$] + AmAc$^-$ (eq 1)

786. OXYGEN ATOM TRANSFER SYSTEMS OF MOLYBDENUM AND TUNGSTEN RELATED TO OXOTRANSFERASES. R. H. Holm, J. P. Donahue, and G. C. Tucci, Department of Chemistry, Harvard University, Cambridge, Massachusetts

Recent developments in the oxo transfer chemistry of molybdenum and tungsten in dithiolene ligand environments are summarized. Topics include the general synthesis of oxo-dithiolene complexes, formulation of new *mono*oxo-Mo(IV,V,VI) systems and their atom transfer reactivity, and comparative kinetics of oxo transfer of molybdenum and tungsten dithiolene complexes.

787. MOLECULAR MECHANICS CALCULATIONS AS A TOOL IN LIGAND DESIGN, R. D. Hancock, IBC Advanced Techologies, American Fork, Utah 84003

The use of Molecular Mechanics (MM) calculations as a tool in ligand design in applications ranging from complexes of Ga(III) and In(III) for biomedical applications, to Fe(III) in the catalytic oxidation of H_2S to S, is described. The problems associated with developing suitable force fields for these metal ions is described. Fe(III) in particular is problematic because of the role of π-bonding in the Fe-O bond. There is an inverse relationship between Fe-N and Fe-O bond length in complexes of Fe(III) with O_3N_3 donor sets. Another area of interest is design of ligands for the complexation of Gd(III) in MRI imaging applications. The number of water molecules that coordinate to the Gd(III) influence its ability to relax the proton signal, and hence the efficiency of the complex for MRI. A method for prediction of the number of water molecules coordinated to the Gd(III) based on MM calculations is presented. A force field for the modeling of complexes of alkali and alkaline earth metal ions is presented. It is shown how calculation of strain energy as a function of metal ion size in the complex can be used to calculate selectivities that ligands display towards metal ions of differing sizes.

788. METAL IONS AND DIOXYGEN IN SWARMER CELL DIFFERENTIATION OF *PROTEUS MIRABILIS*. C. Brakebill, M. L. Kalmer, A. Sutton, and **E. C. Niederhoffer**, Medical Biochemistry, School of Medicine, Southern Illinois University, Carbondale, Illinois 62901

Proteus mirabilis is an important atypical eubacterial kidney pathogen, which has evolved highly developed stress-responses to the immune system of healthy individuals. Iron plays a central role in several of these mechanisms by mediating its own uptake and transport as well as signaling production of various macromolecules by the bacterium. We have observed that iron (in an aerobic environment) is required for swarmer cell differentiation from short (2-μm) swimmer cells to elongated (>20 μm) swarmer cells and that cadmium inhibits this cellular event. This presentation will highlight our understanding of the impact of metal ions and dioxygen on the swarmer cell differentiation of *P. mirabilis*.

789. MOLECULAR INTERACTION OF PYROPHOSPHATE WITH 1,13-DIOXA-4,7,10,16,20,24-HEXAAZACYCLOHEXACOSANE (OBISDIPEN) AND ITS MONO-NUCLEAR AND DINUCLEAR COPPER(II) COMPLEXES, J. B. English, A. E. Martell, R. J. Motekaitis, Dept. of Chemistry, Texas A&M Univ., College Station, Texas 77843-3255, U.S.A., I. Murase, Dojindo Laboratories, Kumamoto, Japan.

The unsymmetrical hexabasic macrocyclic ligand OBISDIPEN forms an assortment of cationic species including mono- through hexaprotonated forms of the macrocyclic ligand in the absence of metal ions as well as mononuclear and dinuclear Cu(II) complexes. These mononuclear and dinuclear Cu(II) complexes also form protonated and hydroxylated species. The cationic hosts combine with certain molecules and anions (guests) which have the requisite size, shape, and donor groups. A bidentate guest molecule or anion can bridge the two metal ions and coordinate them simultaneously. This paper describes the reactions of OBISDIPEN and of its six mononuclear and four dinuclear Cu(II) complexes with the pyrophosphate bridging group to form complexes which exist in a variety of protonated conformations. The binding in these complexes consists of coulombic forces, hydrogen bonds, and coordinate metal-donor bonds. The equilibrium constants were obtained by potentiometric titration at 25.0 ± 0.1 °C and $\mu = 0.1$ M KCl.

790. **A POTENTIOMETRIC STUDY OF METAL CHELATES OF TETRAETHYLENE PENTAMINEHEPTAACETIC ACID AND PENTAETHYLENEHEXAMINEOCTA ACETIC ACID,** P. Letkeman, L. Letkeman, M. Lemaire, Dept. of Chemistry, Brandon University, Brandon, Manitoba, R7A 6A9 Canada, R. J. Motekaitis, A. E. Martell, Dept. of Chemistry, Texas A&M University College Station, Texas 77843-3255, USA, I. Murase, Dojindo Laboratories, Kumamoto, Japan.

The protonation constants of pentaethylenehexamineoctaacetic acid (PHOA) and tetraethylenepentamineheptaacetic acid (TPHA) were determined by potentiometric titration in aqueous solution at an ionic strength of 0.10 M (KNO_3) and at 25.0 °C. The formation constants are reported for Co(II), Ni(II), Cu(II), Zn(II), Cd(II), Hg(II), and Pb(II) complexes of PHOA and TPHA, as well as their related protonated species. The similarities in the stabilities of the PHOA and TPHA are explained in terms of ligand denticity and steric effects. The difference in the stability trend of the 1:1 and 2:1 metal:ligand complexes of the aminopolycarboxylate anions is interpreted on the basis of the probable structures of the complexes in solution.

791. **METAL CHELATION IN THE ACTIVITIES OF NEUROTRANSMITTERS AND NEUROLEPTICS.** Krishna S. Rajan, IIT Research Institute, 10 W. 35 St., Chicago, IL 60616.

On the basis of the observed co-occurrence of significant amounts of Cu(II), Zn(II) and Mn(II) along with neurotransmitter amines such as dopamine (DA) and norepinephrine (NE) in the subcellular organelles of the brain, possible involvement of "heteropolynuclear metal chelation" in the binding, storage and transport activities of these catechol amines has been postulated. Further, chlorpromazine (CPZ), and its metabolites such as 7-(OH)-CPZ, 7,8-di(OH)-CPZ, and others have been found to exhibit strong affinities for binding with brain synaptosomal metals in the combined presence of ATP and the neurotransmitter amines. Results of *in vitro* and *in vivo* studies have indicated the relatively strong neuroleptic activity of these hydroxo-CPZ metabolites. Possible effects of the CPZ's on the cerebral activity of neurotransmitters amines through their chemical (chelation) interactions with co-occurring synaptosomal and synaptic vesicular metal ions is discussed.

792. **METAL CHELATES AND RELATED COMPOUNDS,** A. E. Martell, Department of Chemistry, Texas A&M University College Station, Texas 77843-3255, USA

Fifty years of research on the chemistry of metal chelate compounds are described. The work which began with the reactions of EDTA and related ligands was soon expanded to chelates in general, as indicated in the book with Calvin, "Chemistry in the Metal Chelate Compounds", published in 1952. Many research areas involving metal chelates were then investigated, including vitamin B_6 catalysis and cobalt dioxygen complexes. The work on metal chelates was recently expanded to macrocyclic compounds and cryptates as described in a book with Hancock, "Metal Compounds in Aqueous Solution", published in 1996. The special properties of these coordination compounds have been employed to make major contributions to coordination chemistry, biochemistry, medicinal, environmental, and industrial chemistry. Current and future research involves multinuclear metal complexes for catalysis, chelates of trivalent metal ions and other metal complexes for diagnosis and therapy, and reactions of cobalt dioxygen complexes.

793.

^{1}H AND ^{13}C ANALYSIS OF SUBSITUENT EFFECTS ON THE UNPAIRED ELECTRON SPIN DENSITY DISTRIBUTION IN UNSYMMETRICALLY SUBSTITUTED PARAMAGNETIC MODEL HEME COMPLEXES. Ken Skidmore and Ursula Simonis, Department of Chemistry and Biochemistry, San Francisco State University, San Francisco, CA 94132

Understanding the unpaired electron spin density distribution in paramagnetic heme proteins and their corresponding low-spin iron porphyrin model complexes is important in order to comprehend the mechanisms by which these systems function. In particular, a detailed knowledge of how different subsituents influence the behavior of the unpaired electron spin density in the heme macrocycle is desired. To this end, a series of substituted low-spin iron(III) tetraphenylporphyrin (TPPFe) complexes have been synthesized. These complexes each have a F, Cl, Br, or CF_3 subsituent on one phenyl ring in either the *ortho*, *meta*, or *para* position, and three *para*-OCH_3 substituents on the other phenyl rings. Two-dimensional homo- and heteronuclear ^{1}H and ^{13}C NMR techniques, including pulsed-field gradient COSY, HMQC, and HMBC experiments, have been applied to these paramagnetic systems to determine the unpaired electron spin density distribution at the carbon positions of the porphyrin macrocycle. An analysis of these effects, along with synthetic details and a discussion of the NMR results, will be presented.

794. DESIGN AND SYNTHESIS OF NOVEL DISCOTIC LIQUID-CRYSTALLINE PORPHYRINS. Qing Min Wang and John Arnold, Department of Chemistry, University of California, Berkeley, California, 94720-1460

Molecular design and synthesis of ordering systems at a molecular level to study intermolecular interactions and cooperative phenomena is an active field in molecular materials development. Here we report some novel liquid-crystalline porphyrins containing either two or four flexible triangular DOBOB groups on the porphyrin ring. These materials display nematic discotic mesophase behavior and have been characterized by UV/Vis, ^{1}H NMR and FAB MS. Thermal behavior has been probed using polarized microscopy and DSC. The results of related synthetic modifications will also be described.

794A REACTION CHEMISTRY OF THE $\{[(Me_3Si)_2N]_2V\}^{n+}$ (n = 1, 2, 3) FRAGMENT: SYNTHESIS AND REACTIVITY OF HYDROCARBYL AND OTHER DERIVATIVES. Christopher P. Gerlach and John Arnold, Department of Chemistry, University of California. Berkeley, CA 94720-1460

The synthesis of hydrocarbyls and other derivatives of the $\{[(Me_3Si)_2N]_2V\}^{n+}$ (n = 1, 2, and 3) fragment will be presented. The rich reactivity of these species includes the insertion and oligomerization of isocyanides as well as redox/atom-transfer reactions. The metallation of a -$N(SiMe_3)_2$ ligand via γ-hydrogen elimination is shown to be dependent both on the nature of the R group and the dative ligand L in $[(Me_3Si)_2N]_2VR(L)$ complexes.

795. METAL COMPLEXES OF HETEROSAPPHYRINS AND AMETHYRIN. J. L. Sessler, A. Gebauer, S. J. Weghorn, A. K. Burrell, J. Lisowski, M. C. Hoehner, A. Guba, V. Lynch. Department of Chemistry and Biochemistry, University of Texas at Austin, Austin, TX 78712.

Our program of investigation into the synthesis of new expanded porphyrins has always been driven by our desire to expand the rich metallation chemistry of porphyrins. The larger core size of expanded porphyrins gives them the ability to form metal complexes not observed in regular porphyrins. This reaches from in-plane complexation of large transition metals to the formation of oligo-metallic complexes. Here, a comparison of the metallation chemistry of two expanded porphyrins, heterosapphyrin **1** and amethyrin **2** (with metals like Co, Cu, Zn, and Rh), is presented. These two systems give a first insight into the role of structural composition and aromaticity in the overall metal coordination ability of expanded porphyrins.

1. R = H and/or Et and/or Me
   X = NH, O, S

2. R = Et, Me

796. POSSIBLE INTERMEDIATES IN HEME CATABOLISM. CHARACTERIZATION OF METAL FORMYLBILIVERDINS AND THEIR DEGRADATION PRODUCTS. Richard Koerner, Alan L. Balch, Marilyn M. Olmstead and Pamela M. Van Calcar, Department of Chemistry, University of California, Davis, California 95616.

Coupled oxidation of iron porphyrins produces analogues of heme oxygenase degradation products. We have found that photo-degradation of porphyrins produces formylbiliverdins whose metal complexes oxidize in air to form products identical to those produced by coupled oxidation and heme oxygenase. Several metal complexes (Co, Ni, Cu, In and Fe) of octaethylformylbiliverdin have been synthesized and characterized by single crystal X-ray diffraction, EPR, NMR and electrochemical methods. The oxidative and photo-oxidative reaction products of metalloformylbiliverdins and their relevance to the mechanism of heme degradation by heme oxygenase and coupled oxidation will be discussed.

797. 2,3,7,8,12,13,17,18-OCTAFLUORO-5,10,15,20-TETRAARYLPORPHYRINS AND THEIR ZINC COMPLEXES: FIRST SPECTROSCOPIC, ELECTROCHEMICAL, AND STRUCTURAL CHARACTERIZATION OF A PERFLUORINATED TETRAARYLMETALLOPORPHYRIN. Eric. K. Woller and Stephen. G. DiMagno, Department of Chemistry, University of Nebraska-Lincoln, Lincoln, Nebraska 68588

We report a convenient and general synthesis β-octafluoroporphyrins bearing *meso* tetraaryl substituents, including the first synthesis and full characterization of a perfluorinated tetraarylporphyrin, perfluoro-5,10,15,20-tetraphenylporphyrin. The structural, spectroscopic and electrochemical data indicate that β-octafluoro-*meso*-tetraarylporphyrins are a new class of *planar*, electron-deficient ligands. Particularly impressive is the 0.5 V window over which the formal oxidation potential can be tuned using only aryl substituents. The invariance of the ligand structure with increasingly positive formal oxidation potential is a key advance; electronic effects have been severed from the nonplanar conformations exhibited by all other highly electron-deficient porphyrins.

798.

SYNTHESIS AND CHARACTERIZATION OF A SERIES OF ISOSTRUCTURAL METALLOPORPHYRIN CHALCONITROSYL COMPLEXES M(POR)(NE)X. M=OS,RU E=O,S. **Bryan D. Smith,** D. Scott Bohle*, Chen-Hsiung Hung, Annie K. Powell^, and Sigrid Wocadlo^ Department of Chemistry, University of Wyoming, Laramie, WY, 82071-3838. ^School of Chemical Sciences, University of East Anglia, Norwich, NR4 7TJ, England.

While transition metal chalcocarbonyls have a rich chemistry from CO to CTe, the corresponding chalconitrosyls, excepting NO, have remained poorly developed. Herein we describe: 1) the synthesis and characterization by IR, NMR, differential scanning calorimetry, cyclic voltammetry, and UV-Vis spectroscopy of a new series of isostructural metalloporphyrins containing NE where E=O,S; 2) the remarkable transformation of a thionitrosyl/nitrite complex to a nitrosyl/thiazate complex; 3) the crystal structures of two of these derivatives; and 4) a direct comparison of the M(NE) bonding in these isoelectronic chalconitrosyls.

799. SYNTHESIS, ELECTRON PARAMAGNETIC RESONANCE SPECTROSCOPY, PHOTOPHYSICS, AND ELECTROCHEMISTRY OF HIGHLY-CONJUGATED BIS(PORPHYRIN) COMPOUNDS

Renée Shediac, Paul J. Angiolillo, Mike H.B. Gray, Victor S.-Y. Lin, Michael J. Therien*, *Department of Chemistry, University of Pennsylvania, Philadelphia, Pennsylvania, USA, 19104-6323*

A family of homo- and heterobinuclear *meso*-to-*meso* ethyne-bridged bis(porphyrin) compounds have been synthesized. These species feature variable (porphinato) metal (Cu^{II}, Fe^{II}, Fe^{III}, Zn^{II}) and/or free-base porphyrin components and serve as models in which to probe electron- and energy-transfer events in the strong coupling regime.

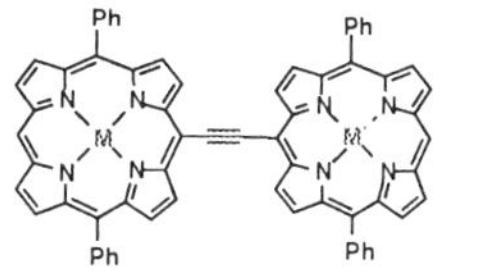

For M = M':
M = Zn^{2+}, H_2, Cu^{2+}, Fe^{2+}, Fe^{3+}

For M ≠ M':
M = Zn^{2+}
M' = H_2, Cu^{2+}, Fe^{3+}

Substantial electronic communication between the chromophores of these bis(porphyrin) compounds is evinced in the spectroscopy of these systems; time-resolved fluorescence, EPR and electrochemical experimental data will be presented.

800.

STUDY OF PROTON COUPLED ELECTRON TRANSFER IN PORPHYRIN SYSTEMS. **Yongqi Deng,** James A. Roberts, Daniel S. Engebretson, Chris K. Chang, and Daniel G. Nocera, Department of Chemistry, Michigan State University, East Lansing, MI 48824

Proton coupled electron transfer (PCET) is fundamental to energy conversion in biology. Porphyrin and chlorin complexes in which proton interfaces are internal or external to the electron transfer pathway have been prepared. It has been found that the electron transfer rate is mediated by the proton motion, especially when the proton interface is strongly coupled in an electronic sense to the electron donor or acceptor. The synthesis and characterization of the complexes will also be presented.

801. FORMATION OF COBALT-CARBON BOND IN ORGANOCOBALT CORRINOIDS: THE REACTION OF COBALT(II) CORRINOIDS WITH ORGANIC HYDROPEROXIDE
<u>X. Shawn Zou</u>, Kenneth L. Brown, Shifa Cheng, Department of Chemistry, Ohio University, Athens, OH 45701.

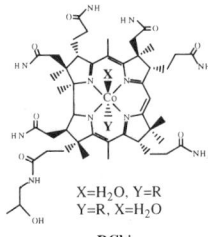

X=H$_2$O, Y=R
Y=R, X=H$_2$O

RCbi

Reactions of cob(II)inamide with 1,1-dimethylpropyl hydroperoxide at various ratios gave mixtures of the diastereomeric α- and β-ethylcobainmides. The compositions of the two diastereomers depends strongly on the ratio of the two reactants, the α-isomer ranging from <2 % when peroxide is in large excess to 87% when Co(II) is in large excess. Under the latter conditions, kinetically controlled products are attained for the reaction of R• with cob(II)inamide. Studies of the temperature dependence of product ratios under these conditions allow determination of the differential activation parameters for the reaction of R• with cob(II)inaminde and two side chain altered analogs and provide significant insight to stereocontrol in this reaction.

802. **DO ORGANOCOBALT PORPHYRINS HAVE AGOSTIC ALKYL GROUPS?** Yang Cao, Jeffrey L. Petersen, and <u>Alan M. Stolzenberg</u>, Department of Chemistry, West Virginia University, Morgantown, WV 26506-6045

Five-coordinate alkyl–cobalt(III) porphyrins have anomalous ^{1}H NMR spectra in which the β-proton peak of the alkyl ligand occurs at higher field than the broadened α-proton peak. Pyridine coordination causes the α-proton peak to sharpen and shift upfield and the β-proton peak to shift downfield. Given these observations, it was proposed that for the five-coordinate complex only, an agostic interaction between the β-protons and cobalt results in the β-protons being closer to the porphyrin and more greatly affected by the ring current than the α-protons. The X-ray structure of $C_2H_5Co(OEP)$ determined in our laboratory establishes that the Co–C–C angle is 118.2°, which is too large for an agostic ethyl group. Chemical shift and coupling constant data for $^{13}CH_3$-cobalt- and $^{13}C_2H_5$-cobalt porphyrins are inconsistent with the presence of an agostic interaction in solution. The spectral shifts upon pyridine coordination appear to be due to changes in the Co–C bonding.

803. PEPTIDE HELIX INDUCTION IN A SELF-ASSEMBLING HEMOPROTEIN MODEL.
Paige A. Arnold, William R. Shelton and <u>David R. Benson</u>, Department of Chemistry, University of Kansas, Lawrence, KS 66045

Formation of 1:1 complexes between cobalt(III) octaethylporphyrin and a water-soluble, cystine-dimerized peptide leads to peptide helix induction. This self-assembled structure is held together by exchange-inert bonds between Co and the imidazolyl side chains of histidine residues in the peptide dimer. Alteration of peptide conformation from random coil to 55% helix is similar to that recorded for a similar system in which two peptides are attached to iron(III) mesoporphyrin IX via amide linkages (Benson, D.R.; Hart, B.R.; Zhu, X.; Doughty, M.B. *J. Am. Chem. Soc.* **1995**, *117*, 8502-8510).

804. STRUCTURE OF MONOMERIC, SOLVENT-FREE METHYLLITHIUM AND METHYLSODIUM. <u>Douglas B. Grotjahn</u>*, B-z. Li, M. D. Allen, and L. M. Ziurys, Department of Chemistry and Biochemistry, Box 871604, Arizona State University, Tempe, AZ 85287-1604

Reaction of $Hg(CH_3)_2$ or $Hg(CD_3)_2$ with alkali metal vapor gives monomeric, solvent-free CH_3M or CD_3M (M = Li or Na), the structures of which have been determined for the first time using high resolution millimeter-wavelength spectroscopy.

805. **KINETICS AND MECHANISMS OF HYDRAZINE OXIDATIONS BY HALOGENS, INTERHALOGENS, AND HYPOHALOUS ACIDS.** Dale W. Margerum, Zhongjiang Jia, Murad Salaita, and Rong Ming Liu, Department of Chemistry, Purdue University, West Lafayette, IN 47907-1393

Rates of reactions of Cl_2, Br_2, I_2, BrCl, ICl, IBr, and HOCl with hydrazine are measured in aqueous acidic solutions. The reactions of Cl_2, Br_2, and BrCl are extremely fast and are measured by pulsed-accelerated-flow (PAF) techniques. The Br_2(aq) reactions are followed by loss of Br_3^- at 266 nm in 0.05 to 1.5 M $HClO_4$ at 25.0° C. The Br_2(aq) kinetics are used to measure the first ionization constant of $N_2H_6^{2+}$, K_{a1} = 1.92 M (μ = 1.0 M). The halogens and halogen compounds react directly with $N_2H_5^+$, because the N_2H_4 concentrations are extremely small in high acid (pK_{a2} = 8.0). The proposed mechanisms proceed via halogen-cation transfer reactions, where the relative rates are BrCl > Cl_2 > Br_2 >> HOCl >> I_2.

806. **MONOMERIC SODIUM AND POTASSIUM THIOLATES AND SELENOLATES - CONTROL OF IONIC ASSOCIATION BY CAREFUL LIGAND AND DONOR CHOICE.** Karin Ruhlandt-Senge, Scott Chadwick and Ulrich Englich, Department of Chemistry, Syracuse University, Syracuse, NY 13244-4100.

Sodium and potassium organochalcogenolates are highly reactive precursors, important for the synthesis of metal organochalcogenolates. In contrast to the relatively well explored lithium analogues, studies concerned with the heavier alkali congeners have been few. Specifically, the correlation between structure and reactivity has not been investigated. We will present syntheses, structural characterizations and reactivity studies of a variety of monomeric sodium and potassium thiolates and selenolates. Interestingly, the formation of contact and solvent separated ion pairs, and consequently, reactivity, can be influenced by careful choice of ligand and donor, examples being the contact ion-pair [K(dibenzo-18-crown-6)(THF)(S-2,4,6-tBu$_3C_6H_2$)] and the solvent-separated [K(18-crown-6)(THF)$_2$][S-2,4,6-tBu$_3C_6H_2$]. We will discuss the ionic association behavior with regard to type and hapticity of the donor, as well as size and electronic properties of the ligands. The independent variation of metal, ligands, donors and chalcogeno atoms will facilitate an understanding of the relationship between coordinative saturation of the metal center (resulting from metal-donor interaction), and steric saturation (achieved *via* steric bulk of the ligand).

807. **MAGNESIUM THIOLATES: AGGREGATION AND RELATION TO FUNCTION.** Scott Chadwick, Karin Ruhlandt-Senge and Ulrich Englich, Department of Chemistry, Syracuse University, Syracuse, NY 13244-4100

Magnesium thiolates are finding widespread application in a variety of chemical sciences—from their use as chalcogenolate transfer agents for redox sensitive transition metal chalcogenolates to asymmetric organic synthesis and as volatile precursors for magnesium doped solid state materials. Despite this emerging wealth of chemical utility, and the potential for new science as well, very little is known with regards to the structures exhibited in magnesium thiolate systems, and even less is known about the factors influencing solid state and solution structure and the relationship to reactivity. We have been systematically investigating the role of organo ligand and donor adjuncts on the solid state and solution structure of magnesium thiolates, with the goal of identifying the factors controlling ionic association and aggregation and ultimately chemical reactivity. We will present the synthesis and characterization of a family of magnesium thiolates, representing a range of aggregation states, including the first structurally characterized solvent separated magnesium thiolate ion pair, [Mg(15-crown-5)(THF)$_2$][SMes*]$_2$, (Mes*=2,4,6-Bu$^t_3C_6H_2$).

808. OLIGOMERIC SODIUM AND POTASSIUM THIOLATES: SYNTHESIS, STRUCTURE AND REACTIVITY. Ulrich Englich and Karin Ruhlandt-Senge, Department of Chemistry, Syracuse University, Syracuse, NY 13244-4100.

Sodium and potassium thiolates are widely used, highly reactive reagents for the synthesis of metal organochalcogenolates, some of which are not available if the corresponding lithium reagents are employed. Despite their importance in synthetic chemistry, the knowledge about the nature of these compounds and their properties remains poor. Specifically, the correlation between association on reactivity has not been explored. We prepared a family of sodium and potassium thiolates under variation of ligands and donors to investigate the dependency of structural characteristics on reactivity. Examples include the one-dimensional, polymeric, single chain derivative [Na(TMEDA)(STrip)]$_\infty$ (Trip = 2,4,6-iPr$_3$C$_6$H$_2$) and the hexameric box shaped [{Na(Et$_2$O)STrip}$_2${NaSTrip}]$_2$. The talk will present a selection of new oligomeric and polymeric sodium and potassium thiolates and will discuss their synthesis, structural and spectroscopic features.

809. SYNTHESES OF MULTIPLY BONDED MAIN GROUP 14 COMPOUNDS. **R.S. Simons**, P.P. Power, Department of Chemistry, University of California, Davis, CA 95616

The stabilization of multiply bonded compounds involving the heavier main group 14 elements (Si, Ge, Sn, and Pb) has been investigated. The bulky σ-bonded aryl 2,6-Mes$_2$C$_6$H$_3$ (Mes = 2,4,6-Me$_3$C$_6$H$_2$) substituent has been employed for this purpose, providing the first example of a stable compound containing a Mo≡Ge triple bond. Results pertaining to our efforts in this area will be presented.

810. A NEW HOMOGENEOUS SUPPORT FOR THE DENSIFICATION OF WELL-DEFINED TRANSITION METAL COMPLEXES. Lawrence R. Sita, Rimo Xi. Department of Chemistry, The University of Chicago, 5735 South Ellis Avenue, Chicago, Illinois, 60637

Thermolysis of [Sn(μ-OSiMe$_3$)(OSiMe$_3$)]$_2$ at 180° C quantitatively provides the novel main group cluster, Sn$_6$(μ$_3$-O)$_4$(μ$_3$-OSiMe$_3$)$_4$ (**I**). The ability and extent of the six tin (II) centers to participate as 2 e σ-donors in the formation of well-defined transition metal complexes has been assessed. This talk will focus on recent results obtained from experiments designed to (1) determine the highest level of complexation that can be achieved for **I**, (2) determine the extent to which electronic effects are communicated through the Sn-O framework of heterobimetallic adducts of **I** and (3) determine if catalytically active adducts can be prepared that possess multiple reaction centers.

811. METATHESIS OF CARBON DIOXIDE AND ISOCYANATES WITH TIN (II) BIS(SILYL)AMIDES. Jason R. Babcock, Rimo Xi, and Lawrence R. Sita, Searle Chemistry Laboratory, Department of Chemistry, The University of Chicago, 5735 S. Ellis Ave., Chicago, IL 60637

Recently, we reported that bis[bis(trimethylsilyl)amino]tin (II) reacts with carbon dioxide to form in high yield [Sn(OSiMe$_3$)(μ-OSiMe$_3$)]$_2$, trimethylsilyl isocyanate, and N,N'-bis(trimethylsilyl)carbodiimide.[1] Extension of this process to the synthesis of a variety of N-substituted, N'-trimethylsilylcarbodiimides from the corresponding isocyanates has also been achieved. This talk will focus on the results obtained from a series of experiments designed to probe the general mechanism, the extents, and the limitations of this novel metathesis process.

1) Sita, L. R.; Babcock, J. R.; Xi, R. *J. Am. Chem. Soc.* **1996**, *118*, 10912.

812.
STRUCTURE OF THE LEAD COMPLEXES WITH S,O-MACROCYCLES FROM 18-CROWN-6 TO HEXATHIA-18-CROWN-6 IN THE CRYSTAL STATE AND IN ORGANIC SOLVENTS.
Alexander Y. Nazarenko, N. Kent Dalley, John D. Lamb, Department of Chemistry and Biochemistry, Brigham Young University, Provo, Utah 84602-5700

As part of a systematic study of hazardous cation recognition using macrocyclic compounds, we report the isolation of several lead picrate complexes with crown-6 ethers with various number of sulphur atoms: Pb(18C6)Picr2, Pb(monothia-18C6)Picr2, Pb(1,10-dithia-18C6)Picr2, Pb(1,10-trithia-18C6)Picr2, Pb(1,4,10,13-tetrathia-18C6)Picr2, Pb(1, 5-dithia-19C6)Picr2, Pb(hexathia-18C6)Picr2. The structures provide information concerning Pb(II) interaction with O and S donor atoms and also features of picrate ion coordination. Coordination numbers vary from 8 to 10. The structural data together with our previously obtained structures of lead macrocyclic complexes enable us to discuss the features of ion recognition of Pb(II) by crown ethers. The molecular structures are compared with the characteristics of the same complexes in organic solvent.

813. **Olefin Metathesis with Silsesquioxanes**

<u>Andrew G. Eklund</u>, Daravong Soulivong and Frank J. Feher
Department of Chemistry, University of California, Irvine, California 92697

General methodology has been developed for the synthesis of highly functionalized silsesquioxanes from readily available olefinic silsesquioxanes. Our methodology is compatible with a wide range of functional groups, and it provides ready access to synthetically useful quantities of functionalized Si/O frameworks. This paper will describe our efforts to utilize olefin metathesis methodology for the large scale production of functionalized silsesquioxanes.

814. **Nucleophilic Substitution at Framework Si Atoms of Incompletely-Condensed Silsesquioxanes**

<u>F. J. Feher</u>, S. H. Phillips, M. A. Esparza, G. T. Lewis and J. M. Spadanuta
Department of Chemistry, University of California, Irvine, California 92697

Over the past several years, incompletely-condensed silsesquioxane frameworks (e.g., **1**) have attracted attention as precursors to a wide variety of interesting molecules. Until now, any chemical modification of **1** has involved reactions which transform SiOH groups into new siloxane (i.e., Si-O-Si) or heterosiloxane (i.e., Si-O-M) linkages. In this paper we report several facile and remarkably selective substitution reactions involving the framework silicon atoms of **1**. For example, the reaction of **1** with HBF4 affords **2** in quantitative yield. Reaction of **2** with MeLi produces **3**, while net hydrolysis affords **4**. Both reactions proceed in nearly quantitative yields with complete retention of stereochemistry at Si.

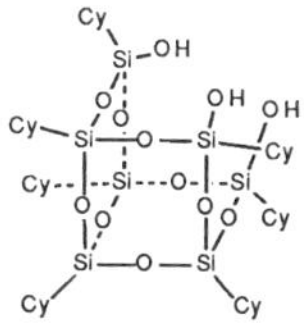

1

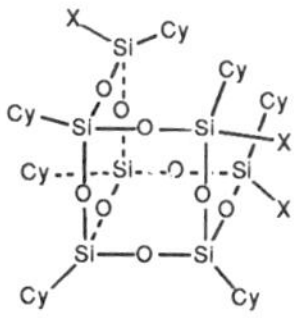

2 X = F
3   Me
4   OH

815. **ESR Studies Of Atomic Hydrogen And Atomic Deuterium Encapsulation Within A Series Of Silsesquioxanes**

Kevin Wyndham and Frank J. Feher
Department of Chemistry, University of California, Irvine, California 92697

Cubeoctameric silsesquioxane **1** was recently shown to encapsulate atomic hydrogen after γ-ray irradiation at room temperature. We have reinvestigated and confirmed this claim. We have also demonstrated that H atoms and D atoms can be encapsulated by a variety of other cubeoctameric frameworks, but there is no encapsulation within larger or smaller [RSiO$_{3/2}$]n frameworks. In the case of **2**, small amounts of H atoms are incorporated within the Si$_8$O$_{12}$ framework, but the major products of γ-irradiation are unstable radicals generated by disruption of the H$_8$Si$_8$O$_{12}$ framework. This paper will describe our efforts to use γ-irradiation as means for preparing silsesquioxane frameworks containing encapsulated small molecules.

1 R = (CH$_3$)$_3$SIO
2 R = H

816. **ELECTRON TRANSFER AT HIGH DRIVING FORCES IN RHENIUM-MODIFIED PROTEINS**; Mahdi M. Abu-Omar, Angelo Di Bilio, Jay R. Winkler, and Harry B. Gray; Beckman Institute, California Institute of Technology, Pasadena, CA 91125.

According to semiclassical theory, ET rates should exhibit a Gaussian dependence on the driving force (-ΔG°). Therefore, in the region where driving forces are greater than the reorganization energy (λ), ET rates are predicted to decrease (the inverted region). Our recent work on the kinetics of long-range ET reactions at high -ΔG° has involved Re-modified proteins. Tricarbonyl (1,10-phenanthroline)(aqua)rhenium(I) is used to label His33 of *cytochrome c*. The MLCT excited state of the rhenium complex is long lived, τ = 50 ns. The model complex, [(phen)Re(CO)$_3$(imidazole)]$^+$, exhibits a reversible reduction wave at -1.30 V and a quasi-reversible oxidation at 1.85 V vs. SCE. The Re(0) to Fe(III) ET reaction was initiated by reductive quenching of the MLCT excited state by ferrocyanide or *p*-methoxyanaline. Although the -ΔG° for the oxidation of Re(0) by *cytochrome c*-Fe(III) is 1.80 eV, which is deep in the inverted region, the observed first-order rate constant for this ET reaction (~ 5 x 10^7 s^{-1}) is two orders of magnitude greater than those of analogous Ru(His33)$^+$ to Fe(III) reactions. Whether this acceleration of the rate is due to a tunneling-energy effect is currently being tested by measuring several highly exoergic ET rates at longer distances.

817. FREE ENERGY PERTURBATION CALCULATIONS OF DURENE CYCLOPHANE HEMES. Marco A. Lopez and Julia Weidler. Department of Chemistry & Biochemistry, California State University, Long Beach, CA 90840-3903

Estimating the contributions of polar and steric effects towards the binding of ligands to heme models systems[1] and hemeproteins[2] has been reported using the AMBER software package. We have now carried out Free Energy Perturbation calculations of a series of durene cyclophane heme model systems.[3] We have reproduced the trends in the M values for the experimental systems and, using the same force field, extended the calculations to include both smaller and larger straps. Our calculations are gas phase with a constant dielectric. Our calculations suggest that the bound CO ligand tilts at the Fe atom. rather than bends at the carbon atom.

(1)Lopez, M. A.; Kollman, P. A. *J. Am. Chem. Soc.* **1989**, *111*, 6212-6222. (2)Lopez, M. A.; Kollman, P. A. *Protein Science* **1993**, *2*, 1975-1986. (3)David, S.; James, B. R.; Dolphin, D.; Traylor, T. G.; Lopez, M. A. *J. Am. Chem. Soc.* **1994**, *116*, 6-14.

818. GUANIDYL PORPHYRINS. J. Daniel Ponce, and M.A. Lopez, Department of Chemistry & Biochemistry, California State University, Long Beach, CA 90840-3903

The discrimination between dioxygen and CO binding to hemeproteins and heme model systems is affected by the polarity of the binding site.[1,2] We are extending this to include a cation in the local binding site. Our work is toward the synthesis of a porphyrin model system having a guanidinium ion at the binding site. We have two approaches to this end: 1) Building a tetra-(2-guanidylphenyl)porphyrin, and 2) building a guanidylphenanthrene porphyrin. We have quantitatively converted 2-nitroaniline to orthonitroguanidyl and are now working with the porphyrins. We will report on our progress to date

(1) David, S.; James, B. R.; Dolphin, D.; Traylor, T. G.; Lopez, M. A. *J. Am. Chem. Soc.* **1994**, *116*, 6-14.
(2) Traylor, T. G.; Koga, N.; Deardurff, L. A. *J. Am. Chem. Soc.* **1985**, *107*, 6504-6510.

819. PROTEIN-BOUND IRON-SULFUR CLUSTERS: REACTIVITY AND MECHANISM OF ASSEMBLY. J. A. Cowan, K. Natarajan, S. Bian, Evans Laboratory of Chemistry, The Ohio State University, 100 West 18th Avenue, Columbus, OHIO 43210, USA.

The role of aromatic core residues in regulating the redox chemistry, solvent accessibility, and stability of the prosthetic $[Fe_4S_4]^{3+/2+}$ cluster in *Chromatium vinosum* high potential iron protein (HiPIP) has been addressed by a combination of site-directed mutagenesis, high field NMR (EXSY) experiments, and variable temperature spectrochemical redox titration measurements. Minimal changes are observed following non-conservative mutation of residues Tyr19, Phe48, and Phe66. Apparently these hydrophobic residues play only a minor role in defining the redox properties of the cluster. Rather, the aromatic core restricts solvent accessibility, and thereby stabilizes the oxidized $[Fe_4S_4]^{3+}$ cluster. The early stages of iron-sulfur cluster assembly and backbone folding have been monitored by NMR, EPR, fluorescence, and MALDI-TOF mass spectrometry experiments. The $[Fe_4S_4]$ cluster assembles through a key structural intermediate that possesses most of the secondary and tertiary structural elements associated with the native protein. This species is a kinetically competent intermediate that contains four equivalents of protein-bound iron and no sulfide, possesses distinctive spectral characteristics, and is most likely relevant to the biosynthetic assembly of other complex clusters in metalloproteins and enzymes.

820. **FACIAL N_3- AND CYSTEINYL THIOLATE-DONOR COMPLEXES OF NICKEL** Patrick J. Desrochers*, Russell W. Cutts, Phillip K. Rice Department of Chemistry, University of Central Arkansas, Conway, AR 72035

Mixed nitrogen- and sulfur-donor coordination geometries are effective models for active sites in some nickel-hydrogenase enzymes. Cysteine is believed to be the source of sulfur-donor atoms binding nickel in these enzymes, making model studies of nickel-cysteine interactions particularly important. A series of $Tp^{XY}Ni$-SR complexes (Tp^{XY} = hydrotris{3(X)-5(Y)-disubstitutedpyrazolylborate}; SR = alkyl and cysteinyl thiolates) were prepared. The characteristic sulfur-to-nickel charge transfer bands of these complexes correlates with the changing thiolate basicity. Mass spectrometric analyses of $Tp^{PhMe}Ni$-SCysEt and $Tp^{MeMe}Ni$-SCysEt suggest a stronger nickel-sulfur bond for the Tp^{MeMe} complex. Unlike $Tp^{PhMe}Ni$-SCysEt, $Tp^{MeMe}Ni$-SCysEt reacts rapidly with one equivalent of molecular oxygen. Kinetics measurements indicate a strong dependence of the reaction rate on oxygen partial pressure. Deactivation of nickel-hydrogenases by dioxygen has been postulated to occur via oxidation of a nickel-cysteinyl-thiolate center in the enzymes.

821. STRUCTURE AND REACTIVITY OF MOLYBDENUM-PTERIN COMPLEXES
Heather L. Kaufmann, Sharon J. N. Burgmayer
Department of Chemistry, Bryn Mawr College, Bryn Mawr, PA 19010.

Tetrahydropterins (H4pterin) react with molybdenum(+6)-dioxo complexes to form stable MoO(H4pterin) complexes under acidic conditions[1]. The new complex [MoOCl(detc)(H4pterin)]Cl is the first example of a molybdenum-pterin complex possessing coordinated sulfur atoms, and its structure has been determined by X-ray diffraction. Differences in reactivity between this complex and other molybdenum-pterin complexes that lack sulfur atoms are being investigated. In parallel with our studies of reactions of reduced tetrahydropterins, we are also examining the reactivity of oxidized pterins with molybdenum. During the course of these investigations, we have confirmed the results reported by Selbin[2] and reputed by Sawyer[3]. The reactivity of these oxidized pterin complexes of molybedum is currently being explored in order to understand its relationship to fully reduced pterin complexes of molybdenum.

[1] Burgmayer, S.J.N. et al. *J. Amer. Chem. Soc.* **1995**, *117*, 5812.
[2] Selbin, J.; Sherrill, J.; Bigger, C.H. *Inorg. Chem.* **1974**, *13*, 2544.
[3] Sawyer, D.T. and Doub, Jr. W. H. *Inorg. Chem.* **1975**, *14*, 1736.

822.
CONSTRUCTION AND CHARACTERIZATION OF AN AZURIN ANALOG FOR THE PURPLE COPPER SITE IN CYTOCHEOME *c* OXIDASE: SPECTROSCOPY AND METAL SUBSTITUTION STUDIES. **Michael T. Hay,** Yi Lu, Department of Chemistry, University of Illinois at Urbana-Champaign, Urbana, Illinois 61801

Purple copper centers, found in cytochrome *c* oxidase and nitrous oxide reductase, are a new class of copper centers in biology. We have prepared a protein analog of a purple copper center from a recombinant blue copper protein (*Pseudomonas aeruginosa* azurin) by replacing the loop containing the three ligands to the blue copper center with the corresponding loop of the Cu_A center in cytochrome oxidase from *Paracoccus denitrificans* (Hay, M. T.; Richards, J. H.; Lu, Y. *Proc. Natl. Acad. Sci. U.S.A.* **1996**, *93*, 461-464). Spectroscopic studies by UV-vis, multifrequency EPR, MCD, RR, EXAFS, and paramagnetic NMR show remarkably similarity of this protein analog to those of other purple copper centers. Mercury and silver derivatives of this protein analog have also been prepared. Spectroscopic studies of this protein analog and its Hg/Ag derivatives will be presented.

823.
A PROTEIN ANALOG APPROACH TO BIOMIMETIC STUDY OF METAL-BINDING SITES IN CYTOCHEOME *c* OXIDASE AND MANGANESE PEROXIDASE. **Yi Lu,** Michael T. Hay, Bryan K.-S. Yeung, Xiaotang Wang, Marjorie C. Ang, and Priscilla D. Massey, Department of Chemistry, University of Illinois at Urbana-Champaign, Urbana, Illinois 61801

Small, stable, and well-characterized proteins have been used as scaffolds for biomimetic studies of metalloenzymes in an effort to elucidate the roles of metal ions in protein structure and function, to expand our knowledge of coordination chemistry, to make new enzymes with novel reactivities, and to apply the engineered proteins in biotechnology. This "protein-analog" approach has allowed us to engineer a purple copper center in azurin that is similar to the Cu_A site in cytochrome *c* oxidase, and a manganese center in cytochrome c peroxidase that closely mimic manganese peroxidase. Spectroscopic studies of both protein analogs will be presented, along with the study of their functional properties.

INOR

824. REDOX REACTIONS BETWEEN NITRIC OXIDE (NO) AND $Cu(II)L_2^{2+}$ COMPLEXES (L= phen, dmp). <u>Dat Tran</u> and Peter C. Ford*, Department of Chemistry, University of California, Santa Barbara, CA 93106.

The metal-NO interaction is thought to play important roles in the bioregulatory functions of NO. Given the importance of Cu(I/II) couples in biological redox systems, we have examined the redox reactions between NO and the Cu(II) complexes $Cu(phen)_2^{2+}$ (**A**) and $Cu(dmp)_2^{2+}$ (**B**; dmp= 2,9-dimethyl-1,10-phenanthroline). In neutral or alkaline aqueous or methanol solutions, **B** ($E_{1/2}$= 0.59 V vs NHE) is reduced by NO to form the corresponding cuprous complex $Cu(dmp)_2^+$ and RONO (R= H, CH_3) at a much faster rate than **A** ($E_{1/2}$= 0.15 V). At lower pH, the cuprous complex $Cu(phen)_2^+$ can be oxidized by nitrite (NO_2^-) in the reverse reaction to yield **A** plus NO. The amount of NO formed has been measured directly by a NO electrode. The kinetics of these reactions, as measured by stopped-flow spectroscopy, will be discussed in the context of the proposed mechanisms.

825. DIRECT THROUGH-BONDING CONTACT SHIFT OF AMIDE N^2H NMR SIGNALS OF NH···S HYDROGEN BOND IN PARAMAGNETIC Fe(II) AND Fe(III) THIOLATO COMPLEXES

<u>Akira Nakamura</u>, Norikazu Ueyama, Taka-aki Okamura, Wei-Yin Sun, and Nami Nishikawa
Department of Macromolecular Science, Graduate School of Science, Osaka University,
Toyonaka, Osaka 560, Japan

^2H NMR technique is an important tool for the detection of contact-shifted signal of high-spin Fe(II) or Fe(III) complexes with the fast relaxation times. For example, an amide NH group involved in NH···S hydrogen bonding in a high-spin Fe(II) peptide complex, $[Fe^{II}(Z\text{-}\mathbf{cys}\text{-}Pro\text{-}Leu\text{-}\mathbf{cys}\text{-}Gly\text{-}Val\text{-}OMe)_2]^{2-}$ (Z = $C_6H_5CH_2OCO$), provides almost undetectably broad ^1H signals, whereas the ^2H NMR spectrum of the amide N^2H-substituted complex, $[Fe^{II}(\text{amide } N^2H\text{-substituted-}Z\text{-}\mathbf{cys}\text{-}Pro\text{-}Leu\text{-}\mathbf{cys}\text{-}Gly\text{-}Val\text{-}OMe)_2]^{2-}$, shows clear N^2H signals at 38.0, 19.5, -4.0 and -20 ppm. Similar NH···S hydrogen bonded ^2H signals for two P-450 model complexes, $[Fe^{III}(OEP)(\text{amide } N^2H\text{-substituted-}Z\text{-}\mathbf{cys}\text{-}Pro\text{-}Leu\text{-}Ala\text{-}OMe)]$ (OEP = octaethylporphyrin) and $[Fe^{III}(OEP)(S\text{-}2\text{-}CF_3CON^2H\text{-}C_6H_4)]$, were observed at -52.8 ppm (CH_2Cl_2) and -38.1 ppm (C_6H_6), respectively. These shifts are mainly caused by scalar-type contact through a partially covalent NH···S hydrogen bond.

826.
STEREOSELECTIVE B-HYDROGEN ELIMINATION FROM NICKEL(II) N-GLYCOSIDE COMPLEXES. **Glenn Smith**, Julie A. Leary, University of California, Berkeley, Department of Chemistry, Berkeley, CA, 94720

Octahedral nickel(II) N-glycoside complexes of glucose, galactose, mannose and talose were analyzed by electrospray ionization-mass spectrometry (ESI-MS). A resulting tricoordinate complex generated from the octahedral species was subjected to collision induced dissociation. A highly stereoselective dissociation pathway involving b-hydrogen elimination and cross-ring cleavages was observed in complexes possessing equatorial C-2 substituents on the carbohydrate. 2H and 13C labeling experiments indicate that the hydrogen on C-2 and a labile proton are involved in the b-hydrogen elimination. Additionally, C-4, C-5 and C-6 are shown to be lost from the monosaccharide as a result of the cross-ring cleavages. The stereoselectivity of this dissociation mechanism is postulated to result from differing geometries of the N-glycoside ligands that possess axial or equatorial C-2 substituents.

827. SELF-ASSEMBLY OF POLYNUCLEAR COMPLEXES INVOLVING TETRADENTATE LIGANDS.
Zhi-Yong Zhang and Marie-Thérèse Youinou, Institut Le Bel, Université Louis Pasteur, 4, rue Blaise Pascal, 67000 Strasbourg, France

As an extension of our work on self-assembly systems, we have investigated some factors responsible for the elaboration of highly organized complexes. Our approach was to modify chemically the ligand in maintaining its coordinating tetradentate properties and therefore its ability to complex d^{10} or d^9 metal ions. The reactions of copper(I) and copper(II) with the 3,5-di(pyrid-2-yl)pyrazole ligand (dpplH) lead to a series of polynuclear complexes. Their characterization by crystallography, NMR or ESR and MS has been carried out. In particular, the spontaneous formation of mixed-valence complexes will be presented.

828.
Self Organization into 2D and 3D Superlattices Differing by their Size. **M.P. Pileni**, F. Billoudet, C. Petit, A. Talef.

Self assemblies made of silver surfide and silver metallic nanoparticles are observed. The domain of monolayer and the number of large aggregate formed strongly dépend on the preparation of the sample. Monolayers organized in a hexagonal network are observed in a very large domain. This is observed with various particle sizes. Large and small aggregates made of nanoparticles differing by their size have been obtained. The particles are highly oriented and form a face centered cubic structure. The size and the shape of the 3D crystals are not yet controlled. This could be due to the preparation mode but also to the quality of the support used.

WITHDRAWN

829. STRUCTURE STABILIZATION IN FAST ION CONDUCTIVITY PHASES. E. A. Secco, Department of Chemistry, St. Francis Xavier University, Antigonish, N.S. CANADA B2G 2W5

Solid state fast-ion conductors have been the focus of much attention over the past few years in view of their potential application in various electrochemical devices. Fast-ion conduction in solids has been considered a paradigm for a structure-property-relation. A number of compounds undergo a structural change at the phase transition accompanied by a sharp jump in ionic conductivity, up to a factor 10^4. We have been successful in some cases to lock-in or preserve the high-temperature high-conductivity structure by the incorporation of various isovalent guest ions. This talk will present our recent efforts, successes and failures, and discuss the basis for these results.

830. STRUCTURAL INVESTIGATION OF THE CHABAZITE LIKE ALUMINOPHOSPHATE PREPARED IN THE PRESENCE OF HF. Anton Meden[1], Alain Tuel[2], Nevenka Rajic[4,3] and Venceslav Kaucic[1,4], [1]Department of Chemistry and Chemical Technology, University of Ljubljana, Slovenia, [2]Institut de Recherches sur la Catalyse, CNRS, Villeurbanne, France, [3]Faculty of Technology and Metallurgy, Belgrade University, Yugoslavia, [4]National Institute of Chemistry, Ljubljana, Slovenia.

Microporous solids are used extensively as heterogeneous catalysts and adsorbents. These materials contain regular, uniformly spaced cages and channels that allow molecules access to large internal surface that enhance catalytic activity and adsorptive capacity. Among these materials, aluminophosphates are of special interest. They have been usually prepared in an aqueous medium, and in a presence of an amine as structure directing agent. Recently, it has been found that an addition of HF to the reaction mixture offers new possibilities in tailoring aluminophosphates. Here we report structural characteristic of an alumino-phosphate with the chabazite like structure prepared in the presence of piperidine and HF. The material exhibits a particular (unexpected) moisture sensitivity after template and fluoride removal. The hydration leads to a deformation of the chabazite framework and it is a reversible process. The structural deformation has been explained by synchrotron powder XRD, solid state NMR, and FTIR data.

831. THE CRYSTAL STRUCTURE OF RHOMBOHEDRALLY DISTORTED SODIUM TITANIUM SILICATE PHARMOCOSIDERITE

M S Dadachov and W T A Harrison Materials Division, Australian Nuclear Science and Technology Organisation, PMB 1 Menai, NSW, 2234, Australia and Chemistry Department, University of Western Australia, Nedlands, WA 6907, Australia

The synthesis and crystal structure of rhombohedrally distorted Sodium Titanium Silicate $Na_4[(TiO)_4(SiO_4)_3]$ $6H_2O$ Pharmacosiderite structure type are reported. The $Na_4[(TiO)_4(SiO_4)_3]$ $6H_2O$ was synthesised by low temperature hydrothermal crystallisation of alkaline titanium silicate gel, prepared by peroxide route. Three dimensional mixed anion framework is built up from four TiO_6 octahedra which share edges using imitation symmetry of 43m $[Ti_4O_{16}]$. These groups are connected along the three directions via Si atom which is four-coordinated. The large cages filled with water molecules are connected with each other via eight membered rings (~5A in diameter) where Na cations are located. Rietveld refinement and difference Fourier map calculation resulted in finding additional Na cations located on 3 fold axes thus causing the distortion of this structure.

Crystal data: $Na_4[(TiO)_4(SiO_4)_3]$ H_2O, F.w 719.8, Rhombohedral, Space group R3m (160) a=7.81239(58) α=88.794(9), V=476.50(6), Z=1, R=0.09 Rw=0.12

832. AB INITIO STRUCTURE DETERMINATION OF $K_2TiSi_3O_9H_2O$ FROM X-RAY POWDER DIFFRACTION DATA

M.S. Dadachov and A.LeBail Materials Division, Australian Nuclear Science and Technology Organisation, PMB 1, Menai, NSW, 2234, Australia and Laboratorie de Fluorures URA CNRS 449 Faculte des Sciences, Universite du Maine, F-72017 Le Mans, France

The crystal structure of $K_2TiSi_3O_9H_2O$ has been solved ab initio from X-ray powder data. The unit cell is orthorhombic (space group $P2_12_12_1$, Z=4) with cell dimensions a= 7.1353(2)A b=9.9073(3)A c=12.9387(4)A V=914.66Å^3. The structural model was obtained from direct methods and difference Fourier maps. The Rietveld refinement converged to final crystal structure and profile indicators Rp=6.94%, Rwp=9.31%. The structure consists of SiO_4 tetrahedra forming infinite $[SiO_3]_n$ 2n-chains extending along the α axis, connected by isolated TiO_6 octahedra thus giving mixed anion framework. The ellipsoidal channels along α axis are filled with exchangeable K cations and water molecules, which can be reversely disassociated.

833. GEOMIMETIC COORDINATION POLYMERS: TOWARDS MOLECULAR QUARTZ. Susan Lopez, Brook Novak and Steven W. Keller, Department of Chemistry, University of Missouri-Columbia, Columbia MO 65211

The development of a new class of three-dimensional coordination polymers will be discussed, in which bent, bifunctional ligands (pyrimidine, 1,3-dicyanobenzene) are used to link tetrahedral metal centers. In so doing, we hope to mimic the structural building blocks of silicate minerals like quartz and zeolites, in which the metal-oxygen tetrahedra corner-share to form M-O-M angles of *ca.* 145°. and thereby create molecular-based extended solids having similar properties (i.e. piezoelectricity, chirality, microporosity). We have structurally characterized, both a three-dimensional, acentric framework, having the formula [Cu(pyrimidine)$_2$]BF$_4$ (similar in connectivity to the feldspar aluminosilicate family), and an *isolated* [Cu$_4$(pyrimidine)$_4$(MeCN)$_8$]$^{4+}$ tetramer, simply by changing the initial reagent stoichiometry. Furthermore, we have preliminary evidence of a templating effect of the extraframework anions. If the BF$_4^-$ ions are replaced with PF$_6^-$, the feldspar-like framework does not form, and a new structure is observed, that is related to quartz, although not chiral. Attempts are underway to explore the non-linear optical properties of these solids, as well as to investigate their microporous nature through anion-exchange reactions.

834. SOLUTION CHEMISTRY OF LANTHANIDE(III) DTPA-BIS-(GLUCOSAMIDE) COMPLEXES: IMPLICATIONS FOR MRI CONTRAST AGENT DESIGN

Kevin M. Schaab and G. R. Choppin. The Florida State University, Dept. of Chemistry, Tallahassee, FL 32306-3006

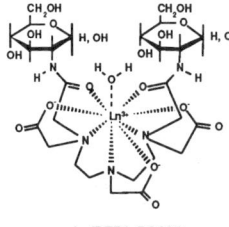

Ln(DTPA-BGAM)

The current generation of contrast agents available or under development for clinical MRI consist of mainly gadolinium complexes of acyclic and macrocyclic aminopolycarboxylates, aminopolyphosphonates, and aminopolyphosphinates. Improved contrast agents will require agents which can remain in the body longer and/or be targeted toward specific tissues. We have undertaken a study of lanthanide complexes of DTPA-BGAM as a possible step toward preparing glucose receptor targeted MRI contrast agents. The effect of the incorporation of the glucose moieties on the thermodynamic stability, kinetic lability, solution dynamics, and biological activity of the lanthanide complexes of this ligand (as measured using a variety of experimental techniques) will be discussed relative to similar compounds which lack the sterically bulky, potentially biologically active glucose moiety.

This research was supported by a contract with Mallinckrodt Medical Inc.

835. SYNTHESIS AND STRUCTURAL CHARACTERIZATION OF POLYMETALLIC AND HETEROPOLYMETALLIC EUROPIUM COMPLEXES. William J. Evans, Michael A. Greci, and Joseph W. Ziller, Department of Chemistry, University of California, Irvine, CA 92697-2025.

The fluorescent properties of europium make it a desirable component in light emitting devices including trichromatic fluorescent lights. Polymetallic europium compounds could be useful precursors to materials containing europium, but few examples of polymetallic europium complexes are known. We have investigated the formation of polymetallic and heteropolymetallic europium complexes from the bulk metal in both liquid ammonia and in alcohols. Well characterized tetra-, penta-, and octametallic europium complexes have resulted as well as a series of heteropolymetallic species containing barium or zirconium. General structural patterns will be discussed as well as the structure of the cubic octametallic complex which has six face-centered oxide ligands. The synthetic results suggest some general principles of reactivity which appear to be applicable to other mixed metal combinations.

836. AMINOTROPONIMINES IN ORGANOLANTHANIDE CHEMISTRY.

Peter W. Roesky
Institut für Anorganische Chemie, Universität Karlsruhe, Karlsruhe, Germany

Aminotroponimines with different substitution patterns on the nitrogen atoms can be introduced into organolanthanide chemistry. The single charged bidentate 10 π-electron ligand system is an analogous to the classical diazadiene ligands. The metal center can be coordinated by one, two or three aminotroponimine ligands whereas diazadienes show a poor coordination chemistry for the lanthanides. The single and double substituted complexes offer a broad pathway for new derivatives, e.g. new homo- and heteroleptic amines of the lanthanides are easily formed. Some of the derivatives discussed in the talk show a similar behavior to cyclopentadienyl or benzamidate compounds.

837. SELF-ASSEMBLED LOW DIMENSIONAL MATERIALS COMPOSED OF BOTH LANTHANIDE AND TRANSITION METAL CENTERS. Jianping Liu, Edward A. Meyers, Sheldon G. Shore, Department of Chemistry, The Ohio State University, Columbus, Ohio 43210

Reactions of $LnCl_3$ (Ln = Sm, Eu) with $NiCl_2$ in the presence of KNCO in DMF produced two types of complexes, monomeric $(DMF)_8Ln_2Ni(NCO)_8$ and a one-dimensional infinite chain complex $\{(DMF)_6Ln_2Ni(NCO)_8\}_\infty$. When $LnCl_3$ (Ln = Sm, Eu) was reacted with $K_2[Pd(CN)_4]$ in DMF, a one-dimensional array $\{(DMF)_{10}Ln_2[Pd(CN)_4]_3\}_\infty$ was formed. In the reaction between $YbCl_2$ and $K_2[Pd(CN)_4] \cdot xH_2O$ in DMF, the complex $\{(DMF)_{10}Yb_3(OH)_4O_{1/2}[Pd(CN)_4]_2\}_\infty$ was formed. This complex contained a one-dimensional chain of $\{(DMF)_{16}Yb_6(OH)_8O[Pd(CN)_4]\}^{6+}_\infty$ and three free $[Pd(CN)_4]^{2-}$ anions served as counter ions. $[(DMF)_{16}Yb_6(OH)_8O]^{8+}$ consists of a Yb_6 octahedron whose center is occupied by an oxygen atom and each trianglur face is capped by an OH^-. The molecular structures and properties of these complexes will be discussed.

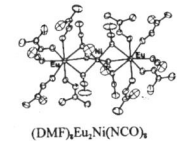

$(DMF)_8Eu_2Ni(NCO)_8$

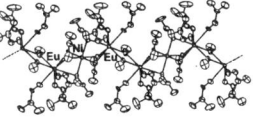

$\{(DMF)_6Eu_2Ni(NCO)_8\}_\infty$

838. SYNTHESIS AND STRUCTURE OF URANYL COMPOUNDS CONTAINING STERICALLY DEMANDING ALKOXIDE LIGANDS. Marianne P. Wilkerson[*,**], Carol J. Burns[*], Robert T. Paine[**], Brian L. Scott[*], [*]Chemical Science and Technology Division, Los Alamos National Laboratory, Los Alamos, NM 87545; [**]Department of Chemistry, University of New Mexico, Albuquerque, NM 87131.

The uranyl aqua ion can exhibit complex hydrolytic equilibria in aqueous solution, often resulting in the formation of polynuclear charged or neutral species. In comparison, metathetical synthesis of uranyl alkoxides in nonprotic solution allows for ligand redistribution and formation of a variety of product structures. The stability of these compounds with respect to redistribution is dependent upon the degree of steric substitution on the alkoxide ligand. We have prepared a series of uranyl alkoxide compounds that show such reactivity to be facile but inhibited in the presence of a strongly coordinating Lewis base. Alternatively, coordination of phenoxide ligands to the uranyl ion may impart variable electronic properties important to the stability of the uranyl product. We report the synthesis and spectroscopic characterization of the monomeric uranyl phenoxide compound, $UO_2(O-2,6-t-Bu_2C_6H_3)_2(THF)_2$. A comparison of this complex with a series of other uranyl phenoxide compounds will also be discussed.

839. STOICHIOMETRIC AND STRUCTURAL STUDIES OF LANTHANIDE COMPLEXES OF $[\alpha\text{-}2\ P_2W_{17}O_{61}]^{10-}$. Judit Bartis[†], Michaela Dankova[†], Robin Rogers[§], Victor Young[¥], Lynn C. Francesconi[†*] Departments of Chemistry, [†]Hunter College of the City University of New York, New York City, New York 10021, [§] University of Alabama, Tuscaloosa, Alabama 35487, [¥]University of Minnesota, Minneapolis, Minnesota 55455.

Investigation of the solution stoichiometry of lanthanide (Ln) complexes of polyoxotungstates can give insight into potential synthetic targets. Only the 1:2 Ln: polyoxotungstate complexes have been isolated in the solid state. Our solution titration data show that both the the 1:1 and 1:2 Ln: $[\alpha\text{-}2\ P_2W_{17}O_{61}]^{10-}$ exist in aqueous solution at pH=4.7 (lithium acetate/acetic acid buffer) for a variety of Ln ions proceeding across the lanthanide series. The speciation was monitored by ^{31}P NMR Spectroscopy. The ^{31}P NMR chemical shifts of the 1:2 Ln: $[\alpha\text{-}2\ P_2W_{17}O_{61}]^{10-}$ species were consistent with the isolated and characterized complexes. The crystal structure of the $K_{17}[Lu(\alpha\text{-}2\ P_2W_{17}O_{61})_2]$ shows the lutetium ion bound to four oxygen atoms of the $[\alpha\text{-}2\ P_2W_{17}O_{61}]^{10-}$ in a square antiprismatic coordination environment. ^{183}W NMR data for the isolated $[Ln(\alpha\text{-}2\ P_2W_{17}O_{61})_2]^{17-}$ complexes in aqueous solution are consistent with the crystal structure data.

840. SOLUTION STOICHIOMETRIC AND STRUCTURAL STUDIES OF LANTHANIDE $[\alpha\text{-1 }P_2W_{17}O_{61}]^{10-}$ COMPLEXES. Judit Bartis, Michaela Dankova, Lynn C. Francesconi*, Department of Chemistry, Hunter College of the City University of New York, New York City, New York 10021.

A number of studies of lanthanide (Ln) complexes of the α-1 isomer of the monovacant $[P_2W_{17}O_{61}]^{10-}$ ligand resulted in inconsistant data and conclusions regarding the stoichiometry of the species. Further no direct structural proof has been provided for the complexes, perhaps due to the rapid isomerization and instability of the $[\alpha\text{-1 }P_2W_{17}O_{61}]^{10-}$ isomer. We report herein a study of the solution stoichiometry and proof of the C_1 symmetry of the lanthanide complexes of the $[\alpha\text{-1 }P_2W_{17}O_{61}]^{10-}$ isomer. To determine solution stoichiometry, we used ion exchange and titration experiments as described here. Accurate initial concentrations of Ln and $[\alpha\text{-1 }P_2W_{17}O_{61}]^{10-}$ were obtained by complexometric titration of lanthanide, using xylenol orange and titration of the ligand using cobalt, monitoring the UV-Visible spectroscopy. Excess Ln counterions were removed by Chelex resin, buffered at pH=4.7. This step was followed by a series of titration experiments. First, a complexometric titration with EDTA to decompose the complex, following the progress of this decomposition with ^{31}P NMR, allowed determination of the lanthanide concentration in the complex. The next titration with cobalt, followed by UV-Visible spectroscopy, allowed determination of the $[\alpha\text{-1 }P_2W_{17}O_{61}]^{10-}$ isomer concentration. These data show the complexes have 1:1 Ln: $[\alpha\text{-1 }P_2W_{17}O_{61}]^{10-}$ stoichiometry. Structural proof of the C_1 symmetry of the complexes was obtained by ^{183}W NMR spectroscopy.

841. SPECTROSCOPIC AND LUMINESCENCE PROPERTIES OF NEW SOLID URANYL PHOSPHONATES. Daniel Grohol and Abraham Clearfield,* Department of Chemistry, Texas A&M University, College Station, Texas 77843

Several recently prepared and structurally characterized solid uranyl phosphonates were studied by means of IR, UV-Vis absorption and steady-state emission spectroscopies as a function of temperature, as well as the kinetics of their excited state. The main objective was determining the decay routes of the excited state of the uranyl ion as a function of the organic group of the phosphonate anion, and as a function of the compounds' structures. At room temperature, some uranyl phosphonates luminesce very strongly, some weakly, whereas others do not emit at all. This different behavior was attributed to a different level of competition between the radiative decay, electron transfer and thermal deexcitation. Temperature dependence of emission also varied significantly from compound to compound. Surprisingly, even luminescence properties of structurally very similar compounds was very different, for which an explanation was sought and suggested.

842. MACROCYCLIC STABILISATION OF NEUTRAL AND CATIONIC LANTHANIDE SPECIES. Timothy J. Woodman and Gerald R. Willey* Department of Chemistry, University of Warwick, Coventry, CV4 7AL UK.

Dehydration of hydrated lanthanide chlorides, $LnCl_3 \cdot xH_2O$ (where $x \geq 6$), with thionyl chloride in the presence of THF generates $LnCl_3(THF)_n$ solvates where Ln = Ce-Lu and n may be 1.5, 2, 3, 3.5 or 4. Each value of n corresponds to a discrete structural type. Recent structural determinations from our laboratory include $DyCl_3(THF)_{3.5}$, $GdCl_3(THF)_4$ and $[PrCl_3(THF)_2]_n$. Total dehydration is not always obtained; for the "mixed" solvates $[LnCl_3(THF)(H_2O)]_n$ (Ln = Nd, La) residual water molecules feature in an extended network of hydrogen bonded interactions. Removal of halide ion(s) from $LnCl_3$ can be effected with $SbCl_5$, $FeCl_2$ or $SnCl_4$ to provide the sequence of cations:-
$$LnCl_3(solvated) \rightarrow LnCl_2^+(solvated) \rightarrow LnCl^{2+}(solvated) \rightarrow Ln^{3+}(solvated)$$
We have identified various members of this sequence following stabilisation *in situ* by either solvent (e.g. $[GdCl_2(THF)_5][SnCl_5(THF)]$) or macrocyclic (crown ether) molecules (e.g. $[DyCl_2(H_2O)_2(12\text{-crown-}4)][SbCl_6]$ and $[DyCl(H_2O)_2(18\text{-crown-}6)][FeCl_4]$).

843.
OPTICAL SENSING VIA ABSORPTION-ENERGY TRANSFER-EMISSION IN LANTHANIDE CYCLODEXTRIN SUPRAMOLECULES. **Wanda K. Hartmann,** Mark A. Mortellaro and Daniel G. Nocera, Department of Chemistry and Center for Fundamental Materials Research, Michigan State University, East Lansing, MI 48824.

Understanding the fundamental pathways of energy transduction from aromatic hydrocarbons to lanthanide ions permits the design of optical sensing schemes based on supramolecular active sites. We have synthesized a cyclodextrin (CD) modified with diethylenetriamine pentaacetic acid (DTPA) to the primary side of A,D-diamino-β-cyclodextrin, thereby permitting Tb^{3+} or Eu^{3+} to be juxtaposed to the hydrophobic interior of the CD. By itself the complex only weakly emits when irradiated with UV light, but when aromatic hydrocarbons are added visible luminescence of the lanthanide is enhanced. Neutral aromatics bind in the cavity permitting absorption-energy transfer-emission process to occur over a fairly short distance. The binding and photophysical properties of theses complexes will be discussed and compared to a previous system.

844. ACTINIDE DICARBOLLIDE COMPLEXES.
Francis M. de Rege, Kent D. Abney, Jon B. Nielsen and Wayne H. Smith.
Los Alamos National Laboratory, Los Alamos, NM 87545.

Several new anionic actinide dicarbollide complexes have been prepared through metathesis, transmetallation, and chemical and electrochemical reduction. The alkyl, imido, and halide complexes will be discussed. These coordinatively unsaturated complexes are easily prepared yet highly reactive. The dicarbollide ligand has been shown to stabilize actinide oxidation states of (III) and (IV), and most recently the unusual (V). The U(V) complex was shown to be stable by cyclic voltammetery and prepared independently by bulk electrolysis. The NMR X-ray, and other characterization data will be presented and compared to bis(cyclopentadienyl) analogs. In addition to the reactivity of these complexes, the application of these systems to advanced nuclear materials will be discussed.

845. URANIUM AND THORIUM ALKYNE COMPLEXES
Benjamin P. Warner and Carol J. Burns
Chemical Science and Technology Division
Los Alamos National Laboratory, Los Alamos, NM, 87545

We have synthesized alkyne complexes of uranium and thorium. Because of the many similarities between uranium and thorium, some differences between analogous complexes can be attributed to the availability of f-electrons in uranium. New routes to actinide-alkyne complexes have been developed. For example, both thorium alkyne complexes and thoracyclopentadienes have been synthesized by the reaction of alkynes with bis(pentamethylcyclopentadienyl)thorium(methy)hydride.

846. THE MAGNETIC PROPERTIES OF NANOPARTICLES AND THEIR CORRELATION WITH THE COORDINATION CHEMISTRY IN SPINEL IRON FERRITES. Z. John Zhang, School of Chemistry & Biochemistry, Georgia Institute of Technology, Atlanta, Georgia 30332-0400

Design and rational synthesis of magnetic nanoparticles with desired properties require a clear understanding of the magnetic properties in the materials at nanometer scale. Magnetic couplings at atomic level fundamentally control the magnetic behaviors of the materials. We are systematically studying the correlation between the magnetic properties of nanoparticles and the coordination chemistry in spinel iron ferrites. The magnetic properties of $MnFe_2O_4$ nanoparticles and their coordination chemistry are being investigated using various physical techniques such as SQUID, Mössbauer spectroscopy, X-ray and neutron scattering. The magnetic behaviors of ferrites nanoparticles and their relationship with cation distribution will be discussed.

847.
A NOVEL APPROACH TO MEASURING DEFECT DENSITY IN MONOLAYER AND MUTILAYER FILMS USING NONLINEAR OPTICAL PROPERTIES. **W. C. Flory-Stary** and G. J. Blanchard, Department of Chemistry, Michigan State University, East Lansing, MI 48824

Monolayer and multilayer thin films are of growing interest in the materials community because of their potential relevance to applications such as optical information storage. Much attention has recently been placed on metal phosphonate chemistry to construct these films because of its structural and thermal stability. In order for these self-assembled multilayers to be applicable as electro-optic devices, they must be essentially defect free. A novel approach to measuring defects on a molecular scale is to utilize the functional chemistry for orientational control in the growth of these films. By constructing these films with polar azo dyes and controlling the orientation at each layer, a multilayer film can be assembled with centrosymmetric bulk ordering. Theoretically, the $\chi^{(2)}$ nonlinear response should be zero, and any second harmonic generation (SHG) observed from this film is a direct result of defects. The results from the SHG measurements on of zirconium phosphonate multilayers will be presented.

848.
USING ZIRCONIUM-PHOSPHONATE MULTILAYER ASSEMBLIES TO EXAMINE PHOTONIC ENERGY TRANSPORT. **J. C. Horne** and G. J. Blanchard, Department of Chemistry, Michigan State University, East Lansing, MI 48824

The ability to store information in a thin film system depends on making a specific, stable modification to the optical response of a given layer of the material. The presence of other optically active constituents in the same layer or in neighboring layers creates the possibility for intra- or interlayer energy exchange processes, which serve to place limits on the utility of these materials for practical applications. Metal-phosphonate organic multilayer films are nearly ideal models for the selective examination of intra- and interlayer energy relaxation processes, because their properties can be controlled at each layer. Characterization of photonic energy exchange within a single layer of a bisphosphonated bithiophene chromophore and transport between layers of energetically overlapped oligothiophene chromophores by steady-state (absorption and emission) and transient (single photon counting) optical spectroscopies will be presented.

849.
SYNTHESIS OF GALLIUM NITRIDE BY SOLID-STATE METATHESIS REACTIONS UNDER PRESSURE
Charles H. Wallace, Lin Rao, Jennifer O'Loughlin, Sang-Ho Kim, James R. Heath, Malcolm F. Nicol and **Richard B. Kaner**.*
Dept. of Chemistry and Biochemistry & the Solid State Science Center
University of California, Los Angeles, CA 90095-1569

Solid-state metathesis (exchange) reactions have been used to produce a wide variety of crystalline refractory materials under ambient conditions. These highly energetic reactions which are driven by the formation of very stable products provide a rapid method to produce high quality refractory materials. In order to test whether the application of pressure to solid-state metathesis reactions could lead directly to pure high pressure/ temperature phases, a high pressure reaction scheme was developed using Bridgman anvils. This is the first time that these highly energetic solid-state metathesis reactions have been studied at high pressures. Polycrystalline gallium nitride has been prepared at static pressures of 45 kbar from gallium triiodide and lithium nitride. The product was characterized by powder X-ray diffraction (XRD) and shown to produce single phase hexagonal GaN (a= 3.190 Å, c= 5.177 Å). The photoluminescence (PL) spectrum at room temperature shows significant luminescence in the blue/UV region. The PL spectra compare well to bulk properties of GaN previously reported and agrees with the direct band gap of 3.39 eV. Other nitrides prepared by this method will be discussed.

850.
SYNTHESIS OF CARBIDES AND CARBIDE/NITRIDE SOLID SOLUTIONS USING DISPLACEMENT REACTIONS
Steven R. Holm and Richard B. Kaner, Department of Chemistry and Biochemistry, Los Angeles, CA 90095
Dept. of Chemistry and Biochemistry & the Solid State Science Center
University of California, Los Angeles, CA 90095-1569

Transition metal carbides are useful primarily for their great hardness. These materials have been produced from a variety of methods starting from the elements which require long reaction times and high temperatures. We have devised two new flexible synthetic routes to these materials by using displacement reactions. Transition metal carbides (e.g. TiC, NbC and WC) can be readily synthesized by combining transition metal halides with lithium acetylide to rapidly form the transition metal carbide through a "double" displacement (metathesis) reaction, or by a combination of transition metal halides with carbon tetrabromide and magnesium powder (single displacement). These reactions are very rapid and exothermic and are driven by the formation of the by-product salt. This synthetic technique can be easily combined with other reactants to produce solid solutions of mixed metal carbides such as $Zr_{1-x}Hf_xC$, as well as carbide and nitride solid solutions, such as $ZrC_{1-x}N_x$.

851. NEW TERNARY MAIN GROUP ELEMENT ANTIMONIDES. **Arthur Mar**, Department of Chemistry, University of Alberta, Edmonton, AB, Canada T6G 2G2

Ternary antimonides $M_xA_ySb_z$ consisting of an electropositive metal M (alkaline earth or transition metal) and a main group element A provide a rich area for exploring the interplay between homoatomic and heteroatomic bonding in stabilizing a structure. Several examples of such compounds will be presented to illustrate this idea. $Ba_2Sn_3Sb_6$ and $SrSn_3Sb_4$ are typical Zintl compounds whose channel structures contain infinite Sb–Sb chains. $ZrSi_{0.7}Sb_{1.3}$ and ZrGeSb adopt the ZrSiS-type structure containing square Si or Ge sheets, while $ZrSn_{0.4}Sb_{1.6}$ reverts to the β-$ZrSb_2$-type structure. Co_3Ge_2Sb possesses an unusual alloy-type structure containing Ge 6-rings and Sb_2 pairs.

852.
LAYER-BY-LAYER GROWTH OF ORIENTED METAL BISPHOSPHONATE MULTILAYERS
Grace A. Neff and Catherine J. Page; Dept. of Chemistry, University of Oregon, Eugene, OR 97403

We are exploring various ways to incorporate nonlinear optical (NLO) activity into multilayer thin films. One method under investigation is the use of asymmetric α,ω-bis(phosphonates) containing NLO chromophores which bind preferentially in one orientation. Preliminary work utilizing Hf and aryl(4-diethylphosphonate)-10-decylphosphonic acid ether on silicon substrates has shown this system to produce multilayer films of ~21 Å/layer with varying densities. However, we have been unable to properly characterize the NLO properties of these films, since the second harmonic (SH) signal generated by the silicon substrate interferes destructively with that generated by the film. This paper will discuss recent results from this same system, using glass as a substrate. X-ray photoelectron spectroscopy (XPS), atomic force microscopy (AFM) and second harmonic generation (SHG) are utilized to characterize these films.

853. SELF-ASSEMBLY OF CYCLIC PEPTIDE BASED NANOTUBES ON GOLD AND SILICON DIOXIDE SURFACES
Jillian M. Buriak and M. Reza Ghadiri, Department of Chemistry, The Scripps Research Institute, BCC 104, 10666 North Torrey Pines Road, La Jolla, CA 92037

Cyclic peptides with alternating D-,L- chirality around the ring self-assemble into hollow, hydrogen-bonded nanotube structures. Functionalization of the amide nitrogens on one face of the cyclic peptides with alkyl thiols and alkyl chlorosilanes allows binding of the peptide to gold and silicon dioxide surfaces, respectively. Bidentate binding of the cyclic peptide to the surface, followed by self-assembly of one or more appropriately functionalized cyclic peptides, yields nanotubes perpendicular to the surface. Application of these nanotube arrays as biosensors will be discussed.

854.
NANOSTRUCTURED SUPERCONDUCTORS. **Peidong Yang** and Charles M. Lieber, Department of Chemistry and Chemical Biology, Harvard University, Cambridge, MA 02138

A chemical approach to introduce columnar defects into high temperature superconductors (HTSCs) which involves growth of MgO nanorods and subsequent incorporation of these rods into HTSCs to form nanrod-HTSC composites has been developed. Electron microscopy analyses of the composites demonstrate that a high density of nanorod columnar defects can be created with orientations perpendicular and parallel to the copper oxide planes. A mechanism of self-organization of nanorods in the inorganic crystal matrix has been proposed to explain these results. These nanorod columnar defects serve as strong pinnig centers for magnetic flux lines. The critical current densities of the nanorod-HTSC composites are enhaced dramatically compared with reference samples. The use of nanorods may thus represent a technologically viable strategy for improving critical currents in large-scale superconductor applications.

855. SYNTHESIS OF SILVER(0)-POLYIMIDE NANOCOMPOSITE FILMS VIA *IN SITU* THERMAL REDUCTION OF SILVER(I) COMPLEXES R.E. Southward[1], D.W. Thompson[1], and A.K. St. Clair[2], William and Mary[1], Wmsburg, VA and Langley Research Center, NASA[2], Hampton, VA 23665

Surface metallized composite films with high specular reflectivity and electrical conductivity have been prepared by soluble silver(I)-beta-diketonate (dik) complexes in a DMAc solution of the poly(amic acid) BTDA-4,4'-ODA. Curing of the Ag(I)-PAA film leads to imidization with Ag-(I) reduction yielding a reflective and conductive Ag surface when the Ag concentration is 5-15% (wt), and the film is cured to 573-613 K. Ag-dik complexes are prepared by reacting acidic dik's (trifluoro-, hexafluoro-, and 3-cyano-acetylacetone) with Ag acetate in DMAc. 3-Cyanoacetylacetone gives an insoluble oligomeric complex with both O and CN coordination which becomes soluble with triarylphosphines. Flexible films with reflectivities of 70-100% are obtained; films retain the mechanical properties of the parent polyimide. Thermal stability in N_2 is near that of the parent. The bulk is not conductive and contains nm-size Ag particles. X-ray shows only Ag reflections and Scherer broadening. The dik ligand has pronounced effects on Ag(I) reduction and migration of silver atoms. Adhesion is outstanding. Characterization was by X-ray, DSC, TGA, XPS, TEM, SEM, and AFM.

856.

HEAT OF FORMATION OF ALPHA AND BETA SI3N4 MEASURED USING HIGH TEMPERATURE CALORIMETRY. **J. Liang** A. Navrotsky, Department of Geosciences and Princeton Materials Institute, Princeton University, Princeton, NJ 08544-1003, and M. Mitomo, National Institute for Research in Inorganic Materials, 1-1 Namiki, Tsukuba, Ibaroki 305, Japan

The standard heats of formation of alpha- and beta-Si3N4 were determined using a Tian-Calvet type high temperature calorimeter. Drop solution technique was used, with alkali borate solvent at 770 C. Rapid dissolution (decomposition/oxidation) of the nitride samples was achieved by mixing the samples with sodium molybdate (3Na2O·4MoO3) in 1:3 (wt%) ratio. The heat of formation of alpha-Si3N4 was determined to be -791.37 ± 10.84 KJ/mol, compatible with previous published determinations; that of beta-Si3N4, -834.22 ± 11.02 KJ/mol, was determined to be ~ 5% more exothermic. The lack of reverse phase transformation, i.e., the transformation from beta to alpha form was explained on this thermochemical basis.

857. LASER-EXCITED LUMINESCENCE STUDY OF AQUEOUS SOLUTION PROPERTIES OF THE Eu^{3+}-DO2A SYSTEM   C. Allen Chang and Yuh-Herng Chen, Institute of Biological Science and Technology, National Chiao Tung University, Hsinchu 300, Taiwan

Excitation spectroscopy (emission monitored at 614.00 nm) of the $^7F_0 \rightarrow {^5D_0}$ transition of Eu^{3+} is used to study the aqueous Eu^{3+}-DO2A complex system where DO2A is a macrocyclic ligand, 1,7-dicarboxymethyl-1,4,7,10-tetraazacyclododecane. At low pH (e.g. pH~6 or below), one major band at 579.32 nm indicates that Eu^{3+} forms a 1:1 species with the ligand DO2A, presumably $Eu(DO2A)(H_2O)_n^+$. As the solution pH increases, a new species at 580.25 nm indicates that the hydrolysis product, $Eu(DO2A)(OH)(H2O)_{n-1}$, is formed. By measuring the lifetimes of each species in H_2O and D_2O, the corresponding number of the inner-sphere coordinated water molecules are 2.96 and 2.64, consistent with the proposed structures (i.e. n = 3). The hydrolysis constant (pK_h) is estimated to be around 8.1±0.3.

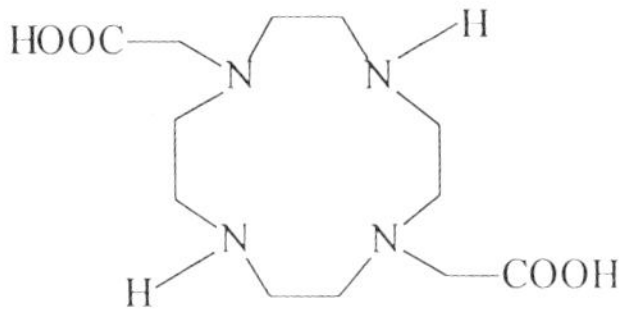

858. PROTON-COUPLED ELECTRON TRANSFER REACTIONS OF RUTHENIUM AMMINE COMPLEXES IN THE LOW-DRIVING FORCE REGIME. Reza Foroughi, H.J. Lee, Jeff C. Curtis. University of San Francisco, San Francisco, CA 94117

Stopped-flow kinetic studies of both simple and proton-coupled, low-driving force electron-transfer reactions such as the one shown below,

$$trans\text{-}(py)Ru(II)(NH_3)_3(4,4'\text{-}bpyH)^{3+} + trans\text{-}(py)Ru(III)(NH_3)_3(2,6\text{-}Me_2pyz)^{3+} \xrightarrow{k(PCET)}$$

$$trans\text{-}(py)Ru(III)(NH_3)_3(4,4'\text{-}bpy)^{3+} + trans\text{-}(py)Ru(II)(NH_3)_3(2,6\text{-}Me_2pyzH)^{3+}$$

indicate that the proton-coupled e.t. reaction actually occurs at a rate which is up to *3 times faster* than simple electron transfer. This rather unexpected result is thought to indicate the importance of stabilization of the precursor complex to electron transfer arising from strong hydrogen bonding between donor and acceptor.

859.

BINDING OF TOXIC METAL IONS BY MODIFIED DENDRIMER COMPLEXES.
Eva R. Birnbaum, Jeffrey G. McDonald, and Nancy S. Sauer, Chemical Science and Technology Division, Los Alamos National Laboratory, MS-J514, Los Alamos, NM, 87545.

We are investigating dendrimer complexes as ligands for selective binding of toxic metal ions. Such systems could potentially be used for remediation of contaminated water and soils. Several generations of PAMAM Starburst Dendrimers have been chemically modified to incorporate new ligands or hydrogen bonding groups at the primary amines. The presence of hydrogen bond donor and acceptor moieties is designed to enhance binding to metallooxy-anions such as arsenate or chromate. Total binding capacity for several metal cations and metallooxyanions by the dendrimers have been determined, as well as relative ion selectivity. Heteronuclear NMR indicates that even at relatively low functionalization, metal ions interact with functional groups added onto the dendrimer periphery. The modes of metal binding with the modified dendrimer complexes have been characterized by proton and metal NMR.

860. RETENTION OF MERCURY ON GOLD ELECTRODES IN POTENTIOMETRIC STRIPPING VOLTAMMETRY EXPERIMENTS Charles M. Watson, Alice E. Bruce, Daniel J. Dwyer, Mitchell R. M. Bruce, University of Maine, Orono, Maine 04469.

$HgCl_2$ in concentrations ranging from 48 nM to 160 µM are reduced onto a 1.0 cm^2 gold foil electrode at − 0.3 V vs. Ag/AgCl reference electrode and subsequently stripped from the electrode at potentials from +0.5 V to + 1.4V. Baseline stripping current levels are obtained at the end of 5 to 10 minute stripping procedures; however, the charge passed during reduction is about 10 times greater than that for the oxidation when using a clean (no residual mercury) electrode. X-ray photoelectron spectroscopic analysis of the gold electrode, after the stripping procedure, clearly shows mercury retention by the gold electrode. The evidence suggests that either most of the reduced mercury does not stay on the electrode, and or that it diffuses into the gold and is not available for oxidation.

861. DEVELOPMENTS IN DINUCLEAR GOLD(I) COMPLEXES WITH PHOSPHOR-1,1-DITHIOLATO TYPE LIGANDS. Werner E. van Zyl and John P. Fackler, Jr., Department of Chemistry and Laboratory for Molecular Structure and Bonding, Texas A&M University, College Station, Texas 77843-3255.

Phosphor-1,1-dithiolato type ligands have found application in diverse technological fields such as antioxidants in motor oils, flotation agents for the separation of heavy metals from ores, and as insecticide derivatives. The syntheses and characterization of new gold(I) dithiophosphinate and dithiophosphonate complexes of the type $[AuS_2PR_2]_2$ and $[AuS_2PR(OR)]_2$ will be presented. Structural studies have indicated that weak intermolecular gold-gold interactions are present in some complexes and absent in others.

862. PALLADIUM AND PLATINUM COMPLEXES OF DIALKYL DIAZIRIDINES. Penny A. Chaloner*, Juliet Cornford, Shaliza S. Dewa, Peter B. Hitchcock and Audrey Wagenaar, School of Chemistry, University of Sussex, Falmer, Brighton BN1 9QJ, UK

3,3-Dialkyldiaziridines react with $[M_2Cl_2(\mu-Cl)_2(PEt_3)_2]$ (M = Pd or Pt) to give bis metallated complexes. The bis metallated species predominates irrespective of the composition of the original reation mixture. The structures of a range of such complexes have been determined, and their dynamic behaviour in solution investigated by multinuclear nmr spectroscopy. When one of the substituents on the diaziridine is CF_3, however, monometallated diaziridine complexes are isolated in the solid state. In solution there is an equilibrium between two diastereomeric monometallated species, and a small amount of a dimetallated diaziridine.

863. STRUCTURAL AND SPECTROSCOPIC INVESTIGATIONS OF POLYNUCLEAR GOLD ORGANOMETALLIC COMPLEXES DISPLAYING AU···AU INTERACTIONS. Jess C. Vickery, Alan L. Balch, and Marilyn M. Olmstead, Department of Chemistry, University of California, Davis, CA. 95616

Trinuclear gold organometallic complexes have been synthesized, structurally characterized, and spectroscopically probed. An intermediate in the formation of these compounds, $Ph_3PAuC(O)NHCH_3$, has also been isolated and characterized. Exciting structures result from self-association in the solid state due to intermolecular Au···Au interactions and hydrogen bonding. Additionally, the remarkable photophysical and electrochemical properties of the archetypical trimer, $[AuC(OCH_3)NCH_3]_3$, and related compounds have been investigated.

864. AZA-CROWNS AS SUPPORTING LIGATION IN F-ELEMENT AND EARLY TRANSITION METAL ORGANOMETALLIC CHEMISTRY. **David J. Berg** and Lawrence Lee, Department of Chemistry, University of Victoria, P.O. Box 3065, Victoria, B.C., Canada V8W 3V6

Early transition metal and f-element organometallic chemistry supported by the deprotonated aza crown ligands 4,13-diaza-18-crown-6 (DAC), aza-18-crown-6 (MAC) and aza-15-crown-5 (15-AC-5) will be discussed. Stable yttrium alkyl complexes such as Y[DAC]R, Y[MAC]R$_2$, and Y[15-AC-5]R$_2$ (R = CH_2SiMe_3, $CH(SiMe_3)_2$, Ph) react cleanly with CO to produce the enolate complexes and react with terminal alkynes to form alkynide complexes which undergo reversible carbon-carbon bond formation to form butatrienediyl complexes. Generation of highly reactive alkyl cations and attempts to prepare an yttrium benzyne complex will also be reported. Syntheses and crystallographic studies of Zr[DAC]R$_2$ and Zr[DAC]R$^+$ (R = CH_2Ph, CH_2SiMe_3, CH_2CMe_3, Me) will be discussed. The cationic species undergo insertion chemistry with CO, alkynes and isonitriles (the latter to yield a vinyl amido cation).

865. DINITROGEN BRIDGED GOLD CLUSTERS, Paul R. Sharp, Hui Shan, Yi Yang, and Alan J. James, Department of Chemistry, University of Missouri, Columbia, MO 65211.

A family of dinitrogen complexes, $[(LAu)_6(N_2)]^{2+}$ **1** (L = a phosphine), with the dinitrogen unit bridging two clusters of three gold atoms, have been prepared from hydrazine and $[(LAu)_3(O)]^+$. Structural characterization of the PPh_2Pr^i derivative shows a nitrogen-nitrogen single bond distance indicative of hydrazido (N_2^{4-}) character for the dinitrogen unit. The complexes can be reduced and protonated to give low (13%, L = PPh_2Me) to quantitative yields (100%, L = $P(p-MeOC_6H_5)_3$) of ammonia, establishing that bonding of dinitrogen to six metal atoms can lead to facile cleavage of the nitrogen-nitrogen bond.

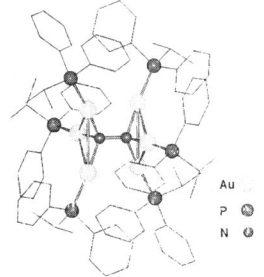

Structure of $[(LAu)_6(N_2)]^{2+}$ **1** (L = PPh_2Pr^i)

American Chemical Society
DIVISION OF MEDICINAL CHEMISTRY

213th ACS National Meeting

San Francisco, CA
April 13-17, 1997

T.J. Perun, Program Chairperson

SUNDAY MORNING

- **Science of Synthesis: A Legacy of William S. Johnson**
  B.M. Trost, Presiding

  Papers 1-5

SUNDAY AFTERNOON

- **Science of Synthesis: A Legacy of William S. Johnson**
  P.A. Wender, Presiding

  No Papers

- **General Oral Papers**
  E. Thorsett, Presiding

  Papers 6-17

SUNDAY EVENING

- **Posters**

  Papers 18-143

MONDAY MORNING

- **General Oral Papers**
  J.J. Wright, Presiding

  Papers 144-151

MONDAY AFTERNOON

- **Calorimetric Techniques in Drug Discovery**
  P. Weber, Presiding

  Papers 152-157

TUESDAY MORNING

- **Dopamine D4 Receptor Ligands**
  D. Wustrow, Presiding

  Papers 158-163

TUESDAY AFTERNOON

- **Patenting Issues for Molecular Diversity**
  J.A. Bristol, Presiding

  Papers 164-166

- **General Oral Papers**
  S.K. Davidsen, Presiding

  Papers 167-178

WEDNESDAY MORNING

- **Strategies for the Treatment of Obesity**   Papers 179-183
  R.R. Wexler, D.W. Robertson, Presiding

WEDNESDAY MORNING AND AFTERNOON

- **Posters**   Papers 184-283
  T.J. Perun, Presiding

WEDNESDAY AFTERNOON

- **Aspects of Chirality in Medicinal Chemistry**   Papers 284-288
  I. Agranat, Presiding

- **General Oral Papers**   Papers 289-300
  N.H. Lin, Presiding

THURSDAY MORNING

- **Gene Knockouts**   Papers 301-304
  G. Johnson, Presiding

- **General Oral Papers**   Papers 305-316
  S.A. Boyd, Presiding

THURSDAY AFTERNOON

- **Antifungal Therapy**   Papers 317-321
  W. Watkins, Presiding

DIVISION OF MEDICINAL CHEMISTRY

001.
PEPTIDE RELATED RESEARCH AS A VEHICLE TOWARDS CHEMICAL AND BIOLOGICAL UNDERSTANDING. Ralph Hirschmann, Department of Chemistry, University of Pennsylvania, 231 South 34th Street, Philadelphia, PA 19104

The paper will describe collaborative research with Professors Amos B. Smith, III and K.C. Nicolaou at the University of Pennsylvania, initiated in 1988, involving the design and synthesis of novel scaffolds, devoid of the amide backbone or of isosteres thereof. These scaffolds successfully mimicked peptidal β-turns and β-pleated sheets as evidenced by the biological activities of these compounds. The research has generated unexpected results. For example, peptidomimetics using a β-D-glucose scaffolds revealed similarities between the somatostatin and the substance P (NK-1) receptors not shown by their endogenous ligands. HIV-1 protease inhibitors were synthesized with Smith, incorporating a pyrrolinone scaffold. Cocrystallization with the protease by Drs. Kuo and Chen (Merck Research Laboratories) provided interesting information which is consistent with biological studies with mutant enzymes by Dr. Donald Heefner (Sepracor). The transport properties of these pyrrolinones are being analyzed in terms of earlier physical concepts (solvation) and - with Professor R.T. Borchardt (University of Kansas) - in terms of the role of efflux systems typified by P-glycoprotein.

002. TOTAL SYNTHESIS OF THE CYCLIC BIPHENYL ETHER PEPTIDES K-13 AND OF4949-III *VIA* $S_N Ar$ MACROCYCLIZATION OF PEPTIDYL RUTHENIUM ARENE COMPLEXES. APPLICATION TO THE DESIGN OF NOVEL PEPTIDEMIMETICS. **Daniel H. Rich**, James Janetka, and George Flentke, School of Pharmacy and Department of Chemistry, University of Wisconsin-Madison, Madison, WI 53706.

The metalloprotease inhibitors, K-13 and OF4949-III, can be synthesized in seven steps by peptide coupling of ruthenium arene p-complexes of aromatic amino acids followed by $S_N Ar$ macrocyclization of the resultant peptidyl ruthenium p-complexes. The route is ammenabl/e to synthesis of compound libraries. The cyclic biphenyl ether portion of these molecules is formed in higher yields from the ruthenium-based $S_N Ar$ cyclizations than cyclizations via amide bond formation. X-ray crystallography and molecular modeling establish that the cyclic biphenyl ether tripeptide system is a mimetic of protease-bound β-sheet structures. Potent HIV protease inhibitors can be constructed from this mimetic.

003. INHIBITION OF CARBOHYDRATE RECOGNITION IN BIOLOGICAL SYSTEMS. C. -H. Wong, Department of Chemistry and the Skaggs Institute of Chemical Biology, The Scripps Research Institute, La Jolla, California 92037

Carbohydrates in biological systems are often associated with many specific recognition and signaling processes leading to important biological functions and diseases. Considerable efforts have been directed toward understanding and mimicking the recognition processes, and developing effective agents to control the processes. The pace of discovery research in glycobiology and development of carbohydrate-based therapeutics, however, has been relatively slow due to the lack of appropriate strategies and methods available for carbohydrate-related research. This lecture will present some of the most recent development in our laboratories regarding the use of chemo-enzymatic strategies to tackle the recognition problem. Highlights include the study of selectin-carbohydrate and aminoglycoside-RNA interactions, glycosyltransferase reactions, and development of agents for the intervention of these recognition processes.

004. NATURAL PRODUCTS SYNTHESIS AND DEVELOPMENT OF PROTECTING GROUPS. Tohru Fukuyama, Faculty of Pharmaceutical Sciences, University of Tokyo, 7-3-1 Hongo, Bunkyo-ku, Tokyo 113, Japan.

An ideal synthesis of a complex molecule would be the one that requires no protecting groups. In practice, however, total synthesis of complex natural products often depends on a successful use of a variety of protecting groups to elaborate molecules with multiple functional groups. A desirable protecting group should survive a number of steps and, when needed, can easily be deprotected without interfering with the rest of the molecule. While mastery of the known protecting groups is essential for synthetic chemists, one would occasionally encounter the cases where a novel protecting group has to be developed to realize one's creative synthetic scheme. Several examples of this lecturer's own experiences in this area will be discussed.

005. AN EXCITING NEW MEDICINAL LEAD: SYNTHETIC, BIOLOGICAL, COMPUTER, AND NMR STUDIES. Paul A. Wender, Department of Chemistry, Stanford University, Stanford, CA 94305 USA

Research in our laboratories is directed at the design and development of new reactions, reagents, and strategies for synthesis and at the application of these studies to problems in biology, medicine, and materials science. In this lecture, we will report for the first time the genesis, rational design, and development of a new class of medicinal leads, based on computer modeling work reported by our laboratory in 1988 (*PNAS*) and recently translated into an active lead.

006. DESIGN, SYNTHESES AND STRUCTURE-ACTIVITY RELATIONSHIP STUDY OF PHOSPHAPEPTIDE INHIBITORS OF GLUTATHIONYLSPERMIDINE SYNTHETASE.
Shoujun Chen[†], Chun-Hung Lin[‡], David S. Kwon[‡], William P. Malachowski[†], Christopher T. Walsh[‡], and James K. Coward[†]
[†]*Departments of Chemistry and Medicinal Chemistry, The University of Michigan, Ann Arbor, MI 48109-1055, and*
[‡]*Department of Biological Chemistry and Molecular Pharmacology, Harvard University Medical School, Boston, MA 02115*

Acting as an antioxidant cofactor found uniquely in parasitic trypanosomes, a conjugate of the tripeptide glutathione (GSH) and the polyamine spermidine, namely trypanothione (N^1, N^8-bisglutathionylspermine, TSH), has attracted special interest as an ideal target for antitrypanosomal drug design. Glutathionylspermidine synthetase (GSPS) catalyzes the ATP-dependent reaction of glutathione (GSH) and spermidine to form glutathionylspermidine (GSP), a key step in the biosynthesis of TSH. As part of recent research aimed at the discovery of new antiparasitic drugs, we have designed and synthesized several polyfunctionalized phosphonates, phosphonamidates and phosphinates with a common structure, **1**, in addition to truncated derivatives (structures not shown), as mimics of the proposed tetrahedral intermediate for the GSPS-catalyzed reaction. The synthesis of these highly functionalized targets and their biological evaluation as potential inhibitors of GSPS will be presented. (Supported by grants: WHO, ID No. 930598 (JKC), and USPHS/NIH, GM-20011 (CTW).)

	R^1	R^2	X
a	H	H	O
b	H	Me	NH
c	H	H	CH_2
d	OH	H	O
e	OH	H	CH_2
f	OAc	H	CH_2

007. SOLUTION PHASE COMBINATORIAL CHEMISTRY. DISCOVERY OF POTENT ANTIMICROBIAL ACTIVITY FROM NOVEL LINEAR PYRIDINOPOLYAMINE LIBRARIES. H. An, B. D. Haly, A. S. Fraser, C. J. Guinosso, L. Wilson-Lingardo, L. M. Risen, J. R. Wyatt and P. D. Cook, Isis Pharmaceuticals, 2292 Faraday Avenue, Carlsbad, CA 92008

A novel mono-*t*-Boc-protected linear pyridinopolyamine scaffold was synthesized by a Schiff base type cyclization of 2,6-pyridinedicarboxaldehyde with mono-protected triamine using Ni^{2+} as a template followed by reductive cleavage and decomplexation in a one-pot procedure. The corresponding orthogonally protected scaffold, required for total deconvolution and future library generation, was also synthesized by the reaction of novel 2-bromomethyl-6-phthalimidomethylpyridine and protected hydroxypropylethylenediamine. A solution phase simultaneous addition of functionalities (SPSAF) combinatorial strategy was used to prepare 13 high purity tertiary amine libraries (total complexity: 1638). All libraries were examined by TLC, purified by chromatographic techniques and confirmed by ESI MS spectral data. Five sublibraries exhibited potent antimicrobial activities; several single compounds derived from these active libraries demonstrated promising antibacterial activity *in vivo*.

008. NEW DNA TARGETING AGENTS FOR CANCER TREATMENT: SYNTHESIS AND BIOLOGICAL EVALUATION OF BORONATED POLYAMINES. J. Cai[1], A. H. Soloway[1], D. M. Adams[2], R. F. Barth[2], J. R. Hariharan[1], I. M. Wyzlic[1] and K. Radcliffe[1], The Ohio State University, College of Pharmacy[1] and Department of Pathology[2], Columbus, OH 43210

In order to develop drugs for Boron Neutron Capture Therapy that may be more effective as tumor targeting agents than compounds currently used clinically, a series of new boron-containing spermidine/ spermine (SPD/SPM) analogs were synthesized. Boron moieties such as *o*-carborane, 2-aminoethyl-*o*-carborane, 2,3-dihydroxypropyl-*o*-carborane were attached to the internal or terminal nitrogens of SPD/SPM via a chain of four methylene groups and in the case of phenylboronic acid, was attached to the internal nitrogen via one methylene group. *In vitro* biological evaluations demonstrate that these substituted SPD/SPM compounds retain the ability to displace ethidium bromide from calf thymus DNA and are rapidly taken up by F98 glioma cells. However, in contrast with SPD/SPM, these polyamine analogs, which have a mono-substituted *o*-carborane group, have much greater toxicity, especially those with terminal *N*-substituted boron compounds. The toxicity of these carboranyl polyamines can be significantly reduced if a hydrophilic groups such as ammonium or hydroxyl entities are attached on the unsubstituted carbon of the carborane cage. These compounds begin to approach SPD/SPM with respect to toxicity.

009.

ANTIBACTERIAL ACTIVITY OF TERTIARY NITROGEN-BASED COMBINATORIAL LIBRARIES PREPARED BY SOLUTION PHASE SYNTHESIS
Pei-Pei Kung, Laura Lingardo, Mike Greig, Jacqueline Wyatt and P. Dan Cook
Medicinal Chemistry, Isis Pharmaceuticals, 2292 Faraday Avenue, Carlsbad, CA 92008

Tertiary amine and oxyamine derived combinatorial libraries were prepared using a solution phase simultaneous addition of functionalities (SPSAF) approach in order to discover novel pharmacophores with broad-spectrum antibacterial activities. The resulting libraries were characterized by ES/MS and their biological activities against *gram positive* and *gram negative* bacteria will be reported. In addition, model studies were conducted prior to library synthesis to demonstrate the use of a simple liquid-liquid extraction purification procedure. The active leads generated from "subtractive deconvolution" of the above libraries along with other acylated amine and oxyamine analogues (prepared utilizing the differential reactivity of amine and oxyamine nucleophiles toward sp^3 electrophiles) will also be discussed.

010. SUPRAMOLECULAR ASSEMBLY OF MODIFIED CEREBROSIDES. A. Goldstein,* M. Gelb, A. Lukyanov,# P. Carlson,# P. Yager,# Departments of Chemistry and Biochemistry, Box 351700 and Center for Bioengineering,# Box 352255, University of Washington, Seattle, WA 98195

Lipid tubules are a recently discovered self-organizing system in which lipids crystallize into organized microstructures (hollow tubules, helices and cochleate cylinders). Naturally occurring cerebrosides are known to form such microstructures and could serve as a core for a continuous drug release system having application in antibacterial, antifungal and antitumor therapy. The microstructure should dissolve or be enzymatic hydroyzed from its ends due to tight crystal lattice packing. The lipidated drugs might themselves be active if the hydrophobic groups do not interfere with binding at the sites of action or function as prodrugs when the hydrophobic moieties are cleaved. Modifications to the sphingosine base and head group replacement of cerebroside has yielded further insight into microstructre formation and provided novel drug delivery cores.

011.

CONFORMATIONALLY CONSTRAINED 7-AZABICYCLOHEPTANE AMINO ACIDS AS PROLINE MIMETICS. W. Han,* J. Pelletier, A. Sahli, L. J. Mersinger, C. A. Kettner, and C. N. Hodge, Du Pont Merck Pharmaceutical Co., P. O. Box 80500, Wilmington, DE 19880

Conformationally constrained 7-azabicycloheptane amino acids (**IV**) were discovered to be effective proline mimetics in enzyme inhibitors. Employing this mimetic, two novel classes of cyclic ureas (**I** and **II**) were designed and proved to be HIV protease inhibitors with moderate to high affinity. In another example, **IV** was incorporated with a boroarginine to give potent thrombin inhibitor **III**. The synthesis and biological activities of classes **I, II,** and **III** will be disclosed.

(I)   (II)   (III)   (IV)

012. ACYLATED NON-α-AMINO ACIDS AS NOVEL AGENTS FOR THE ORAL DELIVERY OF THERAPEUTIC LEVELS OF USP HEPARIN. Andrea Leone-Bay*, Duncan Paton, John Freeman, Christine Lercara, Doris O'Toole, Theresa Rivera, Connie Rosado, Elizabeth Harris, and Robert Baughman, Emisphere Technologies, Inc., Hawthorne, NY 10532

A family of novel compounds (below) that promote the gastrointestinal absorption of USP heparin in rats and primates has been discovered. The delivery agents with heparin were administered either orally or intracolonically in an aqueous propylene glycol solution and caused dramatic increases in both plasma heparin concentrations (anti-factor Xa) and clotting times (APTT). Using one of the most effective delivery agents (n = 7) in this series, an estimated relative bioavailability of 10% can be achieved following oral administration to cynomolgus monkeys. To establish a correlation between the *in vivo* data and an *in vitro* parameter, immobilized artificial membrane (IAM) chromatography and heparin affinity chromatography studies were performed. Heparin affinity chromatography k' values were correlated to the efficiency of oral heparin delivery. A similar correlation was obtained for heparin delivery and log relative k' values on the IAM column.

$X = (CH_2)_n$; n = 1-11
$Y = $ 2-hydroxyphenyl

013. NMR STUDIES OF THE AMYLOID β-PEPTIDE: SOLUTION STRUCTURE OF RESIDUES 1-42
Haiyan Shao, Shu-chuan Jao, Michael G. Zagorski, Department of Chemistry, Case Western Reserve University, Cleveland, OH 44106

The three dimensional solution structure of residues 1-42 of the amyloid β-peptide was determined using NMR spectroscopy, distance geometry, and molecular dynamic techniques. The NMR data used to derive the structure consisted of NOE data and temperature coefficients of the amide NH chemical shifts. The β-peptide is a major component of amyloid deposits in the Alzheimer Disease. The sequential and medium range cross peaks that were found in the NOESY indicate that the peptide folds to form two α-helices connected by a loop while the residues at the N-terminus adopt a predominantly undefined structure. The NH temperature studies establish that the peptide is located at the surface of the micelle.

014. STRUCTURAL STUDIES OF PROTEINS RELATED TO ALZHEIMER S DISEASE.
Kan Ma[1], Leigh Hsu[2], Tsunao Saitoh[2], Michael G. Zagorski[1]. [1]Department of Chemistry, Case Western Reserve University, Cleveland, Ohio 44106. [2]Department of Neurosciences, University of California at San Diego, La Jolla, California 92093

Recent studies demonstrate that the non-Aβ protein component precursor (NACP) binds and promotes precipitation of the β-peptide as amyloid. To provide a molecular basis for these binding processes, we undertook NMR and circular dichroism (CD) studies of the NACP protein as a uniformly ^{15}N labeled sample obtained from an E. coli expression system. The NMR experiments were performed at different temperature (10°-50°C) and pH values (3, 5, 7, and 8). Under all of these conditions, the data establish that the major conformation of the NACP protein is random coil.

Another protein related to AD is the amyloid precursor protein (APP) which generates the β-peptide. The APP is an integral membrane protein with a transmembrane domain near the COOH-terminal side. Recent studies pointed out the membrane is indirectly important to β-peptide generation from the APP. In addition, transport of APP through the secretary pathway is a requisite for β-amyloid formation, and insertion of APP into the membrane is essential for transport. We performed CD and NMR studies of a 43-residue peptide fragment of APP that includes the complete transmembrane domain (residues 684-727). Our data show that in micelle solution the predicted transmembrane domain is largely α-helical.

015.
THE DESIGN AND SYNTHESIS OF POTENTIAL SELECTIVE PROGESTERONE RECEPTOR ANTAGONISTS. **John R. Woolfrey** Mitchell A. Avery. The Department of Medicinal Chemistry; The National Center for the Development of Natural Products, The Research Institute of Pharmaceutical Sciences, School of Pharmacy, University of Mississippi, University, Mississippi 38677

A 3-D quantitative structure-activity relationships (QSAR) study involving compounds with antiprogestational activity was used to develop a progesterone receptor (PR) pharmacophore. Use of comparative molecular field analysis (CoMFA) has indicated regions of favorable and disfavorable steric and electrostatic environments for ligand-receptor binding.The incorporation of glucocorticoid receptor (GR) binding activity of these compounds into the CoMFA model has resulted in observable differences between GR and PR binding. Additional 2-D and 3-D QSAR data was included in preparation of the pharmacophore. The model was used to design steroidal and non-steroidal potential PR antagonists.

016. MECHANISM OF INACTIVATION OF NERUONAL NITRIC OXIDE SYNTHASE BY N^{ω}-ALLYL-L-ARGININE. Henry Q. Zhang,[§] Robert P. Dixon,[§] Michael A. Marletta,[†] and Richard B. Silverman*,[§] [§]Department of Chemistry and Department of Biochemistry, Molecular Biology, and Cell Biology, Northwestern University, Evanston, Illinois 60208 and the [†]Interdepartmental Program in Medicinal Chemistry, University of Michigan, Ann Arbor, Michigan 48109

N^{ω}-Allyl-L-arginine or a viable intermediate N^{ω}-allyl-N^{ω}-hydroxy-L-arginine is shown to be a competitive reversible inhibitor and time-dependent inactivator of bovine brain nitric oxide synthase (nNOS). The enzyme is protected against inactivation by the presence of the substrate, in the absence of O_2, or if $NADP^+$ is substituted for NADPH. Inactivation of nNOS with N^{ω}-allyl-[^{14}C]-L-arginine or N^{ω}-[^3H]allyl-L-arginine is investigated and indicates that only the allyl part of the inactivator is bound to the heme. A mechanism that is consistent with these results is proposed.

017. MEASUREMENTS ON WHOLE BLOOD AND UNDILUTED URINE EMPLOYING NEWLY DEVELOPED SIRE BASED BIOSENSORS. D. Kriz, Department of Pure and Applied Biochemistry, University of Lund, P.O. Box 124, SE-221 00 Lund, Sweden

A new emerging technology, SIRE (Sensors based on Injectable Recognition Elements), has proven to solve many of the problems concerning biosensors. Traditionally, biosensor developing scientists have focused on biomolecules being expensive and thus must be reused (i.e. immobilized). This has often resulted in situations where unstable recognition elements have been bound to stable transducers. This talk will discuss a biosensor based on flow injection of the recognition element applied on small volume (30 μl) whole blood samples and undiluted urine samples. A glucose tolerance monitoring, after a peroral intake, was performed. The versatility of the biosensor was demonstrated by determination of several different metabolites (glucose, lactate, phenyl alanine and galactose). The advantages of the presented biosensor concerning multi analyte analysis and background signal control are focused upon. These features makes it suitable for real time bed side monitoring.

018. EFFECTS OF TAXANES ON β-AMYLOID INDUCED TOXICITY IN PRIMARY NEURONS. R.R. Ranciat, R.E. Ragan, Y. Liu, M. Hepperle, M.L. Michaelis,* and G.I. Georg,* Departments of Medicinal Chemistry and Pharmacology / Toxicology, University of Kansas, Lawrence, KS 66045

Alzheimer's Disease (AD) is a neurodegenerative disorder with no pharmacological treatments to stop or slow the progression of the disease. AD pathology includes extracellular deposits of β-amyloid (Aβ) in senile plaques and the intracellular formation of paired helical filaments (PHF's) that form neurofibrillary tangles. PHF's contain abnormally phosphorylated microtubule associated τ protein, which has been implicated in diminished axonal transport and death of neurons. β-Amyloid induces τ-phosphorylation and loss of microtubule binding. We will present data to demonstrate that microtubule-stabilizing taxanes protect primary cortical neurons against Aβ-induced cytotoxicity. Cell counts in several cultures revealed that paclitaxel/Aβ-treated cultures had an average 97% survival rate compared to controls, whereas the Aβ only cells had a 61% survival rate. Current work focuses on the synthesis and evaluation of taxanes more likely to cross the blood-brain barrier.

019. HIGHLY SELECTIVE INHIBITORS OF PHENYLETHANOLAMINE N-METHYLTRANSFERASE T. M. Caldwell, K. R. Criscione, V. H. Dahanukar, R. K. Jalluri, M. Slavica, and G. L. Grunewald, Department of Medicinal Chemistry, University of Kansas, Lawrence, KS 66045.

In order to probe the role of epinephrine in the central nervous system, a potent and selective inhibitor of phenylethanolamine N-methyltransferase (PNMT) is desired. Most widely studied inhibitors of PNMT such as 7,8-dichloro-1,2,3,4-tetrahydroisoquinoline (SK&F 64139) have strong affinity towards α_2-adrenoceptors (PNMT K_i = 0.22 μM, α_2 K_i = 0.021 μM, selectivity = 0.095). We will present data on 3-hydroxymethyl-7-aminosulfonyl-THIQ which displays the highest PNMT selectivity thus far obtained (PNMT K_i = 0.342 μM, α_2 K_i = 1400 μM, selectivity = 4100). However, this compound is too hydrophilic to penetrate the blood-brain barrier (calculated log P = -1.01). Comparitive Molecular Field Analysis (CoMFA) models of both the PNMT active site and the α_2-adrenoceptor were used to design potent inhibitors of PNMT with low affinity for the α_2-adrenoceptor and with a lipophilicity sufficient to ensure access to the CNS. Biological data for these analogs along with preliminary data on blood-brain barrier penetration will be presented. Sponsored by NIH HL34193.

020. COMPARISON OF LIGAND BINDING AFFINITIES TO RECOMBINANT HUMAN PHENYLETHANOLAMINE N-METHYLTRANSFERASE (hPNMT) WITH BOVINE ADRENAL ENZYME AND CRYSTALLIZATION OF hPNMT. G. L. Grunewald,[1] K. R. Criscione,[1] J. L. Martin,[2] and M. J. McLeish.[3] [1]Department of Medicinal Chemistry, University of Kansas, Lawrence, KS, 66045; [2]Centre for Drug Design and Development, University of Queensland, Brisbane 4072, Queensland, Australia; [3]Victorian College of Pharmacy, Monash University, Parkville 3072, Victoria, Australia.

PNMT catalyzes the final step in the biosynthesis of epinephrine where it is found in discrete brain regions and is thought to play a role in the regulation of blood pressure, release of pituitary hormones, and control of α_2-adrenoceptors. We have amplified the gene for hPNMT using PCR and cloned the hPNMT gene into a pET expression vector. IPTG induction gave high levels of protein in 90% soluble form which was purified to homogeneity. hPNMT was co-crystallized with several known PNMT inhibitors (SK&F 29661, SK&F 64139, and LY 134046) and AdoHcy. Crystallization produced large, well-formed, single crystals that diffract to beyond 3 Å. In order to further characterize hPNMT, a comparison study of ligand binding was performed between hPNMT and partially purified bovine adrenal PNMT.

021.
TRIPEPTIDE AND PENTAPEPTIDE ANALOGUES OF N-TERMINUS ARYL-CARBAMOYL-THIENYLALANYL-LEUCYL-PHENYLALANINE CHEMOTACTIC PEPTIDES. Kirsten L. Overoye, Milind Rajopadhye*, John A. Barrett, Stuart J. Heminway, Barrett Newsome, Timothy R. Carroll, Thomas D. Harris, and Steven R. Johnson. Discovery Research, Radiopharmaceutical Division, The DuPont Merck Pharmaceutical Company, 331 Treble Cove Road, North Billerica, MA 01862

Formyl-Met-Leu-Phe (fMLF) is implicated in the inflammation and infection process and exhibits agonist properties. The synthesis and biological activity of new N-terminus urea substituted Met-Leu-Phe and Phe-dLeu-Phe-dLeu-Phe chemotactic peptides has been recently reported (Higgins, et al. J. Med. Chem. 1996, 39(5), 1013). We now wish to present the SAR of a series of tripeptide and pentapeptide analogues of N-terminus urea Thi-Leu-Phe chemotactic peptides. These peptides have exhibited agonist and non-agonist properties. The synthesis of these peptides and results from the fMLF binding assay and Cytochrome C assay will be presented.

022. SYNTHESIS OF ANALOGS OF SDZ NVI 085: PHARMACOLOGICAL EVALUATION AT CLONED HUMAN α ADRENERGIC RECEPTORS. <u>Yoon T. Jeon,</u>[a] Chao Li,[a] Chi Lou,[a] Douglas Craig,[b] Carlos Forray,[b] Boshan Li,[b] Michelle Iacolina,[b] Theresa A. Branchek[b] and Charles Gluchowski.[a] Department of Chemistry[a] and Pharmacology,[b] Synaptic Pharmaceutical Corporation, 215 College Road, Paramus, NJ 07652

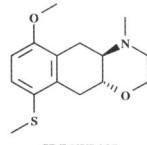

SDZ NVI 085

SDZ NVI 085 has been reported to be a new centrally acting α_1 agonist for the treatment of cerebral ischaemia. As a part of our effort to elucidate therapeutic indications of α_1 adrenergic receptor subtypes and identify novel subtype selective α_1 agonists, we prepared SDZ NVI 085 and several analogs and assayed them at cloned human α adrenergic receptors. Several compounds were found to selectively bind and activate the human α_{1A} adrenergic receptor. The synthesis and SAR for these compounds will be discussed.

023.
DESIGN AND SYNTHESIS OF ALPHA$_2$ ADRENOCEPTOR AGONISTS. <u>Wai C. Wong,</u>* Murali Dhar,* Wanying Sun,* Sriram Tyagarajan,* Carlos Forray,# Pierre J.-J. Vaysse,@ Theresa A. Branchek# and Charles Gluchowski*
*Departments of *Chemistry, #Pharmacology and @Cell Biology*
Synaptic Pharmaceutical Corporation
215 College Road, Paramus, NJ 07652

Alpha$_2$ adrenoceptor agonists have shown promise as agents for the treatment of such conditions as pain, hypertension, glaucoma, cognitive deficiency and migraine. A number of alpha$_2$ agonists have been designed and synthesized based on UK-14304 and medetomidine. Their binding and functional activities at the cloned human alpha$_2$ adrenoceptors will be described.

UK-14304   Medetomidine

024. DESIGN AND SYNTHESIS OF BENZIMIDAZOLE AMINOIMIDAZOLINES AS NOVEL α_2 ADRENERGIC RECEPTOR AGONISTS. <u>Yoon T. Jeon,</u>[a] Chi Lou,[a] Carlos Forray,[b] Vince Jorgensen,[b] Raisa Nagorny,[b] Theresa A. Branchek[b] and Charles Gluchowski,[a] Department of Chemistry[a] and Pharmacology,[b] Synaptic Pharmaceutical Corporation, 215 College Road, Paramus, NJ, 07652

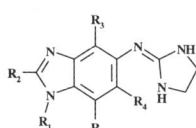

As a part of program directed at the identification of novel subtype selective α_2 adrenergic receptor agonists, we have designed and synthesized a series of benzimidazole aminoimidazolines and evaluated their pharmacology at cloned human α adrenergic receptors. Several compounds were found to have high binding affinity as well as functional efficacy with subtype selectivity at different α_2 receptor subtypes. The synthesis and SAR for these compounds will be discussed.

025. Design, synthesis and evaluation of new α_2/5-HT$_{2A}$ ligands. <u>Robert N. Hanson,</u> Farid A. Mohamed, and Richard C. Deth, Dept. of Pharmaceutical Sciences, Bouve College of Pharmacy and Health Sciences, Northeastern University Boston, MA 02115

Based upon a 3-dimensional model of the interaction of the α_2-receptor and the prototypical ligand, benzobenzamil, we designed and synthesized a series of N-aroyl-N'-

naphthylmethyl guanidines as potential receptor ligands.

Evaluation by receptor binding and isolated tissue assays demonstrated crossover between α_2- and 5-HT receptors. Selectivities and affinities were analyzed based on conformational and molecular modeling studies.

026.

NOVEL 5-AMINO-4-(4-ARYLPIPERAZIN-1-YLMETHYL)PYRAZOLES WITH HIGH AFFINITY FOR 5-HT-1A AND ALPHA-2A RECEPTORS AS POTENTIAL ANTIDEPRESSANTS. **Frank Kerrigan**, Sharon C. Cheetham, Helen Jackson, Claire Martin, Gerry H. Thomas and John P. Watts, Knoll Pharmaceuticals, Research and Development Department, Pennyfoot St., Nottingham NG1 1GF, UK.

Antidepressant drugs are believed to exert their effects by enhancing central 5-HT and/or noradrenaline function. In recent years the focus has been drugs which selectively increase 5-HT function by inhibition of 5-HT reuptake (the SSRIs). This poster presents the design and synthesis of a novel series of 5-amino-4-(4-arylpiperazin-1-ylmethyl)pyrazoles which increase both 5-HT and noradrenaline function via a synergistic combination of agonist action at post-synaptic 5-HT-1A receptors and antagonist action at pre-synaptic alpha-2A adrenoceptors. We also present SAR investigations which enabled the reduction of undesired affinity for alpha-1 adrenoceptors, and in vivo data suggesting high antidepressant potential for several members of the series.

027. STRUCTURE ACTIVITY STUDIES ON A SERIES OF NOVEL BI- AND TRICYCLIC THIENO-PYRIMIDINEDIONES AS α_{1A} ADRENORECEPTOR ANTAGONISTS William A. Carroll, Robert J. Altenbach, Irene Drizin, Karin Tietje, Suzanne A. Lebold, Steven W. Elmore, Fatima Z. Basha, Arthur A. Hancock, James F. Kerwin, Jr., and Michael D. Meyer. Neuroscience Research, Abbott Laboratories, Abbott Park, IL 60064

The identification of multiple subtypes of the α_1 adrenoceptor and the localization of the α_{1A} subtype to the prostate has created the opportunity for the development of uroselective α_1 antagonists for the symptomatic treatment of benign prostatic hyperplasia (BPH). A novel series of α_{1A} selective benz[e]isoindole containing bi and tricyclic thienopyrimidinediones has been synthesized and characterized at subtypes of the α_1 receptor. The structure activity relationships for this series will be presented.

028. STRUCTURE ACTIVITY STUDIES ON A SERIES OF NOVEL QUINAZOLINEDIONES AS α_{1A} ADRENORECEPTOR ANTAGONISTS. Irene Drizin, Michael D. Wendt, William A. Carroll, Robert J. Altenbach, Fatima Z. Basha, Arthur A. Hancock, James F. Kerwin, Jr., and Michael D. Meyer. Neuroscience Research, Abbott Laboratories, Abbott Park, IL 60064

Agents showing selectivity for the α_{1A} subtype of the α-adrenergic receptor have potential utility for the symptomatic treatment of benign prostatic hyperplasia (BPH). A series of substituted quinazolinediones is described which exhibits high affinity and selectivity for the α_{1a} binding site. Extensive structure activity studies demonstrate the importance of aromatic ring substitution for the optimization of potency and selectivity at the α_{1a} site.

$R_1, R_2 = NO_2$, CN, Halogen, Alkyl
Aryl, OAlkyl, Amide, Ester, COOH

029. **SYNTHESIS AND EVALUATION OF CONFORMATIONALLY RESTRAINED AMINO OXAZOLINES AS α_2-ADRENERGIC AGENTS.** S. A. Munk, P. Arasasingham, C. Manlapaz, A. Kharlamb, E. Padillo, E. Runde, D. Hasson, M. Wijono, R. Lai, J. Burke, L. Wheeler, M. Garst; Allergan Pharmaceuticals, Irvine, CA 92612.

We have prepared a series of conformationally restrained analogs of Bay a 6781. The key synthetic step used for the assembly of the agents was a Diels-Alder reaction of cyclopentadiene and an olefin with nascent amine functionality. The agents proved to have high affinity for all three α_2 adrenergic receeotor subtypes. The agents were antagonists in the rabbit *vas deferens* assay.

Bay a 6781

030.

SYNTHESIS OF H-BONDING PROBES OF ALPHA ADRENERGIC RECEPTORS. W. H. Shelver Jooil Kim, College of Pharmacy, North Dakota State University, Fargo, ND 58105

A series of compounds (2-(N-cyanoalkyl,N-methyl)aminomethyl-1,4-benzodioxanes) in which the cyano group was separated from the benzodioxane group by a methylene chain (from 1-6) was synthesized. The receptor binding properties were evaluated using a rat submaxillary gland (alpha- 1A) or rat liver (alpha-1B). In addition the receptor binding to a human alpha-2A receptor was evaluated. The results showed activity markedly increased when the cyano group was separated from the benzodioxane group by four methylene groups. In addition, the compounds were generally more active as ligands to the cloned alpha 2-A receptor.

031.

SYNTHESIS AND ACTIVITY OF CONFORMATIONALLY RESTRAINED HUMAN ß3-ADRENERGIC RECEPTOR AGONISTS. E. R. Parmee, H. O. Ok, A. E. Weber, M. R. Candelore, L. Tota, C. D. Strader, M. H. Fisher, M. W. Baum, G. A. Doss. Merck Research Laboratories, Rahway, NJ 07065.

Conformationally restrained derivatives of human ß3-adrenergic receptor agonists were prepared and it was shown that agonist activity was present only in metabolically labile lactones, typified by L-751,250. Acyclic precursors containing a 4-amidophenylsulfonamide were all potent and selective ß3-agonists and it was shown that a variety of chain lengths and functionality could be tolerated.

L-751,250

032. **SYNTHESIS AND PHARMACOLOGY OF THE FOUR DIASTEREOMERS OF LY215840: A 5HT$_{2A}$/5HT$_7$ RECEPTOR ANTAGONIST,** D.C. Thompson, W.E. Bloomquist, D.W. Robertson and M. L. Cohen, Lilly Research Laboratories, Indianapolis, IN 46285

1-Isopropyl ergoline derivatives possess 5-HT$_{2A}$/5-HT$_7$ receptor affinity. Previous ergolines experienced limitations to their clinical utility prompting a search for derivatives with improved aqueous solubility and greater metabolic stability. This search culminated with the synthesis of the stereoisomers of a 2-hydroxycyclopentyl ergoline amide with the requisite metabolic stability and aqueous solubility required for a clinically effective agent. X-ray crystallography established the absolute configuration of these isomers. 5-HT$_{2A}$ receptor antagonist activity was studied orally (0.1mg/kg) in pithed rats. The two isomers with S configuration at the 1-cyclopentyl carbon atom inhibited 5-HT (0.1mg/kg, iv) - induced pressor responses by 93% and 97%. In contrast, the R isomers were less potent 5-HT$_{2A}$ receptor antagonists. The 5-HT$_7$ receptor affinity paralleled the 5-HT$_{2A}$ receptor antagonist activity of these four isomers. LY215840 possessed the highest 5-HT$_{2A}$ and 5-HT$_7$ receptor antagonist activity. Subsequently, a stereoselective synthesis of LY215840 was developed.

033. **SYNTHESIS OF NOVEL CONFORMATIONALLY RIGID H_3 HISTAMINE RECEPTOR ANTAGONISTS DERIVED FROM ENANTIOMERS OF TRANS-3-[1H-IMIDAZOL-4-YL]-2-CYCLOPROPYLCARBOXYLIC ACID** M. Amin Khan, J. G. Phillips, Clark E. Tedford, S.L.Yates, Leena fadnis, R. Gregory and G. Pawlowski. Gliatech Inc., 23420 Commerce Park Road, Cleveland, Ohio 44122

The conformationally rigid cyclopropane ring was introduced in histamine in place of the flexible chain ethylene. A series of novel, potent and selective H_3 histamine receptor antagonists were prepared from each of the enantiomers of trans-3-[1H-imidazol-4-yl]-2-cyclopropyl carboxylic acid. The H_3-receptor binding affinities were determined from rat cortical membranes, using [^{3}H]-N$^\alpha$-methylhistamine as the ligand. The synthesis, structure activity relationships, and biological activities of these chiral cyclopropyl compounds from these medicinal chemistry studies will be presented.

034. **SYNTHESIS AND STRUCTURE ACTIVITY STUDIES OF A NOVEL SERIES OF GLYCINE ANTAGONISTS: TRICYCLIC BENZAZEPINE ANALOGS.** Kelly A. Brush*, William J. Frazee, Joseph J. Lewis, Laura E. Garcia-Davenport, Linda M. Pullan, Jeffrey M. Goldstein, Departments of Medicinal Chemistry and Pharmacology, ZENECA Pharmaceuticals, A Business Unit of ZENECA Inc., Wilmington, DE 19850-5437.

Glycine antagonists acting at the NMDA receptor site have been studied as potential therapeutic agents in treating a variety of neurological disorders including stroke. We have reported (M.J. Chapdelaine, *Abstracts of Papers*, 211th National Meeting of the ACS, New Orleans, Louisiana; 1996; 109-MEDI) benzazepine 1 as an NMDA antagonist acting at the glycine regulatory site (IC$_{50}$=0.031μM vs. ^{3}H-glycine). In a continuing effort to identify additional antagonists at the glycine site, we have extended that work to include a series of tricyclic benzazepine analogs. It was believed that physical properties such as pK$_a$, log P, and aqueous solubility could be optimized by varying the substitution on the third ring. We wish to report the total synthesis and the SAR surrounding this series of glycine antagonists.

035. **SYNTHESIS AND STRUCTURE ACTIVITY STUDIES OF A NOVEL SERIES OF TRICYCLIC BENZAZEPINE ANALOGS.** Paula A. Payne-Gallimore*, William J. Frazee, Kelly A. Brush, Linda M. Pullan, Jeffrey M. Goldstein and Jitendra B. Patel, Department of Medicinal Chemistry, Zeneca Pharmaceuticals, A Business Unit of Zeneca Inc., Wilmington, DE 19850-5437.

Antagonism of the glycine site of the NMDA receptor has been proposed to modulate the effects of some neurodegenerative diseases. Previous investigations conducted in our lab identified the 3-hydroxy-2,5-dioxo-1-H-[1]benzazepines 1 as selective, potent antagonists of the glycine site. In an effort to diversify the series, tricyclic benzazepine analogs were synthesized in the hopes of discovering compounds with improved physicochemical properties. By varying the functional groups on the third ring, it was hypothesized that the pK$_a$, log P and aqueous solubility could be optimized. Herein, we wish to present the preparation and SAR of this series of glycine antagonists.

036. **SYNTHESIS AND ANTICONVULSANT ACTIVITY OF SUCCINIMIDO [3,4-b]INDANE AND 1,2,3,4,5,6-HEXAHYDRO-1,5-METHANOBENZAZOCINE-2,4-DIONE.** T. D. Greenwood, S. N. Greenwood, S. A. Dandekar, J. M. Tanko, and J. F. Wolfe, Department of Chemistry, Virginia Polytechnic Institute and State University, Blacksburg, VA 24061-0212.

The title compounds, **2** and **3**, were synthesized from **1a** and **1b**, respectively, using alkylation of the lithium enolates of these diamides with *o*-iodobenzyl chloride followed by a dual pathway

(S_{RN}1/aryne) cyclization and then imidization. Both **2** and **3** exhibit activity against scMet- and MES-induced seizures at a level of efficacy significantly above that of their less conformationally restricted analogs, 2-benzylsuccinimide and 2-benzylglutarimide.

037. SYNTHESIS, ANTICONVULSANT EVALUATIONS AND STRUCTURE-ACTIVITY RELATIONSHIP OF N-ACYL-α-AMINOGLUTARIMIDES AND SUCCINIMIDES. M.S. Park[1], J.W.Lee[1], K.C. Son[1], K.I. Jung[1], T.W. Kwon[2], J.W. Choi[1] and D.I. Jung[3] [1]College of Pharmacy, Kyungsung University, Pusan, Korea, 608-736. [2] Dept. of Chemistry, Kyungsung University, [3] Dept. of Chemistry, Dong-A University, Pusan, Korea

A series of (R)- and (S)- N-acyl-α-aminoglutarimides and succinimides, combining the common structures of other anticonvulsants such as N-CO-C-N and imide in a single molecule, were synthesized from the corresponding glutamic acid and aspartic acid and were evaluated their anticonvulsant activities in MES and PTZ tests. All the tested compounds except N-benzylated compounds showed significant anticonvulsant activities in both MES and PTZ tests. In addition, their anticonvulsant activities depended on their stereochemistry, N-acyl groups and N-substituted groups.

n = 1 or 2
R_1 = H, CH_3, C_2H_5, Allyl, isobutyl, benzyl
R_2 = OBz, O-4-nitroBz, O-tert-Bu, OPh, O-4-nitroPh, O-allyl, O-C_2H_5

038.

SYNTHESIS AND SAR OF 3β-SUBSTITUTED ALKYNYLPREGNANES AS ALLOSTERIC MODULATORS OF THE $GABA_A$ RECEPTOR. Ravi B. Upasani*, Richard Carter, Kevin C. Yang, Manuel Acosta-Burruel, James A. McLellan, Derk Hogenkamp, Nancy C. Lan, and Jon E. Hawkinson, CoCensys, Inc., 213 Technology Dr., Irvine, CA 92718.

3β-Phenylethynyl pregnane derivatives have been shown to possess high affinity for the $GABA_A$ receptor (R = MeCOPh, IC_{50} 10 nM). As allosteric modulators of the $GABA_A$ receptor, these compounds have potential uses as anticonvulsants, anxiolytics, and sedative hypnotics. However, these neuroactive steroids are very lipophilic with log P values greater than 6. Therefore, various structural modifications of these molecules were made to lower log P. The different approaches taken to reduce log P, the rationale behind these modifications, and the biological activities of the resulting compounds will be presented.

039. CONTINUED DEVELOPMENT OF BIOPRECURSORS FOR AMINO ACID AMPA ANTAGONISTS: MORPHOLINEDIONES AND α-KETO ACIDS. <u>Satyavijayan N. Danthi</u>, Junge Zhange, Franklin R. Abadie, and Ronald A. Hill, School of Pharmacy, Northeast Louisiana University, Monroe, LA 71209.

Excitatory amino acid neurotransmission pathways are ubiquitous in the mammalian CNS. As a potential means for gaining regioselectivity of action beyond that available with receptor-subtype-selective compounds, we are synthesizing potential bioprecursors of the AMPA antagonist (S)-2'-hydroxy-3',5'-dinitrophenylalanine (1), a triply-charged glutamate analog. The desired characteristics are (1) rapid diffusion into the CNS, and (2) conversion to 1 by biochemical pathways more prevalent in targeted regions of the brain than in others. In this presentation, our progress on the synthesis of morpholinediones (2) and α-keto acid precursors such as 3 will be discussed, as well as our work on the biotransformation of diketopiperazines (4) in rat brain homogenates.

040. A STUDY OF PHARMACOPHORE/RECEPTOR MODELS FOR BENZODIAZEPINE RECEPTOR SUBTYPES AND SAR STUDY VIA A LIGAND-MAPPING APPROACH. <u>Qi Huang</u>, Eric D. Cox, Tong Gan, Xiaohui He, Chunrong Ma, and James M. Cook,* Department of Chemistry, University of Wisconsin-Milwaukee, Milwaukee, WI 53201

The affinities of a number of families of BzR ligands have been determined at 5 distinct recombinant GABA$_A$ receptor isoforms [αxβ3γ2 (x=1,2,3,5,or 6)]. Some of the families, such as β-carbolines, exhibited better selectivity for α1 containing subtypes in comparison to other DS receptor isoforms. *t*-Butyl β-carboline-3-carboxylate(BCCT) displayed both potent affinity (K_i=0.72nM) as well as selectivity(20 times) for the α1β3γ2 isoform. On the other hand, imidazobenzodiazepines exhibited better selectivity for the α5β3γ2 receptor subtype. For instance, the 8-acetylenoimidazobenzodiazepine-3-carboxylic acid ethyl ester displayed potent affinity(K_i=0.49nM) at the α5β3γ2 receptor subtype and was 44 times more selective at the α5 receptor subtype than at other BzR isoforms. An SAR study of the ligands for each receptor subtype has been carried out *via* a ligand-mapping approach. Binding pharmacophores for α1β3γ2 *vs.* α5β3γ2 receptor subtypes have been determined and will be described.

041. A NOVEL FAMILY OF PERIPHERAL BENZODIAZEPINE RECEPTOR (PBR) LIGANDS. <u>Hesheng Zhang</u>,[a] Anthony S. Basille,[b] Phil Skolnick,[b] and R. Ian Fryer,[a] a: Department of Chemistry, Rutgers, The Statae University of New Jersey, Newark NJ 07102, b. Laboratory of Neuroscience, National Institute of Diabetes and Digestive and Kidney Diseases, NIH. Bethesda, Maryland 20892.

We wish to report the synthsis and biological evaluation of a novel type of Peripheral Benzodiazepine Receptor (PBR) ligands, N-(2'-aryl or 2'-alkylcarbony)phenyl-2-(2'-phthalimidyl) acetamides. Using triethylamine as a HBr acceptor and anhydrous methlene chloride as solvent, 2-amino- benzophenones were reacted with bromoacetyl bromide to form the corresponding amides, which in turn were reacted with potassium phthalimide to give the final products. All compounds reported here have been tested against PBR and some have binding IC$_{50}$s in the nanomolar range. A Structure-Activity-Relationship study will also be presented and the structure features of these "ring open" analogs will be compared to other heterocycles with known activity at the PBR.

042.
HIGHLY POTENT AND SELECTIVE HUMAN ADENOSINE A3 RECEPTOR ANTAGONISTS: TRIAZOLOQUINAZOLINE DERIVATIVES. Y.-C. Kim, X.-D. Ji, M. E. Olah*, G. L. Stiles*, K. A. Jacobson, Lab. of Bioorg. Chem., Nat. Inst. of Health, Bethesda, MD 20892, Dept. of Med. and Pharmacol., *Duke Univ. Medical Center, Durham, NC 27710.

Derivatives of the non-selective adenosine antagonist 9-chloro-2-(2-furanyl)[1,2,4]triazolo[1,5-c]quinazolin-5-amine (CGS15943) were prepared in an effort to provide A3 subtype selectivity. A 5-N-benzoyl derivative displayed Ki values of 680 and 273 nM at rat A1 and A2A receptors, respectively, and 3.0 nM at human A3 receptors in radioligand binding assays. A 5-N-phenylacetyl derivative (MRS1220) with a Ki value of 0.65 nM was 470-fold and 80-fold selective for human A3 receptors vs rat A1 and A2A receptors, respectively. This compound also antagonized the effects of an A3 agonist in functional assays (inhibition of adenylate cyclase mediated by human A3 receptors and binding of [35S]GTP-gamma-S).

043.
SYNTHESIS AND PROPERTIES OF 8-SUBSTITUTED-N-6-ALKYL-9-METHYLADENINES. POTENT AND SELECTIVE ADENOSINE ANTAGONISTS. **Ronald J. Wysocki, Jr.**, Gevork Minaskanian, Pauline L. Martin, I. M. Uwaydah and Keith A. Rippel, Discovery Therapeutics Inc., 2028 Dabney Rd Suite E-17, Richmond, VA 23230-3311.

Studies in our laboratories have identified a lead molecule N-0861, N-6-endonorbornyl-9-methyladenine, as a highly selective (100 fold) non-xanthine based A1 adenosine receptor antagonist. Structure activity efforts were initiated to develop a compound with improved potency and aqueous solubility. Strategic placement of a hydroxy group on the N-6 alkyl portion of the molecule was found to increase potency and aqueous solubility. Substantial increases in potency and selectivity were discovered by attachment of different bulky amino groups at the C-8 position of the purine ring of N-0861. The syntheses of these compounds and their pharmacological properties will be presented.

044. SYNTHESIS AND STRUCTURE-ACTIVITY RELATIONSHIPS OF PYRIDINE-MODIFIED ANALOGS OF 3-[[2-((S)-PYRROLIDYL)METHYL]OXY]PYRIDINE, A-84543, A POTENT CHOLINERGIC CHANNEL MODULATOR. Nan-Horng Lin*, David E. Gunn, Yihong Li, Yun He, Hao Bai, Keith B. Ryther, Diana L. Donnelly-Roberts, David J. Anderson, Jeffrey E. Campbell, James P. Sullivan, Stephen P. Arneric and Mark W. Holladay. Neuroscience Research, D-47W, Pharmaceutical Products Division, Abbott Laboratories, 100 Abbott Park Road, Abbott Park, IL 60064-3500.

We have recently identified 3-[[2((S)-pyrrolidinyl)methyl]oxy]pyridine (**1**), A-84543, as a potent cholinergic channel modulator (ChCM). In an effort to probe the influence of substitution on the pyridine ring on binding affinity, analogs of **1** substituted at the 2, 4, 5 and 6- positions of the pyridine moiety were synthesized and tested in vitro for cholinergic channel receptor binding (displacement of [^{3}H](-) cytisine from whole rat brain synaptic membranes) and functional (ion flux) activities. The substituted analogs exhibited Ki values ranging from 0.1 to > 9000 nM compared to a Ki value of 0.2 nM for compound **1**. Among the compounds tested, the 5-butyl substituted A-84543 was the most potent in the cytisine binding assay. The synthesis and structure-activity relationships of these analogs will be presented.

045.
SYNTHESIS AND STRUCTURE-ACTIVITY RELATIONSHIP OF SUBSTITUTED 1-AZABICYCLO[3.2.1]OCTYL-6-OXYPYRAZINES. **R L Simon**, J K Reel, J S Ward, L Merritt, C H Mitch, S C Peters, H E Shannon, F P Bymaster, D O Calligaro, N W DeLapp, P Sauerberg, M J Sheardown, L Jeppesen, C A Whitesitt, Lilly Research Laboratories, Eli Lilly and Company, Indianapolis, Indiana 46285 and Novo Nordisk A/S, Novo Nordisk Park, 2760 Malov, Denmark

A series of 1-azabicyclo[3.2.1]octyl-6-oxy-pyrazines was synthesized and evaluated for the ability to bind to muscarinic receptors as determined by competitive radioligand binding assays. These pyrazines have been demostrated to bind with nM potency to muscarinic receptors without producing the side effects of salivation and tremor in mice usually associated with the m3 receptors. Derivatives linking the 3.2.1 azacycle with 1,2,5-thiadiazoles incorporating an oxygen link also have excellent binding affinity to the muscarinic receptors but are non selective as indicated by the production of salivation and tremor in mice. The most potent analogs were those substituted in the 3 position of the pyrazine by alkylthiols with 2-5 carbons.

046.
1,2,3,4-TETRAHYDRO-2-ISOQUINOLINECARBOXYLATE DERIVATIVES:A NOVEL CLASS OF SELECTIVE MUSCARINIC ANTAGONISTS. III. M. Takeuchi , R. Naito, Y. Yonetoku, K. Ikeda, Y. Isomura , Institute for Drug Discovery Research, Yamanouchi Pharmaceutical Co., Ltd. 21 Miyukigaoka, Tsukuba City, Ibaraki 305, Japan

As a part of searching for novel bladder selective muscarinic M3 antagonists as potential therapeutic agents for urinary incontinence, we have already reported two series of carbamate derivatives. In order to develop more bladder selective M3 antagonists, a series of 1,2,3,4-tetrahydro-2-isoquinolinecarboxylate derivatives was prepared and evaluated for the binding affinities for the muscarinic receptors and for in vivo effects on reflexly- evoked rhythmic contraction of bladder and oxotremorine-induced salivary secretion in rats. Among these compounds, (+)-(1S,3'R)-3'-quinuclidinyl 1-phenyl-1,2,3,4-tetrahydro-2-isoquinolinecarboxylate hydrochloride(YM-53705) showed high affinity for M3 receptor with a Ki value of 12nM and 10-fold selectivity between rhythmic contraction (ID30=0.21mg/kg,i.v.) and salivary secretion (ID50=2.1mg/kg,i.v.). YM-53705 will be expected as a drug for the treatment of urinary incontinence without side-effect such as dry mouth.

047. PHARMACOPHORE IMPROVEMENT OF ALLOSTERIC MODULATORS OF MUSCARINIC RECEPTORS. U. Holzgrabe, H.M. Botero Cid, E. Kostenis, E. Mies-Klomfaß, and K. Mohr, Pharmaceutical Institute, University of Bonn, Kreuzbergweg 26, D-53115 Bonn, Germany

A variety of drugs has been described to retard the dissociation of antagonists, e. g. N-methylscopolamine, from the M_2-acetylcholine receptor. This effect is attributed to an allosteric modulation of the receptor protein which might be exploited in case of organophosphorus poisonings. Recently we developed a pharmacophore model using neuromuscular blockers as well as compounds consisting of a bisquaternary middle chain and aromatic heterocycles, such as phthalimido groups, at both lateral ends. Two positive charges in the middle chain as well as two aromatic areas at both ends arranged in an S-shape conformation were found to be essential for a high allosteric potency. Herein, substituents of increasing size, e. g. alkoxy groups and aryl rings, were added to the phthalimido skeletons, in order to improve the potency and to elucidate whether there is any sterical restriction at the allosteric binding site. In this series a positive correlation was found between the allosteric potency and the lipophilicity of the introduced substituents. Accordingly, an additional hydrophobic interaction is probably the explanation to the high potency of benzylidene substituted phthalimido derivatives.

048. SYNTHESIS AND EVALUATION OF 5-AMINOMETHYL DERIVATIVES OF 3,3-DIETHYL-2(3H)-FURANONE AS SUBTYPE-SPECIFIC MUSCARINIC LIGANDS. Alemayyehu Ahungena, Daniel J. Canney, Department of Pharmaceutical Sciences, School of Pharmacy, Temple University, Philadelphia, PA 19140.

Structurally rigid lactone derivatives of the nonselective muscarinic antagonist benactyzine (**1**) exhibit subtype-specific properties. A series of 5-aminomethyl lactones (**2b**) were prepared and evaluated based on this report. 3,3-diethyl-5-imidazolylmethyl-2(3H)-furanone (**3a**), showed a three-fold selectivity for the M_2 receptor subtype over either M_1 or M_3. To investigate the effect of alkyl-substitution on subtype selectivity, a series of 2'-alkyl derivatives of **3a** were synthesized. The binding properties of the novel lactone derivatives (**3b**) of **3a**, and their thio analogs (**3c**), are described here.

049. 2-FLUORO-4-PYRIDINYLMETHYL ANALOGS OF LINOPIRDINE AS ORALLY ACTIVE ACETYLCHOLINE-RELEASE ENHANCING AGENTS WITH GOOD BRAIN PENETRATION AND DURATION OF ACTION. Richard A. Earl,* R. Zaczek, C. A. Teleha, B. N. Fisher, C. M. Maciag, M. E. Marynowski, A. R. Logue, S. W. Tam, W. J. Tinker, S.-M. Huang and R. J. Chorvat. *The DuPont Merck Research Laboratories; The DuPont Merck Pharmaceutical Co., Experimental Station, P.O. Box 80500 Wilmington, Delaware 19880-0500.*

In an effort to improve the pharmacokinetic and pharmacodynamic properties of the cognition-enhancer linopirdine, a number of core structure analogs were prepared in which the 4-pyridyl pendant group was systematically replaced with 2-fluoro-4-pyridyl. This strategy resulted in the discovery of several compounds with improved activity in acetylcholine-release enhancing assays, *in vitro* and *in vivo*. The most effective compound resulting from these studies, 10,10-bis[(2-fluoro-4-pyridinyl)methyl]-9(10H)-anthracenone (1), is between 10 and 20-times more potent than linopirdine in increasing extracellular hippocampal ACh levels in the rat with a minimum effective dose of 1 mg/kg. In addition to superior potency, 1 possesses an improved pharmacokinetic profile compared to that of linopirdine. The half-life of 1 (2 hrs) in rats is 4-fold greater than that of linopirdine (0.5 hrs). Additionally, a 6-fold improvement in brain penetration over linopirdine has been achieved with 1.

050. SYNTHESIS OF NOVEL MIXED DELTA-MU OPIOID AGONISTS. D. G. Bubacz, A. O. Davis, S. H. Dickerson, P. A. Harris, R. W. McNutt, M. A. Collins and K.-J. Chang. Division of Chemistry, Glaxo Wellcome, 5 Moore Drive, Research Triangle Park, NC 27709.

The synthesis of novel benzhydrylpiperazines with mixed delta-mu opioid agonist activity is described. Eight enantiopure stereoisomers were prepared by several synthetic approaches involving chiral intermediates, including enantiopure derivatives of 2,5-dimethylpiperazine, which were obtained by synthesis from D or L-alanine or by resolution of racemic material. Absolute configurations were determined by x-ray chrystallography of a key intermediate or a selected target. Receptor binding assay results provided a relationship between stereochemistry and activity. The stereoisomer with the highest mu activity produced analgesia in rats and mice.

051.

SYNTHESES AND ACTIVITIES OF DELTA RECEPTOR SELECTIVE OPIOID ANTAGONISTS INCORPORATING 3-AMINOTETRALIN-2-CARBOXYLIC ACID. **Noriyuki H. Kawahata**, Li Zhang, Murray Goodman, Department of Chemistry and Biochemistry, UC San Diego, La Jolla, CA 92093

An antagonist of the delta opioid receptor incorporating 3-aminotetralin-2-carboxylic acid (βAtc) (1) in the sequence Tyr-βAtc-Phe-Phe-OH has been identified. A diastereoselective synthesis of the β-amino acid has been accomplished in order to determine the stereospecificity of the delta opioid receptor for this series of compounds. The biological evaluation and conformational analyses of an initial series of analogs will also be presented.

052. CYCLIC AMINE-BRIDGED ENKEPHALIN ANALOGS. Kevin Shreder, Li Zhang, Trunghau Dang, and Murray Goodman*, Department of Chemistry and Biochemistry, University of California at San Diego, La Jolla, California 92093-0343.

The cyclization of a biologically active linear peptide can yield a derivative with enhanced selectivity, increased potency, and extended biological stability when compared to the parent molecule. To date, there are a variety of structures that can be utilized as "bridges" in the design of cyclic peptides. Commonly used functionalities include amide bonds that result from head-to-tail or mainchain-to-sidechain cyclizations, disulfide bonds, urethanes, and diaryl ethers. We have recently embarked on developing the use of the nitrogen heteroatom as a bridge in the design of amine-bridged cyclic enkephalin analogs such as **1**. The synthesis, biological activity, and conformational analysis of **1** as well as congeners of **1** will be presented here.

053. BIOREVERSIBLE CYCLIC PRODRUGS OF OPIOID PEPTIDES. EVALUATION OF THE RELEASE KINETICS. Huijuan Zhang, Wei Wang, Daxian Shan, and Binghe Wang.* Department of Chemistry, North Carolina State University, Raleigh, NC 27695

Earlier, we reported the development of a coumarin-based prodrug system for amines by taking advantage of the facile lactonization of coumarinic acid and its derivatives. Here we report the application of this prodrug system for the preparation of esterase-sensitive cyclic prodrugs of opioid peptides, [Leu]^5-enkephalin (**1a**) and DADLE (**2b**). These two prodrugs were found to quantitatively release the original peptides with $t_{1/2}$ ranging from 400 to 800 min when incubated in the presence of porcine liver esterase or in plasma.

054. 3,7-DIAZABICYCLO[3.3.1]NONANONES AS KAPPA-OPIOID RECEPTOR AGONISTS. T. Siener, M. Haurand*, W. Englberger*, and U. Holzgrabe, Pharmaceutical Institute, University of Bonn, Kreuzbergweg 26, D-53115 Bonn, *Grünenthal GmbH, Zieglerstr. 6, D-52078 Aachen, Germany

The 2,4-bis-2-pyridine substituted 3,7-dimethyl-3,7-diazabicyclo[3.3.1]nonan-9-one-1,5-diester HZ 2 shows high affinity and selectivity for the kappa (κ)-opioid receptor (K_i=15nM). In mouse and rat the compound produced antinociception in test models such as hot-plate, phenylquinone-induced writhing, Randall-Selitto or tail-flick tests. Antinociception was norbinaltorphimine but not naloxone reversible. Repeated treatment with HZ 2 did not result in naloxone precipitated withdrawal symptoms. Taken together, HZ 2 is a rather selective κ-receptor agonist and characterized by a good peroral bioavailability as well as a long duration of action (Kögel, Holzgrabe et al. Arch. Pharmacol. 354, R43(1996)). In order to establish quantitative structure-activity relationships, the 2-pyridine rings in 2 and 4 position of the skeleton were replaced with methoxy, nitro, hydroxy, chloro and fluoro substituted phenyl rings. With exception of the hydroxy phenyl compound the affinity to the κ-opioid receptor decreases, but was partially shifted to the μ and δ-receptor, respectively. Taken together, the 2-pyridyl substituent turned out to be the best for κ-affinity and subtype selectivity.

055. N-BENZYL-4-AMINOPIPERIDINE PHENYLACETAMIDE ANALOGUES AS POTENT σ_1/σ_2 RECEPTOR LIGANDS. Yunsheng Huang,[1] Brian Whirrett,[1] Steven R. Childers,[2] and Robert H. Mach,[1,2] Department of [1]Radiology and [2]Physiology & Pharmacology, Bowman Gray School of Medicine, Wake Forest University, Winston-Salem, NC 27157

Sigma receptors have been suggested to play an important role in the central nervous system. *In vitro* binding studies have revealed that there are two subtypes of sigma receptors, termed σ_1 and σ_2. The uneven distribution of σ_1 and σ_2 receptors in the central nervous system, as well as in the peripheral tissues, implicate their involvement in a variety of functions. An understanding of the functional role of these receptors has been limited by the shortage of ligands that bind selectively to σ_1 and σ_2 receptors. In the current study, a series of N-benzyl-4-aminopiperidine phenylacetamide analogues were prepared and *in vitro* binding studies for σ_1 and σ_2 receptors were conducted. The structure-activity relationships for the binding of these analogues to both σ_1 and σ_2 receptors will be discussed.

056. HALOGENATED N-SUBSTITUTED 4-(PHENOXYMETHYL)PIPERIDINES AS SIGMA RECEPTOR LIGANDS: EFFECT OF STRUCTURE ON Ki AND *IN VIVO* DISTRIBUTION. R.N. Waterhouse, T.L. Collier, J.C. O'Brien, K. Mardon, Radiopharmaceuticals Division, ANSTO, Lucas Heights, Australia 2234

The development of sigma (σ) receptor probes is a focus of current research due to the implicated involvement of these sites in certain neurological disorders, most notably schizophrenia. A series of thirty-six halogenated 4-(4-phenoxymethyl)piperidines was synthesized in four synthetic steps from 4-hydroxymethyl piperidine using methods previously described (Gilligan *et al. J. Med. Chem.* **1992**, 35, 4344) with modifications. The effects of various N-substituents on the σ-1 and σ-2 dissociation constants were examined. The N-substituents included fluoroalkyl, hydroxyalkyl, iodopropenyl, and selected *ortho*, *meta* and *para* substituted benzyl groups. Also determined were the effects of various moieties on the phenoxy ring; specifically 4-iodo, 4-bromo, 4-nitro, 4-cyano, 3-bromo and pentafluoro substituents. The ranges in the dissociation constants of these compounds for σ-1 and σ-2 receptors were 0.38 - 24.3 nM and 3.9 - 361 nM, respectively. The ratio of Ki (σ-2/σ-1) ranged from 1.19 to 121. Five of the most promising ligands were labeled with ^{123}I or ^{18}F and studied *in vivo* in adult male rats. High uptake, specific binding and good retention of radioactivity was observed in the brain, lung, kidney, heart, muscle and other organs known to possess σ sites. These results indicate that selected 4-(phenoxymethyl)piperidines of this series may provide useful probes for *in vivo* tomographic studies of σ receptors.

057. STRUCTURE-ACTIVITY RELATIONSHIPS OF SESQUITERPENE LACTONES WITH POTENTIAL ANTIMIGRAINE ACTIVITY: Q. Luo[1,2], J.A. Darsey[1,2], R.L. Compadre[1], R.J. Marles[3], C.M.Compadre[1,2], [1]Department of Biopharmaceutical Sciences,University of Arkansas for Medical Sciences, Little Rock, AR 72205, [2]Department of Chemistry, University of Arkansas at Little Rock, AR 72204, [3]Department of Botany, Brandon University, Brandon, MB, Canada.

There is a very limited number of therapeutic alternatives that can be used to treat or prevent migraine headaches. The leaves of the feverfew plant (*Tanacetum parthenium L.*) have been used to treat headaches for over two thousand years. Feverfew contains considerable amounts of parthenolide and other sesquiterpene lactones (SL) with postulated antimigraine effects. In this research we have studied the structural factors responsible for SL with potential antimigraine effects, measured by their ability to inhibit serotonin release from bovine platelets. For this analysis the quantum mechanics parameters for these SL were evaluated using Gaussian 92. Correlation analysis between the structure and the biological activity was evaluated using neural networks and partial least squares methods using comparative molecular field analysis (CoMFA).

058. SYNTHESIS OF HYDROXYL-SUBSTITUTED SULFONYLUREAS AS POTENTIAL SHORT-HALF-LIFE CNS-ACTIVE BLOCKERS OF ATP-SENSITIVE POTASSIUM CHANNELS. Bo Peng, Sonali Rudra, Jeffery K. Bounds, Ahmad A. Adloo, David S. Roane, and Ronald A. Hill, School of Pharmacy, Northeast Louisiana University, Monroe, LA 71209

Certain sulfonylureas exert their hypoglycemic effect by blockade of ATP-sensitive potassium (K^+_{ATP}) channels in pancreatic β-cells. A blocker of CNS K^+_{ATP} channels may have therapeutic potential in stroke, epilepsy, and obesity. The blood-glucose-lowering effect of K^+_{ATP} channel blockers is in pharmacological opposition to some of the centrally-mediated effects of such compounds. Of particular interest to us, Roane and Boyd have found that icv administration of glyburide inhibits feeding in rats, whereas ip glyburide has the opposite effect; the mechanism of this effect is being vigorously studied. We have synthesized a series of hydroxyl-substituted analogs of glyburide, and found that some of these retain potent K^+_{ATP}-channel-blocking activity. Work has begun on converting these compounds to prodrugs optimized for rapid CNS entry and rate-limiting enzymatic cleavage in the periphery.

059. SYNTHESIS AND BIOLOGICAL ACTIVITY OF 1-N-DERIVATIVES OF MELATONIN Alfonso Lira-Rocha[1], Ofelia Espejo[1], Elia B. Naranjo-Rodríguez[1], Cruz Reyez-Vázquez[2], [1]Departamento de Farmacia, Facultad de Química, UNAM and [2]Facultad de Medicina, UNAM, México, D.F., México, 04510

It has been proposed a possible receptor binding site model for melatonin. The way in which it interacts with its hypothetical receptor is through the "northern" part of the molecule. To modify the electronic properties of the interaction zone in the proposed model, a series of 1-substituted melatonin derivatives was prepared as potential melatonin analogs. The synthesis of these compounds and its biological assay on isolated rat ileum will be presented.

060. SYNTHESIS AND BIOLOGICAL CHARACTERIZATION OF LONG-ACTING, ORALLY ACTIVE MORPHOLINE ACETAL HUMAN NK-1 RECEPTOR ANTAGONISTS Jeffrey J. Hale, Sander G. Mills, Malcolm MacCoss, Paul E. Finke, Shuet-Hing L. Chiu, Su Huskey, Margaret A. Cascieri, Sharon Sadowski, Joseph Metzger, George Eiermann, F. David Tattersall*, Nadia Rupniak*, Wayne Rycroft*, Richard Hargreaves* and D. Euan MacIntyre Merck Research Laboratories, P.O. Box 2000, Rahway, NJ 07065 USA *Merck, Sharp & Dohme, Terlings Park, Eastwick Road, Harlow, Essex CM20 2QR, UK

Structural changes were made to the highly potent, orally active human NK-1 receptor antagonist, L-742,694 (J. Med. Chem, **1996**. *39*, 1760), aimed at improving its metabolic profile and enhancing its potency *in vivo* . Compounds of structure **I** in which R_2 = CH_3 could not be prepared using existing methods and required the development of novel synthetic chemistry. Data showing that several of these compounds have an enhanced pharmacologic profile in animal models of peripheral and central NK-1 receptor-mediated events will be presented.

R_1 = H, F, CF_3
R_2 = H, CH_3
R_3 = H, F

061. DEVELOPMENT OF NONPEPTIDIC NEUROTENSIN (8-13) MIMETICS Feng Hong[a], Yuan-Ping Pang[a, *], Bernadette Cusack[b] and Elliott Richelson[b], [a]Neurochemistry and [b]Neuropsychopharmacology Research, Mayo Clinic, 4500 San Pablo Rd., Jacksonville, FL 32224

In order to develop nonpeptidic neurotensin(8-13) ($pR^8R^9P^{10}Y^{11}I^{12}L^{13}$) mimetics with improved properties, such as stability, affinity, selectivity, degree of agonism and antagonism, for the potential treatment of schizophrenia and for pain alleviation, we previously developed the Multiple Template Approach for design of such mimetics. According to this approach, we have synthesized a series of pyrrole-based nonpeptidic mimetics. In the structures of these mimetics (e.g., **1**), the substituted pyrrole moiety mimics the $R^8R^9P^{10}$ fragment, while the 1-adamantyl amino group mimics the $I^{12}L^{13}$ part. Pharmacological studies revealed that these compounds demonstrated submicro molar affinity for the human neurotensin receptors and can serve as new leads for further structural optimization.

062. STRUCTURAL REQUIREMENT FOR BINDING OF ANANDAMIDE-TYPE COMPOUNDS TO THE BRAIN CANNABINOID RECEPTOR. Tzviel Sheskin and Raphael Mechoulam, Natural Products department, Medical Faculty, Hebrew University, Jerusalem 91120, Israel.

In order to establish the structural requirements for binding to the brain cannabinoid receptor (CB_1) we have synthesized numerous fatty acid amides, ethanolamides and some related simple derivatives and have determined their Ki values. A few α-methyl or α,α,-dimethylarachidonoyl-alkylamides were also examined. In the 20:4; n-6 series, the unsubstituted amide is inactive; N-monoalkylation, at least up to a branched pentyl group, leads to significant binding. N,N dialkylation, with or without hydroxylation on one of the alkyl groups, leads to elimination of activity. Hydroxylation of the N-monoalkyl group at the omega carbon atom retains activity. In the 20:x, n-6 series, x has to be either 3 or 4; the presence of only two double bonds leads to inactivation. In the n-3 series, the limited data reported suggest that the derived ethanolamides are either inactive or less active than comparable compounds in the n-6 series. Alkylation or dialkylation of the α carbon adjacent to the carbonyl group retains the level of binding in the case of anandamide; however α-monomethylation or α,α-dimethylation of N-propyl derivatives potentiates binding and leads to the most active compounds seen in the present work (Ki values of 6.9±0.7 to 8.4±1.1 nM). We have confirmed that the presence of a chiral center on the N-alkyl substituent leads to enantiomers which differ in their levels of binding.

063.

IDENTIFICATION AND SYNTHESIS OF A POTENT DUAL INHIBITOR (SP100030) OF NF-κB AND AP-1 MEDIATED GENE EXPRESSION. Robert W. Sullivan[1], Colin G. Bigam[1], Mark J. Suto[1], Lynn J. Ransone[2], Mark E. Goldman[2]. Departments of [1]Medicinal Chemistry and [2]Pharmacology. Signal Pharmaceuticals, Inc. 5555 Oberlin Dr., San Diego, CA 92121.

The transcription factors NF-κB and AP-1 regulate a number of proinflammatory proteins and cytokines. We established high throughput reporter based assays (Jurkat T-cells) to identify inhibitors of these key transcription factors. Using these assays an inhibitor of both NF-κB and AP-1 activation was identified. This compound was also shown to decrease IL-2 and IL-8 levels as well. Through the use of solution phase combinatorial approaches a 10-fold more potent inhibitor was identified (SP100030). The synthesis of these compounds and structure activity relationships will be presented.

064.

SYNTHESIS AND STRUCTURE-ACTIVITY RELATIONSHIP STUDIES OF CONFORMATIONALLY RESTRICTED ANALOGS OF SP100030, A POTENT INHIBITOR OF NF-κB AND AP-1 ACTIVATION. Paul E. Erdman[1], Moorthy S. S. Palanki[1], Mark J. Suto[1], Carla M. Suto[2] and Mark E. Goldman[2], Departments of [1]Medicinal Chemistry and [2]Pharmacology, Signal Pharmaceuticals, Inc., 5555 Oberlin Dr., San Diego, CA 92121 USA

SP100030 is a potent inhibitor of both NF-κB and AP-1 activation. In an effort to further improve the potency and physiochemical properties of this series, a number of conformationally restricted analogs were proposed. The aromatic top portion was positioned in different orientations with respect to the bottom pyrimidine ring. The synthesis of these compounds and their *in vitro* activity will be discussed.

R_2 = Cl, SMe, SO$_2$Me
X = CH$_2$, O, N, S
Y = Cl, CF$_3$

065.

INHIBITORS OF NF-κB AND AP-1 GENE EXPRESSION: STRUCTURE-ACTIVITY STUDIES ON THE PYRIMIDINE PORTION OF SP100030. Moorthy S. S. Palanki[1], Paul E. Erdman[1], Leah Gayo[1], Chella Shevlin[1], Mark J. Suto[1], Lynn J. Ransone[2] and Mark E. Goldman[2], Departments of [1]Medicinal Chemistry and [2]Pharmacology, Signal Pharmaceuticals, Inc., 5555 Oberlin Dr., San Diego, CA 92121 USA

SP100030 was identified as an inhibitor of NF-κB and AP-1 activation. In order to further define the structure-activity relationships, we prepared a series of derivatives of SP100030 with different substituents at the 2- and 4- position of the pyrimidine ring. These compounds were prepared using solution-phase combinatorial methodology. The synthesis of these compounds and their *in vitro* activity will be presented.

066. **AMINO DERIVATIVES OF THE IMMUNOSUPPRESSANT ASCOMYCIN.** Hyun O. Ok[*], John L. Szumiloski, Jeannette E. Brown, Mark T. Goulet, Thomas R. Beattie[#], Mary J. Strauch, Francis J. Dumont and Matthew J. Wyvratt, Departments of Medicinal Chemistry and Immunology, Merck Research Laboratories, P.O. Box 2000, Rahway, New Jersey 07065

C32-Amino derivatives were investigated as replacements for the hydroxyl group at the C32-position of Ascomycin and its C24-deoxy analog. Synthesis and biological evaluation of these amino derivatives are reported herein.

067. **SELECTIVE MODIFICATIONS OF ASCOMYCIN CYCLOHEXYL RING.** Yat Sun Or, Richard F. Clark, Bane Lane, Karl W. Mollison and Jay R. Luly, Abbott Laboratories, Pharmaceutical Products Division, Abbott Park, IL 60064

The related immunosuppressants ascomycin (**1**) and FK-506 (**2**) are subjects of intense chemical research because of their potential applications in organ transplantation and autoimmune disease. We have been interested in studying the effect of northwest cyclohexyl ring modifications on the biological activity of ascomycin.

This paper reports selective modification of cyclohexyl ring without protection of reactive functional groups in **1**. The biological activity of analogues **3-6** will also be discussed.

1 Ascomycin, R = Et
2 FK-506, R = Allyl

068. **2-QUINOLINYLMETHYL-CONTAINING DIARYLPYRAZOLYL HYDROXAMIC ACIDS– INHIBITORS OF 5-LO AND FLAP.** Michael P. Ferro[*], Kathleen A. McCoy, Victor T. Bandurco, William V. Murray, Praful. Lalan, Michael P. Wachter, Dennis C. Argentieri, Monica M. Singer, and Michele Steber, Department of Drug Discovery, The R. W. Johnson Pharmaceutical Research Institute, Route 202, P. O. Box 300, Raritan, New Jersey 08869.

A series of diarylpyrazoles containing the 2-methylquinolinyl functionality (**1-3**) was prepared to assess their potency as inhibitors of leukotriene production. These compounds generally were significantly more potent in the 5-lipoxygenase (5-LO) RBL-1 whole-cell assay than in the corresponding broken-cell assay. Compounds in the series that also contained a methyl hydroxamic acid unit as well as the 2-methylquinolinyl functionality showed potency in both assays. We interpret these results to indicate that 2-methylquinolinyl-containing compounds are inhibiting 5-lipoxygenase activating protein (FLAP), and compounds containing both the 2-methylquinolinyl and the hydroxamic acid functionalities are inhibiting both FLAP and 5-LO. The chemistry and SAR of this series of compounds will be discussed.

069. NEW DUAL INHIBITORS OF INDUCIBLE CYCLOOXYGENASE (COX-2) AND LEUKOTRIENE BIOSYNTHESIS AS POTENTIAL NEW THERAPEUTIC AGENTS FOR RHEUMATOID ARTHRITIS

James A. Sikorski, John J. Talley, Bryan H. Norman, Matthew J. Graneto, Hwang-Fun Lu, Balekudru Devadas, David L. Brown, Gary D. Anderson, Amy W. Veenhuizen, Kelly L. McGarity, Leslie J. Askonas, Carol M. Koboldt, Robert H. Keith, Susan A. Gregory

G. D. Searle R&D, 700 Chesterfield Parkway North, St. Louis, Missouri 63198

Pro-inflammatory products of arachidonic acid metabolism mediated by either the prostaglandin or leukotriene pathways have been implicated in many models of inflammatory disease. Consequently, compounds that inhibit the production of both prostaglandins and leukotrienes may exhibit beneficial anti-inflammatory properties. The recent discovery of an inducible cyclooxygenase (COX-2) has led to potent and selective COX-2 inhibitors with GI-sparing, anti-inflammatory activity. These results suggest that dual inhibitors of COX-2 and either 5-lipoxygenase (5-LO) or LTA$_4$ hydrolase (LTA$_4$H) could be even more effective anti-inflammatory agents. A medicinal chemistry strategy will be presented which combines either a 5-LO or LTA$_4$H pharmacophore with a selective, oxazole-based COX-2 inhibitor. Single molecules establishing a chemical proof of concept will be presented that inhibit either the COX-2 and 5-LO or COX-2 and LTA$_4$H enzymes.

070. NOVEL 1,2-DIARYLCYCLOBUTENES: SELECTIVE AND ORALLY ACTIVE COX-2 INHIBITORS. **R. Frenette,** R. W. Friesen, D. Dubé, Réjean Fortin, S. Prescott, W. Cromlish, G. Greig, S. Kargman, E. Wong, C.-C. Chan, R. Gordon, L. Xu, D. Riendeau, Merck Frosst Center for Therapeutic Research, P.O. Box 1005, Pointe Claire-Dorval, Québec, Canada H9R 4P8.

In the course of research on selective inhibitors of cyclooxygenase-2, a series of novel 2,3-diaryl-2-cyclobuten-1-ones have been synthesized and have been evaluated with respect to their ability to inhibit the isozymes of cyclooxygenase, COX-1 and COX-2. The compound 4,4-dimethyl-2-phenyl-3-[4-(methylsulfonyl)phenyl]cyclobutenone was found to be highly selective for the inhibition of COX-2 and was orally active in the rat paw edema model. The synthesis of the cyclobutenone series will be presented and the regioselectivity observed in the [2+2] cycloaddition step between functionalized stilbenes and keteniminiuim salts will be discussed. The evaluation of a number of analogs with respect to their oral absorption and their efficacy *in vivo*, as determined by the rat paw edema model, will be presented.

071. SYNTHESIS AND STRUCTURE ACTIVITY RELATIONSHIPS OF SULFONYL SUBSTITUTED 4,5-DIARYL THIAZOLES AS POTENT, SELECTIVE COX-II INHIBITORS. Jeff S. Carter*, John J. Talley, D. Joe Rogier, Roland S. Rogers, Stephen R. Bertenshaw, Matthew J. Graneto, Steven Kramer, Paul Collins, Carol M. Koboldt, Yan Y. Zhang, Jaime L. Masferrer, Karen Seibert and Peter C. Isakson. Dept. of Chemistry, Inflammatory Disease Research, Monsanto/Searle, BB-4I, 700 Chesterfield Pkwy. N, Chesterfield MO 63017

A series of sulfone or sulfonamide substituted diarylthiazoles were synthesized and found to be potent and selective inhibitors of cyclooxygenase-II (COX-II). This paper describes the convergent synthesis of diarylthiazoles utilizing the Hantzsch synthesis. By systematically modifying the substituted thiazole, we established the SAR within this series. Analogs included alkyl, aryl and amino 2-position substituents, multiple substitution patterns for the non-sulfone aryl and regioisomeric thiazole pairs in several cases. Among these thiazoles, many exceptionally selective (commonly 2500 to >10,000 fold) and potent (commonly IC_{50} = 7nM-1 uM) COX-II inhibitiors were discovered. In vivo testing in the carrageenan rat paw edema model (CRPE) demonstrated the antiinflammatory activity of this series as shown by CS-179 (Edema ED_{50} = 20 mpk, Hyperalgesia 27 % (20 mpk)).

072. SAR STUDIES OF SUBSTITUTED DI-T-BUTYLPHENOLS AS SELECTIVE COX-2 INHIBITORS
A.D.Sercel, Y. Song, T. Belliotti, V.G. Beylin, D.T. Connor, M.E. Marlatt, R.J. Sorenson, P.C. Unangst, R.B. Gilbertsen, K. Chan, D.J. Schrier, K. Laemont, G.C. Okonkwo, A. Guglietta, D.A. Bornemeier and R.D. Dyer. Departments of Chemistry, Chemical Development, Biochemistry and Immunopathology, Parke-Davis Pharmaceutical Research, Division of Warner-Lambert Co., 2800 Plymouth Road, Ann Arbor, MI 48105

The di-*t*-butylphenol **PD164387** has been identified as a potent and selective PHGS-2 inhibitor. The effects of the phenolic hydroxyl and *t*-butyl groups on potency and selectivity have been systematically investigated. Furthermore, the corresponding pyrimidine analogs have been prepared to study the electronic effects on potency and selectivity. The synthesis, SAR and biological evaluations of this chemical series will be discussed in detail.

PD164387

073. NOVEL DI-T-BUTYLPHENOLS AS SELECTIVE PGHS-2 INHIBITORS AND ORALLY ACTIVE AND SAFE ANTIINFLAMMATORY AGENTS. Y. Song, D.T. Connor, R. Doubleday, R.J. Sorenson, A.D. Sercel, P.C. Unangst, W. Cornell, R.B. Gilbertsen, K. Chan, D.J. Schrier, K. Laemont, G.C. Okonkwo, A. Guglietta, D.A. Bornemeier and R.D. Dyer. Departments of Chemistry, Biochemistry and Immunopathology, Parke-Davis Pharmaceutical Research, Division of Warner-Lambert Co., 2800 Plymouth Road, Ann Arbor, MI 48105

We have investigated the SAR of a series of di-t-butylphenols **1**. **PD138387** (X = S, Y = NHOMe) was identified as a potent, selective and orally active PGHS-2 inhibitor. It had an IC_{50} of 0.98 μM against recombinant human PGHS-2 and inhibited PGHS-2 activity in the J774A.1 cell line with an IC_{50} of 0.22 μM. It was inactive against purified ovine PGHS-1 at 100 μM and did not inhibit PGHS-1 activity in platelets at 20 μM. PD138387 was also orally active *in vivo*. The synthesis, SAR and biological evaluation of this series will be discussed in detail.

PD138387 (X=S, Y=NHOMe)

074. SUBSTITUTED DI-T-BUTYLPHENOLS: A NEW CLASS OF POTENT AND SELECTIVE PGHS-2 INHIBITOR. Y. Song, D.T. Connor, A.D. Sercel, R.J. Sorenson, R. Doubleday, P.C. Unangst, B.D. Roth, W. Cornell, R.B. Gilbertsen, K. Chan, D.J. Schrier, K. Laemont, G.C. Okonkwo, A. Guglietta, D.A. Bornemeier and R.D. Dyer. Departments of Chemistry, Biochemistry and Immunopathology, Parke-Davis Pharmaceutical Research, Division of Warner Lambert Co., 2800 Plymouth Road, Ann Arbor, MI 48105

The SAR of a series of di-*t*-butylphenols **1** has been investigated. **PD164387** (X = N, Y = S and Z = SEt) was identified as a potent, selective and orally active PGHS-2 inhibitor. It had IC_{50} values of 0.14 μM and 100 μM against recombinant human PGHS-2 and purified ovine PGHS-1 respectively. It inhibited PGHS-2 activity in the J774A.1 cell line with an IC_{50} of 0.18 μM, and inhibited PGHS-1 activity in platelets with an IC_{50} of 3.1 μM. PD164387 was also orally active *in vivo*. The synthesis, SAR and biological evaluation of this series will be discussed in detail.

PD164387 (X=N, Y=S and Z=SEt)

075.

SYNTHESIS AND BIOLOGICAL ACTIVITIES OF THE DUAL RECEPTOR ANTAGONISTS OF THROMBOXAN A2 AND PEPTIDOLEUKOTRIENES. M. Yokota, H. Nagaoka, H. Akane, Y. Arakida, Y. Isomura, Institute for Drug Discovery Research, Yamanouchi Pharmaceutical Co., Ltd. 21 Miyukigaoka, Tsukuba City, Ibaraki 305, Japan

Thromboxan A2(TXA2) and peptidoleukotrienes(pLTs) have been thought to play a major role in several inflammatory diseases. This suggests that the dual receptor antagonist of TXA2 and pLTs may be clinically more effective than a selective TXA2 or pLTs antagonist, especially in asthma. We synthesized a novel series of benzanilide derivatives and evaluated for its dual TXA2 and pLTs antagonistic activities. Among these compounds, 3-[(4-tert-butyl-2-thiazolyl)methoxy]-5'-[3-(4-chloro-phenylsulfonyl)propyl]-2'-(1H-tetrazol-5-ylmethoxy)benzanilide sodium salt monohydrate (YM-57158) inhibited the U-46619 and LTD4 induced contraction of guinea pig trachea with a pA2 value of 8.84 and 8.67, respectively. The structure-activity relationships and biological activities of these dual antagonists will be presented.

076. **SAR STUDIES OF THE LEUCINE PORTION OF BIO-1050, A POTENT AND SELECTIVE NANOMOLAR ANTAGONIST OF THE INTEGRIN VLA-4 ($\alpha_4\beta_1$).** Craig N. Zimmerman*, Ko-Chung Lin, Department of Medicinal Chemistry, Biogen Inc., 14 Cambridge Center, Cambridge, MA 02142

The synthesis and thorough SAR studies of the leucine portion of our lead molecule, Bio-1050 (IC_{50} = 30 nM), will be presented. A variety of alkyl as well as nitrogen and sulfur containing sidechain substitutions have been investigated which have resulted in analogs with increased potency.

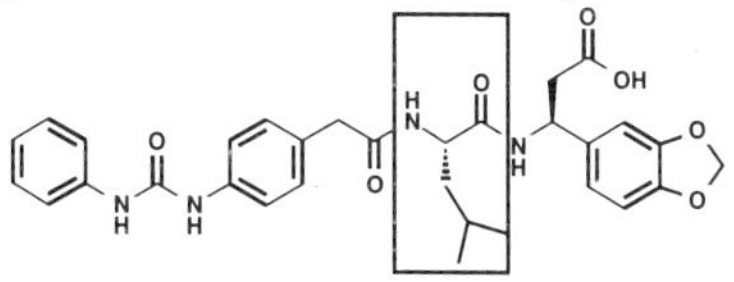

Bio-1050

077. **SAR STUDY ON THE PHENYL ACETYL PORTION OF BIO-1050, A SELECTIVE AND POTENT VLA-4 ($\alpha_4\beta_1$) ANTAGONIST.** Alfredo Castro* and Ko-Chung Lin, Department of Medicinal Chemistry, Biogen Inc., 14 Cambridge Center, Cambridge, MA 02142

We recently described the discovery of the first potent and selective VLA-4 antagonist. Bio-1050 (IC_{50} = 30 nM, cell adhesion assay) is representative of this novel series. In order to establish a structure activity relationship (SAR) of the phenyl acetyl portion of Bio-1050, a series of derivatives were synthesized and evaluated *in vitro* and *in vivo*. The chemistry and the biological results of this SAR study will be presented.

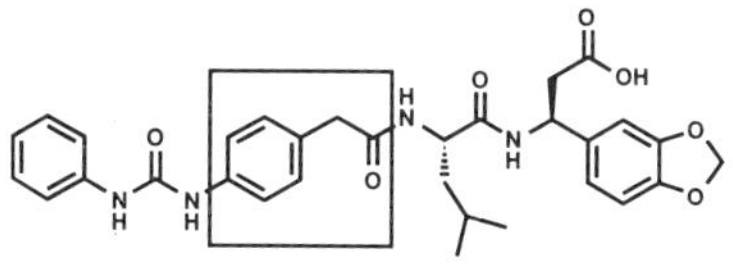

Bio-1050

078. **7-(ALKYLIDENE)CEPHALOSPORIN ESTERS AS ELASTASE INHIBITORS.** John D. Buynak,* A. Srinivasa Rao, Greg Adam, Christa Carver, Bolin Geng, Brian Bachmann, Samer Shobassy, and Stephanie Lackey, Department of Chemistry, Southern Methodist University, Dallas, TX 75275

This research group has recently demonstrated that 7-alkylidenecephalosporins (**1**) and 7-vinylidenecephalosporins (**2**) can be potent β-lactamase inhibitors (*J. Med. Chem.* **1995**, *38*, 1022; *J. Am. Chem. Soc.* **1994**, *116*, 10955). A series of 7-alkylidenecephalosporins and 7-vinylidenecephalosporins, as their benzhydryl esters, have now been tested as inhibitors of human leukocyte elastase. Selected 7-alkylidenecephalosporin esters are found to be potent, irreversible inhibitors of HLE. In contrast to previously reported cephalosporin-based elastase inhibitors, several of these compounds show optimum inhibitory activity as sulfides, rather than as sulfones.

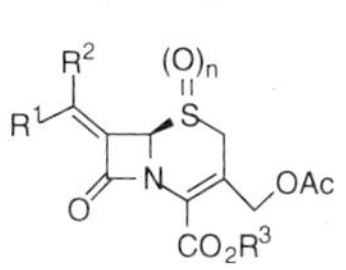

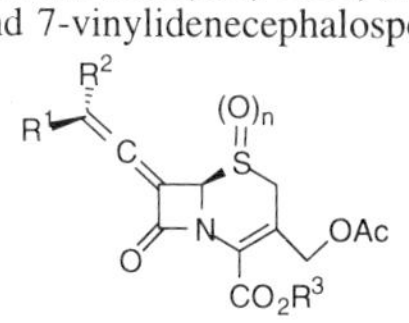

079. **Structure-Based Design and Synthesis of a Series of Hydroxamic Acids with a Quaternary-Hydroxy Group in P_1 as Inhibitors of Matrix Metalloproteinases.** Irina C. Jacobson*, Prabhakar G. Reddy, Zelda R. Wasserman, Karl D. Hardman, Maryanne B. Covington, Elizabeth C. Arner, Patty Welch, John V. Giannaras, Micky D. Tortorella, Robert A. Copeland. The DuPont Merck Pharmaceutical Company, Experimental Station, P. O. Box 80500, Wilmington, Delaware, USA 19880-0500.

Examination of the S_1 area of the active site of pro-stromelysin has led us to the design of novel and potent inhibitors of matrix metalloproteinases containing constrained quaternary-hydroxy group at P_1. Computer modeling indicates the presence of a hydrogen bond between the backbone of ALA 165 and the hydroxyl group at P_1. The synthesis and biological activity of these compounds with variations at P_1', P_2' and P_3' will be described.

080.
TRUNCATED SUCCINAMIDE HYDROXAMATES WITH NANOMOLAR POTENCY AGAINST VARIOUS MMP'S. M. A. Abreo, C. S. Agree, L. Mitchell, S. Zook, S. Margosiak, Agouron Pharmaceuticals Inc, 11099 N. Torrey Pines Road, Suite 130, La Jolla, CA 92037

The SAR of potent non-peptidyl hydroxamic acids having the succinate backbone will be presented.

081. SAR BY NMR AS A NOVEL TOOL FOR LEAD GENERATION: DISCOVERY OF A POTENT SERIES OF NON-PEPTIDE INHIBITORS OF STROMELYSIN (MMP-3).
G. S. Sheppard, S. K. Davidsen, S. W. Fesik, D. H. Steinman, G. M. Carrera, A. S. Florjancic, P. J. Hajduk, S. B. Shuker, E. T. Olejniczak, D. Nettesheim, R. P. Meadows, I. Nagy, P. Marcotte, T. J. Magoc, D. W. Morgan, J. B. Summers, Pharmaceutical Products Division, Abbott Laboratories, Abbott Park, Illinois 60064.

SAR by NMR, a novel NMR-based technique for discovering high affinity ligands (Shuker et al., Science, in press), was applied to the discovery of non-peptide MMP inhibitors. Screening of random small molecules against a truncated form of human stromelysin identified a series of biaryls which bound in the P1' pocket of the enzyme in the presence of acetohydroxamic acid. Linking of the biaryl and hydroxamate fragments with flexible tethers led to molecules having nanomolar IC_{50}s for the enzyme as measured by a thiopeptide enzyme inhibition assay. The effects of substitution on the biaryl nucleus was found to be similar in the linked and unlinked cases.

082. THERMODYNAMIC AND STRUCTURAL INVESTIGATIONS OF INHIBITOR INTERACTIONS WITH STROMELYSIN. R. W. Sarver, P. Yuan, B. Stockman, J. DeZwaan, Pharmacia & Upjohn Inc., Kalamazoo, MI 49001.

Thermodynamics of binding interactions for several inhibitors of stromelysin, a matrix metalloproteinase (MMP), were examined by high sensitivity isothermal titrating calorimetry. Two hydroxamate class MMP inhibitors, PNU-99533 and PNU-143677, formed 1:1 complexes with stromelysin that were both enthalpically and entropically driven. Thermodynamic parameters for another class of MMP inhibitor were differentiated from the hydroxamate type inhibitors. This class showed a two fold increase in enthalpy relative to hydroxamate inhibitors and a negative entropic change. Complexation to the enzyme's catalytic zinc along with hydrogen bond formation in the drug complex was responsible for favorable enthalpic changes. Possible factors contributing to the observed changes in entropy will be presented. Differential scanning calorimetry indicated a 10 to 15°C increase in unfolding temperature for inhibitor complexes compared to the unfolding temperature of native stromelysin which was 75°C. Unfolding was irreversible and followed by protein precipitation.

083. CORRELATION OF PROTEIN DYNAMICS WITH CALORIMETRIC MEASUREMENTS OF STROMELYSIN/LIGAND COMPLEXES. Peng Yuan, Ron W. Sarver, Jack DeZwaan, Brian J. Stockman, Structural, Analytical, & Medicinal Chemistry, Pharmacia & Upjohn, Inc., Kalamazoo, MI 49001

Matrix metalloproteinases including stromelysin are involved in tissue remodeling and connective tissue degradation. Inhibitors of the enzymes may have therapeutic value in the treatment of diseases resulting from imbalanced regulation of these proteins. The complexation of a variety of ligands, representing several chemical templates, to stromelysin was studied using both ^{15}N relaxation and calorimetry. Residues in the active site were found to have varying degrees of mobility depending on the type of ligand. Calorimetry measurements indicated that the driving force of the binding process differs, with one type of ligand associated with favorable enthalpy and entropy changes and the other type with only favorable enthalpy changes. The correlation between calorimetry, structure and dynamics of the various stromelysin/ligand complexes will be discussed.

084. **THE EFFECTS OF A AND C RING MODIFICATION OF OLEANOLIC AND URSOLIC ACID ON THE INHIBITION OF NITRIC OXIDE FORMATION IN MOUSE MACROPHAGES.** Heather Finlay, Gordon W. Gribble, Tadashi Honda, Nanjoo Suh* and Michael B. Sporn,* Departments of Chemistry and Pharmacology and Toxicology,* Dartmouth College and Medical School*, Hanover, NH 03755.

Triterpenoids constitute a large group of C30 pentacyclic compounds derived biosynthetically from the cyclization of squalene. Pharmacological evaluations of two closely related compounds, oleanolic acid and ursolic acid (**1** and **2**) have indicated a potential for use as anti-inflammatory agents. In this context a series of A and C ring modified compounds were synthesized and subsequently assessed by their ability to selectively inhibit inducible nitric oxide synthase in interferon-g activated mouse macrophages. A-ring cleavage and expansion was achieved *via* Schmidt reaction of the related 3-oxo-triterpenoid furnishing the nitrile and the lactam as shown. The nitriles were further functionalized and Baeyer-Villiger oxidation of 3-oxo-compounds yielded the corresponding lactones to complete the series and provide preliminary structure activity relationships for these derivatives.

085. **DISCOVERY OF DIHYDRO-2-AMINO-1,3-THIAZINES AS POTENT AND SELECTIVE iNOS INHIBITORS**
Hongbo Qi[1], Shrenik K. Shah[1], John L. Humes[3], Stephen G. Pacholok[3], Stephan K. Grant[2], Theresa M. Kelly[2], Kenny K. Wong[2], Barbara G. Green[2], Malcolm MacCoss[1]. Departments of Medicinal Chemistry[1], Enzymology[2] and Inflammation Research[3]. Merck Research Laboratories, P. O. Box 2000, Rahway, NJ 07065.

Nitric oxide (NO) is produced from L-arginine by nitric oxide synthases (NOS). Of the three known isoforms of NOS, two constitutive isoforms, neuronal NOS (ncNOS) and endothelial NOS (ecNOS), produce NO which functions as neurotransmitter and vasodilator. The third isoform, inducible nitric oxide synthase (iNOS), generates large amount of NO which acts as a cytotoxic agent in host defense. However, iNOS mediated excess NO production has also been implicated in inflammation. An iNOS selective inhibitor might be useful in treatment of diseases such as arthritis and inflammatory bowel disease. Based on screening leads, 5,6-dihydro-2-amino-4H-1,3-thiazine was found to be a potent inhibitor of NOS (IC_{50} on iNOS=18nM, ecNOS=65nM, ncNOS=29nM). Further modification of this structure lead to the discovery of several compounds with potent and selective iNOS activity. Details of the synthesis of these compounds and the SAR will be presented.

086. **IDENTIFICATION OF NITRIC OXIDE DONORS FROM THE NITROSATION OF HYDROXYGUANIDINES.** **Garry J Southan**, C George[†], L K Keefer[†]. IRSP, SAIC-Frederick and †Laboratory of Comparative Carcinogenesis, NCI-FCRDC, Frederick, MD 21702. ‡Naval Research Laboratory, Washington, DC 20375.

N^G-Hydroxy-L-arginine, an intermediate in the enzymatic oxidation of L-arginine to nitric oxide (NO) by NO synthase, reacts with NO in aerobic aqueous solution to form a species that spontaneously releases nitric oxide. We have isolated analogous NO-donors from the nitrosation of certain hydroxyguanidines in order to elucidate their nature and have obtained a definitive structure by X-ray crystallography. In addition, we show that these NO-donors spontaneously decompose in solution to NO and N_2O with the concomitant formation of, *inter alia*, the corresponding urea and cyanamide. The physiological and pharmacological relevence of these reactions remains to be determined.

087.

DESIGN, SYNTHESIS AND SAR OF BENZENESULFONAMIDE DERIVATIVES AS INHIBITORS OF COMPLEMENT C1r PROTEASE FOR THE TREATMENT OF INFLAMMATORY PROCESSES.

Cuiman Cai, Janet S.Plummer, Sheryl J.Hays, John L.Gilmore, Mark R. Emmerling, Kevin Wang and Juan C. Jaen, Parke-Davis Pharmaceutical Research Division, Warner-Lambert Company, 2800 Plymouth Road, Ann Arbor, MI.48105.

Serine proteases are a group of endopeptidase enzymes that have a serine amino acid in their active center. C1r in the complement system is serine protease and provides a critical and multifaceted defense system in the host defense against infection. Using 2-substituted 4H-3,1-benzoxazin-4-one and benzthiazin-4-one as our template, and following a topliss tree analysis, we synthesized a series of benzenesulfonamide derivatives as potent C1r inhibitors with improved activity. The design, synthesis and structure activity relationship of these improved inhibitors will be presented.

088.

NON-PEPTIDE LIGANDS OF THE HUMAN C5a RECEPTOR: STRUCTURE-ACTIVITY RELATIONSHIP OF A SERIES OF TETRAHYDROIMIDAZOPYRIDINES
Dooseop Kim, Ralph A. Rivero, Stephen E. deLaszlo, Jennifer Tran, Jason Fair, W.T. Ashton, S.M. Hutchins, L. Malkowitz, C. Molineaux, S. Siciliano, M.S. Springer, W.J. Greenlee, and N.B. Mantlo.
Merck Research Laboratories, RY 50G-336, P.O. Box 2000, Rahway, NJ 07065.

C5a, a single chain cationic glycoprotein consisting of 74 amino acid residues generated upon activation of the complement system, is known to be a very effective inflammatory mediator and potentially plays a pivotal role in the pathogenesis of a number of inflammatory diseases. Thus, the blockade of C5a at the receptor level by C5a antagonists could prove effective in the treatment of inflammatory diseases in man. Our efforts in search for non-peptide C5a antagonists led to the discovery of a series of tetrahydroimidazopyridine-base ligands for the C5a receptor. Optimization of this novel series provided a compound with IC_{50} value of 100 nM (30-fold increase in potency over the lead). The synthesis, SAR and functional properties (MPO release from neutrophils) of this class of compounds will be described.

089.

DISCOVERY OF POTENT, SELECTIVE AND ORALLY ACTIVE NONPEPTIDE BRADYKININ B2 RECEPTOR ANTAGONISTS, FR172357 AND FR173657. Yoshito Abe, Hiroshi Kayakiri, Shigeki Sato, Takayuki Inoue, Yuki Sawada, Teruo Oku, Hirokazu Tanaka, Fujisawa Pharmaceutical Co., Ltd. 5-2-3, Tokodai, Tsukuba, Ibaraki, 300-26, Japan

Bradykinin (BK) is an endogenous nonapeptide which is believed to mediate a wide variety of inflammatory diseases. As selective B2 antagonists a number of peptides and recently some nonpeptides have been reported but their clinical use is limited because of their poor oral bioavailability. Seeking orally active nonpeptide B2 antagonists we carried out intentional random screening and found a unique seed compound among the intermediates of angiotensin II antagonists. Extensive investigations revealed the structure activity relationships and led to discovery of a quite novel class of B2 antagonists which were proven to be potent, selective and orally active. The representative compounds such as FR172357 and FR173657 are not only useful probes to elucidate he pathophysiological roles of BK but also promising clinical candidates to treat various inflammatory diseases.

090.

ANTIRHEUMATIC AGENTS. NOVEL METHOTREXATE DERIVATIVES WHICH ARE RESISTANT TO POLYGLUTAMATION. N. Maruyama, H. Matsuoka, N. Ohi, M. Mihara, H. Suzuki, K. Miyamoto, K. Tsuji, N. Kato, T. Akimoto, T. Kuroki, Y. Takeda, K. Yano, Fuji Gotemba Rsearch Laboratories, Chugai Pharmaceutical Co., Ltd., 135, Komakado 1 Chome, Gotemba-city, Shizuoka 412, Japan

Methotrexate (MTX) is reportedly effective for treatment of rheumatoid arthritis (RA) because it undergoes polyglutamation (PG), a process that also causes side-effects. To develop a safer and potent anti-RA agent, we synthesized and tested derivatives which were made resistant to PG by replacing the glutamate of MTX with other amino acids; and potentiated their effects through modification using Oefner's model (Oefner C et al, Eur. J. Biochem., 174, 377-385 (1988)). All tested compounds inhibited in vitro proliferation of human synoviocytes and lymphocytes obtained from RA patients and healthy volunteers, respectively, with varying degrees of potency. Importantly, some derivatives suppressed progression of adjuvant arthritis model (p.o.); N-[4-(2,4-diaminopteridine-6-methyl)-3,4-dihydro-2H-1,4-benzothiazine-7-carbonyl]-L-homoglutamic acid (MX-68) was especially effective. We suggest that MX-68, without undergoing PG, may be of great benefit to RA patients.

091. THE NOVEL POTENT AND SELECTIVE PDE IV INHIBITORS
T. Ukita, M. Sugahara, Y. Terakawa, H. Kikkawa, and K. Wada, Lead Optimization Research Laboratory, Tanabe Seiyaku, Co., Ltd., 16-89, Kashima 3-chome, Yodogawa-ku, Osaka, 532, Japan

PDE IV inhibitor is expected for one of the new promising asthma treatments because of exerting both bronchodilating effect and inactivation of inflammatory cells including eosinophil. The structure of reported PDE IV inhibitors have been classified into two categories; one is Rolipram derived inhibitors, the other is Xanthine and Nitraquazone related inhibitors. We have focussed on the development of a new chemical scaffold (T. Iwasaki et al., J. Med. Chem., 1996, 39, 2696) and recently improved the potency and selectivity of PDE IV inhibitory activity. The synthesis of naphthalene derivatives, structure-activity relationships, PDE IV inhibitory activities($IC_{50} \sim 2 \times 10^{-10}$M), isozyme selectivity, inhibition of bronchoconstriction, and incidence of emesis will be discussed.

092. STABILITY STUDIES OF TRIPTOLIDE AND TRIPTRIOLIDE Yanping Mao, Jian Cai, Xuelian Tao, Jingyu Lai, Peter E. Lipsky. UT-southwestern Medical Center, Dallas Texas 75235-8577

Tripterygium wilfordii Hook. f (TWHF) has been shown to exert therapeutic effects on autoimmune and inflammatory diseases related to the activity of a variety of anti-inflammatory and immunosuppressive diterpenoids. The accelerated stability of two active compounds of TWHF, triptolide (I) and triptriolide (II) was investigated in aqueous solution, at PH 6.9, in a range of 60-90 °C, light protected conditions. The observed degradation rates followed first order kinetics for both compounds. The degradation rate constants ($K_{25°C}$) were obtained by "trend line analysis" and were 8.0843×10^{-5} and 1.8009×10^{-4}, respectively. The times for each compound to degrade 10% ($t_{1/10}$) and 50% ($t_{1/2}$) at 25 °C were 31 days and 204 days for (I), and 24 days and 145 days for (II). HPLC analysis indicated that the decomposition of (I) occurred at the C^{12} and C^{13} oxane moiety and C^{14} hydroxyl. Two of the decomposed products of (I) were identified as (II) and Triptonide (III). A stability test of (I) carried out at room temperature for two years showed that (I) was stable in solid state or dissolved in chloroform.

093.

SYNTHESIS AND PHARMACOLOGY OF NOVEL NON-PEPTIDE ISOXAZOLINE GLYCOPROTEIN IIb/IIIa ANTAGONISTS CONTAINING MODIFIED BASIC MOIETIES.
Donald J.P. Pinto,* John Wityak, Arthur D.Patten, Gary A. Cain, Michael Orwat, Shuaige Wang, Lori L. Bostrom, Susan V. DiMeo, Martin J. Thoolen, Shaker A. Mousa, Adrienne L. Racanelli, Elizabeth A. Hausner, Richard E. Olson and Ruth R. Wexler

The DuPont Merck Pharmaceutical Company, P. O. Box 80500, Wilmington DE 19880-0500

The inhibition of platelet aggregation by selectively blocking the association of fibrinogen with platelet glycoprotein IIb/IIIa has formed the basis of an attractive antithrombotic strategy. We recently reported the chemistry and antiplatelet activity of DMP 754, an orally active isoxazoline GPIIb/IIIa antagonist. This presentation will focus on the synthesis and antiplatelet activity of a series of isoxazolines in which the benzamidine moiety has been replaced or modified.

R = modified benzamidines, piperidyl, amidinopiperidyl, pyridylamidine, benzguanidine, guanidine

094.

SYNTHESIS AND PHARMACOLOGY OF NOVEL N-SUBSTITUTED BENZAMIDINE ISOXAZOLINE GLYCOPROTEIN IIB/IIIA RECEPTOR ANTAGONISTS. Thais M. Sielecki,* Jie Liu, Shaker A. Mousa, Adrienne L. Racanelli, Elizabeth A. Hausner, Martin J. Thoolen, Ruth R. Wexler and Richard E. Olson. *The DuPont Merck Pharmaceutical Company, P.O. Box 80500, Wilmington, DE 19880-0500*

Selective antagonism of the platelet GPIIb/IIIa receptor represents an attractive mechanism for the prevention and treatment of a number of thrombotic disease states. Recently, the antiplatelet activity of the oral GPIIb/IIIa receptor antagonists DMP 754 and DMP 802 were disclosed. The synthesis and biological evaluation of a series of potent N-substituted benzamidine isoxazolines will be presented. The effect of benzamidine substitution on the duration of antiplatelet efficacy in dog will be discussed.

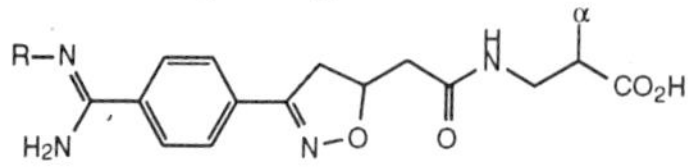

095. NOVEL NONPEPTIDE GPIIb/IIIa RECEPTOR ANTAGONISTS INCORPORATING PIPERIDINE AND OTHER CYCLIC STRUCTURES. Thais M. Sielecki*, Jie Liu, Michael J. Orwat, Donald J. Pinto, George K. Lalka, Shaker A. Mousa, Richard E. Olson, and Ruth R. Wexler, DuPont Merck Pharmaceutical Co., Wilmington, Delaware 19880-0500

In our efforts to discover new anti-thrombotics, we investigated a conformationally constrained series of GPIIb/IIIa receptor antagonists based on the structure of previously reported isoxazoline **XR300**. The effects of ring size and structure, as well as overall chain length, were examined. The synthesis and antiplatelet activity of these compounds, including *in vivo* activity in dog for representative compound **I**, will be discussed.

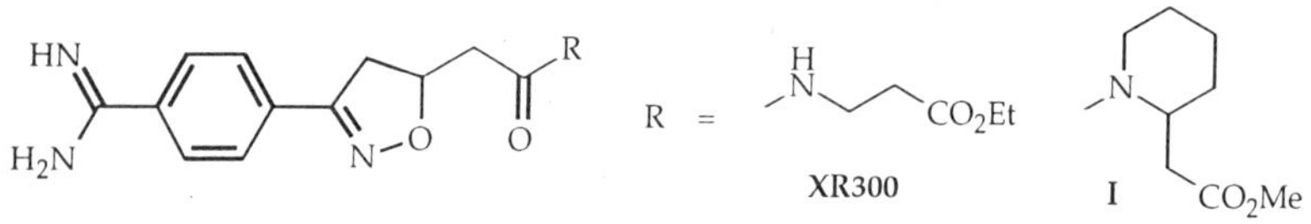

096. GPIIb/IIIa ANTAGONIST ISOXAZOLINYLACETAMIDES HAVING A MODIFIED AMIDE LINKAGE
John Wityak, Douglas G. Batt, A. Ewa Tobin, Shaker A. Mousa, Adrienne L. Racanelli, Elizabeth A. Hausner, Richard E. Olson, and Ruth R. Wexler, The DuPont Merck Pharmaceutical Company, P. O. Box 80500, Wilmington, DE 19880-0500

Antagonism of the glycoprotein IIb/IIIa receptor (GPIIb/IIIa) represents an attractive strategy for the treatment of thrombosis. The synthesis of GPIIb/IIIa antagonist isoxazolinylacetamides bearing various modifications to the amide moiety and the effect of these modifications on the *in vitro* and *in vivo* profile will be presented.

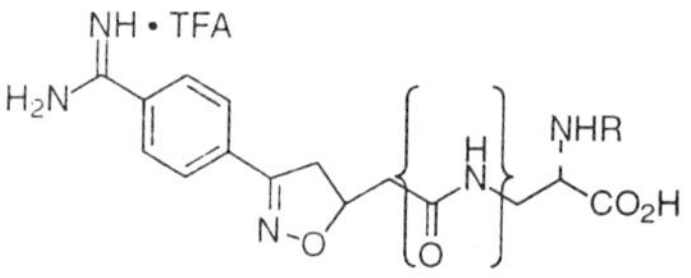

ψ[CSNH], ψ[CONMe], ψ[SO₂NH]

097. **GPIIb/IIIa ANTAGONIST ISOXAZOLINYLACETAMIDES BEARING AN α-PHOSPHORAMIDE MOIETY**
John Wityak, A. Ewa Tobin, Shaker A. Mousa, Adrienne L. Racanelli, Elizabeth A. Hausner, Ruth R. Wexler and Richard E. Olson, The DuPont Merck Pharmaceutical Company, P. O. Box 80500, Wilmington, DE 19880-0500

A new method for the synthesis of phosphoramides has been applied to the preparation of potent GPIIb/IIIa antagonists. The reaction of an amine with trialkyl or trialkenyl phosphites and iodine in the presence of base afforded the phosphorylated products in excellent yield. In vitro and in vivo antiplatelet effects of the resulting phosphoramides will also be discussed.

R = Me, Et, iPr, Bu, Allyl
R' = Me, H

098. DESIGN, SYNTHESIS AND STRUCTURE-ACTIVITY RELATIONSHIP OF A SERIES ARGINAL FACTOR XA INHIBITORS: I. STRUCTURES BASED ON THE (D)-ARG-GLY-ARG TRIPEPTIDE SEQUENCE. Charles K. Marlowe, Uma Sinha, Alice C. Gunn, Alan M. Laibelman, and Robert M. Scarborough. COR Therapeutics, Inc., 256 East Grand Ave., South San Francisco, CA, 94080

Factor Xa is a serine protease of the blood coagulation cascade which plays a pivotal role in linking the intrinsic and extrinsic activation pathways and is responsible for the conversion of prothrombin to active thrombin. Animal studies of factor Xa and thrombin inhibitors have suggested that these different anticoagulant strategies may afford distinct therapeutic indices, with factor Xa inhibitors potentially displaying greater efficacy and safety. A series of transition-state based factor Xa inhibitors utilizing the R-(D)-Arg-Gly-Arg-H tripeptide structure will be presented. The synthesis and in vitro structure-activity relationships obtained by modifications of the R group, the P3 sidechain, the configuration of the P3 residue, and substitution of the P2-Gly residue will be discussed. Potent compounds from this series show both high potency for factor Xa as well as high selectivity against thrombin and other relevant serine proteases.

099. DESIGN, SYNTHESIS AND STRUCTURE-ACTIVITY RELATIONSHIP OF A SERIES ARGININE KETOAMIDE FACTOR XA INHIBITORS: II. STRUCTURES BASED ON THE (D)-ARG-GLY-ARG TRIPEPTIDE SEQUENCE. Charles K. Marlowe, Michael C. Nyman, Alice C. Gunn, Uma Sinha, and Robert M. Scarborough. COR Therapeutics, Inc., 256 East Grand Ave., South San Francisco, CA, 94080

Factor Xa is a serine protease of the blood coagulation cascade which plays a pivotal role in linking the intrinsic and extrinsic activation pathways and is responsible for the conversion of prothrombin to active thrombin. Animal studies of factor Xa and thrombin inhibitors have suggested that these different anticoagulant strategies may afford distinct therapeutic indices, with factor Xa inhibitors potentially displaying greater efficacy and safety. A series of transition-state based factor Xa inhibitors utilizing generic R-(D)-Arg-Gly-Arg-CONHR' structure will be presented. The structure-activity relationships obtained by modifications of the R group, the P3 residue, and R' of the amide funtionality will be discussed. This series of factor Xa inhibitors displayed enhanced activity over a similar aldehyde series however they did not display superior selectivity versus thrombin. Modification of the R' functionality which presumably extends into the prime side of the binding site also displayed some interesting trends.

100. DESIGN, SYNTHESIS AND STRUCTURE-ACTIVITY RELATIONSHIP OF KETOHETEROCYCLIC BASED FACTOR XA INHIBITORS III. STRUCTURES BASED ON THE (D)-ARG-GLY-ARG TRIPEPTIDE SEQUENCE. Bing-Yan Zhu, Hua Yang, Deborah L. Volkots, Charles K. Marlowe, Alice C. Gunn, Paul W. Wong, Uma Sinha, and Robert M. Scarborough. COR Therapeutics Inc., 256 East Grand Ave., South San Francisco, CA, 94080

Factor Xa is a serine protease which plays a pivotal role in the blood coagulation cascade and is responsible for the conversion of prothrombin to active thrombin. Animal studies of factor Xa and thrombin inhibitors have suggested that these different anticoagulant strategies may afford distinct therapeutic indices, with factor Xa inhibitors displaying potentially greater efficacy and safety. A series of transition-state based factor Xa inhibitors utilizing the generic R-(D)-Arg-Gly-Arg-Heterocycle structure will be presented. The synthesis and in vitro structure-activity relationships ....ined by modifications of P4, P3, P2, P1 and heterocyclic motifs will be discussed. This series actor Xa inhibitors showed superior selectivity against a variety of related serine proteases.

101. DESIGN, SYNTHESIS AND STRUCTURE-ACTIVITY RELATIONSHIP OF KETOHETEROCY..C BASED FACTOR XA INHIBITORS IV. STRUCTURES BASED ON THE (D)-PHE-GLY-ARG TRIPEPTIDE SEQUENCE. Bing-Yan Zhu, Wenrong Huang, Arnie Martelli, Alice C. Gunn, Paul W. Wong, Uma Sinha, and Robert M. Scarborough. COR Therapeutics Inc., 256 East Grand Ave., South San Francisco, CA, 94080

Factor Xa is a serine protease which plays a pivotal role in the blood coagulation cascade is responsible for the conversion of prothrombin to active thrombin. Animal studies of factor Xa and thrombin inhibitors have suggested that these different anticoagulant strategies may afford distinct therapeutic indices, with factor Xa inhibitors displaying potentially greater efficacy and safety. A series of transition-state based factor Xa inhibitors utilizing generic R-(D)-Phe-Gly-Arg-Heterocycle structure will be presented. The structure-activity relationship obtained by modifications of P4, P3, P2 and heterocyclic motifs will be discussed. This series of factor Xa inhibitors showed moderate to low selectivity against a variety of related serine proteases.

102. BIARYL SUBSTITUTED ALKYLBORONATED ESTERS AS THROMBIN INHIBITORS. M. L. Quan*, J. Wityak, C. Dominguez, J.V. Duncia, C. A. Kettner, D. E. Duffy, C. D. Ellis, B. Knabb, A.Y. Liauw, J. M. Park, J. B. Santella, M. J. Thoolen, P. Weber, R. R. Wexler. The DuPont Merck Pharmaceutical Company, Experimental Station, P.O. Box 80500, Wilmington, DE 19880-0500

Thrombin is a serine protease which plays an important role in the blood coagulation cascade, and is a target enzyme for new therapeutic agents. Ac-(D)-Phe-Pro-boroArg-OH (DuP 714) was found to be a highly effective thrombin inhibitor. In order to reduce the peptidic nature of DuP 714, we have designed a series of novel biaryl substituted alkylboronated esters as potent thrombin inhibitors. Some of our best compounds such as XM855 have subnanomolar affinity for thrombin.

103. NMR AND MOLECULAR MODELING CHARACTERIZATION OF RGD CONTAINING PEPTIDES. Elsa Locardi, Daniel G. Mullen and Murray Goodman*, Department of Chemistry, University of California at San Diego, La Jolla, CA 92093-0343 and Telios Pharmaceuticals, 11045 Roselle Street, San Diego, CA 92121-1299.

Numerous cell adhesion proteins associated with biochemical processes such as hemostasis, cell migration, metastasis, tissue remodeling and development contain the tripeptide sequence arginine-glycine-aspartic acid (RGD) as the key recognition segment. Our interest has focused on the development of orally active integrin $\alpha_v\beta_3$ antagonists based upon the model of a proprietary pentapeptide pharmacophore, the RGDDV sequence, that exibits subnanomolar potency and high selectivity. The conformational analysis of several cyclic peptides containing the RGDDV pharmacophore and of analogs containing the RGD(Y-OMe)R sequence which has been shown to cross react with integrin $\alpha_{IIb}\beta_3$, were carried out by nuclear magnetic resonance (NMR) spectroscopy and molecular modeling. Comparison of the experimental data in H_2O and DMSO-d_6 suggests that these molecules do not exist in a single, well defined conformation, but as an equilibrating mixture of conformers in fast exchange on the NMR time scale. Despite this inherent flexibility, specific conformational preferences were found which may be related to the biological activity and selectivity of the peptides.

104.
SYNTHESIS AND CHARACTERIZATION OF ORALLY ACTIVE NONPEPTIDE ARGININE VASOPRESSIN V2 RECEPTOR ANTAGONIST. T.Ohkawa, H.Setoi, Exploratory Research Laboratories, Fujisawa Pharmaceutical Co., Ltd. 5-2-3, Tokodai, Tsukuba, Ibaraki, 300-26, JAPAN

Arginine vasopressin (AVP) may contribute to cardiovascular regulation by causing vasoconstriction and stimulating renal water absorption through the V1 and V2 receptors, respectively. Solute-free water diuretics produced by antagonized V2 receptors may be useful in treating disease in which water is retained. During a course of synthetic study, a series of 1,5-benzazepin-4-one derivatives were found to be potent AVP V2 receptor antagonists. These compounds antagonized binding to the human V2 receptors and inhibited the antidiuretic action of exogenously administered AVP in rats. The selected N-[4-[[2,3,4,5-tetrahydro-5-[(4-methylpiperazin-1-yl)carbonylmethyl]-4-oxo-1H-1,5-benzodiazepin-1-yl]carbonyl]phenyl]-4'-methyl-2-biphenylcarboxamide showed excellent oral activity and potential for clinical use. The structure activity relationship study of these compounds and the in vivo activities of representative derivatives will be discussed.

105. SYNTHESIS Of CARIPORIDE MESILATE (HOE 642) AND HOE 694, TWO HIGHLY SELECTIVE NA$^+$/ H$^+$ EXCHANGE (NHE) INHIBITORS. A. Weichert, S. Faber, H.-W. Jansen, W. Scholz and H.-J. Lang, HOECHST MARION ROUSSEL TA Research, Cardiovascular Agents, HOECHST AG, 65926 Frankfurt, Germany

The syntheses of Cariporide mesilate, currently being clinically developed as a protective drug in cardiac ischemia and reperfusion states, and HOE 694 (3-Methylsulfonyl-4-piperidino-benzoylguanidine methanesulfonate) widely used as a pharmacological tool in studies comprising NHE inhibition, are described. Additionally, their selectivity on different NHE subtypes is determined.

Cariporide mesilate

106. BILE ACID TRANSPORT INHIBITORS. SYNTHESIS AND SAR OF 1,5-BENZO-THIAZEPINE-1,1-DIOXIDES. Anthony L. Handlon, Michael C. Lewis, and Chari D. Smith. *Glaxo Wellcome Inc. Research Triangle Park, NC 27709-2700.*

A lowering of serum LDL cholesterol concentrations can be achieved by preventing the reabsorption of bile acids from the gut. This has been demonstrated in the clinic using bile acid binding resins at multi-gram doses. Lewis et al. (J. Lipid Res. *36*, 1995, 1098) have described the first potent non-sterol based inhibitor of the transport protein that is responsible for bile acid reuptake in the small intestine. Here we present the synthesis and SAR of 1,5-benzothiazepine-1,1-dioxide inhibitors of the bile acid transport protein (BAT). One compound in this class, GW577, inhibits human BAT with an IC$_{50}$ of 0.55 µM and prevents the reuptake of bile acid in the rat with an ED$_{30}$ of 0.02 mg/Kg.

GW577

107. SPIROSTANE GLYCOSIDE CHOLESTEROL ABSORPTION INHIBITORS: NON-ANILINE CARBAMATE DERIVATIVES. **Cynthia Eller**, John B. Etienne, Michael P. DeNinno*, Central Research Division, Pfizer Inc., Groton, CT 06340.

Spirostane glycosides have been demonstrated to inihibit cholesterol absorption in rodent and human studies. In previous disclosures we have shown that aryl carbamoyl substituents at the 4" and 6" position of spirostane cellobiosides can greatly improve their potency. The aniline component of these agents raised the concern that cleavage of these groups in vivo would lead to the release of free anilines. Although this risk was perceived to be small due to the compounds' high potency and very low absorption, an effort was made to find non-aniline containing inhibitors. This effort was successful at finding potent inhibitors, (e.g. CP-334,479, ED50 = 0.075 mg/kg), within 2-3 fold the activity of the aryl series. The SAR of these hypocholesterolemic agents will be described.

108. SYNTHESIS AND SAR OF TRIMERIC BILE ACID REABSORPTION INHIBITORS: A NEW APPROACH TO LOWER CHOLESTEROL H. Glombik, K.-H. Baringhaus, G. Böger, A. Enhsen, E. Falk, M. Friedrich, A. Hoffmann, W. Kramer, H.L. Schäfer, S. Stengelin, G. Wess; HMR TA Metabolism Research D-65926 Frankfurt, Germany

Recent attempts in antiatherosclerotic therapy focus on cholesterol lowering agents with new modes of action. With regard to resins and statins real progress is expected from non absorbable bile acid reabsorption inhibitors (BARI) that block the ileal transporter specific for bile acids. This will lead to increased excretion of bile acids, resynthesis from cholesterol in the liver and thus lowering of blood cholesterol by an indirect and non systemic mechanism.
While there is some information available on the size and function of the ileal bile acid transporter, a detailed structure analysis of this transmembrane protein has not been performed. BARI with minimal absorption are designed and synthesized by combining bile acid moieties as recognition units via linkers to trivalent core structures such as Kemp's triacid. Linker chemistry had to be developed for this purpose. A directing effect of the core unit is most important for activity at the primary target as tested in cell and animal models.

109.

SOLID-PHASE PARALLEL SYNTHESIS FOR LEAD OPTIMIZATION. A CONVERGENT MINI-LIBRARY OF NONPEPTIDE GPIIB/IIIA ANTAGONISTS RELATED TO RWJ-50042.
Stephen C. Yabut, William J. Hoekstra,* Bruce E. Maryanoff, Patricia Andrade-Gordon, Michael J. Costanzo, Bruce P. Damiano, Robert Falotico, Barbara J. Haertlein, Jack A. Kauffman, Patricia M. Keane, David F. McComsey, and John A. Mitchell, Drug Discovery,
The R. W. Johnson Pharmaceutical Research Institute, Spring House, Pennsylvania 19477

We have employed solid-phase parallel synthesis for the preparation of over 200 analogues of the nonpeptide GPIIb/IIIa antagonist, RWJ-50042. This involved initial attachment of a amine to a tritylchloride resin, and the subsequent execution of seven chemical steps. A protocol of optimization cycles rapidly afforded a collection of nipecotamide analogues with up to 100-fold enhancement in potency. By generating a convergent compound mini-library, we avoided the synthesis and testing of many compounds predicted to be inferior from the well-behaved structure-activity relationship for this series.

RWJ-50042 (**1**; racemic)
prototype GPIIb/IIIa antagonist

110. SEARCHING FOR DNA BINDER THROUGH OLIGOSACCHARIDE LIBRARY
Masayuki Izumi and Yoshitaka Ichikawa, Department of Pharmacology and Molecular Sciences, Johns Hopkins University School of Medicine, Baltimore, MD 21205

To identify a new class of DNA-binding oligosaccharide, we have developed a novel way to construct an oligosaccharide library. The procedure involves an iodonium ion-catalyzed "non-selective" glycosylation of a glycal derivative (**1**). The resulting 2-iodo oligosaccharides (**2**) will be further modified by dehalogenation, introduction of amino group, etc. to generate a wide variety of oligosaccharide structures.

Further modification
- Dehalogenation
- Replacement with NH_2
- etc.

111. **HIGH SPEED ANALOGUE SYNTHESIS. DERIVATIVES OF CALANOLIDE A, AN ANTI-HIV AGENT, VIA SOLUTION PHASE COMBINATORIAL SYNTHESIS**
David E. Zembower, Pushpa Chandrasekar, Shuyuan Liao, Ze-Qi Xu and Michael T. Flavin, MediChem Research, Inc. 12305 South New Avenue, Lemont, Illinois 60439 USA

(+)-Calanolide A is a naturally occurring inhibitor of HIV-1 reverse transcriptase currently in preclinical development. As part of our program to investigate the structure-activity relationships of this class of compound, we have developed a rapid, parallel synthesis of calanolide analogues modified in the B ring region of the molecule. This procedure afforded calanolide analogues in good isolated yields (52-70%) having sufficient purity (87-94%) for biological evaluation, without the need of any further purification. The synthesis of the common intermediate and the two step solution phase preparation of analogues will be described, as well as the extension of this process to automated parallel synthesis of calanolide libraries and their resulting anti-HIV activity profiles.

(+)-Calanolide A

112. **A Combinatorial Approach to the Phosphatidylinositol-3-phosphates Involved in Intracellular Signal Transduction: Synthesis and Biochemical Studies**
R. Aneja,* A. Parra, C. Stoenescu, W. Xia and S. G. Aneja. Nutrimed Biotech, Cornell University Research Park, Langmuir Laboratory, Ithaca, NY 14850.

The 3-phosphate derivatives of 1D-1-(1´,2´-di-O-fattyacyl-sn-glycero-3´-phospho)-myo-inositols (PtdIns) including phosphatidylinositol-3-phosphate, PtdIns(3)P, and the bis- and tris-phosphate derivatives PtdIns(3,4)P_2 and PtdIns(3,4,5)P_3, have been found in eukaryotic cells, and the occurrence of PtdIns(3,5)P_2 has been suggested. These compounds are putative messengers in cellular signal cascades pertinent to inflammation, cell proliferation, transformation, protein kinesis, and cytoskeletal assembly. Minute quantities are found in cells and therefore synthetic methods are needed to obtain samples for establishing the putative roles.

We report on the first combinatorial approach to the synthesis of phosphatidylinositol-3-phosphates, the characterization of the products as 1D-1-(1´,2´-di-O-fattyacyl-sn-glycero-3´-phospho)-myo-inositol-3-phosphates, and the results of biochemical studies on activation of the calcium independent isoforms of protein kinase C family.

113. **THE SYNTHESIS OF A 4-DIMENSIONAL N-CARBOXYMETHYL PEPTIDE COMBINATORIAL LIBRARY FOR NEW LEAD GENERATION AGAINST METALLOPROTEINASES UTILIZING BOTH MIX AND SPLIT AND INDEXED LIBRARY STRATEGIES**
Nancy J. Kevin*, Craig K. Esser, Kevin T. Chapman
Department of Molecular Design and Diversity
Merck Research Laboratories, P.O. Box 2000, Rahway, New Jersey 07065

N-Carboxyalkyl peptides are known inhibitors of metalloproteinases (Patchett, A. A.; Harris, E. E.; Tristram, M. J.; Wyvratt, M. J. *Nature* **1980**, *288*, 280). An N-carboxymethyl tripeptide combinatorial library patterned after Merck's Enalapril and composed of four subunits (10x10x10x10) was synthesized to generate new leads against metalloproteinases. Using TentaGel S HMB resin as the solid support, a strategy of mix and split, followed by a spatially addressed matrix, was employed to yield 100 mixtures of 200 compounds each.

114. **[1,3]-DIPOLAR CYCLOADDITION REACTIONS ON SOLID SUPPORT AND THE SYNTHESES OF COMBINATORIAL LIBRARIES OF HETEROCYCLIC COMPOUNDS.** Yihan Wang and Stephen R. Wilson* Department of Chemistry, New York University, Washington Square, New York, NY 10003.

[1,3]-Dipolar cycloaddition reactions of azomethine ylides with olefins were explored on solid supports and were utilized in the syntheses of combinatorial libraries of thrombin inhibitors. Various aldehydes, amino acids and dipolarophiles were used as

diversified inputs. The libraries were further functionalized to create second generation of libraries.

115.

SYSTEMATIC STUDY of THE SUBSTITUTED ACTIVE C-TERMINUS of HIRUDIN WITH NATURE AMINO ACIDS. L. Gang, Mu SF, Yun LH, Ding ZK, Sun MJ, Institute of Pharmacology and Toxicology, Beijing Tai-Ping Road 27#, Beijing 100850, P. R. China

Today, using the combinatorial chemistry's strategy for discovery and optimization lead compounds of new drug design is fashionable. Hirudin is a single polypeptide chain protein containing 65 amino acids from *Hirudo Medicinalis*, which is a potent thrombin inhibitor. In this paper, we systematically synthesized C-terminus(desulfo hirudin$^{45-65}$) peptides substituted with 20 natural L-amino acids using the Multipin Method. The peptide library was screened by the fibrinogen clotting assay and amdiolytic hydrolysis assay. Conclude: 1 the desulfo C-terminus(hirudin$^{45-65}$) of hirudin was the potent trifunctional inhibitors of thrombin, which should contain peptide sequence of identifying catalytic site, anion-binding exosite of thrombin and the linker of these two active fragments; 2 the desulfo hirudin$^{45-55}$ replaced by some amino acids is not significantly influence but increase or decrease obviously the anticoagulant potency; 3 the desulfo hirudin$^{56-64}$ is essential, particularly Phe^{56}, Glu^{57}, Ile^{59}, Pro^{60}, Tyr^{63}, Leu^{64} are the key mers for delaying the fibrinogen clotting time and amdiolytic hydrolysis activity of thrombin.

116. COUMARIN-BASED PRODRUGS OF AMINES. STRUCTURAL EFFECTS ON THE RELEASE KINETICS.
Huijuan Zhang, Wei Wang, Ailian Zheng, and Binghe Wang.* Department of Chemistry, North Carolina State University, Raleigh, NC 27695

Earlier, we reported the chemical feasibility studies of a prodrug system by taking advantage of the facile lactonization of coumarinic acid and its derivatives. To further investigate the structural effect on the rates of the esterase-triggered release of the original "drug" entities, several coumarin-based prodrugs of model amines were synthesized. The release rates of these prodrugs were studied using porcine liver esterase in a phosphate buffer (0.05 M, pH = 7.4, 37 °C) and the half lives were found to range from 1.6 to 186 min. The results showed that structural features of both the acyl functional groups (-R) and the model amines (-X) affected the release kinetics.

117.

SYNTHESIS OF ANGIOSTATIC CODRUGS FOR INTRAOCULAR NEOVASCULARIZATION. G. Cynkowska, T. Cynkowski, P.A. Crooks, M. Howard, H. Guo, P. Ashton, College of Pharmacy, University of Kentucky, Lexington, KY 40536

Intraocular neovascularization as a complication in a variety of eye diseases is a leading cause of visual loss. It was demonstrated that a combination of the antiangiogenic steroid, 5 -pregnane-3 17 ,21-triol-20-one (THS) with the antimetabolite 5-fluorouracil (5-FU), has significantly more antiangiogenic activity than either of the components alone. The purpose of this study was to synthesize a variety of codrugs consisting of THS and 5-FU linked at either the 3-position, the 21-position, or both positions via a carbonate link, to permit enzyme-mediated cleavage. Additionally, a number of codrugs of THS and 5-FU with short chain polyethylene glycols (PEGs) linkages, were also synthesized. The main goal of this approach was to improve drug delivery by overcoming solubility problems associated with both THS and 5-FU, and to enhance their permeation through topical membranes. The facile hydrolysis of codrugs was also studied.

118.
SYNTHESIS AND PROPERTIES OF NOVEL PRODRUGS OF CAMPTOTHECIN WITH SHORT CHAIN POLYETHYLENE GLYCOLS AND POLYOXAACIDS. **T. Cynkowski,** G. Cynkowska, P.A. Crooks, H. Guo, P. Ashton, College of Pharmacy, University of Kentucky, Lexington, KY 40536

Camptothecin, a potent natural antitumor drug has limited clinical utility as an anticancer agent due to its toxicity and an extremely poor solubility profile. In an attempt to circumvent the solubility problem and to enhance absorption of the drug, a series of prodrugs of camptothecin linked via the 20-OH group to short chain polyethylene glycols (PEG) or polyoxaacids were prepared and their hydrolysis kinetics studied. The esters of the polyoxaacids were prepared using the DCC/DMAP coupling method, whereas the prodrugs of PEGs were synthesized as carbonates from chloroformates of the corresponding monoprotected PEGs and camptothecin. All the prodrugs studied were much more soluble than camptothecin in both aqueous and organic solutions and underwent hydrolysis to generate camptothecin both in phosphate buffer at pH 7.4 and human serum. The selected prodrugs are being evaluated in in-vivo models.

119.
SYNTHESIS AND EVALUATION OF BIOREVERSIBLE CONJUGATES OF OLIGONUCLEOSIDE PHOSPHOROTHIOATES   Nan-Hui Ho, Dong Yu, Radhakrishnan P. Iyer and Sudhir Agrawal,   Hybridon Inc., One Innovation Drive, Worcester, MA 01605

The design of oligonucleotides conjugated to bioreversible groups (e.g., prodrug) is an important area of research in nucleic acid-based therapeutics. These prodrug oligonucleotides may be used for: (a) tissue-specific targeting, (b) for improving the cellular uptake, (c) for providing enhanced resistance to nuclease-mediated digestion, and (d) providing a slow release depot for the oligonucleotides. We have designed and synthesized a novel prodrug of oligonucleoside phosphorothiate where an acyloxyaryl moiety is site-specifically incorporated into the backbone of the oligonucleotide by using the N-pent-4-enoyl (PNT) as nucleobase protecting group. The bioreversion of the prodrug oligonucleotides to the parent phosphorothioate has been examined using human serum and various esterases.
Our initial results suggest that this bioreversible conjugate is a novel prodrug of the oligonucleotide.

120.
MODIFICATION OF ANCROD WITH POLY(ETHYLENE GLYCOL).
Chengyun Yue, Jeffrey A. Hubbell, Division of Chemistry and Chemical Engineering, Mail Code 210-41, California Institute of Technology, Pasadena, CA 91125

Ancrod is a thrombin-like fibrinogenase from the venom of Akistrodon rhodostoma. It cleaves fibrionopeptides A, AP and AY from fibrinogen. The thus formed clots are not cross-linked and are susceptible to degradation by plasmin. As they lower serum fibrinogen, they appear to be an appropriate alternative to heparin and may be preferable to it in hemorrhagic complications. However, the widespread clinical utility of Ancrod has been limited by immunologic reaction in patients and by availability and thus cost. To reduce the immunogical response of Ancrod and to increase its blood circulation time (thus lower dosage will be required), Ancrod was modified with poly (ethylene glycol) (PEG) in this study. The enzymatic activity, towards both small peptide substrate and fibrinogen, of PEGylated Ancrod at different degree of modification was studied.

121. **LIPID DRUG DELIVERY SYSTEMS FOR METHOTREXATE**, James N. Jacob, Yuanwei Deng, Kumari Batra, Department of Chemistry, University of Rhode Island, Kingston, RI 02881

Natural lipids act as slow release systems for the systemic delivery of anticancer drugs. Methotrexate was used as a model for this study. Several lipid esters of methotrexate were synthesized at the α and γ-carboxyl groups of methotrexate. The in vivo activity was studied again L-1210 cells in mice. There was a three to five fold increase in life span with lipid-methotrexate derivatives compared to methotrexate alone when brain tumor model mice were treated with these drugs. Complete recovery was observed for 87% of lipid methotrexate treated mice after systemic (intraperitoneal or i.p.) administration of L-1210 leukemic cells. Whereas MTX treatment produced only 36% increase in life span, the lipid conjugates increased the life span by over 130% after systemic administration of L-1210 cells. The synthesis of these lipid derivatives and their in vivo animal studies will be presented.

122. **PHYSICAL AND BIOLOGICAL PROPERTIES OF WATER-SOLUBLE PRODRUGS OF MORPHOLINE ACETAL SUBSTANCE P ANTAGONISTS**
Sander G. Mills, Jeffrey J. Hale, Malcolm MacCoss, Conrad P. Dorn, Jr., Paul E. Finke, Richard J. Budhu, Shuet-Hing L. Chiu, Su Huskey, Margaret A. Cascieri, Sharon Sadowski, Robert A. Reamer, James Pivnichny, Dorothy Levorse, Karen Owens, Joseph Metzger, George Eiermann, F. David Tattersall[§], Wayne Rycroft[§], Nadia Rupniak[§], Richard Hargreaves[§], and D. Euan MacIntyre, Merck Research Laboratories, P.O. Box 2000 Rahway NJ, 07065 [§]Merck, Sharp & Dohme, Terlings Park, Eastwick Road, Harlow, Essex CM20, U. K.

Morpholine-based acetals such as L-742,694 have been shown to be highly potent and orally available antagonists of the Neurokinin-1 (NK-1) receptor (J. J. Hale et al. *J. Med. Chem.* **1996**, *39*, 1760). Since such compounds often have limited aqueous solubility (which can complicate the preparation of parenteral formulations), we have prepared novel aminoalkyloxycarbonyl and phosphoryl derivatives of the triazolinone ring that display greatly enhanced solubility in several cases, and which readily revert to the active parent compound *in vivo*. Synthetic routes to such derivatives will be outlined along with detailed physical and biological characterization.

$Q = CO_2CH_2CH_2NH_2 \cdot HCl$, $P(O)O_2^{2-}$

123. **TRIGGERED RELEASE OF CHLORAMBUCIL FROM AN INACTIVE BIOCONJUGATE.**
Ashraf H. Bayomi, Alice M. Mitchell, Ettaya N. Natarajan, Charles B. Grissom and F. G. West, Department of Chemistry, University of Utah, Salt Lake City, UT 84112 & Mohsen A. Aziza and Omar El-Ahmady, Faculty of Pharmacy, Al-Azhar University, Naser City, Cairo - Egypt.

Chlorambucil is widely used in the treatment of chronic lymphocytic leukemia, lymphomas and ovarian carcinoma. A possible solution to the problem of systemic cytotoxicity involves the controlled release of the drug from a nontoxic precursor in the region of the target tumor via some external signal. However, this requires a signal which can penetrate human tissue. Both light (610-700 nm) and ultrasound (20 kHz -50kHz) can penetrate human tissue to depths greater than 10 cm. Since photons of 610 nm red light have an energy of only 47 kcal/mol, light of this wavelength does not possess sufficient energy to break most chemical bonds. However, the extremely weak Co-C bond (to vitamin B_{12} complexes) can be cleaved with light of this wavelength. The high binding capicity of cancer cells for B_{12} also provides specific delivery.

$$Co^{III} - Drug \xrightarrow{h\nu \text{ or }))))} \{Co^{II} + \dot{D}rug\} \xrightarrow{O_2} Co^{III} - OH_2 + Drug\,(released)$$

The bioconjugates have been prepared by: a) reaction of Co(I) (highly nucleophilic) with chlorambucil acid chloride, or b) esterification of β-hydroxyethylcobalt complexes with chlorambucil.

124. **NON-HYDROLYZABLE ANALOGUES OF TAD AND BAD CONTAINIG A FLUORINE ATOM AT THE C2' OF ADENINE NUCLEOSIDE. INDUCTION OF K562 CELL DIFFERENTIATION.** Krystyna Lesiak, Alokes Majumdar, Michael Seidman, Kyoichi A. Watanabe, and Krzysztof W. Pankiewicz. Codon Pharmaceuticals, Gaithersburg, MD 2087

Thiazole-4-carboxamide adenine dinucleotide (TAD) containing a fluorine atom at C2' of the adenine nucleoside in *arabino* configuration (**1**) was found to be nearly as potent inducer of K562 cell differentiation as tiazofurin (TF). Compound **1** could be active by itself or might release TF upon enzymatic hydrolysis. In order to distinguish between these two mechanisms we synthesized non-hydrolyzable analogues of **1** in which the pyrophosphate oxygen was replaced with -CH_2- or -CF_2- group (**2** and **3**, respectively). We also prepared the corresponding non-hydrolyzable, fluoro-substituted analogue **4** of benzamide adenine dinucleotide (BAD). 9-(2-Deoxy-2-fluoro-ß-D-arabinofuranosyl)adenine (F-ara-A) was converted into the 3'-O-tetrahydropyranyl-5'-O-mesyl derivative which was treated with tris(tetrabutylammonium) salt of methylenebisphosphonic acid to give the 5'-methylenebis(phosphonate) analogue **5**. DCC promoted coupling (a novel mechanism will be presented) of **5** with 2',3'-O-isopropylidene-tiazofurin or -benzamide riboside followed by deprotection afforded the desired dinucleotides **2** and **4**, respectively. Similar coupling of tiazofurin 5'-difluoromethylenebis(phosphonate) with protected F-ara-A gave an analogue **3**. Compound **3** as well as ß-CF_2-TAD were found to be potent inducers of K562 cell differentiation whereas **2** and **4** were not.

125. STRUCTURE ACTIVITY STUDY ON LONG CHAIN BISPHOSPHONATES. R.Chen, G.Golomb, and E. Breuer. The Department of Pharmaceutical Chemistry And The Department of Pharmacy, The School of Pharmacy, The Hebrew University of Jerusalem, Jerusalem, Israel 91120.

The bisphosphonates are a class of synthetic compounds used in the treatment of various metabolic bone diseases, including hypercalcaemia of malignancy, Paget's disease, postmenopausal osteoporosis and corticosteroid-induced bone loss. Clinically used bisphosphonates, namely, Etidronate, Clodronate, Pamidronate, and Alendronate, all belong to the geminal P-C-P type. The biological effects of the geminal bisphosphonates are attributed mainly to their strong affinity to calcium phosphate crystals, although direct interaction with osteoclasts and their influence on a variety of biochemical pathways has also been demonstrated. It had been believed that non-geminal bisphosphonates P-$(C)_n$-P with n≥2 were inactive. It was first recognized in our laboratory that long chain P-$(C)n$-P type bisphosphonates are also capable to modulate calcium related disorders, provided that the molecules contain donor groups adjacent to the phosphonic functions. In order to elucidate the roles of various structural features in biological activity we synthesized long - chain bisphosphonates having hydroxy, nitro, amide, carboxy, hydrazono groups adjacent to the phosphonic functions. The ability of the newly synthesized compounds to interfere with calcium metabolism was evaluated in comparison to compounds of confirmed activity in an *in vitro* model.

126. STUDY OF THE MECHANISM OF INHIBITION OF EPOXIDE HYDROLASES BY CHALCONE OXIDES. C. Morisseau[1], G. Du[1], J.W. Newman[1], Y. Nakagawa[2], J. Zheng[1] & B.D. Hammock[1]. 1 Departments of Entomology and Environmental Toxicology, University of California Davis, CA 95616; 2 Department of Agricultural Chemistry, Kyoto University, Kyoto 606-01, Japan.

Metabolism of drugs and xenobiotics is among the important factors in determining the biological and toxicological effects of exposure. Many mutagens and carcinogens are degraded by the soluble and microsomal epoxide hydrolases. Conversely, the diol resulting from the hydrolysis of leukotoxin by an epoxide hydrolase is the metabolite responsible for the toxicity of this compound in cell culture. If production of leukotoxin diol results in the clinical symptoms of ARDS, inhibition of the epoxide hydrolase could reduce symptoms. In this study, we report (1) the quantitative analysis of the structure-activity relationship for about forty inhibitors (chalcone oxide derivatives) of soluble epoxide hydrolases, (2) the kinetic study of their action, and (3) the determination of the structure of the enzyme-inhibitor complex. These results provide an understanding of the mechanism of inhibition permitting the design of therapeutic drug or pro-drug for the treatment of ARDS.

127. SYNTHESIS AND STUDY OF NEW SELENO-ORGANIC MIMICS OF GLUTATHION PEROXIDASE   I. Erdelmeier, C. Tailhan-Lomont, M. Moutet, J.-C. Yadan, OXIS International S. A., F - 94385 Bonneuil-sur-Marne, France

Studying the synthesis of new selenium-nitrogene-heterocycles as a part of our medicinal chemistry program for the treatment of oxidative stress-related diseases, we uncovered a mild, copper(I)-assisted reaction of selenocyanate with suitably substituted non-activated aromatic bromides. This paper will discuss the scope and limitations of the new copper(I)-assisted synthesis of arylisoselenazolines and azines in terms of (1) the role of the metal salt and (2) the compatibility with amine-, amide-, ester- and nitro-substituents. Like the natural selenium containing enzyme Glutathion Peroxidase, these new low-molecular-weight mimics are able to catalyze the thiol-mediated reduction of hydrogen peroxide and organic peroxides, as will be shown by the comparison of their GPx-activities with the sulfur-analogues and the reference compound Ebselen. Direct competition experiments between the lead compound 4,4-dimethylbenzisoselenazine and Ebselen will be presented to illustrate the significantly higher reactivity as well with thiols as with peroxides, modelling isolated reaction steps of the catalytic cycle.

128.
X-RAY STRUCTURAL STUDIES OF ALLOSTERIC EFFECTORS-HEMOGLOBIN COMPLEXES.
Carmen M. Moure and Donald. J. Abraham, Department of Medicinal Chemistry, Richmond, Virginia, 23298.

Structure-activity studies have been conducted on a series of compounds that behave as allosteric effectors of hemoglobin. Their allosteric activity cannot be explained solely from their affinity binding constants. Previous crystallographic studies (low resolution) showed that these effectors bind at the same site in the hemoglobin central water cavity in its deoxy state. The differences in activity could be due to various interactions with surrounding hemoglobin residues. To test this, the X-ray structures of the effector-hemoglobin complexes have been solved and refined at 2 angstrom resolution. The two most active compounds in the series show additional binding sites in the cavity, suggesting a possible explanation for their increased activity. Solution binding studies are being conducted to determine if the new binding sites are also present in solution. Results will be interpreted with respect to the allosteric theory of hemoglobin.

129.
STRUCTURAL REQUIREMENTS OF BENZAMIDE RETINOIDS FOR THE ACTIVATION OF RETINOIC ACID RECEPTORS . **G. Villeneuve**, P. Jones, C-P. Fei, K. T. Nwe, B. Gour, J. Desmarte, D. Martin, T. H. Chan, and B. Leyland-Jones, Department of Chemistry and Department of Oncology, McGill University, Montréal, Québec H3A 2K6

We have determined the ability of two structurally different series of benzamide retinoids to induce differentiation of human and mouse premature cells. The first series is exemplified by 4-[(5,6,7,8-tetrahydro-5,5,8,8-tetramethyl-2-quinoxalinyl)-carboxamido benzoic acid while the second is based on the structure of 4-(b-ionylideneacetamido) benzoic acid. The structure-activity relationship was studied in parallel in both series with particular attention to amide bond modification and effects of substituents on the benzoate ring. The results have been rationalized by conformational studies in the solid state (X-ray crystallography), in solution (NMR), and by molecular modeling. The synthesis, biology, and conformational analysis will be presented and related to that of all-trans-retinoic acid.

130. **SYNTHESIS AND BIOLOGICAL ACTIVITY OF POTENT RXR SPECIFIC RETINOIDS.** **Vidyasagar Vuligonda**,* Peter J. A. Davies, Devendra M. Kochar and Roshantha A. S. Chandraratna.* **Retinoid Research, 2525 Dupont Drive, Allergan Pharmaceuticals, Irvine, CA 92612.**

A series of potent retinoid X receptor (RXR) specific retinoids were designed on the premise that a *cis* substituted cyclopropyl group is a suitable isostere for the 9-*cis* double bond of 9-*cis* RA. These compounds were synthesized employing a stereo/enantioselective cyclopropanation as the key reaction. Compound **1** (racemic) showed high affinity binding and transactivation at the RXRs and weak RAR activity. The biological activity of individual enantiomers of **1** will be discussed. In addtion, 9-*trans* cyclopropyl locked isomers of compound **1** will be described.

131. ETHNOBOTANICAL-DIRECTED DISCOVERY OF THE ANTIHYPERGLYCEMIC PROPERTIES OF CRYPTOLEPINE: ITS ISOLATION FROM *CRYPTOLEPIS SANGUINOLENTA*, SYNTHESIS, AND *IN VITRO* AND *IN VIVO* ACTIVITIES Donald E. Bierer, Diana M. Fort, Larisa G. Dubenko, Christopher D. Mendez, Jian Luo, Michael J. Reed, Patricia Peterli-Roth, R. Eric Gerber, Joane Litvak, Nancy Waldeck, Reimar C. Bruening, Ben K. Noamesi, Richard F. Hector, Thomas J. Carlson, and Steven R. King SHAMAN Pharmaceuticals, 213 East Grand Ave., South San Francisco, CA 94080.

The enormous biodiversity of the tropical rainforest has provided indigenous cultures around the globe with a deep and diversified knowledge about the use of plants for medicinal purposes. As part of our effort to access and record these oral healing traditions before they disappear, we were interested in the bioactive constituent present in *Cryptolepis sanguinolenta* responsible for its observed use for treating associated symptoms of diabetes. *In vivo*-guided fractionation led to the discovery of cryptolepine as an active component. While a number of bioactivities for cryptolepine have been reported, this is the first report demonstrating that cryptolepine can lower blood glucose in rodent models of type II diabetes. The isolation, synthetic approaches, and antihyperglycemic activities of cryptolepine will be reported.

132. ANTIHYPERGLYCEMIC ACTIVITIES OF CRYPTOLEPINE ANALOGUES: AN ETHNOBOTANICAL LEAD STRUCTURE ISOLATED FROM *CRYPTOLEPIS SANGUINOLENTA* Larisa G. Dubenko, R. Eric Gerber, Joane Litvak, Patricia Peterli-Roth, Barbara Sloan, Nancy Waldeck, Jian Luo, Michael J. Reed, and Donald E. Bierer SHAMAN Pharmaceuticals, 213 East Grand Ave., South San Francisco, CA 94080.

Cryptolepine (**1**), a member of the quindoline family of alkaloids, is among a rare class of natural products whose synthesis was reported prior to its isolation from nature. We recently reported, via the patent literature, the discovery of cryptolepine's antihyperglycemic properties. As part of a medicinal chemistry program designed to optimize natural product lead structures originating from our ethnobotanical and ethnomedical field research, we undertook a synthetic analogue program about the cryptolepine nucleus. We will report the synthesis and antihyperglycemic activities of a series of cryptolepine analogues (**2**).

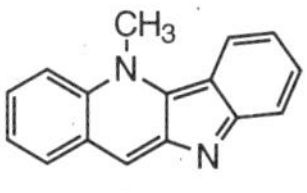

133. SYNTHESES OF 2-(3,5-DIMETHYL-4-METHOXYPYRIDYL)BENZOTHIAZOLIDINE DERIVATIVES AS A POTENTIAL GASTRIC H^+/K^+-ATPASE INHIBITORS. S-H.Yoon, S. K. Seo, Y. H. Lee, S. Y. Choi and S. K. Hwang, Department of Applied Chemistry, Ajou University, Suwon 442-749, KOREA

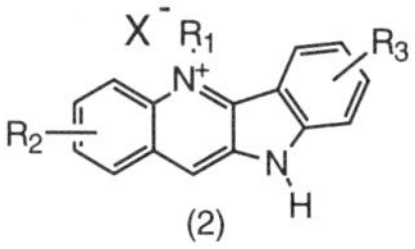

Since the long lasting antisecretory effect of omeprazole results from the formation of tightly bound S-S bond between the thiophilic intermediates and the thiol group in H^+/K^+-ATPase, we considered the benzothiazolidine moiety as a different kind of thiophile, which would afford less tightly bound enzyme inhibition. In modifying the structure of omeprazole, we substituted the sulfinylbenzimidazole moiety of omeprazole with benzothiazolidine one. The syntheses and the pharmacological activities of 2-(3,5-dimethyl-4-methoxypyridyl)benzothiazolidine derivatives as a new class of gastric H^+/K^+-ATPase inhibitors will be presented in detail.

134. SYNTHESIS AND RECEPTOR BINDING STUDIES OF COMPETITIVE AND AFFINITY LABELS OF MELATONIN RECEPTORS. P.K. Li, G.H. Chu, T.N. Parekh, P.A. Witt-Endetby. Graduate School of Pharmaceutical Sciences, Duquesne University, Pittsburgh PA 15282.

The pineal hormone melatonin regulates the variety of functions such as circadian rhythms, seasonal reproduction, visual function and cardiovascular physiology. These effects of melatonin may be mediated through melatonin receptor subtypes (ML1A and ML1B). Thus, potent agonists or antagonists of for melatonin receptors are important to investigate the roles of melatonin in the variety of physiological functions. In addition, a non-reversible affinity label for the melatonin receptor would be useful in purification of the receptor proteins. Recently, it was reported that N-[2-(2-methoxynaphthyl)ethyl]propionamide (structure shown below, R=Me) has the same affinity to the melatonin receptor as melatonin. It was postulated that the methoxy group binds to the hydrophobic pocket at the receptor site. Thus, this study is to focus on designing analogs to probe the size of the hydrophobic pocket and to synthesize affinity labels to alkylate the receptor at the pocket (structures shown below). Most of the compounds have affinities to melatonin receptors at pM range. In addition, preliminary results also indicated that the affinity labels showed time-dependent covalent labeling of melatonin receptor.

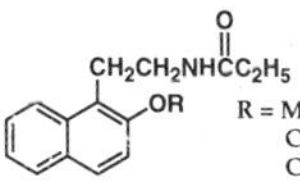

135. SYNTHESIS AND BONE ANABOLIC EFFECTS OF THE ETHYL ESTER OF BW245C AND ANALOGS, SPECIFIC PROSTAGLANDIN D (DP) RECEPTOR AGONISTS. Mitch A. deLong, Cynthia J. Miley, Jack A. Amburgey, Suzette L. Roof, Sherry Pierce, and Mark Lundy, Miami Valley Labs, Procter & Gamble Pharmaceuticals, Cincinnati, OH 45242

Prostaglandins PGE_2 and PGE_1 are potent stimulators of bone formation *in vivo*. Prostaglandin PGD_2 has recently been reported to cause bone growth, but PGD_2 has also been shown to dehydrate rapidly to PGJ_2 *in vivo*. To investigate the bone anabolic effects of the activation of the Prostaglandin D (DP) receptor, we synthesized an analog of the DP-selective compound BW245C, which is not subject to degradation to a J-series prostaglandin. Then we determined that, in osteoclast-like cells (bone-forming cells) the mRNA of the DP receptor was expressed. Finally, using our DP-specific analogs, we demonstrated that selective stimulation of the DP receptor causes an anabolic response in an *in vivo* model of bone growth, the rat local injection model. This implies that the bone growth reported for PGD_2 may be due to its ability to stimulate the prostaglandin DP receptor.

136. FACILE FORMATION OF SUPERELECTROPHILE IN ENZYMATIC REACTIONS. Brajeswar Paul, Grace Cancer Drug Center, Roswell Park Cancer Institute, Buffalo, NY 14263.

We postulated a new concept on the mechanism of action for group transfer reaction of S-adenosyl-L-methionine (AdoMet; Abstr., 211th Nat. Mtg., Am. Chem. Soc., New Orleans, LA, 1996 MEDI 224 and ORGN 214) based on non-bonded atomic contact between the electron pair of the sulfonium and adenosyl C8-H (acidic). This contact makes the sulfonium cation super electron deficient (superelectrophile) and triggers =S—C bond cleavage. Similarly in pyridoxal phosphate the close proximity between phenolic 3-OH (acidic) and the electron deficient 4-CHO (distance 1.6A°) results in a non-bonded atomic contact and enhances the reactivity of the aldehyde group. This facile formation of superelectrophile in enzymatic reactions by non-bonded atomic contact between an acidic hydrogen and an electrophile is in contrast with the report of its formation in superacidic media (HF-SbF_5 etc.) and sometimes under forcing conditions (Angew. Chem. Int. Ed (1993), 32, 767-788). Recently, facile dealkylation via dithiadications was reported (Tetrahedron Lett. (1996) 37, 667-670). This illustrates how the enzymatic reactions are carried out at ambient temperature by fixation of two reacting groups in close geometrical relationship. (Supported by NIH grant CA16056.)

137.
SOLUTION STRUCTURES OF CONSTRAINED β-AMINO ACID OLIGOMERS. Xiaolin Huang[†], Daniel Appella[‡], Laurie Christianson[‡], Daniel Klein[‡], Samuel H. Gellman[‡]*, and Joseph J. Barchi, Jr.[†]*, Laboratory of Medicinal Chemistry[†], Division of Basic Sciences, National Cancer Institute, Bethesda, Md 20892 and Department of Chemistry[‡], University of Wisconsin, Madison, WI 53706

The structures of oligomeric β-amino acids where the two backbone carbon atoms of the amino acid unit were constrained into a cyclopentane ring (compounds **1** and **2**) were studied by 500 MHz NMR spectroscopy and molecular dynamics simulations. It was shown that both **1** and **2** adopted well-defined structures in CD_3OH, DMSO-d_6 and pyridine-d_5 solutions by NOE, ROE and amide exchange data. The data shows that the compounds fold into regular helices defined by 12-membered ring hydrogen bonds between backbone carbonyls and NH groups. The details of the structures and the implications for the design of polymers with the ability to fold into defined secondary structures will be discussed.

1 n = 5
2 n = 7

138. The Syntheses of Gamma Turn Mimetics Containing an Acyl Hydrazine Moiety. Mark D. Ferguson, Joseph Meara, Hiroshi Nakanishi, Min Lee, Michael Kahn, Molecumetics, 2023 120th Avenue NE, Suite 400, Bellevue, WA 98005

Nature has developed several protein secondary structures to account for the molecular recognition events that occur. These secondary structures include beta sheet, beta turn, alpha-helices, and gamma turns. Gamma turns are characterized by a 3 to 1 hydrogen bond. Malformins are naturally occuring pentapeptides derived from Aspergillus niger. It has been claimed that these peptides raised the level of NGF 3-6 times in cell culture supernatants at a concentration of 5mM. Molecular modelling indicates that the use of a seven membered ring is a good mimic of the gamma turn found in these peptides. The design, syntheses, and applications of gamma turn mimetics containing an acyl hydrazine moiety will be discussed.

139. **SYNTHESIS OF TESTOSTERONE DERIVATIVES TO REVEAL THE IMPORTANCE OF THE α, β - UNSATURATED STRUCTURE TO OBTAIN ANTIBODIES.** Y. GUO, Department of Organic Chemistry, bioMérieux, 69280 Marcy, France

The aim of this study was to synthesize steroidal immunogens and then to obtain, from such immunogens, antibodies which are essential to the hormone diagnostic research. Two novel Testosterone derivatives that have no α,β-unsaturated structure in the ring A were synthesized and used to immunization. Result comparison between these two derivatives and other known compounds possessing the α,β-unsaturated structure showed that such a stucture was necessary to prepare antibodies, anti-Testosterone against Position 3 immunogens, with animals. Synthesis methods of the derivatives were discussed.

140. BIOTIN-ANALOG CONJUGATED ANTIBODIES FOR POSITIVE CELL SELECTION AND RELEASE. <u>Yu-An Chang</u>*, Bill Lake, Roy Guillermo, Fahad Al-Abdaly, and Bill Scouten@, Baxter Healthcare Corporation, Immunotherapy Division, 9 Parker, Irvine, CA 92618, @Utah State University, Biotechnology Center, Logan, UT 84322

Several cell selection techniques have been studied and developed for cancer adjunctive therapies, such as CD34+ stem cells from peripheral blood, cord blood, or bone marrow. Positive cell selection technologies are the foundation of future cell therapies. In order to recover the captured target cells, biotin analog conjugated antibodies were investigated and developed into an efficient positive cell selection system. It has been demonstrated that this system is capable of producing not only high recovery yield (78%) of the desired cells such as CD34+ cells but also with very high purity (80%).
Chemistries of the preparation of biotin analog conjugated anti-bodies, factors which are crucial for the optimization of this system, and the cell selection results will be presented.

141. **SYNTHESIS AND EVALUATION OF WATER SOLUBILIZED BIFUNCTIONAL AND TRIFUNCTIONAL BIS-MALEIMIDE PROTEIN CROSS-LINKING REAGENTS.** D. Scott Wilbur, <u>Pradip M. Pathare</u>, Donald K. Hamlin, S. Ananda Weerawarna, Robert L. Vessella, Patrick S. Stayton, and Richard To. Depts. of Radiation Oncology and Urology, and Center for Bioengineering, University of Washington, Seattle, WA 98195.

Several new bis-maleimide protein cross-linking reagents have been synthesized for studies involving cross-linked monoclonal antibodies and a streptavidin mutant. Because of difficulties with water solubility of some bis-maleimides prepared, water solubilizing linkers were employed in subsequent derivatives. A bifunctional bis-maleimide cross-linking reagent, **1**, was readily prepared by reaction of 4,7,10-trioxa-1,13-tridecanediamine with N-methoxycarbonylmaleimide. Trifunctional bis-maleimide cross-linking reagents, **2** and **3**, were prepared in syntheses employing aminoisophthalic acid or benzene tricarboxylate. Cross-linking reagents **2** and **3** were substituted with R = *p*-iodobenzamide; R = *p*-tri-*n*-butylstannylbenzamide; or R = biotinamide. Protein cross-linking reactions using these compounds will be described.

142. CHARACTERIZATION OF THE THERMAL DENATURATION OF RECOMBINANT HUMAN GROWTH HORMONE. Isabel Gomez-Orellana, Sam Milstein and Duncan. R. Paton. Emisphere Technologies, Inc. 15 Skyline Drive, Hawthorne, NY 10532.

We have previously reported the development of a new technology for the oral delivery of protein drugs. The studies included in these reports have shown that a series of small organic molecules can promote the delivery of a protein to the bloodstream when the protein and a delivery agent are administered together. These results have led to much speculation on the mechanism by which delivery is effected, including penetration enhancement, enzyme inhibition, and/or the formation of intermolecular complexes. The studies to date have not shown untoward membrane damage, and the delivery agents are generally poor inhibitors of proteases.

The possibility of a delivery agent/protein interaction coupled to a partial unfolding of the protein is particularly interesting because of the growing evidence that protein penetration and translocation across membranes involves changes in protein conformation. We have used differential scanning calorimetry to study the thermal unfolding of recombinant human growth hormone (rhGH) and the effect of the delivery agents on the population distribution of the protein conformations. Our results show that the delivery agents stabilize an intermediate conformation of rhGH, suggesting the involvement this intermediate in the delivery mechanism.

143. LYMPHOMA-SELECTIVE ANTIBODY LYM-1 RECOGNIZES A DISCONTINUOUS EPITOPE ON THE LIGHT CHAIN OF HLA-DR10. Larry M. Rose, Angelo H. Gunasekera, Sally J. DeNardo, Gerald L. DeNardo, Claude F. Meares, Department of Chemistry, University of California, Davis, California, 95616.

The selectivity of Lym-1 for malignant B lymphocytes makes this monoclonal antibody a promising candidate for the delivery of toxic agents to malignant B cells. The original immunogen used for the development of Lym-1 was Raji Burkitt's lymphoma cell nuclei. The Lym-1 antigen was characterized at that time as a polymorphic HLA-DR variant. We prepared an affinity column using immobilized Lym-1 to isolate the Lym-1 antigen from Raji cell lysate. Immunological characterization of the immunoaffinity-purified Lym-1 antigen on Western blots led to the conclusion that the antigen is the beta chain of HLA-DR10. This was confirmed by Edman sequencing of the isolated polypeptide chain. Western blots further show that the Lym-1 epitope is only recognized if the beta chain disulfide bonds are intact. These results imply that Lym-1 binds a discontinuous epitope on the beta chain of HLA-DR10.

144. **POTENT CYCLIC UREA HIV PROTEASE INHIBITORS WITH INDAZOLE P2/P2' GROUPS.** James D. Rodgers, Haisheng Wang, Barry L. Johnson, Patrick Y. Lam, Yu Ru, George V. De Lucca, Ui Tae Kim, Soo S. Ko, George L. Trainor, Ron M. Klabe, Beverly C. Cordova, Lee T. Bacheler, Susan Erickson-Viitanen, Gilbert N. Lam and Chong-Hwan Chang. Dupont Merck Pharmaceutical Company, P.O. Box 80500, Experimental Station, Wilmington, DE 19880

We have discovered that cyclic ureas containing indazole P2/P2' groups are extremely potent inhibitors of HIV protease. In addition, the potent enzyme inhibitory activity associated with the indazole group translates exceptionally well to antiviral activity in the cell-based assay. The symmetrical bis-indazole XR835 is an extremely potent inhibitor of HIV protease (K_i = 0.018 nM) with excellent antiviral activity (IC_{90} = 8 nM). This presentation will discuss potent cyclic urea HIV protease inhibitors with indazole P2/P2' groups as potential clinical candidates.

XR835 K_i = 0.018 nM
IC_{90} = 8 nM

145. ABT-627 AND BEYOND: THE DISCOVERY OF ET_A-SELECTIVE, ET_B-SELECTIVE, AND BALANCED ENDOTHELIN ANTAGONISTS. Thomas W. von Geldern, Pharmaceutical Products Division, Abbott Laboratories, Abbott Park, IL 60064-3500

Pharmacophore analysis led us to identify several critical features common to small-molecule endothelin antagonists, and to incorporate these features into a novel framework, a 1,2,4-trisubstituted pyrrolidine-3-carboxylic acid. The ease of synthesis and ready modification of this framework allowed rapid optimization of biochemical and pharmacokinetic properties, and resulted in clinical candidate ABT-627. Additional structure-activity profiling has identified subtle structural modifications which have dramatic effects on receptor binding selectivity. Followup studies have led to the development of both ET_B-selective and balanced antagonists, as well as to compounds with greater ET_A-selectivity than ABT-627.

146. A Novel Series of ET_A Selective Antagonists

Jia-Ning Xiang, Steven T. Atkinson, Dimitri Gaitanopoulos, Deborah Bryan, Ponnal Nambi, Eliot H. Ohlstein, Joseph Weinstock and John D. Elliott

SmithKline Beecham Pharmaceuticals, King of Prussia, Pennsylvania 19406 U.S.A.

The endothelins are a family of three potent vasoconstrictor peptides (ET-1, 2 and 3) encoded in the human genome and each composed of 21 amino acids, with disulfide linkages between cysteine residues 1,15 and 3,11. The ETs elicit their effects through binding to receptors of the G-protein coupled seven-transmembrane spanning superfamily, and two receptor subtypes (ET_A and ET_B) have been fully characterized from human tissues through molecular cloning and expression. Recently our laboratory has disclosed the design and synthesis of a series of indane-based dual antagonists of the ET_A and ET_B receptors. Herein we describe the results of structure-activity relationship studies on a new series of potent ET_A selective antagonists based upon the 2-benzyl-3-imidazolylpropenoic acid framework.

147. DISCOVERY OF ASCOMYCIN-DERIVED IMMUNOSUPPRESSANT A-86281. A NOVEL AGENT WITH POTENT TOPICAL BUT WEAK SYSTEMIC ACTIVITY.

Jay R. Luly, Yat Sun Or, Megumi Kawai, Rolf Wagner, Paul E. Wiedeman, Richard Clark, Indrani Gunawardana, Hana Kopecka, Teresa Rhoades, Cynthia Henry, Noah Tu, Nwe BaMaung, Ben C. Lane, Kennan Marsh, Yung-Wu Chen, Gin C. Hsieh, George W. Carter, Karl W. Mollison. Abbott Laboratories, Pharmaceutical Products Division, Immunoscience Research, Abbott Park, IL 60064

The immunosuppressant tacrolimus (FK506), an ascomycin analogue, shares with cyclosporine a similar mechanism of action, good oral efficacy in transplantation and inflammatory/autoimmune diseases, and, unfortunately, similar major toxicities. In an effort to minimize potential side effects of this class, we synthesized several series of ascomycin analogues to identify compounds with potent topical immunosuppressive activity, but less systemic immunosuppression following absorption. *In vitro*, A-86281 (ABT-281) potently inhibits T cell activation, and *in vivo* shows topical potency comparable to FK506 (ED50's 0.9% and 0.6%, respectively) in a swine contact hypersensitivity model. In contrast, A-86281 displays a more rapid disappearance from the circulation, and is a markedly weaker systemic immunosuppressant in multiple animal models. The A-86281 profile of potent topical but weak systemic activity should be advantageous for minimizing systemic immunosuppression and consequently the mechanistically-related toxicities associated with immunophilin-dependent immunosuppressants.

148.

THE EVOLUTION AND SYNTHESIS OF ELETRIPTAN - A SELECTIVE "5-HT1D-LIKE" RECEPTOR PARTIAL AGONIST WITH RAPID ORAL ABSORPTION. Martin J. Wythes, David Brown, Paul Butler, Michael D. Closier, Roger P. Dickinson, Peter Ellis, Paul Gupta, Keith James, John E. Macor, Paul Morgan, David J. Rance, James W. Rigby, Nicholas B. Shepperson. Departments of Discovery Chemistry, Discovery Biology and Drug Metabolism, Pfizer Central Research, Sandwich, Kent. CT13 9NJ, U.K.

The chemical evolution and synthesis of eletriptan, a selective "5-HT1D-like" receptor partial agonist with rapid oral absorption, will be described for the first time. Eletriptan is currently in Phase III clinical trials for migraine. Our preclinical objectives were met by optimising lipophilicity and stability within a series of indoles bearing relatively lipophilic sulphone moieties at C-5, and conformationally restricted 2-(R)-pyrrolidinylmethyl groups at C-3. The key steps along the evolutionary pathway of eletriptan will be discussed, encompassing synthesis, in vitro SAR, physicochemical/pharmacokinetic parameters, and in vivo profiles. Clinical data show that in man eletriptan is rapidly orally absorbed, it has a relatively long half-life, good bioavailability, and it rapidly relieves migraine headaches following oral administration.

149. 1997 ACS Award for Team Innovation sponsored by The Corporation Associates

"The Discovery of COZAAR®†, Losartan Potassium. The First Angiotensin II Antagonist for the Treatment of Hypertension"

Awardees: John V. Duncia, David J. Carini, and Pancras C. Wong
The Du Pont Merck Pharmaceutical Company, P.O. Box 80500, Wilmington, DE 19880-0500

COZAAR®, losartan potassium, otherwise known as DuP 753, and MK-954, is the first high blood pressure drug with a novel mechanism of action to be approved and introduced to the world-wide marketplace in over a decade. It was first synthesized in 1986 at DuPont and represents the first potent, orally active, nonpeptide angiotensin II antagonist.

A brief history behind its discovery as well as various issues encountered in its preclinical development will be presented. †Registered trademark of the E.I. du Pont de Nemours Co., Wilmington, DE.

COZAAR®, losartan potassium, DuP 753, MK-954

150.

THE DEVELOPMENT OF COZAAR® (LOSARTAN POTASSIUM)†, THE FIRST ANGIOTENSIN II RECEPTOR ANTAGONIST FOR HYPERTENSION.
MICHAEL R. GOLDBERG, RONALD S. EYDELLOTH; MERCK RESEARCH LABORATORIES, WEST POINT, PA 19486.
The AT_1-selective angiotensin II receptor antagonist, losartan, has been studied extensively in animals and humans to investigate its safety, efficacy, specificity of action, and long-term effects. Critical issues included long-term effects in animal toxicity studies, clinical differentiation from inhibitors of angiotensin converting enzyme, and effects on endogenous mediators of the renin-angiotensin-aldosterone system. Preclinical Safety studies and the results of clinical studies which address these issues will be presented.

†Trademark owned by E.I. duPont deNemours & Company.

151. SCIENCE, ART AND DRUG DISCOVERY - A PERSONAL PERSPECTIVE
Simon F Campbell, Pfizer Central Research, Sandwich, Kent, UK, CT13 9NJ

Tremendous progress has been made over the last 50 years in developing rational approaches to drug design as the scientific principles underlying many biological processes have become better understood. On the other hand, the three-dimensional shapes of most proteins targeted by medicinal chemists have not been defined, and there is a fine art in imagining how biologically-active molecules exert their effects within an unknown environment. This Award Address will provide a personal perspective on the discovery of some novel therapeutic agents which have emerged from our Sandwich laboratories over the past twenty-five years, and will illustrate synergies between rational and intuitive approaches. Particular attention will be paid to amlodipine (Norvasc^R) and doxazosin (Cardura^R) which have proved major clinical and commercial successes, but other agents at earlier stages in development will also be discussed. Finally, some thoughts on how the drug discovery process might evolve over the next decade will be offered.

152. THERMODYNAMIC SOLVENT ISOTOPE EFFECTS AND MOLECULAR HYDROPHOBICITY.
Eric J. Toone, Department of Chemistry, Duke University, Durham NC 27708-0346

The thermodynamic solvent isotope effect, or enthalpy of transfer from light to heavy water, has long been recognized as a useful tool for understanding the molecular basis of hydrophobicity. The recent development of commercial microcalorimeters has led to renewed interest in the field in the context of biologically relevant processes. We will present recent results aimed at understanding the role of structure in determining overall hydrophobicity, the role of hydrogen bonding in the stabilization of non-covalent complexes, and the underlying molecular origin of the hydrophobic effect

153. CALORIMETRY AS A TOOL FOR LOOKING AT STRUCTURAL CHANGES AND LINKED BINDING. Kenneth P. Murphy, Department of Biochemistry, University of Iowa, Iowa City, IA 52242

The ability to use structural information for the prediction of the thermodynamics of a protein-ligand interaction is a long-standing goal of biophysical chemistry. Recently, we have developed an empirical approach which is able to calculate accurately the thermodynamics of protein unfolding (i.e. $\Delta H°$, $\Delta S°$, and ΔC_p) based on changes in accessible surface area. This approach is also being applied to predicting the thermodynamics of protein-ligand interactions. Given the close relationship between energetics and changes in surface area, the converse is also feasible; it is possible to estimate structural changes from well-determined thermodynamics of binding. This talk will present the basic approach and provide examples of its application, as well as presenting possible pit-falls in the application of these calculations, especially those that can arise from linked-binding events.

154.

MICROCALORIMETRIC STUDIES OF THE BINDING OF SMALL MOLECULES TO TARGET PROTEINS. James A. Thomson, Department of Biophysics, Agouron Pharmaceuticals, Inc., San Diego, California 92121

The binding of several small molecules to FK506 binding protein has been studied by isothermal titration calorimetry. Owing to the low solubility of the inhibitors sample solutions were prepared by addition of small aliquots of inhibitor stock solutions in dimethyl sulfoxide or methanol to an appropriate buffer. To minimize the heat of dilution a comparable amount of dimethyl sulfoxide or methanol was added to the protein solution. The consequences of the addition of the cosolvents will be presented.

155. DETERMINATION OF TIGHT LIGAND-PROTEIN AFFINITIES BY THERMAL STABILITY PERTURBATION. M. L. Doyle, Department of Macromolecular Sciences, SmithKline Beecham Pharmaceuticals, King of Prussia, PA 19406-0939

One of the virtues of thermodynamics is the pathway-independence of state functions such as the Gibbs energy of binding. This enables free energy changes that are difficult/impossible to measure to be determined indirectly by constructing an experimentally accessible thermodynamic cycle relating reactants and products. It has been known for some time (reviewed recently by Brandts & Lin (1990) Biochemistry 29, 6927) that ligand binding affinities of proteins can be determined indirectly from protein thermal stability measurements in the presence and absence of ligand. This approach is especially useful for extremely tight binding which is otherwise difficult to measure. Here we apply this strategy using calorimetery to characterize the following sub-picomolar affinities: 1) a protease-proteinaceous inhibitor interaction, and 2) interactions of a bacterial tRNA synthetase with a series of drug candidate compounds.

156. EXPLORING ALLOSTERIC MULTI-VALENT MECHANISMS FOR RECEPTOR DIMERIZATION THROUGH MICROCALORIMETRY.
Michael W. Pantoliano & Barry A. Springer, 3-Dimensional Pharmaceuticals, Inc., Exton, PA 19341.

Many cytokines and growth factors display multi-valent binding sites for their respective cell surface receptors, and initiate transmembrane signaling events through the subsequent dimerization or higher order aggregation of these receptors. These binding reactions are often comprised of complex allosteric binding mechanisms that are difficult to study using conventional methods. Isothermal titrating calorimetry (ITC) in combination with site directed mutagenesis, however, were found to be useful tools for dissecting out the individual binding interactions for the three primary reactants of the fibroblast growth factor (FGF) system: basic FGF (bFGF), the FGF receptor, FGFR1, and the cofactor, heparin/heparan sulfate (HS). These studies suggested that bFGF and HS participate in a concerted bridge mechanism for the dimerization of FGFR1. Moreover, ITC assisted in the characterization of the secondary FGFR binding site on bFGF, thereby allowing a quantitative means for assessing the future redesign of bFGF into a mono-valent receptor antagonist.

157. THE MULTIPLE ROLES OF CALORIMETRY IN A STRUCTURE-BASED DRUG DESIGN PROGRAM, P. R. Connelly, Connelly Scientific, Lower Burrell, PA 15068

Microcalorimetric methods have played important roles in three key areas of structure-based drug design technology: theoretical lead compound generation, lead compound optimization, and assay technology. Relevant to the areas of lead generation and optimization, I will first discuss the utilization of calorimetric and computational methods in dissecting the energetics of formation of hydrogen bonds and hydrophobic interactions in a small molecule / protein binding reaction. Secondly, a demonstration of the utility of calorimetric analysis for revealing novel properties of therapeutically important protein targets will be detailed through an account of the discovery of the allosteric properties the immunosuppressive target, inosine monophosphate dehydrogenase. The implications of this discovery for the development of immunosuppressants that overcome the "Differential Molecular Recognition Problem" will be discussed. Finally, the topic of Field Gradient Analysis, a new patented, calorimetric assay technology that can be used to measure the thermodynamic and kinetic properties of biological macromolecules, will be introduced.

158. ANTIPSYCHOTICS BLOCK TYPE 2 AND TYPE 4 DOPAMINE RECEPTORS

Philip Seeman, Pharmacology Department, Medical Science Building, Room 4344, University of Toronto, Toronto, Canada M5S 1A8

Hallucinations are blocked when any antipsychotic drug (clozapine included) occupies >70% of D2 dopamine receptors in the brain. Antipsychotics also block D2 in the brain motor-controlling regions, yielding Parkinsonism. Atypical antipsychotics (clozapine, olanzapine) elicit little Parkinsonism. To develop atypical antipsychotics, it is helpful to have accurate dissociation constants (K) at various receptors. We found that the inhibition constant (Ki) of a competing antipsychotic depended on the radioligand partition coefficient. After obtaining antipsychotic Ki's using different radioligands, we extrapolated to "zero" ligand partition, yielding the real (or radioligand-independent) K of the antipsychotic. Clozapine's K of 1.6 nM at D4 agreed with that for [^{3}H]clozapine at D4. The atypical neuroleptics, remoxipride, clozapine, perlapine, seroquel and melperone, have low affinity for the D2 receptor with K = 30-90 nM, making these drugs displaceable by endogenous dopamine. Typical neuroleptics have K = 0.3-5 nM at D2, making them less displaceable by dopamine. Omitting the five antipsychotics with low affinity for D2, a relation was found between the antipsychotic doses (20 drugs) for rat catalepsy and the D2/D4 ratio of the K values. Thus, atypical neuroleptics have either low affinity for D2 or are selective for D4. D4 antagonist/agonist action may clinically supplement D2 block for antipsychotic action.

159. 2-PHENYL-4(5)-(1-[4-(PYRIMIDIN-2-YL)PIPERAZIN-1-YL])METHYL IMIDAZOLE AND RELATED COMPOUNDS AS SELECTIVE ANTAGONISTS AT CLONED HUMAN D4 RECEPTORS. Andrew Thurkauf, Neurogen Corporation, 35 NE Industrial Road, Branford, Connecticut 06405.

It has been suggested that the atypical nature of clozapine is due, at least in part, to its affinity for dopamine D4 receptors. In contrast, the benzamide remoxapride possesses a behavioral profile characteristic of an atypical antipsychotic yet has very low D4 affinity and modest D2 affinity relative to other related benzamides such as sulpiride. The unique profile of remoxipride led us to a reexamination of the benzamides as a starting point for the identification of new atypical antipsychotic agents. This talk will describe the identification of the title compounds as D4 selective ligands through the progressive and systematic modification of the benzamides.

160. QUINOLINYLMETHYL-PIPERAZINES HAVING POTENT AND SELECTIVE AFFINITY FOR THE DA D4 RECEPTOR. D. Wustrow, T. Belliotti, J. Blankly, T. Heffner, R. Kennedy, T. Pugsley, L. Wise. *Parke-Davis Pharmaceutical Research Division, Warner-Lambert Company, Ann Arbor MI, 48105.*

Unlike the dopamine (DA) D2 receptors which are present in many brain regions, the DA D4 receptors are located primarily in the hippocampus and cortical regions. These brain areas are thought to be involved in the etiology of schizophrenia. To understand the function of DA D4 receptors in the CNS and their possible role in schizophrenia, selective ligands for this receptor are needed. This talk will highlight the use of a combination of parallel synthesis and traditional medicinal chemistry techniques to optimize the mass screen lead PD 089232. The neurochemical and behavioral effects of selective D4 antagonists in this series will be discussed as it relates to the possible role of the D4 receptor in the CNS and the potential utility of D4 antagonists as antipsychotic agents.

PD 089232

161. **SYNTHESIS, SAR AND PRECLINICAL PROFILE OF PYRIDO[1,2-a]PYRAZINE DOPAMINE D4 RECEPTOR ANTAGONIST CP-293,019.** M. A. Sanner, A. R. Dunaiskis, T. A. Chappie, A. Fliri, S. H. Zorn, C. G. Johnson, E. R. Jackson, J. M. Morrone, P. A. Seymour, M. J. Majchrzak, T. F. Seeger, D. G. Costello, S. Faraci, J. L. Collins, J. R. de Wet, J. P. Holland, J. S. Lee, D. B. Duignan, J. F. Blake, H. Rollema. Pfizer Central Research, Eastern Point Rd., Groton, CT 06340

The discovery that the atypical antipsychotic clozapine has high affinity for the D4 receptor touched off a new phase of antipsychotic research. Starting from a moderately selective screening hit, analoging in a series of pyrido[1,2-a]pyrazines led to CP-293,019, a potent, selective human D4 receptor antagonist (D4 Ki = 3.4 nM, D2 Ki > 3,310 nM). Microdialysis and c-fos experiments indicate CP-293,019 activity in the prefrontal cortex but not striatum. The combination of *in vitro* and *in vivo* properties suggest that a dopamine D4 receptor antagonist such as CP-293,019 may be an effective antipsychotic free of extrapyramidal side effects.

CP-293,019

162. HETEROARYL PIPERIDINES AND PIPERAZINES AS SELECTIVE hD4 ANTAGONISTS Michael Rowley, Merck Sharp and Dohme, Terlings Park, Harlow, Essex CM20 2QR, UK

Schizophrenia is a mental illness for which there is still a great need for novel drug therapy. Classical neuroleptics, which are presumed to act by antagonism of dopamine D2 receptors, are useful for the treatment of the positive symptoms of schizophrenia, but cause major movement disorders which are thought to be due to blockade of D2 receptors in the striatum. With the recent cloning of 5 dopamine receptor subtypes, and the finding that the atypical neuroleptic Clozapine has higher affinity for hD4 than hD2 receptors, we became interested in selective hD4 antagonists as possible novel neuroleptics. We have described the development of piperidinyl piperidines as high affinity selective hD4 antagonists, but these compounds suffer from affinity at a number of ion channels. In order to address this problem, various approaches were taken. The most successful was to reduce the pKa of the basic nitrogen with a series of piperazinylpyrazoles. Further conformational restraints helped to restore the high affinity.

163. S-(-)-4-[4-[2-(ISOCHROMAN-1-YL)ETHYL]PIPERAZIN-1-YL]BENZENESULFONAMIDE, A HIGH AFFINITY, SELECTIVE ANTAGONIST FOR THE HUMAN DOPAMINE D_4 RECEPTOR. Ruth E. TenBrink, Carol L. Bergh, J.Neil Duncan, Douglas W. Harris, Rita. M. Huff, Robert A. Lahti, Barry S. Lutzke, Iain J. Martin, John C. Sih, and Martin W. Smith

Overactivity of the dopamine system in the brain has been proposed a factor in disease states such as schizophrenia. Current drugs such as haloperidol and clozapine, which function as dopamine antagonists, serve to block the altered thought processes associated with psychotic diseases. Clozapine is an atypical antipsychotic (lacking in motor side effects) which under therapeutic conditions (measured plasma or CSF levels) binds at a higher percentage of dopamine D4 than dopamine D2 receptors. This report summarizes the synthesis and structure/activity relationships of a series of isochromanyl aryl piperazines with high affinity and selectivity for the dopamine D4 receptor. The key intermediates 1-(2-

chloroethyl)isochroman and ethyl 2-(isochroman-1-yl)acetate were prepared via a Lewis acid-mediated ring closure. Resolution of 2-(isochroman-1-yl)acetic acid was achieved via salt formation with R-(+)-α-methylbenzylamine or by a lipase-mediated hydrolysis of ethyl 2-(isochroman-1-yl)acetate. (S)-(-)-4-[2-(Isochroman-1-yl)ethyl]piperazin-1-yl]benzenesulfonamide (PNU-101387) exhibits nanomolar affinity and high selectivity for the human D4 dopamine receptor. PNU-101387G (methanesulfonate salt) is currently undergoing clinical trials for the treatment of schizophrenia.

PNU-101387G

164.
PATENT THE WHEAT, NOT THE CHAFF: MEETING CHALLENGES IN OBTAINING PATENT PROTECTION FOR INVENTIONS FOR COMBINATORIAL CHEMISTRY. Caldwell, John W., Woodcock Washburn Kurtz, Mackiewicz & Norris, 1 Liberty Place, 46th Floor, Philadelphia, PA 19103

Utility, enablement and appropriateness of claim scope must be demonstrated to the United States Patent and Trademark Office in order to be accorded patent protection for a combinatorial process or scheme for obtaining or exploiting chemical diversity. Libraries, diversity methods and eventual therapeutic regimes are often met with skepticism. Some approaches can succeed - some do not. Not all types of inventions have business significance. owning a legal and business tool without having a use for the tool is impractical.

165.
STRATEGIES FOR OBTAINING PATENT PROTECTION FOR LIBRARIES OF DIVERSE COMPOUNDS. Seidman, Stephanie, Brown Martin Haller & McClain, 1660 Union Street, San Diego, CA 92101

A fundamental tenet of the US patent laws is the requirement that the patent teach how to make and use the subject matter that is claimed. Since libraries by their nature contain a large number of compounds, obtaining patents that protect these libraries is challenging. Strategies for protecting libraries are presented

166. OBTAINING PATENT PROTECTION FOR PHARMACEUTICAL INVENTIONS DERIVED THROUGH MOLECULAR DIVERSITY TECHNIQUES. Barry Elledge, Ph.D., J.D., Intellectual Property Law Office, 466 Monroe Drive, Palo Alto, CA 94306

Combinatorial chemistry and other molecular diversity techniques are frequently employed to identify new pharmaceutical lead compounds. However, these compounds often present particular difficulties when patent protection is sought. In comparison with molecules generated by conventional pharmaceutical synthesis programs, the products of combinatorial chemistry research protocols are likely to be poorly characterized: their structure frequently is presumed rather than established instrumentally; synthetic products are unpurified mixtures; concentrations of the active species are not determined; bioactivity measurements are performed on the impure mixtures; and biological activity often is deduced from receptor binding rather than biological assays. Moreover, often only a small number of active species of diverse structure are identified. This limited range of information restricts the scope of patent protection, both for particular active species and for the library of compounds as a whole. Strategies to maximize the available patent protection will be discussed.

166A   U.S. PATENT OFFICE POSITION ON COMBINATORIAL CHEMISTRY. John Kite, III, U.S. Patents & Trademarks Office, Arlington, VA 22202

(No Abstract Available)

167. SYNTHESIS AND BIOLOGICAL EVALUATION OF HIMBACINE AND ANALOGS. Samuel Chackalamannil,* Robert J. Davies, Darío Doller Yuguang Wang, Theodros Asberom, Daria Leone, Robert McQuade, and Vilma Ruperto, *Schering-Plough Research Institute, 2015 Galloping Hill Road, Kenilworth, NJ 07033* ; Andrew T. McPhail, *P.M. Gross Chemistry Laboratory, Duke University, Durham, NC 27708-0346*

Himbacine (**1**) is a tetracyclic piperidine alkaloid isolated from the bark of the Australian pine tree of *Galbulimima* species. It has attracted considerable attention due to its promising biological property as a selective muscarinic receptor antagonist. Enhancement of synaptic acetylcholine levels by selective inhibition of presynaptic muscarinic receptors is a promising therapeutic approach for the treatment of senile dementia associated with Alzheimer's disease. Himbacine is a potent inhibitor of the muscarinic receptor of M_2 subtype with 10 to 20-fold selectivity toward other receptors. In the context of our efforts to develop potent and selective muscarinic receptor antagonists, we have developed a general and practical approach for the synthesis of himbacine analogs. We wish to report the syntheses of (+)-himbacine (**1**) and a number of its analogs using a highly convergent and practical approach. Additionally, a systematic evaluation of the structure-activity relationship and selectivity of these compounds against various muscarinic receptors will be presented.

168. ARYLSULFONAMIDE AGONISTS OF THE HUMAN β3 ADRENERGIC RECEPTOR FOR THE TREATMENT OF OBESITY. A. E. Weber, R. J. Mathvink, J. E. Hutchins, L. M. Perkins, E. R. Parmee, H. O. Ok, M. R. Candelore, P. P. Vicario, M. A. Cascieri, C. D. Strader, G. J. Hom, M. J. Forrest, D. E. MacIntyre, R. F. Alvaro, J. Kao, M. J. Wyvratt, Jr., M. H. Fisher, *Merck Research Laboratories, Rahway, NJ 07065*

Activation of β3 adrenergic receptors (β3 ARs) on the surface of adipocytes leads to an increase in metabolic rate, thus stimulation of these receptors with a selective agonist is an attractive approach for the treatment of obesity. Using a cloned human receptor assay, we have identified a family of arylsulfonamides, typified by L-755,507, which are potent and selective β3 AR agonists. The *in vitro* and *in vivo* activity of these derivatives as well as their pharmacokinetic profile will be discussed.

169. SYNTHESES AND β-ADRENERGIC ACTIVITIES OF R- AND S-FLUORO-NAPHTHYLOXYPROPANOLAMINES. Adeboye Adejare,* Stacey A. Deal and Sophie S. Sciberras, *Division of Pharmaceutical Sciences, School of Pharmacy, University of Missouri-Kansas City, Kansas City, MO 64110*

Many biogenic amines where an aromatic proton is substituted with fluorine have exhibited pharmacological properties that are dependent on the position of fluorine on the aromatic ring. For example, 6-fluoronorepinephrine is selective for α adrenergic receptors whereas the 2-fluoro is selective for β-receptors. Furthermore, we have reported syntheses of fluorophenoxypropanolamines and their selectivities as β-receptor agonists. The current report is on syntheses of an enantiomeric pair that are expected to be antagonists. Starting with fluoronaphthalene, the steps involved introduction of the 4-hydroxy group by Friedel-Crafts acylation followed by Baeyer Villiger oxidation. The side chain for each enantiomer was introduced using the Sharpless epoxidation technique followed by opening of the side chain epoxide with t-butylamine. HPLC methods were used to characterize %ee of the enantiomers. Binding activities of the compounds on β-adrenergic receptors were evaluated. Partial support: NIMH.

170. SELECTIVE DOPAMINE D_4 RECEPTOR LIGANDS. SYNTHESIS AND SAR OF 1-ARYL-4-ARYLETHYLPIPERAZINE DERIVATIVES. Kerry A. Cleek[•], Martin G. Smith[†], Charles F. Lawson[†], Siusaidh K. Schlachter[†], John C. Sih[‡], Connie G. Chidester[•], Chiu-Hong Lin[•], [•]*Structural, Analytical & Medicinal Chemistry,* [†]*Chemical & Biological Screening,* [‡]*Chemical Process R&D, Pharmacia & Upjohn, Kalamazoo, MI 49001*

Classical antipsychotic agents used in the treatment of schizophrenia suffer from serious, sometimes irreversible extrapyramidal side-effects (EPS) which have been attributed to indiscriminate binding. Due to the high localization of the dopamine (DA) D_4 receptor in areas of the brain associated with cognitive and emotional functions (such as the cortex and limbic regions), a selective DA D_4 antagonist may afford an atypical neuroleptic devoid of EPS and other adverse side effects that current therapies possess. The compound PNU-96970 (**I**, n = 2; X = O; R = OCH_3) was identified as a potent D_4

preferring ligand. In order to investigate the SAR of this molecule, compounds which varied the benzoxan moiety as well as the substituted phenyl piperazine (**I** and **II**) were synthesized and evaluated. Racemic analogs that demonstrated potent and selective D_4 activity were synthesized in enantiomerically pure form utilizing an enzymatic resolution of an ester intermediate in >99% ee. The (S)-enantiomer PNU-106161 (**I**, n = 2; X = CH_2; R = $SONH_2$) was identified as a potent and selective DA D_4 ligand demonstrating a Ki of 8 nM at the DA D_4 receptor and Ki's >1184 nM for other monoaminergic receptors tested.

171. SYNTHESIS AND BIOLOGICAL EVALUATION OF FLUORINE-18 LABELED RS-15385-197 ANALOGS: POTENT AND SELECTIVE ALPHA-2 ADRENERGIC RECEPTOR RADIOLIGANDS FOR PET. Joel D. Enas*, Robin D. Clark[†], Henry F. VanBrocklin and Thomas F. Budinger, Center for Functional Imaging, Lawrence Berkeley National Laboratory, University of California, Berkeley, CA 94720 and [†]Roche Bioscience, Palo Alto, CA 94304

Aberrations in the α_2-adrenergic receptor system have been implicated in a number of disease states including hypertension, drug abuse, depression, and neurodegenerative disorders such as Alzheimer's Disease. RS-15385-FP (**1**) and RS-15385-FPh (**2**) are analogs of the α_2-adrenergic receptor antagonist RS-15385-197 which display a high receptor binding affinity (K_i = 0.2 and 0.5 nM, respectively) as well as a high degree of α_2/α_1 selectivity (7000:1 and 2000:1, respectively). We synthesized [F-18]-**1** starting from the corresponding hydroxypropyl sulfonamide which was converted to the mesylate, followed by nucleophilic displacement with [F-18]fluoride. [F-18]-**2** was synthesized by fluoro-for-nitro exchange on the corresponding nitrophenyl derivative which was produced in two steps from the hydroxypropyl sulfonamide. In vivo distribution studies in rats and PET studies in monkeys demonstrate uptake in α_2-adrenergic receptor rich regions of the brain, particularly the locus coeruleus.

172. NEUROIMMUNOPHILINS AS NOVEL TARGETS FOR CNS DRUG DISCOVERY G.S. Hamilton,* D. T. Ross, H. Valentine, H. Guo, M.A. Connolly, S. Liang, C. Ramsey, J. Li, W. Huang, L. Wei, P. Howorth, R. Soni, M. Fuller, H. Sauer, A. Nowotnick, P.D. Suzdak and J.P. Steiner, Guilford Pharmaceuticals, Inc., 6611 Tributary Street, Baltimore, MD 21224.

The immunophilins FKBP and cyclophilin are "receptors" for the immunosupressant drugs FK506 and cyclosporin A, respectively. Recently it has been discovered that that immunophilins are enriched 10-40 fold more in the brain than in the immune tissues. Within neural tissues, they influence neuronal process extension, and may have potential therapeutic applications in neurodegenerative disorders. We have prepared novel small molecule neuroimmunophilin ligands which possess neurotrophic actions, but are devoid of immunosuppressive activity. Compounds such as GPI 1046 potently promote neurite outgrowth in these cultures at picomolar concentration. We have evaluated the in vivo efficacy of these compounds in an animal model of Parkinson's Disease. Using N-methyl-4-phenyltetrahydropyridine (MPTP) to induce lesioning of dopaminergic neurons, we found that systemic treatment of lesioned mice with GPI 1046 resulted in striking recovery of tyrosine-hydroxylase-positive innervation of the striatum. More importantly, GPI 1046 was effective in causing such restoration even when administered eight days following MPTP lesioning, suggesting that the actions of these compounds is not merely protective but also regenerative. As such, systemically active small molecules promote neuronal recovery of degenerated neurons and define a novel class of therapeutic agents for the treatment of neurodegenerative disorders, such as Parkinson's Disease.

173. SAR OF A SERIES OF NUCLEOSIDE ANALOGS AS ADENOSINE KINASE INHIBITORS M. Cowart, M. J. Bennett, G. Gfesser, S. S. Bhagwat, K. L. Kohlhaas, K. M. Alexander, C. Zhu, D. Britton, J. Lynch, H. Yu, K. Gunther, J. F. Kerwin Jr., E. A. Kowaluk *Neuroscience Research, Pharmaceutical Products Division, Abbott Laboratories, Abbott Park, Illinois 60064-3500*

One approach to the development of neuroprotective agents is to take advantage of natural protective effects of endogenously released adenosine by inhibiting adenosine kinase (AK), a key route of adenosine metabolism in the brain. More than 60 carbocyclic nucleosides were synthesized using a palladium mediated coupling or the Mitsunobu reaction as the key step. The most interesting compounds were A-127157 and A-134974, both with picomolar affinity for adenosine kinase, and inactive at other sites of adenosine interaction. A-134974 was found to inhibit chemically induced seizures in the rat, and to reduce brain infarct volume in an animal model of stroke damage at 3 μmol/kg i.p.

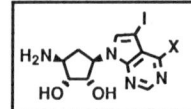

assay	A-127157 (X=Cl) IC_{50} (μM)	A-134974 (X=NH_2) IC_{50} (μM)
AK (rat brain cytosol)	0.000035	0.000080
AK (IMR cells)	0.015	0.043
A_1,A_{2a} receptor binding	>100	>100
in vivo seizure inhibition	30 μmol/kg	4.5 μmol/kg

174. DESIGN AND SYNTHESIS OF THE ANTICONVULSANT Co 2-1068: A NOVEL NEUROACTIVE STEROID MODULATOR OF THE GABA$_A$ RECEPTOR. Ravi B. Upasani*, Kevin C. Yang, Jon E. Hawkinson, Manuel Acosta-Burruel, James A. McLellan, and Richard Carter, CoCensys, Inc., 213 Technology Dr., Irvine, CA 92718.

3β-(4-Acetylphenyl)ethynyl pregnane exhibits high affinity for the GABA$_A$ receptor (R = H, IC$_{50}$ 10 nM). Structural modification of this potent GABA$_A$ receptor modulator resulted in a long-acting, water-soluble, orally-bioavailable neuroactive steroid prodrug, Co 2-1068 (R = O-hemisuccinate Na salt), that exhibits potent anticonvulsant activity across a range of animal procedures with ED$_{50}$ values of 3 mg/kg, p.o., against PTZ-induced seizures, 4.5 mg/kg p.o. against MES-induced seizures in mice, and 3.1 mg/kg, p.o., against Stage-5 corneal-kindled seizures in rats. The design, synthesis and the biological activities of this highly active neuroactive steroid will be presented.

175. EVIDENCE FOR THE CONSERVATION OF CONFORMATIONAL TOPOGRAPHY AT FIVE MAJOR GABA$_A$/BENZODIAZEPINE RECEPTOR SUBSITES. POTENT AFFINITIES OF THE (S)-ENANTIOMERS OF FRAMEWORK-CONSTRAINED 4,5-SUBSTITUTED PYRROLOIMIDAZOBENZODIAZEPINES. Qi Huang,[1] Ruiyan Liu,[1] Ruth M. McKernan,[2] and James M. Cook,[1]* [1]. Department of Chemistry, University of Wisconsin-Milwaukee, Milwaukee, WI 53201; [2]. Merck, Sharp & Dohme Research Laboratories, Harlow, Essex CM20 2QR, UK

The synthesis and *in vitro* affinities of a series of optically active pyrrolo or azetidinylimidazobenzodiazepines at five recombinant GABA$_A$/BzR subtypes expressed from human cell lines are described. Examination of the *in vitro* binding data indicates that the (S)-enantiomers of these ligands are much more potent *in vitro* than their corresponding (R)-isomers. The binding affinities of these framework constrained (rigid) ligands provide evidence for the first time which indicates the conformational preference for BzR ligands is highly conserved at these five recombinant receptor subsites. The configurational preference for the (S)-enantiomers in both series suggests that pharmacophoric descriptors H$_1$, H$_2$ and L$_1$ are the same in these five binding subsites in agreement with previous modeling studies. Data from the present study should be useful for understanding the conformational requirements of GABA$_A$/BzR subtypes as well as providing some insight into the conservation of binding site topography across a series of BzR binding subsites.

176. UNEXPECTED INACTIVATION MECHANISMS OF DIHYDROHETEROAROMATIC AMINO ACID ANTICONVULSANT AGENTS. Richard B. Silverman and Mengmeng Fu, Department of Chemistry, Northwestern University, Evanston, IL 60208

4-Amino-4,5-dihydro-2-thiophene carboxylic acid (**1**) and 4-amino-4,5-dihydro-2-furan carboxylic acid (**2**) were rationally designed mechanism-based anticonvulsant agents based upon the known mechanism of the dihydroaromatic γ-aminobutyric acid aminotransferase (GABA-AT) inactivators gabaculine (**3**) and isogabaculine (**4**). It was believed that **1** and **2**, when incubated with GABA-AT, would be converted to aromatic pyridoxamine-5'-phosphate adducts. Our study of **1** and **2**, however, has shown that both of these compounds inactivate pig brain GABA-AT via mechanisms different from the proposed pathway.

177. SYNTHESIS OF 4-ARYL KAINOIDS, G. A. Kraus, Department of Chemistry, Iowa State University, Ames, Iowa 50011

The observation by Shirahama that **1a** and **1b** have potent neurophysiological activity has spawned intense synthetic attention toward C-4 aryl analogs. Control of the C-3/C-4 stereochemistry is important, since the isomer bearing the opposite stereochemistry exhibits little biological activity. Most of the synthetic routes to C-4 aryl analogs begin from 4-hydroxyproline. As part of a program to better understand the interplay between the structure and activity of kainoids, we report a direct route to **1a** from the dimethyl ester of a-keto glutaric acid (**2**) and nitrostyrene **3**.

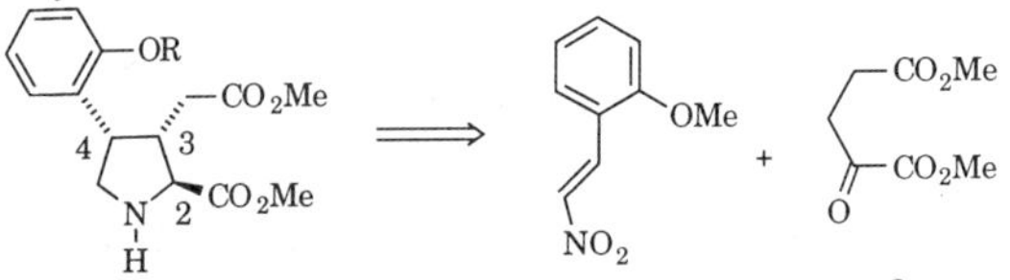

178. NOVEL CLASS OF INDOLE-2-CARBOXYLATES AS POTENT GLYCINE SITE NMDA ANTAGONISTS. R. Di Fabio*, G. Araldi, S. Giacobbe, G. Pentassuglia, A. Bozzoli, Z. Cimarosti, P. Maragni, T. Rossi, A. Reggiani.
Medicines Research Centre, *GlaxoWellcome S.p.A.*, Via A. Fleming 4, 37100 Verona (Italy)

Antagonists of the glycine binding site associated with the NMDA receptor have been proven to be potent neuroprotective agents in animal models of stroke, with a cleaner side effect profile with respect to both competitive antagonists and channel blockers. After the identification of GV150526, the indole-2-carboxylate currently in clinical development for treatment of stroke, as part of a broader SAR investigation of the C-3 position of the indole nucleus, we designed the novel class of conformationally restricted analogues shown in the figure. The synthesis and the SAR studies of this new series of indole derivatives will be presented.

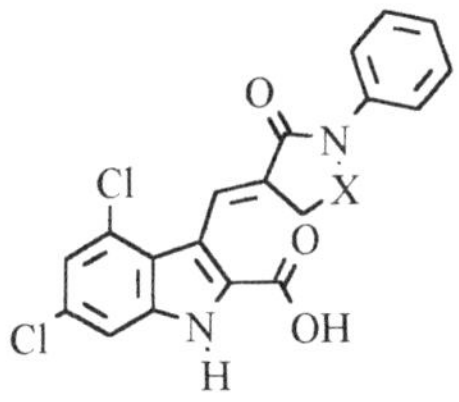

179. NEW TREATMENT OPTIONS IN OBESITY - F. Xavier Pi-Sunyer, M.D., Columbia College of Physicians and Surgeons, New York, N.Y. 10025.

New pharmacologic treatment options are emerging for obesity. Currently available anti-obesity agents include phetermine, d,l fenfluramine, and dexfenfluramine. These agents act on brain neurotransmitters in order to achieve appetite control. Phentermine affects the adrenergic system and the fenfluramines affect the serotoninergic system. Clinical trials with dexfenfluramine and with the combination of phentermine and fenfluramine have been found to be effective in enhancing weight loss over that occurring with placebo. for periods as long as a year. Potential side effects of all three drugs include primary pulmonary hypertension, which, although very rare, is lethal. Two other weight loss drugs, sibutramine and orlistat, are undergoing clinical trials. Leptin is a peptide released from fat cells whose central hypothamic effect is to inhibit food intake and enhance energy expenditure. It has now begun clinical trials. Other potential agents that are being developed include leptin agonists, neuropeptide Y antagonists, cholecystokinin agonists, and urocortin.

180. PROGRESS TOWARDS SELECTIVE, FULL-AGONISTS OF THE HUMAN ß$_3$-ADRENERGIC RECEPTOR. R. L. Dow, T. T. Chou, T. V. Olson, S. R. Schneider, D. M. Hargrove, D. K. Kreutter and S. Yee Central Research Division, Pfizer, Inc., Groton, CT 06340

Augmentation of basal energy expenditure through selective stimulation of thermogenesis has the potential to serve as an effective weight loss therapy. The identification of a novel member of the ß-adrenergic receptor family (ß$_3$), which upon activation leads to enhanced lipolysis and thermogenesis has stimulated significant drug discovery activity. Early efforts in this area lead to the discovery of agents which possessed excellent activity in rodent models but failed in human trials due to poor efficacy and ß$_1$-/ß$_2$-mediated side effects. Subsequent cloning efforts on the ß$_3$ adrenergic receptor revealed these compounds to be weakly efficacious in activating the human receptor, while being full-agonists of rodent receptors. This presentation will detail our efforts to identify agonists of the human ß₃ receptor with enhanced efficacy, based on a series of compounds (I) in which the key acid pharmacophore is conformationally-restricted.

$$Ar\overset{OH}{\underset{}{\diagdown}}\overset{H}{\underset{R'}{N}}\diagdown X\text{—}\diagup\text{—}CO_2H$$

I

181. CCK AND FEEDING INHIBITION: THE QUEST FOR HIGHLY SELECTIVE SMALL MOLECULE MIMETICS. William F. Michne, Department of Chemistry, Astra Arcus USA, P.O. Box 20890, Rochester, NY 14602

Cholecystokinin (CCK) is a 33-amino acid peptide with multiple functions in both the central nervous system (via CCK_B receptors) and the periphery (via CCK_A receptors). Of particular interest to the management of obesity is the role of CCK in the mediation of satiety via the A-receptor subtype. The carboxy terminal octapeptide (CCK-8) is fully active in this regard, but is lacking in receptor selectivity, metabolic stability, and oral bioavailability. Previous work has shown that smaller peptides can be obtained which are potent and highly selective full agonists, but which lack oral bioavailability. Detailed NMR studies of one such peptide have resulted in the determination of a solution structure, and further SAR studies have established the biological relevance of the structure. Comparison of this peptide structure with that of a reported small molecule partial agonist led to the design of related compounds which are full agonists at the A-receptor, with superior selectivity and bioavailability properties.

182. REGULATION OF LEPTIN GENE EXPRESSION Johan Auwerx, LBRE, Institut Pasteur de Lille, 59019 Lille, France

To understand leptin's role as a satiety factor, it is crucial to delineate the factors involved in *ob* gene regulation. The expression of the *ob* gene is controlled by the nutritional status. Fasting reduces, whereas food intake increases its expression, an effect accounted for by changes in plasma insulin levels. Glucocorticoids, endotoxin and cytokines, have been shown to induce *ob* gene expression, whereas cAMP treatment, β3 adrenergic-receptor agonists, and antidiabetic thiazolidinediones down-regulate leptin gene expression. In obesity leptin levels are increased. Leptin gene expression is furthermore controlled by transcription factors of the C/EBP family and by PPARγ. C/EBPα induces leptin expression, an effect mediated by a C/EBP site in the *ob* gene promoter. In addition, the expression of the *ob* gene is reduced by PPARγ and its activators both *in vivo* and *in vitro*, via a direct effect on the *ob* gene promoter.

183. PPARγ LIGANDS: POTENTIAL THERAPEUTICS FOR DIABETES AND OBESITY.
Jeff E. Cobb, Glaxo Wellcome Inc., 5 Moore Drive, RTP, NC 27709

Peroxisome Proliferator Activated Receptor γ (PPARγ) is a member of the nuclear receptor super family of ligand-activated transcription factors. It is most abundant in adipose tissue and plays a key role in adipocyte differentiation and in the expression of a number of genes involved in fuel metabolism. Endogenous activators of PPARγ include a number of fatty acids and 15-deoxy-$\Delta^{12,14}$-PGJ₂. Recently, PPARγ was shown to be the biochemical target for the thiazolidinedione class of antidiabetics. These compounds bind to and activate PPARγ. Thiazolidinediones improve insulin sensitivity in Type 2 diabetics while lowering plasma glucose, insulin, and triglycerides. While current research is focused on the antidiabetic applications of PPARγ agonists, modulation of PPARγ activity may also have application in the treatment of obesity. In particular, PPARγ's effects on white and brown fat differentiation, and on the expression of the uncoupling protein and *ob* genes suggest avenues for future research.

184. SYNTHESIS AND STRUCTURE-ACTIVITY RELATIONSHIP OF TRIMETHOXYBENZYL PIPERAZINES: NOVEL DOPAMINE D4 ANTAGONISTS
Terri S. Purchase, Shelly A. Glase, Hyacinth C. Akunne, Thomas G. Heffner, Robert G. MacKenzie, Thomas A. Pugsley, and Lawrence D. Wise, Departments of Chemistry and Therapeutics, Parke-Davis Pharmaceutical Research, Division of Warner-Lambert Company, 2800 Plymouth Road, Ann Arbor, MI 48105

Screening of our chemical library revealed that PD 35680 potently and selectively bound to dopamine (DA) D4 receptors. A series of analogs was prepared and evaluated as potential DA D4 antagonists. In binding studies, compounds of this series were found to be selective for D4 over D2 and D3 human receptors expressed in Chinese hamster ovary (CHO) K1 cells, and were found to be antagonists as they blocked quinpirole stimulation of D4 receptor mitogenesis. They were also found to increase dopamine turnover in vivo, as would be expected of a DA antagonist. The synthesis and binding profile of these compounds will be discussed.

185. BISARYLMETHYLAMINES, A NEW CLASS OF D4 LIGANDS: STRUCTURE ACTIVITY RELATIONSHIPS. M. Arlt, H. Boettcher, G. D. Bartoszyk, H. E. Greiner, C. A. Seyfried, Merck KGaA, Frankfurter Strasse 250, D-64271 Darmstadt, Germany, J. J. Berthelon, M. Brunet, J. J. Zeiller, Lipha Centre Recherche Lacassagne, 69003 Lyon, France

Typical antipsychotics are mainly dopamine antagonists. Recently, it has been found that the atypical antipsychotic Clozapine displays a certain selectivity towards a dopamine receptor subtype, namely the D4 receptor. In addition it has been proposed that the D4 receptor is upregulated in schizophrenic brains. These findings have caused a lot of interest in D4 receptor ligands. We have screened our in house collection and have identified bisarylmethylamines as ligands for the D4 receptor with affinities in the nanomolar range. Structure activity relationships for this class of compounds will be presented.

186. LU 111995: A NOVEL D_4 / $5-HT_2$ RECEPTOR ANTAGONIST AND POTENTIAL NEW ANTIPSYCHOTIC. Gerd Steiner, Alfred Bach, Siegfried Bialojan, Gerd Greger, Thomas Höger, Gerhard Klebe, Rainer Munschauer, Hans-Jürgen Teschendorf, Martin Traut and Liliane Unger. Main Laboratory BASF AG, D-67056 Ludwigshafen, Germany and Knoll AG, D-67008 Ludwigshafen, Germany

All classical antipsychotic agents act as antagonists at the dopamine D_2 receptor. However, their clinical efficacy is hampered by extrapyramidal side effects. In contrast, atypical antipsychotics such as clozapine induce fewer side effects and are often effective in nonresponders to classical agents and in attenuating the negative symptoms of schizophrenia. A common feature of most of these atypical antipsychotics is a combination of D_2- and $5-HT_2$-antagonism. 5 distinct dopamine receptors have been revealed, D_1-D_5, of which the D_4 receptor is of particular interest, because clozapine binds to this receptor with the highest affinity over all other dopamine receptor subtypes.
Our aim has been to substitute the conformationally flexible amino side chains, present in most atypical neuroleptics, with new more rigid 3-azabicyclo-[3.2.0]heptane building blocks synthesized by an intramolecular 2+2 photocycloaddition reaction. The chemistry, including X-ray analysis of the absolute configuration of the bicyclic side chain, and SAR of a series of compounds will be discussed. LU 111995 ((+)-(1S,5R,6S)-exo-3-[2-[6-(4-fluorophenyl)-3-azabicyclo[3.2.0]heptane-3-yl]-ethyl]-1H,3H-quinazoline-2,4-dione) has been selected for development based on its D_4-versus D_2 selectivity and its high D_4 and $5-HT_2$ receptor affinities. In vivo pharmacology as well as pharmakokinetic data will be presented. LU 111 995 was well tolerated in a double blind placebo controlled study with 12 difficult to treat chronic schizophrenic patients.

187. CHROMENO[3,4-C]PYRIDINE-5-ONES: SELECTIVE DOPAMINE D_4 RECEPTOR ANTAGONISTS AS POTENTIAL ANTIPSYCHOTIC AGENTS. **Steven R. Miller**, David T. Connor, Thomas G. Heffner, Thomas A. Pugsley, Paul C. Unangst, and Lawrence D. Wise. Parke-Davis Pharmeceutical Research, Division of Warner-Lambert Co., 2800 Plymouth Rd. Ann Arbor, MI 48105

Existing antipsychotic drugs such as haloperidol are believed to act by nonselective blockade of dopamine receptors. The dopamine D_4 receptor subtype is thought to be concentrated in limbic areas of the brain, while D_2 receptors are present in both striatal and limbic regions. The striatum is believed to be responsible for motor control, whereas the limbic region controls thought and emotion. Thus, a selective D_4 receptor antagonist may demonstrate antipsychotic efficacy with a reduced incidence of extrapyramidal side effects. Screening of the Parke-Davis compound library identified a series of chromeno[3,4-c]pyridine-5-ones as selective D_4 receptor antagonists. The synthesis and binding profile of this series are presented.

188. SYNTHESIS AND EXAMINATION OF 1-ARYL-4-(PHTHALAN, CHROMAN, AND BENZOXAPINE) ETHYLPIPERAZINE DERIVATIVES AS DOPAMINE D_4 RECEPTOR LIGANDS. <u>Chiu-Hong Lin</u>[*], Kerry A. Cleek[*], Michael D. Ennis[*], Robert L. Hoffman[*], Martin W. Smith[†], Charles F. Lawson[†], Siusaidh K. Schlachter[†], Connie G. Chidester[*], [*]Structural, Analytical & Medicinal Chemistry, [†]Chemical & Biological Screening, Pharmacia & Upjohn, Kalamazoo, MI 49001

The discovery of the dopamine D_4 receptor has elicited considerable interest in finding a selective antagonist which might be useful as an antipsychotic agent with fewer adverse side effects. To search for a selective D_4 receptor ligand, 1-aryl-4-(phthalan, chroman and benzoxapine)ethylpiperazine derivatives (**I**, **II**, and **III**) were synthesized and evaluated. The 1-aryl moiety was varied by *para*-substitution as shown by R_2. The analogs were evaluated in binding assays which identified PNU-103921, (**II** where $R_2 = SO_2NH_2$) as a very selective D_4 receptor ligand. The enantiomers of this racemic analog were then synthesized via an enzymatic resolution of a key intermediate. The activity was found to reside in the (*R*) enantiomer (PNU-106675) with a *Ki* = 23 nM for the D_4 receptor and *Ki*'s > 1900 nM for the rest of the receptors. The (*S*) enantiomer (PNU-106723) was 20-fold less potent than the (*R*) enantiomer.

189. **A NOVEL SERIES OF 1-ARYLOXY-3-(4-ARYLOXYPIPERIDINYL)-2-PROPANOLS AS POTENT, SELECTIVE DOPAMINE D4 RECEPTOR ANTAGONISTS.**
<u>Tracy F. Gregory</u>, Jonathon L. Wright, Hy C. Akunne, Thomas G. Heffner, Robert G. MacKenzie, Thomas A. Pugsley, Seth J. Vander Meulen and Lawrence D. Wise. Departments of Chemistry and Therapeutics, Parke-Davis Pharmaceutical Research, Division of Warner-Lambert Company, 2800 Plymouth Rd., Ann Arbor, MI 48105.

PD 141977

A selective dopamine (DA) D4 antagonist might provide antipsychotic efficacy without causing neurological side effects. Screening of our chemical library revealed that PD 141977 bound to the human (h) D4.2 receptor selectively with a Ki of 26 nM. The selectivity over the h D2L receptor (Ki 1438 nM) was 55-fold and the h D3 receptor (Ki 608 nM) was 23-fold. In a functional assay this compound acted as an DA D4.2 antagonist by blocking the stimulation of mitogenesis caused by the DA agonist quinpirole. Despite the chemical similarity of PD 141977 to propranolol, it shows only weak affinity for rat brain ß-adrenergic receptors (ß1 3706 nM, ß2 1041 nM). Subsequent studies have led to a novel series of 1-aryloxy-3-(4-aryloxypiperidinyl)-2-propanols with both higher selectivity and more potent affinity for DA h D4.2 receptors. SAR studies to optimize the selectivity of h D4.2 over h D2L receptors and potency will be presented.

190. 8-(BROMOMETHYL)BENZYL-6,7-DIHYDROXY-2-AMINOTETRALIN DERIVATIVES AS POTENTIAL DOPAMINE D2-LIKE RECEPTOR ALKYLATING AGENTS. <u>Debasis Ghosh</u>,[a] David E. Nichols,[a] Clifford R. Peck,[b] and Richard B. Mailman.[b] [a]School of Pharmacy, Purdue University, W. Lafayette, IN 47907 and [b]School of Medicine, University of North Carolina, Chapel Hill, NC 27599

In the process of preparing 8-benzyl derivatives of 6,7-dihydroxy-2-aminotetralin-type dopaminergic ligands, a novel rearrangement during the oxygen ether cleavage step led to derivatives with a 2'-bromomethyl moiety attached to the 8-benzyl group. The reactivity of benzyl bromides toward nucleophiles suggested the possibility that these compounds might interact with neucleophilic residues in the dopamine receptor such as serine, cysteine, or histidine, leading to irreversible receptor alkylation. The most selective of these compounds, possessing an N,N-dipropyl group, had excellent affinity for D2-like receptors. Incubation of rat striatal homogenates with this compound showed a time-dependent loss of Bmax for D2 receptor ligand binding, providing evidence that this molecule was capable of inactivating the ligand binding domain of the receptor. The syntheses, mechanism of the novel bromomethyl rearrangement reaction, and inactivation kinetics of these compounds will be presented.

191. 3α-(SUBSTITUTED PHENYL)TROPANE-2β-CARBOXYLIC ACID ESTERS: LIGANDS WITH UNEXPECTEDLY HIGH AFFINITY AND SELECTIVITY AT THE DOPAMINE TRANSPORTER. Christopher R. Holmquist, Kathyrn I. Keverline-Frantz, Phillip Abraham, John W. Boja, Michael J. Kuhar and F. Ivy Carroll, Chemistry and Life Sciences, Research Triangle Institute, Research Triangle Park, NC 27709, and National Institute on Drug Abuse, Addiction Research Center, Baltimore, MD 21224

3α-(4-Substituted phenyl)tropane-2β-carboxylic acid esters (**1**) are a new class of ligand which possess interesting activity at the cocaine receptor site. In comparison with analogous 3β-(4-substituted phenyl)tropane-2β-carboxylic acid esters (**2**), most analogs of **1** show a marked increase in selectivity for the dopamine transporter (DAT) relative to the 5-HT transporter with only a modest drop in potency at the DAT. These results were unexpected as allococaine (**3**) is 60-fold less potent than cocaine (**4**) at the DAT. The first comprehensive comparison of esters **1** and **2** will be presented along with the synthesis of **1** and the results of modeling and NMR studies seeking to explain the high binding affinity of **1**.

192. SYNTHESIS AND BIOLOGICAL ACTIVITY OF NOVEL 6-ALKYL-3-BENZYL-2-(METHOXYCARBONYLMETHYL)TROPANE DERIVATIVES. Stacey A. Lomenzo[1], Sari Izenwasser[2], Jonathan L. Katz[2], Mark L. Trudell[1]. [1]Department of Chemistry, University of New Orleans, New Orleans, LA 70148 and [2]NIDA Division of Intramural Research, P. O. Box 5180, Baltimore, MD 21224.

A series of 6-alkyl-3β-benzyl-2-(methoxycarbonylmethyl)tropane analogs were synthesized and evaluated as cocaine binding-site ligands at the dopamine transporter. The K_i values for displacement of [^{3}H]WIN 35,428 for the 6α-alkyl-3β-benzyl-2β-(methoxycarbonylmethyl)tropane analogs **1** (211– 8,579 nM) were two- to four-fold more potent than the corresponding 6α-alkyl-3β-benzyl-2α-(methoxycarbonylmethyl)tropane analogs **2** (531–19,658 nM). In general, the binding affinity of the 6α-alkyl derivatives decreased as the steric bulk of the 6α-alkyl group increased. The chemical syntheses and *in vitro* binding data for compounds **1** and **2** will be presented.

193. HIGHLY SELECTIVE ENDOTHELIN ANTAGONISTS FOR ET$_A$ RECEPTOR: SYNTHESIS AND STRUCTURE-ACTIVITY RELATIONSHIPS Gang Liu, Kenneth J. Henry, Bruce G. Szczepankiewicz, Martin Winn, Steven A. Boyd, Robert A. Mantei, James Wasicak, Thomas W. von Geldern, Jinshyun R. Wu-Wong, William Chiou, Douglas Dixon, and Terry J. Opgenorth. Pharmaceutical Products Division, Abbott Laboratories, Abbott Park, IL 60064-3500

Endothelin-1 acting through the ET$_A$ receptor has been implicated in the pathophysiology of a number of diseases, including renal failure, cerebral vasospasm, pulmonary hypertension, and angioplasty-induced restenosis. Selective antagonism of ET$_A$ receptor would offer beneficial therapeutical profiles. We have expanded upon the potent and selective endothelin-A antagonist, A-127722 (I), and discovered potent compounds (II) with exceptional selectivity for ET_A receptor. This paper discusses the SAR of this series of ET_A antagonists and methods for their synthesis.

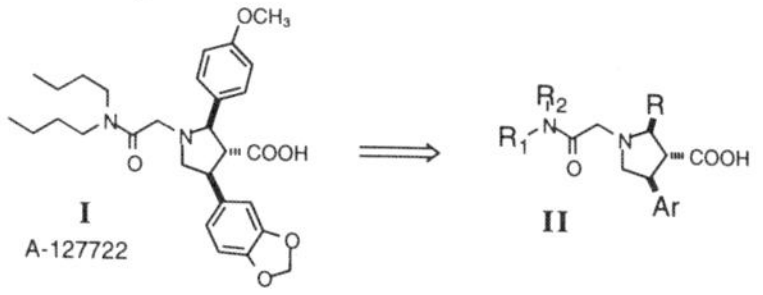

194. DISCOVERY OF BALANCED ENDOTHELIN-A/B ANTAGONISTS Hwan-Soo Jae, Martin Winn, Thomas W. von Geldern, Ruth Wu-Wong, William Chiou, Douglas Dixon, and Terry J. Opgenorth. Pharmaceutical Products Division, Abbott Laboratories, Abbott Park, IL 60064-3500

We have developed a very potent and selective Endothelin-A antagonist, A-127722, in which an N,N-dibutyl acetamide is attached to the 1-position of a 2,4-diaryl pyrrolidine-3-carboxylic acid (I). We have now discovered that when the nitrogen substituent is N'-alkyl, S-alkylsulfonamidoethyl (II), the resultant compounds have high affinity for both ET_A and ET_B receptors. This paper discusses the SAR of this series of balanced antagonists and methods for their synthesis.

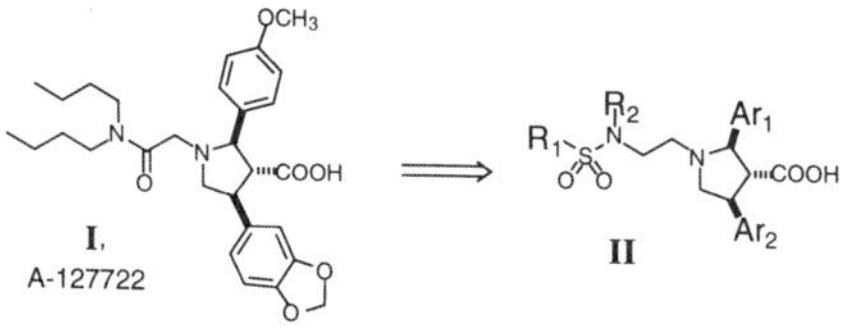

195.
PD 163070, A POTENT ENDOTHELIN-A SELECTIVE ANTAGONIST WITH LONG DURATION OF ACTION. C. Lee, X.-M. Cheng, J. Repine, L. Bradford, E. Reyner, B. Stewart, H. Hallak, K. M. Welch, D. Walker, S. Haleen, A. Doherty. Parke-Davis Pharmaceutical Research Division, Warner-Lambert Company, Ann Arbor, MI 48105

Structure-activity relationship studies in the butenolide series, represented by PD 156707, led to the discovery of PD 163070, a potent endothelin-A selective antagonist with long duration of action. In rat, the dimethylamino analog was shown to retain inhibitory activity against an endothelin challenge for up to 16 h after 10 mg/kg intravenous or oral administration. Metabolism studies of PD 163070 will be discussed.

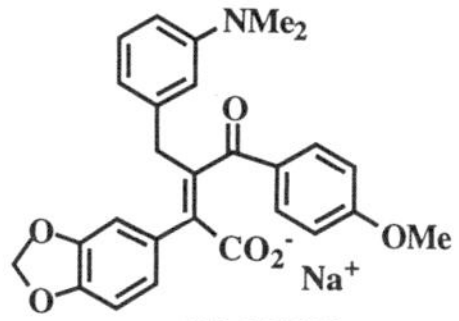

PD 163070
$hET_A = 3$ nM
$hET_B = 2500$ nM

196. THE TOTAL SYNTHESIS OF THE NATURAL PRODUCT ENDOTHELIN CONVERTING ENZYME (ECE) INHIBITOR, WS75624 B. M. A. Massa[†], W. C. Patt[†], K. Ahn[‡], A. Sisneros[‡], S. Herman[‡], Medicinal Chemistry Department[†], Biochemistry Department[‡], Parke-Davis Pharmaceutical Research Division, Warner-Lambert Company, 2800 Plymouth Road, Ann Arbor, MI 48105

The first synthesis of WS75624 B, an endothelin converting enzyme (ECE) inhibitor isolated from a fermentation broth of *Saccharothrix* sp. No. 75624, will be presented. The 14 step synthesis of WS75624 B utilized kojic acid to prepare a bromoacetyl pyridine intermediate. The thiazole ring was formed by condensation of this intermediate with a thioamide in the penultimate step. Employing various thioamides led to a series of 2-alkyl-thiazol-4-yl compounds. The ECE inhibition data for these analogs will be discussed and compared to the natural product.

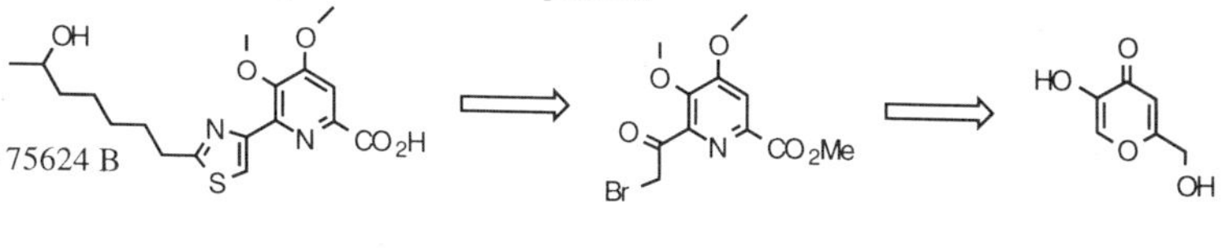

197. RAS FARNESYL TRANSFERASE INHIBITORS: SYNTHESIS AND STRUCTURE-ACTIVITY RELATIONSHIPS OF A NEW SERIES OF HISTIDINE DERIVATIVES
Quin, III, J., Kaltenbronn, J. S., Leonard, D., Shuler, K., Sebolt-Leopold, J., Gowan, R., Scholten, J., Zimmerman, K., and Doherty, A. M.; Parke-Davis Pharmaceutical Research, Division of Warner Lambert Company, Ann Arbor, MI. 48105

Mutant ras proteins in the cells are linked to a number of cancers due to their inability to regulate cell division, allowing uncontrolled cell growth and differentiation. Inhibiting the processing of these mutant ras proteins has the potential for the treatment of cancer, particularly cancers of the colon and pancreas. Screening of the Parke Davis compound library identified **1** as a novel inhibitor of ras farnesyl transferase. This bis-(1-naphthylmethylacetyl) derivative of histidine methyl ester was selective for inhibition of ras farnesyl transferase (FTP) over geranyl-geranyl transferase (GGPT) with IC_{50}s of 0.27 μM and 20.4 μM respectively. Compound **1** also inhibited the processing of H-ras in Me 12/+ cells with a minimum effective dose (MED) of 2.5 μM. We will describe the synthesis and structure activity relationships (SAR) of a series of potent ester and amide derivatives based on this lead structure.

198. BENZOCYCLOHEPTAPYRIDINE FARNESYL PROTEIN TRANSFERASE INHIBITORS: EFFECTS OF SUBSTITUTION AT POSITION 5 AND 6 ON ENZYME INHIBITION

Jocelyn D. del Rosario[‡], Stacy W. Remiszewski[‡]*, W. Robert Bishop[†], Ronald J. Doll[‡], Linda James[†], Paul Kirschmeier[†], Mehdi Nafissi[‡], F. George Njoroge[‡], Joanne Petrin[†], Patrick Pinto[‡]

Departments of Chemistry[‡] and Tumor Biology[†], Schering-Plough Research Institute
2015 Galloping Hill Road, Kenilworth, NJ 07033

The novel non-peptidic farnesyl protein transferase inhibitor SCH 44342 was found to undergo oxidative metabolism at position 5 or 6, the bridgehead. In order to determine the effect of that oxidation on the biological activity the synthesis of 5 and 6 oxygenated analogs of SCH 44342 was achieved. Additionally, the preparation of compounds modified at position 5 and/or 6 has been accomplished. The SAR for the compounds reveals that modifications at the bridgehead have a dramatic effect on enzyme inhibition.

199. INHIBITORS OF FARNESYL PROTEIN TRANSFERASE. PART 1. 4-AMIDO, 4-CARBAMOYL AND 4-CARBOXAMIDO DERIVATIVES OF 1-(8-CHLORO-6,11-DIHYDRO-5H-BENZO[5,6]CYCLOHEPTA[1,2-b]PYRIDIN-11-YL)PIPERAZINE. R. R. Rossman, A. K. Mallams*, R. J. Doll, V. Girijavallabhan, A. K. Ganguly, J. Petrin, L. Wang, R. Patton, W. R. Bishop, P. Kirschmeier, J. Catino, M. S. Bryant, K.-J. Chen, W. A. Korfmacher, C. Nardo, S. Wang, A. A. Nomeir, C.-C. Lin, M. Liu, Z. Li, P. Chen, S. Lee and I. King, Antiinfectives and Tumor Biology Research, Schering-Plough Research Institute, Kenilworth, NJ 07033.

The discovery of Sch 44342, a member of a new class of selective, non-peptidic, non-sulfhydryl inhibitors of ras protein farnesylation in these laboratories, led to the synthesis of a series of novel 4-amido, 4-carbamoyl and 4-carboxamido derivatives of 1-(8-chloro-6,11-dihydro-5H-benzo[5,6]cyclohepta[1,2-b]pyridin-11-yl) piperazine. The synthesis and SAR of these new piperazine derivatives will be described. This led to the selection of two amido derivatives, namely the 3- and 4-pyridylacetyl analogs (Sch 54419 and 54429 respectively) for evaluation as lead structures. Both were orally active, but were found to be rapidly metabolized *in vivo*. This resulted in the synthesis of a variety of new compounds that would not be readily metabolized, such as the N-oxides (Sch 57665 and 59413) and the 4-N-methyl and 4-N-carboxamidopiperidylacetyl derivatives (Sch 58006 and 59836 respectively).The *in vitro* and *in vivo* activity of these novel, orally active, antitumor agents will be described.

200. NOVEL CONFORMATIONALLY EXTENDED NAPHTHALENE-BASED INHIBITORS OF FARNESYLTRANSFERASE. C. J. Burns, J.-D. Guitton, B. Baudoin, A. Lebrun, Y. LeLievre, M. Duchesne, F. Parker, N. Fromage and A. Commercon, Rhone-Poulenc Rorer S. A. - Centre de Recherches de Vitry-Alfortville, 13, Quai Jules Guesde - B. P. 14 - 94403 Vitry-sur-Seine, France

Two series of very potent conformationally-fixed, naphthalene-based inhibitors of farnesyltransferase have been prepared and will be presented. These inhibitors were designed based on the molecular modeling premise that the active conformation of the inhibitory CAAXbased peptides and pseudopeptides is "extended" rather than "turned". The 1,5- and 1,6-substituted naphthalene scaffolds unambiguously impart an extended conformation to the appended amino acids Met and Cys. These new prototypes, particularly the 1,6-naphthalene-substituted series, exhibit a considerable increase in cellular activity both in terms of ras processing and in clonogenic assays of tumor cells (activated Ha-ras and Ki-ras).

201. SYNTHESIS AND TYROSINE KINASE INHIBITORY ACTIVITY OF A SERIES OF 2-AMINO-8H-PYRIDO[2,3-d]PYRIMIDINES
Z. Wu, D. Boschelli, S. Klutchko, G. Lu, R. Panek, A. Kraker, A. Doherty. Parke-Davis Pharmaceutical Research, Division of Warner-Lambert Company, 2800 Plymouth Road, Ann Arbor, MI 48105

Screening of a compound library lead to the identification of PD 090560 as a tyrosine kinase inhibitor. It was recently reported that an analog, PD 166285, had increased activity with IC_{50}'s of 0.078, 0.042, and 0.0088 µM against PDGF, FGF, and c-src respectively. We present here an SAR study where the groups at the 2, 6, 7 and 8 positions of the pyrido[2,3-d]pyrimidine were varied. The synthesis and tyrosine kinase inhibitory activity of these compounds will be discussed.

R^2 = H, Me, Ph, Ph-4-O-(CH$_2$)$_2$-NEt$_2$, 4-pyridyl
R^6 = Ph, 3-thiophene, 2-thiophene, biphenyl, 2,6-dichlorophenyl
R^8 = alkyl, substituted alkyl
X = O, NH

R=H: PD 090560
R=Ph-4-O-(CH$_2$)$_2$-NEt$_2$: PD 166285

202. SYNTHESIS AND BIOLOGICAL EVALUATION OF NOVEL PROTEIN KINASE C (PKC) INHIBITORS. Kelvin K. C. Sham, Robert A. Daines, William D. Kingsbury, Glenn A. Hofmann, Michael R. Mattern, Ponnal Nambi
Departments of Medicinal Chemistry, Biomolecular Discovery, and Renal Pharmacology, SmithKline Beecham Pharmaceuticals, P.O. Box 1539, King of Prussia, PA 19406.

Protein kinase C (PKC) is a serine/threonine kinase which plays an important role in a number of physiological functions and pathological states. Since activators of PKC, such as phorbol esters, are known to cause intense inflammation and tumor promotion, compounds which are able to regulate PKC may be useful therapeutic agents for the treatment of inflammatory and proliferative diseases. We have prepared a series of phenylmethyl hexanamides and benzamides which are novel PKC inhibitors that interact with the regulatory domain of the enzyme. Synthesis and structure-activity relationships of selected compounds will be presented.

203. Stereochemical probes for the estrogen receptor: synthesis and receptor binding of (17α,20E/Z-21-phenyl-19-norpregna-1,3,5(10),20-tetraene-3,17β-diols. Robert N. Hanson, Lee W. Herman, Rita Fischi, and Elio Napolitano, Dept. of Pharmaceutical Sciences, Northeastern University, Boston, Massachusetts 02115

Previous studies from our laboratory using 17α-E, and 17α-Z-halovinyl and phenylthiovinyl estradiols demonstrated a marked preference for the Z-stereochemistry and a significant steril tolerance for the Z-vinyl substitutent. To further explore the extent of that stereochemical preference and steric tolerance we have prepared stereoselectively the 17αE- and 17αZ-phenylvinyl estradiols (E- and Z-styrylestradiols). The results, in addition to demonstrating a facile preparation of the target compounds. supported the previously observed stereochemical and steric effects. The relative binding effinities for the Z isomer were 3-4-fold greater than the E isomer at both 4°C and 25°C and only one-half to one-fourth those of estradiol under similar conditions. The developing model for ligand-accessible space within the estrogen receptor suggests that Z-phenylvinyl estradiols may provide interesting and useful probes for mapping the receptor.

204. **DESIGN, SYNTHESIS, AND PHARMACOLOGICAL EVALUATION OF A NOVEL SERIES OF BENZOTHIOPHENE DERIVED SELECTIVE ESTROGEN RECEPTOR MODULATORS (SERMs)** Alan D. Palkowitz*, Andrew L. Glasebrook, Henry U. Bryant, Kenneth J. Thrasher, Lorri L. Short, Pamela K. Shelter, Kenneth L. Hauser, D. Lynn Phillips, Harlan W. Cole, Masahiko Sato, and Preston C. Conrad
Lilly Research Laboratories, Indianapolis IN, 46285

Raloxifene, [2-(4-hydroxyphenyl)-6-hydroxybenzo[b]thien-3-yl][4-[2-(1-piperidinyl)ethoxy]phenyl]methanone hydrochloride, is a 2-arylbenzothiophene derivative that is representative of a newly described class of compounds known a Selective Estrogen Receptor Modulators (or SERMs). Characteristic of raloxifene is its ability to potently mimic the actions of estrogen on bone tissue and serum lipids, while effectively antagonizing the stimulatory actions of estrogen in the breast and uterus. This unique profile has suggested numerous safe uses for such compounds in estrogen dependent diseases such as osteoporosis, breast cancer, and gynecological disorders. As part of ongoing SAR studies, we have identified a novel series of heteroatom modified benzothiophene derivatives that demonstrate log-fold increases in potency over raloxifene in both *in vitro* and *in vivo* assays used to evaluate the SERM profile. Reported in this presentation will be the design, synthesis, and pharmacology of these novel agents.

205. **THE SYNTHESIS AND SAR OF RALOXIFENE DERIVATIVES IN THE DIARYLKETONE REGION** Brian S. Muehl*, Charles D. Jones, Mary D. Adrian, Henry U. Bryant, Harlan W. Cole, Andrew L. Glasebrook, David E. Magee, David L. Phillips, Lorri L. Short, and Pamela K. Wilson
Lilly Research Laboratories, Indianapolis IN 46285

A variety of modifications in the diarylketone region of Raloxifene was synthesized by two primary routes. Ether and thioether modifications were made by reacting the carbinol of bis-(t-butyldimethylsilyl) protected Raloxifene with an alcohol or thiol in the presence of trifluoroacetic acid. A variety of amine modifications were synthesized by displacing the mesylate of the carbinol of the protected Raloxifene with the desired amine. All modifications resulted in decreases in both *in vitro* and *in vivo* activities. While this position shows a limited response to the size of the modification, no consistent correlation is seen between the activity and the electronics of the modifications. Methyl ether and n-methylamino modifications at this position in the benzofuran analog of Raloxifene resulted in a greater decrease in *in vivo* activity than in seen with the same modifications in the benzothiophene system.

206. **SYNTHESIS AND BIOLOGICAL EVALUATION OF NOVEL NON-STEROIDAL INHIBITORS OF STEROID 17,20 LYASE.** T.Yoden, M.Okada, E.Kawaminami, I.Kinoyama, Y.Ideyama, and Y.Isomura, Institute for Drug Discovery Research, Yamanouchi Pharmaceutical Co.,Ltd. 21 Miyukigaoka, Tsukuba, Ibaraki 305, Japan

Steroid 17,20 lyase, which is a cytochrome P-450 enzyme, is responsible for the last step in androgen biosynthesis. Therefore, inhibition of this enzyme is expected to be useful for the treatment of androgen-dependent diseases such as prostate cancer. In order to find a potent non-steroidal inhibitor of steroid 17,20 lyase, we have focused on the azole type inhibitors of P-450 and have synthesized a series of imidazole derivatives. Among them, we found that compounds having a tricyclic substituent showed potent inhibitory activities. SAR studies led us to finding 2-(1H-imidazol-4-ylmethyl)-9H-carbazole monohydrochloride monohydrate (YM116) with an IC50 value of 6.8 nM(rat testicular microsome). The synthesis and SAR of these compounds will be presented.

207. **SAR STUDY OF TAXOIDS FOR NO AND TNFα PRODUCTION BY MURINE MACROPHAGES** Iwao Ojima*, **Zhuping Ma**, Teruo Kirikae[§], Scott D. Kuduk, John C. Slater, Craig S. Takeuchi, Pierre-Yves Bounaud, Fumiko Kirikae[§], and Masayasu Nakano[§]. Department of Chemistry, State University of New York at Stony Brook, Stony Brook, NY 11794. [§]Department of Microbiology, Jichi Medical school, Minamikawachi-machi, Tochigi-ken, 329-04, Japan

Recent studies have revealed that paclitaxel effects macrophages cycle-independently and possesses LPS-like effects on murine macrophages. Paclitaxel, as well as LPS, can activate macrophages and induce TNFα and NO production. In the present SAR study of a series synthetic taxoids, we found that (i) the benzoylamino at C-3' position is the most important site to activate LPS-responsive mice, (ii) a phenyl group at C-3' position is not a requisite for the activation but the acetyl group at C-10 has distinct effects on the activity, and (iii) there is no correlation between the cytotoxity and macrophage-activating ability of these taxoids.

208. NEW TAXANES AS HIGHLY EFFICIENT REVERSAL AGENTS FOR MULTI-DRUG RESISTANCE IN CANCER CELLS
Iwao Ojima*[1], **Pierre-Yves Bounaud**[1], Craig Takeuchi[1], Paula Pera[2], Jean M. Veith[2], and Ralph J. Bernacki*[2]. [1]Department of Chemistry, State University of New York at Stony Brook, Stony Brook, New York 11794-3400, [2] Department of Experimental Therapeutics, Roswell Park Cancer Institute, Elm & Carlton Streets, Buffalo, New York, 14263.

Multi-Drug Resistance is a serious problem in cancer chemotherapy. It is mainly associated with amplification of P-glycoprotein (Pgp), a membrane phosphoglycoprotein, that acts as a drug-efflux pump to maintain low intracellular cytotoxic drug levels. A series of new non-cytotoxic taxanes featuring one hydrophobic side-chain at R_1, R_2 or R_3 were synthesized and tested for reversal activity in drug-resistant tumor cell lines. In combination with anticancer agents like paclitaxel or doxorubicin, several taxanes were found to help those drugs to recover their cytotoxicity up to 99.8%.

209. SYNTHESIS AND EVALUATION OF NEW TAXANE-BASED REVERSAL AGENTS
Iwao Ojima*[1], **Pierre-Yves Bounaud**[1], Paula Pera[2], Jean M. Veith[2], and Ralph J. Bernacki*[2]. [1]Department of Chemistry, State University of New York at Stony Brook, Stony Brook, New York 11794-3400, [2] Department of Experimental Therapeutics, Roswell Park Cancer Institute, Elm & Carlton Streets, Buffalo, New York, 14263.

Cancer chemotherapy is often limited by the development of clinical resistance to antitumor drugs. Drug-resistant tumor cells were found to overexpress P-glycoprotein, a drug-efflux pump in charge of maintaining low intracellular drug concentrations. Several non-cytotoxic taxane-based reversal agents were synthesized, featuring two hydrophobic groups, R_1 and R_2, at C-7 and C-10 positions. The cytotoxicity of paclitaxel against drug-resistant tumor cell lines was found to be restored dramatically (up to 99.7%) by a combined treatment with those taxane reversal agents.

210.
SYNTHESIS AND CYTOTOXICITY OF ANDROSTANYL UREAS AS DOLASTATIN 10 MODEL CANDIDATES. **P.K.Subramanian(1)**, C.C.Tseng (1), and R. Vishnuvajjala (2), (1) Ash Stevens Inc., Detroit Research Park, Detroit, Michigan 48202; (2) Pharmaceutical Resources Branch, National Institute of Health, Bethesda, Maryland 20892.

Dolastatin 10 is a chiral natural product first isolated from sea hare Dolabella auricularia, and exhibits remarkable antineoplastic activity. Multi-step total synthesis was developed because of its limited availability from natural source. Several analogues with structural and chiral modifications have been prepared and their biological properties assessed. In the present study, the thiazolyl steroids were made as simpler models of this very potent compound. The synthesis of steroidal ureas as well as their cytotoxicities in a variety of cancer cell cultures will be described.

211. LINKED ACRIDINE DERIVATIVES WITH S-NITROSO-β-AMINOTHIOLS A[ND] THEIR RADIATION SENSITIZATION PROPERTIES. D. Christodoulou[1], J. Chu[1], W. DeG[raf]f[2], D.A. Wink[2]. [1]Chemistry Section, National Cancer Institute, Frederick Cancer Research and Develo[pme]nt Center, Frederick, MD 21702 and [2]Radiation Biology Branch, National Cancer Institute, Bethesda, M[D 20]892.

Nitric oxide (NO) is an endogenously generated bioregulatory molecule in mammalian [cell]s. Among the pathophysiological functions of NO, its cytotoxic effects, impact on reactive oxygen species a[nd th]e immune system, have been a controversial and active topic in current literature. Following up studies o[n se]nsitization of hypoxic cells to ionizing radiation by NO, we have designed compounds with potential medicinal applications as radiosensitizers. The focus of the present study is, compounds based on certain S-nitroso-β-aminothiols, since S-nitrosoglutathione (GSNO) and S-nitroso-N-acetylpenicillamine (SNAP) have been effective radiosensitizers among other NO release molecules. To increase NO delivery, we have synthesized linked derivatives of cysteine, cysteamine and penicillamine with acridine, a known DNA intercalator. Several products contain an S-nitroso-β-aminothiol linked on the 9-position of the acridine nucleous, either directly or via an α,ω diamine linker, -NH(CH$_2$)$_n$NH-, n = 1-4. The stability of this class of compounds, their NO release properties, and their in vitro radiosensitization properties will be discussed.

212. **SYNTHESIS AND IN VITRO ANTICANCER ACTIVITY OF SOME PERMA[N]ENTLY CHARGED DERIVATIVES OF TAMOXIFEN.** Emil Pop, Bogumila Rachwal, Stanislaw Rachwal, Anat Biegon, Sarina Striem and Marcus E.Brewster, Pharmos Corporation, 2 Innovation Drive, Alachua, FL 32615 and Rehovot Israel.

Permanently charged derivatives of tamoxifen were synthesized by quaternization of the tertiary nitrogen with aliphatic (methyl, ethyl, propyl, 2-methyl-2-butenyl, cyclohexyl, etc.) and aromatic (benzyl, fluoro-, chloro-, bromo-, methyl-, methoxy-, nitro-benzyl, naphthylmethyl, etc.) halides . The anticancer activity of the novel derivatives was tested by determination of their potency to inhibit EJ or Panc-1 cell growth as well as the phorbol ester binding to EJ cells. Cell viability was assessed using the XTT method. Aromatic derivatives proved to be more active than aliphatic ones. The naphthylmethyl analog was more potent than the benzyl . In the benzyl series, methyl (o, m, p) substituents potentiated inhibition of both EJ and Panc-1 cells and phorbol ester binding, while fluoro derivatives were the least active. In the aliphatic series, the 2-methyl-2-butenyl derivative was more active then the ethyl, while the activity of the cyclohexyl derivative was very low. A good correlation between the ability to inhibit EJ and Panc-1 cell growth and to block phorbol ester binding was noticed.

213. REACTIVITY OF CYTOTOXIC AZIRIDINYL POLYAMINE ANALOGUES WITH DNA AND 4-(p-NITROBENZYL)PYRIDINE. F.A. Rogers, Y. Li, J.L. Eiseman, D.L. Sentz, S. Pan, L.-T. Hu, M.J. Badrian, M.J. Egorin, and P.S. Callery, Schools of Pharmacy and Medicine, and University of Maryland Cancer Center, University of Maryland at Baltimore, Maryland 21201

The diaziridinyl spermine analogue, 1,12-diaziridinyl-4,9-diazadodecane (**1**), and two methylated analogues, **2** and **3**, were synthesized as DNA targeted alkylating agents. Compound **1** was approx. 10-fold more cytotoxic against L1210 murine leukemia cells and against an NCI panel of 60 cell lines than were **2** and **3**. The *in vivo* toxicity of **2** and **3** was significantly lower than that of **1** in female CD$_2$F$_1$ mice. DNA reactivity of pBR322-derived DNA with **1** was observed at a 10-fold lower concentration than that for **2** or **3**. Rates of formation of adducts formed with 4-(p-nitrobenzyl)pyridine were similar for **1**, **2**, and **3**. Methylation of polyamine **1** results in aziridine analogues with decreased biological activity.

1 R^1,R^2 = H
2 R^1 = CH$_3$, R^2 = H
3 R^1 = H, R^2 = CH$_3$

214. COVALENT MODIFICATION OF N3 OF GUANINE BY (+)-CC-1065 RESULTS IN PROTONATION OF THE CROSS-STRAND CYTOSINE. H.-J. Park, L. H. Hurley, Drug Dynamics Institute, College of Pharmacy, University of Texas at Austin, Austin, Texas 78712

The antitumor antibiotic (+)-CC-1065 can alkylate N3 of adenine in defined sequences of duplex DNA, resulting in an extra positive charge that is delocalized over the covalently modified adenine. Recently, it has also been found that (+)-CC-1065 can alkylate N3 of guanine in select oligomers. A High field proton NMR analysis coupled with restrained molecular dynamics was conducted on the (+)-CC-1065-d[GCGCAATTG*CGC]2 adduct (* indicates the drug alkylation site). The exchangeable imino and amino proton assignments reveal that the cross-strand cytosine is protonated due to drug modification on N3-guanine. This study suggests a novel mechanism for drug alkylation on N3-guanine involving the protonation of cross-strand cytosine. Finally, the cross-strand protonated cytosine may be deaminated to uracil with the greater probability, and this may contribute to the mutagenic potential of drugs that alkylate N3-guanine.

215. DESIGN AND SYNTHESES OF STEREOCHEMICALLY DEFINED BISDIOXOPIPERAZINE ANALOGS AS TOPOISOMERASE II INHIBITORS. Shabana S. Insaf, Donald T. Witiak*, Division of Medicinal Chemistry, University of Wisconsin, Madison, WI 53706.

Recently, DNA topoisomerases have been recognized as viable chemotherapeutic targets. These enzymes catalyze the interconversion of different topological forms of DNA. Bis(2,6-dioxopiperazine)s are micromolar inhibitors of topoisomerase II. Structurally, these molecules are cyclized imide derivatives of ethylenediaminetetraacetic acid. Structure activity studies have established the relevance of the central linker chain and intact imide ring systems. Substitutions on the linker and/or imide nitrogens has been thoroughly investigated. In order to probe the effect of substitution on the ring systems, bicyclic and chiral bisdioxopiperazine analogs were developed. This poster will discuss the design principles and synthetic strategies used.

216. THE NOVEL BIS(BENZOXAZOLE) CYTOTOXIC NATURAL PRODUCT UK-1 IS A MAGNESIUM ION-DEPENDENT DNA BINDING AGENT AND INHIBITOR OF HUMAN TOPOISOMERASE II. Sean M. Kerwin, Mark R. DeLuca, and Yan Kwok Department of Medicinal Chemistry, College of Pharmacy, University of Texas, Austin, Texas 78712

We describe the chemical and biochemical properties of a structurally unique cytotoxic bis(benzoxazole) natural product, UK-1. Synthetic UK-1 was prepared and found to bind a variety of di- and tri-valent metal ions. UK-1 binds double-stranded DNA in a metal ion-dependent fashion. Significantly, we have found that UK-1 is a potent inhibitor of human topoisomerase II. This work provides a rational for the origin of the cytotoxicity of this compound, and may lead to the design of more active agents of potential use in cancer chemotherapy.

UK-1

217. SYNTHESIS, METAL ION BINDING, AND DNA CLEAVAGE STUDIES OF BIS(PROPARGYLIC) SULFONE CROWN ETHERS. Sean M. Kerwin, Mark M. McPhee Department of Medicinal Chemistry, College of Pharmacy, University of Texas, Austin, Texas 78712

A new class of crown ethers possessing a bis(propargylic) sulfone moiety has been designed as potential ion-selective DNA cleavage reagents. We have prepared a homologous series of these crown ethers (**1**, n = 1 - 4). These crown ethers bind alkali metal ions with varying affinity. In addition, these crown ethers cleave double stranded DNA, apparently through the intermediacy of an electrophilic species generated from **1** in basic aqueous solutions.

1

218.

TEMPLATE DIRECTED DESIGN AND DEVELOPMENT OF NEW QUINOBENZOXAZINE ANALOGS AS ANTI-CANCER AGENTS. **Q. Zeng,** Y. Kwok, S. M. Kerwin, L. H. Hurley. Drug Dynamic Institute, College of Pharmacy, University of Texas, Austin, TX 78712

Quinobenzoxazines are synthetic fluoroquinolone analogs that show excellent antineoplastic activity. They are catalytic inhibitors of topoisomerase II. In studying their mechanism of function, we proposed a model of novel 2:2 quinobenzoxazine:Mg2+ complex assembled on DNA. Based on this model, a series of compounds were designed and synthesized. Topoisomerase II decatenation, B16 cell growth inhibition, DNA unwinding, DNaseI footprinting assays were used to evaluate these agents. Two types of the agents are comparable or better topoisomerase II inhibitors than the parent quinobenzoxazine A-62176. The results indicate a general trend for this class of agents that higher DNA binding affinity leads to higher potency of topoisomerase II inhibition and therefore stronger cytotoxicity against tumor cells. These results provide further evidence to support our understanding of interaction between DNA and quinobenzoxazines.

219. SYNTHESIS AND CYTOTOXICITY OF 3,5-ISOXAZOLIDINEDIONES AND 2-ISOXAZOLINE-5-ONES IN MURINE AND HUMAN TISSUE CULTURED CELL LINES, R. A. Izydore, I. H. Hall, X. Zhou, D. L. Daniels, T. Woodard, E. Tse, and R. A. Muhammad, Department of Chemistry, North Carolina Central University, Durham, NC 27707 and Division of Medicinal Chemistry and Natural Products, University of North Carolina, Chapel Hill, NC 27599

A series of 3,5-isoxazolidinediones was synthesized and evaluated for cytotoxic activity. Twenty six derivatives were prepared by reacting 2-unsubstituted 3,5-isoxazolidinedione with halo ketones, esters, amides and nitriles in the presence of potassium hydroxide and tetrabutylammonium bromide or by the condensation of malonyl chlorides and benzohydroxamic acids. In four instances the isolated products were 2-isoxazoline-5-ones. The isoxazolidines were evaluated for cytotoxic activity in the Ehrlich ascites carcinoma screen in CF-1 mice at 8 mg/kg/day. A number of the compounds showed greater than 80% inhibition of tumor growth. The compounds were also evaluated in the L1210 lymphoid leukemia screen and in cultured cell lines derived from human tumors. These included the Tmolt₃ T cell leukemia, HeLa suspended cell uterine carcinoma, HeLa solid carcinoma, colon adenocarcinoma SW-480, HCT-8 ileum carcinoma, osteosarcoma TE 418, lung MB9812 and A549, skin A431, KB nasopharynx and brain glioma screens. Selected compounds showed potent activities in a number of these screens. Four of the componds were selected to examine the biochemical mode of action in L1210 leukemia cells over 60 minutes at concentrations of 25-100 uM. Several of the rate-limiting enzymes were significantly inhibited in a concentration dependent manner. The 3,5-isoxazolidinediones and 2-isoxazoline-5-ones are potent cytotoxic agents that have activities and modes of action that are similar to other five-membered heterocyclic systems that we have studied.

220.

STRUCTURAL FEATURES OF A SIX NUCLETIDE HAIRPIN LOOP. **Matthew A. Fountain** Thomas R. Krugh, Martin Serra, Douglas H.Turner, University of Rochester, Rochester, NY, 14627

The hairpin loop GUAAUA occurs frequently in ribosomal RNA. The structural features of the r(GGCGUAAUAGCC) hairpin have been determined by NMR and molecular modeling. The U5 base is stacked on the sheared base pair formed by G4 and A9 and can initiate a uridine turn similar to that observed in the anticodon loop of tRNA. The A6, A7, and U8 bases can stack on one another with their hydrogen-bonding surfaces exposed to the solvent, suggesting that they are available for tertiary interactions or protein recognition in rRNA.

221.

INFLUENCES OF STRUCTURE AND REACTIVITY ON THE BINDING OF QUINONE METHIDES TO DNA. **Mark A. Lewis,** Minfei Qian, and John A. Thompson, Department of Pharmaceutical Sciences, University of Colorado Health Sciences Center, Denver, Colorado 80262

Peroxidase- or cytochrome P450-catalyzed oxidations of compounds bearing a 4-methylphenol moiety result in the formation of electrophilic p-quinone methides (QMs) capable of alkylating DNA bases. A series of QMs with varying substituents in the 2- and 6-positions were investigated and those with bulky alkyl groups preferentially alkylated the N6 position of dA and secondarily the N2 position of dG. On the other hand, QMs with small substituents alkylated dA and dG to a similar degree with the latter adducts distributed among the N2, N1 and N7 positions of guanine.

Reactions between QMs and individual deoxynucleosides demonstrated that alkylation regioselectivity is not greatly influenced by the intrinsic electrophillicity of the QM. In each case N1-dG, N2-dG, N7-dG, and N6-dA adducts were formed in similar ratios. Steric accessibility, therefore, is the main factor which influences the preferred DNA alkylation sites of simple QMs.

222. TOTAL SYNTHESIS AND INVITRO ACTIVITY OF CRYPTOPHYCINS AND THEIR 16-(3-PHENYLACRYLOYL) DERIVATIVES. D. Nguyen, R. Rej, B. Go, S. Fortin, J.-M. de Muys, J.-F. Lavallée*; BioChem Therapeutic Inc., Laval, Quebec, Canada H7V 4A7

Cryptophycin A, a cyclic depsipeptide isolated fron the blue-gree algae (cyanobacterium) Nostoc sp. GSV 224, has shown excellent activity against solid tumors implanted in mice. The benzylic epoxide, which was shown to be very important for biological activity, is also fairly unstable under both acidic and alkaline conditions. The high doses needed to observe in vivo activity might be a result of this instability. In order to solve this problem while preserving the electrophilic character of the benzylic position, the 16-(3-phenylacryloyl) derivatives have been proposed to give promising analogs of the natural product. An efficient and convergent total synthesis was developed to prepare Cryptophycins A, B, C, D and their 16-(3-phenylacryloyl) derivatives. The in vitro biological activity data will be presented.

223. LACTONE-ADDUCTS OF (E)-RAC-(2'-BUTEN-1',4'-OLIDO)-[3',4':1,2]-6-FLUORO-2-METHYL-3-(P-METHYLSULFONYLBENZYLIDENE)INDAN AND THEIR GROWTH INHIBITORY ACTIVITY.
Gerhard J. Sperl, Ph.D.[+], Rifat Pamukcu, MD[#], David S. Alberts, MD[^], Klaus Brendel, Ph.D.[%], Paul H. Gross, Ph.D.[+*]. University of the Pacific, Stockton, California, 95211: Department of Chemistry[+*]; Cell Pathways, Inc.[#], Aurora, Colorado; Arizona Cancer Center[^]: Tucson, Arizona; University of Arizona, Tucson, Arizona: Department of Pharmacology[%].

Products from the β-addition of S, N, O and C nucleophiles to the recently synthesized $\Delta^{\alpha,\beta}$ butenolide are reported, along with their growth inhibitory activity data against human colon adenocarcinoma cell line HT-29.

R1, R2 = π-bond ⟶ R1, R2 various substituents

224. SYNTHESIS OF RADIOACTIVE SULINDAC-SULFONE-LACTONE VIA SULFONIUM SALT AND ^{14}C ALKYLHALIDE.
Gerhard J. Sperl, Ph.D.[+], Rifat Pamukcu, MD[#], Klaus Brendel, Ph.D.[%], Paul H. Gross, Ph.D.[+*]. University of the Pacific, Stockton, California, 95211: Department of Chemistry[+*]; Cell Pathways, Inc.[#], Aurora, Colorado; University of Arizona, Tucson, Arizona: Department of Pharmacology[%].

A new method of radioactive labeling via sulfonium salts and ^{14}C alkylhalides is reported, along with a novel synthesis of (E)-rac-(2'-buten-1',4'-olido)-[3',4':1,2]-6-fluoro-2-methyl-3-(4-[^{14}C-methyl]-sulfonylbenzylidene)-indan, a drug that showed very high activity, when tested in various tumor cell lines.

225.

DESIGN AND SYNTHESIS OF SELECTIVE INHIBITORS OF CATHEPSIN B AS NOVEL ANTI-METASTATIC AGENTS. **Stephanie Michaud** T.-H. Chan; J. S.Mort; J. DeMarte; B. J. Gour-Salin; Dept. of Chemistry, McGill University, 801 Sherbrooke St. W., Montreal, PQ, H3A 2K6, Canada

In recent years, it has become apparent that the cysteine protease cathepsin B plays a crucial role in cancer metastasis. As such, much research has focused on the development of specific inhibitors both in order to dissect out the exact role of the enzyme in normal biological processes and to provide compound 'leads'. Since many different cysteine proteases exist in mammalian cells, high selectivity of such inhibitors is essential. Cathepsin B varies from the other cysteine proteases in its class in that it has both endo- and exo-peptidase activity. This particular activity is attributed to the presence of an additional structural feature found only in cathepsin B: the occluding loop. We have used this structural feature of cathepsin B as a unique template for designing epoxysuccinyl dipeptides as highly selective inhibitors of cathepsin B. We will present the synthesis of these inhibitors, kinetic data and the results of our in vitro assays.

226.
SYNTHESIS AND STRUCTURE ACTIVITY-RELATIONSHIPS OF A SERIES OF 1,1-DIARYL INHIBITORS OF TUMOR NECROSIS FACTOR-α.
Mary G. Shire, Laura G. Corral, Lu-Min Wong, David I. Stirling and George W. Muller.
Celgene Corporation, Warren, NJ 07059.

Tumor necrosis factor-alpha (TNFα) is a key proinflammatory cytokine released by monocytes /macrophages in response to immunostimulants. Elevated levels of TNFα have been implicated in a number of disease states including septic shock, cachexia, rheumatoid arthritis, GvHD, asthma, ARDS and AIDS. The development of inhibitors of TNFα is an intensive area of research. Monoclonal antibodies to TNFα have been shown to be efficacious in clinical trials in the treatment of rheumatoid arthritis. Other approaches to the inhibition of elevated levels of TNFα include the use of glucocorticoids, cAMP elevating agents, adenosine agonists, TNF convertase inhibitors and CSAID cytokine inhibitors. A series of 1,1-diaryl compounds which inhibit TNFα in LPS-stimulated PBMCs have been prepared. The structure activity-relationships and the mechanism of action of these compounds will be presented.

227. INHIBITION OF TUMOR NECROSIS FACTOR ALPHA (TNFα) AND INTERLEUKIN-1β (IL-1β) PRODUCTION BY NOVEL 9-DEAZA-9-PHOSPHAPURINES
G. R. Revankar, T. S. Rao, J. O. Ojwang, R. F. Rando, and T. L. Wallace
Aronex Pharmaceuticals, Inc., 9391 Grogans Mill Road, The Woodlands, TX 77380

Cytokines such as tumor necrosis factor alpha (TNFα) and interleukin-1β (IL-1β) have been implicated in inflammatory processes and in the pathogenesis of many diseases including tumors in humans. Agents which can inhibit the production or maturation of TNFα and IL-1β in these indications may have excellent therapeutic potential. We designed and synthesized novel 1,3-azaphospholo[4,5-d]-pyrimidines (9-deaza-9-phosphapurines) as inhibitors of cytokines. 7-Amino-1H-1,3-azaphospholo[4,5-d]-pyrimidine (1), and several of its derivatives exhibit superior activity against TNFα and IL-1β production in the lipopolysaccaride (LPS) or a mixture of phorbol myristate acetate and interferon gamma (PMA/IFN$_\gamma$) stimulated human monocytic THP-1 cells. Compound 1 also inhibits the production of TNFα and IL-1β in peripheral blood mononuclear cells (PBMC) and is not toxic to mice in a five-day toxicity study. Detailed synthesis of these compounds and cytokine inhibition results are the subject of this presentation.

228. DESIGN RATIONALE AND SYNTHESIS OF NON-NATURAL ANALOGUES OF THE CATIONIC AMINO ACIDS ARGININE AND LYSINE. **Thomas A. Dix**, Kevin J. Kennedy, Joseph T. Lundquist, Tiberiu Simandan, Kyle P. Kokko, and Craig C. Beeson§, Department of Pharmaceutical Sciences, Medical University of South Carolina, 171 Ashley Avenue, Charleston, SC, 29425-2303 and §Department of Chemistry, University of Washington, Seattle, WA, 98195-1700.

A series of 34 non-natural isosteric analogues of the cationic, ion-pairing, natural amino acids arginine and lysine have been synthesized. The design of these molecules is based on the concept of steric inhibition of solvation, in that judicious placement of alkyl groups can destabilize aqueous ion solvation and favor ion-pairing (see - Beeson and Dix (1993) *J. Am. Chem. Soc.* **115**, 10275). Thus, when the residues are substituted for the natural amino acids in biologically active peptides, enhanced ion-pairing of the peptides to their receptors to increase the peptides' biological activities should result. The increased lipophilicity of the non-natural residues also is expected to improve pharmacokinetic parameters and agonist/antagonist behaviors of peptides in many cases. The non-natural analogues have been protected for routine inclusion in automated Merrifield peptide synthesis protocols.

229.

SYNTHESIS OF IMIDAZOLE BASED ARGININE SURROGATES. S. Lévesque, M.A. Siddiqui, Biochem Therapeutic, Laval, Quebec, Canada, H7V 4A7

Arginine is an important amino acid which as a part of bioactive peptides and proteins participates in important interactions. The guanidino group of arginine plays a crucial role in binding to many receptor's active sites, however this highly charged function contributes to impermeability of ligands/drugs to cells. In order to improve the cell permeability properties of arginines, several of its surrogates, incorporating imidazole as guanidine isostere, are synthesized. The novel methodologies to prepare imidazole based arginine surrogates will be presented.

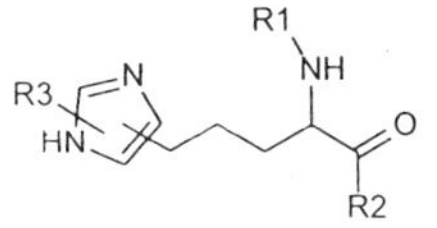

R1,R2:Protecting group
R3:H, NH_2, NO_2

230. ASYMMETRIC SYNTHESES AND NEUROPHARMACOLOGICAL ASSESSMENTS OF METHYLPHENIDATE ENANTIOMERS AND THEIR PARA-SUBSTITUTED DERIVATIVES
Dung L. Thai†, Donald E. Bierer§, and James M. Perel†
†Departments of Pharmacology and Psychiatry, University of Pittsburgh, Pittsburgh, PA 15213
§Shaman Pharmaceuticals, Inc., 213 East Grand Avenue, South San Francisco, CA 94080

The individual enantiomers of methylphenidate (**1**) were prepared in 6 steps and 10-28% overall yield from optically pure pipecolic acid. Key steps in the synthesis included aryl organometallic addition to a carboxy-activated N-Boc-pipecolate, Wittig olefination of an α-amino ketone, and hydroboration of a chiral olefin. An investigation of the level of 1,2-asymmetric induction in the hydroboration of a chiral allylic alkene intermediate using various borane reagents was also conducted. Oxidation of the resulting alcohol, esterification, and deprotection afforded pure **1**. From (d)-pipecolic acid, the (2R, 2'R) and (2S, 2'R) enantiomers of **1** were obtained in >99% optical purity while the (2S, 2'S) and (2R, 2'S) enantiomers were derived from (l)-pipecolic acid in 96% optical purity. The versatility of this methodology is demonstrated with the synthesis of the (2R, 2'R) and (2S, 2'S) enantiomers of p-bromo and p-methoxy analogues in similar yields and enantiomeric purities.

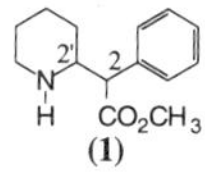

231. A NEW SYNTHETIC APPROACH TO HEXAHYDROISOQUINO[3,2-b][3]BENZAZEPINES.
D. Zlotos, W. Meise, Pharmaceutical Institute, University of Bonn, Kreuzbergweg 26 , D-53115 Bonn, Germany.

"Iso-B-homoberbines" **6** are of attracting interest as phosphodiesterase inhibitors. The synthesis started off from alkylation of the ß-ketoester **1** (NaH, benzylchloride) and **1'** resp. (dianion approach: 2 LDA, benzylchloride). Subsequent demethoxycarbonylation with LiI in DMF gave the 3-benzyl-2-tetralones **3** which can be rearranged to the lactones **4**. From a following reduction using borane and a final Pictet-Spengler cyclisation the homoberbine **6** can be obtained. The conformation of **6** is characterized by means of IR und NMR methods.

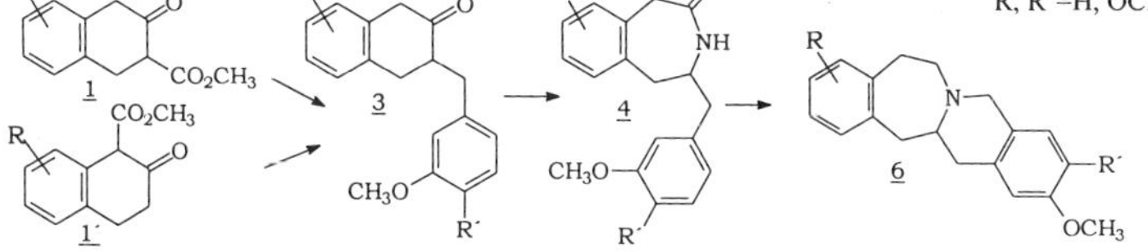

R, R'=H, OCH_3

232.
TITLE. **CHIRAL SEPARATIONS OF LOBELINE ANALOGS USING HIGH PERFORMANCE CAPILLARY ELECTROPHORESIS AND DERIVATIZED CYCLODEXTRINS AS CHIRAL ADDITIVES.** Christopher R.McCurdy*, Thirunellai G.Venkateshwaran, JamesT.Stewart, and J.Warren Beach. Department Of Medicinal Chemistry, College of Pharmacy, University of Georgia, Athens, GA: 30602-2352.

Abstract : Several Lobeline analogs have been analyzed using High Performance Capillary Electrophoresis (HPCE). The chiral separation of the enantiomers was achieved using derivatized-B-cyclodextrins as chiral additives. The effect of the additive concentration on the enantiomeric separation was investigated. The effect of other parameters such as temperature, organic modifier concentration, pH of buffer solution and type of additive used has also been investigated. The addition of methyl-B-cyclodextrin, dimethyl-B-cyclodextrin and hydroxypropyl-B-cyclodextrins worked efficiently for the enantiomeric separation of most compounds. However, in the case of one of the analogs, separation was not achieved under any of the conditions

233. A NEW AND EFFICIENT ROUTE IN PEPTIDE NUCLEIC ACIDS (PNA) BACKBONE SYNTHESIS. Kelly Teng and Kurt E. Weiler, Department of Medicinal Chemistry, Isis Pharmaceuticals, 2280 Faraday Ave, Carlsbad, CA 92008

A convenient and general two steps procedure for the synthesis of PNA backbone **3** is described. The procedure involved the coupling of two α-amino acid derivatives, alcohol **1** and sulfonamide **2**, by Mitsunobu reaction, followed by facile *N*-sulfonyl protecting group removal with $PhSH/K_2CO_3$.

234. UTILIZATION OF A BENZOYL MIGRATION TO EFFECT AN EXPEDITIOUS SYNTHESIS OF THE PACLITAXEL C-13 SIDE CHAIN. Hu, Zhiyong and Erhardt, Paul W. Center for Drug Design and Development/Department of Medicinal and Biological Chemistry, The University of Toledo College of Pharmacy, Toledo, OH 43606-3390

In order to prepare simplified analogues of the anticancer drug paclitaxel, we needed gram quantities of paclitaxel's C-13 side chain. Of the reported routes, the Sharpless et al. asymmetric dihydroxylation method seemed to be the most promising (J. Org. Chem., 59, 5104 (1994)). However, during this procedure an alcohol is protected as its acetyl ester and when an adjacent position is elaborated to an amine, the acetyl spontaneously migrates to form the acetamide. This necessitates removal of the acetyl as a distinct step so that the amine can be subsequently reconverted to a desired benzamide. We have instead utilized a benzoyl to initially protect the hydroxy group and, after observing a similar migration, have thus received the desired product directly. Although this possibility has also been alluded to by Nicolaou et al.(Angew. Chem. Int. Ed. Engl., 33, 15 (1994)) with regard to a pending Sharpless patent, no experimental details or physical properties for the actual intermediates have been reported. Our results for this specific process will be conveyed in detail along with our preliminary findings associated with extending the process to include novel heterocyclic systems.

235. SYNTHESIS OF STERIC PROBES TO EXAMINE THE CHEMICAL BEHAVIOR OF POTENTIAL BENZYLAMINE-RELATED CHIRAL AUXILIARY SYNTHETIC REAGENTS Ni, Yidong and Erhardt, Paul W. Center for Drug Design and Development/Department of Medicinal and Biological Chemistry, The University of Toledo College of Pharmacy, Toledo, OH 43606-3390

In order to confidently employ benzylamine-related systems as practical chiral auxiliary synthetic reagents, we needed to assess the ease and stereoselectivity with which debenzylation can occur when hydrogenolysis is attempted in a variety of steric and asymmetric environments. To first define the general scope for such debenzylations, a family of six probe molecules was prepared having incremental elements of bulk. The synthesis and characterization of these steric probes, along with the unanticipated results from kinetic studies pertaining to their hydrogenolyses, will be conveyed.

236. **BISLITHIATION OF DIARYL COMPOUNDS: FACILE SYNTHESIS OF TRICYCLIC PHARMACEUTICALS BASED ON HETEROCYCLES FUSED TO TWO BENZENE RINGS.** Anu Chopra, Donna C. Dorton and Craig A. Ogle,* *Department of Chemistry, University of North Carolina at Charlotte, Charlotte, NC 28223*

Tricyclic compounds such as substituted acridans, dibenzo[b,f][1,4]diazepines and dibenzo[b,f][1,4]oxazepines exhibit various types of pharmacological activity. These compounds contain a heteroatom in the central ring and an alkamine chain or a nitrogen-containing heterocycle attached to the carbon atom in the central ring. This class of compounds form a significant part of one of the most extensive and well-studied group of drugs known as tricyclic pharmaceuticals and have found extensive use in treatment of various mental and physical disorders. A facile preparation of tricyclic pharmaceuticals based on the acridan and dibenzo[b,f][1,4]oxazepine skeletons will be described. Our approach allows for the cyclization of the central ring and the introduction of the side chain in one step. These tricyclic compounds were prepared by utilizing the reaction between dilithiodiaryl compounds and appropriately substituted ester moieties.

237. **PREPARATION OF ENANTIOMERICALLY PURE PRECURSORS FOR THE SYNTHESIS OF [C-11]McN5652X AND [C-11]McN5652W.** Y. Huang, K. Mahmood, N. Simpson, S. Mason, C. Mathis,* PET Facility, Department of Radiology, University of Pittsburgh, Pittsburgh, PA 15213.

The serotonin reuptake complex ligands McN5652X and McN5652W (the active [(+)-] and inactive [(-)-] enantiomers) have been radiolabeled and utilized in positron emission tomography (PET) imaging studies [Szabo et al., Synapse (1995)]. Suehiro and coworkers recently reported the synthesis of the thio ester precursors **2a** and **2b** for radiolabeling with high specific activity [C-11]methyl iodode [Nucl. Med. Biol. (1995)]; however, low isolated yields of **2a** and **2b** were obtained as a result of isomerization. New, high yielding routes are reported here for the synthesis of **2a** and **2b** from **1a** and **1b**, respectively. No isomerizations or losses of enantiomeric excess were obtained for the transformations of **1a** to **3a** and **1b** to **3b**, providing high yielding routes for the stereospecific syntheses of **3a** and **3b**.

1a: (+)-enantiomer
1b: (-)-enantiomer
2a, 2b: R = alkyl or aryl
3a: [C-11]McN5652X
3b: [C-11]McN5652W

238. **SYNTHESIS AND USE OF *N1'-(p*-FLUOROBENZYL) NALTRINDOLE.** E. Akgün[a], D.A. Rottenberg[a], P.S. Portoghese[b]. a) PET Imaging Service, VAMC, Minneapolis, MN 55417; b) Department of Medicinal Chemistry, University of Minnesota, Minneapolis, MN 55455.

In an effort to design δ_2-selective ligands for PET imaging, *N1'-(p*-fluorobenzyl)naltrindole (FBNTI) was synthesized by reacting of 3-*O*-benzyl-protected naltrindole with *p*-fluorobenzyl bromide, followed by deprotection. New ligand binds strongly to δ-receptor. Results of these studies will be discussed.

Ar : C_6H_4F (*p*)

FBNTI

239. REDUCTION OF THE (+)- CAMPHOR OXIME WITH BORANE•THF COMPLEX, L. D. Vazquez, E. Garcia, M. De Jesus, M. Ortiz, Department of Chemistry, University of Puerto Rico-Humacao Campus, CUH Station, Humacao PR 00791

Amines are important organic compounds due to their biological activity and as precursors for a large variety of pharmaceutical products. Our research objective is to develop new methods for the synthesis of chiral primary amines. Our recent study involves the camphor oxime synthesis, and its subsequent reduction with BH_3 in THF to prepare the chiral primary amine. Preliminary results indicate the formation of mainly the two amino derivates **1** and **2**.

240. THE REDUCTIVE REARRANGEMENT OF O-SILYL SUBTITUTED ACETOPHENONE OXIMES WITH BORANE•THF COMPLEX. M. Ortiz, D. Figueroa and J. López. University of Puerto Rico, Humacao P.R. 00791

We have synthesized O-Silyl subtituted acetophenone oximes (**1**) as precursors for the enantioselective synthesis of chiral α-primary amines. Preliminary studies on the reduction of the O-Silylated oxime ethers with Borane•THF complex, however indicated a migration of the aryl moiety and a subsequent reduction of the intermediate. The ratio of **2** and **3** was found to be dependent on the size of the substituents on silicon.

241.   STERIC EFFECTS ON THE SYNTHESIS OF PROMAZINE DERIVATIVES
C. GARCÍA[1], L. E. PIÑERO[1], C. MATEO[2] AND D. MARALET[2]; University of Puerto Rico - Humacao, Department of Chemistry ; Humacao, Puerto Rico 00791

Promazine derivatives have numerous pharmacological activities, the best known being their action as tranquilizers. Chlorpromazine (CPZ), the most important representant of these drugs, is actually being used in the United States to treat thousands of psychiatric patients annually. The molecular photochemical mechanisms for the photosensitizing ability of CPZ are still unknown, since very little is known about their structure-activity relationship. We have synthesized several non available 2-substituted promazine derivatives by thionation, Smile's rearrangement and Ulmann's reaction using several catalysts. The yield of the corresponding phenothiazine intermediate is determined by the torsion angle of the compound.

This Work is supported by
1) NIH-MBRS Grant GM 8216-11 to Humacao University College and
2) NSF-AMP Grant 539570

242.   Trifluoromethylthiolation of Enamines: S. Munavalli[1], D. K. Rohrbaugh[2], G. W. Wagner[1], F. J. Berg[2], D. I. Rossman[2] and H. Dupont Durst[2]

[1]Geo-Centers, Inc., Ft. Washington, MD 20774 and [2] U. S. Army, Edgewood Research Development and Engineering Center, Aberdeen Proving Ground, MD 21010

The use of enamines in carbon-carbon bond formation has become an established classical method in organic synthesis. The ease of enamine synthesis, regiospecific C-alkylation and facile regeneration of 2-substituted carbonyl compounds have transformed the enamines into versatile synthetic intermediates. Introduction of fluorine and fluorine containing moieties such as the trifluoromethyl and trifluoromethylthio groups has been reported to considerably enhance the biological potency of the parent compounds. The reaction of cyclic enamines (**1** and **2**) with N-trifluoromethylthiophthalimide (**3**) has been found to regiospecifically introduce the trifluoromethylthio function and to furnish 2-(trfluoromethylthio)ketones (**4** and **5**) in excellent yields (>80%). However, the treatment of **2** with trifluoromethylsulfenyl chloride (**6**) gives a complex mixture of **8** compounds including **5**. The extension of this reaction to biologically active compounds and the spectrometric characterization of the various products are presented in this paper.

243. INHIBITORS OF THYMIDYLATE SYNTHASE: SYNTHESIS AND BIOLOGICAL ACTIVITIES OF 2-AMINO-4-OXO-5-SUBSTITUTED-PYRROLO[2,3-d]PYRIMIDINES. Aleem Gangjee*, Anup P. Vidwans, Eifatih Elzein, Division of Medicinal Chemistry, Graduate School of Pharmaceutical Sciences, Duquesne University, Pittsburgh, PA 15282.

Thymidylate synthase (TS) is a crucial enzyme that catalyses the reductive methylation of dUMP to dTMP. This is the exclusive denovo source of dTMP, hence inhibition of TS, in the absence of salvage, leads to "thymineless death" of the cell. As part of our ongoing search for potent inhibitors of TS as potential antitumor agents, as well as selective inhibitors of organisms that cause opportunistic infections in AIDS patients we report here the synthesis and biological activities of a series of 2-amino-4-oxo-5-substituted-pyrrolo[2,3-d] pyrimidines with a variety of substitutions on the phenyl ring of structure **1**.

244. THE SYNTHESIS OF 6-SUBSTITUTED 2-METHYL-4-OXOPYRIDO[3,2-d]PYRIMIDINE ANALOGUES AS THYMIDYLATE SYNTHASE INHIBITORS. Aleem Gangjee*, Yuanming Zhu, Division of Medicinal Chemistry, Graduate School of Pharmaceutical Sciences, Duquesne University, Pittsburgh, PA 15282.

The quinazoline antifolate analogue **1** (tomudex) is an inhibitor of thymidylate synthase which is an approved antitumor drug in Europe and is in phase III clinical trials in the USA. We are interested in the design and synthesis of 2-methyl-4-oxopyrido[3,2-d]pyrimidine classical and nonclassical antifolates **4-5** which are N5 isosteres of the quinazoline ring in **1**. The synthesis was accomplished via the cyclization of pyridine N-oxide **2** which afforded a key intermediate **3** with a 6-chloro moiety. Nucleophilic displacement of this chloro allows the synthesis of a variety of classical and nonclassical pyrido[3,2-d]pyrimidine analogues represented by **4** and **5**.

3. R = Cl
4. R = 4'-SPh-CO-Glu^L
5. R = 4-pyridinyl

245. SYNTHESIS AND BIOLOGICAL ACTIVITY OF 5-DESMETHYL-2,4-DIAMINOPYRIDO[2,3-d]PYRIMIDINE DIHYDROFOLATE REDUCTASE INHIBITORS.
Aleem Gangjee[*1], Ona Adair[1] and Sherry F. Queener[2], [1]Division of Medicinal Chemistry, Graduate School of Pharmaceutical Sciences, Duquesne University, Pittsburgh, PA 15282 and [2]Department of Pharmacology and Toxicology, Indiana University School of Medicine, Indianapolis, IN 46202.

SAR established in our laboratory, as well as other laboratories, indicates that a methyl group in the 5-position of the 2,4-diaminopyrido[2,3-d]pyrimidine ring can effect both potency and selectivity against *Pneumocystis carinii* (pc) and *Toxoplasma gondii* (tg) dihydrofolate reductase (DHFR). Compounds of the general structure **1** have been synthesized in this study to investigate the effect of 5-desmethyl analogs with various side chains on the potency and selectivity against pcDHFR and tgDHFR. The synthesis, as well as the inhibitory activities against pcDHFR, tgDHFR and mammalian DHFR, will be presented.

246. NITROGEN BRIDGE HOMOLOGS OF TRIMETHOPRIM AS INHIBITORS OF DIHYDROFOLATE REDUCTASES. Aleem Gangjee*, Kshitij Patkar, Division of Medicinal Chemistry, Graduate School of Pharmaceutical Sciences, Duquesne University, Pittsburgh, PA 15282 and Sherry F. Queener, Department of Pharmacology and Toxicology, Indiana University School of Medicine, IN 46202

Trimethoprim (TMP, **1**) is the first line agent in combination with sulfamethoxazole for the treatment of the opportunistic infections caused by *Pneumocystis carinii* (pc). TMP is a selective but not a potent inhibitor of dihydrofolate reductase from this organism or from other organisms like *Toxoplasma gondii* (tg) which cause opportunistic infections in patients with AIDS. In order to preserve selectivity and increase potency, 5N-substituted, 2 atom-bridge homologs of TMP, were designed. The synthesis and biological evaluation of a variety of substituted analogues of general structure **2** will be reported.

247. SYNTHESIS AND BIOLOGICAL ACTIVITIES OF HOMOLOGS OF TRIMETHOPRIM AS INHIBITORS OF DIHYDROFOLATE REDUCTASE FROM *PNEUMOCYSTIS CARINII* AND *TOXOPLASMA GONDII*. Aleem Gangjee*, Mohit Kothare, Anil Vasudevan, Division of Medicinal Chemistry, Graduate School of Pharmaceutical Sciences, Duquesne University, Pittsburgh, PA 15282 and Sherry F. Queener, Department of Pharmacology and Toxicology, Indiana University School of Medicine, IN 46202.

Trimethoprim (TMP, **1**) in combination with sulfamethoxazole, is the first line regimen approved for the treatment of *Pneumocystis carinii* infections. In our quest towards developing a single selective agent capable of inhibiting dihydrofolate reductase from organisms causing opportunistic infections in patients with AIDS, we report here the synthesis and biological evaluation of compounds of general structure **2**, with a variety of substituted aromatic rings as homologs of TMP.

248. WATER-SOLUBLE ECHINOCANDIN ANALOGS 1. - DESIGN AND SYNTHESIS OF PHOSPHATE, PHOSPHONATE AND AMINO ACID ESTERS OF LY303366 - Peter S. Borromeo, James A. Jamison, Michael J. Rodriguez, William W. Turner, Vasu Vasudevan. Infectious Diseases Research, Eli Lilly & Co., Lilly Corporate Center, Indianapolis, IN 46285

LY303366 (R^1 = OH, R^2 = H), presently in Phase II clinical trials, is a potent antifungal compound, especially against *Candida spp.* Part of our continuing support effort has been directed toward increasing water-solubility for improved IV formulations. Modifications at the phenolic hydroxyl (R^1 = OH) have been made utilizing phosphate, phosphonate and amino acid moieties. Synthesis and biological data on these compounds will be presented.

249. WATER-SOLUBLE ECHINOCANDIN ANALOGS 2. - DESIGN AND SYNTHESIS OF POLAR HEMIAMINAL ETHERS. James A. Jamison, Lisa M. LaGrandeur, Michael J. Rodriguez, William W. Turner. Infectious Diseases Research, Eli Lilly & Co., Lilly Corporate Center, Indianapolis, IN 46285

LY303366, a semisynthetic derivative of the naturally occuring Echinocandin B class of antifungal agent, has recently entered clinical trials. Ongoing chemistry on the LY303366 molecule is directed towards incorporation of functional groups that were chosen for their polarity intending to impart water solubility. In this vein, several ethers with amine, carboxyl and phosphate ester functions at the terminus were synthesized. These analogs maintained activity compared to alkyl ethers that were synthesized previously. Synthetic schemes as well as biological results will be presented.

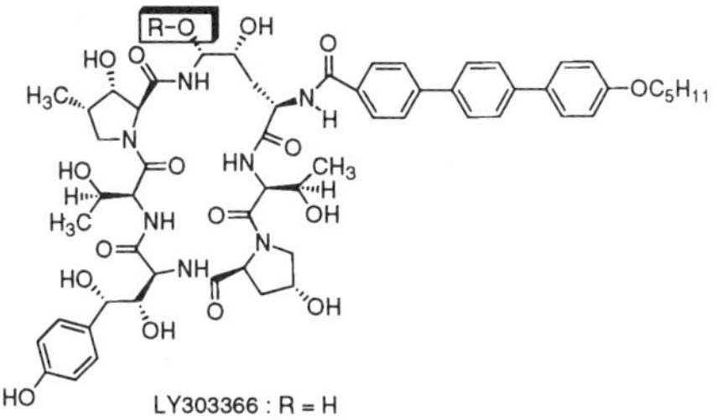

LY303366 : R = H

250. SYNTHESIS AND *IN VITRO* ANTIMICROBIAL ACTIVITY OF 3-KETO 16-MEMBERED MACROLIDES DERIVED FROM TYLOSIN. Lawrence C. Creemer[1], Herbert A. Kirst[1], and John E. Toth[2]. 1)Research and Development, Elanco Animal Health 2001 W. Main Street, Greenfield, Indiana, 46140-0708, 2)Lilly Research Laboratories, Eli Lilly Corporate Center Indianapolis, Indiana, 46285

Recent reports have described 14-membered ketolides derived from erythromycin that exhibit potentially useful antimicrobial activity against macrolide resistant bacteria. To determine if 16-membered macrolides possess analogous activity, 3-keto-5-O-mycaminosyl-tylonolide, 3-keto-desmycosin, and other 3-keto derivatives of tylosin having the general structure below were synthesized.

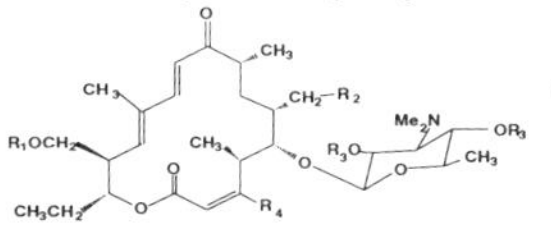

Where:
R_1 = H, Ac, β-D-mycinosyl, 4-O-acetyl-β-D-mycinosyl
R_2 = CHO, CH(OEt)$_2$
R_3 = H, Ac
R_4 = OH

These derivatives were synthesized by protection of susceptible functional groups, oxidation with DCC / pyridine / TFA in DMSO / benzene (1:1), and deprotection. The resulting 3-keto products adopted the enolic tautomer preferentially in chlorinated solvents and exclusively in polar solvents such as EtOH. This effect was demonstrated by integration of the enolic OH proton in the NMR spectrum and measurement of the enolic chromophore in the UV spectrum in EtOH. The *in vitro* antimicrobial activity of these derivatives was greatly reduced compared to their unoxidized parent structures. This preference for the enolic tautomer in 16-membered macrolides is not seen with the 14-membered ketolides. The lack of antimicrobial activity of these enolic derivatives parallels that found with related 2,3-anhydro (R_4 = H) derivatives in the 16-membered macrolide series.

251.

GLYCOPEPTIDE DIMERIZATION: SYNTHESIS AND STRUCTURE ACTIVITY RELATIONSHIP OF COVALENT GLYCOPEPTIDE DIMERS. Douglas R. Stack, Richard C. Thompson, Thalia I. Nicas, Deborah L. Mullen, Thomas F. Butler, Eli Lilly and Company, Indianapolis, IN 46285.

The antibacterial activity of certain glycopeptides has been shown to be influenced by their ability to form a non-covalent dimeric species. It has been reported that dimerization enhances the ability of the glycopeptide to bind to its bacterial cell wall target as well as effectively increasing the concentration of the drug at its target. In this study we investigated the potential advantage of pre-forming dimeric glycopeptide derivatives through the use of covalent linkers. We have successfully prepared a series of covalent glycopeptide dimers utilizing a variety of tethers for attachment. Evaluation of the antibacterial activity of these dimeric species revealed that certain dimers did indeed exhibit a significant improvement in activity as compared to the parent monomeric glycopeptide. Herein we report the synthesis and SAR study of covalent glycopeptide dimers, and show that this is a valid approach to the preparation of glycopeptide derivatives with potent activity against vancomycin-resistant enterococci as well as other Gram-positive organisms.

252. MOLECULAR RECOGNITION IN β-CYCLODEXTRIN-CATALYZED HYDROLYSIS OF PENICILLINS. Yu-Chien Wei, Ching-jer Chang, Department of Medicinal Chemistry and Molecular Pharmacology, Purdue University, West Lafayette, IN. 47907

The hypothesis that β-CD may serve as the biomimetic model of penicillinase was first proposed by Schwartz. This prompted us to study the molecular recognition between β-CD and penicillin since the insight of this specific molecular recognition may serve as the bases for biomimetic catalysis. Detailed kinetic studies, ^{1}H NMR and FAB-MS analysis indicated that a stable complex was formed between β-CD and penicillin V (1:1) in the solid state and in solution. Further molecular modeling study showed that the most stable complex was formed through the recognition between the β-face of penicillin V and the secondary hydroxyl face of β-CD. This molecular recognition was further supported by the isolation and characterization of O^2-penicilloyl-β-CD tetrahedral transient intermediate and O^2-penicilloyl-β-CD (acyl-β-CD) intermediate.

253. SYNTHESIS APPROACH AND BIOLOGICAL STUDY OF MADUMYCIN II
Arun K. Ghosh* and Wenming Liu
Department of Chemistry, University of Illinois at Chicago, 845 West Taylor Street, Chicago, IL 60607

Madumycin II, a streptogramin antibiotic, has shown several very important biological properties. We have developed a synthetic pathway for the southwestern portion of the molecule by using a novel asymmetric synthesis, followed by a Wittig reaction and a Horner-Emmons olefination reaction as key steps. The northeastern portion was synthesized by reacting cis-2-butene, (-)-B-Methoxydiisopinocampheylborane with isobutyraldehyde, followed by ozonolysis, Horner-Emmons olefination. Then LiOH aqueous THF hydrolysis gave the Hydroxy-acid. We will present synthesis and some of our on-going biological studies of Madumycin II and its variants.

254. DISCOVERY OF SMALL MOLECULE INHIBITORS OF A NOVEL THERAPEUTIC TARGET: SELF-SPLICING GROUP I INTRON RIBOZYME. **Mei Cui**, Shannon M. Lemrow and Houng-Yau Mei. BioOrganic Chemistry, Parke-Davis Pharmeceutical Research, Division of Warner-Lambert Co., 2800 Plymouth Rd. Ann Arbor, MI 48105

Group I introns have been found to undergo *in vitro* self-splicing reactions in the absence of any proteins. These introns (ribozymes) have been found in various microorganisms including pathogenic *Pneumocystis carinii* but not in higher eukaryotes. We have studied catalytic group I intron RNA as a potential therapuetic target. High-throughput screening methodologies have been established to search for inhibitors of self-splicing *P. Carinii* introns. Small molecule inhibitors that specifically interfere with the catalytic functions of *P. Carinii* introns have been identified. These inhibitors represent potential pharmacological agents for infectious diseases caused by group I intron-containing microbials.

255.
SYNTHESIS AND EVALUATION OF NEW ANTIMALARIAL AGENTS.
[1]Lindsay A. Slater, [1]David J. Robins, [2]Stephen R. Phillips, [2]Fiona A. McMonagle, [1]Department of Chemistry, [2]Division of Infection and Immunity, Joseph Black Building, University of Glasgow, Glasgow, Scotland, G12 8QQ.

Malaria is by far the world's most important tropical parasitic disease, infecting 300-500 million people each year, and killing more people than any other communicable disease except tuberculosis. Due to the increasing prevalence of resistant strains of parasite to established chemotherapy new drugs are desperately needed. Polyamines are polycationic compounds some of which are naturally occurring and ubiquitous in nature. Parasitic protozoa are sensitive to inhibitors of polyamine biosynthesis and we have investigated analogues of putrescine, spermidine and spermine for activity against *P. falciparum in vitro*. Chain length and composition and nitrogen substitution patterns have been varied to establish structure-activity relationships. Results show a general increasing activity as the number of nitrogen atoms and the hydrophobic character are increased. *In vivo* tests are on going. The *P. falciparum* parasite has AT rich DNA and we are incorporating AT specific binding agents into our molecules to improve activity.

256. STRUCTURE AND ACTIVITY RELATIONSHIPS OF 4-AMINOQUINOLINES (4-AQs) AGAINST DRUG-RESISTANT *P. falciparum*. D. De, F. M. Krogstad, L. D. Byers and D. J. Krogstad. Depts. of Tropical Medicine and Chemistry, Tulane University, New Orleans, LA 70112.

We have recently shown that modification of the diaminoalkane side chain at the 4-position of the AQ ring produces 4-AQs which are active against chloroquine-susceptible and -resistant *P. falciparum*. In the studies reported here, we examined the effect of substitution for the Cl atom at the 7-position of the AQ ring on biologic (antiplasmodial) activity. The compounds examined were synthesized by condensation of *m*-substituted aniline with diethyl ethoxymethylene malonate to produce 7-substituted-

4-chloroquinoline, nucleophilic substitution of the Cl at the 4-position with diaminoalkanes, and regioselective reductive alkylation with $NaBH_4$-CH_3COOH to convert terminal primary amine into a diethylamino group. This strategy yielded a series of 4-AQs with different substituent at the 7-position of the AQ ring and the length of the diaminoalkane side chain varies from 2-12 carbons. A number of these 4-AQs are equally active against chloroquine-susceptible and -resistant *P. falciparum in vitro* (IC50s, 5-20 nM). We will discuss the synthesis and antiplasmodial activity of these compounds.

257. **THE SYNTHESIS OF SELECTIVE NEW INHIBITORS OF TRYPANOSOMAL GLYCERALDEHYDE-3-PHOSPHATE DEHYDROGENASE VIA RATIONAL PROTEIN STRUCTURE-BASED DRUG DESIGN.** <u>A. M. Aronov</u>, C. L. M. J. Verlinde, W. G. Hol, M. H. Gelb*, Departments of Chemistry and Biological Structure, University of Washington, Seattle, Washington 98195

As part of a project targeting glyceraldehyde-3-phosphate dehydrogenase (GAPDH) for rational structure-based design of drugs against trypanosomiasis and leishmaniasis, a new series of adenosine analogs was designed, synthesized and tested. The design was based mainly on structural data from a number of NAD:GAPDH complexes which have been reported earlier. Several substitution alternatives have been explored. Successive rounds of modelling and screening based on the use of multisubstituted matrices provided an overall greater than 10^4-fold affinity improvement versus adenosine, with at least 30-fold higher potency toward trypanosomal GAPDH over the human enzyme. The inhibitors have been shown to act in a competitive manner. Ways to increase the number of substituents and retain solubility are being discussed.

258. **THE IMPORTANCE OF THE 4'-HYDROXYL HYDROGEN FOR THE ANTI-TRYPANOSOMAL PROPERTIES OF (+)-7-DEAZA-5'-NORARISTEROMYCIN.** <u>Katherine L. Seley</u>* and Stewart W. Schneller, Department of Chemistry, Auburn University, Auburn, AL 36849.   Cyrus J. Bacchi, Donna Rattendi and Schennella Lane, Haskins Laboratories, Pace University, New York, NY 10038.

We have recently found that (+)-7-deaza-5'-noraristeromycin (**1**) is a potent inhibitor of the growth of *Trypanosomal brucei*. As part of a plan to improve upon this activity and to determine the structural features of **1** that are necessary for its anti-trypanosomal properties, the methylated derivative **2** became a target compound. This compound lacks the 4'-hydroxyl hydrogen. The synthesis of **2** and its anti-trypanosomal properties will be presented. This investigation received financial support from the UNDP/World Bank/WHO Special Programme for Research and Training in Tropical Medicine; Grant #930041 (S.W.S.) and Grant #950594 (C.J.B.).

1, R=H
2, R=Me

259. **SYNTHESIS AND ANTHELMINTIC ACTIVITY IN EXPERIMENTAL TRICHINELLOSIS OF 1-METHYLTRICLABENDAZOLES**, <u>Alicia Hernández-Campos</u>,[1] Rafael Castillo,[1] Raúl Morales H.,[2] Lilian Yepes M.,[2] Pablo Basilio E.,[2] Roberto Cedillo R,[2] Francisco Hernández-Luis,[1] and Onofre N.H.,[2] [1]Departamento de Farmacia, Facultad de Química, UNAM. C.U., México D.F. México, 04510, [2]Unidad de Investigación Médica en Enfermedades Infecciosas y Parasitarias, IMSS. México D.F. México.

Triclabendazole (TBC), 6-Chloro-5-(2,3-dichlorophenoxy)-2-methylthiobenzimidazole is a potent fasciolicide with better bioavailability than the benzimidazole carbamates. Structurally, it differs from the later in having a 2-methylthio group instead of a 2-methylcarbamate group. In order to increase the liposolubility of TBC and to know the role of the hydrogen at position 1 in the anthelmintic activity, we have synthesized two TBC derivatives: 6-chloro-5-(2,3-dichlorophenoxy)-1-methyl-2-methylthiobenzimidazole (1,5TBC) and 5-chloro-6-(2,3-dichlorophenoxy)-1-methyl-2-methylthiobenzimidazole (1,6TBC), and the mixture of both. These compounds were tested in a novel model of experimental trichinellosis in mice. The results of these studies showed that 1,5TBC is more active than TBC, 1,6TBC and the mixture 1,5TBC-1,6-TBC. Details of the synthesis and biological assays will be presented.

260. SYNTHESIS AND GIARDICIDAL ACTIVITY OF 1-METHYLBENZIMIDAZOLES. Rafael Castillo,[1] Remedios Sánchez,[1] Roberto Cedillo R,[2] Amparo Tapia,[2] Bibiana Chávez,[2] Alicia Hernández-Campos,[1] Francisco Hernández-Luis,[1] Lilian Yepes M.,[2] and Onofre N.H.,[2] [1]Departamento de Farmacia, Facultad de Química, UNAM. C.U., México D.F. México, 04510, [2]Unidad de Investigación Médica en Enfermedades Infecciosas y Parasitarias, IMSS. México D.F. México.

A series of 1-methylbenzimidazole derivatives of general formula shown below, has been synthesized and tested in vitro against *Giardia lamblia*. The synthesized compounds shown a giardicidal activity due to alteration at cellular levels different than those of metronidazole, albendazole, and mebendazole. The synthesis, biological assays, and ultrastructural studies will be presented.

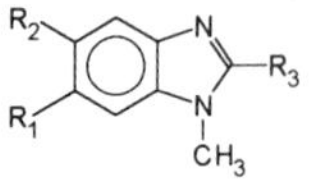

R_1 or R_2 = H or Cl
R_3 = NH_2, CH_3, or SCH_3

261. SYNTHESIS AND PLASMA ENZYMATIC HYDROLYSIS STUDIES OF THREE ALBENDAZOLE CARRIER PRODRUGS, Francisco Hernández-Luis, Rafael Castillo, Alicia Hernández-Campos, Manuel Morales H., Helgi Jung, Departamento de Farmacia, Facultad de Química, UNAM, México D.F. México, 04510

Albendazole (ABZ) is a potent, broad spectrum anthelmintic of the benzimidazole carbamate family of compounds. Due to its poor solubility and first pass metabolism, ABZ has been restricted to be used in the treatment of intraintestinal parasitosis. However, recent findings show that ABZ is effective in the therapy of neurocysticercosis, requiring for this high doses and long treatments. With the purpose of increasing the solubility and absorption of ABZ, we have synthesized three carrier prodrugs of this compound: 1-[(4-aminomethylbenzoyloxy)methyl]albendazole, 1-(ethoxycarbonyl)albendazole, and 3-(ethoxycarbonyl)albendazole. The synthesis, isolation, purification, spectroscopic identification, as well as solubility at different pH values, distribution coefficient, and plasma hydrolysis of these compounds will be presented.

262. PREPARATION OF TWO PRODRUGS (BIOPRECURSORS) OF ALBENDAZOLE AND EVALUATION OF THEIR ACTIVITY AGAINST TRICHINELLA SPIRALIS, Francisco Hernández-Luis,[1] Lilian Yepes-Mullia,[2] Raúl Morales H.,[2] Rafael Castillo,[1] Gabriel Martinez,[1] Roberto Cedillo R,[2] Alicia Hernández-Campos,[1] and Onofre N. H.,[2] [1]Departamento de Farmacia, Facultad de Química, UNAM. C.U., México D.F. México, 04510, [2]Unidad de Investigación Médica en Enfermedades Infecciosas y Parasitarias, IMSS. México D.F. México.

Some attempts to increase the solubility of albendazole (ABZ), a potent, broad spectrum anthelmintic, include the preparation of ABZ prodrugs. In this regard, two ABZ prodrugs, N-methoxycarbonyl-N'-[(2-nitro-4-propylthio)phenyl]thiourea (2) and N-methoxycarbonyl-N'- [(2-nitro-5-propylthio)phenyl]thiourea (3), have been synthesized in our laboratory and tested in experimental trichinellosis. Anthelmintic activity of 2 and 3 in mice infected with Trichinella spiralis showed that 3 is more active than 2. It was determined that 3 has a similar efficacy than ABZ against T. spiralis pre-adult, adult, and female fecundity, but it was not as effective as ABZ against the muscle stage of the parasite. The synthesis and experimental details of the bioassays will be presented.

263. **ROBUSTAFLAVONE, A NATURALLY OCCURRING BIFLAVANOID, IS A POTENT INHIBITOR OF HEPATITIS B VIRUS REPLICATION**
Yuh-Meei Lin, Herbert M. Anderson, Ralph M. Schure, Brent E. Korba[†], Michael T. Flavin and David E. Zembower, MediChem Research, Inc., 12305 South New Avenue, Lemont, Illinois 60439 USA, and [†]Division of Molecular Virology and Immunology, Georgetown University Medical Center, 5640 Fishers Lane, Rockville, Maryland 20852 USA

Hepatitis B virus (HBV) has been designated the ninth leading cause of death by the World Health Organization. Though several highly effective vaccines are available, HBV remains the most significant viral pathogen infecting man. There are currently no effective drugs available which have been approved for treatment of HBV infection. Our program to discover small molecule drug leads from plant sources identified robustaflavone, a biflavanoid isolated from *Rhus succedanea*, as a potent inhibitor of HBV replication *in vitro*, with an effective concentration (EC_{50}) of 0.25 µM, and a therapeutic index (IC_{50}/EC_{90}) of 153 in chronically infected 2.2.15 cells. An improved isolation protocol and additional biological studies will be presented.

264. **ANALYSIS OF THE BINDING MODES OF SUBSTRATES AND INHIBITORS OF THE HERPES SIMPLEX VIRUS TYPE I THYMIDINE KINASE (HSV-I TK) USING 3D QSAR AND MOLECULAR SURFACE PROPERTIES:** T.A. Hinds, R.R. Drake, C.M. Compadre, Departments of Biochemistry and Molecular Biology and Biopharmaceutical Sciences, University of Arkansas for Medical Sciences, Little Rock, AR 72205.

HSV-I TK is an important pharmacological target in antiherpetic therapy. Additionally this enzyme is used in combination with specific nucleoside drugs as a suicide gene in the treatment of cancer and AIDS. We have examined the putative binding of different types of nucleoside analogs using electrostatic and steric contour maps generated by CoMFA analysis and lipophilic, electrostatic and hydrogen bonding potential surfaces created using MOLCAD. The results indicate that the combination of a hydrogen bond donor and acceptor at the 3 and 4 positions, and hydrophobicity at the 5 position of the heterocyclic moiety in pyrimidines is critical for binding. Purine acyclic nucleosides bind by mimicking pyrimidine furanosides. We were also able to rationalize the difference in the activity of β-L-thymidine, α-D-thymidine and the natural substrate β-D-thymidine based on different binding modes.

265. PEPTIDOMIMETIC INHIBITORS OF HCMV PROTEASE. **Jeff O'Meara**, A. Abraham, M. Bailey, A. Bhavsar, P. Bonneau, J. Bordeleau, Y. Bousquet, C. Chabot, J.-S. Duceppe, G. Fazal, S. Goulet, C. Grand-Maître, I. Guse, T. Halmos, P. Lavallée, E. Malenfant, W. Ogilvie, R. Plante, C. Plouffe, M. Poirier, M.-A. Poupart, F. Soucy, C. Yoakim, R. Déziel. *Bio-Méga/Boehringer Ingelheim Research, Inc., 2100 Cunard Street, Laval, Quebec, Canada H7S 2G5*

The Human Cytomegalovirus (HCMV) is a highly prevalent member of the herpesvirus family which is responsible for opportunistic infections in immunocompromised individuals including organ transplant recipients, cancer patients and AIDS sufferers. Current treatments for HCMV infections include DNA polymerase inhibitors which show severe side effects when used alone or (in the case of AIDS) in combination with AZT. Therefore there exists a need for alternative anti-HCMV therapies. All members of the herpesvirus family express a protein late in the virus life cycle that appears to function as a self assembling scaffold during the maturation of the viral capsid. This assembly protein must be processed to remove a short segment of the C-terminus in order to permit the entry of viral DNA and produce an infectious virus particle. Recently it has been shown that this processing is mediated, at least in part, by a protease which is encoded by the virus. This suggests that specific inhibitors of this protease would have therapeutic value. The development of inhibitors of HCMV protease inhibitors based on the amino acid sequences of the peptide cleavage sites and showing sub micromolar potency *in vitro* is described. Optimization of the amino acid residues of these inhibitors has produced a 70 fold increase in inhibitor potency relative to the natural peptide sequence. This study also proposes possible roles for the various amino-acids of the resulting inhibitor sequence and provides a firm foundation for the design of second generation inhibitors.

266. **SYNTHESIS AND IN VITRO ANTIVIRAL ACTIVITY OF UNSATURATED 3-ALKYL DERIVATIVES OF 5-AMINOTHIAZOLO[4,5-*d*]PYRIMIDINE-2,7(3*H*, 6*H*)-DIONE.** Arthur F. Lewis[1], Ganapathi R. Revankar[1], Joshua O. Ojwang[1], Shawn D. Mustain[1], Robert F. Rando[1], John H. Huffman[2], and Erik De Clercq[3], [1]Aronex Pharmaceuticals, Inc., 9391 Grogan's Mill Road, The Woodlands, TX 77380, [2]Institute for Antiviral Research, Utah State University, Logan, UT 84322, and [3]The Rega Institute for Medical Research, Katholieke Universiteit Leuven, B3000 Leuven, Belgium.

5-Amino-3-[(*Z*)-2-penten-1-yl]thiazolo[4,5-*d*]pyrimidine-2,7(3*H*, 6*H*)-dione (AR-132) is a potent inhibitor of human cytomegalovirus (HCMV) in cell culture. To explore the therapeutic potential of unsaturated alkyl derivatives related to AR-132, a study was designed to define (1) the chain length (3-6 carbon) of the 3-substituent, (2) the geometry (*Z* or *E*) about the unsaturation, (3) the type (alkene, alkyne, or allene) of unsaturation, (4) the degree (monoene, diene) of unsaturation, and (5) the degree of branching (α, β, γ) of the 3-substituent which will result in the more potent antiviral agent. The structure activity relationships resulting from this study will be presented.

AR-132

267. **SYNTHESIS AND IN VITRO ANTIVIRAL ACTIVITY OF 3-HYDROXYALKENYL AND RELATED DERIVATIVES OF 5-AMINOTHIAZOLO[4,5-*d*]PYRIMIDINE-2,7(3*H*, 6*H*)-DIONE.** Arthur F. Lewis[1], Ganapathi R. Revankar[1], Joshua O. Ojwang[1], Shawn D. Mustain[1], Robert F. Rando[1], John H. Huffman[2], and Erik De Clercq[3], [1]Aronex Pharmaceuticals, Inc. 9391 Grogan's Mill Road, The Woodlands, TX 77380, [2]Institute for Antiviral Research, Utah State University, Logan, UT 84322, and [3]The Rega Institute for Medical Research, Katholieke Universiteit Leuven, B3000 Leuven, Belgium.

The finding of potent, in vitro inhibition of human cytomegalovirus (HCMV) by 5-amino-3-[(*Z*)-2-penten-1-yl]thiazolo[4,5-*d*]pyrimidine-2,7(3*H*, 6*H*)-dione (AR-132) prompted the study of certain related derivatives which are hydroxylated in the alkenyl chain. 5-Amino-3-[(*Z*)-4-hydroxy-2-buten-1-yl]thiazolo[4,5-*d*]pyrimidine-2,7(3*H*, 6*H*)-dione was found to have significant anti-HCMV activity. A comparative study of this compound and related derivatives to AR-132 will be presented.

AR-132

268. **SYNTHESIS AND ANTIVIRAL ACTIVITY OF ALKOXYPROPYL ESTERS OF GANCICLOVIR MONOPHOSPHATE IN HCMV- AND HSV-INFECTED CELLS**
James R. Beadle, Ganesh D. Kini, Kathy A. Aldern and Karl Y. Hostetler
Department of Medicine, VA Medical Center, San Diego, CA 92161, and the University of California, San Diego, La Jolla, CA, 92093

Ganciclovir has proven to be an effective treatment for cytomegalovirus disease in humans (HCMV). However, the oral bioavailability of ganciclovir is only 6 to 9 percent and the plasma half-life is short, 2.9 hours. In an attempt to develop a prodrug of this agent with improved pharmacokinetics, we synthesized 1, a ganciclovir monophosphate-lipid conjugate, and tested it for inhibitory activity against HCMV and herpes simplex virus (HSV-1) in DNA-reduction assays. Conjugate 1 is 6-fold more potent than ganciclovir in the HCMV assay (IC$_{50}$ = 0.3 μM). Ganciclovir and 1 are equally active against HSV-1. The synthesis and evaluation of 1 will be presented.

1

269. THE DESIGN, SYNTHESIS, AND EVALUATION OF A SERIES OF ACYCLIC TRICIRIBINE ANALOGS FOR THEIR ANTIVIRAL AND ANTINEOPLASTIC ACTIVITY.
Anthony R. Porcari, John C. Drach, Linda L. Wotring, and Leroy B. Townsend. Department of Medicinal Chemistry, College of Pharmacy and Department of Biologic and Material Sciences, School of Dentistry, The University of Michigan, Ann Arbor, Michigan, 48109-1065.

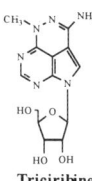
Triciribine

Studies of triciribine as an antineoplastic agent have been on going since it was first synthesized in 1971 by Schram and Townsend. Recently, Drach and Kucera found triciribine to exhibit potent *in vitro* activity against HIV. The active form of triciribine is believed to be the 5'-monophosphate and its mechanism of action is proposed to be inhibition of DNA or protein synthesis. In our attempts to further explore the antineoplastic and antiviral activity of triciribine, we have synthesized acyclic analogs and evaluated them for antineoplastic and antiviral activity as well as their ability to be phosphorylated. These analogs have proved to be valuable in determining the rigidity requirements of the ribosyl moiety of triciribine. This study was supported by NIH Training Grant 5-T32GM07767 and Research Grants 5-RO1-AI-33332 and 5-RO1-AI-36872 from the National Institute of Allergy and Infectious Diseases.

270. SYNTHESIS AND ANTIVIRAL EVALUATION OF ACYCLIC UNSATURATED URACIL NUCLEOSIDES. Ragine Rajaratnam, Shashi Phadtare, College of Pharmacy, Xavier University of Louisiana, New Orleans, Louisiana 70125.

Acyclic C_5- & C_6-substituted uracil and thymine nucleoside analogues have been reported to exhibit promising anti-HIV activity against HIV-1. We have synthesized a new series of C_5- & C_6-thiophenyl (-SPh) substituted acyclic unsaturated uracil nucleosides as potential antiviral agents. The thiophenyl (-SPh) substituent at the C_5 or C_6-position of uracil was introduced by reacting the corresponding chlorouracil with thiophenol in ethylene glycol. The alkylation of these substituted uracil base at N^1-position was achieved by reacting 1,4-dichloro butene or 1,4-dichlorobutyne in DMSO/K_2CO_3 at room temperature. The 4'-chloro compound was hydrolyzed in aqueous acidic medium to acyclic unsaturated 4'-hydroxy thiophenyl uracil nucleosides. In this presentation, the synthesis, *in vitro* anti-HIV evaluation and SAR-studies of these C_5- & C_6-thiophenyl uracil nucleosides will be presented. This research is supported by a grant from National Institute on Drug Abuse (NIDA).

271.

ANTIVIRAL OLIGO AND POLYRIBONUCLEOTIDES CONTAINING SELECTED TRIAZOLO[2,3-a]PURINES
Mayoka G. Tutonda,[*] Robert W. Buckheit, Jr.,[f] and Arthur D. Broom[*]
[*]Medicinal Chemistry, 308 Skaggs Hall, University of Utah, Salt Lake City, UT 84112
[f]Southern Research Institute, Frederick Research Center, Frederick, MD 21701

This laboratory demonstrated earlier that poly(1-methyl-6-thioguanylic acid) PMTG and poly(1-methyl-6-thioinosinic acid) PMTI were potent inhibitors of HIV replication and cytopathicity. These findings prompted us to explore the parameters responsible for this activity. The working hypothesis has been that amphiphilic character (hydrophobic base/hydrophilic backbone) and the ability to form a highly ordered, non-hydrogen bonded array in solution are prerequisites for antiviral activity. Synthesis, characterization and biological evaluation of oligo and polyribonucleotides containing triazolo[2,3-a]purines further testing the hypothesis will be discussed.

X=O R=H
X=O R=CH₃
X=S R=CH₃

272. ANTISENSE OLIGODEOXYRIBONUCLEOTIDE CONTAINING A NOVEL 2'-O MODIFIED ADENOSINE. Seema P. Joglekar and Arthur D. Broom. Medicinal Chemistry, 308 Skaggs Hall, University of Utah, Salt Lake City, UT 84112

Antisense oligodeoxyribonucleotides (ODNs) offer potential to block expression of specific genes within cells. Inhibition of gene expression relies on the ability of an ODN to bind a complementary messenger RNA sequence. Chemical modifications in a nucleoside can enhance potency and stability of the antisense ODNs. One such modified nucleoside (2'O-(3',5'-diacetylthymidin-5-yl)adenosine) has been synthesized in this laboratory. This modified nucleoside has been incorporated into an oligodeoxyribonucleotide sequence (poly dA, 12mer) using solid phase DNA synthesis. Synthesis of the above nucleoside will be reported. T_m and NMR data on the polymer will be discussed. This incorporated modified nucleoside will provide the basis for further modifications.

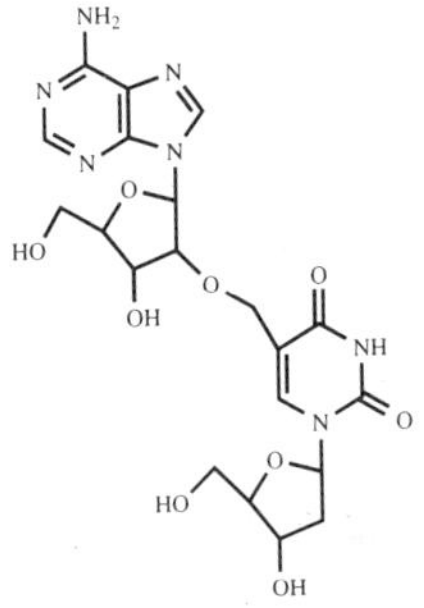

273. Serinol-Terpyridine Containing Oligonucleotides for Ribozyme Mimics. Andrew T. Daniher, James K. Bashkin*, Department of Chemistry, Washington University, St. Louis, MO, 63130.

Terpyridine complexes of metal ions are well known RNA transesterification and hydrolysis catalysts. We have previously prepared sequence-specific RNA transesterification agents, or ribozyme mimics, by incorporating this active complex at C-5 of deoxyuridine into oligonucleotides. This mimic cleaved its target strand (28% at 45°C and 72h). To improve the efficiency of cleavage, we designed mimic **1**. **1** allows the terpyridine ligand to be incorporated into the oligo as a non-nucleoside monomer. This introduces greater flexibilty into the target strand because of the absence of a base-pair. Site-directed cleavage with RNase H demonstrates that these modified oligomers can hybridize with its RNA target sequence. Improved site-specific RNA hydrolysis will be discussed.

5' AAC CCU UUA GAG ACU AUG UAG AC 3'
3' GA AAT CTC TGA XAC ATC 5'

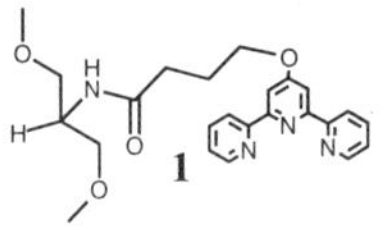

274.
HOMOPHTHALIMIDES ARE INHIBITORS OF HUMAN RHINOVIRUS 3C PROTEASE. L.N. Jungheim*, **J.D. Cohen**, Q.M. Wang, R.B. Johnson, E.C. Villarreal, M. Wakulchik, R.J. Loncharich, Eli Lilly & Company Infectious Diseases Research, Lilly Corporate Center, Indianapolis, IN. 46285

Rhinoviruses are the causative agents in the majority of "common cold" episodes. The Rhinovirus 3C protease is an enzyme involved in the co-translational processing of the viral genome, and is responsible for making eight proteolytic cleavages of the precursor polyprotein which are necessary for the virus to replicate. A high throughput screen was run in an effort to discover inhibitors of Rhinovirus 3C protease. Among the more interesting "hits" were homophthalimides related to LY046601, shown below. Modifications of the group at nitrogen were made, and new acylations and alkylations α to carbonyl were explored, leading to compounds which exhibited enzyme inhibition in the low micromolar range. These compounds were shown to have tight/covalent binding to the 3C protease enzyme by LC/MS experiments. Calculations showing inhibitor/enzyme fit will also be presented.

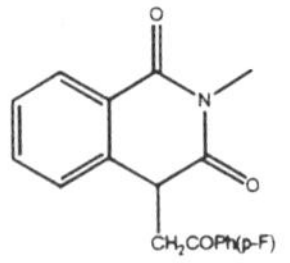

IC_{50} = 12.8 μg/mL

275. NEW AROMATIC INHIBITORS OF INFLUENZA NEURAMINIDASE. Venkatram R. Atigadda[1], Y. S. Babu[3], S. Bantia[3], P. Chand[3], N. Chu[3], J. A. Montgomery[3], D. A. Walsh[3], M. Luo[2] and Wayne J. Brouillette[1*], [1]Department of Chemistry and [2]Center for Macromolecular Crystallography, University of Alabama at Birmingham, Birmingham, AL 35294, [3]BioCryst Pharmaceuticals, Inc., 2190 Parkway Lake Dr., Birmingham, AL 35244.

Neuraminidase plays a critical role in the life cycle of influenza virus and is a target for new therapeutic agents. The most potent inhibitors of neuraminidase are derivatives of 2-deoxy-2,3-didehydro-N-acetyl neuraminic acid (DANA). We have developed benzoic acid analogs of DANA, and the most potent reported to date was BANA 113 (IC_{50} = 0.01mM). Here we will report the synthesis and biological activity of analogs of BANA 113 containing replacements for the N-acetyl group.

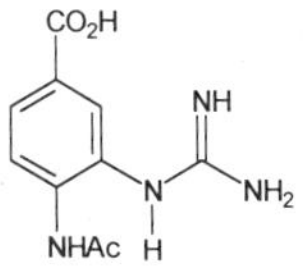

276. SUBSTITUTED BENZOIC ACIDS AS INHIBITORS OF INFLUENZA NEURAMINIDASE. F. J. Duarte[1], Y. S. Babu[3], S. Bantia[3], P. Chand[3], N. Chu[3], J. A. Montgomery[3], D. A. Walsh[3], M. Luo[2] and Wayne J. Brouillette[1*], [1]Department of Chemistry and [2]Center for Macromolecular Crystallography, University of Alabama at Birmingham, Birmingham, AL 35294, and [3]BioCryst Pharmaceuticals, Inc., 2190 Parkway Lake Dr., Birmingham, AL 35244.

Neuraminidase (NA) is one of the two major suface glycoproteins of influenza virus and is necessary for infection of the virus. Our laboratory has shown that benzoic acid derivatives, such as BANA 113, can moderately inhibit the NA activity. We have synthesized new substituted benzoic acids by replacing the N-acetyl group in BANA 113 with various substituents. Some modifications permit further interactions with the protein. The synthesis and biological activities for selected benzoic acid derivatives will be presented.

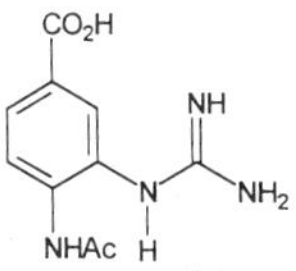

277. FLUORO SUBSTITUTED AROMATIC INHIBITORS OF INFLUENZA NEURAMINIDASE. Shoukath M. Ali,[1] Y.S. Babu,[2] S. Bantia,[2] P. Chand,[2] N. Chu,[2] J. A. Montgomery,[2] D. A. Walsh,[2] M. Luo,[2] and W. J. Brouillette,[*1] [1]Department of Chemistry, and [3]Center for Macromolecular Crystallography, University of Alabama at Birmingham, Birmingham, AL 35294, and [2]Biocryst Pharmaceuticals, 2190 Parkway Lake Drive, Birmingham, AL 35244.

Influenza virus neuraminidase plays a major role in releasing progeny virus from the surface of infected cells and hence serves as a potential target for the development of influenza drugs. Previously, based on the x-ray crystal structure of neuraminidase, we synthesized benzoic acids (*J. Med. Chem.* **1995**, *38*, 3217-3225) with moderate inhibitory activity. The most potent compound in this series was BANA 113 (IC_{50} 0.01 mM). This result prompted us to investigate various substituted benzoic acids. In this presentation, approaches to the synthesis of fluorobenzoic acids and their biological activities will be presented.

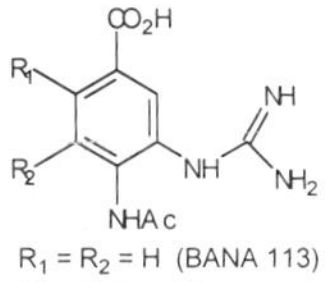

$R_1 = R_2 = H$ (BANA 113)
$R_1 = F, R_2 = H$;
$R_1 = H, R_2 = F$

278. **CYCLIC HIV PROTEASE INHIBITORS: NONSYMMETRIC CYCLIC UREAS WITH EXCELLENT ORAL BIOAVAILABILITY AND LOW DRUG RESISTANCE PROFILES.** P. Y. S. Lam, J. D. Rodgers, R. Li, Y. Ru, P. K. Jadhav, C.-H. Chang, C. G. Clark, S. P. Seitz, L. T. Bacheler, G. N. Lam, S. Erickson-Viitanen, G. L. Trainor and P. S. Anderson. The DuPont Merck Pharmaceutical Co., P.O.Box 80500, Wilmington, DE 19880-0500.

Due to the long-term potential drug resistance problems associated with HIVPR mutations, a main emphasis in the search for the next generation of structurally different HIVPR drugs is in the area of cyclic ureas. Cyclic ureas tend to have low solubility due to the rigidity of the preorganized structure and the intermolecular diol/urea bidentate chelation in the crystal lattice. This presentation describes the use of nonsymmetric cyclic ureas with a floppy alkyl P2 to break up the crystal packing and to increase solubility and thus improve the pharmacokinetics to contribute to the discovery of potential clinical candidates. Inhibitors containing heterocyclic amides with flat drug resistance profiles against the known clinical mutants will also be presented.

P2' = hydrogen-bonding and/or solubilizing groups

279. HETEROARYLMETHYL CARBOXAMIDOBENZYL SUBSTITUTED CYCLIC UREAS AS HIV PROTEASE INHIBITORS. S.R. Sherk, T.P. Selby, DuPont Merck Pharmaceutical Company, Wilmington, DE 19880

An extremely potent subclass of substituted cyclic urea HIV protease inhibitors are those which contain heteroaryl carboxamides substituted at the P2/P2' positions of the molecule. However, these compounds generally suffer from poor bioavailability due most likely in part to the hydrolytic instability of the heteroaryl amide bond functionality. In an effort to find more hydrolytically stable carboxamide analogs we have investigated a number of heteroaryl methyl derivatives. The synthesis, biological activity and SAR studies of these compounds will be presented.

280. **POTENT HIV PROTEASE INHIBITORS CONTAINING A NOVEL (HYDROXYETHYL)AMIDOSUCCINOYL ISOSTERE.** P. L. Beaulieu*, D. Wernic, A. Abraham, P. C. Anderson, T. Bogri, Y. Bousquet, G. Croteau, I. Guse, D. Lamarre, F. Liard, W. Paris, D. Thibeault, L. Tong[#] and S. Pav[#]. Bio-Méga/Boehringer Ingelheim Research Inc., 2100 Cunard Street. Laval (Québec), Canada, H7S 2G5 and [#] Boehringer Ingelheim Pharmaceuticals Inc., Ridgefield, CT 06877, USA.

A series of HIV protease inhibitors containing a novel (hydroxyethyl)amidosuccinoyl transition state isostere has been synthesized. These peptidomimetic structures (**1**) inhibit viral protease activity at low nanomolar concentrations (IC_{50} <10 nM for HIV-1). Several compounds inhibited HIV-1 replication in cell culture assays with EC_{50} = 3.7-35 nM. The synthesis and biological properties of these compounds will be presented.

281. NONNUCLEOSIDE INHIBITOR POTENCY AND POLYMERASE EFFICIENCY AGAINST DRUG-RESISTANT STRAINS OF HIV-1 REVERSE TRANSCRIPTASE: CORRELATION OF COMPUTER MODELING AND EXPERIMENTAL DATA. Marilyn B. Kroeger Smith, Paul L. Boyer, Stephen H. Hughes, Christopher J. Michejda, Richard H. Smith, Jr., ABL-Basic Research Program, NCI-Frederick Cancer Research and Development Center, Frederick, MD 21702

Successful clinical treatment of HIV-1 infection by targeting reverse transcriptase (RT) is difficult due to the emergence of drug-resistant strains of the virus following dosage with inhibitors. One current strategy to combat this problem involves therapy which combines drugs from more than one class, focusing on compounds that inhibit variant strains of RT having low polymerase activity. In order to screen nonnucleoside inhibitors for this purpose, a computer model has been developed which allows for determination of drug-protein interaction energy in modified RTs. The energy values were found to correlate with inhibitor potency against various drug-resistant strains of RT, as measured by EC_{50} values. In addition, the distance between a residue in the primer-grip region of RT and one in the dNTP binding site in the uncomplexed, modified RTs was found to be related to the measured polymerase activity. This method thus provides both for rapid assessment of nonnucleoside inhibitor potency against drug-resistant variants and, in addition, yields structural information relating to the mechanism of enzyme action in mutant strains of HIV-1 RT. Research sponsored by the National Cancer Institute, DHHS, under contract with ABL.

282. HYDRAZIDE-CONTAINING INHIBITORS OF HIV-1 INTEGRASE

He Zhao,[†] Nouri Neamati,[‡] Sanjay Sunder,[‡] Huixiao Hong,[†] Shaomeng Wang,[†] George W. Milne,[†] Yves Pommier [‡] and Terrence R. Burke, Jr.[†]
[†]*Laboratory of Medicinal Chemistry*, [‡]*Laboratory of Molecular Pharmacology, Division of Basic Sciences, National Cancer Institute, NIH, Bethesda, MD 20892*

While clinically effective inhibitors of HIV reverse transcriptase and protease are already in use, efforts to target the viral integrase have yet to yield comparably efficacious agents, despite the fact that the integrase is a critical enzyme in the viral life cycle. In order to discover new integrase inhibitors as potential anti-AIDS therapeutics, we have prepared a large number of aroyl hydrazides and have found that several of these afford effective inhibition of HIV-1 integrase preparations at low micromolar concentrations. These hydrazides represent a new class of integrase inhibitors which are unusual in not requiring the ortho bis-hydroxy "catechol" functionality which is common to many other inhibitors. The synthesis and biological evaluation of several aroyl hydrazides will be presented.

283. LIPID PRODRUG OF A FOSCARNET-AZT CONJUGATE: SYNTHESIS AND ANTIVIRAL ACTIVITY AGAINST PFA- AND AZT-RESISTANT STRAINS OF HIV

Ganesh D. Kini, James R. Beadle, Kathy A. Aldern, John W. Mellors,*
Douglas D. Richman and Karl Y. Hostetler
Department of Medicine, VA Medical Center, San Diego, CA 92161 and the University of California, San Diego, La Jolla, CA 92093-0676 and the Department of Medicine, University of Pittsburgh,* Pittsburgh, PA 15261

Synergistic activity of AZT and PFA against drug susceptible (wild-type) HIV has been reported. Of note is the persistent activity of these agents in combination against AZT or PFA resistant mutants. The syntheses and highly improved antiviral activity of lipid prodrugs of foscarnet (1-O-octadecyl- *sn*-glycero-3-phosphono-formate) and AZT (1-(octadecyloxy)-propane-3-phospho-AZT) have been previously reported by our laboratory.

We have now synthesized **1**, a novel lipid-PFA-AZT conjugate, and screened it for antiviral activity against several isolates of HIV in vitro. Preliminary results indicate that conjugate **1** shows significant activity against wild-type and mutant HIV strains that are highly resistant to AZT and PFA individually. The synthesis and antiviral activity data will be presented.

284. CHIRALITY IN MEDICINAL CHEMISTRY: AN INTRODUCTION

Israel Agranat, Institute for Advanced Studies at The Hebrew University of Jerusalem, Jerusalem 91904, Israel

The notion of Chirality has emerged as a major theme in drug design and discovery. Many of the new chiral drugs are being developed as single enantiomers. Chiral recognition is one of the most significant topics of contemporary stereochemical studies and is an important component of pharmacological events. Stereoisomer distinction abounds in the fields of Medicinal Chemistry and Pharmacology. The tragedy of thalidomide (racemic 2-(2, 6-dioxo-3-piperidnyl-1H-isoindole-1,3 (2H)-dione), approved and used (in Europe) for nausea associated with pregnancy, had a major impact on the regulatory environment. The tracing of the teratogenicity of thalidomide to the (S)-(-)-enantiomer is still a controversial issue. Thalidomide is effective in the treatment of ENL, a complication of lepromatous leprosy and is used as an experimental drug in the treatment of a variety of diseases with an autoimmune character. Thus, the incident of thalidomide proved to be not only a disaster, but also an opportunity.

285. CHIRALITY AND DRUG DESIGN. P. S. Anderson, Chemical & Physical Sciences, The DuPont Merck Pharmaceutical Company, Experimental Station, P.O. Box 80500, Wilmington, Delaware 19880-0500

A key step in the drug discovery process is the optimization of a lead molecule for pharmacological potency and selectivity. The tools of structural biology can be used to facilitate this step by identifying and focusing attention on important non-covalent interactions between the drug candidate and its macromolecular target. Stereochemical detail is frequently important for obtaining proper alignment of complimentary features in the small molecule ligand and its target structure. For this purpose, X-ray crystallography has been used to select stereochemically-based tactics for optimizing inhibitors of carbonic anhydrase and HIV protease.

286. STEREOCHEMISTRY IN DRUG METABOLISM: MECHANISMS AND BIOLOGICAL SIGNIFICANCE. John Caldwell, Pharmacology and Toxicology, Imperial College School of Medicine at St. Mary's, London W2 1PG, U.K.

Differences between stereoisomeric forms of numerous drugs in terms of their absorption, distribution, metabolism and excretion are common. The pharmacokinetic importance of drug stereochemistry depends upon the mechanism under consideration: this has little influence on passive processes such as diffusion across membranes, the major mechanism of absorption, distribution and renal elimination, but when the drug interacts with an enzyme or a transporter system, then discrimination is common. This is particularly the case with the Phase I and Phase II metabolic enzymes, whose stereoselectivity can provide information about mechanisms of action and enzymic structure. In general, there exist better relationships between effect and plasma concentration, than with the external dose offered. The determination of the molecular species responsible for the activity of interest, be it parent compound, a metabolite, the active stereoisomer etc. is thus of paramount importance. Failure to take inter- and intra-species pharmacokinetic variation into account can result in delay and/or failure during development or make an approved drug vulnerable in the marketplace.

287. STEREOCHEMISTRY, A KEY PARAMETER IN THE DESIGN OF NOVEL TYPES OF EXCITATORY AMINO ACID RECEPTOR LIGANDS. P. Krogsgaard-Larsen, T.N. Johansen, B. Ebert & E. Falch, Department of Medicinal Chemistry, The Royal Danish School of Pharmacy, DK-2100 Copenhagen, Denmark

The excitatory amino acid (EAA) receptors are subdivided into NMDA, AMPA, kainic acid, and metabotropic EAA receptors. Whereas all AMPA agonists, including AMPA, ATPA, APPA, 2-Py-AMPA, and Br-HIBO, show (S)-configuration, the corresponding (R)-enantiomers show widely different pharmacological effects. Thus, (R)-AMPA is inactive, (R)-2-Py-AMPA is an AMPA antagonist, and (R)-Br-HIBO shows AMPA agonist enhancing effects. Analyses of the AMPA agonist and antagonist effects, respectively, of (S)- and (R)-APPA have led us to introduce the new pharmacological concept, functional partial agonism, which may have therapeutic applicability. Molecular mechanisms will be discussed.

288. IS THERE A PREFERABLE WAY LEADING TO ENANTIOMERICALLY PURE COMPOUNDS? D. Arlt, Bayer AG, Central Research, D-51368 Leverkusen, Germany

Among the enantiomerically pure drugs not derived from natural products 65 % are obtained by classical resolution. Classical resolution becomes highly attractive where it can be combined with in-situ racemization in an asymmetric transformation process. Particularly advantageous to obtain an enantiomerically pure compound is a separation process by direct crystallization, but only in a few cases such advantageous variations of resolution methods ca be realized.
Main topic of the lecture will be the development of a large set of CSP's designed for the separation of racemates by chromatography, espe-

cially of drugs or drug intermediates in all stages of research and development. Also an example for a production scale separation will be given. A further topic will be the possibility to use this separation technology for the production of optically active ligands to prepare enantioselective catalysts.

289. SYNTHESIS AND EVALUATION OF NOVEL INHIBITORS OF THROMBIN. Terry A. Lyle, Merck Research Laboratories, Department of Medicinal Chemistry, West Point, PA 19486

This presentation will cover the synthesis and characterization of novel thrombin inhibitors based on a D-Phe-Pro-X template. Progress toward the goal of a selective, orally bioavailable inhibitor will be presented. A number of structural features that provide increased selectivity with respect to related serine proteases have been discovered, and will be discussed with the aid of crystal structures of selected inhibitors bound to the active site of thrombin. The impact of a variety of structural changes on oral bioavailability will also be addressed, as well as the performance of several compounds in an animal model of arterial thrombosis.

290. DESIGN OF MACROCYCLIC THROMBIN INHIBITORS. Michael N. Greco,*,† Eugene T. Powell,† Leonard R. Hecker,† Patricia Andrade-Gordon,† Jack A. Kauffman,† Joan M. Lewis,† Venkatapathy Ganesh,‡ Alexander Tulinsky,‡ and Bruce E. Maryanoff,† Drug Discovery,† The R. W. Johnson Pharmaceutical Research Institute, Spring House, Pennsylvania 19477, and the Department of Chemistry,‡ Michigan State University, East Lansing, Michigan 48824

Since thrombin is a trypsin-like serine protease with a central role in the bioregulation of thrombosis and hemostasis, selective active-site-directed inhibitors represent potentially useful therapeutic agents for the management of thrombotic disorders. By following a protein structure-based protocol, we have designed potent, macrocyclic active-site inhibitors of thrombin. We plan to discuss structure-function issues relating ring size and P3/P1' modifications to enzyme inhibition. Chemical synthesis, in vitro biochemical evaluation, and details of the X-ray crystal structure of a complex between 1 and thrombin will also be presented.

291. DISCOVERY OF NOVEL POTENT AND SELECTIVE α_{1A} ADRENOCEPTOR ANTAGONISTS FOR THE SYMPTOMATIC TREATMENT OF BENIGN PROSTATIC HYPERPLASIA (BPH) Michael D. Meyer, Robert J. Altenbach, Fatima Z. Basha, William A. Carroll, Irene Drizin, Paul P. Ehrlich, Steven W. Elmore, Suzanne A. Lebold, Kevin B. Sippy, Karin Tietje, Michael D. Wendt, Arthur A. Hancock, Michael E. Brune, Steven B. Buckner, and James F. Kerwin, Jr. Neuroscience Research, Abbott Laboratories, Abbott Park, IL 60064

α_1-Adrenoceptor antagonists such as terazosin have become widely recognized as an effective symptomatic treatment of BPH. The currently available α_1-antagonists, first developed as antihypertensive drugs, frequently demonstrate limiting cardiovascular effects in the treatment of BPH. Recent advances in molecular biology have shown that the α_{1A} subtype predominates in prostatic tissue. Consequently α_{1A} selective antagonists may be useful clinically for the treatment of BPH with fewer cardiovascular side effects. Herein, we report structure activity studies on a series of novel benz[e]isoindoles which exhibit a high level of in vitro selectivity for the α_{1A} binding site, and in vivo selectivity for relaxation of prostatic smooth muscle. From this series, A-131701 was selected for more extensive evaluation. In vitro and in vivo characterization of this agent will be presented.

292. **STUDIES OF THE MOLECULAR MECHANISM OF TNF-α INHIBITION BY TETRAFLUOROPHTHALIMIDES**
Satomi Niwayama, Benjamin E. Turk, Christine Loh and Jun O. Liu*
Center for Cancer Research, Massachusetts Institute of Technology, Cambridge, MA 02139
We recently reported a series of potent inhibitors of LPS-induced TNF-α production in human monocytes with the following tetrafluoro derivatives being the most potent *(J. Med. Chem.*, **1996**, 3044). Progress on the elucidation of the molecular mechanisms of action of F_4CC-1104 will be presented. Studies on the enantioselectivity of action of tetrafluorophthalimides will also be described.

tetrafluorothalidomide
IC_{50} = ~400nM

tetrafluoro[3-phthalimido-3-(3,4-dimethoxyphenyl)]propionate
(F_4CC-1104)
IC_{50} = ~100nM

293.

DISCOVERY AND CHARACTERIZATION OF POTENT, SMALL MOLECULE IL-8RB ANTAGONISTS. **K. L. Widdowson**, D. F. Veber, J. D. Elliott, H. Nie, M. C. Rutledge, A. J. Jurewicz, R. P. Hertzberg, J. M. Lee, J. J. Foley, S. Sarau, W. M. Potts, A. D. Ayrton, J. A. Lee, W. W. Holl, P. R. Young and J. R. White , Medicinal Chemistry Department, SmithKline Beecham Pharmaceuticals, 709 Swedeland Rd., King of Prussia, PA, 19406

Recent work in our laboratory has resulted in the discovery and characterization of a series of N-(2-hydroxy phenyl) N'-phenyl ureas as potent, selective Interleukin-8RB antagonists. The potency of these compounds correlates well with their acidity. The position of the phenol on the phenyl ring also appears to be critical. The in vivo and in vitro characterization of these molecules will also be discussed.

294. **HISTIDYL-(N-BENZYLGLYCINAMIDES) AS RAS FARNESYLTRANSFERASE INHIBITORS POSSESSING ANTITUMOR ACTIVITY IN MICE.** D.J. McNamara[1], E.M. Dobrusin[1], D.M. Leonard[1], K.R. Shuler[1], K.M. Smith[1], J.S. Kaltenbronn[1], J. Quin III[1], S. Bur[1], C.E. Thomas[1], J.D. Scholten[2], K.K. Zimmerman[2], J.S. Sebolt-Leopold[2], R.C. Gowan[2], M.P. Latash[2], W.R. Leopold[3], S.A. Przybranowski[3], and A.M. Doherty[1]. Departments of Chemistry[1], Biochemistry[2], and Cancer Research[3], Parke-Davis Pharmaceutical Research, Division of Warner-Lambert Co., 2800 Plymouth Road, Ann Arbor, MI 48105

We have previously disclosed that PD152440, CbzHis-N-[2-(benzyloxy)ethyl]-N2-[(4-benzyloxy)benzyl]glycinamide, inhibits the farnesylation of Ras by the isolated enzyme farnesyltransferase (FT) with an IC_{50} of 0.26 μM, inhibits the farnesylation of Ras in cultured, H-Ras transformed, NIH 3T3 fibroblast cells (H61) at 1 μM, and inhibits H61 colony formation in soft agar with an IC_{50} of 14.3 μM. We here report a SAR study of PD152440, in which, by systematically varying the C-terminal, N-benzyl, and N-terminal moieties, we have developed novel compounds, highlighted by PD169451, with considerably enhanced potencies. PD169451 inhibits FT with an IC_{50} of 7 nM, is selective for FT compared to the closely related geranylgeranyl transferase, inhibits the farnesylation of Ras in H61 cells at 10-50 nM, and inhibits H61 colony formation in soft agar with an IC_{50} of 210 nM. Furthermore, PD169451 inhibits the growth of subcutaneously implanted H61 tumors in mice by 88% when administered intraperitoneally on days 1-14 at 150 mg/kg.

295. 9-ARYL- AND 9-CYCLOALKYLHETEROATOM-SUBSTITUTED DERIVATIVES OF 9-DEAZAGUANINE, INHIBITORS OF PURINE NUCLEOSIDE PHOSPHORYLASE (PNP). Y. Sudhaker Babu, Arthur J. Elliott, John A. Montgomery, Philip E. Morris, Jr., Sandra L. Petty, David A. Walsh, and Carl H. Williams, BioCryst Pharmaceuticals, Inc., 2190 Parkway Lake Drive, Birmingham, AL 35244

Purine Nucleoside Phosphorylase (PNP) is a key enzyme in T-cell metabolism. Using the three-dimensional crystal structure of PNP, we designed (*J. Med. Chem.* **1993**, *36*, 55-69) certain 9-substituted-9-deazaguanine derivatives (X = CH_2) and have shown them to be potent PNP inhibitors. One of these, peldesine (BCX-34), is currently in Phase III clinical trials for the treatment of cutaneous T-cell lymphoma (CTCL) and psoriasis. In this report, the syntheses, *in vitro* inhibitory activity against PNP, and the PNP:inhibitor crystal structures for some novel 9-deazaguanine derivatives (X = O, S, SO_2, and NH) will be discussed.

Y = H, NH_2
X = O, S, SO_2, NH
R = aryl, cycloalkyl

296. Regioisomeric Azabenzothiopyranoindazoles. Synthesis and Antitumor Evaluations. A. Paul Krapcho[1], Simon N. Haydar[1], Miles P. Hacker[1], Starlan Truong-Chiott[1] and Ernesto Menta[2]. [1] University of Vermont, Burlington, VT 05405 and [2]Boehringer Mannheim Italia Research Center, I-20052, Monza, Italy.

A number of benzothiopyranoindazoles have been synthesized in order to avoid enzymatically mediated radical cycling and specific chemotypes exhibited broad-spectrum antitumor activities. At part of a program to develop new chemotypes, we have replaced the carbon atom of the benzothiopyranoindazole skeleton at positions 7, 8 and 9 with a nitrogen atom leading to the 7-aza-, 8-aza- and 9-azabenzothiopyranoindazoles. The synthesis of the 10-aza regioisomer is currently under investigation. The antitumor activities, which are highly dependent on the position of the nitrogen atom and the side arm structure (R and R_1), will be discussed. Several promising candidates for preclinical trials have been identified.

297. LACTONES OF NSAID DERIVATIVES WITH INCREASED GROWTH INHIBITORY ACTIVITY. Gerhard J. Sperl, Ph.D.[+], Rifat Pamukcu, MD[#], David S. Alberts, MD[^], Klaus Brendel, Ph.D.[%], Paul H. Gross, Ph.D.[+*]. University of the Pacific, Stockton, California, 95211: Department of Chemistry[+*]; Cell Pathways, Inc.[#], Aurora, Colorado; Arizona Cancer Center[^]: Tucson, Arizona; University of Arizona, Tucson, Arizona: Department of Pharmacology[%]

A series of 3-indenyl acetic acids has been synthesized. Halolactonization of these and other related β,γ unsaturated acids led to $\Delta^{\alpha,\beta}$-butenolides, which had up to 3500 fold increased growth inhibitory activity against human colon adenocarcinoma cell line HT-29.

298. SYNTHESIS AND ANTITUMOR PROPERTIES OF PYRIMIDOAZEPINE-BASED TETRAHYDROFOLIC ACID DERIVATIVES. Mark W. Read, Michael L. Miller and Partha S. Ray,* Department of Chemistry, The University of Memphis, Memphis, Tennessee 38152

Despite the many hundreds of folic acid analogs which have been prepared over the past several decades, as potential antitumor agents, there have been no reports of pyrimidoazepine based folic acid derivatives. We have initiated a program aimed at the synthesis of such compounds as potential inhibitors of the enzyme glycinamide ribonucleotide formyltransferase (GARFT) which catalyzes the transfer of the formyl group from 10-formyl-5,6,7,8-tetrahydrofolic acid to the amino group of glycinamide ribonucleotide (thus, introducing the C-8 carbon of the purine ring) during *de novo* purine biosynthesis. In this paper we report on the synthesis of these novel pyrimidoazepine-based tetrahydrofolic acid derivatives using 1,3-dipolar cycloaddition and dienolate chemistry. We will also discuss the antitumor properties of our targets.

299. SYNTHESIS, ANTITUBULIN ACTIVITY, AND ANTICANCER ACTIVITY OF MORE POTENT ANALOGS IN THE 2-ALKOXYESTRADIOL SERIES. Mark Cushman,[*,†] Hu-Ming He,[†] John A. Katzenellenbogen,[‡] Ravi K. Varma,[§] Ernest Hamel,[∥] Chii M. Lin,[§] Pascal Verdier-Pinard,[∥] Siya Ram,[⊥] and Yesh Sachdeva.[⊥] [†]Department of Medicinal Chemistry and Molecular Pharmacology, School of Pharmacy and Pharmacal Sciences, Purdue University, West Lafayette, Indiana 47907; [‡]Department of Chemistry, University of Illinois, Urbana, Illinois 61801; [§]Drug Synthesis and Chemistry Branch, National Cancer Institute, Bethesda, Maryland 20852; [∥]Laboratory of Drug Discovery Research and Development, National Cancer Institute, Frederick, Maryland 21702; and [⊥]Pharm-Eco Laboratories, Inc., 128 Spring Street, Lexington, Massachusetts 02173

A new series of estradiol analogs was synthesized in an attempt to improve on the anticancer activity of 2-methoxyestradiol, a naturally occurring mammalian tubulin polymerization inhibitor. The compounds were evaluated as inhibitors of tubulin polymerization, as well as for cytotoxicity in human cancer cell cultures. The most potent of the new compounds in both of these assays proved to be 2-(2',2',2'-trifluoroethoxy)-6-oximinoestradiol, followed closely by 2-ethoxy-6-oximinoestradiol and 2-ethoxy-6-methoximinoestradiol, all of which were extremely cytotoxic in cancer cell cultures while lacking any significant affinity for the estrogen receptor. The cytotoxicities of the compounds correlated in general with their abilities to inhibit tubulin polymerization, thus supporting the inhibition of tubulin polymerization as the mechanism of the cytotoxic effect.

300. NUCLEOSIDE SILYL ETHERS INHIBIT *RAS*-P21 NUCLEOTIDE EXCHANGE

Jocelyn D. del Rosario[‡], Stacy W. Remiszewski[‡,*], Barr Bauer[‡], Joan Brown[‡], Donna Carr[†], Ronald J. Doll[‡], Linda James[†], Paul Kirschmeier[†], Mark Snow[‡]

Departments of Chemistry[‡] and Tumor Biology[†], Schering-Plough Research Institute
2015 Galloping Hill Road, Kenilworth, NJ 07033

The guanine nucleotide binding protein *ras*-p21 has been implicated in up to 40% of all human cancers. In response to extracellular signals the protein cycles from an inactive GDP-bound state to an active GTP-bound state, signaling cell growth, by exchanging GDP for intracellular GTP. In mutated oncogenic protein this exchange occurs spontaneously, resulting in uncontrolled cell growth. Nucleoside silyl ethers were found to inhibit this nucleotide exchange process. Several silyl and carbon ether analogs have been prepared and tested, as well as ribose and heterocycle analogs. Based on the SAR of the ribose and the heterocycle a binding mode for the silyl ethers is proposed.

301 GENETICALLY ENGINEERED MICE AS IMPORTANT NEW TOOLS FOR DRUG DISCOVERY. S. A. Lira, Department of Immunology, Schering-Plough Research Institute, Kenilworth, NJ 07033

Recent advances in molecular embryology and gene regulation now allow us to modify genetic information at the level of the organism. This presentation will review recent progress in this technology and illustrate how it can be applied to drug discovery.

302. TrkB SIGNALING IS REQUIRED FOR POSTNATAL SURVIVAL OF CNS NEURONS AND PROTECTS HIPPOCAMPAL AND MOTORNEURONS FROM AXOTOMY-INDUCED CELL DEATH. I. Silos-Santiago. Department of Molecular Oncology. Bristol-Myers Squibb. Pharmaceutical Research Institute, Princeton, NJ 08543.

Newborn mice carrying targeted mutations in genes encoding neurotrophins or their signaling receptors display severe neuronal deficits in the peripheral, but not in the central nervous system. However, trkB (-/-) mice that survive after the first postnatal week have a significant increase in apoptotic cell death in cortical layers II-III and V-VI, although the most affected region in the brain is the dentate gyrus of the hippocampus. Furthermore, axotomized hippocampal and motorneurons of trkB (-/-) mice have significantly lower survival rates than those of the wild type littermates. These results suggest that neurotrophin signaling through trkB receptors play a role in the survival of CNS neurons during postnatal development, and also indicate that trkB receptor signaling protects CNS neurons from axotomy-induced cell death.

303. ENDOTHELINS: FROM VASCULAR BIOLOGY TO DEVELOPMENTAL GENETICS
Masashi Yanagisawa, M.D., Ph.D., Howard Hughes Medical Institute and Department of Molecular Genetics, University of Texas Southwestern Medical Center, Dallas, TX 75235-9050, U.S.A.

Endothelins are a family of small peptide ligands (ET-1, ET-2 and ET-3) that act on G protein-coupled heptahelical receptors (ETA and ETB receptors). ET-1 was originally identified as a powerful vasoconstrictor factor derived from the vascular endothelium. Mature, 21-residue endothelins are produced from biologically inactive, 38-41 residue intermediates, termed big endothelins, via a proteolytic cleavage catalyzed by the recently identified membrane-bound metalloproteases, endothelin converting enzyme-1 and -2 (ECE-1 and ECE-2). Deregulation of endothelins has been implicated in a number of disorders involving abnormal vascular tone.

As an approach to better understanding the physiological role of the endothelin system, we have been systematically knocking out its individual molecular components in mice. Unexpected results from gene targeting experiments have provided new insights into important roles of the endothelin pathway in development of neural crest-derived tissues. Findings in the knockout mice have directly led to the discovery that defects in the human ETB and ET-3 genes cause Hirschsprung disease (congenital megacolon) and related neurocristopathies. Endothelins emerge as the first examples of G protein-coupled cell-cell signaling molecules that are involved in mammalian embryonic development. Experiments are underway utilizing these mutant mice to genetically dissect the role of endothelins in adult homeostasis, including the regulation of blood pressure and respiration.

304. INSIGHTS INTO THE PHYSIOLOGICAL ROLE OF DOPAMINE THROUGH GENE TARGETING OF THE DOPAMINE TRANSPORTER. Marc G. Caron, Howard Hughes Medical Institute Laboratories, Dept. of Cell Biology, Duke Univ. Med. Ctr., Durham, NC 27710.

Dopamine (DA) is implicated in the control of several important physiological paradigms in the central nervous system (CNS) and periphery. These actions of DA are mediated through 5 distinct G protein-coupled receptor subtypes. Recent studies have used genetic deletion of several of these receptor genes to examine their specific involvement in these paradigms. In the CNS, the dopamine transporter (DAT) regulates the temporal and spatial activity of DA by its reuptake into presynaptic terminals. To assess the contribution of DAT in dopaminergic transmission, we created a mouse line in which the DAT gene has been deleted. These mice display marked spontaneous hyperlocomotion due to the fact that DA spends 300 times longer at the synapse of DAT lacking mice. DAT knockout mice do not respond to cocaine or amphetamine. Most remarkably, the magnitude of the adaptive changes that occurred in an attempt to dampen the hyperdopaminergic tone of these mice (decreased in DA levels, tyrosine hydroxylase, D1 and D2 receptors) were all decreased to an extent never previously observed with any pharmacological or surgical manipulations. In addition, the absence of the DAT revealed that DA plays an important role as an antiproliferative agent on pituitary lactotrophs and also controls growth hormone and somatic growth through hypothalamic sites. The results indicate that the DAT is a crucial component in the maintenance of dopaminergic tone in both basal ganglia and hypothalamus. Using this approach we now document a physiological role for DA on growth and development which had not been widely appreciated.

305. DESIGN OF PROTEASE INHIBITORS FROM STRUCTURES OF SUBSTRATE PRODUCTS. THE EXAMPLE OF BUTABINDIDE, AN INHIBITOR OF THE CHOLECYSTOKININ (CCK-8)-INACTIVATING PEPTIDASE. C.R. Ganellin, P.B. Bishop, R.B. Bambal, S.M.T. Chan, A.N.J. Moore, P. Bourgeat[†], C. Rose[†], F. Vargas[†] and J.C. Schwartz[†] Department of Chemistry, University College London, 20 Gordon Street, London WC1H 0AJ, England. [†]Unité 109 de Neurobiologie et Pharmacologie, Centre Paul Broca de l'INSERM, 2ter rue d'Alésia, 75014 Paris, France.

Tripeptidyl peptidase II cleaves the neurotransmitter CCK-8 at the Met—Gly bond to give Asp·Tyr·Met·OH + Gly·Trp·Met·Asp·Phe·NH$_2$. The novel TPPII inhibitor, butabindide, [C. Rose et al, Nature (London) 1996, 380, 403], was discovered via lead generation followed by optimization of structure with respect to activity. Comparison with the substrate products shows that, in retrospect, butabindide could have been designed from Asp·Tyr·Met·OH by removing polar groups (CO$_2$H, NH, OSO$_3$H, SMe). It is likely that polar groups facilitate the dissociation of hydrolysis products from the enzyme thereby freeing the active site to promote enzyme recovery. This suggests a *general approach to the design of protease inhibitors*: i.e. remove polar groups from the substrate products and increase affinity through optimization of hydrophobic binding. Structure-activity relationships for these comparisons will be presented. We thank Pharmacia and Upjohn Inc. for financial support.

306. STRUCTURE ACTIVITY RELATIONSHIP OF CYCLIC UREAS WITH ASYMMETRIC P2/P2' SUBSTITUENTS: POTENT INHIBITORS OF HIV-1 PROTEASE. George V. De Lucca, Ui T. Kim, Emeka Akamike, Patrick Lam, Charles Clark, James Rodgers, Haisheng Wang, Soo Ko, Anurug Srivastava, Keichi Tanabe, Ron Klabe, Beverly, Lee Bacheler and Susan Erickson-Viitanen. Dupont Merck Pharmaceutical Company P.O. Box 80500 Experimental Station, Wilmington, DE 19880-0500

Cyclic ureas are a class of potent HIV-1 Protease inhibitors exemplified by the C_2 symmetric clinical candidate DMP450. Our goal has been to develop the next generation of cyclic urea HIVPR inhibitor with a ten fold increase in potency while maintaining the excellent pharmacokinectics of DMP450. This paper will discuss the factors that effect potency, protein binding, solubility, bioavailability, and resistance profile in the cyclic urea series and our efforts to juggle all these factors to obtain very potent (SE063 IC_{90} = 11 nM) compounds with excellent oral bioavailability.

DMP450
IC_{90} = 125 nM
Dog F % = 80

SE063
IC_{90} = 11 nM
Dog F % > 95

307. DESIGN AND SYNTHESIS OF NOVEL CONFORMATONALLY RESTRICTED HIV PROTEASE INHIBITORS. F.G. Salituro, C.T. Baker, J.T. Court, D.D. Deininger, E.E. Kim, B. Li, P.M. Novak, B.G. Rao, S. Pazhanisamy, W.C. Schairer, R.D. Tung, Vertex Pharmaceuticals, 130 Waverly Street, Cambridge MA, 02139

During the past several years we have been engaged in a research program aimed at the development of structurally novel HIV protease inhibitors, with the goal of complementing the resistance profile of our current clinical candidate (VX-478, currently under development by Glaxo Wellcome as 141W94 and by Kissei Pharmaceuticals in the Far East as KVX-478). Early results in this program, driven in part by combinatorial chemistry approaches, yielded a set of promising basis compounds with low nM Ki activity. Structural and computational analysis indicated opportunities for activity enhancement through conformational restriction. Incorporation of a variety of such changes resulted in the development of several classes of novel, highly potent inhibitors. In this presentation we will discuss the design, synthesis and enzymological analysis of these compounds.

308.
DESIGN, SYNTHESIS, AND ELABORATION OF A NONPEPTIDE SCAFFOLD USEFUL IN BUILDING A PROTEASE INHIBITOR LIBRARY. **Nha Huu Vo,** C. Nicholas Hodge, and Charles J. Eyermann The DuPont Merck Pharmaceutical Company, Chemical and Physical Sciences, Experimental Station, Wilmington, Delaware 19880-0500

The design of protease inhibitors has emerged as an important strategy for developing new therapeutics. While it is facile to access potent inhibitors by replacing the substrate's scissile bond with a transition-state mimetic, the de novo design of potent nonpeptide inhibitors is still a significant challenge. Using structure-based design techniques, we have designed a series of scaffolds which may be useful for building protease inhibitor libraries. We have selected one of these scaffolds which is amenable to combinatorial synthesis. Details of the design process, the solid-phase synthesis, and biological results will be presented.

309. SEGMENTAL STABILITY OF $CD8^+$ CTL gp41 EPITOPE PEPTIDES OF HIV TYPE I: PEPTIDE CONFORMATIONAL ANALYSIS USING PROTON NUCLEAR MAGNETIC RESONANCE SPECTROSCOPY AND MOLECULAR MODELING. John K. Young, Joseph J. Gambino, Francis A. Ennis and George E. Wright Department of Pharmacology and Molecular Toxicology, University of Massachusetts Medical Center, Worcester, MA 01655

NMR and molecular modeling have been used to determine the three-dimensional structure in 50% trifluroethanol-water of two HIV-1 peptide epitopes: HIV-140c (RLRDLLLIVTR) and WMJ2 (RLRDLLLIVKR). These peptides represent the most potent and weakest, respectively, among several variants in sensitizing autologous transformed B cells for Lysis by this $CD8^+$ CTL clone. Based on NOE distance constraints, the calculated conformation of HIV-140c was α-helical from R1 to L7 with a type III β-turn at the C-terminus. WMJ2 possessed a similar but less stable α-helical structure from R1 to L7, but the β-turn at the C-terminus was lost. The presence of a weak $d_{\alpha N}(i,i+2)$ NOE from R11 to V9 in the NMR spectrum of WMJ2 suggest that a population of the conformers present in solution may contain this reverse turn. We believe that the significant drop in biological activity of WMJ2 compared to HIV-140c is a result of the former's reduced stability at the C-terminus.

310.
SYNTHESIS, STRUCTURE AND ANTIBODY BINDING OF GLYCOPEPTIDES FROM IMMUNOGENIC DOMAINS OF HIV-1 GP120. Joseph J. Barchi, Jr.[†]*, Xiaolin Huang[†], Feng-Di T. Lung[†], Peter P. Roller[†], Robert R. Garrity[‡] and Jeff Muschik[‡], Laboratory of Medicinal Chemistry[†], Division of Basic Sciences, National Cancer Institute, Bethesda, Md 20892 and Laboratory of Tumor Cell Biology[‡], Frederick Cancer Research and Development Center, Division of Basic Sciences, National Cancer Institute, Frederick, Md 21702.

We have prepared glycosylated analogues of peptides from antigenic domains of gp120, in particular the V3 loop, and studied their solution conformations by NMR. Modification by O-glycosylation, with sugars attached to serine and threonine residues flanking the immunodominant tip ot the loop, was shown to profoundly effect both the local conformation and the antibody binding properties of V3 peptides. A comparison of the structural and antibody binding features of a series of glycopeptides will be presented as well as the implications of their use as immunogens in the design of novel vaccine constructs for the prevention of HIV-1 infection.

311. MOLECULAR ENGINE FOR DRUG DELIVERY.
James Summerton, Paula Matthews, Mike Partridge, Doreen Weller, and Dwight Weller, Antivirals, 4575 SW Research Way, Corvallis, OR 97333.

A molecular engine was constructed from a novel polypeptide which exists as a highly water soluble random coil at high pH and a water insoluble alpha-helix at low pH. Using fluorescent tags, the engine can be shown to pass from the low pH environment of the endosome into the higher pH cytosol of cultured cells. The talk will focus on the design and properties of the engine as they relate to the transport of substances across the cell membrane.

312. SYNTHESIS AND ANTIVIRAL ACTIVITY OF Z- AND E-2-HYDROXYMETHYLCYCLO-PROPYLIDENEMETHYLPURINES AND -PYRIMIDINES.
Y.-L. Qiu[a], M. B. Ksebati[b], J. C. Drach[c], E. R. Kern[d] and J. Zemlicka[a], [a]Karmanos Cancer Institute, Wayne State University School of Medicine, Detroit, MI 48201, [b]Central Instrumentation Facility, Wayne State University, Detroit, MI 48202, [c]School of Dentistry, University of Michigan, Ann Arbor, MI 48019, [d]Department of Pediatrics, University of Alabama School of Medicine, Birmingham, Alabama 35294.

Reaction of nucleic acid bases or suitable precursors with E/Z ethyl 2-bromo-2-bromomethylcyclopropane carboxylates using K_2CO_3 in DMF at 100°C afforded products of alkylation/elimination sequence, E/Z-esters 1. Reduction with DIBALH in THF led to the title compounds 2 and 3. The isomeric assignment relied on NMR spectroscopy (NOE). Compounds 2 (B = Gua, Ade

B = nucleic acid base

and Cyt) are strong inhibitors of human cytomegalovirus (HCMV) effective also against other herpesviruses. Analogues 2 (B = Gua and Ade) were also active *in vivo* against murine CMV. Supported by NIH grants RO1-CA32779, U19-AI31718, RO1-AI33332 and NO1-AI65290.

313. STRUCTURE-BASED DESIGN OF PROTEIN-TYROSINE PHOSPHATASE 1B INHIBITORS.
Terrence R. Burke, Jr.[‡]*, Zhu-Jun Yao[‡], Bin Ye[‡], Shaomeng Wang[†], and Zhong-Yin Zhang[§]. [‡]*Laboratory of Medicinal Chemistry, Bldg. 37, Rm 5C06, Division of Basic Sciences, NCI, National Institutes of Health, Bethesda, MD 20892 and [†]Georgetown University Medical Center, Washington, DC 20007 and [§]Albert Einstein College of Medicine, Bronx, NY 10461.*

Although protein-tyrosine phosphatases (PTPs) constitute important components of protein-tyrosine kinase signalling cascades, little has been reported on the design and synthesis of PTP inhibitors.

Protein-tyrosine phosphatase 1B (PTP1B) is a widely expressed non-receptor PTP which serves as a model for the structure-based design of PTP inhibitors. Using the X-ray structure of the difluorophosphonate **1** bound within the catalytic site of PTP1B, we identified structural modifications which could potentially afford enhanced binding interactions. Several of these rationally designed inhibitors were subsequently prepared and shown to exhibit up to 14-fold higher PTP1B inhibitory potency than the parent **1**. An overview of this work will be presented.

314 THE STRUCTURE AND FUNCTION OF PROTEIN TYROSINE PHOSPHATASES AND THEIR ROLE IN INSULIN SIGNALING. Mike Visnick, Department of Metabolic Diseases, Hoffmann-La Roche Inc., 340 Kingsland Street, Nutley, NJ. 07110.

Protein Tyrosine Phosphatases (PTPases) represent an important class proteins which modulate cellular function through their influence on the intracellular signaling cascade of growth factor, cytokine and hormone receptors. PTPases working together with protein tyrosine kinases control the phosphorylation state of key tyrosine residues on various target proteins and thereby carefully regulate protein function. In certain cases such as the non-receptor kinase p56lck, dephosphorylation of a specific tyrosine residue activates the kinase domain and serves to initiate a cascade leading to T-cell activation. In contrast, tyrosine dephosphorylation at critical sites of the insulin receptor down regulates its kinase domain which leads ultimately to insulin desensitization in key target tissues. Thus PTPases have been shown to occupy key roles in both signal initiating as well as terminating events in response to specific biological effectors. This lecture will present an overview of the structure and function of several PTPases involved in physiologically relevant processes as well as a more detailed analysis of the PTPases of importance in insulin signaling and hence of potential therapeutic significance in non insulin dependent diabetes mellitus (NIDDM).

315. DESIGN AND SYNTHESIS OF PEPTIDE-BASED VLA-4 ANTAGONISTS LEAD TO ASTHMA DRUG CANDIDATES, Ko-Chung Lin,* Craig N. Zimmerman, Alfredo Castro, Wen-Cherng Lee, Charles E. Hammond, Humy S. Ateeq, Sherry H. Hsiung, Lillian T. Chong, Yusheng Liao, Ronald G. Almquist, Sandhya Kalkunte and Steven P. Adams, Department of Medicinal Chemistry, Biogen Inc., Cambridge, MA 02142

In this presentation, SAR studies of the first VLA-4 small molecule antagonists exhibiting *in-vitro* potency and *in-vivo* efficacy will be discussed. Eosinophils and T lymphocytes recruited into airway tissues are involved in inflammation in asthma. Both of these cells express the integrin $\alpha_4\beta_1$, very late antigen-4 (VLA-4), a key cell surface receptor that mediates cellular adhesion and activation through a variety of cell-cell and cell-matrix interactions. VLA-4 serves as a receptor for the cytokine-inducible endothelial cell surface protein, vascular cell adhesion molecule-1 (VCAM-1), as well as to the extracellular matrix protein fibronectin (FN). The LDV amino acid sequence of the alternatively spliced connecting segment 1 (CS1) region of FN, was shown to be recognized by VLA-4. Although the affinity of LDV towards VLA-4 is quite low (IC_{50} ~5,000 µM in cell adhesion assay (CAA), it provided precedence that small molecules could block VLA-4 / ligand interactions. A hypothetical model of our inhibitor/VLA-4 interactions was formulated and it resulted in the design and synthesis of a series of compounds with IC_{50} in the sub-nanomolar range. Furthermore, *in vivo SAR* studies in mice and sheep have led to the discovery of a series of potent drug candidates for clinical evaluation in asthma.

316. EFFECT OF NUCLEOSOME STRUCTURE ON MUSTARD CROSS-LINKING OF DNA. **Julie T. Millard**, Nathan Leudtke, Rebecca J. Spencer, Paul B. Hopkins; Departments of Chemistry, Colby College, Waterville ME 04901 and University of Washington, Seattle WA 98195.

Mechlorethamine (*N*-methylbis(2-chloroethyl)amine) is the structurally simplest of the nitrogen mustards. The antitumor activity of this agent has been attributed to its ability to form DNA interstrand cross-links. Such cross-links result in considerable bending of the DNA (Rink & Hopkins [1995] *Biochemistry 34*, 1439). Previously we found that in DNA oligomers, mechlorethamine forms cross-links through the distal deoxyguanosine residues at the sequence 5'-GNC (N=G or C) in preference to 5'-GC and 5'-CG sequences (Millard, Raucher, & Hopkins [1990] *J. Am. Chem. Soc. 112*, 2459). We have extended our studies to a 154 base pair restriction fragment of the *Xenopus borealis* 5S RNA gene and found that the 5'-GNC sequence preference is preserved in this longer DNA. Moreover, upon reconstitution of this

DNA with chicken histones to form uniquely positioned nucleosomes, the 5'-GNC sequence remains the preferred target. Likewise, the aromatic mustards chlorambucil and melphalan also display the 5'-GNC consensus sequence in both free and nucleosomal DNA.

317.

THE DISCOVERY OF VORICONAZOLE - A NOVEL, BROAD-SPECTRUM TRIAZOLE ANTIFUNGAL. Andrew S.Bell, Catherine Burt, Roger P. Dickinson, and Ken Richardson, Department of Discovery Chemistry, Pfizer Central Research, Sandwich, Kent, CT1 3 9NJ, U.K.

Fluconazole is the agent of choice for the treatment of infections due to Candida and Cryptococcus spp. However, it is only weakly active against Aspergillus spp., due to low potency against its target enzyme, lanosterol 14alpha-demethylase. Based on molecular overlays of fluconazole and lanosterol, enzyme potency was improved by introduction of an additional group on the propanol backbone of fluconazole and replacement of one triazole by other heterocycles. In vivo activity and physicochemical properties were optimized, resulting in the fluoropyrimidine derivative UK-109,496 (voriconazole), which was selected for further investigation. The efficacy of voriconazole in animal infection models and in early clinical studies will be presented.

318.

RECENT STUDIES ON AZOLES; DISCOVERY AND SYNTHESIS OF SCH 56592, A BROAD SPECTRUM ORALLY ACTIVE ANTIFUNGAL AGENT. A.K. Saksena, V.M. Girijavallabhan, R.G.Lovey, R.E.Pike, H.Wang, Y.T.Liu, P.Pinto, F. Bennett, N. Patel, A.K. Ganguly, Scering-Plough Research Institute, Kenilworth, NJ 07033, USA

As part of an extensive search for orally effective agents, we developed a series of novel tetrahydrofuran based azole antifungals having broad-spectrum activity. Close attention to stereo and regiochemical requirements for optimal oral activity led to discovery of 2,2,4-cis-trisubstituted tetrahydrofurans having 2R absolute configuration at the benzylic carbon. A follow-up on these observations led to Sch 51048 as a potent antifungal having superior activity over existing agents against a variety of systemic infections in normal and immunocompromised animal models. In an effort to further improve oral absorption properties of Sch 51048, synthesis of its novel analogs having polar side chains was undertaken. While introduction of relatively basic or acidic side chains resulted in loss of activity, incorporation of hydroxyl groups led to analogs with greatly improved *in vivo* activity. Practical stereo and enantioselective synthesis of the 2,2,4-cis-trisubstituted tetrahydrofuran portion and methodology for synthesis and introduction of hydroxylated side chains with defined absolute and relative stereochemistry will be described. From biological evaluation of a variety of hydroxylated analogs, Sch 56592 displayed the best profile in terms of its superior bioavailability upon oral administration in several animal species. Chemistry leading to the discovery of Sch 56592 and other highly active analogs will be discussed. Based on its superior broad-spectrum activity over Sch 5I048, itraconazole, fluconazole and amphotericin B in a number of infection models, Sch 56592 was selected for clinical evaluation.

319.

ECHINOCANDIN ANTIFUNGALS FROM THE LAB TO THE CLINIC. William W. Turner, Eli Lilly & Co., Lilly Corporate Center, Indianapolis, IN 46285

With the initial isolation of the echinocandin natural product class over 20 years ago, the steady efforts of several research groups have finally resulted in clinical candidates to treat serious, systemic fungal infections. These cyclic lipopeptide natural products have significant fungicidal anti-*Candida* activity as well as anti-*Pneumocystis* activity. Modifications have reduced toxicity, increased antifungal potency, and increased water solubility. The increases in potency have also effectively broadened the spectrum to include all *Candida* species and the pathogenic *Aspergillus* species. Effective oral uptake in humans has now been demonstrated giving the possibility of oral as well as IV routes of administration. Chemistry and SAR studies which led to these improvements will be discussed.

320.

NEW DEVELOPMENTS IN ANTIFUNGAL AGENTS. J.M. Balkovec, R.M. Black, F.A. Bouffard, J.F. Dropinski, M.L. Hammond, Department of Medicinal Chemistry, Merck Research Laboratories, Rahway, New Jersey, 07065-0900

The incidence of serious fungal infections has increased over the past two decades due mainly to an increase in the immunocompromised patient population. In response to this problem, several groups have developed agents with novel mechanisms of action. Antifungal targets that affect the integrity of the cell wall have proven a fruitful area of research. We have described derivatives of pneumocandin B_0, a cyclic lipopeptide related to the echinocandins, that inhibit the synthesis of β-1,3-glucan. Cationic derivatives of the natural product possess activity against Candida and Aspergillus pathogens. The studies leading to the selection of a promising new clinical agent, L-743,872, will be, presented along with the compound's in vitro and in vivo activity.

Pneumocandin B_0 X = O   Y = OH
L-743,872   X = H,H   Y = NHCH$_2$CH$_2$NH$_2$

321.

AMINO-KETONES AS INHIBITORS OF FUNGAL 2,3-OXIDOSQUALENE-LANOSTEROL CYCLASE. S.Jolidon, A.Polak-Wyss, P.G.Hartman & P. Guerry, Pharma Division, Prelinical Research, F. Hoffmann-La Roche Ltd., CH-4070 Basel, Switzerland.

2,3-Oxidosqualene-Lanosterol Cyclase (OSC) is an important enzyme in the biosynthesis of the essential fungal cell wall constituent ergosterol. Whereas many other enzymes of this pathway have been intensively studied as targets for antifungals, OSC has been somewhat underexploited. We decided to aim at this target using known inhibitors as a starting point. Simple mechanistic considerations led to the design of bifunctional inhibitors containing a tertiary amino group and a ketone. These compounds should have the capacity to interact with two different enzyme sites, in contrast to earlier inhibitors. A cell-free assay allowing accurate determination of IC$_{50}$-values revealed very potent enzyme inhibitors in this series. These compounds represent the first OSC-inhibitors which also show an effective *in vitro* activity against a wide range of medically important fungi. Unfortunately, the amino-ketones proved to be almost inactive against systemic fungal infections in mice models. This lack of activity is probably due to the high lipophilicity and thus high protein binding of the inhibitors. However, the title compounds proved to be very effective as cholesterol lowering agents in mammals.

American Chemical Society
DIVISION OF NUCLEAR CHEMISTRY & TECHNOLOGY

213th ACS National Meeting

San Francisco, CA
April 13-17, 1997   P.E. Haustein, Program Chairperson

SUNDAY MORNING

- **Award Symposium in Honor of Peter J. Armbruster - I**
G. Friedlander, Presiding   Papers 1-5

SUNDAY AFTERNOON

- **Award Symposium in Honor of PeterJ. Armbruster - I**
Cluster, Fusion, and Fission
G. Herrmann, Presiding   Papers 6-11

- **Radioactive Beams, Radioactive Atoms in Traps, Exotic Nuclei - I**
J. Cerny, Presiding   Papers 12-18

MONDAY MORNING

- **Chemical and Nuclear Properties of the Heaviest Elements - I**
K.E. Gregorich, Presiding   Papers 19-26

- **Radioactive Beams, Radioactive Atoms in Traps, Exotic Nuclei - I**
D.J. Morrissey, Presiding   Papers 27-32

MONDAY AFTERNOON

- **Chemical and Nuclear Properties of the Heaviest Elements - I**
R.W. Lougheed, Presiding   Papers 33-40

- **Radioactive Beams, Radioactive Atoms in Traps, Exotic Nuclei - I**
E. Zganjar, Presiding   Papers 41-47

TUESDAY MORNING

- **Nuclear Data Resources on the World Wide Web I**
J. Dairiki, Presiding   Papers 48-53

- **Radioactive Beams, Radioactive Atoms in Traps, Exotic Nuclei I**
  W. Loveland, Presiding   Papers 54-59

TUESDAY AFTERNOON

- **Nuclear Data Resources on the World Wide Web - I**
  R.A. Meyer, Presiding   Papers 60-65

- **Radioactive Beams, Radioactive Atoms in Traps, Exotic Nuclei - I**
  D.J. Vieira, Presiding   Papers 66-71

WEDNESDAY MORNING

- **Nuclear Data Resources on the World Wide Web - I**
  R. MacFarlane, Presiding   Papers 72-77

- **Nuclear Decay Probes in Chemical Physics and Materials Science - I**
  D.G. Fleming, Presiding   Papers 78-81

WEDNESDAY AFTERNOON

- **Nuclear Data Resources on the World Wide Web - I**
  D. Brenner, Presiding   Papers 82-87

- **Nuclear Decay Probes in Chemical Physics and Materials Science - I**
  R.S. Hayano, Presiding   Papers 88-90

THURSDAY MORNING

- **Additional Aspects of Nuclear Science**
  P.E. Haustein, Presiding   Papers 91-100

- **Nuclear Decay Probes in Chemical Physics and Materials Science - I**
  R.F. Kiefl, Presiding   Papers 101-104

THURSDAY AFTERNOON

- **Nuclear Decay Probes in Chemical Physics and Materials Science - I**
  R.H. Heffner, Presiding   Papers 105-109

DIVISION OF NUCLEAR CHEMISTRY & TECHNOLOGY

001.
WITH RECOIL SEPARATORS TO THE LIMITS OF NUCLEAR STABILITY. **G. Muenzenberg**
Gesellschaft fuer Schwerionenforschung mbH, Planckstrasse 1, D-64220 Darmstadt, Germany

Abstract Not Available

002.
AWARD ADDRESS: 100 YEARS OF RADIOACTIVITY - FROM POLONIUM TO ELEMENT 112.
P. Armbruster Gesellschaft fuer Schwerionenforschung mbH, Planckstrasse 1, D-64220 Darmstadt, Germany

ABSTRACT NOT AVAILABLE

003. A REVIEW OF THE INSTRUMENSTATION USED FOR THE DISCOVERY OF NEW ELEMENTS DURING THIS CENTURY* A. Ghiorso LBNL. 1 Cyclotron Rd., Berkeley, Ca. 94720

Instrumentation, of necessity, has played a dominant role in the search for new elements during this century and has gone through many evolutionary changes. It is the purpose of this paper to review the development of the techniques for identifying new atoms from the beginnings, when optical spectroscopy was the only tool available and milligrams of material were needed, to the present time when a transactinide element can be singled out with certainty even though it might take weeks to produce one single atom.
*This work was supported by the Director, Office of Energy Research, Division of Nuclear Physics of the office of High Energy and Nuclear Physics of the US Department of Energy under contract DE-AC0376SF00098.

004. SYNTHESIS AND STABILITY OF HEAVIEST ELEMENTS
Sigurd Hofmann, GSI, Planckstrasse 1, D-64291 Darmstadt, Germany

The synthesis of the heaviest elements by fusion of lighter nuclei is more and more hindered by the increasing repulsive Coulomb forces of the reacting system of nucleons. The presently known heaviest element has atomic number Z=112 and atomic weight A=277. The reaction cross-section for fusion of the ^{70}Zn projectiles with the ^{208}Pb targets is 1 pb. The projectile energy had to be chosen extremely low, so that projectile and target nuclei come to rest when, in a central collision, the diffuse nuclear surfaces are just in contact. Investigation of such extremely rare reaction processes could be achieved by continuously improved experimental sensitivity using high current beams, effective separation techniques and sophisticated detection methods. The lifetime of the heaviest nuclei is determined by shell-structure effects, which result in a stability high enough, so that the nuclei decay by emission of alpha particles and not fission. Even higher stability and longer lifetime is predicted by theoretical models for nuclei closer to the magic neutron number N=184. Presently, the most interesting research topic is to examine a possible correlation of stability and reaction cross-section, in order to find an optimum road for production of superheavy nuclei.

005. PREDICTED PROPERTIES OF HEAVY ELEMENTS, A. Sobiczewski, Soltan Institute for Nuclear Studies, Warsaw, Poland and GSI, Darmstadt, Germany

Theoretical predictions for the properties of nuclei of heaviest elements are presented. Such properties as mass, modes of decay and respective half-lives are discusssed. Much attention is paid to effects of the shell structure of a nucleus in these properties. Even-even nuclei with the proton number $Z=82$-120 and the neutron number $N=126$-190 are considered. Two regions of increased stability, due to large shell effects, are predicted: one around the nucleus 270-108 (270-Hs) (deformed superheavy nuclei) and the other around the nucleus 298-114 (spherical superheavy nuclei). Alpha decay is expected to be the dominating decay mode for nuclei around the centers of both regions. Predicted half-lives are long enough to study the properties, both physical and chemical, of many of these nuclides, if they are synthesized.

006. SHELLS IN CLUSTERS AND NUCLEI.
S. Bjørnholm, The Niels Bohr Institute, University of Copenhagen, Blegdamsvej 17, DK-2100 Copenhagen, Denmark

The conduction electrons in small metallic particles are organized in much the same way as the nucleons in the nucleus. These particles therefore form a model system where analogies to nuclear structure, in particular the occurrence of shells and magic numbers, can be explored in a new and wider context. The particle number can easily exceed the nuclear limit of $A \approx 300$, and this makes it possible to exhibit a higher periodicity in the shell oscillations, the supershell beat mode, which finds a simple explanation in terms of triangular and square orbits of a classical particle moving with Fermi velocity in the effective mean field of the spherical particle.

Metal particles may be charged by removing electrons, and this may lead to decay by fission. The possible analogies and differences compared to nuclear fission are presently being studied.

007. THE DYNAMICS OF NUCLEAR FUSION AND FISSION. Wladyslaw J. Swiatecki, Nuclear Science Division, Lawrence Berkeley National Laboratory, One Cyclotron Road, Berkeley, CA 94720

Attempts to formulate a collective dynamics of the fission and fusion of nuclei will be reviewed, with emphasis on the hindrance to the fusion of very heavy systems.

008. SHELL EFFECTS IN FUSION OF HEAVY NUCLEI. P. Möller, P. Moller Scientific Computing and Graphics, Inc., P. O. Box 1440, Los Alamos, NM 87544 and J. R. Nix, Theoretical Division, Los Alamos National Laboratory, Los Alamos, NM 87545.

The spontaneous-fission properties of Fm isotopes undergo dramatic changes between ^{256}Fm and ^{258}Fm. The fission fragments of the former isotope are mass asymmetric with kinetic energies of about 200 MeV, whereas the fission fragments of the latter isotope are symmetric with kinetic energies of about 235 MeV. This rapid change occurs because the division into nearly doubly magic fragments near ^{132}Sn becomes possible and opens up new valleys in the fission potential-energy surface. In the cold-fusion reactions leading to the heaviest elements, the nearly doubly magic targets and/or projectiles may give rise to important features associated with this magicity. Cold fusion is thought to favor heavy-element formation because it leads to low excitation energies of the compound nuclei. We investigate how near-magic targets and projectiles may lead to persistent survivability of the shells in the fusion valley as the ions merge, in addition to their effect on the compound-nucleus excitation energy. This work was supported by the U. S. Department of Energy.

009. FISSION OF RELATIVISTIC NUCLEI AND THE PRODUCTION OF NEW ISOTOPES. M.S. BERNAS, Institut de Physique Nucléaire, ORSAY F-91400

Collisions of U-ions accelerated at 750 A.MeV - 83% of light velocity - with Pb and Be targets were investigated. The resulting fragments were on-line analyzed using magnetic separation with the fragment separator, energy-loss and time-of-flight measurements, event by event.
The reverse kinematics at large velocity provides an efficient collection

of fragments originating from Coulomb and nuclear fission as well as from fragmentation. For each tuning of the FRS, some hundred isotopes are unambiguously identified. The production yields are measured independantly from chemical and radiochemical properties or from ß-decay life-time since the identification takes place in 0.2µs. The method is highly sensitive; cross-sections as low as 0.3nb (i.e. $3.10^{-34}cm^2$) were measured for the doubly magic isotope of ^{78}Ni, observed for the first time. Over two beam time periods, more than a hundred new neutron rich isotopes were identified.

010. FISSION OF EXOTIC NUCLEI. K.-H. Schmidt, GSI Darmstadt, Planckstrasse 1, D-64291 Darmstadt

The radioactive-beam facility BRENDA of GSI Darmstadt offers the unique possibility to produce isotopically separated secondary beams of short-lived fissile nuclei by fragmentation of a 1 A GeV 238U beam. These radioactive beams have been used to extend studies on nuclear fission to previously unexplored nuclei. Fission in-flight was induced in a secondary lead target at average secondary-beam kinetic energies of about 400 A MeV. By measuring the velocities and the energy losses of both fission fragments their nuclear charges and their kinetic energies could be determined. From the data, fission-fragment charge distributions and total kinetic energies for electromagnetic-induced fission in lead at an average excitation energy of 11 MeV were extracted. The transition from asymmetric to symmetric fission around 226Th is systematically covered. Variations of the total kinetic energies and of the even-odd effect in the charge yields are discussed in terms of the scission configurations and the thermal excitations in different fission modes.

011. HYBRID REACTORS FOR ENERGY PRODUCTION AND WASTE INCINERATION
H. Nifenecker, Institut des Sciences Nucléaires, 53, avenue des Martyrs 38026 - Grenoble Cedex - France

Nuclear energy meets and will increasingly meet considerable societal opposition if at least two prerequisite are not met :
1) Complete impossibility of a major accident. This requires non-critical systems as well as passive cooling of the residual heat after shut-down of the reactor. 2) Stabilization and, better, decrease of the waste inventory. Hybrid system where an accelerator is associated to a sub-critical neutron multiplying assembly seems to be able to meet these aims. The physics of these systems will be briefly explained. Afew proposed systems will be described with special reference to the Fast Energy Amplifier proposed by C. Rubbia. Prospects for building a first test system will be discussed.

012.
SHELL-MODEL CALCULATIONS LEADING TO THE DRIP LINE. **B.A. Brown,** Department of Physics and Astronomy, Michigan State University, East Lansing, Michigan 48824

Theoretical calculations which combine the Hartree-Fock mean-field with configuration-mixed shell-model correlations have been extremely successful in describing the properties of nuclei near the valley of stability. The hamiltonians which have been derived from such studies can be used to predict the properties of nuclei out to the proton and neutron drip lines which are the subject of recent and planned experimental investigations. I will review the results of such extrapolations, their comparison to recent experimental data, and the problems and goals for future theoretical work.

013. MASS MEASUREMENTS OF SHORT LIVED NUCLEI AT GSI,
H. Wollnik[1], K. Beckert[2], T. Beha[3], F. Bosch[2], H. Eickhoff[2], B. Franzke[2], Y. Fujita[4], H. Geissel[2], M. Hausmann[2], H. Irnich[2], H. C. Jung[1], Th. Kerscher[3], O. Klepper[2], H.-J. Kluge[2], C. Kozhuharov[2], G. Kraus[2], K. E. G. Löbner[3], G. Münzenberg[2], Yu. Novikov[5], F. Nickel[2], F. Nolden[2], Z. Patyk[2], T. Radon[2] H. Reich[2], C. Scheidenberger[2], B. Schlitt[2], W. Schwab[2], M. Steck[2], K. Sümmerer[2]; [1] Universität Gießen, [2] Gesellschaft für Schwerionenforschung, Darmstadt, [3] Universität München, [4] Osaka University, [5] St. Petersburg Nuclear Physics Institute

High-accuracy mass measurements of proton-rich isotopes in the range of $60 \leq Z \leq 84$ were performed at the GSI using the technique of Schottky spectrometry by which the frequency of rotation in the "experimental storage ring" (ESR) was determined for cooled projectile fragments. Such fragments were produced by ^{209}Bi ions at $930 A \cdot$MeV, separated by the recoil mass separator FRS, and stored in the ESR, where the momentum spread was reduced via electron cooling to $\Delta p/p \simeq 10^{-6}$. Determining the frequency of rotation for ion bunches mass resolving powers of 350000 and mass accuracies of 100 keV were achieved for ions in the region of $A \approx 200$.

014. RADIOACTIVE ION BEAMS AND PENNING TRAPS AT ISOLDE
G. Bollen, PPE-ISOLDE, CERN, CH-1211 Geneva 23

The ISOLTRAP system installed at the PS-Booster ISOLDE facility at CERN/Geneva is a Penning trap mass spectrometer particularly designed for the investigation of unstable isotopes. More than 100 radioactive isotopes of the alkali, alkali earth, earth alkali elements and of mercury, most of them far from stability with half-lives down to the 1 second region, have been determined. An accuracy of typically dm/m = 1×10^{-7} has been obtained for all of the investigated isotopes. The accurate data have a large impact on the nuclide chart of masses. The ISOLTRAP results helped to reveal a number of wrong mass values, to clarify a large number of inconsistencies between other experimental data and to set a firm anchor ground for many Q-values links to different isotopes of other elements. At ISOLDE another Penning trap system is presently under construction which plays an important role within the REX-ISOLDE project. In this project a novel, cost effective and efficient scheme will be employed for the post-acceleration of the existing low-energy ISOLDE beams up to an energy of 2.2 MeV/u. First experiments aim for the investigation of very neutronrich isotopes close to N=20 and N=28.

015. New Proton Radioactivities and Radioactive Beams at ATLAS*
Cary N. Davids, Physics Division, Argonne National Laboratory, Argonne, IL 60439

In experiments using the Fragment Mass Analyzer at ATLAS and a double-sided silicon strip detector placed behind the focal plane, a series of new proton emitters have been discovered with $73 \leq Z \leq 83$. Measurements of the energies and half-lives of the protons allow the angular momentum of the emitted proton to be determined, providing nuclear structure information on nuclei situated several neutrons beyond the proton drip line. Combining the present data with earlier work, a consistent picture of the proton emission rates has been obtained, using simple barrier penetration and a low-seniority shell model calculation.

Progress at ATLAS in the area of radioactive beams will also be reported.

*Work supported by the U.S. Department of Energy, Nuclear Physics Division, under Contract W-31-109-ENG-38.

016. EVIDENCE FOR COLLECTIVITY IN THE LIGHT Po ISOTOPES: THE α-DECAY OF ^{190}Po
J. C. Batchelder, UNIRIB, Oak Ridge Associated Universities, Oak Ridge Tn, 37831

The α-decay properties of ^{190}Po were investigated through the use of a fragment mass analyzer in conjunction with a double-sided Si strip detector. The isotope was produced via the ^{96}Mo(^{96}Mo,2n) reaction, and its α-decay energy and $T_{1/2}$ were measured as 7529(15) keV and $2.4^{+0.4}_{-0.3}$ ms respectively. The resulting reduced width is nearly identical to that of the ^{192,194}Po isotopes. This is believed to result from significant mixing between the ground state $\pi(2p)$ and the low-lying 0^+ $\pi(4p-2h)$ intruder state in the Po parent. This provides further evidence for collectivity in the light Po isotopes. In addition, ^{191}Po was unambiguously identified and the ^{186}Pb α-decay branch was determined experimentally for the first time.

UNIRIB is a consortium of universities, State of Tennessee, Oak Ridge Associated Universities, and Oak Ridge National Laboratory and is partially supported by the U.S. Department of Energy under contract No. DE-AC05-76OR00033 with the Oak Ridge Associated Universities.

017. NEW ALPHA EMITTERS AND PLANNED CHARGED-PARTICLE DECAY STUDIES USING THE HRIBF RECOIL SEPARATOR. Kenneth S. Toth, Oak Ridge National Laboratory*, Oak Ridge, TN 37831

With the use of beams from the Holifield Radioactive Ion Beam Facility (HRIBF) at Oak Ridge National Laboratory it should be possible to identify many new isotopes near and beyond the proton drip line. In these investigations, nuclei produced in fusion evaporation reactions will be separated from the incident ions and dispersed in mass/charge with a recoil mass separator and implanted in double-sided Si strip detectors which will be used in conjunction with a gas-filled position sensitive avalanche counter for the study of proton and alpha-particle radioactivity. The present talk will illustrate these possibilities by describing the discovery of Pt, Pb, and Po alpha emitters at the Argonne National Laboratory ATLAS facility with the use of a fragment mass analyzer. It will then focus on experiments looking for new ground-state proton emitters in the mass region between Cs and Tm that will be carried out with the first radioactive beams to become available at HRIBF.

*Research sponsored by the Oak Ridge National Laboratory managed by Lockheed Martin Energy Research Corporation for the U.S. Department of Energy under contract number DE-AC05-96OR22464.

018. NEW MEDIUM-MASS PROTON-RICH ISOTOPES AND THEIR DECAYS. B. Blank, C.E.N. Bordeaux, F-33170 Gradignone, France

In experiments at GANIL and GSI, we searched for new proton-rich isotopes in the Cr-to-Mo region. These isotopes are of special interest for modeling the astrophysical rp process and for the search for the yet unknown two-proton radioactivity. The studies allowed to observe 11 new isotopes. The odd-Z isotopes Ga60, As64, and Y77 are of prime importance for the rp process, whereas Fe45 is according to theoretical predictions the best candidate for the two-proton radioactivity. The decay studies of isotopes close to the drip line in the Se-to-Mo region show the influence of deformation on the beta-delayed proton emission. E.g., a change of a factor of ten in the proton-emission probability between Se67 and Kr71 is interpreted as a sudden change in deformation.

019. CHROMATOGRAPHIC STUDIES OF Rf (ELEMENT 104) WITH TRIBUTYLPHOSPHATE (TBP)
W.Bruechle, E. Jaeger, M. Schaedel, B. Schausten, Gesellschaft fuer Schwerionenforschung, D64220 Darmstadt,Germany;R. Guenther, J.V. Kratz, W. Paulus, A. Seibert, P. Thoerle, S. Zauner, Inst. fuer Kernchemie Universitaet Mainz, D55099 Mainz, Germany; D. Schumann, Inst. fuer Analytische Chemie Techn.Universitaet D01062 Dresden, Germany

Liquid- liquid extractions of Rf in the system HCl/TBP (Benzene) were carried out by different authors in dependence on concentrations of chloride, TBP, and hydrogen. We transfered this system to the Automatic Rapid Chemistry System ARCA, which can perform chromatographic separations within seconds. ^{261}Rf was produced in the reaction ^{248}Cm (98MeV ^{18}O, 5n) ^{261}Rf at the Philips Cyclotron of Paul Scherrer Institute (Villigen, CH). The produced activities were transported with a He(KCl)- jet to the apparatus, where small columns filled with TBP on an inert support (Voltalef) were used to separate a hafnium- fraction (8M HCl) and a zirconium- fraction (2M HCl). The samples were dried and measured by α- spectroscopy. After 1140 experiments 30 α- decays with the energy of ^{261}Rf or ist daughter ^{257}No were detected, including 5 correlated mother- daughter- decay chains. The distribution between the two chemical fractions indicates that Rf behaves intermediate between Zr and Hf.

020.
AQUEOUS CHEMISTRY OF ELEMENT 105. **J. V. Kratz** Institut fuer Kernchemie, Universitaet Mainz, D-55099 Mainz, Germany

ABSTRACT NOT AVAILABLE

NUCL

021. FIRST AQUEOUS CHEMISTRY WITH SEABORGIUM. <u>Matthias Schädel,</u> Gesellschaft für Schwerionenforschung, D-64220 Darmstadt, Germany, for a Darmstadt–Mainz–Dresden–Krakow–Berkeley–Tokai–Collaboration.

For the first time, chemical separations of element 106 (Seaborgium, Sg) were performed in aqueous solutions. The isotopes ^{265}Sg and ^{266}Sg were produced in the ^{248}Cm + ^{22}Ne reaction at a beam energy of 121 MeV. The reaction products were continuously transported by a He(KCl)-jet to the computer-controlled liquid chromatography system ARCA. In 0.1 M HNO$_3$/ 5×10^{-4} M HF, Sg was found to be eluted within 10 s from cation-exchange columns together with the hexavalent Mo- and W-ions, while hexavalent U-ions and tetravalent Zr-, Hf-, and element 104 ions were strongly retained on the column. Element 106 was detected by measuring correlated α-decays of the daughter isotopes 78-s ^{261}104 and 26-s ^{257}102. The chemical results show that the most stable oxidation state of Sg in aqueous solution is +6, and that like its homologs Mo and W, Sg forms neutral or anionic oxo- or oxohalide-compounds under the present condition. In these first experiments, Sg exhibits properties very characteristic of group 6 elements, and does not show U-like properties.

022. SEARCH FOR ^{264}Sg.* <u>C. A. McGrath,</u> K. E. Gregorich, D. M. Lee, M. R. Lane, E. R. Sylwester, D. A. Shaughnessy, M. B. Hendricks, D. A. Strellis, P. A. Wilk, D. C. Hoffman, Nuclear Science Division, Lawrence Berkeley National Laboratory, Berkeley, California 94720.

Recently, a search was undertaken at the 88-Inch Cyclotron at the Lawrence Berkeley National Laboratory for the new isotope ^{264}Sg. This nucleus is predicted to decay with equal probability by spontaneous fission(SF) and by alpha emission with E$_\alpha$ ≈ 9 MeV and with a half life between 0.5 and 2 seconds. The ^{260}Rf daughter of this alpha decay undergoes SF with a reported half life of 20 ms. We tried to produce ^{264}Sg via the ^{250}Cf(^{18}O,4n)^{264}Sg reaction at 111 MeV and to correlate parent and daughter decays using our rotating wheel system. Only 3 α-SF correlations were detected, while our calculations predict 2 or 3 random correlations of unrelated background activity. This result permits us to set an upper limit on the cross section for this reaction. If the α-decay branch is ≥25% and the half life is between 0.5 and 2 seconds, the cross section is ≤100 pb at 111 MeV. Other production methods are being considered.

*This work was supported by the Director, Office of Energy Research, Division of Nuclear Physics of the Office of High Energy and Nuclear Physics of the U. S. Department of Energy under Contract DE-AC03-76SF00098

023. PRODUCTION CROSS SECTION OF ^{261}Ha FROM ^{250}Cf(^{15}N,4n) AND ^{243}Am(^{22}Ne,4n).* <u>M.R. Lane</u>, K. E. Gregorich, B. M. Wierczinski, D. M. Lee, M. B. Hendricks, C.A. McGrath, D. A. Shaughnessy, D. A. Strellis, E. R. Sylwester, P. A. Wilk, and D. C. Hoffman, Nuclear Science Division, Lawrence Berkeley National Laboratory, MS 70A-3307, Berkeley, California 94720.

The production cross sections of 1.8-sec ^{261}Ha produced at the LBNL 88-Inch Cyclotron from 85-MeV ^{15}N projectiles on ^{250}Cf and from 116-MeV ^{22}Ne projectiles on ^{243}Am, determined by measuring alpha particles and spontaneous fission fragments using our MG Rotating Wheel System, will be discussed.

024. SISAK 3 - FIRST APPLICATION TO STUDY THE NUCLEAR AND CHEMICAL PROPERTIES OF ELEMENTS 104 (RUTHERFORDIUM), 105 (HAHNIUM), AND 106 (SEABORGIUM) <u>Birgit Wierczinski,</u> Gunnar Skarnemark, Department of Nuclear Chemistry, Chalmers University of Technology, 412 96 Göteborg, Sweden

Solvent extraction studies of short-lived elements is a powerful method to study the nuclear and chemical properties of very short-lived nuclides. Since many years SISAK 3, a system that is capable of performing on-line liquid-liquid extractions, was successfully applied to study the properties of short-lived β-decaying nuclides. Due to its ability to detect even nuclides with half-lives down to one second, several adjustments were made to enable the system to investigate even the properties of transactinide elements. First attempts to investigate the properties of transactinides revealed unexpected problems, such as high β-background, interference of polonium, produced in the production reactions, and adsorption of the nuclides to the surface of several parts of the experimental set-up.

This paper will present the results obtained during the experiments to investigate the properties of elements 104 (rutherfordium), 105 (hahnium), and 106 (seaborgium), which were produced at Paul Scherrer Institut, Switzerland, Lawrence Berkeley National Laboratory, USA, and Gesellschaft für Schwerionenforschung, Germany, respectively.

025. IMPROVED HALF-LIFE MEASUREMENT OF ^{224}Pa AND ITS ^{209}Bi(^{18}O,3n)^{224}Pa PRODUCTION CROSS SECTION. P. A. Wilk, K. E. Gregorich, M. B. Hendricks, M. A. Lane, D. M. Lee, C. A. McGrath, D. A. Shaughnessy, D. A. Strellis, E. R. Sylwester, D. C. Hoffman, Nuclear Science Division, Lawrence Berkeley National Laboratory, MS 70A-3307, Berkeley, California 94720 and Chemistry Department, University of California, Berkeley, California 94720

The half-life of ^{224}Pa was determined to be 860±20 ms by measuring its alpha decay using our rotating wheel system at the Lawrence-Berkeley National Laboratory 88-Inch Cyclotron. The new value is much more precise but is consistent with a previously reported half-life of 950±150 ms. The ^{224}Pa was produced via the ^{209}Bi(^{18}O, 3n) reaction with a production cross section of 0.5±0.1 mb using 111 MeV ^{18}O^{5+} projectiles. No cross section for this reaction has been reported before.

026. ELECTRONICS FOR LIQUID SCINTILLATION ALPHA PARTICLE SPECTROSCOPY*
K.E. Gregorich, Lawrence Berkeley National Laboratory, MS-88, Berkeley, CA 94720

Recent developments in liquid scintillation counting now make alpha-particle pulse-height spectroscopy possible with energy resolution as good as 200 KeV FWHM. Since the light output per MeV of decay energy is much greater for beta particles than for alpha particles, it is necessary to use pulse shape discrimination based on the differing decay times of the alpha- and beta-light pulses to produce a clean alpha energy spectrum. In recent on-line accelerator experiments we have used this technique for the study of the chemical properties of the transactinide elements. In these experiments, the beta-decay of large amounts (>1 kHz) of background activities caused pileup pulses which passed the pulse shape discrimination circuit. In subsequent experiments an additional pulse shape discrimination technique was used which allows the measurement of alpha decay rates as small as a few counts per hour in samples with beta background rates of over 1 kHz.

*This work was supported by the Director, Office of Energy Research, Division of Nuclear Physics of the office of High Energy and Nuclear Physics, and the Chemical Sciences Division of the Basic Energy Sciences of the US Department of Energy under contract DE-AC0376SF00098.

027. PHYSICS OF RADIOACTIVE NUCLEAR BEAMS: THEORETICAL PERSPECTIVE, W. Nazarewicz, Department of Physics and Astronomy, University of Tennessee, Knoxville, Tennessee 37996 and Oak Ridge National Laboratory*, Oak Ridge, TN 37831-6373

Many of the current limitations in nuclear physics studies promise to be lifted by the field of radioactive nuclear beam experimentation. In this presentation, I shall concentrate on some theoretical aspects of physics of the drip-line nuclei. Experiments with radioactive nuclear beams will make it possible to look closely into many exciting aspects of the nuclear many-body problem. Although an experimental excursion into new territories of the chart of the nuclides will offer many excellent opportunities for traditional nuclear structure, there are also many unique features of exotic nuclei that define "new" physics far from stability.

028.
NEW REGION OF DEFORMATION VIA INTERMEDIATE ENERGY COULOMB EXCITATION. **T. Glasmacher,** Department of Physics and Astronomy, Michigan State University, East Lansing, MI 48824

Fast exotic beams of neutron rich sulfur and silicon isotopes with intensities as low as 15 particles/second have been excited in the virtual photon field of a gold target. The deexcitation photons have been measured in an array of position sensitive NaI(Tl) detectors. These measurements reveal the locations of the first excited states and the strengths of the E2 transitions from these states to the beta-unstable ground states. The data provide experimental evidence for a new region of deformation in the neutron rich sulfur isotopes.

NUCL

029. LOW ENERGY COULOMB EXCITATION OF RADIATIVE BEAMS IN INVERSE KINEMATICS. R. F. Casten, Department of Physics, Yale University, New Haven, Connecticut 06520-8124

The advent of radioactive nuclear beams (RNBs) opens a new era in nuclear structure physics. A number of exciting opportunities for the study of exotic nuclei present themselves. However, RNB intensities will be orders of magnitude lower than beam intensities traditionally used in nuclear physics investigations. Moreover, it will normally be the radiative beam nuclei themselves (or their N/Z neighbors) that are of interest: this fact focuses attention on the use of inverse kinematics. To survey the basic structure of new nuclei accessible with RNBs requires a highly efficient technique suitable for disclosing the basic observables (lowest energy levels and B(E2) values) that are structural signatures. Low energy Coulomb excitation is an ideal technique for this. New ideas, instruments and recent results in its exploitation will be discussed.

030.
NEW TYPES OF HIGH-EFFICIENCY MAGNETIC SEPARATORS FOR RNB AND OTHER REACTION STUDIES. **F.D. Becchetti,** Department of Physics, University of Michigan, Ann Arbor, Michigan 48109-1120

It has been shown that the efficient production and utilization of secondary readioactive nculear beams (RNB) and the analysis of heavy-ion reaction products can both benefit by exploiting devices based on large-solid-angles solenoids. We (and others) have recently developed multi-solenoid devices which overcome some of the previous limitations on beam purity. Likewise, new types of radial electric lenses and de-focussing solenoid elements have been designed which, if proven practical, should result in both high purity and good mass resolution while retaining very large collecton efficiency. New types of RNB target-detector geometries may also be utilized to further improve the overal efficiency, viz. to x100 to x1000 that of a more conventional device. A new double-solenoid syste m now under construction at the U. Notre Dame as a joint U. Michigan-U. Notre Dame NSF-ARI Grant project will test many of these concepts.

031.
HEAVY-ION REACTION STUDIES WITH ISOSPIN SELECTED BEAMS. **Sherry J Yennello**, Texas A&M University, College Station, Texas, USA 77843-3366

The ability to vary the isospin of projectiles through the use of unstable beams has expanded the playing field for nuclear reaction mechanism studies. The first of these experiments was a study of the degree of equilibration of heavy-ion reactions. Recently work has been done on the isospin dependence of the collective flow. Work is currently underway to understand the Coulomb contributions to fragmentation. This talk will give an overview of these recent heavy-ion studies that have added new insight to traditional reaction mechanism studies by using the isospin degree of freedom.

032. FUSION ENHANCEMENT WITH NEUTRON-RICH RADIOACTIVE BEAMS
K.E. Zyromski[1], W. Loveland[1], G.A. Souliotis[1], D.J. Morrissey[2], C.F. Powell[2], O. Batenkov[3], K. Aleklett[4], R. Yanez[4], I.B. Forsberg[1], M. Sanchez-Vega[4], J.R. Dunn[1], and B.G. Glagola[5]; [1]Oregon State University, Corvallis OR 97331; [2]Michigan State University, E. Lansing MI 48824; [3]Khlopin Radium Institute, St. Petersburg, Russia; [4]Uppsala University, Uppsala, Sweden; [5]Physics Division, Argonne National Laboratory, Argonne IL 60439

We have measured the fusion-fission excitation functions for the ^{32,38}S + ^{181}Ta reactions. The radioactive ^{38}S was produced by projectile fragmentation of ^{40}Ar at NSCL-MSU, and separated in the A1200 fragment separator. Time-of-flight energy was used to determine coincident fission fragments corresponding to full momentum transfer. The threshold energies were measured to be 137.5 $\pm$ 1.0 MeV for the ^{32}S reaction, and 130.7 $\pm$ 2.0 MeV for the ^{38}S reaction. These values are in agreement with the predictions of various systematics, and the shift in the fusion threshold with the n-rich projectile may be of interest in the synthesis of the heaviest nuclei.

033. EVIDENCE FOR RELATIVISTIC EFFECTS IN THE CHEMISTRY OF ELEMENT 104

H.W. Gäggeler[1,2], B. Eichler[2], A. Türler[2], D.T. Jost[2], S. Hübener[3], A. Yakushev[4], I. Zvara[4], S. Timokhin[4], V. Lebedev[4], [1]Universität Bern, CH-3012 Bern, Switzerland, [2]Paul Scherrer Institut, CH-5232 Villigen, Switzerland, [3]Forschungszentrum Rossendorf, D-01314 Dresden, Germany, [4]Flerov Laboratory for Nuclear Reactions, Dubna, Russia

On the basis of thermodynamic extrapolations, the first transactinide element 104 is expected to form volatile chlorides and bromides of similar volatility if compared to those of the homologuous element Hf. The α-decaying 78 s isotope ^{261}104 was produced at the 4m cyclotron of FLNR/Dubna in 0.5 pµA ^{18}O bombardments of a 1 mg/cm² thick ^{248}Cm target, containing small amounts of Gd. Via fusion reactions short-lived isotopes of Hf and of ^{261}104 were produced. With the On-Line Gaschemistry Apparatus OLGA the retention times in quartz columns were investigated using chlorinating and brominating gases. The results showed that element 104 forms chlorides and bromides of higher volatility if compared to those of Hf, in agreement with relativistic calculations. In addition, as expected, element 104 chloride was found to be more volatile than element 104 bromide.

Work supported by INTAS & Swiss National Science Foundation

034. FIRST CHEMICAL SEPARATION AND IDENTIFICATION OF SEABORGIUM

A. Türler[1], B. Eichler[1], H.W. Gäggeler[1,2], K.E. Gregorich[3], D.C. Hoffman[3], S. Hübener[4], D.T. Jost[1], M. Schädel[5], A. Yakushev[6], [1]Paul Scherrer Institute, CH-5232 Villigen PSI, Switzerland, [2]Department of Chemistry and Biochemistry, University of Bern, 3012 Bern, Switzerland, [3]Lawrence Berkeley National Laboratory, Berkeley, California 94720, USA, [4]Forschungszentrum Rossendorf, D-01314 Dresden, Germany, [5]Gesellschaft für Schwerionenforschung, D-64291 Darmstadt, Germany, [6]Flerov Laboratory of Nuclear Reactions, Dubna, Russia

The chemical properties of element 106 (Seaborgium, Sg) were successfully studied using the On-line Gas Chromatography Apparatus (OLGA III). After chemical separation of Sg in the form of volatile oxichlorides the nuclides ^{265}Sg and ^{266}Sg were unambiguously identified and their half-lives were determined for the first time. The Sg nuclides were produced from the ^{248}Cm(^{22}Ne, 4,5n)^{266,265}Sg reaction at the GSI Darmstadt UNILAC accelerator. Simultaneously, short-lived W nuclides were produced from a small admixture of ^{152}Gd to the Cm target material. As predicted by relativistic calculations and by extrapolations of chemical properties, it was demonstrated that Sg oxichlorides are indeed less volatile than their lighter homologue W- and Mo-oxichlorides.

035. ON-LINE GAS PHASE CHROMATOGRAPHY OF THE BROMIDES OF THE GROUP 4, 5, AND 6 ELEMENTS.[*] E.R. Sylwester, K.E. Gregorich, D.M. Lee, Y.H. Chung, M.B. Hendricks, M.R. Lane, C. Laue, C.A. McGrath, D.A. Shaughnessy, D.A. Strellis., D.C. Hoffman. Nuclear Science Division, MS 88/136, Lawrence Berkeley National Laboratory, Berkeley, California 94720.

Gas phase chromatography has been used to determine the volatility of bromides of the group 4, 5, and 6 elements, including the transactinides 104 (Rf) and 105 (Ha). The Heavy Element Volatility Instrument (HEVI) was used to measure the volatilities of the bromides of short-lived isotopes of these elements. Adsorption enthalpy values were calculated from the observed volatilities using a Monte Carlo program. The values for RfBr$_4$ and HaBr$_5$ are similar to ZrBr$_4$ and NbBr$_5$ rather than to HfBr$_4$ and TaBr$_5$. This deviates from a simple extrapolation from periodic table trends and may indicate the influence of relativistic effects on the electronic level structure of these transactinides. The group 6 element Seaborgium (Sg) has not yet been studied due to the low production cross-section.

[*]This work was supported in part by the Director, Office of Energy Research, Division of Nuclear Physics of the Office of High Energy and Nuclear Physics of the US Department of Energy under contract DE-AC03-76SF00098

036. ELECTRON-CAPTURE DELAYED FISSION IN ^{246}ES.* D.A. Shaughnessy, K.E. Gregorich, M.B. Hendricks, M.R. Lane, D.M. Lee, C.A. McGrath, D.A. Strellis, E.R. Sylwester, P.A. Wilk and D.C. Hoffman, Lawrence Berkeley National Laboratory, MS 70A-3307, Berkeley, CA 94720.

We have extended our systematic study of electron-capture delayed fission (ECDF) in neutron-deficient isotopes to ^{246}Es. The ^{246}Es was produced at the 88 Inch Cyclotron at Lawrence Berkeley National Laboratory via the ^{249}Cf(p, xn)$^{250-x}$Es reaction with 37 MeV protons. There were 19 ^{249}Cf targets used simultaneously in our light ion multiple (LIM) target system. Alpha particles and fission fragments were detected in our rotating wheel system. In some experiments, TTA extractions were performed to remove interfering activities. The chemically separated samples were positioned between a solid-state particle detector and two x-ray detectors. This configuration enabled us to look for fissions in coincidence with K x-rays following electron-capture. Our measured production cross section of 13 ± 5 μb for ^{246}Es was much lower than the cross section predicted by a neutron evaporation code. The probability of delayed fission was determined from the number of x-ray/fission coincidences measured.
*This work was supported by the Director, Office of Energy Research, Division of Nuclear Physics of the Office of High Energy and Nuclear Physics of the U.S. Department of energy under Contract DE-AC03-76SF00098.

037. IN-FLIGHT SEPARATION AND DETECTION METHODS. V. Ninov, P. Armbruster, S. Hofmann, F.-P. Heßberger, H. Folger, G. Münzenberg Gesellschaft für Schwerionenforschung mbH, Max-Planck Str. 1, 64291 Darmstadt K.E. Gregorich, C.A. McGrath LBNL, 1 Cyclotron Rd., Berkeley, Ca. 94720

Since accelerators providing high intensity beams of more than 10^{12} particles/second recoil spectrometers making use of the reaction kinematics became the most important tools for studies of nuclear matter far beyond the valley of β-stability. The major goals of recoil separators were the discovery of the heaviest elements as well as connected with RIB's the synthesis of the doubly magic nuclei ^{100}Sn and ^{78}Ni, giving new insights exploring nuclei at the proton and neutron drip lines. In the presentation a general survey about the well established recoil separator techniques like electro-magnetic separators, gas-filled separators as well as fragment separators and their applications will be given. Furthermore detection techniques used for the study of the decay properties as half-lives and decay modes of the synthesized particles will be reviewed. Future possibilities connecting recoil separators with trapping techniques will be discussed.

038.
THE DISCOVERY OF ELEMENT 110 AND THE VERIFICATION OF A CLOSED NEUTRON SHELL AT N=162. J.F.Wild, R.W. Lougheed, and K.J. Moody, Lawrence Livermore National Laboratory, P.O. Box 808, L-231, Livermore, CA 94551; Yu.A. Lazarev, Yu.V. Lobanov, Yu.Ts. Oganessian, V.K. Utyonkov, F.Sh. Abdullin, A.N. Polyakov, J. Rigol, I.V. Shirokovsky, Yu.S. Tsyganov, S. Iliev, V.G. Subbotin, A.M. Sukhov, G.V. Buklanov, B.N. Gikal, V.B. Kutner, A.N. Mezentsev, and K. Subotic, Joint Institute for Nuclear Research, Dubna, Russia, 141980.

In this paper, we discuss the experiments leading up to and including the discovery of the isotope 110-273, and the importance of this nuclide in deducing the decay properties of nuclei in this region through the establishment of a closed neutron shell at N=162. This shell and the closed proton shell at Z=108 are responsible for an enhancement in stability against spontaneous fission decay of several orders of magnitude. This research was performed partially under the auspices of the US Department of Energy by the Lawrence Livermore National Laboratory under Contract no. W-7405-Eng-48.

039.
TOWARDS SPHERICAL SHELLS Z=114 AND N=178-184. **V.K. Utyonkov,** Yu.Ts. Oganessian, Yu.V. Lobanov, F.Sh. Abdullin, A.N. Polyakov, I.V. Shirokovsky, Yu.S. Tsyganov, S. Iliev, V.G. Subbotin, A.M. Sukhov, Joint Institute for Nuclear Research, Dubna, Russia, 141980, R.W. Lougheed, J.F. Wild, K.J. Moody, N. J. Stoyer, University of California, Lawrence Livermore National Laboratory, Livermore, California, 94551

The preparation of experiments designed to produce spherical superheavy nuclides with Z=114 and N=174-176 via the complete fusion reaction 244Pu+48Ca using the Dubna gas-filled recoil separator is described. These experiments are a continuation of a series of Dubna-Livermore experiments which led to the identification of the heavy nuclides with Z=104-110. The sensitivity of these experiments is expected to exceed by a factor of 100-1000 previous attempts to produce spherical superheavies in 48Ca-induced fusion-evaporation reactions on actinide targets, providing the opportunity to search for superheavy nuclides with halflives from microseconds to several days at cross section levels down to 0.1 picobarn.

040. THE BERKELEY GAS-FILLED SEPARATOR* K.E. Gregorich, Lawrence Berkeley National Laboratory, MS-88, Berkeley, CA 94720

The Berkeley Gas-filled Separator (BGS), a separator for recoiling compound nuclei, is being constructed at the Lawrence Berkeley Laboratory 88-Inch Cyclotron. It will take advantage of the high intensities of heavy ion beams at the 88-Inch Cyclotron for studies of production and decay properties of nuclides with cross sections as small as picobarns. The large angular acceptance (45msr), large bend angle (70°), and microsecond separation times, together with a momentum acceptance of over 50%, will provide an efficient and selective separation of compound nucleus products from beam particles and other reaction products. The BGS experimental program will include (but is not limited to) heavy elements, gamma-ray studies at the focal plane, nuclides near the proton drip-line, secondary reactions, transfer reactions, and even the use of the BGS as a preseparator for chemical studies. First operation of the BGS is planned for late in 1997.
*This work was supported by the Director, Office of Energy Research, Division of Nuclear Physics of the office of High Energy and Nuclear Physics of the US Department of Energy under contract DE-AC0376SF00098.

041.
RECENT RADIOACTIVE BEAM EXPERIMENTS IN NUCLEAR ASTROPHYSICS.
M. Wiescher, J. Goerres, Department of Physics, University of Notre Dame, Notre Dame, IN 46616

Radioactive beam experiments deliver important input parameters for the understanding and interpretation of nucleosynthesis processes in explosive stellar scenarios. Two examples of recent measurements will be presented to elucidate the various experimental quests in nuclear astrophysics and the experimental approaches on todays available radioactive beam facilities. As first example we will present a recent measurement of low energy resonances in 18F(p,alpha) at the radioactive beam facility at Louvain-la-Neuve (Belgium). This reaction is important for the understanding of the energy generation in the hot CNO cycles. As a second example we will present the results of a recent lifetime measurement of 44Ti using the NSCL/MSU A1200 fragment separator facility. This result is important for the interpretation of nucleosynthesis in the alpha-rich freeze out in supernovae models.

042.
REACTION STUDIES ON SHORT-LIVED NUCLEI FOR BIG BANG NUCLEOSYNTHESIS, R.N. Boyd, Department of Physics, Ohio State University Columbus, OH

Recent difficulties with the standard model of big bang nucleosynthesis emphasize the importance of studying alternate models. Some of those models, e.g., the inhomogeneous models, require information about reactions on short-lived nuclei, e.g., 8Li(4He,n)11B, 15C(n,gamma)16C, and 18N(n,gamma)19N. A comparison of the observed primordial abundances to the predictions of the standard and inhomogeneous models will be presented, In addition, the experimental studies that have contributed to the state of our knowledge of these and other reactions, all involving the use of radioactive ion beams, will be discussed.

043.
MEASUREMENT OF THE ^{7}BE(p, GAMMA) CROSS SECTION AT LOW ENERGIES.
D.M. Moltz, J.D. Powell, A. Rice, M.W. Rowe and J. Cerny, Dept. of Chemistry and Lawrence Berkeley National Laboratory, Berkeley, CA 94720, and V. Hansper, E. Wulf, H.R. Weller, A.E. Champagne, Triangle Universities Nuclear Laboratory, Durham, NC 27708, and M. Hofstee, Dept. of Physics, Colorado School of Mines, Golden CO.

The solar neutrino problem has generated enormous interest over the past 20 years. The problem can best be summarized as the lack of agreement between solar models and three primary types of experiments measuring the solar neutrino flux. The situation is complicated further by the lack of agreement between the three experiments, all of which sample different parts of the neutrino spectrum. One of the most enduring problems with this non-agreement is the lack of a consistent value for the cross section of the ^{7}BE(p,gamma) ^{8}B reaction extrapolated to low energies (the S_{17} factor). Although much work on this reaction cross section has concentrated on getting the correct cross section at higher energies (about 1 MeV) via various methods, little work has been done on actually measuring the cross section below the current lower limit of 117 keV set by Filippone *et al* in 1983. In this presentation, we will detail the problem, the current status of experiments in progress and our plans for measuring the cross section at lower energies.

044. EXPERIMENTS OF INTEREST TO NUCLEAR ASTROPHYICS USING 17F, 18F AND 56NI BEAMS FROM THE ATLAS ACCELERATOR. K. E. Rehm, Argonne National Laboratory, Argonne, IL 60439

First experiments with radioactive beams of 17F, 18F and 56Ni have been performed at the superconducting accelerator ATLAS at Argonne National laboratory. The experiments address several questions related to the hot CNO cycle and the breakout into the rp process, in particular the production of 19Ne via the 18F(p,gamma) and of 17F via the 14O(4He,p) reactions. The beams were produced by using either the two-accelerator method (18F, 56Ni) or (for 17F) by bombarding a hydrogen (deuterium) target with 17O or 16O, respectively. Special high-efficiency detection techniques were developed to perform experiments with low beam intensities and sometimes considerable isobar impurities. Planned measurements with other radioactive ion beams will be discussed.

045.
THE S800 SPECTROGRAPH. **B.M. Sherrill,** Department of Physics, Michigan State University, East Lansing, Michigan 48824

The S800 is a high resolution, large solid angle spectrogaph that was recently completed at the National Superconducting Cyclotron Laboratory. The solid angle is up to 20 msr (7 degrees by 10 degrees) which makes it well suited to work using radioactive beams. The energy resolution of 1 part in 10,000 provides adequate energy resolution even for the high energy beams availible at the NSCL. The design of the spectrograph is based on superconducting magnets and ray reconstruction techniques. The design and first operation will be described along with early results from the first experiments. Other examples of the upcoming scientific program and futurre possibilities will also be presented.

046.
WEAK INTERACTIONS IN NUCLEI. **J.C. Hardy,** E. Hagberg, V.T. Koslowsky, G. Savard, I.S. Towner, AECL Research, Chalk River Laboratories, Chalk River, Ontario, Canada K0J 1J0

Abstract not available.

047. GENERATION OF RADIOACTIVE BEAMS BY INTERMEDIATE ENERGY FRAGMENTATION OF HEAVY NUCLEI, G. A. Souliotis[a], W. Loveland[a], G. J. Wozniak[b], K. Hanold[b], D. J. Morrissey[c], K. E. Zyromski[a], K. Aleklett[d], J. O. Liljenzin[e], Oregon State University[a], Lawrence Berkeley Laboratory[b], Michigan State University[c], Uppsala University[d], Chalmers University of Technology[e].

New developments in the fragmentation of heavy nuclei at relativistic energies have led to the formation of several new neutron-rich nuclei and new possibilities for their study. We wish to point out that fragmentation of heavy nuclei at intermediate energies also appears to hold promise as a new, novel source of radioactive beams. We have studied the fragmentation of 20 MeV/nucleon ^{197}Au and ^{238}U by C, Al, Ti, Zr, Au and ^{208}Pb using the A1200 fragment separator at MSU. By careful calibration of the pulse height defect, plasma delay effects, energy losses in the spectrometer, etc with numerous calibration beams, we have been able to achieve unit resolution for Z, A and q for each reaction product. Study of the fragmentation of ^{197}Au revealed several new n-deficient nuclei (A=150-200) while study of the ^{238}U - ^{208}Pb reaction led to the formation of several new n-rich nuclei (A=85-130). The latter reaction proceeds by deep inelastic scattering and because of the symmetric fragment mass distribution, complements the work at relativistic energies. The implication of these results for new radioactive beam facilities will be discussed.

048. THE TABLE OF ISOTOPES EARLY STORY. Glenn T. Seaborg, Associate Director-at-Large, Lawrence Berkeley National Laboratory, Berkeley, California, 94720

The first complete Table of Isotopes, entitled "A table of induced radioactivities," consisting of 13 pages including references, was published by John J. Livingood and me in the January, 1940 issue of Reviews of Modern Physics; this publication also included a table "Stable Isotopes of the Elements" but didn't include the natural radioactivities. This was soon followed by a similar "Complete list of induced radioactivities" included in my extensive article "Artificial Radioactivity" published in the August 1940 ussue of Chemical Reviews. Then during World War II, working at the wartime Metallurgical Laboratory of the University of Chicago (but with comouflaged identification with U.C. Berkeley) I published in the January 1944 issue of Rev. Mod. Phys. a 32-page "Table of Isotopes" which included the stable isotopes and natural radioactivities. My "Table of Isotopes" in Rev. Mod. Phys. was updated in 1948 (83 pages), 1953 (182 pages), 1959 (319 pages) with Isadore Perlman, Jack M. Hollander and Donald Strominger as various co-authors.

049. NUCLEAR DATA: FROM LIVINGOOD AND SEABORG TO THE 21ST CENTURY. C. M. Lederer, University of California Energy Institute and Department of Nuclear Engineering, Berkeley, CA 94720-5180

The first compilations of radioisotopes helped nuclear chemists and physicists keep track of the rapidly growing number of artificially produced radioactive species. Although the focus of research has shifted from the discovery of isotopes to the understanding of their properties, the fundamental goal of the nuclear data program remains the same: to provide information in forms that support and inspire the creation of new knowledge and applications. As nuclear data programs evolved, both the evaluation process and the techniques for managing data have changed. Enormous growth in the amount of data forced the nuclear data business to change from a sideline of research to a dedicated program and, in the past few decades, a worldwide network of centers in which evaluators share the effort, increasingly along lines of specialization. Computers have become essential tools to handle and evaluate the rapidly expanding mass of data. Computers also play a growing role in the way scientists and engineers access and use nuclear data. Nuclear data are available today at the click of a mouse. Ahead lie developments that will make the response to that mouse click dramatically more useful.

050. NATIONAL AND INTERNATIONAL NUCLEAR DATA ACTIVITIES. C.L. Dunford, National Nuclear Data Center, Department of Advanced Technology, Brookhaven National Laboratory, Upton, New York 11973-5000

Compiled and evaluated nuclear data are generated through national and international collaboration. The National Nuclear Data Center (NNDC) at Brookhaven National Laboratory is the focal point for the U.S. nuclear data activities and the

U.S. repository for data generated in the U.S. program or obtained through international collaboration. The U.S. activity has been carried out through networks of nuclear structure and nuclear reaction data evaluators. Bibliographic and experimental data compilation is carried out by the NNDC within an international collaboration of nuclear data centers. In the near future, all U.S. nuclear data activity will be coordinated in a single body with the three existing networks for reaction data, structure data and dissemination forming working groups within the new U.S. network. The NNDC also serves as the point of contact with international nuclear data activities and as the repository and primary data dissemination point for the United States.

051. EXPLORING THE ISOTOPES, R. B. Firestone, S. Y. F. Chu, and H. Nordberg, Lawrence Berkeley National Laboratory, Isotopes Project, Berkeley, CA 94720

The explosion in computer technology has led to a revolution in the way nuclear structure and decay data is being disseminated. For nearly sixty years the *Table of Isotopes* and later the *Nuclear Data Sheets* were primary sources of information for nuclear researchers. In 1996, the 8th edition of the *Table of Isotopes* (John Wiley and Sons, Inc.) was published. This two-volume, 3100 page book signaled the end of the era of publishing nuclear data comprehensively on paper. The CD-ROM version that comes with the 8th edition provides the same information spread over 14,000 "pages" for better readability on the computer screen. With Adobe Acrobat technology to view the data, over 100,000 hypertext links to navigate through it, and nearly unlimited expansion possibilities, the CD-ROM becomes the format of choice for "hard copy" publishing of nuclear data. Even a CD-ROM cannot be updated continuously to provide the most current data. We are therefore developing an Internet version of the *Table of Isotopes* (http://isotopes.lbl.gov/isotopes/toi.html) to supplement the CD-ROM. The website provides home pages focusing on nuclear structure data, radioactive decay data, nuclear astrophysics data, capture gamma data, and atomic masses. The nuclear structure (ENSDF) and reference (NSR) databases are available on both the CD-ROM and Internet. We have developed the Isotopes Explorer application (previously known as VuENSDF) to view these databases and display level schemes, tables, plots, references, and nuclear charts. The databases can be searched and derived quantities can be calculated. The Isotopes Explorer is a 32-bit Windows application that can download information directly from the CD-ROM, Internet, or user generated files. Information about the Isotopes Explorer can be found at http://isotopes.lbl.gov/isotopes/vuensdf.html.

052. WEB-BASED INFORMATION SYSTEMS FOR NUCLEAR PHYSICS, Peter Ekström, Department of Physics, Lund University, Sölvegatan 14, SE-223 62 Lund, Sweden

The development of the Internet, both in capacity and in browser/server technology, will transform the World Wide Web into a general and easy-to-use global information system. The most important features of the Web (and Intranets) today are **decentralisation**, *i.e.* information is stored anywhere, and **platform independence**, *i.e.* the information can be retrieved in the same way from most computer platforms using an advanced Web browser. In this talk I will discuss recent developments of Web-based client-server database systems for literature references and for nuclear structure and decay data. The client consists of an HTML page with frames, forms and javascript code, and the server is a tailor-made database program that satisfies client requests by returning HTML-formatted pages.

053.
THE MACNUCLIDE NUCLEAR DATABASE MANAGEMENT SYSTEM. **Craig A. Stone** Robert Sutton, Department of Chemistry, San Jose State University, San Jose, CA 95192-0101, Min Li, National Nuclear Data Center, Brookhaven National Laboratory, Upton, NY 11973-5000 and San Jose State University

ABSTRACT NOT AVAILABLE

054. FIRST EXPERIMENTS AT THE HRIBF. J. D. Garrett, Physics Division, Oak Ridge National Laboratory, Oak Ridge, Tennessee 37831

The Holifield Radioactive Ion Beam Facility (HRIBF) at Oak Ridge National Laboratory (ORNL) has produced its first accelerated beams of radioactive ^{69,70}As and is now authorized to start its experimental program. The Program Advisory Committee (PAC) has approved research for 72 eight-hour shifts with radioactive arsenic beams and 106 shifts of associated research with stable beams. These experiments are expected to start in January 1997. This experimental program will be reviewed, as will the most recent developments in producing beams of other radioactive ions at the HRIBF. The long term possibilities for radioactive ion beam research at ORNL associated with both the present facility and the National Spallation Neutron Source (NSNS) also will be discussed. Research sponsored by the Oak Ridge National Laboratory, managed by Lockheed Martin Energy Research Corporation for the U.S. Department of Energy under contract number DE-AC05-96OR22464.

055. RESEARCH WITH RADIOACTIVE BEAMS AT TRIUMF. John M. D'Auria, Department of Chemistry, Simon Fraser University, Burnaby, B.C. Canada, V5A IS6

Over the last 10 years, a program using radioactive beams from the thick target on-line isotope separator, TISOL, has been in progress at the high intensity, intermediate energy proton cyclotron facility, TRIUMF. This program has included studies in fundamental interactions, and nuclear astrophysics. In 1995 a second facility, ISAC, was funded and is presently being constructed. Based upon the ISOL method, this facility will produce a wide range of high intensity beams including beams accelerated to energies from 0.15 to 1.5 MeV/u. The program is particularly aimed at reaction studies in explosive nucleosynthesis, solar neutrino flux questions, fundamental interactions (using a neutral atom trap), and material science. A summary of the entire program (present and future) will be presented along with a status report on the progress of the ISAC facility.

056. PLANNED DEVELOPMENT OF A RADIOACTIVE BEAM CAPABILITY AT THE LBNL 88-INCH CYCLOTRON. P.E. Haustein, Chem. Dept, BNL, Upton, NY 11973, D. M. Moltz, E. B. Norman, J. Powell and J. Cerny, Nuclear Science Division, LBNL, Berkeley, CA 94720

Planned development of low-Z, proton-rich, radioactive beams (^{11}C, ^{13}N, ^{14,15}O, and ^{18}F) at the 88" Cyclotron of the Lawrence Berkeley National Lab is described. Based on the "coupled cyclotron method", isotopes produced by (p,n) and (p,a) reactions at a high-current (30 mA), low-energy (10 MeV) medical cyclotron will be transferred ~300 meters by high-speed gas-jet transport to the ECR ion-source at the 88" Cyclotron. Important features of this approach are its low cost, use of simple and well tested technology, applicability to nearly all elements, and avoidance of lengthy (chemical or physical) isotopic release delays at the production target. Developmental progress is reported for various operational components. Based on conservative estimates, e.g. 1% ECR ion-yield, extracted radioactive ion beams are projected to exceed 10^6 ions/sec. Experiments which will use these beams include studies of the scattering of mirror nuclei, single and mutual excitation in inelastic scattering and single nucleon transfer reactions.

Research supported by USDOE contracts DE-AC02-76CH00016 and DE-AC03-76SF00098.

057. MEASUREMENT OF THE PROPERTIES OF NUCLEI FAR FROM STABILITY AT GANIL AND THE SPIRAL PROJECT. W.Mittig Ganil, B.P.5027 14021 Caen Cedex

A brief review of current research at GANIL in the domain of exotic nuclei will be given: mass-measurements, decay properties, elastic and inelastic scattering and charge exchange reactions. These results will be illustrated by exemples, and the comparison with theory will be discussed. The use of secondary beams from fragmentation for the study of nuclear structure has already become a powerful tool for the exploration of nuclear structure in unusual conditions.These possibility will be enormously enlarged by the acceleration of radioactive ISOL beams. The future facility SPIRAL, presently under construction at GANIL, wil be described. It will use high intensity primary heavy ion beams to produce radioactive species in a thick target. After ionisation, these nuclei will be accelerated by a new cyclotron and will be transported to the experimental areas.

058.
EXPERIMENTS WITH EXOTIC BEAMS, TOWARDS THE FUTURE AT THE NSCL. D.J. Morrissey, Department of Chemistry, Michigan State University, East Lansing, MI 48824

The current research program at the National Superconducting Cyclotron Laboratory (NSCL) relies heavily on exotic beams produced by the K1200 cyclotron and the A1200 fragment separator. The program ranges from accelerator mass spectroscopy of atmospheric krypton isotopes, through production of nuclei at the limits of stability for decay spectroscopy, to production of intense beams for secondary reactions. A sampling of recent projects will be described. In addition, the major upgrade of the capabilities of the NSCL, recently approved by the National Science Foundation, will be described. This upgrade will provide primary and exotic beams with up to three orders of magnitude higher intensity and open the way for the next generation of experiments.

059.
RECENT RADIOACTIVE NUCLEAR BEAM STUDIES AT RIKEN AND THE RI BEAM FACTORY PROJECT. I. Tanihata, RIKEN, 2-1 Hirosawa, Wako, Saitama 351-01, Japan

Recent structure studies of neutron-rich nuclei at RIKEN Ring Cyclotron Facility will be reviewed. Particular interests is placed on the elastic and inelastic scatterings of protons and quasi-free nucleon-nucleon scatterings, (p, 2p), (p, pn). These studies show the low-energy excitation of halo and also provide information on ls coupling strength in neutron-rich nuclei. The RI Beam Factory Project, where high-energy radioactive beam over the whole range of elements would be supplied, has been approved to start its construction. A brief introduction to the planned facility will be presented.

060. INTERNATIONAL PROJECT FOR EVALUATION OF DECAY DATA, R. G. Helmer, Idaho National Engineering Laboratory, Lockheed Martin Idaho Technologies Co., Idaho Falls, Idaho 83415-2114, USA, and E. Browne, Isotopes Project, Lawrence Berkeley National Laboratory. Berkeley, California 94720

With the hope of reducing the redundancy of the many existing files of decay data and improving the quality of the available data, an international collaboration was initiated in 1994 to carry out cooperative evaluations for radionuclides of importance in various applications. The objective of this effort is to provide an authoritative and widely accepted set of decay data for these radionuclides. The participants in this collaboration are M-M. Bé, Laboratoire Primaire des Rayonnements Ionisants (LPRI), France; E. Schönfeld, Physikalisch-Technische Bundesanstalt (PTB), Germany; T. D. MacMahon, Centre for Analytical Research in the Environment (CARE), United Kingdom; J. M. Los Arcos, Centro de Investigaciones Energeticas Mediombientales y Tecnologicas (CIEMAT), Spain; J. K. Tuli, Brookhaven National Laboratory (BNL); and the authors. The evaluations will be prepared both in a form similar to that used in the Table de Radionucléides formerly published by LPRI, 1982 - 1987, and that for the Evaluated Nuclear Structure Data File (ENSDF). The results will also be submitted to the Nuclear Data Section of the International Atomic Energy Agency for possible inclusion in their decay-data database.

Supported by the US Department of Energy under contracts DE-AC07-94ID13223 for INEL and DE-AC03-76SF00098 for LBNL.

061. EVALUATION OF DECAY DATA: QUALITY AND UNIFORMITY. E. Browne, Lawrence Berkeley National Laboratory, Berkeley, CA 94720; R. G. Helmer, Idaho National Engineering Laboratory, Lockheed Idaho Technologies Co., Idaho Falls, ID 83415-2114

The various uses of radionuclides in applied fields require the availability of evaluated nuclear data of good quality. The main objective of the international Decay Data Evaluation Project is to provide such information. Since this Project includes evaluators from several institutions, it has been necessary to agree on the application of well-defined procedures and the use of other sources of data to insure uniform evaluations. Some of these additional sources are the Rosel et al. theoretical internal-conversion coefficients, and the recent evaluation of the atomic fluorescence yields by Schonfeld and Janssen. A major evaluation issue has always been the treatment of discrepant values; we have chosen to use the *Limitation of Relative Statistical Weights* method, which had been previously adopted by a Coordinate Research Programme of the International Atomic Energy Agency. We have tried to generate realistic uncertainties and to document all steps in the evaluations.

Supported by the US Department of Energy under contracts number DE-AC03-76SF00098 for LBNL, and DE-AC07-94ID13223 for INEL.

062. DESIGN OF NEW ELECTRONIC EDITION OF GAMMA-RAY SPECTRUM CATALOGUES FOR INTERNET PRESENTATION, R. L. Heath, Idaho National Engineering Laboratory, Lockheed Martin Idaho Technologies, Co., Idaho Falls, Idaho 83415-2211

New editions of the original Gamma-ray Spectrum Catalogues are being prepared for publication in electronic format. The Catalogues will be available on CD-ROM and as an Internet resource from the INEL. All material is formatted in Adobe Acrobat. Additions to the original content of the Catalogues include integrated decay schemes, and tables of related decay data from the ENSDF file. Spectra representing the response of large-volume Ge detectors, alpha-particle spectra, prompt neutron capture and inelastic scattering gamma-ray spectra, and gross fission product spectra are being developed for future addition. All numerical data are available in relational database format with an advanced graphic user interface (GUI), designed for retreival of both graphics and data for general laboratory use. The major emphasis of the presentation will be on technology applied in the design of material for screen presentation, incorporating color, advanced database technology, and graphics formatting to overcome bandwidth limitations on the internet. The first edition of the Catalogues in CD-ROM format is currently in the beta test phase. The web site content presently available will be described. This work is supported by the U. S. Department of Energy under contract DE-AC07-94ID13223.

063. THE RECENT RENAISSANCE IN HEAVY ELEMENT RESEARCH*. Darleane C. Hoffman and Diana M. Lee, Nuclear Science Division, 70A-3307, Lawrence Berkeley National Laboratory, University of California, Berkeley, CA 94720.

New information concerning the nuclear properties of the heaviest elements generated by the recent renaissance in heavy element research will be briefly discussed. Some of the data needs and compilations that would be especially helpful to researchers in this field will be outlined. A few examples are: gamma- and alpha-decay data cross-indexed by energy, intensity, other decay modes, metastable states, parent-daughter relationships, and isotope; spontaneous fission partial half-lives and fragment properties; neutron and charged-particle reaction cross sections. Particularly helpful would be the availability of working libraries or sub-databases in a format which can readily be edited and/or used for other application programs.

064. EVALUATION OF NUCLEAR STRUCTURE DATA FOR THE ACTINIDES. Y. A. Akovali, Oak Ridge National Laboratory,*Oak Ridge, TN 37831

Evaluated experimental data pertaining to the nuclear structure of all nuclei are published in Nuclear Data Sheets and in Table of Isotopes and are stored in the Evaluated Nuclear Structure Data Files. These files may be utilized for the systematics of nuclear properties. These systematic studies provide valuable tools for understanding the nuclear structure, and they are utilized routinely by the nuclear structure evaluators in evaluating various data, resolving discrepancies and adopting nuclear properties. One such systematics is the study of alpha decay, which is the predominant decay mode in the actinide region. In the present work, alpha-decay data from doubly-even nuclei throughout the periodic table are reviewed and evaluated. From these data, nuclear radius parameters are calculated by using the Preston formula for alpha-decay probabilities. The radius parameters for each element behave rather regularly as a function of neutron number. Any irregularity or large departure from this behavior indicates probable incorrect input data. This systematic behavior can also be used in estimating partial half-lives.

*Managed by Lockheed Martin Energy Research Corporation for the U.S. Department of Energy.

065. INTERNET AND THE NATIONAL NUCLEAR DATA CENTER. T.W. Burrows, National Nuclear Data Center, Department of Advanced Technology, Brookhaven National Laboratory, Upton, New York 11973-5000

The National Nuclear Data Center (NNDC) has been providing electronic access to the nuclear data bases it maintains since 1986. Originally the access was via modem and DecNet; primary access now is through the Internet. After the Internet became widely available to the scientific community, retrievals from the NNDC's data bases have risen almost exponentially with an estimated 124,000 retrievals for 1996. Primary access is still through TELNET but there is substantial activity through the World Wide Web (W3). A current goal of the Center is to provide W3 access to all of its major data bases by the end of 1997. W3 access to the NNDC data bases will be demonstrated along with demonstrations of "work in progress." The Center has also taken the initiative to develop and maintain new data bases covering the "frontiers" of nuclear science.

066.
PROSPECTS IN RADIOACTIVE ATOM TRAPPING: ^{21}NA AS AN EXAMPLE. S.J. Freedman, B. Fujikawa, G. Gwinner, M. Rowe, S.Q. Shang, and P. Vetter, Physics Department and Lawrence Berkeley National Laboratory, University of California, Berkeley, CA 94720

Recent demonstrations that laser trapping and cooling techniques are applicable to exotic short-lived radioactive atoms are providing new opportunities for better measurements of important fundamental constants and more sensitive searches for physics beyond the presently established Standard Model of the fundamental interactions. The application of laser techniques to control the spatial extent and degree of polarization of essentially massless radioactive sources in beta-decay experiments should allow studies of the symmetries of the weak interactions with unprecedented precision. Experiments with rare atoms having particular sensitivitics to subtle weak-interaction modifications to atomic structure should be practical with laser manipulation techniques. We review some of the prospects in this emerging area of experimental research and we describe an example of an ongoing experiment to study the beta decay of the short-lived isotope ^{21}Na.

067.
BETA-NEUTRINO CORRELATION MEASUREMENTS IN A LASER TRAP. J.A. Behr, A. Gorelov, T. Swanson, O. Hausser, M. Trinczek, D. Melconian, M. Dombsky, J.M. D'Auria, D. Asgeirsson, Simon Fraser U.; J. Dilling, U. Heidelberg; U. Giesen, G. Roy, U. Alberta; K.P. Jackson, C.D.P. Levy, L. Buchmann, TRIUMF; J. Deutsch, Louvain-la-Neuve; W.P. Alford, U. Western Ontario.

We have begun a program to test weak interaction symmetries by trapping in a Zeeman-optical trap isotopes produced copiously at the TRIUMF cyclotron's on-line separator TISOL. We first plan to search for non-Standard Model scalar

contributions in the beta-neutrino correlation in the 0+ to 0+ Fermi decay of 38mK. We have trapped 38mK and 37K in a vapor-cell ZOT, and transferred them to a second ZOT which houses the nuclear detectors. First measurements of beta-recoil coincidences will be presented.
Supported in part by NSERC.

068.
38K, 37K LASER TRAPPING AND BETA DECAY MEASUREMENTS. **P.A. VOYTAS** R.S. Williamson, III, B.P. Schwartz, T.G. Walker, P.A. Quin, Department of Physics, University of Wisconsin-Madison, Madison, WI 53706

Recent progress in laser trapping and target development has brought about the realization of on-line laser trapping. Our approach uses a unique production and transport technique which produces neutral atom fluxes of >3x10^5 38K/sec or >3x10^4 37K/sec. Using stable potassium we have developed a primary laser trap with a capture efficiency of 1x10^-4 from a thermal beam. We are developing a pre-trap collection stage which should increase the overall capture efficiency by a factor of up to 100 by capturing from a vapor cell and delivering the atoms as a slow beam to the primary trap. Capture from the slow beam should provide an essentially background-free environment in which to perform precise beta-decay measurements. Progress and planned measurements will be presented.

069. PROGRESS TOWARDS TRAPPING ^{82}Rb IN AN OPTICAL TRAP.
R. Guckert[1,2], V. D. Sandberg[1], D. Tupa[1], D. J. Vieira[1], and X. Zhao[1]
[1] Los Alamos National Laboratory, Los Alamos, NM 87545, USA
[2] II. Phys. Institute, Justus Liebig University, 35392 Giessen, FRG

The efficient coupling of an optical trap to a mass separator is being developed to undertake high-precision electroweak interaction measurements. The use of an ion implantation and subsequent heated foil release method is being pursued as a suitable way of introducing radioactive samples into the ultra high vacuum region of the optical trap without gas loading. The batchwise trapping and cooling of one million ^{82}Rb atoms is envisioned with overall efficiencies of 1% or more. The cloud of cold radioactive atoms (~1 mm in diameter) will than be spin polarized and transferred to a second optical trap, repolarized, and loaded in a magnetic trap in which a beta-asymmetry measurement will be performed. The latest developments towards undertaking this experiment will be presented.

070.
ATOMIC LIFETIME MEASUREMENTS WITH TRAPPED RADIOACTIVE FRANCIUM ATOMS.
G. D. Sprouse, L.A. Orozco, J.E. Simsarian, W. Shi and W.Z. Zhao,
Physics Department, State University of New York, Stony Brook, NY, 11794-3800

We have produced 210Fr(3 minute) in a nuclear reaction and trapped about 10,000 atoms in a magneto-optic trap. Because the atoms are confined in both space and in momentum space, they make an ideal sample for further study of the atomic properties of Fr. We will report on our program to measure lifetimes and energy levels which are important for the use of Fr as a laboratory for fundamental physics measurements.

NUCL

071.
AN EFFICIENT TRAP OF ^{221}FR ATOMS. Z.T.Lu, K.L.Corwin, K.R.Vogel, C.E.Wieman, JILA, University of Colorado, Boulder, CO 80309. T.P.Dinneen, J.A.Maddi, H.Gould, LBNL, Berkeley, CA 94720

Francium, the heaviest alkali, is the ideal candidate for the next generation of experiments that study atomic parity non-conservation(PNC) effects or search for a CP-violating permanent atomic electric dipole moment(EDM). Since all francium isotopes are radioactive and have half-lives of 22 minutes or less, a highly efficient atom trap is needed to provide a sufficient sample for PNC or EDM experiments. Using an improved vapor cell magneto-optical trap, we have demonstrated an efficient trap of ^{221}Fr($t_{1/2}$=5m) atoms. The 4cm cube cell has a geometry optimized for a high trap loading rate and its walls are coated with dryfilm to slow the loss of francium atoms to the walls. Produced from the decay of ^{225}Ac($t_{1/2}$=10d), the francium atoms are efficiently transferred into the cell using a novel orthotropic oven. We have also measured the atomic hyperfine structure of the excited $7P_{3/2}$ and $7P_{1/2}$ levels of ^{221}Fr. Work supported by NSF, ONR, and U.S.DOE under contract number DE-AC-76SF00098.

072.
Nuclei in the Cosmos
by Stanford Woosley

Few scientific disciplines have the broad appeal and the truly interdisciplinary nature of nuclear astrophysics. Laboratory data and nuclear theory, our best understanding of the microscopic, are applied to understanding the evolution of matter under nature's most extreme conditions of temperature, density, and energy. Most of this nucleosynthesis occurs in stars and much of it in the more massive stars, greater than 8 solar masses, which become Type II and Ib supernovae. An important additional component comes from red giant stars and those white dwarfs that explode as Type Ia supernovae. Models for these events and the resulting nucleosynthesis will be reviewed and compared to what makes up the present Galaxy. The central role played by nuclear data will be emphasized and some examples of important data presented.

073. NUCLEAR DATA CAPABILITIES & OPPORTUNITIES FOR NUCLEAR ASTROPHYSICS.*
M. S. Smith, Physics Division, Oak Ridge National Laboratory, Bldg. 6010, MS-6354, P. O. Box 2008, Oak Ridge, TN 37831-6354

The nuclear data needs of the nuclear astrophysics community will be reviewed, and the strong overlap of these needs with the capabilities of the nuclear data community will be discussed. Additionally, some ongoing efforts in nuclear data evaluation relevant for nuclear astrophysics will be described. Special attention will be given to emerging electronic means of disseminating nuclear data.
*Research sponsored by the Oak Ridge National Laboratory, managed by Lockheed Martin Energy Research Corp. for the U. S. Department of Energy under contract number DE-AC05-96OR22464.

074. ENERGY LEVELS OF LIGHT NUCLEI, A=3-20, AVAILABLE FROM TRIANGLE UNIVERSITIES NUCLEAR LABORATORY ON THE WORLD WIDE WEB. G.M. Cheves, C.-W. Sheu, H.R. Weller, Duke University and TUNL, Durham, NC 27708 and D.R. Tilley, North Carolina State University and TUNL, Raleigh, NC 27695

The Triangle Universities Nuclear Laboratory (TUNL) Nuclear Data Evaluation Group provides an extensive collection of nuclear data information for A=3-20 on the World Wide Web. Currently included are: (1) Energy Level Diagrams for A=4-20. (2) Abridged versions of TUNL's most recent evaluations, "Energy Levels of Light Nuclei" for A=3, 16-20, and 5 (preliminary). (3) Adopted levels and decay data in ENSDAT style for A=3-20 nuclei as well as reaction data for A=18, 19. This format is similar to the Nuclear Data Sheets style for higher mass nuclei and is based on information from the NNDC Evaluated Nuclear Structure Data Files (ENSDF). (4) The A=1-20 portion of the Table of Isotopes, which was provided to TUNL by the Berkeley Isotopes Project. (5) Information about the current status of A=3-20 "Energy Levels of Light Nuclei" reviews by Fay Ajzenberg-Selove and by the TUNL group. (6) Abridged versions of Fay Ajzenberg-Selove's most recent evaluations for A=5-15. ** Work supported by DOE Contract Nos. DEFG05-91-ER40619 and DEFG05-88-ER40441.

075. RADIOACTIVE BEAMS - THE RESEARCH PERSPECTIVE. J. D. Garrett, Physics Division, Oak Ridge National Laboratory, Oak Ridge, Tennessee 37831

The advent of accelerated beams of radioactive ions at a variety of nuclear physics facilities throughout the world are producing a renewed interest in the structure of nuclei at the extremes of nuclear stability! This talk will address this frontier of nuclear structure data. Not only will the prospects for new data at the very limit of nuclear stability be described, the applications of nuclear data for planning the next generation of radioactive beam accelerators also will be considered. Research sponsored by the Oak Ridge National Laboratory, managed by Lockheed Martin Energy Research Corporation for the U.S. Department of Energy under contract number DE-AC05-96OR22464.

076. NUCLEAR DATA FOR RADIOACTIVE ION BEAMS
M.B. Chadwick, Theoretical Division, Los Alamos National Laboratory, Los Alamos, NM 87545

Radioactive Ion Beam (RIB) facilities that are under development, and that have recently come online, enable the study of nuclei far from stability to probe the nature of the nuclear force, as well as determining nuclear reaction cross sections important in astrophysical nucleosynthesis. To optimize the design of these facilities, nuclear data are needed for (p,xn) reactions producing proton-rich nuclides, and fission reactions producing neutron-rich fragments. In many cases measurements of these cross sections do not exist, and therefore one must rely on nuclear model calculations. I shall describe such calculations which include direct, preequilibrium, and equilibrium theories to describe how the projectile's energy is distributed in the reaction, and how nuclear break-up occurs. These calculations are benchmarked against measurements where they exist, and can be used to determine suitable target materials for RIB production.

077. NUCLEAR DATA SYSTEM-2000. **D. A. Resler** R. M. White, Nuclear Data Group, L-59, Lawrence Livermore National Laboratory, Livermore, CA 94550

078. ANTIPROTONIC HELIUM AND NEW ASPECTS IN THE ATOMIC CAPTURE PROCESS. R.S. Hayano, Department of Physics, University of Tokyo, Tokyo 113, Japan.

Antiprotonic helium atom is an exotic 3-body metastable system consisting of an antiproton (with a large principle quantum number around 38 and an angular momentum larger than about 35), an electron (in 1s state) and a helium nucleus. About 3% of antiprotons stopped in helium media (solid, liquid, gas) form antiprotonic helium atoms, which survive for several microseconds. It has recently become possible to study its formation, structure and decay by using a laser spectroscopy technique. The resonance wavelengths now agree within a few ppm with the results of recent 3-body calculations, while from the resonance intensities we can study the (n,l) populations of antiprotons immediately after atomic capture.

079. SLOWING DOWN AND THERMALIZATION OF THE PROTON, MUON, AND POSITRON IN GASES. Masayoshi Senba, Department of Physics, Dalhousie University, Halifax, NS, B3H 3J5.

While slowing down in gases, positively charged particles go through roughly three stages: Bethe-Bloch, cyclic charge exchange, and thermalization regimes. Whether the projectile (p^+, μ^+ or e^+) thermalizes as a positively charged species or as a bound state with an electron depends on the subtle competition between charge exchange and energy moderation at the end of the cyclic charge exchange regime. In radiation chemistry/physics, including positron and μSR experiments, it is important to know (1) how fast the energetic particle loses kinetic energy, (2) at what energy cyclic charge exchange sets in, (3) at what energy the final product forms and (4) how long it takes for thermalization. Based on muon slowing down times and on recent work on hot atom reaction yields for T and Mu in H_2, I will discuss these questions (1)-(4), emphasizing effects of the projectile mass ($M_p : M_\mu : M_e = 1 : 1/9 : 1/1836$) on slowing down and thermalization.

080. AN APPROACH TO COMPREHENSIVE UNDERSTANDING OF THE ORE AND THE SPUR MODELS OF POSITRONIUM FORMATION. Yasuo Ito, Research Institute for Nuclear Science and Technology, The University of Tokyo, Ibaraki 319-11, Japan.

A way is proposed to overcome the existing, apparently inconsistent, two models of Ps formation: the Ore model and the spur model. The basic idea lies in admitting that the two models are possible but that Ps formation must take place inside the terminal positron spur, which is in fact a short track including many ion pairs. A Ps formed by the Ore model must get into a highly oxidizing environment of the short track and is efficiently converted to a bare positron. This view gives precedence to the spur model in most condensed phases, but it also allows for existence of two o-Ps states or predicts that in special cases Ps can survive such an oxidizing atmosphere. It is thus aimed to put an end to the long-lasted conflicting arguments on the validity of the two models.

081. CHEMICAL REACTIONS AND THERMALIZATION OF THE POSITRONIUM "ATOM" IN CONDENSED MATTER
Hermann Stoll, Max-Planck-Institut für Metallforschung, D-70569 Stuttgart, Germany

Chemical reactions of positronium "atoms" (Ps) can be investigated in the time domain on a ns time scale by correlated measurements of the individual positron (e^+) lifetimes (= positron ages) and the Doppler shifts ΔE ($\rightarrow$ momentum of the annihilating positron-electron pairs) of the energy of one of the annihilation quanta ("Age-Momentum Correlation, AMOC") in a so-called $\beta^+\gamma\Delta E$ triple-coincidence set-up at a relativistic (MeV) positron beam. Examples discussed are the oxidation, complex formation, and spin conversion of Ps as well as the inhibition of Ps formation and the formation of bound states of e^+ with various ions. The full analysis of the two-dimensional AMOC data allows us to separate different chemical reactions, e.g., Ps spin conversion and Ps complex formation in the system HTEMPO/benzene, even if they take place simultaneously. Beam-based AMOC measurements enabled us to obtain first quantitative data on the slowing-down process of Ps. Results will be given for the kinetic energy of Ps at its formation and the Ps slowing-down time in different solids and liquids.

082 GAMMA-RAY SPECTROSCOPY WITH GAMMASPHERE
A.O.Macchiavelli, Lawrence Berkeley National Laboratory, Berkeley, CA 94720

Gamma-ray spectroscopy techniques have played a major role in our understanding of the atomic nucleus. Today, the new generation of γ-ray detector arrays (GAMMASPHERE in the United States, Eurogam and GA.SP. in Europe) is providing new insights into the understanding of the nuclear structure at extreme conditions, due to the large improvement in the *resolving power* of these arrays.

GAMMASPHERE, a national US facility, is currently in operation at the 88-inch Cyclotron of the Lawrence Berkeley National Laboratory and over 70 experiments have been carried out in its final configuration. In this talk I will present a short review of some of the technical aspects of the array and discuss a few cases among the broad range of nuclear physics topics that can be addressed with such a detector system. These will include: (1)Decay out of superdeformed bands, (2) Lifetime measurements of *magnetic rotation* bands, (3) Coulomb excitation studies of ^{235}U and (4) A search for double octupole-phonons in ^{208}Pb.

083. HIGH-SPIN NUCLEAR STRUCTURE DATA ON THE INTERNET. B. Singh, Department of Physics and Astronomy, McMaster University, Hamilton, Ontario L8S 4M1, Canada & Nuclear Science Division, LBNL, Berkeley, Ca 94720.

The study of nuclear structure at fast nuclear rotations, using fusion-evaporation reactions, started in the early sixties but since the experimental observation of superdeformation about a decade ago it has become one of the most pursued research topics in nuclear physics. Large gamma-ray detector arrays GAMMASPHERE, EUROGAM, and GASP were developed during the last few years and these continue to produce a wealth of new information about the properties of nuclei at high spins, including superdeformation. It is considered vital to compile, evaluate and systematize published data on many thousands of levels and gamma rays and associated nuclear bands obtained in such studies and make these available to the research community in conveniently retrievable and modern formats. This talk will describe the numerical, bibliographic and other high-spin related databases that are already accessible via INTERNET. Present limitations and ways to improve the current status and display of such databases will also be discussed.

084.
RELATIVISTIC HEAVY ION COLLISIONS: A FIELD ON THE THRESHOLD. **D. Cebra**
Department of Physics, University of California, Davis, CA 95616

ABSTRACT NOT AVAILABLE

085. COMPILATION OF RELATIVISTIC HEAVY-ION DATA. M.R. Bhat, National Nuclear Data Center, Department of Advanced Technology, Brookhaven National Laboratory, Upton, New York 11973-5000

Data from relativistic heavy-ion interactions measured at the AGS have been compiled into the data bases at the National Nuclear Data Center (NNDC), and are available for online access on INTERNET or ESNET. This work is part of a nuclear data program in support of basic research funded by the Division of Nuclear Physics of the U.S. Department of Energy. The goal of this effort is to have a complete database of high energy heavy-ion interactions measured at all experimental facilities including the Relativistic Heavy Ion Collider (RHIC) when it is operational. Data that have been reduced to well-defined physical quantities and not raw data will be in the database. The archived data listings and plots may now be accessed on the WWW. Further improvements in this data resource and additional features will be implemented based on user feedback.

086. A NUCLEAR INFORMATION SERVICE. R. E. MacFarlane, Theoretical Division, Los Alamos National Laboratory, Los Alamos, NM 87545

The T-2 Nuclear Information Service (http://t2.lanl.gov/) came up in September of 1993 and has grown ever since in both content and usage. Its purpose was to feature the work of our Group and to pioneer methods for delivering nuclear information through the Web. One popular interface is the Nuclear Data Viewer, which can be used interactively to prepare high-quality custom plots from evaluated and experimental nuclear data. Another interface converts files in the standard ENDF format into easy-to-read HTML pages with help links. This provides access for nonexperts. A third interface uses a graphical Map to the Nuclides to provide point-and-click access to information on a particular nuclide; for example, cross sections, decay data, or calculated nuclear masses. Other features of our Service include a dedicated area for nuclear astrophysics, publications from our group, and an area on computer codes that we use in our work.

087. **TECHNOLOGIES FOR NETWORK ACCESS AND UPDATING OF NUCLEAR DATABASES**. **Robert Sutton** Erik Miyake, Tracy L. M. Langlands, Scientific Digital Visions, Inc., 2 N 2nd St. Suite 1215, San Jose, CA 95113; Craig A. Stone, Department of Chemistry, San Jose State University, San Jose, CA 95192-0101

NO ABSTRACT AVAILABLE

088. PROBING MATTER WITH RADIOACTIVE NUCLEI, Robert F. Kiefl, Canadian Institute for Advanced Research, TRIUMF and Physics & Astronomy Department, University of British Columbia, 6224 Agricultural Road, Vancouver, B.C. V6T 1Z1

Magnetic spin resonance of stable nuclei (NMR) is a powerful method for probing internal magnetic fields and fundamental excitations in condensed matter. However, conventional NMR is primarily a bulk probe since about 10^{19} nuclear spins are required to generate a signal. High intensity proton accelerators such as TRIUMF in Vancouver can be used to generate intense beams of radioactive nuclei or muons which can be implanted into materials of interest. Radioactive nuclear spins are studied using methods closely related to NMR but with an enormous increase in sensitivity i.e. about 10^{10} fewer spins are required. In this talk I will describe how these techniques work and then give a few examples of how they can be used to probe ultra thin structures and interfaces.

089. MUON SPIN ROTATION STUDIES OF UNCONVENTIONAL SUPERCONDUCTORS: FULLERIDES, CUPRATES AND HEAVY FERMION SYSTEMS. *Robert H. Heffner, MS K764, Los Alamos National Lab, Los Alamos, N.M. 87545 USA.*

In 1957 the Bardeen-Cooper-Schrieffer (BCS) theory successfully provided a microscopic explanation of superconductivity in simple metals based on an electron-phonon interaction which produces a pairing force between electrons in a relative zero-orbital-angular-momentum, spin-singlet state. Since that time new classes of superconductors have been discovered exhibiting higher angular momentum pairing states, together with new pairing interactions. This talk will examine the role that muon spin rotation (μSR) has played in elucidating some of the unconventional properties of these superconductors. I will begin with an introduction to the superconducting order parameter and pairing mechanism and then discuss the parameters characterizing the superconducting state, such as the coherence length, magnetic penetration depth, critical fields, etc. Following a brief introduction to μSR, three classes of new superconductors will be explored: fullerides, heavy fermions and high-temperature copper oxides. Of particular interest is the interplay between superconductivity and magnetism in the latter two classes of materials. Prospects for future research will then be discussed.

090. ^{13}N, THE NUCLEAR PROBE TO THE PROCESSES OF NO, NO_2 AND HONO ON AEROSOL SURFACES. D. T. Jost, M. Ammann, M. Kalberer, E. Rössler, D. Piguet, U. Baltensperger, H.W. Gäggeler; Paul Scherrer Institut, CH-5232 Villigen, Switzerland

^{13}N (β^+, $T_{1/2}$ = 9.96 min) is used as in situ tracer in the adsorption and conversion processes of NO_2 on carbonaceous aerosol surfaces. ^{13}N is continuously produced in the reaction $^{16}O(p,\alpha)^{13}N$ with an incident energy of 12.8MeV in a gas target (480 ml/min He +20 ml/min O_2). The mixture of ^{13}N compounds produced in the target is converted in a Mo-converter to NO and subsequently oxidized over Cr_2O_3 to yield NO_2. The NO_2 is brought in contact with the aerosol in a temperature and humidity controlled flow tube reactor. Reaction products are collected in subsequent selective denuders (NaCl for HNO_3, Na_2CO_3/glycerol for HONO, triethanolamine for NO_2 and Co_2O_3 for NO, and a filter for any activity attached to the aerosol particle) and the activities are measured continuously in the traps. Sticking coefficients for NO_2 on various carbonaceous aerosol was found in the range of 2.4 to $4.0 \cdot 10^{-4}$. Reduction to NO and its desorption was measured. In addition, the formation of HONO on the particle surface was for the first time observed in situ.

091. FREEZEOUT PROPERTIES IN RELATIVISTIC HEAVY-ION COLLISIONS. Scott Chapman and J. Rayford Nix, Theoretical Division, Los Alamos National Laboratory, Los Alamos, New Mexico 87545.

We use a new realistic expanding source model for relativistic heavy-ion collisions to perform a detailed analysis of invariant π^+, π^-, K^+, and K^- one-particle multiplicity distributions and π^+ and K^+ two-particle correlations in nearly central collisions of Si + Au at $p_{\rm lab}/A = 14.6$ GeV/c. By minimizing χ^2 with a total of 1416 data points for the six types of data considered, we determine the values of nine independent freezeout properties, along with their uncertainties at 99% confidence, that characterize the gross properties of the source during its freezeout from a hydrodynamical fluid into a collection of noninteracting, free-streaming hadrons. These nine freezeout properties are the central baryon density n, nuclear temperature T, transverse velocity $v_{\rm t}$, longitudinal velocity v_{ℓ}, source velocity $v_{\rm s}$, transverse radius $R_{\rm t}$, longitudinal proper time $\tau_{\rm f}$, width in proper time $\Delta\tau$, and pion incoherence fraction λ_π. The first five properties are determined primarily by one-particle multiplicity distributions, and the last four properties are determined primarily by two-particle correlations. This work was supported by the U. S. Department of Energy.

092. THE DECAY OF ^{137}Cs AND THE FIRST EXCITED STATE OF ^{137}Ba. B. K. Wagner, P. E. Garrett, Minfang Yeh, and S. W. Yates, Departments of Chemistry and Physics & Astronomy, University of Kentucky, Lexington, Kentucky 40506-0055

^{137}Cs/^{137m}Ba sources are among those most frequently used for γ-ray calibration. The well-known 661.6 keV γ-ray de-excites a 153-s isomer, the second excited state of ^{137}Ba, while a lower-lying 1/2$^+$ state has been observed in a variety of nuclear reactions. In a recent study of the γ-rays emitted in the decay of ^{137}Cs, a 283.4 keV γ-ray was reported and was suggested as evidence for a previously unobserved weak β$^-$ decay branch to the 1/2$^+$ state; however, the discrepancy between the energy of the observed γ-ray and that accepted for the energy of the first excited state (279.96 keV) is disturbing. To resolve this question, we initiated an (n,n'γ) study of ^{137}Ba. The energy of the first excited state was determined to be 283.5 keV, in accord with other recent measurements, and no evidence for a doublet of states near this energy was found. In view of these conclusions, a re-examination of the decay of ^{137}Cs was pursued with a large-volume Compton-suppressed HPGe detector. A 283.5 ± 0.1 keV γ ray was observed with an absolute intensity of 6.1(1.0) x 10^{-6}, in excellent agreement with recent results, and is attributed to population of the first excited state of ^{137}Ba through a weak β$^-$ decay branch.

093. SPIN DISTRIBUTIONS OF SPONTANEOUS FISSION FRAGMENTS. D.A. Strellis[1,2], G.M. Ter-Akopian[3], A.V. Daniel[3], G.S. Popeko[3], Yu.Ts. Oganessian[3], J.O. Rasmussen[1], R. Aryaeinejad[4], S. Asztalos[1,2], B.R.S. Babu[5], J. Becker[6], L. Bernstein[6], K. Butler-Moore[4], S.Y. Chu[1], J.D. Cole[4], Y.X. Dardenne[4,6], R. Donangelo[7], M.W. Drigert[4], R. Firestone[1], T.N. Ginter[5], K.E. Gregorich[1], H. Griffin[8], J.H. Hamilton[5], J. Kormicki[5], I.Y. Lee[1], R.W. Lougheed[6], W.C. Ma[9], A.O. Macchiavelli[1], R. McLeod[1], M.F. Mohar[1], K.J. Moody[6], S. Prussin[2], A.V. Ramayya[5], M.A. Stoyer[6], P. Varmette[9], M.G. Wang[10], S.J. Zhu [5,10,11]

[1]Lawrence Berkeley National Laboratory, Berkeley, California 94720
[2]Nuclear Engineering Department, University of California, Berkeley, California 94720
[3]Joint Institute of Nuclear Research, Dubna 141980, Russia
[4]Idaho National Engineering Laboratory, Idaho Falls, Idaho 83415
[5]Department of Physics, Vanderbilt University, Nashville, Tennessee 37235
[6]Lawrence Livermore National Laboratory, Livermore, California 94550
[7]University Federal Rio de Janeiro, BR 21945-470 Rio de Janeiro, Brazil
[8]University of Michigan, Ann Arbor, MI 48109-1120
[9]Department of Physics, Mississippi State University, Mississippi 39762
[10]Physics Department, Tsinghua University, Beijing, People's Republic of China
[11]Joint Institute For Heavy Ion Research, Oak Ridge, Tennessee 37831

We have repeated and extended earlier studies of the GANDS93 collaboration on the Yrast population patterns of spontaneous fission using the newer data set from Gammasphere taken with ^{252}Cf in late 1995.

094. Gamma ray spectroscopy of odd-Z nuclei from ^{252}Cf at GAMMASPHERE†.

Early studies of prompt gamma rays from spontaneous fission were successful mainly in working out the level schemes of even-Z nuclei, since a majority of the population of even-even nuclei passes through the first excited $2^+ \to 0^+$ of the even-even fragment. Now with the more powerful current generation of detector arrays it is feasible to study the odd-Z nuclei. Especially helpful in our last ^{252}Cf run at GAMMASPHERE was the addition of two small Ge x-ray detectors. By gating on the odd-Z x-rays we have generated 3-dimensional gamma energy histograms strongly enhanced in the gated Z and 98-Z fission complement. We can thus build on the known level schemes to extend them to higher spins.

†S. Asztalos[1,2], G.M. Ter-Akopian[3], A.V. Daniel[3], G.S. Popeko[3], Yu.Ts Oganessian[3], J.O. Rasmussen[1], R. Aryaeinejad[4], B. R. S. Babu[5], J. Becker[6], L. Bernstein[6], K. Butler-Moore[4], S. Y. Chu[1], J. D. Cole[4], Y. X. Dardenne[4,6], R. Donangelo[7], M. W. Drigert[4], R. Firestone[1], T. N. Ginter[5], K. E. Gregorich[1], H. Griffin[8], J. H. Hamilton[5], J. Kormicki[2], I.Y. Lee[1], W. Lougheed[6], W. C. Ma[8], A. O. Macchiavelli[1], R. McLeod[1], M. F. Mohar[1], K. J. Moody[6], S. Prussin[2], A. V. Ramayya[5], M. A. Stoyer[6], P. Varmette[9], M. G. Wang[10], J. F. Wild[6], S. J. Zhu[5,10,11]
[1]Lawrence Berkeley National Laboratory, Berkeley, California 94720
[2]Nuclear Engineering Department, University of California, Berkeley, California 94720
[3]Joint Institute of Nuclear Research, Dubna 141980, Russia
[4]Idaho National Engineering Laboratory, Idaho Falls, Idaho 83415
[5]Department of Physics, Vanderbilt University, Nashville, Tennessee 37235
[6]Lawrence Livermore National Laboratory, Livermore, California 94550
[7]University Federal Rio de Janeiro, BR 21945-470 Rio de Janeiro, Brazil
[8]University of Michigan, Ann Arbor, MI 48109-1120
[9]Department of Physics, Mississippi State University, Mississippi 39762
[10]Physics Department, Tsinghua University, Beijing, People's Republic of China
[11]Joint Institute For Heavy Ion Research, Oak Ridge, Tennessee 37831

095. STABILITY OF LANTHANIDE AND ACTINIDE COMPLEXES OF A HEXADENTATE NITROGEN DONOR. Lester R. Morss, Chemistry Division, Argonne National Laboratory, 9700 S. Cass Ave., Argonne, Illinois 60439, and Dale D. Ensor, Department of Chemistry, Tennessee Technological University, Cookeville, TN 38505.

This study is part of a research project whose objective is to develop trivalent lanthanide and actinide complexes with soft-donor ligands, to compare their bonding properties, and to enhance the bonding differences of 4f and 5f ions coordinated to soft donors. The stability constants of 1:1 complexes of several M^{3+} (M= lanthanide or actinide) ions with the hexadentate ligand tetrakis(2-pyridyl-methyl)-1,2-ethylenediamine (TPEN) have been determined by potentiometry or by distribution measurements. The La^{3+} and Tb^{3+} complexes are known from crystal structures to be 10-coordinated to six TPEN nitrogens and to four oxygens from two nitrates. Of particular interest is the fact that in these complexes the soft donor nitrogen coordinates to f-element ions in the presence of hard donor oxygens of water. The stability constants will be compared with those of other hexadentate ligands with oxygen donors. This work was sponsored by the U.S. Department of Energy, Office of Basic Energy Sciences, Division of Chemical Sciences, under Contract W-31-109-ENG-38.

096.
SEPARATION OF 210-BI AND 210-PO BY USING ION-EXCHANGE MEMBRANE FILTERS: ENVIRONMENTAL APPLICATIONS. J.S. Gaffney, K.A. Orlandini, N.A. Marley, and M.M. Cunningham, Environmental Research Division, Bldg. 203, Argonne National Laboratory, Argonne, Illinois 60439

Ion-exchange membrane filters have been examined as a simple and inexpensive method for the simultaneous isolation and counting of 210Bi and 210Po in environmental samples. The procedure enables the determination of 210Pb in soils, waters, and aerosol samples through low level beta counting of the 210Bi. Concurrent measurement of the 210Po by alpha counting allows the 210Po/210Pb disequilibria to be determined and used for age determinations. Data comparing this method to more conventional methods will be presented, along with some examples of applications in atmospheric science. The advantages and limitations of this method over the current separation and counting of the naturally occuring radon daughters will be discussed.

097. SYNTHESIS OF F-18-LABELED 5-FLUORO-2'-DEOXY-2'-FLUORO-1-β-ARABINO-FURANOSYLURACIL ([^{18}F]-FFAU) FOR PET IMAGING STUDIES. M. M. Alauddin, J. D. Fissekis and P. S. Conti. University of Southern California, PET Imaging Science Center, Los Angeles, CA 90033.

5-Fluoro-2'-deoxy-2'-fluoro-1-β-arabinofuranosyluracil (FFAU) displays antiviral activity against simplex herpes virus-1 with moderate cytotoxicity in cell culture. Cytotoxic action involves conversion to the corresponding nucleotide, FFAUMP, and subsequent inhibition of dTMP synthase. Most recently tritiated FFAU has been evaluated for catabolism and biodistribution *in vivo* in mice as a potential agent for imaging of dTMP synthase activity using PET. Here, we report a F-18 labeled synthesis of FFAU, the label being in the base moiety. 3', 5'-O-bis-(tetrahydropyranyl)-2'-deoxy-2'-fluoro-1-β-arabinofuranosyluracil was used as the precursor in this synthesis. Carrier added labeling was performed by a modification of literature methods for the preparation of [^{18}F]FUdR. The radiolabeled product was purified by HPLC, using either an ion exchange column or a C$_{18}$ reverse phase column, and characterized by NMR (^{1}H and ^{19}F) spectroscopy and mass spectrometry. Chemical yield was quantitative, producing mCi levels of labeled product suitable for *in vivo* animal and human studies. Radiochemical purity of the product was >99%, and specific activity was 450 mCi/mmol.

098. SYNTHESIS OF 9-(4-[^{18}F]-FLUORO-3-HYDROXYMETHYLBUTYL)GUANINE ([^{18}F]FHBG) FOR IMAGING OF VIRAL INFECTION AND GENE THERAPY USING PET. M. M. Alauddin and P. S. Conti. University of Southern California, PET Imaging Science Center, Los Angeles, CA 90033.

The 4-fluoro-analogue of ganciclovir, FHPG, is selectively monophosphorylated by viral kinase, and is under investigation for *in vivo* imaging assessment of selective accumulation in infected cells or tumor cells transfected for gene therapy (J. Nucl. Med. 1996, 53P). Penciclovir also demonstrates antiviral activity against herpes simplex virus and varicella zoster virus based on selective monophosphorylation by viral kinase, and display longer intracellular retention. A new fluoro-analogue of penciclovir, FHBG, is also being considered as an imaging agent for gene transfection. Here, we report a chemical and radiochemical synthesis of FHBG. Penciclovir was converted to N^2,-O-bis(methoxytrityl)-4'-tosyl-derivative by treatment with methoxytrityl chloride followed by tosylation. The tosylate was fluorinated following our reported method for FHPG, with subsequent deprotection using HCl. All intermediates and the final product FHBG were characterized by NMR (^{1}H and ^{19}F) spectroscopy and mass spectrometry. Radiosyntheses were performed using [^{18}F]-KF, with average radiochemical yield being 7% in seven runs. Product could be co-eluted with the authentic sample and was purified by HPLC using a C$_{18}$ column.

099. FeCl$_3$- doped Polypyrroles probed by Positron Annihilation Technique, V. S. Subrahmanyam et al., Saha Institute of Nuclear Physics, 1/AF Bidhannagar, Calcutta-700 064, India.

FeCl$_3$ doped polypyrrole samples have been synthesised at different temperatures ranging from 308K and below. The induced defects in the doped structure were analysed by positron lifetime and Doppler broadening of the annihilation radiation. The important feature of the study is the dependence of the defect concentration on the synthesis temperature. This has been indicated by the varying intensity I$_2$ of the intermediate lifetime component originating from a state, with ~ 385 ps. The magnitude of the conductivity measured in the samples was found to increase monotonically with I$_2$ (which is proportional the defect concentration). Samples synthesised at lower temperatures show higher 'S' parameter characteristics and also increased trend of conductivity.

100. MECHANISMS OF NUCLIDE DECAYS T. Garcia, W. Slaughter, and E. Stedman RESEARCH DATA ANALYSIS, Box 324, Redlands, CALIFORNIA 923373-0121

To reach superheavy elements (proton sequential count) this present fairyland effort must match Periodic Chart chemist, iff we presume that proton-electron correspondence will indeed continue (number, valences directions, and depth with this proton vine being supported by its neutron vine as illustrated with a few light element examples). With well known past ^{209}Bismuth detrimental core contraction, how far can actual chemistry go with the only evidence apparently being alpha decays and are these at so called ground state well enough to have a nice beta decay example to at least claim a success? Are the experimental idealists aware of the eight beta decays and what energetics occur when such leptons are ejected to have an isobar mechanism? As one guess, should we forget any electron capture with the presumed lack of any proton-proton clustering concentrations or also too far apart except for alpha decays? Are any of these superheavy elements going to provide present civilization with any practicality unless much more basic scientific value is discovered enroute and during any achievements? Instead, let us head toward preparing some nuclides (as atoms) that can be kept in "bottles" in much larger quantities to show that supposed $\frac{1}{2}$-life values are incorrect or far more stable. This means that selections of shapes ought to have a much higher priority as will be illustrated via real models.

101. MODELLING HYDROGEN IN MATERIALS WITH ITS PSEUDO-ISOTOPE MUONIUM
Stephen F.J. Cox, Rutherford Appleton Laboratory, Oxfordshire, OX11 0QX UK and University College London, WC1E 6BT UK.

Many of the puzzles concerning hydrogen in materials, and its effect on their physical or functional properties, reflect the lack of atomistic models for the sites which protons, hydrogen atoms and hydride ions occupy or the dynamical processes they undergo. This is just the sort of information which is obtained by using implanted positive muons to mimic proton activity, or muonium as a model for hydrogen. The special properties of muon production and decay permits their study by an unusual but sensitive spectroscopy, developed from techniques in sub-atomic physics, which now finds widespread use in materials science. Its application has greatly illuminated many problems of hydrogen in metals and in semiconductors has discovered an unexpectedly complex interplay between crystallographic location and charge state, with important implications for hydrogen transport and electronic activity.

102.
µSR STUDIES OF BUCKYBALL RADICALS. Paul W. Percival, Department of Chemistry and TRIUMF, Simon Fraser University, Burnaby, B.C., Canada V5A 1S6

Irradiation of solid fullerenes with muons results in two types of muonium adduct: fullerenyl radicals, such as MuC_{60} and MuC_{70} (five isomers), in which the muonium atom is attached to the outside of the carbon skeleton; and encapsulated muonium, $Mu@C_{60}$ or $Mu@C_{70}$, formed from energetic atoms which penetrate the cage. All these species may be detected by muon spin rotation (µSR) or radio-frequency muon spin resonance (RF-µSR). They are easily distinguished by their muon–electron hyperfine coupling constants ($A\mu$). Encapsulated muonium has a value of $A\mu$ close to that of the free atom (4463 MHz), indicating that it is unbonded. In contrast, the fullerenyl radicals have order-of-magnitude smaller hyperfine constants typical of organic radicals. The hyperfine constants depend on the spin distribution of the unpaired electron in the radicals. This has been investigated by measuring 13C hyperfine constants in $Mu^{13}C_{60}$ and $Mu^{13}C_{70}$.

103. MuSR STUDIES OF MUONIUM RELAXATION BY DIOXYGEN AND ETHYLENE ADSORBED ON ULTRA-PURE SILICA POWDER. Robert F. Marzke, Department of Physics and Astronomy, ASU, P.O. Box 871504, Tempe, AZ 85287-1504; Donald G. Fleming, Department of Chemistry, University of British Columbia, Vancouver, BC V6T 1Z1, Canada; Paul W. Percival, Department of Chemistry, Simon Fraser University, Burnaby, BC V5A 1S6, Canada; Masayoshi Senba, TRIUMF, 4004 Wesbrook Mall, Vancouver, BC V6T 2A3, Canada

Extremely low (<5 ppb) transition metal impurity levels in a finely divided SiO_2 powder (EM Optipur, Merck) make it ideally suited for wide-ranging studies of muonium (Mu) relaxation by gases adsorbed onto a clean silica substrate. We have performed transverse and longitudinal-field MuSR measurements of muonium relaxation by O_2 and C_2H_4 at very low surface loadings (4-12 molecules per 10^6 SiO_2), from 5 to 300 K. Our findings elucidate the behavior of Mu on the silica surface, and show the prominent role of chemical reactions between Mu and the adsorbate. We also observe the effect of spin exchange upon relaxation in the case of O_2, in both LF and TF experiments. TF rate temperature dependences for both adsorbates are similar, but their LF rate field dependences differ strongly.

104. RECENT ASPECTS OF POSITRONIUM INTERACTIONS IN GASES AND SOLIDS. Toshio Hyodo, Department of Basic Sciences, Graduate School of Arts and Scieces, University of Tokyo at Komaba, Tokyo 153, Japan

Positronium(Ps) atoms formed in free space between the grains of silica aerogel or fine powders are used to study their interactions with molecules. The momentum transfer cross sections and the cross section for the Ps spin conversion are determined. Information about the thermalization of Ps through collisions with the silica particles and the gas molecules is important for the high precision measurement of ortho-Ps lifetime to be compared with QED theory. The Ps atoms are also used to detect paramagnetic radicals on the surfaces of the silica particles. Ps formed in some alkali halides is free at low temperatures and self-trapped at high temperatures. The transition between the free and the localized states occurs through tunneling at low temperatures as well as clasically at higher temperatures. It is interesting that the {200} Fourier components of the free Ps wave functions in alkali halides are much larger than the other high momentum components. Detection of spin-crossover phase transion with Ps and the effect of phonon scatterings on the Ps momentum distribution are also discussed.

105. DEPTH PROFILING OF DEFECTS IN SEMICONDUCTORS AND THIN FILMS. K.G. Lynn, Department of Physics and MME, Washington State University - Pullman, WA 99164-2711.

Defects play a critical role in determining the electrical properties of semiconductors and thin films, notably in controlling and understanding the next generation of semiconductor electronic devices. Positron beams are one of the newer tools available for the characterization of defects and their environments. This probe is capable of non-destructively depth profiling defects of various materials, quantitatively determining defect concentrations as a function of depth in some samples. Positrons can provide information that is both unique and complementary to that extracted from more standard techniques. An overview of the physical measurements performed with this technique will be provided with an emphasis on recent developments that extend the information obtainable. In particular, results will be presented showing quantitative comparisons between theory and defect-specific experiments on metals and semiconductors. Results on molecular beam epitaxy and impurity-defect complexes in both amorphous and crystalline Si, and on optically excited defects using positron beams will be examined.

106. ANALYTICAL APPLICATIONS OF POSITRON BEAM LIFETIME SPECTROSCOPY. Richard H. Howell, Lawrence Livermore National Laboratory, Livermore, CA 94550.

We are developing a defect analysis capability based on two positron beam lifetime spectrometers: the first uses a 3 MeV electrostatic accelerator and the second on the high current beam at our electron linac. The 3 MeV beam is now being used for bulk sample analysis and analysis of samples encapsulated in controlled environments. At our high current beam we are developing a low energy, microscopically focused, pulsed positron beam for small sample features. This beam will enable defect specific, 3-dimensional maps of defect concentrations with sub-micron location resolution. We will describe both instruments and discuss uses of these instruments in studying defect distributions in materials. This work was performed under the auspices of the US Department of Energy by LLNL under contract No. W-7405-ENG-48.

107.
ETLA - ELASTIC THERMALISATION LIFETIME ANALYSIS OF CROSS LINKED PMMA.
C. Dauwe, B. Van Waeyenberge, D. Segers, T. Van Hoecke. Dept. Subatomic and Radiation Physics, University of Ghent, Proeftuinstraat 86, B-9000 Gent, Belgium
Du Prez F., Dept. of Organic Chemistry, Krijgslaan 281, B 9000 Gent, Belgium

Positronium can be formed in a variety of "hole" type sites such as in the free space between fine powders, in microporous materials and even in the free volume sites of polymers. We present a recently developed model for the interpretation of the lifetime spectra from such o-Ps which is formed with an epithermal initial energy and which thermalises during its annihilation time. The relevant parameters of the analysis are the o-Ps asymptotic decay rate, the o-Ps intensity, the initial energy of the Ps and an energy-relaxation parameter. Positron lifetime spectra were obtained from PMMA with different amounts of cross-linking. The energy relaxation parameter was found to relate to the cross-linking while all other parameters are constant.

108. ATOMIC CLUSTERS AND POSITRON CAPTURE SITES IN AMORPHOUS AND UNDERCOOLED METALLIC STRUCTURES: K. Krištiaková, J. Krištiak, Institute of Physics, Slovak Academy of Sciences, 842 28 Bratislava, Slovakia

The presence and evolution of two phenomena crucial for the comprehension of processes taking place in noncrystalline metallic alloys, namely the nearest neighbour arrangement of atoms, i. e. clustering or local atomic ordering, and the density of voids or empty spaces between atoms in such structures, have been determined by positron probing (PAL). The observations provide the insight into the microstructural processes taking place via determination of the evolution of the positron capture site density and distribution. The observed shifts in positron lifetime distributions are analyzed in terms of relaxation of free volume on samples of NiZrAl in amorphous and undercooled liquid states. Atomic clustering has been analyzed on samples of rapidly quenched FeCoB alloys with different thermal history in the melt.

109.

1D-ACAR SPECTROSCOPY OF OXYGEN-RELATED CENTERS IN SILICON. **N. Yu Arutyunov** Institute of Electronics of the Uzbek Academy of Sciences, 700143 Tashkent, Uzbekistan

ABSTRACT NOT AVAILABLE.

American Chemical Society
DIVISION OF ORGANIC CHEMISTRY

213th ACS National Meeting

San Francisco, CA
April 13-17, 1997

S.S.Hall, Program Chairperson

SUNDAY MORNING

- **Recent Advances in Organopalladium Chemistry**
  E-I. Negishi, Presiding — Papers 1-7

- **Porphyrins and Polyaromatics**
  C.M. Muzzi, Presiding — Papers 8-19

- **Enzyme-Mediated Reactions: Biocatalysis, Chemoenzymic Synthesis, and Enzymology**
  R.L. Polt, Presiding — Papers 20-31

SUNDAY AFTERNOON

- **Science of Synthesis: A Legacy of William S. Johnson**
  P.A. Wender, Presiding — Papers 32-37

- **Recent Advances in Organopalladium Chemistry**
  R.C. Larock, E-I. Negishi, Presiding — Papers 38-43

- **Synthesis of Organic Light Emitting Materials
  Organic Materials and Devices in Display Technology**
  B.R. Hsieh, C.H. Chen, Presiding — Papers 44-50

- **Peptidomimetics, Isosteres, Surrogates, Mimics, and Artificial Biomolecules and Processes**
  L.O. Weigel, Presiding — Papers 51-60

SUNDAY EVENING

- **Posters**
  S.S. Hall, Presiding — Papers 61-130

MONDAY MORNING

- **R. F. Hirschmann Award in Peptide Chemistry**
  R. Johnson, Presiding — Papers 131-134

- **Recent Advances in Organopalladium Chemistry**
  J-E. Backvall, E-I. Negishi, Presiding — Papers 135-140

- **Design and Optimization in Organic Synthesis**
  V. Snieckus, S.H. Bertz, Presiding — Papers 141-145

- **Stereoselective Syntheses**
C.J. Forsyth, Presiding						Papers 146-156

- **Heterocycles, Physical Organic, Nuclear Magnetic Resonance, and Photochemistry**
S.S. Hall, Presiding						Papers 157-229

MONDAY AFTERNOON

- **ACS Award for Creative Work in Synthetic Organic Chemistry**
W.R. Roush, Presiding						Papers 230-233

- **Photochemistry, Optics, Sensors, and Switches**
S.A. Fleming, Presiding						Papers 234-244

- **Strain, Mechanisms, and Theory**
W.T.G. Johnson, Presiding						Papers 245-255

- **Carbenes, Cycloadditions, and Carbocations**
A.P. Marchand, Presiding						Papers 256-267

MONDAY EVENING

- **Sci-Mix**
S.S. Hall, Presiding						Papers 61-130

TUESDAY MORNING

James Flack Norris Award in Physical Organic Chemistry
P.W. Jennings, Presiding						Papers 268-271

- **Merging of Solid-Phase Organic Synthesis and Molecular Diversity for the Generation of Combinatorial Libraries**
Perspective, Solid-Phase Organic Synthesis Examples, Methods of Analyses and Alternative Techniques
J.S. Kiely, M.J. Green, Presiding					Papers 272-278

- **Design and Optimization in Organic Synthesis**
S.H. Bertz, V. Snieckus, Presiding					Papers 279-283

- **New Reactions and Methodologies**
A.R. Howell, Presiding						Papers 284-294

- **Porphyrins, DNA, Enzymology, Peptides, Structural Mimetics, Marine Natural Products and Polymers**
S.S. Hall, Presiding						Papers 295-344

TUESDAY AFTERNOON

- **Ernest Guenther Award in the Chemistry of Natural Products**
S.D. Burke, Presiding						Papers 345-348

- Boron Neutron Capture Therapy, X-Ray and Neutron Diffraction, Nuclear Magnetic Resonance, Mass Spetrometry, and UV Spectroscopy Techniques
G.W. Kabalka, Presiding
Papers 349-359

- Oligonucleotides, Nucleosides, Nucleases and Synthases
K.D. Turnbull, Presiding
Papers 360-370

- Host-Guest, Molecular Recognition, Self Assembly, and Donor-Acceptor Systems
M.D. Distefano, Presiding
Papers 371-379

WEDNESDAY MORNING

- Merging of Solid-Phase Organic Synthesis and Molecular Diversity for the Generation of Combinatiorial Libraries
Methods of Analyses and Alternative Techniques, and Combinatorial Libraries Now and Tomorrow
J.S. Kiely, M.J. Geen, Presiding
Papers 380-386

- Synthetic Receptors for Optical Chemosensors
Biosensing and Biosensors: Chemistry, Recognition, and Process Control
T.W. Bell, Presiding
Papers 387-391

- Design and Optimization of Organic Synthesis
V. Snieckus, S.H. Bertz, Presiding
Papers 392-396

- Macrocycles: Synthetic Strategies and Total Synthesis of Cyclic Peptides and Macrolides
P.L. Toogood, Presiding
Papers 397-408

- Palladium and Other Metal-Mediated Reactions, Asymmetric Reactions, Stereoselective Syntheses, and Cycloadditions
S.S. Hall, Presiding
Papers 409-465

WEDNESDAY AFTERNOON

- ACS Award in Pure Chemistry
P. DeShong, Presiding
Papers 466-469

- Synthetic Receptors for Optical Chemosensors
Biosensing and Biosensors: Chemistry, Recognition, and Process Control
T.W. Bell, Presiding
Papers 470-474

- Oligosaccharides, Cell-Surface Molecules, and Peptides
C.R. Bertozzi, Presiding
Papers 475-485

- Aromatics and Heterocycles
V.V. Zhdankin, Presiding
Papers 486-496

- Posters
S.S. Hall, Presiding
Papers 497-568

THURSDAY MORNING

- **Solid and Solution-Phase Synthesis, Techniques, and Combinatorial Library Generation**
  J. Gervay, Presiding
  Papers 569-579

- **Metal-Mediated Reactions and Syntheses**
  J.B. Sheridan, Presiding
  Papers 580-591

- **Alkaloid Syntheses: New Methodologies and Strategies**
  D.A. Horne, Presiding
  Papers 592-604

- **Asymmetric Reactions and Methodology**
  D.V. McGrath, Presiding
  Papers 605-616

THURSDAY AFTERNOON

- **Stereoselective Syntheses**
  J.W. Leahy, Presiding
  Papers 617-628

- **Palladium and Other Metal-Mediated Reactions and Syntheses**
  B.C. Soderberg, Presiding
  Papers 629-638

- **Radicals and Their Applications in Selective Synthesis**
  M.P. Sibi, Presiding
  Papers 639-650

- **Tandem and Environmentally Benign Reactions and Processes**
  J.A. Soderquist, Presiding
  Papers 651-659

DIVISION OF ORGANIC CHEMISTRY

001. NEW DEVELOPMENTS IN Pd-CATALYZED FORMATION OF CARBON-CARBON AND CARBON-HETEROATOM BONDS AND THEIR APPLICATIONS TO NATURAL PRODUCTS SYNTHESIS. Ei-ichi Negishi, Department of Chemistry, Purdue University, W. Lafayette, Indiana 47907-1393.

Recent developments made in the areas of Pd-catalyzed reactions involving (a) **cyclic carbopalladation,** (b) **cyclic acylpalladation,** (c) **α-alkenylation,** and (d) **tandem cross coupling–lactonization** will be discussed with emphasis on their applications to the synthesis of various natural products including **rubrolides, goniobutenolide A,** and **nakienones A and B.**

002. PALLADIUM(0) CARBENES AS INTERMEDIATES IN ORGANOMETALLIC CHEMISTRY. Carl A. Busacca, Azad Hossain, and Yong Dong, Department of Chemical Development, Boehringer Ingelheim Pharmaceuticals, Inc., Ridgefield, CT 06877

The Stille reaction is a powerful synthetic tool for carbon-carbon bond construction. In certain substituted vinyl stannanes, however, cine rather than ipso regioisomers may form. This same behavior has been observed for palladium catalyzed coupling of other vinyl metal species including silicon, germanium, mercury and boron. The mechanistic implications of deuterium labeling studies, crossover experiments, and Tin-119 NMR data will be discussed. The data supports the intermediacy of a palladium(o) carbene in these processes. Several alternative explanations, including anti beta elimination of palladium hydride will be shown to be inconsistent with the experimental results.

003. PALLADIUM-CATALYZED AMINATION AND ETHERATION OF ARYL HALIDES AND TRIFLATES: FUNDAMENTALS AND METHODOLOGY.

John F. Hartwig, Department of Chemistry, Yale University, P.O. Box 208107, New Haven, CT 06520-8107

Palladium-catalyzed reactions that form amines, ethers, and sulfides from aryl halides and triflates will be discussed. New synthetic methods will be presented, with an emphasis on how the catalytic amination and etheration reactions occur and how this information has lead to catalyst development. Primary reactions involved in the catalytic cycle, including N-H activation of alkylamines, reductive elimination of amines and ethers, and β-hydrogen elimination from amides and alkoxides, have been observed directly. Information on these reactions will be presented, in addition to the information on synthetic methodology.

004. THE SOLID PHASE SYNTHESIS OF TRI-SUBSTITUTED INDOLES. J. W. Ellingboe, M. D. Collini, Chemical Sciences Division, Wyeth-Ayerst Research, Pearl River, New York 10965

The solid phase synthesis of small organic molecules has recently become important because of the utility of this methodology for producing combinatorial libraries. While the formation of amide bonds on a solid support has been optimized, other types of reactions have not been studied as extensively. Carbon-carbon bond formation and reactions in which a heterocycle is formed are of particular importance for the synthesis of small organic molecules used in drug discovery research. Palladium catalyzed processes have been used for both of the above reaction classes, and therefore appeared to be a fruitful area for extension to solid phase synthesis. This talk will focus on the Pd catalyzed coupling of terminal alkynes with aryl iodides on a solid support. The extension of this reaction to the synthesis of indoles will also be discussed.

005. PALLADIUM-CATALYZED AROMATIC CARBON-HETEROATOM BOND FORMATION: SYNTHETIC AND MECHANISTIC RESULTS. John Wolfe, Seble Wagaw, Jean Francois Marcoux, Ross Widenhoefer, Annita Zhong and <u>Stephen L. Buchwald</u>,* Department of Chemistry, Massachusetts Institute of Technology, Cambridge, MA 02139

Progress in our palladium-catalyzed methodology for the conversion of aryl halides to aniline and aryl ether derivatives will be reported.

006. A NOVEL MODE OF PALLADIUM-CATALYZED CYCLOISOMERIZATION OF 1,5-DIENE SYSTEMS.
<u>Takeshi Nakai</u>, A. Nagasawa, M. Harada, A. Takada, M. Sugiura, and K. Tomooka
Department of Chemical Technology, Tokyo Institute of Technology, Meguro, Tokyo 152, Japan.

During the course of studies on Pd-catalysis of the Claisen and Cope rearrangements, we have recently discovered an unprecedented mode of Pd-catalyzed cycloisomerization of 1,5-diene systems. For example, exposure of 1,5-diene **1** to a catalytic amount of $PdCl_2(PhCN)_2$ gives rise to cyclopentene **2** as the ultimate product. The scope and limitation of the new type of cycloisomerization will be presented, and mechanisms thereof will also be discussed.

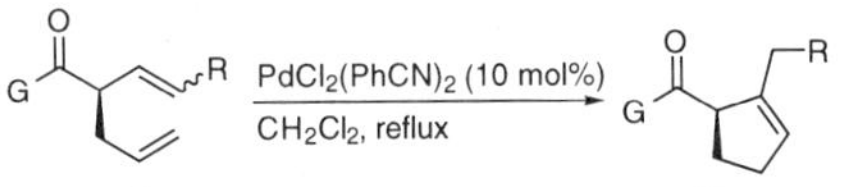

$$\text{1} \xrightarrow[\text{CH}_2\text{Cl}_2, \text{ reflux}]{\text{PdCl}_2(\text{PhCN})_2 \text{ (10 mol\%)}} \text{2}$$

G= Oppolzer's chiral sultam or OEt

007. DEVELOPMENT OF PALLADIUM(IV) CHEMISTRY AND THE OXIDATION OF ORGANOPALLADIUM(II) COMPLEXES BY WATER AND HALOGENS. **A. J. Canty**, Chemistry Dept, University of Tasmania, Hobart, Tasmania, Australia 7001

The development of organopalladium(IV) chemistry will be reviewed followed by an assessment of the potential roles of Pd(IV) intermediates in organic synthesis and catalysis, and recent advances demonstrating that water is able to oxidise Pd(II) to Pd(IV), and that oxidation by halogens is possible without cleavage of Pd-C bonds. Isolation of the first examples of diorganopalladium(IV), hydroxo and aquapalladium(IV) complexes is described, together with hydrogen-bonding networks such as "$Pd^{IV}(OH_2)^+...^-OC_6F_5$" interactions in a tris(pyrazol-1-yl)borate complex containing a pallada(IV)cyclopentane group $[Pd(CH_2CH_2CH_2CH_2)(OH_2)\{(pz)_3BH\}.OC_6F_5]_2$. Oxidation and alkyl transfer occur together in some reactions, e.g.

$2 [Pd^{II}Me_2\{(pz)_3BH\}]^- + 2 H_2O + PPh_3 \rightarrow$

$Pd^{IV}Me_3\{(pz)_3BH\} + Pd^{II}Me\{(pz)_3BH\}(PPh_3) + H_2 + 2 OH^-$

008.

SYNTHESIS OF NOVEL DODECASUBSTITUTED PORPHYRINS VIA THE SUZUKI COUPLING REACTION. **Cinzia M. Muzzi** Craig J. Medforth, Kevin M. Smith, Department of Chemistry, University of California, Davis, CA, 95616

Recently, interest has centered on the possible functional role of the nonplanar conformational distortions seen for some tetrapyrroles in biological systems. One way the effect of these nonplanar distortions has been investigated is by studying sterically crowded dodecasubstituted porphyrins. These porphyrins can be prepared from the acid catalyzed condensation of appropriately substituted pyrroles and benzaldehydes; however, this route is limited by difficulties in the synthesis of substituted pyrroles. Fortunately, large quantities of brominated dodecasubstituted porphyrins can be easily prepared and then coupled with aryl boronic acids via a palladium catalyzed (Suzuki) reaction. This eliminates the need to synthesize substituted pyrroles and permits the synthesis of many novel dodecaarylporphyrins.

009.

MECHANISTIC STUDIES ON THE OXIDATIVE CYCLIZATION OF A,C-BILADIENES: NEW INSIGHTS INTO THE ROLE OF THE METAL ION. **J.J. Lin,** R.T. Holmes, K.M. Smith. Department of Chemistry, U.C. Davis, Davis, California 95616.

The role of the metal in metal assisted oxidative cyclization will be discussed. Insights from previous experimental results have implied that the role of the metal serves as a chelator as well as an oxidant. Further exploration suggests that the metal is not primarily important in aligning the tetrapyrroles into proper conformation prior to cyclization. The metal probably inserts after formation of the dihydroporphyrin intermediate and serves to facilitate the elimination of the 1-substituent to produce porphyrin.

010.

PORPHYRINYL BETA-KETO ESTER RADICALS: SYNTHESIS AND CHEMISTRY. **R. T. Holmes,** J. J. Lin, and K. M. Smith, Department of Chemistry, U.C. Davis, Davis, California, 95616

Oxophlorins have long represented a class of porphyrinoid macrocycles with the capacity to form stabilized radicals. These radicals have displayed the unique ability to reversibly self assemble as dimers or, in some cases, react with molecular oxygen and form interesting monomers. Porphyrins, although possessing the ability to form stabilized radicals themselves, have failed to display chemistry analogous to that of oxophlorins. We can now report that covalently linked porphyrinoid dimers and oxygenated monomers have been prepared from porphyrinyl beta-keto esters. The syntheses of these molecules will be discussed. Successful characterization of by-products provides good insight into mechanistic details. The isolation of a unique peroxide dimer reveals mechanistic detail concerning both porphyrin radical chemistry and oxophlorin radical chemistry.

011.

DIHYDROPORPHYRIN SYNTHESIS: NEW METHODOLOGY. **Kalyn M. Shea,** Laurent Jaquinod, Richard G. Khoury, Kevin M. Smith, Department of Chemistry, University of California, Davis, CA, 95616

The selective formation of either cyclopropanochlorins or functionalized trans-chlorins by reaction of 2-nitro-5,10,15,20-tetraphenylporphyrins with "activated" methylene compounds has been achieved and confirmed by X-ray crystallography. Reaction control is accomplished via sequential Michael addition, followed by inter- or intra-molecular nucleophilic displacement of a nitro group. Steric as well as thermodynamic effects have been found to govern the observed selectivity. Compounds containing bulky methylene groups are found to favor intramolecular formation of the cyclopropyl annulated chlorins at ambient temperature. In contrast, trans-chlorin formation is favored at higher temperatures. Modification of the ß, and/or meso-phenyl chlorin positions using regioselective dibromination is applied to the synthesis of new photodynamic therapy (PDT) reagents, as well as to the preparation of highly nonplanar chlorins.

012.
NEW PHOTOSENSITIZERS FOR USE IN PHOTODYNAMIC THERAPY.
M. M. RESSLER, R. K. PANDEY, AND K. M. SMITH, DEPARTMENT OF CHEMISTRY, U. C. DAVIS, DAVIS, CALIFORNIA 95616

PHOTODYNAMIC THERAPY (PDT) IS A NEW METHOD OF CANCER TREATMENT UTILIZING THE SELECTIVE RETENTION OF A PHOTOSENSITIZER IN A TUMOR AND ACTIVATING LIGHT TO ELICIT AN EFFICIENT PHOTOTOXIC REACTION AND DESTROY THE TUMOR CELL. IN THE LAST FEW YEARS A NUMBER OF PHOTOSENSITIZING DYES SUCH AS PORPHYRINS, CHLORINS, PURPURINS, AND BENZOPORPHYRIN DERIVATIVES HAVE BEEN REPORTED AS POTENTIAL PHOTOSENSITIZERS IN PDT. TO FURTHER THIS RESEARCH, A SERIES OF ALKYL ETHER BENZOPORPHYRINS AND PURPURINS IS BEING SYNTHESIZED TO DETERMINE THE EFFECTS OF INCREASED CARBON-CHAIN LENGTHS AT THE ETHER LINKAGE IN THESE PHOTOSENSITIZERS. CURRENT WORK IN PROGRESS WILL BE PRESENTED DETAILING THE SYNTHETIC STRATEGY.

013. SYNTHESIS AND SPECTROSCOPY OF EDGE-OVER-EDGE Zn(II) PORPHYRIN AND BACTERIOCHLORIN DIMERS. Jayasree Vasudevan, Spencer Knapp,* Robert T. Stibrany, Jean Bumby, Joseph A Potenza, Tom Emge, and Harvey J. Schugar,* Department of Chemistry, Rutgers University, New Brunswick, NJ 08903 USA.

We report UV-vis and NMR spectroscopic analysis of a series of bacterial photosystem special pair model dimers whose extent and direction of porphyrin or bacteriochlorin ring overlap is controlled by positioning a Zn-ligating amino tether at a *meso*- or β-position.

014. NOVEL AROMATIC STRUCTURES BY CONCEPTUAL RING TRANSPLANTATION OF THE PORPHYRIN MACROCYCLE. Timothy D. Lash, Department of Chemistry, Illinois State University, Normal IL 61790-4160.

The "3 + 1" approach to porphyrin synthesis (Lash, T.D. *Chem. Eur. J.* **1996**, *2*, 1197) has been applied to the preparation of novel 18 π electron bridged annulenes **1-6**. These structures are formally related to the porphyrins by the conceptual replacement, or ring transplantation, of a single pyrrole ring. This approach allows the straightforward synthesis of many unique porphyrinoid structures of both theoretical and practical significance.

1 a. X = CH; b. X = N   2 a. X = CH; b. X = N   3 Tropiporphyrin   4 Azuliporphyrin   5 Benzocarbaporphyrin   6

015. THE AZULIPORPHYRIN-CARBAPORPHYRIN CONNECTION. Sun T. Chaney and Timothy D. Lash, Department of Chemistry, Illinois State University, Normal IL 61790-4160.

Novel porphyrin analogs with benzene or pyridine subunits have recently been prepared by the "3 + 1" methodology (Lash, T.D.; Chaney, S.T. *Chem. Eur. J.* **1996**, *2*, 944). We have now extended this chemistry to the synthesis of the first porphyrin analog with an azulene subunit. The new system, azuliporphyrin (**1**), is formally cross-conjugated although proton NMR spectroscopy shows a weak porphyrin-like ring current. This indicates that the dipolar tropylium/carbaporphyrin canonical form **2** makes a significant contribution to this structure. In the presence of pyrrolidine, a carbaporphyrin species is formed due to nucleophilic addition onto the seven-membered ring. This observation indicates that the chemical modification of this fascinating new porphyrinoid system will give related carbaporphyrin structures.

016. TWO NEW POLYHETERO[9](9,10)ANTHRACENOPHANES. Bethany Blackwell[†], Sarah Ngola[†], Heidi Peterson[†], **Stuart Rosenfeld**[†], Cynthia White-Tingle[†], Jerry P. Jasinski[‡], Yu Li[‡], and Jonathan E. Whittum[‡], Departments of Chemistry, [†]Smith College, Northampton, MA 01063 and [‡]Keene State College, Keene, NH 03431

We have prepared and characterized two new polyhetero[9](9,10)anthracenophanes **1** and **2**. Although molecular mechanics calculations (MMX) suggest that the tautomeric partners **3** and **4** are more stable than **1** and **2**, X-ray crystal structures and nmr spectra of these compounds demonstrate unambiguously that it is the anthracene tautomer that is favored in both solid state and solution. At room temperature, both compounds are conformationally mobile on the nmr time scale and for both, reversible broadening consistent with the interconversion of conformational enantiomers is observed at low temperatures.

017. DEHYDROBENZOANNULENES REVISITED: A NEW LOOK AT AN OLD MOLECULAR SYSTEM. Michael L. Bell, Stephen C. Brand, Jamieson J. English, Charles A. Johnson, Joshua J. Pak, and Michael M. Haley,* Department of Chemistry, University of Oregon, Eugene, OR 97403-1253

Calculations suggest polymeric phenylacetylene networks have great potential as advanced materials characterized by exceptional properties like thermal stability, chemical inertness, non-linear optical activity, and electrical conductivity. Dehydrobenzoannulenes (*e.g.*, **1-6**) represent possible precursors to these densely cross-linked carbon-rich solids. Accordingly, we will present a versatile new synthetic route that permits easy preparation of novel α,ω-poliynes, thus leading to dehydrobenzoannulenes of varying topologies previously available only in low yield (**3**, **6**) or altogether inaccessible by traditional copper-mediated dimerization/cyclooligomerization routes (**1, 2, 4, 5**).

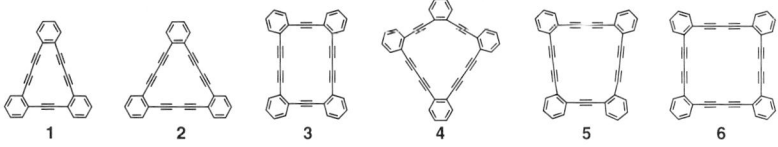

018. A NEW ITERATIVE SYNTHESIS OF [n]PHENACENES, A FAMILY OF NOVEL GRAPHITE RIBBONS. Frank B. Mallory, Kelly E. Butler, and Emilie J. Brondyke, Department of Chemistry, Bryn Mawr College, Bryn Mawr, Pennsylvania 19010, and Clelia W. Mallory, Department of Chemistry, University of Pennsylvania, Philadelphia, Pennsylvania 19104

We have recently synthesized several alkyl-substituted [7]phenacenes [*Tetrahedron Lett.* **1996**, *40*, 7173-7176] and an alkyl-substituted [11]phenacene. The alkyl substituents were incorporated along the aromatic framework to confer solubility on these otherwise intractable compounds. We now report a new five-step iterative strategy for the synthesis of *tert*-butyl-substituted [n]phenacenes. In each five-step iteration, the two key steps are Wittig coupling of a benzylic phosphonium salt with an aromatic aldehyde, and photocyclization of the resulting stilbene derivative. This new strategy should lead efficiently to systems with much larger values of n than we have achieved previously. These molecules, which are related structurally to graphite as ribbons are to sheets, may have interesting and useful properties as materials.

019. PREPARATION AND OXIDATIVE POLYMERIZATION OF 4-SUBSTITUTED-2,2'-BITHIOPHENES. Seth C. Rasmussen, Jason C. Pickens, and James E. Hutchison, Department of Chemistry and Material Science Institute, University of Oregon, Eugene, OR 97403.

The functionalization of polythiophene at the β positions of the repeat units has been shown to be a very powerful method for controlling the physical, electrical, and optical properties of the polymer. Such derivatization has produced a class of conjugated polymers exhibiting good solubility, processibility, environmental stability, and electrical and optical activity. However, complicating steric and electronic effects introduced through the addition of sidechains are not always desirable. In order to produce derivatized polythiophenes with a decreased density of sidechains, we are currently investigating the oxidative polymerization of monosubstituted bithiophenes. The resulting polymers should display the desirable properties imparted by sidechain incorporation while reducing any undesirable steric and electronic effects. The synthesis and characterization of these novel bithiophene "monomers" will be presented. Conditions for the chemical and electrochemical polymerizations will be discussed along with electrochemical, spectroscopic, and conductivity properties of the resulting polymers.

020. CONFORMATIONAL AND ELECTRONIC CONTROL OF FLAVINS REDOX CHEMISTRY
Justin J. Hasford and Carmelo J. Rizzo*;
Department of Chemistry; Vanderbilt University; Nashville, TN 37235-1822

As models for flavin coenzyme mediated reactions, a series of conformationally biased flavins have been synthesized and their redox properties studied by cyclic voltammetry. These studies show that the redox chemistry of flavins can be driven by conformational considerations. In the course of this work we have also observed a linear free energy substituent effect which allows for the accurate prediction of flavin reduction potential.

021. EXPLORING THE CATALYTIC MECHANISM OF PHOSPHOLIPASE C. Paul J. Hergenrother, Mark R. Spaller, and Stephen F. Martin*, Department of Chemistry and Biochemistry, University of Texas at Austin, Austin, Texas 78712

With three active site zinc atoms, the phosphatidylcholine-preferring phospholipase C from *Bacillus cereus* (PLC$_{Bc}$) possesses an unusual molecular architecture. Inspection of the crystal structure of the enzyme complexed with a substrate analogue inhibitor reveals three amino acid side chains positioned to serve as a general base in the PLC_{Bc} catalyzed hydrolysis of phospholipids. Using site-directed mutagenesis, these residues (Glu4, Asp55, Glu146) have been systematically altered and the resulting mutant proteins examined. Comparing the wild-type and mutant pH vs. activity profiles, structures, and thermostabilities indicates that Asp55 is an excellent candidate for the general base. The mechanism behind the upward pK_a shift of this residue and its indispensable role in catalysis has been investigated. The role that the three zinc atoms fulfill has also been investigated in the context of enzyme stability, substrate binding, and catalysis. These results lend insights into the mechanism of enzymatic hydrolysis of phosphodiesters.

022. **SYNTHESIS AND APPLICATIONS OF ADVANCED COENZYME A ANALOG SYNTHONS.** Kurt W. Vogel, Richard T. Bibart, and Dale G. Drueckhammer*. Department of Chemistry, Stanford University, Stanford CA 94305.

Analogs of coenzyme A (CoA, **1**) and its acyl derivatives have proven useful in the study of enzyme structure and mechanism. Our lab recently reported the synthesis of CoA synthon **2** via a combination of enzymatic and non-enzymatic reactions from which CoA analogs modified in the cysteamine region of the molecule could be readily prepared. Here we report the synthesis of CoA synthons **3** and **4**. Synthon **3** allows for modifications in the ß-alanine region of the molecule, while synthon **4** allows for the synthesis of other CoA analogs not previously available. Applications of CoA analogs available via this new methodology will be discussed.

1. X=NH, Y=NH, Z=SH
2. X=NH, Y=S, Z=CH$_3$
3. X=S, Y=NH, Z=CH$_3$
4. X=NH, Y=NH, Z=CH$_2$SC(O)CH$_3$

023. CONVERSION OF HYDROXYNITRILES TO LACTONES BY MICROBIAL HYDROLYSIS
Stephen K. Taylor, Lloyd J. Simons, James R. Vyvyan, and Noelle K. Wood
Department of Chemistry, Hope College, Holland, MI 49422-9000

The 4- and 5-hydroxynitriles **1** are readily available and in principle could be used in a general synthesis of lactones **2** (Equation 1). Unfortunately, the chemical hydrolysis of nitriles requires extended times at high temperature under strongly acidic or basic conditions. These harsh conditions have forestalled the development of this approach. We have developed the use of *Rhodococcus rhodochrous* to hydrolyze hydroxynitriles **1** to lactones **2** under extremely mild conditions--pH 6-7 at 30 °C--in a few minutes to a few hours. We have demonstrated this approach in the synthesis of several lactones, including the *Trogoderma* beetle pheromone in moderate enantiomeric excess. Our latest results and the development of this microbial hydrolysis will be presented.

024. **APPLICATION OF MICROBIAL DIHYDROXYLATION TO THE PREPARATION OF ENANTIOMERICALLY PURE ISOQUINOLINE DERIVATIVES.** T. Hudlicky, **M. A. A. Endoma**, and G. Butora, Department of Chemistry, University of Florida, Gainesville, FL 32611

A practical route was developed to prepare the highly functionalized derivatives **5a** and **5b** starting from homochiral diol **2**, available from microbial oxidation of phenethyl bromide. The key step in the synthesis was intramolecular addition of a trisubstituted olefin to a cyclic *N*- acyliminium ion formed *in situ* by acid treatment of hydroxycarbamates **4a** and **4b**.

R = H, alkylcarbonyl
5a, X = CH$_2$
5b, X = O

4a, X = CH$_2$
4b, X = O

025. HORSERADISH PEROXIDASE-CATALYSED OXIDATIVE COUPLING OF PHENOLIC PHENYLPROPENOIDS TO GIVE ENANTIOMERIC β-5 PHENYLCOUMARANS, Ezio Bolzacchini, Simone Meinardi, Marco Orlandi, Bruno Rindone, Gosta Brunow*, Harry Setala*, Jussi Sipila*, Dipartimento di Scienze Dell'Ambiente e del Territorio, Universita' di Milano, Via Emanueli 15, 20126 Milano, Italia, *Laboratory of Organic Chemistry, P.O. Box. 55, Fin-00014, University of Helsinki, Finland. Phenylcoumarans are an important group of natural lignans. The oxidative coupling of phenols catalyzed by peroxidases is a very attractive method for the preparation of phenylcoumarans from phenolic cinnammate derivatives. The enzymatic method takes advantage from mild reaction conditions and fast reaction rates. Two stereogenic carbons are formed in this reaction and diastereoselection to give the trans stereochemistry in usually obtained. Enantioselection is obtained in this enzyme-catalyzed reaction by reacting phenylpropenoids bound to a chiral auxiliary to the horseradish peroxidase (HRP)-catalyzed oxidative dimerization with hydrogen peroxide.

026. PRACTICAL, STEREOSELECTIVE CHEMO-ENZYMATIC SYNTHESIS OF BILA2157BS, A POTENT RENIN INHIBITOR: **P. L. Beaulieu***, M. Bailey, C. Beaulieu, C. Boucher, J.S. Duceppe, J. Gillard, P. Lavallée, B. Simoneau, B. Thavonekham, S. Valois, D. Wernic. Bio-Méga/Boehringer Ingelheim Research Inc., 2100 Cunard Street, Laval (Québec), Canada, H7S 2G5. BILA 2157 BS (**1**) is a potent and orally active (in animal models) renin inhibitor. In support of preclinical studies, we have developed an efficient stereoselective synthesis of this compound. The key step is an enzymatic resolution providing succinic acid derivative **2** in >98% ee.

027. CHEMOENZYMATIC SYNTHESIS OF THE MORPHINE SKELETON: STEPWISE VERSUS TANDEM RADICAL CYCLIZATION.
Gabor Butora, Tomas Hudlicky, Andrew G. Gum, Stephen P. Fearnley, Michele R. Stabile and Khalil A. Abboud
Department of Chemistry, University of Florida, Gainesville, FL 32611

A short synthesis of the morphinan skeleton has been accomplished. The key steps involved enzymatic dihydroxylation of 2-(2-bromoethyl)bromobenzene, a stepwise vinyl and aryl radical cyclizations and an acid-catalyzed closure of an aldehyde to form the C_{10}-C_{11} bond.

028. ADVANCED INTRAMOLECULAR DIELS-ALDER STUDY TOWARD THE SYNTHESIS OF (-)-MORPHINE
Gabor Butora, **Andrew G. Gum**, Tomas Hudlicky and Khalil A. Abboud
Department of Chemistry, University of Florida, Gainesville, FL 32611

A tricyclic ring system containing all 5 chiral centers of the natural (-)-morphine skeleton has been synthesized in 11 steps. The key steps in the synthesis were an enzymatic oxidation of (2-bromoethyl)benzene to the corresponding *cis*-dienediol and a thermal [4+2] intramolecular Diels-Alder cycloaddition. This result demonstrated that the stereochemistry at the C_9 and C_{14} of (-)-morphine could be properly set in one chemical step.

029.

ENZYMES WITH TRANSITION METALS- POWERFUL METHODOLOGY FOR ASYMMETRIC SYNTHESIS. **Judith A. Howarth** and Jonathan M.J. Williams, Department of Chemistry, University of Bath, Claverton Down, Bath BA2 7AY

Both enzymes and transition metals have been used for the preparation of enantiomerically enriched products. It was anticipated than an enzyme could be used to enantioselectively hydrolyse one enantiomer of an allyl acetate. At the same time, a transition metal would be employed to racemise enantiomerically enriched acetate, but not the product. Thus the enantiomer favoured by the enzyme would be replenished, and so both enantiomers of allylic acetate could be converted into one enantiomer of allylic alcohol.

030.

ENZYME MEDIATED, PALLADIUM CATALYSED NUCLEOPHILIC SUBSTITUTION OF ALLYLIC ALCOHOLS. **Elizabeth M. Davies** and Jonathan M.J. Williams, Department of Chemistry, University of Bath, Claverton Down, Bath, BA2 7AY, UK.

Palladium catalysed nucleophilic substitution of allylic acetates has been well studies over the past 30 years. A novel method has been developed which allowed the less volatile allylic alcohols to be used instead of the acetates. The method utilised an enzyme to carry out the acylation in situ alongside the palladium catalysed substitution.

031. DEVELOPMENT OF MICRO-METALLO ENZYMES FOR USE AS ASYMMETRIC EPOXIDATION CATALYSTS. B. D. Dangel and R. L. Polt*, Department of Chemistry, University of Arizona, Tucson, Arizona 85721

Chiral C_2-symmetric ligands have been synthesized from readily available amino acids. Coupling of L-Phe to phenylene diamine with subsequent formation of the Schiff base has led to a chiral ligand with C_2-symmetry based on X-ray data. Complexation of the ligand to nickel (II) bromide resulted in a model compound for use as an epoxidation catalyst. This complex possesses a screw axis centered on the metal (tetrahedral distortion due to "stacked" $Ph_2C=N$ groups).

032. NEW DIRECTIONS FOR ORGANIC SYNTHESIS. Paul A. Bartlett, Department of Chemistry, University of California, Berkeley, CA 94720-1460

Developments in organic synthesis have long been stimulated by the desire to prepare compounds of biological activity. Whereas compounds isolated from natural sources provided the first targets, bioorganic concepts and structure-based approaches that have emerged in recent years have led to the design of novel structures and ring systems that are decidedly "unnatural". Now, the explosion of interest in combinatorial libraries provides another opportunity to develop new strategies and methods for organic synthesis. A number of projects from our own work that illustrate these points will be described, including 1) the structure-based design of macrocyclic inhibitors of aspartic and zinc peptidases and qualitative and quantitative assessments of the efficacy of this strategy, and 2) the development of novel ring systems and heterocyclization processes appropriate for combinatorial library design.

033. SYNTHESIS AND CHEMISTRY OF DISPIROACETALS AND OTHER 1,2- DIACETALS. Steven V. Ley, Department of Chemistry, University of Cambridge, Lensfield Road, Cambridge CB2 1EW, UK

The spiroacetal motif occurs in a very wide range of biologically active molecules from insect pheromones to ionophores and antiparasitic agents. This unit imparts chirality, provides a rigid scaffolding for functional groups and contributes to the overall shape of the systems. New vicinal diol protecting groups have been developed using dispiroacetals. These methods may also be used to effect desymmetrization of meso-polyols for resolution and in asymmetric synthesis. Recently we have discovered the use of 1,2-diacetals as alternative agents for selective ketalisation of carbohydrates. These new studies open the way to the synthesis of large glycosidic arrays and multi antennary oligosaccharide structures. Application of these methods to complex molecule synthesis will be presented.

034. SYNTHETIC STUDIES ON ANNONACEOUS ACETOGENINS. James A. Marshall, Department of Chemistry, University of Virginia, Charlottesville, VA 22901.

The C_{35}-C_{38} polyketide natural products from the Annonacea family of tropical and subtropical trees and shrubs exhibit a wide range of biological activities. A bidirectional approach to certain bis-tetrahydrofuran representatives of this family will be described.

035. THE IMPACT OF GENOMICS ON MEDICINAL CHEMISTRY - FROM GENES TO COMBINATORIAL LIBRARIES, Brian W. Metcalf, Chemical & Cellular Sciences, SmithKline Beecham Pharmaceuticals, King of Prussia, PA 19406

One of SB's most successful recent efforts to drug design, that of the discovery of potent endothelin antagonists will be discussed. The approaches used will then be put into context in the changing environment which demands simulataneous attacks on multiple related targets arising from the superfamily of G protein-coupled receptors which includes the endothelin receptors. Two examples of the exploitation of a human gene data base will be discussed. The discovery of Cathespin K, an osteoclast specific cysteine protease, by random sequencing of an osteoclast cDNA library will be described. Provision of the recombinant protease then sets the scene for structure and mechanism based approaches to protease inhibition. On the other hand, the identification of over seventy novel orphan G protein-coupled receptors demands an alternative paradigm in drug discovery. Approaches to an opportunity of this magnitude will be presented.

ORGN

036. METAL CATALYZED ENE-TYPE REACTIONS IN THE SERVICE OF NATURAL PRODUCTS' CHEMISTRY. Barry M. Trost, Department of Chemistry, Stanford University, Stanford, CA 94305-5080

The all-carbon Alder ene reaction has had limited applications in the synthesis of complex molecules, in part, because of limitations posed by the required conditions. A catalyzed version may overcome some of these limitations. Furthermore, catalysis opens up prospects for molecular modifications not possible in the thermal process. A palladium catalyzed intramolecular version is explored in the context of the synthesis of vitamin D analogues. A ruthenium catalyzed intermolecular version is explored in the context of the synthesis of acetogenins.

037. WILLIAM S. JOHNSON: GENTLEMAN CHEMIST. Carl Djerassi, Department of Chemistry, Stanford University, Stanford, CA 94305-5080

Since all other contributors to this symposium will honor William S. Johnson, the chemist, through references to his chemical contributions and to those of his former students and colleagues, it seemed appropriate to also honor his human side. A brief distillate of personal reminiscences spanning an acquaintenceship of five decades will be presented as homage to William Johnson, the gentle man.

038. PALLADIUM - CATALYZED CYCLIZATION AND ANNULATION
Richard C. Larock, Department of Chemistry
Iowa State University, Ames, Iowa

Two major applications of palladium in organic synthesis will be discussed: (1) palladium(II)-catalyzed cyclization and oxidation, and (2) palladium(0)-catalyzed annulation of alkenes, dienes and alkynes. This chemistry provides novel new routes to a wide variety of carbocycles and heterocycles of potential pharmaceutical interest.

039. Pd-CATALYZED NEW BENZANNULATION OF CONJUGATED ENYNES
Y. Yamamoto, Department of Chemistry, Graduate School of Science, Tohoku University. Sendai 980-77, Japan

Conjugated enynes cyclodimerized in the presence of Pd(0) catalysts to give the corresponding 1,4-disubstituted aromatic compounds in good to high yields. A substituent at C-1 and /or C-4 position of enynes did not hamper aromatization. Enynes substituted at C-2 position underwent cyclization very smoothly and functional groups such as hydroxy and carbonyl could be introduced at the alkyl chain attached to C-2 position. Bis-enynes cyclized intramolecularly to give the corresponding para-cyclophanes in good to high yields. Cross-coupling between conjugated enynes and certain acetylenes proceeded smoothly to give the substituted aromatic compounds in good yields.

040.

PALLADIUM CATALYSED POLYCOMPONENT MOLECULAR QUEUING CASCADES. R. Grigg, School of Chemistry, Leeds University, Leeds, LS2 9JT, U.K.

Based on relative rate data acquired from our extensive studies on palladium catalysed cyclisation-anion capture processes we will discuss the development of a series of polycomponent cascades involving mono- or poly-cyclisation with incorporation of a wide variety of additional substrates including carbon monoxide, allene, amines etc. in a well ordered "molecular queue". By appropriate design these cascades can be initiated by a cyclisation step or cyclisation can be deferred to allow incorporation of, for example, carbon monoxide. Applications of these queuing cascades to alkaloid synthesis will be described.

041. The Preparation of Chiral Intermediates Using the Fundamentals of Organopalladium Chemistry. William M. Clark, SmithKline Beecham Pharmaceuticals, Synthetic Chemistry Department, P.O. Box 1539, King of Prussia, PA 19406

The development of practical inexpensive methods for the synthesis of chiral intermediates is still a challenge within the pharmaceutical industry. This presentation will describe some of the efforts at SmithKline Beecham to utilize palladium chemistry to synthesize compounds on large scale in a catalytic enantioselective manner.

042. RECENT DEVELOPMENTS IN THE USE OF INTRAMOLECULAR HECK REACTIONS IN NATURAL PRODUCTS TOTAL SYNTHESIS. Larry E. Overman, Department of Chemistry, 516 Physical Sciences 1, University of California, Irvine, CA 92697-2025

During the past decade the intramolecular Heck reaction has emerged as a uniquely powerful tool for constructing complex carbocycles and heterocycles. This lecture will describe recent total synthesis investigations in our laboratories in which intramolecular Heck reactions play a central strategic role.

043. *SYN-ANTI* DICHOTOMY IN THE Pd(0)-CATALYZED ALLYLIC SUBSTITUTION. P. Kočovský, Department of Chemistry, University of Leicester, Leicester LE1 7RH, U.K.

The Pd(0)-catalyzed allylic substitution is known to occur via η^3-complexes, which arise from the allylic substrates in an *anti*-fashion (**1** → **3**). The subsequent reaction with stabilized C-nucleophiles (e.g., malonate) again proceeds with an *anti*-mechanism, giving a product of an overall retention of configuration. We have now found that the steric course of the first step can be completely reversed by a suitably positioned coordinating group (**2** → **4**). This stereochemistry control appears to be more general than that relying on the coordinating leaving group ($Ph_2PCH_2CO_2^-$ or Cl^-) reported previously (*JACS* **1989**, *111*, 4981; *JACS* **1990**, *112*, 2813).

044. CHEMISTRY AND DEVICE TECHNOLOGY IN POLYMERIC LED'S. H. Bässler, A. Greiner, **W. Heitz**, J. H. Wendorff, Philipps Universitaet Marburg, Fb Physikalische Chemie, Polymere, Marburg, Germany, D-35032

Poly-p-phenylene (PPP), poly(phenylenevinylene) (PPV), and poly(phenylene-ethinylene) (PPE) show photoemission similar to inorganic materials if a voltage is applied. By structural variations processable polymers of this type are obtained. Blend with polymers like polycarbonate show a relative quantum efficiency two orders of magnitude higher than the pure polymer. This is a guide to construct systems where the active material is incorporated chemically or physically. Soluble PPP, PPV, and PPE are mostly prepared by metal catalyzed reactions. Side reactions cause defect structures in the polymer chain. These defects are critically examined and the consequences for the optical properties demonstrated. Polymers are usually immiscible. This can be used to create white color emission. Using conventional orientation technology of liquid crystal displays oriented LED's can be fabricated on commercial polyimide which behaves as hole carrier.

045. NEW ORGANIC LIGHT EMITTING MATERIALS. Yves Geerts, Uwe Keller, Ullrich Scherf, Monika Schneider and **Klaus Müllen**, Max-Planck-Institut für Polymerforschung, Ackermannweg 10, D-55128 Mainz, Germany

There is still considerable progress in the synthesis and processing of organic materials (emitters, hole and electron transporting layers) for light emitting diodes LED's. We report on novel polymeric, blue emitters, also with an improved support to the electron injection process, based on ladder-type (poly-para-phenylene)s LPPP's and related copolymeric step-ladder structures. LED's consisting of highly photostable organic dyes (terrylene biscarboximides, quinacridones) as emissive materials represent another attractive goal. Some of these chromophors form stable columnar liquid crystalline mesophases (dyes with the perylene sceleton) or hydrogen bonded supramolecular aggregates (quinacridone based chromophors) and, therefore, open possibilities to arrange the molecules in highly ordered emissive layers. A very promising novel approach is the fabrication of emitting materials based on inorganic/organic hybride systems, composed of organic chromophors (emitters) covalently bonded onto a silicon-based sol/gel matrix.

046. Red Fluorescent Dopants For Organic Electroluminescent Devices. **Chin H. Chen**, Jianmin Shi and C.W. Tang, Imaging Research and Advanced Development, Eastman Kodak Company Rochester, NY 14650-2110

In organic electroluminescent (EL) devices, dopants play an important role in the improvements of device efficiency, operational stability and color fidelity. The function of the dopant is to effect efficient EL excitation energy transfer from the host matrix, resulting in emission from the dopant molecules. This presentation will focus our recent studies on the design and synthesis of novel red dopants based on the DCM [4-(dicyanomethylene)-2-methyl-6-styryl-4H-pyran] class of molecules and their luminescence properties as dopants in the Alq host. The incorporation of steric groups in DCM such as tetramethyljulolidine has produced highly red fluorescent dyes. The steric bulkiness has been found to be particularly effective in suppressing the onset of concentration quenching due to intermolecular interactions between adjacent dopants. Material issues such as chemical and thermal stabilities and their dependence on the molecular structures and other device characteristics such as chromaticity shift, carrier trapping, EL operational stability will be discussed.

ORGN

047. OLIGO(DIAZAPHENYLS) AND POLY(DIAZAPHENYLENES)-NOVEL MATERIALS FOR LEDS AND FLUORESCENCE LIQUID LIGHTGUIDES. **R. Gompper**, J. Einsiedler, O. Freundel, , C. Harfmann, H.-J. Mair, and G. Nath. Institut fuer Organische Chemie, University of Munich, D-80333 Muenchen, Germany

A methodology has been developed which repeatedly uses the condensation of vinamidinium salts with acetamidine to form 2-methylpyrimidines that in turn can be con=1Fverted in new vinamidinium salts. By reaction of various vinamidinium salts with amidines oligo(diazaphenyls) can be prepared in high yields. They are thermally stable, can be reversibly reduced at low potentials and dissolved in TFA. They show (mainly blue) fluorescence (with high quantum yields in the solid state) and electroluminescence and can be used in fluorescence liquid lightguides. Oligo(diazaphenyls) containing pyridine, terpyridine, anthracene, pyrene, thiophene, phenothiazine, dihydrophenazine, flavine, spirobifluorene and porphyrine rings can be obtained as well.

048. NEW LUMINESCENT POLYMERS FOR LEDs AND LIGHT EMITTING ELECTRO-CHEMICAL CELL (LECS). **A.B. Holmes**, B.-S. Chuah, D.-H. Hwang, X.-C. Li, S.C. Moratti, S.T. Kim, J.C. DeMello and R.H. Friend, Departments of Chemistry & Physics, University of Cambridge, Lensfield Road, Cambridge CB2 1EW, & Madingley Road, Cambridge CB3 0HE, U.K.

The dehydrohalogenation condensation polymerization has been used to prepare polymers **1** and **2**. Polymer **1** exhibits a remarkably high photoluminescence quantum efficiency and can be used as a solution processable green emissive layer in single layer electroluminescent (EL) devices. The copolymers **2** emit at longer wavelengths as the proportion of y increases. Accompanied with this is a reduced efficiency and turn-on voltage in EL devices. A LEC was prepared using polymer **1** blended with polyethyleneoxide-lithium triflate.

049. A GENERAL APPROACH TO PRECURSORS FOR POLY(ARYLENE VINYLENE) DERIVATIVES: SYNTHESIS AND BEHAVIOUR IN ORGANIC LED'S. **D.J. Vanderzande D.**, A.C. Issaris, M.J. Van Der Borght, A.J. Van Breemen, M. M. de Kok, and J. M. Gelan, Research group of Organic and Polymeric Chemistry, Institute of Material Research (IMO), Limburg University, Universitaire Campus, B-3590 Diepenbeek, Belgium.

A new precursor route for poly(arylene vinylene) derivatives will be presented. In this way non-ionic precursor polymers are obtained which show an enhanced thermal stability and are soluble in organic solvents. This enables a thorough structural characterisation and the study of the mechanism. Experiments have been performed to differentiate between a radical or an ionic polymerisation mechanism. Furthermore the scope and limitations for this new route was explored. The first results indicate that the versatility exceeds that of the Wessling route, e.g. the synthesis of poly(p-biphenylene vinylene) and poly(2,6 naphthylene vinylene) could be accomplished in our case, where it failed in the case of the Wessling route.

050. AROMATIC POLYMERS WITH 1,3,5-TRIAZINE UNITS AS HOLE BLOCKING/-ELECTRON TRANSPORT MATERIALS IN LEDS, R. Fink, C. Frenz, M. Thelakkat, **H.-W. Schmidt**, Makromolekulare Chemie I, Bayreuther Institut für Makromolekulforschung (BIMF), 95440 Bayreuth, Germany

In multilayer LED devices the chemical and physical properties of a hole transport layer (HTL), an emission layer (EML), and an electron transport layer (ETL) have to be optimized and adjusted. In this paper we report the synthesis of polymeric 1,3,5-triazines as hole blocking/electron transport layer in multilayer LEDs. Various aromatic polyethers containing the triazine units have been prepared. Due to the high electron affinity and structural symmetry of the triazine unit, electron injection and transport should be favored. The lower reduction potential and the higher oxidation potential results in a decreased barrier for electron injection and an increased barrier for holes. This was confirmed in a two-layer device using a triazine polymer as ETL and PPV as HTL and EML.

051. Design and Synthesis of Pyrrolinone Based Non-Peptidal Peptidomimetics: From β-Sheet Mimetics to an Orally Active HIV-1 Protease Inhibitor.
Amos B. Smith III,[1*] David A. Favor,[1] Mark C. Guzman,[1] Alexander Pasternak,[1] Andrew B. Benowitz,[1] Joel R. Huff,[2] Lawerance Kuo,[2] Paul L. Darke,[2] Zhongguo Chen,[2] Emilio A. Emini,[2] Paul A. Sprengeler,[1] and Ralph Hirschmann[1*]

[1]*Department of Chemistry, University of Pennsylvania, Philadelphia, PA 19104, and* [2]*Departments of Molecular Biology and Virus and Cell Biology, Merck Research Laboratories, West Point, PA 19486*

This paper will describe collaborative research directed at the design and synthesis of 3,5-linked polypyrrolinones as mimics of β-sheets. We have been able to demonstrate that these novel polymers adopt an extended sheet structure both in the solid state and in solution. The pyrrolinone scaffold was then incorporated into known peptidal based inhibitors of aspartic acid proteases (renin and HIV-1 protease). The culmination of these efforts was the orally active inhibitor **1**. The X-ray structure of **1** co-crystallized with the HIV-1 protease will also be discussed.

052. NEW APPROACHES TO THE ASYMMETRIC SYNTHESIS OF DIPEPTIDE ISOSTERES FOR HIV-1 PROTEASE INHIBITORS
Iwao Ojima*, Hong Wang, Tao Wang, and Edward Ng, Department of Chemistry, State University of New York at Stony Brook, Stony Brook, NY 11794-3400.

A new asymmetric synthetic route based on the β-lactam synthon method has been developed to obtain a variety of non-protein amino acid dipeptide isosteres for potential HIV-1 protease inhibitors. Various enantiomerically pure β-lactams (**1**) were prepared through highly stereoselective cyclocondensation of ester enolates with imines. Ring opening of these β-lactams led to two key intermediates **2** and **3**, which were then converted, in a few steps, to dihydroxyethylene, hydroxyethylene and other dipeptide isosteres.

053. DIPEPTIDE SURROGATES CONTAINING ASPARAGINE-DERIVED TETRAHYDRO-PYRIMIDINONES. PREPARATION, STRUCTURE, AND USE IN SOLID PHASE SYNTHESIS.
Joseph .P. Konopelski and **Lubov K. Filonova**, Department of Chemistry, University of California, Santa Cruz, CA 95064

Novel dipeptide surrogates Fmoc-Xxx-(cyclo)Asn-OBu^t (**1**) are prepared from asparagine, aromatic aldehydes, and Fmoc-protected amino acid chlorides. The tetrahydropyrimidinone structures exist in a single preferred ring conformation, as observed in a single crystal X-ray analysis of one of the derived carboxylic acids. As with other molecules with similar substitution patterns, the six-membered ring is held in a boat conformation. These dipeptide building blocks can be deprotected at the C-terminal and/or N-terminal positions by standard protocols. In addition, they function as ready participants for solid phase peptide synthesis. Finally, the constrained asparagine residue is transformed to the native amino acid within a polypeptide framework, allowing for the direct interchange of a restricted system of known conformation and configuration to the native state under mild conditions.

054.

DESIGN AND SYNTHESIS OF DIPEPTIDE SECONDARY STRUCTURE MIMETICS.
Masakatsu Eguchi, Hiroshi Nakanishi, Min S. Lee, and Michael Kahn, Molecumetics Ltd., 2023 120th Ave. N.E., Suite 400, Bellevue, WA 98005 and Department of Pathobiology, University of Washington, Seattle, WA 98195

Folded structures in biologically active proteins or peptides often play an important role in their interactions with biological systems. Mimetics of such local structure serve as a good probe for drug design and discovery efforts. We have synthesized 6,5- and 5,5-bicyclic dipeptides ((3S)-3-(t-butyloxycarbonyl)amino-1,5-diaza[3.3.0]octan-2-one-8-carboxylic acid and (3S)-3-(t-butyloxycarbonyl)amino-1,6-diaza[4.3.0]nonan-2-one-9-carboxylic acid) from commercially available Boc-Ser-OMe and Boc-Asp-OBn in good yield using 1,3-dipolar cycloaddition as the key step. Conformational analysis predicted that the torsional angles of these compounds are similar to those of a type II' β-turn conformation. The Fmoc derivatives of these templates were successfully applied to solid phase peptide synthesis.

055. DESIGN, SYNTHESIS, AND CONFORMATIONAL ANALYSIS OF A PROPOSED β-TURN MIMIC. Brian E. Fink, Phil R. Kym and John A. Katzenellenbogen, Department of Chemistry, University of Illinois, Urbana, Illinois 61801

In connection with our interest in the receptor-bound conformation of biologically active peptides, we have designed an external β-turn mimic based on the 2-aza-cyclodec-6-enone ring system **1**. Molecular modeling suggests this system should adopt the ϕ and φ torsion angles of a type I β-turn. We have synthesized this ten-member lactam through the ring closing olefin metathesis of dipeptide **2** derived from 2-aminohex-5-enoic acid **3**. We have prepared a number of peptide-like derivatives and their evaluation as potential β-turn mimics through X-ray analysis, NMR solution studies and molecular modeling will be presented.

056. SYNTHESIS OF A NEW TYROSINE ANALOGUE HAVING X_1 AND X_2 ANGLES CONSTRAINED TO VALUES OBSERVED FOR AN SH2 DOMAIN-BOUND PHOSPHOTYROSYL RESIDUE.
Terrence R. Burke, Jr. and Bin Ye. *Laboratory of Medicinal Chemistry, Bldg. 37, Rm 5C06, Division of Basic Sciences, NCI, National Institutes of Health, Bethesda, MD 20892*

Synthesis is reported of a new tricyclic amino acid, (±)-(rel-2S,3R)-2-carboxy 1,2,3,4,5,6-hexahydro-8-hydroxy-1,5-methano-3-methyl-3-benzazocine (**3**), which contains within its structure the elements of a tyrosine moiety having χ_1 and χ_2 angles (constrained to values observed for a phosphotyrosyl (pTyr) residue **1** bound to the p56[lck] SH2 domain . Additionally, the ϕ angle of N acylated **2** correlates well with the ϕ angle of the SH2 domain-bound pTyr residue. Compound **3** represents a unique, highly constrained amino acid which may be of value in signal transduction studies.

057. STEREOSELECTIVE SYNTHESIS OF A CONFORMATIONALLY RESTRICTED β-HYDROXY-γ-AMINO ACID. D. Xiao, M. M. Joullié, Department of Chemistry, University of Pennsylvania, Philadelphia, PA 19104

β-Hydroxy-γ-amino acids are a class of non-proteinogenic amino acids having potential use in drug discovery. In conjunction with our research on didemnins, we designed a generic cyclic β-hydroxy-γ-amino acid in which the requisite carboxyl, hydroxyl and amino groups are positioned on a conformationally stable 6-membered ring. We now present a stereoselective synthesis of the (S,S,S) cyclohexane β-hydroxy-γ-amino acid which will be used to generate a new didemnin B analog.

058.

PEPTIDE BOND FORMATION AT THE AIR-WATER INTERFACE
Jaya Singh and John S. Oliver
Department of Chemistry, Brown University, Providence, RI 02912

We have demonstrated amide bond formation at the air-water interface. Proximity and orientation between reactive groups provide the driving force for the reaction. Two amphiphiles that incorporate a C_{16} hydrophobic tail and a glycine derivative as the reactive hydrophilic head group were designed. The nucleophilic amphiphile (**1**) was constituted by the hexadecyl amide of glycine and the hexadecyl thioester of N-acetyl glycine served as the electrophilic amphiphile (**2**). Upon compression of the monolayer on a Langmuir-Blodgett (LB) trough, nucleophilic attack of the amino group at the carbonyl of the thioester resulted in the formation of the dipeptide (**3**). This product was isolated and characterized by TLC, ^{1}H NMR and HRMS. To study polypeptide formation, the hexadecyl thioester of glycine was synthesized. Compression on a LB trough resulted in the formation of several [glycine]$_n$ -polypeptides (n = 2-15). The products of the polymerization reaction were characterized by amino acid analysis and electrospray-MS. Implications for prebiotic synthesis of peptides will be highlighted and discussed.

059. NEOGLYCOPHOSPHOLIPIDS WITH ALKYL SPACERS: THEIR SYNTHESIS VIA AN IMPROVED REDUCTIVE AMINATION AND MONOLAYER PROPERTIES
Lijun Sun and Elliot L. Chaikof* Laboratory for Biomolecular Materials Research and Department of Surgery, Emory University School of Medicine, Atlanta, GA 30322

Glycosidation of activated saccharides (**1**) with alkyldiols followed by oxidation with PDC afford aldehydes **2** in overall yields of 50-70%. An improved reductive amination was used to conjugate **2** with distearoyl phosphoethanolamine (DSPE), which has significant improvement in both reaction yields (80-95%) and reaction efficiency (1.3-1.5 eq. of DSPE; 0.5-1.0 h). Amphiphiles **3**, obtained in 62-82% yields by hydrogenation of the reductive amination products, revealed interesting monolayer behavior at an air-water interface.

a: HO(CH$_2$)nOH (n=3, 5, 7, 10, 16); b: PDC; c: DSPE/NaBH$_3$CN; d: H$_2$, 10% Pd-C

060. FACLITATED TRANSPORT OF ZWITTERIONIC PHENYLALANINE THROUGH PLASTICIZED CELLULOSE TRIACETATE MEMBRANES
Tracey A. Munro and Bradley D. Smith* Department of Chemistry and Biochemistry, University of Notre Dame, Notre Dame, Indiana 46556, USA.

The facilitated transport of amino acids through lipophilic membranes has application in various analytical and purification processes. We have discovered that plasticized cellulose triacetate membranes containing quaternary ammonium salts are selectively permable to phenylalanine at neutral pH. The membranes are quite stable and maintain high fluxes (> 10^{-5} mol/m^{2}s) over many weeks. Mechanistic studies suggest a fixed-site jumping transport mechanism where the ammonium salt acts as fixed "stepping stones" and the zwitterionic phenylalanine permeates through the membrane by jumping from one fixed-site to another.

061. TITANIUM ISOPROPOXIDE CATALYZED REDUCTIVE AMINATION REACTIONS ON SOLID SUPPORT. John C. DiCesare, Celesta E. McGee, Holly B. Black and Wendy E. Rasmussen, Department of Chemistry, The University of Tulsa, Tulsa, Oklahoma 74104,

As part of a program directed toward the synthesis of ligands for a combinatorial library, we were interested in preparing amines derived from the reductive amination of hindered ketones and/or weakly basic amines. A key step in the planned synthetic pathway is the reductive amination of an aryl amine and a piperidone derivative. Mattson and co-workers (*J. Org. Chem.*, **1990**, *55*, 2552) reported that the amines could be accessed by using a modified Borch reduction using titanium (IV) isopropoxide as a catalyst. A thorough understanding of the reaction mechanism is beneficial in determining the optimum reaction conditions needed for solid phase combinatorial synthesis. Results from the mechanistic study were utilized to adapt the modified reductive amination procedure to the solid phase synthesis used in combinatorial chemistry.

R = H or Polystyrene resin

062. SYNTHESIS OF TERTIARY AMINE COMBINATORIAL LIBRARIES FROM NOVEL PIPERAZINYL SUBSTITUTED POLYAZAPYRIDINOPHANE SCAFFOLDS. B. D. Haly, H. An, R. H. Springer and P. D. Cook, Isis Pharmaceuticals, 2292 Faraday Ave., Carlsbad, CA 92008

Two novel asymmetric mono-*t*-Boc-protected piperazinyl polyazapyridinophanes were synthesized as scaffolds for solution phase combinatorial chemistry drug discovery. The key intermediate *t*-Boc-piperazinyl pyridine-2,6-ditosylate was cyclized with tris(2-nitrobenzenesulfonyl)triamines to give the corresponding orthogonally protected piperazinyl polyazapyridinophanes in high yields. These polyazapyridinophane derivatives were selectively deprotected by thiophenol to afford the desired mono-*t*-Boc-protected piperazinyl polyazaphane scaffolds. Twelve libraries (complexity: 12000) were generated from these novel scaffolds and 10 benzylic bromides by using our solution phase simultaneous addition of functionalities (SPSAF) combinatorial strategy.

063. SOLID PHASE SYNTHESIS OF PEPTIDYL TRIFLUOROMETHYL KETONES. Marc-André Poupart, Gulrez Fazal, Sylvie Goulet and Ly Thy Mar; Bio-Méga/Boehringer Ingelheim Research Inc., 2100 Cunard Street, Laval, Québec, Canada, H7S 2G5.

A solid phase process for the synthesis of peptidyl trifluoromethyl ketones **1** was developed. The methodology is suitable for the preparation of different P_1 side chain analogues and can therefore be adapted to the synthesis of inhibitors of various serine proteases. Details of the process as well as the scope and limitations of the methodology will be presented.

064. AN ESTER ENOLATE CLAISEN/RING CLOSING METATHESIS ROUTE TO FUNCTIONALIZED CARBOCYCLES AND HETEROCYCLES. John F. Miller, Andreas Termin, Kevin Koch, and <u>Anthony D. Piscopio</u>*, *Department of Chemistry, Amgen Inc., Boulder, CO 80301.*

Functionalized carbocycles and heterocycles have been prepared using variations of the ester enolate Claisen rearrangement and ring closing metathesis (RCM) protocol as consecutive key steps. The manifold has been demonstrated on solid phase wherein the R_1 substituent serves as the point of resin attachment. Following [3,3] sigmatropic rearrangement and derivatization of the resulting α,ω dienes, cyclative cleavage *via* RCM affords the desired solubilized products. In addition to solid phase heterocyclic synthesis, the application of this methodology to the solution phase synthesis of fused bicyclic and spirocyclic molecules will be presented.

X = O, S, CH2; Y = allyl group
X = N; Y = (H)*t*-Boc
R1 = H, alkyl or ●〜〜

X = O, SO2, CH2, N-*t*-Boc
Z = OH, OR', NR"2
R1 = H, alkyl or ●〜〜

X = O, SO2, CH2, N-*t*-Boc
Z = OH, OR', NR"2

065. SOLUTION/SOLID PHASE EXTRACTION METHODS FOR REMOVING UNREACTED ELECTROPHILES FROM SOLUTION PHASE COMBINATORIAL REACTION MIXTURES
Andrew M. Kawasaki[§], Michael E. Jung[¥], P. Dan Cook[§]
[§]ISIS Pharmaceuticals, 2292 Faraday Ave, Carlsbad, CA 92008. [¥]Dept. Chemistry and Biochemistry, UCLA, Los Angeles, CA 90024.

Solution phase organic reactions often require effective methods for removal of excess reactants. In this work we present an efficient method for the removal excess electrophiles by use of a combination of solution and solid phase extraction procedures. Hence, excess electrophiles were quenched with 2-mercaptoethanesulfonic acid, sodium salt. On workup the bulk of the quenched reagents partitioned into the aqueous layer as thioether sulfonates. On occasion, the quenched reagent remained lipophilic enough to partition into the organic layer, in which case, a minimal amount of anion exchange resin is used to remove the contaminant. Typically the products after workup are of sufficient purity for biological assays sans chromatography. Hence, an efficient, economical method has been developed for the removal of excess electrophiles from a reaction mixture. The procedure is especially well-suited for solution-phase combinatorial chemistry.

066. ARYL AND ALKYL SULFONAMIDE "SAFETY-CATCH" LINKERS FOR SOLID-PHASE SYNTHESIS. <u>Bradley J. Backes</u>, John Phillips, and Jonathan A. Ellman, Department of Chemistry, University of California, Berkeley, CA 94702

The stability of the acylsulfonamide linkage towards strongly acidic and strongly basic reaction conditions makes it a superior carboxylic acid linker for solid-phase organic synthesis. We have used sulfonamide linker **1** for the synthesis of substituted aryl acetic acid derivatives in which enolate alkylation and Suzuki cross-coupling reactions were demonstrated. A new activation method to prepare a highly reactive acylsulfonamide for nucleophilic cleavage has been developed that allows for the use of limiting quantities of nucleophile to provide pure products in solution. In order to extend our activation protocol to aryl carboxylic acids and carboxylic acids that posses α-electron withdrawing groups such as amino acids, alkyl sulfonamide linker **2** is required. Additional synthesis methods using sulfonamide linkers will be reported.

067. COMBINATION OF STRUCTURE-BASED DESIGN AND COMBINATORIAL CHEMISTRY FOR THE DEVELOPMENT OF CATHEPSIN D INHIBITORS. Ellen K. Kick,[†] Diana C. Roe,[‡] A. Deoffrey Skillman,[‡] Guangcheng Liu,[†] Todd J. A. Ewing,[‡] Yaxiong Sun,[‡] Irwin D. Kuntz,[‡] Jonathan A. Ellman.[†] [†]Department of Chemistry, University of California, Berkeley, CA 94720. [‡]Department of Pharmaceutical Chemistry, University of California, San Francisco, CA 94143.

Two small organic molecule libraries targeting human cathepsin D were synthesized on solid support by the display of functionality about the hydroxyethylamine isostere. The building blocks for the directed library were selected using the structure-based design based on the crystal structure of cathepsin D complexed with the peptide-based natural product pepstatin. The building blocks for the diverse library were selected using a clustering program designed to maximize diversity. The screening of the libraries for inhibitory activity toward cathepsin D provided a "hit rate" of 6-7% at 1 μM inhibitor for the directed library and a "hit rate" of 2-3% at 1 μM inhibitor for the diverse library. Detailed characterization of several of the most active compounds from the libraries showed that the diverse library is 4-5 fold less potent that the directed library. A second generation library was made that rapidly resulted in the identification of a number of potent inhibitors with Ki ≤ 15 nM. These general approaches can be applied successfully to many other enzyme targets.

068.

BUILDING BLOCK SYNTHESIS FOR COMBINATORIAL BALANOL LIBRARIES L. O. Lyngsø and E. V. Byrgesen and J. Nielsen, Department of Organic Chemistry, Building 201, The Technical University of Denmark, DK-2800 Lyngby, Denmark

We are engaged in combinatorial solid-phase synthesis of selected natural product structures. A such example is balanol which is a very potent protein kinase C inhibitor. From our retro-synthetic analysis, balanol is divided into three main synthons: A mono allylated benzophenone diacid, an amino-protected cyclic aminoalcohol and a benzoic acid. For parent balanol these building blocks are 4-(2-carboxy-6-hydroxybenzoyl)-3,5-dihydroxy-benzoic acid monoallyl ester, (3R,4R)-3-Fmoc-amido-1-(2-nitrophenylsulfonyl)azepan-4-ol and 4-hydroxy-benzoic acid. We have synthesized a series of substituted mono-allylated benzophenone diacids and substituted amino-protected cyclic aminoalcohols including the building blocks for genuine balanol for the solid-phase synthesis of balanol libraries.

069. **DIVERGENT SYNTHESES OF LIGANDS FOR HIGH THROUGHPUT SCREENING OF CATALYST LIBRARIES**

Kevin Burgess, Alex Porte, and Mark Powell
Chemistry, Texas A & M University, College Station, TX 77843-3255.

In previous work we have introduced parallel formation and automated screening of relatively large collections of potential catalysts for asymmetric transformations. Preparation of, for instance, a 96-member catalyst library requires a relatively large number of ligands, usually more than are commercially available. Our response to this problem is to develop *divergent syntheses of optically active ligands*. Such syntheses involve a large scale preparation of an advanced intermediate which is then transformed into a series of ligands at the end of the sequence. This contribution will illustrate this idea for the preparation of C_3 symmetric phosphines and a new class of oxazoline ligands.

070.

PRODUCT SPECIFIC RELEASE IN COMBINATORIAL CHEMISTRY. Beverley J. Hall*, John D. Sutherland* and Adrian J. Pipe**, *Dyson Perrins Laboratory, South Parks Road, Oxford, OX1 3QY, UK, **GlaxoWellcome Medicines Research Centre, Gunnels Wood Road, Stevenage, Herts, SG1 2NY, UK

In combinatorial synthesis release of products from the solid phase is key to most screening strategies. A variety of cleavable linkers have been developed which are inert to the reaction conditions employed in a sequence and are then cleaved by orthogonal chemistry. It would be extremely useful if a means of cleaving only the final product from a successful sequence of reactions could be devised, so that cleavage products are not contaminated with intermediates and failure sequences. This would allow reactions which are not yet optimised for the solid phase or which are intrinsically low yielding to be employed. We are interested in developing solid phase routes to carbonyl compounds for screening as potential drug leads. The results of these investigations will be reported, including a successful seven-step synthetic route on solid support, and the use of acylated oxazolidinones towards generation of polyketide libraries.

071. A NEW SYNTHESIS OF ARYL GLYCINES FROM ARYL BORONIC ACIDS
Nicos A. Petasis,* Andrew Goodman and Ilia A. Zavialov, *Department of Chemistry and Loker Hydrocarbon Research Institute, University of Southern California, Los Angeles, CA 90089-1661*

Aryl boronic acids (**1**) react with the adducts of amines (**2**) and α-keto acids (**3**) to give aryl glycine derivatives (**4**). This reaction works with substituted aryl as well as hetero aryl boronic acids. Several examples of this process with various amines will be reported, including solid phase applications.

072. DEVELOPMENT OF TFP RESIN FOR COMBINATORIAL LIBRARY SYNTHESIS.
Hui Shao, Shaojing Tong and Arlindo Castelhano* Cadus Pharmaceutical Corporation, 777 Old Saw Mill River Road, Tarrytown, NY 10591

Recently, a considerable amount of attention in pharmaceutical industry has been focused on the use of solid phase organic synthesis for the generation of combinatorial libraries of non-oligomeric, small organic compounds for high throughput screening(HTS) in order to speed up the drug discovery process. One of the key elements in solid phase organic synthesis is to design a proper linker on solid support in order to perform organic reactions. Most of the linkers available were developed based on solid phase peptide synthesis. This presentation will discuss our recent effort on the development of a new resin with a unique linker which enable the execution of amidation and cyclocleavage reactions in a versatile and cost effective manner. The preparation of this resin and its use in combinatorial library synthesis will be addressed.

073. DOWEX 1-SUPPORTED RHODIUM AND PLATINUM CONTAINING ION PAIRS AS SELECTIVE AND RECYCLABLE CATALYSTS FOR VARIOUS HYDROGEN TRANSFER PROCESSES. [1]**Merav Setty-Fichman**, [2]Yoel Sasson and [1]Jochanan Blum. [1]Department of Organic Chemistry and [2]Casali Institute of Applied Chemistry, Hebrew University, Jerusalem 91904, Israel.

The immobilized ion pair [polystyrene-CH_2NMe_3]$^+$[$RhCl_4(H_2O)_n$]$^-$, prepared by room temperature interaction of commercial Dowex 1® with $RhCl_3·3H_2O$, was shown to catalyze at 78°C in 92% aq EtOH, the isomerization of allylarenes by an order of magnitude faster than the soluble [$(C_8H_{17})_3NMe_3$]$^+$[$RhCl_4(H_2O)_n$]$^-$ catalyst under PTC conditions. The reaction rate proved to depend on the electronic nature of the substrate, the degree of crosslinking of the support, the size of the catalysts beads, and the stirring rate. The supported ion pair was shown to be leach-proof and recyclable in numerous runs without loss of activity. The catalyst promoted also selective disproportionation of 1,3-cyclohexadiene to benzene and cyclohexene free of any side products. The ion pairs [polystyrene-$CH_2NMe_3]_2^+[PtCl_6]^{2-}$, generated from Dowex 1 and $PtCl_4$, was found to be an efficient hydrogenation catalyst for a variety of alkenes, alkynes and other unsaturated substrates in THF at 30-35°C under 50-600 psi H_2. Also this polymer-bound catalyst was found leach-proof and recyclable. The mechanisms of the various hydrogen transfer processes have been studied and will be discussed.

074. **METAL CATION TRANSPORT STUDIES COMPARING DB18C6 WITH N,N,N',N'-TETRAKIS (n-PROPYL)-2,3-NAPHTHALENEDIOXYDIACETAMIDE (NPR). Grace B. Borowitz**, Irving J. Borowitz, Brian Benoff, and Paul Miller, Department of Chemistry, Ramapo College of New Jersey, Mahwah, New Jersey 07430

Our previously synthesized N,N,N',N'-tetrakis (n-propyl)-2,3-naphthalenedioxydiacetamide (**NPr**) and related diacetamides bind cations in methanol in the order: $Ca^{2+} > Sr^{2+} > Ba^{2+} > Mg^{2+} >> Na^+$, K^+. They also extract Group II cations >Group I cations from water into dichloromethane. In contrast, our diacetamides including **NPr** were found by W. Simon *et al* (ETH Zurich) to sense $Na^+ >> Ca^{2+}$ when incorporated into ion-selective electrodes with a low dielectric constant solvent. In an effort to understand this selectivity reversal, we studied the three phase transport of cations with **NPr** in a bulk liquid membrane system using dichloromethane. We now report that **NPr** transports thiocyanates in the order K^+ (7.2) > Ca^{2+} (6.6) > Ba^{2+} (5.8) >Na^+ (1). **DB18C6** transports KSCN 20.5 times faster than does **NPr**. The results will be discussed.

075. **POLYMERS CONTAINING CALIX[4]ARENES. Michael T. Blanda*** and Eba Adou Department of Chemistry, Southwest Texas State University, San Marcos, TX 78666.

A series of copolymers have been synthesized which contain calix[4]arene molecules as part of the polymer backbone or pendant to the polymer backbone. Both types of polymers have the calixarene moieties incorporated via the upper rim of the calixarene which is held in the cone conformation. This leaves the lower rim of the calixarene unhindered and available to complex metal ions. In addition to metal ions, the calixarene-containing polymers also have the potential to bind neutral molecules within the cavity of the calixarene. Preparation of the pendant calixarene polymers was achieved by free radical copolymerization reactions between a monofunctionalized calixarene monomer and methylacrylate, acrylonitrile and styrene using AIBN as the initiator. In the preparation of the copolymers in which the calixarene moieties are part of the polymer backbone, difunctionalized calixarene alcohols and amines were condensed with a variety of diacid chlorides to produce poly-esters and amides, respectively. The syntheses of the monomers, polymer composition and physical properties will be discussed.

076.
ENANTIOMERIC DISCRIMINATION OF FERROCENECARBOXYLIC ACID DERIVATIVES OF AMINO ACIDS IN AQUEOUS CYCLODEXTRINS. **B. S. Jursic**, Department of Chemistry, University of New Orleans, New Orleans, Louisiana 70148

Amides between ferrocenecarboxylic acid and amino acids were use to explore ferrocene as a core group for binding to a cyclodextrin cavity. It was determined that the binding energy is sufficient to cause spectroscopic differences between enantiomeric amides of aminoacids. The best results were obtained with gamma-cyclodextrine as host molecule. The preparation of the ferrocene amides, as well as results of their epectroscopic discrimination in aqueous cyclodextrines will be presented.

077.

ENANTIOMERIC DISCRIMINATION AND CYCLOADDITION REACTION IN AQUEOUS CYCLODEXTRIN. EXAMPLE OF THE CYCLOADDITION REACTION BETWEEN ANTHRACENE AND MALEIMIDE DERIVATIVES OF AMINO ACIDS. **B. S. Jursic** and Alexander Echevarria Department of Chemistry, University of New Orleans, New Orleans, Louisiana 70148

The cycloadduct between anthracene and maleimide derivatives of amino acids were prepared without and with cyclodextrin catalysis. The rates of the formation of the enantiomeric cycloadducts were different in aqueous cyclodextrins due to the difference in the binding constant. These differences were also reflected in spectroscopic discrimination of enantiomeric cycloadducts in aqueous cyclodextrins.

078. INCLUSION COMPLEXES OF β-CYCLODEXTRIN WITH SULPHANILAMIDE DRUGS: PREPARATION AND THERMAL STUDIES, **Lakshmanan Venkatasubramanian**, Department of Plant Sciences, University of Cambridge, Cambridge CB2 3EA, UK.

Inclusion complexes of β-Cyclodextrin with different sulphanilamide drugs namely sulphapyridine, sulphadiazine, sulphamerazine, sulphamethazine and sulphathiazole were prepared. All these complexes along with pure β-Cyclodextrin showed a decrease of 10-12% weight between 30-100 °C which is due to the loss of water molecules of crystallisation for 1:1 inclusion complexes of β-cyclodextrin with sulphanilamide drugs. Following this loss of weight these complexes showed an onset of decomposition at 250-260 °C. DSC studies showed an endothermic peak at about 100 °C indicating the loss of water molecules from these complexes. Other endothermic peaks observed were due to the phase change of the dehydrated complexes followed by melting with concomitant decomposition. All these sulphanilamide complexes showed an increase in the decomposition temperature compared to the melting temperatures of the pure sulphanilamide drugs. The stability of the inclusion complex of the respective sulphanilamide drug with β-cyclodextrin is predominantly due to the hydrophobic forces which help to hold the guest drug molecule inside the macrocyclic cavity.

079.

ANION RECOGNITION BY HYDROGEN BONDING: A READILY AVAILABLE NEUTRAL ACYCLIC HALIDE BINDING AGENT. **Konstantinos Kavallieratos,** Susan R. de Gala, David J. Austin and Robert H. Crabtree*, Department of Chemistry, Yale University, New Haven, CT 06511-8118.

Isophthalic diamides, m-$C_6H_4(CONHAr)_2$ (Ar = Ph, **1**; p-n-BuC_6H_4, **2**), readily synthesized on a large scale, have been shown to bind halide ions with an association constant K_a (for **2**) of 6.1×10^4 M^{-1} (Cl$^-$), 7.1×10^3 M^{-1} (Br$^-$) and 5.0×10^2 M^{-1} (I$^-$) determined by ^{1}H-NMR spectroscopy. The structure of the complex [PPh$_4$]$_2$[**1**.Br][Br].CH$_2$Cl$_2$, was determined by X-ray crystallography and shows a 1:1 complexation of bromide ion exclusively via rarely structurally characterized N-H...Br hydrogen bonds and a [PPh$_4$]Br of crystallization. The acyclic diamide adopts an unusual *syn-syn* non planar binding conformation. The presence of hydrogen bonding in the 1:1 complex was confirmed in solution by FT-IR and NMR spectroscopy. The anion binding properties of this and other simple artificial receptors are expected to allow applications in a wide variety of situations.

080. STUDIES TOWARDS CALCIUM ION INDICATORS EMPLOYING STABLE RADICAL REPORTER GROUPS. K. H. Gibson, J. M. Peach, **J. Skidmore**, Dyson Perrins Laboratory, University of Oxford, South Parks Road, Oxford, OX1 3QY, UK.

Calcium ion flux is a controlling factor in many biological processes; probes facilitating the measurement of calcium ion concentration *in vivo* are an important tool in the armoury of molecular biologists. Current probes are based on fluorescent or nmr reporting groups. Whilst these have allowed considerable advances in the study of cellular calcium they are hampered by a lack of three dimensional imaging capability and poor sensitivity respectively. This poster presents ongoing studies towards calcium ion probes utilising stable radicals connected to a suitable ionophore. It is hoped that these will show a change in esr on binding thus allowing calcium concentration to be measured. Additionally, utilisation of radio-frequency esr should allow imaging of calcium ion concentration. Included in the report are synthetic approaches to a number of potential probes which have allowed some of the criteria for a successful probe to be determined.

081. **SELF-ASSEMBLED RECEPTOR FOR DICARBOXYLATES *VIA* METAL TEMPLATION**
Heidi Chen, Andrew D. Hamilton*, University of Pittsburgh, Pennsylvania 15260

A self-assembled receptor that utilizes the advantages of metal templating and hydrogen bonding interactions has shown strong binding for dicarboxylates. Binding was studied by ^{1}HNMR titrations in 10% acetonitrile/chloroform. The synthesis and binding properties of these self-assembled receptors will be discussed as well as an extension of this approach to metal-catecholate complexes.

082. **THE DESIGN AND SYNTHESIS OF A PREORGANIZED TRIPHOSPHONATE IONOPHORE** Jeremy J. Hans and Steven D. Burke*, *Department of Chemistry, University of Wisconsin-Madison, Madison, WI 53706.*

The design, synthesis and study of preorganized triolide ionophore **2** is described. Design features of **2** include conformational control elements that help define the overall shape and rigidity of this macrocycle. The synthesis of monomer unit **1** includes a dioxanone-to-dihydropyran Claisen rearrangement to form the dihydropyran ring and an Arbuzov reaction to install the phosphonate residue. Iterative esterification and macrolactonization produces triolide **2**. The X-ray crystal structure of a Na$^+$ complex and cation binding data of **2** are also reported.

083. A NEW CLASS OF DNA MINOR GROOVE BINDERS THAT RIVAL THE NATURAL PRODUCT DISTAMYCIN IN SEQUENCE-SELECTIVITIES AND BINDING AFFINITIES. B. Gong* and Y. Yan, Department of Chemistry, The University of Toledo, Toledo, OH 43606

New DNA minor groove binders with high sequence-selectivities and binding affinities have been discovered. The designed compounds represent different combinations of the meta- and para-substituted benzene rings, which represent different curvatures (shapes) of potential groove binders that may fit into the minor groove of a DNA molecule. The molecules also bear amide, urea, amino and alkyl groups that can introduce hydrogen bonding, electrostatic and Van der Waals interactions respectively. Among the eight compounds screened, two showed sequence-selectivities and affinities that rival the natural product distamycin. Viscometric analysis indicated that these ligands are indeed groove binders. This new class of minor groove binders not only offer the opportunities for studying the molecular recognition between DNA and small molecules, but also lay the foundations for designing a new generation of minor groove binders, and for the discovery of duplex DNA or RNA binders through combinatorial approaches.

084. HYDROGEN BONDED, SELF-COMPLEMENTARY MOLECULAR DUPLEXES FROM SIMPLE BUILDING BLOCKS: MIMICKING DNA ASSEMBLY. B. Gong, Department of Chemistry, The University of Toledo, Toledo, OH 43606

A new class of peptides based on the readily available building blocks, m-aminobenzoic acid (MABA) and the amino acid sarcosine, have been designed and synthesized. The peptides are designed in such a way that their backbones are self-complementary in terms of the arrangement of hydrogen bond donors and acceptors. As shown by ^{1}H-NMR studies, these self complementary peptides form duplexes in organic solvents. Properties such as positive cooperativity and "base pair breathing" that are characteristic of DNA duplex formation are also demonstrated by these peptide molecules. This class of complementary molecular strands provide a new example of simple models that mimic DNA assembly processes. More importantly, the same design described here can be extended to the design of molecular duplexes with different sequences, which will result in a new class of information-carrying molecules.

085. MOLECULAR RECOGNITION OF BIOLOGICAL POLYANIONS AND ZWITTERIONS BY SAPPHYRIN-BASED RECEPTORS. Andrei Andrievsky, Jonathan L. Sessler,* and Vladimír Král. Department of Chemistry and Biochemistry, University of Texas at Austin, Austin, TX 78712.

Sapphyrin, a unique pentapyrrolic "expanded porphyrin", acts as an efficient anion binding agent in solution and in the solid state. Here we report its successful use as a building block in the construction of several conjugated and oligomeric receptors comprised of natural ionophores, oligopeptides and polyamines. These multidentate host molecules display excellent binding and transport of important biological polyanions (*e.g.*, nucleotide phosphates, oligonucleotides, and DNA) and zwitterions (*e.g.*, amino acids). They have been specifically designed to complement their prospective guests' shapes, charges and recognition sites. Synthetic incorporation of chiral moieties into receptor structures allowed for enantioselective differentiation of natural substrates.

R^1, R^2 = Sapphyrins, Distamycin, Lasalocid, Polyamines, etc.

086. SELF-ASSEMBLED MONOLAYERS OF 7-(10-THIODECOXY)COUMARIN ON GOLD: SYNTHESIS, CHARACTERIZATION, AND PHOTOCHROMIC BEHAVIOR Weijin Li, Vincent Lynch, and Marye Anne Fox, Department of Chemistry and Biochemistry, the University of Texas at Austin, Austin, Texas 78712.

Self-assembled monolayers (SAMs) of 7-(10-thiodecoxy)coumarin on polycrystalline gold, upon irradiation at 350 nm, produce [2+2] photodimers whose reflection FTIR absorption and surface fluorescence emission differ appreciably from the initial monolayer. The photodimerization can be reversed by a 254 nm irradiation as demonstrated by fluorescence spectroscopy, reflection absorption FTIR and sessile drop contact angle measurements. Our objective is to study the photochemical reversibility of a chemical system on a solid metal surface and to demonstrate that emission of an attached group can be used for surface patterning in a write-erase device.

087.

SYNTHESIS AND CHARACTERIZATION OF TERMINALLY FUNCTIONALIZED N-ALKANETHIOLS
. **Mamoun M. Bader**, Department of Chemistry, United Arab Emirates University P. O. Box 17551, Al-Ain, United Arab Emirates; Michael D. Ward, Lynn M. Frostman, **Department of Chemical Engineering and Materials Science,University of Minnesota, 421 Washington Ave. S.E., Minneapolis, MN 55455**

Abstract: Synthesis and characterization of a series of terminally functionalized n-alkanethiols with the general formula X-$(CH_2)_n$-SH are reported, where X= OH, NO_2, CO_2H, CO_2Me, CO_2Et, p-NO_2-Ph, p-NH_2-Ph, and n has a value between 5 and 12. These compounds can be used for the synthesis of self-assembled monolayers on gold surfaces, which provides a simple approach to chemically modified surfaces with controlled interfacial structure.

088. **HETERONUCLEAR POLYACETYLENIC MACROCYCLES AND MOLECULAR CAGES.** James C. Bohling and Lawrence T. Scott, *Department of Chemistry, Boston College, Merkert Chemistry Center, Chestnut Hill, MA 02167.*

In our quest to expand the world of multidimensional polyacetylenic molecules, we have assembled a group of building blocks with various geometries and properties. Syntheses of cages and macrocycles containing multiple enediynes linked by trivalent phosphorus bridgeheads will be presented, and the chemistry of these novel heterocycles will be discussed.

089. **SYNTHESIS OF A FUNCTIONALIZED CHIRAL MOLECULAR TWEEZER.** Michael Harmata and Sriram Tyagarajan, Department of Chemistry, University of Missouri-Columbia, Columbia, Missouri 65211

The synthesis of the illustrated molecular tweezer will be presented.

090. AN EFFICIENT PROCEDURE FOR THE SYNTHESIS OF UNIQUE MONOSUBSTITUTED BIPYRIDINES. **Hephzibah J. Kumpaty,**[a] Charles A. Panetta,[b] Norman E. Heimer,[c] [a]Department of Chemistry, The University of Wisconsin-Whitewater, Whitewater, Wisconsin, 53190, [b]Department of Chemistry, The University of Mississippi, University, Mississippi, 38677, [c]Department of Chemistry, U.S. Air Force Academy, USAFA, CO 80840-6230

Our goal was to synthesize novel 3-, 4-, 5-, and 6-monosubstituted disulfide functionalized 2,2'-bipyridines and the corresponding ruthenium complexes. These molecules were synthesized so as to examine their monolayer forming abilities and to perform various electrochemical studies. The key step in syntheses was the construction of the unsymmetrical bipyridine unit. Eleven new bipyridines were successfully synthesized in reasonable yields using palladium catalyzed cross-couplings between 2-trimethylstannylpyridine and the appropriately substituted 2-chloropyridines. All new intermediates and the final target compounds have been characterized by spectral and elemental analyses. Presented herein are the synthetic details of these reactions and their subsequent conversion of bipyridines to the ruthenium complexes.

091. **Photoresponsive Dendrimers** Denise M. Junge and Dominic V. McGrath* Department of Chemistry, University of Connecticut, Storrs, CT 06269-4060

The azobenzene moiety undergoes photoinduced, reversible *cis-trans* isomerization in which the strucural change is considerably large. Benzyl aryl ether dendrimers up to the second generation have been prepared with photoresponsive 4,4'-azobenzene central linkers. Photoirradiation of these dendrimers with 350 nm light resulted in the isomerization of the central linker from the extended *trans* to the more compact *cis* form. Back isomerization occurred thermally over a period of approximately 5 hours in the dark or in 10 minutes by exposing the dendrimers to visible light.

092.

CONFORMATION AND FUNCTION STUDY OF THE PHOTOSWITCHABLE BIS(CROWN ETHER) IN MEMBRANE DISRUPTION PROCESSES. **Binqi Zeng**, T. M. Fyles, Department of Chemistry, University of Victoria, Victoria, B.C., V8W 3P6, Canada

Controlled release of drugs from liposomes has potential applications in liposomes as drug carriers for chemotherapy. Bis(crown ether)s are linear bola-amphiphiles which are capable of acting as molecular recognition membrane disruption agents. The conformation changes of the bis(crown ether) with 77'-thioindigo as photo antenna can be regulated by visible light. The U-shaped conformation is proven to be the active form in membrane disruption by fluorescence self-quenching (FSQ) method. The barium recognition controlled membrane disruption by the bis(crown ether) is confirmed by forming 1:1 'sandwich' complexes which create the required U-shaped conformation in the linking chain.

093. SYNTHESIS AND ANALYSIS OF LIQUID CRYSTALLINE DIPHENYLACETYLENE DERIVATIVES. Aline F. Miller, Matthew Scheideman, Jessica L. Lavoie, James Pape, and Eric Scharrer*, Department of Chemistry, Franklin and Marshall College, P.O. Box 3003, Lancaster, PA 17604-3003.

We have initiated investigations into the liquid crystalline behavior of monosubstituted diphenylacetylene derivatives, **1**. A series of para substituted ether derivatives have been prepared and using polarizing microscopy, it was determined that derivatives with R = $C_{12}H_{25}$, $C_{14}H_{29}$, $C_{16}H_{33}$, $C_{17}H_{35}$, and $C_{18}H_{37}$ were mesomorphic. Several of the para and meta substituted ester derivatives also exhibited liquid crystalline behavior. The synthesis and liquid crystal properties of these molecules will be discussed.

X = OR, CO_2R

1

094.

RED LIGHT EMITTING "PUSH-PULL" DISUBSTITUTED POLY(1,4-PHENYLENEVINYLENES). **Ronald M. Gurge** Paul M. Lahti, Ananda Sarker, Bin Hu, and Frank E. Karasz, Chemistry Department, University of Massachusetts, Amherst, MA 01003.

During the past decade, significant advances in LEP (light emitting polymer) technologies have lead to the realization of the marketability of these materials in LED's (light emitting displays). PPV [poly(p-phenylene vinylene)] and its derivatives, synthesized by soluble precursor routes, have been among the most studied LEP's to date. PPV's are advantageous because of their processibility, high thermal stability, efficiency and color tunability. In many cases it is possible to synthesize polymers that emit across the blue to red spectral regions. This poster will present the synthesis and characterization of some new "push-pull" substituted PPV's by a modified precursor route. These polymers have been film cast from solution to make single layer and multi-layer LED's which have red light emission in the 620 nm region. The convenient synthesis and processibility of these polymers make them candidates for emissive layers in Organic LED's.

095. **POLYCATIONS: SYNTHESES OF POLYAMMONIUM STRINGS AS ANTIBACTERIAL AGENTS.** Robert Engel, Jeanne Fabian, Tessie October and Alice Cherestes, Department of Chemistry and Biochemistry, Queens College, C.U.N.Y., 65-30 Kissena Boulevard, Flushing, NY 11367

A series of polyammonium species, each bearing two to ten cationic sites within the covalent structure, have been synthesized. For these species quaternary ammonium sites are located in a linear array at regular intervals along a "string" backbone with free-floating anions. Syntheses of the "string" species are accomplished by initial quaternization of 1,4-diazabicyclo[2.2.2]octane (dabco) using linear ω-halo-1-alkanols. Selectivity for mono- or di-substitution is accomplished through solvent choice. Diols from disubstitution are converted to the corresponding α,ω-dichlorides and elaboration is accomplished by reaction of the appropriate monosubstituted species. Continued linear elaboration of the polycationic strings is accomplished by repetition of the previously noted reaction sequence. The resultant polycationic species exhibit antibacterial activity using mutant strains of *E. coli*.

096. **SYNTHESIS, CONFORMATIONAL ANALYSIS AND ACTIVITY OF *CYCLO*[IAA-ASP-IAA-ASP] AS A PROTEASE MIMETIC.** Eryk Thouin and William D. Lubell*, Département de chimie, Université de Montréal, C. P. 6128, Succ. A, Montréal, Canada H3C 3J7

In detailing the synthesis, conformational analysis and activity of cyclic peptide **1**, our presentation will discuss the merits of using natural and conformationally rigid amino acids in order to prepare cyclic peptide libraries that may mimic the structural aspects and catalytic activities of various enzyme active sites.

cyclo[Asp-IAA-Asp-IAA], **1**

097. **APPLICATIONS OF UNNATURAL OLIGOSACCHARIDE BIOSYNTHESIS TO SELECTIVE DRUG DELIVERY.** Kevin J. Yarema, Lara K. Mahal and Carolyn R. Bertozzi*, Department of Chemistry, University of California, and Materials Sciences Division, Lawrence Berkeley National Laboratory, Berkeley, CA 94720

Cell surface antigens are attractive targets for the selective delivery of toxins to diseased cells. Most efforts toward the development of cell-specific drug targeting agents have focused on the use of monoclonal antibodies that recognize a particular antigen, an approach that suffers from severe limitations. We have recently demonstrated that unnatural epitopes can be displayed on cells using the biosynthetic pathways for cell-surface oligosaccharides. This discovery has now been extended to the development of a novel strategy for selective drug targeting. We installed ketone groups on cells through unnatural sialic acid biosynthesis, and selectively modified the cells with biotinamidocaproyl hydrazide under physiological conditions. Synthetic ricin-avidin conjugates killed the biotinylated cells with remarkable selectivity. Applications of this and related approaches to anti-cancer therapy will be presented.

098.

SYNTHESIS AND TRIGGERING OF A SPIROCYCLIC CALICHEAMICINONE ANALOG. David K. Moss, John D. Spence and Michael H. Nantz, Department of Chemistry, University of California, Davis, California 95616

A triggering mechanism for an enediyne analog of calicheamicinone has been developed and tested. The mode of activation resembles the mechanism that is inherent to the calicheamicins. An intramolecular Michael addition is planned to rehybridize a Csp2 center that lies within the 10-membered enediyne core. The key step has been demonstrated by the formation of 2 on deprotection of 1. The synthesis of 2 and progress toward enediyne 3 will be presented.

099.

ANTIBODY-CATALYZED [2,3]-SIGMATROPIC REARRANGEMENTS. **Alexander Flohr,** Zhaohui S. Zhou, Donald Hilvert, The Scripps Research Institute, 10550 North Torrey Pines Road, La Jolla, Clifornia 92037

Several antibody-catalysts for [2,3]-sigmatropic rearrangements are described. The antibodies were elicited with N-aryl-3-methoxyphenyl proline derivatives and accelerate syn eliminations of primary or secondary alkyl arylselenoxides as well as rearrangements of allylic sulfoxides. Kinetic data and investigation of the substrate selectivities will be presented.

100. IN VITRO SELECTION OF RNA ENZYMES FOR THE DIELS-ALDER REACTION. **Paramjit S. Arora** and James S. Nowick*, Department of Chemistry, University of California, Irvine, CA 92697-2025.

RNA enzymes are postulated to have catalyzed all chemical reactions in the earliest living cells. RNA has been shown to efficiently catalyze reactions involving phosphoryl group transfers; however, there are no examples of RNA enzymes for carbon-carbon bond forming reactions. We are investigating the potential of RNA molecules, selected from a pool of random RNA sequences, to catalyze a Diels-Alder reaction. Two different methods are being used to develop RNA enzymes: (a) the transition state analog (TSA) approach that has been successfully exploited for generation of catalytic antibodies, and (b) the direct selection approach that has shown more promise than TSA strategy for selection of catalytically active nucleic acids. The results of our studies with the TSA-selected RNAs and the progress of our efforts toward the direct selection of RNA enzymes will be presented.

101. O^2-ARYL DIAZEN-1-IUM-1,2-DIOLATES AS NITRIC OXIDE RELEASING AGENTS. **Aloka Srinivasan,** †Joseph E. Saavedra, Melissa Booth, Larry K. Keefer, Lab. of Comparative Carcinogenesis, NCI-FCRDC, and †IRSP, SAIC Frederick, Frederick, MD 21702.

As an approach to delivering the multifaceted bioregulatory agent nitric oxide to specific bodily sites without adversely affecting other tissues, diazeniumdiolates have been synthesized with the terminal oxygen protected by electron deficient aryl groups such as 2,4-dinitrophenyl, purines and pyrimidines. Here, we report the synthesis of selected examples and their reactions with various nucleophiles including glutathione to regenerate nitric oxide.

102. CYCLOPROPANE-DERIVED PEPTIDOMIMETICS. DESIGN, SYNTHESIS, EVALUATION AND STRUCTURE OF NOVEL HIV-1 PROTEASE INHIBITORS
Michael C. Hillier, Gordon O. Dorsey, Todd H. Gane, Stephen F. Martin*
Department of Chemistry, University of Texas at Austin, Austin, TX 78712

Modeled after symmetrical diamino diol HIV-Protease inhibitors developed at Abbott laboratories, analogs **1-5** were assembled to determine the efficacy of the cyclopropyl subunit to enforce side-chain orientation as well as "extended" secondary structure. The synthesis of these compounds will be presented in addition to biological and structural data.

103. SYNTHESIS OF CADA ANALOGS AS POTENTIAL ANTIVIRAL AGENTS
Qi Jin, Andrej Sodoma and Thomas W. Bell, Department of Chemistry, University of Nevada, Reno, NV 89557-0020

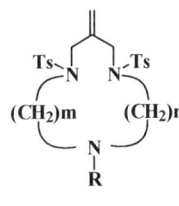
CADA Analogs

Cycloazadisulfonamide (CADA) **1** (n = m = 3, R = Ph) was found to display anti-HIV activity (EC_{50} ~2 µM) and relatively low cytotoxicity (IC_{50} > 200 µM, TI > 100) in vitro. However the low water solubility of CADA (8 µM) limits studies on its mechanism of action. Many CADA analogs have been made in order to examine specific structure-activity relationships, to improve water solubility and to discover new lead compounds. This poster will introduce the synthesis of CADA analogs which have either different functional groups at the bottom nitrogen atom (R = R'CO-, R"CH_2-) or various macroring sizes (m = 2, 3, 4, n = 2, 4).

104.
NEW ANALOGS OF CADA HAVING ACTIVITY AGAINST THE HUMAN CYTOMEGALOVIRUS (HCMV). **Meinrado F. Samala**, Athanasios Glekas, Andrej Sodoma, Thomas W. Bell, Stephen C. St. Jeor and Albert van Geelen, Departments of Chemistry and of Microbiology, University of Nevada, Reno, Nevada 89557-0020

CADA, **1**, is a promising new antiviral compound that has been shown by the National Institutes of Health (NIH) to have anti-HIV activity (EC_{50} ~ 0.2 - 2.0 µM). New analogs, in linear and cyclic forms, are being synthesized to help establish the mechanism of action of CADA against HIV. Meanwhile, we have discovered two new analogs (ARB 95-213 and ARB 95-214) that exhibit activity against the human cytomegalovirus (HCMV). HCMV is a ubiquitous member of the herpes family of viruses and is associated with high incidence of mortality in HIV patients. The syntheses of these new potential drugs as well as their anti-HIV and anti-HCMV activities will be presented.

105. **ENANTIOSPECIFIC SYNTHESIS OF NOVEL ANALOGS OF GLYCOSIDASE INHIBITORS. M.J. Blanco** and F.J. Sardina. Dpto de Química Orgánica. Universidad de Santiago. 15706 Santiago de Compostela. Spain

Glycosidases are key enzymes in the biosynthesis and processing of glycoproteins, which are vital for normal cell-cell communication. Glycosidase inhibitors have become important as potential antibacterial, antiviral and antitumoral agents. Many polyhydroxylated pyrrolidines show potentially useful activity due to their structural resemblance to sugars and their resultant ability to act as glycosidase inhibitors. We present stereoselective syntheses of galactosidase inhibitor (1) analogs for studies of structure-activity relationships.

106. THE REGIO AND STEREOSPECIFIC SYNTHESIS OF POTENT CARBOCYCLIC INFLUENZA NEURAMINIDASE INHIBITORS: GS4071 AND ANALOGUES. L. Zhang, M. A. Williams, W. Lew, H. Liu, C. U. Kim. Gilead Sciences Inc., 353 Lakeside Dr. Foster City, CA 94404.

A series of transition state analogues based on a carbocyclic scaffold was synthesized as inhibitors of influenza neuraminidase. Starting from commercially available quinic acid (**1**), N-Trityl protected aziridine (**2**) was prepared as a key intermediate in a regio- and stereospecific manner. Aziridine **2** was opened with various linear, branched, cyclic and aromatic alcohols under Lewis acid catalysis followed by reduction of the azide group and ester hydrolysis to afford the corresponding amino acids **3**. Structure-activity studies of a series of carbocyclic analogues identified GS4071 (**3**, R=3-pentyl) as a potent inhibitor of influenza neuraminidase. The 3-pentyl sidechain of GS4071 interacts in a novel hydrophobic binding mode in the neuraminidase active site. Hydrophobic interactions which appeared to be optimal for the 3-pentyl group were confirmed by X-ray crystallographic analysis.

107. **SYNTHESIS OF THE SERINE-THREONINE PHOSPHATASE INHIBITOR MICROCYSTIN LA, AND PROGRESS TOWARDS THE SYNTHESIS OF SELECTED ANALOGS.** John M. Humphrey, **James B. Aggen**, and A. Richard Chamberlin*; Department of Chemistry, University of California, Irvine, California 92697

Protein kinases and phosphatases mediate reversible protein phosphorylation, which is a major control element of the cell. There is a diverse group of toxic natural products that inhibit certain phosphatases, thereby disrupting normal biochemical pathways. These toxins can be useful for dissecting the individual biochemical pathways associated with these enzymes. Here we describe the first total synthesis of one such toxin, the cyclic heptapeptide microcystin LA. The synthesis features a convergent route that is amenable to analog preparation in the search for more selective phosphatase inhibitors.

108. NOVEL PROTEIN PHOSPHATASE INHIBITORS BASED UPON OKADAIC ACID. Craig J. Forsyth,* Valerie A. Frydrychowski, Steven F. Sabes, and Rebecca A. Urbanek, Department of Chemistry, University of Minnesota, Minneapolis, MN 55455.

Okadaic acid is a marine natural product that potently inhibits protein serine / threonine phosphatases 1 and 2A. The design, synthesis, and assay of protein phosphatase 2A (PP2A) inhibitory activity of novel okadaic acid structural variants has been used to elucidate further which structural features are essential for phosphatase inhibition. This work has led to the identification of structurally simplified potent inhibitors of PP2A. The details of this work will be presented.

Okadaic Acid

109. **Stereostructure of (-)-Chloropeptin I, a Novel Inhibitor of gp120-CD4 Binding, Via High-Temperature Molecular Dynamics, Monte Carlo Conformational Searching, and NMR Spectroscopy**
Hiroaki Gouda,[†] Keiichi Matsuzaki,[†] Haruo Tanaka,[*†] Shuichi Hirono,[†] Satoshi Ōmura,[*¥] John A. McCauley,[‡] Paul A. Sprengeler,[‡] George T. Furst,[‡] and Amos B. Smith, III[*‡]

[†]School of Pharmaceutical Sciences, Kitasato University, Minato-ku, Tokyo 108, Japan; [¥]Research Center for Biological Function, The Kitasato Institute, and School of Pharmaceutical Sciences, Kitasato University, Minato-ku, Tokyo 108, Japan; [‡]Department of Chemistry, Laboratory for Research on the Structure of Matter, and Monell Chemical Senses Center, University of Pennsylvania, Philadelphia, PA 19104

The relative and absolute stereochemistry of chloropeptin I (1), a potent inhibitor of the binding between the gp120 envelope glycoprotein of HIV-1 and CD4, has been elucidated. We previously established the planar structure of this novel chlorinated hexapeptide. The C_α absolute configurations of three amino acids were readily assigned following acidic hydrolysis. The stereochemistry of the remaining three linked residues were determined employing two computational methods in conjunction with NMR analysis.

110. **(+)-THIAZINOTRIENOMYCIN E: RELATIVE AND ABSOLUTE STEREOCHEMISTRY**
Amos B. Smith III,[*1] Joseph Barbosa,[1] Nobuo Hosokawa,[2] Hiroshi Naganawa[2] and Tomio Takeuchi[*2]
[1]Department of Chemistry, Monell Chemical Senses Center, Laboratory for Research on the Structure of Matter, University of Pennsylvania, Philadelphia, Pennsylvania 19104 and [2]Institute of Microbial Chemistry, Tokyo 141, Japan

In 1995, Takeuchi and coworkers at the Institute of Microbial Chemistry (Tokyo) revealed the isolation and planar structures of the thiazinotrienomycins, five novel ansamycin antibiotics which displayed significant *in vitro* cytotoxic activity against several human cancer cell lines Structurally, the thiazinotrienomycins represent heterocyclic congeners of the trienomycins and mycotrienins (See Smith, A. B., III et. al. *J. Am. Chem. Soc.* **1996**, *118*, 8308 and 8316). This presentation will focus on the determination of the relative and absolute stereochemistry of thiazinotrienomycin E, the most abundant isolate, utilizing a degradation and chemical correlation approach.

111.

NEW NORTERPENE PEROXIDES FROM INDO-PACIFIC LATRUNCULIIDS . **S. Sperry**, P. Crews, Department of Chemistry and Biochemistry, University of California, Santa Cruz, California 95064

For several years, marine sponges belonging to the family Latrunculiidae have been known to produce norterpene cyclic peroxides. These compounds often display biactive properties. In our studies of Diacarnus ?spinopoculum we have isolated several compounds of this class, including six which have not been previously described. Structure elucidation was based on analyses of spectral data. One of the new compounds, nuapapuin B Me-ester, has demonstrated cytoxicity in NCI's 60 cell-line screen and is presently undergoing in vivo evaluation.

112. BIOMIMETIC SYNTHESIS OF XESTOSPONGIN A.   J. E. Baldwin, **C. R. Firkin** and R. C. Whitehead, Dyson Perrins Laboratory, South Parks Road, Oxford, OX1 3QY, UK.

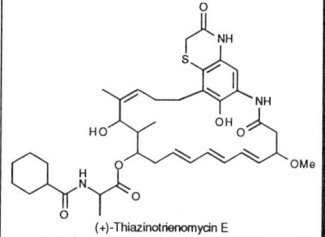

Xestospongin A (1) is a bis-quinolizidine with an axis of symmetry isolated from marine sponge that has been shown to possess vasodilative activity *in vivo*. It is proposed that this and other related marine alkaloids are biosynthesized *via* dimerization of a 3-alkyl pyridine unit. Previous studies have shown that a stereospecific synthesis of (1) may be based around one asymmetric centre, C2. A potential monomer (2) was synthesized and a solution of this heated in acetone. The desired dimeric pyridinium salt (3) was shown to be the major product. A small amount of trimeric material was also produced but no higher order polymers could be detected. The excellent selectivity of dimerization adds credence to the proposed biosynthesis of (1). Investigations are underway into completion of the synthesis of xestospongin A.

113.

STUDIES TOWARDS THE TOTAL SYNTHESIS OF NAGSTATIN. K. Tchabanenko, J.Robertson, The Dyson Perrins Laboratory, South Parks Road, Oxford, OX1 3QY, U.K.

Nagstatin, being an inhibitor of a wide range of enzymes, whose activity increases in several diseases such as diabetes mellitus, leukaemia and cancer, may aid in understanding the mechanisms or the processes of the above diseases. The aim of this study is to design an effective and versatile synthetic pathway towards Nagstatin and related amino sugars starting with commercially available reagents using desymmetrisation methodology in the early steps of the synthesis. The approaches towards the amino sugar core structure of Nagstatin include Aza-Diels-Alder, asymmetric amino-hydroxylation and desymmetrisation of a meso epoxide. The data on a cis-selective Horner-Emmons reaction will also be presented.

114. TOTAL SYNTHESIS OF MICROCOLIN B
Merritt B. Andrus,* Wenke Li and Robert F. Keyes, Department of Chemistry, Purdue University, West Lafayette, IN 47907

Microcolins are new potent immunosuppressants. A convergent, asymmetrical total synthesis of microcolin B will be presented, which features an efficient preparation of chiral α,γ-dimethyl acid **1** from readily available **2** and a novel mixed imide synthesis using pentafluorophenol ester **3** and lactam **4** in the preparation of prolyl 2-pyrrolinone portion of the molecule. A formal synthesis of microcolin **A** will also be discussed.

115. STEREOCONTROLLED SYNTHESIS OF BIOSTABLE LIPOXIN ANALOGUES
Nicos A. Petasis,*[†] Valery V. Fokin,[†] Irini A. Zanze,[†] Saeed Binaei[†] and Charles N. Serhan*[‡]
[†]Department of Chemistry and Loker Hydrocarbon Research Institute, University of Southern California, Los Angeles, CA 90089-1661; [‡]Center for Experimental Therapeutics and Reperfusion Injury, Brigham & Womens Hospital and Harvard Medical School, Boston, MA 0211.

The combined action of various lipoxygenases on arachidonic acid leads to the formation of the lipoxins (**1, 2**), which play a key role in cell-cell interactions and the regulation of inflammatory signals. It was recently found that aspirin triggers the formation of two new lipoxins with a 15-epi configuration (**3, 4**), which have different properties than the native compounds. In order to study the bioactions of these lipoxins we have designed a series of new analogues with different R-groups, which have increased biostability. The stereocontrolled synthesis of these compounds will be reported.

1, LXA$_4$: R = (CH$_2$)$_3$Me X = OH, Y = H
3, 15-epi-LXA$_4$: R = (CH$_2$)$_3$Me X = H, Y = OH
2, LXB$_4$: R = (CH$_2$)$_3$Me X = OH, Y = H
4, 15-epi-LXB$_4$: R = (CH$_2$)$_3$Me X = H, Y = OH

116. ROUTE TO THE SYNTHESIS OF MULTI-KILOGRAM BATCHES OF ONTAZOLAST.
Jana Vitous, Victor Fuchs, Karl Grozinger, Palayakotai Raghavan, Lana Smith. Boehringer Ingelheim Pharmaceuticals, Inc. Ridgefield, CT 06877-0368, Department of Chemical Development.

From a series of benzoxazolamine analogs that inhibit leukotriene (LT) biosynthesis, ONTAZOLAST, (S)-N-[2-Cyclohexyl-1-(2-pyridinyl)ethyl]-5-methyl-2-benzoxazolamine, is in development as a topically administered treatment for asthma (IC_{50} 0.001 µM). ONTAZOLAST inhibits LT biosynthesis by affecting the availability of arachidonic acid (acting at the level of arachidonic acid release). The Chemical Development group at BIPI began extensive studies to develop a safe, efficient, and economic route for the production of bulk quantities of Ontazolast, the results of which are disclosed herein.

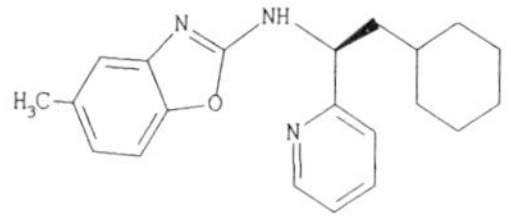

ONTAZOLAST

117.

Synthetic Studies Towards the Welwitindolinone Alkaloids: Efficient Construction of an Oxindole Core

John L. Wood, <u>Alexandra A. Holubec</u>, and Brian M. Stoltz
Sterling Chemistry Laboratory, Department of Chemistry, Yale University
New Haven, Connecticut 06520-8107

Abstract. Recent efforts toward a total synthesis of N-methyl welwitindolinone C isothiocyanate (**1**) have resulted in the development of an efficient means of converting isatin (**3**) to a suitably functionalized oxindole synthetic intermediate (**2**). The evolution of this synthetic strategy as well as efforts to advance **2** to **1** will be described.

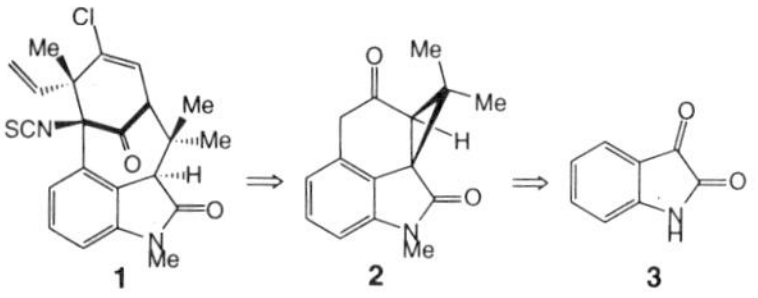

118. HIGHLY REGIOSELECTIVE MITSUNOBU REACTION AND BRESLOW REMOTE FUNCTIONALIZATION OF STEROIDS: PREPARATION OF 14β,15-KETOSTEROIDS. Michael E. Jung* and <u>Ted W. Johnson</u>, Department of Chemistry and Biochemistry, University of California, Los Angeles, CA 90095

A practical, efficient synthesis of **2** from 3,6-cholestanediol **1** has been achieved in 7 steps utilizing Breslow's remote functionalization methodology. Synthesis of **2** is based on a regioselective Mitsunobu reaction on diol **1**, followed by remote functionalization of C14, and finally installation of the 15-keto functionality via epoxide opening and oxidation. Other 14β,15-ketosteroids will also be synthesized and studied for their potential anti-histamine properties.

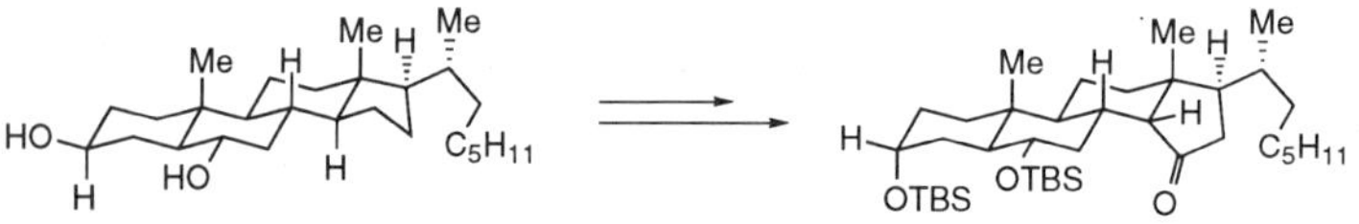

119.

RING EXPANSIONS IN THE SYNTHESIS OF MEDIUM RING DILACTAMS. <u>Ian N.Houson</u>*, J.M. Brown*, S.P.Watson**. *Dyson Perrins Laboratory, Oxford University, South Parks Rd, Oxford, OX2 7AT, UK. **Glaxo Wellcome Medicines Research., Gunnels Wood Rd, Stevenage, SG1 2NY, UK.

We present an overview of the ring expansion route to eight and nine membered cyclic dilactams. Ring expansions of cyclic amides by a pendant amine to form cyclic polyaminoamides has been well documented. However, there are relatively few reports of this reaction being applied to the formation of cyclic amides via the use of an acylated amide. We have further explored the transamidative ring reactions and have developed a procedure which allows synthesis of eight or nine membered rings in almost quantitative yield. N-tBoc protected beta- or gamma-amino acids are coupled to cyclic amides, deprotected and then cyclised under mildly basic conditions. Further studies are in progress to generalise and extend the reaction.

120. **2-METHYLENEOXETANES: VERSATILE SCAFFOLDS FOR HIGHLY FUNCTIONALIZED CARBONYLS.** L. M. Dollinger, and A. R. Howell,* Department of Chemistry, University of Connecticut, Storrs, CT 06269

2-Methyleneoxetanes **1**, containing a pendant alkene, should provide a versatile scaffold for ring closing metathesis reactions, resulting in highly functionalized carbocyclic systems **2**. We have recently described the conversion of β-lactones **3** to 2-methyleneoxetanes. Further, current efforts in our group have demonstrated that ethyl acetoacetate can be readily transformed by a straightforward sequence to highly functionalized β-lactones. The protocol we have developed, the synthetic versatility of the sequence and useful transformations of **2** will be discussed.

121. **FIRST ENANTIOMERIC RESOLUTION OF A SULFENAMIDE WHICH OWES ITS CHIRALITY TO THE S-N CHIRAL AXIS.**
M. Ben-David Blanca, D. Kost, and E. Maimon, Department of Chemistry, Ben Gurion University of the Negev, Beer-Sheva 84105, Israel.

The sulfenamide **1**, possessing an electron withdrawing 2,4-dinitrophenyl group on sulfur and severe steric crowding was prepared. The compound is chiral at the ground state due to the S-N chiral axis. Racemization is effected by rotation about the S-N bond. The free energy barrier for this rotation was measured by NMR to be 21.0 kcal mol^{-1}. At ambient temperature the racemization rate was too fast to allow resolution. However, pure enantiomers were resolved at 0°C by HPLC on a chiral column.

122. **A PRACTICAL ASYMMETRIC BAYLIS-HILLMAN REACTION AND USE OF THE SCALEMIC PRODUCTS**
Sara Rafel, Linda J. Brzezinski, Michael Piber and James. W. Leahy*
Department of Chemistry, University of California, Berkeley, CA 94720-1460

While the Baylis-Hillman reaction is a useful method for the construction of carbon-carbon bonds, it suffers from protracted reaction times and the generation of racemic products. We have developed methods that allow for the formation of these products in a significantly more rapid fashion and with absolute stereocontrol. The details of this work will be discussed, as will some of the uses of this new methodology.

123. **CHIRAL CONTROL OF THE MICHAEL REACTION ON FUNCTIONALIZED CYCLOALKANONES BY CHIRAL AUXILIARIES.** Ivan Jabin and Michel Pfau, Chem. Dept., ESPCI, Paris V, France, & N. D. Heindel, Dept of Chem., Lehigh Univ., Bethlehem, PA. 18015

Use of a chiral auxiliary to direct substitution on a pendant moiety is a valuable tool in construction of chiral natural products. We have probed structural effects on stereochemical control in the Michael reaction where a chiral imine directs attack onto a cycloalkanone scaffold. When the functionalization is at the α' or β position, the corresponding chiral imine displays low reactivity to electrophilic olefins. However, with functionalization in γ, the chiral imine has led to chiral polyfunctional building blocks in good yields and high enantioselectivities. We have exploited this chemistry to prepare a compound with C_2 symmetry, a model construct of the CD steroid ring junction, and the first enantioselective synthesis of (-)Polywood^R, an important compound in perfumery. We have also applied this method to the enantioselective synthesis of (-)TMD, an inhibitor of cholesterol biosynthesis.

124.

NOVEL SYNTHETIC APPLICATIONS OF CATECHOL BORANE DERIVATIVES Mark W. Hooper, Elena Fernandez, John M. Brown. Dyson Perrins Laboratory, South Parks Rd. OXFORD OX1 3QY, U.K.

A novel approach to the work-up of catechol borane derivatives is described. Catechol borane is widely used in catalytic hydroboration. However, the synthetic application of the resulting catechol borane adducts has been limited to basic H2O2 work-up leading to the product alcohols. This work describes the conversion of the catechol borane derivatives to dialkylborane derivatives, using MeMgCl or ZnEt2, therefore activating the boranes towards the standard work-up reagents. This is illustrated by catalytic hydroboration/amination of vinylarenes, giving primary amines of up to 98% e.e.. The activation of catechol borane adducts offers the possibility of extending the scope of catalytic hydroboration to the synthesis of a wide range of functionalities.

125. BICYCLIC ORTHOESTER AS A NEW CHIRAL AUXILIARY FOR ASYMMETRIC ALLYLATION: APPLICATION TO SYNTHESIS OF α-TERTIARY HYDROXY ACIDS. **S. Prescott**, S. Gingras and D. Dubé, Merck Frosst Centre for Therapeutic Research, P. O. Box 1005, Pointe Claire-Dorval, Québec, Canada H9R 4P8.

A simple and efficient method for the preparation of α-keto orthoesters was developed from enantiomerically pure dimethyl tartrate. We wish to demonstrate the utility of the bicyclic orthoester in the diastereoselective nucleophilic addition of allyl reagents to ketones to obtain, after hydrolysis, optically pure tertiary alcohols. High levels of asymmetric induction are achieved with various alkyl and aryl ketones. The effect of Lewis acids and solvents are described.

126.

α-HYDROXY-β-AMINOPHOSPHONATES VIA THE SHARPLESS CATALYTIC ASYMMETRIC AMINOHYDROXYLATION. **Allen A. Thomas**, K. Barry Sharpless, The Scripps Research Institute, Department of Chemistry, 10550 N. Torrey Pines Rd., La Jolla, CA, 92037.

Phosphonates have recently attracted attention as amino acid mimetics for use as inhibitors and as antigens in antibody production. Reliable and inexpensive methods for their synthesis would be beneficial. Using the Sharpless catalytic asymmetric aminohydroxlation (AA), α-hydroxy-β-aminophosphonates of the general type **1** were prepared with high enantioselectivity from the corresponding olefins. The phosphonates **1** were further elaborated to give the α-aminophosphonates **2**.

127.

CATALYTIC AMINOHYDROXYLATION OF α,β-UNSATURATED AMIDES. **A. Erik Rubin** K. Barry Sharpless, The Scripps Research Institute, Department of Chemistry BCC-315, 10550 N. Torrey Pines Rd., La Jolla, CA, 92037

α,β-Unsaturated amides afford *only* racemic products from the Sharpless asymmetric aminohydroxylation (AA) reaction but represent one of the few olefin classes which exhibit excellent rates and yields of the aminohydroxylated products (**1** and **2**) in the absence of external ligands. The scope of the reaction was examined and the hydroxysulfonamides were elaborated to give aziridines **3**. The reactivity of these aziridines with external nucleophiles was also investigated.

128. MECHANISTIC STUDY OF THE JACOBSEN ASYMMETRIC EPOXIDATION OF INDENE. **Ji Liu**, David L. Hughes, George B. Smith, George C. Dezeny, Chris H. Senanayake, Robert D. Larsen, Thomas R. Verhoeven, and Paul J. Reider, RY80Y-245, Merck Research Laboratories, P. O. Box 2000, Rahway, NJ 07065

The asymmetric epoxidation of indene using aq. NaOCl, catalyzed by Jacobsen's chiral manganese salen complex, provides indene oxide in 90% yield and 85-88% enantioselectivity. The axial ligand, 4-(3-phenylpropyl)pyridine N-oxide (P3NO), increases the rate of epoxidation without affecting enantioselectivity, and also stabilizes the catalyst. These two effects afford a reduction in catalyst loading to <1%.

129. SYNTHESIS, ISOMERISM, AND DECOMPOSITION BEHAVIOUR OF PALLADA(IV)CYCLOPENTANE COMPLEXES. **A. J. Canty** and J. L. Hoare, Chemistry Dept, University of Tasmania, Hobart, Tasmania, Australia 7001

The oxidative addition of RX (R=CH$_3$, CD$_3$, CF$_3$, Et, Bz; X = I, Br) to Pd(CH$_2$)$_4$(bpy) gives isolable Pd(IV) products. For R = C$_1$(CH$_3$, CD$_3$, CF$_3$) the Pd(IV) complexes occur as a mixture of isomers **A** (80%) and **B** (20%), but for R=Et and Bz isomer **B** occurs alone. **A** and **B** do not interconvert in solution, and decomposition is more facile from **A**. For R=C$_1$ pentane is the major decomposition product, with pentenes, butenes and iodopentane as minor products. For R=Et, hexane is the major product, but for R=Bz butenes are the major products. Reductive elimination from Pd(IV) appears to be facile for C$_1$ and Et, but does not occur for R = Bz.

130. RUTHENIUM-CATALYZED REDOX ISOMERIZATION OF PROPARGYL ALCOHOLS **Robert C. Livingston** and Barry M. Trost*, Department of Chemistry, Stanford University, Stanford, CA 94305-5080.

One common approach to enals and enones is the sequential reduction and oxidation of readily accessible propargyl alcohols. The internal redox isomerization of propargyl alcohols by indenylruthenium complex **1**, in conjunction with ammonium hexafluorophosphate salts and indium trichloride as cocatalysts, effects this transformation in one step and with excellent chemoselectivity to give exclusively the *trans* product in good yields. Tandem isomerization and conjugate addition of an internal nucleophile allows preparation of heterocycles directly from propargyl alcohol precursors.

131. SOLID PHASE SYNTHESIS AND ON-RESIN CYCLIZATION OF CYCLOSPORINS. Daniel H. Rich,*# Prakash Raman,* Yvonne M. Angell#, George R. Flentke#. Department of Chemistry* and School of Pharmacy,# University of Wisconsin-Madison, 425 N. Charter St., Madison WI 53706.

For over 15 years cyclosporin analogs have been synthesized in solution, and not on solid phase using the Merrifield method, in spite of the obvious advantages of the latter approach. The difficulties of synthesizing CsA peptides on solid supports are caused by the numerous sterically hindered, N-methylated amino acids in the CsA structure. By exhaustive evaluation of coupling reagents, we found that the azobenzotriazole reagents, HATU and/or HOAt/DIPCDI, successfully coupled hindered N-methylated amino acids and enabled us to synthesize analogs of the sterically hindered CsA 8-11 sequence ((D)-Ala^8-MeLeu^9-MeLeu^{10}-MeVal^{11}-MeBmt^1--) on the solid phase resin.[1] We now report the synthesis and on-resin cyclization of a series of cyclosporins. Approaches to the combinatorial synthesis of cyclosporin libraries will be described.

1. Y. M. Angell, T.L. Thomas, G. R. Flentke, and D.H. Rich, J. Amer. Chem. Soc. 1995, 117, 7279-7280.

132.

TOWARDS AN UNDERSTANDING OF B-SHEET STRUCTURE: FROM AMYLOID TO DESIGNER B-SHEETS. **Jeffery W. Kelly** Department of Chemistry, Texas A&M University, College Station, Texas 77843

This talk will focus on our current understanding of the biophysical mechanism of amyloid fibril formation. Mechanistic insights coming from these studies have been used to develop a new small molecule approach to intervene in human amyloid disease, an approach which is now undergoing clinical evaluation. The second part of this talk will focus on the progress made in developing templated b-sheet structures. Recent work involving the folding of Gly-Ala sequences into monomeric sheets provide further insights into the mysteries of b-sheet structure.

133. DESIGN AND CHARACTERIZATION OF INDEPENDENTLY FOLDED POLYPEPTIDE DOMAINS. **Barbara Imperiali**, Division of Chemistry and Chemical Engineering, California Institute of Technology, Pasadena, CA 91125.

The polypeptide architecture, defined by a network of noncovalent interactions, shows remarkable versatility of structure and function and represents an enticing target for synthetic manipulation. Our approach to *de novo* protein synthesis is based on the assembly of "compact" supersecondary structural motifs (ca. 20-40 amino acids in length). This limited size makes the polypeptides accessible through solid phase peptide synthesis, permitting incorporation of a wide array of natural and synthetic amino acids and allowing for comprehensive solution state structure evaluation by 2-D NMR and CD techniques.

In this presentation I will describe our recent progress in the assembly of independently folded polypeptide motifs. While small proteins or protein domains generally require disulfide bridges or metal sites for their stabilization, a strategy for circumventing these cross-linking restraints has been developed. Recent studies on a metal ion independent, 23-residue, ββα motif (BBA1), derived from the zinc fingers domains highlight the application of this approach. Additionally, progress in the assembly of functional polypeptide motifs that incorporate a new class of coenzyme-amino acid chimera residues will be described.

134. PEPTIDE RESEARCH: AN INTERDISCIPLINARY CHALLENGE. **Murray Goodman**, Department of Chemistry and Biochemistry, University of California, San Diego, San Diego, California, 92093-0343.

Peptides come in many forms and are enormously varied in structure and function. The development of drugs, immunomodulating agents, antibiotics and neuropeptides, among others depend on research by organic chemists, biochemists, biophysicists, pharmacologists and molecular biologists. It is in this interdisciplinary vein that we commenced our research many years ago. We developed synthetic methodologies, investigated conformations of oligopeptides in solution and studied structure-bioactivity relationships of peptides and peptidomimetics. In this symposium, our integrated program will be explored involving the interplay of synthesis, spectroscopy and computer simulations together with biological studies. The synthesis of novel building blocks will be described as well as their incorporations into opioids and somatostatin as target drug families and triple helical collagen mimetics. Using our approach to peptidomimetics we are building a bridge to the twenty-first century.

135. PALLADIUM-CATALYZED OXIDATION OF ALLENES AND CONJUGATED DIENES J. E. Bäckvall, Department of Organic Chemistry, University of Uppsala, Box 531, S-751 21 Uppsala, Sweden

The palladium-catalyzed 1,4-oxidation of conjugated dienes proceeds via 4-substituted $(1,2,3)$-η^3-allyl-palladium complexes. Nucleophilic attack on these π-allyl intermediates and subsequent reoxidation of palladium gives the product. There is an inherent problem to obtain carbon-carbon bond formation in oxidation reactions due to the ease of oxidation of carbon nucleophiles. We were recently able to overcome this problem in the palladium-catalyzed 1,4-oxidation of conjugated dienes and have obtained carbon-carbon bond formation via three different approaches: (i) via vinylpalladation (ii) via nucleophilic attack by an allylsilane and (iii) via nucleophilic attack by a stabilized carbon nucleophile.

We recently developed a related palladium-catalyzed oxidation of allenes. This reaction proceeds via nucleophilic attack on an allene-palladium complex to produce a 2-substituted $(1,2,3)$-η^3-allyl-palladium complex, which is oxidized in an analogous manner as that of the 1,3-diene reaction. In this way an overall 1,2-addition to the allene is obtained.

136. ASYMMETRIC SYNTHESES OF NOVEL AMINO ACID DERIVATIVES VIA PALLADIUM CATALYZED ENANTIOSELECTIVE REACTIONS. Barry M. Trost, Department of Chemistry, Stanford University, Stanford, CA 94305-5080.

Peptidomimetics frequently include unusual amino acid analogues and surrogates. New strategies for synthesizing such systems asymmetrically employing palladium catalyzed enantioselective alkylations have been developed. In one, deracemization of vinyl epoxides with regioselective formation of the new bond at the allylic position proximal to the alcohol substituent can lead to asymmetric formation of a quaternary center. In the second version, enantioselective alkylation of azlactones for creation of quaternary centers asymmetrically is explored.

137. **PALLADIUM-CATALYZED HYDROARYLATION OF PROPIOLAMIDES: A STEREO- AND REGIOSELECTIVE SYNTHESIS OF TRISUBSTITUTED α, β-UNSATURATED AMIDES.** Lynne A. Hay, Thomas M. Koenig, Francis O. Ginah, James D. Copp and <u>David Mitchell</u>,* Chemical Process R&D, Lilly Research Laboratories, A Division of Eli Lilly and Company, Indianapolis, IN 46285-4813.

Unlike most propiolates, propiolamides undergo the Heck reaction with a variety of aryl halides or triflates in good to excellent yields. These resulting arylpropiolamides, in the presence of palladium (0), react in a stereo- and regioselective hydroarylation reaction to provide trisubstituted α, β-unsaturated amides. By judicious choice of the aryl coupling partner and reaction sequence, either the Z or E stereoisomer can be independently prepared.

138. **ROLE OF HALIDE ION IN PALLADIUM(II) CATALYZED CARBON-CARBON COUPLING REACTIONS.** <u>X. Lu</u>,* Z. Wang and Z. Zhang, Shanghai Institute of Organic Chemistry, Chinese Academy of Sciences, Shanghai 200032, China

Vinyl palladium species usually react with substituted olefins to give vinylation products through β-hydride elimination of the insertion intermediate. We found that the presence of excess halide ions may suppress the β-hydride elimination and enhance the protonolysis of the mesomeric palladium enolates (Y=COR) or the β-heteroatom elimination (Y=CH_2X).

139. **FACILE CONSTRUCTION OF COMPLEX ORGANIC SKELETONS BY PALLADIUM-CATALYZED CASCADE REACTIONS.** <u>Armin de Meijere</u>, Institut für Organische Chemie der Georg-August-Universität Göttingen, Tammannstrasse 2, D-37077 Göttingen, Germany

The Heck reaction and mechanistically analogous carbopalladations are among the most important modern C-C bond forming processes. Several such cross coupling steps can be combined to occur in a reaction cascade which may also involve other thermal rearrangements or cycloadditions in between or at the end. Such transformations of appropriately substituted open-chain or monocyclic precursors yield complex oligocyclic all carbon or heteroatom containing skeletons (e. g. **1** → **2**). The increase in molecular complexity in such transformations during a single synthetic operation can be quite remarkable. The basic principles[1] will be discussed and recent examples presented.

[1] A. de Meijere, F. E. Meyer, *Angew. Chem. Int. Ed. Engl.* **1994**, *33*, 2379–2411.

140. **Syntheses of Indoles *via* a Palladium-catalyzed Annulation Between Iodoanilines and Ketones** Cheng-yi Chen*, David R. Lieberman, Robert D. Larsen, Thomas R. Verhoeven, and Paul J. Reider, Department of Process Research, Merck Research Laboratories, Division of Merck & Co., Inc., P. O. Box 2000, Rahway, New Jersey 07065, USA

Abstract: The indole nucleus is a common and important feature of a variety of natural products and medicinal agents. Due to their potent biological activity these compounds continue to be attractive synthetic targets. A new and efficient method for indole synthesis using a palladium-catalyzed annulation between o-iodoanilines and ketones will be described.

141. TOWARD THE IDEAL SYNTHESIS. **Paul A. Wender**, Department of Chemistry, Stanford University, Stanford, California 94305, USA

One of the most significant challenges in synthesis is the development of new reactions and strategies that allow for the conversion of simple starting materials into structurally complex targets in a minimum number of steps, ideally one. Encompassing all of the objectives of selectivity and economy, such ideal syntheses would provide the target molecule in 100% yield and in a cost effective, operationally simple, safe, and environmentally acceptable fashion – allowing for the realization of the full potential of synthesis and its role in chemistry, biology, medicine, environmental science, and materials science. In this lecture, the two fundamental approaches to the ideal synthesis will be discussed from a theoretical viewpoint based on complexity and connectivity analysis and exemplified by studies on new reactions and strategies for the synthesis of new materials and medicinal leads.

142. **THE MULTICOMPONENT REACTIONS (MCR'S) AND THEIR LIBRARIES - PERSPECTIVES AT A PROGRESS IN ORGANIC CHEMISTRY AND ITS APPLICATIONS.** Ivar K. Ugi, Alexander Doemling, Bernhard Gruber Organ.-Chemie u. Biochemie I - Lichtenbergstr. 4, D- 85747 Garching

For a whole century the isocyanides were then not well available, and their chemistry was the part of a rather empty field. At the same time the classical MCR's were introduced, which are generall α–aminoalkylations of nucleophiles and their secondary reactions with further components. From 1958 on, the isocyanides could be prepared well, and one year later the four component reaction of the isocyanides, the Ugi-4CR, was introduced, and since 1993 the 7CR's and even higher MCR's became available, which have advantages in preparative chemistry and its libraries. Already in 1961 the MCR 'libraries' were introduced. Simultaneously the advantageous mathematically oriented computer supported methods were developed and investigated. At the conference „Cambridge Healthtech Institute", La Jolla, CA, on Jan 23-25, 1995, a poster of these results has started the activity of MCR libraries of the still growing activities of the MCR's and their libraries.

143. SYNTHESIS OF ORGANOFLUORINE COMPOUNDS WITH REAGENTS DERIVED FROM HYDROFLUORIC ACID. **T. Hiyama**, Research Laboratory of Resources Utilization, Tokyo Institute of Technology, Nagatsuta, Midori-ku, Yokohama 226, Japan

Organofluorine compounds are the key for the future biological and materials sciences. To introduce fluorine atom into organic substrates, highly reactive molecular fluorine and the reagents derived from it are often used. In view of economy, however, it is desirable to utilize agents close to natural resources. In this sense, hydrofluoric acid and its derivatives should be the reagents of choice. We observed that 1 M hydrofluoric acid in combination with N-halosuccinimide and a phase-transfer catalyst n-Bu$_4$NF could convert olefins into the halofluorination products. The active nucleophilic species was assumed to be n-Bu$_4$NH$_2$F$_3$. The reaction using a halonium ion oxidant and the HF-derived reagent allows us to convert various types of C-S bonds into C-F bonds, and we call it *Oxidative Desulfurization-Fluorination Reaction*. Examples of the reaction will be presented and discussed.

144. PROCESS R&D DIRECTED TOWARD THE SYNTHESIS OF CYCLIC UREA DIOLS AS INHIBITORS OF HIV PROTEASE. **Pat N. Confalone**, Chemical Process R&D, The Dupont Merck Pharmaceutical Company, Wilmington, DE 19980

A number of syntheses of cyclic urea diols of general structure **4** from the chiral pool starting material (L)-tartaric acid (**1**) are described. Considerations such as overall yield, reproducibility, cost, thermochemistry, and environmental in the context of developing a commercial synthesis are presented.

Installation of the key R,S,S,R stereochemistry in the cyclic urea system, reductive amination of the diamine precursor **2** vs N-alkylation of the urea **3**, and cylization protocols to yield the 7-membered cyclic urea are discussed.

145. Enzymes, Electrochemistry, and Efficiency in Iterative Design for Complex Molecules.
Tomas Hudlicky, Department of Chemistry, University of Florida, Gainesville, FL 32611

Tandem enzymatic and electrochemical methods are exploited to decrease further the length of a given synthetic sequence, with a greater part of the sequence performed in either water or alcohols, frequently without isolation. The emphasis is on iterative and self-correcting multi-generation design of a given target. Progress toward practical, large scale syntheses of morphine **1**, pancratistatin **2**, and selected carbohydrates will be presented.

146. TOTAL SYNTHESIS OF STIPIAMIDE USING A TIN CUPRATE ACETYLENE ADDITION REACTION, AND A NEW MULTI-DRUG RESISTANCE REVERSAL AGENT, 6,7-DEHYDRO-STIPIAMIDE
Merritt B. Andrus* and Salvatore D. Lepore
Purdue University, Department of Chemistry, West Lafayette, IN 47907

(-)-Stipiamide has been synthesized in a direct, convergent manner using a tin cuprate acetylene addition reaction and a Stille tin coupling as key steps for the assembly of the challenging *E,E,Z,E,E* - polyene array. Stipiamide, a polyene antibiotic of wide ranging biological activity, has recently been found to contravene P-glycoprotein-mediated multidrug resistance. The synthesis of stipiamide demonstrates the utility of the tin cuprate acetylene addition reaction. and provides a starting point for the design of new compounds that inhibit P-glycoprotein. In particular, assays of adriamycin resistant cells using 6,7-dehydro-stipiamide show MDR reversal activity to be at least 10 times better than verapamil (a widely used MDR reversal agent).

147. Stereoselective Synthesis of *des*-C Analogues of 1α,25-Dihydroxyvitamin D$_3$
Salvador G. Alvarez, Tomohiro Yamazaki, and Isao Shimizu*
Department of Applied Chemistry, School of Science and Engineering, Waseda University
3-4-1 Ohkubo, Shinjuku-ku, Tokyo 169 JAPAN, Email: shimizui@cfi.waseda.ac.jp

As part of our research program directed toward the design and synthesis of novel analogues of vitamin D$_3$ (**1a**) for detailed pharmacological characterization, we report the stereoselective synthesis of *des*-C-1α,25-dihydroxyvitamin D$_3$ (**2**), a novel analogue of vitamin D$_3$ (**1a**). The key step involved the reductive cleavage of **6** to give **5** which corresponds to the D-ring synthon with the proper stereochemistry at C-21 (Shimizu, I.; et al. *J. Org. Chem.* **1993**, *58*, 1483). Elaboration of **5** afforded the D-ring **3**. A-ring synthon **8** was readily prepared by our published procedure (Shimizu, I.; et al. *J. Org. Chem.* **1993**, *58*, 2523). Coupling of **8** with **3** followed by desilylation gave **2**. Synthesis of **2** and other analogues will be discussed.

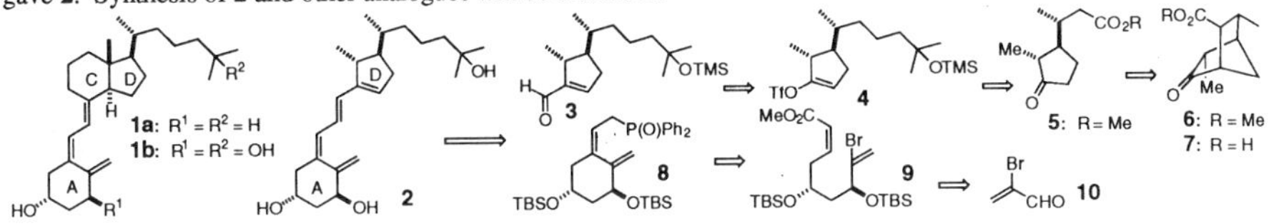

1a: R^1 = R^2 = H
1b: R^1 = R^2 = OH
5: R = Me
6: R = Me
7: R = H

148. STEREOSELECTIVE SYNTHESIS OF WOODY-AMBER, POLYSUBSTITUTED DECALINS.
G. Fráter, U. Müller, C. Nussbaumer, Givaudan-Roure Research Ltd, 8600 Dübendorf, Switzerland

The acid catalyzed cyclization of the Diels-Alder adduct of myrcene and (E)-4-methyl-pent-3-en, 2-one, furnished a minor side product **1** with a powerful woody-amber odor. **1** has been stereoselectively synthesized from α-ionone, the key step being the lactone-enolate Claisen rearrangement of **2**. **1** displays a striking stereoselectivity of the odor strength; the enantiomers of **1** differ in odor threshold by a factor of at least 10^4. The double bond isomer and epimer of **1**, i. e. **3** has been dia- and enantioselectively prepared from homomyrcene and is as powerful as **1**.

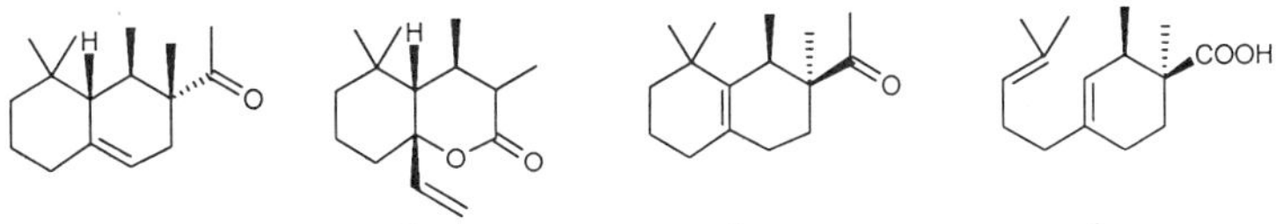

149.

NOVEL USE OF AN INTRAMOLECULAR ATOM TRANSFER REACTION OF A TRIMETHYLENEMETHANE (TMM) DIYL IN THE TOTAL SYNTHESIS OF CONFERTIN. R. D. Little, M. K. Schwaebe and **A. L. Allan**, Department of Chemistry, University of California, Santa Barbara, Santa Barbara, CA. 93106

The intramolecular atom transfer reaction of the trimethylenemethane (TMM) diyl provides an efficient route to functionalized [5.3.0] and [6.3.0] systems. Application of this method will be highlighted in our efforts toward the total synthesis of the pseudoguaianolide confertin.

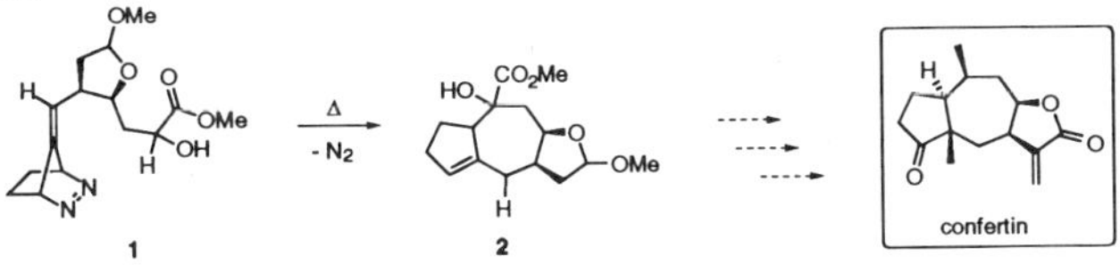

150. **ENANTIOSELECTIVE TOTAL SYNTHESIS OF (–)-KUMAUSALLENE.**
P. Andrew Evans* and Jamie D. Roseman Lammot du Pont Laboratory, Department of Chemistry and Biochemistry, University of Delaware, Newark, DE 19716.

The first enantioselective total synthesis of the halogenated sesquiterpene (-)-kumausallene will be described. The synthesis utilizes an intramolecular acyl radical cyclization to form the key cis-2,5-disubstituted tetrahydrofuran-3-one which is then further elaborated into the natural product.

151. **SINGLET OXYGENATION/RADICAL REARRANGEMENT AS A GENERAL APPROACH TO 1,4-DIOXYGENATED PEROXIDE SUBUNITS: ASYMMETRIC TOTAL SYNTHESES OF CHONDRILLIN AND PLAKORIN** Patrick H. Dussault and Kevin R. Woller
Department of Chemistry, University of Nebraska–Lincoln, Lincoln, NE 68588-0304

The allylic hydroperoxides derived from singlet oxygenation of allylic alcohols undergo stereoselective radical rearrangement to furnish single enantiomers of 4-peroxy-2-en-1-ols. The strategy is illustrated with the first asymmetric total syntheses of the marine natural products chondrillin and plakorin.

152. **PROGRESS TOWARDS THE SYNTHESIS OF ZARAGOZIC ACID C: USE OF REVERSIBLE KETALIZATION TO CONTROL THE STEREOCHEMISTRY OF THE 2,8-DIOXABICYCLO[3.2.1]OCTANE CORE.** Sayee G. Hegde and David C. Myles*, Department of Chemistry and Biochemistry, University of California, Los Angeles, 90095-1569.

The zaragozic acids (ZAA)/squalestatins act as cholesterol-lowering agents by inhibiting squalene synthase (IC_{50} = 12.5 nM). Several elegant total syntheses have been reported. The reversible ketalization of tetrahydroxyketone **6** to obtain the desired stereochemistry and shape of the emerging ketal has been studied. The application of this *novel* approach in the synthesis of zaragozic acid C will also be discussed.

153. **AN EFFICIENT AND FLEXIBLE TOTAL SYNTHESIS OF OKADAIC ACID.**
Craig J. Forsyth,* Steven F. Sabes, and Rebecca A. Urbanek, Department of Chemistry, University of Minnesota, Minneapolis, MN 55455.

An efficient and flexible total synthesis of the marine natural product okadaic acid has been developed. This involves the sequential coupling of 3 fragments, representing carbons C1-C14, C15-C27, and C28-C38 of the natural product. Direct addition of a $CeCl_3$-modified C28 anion to a C27 β,γ-unsaturated aldehyde lead succinctly to a C15-C38 intermediate. Condensation of a C14 aldehyde with a C15 ketophosphonate, followed by CBS / BH$_3$ reduction of the resultant C16 ketone, spiroketalization, and deprotection completed the synthesis in ca. 27 steps in the longest linear sequence.

Okadaic Acid

154. THE FIRST ENANTIOSELECTIVE SYNTHESIS OF A DAPHNANE DITERPENE: THE TOTAL SYNTHESIS OF RESINIFERATOXIN. Paul A. Wender,* Cynthia D. Jesudason, Hiroyuki Nakahira, Norikazu Tamura, Anne Louise Tebbe, Yoshihide Ueno, Department of Chemistry, Stanford University, Stanford, California 94305

Resiniferatoxin, a naturally occurring daphnane diterpene, was identified on the basis of its extraordinary irritant activity. Resiniferatoxin displays a structural similarity to capsaicin, which is the major active constituent of common red pepper and acts as a ultrapotent capsaicin analog. The first enantioselective total synthesis of resiniferatoxin will be described.

Resiniferatoxin   Capsaicin

155. TAXANES FROM PINENE: AN UNEXPECTED AND HIGHLY TORQUOSELECTIVE NAZAROV CYCLIZATION. Heidi R. Vollmer and F. G. West, University of Utah, Department of Chemistry, Salt Lake City, UT 84112.

Dienone **1** can be easily prepared via three efficient steps from pinene. Although related substrates could be converted to the *exo* cyclopentenone product under standard Nazarov conditions, **1** underwent conversion to *endo* cyclopentenone **2** with complete torquoselectivity. Product **2** contains all of the carbons of the taxane AB skeleton. The origins of this selectivity and approaches toward carrying **2** to taxane natural products will be discussed.

156. THE PINENE PATH TO TAXANES: A CONCISE STEREOCONTROLLED SYNTHESIS OF TAXOL®. Paul A. Wender,* Neil F. Badham, Simon P. Conway, Paul E. Floreancig, Timothy E. Glass, Christian Gränicher, Jonathan B. Houze, Jan Jänichen, Nancy E. Krauss, Daesung Lee, Daniel G. Marquess, Paul L. McGrane, Wei Meng, Thomas P. Mucciaro, Michel Mühlebach, Michael G. Natchus, Holger Paulsen, David B. Rawlins, Jeffrey Satkofsky, Anthony J. Shuker, James C. Sutton, Richard E. Taylor, and Katsuhiko Tomooka. Department of Chemistry, Stanford University, Stanford, California 94305-5080

Taxol® has served as a stimulus for the development of fundamentally new advances in chemistry, biology, and medicine. In connection with efforts to improve the availability of Taxol® and its precursors, to elucidate its novel mode of action, and to identify clinically superior agents, this study describes a stereocontrolled synthesis of a highly versatile taxane precursor which provides concise access to Taxol® analogs and conversion of this precursor into Taxol® itself. The strategy is based on pinene, an abundant and inexpensive building block. The use of a conformational control element for functionalization of the B-ring as well as the development of a late-stage intramolecular aldol closure for formation of the C-ring of taxanes is also described. This strategy results in a uniquely concise total synthesis of Taxol® in 37 steps and provides even more concise access to key analogs.

157.

SIMPLE SYNTHETIC PROCEDURES FOR TRANSFORMATION OF ALKYL AND ARYL HALIDES INTO 2-ALKYL AND 2-ARYLTHIO-1,3,4-OXADIAZOLES. **B. S. Jursic** and Blaise LeBlanc, Department of Chemistry, University of New Orleans, New Orleans, Louisiana 70148

An efficient method for the preparation of 5-alkylthio and 5-arylthiotetrazoles and 2-alkylthio and 2-arylthio-1,3,4-oxadiazoles will be presented. The procedure begins with alkyl and activated aryl halides or activated benzene derivatives. These were transferred to thiocyanates through the [3+2] cycloaddition of the azide anion to thiocyanates under PTC conditions. The 5-thioalkyl and 5-thioaryltetrazoles were prepared and then transformed into corresponding oxadiazoles.

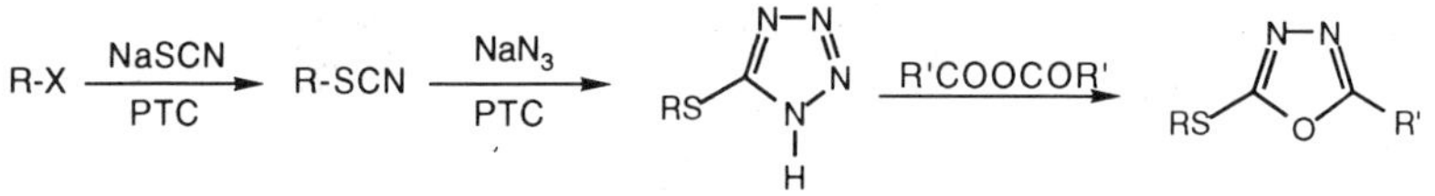

158.

PREPARATION OF MALONIC ACID AMIDE DERIVATIVES THROUGH DECARBONYLATION OF CORRESPONDING BARBITURIC ACIDS. **B. S. Jursic** and Heather Boyle, Department of Chemistry, University of New Orleans, New Orleans, Louisiana 70148

The preparation of malonic acid amides is not a straightforward synthetic process. On the other hand, preparation of a wide variety of barbituric acids is well established. A very efficient way of preparing malonic amides is through decarbonylation of barbituric acid in basic media. The procedures are simple and the yields of this preparation are very high.

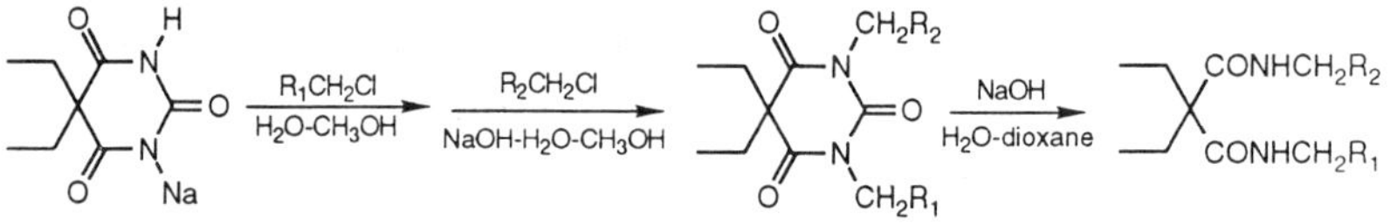

159. SYNTHESIS OF STEROIDAL THIAZOLIDINONES

D. RAMESH and K. VENKATESHWER RAO
DEPT. OF CHEMISTRY, OSMANIA UNIVERSITY
HYDERABAD, A.P. INDIA

Extranucleohetero steroids with thiazolidinone ring condensed to the steroidal system are endowed with variety of biological activities. The steroidal ketone 17-0x0-5-androsten - 3 - ßyl acetate on reaction with hydrazine hydrate gave 17-hydrazone derivative which on base catalysed condensation with benzaldehyde furnished schiff's base. The subsequent reaction of schiff's base with thioglycolic acid afforded thiazolidinone derivative. The structure of the compound was confirmed by spectral data.

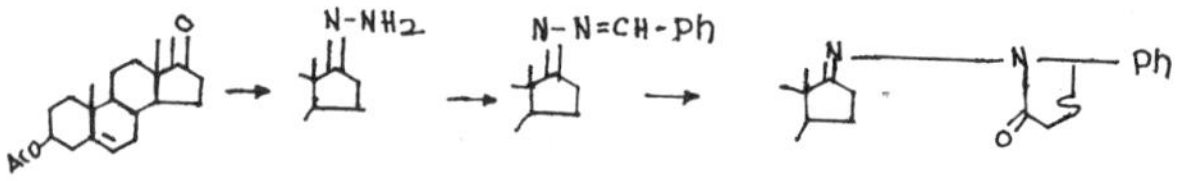

160. PREPARATION OF NEW BENZIODAZOLES
Ruslan M. Arbit, Marc McSherry, Brian Mismash, and Viktor V. Zhdankin*
Department of Chemistry, University of Minnesota - Duluth, Duluth, MN 55812.

Only few examples of hypervalent iodine derivatives with I-N bonds were reported in the literature. Most of these compounds lack stability and are sensitive to moisture. Nevertheless, some of them are useful reagents for organic chemistry. A variety of new, stable derivatives of benziodazole can be prepared from a readily available acetate **1** by ligand exchange on iodine. The preparation, structure and chemistry of new benziodazoles **2,3** will be discussed.

R = p-Tol, Me, CF$_3$ X = N$_3$, CN, OAlk, NHC(O)Alk

161.   THE REACTION OF N-PHENYLSULFONIMIDOYL CHLORIDE WITH TRIMETHYLSILYLETHENE. A NEW ROUTE TO 2-ALKENYLANILINES.
Michael Harmata*, Darin E. Jones, Mehmet Kahraman and Susan Weatherwax, Department of Chemistry, University of Missouri-Columbia, Columbia, Missouri 65211

Benzothiazine **1** is available from the reaction of N-phenyl-S-(4-methylphenyl)sulfonimidoyl chloride with trimethylsilylethene. This compound can be deprotonated, alkylated to give benzothiazines **2**. Treament with fluoride and subsequent hydrolysis leads to 2-alkenylanilines in fair to good yields.

162. NOVEL REACTION OF AMINOPYRIDINES WITH GLYOXAL AND FORMALDEHYDE; SYNTHESIS OF 4,8-Di(N-AMINOPYRIDYL) 2,8- DIOXA 4,8-DIAZABICYCLO[3.3.0]OCTANE, 6,8-Di(N-AMINO-PYRIDYL) 2,4-DIOXA 6,8-DIAZABICYCLO[3.3.0]OCTANE AND X-RAY STRUCTURAL STUDY OF RELATED 1,3-Di(N-AMINOPYRIDYL) 4,5-DIHYDROXY IMIDAZOLIDINE. S. M. F. Farnia, A. Kakanejadifard and F. Raofi, Department of Chemistry, University of Tehran, Tehran, Iran

Condensation of 2-aminopyridine with glyoxal proceeds with high selectivity to a mixture of meso and dl diol to be easily transformed into the corresponding bicyclooctanes with formaldehyde and acetonitrile as solvent. In water, however, the reaction selectively produces imidazolidine dihydroxy compound. Based on NMR analysis, the major diastereomers were assigned as syn derivatives. X-ray crystal structure of dihydroxy imidazolidine shows a planar configuration for the ring system, in line with two anomeric effects, first a strong $n_N \rightarrow \sigma^*_{C-O}$ interaction and second a weak $n_O \rightarrow \sigma^*_{C-N}$ one. A network of hydrogen bonding between pyridyl nitrogens and hydroxy hydrogens forms a pattern in a chair-conformation.

163.

SYNTHESES OF 2-AMINOTRIMETHYLIMIDAZOPYRIDINE (TMIP) ISOMERS. **Todd K. Tochimoto,** James E. Bupp, Mary J. Tanga . SRI International 333 Ravenswood Ave, Menlo Park, CA 94025

Heterocylic amines are endogenous chemicals found in cooked and heat processed protein-rich foods that constitute a significant health hazard. To assess the risk that consumption of these mutagens pose to humans, it is essential to isolate, identify, and synthesize these compounds. While some of the heterocyclic amines have been identified and synthesized, one that contributes 10-15% of the total mutagenic activity of fried meat samples has only been identified by mass spectra to have a molecular weight of 176. From the available data, the mutagenic compound was determined to be one of the twelve isomers of 2-aminotrimethylimidazopyridine (TMIP). The syntheses of some of the isomers of this potential food mutagen will be presented.

164. **SYNTHESES OF BENZO[k]FLUORANTHENES VIA O-QUINODIMETHANES.**
James E. Bupp, Mary J. Tanga. SRI International 333 Ravenswood Ave, Menlo Park, CA 94025

A very important class of polycyclic aromatic hydrocarbons (**PAHs**) is the benzo[k]fluoranthenes. The carcinogenic and mutagenic properties of these compounds make them interesting and important synthetic targets. Benzo[k]fluoranthenes are environmental carcinogens which have been detected in automobile exhaust, cigarette smoke, soil, drinking water, marine sediments and broiled and smoked foods. Utilization of o-quinodimethanes in Diels Alder reactions is a highly efficient way to produce polycyclic molecules. The synthesis of benzo[k]fluoranthenes based upon trapping o-quinodimethanes generated from sultines is described.

165. SYNTHESIS OF TRIAZOLOPYRIDAZINES AS PRECURSORS FOR NOVEL HIGH DENSITY INSENSITIVE ENERGETIC MATERIALS. K. C. V. Ramanaiah and Mark L. Trudell, Department of Chemistry, University of New Orleans, New Orleans, LA 70148.

The synthesis of triazolopyridazines **1** and **2** have recently been completed. The compounds **1** and **2** have been envisaged as useful precursors for the preparation of novel energetic materials (**3**). The synthesis of **1** and **2** as well as progress toward the preparation of **3** will be reported.

166. SYNTHESIS OF NEW DIPYRIDOTETRAAZAPENTALENE DERIVATIVES AS POTENTIAL HIGH DENSITY INSENSITIVE ENERGETIC MATERIALS. Devan Balachari and Mark L. Trudell, Department of Chemistry, University of New Orleans, New Orleans, LA 70148.

The synthesis of two new tetranitrodipyridotetraazapentalene derivatives **3** and **4** have been prepared from the corresponding pyridotriazoles **1** and **2**. Arylation of **1** and **2** with 2-chloro-3-nitropyridine, followed by reductive cyclization in triethylphosphite and nitration (HNO_3/H_2SO_4, 80 °C) furnished **3** and **4** in good overall yields. The details of the syntheses and physical properties of **3** and **4** will be presented.

167. SYNTHESIS AND STRUCTURE OF NEW FURAZAN DERIVATIVES. Alexander K. Zelenin, Edwin D. Stevens, Mark L. Trudell, Department of Chemistry, University of New Orleans, New Orleans, LA 70148.

Diaminofurazan **1** has been shown to be a useful precursor for the synthesis of energetic furazan derivatives as well as bioisosteres of urea derivatives. An improved two-step synthesis of **1** will be described. In addition, the synthesis of the new energetic material **2** as well as aryl furazans (**3**) and diaryl furazans (**4**) will be reported.

168. [4 + 2]CYCLOADDITION REACTION OF PYRROLES WITH 3-BROMOPROPIOLATES.
Chunming Zhang and Mark L. Trudell, Department of Chemistry, University of New Orleans, New Orleans, Louisiana 70148.

3-Bromopropiolates (**1**) underwent a smooth [4 + 2] cycloaddition reaction with *N*-acyl pyrroles (**2**) to afford *N*-acyl-7-azabicyclo[2.2.1]heptadiene derivatives (**3**) in good yields. The 7-azabicycloheptadienes (**3**) could then be converted into highly functionalized derivatives (**4**) *via* a variety of coupling reactions.

169. PREPARATION AND ALKYLATION OF 3-(PHENYLSULFONYL)-2-ALKYL-2,3-DIHYDRO-ISOINDOL-1-ONES Frederick A. Luzzio* and Deanna Piatt-Zacherl
Department of Chemistry, University of Louisville, Louisville, Kentucky 40292

3-(Phenylsulfonyl)-2,3-dihydroisoindol-1-ones (**1**) were prepared by reduction of a series of N-substituted phthalimides with aluminum amalgam followed by phenylthiation of of the resulting hydroxylactams. The 3-(phenylthio)-2,3-dihydroisoindol-1-ones (**2**) were oxidized to the corresponding sulfones with *m*-chloroperbenzoic acid. The sulfones were deprotonated by means of sodium hydride/DMSO and the resulting anions were treated with alkyl halides giving alkylated products.

(**1**) R_1=alkyl, R_2=phenylsulfonyl

(**2**) R_1=alkyl, R_2=phenylthio

170. TRIMETHYLSILYL DERIVATIVES AS CARBANION EQUIVALENTS: A CONVENIENT ROUTE TO 4,4'-*BIS*-(HALOMETHYL)-2,2'-BIPYRIDINES
Jaydeep J. S. Lamba, Natia R. Anastasi, **Cassandra L. Fraser*** Department of Chemistry, University of Virginia, McCormick Rd, Charlottesville, VA 22901

One of the most widely used ligands in inorganic chemistry is bipyridine (bpy) and its various derivatives. From foundational studies in coordination chemistry to present day research in biological electron transfer and supramolecular chemistry, bpy has played a central role. In the course of trying to develop clean, high yield routes into 4,4'-*bis*-(halomethyl)-2,2'-bipyridines, it was discovered that these widely used compounds could be readily accessed in high yield by reaction of 4,4'-*bis*-[(CH$_3$)$_3$SiCH$_2$-]-2,2'-bpy with BrCF$_2$CF$_2$Br and Cl$_3$CCCl$_3$ in the presence of n-Bu$_4$NF/SiO$_2$. More profoundly perhaps, these results demonstrate a new mild way of masking and unmasking carbanions for subsequent elaboration. Investigation of the scope of this reaction of various silyl derivatives with electrophiles will be described.

171. FORMATION OF A CARBON-NITROGEN BOND VIA A 5-EXO-TRIG RADICAL CYCLIZATION ONTO THE NITROGEN OF AN IMIDATE ESTER
Cynthia K. McClure*, Jeffrey S. Link, Anthony J. Kiessling[†]
Montana State University, Department of Chemistry and Biochemistry, Bozeman, MT 59717
[†]Salisbury State University, Department of Chemistry and Biochemistry, Salisbury, MD 21801

Treatment of imidate ester **1** with various radical initiators has shown to undergo cyclization through a 5-exo-trig radical cyclization. Results of these cyclizations show that when imidate ester **1** is treated with Bu$_3$SnH and AIBN, under standard conditions, the cyclized product **2** (33%) and the aryl-bromide reduced form of **1** (23%) were the two major products isolated with no 6-endo cyclization product detected. Results of the cyclization of **1** using $((CH_3)_3Si)_3SiH$, BEt_3, and O_2 as the radical initiator, were similar, with the aminal **3** and the aryl-bromide reduced form of **1** being the only two products detected by GC/MS.

172. **DIRECT CONVERSION OF HYDROXYQUINOLINES TO AMINOQUINOLINES**
  John J. Weidner and Norton P. Peet, Medicinal Chemistry, Hoechst Marion Roussel, 2110 E. Galbraith Rd., Cincinnati, OH 45215

A concise, one-pot, two-step route to aminoquinolines and related heterocycles has been developed, as exemplified by the conversion of 8-hydroxyquinoline (**1**) to 8-aminoquinoline (**2**).

The key step is the Smiles rearrangement and *in situ* hydrolysis of an 2-aryloxyacetamide, resulting in the direct conversion of a heterocyclic aromatic alcohol to its corresponding heterocyclic aromatic amine.

173. **DIHYDROPYRONES AS DIENOPHILES IN THE DIELS-ALDER REACTION.**
  Deqi Chen, Nancy I. Totah,* Department of Chemistry, University of Iowa, Iowa City, IA 52242

A new strategy for the preparation of oxydecalin systems of type **1** is described via the Diels-Alder reaction of functionalized dihydropyrone dienophiles. The preparation of several dienophiles of this type is described. The use of these substrates in both inter- and intramolecular Diels-Alder reactions is reported.

174. **SYNTHESIS OF 2,6-DIBENZOYL-5-METHYL-3-STYRYL BENZO [1, 2-b: 5, 4-b'] DIFURANES UNDER PHASE TRANSFER CATALYSIS.**
  P. Sampath Rao and D. Ashok, Department of Chemistry, P.G. College of Science, Saifabad Hyderabad 500 004, INDIA.

Synthesis of 2,6-dibenzoyl-5-methyl-3-styryl benzo [1, 2-b: 5, 4-b'] difuranes have been synthesized with a view to study their ease of formation and evaluate biological activity. These bifuranes have been synthesized by condensing substituted resorcinol with α-bromoacetophenone under two different conditions: (1) refluxing in acetone with K_2CO_3 as a base and (2) with phase transfer catalysis (PTC) which is advantageous over first method with respect to reaction conditions, yield and purity of the products.

175. **ACID PROMOTED REARRANGEMENT-CYCLIZATION OF *t*-BUTYL STYRYL SULFOXIDES AS A NEW METHODOLOGY FOR THE SELECTIVE SYNTHESIS OF BENZO[*b*]THIOPHENES.** Tony Y. Zhang*, John C. O'Toole, James Aikins, and Kevin A. Sullivan, Chemical Process R&D, Lilly Research Laboratories, Eli Lilly and Company, Lilly Corporate Center, Indianapolis, IN 46285-4813

t-Butyl styryl sulfoxides were found to undergo facile fragmentation-cyclization to give benzo[*b*]thiophenes upon being heated in acidic media. Scope, limitation and mechanistic study of this new benzothiophene synthesis will be presented.

176. **STRUCTURE AND BONDING IN IMIDAZOLIUM CATIONS: NMR AND SINGLE CRYSTAL X-RAY STUDIES** R. J. Morgan*[1], R. J. Donovan, Clifford George[2] and J. P. Jasinski[3]
[1] *Department of Chemistry, Long Island University, Brooklyn N.Y, 11201.*
[2] *Laboratory for the Structure of Matter, Naval Research Laboratory, Washington D.C. 20375.*
[3] *Keene State College, Keene, New Hampshire 03431.*

Novel heteroaromatic systems 1-4 have been prepared and characterized by NMR and single crystal X-ray crystallography. The NMR and X-ray data provide evidence in support of the suggested aromatic structures 1-4, and also provides insight into the bonding and charge distribution in these interesting cationic systems.

177. **DILITHIATON OF PYRAZINES: A FACILE ONE-POT SYNTHESIS OF MULTI-SUBSTITUTED PYRAZINES.** Weimin Liu, Dean S. Wise, and Leroy B. Townsend, Department of Chemistry, College of Literature, Science, and the Arts; Department of Medicinal Chemistry, College of Pharmacy, The University of Michigan, Ann Arbor, MI 48109-1065

As a part of our ongoing research in the organo-lithium chemistry of pyrazines, a dilithiation-sequential-addition method has been developed, which allows the introduction of two different functional groups onto the pyrazine ring in one pot. When alkyl or aryl disulfides are used as electrophiles, a secondary displacement also occurred. Thus, a maximum number of substitutions can be achieved via lithiation and nucleophilic displacement. This method has been used in the synthesis of pyrazine C-nucleosides. Supported by Grant R01-CA56842 from NCI and R01-AI36872 from NIAID, NIH.

E_1, E_2 = halogens, disulfides, carbohydrate lactone; R_1, R_2 = halogens, carbohydrate, alkyl/phenylthio; R_3 = alkyl/arylthio

178. SCHIFF BASE DERIVATIVES OF WEAKLY BASIC AMINES WITH ACROLEIN AND RELATED ALDEHYDES, Hui Liu, N. Aratari, P.G. Rasmussen*, Dept. of Chem., Univ. of Mich., Ann Arbor MI 48109-1055

The Schiff base reaction has found extensive use in synthesis but the imine products which result are often subject to hydrolysis. If electron withdrawing groups are present, the imine linkage is stabilized. However amines carrying such groups can be reluctant to participate in the Schiff base reaction. In some cases this can be overcome by increasing the severity of the conditions. However we are interested in preparing Schiff base derivatives of acrolein, and acrolein derivatives for investigation of the products as monomers. These aldehydes do not permit harsh conditions. In this presentation we describe the preparation, characterization and polymerization of mono-anil monomers derived from weakly nucleophilic amines, such as 2,5-diamino-3,4-dicyanothiophene and 2-amino-1-methyl-4,5-dicyanoimidazole.

179. THERMAL GENERATION OF NONSTABILIZED CYCLIC AZOMETHINE YLIDES. A NOVEL ROUTE TO INDOLIZIDINES. William H. Pearson* and Yuan Mi, Department of Chemistry, University of Michigan, Ann Arbor, MI 48109-1055.

It has been demonstrated that the imines **1** are precursors of nonstabilized azomethine ylides under thermal conditions. Their cycloadditions with both activated and unactivated dipolarophiles produce indolizidines in moderate to good yields via a double-cyclization strategy. The indolizidines can also be obtained *directly* from the corresponding amines, aldehydes and dipolarophiles.

180. SYNTHESIS OF NOVEL HETEROCYCLES FROM 2-AMINO-3-CYANOMETHYL-SULFONYL-4,5-DIMETHYLFURAN. Chad E. Stephens and J. Walter Sowell Sr., College of Pharmacy, Division of Medicinal Chemistry, University of South Carolina, Columbia, SC 29208

The versatile heterocycle 2-amino-3-cyanomethylsulfonyl-4,5-dimethylfuran (**I**) is synthesized by base-catalyzed condensation of acetoin with sulfonyldiacetonitrile in methanol. Subsequent reaction of **I** with various ortho esters or phosgene iminium chloride leads to a number of novel furo[3,2-*b*] [1,4]thiazine 1,1-dioxides. Condensation of **I** with benzaldehyde yields the dihydrofurothiazine, while condensation with substituted benzaldehydes leads to the uncyclized α-[(2-amino-4,5-dimethyl-3-furanyl)sulfonyl]cinnamonitriles. Reduction of the 2-nitrocinnamonitrile with iron/acetic acid results in an intramolecular triple-cyclization in which the furan ring is opened and reclosed as a pyrrole to give a novel tetracyclic system, namely pyrrolo[2',3':5,6][1,4]thiazino[3,2-*b*]quinoline 4,4-dioxide.

181. MICHAEL ADDITIONS TO 3-NITROINDOLES. INVESTIGATION INTO THE SYNTHESIS OF PYRROLO[2,3-*b*]INDOLES AND PYRROLO[3,4-*b*]INDOLES. Erin T. Pelkey and Gordon W. Gribble, Department of Chemistry, Dartmouth College, Hanover, NH 03755.

We recently discovered a novel rearrangement leading to pyrrolo[2,3-*b*]indole **1**, which occurs upon treatment of **2** with ethyl isocyanoacetate and DBU. This rearrangement can be circumvented by changing the indole protecting group from benzenesulfonyl to benzyl. Thus, treatment of **3** with ethyl isocyanoacetate and DBU gave pyrrolo[3,4-*b*]indole **4**. The mechanism leading to the two regiochemical outcomes will be discussed.

182. MULTIPLE SYNTHETIC ROUTES TO 3-SUBSTITUTED-5-PHENYL BIS-HYDANTOINS. Kevin S. Emerick, David J. Mustra, Kent VanTyle, and O. LeRoy Salerni, Department of Chemistry and College of Pharmacy and Health Sciences, Butler University, Indianapolis, Indiana 46208

In the course of our studies directed toward the synthesis of antiseizure agents, we had occasion to synthesize a number of bis-hydantoins and bis-thiohydantoins having a strategically placed aryl group. We wish to report two synthetic routes to the target compounds. Pathway A involves the reaction of N,N'-[(bis-α-phenyl-α-t-butylamino carbonyl) methyl]ethylenediamine with isocyanates or isothiocyanates to form 1,2-bis(5-aryl-2,4-imidazolinedione-1-yl)ethanes or their thio analogs. Pathway B utilizes the conversion of N,N'-[(bis-α-phenyl-α-cyano)methyl]ethylenediamine to the hydantoin derivatives.

Pathway A: Z = CONHC(CH$_3$)$_3$
Pathway B: Z = CN

R-N=C=X
X = O or S

Details of the two synthesis pathways are discussed, including a novel fragmentation to 1-(2-aminoethyl)-3-methyl-5-phenyl-2,4-imidazolinedione. Mass spectroscopy and nmr data that provides unequivocal proof of structure is also presented.

183. BASE CATALYZED CONDENSATION OF 1-INDANONE WITH 4-PYRIDINE-CARBOXALDEHYDE. **Hesheng Zhang** Julia C. Pinto,*and R. Ian Fryer Department of Chemistry, Rutgers, The State University of New Jersey, Newark, New Jersey 07102, *Present address: Leukosite Inc., Cambridge, MA 02142.

While attempt to prepare substituted 2-(4'-pyridinylmethalene)-1-indanone by the condensation of the 1-Indanone with 4-pyridinecarboxaldehyde, the major product isolated was a spiro-pentacyclic system by an unexpected double condensation reaction. This reaction was initiated by stirring 1-indanone analogs with an equivalent amount of 4-pyridinecarboxaldehyde in the presence of a catalytic amount of potassium hydroxide in ethanol. The structures of the spiro-pentacyclic compounds were verified by spectra and confirmed by X-ray analysis. A mechanism was also proposed and was proved throuth the synthesis of intermediates.

184. LITHIUM BROMIDE ASSISTED ELECTROPHILIC RING-OPENING OF THF. Craig A. Ogle* and **Anissa Slusher**, Department of Chemistry, University of North Carolina at Charlotte, Charlotte, NC 28223.

The electrophilic ring-opening of THF is promoted utilizing LiBr. The following electrophiles were used: acetyl chloride, sulfuric acid, t-butyldimethylsilylchloride, and benzyl chloride. These reactions appear to go to completion for the silane and acid chloride. When benzyl bromide is used as the electrophile, an equilibrium is reached and the reaction gives poor yields. When sulfuric acid is used as the electrophile, the reaction goes nearly to completion, however the product, 4-bromo-1-butanol, is unstable and the reaction reverses slowly giving HBr and THF.

1a, E=H; 1b, E=Ac; 1c, E=CH$_2$Ph
1c, E=Si-t-Bu(Me)$_2$

185. THE SYNTHESIS OF THREE NITROGEN/SULFUR-CONTAINING POLYCYCLIC HETEROCYCLIC NOVEL RING SYSTEMS VIA PHOTOCYCLIZATION. Jiann-Kuan Luo, Maria-Paz Cabal, Ronald F. Federspiel and Raymond N. Castle, Department of Chemistry, University of South Florida, Tampa, Florida 33620-5250

Three previously unknown heterocyclic ring systems, naphtho[2',1':4,5]thieno[2,3-c]naphtho[2,1-f]quinoline (1), naphtho[2',1':4,5]thieno[2,3-c]naphtho[2,1-f][1,2,4]triazolo[4,3-a]quinoline (2) and naphtho-[2',1':4,5]thieno[2,3-c]naphtho[2,1-f][tetrazolo[1,5-a]quinoline (3) were synthesized via oxidative photo-cyclization of 3-chloro-N-(2-phenanthryl)naphtho[1,2-b]thiophene-2-carboxamide (6). Chlorination of the lactam **7** with phosphorus oxychloride followed by substitution with hydrazine afforded 6-hydrazinonaphtho[2',1':4,5]thieno[2,3-c]naphtho[2,1-f]quinoline (**10**). The unsubstituted ring system **1** was obtained by treatment of hydrazine derivative **10** with 10% copper sulfate solution. When the hydrazine derivative **10** was allowed to react separately with trimethyl orthoformate and nitrous acid, the triazole **2** and tetrazole **3** were obtained, respectively.

The structure confirmation of compound **1** was accomplished by a total assignment of its ^1H and ^{13}C nmr spectra utilizing the concerted two-dimensional nmr spectroscopic methods.

186.

KETO-ENOL TAUTOMERISM OF 4-METHYLDIHYDROFURO[2,3]COUMARIN-9-ONE; SYNTHESIS OF THE FIRST DIFUROCOUMARIN. E.A. Carberry, Southwest State University, Marshall, MN 56258; V.F.Traven, R.V. Rozhkov, D.V.Kravtchenko, T.A.Chibisova, D.Mendeleev University of Chemical Technology of Russia, Moscow 125190, Russia

Even though many furocoumarins are useful phototherapeutic compounds, synthesis of difurocoumarins has not been reported yet. We have studied keto-enol transformations of 4-methyldihydrofuro[2,3-h]coumarin-9-one (1) and its condensation reactions at position 8, including the dimerization reaction. Compound (1) has been acetylated both with 8-acetyl-9-acetoxy-4-methylangelicin and with 9-acetoxy-4-methylangelicin formation. Compound (1) reacts with aldehydes and ketones via aldol condensation. In accordance with electron absorption spectra of the compound (1) in CCl4-CH3OH mixtures, it exists in two tautomeric forms (keto and enol). MNDO calculations show that the keto form is definitely more stable, even though the enol form is stabilized due to aromaticity of the furane ring. Aldol self-condensation of (1) has also been found providing the formation of first difurocoumarin, 9-(4'-methyldihydrofuro[2',3']coumarin-9'-on-8'-yl)-4-methylangelicin.

187.

ONE-POT SYNTHESIS OF 3-CYANOCOUMARINS FROM 3-CARBONYL SUBSTITUTED COUMARINS. E.A. Carberry and J.D. Beebout, Southwest State University, Marshall, MN 56258; M.R. Salem, Al-Bouth Univeristy, Homs, Syria; R.V. Rozhkov, V.D. Dimitrova, A.L. Sedov, M.P. Nemeryuk and V.F. Traven, D.Mendeleev University of Chemical Technology of Russia, Moscow 125190, Russia.

Our investigation of the reaction of 3-acetyl-, 3-ethoxycarbonyl coumarins with cyanoacetohydrazine has yielded unusual results. These reactions produced corresponding 3-cyanocoumarins. We suggest that the reaction starts with a Micheal addition of the nucleophile to position 4 of the coumarin and results in the 4-substituted 3,4-dihydrocoumarin. Then following pyron ring opening and recyclizaton, the elimination of the hydrazine derivative occurs. This last step is a Michael retroreaction. Molecular peaks of side products of the reaction have been detected in the mass spectra of the reaction mixture. Yield dependence of the reaction on different substitutions of coumarin and cyanoacetohydrazine derivatives have been investigated, resulting in a range of yields from 32-76%.

188. AN EFFICIENT SYNTHESIS OF 3-HYDROXYBENZ[c]PHENANTHRENE AND 12-HYDROXYBENZO[g]CHRYSENE. Subodh Kumar, Environmental Toxicology & Chemistry, Great Lakes Center, SUNY College at Buffalo, Buffalo, NY 14222

3-Hydroxybenzo[c]phenanthrene and 12-hydroxybenzo[g]chrysene are precursors in the synthesis of the bay-region diol epoxides which are important derivatives for studying mechanism of carcinogenesis of the parent hydrocarbons. A short and efficient synthesis of these phenols from readily available starting materials, which involves the palladium-catalyzed cross coupling reaction will be presented (Scheme).

189. **A NEW AND CONVENIENT APPROACH TO THE SYNTHESIS OF DIHYDRODIOLS AND DIOL EPOXIDES OF CHRYSENE.** Subodh Kumar, Environmental Toxicology & Chemistry, Great Lakes Center, SUNY College at Buffalo, Buffalo, NY 14222

Palladium catalyzed cross-coupling reaction was investigated for the synthesis of racemic and optically pure dihydrodiol and diol epoxide derivatives of chrysene from readily available starting materials (Scheme). The details of the synthesis of various 2-arylnaphthalenes and their conversion to dihydrodiol and diol epoxide derivatives of chrysene will be presented.

190. **THE SYDNONE RING AS AN *ORTHO*-DIRECTOR OF LITHIATION.** Douglas M. Krein and Kenneth Turnbull*, Department of Chemistry, Wright State University, Dayton, OH 45435

Disubstituted sydnones (*cf.* **2**) have been prepared, generally in excellent yield, by dilithiation of 3-(2-phenyl)sydnone (**1**) with *n*-BuLi / TMEDA and subsequent reaction with electrophiles. Selective reaction at the *ortho*-aryl position to form the acyl species **3** is possible under the same conditions using Weinreb's amides as the electrophiles. These processes represent the first utilization of the sydnone ring as an *ortho*-director of lithiation.

191.

CONVERSION OF HEXAHYDROTRIAZINES TO IMIDAZOLINES
J. Thomas Ippoliti*, Silas Blackstock[†] and Jeffrey Evanson, Department of Chemistry, University of St. Thomas, St. Paul, MN 55105, [†]Department of Chemistry, University of Alabama, Tuscaloosa, AL 35487

The isolation and characterization of an imidazoline obtained from the reaction of ammonia and an aldehyde is reported. The stereochemistry of the imidazoline is confirmed using x-ray diffraction.

192. **PREPARATION OF CYCLIC GUANIDINE ADDUCTS FROM 2-AMINOIMIDAZOLE. SYNTHESIS OF THE $C_{11}N_5$ MARINE METABOLITES: OROIDIN AND THE DISPACAMIDES.** Anne Olofson, Kenichi Yakushijin, and David A. Horne*. Department of Chemistry, Columbia University, New York, NY 10027

Cyclic guanidine adducts of the generalized type **A** form the basis of a number of structurally complex marine natural products. Such metabolites include, palau'amine, the phakellins, the agelaspongins, and the gonyautoxins (e.g. saxitoxin). In each case, a 5-membered ring cyclic guanidine unit forms the alkaloid core. We report the synthesis and isolation of cyclic guanidine adducts **A** directly from 2-aminoimidazole (AI) and describe their subsequent rearrangements. The synthesis of the $C_{11}N_5$ marine metabolites oroidin and the dispacamides is illustrative.

A

oroidin R=Br

dispacamides R=H or Br

193. **OXIDATIVE DIMERIZATION OF 2-AMINOIMIDAZOLES. BIOMIMETIC SYNTHESIS OF THE $C_{10}N_6$ MARINE PIGMENT PARAZOANTHOXANTHIN A.** Ying-zi Xu, Kenichi Yakushijin, and David A. Horne* Department of Chemistry, Columbia University, New York, NY 10027

The oxidative dimerization of 2-aminoimidazoles (AIs) by molecular bromine has been investigated. In particular, a biomimetic synthesis of the tricyclic $C_{10}N_6$ marine pigment, parazoanthoxanthin A (**1**) has been achieved in a single step from the oxidative dimerization of a C_5N_3 AI precursor. Both oxidative homo- and heterodimerization of AI derivatives proceed efficiently. The results are suggestive of a common bromonium ion pathway in the biosynthesis of AI containing metabolites.

"C_5N_3" [ox] "$C_{10}N_6$"
parazoanthoxanthin A

194. **RATIONAL DESIGN AND SYNTHESIS OF COMPONENTS FOR A BARBITURATES SCREENING IMMUNOASSAY** Jonathan Grote, Maciej Adamczyk, Robert Dubler, Charles Harrington, and Jeanine Douglas, Divisional Organic Chemistry, Dept 9NM, Abbott Laboratories, Abbott Park, IL 60048-3500

Barbiturates are central nervous system depressants which are frequently administered on a therapeutic basis as sedatives, hypnotics, and anticonvulsants. These 5,5-disubstituted 2,4,6-pyrimidinetriones have been known for several decades, having been introduced into therapeutic use by Fisher and von Mering in the early 1900s. Small structural variations in the 5-substituents cause substantial changes in the drug's physiological effects and duration: the most frequently abused barbiturates have alone or in combination with other drugs often been used to commit suicide. Therefore, rapid determination of the presence or absence of barbiturates in a comatose patient prior to emergency medical treatment can be the difference between life and death. We present here the design of a hapten and tracer which reflect the structural properties of a variety of barbiturates. Hapten conjugation allowed for development of an antisera which demonstrated good recognition of several different barbiturates in a fluorescence polarization immunoassay format. The relationship between the immunogen architecture, the characteristics of the resulting antisera, tracer structure, and the chemical structure of the analyte barbiturates will be presented.

195. **SYNTHESIS OF NONLINEAR OPTICAL DONOR-ACCEPTOR STILBENES BEARING PHOSPHONATE ESTER ELECTRON WITHDRAWING GROUPS.** Kevin D. Belfield*, Katie Schafer, Chandra Chinna, Suresh Sriram, and Ousama Najjar, Department of Chemistry, University of Detroit Mercy, P.O. Box 19900, Detroit, MI 48219-0900

An efficient synthesis of novel nonlinear optical chromophores will be presented. The phosphonate ester group is an effective pi electron acceptor group in donor acceptor substituted stilbene derivatives for nonlinear optics, photorefractive materials, and other electro-optic applications. It has the added advantages of being easily synthesized and functionalized for further fabrication into crosslinked films, for use as dopants with strong hydrogen bonding characteristics in host-guest self-assembled macromolecular composites, and can be incorporated as a main chain component in fully functionalized photorefractive polymeric materials. The synthesis will show the ease and utility of Ni-catalyzed Arbuzov reactions and the powerful Pd-catalyzed Heck reaction.

196. AB INITIO MOLECULAR ORBITAL STUDIES ON THE CATALYTIC MECHANISM OF THE PENICILLIN-BINDING PROTEINS AND BETA-LACTAMASES. **Javier Gonzalez** and Edward Turos, Departamento de Quimica Organica e Inorganica, Universidad de Oviedo, 33071-Oviedo, Spain and Department of Chemistry, University of South Florida, 4202 E. Fowler Avenue, Tampa, Florida 33620-5250

Despite all the evidences derived from the crystallographic studies, the detailed mechanism of the interaction of penicillin-binding proteins and beta-lactamases with beta-lactam antibiotics is still under discussion. Here we present the results obtained in a theoretical study of the initial step of the catalytic mechanism of these enzymes. Ab initio calculations, at the HF/6-31G* and MP2/6-31G*//6-31G* levels of theory were carried out on a model for the base-catalyzed nucleophilic addition of serine to a beta-lactam. According to this study a reaction pathway exists in which the nucleophilic attack of the oxygen of the serine 70 residue to the carbonyl group of the beta-lactam is concerted with the hydrogen transfer from the OH group to the nitrogen atom of the Lysine 73 residue, which acts a general base.

197. STUDY OF THE INTERACTIONS BETWEEN CARBOXYPEPTIDASE A AND ITS SUBSTRATES USING ENZYME KINETICS AND MONTE CARLO/ENERGY MINIMIZATION PROCEDURES. **Guiqing Liang**, Amanda Jones, and John F. Sebastian, Department of Chemistry, Miami University, Oxford, Ohio, 45056

The kinetics of tripeptide substrates of carboxypeptidase A (CPA) were studied for the purpose of examining the interactions of P_1/S_1. It was found that increasing the apolar size of the side-chain group of amino acid in the P_1 site resulted in an increase of k_{cat}/K_S. Monte Carlo techniques along with energy minimization (MC/EM) were applied to conformational searches of 8 substrates "docked" to the active site of CPA. The computational data were consistent with the kinetic data in which those substrates that bound to CPA with more hydrogen bonds in P_1'-P_2/S_1'-S_2 have relative lower K_S values. The results suggest that MC/EM procedure is a useful tool to explore the interactions between CPA and its substrate (P_1'-P_2/S_1'-S_2) in the ground state.

198. AM1 STUDY OF THE THERMODYNAMICS OF SIMPLE KETO-ENOL TAUTOMERISM, <u>Brian H. Nordstrom</u>, Physical Science Department, Embry-Riddle Aeronautical University, Prescott, AZ 86301-3720.

Gas-phase equilibrium constants for keto-enol tautomerism have been calculated semiempirically. A number of simple ketones and their enol tautomers have been investigated. Selected examples are presented in this paper. Structures are drawn using PC Model, and then the lowest-energy conformers are found using the steric energy minimization program GLOBAL-MMX (GMMX). AM1 calculations are done on all the lowest-energy conformers of each ketone and enol yielding heats of formation, ΔH_f^o, and molar entropies, S^o. Using the Boltzmann distribution of conformers of each compound, weighted averages of these thermodynamic functions are calculated. Finally, equilibrium constants are calculated using the equation
$$\Delta H^o - T \Delta S^o = - RT \ln K_{eq}.$$

199.

LOW-BARRIER-HYDROGEN BONDS: AN AB INITIO INVESTIGATION. Michael A. McAllister, A. Ganeshkumar, Yongping Pan, & Jay Smallwood. Department of Chemistry, University of North Texas, P.O. Box 5068, Denton, TX, 76203.

High-level ab initio and DFT molecular orbital calculations have been used to investigate the physical properties of several systems which are believed to contain low-barrier-hydrogen-bonds. Systems investigated include: formic acid/formate; enol/enolate; and hydrogen maleate. In each of these systems it is found that the hydrogen bond formed is extraordinarily short and strong, with a calculated enthalpy of activation for proton transfer from donor to acceptor which is less than the zero point vibrational energy available to the system. Several perturbations to these systems were studied. Forcing a miss-match of pKa's between donor and acceptor, via the use of substituents, causes the strength of the hydrogen bond to decrease significantly. Microsolvation of the hydrogen bonded complex, on the other hand, does not affect the strength of the low-barrier-hydrogen-bond.

200. COMPUTATIONAL INVESTIGATIONS OF CYCLOBUTYLIDENE. Robert J. McMahon and Louise L. Stracener, Department of Chemistry, University of Wisconsin, Madison, Wisconsin, 53706

Cyclobutylidene (**1**) has been investigated computationally using a variety of methods, including density functional theory, complete-active-space self-consistent-field method, and coupled-cluster theory. The singlet carbene was found to have a bent ring structure, while the triplet species is planar. The singlet-triplet energy gap is very small; there is disagreement at the highest levels of theory regarding the ground-state multiplicity. The products of 1,2-H migration (cyclobutene, **2**) and ring contraction (methylenecyclopropane, **3**) were also examined; 1,2-H migration appears to be slightly more exothermic than ring contraction.

1   2   3

201.
PROBING THE ISOMERIC DISTRIBUTION OF METHYL ALLENYL ANION IN THE GAS PHASE USING ITS REACTION WITH CARBON DISULFIDE. **R. Gareyev**, V. M. Bierbaum, C. H. DePuy, Department of Chemistry and Biochem., University of Colorado, Boulder, Colorado 80309-0215

The abstraction of a proton from methyl allene may proceed at two different sites in the allene system: vicinal to the methyl group or terminal, yielding two isomeric anions. Exclusive formation of the former of the two anions is expected if a proton is abstracted from dimethyl acetylene with consequent rearrangement of π-bonds. The actual distribution in the gas phase of the two isomeric anions obtained by deprotonation of both precursors was studied by allowing the mixture of anions to react with CS2. In a multistep addition-rearrangement-dissociation reaction this reagent yields different products with the two isomers of the anion studied thus allowing for the assay of the mixture. The influence of the basicity of the deprotonating agent on the distribution of anions was also investigated. Reaction with CS2 may be a useful test for the determination of the reactive site in substituted allenyl anions and other anions with cumulated double bonds.

202. THE RELATIONSHIP BETWEEN COMPLEX INDUCED PROXIMITY EFFECTS OF ORGANOLITHIUM REAGENTS AND RELATIVE ANION STABILITIES RESPONSIBLE FOR VINYLIC VS ALLYLIC DEPROTONATION OF CYCLIC VINYL ETHERS: AN AB INITIO STUDY **Trevor D. Power**, John F. Sebastian, Department of Chemistry, Miami University, Oxford, Ohio, 45056

The relative energies of allylic and vinylic anions of several vinyl ethers were determined by ab initio calculations at the Hartree-Fock and Møller Plesset 2 (single point) levels using basis set 6-31++G(d,p) in an attempt at explaining experimental results concerning allylic vs vinylic deprotonation. A general trend with cyclic vinyl ethers has been discovered where the stability of the allyl anion over the vinyl anion in terms of ring size is 8≈7>6>5≈4. Generally speaking, optimized vinyl anions exhibit a vinyl angle compression whereas optimized allyl anions exhibit an allyl angle expansion. Additionally, transition state structures are examined that invoke a Complex Induced Proximity Effect (CIPE) for the formation of the same products. Structures of the lithiated species (solvated), allylic and vinylic complexes will be described.

203. INTRAMOLECULAR B-N COORDINATION IN BORATABENZENE COMPLEXES, A.J. Ashe, III*, J.W. Kampf, and J. R. Waas, Department of Chemistry, The University of Michigan, Ann Arbor, MI 48109.

The crystal structure of tricarbonyl[1-(3-dimethylaminopropyl)boratabenzene]manganese(I) shows weak intramolecular B-N coordination (**B**). However in a solution of toluene-d_8 **B** is in mobile equilibrium with the ring opened isomer **A**. Using ^{11}B NMR spectroscopy the equilibrium constants for **B**→**A** have been measured over the temperature range -35°C to 48°C, allowing evaluation of $\Delta H°$ (6.0 kcal/mol) and $\Delta S°$ (23 cal/mol K).

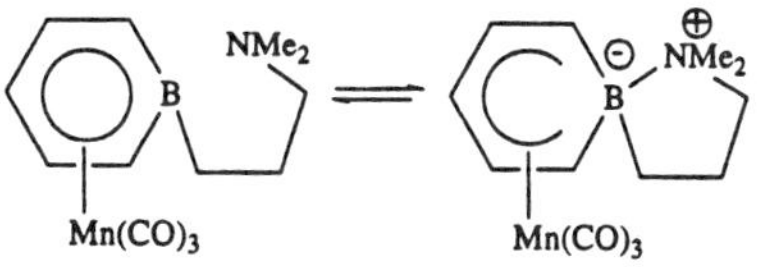

204. CARBON-FLUORINE BOND ACTIVATION BY ARYL CARBOCATIONS: CONCLUSIVE INTRA-MOLECULAR FLUORIDE SHIFTS BETWEEN CARBON ATOMS IN SOLUTION AND THE FIRST EXAMPLES OF INTERMOLECULAR FLUORIDE ION SHIFTS **Dana Ferraris**, Christopher Cox, and Thomas Lectka, Department of Chemistry, Johns Hopkins University, Baltimore, MD 21218

The chemistry of selective C-F bond activation is undergoing a surge of interest due to the scientific and commercial importance of fluorocarbons. As part of a new approach to C-F bond activation, we report the use of *carbocations* to abstract fluoride from organic molecules through a three-center transition state or intermediate in condensed media. Although cationic fluoride transfers involving hypervalent [C-F-C]+ interactions are documented in the gas phase, reports of such intermolecular transfers in solution are virtually nonexistent, and the published cases of intramolecular shifts are fraught with controversy. In fact, the question of whether Wagner-Meerwein-type rearrangements of fluoride could ever occur in solution has been a source of continuing debate. In this poster, we describe intramolecular fluoride shifts between carbon atoms in solution and in the solid state, and the first documented examples of analogous *intermolecular* fluoride transfers between an aryl cation and its counterion.

205. HYDROGEN ATOM SHIFTS IN ARYL RADICALS. Michele A. Brooks and Lawrence T. Scott, *Department of Chemistry, Boston College, Merkert Chemistry Center, Chestnut Hill, MA 02167.*

Hydrogen atom 1,2-shifts in aryl radicals have not been previously reported, to the best of our knowledge. We have found evidence for such a rearrangement in the radical derived from 2-bromobenzo[c]phenanthrene by flash vacuum pyrolysis at temperatures ranging from 950-1100 °C. Deuterium labeling experiments are consistent with the mechanism shown below.

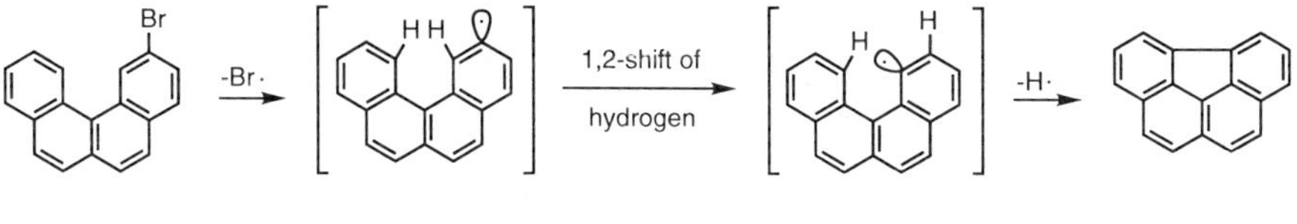

206. CONFORMATIONAL ANALISIS OF 5-SUBSTITUTED 1,3-DIOXANES: EFFECT OF LITHIUM BROMIDE ADDITION. E. Juaristi,[a] F. Díaz,[a] G. Cuéllar,[a] and H.A. Jiménez-Vázquez,[b] (a) Depto. Química, CINVESTAV-IPN, Apdo. 14-740, 07000 México, D.F., México. (b) Depto. Química Orgánica, ENCB-IPN, 11340 México, D.F., México.

The position of equilibria, established by means of BF_3, between diastereomeric cis- and trans-5-substituted-2-phenyl-1,3-dioxanes, in solvent THF and in the presence of 1 to 10 equivalents of LiBr has been determined. The observed $\Delta G°$ values show that the addition of salt to the reaction medium influences the position of equilibrium. Lithium bromide effects on the conformational behavior are discussed in terms of hydrogen-bond and electrostatic intra- or intermolecular perturbations in the heterocycles, which may model salt effects in physiological events.

X= CO_2H, CO_2CH_3, $CONHCH_3$, CH_2OH, OH, $OCOCH_3$, NO_2, $NHCOCH_3$

207. A STABILITY STUDY OF MOLYBDENUM CARBENE COMPLEXES. Daniel F. Harvey* and Dina M. Sigano. Department of Chemistry and Biochemistry - 0358, University of California, San Diego, La Jolla, CA 92093-0358.

Over the past 30 years Fischer carbene complexes have been shown to participate in a variety of synthetically useful transformations. Their stability and reactivity have been shown to vary dramatically depending on (i) the metal employed, (ii) the ligands on the metal, (iii) the substituents on the carbene carbon, and (iv) the reaction environment. In an effort to better understand what governs their decomposition, we sought to compare and quantify the stability of a series of alkyl alkoxy molybdenum carbene complexes. The results of this investigation will be presented.

208. REACTIVITY AND SELECTIVITY OF N-ACETYL-4-STILBENYLNITRENIUM ION, Kelly J. Kayser and Michael Novak*, Department of Chemistry, Miami University, Oxford OH 45056.

Decomposition of the carcinogen N-Acetyl-N-(Sulfonatooxy)-4-aminostilbene in 5% CH_3CN/H_2O, $\mu=0.5$, at $20°C$ apparently occurs via rate-limiting N-O bond heterolysis to generate the N-acetyl-4-stilbenylnitrenium ion. The hydrolysis rate constant, k_o, is pH independent with an average value of $(1.89\pm.02) \times 10^{-2}$ s^{-1}. This rate constant falls on the previously determined correlation line for a plot of log k_o vs. σ^+ for a series of N-sulfonatooxy-N-acetyl-N-arylamines, all of which undergo hydrolysis via a common N-O bond heterolysis mechanism. Trapping of the nitrenium ion in aqueous solution results in the formation of four major hydrolysis products which have been tentatively identified. The nitrenium ion is trapped very inefficiently by N_3^- ($k_{az}/k_s < 1$ M^{-1}). This is quite surprising since nitrenium ions derived from other carcinogenic N-arylhydroxamic acid esters react quite selectively with N_3^-. The apparently aberrant reactivity of this nitrenium ion, and the consequences of this reactivity pattern for the mechanism of carcinogenesis by 4-aminostilbene derivatives, will be discussed.

209. HYDROLYSIS MECHANISMS OF ESTER DERIVATIVES OF N-HYDROXY-2-AMINO-5-PHENYLPYRIDINE AND ITS ANALOGUES. Lulu Xu and Michael Novak*, Department of Chemistry, Miami University, Oxford, OH 45056

2-Amino-5-phenylpyridine(Phe-P-1) and its analogues represent a group of carcinogenic and mutagenic heterocyclic amines found in cooked meats[1,2]. In order to explore the mechanism of their carcinogenicity and mutagenicity we have synthesized pivalic acid esters of the N-hydroxy-2-amino-5-phenyl-pyridine, **1**, N-hydroxy-2-amino-3-methyl-5-phenylpyridine, **2** and N-hydroxy-3-amino-6-phenylpyridine, 3. Hydrolysis rate/pH profiles determined by UV spectrophotometric methods in pH range 1.5-7.5 show that the decomposition rates of all three esters are pH dependent. The data indicate that only the neutral esters are subject to hydrolysis in the pH range examined. The pK_a values for the esters 1, 2 and 3 are 2.9, 3.4 and 4.1 respectively, determined by both kinetic and spectrophotometric methods. The products derived from hydrolysis of 3 are primarily those expected from a singlet nitrenium ion derived from N-O bond heterolysis. The products derived from 1 are 5-phenyl-2-aminopyridine and dimers. These products suggest a mechanism involving either a nitrene or triplet nitrenium ion.

210. SYNTHESIS AND RELATIVE DIATROPICITY OF NOVEL AROMATIC *CIS*- AND *TRANS*- THIA[13]ANNULENES

Reginald H. Mitchell, Ji Zhang, Department of Chemistry, University of Victoria, P.O.Box 3065,Victoria, B.C. Canada, V8W 3V6

4-Bromo-*trans*-9b,9c-dimethyl-9b,9c-dihydrophenyleno[1,9-bc]thiophene **2** and 4-Bromo-*cis*-9b,9c-dimethyl-9b,9c-dihydrophenyleno[1,9-bc]thiophene **3** were successfully synthesised and isolated in 11 steps, starting from 2,4-Dimethyl 2-amino-3-methylthiophene-2,4-dicarboxylate **1**. The bromo substituent in the 4-position stabilizes the *trans*- thia[13]annulenes **2** and allow us to access the first *cis*- bridged thia[13]annulene **3**. The thia[13]annulene **2** is notably diatropic and shows about 37% of the ring current of the parent bridged [14]annulene **4**.

211. SOLVENT EFFECTS ON THE NUCLEOPHILIC RING CLEAVAGE OF SPIROACTIVATED CYCLOPROPANES.

M. A. McKinney*, S. S. Templin** and K. G. Kremer*, *Department of Chemistry, Marquette University, Milwaukee, WI 53233. **Department of Chemistry, Cardinal Stritch College, Milwaukee, WI 53217.

Kinetic data for the reactions of 6,6-dimethyl-1-phenylspiro[2.5]octan-4,8-dione (1a) and 6,6-dimethyl-1-phenyl-5,7-dioxaspiro[2.5]octan-4,8-dione (2a) with pyridine were obtained in various solvents. The data for these S_N2 reactions were compared to data for the Menschutkin Reaction. The reaction rates increased with increasing solvent polarity as expected for S_N2 reactions. But solvent polarity alone has been shown to be an incomplete analysis of solvent effects. The data were interpreted using recently developed methods of multiparameter analysis in which four independent solvent parameters were correlated with the reaction rates.

2a: x = O ; 1a: x = CH$_2$ 1b, 2b

212. THE MECHANISM OF THE THERMAL REARRANGEMENT OF CYCLOPROPYL DICARBONYL COMPOUNDS TO DIHYDROFURANS.

Mary K. Budowle, Glen R. Frazee, Sandeep Shiroor, and Michael A. McKinney, Department of Chemistry, Marquette University, Milwaukee, WI 53201-1881.

The mechanism of the thermal rearrangement of cyclopropyl dicarbonyl compounds to dihydrofurans can be compared to that of the well-studied rearrangement of vinyl cyclopropanes to cyclopentenes. We studied the epimerization of E and Z-2 as well as their rearrangements and the rearrangements of 1 with various aryl substituents. Solvent effects, activation parameters, stereochemistry, and substituent effects on the rearrangement will be discussed. A mechanism consistent with the data obtained will be presented.

1: X=Y=CH$_2$ 2: X=O,Y=CH$_2$ 3: X=CH$_2$, Y=O

213. **DHP-A BETTER NMR PROBE FOR THE LONG DEBATED MILLS-NIXON EFFECT?**
Reginald H. Mitchell* and Danny Y.K. LAU (e-mail: LYK@UVIC.CA)
Department of Chemistry, University of Victoria, PO Box 3065, Victoria, B.C. Canada, V8W 3V6

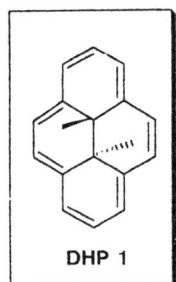

DHP 1

Benzene has been used as a probe to study the long debated Mills-Nixon (MN) effect which is about *the perturbation of aromatic π-electrons on small ring annelation*. However, benzene as a NMR probe in benzocycloalkenes or benzocycloalkenones gave inconclusive results to comment on whether the MN effect exists or not. This may due to geometry distortion, rehybridization, steric compression and hyperconjugation, which make the NMR results complicated for the annelated benzene derivatives. Dimethyldihydropyrene (DHP) **1** has been known as a **sensitive** NMR probe which responds to ring annelation. By fusing cycloalkene or cycloalkenone rings onto DHP, we may be able to comment on the MN effect by monitoring the chemical shift of its internal methyl protons which depends on its ring current and less seriously on those factors mentioned in annelated benzene derivatives. The MN effect probed by cycloalkene and cycloalkenone-annelated DHP will be discussed.

214. TRIMETHYLTRIS(TRIFLUOROPROPYL)CYCLOTRISILOXANE: NUCLEAR MAGNETIC RESONANCE ASPECTS. Dong Lin, Timothy R. Willson, Muhammad E. Haque, and Robert V. Honeychuck, Department of Chemistry, George Mason University, Fairfax, Virginia 22030

Cyclic fluorinated siloxane trimer 2,4,6-trimethyl-2,4,6-tris(3,3,3-trifluoropropyl)cyclotrisiloxane (F_3) has been studied via solution phase ^1H, ^{13}C, ^{19}F, and ^{29}Si NMR on a single instrument using several pulse sequences for ^{13}C and ^{29}Si. Two forms of F_3 were used: a white solid obtained at room temperature, and a colorless liquid isolated at -78 °C. The solvents used were chloroform-d and acetone-d_6. The possible relationship of the spectra to the stereoisomeric forms of F_3 is discussed. The decoupled ^{13}C spectra in chloroform-d reveal a pattern of low complexity in the solid isolated at room temperature, and increased numbers of peaks in the liquid separated at low temperature. Since F_3 is a monomer used in production of fluorinated siloxane polymers, and since it could conceivably undergo a redistribution reaction to the corresponding cyclic tetramer, a brief comparison to these fluorinated Si-O species is provided.

215. **NOVEL INSIGHTS INTO THE MECHANISM OF METHANE CARBONYLATION IN SUPERACIDS USING HIGH-PRESSURE NMR SPECTROSCOPY.** Peter J. F. de Rege,[a] John. A. Gladysz,[b] István. T. Horváth,[a]
[a]Exxon Research and Engineering Company, Annandale, NJ 08801 and
[b]Department of Chemistry, University of Utah, Salt Lake City, UT 84112.

The facile activation of methane in the liquid superacids HSO_3F/SbF_5 and HF/SbF_5 leading to C_n-carbocations (n ≥ 4) was discovered by Olah et. al.. High-pressure (HP-) NMR confirms Olah's results, and further reveals the formation of a number of oxygenated species in the case of HSO_3F/SbF_5. Hogeveen et. al. reported the exclusive formation of acylium ion, CH_3CO^+, from the reaction of CH_4, CO and SbF_5. HP-NMR shows that methane carbonylation occurs in HSO_3F/SbF_5 and HF/SbF_5, and suggests that the similar reactivity observed for SbF_5 results from the presence of HF/SbF_5. Remarkably, the $CO/HF/SbF_5$ system exhibits previously unreported spectral features under CO pressure which are attributed to *the first spectroscopic observation of the formyl cation*, **HCO^+**, in a condensed phase.

216. NMR STUDIES OF ROTATIONAL BARRIERS IN PROTONATED PUSH-PULL ENAMINES **Tammy J. Dwyer**, Charles M. Buxton, Scott Kirkowski, Department of Chemistry, University of San Diego, San Diego, CA 92110

We have used NMR to investigate the site of protonation and the effect of protonation on the C-N and C=C rotational barriers in a series of push-pull enamines, in which one or more electron-donating groups are positioned at one end of the C=C bond and one or more electron-accepting groups' at the other. In preliminary studies we have obtained compelling evidence for the protonation of these enamines in an uncommon position with attendant formation of a keteneimine functional group. Protonation of 1,1-bis(dimethylamino)-2-cyano-2-(4-X-phenyl)-substituted ethylenes (**1**) with one equivalent of trifluoroacetic acid yields exclusively cyano N-protonated **1** (keteneimine) when X=NO$_2$, and a mixture of the keteneimine and C-protonated **1** when X = F and CH$_3$. The rotational barriers in these protonated forms have been measured and are also reported.

1 X = NO$_2$, F, CH$_3$

217. SYNTHESIS OF 13C, 19F LABELED RETINAL AND SOLID STATE 19F NMR. <u>M. Groesbeek</u> & S.O. Smith, Department of Molecular Biophysics and Biochemistry, Yale University, New Haven, CT 06520-8114, U.S.A.

We have investigated methods to monofluorinate the methyl positions 19 and 20 in retinal, and we found a general, mild strategy that avoids side reactions, proceeds in high yields and is compatible with current carbon-13 labeling strategies. Key in the new sequence is the conversion of N-methoxy-N-methyl amides via the α-chloromethyl ketone to the corresponding γ-hydroxynitrile, which can be fluorinated by DAST. The synthetic results will be presented as well as solid-state NMR data of rhodopsin, reconstituted with the novel retinals.

218. NEW RATIOABLE CALCIUM INDICATORS FOR FLUORESCENCE AND ^{19}F NMR DETECTION. <u>Jun-Rui Yang</u>, Karen A. LeCompte and Margaret E. Langmuir, Covalent Assoc., Inc., 10 State St., Woburn, MA 01801; David A. Forsyth, Dept. of Chemistry, Northeastern University, Boston, MA 02115

The syntheses of three new calcium indicators are presented. The absorption maxima of these indicators are above 500 nm and the emission maxima beyond 600 nm. Their absorption and emission maxima change upon calcium binding making them good candidates for ratiometric fluorescence detection. Their pK$_a$s are all below pH 7 so that binding to calcium at physiological pH is possible. Their calcium complex dissociation constants fall in the range of 0.67 to 1.46 μM. It is thus possible to use these ionophores to study cytosolic calcium ion concentration spikes in certain cells which cannot be studied using indo-1 or fura-2 whose dissociation constants are in the 0.22 to 0.25 μM range. ^{19}F NMR data are reported for one of the indicators which show that the indicator is highly selective for Ca^{2+} in the presence of Mg^{2+} and K$^+$ and has a 6 ppm downfield shift on binding calcium compared to the tetra anion at pH 6.9.

219. Deuterium Isotope Effects on Resonance Raman Spectra of the N-Ethyl-4-Phenylpyridinium Iodide Complex. <u>David Wiedenfeld</u>, Jay R. Winkler, and Harry B. Gray, Beckman Institute, California Institute of Technology, Pasadena, California 91125.

Absorption spectra and absolute resonance Raman cross sections of the natural abundance N-ethyl-4-phenylpyridinium iodide charge-transfer complex and three different deuterated variants of this complex were measured. At least five modes are enhanced in the resonance Raman spectrum of the natural abundance derivative; the dominant mode is 1565(18) cm^{-1}. Spectral modeling led to estimates of single-mode reorganization energies upon pyridinium photoreduction. The implications of these results for weakly-coupled intramolecular electron-transfer reactions in related charge-transfer complexes will be presented.

220. PHOTOCHEMISTRY OF 7,7-DISUBSTITUTED 3,4,4a,5,6,7-HEXAHYDRO-4a-METHYL-1(2H)-NAPHTHALENONES, Richard A. Bunce, R. Shawn Childress and Elizabeth M. Holt, Department of Chemistry, Oklahoma State University, Stillwater, OK 74078-3071.

Naphthalenones **1** and **2** have been synthesized and their photochemistry studied. Both molecules incorporate a rigid s-cis enone extended over two rings. In benzene, both were photoinert, but in alcohol solvents, hydrogen-bonded enols resulting from solvent addition to the enone double bond were isolated as the only products. Rearrangements anticipated for γ,γ-disubstituted enones, acyl migration for **1** and Type A rearrangement for **2**, were not observed. The addition products were formed in synthetically useful yields and with high regio- and stereospecificity. Mechanisms accounting for product formation and product stability will be presented.

1 (R, R = Ph)
2 (R, R = Me)

221. 2-BENZOYL BENZOIC ACID: A PHOTOLABILE MASK FOR ALCOHOLS AND THIOLS. P.B. Jones, M.P. Pollastri, and N.A. Porter, Department of Chemistry, Duke University, Durham, NC 27708

Photolysis of 2-benzoyl benzoate esters of primary and secondary alcohols in the presence of a hydrogen donor or electron donor produces the corresponding alcohol in high yield. The fate of the benzoate is dependent on the conditions used for the reaction. In 2-propanol, the ketyl radical that derives from the photoreduction dimerizes, to afford the benzpinacol product 3,3'-diphenyl-biphthalidyl. In the presence of amines the product is 3-phenyl phthalide. While the photoproduct of the benzoate:2-propanol reaction results from anticipated free radical chemistry, the amine promoted reaction appears to result from a second, "dark", electron transfer process. We conclude that 2-benzoyl benoic acid is an effective photolabile protecting group for primary and secondary alcohols and preliminary studies indicate that thiols can be protected in an analogous way. Studies on the effect of benzophenone substituents and reaction solvent on the benzhydrol:benzpinacol product ratio provide mechanistic insight into the process.

222. A PHOTO-REMOVABLE PROTECTING GROUP FOR AMINES. Ailian Zheng and Binghe Wang.* Department of Chemistry, North Carolina State University, Raleigh, NC 27695

Our continuing interest in the development of new protecting groups for amines has led us to design a photo-removable protecting group. This design takes advantage of the photoisomerization and the ensuing lactonization of o-hydroxy-*trans*-cinnamic acid and its derivatives. We have found that both the protection and deprotection can be carried out in high yields for a variety of amines. With all the model compounds we have studied, the deprotection reaction was accomplished using a low intensity (4 Watts) UV lamp without any undesirable side reactions.

223. A PHOTOCHEMICAL APPROACH TO THE SYNTHESIS OF COUMARIN-BASED PRODRUGS OF AMINES. Ailian Zheng, Huijuan Zhang, and Binghe Wang.* Department of Chemistry, North Carolina State University, Raleigh, NC 27695

Earlier, we reported the development of a coumarin-based prodrug system (**4**) for amines (Wang, *et al.* Bioorg. Med. Chem. Lett. **1996**, 6, 945). The design takes advantage of the facile lactonization of coumarinic acid and its derivatives. However, the original synthetic route was lengthy. Here we report a new synthetic approach to these prodrugs by taking advantage of the photoisomerization of o-hydroxy-trans-cinnamic acid. Using this approach, we have synthesized model amides (**4**) of several amines with different structural features.

224. **SYNTHETIC APPROACHES TO THE PREPARATION OF NOVEL COUMARIN-BASED CYCLIC PRODRUGS OF PEPTIDES.** Daxian Shan, Wei Wang, Huijuan Zhang, Ailian Zheng, and Binghe Wang.* Department of Chemistry, North Carolina State University, Raleigh, NC 27695

Earlier, we reported the development of a coumarin-based prodrug system for amines by taking advantage of the facile lactonization of coumarinic acid and its derivatives. Such a system can also be used for the preparation of cyclic prodrugs of peptides and peptide mimetics. We have developed two approaches to the synthesis of such cyclic prodrugs. The first method started with coumarin and required reductive opening of the lactone ring. The second method was carried out by taking advantage of the photo isomerization of trans-cinnamic acid into to its cis-isomer.

1. peptide = -HN-Tyr-Gly-Gly-Phe-Leu-
2. peptide = -HN-Tyr-Gly-D-Ala-Phe-D-Leu-

225. **FORMATION OF COMPLEX BICYCLIC PHOTOADDUCTS VIA TANDEM 4-PYRONE OXYALLYL-ZWITTERION TRAPPING AND NORRISH TYPE II REARRANGEMENT.** Mike Fleming, Rehan Basta, Scott Mitchell, P. V. Fisher and F. G. West, University of Utah, Department of Chemistry, Salt Lake City, UT 84112.

Prior investigations from our labs have shown that 4-pyrones undergo facile photorearrangements via oxyallyl-zwitterion trapping to form highly functionalized cyclopentenone products (**A**). Interestingly, continued irradiation of compound **A**, in the presence of a sensitizer, initiates Norrish type II chemistry to give a complex bicyclic photoadduct of the type **B**. Overall, this process creates two new C-C bonds and up to five stereocenters. The scope and limitations of this transformation will be discussed in detail.

226. **SURPRISING PHOTOCHEMICAL AND THERMAL ISOMERIZATIONS OF A CYCLIC 1,2,3-BUTATRIENE.** Yiyuan Chen, Zhongxin Ma, Sina Ijadi-Maghsoodi and Thomas. J. Barton,* Ames Laboratory (USDOE) and Department of Chemistry, Iowa State University, Ames, Iowa 50011

Photolysis of cyclic 1,2,3-butatriene **1** exclusively affords cyclopropane **2** for which an unprecedented initial isomerization to a methylenecyclopropanylidene is proposed. Thermolysis of **1** produces butadiene **3** in 83% yield. Formation of **3** is an apparent disallowed S, A [2+2] cycloaddition but we will argue that it is actually an allowed [4+2] cycloaddition.

227. THE ABSOLUTE REACTIVITY OF SINGLET PENTAFLUOROPHENYL NITRENE WITH REPRESENTATIVE ALKENES. **Hongbin Zhai** and Matthew S. Platz*, Department of Chemistry, The Ohio State University, 120 West 18th Avenue, Columbus, OH 43210

Unlike the parent system, polyfluorinated aryl nitrenes can add to alkenes owing to the so-called fluorine effect. In this study, the relative reactivity of singlet pentafluorophenyl nitrene (^{1}PFPN) towards representative alkenes (styrene, 1-hexene, cyclohexene, 2-methyl-2-butene, and tetramethylethylene) was determined by laser flash photolysis. The absolute rate constants for the trapping of ^{1}PFPN were deduced for alkenes assuming that $k_{pyridine}$ of ^{1}PFPN is 3×10^7 $M^{-1}s^{-1}$ in methylene chloride and acetonitrile. The nitrene reacts faster with more highly substituted alkenes. But the selectivity is quite low and varies only slightly with solvent polarity. The selectivity of ^{1}PFPN resembles that of singlet phenylcarbene with alkenes. Moreover, the selectivity of ^{1}PFPN towards tetramethylethylene and pyridine does not vary with temperature in methylene chloride. Thus, as per many carbenes, the reactions of ^{1}PFPN must have early transition states and the selectivity differences are due to entropic rather than to enthalpic factors.

228. LASER FLASH PHOTOLYSIS STUDIES OF RADICAL FRAGMENTATIONS. Pavel A. Simakov, Martin Newcomb, John H. Horner, Felix N. Martinez. Department of Chemistry, Wayne State University, Detroit, MI 48201, USA.

The direct observation of radical fragmentations has fundamental importance for organic chemistry. This work is an attempt to measure directly rate constants of decarboxylation and decarbonylation reactions of carbon centered radicals. In order to satisfy the requirements for laser flash photolysis, several classes of radical precursors has been investigated, including: N-(alkoxyoxalyloxy)pyridine-2-thiones (PTOC esters), dialkoxy iodoarenes, 2-alkoxysulfanyl-benzothiazoles and pivaloketones. The radical of interest was produced by irradiating the radical precursor with 266 or 355 nm light in the laser flash; fragmentation reaction rate constants were calculated from the observation of UV absorbance of the fragments. The "reporter group" approach applying ring opening of a 2-substituted-cyclopropylcarbinyl radical was used in investigations of the fragmentation reactions leading to primary alkyl radicals. Fragmentation kinetics and temperature dependence will be discussed.

229. NON-ALTERNATE CYCLOPENTENE FUSED POLYCYCLIC AROMATIC HYDROCARBONS: A NEW ADDITION
B. F. Plummer, D. C. Goodman, S. A. Faiz
Department of Chemistry, Trinity University, San Antonio, TX 78212

Non-alternate, cyclopentene-fused polycyclic aromatic hydrocarbons (CPAH) often exhibit different behavior than alternate PAH. The phenyl substituents in 1,4-diphenylbenz[1,2-a]aceanthrylene, **1**, cause distortion of **1** from planarity. The photochemistry of **1** has been examined and the novel [2π + 2π] cycloaddition products, **2**, characterized. An analysis of their structures and an interpretation of the photochemistry of **1** will be presented.

230. RECENT PROGRESS IN TOTAL SYNTHESIS OF NATURAL PRODUCTS. Clayton H. Heathcock, Department of Chemistry, University of California, Berkeley, CA 94720.

This lecture will deal with recent work in the total synthesis of complex natural products having promising bio-activity.

231. DEVELOPMENT AND APPLICATION OF RHODIUM CARBENOID CHEMISTRY IN THE SYNTHESIS OF NATURAL PRODUCTS

John L. Wood
Sterling Chemistry Laboratory, Department of Chemistry, Yale University
New Haven, Connecticut 06520-8107

Abstract. Our recent efforts to exploit the rich chemistry of rhodium carbenoids in natural products total synthesis will be presented in the context of syntheses directed toward K252a, staurosporine, N-methyl welwitindolinone C isothiocyanate, and syringolide 1 and 2.

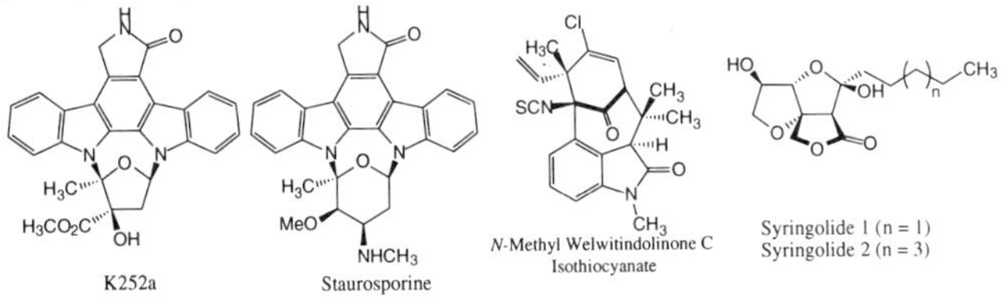

K252a

Staurosporine

N-Methyl Welwitindolinone C Isothiocyanate

Syringolide 1 (n = 1)
Syringolide 2 (n = 3)

232. PROGRESS AND PROBLEMS IN TOTAL SYNTHESIS.
G. Stork, Department of Chemistry, Columbia University, New York, NY 10027.

Recent progress and problems in total synthesis will be presented.

233. SYNTHESIS OF ARCHITECTURALLY COMPLEX NATURAL PRODUCTS.
Amos B. Smith, III
Department of Chemistry, Laboratory for Research on the Structure of Matter, and Monell Chemical Senses Center, University of Pennsylvania, Philadelphia, PA 19104

Progress towards the total synthesis of spongistatins 1 and 2, potent antitumor agents, will be discussed. Central to this synthetic venture is the development of an efficient, "one pot" multicomponent linchpin coupling of 2-lithio-2-trialkylsilyl-1,3-dithiane with a variety of electrophiles via controlled Brook rearrangement.

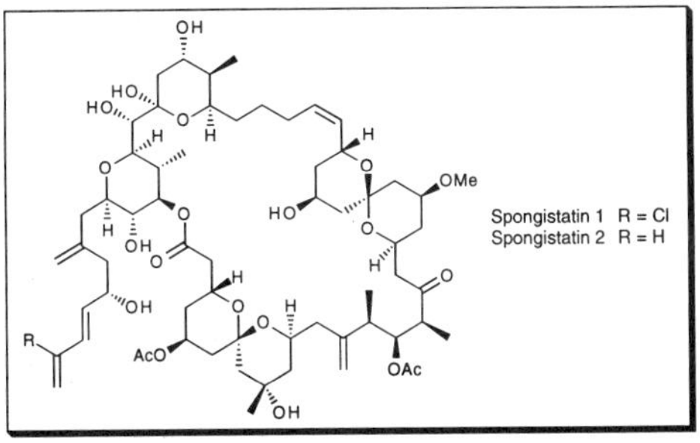

Spongistatin 1 R = Cl
Spongistatin 2 R = H

234.
SHORT LIVED 1,5-BIRADICALS FORMED FROM TRIPLET 1-ALKOXY- AND 1-(BENZYLOXY)-9,10-ANTHRAQUINONES. **Ronald L. Blankespoor,** Robert P. Smart, Timothy J. Peelen, Donald L. Ward, Department of Chemistry, Calvin College, Grand Rapids, MI, 49546

The cyclopropylmethyl and trans-2-phenylcyclopropylmethyl radical clocks were used to estimate the lifetimes of triplet state biradicals formed from 1-alkoxy-9,10-anthraquinones by photoexcitation. Irradiation (350 nm) of 1-(cyclopropylmethoxy)-2-methyl-9,10-anthraquinone gave cyclopropanecarboxaldehyde and 1-hydroxy-2-X-9,10-anthraquinone (3) in high yields with no rearranged products. In contrast, irradiation of 1-{(trans-2-phenylcyclopropyl)-methoxy}-2-benzyl-9,10-anthraquinone under similar conditions produced only small amounts of 3 and trans-(2-phenylcyclopropyl)carboxaldehyde, and a major product (60%) resulting from rearrangement of the 1,5-biradical to a homoallylic 1,8-biradical. Using a rate constant of ca. 1.0e10 s-1 for the ring opening of an oxygen substituted phenylcyclpropylmethyl radical, a lifetime of 1-2 ns is obtained for this 1,5-biradical, which is much shorter than those from benzophenones and acetophenones.

235. **BORON ASSISTED PHOTOCHEMICAL HYDROGEN ATOM ABSTACTIONS**
Steven A. Fleming and Cara L. Bradford, Department of Chemistry and Biochemistry, Brigham Young University, Provo, UT 84602

A variety of boron templates have been synthesized and used to coordinate chromophores in order to facilitate intermolecular photochemical reactions. This methodology has been applied to 2+2 photocycloadditions and radical coupling procedures. Irradiation of acetophenone in the presence of tetrahydrofuran with added boron reagent for complexing provides a good yield of a diastereomeric mixture the coupled alcohol (**1**) shown below. Use of a tosyl protected amino borane as the template coordinating cinnamyl alcohol and tetrahydrofuran results in formation of acetal **2**, presumably via radical cleavage, hydrogen abstraction, and electron transfer. A mechanism for selectivities and formation will be presented.

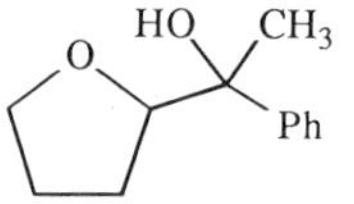

236. FLUORESCENCE QUENCHING OF PYRENE LABELLED POLY(ARYL ETHER) MONODENDRONS
James E. Hanson, Jeanne M. Riley, and Wyatt R. Murphy, Jr.
Department of Chemistry, Seton Hall University, South Orange, NJ 07079-2694

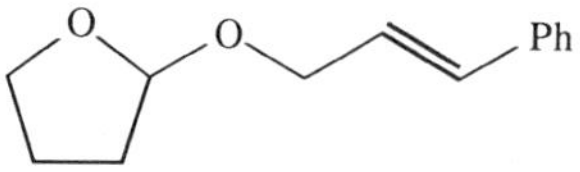

Little is known about the solution structure of dendritic polymers; fluorescence techniques have the potential to elucidate many of the details of dendrimer structure and dynamics. A series of poly(aryl ether) monodendrons labelled with 1-oxo-pyrene at the focal point have been synthesized and the fundamental photophysical data for these molecules have been reported previously. Molecular oxygen ($^3O_2$) is a good quencher for the fluorescence of polynuclear aromatic hydrocarbons such as pyrene: it quenches efficiently by a contact mechanism. Fluorescence quenching by oxygen was used to probe the permeability and microstructure of the labelled dendritic polymers. The native lifetimes of the pyrene labels were essentially identical for all generations of dendrimers, but when measured in the presence of oxygen, the smaller dendrimers were quenched more efficiently. This may be a result of increased shielding by the larger dendrimers or a result of slower diffusion for the larger molecules. More detailed studies and mathematical modelling should be able to distinguish between these two possible mechanisms.

237. PHOTOGENERATION AND CHEMISTRY OF NON-KEKULÉ M-QUINONE METHIDES: α,α-DIPHENYL-M-QUINONE METHIDE. **Li Diao** and Peter Wan, Department of Chemistry, University of Victoria, POBox 3065, Victoria, B.C., CANACA V8W 3V6

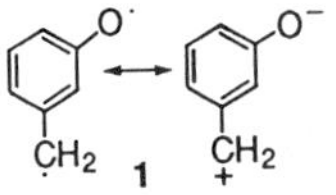

Non-Kekulé molecules, for example *m*-quinone methides (*m*-QMs, **1**), have been of theoretical and experimental interest for many years (Borden, W.T.; Iwamura, H.; Berson, J.A. *Acc. Chem. Res.* **1994**, *27*, 109). However, experimental investigation of these molecules has been hampered due to the limited number of general routes to these molecules. We have recently reported an efficient and general method for the photogeneration of *o*, *m*, and *p*-QMs, by photolyzing the corresponding hydroxybenzyl alcohols (Diao, L.; Cheng, Y.; Wan, P. *J. Am. Chem. Soc.* **1995**, *117*, 5369). In this work, we present results for the photogeneration and subsequent chemistry of α,α-diphenyl-*m*-QM, by irradiating its precursor (*m*-hydroxyphenyl)diphenyl methanol in aqueous solution. Two transients are formed using nanosecond laser flash photolysis (LFP), namely the carbocation in acidic solution with a 6 μs lifetime, and α,α-diphenyl-*m*-QM with a 35 μs lifetime in basic solution. The latter intermediate gives rise to polymer formation (in basic solution) which we have characterized. The mechanism of photogeneration and chemistry of this interesting *m*-QM will be presented based on studies using LFP and quantum yield measurements at different pHs.

238. PHOTOINDUCED ELECTRON TRANSFER OF FLUORESCEIN DERIVATIVES IN AQUEOUS POLYVINYLPYRROLIDONE. **Xiaohua Qian** and Guilford Jones, II, Department of Chemistry, Boston University, Boston, MA 02215

Photoinduced electron transfer between model electron donor-acceptor systems in the microdomain of synthetic polymers such as aqueous polyvinylpyrrolidone (PVP) has been studied. The conformation of the polymer plays an important role in regulating the rate of intramolecular electron transfer (ET). Two model compounds (DNBF and DTAF) having a common electron donor moiety (xanthene group) and different electron acceptor groups (dinitrobenzoylamido and dichlorotriazinylamino) were employed. For both systems, the very efficient fluorescence emission of the fluorescein moiety was found to be quenched via intramolecular photoinduced ET process to the electron acceptor groups and the rate of electron transfer is found to be depend on the driving force, the solvents and the property of the polymer interior under circumstances in which the electron donor-acceptor ligands are bound into a polymer microdomain.

239.
SYNTHESIS OF NOVEL THIOPHENE BASED DYES FOR NONLINEAR OPTICAL APPLICATION. A. Abbotto, S. Bradamante, A. Facchetti and G. A. Pagani, Department of Chemistry, University of Milano, via Golgi 19, I-20133 Milano, Italy.

Push-pull molecules where a strong intramolecular charge transfer occurs from the donor to the acceptor end group have become very attractive as non linear optical (NLO) materials for potential electrooptical applications. We report the synthesis of new dyes consisting of an N-alkylpyridinium ring (the acceptor) bonded to a dicyanomethanide functionality (the donor) through a conjugated spacer group containing a variable number of thiophene rings. The NLO properties have been monitored through the solvatochromic effect. The solvatochromic response of the new chromophores is among the highest so far reported in the literature.

240. POLYTHIOPHENE/POLYPEPTIDE HYBRIDS: SYNTHESIS AND POLYMERIZATION STUDIES OF BITHIOPHENE DERIVATIZED POLY-L-TYROSINE. Seth C. Rasmussen, Jason C. Pickens, and James E. Hutchison, Department of Chemistry and Materials Science Institute, University of Oregon, Eugene, OR 97403.

Conjugated polymers are of interest for applications in the field of electronic and electrooptical devices. To optimize these systems, routes for producing structurally defect-free and highly oriented polymeric systems must be developed. Our approach to controlling the orientation, supermolecular arrangement and dimensionality of conducting polymers utilizes rigid-rod, α-helical polypeptides to "template" the oxidative polymerization of the conjugated polymer. Our method utilizes bithiophene "monomers" covalently bound to the liquid crystalline polypeptide which holds them in position for polymerization. Attachment of the monomers to the "template" potentially directs the regiochemistry and sequence of monomer couplings, degree of polymerization, and polydispersity of the resulting bound polymers. Synthesis of the bithiophene "monomer" and of bithiophene derivatized poly-L-tyrosine have been accomplished and will be reported. Chemical and electrochemical polymerization studies of this loaded "template" are currently being investigated and will be presented. Spectroscopic and electrochemical data of the bithiophene-derivatized polytyrosine and of the resulting polythiophene/polytyrosine hybrids will also be discussed.

241. CHIRAL RESOLUTION OF 10, 10'-DIHYDROXY-9, 9'-BIPHENANTHRYL AND ITS APPLICATION IN THE COLOR PATTERNING OF REFLECTIVE CHOLESTERIC DISPLAYS, **Yuhui Lin** and L.-C. Chien*, Liquid Crystal Institute, Chemical Physics and NSF ALCOM Center, Kent State University, Kent, OH 44242

Two methods for the resolution of 10, 10'-dihydroxy-9, 9'-biphenanthryl were developed. The first involved the salt formation with (-)-strychnine via its cyclic diester with phosphoric acid. The second method was achieved by the formation of diastereomers (R)- and (S)-9, 9'-biphenanthryl-10, 10'-diyl N-((-)-α–methylbenzyl)phosphoro-amidates. (+)-10, 10'-dihydroxy-9, 9'-biphenanthryl was utilized in the study of color patterning of reflective cholesteric displays. A color pattern consisting three primary colors, red, green, and blue, was achieved by irradiating the display cell containing the optical active 10, 10'-dihydroxy-9,9'-biphenanthryl with different doses of UV light.

242.

EXCITED STATE CHARGE TRANSFER IN 2,3,5,6-TETRAARYL BENZO[1,2-B:4,5-B']
DIFURANS: EVIDENCE FOR TANDEM TICT FORMATION. **M. A. Meador** M. Abdul-Aziz, D. Hardy-Green, W. Youngs, J.V. Auping, A.Hess, J.D. Kinder, Polymers Branch, NASA Lewis Research Center, Cleveland, OH 44135; Department of Chemistry, Case Western Reserve University, Cleveland, OH 44106; Department of Chemistry, The University of Akron, Akron, OH

Excited state charge transfer in donor/acceptor substituted 2,3,5,6-tetraarylbenzo[1,2-b;5,4-b'] difurans, 1,was investigated in solvents of varying polarity. These compounds have high fluorescence quantum yields (from 0.20 to 0.98 depending upon solvent and substituent). They also exhibit solvatochromic behavior similar to TICT dyes, e.g., p-(N,N-dimethylamino) benzonitrile (Grabowski, Z.R.; et al Nouv. J. Chemie 1979, 3,443). However, unlike most TICT systems donor and acceptor are not conjugated with one another. A mechanism involving two cascading "TICT-like" charge transfers, similar to that recently described for p-(N,N-dialkylamino)terphenyls and -quaterphenyls (Foley, M.J.; Singer, L.A. J. Phys. Chem. 1994, 98, 6430), is proposed for 1.

243.

PHOTORESISTANT ANTIBODY-SENSORS USING ACRIDONE AS FLUORESCENT LABEL. **J.-L. Reymond**, N. Bahr, E. Tierney, Department of Molecular Biology, BCC-550, The Scripps Research Institute, la Jolla, CA 92037

Acridone is a highly photoresistant fluorophore that can be used to build selective fluorescent sensors using antibodies. We show that anti-analyte antibodies quench the fluorescence of acridone-labeled analytes by up to 98 %. The fluorescence signal (lem = 445 nm) of the resulting homogeneous immunosensors is highly sensitive and selective for the analyte. Most importantly, these sensors are completely stable to prolonged irradiation with UV light (356 nm), which suggests that acridone-based sensors might be useful for imaging application when very long measurement times are required.

244. **A CHIROPTICAL REDOX SWITCH BASED ON TRIPODAL LIGAND/ COPPER COORDINATION COMPLEXES**, James W. Canary* and Steffen Zahn *Department of Chemistry, New York University, New York, NY 10003*

The preparation and characterization in solution of a coordination complex that displays several physical properties consistent with behavior as a reversible chiroptical redox switch are reported. The ligand, bis(2-(quinolyl)methyl)-1-((2-quinolyl)ethyl)amine (α-MeTQA), was prepared by asymmetric synthesis. Complexes of α-MeTQA with Cu^I and Cu^{II} display bisignate circular dichroism (CD) spectra with very large amplitudes. The shapes of the CD signals agree with previously published analyses of the conformational behavior of closely related coordination complexes. The amplitude of the spectrum of the $Cu^{II}(L)$ complex is dramatically larger than that of corresponding $Cu^I(L)$ complex. The CD spectra can be rapidly and reversibly interconverted by adding chemical oxidants or reductants to the solutions. Addition of certain coordinating anions enhances the magnitude of the effect.

245. WHY ARE METHYLENECYCLOPROPANE AND CYCLOPROPENE MORE STRAINED THAN CYCLOPROPANE? William T. G. Johnson and Weston Thatcher Borden, Department of Chemistry, Box 351700, University of Washington, Seattle, Washington 98195-1700

Ab Initio calculations have been performed in order to investigate why the introduction of each trigonal center into a cyclopropane ring results in an increase in strain energy of 12 - 14 kcal/mol. Our calculations find that, as is commonly believed, introduction of a trigonal center into a three-membered ring does create additional angle strain. However, our computational results show that the major source of the additional "strain" that results from the introduction of each trigonal center into cyclopropane is not an increase in angle strain but the loss of a very strong cyclopropane C-H bond.

246. **TRIS(TRIISOPROPYLSILYL)SILANE AND BIS(TRIISOPROPYLSILYL)SILYLENE.**
P. P. Gaspar, A. M. Beatty, T. Chen, T. Haile, W. T. Klooster, T. F. Koetzle, D. Lei, T. S. Lin, Departments of Chemistry, Washington University, St. Louis, MO 63130-4899 and Brookhaven National Laboratory, Upton, NY 11973-5000

It has finally been possible to test the strategy presented in 1991 (Grev, Schaefer, Gaspar, *J.Am.Chem.Soc.* **1991**, *113*, 5638) for the design of a ground-state triplet silylene: attaching to the divalent silicon atom groups sufficiently bulky to open the R-Si-R angle to a value such that the triplet silylene is lower in energy than the singlet, and lowering the electronegativity of the substituents to minimize the "crossover angle" above which the ground state is a triplet. Tris(triisopropylsilyl)silane has been synthesized as a precursor to $(i\text{-}Pr_3Si)_2Si:$, and both thermal and photochemical dissociation to the silylene has been successful, and the multiplicity of the ground state will be reported. X-ray and neutron diffraction studies revealed novel features of $(i\text{-}Pr_3Si)_3SiH$ such as the near coplanarity of the four silicon atoms and the length of its Si-H bond. This work receives support from National Science Foundation Grant No. CHE-9632897.

247. **DETECTION AND STRUCTURE-REACTIVITY STUDIES OF TRANSIENT INTERMEDIATES IN THE NUCLEOPHILIC SUBSTITUTION REACTIONS OF ACTIVATED VINYLIC SYSTEMS.** Claude F. Bernasconi,*[a] **Rodney J. Ketner**,[a] and Zvi Rappoport,[b] [a]Department of Chemistry and Biochemistry, University of California, Santa Cruz, California 95064, [b]Department of Organic Chemistry, Hebrew University, Jerusalem, Israel, 91904

The tetrahedral intermediates in the nucleophilic substitution reactions of several activated vinylic substrates with thiolate ion nucleophiles have been observed spectroscopically. The substrates studied are $Ph(X)C=C(CO_2)_2C(CH_3)_2$ (X = OMe, SMe); $Ph(MeO)C=C(CN)_2$; α-thiomethylbenzylideneindandione; $Ph(MeS)C=C(CO_2Me)NO_2$; and $Ph(X)C=C(Ph)NO_2$ (X = OMe, OCH_2CF_3). These observations confirm the hypothesis that nucleophilic vinylic substitution reactions take place by an addition-elimination mechanism (1). Rate constants for the formation

$$Nu^- + \underset{LG}{\overset{Ph}{>}}=\underset{Y}{\overset{X}{<}} \underset{k_{-1}}{\overset{k_1}{\rightleftharpoons}} Ph-\underset{LG}{\overset{Nu}{\underset{Y}{\overset{X}{|}}}} \overset{k_2}{\longrightarrow} \underset{Nu}{\overset{Ph}{>}}=\underset{Y}{\overset{X}{<}} + LG^- \quad (1)$$

and breakdown of these tetrahedral intermediates are to be reported. The factors that influence the intermediate's stability and lifetime will be discussed.

248. **SPECTRAL EVIDENCE FOR SINGLE ELECTRON TRANSFER IN NUCLEOPHILIC ALIPHATIC SUBSTITUTION OF A CARBANION BY METHYL IODIDE.** <u>Laren M. Tolbert</u>, **Joanne Bedlek, Michael Terapane and Janusz Kowalik, School of Chemistry and Biochemistry, Georgia Institute of Technology, Atlanta, GA 30332.**

9-Mesitylfluorenyl anion (**9MsF-**) is unreactive toward methyl iodide a low temperatures (<-78°C). Above 60°C, the spectrum of **9MsF-** in the presence of MeI is replaced with that of the corresponding 9-mesitylfluorenyl radical (**9MSF·**). This spectral change is accompanied by the formation of 9-methyl-9-mesitylfluorene in low yield. These observations are consistent with the oxidation of **9MsF-** by methyl iodide and represent the first observation of the oxidation of a carbanion by methyl iodide. The formation of the nucleophilic substitution product by single electron transfer thus represents a tenable pathway for product formation.

249. SYMMETRIES OF HYDROGEN BONDS IN SOLUTION. Charles L. Perrin and Jennifer B. Nielson, Dept. of Chemistry, Univ. Calif. San Diego, La Jolla, CA 92093-0506.

Is a hydrogen bond symmetric (single-well potential, O⋯H⋯O, one structure) or asymmetric (double-well potential, O-H⋯O ↔ O⋯H-O, pair of tautomers)? The NMR method of isotopic perturbation can distinguish these in monoanions of mono-^{18}O-labeled dicarboxylic acids. According to observed ^{13}C isotope shifts (ISs) in aqueous solution, these hydrogen bonds are asymmetric (Perrin, *Science*, **1994**, *266*, 1665). The results on hydrogen maleate and phthalate were especially surprising since they show symmetric hydrogen bonds in crystals. The difference had been attributed to the disorder of aqueous environment, and it was thought that these hydrogen bonds are symmetric in nonpolar solvents. Yet further, more extensive studies show that even in nonpolar solvents these hydrogen bonds are asymmetric. The observed ^{18}O-induced isotope shifts, especially at the *ipso* carbons, demonstrate that a wide variety of these monoanions exist as two equilibrating tautomers in all solvents. In support of this claim, "primary" ISs, which had been diagnostic for a single-well potential, were reinvestigated.

250. CONFORMATIONAL ANALYSIS OF SINGLET-TRIPLET STATE MIXING IN ORGANIC DIRADICALS SEPARATED BY A PARTIALLY CONJUGATED SPACER.
Andrei G. Kutateladze, William A. McHale, Department of Chemistry and Biochemistry, University of Denver, Denver, CO 80208

A systematic *ab initio* casscf study has been undertaken to elucidate the conformational dependence of spin-orbit coupling (soc) in triplet organic diradicals of general type ·CCXC· and ·CCXZ·, where X is a fragment (lone pair or multiple bond) capable of conjugation with one of the diradical centers. It is found that the overall problem of conformational dependence of soc values in such triplet species can in general be reduced to a fairly well understood case of soc dependence in 1,3-propanediyl. A simple "empirical" fitting equation for accurate evaluation of spin-orbit coupling matrix elements in various triplet organic species will be presented and the prospects of a "molecular mechanics"-like approach to treating the relativistic corrections to the Hamiltonian will be discussed.

251.

N-NITROSIMINIUM ION INTERMEDIATES IN THE DECOMPOSITION OF α-ACETOXYNITROSAMINES. Hongliang Cai, James C. Fishbein, Department of Chemistry, Wake Forest University, Winston-Salem, N. C. 27109

α-Acetoxynitrosamines are widely used to understand the molecular basis of dialkylnitrosamine carcinogenicity and mutagenicity. A study of the equilibrium and rate constants, product structure, isotope effects and structural-reactivity correlations in the decomposition of α-acetoxynitrosamines **1** will be presented. It is concluded that the pH-independent decomposition of α-acetoxynitrosamines involves the formation of N-nitrosiminium ion intermediates **2**. Differential effects of reactivity, mechanisms, and equilibrium of R_1 and R_2 will be discussed.

252. The Mechanism of Catalytic Lactic Acid Conversion to 2,3-Pentanedione: Correlation of Experimental and Molecular Modeling Data

Radu Craciun*, Scott M. Perry+, Dennis J. Miller+, and James E. Jackson*; Departments of *Chemistry and +Chemical Engineering, Michigan State University, East Lansing, MI 48824

In recent reports we described the discovery (*J. Catal.* **1994**, *148*, 252-260) and optimization (*Ind. Eng. Chem. Res.* **1995**, *34*, 974-980) of a new catalytic process: the condensation of lactic acid to form 2,3-pentanedione over bases on silica/alumina:

$$2\ CH_3CHOHCOOH \longrightarrow CH_3C(O)C(O)CH_2CH_3 + CO_2 + 2\ H_2O$$

We now present condensed-phase reaction studies, variable temperature-MS (VT-MS), diffuse reflectance IR (DRIFTS), deuterium labeling studies, and molecular orbital calculations at semiempirical PM3 and ab initio 3-21G levels of theory, with solvation effects included via the PM3 model. The consensus mechanism begins with a Claisen condensation, followed by decarboxylation and dehydrative ketonization of the resultant enediolate.

253. PHYSICAL ORGANIC CHEMISTRY OF FISCHER CARBENE COMPLEXES. KINETICS OF HYDROLYSIS OF $(CO)_5M=C(OR)Ph$ (M = Cr, W; R = Me, Et) AND $(CO)_5Cr(OMe)CH=CHPh$ IN AQUEOUS ACETONITRILE. C.F. Bernasconi, F.X. Flores, and **K.W. Kittredge**, Department of Chemistry and Biochemistry, University of California, Santa Cruz, CA 95064

A study of the hydrolysis of the title compounds in 50% MeCN-50% water (v/v) at 25°C is reported. The reaction occurs in two stages; the first is the formation of $(CO)_5M=C(O^-)R'$ (R' = Ph or CH=CHPh) while the second, much slower stage is the formation of R'CH=O and $(CO)_5MOH^-$. This presentation focuses on the first stage. The results lead to insights regarding the following points. (1) Effect of changing the metal (Cr vs. W), the alkoxy group (MeO vs. EtO) and the R' group (Ph vs. CH=CHPh) on reactivity. (2) Relative reactivity of Fischer carbene complexes vs. carboxylic esters. (3) Reasons why a tetrahedral intermediate is not detectable even though the equilibrium for its formation is probably favorable at high pH. (4) Reasons why the hydrolysis of Fischer carbene complexes containing an acidic proton such as $(CO)_5Cr=C(OMe)CH_3$ follow an entirely different mechanism.

254. COMPUTATIONAL STUDY OF THE STABILIZATION OF 2-FERROCENYLMETHYL CATION. Kathleen V. Kilway[†] and Andrew Streitwieser
Department of Chemistry, University of California, Berkeley, CA 94720-1460
[†]Department of Chemistry, University of Missouri, Kansas City, MO 64110-2499

2-Ferrocenylmethyl cation was studied using density functional theory (Becke3LYP). Within designated symmetry constraints, geometries were optimized with a quasi-relativistic effective core potential for iron and standard double zeta basis sets (6-31G(d,p)) with a set of d polarization functions for carbon and p polarization functions for hydrogen atoms. The resulting structure is a hybrid of the previously proposed structures where the α-carbon is bent out of the cyclopentadienyl plane towards the iron center with no ring slippage. The formation of the ferrocenyl vs. benzyl cation is favored by over 24 kcal/mol. This stabilization is due to the sharing of the positive charge by the cyclopentadienyl rings, effectively spreading the charge throughout the ferrocenyl moiety.

255. THE EFFECTIVE COMBINATION OF MOLECULAR MODELING CALCULATIONS AND REFRACTIVE INDEX MEASUREMENTS TO STUDY ENTHALPIES OF FORMATION AND HYDRATION EQUILIBRIA RESPECTIVELY, FOR A SERIES OF CYCLOKETONES. Madeline Adamczeski*, Keiko Tomioka, **Zhiwen Zhang**, Karen Byrne. Department of Chemistry, American University, N.W., Washington, D.C. 20016-8014

We wished to experimentally determine the hydration equilibrium constants for a series of cyclic ketones. From refractive indices of each pure carbonyl compound and two corresponding aqueous solutions of different compositions, the equilibrium constants were obtained. The hydration equilibria for respectively, acetone, cyclobutanone, cyclopentanone, cyclohexanone, cycloheptanone, and cyclooctanone were fitted by an empirical formula which relates the refractive indices to temperature and density. It is noteworthy to mention that the hydration equilibrium constant for cycloheptanone was not previously published. Molecular mechanics calculations were employed to assist our understanding of equilibria and predict the enthalpies of hydration of the aforementioned carbonyl compounds. Formaldehyde and cyclopropanone were also included in the theoretical calculations. We anticipated that analysis of the theoretical calculations for these hydration reactions could: (a) support our experimental results, (b) enhance our current understanding about the individual factors that govern the equilibrium, and (c) be used when experimental values are lacking.

256. **GENERATION AND TRAPPING OF N-SUBSTITUTED-3-AZETIDINYLIDENECARBENES**
Alan P. Marchand,* K. C. V. Ramanaiah, and Simon G. Bott, Department of Chemistry, University of North Texas, Denton, TX 76203; John C. Gilbert,* Department of Chemistry, University of Texas at Austin, Austin, TX 78712; Steven Kirschner,* Department of Chemistry, Austin Community College, Riverside Campus, Austin, TX 78741.

The reactive intermediates produced via base-promoted reactions of **1a** and **1b** with diethyl Diazomethylphosphonate (DAMP) have been shown to be vinylidenecarbenes rather than the corresponding cycloalkynes. The results of semi-empirical MO calculations for ring expansion of azetidinylidenecarbenes to azacyclopentynes will be presented and discussed.

257. **DISTINGUISHING "TRUE" CARBENE FROM EXCITED DIAZO REACTIVITY USING TRIPLET SENSITIZATION.** Krista R. Motschiedler and Miguel Garcia-Garibay*. Department of Chemistry and Biochemistry, University of California, Los Angeles, CA 90095.

Irradiation of 1,2-diarydiazoethenes in methanol gives rise to stilbenes and ethers by intramolecular 1,2-H shifts and intermolecular reaction with methanol, respectively. Although these products are known to originate from the singlet carbene, stilbenes may also form from the singlet excited state of the diazo precursor. In order to determine the relative contributions of carbene vs. excited state diazo reactivity, the results of inter- and intramolecular triplet sensitization were compared with those obtained from direct photolysis. It was found that the majority of 1,2-H shift products from direct photolysis originate from the carbene rather than the excited diazoethane.

258. **COMPUTATIONAL STUDIES ON THE MECHANISMS AND KINETICS OF ALKYLHALOCARBENE REACTIONS.** Amy E. Keating, Miguel A. Garcia-Garibay* and K.N. Houk*, Department of Chemistry and Biochemistry, University of California, Los Angeles, CA 90095-1569

We have used ab initio and density functional computational methods to investigate the reactivity of various alkylhalo and dihalocarbenes. Results indicate that 1,2-H shifts in the free carbenes to give olefin products are sensitive to the conformation of the carbene, in agreement with the high Z/E selectivities observed experimentally. We have also investigated reaction trajectories for cyclopropanation with several olefins, and we show that a delicate balance between the barriers for carbene isomerization, rearrangement and cycloaddition leads to a kinetic scheme which can explain provocative experimental results.

259. **VINYL OXOCARBENIUM IONS IN INTERMOLECULAR 4+3 CYCLOADDITION REACTIONS.** Michael Harmata*, and Darin E. Jones, Department of Chemistry, University of Missouri-Columbia, Columbia, Missouri 65211

The preparation of bifunctional conjunctive reagents **1** and their Lewis acid-mediated reaction with dienes to produce 4+3 cycloadducts will be presented.

260. **DiazoDecomposition Applied to Intramolecular Aziridination Reactions**
Dennis L. Wright and Mark C. McMills*
Department of Chemistry and Program in Molecular and Cellular Biology
Ohio University
Athens, Ohio 45701

Cyclic and acyclic diazoamide precursors for diazodecomposition have been prepared with tethered imino C=N bonds for an approach to 1,8-diazabicyclo[3.2.1]octane containing natural products such as quinocarcin. Intramolecular aziridination has been found to be the dominant behavior as opposed to the other possible modes of reactivity such as ylide formation or C-H insertion.

261. **INTRAMOLECULAR AZOMETHINE YLIDE CYCLOADDITION APPROACH TO THE HEXACYCLIC BISGUANIDINE (−)-PALAU'AMINE**
Larry E. Overman,* John E. Tellew, and William C. Trenkle
Department of Chemistry, University of California, Irvine, CA 92697-2025.

Studies directed toward the total synthesis of palau'amine will be discussed.

262. **ASYMMETRIC DIELS-ALDER REACTION EMPLOYING CHIRAL 1-AMINO-3-SILOXY-1,3-BUTADIENES: APPLICATION TO ENATIOSELECTIVE SYNTHESIS OF (−)-α-ELEMENE.** Sergey A. Kozmin, Viresh H. Rawal*, Department of Chemistry, University of Chicago, Chicago, Illinois 60637

The asymmetric Diels-Alder reaction is a powerful tool for the synthesis of enetiomerically-enriched, complex molecules. We report here the first preparation of chiral 1-amino-3-siloxy-1,3-butadienes and show their usefulness in Diels-Alder reaction. These dienes exhibit excellent reactivity and enantioselectivity in [4+2] cycloadditions with various dienophiles. The synthetic potential of this methodology is exemplified by a concise, enantioselective synthesis of (-)-α-elemene.

263. DIELS-ALDER REACTIVITY OF A KETOVINYLPHOSPHONATE. EMPIRICAL AND THEORETICAL OBSERVATIONS. STUDIES TOWARD THE SYNTHESES OF PHOSPHONATE ANALOGS OF INOSITOLS. Cynthia K. McClure*, Karl B. Hansen[†], Keith J. Herzog[†], Ross J. Fischer, Jeffrey S. Link, Montana State University, Department of Chemistry and Biochemistry, Bozeman, MT 59717
[†]University of Delaware, Department of Chemistry and Biochemistry, Newark, DE 19713

The Diels-Alder reactivity of the ketovinylphosphonate **1** with various dienes has been investigated under thermal and Lewis acid assisted conditions, with the ultimate targets being phosphonate analogs of inositol phosphates. The dienes studied were cyclopentadiene, furan, 2-silyloxyfuran, (E)-1-acetoxy-1,3-butadiene, and Danishefsky's diene. We have also performed *ab initio* calculations on the phosphonic acid analogue of **1** using a model Lewis acid (H$^+$). The resulting LUMO levels of these model dienophiles were compared, and will be discussed.

264. NOVEL CYCLOISOMERIZATION PATHWAYS FOR ENYNE ALLENES. Thomas Gillmann*, Stefan Heckhoff, and Wilhelm Dolle, *Institut für Pharmazeutische Chemie der Philipps-Universität Marburg, Marbacher Weg 6, D-35032 Marburg, Germany.*

Since the discovery of the enediyne antitumor agents the closely related enyne allenes have been studied primarily with the aim to design simple mimics of their natural prototypes that would be useful as DNA-cleaving agents. In addition, these π-electron systems are increasingly used for sequential radical cyclization reactions, too. We and others have recently discovered a novel mode of cycloisomerization of enyne allenes that is characterized by the formation of five-membered instead of six-membered rings.

Herein we report on cycloisomerization experiments with a series of enyne allene esters and discuss substituents effects that direct these reactions to either of the two modes.

265. STEREOCHEMISTRY OF THE THERMAL ISOMERIZATION OF 1-VINYL-7-EXO-PHENYLBICYCLO[4.1.0]HEPTANE TO 7-PHENYLBICYCLO[4.3.0]NON-1(9)-ENE. John E. Baldwin and Samuel J. Bonacorsi, Jr., Department of Chemistry, Syracuse University, Syracuse, NY 13244

Thermal isomerization of 1-(2'-(E)-deuterioethenyl)-7-*exo*-phenylbicyclo[4.1.0]heptane gives two 7-*exo*-phenylbicyclo[4.3.0]non-1(9)-enes, the *suprafacial, retention* and *antarafacial, retention* products shown, in a ratio of 86:14. The substantial antarafacial component is noteworthy, for the rearranging hydrocarbon cannot rotate about the C1-C6 bond following C1-C7 bond cleavage.

sr product, 86 %   *ar* product, 14 %

266. PROGRESS TOWARDS DIPHENYLPENTAPRISMANE. William R. Sponholtz, III[a], Frank L. Switzer[a], Gordon W. Gribble[a], Marianne P. Byrn[b], and Charles E. Strouse[b], Departments of Chemistry, a: Dartmouth College, Hanover, NH 03755; b: University of California, Los Angeles, CA 90095.

Mesoionic "munchnone" **1** (oxazolium-5-olate) reacts smoothly with 1,3,5,7-cyclooctatetraene (COT) (**2**) in a tandem 1,3-dipolar cycloaddition reaction to give the novel N-benzyl-9,11-diphenyl-10-azatetracyclo[6.3.0.0^{4,11}0.^{5,9}]undeca-2,6-diene (**3**) ring system in low yield. Intramolecular [2+2] photocyclization affords in high yield the desired polycyclic caged amine **4**, N-Benzyl-3,5-diphenyl-4-azahexacyclo[5.4.0.0.^{2,6}0.^{3,10}0.^{5,9}0.^{8,11}]undecane. These unprecedented reactions, if further exploited successfully, will afford an exceptionally concise and efficient synthesis of diphenylpentaprismane (**5**).

267. **A Novel Pathway for Carbocation Generation Through Carboxylic Acid Decarboxylation**
Pavel A. Krasutsky *, Hong Wang, Xiaoling Zhang, Robert M. Carlson
Department of Chemistry, University of Minnesota-Duluth, Duluth, Minnesota 55812

The decarboxylation of carboxylic acids with trifluoroacetic anhydride and > 95% H_2O_2 proceeds at room temperature. Products of this reaction clearly demonstrates the intermediate formation of carbocations under these conditions which could undergo to typical reactions: collapse of ion pairs (**c**); elimination-epoxydation-ring opening (**e**); onium rearrangements (**o**); carbocation internal rearrangements (**i**). The formation of unstable asymmetric acyl trifluoroacetyl peroxides from bis(trifluoroacetyl)peroxide and the corresponding carboxylic acid is proposed. The details of the reaction between H_2O_2 and TFAAn are also clarified.

$$RCOOH \xrightarrow{(CF_3COO)_2} R-\overset{O}{\underset{\|}{C}}-O-OCOCF_3 \xrightarrow{(e) \quad (i)} R^+|CO_2|OCOCF_3^- \xrightarrow{(c)} ROCOCF_3 + CO_2$$
(o)

268.

IONIC AGGREGATES FOR SUPRAMOLECULAR CONTROL OF CHEMICAL REACTIONS.
Craig S. Wilcox, Paul J. Smith, John Stephens, Eu-nil Kim, Jaemoon Yang, and Daniel J. Soose; Department of Chemistry, Chevron Science Center, University of Pittsburgh, Pittsburgh, PA 15260

Organized assemblies of ionic groups that are close to reaction centers can strongly affect the rate and stereochemical outcome of chemical processes. Ionic groups can also provide a powerful organizing force in aggregate self-assembly. These intramolecular electric field effects and intra-aggregate electric field effects provide a means of predictably controlling reactivity. Examples of such ion field effects will be presented and an initial approach to quantitative analysis of these results will be described.

269. **SELF-REPRODUCING MOLECULAR SYSTEMS**
M. Reza Ghadiri, Departments of Chemistry and Molecular Biology and the Skaggs Institute for Chemical Biology, The Scripps Research Institute, La Jolla, California 92037

Living systems are autonomous self-reproducing entities that operate based on molecular information originating and changing through natural selection. In a living system, a complex blend of molecular information-transfer processes creates a coherent self-organized multicomponent catalytic chemical network—a "molecular ecosystem"— that not only interacts with, and adapts to the changes in the environment, but one that can also bring about emergent properties far greater than the simple sum of the chemical constituents of the system. Therefore, in order to understand and ultimately mimic the properties of living molecular systems, it is essential to construct and study the characteristics of simpler, but informationally-rich, multicomponent autocatalytic chemical systems. In this lecture, within the context of (auto)catalytic peptide-based systems, we will discuss the peptide self-replication process and the construction of self-organized multicomponent autocatalytic networks that display some of the most basic properties of living molecular systems such as selection, adaptation, and the acquisition of new functions.

270. CHEMICAL MODELS OF PROTEIN β-SHEETS. James S. Nowick, Department of Chemistry, University of California, Irvine, CA 92617-2025.

We have set out to learn about protein β-sheets by developing chemical models that mimic their structures and hydrogen-bonding patterns. In these models, peptidomimetic templates induce β-sheet structure in attached peptide strands. This talk will describe ongoing studies of oligourea templates designed to hold peptide strands in proximity, β-strand mimics designed to induce a β-sheet conformation in adjacent peptide strands, and β-sheet models containing these structural elements.

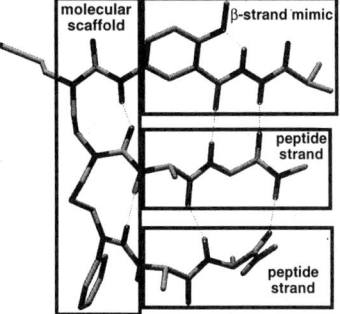

271. HOST-HOSTAGE RELATIONSHIPS OF MOLECULAR CAPSULES. J. Rebek, Jr., The Skaggs Institute for Chemical Biology, The Scripps Research Institute, 10550 North Torrey Pines Road, La Jolla, CA 92037

Molecular assemblies provide a means of evaluating intermolecular forces and chemical information involved in recognition processes. Self-complementary molecules are especially useful in this regard, and here are decribed synthetic structures inspired by, among other things, sports equipment. These molecules dimerize through hydrogen bonding to generate pseudo-spherical capsules. Their assembled states are capable of reversible encapsulation of smaller molecules of complementary size and shape. The unusual thermodynamic parameters observed for the encapsulation process augur well for the use of these capsules as reaction chambers.

272. COMBINATORIAL CHEMISTRY IN PERSPECTIVE. **Robert S. Root-Bernstein,** Department of Physiology, Michigan State University, East Lansing, MI 48824

Historical, philosophical, scientific, managerial and organizational perspectives will be used to analyze the challenges posed by combinatorial chemistry: What problem was combi-chem invented to solve? How successful has it been and can it possibly be? And what are the dangers of succeeding? In order to address these questions, some basic questions must be answered: What *is* diversity? Should chemists try to explore all of chemical diversity space or can subsets of that space, such as pharmaceutical or materials science diversity, be defined and explored more fruitfully? Should we explore or optimize? How can these strategies best be managed? Since new technologies change habits of work and thought, they require new organizational and managerial approaches. Technologies also uniformatize work, eliminating spontaneity and serendipity – traditional sources of discoveries that still need to be fostered. Finally, combi-chem introduces new challenges for adjacent fields such as analytical and computational chemistry, biological screening, and materials sciences. How can the integration of these fields be accomplished quickly and functionally and what is at stake in this process?

273. NOVEL CYCLIZATION REACTIONS FOR COMBINATORIAL CHEMISTRY. Paul A. Bartlett, Jason G. Lewis, Donna D. Johnson, Ryan P. Trump, Department of Chemistry, University of California, Berkeley, CA 94720-1460

Desirable criteria for synthetic routes to combinatorial or parallel libraries are 1) short sequences amenable to solid phase, 2) starting materials available with diverse functionality, 3) no more than one variable introduced in any step, and 4) target structures that are cyclic or conformationally constrained. Since amino acids (both natural and unnatural) are among the most diverse starting materials available, they are attractive input materials for library synthesis. In contrast to their oligomerization into conventional peptide libraries, we have incorporated these starting materials into short sequences that afford novel fused and spirocyclic bi- and tricyclic compounds. The development of this chemistry and its adaptation to solid phase library synthesis will be described.

274. THE SOLID PHASE SYNTHESIS OF TERTIARY AMINES. David R. Barn, Angus R. Brown, Zoran Rankovic, David C. Rees and **J. Richard Morphy**, Medicinal Chemistry Department, Organon Laboratories Ltd., Newhouse, ML1 5SH, Scotland, U.K.

A range of tertiary amines was constructed on "REM resin", starting from secondary amines, primary amines and resin-bound "ammonia". REM methodology involves three essential steps: 1) coupling of the starting amine to the resin via a Michael addition, 2) quaternisation with an alkyl halide and 3) base catalyzed cleavage of the product via Hofmann elimination. The resin linker is compatible with a range of chemistries, including acid and base sensitive protecting group strategies and palladium catalyzed couplings. The tertiary amine products are obtained in consistently high purity (95% or greater) and can be used in biological assays without purification. Reactions on REM resin can be monitored by a combination of FT-IR and gel phase ^{13}C NMR. After cleavage of the product, REM resin is regenerated and can be recycled in an automated synthesis of larger quantities (>0.5g) of pure tertiary amine.

275. PREPARATION OF TRICYCLIC TETRAHYDROQUINOLINE LIBRARIES VIA SOLID-PHASE ORGANIC SYNTHESIS. **Thomas K. Hayes**, John S. Kiely, Behrouz Forood and Thomas P. Brady, Combinatorial and Exploratory Chemistry, Houghten Pharmaceuticals, Inc., San Diego, CA 92121

Adaptation of a putative hetero Diels-Alder cyclization chemistry to solid-phase has been performed. This multicomponent reaction between solid bound anilino derivatives with aldehydes and a dienophile has been utilized in the generation of tricyclic tetrahydroquinoline derivatives. Integrating this chemistry in to various synthetic schemes has lead to the construction of several libraries of related compounds.

276. COMBINATORIAL LIBRARIES OF BALANOL ANALOGUES. **J. Nielsen** and L. O. Lyngsø, Department of Organic Chemistry, Building 201, The Technical University of Denmark, DK-2800 Lyngby, Denmark.

We are engaged in combinatorial solid-phase synthesis of selected natural product structures. Using the potent protein kinase C inhibitor balanol as a structural template, we have designed combinatorial libraries which are expected to elucidate structure-activity relationships effectively. From our retro-synthetic analysis, balanol is broken up in three main synthons:
♦ A mono-allylated aromatic diacid. For parent balanol, this synthon or building block is a complex, substituted benzophenone structure [4-(2-carboxy-6-hydroxybenzoyl)-3,5-dihydroxy-benzoic acid allylester].
♦ A Fmoc-protected cyclic aminoalcohol. For parent balanol, this building block is the *trans*-3-Fmoc-amino-4-hydroxy-hexahydroazepine.
♦ A benzoic acid. For parent balanol, this building block is protected 4-hydroxybenzoic acid. Procedures for the synthesis of all 3 building blocks have been implemented and will be presented. Likewise, application of a diverse set of building blocks in a combinatorial experiment will be demonstrated.

277. NEW NMR TOOLS FOR COMBINATORIAL CHEMISTRY INCLUDING SOLID-PHASE-SYNTHESIS-RESIN NMR AND HIGH-THROUGHPUT SPECTROSCOPY. **Paul A. Keifer**, Varian NMR Instruments, Palo Alto, California USA.

Two different technologies have evolved that improve the capability of NMR to monitor the progress of reactions, and characterize their products, at any stage of a combinatorial-chemistry program. First, we developed the techniques which are now being used to obtain high-resolution 1H NMR spectra of compounds still bound to solid-phase synthesis (SPS) resins; this includes the use of magic-angle-spinning (MAS) probes optimized for true high-resolution spectroscopy (Nano•nmr probes). We have used these Nanoprobes to determine how the quality of SPS-resin NMR spectra are influenced by resin, tether, and linker structures, and by different solvents, temperatures, nuclei, and experimental conditions, and have also demonstrated that the Nanoprobes can obtain NMR data from single beads of SPS resin (90 µ). Secondly, we have developed a complementary solution-state-NMR analysis method to rapidly (and automatically) obtain 1H NMR spectra on microgram quantities of sample. This fully automated "High-Throughput Spectroscopy" (HTS) method can analyze 100-nanomole-sized microtiter-plate samples in < 3 minutes per sample, even without deuterated solvents.

278. COMBINATORIAL LIBRARIES FOR DRUG DISCOVERY: THEORETICAL EVALUATION OF DECONVOLUTION STRATEGIES. Danielle A.M. Konings, Jacqueline R. Wyatt, David J. Ecker and **Susan M. Freier**, Department of Microbiology, Southern Illinois University, Carbondale, IL 62901 and Isis Pharmaceuticals, 2292 Faraday Avenue, Carlsbad, CA 92008

Synthesis and testing of complex mixtures maximizes the number of compounds that can be prepared and tested in a combinatorial library. When mixtures of compounds are screened, however, the identity of the compound(s) selected may depend on the deconvolution procedure employed. Previously, we developed a model system for evaluation of deconvolution procedures and used it to compare pooling strategies for iterative and non-iterative deconvolution [Freier et al, *J. Med. Chem.* **38**, 344-352]. We have now extended the model studies to include simulations of procedures with overlapping subsets such as subtractive pooling [Carell et al, *Angew. Chem. Int. Ed. Engl.* **33**, 2061-2064], bogus coin [Blake and Litzi-Davis, *Bioconjugate Chem.* **3**, 510-513] and orthogonal pooling [D'Prez et al, *J. Am. Chem. Soc.* **117**, 5405-5406]. These strategies required synthesis and testing of fewer subsets than did the more traditional non-overlapping iterative strategies. The compounds identified, however, were not the most active compounds in the library and were substantially less active than those identified using more traditional strategies.

279. RUTHENIUM CATALYZED REACTIONS FOR ATOM ECONOMY. **Barry M. Trost**, Department of Chemistry, Stanford University, Stanford, CA 94305-5080

Enhancement of synthetic efficiency requires maximizing our effectiveness in utilizing our raw materials and minimizing waste. A beginning toward this goal is to use reactions that generate minimal stoichiometric by-products and, in the ideal, are simple additions. An examination of a semi-rational approach for the invention of new addition reactions based upon ruthenium catalysis is presented.

280. THE SYNGEN PROGRAM FOR GENERATION OF EFFICIENT SYNTHESES. J. B. Hendrickson Department of Chemistry, Brandeis University, Waltham, MA 02254-9110

The SYNGEN program is designed to generate all the shortest synthetic routes to a given, input target molecule, and order them by cost. SYNGEN follows a clear logical protocol which focuses sharply on generating all routes with the fewest steps systematically, from a catalog of some 6000 Aldrich starting materials.

The stringent criteria require that each step in the route make at least one skeletal bond of the target, i.e., no refunctionalization steps to repair intermediate functional groups. Each generated reaction in each route is checked automatically against a literature database of some half-million reactions to ascertain its validity, by the frequency with which close precedents occur in this literature, and their average yield. The talk will summarize the present state of the SYNGEN program.

281. **Liquid Phase Synthesis of the Pulmonary Surfactant Polypeptide KL-4**
Cynthia A. Maryanoff, Chemical Development Department. The R. W. Johnson Pharmaceutical Research Institute, Spring House, PA 19477

The synthetic 21-amino acid polypeptide KL-4 [$(KL_4)_4K$] was developed as a pulmonary surfactant to be used in treatment of respiratory distress syndrome (RDS) in infants (IRDS) and in adults (ARDS). KL-4 is designed to contain hydrophobic (Leucine) and hydrophilic (Lysine) regions to mimic the structure of the natural surfactant protein B (SP-B). The liquid phase synthesis of KL-4 is a challenging task. The protection and deprotection patterns, the possibility of diastereomer formation in every coupling step, the solubility (or lack of it) of the different fragments and the isolation of intermediates are among the potential problems that face such a synthesis. This synthesis will be discussed.

282. **CONCEPTUALLY SIMPLE, COMPLETELY GENERAL NEW COMPLEXITY INDICES. RIGOROUS TOPOLOGICAL APPROACH TO STRATEGIC BONDS**

Steven H. Bertz, Complexity Study Center, Mendham, New Jersey 07945

The total number of subgraphs $N_T(G)$ of a graph G is a robust measure of its topological complexity. This new complexity index can be applied to finding strategic bonds by calculating it for the target and all possible precursor structures. The bonds that give the greatest simplification when disconnected in the retrosynthetic direction are the topological strategic bonds. A similarity index such as $SI(j,k)$ can also be used for this purpose by finding the bondset of the target T that gives the least similar precursor structure T'. For example, the most efficient 1-bond disconnection in the case of the 6-membered ring corresponds to the ring-closure step of the Robinson Annulation. The best 2-bond disconnections correspond to the reflexive Diels-Alder Reaction and the (yet-to-be-developed) "bis-allyl coupling." The companion index $N_S(G)$, the number of kinds of subgraphs in graph G, is a good measure of structural complexity and is well suited for the study of molecular diversity.

283.
GROWTH IN DEMAND FOR CHIRAL DRUG SUBSTANCES. Gerhard Beck, Chemical Development, Hoechst Marion Roussel, Frankfurt, Germany, 65926

A wide variety of biological functions emerge through molecular recognition which requires strict matching of chirality. Enzymes and receptor sites in biological systems have the ability to differentiate between two enantiomers of a parent compound. Therefore nowadays chemical synthesis of optically pure compounds plays an important role in the development of new drugs, agrochemicals and intermediates. Synthetic chemists have discovered a variety of methods to complement biological processes. The most ambitious technique is to synthesize a homochiral product from non-chiral starting materials. But of course, also more traditional methods, like adaptation of naturally occuring chiral pool compounds or simply chemical or biological resolution of racemates are important methods, especially for large scale synthesis. An overview of the syntheses of chiral intermediates and drug substances, which have been performed during the last years in our laboratories and pilot plant will be given. Special emphasis will be taken on asymmetric synthesis with chiral metal catalysts. Future needs and perspectives for asymmetric large scale synthesis will be given.

284. 2-METHYLENEOXETANES: PROMISING INTERMEDIATES IN ORGANIC SYNTHESIS. A. R. Howell, A. J. Ndakala, and G. Wang, Department of Chemistry, University of Connecticut, Storrs, CT 06269

The reactivity of 2-methyleneoxetanes **1** has been largely unexploited. The unique combination of functionalities--a reactive oxetane, an enol cyclic ether and an exocyclic double bond--offers intriguing possibilities for further manipulation. Recently, we have demonstrated the conversion of **1** to 1,5-dioxaspirohexanes **2** and will discuss the reactivity of this novel class. In addition, a 2-methyleneoxetane with a remote hydroxy group undergoes cyclization to a fused ketal **3** by an iodoetherification procedure. The synthetic protocol and utility of these compounds will also be discussed.

285. SYNTHESIS OF MEDIUM-RING CYCLOALKENONES BY SILICON-PROMOTED RING EXPANSION REACTION. Jih Ru Hwu, K.-L. Chen, and K.-T. Yong, Department of Chemistry, National Tsing Hua University, Hsinchu, Taiwan 30043, R.O.C. and Institute of Chemistry, Academia Sinica, Taipei, Taiwan 11529, R.O.C.

A new and efficient method was developed for the synthesis of cycloalkenones of medium size. Oxidative ring expansion of γ-silylbicycloalkanols with ceric ammonium nitrate (2.0 equiv) in aqueous acetonitrile at 85 °C gave cycloalkenones with 8- to 11-membered ring in 73–92% yields. In contrast, performance of the same type of reaction on a non-silylated bicycloalkanol afforded a mixture of *cis*- and *trans*-cyclonon-5-enones in poor yields (16% overall). These results indicate that silicon promoted ring expansion and specified the position of C=C in the products.

286. **UNEXPECTED AND SELECTIVE OXIDATIONS USING ALUMINA SUPPORTED PERMANGANATE.**
Clifford E. Harris, Will Chrisman, Chris McBride, Brady Taylor, and Bakthan Singaram, Department of Chemistry and Biochemistry, University of California, Santa Cruz, CA 95064.

Potassium permanganate supported on neutral alumina acts as a strong, selective oxidizing agent. Treatment of β,β- disubstituted enamines with this reagent in acetone gives a heterogeneous oxidative cleavage of the enamine carbon-carbon double bond.. The unexpected aromatization of 1, 4-dienes was observed when we oxidized an enol ether, 1-methoxy-1,4-cyclohexadiene. Treatment of ketoximes with the supported permanganate provides a facile deoximation to afford ketones.

287. **A NEW MILD CONVERSION OF AMIDES TO ESTERS USING TRIFLIC ANHYDRIDE AND PYRIDINE**

André B. Charette* and Peter Chua

Département de Chimie, Université de Montréal, Montréal, Québec, Canada, H3C 3J7

Abstract: A new mild method to convert secondary and tertiary amides to esters will be presented. Using a stoichiometric amount of triflic anhydride and an excess of pyridine followed by an alcohol, amides can be converted to the corresponding esters at temperatures below zero. The esters can then be further hydrolysed allowing access to the corresponding acid. Good to excellent yields are obtained for all secondary amides and for less sterically hindered tertiary amides. The methodology is applicable to both aliphatic and aromatic amides.

288. **CONTROLLED REDUCTION OF TERTIARY AMIDES USING DIALKYLBORANES TO THE CORRESPONDING ALDEHYDES OR AMINES.** **Gayane Godjoian**, Bakthan Singaram, Department of Chemistry and Biochemistry, University of California, Santa Cruz, CA 95064.

Several tertiary amides were reduced using one or two equivalents of various dialkylboranes, such as 9-BBN, dicyclohexylborane(Chx_2BH), or disiamylborane(Sia_2BH). The tertiary amides used in this study had alkyl substituents of varying steric requirements at the nitrogen atom. It was found that the sterics requirements of dialkylboranes played a major role in the determination of the nature of the reduction products. Thus, tertiary amides can be reduced to either the corresponding amine or to the corresponding aldehyde. The composition of the reduction products will be presented.

$$R\text{-}N(R')(R'') \xleftarrow{\text{9-BBN}}_{\text{THF, 25 °C}} R\text{-C(O)-}N(R')(R'') \xrightarrow{\text{Sia}_2\text{BH / Chx}_2\text{BH}}_{\text{THF, 25 °C}} R\text{-CHO}$$

289. **REDUCTION OF SUBSTITUTED NAPHTHALENES WITH 2-PROPANOL AND RANEY® NICKEL.** **Benjamin H. Gross** and Robert C. Mebane, Chemistry Department, University of Tennessee at Chattanooga, Chattanooga, Tennessee 37403.

The reduction of naphthalene to 1,2,3,4-tetrahydronaphthalene by catalytic transfer hydrogenation utilizing 2-propanol and Raney nickel has been reported. However, no subsequent investigation has focused on the influence of a substituent on the ring which more readily undergoes reduction. As a continuing interest in catalytic transfer hydrogenation involving the Raney catalysts, we have conducted the reduction of sixteen naphthalenes with substituents which include methyl, methoxy, acetyl, hydroxy, acetoxy, carbomethoxy, fluoro, and acetamido. These compounds were heated at reflux with 2-propanol and stirred with the catalyst. The progress of the reactions was monitored by gas chromatography and mass spectrometry. Proton nuclear magnetic resonance was utilize in the identification of products where authentic samples were not available. Our results indicate that unsubstituted rings generally undergo hydrogenation more readily than substituted rings; however, significant differences were observed in rates of hydrogenation and degree of hydrogenolysis.

290. **A USEFUL MODIFICATION OF THE BIRCH REDUCTION WITH LITHIUM IN AMINES.** M. E. Garst, Allergan Pharmaceuticals, 2525 Dupont Drive, Irvine, California 92715, L. J. Dolby, **S. Esfandiari**, N. C. Chamberlain, N. A. Fedoruk, Organic Consultants, Inc., 132 East Broadway, Eugene, Oregon 97401

A modification of the Birch reduction, using lithium, n-propylamine and ethylenediamine has been applied to several reductions: debenzylation of N-benzylamides, reduction of anisole to 1-methoxy-1,4-cyclohexadiene, reduction of naphthalene to isotetralin, reduction of ketoximes and aldoximes to amines, conversion of thioanisole to thiophenol, cleavage of a sulfonate ester to the alcohol, and cleavage of a p-toluenesulfonamide to the amine. The reactions were carried out without any added cosolvent, but in some cases added t-butanol was beneficial. Preliminary experiments demonstrated that the reaction is very slow in the absence of ethylenediamine.

291. α-ALKYL-α-AMINOSILANES BY HYDROSILYLATION OF N-ALKENYL CARBAMATES. Gregory W. Hewitt, Scott McN. Sieburth* and Farah Sandaire, Department of Chemistry, State University of New York at Stony Brook, Stony Brook, New York 11794-3400

Alpha-amino silanes (**4**) are important synthetic reagents and can be found in a variety of biologically active molecules. Few methods are available for the preparation of these molecules with substitution between the two heteroatoms. Catalytic hydrosilylation of the readily available N-alkenyl carbamates **2** offers a versatile approach with the potential for asymmetric synthesis. Hydrosilylation of **2** using rhodium catalysts proceeds with complete regioselectivity. Rates of reaction are related, in part, to the nature of R^1, R^2 and R^3.

$R = PhCH_2$, t-Bu $R_1, R_2 = H$, alkyl, Ph $R_3 = Me$, Ph, EtO

292. **METHOXYMETHYL METHYL SULFATE (MOM-MSf) - A NEW POWERFUL METHOXYMETHYLATING REAGENT.** Mikhail Lebedev, Department of Chemistry, Simon Fraser University, Burnaby, B.C., V5A 1S6, Canada

The highly versatile methoxymethylating reagent, methoxymethyl methyl sulfate **1**, was synthesized by the reaction of dimethoxymethane and sulfur trioxide. Reactions of reagent **1** with alcohols and carboxylates led to the corresponding methoxymethyl methyl ethers and esters in good yields, while alkenes, alkynes and dienes formed alkyl, vinyl, and allyl sulfates, respectively. Electrophilic strength of new reagent **1** was evaluated on the basis of its reactions with aromatic hydrocarbons and charge distribution in its global minimum conformation found by *ab initio* calculations.

293. TRIMETHYLSILYL DERIVATIVES AS FINAL NUCLEOPHILES IN THE TANDEM SEQUENCE OF AN ArSCl INITIATED Ad$_E$ REACTION RESULTING IN THE SYNTHESIS OF POLYFUNCTIONAL COMPOUNDS. **Margarita I. Lazareva**, Ron Caple, and William A. Smit, Chemistry Department, University of Minnesota-Duluth, Duluth, MN 55812

A convenient synthetic scheme for assembling complex molecules from simple precursors based on the Lewis acid mediated sequence of reactions ArSCl + vinyl ether-I + vinyl ether-II + trimethylsilyl derivatives has been developed. The use of silyl derivatives (like trimethylsilyl vinyl ethers, allylsilanes, trimethylsilyl ketene acetals) in the final step of the sequence enables us to introduce a wide variety of functional groups to a molecule.

294. **ALKENYLATION AND ALKYNYLATION OF C-H BONDS VIA FUNCTIONALIZED TRIFLONES.** Jason S. Xiang, J. Gong, W. Jiang, and P. L. Fuchs, Department of Chemistry, Purdue University, West Lafayette, Indiana 47907.

Aryl and alkyl β-heteroatom trisubstituted vinyl triflones react with tetrahydrofuran and cyclohexane to undergo trifluoromethyl radical-mediated C-H functionalization reactions to afford E and Z β-heteroatom trisubstituted olefins. Most reactions proceed with both high yield and high stereospecificity (retention of configuration). Beta-substituents which have been employed in this study are: iodo, bromo, fluoro, benzoate, ethylcarbonate, and phthalimide. Beta-substituents bearing powerful electron-releasing groups such as alkoxy or amino render the vinyl triflone unreactive.

Silicon (TIPS) substituted ethynyl triflone has been prepared for the first time. This reagent reacts with ethers and hydrocarbons to provide 1-silyl 2-substituted alkynes. Application of the above reagents in model studies directed toward the construction of spiroketals of biological significance will be discussed.

295. **COMPUTER SIMULATIONS OF SYNTHETIC ANALOGS OF COPROPORPHYRINOGEN-III.** Jingyuan He, Marjorie A. Jones, Jean M. Standard, and Timothy D. Lash, Department of Chemistry, Illinois State University, Normal IL 61790-4160.

In the heme biosynthetic pathway, coproporphyrinogen-III (copro'gen III, **1**) is converted to protoporphyrinogen-IX by Coproporphyrinogen Oxidase (C.O.) through two sequential oxidative decarboxylations. The A then B ring propionate groups are oxidatively decarboxylated to vinyl groups. The roles of the C and D ring propionate groups in substrate recognition are not well understood. Six analogs of copro'gen III with changes at positions 13 and/or 17 (**2** to **7**, respectively) have been simulated using BIOGRAF 3.1 on a Silicon Graphics workstation to evaluate conformational effects. Analog **2** exhibits the same lowest energy conformation as the authentic substrate; all the others are different from the authentic substrate. These results correlate with previous kinetic data for synthetic samples of 2-7 which indicate that the substrate recognition by C.O. involves both substituent sequence and conformational effects. (Supported by NIH Grant No. 1 R15 DK45206-1)

P = $CH_2CH_2CO_2H$
Me = CH_3
Et = CH_2CH_3
Pr = $CH_2CH_2CH_3$
Bu = $CH_2CH_2CH_2CH_3$

1. $R^1=R^2=P$
2. R1=P; R2=Et
3. R1=Et; R2=P
4. $R^1=R^2=Me$
5. $R^1=R^2=Et$
6. $R^1=R^2=Pr$
7. $R^1=R^2=Bu$

296. **MONOVINYL PORPHYRINOGEN PROBES FOR THE ACTIVE SITE OF COPROPORPHYRINOGEN OXIDASE.** Anna-Sigrid I.M. Keck, Marjorie A. Jones, Ukti N. Mani and Timothy D. Lash, Department of Chemistry, Illinois State University, Normal, IL 61790-4160

Me = CH_3
V = $CH=CH_2$
P = $CH_2CH_2CO_2H$

1 $R_1 = R_2 = R_3 = P$; $R_4 = Me$
2 $R_1 = V$; $R_2 = R_3 = P$; $R_4 = Me$
3 $R_1 = V$; $R_2 = R_4 = P$; $R_3 = Me$

The sixth enzyme in the protoheme biosynthetic pathway, Coproporphyrinogen Oxidase (C.O.), metabolizes Coproporphyrinogen-III (**1**) to Protoporphyrinogen-IX by the oxidative decarboxylation of the A and B ring propionate groups. To increase our understanding of what specific substituent sequence on the substrate is required for this mitochondrial enzyme to recognize a porphyrinogen substrate, different synthetic porphyrinogens were incubated with a crude enzyme preparation derived from chicken blood hemolysates. Although C.O. is highly specific for the authentic substrate, the enzyme also metabolized the two different synthetic monovinyl porphyrinogen analogs, Harderoporphyrinogen-III (**2**) and Harderoporphyrinogen-I (**3**) to divinyl products. The type I isomer was further metabolized to a trivinyl porphyrinogen, which was consistent with a proposed model [Lash et al., *BioMed. Chem. Lett*. **1994**, *4*, 1607] for the active site of this important enzyme. (Research supported by NIH Grant No. 1 R15 DK 45206-1)

297. **SYNTHESIS AND SPECTROSCOPIC PROPERTIES OF CARBAPORPHYRINS.** Michael J. Hayes and Timothy D. Lash, Department of Chemistry, Illinois State University, Normal IL 61790-4160.

Aromatic porphyrinoids with internal CH units have only recently been reported in the literature (Lash, T.D. *Angew. Chem. Int. Ed. Engl.* **1995**, *34*, 2533). Using the "3 + 1" methodology (Lash, T.D. *Chem. Eur. J.* **1996**, *2*, 1197), we have synthesized the novel carbaporphyrins 1-3. Carbaporphyrins have porphyrin-like electronic spectra and exhibit strong macrocyclic ring currents in proton NMR spectroscopy. However, no unusual tautomers or conformers could be detected in solution, as had previously been claimed by other workers (*Angew. Chem. Int. Ed. Engl.* **1996**, *35*, 1820).

298. TROPIPORPHYRIN AND AZEPIPORPHYRIN: THE FIRST PORPHYRIN ANALOGS WITH SEVEN-MEMBERED RING SUBUNITS. Joseph C. Petryka, Sun T. Chaney and Timothy D. Lash, Department of Chemistry, Illinois State University, Normal IL 61790-4160.

Acid catalyzed condensation of dialdehydes with tripyrranes provides a superior route to porphyrin analogs (Lash, T.D. *Angew. Chem. Int. Ed. Engl.* **1995**, *34*, 2533). Condensation of tripyrrane **1** with cycloheptatrienedialdehyde **2a** in the presence of 5% TFA in dichloromethane gave the cycloheptatrienyl porphyrin analog "tropiporphyrin" (**3a**) in 23% yield. Proton NMR and uv-vis spectroscopy demonstrate that this is an aromatic 18 π electron system. The synthesis of the aza-analog "azepiporphyrin" (**3b**) is presently under investigation.

299. NEW PSORALEN-PORPHYRINS WHICH INTERCALATE INTO DNA. Xiao Wang, Olga Fedorova[†], William R. Trumble and Leszek Czuchajowski, Department of Chemistry, University of Idaho, Moscow, ID 83844-2343; [†]Institute of Bioorganic Chemistry, Novosibirsk 630090, Russia.

Psoralen was joined at the C(4') center through -CH_2-NH-$(CH_2)_n$-NH- bridges (n=2, 4, 6) with the para position of tetrafluorophenyl meso-substituent of porphyrin. Three other meso-substituents were represented by 4-N-methylpyridinium rings. Each of these tethered structures is able to interact with DNA double strands due to the specificity of its terminal units, causes a bathochromic shift of the porphyrin Soret band and exhibits the remarkable hypochromicity. The association constant of psoralen-porphyrin with calf thymus DNA was calculated from the titration curve: K = $(2.33 \pm 0.36) \times 10^7$ M^{-1} when n=4. The appearance of two isobestic points in the UV-visible spectra suggests the formation of one type of complex, more stable than those formed by the respective detached porphyrin unit.

300. THE SYNTHESIS OF ISOMERIC DITHYMIDYL-PHOSPHORUS(V)-MESO-TETRAPHENYL-PORPHYRINS AND THEIR INTERACTION WITH DNA. Gerard Kian-Meng Goh, Olga Fedorova[†], and Leszek Czuchajowski, Department of Chemistry, University of Idaho, Moscow, ID 83844-2343; [†]Institute of Bioorganic Chemistry, Novosibirsk 630090, Russia.

The synthesis of three new analogs of dinucleosides containing thymidine units connected axially to the P-center of phosphorus(V)-meso-tetraphenylporphyrin was accomplished. The compounds are 5'-O, 5'-O-dithymidylphosphorus(V)-meso-tetraphenylporphyrin, 5'-O,3'-O-dithymidylphosphorus(V)-meso-tetraphenylporphyrin and 3'-O, 3'-O-dithymidylphosphorus(V)-meso-tetraphenylporphyrin. In the presence of calf thymus DNA the Q bands in the electronic spectra of these compounds became red-shifted in aqueous solution of low ionic strength. These shifts disappear at high ionic strength (~ 0.1 M) which may be due to electrostatic interaction primarily between positively charged P(V) centers of porphyrins and negatively charged phosphate groups of DNA. The new analogs will provide a lead towards the synthesis of oligo-porphyrinyl DNA analogs in which numerous phosphorus(V) porphine units are axially connected through the 5'-O-thymidine-3'-O-bridges.

301. SITE-SPECIFIC PHOTOMODIFICATION OF DNA BY PORPHYRINYL OLIGO-NUCLEOTIDE CONJUGATES SYNTHESIZED VIA SOLID PHASE H-PHOSPHONATE APPROACH. Handong Li, Olga Fedorova, William Trumble, T. Rick Fletcher, and Leszek Czuchajowski, Department of Chemistry, Department of Microbiology, Molecular Biology and Biochemistry, University of Idaho, Moscow ID 83843 and Institute of Bioorganic Chemistry, Novosibirsk 630090, Russia.

Introduction of DNA-cleaving functional groups into an antisense oligonucleotide generates site-directed damage on the target nucleic acids through chemical or photochemical reactions. In our study the meso-tris (4-pyridyl) (ω-hydroxyhexamethylenamidophenyl)porphyrin was converted to its H-phosphonate derivative and conjugated using solid phase synthesis with the 5'-hydroxyl group of deoxyribonucleotides d(TCTTCCCA) and d(T)$_{12}$. These conjugates were transformed into its N-methyl-pyridiniumyl-porphyrin analogs. A 532 nm laser beam was utilized to photoactivate both types of the conjugates in the presence of the target 22-mer and 16-mer oligonucleotides. Photoactivation of porphyrin-oligonucleotide conjugates resulted in site-specific DNA modification characterized by a main reaction site size of ~5 bases.

302. HOECHST 33258 DERIVATIVES TETHERED TO OLIGO(DEOXYNUCLEOTIDES): DUPLEX STABILIZATION AND FLUORESCENCE CHARACTERISTICS. KRISTIN WIEDERHOLT, Jordi Robles, Sharanabasava B. Rajur, and Larry W. McLaughlin, Merkert Chemistry Center, Boston College, Chestnut Hill, MA 02167.

We have synthesized two related analogues of the fluorescent dye Hoechst 33258 in which the terminal phenol moieties have been employed to permit their covalent conjugation to oligo(deoxynucleotides). Conjugation of the Hoechst derivatives to the DNA sequences employs a cystamine or a hexa(ethylene glycol) linker tethered to an internucleotide or terminal phosphorus. We have prepared a series of DNA-Hoechst conjugates and studied their thermal stabilization and fluorescence characteristics

303. OLIGONUCLEOTIDE ANALOG ARRAYS FOR ENHANCED HYBRIDIZATION. Jacqueline A. Fidanza and Glenn H. McGall, Affymetrix, Inc., 3380 Central Expressway, Santa Clara, CA 95051

Nucleic acid analogs have been used extensively to modulate hybridization to target sequences. We are investigating the use of oligonucleotide analog probe arrays to enhance target hybridization while maintaining specificity. A:T and G:C rich probe arrays display different hybridization thermodynamics under a given set of conditions. Our work has focused on incorporation of 2,6-diaminopurine to enhance the A:T base pairing and 7-deaza-2'-deoxyguanosine to disrupt 2° structure associated with runs of G. The incorporation of 2'-O-methyl-ribonucleoside analogs within arrays is also under investigation. Analog arrays should improve binding to and therefore detection of both DNA and RNA targets.

304. CATALYTIC DNGS FOR PHOSPHODIESTER HYDROLYSIS. Dev P. Arya and Thomas C. Bruice, Department of Chemistry, University of California at Santa Barbara, CA, 93106.

The novel structural modification in DNG (deoxyribonucleic guanidine) is the replacement of the negatively charged phosphate bridge, found in DNA, with a positively charged achiral 5'-3' guanidyl linkage. The development of deoxyribonucleic guanidines as stable carriers for catalysts capable of hydrolyzing phosphate backbones of nucleic acids will be presented. The use of deoxyuridyl DNGs for the attachment of these catalysts will be discussed.

305.

NOVEL NON-NUCLEOSIDIC PHOSPHORAMIDITES FOR OLIGONUCLEOTIDE MODIFICATION AND LABELING. S. SU, R. S. Iyer, S. K. Aggarwal, K. L. Kalra, BioGenex Laboratories, 4600 Norris Canyon Road, San Ramon, California 94583

A series of novel labeled phosphoramidites with a new non-nucleosidic backbone based upon cyclohexyl-4-amino-1,1-dimethanol have been synthesized. The new backbone has been obtained by hydroboration of inexpensive 3-cyclohexene 1,1-dimethanol followed by reaction with hydroxylamine- O-sulfonic acid. This new backbone has been used to synthesize biotin, fluorescein and amino labeled phosphoramidites. The natural 3-carbon atom internucleotide phosphate distance in DNA/ RNA is retained in the structure which will not affect the hybridization and annealing properties of the duplex. The reporter group is kept outside of the nucleotide chain by rigid cyclohexane ring resulting in high coupling efficiency and yields. Synthesis and the use of these compounds for oligonucleotide labeling will be discussed.

306. REGIOSELECTIVE SYNTHESIS OF THYMIDINE DINUCLEOTIDE VANADATE ANALOG, Long Mao and **Krishnan P. Nambiar**[*], Department of Chemistry, University of California, Davis, CA 95616.

Organovanadium compounds are receiving considerable attention due to their biological activities such as glucose transport and phosphatase inhibition. Vanadates are transition state analogs of phosphate ester hydrolysis and hence well suited for structural investigation of DNA polymerases and reverse transcriptases. In this report, we describe an efficient method for the synthesis of thymidine dinucleotide vanadate analog and its characterization. 5'protected thymidine is first reacted with VCl_3 and the resulting 3'-dichlorovanadium derivative is reacted with 3'-protected thymidine to yield the trivalent vanadium dinucleotide which is then oxidized with hydrogen peroxide to yield the pentavalent Vanadium dinucleotide.

307. **AMIDE BOND FORMATION CATALYZED BY NUCLEOSIDES IN A NON-POLAR MEDIUM. EVIDENCE FOR BIFUNCTIONAL CATALYSIS**

Christian Melander and David A. Horne*
Department of Chemistry, Columbia University, New York, NY 10027

The intrinsic catalytic abilities of the nucleosides, adenosine, cytidine, guanosine, uridine, and pseudouridine (Ψ) were examined with respect to their abilities to catalyze amide bond formation. Rate enhancements of 24% and 29% were observed with G and Ψ, respectively, whereas C showed a 210% increase. Less than 10% rate enhancement was observed with A and U. The relatively large rate increase with C points to a *bifunctional* mode of catalysis wherein a greater degree of stabilization of the ionic transition state is achieved. The implications for peptidyl transferase activity manifested solely by RNA and a possible role for the modified base Ψ will discussed with respect to protein synthesis. These findings support the proposal that nucleic acids may have played an important role in the development of modern day protein synthesis and that early nucleic acids may, in fact, have recruited proteins for desolvation purposes in generating hydrophobic cores in facilitating nucleic acid mediated catalysis.

308. **RNA SUBSTRATE RECOGNITION BY E. COLI PSEUDOURIDINE SYNTHASE (rluA) WITH DUAL SUBSTRATE SPECIFICITY.** David A. Horne*[1], Kirsten Rood[1], Elie Levine[1], and James Ofengand[2]. [1]Department of Chemistry, Columbia University, New York, NY 10027. [2]Department of Biochemistry and Molecular Biology, University of Miami, Miami, FL 33101.

The *rluA* gene product of *E. coli* is a pseudouridine synthase with dual specificity that catalyzes the rearrangement of uridine to pseudouridine in positions Ψ746 of E. coli 23S ribosomal RNA and Ψ32 in tRNAs. Unlike pseudouridine synthase I in which the 3-D architecture is important for activity, Ψ746 is able modify simple stem-loop structures. This was shown for stem-loop structures of cysteine and leucine tRNAs which contain pseudouridine at position 32. Ψ746 also recognizes and modifies the stem-loop structure of phenylalanine tRNA, but does not modify a single stranded construct. The overall efficiency for modification of simple stem-loop structures is about 6-fold less than full length tRNA. Bases within the stem do not seem to be specifically recognized and necessary for modification, but 2 (U33 and A36) within the loop are required. A chimeric RNA containing the tRNA stem and the 23S loop (F+G) is a substrate for the enzyme.

309.

DEVELOPMENT OF A DNA ALKYLATING REAGENT. Tony L. Hudgens, Kenneth D. Turnbull*, Department of Chemistry & Biochemistry, University of Arkansas, Fayetteville, Arkansas 72701

The progress in the synthesis of an alkylating reagent designed to chemo-selectively alkylate the phosphodiester backbone of nucleic acids via a quinone methide will be presented. The focus will be on the synthesis and reactions of model compounds prepared for optimizing the design of this alkylating reagent.

310. A MIMIC OF CAP BENT DNA. Felicia A. Etzkorn, Vladimir V. Kalashnikov, Allison M. Hager, Department of Chemistry, University of Virginia, Charlottesville, VA 22901

DNA bending is thought to be important in a number of critical biological processes such as transcriptional activation, recognition of damage and repair. We have undertaken the design and synthesis of a mimic of the bent conformation of the catabolite activating protein (CAP) DNA binding site found in the X-ray crystal structure of the CAP/DNA complex (Schultz, S. C.; Shields, G. C.; Steitz, T. A. *Science* **1991**, *253*, 1001-1007). The mimic was designed with a linker between two ends of double-stranded DNA, much as a string bends the bow of a bow and arrow. The design criteria were that the linker should be flexible, water-soluble, uncharged and easily synthesized. The final *successful* design was based on tetraethylene glycol linked to the DNA phosphate backbone via an aqueous-stable phosphoramide bond. The mimic included the native asymmetric CAP sequence, instead of the X-ray structure palindromic sequence, to preclude hairpin formation. The mimic was folded into the monomeric bent conformation by dilution and annealing. The bent mimic was compared with a straight DNA control by circular dichroism (CD), polyacrylamide gel electrophoresis (PAGE), and matrix-assisted-laser-desorption time-of-flight mass spectrometry (MALDI-TOF).

311. ORGANOMETALLIC PHOTONUCLEASES: A NOVEL CLASS OF DNA CLEAVING AGENTS. Debra L. Mohler,* David R. Dain, Angela D. Kerekes, and William D. Nadler, Department of Chemistry, West Virginia University, Morgantown, WV 26506

As part of an effort to develop organometallic compounds as new agents for the modification of oligonucleotides, we have begun to study the reactivity of cyclopentadienyl (Cp) metal complexes towards DNA. Thus, irradiation of $CpW(CO)_3CH_3$ (**1**) at various concentrations in the presence of pBR322 DNA resulted in single-strand cleavage at 6 molecules/base pair. The mechanism of strand scission and the attachment of DNA recognition elements to the complex will be discussed.

$$CpW(CO)_3CH_3 \xrightarrow[-CO]{h\nu} CpW(CO)_2 + \dot{C}H_3$$

1

— intact (form I) DNA
— cleaved (form II) DNA

312. **MECHANISTIC STUDIES ON MODEL OXETANES OF (6-4) PHOTOPRODUCT PHOTOLYASE.** Gautam Prakash and Daniel E. Falvey*, Department of Chemistry and Biochemistry, University of Maryland, College Park, MD 20742.

The DNA repair enzyme, *(6-4) Photoproduct Photolyase*, specifically targets UV light induced (6-4) dimers between pyrimidines. The mechanism proposed for this enzyme rests on its ability to form an oxetane intermediate in the repair pathway. Unfortunately, the putative oxetane intermediate is unstable above -80 °C, and therefore is unhelpful in evaluating the mechanism. To address this problem, we have synthesized a stable model oxetane adduct on thymine, and have shown that this model oxetane can be made to revert back to its constituent halves in the presence of light, and either electron donor or electron acceptor molecules. We have also synthesized, and split, a model oxetane on cytosine, which suggests that the mechanism is similar for pyrimidines in general. In addition, using the technique of Laser Flash Photolysis, both the splitting, and the formation reactions of a few of the above oxetanes were studied. The results provide kinetic data for both reactions, and lend further support to the proposed mechanism.

313.

PHOTOACTIVE ANALOGS OF FARNESYL- AND GERANYLGERANYL PYROPHOSPHATE: APPLICATION TO STUDIES OF STRUCTURE AND FUNCTION OF PROTEIN PRENYL-TRANSFERASES. **Mark D. Distefano,** Rebecca L Edelstein, Tammy C. Turek, Igor Gaon and Valerie A. Weller. Department of Chemistry, University of Minnesota, Minneapolis, MN 55455

Farnesyl and geranylgeranyl pyrophosphate are involved in a large number of cellular processes including the prenylation of transforming mutants of Ras proteins implicated in cancer. Photoactive analogs can provide useful information about enzyme active sites that bind isoprenoids. We have synthesized two classes of photoactive prenyl pyrophosphate analogs to study the structure and function of prenylating enzymes. Benzophenone-containing analogs are competitive inhibitors of farnesyl transferase, they inactivite the enzyme upon photolysis and they label the beta subunit; peptide mapping experiments are in progress. Diazoester analogs are substrates for the transferase. This has allowed us to prepare photoactive Ras protein which is being used to study the interactions between prenylated Ras and other proteins and/or lipids.

314. SYNTHESIS OF β-FLUORO-S-ADENOSYLMETHIONINE: A POTENTIAL AFFINITY LABEL OF S-ADENOSYLMETHIONINE-DEPENDENT ENZYMES. Caroline Leriche & Hung-wen Liu, Department of Chemistry, University of Minnesota, Minneapolis MN 55455

Ethylene, one of the simplest organic molecules with biological activity, is a plant hormone that controls many aspects of plant growth and development and whose synthesis is strictly regulated. Induction of this plant hormone by a variety of inducers is due to *de novo* synthesis of ACC synthase which catalyzes the rate-limiting step in the conversion of AdoMet to 1-aminocyclopropane-1-carboxylate (ACC).
In order to study the catalytic mechanism of this enzyme, an analog of AdoMet, β-fluoro S-adenosylmethionine (**1**), was synthesized. This possible affinity label will be a tool to trap and identify active-site residues which may be involved in the catalysis of ACC synthase.
Analog **1** will be also important for the study of SAM-dependent enzymes.

315.

RESCUE OF AN INACTIVE MUTANT OF CHORISMATE MUTASE BY ADDITION OF CHEMICAL ADDITIVES. **C. Grisostomi,** P. Kast, D. Hilvert, Department of Chemistry, The Scripps Research Institute, BCC-377, 10550 North Torrey Pines Road, La Jolla, California 92037

The specific chemical requirements of Bacillus subtilis chorismate mutase (BsCM) at position 90 were examined by chemical rescue experiments. BsCM R90G showed a kcat value 5 orders of magnitude lower than for the wild type. This result suggests the involvement of Arg90 in the catalytic step, by stabilization of a partial negative charge on the ether-oxygen in the transition state rearrangement from chorismate to prephenate. Methylguanidine, but none of the primary amines tested, rescued the Arg90Gly variant, by increasing the rate of the rearrangement by one order of magnitude, adequate for the stabilization of a partial, rather than a full, negative charge. Saturation was observed at high concentrations of the additives. Apparently, the positive charge needs to be correctly positioned as well as polarizable.

316. YEAST C-METHYLTRANSFERASE IN UBIQUINONE BIOSYNTHESIS. Robert J. Barkovich, Andrey Shtanko[a], Jennifer A. Shepherd, Peter T. Lee, David C. Myles, Alexander Tzagoloff[a] and Catherine F. Clarke, Department of Chemistry and Biochemistry, University of California, Los Angeles, CA 90095 and the [a]Department of Biological Sciences, Columbia University, New York, NY 10027

Ubiquinone (coenzyme Q or Q) is a lipophilic metabolite that functions in the electron transport chain in the plasma membrane of prokaryotes and in the inner mitochondrial membrane of eukaryotes. Q deficient mutants of *Saccharomyces cerevisiae* fall into eight complementation groups (*coq1-coq8*). Yeast mutants from the *coq5* complementation group lack Q and as a result are respiratory defective and fail to grow on nonfermentable carbon sources. A nuclear gene, designated *COQ5* was isolated from a yeast genomic library based on its ability to restore growth of a representative *coq5* mutant on media containing glycerol as the sole carbon source. A gene fusion expressing a biotinylated form of Coq5p retains function, as assayed by the complementation of the *coq5* mutant, and is located in mitochondria. The synthesis of two farnesylated analogs of intermediates in the ubiquinone biosynthetic pathway is reported. These reagents have been used to develop *in vitro* C-methylation assays with isolated yeast mitochondria. These studies show that Coq5p is required for the C-methyltransferase step that converts 2-polyprenyl-6-methoxy-1,4-benzoquinone to 2-methyl-3-polyprenyl-5-methoxy-1,4-benzoquinone.

317.
AUTOCATALYSIS OF AMIDE ISOMERIZATION. **Christopher Cox** and Thomas Lectka, Department of Chemistry, Johns Hopkins University, Baltimore, Maryland 21218

The catalysis of amide bond isomerization (AI) by Brönsted acids is a well-documented reaction that proceeds through a putative N-protonated intermediate. However, intramolecular general acid-catalyzed AI, termed autocatalysis, is a much less-studied but likely biologically-relevant process in which hydrogen bond donation to the amide nitrogen (N_a) through a correctly aligned cyclic intermediate replaces discrete N-protonation. Autocatalysis has been proposed to play a key role in the folding of several proteins including dihydrofolate reductase, and Karplus et al. have proposed in an intriguing theoretical study that autocatalysis plays an important role in the mechanism of cyclophilin and FKBP-induced folding. The authors also predicted that the effect should be general and measurable in model prolines; however, experimental conformation of these proposals has yet to appear. We wish to report experimental confirmation of autocatalysis in model systems, including crystallographic and spectroscopic evidence for an unusual hydrogen bond between the side chain and the prolyl N_a in a novel *cis*-proline peptidomimetic.

318. DIOXYGEN OXIDATION OF 4A,8A-DIAZA-2,6-DIOXA-3,4,7,8-TETRAHYDRO-4,4,8,8-TETRAMETHYLANTHRACENE-1,5-DIONE (DDTTA). David J. R. Brook, Bruce Noll and Tad H. Koch, Department of Chemistry and Biochemistry, University of Colorado, Boulder, CO 80302.

The stable 1,4-dihydropyrazine, DDTTA, is oxidized by dioxygen in various organic solvents to give mixtures of 5,6-dihydro-2-oxo-5,5-dimethyl-1,4-oxazine-3-carboxaldehyde (**1**), and a second product that was previously ascribed a dioxetane structure. The latter was fully characterized by X-ray crystallography and found to be the diol, 4a,8a-diaza-9,9a-dihydroxy-2,6-dioxa-3,4,7,8,9,9a-hexahydro-4,4,8,8-tetramethylanthracene-1,5-dione (**2**). The oxidation rate and product ratios are highly solvent dependant. Trapping experiments, reaction stoichiometry, and kinetic measurements are all consistent with a hydroperoxide intermediate that reacts with DDTTA to give the diol **2** or with **2** to give the aldehyde **1**. Both DDTTA and the intermediate hydroperoxide are significantly less reactive than biologically active 1,4-dihydropyrazine counterparts.

319. INVESTIGATION OF PLANT CHOLINESTERASE ACTIVITY IN CHOLINE ESTER AUXIN CONJUGATES. R.A. Fluck, P.A. Leber, J.D. Lieser, Departments of Biology and Chemistry, Franklin & Marshall College, Lancaster, PA 17604-3003.

Based on bioassays of pea stem segments, it has been proposed that auxins are responsible for plant cell elongation and are stored in the form of choline esters. It has also been proposed that in response to environmental stimuli, plant cholinesterase (ChE) hydrolyzes choline esters thus generating active auxins. These active auxins then stimulate growth by causing plant cells in the stem to elongate. Studies conducted in support of this proposal have yielded the following results:
I. The Steglich esterification has proven to be an effective method of synthesizing indole-3-acetylcholine iodide (IATC).
II. A protocol for extracting plant ChE from light grown pea roots has been proven effective by colorimetric assays in which acetylthiocholine chloride (ATC) was used as the substrate. Animal acetylcholinesterase (AChE) and butyrylcholinesterase (BChE) enzymes were also subjected to colorimetric assays to determine their viability as potential models for plant enzyme behavior. Thus far, variations in IATC concentration have not had a significant impact on either the plant or animal enzyme studies.

320. INVESTIGATIONS OF A THEORETICAL MECHANISM FOR REGULATING PLANT GROWTH. R.A. Fluck, P.A. Leber, S.K. Szczerbicki, J.G. Varnes, Departments of Biology and Chemistry, Franklin & Marshall College, Lancaster, PA 17604-3003, and J.D. Cohen, USDA/ARS Agricultural Research Center, Beltsville, MD 20705-2350.

A theoretical mechanism for regulating the growth of plants based on a pea stem segment bioassay suggests that plant growth regulators responsible for cell elongation (auxins) are stored as choline esters in plants. It is postulated that, in response to environmental stimuli, these choline conjugates are hydrolyzed by the enzyme cholinesterase (ChE) contained in cell walls or membranes to the active growth-promoting free auxins. A methodology for extracting unlabeled indole-3-acetylcholine (IAC), the choline ester of the potent auxin indole-3-acetic acid (IAA), from various plants has been developed using the labeled (5-^{3}H- and $^{13}C_6$-IAC) conjugate as a marker. To date, this methodology has been tested on plant extracts from *Pisum sativum* L. (pea) and *Brassica oleracea capitata* (cabbage); our preliminary results support the conclusion that IAC is endogenous in pea plants.

321. **MODELS FOR NITRIC OXIDE SYNTHASE**

J. T. Groves* and **C.-Y. Wang,** Department of Chemistry, Princeton University, Princeton, NJ 08544

A model approach to the second step of the nitric oxide synthase (NOS) reaction will be described. NOS carries out an unusual 3-electron oxidation of N-hydroxyarginine to NO and citrulline. We have demonstrated that an iron(III)porphyrin can catalyze a similar oxidation of the N-hydroxyguanidine group to form NO and the corresponding urea under aerobic conditions. Mechanistic studies including ^{1}H-NMR and ^{18}O-labeling experiments support a radical reaction pathway.

322. SYNTHESIS OF A LIBRARY OF ARTIFICIAL β-SHEETS FOR DETERMINING AMINO ACID PROPENSITIES FOR β-SHEET FORMATION. Shabana S. Insaf, James S. Nowick*, Department of Chemistry, University of California, Irvine, CA 92697-2025.

We have designed a flexible oligourea molecular scaffold to nucleate and stabilize β-sheet structures (Nowick, J. S.; Powell, N. A.; Martinez, E. J.; Smith, E. M.; Noronha, G. *J. Org. Chem.* **1992**, *57*, 3763). This oligourea scaffold holds β-strands at hydrogen bonding distances. The amino acid contributions to β-sheet stability exhibit context dependence. Therefore, we decided to explore the propensity scale of various amino acids for stabilizing our template-based β-sheet design system. This poster describes a combinatorial library [**1**, R = H, Me, *i*Pr or *i*Bu and R' = H, Me, *i*Pr or *i*Bu] created to explore the effect of amino acid variations on the thermodynamic stability of artificial β-sheets described.

323. A TALE OF THREE β-STRAND MIMICS. **James H. Tsai**, Amy Sue Waldman, Mason Pairish, and James S. Nowick*, Department of Chemistry, University of California, Irvine, CA 92697-2025.

We are interested in combining β-strand mimics, other peptidomimetic templates, and peptide strands to create compounds that mimic the structure and hydrogen-bonding patterns of β-sheets. We designed and synthesized three β-strand mimics with the appropriate hydrogen-bonding groups. These mimics induce β-sheet structure in adjacent peptide strands. From these mimics, we hope to gain insight into protein folding, enzyme-substrate interactions, and β-amyloid deposition in Alzheimer's disease. This poster will discuss the synthesis and the structural studies of these three β-strand mimics.

324. β-TURN MIMICS BASED ON DIAMINE TEMPLATES. **Michael J. Soth** and James S. Nowick, Department of Chemistry, University of California, Irvine, California 92697.

The design and development of potential β-Turn mimics based on amino acid-derived diamines will be described. Unlike most turn mimics in which the turn is enforced by an internal covalent linkage, these mimics are simply biased towards turn formation by a nine-membered hydrogen-bonded ring. The synthesis incorporates four variable sidechains onto a general backbone structure, and was designed to be used for the creation of a combinatorial library. The incorporation of an aza-amino acid into the backbone appears to facilitate hairpin turn formation.

325. TURN MIMICS FOR THE HELIX-TURN-HELIX DNA-BINDING PROTEIN MOTIF. Jeremy M. Travins and Felicia A. Etzkorn, University of Virginia, Department of Chemistry, McCormick Rd., Charlottesville, VA 22901

The goal of this project is to mimic the helix-turn-helix (HTH) motif of DNA-binding proteins by constraining the native conformation of this structural motif. The target sidechain-linked tripeptide **2** was synthesized in 9 steps from benzene 1,3-dialdehyde, amino acid synthons **6** and **8** (Schmidt, U.; Meyer, R.; Leitenberger, V.; Griesser, H.; Lieberknecht, A. *Synthesis* **1992**, 1025-1030), and benzyl-glycine.

326. TYPE VI β-TURN MIMETIC DESIGN; SYNTHESIS OF 5-*TERT*-BUTYLPROLINE ANALOGUES OF THE FKBP SUBSTRATE ALA-LEU-PRO-PHE. Liliane Halab, Francis Gosselin and William D. Lubell*, Département de chimie, Université de Montréal, C. P. 6128, Succ. A, Montréal, Canada H3C 3J7

We are using 5-*tert*-butylprolines (5-tBuPros) in order to prepare Xaa-Pro amide *cis*-isomer surrogates that may serve as conformationally rigid type VI β-turn mimetics.[1,2] Since the peptidyl prolyl isomerases (PPIases), which isomerize X-Pro amide bonds and accelerate the folding of particular proteins, have been proposed to bind preferably to peptides possessing type VI β-turn conformations, we are synthesizing type VI β-turn mimetics in order to furnish novel PPIase inhibitors. We will present syntheses of analogues of

the FKBP tetrapeptide substrate (Ala-Leu-Pro-Phe) that is proposed to bind to FKBP as a type VI β-turn. In particular, by detailing our work on the incorporation of 5-*tert*-butylproline into di- and tri-peptides using both Boc and Fmoc methodologies, we will present a general synthetic strategy for preparing libraries of Xaa-Pro amide *cis*-isomer surrogates and type VI β-turn mimetics.
1. Beausoleil, E.; Lubell, W. D. *J. Am. Chem. Soc.* In Press.
2. Beausoleil, E.; L'Archevêque, B.; Bélec, L.; Atfani, M.; Lubell, W.D. *J. Org. Chem.* In Press.

327. SYNTHESIS AND BIOLOGICAL EVALUATION OF β-TURN MIMETICS THAT INCORPORATE i+1, i+2, AND i+3 SIDE-CHAINS. Andrew J. Souers, Alex A. Virgilio, and Jonathan A. Ellman. Department of Chemistry, University of California, Berkeley, CA 94720

The β-turn motif (**1**) plays an essential role in many molecular recognition events in biological systems. Here, we describe the solid phase synthesis of β-turn mimetics (**2**) that incorporate the i+1, i+2, and i+3 residues from a readily available and diverse set of precursors. Large libraries of these derivatives have been prepared, and biological evaluation has resulted in the identification of several potent small molecule ligands. In particular, turn mimetics that bind to specific somatostatin receptor subtypes with IC_{50} values < 80 nm have been identified, and resynthesis on a larger scale has allowed for further evaluation.

328.
POLYPEPTIDES OF CYCLIC BETA-AMINO ACIDS THAT FORM STABLE HELICES. **D. H. Appella**, L. A. Christianson, I. L. Karle, D. R. Powell, and S. H. Gellman, Department of Chemistry, University of Wisconsin, Madison, WI, 53706

The synthesis and structural characterization of polypeptides made from trans-2-aminocyclohexanecarboxylic acid (trans-ACHC) or trans-2-aminocyclopentanecarboxylic acid (trans-ACPC) will be presented. Molecular mechanics and molecular dynamics calculations predicted that polypeptides of trans-ACHC would adopt a stable helical structure characterized by intramolecular 14-membered ring hydrogen bonds. Crystal structures of a tetramer and a hexamer both show the predicted helical structure. Amide proton H/D exchange data also indicate that the hexamer adopts a very stable internally hydrogen bonded conformation in methanol. Calculations also predicted that polypeptides of trans-ACPC would adopt a helical structure with intramolecular 12-membered ring hydrogen bonds. Experimental results on this system will be presented.

329. THE PREPARATION OF A PEPTIDE CONTAINING AN E-DEHYDRO-PHENYLALANINE RESIDUE. Francis M Rossi[a], Christopher Feng[a], and William A. Shirley[b], a. Department of Chemistry, Santa Clara University, Santa Clara CA 95053, and b. Department of Molecular Biology MB-19, The Scripps Research Institute, 10666 N. Torrey Pines Rd. La Jolla CA 92037.

The major product (**1**) of the condensation of benzaldehyde with a peptide containing a phosphonoglycine residue was found to contain an E-dehydrophenylalanine residue. Characterization of **1** by NMR spectroscopy indicated that it has a different backbone conformation than the previously described Z-isomer.

330.

THE STRUCTURES OF A PEPTIDE CONTAINING A Z- AND E-DEHYDROAMINO ACID RESIDUE. William A. Shirley, Francis M. Rossi, a. Department of Molecular Biology, The Scripps Research Institute, La Jolla, CA 92037, and b. Department of Chemistry, Santa Clara University, Santa Clara, CA 95053

The structure of small peptides of biological interest with a Z-dehydrophenylalanine residue is known from solution and solid state studies. The recent synthesis of the E-isomer of dehydrophenylalanine and the characterization of a peptide containing this residue by NMR spectroscopy indicates that it, the peptide, possesses a different backbone conformation from that of a peptide with the Z-isomer. Preliminary data indicates that peptides with the Z-isomer form beta-turns while those with the E-isomer may form gamma-turns. We investigate the relative energy of these two conformations for each of the isomers using molecular modeling and molecular dynamics.

331.

SYNTHESIS OF RIGID HYDROPHILIC ORGANIC SCAFFOLDS FOR USE IN DE NOVO PROTEIN SYNTHESIS. A. R. Mezo and J. C. Sherman,[*] Department of Chemistry, University of British Columbia, Vancouver, B.C., Canada, V6T 1Z1.

We are designing and synthesizing de novo proteins using tetrathiol cavitands as rigid organic scaffolds. Previously, cavitands with methyl pendent groups (R = CH_3) have been used for de novo protein synthesis, but were later determined to be too hydrophobic for their study in aqueous solution. We are now incorporating hydrophilic moieties such as hydroxyls and phosphates into the pendent groups of the cavitand to help reduce the hydrophobicity of the cavitand. The synthesis of these hydrophilic tetrathiol cavitands will be presented.

R = CH_3
R = $(CH_2)_3OH$
R = $(CH_2)_3OPO_3H_2$

332. SYNTHESIS OF COMPLEX PEPTIDE-CONJUGATES BY USE OF ENZYME-CLEAVABLE PROTECTING GROUPS. **Stefanie Gabold**, Dagmar Sebastian, Axel Heuser, Sonja Schulze, Herbert Waldmann, University Karlsruhe, Germany

Peptide-conjugates e.g. phosphopeptides are found to play an important role in signal transduction, whereas nucleopeptides are involved in viral replication. To study these biological phenomena associated with peptide conjugates, characteristical partial structures of the complex parent proteins are required. Due to the multifunctionality and the pronounced acid- and base-senstivity of the peptide-conjugates their chemical synthesis is very difficult and needs the application of a variety of protecting groups, that can be removed under mild conditions. We now report that such sensitive and multifunctional peptide conjugates can advantageously be built up under mildest conditions (pH 6.5-7.0, 25-37°C) by employing enzymatically cleavable blocking groups. We want to present the synthesis of a phosphopentapeptide from the human raf-protein and of various nucleo-peptides using the penicillin G acylase sensitive phenylacetamidogroup and lipase-cleavable protecting groups such as the heptylester.

333.

PROTECTION OF AMINO ACID BY THIOSUBTILISN CATALYZED ACETYLATION

Dar-Fu Tai, Shu-Li Wang
Department of Chemistry, National Dong-Hwa University, Hwalien, Taiwan, ROC.

Thiolsubtilisin, an artificial thiol-protease, is chemically modified from subtilisin by replacing serine-221 with cysteine at the active site. Adsorption of thiolsubtilisin on cellulose or celite stabilized the enzyme and was able to reuse. The suitable pH for immobilization is among 6 to 6.5. Immobilized thiolsubtilisin catalyzed *N*-acetylation of free amino acids by using ethyl acetate as the acetylating reagent to form *N*-acetyl amino acids in organic solvent.

334. CONSTRAINED *CIS*-PROLINE DIPEPTIDES AS BIOCHEMICAL PROBES.

Timothy P. Curran*, Jason F. Hall and Patrick M. McEnaney
Department of Chemistry, College of the Holy Cross, Worcester, MA 01610-2395

It has been speculated that *cis-trans* proline isomerization plays an important role in many biochemical events. As such, molecules that constrain the proline amide to either the *cis*- or *trans* -amide conformation could be used to probe the relationship between proline amide conformation and biochemical activity. We have prepared four bicyclic dipeptides (**1-4**) that constrain the proline amide to the *cis*-amide conformation. In this presentation the synthesis and characterization of **1-4** will be given. In addition, preliminary results from the use of **1-4** as biochemical probes will be presented.

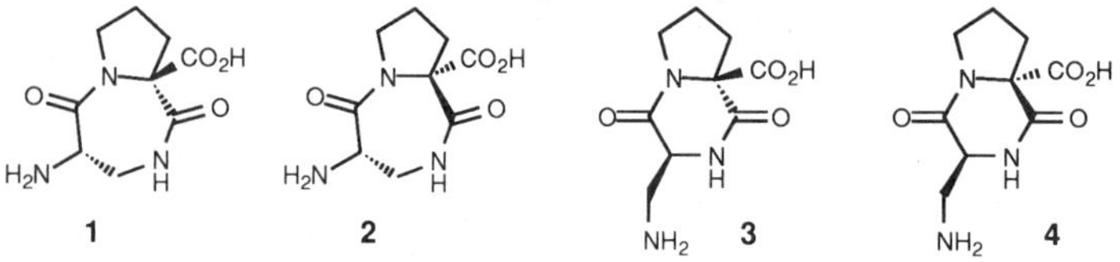

1 **2** **3** **4**

335.

PREPARATION OF AROMATIC DERIVATIVES OF SMALL, POLAR PEPTIDES AS POTENTIAL ANTICANCER COMPOUNDS. **B. S. Jursic** and Ingrid P. Buchler, Department of Chemistry, University of New Orleans, New Orleans, Louisiana 70148

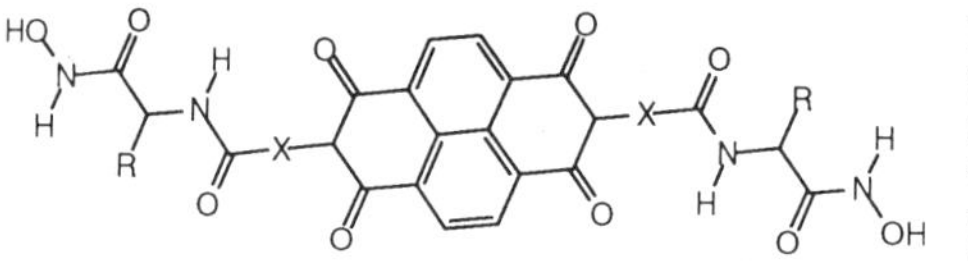

It was demonstrated that small polar peptide units separated by certain number of methylene groups have anticancer activity. Similar compounds are capable to activate Protein Kinase C (PKC) that is responsible for production of messenger molecules necessary for cell differentiation. It is also known that flat aromatic compounds can intercalate DNA and in this way stop its replication. We do believe that there is a complexation between PKC and DNA. The synthesis of some of these compounds will be presented and their activity will be discusses.

336.

PREPARATION OF ALIPHATIC HYDROXYUREA DERIVATIVES AS POTENTIAL ANTICANCER COMPOUNDS. **B. S. Jursic** and Rajalakshmi Nair, Department of Chemistry, University of New Orleans, New Orleans, Louisiana 70148

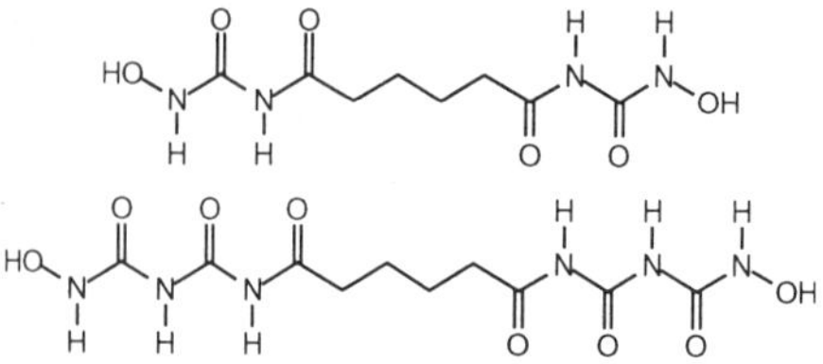

It well known that hydroxamic acid derivatives with a certain balance polarity in the molecule can induce differentiation of the cancer cells. On the other hand, hydroxy urea can be used for a long time to suppress white blood cell growth for leukemia patients. We will present the preparation of the organic compounds that are a combination of these groups with some of their biological activities.

337. ISOLATION, CHARACTERIZATION AND SYNTHESIS OF THE (E) AND (Z) CONFIGURATIONAL ISOMERS OF THE NOVEL 7-METHYL-6-OCTADECENOIC ACID FROM A MARINE BACTERIUM OF THE GENUS *VIBRIO ALGINOLYTICUS*. Néstor M. Carballeira*[a], Anthony Sostre[a], and Thomas Tosteson[b]. [a]*Department of Chemistry, University of Puerto Rico, P.O. Box 23346, San Juan, Puerto Rico 00931-3346*, [b]*Department of Marine Sciences, University of Puerto Rico-Mayaguez, Lajas, Puerto Rico 00667*

Both stereoisomers of 7-methyl-6-octadecenoic acid have been identified in a marine bacterium of the genus *Vibrio alginolyticus*, isolated from the Caribbean sponge *Pandaros acanthifolium*. The synthesis of the (E) and (Z) configurational isomers was accomplished using the well known Wittig reaction. The total synthesis of these isomers as well as the identification of other fatty acids in this bacterium will be presented.

338. CHARACTERIZATION AND SYNTHESIS OF NOVEL FATTY ACIDS FROM A NEW STRAIN OF *VIBRIO ALGINOLYTICUS* FROM THE BLACK SEA. Néstor M. Carballeira*[1], Anthony Sostre[1], Kamen Stefanov[2], Simeon Popov[2], and Atanas Kujumgiev[3], [1]*Department of Chemistry, University of Puerto Rico, P. O. Box 23346, San Juan, Puerto Rico 00931*, and [2]*Institute of Organic Chemistry with Center of Phytochemistry, and* [3]*Institute of Microbiology, Bulgarian Academy of Sciences, Sofia 1113, Bulgaria.*

As part of our continuing program in the search of novel marine antimicrobial fatty acids, we identified the new fatty acids 9-methyl-10-hexadecenoic acid (**1**) and 11-methyl-12-octadecenoic acid (**2**) in a new strain of *Vibrio alginolyticus* associated with the alga *Cladophora coelothrix* from the Black Sea. Linear alkylbenzene fatty acids, such as 10-phenyldecanoic, 12-phenyldodecanoic, and 14-phenyltetradecanoic acid were also identified in this bacterium. The complete characterization of these novel fatty acids, as well as efforts towards the total synthesis of **1** and **2** will be presented.

339.

THE IDENTIFICATION AND STRUCTURE ELUCIDATION OF CYCLIC PEROXIDES FROM THE MARINE SPONGE *PLAKORTIS SP.*

Blaine Harrison and Phillip Crews

Department of Chemistry and Bio-chemistry, University of California, Santa Cruz, Santa Cruz, CA 95064

Abstract: The genra *chondrosia* and especially *plakortis* of marine sponges has been a rich source of cyclic peroxides and their open ring analoges. Here we report on the collection of a sponge identified as *Plplakortis sp.* (collected in Papua New Guinea) that was a source of the new cyclic peroxides dihydroplakortin B ethyl ester (**1**), plakortin B ethylester (**2**), and plakortin B methylester (**3**). The structures of these compounds was established from spectroscopic data and the relative stereochemistry determined by comparison to known plakortin analogues. These compounds appear to be derived biosynthetically from simple carboxylic acids via the polyketide pathway.

340.

NEW CORIOLIN SESQUITERPENES PRODUCED BY THE WATER CULTURED FUNGI SEPARATED FROM A HALICLONA INDO-PACIFIC SPONGE. **Yangsheng Wanggui,** Bethel Borgeson, Phillip Crews, Department of Chemistry and Biochemistry, Unversity of California, Santa Cruz, Santa Cruz, CA, 95064

During the recent years we have begun the exploration of marine derived heterotropic microorganisms, especially marine fungi, as new source of bioactive natural products. Although this research field is still in its infancy, some remarkable compounds, ranging from alkaloids, terpenes to polyketide, have been isolated from marine media cultured fungi separated from Indo-Pacific sponges. Four new cyclic sesquiterpenes, coriolins D, E, F and G, have been isolated from an unidentified marine media cultured fungi which were separated from an Indo-Pacific sponge Haliclona sp6. Coriolin D and G were found to be antimicrobially (Bacillus subtilus) active. We will present the discussion about the isolation and structure determination of these new sesquiterpenes.

341.

GEOGRAPHIC VARIATION IN THE TROPICAL MARINE SPONGE JASPIS ?JOHNSTONI: AN UNEXPECTED SOURCE OF TWO NEW TERPENOIDS. **Leanne Murray,** Amanda Johnson and Phillip Crews, Department of Chemistry and Biochemistry, Unversity of California, Santa Cruz, Santa Cruz, CA 95064

Diversity in the chemistry of Jaspis sp. sponges has been well documented in the literature with reports of macrolides, styryl sulphates, nucleosides, purine bases and the cyclic depsipeptide jasplakinolide. Our research group has had the opportunity to collect Jaspis ?johnstoni from numerous locations throughout the South Pacific including Indonesia, Vanuatu, Fiji, Papua New Guinea and the Solomon Islands. Subsequent investigation of the chemistry of these specimens has invariably led to jasplakinolide being isolated as the major component. However, a recent collection of Jaspis ?johnstoni from Madang Harbor in Papua New Guinea unexpectedly led to the isolation of two new terpenoids jaspaquinol and jaspic acid.

342. A NEW PROCESS FOR THE RECOVERY AND RECYCLE OF 4,4'(HEXAFLUOROISOPROPYLIDENE)BIS(PHTHALIC ANHYDRIDE) [6-F DA] FROM FLUORINATED COMPOSITE MATERIALS. Joseph R. Wetzel* and Andrew J. Caruso, Chemical Process Technology Laboratory, General Electric Corporate Research and Development, One Research Circle, Niskayuna, NY 12309

AFR-700B, is a polyimide, which in its cured state is a thermoset with many interesting properties. However, its high cost limits its use to extremely specialized cases such as those found in stealth avionics and aerospace applications. Much of this polyimide's cost is associated with a key monomer, 6-F dianhydride, which is commercially available at a cost of $400-500 per pound. We have discovered an economically and environmentally attractive, two step recycle process which yields pure 6-F dianhydride monomer. The recycle process, involving scrap AFR-700B composite parts, a novel medium pressure reaction, and the formation of a new copolyimide from recovered 6-F dianhydride will be discussed.

343. STUDIES DIRECTED TOWARDS SYNTHESIS OF CONDUCTIVE POLYMERS FROM ENERGETIC MATERIAL. **Sreenivasa R. Eturi**, A. Bashir-Hashemi, Geo-Centers, Inc.,762 Rt 15 S, Lake Hopatcong, NJ 07849 and Sury Iyer, ARDEC, Picatinny Arsenal, NJ 07806

Chemical conversion of surplus energetic material into useful high value industrial products will be discussed. Our research in this area is focused on the conversion of energetic material such as TNT into conductive polymers. Further conversion of TNT into nonlinear optical material and substituted heterocyclic compounds (e.g. indazole and indole) will also be presented.

344. SYNTHESIS AND CHARACTERIZATION OF POLY[*ortho*-PHENYLENE ETHYNYLENE]
David. A. Shultz* and Martha G. Hollomon
Department of Chemistry, **North Carolina State University**, Raleigh, NC 27695-8204

As part of an ongoing research project to prepare organic polymers with stable, paramagnetic repeat units, we have designed poly[*ortho*-phenyleneethynylene] with paramagnetic, chelating ligands as part of the repeat unit. As with the poly[*ortho*-phenylenevinylene] design, proposed by Lahti, this design will have spins of which are high-spin coupled along the polymer backbone even in the presence of defects. Our polymers will be among the first organic polymers having high-spin coupling along the main chain, and also containing structural units for controlling interchain interaction. Extended solids prepared from this polymer may be thought of as infinitely crosslinked polymers having a main-chain paramagnetic repeat unit. This paper will discuss the synthesis and characterization of **poly[14(catH2)]**. The electronic and magnetic properties of **poly[14sq]** will be discussed.

345. HOW NATURE SYNTHESIZES VITAMIN B_{12}. A SURVEY OF THE LAST FOUR BILLION YEARS. **A.I. Scott**, Department of Chemistry, Texas A&M University, College Station, Texas 77843-3255

The acquisition and sequencing of the genes encoding the enzymes for vitamin B_{12} biosynthesis in *Salmonella typhimurium* and *Pseudomonas denitrificans* have dramatically altered the direction of research on the pathway from uro'gen III to the corrinoids. Using a combination of molecular biology, organic chemistry and NMR spectroscopy, logical progression along the sequence has been made. Recent work from our laboratory is focused on the discovery and specificities of the methyl transferases connecting uro'gen III with cobyrinic acid, the temporal resolution of cobalt insertion and a comparison of the anaerobic and aerobic pathways in *S. typhimurium* and *P. denitrificans* respectively. The implication of two parallel routes to corrins in these bacteria will be discussed. The complete repertoire of gene products can be combined to reconstruct the biosynthetic pathway *in vitro*. Of great interest is the mechanism of ring contraction leading from porphyrinoids to corrins. The details of this process will also be discussed in the light of recent discoveries in the semi-anaerobic organism, *Propionibacterium shermanii*.

346.
RECENT STUDIES ON RETINAL PIGMENTS. Koji Nakanishi Department of Chemistry, Columbia University, New York, NY 10027

The structure of the fluorophore A2E which accumulates in old age eyes and is associated with macular degeneration, an incurable eye disease that may lead to blindness, has been determined (1). The structure consists of two retinal moieties condensed with an ethanolamine group. Its synthesis has been completed (R. Ren et al., submitted). Its properties in relation to age-related blindness and other aspects will be presented.
Photoaffinity studies indicate that the C-3/C-4 of the retinal ionone moiety is located close to the mid-section of helix F of rhodopsin (2). The absolute sense of twist around C-12/C-13 in the 11-cis-retinal chromophore within the rhodopsin binding site, as determined by exciton coupled CD, will be presented (3). The chromophore is known to be twisted around the C/6/C-7 single bond as well. We plan to study this by measuring the fluorine distance in a retinal analog by solid state F-NMR (K. Monde, L. Gilchrist, K. Nakanishi, CA. McDermott, in prep.).
1. N. Sakai, et al., J. Am. Chem. Soc., 118, 1559-1560 (1996).
2. H. Zhang, et al., J. Am. Chem. Soc ., 116, 10165-10173, (1994).
3. J. Lou, Q. Tan, B. Borhan, E. Karnaukhova, N. Berova, K. Nakanishi, unpublished.

347. CHEMISTRY OF FUNGAL ANTAGONISM AND DEFENSE. **James B. Gloer,** Department of Chemistry, University of Iowa, Iowa City, IA 52242

Studies of the chemical ecology of plants, insects, and marine organisms are widely recognized as valuable guides to the discovery of bioactive natural products. However, relatively little is known about whether applications of microbial ecology might be similarly useful. Many antagonistic and defensive interactions are known to occur among the fungi, and we view such interactions as leads to the discovery of new biologically active metabolites. For example, antagonism between species of naturally-competing fungi occurs in many ecosystems. Our studies of antagonistic species from selected habitats have led to the isolation of a variety of new antifungal agents. In another project, we have found that certain important fungal bodies (sclerotia, ascostromata) contain unique metabolites that help to protect them from attack by fungivorous insects. Many of these compounds exhibit effects on agriculturally important insects and/or show medically relevant bioactivities. Results from these projects will be presented.

348. LESSONS FROM THE DEEP. Kenneth L. Rinehart, 454 Roger Adams Laboratory, University of Illinois, Urbana, IL 61801

In the past 40 years the ocean has presented natural products chemists a cornucopia of new structure types and biologically active molecules. Examples of the former from our laboratory are nickel-containing porphyrins and families of novel guanidines, steroidal sulfates and sphingosines. The list contains toxins as well as chemotherapeutic agents, of special concern being hepatotoxic cyclic peptides implicated in numerous recent deaths in Brazil. Pharmaceutical marine-derived natural products include the didemnins, cyclic depsipeptides with antitumor, antiviral and immunosuppressive activities and ecteinascidins, bis(tetrahydroisoquinolines). Didemnin B has undergone extensive clinical trials but recent structure-activity relationship studies demonstrate greater activities for didemnin M and dehydrodidemnin B. The ecteinascidins show potent antitumor properties, in xenografts of human lung, ovarian, breast, and kidney tumors on nude mice as well as in leukemia and melanoma, and ecteinascidin 743 (ET 743) is currently in clinical studies in four countries. Marine-derived compounds also provide examples of new pharmacophores to be developed in synthetic studies.

349.

SYNTHESIS OF 1-AMINO-3-[2-(1,7-DICARBA*CLOSO*DODECA-BORAN(12)-1-YL)ETHYL]CYCLOBUTANECARBOXYLIC ACID: A POTENTIAL BNCT AGENT. Rajiv R. Srivastava, Robert R. Singhaus and George W. Kabalka*, Department of Chemistry and Radiology, University of Tennessee, Knoxville, TN 37996-1600.

The synthesis of 1-amino-3-[2-(1,7-dicarba*closo*dodecaboran(12)-1-yl)ethyl]cyclobutanecarboxylic acid was achieved in four steps starting from *m*-carborane. This is the first report on the synthesis of *m*-carboranyl amino acid. The title compound is currently being evaluated for the potential use in boron neutron capture therapy (BNCT) of cancer.

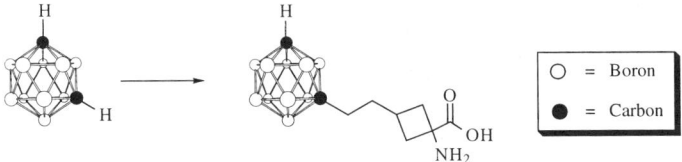

350.

INTERMOLECULAR PROTON TRANSFER ALONG THE AMIDE HYDROGEN BOND IN CRYSTALLINE N-(4-CHLOROPHENYL)FORMAMIDE. A.G. Sykes P. Hrvatin, T.J.R. Weakley, and V.G. Young, Department of Chemistry, University of South Dakota, Vermillion, SD 57069

A fundamental question remains regarding whether hydrogen has two nondegenerate configurations within the amide hydrogen bond. If a double well potential governing proton transfer exists along the N-H...O hydrogen bond, then an iminol tautomer is an alternate configuration for the familiar amide form. Previous studies of the structures of crystalline anilides indicate that a non-degenerate configuration for the N-H...O amide proton has not been observed. We report, here, the first crystallographic evidence of solid-state proton transfer in the amide hydrogen bond in crystalline N-(4-chlorophenyl)formamide. X-ray diffraction studies at room temperature and 173 K and accompanying neutron diffraction results will be presented.

351. 9-SUBSTITUTED 9-(*o-t*-BUTYLPHENYL)FLUORENE ROTOMERS: REACTIVITY, DYNAMIC NMR, X-RAY CRYSTAL STRUCTURE, AND THEORETICAL STUDIES. Cal Y. Meyers, Yuqing Hou, Hisham G. Lutfi, Dept. of Chemistry, Paul D. Robinson, Dept. of Geology, Southern Illinois University, Carbondale, IL 62901; Howard E. Dunn, Jeff W. Seyler, Dept. of Chemistry, University of Southern Indiana, Evansville, IN 47712.

9-X-9-(*o-t*-Butylphenyl)fluorenes in which X = MeS (**1**) or Me (**2**) are observed in solution at 25 °C (^{1}H NMR) and crystalline form (X-ray) solely as their *ap* rotamers. When X = H (**3**), OH (**4**), OEt, Oallyl (**5**), OAc or F, these fluorenes were observed solely as their *sp* rotamers. Calculations support the relatively greater stability of **1** as its *ap* rotamer, but also support the X-ray data indicating the 6'-H···S distance is 0.51 Å shorter than the sum of these atoms' van der Waal's radii, which leads to the apparent H-bonding exhibited in the extraordinary deshielding of the Ph 6'-H (δ9.0). In contrast, 6'-H in *ap*-**2** resonates at δ8.1. Sterically crowded *ap*-**1** homolytically dissociates slowly in crystalline form at 25 °C and in DMSO the resulting CH$_3$S· and 9-Ar-Fl· leads to the formation of *sp*-**3** and *sp*-3-MeS-**3**, but *t*-Bu rotation is not inhibited even at -85 °C.

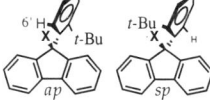 α-Hydride transfer from Et$_2$O, THF, (PhCH$_2$)$_2$O or AcOEt cleanly converts the 9-(*o-t*-butylphenyl)fluorenyl cation (**6**) to **3**, which is also produced by base induced E2 elimination of *sp*-**5**. The bulky *t*-Bu allows reduction of cation **6** to *sp*-**3** with H$_3$PO$_2$/I$_2$/HOAc, and oxidation of the 9-anion with O$_2$ to the blue 9-radical and *sp*-**4**, without also forming a fluorenyl peroxide (*cf.* the *i*-Pr analog).

352. 9-SUBSTITUTED 9-(o-i-PROPYLPHENYL)FLUORENE ROTAMERS: REACTIVITY, DYNAMIC NMR, X-RAY CRYSTAL STRUCTURE AND THEORETICAL STUDIES. Cal Y. Meyers, Yuqing Hou and Delano Scott, Dept. of Chemistry, and Paul D. Robinson, Dept. of Geology, Southern Illinois University, Carbondale, IL 62901, and Howard E. Dunn and Jeff W. Seyler, Dept. of Chemistry, University of Southern Indiana, Evansville, IN 47712.

A dynamic ^{1}H NMR study of ap-9-hydroxy-9-(o-i-propylphenyl)fluorene (1) indicated that rotation of the i-propyl group is restricted. Reduction of 1 with H_3PO_2/I_2 in HOAc in the presence of air produced the ap and sp rotamers of 9-(o-i-propylphenyl)fluorene (2) in a 29:71 ratio (^{1}H NMR), in close agreement with calculations, and 25% of ap-bis[9-(o-i-propylphenyl)fluorenyl] peroxide (3). Under argon only 2 was formed. When the reagent was H_3PO_2 alone, only 2 was produced even in the presence of air. In the presence of I_2 (HI), an SET route is followed; with H_3PO_2 alone, a hydride-transfer mechanism is proposed. Compound 2 crystallizes only as its sp rotamer. Treatment of lithiated 2 with O_2 in THF led to 1 and 3. Intramolecular H-bonding of the 6'-H to O in 1 and 3 and to Cl in ap-9-chloro-9-(o-i-propylphenyl)-fluorene apparently is responsible for its exceptional deshielding (as low as δ8.6) not observed in 2 or ap-9-methyl-9-(o-i-propylphenyl)fluorene.

353. CONFORMATIONAL ANALYSIS OF 1-ALKOXYMETHYL-5(R)-METHYL-2-PYRROLIDINONE DERIVATIVES. DETERMINATION OF THE ABSOLUTE STEREOCHEMISTRY OF ALCOHOLS
Shamil K. Latypov,*,# Ricardo Riguera,‡ Michael B. Smith,† and Jana Polivkova†

#The Institute of Organic and Physical Chemistry of Russian Academy of Sciences, Kazan, 420083, Tatarstan Republic, the Russian Federation. ‡Departamento de Quimica Organica, Facultad de Quimica, Universidad de Santiago de Compostela, 15706, Santiago de Compostela, Espana. †Department of Chemistry University of Connecticut, 215 Glenbrook Road, Room 151 Storrs, Connecticut 06269-4060, U.S.A.

This work identifies parameters that will allow a prediction of the sense of non-equivalence in derivatives prepared by reaction of chiral racemic and chiral non-racemic alcohols with 1-chloromethyl-5R-methyl-2-pyrrolidinone. The factors governing NMR non-equivalence (n.e) (Δδ) in these derivatives was established and the NMR parameters were determined. When compared with parameters derived from calculations of the conformations of the derivatives, we were able to establish the absolute stereochemistry.

354. CONFORMATIONS OF CROWN THIOETHERS. **V.V.Samoshin,** Moscow State Academy of Fine Chemical Technology, Moscow, 117571, Russia; E.I.Troyansky, Institute of Organic Chemistry, Russian Academy of Sciences, Moscow, B-334, Russia

Stereochemical studies of sulfur- and oxygen-containing macrocycles by means of NMR measurements, X-ray crystal structure analysis and molecular mechanics calculations revealed new conformational regularity and peculiarities. Conformationally rigid structural fragments (a fused cyclohexane moiety, ester moieties) restrict a number of possible conformers, thus facilitating the analysis of conformational structure and distribution for such complicated objects as macroheterocycles in solution.

355.
CONCENTRATION DEPENDENCE OBSERVED IN THE CHEMICAL SHIFTS OF AROMATIC AND OTHER PROTONS Abhijit Mitra*, Pamela J. Seaton**, *Department of Chemsitry and Biochemistry, Manhattan College/College of Mount St. Vincent, Riverdale NY, 10471, ** Department of Chemistry, Univ. North Carolina - Wilmington, Wilmington, NC 28403.

Our studies show that the proton NMR spectra of a wide variety of heterocyclic and nonheterocyclic aromatic compounds show concentration dependent chemical shift changes. An important observation is that different protons shift at different rates, resulting in crossing or coalescence of the signals, thus altering the spectral pattern. The extent of these concentration dependent changes also differ with solvents. In the H-NMR spectrum of quinoline in chloroform-d, the chemical shift change of 4-H is largest and crosses the 8-H at ~0.5M. For isoquinoline, the chemical shift changes are also substantial and in cyclohexane-d12 the 4-H crosses the 6-H at high concentration (~4M) and the 7-H at lower concentration (~0.1M). The concentration dependent chemical shift changes of naphthalene are small. However, 1-fluoronaphthalene showed larger changes and, as in quinoline, the changes in chemical shift of 4-H and 8-H are most significant.

356.
CONCENTRATION DEPENDENT CHEMICAL SHIFT CHANGES: A STRUCTURE ELUCIDATION PROBLEM Pamela J. Seaton* Abhijit Mitra **, *Department of Chemistry, Univ. North Carolina - Wilmington, Wilmington, NC 28403; ** Departments of Chemistry and Biochemistry, Manhattan College / College of Mount St. Vincent, Riverdale, NY 10471

Our recent research involving proton NMR spectroscopy of a variety of heterocyclic and non-heterocyclic aromatic systems, commonly present in natural products, show concentration dependent chemical shift changes, leading to changes in the spectral pattern. For example, for indoles in deuterochloroform, the 2-H and 4-H signals shift dramatically with concentration, resulting in crossover of the signals. In deuteromethanol, the 2-H/7-H signal of acridine coalesces with the 4-H/5-H signal at high concentration and with the 1-H/ 8-H signal at low concentration. Concentration dependent changes for beta-carboline, naphthoquinone and flavone systems were also studied. Spectral pattern alterations have far reaching implications since NMR has been extensively used as one of the tools for the structure elucidation of new natural products.

357. Correlation of ^{13}C-^{1}H Coupling Constants with Electronic Structure in Bi- and Polycycloalkanes: A PM3 and HF/6-31G* Analysis

Liliana Craciun and James E. Jackson; Department of Chemistry, Michigan State University, East Lansing, MI 48824

Muller-Pritchard type ($J_{C-H} = a \times \% \, s_C$) and related expressions are explored for the prediction, from standard quantum chemical models, of one-bond C-H spin-spin coupling constants, $^{1}J_{13C-1H}$, in a series of bi- and polycyclics. Correlations of experimental $^{1}J_{13C-1H}$ with quantities computed from NBO analyses of PM3 and HF/6-31G* wavefunctions//geometries are examined for 39 aliphatic hydrocarbons (>150 C-H sites; J range >100 Hz). Experimental vs. calculated coupling constants are best fit when the model includes contributions from atomic charges (q_H and q_C) along with s-character at carbon (% s_C). Previously used geometrical measures of hybridization are also discussed.

That $^{1}J_{13C-1H}$ depends on carbon orbital hybridization is part of the canon of organic chemistry. Numerous equations have been proposed based on modest data sets and various measures of hybridization. By using common computational chemistry methods for a large data set, we offer both a predictive tool for the practicing chemist, and insights into the validity of hybridization-based interpretations of coupling.

358. USE OF HYDROGEN/DEUTERIUM EXCHANGE TO DIFFERENTIATE FRAGMENT IONS FROM PSEUDO-MOLECULAR IONS IN MASS SPECTROMETRY. Adeboye Adejare* and Paul W. Brown, Division of Pharmaceutical Sciences, School of Pharmacy, University of Missouri-Kansas City, Kansas City, MO 64110

Fragmentation of protonated molecules can occur for some compounds even when using gentle electrospray ionization conditions. For many compounds, the fragment ions generated during desolvation (in-source dissociation) cannot be distinguished from protonated molecules (pseudo-molecular ions) such as an impurity, decomposition product, metabolite, or reaction by-product which have the same structure. This application of hydrogen/deuterium (H/D) exchange allows fragment ions formed from in-source dissociation to be distinguished from pseudo-molecular ions using hydrogen/deuterium exchange. This technique is based on fragmentation processes which involve neutral losses arising from cleavage of carbon-heteroatom bonds. In this situation, the fragment ion does not have all active hydrogens exchanged and can be differentiated from a fully deuterium exchanged molecule by one dalton. The method being reported will be useful for some organic structure elucidation or analysis and may be more suitable than chromatographic methods in some cases.

359. SYNTHESIS, UV-SPECTRA AND SOME ELECTROCHEMICAL PROPERTIES OF NOVEL 3,3'-BRIDGED-2,2'-BITHIENYLS. Hans Zimmer, Harry B. Mark, Jr., and Mike Wehmeier, Department of Chemistry, University of Cincinnati, PO Box 210172, Cincinnati OH 45221-0172.

A convenient synthesis of a number of novel N-3,3'-bridged-bis-2,2'-thienyls as potential building blocks for novel conducting polymers is reported.
Although their UV-spectra correlate to the degree of coplanarity of these species, their oxidation potentials did not correlate. Also, these compounds did not show any tendency to polymerize as shown by their cyclic voltammograms. This behavior is in contrast to results found earlier by us for similar -$(CH_2)_n$- bridged 2,2'-bithienyls. This finding is explained by nitrogen and sulfur adsorption orientation of the N-bridged species on the electrode surface prior to electron transfer.

360. DEVELOPMENT OF A REAGENT FOR NUCLEIC ACID PHOSPHODIESTER LABELING. **Kenneth D. Turnbull***, Qibing Zhou, Tony L. Hudgens, Department of Chemistry and Biochemistry, University of Arkansas, Fayetteville, Arkansas 72701

Progress in the development of a reagent to be used for site-specific phosphodiester labeling of DNA will be discussed. We are systematically optimizing the various components of a reagent designed to: (1) site-specifically recognize DNA and chemoselectively bind to the phosphodiester, (2) crosslink to the target DNA through photo-initiated quinone methide formation and phosphate alkylation, (3) release the delivery oligonucleotide *in situ*, and (4) produce a covalently stable phosphotriester modified DNA carrying an independently tethered reporter group, drug agent, or protein conjugate.

361. **BIOCONJUGATE SYNTHESIS USING PROTECTED OLIGONUCLEOTIDES.** Dustin L. McMinn, Tracy J. Matray, and Marc M. Greenberg*, Department of Chemistry, Colorado State University, Fort Collins, Colorado 80523.

Oligonucleotide bioconjugates are useful as diagnostic tools, mechanistic probes and as potential therapeutic agents. However, many synthetic methods for bioconjugate formation suffer from low yields, nonspecific covalent modifications, and undesirable reaction conditions. We wish to report general methodology for the synthesis of bioconjugates (e.g. **1**) that produces high yields of homogeneous products rapidly, and under mild reaction conditions. Reactions are carried out using oligonucleotides containing 3'-terminal alkyl amines, but retain all standard protecting groups. The protected oligonucleotides are synthesized using a previously reported photolabile solid phase oligonucleotide synthesis support (*Tetrahedron* **1996**, *52*, 3827.).

362. PROOFING OF PHOTOLITHOGRAPHIC DNA SYNTHESIS METHODS. FABRICATION OF DNA MICROCHIPS

Michael C. Pirrung* and Lara Fallon, Department of Chemistry, P. M. Gross Laboratory, Duke University, Durham, NC 27708-0346; Glenn McGall, Affymetrix, 3380 Central Expressway, Santa Clara, CA 95051.

We have evaluated in a microchip format the dimethoxybenzoincarbonate-based phosphoramidite solid-phase DNA synthesis method we previously developed. Easy-off base protecting groups were used to enable their removal without cleavage of the oligonucleotide from the chip. Rates of photolysis under both 310 and 365 nm irradiation of the DMBOC-protected nucleotides while attached to the surface were determined by quantitative fluorescence scanning. Solvent effect studies using a similar protocol showed that the deprotection of the DMBOC-T occurs fastest in dioxane or without solvent. The coupling efficiencies of these amidites in the synthesis of homopolymers were also determined. Sixteen decanucleotides of the sequence 5'-AAXTAXCTAC-chip were prepared by way of these amidites in a 4 × 4 array, where the positions marked by X comprise all combinations. The array was hybridized at 23 °C with a target deoxyeicosanucleotide of the sequence fluorescein-5'-CTGAACG**GTAGCATCTT**GAC. Fluorescence imaging demonstrated high-fidelity hybridization to the perfectly-matched probe.

363. DNA CATALYZED REDUCTIVE AMINATION
Z-Y, J, Zhan and D. G. Lynn*
Searle Chemistry Laboratory, The University of Chicago, Chicago, IL 60637

Here we report that DNA (dGCAACG) catalyzes the reductive amination of the 3'-aldehyde-modified DNA trimer (5'-HO-dCGT-CH$_2$CHO) with 5'-amine-modified DNA trimer (5'-H$_2$N-dTGC). This reaction was developed to employ thermodynamic control on the templating DNA to enhancing the fidelity of the coupling reaction. As a result of destabilization of the amine product on the template, conditions have been developed which allow for turnover numbers of 10 to 100 depending on the nature of the reducing environment. The reaction is first order in template and suggests a general strategy by which the information encoded in DNA may be recorded into different molecules.

$$5'\text{-}^+\!H_3N\text{-}dTpGpC$$
$$+\qquad\qquad [H]\qquad\longrightarrow\qquad ^+H_2N\sim$$
$$3'\text{-}CHO\text{-}dTpGpC5'\qquad [T]$$

$v_T/v = 1.2 \times 10^2$

[T]: dGCAACG
v_T: Initial rate with template
V: Initial rate without template

364. A REDOX PATHWAY LEADING TO THE ALKYLATION OF DNA BY THE ANTHRACYCLINE, ANTI-TUMOR DRUGS, ADRIAMYCIN AND DAUNOMYCIN. D.J. Taatjes, G. Gaudiano, K. Resing, T.H. Koch, Department of Chemistry and Biochemistry, University of Colorado, Boulder, Colorado 80309-0215 and Istituto di Medicina Spermentale, CNR, Viale Marx, 15-43, 00137 Rome, Italy

Reaction of the anthracycline, antitumor drugs, adriamycin and daunomycin, with the self-complementary DNA oligonucleotide GCGCGCGC, (GC)$_4$, in the presence of the reducing agent, dithiothreitol; the oxidizing agent, hydrogen peroxide; or the alkylating agent, formaldehyde, gives a similar mixture of DNA-drug adducts. An adduct structure with each anthracycline intercalated at a 5'-CpG-3' site and covalently bound via a methylene group from its 3'-amino group to a 2-amino group of a deoxyguanosine is proposed based upon electrospray MS data, UV-Vis absorption data, and a relevant crystal structure. The results suggest a pathway to the inhibition of transcription by reductively activated adriamycin and daunomycin. Reductive activation in the presence of oxygen yields hydrogen peroxide; hydrogen peroxide oxidizes constituents in the reaction mixture to formaldehyde, and formaldehyde couples the drug to DNA. Formaldehyde results from hydrogen peroxide oxidation of Tris (tris-(hydroxymethyl)aminomethane), present in transcription buffer, and spermine, a polyamine commonly associated with DNA *in vivo*, presumably via the Fenton reaction.

365. STEREOCONTROLLED, DE NOVO SYNTHESES OF β-2'-DEOXYNUCLEOSIDES, α-2'-DEOXYNUCLEOSIDES AND β-2',3'-DIDEOXYNUCLEOSIDES
Jason R. Buck, Minnie Park, Daniel Prudhomme, Zhiwei Wang and **Carmelo J. Rizzo***
Department of Chemistry; Vanderbilt University; Nashville, TN 37235-1822

We have recently developed a stereocontrolled, de novo synthesis of β-2'-deoxypyrimidines. The key step of this sequence is the use of a *m*-trifluoromethylbenzoyl group at the 2-position of ribose as both a directing group for Vorbruggen glycosidation and a deoxygenation precursor. We have now extended this strategy to the synthesis of β-2'-deoxypurines, α-2'-deoxynucleosides and β-2',3'-dideoxynucleosides.

β-2'-Deoxynucleosides   α-2'-Deoxynucleosides   β-2',3'-Dideoxynucleosides

366. SYNTHESIS AND CHARACTERIZATION OF THE AFLATOXIN B1 FORMAMIDO-PYRIMIDINE ADDUCT IN DUPLEX DNA. **Zhengwu Deng,** Hui Mao, Fang Wang, Michael P. Stone, Thomas M. Harris, Department of Chemistry and Center in Molecular Toxicology, Vanderbilt University, Nashville, Tennessee 37235

Aflatoxin B1 (AFB1) is one of the most potent chemical carcinogens. Metabolic activation yields the 8,9 exo epoxide which forms adducts in DNA. The main adduct involves reaction at the N7 position of deoxyguanosine. This cationic adduct is labile in aqueous solution and can readily depurinate. Under basic conditions the imidazole ring of guanine is attacked yielding the AFB1-formamidopyrimidine derivative which exists as a mixture of conformers/geometrical isomers. In the present study, the AFB1-formamidopyrimidine adduct has been constructed in duplex DNA. A single stable conformation has been obtained which has made it possible to characterize the structure by 2D NMR and other physical methods.

367. SYNTHESIS OF HIGH-DENSITY OLIGONUCLEOTIDE ARRAYS FOR HYBRIDIZATION-BASED SEQUENCE ANALYSIS. Glenn H. McGall, A. Dale Barone, Jacqueline A. Fidanza, Jody E. Beecher, Martin J. Goldberg, Nam Ngo, Thadeus S. Block, Affymetrix, Inc., 3380 Central Expwy., Santa Clara, CA 95051.

Hybridization to large arrays of DNA probes is proving to be a powerful technique for large-scale DNA and RNA sequence analysis. As the application of this technology grows, one of the primary challenges will be to increase the density of sequence information encoded in these arrays. This presentation will discuss recent advances in the chemistry and methods used for high-density ($>10^6$ sequences/cm^2) array fabrication which integrate solid-phase oligonucleotide synthesis with photolithographic techniques adapted from the microelectronics industry.

368.

NOVEL NON-NUCLEOSIDIC PHOSPHORAMIDITES FOR MULTIPLE LABELING OF OLIGONUCLEOTIDES AND SYNTHESIS OF BRANCHED DNA. **R. S. IYER,** S. SU, A. INAMDAR, K. L. KALRA, BioGenex Laboratories, 4600 Norris Canyon Road, San Ramon, California 94583

Novel symmetric and asymmetric branching phosphoramidites that can be used to synthesize branched oligonucleotides and to amplify the signal intensity of DNA probes have been developed. The detection sensitivity of oligonucleotide probes is increased by the introduction of multiple labels such as biotin or fluorescein onto the oligonucleotides. Arranging the labels non-linearly at the 5'-end of an oligonucleotide chain is particularly desirable as it minimizes steric hindrance and increases the detectability of the labels. Branching phosphoramidites that result in the formation of DNA 'forks' and 'combs' have been used successfully for this purpose. Poly-labeled branched DNA multimers constructed using these reagents showed significantly increased signal intensity relative to singly labeled probes as determined by in situ hybridization and microtiter plate assays.

369. ORGANIC NITRATES AND NITRITES AS NOVEL ARTIFICIAL NUCLEASES. Jih Ru Hwu, S.-C. Tsay, K. A. Vyas, S.-F. Yu, and H. V. Patel, Department of Chemistry, National Tsing Hua University, Hsinchu, Taiwan 30043, Republic of China and Development Center for Biotechnology, Taipei, Taiwan 10671, Republic of China

A new method was designed for the single-strand scission of DNA in vitro involving the use of phthalimidoalkyl nitrates and nitrites as the source of $NO_2\cdot$ and $NO\cdot$, respectively. Photolysis of phthalimidoalkyl nitrates and nitrites in a buffer at pH 6.0 containing 10% EtOH with 312 nm UV light induced a homolytic fission of the RO–N bond to generate the corresponding alkoxyl radicals accompanied by $NO_2\cdot$ and $NO\cdot$, respectively. These radicals functioned as efficient DNA cleavers and low energy UV light was used as the trigger.

370. **FORMATION OF PSEUDOURIDINE IN TRANSFER RNA CATALYZED BY E. COLI PSEUDOURIDINE SYNTHASE I.** Xuemei Zhao, Joe Chihade, and David A. Horne*. Department of Chemistry, Columbia University, New York, NY 10027.

The last step in the biosynthesis of cellular RNAs is the formation of modified bases at the polynucleotide level. *E. coli* pseudouridine synthase I (PSUI) catalyzes the rearrangement of uridine residues to pseudouridine (Ψ) in positions 38, 39, and 40 of various tRNAs. We have investigated both the RNA recognition and the mechanism of the U to Ψ rearrangement. Kinetic data will be presented in conjunction with site-directed mutagenesis studies that refutes the previously postulated mechanism involving covalent cysteine intermediates. Based on these studies, a new mechanistic proposal will be advanced.

371.

HOST-GUEST CHEMISTRY IN A PROTEIN CAVITY: TUNING THE SELECTIVITY AND RATE OF PYRIDOXAMINE PROMOTED REDUCTIVE AMINATION USING SITE DIRECTED MUTAGENESIS. **Mark D. Distefano,** Hao Kuang, Aram Mazhary, Jeramia Ory and Leonard J. Banaszak. Department of Chemistry, University of Minnesota, Minneapolis, MN 55455

Lipid binding proteins are members of a class of small proteins (131 residues) with a simple architecture that consists of two orthogonal planes of beta-sheet secondary structure and a large sequestered cavity. We are using this system to prepare artificial enzymes and recently reported a conjugate that incorporates a pyridoxamine moiety that reductively aminates alpha keto acids to the corresponding amino acids. Here we describe the use of three mutant proteins that allow the position of pyridoxamine attachment to be moved within the cavity. One mutant conjugate reductively aminates with opposite enantioselectivity (relative to the original system), a second mutant reacts much more rapidly, and a third mutant excludes branched keto acids. A x-ray crystal structure of one of the conjugates has been solved to 2.6 Å resolution.

372. UNSYMMETRICAL DISULFIDE-FUNCTIONALIZED BIPYRIDINES AND THEIR RUTHENIUM DERIVATIVES. **Charles A. Panetta**,[a] Hephzibah Jayasheela Kumpaty,[a] Montray C. Leavy,[a] Charles L. Hussey,[a] Norman E. Heimer,[b] [a]Department of Chemistry, The University of Mississippi, University, Mississippi, 38677, [b]Department of Chemistry, U.S. Air Force Academy, USAFA, CO 80840-6230

We have synthesized eleven unique 2,2'-bipyridines substituted on the 3-, 4-, 5-, and 6- positions with chains ending with disulfide functional groups. The chains varied from four to thirteen atoms in length between the disulfide and the bipyridine nucleus and incorporated ester or amide linkages. Eight of these were converted to their redox-active ruthenium(II) complexes and seven of the latter were chemisorbed on gold and antimony-doped tin oxide (ATO) substrates and were electrochemically evaluated. The best surface coverages on ATO were obtained with those complexes with the longer side chains: the long alkyl tethers probably facilitated surface packing. The longer side chain on the 4-position also had the highest electron-transfer rate. The formal potentials of the surface-confined complexes did not vary appreciably with the position and nature of the side chain.

373. **MOLECULAR CLEFTS DERIVED FROM KAGAN'S ETHER. SYNTHESIS AND SOLID STATE INCLUSION COMPLEXES OF A CHIRAL MOLECULAR TWEEZER.** Michael Harmata[†], Mehmet Kahraman[†], Sriram Tyagarajan[†], Charles L. Barnes[†] and Christopher J. Welch[§], [†]Department of Chemistry, University of Missouri-Columbia, Columbia, Missouri 65211 and [§]REGIS Technologies, Inc. 8210 Austin Avenue, P.O. Box 519, Morton Grove, Illinois 60053

The synthesis of **1** and a discussion of some of its solid state inclusion complexes will be presented. The chromatographic behavior of **1** and various congeners on a chiral stationary phase will also be discussed.

374. SOLVENT-ION PAIR CHIRAL RECOGNITION. **Cheryl D. Stevenson**, Rosario M. Fico, Jr., Tyler D. Schertz, Wayne E. Zeller, and Jim M. Graham, Department of Chemistry, Illinois State University, Normal, IL 61790

^{7}Li-NMR spectra of solutions of the lithium salts of the R and S chiral isomers of the sec-butoxy-cyclooctatetraene dianion (C_8H_7-$OC_4H_9^{2-}$) in S,S(+)-2,3-dimethoxy-1,4-bis(di-methylamino)butane (SS-DDB) were found to be remarkably different. The lithium cations in both solutions exist as solvent separated and contact ion pairs with C_8H_7-$OC_4H_9^{2-}$. However, the concentration of the contact ion pair is much greater for the R isomer of C_8H_7-$OC_4H_9^{2-}$ than for S isomer. The steric interactions involving the chiral solvent and the R and S sec-butoxy groups results in the solvent being much more capable of partially separating Li^+ from the S isomer of C_8H_7-$OC_4H_9^{2-}$ than from the R isomer. Analogous results were obtained with the menthol system connected to cyclooctatetraene.

375. DESIGN OF ION SENSORS BASED ON SAM SUPPORTED PEPTIDE NANOTUBES

Kianoush Motesharei and M. Reza Ghadiri*
*Department of Chemistry & Molecular Biology
The Scripps Research Institute
La Jolla, California 92037*

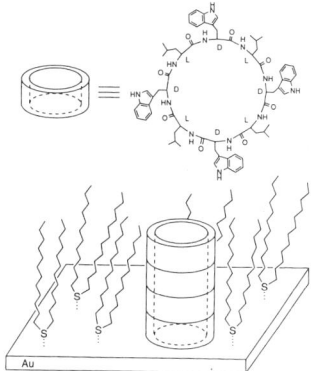

This talk will focus on design, construction, and characterization of cyclic peptide based ion channels in self-assembled organosulfur monolayers. Cyclic peptides consisting of alternating units of D-Leu and L-Trp were incorporated into organosulfur monolayers on gold surfaces through a self-assembly process in order to form hollow tubular structures of fixed diameter. The resulting peptide incorporated monolayer surfaces were then examined for their channeling activity and selectivity using cyclic voltammetry and impedance spectroscopy.

376. TEMPERATURE AND CONCENTRATION-DEPENDENT BINDING OF A NAPHTHYL SUBSTITUTED β-CYCLODEXTRIN WITH HYDROPHOBIC GUESTS Shelli R. McAlpine and Miguel A Garcia-Garibay,* Department of Chemistry and Biochemistry, University of California Los Angeles, Los Angeles, California 90095-1569

An investigation of temperature and concentration effects on a naphthyl substituted β-cyclodextrin will be discussed. The self-aggregation of this derivative and the effects of aggregation on guest binding will be addressed.

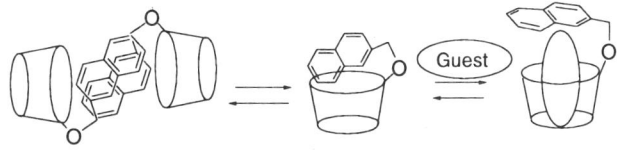

377.

DOUBLE DIELS-ALDER MACROANNULATION OF BISDIENES BY BISDIENOPHILES: SYNTHESIS OF MACROCYCLIC CYCLOPHANE BELTS AS PRECURSORS TO CYCLACENES AND RELATED MOLECULAR HOOPS. **Robert M. Cory,** Anton J. Dikmans, Cameron L. McPhail, Darren Carrozzella and Carl J. Scott, Department of Chemistry, University of Western Ontario, London, Ontario, CANADA N6A 5B7. Fax: (519) 661-3022. E-mail: rcory@uwo.ca

Rapid convergent synthesis of belt-shaped cyclophanes consisting of linearly fused carbocyclic six-membered rings is accomplished via double Diels-Alder cycloaddition of diquinones to bisdienes bearing pendant solubilizing hexyl groups . Optimization of macrocyclization over polymerization of the presumed initial adducts is controlled by temperature: higher temperatures favoring the former and lower temperatures the latter. A hypothetical explanation for this will be discussed in terms of transition state theory. Subsequent transformations of the double Diels-Alder adducts to advanced precursors to cyclacenes and cyclacene quinones, and a second generation synthesis involving phenylthio-substituted diquinones as the bisdienophiles will also be described.

378. **ENERGY TRANSFER BETWEEN HEMICARCERAND AND BIACETYL GUEST.** Ileana Balanescu, Angeles Farran, Kurt D. Deshayes Center for Photochemical Sciences and Department and Chemistry, Bowling Green State University, Bowling Green, OH 43403

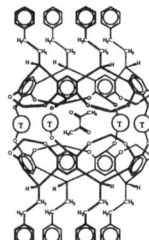

We report intramolecular energy transfer from two different types of hemicarcerand hosts to incarcerated biacetyl (2,3-butanedione). Phosphorescence quantum yields studies demonstrate energy is transferred from hemicarcerand to biacetyl through a triplet-triplet energy transfer mechanism with greater than 90% efficiency.

379. ELECTRON TRANSFER TO INCARCERATED ACCEPTOR SYSTEMS: INFLUENCE OF THE INTERVENING MEDIA, David W. Place, Kurt D. Deshayes, John Miller*, Rich Marasas Jr.*, Center for Photochemical Sciences, Bowling Green State University, Bowling Green, OH 43403, *Chemistry Division, Argonne National Laboratory, Argonne, IL 60439

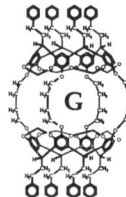

We have encapsulated two electron acceptors (G), biacetyl and p-benzoquinone, within hemicarcerand **1**. Quenching curves of E vs. log kq reveal, in both cases, a leveling effect at 2.2×10^9 $M^{-1}s^{-1}$ for **1.BQ** and 1.0×10^8 $M^{-1}s^{-1}$ for **1.BA** at high driving force. These rates are approximately one order of magnitude lower than diffusion controlled rates measured with free p-benzoquinone and biacetyl, respectively. The major influence in the results has been attributed to the intervening media, hemicarcerand **1**.

380. STRATEGIES FOR PARALLEL PURIFICATION OF COMBINATORIAL SYNTHESIS PRODUCTS. **John C. Hodges**, R. John Booth, Mark W. Creswell, Michael W. Wilson and Joseph S. Warmus, Chemistry Department, Parke-Davis Pharmaceutical Research, 2800 Plymouth Road, Ann Arbor, Michigan, 48105

It is a popular notion that solid supported synthetic methods are essential for the efficient practice of combinatorial chemistry. In pharmaceutical discovery applications, the typical synthetic target chosen for small molecule combinatorial libraries has two to four sites of structural diversity and can be prepared by a short synthesis wherein each synthetic step is accomplished at high yield. In such a setting, the benefits of solid supported synthesis can be outweighed by its limitations. We have found that for such short and high yielding syntheses, strategic planning for parallel purification makes it possible and often preferable to use traditional solution phase synthetic methods and solid supported reagents in the generation of small molecule libraries. High throughput purification methods as applied to a variety of organic reactions and combinatorial synthesis paradigms will be presented.

381. RAPID MICROSCALE SYNTHESIS: A NEW TOOL FOR THE MEDICINAL CHEMIST. **James R. McCarthy**, Yun Feng Xie, Philip E. Erickson, Richard F. Lowe, Thomas R. Webb, Dimitri E. Grigoriadis, Errol B. De Souza and Jeffrey P. Whitten. Neurocrine Biosciences, Inc. 3050 Science Park Road, San Diego CA 92024

Recently there has been a growing interest in the use of automated synthesis for the preparation of small organic molecules to facilitate the drug discovery process. Solid-phase synthesis is playing a major role in this process particularly for the preparation of large libraries for screening to obtain active molecules or "hits". A new solution phase robotics-driven synthesis method called rapid microscale synthesis (RMS) will be presented that compliments solid phase techniques. Examples will be presented of the utility of RMS to the medicinal chemist for the preparation of milligram quantities of several hundred individual analogs of a biologically active molecule. Analogs were prepared by the robot in a relatively short period of time and molecules with dramatically increased potency were obtained.

382. Diverse Scaffolds for Lead Generation. **Philip F. Hughes**, Sphinx Pharmaceuticals, A Division of Eli Lilly & Co., Durham, NC 27707.

To maximize the efficiency of high throughput screening for lead generation, it became necessary to provide a large number of diverse compounds. One approach used at Sphinx was to develop a general method for the solid phase synthesis of individual bis-amides derived from sequential acylation of suitably protected diamine scaffolds. Compound diversity arose from the collection of acylating agents (diversity elements) and a variety of diamine scaffolds. Scaffolds were chosen to display nitrogens with a diversity of geometric and spatial arrangements. Alternative chemistries for the coupling of diversity elements were also utilized. The methodology, which relies on minimal investment in hardware and automation was used to construct a library of over 30,000 compounds. The considerations for this approach, the methodology used in library construction, computer diversity analysis of diversity elements and examples of chemistry used will be discussed.

383. NEW LINKERS FOR GENERATING SUBSTITUTED AMIDES AND UREAS FROM SOLID SUPPORTS. **C.P. Holmes**, Affymax Research Institute, 4001 Miranda Ave., Palo Alto, CA 94304

We have been developing two new linkers which lead to substituted amides and ureas in order to extend beyond the carboxamides formed from commercially available amine linkers. Displacement of a benzylic bromide from a photolabile ortho-nitrobenzyl linker with primary amines leads to support-bound secondary amines which can be subsequently acylated with acids or isocyanates to give the amide or urea, -respectively. Mild photolysis liberates the compounds from the support suitable for direct biological assay. An alternative approach has also been developed in an acid cleavable format. Reductive amination of an aldehyde linker with primary amines affords the analogous secondary amine-supports, which are then acylated as before. Cleavage with TFA liberates the compounds from the support. Comparison of these two complementary linkers for use in combinatorial chemistry will be discussed.

384. HETEROCYCLIC POSITIONAL SCANNING COMBINATORIAL LIBRARIES. R. A. Houghten, Torrey Pines Institute for Molecular Studies & Houghten Pharmaceuticals, Inc., 3550 General Atomics Court, San Diego, CA 92121

The simultaneous multiple synthesis method (*PNAS* 82:5131, 1985) has been expanded to include micro-"T-bags". When combined with our original combinatorial library approaches (*Nature* 354:84, 1991), we are now able to synthesize libraries of thousands to millions of individual heterocycles in arrays or mixtures, respectively. Heterocyclic combinatorial libraries in positional scanning format (*Biotechniques* 13:901, 1992) have been prepared using the "libraries from libraries" approach (*PNAS* 91:11138, 1994). Upon screening these libraries, highly active individual heterocycles can be identified in as little as 24 hours. A range of heterocyclic (i.e., isoquinolines, quinazolinones, hydantoins, cyclic ureas, indoles, etc.), small molecule acyclic (i.e., acyl amines, guanidines, etc.), peptidomimetic (i.e., di-, tri- and tetra-) and peptide libraries have been prepared. Case studies illustrating the synthesis and identification of individual, highly active heterocyclic compounds having agonist or antagonist activities will be presented.

SOLID-PHASE SYNTHESIS OF P1 PEPTIDOMIMETIC LIBRARIES
DIRECTED TOWARD ASPARTYL PROTEASE INHIBITION. Roland E. Dolle,
Timothy Herpin, and Yvonne Shimshock; Department of Chemistry, Pharmacopeia Inc.,
101 College Road East, Princeton, NJ 08540

Aspartic acid proteases, including HIV protease, cathepsin D and plasmepsin, are a family of enzymes of current therapeutic interest. Previously reported library designs directed toward their inhibition have required the individual synthesis of orthogonal protected P1 scissile bond mimetics in solution followed by attachment to the solid support (Kick, E. K.; Ellman, J. A. *J. Med. Chem.* **1995** *38*, 1427; Wang, G. T.; et. al. *J. Med. Chem.* **1995** *38*, 2995). In our laboratories, we have taken a different approach in that chemistries have been discovered which allow the direct *combinatorial* synthesis of P1 mimetics on the solid phase. The synthetic details of encoded libraries produced in this fashion will be presented.

386. COMBINATORIAL CHEMISTRY AT CROSS ROADS. **Walter H. Moos**, Chiron Corporation, Technologies Division, 4560 Horton Street, Emeryville, CA 94608

Combinatorial chemistry and related approaches to generating and screening molecular diversity are modern day additions to the medicinal chemistry toolbox. Up to the 1960s, most drug discovery utilized natural products sources or one-at-a-time chemical synthesis to identify and optimize leads. Among the many advances over the last several decades, solid phase synthesis techniques, sensitive and high performance chromatographies and spectroscopies, automation and miniaturization of in vitro assays, and molecular and cellular biology have paved the way for new paradigms in pharmaceutical research. Combinatorial discovery is taxing preclinical development throughput, thus the application of combinatorial principles to later stage bottlenecks will become the focus of a number of laboratories. Early stories of the field's contributions to development pipelines are emerging. Combinatorial discovery is unquestionably faster and cheaper than older methods; better may take more time.

387. PRACTICAL CHALLENGES AND OPPORTUNITIES FOR CHEMOSENSOR DEVELOPMENT.
S. Bambot, R. Fraatz, H. He, M. Mortellaro and **James Tusa**, AVL Scientific Corporation, Roswell, Georgia 30076

Chemosensors and Biosensors have been utilized for many years in whole blood benchtop analyzers in the hospital lab setting and are just now appearing in single-use disposable sensors run on small "Point of Care" systems. The major commercial successes of such systems include glucose sensing, blood gas and blood electrolyte sensing. Desirable characteristics for sensors measuring these and other key analytes are described for bedside utility in the hospital critical care setting, with specific requirements for responsivity, specificity, and speed. Ultimately, the simplicity and stability afforded by such sensors, especially optically interrogated chemosensors, are the driving attributes for their commercial attractiveness and success. Properly configured, sensors with these attributes can achieve the low cost, manufacturability, reliability, and stable calibration necessary for product success.

388. MOLECULAR RECOGNITION OF ANALYTES IN CLINICAL CHEMISTRY ASSAYS.
B.P. Czech, Diagnostics Division, Bayer Corporation, Tarrytown, New York 10591

Most automated clinical diagnostic assays are based on spectrophotometric determinations. In the past 25 years many attempts have been made to apply molecular recognition by artificial receptors to the measurement of analytes in physiological fluids. Despite the multitude of "host" compounds synthesized to date, only a handful of successful commercial applications were realized. Spectrophotometric assays for the determination of sodium and potassium in blood, based on chromogenic cryptahemispherands, serve as distinct examples of such products. Also, work has been completed to introduce the first practical colorimetric reagent for measurement of lithium in blood. Two cryptand chromoionophores, which exhibit unprecedented selectivity for lithium, have been identified. Other efforts aimed at an improved calcium assay produced a series of potentially useful calcium chromoionophores. A collaborative effort led to the first chromogenic receptor with significant specificity toward creatinine, an important indicator of renal function. A summary of these efforts and insights into commercialization aspects will be discussed.

389. MOLECULAR CONTAINER COMPOUNDS.
Donald J. Cram, Department of Chemistry and Biochemistry, University of California, Los Angeles, CA 90095-1569

We report the design, synthesis and chemistry of host compounds having closed surfaces and hollow cavities large enough to form complexes by enclosing guest molecules as small as H_2O or as large as [3.3]paracyclophane or ferrocene. Guests in some cases are introduced into these hosts at high temperature to form, usually, one-to-one complexes stable at ambient temperature. In other cases complexes are formed by encapsulation of guests during shell closures. Many reactions have been carried out in the inner phase of these hosts, including thermal and photochemical rearrangements, alkylations, eliminations, additions, Diels-Alder reactions, acid-base reactions, oxidations and reductions. The shells of these hosts have been used to stabilize cyclobutadiene, benzyne and benzocyclopropenone. Complexes soluble in water are described.

390. ARTIFICIAL RECEPTORS USEFUL FOR AQUEOUS SUGAR SENSING
Seiji Shinkai, Department of Chemical Science & Technology, Faculty of Engineering, Kyushu University, Fukuoka 812, Japan

For the development of new receptor molecules that can precisely recognize sugar molecules, we synthesized a number of diboronic acids. Since one boronic acid can react with two OH groups (one diol group) to form a boronate ester, one diboronic acid can immobilize two diol units to form a sugar-containing macrocycle. The selectivity can be tuned by the relative spatial position of the two boronic acids and the complexation event can be read-out by CD spectroscopy. When a boronic acid group is combined intramolecularly with an aminomethylfluorophore, the complexation event can be conveniently read-out by fluorescence spectroscopy. This is a novel application of PET (photoinduced electron transfer) sensor: the sugar-binding changes the strength of the B⋯N interaction, which eventually changes the fluorescence quenching efficiency of the amine. We have demonstrated using a chiral 1,1'-binaphthyl group as a fluorophore that even the chiral recognition of sugars is possible. These abundant examples support the superiority of boronic-acid-based covalent-bond recognition over hydrogen-bond-based noncovalent-bond recognition for sugars in water.

391. ENHANCED SENSITIVITY IN SENSORY MATERIALS: CONDUCTING POLYMER-BASED POLY-RECEPTOR ASSEMBLIES. **Timothy M. Swager,** Department of Chemistry, Massachusetts Institute of Technology, Cambridge, Massachusetts 02139

We will present a number of approaches to the design and synthesis of conducting polymer-based sensory materials. Conducting polymers are ideal sensory materials since their conductivity and photophysical properties are very sensitive to chemical composition and electronic perturbations. We have synthesized polythiophenes which display ion specific changes in their band gap. Other approaches to ion sensory materials make use of the electronic perturbation induced by ion complexation. We have also developed polymers which contain electron rich macrocycles capable of hosting electron poor organic compounds. These charge transfer complexes are a direct result of the macrocyclic structure and are not observed in non-macrocyclic analogs. In one system we have demonstrated a novel chemoresistive effect in which the conductivity is lowered by paraquat. We have also demonstrated how a fluorescence-based chemosensory response can be enhanced (amplified) by migration of the photogenerated exciton to a complexed site.

392. **Efficient Domino-Type Transition Metal Catalyzed Synthesis of Natural Products**
Lutz F.Tietze, Thomas Nöbel, Maurus Spescha, Wilm Buhr, Hartmut Schirock.
Institut für Organische Chemie der Universität Göttingen, Tammannstr. 2, 37077 Göttingen, FRG

Domino reactions[1] are suitable for the construction of complex molecules using simple starting materials in a highly efficient way. In the lecture I shall present novel multiple transition metal catalyzed transformations for the synthesis of anticancer agents, antibiotics and steroids. As one example the synthesis of the estrone derivative 3 from 1 and 2 using two succesive Pd-catalyzed transformations will be discussed[2].

[1] a) L. F. Tietze, *Chem. Rev.* **1996**, 115. b) L. F. Tietze, U. Beifuss, *Angew. Chem. Int. Ed. Engl.* **1993**, *32*, 131. [2] L. F. Tietze, T. Nöbel, M. Spescha, *Angew. Chem. Int. Ed. Engl.* **1996**, *5*, 2259.

393. THE USE OF DESIGNED EXPERIMENTS FOR THE DEVELOPMENT OF ROBUST MANUFACTURING PROCESSES FOR β-KETOCARBOXAMIDES. R. R. Valente*, L. F. Valente, J. O. Rosser, D. R. Frank, Chemicals Development Division, Eastman Kodak Company, Rochester, New York 14652

α-Substituted-β-ketocarboxamides are widely used as yellow-dye-forming image couplers in photographic systems. The typical synthetic sequence is composed of an amidation of a β-ketoester followed by halogenation and displacement with an appropriate nucleophile. We plan to describe the use of a statistically designed set of experiments to efficiently optimize the halogenation and displacement sequence. Additionally, we will describe the sytematic approach taken to monitor and control the process to manufacture the α-unsubstitued precursor.

394. Chiral Dendrimers. Direct Glycosidation of Glycals to Give Linear and Dendritic Glycopolymers. Dotsevi Y. Sogah, Charles J. Brandenburg and Pawel Klosinski. Department of Chemistry, Baker Laboratory, Cornell University, Ithaca, NY 14853.

Carbohydrates constitute one of the most structurally diverse group of naturally occurring polymers. They have been implicated in many biological processes such as cell surface recognition. As materials they have been used in food additives, adhesives, textiles, detergents, cosmetics and pharmaceuticals, and enhanced oil recovery. This presentation will deal with direct glycosidation of glycals under catalysis by Ph3P.HBr or ZrCl4 as a route to materials of well-controlled architectures. The method is readily adaptable to the synthesis of both linear and dendritic structures. Thus, both divergent and convergent (with modifications) approaches to chiral dendrimers have been demonstrated and will be discussed. The presentation will concentrate on methodology rather than properties of the materials.

395. A TOOLBOX FOR THE DESIGN OF ORGANIC SYNTHESIS, J. Gasteiger, W.D. Ihlenfeldt, R. Fick, M. Pfoertner, Computer-Chemie-Centrum, Universität Erlangen-Nürnberg, Nägelsbachstr. 25, D-91052 Erlangen, Germany

The design of the synthesis of an organic compound by a chemist is not a straight-forward process. Many different types of considerations and searches are used in an interwoven manner. A variety of search methods incorporated into the WODCA system have been designed to support this multi-faceted approach.[1] Emphasis will be put onto the methods directed to find appropriate starting materials and those to define strategic bonds in the target molecule. Examples of the application of the WODCA system will be given.

[1] W.D. Ihlenfeldt, J. Gasteiger, Angew. Chem. Int. Ed. Engl. $\underline{34}$, 2613 (1995)

396. CARBON—CARBON BOND FORMATION—OUTDATED GRAIL OF ORGANIC SYNTHESIS? **K. Barry Sharpless**, Department of Chemistry, The Scripps Research Institute, La Jolla, California 92037

Organic chemists should consider synthetic strategies that don't depend so heavily on C—C bonds: quicker and more modular routes exist to structurally and chemically diverse molecules, for example interconnecting islands of small molecules with heteroatoms bridges.

397. SYNTHESIS OF MOTUPORIN AND RELATED PEPTIDES. **Peter L. Toogood**, Hong Yong Kim and Raghu Samy, Department of Chemistry, University of Michigan, Ann Arbor, MI 48109-1055.

Motuporin is a naturally occurring cyclic peptide that inhibits protein phosphatase-1 with $IC_{50} < 1$ nM. It is potentially a useful tool for studying the role of phosphatases in cells. To provide sufficient material for structure-function and biochemical studies a highly convergent synthesis of motuporin was designed that should provide access to analogs of this peptide and related natural products. The execution of this synthetic plan will be described, including new chemistry for preparation of the characteristic non-proteinogenic amino acid Adda.

398. SYNTHETIC STUDIES ON A CYCLIC PEPTIDE MODEL OF CFG RINGS OF TEICOPLANIN. Anthony J. Pearson, **Penglie Zhang**. Department of Chemistry, Case Western Reserve University, Cleveland, Ohio 44106

Teicoplanin is a heptapeptide antibiotic that belongs to the vancomycin family. Synthetic studies on a 14-membered CFG rings model will be presented. Two approaches are described. The first one involves a ruthenium-mediated cycloetherification. The second approach calls for ruthenium-assisted diaryl ether bond formation followed by cycloamidation.

399. **SYNTHESIS OF THE TRIPEPTIDE MOIETY AND C(1)-C(5) AND C(6)-C(11) SUBUNITS OF (-)-ANTILLATOXIN**
Duncan Wardrop and James D. White,* Department of Chemistry, Oregon State University, Corvallis, Oregon, 97331.

Antillatoxin (1) is a highly ichthyotoxic cyclic lipopeptide recently isolated from the marine cyanobacterium *Lyngbya majuscula*. The tripeptide **2** has been prepared from the constituent amino acids. The synthesis of **3** from enyne **4** and (*R*)-3-hydroxy-2-methyl-propionate (**5**) will also be discussed.

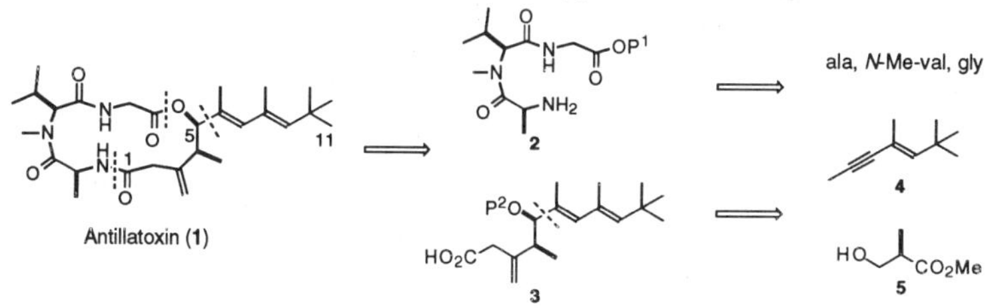

400. **ASYMMETRIC HORNER-WADSWORTH-EMMONS REACTIONS WITH ALDEHYDE SUBSTRATES: SCOPE AND SYNTHETIC APPLICATIONS. T. Rein**, Department of Organic Chemistry, Technical University of Denmark, DK 2800 Lyngby, Denmark

Results from reactions between *meso*-dialdehydes (e.g., **2**) and chiral phosphonate reagents such as **1** will be described. Possibilities for using **3**, and products from similar asymmetric Horner-Wadsworth-Emmons reactions, as building blocks in natural product synthesis (e.g., in an approach to the cytotoxic macrolide dictyostatin 1) will also be discussed.

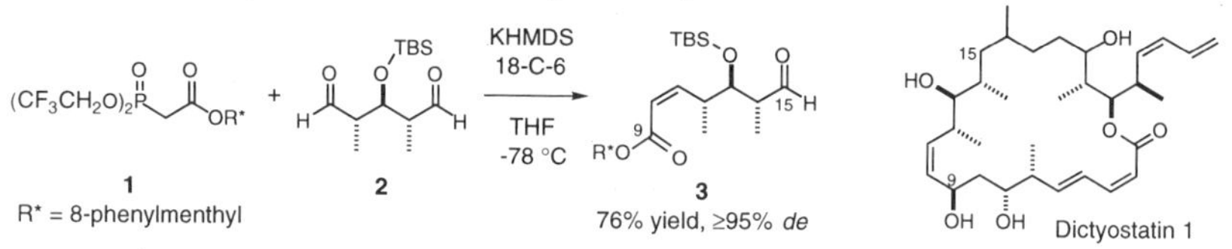

401. ASYMMETRIC HORNER-WADSWORTH-EMMONS CONDENSATIONS OF *MESO*-DIALDEHYDES: METHODOLOGY DEVELOPMENT AND SYNTHETIC APPLICATIONS **J. S. Tullis**, P. Helquist, T. Rein, Department of Chemistry and Biochemistry, University of Notre Dame, Indiana 46556 and Department of Organic Chemistry, Technical University of Denmark, DK-2800 Lyngby, Denmark

A specific focus of our efforts is the development of asymmetric Horner-Wadsworth-Emmons (HWE) reactions and the application of these methods in the synthesis of new cytotoxic agents of potential use as anticancer agents. The underlying principle is the ability of the chiral phosphonate reagent to distinguish between enantiotopic aldehyde units in the *meso*-dialdehyde. As shown in the figure below, alkene geometry and enantioselectivity are controlled by choice of R_1 and R^*, respectfully.

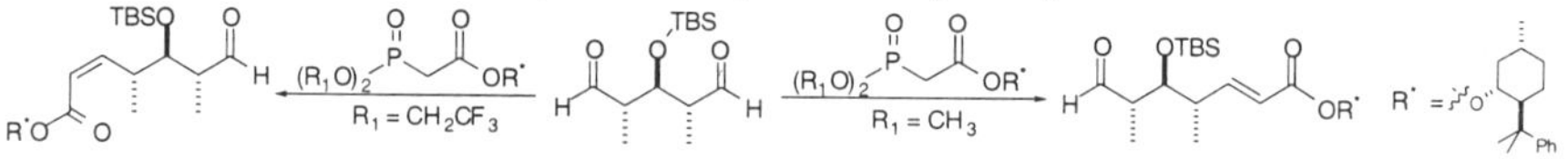

Asymmetric HWE condensations of *meso*-dialdehydes very directly generate chiral moieties that are seen in a number of cytotoxic natural products currently under investigation in our laboratory, such as superstolide A, doliculide, the sphinxolides and the iejimalides.

402.

SYNTHESIS OF THREE SUBUNITS OF THE MACROLIDE RHIZOXIN. **Christine S. Nylund**, Neal J. Green, Mark Holoboski and James D. White,* Department of Chemistry, Oregon State University, Corvallis, Oregon 97331.

Rhizoxin (**1**) is a 16-membered macrolactone that possesses antibacterial, antifungal and antitumor activity. A convergent route toward this macrolide will be described that features syntheses of pyran **2** via a stereocontrolled radical cyclization, mesitylate **3** from a chirality-transfer sequence, and vinylstannane **4** from methyl 2-methyloxazole-4-carboxylate.

403. SYNTHESIS AND BIOLOGICAL EVALUATION OF BACKBONE ANALOGS OF AMPHOTERICIN B: Bruce N. Rogers and Scott. D. Rychnovsky, University of California, Irvine, CA 92697-2025

Polyene macrolide antibiotics such as amphotericin B are an important class of biologically active natural products currently used in the treatment of serious systemic fungal infections. The synthesis of the first polyene macrolide antibiotic containing a rigid non-polyene backbone has been accomplished. The sterol recognition surface of amphotericin B has been modified in an effort to better understand the role of the polyene backbone. Degradation, synthesis, and *in vitro* activity will be discussed.

404. SYNTHETIC STUDIES ON THIOSTREPTON ANTIBIOTICS. Yongchun Shen, Marco A. Ciufolini, Department of Chemistry, Rice University, Houston, TX 77251

Thiostrepton antibiotics are potent antitumor agents that appear to express their activity through strong inhibitors of protein synthesis. More recently, they have been found to induce expression of various genes of unknown function in Streptomyces species. Micrococcin P1 is a member of this family of antibiotics. Progress toward the total synthesis of this complex natural product will be discussed.

405. SYNTHETIC STUDIES TOWARDS THE PHORBOXAZOLE NATURAL PRODUCTS. Craig J. Forsyth,* Feryan Ahmed, Russell D. Cink, Arundhati Deo, and Chi Sing Lee. Department of Chemistry, University of Minnesota, Minneapolis, MN 55455.

Phorboxazoles A and B are phenomenally potent cytostatic agents with an, as yet, undefined mode of action. We have initiated a convergent approach to the total synthesis of the phorboxazoles that features the independent assembly of 3 fragments, representing carbons C1-C17, C18-C30 [*Tetrahedron Lett.* **1996**, *37*, 6449.] and C31-C46 of the natural product. The stereoselective syntheses of each fragment, as well as progress towards their convergent couplings will be described.

406. **A SYNTHETIC APPROACH TO THE POTENT ANTITUMOR MACROLIDE CRYPTOPHYCIN 1**

Kevin M. Gardinier and James W. Leahy*
Department of Chemistry, University of California, Berkeley, CA 94720-1460

Cryptophycin 1 has received much attention recently due to its impressive antimitotic activity, both in vivo and in vitro, and unusually high potency against multi-drug-resistant tumors. We present here our efforts to a convergent and stereospecific total synthesis of the title compound and analogs.

Cryptophycin 1

407. **TOTAL SYNTHESIS OF (-)-MACROLACTIN A**

Amos B. Smith, III* and Gregory R. Ott

Department of Chemistry, Laboratory for Research on the Structure of Matter, and Monell Chemical Senses Center, University of Pennsylvania, Philadelphia, PA 19104, U.S.A.

The total synthesis of macrolactin A, an effective anti-HIV-1 agent in vitro, will be presented. Our successful approach is both concise and versatile, exploiting Stille cross-couplings for stereospecific construction of the three diene moieties and closure of the 24-membered macrocyclic ring.

MACROLACTIN A (1)

A, X = β-H, α-OH, R = H
B, X = β-H, α-OH, R = β-glucosyl
C, X = β-H, α-O-β-glucosyl, R = H
D, X = β-H, α-OH, R = R(D)
E, X = O, R = H
F, X = O, R = H, 16, 17 dihydro

$R(D) = $ —$O_2C(CH_2)_2CO_2H$

408. **STRUCTURE-ACTIVITY RELATIONSHIPS OF THE DIDEMNINS.** Alexandra J. Katauskas and Kenneth L. Rinehart, Roger Adams Laboratory, University of Illinois, Urbana, Illinois 61801

The didemnin class of biologically active cyclodepsipeptides, isolated from a marine tunicate, has shown considerable antitumor, antiviral, and immunosuppressive activities. Didemnin B and most other natural didemnins contain a common macrocycle and differ only in the composition of the side chain. Structural modifications were introduced in the side chain to afford several didemnin analogues for structure-activity relationship studies. We also present synthetic studies directed toward a modified macrocycle which possesses an amide bond in place of an ester bond. The instability of this C-O bond leads us to believe its replacement by a C-N bond should improve the stability of the compound, and thus, provide a more active analogue. Of the compounds synthesized, glutaminyldidemnin B is the most active against the L1210 (murine leukemia) cell line, at the 0.1 ng/mL level. The L1210 activity of the benzyloxycarbonyl (Cbz) derivative of the amino hip analog is significantly greater than that of Cbz-Didemnin A and compares favorably to didemnin B.

409. ORGANOPALLADIUM(III) CHEMISTRY? : ELECTROCHEMICAL STUDIES.
A. J. Canty,[a] A. M. Bond,[b] J. B. Cooper,[c] V. Tedesco,[b] P. R. Traill,[a] D. M. Way,[c] Chemistry Depts, University of Tasmania,[a] Monash[b] and La Trobe[c] Universities, Australia

X-ray studies of dimers such as **1** (L = γ-picoline) show that the palladium square planes are adjacent and ideally suited to potential formation of Pd(III)Pd(III) dimers with Pd-Pd bonds. Cyclic voltammetry of **1** and relatives show chemically reversible 1e⁻ oxidation, and a second irreversible 1e⁻ oxidation at more positive potentials, consistent with oxidation at the metal centres although oxidation of the organic ligand cannot be discounted. In the presence of additional ligands, the first oxidation becomes a 2e⁻ process, perhaps leading to a Pd(III)Pd(III) dimer with a Pd-Pd bond where the ligand, solvent or electrolyte occupies axial coordination sites, e.g. a neutral ligand L' in **2**.

410. **HYDROGEN-BONDING IN ORGANOPALLADIUM(IV) COMPLEXES: Pd(OH)...HOAr AND Pd(OH$_2$)$^+$...$^-$OAr INTERACTIONS. A. J. Canty**, H. Jin, B.W. Skelton and A. H. White, Chemistry Departments, University of Tasmania, Hobart, and University of Western Australia, Perth, Australia

The interaction between coordinated ligands and solvent molecules is an important phenomenon in organometallic chemistry, and there have been extensive reports on the interaction between coordinated alkoxide and alcohols in organopalladium(II) chemistry, Pd(OR)...HOR. In the relatively new field of organopalladium(IV) chemistry such interactions have yet to be explored, but we have found that stable pallada(IV)cyclopentane complexes are accessible which have coordinated hydroxo and aqua ligands indicating that an aqueous solution chemistry for palladium(IV) is feasible. These complexes form interesting hydrogen bonding networks in the solid state, involving Pd(OH)...HOAr and Pd(OH$_2$)$^+$...$^-$OAr interactions in complexes formulated as Pd(CH$_2$)$_4$(OH){(pz)$_3$BH}.2HOAr and [Pd(CH$_2$)$_4$(OH$_2$)(pz)$_3$BH}. OC$_6$F$_5$]$_2$.

411. **SYNTHESIS OF 2-ARYLTRYPTAMINES WITH PALLADIUM CATALYZED CROSS-COUPLING OF 2-BROMOTRYPTAMINES AND ARYLBORONIC ACIDS. Lin Chu**, Michael H. Fisher, Mark T. Goulet, and Matthew J. Wyvratt, Department of Medicinal Chemistry, Merck Research Laboratories, PO Box 2000, Rahway NJ, 07065

The 2-aryltryptamine (5) moiety is present in many natural products and medicinal chemistry targets with various biological activities. Our interest in this class of compounds led to the investigation of a versatile process to introduce a wide range of aryl substituents onto the 2-position of the indole, and which requires no protection of the indole nitrogen. We wish to report here a high-yielding aryl-indole coupling methodology, which employs a modification of the Suzuki reaction with Pd(0) as the catalyst and bromide **2** as the key intermediate.

412. DATABASE RELATED TO PALLADIUM-CATALYZED ORGANIC CHEMISTRY
J-L. MALLERON, A. JUIN, L. BAUDRY, P. DESMAZEAU, C. GIRARDET, C. M'HOUMADI, T-P-H. NGUYEN, Rhône-Poulenc Rorer S.A. Central Research, Centre de Recherche de Vitry-Alfortville 13 Quai Jules Guesde, B.P. 14, 94403, Vitry-sur Seine Cedex, France

Creation of a database on organic chemistry catalyzed by palladium salts. Most of the mechanisms are identified and presented (at present, about 87 types of mechanisms are described). They are described with respect to data of the literature. The various chemical parameters of these reactions are listed. References are chosen depending on the nature of the mechanism and on the nature of the chemical parameters. ISIS software from Molecular Design Limited (MDL) and EXCEL software (Microsoft) are used. The general mechanism as the catalytic cycle and the tables of the chemicals parameters can be obtained for each reference related to the REACTION NAME or the REACTION NUMBER. Searching can be performed by substructure, by mechanism, for asymmetric catalysis, for carbonylation reactions (under atmospheric pressure or not), by ligand, catalyst, solvent, author, year ... so far 4200 reactions have been inputted. This database and a handbook related to palladium chemistry have been published in November 1996 by Academic Press (database is updated twice per year).

413. **MECHANISTIC STUDIES OF THE PALLADIUM-CATALYZED ISOMERIZATION OF EPOXIDES TO CARBONYL COMPOUNDS. Sanjitha Kulasegaram** and Robert J. Kulawiec[*], Department of Chemistry, Georgetown University, Washington, DC 20057-1227.

The isomerization of epoxides to carbonyl compounds occurs in the presence of the zerovalent Pd catalysts Pd(OAc)$_2$/ PBu$_3$, Pd(OAc)$_2$/ PPh$_3$ or Pd(PPh$_3$)$_4$. This isomerization is chemo- and regioselective for aliphatic and aryl-substituted epoxides, giving aldehydes or ketones in almost quantitative yields. We have previously suggested a mechanism involving a hydrido-palladium enolate species which is formed after oxidative addition and β-hydride elimination; subsequent reductive elimination forms the carbonyl compound. The results of isotopic labeling, solvent variation and kinetic studies will be presented, and their implications for the mechanism and Pd-catalyzed epoxide isomerization will be discussed.

$$Ar\text{-epoxide-}R \xrightarrow[t\text{-BuOH, reflux}]{Pd(OAc)_2/ PBu_3} Ar\text{-}CH_2\text{-}C(O)\text{-}R$$

Ar = 2-naphthyl; R = $C_8H_{16}COCH_3$, $C_9H_{18}CN$, $C_8H_{16}CO_2Et$, $C_9H_{18}OH$

414.

THE APPLICATION OF NOVEL LIGANDS IN A PALLADIUM CATALYSED ASYMMETRIC ALLYLIC SUBSTITUTION REACTION. **Simon J. Sesay,** Jonathan M. J. Williams Loughborough University, Loughborough, Leicestershire, LE 11 3TU. U. K.

The use of homochiral phenyloxazolines as ligands in metal catalysed reactions has received a great deal of attention[1]. We are currently designing homochiral imine ligands that contain a similar backbone to the homochiral 2-diphenylphosphinophenyloxazoline.
The poster will describe the synthesis and application of these ligands in palladium catalysed allylic substitution reactions.The initial levels of enantioselectivity of the substituted product are encouraging (50- 76% e.e.) , at present we are optimising the reaction conditions and ligand design to increase selectivities still further.

Ref. 1 C.F. Frost, J. Howarth, J.M.J. Williams; Tetrahedron Asymmetry, 1992, 3, 1089-1122

415.

HOMOCHIRAL LIGANDS IN METAL CATALYSED ASYMMETRIC SYNTHESIS. **L.M.Newman** Loughborough University, Loughborough, Leicestershire. LE11 3TU; J.M.J. Williams, Bath University; R. McCague, Chiroscience Ltd, Cambridge

The use of homochiral acetal and oxazoline ligands in asymmetric synthesis have shown to be of great synthetic importance. We would like to report the synthesis and application of some novel acetal and oxazoline ligands. We are currently studying the use of acetal based ligands in organometallic addition to aldehydes and rhodium (I) catalysed hydrosilylation of ketones. These reactions at present have shown only modest enantiomeric excesses <60%. More successfully however have been the rhodium (I) catalysed hydrosilylation of ketones and palladium (0) catalysed allylic substitution reactions using homochiral oxazoline ligands. These reactions have shown excellent enantiomeric excesses of <86% and <97% respectively. Optimisation of these results has yet to be completed, hence an improvement of enantioselectivity is expected.

416. ASYMMETRIC ALKYLATION OF ALLYLIC *GEM*-DICARBOXYLATES.
Barry M. Trost* and Chul Bom Lee, Department of Chemistry, Stanford University, Stanford, CA, 94305-5080

Enantioselective alkylations of allylic *gem*-dicarboxylates with soft nucleophiles have been investigated. The complex of palladium and the Trost ligand **A** efficiently catalyzes the asymmetric process providing allylic carboxylates as the major or only product in high *ee*. The product obtained, being a substrate for palladium catalysis, has been successfully employed for second allylic alkylations with complete transfer of chirality. This palladium catalyzed alkylation reaction constitutes the equivalent of a stabilized nucleophile addition to a carbonyl group with high asymmetric induction.

85-95% ee

A

417. NEW CHIRAL AMINE-IMINE LIGANDS CONTAINING *TRANS*-2,5-DISUBSTITUTED PYRROLIDINES James A. Sweet, Jennifer M. Cavallari, John Baluyut, and Dominic V. McGrath* Department of Chemistry, University of Connecticut, Storrs, CT 06269-4060

The C_2-symmetric *trans*-2,5-disubstituted pyrrolidine moiety (**1**) has seen wide use as a chiral auxiliary for asymmetric synthesis. Despite this, **1** has scarcely been used in ligands for transition metal-catalyzed asymmetric synthesis. We present the modular preparation of a new class of chiral *N,N*-chelating ligands which contain **1** linked to a pyridine ring. The use of these ligands in palladium-catalyzed allylic alkylations will be presented.

418. PALLADIUM-MEDIATED SUBSTITUTION OF CHLORINE IN 2-CHLOROQUINOLINES
James W. Mitchell, Frank Roschangar, Marco A. Ciufolini
Department of Chemistry, Rice University, Houston, TX. 77251

We found that unlike ordinary aryl chlorides, 2-chloroquinolines readily participate in Castro-Stevens, Stille, Suzuki, & carbonylative reactions under the catalytic influence of zerovalent palladium complexes. These processes furnish diverse quinoline derivatives that are otherwise difficult to obtain through classical syntheses (Friedländer, etc.). The starting chloroquinolines are especially readily available through the Meth-Cohn reaction, which is particularly suitable for the creation of 2-substituted quinolines. As a result, facile access to the 2,3-disubstitution pattern in these heterocycles becomes possible. Quantities of various synthetic intermediates for bioactive natural products have been prepared in a concise and efficient fashion thanks to the new chemistry.

419. REACTIONS OF TRIALKYLSILYL DIENOLETHERS WITH PALLADIUM(II) SALTS: FORMATION OF NOVEL η^3-ALLYL PALLADIUM COMPLEXES AND 4-ACYLOXY SUBSTITUTED ALKENALS. Björn C. Söderberg,* Angela K. Berry, and Paula. C. Jones, Department of Chemistry, West Virginia University, Morgantown, WV 26506-6045.

Reaction of 1-trimethylsilyloxy-1,3-butedienes with bis(acetonitrile)palladium dichloride in acetonitrile affords η^3-allylpalladium complexes having a formyl group attached to one of he terminal carbons of the allylic system. In contrast, using palladium diacetate as the palladium(II) source gives *trans*-4-acetoxy substituted 2-butenals. Esters of other acids can be obtained by addition of an excess of the appropriate acid salt to the reaction. The isolated η^3-allylpalladium complexes can be transformed into 4,5-dihydrofuranes by reaction with methyl acetoacetate in the presence of triethylamine.

420. SYNTHESIS AND CHARACTERIZATION OF ANNULENE SEMIQUINONE LIGANDS
David. A. Shultz* and Martha G. Hollomon
Department of Chemistry, **North Carolina State University,** Raleigh, NC 27695-8204

We have explored the Siegrist reaction as a method for the preparation of novel annulene semiquinone ligands. These ligands might provide a basis for novel extended solids. We also have evidence of annulene formation during Pd-catalyzed Heck polymerization of a bromo-styrene derivative. This paper will describe the synthesis and characterization of these liqands.

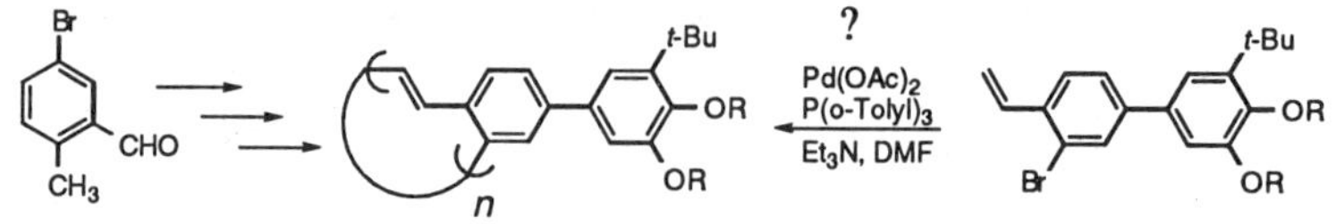

421.

PREPARATION OF CYCLOPHANES VIA A PALLADIUM CATALYZED COUPLING REACTION.
William R. Kwochka, Beverly B. Smith, Department of Chemistry and Physics, Western Carolina University, Cullowhee, NC 28723.

The examination of simple cyclophane structures has contributed greatly to areas of study as diverse as molecular and cationic recognition to fundamental principles of spectroscopy. Palladium(0) catalyzed coupling reactions have long been known to be extremely mild and versatile methods of constructing carbon-carbon bonds. We now wish to report the palladium catalyzed coupling of bis-trialkylboranes with meta-dibromobenzene (the Suzuki coupling) to form several carbon-carbon bonds in a single reaction vessel selectively providing either [n]- or [n.n]metacyclophanes. [n.n]Paracyclophanes can be prepared in this manner as well.

422.

DESIGN AND SYNTHESIS OF SUBSTITUTED TRIS-BIPYRIDYL RUTHENIUM COMPLEXES.
Lynne A. Miller and Robert P. Dixon, Department of Chemistry, Southern Illinois University, Edwardsville, IL 62026-1652

Previous research has shown that tris-bipyridyl ruthenium (II) and rhodium (II) complexes have the potential to bind and cleave DNA with isoform specificity (A, B and Z forms). Our research proposes that substituted tris-bipyridyl complexes (shown below) containing hydrogen bond donating and accepting functionalities (as well as moieties capable of electrostatic interaction) will bind and cleave DNA not only with isoform specificity, but also with sequence specificity. In addition, these tris-bipyridyl complexes will be tested for their ability to detect and destroy various insecticides and chemical warfare agents in the phosphate and phosphonate families.

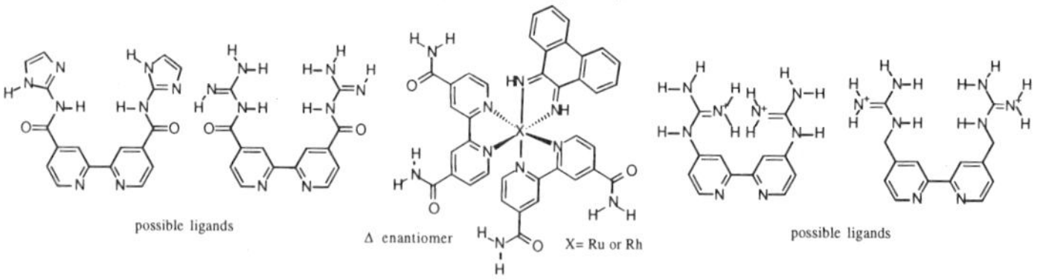

423. DENDRITIC-FERROCENE DERIVATIVES. C. Cardona, A. E. Kaifer, Department of Chemistry, University of Miami, Coral Gables, Florida, 33124

Novel dendritic-ferrocene derivatives were synthesized and their electrochemical behavior was analyzed in aqueous media. These globular molecules are characterized by containing a redox active core with a surrounding structurally rigid and sterically compact dendritic matrix and the radial distance between the ferrocene residue and the surface of the molecule can be controlled by adding subsequent generations using Newkome's polypeptide building blocks.

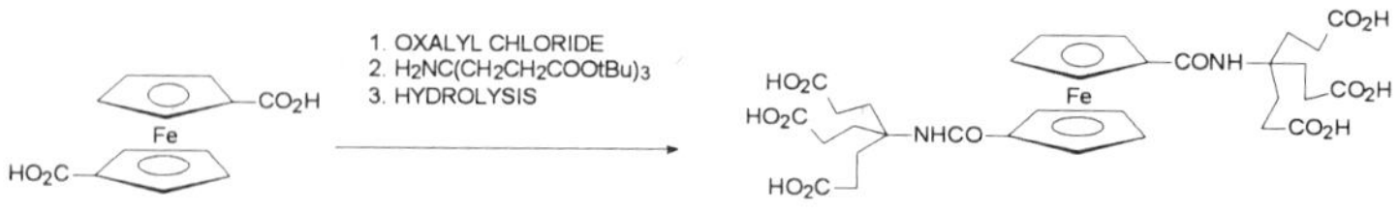

Previous studies done on similar molecules have shown that the electrochemical behavior becomes irreversible as the electroactive core becomes deeply buried in the dendritic envelope. Our aim is to elucidate quantitative information on the distance dependence in the electron transfer process of this type of molecules, which potentially can model redox active proteins.

424.
A NEW SYNTHESIS OF CHIRAL ANSA-TITANOCENES FROM 2-(HYDROXYMETHYL)INDENE.
Hasan Palandoken, Shawn R. Hitchcock, Michael, H. Nantz*, Department of Chemistry, University of California, Davis, CA, 95616

ansa-Titanocenes that contain a 1,2-bis(2-indenyl)ethane framework have shown promise as epoxidation catalysts. We have developed a new synthesis of these complexes from 2-(hydroxymethyl)indene. A titanium-mediated reductive coupling of allylic alcohols constitutes the key step, and provides the ansa-ligands in improved yields relative to existing methods. Yield optimization and synthesis details will be presented.

425. ALLENATION OF CARBONYL COMPOUNDS WITH ALKENYL TITANOCENES
Nicos A. Petasis,* and <u>Yong-Han Hu</u>, *Department of Chemistry and Loker Hydrocarbon Research Institute, University of Southern California, Los Angeles, CA 90089-1661*

Alkenyl titanocene derivatives (**2**) prepared in situ from alkenyl magnesium bromide and titanocene dichloride (**1**), react with aldehydes and ketones to give the corresponding allene derivatives (**4**) in good yields. Presumably, this carbonyl allenation involves a titanocene alkenylidene intermediate (**3**), which could be trapped with bis(trimethylsilyl)acetylene, forming the corresponding titanacyclobutene adduct.

R= Me, CH$_2$SiMe$_3$, alkenyl

426. RADICAL ADDITIONS TO CHIRAL Co-COMPLEXED CONJUGATED ENYNES.
Sarkhadoun Yadegar, Todd Monroe, and Gagik G. Melikyan,* Department of Chemistry, California State University, Northridge, CA 91330

Diastereoselectivity of Mn(III)-induced radical additions to chiral racemic Co-complexed enynes **1** has been examined. Results on mechanistic studies and the effect of bulkiness of the axial ligand L and substituent R upon the ratio of stereoisomers **2** and **3** will be presented.

427. **RHENIUM-CATALYZED OXIDATION OF THIOLS AND DISULFIDES WITH SULFOXIDES.** **Jeffrey B. Arterburn**, Sherry L. Nelson, Marc C. Perry, Mylena Holguin. Dept. of Chemistry and Biochemistry, New Mexico State University, Las Cruces, New Mexico 88003

The oxidation chemistry of organosulfur compounds is of tremendous biological and industrial importance. Higher oxidation products of disulfides, thiosulfinates and thiosulfonates, have been identified in natural products with antibiotic properties and DNA-cleaving ability, and simple derivatives are also useful reagents. The oxidation of disulfides provides a direct route to these compounds, but their sensitivity requires mild, selective methods. While most transition metal-catalyzed oxidations use peroxides, iodoso benzene, or N-oxides as oxidants, we have investigated oxygen atom transfer reactions from sulfoxides with the rhenium(V) oxo complex $ReOCl_3(PPh_3)_2$ (**I**). Complex (**I**) catalyzes two important reactions with sulfoxides in this context: 1) the oxidation of thiols to disulfides with methyl sulfoxide; and 2) the oxidation of disulfides to thiosulfinates and thiosulfonates with phenyl sulfoxide. This chemistry is compatible with many organic functional groups, does not require the exclusion of air or moisture, and provides cyclic and acyclic disulfides, thiosulfinates and thiosulfonates. The structural diversity of compounds accessible with this chemistry, and our mechanistic hypothesis will be the subjects of this presentation.

428. PREPARATION OF 1,4-DIOXENES FROM α-DIAZO-β-KETOESTERS
John B. Brogan, Ramona Hilgenkamp, and <u>Charles K. Zercher</u>*, Department of Chemistry, University of New Hampshire, Durham, NH 03824

Both intramolecular and intermolecular methods for the formation of dioxenes have been developed. Both of these methods involve the catalytic decomposition of α-diazo-β-ketoesters in the presence of diols or diol equivalents. Formal OH-insertion followed by dehydration generates the 1,4-dioxenes. A wide variety of α-diazo-β-ketoesters and diols have been used for this study. The general nature and the mechanism of the two complementary methods will be discussed.

429. **REDIRECTED DIRECTED METALATION XIX. THE METHYLANISOLES, IN PARTICULAR o-METHYLANISOLE.** <u>D. W. Slocum</u>, C. Friesen, D. Smith, Department of Chemistry, Western Kentucky University, Bowling Green, KY 42101.

Directed-ortho metalation (DoM) is a very useful alternative to electrophillic aromatic substitution (EAS) for the synthesis of substituted aromatic compounds. DoM is highly regiospecific providing metalation in the ortho-position. However, lateral (α-) metalation of an alkyl side-chain and ortho-metalation of a second ortho-position can compete when such opportunities are available. Investigation of the three methylanisoles was therefore undertaken to determine the interplay between these factors and to discover metalation conditions to achieve selectivity. This goal was uniquely achieved for p-methylanisole where conditions were discovered which allowed >85% ortho-metalation as determined by GC analysis with virtually no competing lateral metalation.[1] Metalation conditions were found that permitted significant discrimination between 2- and 6-position metalation of m-methylanisole. Most intriguing, a regiospecific α-metalation of o-MA has been discovered in direct contradiction to literature assessments. Evidence for the generation of a unique α-, o-dilithio intermediate for o-MA will also be presented. [1. D. W. Slocum, D. Reed, F. Jackson, III and C. Friesen, JOM, 512, 265 (1996).]

430. REDIRECTED DIRECTED METALATION XX. THE DIMETHOXYBENZENES, IN PARTICULAR o-DIMETHOXYBENZENE. <u>D. W. Slocum</u>, J. Ray and P. Shelton, Department of Chemistry, Western Kentucky University, Bowling Green, KY 42101

As an alternative to electrophillic aromatic substitution for the preparation of substituted aromatic compounds, directed ortho-metalation (DoM) has no peer. DoM is highly regiospecific directing metalation exclusively to the ortho-position. However, when two (or more) directing groups are available, competition between mono- and dilithiation can take place which often produces mixtures of the mono- and disubstituted products. This situation is examined for p-, m-, and o-dimethoxybenzene (DMB). Each presents a unique confrontation with the specific generation of the mono- or dilithio intermediate. Use of our metalation methodologies of catalytic TMEDA and THF in hydrocarbon solvents has provided insight and advances in our ability to affect these metalations. Specific monometalation of o-DMB in >85% GC yield can be accomplished using a combination of THF/cyclohexane. A unique observation is that m-DMB undergoes 50% mono metalation in neat n-hexane and p-DMB about 25% metalation whereas o-DMB and n-BuLi produces an immediate precipitate. Progress towards efficient generation of a dilithio intermediate in each of these systems will be discussed.

431. GLYCOLATE CLAISEN-OLEFIN METATHESIS FOR THE SYNTHESIS OF FUNCTIONALIZED DIHYDROPYRANS
Raymond A. Ng, James A. Morrison, and Steven D. Burke*, *Department of Chemistry, University of Wisconsin, Madison, WI 53706.*

The development of a glycolate Claisen/ring-closing metathesis sequence for accessing functionalized dihydropyrans is described. Combination of a Claisen rearrangement to set the desired stereocenters and the ring-closing metathesis as a highly efficient C-C bond forming reaction offers a convergent route to functionalized dihydropyrans.

432. TUNGSTEN CARBONYL-INDUCED ALKYNOL CYCLIZATIONS IN CARBOHYDRATE SYNTHESIS. **Hugh Y.H. Zhu**, Frank E. McDonald*, Department of Chemistry, Northwestern University, Evanston, IL 60208

This presentation will describe applications of tungsten carbonyl-induced alkynol cyclizations to carbohydrate synthesis. The six-membered ring oxacarbenes **2** are versatile intermediates for preparing glycals **3** and thiopyranosides **4**, thus providing a novel strategy for the synthesis of monosaccharides and oligosaccharides from acyclic starting materials.

433. RAPID ACCESS TO [4.3.1]PROPELLANES VIA MOLYBDENUM FISCHER CARBENE COMPLEXES. Daniel F. Harvey* and Thomas M. Bertolini. Department of Chemistry and Biochemistry - 0358, University of California, San Diego, La Jolla, CA 92093-0358.

Several molybdenum Fischer carbene complexes were employed in the cyclization of dimethyl 6-methylene-7-octen-1-yne-4,4-dicarboxylate to give 3-alkyl substituted tricyclo[4.3.1.0]decadienes in good yields. Pentacarbonyl(butylmethoxycarbene)molybdenum(0) afforded the highest yields of propellane (54%) while the analogous chromium carbene complex gave no reaction. The experimental details and mechanism of this reaction will be presented.

434. MODIFIED TITANIUM CATALYSTS FOR THE ENANTIOSELECTIVE RING OPENING REACTIONS OF MESO EPOXIDES. Alecia Eppley, Nancy I. Totah,* Department of Chemistry, University of Iowa, Iowa City, IA 52242

Chiral organotitanium complexes prepared from functionalized 1-7-dioxaspiro-[5.5]undecane ligands can be used to promote the enantioselective addition of trimethylsilyl azide to meso epoxides under mild conditions. These reactions proceed with modest to good levels of enantioselectivity. The influence of *in situ* modification of the chiral titanium complexes on reaction rate and enantioselectivity will be discussed.

435. Asymmetric Synthesis Catalyzed by Transition Metal Complexes with Chiral Tridentate Ligands. Xumu Zhang*, Yutong Jiang, Qiongzhong Jiang, Guoxin Zhu, James Longmire, Michael Terry, Betsy Pierce and Ted Webb, Department of Chemistry, The Pennsylvania State University, University Park, PA 16802

Families of chiral tridentate ligands are used to promote enantioselective reactions of unfunctionalized substrates (i.e., ketones, imines, olefins). Many asymmetric reactions are explored basing on these new ligand systems (e.g., **1, 2**). A general strategy based upon predictable steric and electronic effects of catalytic systems bearing these chiral ligands is discussed. Good to excellent enantioselectivities have been achieved in asymmetric hydride transfer, hydrosilylation and allylic alkylation reactions.

436. Asymmetric Catalysis Based on New Chiral Cyclic Phosphine Ligands. Xumu Zhang*, Guoxin Zhu, Ping Cao, and Qiongzhong Jiang Department of Chemistry, The Pennsylvania State University, University Park, PA 16802

A new chiral cyclic 1,4-bisphosphine, (2R, 2'R)-bis(diphenylphosphino)-(1R, 1'R)-dicyclopentane (abbreviated (R, R)-BICP) was synthesized. Rhodium complexes of this ligand are excellent catalysts for asymmetric hydrogenation of α-(acylamino)acrylic acids giving up to 99% ee. The key feature of this ligand is that it contains two cyclopentane rings in its backbone which can restrict its conformational flexibility leading to high enantioselectivity in asymmetric reactions. The enantioselectivies obtained with this ligand in asymmetric hydrogenation are comparable or higher than the results reported with other chiral bisphosphine ligands.

BICP

437. Asymmetric Catalysis Based on Novel Chiral Monophosphines Xumu Zhang*, Guoxin Zhu, Zhaogen Chen, Qiongzhong Jiang, Dengming Xiao, Ping Cao and Cheng Guo Department of Chemistry, The Pennsylvania State University, University Park, PA 16802

A new family of chiral monophosphines, 2,5-dialkyl-7-phenyl-7-phosphabicyclo[2.2.1]heptanes, were synthesized. The palladium complexes of these ligands are excellent catalysts for asymmetric allylic alkylation with up to 99 % ee being observed. These ligands contain a rigid fused bicyclic ring structure which limits their conformational flexibility and affords a well defined chiral environment around the palladium. Other enantioselective reactions such as [3 + 2] cycloaddition are also explored with these chiral monophosphines.

R = Me, ^{i}Pr, etc.

438. Asymmetric Catalysis Based on Chiral Phosphine Ligands with Aromatic Backbones
Xumu Zhang*, James Longmire, and Guoxin Zhu
Department of Chemistry, The Pennsylvania State University, University Park, PA 16802

A family of chiral phosphine ligands with aromatic backbones are used in transition metal catalyzed enantioselective reactions. The common feature of these ligands are that they contain ridig aromatic backbones which could be used to restrict conformational flexibility of the ligands. The bite angle of P-M-P is changed systematically depending on the group of aromatic backbones. A number of asymmetric reactions have been explored using these chiral ligands.

439. DEVELOPMENT OF A CHIRAL C_3 SYMMETRIC LIGAND SYSTEM FOR LANTHANIDE LEWIS ACID CATALYSIS. **Laurel L. Clouston*** Claude Spino**, David J. Berg*, * University of Victoria, Chemistry Department, PO Box 3065, Victoria, BC, Canada, V8W 3P6, ** Université de Sherbrooke, Département de Chimie, Sherbrooke, QC, Canada, J1K 2R1

Multidentate, fluorinated tris-chelate complexes of yttrium have been shown to be air-stable Lewis Acid Catalysts for the Inverse Demand Hetero Diels-Alder Reaction. Current developments focus on a modular approach for the synthesis of chiral C_3 ligand systems. Progress will be discussed in the case where phloroglucinol (1,3,5 - trishydroxybenzene) provides a benzene cap and ether links, while the chiral component is achieved through incorporation of camphor.

440. Asymmetric Cleavage of Achiral Acetals Mediated by Chiral Lewis Acids. **L. A. Newitt** and P. G.Steel*Department of Chemistry, University of Durham, South Road, Durham, UK, DH1 3LE

Recent years have seen intense interest in the use of acetals as substrates for C-C bond forming reactions. High stereocontrol in these reactions has been induced by the use of chiral acetals to direct Lewis acid complexation. Unfortunately the chiral information of the original acetal is destroyed when the newly created chiral centre is liberated. This poster will describe studies towards the enantioselective cleavage of achiral acetals mediated by chiral Lewis acids, as a means of overcoming this problem. Initial investigations of 2-substituted dioxane cleavage, with a number of Lewis acids complexed to C2 -symmetric chiral ligand, afforded only moderate success. The formation of an oxacarbenium ion intermediate was suggested to explain the absence of stereochemical control. This was verified in a mechanistic study examining the reduction of t-butylspirodioxane with silane-Lewis acid systems and comparison with the corresponding reactions of t-butylcyclohexanone. Current investigations are focused on the desymmetrisation of *meso* systems, utilising more reactive nucleophiles and stronger chiral Lewis acids.

441. ASYMMETRIC CATALYTIC HYDROBORATION REACTIONS. Mark J. Burk, **Charles A. Di Tusa**, William F. Kiesman, Department of Chemistry, Duke University, Durham, N.C. 27708

The application of homogeneous chiral catalysts to the asymmetric hydroboration of aryl olefins will be described. Cationic rhodium complexes containing chiral bis(phospholane) ligands, including 1,2-bis(phospholano)-

benzene (DuPHOS) Rh(I) and 1,2-bis(phospholano)ethane (BPE) Rh(I) have exhibited excellent reactivity and good enantioselectivities in the catalytic asymmetric hydroboration of styrene and substituted styrenes.

442. ASYMMETRIC SYNTHESIS OF SUBSTITUTED CYCLOBUTANES VIA INTRAMOLECULAR CYCLIZATION OF 1-CHLOROALKYLBORONIC ESTERS.
William C. Hiscox, Donald S. Matteson, Hon-Wah Man, Department of Chemistry, Washington State University, Pullman, Washington 99164-4630

A highly stereoselective and efficient synthesis of substituted cyclobutanes has been developed which involves an intramolecular cyclization of the appropriately substituted DICHED 1-chloro-4-cyanobutylboronic ester (DICHED = dicyclohexylethanediol, (R,R)-isomer shown). The story of this discovery will be presented, and the scope of the reaction and its application to natural products synthesis will be discussed.

443. DESIGN, SYNTHESIS, AND ASYMMETRIC REDUCTION OF BRIDGED NADH MODELS HAVING PARAPYRIDINOPHANE STRUCTURE
Nobuhiro Kanomata* and Tadashi Nakata
The Institute of Physical and Chemical Research (RIKEN), Wako, Saitama 351-01, Japan

Bridged NADH models, **1** and **2**, were synthesized from the corresponding bridged nicotinate intermediate prepared by the reaction of formyl-substituted (vinylimino)phosphorane with methyl propiolate. Biomimetic reaction with these model compounds effected enantioselective and stereospecific reduction of benzoylformate to mandelate with 99-97% ee.

444.

ENANTIOSELECTIVE ACYLATIONS OF ALCOHOLS CATALYZED BY CHIRAL PHOSPHINES.
Olafs Daugulis, Edwin Vedejs*, and Steven T. Diver, Department of Chemistry, University of Wisconsin-Madison, 1101 University Ave., Madison, WI 53706

The combination of acid anhydrides and a catalytic amount of chiral phosphines is shown to effect the enantioselective acylation of certain secondary alcohols with selectivities (s) up to 15. These are the first examples of enantioselective acylation using chiral phosphine catalysis, and the best results to date for any chemical system using an achiral acyl donor as the stoichiometric reagent under conditions of catalytic turnover of catalyst.

445.

THE ASYMMETRIC SYNTHESIS OF C2 SYMMETRIC 1,6-DIEPOXIDES. **Justin K. Wyatt** and Michael H. Nantz, Department of Chemistry, University of California, Davis, California 95616

The synthesis of diepoxide 2 from diene 1 (R* = chiral auxiliary) was examined. Matched and mismatched double stereodifferentiation reactions were studied to determine conditions for optimal formation of enantioenriched (R, R)-diepoxide 2. It is intended to use diepoxide 2 in a synthesis of phorboxazole ring A.

446. ASYMMETRIC SYNTHESIS OF NOVEL STRUCTURAL MIMICS OF SULPHIDOLEUKOTRIENES. Timo V. Ovaska, George H. Voynov, Nicole McNeil, and Joel A. Hokkanen, Department of Chemistry, Connecticut College, New London, CT 06320.

Sulphidoleukotrienes LTC_4, LTD_4, and LTE_4 have been implicated in the pathophysiology of asthma and related disorders. Thus, agents capable of interfering with binding of leukotrienes or preventing their biosynthesis are attractive therapeutic targets. This presentation will focus on an efficient asymmetric route toward a new class of structural mimics of the sulphidoleukotrienes with potential LTD_4 antagonist activity.

447. The Stereocontrolled Synthesis of α-Substituted α-Amino Acid Derivatives.
Murray Goodman and Jinfang Zhang, Department of Chemistry and Biochemistry, University of California, San Diego, La Jolla, California 92093-0343

In our continuing research on the design and synthesis of the conformationally constrained peptidomimetics, we have developd an efficient synthetic route for the synthesis of selected α-substituted α-amino acid derivatives, such as α-alkylated tyrosine and a fully protected (R)-α-methyltrytophan which can be readily used in solid phase and combinatorial synthesis. Our synthetic approach for the construction of the chiral quarternary α-carbon relies on the stereoselective alkylation of salicylaldehyde-derived oxazolidinones recently reported by Zydowsky and coworkers(J. Org. Chem. 1990, 55, 5437-5439). The structures and stereochemistry of the products were unambiguously characterized by NMR studies and X-ray crystallography. Details of the syntheses of these building blocks will be presented.

448. IMPROVED ENANTIOSELECTIVITY IN THE HYDROCYANATION OF ALDEHYDES WITH *CYCLO*[(R)-HIS-(R)-PHE] USING CATALYTIC AMOUNTS OF OPTICALLY ACTIVE AROMATIC CYANOHYDRINS.
Eugene F. Kogut, Jason C. Thoen, and Mark A. Lipton, Purdue University, West Lafayette, IN, 47907.

The cyclic dipeptide *cyclo*[His-Phe] has been shown to catalyze the formation of cyanohydrins from some aromatic aldehydes and hydrogen cyanide with very high enantioselectivity (up to 97% ee), but with lower selectivity for aliphatic aldehydes It has been shown for the reaction with meta-phenoxy-benzaldehyde that the reaction exhibits enantioselective autoinduction; the presence of product results in a greater optical purity of the product formed, implying that the product is part of the true catalytic species. This phenomenon has been utilized to obtain increased ee's for several cyanohydrins by seeding the hydrocyanation reaction with aromatic cyanohydrins of high optical purity. These seeded reactions give products with higher optical purity than control reactions, in some cases increasing ee's by more than 20%.

449. Synthetic approaches to enantiomerically pure silacycles. **C. Douglas** and P.G. Steel*, Department of Chemistry, University of Durham, South Road, Durham, UK, DH13LE.

This poster describes various routes towards the asymmetric synthesis of enantiomerically pure silacycles, with applications as silicon based bifunctional chiral auxiliaries and resolving agents. The synthesis of silacycles has been explored through three distinct synthetic strategies, silacyclohexanone formation, heterosilacycle formation and the silylfunctionalisation of dienes. Following literature precedent, hydroboration-redistribution of divinylsilanes affords a range of silacyclohexanones. Asymmetric hydroboration has been employed with α-substituted divinylsilanes to afford scalemic silacyclohexanones in moderate yields, over five steps. Work is currently underway towards optimisation of these procedures for applications in asymmetric synthesis. Heterosilacycles can be accessed via an asymmetric hydroboration-oxidation-cyclisation sequence. For example, silapyrans may be obtained in moderate yields by Mitsonobu cyclisation of diols, obtained by asymmetric hydroboration-oxidation of α-substituted divinylsilanes. The silylfunctionalisation strategy is currently being explored via asymmetric hydrosilylation of appropriate dienes.

450.
PRINS DESYMMETRIZATION OF A C$_2$-SYMMETRIC DIOL: APPLICATION TO THE SYNTHESIS OF 17-DEOXYROFLAMYCOIN.
Yueqing Hu, Guang Yang, Uday R. Khire and Scott D. Rychnovsky*, Department of Chemistry, University of California, Irvine, CA 92697

Prins cyclizations have been used to prepare a variety of monooxygenated rings. Mixed acetals derived from unsaturated alcohols are the most common presursors. An unusual Prinz cyclization-desymmetrization of a symmetric diol acetal was employed as a key step to construct a tetrahydropyran segment. Alkylations followed by manipulations of polyene moiety finished the total synthesis of 17-deoxyroflamycoin.

451. **STUDIES TOWARDS THE SYNTHESIS OF PIPERAZINOMYCIN. SYNTHESIS OF SELECTIVELY PROTECTED L-DOPA DERIVATIVES.** Michael E. Jung* and **Tsvetelina I. Lazarova**, Department of Chemistry and Biochemistry, University of California, Los Angeles, California 90095.

The advanced intermediate **4** in the synthesis of piperazinomycin (**1**) has been prepared in 9 steps (12% overall yield), using a novel S$_{RN}$1 reaction of a 3-iodotyrosine derivative as the key step. The mono-protected L-DOPA derivative **8b** has been synthesized in 5 steps (33% overall yield), using a Reimer-Tiemann formylation and a Dakin oxidation as the key steps.

452. STEREOSELECTIVE SYNTHESIS OF 2-ALKYL KETOMETHYLENE AND HYDROXYETHYLENE DIPEPTIDE ISOSTERES. Junhua Tao; Robert V. Hoffman*. Department of Chemistry and Biochemistry, New Mexico State University, Box 30001, Department 3C, Las Cruces, NM 88003

A simple and general protocol for the diastereoselective synthesis of ketomethylene and hydroxyethylene dipeptide isosteres which have an alkyl group at C-2 is reported. The alkyl group is introduced by chiral alkylation using a scalemic 2-triflyloxy ester. The alkylation takes place with inversion of configuration and is compatible with a variety of groups. This methodology is thus well suited for the preparation of a wide variety of carbapeptides.

453. A NOVEL SYNTHETIC ROUTE TOWARDS THE AUSTRALINES. Emma L. Trown, Department of Chemistry, Cardiff University, P.O. Box 912, Cardiff CF1 3TB, United Kingdom.

We have recently shown that the protected hydroxy aldehyde **1** can be obtained from the hydroxy ester, itself available in highly enantiomerically enriched form from a yeast reduction of 3-keto proline esters. The objective of this project is to examine methods for the homoligation of this intermediate to a range of natural and non-natural products. The theme of the poster will be centred around the use of this chemistry in the elaboration of members of the australine family **3** *via* cyclisations of various hydroxy epoxides (e.g. **2**).

454. ZARAGOZIC ACID CORE PREPARED USING OXONIUM YLIDES
John B. Brogan and Charles K. Zercher* University of New Hampshire, Durham, NH 03824

We have developed a method which provides rapid entry into highly functionalized 2,8-dioxabicyclo[3.2.1]octane and 1,9-dioxabicyclo[3.3.1]-nonane skeletons. The method involves intramolecular exposure of a carbenoid to a functionalized ketal.

The 2,8-dioxabicyclo[3.2.1]octane skeleton formed above is remarkably similar to the skeleton found in Zaragozic Acid.

455.

ENANTIOSPECIFIC TOTAL SYNTHESIS OF (+)-AJMALINE AND ALKALOID G *via* THE ASYMMETRIC PICTET-SPENGLER REACTION. Jin Li and J. M. Cook, Department of Chemistry, University of Wisconsin-Milwaukee, Milwaukee, WI. 53201

D-(+)-tryptophan **1** has been converted into the cardiovascular alkaloid (+)-ajmaline **2** *via* the *trans* transfer of chirality in the asymmetric Pictet-Spengler reaction. Alkaloid G can also be prepared by this route.

456.

PREPARATION OF(2S, 3S)-N-PROTECTED-3-AMINO-1,2-EPOXY-4-PHENYLBUTANE-A KEY INTERMEDIATE IN HYDROXYETHYLAMINE-BASED HIV PROTEASE INHIBITOR 141W94
Michelle J. Ferry Fugett Rose O'Mahoney Glaxo Wellcome, 5 Moore Dr., RTP, N.C. 27709

N-protected-3-amino-1,2-epoxy-4-phenylbutanes are key intemediates in many syntheses of hydroxyethylamine-based HIV protease inhibitors. This novel and economical route starts with the readily available N-protected L-phenylalanine. The starting L-phenylalanine is converted to the methyl ketone via the Weinreb amide, then brominated and reduced to the alkylbromohydrin. The bromohydrin is treated with base to afford the desired epoxide.

457. SYNTHESIS AND RESOLUTION OF *trans*-2-(1-ARYL-1-METHYLETHYL) CYCLOHEXYLAMINES. Kalavathi Bodige, Wen-Yee Lee, James M. Salvador, Department of Chemistry, The University of Texas at El Paso, El Paso, Texas 79968

The racemic title amines (aryl = phenyl, 4-t-butylphenyl, 2-naphthyl) were synthesized by opening N-(diphenylphosphinoyl)-7-azabicyclo[4.1.0]heptane with the corresponding α-potassium isopropylarene and hydrolysing the resulting phosphinamides. *Candida rugosa* lipase was used to enantioselectively acylate the (-)-title amines with methyl methoxyacetate.

458. THE DISCOVERY OF THE [2$_s$+2$_a$] REACTION OF DISILANES AND ACETYLENES AND THEIR APPLICATIONS. Zhongxin Ma, Thomas J. Barton, Sina Ijadi-Maghsoodi, Jibing Lin, and Mark Gordon Ames Laboratory, USDOE and Department of Chemistry, Iowa State University, Ames, Iowa 50011

A concerted [2$_s$+2$_a$] intramolecular addition reaction was discovered between disilanes and acetylenes in unimolecular processes. The reaction was performed with the 5,5,6,6-tetramethyl-3,3,8,8-tetraphenyl-4,7-dioxa-5,6-disilyl-cyclooctyne (**1**) under an argon flow at 600°C to give product 1,1,4,4-tetramethyl-3,3,6,6-tetraphenyl-2,5-dioxa-1,4-disilyl-dicyclo[3,3,0]octa-7-ene (**2**). *Ab initio* calculations were performed at the MP4/6-31G** level. The transition state structure was found and the activation energy was calculated to be 34 kcal/mol. The first example of a 3,3'-disilyl-indeno[2,1-*a*]indene (**4**) was synthesized from dibenzo-5,6-disilylcyclooctyne(**3**) in the same manner.

459. A SYMPHONY OF VARIATIONS ON THE DIELS-ALDER CYCLOADDITION Alexander Z. Bradley, Richard C. Burrell, Kimberly J. Daoust and Richard P. Johnson, Department of Chemistry, University of New Hampshire, Durham, NH 03824

Diels–Alder cycloadditions of enynes or diynes that should result in strained intermediates have been investigated by molecular orbital calculations and by flash vapor thermolysis experiments. In each case, results are consistent with [2+4] cycloaddition to give the strained cumulene or benzyne, followed by secondary reactions, which include aromatization, retro–cycloaddition, electrocyclic ring opening, and hydrogen abstraction. Ab initio MO calculations have been used to predict reaction energetics.

460. PREPARATION AND DIELS-ALDER REACTIVITY OF 1-AMINO-3-SILOXY-1,3-BUTADIENES. Jacob M. Janey, Sergey A. Kozmin, Viresh H. Rawal*, Department of Chemistry, University of Chicago, Chicago, Illinois 60637

The use of heteroatom-substituted dienes has greately expanded the scope of the Diels-Alder reaction, providing access to complex, highly functionalized products. We have developed an efficient and practical method for the preparation of 1-amino-3-siloxy-1,3-butadienes, a new class of dienes that exhibits excellent reactivity in [4+2] cycloadditions with various dienophiles. The scope of this methodology and its application to organic synthesis will be discussed.

461. SILICON TETHERED ENE CYCLISATIONS. J. Robertson and G. O'Connor, Dyson Perrins Laboratory, South Parks Road, Oxford, OX1 3QY, U.K., D.S. Middleton, Discovery Chemistry, Pfizer Central Research, Sandwich, Kent, CT 13 9NJ, U.K.

Silicon tethered cycloadditions, radical reactions and hydro- and bis-silylation of double bonds have become the subject of intense investigation by synthetic chemists. In connection with our efforts to develop stereoselective syntheses of carbohydrate and amino sugars from non-carbohydrate precursors we considered the synthetic possibilities offered by intramolecular ene reactions in which the reacting components are linked by a silicon tether. Not only would we predict high levels of regio- and stereo- control, but subsequent oxidative removal of the tether would allow additional oxygen functionality to be introduced at the chain termini. We will report the results of our investigation into the Type I and Type II Si-O and Si-C tethered carbonyl ene cyclisations and of their oxidative cleavage to polyhydroxylated compounds.

462.

FACILE 1,3-DISUBSTITUTED CYCLOBUTANE FORMATION VIA RADICAL CYCLIZATIONS, THE *GEM*-DIALKOXY AND *GEM*-DICARBOETHOXY EFFECTS. Michael E. Jung* and **Rodolfo Marquez**, Department of Chemistry and Biochemistry, University of California, Los Angeles, California 90095

New studies in four membered ring formation, taking advantage of the *gem*-dialkoxy and *gem*-dicarboethoxy effects, are described. The use of these effects allows for the easy preparation of substituted cyclobutane ring systems is also ilustrated.

X=CO_2Et, OCH_2CH_2O, $OCH_2CH_2CH_2O$

463. **MECHANISMS OF REARRANGEMENT REACTIONS OF AZIRIDINYLBENZALDOXIMES.**
Martha Hotema, Delphine Nwoko, and James E. Johnson, Department of Chemistry and Physics, Texas Woman's University, Denton, TX 76204.

Reactions of 2-methyl- and 2,2-dimethylaziridine with benzohydroximoyl chlorides [ArC(Cl)=NOH] give aziridinylbenzaldoximes (**1**). It has been found that the aziridine ring in these compounds undergoes ring opening in HCl-dioxane solution to give (**2**). Cyclization of **2** gives either a 1,2,4-oxadiazine (**3**, sodium hydroxide in water)

or a 4,5-dihydro-1,2,4-oxadiazole (**4**, sodium hydride in dioxane). The mechanisms of these ring closures have been investigated using deuterium labeled **2** [ArC(NHCD$_2$CR^{1}R^{2}Cl)=NOH]. (Supported by NIH-MBRS Grant No. GM08256 and NSF-RUI Grant No. CHE-921-1435)

464. **THE NATURE OF CYCLOPENTYNE FROM DIFFERENT PRECURSORS.**
John C. Gilbert, Everett G. McKinley, and Duen-Ren Hou, Department of Chemistry and Biochemistry, The University of Texas at Austin, Austin, Texas 78712

Cyclopentyne (**1**), as generated from three different precursors, affords both [2+2]- and [2+4]-cycloaddition products with spiro[4.2]hepta-1,3-diene (Eq. 1). The ratio of these cycloadducts is strongly precursor-dependent, favoring the [2+4] product when diazomethylenecyclobutane is the source of **1**, and the [2+2] isomer when **1** is derived from the dibromides. Moreover, the ratio is temperature-*independent* with the former precursor, but temperature-*dependent* with the latter two. Mechanistic ramifications of these and related observations will be presented. Research supported by the Robert A. Welch Foundation (Grant F-815) and THECB (Grant ARP-455).

465.

CARBENE CATION RADICALS: GENERATION, REACTIVITY AND UTILITY. Darren G. Stoub Joshua L. Goodman*, Department of Chemistry, University of Rochester, Rochester, NY 14627

A fundamental focus among organic chemists has been understanding the reactivity of organic compounds. Toward this end, the chemistry of neutral carbenes has been extensively studied. We have sought to understand the effect of carbene oxidation; that is, investigate the chemistry of carbene cation radicals. Product and mechanistic studies from the chemical oxidation of the diaryl diazomethanes in the presence of olefins, as well as absolute reaction rates measured via a double laser flash photolysis method have revealed the novel electrophilic reactivity and synthetic utility of these compounds.

ENANTIOSELECTIVE C–C BOND FORMING REACTIONS.
466. **D. A. Evans**, Department of Chemistry & Chemical Biology, Harvard University, Cambridge, MA 02138

This lecture will address the progress that is being made in the design of chiral divalent metal catalysts for cycloaddition, aldol and Michael reactions. Issues pertaining to the structure of the catalyst-substrate complexes, reaction scope, and the problems of product inhibition will be discussed.

467. **STEREOCONTROLLED CARBON-CARBON BOND FORMATION VIA TORQUOSELECTIVE NAZAROV CYCLIZATIONS.** F. G. West, Department of Chemistry, University of Utah, Salt Lake City, UT 84112.

The Nazarov Reaction has been employed extensively for the introduction of new cyclopentenone rings. The key step involves carbon-carbon bond-formation via conrotatory electrocyclization of a pentadienyl cation, with creation of up to two new stereocenters. Typically, some or all of this stereochemical information is lost in subsequent deprotonation or desilylation steps; however, the "Interrupted Nazarov Reaction" preserves these stereocenters through the intervention of a nucleophilic trapping event. This talk will focus on recent advances in asymmetric Nazarov Cyclizations via either chiral auxiliaries or chiral Lewis acids.

468.

NEW SYNTHETIC METHODOLOGY FOR ALKALOID TOTAL SYNTHESIS. Steven M. Weinreb, Department of Chemistry, Penn State University, 152 Davey Lab, University Park, PA 16802.

A description of recent studies on development of new synthetic methodology will be presented. Application of this chemistry to construction of alkaloids will also be described.

469. DESIGN, DEVELOPMENT, AND APPLICATIONS OF TRANSITION–METAL COMPLEXES IN CATALYTIC ASYMMETRIC SYNTHESIS. Erick M. Carreira, Arnold and Mabel Beckman Laboratory of Chemical Synthesis, Division of Chemistry and Chemical Engineering, California Institute of Technology 164–30 Crellin, Pasadena, CA 91125.

Using the tools and methods of coordination chemistry, molecular recognition and organic synthesis we have prepared transition–metal catalysts and reagents which are useful in the enantioselective and diastereoselective synthesis of optically active compounds. The latest developments in our laboratory regarding the preparation and use of chiral catalysts and atom–transfer reagents will be highlighted. Emphasis on the relationship between the structure and reactivity will be discussed. Additionally, the presentation will include applications to the synthesis of natural products as well as important, useful building blocks for asymmetric synthesis.

470. **CHEMICAL COMMUNICATION IN WATER USING FLUORESCENT CHEMOSENSORS.** Anthony W. Czarnik, Department of Chemistry, Ohio State University, Columbus, OH 43210 [current address: IRORI Quantum Microchemistry, San Diego; aczarnik@irori.com]

The modern era in fluorescence commenced with the work of George Stokes, who in the mid-19th century first reported that fluorescence emission occurs at a longer wavelength than excitation. This wavelength shift, termed the Stokes shift, provides the primary conceptual basis for the enormous sensitivity of fluorimetric analyses. Stokes also outlined the relationship between the concentration of a fluorophore and observed fluorescence intensity, describing the quenching of fluorescence at high concentration and by the presence of 'foreign substances'. Neither of these effects occurs in absorption spectroscopies, and led Stokes in 1852 to propose that fluorescence be used for the detection of organic substances.

Compounds incorporating a binding site, a fluorophore, and a mechanism for communication between the two are called *fluorescent chemosensors*. This talk will survey our OSU group's efforts in the design and discovery of such compounds over the past ten years.

471. LUMINESCENT SENSING WITH LUMOPHORE-FUNCTIONALIZED RECEPTORS. **A. P. de Silva**, H. Q. N. Gunaratne, T. Gunnlaugsson, A. J. M. Huxley, C. P. McCoy and T. E. Rice. School of Chemistry, Queen's University, Belfast BT9 5AG, Northern Ireland.

Modular construction of luminescent sensors as 'lumophore-spacer-receptor' systems according to the PET(Photoinduced Electron Transfer) signalling principle has several endearing features. 1. Receptors for a given analyte with established selectivity profiles can be converted into luminescent sensors without losing the binding characteristics. 2. The lumophore can be chosen according to the end-user's needs of excitation & emission wavelengths and emission lifetime. 3. Compatibility of lumophore-receptor pairs can be evaluated by checking PET thermodynamics according to electrochemical data for the individual components, before the synthesis of the sensor system is attempted. 4. Special sensing needs can be addressed by grafting additional modules with appropriate functionalities.

Recent examples from our laboratory will be used to illustrate these features.

472. LUMINESCENT SENSORS WITH AND FOR TRANSITION METALS. **Luigi Fabbrizzi**, Dipartimento di Chimica Generale, Università di Pavia, I-27100 Pavia, Italy

A two-component approach has been followed in order to design luminescent chemosensors for transition metals. Substrate recognition by a receptor moiety is signalled through the quenching of an adjacent luminescent subunit, which takes place via either an electron transfer or an energy transfer process involving the metal center. The photophysically inactive Zn(II) ion has been incorporated into a variety of concave receptors suitable for the inclusion of anions or ionizable substrates. Recognition is based on the metal-ligand interaction and is signalled through the quenching of an appended luminescent subunit. The quenching mechanism is assigned to an electron transfer process from the anion to the proximate lumophore. Some selective Zn(II) containing polyamine systems, both cyclic and non-cyclic, will be presented: these display sensing tendencies towards amino acids.

473. PEPTIDE PLATFORMS FOR METAL CATION SENSING. **Barbara Imperiali**, Division of Chemistry and Chemical Engineering, California Institute of Technology, Pasadena, CA 91125.

Recent efforts have been directed toward the production of fluorescent chemosensors for trace quantities of divalent transition metal cations. Chemosensors for divalent metals require two components to function - a metal ion recognition and binding unit and a signaling mechanism. For example, to effect highly selective binding to divalent zinc ions, a chemically synthesized peptide patterned after the zinc finger peptides has been employed. To this template, fluorophores that have solvent-dependent emission properties have been attached. Upon binding zinc, the peptide folds and the change in environment experienced by the pendant fluorophore is detected by fluorescence spectroscopy. A fluorescent chemosensor for divalent zinc sensitive to nanomolar concentrations of divalent zinc has been prepared. This chemosensor allows sensitive measurements of free zinc ion to be made but is oxidatively labile; recent efforts to address this problem will be presented. In addition, a peptide-based chemosensors with selectivities for other transition metal ions will be introduced.

474. **FLUORESCENT CHEMOSENSORS OF INTRACELLULAR SIGNAL TRANSDUCTION: APPLICATIONS TO PHARMACEUTICAL SCREENING. R.Y. Tsien**, R. Heim, A.B. Cubitt, G. Zlokarnik, J.E. Gonzalez. Howard Hughes Med. Inst., UCSD, La Jolla, CA 92093, and Aurora Biosciences Corp., 11149 N. Torrey Pines Rd., La Jolla, CA 92037.

Fluorescent readouts of intracellular signal transduction, together with engineering of responsive cell lines and miniaturized parallel assays, provide some promising approaches to ultra-high-throughput screening of pharmaceutical libraries. a) Proteins can be fluorescently tagged *in vivo* by fusion to Green Fluorescent Protein (GFP). Mutagenesis, guided in part by the crystal structure, has improved GFP and yielded new colors, which reveal protein-protein interaction and protease activity by fluorescence resonance energy transfer (FRET). b) Gene expression can be visualized in single living cells by using β-lactamase as a reporter enzyme. It cleaves novel membrane-permeant substrates and changes their fluorescence from green to blue by disrupting FRET. Because of its enzymatic amplification and nondestructive assay, β-lactamase has major advantages over existing reporter genes. c) A fast and sensitive optical readout of membrane voltage results from FRET between a fluorescent gating charge and another fluorophore attached to one side of the membrane.

475. REMODELLING GLYCOFORMS ON PROTEINS AND CELLS BY CHEMOSELECTIVE LIGATION. Carolyn R. Bertozzi, Department of Chemistry, University of California, and Materials Sciences Division, Lawrence Berkeley National Laboratory, Berkeley, CA 94720.

The chemoselective ligation strategy has found tremendous success in the synthesis of macromolecular protein conjugates. At the heart of this approach is the introduction of mutually and uniquely reactive functional groups onto unprotected fragments and the coupling of these fragments in an aqueous environment. We have extended the principle of chemoselective ligation to the generation of glycoproteins and cells with tailor-made glycoforms. New enzymatic and biosynthetic methods for introducing uniquely reactive functional groups onto protein-bound oligosaccharides will be presented, along with applications to glycoform engineering.

476. SYNTHETIC INVESTIGATIONS DIRECTED TOWARDS LANDOMYCIN A: ASSEMBLY OF THE HEXASACCHARIDE FRAGMENT. Yu Guo, Jianjun Lu and Gary Sulikowski, Department of Chemistry, Texas A&M University, College Station, TX 77843

Progress towards the synthesis of the hexasaccharide fragment of landomycin A (a.k.a. NSC 639187) will be described. The stereocontrolled construction relies on the combined application of glycosyl tetrazoles and phosphites to introduce alpha and beta glycosidic linkages, respectively.

Landomycin A

477. 6A,6B-, 6A,6C-, AND 6A,6D-DIHYDROXY PERMETHYLATED β-CYCLODEXTRINS: A NOVEL METHOD LEADING TO BIFUNCTIONAL ENZYME MIMICS. Zhen Chen, Jerald S. Bradshaw* and Milton L. Lee, Department of Chemistry & Biochemistry, Brigham Young University, Provo, UT 84602

A,B - DIOL   A,C - DIOL   A,D - DIOL

The title compounds have been prepared through a convenient method as shown in the scheme (tert-butyldimethylsilylation and the permethylation steps are in one pot). The permethylated CD diols were separated and the structures proved by ^{1}H NMR and ^{13}C NMR spectroscopies and HRMS and elemental analyses. The ^{13}C NMR spectra exhibited five signals for the carbon-6s containing OCH$_3$ and two signals for carbons-6s containing OH. The permethylated CD diols were also converted to known disubstituted permethylated CD derivatives. The primary 6A,6B-; 6A,6C-; and 6A,6D-bifunctionalized β-CDs are now available by the method described. The details of the synthesis, purification, and characterization of these compounds will be described.

478. **A Unified Approach to Unambiguous Synthesis of the Phosphatidylinositol-3-phosphates Involved in Intracellular Signal Transduction**
R. Aneja,* A. Parra, C. Stoenescu, W. Xia and S. G. Aneja. Nutrimed Biotech, Cornell University Research Park, Langmuir Laboratory, Ithaca, NY 14850.

The 3-phosphate derivatives of 1D-1-(1´,2´-di-O-fattyacyl-sn-glycero-3´-phospho)-myo-inositols (PtdIns) including phosphatidylinositol-3-phosphate, PtdIns(3)P, and the bis- and tris-phosphate derivatives PtdIns(3,4)P_2 and PtdIns(3,4,5)P_3, have been found in eukaryotic cells, and the occurrence of PtdIns(3,5)P_2 has been suggested. These compounds have been demonstrated as activators of protein kinase C isoforms δ, ϵ, and η, and are putative messengers in cellular signal cascades pertinent to inflammation, cell proliferation, transformation, protein kinesis, and cytoskeletal assembly. Minute quantities are found in cells and therefore synthetic methods are needed to obtain samples for establishing the putative roles.

We report on a unified approach to unambiguous synthesis of the complete series of the phosphatidylinositol-3-phosphates and illustrate by the synthesis of 1D-1-(1´,2´-dihexadecanoyl-sn-glycero-3´-phospho)-myo-inositol-3,4,5-trisphosphate.

479. **APPLICATION OF ASYMMETRIC AMPLIFICATION VIA ACHIRAL AUXILIARIES FOR THE SYNTHESIS OF DEOXY-CARBOHYDRATES**
Malinda E. Pauly, Wallace C. Pringle, and Craig A. Merlic* Department of Chemistry and Biochemistry, University of California, Los Angeles, California 90095-1569

The synthesis of deoxy- and deuterium-labeled carbohydrates in high ee's using a new method of asymmetric amplification via achiral auxiliaries is described. The method avoids classical derivatization with stoichiometric chiral auxiliaries. Instead, achiral auxiliaries are employed in catalytic asymmetric dihydroxylation reactions. Subsequent cleavage provides sugar products with ee's typically greater than 99%.

480. **A SELF-REPLICATING PEPTIDE**
David H. Lee, Juan R. Granja, Jose A. Martinez, Kay Severin, M. Reza Ghadiri*
Depts of Chemistry and Molecular Biolgy, The Scripps Research Institute
La Jolla California 92037

The production of amino acids and proteinoid species under likely prebiotic conditions have long been demonstrated. Because of the central importance of molecular self-replication in the origin of life, the feasibility of peptide self-replication was investigated. This talk will decribe the first example of a self-replicating peptide. A 32-residue a-helical peptide based on the leucine zipper motif of GCN4 is shown to act autocatalytically in templating its own synthesis by accelerating the thioester-promoted amide bond condensation of 15- and 17-residue constitutional peptide fragments in neutral and dilute aqueous solutions. The self-replicating process displays parabolic growth and square-root law kinetics in the initial rates of product formation.

481. **INVESTIGATION OF END CAPPING PROPENSITIES OF NOVEL FUNCTIONAL GROUPS. Krishnan P. Nambiar *,** Hemender K. Reddy, Selvasekaran Janardhanam and Ling Li, Department of Chemistry, University of California, Davis, CA 95616.

We and others have recently shown that α-helical structures in short peptides can be stabilized by the introduction of amino acids at the helical termini that are capable of hydrogen bonding with the unpaired CO and NH groups located at the helical termini. In this report we describe the synthesis of peptides incorporating several novel functional groups at the amino terminus to test their efficacy in stabilizing α-helical structures via end capping. The end capped peptides show remarkable helicity in aqueous solution. The trend in α-helix inducing ability was found to be sulfonate > carboxylate > Phosphate > Phosphonate > hydroxyl > nitro > acetyl. Our results demonstrate the feasibility of designing and synthesizing novel molecules for incorporation into peptides at the helical termini for inducing helical folding and stabilizing the folded structure via end capping.

482. STABILIZATION OF β-RIBBON STRUCTURES IN PEPTIDES USING STRAIN-FREE DISULFIDE BRIDGES and TURN STRUCTURES. **Krishnan P. Nambiar**[*], Selvasekaran Janardhanam, Devan Balachari and Ling Wang, Department of Chemistry, University of California, Davis, CA 95616.

The effect of a disulfide cross link between two peptide chains on the stability of β-ribbon secondary structures formed by these peptides has been investigated. We designed and synthesized four sets of 9-residue peptides incorporating cysteine, homocysteine, (S)-α-amino-δ-mercaptopentanoic acid and (S)-α-amino-ε-mercaptohexanoic acid respectively. Comparison of the CD data clearly show that the dimers containing a strain-free disulfide bridge formed between the longer side chains of higher homologs of cysteine show dramatically higher β-ribbon character as compared to the dimer with cystine disulfide bond. We also investigated the effect of incorporating a turn structure in promoting β-ribbon character in peptides. We show that the tripeptide β-Alanine-3-amino-5-nitrobenzoic acid-Sarcosine when incorporated in the middle of a peptide induces it to adopt β-ribbon structure.

483. STERIC EFFECTS OF 5-*TERT*-BUTYLPROLINE ON THE CONFORMATION OF POLYPROLINE. Eric Beausoleil and William D. Lubell*, Département de chimie, Université de Montréal, C. P. 6128, Succ. A, Montréal, Canada H3C 3J7

The left handed polyproline II (PPII) helix is now recognised as an important peptide conformation in biological systems; however, the related right handed polyproline I (PPI) helix is much less well studied due to its conformational lability. In this presentation, we report a new method to control the conformation of the polyproline helix designed to augment the population of PPI. In particular, we are incorporating 5-*tert*-butylproline (5-tBuPro) into polyprolines to create steric interactions that disfavor PPII and thereby increase the population of PPI. Using 5-*t*-BuPro to bias *cis*-amide conformation, we are striving to synthesize PPI for study in solution by NMR and in the crystal state by X-ray. We have synthesized different length polyprolines and have incorporated 5-tBuPros at various points along the prolyl peptide chain. By using both NMR and CD spectroscopy to measure the PPII to PPI ratio, we are identifying the requirements to induce a PPI helix using 5-tBuPro. In summary, our presentation will feature the synthesis of 5-tBuPro, its incorporation into a library of N-Ac-(Pro)$_n$-NH$_2$ (n = 3-6) oligomers as well as the influence of the position of a 5-*tert*-butylprolyl residue on the PPI/PPII equilibrium.

484. **STRUCTURE OF PHOSPHINE CONTAINING PEPTIDE LIGANDS**
Scott R. Gilbertson,*[1] Guohua Chen,[1] Jeff Kao,[1] Alicia Beatty[1] and Charles Campana,[2]
[1]Department of Chemistry, Washington University, Campus Box 1134, St. Louis, Missouri 63130-4899. [2]Siemens Energy and Automation Inc. 6300 Enterprise Lane, Madison, Wisconsin 53719-1173

The structure of a series of phosphine containing peptides will be reported. The X-ray structure of a phosphorus containing dodecapeptide will be discussed. Analysis of the solution structure of similar phosphorus containing peptides before and after coordination of rhodium is also reported. In both the solid state and in solution, the peptides were found to exist as a mixture of α-helical and 3$_{10}$ helical conformations. Coordination of rhodium to the i, i+4 orientated phosphine groups appears to alter the conformational preference.

485.

A NOVEL CYTOTOXIC HEXAPEPTIDE, CYCLOCINIAMIDE A, FROM THE MARINE SPONGE *PSAMMOCINIA* William D. Clark and
Phillip Crews Department of Chemistry and Biochemistry and Institute for Marine Sciences University of California, Santa Cruz, California 95064

Marine sponges continue to be rich sources of novel peptide constituents that do not appear to be metabolites of the sponge. A novel cytotoxic cyclic peptide, cyclociniamide A (1), was isolated as a minor constituent from the marine sponge *Psammocinia* collected in Papua New Guinea. Previously, *Psammocinia* has yielded only acyclic sesterterpene compounds containing both terminal tetronic acid and furan moieties. The structure elucidation of cyclociniamide A (1) was carried out using the data derived from a variety of one-dimensional and two-dimensional ^{1}H and ^{13}C NMR techniques. The absolute stereochemistry was assigned as 4R, 7S, 11R, 14S based on chiral TLC analysis and NOESY and molecular modeling data. Cyclociniamide A (1) exhibited potent biological activity in the brine shrimp assay.

486. BLENDING DIRECTED *ortho* AND REMOTE METALATION AND CROSS COUPLING IN NATURAL PRODUCT SYNTHESIS. GILVOCARCIN-V, KINAFLUOREONE, AND PHENANTHROVIRIDIN. James, C.; Britton, R.; Mohri, I.; Zhao, B.-p.;Iwemma Bakker, W.; Stefinovic, M.; **Snieckus**, V. Guelph-Waterloo Centre for Graduate Work in Chemistry, University of Waterloo, Ontario, Canada N2l 3G1

Current methodological developments in ortho and remote aromatic metalation and cross coupling reactions (*J. Org. Chem.* **1991**, *56*, 1683, *Synlett* **1994**, 349) in our laboratories have set the conceptual foundations for the synthesis of diverse aromatic and heteroaromatic natural products. This presentation will deal with the total synthesis of defucogilvocarcin V (**1**), kinafluorenone (**2**), and phenanthroviridin (**3**), all antibiotic agents derived from various strains of *Streptomyces*.

487. ### A Novel High Yield Synthesis of Substituted Isoindolequinones
Manisha Chakraborty, David B. McConville, Gerald F. Koser and Wiley J. Youngs. Chemistry Department, University of Akron, Akron, OH 44325-3601

A general, convenient, high yield, one step synthesis of heterocyclic isoindolequinones is described. Primary amines react with 2,3-di(trimethylsilyl)ethynyl 5,6-dimethyl hydroquinone(**1**) or its desilylated derivative(**2**) at room temperature in argon to give substituted isoindolequinones(**3**) in moderate(48%) to good(65%) yield using the corresponding amine as solvent. Presence of a catalytic amount of Pd(PPh$_3$)$_2$Cl$_2$ increases the yield about 10%. Diisopropyl amine reacts with substrate **2** on reflux producing **3** in low yield(32%). The reaction of **1** and primary amines gives almost quantitative yields of **3** in methanol at room temperature.

488. A NOVEL ASYMMETRIC SYNTHESIS OF SUBSTITUTED INDENES. **K. D. Lovasz**, J. K. Morris, S. L. VanderVelde, T. M. Judge, J. Tulinsky, R. B. Gammill, S. A. Mizsak, and K. R. Romines, Structural, Analytical and Medicinal Chemistry Research, Pharmacia & Upjohn, Kalamazoo, MI 49001

An efficient asymmetric synthesis of 1,2,3-substituted indenes has been developed. The asymmetric center at C-1 of the indene is established by the successful addition of an organocuprate to a chiral unsaturated acylimide. This selectively forms the asymmetric center and allows alteration of the stereochemistry simply by choice of the chiral auxiliary. The indene is formed in one step under mild Lewis acid conditions and low to ambient temperatures. This synthesis offers complete control of the asymmetric center at C-1, a single, carbocyclic annulation step, and the preparation of complex indene derivatives.

489. SYNTHESIS OF NOPINONE AND VERBENONE DERIVED CHIRAL ANNULATED INDENES
Chao Liu and <u>John R. Sowa, Jr.</u>*, Department of Chemistry, Seton Hall University, South Orange, NJ 07079

Recently there has been considerable interest in asymmetric synthesis using chiral transition metal catalysts. Our interest in this area is in the design of new chiral ligands. In this presentation, we show the general synthesis of chiral annulated indenes that are derived from enantiomerically pure chiral bicyclic terpenes: nopinone and verbenone.

490. A UNIFIED THEORY OF THE REACTIVITY OF 5-SUBSTITUTED TOCOPHEROLS
T. **Rosenau** and W. D. Habicher, TU Dresden, Institut für Organische Chemie, Mommsenstr. 13, D - 01062 Dresden, Germany

Tocopherol derivatives, especially 5-substituted tocopherols, are used as auxiliaries in synthesis and for medical purposes, for example as lipophilic drug carriers. The chemical behavior of those tocopherol (vitamin E) derivatives has not been well understood, but can now be described by the theory presented. The reactivity in the tocopherol system is determined almost exclusively by the substituent at carbon atom 5. This is proved with the help of various 5-substituted tocopherols carrying C_1, C_2 and C_3 units at this position. The electronic effects of the substituent, and thus the polarization of C-5a, is the crucial factor that influences the whole chromanol system of tocopherols, and determines the reactivity in an almost unparalleled manner. It is demonstrated that the chemical reactions of tocopherol derivatives with the huge variety of possible products can be reduced to two basic pathways. These reaction paths involve either the *para*-tocopheryl quinone or the 5a-*ortho*-quinone methide as initial intermediates. In the present paper, the synthesis and chemical properties of differently substituted tocopherols are described. The data obtained are integrated into a theory on the reactivity of tocopherol derivatives in dependence on the 5-substituent.

491.

THE SYNTHESIS OF 2-PYRONES FROM CYCLOBUTENDIONE PRECURSORS. **P. A. Mingo**, S. Zhang, L. S. Liebeskind, Department of Chemistry, Emory University, Atlanta, Georgia 30322

The synthesis of 5-silyloxy-2-pyrones was achieved via the addition of silyl-protected cyanohydrins to 3-cyclobuten-1,2-diones. Deprotonation of the cyanohydrin with lithium hexamethyldisilazide followed by addition of the resulting anion to the cyclobutendione generates a 4-alkoxide intermediate. Silyl migration and subsequent displacement of cyanide gives a 4-acylcyclobutenone. This intermediate undergoes a 4π-electrocyclic ring opening followed by a 6π-electrocyclic ring closure to afford the desired 2-pyrone in good to excellent yield.

492. **ALDOL-INSERTION REACTION SEQUENCE FOR THE SYNTHESIS OF TETRAHYDROFURANS.** Michael A. Calter, Priyantha Sugathapala and Cheng Zhu, Department of Chemistry, Virginia Tech, Blacksburg, Virginia, 26021-0212

The aldol reactions of the dicholorophenylborane enolates of α-diazo-β-ketoesters with aldehydes afford aldol adducts in good to excellent yields. Cylization of these adducts to (2H)-3-keto-dihyrofurans occurs readily in the presence of rhodium (II). Further reactions of the ketofurans yield a variety of substituted di- and tetrahydrofuran derivatives.

493. **DIMERS OF 3-METHYLINDOLE WITH ARYL AND ALKYL SULFONIC ACIDS.** Wayne M. Stalick, George W. Mushrush and Sami Faour, Department of Chemistry, George Mason University, Fairfax, Virginia 22030

When 3-methyl indole is reacted with dodecyl benzene sulfonic acid (DBSA) in the presence of a hydrocarbon solvent, two distinctly different compounds are produced. One of these is a white, sharp melting solid and the other is a gum which is relatively insoluble in most solvents. Elemental analysis of the white solid shows that it contains one DBSA per two 3-methyl indole units. ^{1}H NMR and ^{13}C NMR spectra were complex and even though consistent with the analysis, were inconclusive. To simplify the spectra p-ethyl benzene sulfonic acid, p-toluene sulfonic acid, ethane sulfonic acid and methane sulfonic acid were all reacted with 3-methyl indole. Each of these acids produced white solids with one sulfonic acid residue per two 3-methyl indole units. The resultant NMR and FTIR spectra of these compounds were clear and difinitive. The spectral data collected, along with the proposed structures, will be discussed along with the results of an X-ray crystallograpic study.

494. **INVESTIGATION OF THE REACTION OF PENTACOVALENT OXAPHOSPHOLENES WITH ISOCYANATES. SYNTHETIC AND MECHANISTIC ASPECTS**
Cynthia K. McClure* and Baozhong Cai.
Montana State University, Department of Chemistry and Biochemistry, Bozeman, MT 59717

The reaction of pentacovalent oxaphospholene **1** with isocyanates gave novel pyrimidine derivatives **4** via new heterocyclic P(V)s **2** and **3** as key intermediates. The mechanism was supported by kinetic NMR data and crossover experiments.

495. ELECTROCHEMICAL INVESTIGATION OF PYRIDINIUM ANTHRAQUINONES. Andrew S. Koch and Jean L. Raphael, Department of Chemistry, Williams College, Williamstown, MA 01267

Mono- and bis-pyridinium anthraquinones, anthraquinones substituted with a cationic pyridinium moiety via the pyridine nitrogen, have been prepared and an electrochemical investigation has been carried out. Both the position and the number of pyridinium substituents were used to vary the reduction potential of the anthraquinone ring. It was found that effects due to the position of the pyridinium moiety were very small, while the number of pyridinium moieties had an effect on the quinodal reduction potential on the order of 150 mV. In all cases the reduction potentials of the anthraquinones were well below the reduction potential of the pyridinium moiety.

496. IMIDAZOLIUM CATIONS: CARBON ADDUCTS OF 2,2'-BIPYRIDINES
R. J. Morgan[1]*, R. J. Donovan, E. Rodriguez[1] and M. Forester[1].
[1]*Department of Chemistry, Long Island University, Brooklyn N.Y, 11201.*

The addition of chloroiminium salts using Vilsmeier-Hack methodology can be used to add carbon across the nitrogen atoms in diimine compounds related to 2,2'-bipyridine. In most instances, the procedure offers good to excellent yields of novel fluorescent dyes (I, II and III). These imidazolium cations possess interesting structures and useful physical properties. Their ease of preparation, photophysical properties, water solubility and large Stoke's shift make them ideal candidates for a variety of applications.

IIIa R = H
IIIb R = CO_2H
IIIc R = CO_2CH_3

497. SYNTHESIS OF POLYCYCLIC AROMATIC COMPOUNDS IN SUPERACID.
Douglas A. Klumpp[a], Donald N. Beak[b], G. K. Surya Prakash[b], George A. Olah[b], [a]Department of Chemistry, California State Polytechnic University, Pomona, CA 91768, and [b]Loker Hydrocarbon Institute and Department of Chemistry, University of Southern California, Los Angeles, CA 90089.

Acenaphthenequinone is converted to aryl pinacols by reaction with aryl Grignard reagents and the aryl pinacols react in triflic acid to give polycyclic aromatic compounds. Pinacols **1** and **2** upon reaction in triflic acid gives **3** and **4**, respectively, by a dehydrative cyclization process. In studies using acid systems of varying strength, **3** is only produced from **1** in media more acidic than H_o -11. This is consistent with the formation of a dicationic intermediate leading to the formation of **3**. The results of these and related studies will be presented.

498. Generation of Four Five-Membered Rings in a One Pot Process. Studies Toward the Synthesis of Dicyclopenta[a,e]pentalene via the Tandem Pauson-Khand Reaction. Scott G. Van Ornum, M. Sreenivasa Reddy, and J. M. Cook, Department of Chemistry, University of Wisconsin-Milwaukee, Milwaukee, Wisconsin 53201

There has been a long standing interest in our laboratory in the synthesis of strained polyquinanes and polyquinenes. Cyclopentapentalenes (see **3**) which may exhibit Huckel stability are of particular interest in regard to bonding and stabilization in organic molecules. The Pauson-Khand reaction has proven to be an effective means to generate fused five-membered rings. In this report we present the synthesis of **2** from **1** and recent attempts to cyclize this to the tetraquinane ring system under the conditions of the Pauson-Khand reaction.

499. THE USE OF INTRAMOLECULAR FRIEDEL-CRAFTS REACTIONS TO PREPARE POLYCYCLIC SYSTEMS WITH ANGULAR METHYL GROUPS. George Majetich, Paul Spearing and Jing Fang. *Department of Chemistry*, The University of Georgia, Athens, Georgia 30602.

Treatment of dienone **1** with $TiCl_4$ at -78 °C for five minutes produces enone **2** in good yield; note that cyclization also creates a new quaternary center at the ring fusion. In contrast, the use of boron trifluoride etherate at elevated temperatures produces hydrophenanthrene **3**, undoubtedly via a "dienone-phenol rearrangement" mechanism. Several substrates have been synthesized in order to establish the utility of this new method for preparing other fused polycyclic systems having an angular methyl group.

500.

A CONVENIENT STEREOSELECTIVE SYNTHESIS OF 17-β-7-(α-3'-CARBOXYPROPYL)ESTRADIOL: PREPARATION OF AN ESTRADIOL FLUORESCENT PROBE
Don Johnson, Maciej Adamczyk and Rajarathnam E. Reddy, Divisional Organic Chemistry (9NM, AP20) Abbott Laboratories, 100 Abbott Park Road, Abbott Park, IL 60064

Synthesis of appropriately labeled estradiol probes is critical for the understanding, diagnosis, and treatment of estrogen related illnesses.
Although it has been shown that probes derived via functionalization at the C7 position of estradiol show superior binding to estrogen receptors and antibodies, and also exhibit significant antiestrogenic activity, published synthetic methods for 7-substituted estradiol derivatives involve use of controlled starting materials and hence require strict adherence to stringent regulatory procedures.
In this poster, we describe a stereoselective synthesis 17-β-7-(α-3'-carboxypropyl) estradiol from a noncontrolled substance, 6-oxoestradiol and its application to the preparation of a fluorescent probe.

501. SYNTHESIS AND STRUCTURAL STUDIES BY NMR OF SOME STEROIDS RELATED TO LOTEPREDNOL ETABONATE. Stanislaw Rachwal, Pascal Druzgala, Zong-Zheng Liu, Marcus E. Brewster, Emil Pop. Pharmos Corporation, 2 Innovation Drive, Alachua, FL 32615

Several derivatives of Lotoprednol Etabonate (1), a soft corticosteroid antiinflammatory drug, are formed during the preparation and sterilization processes. Some of these compounds, more abundant contaminants of 1, have been identified, synthesized, and fully characterized by NMR spectroscopy.

502. SYNTHESIS AND NMR CHARACTERIZATION OF 1-(2-CHLOROETHYL)-1,4-DITHIANIUM SALT FOUND IN MUSTARD HEELS. F.-L. Hsu, G. Hondrogiannis and Y.-C. Yang, U.S. Army Edgewood Research, Development & Engineering Center, Aberdeen Proving Ground, Maryland 21010-5423

Mustard heels are insoluble materials deposited in sulfur mustard (**1**) agent storage ton container after a long period of time and 1-(2-chloroethyl)-1,4-dithianium chloride (**2**) was proposed as the structure based on the NMR analysis. The synthesis of **2** and its analogs was investigated by the reactions of pentamethylene sulfide and 1,4-dithiane with 1-chloro-ω-iodo-alkanes, I-$(CH_2)_n$-Cl (n = 2, 3) and was accomplished in high yields (70-80%) as the tetrafluoroborate salts under reflux conditons in nitromethane in the presence of $AgBF_4$. Structural identification of the sulfonium salts was accomplised by ^1H and ^{13}C NMR spectroscopies.

503. SELECTIVE REDUCTIONS OF ALDEHYDES, KETONES, ESTERS, AND EPOXIDES IN THE PRESENCE OF A NITRILE USING LITHIUM *N,N*-DIALKYLAMINOBOROHYDRIDES. **Christopher J. Collins**, Gary B. Fisher, Adeena Reem, and Bakthan Singaram*, Department of Chemistry and Biochemistry, University of California, Santa Cruz, Santa Cruz, CA 95064

Lithium *N,N*-dialkylaminoborohydrides (LiR_2NBH_3) selectively reduced aldehydes, ketones, esters, and epoxides in the presence of aliphatic or aromatic nitriles. The reactive functionality was reduced in 85-99% yield with no detectable reduction of the nitrile.

504. REDUCTION OF ALKYLCYCLOHEXANONES TO THE CORRESPONDING ALCOHOLS USING VARIOUS SUBSTITUTED LITHIUM BOROHYDRIDES **Gayane Godjoian**, Bakthan Singaram, Department of Chemistry and Biochemistry, University of California, Santa Cruz, CA 95064, USA

Aminoborohydrides ($LiABH_3$) are powerful reducing agents which reduce various alkylcylohexanones, such as 4-*tert*-butylcyclohexanone, to give predominantly *trans*-alcohol. This selectivity can be altered by pre-mixing $LiABH_3$'s with trisiamylborane to get essentially pure *cis*-alcohol. Reduction of alkylcylohexanones with lithium B-aminodialkylborohydrides will also be presented.

505. REDUCTIVE PHOTOCHEMICAL CARBOXYMETHYLATION OF OLEFINS WITH α-ALKYLTHIOACETATES. Lisa X. Deng, **Andrei G. Kutateladze**, Department of Chemistry and Biochemistry, University of Denver, Denver, CO 80208.

Radical photochemical reactions in reducing media (X• + C=C → X-C-C• → X-C-C-H) have been studied in depth over the years and proved useful for various synthetic procedures. We now report our mechanistic and synthetic study of the photochemically initiated reductive carboxymethylation of olefins with α-(alkylthio)acetates which (i) is a carbon-carbon bond forming reaction, (ii) does not require a hydrogen-rich solvent or an additional reducing reagent for its completion and (iii) is experimentally simple, i.e. does not require degassing, is indifferent to the presence of water or polar functional groups, etc. An example of the synthesis of an ω-hydroxy-substituted fatty acid is shown below.

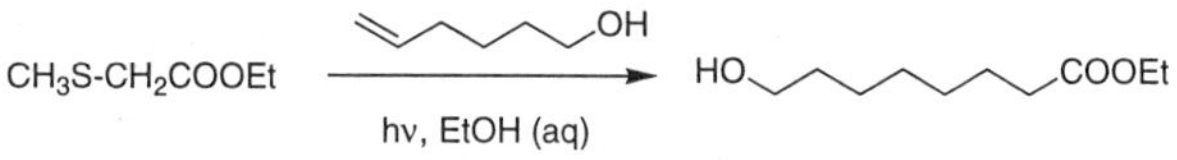

$$CH_3S\text{-}CH_2COOEt + \text{(alkene-OH)} \xrightarrow{h\nu,\ EtOH\ (aq)} HO\text{-}(CH_2)_n\text{-}COOEt$$

506. CHEMISTRY OF N-ALKYL-O-ACYL HYDROXAMIC ACID DERIVATIVES. Andrew J. Clark*, Joanne L. Peacock, Department of Chemistry, University of Warwick, Coventry, CV4 7AL, England UK.

N-alkyl-O-benzoyl hydroxamic acids (**1**) undergo rearrangement to give α-benzoyloxy amides (**2**) under basic conditions. Alternatively reaction with tributyltin hydride furnishes the corresponding β-lactams (**3**) via a 4-exo amidyl radical cyclisation. In addition to cyclisation (**3**), some reduction (**4**) and rearranged compounds (**2**) are also detected.

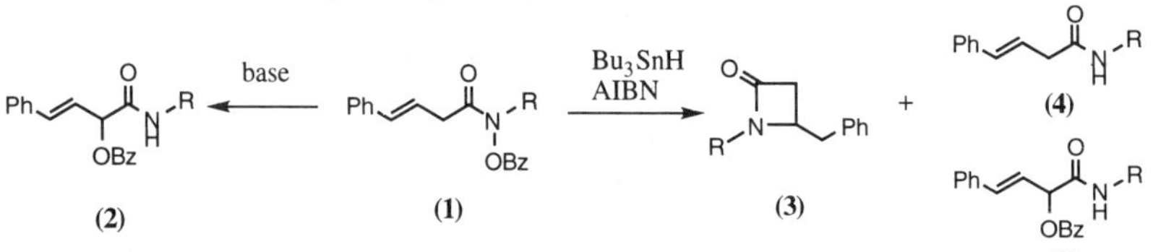

507. THE COUPLING REACTION OF ALKENES WITH α-BROMO CARBOXYLIC ACID DERIVATIVES USING NICKEL BORIDE AND BOROHYDRIDE EXCHANGE RESIN IN METHANOL. Meyoung Ju Joung, Dong Won Lee, Jin Hee Ahn†, Ki Hong Lee†, and **Nung Min Yoon***, Department of Chemistry, Sogang University, Seoul 121-742, Korea, †Jinro Central Research Institute, Yong In 449-910, Korea

The coupling reaction of alkenes with α-bromo carboxylic acid derivatives was studied in the presence of excess sodium iodide using nickel boride (cat.) and borohydride exchange resin in methanol. In the reactions with excess (2.5 equiv) α-bromo carboxylates, monosubstituted and 1,2-disubstituted alkenes gave the corresponding γ-iodo carboxylates, halogen atom transfer reaction products, whereas 1,1-disubstituted and trisubstituted alkenes gave the corresponding coupling products, unsaturated carboxylates. Reactions of norbornene and 1,1-disubstituted alkenes gave good to excellent yields (66-91%) of the corresponding coupling products, whereas those of 1,2-disubstituted and monosubstituted alkenes gave poor to moderate yields (15-43%).

508. A NEW COUPLING REACTION OF ALKYL IODIDES WITH ELECTRON DEFICIENT ALKENES USING NICKEL BORIDE AND BOROHYDRIDE EXCHANGE RESIN IN METHANOL. Tae Bo Sim, Jaesung Choi, Meyoung Ju Joung, and **Nung Min Yoon***, Department of Chemistry, Sogang University, Seoul 121-742, Korea

The radical addition reaction of alkyl iodides with α,β-unsaturated esters, nitriles and ketones proceeds in moderate to excellent yields (54-95%) using nickel boride(cat.)-borohydride exchange resin in methanol. Compared with tributyltin hydride method, this method has an advantage of simple work-up, since Ni_2B-BER can be removed readily by filtration.

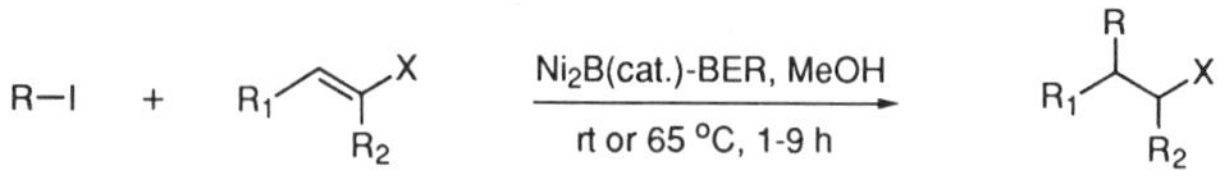

$$R\text{-}I + R_1\text{-}CH=CR_2\text{-}X \xrightarrow[\text{rt or 65 °C, 1-9 h}]{Ni_2B(cat.)\text{-}BER,\ MeOH} R_1\text{-}CHR\text{-}CR_2\text{-}X$$

R_1 = H, Me, Ph R_2 = H, Me   X = CO_2R, CN, COR

509.

SIGMATROPIC SHIFTS IN ALLENE OXIDE REARRANGEMENTS. FIRST GENERAL ROUTE TO ALIPHATIC [3,4] SHIFTS. Ihsan Erden, Fu-Pei Xu, Wei-Guo Cao, Department of Chemistry and Biochemistry, San Francisco State University, San Francisco, California, 94132

[3,3] Sigmatropic rearrangements are among the most extensively studied intramolecular shifts in organic chemistry. Their [3,4] counterparts were first observed in 1,4-cyclohexadienol-benzene rearrangements alongside competing [1,2] and [3,3] shifts. Aliphatic examples of [3,4] sigmatropic shifts have remained virtually unknown, except for one isolated case. Dewar's prediction that aliphatic [3,4] shifts should proceed with ease if suitable systems became available has now been borne out in our experiments. Extension of our saturated fulvene endoperoxide-allene oxide methodology to suitably substituted analogs has uncovered the first general route to acyclic [3,4] shifts. The scope and limitations of these reactions as well as application to other intramolecular shifts will also be presented.

510.

A MILD DEPROTECTION OF N-ACETYL AMINO ACIDS. Mark J. Burk, **John G. Allen**, Judy A. Straub, Paul M. Gross Chemical Laboratory, Department of Chemistry, Duke University, Durham NC 27708-0348

A convenient protocol has been developed for the conversion of N-acetamido amino acids to the corresponding N-t-butoxycarbamate (Boc) derivatives. The Boc-protected amino acids were obtained in high yield without racemization.

511. **LEWIS ACID-DEPENDENT PRODUCT FORMATION IN THE INTERRUPTED NAZAROV CYCLIZATION.** Cindy C. Browder and F. G. West, Department of Chemistry, University of Utah, Salt Lake City, Utah 84112

Treatment of divinyl ketones, such as trienone **1**, with Lewis acids results in Nazarov cyclization with concomitant trapping of the oxyallyl intermediate by the pendant π-nucleophile. In the case of trienone **1**, variation of the Lewis acid allowed for formation of different polycyclic systems, including bicyclic products **2** and tricyclic [3+2]-type products **3**.

512. **TORQUOSELECTIVE NAZAROV CYCLIZATIONS OF FACIALLY BIASED DIENONES.** Timothy D. White, Miles G. Egan and F. G. West, Department of Chemistry, University of Utah, Salt Lake City, UT 84112

Dienones **1** were synthesized and subjected to various Lewis acids under Nazarov cyclization conditions. A silicon directed cyclization (R=TMS) produced α, β-unsaturated ketone **2** in good yield as a 9:1 mixture of exo and endo diastereomers. When R=Ph, treatment with TiCl$_4$ furnished tricycle **3** as a single diastereomer, presumably via chloride trapping of the oxyallyl intermediate. The origins of the high torquoselectivity seen in these cases will be discussed.

513. STEREOCHEMICALLY CONTROLLED EPOXIDE OPENING OF 6-CYANO-3,4-EPOXY-2,2-DIMETHYLCHROMANE BY METALLO-ANILINE DERIVATIVES.
Jeffrey T. North,* David R. Kronenthal, Annie J. Pullockaran, Chemical Process Research, Bristol-Myers Squibb, P.O.Box 4000, Princeton, NJ 08543

The reaction of metallated aniline **1** with epoxide **2** was found to be highly counter-ion dependent. Employment of the lithium salt of **1** produced the expected *trans*-aminol **BMS 191,095**. Conversely, the use of magnesium as the counter-ion resulted in a syn-opening of the epoxide to produce the *cis*-aminol **BMS 199,080**. The reaction of the lithium and magnesium salts of *N*-methylanilines with **2** was also examined and the results of both studies will be discussed.

514.
COMPLEMENTARY ROUTES TO (*E*)- AND (*Z*)-TETRASUBSTITUTED β-ALKYNYL ACRYLONITRILES. Sean M. Kerwin and **Wendi M. David** Department of Medicinal Chemistry, College of Pharmacy, University of Texas, Austin, Texas 78712

Tetrasubstituted β-alkynyl acrylonitriles are relatively rare. We have investigated a number of routes in order to stereoselectively produce both the (*E*)- and (*Z*)-stereoisomers of these compounds. We have found that *C,N*-dialkynyl imines undergo a facile, stereoselective thermal rearrangement to (*Z*)-β-alkynyl acrylonitriles. This rearrangement may take place through an aza-Bergman–retro-aza-Bergman cyclization process. The corresponding (*E*)-β-alkynyl acrylonitriles can be obtained through a palladium catalyzed coupling of the appropriate enol triflate and an alkynyltrialkyl stannane.

515. BENZYL ESTER AMIDATIONS CATALYZED BY *PSEUDOMONAS CEPACIA* LIPASE
Jonathan Grote and Maciej Adamczyk, Divisional Organic Chemistry, Dept 9NM, Abbott Laboratories, Abbott Park, IL 60048-3500

Regioselective transformation of polyfunctional compounds is a challenging problem in organic synthesis, especially in the cases where a structure that is sensitive to acid, base, oxidation, or reduction limits the choice of reagents needed to accomplish a particular transformation. In recent years, lipases have become attractive as biocatalysts capable of achieving regioselective reactions under extremely mild (near neutral pH) conditions; they can be used in a wide variety of organic solvents and do not require a coenzyme for activity.

Although lipases have been known for some time as catalysts useful for regioselective transesterification or hydrolysis, studies regarding their use as ester *amidation* catalysts are limited. With the lipase from *Pseudomonas cepacia*, we have found that amidation reactions are catalyzed with *benzyl* esters, producing high yields of amides with a variety of different amines. Furthermore, regioselective reactions have been observed for a variety of different diesters and diamines.

516. POLYETHYLENE GLYCOL AS AN EFFECTIVE CATALYST IN ADDITION OF HYDROGEN BROMIDE TO OLEFINS. Tomoo Matsuura and Kiyoshi Watanabe, Nippon Zeon Co., Ltd., R&D Center, 1-2-1 Yako, Kawasaki-ku, Kawasaki-shi, 210 Japan

Addition of hydrogen halides to olefins is one of the important reactions in organic synthesis. Some catalysts, e.g., metal halides and quaternary ammonium salts are known. However, metal halide catalysts sometimes cause side reactions that depress the yield of the corresponding adduct, and quaternary ammonium salts are not effective when HBr gas is used. We wish to report herein that hydrobromination is accelerated in the presence of polyethylene glycol (PG). The reaction proceeds by a free-radical pathway, and the rate of HBr transfer to the organic phase is accelerated in the presence of PG. Thus PG is effective even when HBr gas is used in place of hydrobromic acid as a HBr source.

517.
REACTIVITY UMPOLUNG USING PHOSPHINE CATALYSIS: APPLICATION TO C-N BOND FORMATION . Barry M. Trost* and Gregory R. Dake, Department of Chemistry, Stanford University, Stanford CA 94305-5080.

Through Michael addition, phosphines can impart a "reactivity umpolung" to ynoate acceptors. The use of nitrogen nucleophiles in the phosphine-catalyzed reaction will be presented. 2-Alkynoates react at the C_α position with nitrogen nucleophiles to provide dehydroamino acid derivatives. Methyl 2-butynoate reacts at its C_γ position with nitrogen nucleophiles such as hydroxamic acid esters. The ability to form 1:1 adducts between electron deficient alkynes and nitrogen nucleophiles using catalytic phosphine constitutes a synthetically efficient process.

518. CATIONIC COUPLING OF 4-ACETOXY-1,3-DIOXANES WITH NUCLEOPHILES: SYNTHESIS OF *ANTI* 1,3-DIOL SYNTHONS
S. D. Rychnovsky,* N. A. Powell, University of California, Irvine, CA 92697

4-Acetoxy-1,3-dioxanes, readily prepared from 3-hydroxycarboxylic acids, have been shown to undergo cationic coupling with a wide variety of nucleophiles to produce *anti* 1,3-dioxanes in excellent diastereoselectivities.

519. FORMATION OF SUBSTITUTED CYCLIC OLEFINS THROUGH THE RING CLOSING METATHESIS (RCM) OF DIENES WITH 2,2-DISUBSTITUTED DOUBLE BONDS
Thomas A. Kirkland and Robert H. Grubbs,* Arnold and Mabel Beckman Laboratories of Chemical Synthesis, Division of Chemistry and Chemical Engineering, California Institute of Technology, Pasadena, CA 91125.
The ruthenium alkylidene $Cl_2(PCy_3)_2Ru(CHPh)$ has been shown to be active for the ring closing metathesis (RCM) of unsubstituted dienes. In order to examine the synthetic versatility of this alkylidene, model dienes containing 2,2-disubstituted double bonds were ring closed to yield

cyclics containing variously substituted double bonds. Five, six and seven membered cyclics were formed with trisubstituted double bonds. Formation of a cyclic containing a tetrasubstituted double bond was unsuccessful. Although some steric tolerance was shown through the RCM of an isopropyl substituted diene, no RCM was observed with a *tert*-butyl substituent. RCM is not observed with either electron donating or electron withdrawing substituents in conjugation with the double bond; however the reaction proceeds with substitutents containing functional groups which do not alter the electronics of the double bond. In similar studies, the molybdenum alkylidene $[(CF_3)_2(CH_3)CO]_2(NAr)Mo(CMe_2Ph)$ was shown to be more active for the RCM of many of these substrates.

520. SYNTHESIS OF 1,1'-IMINODICARBOXYLIC ACIDS FROM ORGANOBORONIC ACIDS
Nicos A. Petasis,* Ilia A. Zavialov and Valery V. Fokin
Department of Chemistry and Loker Hydrocarbon Research Institute
University of Southern California, Los Angeles, CA 90089-1661

Reaction of α-amino acids (**1**) with glyoxylic acid and various organoboronic acids (**2**) gives 1,1'-iminodicarboxylic acid derivatives (**3**) with good yields and good to excellent diastereoselectivities. The reaction also works with peptides to form the corresponding N-carboxyalkyl peptide derivatives.

521. A PRACTICAL SYNTHESIS OF UREAS FROM PHENYL CARBAMATES.
Bounkham Thavonekham, Bio-Mega/Boehringer Ingelheim Research Inc., 2100 Cunard Street, Laval, Quebec, H7S 2G5 Canada.

Ureas derivatives are important in the pharmaceutical area. Here we report a synthetic method for their preparation. We explore the potential of phenyl carbamates to serve as intermediates for the preparation of N,N'- unsymmetrically substituted ureas. The carbamates are treated with a stoichiometric amount of amine in DMSO at ambient temperature, generating the ureas in high yield. The reaction is simple, safe, fast, inexpensive and easily scaled up. The scope and limitations of the reaction will be presented.

R_1 = Aryl, Aliphatic
R_2 = ester, ketone, CN, OMe ...

(74-95.5%)

522. BASE-PROMOTED SYNTHESIS OF UNSYMMETRIC UREA AND HYDANTOIN DERIVATIVES VIA α-LACTAMS. M. Madhava Reddy; Robert V. Hoffman.* Department of Chemistry and Biochemistry, New Mexico State University, Box 30001, Department 3C, Las Cruces, NM 88003

Unsymmetric ureas have been found to be useful as cholinesterase inhibitors and peptidase inhibitors. A general synthesis of urea and hydantoin derivatives from *O*-sulfonylated hydroxamic acids is described. Sodium hydride was used as the base to form an α-lactam intermediate which subsequently reacts with amines or hydrazines to give urea derivatives and hydantoins, respectively.

523. NEW METHODS FOR SYNTHESIS OF 3-FLUORO 2-KETO ESTERS AND OXAZOLIN-2-ONE-4-CARBOXYLATES. John F. Okonya; M. Catherine Johnson; Robert V. Hoffman.* Department of Chemistry and Biochemistry, New Mexico State University, Box 30001, Department 3C, Las Cruces, NM 88003

Densely functionalized compounds occur very commonly in natural products of varied biological activity. Several examples of 1,2,3-trifunctionalized compounds are useful as peptidomimetics. New methods for the synthesis of 3-fluoro 2-keto esters and oxazolin-2-one-4-carboxylates have been developed. The fluoro keto esters were obtained from α-keto esters by electrophilic fluorination with recently introduced fluorinating agent, Selectfluor.® The oxazolinones were prepared by the condensation reactions of 3-nosyloxy 2-keto esters with methyl carbamate.

524. A SHORT, EFFICIENT SYNTHESIS OF MONOFLUORO KETOMETHYLENE PEPTIDE ISOSTERE CORE UNITS. James Saenz; Robert V. Hoffman,* Department of Chemistry and Biochemistry, New Mexico State University, Box 30001, Department 3C, Las Cruces, NM 88003

Monofluoro ketomethylene peptide isostere core units (six examples) were prepared by a four step sequence from carboxylic acids. Little purification of intermediates is required and reasonable overall yields (30-60%) were obtained. The introduction of fluorine is accomplished by electrophilic fluorination using Selectfluor.™

525. A NOVEL SYNTHESIS OF S-METHYL THIOLCARBAMATES FROM N-ALKYL CARBONIMIDODITHIOIC ACID ESTERS CATALYZED BY H-MORDENITE AND H-Y ZEOLITES Indrasena T. Reddy and S. Rajappa. Division of Organic Chemistry (Syn.), National Chemical Laboratory, Pune 411 008, INDIA. FAX: +91-212-335153

Two different zeolite catalysts are used for the conversion of dimethyl carbonimidodithioates derived from various primary amines, amino acid esters and sulfonamides to corresponding S-methyl thiolcarbamates. Zeolites catalyze only the derivatives of amines and amino acid esters but not the sulfonamides, Elecronic environment around the reactive centre (imine) may be the reason. AM1 calculations are proved this effect. Details about the reaction mechanisms and thoeretical data will be discussed. We thank CSIR, New Delhi for financial support.

$$MeS\text{-}C(=N\text{-}R)\text{-}SMe \xrightarrow{\text{zeolite, toluene, reflux, 24 h}} MeS\text{—}CO\text{—}NH\text{—}R$$

where '=N-R' = amines and amino acid esters

526.

NEW REAGENTS FOR THE STEREOSELECTIVE SYNTHESIS OF CIS-α,β-UNSATURATED CARBOXYLATES AND NITRILES, Tony Y. Zhang*, John C. O'Toole, James M. Dunigan and Veronica Stewart, Chemical Process R&D, Lilly Research Laboratories, Eli Lilly and Company, Lilly Corporate Center, Indianapolis, IN 46285-4813

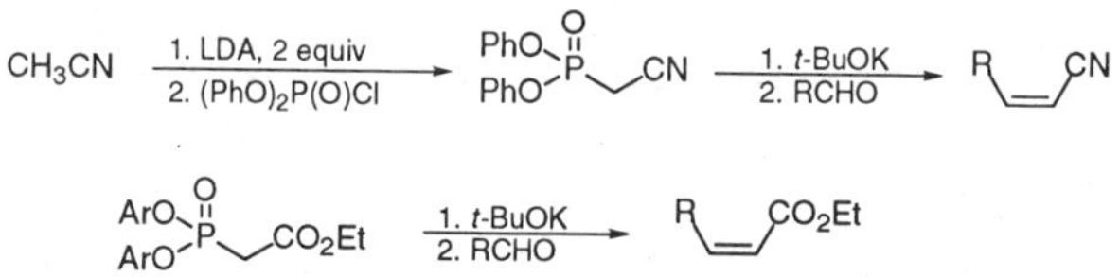

Diphenyl cyanomethylenephosphonate was prepared in an efficient manner from acetonitrile and was found to be a selective reagent for the synthesis of cis-alkenenitriles. For the carboxylates series, steric and electronic effects of the aryloxy (ArO-) group on the stereoselectivities of the olefination reaction were investigated.

527. **REACTION OF ORGANOLITHIUMS WITH FLUORO-N,N-DIALLYL-ANILINES: A BENZYNE–MEDIATED ANIONIC CASCADE.** Matthew W. Carson and William F. Bailey, Department of Chemistry, University of Connecticut, Storrs, CT 06269-4060.

Regioselective ortho-lithiation of either 2-fluoro- or 3-fluoro-N,N-diallylaniline initiates an unprecedented anionic cascade leading, as shown below, to an N-allyl-3,4-disubstituted indoline. The transformation involves loss of LiF from the ortho-lithiated species, regioselective intermolecular addition of the organolithium to the benzyne intermediate, and cyclization of the resulting [2-(N,N-diallylamino)phenyl]lithium.

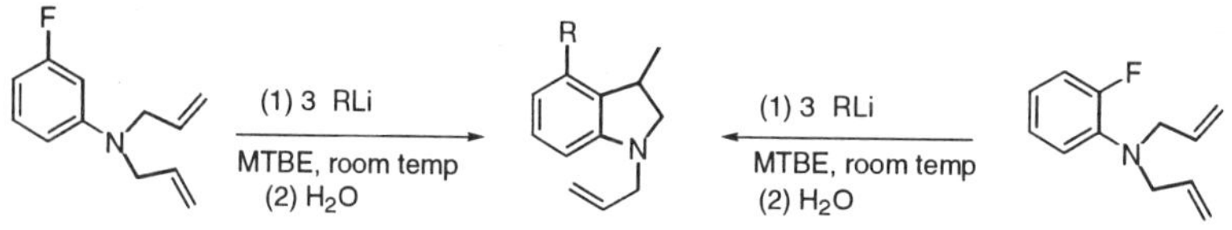

528. 1-OXO-2-CYCLOALKENYL-2-CARBONITRILES: FACILE CONJUGATE ADDITION OF GRIGNARD REAGENTS. **Fraser F. Fleming,*** Yifang Pu, Vaqar Sharief, and Faith Tercek; Department of Chemistry and Biochemistry, Duquesne University, Pittsburgh, PA 15282.

Several 1-oxo-2-cycloalkenyl-2-carbonitriles have been prepared as highly polarized alkenes that are ideal partners for Diels-Alder and Michael addition reactions. We have found that Grignard reagents exhibit a high propensity to react conjugately with 1-oxo-2-cycloalkenyl-2-carbonitriles and have performed a number of conjugate addition-silylation reactions to generate substituted nitriles (3). The advantages and limitations of this novel conjugate addition methodology will be presented.

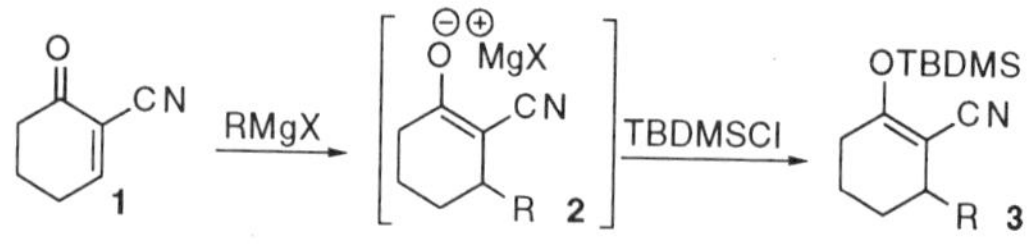

529. **STEREOCHEMISTRY OF ArSCl INITIATED STEPWISE COUPLING OF TWO VINYL ETHERS WITH SILICON AND TIN CONTAINING CARBON NUCLEOPHILES.** **Yury K. Kryschenko**, Margarita I. Lazareva, Karl Pracht, Ron Caple, and William A. Smit, Department of Chemistry, University of Minnesota-Duluth, Duluth MN 55812

The four component reaction sequence (ArSCl, VE-I, VE-II, Nu_c) is a promising way to make complex organic molecules. However, the validity of a synthetic protocol depends highly on ability to control its streochemical outcome. The use of cyclic vinyl ethers in this sequence leads to predominant or exclusive formation of only one diatereomer. For example, **1** is formed as 6:1 mixture of the two diastereomers, R^*,R^*,S^*,S^* predominant.

530. NOVEL SYNTHESIS AND REACTIONS OF HYDROXYLAMINES Mathew P. Leese
Department of Chemistry, Cardiff University, P.O. Box 912, Cardiff CF1 3TB, United Kingdom.

During the course of our studies on the *retro*-Cope elimination we have found it expeditious to synthesise the precursor hydroxylamines *via* the Mitsunobu reaction of mixed carbonate/carbamates **1** with alcohols. To this end we have evaluated a range of nucleophiles and subsequent deprotection of their Mitsunobu products **2**. We have utilised the *retro*-Cope reaction to elaborate a number of heterocyclic systems; N-hydroxy isoindolines **3** have been found to form quantitatively at room temperature, an investigation into the generality of this synthesis and the interactions of the products with metals has been initiated.

531. **Trifluoromethylthiolation of Trimethylsilyl Enol Ethers**: S. Munavalli[1], D. K. Rohrbaugh[2], G. W. Wagner[1], F. J. Berg[2], D. I. Rossman[2] and H. Dupont Durst[2]

[1]Geo-Centers, Inc., Ft. Washington, MD 20774 and [2] U. S. Army, Edgewood Research Development and Engineering Center, Aberdeen Proving Ground, MD 21010

Trimethylsilyl enol ethers have attracted considerable attention as highly useful synthons. The popularity of the trimethylsilyl enol ethers squarely rests on the fact that the bulky silyl moiety permits regio- and stereoselective reactions. In continuation of our interest in the chemistry of the trifluoromethylthio group, the reaction of trimethylsilyl enol ethers (1-4) with trifluoromethylsulfenyl chloride (5) has been found to furnish 2-(trifluoromethylthio)carbonyl compounds (6-10) accompanied with other minor products. The synthesis and the spectral characterization of the various products formed in this reaction are presented in this paper.

532. **SILYLACETIC ESTERS: ENOLATE REACTIONS AND POLYOL PREPARATION.** Mary M. Mader, Department of Chemistry, Grinnell College, Grinnell, IA 50112

The aldol reaction of α-silylacetic esters is described. Conditions have been found that minimize the subsequent Peterson elimination of the aldolate intermediate, and the β-hydroxy-α-silylester is isolated in good yield. The reaction is also applicable to imines. Work is in progress to convert the aldol products to diols and triols using known methodology.

$$EtO_2CCH_2SiMe_2Ph \xrightarrow[\substack{\text{a. LDA, -78}^\circ \\ \text{b. MgBr}_2\cdot\text{OEt}_2\text{, -78}^\circ\text{ (60 min)} \\ \text{c. R'CH=O, -78}^\circ\text{ (60 min)} \\ \text{d. DIBALH}}]{} \underset{\substack{\text{HO}\\\text{SiMe}_2\text{Ph}}}{\text{OH}\atop\text{R'}} \xrightarrow[\text{2. [O]}]{\text{1. (PhCO)}_2\text{O}} \underset{\substack{\text{BzO}\\\text{OH}}}{\text{OBz}\atop\text{R'}}$$

533. PREPARATION AND CHARACTERIZATION OF TRANS-1,4-DIAZIDO-1,4-DINITROCYCLOHEXANE AND EXO-2,5-DIAZIDO-ENDO-2,5-DINITRONORBORNANE: STABLE GEMINAL AZIDO-NITRO COMPOUNDS. J. J. Struckhoff Jr., G. K. S. Prakash, K. Weber, A. Schreiber, R. Bau, and G. A. Olah, Loker Hydrocarbon Research Institute and Department of Chemistry, University of Southern California, Los Angeles, California 90089-1661

The preparation and full characterization of compounds with geminal azido-nitro groups is described. This novel functinality is of potential interest for energetic materials. Using potassium ferricyanide as an oxidant in a modified Kornblum procedure, starting from the corresponding secondary nitro derivative, geminal azido-nitro compounds can be synthesized in high yields. Using this procedure trans-1,4-diazido-1,4-dinitrocyclohexane (6b) and exo-2,5-diazido-endo-2,5-dinitronorbornane (8b) were prepared and found to be stable crystalline compounds, which were fully characterized by ^{1}H and ^{13}C NMR, IR, and X-ray crystallography.

534. SYNTHESIS OF [3]DENDRALENE DERIVATIVES.
Wencheng Lou and Tze-Lock Chan*, Department of Chemistry,
The Chinese University of Hong Kong, Shatin, N.T., Hong Kong

[3]Dendralenes are acyclic hydrocarbons possessing the 3-methylene-1,4-pentadiene structural unit. These cross-conjugated systems serve as important vehicles for the comparison of resonance and molecular orbital theories and as unique twofold diene components in "diene-transmissive" Diels-Alder reactions. Although scattered examples of this class of molecules have been reported, a general synthetic pathway for them was lacking. We have developed a viable route to both symmetrical and unsymmetrical [3]dendralenes of the type 1 from diols 2 *via* bissulfones 3 using our one-flask Ramberg-Bäcklund reaction as the final step. Details of the reactions involved will be presented.

535. EFFICIENT SYNTHESES OF 2-ISOPINOCAMPHEYL- AND RELATED 2-TERPENYL-1,3,2-DIOXABOROLES FROM 1,2-DIKETONES AND 2-HYDROXYKETONES
P. V. Ramachandran, Z. H. Lu, and Herbert C. Brown* H. C. Brown and R. B. Wetherill Laboratories of Chemistry, Purdue University, West Lafayette, IN 47907-1393

Efficient syntheses of synthetically important chiral 1,3,2-dioxaboroles have been developed. The reduction of 1,2-diketones with *B*-chlorodiisopinocampheylborane, or a reaction of α-hydroxyketones with isopinocampheyldichloroborane, provides rapid, convenient syntheses of chiral enediolboronates of the type, 2-isopinocampheyl-1,3,2-dioxaboroles. The latter procedure has been utilized to synthesize a series of 2-terpenyl-1,3,2-dioxaboroles from terpenyldichloroboranes.

536. REGIOSELECTIVE OPENING OF 3,4-EPOXY ALCOHOLS WITH ALKYNYL ALANES-CONFORMATIONAL SUDIES L. Castro-Rosario, J. A. Prieto*, R. Tirado, A. Sanchez-Santos and I. Kock. Department of Chemistry, University of Puerto Rico, Río Piedras

The preparation of polypropionates continues to be of great interest. Their broad range of biological activity and extensive array of contiguous stereogenic centers makes them formidable targets. We have been developing a stereoselective reiterative approach for their synthesis based on oxirane chemistry. As part of this study we had to prepare and cleave several hindered 3,4-epoxy alcohols. Alkynyl alanes proved to be effective reagents for this task showing a preference for attack on the less hindered side. This preference was sometimes compromised by the stereochemical relationship between the groups present at the reacting epoxides A study on the scope and limitations of this reaction will be presented.

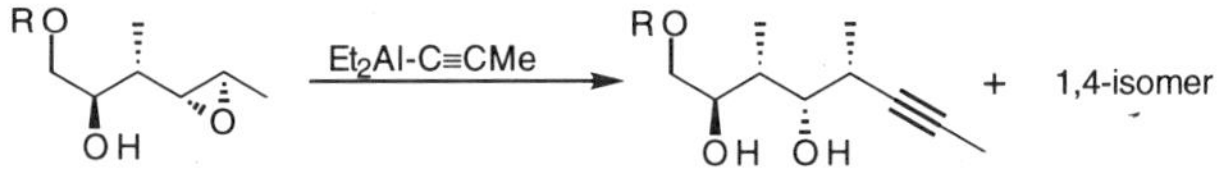

Support provided by the Department of Education-Patricia Harris Fellowship and NIH MBRS-RR08102

---

537. **THE SYNTHESIS OF HIGHLY FUNCTIONALIZED BICYCLO[3.2.0]HEPTANONES FROM 3-HOMOALLYLCYCLOBUTENONES. A TOTAL SYNTHESIS OF (±)-PRECAPNELLADIENE.** James M. MacDougall, Philip S. Turnbull, Sharad K. Verma and Harold W. Moore, Department of Chemistry, University of California at Irvine, Irvine, CA 92717.

A novel synthetic route to highly functionalized bicyclo[3.2.0]heptanones from 3-homoallylcyclobutenones is reported. The methodology is applied to the total synthesis of the sesquiterpene natural product (±)-precapnelladiene.

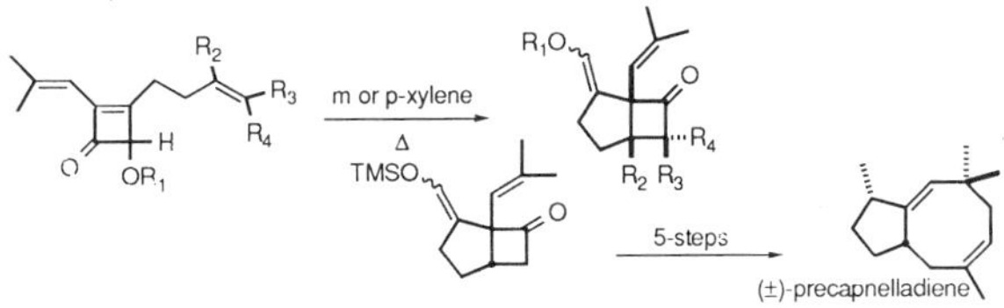

---

538. **AZA-ANNULATION/IONIC REDUCTION OF 8-SUBSTITUTED-2-TETRALONES DOES NOT AFFORD *TRANS*-BENZO[f]QUINOLINONES**

John Brennan, Perry C. Heath, Gregory A. Stephenson, **Leland O. Weigel**
Chemical Process Research and Development
Eli Lilly and Company, Indianapolis, IN 46285-4813

Aza-annulation of 8-substituted-2-tetralones (**1**, R = OMe, Br) followed by ionic reduction afforded **2** and **3** as the major products and not **4** as reported (R = OMe, *Synthesis*, **1993**, 1151). We will present corrected product distribution analyses and single crystal X-ray data to support our assignments as well as a model to rationalize the selectivity observed.

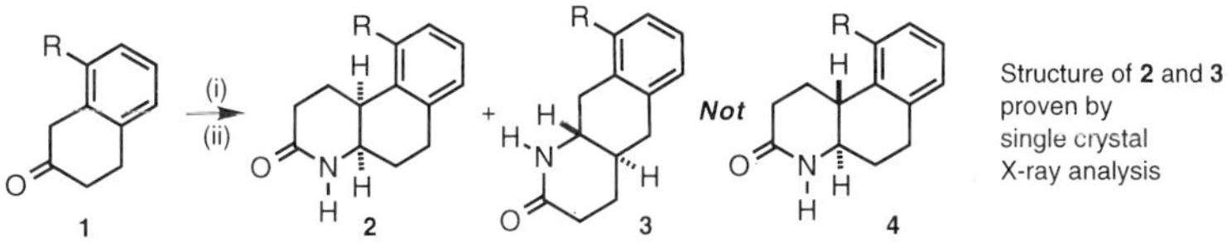

Structure of **2** and **3** proven by single crystal X-ray analysis

(i) Pyrrolidine, -H_2O; CH_2=CHCONH$_2$, 84-91 °C (ii) Et$_3$SiH, TFA, 15-80 °C

539.

A DIASTEREOSELECTIVE SYNTHESIS OF TRANS, TRANS-OCTAHYDRONAPHTHOQUINOLIZINE I'ai Mi David I. Schuster, Department of Chemistry, New York University, New York City, New York 10003-6603

The rigid tetracyclic systems of octahydronaphthoquinolizines (OHNQ) are close structural analogues of compounds which are known to be biologically active due to specific interactions at neuroreceptor sites in the central nervous system. In particular, OHNQ's show specific affinity for sigma receptor binding sites. Sigma receptors in the central nervous system and the periphery have been implicated in a variety of physiological processes, and may represent links between the nervous, endocrine and immune systems. The approaches to this system which have been reported previously produce a mixture of diastereomers with a low percentage of the biologically most active trans, trans isomer. A facile approach involving stereospecific reduction of an enamine moiety in a tetracyclic system has made it possible to synthesize diastereospecifically the biologically active octahydronaphthoquinolizine with trans, trans stereochemistry.

540. DIASTEREOSELECTIVE END DIFFERENTIATION OF PSEUDO C_2-SYMMETRIC 1,3,5-TRIOLS. Jennifer A. Shepherd, Cecile S. McKnight, and David C. Myles*, Department of Chemistry and Biochemistry, UCLA, Los Angeles, CA 90095-1596, USA.

Oxidative cyclization of pseudo C_2-symmetric *para*-methoxybenzyl (PMB) ethers to *para*-methoxybenzylidene (PMP) acetals gives highly diastereoselective end differentiation. For example, when R = CH(OH)CO$_2$Et and R' =H, a **91:9** diastereoselectivity is achieved.

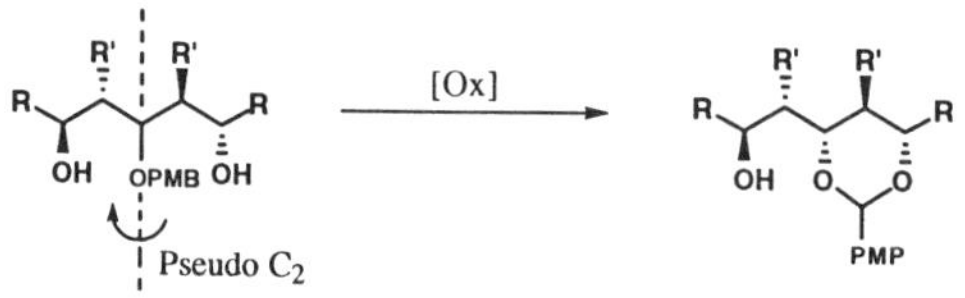

541.

A Novel Reaction of Cyanogen Iodide with Cyclic Tertiary Amines. B. H. Lee*, M. F. Clothier and D. A. Pickering. Animal Health Discovery Research, Pharmacia & Upjohn Inc., 7000 Portage Road, Kalamazoo, MI 49001-0199

It has recently been shown that cyanogen iodide reacts with the tertiary amine ring of marcfortine A via an interesting mechanism to give cyano and iodocyano substituted products. We have now extended this reaction to various cyclic tertiary amines, the results of which will be presented.

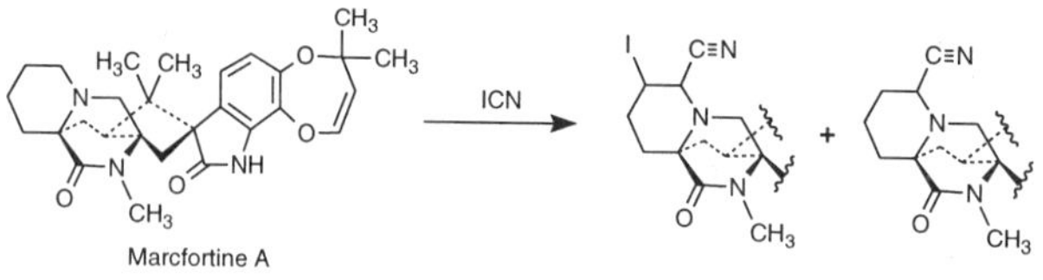

542.

Hydroxylation of Marcfortine A *via* Biotransformation and Organic Synthesis. B. H. Lee*, and M. F. Clothier. Animal Health, Pharmacia & Upjohn Inc., 7000 Portage Road, Kalamazoo, MI 49001-0199

Utilizing selected isolated, we successfully hydroxylated marcfortine A at carbon atoms 5, 10, 12, 14, 15, 16, 24 and 27. In order to obtain useful amounts of some of these hydroxy derivatives and to provide unequivocal proof of structure, we hydroxylated marcfortine A at C14, C15 and C16 by semi-synthetic procedures. These semi-synthetic derivatives were identical in all respects to those obtained from biotransformation experiments.

Marcfortine A

543. **Studies Toward Pancratistatin: Cycloadditions of a (-)-*trans*-6-Amino-2,4-cyclohexadien-1-ol Derivative.** Paul I. Higgs and Tomas Hudlicky.
Department of Chemistry, University of Florida, Gainesville FL 32611, USA.

The optically pure amino alcohol **1** has been prepared from the halogenated aziridines **2**. Successful Diels Alder cycloadditions were accomplished with *p*-benzoquinone and benzyne. Details of further utility of synthon **1** will be presented.

544. **CYCLIZATIONS OF UNSYMMETRICAL BIS-1,2-(3-INDOLYL)ETHANES: SYNTHESIS OF (-)-TJIPANAZOLE F1.** Eric J. Gilbert, David L. Van Vranken*, Department of Chemistry, University of California, Irvine, CA 92697, USA.

The cyclization of bis-1,2-(3-indolyl)ethane **1** can be controlled using thermodynamic or kinetic conditions to provide either tetrahydroindolo[2,3-a]carbazole **2** or **3** as the preferred product. This methodology was used in the first synthesis of (-)-tjipanazole F1.

545. **THE DIYL TRAPPING REACTION: A VIABLE ROUTE TOWARD THE SYNTHESIS OF APHIDICOLIN.** R. D. Little, M.M. Ott, A. Matzeit, J. Dickhaut and Z. Tesfai, Department of Chemistry, University of California, Santa Barbara, Santa Barbara, CA 93106

A synthetic route has been devised that converges at a key intermediate en route to a formal total synthesis of aphidicolin (**1**). The main features include the regioselective intramolecular diyl trapping reaction of that creates the C-D bicyclic ring system, the oxidative cleavage of a trisubstituted olefin to begin construction of the A-B rings, and a highly efficient Shapiro reactions that affords the key intermediate towards the synthesis of the tetracyclic system. The details of this approach as well as the progress made towards the synthesis of aphidicolin (1) will be described.

546. CONVENIENT SYNTHESIS OF TAVACPALLESCENSIN AND OCCIDOL. Tse-Lok Ho* and Y.-j. Lin, Department of Applied Chemistry, National Chiao Tung University, Hsinchu, Taiwan, ROC.

The access to both tavacpallescensin (1) and occidol (2) from a common intermediate (3) demonstrates the effectiveness of symmetry considerations in synthetic design.

547. SYNTHESIS OF DIMETHYLNAPHTHGERANINE E. Dario Gomez, Michael Winters, Thomas J. Onofrey and Harold W. Moore, Department of Chemistry, University of California at Irvine, Irvine, CA 92717.

Reported is the synthesis of dimethylnaphthylgeranine E. Key to the synthesis are the utility of thermally induced ring expansion of 4-arylcyclobutenones and a novel photo-annulation reaction of quinones for the construction of the pyranonaphthoquinone nucleus found in the natural product.

548. NEW METHODS FOR THE PREPARATION OF HYDROPHENANTHRENES: THE TOTAL SYNTHESES OF (±)-NIMBIDIOL, (±)-HINIKIONE, AND (±)- ARUCADIOL . George Majetich, Shuang Liu and Jing Fang. *Department of Chemistry*, The University of Georgia, Athens, Georgia 30602.

Functionalized hydrophenanthrenes can be prepared using a cyclialkylation-based strategy. These annulation are highly dependent on the directing effects of the arene substituents and on conformation considerations. However, the introduction of steric hindrance, either in the form of a trimethylsilyl moiety or an alkyl group, has been found to facilitate ring closure. This methodology has been used to synthesize arucadiol and several other related diterpenes.

549. STUDIES DIRECTED TOWARD THE SYNTHESIS OF EPIBATIDINE AND RELATED ANALOGS. Neville P. Payri and Mark L. Trudell, Department of Chemistry, University of New Orleans, New Orleans, LA 70148.

The [4 + 2]cycloaddition reactions of N-acyl pyrroles (1) with 1,3-allenic esters (2) has been shown to proceed in high yield to afford the corresponding cycloaddition adducts (3). The application of this approach to the synthesis of epibatidine 4 and related analogs will be presented.

550. REGIOSPECIFIC SYNTHESIS OF 6-ALKYL TROPAN-2-ONES. Stacey A. Lomenzo[1], Amy L. Bradley[1], Naijue Zhu[2], Cheryl L. Klein[2], Mark L. Trudell[1]. [1]Department of Chemistry, University of New Orleans, New Orleans, LA 70148 and [2]Department of Chemistry, Xavier University of Louisiana, New Orleans, LA 70125.

A new synthetic approach for the regiospecific alkylation of the 6-position of the tropane ring system has been developed. Alkylation, desulfonylation and deprotection of sulfones 1 and 2 furnished a series of 6-*endo*-alkyl tropan-2-one derivatives 3 and 4 (R = Me, Et, Pr, Bu, Bn) stereoselectively in good yields.

551. STUDIES DIRECTED TOWARD THE TOTAL SYNTHESIS OF HAPLOMYRTIN. Thomas A. Gilmore and William A. Feld, Department of Chemistry, Wright State University, Dayton, OH 45435.

The diverse structures of natural lignans, a category of plant phenols, has resulted in the frequent review of their chemistry. The arylnaphthalene lignans include Haplomyrtin whose structure has been thoroughly characterized but for which a total synthesis has yet to be reported. The construction of the arylnaphthalene nucleus associated with Haplomyrtin has been carried out employing vanillin and piperonal as starting materials. The reaction sequence includes protection of a halogenated vanillin, a lithium metallation followed by coupling with piperonal, *in-situ* formation and DMAD Diels-Alder reaction of a benzofuran and subsequent deprotection followed by regiospecific reduction.

Haplomyrtin

552.
PROGRESS TOWARDS THE TOTAL SYNTHESIS OF (-)TECOMANINE. **Denise A. Ockey** and Neil E. Schore Department of Chemistry, University of California, Davis, Ca. 95616

The Pauson-Khand reaction has been proven to be a valuable method of obtaining complicated cylopetenone systems. Recent advances in the chiral auxiliary mediated Pauson-Khand reaction has provided an efficient route to the asymmetric synthesis of natural products that contain five membered rings. Using this approach the total synthesis of (-)-Tecomanine has been undertaken. The synthesis proceeds through an intermediate which can be obtained in a few steps by using straight forward chemistry. Pauson-Khand cyclization of the intermediate provides the basic framework of (-)- Tecomanine.

553. **PROGRESS TOWARD THE TOTAL SYNTHESIS OF THE HENNOXAZOLES**
Eric J. Zylstra, Miles Wan-Li She, and James W. Leahy*
Department of Chemistry, University of California, Berkeley, CA 94720-1460

The hennoxazoles are a family of bis-oxazole-containing marine natural products isolated by Scheuer *et. al.* in 1991. Hennoxazole A, the most abundant of these compounds, displays antiherpetic and analgesic properties. We will describe our efforts toward the enantiospecific total synthesis of the hennoxazoles by a highly convergent pathway.

554. **AN ATOM-ECONOMICAL STRATEGY FOR THE SYNTHESIS OF THE UNIQUE ALKALOID LEUCETTAMOL A**
Katharine McElhone Greene and James W. Leahy*
Department of Chemistry, University of California, Berkeley, CA 94720-1460

Leucettamol A (1) was reported in 1993 as an isolate from the sponge *Leucetta microraphis*. The material was reported to show no optical activity, an uncommon finding among natural products containing non-epimerizeable stereocenters. Progress toward the total synthesis of 1 via the retrosynthetic strategy outlined below will be discussed.

555. **THE SYNTHESIS OF C(6) AND C(7)-ALKYLATED SWAINSONINE ANALOGS VIA THE REDUCTIVE DOUBLE CYCLIZATION OF AZIDO-EPOXY ESTERS.**
William H. Pearson and Erik J. Hembre, Department of Chemistry, University of Michigan, Ann Arbor, MI. 48109-1055

The reductive double cyclization of azido-epoxy esters has been exploited to prepare a series of novel C(6) and C(7) ring-alkylated analogs of the anti-cancer agent (-)-swainsonine (1). The synthesis and mannosidase inhibitory activity of these compounds will be presented.

1 (-)-Swainsonine ($R^1=R^2=H$)
2 ($R^1 = H$, $R^2 =$ alkyl)
3 ($R^1 =$ alkyl, $R^2 = H$)

556. A FORMAL TOTAL SYNTHESIS OF (±)-GEPHYROTOXIN VIA THE INTRAMOLECULAR SCHMIDT REACTION. Wen-kui Fang and William H. Pearson*, Department of Chemistry, The University of Michigan, Ann Arbor, MI 48109-1055

Treatment of secondary azido bromoalkene 1 with triflic acid followed by reduction of the resultant iminium ions with sodium borohydride and transformation of the bromide into hydroxyl group provided the known tricyclic aniline 2 as the major product in overall 72% yield. Compound 2 had previously been converted to (±)-gephyrotoxin.

557. SYNTHESIS OF TWO SUBUNITS OF THE MARINE TOXIN POLYCAVERNOSIDE A
Lonnie A. Robarge, Duncan J. Wardrop, and James D. White*, Department of Chemistry, Oregon State University, Corvallis, Oregon, 97331

Polycavernoside A (1) is a highly toxic macrolide recently isolated from the red alga *Polycavernosa tsudai*. The southern fragment 2 has been prepared by a scheme incorporating an enantioselective microbial reduction and a stereoselective palladium (II) catalyzed alkoxycarbonylation. Details of the synthesis of 2 and progress towards the northern fragment 3 from R-(-)-pantolactone will be presented.

558. SYNTHESIS OF PODOPHYLLOTOXIN A-RING PYRIDAZINE ANALOGUE
Emmanuel Bertounesque,[a] Thierry Imbert,[b] and Claude Monneret*,[a]
[a]UMR 176, Institut Curie, Section Recherche, 26 rue d'Ulm, F-75231 Paris Cedex 05, France.
[b]Pierre Fabre Médicament, C.R.P.F., 17 avenue Jean Moulin, F-81106 Castres, France.

We have developed a Stille approach to N-heterocyclic A-ring podophyllotoxin derivatives exemplified by a nine-step synthesis of A-ring pyridazine podophyllotoxin 2 from (-)-podophyllotoxin 1.

559. SYNTHETIC STUDIES DIRECTED TOWARD THE RHIZOXINS: SYNTHESIS OF THE C10-C20 SUBUNIT
Gary E. Keck, Dhileepkumar Krishnamurthy, Ken A. Savin, and Victor J. Cee
Department of Chemistry, University of Utah, Salt Lake City, Utah 84112

The C10-C20 subunit of rhizoxin has been synthesized. The key steps of this synthesis are an asymmetric stannane addition to aldehyde **1** and an aldol condensation of the corresponding ketone with aldehyde **2**. The use of these three simple components provides a very short, convergent synthesis of the C10-C20 subunit.

560. PROGRESS DIRECTED TOWARD THE TOTAL SYNTHESIS OF RHIZOXIN: ENANTIOSELECTIVE SYNTHESIS OF C12 TO C26 SEGMENT. Jian Hong and Steven D. Burke*, Department of Chemistry, University of Wisconsin-Madison, Madison, WI 53706.

Rhizoxin is a novel type of 16-membered macrolide isolated from *Rhizopus Chinensis*. It exhibits antimicrobial, antifungal and *in vivo* antitumor activity. Phase II clinical trials for lung cancer and breast cancer are currently underway. The construction of a C12-C26 segment was achieved in an efficient manner by coupling of the oxazole chromophore side chain with the central portion *via* a HWE reaction. The central portion was prepared *via* a chelation-controlled Ireland-Claisen rearrangement.

561. PROGRESS TOWARDS THE TOTAL SYNTHESIS OF AT2433-B$_2$
John D. Chisholm and David L. Van Vranken*
Department of Chemistry, University of California, Irvine, CA 92697

A simple and efficient route to indolocarbazole *N*-glycosides possessing a hexacyclic aromatic core has been developed. This method takes advantage of non-aromatic precursors to overcome the lack of nucleophilicity of indole nitrogens. The mild glycosylation conditions allow the use of unactivated carbohydrates, proceed with excellent control of stereochemistry, and allow the use of disaccharides. The synthesis of a protected form of the antitumor alkaloid AT2433-B2 has been accomplished using this methodology.

562. TOTAL SYNTHESIS OF THIARUBRINE C, A DNA-CLEAVING NATURAL 1,2-DITHIIN. Yamin Wang and Masato Koreeda,* Department of Chemistry, The University of Michigan, Ann Arbor, Michigan 48109-1055

Thiarubrine C (**1**), a naturally occurring 1,2-dithiin, possesses potent antibacterial and antifungal activity. Unlike other natural thiarubrines, it has been shown to cleave DNA-strand. The first total synthesis of thiarubrine C has been achieved in 10 steps starting from 2,4-hexadiyne-1,6-diol. The final steps in the synthesis include two sequential palladium(0)-catalyzed cross-couplings, followed by fluoride anion-mediated deprotection and oxidative 1,2-dithiin ring formation. The details of the total synthesis of thiarubrine C as well as its mode of action for DNA cleavage will be discussed.

* To whom correspondence should be addressed. Fax: 1-313-764-7371. E-mail: koreeda@umich.edu

563. TOTAL SYNTHESIS OF TAUTOMYCIN AND TAUTOMYCIN ANALOGS
James E. Sheppeck II, **Wen Liu**, and A. Richard Chamberlin*
Department of Chemistry, University of California, Irvine, California, 92717

Tautomycin is a very potent protein phosphatase inhibitor. Protein phosphatases along with protein kinases are involved in a variety of cellular processes such as cell division, gene expression, neurotransmission, and many second messenger and signal transduction pathways. We have achieved the total synthesis of tautomycin through a highly convergent and efficient synthetic route. The strategy used for total synthesis makes available analogs for structure and activity studies.

564. PP1 AND PP2A AND THE OKADAIC ACID CLASS OF INHIBITORS. **Carla-Maria Gauss**, James E. Sheppeck, and A. Richard Chamberlin, Dept. of Chemistry, U.C. Irvine, Irvine, CA 92697

There is considerable interest in the elucidation of signaling pathways in the central nervous system, both between neurons, via neurotransmitters, and within them. The elements responsible for transmitting signals from membrane bound receptors to other destinations within the neuron comprise a complex system that is only partially understood. This project focuses on one aspect of that system, the reversible phosphorylation of Dopamine Cyclic AMP Regulated Phosphoprotein (DARPP) and its isoform Inhibitor-1 (I-1). These "molecular switches" are under the immediate control of protein kinase A and protein phosphatases (PPs), the activities of which can in turn be modulated indirectly by extracellular events. There is a structurally diverse group of natural toxins which exert their cytotoxic effects by inhibiting PP2A and/or PP1, thereby preventing the dephosphorylation of DARPP and disrupting the normal signaling pathways. These toxins are collectively known as the Okadaic Acid class of inhibitors which includes okadaic acid, calyculin, microcystin LR, and tautomycin. They are competitive inhibitors; i.e., their PP1/PP2A binding sites overlap. A recently published X-ray structure of a PP1-microcystin LR complex provides a unique opportunity to explore binding interactions of all of these inhibitors with both PP1 and PP2A, and to develop new selective inhibitors based on these natural products.

565.
SYNTHESIS OF NOVEL, ACHIRAL, AND REVERSIBLE INHIBITORS OF CARNITINE PALMITOYLTRANSFERASE I. **John A. Brinkman**, Goli B. Naderi, Robert C. Anderson, Edwin B. Villhauer,. Department of Metabolic Diseases, Preclinical Research, Sandoz Research Institute, Sandoz Pharmaceutical Corporation, East Hanover, New Jersey 07936.

Inhibition of carnitine palmitoyltransferase I (CPT I) has been recognized as a mechanistic approach of potential clinical value in the treatment of non-insulin-dependent diabetes mellitus (NIDDM). Our own efforts in this area have resulted in the creation of the first reported achiral inhibitors of CPT I. Approaches to couple a sterically hindered, very water-soluble aminocarnitine analog with very water-insoluble electrophiles provide high yields of the amide- and urea-based inhibitors.

$X = CH_2, NH$

566.

Synthesis of LY333531, an Isozyme Selective Inhibitor of Protein Kinase Cβ (PKCβ): Part I - Chemistry of Bisindolylmaleimides. Margaret M. Faul,* Christine A. Krumrich and <u>Leonard L Winneroski</u>. Lilly Research Laboratories, A Division of Eli Lilly and Company, Indianapolis, IN 46285-4813.

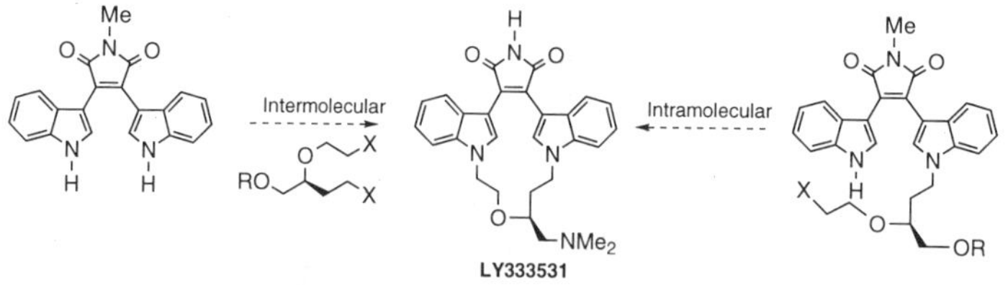

This talk will describe and compare the inter- and intramolecular approaches to the synthesis of 14-membered macrocylic bisindolyimaleimide **LY333531**. Formation of other amino substituted bisindolylmaleimides will also be discussed.

567. Synthesis of LY333531, an Isozyme Selective Inhibitor of Protein Kinase Cβ (PKCβ): Part II - Approaches to the Synthesis of the 6 Atom Bridge. Margaret M. Faul,* <u>Christine A. Krumrich</u> and Leonard L Winneroski. Lilly Research Laboratories, A Division of Eli Lilly and Company, Indianapolis, IN 46285-4813.

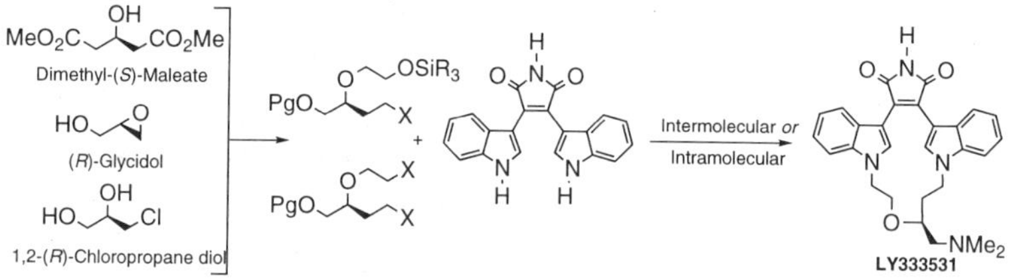

This talk will describe a number of approaches to the synthesis of the 6 atom chiral linker, required for the synthesis of **LY333531**, from dimethyl-(S)-maleate, (R)-glycidol and 1,2-(R)-chloropropane diol. The routes will be compared and the optimal synthesis of **LY333531** described.

568.

SYNTHESIS OF 2,2-DIFLUORO-1-AMINOCYCLOPROPANE-1-CARBOXYLIC ACID: AN UNUSUAL SUBSTRATE OF L-LACTATE DEHYDROGENASE. <u>K. Li</u>, W. Du, H. -W. Liu, Department of Chemistry, University of Minnesota, Minneapolis, MN 55455

A simple procedure for the preparation of 2,2-difluoro-1-aminocyclopropane-1-carboxylic acid was developed.

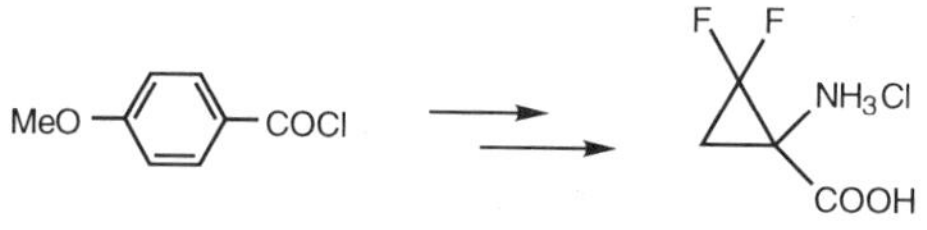

This compound which lacks the keto moiety was found to be an unusual substrate of L-lactate dehydrogenase. Details regarding the synthesis and kinetic studies will be presented.

569. PHOTOACOUSTIC FTIR SPECTROSCOPY, A NON-DESTRUCTIVE METHOD FOR SENSITIVE ANALYSIS OF SOLID-PHASE ORGANIC CHEMISTRY.
Francis Gosselin, Mauro Di Renzo, Thomas H. Ellis* and William D. Lubell*, Département de chimie, Université de Montréal, C. P. 6128, Succ. A, Montréal, Canada H3C 3J7

Growing use of combinatorial chemistry for drug discovery has accelerated the development of analytical techniques for studying organic chemistry on solid phase. In contrast to traditional FT-IR, we have found that photoacoustic FT-IR (PA-FTIR) requires no sample preparation and gives spectra of superior quality.[1] In PA-FTIR, a sensitive microphone measures an acoustic wave created by absorbed radiation diffusing as heat through the sample towards a boundary layer of gas. By detecting only absorbed radiation, PA-FTIR spectroscopy eludes the effects of light scattering and reflection. We will present the use of PA-FTIR to effectively monitor the modification of solid supports as well as syntheses of polymer-bound organic compounds without material loss.[1]

1. Gosselin, F.; Di Renzo, M.; Ellis, T. H.; Lubell, W. D. *J. Org. Chem.* In Press.

570. SOLID-PHASE SYNTHESIS OF 5-tBuPro^7-OXYTOCIN. EXAMINATION OF THE Cys-5-tBuPro AMIDE BOND STABILITY. Laurent Bélec, Benoît L'Archevêque and William D. Lubell*, Département de chimie, Université de Montréal, C. P. 6128, Succ. A, Montréal, Canada H3C 3J7

We are exploring the relationship between X-Pro amide conformation and peptide bioactivity through the use of 5-alkylprolines that function as rigid *trans*- and *cis*-rotamer surrogates.[1-3] A potent oxytocin antagonist possessing the Cys^6-Pro^7 amide locked in the *cis*-rotamer by a lactam cyclization provides evidence that the *cis*- and *trans*-rotamers interact at the receptor to produce contrasting biological responses.[2] Replacement of Pro^7 in oxytocin with 5-*tert*-butylproline (5-tBuPro) should increase the Cys^6-Pro^7 *cis*-amide rotamer population and thereby provide means to study the hypothesis that the *cis*-rotamer favors antagonism and the *trans*-rotamer is necessary for agonist activity. Our progress in the synthesis, conformational analysis and biological analysis of 5-tBuPro^7-oxytocin will be presented.

1. Beausoleil, E.; Lubell, W. D. *J. Am. Chem. Soc.* In Press.
2. Beausoleil, E.; L'Archevêque, B.; Bélec, L.; Atfani, M.; Lubell, W.D. *J. Org. Chem.* In Press.
3. Lombart, H.-G.; Lubell, W. D. *J. Org. Chem.* In Press.
4. Smith, D.D.; Slaninova, J.; Hruby, V. *J. Med. Chem.* **1992**,*35*, 1558.

571.
RING OPENING CROSS-METATHESIS ON SOLID SUPPORT. **Gregory D. Cuny,** Jingrong Cao and James R. Hauske, Sepracor, Inc., 111 Locke Dr., Marlborough, MA 01752

Recently, ring closing-metathesis has emerged as an effective means for the synthesis of macrocycles, carbocycles and heterocycles. However, the application of *inter*molecular ring opening cross-metathesis (ROM) for the convergent synthesis of small organic molecules has remained relatively unexplored due in part to competing polymerization reactions and to lack of regioselectivity. We now report that the application of solid-phase synthesis techniques to ROM has been found to be beneficial in preventing polymerization. In addition, enhanced regioselectivity was observed for some resin bound substrates compared to solution phase conditions. Specifically, we will report (1) ROM of resin bound bicyclic alkenes with terminal aryl olefins, (2) utilization of solid-phase ROM to produce a combinatorial library of several thousand highly substituted cyclopentane derivatives in a stereospecific fashion, and (3) regioselective ROM for a resin-bound bicyclic alkene with electron rich terminal aryl olefins.

572.
CONVERGENT PARALLEL SYNTHESIS. **Carmen M. Baldino,** David S. Casebier, Justin Caserta, Gregory Slobodkin, Cheng Tu, and David L. Coffen, ArQule, Inc., 200 Boston Avenue, Medford, MA 02155

The considerable gain in efficiency achievable by introducing the principle of convergence into automated parallel synthesis is illustrated with a seven-step synthesis of a sixteen-hundred-compound array of potential serine protease inhibitors, conducted on a 50 µmole scale (ca. 20 mg.

of each compound).

573.
SELF-ASSEMBLED, SELF-AMPLIFYING COMBINATORIAL LIBRARIES AS A NEW METHOD FOR LIGAND DISCOVERY: APPLICATION TO DNA-BINDING COMPOUNDS. **Bryan Klekota, Mark H. Hammond, and Benjamin L. Miller***, Department of Chemistry, University of Rochester, Rochester, New York 14607

Processes such as the polymerase chain reaction and in vitro selection of RNA aptamers have demonstrated the power of selection and amplification methods for the identification of biopolymeric ligands to receptors. As part of an effort to study analogous selection and amplification processes for nonbiopolymeric materials, we have developed a self-assembled combinatorial library approach to ligand synthesis that employs a solid-supported affinity reagent to select tight-binding compounds from an equilibrating mixture of coordination compounds. Results of the application of this method to the identification of novel DNA-binding compounds will be presented.

574.
SOLID PHASE SYNTHESIS OF A CHEMICAL DIVERSITY LIBRARY DESIGNED TO ELUCIDATE BINDING STRUCTURAL INFORMATION. **Joshua Odingo,** Bradley Sharpe and David Oare, Department of Bioorganic Chemistry, Genentech Inc., 460 Point San Bruno Boulevard, South San Francisco, CA 94080-4990.

A general approach utilizing solid phase organic chemistry techniques for generation and application of molecular diversity to rationally probe ligand-receptor binding structural requirements will be presented. This approach to lead compound identification incorporates all available information into a set of compounds that probe the topological disposition of key binding residues or functionalities for optimal substrate binding. Its integration with computational approaches to provide a powerful structural tool will be discussed. The synthetic reaction sequence involved a Palladium-catalyzed condesation of aryl iodides with acetylenic amines and a conversion of amines to guanidines. Details regarding the chemistry development and scope of the entire process will be discussed

575. PREPARATION OF A COMBINATORIAL LIBRARY OF TRICYCLIC NITROGEN HETEROCYCLES ON SOLID SUPPORT VIA DIELS-ALDER REACTION OF FURAN DIENES. K. Paulvannan and Jeffrey W. Jacobs, Affymax Research Institute, 3410 Central Expressway, Santa Clara, CA 95051.

An efficient methodology has been established on solid support to prepare tricyclic nitrogen heterocycles by the intramolecular Diels-Alder reaction of furan dienes with various dienophiles under mild condition (R_2 = H; 22 °C: R_2 = Me, Et; 85 °C). The methodology development and preparation of a combinatorial library of 2500 tricyclic lactams will be discussed.

576. SOLID PHASE SYNTHESIS OF COMBINATORIAL LIBRARIES USING ANHYDRIDES AS TEMPLATES John Perumattam, Sarvajit Chakravarty, Glenn McEnroe, Scios Inc., 820 West Maude Av. Sunnyvale, CA 94086

ABC type libraries are rapidly generated using readily available components such as anhydrides (A), amine nucleophiles (B), and resin-bound amines (C). Symmetrical diamines are reacted with chlorotrityl resin where one amine is protected as result of attachment to the resin leaving the other amine available for coupling reaction with various acids. The diverse acids are prepared by the reaction of anhydrides with amines as reported earlier.[1] A method is developed for the simultaneous synthesis of hundreds of amino compounds using array synthesis which results in a single well-defined compound per well..

1. John Perumattam, et al. 212th ACS meeting, August 25-29, 1996, Orlando, Florida, Paper # 86.

577. SOLID-PHASE SYNTHESIS OF PROSTAGLANDINS.
Lorin A. Thompson, Frederick L. Moore, Young-Choon Moon and Jonathan A. Ellman. Department of Chemistry, University of California at Berkeley, 94720

Many current libraries are assembled using simple and well-precedented reactions such as amide bond formation or one-pot heterocycle synthesis. However, the development of specific and high affinity ligands to a receptor target often requires the preparation of more complex structures. Prostaglandins are complex and delicate compounds, and the efficient and general synthesis of these molecules continues to provide a standard for demonstrating the versatility of new synthesis methods. This report describes solid-phase methods for the preparation of a variety of prostaglandins. A key challenge in these efforts is to incorporate a wide range of functionality that ideally would be obtained from readily accessible building blocks.

578. SYNTHESIS OF NOVEL POLYAZAPHANE LIBRARIES WITH POTENT ANTI-MICROBIAL ACTIVITIES BY SOLUTION PHASE SIMULTANEOUS ADDITION OF FUNCTIONALITIES (SPSAF). H. An, L. L. Cummins, R. H. Griffey, R. Bharadwaj, L. Wilson-Lingardo, B. D. Haly, L. M. Risen, J. R. Wyatt and P. D. Cook, Isis Pharmaceuticals, 2292 Faraday Avenue, Carlsbad, CA 92008

Three novel, asymmetric polyazaphanes have been synthesized in high yields by an efficient cyclization of 2,6-bis(bromomethyl)pyridine with novel protected triamines. Selective deprotection of orthogonally protected polyaza-phanes by thiophenol provided mono-*t*-Boc-protected polyazaphane scaffolds suitable for SPSAF studies. Model studies of small libraries of the scaffold using CZE analysis indicated that simultaneous addition of 10 benzylic bromides would result in libraries containing approximately equimolar amounts of all possible compounds. Sixteen high quality tertiary amine libraries (total complexity: 1600) were generated by this procedure. All libraries were monitored by TLC, purified by chromatographic techniques, and comfirmed by ESI MS spectral data. Four first round sublibraries exhibited antimicrobial activities. Twenty single compounds were synthesized; several demonstrated potent antibacterial activities.

579. A GENERAL STRATEGY FOR THE SYNTHESES OF SACCHAROPEPTIDES APPLICABLE TO A SOLUTION PHASE COMBINATORIAL APPROACH. P.S. Ramamoorthy and Jacquelyn Gervay,* Department of Chemistry, The University of Arizona, Tucson, Arizona 85721

Combinatorial chemistry has become an important tool in the pharmaceutical industry for developing new lead compounds possessing desired biological activity. Carbohydrates are biologically important molecules and provide an untapped source for generating combinatorial libraries. By appropriate modification, monosaccharides can be converted to sugar amino acids that are suitable for coupling with other classes of molecules. In this report, we outline a route to saccharopeptides derived from sialic acid that is amenable to a solution phase combinatorial approach. Peracetylated sialic acid was coupled to glycine, serine and alanine benzyl esters using standard peptide coupling procedures. The benzyl esters can be subject to hydrogenolysis to yield free acids that can be coupled again. The ease of reactions coupled with the minimal purification necessary make this applicable to combinatorial chemistry.

580. **METAL-MEDIATED [6π + 2π],*homo*[6π + 2π] ALKYNE CYCLOADDITIONS.**
John B. Sheridan, W. Chen, H-J. Chung, K. Chaffee, Department. of Chemistry, Rutgers, The State University of New Jersey, University Heights, Newark, NJ 07102.

One or two alkynes have been sequentially added to coordinated 6π-manifolds via photoassisted metal-promoted [6π + 2π],*homo*[6π + 2π] cycloadditions, forming novel tri- and tetracyclic species. In the case of cyclohexadienyl precursors, a [5 + 2],*homo*[5 + 2] process (i) ($ML_x = Mn(CO)_3$) gave complexes of the tricyclic ligand (**1**). For trienes, a [6 + 2],*homo*[6 + 2] coupling (ii) ($ML_x = Cr(CO)_3$) gave the metal-free tetracycles (**2**).

581. **OXIDATIVELY INDUCED REDUCTIVE ELIMINATION OF (PENTENEDIYL)IRON COMPLEXES: MECHANISM OF AN UNUSUAL STEREOCHEMICAL OUTCOME**
William A. Donaldson* and Young K. Yun, Department of Chemistry, Marquette University, P.O. Box 1881, Milwaukee, WI 53201-1881

The oxidatively induced decomplexation of (pentenediyl)iron complexes **1** leads to the formation of vinylcyclopropane carboxylates. The stereochemistry of the vinyl substituent relative to the other two substituents on the cyclopropane ring depends on the nature of the group R, as well as the conditions for oxidative decomplexation. A mechanistic rationale will be proposed which is further supported by stereochemical labelling experiments.

582.

THE SUZUKI-MIYAURA COUPLING: THE TRANSMETALLATION STEP. **John A. Soderquist** Anil Rane and Karl Matos, Department of Chemistry, University of Puerto Rico, Rio Piedras, Puerto Rico, 00931-3346

Extensive studies on the stereochemistry of alkyl group transfer from boron to palladium have revealed that the transmetallation occurs with complete retention of configuration with respect to carbon. For this base-catalyzed process, the Lewis acidity of the boron plays an important role with more oxygenated boranes undergoing coupling more slowly or not at all. Our results suggest that the process proceeds through a four-centered transition state (**I**) which can be reached either by the electrophilic attack of Pd(II) on the oxygen of the hydroxyborate, or for more oxygenated boranes, through the nucleophilic attack of a hydroxypalladium (**II**) on the neutral organoborane.

583. ENANTIOSELECTIVE SYNTHESIS OF THE C11-C24 SEGMENT OF MACROLACTIN A VIA ORGANOIRON METHODOLOGY. William A. Donaldson* and Vadapalli Prahlad, Department of Chemistry, Marquette University, P.O. Box 1881, Milwaukee, WI 53201-1881

Macrolactin A (**1**), a polyene macrolide with antitumor and antiviral activity, has been of recent synthetic interest. The enantioselective synthesis of the C11-C24 segment **2** will be described. This approach utilizes the ability of a (tricarbonyl)iron adjunct to control the formation of both the C15 and C23 stereocenters, *nine carbons remote from each other*.

584. A SYNTHETIC APPROACH TO AMPHIDINOLIDE A: APPLICATION OF NOVEL METAL-MEDIATED METHODS. Robert E. Maleczka, Jr. Lamont R. Terrell, Deanna J. Clark, Elena S. Schoenberger, and Carolyn M. Horvath, Department of Chemistry, Michigan State University, East Lansing, MI 48824

A highly convergent approach to the anti-leukemic macrocycle, amphidinolide A, as well as our related studies aimed at the development of new metal mediated processes will be presented.

585. PREPARATION OF PSEUDO-STEROIDAL STRUCTURES VIA CASCADE REACTIONS INITIATED BY INTERRUPTED NAZAROV CYCLIZATIONS. John A. Bender and F. G. West, Department of Chemistry, University of Utah, Salt Lake City, UT 84112.

The Nazarov cyclization is a well-known and convenient method for the preparation of cyclopentenone ring systems from acyclic cross-conjugated dienones. When initiated with Lewis acids, the mechanism is believed to proceed through an oxyallyl zwitterion which we have efficiently trapped with pendant nucleophiles to give complex polycyclic ring systems. The interrupted Nazarov cyclization has been further elaborated to initiate cascade reactions which produce three rings and up to six stereocenters in high yields with excellent diastereoselectivity.

586. POLYCYCLIC MOLECULES VIA TANDEM REDUCTIVE CYCLIZATION -INSERTION REACTION SEQUENCES. William E. Crowe, An T. Vu, John Ginn, Department of Chemistry, Emory University, Atlanta, GA 30332

Simple acyclic substrates, such as 1, can be converted to a variety of complex, polycyclic products such as 2-6 via a sequence of carbon-carbon bond forming reactions starting with titanium-mediated reductive cyclization. Products 3-6 result from an unusual cyclopentadienyl reductive elimination reaction.

587. NOVEL SYNTHETIC ROUTES TO CYCLOPENTANOID AND CARBAPENEM SKELETONS. Jiawang Zhu, Ephraim S. Vidal, Dora Fracchiolla Kass, and Iwao Ojima*. Department of Chemistry, State University of New York at Stony Brook, Stony Brook, New York 11794-3400.

New developments in the rhodium complex catalyzed silylcarbocyclizations of diynes and enynes will be reported. The bicyclocarbonylation of 4-substituted hepta-1,6-diynes such as **1**, with $HSiMe_2Bu$-t in the presence of $Rh(acac)(CO)_2$ under carbon monoxide (50 atm), gives **2** in good yield. The reaction of azetidinone **3** with $HSiMe_2Ph$ catalyzed by $Rh(acac)(CO)_2$ under nitrogen atmosphere affords **4** with carbapenem moiety.

588.

A DIASTEREOSELECTIVE SYNTHESIS OF BENZOPYRANS USING A NOVEL INTRAMOLECULAR NICHOLAS REACTION. C. Muller, E. Tyrrell*, School of Applied Chemistry, Kingston University, Penrhyn Road, Kingston, Surrey, KT1 2EE, England.

As a class of heterocyclic compounds benzopyran derivatives exhibit a range of physiological activities such as potassium channel modulation and adrenoreceptor antagonist activity. Our interest in the chemistry of organocobalt clusters has focused our attention towards its use in heterocyclic synthesis. In the process, we have developed a novel intramolecular cyclisation reaction (Nicholas reaction) between an organocobalt stabilised cation and a trisubstituted alkene to afford a new family of benzopyran derivatives. This efficient cylisation step was further optimised by the elaboration of a one pot process involving: a) the cobalt complexation of the teminal alkyne, b) the Lewis acid mediated intramolecular cyclisation reaction, and c) the cobalt decomplexation of the resulting product.

589. SELECTIVE C-C BOND FORMATION ON THE FIRST KETENE-ALKYNE COMPLEXES. H. C. Lo and Douglas B. Grotjahn*, Department of Chemistry and Biochemistry, Box 871604, Arizona State University, Tempe, AZ 85287-1604

Chemo-, regio-, and stereoselective coupling of diphenylketene with internal alkynes on Ir(I) at 60 °C gives iridabenzopyrans **2** (R^3 = Ph, 41-55%). In these reactions, the fragment $ClIr[P(i-Pr)_3]_2$ chemoselectively forms the first ketene-alkyne complexes **1** by loss of $P(i-Pr)_3$. Spectral data for **1** point to

η^2-arene coordination, explaining facile regiospecific C-H activation of the ketene phenyl substituent and subsequent alkyne and ketene insertion chemistry leading to **2**. Transformations of the π-system in **2** leading to **3** and demetallation of **2** leading to **4** will be presented.

590. CATALYTIC ENANTIOSELECTIVE ALLYLIC OXIDATION OF OLEFINS WITH COPPER(I) CATALYSTS AND NEW PERESTER OXIDANTS. Merritt B. Andrus* and Xi Chen, Purdue University, Department of Chemistry, 1393 H.C. Brown Building, West Lafayette, IN 47907

Asymmetric allylic oxidation of olefins using catalytic copper(I) bisoxazoline complexes with peresters in acetonitrile at low temperature gives allylbenzoate ester products in good enantioselectivity. New electron deficient peresters have been synthesized and used as oxidants to now increase the rate and selectivity of this important reaction. Dependency on the nature and stoichiometry of the copper(I) salt will also be discussed together with the use of new ligands.

591. FIRST ENANTIOSELECTIVE CATALYSTS FOR THE REARRANGEMENT OF ALLYLIC IMIDATES TO ALLYL AMIDES. Michael Calter, T. Keith Hollis, Larry E. Overman*, Joseph Ziller, G. Greg Zipp, Department of Chemistry, 516 Physical Sciences I, University of California, Irvine, CA 92697-2025.

The first enantioselective catalysts for the 1,3-transposition of allylic imidates to allylic amides will be reported. The first generation catalysts are based on enantiopure Pd diamine complexes (**1**) and a number of structural analogs. The second generation catalysts are based on cyclopalladated complexes (**2**). The isolation and characterization of these complexes and the results of catalysis will be discussed.

592. SYNTHESIS OF $C_{11}N_5$ MARINE ALKALOIDS: (±)-HYMENIN, STEVENSINE, HYMENIALDISINE, AND DEBROMOHYMENIALDISINE. Ying-zi Xu, Kenichi Yakushijin, and David A. Horne*, Department of Chemistry, Columbia University, New York, NY 10027

The synthesis of $C_{11}N_5$ marine sponge alkaloids (±)-hymenin (**1**), stevensine (**2**), hymenialdisine (**3**), and debromohymenialdisine (**4**) is described. The key steps in the synthesis centered around the generation of novel azafulvenium ions and their regioselective heterodimerization with 2-aminoimidazole (AI). A rarely used protodebromination/oxidation strategy was employed to selectively generate the desired α-bromo substitution pattern seen in hymenialdisine (3). In addition, the AI moiety was shown to be a useful precursor to the glycocyamidine unit found in 3 and 4, which suggests that AI derived natural products may be the biogenic forerunners to glycocyamidine metabolites.

Hymenin (1) Stevensine (2) Hymenialdisine (3) R=Br
Debromohymenialdisine (4) R=H

593. SYNTHETIC EFFORTS TOWARD THE TOTAL SYNTHESIS OF MADANGAMINE A.
Norbert Matzanke, Robert Gregg, and Steven M. Weinreb*, Department of Chemistry, The Pennsylvania State University, University Park, PA 16802-6300

Tricyclic compound **2** constitutes an advanced intermediate in the total synthesis of the marine alkaloid madangamine A (**1**). Synthesis of compound **2** has been achieved via high pressure Diels-Alder reaction of enone **4** and stereoselective aza-Cope rearrangement of intermediate **3**.

594. SYNTHESIS OF THE OCTAHYDRO-8B-AZAACENAPHTHYLENE RING SYSTEM, A PORTION OF THE DIMERIC COCCINELLID ALKALOIDS, James C. Shattuck, Department of Chemistry, University of Hartford, 200 Bloomfield Ave., W. Hartford, CT 06117, Jerrold Meinwald, Department of Chemistry, Baker Laboratory, Cornell University, Ithaca, NY 14853.

The "dimeric" coccinellid alkaloids exochomine, chilocorine A, and chilocorine B represent a challenging set of synthetic targets. An interesting structural feature of these alkaloids is the novel 3,4-disubstituted octahydro-8b-azaacenaphthylene ring system. Representatives of this ring system have been synthesized and functionalized for possible use in a synthesis of the "dimeric" coccinellid alkaloids.

595. ENANTIOSPECIFIC TOTAL SYNTHESIS OF TRYPROSTATIN A. Tong Gan and James M. Cook*, Department of Chemistry, University of Wisconsin-Milwaukee, Milwaukee, WI 53201.

The 3-methyl-6-methoxyindole **1** was converted into tryprostatin A (**3**) via a regiospecific bromination process. The 2-isoprenyl-6-methoxytryptophan **2** served as a key intermediate in this route.

596. A NOVEL ATOM TRANSFER APPROACH TO THE INDOLO[2,3-*a*]QUINOLIZIDINE RING SYSTEM. Heidi L. Hassinger and Gordon W. Gribble, Department of Chemistry, Dartmouth College, Hanover, NH 03755.

In recent years radical reactions have found utility in the formation of various ring systems leading to the elegant, convergent synthesis of numerous natural products. A hydrogen atom transfer reaction will be employed in a novel approach to the indolo[2,3-*a*]quinolizidine ring system (**III**) from the corresponding indole amide (**II**).

597. A NEW SYNTHETIC ROUTE TO 2-PYRIDONES AND ITS APPLICATION TOWARD THE SYNTHESIS OF INDOLIZIDINE ALKALOIDS. **Scott M. Sheehan** and Albert Padwa, Department of Chemistry, Emory University, Atlanta, GA 30322

Treatment of 1-(benzenesulfonyl-diazo-acetyl)-pyrrolidin-2-one with a catalytic quantity of rhodium(II) acetate resulted in the formation of an isomünchnone dipole which underwent bimolecular trapping with various dipolarophiles in high yield. The initially formed cycloadducts were not isolable or observed, as they all readily underwent ring opening to give the 3-hydroxy-2(1*H*)-pyridone ring system. The 3-hydroxy-2(1*H*)-pyridones were readily converted to the corresponding triflates which function as suitable substrates in various types of palladium-catalyzed cross coupling reactions. An application of the method to the synthesis of the indolizidine alkaloid (±)-ipalbidine was carried out in eight steps in 17% overall yield.

598. DIELS-ALDER REACTIONS OF PHENACYL OXAZOLONE DERIVATIVES
Michael B. Smith,[*] Yun-hui Zhang, Mohammed E. Azab, Mounir A.I. Salem, and Dongping Fan Department of Chemistry, University of Connecticut, 215 Glenbrook Road, Room 151, Storrs, Connecticut 06269-4060, U.S.A. [smith@nucleus.chem.uconn.edu] and the Faculty of Science, Ain Shams University, Cairo, Arab Republic of Egypt

We have prepared several phenacyl oxazolone derivatives for two purposes. The first is to examine the intramolecular variation of this reaction. Although cycloaddition occurs, it is accompanied by decarboxylation. We therefore examined the intermolecular version of the reaction to find a suitable Lewis acid to use as a catalyst. Catalysis of this reaction is problematic, however, and we are examining non-catalysis conditions to determine the viability and regioselectivity of the reaction.

599. NEW METHODOLOGY TOWARDS (+)-LYCORICIDINE: THE CONDURITOLAMINE APPROACH Jason Chruma, Robin Polt,[*] Dalibor Sames Department of Chemistry, University of Arizona, Tucson, AZ 85721

Our synthetic methodology of the anti-viral and anti-tumor agent (+)-lycoricidine (**1**) from D-serine, which is focused around the synthesis of the conduritolamine **3** via Polt-Peterson reduction-alkenylation of **2**, followed by Sharpless dihydroxylation, and McMurray coupling will be discussed.

600. **NEW METHOD FOR THE SYNTHESIS OF (-)-SWAINSONINE AND ITS ANALOGS.** Hossein Razavi and Robin Polt*, Department of Chemistry, University of Arizona, Tucson, Arizona 85721

Synthesis of indolizidine alkaloid **2** has been achieved from TBDMS-protected methyl D-serinate benzophenone Schiff base **1**. Tandem reduction-alkenylation followed by stereoselective, substrate directed, dihydroxylation of the resulting β-amino alcohol gave the requisite adduct which was converted to the corresponding trihydroxylated pyrrolidine. Subsequent Swern oxidation of the pyrrolidine and selective homoallylation afforded the required 3-carbon homologue which was cyclized to give the desired product.

601. **TOTAL SYNTHESIS OF (-)-INDOLIZIDINE 167B.** Robert M. Henry, Steven R. Angle*, and J. Guy Breitenbucher, Department of Chemistry, University of California, Riverside, California 92521-0403.

The total synthesis of (-)-Indolizidine 167B from D-norvaline derivative **2** will be presented. The key step of the synthesis is a conformationally restricted Claisen rearrangement of silyl ketene acetal **1**.

602. **STEREOSPECIFIC TOTAL SYNTHESES OF MICHELLAMINES A, B, AND C.** Velaparthi Upender, Peter D. Hobbs, and Marcia I. Dawson*, SRI International, Menlo Park, CA 94025.

The three anti-HIV dimeric alkaloids michellamines A, B, and C were synthesized stereospecifically by the individual palladium-catalyzed cross-coupling of the appropriately 6-substituted tetrabenzylated korupensamine A and B derivatives, followed by deprotection.

603. **TOTAL SYNTHESES OF *MESO*-CHIMONANTHINE AND *MESO*-CALYCANTHINE: A REMARKABLE SAMARIUM MEDIATED ALKYLATION.** J. T. Link and Larry E. Overman*, Department of Chemistry, University of California at Irvine, Irvine, California 92697-2025

Psycholeine, a natural product comprised of an achiral hexacyclic core with two pendant pyrroloindoline units, was reported to be the first non-peptide antagonist of the somatostatin family of receptors. The core, called *meso*-calycanthine, has been concisely synthesized via the bis(pyrroloindoline) alkaloid *meso*-chimonanthine. The two key quaternary centers present in *meso*-chimonanthine and *meso*-calycanthine were steroselectively assembled in a novel samarium mediated reductive alkylation. Further advances toward a total synthesis of psycholeine will also be described.

604. APPLICATION OF THE COBALT-MEDIATED [2+2+2]-CYCLOADDITION TO THE SYNTHESIS OF THE OPIOID ANALGESICS MORPHINE, CODEINE, AND THEBAINE.
Monika Knopp, Michael J. Malaska, Ellen David, Dolores Perez-Meiras, Brian A. Siesel, K. Peter C. Vollhardt, Department of Chemistry, Universitiy of California at Berkeley, Berkeley, CA 94720.

The cobalt-mediated [2+2+2]-cycloaddition of three unsaturated functionalities opens up a highly efficient and flexible entry to a large number of complex polycyclic systems with excellent chemo-, regio- and stereoselectivity. The usefulness of this methodology in the synthesis of natural and bioactive products has already been demonstrated in several applications. The opioid analgesics morphine, codeine, and thebaine are likely ideal candidates, too. Indeed, treatment of a 4-(3-alkynyl)benzofuran with an external alkyne using the cobalt-cycloaddition conditions as the key step affords access to four of the five rings of the morphinan frame in good yields (53-72 %) and excellent stereoselectivity (>98%). The relative configuration of the Co-complexes could be proven by X-ray structure determination. Current efforts are directed towards an extension of this method to 3-substituted benzofurans. This approach should provide tetracyclic precursors containing a quaternary stereogenic center at C-13 of the morphinan framework with the functionality necessary for the further elaboration to the natural product.

605. DIASTEREOSELECTIVE REACTIONS OF CHIRAL, NITROGEN-SUBSTITUTED OXYALLYL CATIONS. Michael A. Walters and <u>Debra J. Sponholtz</u>, Chemistry Department, Dartmouth College, Hanover, NH 03755

Hetero-substituted allyl cations have been shown to be extremely valuable intermediates in organic synthesis. We have successfully generated α-nitrogen stabilized oxyallyl cations utilizing different protecting groups on nitrogen and have studied the [4+3] cycloadditions of these compounds with cyclic dienes. As an outgrowth of this work, we have recently discovered that chiral N-substituted oxyallyl cations can be generated by employing N-linked oxazolidinone chiral auxiliaries as part of the oxyallyl-cation system. These reactive intermediates undergo diastereoselective cycloadditions with furan.

606. CHIRAL ALKOXIDE-MEDIATED ASYMMETRIC ALDOL REACTION OF ARYLACETONITRILES: AN ASYMMETRIC TRANSFORMATION OF THE FIRST KIND.
Paul R. Carlier,* Weldon W-F. Lam, Department of Chemistry, University of Science and Technology, Clear Water Bay, Kowloon, Hong Kong.

Lithiated arylacetonitriles undergo *anti*-selective aldol reaction with aldehydes. The resulting beta-hydroxy nitriles are useful precursors for synthesis of gamma-amino alcohol monoamine reuptake inhibitors. To develop an asymmetric version of the nitrile aldol reaction, a wide range of diamine and aminoalkoxide ligands were screened: the lithium alkoxide of ephedrine proved optimum.

Ar-CN → 1) 1.2 eq. (-)-ephedrine, 2.2 eq. n-BuLi; 2) t-BuCHO, -78°, 24 hrs; 3) NH₄Cl (aq.) → Ar-C(OH)(CN)-C(Me)₃ 31-86% ee

Ar-CN → 1) 1.2 eq. (-)-ephedrine, 1.9 eq. n-BuLi, -78 °C, 24hrs; 2) NH₄Cl (aq.) → racemic

Detailed investigation revealed that asymmetric induction in this reaction is thermodynamically controlled, and is therefore the product of an "asymmetric transformation of the first kind."

607. ASYMMETRIC ALLYLATION AND ALDOL REACTIONS CATALYZED BY CHIRAL LEWIS BASES. Scott E. Denmark*, Xiping Su, Stephen B. D. Winter, Ken-Tsung Wong, and Robert A. Stavenger, Department of Chemistry, University of Illinois, Urbana, IL 61801.

Chiral, non-racemic phosphoramides have been found to catalyze the addition of allyltrichlorosilane to aromatic aldehydes in moderate enantiomeric excess. The phosphoramides are also effective catalysts for the reactions of aldehydes with trichlorosilyl ketene acetals and trichlorosilyl enol ethers. Very high enantiomeric excesses (up to 97% e.e.) have been achieved with phosphoramide 1:

cyclohexene-OSiCl₃ + naphthalene-CHO →(1, 0.1 equiv, CH₂Cl₂, -78°C, 2h then, NaHCO₃, H₂O)→ product, 97% e.e., 94% yield

1 = Ph,Me-N-P(=O)(N(Ph)Me)(pyrrolidine)

The complexation of phosphoramides with Lewis acids has been studied by both NMR spectroscopy and X-ray crystallography. The insights provided by these studies for the enantioselectivity of these reactions will be presented.

608. HIGHLY SELECTIVE ALDOL REACTIONS BETWEEN A β-KETOIMIDE AND AN ACETYLENIC KETONE IN THE ASYMMETRIC SYNTHESIS OF NON-PEPTIDIC HIV PROTEASE INHIBITORS. Susan L. Vander Velde, Thomas M. Judge, Kristine D. Lovasz, Jeanette K. Morris, Karen R. Romines, Fusen Han, Ronald B. Gammill, Pharmacia & Upjohn, Inc., Kalamazoo, MI 49001

Inhibition of the HIV protease enzyme is a promising therapeutic strategy for treatment of escalating HIV infection. A new class of low molecular weight HIV protease inhibitors has been developed at Pharmacia & Upjohn, 4,5-dihydro-α-pyrones. The asymmetrical synthesis of the clinical candidate PNU-140690 is described in this paper. The first of two stereogenic centers in this molecule is established via addition of an organocuprate to a chiral imide. Either enantiomer of the chiral imide may be used to generate the C3α center based on the choice of organocuprate. The second center is installed with high selectivity via an aldol reaction between a methyl ketone, in either diastereomeric series, and an appropriate unsymmetrical, acetylenic ketone. Through an extensive survey of the possible aldol substrates, a combination which cleanly afforded the desired C6 stereocenter was identified.

PNU-140690 (3R,6R)

609. B,B-DIHALOTERPENYLBORANES FOR THE DIASTEREO- AND ENANTIOSELECTIVE SYNTHESIS OF SYN-ALDOLS

P. V. Ramachandran, W. C. Xu, and Herbert C. Brown* H. C. Brown and R. B. Wetherill Laboratories of Chemistry, Purdue University, West Lafayette, IN 47907-1393

B,B-dihaloterpenylboranes, readily synthesized from the corresponding terpenes, in the presence of i-Pr₂NEt converts 3-pentanone to ≥ 99% Z-enolates, converted by aldehydes to pure syn-aldols in 10-74% enantiomeric excess (ee). One of these, B,B-dibromo-2-isocaranylborane, provides 58-74% ee for the syn-cross aldols derived from the reaction of the enolate of 3-pentanone with a series of aldehydes.

(–)-IpcBCl₂ (+)-EapBCl₂ (–)-2-CarBCl₂ (–)-4-CarBCl₂ (–)-LgfBCl₂ (+)-MyrBCl₂ (–)-2-CarBBr₂

610. **REMOTE ASYMMETRIC INDUCTION IN CONJUGATE ADDITION REACTIONS.**

Jane Betty Goh, Dennis C. Liotta*
Emory University, Atlanta, GA 30306

Conjugate addition of various nucleophiles to α-N,N-dibenzylamino enones are studied. Diastereoselectivies were low for enolate and nitrile additions (0 to 48% de) but Gilman reagent (Me_2CuLi), in the presence of chlorotrimethylsilane (TMSCl), add to give high diastereoselectivities (up to 92 % de).

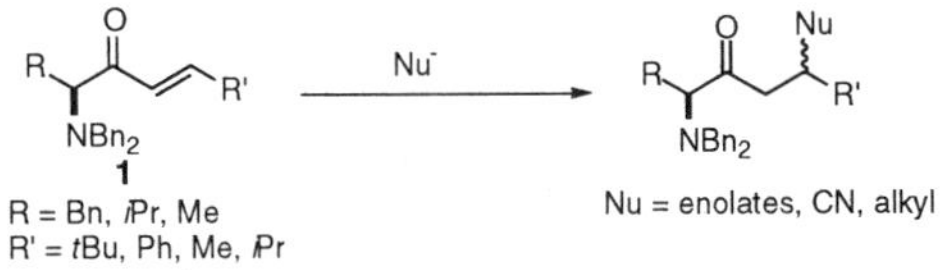
R
O
R'
NBn2
1
Nu⁻
O
Nu
R
R'
NBn2
R = Bn, iPr, Me
R' = tBu, Ph, Me, iPr
Nu = enolates, CN, alkyl

611. **α-LITHIATION OF *N*-BOC-*N*-(*p*-METHOXYPHENYL)ALLYLAMINES MEDIATED BY (-)-SPARTEINE.** **G. A. Weisenburger** and Peter Beak, Department of Chemistry, University of Illinois at Urbana-Champaign, Urbana, IL 61801

The lithiation and subsequent electrophilic substitution of the cinnamylamine **1** in the presence of (-)-sparteine (**2**) affords enecarbamates **3-6** with enantiomeric ratios greater than 96:4 in good yields. A transmetalation-substitution sequence can provide the other enantiomer of **3-6**. The high enantioselectivity and regioselectivity is not limited to cinnamate derivatives. Experiments suggest that the enantiodetermining step in the sequence is an asymmetric deprotonation.

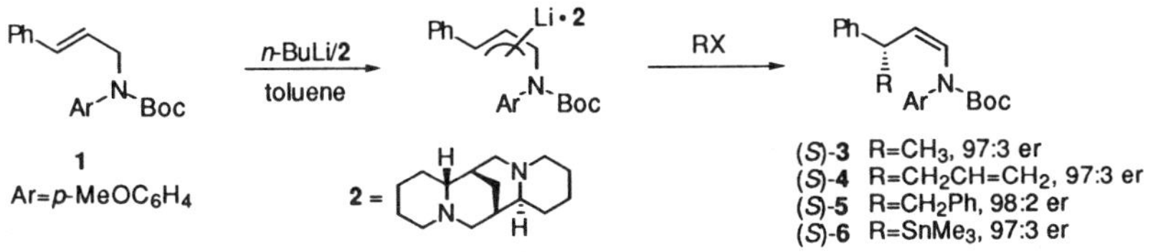
Ph
Ar
N
Boc
1
Ar=p-MeOC6H4
n-BuLi/2
toluene
Li • 2
RX
2 =
H
N
N
H
R
(S)-3 R=CH3, 97:3 er
(S)-4 R=CH2CH=CH2, 97:3 er
(S)-5 R=CH2Ph, 98:2 er
(S)-6 R=SnMe3, 97:3 er

612. CESIUM CARBOXYLATE MEDIATED HYDROXYL AND EPOXIDE INVERSIONS IN POLYPROPIONATE SYNTHESIS. J. A. Prieto*, D. O. Arbelo and L. Castro-Rosario, Department of Chemistry, University of Puerto Rico, Rio Piedras, PR 00931-3346

The development of general methodologies for the preparation of polypropionates is still regarded as a very challenging endeavor for the synthetic organic chemist. Our approach towards the stereoselective elaboration of these systems is based on oxirane chemistry. As part of this studies we confronted the need to develop hydroxyl and epoxide inversion protocols on our hindered polyfunctional systems. This goal was successfully achieved on both instances using Cesium carboxylate reagents. The details of this study will be presented

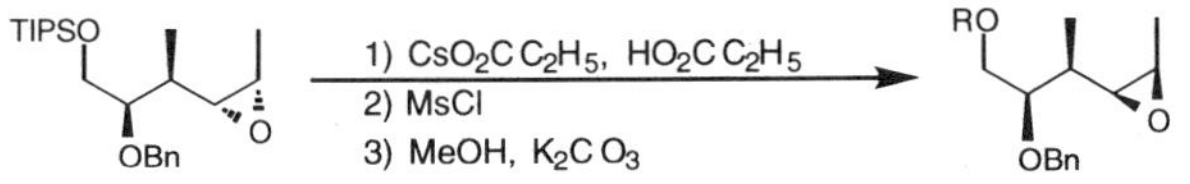
TIPSO
OBn
O
1) CsO2C C2H5, HO2C C2H5
2) MsCl
3) MeOH, K2C O3
RO
O
OBn

Support provided by the Department of Education-Patricia Harris Fellowship and NIH MBRS-RR08102

613. Rh-DUPHOS CATALYZED ENANTIOSELECTIVE HYDROGENATION OF ENOL ESTERS. SYNTHESIS OF HIGHLY ENANTIOENRICHED α-HYDROXY ESTERS.

Mark J. Burk, **Christopher S. Kalberg,** Antonio Pizzano, Paul M. Gross Chemical Laboratory, Duke Uniersity, Durham NC 27708

Recently, catalytic asymmetric hydrogenation has proven very successful in the synthesis of enantiomerically pure novel amino acids. The extention of this methodology to other classes of biologically relevent compounds such as α-hydroxy acids has met with only limited success. The EtDuPHOS-Rh catalyzed hydrogenation of a wide range of enol esters with excellent levels of enantioselectivity will be discussed (Eq. 1). Deprotection then yields the desired hydroxy esters without racemization.

$$'R\text{-}C(OBz)=CH\text{-}CO_2R \longrightarrow 'R\text{-}CH(OBz)\text{-}CH\text{-}CO_2R \longrightarrow 'R\text{-}CH(OH)\text{-}CH\text{-}CO_2R \quad (1)$$

614. **CHIRAL π-COMPLEXES OF HETEROCYCLES WITH TRANSITION METALS AND THEIR USE IN ASYMMETRIC CATALYSIS.** J. Craig **Ruble** and Gregory C. Fu,* Department of Chemistry, Massachusetts Institute of Technology, Cambridge, MA 02139

Although a wide array of reactions are known to be catalyzed by nucleophiles, there has been surprisingly little progress in the development of chiral nucleophilic catalysts that are both versatile and highly enantioselective. We are exploring the possibility that π-complexation of a heterocycle to a transition metal may be an especially effective approach to the development of chiral analogues of nucleophilic catalysts such as DMAP and imidazole. Initial results with iron complex **1** have been particularly promising.

615. **CATALYTIC ASYMMETRIC EPOXIDATION OF OLEFINS VIA DIOXIRANES** Yong Tu, Zhi-Xian Wang, Michael Frohn, Yong Tang, and <u>Yian Shi</u>*, Department of Chemistry, Colorado State University, Fort Collins, Colorado 80523.

Chiral dioxiranes have appeared to be promising reagents for asymmetric epoxidation, particularly for *trans*-olefins bearing no allylic alcohol groups. We wish to report a highly effective asymmetric epoxidation method using Oxone as oxidant and a fructose-derived ketone **1** as catalyst. The olefin substrates include *trans*- and trisubstituted olefins that can bear a wide range of functional groups. The enantiomeric excesses range from 70% to >95%. The epoxidation is believed to proceed via a spiro transition state, which provides a model for predicting the stereochemical outcome of the reaction.

616.

REEVALUATION OF THE KINETICS OF THE CATALYTIC ASYMMETRIC DIHYDROXYLATION OF ALKENES. **Derek W. Nelson**, K. Barry Sharpless*, Department of Chemistry BCC-315, Department of Chemistry, The Scripps Research Institute, 10550 North Torrey Pines Road, La Jolla CA 92037

Stoichiometric kinetic studies on the ligand-accelerated osmylaton reaction in conjunction with turnover studies on the catalytic Asymmetric Dihydroxylation (AD) have revealed that the apparent Michaelis-Menten behavior recently reported does not result from the proposed 20-electron, alkene-osmium-ligand complex. The specifics of the kinetic studies and latest refinements in the model for reactivity at the metal-oxo functions will be discussed.

617. NEW STEREOSELECTIVE SYNTHESIS OF α,α-DISUBSTITUTED-α-AMINO ACIDS

André B. Charette* and Christophe Mellon

Département de Chimie, Université de Montréal, Montréal, Québec, Canada, H3C 3J7

Abstract: A new stereoselective synthesis of quaternary amino acids will be presented. The diastereoselective sequential addition of Grignard and organocerium reagents to a chiral α-alcoxy nitrile constitutes the key transformation of this methodology. After protection of the amine, the auxiliary is cleaved and the product oxidized to the N-protected α-amino acid in good yields. A wide variety of α,α-disubstituted-α-amino acids can be prepared by this method, including arylglycines. The auxiliary is readily prepared in either enantiomeric form in only 4 steps from commercially available material.

618. ASYMMETRIC SYNTHESIS OF γ,δ-UNSATURATED AMINO ACIDS. Mark J. Burk, **John G. Allen**, William F. Kiesman, Paul M. Gross Chemical Laboratory, Department of Chemistry, Duke University, Durham NC 27708-0348

Two flexible and complementary strategies were developed to synthesize a variety of natural and unnatural γ,δ-unsaturated α-amino acids. Dienamide esters were prepared by the Horner-Emmons olefination of unsaturated aldehydes or by a novel palladium catalyzed Suzuki cross-coupling reaction. Catalytic asymmetric hydrogenation using the Rh-DuPHOS system gave high yields of γ,δ-unsaturated amino acids with excellent enantioselectivity.

619. ASYMMETRIC SYNTHESIS OF α-METHYL ASPARAGINE. Joseph P. Konopelski and **Todd A. Ritsema**, Department of Chemistry, University of California, Santa Cruz, CA 95064

Methyl ester **1**, available in enantiomerically pure form from the amino acid asparagine *via* a one-pot cyclization/protection sequence, can be effectively deprotonated with LDA/DMPU/LiCl. Treatment with MeI affords corresponding alkylated adduct **2** in diastereomerically pure form, from which both α-methyl asparagine and α-methyl aspartic acid can be obtained.

620. SYNTHESIS, CIRCULAR DICHROISM SPECTRAL ANALYSIS AND BIOLOGICAL ACTIVITY OF A BIOACTIVE GRAMICIDIN S ANALOGUE POSSESSING TWO INDOLIZIDINONE AMINO ACIDS. Henry-Georges Lombart and William D. Lubell*, Département de chimie, Université de Montréal, C. P. 6128, Succ. A, Montréal, Canada H3C 3J7

In the search for new agents to combat bacterial strains exhibiting resistance to traditional antibiotics, Gramicidin S [GS, *cyclo*-(Val-Orn-Leu-D-Phe-Pro-Val-Orn-Leu-D-Phe-Pro)] is an attractive target for development because of its antibiotic properties against both Gram-positive and Gram-negative bacteria.

Using (3S, 6S, 9S)-indolizidinone amino acid (**1**), we synthesized [IAA$^{4-5,4'-5'}$]-GS (**2**) which exhibited similar CD spectra and antibacterial activity to that of GS. Furthermore, **2** exhibited 8 times lower hemolytic activity than GS. The synthesis of **1** and **2**, and the conformation and bioactivity of **2** will be presented.

(3S, 6S, 9S)-Indolizidinone Amino Acid (IAA)

	R	R'
IAA:	H	OH
IAA-**1**:	Boc	OH
2:	Cyclo-[Val-Orn-Leu-IAA-Val-Orn-Leu-IAA]	

621. HIGHLY EFFICIENT SYNTHESIS OF (D)-SERINOL FROM AZIRIDINE-2-METHANOL DERIVATIVES. Soo-Kyung Choi, and <u>Won Koo Lee</u>*, Department of Chemistry, Sogang University, Seoul 121-742, Korea

The C(3)-N bond of N-(S)-α-methylbenzylaziridine-2(S and R)-methanol benzyl ether, **1** and **2**, is regioselectively cleaved by an oxygen nucleophile to give 2-amino-1,3-propanediols. Selective deprotection of an oxygen protecting group and slight modification provides (D)-serinols as protected forms in high yields.

622. STEREOSELECTIVE SYNTHESIS OF (1S, 2R) - TRANS -2- (1H-IMIDAZOL-4-YL) CYCLOPROPYL AMINE: KEY INTERMEDIATE FOR THE PREPARATION OF NOVEL H$_3$ HISTAMINE RECEPTOR AGENTS

M. Amin Khan, James G. Phillips, Clark E. Tedford, and K. Kirschbaum,* Gliatech Inc., 23420 Commerce Park Rd., Cleveland, Ohio 44122, and * Department of Chemistry, University of Toledo, Toledo, Ohio, 43606.

A procedure for the preparation of the H$_3$ histamine receptor agonist (1S, 2R) – trans-2-(1H-imidazol-4-yl) cyclopropyl amine is described. the key step in the synthesis is a stereoselective cyclopropanation of the trans -3-(1-triphenylmethyl-imidazol-4-yl) acryloyl derivative of Oppolzer's chiral sultam, (1R)-(+)-2,10- camphorsultam, using sulfur ylide chemistry. The mixture of diastereomeric trans cyclopropane sultams was separated and purified by routine column chromatography. X-ray crystallography provided the absolute configuration of the sultam intermediate, and that of the (1S, 2R) -trans -2-(1H-imidazol-4-yl) cyclopropyl amine.

623. ENANTIOPURE α-HYDROXY-β-LACTAMS *VIA* STEREOSPECIFIC GLYCOSYLATION.
<u>Bimal K. Banik</u>, Maghar S. Manhas and Ajay K. Bose, Department of Chemistry and Chemical Biology, Stevens Institute of Technology, Hoboken, NJ 07030.

Iodine catalyzed Ferrier rearrangement involving a *trans*- α-hydroxy-β-lactam (+/-)-**1** and D-glucal triacetate failed although the *cis* isomer reacted readily. The rearrangement, however, was successful with L-rhamnal diacetate **2** and led to the exclusive formation of **3a** and **3b**, two isomeric glycosides - shown to be α-glycosides by the ^{1}H NMR spectra of **4** (dihydro **3a**). Hydrolysis of **3a** and **3b** produced enantiomeric forms of *trans* 1-*p*-anisyl-3-hydroxy-4-phenyl-2-azetidinone (**1**) which were efficient synthons for diastereomers of Taxol® and analogs.

624. **CONCISE, STEREOSELECTIVE SYNTHESES AND NOVEL TRANSFORMATIONS OF β-LACTONES: APPLICATIONS TO NATURAL PRODUCT SYNTHESIS.** Daniel Romo,* Hong Woon Yang, and William D. Schmitz, Dept. of Chemistry, Texas A&M University, College Station, TX 77843-3255 USA

β-lactones (2-oxetanones) are versatile synthetic intermediates, are found in several natural products, and are useful as monomers for the synthesis of biodegradable polymers. Thus, there is growing interest in the development of concise and stereoselective routes to these strained heterocycles. We have initiated a program directed towards the development of direct and stereoselective syntheses of β-lactones in efforts to extend the utility of these systems as intermediates in natural product synthesis and to find direct methods for the synthesis of β-lactone natural products. A highly diastereoselective, tandem Mukaiyama aldol-lactonization approach to β-lactones and its use in a concise synthesis of (-)-panclicin D and the synthesis of the γ-butyrolactone segment of okinonellin B will be presented.

625. **PROGRESS TOWARD THE TOTAL SYNTHESIS OF RHIZOXIN**

Jennifer A. Lafontaine, David P. Provencal and James. W. Leahy*
Department of Chemistry, University of California, Berkeley, CA 94720-1460

A concise synthetic approach to the C(1) - C(9) fragment of the antitumor macrolide rhizoxin (**1**) via a three-fold pseudosymmetric intermediate is described. The preparation (from readily available γ-butyrolactone) includes both an asymmetric allylation and an asymmetric aldol addition. The pseudosymmetry proved useful in the realization of the target. Coupling strategies toward unification of aldehyde **2** and sulfone **3** will also be described.

626.

SYNTHESIS OF TWO PROPOSED INHIBITORS OF PARA-AMINOBENZOIC ACID SYNTHASE.
Kimberly D. Stigers, Roger Mar-Tang, Paul A. Bartlett, University of California at Berkeley, College of Chemistry, Berkeley, CA, 94720

The Shikimate Pathway is a metabolic pathway in plants, bacteria, and fungi that produces the aromatic amino acids and several secondary aromatic metabolites. Because this pathway is not present in animals, it is an attractive target for the development of herbicides and anti-bacterial drugs. para-Aminobenzoic acid synthase (PABS) is an enzyme acting late in the Shikimate

Pathway. In spite of significant homology to two other shikimate enzymes (isochorismate synthase, IS, and anthranilate synthase, AS), recent work has demonstrated that PABS operates through a unique mechanism (Kozlowski, M.C; Tom, N.J; Seto, C.T; Sefler, A.M; Bartlett, P.A. J. Am. Chem. Soc. 1995, 117, 2128-2141). Two potential inhibitors were designed as transition state analogs to explore the mechanism of PABS. The synthesis of these inhibitors has been completed and biological assays against PABS, IS, and AS are underway.

627. TOTAL SYNTHESIS OF ANTIMICROBIAL ISO BRANCHED MARINE FATTY ACIDS. Néstor M. Carballeira*, and Elba D. Reyes, *Department of Chemistry, University of Puerto Rico, P. O. Box 23346, San Juan, Puerto Rico 00931*

As part of our program in novel marine antimicrobial fatty acids the acids (5Z,9Z)-14-methyl-5,9-hexadecadienoic acid (**1**), recently identified by us in the Caribbean gorgonian *Eunicea succinea*, and the known longer-chain analog (5Z,9Z)-24-methyl-5,9-hexacosadienoic acid (**2**), isolated for the first time from the sponge *Petrosia ficiformis*, have been synthesized for the first time in our laboratory through a common synthetic route. To introduce the *iso* ramification, commercially available 4-methyl-1-pentanol was used as the starting material for the two synthetic targets. The key steps in the synthesis of the *cis* Δ5,9 acids **1** and **2** was a combination of alkyne-bromide coupling and Wittig reaction.

1, x= 1; **2**, x= 11

628. STEREOSELECTIVE INSERTION REACTIONS OF FORMAMIDES INTO SILIRANES AND APPLICATIONS TO ORGANIC SYNTHESIS. K. A. Woerpel* and Jared T. Shaw, Department of Chemistry, University of California, Irvine, CA 92697-2025

Although the carbonyl groups of unactivated formamides are typically poor electrophiles for carbon-carbon bond formation, they react readily with siliranes. For example, treatment of **1** with *N*-formylpyrrolidine provided the insertion product **2** as a single stereoisomer and in high yield. The formamide insertion reaction presents opportunities for stereoselective synthesis when carbon nucleophiles such as silylenol ethers are added to the resulting oxasilacyclopentane electrophile. Observations regarding the insertion reaction and examples of nucleophilic additions will be presented.

629. REGIOSELECTIVITY IN THE PALLADIUM-CATALYZED ADDITION OF CARBON NUCLEOPHILES IN DIHYDROPYRAN DERIVATIVES. Marc-Raleigh Brescia, Yvonne C. Shimshock and Philip DeShong, Department of Chemistry and Biochemistry, University of Maryland, College Park, Maryland 20742

The regioselectivity of Pd-catalyzed malonate or sulfonylacetate addition in dihydropyran derivatives is highly dependent upon the substitution pattern of the dihydropyran nucleus and is governed by electronic, rather than steric factors. In certain instances, subtle steric features also play a role in controlling regioselectivity by altering the conformation of the intermediate η^3-allyl Pd-complex.

630. DESIGN AND SYNTHESIS OF NOVEL REAGENTS FOR AMINO ACID DETECTION. O. Petrovskaia, M. M. Joullié, D. B. Hauze, Department of Chemistry, University of Pennsylvania, Philadelphia, Pennsylvania 19104-6323

A plethora of compounds for amino acid detection on solid surfaces has been evaluated by forensic scientists due to the importance of latent fingerprint development. A sensitive developing reagent should form a product with high absorption and/or fluorescence in the visible range. In order to obtain high contrast the reagent itself should be colorless and non-fluorescent. It should also be selective and highly reactive at ambient conditions. We have synthesized a series of substituted 1,2,3-indanetriones and p-dimethylaminocinnamaldehyde analogs in order to evaluate their potential as fingerprint developing reagents. A variety of palladium-catalyzed coupling conditions was employed. We also report 1,2-indanediones as a new class of potential reagents for amino acid detection. Studies of the reaction mechanism between the 1,2-indanediones and amino acids are currently underway.

631.
SYNTHESIS OF 9-ALKYLIDENE-9H-FLUORENES BY A NOVEL PD-CATALYZED DOMINO REACTION OF ARYL IODIDES AND INTERNAL ALKYNES
Richard C. Larock* and Qingping Tian
Department of Chemistry, Iowa State University, Ames, Iowa 50011

In the presence of a palladium catalyst and an appropriate base, the reaction of iodobenzene and diphenylacetylene affords 9-benzylidene-9H-fluorene, a totally different product from those previously reported by Heck and Dyker. This methodology has been extended to other aryl iodides and a variety of internal alkynes. The reaction apparently proceeds by arylpalladium addition to the alkyne, a novel palladium rearrangement, and subsequent aryl-aryl coupling.

632.

LIGAND EFFECTS ON THE PALLADIUM-CATALYZED ARYLAMINATION OF OPTICALLY ACTIVE α-CHIRAL AMINES. Seble Wagaw, Roger A. Rennels, and Stephen L. Buchwald,* Department of Chemistry, Massachusetts Institute of Technology, Cambridge, MA 02139

The effect of the choice of ligand on the palladium-catalyzed coupling of optically pure α-chiral amines with aryl bromides will be discussed. While $L_nPd/P(o$-tolyl$)_3$ successfully catalyzes the intramolecular arylamination of optically pure α-chiral amines, intermolecular coupling reactions with this catalyst system gives racemized products. A mechanism for the observed racemization will be proposed. In contrast to results obtained with $L_nPd/P(o$-tolyl$)_3$, $L_nPd/(\pm)$-BINAP catalyzes intermolecular N-arylations of α-chiral amines to give coupled products in good yields and with no loss in optical purity. A three step synthesis for (S)-N-acetylindoline-2-carboxylate methyl ester employing a Pd/P(o-tolyl)$_3$ catalyzed intramolecular arylamination will also be described.

633. ON THE NATURE OF THE PALLADIUM-CATALYZED ELIMINATION OF ALLYLIC CARBONATES TO FORM 1,3-DIENES
James M. Takacs, Edward C. Lawson, and Francis Clement, Department of Chemistry, University of Nebraska-Lincoln, Lincoln, NE 68588-0304.

Investigations into deuterium isotope effects and stereospecificty in the palladium-catalyzed elimination of allylic carbonates using a catalyst system recently introduced by Tsuji and coworkers (0.1 equiv. [Pd(OAc)$_2$/PBu$_3$], THF, 25 °C; Mandai, T.; Matsumoto, T.; Tsuji, J.; Saito, S. *Tetrahedron Lett.*

1993, *34*, 2513-16) will be described. Large isotope effects are observed, and as illustrated below, the elimination is stereospecific and overall syn with respect to loss of the elements H-O$_2$COR to form the 1,3-diene product. The results are not consistent with a commonly accepted mechanism, wherein hydrogen is lost via β-hydride elimination, instead support the stereospecific base promoted *anti*-elimination of an allylpalladium intermediate.

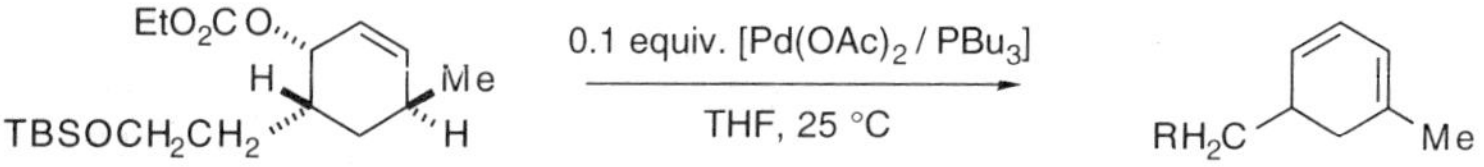

634.
PALLADIUM CATALYZED ALKYNE DIMERIZATIONS IN NATURAL PRODUCT SYNTHESIS. **Matthias C. McIntosh** † and Barry M. Trost,* Stanford University, Stanford, CA 94305-5080
† present address: Department of Chemistry and Biochemistry, University of Arkansas, Fayetteville, AR 72701

The dimerization of alkynes to 1,3-enynes under transition metal catalysis has been known for many years, but has been of little synthetic value due to a lack of chemoselectivity in cross dimerizations. In recent years, however, Trost et al have shown that dimerization of a diverse range of alkynes using a palladium-phosphine catalyst proceeds with high chemo-, regio- and stereoselectivity. Applications of alkyne dimerization in the total synthesis of cleviolide, syributin 1, and 7,7-C-didehydro-6-hydroxy-6,7-dihydrocaulerpenyne will be described.

635. TITANOCENE-MEDIATED OR CATALYZED CONVERSION OF ENONES TO γ-BUTYROLACTONES. <u>Natasha M. Kablaoui</u>, Frederick A. Hicks, and Stephen L. Buchwald*, Massachusetts Institute of Technology Department of Chemistry, Cambridge, MA 02139

A method for the diastereoselective transformation of enones to γ-butyrolactones using Cp$_2$Ti(PMe$_3$)$_2$ will be described. This heteroatom-containing variant of a Pauson-Khand-type reaction proceeds through carbonylation of an oxatitanacycle followed by reductive elimination to produce γ-butyrolactones. Depending on the substrate, either a stoichiometric or a catalytic amount of the metal can be used to affect the transformation. Substrates which contain an aryl ketone moiety are efficiently transformed using a catalytic amount of the metal complex. Our results are consistent with the view that the key step in this catalytic cycle is formation of a charge transfer complex or involves reversible electron transfer between the catalyst and the substrate.

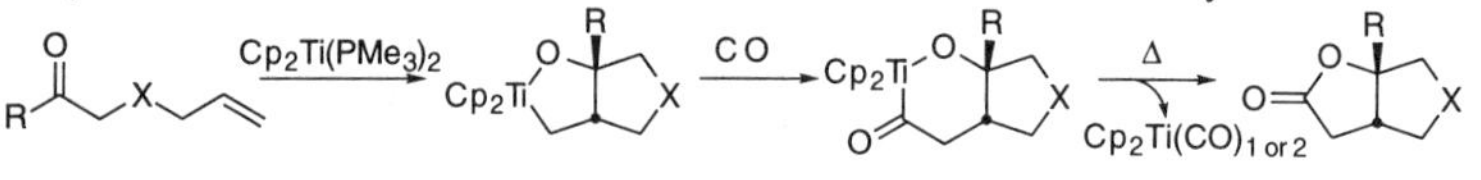

636. AN ENANTIOSELECTIVE TITANIUM-CATALYZED SYNTHESIS OF CYCLOPENTENONES. <u>Frederick A. Hicks</u>, Natasha M. Kablaoui, Stephen L. Buchwald*, Massachusetts Inst. of Technology Dept. of Chemistry, Cambridge, MA 02139

The conversion of enynes to cyclopentenones catalyzed by Cp$_2$Ti(CO)$_2$ will be described. This system produces cyclopentenones in excellent yields (75 - 95%) and displays functional group tolerance unprecedented in Group 4 metallocene chemistry. An enantioselective version of the cyclization has been developed utilizing (*S,S*)-

(EBTHI)Ti(CO)$_2$, which produces cyclopentenones in good yields (70 - 94%) and generally good to excellent ee's (72 - 96%). A rationale for the observed enantioselectivity will be discussed.

637.
COBALT MEDIATED, EXO-SELECTIVE DIELS-ALDER REACTIONS WITH CYCLOHEXENONES. B. Matthew Richardson Mark E. Welker, Department of Chemistry, Wake Forest University, Winston-Salem, NC, 27109-7486

Decalone systems form the core of many diterpenoid natural products, notably eremophilane and clerodane families. This work entails a facile and selective method of forming a cis-decalone framework. 2-Cobaloxime-E-1,3-pentadienyl complexes (cobaloxime = pyridine (dimethylglyoxime)2 cobalt (III)) react with a variety of cyclohexenone dienophiles under thermal and Lewis acid catalyzed methods to give decalone products. These products contain a cis-fused ring juntion and are resultant from an exo-transition state. Decalone products are produced in high yields and are easily cleaved from the metal template.

638. REARRANGEMENTS OF CYCLOPROPENES INDUCED BY Cp*ReO$_3$. **Kevin P. Gable** and Abdullah AbuBaker, Department of Chemistry, Oregon State University, Corvallis, OR 97331-4003.

Several cyclopropenes give different reactivity with Cp*ReO$_3$, depending on structure. Alkyl substituents (Me, cyclohexyl) lead to oxidative cleavage of the ring in a stoichiometric reaction with the rhenium compound. The reaction, surprisingly, requires elevated temperatures (70-90°C). This process can be rationalized based on an initial alkene-metal interaction, perhaps followed by formation and fragmentation of a metallaoxetane. However, phenyl substituents lead to rearrangement to indenes, a process which is catalytic in the metal. This is explained as a Lewis acid-catalyzed phenomenon based on comparison with the behavior of other transition metal compounds and Lewis acids.

639. PRACTICAL AND EFFICIENT ENANTIOSELECTIVE CONJUGATE RADICAL ADDITIONS. **Mukund P. Sibi** and Jianguo Ji, Department of Chemistry, North Dakota State University, Fargo, North Dakota 58105-5516

Enantioselective conjugate radical additions to **1** proceed with >97%ee using MgI$_2$ in combination with a chiral bisoxazoline ligand derived from cis-2-aminoindanol. The reactions provide high enantioselectivity (up to 95% ee) using 5-30 mol% of the chiral Lewis acid at room temperature. Details of these reaction protocols and extension of the enantioselective process to more complex systems will be presented.

640. EFFECT OF LANTHANIDE LEWIS ACID AND ADDITIVES ON ACYCLIC DIASTEREOSELECTIVITY IN PROCHIRAL RADICAL ADDITION TO PROCHIRAL OLEFINS. S. V. Chandramouli, Mukund P. Sibi*, Craig P. Jasperse, Megan Kirsch, Sarah Steffl, and Nathan Dodder, Department of Chemistry, North Dakota State University, Fargo, North Dakota 58105-5516

Intermolecular addition of neutral prochiral radicals to olefins has not been systematically studied. This work reports the Lewis acid-mediated addition of prochiral radicals to prochiral olefins (eqn 1). The dependence of diastereoselectivity on Lewis acid, additives, and subtituents in **1** and **2** will be discussed.

$$\underset{\mathbf{1}}{X\overset{O}{\underset{}{\diagdown}}R} \quad \xrightarrow{\underset{\mathbf{2}}{R_1\diagdown OR_2}} \quad \underset{\mathbf{3}\ syn}{X\overset{O}{\underset{}{\diagdown}}\overset{R}{\underset{OR_2}{\diagdown}}R_1} \quad + \quad \underset{\mathbf{4}\ anti}{X\overset{O}{\underset{}{\diagdown}}\overset{R}{\underset{OR_2}{\diagdown}}R_1} \quad (1)$$

641.

SYNTHESIS OF 4-(METHYL)PHENYLSULFONYL-SUBSTITUTED DIENES VIA RADICAL CYCLIZATION OF DIYNES Craig L. Shering*, Stephen Caddick* and Sjoerd N. Wadman**. * The Chemistry Laboratory, University of Sussex, Falmer, Brighton BN1 9QJ, United Kingdom. ** Glaxo-Wellcome, Medicines Research Centre, Gunnels Wood Road, Stevenage, Herts SG1 2NY, United Kingdom.

Radical mediated cyclizations have emerged as a powerful tool for organic synthesis. Addition of a free-radical to a multiple bond is an effective method for the production of carbon-centred radicals, which allows the incorporation of useful functionality. The synthetic utility of this type of approach using dienes and acetylenic olefins has been shown but diynes have attracted little attention. Herein, we demonstrate that the radical cyclization of diynes can be used successfully in the synthesis of two new classes of arylsulfonyl-substituted diene system.

642.

THE THERMAL INTRAMOLECULAR CYCLIZATION OF A SUBSTITUTED BICYCLO[2.1.0]PENTANE: FORMATION OF A NEW NONSTABILIZED INTERCEPTIBLE BIRADICAL INTERMEDIATE. **Kate Redmond** and Barry K. Carpenter, Department of Chemistry, Baker Laboratory, Cornell University, Ithaca, NY 14853

5-(5-Cyano-4-pentynyl)bicyclo[2.1.0]pentane was synthesized as a mixture of exo and endo epimers. Upon thermolysis at 160°C, the endo epimer rearranged to give a biradical intermediate that was able to rapidly undergo an intramolecular hydrogen atom abstraction to give a stable, closed shell product. The exo epimer, however, rearranged to give a biradical that was too conformationally constrained to undergo intramolecular reaction. This long-lived didehydromethyleneindane biradical is a new addition to the interesting and important class of nonstabilized but long-lived biradicals headed by 1,4-didehydrobenzene.

643.

AZULENYL NITRONES: EVIDENCE FOR ENHANCED FREE RADICAL SCAVENGING AS A RAMIFICATION OF STRUCTURE AND OXIDATION POTENTIAL. **David A. Becker*** and Reina Natero, Department of Chemistry, Florida International University, University Park, Miami, Florida 33199. Luis Echegoyen* and Raphael C. Lawson, Department of Chemistry, University of Miami, Coral Gables, Florida 33124.

Nitrones are currently of interest as antioxidant therapeutics. The azulene derivative (Z)-3-{{(1,1-dimethylethyl)oxidoimino}methyl}-8-methyl-5-(1-methylethyl)-1-Azulenecarboxylic acid, ethyl ester, **1**, is the first representative of a class of novel chromotropic nitrone spin traps. Cyclic voltammetric studies of **1** reveal that one-electron oxidation of **1** is unusually facile and both electrochemically and chemically reversible. The results of radical scavenging competition studies involving **1** and other nitrones will be presented.

644. EXPERIMENTAL CHARACTERIZATION OF THE 2,6- AND 3,5-DIMETHYLENE-PYRIDINIUM BIRADICALS. <u>Kami K. Thoen</u>, Eric D. Nelson and Hilkka I. Kenttämaa, Department of Chemistry, Purdue University, West Lafayette, IN 47904-1393

The 2,6- (**1**) and 3,5-dimethylenepyridinium (**2**) distonic biradical ions have been generated and their reactivities characterized in the gas phase. The biradicals were found to yield qualitatively different products in reactions with several neutral reagents. The differences in reactivity were attributed to different ground-state multiplicities predicted earlier for these isomeric species. The closed-shell singlet biradical **1** reacts with 2,3-dimethoxy-1,3-butadiene by adduct formation and by loss of one or two methanol molecules from the adduct. In sharp contrast, the triplet biradical **2** only exhibits formation of a stable adduct. Similarly, *tert*-butylisocyanide reacts with **2** by exclusive adduct formation. However, the reaction of **1** yields an adduct and an ion that arises from abstraction of HCN. Most of the products observed for **1** in the above reactions are known to be formed for another closed-shell singlet biradical also, a *meta*-benzyne with an inert, positively charged substituent. Hence, this reactivity seems to be diagnostic of a closed-shell singlet electronic ground state. The same products were also observed for an analogous benzyl cation, 3-methylene-5-methylpyridine. In general, the triplet biradical **2** was found to be significantly less reactive than either the singlet biradical **1** or related monoradicals.

645. **Generation and Reactivity of N-Phenylpyridinium Biradicals.**
<u>E. D. Nelson</u>, K. K. Thoen, and H. I. Kenttämaa,
Department of Chemistry, Purdue University, West Lafayette, IN 47907-1393

The kinetics of biradical reactions are often difficult to examine due to the transient nature of biradicals in solution and interfering reactions from impurities. These problems can be circumvented by attaching an inert pyridinium charge site to the biradical, which allows mass spectrometric isolation and detection of the biradical ion in the high vacuum environment of an FT-ICR ion trap mass spectrometer. These gas-phase experiments allow the measurement of intrinsic reaction rates, uncomplicated by the presence of solvent molecules. We present here the gas-phase synthesis and reactivity of *N*-phenylpyridinium biradicals. Some of these biradicals (*e.g.*, those with each radical site on a different ring) display reactivity similar to that of analogous N-phenylpyridinium monoradicals, implying a triplet ground state. Others (*e.g.*, the *meta*-benzynes) undergo reactions similar to the benzyl cation, suggesting a singlet ground state.

646. EXPLOSION AND ION ASSOCIATION CHEMISTRY OF THE ANION RADICALS OF TNT, DNT, AND TRINITROBENZENE. **Matthew L. Batz**, Paul M. Garland, Richard C. Reiter, Michael D. Sanborn, and Cheryl D. Stevenson

EPR analysis shows that the anion radical of 2,6-dinitrotoluene (DNT) in liquid NH_3 exists with the counterion (either K^+ or Na^+) associated with one of the two nitro groups. This tight association ($-NO_2^{\bullet-}$, M^+) persists after solvent removal, and it renders the anion radical very susceptible to loss of metal nitrite. Slight agitation of the solid salt of $DNT^{\bullet-}$ leads to detonation, and formation to KNO_2 and polymer (in the solid phase) and CH_4, HCN, H_2, and N_2O (in the gas phase). Trapping experiments suggest that the methane comes from carbenes, and it is suggested that the HCN comes from an anthranil radical intermediate, and the ease, of $C-NO_2^{\bullet-}$, M^+ bond rupture increases with the degree of nitration.

647. STEREOSELECTIVE TRAPPING OF PROCHIRAL CARBON RADICALS WITH CHIRAL NITROXYL RADICALS. Rebecca Braslau,* Leland C. Burrill II, Michael Siano, Vladimir Chaplinski and Patrick W. Papa, Department of Chemistry and Biochemistry, University of California, Santa Cruz, California 95064

Optically active nitroxyl radicals can react differentially between the two enantiotopic faces of prochiral carbon radicals. Several chiral nitroxyl radicals have been prepared and stoichiometric generation methods of prochiral radicals have been developed. Reactions in which optically active nitroxyl radicals react with transient prochiral carbon radicals will be discussed.

648. A NOVEL ROUTE INTO THE [6.3.0] RING SYSTEM VIA RADICAL INITIATED VINYLCYCLOPROPYL RING OPENING. Georgia L. Carroll and R. Daniel Little, Department of Chemistry, University of California, Santa Barbara, Santa Barbara, CA 93106.

Deazetation of diazene **1** affords a TMM diyl **2** that undergoes vinylcyclopropyl ring opening to produce the [6.3.0] carbocycle **4** in preference to the [4.3.0] product **5**. In addition, this process is potentially useful for the formation of [5.3.1] systems, as found in the anti-tumor agent taxol. Studies of this process, and efforts towards the [5.3.1] taxane skeleton will be presented.

649. COMPLETELY DIASTEREOSELECTIVE RADICAL REACTIONS USING ARENECHROMIUM TRICARBONYL COMPLEXES. Craig A. Merlic* and Joseph C. Walsh, Department of Chemistry and Biochemistry, University of California, Los Angeles California, 90095-1569.

Complete control of diastereoselectivity in radical reactions is achieved through the use of chromium tricarbonyl complexed arenes. When chromium-complexed indanone **1** is reacted with samarium(II) iodide, methyl acrylate, and t-butanol; a single, diastereomerically pure spiro-lactone **2** is produced with none of the other diastereomer detected. This reaction demonstrates the power of metal-templated radical reactions.

650. DIASTEREOCONTROL IN ORGANOMETALLIC RADICAL CHEMISTRY: RECENT RESULTS. **Gagik G. Melikyan**,* Sarkhadoun Yadegar, Asatour Deravakian, Steven Myer, and Todd Monroe, Department of Chemistry, California State University, Northridge, CA 91330

Diastereocontrol in organometallic radical reactions will be discussed including: 1) Mn(III)-mediated radical cycloaddition to chiral racemic $Co_2(CO)_5L$ complexes of conjugated enynes, and the effect of bulkiness of the axial ligand L ($40Å^3$-$700Å^3$) and substrate structure upon stereoselectivity of cyclization; 2) a novel one-step and highly diastereoselective dimerization of Co-complexed propargyl alcohols (de 86-94%) induced by tandem action of various hydride ion donors and Lewis acids; and 3) inter / intramolecular coupling of electrochemically generated Co-complexed propargyl radicals. The scope, synthetic potential and limitations of the reactions will be considered together with mechanistic studies.

651. A SIMPLE ONE-POT ASYMMETRIC ENOLBORATION-ALDOLIZATION-REDUCTION PROCEDURE FOR THE SYNTHESIS OF *ANTI*-1,3-DIOLS IN VERY HIGH ENANTIOMERIC EXCESS
P. V. Ramachandran, W. C. Xu, Z. H. Lu, and Herbert C. Brown* H. C. Brown and R. B. Wetherill Laboratories of Chemistry, Purdue University, West Lafayette, IN 47907-1393

The enolization of ketones, at 0 °C, with B-chlorodiisopinocampheylborane in the presence of Et_3N, in CH_2Cl_2, followed by treatment with aldehydes, at –78 °C, provides the aldolates. These, upon warming to 0 °C, undergo an intramolecular reduction to afford *anti*-1,3-diols in very high enantiomeric excess.

652. UNSATURATED NITRILES: A TANDEM OZONOLYSIS-ALDOL SYNTHESIS OF HIGHLY ACTIVATED MICHAEL ACCEPTORS. **Fraser F. Fleming**,* Aiju Huang, Vaqar Sharief, and Yifang Pu, Department of Chemistry and Biochemistry, Duquesne University, Pittsburgh, PA 15282.

A tandem ozonolysis-aldol sequence has been used to prepare several 1-oxo-2-cycloalkenyl-2-carbonitriles. The synthesis of **4** is representative of this exceptionally mild transformation that occurs in 98% yield with no added acid or base. This presentation will describe the synthesis of several of these highly activated Michael acceptors and their propensity to undergo conjugate addition reactions.

653.

A NEW PREPARATION OF ESTERS FROM CARBONYL COMPOUNDS FOLLOWING LITHIUM ALUMINUM HYDRIDE REDUCTION. Zhengming Guo Robert D. Sindelar The Department of Medicinal Chemistry; The National Center for the Development of Natural Products, The Research Institute of Pharmaceutical Sciences, School of Pharmacy, University of Mississippi, University, Mississippi 38677

Abstract: A modified work-up procedure for the lithium aluminum hydride reduction of esters and other carbonyl compounds will be described. The one pot procedure converts an ester to its "homo-reverse ester" and other carbonyl compounds to esters. The key step is trapping the resulting intermediate aluminum alkoxide during reflux with an ester.

654. NEW ASPECTS OF THE BORONIC ACID MANNICH REACTION

Nicos A. Petasis,* Ilia A. Zavialov, Andrew Goodman, Irini A. Zanze and Valery V. Fokin
Department of Chemistry and Loker Hydrocarbon Research Institute
University of Southern California, Los Angeles, CA 90089-1661

The boronic acid Mannich reaction is a three-component process recently discovered in our laboratories which involves the condensation of an amine (2) with a carbonyl compound (3) and an alkenyl or aryl boronic acid (1). Good to excellent diastereoselectivities can be obtained when certain chiral components are used in the reaction. Some new synthetic and mechanistic aspects of this process will be described.

655. *N*-UNSUBSTITUTED β-LACTAMS *VIA* AN ENVIRON-FRIENDLY APPROACH: ^{15}N-LABELED TAXOL SYNTHONS. Ajay K. Bose, Sang-Hyun Park, M. Jayaraman, Ernest W. Robb, Maghar S. Manhas, Department of Chemistry and Chemical Biology, Stevens Institute of Technology, Hoboken, NJ 07030.

Study of an old method (Wells and Tarwater, 1969) has led to a short, efficient and eco-friendly approach to *N*-unsubstituted α-hydroxy-β-lactams (3, R = OH). Reaction of an aromatic aldehyde, NH_4Cl, strong NH_4OH and ^{i}PrOH leads to the separation of the crystalline hydrobenzamide 1 which is treated with $ROCH_2COCl$ (R = Ac, Me, Bn) and NEt_3 in toluene solution. Two *cis*-β-lactams (2) are produced which are hydrolyzed by stirring with silica gel to give *cis*-β-lactam 3 (R = OAc, Ar = Ph) in 63 - 88% yield. Using $^{15}NH_4Cl$, ^{15}N-labeled 3 (R = OAc, Ar = Ph) was easily obtained. α-Hydroxy-β-lactams (3, R = OAc) can be resolved into both enantiomers *via* the Ferrier reaction with glucal triacetate and iodine (Banik *et al*.1994) to serve as synthons for Taxol® and analogs. ^{15}N-labeled compounds derived from 3 can be used for metabolic studies and for use as an internal standard for rapid mass spectral quantitation.

656. ORGANIC REACTIONS IN WATER: ON THE REGIO- AND STEREO-SELECTIVITIES OF PROPARGYLATION-ALLENYLATION REACTIONS OF CARBONYL COMPOUNDS, Chao-Jun Li and Xiang-Hui Yi, Department of Chemistry, Tulane University, New Orleans, LA 70118

The regio- and stereoselectivities of metal-mediated propargylation-allenylation of carbonyl compounds in water have been investigated. A dramatic change of regio-selectivity was observed in the metal-mediated propargylation-allenylation reactions of aldehydes with various propargyl bromides. The product distribution is strongly affected by the presence of an α-hydroxyl group on the propargyl bromide moiety. While the reaction of carbonyl compounds with non-hydroxylated alkylated propargyl bromide generated allene as a single product, similar reactions with hydroxylated propargyl bromide significantly increased the propargylation product. Potential synthetic application of the reactions will also be discussed.

657. HIGHLY SELECTIVE SURFACE–MEDIATED OXIDATION OF SULFIDES, SULFOXIDES AND AMINES. John D. Fields and Paul J. Kropp, Department of Chemistry, University of North Carolina, Chapel Hill NC 27599–3290.

$(CH_3)_3COOH$ mediated by silica gel effectively oxidizes sulfides, sulfoxides and amines with selectivities often difficult to otherwise achieve. These reactions occur quickly under mild conditions while obviating the need for metal catalysts. Sulfide oxidation occurs via an electrophilic oxidation promoted by double hydrogen bonding of $(CH_3)_3COOH$ to isolated silanol groups. Synthetic viability and mechanistic considerations will be discussed.

658. "NO-SOLVENT" ORGANIC REACTIONS INDUCED BY TELLURIDE ION AND ULTRASOUND Qinyu Xu and Donald C. Dittmer, Department of Chemistry, Room 1-014 CST, Syracuse University, Syracuse, NY 13244

Elemental tellurium is reduced in the solid state by ultrasonication with KOH and rongalite.

$$Te + HOCH_2SO_2Na + KOH \xrightarrow{)))} K_2Te + HOCH_2SO_3Na + H_2O$$

The use of alumina or other inert solid (eg molecular sieves) aids in keeping the mixture free-flowing and less sticky (due to KOH and H_2O). Addition of organic substrate with further sonication results in the regeneration of black elemental tellurium (which may be reduced by excess rongalite) and formation of the desired product.

659. **MICROWAVE-ASSISTED REACTIONS ON CLAY AND SUPPORTED REAGENTS UNDER SOLVENT-FREE CONDITIONS.**
Rajender S. Varma,*[1] Rajesh K. Saini,[1] Rajender Dahiya[1] and Manju Varma[2]
[1]Department of Chemistry and Texas Regional Institute for Environmental Studies (TRIES), Sam Houston State University, Huntsville, Texas 77341-2117; [2]Present address: Genosys Biotechnologies, 1442 Lake Front Circle, Suite 185, The Woodlands, TX 77380-3600.

A variety of organic reactions in presence of montmorillonite K 10 clay or 'doped' clay (clayfen) are accelerated by microwave (MW) irradiation under solvent-free conditions. Several examples from this environmentally benign approach will be presented that utilize neat reactants in open containers under MW activation. The solvent conservation and the elimination of the hazardous and toxic reagents will be discussed for these synthetically useful protocols that encompass condensation, oxidation, and cyclization reactions. A convergent strategy for the use of these valuable intermediates (enones, enamines, and nitroalkenes) in the synthesis of heterocycles will be elaborated. [Supported by Texas ARP in chemistry (Grant # 003606-023), and at TRIES by Office of Naval Research/SERDP (Grant # N00014-96-1-1067)].

American Chemical Society
DIVISION OF PETROLEUM CHEMISTRY

213th ACS National Meeting

San Francisco, CA
April 13-17, 1997   J. G. Reynolds, Program Chairperson

SUNDAY MORNING

- **Tutorial: The Chemistry and Refining of Petroleum I: Crude Oils**
J.G. Speight, Presiding   Papers 1-4

- **Advances and Applications of Computational Chemical Modelling to Heterogenous Catalysis - Metal Sulfides and Homogenous Catalysis**
M. Neurock, R.A. van Santen, Presiding   Papers 5-10

SUNDAY AFTERNOON

- **Tutorial: The Chemistry and Refining of Petroleum II: Heavy Crude Oils**
J.G. Speight, Presiding   Papers 11-14

- **Advances and Applications of Computational Chemical Modelling to Heterogenous Catalysis - Metals and Zeolites**
M. Neurock, Presiding   Papers 15-20

MONDAY MORNING

- **George A. Olah ACS Award in Petroleum Chemistry to Roby Bearden, Jr.**
I.A. Wiehe, Presiding   Papers 21-24

MONDAY AFTERNOON

- **Advances in Catalysis and Processes for Heavy Oil Conversion Fundamental Aspects of Heavy Oil Upgrading**
C. Song, P.O'Connor, Presiding   Papers 25-32

- **Advances and Applications of Computational Chemical Modelling to Heterogenous Catalysis - Zeolites**
R.A. van Santen, Presiding   Papers 33-38

- **George A. Olah ACS Award in Petroleum Chemistry to Roby Bearden, Jr.**
I.A. Wiehe, Presiding   Papers 39-43

TUESDAY MORNING

- **Advances in Catalysis and Processes for Heavy Oil Conversion**
  Resid Hydroconversion
  T. Takatsuka, T.J. Gardner, Presiding   Papers 44-51

- **Advances and Applications of Computational Chemical Modelling to Heterogenous Catalysis - Molecular Simulation and Statistical Methods**
  M. Neurock, Presiding   Papers 52-57

- **Catalytic Combustion**
  E. Kikuchi, Presiding   Papers 58-64

TUESDAY AFTERNOON

- **Advances in Catalysis and Processes for Heavy Oil Conversion**
  Removal of S,N, and Metals from Heavy Oils
  C. Song, T.Takatsuka, Presiding   Papers 66-73

- **Advances and Applications of Computational Chemical Modelling to Heterogenous Catalysis - Metal Oxides**
  R.A. van Santen, Presiding   Papers 74-79

- **Catalytic Combustion**
  R.A. Dalla Betta, Presiding   Papers 65,80-86

WEDNESDAY MORNING

- **Advances in Catalysis and Processes for Heavy Oil Conversion**
  Thermocracking, Hydrocracking, and Coprocessing
  T.J. Gardner, C. Song, Presiding   Papers 87-94

- **Advances and Applications of Computational Chemical Modelling to Heterogenous Catalysis - Metal Sulfides and Homogeneous Catalysis**
  M. Neurock, Presiding   Papers 95-101

- **Worldwide Perspectives on the Manufacture, Characterization and Application of Lubricant Base Oils I - New Products and Processes**
  A.J. Stipanovic, Presiding   Papers 102-109

WEDNESDAY AFTERNOON

- **Advances in Catalysis and Processes for Heavy Oil Conversion**
  Catalytic Cracking of Heavy Oils
  P. O'Connor, T. Takatsuka, Presiding   Papers 110-117

- **Advances and Applications of Computational Chemical Modelling to Heterogenous Catalysis - Future Directions**
  M. Neurock, R.A. van Santen, Presiding   Papers 118-120

- **Worldwide Perspectives on the Manufacture, Characterization and Application of Lubricant Base Oils II**
  Oxidation/Degradation Chemistry
  M.P. Smith, Presiding   Papers 121-127

THURSDAY MORNING

- **Advances in Catalysis and Processes for Heavy Oil Conversion**
  New Approaches to Catalyst and Process Development
  P. O'Connor, C. Song, T. Takatsuka, Presiding   Papers 128-135

- **Worldwide Perspectives on the Manufacture, Characterization and Application of Lubricant Base Oils III**
  Synthetic and Biodegradable Lubricants
  M.W. Wilson, Presiding   Papers 136-139

- **Base Oil Characterization, Analysis and Performance Prediction**
  M.W. Wilson, Presiding   Papers 140-144

THURSDAY AFTERNOON

- **Advances in Catalysis and Processes for Heavy Oil Conversion**
  Feed Pretreatments, Product Stabilization and Chemicals
  C. Song, P. O'Connor, T. Takatsuka, Presiding   Papers 145-149

- **Worldwide Perspectives on the Manufacture, Characterization and Application of Lubricant Base Oils IV**
  Other New Concepts in Base Oil and Lubricant Technology
  C.A. Migdal, Presiding   Papers 150-156

DIVISION OF PETROLEUM CHEMISTRY, Inc.

001. PETROLEUM RESIDUA AND HEAVY OIL CONSTITUENTS. J. F. Schabron, Western Research Institute, 365 N. 9th St., Laramie, WY 82070

      Petroleum residua, either obtained by atmospheric or vacuum distillation, range from viscous liquids to near solid materials. Heavy oils are obtained by enhanced recovery processes such as steam flooding. The chemical composition of these materials is complex, and can lead to challenges in selecting and designing optimal upgrading schemes. Compositional considerations include polarity and molecular size. The types of polynuclear aromatic constituents, including their size and degree of condensation determine properties which affect processability. Asphaltenes, defined by precipitation in an excess of a hydrocarbon solvent such as heptane are concentrated in residua and heavy oils. Heteroatoms (N,S,O) and metals (Ni,V,Fe) are concentrated in the asphaltenes. Characterization and evaluation schemes should be designed to provide insights as to the processability of these materials.

002. **Metals and Heteroatoms in Heavy Crude Oils.** John G. Reynolds, University of California, Lawrence Livermore National Laboratory, P. O. Box 808, L-369, Livermore CA 94551.

In the conversion of crude oil to transportation fuels, heteroatoms (principally V, Ni, Fe, S, and N) generally cause processing problems, particularly when utilizing heterogeneous catalysts for upgrading, and have been the focus of much applied research over the years. Knowledge of the structure of the compounds which contain these elements as well as the behavior of these elements during upgrading is essential for process improvements and flexibility. This presentation will summarize the structural data on heteroatomic compounds found in petroleum and related feedstocks, and will review reaction mechanisms thought to occur during widely used processing such as hydrotreating and catalytic cracking.

This work was performed under the auspices of the U. S. Department of Energy by the Lawrence Livermore National Laboratory under Contract No. W-7405-ENG-48

003. ASPHALTENES AND THE STRUCTURE OF PETROLEUM. James G. Speight, Western Research Institute, 365 North 9th Street, Laramie, WY 82070-3380

Understanding the complex nature of the asphaltene fraction of petroleum needs the recognition that the constituents are multidimensional in space in contrast to the two-dimensional aspects often represented, for convenience, on paper. Thus, a key to understanding crude oil behavior during refining operations is to develop an understanding of the principal precursors (the asphaltenes and other polar constituents) and the relationships between the asphaltenes and the other constituents of petroleum feedstocks.

This presentation summarizes asphaltene character and the interrelationships of the constituents of petroleum which can then be used to explain various behavioral aspects including the onset of coke formation, the events that lead to sediment formation in cracked and visbroken feedstocks, as well as catalyst deactivation, and related phenomena.

004. **THERMAL REACTIVITY OF HEAVY OILS**
I. A. Wiehe, Exxon Corporate Research, Route 22 East, Annandale, NJ 08801-0998

The key in understanding the thermal reactivity of heavy oils is to recognize that the thermal chemistry of these complex mixtures of hydrocarbons is usually coupled with phase change. The phase change may be evaporation of volatile liquids in the reactant heavy oil or of the volatile liquid product or it may be the liquid - liquid phase separation of asphaltene aromatic cores to form coke. Therefore, to measure reactivity directly in open reactors, one must select conditions to minimize evaporation of the reactant while allowing volatile liquid products to escape the reactor. On the other hand, one might do the thermal reaction in a closed reactor and distill the volatile liquids out of the reactor before exposing it to solvents. In comparing the coke formation for these two reactors, the coke forms earlier in the closed reactor but less forms than in the open reactor at long times, as is predicted by the liquid - liquid phase separation mechanism. As a result, the series kinetic model, including rate constants and the stoichiometric coefficients, are shown to be independent of reactor type; the first order rate constants depend on aliphatic sulfur concentration; and the stoichiometric coefficients are independent of temperature.

005. TRANSFORMATION OF SMALL HYDROCARBON MOLECULES ON A PD(111) SURFACE: A QUANTUM CHEMICAL STUDY. J.F. Paul and P. Sautet, Institut de Recherches sur la Catalyse, Villeurbanne and Ecole Normale supérieure, 69364 Lyon, FRANCE

The chemisorption of CHx (x=0-3) and H on a Pd(111) surface has been studied with Density Functional periodic calculations. The carbon atom prefers a hollow site, while the CH_2 is preferencially in a bridge and the CH_3 on top of a Pd atom, in simple agreement with the free valance. The binding energy for the CH_x fragments on Pd(111) decreases from C to CH_3. On the surface the C-H bond cleavages are only slightly endothermic and the most stable surface species is CH_3. The Preferred binding mode for ethylene is di-σ with a small binding energy, the π mode is only slightly less stable. The transformation of ethylene on the surface is discussed.

006.
MAKING AND BREAKING CHEMICAL BONDS AT METAL SURFACES
J. K. Nørskov[1], B. Hammer[1,2], Y. Morikawa[2], P. Kratzer[1], and J. J. Mortensen[1]
[1]CAMP, Department of Physics, Technical University of Denmark, DK-2800 Lyngby, Denmark
[2]JRCAT, 1-1-4 Higashi, Tsukuba, Ibaraki 305, Japan.

Density functional calculations for H_2, CO, N_2, and CH_4 adsorption and dissociation on a number of metal and alloy surfaces surfaces are presented. Equilibrium structures, chemisorption energies, and activation energies are shown to compare well with experiment provided non-local corrections (GGA) to the local density approximation are included. We study the effect of alloying, bimetal overlayer structures, and geometrical effects including the role of steps and reconstructions. Based on the self-consistent density functional calculations we develop a model singling out some important factors determining the ability of a metal surface to make and break chemical bonds.

007.
ELECTRONIC STRUCTURE INVESTIGATIONS OF METAL SPECIES SUPPORTED BY IONIC SUBSTRATES. **Notker Rösch,** Theoretische Chemie, TU München, 85747 Garching, Germany

The bonding of transition metal atoms at various sites of the magnesium oxide surface has been studied with the help of density functional cluster models. Two different types of bonding to surface oxygen atoms of the regular surface MgO(001) can be discriminated, a rather weak one and a more covalent polar one. The preferred adsorption site of Pd atoms is discussed in the light of recent experimental findings. The interaction of small transition metal clusters with various sites of MgO(001) has been investigated to quantify the competition between metal-metal and metal-substrate interaction. Also, models describing metal atoms and clusters in zeolite cages have been calculated, addressing the question of electron-deficient metal species (Pt in mordenite, Pd clusters in faujasites).

008.

ADSORPTION OF O_2 ON Ag(110). D.M. Bird and P.A. Gravil, School of Physics, University of Bath, Bath BA2 7YA, United Kingdom

The generalised gradient approximation (GGA) is widely used in calculations based on density functional theory of reaction pathways, activation barriers, etc. There have been some recent indications that activation barriers calculated using GGA are underestimated. We report an extensive series of calculations of the adsorption of O_2 on Ag(110) within a plane-wave, pseudopotential approach and using a supercell geometry. This system has been chosen both for its catalytic interest and because it provides a well-characterised example of precursor-mediated molecular dissociation. Molecular beam experiments indicate that adsorption into a molecularly chemisorbed state is activated, with a barrier of order 0.2eV. On the contrary, our calculations show non-activated pathways for adsorption and a set of calculations for eight high symmetry sites and orientations in the surface unit cell indicate that the surface appears generally attractive to an incoming O_2 molecule, rather than repulsive, as indicated by experiment. The implication is that the GGA is failing to give a barrier in this case, the consequences of which will be discussed.

009. THEORETICAL STUDIES OF SURFACE REACTIONS ON METALS. J. L. Whitten, Department of Chemistry, North Carolina State University, Raleigh, N.C. 27695

The goal of this research is the development and application of theoretical techniques that will provide a molecular level understanding of surface processes, especially reaction mechanisms and energetics. Electronic structures are described by an *ab initio* embedding formalism that permits an accurate determination of reaction energetics and adsorbate structure. An overview of the theoretical method of embedding of clusters in an extended lattice, the expansion of wavefunctions by configuration interaction and the inclusion of relativistic effects for heavy atoms is presented. Applications to several catalytic and photochemical reactions on nickel and platinum surfaces are reported. Topics to be discussed include: mechanistic studies of ethylene production and competing pathways on platinum catalysts, specifically α vs. β hydrogen elimination, and the investigation of photochemical reactions of methane on platinum.

010. TOWARD UNDERSTANDING THE MECHANISM FOR THE SELECTIVE HYDROGENATION OF MALEIC ANHYDRIDE TO TETRAHYDROFURAN OVER PALLADIUM
Matthew Neurock, P.S. Venkataraman, Department of chemical engineering, University of Virginia, Charlottesville, VA 22903, USA, and George W. Coulston, DuPont Central Research and Development, Experimental Station, Wilmington, DE 19880-0262

Tetrahydrofuran is produced commercially by the hydrogenation of maleic anhydride over heterogeneous, bimetallic Pd-Re catalysts. The general overall reaction pathway involves the successive hydrogenation of the unsaturated carbon-carbon bond of maleic anhydride (MA) to form succinic anhydride (SA), and the subsequent hydrogenation of the carbonyl moieties to form γ-butyrolactone (GBL) and Tetrahydrofuran (THF). While the general pathway is known, very little is known about the governing mechanism at the molecular level.

First principle density functional calculations are carried out in conjunction with ultrahigh vacuum (UHV) temperature-programmed -desorption (TPD) studies to examine the chemisorption and reactivity of MA and its hydrogenation reaction intermediates on well-defined model surfaces. We study the binding of MA, SA, GBL and THF on model Pd(111) surfaces and explore possible reaction mechanisms via detailed reaction-coordinate calculations. Specifically, we analyze the hydrogenation of MA to SA and the deleterious ring-opening reactions.

Comparisons are made with experimental UHV TPD/HREELS spectra wherever data is available. The results of this work are expected to provide adsorption and kinetic parameters for detailed reaction models.

011.   HEAVY OIL UPGRADING PROCESSES:
M. Rashid Khan
Texaco Fuels & Lubricants Technology Department
P O Box 509
Beacon NY 12508

This paper presents an overview of the current state of the art of upgrading technology. The review will discuss upgrading in terms of the type of technology and the chemistry involved. The need to convert heavy oil and residue is growing. Technologies for upgrading heavy oils can be broadly divided into hydrogen addition and carbon rejection processes. Hydrogen addition processes involve reacting heavy oils with an external source of hydrogen which result in an overall increase in H/C ratio. Carbon rejection, on the other hand, redistributes hydrogen among various components resulting in fractions with increased H/C and fractions with lower H/C. The paper will demonstrate that most upgrading processes can be classified within these broad ranges.

012.   REFINING OF HEAVY OILS: TAR SAND BITUMEN
Murray R. Gray
Department of Chemical and Materials Engineering
University of Alberta
Edmonton, Alberta T6G 2G6
Canada

Current Canadian production of heavy oil and bitumen is approximately 410,000 bbl/d, with 68% of this production from the Athabasca tar sands. The profitability of these operations has led to announcements of $2.8 billion (US) in new investment by 2006 in the Athabasca deposit alone to increase production, and intense scrutiny of the various alternative refining strategies. The mining and extraction of the bitumen from the tar sands gives an unusual feed stock with a high content of mineral solids. The chemical and physical properties of the bitumen, as well as the combination of mining and refining at a single remote site, has led to a heavy reliance on coking processes. This paper will review the current technology and discuss future development in light of the unique circumstances of Northern Alberta.

013.   HYDROPROCESSING OF HEAVY OILS AND RESIDS
Geoffrey E. Dolbear, G. E. Dolbear and Associates, 23050 Aspen Knoll Dr., Diamond Bar, CA 91765

Heavy oils and resids are rich in nickel, vanadium, sulfur, nitrogen, and the polynuclear aromatics that lead to coke. Hydrogenation provides an effective means of reducing or removing these impurities. Hydrogenation is applied in several ways, depending on the refiner's needs. High severity removes metals and sulfur, and converts the heavy ends, providing an excellent crude. Coke yields from the unconverted bottoms are low, an advantage for coking and FCC. Low sulfur fuel oil can be made by treating at lower severity with a controlled pore catalyst.
Catalysts for these processes are patterned after HDS, relying on CoMoS and related systems. To deal with deposition of metals, catalysts are designed with the majority of their porosity in wide (>150Å) pores. These larger pores also enhance diffusion of large asphaltene molecules into the catalyst particle.
Processes for hydrotreating heavy oils and resids usually operate at high pressures (>1500 psi). These are needed to combat coke deposition and to saturate aromatics. High temperatures (>700°F) provide thermal cracking to reduce MW, but this creates a problem with thermodynamics since aromatics saturation equilibria tend to favor the aromatics at these temperatures. Both fixed and ebullated bed processes are used successfully. Typical fixed bed systems must treat an atmospheric resid for viscosity reasons, while ebullated bed systems rely on product liquid to solve this problem. Feeds are often treated in stages, with a different catalyst in each stage, removing metals and sulfur in successive reactors. This is because once the metals have been removed, more aggressive catalysts can be used for further reaction.

014.   PETROLEUM PRODUCTS AND STABILITY. George W. Mushrush, George Mason University, Fairfax, VA 22030; Naval Research Laboratory, Fuels Section, Washington, DC 20375 and James G. Speight, Western Research Institute, Laramie, WY 82070.

Middle distillate fuels continue to have instability problems. Over the past few years, the quality of crude oils has decreased and if synfuels are ever used in quantity, stability and chemical incompatibility in these fuels will be especially troublesome for both refiners and consumers. Specific chemical moities and reaction pathways that initiate the reactions that lead to fuel instability and incompatibility are difficult to isolate. Many diverse and competing mechanisms could be used to explain the observed instability: free radical hydroperoxide initiated oxidation mechanisms, electron transfer initiated oxidation mechanisms, and soluble macro-molecular oxidation mechanisms. In a mixture as complicated as a middle distillate fuel, these reactions in addition to many others areprobably simultaneously occurring. Results from our laboratory, indicate that the sedimentation process can be modeled by considering acid catalyzed condensation reaction initiated by a hydroperoxide induced oxidation process. Research supporting these various mechanisms will be presented.

015. Ab initio cluster model comparative study of atomic oxygen and sulfur chemisorption on Pt(111) surfaces. Relevance to heterogenous catalysis
J. M. Ricart[a], A. Cloteta and F. Illas[b]
a) Departament de Qu°mica, Facultat de Qu°mica, Universitat Rovira i Virgili, Páa. Imperial TÖrraco 1, 43005 Tarragona, Spain
b) Departament de Qu°mica F°sica, Universitat de Barcelona, C/Mart° i Franquäs, 1, 28028 Barcelona Spain

Chemisorption of atomic oxygen and sulfur on Pt(111) has been studied by means of the ab initio cluster model aproach. For both adsorbates, we consider chemisorption on the three fold open site of the Pt(111) surface which is represented by a Pt25 cluster model having 12 atoms in the first layer, 6 in the second and 7 in the third. Ab initio molecular orbital Hartree-Fock and explicitly correlated wavefunctions are used to obtain a reliable estimate of the interaction energy, vibrational frequency for the normal mode of the adsorbate above the surface, and the equilibrium geometry. The chemisorption bond is analyzed using different theoretical techniques including the Constrained Space Orbital Variation, CSOV, method, the analysis of dipole moment curves and the use of projections operators. The influence of electronic correlation effects is analyzed using Multireference Configuration Interaction, MRCI, techniques and, also, by using correlation functionals.

016. VARIATIONAL DENSITY FUNCTIONAL THEORY USING THE ATOM-IN-JELLIUM REFERENCE SYSTEM
Andrew E. DePristo, Liqiu Yang and Thiele L. Wetzel, Ames Laboratory, Iowa State University,á Ames, IA 50011.

We present a new variational extension of the corrected effective medium (CEM) theory. The first new part is the use of a linear combination of spherical atomic densities from full self-consistent field density functional theory (SCF-DFT) calculations on atoms and ions. This incorporates large changes from the summed atomic densities due to atomic configuration changes, e.g., {Ni, Ni+, Ni2+} or {O, O-, O2-} or {Pd(5d10 6s0), Pd(5d9 6s1)}, and thus is capable of describing large electron transfer and configuration effects. The second new part is the use of a first order perturbation correction for non-additive atomic densities. This includes directional bonding effects. The energy is minimized with respect to the combination of these atomic densities, and is thus variational. If this method is sufficiently accurate, it will provide capabilities for treating some of the most important systems in catalysis, e.g., zeolites and multimetallic clusters adsorbed on oxide supports. Initial tests for energies and structures reported in this talk will be chosen from bulk metals (bcc, hcp, fcc), oxides and small molecules.

á This work was supported by the Division of Chemical Sciences, Office of Basic Energy Sciences of the U. S. Department of Energy through the Ames Laboratory, which is operated for the U. S. DOE by Iowa State University under Contract No. W-7405-Eng-82.

017. DFT STUDIES OF OXAMETALLACYCLES FOR GROUPS VIII, IB, AND IIB TRANSITION METALS. M. Mavrikakis, D. Doren, M. A. Barteau, Center for Catalytic Science and Technology, University of Delaware, Newark, DE 19716

Ethylene oxide is among the most useful intermediates in chemical synthesis and silver is the most efficient catalyst for ethylene epoxidation. Our goal is to test the hypothesis that oxametallacycles are ethylene epoxidation intermediates. We have used the Amsterdam Density Functional (ADF) program to calculate the fully optimized structures for simple oxametallacycles. The electronic structure of oxametallacycles generated by the insertion of 1 to 4 metal atoms in the C-O bond of the ethylene oxide molecule for 14 different transition metals has been calculated. The thermodynamic stability of these complexes decreases from left to right of the periodic table, and from top to bottom of each group of metals. Our results suggest that Ag gives a moderately stable intermediate (ca. 10 kcal/mol), which is considerably less stable than the corresponding structures for most of the rest of the metals studied. This may explain the formation of the less thermodynamically favorable C_2H_4O isomer, ethylene oxide rather than acetaldehyde, on Ag. Our calculations of the relative stability of oxametallacycles versus acetaldehyde on various metal surfaces are consistent with experimental evidence that an oxametallacycle is preferred on a Rh(111) surface, whereas η^2-CH_3CHO is preferred on a Pd(111) surface.

018. ADVANCES IN AB INITIO MODELLING OF ZEOLITE CATALYSTS - FROM ACTIVE SITE MODELS TO FULLY PERIODIC STRUCTURES. Joachim Sauer, Max-Planck-Gesellschaft, Arbeitsgruppe Quantenchemie an der Humboldt-Universität, D-10117 Berlin/Germany

Cluster models of active sites of zeolite catalysts made quantum chemical ab initio calculations possible. The contribution of this approach to the understanding of the local structure, the spectra and the reactivity of zeolitic Brønsted sites is reviewed. The cluster approach neglects the electrostatic crystal potential and the strcutural constraints imposed by the silicate framework. I.e. it cannot reveal reactivity differences between different zeolite frameworks or between different crystallographic sites of a given framework. Recent developments allow to overcome such limitations. (i) The ab initio cluster model is extended by embedding techniques which describe the periodic environment by a shell model potential. (ii) Periodic techniques which use plane wave basis sets within the density functional method (e.g. Car-Parrinello) become applicable to zeolite catalysts with not too large unit cells. Reliable predictions of absolute acidities and subtle acidity differences between different framework types can be made.

019. RELATIONS BETWEEN THE STRUCTURE OF A ZEOLITE AND ITS ADSORPTION PROPERTIES. A. Goursot, V. Vasilyev, A. Arbuznikov, UMR CNRS 5618, Ecole de Chimie, 8 rue de l'Ecole Normale, 34296 Montpellier CEDEX 5, France,

Gaseous molecules adsorb in zeolites through their interaction with the accessible extra-framework cations, located in the supercages. The adsorption of nitrogen in Na LSX zeolites has been studied for different experimental and modified structures. The adsorption of these molecules, in turn at each of the 64 accessible cationic site, has been studied quantum chemically, using a density functionnal method point charges. These calculations have shown that only a few cations are efficient for initial adsorption and that they are always site III cations. Their efficiency depends both on the framework geometry and on their location in the supercages. Moreover, the study of modified structures leads to the conclusion that the framework geometry is a key factor determining the adsorption properties of a zeolite.

020. METHANOL IN MICROPOROUS MATERIALS FROM FIRST PRINCIPLES. Julian D. Gale, Department of Chemistry, Imperial College, South Kensington, SW7 2AY, U.K, Rajiv Shah and Michael C. Payne, Cavendish Laboratory (TCM), University of Cambridge, Madingley Road, Cambridge CB3 0HE, U.K, and Ivan Stich, Joint Research Center for Atom Technology, 1-1-4 Higashi, Tsukuba, Ibaraki 305, Japan.

Recently there has been an increasing interest in the use of planewave techniques to tackle chemical problems. In this presentation the application of such methods to the problem of methanol adsorption and reactivity in zeolites will be explored as an alternative to conventional finite cluster models. The nature of the isolated Brønsted acid sites in both aluminosilicates and silicoaluminophosphates will be compared, as well as addressing the question of whether methanol is adsorbed as a neutral or ion pair complex and how this depends on framework structure/chemical composition. Results from both static energy minimisation and first principles dynamics will be presented, as will calculations of the stability of possible reaction intermediates.

021. MICROCAT-RC™: TECHNOLOGY FOR THE HYDROCONVERSION UPGRADING OF PETROLEUM RESIDUA. R. Bearden, Jr.,Exxon Research and Development Laboratories, Baton Rouge, Louisiana   70821

MICROCAT-RC™ is a process for converting residual, essentially non-volatile (566+°C) fractions of petroleum into a high yield (> 100 volume %) of distillable product, which is comprised of naphtha, distillate fuels and vacuum gas oil. It is a slurry-catalyst hydroconversion process that is based in large part on controlled thermal cracking in the presence of hydrogen and small amounts of finely-divided hydrogen transfer catalyst. Catalysts are typically prepared in a portion of process feed from oil-soluble or oil-dispersible metal compounds and consist of micron-size composites of nanoscale, metal-sulfide catalyst sites stabilized by a hydrocarbonaceous matrix derived from the feed. This presentation will cover several of the fundamental components of MICROCAT technology including (1) chemistry of catalyst formation, (2) role of catalyst in conversion and influence on reaction kinetics and (3) methods used to detect and to suppress formation of mesophase coke, a material that can have an adverse effect on process operability.

022. UPGRADING HEAVY OIL USING SLURRY PROCESSES.
A. Del Bianco, ENIRICERCHE S.p.A, Via Maritano 26, 20097 San Donato M.se, ITALY

Slurry processes offer an interesting solution to convert heavy feedstocks, because they combine the advantages of the C-rejection technologies in terms of flexibility with the high performances peculiar of the H-addition processes. The origin of these processes can be found in the Bergius-Pier technology (1920-1930) for the conversion of heavy oils and coal into distillates. Whereas the Bergius-Pier technology did not employ any catalyst, more recently small amounts of inexpensive "additives" or finely dispersed hydrogenation catalysts are used to inhibit coke formation physically interfering with the coalescence of mesophase. Furthermore, the use of high active catalysts can also contribute to reduce the severity of the process and to improve the quality of the products. For this reason, most of the research activity carried out in the last decade in the field of slurry processes dealt with the identification of more effective/less expensive catalysts, as well as the technological problems connected with their use.
In the present paper we will discuss extensively both the subjects, with the aim to describe the path from the fundamental chemical aspects of catalytic dispersed systems to the development and initial commercialization of slurry processes.

023. REACTION MECHANISMS DURING RESIDUE HYROPROCESSING WITH FeS CATALYSTS. M. Ternan, P. Rahimi, D. Liu, and D.M. Clugston, National Centre for Upgrading Technology, Devon, Alberta, T0C 1E0, Canada

The difference between molecules that are converted to distillates during coprocessing and those that are not has been studied. (1) The molecules that were converted had larger H/C atomic ratios and smaller N/C atomic ratios than the feedstock molecules. (2a) S,V, and Ni heteroatoms could be removed without causing much change in molecular weight. (2b) It was not possible to increase the H/C atomic ratio or decrease the N/C atomic ratio without decreasing the molecular weight. (3) Generally, the unconverted +525°C residue molecules became smaller as the processing severity increased. (4) The unconverted resid molecules retained their side chains at mild processing severities (425°C). When the processing severity increased (450°C) side chain carbon began to be removed. (5) If there was insufficient reaction time to provide enough hydrogen, then molecules which were larger than the feedstock molelules were formed. Conversion of residue to distillate molecules appeared to be limited by hydrogen addition (Energy & Fuels $\underline{9}$, 1011, 1995).

024. **EXXON'S MICROCAT COAL LIQUEFACTION (MCL) PROCESS.** P. S. Maa, S. C. Mraw, L. R. Veluswamy, M. Y. Wen, Exxon Research & Development, Baton Rouge, LA 70821; G. A. Melin, D. C. Spencer, Exxon Engineering, Florham Park, NJ 07932

MICROCAT-CL is a proprietary Exxon process that uses a highly-dispersed slurry catalyst in the liquefaction reactor. The high yields of raw coal liquids are subsequently upgraded to a range of premium transportation fuels. The process flow-plan will be described, and examples given of conversion and yields under a variety of conditions for various coals. In keeping with the special symposium honoring Dr. Bearden, the behavior of the microcatalyst itself under liquefaction conditions will be highlighted, including some ideas for altering the chemistry pathways of the current microcatalyst.

025. A STRUCTURAL ATTRIBUTE REACTION MODEL OF COMPLEX FEEDSTOCKS. D. M. Campbell and M. T. Klein, University of Delaware, Newark, Delaware 19713

The value of representing a complex petroleum mixture in terms of a small (10-100) number of molecules motivated the development of a hybrid Monte Carlo-quadrature molecular structure building technique. The focus of this paper is the incorporation of quantitative pyrolysis data in constraining this representation. To this end, a reaction model of the reactive attributes comprising a series of differential equations representing mass balances for the reactive attributes has been developed. Construction of a representative set of molecules using the attribute concentrations allows a comparison to analytical properties of an actual reaction mixture.

These techniques have been applied to petroleum resid and gas oil to obtain a small molecular representation which represents the initial feedstock well. Often the small quadrature representation can match the experimental results as well or better than a larger Monte Carlo representation. Reaction data provide much more detail and therefore a statistically more unique, and often larger, set of molecules.

026. DEVELOPMENT OF ADVANCED EVALUATION METHOD FOR CRUDE OIL.

N. Yoshimoto and H. Kato, Petroleum Research Laboratory, Mitsubishi Oil Co., Ltd., 4-1 Ohgimachi, Kawasaki-ku, Kawasaki, 210 JAPAN

A quick and useful evaluation method for crude oil has been developed. Not only yields and properties of the fractions of the refinery topper, but also yields and properties of the products through secondary process (e.g. hydrodesulfurization of gas oil) can be estimated by this advanced evaluation method in place of TBP distillation. This method is valid for both the early selection of proper crude oil and the running efficiency.

The development for crude oil distillation method by SFC (supercritical fluid chromatography) is reported. The fraction yields of crude oil were determined from applied simulated distillation using sensitivity factors of the fractions. Although the sensitivity factor of vacuum residue was different from every crude oil, this could be explained by carbon residue content of crude oil. Good correlation between the results of TBP and those of this method could be shown for many kind of crude oil.

027. SELECTIVE BIARYL CLEAVAGE BY HYDROGEN ATOM
Bruce R. Cook[a], Bernadette B. Wilkinson
Exxon Research and Development Laboratories; P.O. Box 2226, Baton Rouge, LA 70821-2226
[a]Author's current address: Exxon Reserach and Engineering, Annandale, NJ 08801

The selective cleavage of strong bonds, such as biaryl bonds, is a key goal for the efficient conversion of heavy hydrocarbon resources. Biaryl cleavage is a key limiting reaction in the catalytic and thermal center ring cracking of multi-ring aromatics, such as phenanthrenes and dibenzothiophenes. Hydrogen atoms have been proposed as active agents for faciliting both biaryl cleavage and dealkylation reactions under high temperature hydrogenolysis conditions. However, no studies have reported selectivities in systems where both reactions can take place. This paper examines the relative propensity for hydrogen atom to induce biaryl cleavage vs. dealkylation in a series of symmetric dimethylbiphenyls. In these studies co-hydrogenolysis of 2,2,3,3-tetramethylbutane was utilized as a source for hydrogen atoms. Hydrogen atom induced biaryl cleavage was found to occur at rates competitive with demethylation, with the biaryl cleavage reaction exhibiting a 20 kJ/mole higher activation energy versus demethylation. The relative rate of biaryl cleavage versus demethylation was found to be highly dependent on the rate of formation and stability of the corresponding ipso hydrogen atom adducts.

028. CATALYSIS OF IRON SULFATES AS A HYDROCONVERSION CATALYST OF HEAVY MATERIALS.
E. Ogata, X.-Y. Wei, K. Horie, A. Nishijima, I. Saito, and K. Ukegawa; Department of Chemistry and Biotechnology, Graduate school of Engineering, TheUniversity of Tokyo;
7-3-1, Hongo, Bunkyo-ku, Tokyo 113, Japan

Catalysis of iron sulfates in the hydroconversion of 1-methylnaphthalene (1-MN) was carried out in the development of hydroconversion catalyst of heavy hydrocarbon materials. The catalytic activities of iron sulfates in the absence of sulfur was very low in 1-MN hydroconversion. Hydrogenation activity of ferrous sulfate dramatically increased with addition of sufficient solid sulfur. However, the hydrogenation activity of ferric sulfate did not increase with addition of sulfur, although demethylation was accelerated by sulfur. It was indicated from the XRD analysis of catalysts recovered after the reaction that ferrous sulfate was transformed to pyrrhotites above 350°C in the presence of excess sulfur for the stoichiometry. These pyrrhotites from ferrous sulfate showed very high activity in 1-MN hydrogenation. Ferric sulfate was not transformed easily to pyrrhotites, but ferric sulfate was transformed to pyrrhotites in the presence of a large excess of sulfur, demethylation was promoted and no hydrogenation occurred.

029. HYDROGEN TRANSFER IN THE HYDROCONVERSION OF A VACUUM RESIDUE. AN ANALYTICAL APPROACH. R. Bacaud, L. Rouleau, Institut de Recherches sur la Catalyse, CNRS, 2 Avenue Albert Einstein, 69626 Villeurbanne Cedex, France, and V.L. Cebolla, L. Membrado, Departamento de Procesos Químicos, Instituto de Carboquímica, 50015 Zaragoza, Spain

Analytical evaluation of petroleum based materials and processed feeds is a complex task relying upon a compromise between tedious in-depth characterizations and fast responding tools for process control. In the present paper, a large number of hydroprocessed vacuum residues, obtained either under catalytic or thermal conditions, have been submitted to the following analytical techniques: Simulated Distillation, coupled Simdis/MS, UV spectroscopy, ^{13}C NMR, quantitative Thin Layer Chromatography/FID, Vapor Phase Osmometry. A confrontation of analytical data in the light of correlations with hydrogen transfer evaluation is proposed, which accounts for observed variations in aromatic content. Conradson Carbon Residue largely influences the results obtained with some of the examined techniques. Apparent discrepancies are rationalized and a strategy for a comprehensive analytical evaluation of hydroprocessed feeds is proposed.

030. DEVELOPMENT OF HEAVY OIL HYDROCRACKING CATALYSTS -CATALYST CHARACTERIZATION WITH TEMPERATURE PROGRAMMED-DESORPTION, -REDUCTION AND -SULFIDING-
E. Iwamatsu‡, E. Hayashi‡, Y. Sanada‡ and T. Yoneda‡
Shakeel Ahmed‡‡, S. Ali‡‡, A. Lee‡‡ and H. Hamid‡‡
‡Advanced Catalysts Research Laboratory, Petroleum Energy Center, Kanagawa, Japan
‡‡Research Institute, King Fahd University of Petroleum & Minerals, Dhahran 31261, Saudi Arabia

As an acid function of catalyst for hydrocracking, Al2O3 pillared montmorillonite catalysts were prepared and characterized by means of temperature programmed desorption(TPD) with NH3, temperature programmed reduction(TPR) and temperatue programmed sulfiding(TPS) with H2S. The correlations between the characteristics of catalysts thus obtained and activities of cumene cracking were studied. Commercial hydrocracking catalysts were also characterized by means of temperature programmed methods and pulse reaction of cumene. TPD, TPR profiles and cumene cracking activity were changed by sulfiding of Al2O3 pillared montmorillonite. It is suggested that heat treatment and S species adsorbed strongly on the surface influenced upon acid property, thefore, cumene cracking activity was changed.

031. DISTRIBUTION OF HETEROATOMS IN ASPHALTENES SEPARATED FROM KHAFJI RESIDUE BEFORE AND AFTER HYDROTREATMENT AS STUDIED BY GPC FRACTIONATION. Y. Koinuma, S. Kushiyama, R. Aizawa, S. Kobayashi, I. Uemasu, K. Mizuno and Y. Shimizu, National Institute for Resources and Environment, 16-3 Onogawa, Tsukuba, Ibaraki 305 Japan

Asphaltenes are complex mixtures consisting of hydrocarbon molecules of so wide a range of molecular size that fractionation into narrower cuts is desirable to better understand their structure and reactivity. In this study, the asphaltenes separated from Khafji residue and its hydrotreated product were fractionated into nine cuts by preparative gel permeation chromatography. The results of the analysis showed that, for the feed asphaltenes, the distributions of sulfur, nitrogen, vanadium, nickel and H/C ratio are all fairly uniform over the whole range of molecular size which corresponds to the VPO average molecular weight range of about 30,000 to 800. In contrast, for the product asphaltenes, the smaller the molecular size, the lower were the concentrations, except for nitrogen and H/C ratio. The results of NMR analysis and the fractionation of resins and aromatics in maltenes will also be reported.

032. High-Temperature Simulated Distillation Study on Catalytic Upgrading of Atmospheric and Vacuum Resids.
Boli Wei, K. Madhusudan Reddy and Chunshan Song*
Fuel Science Program and Laboratory for Hydrocarbon Process Chemistry, The Pennsylvania State University, 209 Academic Projects Building, University Park, PA 16802.

Simulated distillation by gas chromatography (SimDis) has been used to measure boiling point distribution of petroleum fractions for years. However, SimDis method is not established for analyzing heavy oils and resids with final boiling point above 538°C (1000°F). This paper presents our work on high-temperature SimDis (H-T SimDis) analysis of two petroleum resids (atmospheric and vacuum resids) and their upgrading products from catalytic hydrotreating in batch reactors. A 6 m x 0.53 mm high-temperature megabore capillary column was used for the H-T SimDis analysis. Effects of GC conditions were examined using four different temperature programs and two detection temperatures. Best conditions were identified for heavy hydrocarbons in the resids with boiling point up to about 750°C (1382°F). The resids were hydrotreated under 6.9 MPa H_2 in the temperature range of 350-450°C with or without catalyst. Three Mo-based supported and unsupported catalysts were tested and the products were analyzed by both H-T SimDis and sulfur analyzer. The SimDis results indicate that through proper hydrotreating the heavy resids can be upgraded to light oils. Different Mo catalysts showed different behavior for molecular weight reduction and for desulfurization of the resids.

033. DENSITY FUNCTIONAL THEORY STUDIES OF NMR ^{29}SI CHEMICAL SHIFTS IN ZEOLITES. John B. Nicholas, Environmental Molecular Sciences Laboratory, Pacific Northwest National Laboratory, Richland, WA 99352, USA, jb_nicholas@pnl.gov

Solid State NMR is a powerful technique by which a detailed picture of the crystalline symmetry of zeolites and the local environment of particular atoms can be obtained. We present theoretical predictions of ^{29}Si chemical shifts for atoms in siliceous zeolites. A variety of clusters representing particular sites in the MFI framework are used as models. Each cluster was optimized using DFT at the SVWN/DZVP2 level of theory. Predictions of the NMR data were then obtained using the GIAO formalism with both the SVWN and BLYP exchange-correlation functionals and TZVP+ basis sets. The theoretical results accurately predict changes in ^{29}Si chemical shift as a function of the local geometry and can be used to identify particular sites within the zeolite. The influence of other structural factors, such as framework Al and bridging hydroxyl groups has also been investigated. The factors relating to the accuracy of the calculations will be discussed.

034. THEORY, SPECTROSCOPY AND KINETICS OF zEOLITE CATALYSED REACTIONS, Rutger A. Van Santen, Schuit Institute of Catalysis, Eindhoven University of Technology, P.O. Box 513, 5600 MB Eindhoven, The Neterlands.

Quantum-chemical DFT cluster calculations on models of zeolite proton calalysed reactions show that protonated intermediates usually are transition states. There is rarely a direct relation between groundstate and transition state structures. Associative reaction mechanisms are preferred, that do not proceed via intermediate alcoxy formation. Pico-second infrared laser spectroscopy experiments have been used to analyse the spectral features of adsorbed methanol clusters. The relaxation times of protons provide information on the degree of hydrogen bonding and site in homogeneity. The overall kinetics of a reaction of the hydroisomerisation of hexane using quantum-chemical cluster data. One then requires also additional or experimental data on the heats of adsorption of reactants of products. Predictions on the degree of pore filling at reaction conditions have been using Position Profiling measurements.

035.
FIRST-PRINCIPLES MOLECULAR DYNAMICS STUDY OF SMALL MOLECULES IN ZEOLITES. **Karlheinz Schwarz** Ernst Nusterer, Theoretical Chemistry Group, Vienna University of Technology, A-!060 Vienna, Austria, and Peter E. Blöchl, IBM Zurich Research Laboratory, Säumerstr.3, CH 8803 Rüschlikon, Switzerland.

The key in the understanding of zeolite catalysis is the interaction of the reactants with the inorganic framework. We performed molecular dynamics (MD) simulations using the Projector Augmented Wave (PAW) method in order to study the adsorption and reaction of small molecules (methanol and water) in a zeolite (Si-rich sodalite). The MD at finite temperature was essential to find the global minimum and to determine the proton transfer from the acid site which occurs only at high coverage. At low coverage a proton exchange reaction was observed during an MD run of 10 psec.

036. NOx DECOMPOSITION PATHWAYS IN Cu-ZSM-5: INSIGHTS FROM QUANTUM AND STATISTICAL MECHANICAL CALCULATIONS. A. K. Chakraborty, Dept. of Chemical Engineering, Univeristy of CA, Berkeley, CA.

The efficient catalytic decomposition and reduction of nitrogen oxides presents a challenge, overcoming which will lead to significant environmental benefits. Cu-ZSM-5 has been found to be the most active zeolite catalyst for this purpose. However, its use is hampered due to many reasons (e.g., steam inhibition). We have studied the thermochemistry of nitrogen oxide decomposition pathways in Cu-ZSM-5 using quantum Density Functional theory and rudimentary statistical mechanics. We first characterize the catalytic sites in Cu-ZSM-5. We then study an enormous number of possible reactions, and propose a reaction pathway that is thermodynamacally feasible. This pathway allows consistent interpretation of a vast body of experimental results. Insight into the reasons underlying the ability of Cu species in ZSM-5 to catalyze Nox decomposition will also be provided.

037.
THEORETICAL STUDIES OF HYDROCARBON CRACKING IN ZEOLITES. A. Redondo, P. J. Hay, Los Alamos National Laboratory, Los Alamos, New Mexico

We have carried out *ab initio* (Hartree-Fock and Moller-Plesset perturbation theory, MP2) and semiempirical (MOPAC, PM3) quantum mechanical calculations of olefin and paraffin cracking in acid zeolites. For definiteness, we have chosen as examples 1-pentene and *n*-pentane in faujasite. The zeolite is represented by a small cluster terminated with OH bonds. After a brief review of the cracking mechanisms we will consider then the thermochemistry of the gas phase reactions for the olefin and the paraffin. These will then be compared with the corresponding calculations for the hydrocarbons interacting with the zeolite cluster. Differences in the reaction mechanisms will be discussed.

038.
VALIDATING SIMULATIONS OF NON-FRAMEWORK SPECIES LOCATIONS IN ZEOLITES V.V. GULIANTS, J.T. MULLHAUPT, Praxair Inc., P.O. Box 44, Tonawanda, NY 14151, J.M. NEWSAM, A.M. GORMAN, C.M. FREEMAN, Molecular Simulations Inc., 9685 Scranton Road, San Diego, CA 92121

A grid-based cation-insertion algorithm described recently and combined Monte Carlo docking and molecular dynamics method have been applied to simulating the locations of non-framework species in zeolites. The new grid-based method correctly identifies and populates preferred non-framework cation locations in various frameworks with the non-framework species. The cation locations were predicted in dehydrated zeolite adsorbents and catalysts, such as Na88X, Ca48X, mixed cation zeolite 3A, and Cu-mordenite using only suitable framework structure models. The locations of benzene molecules in the supercage of Ca48X were predicted via combined Monte Carlo docking and molecular dynamics method. These examples demonstrate that the new grid-based insertion technique and the Monte Carlo docking procedure can provide reasonable starting models to aid in determining crystal structures of zeolites containing non-framework species.

039. THE ROLE OF POLAR AROMATICS IN RESIDUUM HYDROCRACKING, Barry B. Pruden, Department of Chemical and Petroleum Engineering, University of Calgary, Calgary, Alberta T2N 1N4, and N.Kelly Benham, Petro-Canada, Calgary, Alberta T2P 3E3.

Recent improvements in residuum upgrading are based on the prevention and control of mesophase, a pre-coke material. Mesophase results from the precipitation and condensation of asphaltenes, which can occur when insufficient resins (polar aromatics) are present to either solubilize the asphaltenes or hold them in colloidal suspension. Increasing the polar aromatics in the reactor through recycle is the basis for the technological improvements described in this paper This technique, applied in the CANMET Hydrocracking Process, had led to higher conversions through higher operating temperatures and higher hydrogen transfer rates to the asphaltenes; lower additive requirements through recycle of pitch, maintaining the additive in a fine particle size; and the ability to convert to a 700-°F product slate.

040. MOLECULAR TRANSFORMATION DURING HEAVY OIL UPGRADING

P.M. Rahimi[1], V. Nowlan[2], H. Dettman[1] and A. Delbianco[3]
National Centre for Upgrading Technology, One Oil Patch Drive, Devon, Alberta, Canada T0C 1E0
[2] Syncrude Research, 9421 17th Ave, Edmonton, Alberta, Canada T6N 1H4
[3] Eniricerche SpA, San Donato Milanese, Italy

Conversion of heavy oils to distillate products requires an increase in H/C ratio, reduction in molecular weight and lowering of the boiling point of the feedstocks. Understanding the chemistry of thermal and catalytic processes is the first step in improving product yields and quality. The lack of progress in understanding the chemistry of heavy oils/bitumen conversion is due partly to the molecular complexity of these materials and partly to the occurrence of simultaneous reactions under a given condition. The use of model compounds in studying reaction mechanisms such as hydrogen transfer or cleavage of strong C-C bonds has been helpful but its applicability in understanding heavy oil conversion has not yet been fully exploited.

In this study we use fractions of heavy oils that were separated by liquid chromatography, and well characterized, as model compounds to investigate the conversion chemistry of complex hydrocarbon mixtures. The major reaction paths leading to the observed productsare identified by analyzing fractionated feedstocks before thermal reaction as well as the reaction products using various analytical techniques such as MCR, molecular weight, and NMR.

041. MECHANISMS OF COKE AND SEDIMENT RESULTING FROM RESID CONVERSION
I. A. Wiehe, Exxon Corporate Research, Annandale, NJ 08801-0998

The conversion of petroleum atmospheric and vacuum resids in visbreaking and in hydroconversion processes are usually limited by either the onset of coke formation in the reactor or by carbonaceous sediments in the liquid products. Coke and sediment result because two of the major products of conversion, volatile liquids and asphaltene aromatic cores, are insoluble in each other. By combining kinetic modeling of the thermal conversion of resids with optical microscopy and analytical data on the products, a detailed mechanism of coke and sediment formation is revealed. Coke results when asphaltene cores phase separate at thermal reaction conditions to form a liquid crystalline phase that coats metal surfaces as a liquid film before oligimerizing into a solid. On the other hand, toluene soluble sediments are formed when asphaltene cores become insoluble on cooling below thermal cracking temperatures to form oil insoluble, isotropic solids. Unlike coke, sediments can be redissolved by heating, by adding a solvent, or by removing the nonsolvents, the volatile liquids.

042. HYDROPROCESSING CATALYST DEACTIVATION: MECHANISMS AND CONTROL

V.J. Nowlan[1], C.A. McKnight[1], and R.A. McFarlane[2]
[1]Syncrude Research, 9421 17th Ave., Edmonton, Alberta, Canada T6N 1H4
[2]National Centre for Upgrading Technology, One Oil Patch Drive, Devon, Alberta, Canada T0C 1E 0

An investigation of catalyst deactivation under hydroprocessing conditions with Athabasca bitumen was performed using a series of time on stream experiments. Porosity measurements showed that up to 9 days age, the catalysts can be considered to be porous but the 18 day old catalyst is essentially non-porous, having a surface area of 6.9 m^2/g and pore volume of 0.024 cm^3/g compared to the fresh catalyst values of 272.5 m^2/g and 0.534 cm^3/g, respectively. When low temperature ashing was used to remove 95% of the coke deposited on the 1 day old catalyst, the resulting pore volume and pore size distribution indicated that deposited metal sulphides were also contributing to pore-mouth plugging.

The adsorptive capacity of the catalyst for NO decreased dramatically with age. This capacity was diminished by 40% after 1 day, 53% after 9 days and 77% after 18 days. The rapid drop-off in NO uptake corresponded to a steep decline in sulphur removal and hydrogen consumption activity.

From the performance data, it can be seen that the behaviour of the pitch and CCR conversions is different from the other hydroprocessing reactions. The behaviour of the sulphur, nitrogen, vanadium and nickel deactivation curves can be attributed to either of two different mechanisms for each reaction or to deactivation caused by some component of the feed that will poison the catalyst directly, but whose products are not catalyst poisons.

043. SODIUM DESULFURIZATION OF BITUMEN: A NOVEL ROUTE TO CONTAMINANT REMOVAL.

R.D. Myers, Imperial Oil Resources Limited, 3535 Research Rd. N.W. Calgary, AB, Canada, T2L 2K8; R. Bearden Jr., Exxon Research and Development Laboratory, 4045 Scenic Highway, Baton Rouge, LA, 70805; G. Brons, Exxon Research and Engineering, Route 22 East, Annandale, NJ, 08801-3059; P.A. Fletcher, AEA Technology plc, 429 Harwell, Didcot, Oxn. OX11 0RA.

Bitumen contains high concentrations of contaminants such as sulfur, nitrogen and metals (Ni and V). The presence of these contaminants and the related high resid content cause the value of bitumen in the marketplace to be significantly discounted relative to light sweet oil. It is well known that alkali metals, particularly sodium, are effective in removing sulfur and metal contaminants from hydrocarbon streams. Although sodium is much more efficient and selective than hydrogen, it is also more expensive. Thus any upgrading scheme utilizing sodium as a desulfurizing agent requires an economic method for sodium regeneration. This paper will present bitumen desulfurization results obtained using a stirred autoclave reactor at Imperial Oil Resources Limited. Although previous work showed that a sodium:sulfur ratio of at least 2:1 was required for complete desulfurization, we have shown that hydrogenolysis of the initially formed intermediate sodium salt, [R-S⁻Na⁺] to form R-H and NaSH may occur. This paper will also present the results of a program conducted by AEA Technology to design and construct a prototype flowing sodium regeneration cell which is compatible with sodium desulfurization chemistry. Critical flowing salt cell design features and experimental data associated with high current density cell operation will be presented.

044. RECENT DEVELOPMENTS IN FIXED BED CATALYTIC RESIDUE UPGRADING.

S. Kasztelan, V. Harlé, S. Kressmann, F. Morel, Institut Français di Pétrole, France

Atmospheric or vacuum residue can be hydrotreated and converted into valuable distillates using high temperature, high pressure, low contact time and fixed bed hydroprocessing units. This paper will review the recent developments made, both on processes and catalysts to improve the run length and the level of conversion. This includes the development of multiple swing reactors, the use of improved fluid distribution, catalyst loading, guard bed materials and sophisticated association of catalysts using particle size, activity, pore size and shape grading.

045. SYNERGISTIC EFFECT OF CATALYST COMBINATION ON RESID HYDROCONVERSION.

Y.Mizutani, A.Nishizawa, Y.Yamamoto, S.Takehara and H.Yamazaki, COSMO Research Institute, Satte, Saitama 340-01, Japan

The synergistic effect of catalyst combination with various pore sizes has been reported to make the optimum catalyst system for Resid Hydroconversion Process. This optimum catalyst system shows the high stability at high reaction temperature to achieve high conversion. The mechanism on this synergistic effect of catalyst combination has been investigated. This talk will discuss recent developments and application on Resid Hydroconversion catalyst. Included in the set of experimental results on reactivities and properties of intermediate products, sampled between reactors, etc, will be probes of sequential conversion of large moleculars, such as the asphaltene, by catalysts with the different pore size, and the increased HDS reactivity.

046. **THE SHELL RESIDUE HYDROCONVERSION PROCESS: DEVELOPMENT AND ACHIEVEMENTS.** B. Scheffer, K.W. Röbschläger (Shell International Oil Products, 1030 BN Amsterdam) and F.C. de Boks (Shell Netherlands Refinery, Pernis)

The Shell HYCON process converts high sulphur, high metal content vacuum residues into atmospheric and vacuum distillates. The HYCON process is equipped with catalyst bunkering facilities that make on-line catalyst replacement possible. The first unit was built at Pernis refinery and was started up in 1989.
Initial teething problems were resolved by some modifications. The process performance (in terms of sulphur/metals removal and conversion into distillates) is now in line with predictions, and the catalyst bunkering system works well.
The third HYCON cycle lasted over 9000 hours. The throughput has been increased to well above design value in 1995 and a high conversion of vacuum residue into distillates has been consistently achieved.
The technical achievements since the start-up of the HYCON unit are described in the paper.

047. The New Concept of the Asphaltene Conversion Catalyst. S. Inoue, T. Takatsuka and Y. Wada, Environmental Technology Center, Chiyoda Corporation, 3-13 Moriya-cho, Kanagawa-ku, Yokohama 221, Japan

A clear understanding of the chemistry of asphaltenes cracking is the basis for a good perception of the mechanism of bottoms cracking. Asphaltene molecules are hard to crack over catalyst because they are so big and have multiple stacked, porphirin-like structures, containing high amounts of heteroatoms, which easily deactivate the catalysts. So far, the first stage of reaction is usually not given much attention. It is mostly regarded as a simple "Metal take-up zone", guarding the subsequent hydrotreating catalyst from undue fouling by these metals, or rather their sulfides.
A catalyst with a high pore size and still good strength is required there, rather than one with high hydrotreating activities. Catalyst supported by sepiolite and its modification is a good candidate for the target.

048.
EFFECT OF THE ADDITION OF LIGHT HYDROCARBON SOLVENTS ON THE CATALYTIC HYDROTREATMENT OF PETROLEUM HEAVY FRACTION. M. Nomura, W. Trisunaryanti, R. Shiba, and M. Miura, Department of Applied Chemistry, Faculty of Engineering, Osaka University, Suita, Osaka 565, Japan

Hydrotreatment of a petroleum asphalt fraction was examined in the presence of a series of light hydrocarbon solvents in order to obtain basic information of the solvent effect. The treatment was conducted with a Ni-Mo-Al2O3 catalyst at 673 K under 50 kg/cm2 of hydrogen in a batch autoclave. The effectiveness order of the solvents tested in terms of desulfurization as well as cracking of heavier components in the feed was found to follow the sequence cyclohexane > benzene > tetralin > decalin > 1-methylnaphthalene. This would suggest that the supercritical nature of the lighter solvents, cyclohexane and benzene, under the conditions employed is relatively more influential than the hydrogen-donating ability of tetralin for the up-grading.

049. **THERMAL AND CATALYTIC HYDROCRACKING OF SUPERCRITICAL FLUID-EXTRACTED VACUUM RESIDUE**
Murray R Gray[1], Keng Chung[2] and Xiao-Feng Qian[1]
1. Department of Chemical and Materials Engineering, University of Alberta, Edmonton, Alberta. T6G 2G6
2. Syncrude Canada Ltd., 9421 - 17 Ave, Edmonton, Alberta. T6N 1H4

A series of ten fractions of Athabasca vacuum residue, prepared by supercritical fluid extraction with pentane, were reacted in a microbatch reactor at 430°C. Three reaction conditions were used: hydrogen gas at 10 MPa, hydrogen gas with Ni/Mo on alumina catalyst and nitrogen gas at 0.24 MPa. Analysis included boiling point distribution by simulated distillation, yield of coke in the reactor and on the alumina catalyst, and sulfur content. The data provide insight into reactivity and coking propensity as a function of the molecular weight of the residue fraction.

050. VACUUM RESIDUE HYDROPROCESSING SIMULATION: THE KEY FOR A BETTER UNDERSTANDING OF THE PROCESS. Capistrano Nobre de Abreu, Reginaldo Guirardello (UNICAMP), Carlos de Medeiros (PETROBRAS)

PETROBRAS is facing a new challenge in its history that is to meet society requirements of light products to be obtained from an increasing heavy oil supply for our deep water oil feeds. Another reality is the increased strictness in the specification concerning quality and environmental aspects.
To cope with this, PETROBRAS implemented the PROTER - Program for the Development of Strategical Refining Technologies - which aims to optimize the existent technology, develop new refining schemes and implement new processes able to solve our problems.
Hydrorefining processes will play an important role in this context, more specifically the vacuum residue hydroprocessing that is able to convert heavy feed in high quality products with low amount of sulphur and nitrogen.

051. KINETICS OF RESIDUE HYDROTREATING REACTIONS. Jie Chang, Jiansheng Liu, Dadong Li, Research Institute of Petroleum Processing (RIPP), P. O. Box 914-17, Beijing, 100083, P. R. China

Four kinds of typical residues from Middle East and China have been hydrotreated in a pilot plant, which has three fixed bed reactors in series. Many series of commercial catalysts from CHEVRON, UNOCAL and RIPP have been employed in this research. The process conditions are pressure of 12.0~16.0 mpa, temperature of 360~400 °C, LHSV of 0.15~1.0 h^{-1}, H_2/Oil ratio of 400~1000 Nm^3/m^3. The feedstocks include not only Mideast original residues, but also Chinese original residues, such as Shengli residue, Xinjiang residue and Liaohe residue as well. The kinetic parameters of residue hydrotreating reactions are calculated in terms of pore diffusion resistance and correlated with some special properties of oils and catalysts. It supplies important basic data for the selection of catalysts, the design of composite catalyst beds and the optimum of process conditions.

052. ATOMISTIC SIMULATIONS OF ADSORPTION IN ZSM-5. Alexis T. Bell, Department of Chemical Engineering, University of California, Berkeley, CA 94720-1462, USA and Doros N. Theodorou, Department of Chemical Engineering, University of Patras, GR26500 Patras, Greece

The adsorption of atoms and molecules in ZSM-5 can be simulated successfully using an atomistic representation of adsorbate-zeolite and adsorbate-adsorbate interactions and the principles of statistical mechanics. The calculation of Henry's constants can be achieved by Monte Carlo integration and complete adsorption isotherms can be determined by Grand Canonical Monte Carlo. Special techniques must be adopted to achieve computational efficiency for cases involving tightly fitting or long-chain molecules. The extent to which the adsorption of C_1-C_{20} alkanes can be simulated will be shown. Recent applications of Monte Carlo simulations to describe the orientational distribution of ethane in ZSM-5 will be described and compared with information derived from ^2H NMR experiments.

053.

APPLICATION OF INTEGRATED COMPUTATIONAL CHEMISTRY SYSTEM TO THE DESIGN OF INORGANIC MEMBRANES. Hiromitsu Takaba, Koichi Mizukami, Momoji Kubo, Adil Fahmi, and Akira Miyamoto, Department of Molecular Chemistry and Engineering, Faculty of Engineering, Tohoku University, Aoba, Aramaki, Aoba-ku, Sendai, 980-77, Japan.

Microporous inorganic membranes have been investigated as possible bifunctional systems where the membrane is involved in the catalytic reaction besides separation processes, because of their thermal and chemical stabilities relative to organic membranes. The separation mechanism, which depends on the gas pressure and temperature, is not yet well understood at the molecular level. Therefore, molecular modeling of transport mechanisms is necessary to develop a new efficient membrane reactors. We will report the MD simulation results of the transport mechanism of gases (CO_2, He, Ar, N_2, C_4H_{10}) on some inorganic membranes, (1) ceramics having cylindrical pore, (2) ZSM-5 type silicalite, and (3) amorphous silica.

054.

DIFFUSION AND REACTION IN ZSM-5 AND COMPOSITE CATALYSTS - METHANOL-TO-OLEFINS. F.J.Keil, J. Hinderer, A.R. Garayhi

Multicomponent diffusion and reaction in HZSM-5 and a composite catalyst is modeled. The reaction Methanol-To-Olefins (MTO) is taken as an example. Diffusivities of the molecules within a three-dimensional network of HZSM-5 is computed by a Monte Carlo approach. Some results obtained by Molecular Dynamics are also given. The results are compared to adsorption and NMR measurements. Diffusivities in the amorphous phase of a composite catalyst is described by the dusty-gas model. Any type of common reaction Kinetics is included in the amorphous phase. With the aid of kinetic parameters found by experiments, MC simulations of the MTO reaction are executed.

055.

PREDICTION OF SORPTION AND DIFFUSION OF HYDROCARBONS IN ZEOLITES THROUGH NEW, HIERARCHICAL MOLECULAR SIMULATION TECHNIQUES
Leonidas Gergidis and Doros N. Theodorou, University of Patras, GR-26500 Patras, Greece; Edward J. Maginn, University of Notre Dame, Notre Dame, IN 46556, USA; and Alexis T. Bell, University of California, Berkeley, CA 94720-1462, USA.

Computer simulations offer the possibility of predicting the sorption equilibria and transport rates of organic molecules in zeolites from miscroscopic structure and interactions. This presentation will focus on examples of new molecular simulation strategies developed to overcome the large computer time requirements of conventional techniques and thereby address real-life zeolite/sorbate systems. The first example concerns the prediction of transport (as opposed to self-) diffusivities for pure and mixed small alkanes in the zeolite silicalite through a combination of "color field" nonequilibrium molecular dynamics and grand canonical Monte Carlo simulations. The second example concerns the prediction of the sorption equilibria, siting and conformation of long (C6 to C25) alkanes in silicalite. For this problem, a configurational bias Monte Carlo integration approach has been developed, which also provides a basis for systematic coarse-graining of the representation of the sorbed chain molecules and subsequent study of their long-time dynamics.

056.

MIXED QUANTUM - CLASSICAL SIMULATIONS OF CHEMICAL DYNAMICS AT SURFACES.
John C. Tully, Department of Chemistry, Yale University, New Haven, CT 06520

Conventional molecular dynamics, based on the Born-Oppenheimer separation of electronic and nuclear motion and classical mechanical treatment of nuclear motion, is inadequate to describe surface chemical processes such as charge transfer, photochemistry, chemistry at metal surfaces and reactions involving hydrogen atoms for which quantum mechanical effects such as zero-point motion and tunneling have important consequences. We are developing methods to extend the molecular dynamics approach to incorporate both electronic transitions and quantum atomic motion. We outline the challenges in combining classical and quantum dynamics, and present two applications. The first application is to chemistry at metal surfaces, where electron-hole pair transitions contribute significantly to energy transfer. The second is to inclusion of tunneling effects in surface chemical reactions. This work was carried out in collaboration with M. Head-Gordon, S. Hammes-Schiffer and J. Kindt.

057. THE EFFECTS OF ADSORBATES ON THE DYNAMICS OF ADSORPTION ETHANE ON PT(111), J.A. Stinnett and R.J. Madix, Department of Chemical Engineering, Stanford University, Stanford, CA 94305

The adsorption probabilities, α, of ethane on Pt(111) covered with monolayers of ethane, ethylidyne and sulfur were determined experimentally. In all cases significant increases in α relative to the clean surface were observed. Furthermore, whereas α showed a weak dependence on the angle of incidence on clean Pt(111), on the adsorbate-covered surface, no dependence was observed. These effects were examined further for a sulfur adlayer by stochastic trajectory simulations. The simulations reveal that the origin of both the increase in α and the independence on incident angle results from the corrugation of the surface introduced by the adsorbate, which causes substantial interconversion of the parallel and normal momentum components of the incident species.

058. METAL DISPERSION AND CATALYTIC ACTIVITY IN Pd CATALYSTS FOR METHANE COMBUSTION R.A. Dalla Betta, D.G. Loffler and S. Magno, Catalytica Combustion Systems, Inc. 430 Ferguson Dr., Mountain View, CA 94043

Characterization of Pd-based catalysts for methane combustion frequently include measuring metal dispersion using either H2 or CO chemisorption. In spite of uncertainties derived from the formation of palladium oxides under reaction conditions, those techniques generally result in reproducible results. Yet, the correlation of catalytic activity with physical properties of the catalyst is not well established. Literature reports suggest that, under reaction conditions, bulk palladium oxide is formed. Oxidation breaks apart the metal crystallites, so that most of the oxide participates in the reaction. This process may be reversed when the sample is reduced prior to measuring hydrogen uptake.

We have prepared a series of Pd-based catalysts for methane combustion with different metal loading. The catalytic activity of the samples was correlated with metal dispersion measured using CO and H2 chemisorption. The results suggest a linear correlation between hydrogen chemisorption and catalyst activity for samples with relatively high metal dispersion. This correlation seems to break down for poorly dispersed samples.

059. KINETIC CHANGES DURING THE ACTIVATION OF Pd METHANE OXIDATION CATALYSTS. Carl R. F. Lund, Sarves Peri, Saley Harou-Kouka. Department of Chemical Engineering, SUNY - Buffalo, Buffalo, NY 14260-4200.

Methane combustion was studied over supported Pd catalysts under conditions of high methane concentration and low oxygen concentration. The catalysts undergo activation under reaction conditions. A change in the kinetics was observed to accompany the activation process. A mechanistic model was developed wherein methane undergoes deep dehydrogenation on the surface of the catalyst and oxygen subsequently scavenges the resulting species. The mechanistic model was capable of describing the kinetics of the reaction. It was also consistent with transient kinetic experiments and offers a description of possible mechanistic changes that occur during the activation process.

060. CATALYTIC COMBUSTION OF LOW-HEATING VALUE FUEL-GASES: IGNITION OF CO, H_2 AND CH_4. E.M. Johansson and S.G. Järås, KTH - Royal Institute of Technology, Department of Chemical Engineering and technology, S-100 44 Stockholm, Sweden.

There is an increasing interest in catalytic combustors fuelled by low-heating value gases, with a LHV of 5-7 MJ/Nm^3, because catalytic combustion could be advantageous compared to flame combustion with respect to stable combustion of LHV-gases and low conversions of fuel-N to NO_x. In the present European project, catalytic combustion of gasified wood for gas turbine applications is studied. We are using a synthetic gas mixture of H_2, CO, CO_2, H_2O, N_2, NH_3 and C_2H_4, which is mixed with air at atmospheric pressure and combusted over washcoated cordierite monoliths. The objective of the presentation is to discuss how the ignition of CO, H_2 and CH_4 over washcoats, such as Cu, Mn or Fe doped La-hexaaluminate catalysts, is influenced by fuel gas composition. Recent results show (1) that CO and H_2 enhance the combustion of CH_4, (2) coadsorption phenomena control light-off and (3) choice of transition metal influences the ignition strongly.

061.

CATALYTIC PROPERTIES OF TRANSITION-METAL-SUBSTITUTED HEXAALUMINATE FOR HIGH TEMPERATURE COMBUSTION, K. Eguchi, H. Takahara, H. Inoue, K. Sekizawa, and H. Arai, Graduate School of Engineering Sciences, Kyushu University, Kasuga 816 Japan

Cation-substituted hexaaluminate compounds, $AMAl_{11}O_{19}$ (A = Ba, La, and Sr; M = Co, Cr, Cu, Fe, Mn, and Ni), were investigated. Mn and Cu are the excellent substituents in enhancing the activity for the hexaaluminates with different A cations. The difference in catalytic activity among A cation species depends on the valence of A cation. Structural difference between magnetoplumbite and β-alumina types does not show large effect for the activity. Although Mn and Cu were also effective substituents for enhancing catalytic activity in Ba-based hexaaluminate compounds, their activities were low as compared with the La-based catalysts. These results indicate that the redox cycle of transition metal in hexaaluminate lattice and catalytic activity appear to be affected with the electronic or structural effect of the large cation in the mirror plane. Activation energy for methane oxidation indicates that Mn is the most effective substituent in this series.

062.

IN-SITU STUDIES OF SPATIAL PATTERNS IN A PACKED-BED CATALYTIC COMBUSTOR. S. Marengo and P. Comotti, Stazione Sperimentale per I Combustibili, S. Donato Mil., Italy 20097

NO ABSTRACT AVAILABLE

063. MODELING AND DETERMINATION OF KINETIC PARAMETERS FOR CATALYTIC COMBUSTION OF METHANE. D. Papadias, E.M. Johansson, L. Edsberg* and P. Björnbom, KTH - Royal Institute of Technology, S-100 44 Stockholm, Sweden, Dep. of Chemical Engineering and Technology, *Dep. of Numerical Analysis and Computing Science

In earlier work we have used one-dimensional models for various studies of catalytic combustion. In this paper we intend to give a status report on our efforts of extending this work to combine a more rigorous, two-dimensiona, model of a monolith channel of a catalytic combustor with experimental data.

For determination of kinetic data for catalytic combustion, an isothermal annular reactor with a washcoated rod is used according to McCarty. Steady state experiments at different temperatures are to give activation energies and preexponential factors for the catalytic reactions. These kinetic parameters are to be used to predict experimental results on an adiabatic monolithic catalytic combustor using our two-dimensional model.

064.

KINETICS OF HIGH TEMPERATURE METHANE COMBUSTION BY METAL OXIDE CATALYSTS
Jon G. McCarty, Yun-feng Chang, and Victor L. Wong, SRI International, 333 Ravenswood Avenue, Menlo Park, CA, 94025-3493 and E. Magnus Johansson, KTH, Teknikringen 42, S-100 44 Stockholm, Sweden

The kinetics of the catalytic combustion of methane by supported palladium oxide and other supported and unsupported transition metal oxide catalysts were examined for several oxygen partial pressure levels using temperature programmed reaction (TPR). By preparing the catalysts as thin (≈10 μm) coatings on an alumina tube and conducting the experiments with high annular flow rates of dilute methane and oxygen in helium, the rate measurements were extended up 900 °C without significant contribution by gas phase reactions. In the case of supported PdO, ramp and hold temperature-time transient techniques were used to investigate the hysteresis in rate at the transition between metallic Pd and PdO. First row transition metal oxides from chromia to copper oxide were supported with calcined lanthanum aluminate peroskite and also examined unsupported in a fixed bed microreactor. With the significant exception of supported PdO, the transition metal oxides show Arrhenius kinetics over three orders of magnitude in the effective first-order rate constant for combustion of dilute methane.

065.

STRUCTURE AND CATALYTIC PROPERTIES OF PdO_x/ZrO_2 CATALYSTS FOR METHANE OXIDATION AT LOW TEMPERATURES. E. Fujimoto, F. H. Ribeiro, M. Avalos-Borja, and E. Iglesia. Department of Chemical Engineering, University of California, Berkeley, Berkeley, CA 94720

NO ABSTRACT AVAILABLE

066. INFLUENCE OF HETERO-ATOM CONTAINING COMPOUNDS ON THE HYDRO-METALLISATION OF VANADYL-TETRAPHENYLPORPHYRIN, J.P. Janssens, A.D. van Langeveld, S.T. Sie and J.A. Moulijn, Delft University of Technology, the Netherlands

Oil residua contain large amounts of hetero-atoms, such as sulfur, nitrogen, oxygen and metals. The metals, mainly vanadium and nickel, can be removed by catalytic hydrotreatment. The presence of sulfur, nitrogen and oxygen containing compounds in oil residua can influence the rate at which these metals are removed, since they can adsorb competitively on the active site of the catalyst and inhibit the hydrogenation and ring cleavage reactions of organo-metallic compounds.
In the present work the influence of characteristic hetero-atom containing compounds, i.e., quinoline, ammonia, benzofuran, water, dibenzothiophene, and anthracene on the hydrodementalisation (HDM) of vanadyl-tetraphenylporphydrin (VO-TPP) has been investigated. The added amount of hetero-atoms has been chosen near to that occurring in oil residua.

067. DEPOLYMERIZATION AND DEMETALLATION TREATMENTS OF ASPHALTENE IN VACUUM RESIDUE. K. Sakanishi, N. Yamashita, D.D. Whitehurst, I. Mochida Institute of Advanced Material Study, Kyushu University, Kasuga, Fukuoka 816, Japan

The NiMo sulfides supproted on carbon nanoparticles of hollow structure was effective for the complete hydroconversion of asphaltene(hexane insoluble fraction) in a vacuum residue(VR) into maltene(hexane soluble) fraction with the aid of hydrogen-donating solvents. The ESR spectroscopy showed that metallic compounds in the asphaltene was selectively converted to polar maltene fraction by hydrogenative decoagulation under the mild catalytic hydrotreatment. The modified cation exchange resins under ultrasonic irradiation at 50 ℃ was also effective for the decoagulation of asphaltene to the hexane soluble fraction without hydrogen consumption. Based on the structural analyses of asphaltene and maltene fractions before and after the treatments, the decoagulation, depolymerization, and demetallation schemes of the asphaltene in VR are discussed, referring the roles of carbon-supported NiMo sulfides and ion-exchange resin.

068. Development of New Residue Hydrotreating Catalysts. C.J.J. den Ouden*, Shell International Chemicals BV, Shell Research and Technology Centre, Amsterdam, Badhuisweg 3, 1031 CM Amsterdam, The Netherlands; O.K. Bhan, Shell Development Company, Westhollow Technology Center, 3333 Hwy. 6 South, Houston, Texas 77082, USA; S. Boardman, Criterion Catalyst Company Ltd., 1650 Parkway, The Solent Business Park, Whiteley, Fareham, PO15 7AH, England; R.D. Street, and S.E. George, Criterion Catalyst Company, 16825 Northchase Drive, Two Greenspoint Plaza, Houston, Texas, 77060, USA

With the ever increasing severity of residue hydroprocessing operating conditions in terms of tighter product specifications and lower quality residue feedstocks, the development of improved residue hydrotreating catalysts is an ongoing process. In this paper, the development of a new hydrodesulfurization (HDS) catalyst (RN-450) with both high stability and high activity will be discussed. In addition, other novel prototype catalysts and their role in residue hydroprocessing will be introduced and discussed. Factors affecting resid unit performance including catalyst choice, feedstocks, chemistry and unit design considerations will be discussed as well. Case studies will be used to illustrate the optimum combination of these factors.

069. TPR and TPD Studies and Carbazole Hydrodenitrogenation Activity of Supported Molybdenum Nitrides. Masatoshi Nagai, Yosuke Goto, Osamu Uchino, Shinzo Omi, Graduate School of Bio-applications and Systems Engineering, Tokyo University of Agriculture & Technology, Koganei, Tokyo 184, Japan.

The activity and selectivity of the supported Mo nitrides for the HDN of carbazole were studied. The effects of catalyst preparation and pretreatment on the HDN activity of the nitrided Mo/Al_2O_3 catalyst were discussed. The surface composition and structure of the nitrided 12.5% and 97.1% Mo catalysts have been studied using TPR and TPD with the XRD analysis. The nitrided catalysts were prepared by the temperature-programmed reaction of Mo/Al_2O_3 with NH_3 at temperatures of 773, 973, and 1173 K and then cooled to room temperature in flowing NH_3 or He gas. The NH_3-cooled 12.5% Mo/Al_2O_3 catalysts were more active than the He-cooled catalysts for both hydrogenation and C-N hydrogenolysis in the HDN of carbazole. The activities of the NH_3- and He-cooled catalysts increased with increasing nitriding temperature. From the results, Mo metal was formed, together with γ-Mo_2N on the surface of the nitrided Mo/Al_2O_3 catalysts at 1173 K.

070. DEEP GAS OIL HDS ACTIVITIES OF NiW BASED CATALYSTS.

A.D. van Langeveld, H.R. Reinhoudt, R. Troost, S.T. Sie and J.A. Moulijn
Delft University of Technology, PO Box 5045, 2600 GA, Delft, The Netherlands
(A.D.vanLangeveld@stm.tudelft.nl).

The performance of NiW based sulfidic hydrotreating catalysts strongly depends on their degree of sulfidation. The genesis of the active phase under temperature programmed sulfidation was followed by FTIR of adsorbed CO and NO, and quasi in situ XPS analysis. Micro crystalline Ni appeared to sulfide essentially up to 613 K, whereas nickel in nickelaluminate and nickeltungstate are still in the oxidic state. Furthermore, it was concluded that calcining at 823 K favored the interaction between the WO3 and the alumina, thus reducing the sulfidability of the tungsten present. The catalytic performance in the hydrodesulfurization of thiophene, dibenzothiophene and 4-ethyl, 6-methyl dibenzothiophene will be discussed for the catalysts in various stages of sulfidation. For dibenzothiophene, optimum catalyst performance was found upon mild sulfiding at 613 K or lower temperatures. In contrast, the highest activity for thiophene HDS was observed after sulfiding at 823 K. In the view of these results, the HDS of 4-ethyl, 6-methyl dibenzothiophene will be discussed.

071. THE EFFECT OF INDIVIDUAL GASOIL COMPOUNDS ON THE DEEP HDS PERFORMANCE OF ASA SUPPORTED NOBLE METAL CATALYSTS.

H.R. Reinhoudt, S. van Schalkwijk, A.D. van Langeveld, S.T. Sie and J.A. Moulijn
Delft University of Technology, PO Box 5045, 2600 GA, Delft, The Netherlands
(H.Reinhoudt@stm.tudelft.nl).

In the strive towards a cleaner environment, new legislation has been adapted for the reduction of the sulfur content in diesel fuel. However, this reduction involves the conversion of 4,6 alkylated dibenzothiophenes which requires different catalytic functions than present in traditional HDS catalysts. To realize the reduction in an economically attractive way, the development of catalysts with a high activity for 4,6 alkylated DBT,s would be highly beneficial. It will be shown that ASA based noble metal catalysts are very active for the conversion of 4,6 alkylated DBT,s under model reaction conditions. To bridge the gap between model testing and real feed processing, a detailed study was undertaken to quantify the effect of individual diesel fuel components on the deep HDS performance of ASA based catalysts. It appeared that basic nitrogen compounds like quinoline have a large influence on the Pt/ASA performance. A model for the different active sites on Pt/ASA is proposed. The results are used to improve the catalytic performance of the ASA catalysts in deep HDS processing.

072. CATALYTIC ACTIVITY OF COBALT OXIDE LOADED ON HIGH SURFACE SUMECTITE FOR THIOPHENE HYDRODESULFURIZATION

E. Hayashi,* E. Iwamatsu,* Y. Yamamoto,* Y. Sanada,* and T. Yoneda*
M.E. Biswas,[†] S. Ahmed Ali,[†] A. Lee,[†] and H. Hamid[†]

* Advanced Catalysts Research Laboratory, Petroleum Energy Center, Kanagawa, Japan
† The Research Institute, King Fahd University of Petroleum and Minerals, Dhahran 31261, Saudi Arabia

Catalysts of cobalt oxide loaded on sumectite having high surface area are prepared and tested on thiophene hydrodesulfurization activity (HDS). Sumectite clays (montmorillonite, saponite, porous saponite, hectorite and stevensite) were used as supports. The catalysts were tested on HDS activity for thiophene by means of pulse reaction. Co- porous saponite catalyst in the series shows a highest thiophene HDS activity so far studied. It appears that the kind of sumectite and the structure have a strong influence on the catalytic activity. The catalysts were characterized by Temperature - Programmed Sulfiding (TPS) method and ESR. Reduction and sulfiding conditions of loaded Co were discussed.

073. Mesoporous Zeolite-Supported Co-Mo Catalyst for Hydrodesulfurization of Petroleum Resids.
K. Madhusudan Reddy, Boli Wei and Chunshan Song*
Fuel Science Program and Laboratory for Hydrocarbon Process Chemistry, The Pennsylvania State University, 209 Academic Projects Building, University Park, PA 16802.

Mesoporous zeolites with MCM-41 type structure with different SiO_2/Al_2O_3 ratios were synthesized according to the procedure described elsewhere (K. M. Reddy and C. Song, Catal. Lett., 1996, 36, 103). MCM-41 supported Co-Mo catalysts were prepared by co-impregnation of $Co(NO_3)_2 \cdot 6H_2O$ and $(NH_4)_6Mo_7O_{24}$ followed by calcination and sulfidation. Commercial Al_2O_3 supported Co-Mo (Shell 344) was sulfided and tested for comparison purpose. In our earlier work, the hydrodesulfurization of a model fuel containing 3.5 wt% sulfur as dibenzothiophene in n-tridecane (C. Song and K. M. Reddy, ACS Div. Petr. Chem. Prepr., 1996, 41(3), 567) Co-Mo/MCM-41 catalyst was found to be more active than Co-Mo/Al_2O_3 catalyst due to higher hydrogenation and hydrocracking activities. In this work, we aim at exploring the potential of mesoporous zeolite-supported Co-Mo catalysts for hydrodesulfurization of petroleum resids. Two types of petroleum resids, atmospheric and vacuum resids, were used as feedstocks. Catalysts were examined at 350-450°C under higher H_2 pressure (6.9 MPa) with and without solvent at varying catalyst amounts. The preliminary results indicate that Co-Mo/MCM-41 type catalysts have potential for hydrodesulfurization of petroleum resids.

074. CLUSTER STUDIES ON ADSORPTION AND REACTIONS AT THE V_2O_5 (010) SURFACE. M. Witko[1], K. Hermann[2], R. Tokarz[2], and A. Michalak[3], [1]Institutue of Catalysis and Surface Chemistry, PAS, ul. Niezapominajek, 30-239 Krakó, Poland, [2]Fritz-Haber-Institut der MPG, Faradayweg 4-6, D-14195 Berlin, Germany, [3]Faculty of Chemistry, Jagiellonian University, ul. Ingardena 3, 30-060 Kraków, Poland.

The electronic structure of vandium pentoxide, V_2O_5, is discussed based upon the cluster model using ab initio DFT and semiempirical INDO-type methods. No influence of the second layer on the surface properties is found. Results of the adsorption of hydrogen (a probe reaction to model the first step in the selective oxidation of hydrocarbons) at the structurally different oxygen sites, are compared with the adsorption/activation of hydrocarbons at the V_2O_5 (010) surface. Among the different oxygen centers those bridging two bare V atoms are most nucleophilic sites. Hydrogen binds to all inequivalent oxygen sites with the strongest binding occurring for O bridging two bare V atoms. Oxidation of propene and toluene on V_2O_5 (010) to yield aldehyde species proceeds through the formation of C-O bond with the bridging oxygen, abstraction of two H atoms from CH_3 group and formation of two OH surface groups.

075.
REACTIVITY OF CO AND O_2 INTERACTING WITH MgO OXYGEN VACANCIES.
G. Pacchioni and A. M. Ferrari, Dipartimento di Scienza dei Materiali, via Emanueli 15, Milano, Italy

The electronic structure of oxygen vacancies on the MgO surface (F_s centers) and their interaction with O_2 and CO has been studied by means of *ab initio* Hartree-Fock cluster wavefunctions. Missing oxygen atoms or ions on the MgO surface result in the formation of charged and paramagnetic defects exhibiting a pronounced reactivity. O_2 interacts with the electrons trapped in the surface defect with formation of the superoxide anion, O_2^-. CO interacts with F_s centers forming complex surface species of general formula $C_nO_{n+x}^{2-}$. The precursor of these species is the CO^- radical anion. Adsorbed CO^- is much less stable than O_2^- and easily react to form more stable species. The reactivity of F_s centers is dominated by electrostatic effects and the different stability of O_2^- and CO^- surface complexes is due to the different electron affinity of the neutral molecules. Comparsion of experimental and computed EPR and IR spectra show that the O_2^- molecule is adsorbed parallel to the surface while CO^- lies normal to the F_s center with the O-end oriented toward the surface.

076.
HARTREE-FOCK PERIODIC STUDY OF THE CHEMISORPTION OF SMALL MOLECULES ON TiO_2 AND MgO SURFACES. J.Ahdjoudj, A. Markovits and C. Minot Laboratoire de Chimie Théorique, UPMC, 75252 Paris, Cédex05, France

Adsorption of molecules on TiO_2 and MgO depend on their acidic (basic) properties in the gas phase. When molecularly adsorbed, the site for adsorption is the metal and the E_{ads} correlate with the proton affinities. On MgO, CO_2 molecules that ordinary are considered as acidic probe molecules, are absorbed on Mg sites. HSBA theory allow to compare TiO_2 with MgO. In contrast H_2O is absorbed on MgO by H-bond, revealing the cleavage for MeOH and acidic cleavage for MeSH. At high coverage, stabilization through H-bonds favors the vacinity of adsorbates on the surface (islands of absorbates and coadsorption effects). Van der Waals interactions explain the chage of orientation (CO_2) with the coverages. Lateral interactions are more important on MgO than on TiO_2.

077. MODEL STUDIES OF PARTIAL OXIDATION PROCESSES ON OXIDIZED MO(110) AND CO-MO(110) C. M. Friend, D. A. Chen, K. T. Queeney, and M. Chen, Department of Chemistry, Harvard University, Cambridge, MA

This work addresses issues important to developing a fundamental understanding of the interplay between geometric structure and chemical reactivity in heterogeneous oxidation chemistry. The main objective of this project is to identify factors that control partial oxidation reactions. Experimental studies of oxidation reactions have been performed on oxidized Mo(110) and mixed Co-Mo(110)-O phases in order to understand the role of specific types of oxygen in dictating the oxidation processes and to examine how the functionality of the Co and Mo combine in oxidation processes. Theoretical studies of Mo(11)) slabs with different oxygen phases using the method developed by Kaxiras, et al. are also underway.

078.
Structure, Bonding, and Reactivity at Molybdenum Oxide (MoO_3) Surfaces. K. Hermann[1], M. Witko[2], and A. Michalak[3]; [1]Fritz-Haber-Institut der MPG, D-14195 Berlin (Germany), [2]Institute of Catalysis, PAN, 30239 Cracow (Poland), [3]Faculty of Chemistry, Jagiellonian University, 30060 Cracow (Poland)

The electronic structure and bonding near different oxygen sites at MoO_3(010) and (100) surfaces is reviewed on the basis of ab initio density functional theory (DFT-LCGTO) cluster calculations. The clusters are chosen as finite sections of the ideal MoO_3 surface where cluster embedding is achieved by bond saturation with hydrogens yielding clusters up to $Mo_7O_{30}H_{18}$. Resulting charge density distributions and binding properties are analyzed by populations, bond orders, and electrostatic potential maps. Interatomic binding at the surface is determined by both ionic and covalent contributions with clear distinctions between the terminal oxygens and the different bridging surface oxygens. The difference in electronic structure between the (010) and (100) surface is found to be mainly due to the different atom arrangement while local atom charging and binding properties seem surface independent. The electronic surface parameters influence the behavior and reactions of adsorbed molecules as will be shown for H, OH, and C_3H_5 adsorbates.

079. **On The Mechanism and Energetics of Metal Dusting at Transition Metal Oxides**
Hansong Cheng, David Reiser, and Sheldon Dean, Jr., Air Products and Chemicals, Inc., 7201 Hamilton Boulevard, Allentown, PA 18195-1501

Metal dusting is one of the most catastrophic corrosive phenomena that frequently occurs in many industrial processes. Its dominant chemical processes include chemisorption of carbon monoxide at alloys or metal oxides and the subsequent deposition of carbon on these materials resulting from the interaction between the adsorbed species and the gas steams through the Boudouard and steam-carbon reactions at high temperature. In this presentation, we report a density functional study on possible mechanisms of metal dusting at transition metal oxides. Minimum energy paths are calculated and the structural changes of the reaction species along the paths are examined. The results suggest that dusting is unfavorable energetically at the transition metal oxides, in agreement with practical observations. Population analysis is then performed to analyze the electron transfer process at the gas-surface interface and explains why CO dissociation at the oxide surfaces is difficult and these materials are dusting resistant.

080. CATALYTIC PROPERTIES OF Pd/Al_2O_3 CALCINED AT HIGH TEMPERATURES FOR METHANE COMBUSTION. E. Kikuchi[1], T. Matsuda[2], N. Takahashi[2], [1] Department of Applied Chemistry, Waseda University, Okubo, Shinjuku, Tokyo 169, [2] Department of Materials Science, Kitami Institute of Technology, Kitami, Hokkaido 090.

A series of palladium catalysts supported on Al_2O_3 which was calcined at temperatures in the range from 900 to 1600℃ were tested for methane combustion. Dispersion of palladium decreased with increasing calcination temperature of Al_2O_3 support. The catalytic activity of Pd/Al_2O_3 was not related to dispersion of palladium, and palladium supported on Al_2O_3 calcined at 1600℃ exhibited the highest activity among the catalysts tested. The studies of temperature programmed oxidation of Pd and temperature programmed reaction of methane with PdO showed that the oxidation temperature of metallic Pd to PdO was lowered by increasing calcination temperature of Al_2O_3 support, although no appreciable difference was observed in the reaction of PdO and methane. We conclude from these results that the catalytic activity of Pd/Al_2O_3 for methane combustion was strongly affected by the rate of oxidation of metallic Pd to PdO.

081.
CATALYTICALLY SUPPORTED COMBUSTION - HETEROGENEOUS REACTION MODELLING
C. Niehörster Lehrstuhl für Technische Thermodynamik, RWTH-Aachen, D-52062 Aachen, Germany

Reducing pollutant emissions from combustion is of high environmental concern. Improving combustion processes in radiant burners is achieved using catalytically supported combustion. Here the stabilization of the flame is mainly dependent on radiant heat transfer from the burner surface. Catalytically coating of the surface leads to lower gas-phase temperatures at higher surface temperatures. Low gas-phase temperatures and low gas-phase fuel conversion give almost zero pollutant emissions. The modelling of such a system is dependent on the coating and heterogeneous kinetics, including homogeneous-heterogeneous interactions. This work deals with kinetic modelling of the most important heterogeneous reactions. Simulations are presented for one-dimensional modelling using reduced heterogeneous and detailed homogeneous reaction kinetics. Additionally the surface-region is modelled as a channel flow including detailed heterogeneous reaction modelling. As the final result a four-step reaction mechanism including radical desorption and adsorption is presented. A comparison with laser-optical measurements shows very good agreement.

WITHDRAWN

082. THE CATALYTIC COMBUSTION OF VOLATILE CHLORO-ORGANIC COMPOUNDS USING URANIUM OXIDE CATALYSTS. Graham J. Hutchings, Catherine S. Heneghan, Stuart H Taylor, Leverhulme Centre for Innovative Catalysis, Department of Chemistry, University of Liverpool, PO Box 147, L69 3BX, UK; Ian D. Hudson, BNFL, Company Research Laboratory, Springfields Works, Salwick, PR4 0XJ, UK

Chloro-organic compounds are known to be particularly stable and conventional catalysts for their destruction, based on supported noble metal systems, often deactivate rapidly. It is therefore desirable, to discover new catalysts which demonstrate high activity and are resistant to deactivation. We have found that catalysts based on uranium oxide are particularly effective for the oxidative destruction of chloro-organic compounds at industrially relevant temperatures and flow rates. Chlorobenzene and chlorobutane were used as typical examples of VOCs. Chlorobenzene conversion (1% in air) over U_3O_8 was 99.7% at 350°C and GHSV = 70,000 h^{-1}, with CO_X the only products detected. Similar results were also obtained for the destruction of chlorobutane over U_3O_8. Comparison with the metal oxide Co_3O_4, a known highly active combustion catalyst, showed that U_3O_8 was considerably more active under our experimental conditions. It has also been shown that U_3O_8 was an effective catalyst for the destruction of lower levels of chlorobenzene converting 99.9% of a 1200 ppm feed at 400°C. Time on line studies indicated that the catalyst was not readily deactivate as virtually complete conversion (99.9%) of a chlorobenzene stream (1200 ppm in air) was maintained at 450°C for over 400 hours continuous use. These studies have demonstrated that under commercially realistic operating conditions catalysts based on uranium oxide show high activity for the oxidative destruction of chloro-organics with high efficiency. These findings are clearly important for the development of new VOC destruction catalysts.

083. CATALYZED COMBUSTIONS OF CHLORINATED HYDROCARBONS. K. Griesbaum, D. Hönicke, A. Haas, F. Schwerdtner and A. Khemili, Engler-Bunte-Institut, Bereich Petrochemie, Universität Karlsruhe (TH), D-76128 Karlsruhe, Germany

Our goal was the total oxidation of an industrially occurring mixture of chlorinated hydrocarbon waste products with the concurrent formation of a maximum amount of Cl_2 rather than HCl. The reactions were carried out with air on a $CuO-Cr_2O_3/\gamma-Al_2O_3$ catalyst at 500 °C. A point of major concern was the performance of the catalyst with time-on-stream. In model experiments with individual chlorinated substrates, we could demonstrate that the stability of the catalyst depends decisively on the ratio of hydrogen to chlorine in the feed, independent of the carbon skeleton of the chlorinated substrates. Substrates containing H:Cl in a ratio of greater than one could be totally oxidized over a long period of time to give chlorine and HCl in the proportions corresponding to the Deacon equilibrium. Substrates containing H:Cl in a ratio smaller than one led to rapid deactivation of the catalyst. Based on analyses of the spent catalysts, an explanation for this behavior will be advanced.

084.
ELECTROCHEMICAL PROMOTION OF PD-CATALYZED OXIDATION OF CO BY NO ON YTTRIA STABILIZED ZIRCONIA Soonho Kim Gary L. Haller, Department of Chemical Engineering, Yale University, New Haven, CT 06520-8286

The Pd-catalyzed oxidation of CO by NO exhibits strong electrochemical promotion (EP) by electrochemically pumped oxygen ion from yttria-stabilized zirconia (YSZ) under appropriate conditions of temperature, gas composition, and catalyst potential. The Pd catalyst film was deposited on the face of an oxygen ion conducting YSZ disc, and the Ag counter and reference electrodes were deposited on the other side of the disc. Both the catalyst film and the Ag counter and reference electrodes were exposed to the reacting NO-CO mixture. The observed increase in the rate of carbon dioxide production was typically a factor of 1000 higher than the rate of oxygen ion removal from the catalyst. The EP system provides direct information about the nature of the catalyst and the reaction mechanism under conditions of elevated pressure.

085. CONTROLLED SYNTHESIS OF PLATINUM-RUTHENIUM CATALYTIC SURFACES FOR OXIDATION OF METHANOL IN FUEL CELLS. S. H. Bergens, C. E. Lee, Y. Xing, P. B. Teige, Department of Chemistry, University of Alberta, Edmonton, Alberta, Canada, T6G 2G2

Platinum black effected the reduction of $[Ru(COD)(\eta^3-C_3H_5)_2]$ (**1**, COD is 1,5-cyclooctadiene) by dihydrogen (pressure H_2 = 1 atm, T = -10°C) in hexane to generate ruthenium(0), cyclooctane, propane, n-octane, and octahydropentalene. The ruthenium(0) was adsorbed by the surface of platinum. The amount of **1** consumed equaled the amount of reduced COD generated at all stages of the reaction, allowing real-time monitoring of the equivalents ruthenium deposited on platinum. Reactivity studies and cyclic voltammetry indicated that complete coverage of platinum by ruthenium was achieved after reduction of 1.8 equivalents **1**. Reductions of **1** were interrupted after deposition of sub-monolayer amounts of ruthenium and the resulting surfaces were evaluated as electrocatalysts for oxidation of methanol. A platinum surface covered with 0.1 equivalents ruthenium was up to 14 times more active than bare platinum for potentiodynamic oxidation of methanol. A surface covered with 0.8 equiv ruthenium was operated for 1.5 h with negligible dissolution of ruthenium.

086. CATALYTIC DECOMPOSITION OF TRIMETHYLAMINE OVER GOLD SUPPORTED ON IRON BASED OXIDES. Atsushi Ueda, Ryoichi Taniguchi*, Masashi Azuma*, and Masatake Haruta, Osaka National Research Institute, AIST, Midorigaoka 1-8-31, Ikeda, Osaka 563, JAPAN, *Department of Applied Chemistry, Osaka Institute of Technology, Omiya 5-16-1, Asahi-ku, Osaka 535, JAPAN.

Gold exhibits good activity for the oxidative decomposition of $(CH_3)_3N$ when it is supported on ferric oxide and ferrites with high dispersion. The crystallite size of gold supported on metal oxides is 3-5nm. In the case of $(CH_3)_3N$ decomposition over $Au/NiFe_2O_4$, which is especially active and selective into N_2 formation, oxidation to CO_2 starts even at 373K accompanied by N_2O formation and above 423K N_2 formation gradually prevails with increasing temperature. Since $(CH_3)_3N$ is one of typical odors, the supported gold catalyst appears to be the best candidate for its removal. The genesis of the unique catalytic behavior of supported gold will be discussed to emphasize the role of perimeter between ultrafine gold particles and metal oxide supports.

087. HYDRO-THERMAlCRACKING OF HEAVY OILS AND ITS MODEL COMPOUNDS, I.Nakamura, M.G. Yang and K. Fujimoto, Department of Applied Chemistry, School of Engineering, The University of Tokyo, 7-3-1 Hongo, Bunkyo-ku, Tokyo 113 Japan

Liquid phase cracking of vacuum gas oil (VGO) was conducted with Ni-Mo catalysts supported on non-acidic supports such as active carbon (AC) at higher reaction temperature (713 K) and under lower hydrogen pressure (3.0-8.0 MPa) than typical hydrocracking process using metal supported acid catalyst, which is termed as hydro-tharmalcracking process. In the Ni-Mo/AC system, the yield of middle distillate increased from 44 to 49 wt% and naphtha yield was suppressed (from 22 to 14 wt%) compared to the thermal cracking, whereas the conversion level decreased slightly (from 66 to 64 wt%). Although acidic catalyst such as Ni-Mo/USY gave higher conversion level than that of Ni-Mo/AC, the yield ratio of middle distillates to naphtha for the new process was by two times higher than that for the acid-catalyzed hydrocracking at the same conversion level. The hydro-tharmalcracking or thermal cracking of n-dodecylbenzene gave toluene as the major aromatic product, whereas the hydrocracking with Ni-Mo/USY gave benzene. The reaction mechanism of the process was assumed to be composed of the thermal cracking of hydrocarbon molecules via the free radical chain mechanism and the hydro-quenching of free radicals with the catalysts.

088. PYROLYSIS MECHANISM OF HEAVY OIL MODEL COMPOUNDS. H. Kawai and F. Kumata, Petroleum Energy Center, KSP D-1237, 3-2-1, Sakato, Takatsu-ku, Kawasaki, Kanagawa 213, Japan

During the hydroprocessing of petroleum heavy oil, pyrolysis occurs concurrently with the main catalytic reaction. Generally, in case of heavier feedstocks, the influence of pyrolysis is more overwhelming. Asphaltene closely relates to pyrolysis, but the reaction mechanism is not elucidated sufficiently. In this study, 1,2-diphenylethane(DPE) and dibenzylsulfide(DBS), which represent a part of asphaltene structure, were used as model compounds under various hydrogen partial pressures. DBS pyrolysis easily happened at 400°C while DPE pyrolysis hardly decomposed. The results showed that an increase of hydrogen partial pressure generated more hydrogen radical, and the hydrogen radical attacked at aromatic ring of DPE and sulfur atom of DBS. The weaker C-S bond of DBS was affected more by the hydrogen radical than Ph-C bond of DPE. Furthermore, during pyrolysis of DBS, product ratio of stabilization to toluene and dimerization of benzyl radical was constant. Finally, Kuwait Vacuum Residue was used to compare with the model compounds, and the results showed the same trend concerning the effects of hydrogen partial pressure.

089. NEOPHASE INCEPTION AND COKE FORMATION IN THERMAL CONVERSION OF SHENGLI VACUUM RESIDUE. Shenghua Li, Logistical Engineering College, Chongqing 630042, CHINA; Chenguang Liu, Guohe Que, Wenjie Liang and Yajie Zhu, University of Petroleum, Shandong 257062, CHINA.

Neophase formation in the thermal conversion of Shengli vacuum residue is elucidated in view of its colloid form structure disclosed by the freeze-fracture replication transmission electron microscopy. Together with the comprehensive time sequential approach to the thermal reaction system of Shengli vacuum residue, as well as its real-time optical microscopic tracking conducted and reported previously by the authors, it is indicated that there exist close relations between the onset points and morphologies of neophases and the induction period of coke formation and morphological compositions of coke.

Keywords: neophase inception, coke formation, thermal resid conversion

090. MODELLING HEAT EFFECTS OF VGO HYDROCRACKING, C.S. Laxmi Narasimhan, M. Sau & R.P. Verma, Indian Oil Corporation Ltd, India

Hydrocracking, which is employed in petroleum refining industry to heavy gas oils and resids to useful products, is a highly exothermic process involving simultaneous cracking and hydrogenation of complex hydrocarbon mixtures. The reaction is carried out in a trickle bed reactor involving complex kinetics, heat & hydrodynamics effects.
Heat effects are critical for stable hydrocracking operation, catalyst life, process design, optimization & control. Therefore, it is quite important to accurately model the heat effects of hydrocracker.
As kinetics is always coupled with temperature effects, a novel approach has been developed to model the combination of heat effects with kinetics of hydrocracking of complex petroleum mixture using continuum theory of lumping.
The resulting coupled integro-differential equations are solved numerically to obtain both temperature & concentration profiles. The model prediction are compared with reported experimental data from literature and are found to be in good agreement.
The monitoring of a commercial VGO-Hydrocracker plant using the model has beee discussed.

091. HYDROCRACKING ACTIVITY AND SELECTIVITY OF VGO.1. EFFECT OF METAL LOADING TO Y-ZEOLITE FOR THE CATALYTIC ACTIVITIES. T. Isoda, K. Kusakabe, S. Morooka, Department of Chemical Science and Technology, Kyushu University, 6-10-1, Hakozaki, Higashi-ku, Fukuoka 812-81, Japan

Hydrocracking, hydrodenitrogenation (HDN), and hydrodesulfurization (HDS) of vacuum gas oil (VGO) were examined over Ni, Co, or Fe loaded Y-zeolites dealuminated to different extents at 300 - 380 °C under 5 MPa H_2 pressure. Ni loaded zeolite of Si/2Al = 16.0 (Ni-HY-A) enhanced the hydrocracking of paraffins in VGO at least the gas yields. Extensive dealuminated Ni-HY-B zeolite of Si/2Al = 50.0 provided large amount of middle distitrate than that of Ni-HY-A, and supressed the gas and coke formation. Both zeolites also exhibited an excellent HDN and HDS activity among the metal-Y zeolites examined. HY-A zeolite of metal free enhanced the gas, coke and hexane insoluble products at higher reaction temperatures. Such a higher catalytic activity of Ni-HY zeolites originate from optimum acidity and higher hydrogenation activity. The Si/2Al ratio and metal loading are the controlling factors for the catalytic performances of novel zeolite.

092. HYDROCRACKING ACTIVITY AND SELECTIVITY OF VGO.2. PRODUCT DISTRIBUTION IN THE HYDROCRACKING OVER Ni LOADED DEALUMINATED Y-ZEOLITE. T. Isoda, K. Kusakabe, S. Morooka, Department of Chemical Science and Technology, Kyushu University, 6-10-1, Hakozaki, Higashi-ku, Fukuoka 812-81, Japan

The product oils from VGO over Ni loaded Y-zeolite in the different extent of dealumination (Si/2Al = 16.0 and 50.0) and Y zeolite of Ni free (Si/2Al = 16.0) were separated several fractions by alumina liquid chromatograph and HPLC, and was analyzed by GC-FID to clarify the effect of hydrogenation and acidity in the zeolite for the products selectivity. VGO contained n-paraffins was the major product in the feed, of which content was 45 wt %. The second major was three and four rings aromatics, and one and two rings aromatics and polars were minor. Ni-HY zeolites exhibited higher activity for the removal of polars and four rings aromatics with less coke and gas yield. Straight paraffins of which carried 20 to 32 of the carbon numbers were completely cracked regardless of the Ni-HY zeolites examined, whereas the selectivities for the product such as branched paraffins were different. HY of Ni free zeolite exhibited significant amount of coke, polars, four rings aromatics, and gas yield, while yield of n-paraffins were markedly reduced through the deep hydrocracking. The controle for the acidity and hydrogenation activity in the zeolite will be the key issue to provide for the flexibility in the product oil through the hydrocracking from heavy feed.

093. Catalytic Reactions of High Density Polyethylene and Commingled Post Consumer Plastic for producing Distillate Fuels.
W. Ding, J. Liang, and L. L. Anderson, 3290 MEB, Department of Chemical & Fuels Engineering, University of Utah, Salt Lake City, UT 84112.

Catalytic reactions of high-density polyethylene (HDPE) and commingled post-consumer plastic waste (CP#2) for producing clean distillate fuels were systematically studied over hydrogenation-hydrocracking catalysts NiMo/zeolite and NiMo/SiO$_2$-Al$_2$O$_3$. The reaction parameters, reaction temperature, H$_2$ pressure, reaction time, ratio of catalyst to feed, were found to have dramatic effects on conversion and product distribution of HDPE and CP#2. At 375 °C, 1000 psig H$_2$ (initial), 40 wt% catalyst, at reaction time of 60 min., catalytic reactions over NiMo/zeolite resulted in over 90% conversion for both HDPE and CP#2 although about 50% gases were produced. The liquid products contained C$_4$-C$_{12}$ hydrocarbons with mostly C$_5$-C$_7$. For HDPE, nearly 100% conversion with about 70% liquid was obtained from reactions over NiMo/SiO$_2$-Al$_2$O$_3$ under the same reaction conditions. A reaction mechanism is proposed based on detailed product analyses. Thermal reactions at higher reaction temperatures (480-525 °C), shorter residence time (5-10 min.) were also performed.

094. THE USE OF COAL LIQUEFACTION CATALYSTS FOR COAL/OIL COPROCESSING AND HEAVY OIL UPGRADING. A.V. Cugini and D. Krastman, Pittsburgh Energy Technology Center, Pittsburgh, PA 15236-0940 and T. J. Gardner, Sandia National Laboratories, Albuquerque, NM 87185-0709

The catalytic hydrogenation of coal and model solvents using dispersed and/or supported catalysts has been the focus of several recent studies at PETC. The major findings include the need for catalyst activation (transformation from precursor to active state), the effect of catalyst dispersion on solvent hydrogenation and coal conversion, the potential of iron catalysts, and the benefits of novel supported catalysts (including a supported hydrous titanate catalyst developed at Sandia National Laboratories). This study extends the use of these catalysts to coal-oil coprocessing and heavy oil upgrading applications. The results indicate the potential for coal liquefaction catalysts for use in coprocessing and heavy oil upgrading. High conversions of coal-oil mixtures were observed at 1000 psig H$_2$ with catalyst additions as low as 100 ppm Mo. Similar results were observed in heavy oil systems. The catalysts tested in this study include molybdenum and iron based catalysts.

095. SULFUR BONDING IN MoS$_2$ AND Co-Mo-S STRUCTURES
L. S. Byskov[1], B. Hammer[1], J. K. Nørskov[1], B. S. Clausen[2], and H. Topsøe[2]
[1]CAMP, Department of Physics, Technical University of Denmark, DK-2800 Lyngby, Denmark
[2]Haldor Topsøe Research, Nymøllevej 55, DK-2800 Lyngby, Denmark.

The structure and bonding in MoS$_2$ clusters with and without Co doping is studied theoretically using self-consistent density functional theory with a non-local exchange-correlation energy. The clusters model the Co-Mo-S catalysts used extensively in hydrotreating. We study in detail the structure and bond energies of MoS$_2$ as a function of the amount of sulfur. This leads us to a model of the active catalyst which is in excellent agreement with available experimental data. We also study the energy required to form sulfur vacancies, which are believed to be the active sites for many hydrotreating reactions. Substituting some Mo atoms at the edge by Co leads to a significant lowering of the sulfur bond energy. This will lead to an increase in the concentration of active sites for the reactions and thus explains the promoting effect of Co in accordance with the Bond Energy Model [J. K. Nørskov, B. S. Clausen and H. Topsøe, *Catal. Lett.*, **13**, 1 (1992)].

096. TRANSITION METALS TO SULFUR BINDING ENERGIES RELATIONSHIP TO CATALYTIC ACTIVITIES IN HDS: BACK TO SABATIER WITH FIRST PRINCIPLE CALCULATIONS. Herve Toulhoat, Pascal Raybaud, Slavik Kasztelan, Institut Français du Pétrole BP311 92852 Rueil-Malmaison FRANCE, Georg Kresse and Jurgen Hafner, Teschnische Universität Wien, Wiedner Hauptstrasse 8-10 A-1040 AUSTRIA

We have undertaken systematic calculations of Transition Metal Sulfides bulk crystal structures, electronic and energetic properties at the first principles level (DFT, GGA, PW-USPP, PBC, implemented in the Vienna Ab initio Simulation Package, VASP). Relaxed cell parameters and ionic positions showed an excellent agreement with experimental values. Computed and experimental cohesive energies agreed within 3%. We re-defined the metal-sulfur (M-S) bond strength as the cohesive energy per metal-sulfur bond: we show that all experimental HDS activities (Pecoraro and Chianelli, 1981) fit nicely on a single volcano master curve when plotted against this simple energetic parameter. Metallic (i.e. zero gap) ionic sulfides consistently exhibit the weakest M-S bonds and semi-conductor iono-covalent sulfides the strongest. However, the Sabatier principle suggests a simple kinetic interpretation of this master curve. This new interpretation also accounts for the well known synergetic effects in mixed sulfides and therefore opens new prospects for exploratory applied research.

097. The Development of an All-Electron Atom-in-Molecule SCF Method and Its Application to Heterogeneous Catalysis and Other Extended Systems.
Jack A. Smith, Union Carbide, Catalyst Skills Center, P.O. Box 8361, South Charleston, WV

A non-empirical all-electron SCF method is presented which exploits an Atom-in-Molecule picture and allows for a localized treatment of large and extended systems. The method is derived by projecting out atomic (single-centered) subspaces from within a molecular or periodically extended system and iteratively solving perturbed atomic-like subsystems until a self-consistent Atom-in-Molecule picture is obtained. The partitioning leads to a familar non-local energy-dependent pseudopotential formalism, but the use of pseudometric and truncated (renormalized) pseudoeigenfunctions keeps the problem energy-independent and locally tractable. This approach is outlined for surface chemisorption, but its application to polymers, solvation, and other extended systems can also be discussed.

098. THEORETICAL STUDIES OF HOMOGENEOUS CATALYTIC REACTIONS: Pt (0) CATALYZED ALKENE AND ALKYNE DIBORATION AND DIIMINE-Ni AND -Pd CATALYZED ETHYLENE POLYMERIZATION AND CARBON MONOXIDE/ETHYLENE COPOLYMERIZATION. Keiji Morokuma, Qiang Cui, Mats Svensson, Djamaladdin G. Musaev, Robert D. J. Froese and Toshiaki Matsubara, Cherry L. Emerson Center for Scientific Computation and Department of Chemistry, Emory University, Atlanta, Georgia 30322

In this talk we will present applications of DFT and ab initio MO calculations to three catalytic reactions of transition metal complexes: (1). the mechanism of Pt(0)-catalyzed alkene and alkyne diboration reactions, (2). the mechanism of the diimine-Ni and -Pd catalyzed ethylene polymerization, and (3). the mechanism of Pd(II) and Ni(II)-catalyzed alternating copolymerization of carbon monoxide with ethylene. The most favorable pathway of the catalytic diboration reaction involves (i) oxidative addition of B-B to the M(0) complex, (ii) dissociation of one of phosphines, (iii) coordination of alkyne to M, (iv) insertion of alkyne into the M-B bond, (v) isomerization of the resultant complex accompanied by readdition of phosphine, and (vi) reductive elimination of alkenyldiboron product. For olefin polymerization reaction (2), chain initiation and propagation pathways as well as various chain transfer and termination pathways have been studied. A comparison has been made for the potential energy profiles between the Ni(II)- and Pd(II)-catalyzed and zirconocene-catalyzed ethylene polymerization.

099. RHODIUM CATALYZED HYDROGENATION Of CO_2 INTO FORMIC ACID: THEORETICAL STUDIES. A. Dedieu*, F. Hutschka, Laboratoire de Chimie Quantique, UPR 139 du CNRS, Université Louis Pasteur, Strasbourg, France

We present here a theoretical study - based on MP2 and QCISD(T) calculations - of the hydrogenation of CO_2 to formic acid homogeneously catalyzed by rhodium phosphine complexes.[1,2] A striking feature revealed by these studies is an alternative to the classical oxidative addition/reductive elimination sequence for the reaction of the formate intermediate with dihydrogen. This alternative involves a σ bond metathesis reaction between H_2 and the rhodium formate intermediate. The factors which are responsible for a low energy barrier, including the effect of an external base[3] are delineated. Reactions occuring within a square planar and an octahedral environment are compared. We extend the scope of our discussion to related reactions occuring in heterogeneous conditions.

(1) Hutschka, F.; Dedieu, A.; Leitner, W. *Angew. Chem. Int. Ed. Engl.* **1995**, *34*, 1742.
(2) Hutschka, F.; Dedieu, A.; Eichberger, M.; Fornika, R.; Leitner, W. *submitted for publication*
(3) Hutschka, F.; Dedieu, A. *submitted for publication*

100. TOWARDS MORE REALISTIC MOLECULAR MODELING OF HOMOGENEOUS CATALYSIS BY DENSITY FUNCTIONAL THEORY: COMBINED QM/MM AND *AB INITIO* MOLECULAR DYNAMICS. Tom K. Woo, Peter M. Margl, Liqun Deng, Tom Ziegler, Department of Chemistry, University of Calgary, Calgary, AB, T2N 1N4, CANADA, email: tkwoo@zinc.chem.ucalgary.ca.

Both combined Quantum Mechanics/Molecular Mechanics (QM/MM) and *ab initio* molecular dynamics methods are fast emerging as viable computational molecular modeling tools. In the combined QM/MM method part of the system, say the active site, is treated quantum mechanically whereas the remainder of the system is treated with a molecular mechanics force field. With the *ab initio* molecular dynamics methods, the system is simulated at a finite temperature. Here, no empirical force field is utilized. Rather, the forces at each time step are determined with a full electronic structure calculation at the density functional level. Both methods have been applied to study homogeneous catalytic processes. We will present a QM/MM study of the Brookhart Ni(II) diimine olefin polymerization catalyst (**1**). We will also present a study of the Ti(IV) constrained geometry olefin polymerization catalyst(**2**) where free energy barriers are determined at the *ab initio* molecular dynamics level for several chain termination and long chain branching mechanisms.

101.
INSIGHTS INTO HOMOGENEOUS CATALYSIS FROM FIRST PRINCIPLES THEORY: DENSITY FUNCTIONAL STUDIES OF INORGANIC COBALT COMPLEXES. **D.J. Doren** and R. Konecny, Department of Chemistry and Biochemistry, University of Delaware, Newark, Delaware 19716

Density functional theory (DFT) has been used to predict geometries, energies and spin states of a series of inorganic cobalt complexes containing the geometry-constraining Tp (tris-pyrazolyl borate) ligand. For example, the carbonyl, TpCoCO, is a stable 16-electron compound with an unusual asymmetric geometry. This molecule is the structural and electronic analog of a proposed (unstable) intermediate in C–H bond activation. DFT studies of TpCoCO permit detailed tests of DFT theory and a rationale for the structure that applies to unstable species as well. More recent work has focused on reactive molecules, including and bis-(TpCo)-dioxygen complexes (with 36 heavy atoms) that may be useful in catalyzing alkane oxidation. DFT provides a useful guide to the feasibility of proposed oxidation mechanisms.

102. **SELECTED COMMERCIAL ISSUES IMPACTING THE REQUIREMENTS FOR LUBRICANT BASESTOCKS.** W. R. Downey Jr., Kline & Company, Inc., 165 Passiac Avenue, Fairfield, NJ 07004.

The worldwide market for basestocks is being impacted by several commercial factors: changing supply capacity and capability, shifting performance requirements in the finished lubricant product categories and new manufacturing technologies. These factors are going to shape the number, size and capabilities of the basestock suppliers. It will also determine the companies will continue to serve this changing business; selected exits are expected.

103. **PETRO CANADA'S LUBRICANT MANUFACTURING EXPANSION BASESTOCKS FOR THE NEXT MILLENIUM**

T. Cully, T. R. Lynch, and H. G. Marr
Petro Canada Lubricants

Petro Canada's new lubricant base oil manufacturing facility is now in operation, making our company a world class basestock manufacturer. This plant, using the latest in catalytic dewaxing and high pressure hydrofinishing technologies, will produce high quality basestocks with unique properties. This paper will describe the strategic objectives underlying this investment, and the economic and performance benefits from these advanced basestocks.

104. OPTIMIZING CBO/UCBO BASE OIL BLENDS TO MEET THE PERFORMANCE DEMANDS OF FUTURE PCMO/HDMO LUBRICANTS.
J.N. Ziemer, B.K. Lok, and J. Lopez, Chevron Global Lubricants, Richmond, CA 94802

Unconventional base oils (UCBO's) with very high Viscosity Indexes have properties which make them uniquely suited to meet future lubricant requirements for high oxidation stability and low volatility. With Chevron's all catalytic Isocracking, Isodewaxing and lube hydrofinishing technologies, a wide range of CBO's and UCBO's can be manufactured with different viscosity's, V.I.'s, and volatility's. In order to determine the optimum CBO/UCBO blend ratio for making cost effective base oils that meet future PCMO/HDMO specifications, we have utilized computer models and experimental blend studies to allow us to predict finished product kinematic and Cold Cranking Simulator viscosity's, and Noack and D2887 volatility's. This paper will discuss the sensitivity of different variables on finished product volatility.

105. **KINETIC METHOD FOR MEASURING THE PERFORMANCE OF LUBE HYDROFINISHING CATALYSTS**

T. R. Lynch, P. D. Mack, H.G. Marr, and A. Pidutti,
Petro Canada Lubricants

Petro Canada has had nearly 20 years operating experience with manufacturing lubricants by hydrocracking vacuum gas oils, followed by high pressure hydrofinishing of the dewaxed intermediate product to produce extremely high quality, crystal clear basestocks. This paper addresses some of the details of the hydrofinishing step, whose purpose is to improve the appearnace and oxidative stability of the hydrocracked products. Our experience has shown that the stability of the hydrofinished stock is a function of the feed quality to the hydrofinisher and the capability of the hydrofinishing catalyst itself. We have developed a simple methodology based on 1st order kinetics for measuring the performance of the catalyst. This method has been used to track the performance of our catalyst in our production unit in commercial use and to evaluate competitive catalysts under pilot plant conditions.

106. **Composition and Performance of a Hydroisomerized Wax Base Oil.** J.A. Patel, M.P. Smith, J.R. Powers, J.R. Whiteman and G.F. Prescott, Texaco Fuels and Lubricants Technology Department, P.O. Box 509, Beacon, NY, 12508-0509, and Star Enterprise, Port Arthur, TX.

In response to an emerging demand for higher performance base oils, two high viscosity index (VI) hydroisomerized basestocks were developed and commercialized in 1990. These oils, tradenamed TEXHVI-70 and TEXHVI-100, meet the API Group III definition (>120 VI and <0.03 %sulfur) and were developed to satisfy critical low temperature viscosity and volatility performance requirements in several lubricant product areas. Manufacturing process, physical and chemical test data, and compositional make-up will be outlined. Computer model predictions and actual ASTM Sequence IIIE and VE Engine Test performance data and potential product applications will also be discussed.

107. **MOHAWK - CEP RE - REFINING PROCESS PRODUCES HIGH QUALITY LUBE BASE OILS**, L. M. Magnabosco, MAGNA Associates, Lake Forest, CA 92630 and K. C. Khurana, Chemical Engineering Partners, Newport Beach, CA 92660

The Mohawk - CEP Process is based upon vacuum distillation and thin film evaporation, followed by multi-stage hydrotreating at moderate to high severity. The basic Mohawk treatment addresses and properly deals with the fundamentals involving the chemistry of used oil. This treatment results in a reaction mixture that is begnin with respect to corrosion and fouling that have plagued all other waste lube oil reclamation processes. Catalytic hydroprocessing has been improved dramatically by a number of improvements and advancements that have been incorporated into the Mohawk-CEP process package. These improvements include proper identification of hydrotreating feed constituents so as to permit selection of optimum hydroprocessing reaction conditions that minimize catalyst deactivation and enhance imparting desirable lube base stock characteristics. Furthermore, incorporation of a patented hydrotreating feed poison reduction process step has resulted in reduced catalyst consumption of only 0.8 lb. catalyst/metric ton of lube base stock produced vs. 2.5 to 4.0 lb. catalyst/metric ton of lube produced before. The lube base stocks produced with the Mohawk-CEP process are of equal or better quality than virgin. The Mohawk-CEP process represents the only proven re-refining technology that works and is economically viable.

108. **MASS SPECTRAL INVESTIGATION OF WHITE OILS MANUFACTURED BY DIFFERENT PROCESSES.** Michael T. Cheng, Analytical Sciences Unit, Chevron Research and Technology Company, Richmond, California 94802.

Structure - activity relationship has been explicitly applied to many fields, such as in the development of new drug and pesticide. It is also important in the petroleum related products. It is more difficult in petroleum applications because one usually deals with complex mixtures. Before one can obtain the structure - activity relationship, one needs to obtain some way to describe the structures. In this paper we will present an approach in the structural characterization of key compounds in three different white oils. All three oil samples were treated with silicalite to separate the compounds by size and shape. And we will discuss the GCMS analysis of these fractions.

109. **PRODUCT APPROVALS FOR LUBRICANTS BASED ON RE-REFINED BASE STOCKS.** Michael P. Smith, Greg Hutchison and Kirk McNaughton, Safety-Kleen Corp., Elgin, IL 60123

A wide variety of premium finished lubricants have been formulated with high quality re-refined base stocks. These state-of-the-art base oils are produced by many of the same processes used to refine crude oils including vacuum distillation and hydrotreating. The resultant paraffinic base oils meet or exceed customer, industry, and government standards for product toxicology, consistency and quality. The high quality re-refined base oils are then blended with premium lubricant additives to produce finished lubricants. The quality of the re-refined base oils will be demonstrated by reviewing product approval data on a wide range of industrial and automotive lubricants.

110. CATALYST TECHNOLOGY FOR RESID FCC: EVOLUTION AND FUTURE TRENDS.
P.O'Connor *), J.P.J.Verlaan and S.J.Yanik, *) Akzo Nobel Catalysts
R&D Centre Amsterdam, The Netherlands.

The evolution of resid FCC catalyst technology over the years is discussed.
Present and future trends are elucidated.

Resid FCC Catalysts today are designed making use of the following features:
- High activity, selectivity and metal resistant zeolites
- High accessible catalyst architecture, for optimal site utilization, Conradson Carbon conversion and easy stripping.
- Special designed metal-support interaction to reduce the detrimental effects of metal contaminants.

111. Application of HGMS Technology to the Petroleum Refining Industry. M. Ushio, Central Technical Research Laboratory. Nippon Oil Co., Yokohama 231, Japan

High gradient magnetic separation (HGMS) technology was applied in two petroleum refining fields as the first attempt in the world. One application was the magnetic separation of FCC equilibrium catalyst. A pilot test was successfully conducted at 3700 BD unit. More than 20 wt% reduction of catalyst consumption could be observed by integration of HGMS with the FCC unit while maintaining the same metal and activity level of the equilibrium catalyst as without HGMS. The other application was the magnetic separation of small iron contaminant in RDS feed. We found that small iron particles in the RDS feed cause a ΔP problem of the RDS catalyst bed, but they can be removed efficiently by the magnetic separator we developed. This technology has already been commercialized in our two refineries. The commercialized units work well and help to solve the ΔP and solidification problems of the RDS catalyst bed.

112. CATALYTIC CRACKING OF NARROW FRACTIONS SEPARATED BY SUPERCRITICAL SOLVENT EXTRACTION FROM A VACUUM RESIDUUM. Yaofang Liu, Renan Wang, Jiujing Yang, Genlin Niu ,The National Key Laboratory for Heavy Oil Process, University of Petroleum, Changping, Beijing 102200, China.

A supercritical extraction and fractionation apparatus was developed. With n-heptane as the solvent, the vacuum residuum from Shengli crude oil was separated in the apparatus into 15 narrow fractions, each of them being around 5%m of the residuum. The distributions of SARA components, molecular weight, carbon residue,H/C, metal contents, et al., were investigated as well as the structural parameters. The 13 front fractions were respectively cat-cracked in a fixed- fluidized bed reactor. It is indicated that the fractions were almost separated based on their molecular weight and that the extract oil with less than 60% extraction yield was good in cat-cracking.

113. ASPHALTENE CRACKING BEHAVIORS IN RFCC FEED PRETREATMENT RELATING TO THE RFCC PERFORMANCE
K. Fujita, S.Abe and Y. Inoue, Nippon Ketjen Co., Ltd.,Tokyo, Japan

The roles of resid HDS units have been diversified from the time when sulfur removal was the only and most important requirement. Today, to supply the higher quality feed production to the RFCC units, combined functions are required as well such as decomposition of Conradson Carbon Residue, removal of metals (V+Ni), denitrofication and deeper hydrogenation of oils which relates to the RFCC performance. The RFCC feed, which is processed by a new catalyst system of new functional properties for asphaltene molecules, led to higher gasoline and middledistillates yields at the same coke level with less bottoms fraction.

The degree of cracking in asphaltenes affects the reacting and initial deactivation rate of the pretreatment catalyst system. On the other hand, RFCC, as curie point pyroriser study showed, the higher H/C rates of asphaltenes had better RFCC crackability.

114. NOVEL COMBINATION PROCESS OF HEAVY OIL CRACKING USING ADDITIVE CATALYST.
Koichi Kato. Nikko Consulting & Engineering Co.,Ltd.

A new combination process of heavey oil cracking has been introduced and proven to be commercially feasible in several bench scale test runs by Japan Energy Corporation. Japan. "Succeed Process" is a combination of slurry phase hydrocracking and delayed coker. The first section which is main part of "Succeed" uses tubular reactor and small amounts of disposable additive catalyst so that the catalyst does not cause any problem not only on the reaction but in the quality of the product in the consecutive second coking section. For the vacuum residue feedstock. the first section gives around 65wt% conversion (not too high) and the second covers around 55wt% on the unconverted of the first section. therefore resulting in the total conversion of 85wt%. The Succeed catalyst is suitable for use in overcoming coking problems and in conducting a conversion of heavy oil feedstocks into lighter and more valuable oil products.
A 3500 BPSD demonstration plant work will be carried out in 1998.

115. THERMODYNAMIC AND KINETIC MODELING OF FCC PROCESS
M.M.Sugungun, I.M.Kolesnikov, S.I.Kolesnikov
State Gubkin Academy of Oil and Gas, Moscow, RUSSIA.

Catalytic cracking of petroeum fractions is usually carried out in a reactor block with a somewhat complicated hydrodynamic regime. The reactor block can in the actual sense be considered as a combination of two different reactors. The riser is a near plug-flow ideal displacement of the catalyst and the reaction mixture, while the main reactor vessel is considered as an ideal mixing CSTR.
Various researchers have modeled the FCC process with different approach, exact kinetic parameters are not however established. A new kind of approach to modeling the process was proposed that allows the computation of the kinetic parameter to some certain accuracy. Cracking reactions are considered to be a five-stage process. Model adequacy was verified on a commercial FCC unit.

116. VACUUM GAS OIL AND BITUMEN MIX CATALYTIC CRACKING
I.M.Kolesnikov, Petra Saidel, M.Yu.Kilyanov, S.I.Kolesnikov
State Gubkin Academy of Oil and Gas, Moscow, RUSSIA.

Vacuum gas oil and or a mixture of gas oil and mazut are usually subjected to catalytic cracking with a zeolite catalysts. Special mazut cracking commonly termed as residue cracking allows an increase im the conversion of heavy feed to a more lighter products for the production of unsaturated hydrocarbons for synthesis of various petrochemical products, high octane gasoline and different grades of distillates.
The effects of deasphalt bitumen addition to vacuum gas oil cracking were discussed in this paper.

117. CRACKING CATALYSTS ACTIVATION WITH METALLORGANIC SILOXANES
I.M.Kolesnikov, V.M.Vinogradov, S.I.Kolesnikov, A.V.Yablonsky
State Academy of Oil and Gas, Moscow, RUSSIA.

Cracking catalysts are exposed to extreme reactor hydrodynamic thermal conditions and heavy metal contamination on processing a heavy feed. Under these conditions the catalyst deactivates with subsequent reduction of the catalysts activity which might be partialy regain on regeneration.
Vanadium and Nickel oxides upon deposition on the catalysts surface and within the pores deactivates the catalyst irreversibly by poisoning the catalysts active centers thereby decreasing gasoline selectivity.
Catalysts treatment with a suitable metallorganic siloxane was found to reduce metal contamination, increase and retain the catalysts activity and enhance selectivity. Russian and foreign catalysts are treated with Alumino Phenylsiloxane at various concentration.

118.

FUTURE DIRECTIONS IN MODELLING HETEROGENEOUS CATALYSIS WITH DENSITY FUNCTIONAL THEORY AND OTHER TOOLS **Dennis R. Salahub**. Département de chimie, Université de Montréal, C.P. 6128, Succursale centre-ville, Montréal, Québec H3C 3J7, Canada

The conference organizers asked me to give a talk on "Future Directions in Computational Chemical Modeling to Heterogeneous Catalysis". This is, of course a "Mission: Impossible" kind of assignment. Nevertheless, I will gaze into my own biased crystal ball and try to extrapolate a few trends in what I would see as fruitful directions. The examples will all involve Density Functional Theory because we know the strengths, and the limitations, of the DFT techniques best. However, the future will definitely be cosmopolitan; combining DFT with other methods will represent a strong subtheme. The talk will focus on:

- The problem: structure , properties, dynamics and reactions of complex catalysis models
- Choice of model: clusters?, embedded clusters?, slabs?
- Choice of methodology, Kohn-Sham DFT
- Combining quantum chemistry with dynamics
- "Best bets" for the future.

119.

Computational Design of "Green" Process Technologies David A. Dixon, Jeffrey A. Nichols, Maciej Gutowski, Donald R. Jones, and John B. Nicholas, Environmental Molecular Sciences Laboratory, Pacific Northwest National Laboratory, Battelle Blvd, P.O. Box K1-83, Richland WA, 99352 and Matthew Neurock, Department of Chemical Engineering, University of Virginia, Charlottesville, VA, 22903

The challenges in the optimal design of industrial chemical processes arise from the need to minimize the environmental impact of the process and waste stream by optimizing the usage of raw materials and the utilization of energy.The development of new theoretical methods, new algorithms, and new software which incorporates these advances on new parallel computer architectures promises to lead to a fundamental new way to design chemical processes, including the design of catalysts. We will describe the development and use of the new computational chemistry software system, NWChem, for the prediction of the properties of catalysts of chemical systems of interest to the environmental restoration effort of the Department of Energy and to the chemical industry. Examples will include applications to VPO and to zeolites.

120. HETEROGENEOUS CATALYSIS: LOOKING FORWARD WITH MOLECULAR SIMULATION,
J. W. Andzelm, A. E. Alvarado-Swaisgood, F. U. Axe, M. W. Doyle, C. M. Freeman, A. M. Gorman, J. F. Harris, C. M. Kölmel, S. M. Levine, P. W. Saxe, M. A. van Daelen, E. Wimmer & J. M. Newsam, Molecular Simulations Inc., 9685 Scranton Road, San Diego, CA 92121 USA

Some of the areas in which we anticipate, over the next 5 years, notable advances in the application of molecular simulation to problems in heterogeneous catalysis are considered, in the context of recent progress to date. The areas specifically considered are • Expanding access to methods, • Quantitative Structure-Property Relationships, • Building structural models to focus or pre-screen experiments, • Confidence in predicting local and extended structurem, • Reaction mechanisms, barriers and kinetics and • Data for Process Models. In each of these areas, we indicate why we consider the topic significant, provide reference to topical work and suggest opportunities for future developments.

We thank the various member representatives of the MSI Catalysis Consortium for their ongoing input, ideas, guidance and advice.

121. **CONCEPT OF OPTIMAL AROMATICITY IN BASE OIL OXIDATIVE STABILITY REVISITED.**
Jinichi Igarashi, Toshio Yoshida and Harumichi Watanabe, Central Technical Research Laboratory, Nippon Oil Co., Ltd., Yokohama, Japan 231

Before the recent commercial emergence of "non-conventional" severely hydrocracked and wax isomerate lubricant base oils, it was suggested that it was inadvisable to use these over-refined base oils without "natural antioxidants" because of their tendency to "break down" very rapidly after all synthetic antioxidants are consumed. Thus it was previously pointed out that the best oxidative stability of base oils is obtained when aromatic and sulfur compounds are present in the oil in optimal concentrations in concert with Fuchs' concept of "optimal aromaticity". In this paper, we explain the kinetics and mechanism of the optimal aromaticity phenomenon and show that the optimal aromaticity still plays a very important role in the advanced lubricants from the non-conventional base oils with certain synthetic antioxidants.

122.
OXIDATION BEHAVIOUR OF BASE OILS AND THEIR CONSTITUTING HYDROCARBON TYPES, Himmat Singh & (Ms) Sarika Swaroop, Indian Institute of Petroleum, Dehradun 248 005, INDIA

Hydrocarbon types of a base oil matrix behave differently under oxidizing environment. This aspect has been investigated by separating a number of HVI base oils into saturates (I & II) and aromatics using adsorption chromatography. It is observed that saturates play a key role in lube oxidative degradation while aromatics have a moderating role. The propensity and magnitude of oxidation of different aromatic types with ring substitution have been discussed. It is shown that the rate of oxidation is affected by the presence of type and amount of aromatics in the overall hydrocarbon blends. Differences in the oxidation profiles of blended base oils in respect of their separated fractions and molecular parameters have been discussed.

123. THE INFLUENCE OF HINDERED PHENOLIC AND AROMATIC AMINE ANTIOXIDANTS ON THE STABILITY OF BASE OILS. C. A. Migdal, Uniroyal Chemical Company, Inc., Middlebury, CT 06749

Base oils are continuously being improved through processing and synthesis in order to meet new performance and regulatory requirements of lubricants. This study was conducted to measure the oxidative stability of base oils representative of the American Petroleum Institute's Group I-IV categories . A test method was developed using Pressure Differential Scanning Calorimetry (PDSC) to measure the oxidation induction times of the oils. The PDSC results are discussed and compared to oxidation induction times measured using the Rotary Bomb Oxidation Test (RBOT, ASTM D 2272). The base oils were tested with and without alkylated diphenylamine and/or hindered phenolic antioxidants present.

124. EFFECT OF COMPOSITION ON THE DEGRADATION BEHAVIOUR OF BASE OIL. A.Adhvaryu; D.C. Pandey; I.D. Singh, Indian Institute of Petroleum, Dehradun,248005, INDIA

Of late, FTNMR has developed as a highly specialised tool to study molecular composition and transformation in hydrocarbon systems. In many cases, today, this technique provides information on changes in the molecular level that reveal new vistas of molecular participation in product formation. The present talk covers a novel approach to study additive response in the light of molecular composition of the base fluid using NMR technique. Quantitative data involving 1H and 13C NMR have been used to study changes in carbon skeleton during oxidative degradation. DSC studies under oxidising environment is done on both neat and additive doped base oil. Order of reaction and several kinetic parameters were computed and variations explained in terms of composition of base oils. The data obtained on temperature and time of start of degradation were used to explain additive response.

125.

EFFECT OF ALCOHOL FUELS ON LUBRICANTS : OXIDATIVE AND LUBRICITY PROPERTIES. Rajesh J. Shah, Koehler Instrument Company, Bohemia, NY 11716., J. Larry Duda & Joseph Perez, Penn State University, University Park, PA 16802

Energy and environmental concerns have created a major thrust in the US towards alternative fuels for internal combustion engines. One of the important issues to be resolved for the more ubiquitous use of alcohol fuels is the effect of these fuels on the performance of the engine and its lubricating systems. The potential lubricating problems can be severe and quite diverse depending upon the specific characteristics of the alternative fuel chemistry, composition of the base lubricant and its additives.

In this work a bench test protocol has been developed to study the effects of various alcohol fuels on different lubricants. The test procedure consists of a series of oxidative and wear tests on lubricant samples extracted with alcohol fuels, followed by Atomic Absorption to get a better perspective of the distribution of additives as opposed to the base oils. The test procedure and the results obtained on two typical lubricants shall be discussed. This test procedure has been designed to act as a precursor evaluation technique at the lab scale for lubricants to be used with alternative fuels, before expensive engine tests are conducted.

126. EXAMINATION OF TERTIARY ALKYL PRIMARY AMINES AS LUBRICANT ADDITIVES. Rajiv M. Banavali and Shyam B. Karki, Rohm & Haas Company, Deer Park, TX 77536

Branched tertiary alkyl primary amines possess excellent oil solubility, thermal and oxidative stability, strong basicity, and fluidity over a wide temperature range. These properties make them well suited for enhancing the performance of base oils, particularly in dispersancy, corrosion inhibition, antioxidant, demulsibility, and EP/AW applications. The amines are also excellent solubilizers of other lubricant additives for base oils. The multifunctional aspects of these amines, as well as their ability to improve the function of other additives, can be exploited fully in a variety of formulations. We have designed studies to evaluate the use of these amines in lubricant formulations with respect to the aforementioned properties. We will present the results of our ongoing program.

127. **TURBINE AND HYDRAULIC FLUIDS: OPTIMIZING OXIDATION PERFORMANCE.** *Bernard C. Roell, Jr., Carlos L. Cerda de Groote and Margaret Lemmon, The Lubrizol Corporation, Wickliffe, Ohio 44092-2298.*

Oxidation characteristics of turbine and hydraulic formulations are dependent on base oil quality, package components and the interaction between the base oil and the components in a formulation. Oxidation performance is commonly measured by the Rotary Bomb Oxidation Test (ASTM D2272) and Turbine Oxidation Stability Test (ASTM D943).

These screen tests are often used in the lubricant industry to evaluate oxidation performance of a formulation. Two complementary studies were conducted. The objectives were to study chemical composition effects on oxidation performance and to find additive formulations which maximize oxidation life across three oils having different chemical composition. In the first study, only the level of oxidation inhibitor was varied. In the second study metal deactivator level, total oxidation inhibitor level and the ratio of two oxidation inhibitors were varied using a statistical design. The relative importance and direction of additive component effects on oxidation performance is different for various base oil types. Optimizing additive performance across a wide variety of base oils requires a delicate balance of these components.

128. Product Absorptivity On Tetrachloroaluminate Residuum Upgrading Catalysts -- Effect of Catalyst Type And Product Polarity. M.A. Plummer, Marathon Oil Company, Littleton, CO and S.M. Plummer, Cornell University, Ithaca, NY

Previous work investigated tetrachloroaluminate catalysts for converting petroleum and shale oil residuums into lower molecular weight aromatic products. Unfortunately, product recovery was limited by its absorptivity on the catalyst. This study quantifies catalyst parameters that affect the level of product absorptivity. The absorptivities of five hydrocarbons on three tetrachloroaluminate catalysts were measured. Increasing absorptivity levels were obtained with increasing hydrocarbon polarity and with decreasing size of the catalyst cation. Using molecular modeling, adsorptivity level was also shown to be a function of ion charge.

HEAVY OIL CONVERSION ASSISTED BY CARBON RELATED MATERIALS.
I. Mochida, K.Sakanishi, H. Hasuo, F. Hamdy, N.Yamashita, T. Nagamatsu,
Institute of Advanced Material Study, Kyushu University, Kasuga, Fukuoka 816, Japan

The catalytic activities of various carbon-supported binary metal sulfides were investigated in order to elucidate the roles of metals and carbon species in the hydrogenation(Hy) of 1-methyl-naphthalene(1-MN), hydrodesulfurization(HDS) of dibenzothiophenes(DBT) and gas oil, hydrotreatment of petroleum vacuum residue(VR), and coal liquefaction. NiMo sulfides supported on Ketjen Black(KB), Black Pearl 2000(BP2000), or DIAHOPE active carbon(D-AC) of large surface areas above 1000 m^2/g exhibited the higher activity for the hydrogenation of 1-MN than the same sulfides on active cokes of lower surface areas. The carbon-supported CoMo and NiMo catalysts showed the higher activities for the HDS and hydro-conversion of asphaltene fraction in VR than commercial alumina-supported CoMo and NiMo catalysts, respectively. The sulfides on carbon supports were analyzed to find a clue to their higher activity.

130. MILD UPGRADING OF MIDWAY SUNSET CRUDE OIL FROM SAN JOAQUIN VALLEY CALIFORNIA BY AQUEOUS PYROLYSIS -- CATALYSIS AND MODELLING. John G. Reynolds and Charles B. Thorsness, University of California, Lawrence Livermore National Laboratory, P. O. Box 808, L-369, Livermore CA 94551.

Midway Sunset crude oil and well-head oil were treated at elevated temperatures in a closed system with the presence of water. Mild to moderate upgrading, as measured by increase in API gravity, was observed at 400°C or above. Reduced pressure operation exhibited upgrading activity comparable to upgrading under normal aqueous pyrolysis conditions. The use of catalytic additives provided additional upgrading. Several different types of catalysts were examined, such as metal complexes, aerogels, and nitrides. The best of the minimum set of catalysts (Ni, Co, Cu, Mo, Zn, and Ca as carboxylate and/or inorganic salts) tested was Co(II) 2-ethylhexanoate. Fe(III) 2-ethylhexanoate also showed some reasonable activity under certain conditions. In addition, a process model to describe the behavior of the pressure and upgrading was also developed to account for the reduced pressure operation.

This work was performed under the auspices of the U. S. Department of Energy by the Lawrence Livermore National Laboratory under Contract No. W-7405-ENG-48

131. HYDROCRACKING STUDIES USING DISPERSED CATALYSTS PREPARED IN REVERSE MICELLES Kevin J. Smith, Apinya Duangchan and Andrew G. Hall, Department of Chemical Engineering, University of British Columbia, Vancouver, BC, Canada, V6T 1Z4.

A series of Fe, Co and Mo colloids, prepared in the reverse micelles of water-in-oil microemulsions, have been characterized and tested for the hydrocracking of diphenylmethane (DPM) and Cold Lake residue. Particle size of the metal colloids varied in the range 3-16 nm, depending on the microemulsion composition and the metal. Sulphided metals, also prepared in the reverse micelles using hydrogen sulphide, had larger particle size than the reduced metals. Batch hydrocracking of DPM showed that the Co dispersed in the microemulsion was more active than the Fe. The catalysts were also effective for reducing coke yield during the batch hydrocracking of Cold Lake residue at 430°C for 1 hour, with coke yield increasing in the order Mo (5.4 wt% of feed) < Co (6.0%) < Fe (7%) whereas the coke yield from thermal cracking was 7.4%. For the Mo catalysts prepared in different water-in-oil microemulsions, coke yield decreased with the type of oil, decalin having a lower coke yield (5.8%) than toluene (12.2%) or hexane (19.6%).

132. UPGRADING ACTIVITY OF RECYCLED Mo-BASED DISPERSED CATALYSTS.
A. Del Bianco*, N. Panariti*, M. Marchionna[#], * ENIRICERCHE S.p.A
[#] SNAMPROGETTI S.p.A, Via Maritano 26, 20097 San Donato M.se, ITALY

Among the various options available for upgrading bottoms materials, slurry processes have excellent potential, particularly on high-metals feedstocks. Nevertheless, main limitations to their commercialization refer the high investment and operating costs due to the severe operating conditions and/or the catalyst make-up.
The use of metals with higher catalytic activity, such as molybdenum, may favor the process of hydrogen uptake thus promoting the upgrading reactions. However, when considering the high price of this metal, a *once-through* scheme can be considered only for processes employing very low molybdenum concentrations. For higher concentrations, the recycle of the catalyst is mandatory and this step requires a proper separation unit and a complex/expensive regeneration unit. In order to overcome this problem, a process solution based on the recycle (without regeneration) of substantial amounts of molybdenum-based dispersed catalysts has been successfully developed.

133. PILLARED RECTORITE MICROSPHERIC CATALYST for CRACKING HEAVY OIL.
J.guan Z.Yu Z.Chen Q. Liu and X.Wang. Research Institute of Petroleum Processing, China Petro-Chemical Corporation. Beijing 100083. P.R. China

To convert more heavy oil into gasoline and light cycle oil in FCC process, a class of novel pillared rectorite (PIR) microspheric catalyst was provided. The catalyst was prepared according to procedures of mixing, spray drying to form microspheric shape, pillaring reaction and calcination. It has two dimensional structures with basal spasing of 27-32Å attrition index of 2.2wt% and apparent bulk density of 0.76. Under same evaluation condisions its activity for cracking heavy oil is 3.6 hihger and total yields of gasoline and light cycle oil are 11m% higher than that of commercial cracking catalysts containing REHY after treatment at 800°C for 4hrs with 100% steam. The novel PIR catalyst with good attrition resistant index has high activity hydrothermal stability and good selectivity for cracking heavy oil that is able to meet requirements of commercial FCC process.

134. EVALUATION OF CATALYSTS BASED ON HETERONUCLEAR HYDROXIDES FOR HEAVY OIL CONVERSION. A.N. Kopytovich, A.A. Chernik, Department of Chemistry and Technology of Electrochemical Manufactures, Byelorussian State Technological University, Sverdlov Str. 13a, Minsk, 220030, Republic of Belarus

At present the intensification of heavy oil conversion is connected with using of proton-effective catalysts. It was found out that hydroxide and oxide catalysts, received on basis of jointly precipitate metal ions in a number of cases display very high catalytic activity. Thus the activity of such catalysts is not equal for total activity of mixed hydroxides or oxides. Assumption that heteronuclear hydroxides will be formed at joint precipitation of metal hydroxides in a number of cases was stated. These hydroxides are new compounds and can show high catalytic activity owing to features of their structure. Highly effective oxide catalysts can be received by thermal decomposition on the basis of these heteronuclear hydroxides.

135. DEVELOPING STRATEGIC HEAVY OIL REFINING TECHNOLOGIES. Albertino Machado de Carvalho, PETROBRAS

The high competitiveness observed nowadays in the oil industry and the growing of market globalization and desregulation are making technology to be the main business leverage, encancing the competitive differential between companies.
For the next years PETROBRAS will need to match the increase in the offer of Brazilian heavy oil, coming from deep water oil feeds with the needs of society, that demands light products with more rigid specifications concerning quality and environmental aspects.
In this scenario PETROBRAS has decided to make investments in the main segments of its business. The Program for the Development of Strategical Refining Technologies (PROTER) was conceived to the refining area. Starting from the increasing participation of Brazilian Crudes, its conciliates supply and demand by developing new processes and also increasing the use of our technologies for improved processing of heavy residues.

136. POLYOLESTER LUBRICANTS FOR USE IN ENVIRONMENTALLY FRIENDLY REFRIGERATION APPLICATIONS. S. T. Jolley, The Lubrizol Corp., Wickliffe, Ohio 44092

The discovery that fully halogenated, chlorine containing refrigerants (CFC's) are a major contributor to the depletion of the earth's ozone layer resulted in their replacement by hydrofluorocarbon (HFC) refrigerants. One unexpected side effect to the use of HFC refrigerants turned out to be the need for a new lubricant for use in the compressors used to operate refrigeration systems. Mineral oil lubricants, which worked well with CFC systems because of their mutual miscibility, proved less effective in HFC containing systems where mineral oils are immiscible with this refrigerant type. Although several types of polar, synthetic lubricant have been identified which work in these systems, polyolesters have gained the widest acceptance by this industry as useful lubricants to replace mineral oils. This paper will detail the development of this new use for polyolester lubricants, the physical characteristics necessary for their use and some of the challenges that are yet to be overcome for their long-term successful implementation.

137. BIODEGRADABLE GEAR OILS. D. E. Weller, J. M. Perez, R. E. Keay, Department of Chemical Engineering, Pennsylvania State University, University Park, Pennsylvania 16802

With many countries increasing governmental restrictions on lubricants, the need for biodegradable lubricants is increasing. This study was performed to develop a more biodegradable lubricant for marine main reduction gears applications. These biodegradable lubricants will only gain acceptance if they can either perform as well or better than the currently used mineral oils, or governmental restrictions totally require their use. Some possible replacements for currently used mineral oils discussed will be polyalphaolefins (PAO), vegetable oils and PAO/vegetable oil blends. In this study, a comparison of some currently important requirements of gear oils will be made between mineral oils and more biodegradable lubricants. Some important lubricant characteristics discussed include viscosity/temperature behavior, wear and scuffing performance, corrosion prevention, low temperature performance, and seal compatibility.

138.
Suitability of Basestocks for Biodegradable Lubricants.
S.J. Asadauskas, J.M. Perez. and J.L. Duda, 133 Fenske Laboratory, Chemical Engineering Department, The Pennsylvania State University, University Park, PA 16802

ABSTRACT. Some of the advantages and disadvantages of basestocks for biodegradable lubricants are discussed. The focus is on the selection of base fluids as to their suitability as biodegradable lubricants. When considering a wide variety of base fluids, properties such as viscosity, hydrolytic stability, seal compatibility or heat conductivity may be as important as volatility, low temperature properties and solvency. A knowledge of the base stock properties prior to establishing thermal oxidative stability, additive requirements and additive susceptibility is required. The priority as to which properties are most significant depend on the application and required performance level. These factors limit chances of developing a universal basestock to satisfy biodegradability requirements and to replace mineral oils in lubricants. In each case the most appropriate base fluid should be carefully selected. Properties of five of the major biodegradable basestocks are discussed.

139.
CHARACTERIZATION OF SYNTHETIC LUBRICANTS IN BASE OILS AND FINISHED MOTOR OILS BY FAST SUPERCRITICAL FLUID CHROMATOGRAPHY
Evan N. Chen, Jr., Texaco, Inc., P.O. Box 509, Beacon, NY 12508

In the petroleum industry, synthetic lubricants, such as poly-alpha-olefins (PAO's), are widely used for crankcase engine, gear, and automotive lubrication. Although PAO's are available in viscosity grades from 2 cST to 100 cST (at 100 °C), the 4 and 6 cST are generally used for automotive-type applications due to their low manufacturing costs compared to the higher viscosity grades. Characterization of the amount and type of PAO in a motor oil has been performed by high temperature gas chromatography (HTGC). Unfortunately, HTGC methods often use oven temperatures in excess of 400°C thus requiring specialty glass or aluminum-clad columns for the analysis. In addition, the long 80 minute analysis time decreases sample throughput and productivity. For this reason, a fast supercritical fluid chromatography (SFC) method has been developed to quantitatively analyze 2, 4, and 6 cST PAO's in mineral oils and motor oils utilizing low temperature conditions (125°C) and fast analysis times (<12 min.).

140. CARCINOGENIC ASSESSMENT OF LUBRICANT BASE OILS BY PROTON NMR. C. S. Nessel[1], D. T. Coker[2], A. G. King[2], and D. L. Mumford[3]. [1]Exxon Biomedical Sciences, Inc., East Millstone, NJ 08875, [2]Exxon Research Centre, Abingdon, OX13 6AE,UK, [3]Exxon Company, International, Florham Park, NJ 07932.

The need for rapid, scientifically-sound tests for the carcinogenic evaluation of lubricant base oils is increasing worldwide. A research program was initiated to evaluate the use of Bay Region Hydrogen (BRH) measurement by Proton NMR for carcinogenic assessment of base oils. Proton NMR can identify and quantify protons in locations on polynuclear aromatic hydrocarbon molecules which are associated with carcinogenic activity ("bay regions"), and is potentially applicable to a broad range of petroleum products. Approximately 60 base oils and related materials were tested. All known and suspected noncarcinogenic oils had BRH measurements of less than 600 ppm. Carcinogenic base oils contained greater than 1500 ppm. BRH measurements on straight-run and cracked gas oils supported these results. These data suggest that BRH measurement by Proton NMR may be able to discriminate carcinogenic and noncarcinogenic base oils with a threshold for carcinogenicity of approximately 1000 ± 400.

141.

NMR BASED CHARACTERIZATION OF LUBE BASE OILS, Himmat Singh, A. Adhvaryu, I.D. Singh and G.S. Chaudhary, Indian Institute of Petroleum, Dehradun 248 005, INDIA

Nuclear Magnetic Resonance Spectroscopy, of late, has emerged as a highly potential technique to study the molecular composition of a base oil matrix. The usefulness of the technique lies in its capability to explain the chemistry involving molecular changes in various extraction and refining stages. This paper deals with the application of this technique to understand the variation of gross properties of the base matrix. The variation in Wt. % S, VI, CCR, KUOP etc. is explained at the molecular level using high resolution ^{1}H and ^{13}C NMR method. A single spectra can be utilised to generate information on different properties of base oils used for their specification.

142.

INFLUENCE OF REFINING ON BASE OIL COMPOSITION, Himmat Singh, A. Adhvaryu, I.D. Singh and G.S. Chaudhary, Indian Institute of Petroleum, Dehradun 248 005, INDIA

Today with the rapid advancement of technology in the lube refining sector, crude type and source, have ceased to be the deciding factor in governing the product quality. Certain alterations/additions in the refining sequence of the distillate improves yield and essential properties of finished base oils. In the present paper, the effect of hydrotreatment of lube distillates prior to solvent extraction on base oil quality has been studied. NMR and FTIR techniques have been used to identify the structural changes in the base oils obtained with and without hydrotreating step. It is observed that hydrotreating has modified the functional groups in the hydroxyl region while in carbonyl region the type of >C=O functionalities have increased. The NMR data on base oils support the reduction in condensed aromatic carbon and increase in naphthenic carbons. The changes observed in base oil properties are explained in light of structural/molecular parameters.

143.

DETERMINATION OF PHYSICO CHEMICAL PROPERTIES AND CARBON TYPE ANALYSIS OF BASE OILS USING MID-IR SPECTROSCOPY AND PARTIAL LEAST SQUARES REGRESSION ANALYSIS. M.I.S.Sastry, Anju Chopra, A.S.Sarpal, S.K.Jain, S.P.Srivastava and A.K.Bhatnagar. Indian Oil Corporation Limited, Research & Development Centre, Sector 13, Faridabad 121 007, Haryana, INDIA.

The paper highlights an experimental and computational protocol for the simultaneous determination of physico-chemical properties (Viscosity index and pour point) and carbon type analysis (C_a; C_p; C_n; and Iso paraffin carbon content, I_p) of base oils from their IR spectral features. IR spectra of 50 base oils of different origin and processing schemes, are recorded using HATR accessory having ZnSe windows. Partial least squares (PLS) technique has been applied to construct the mathematical models which correlate the IR spectral features with the experimentally determined values (using NMR and standard ASTM tests). The regression coefficients are found to be in the range 0.85-0.95.

144. BIAS IN THE IP 346 METHOD FOR POLYCYCLIC AROMATICS IN BASE OILS AND IN ASTM D2007 METHOD FOR HYDROCARBON TYPE DETERMINATION
Bhajendra N. Barman, Texaco Inc., Fuels and Lubricants Technology Department
P.O. Box 1608, Port Arthur, TX 77641

Thin-layer chromatography with flame-ionization detection (TLC-FID) was used to examine the composition of polycyclic aromatic (PCA) and non-PCA fractions from lubricant base stocks obtained by the IP 346 method. TLC-FID chromatograms of the original base oil, and PCA and non-PCA fractions indicate that liquid-liquid extraction procedures do not provide pure fractions. Aromatics with multiple rings and polar materials were found to be present in the non-PCA fractions while PCA fractions were contaminated with saturates and both one- and two-ring aromatics. Such cross-contamination can result in overestimation of PCAs at low concentrations, or their underestimation at high concentrations.

Using aromatics, saturates and polar fractions obtained by ASTM D2007 open-column clay-gel chromatography and extracts from adsorbents, cross-contamination of hydrocarbon types and incomplete recovery of samples were also ascertained by TLC-FID. Inaccuracies in ASTM D2007 results were found to be worse for very high boiling distillates, heavy alkylbenzenes and highly paraffinic base stocks.

145. SOLVENT EXTRACTION OF HEAVY COKER GAS OIL. Y.Wang, Research Institute of Petroleum Processing, Beijing 100083, P.R.China

Heavy coker gas oil (HCGO) is usually used as a blended component of FCC feedstock. However, the blended ratio is limited due to its high contents of basic nitrogen compounds (BNC) and resin, which deactivate the activity of cracking catalyst and result in low conversion and high coke deposition. Pilot plant results showed that solvent extraction could solve the problem, which had two advantages, (1) the process could reduce the BNC and resin contents of HCGO, (2) the extract of HCGO could be used as a feedstock to produce needle coke. More preferably, a lubricant extract without solvent recovery was used as a solvent instead of the fresh solvent to extract the HCGO for reducing its BNC and resin contents.

146. STABILIZATION OF FUELS PRODUCED BY HIGH TEMPERATURE CRACKING. Matthew E. Gande, Ciba Additives, 540 White Plains Rd., Tarrytown, NY 10951

When naphtha, gas oil, or ethane/propane are cracked to produce ethylene and propylene, a higher molecular weight component is also produced. While this material has some properties that allow it to be used as gasoline, it is extremely unstable, and is prone to forming gums during processing and storage. In this paper we discuss the use of amine derivatives to impart stability to the pyrolysis gasoline thereby increasing the ease of handling, storage, and use of the fuel. While the amine derivatives investigated in this study were effective when used alone, a synergistic effect was observed when used in conjunction with co-stabilizers.

147. EFFECT OF HEAVY ASPHALTENE ON STABILITY OF RESIDUAL OIL. M.Tojima, S.Suhara, M.Imamura, A.Furuta, JGC Corporation, Kinu-ura Research Center, 2-110 Sunosaki-cho Handa Aichi Pref. 475, Japan

An asphaltene fractionation method was developed in order to investigate the effect of asphaltene on residual oil stability. Using the binary solvent system, asphaltenes can be separated into fractions according to their solubility. The fractionation method and the Heithaus stability evaluation method were applied to hydroprocessed residue. The fraction with the lowest solubility in asphaltenes was found to consist of the most highly condensed polyaromatics (heavy asphaltenes). Further, it was observed that these heavy asphaltenes were condensed by hydroprocessing at high temperatures above 400°C. Then the peptizability of asphaltene defined by the Heithaus method decreased with the change in heavy asphaltene structure. It was concluded that the condensation of heavy asphaltenes would be a cause of residue destabilization.

Thus, the application of this asphaltene fractionation method and Heithaus method is effective in understanding the cause of destabilization of residual oil during hydroprocessing.

148. A NEW HYDRODEALKYLATION PROCESS WITH FLUID-BEDS TO PRODUCE HIGH PURITY NAPHTHALENE AND METHYLNAPHTHALENES FROM HEAVY AROMATIC OILS. T. TSUTSUI, O. Kubota, S. Okada, K. Sato and T. Takeuchi, Fuji Oil Co., Tokyo, Japan

Recently, the significance of chemical grade naphthalene and β-methylnaphthalene as base materials for synthesis has been increasing, and it seems important to establish a rational process to dealkylate heavy aromatic oils. In this presentation, a new hydrodealkylation process with a fluid-bed reactor and its reaction performance in dealkylation of LCO are described. With this process, even a heavy aromatic oil containing substantial amount of poly-cyclic aromatics, polyalkylated aromatics and impurities such as sulfur and nitrogen compounds can be directly processed, and dealkylation with sufficient desulfurization and denitrogenation proceeds stably by a selected catalyst in the presence of coke deposited on it. This process and concept are also applicable to dealkylation of heavy polyalkylated-benzenes, coal tar or other heavy aromatic oils. The fluid-bed reaction analysis for a consecutive dealkylation reaction and the catalyst reactivity under coke deposition are described and discussed.

149. EFFECT OF Si/Al RATIO IN ZSM-5 ZEOLITE ON CATALYST DEACTIVATION IN REACTION OF CHLOROTOLUENE. T.TAKAHASHI and T.KAI, Department of Applied Chemistry and Chemical Engineering, Kagoshima University, Kagoshima 890, JAPAN

When the reaction of chlorotoluenes was carried out over a zeolite, isomerization and demethylation simultaneously proceeded. The selectivity and catalyst deactivation rate were strongly dependent upon the concentration of strong acid and the size of micropores in zeolites, that is, the demethylation mainly proceeded with rapid deactivation over HY and H-mordenite which had high concentration of strong acid sites and large micropore size. On the other hand, the isomerization mainly proceeded over a HZSM-5. These results suggest that hydrogen evolved with coke formation was played an important role for the selectivity. In the present study, the reaction of chlorotoluenes was carried out over HZSM-5 zeolites with different Si/Al ratio to elucidate the effect of strong acid site concentration on the selectivity and catalyst deactivation rate. The concentration of strong acid sites decreased with the Si/Al ratio and the deactivation rate also decreased with the ratio. Whereas, the selectivity to the isomerization increased with the ratio. The coke contents deposited on the zeolite surface changed from 8 wt%(Si/Al=50) to 0.3 wt%(The ratio=300). These results indicate that the concentration of strong acid sites is the most important factor to decide the selectivity and catalyst deactivation rate.

150. LIGAND EXCHANGE IN ENGINE OIL ADDITIVE AND BASE STOCK SYSTEMS CONTAINING MOLYBDENUM DIALKYLDITHIOCARBAMATE. R. K. Jensen, M. D. Johnson, S. Korcek and M. J. Rokosz. Ford Motor Company, P. O. Box 2053, MD 2629/SRL, Dearborn, Michigan 48121.

Molybdenum dialkyldithiocarbamate, Mo(dtc)$_2$, is used in advanced fuel efficient engine oils as a friction reducing additive. Investigation of the frictional properties of oils containing Mo(dtc)$_2$ during aging in engines showed that the friction reducing activity of these oils is gradually depleted with mileage accumulation and is dependent on the presence of other additives such as zinc dialkyldithiophosphate, Zn(dtp)$_2$, hindered phenols and sulfur compounds. From these studies, it is clear that interactions of antifriction, antiwear and antioxidant additives and their reactions with free radicals and other oxidation products during aging play an important role in optimizing performance of engine oils. This work describes results of our investigations of interactions of Mo(dtc)$_2$ with other additives in additive systems containing Mo(dtc)$_2$, Zn(dtp)$_2$, copper carboxylate and hindered phenol in a model hydrocarbon lubricant and various base oils at 160°C in the presence of simulated engine blowby gas. Our results indicate that Mo(dtc)$_2$ undergoes ligand exchange reactions with other metal containing additives and that the extent of these exchange reactions are significantly influenced by oxidation reactions at elevated temperatures as well as by composition of the medium in which the exchange takes place. This has important implications for formulating fuel efficient engine oils as well as for development of guidelines for base oil interchangeability.

151. **TEOST Study of Base Oil Blends.** T.W. Selby and J. J. Richardson, Savant, Inc., 4800 James Savage Road, Midland, MI 48642.

The Thermo-Oxidation Engine Oil Simulation Test (TEOST) device has been developed to study the deposit characteristics of engine oils in the turbocharger using a special Protocol 33C involving cyclic exposure of the oil to high temperature from 200° to 480°C. This same approach has been found interesting in thermal conductivity studies. This paper presents results using the TEOST Protocol 33C on a variety of base oils blended with a chosen additive system.

152. **Film Formation Behavior of Lubricant Base Stocks in Concentrated Contacts.** S. Gunsel and H.A. Spikes, Pennzoil Products Company, P.O. Box 7569, The Woodlands, TX 77387 and Imperial College, London, England.

The film formation behavior of lubricant base stocks was studied in rolling, concentrated contacts using ultra-thin film interferometry. By using this technique, it was possible to measure thin films formed by low viscosity (4 cst @ 100°C) base stocks in mixed elastohydrodynamic and boundary regimes. Film thickness measurements were conducted on a series of lubricant base stocks processsed via different refining technologies. These included severely hydrotreated, wax isomerized, solvent extracted / hydrotreated and conventionally solvent extracted base stocks. The effects of base stock composition and processing technology on film-forming properties were investigated over a temperature range of 25- 170°C. Effective pressure / viscosity coefficients were calculated for each base stock. It was found that highly refined base stocks (i.e. hydrotreated and wax isomerized) behaved as expected from the classical EHD theory while conventionally solvent extracted base stocks produced thicker films in the contact than predicted from the EHD theory under low speeds and thin film conditions. The formation of these residual films in the contact was attributed to polar compounds in the solvent extracted base stocks. The residual films were approximately 10-15 nm thick.

153. INVESTIGATION OF LOW TEMPERATURE CRYSTALLIZATION PROCESSES IN BASE OILS UTILIZING A PERDEUTERATED PROBE MOLECULE AND ^{2}H NMR. John C. Edwards, Arthur J. Stipanovic*, Paul J. Giammatteo, Michael P. Smith Texaco Fuels and Lubricants Technology Department, P.O. Box 509, Beacon, NY 12508

The low-temperature performance characteristics of a motor oil is highly dependent on the extent, and type, of micro-crystallite formation in the paraffin components of the product. We have utilized solid-state ^{2}H NMR to observe the formation of crystallites in various base oils by observing the behavior of 1 wt% of a perdeuterated probe molecule blended into a number of different classes of base oil. Different cooling cycles were utilized, one involving a fairly rapid cycle (+20 to -35° C in 15 minutes) while the other mimicked the cooling cycle used in an ASTM mini-rotary viscometer (MRV) base oil performance test (+80 to -35°C in 51 hours). The effects of these different cooling cycles on the amount, and type (whether orthorhombic or hexagonal crystal-packing), of crystalline domain will be presented, along with the solid/liquid, and mobility profiles that can be obtained from the data for the different base oils. Finally, the effect of addition of pour-point depressant additives will be discussed.

154. DEVELOPMENT OF A MICROREACTOR BENCH TEST FOR LUBRICANT EVALUATION, P. A. Gabilondo, J. M. Perez, W. A. Lloyd, Department of Chemical Engineering, The Pennsylvania State University, State College, Pennsylvania 16803.

Deposit formation and viscosity increase are the major causes of lubricant failure in engine tests, such as the IIIE sequence test. Oxidation in the piston ring/cylinder area contributes significantly to lubricant degradation. The Penn State Micro Oxidation Test (PSMO) can be used to evaluate automobile lubricants by a thin film oxidation mechanism. This bench test, coupled with Gel Permeation Chromatography (GPC), determines the amount of polymerization oxidation products, deposit forming tendencies and volatility of the lubricants evaluated. A modified version of the PSMO, the Portable Penn State Micro Oxidation test (PPSMO), is under development to decrease costs and increase the availability of this test. The PPSMO successfully evaluates the less volatile vegetable oils for their resistance to oxidative degradation. Further studies are needed to confirm the capabilities of this test in evaluating oxidative stabilities of the more volatile basestocks such as hydroisomerized waxes.

155. Investigation of Wear Properties of Several Commercial Greases
Ryan B. Adams and Joseph M. Perez, 133 Fenske Laboratory, Chemical Engineering Department, Pennsylvania State University, University Park, PA 16802

Abstract

Several commercial greases were evaluated for their wear protection at moderate and extreme pressure conditions. Candidate greases contained both synthetic and mineral oil base fluids. The motivation behind the study was to increase the lifetime and load carrying capacity of certain gears. The application puts the grease under severe wear conditions for brief periods of time. The time between these cycles varies and the grease is replaced by a pump when required. The current grease has a clay thickener with a polyalphaolefin (PAO) base fluid. It does not contain any extreme pressure (EP) additives which is the main reason for the high wear currently being observed. All the replacement candidates contain EP additives.

156. Having Fun With Base Oils: Predicting Properties Using Neural Networks. A.J. Stipanovic, G.P. Firmstone and M.P. Smith, Texaco Fuels and Lubricants Technology Department, P.O. Box 509, Beacon, New York 12508-0509, U.S.A.

Lubricant base oils are composed of a complex mixture of aromatic, paraffinic and naphthenic molecules, along with sulfur and nitrogen containing compounds, that also range in molecular size and other characteristics. As a result, it is often difficult to develop direct correlations between the physical and chemical characteristics of base oil and its compositional constituents. Although linear regression analysis and more sophisticated partial least squares (PLS) techniques have been quite useful in this regard to predict engine test results and oxidation test data,, Neural Network (NN) technology offers exceptional promise in helping to relate base oil composition directly to performance in many automotive and industrial oil applications, especially for properties that exhibit a non-linear dependence on composition or other variables. In this study, as an example, the ability to predict viscosity index from base oil composition and viscosity was compared using NNs and more traditional statistical methods and better predictability was obtained with the former method.

American Chemical Society
DIVISION OF PHYSICAL CHEMISTRY

213th ACS National Meeting

San Francisco, CA
April 13-17, 1997   E.B. Stechel, Program Chairperson

SUNDAY MORNING

- **Frontiers in Electronic Structure Theory**
  Jan Almlof and his Science
  R.J. Bartlett, M. Head-Gordon, Presiding   Papers 1-5

- **Chemistry of Single Molecules**
  High-Resolution Spectroscopy of Single Molecules in Solids
  W.E. Moerner, Presiding   Papers 6-11

- **Kinetics of Growth on Surfaces**
  R.S. Williams, Presiding   Papers 12-16

- **Symposium Honoring the Memory of B.E. Bent**
  G.W. Flynn, Presiding   Papers 17-21

SUNDAY AFTERNOON

- **Frontiers in Electronic Structure Theory**
  Linear Scaling Alorithms and Massively Parallel Computing
  M.Head-Gordon, Presiding   Papers 22-26

- **Chemistry of Single Molecules**
  Single-Molecule Microscopy and Spectroscopy
  X.S. Xie, Presiding   Papers 27-31

- **Kinetics of Growth on Surfaces**
  R.S. Williams, Presiding   Papers 32-36

- **Symposium Honoring the Memory of B.E. Bent**
  B.E. Koel, Presiding   Papers 37-42

SUNDAY EVENING

- **Posters**
  W.E. Moerner, X.S. Xie, Presiding   Papers 43-167

MONDAY MORNING

- **Frontiers in Electronic Structure Theory**
  Applications Frontiers
  M.S. Gordon, Presiding — Papers 168-173

- **Chemistry of Single Molecules**
  Single Molecule Detection in Analytical and Electro-Chemistry
  U.P. Wild, Presiding — Papers 174-179

- **Kinetics of Growth on Surfaces**
  R.S. Williams, Presiding — Papers 180-186

- **Debye Award Symposium: Dynamics of Complex Systems**
  G.W. Flynn, Presiding — Papers 187-190

MONDAY AFTERNOON

- **Frontiers in Electronic Structure Theory**
  Applications Frontiers
  A. Wilson, Presiding — Papers 191-195

- **Chemistry of Single Molecules**
  Probing Single Biomolecules I
  P.S. Weiss, Presiding — Papers 196-199

- **Kinetics of Growth on Surfaces**
  R.S. Williams, Presiding — Papers 200-205

- **Debye Award Symposium: Dynamics of Complex Systems**
  P.F. Barbara, Presiding — Papers 206-210

- **Nanostructured Materials: Clusters, Composites & Thin Films**
  M. Moskovits, Presiding — Papers 211-215

MONDAY EVENING

- **Sci-Mix**
  E.B. Stechel, Presiding — Papers
  70,110,124,141,157,161,162,165,166,167,
  216,267,316,326,331,338,347,351,363,386,
  396,398-400,404,405,408,409,457

TUESDAY MORNING

- **Frontiers in Electronic Structure Theory**
  Advances in Electron Correlation
  S. Langhoff, Presiding — Papers 217-222

Chemistry of Single Molecules
Probing Single Biomolecules II
W.P. Ambrose, Presiding   Papers 223-228

- **Orientation and Alignment in Chemical Processes**
Stereochemistry at Interfaces
R.E. Miller, Presiding   Papers 229-235

- **Nanostructured Materials: Clusters, Composites & Thin Films**
J. Jellinek, Presiding   Papers 236-242

TUESDAY AFTERNOON

- **Awards Symposium**
E.B. Stechel, Presiding   Papers 243-247

WEDNESDAY MORNING

- **Frontiers in Electronic Structure Theory**
Density Functional Theory
M. Levy, Presiding   Papers 248-253

- **New Concepts in Surface Chemistry: Diffusive Motion of Atoms and Molecules on Surfaces**
H. Jonsson, Presiding   Papers 254-256

- **Orientation and Alignment in Chemical Processes**
Small Molecules and Clusters
P.M. Felker, Presiding   Papers 257-263

- **Nanostructured Materials: Clusters, Composites & Thin Films**
S.E. Webber, Presiding   Papers 264-270

- **Chemistry of Single Molecules**
Probing Single Molecules with STM/AFM
J.K. Trautman, Presiding   Papers 271-275

WEDNESDAY AFTERNOON

- **Frontiers in Electronic Structure Theory**
Ab-Initio Dynamics
W.H. Miller, Presiding   Papers 276-281

- **New Concepts in Surface Chemistry: Diffusive Motion of Atoms and Molecules on Surfaces**
H. Jonsson, Presiding   Papers 282-284

- **Orientation and Alignment in Chemical Processes**
  Chemical Control and Alignment
  A.M. Wodtke, Presiding   Papers 285-289

- **Nanostructured Materials: Clusters, Composites & Thin Films**
  U. Kreibig, Presiding   Papers 290-295

WEDNESDAY EVENING

- **Posters**
  E.B. Stechel, Presiding   Papers 296-438

THURSDAY MORNING

- **Frontiers in Electronic Structure Theory**
  Excited States
  S.P. Karna, Presiding   Papers 439-443

- **New Concepts in Surface Chemistry: Diffusive Motion of Atoms and Molecules on Surfaces**
  H. Jonsson, Presiding   Papers 444-446

- **Orientation and Alignment in Chemical Processes**
  Stereochemical Reactions
  D.J. Nesbitt, Presiding   Papers 447-453

- **Nanostructured Materials: Clusters, Composites & Thin Films**
  K.L. Rowlen, Presiding   Papers 454-460

THURSDAY AFTERNOON

- **Frontiers in Electronic Structure Theory**
  Properties
  H. Partridge, Presiding   Papers 461-465

- **New Concepts in Surface Chemistry: Diffusive Motion of Atoms and Molecules on Surfaces**
  H. Jonsson, Presiding   Papers 466-471

- **Orientation and Alignment in Chemical Processes**
  Orientation and Alignment in Photochemistry
  M. Shapiro, Presiding   Papers 472-476

- **Nanostructured Materials: Clusters, Composites & Thin Films**
  V.M. Shalaev, Presiding   Papers 477-482

DIVISION OF PHYSICAL CHEMISTRY

001. JAN ALMLOF AND HIS SCIENCE, Peter R. Taylor, San Diego Supercomputer Center, and Department of Chemistry and Biochemistry, University of California, P.O. Box 85609, San Diego, California 92186-9784

Jan Almlof explored many areas of quantum chemistry during his career, but his primary goals were always to obtain the right answer for the right reason, and to find ways of extending quantum-chemical methodology so that questions could be answered in this way for larger and larger molecules. We shall review a number of his contributions, both methodological and applications, in this retrospective.

002
THE ACCURATE TREATMENT OF CORRELATION EFFECTS IN COUPLED-CLUSTER THEORY. T. Helgaker and W. Klopper, Department of Chemistry, University of Oslo, N-0315 Oslo, Norway H. Koch and Poul Joergensen, Department of Chemistry, Aarhus University, DK-8000 Aarhus C, Denmark J. Noga, Institute of Inorganic Chemistry, Slovak Academy of Sciences, SK-84236 Bratislava, Slovakia

This talk discusses the convergence of the coupled-cluster electronic energy and molecular properties with respect to the size of the atomic-orbital basis set and the correlation treatment. High accuracy in the calculated properties puts severe demands on the basis set and methods that address the basis-set problem are discussed, in particular the use of integral-direct methods and multiple basis sets.

003.

THE AB INITIO COMPUTATION OF FERROCENE. Hans P. Lüthi, Swiss Center for Scientific Computing, ETH Zürich, CH-8092 Zürich, Switzerland

Ferrocene is a molecule that has challenged experimental and theoretical chemistry in many ways. We will review the inspiring history of its *ab initio* computation, an area of research where Jan Almlöf was a major contributor. Some of the important lessons learned, along with a quantitative analysis of how accurate a property such as the metal–ligand binding energy of a transition metal compound can be evaluated using today's most advanced *ab initio* methods, will be presented.

004. APPLYING LINEAR SCALING ELECTRONIC STRUCTURE METHODS TO LARGE MOLECULES. Gustavo E. Scuseria, Chemistry Department, Rice Quantum Institute, and Center for Nanoscale Science and Technology, 6100 Main Street, MS 60, Rice University Houston, Texas 77005-1892

This presentation will address our recent implementation of fast, linear scaling methods (tight-binding, semiempirical, density functional, and Hartree-Fock) and its application to very large molecular systems. Our recent implementation of the Gaussian very Fast Multipole Method (GvFMM) together with very fast numerical quadratures that achieve linear scaling has made it possible to carry out DFT geometry optimizations in very large molecules. In particular, we will discuss results obtained on graphene-sheets containing hundreds of atoms using Hartree-Fock and DFT methods. A conjugate gradient density matrix search (CGDMS) technique that avoids the diagonalization step has been implemented for AM1, tight-binding (TB), and both Hartree-Fock and DFT methods. Applications with AM1 to biological molecules containing thousands of atoms will be discussed. Using tight-binding, we have fully optimized the geometry of carbon clusters containing up to 8,640 atoms, and compared their geometrical shape with that observed in TEM images.

005.

DENSITY BASED METHODS IN ELECTRONIC STRUCTURE THEORY: FROM LINEAR FOCK BUILDS TO ENERGY MINIMIZATION. **C.A. White** and M. Head-Gordon, Department of Chemistry, University California at Berkeley, Berkeley, CA 94720

The treatment of large molecules using ab initio techniques is limited by two bottlenecks. The first concerns the long range nature of the Coulomb interaction between electrons. The second concerns the manipulation (diagonalization, multiplication and storage) of large matrices. By formulating theories based upon the molecular density matrix rather than atomic or molecular orbitals, it is possible to ameliorate the effects of these bottlenecks. Within this talk, I will present methods which address both the Coulomb and diagonalization bottlenecks. The Coulomb portion will focus upon recent advances concerning the speed and efficacy of our Continuous Fast Multipole Method (CFMM), a linear scaling, O(N), method for the formation of the Coulomb interaction matrix. The diagonalization portion of the talk will focus upon a method of directly solving for the density matrix in self consistent field theories.

006. NEW HOST-GUEST SYSTEMS FOR SINGLE MOLECULE SPECTROSCOPY. F. Jelezko, B. Lounis and M. Orrit, C.P.M.O.H., CNRS et Université Bordeaux I, 33405 Talence, France

These last few years, we have been looking for new molecular host-guest systems giving strong and narrow single molecule lines in low temperature solids. We report on two new highly fluorescent aromatic compounds for single molecule spectroscopy, dibenzanthanthrene (DBATT), and dibenzoterrylene (DBT) studied either in an n-hexadecane matrix or in a naphthalene crystal. The very weak triplet yields and short triplet lifetimes make these molecules nearly perfect optical two-level systems, with negligible buildup of triplet population. DBATT is used for pump-probe measurements under heavy illumination. DBT is absorbing in the near infrared (around 750 nm), and can be studied with diode lasers.

007. REVERSIBLE, LIGHT-INDUCED FREQUENCY JUMPS OF TERRYLENE IN p-TERPHENYL: SINGLE MOLECULE OPTICAL SWITCHING. Th. Basché, Institut für Physikalische Chemie, Sophienstrasse 11, 80333 München, F.R.G.

The controlled optical manipulation of a single absorber in a solid is one fascinating goal of single molecule spectroscopy. Single terrylene molecules in site X_1 of a p-terphenyl crystal studied at low temperatures exhibit an interesting property: In very low concentration crystals all absorbers from that site undergo light-induced frequency jumps by which they increase their excitation frequency by 843 ± 3 GHz. These spectral jumps are fully reversible with only about 0.4 % variation in jump width from one sample to another. Aside from this new spectral position there are additional positions for the absorbers which differ from molecule to molecule. By raising the temperature above 40 K all molecules can be reset to their original X_1 absorption frequency. Spectral jumps of a given molecule can be investigated over several weeks allowing a detailed look at its light driven spectral dynamics.

008. SPECTRAL LINE SHAPES OF INDIVIDUAL CHROMOPHORES IN GLASSES. J. L. Skinner, Department of Chemistry, University of Wisconsin, Madison WI 53706

Hole burning and photon echo spectroscopy of dilute chromophores in low-temperature glasses provide information about the chromophore's fluctuating transition frequency. This is usually interpreted in terms of the two-level system (TLS) model. From the waiting time and temperature dependence of the hole shape or echo decay we have learned about the distribution of TLSs in glasses. The line shapes of individual chromophore molecules are also sensitive to the same TLS-induced fluctuating frequency. In principle, each chromophore has a different line shape, since each chromophore has its own unique TLS environment. Unlike hole burning or echo experiments, current single molecule line shape experiments have almost no dynamic range. Nonetheless, the distribution of line shapes contains information about the distribution of TLSs that cannot be obtained from hole burning or echo experiments, each of which average over many chromophores. The theoretical framework for single molecule line shapes is discussed, and theoretical and simulation results for the distribution of line shapes is presented. The results are compared to experiments on terrylene in polymer glasses.

009. OPTICAL SPECTROSCOPY AND MICROSCOPY OF SINGLE TERRYLENEDIIMIDE (TDI) MOLECULES BETWEEN 1.4 K AND ROOM TEMPERATURE. C. Bräuchle, S. Mais, T. Basché, K. Müllen+, W. Göhde#, H. Fuchs#, Inst. Physikalische Chemie, University Munich, +MPI Polymerforschungs Mainz, #Physikalisches Institut, University Münster, Germany

For single molecule spectroscopy (SMS) highly stable fluorophores with functional groups are of interest since they can be attached as SMS-markers to systems like polymer chains or biomolecules. With low temperature SMS (1.4 K) and a temperature variable combined confocal and nearfield optical microscope we investigated fluorescence excitation and emission properties as well as intra (triplet) and intermolecular (spectral diffusion) dynamics of single TDI molecules at temperatures between 1.4 K and room temperature. Further more, energy transfer between a pair of two covalently linked chromophores (Perylene-TDI) will be discussed.

010.
PHOTOLUMINESCENCE SPECTROSCOPY OF SINGLE CDSE NANOCRYSTALLITE QUANTUM DOTS. **Stephen A. Empedocles** David J. Norris, Moungi G. Bawendi, MIT, 77 Massachusetts Ave, Cambridge, Ma, 02139

We use the techniques of single molecule detection to reveal the lineshape of single CdSe Nanocrystallite quantum dots. Using far-field epifluorescence microscopy, we are able to image and obtain fluorescence spectra that are more than 50x narrower than what can be achieved using ensemble techniques. Resolution limited linewidths as narrow as $120\mu eV$ demonstrate the true zero-dimensional nature of this unique physical system, reinforcing their description as "artificial atoms". In addition, light driven spectral diffusion is observed, resulting in spectral shifts as large as 80meV over several minutes.

011. HOLE BURNING AND SINGLE MOLECULE LINESHAPES. R. Silbey, Massachusetts Institute of Technology, Department of Chemistry, 77 Massachusetts Avenue, 18-390, Cambridge, MA 02139-4307.

The shape of a hole burned line of a chromophore in an amorphous glass is related to the distribution of line shapes and positions of single molecules in the same host. In this talk, we consider whether the same microscopic model can succesfully predict the hole line shape as well as the distribution of line shapes and positions of single molecule lines. We use the standard two level system model and, in addition, consider a number of new proposals for variations on this model.

012. ATOMIC FORCE MICROSCOPY STUDIES OF SOLUTION CRYSTAL GROWTH MORPHOLOGY AND DYNAMICS, J.J. De Yoreo, T.A. Land, and J.D. Lee, Lawrence Livermore National Laboratory, Livermore, CA 94550, A.J. Malkin, Yu. G. Kuznetsov and A. McPherson, University of California at Riverside, Riverside, CA, 92521

In recent years, the atomic force microscope has been used to investigate the growth of a number of single crystal surfaces from solutions at the 10nm to 100μm length scale both in real time and ex situ. This talk will discuss the results from studies on inorganic salts, simple organic molecules and macromolecular systems where the size of the growth units varies from a few Å to over 10nm. It will compare the evolution of surface morphologies with increasing supersaturation, the modes of crystallization, step kinetics and defect formation. Results will be analyzed within the framework of BCF-type models to derive the rate controlling thermodynamic and kinetic parameters of growth. This work was performed under the auspices of the U.S. Department of Energy by Lawrence Livermore National Laboratory under contract No. W-7405-ENG-48.

013. IN-SITU TEM STUDY OF STRAIN-INDUCED ORGANIZATION IN EPITAXY. J. M. Gibson, Depts. of Physics and Materials Science, University of Illinois, Urbana, IL 61801

It is now recognized that strain associated with non-uniform epitaxial growth can give rise to long-range interactions which can lead to useful self-organization during epitaxy. The transmission electron microscope (TEM) can directly observe and measure these strain fields beneath the substrate surface, and simultaneously observe the microstructure of the deposit. Using an ultra-high vacuum in-situ TEM, we can visualize these effects directly during epitaxial growth. We have measured strain associated with monolayer steps on $Ge(x)Si(1-x)$ films, and examined strain-induced ordering of islands during metal silicide growth on silicon. In particular, we discuss the possiblity of using self-organization of pinholes in $CoSi(2)$ layers on Si for a novel high-speed naturally-patterned permeable-base transistor.

014. ISLAND FORMATION DURING GROWTH OF Ge ON Si(001) BY CHEMICAL VAPOR DEPOSITION T.I. Kamins, E.C. Carr, R.S. Williams, and S.J. Rosner, Hewlett-Packard Laboratories, Palo Alto CA 94303-0867

Ge layers were grown on Si(001) by chemical vapor deposition in the viscous-flow regime in a single-wafer, Si epitaxial reactor using germane in a hydrogen carrier gas. The deposited layers were examined by atomic force microscopy and transmission electron microscopy after air exposure. In this lattice mismatched system Stranski-Krastanow growth is observed, with four stages of growth: (1) uniform layer formation up to about 4 monolayers of Ge; (2) square-based, short islands with a constant aspect ratio, which grow until the base reaches a critical size; (3) dome-shaped islands, which retain the same diameter but grow taller until the height also reaches a limiting size, probably related to the strain energy in the system; (4) large, faceted islands containing defects. The island density increases with decreasing deposition temperature. Under some conditions, the island density remains constant while the square-based islands transform into dome-shaped islands as more Ge is added.

015. Self-assembled quantum dot formation on semiconductor surfaces, A. -L. Barabasi, Department of Physics, University of Notre Dame, Notre Dame, IN 46556

Heteroepitaxial growth of highly strained structures offers the possibility to fabricate islands with very narrow size distribution, coined self-assembling quantum dots (SAQD). In spite of the high experimental interest, the mechanism of the SAQD formation is not well understood. This talk will discuss recent theoretical results on the formation and the stability of the islands. Thermodynamic approaches will be presented that predict the equilibrium island sizes and densities, as well as the nature and the magnitude of the critical thickness needed to be deposited for SAQD formation. The nonequilibrium aspects of the growth process are captured by Monte Carlo simulations, that provide the dynamics of the island formation. The results will be compared with experimental results on SAQDs observed in InAs and InP grown on GaAs and CdSe grown on ZnSe.

This work was done in collaboration with I. Daruka and J. Furdyna.

016.

KINETICS OF LARGELY LATTICE-MISMATCH EPITAXY. Yong Chen, Hewlett-Packard Laboratories, 3500 Deer Creek Road, Palo Alto, CA 94304, and Jack Washburn, Lawrence Berkeley National Laboratory, University of California, Berkeley, CA 94720

The kinetics of island nucleation, growth, and dislocation formation in largely lattice-mismatch heteroepitaxy are analyzed theoretically. It is shown that 2D platelets tend to transform to 3D islands as they exceed a certain critical size. During island growth, the increase of the strain concentration at the island edge makes it increasingly difficult for adatoms to reach the island, which leads to the formation of homogeneously sized islands. The high strain concentration at the island edge is eventually relieved by growing-in dislocations.

017. HYDROCARBON FILMS FROM METHYL RADICALS, Peter C. Stair, Department of Chemistry, Northwestern University, Evanston, IL 60208

We have used a gas phase source of methyl radicals based on the thermal decomposition of azomethane to study the adsorption and surface chemistry of methyl radicals on molybdenum, platinum and nickel surfaces. Measurements of the methyl radical sticking probability as a function of coverage have revealed a surface reaction limited adsorption of methyl radicals which takes place at high surface coverages, >1 monolayer, on clean and oxidized molybdenum and nickel but not on platinum. In this regime the uptake of carbon from impinging gas phase methyl radicals continues without any evidence of saturation even at a coverage of 10 monolayers of carbon. This surface carbon is in the form of alkyl hydrocarbon chains as revealed by XPS and TPD measurements. C_4 hydrocarbons have been observed by TPD. We tentatively attribute the growth of these surface hydrocarbons to a methyl insertion mechanism analogous to the chain growth reaction in Fischer-Tropsch synthesis. However, an alternative mechanism involving sequential hydrogen abstraction followed by radical addition is another possibility.

018. ACTIVATION OF C5-C8 CYCLOALKANES VIA ELECTRON INDUCED DISSOCIATION (EID) OF MULTILAYERS ON Pt AND Pt-Sn ALLOY SURFACES. Y.-L. Tsai and B. E. Koel, Department of Chemistry, University of Southern California, Los Angeles, CA 90089-0482.

We have previously reported (Surface Sci. 292 (1993) L803) that low energy-electron induced dissociation (EID) of cyclohexane multilayers can be used to form cyclohexyl species on a Pt(111) surface. This method can be extended to multilayers of other hydrocarbons containing a single type of C-H bond in the molecule in order to prepare and investigate the chemistry of monolayer coverages of a single type of adsorbed intermediate on reactive metal surfaces under UHV conditions. While Pt(111) has sufficient reactivity to dehydrogenate the C5-C8 cycloalkanes during heating in UHV, these molecules are mostly reversibly adsorbed on Pt-Sn alloys and very little dehydrogenation occurs. We have now used EID to activate cyclic-C_5 to C_8 alkanes on Pt(111) and two ordered Pt-Sn surface alloys in order to probe the chemistry of cycloalkanes and cycloalkyl species on Pt-Sn alloys. Dehydrogenation of adsorbed cycloalkyl groups on Pt-Sn alloys leads to large increases in the yield of unsaturated gas phase products desorbing from the surface compared to Pt(111).

019. REACTION TRANSITION STATES AT METAL SURFACES. Andrew J. Gellman, A. Paul, Q. Dai, J. Meyers, S. Street, Department of Chemical Engineering, Carnegie Mellon University, Pittsburgh, PA 15213

Measurements of reaction barriers in series of fluorine substituted reactants adsorbed on metal surfaces have revealed some of the characteristics of the transition states for these elementary processes. These measurements have been coupled with the use of vibrational spectroscopy to determine adsorbate orientations and computational methods to calculate reaction energetics. In concert these methods have generated a detailed picture of the reactions of a number of adsorbed species such as alkyl groups, alkoxides, and phenyl groups on metals surfaces such as Cu and Ag. These studies have often taken queues from Professor Brian Bent's work while at the same time attempting to add to the understanding of surface chemistry pioneered in his laboratory.

PHYS

020. THE CATALYTIC CHEMISTRY OF SMALL HYDROCARBONS ON MOLYBDENUM OXIDES. W.T. Tysoe, Department of Chemistry and Laboratory for Surface Studies, University of Wisconsin-Milwaukee, WI 53211

The relative activity for the metathesis of olefins is investigated using various model oxides prepared in an ultrahigh vacuum chamber and tested using an isolable catalytic reactor where it is shown that both MoO_2 and MoO_3 provide the most active catalyst for reaction above ~650 K. It is also shown that the activation energy for reaction above this temperature is ~65 kcal/mol. However, another reaction regime is found below ~650 K where the activation energy decreases to ~6 kcal/mol and the absolute activity is close to that for alumina-supported molybdena. Reaction of ethylene at high temperatures shows the formation of higher hydrocarbons where the product yield is well described by a Schulz-Flory distribution. This suggests that, rather than proceeding via the commonly proposed carbene-metallacycle pathway, reaction occurs via a dissociative/associative mechanism at high temperatures. This chemistry can also be probed in ultrahigh vacuum by grafting carbene species onto various model oxide catalyst surfaces by thermally decomposing iodine-containing precursors.

021. CHANGE IN REACTION PATHWAY INDUCED BY DEUTERATION: THERMAL DECOMPOSITION OF NEOPENTYL GROUPS ON Pt(111) SURFACES, Ton V. W. Janssens and Francisco Zaera, Department of Chemistry, University of California, Riverside, CA 92521.

Most alkyl groups adsorbed on metal surfaces undergo a facile β-H elimination reaction to form the corresponding alkene and surface hydrogen, and as a consequence much less is known about other possible reactions such as α- or γ-H eliminations or C-C-bond-scission steps in those systems. The thermal decomposition of neopentyl groups, which do not contain β hydrogen atoms, can provide more information about those less common processes. In this talk we report results from both temperature-programmed desorption and infrared experiments with non-deutero-, α-deutero, γ-deutero-, and per-deutero-neopentyl groups adsorbed in ultra-high vacuum on Pt(111) surfaces, which were produced in situ by thermal decomposition of the corresponding neopentyl iodides. It was found that a hydrogen, not deuterium, atom was eliminated first from both α- and γ-deutero-neopentyl groups, indicating that γ-H elimination takes place in the former case while α-H elimination occurs in the latter system instead. This change in decomposition mechanism is attributed to the kinetic isotope effect associated with the deuterium substitution, which implies that the rates of the α- and γ-H elimination reactions in non-deutero-neopentyl groups are of the same order of magnitude. The results on platinum will be compared with those reported previously on Ni(100)

022. ELECTRON CORRELATION AT LINEAR COST. Peter M. W. Gill, Department of Chemistry, University of Cambridge, Cambridge CB2 1EW, England

The last decade has seen tremendous advances in the treatment of the Coulomb problem and the resulting algorithms (CFMM, KWIK, COPs,...) now enable us to perform DFT and HF calculations on very large chemical and physical systems. The next step is to attack the correlation problem. This can be approached using DFT but the present correlation functionals are imperfect and it is not yet clear how to improve them systematically. Alternatively, one can develop better computational routes (e.g. via localization) to the CI-based methods. A third possibly is to develop an entirely new approach to correlation. Recent advances in these three areas will be surveyed and compared.

023.

SCF THEORY FROM O(N^4) TO O(N): THE LEGACY OF JAN ALMLOF. **Matt Challacombe,** Eric Schwegler, Chris White, Benny Johnson, Peter Gill, and Martin Head-Gordon, Department of Chemistry, UC Berkeley, Berkeley, California, 94720

The seminal contributions of Jan Almlof to reducing the computational complexity of SCF theory will be reviewed, from direct methods to linear scaling algorithms. The latest advances in linear scaling methods will be introduced, including rigorous methods for calculating the exchange matrix and hierarchical multipole methods for computation of the Coulomb matrix with and without periodic boundary conditions. Large scale HF calculations will be used to illustrate the cpu dependence of these methods for a variety of insulating systems, basis sets, and levels of precision. The performance of advanced methods for the evaluation of the exchange correlation matrix will be assessed, and the crossover of linear scaling HF and HF/DFT Fock builds with state of the art eigensolvers will be examined.

024.

LINEAR SCALING, LINEAR RESPONSE, AND DISTORTED GRIDS IN DENSITY FUNCTIONAL ALGORITHMS. E. B. Stechel , Sandia National Laboratories, MS 1421, Albuquerque, NM 87185-1421.

The rapid pace of advances in computational power and the advent of parallel processing demands equally innovative advances in algorithmic development. Computational materials science now holds the promise of accurate and realistic calculations on complex materials. However, with tried and true traditional algorithms, accurate quantum mechanical calculations of properties of such materials are limited to system sizes of less than a couple hundred atoms. This is due to the unphysical fast scaling with size, typically N^3 or greater. Linear response (or second derivatives) is N^4. This presentation will discuss some recent advances towards the simultaneous goals of accuracy and efficiency for large systems, including linear scaling ground state calculations, linear scaling-linear response and non-uniform grids. Our linear scaling approach is based on invariant subspaces that are spanned by localized non-orthogonal orbitals. Throughout, we emphasize the variational principle and a principle which we refer to as "faithful representations."

025.

THE PARALLEL CONTINUOUS FAST MULTIPOLE METHOD. **Benny G. Johnson** and Richard L. Graham, Q-Chem, Inc., 317 Whipple St., Pittsburgh, PA 15218, and Cray Research, Inc., 655 Lone Oak Dr., Eagan, MN 55121

The Continuous Fast Multipole Method (CFMM) is a powerful new method capable of evaluating the electronic Coulomb interactions in DFT in only linear work in system size for large molecules. Due to its greatly enhanced efficiency over standard methods, the CFMM is rapidly becoming established as a mechanism for studying molcules by DFT which are much larger than previously possible. We have recently demonstrated that the CFMM is quite suitable for parallelism as well; combined with parallel computing, the ability to carry out fast calculations on large molecules is magnified. For example, we have recently carried out LSDA/3-21G calculations on a Conus peptide having 352 atoms (1960 basis functions). On 32 nodes of a Cray T3E machine, a complete SCF calculation on this molecule required less than 30 minutes.

026. BIGGER, BETTER AND FASTER -- THE USE AND IMPACT OF PARALLEL COMPUTATION IN THEORETICAL CHEMISTRY. Robert J. Harrison, Mail Stop K1-90, Pacific Northwest National Laboratory, P.O. Box 999, Richland, WA 99352

I will examine some of the motivations for using both modestly-and massively-parallel computers in computational chemistry and compare our hopes and expectations with obained results. Particular attention will be paid to the algorithmic and programming strategies required to scale computational chemistry methods to massive parallelism, and also how to use of all available resources (memory, disk, cpu and communications bandwidth) often give rise to highly super-linear scaling. This will be illustrated with results for large semi-direct SCF and MP2 gradient calculations. Such results have significant implications for the choice of machine configuration and assignment of resources for maximum total system throughput. This work was performed under the auspices of the High Performance Computing and Communications Program, and the Environmental Molecular Sciences Laboratory construction project, U.S. department of Energy under contract DE-AC6-76RLO 1830 with Battelle Memorial Institute which operates the Pacific Northwest National Laboratory, a multiprogram national laboratory.

027. SINGLE MOLECULE MICROSCOPY. Urs. P. Wild, Hermann Bach, and Alois Renn, Physical Chemistry Laboratory, Swiss Federal Institute of Technology, ETH Zentrum, CH-8092 Zurich, Switzerland.

We have developed an experimental scheme to observe frequency shifts of single molecule absorption lines induced by a controlled manipulation in the surrounding in the solid state at low temperatures. In a mixture of terrylene and triphenylene in the matrix n-octane many single terrylene molecules are observed in parallel with a fluorescence microscope while triphenylene molecules are excited to their long living triplet state. Frequency shifts are induced upon excitation and relaxation of triphenylene resulting in increased spectral dynamics of terrylene molecules during the relaxation period.

028.

Discreet Jumps in Fluorescence Intensity Observed from Single Luminescent Conjugated Polymer Molecules. David A. Vanden Bout, W.T. Yip, Dan-K. Fu, Timothy W. Swager, and Paul F. Barbara. University of Minnesota Chemistry Department, 207 Pleasant St. SE, Minneapolis, MN 55455.

Scanning confocal microscopy is used to observe fluorescence from individual molecules of a luminescent conjugated polymer (a modified copolymer of poly-p-phenylene-vinylene and poly-p-pridylene-vinylene). Despite their large molecular weight (Mn=22,000), the molecules behave like single chromophores. The fluorescence intensity observed from individual molecules makes discreet jumps between a small number of intensity levels before abruptly photobleaching. Histograms of the fluorescence intensity reveal three states: an "on" state of full fluorescence, an "intermediate" state, and an "off" state where no fluorescence is observed. Transitions from the "on" state to the lower fluorescence levels are photo driven, while the transitions out of the "off" state are independent of the excitation rate. Single molecule fluorescence spectra show that the molecules exhibit no spectral diffusion either spontaneously or as the result of intensity jumps. Experiments are continuing to determine the exact nature of the various optical states.

029. ROOM TEMPERATURE SINGLE MOLECULE MICROSCOPIES
W. Patrick Ambrose, Peter M. Goodwin, J. Enderlein, David J. Semin, and Richard A. Keller, Los Alamos National Laboratory, M888, Los Alamos, NM 87545

We have developed three capabilities to image the locations of and interrogate immobilized single fluorescent molecules: near-field scanning optical, confocal scanning optical, and wide-field epi-fluorescence microscopy. Each microscopy has its own advantages. Near-field illumination can beat the diffraction limit. Confocal microscopy has high brightness and temporal resolution. Wide-field has the quickest (parallel) imaging capability. With confocal microscopy, we have verified that single fluorescent spots in our images are due to single molecules by observing photon antibunching. Using all three microscopies, we have observed that xanthene molecules dispersed on dry silica curiously exhibit intensity fluctuations on millisecond to minute time scales. We are exploring the connection between the intensity fluctuations and fluctuations in individual photophysical parameters. The fluorescence lifetimes of Rhodamine 6G on silica fluctuate. The complex nature of the intensity and lifetime fluctuations is consistent with a mechanism that perturbs more than one photophysical parameter.

030.
SINGLE-MOLECULE SPECTROSCOPY AND CHEMICAL DYNAMICS AT ROOM TEMPERATURE. H. Peter Lu and X. Sunney Xie, Pacific Northwest National Laboratory, P.O.Box 999, Richland, WA 99352

We present an analysis of single-molecule spectral trajectories, which provides detailed information about the potential energy surface and single-molecule dynamics in a particular environment. Recording emission spectra for single sulforhodamine 101 molecules on a quartz surface with a time resolution as high as 70 ms, we show spontaneous spectral fluctuations on two distinctly different time scales: hundreds of milliseconds and tens of seconds. We also observe photoinduced spectral fluctuation, a phenomenon intrinsic to repetitive photoexcitations in single molecule experiments. The single-chromophore spectral analysis allows us to conduct detailed investigation of conformational dynamics and energy landscapes for a specific active site in a protein. Measurements of single-molecule chemical reaction kinetics are demonstrated for interfacial electron transfer from excited cresyl violet molecules to the conduction band of indium-tin-oxide. In this system, each single molecule exhibits a single exponential electron transfer kinetics. A wide distribution of site-specific electron transfer rates is observed. The physical origin of multiexponential kinetics of electron transfer in this system is revealed to be the inhomogeneity of molecular interactions on the semiconductor surface. The single-molecule approach has thus revealed phenomena previously hidden in room-temperature experiments performed on large ensembles of molecules.

031. SINGLE MOLECULE STUDIES: FROM CHEMISTRY TO BIOLOGY WITH A POSSIBLE APPLICATION IN GENOMICS. Jay Trautman, SEQ LTD, Sarnoff Research Center, 201 Washington Road, Princeton, NJ 08540.

Over the past few years methods have been developed to manipulate, observe, and generally characterize single molecules. In principle, one can combine these capabilities to make a DNA sequencer, wherein a single molecule of DNA is digested by an single exonuclease and the single nucleotides are detected and identified in order, post-cleavage. The devil is, of course, in the details. For example, my colleagues and myself have worked primarily on optical studies of single molecules and quantum dots at visible wavelengths, whereas native nucleotides fluoresce in the UV. That work on single species in the visible will be covered as well as our progress in the UV. Other aspects of the single molecule studies including single enzyme turnover kinetics and environmental effects on nucleotide photophysics will also be discussed.

032.
LOW TEMPERATURE EPITAXIAL GROWTH OF SEMICONDUCTORS AND CERAMICS. **David G. Cahill**, J. E. Van Nostrand, B. W. Karr, I. Petrov, and J. E. Greene, Department of Materials Science, University of Illinois, Urbana, IL, 61801

We use ultra-high vacuum scanning tunneling microscopy (STM) to study to the evolution of rough surface morphologies during vapor phase crystal growth of Ge and TiN at low temperatures; Ge is grown on Ge(001) substrates using molecular beam epitaxy at $60<T<230$ C and TiN(001) is grown on MgO(001) using reactive magnetron sputtering at $650<T<750$ C. For both materials, we observe nearly periodic, nanometer-scale surface roughness (i.e. growth mounds) produced by asymmetric kinetics for the attachment of adatoms at ascending versus descending steps. By analyzing the evolution of the mound geometry, we have determined that this asymmetry on Ge(001) is small, consistent with the observed slow cross-over between smooth and rough growth. For TiN(001), the geometry of growth mounds is quickly established and stabilized, indicating a relatively strong asymmetry for adatom attachment.

033. REAL-TIME MEASUREMENTS OF THIN FILM GROWTH, E. Chason, J.A. Floro, T.M. Mayer, B.K. Kellerman[a] and D.P. Adams, Sandia National Laboratories, Albuquerque, NM, [a]MEMC Electronic Materials, St. Peters, MO

Real-time monitoring techniques enable us to measure the evolution of surface morphology during thin film growth. We present results from growth studies using two in situ diagnostics: X-ray reflectivity (XRR) and multi-beam optical stress monitoring. XRR measurements of film thickness and roughness during chemical vapor deposition of Fe on Si(001) provide insight into the kinetics of precursor dissociation and Fe cluster nucleation. Stress measurements during heteroepitaxial growth of SiGe alloys on Si(001) demonstrate how films can relieve misfit strain by changing the surface morphology without introducing dislocations into the film. This work was supported by the United States Department of Energy under Contract DE-AC04-94AL85000. Sandia is a multiprogram laboratory operated by Sandia Corporation, a Lockheed Martin Company, for the United States Department of Energy.

034. EPITAXIAL GROWTH OF Ge ISLANDS ON Si (001). G. Medeiros-Ribeiro, D. Ohlberg, T. Kamins, A. Bratkovski, and R. S. Williams, Hewlett Packard Laboratories, 3500 Deer Creek Road, Bldg. 26 Palo Alto, CA 94304

Much attention has been paid in the past years to coherently strained systems as candidates for the synthesis of zero dimensional objects. Growth in the Stranski-Krastanov mode allows us to produce such structures. In this talk we will present the Ge:Si (001) epitaxial system studied by scanning tunneling microscopy. With this technique we were able to observe various stages of the epitaxy, from strained and fractured epitaxial layers to the formation of islands. These could be classified under three different categories - pyramids, domes and superdomes. The first two families were coherently strained to the substrate whereas the last included misfit dislocations. We discuss the shape transition with a simple theoretical model that takes into account energy considerations. Finally, we discuss the spatial distribution of the islands where we observed some degree of ordering.

035. KINETICS OF METAL FILM GROWTH AND ISLAND THICKENING ON OXIDES: Cu / ZnO(0001)-Zn AND Au / TiO$_2$(110). Jun Yoshihara, Steven C. Parker, Ann W. Grant and Charles T. Campbell, Department of Chemistry, Box 351700, University of Washington, Seattle, WA 98195-1700.

The vapor deposition of late transition metals on oxide single crystal surfaces at low temperatures can lead to a growth mode wherein two-dimensional islands (one atom thick) first form and then grow laterally to cover a significant fraction of the surface. After a critical coverage is reached, these islands instead grow thicker, while their lateral growth is greatly suppressed. Annealing causes the islands to irreversibly thicken, thus exposing more metal-free oxide surface. The factors (flux, temperature, defect density, etc.) which influence the critical coverage and the thermal thickening kinetics have been explored using LEIS, XPS, LEED spot widths, and selective chemisorption.

036.
GROWTH OF SELF-ASSEMBLED MONOLAYERS. **P. Fenter** Princeton Materials Institute, Princeton University, Princeton NJ 08544

The growth of self-assembled monolayers has been widely studied, yet the molecular level processes involved in the self-assembly of these systems have yet to be elucidated. I describe recent results on the widely studied decanethiol/Au(111) system using vapor phase deposition in which we obtain a direct measure of the structure, order and coverage during the growth, by using a range of powerful and complementary techniques (grazing incidence x-ray diffraction, He atom diffraction, x-ray photo-electron spectroscopy and x-ray standing waves). Our results reveal a number of unexpected features, including the existence of multiple self-assembly mechanisms as a function of the growth conditions (due to the adsorption energetics), and a complex coverage-temperature phase diagram which has an important impact upon the monolayer evolution. This work was performed in collaboration with A. Eberhardt, F. Schreiber, T.Y.B. Leung, D. Lavritch, P. Schwartz, S. Wetterer, M. Bedzyk, L. Berman, P. Eisenberger, G. Scoles

037. SYNTHESIS, IDENTIFICATION AND MECHANISM FOR FORMATION OF ETHYLIDYNE ADSORBED ON Ni(111). S.T. Ceyer, T.R. Trautman, T. Burgi, K. Haug and A. Utz, Department of Chemistry, Massachusetts Institute of Technology, Cambridge MA 02139

The synthesis of ethylidyne on a Ni(111) surface by the reaction of gas phase H atoms with adsorbed ethylene, acetylene and ethane has been demonstrated. Four isotopomers, CCH$_3$, CCD$_3$, ^{13}C^{13}CH$_3$, ^{13}C^{13}CD$_3$, have been synthesized and their vibrational spectra have been measured by high resolution electron energy loss spectroscopy as a function of scattering angle. A detailed symmetry analysis and a normal modes calculation has enabled the identification of additional features in the bivrational spectra as arising from a small amount of acetylene coadsorbed with ethylidyne. Acetylene is shown to be the product of a dynamic equilibrium between ethylidyne and acetylene in the presence of gas phase hydrogen atoms. The mechanism of ethylidyne formation has also been investigated by studying the kinetics of the reactions of gas phase H atoms with adsorbed ethylene, acetylene or ethane and by identifying the dissociation products resulting from the interaction of <100 eV electrons with adsorbed ethane, ethylene and acetylene.

038. DIRECT ORGANOMETALLIC SYNTHESIS: β-HYDRIDE ELIMINATION AND METAL ETCHING REACTION ON ALUMINUM. S.P. Lokare*, E.L. Crane* L.H. Dubois**, and R.G. Nuzzo*, *School of Chemical Sciences, Department of Materials Science and Engineering, and the Frederick Seitz Materials Research Laboratory, University of Illinois, Urbana, IL 61801, **Defense Advanced Research Projects Agency, Arlington, VA 22203

The thermal decomposition and subsequent reactions of isobutyl iodide on Al(111) lead to a complex mixture of isobutene and dihydrogen (products formed via a β-hydride elimination process of surface-bound alkyl groups derived from the dissociation of the C-I bond) and volatile organometallic etching products (including diisobutyl aluminum iodide, diisobutyl aluminum hydride, methyl aluminum dihydride, and AlI_x, x=1–3). The activation energy of the β-hydride elimination process was measured to be ~1.1 eV using the collision induced reaction of hyperthermal atom beams (a value in excellent agreement with that obtained from an analysis of the thermally activated kinetics of the reaction). A detailed mechanistic picture of these complex surface reaction processes will be presented.

039. SPATIAL PATTERNING OF ADSORBED MONOLAYERS BY A PHYSISORBED TEMPLATE. G.A. Reider, P.A. Williams, U. Hoefer, L. Li, T. Suzuki, and T.F. Heinz, Department of Physics, Columbia University, New York NY 10027

In this paper, we present a new method to achieve spatial patterning of adsorbed monolayers. The scheme involves covering the surface with a physisorbed monolayer. This layer is then patterned by means of laser-induced thermal desorption to form a template. The surface is subsequently exposed to the desired chemisorbed species. A complementary patterning of the surface by the chemisorbed adsorbate will result provided that the sticking coefficient of the chemisorbed species is significantly reduced by the presence of the physisorbed molecules of the template layer. The final patterned chemisorbed layer is obtained by allowing the temperature of the sample to rise to the point where the physisorbed molecules of the template desorb. The method provides a general approach to the patterning of strongly bound and labile molecules for which direct patterning by laser-induced thermal desorption may not be suitable. It may be considered as an ultraclean monolayer photoresist. We have demonstrated the method by preparing a grating structure of adsorbed H on a Si(111)7x7 surface. The physisorbed template layer used in the experiment was Xe and patterning was achieved by two interfering laser beams irradiating the surface. The presence of the grating structure was verified by means of optical diffraction. As will be discussed, this technique should be useful for a variety of studies of surface dynamics.

040. OXIDATION OF GaAs(110) WITH NO_2: A MOLECULAR BEAM STUDY. M.J. Cardillo, C.C. Bahr, A. vom Felde*, and S.K. Buratto[†], Bell Laboratories, Murry Hill, New Jersey 07974; *IBM East Fishkill, 33A Building 630-1, 1580 Route 52, Hopewell Junction, New York 12533; [†]Department of Chemistry, University of California, Santa Barbara, California 93106-9510

I describe time-resolved modulated molecular beam measurements of the adsorption of NO_2 on the GaAs(110) surface. These measurements provide a quantitative description of the gas-surface interaction: thermal NO_2 has a nearly unity sticking probability, dissociating to form NO (which quickly desorbs) and a surface oxide of ~1/3 ML saturation coverage. The NO_2 dissociation probability is strongly controlled by the defect density, the oxygen coverage, and the surface temperature. We have characterized the competition between the processes leading to oxygen deposition, i.e. desorption and diffusion of NO_2 to reactive defects. The barrier energies $E_{diff} = 6 \pm 2$ kcal/mol and $E_{des} = 9 \pm 2$ kcal/mol provide an interesting example of the transition from defect-dominated surface chemistry, typical of low Miller index metal surfaces to defect-insensitive diffusion-limited chemistry, which we argue is generic to corrugated semiconductor surfaces.

041. ATOMIC LAYER GROWTH USING BINARY REACTION SEQUENCE CHEMISTRY. S.M. George, A.W. Ott, and J.W. Klaus, Department of Chemistry and Biochemistry, University of Colorado, Boulder, Colorado 80309-0215

Sequential ABAB... surface chemical reactions can be employed for atomic layer controlled deposition. We have examined the binary reactions: $2Al(CH_3)_3 + 3H_2O \rightarrow Al_2O_3 + 6CH_4$ for Al_2O_3 deposition; $SiCl_4 + 2H_2O \rightarrow SiO_2 + 4HCl$ for SiO_2 deposition; and $2In(CH_3)_3 + 3H_2O \rightarrow In_2O_3 + 6CH_4$ for In_2O_3 deposition. Each binary reaction was divided into two separate half-reactions. For example, the two half-reactions for Al_2O_3 deposition are:

(A) $AlOH^* + Al(CH_3)_3 \rightarrow AlOAl(CH_3)_2^* + CH_4$
(B) $AlCH_3^* + H_2O \rightarrow AlOH^* + CH_4$

where the asterisks designate the surface species. The half-reactions were then performed sequentially and repetitive ABAB... cycling can produce layer-by-layer controlled growth. Extremely flat and conformal Al_2O_3 and SiO_2 films can be deposited with growth rates of ~1Å/AB cycle. Temperature-dependent studies of Al_2O_3 and SiO_2 deposition argue that atomic layer controlled growth is a direct consequence of the binary surface chemical reactions. In addition, investigations of deposited In_2O_3 films reveal that the surface morphology deteriorates if the coverages of the surface species are too low.

042. THE SURFACE CHEMISTRY OF ATOMIC LAYER EPITAXY OF COMPOUND SEMICONDUCTORS. R.M. Osgood, Jr., Columbia Radiation Laboratory, Columbia University, New York, NY 10027

Atomic layer controlled chemistry has been of interest because it has the potential to define precisely the thickness of thin layers of semiconductors. Atomic layer removal or etching was, for example, an important objective of the Bent group at Columbia. This work and others has led us to consider a more surface-chemistry-oriented approach to obtaining reactive atomic growth of bilayers of II-VI materials. Our results show that UHV surface reactions can be used to obtain room-temperature self-limiting growth of epitaxial layers. In particular, we present an experimental study of layer-by-layer growth of CdS on ZnSe(100) using a variety of HV surface probes including LEIS, Auger, XPS and LEED. The results show clear evidence of atomic layer epitaxy based on room- temperature sequential surface reactions of $(CH_3)_2Zn$ and H_2S. This work is supported by the NSF (grant DMR-92-25134) and the DOE (grant DE-FG02-90ER14104).

043. SLOW DYNAMICS OF SINGLE MOLECULES IN A ROOM TEMPERATURE POLYMER MATRIX. **James D. Chesko**, H. Peter Lu, X. Sunney Xie. Pacific Northwest National Laboratory, Environmental Molecular Sciences Laboratory, P. O. Box 999, MS K2-14, Richland, WA 99352

Trajectories of single-molecule emission intensity and dipole orientation were measured with a time resolution up to 10 ms for dye molecules in poly(methylmethacrylate) matrix at room temperature. The intensity autocorrelation functions showed a dynamic process occurring on a rather slow time scale, several hundred milliseconds. Although we observed a small propensity for dipole reorientation, it is too infrequent to account for these intensity fluctuations. The intensity fluctuations on this time scale were proven to be spontaneous (independent of the excitation intensity), and were due to the spectral fluctuations that we observed previously. The higher time resolution in the intensity trajectories allowed us to observe multi-exponential decays in the intensity autocorrelation functions, which indicated complex intermolecular interactions and dynamics within the polymer matrix.

044. SPECTROSCOPY OF SINGLE MOLECULES PERFORMED WITH CONFOCAL MICROSCOPY AT LOW AND ROOM TEMPERATURES. Ludovic Fleury, Universitaet Chemnitz-Zwickau, Institut of Physics, Department of Optical Spectroscopy and Molecular Physics, D 09107 Chemnitz, Germany

Confocal microscopy at liquid helium temperatures has been applied to an interface structure of terrylene in hexadecane on a quartz surface. Upon exploiting the axial resolution of the confocal microscope it becomes possible to select individual terrylene molecules with different distances to the quartz surface. A drastic order of magnitude increase in the homogeneous linewidth of the molecules close (~100 nm) to the quartz surface is found together with an increased tendency towards spectral diffusion. On the other hand, room temperature confocal microscopy of a variety of dye molecules on surfaces and in thin polymer films has been carried out. The dynamics of these systems is characterized by measuring the autocorrelation function g(2) (t) of the fluorescence intensity. The type of dynamics observed and the time scale on which it takes place strongly depends on the system. Sulfordodamine 101 in polyvinylbutyrol as an example shows a dynamic process on a time scale of roughly 100 ms as it also has been reported recently by means of fluorescence spectra of this system.

045. ROOM TEMPERATURE STUDY OF SINGLE MOLECULE DYNAMICS ON A SURFACE. T. Ha, Th. Enderle, P.R. Selvin, D.S. Chemla and S. Weiss, Molecular Design Institute, Materials Sciences Division, MS 2-300, Lawrence Berkeley National Laboratory, Berkeley, CA 94720

We report on a variety of dynamical effects in the laser induced fluorescence (LIF) of single molecules adsorbed on a air-glass interface. The molecules were excited by a diffraction limited spot of a far field scanning optical microscope. Spectral jumps, often correlated with dark states, were observed in two different experimental schemes: emission spectrum and two-channel time trace. A small number of single molecules underwent rotational jumps as a result of desorption and readsorption from/to the surface. These events were observed by modulating the polarization of the excitation light. Quantum jumps of single molecules to the triplet state were resolved. A strong inhomogeneity in quantum jumps parameters was found among different molecules. Careful characterization of the intrinsic dynamical behavior of single molecule LIF is necessary before it can be applied to biological studies.

* Supported by the Laboratory Directed Research and Development Program of Lawrence Berkeley National Laboratory under U.S. Department of Energy, Contract No. DE-AC03-76SF00098 and by the Office of Naval Research, Contract No. N00014-95-F-0099.

046.

ON THE ORIENTATION OF PENTACENE IN P-TERPHENYL. **E. Heinecke** G. Knüpffer, A. Hese, Institut für Strahlungs- und Kernphysik, Technische Universität Berlin, Berlin, Germany, D-10623

Pentacene in p-terphenyl is one of the best investigated systems used in single molecule spectroscopy. But although a lot of data about this system is available, the correlation between the 4 spectral sites O1 to O4 and the 4 crystalline sites P1 to P4 is up to now somewhat ambiguous. Measurements for all 4 sites will be presented, which show a strong correlation between O1/O4 on one hand and O2/O3 on the other hand. This pairing is not only manifested by the same orientation of the projection of the molecular transition dipole moment onto the crystal plane, but also by the structure within the line shapes of the well resolved sites. The results agree well with those obtained by Köhler et. al. by ODMR measurements.

047.

STARK EFFECT OF PENTACENE. **E. Heinecke** Ch. Müller, D. Hartmann, G. Knüpffer, A. Hese, Institut für Strahlungs- und Kernphysik, Technische Universität Berlin, Berlin, Germany, D-10623

Single molecule spectroscopy is an excellent method to show, that qualities of an impurity molecule can be changed by a crystalline surrounding. The different lifetimes and ISC rates of pentacene in p-terphenyl in the 4 spectral sites are good examples. We investigated the Stark effect of pentacene under different conditions. Measurements on "free molecules" in a molecular beam and on single molecules in a crystal matrix have been done. The results show a large change of the polarizability induced by the surrounding.

048. NEAR-FIELD SCANNING OPTICAL MICROSCOPY STUDIES OF PROTEIN CHANNELS. M. Ann Lee and Robert C. Dunn, Department of Chemistry, University of Kansas, Lawrence, Kansas 66045

The introduction of atomic force microscopy (AFM) and related scanning probe techniques have opened new areas of high resolution investigation for biological processes *in situ*. Recently, near-field scanning optical microscopy (NSOM) has emerged as a scanning probe technique capable of optically probing biological structures at the nanometric scale. NSOM uses a nanometric light source positioned close to a sample surface to obtain high resolution fluorescence images that go beyond the diffraction limit normally encountered in optical microscopy. Moreover, the fluorescence sensitivity of the technique allows for the detection and study of single molecules. The low detection limits and high spatial resolution have opened many unique opportunities to explore biological processes *in situ*.

We have embarked on studies using both NSOM and AFM in a complimentary fashion to study the location and dynamics of protein channels located in cellular membranes. The goal is to study the conformational dynamics of a single voltage-gated ion channel in real time as it opens and closes. Progress towards these goals will be reported.

049. SINGLE MOLECULE METHODOLOGIES FOR DNA ANALYSIS. William A. Lyon and Shuming Nie, Department of Chemistry, Indiana University, Bloomington, IN 47405

The ability to detect single fluorescent molecules and to manipulate genomic DNA has opened many new opportunities in DNA analysis and fundamental studies. One exciting research area is to use DNA-binding proteins and synthetic oligonucleotides as sequence-specific probes. A wide variety of proteins exist that are capable of highly specific binding to unique DNA sequences via electrostatic, hydrophobic, and hydrogen bonding interactions. Also, hybridization methods based on Crick-Watson base pairing are available for detecting specific nucleotide sequences in genomic DNA. In fact, oligonucleotide probes of 17 bases or longer provide enough sequence combinations ($4^{17} = 1.7 \times 10^{10}$) that their complementary sequences will occur no more than once in the diploid human genome (6×10^9 bp). Our approach involves stretching and fixing genomic DNA molecules on a chemically modified surface, and using fluorescently labeled DNA-binding proteins or oligos to target specific sites on the stretched DNA. Individual probe molecules are measured by an integrated laser-induced fluorescence and tapping-mode atomic force microscope. The instrumentation as well as results obtained from prokaryotic and eukaryotic DNA will be discussed in detail.

050.

QUANTUM OPTICAL INTERACTIONS BETWEEN A SINGLE MOLECULE AND A SPHERICAL MICROCAVITY. **D.J. Norris,** M. Kuwata-Gonokami,* and W.E. Moerner, Department of Chemistry and Biochemistry, University of California San Diego, La Jolla, CA 92093-0340, *Department of Applied Physics, Faculty of Engineering, University of Tokyo, Tokyo 113, Japan

The recently demonstrated ability to optically detect a single molecule embedded in a solid host material allows the possibility to study the interactions between a single molecule and an optical cavity. When the molecular transition is in resonance with a particular mode of a low loss microcavity, the optical behavior of the molecule should be strongly modified, including enhancement of its spontaneous emission rate. Such an experiment provides a means to control the behavior of a single molecule in a positive way. In addition, novel quantum physics can be addressed. We present an experimental investigation of this situation, by detecting the emission from a single molecule, terrylene in a p-terphenyl crystal, coupled to a low loss mode of a ($Q \sim 10^6 - 10^7$) spherical microcavity.

051.

TWO-PHOTON AND ONE-PHOTON SPECTROSCOPY ON A SINGLE QUANTUM SYSTEM. **T. Plakhotnik,** D. Walser, A. Renn, and U. P. Wild, Physical Chemistry Laboratory, Swiss Federal Institute of Technology, CH-8092 Zürich, Switzerland

Two-photon fluorescence excitation (TPE) spectra of single diphenyloctatetraene (DPOT) molecules trapped in an n-tetradecane matrix were measured at cryogenic temperatures. The purely electronic $1^1A_g \rightarrow 2^1A_g$ transition at 444 nm was excited by an IR cw single mode laser at 888 nm. The same system was also studied with one-photon excitation (OPE) at 444 nm. We report on a number of unusual effects: (I) single molecule frequency shifts proportional to the IR power in TPE spectra, (II) a two to threefold single molecule linewidth increase in TPE spectra compared to OPE, (III) single molecule lines "erasing" under IR illumination yielding a difference between the number of molecules observed in TPE and OPE spectra. Though many nonlinear optical and thermal effects have been considered, most of the observations were explained in the frame of C-H band excitations followed by energy exchange between acoustic phonons and a local vibration in the "DPOT-matrix" system. Additionally, an ac Stark effect could contribute up to 30% to the shifts. Shpol'skii sites of DPOT with broken and unbroken symmetry are also discussed.

052. FLUORESCENCE IMAGING OF SINGLE MOLECULES AND PHOTOSYNTHETIC MEMBRANES WITH TWO-PHOTON EXCITATION. E. J. Sanchez, L. Novotny, and X. S. Xie, Pacific Northwest National Laboratory, Environmental Molecular Sciences Laboratory, P. O. Box 999, K2-14, Richland, WA 99352

We report the imaging of single-molecule fluorescence induced by two-photon excitation in ambient conditions. Using an inverted fluorescence microscope, we obtained the two-photon images of different single fluorophores (Rhodamine B, Sulforhodamine 101, Coumarin 535 on poly-methyl methacrylate films) and biological membrane fragments by raster scanning the sample with respect to a diffraction limited focus of a mode-locked Ti:sapphire laser beam. The signal to background ratio was as high as 50:1 and the full width at half maximum (250nm) of a single-molecule peak was significantly shorter than that for one photon excitation. With its high sensitivity and simplicity, the two-photon experiment offers a valuable approach for spectroscopic studies on individual immobilized molecules.

053. SINGLE MOLECULE SPECTROSCOPY IN LOW TEMPERATURE SOLIDS USING GRADIENT INDEX LENS. Martin Vacha, Yi Liu, and Toshiro Tani, Electrotechnical Laboratory, 1-1-4 Umezono, Tsukuba, 305 Ibaraki, Japan, Hiroki Nakatsuka, University of Tsukuba, 1-1 Tennoh-dai, Tsukuba, 305 Ibaraki, Japan

One of the crucial requirements for successful detection of fluorescence of single dye molecules embedded in a solid matrix is the use of high N.A. optics for collection and focusing of the signal. So far, this has been achieved by using parabolic or elliptical mirrors or microscope objectives. We present the use of radial gradient index micro-lens (selfoc lens) as a sample substrate and fluorescence collector. The presented design shows characteristics superior to those of a microscope objective of comparable parameters. The performance in various illumination-detection configurations will be demonstrated.

054.

DENSITY MATRIX-BASED SCHEMES IN HARTREE-FOCK AND DENSITY FUNCTIONAL THEORY TO ACHIEVE LINEAR SCALING FOR ENERGIES, GRADIENTS, AND FORCE CONSTANTS. **Christian Ochsenfeld** and Martin Head-Gordon, Department of Chemistry, University of California at Berkeley, Berkeley, CA 94720

A Self-Consistent Field (SCF) algorithm within a density matrix-based formulation is presented, which avoids the conventional density update scheme using an $O(M^3)$ diagonalization step. Our new method opens the way to achieve linear scaling of the computational effort with molecular size M. Further, we reformulate the Coupled Perturbed SCF (CPSCF) theory within such a fully local density matrix-based scheme. The sparsity of the derivative density matrix allows to solve these $O(M)$ CPSCF equations necessary for the calculation of vibrational frequencies of a molecule scaling only linearly with system size. An efficient implementation of these methods is discussed and a comparison to conventional algorithms is presented.

PHYS

055.

DIRECT DETERMINATION OF POLARIZED ATOMIC ORBITALS IN SELF-CONSISTENT FIELD CALCULATIONS. **Michael S. Lee** Martin Head - Gordon, Department of Chemistry, University of California, Berkeley, CA 94709

We introduce a method for simultaneous variational optimization of a minimal basis set and the one-particle density matrix in a self-consistent field calculation. This procedure significantly reduces the number of degrees of freedom in large basis set calculations and still preserves the density matrix as an exact projector. Compact representations of the one-particle density matrix such as the one used in this scheme offer considerable improvement in time and memory scaling, while introducing only small errors. Furthermore, polarized atomic orbitals provide a clear picture of atomic orbital contributions to molecular orbitals.

056. ELMO: LOCALIZED NON-ORTHOGONAL ORBITAL HARTREE-FOCK METHOD FOR TRUELY LARGE SYSTEMS. Marc Couty, Laboratoire de Chimie Théorique, Université Paris XII, M2 Bois de l'Etang, rue Galilée 77240 Champs sur Marne (France); Craig A. Bayse and Michael B. Hall, Department of Chemistry, Texas A&M University, College Station, Texas 77843-3255

A new orbital optimization method, ELMO, is derived in a localized non-orthogonal formalism. The method is based on a quasi Newton-Raphson algorithm in which an approximate diagonal-blocked Hessian matrix is calculated through the Fock matrix. During each iteration the Hessian matrix inverse is updated by a variable metric updating procedure to account for the intrinsically small coupling between the orbitals. The method requires Fock matrix construction which requires on the order of n^3 operations. The updated orbitals are obtained with approximately n^2 operations and no other n^3 processes such as matrix diagonalization, matrix multiplication or orbital orthogonalization are employed. The use of localized orbitals allows for creation of high quality initial "guess" orbitals from optimized molecular orbitals of small systems and thus reduces the number of iterations to converge. The small number of variables, even for a very large system, and the limited number of operations makes ELMO a method of choice to study truly large systems. The delocalization effects are accounted in a perturbative treatment which allows the accurate calculation of the total energy with a limited number of operations. This extension, referred to as ELMO(PT), is a variational method that reproduces the Hartree-Fock energy with an error of less than 2 kcal/mol for a reduced total cost compared to standard Hartree-Fock methods.

057.

LINEAR SCALING COMPUTATION OF HF AND HF/DFT FOCK MATRICES. **Matt Challacombe** Eric Schwegler, Chris White, Benny Johnson, Peter Gill, and Martin Head-Gordon, Department of Chemistry, UC Berkeley, Berkeley, California, 94720

Abstract. Hybrid Hartree-Fock/density functional theories (HF/DFTs) are emerging as highly competitive models for the accurate simulation of weak interactions such as hydrogen bonding. HF/DFT calculations are currently limited by calculation of the Coulomb (J), exact exchange (K), and exchange-correlation matrices (Kxc). In this talk, I will introduce linear scaling methods for computing J and K. Hierarchical multipole methods for computing J with and without periodic boundary conditions will be discussed. I will also describe a rigorous method for computing K that fully exploits the structure of an insulating system's density matrix. The performance of advanced methods for the evaluation of Kxc will also be assessed. Large scale calculations will illustrate the cpu dependence of these methods for a variety of insulating systems, basis sets, and levels of precision.

058. DEPENDENCE OF CORRELATION ENERGY OF MOLECULES ON THE ELECTRON CONTENT AND PARTIAL CHARGES, Sándor Kristyán, Department of Chemistry, Emory University, Atlanta, GA 30322, U.S.A.

An amazingly simple quasi-linear relationship was introduced [S.Kristyan, Chem.Phys.Lett. 247(1995)101 and 256(1996)229] between the number of electrons, N, participating in a molecular system and the correlation energy, Ecorr. Namely, Ecorr/(N-1) is between -0.035 and -0.045 hartree in case of any molecular system with any, equilibrium or nonequilibrium, geometry. In this work this relationship is used and developed (Ecorr =

$\Sigma_{(A=1,M)}$Ecorr(Z_A-(partial charge on A),Z_A) where Ecorr(N_A, Z_A) is the M atomic Ecorr (N_A electrons with Z_A nuclear charge) in the molecule from CI calculation) to estimate correlation energy rapidly in ab initio calculations and unlike the relationship above, it reaches the chemical accuracy 1 Kcal/mol. The only input is the partial charges on atoms in the molecule from the molecular wave function (available in packages like Gaussian94/DFT, Biosym, etc.), easy to obtain after HF-SCF calculations. The method is compared to the well known B3LYP, MP2, RCCSD and G2M methods.

059.

RECENT ADVANCES IN LOCAL CORRELATION. **P.E. Maslen,** M.S. Lee and M. Head-Gordon, Dept. of Chemistry, University of California, Berkeley, CA, 94720

We describe recent advances in the calculation of electron correlation for large molecules. To avoid the costly transformation of basis from atomic orbitals to molecular orbitals, we have derived equations for the MP2, CCSD and CCSD(T) energies which are valid in an arbitrary non-orthogonal basis. We have reduced the cost of MP2 by solving these equations in a localised basis and introducing a new local-correlation ansatz. Our new algorithm offers several improvements over the original local correlation method due to Saebo and Pulay. The new algorithm is considerably faster, the definition of the local correlation space has been greatly simplified, and the local ansatz does not introduce discontinuities into the potential surface.

060.

THE DEVELOPMENT OF CORRELATION CONSISTENT BASIS SETS FOR THIRD ROW ATOMS. **Angela K. Wilson,** David E. Woon, Kirk A. Peterson, Thom H. Dunning, Jr., Pacific Northwest National Laboratory, Environmental Molecular Sciences Laboratory, P.O. Box 999, Richland, WA 99352

We are extending the correlation consistent basis sets to the third row atoms, Ga-Kr. We shall present the approach used to develop the sets, along with preliminary benchmark results. The influence of the 3d-electrons on the properties of molecules containing third row atoms will be discussed. Finally, we will discuss the differences expected as we move left in the periodic table to the transition metals.

061. AN APPROXIMATE FORMULA FOR THE INTERMOLECULAR PAULI REPULSION BETWEEN CLOSED SHELL MOLECULES: APPLICATION TO THE EFFECTIVE FRAGMENT METHOD. Jan H. Jensen and Mark S. Gordon, Department of Chemistry, Iowa State University, Ames, IA 50011-3111

The exchange repulsion formula proposed by Murrell and co-workers [*Proc. Roy. Soc. (London)*, **1965**, *A284*, 566; *J. Chem. Phys.* **1967**, *47*, 4916] is considered in detail. Potentially important terms missing in the Murrell formalism are identified and evaluated for the water dimer using several basis sets. Insights into the contributing terms are obtained by using localized molecular orbitals. The results point toward a relatively simple expression for intermolecular exchange repulsion, based on the isolated wavefunctions of the two overlapping species. This expression is incorporated into the Effective Fragment Potential method [*Modeling the Hydrogen Bond*, ACS Symposium Series, number 569, Chapter 9; *J. Chem. Phys.* submitted].

062. APPROXIMATE SECOND ORDER METHOD FOR ORBITAL OPTIMIZATION OF SCF AND MCSCF WAVEFUNCTIONS. Galina Chaban, Michael W. Schmidt, and Mark S. Gordon, Department of Chemistry, Iowa State University, Ames, IA 50011

A quasi-Newton method involving a diagonal guess orbital Hessian with iterative updates has been proposed recently by Almlof for the optimization of closed shell SCF wavefunctions. The technique is extended in the present work to orbital optimization for more general wavefunctions, including ROHF, GVB, and MCSCF. The method is faster than the standard diagonalization techniques used for RHF and GVB. Its convergence properties (number of iterations) are comparable to that of DIIS accelerated SCF methods, but the time needed for solving the HF equations is about three times less. In addition, the method eliminates the traditional diagonalization step (except for the final orbital canonicalization) which is not easily parallelizable. For MCSCF, the method presented here requires more iterations than an exact second order program, but since each iteration is substantially faster, leads to a more efficient overall program. In addition, the memory requirements for approximate second order method are substantially less than those for exact second order methods.

063. GRID-FREE DENSITY FUNCTIONAL THEORY. Kurt Glaesemann and Mark S. Gordon, Department of Chemistry, Iowa State University, Ames, Iowa 50011

The grid-free density functional theory (DFT) approach developed by Jan Almlöf is implemented in the *ab initio* program of GAMESS. The grid-free approach is based on a matrix representation of the functionals instead of the commonly used grid based representation of the functionals. The matrix representation relies on basis set completeness to represent the functionals. Both results from local and gradient corrected functional are presented with comparisons to results from grid-based codes. Analytic energy gradients are also implemented.

064.

SELF-CONSISTENT SCREENING CALCULATIONS OF THE SURFACE PHONON DISPERSION OF METALS. **Andrew A. Quong** Computational Materials Science Department, Sandia National Laboratories, Livermore CA 94551

Experimental techniques such as HREELS and He-atom scattering have been used to measure the surface-phonon-dispersion curves of a wide range of materials. Fitting the experimentally determined curves yields force constants that are significantly different than their bulk counterparts. The problem with the fitting is that very different values of surface force constants can be obtained depending on the assumptions made concerning the range and nature of the surface force constants. I will present a numerically efficeint method that utilizes the results of linear-response theory and density functional theory that yields force-contants and phonon dispersion curves that are in excellent agreement with experiment. The computational results demonstrate that the screening at the surface is very complicated and that it is much harder to apply simple models to the surface than to the bulk.

065.

COMPACT AB INITIO WAVEFUNCTIONS FOR EXCITED ELECTRONIC STATES: METHODS APPROPRIATE FOR "ON-THE-FLY" DYNAMICS AND FOR EXCITED STATES PROMOTING MORE THAN ONE ELECTRON. **C. David Sherrill,** Anna I. Krylov, and Martin Head-Gordon, Dept. of Chemistry, University of California, Berkeley, CA, 94720-1460

Current single-reference excited state methods such as configuration interaction with singles (CIS) or equation-of-motion coupled-cluster singles and doubles (EOM-CCSD) are applicable primarily to states which are single excitations from the reference. Furthermore, such approaches assume the presence of a dominant single-determinant reference. Unfortunately neither of these criteria are necessarily satisfied during chemical reactions. We consider the minimal elements of a theory capable of avoiding these assumptions and yet efficient enough to be applicable to large molecules or to "on-the-fly" dynamics methods which would require very large numbers of wavefunction evaluations.

066.

MULTIREFERENCE CISD CHARACTERIZATION OF POTENTIAL ENERGY SURFACES OF SMALL MOLECULES WITH ANALYTIC GRADIENTS: BENCHMARK CALCULATIONS. **G.S. Kedziora** and R. Shepard, Chemistry Division, Argonne National Laboratory, Argonne, IL 60439.

Benchmark calculations that characterize interesting regions of potential energy surfaces of ground and excited electronic states of polyatomic molecules will be presented. Excited-state potential minima and reaction paths on ground and excited state potential energy surfaces usually require multireference methods to obtain accurate and reliable results. The COLUMBUS programs are a suite of multireference electronic structure programs that have the ability to calculate analytic gradients of MRCISD wavefuctions with little extra cost over the energy calculation. We can use these gradients to automatically find stationary points and reaction paths of difficult electronic structure problems of real molecules, including excited states. Comparisons with experimental data and timings will be shown.

067.

NEW METHODS FOR ABINITIO EXCITATION ENERGY CALCULATIONS. AN APPROACH FOR NEAR DEGENERACY PROBLEM OF SINGLE REFERENCE CONFIGURATION INTERACTION METHOD. **Manabu Oumi** David Maurice, Martin Head-Gordon, Department of Chemistry, University of California, Berkeley, CA 94720

We have developed a new theory to approach the problem of CIS (Configuration Interaction Singles) excited states and CIS(D) (CIS with Doubles correction) excitation energy which occurs when CIS results are near degenerate. Since CIS(D) is a non-degenerate perturbation theory for CIS results, when CIS is degenerate it is not formally valid. CIS states which are close to each other are not as reliable as the other states. We have solved this problem by diagonalizing the singles block of the response matrix to second order. This requires an iterative procedure which didn't exist in CIS(D), but the actual iteration doesn't require many cycles when we start with CIS eigenvectors. The computation scales N^5 of the molecular size which is the same as CIS(D), but improved reliability is achieved in cases with near-degeneracies. Several examples are presented.

068. FIXED-SAMPLE OPTIMIZATION USING A PROBABILITY DENSITY FUNCTION. Robert N. Barnett, Zhiwei Sun, and William A. Lester, Jr., Chemical Sciences Division, Lawrence Berkeley Laboratory and Department of Chemistry, University of California, Berkeley, California 94720-1460.

We consider the problem of optimizing parameters in a trial function that is to be used in fixed-node diffusion Monte Carlo calculations. We employ a trial function with a Boys-Handy correlation function and a one-particle basis set of high quality. By employing sample points picked from a positive definite distribution, parameters that determine the nodes of the trial function can be varied without introducing singularities into the optimization. For CH as a test system, we find that a trial function of high quality is obtained and that this trial function yields an improved fixed-node energy. This result sheds light on the important question of how to improve the nodal structure and, thereby, the accuracy of diffusion Monte Carlo. (The work was supported by the Director, Office of Energy Research, Office of Basic Energy Sciences, Chemical Sciences Division of the U. S. Department of Energy under Contract No. DE-AC03-76SF00098.)

069.

MODIFIED AUXILIARY-FIELD MONTE-CARLO: A NOVEL APPROACH FOR ELECTRONIC STRUCTURE. **Naomi Rom**, David Charutz, Roi Baer, and Daniel Neuhauser, Dept. of Chemistry, University of California, Los Angeles, CA 90095-1569.

Auxiliary-Field Monte-Carlo (AFMC) is an approach for calculating structure and dynamics of many-body quantum system with two-body interactions, converting the difficult calculation into an ensemble of simple one-particle problems.

We present a new (exact) modification of AFMC which allows very accurate calculations for molecules, as exemplified in an all-electron calculation of Ne in a 6-31G** basis. The method scales very favorably with the number of electrons, and allows the extraction of excited states.

070. **DOUGLAS-KROLL HAMILTONIAN: A SIMPLIFIED IMPLEMENTATION**
Sutjano Jusuf, Kevin Lan, Jan E. Almlöf
Department of Chemistry, University of Minnesota, Minneapolis, MN 55455, USA

The current method of implementing the Douglas-Kroll (DK) scheme of relativistic electronic structure calculation often result in resource-demanding transformations for the two-electron terms. To avoid these transformations, we proposed expansions of the DK operators with GTOs in momentum space. This approach termed as the Taylor-Series-Equivalence (TSE) can be characterized as fitting the zero-th and first order derivatives at various centers of expansion. For high nuclear charge systems, TSE required an extensive number of GTOs for an accurate expansion of the one-electron terms and thus became less viable in comparison to conventional technique. However, the two-electron terms with no direct dependence on nuclear charge have been shown to have negligible effect in earlier studies. On that basis, It may be adequate to evaluate them with a relatively lower accuracy of TSE. In our study, the TSE approach was not applied to the evaluation of the two-electron terms as several details still needed further investigation.

071. **An MI/FTIR Investigation of Reactions of Tungsten Atoms Produced Using A Pneumatically Powered Mechanical Translator-Rotator for Direct Laser Vaporization with Small Molecules**
Earle G. Stone, Chris Hamby, and Stephan B.H. Bach, Chemistry Program, Division of Earth and Physical Sciences, The University of Texas at San Antonio, San Antonio, Texas 78249-0663

A completely mechanical translator-rotator system has been designed, constructed and proven for use in the direct laser vaporization (DLV) of metals and the study of their reactions with small molecules using Matrix Isolation/ Fourier Transform Infrared Spectrometry (MI/FTIR). The system was proved by duplicating work involving titanium and carbon monoxide previously performed by Lester Andrews *et al*. The translator-rotator has applications wherever a reproducible stream of a material is required using the DLV technique. Key features of the new translator-rotator design are the employment of an air ratchet to provide power for the translator-rotator mechanism, and the elimination of magnetic relays and electrical limit switches through the use of an all mechanical gear and slot mechanism. The reproduced work and preliminary data on tungsten reacted with CO, H_2O, and N_2 in argon matricies will be presented.

072. BRANCHING RATIOS OF THE NH_2 + NO_2 REACTION, Ned F. Lindholm and John F. Hershberger, Department of Chemistry, North Dakota State University, Fargo, ND 58105

The NH_2 + NO_2 reaction is an important step in the thermal de-NO_x process, and has several possible exothermic channels. We have measured the product branching ratio of this reaction at room temperature. NH_2 radicals were formed by photolysis of ICN followed by the CN + NH_3 reaction. NO and N_2O products were detected using time-resolved infrared laser spectroscopy. NO + H_2NO was found to be the dominant product channel, with a branching ratio of 0.73±0.15. N_2O + H_2O is a comparatively minor channel with a branching ratio of 0.27±0.1. To measure the OH yield, CO was added in excess. No CO_2 from the reaction OH+CO→CO_2 + H was detected, indicating the the 2OH + N_2 product channel is not significant.

073. DEPLETION KINETICS OF NIOBIUM ATOMS IN THE GAS PHASE. Roy E. McClean,[*] Mark L. Campbell, and Erica J. Kölsch, Chemistry Department, United States Naval Academy, Annapolis, MD 21402

The gas phase depletion kinetics of Nb(a^6D_J, a^4F_J) in the presence of O_2, SO_2, CO_2, N_2O, and NO are reported. Niobium atoms were produced by the 248 nm photodissociation of $Nb(C_5H_5)(CO)_4$, and detected by laser-induced fluorescence. The ground term of Nb ($4d^45s^1$ a^6D_J) reacts at or above the collision rate with all of the aforementioned oxidants. The first excited term, Nb($4d^35s^2$ a^4F_J), is not as reactive with these oxidants. Results are interpreted in terms of long range attractions and valence interactions.

074. INTERNAL ENERGY DISTRIBUTIONS OF SO AND CS PRODUCTS FROM O(^{3}P) + CS_2 REACTIONS. Xirong Chen, Yuying Cheng, Jiande Han and Brad R. Weiner, Dept. of Chemistry, University of Puerto Rico, P.O. Box 23346, San Juan, P.R. 00931

The rotational and vibrational population distributions of both CS and SO products from O(^{3}P) + CS_2 reaction have been measured. The O(^{3}P) atoms are generated by 355 nm photolysis of NO_2. The SO($X^3\Sigma^-$,v"=0 - 5) and CS($X^1\Sigma^+$,v"=0 - 3) were observed directly by using laser induced fluorescence spectroscopy on the ($B^3\Sigma^-$ - $X^3\Sigma^-$) and ($A^1\Pi$ - $X^1\Sigma^+$) transitions respectively. The experimental vibrational state distributions of SO and CS were compared with those calculated by a Franck-Condon model. An energy balance analysis including the translational energy and the internal energies of SO and CS products account for nearly all (>90%) of the available energy. These energy disposal results can be explained in terms of a direct S atom abstraction mechanism.

075. KINETIC STUDY OF THE REACTION OF Ta($a^4F_{3/2}$) WITH O_2, CO_2, N_2O, NO and SO_2 FROM 298 to 573 K. Mark L. Campbell and Kelli L. Hooper, Chemistry Department, United States Naval Academy, Annapolis, MD 21402

The gas phase reactivities of the ground state of tantalum ($a^4F_{3/2}$) with O_2, CO_2, N_2O, NO and SO_2 from 298 - 573 K are reported. Tantalum atoms were produced by the photodissociation of tantalum cyclopentadiene tetracarbonyl and detected by LIF. The bimolecular rate constants for the reactions of tantalum with O_2, CO_2 and N_2O were temperature dependent and are described in Arrhenius form by $(1.7\pm0.1) \times 10^{-10}$ exp(-7.8 ± 0.2 kJ/mole/RT), $(7.1\pm0.5) \times 10^{-11}$ exp(-13.6 ± 0.3 kJ/mole/RT), and $(1.3\pm0.1) \times 10^{-10}$ exp(-27.6 ± 0.5 kJ/mole/RT) cm^3s^{-1}, respectively, where the uncertainties are $\pm\sigma$. The rate constants for the reactions of Ta($a^4F_{3/2}$) with NO and SO_2 were fairly temperature insensitive. The room temperature rate constants are 5.7×10^{-11} and 6.1×10^{-11} cm^3s^{-1} for NO and SO_2, respectively. The disappearance rates in the presence of all the oxidants are independent of total pressure indicating termolecular processes are unimportant.

076. KINETICS OF CN AND NCO REACTIONS WITH MAIN GROUP HYDRIDES. Michael A. Edwards and John F. Hershberger, Department of Chemistry, North Dakota State University, Fargo, ND 58105

The gas phase kinetics of CN and NCO reactions with heavier main group hydrides were studied using the flash photolysis/infrared absorption technique. CN radicals were produced by 248 nm excimer laser photolysis of ICN precursor molecules and thermalized by SF_6 buffer gas. NCO radicals were created by the reaction of CN with molecular oxygen. CN and NCO were then detected by time-resolved infrared absorption spectroscopy using tunable diode lasers. Rate constants for reactions of CN and NCO with SiH_4 and GeH_4 were measured under pseudo-first order conditions. Information on product channels will also be presented.

077. KINETICS, DYNAMICS AND THEORETICAL
STUDIES OF THE Al + CS$_2$ AND Al + OCS REACTIONS
Miriam Pérez, Yasuyuki Ishikawa and Brad R. Weiner
Department of Chemistry, University of Puerto Rico, Rio Piedras PR 00931

The temperature (295-385 K) and pressure (5-50 torr) dependence of the rate constants of the Al + OCS and Al + CS$_2$ reactions have been measured by time-resolved laser induced fluorescence spectroscopy, by monitoring the disappearance of the Al atoms. The rate expression for both of the reactions was determined. The former reaction was found to be independent on total pressure, while the rate constant for Al + CS$_2$ shows a strong total pressure dependence suggesting the formation of an AlCS$_2$ collision complex. The vibrational and rotational state distributions of the AlS product were measured.

Ab initio correlated calculations have been performed to determine the structures of the Al-CS$_2$ and Al-OCS collision complexes. The Gaussian-2 theoretical procedure, based on the *ab initio* correlated methods, is applied to the reactants, complexes and the products in order to investigate their relative stability.

078. **MATRIX ISOLATION STUDY OF THE REACTIONS OF CHROMYL CHLORIDE: INTERMEDIATES IN THE REACTION OF CrCl$_2$O$_2$ WITH CH$_3$OH**, Bruce S. Ault, Department of Chemistry, University of Cincinnati, Cincinnati, OH 45221

Pyrolytic merged jet deposition into argon matrices at 14 K has been employed to examine the gas phase reaction of CrCl$_2$O$_2$ with methanol and small hydrocarbons. With the reaction zone held between room temperature and 100°C, a very reactive intermediate with absorptions near 670, 950, 1150 and 1425 cm^{-1} was identified. The spectra indicate that this species has an intact CH$_3$O group and appears to be bonded to or interacting with one of the oxygens of CrCl$_2$O$_2$. Above 100°C, these bands diminished in intensity, and were completely destroyed by 175°C. In their place, strong bands due to CH$_2$O and HCl grew in. At yet higher temperatures, these bands were reduced and then destroyed, and strong bands due to CO and CO$_2$ were seen. The use of isotopic labeling allowed the tracking of the carbon and oxygen atoms of labeled CH$_3$OH through the reaction scheme and provided evidence as to the mechanism of this oxidation reaction.

079. MEASUREMENT OF THE DIFFERENTIAL CROSS SECTION FOR THE H + D$_2$(v=0, J) $\rightarrow$ HD(v'=4, J'=3) + D REACTION AT E$_{col}$ = 2.2 ± 0.1 eV

Hao Xu, Brian D. Bean, Felix Fernandez-Alonso, Neil E. Shafer-Ray, Russell J. Low, and Richard N. Zare
Department of Chemistry, Stanford University Stanford CA, 94305 USA

A 95:5 mixture of D$_2$ and HI is expanded into a vacuum chamber. Fast H atoms are generated by the photolysis of HI at 212.97 nm, and are allowed to react with D$_2$ for 20 ns. The resulting HD (v'=4, J'=3) product is detected via the Rydberg-tagging time-of-flight spectroscopy. The time-of-flight (TOF) profile can be correlated to the differential cross section using the laws of cosines. The instrument is calibrated with the HD (v'=4, J'=3) photoproduct of the HDS photolysis at 130.6 nm. The resulting differential cross section shows structures unaccounted for in the QCT calculations by Aoiz et al, while matches qualitatively with the Quantum Mechanical calculations by Wu and Kuppermann. Several experimental advances have to be made to make this measurement possible. Those advances include: (1) the use of molecular-Rydberg-tagging, and(2) the removal of photoelectrons by a small bias voltage to suppress the devastating electron-impacted ionization background.

080.
QUANTUM MECNAHICAL THERMAL RATE CONSTANT CALCULATION FOR CL + H2 -> CLH + H REACTION BASED ON FLUX-FLUX AUTOCORRELATION FUNCTION FORMALISM .
Haobin Wang, Ward H. Thompson, and William H. Miller, Department of Chemistry, University of California, and Chemical Science Devision, Lawrence Berkeley National Laboratory, Berkeley, CA, 94720

Recent development of direct (i.e. without first solving the state-to-state reactive scattering problem) thermal rate constant calculation based on flux-flux autocorrelation formalism is applied to calculate the rate constants for Cl + H2 -> ClH + H reaction. The key features in this method are recognition of the low rank of Boltzmannized flux operator and the use of absorbing boundary condition. We calculate thermal rate constants ranging from 200 K to 1500 K, using the J-shifting and Centrifugal Sudden approximations to include the nonzero total angular momentum. The results are compared with experiment and previous theoretical calculations.

081. PYROLYTIC REACTIONS OF HEXAFLUOROPROPYLENE OXIDE. David M. Jollie and Philip G. Harrison, Department of Chemistry, University of Nottingham, University Park, Nottingham, U.K., NG7 2RD.

The mechanism for the pyrolytic decomposition of hexafluoropropylene oxide is well known. Component reactions comprise an initial step of decomposition to difluoromethylene, CF_2, and trifluoroacetyl fluoride accompanied by further secondary reactions of these primary products. However, the rate constants and activation energy for this reaction have not been previously determined. In this paper the simple pyrolysis of hexafluoropropylene oxide is studied by in situ gas phase infrared spectroscopy, and rate constants are determined for the first order reaction. An activation energy is also calculated from the Arrhenius equation. The pyrolysis of hexafluoropropylene oxide under atmospheres of bromine, oxygen and carbon dioxide, respectively, is also studied in order to determine the spin state of the thermally produced difluoromethylene from the products observed.

082.
SIFT STUDIES OF ISOMERIC PRODUCTS OF ION-MOLECULE ASSOCIATION REACTIONS. **Neyoka D. Fisher,** Nigel G. Adams, Department of Chemistry, University of Georgia, Athens, GA 30602
A Selected Ion Flow Tube (SIFT) has been used to study association reactions of interstellar importance and to determine the isomeric identities of association product ions generated in different ways. This also yields information about the reaction mechanisms and probes the potential energy surfaces. The identities of the ions were determined from their reactivities with a series of neutral species. The systems studied were 1) CH_5O^+, generated from CH_3^+ with H_2O, and protonated methanol, along with 2) $CH_3O_2^+$, generated from CH_3^+ with O_2, and protonated formic acid; the latter study is compared with $CH_3O_2^+$ generated in the $O_2^+ + CH_4$ reaction[1]. The $C_2H_5O_2^+$ system is under investigation. Rate coefficient and product ion distributions were determined to investigate the effect of structure on reactivity.
1. Van Doren, J.M.; Barlow, S.E.; DePuy, C.H.; Bierbaum, V.M.; Dotan, I; Ferguson, E.E. *J. Phys. Chem.* **90**, *1986*, 2772.

083.

SEMICLASSICAL TRANSITION STATE THEORY CALCULATIONS OF THERMAL RATE CONSTANTS FOR THE OH + H2 --> H20 + H REACTION. **Kathy L. Sorge** and William H. Miller, Dept. of Chemistry, University of California, Berkeley, and Chemical Sciences Division, Lawrence National Berkeley Laboratory, Berkeley, California 94720

Using two variations of semiclassical transition state theory, the thermal rate constants are calculated. The first variation consists of changing the variable of integration from energy to the barrier penetration integral, theta. This elimates the need to invert the relationship between energy and theta. The second variation relies on the fact that the energy is quadratic in theta. Thus, the semiclassical transmission coefficient is isomorphic to that of the Eckart barrier, and can be replaced with the exact quantum mechanical transmission coefficient for the Eckart barrier. Rate constants are calculated using the above methods and are in good agreement with experiment.

084. SEPARABLE BASIS ANALYSIS OF THE THREE-BODY PROBLEM AS APPLIED TO THE O+HCl--OH+Cl REACTION, Bill Poirier and William H. Miller, Dept. of Chemistry, University of California, and Chemical Sciences Division, Lawrence National Berkeley Laboratory, Berkeley California 94720

The evaluation of quantum scattering quantities for three-body reactive systems is explored in conjunction with the optimal separable basis methodology, which is utilized in two different ways. First, results are obtained for the zero angular momentum case using a three-dimensional DVR Hamiltonian and optimized preconditioning. For the higher values of J, the helicity-conserving and J-shifting approximations are employed, after having first minimized the coriolis coupling via another application of the optimal basis method. Fixed-energy cumulative reaction probabilities and thermal rate constants are presented for the O+HCl reactive scattering system.

085. O ATOM ADDITION TO ALKYNES. S. Jonnalagadda, C. Clohessy, J. Garrido, S. Chan and K. A. Singmaster, Department of Chemistry, San Jose State University, San Jose, California 95192-0101

O atom addition to alkynes has been investigated in a cryogenic matrix environment. We initiated our studies with two simple alkynes, hexafluoro-2-butyne and dimethylacetylene. Matrices containing the alkyne of interest and ozone, our O atom source, are photolyzed with a variety of sources to produce different excited states of the O atom. 642 nm photolysis of $C_4F_6/O_3/Ar$ matrix resulted in the formation of bis(trifluoromethyl)ketene, perfluoro-2,3-butanedione and bis(trifluoromethyl)oxirene. The oxirene was shown to photodecompose with 420 nm light to produce the ketene. Although the oxirenes isomeric ketocarbene was not directly observed, it was trapped as a ketoketene when photolysis was performed on a matrix containing CO. Dilution studies suggest that the butanedione is also a trapping product of the ketocarbene (two O atom addition). Studies on dimethylacetylene have just been initiated. So far we have been able to identify the final products of the reaction. These are dimethylketene, 2,3-butanedione and methyl vinyl ketone. This work is supported by the Petroleum Research Fund and NIH-MBRS Program.

086. PHOTOCHEMISTRY OF 4-METHYLENE-1,3-DIOXOLAN-2-ONE. M. Coughlan, V. Maraschin and K. A. Singmaster, Department of Chemistry, San Jose State University, San Jose, California 95192-0101

Cryogenic matrices serve as an excellent environment for trapping and detecting photolysis products of a broad spectrum of compounds. Pursuing our interest in small cyclic molecules we speculated that 4-methylene-1,3-dioxolan-2-one could serve as a precursor for allene oxide, a molecule first detected as an intermediate in the addition of O atoms to allene. The dioxolanone was synthesized as documented by Trost and Chan (*J.Org.Chem.* 1983). A matrix containing the molecule was photolyzed using a 75 W Xe arc lamp. When the dioxolanone was exposed to full arc radiation CO_2, CO, ethylene, formaldehyde and ketene were observed as the photoproducts. When 200 nm light was used the main products observed were CO_2, ketene and cyclopropanone. Cyclopropanone was identified by comparison to previous spectra and by it's photochemistry. Photolysis of cyclopropanone with 260 - 380 nm light resulted in the formation of CO and ethylene. Unfortunately allene oxide could not be detected. This work is supported by the Petroleum Research Fund and NIH-MARC Program.

087. PHOTOCHEMISTRY OF THE TETRAMETHYLETHYLENE-OZONE CHARGE-TRANSFER COMPLEX. S. Chan, V. Maraschin and K. A. Singmaster, Department of Chemistry, San Jose State University, San Jose, California 95192-0101

Charge transfer complexes between alkenes and ozone have only been detected in cryogenic matrices. As expected, the charge transfer band maximum is dependent on the alkene's ionization potential. Most of these complexes exhibit photochemistry, however, due to ozone's rich electronic spectra, it is difficult to distinguish between photoproducts of the CT complex and photoproducts resulting from ozone photolysis. The tetramethylethylene-ozone CT complex is the simplest complex for which the CT band maximum (at 500 nm) does not overlap an ozone absorption. We have investigated the photochemistry of this complex. Photolysis of the complex with very low intensity 514 nm light resulted in rapid loss of the complex and growth of three primary photoproducts. These have been identified as 2,2,3,3-tetramethyloxirane, 3,3-dimethyl-2-butanone and isopropenyl isopropyl ether. This chemistry is identical to the chemistry observed in the addition of $O(^3P)$ to tetramethylethylene. This work is supported by the Petroleum Research Fund.

088.

PHOTODISSOCIATION DYNAMICS OF THIOPHOSGENE AT 193 AND 248NM. **Harry Gomez** Yasmín Pedrogo, K. Ravichandran and Brad R. Weiner Department of Chemistry, University of Puerto Rico, San Juan, P.R. 00931

The photodissociation dynamics of thiophosgene ($CSCl_2$) at 248 and 193nm have been studied using laser induced fluorescence spectroscopy. The nascent CS ($X^1 \Sigma^+$) rovibrational state distributions were measured by probing the ($A^1 \Pi - X^1 \Sigma^+$) transition for both photolysis wavelengths. Results were compared to a three body impulsive model which suggests a mechanism via a CSCl intermediate consistent with previous studies.

089.

PHOTODISSOCIATION OF KETENE: DISTRIBUTIONS AND RATE CONSTANTS. **E. A. Wade**, A. Mellinger, H. Clauberg, M.A. Hall, and C.B. Moore, Department of Chemistry, U.C. Berkeley, Berkeley, CA 94720

Abstract

Ketene (CH_2CO) was photolyzed in a supersonic jet by a supersonic jet by a tunable pulsed laser, and the vibrationally excited carbon monoxide (CO(v=1)) was detected by vacuum ultraviolet (VUV) laser-induced fluorescence (LIF). Ketene can dissociate along two channels, to produce singlet and triplet methylene ($CH_2(^1A_1)$ and $CH_2(^3B_1)$). Rotational distributions were obtained at several energies up to 1720 cm^{-1} over the threshold for production of singlet methylene and CO(v=1), and used to determine the singlet yield. Up to 500 cm-1 over the same threshold, vibrational branching ratios for CO(v=1) and CO(v=1) were also determined. The rotational and vibrational distributions and the singlet yield were then used to determine rate constants for dissociation along the singlet surface. These measurements will be discussed and compared with several statistical models, including phase space theory (PST), separate statistical ensembles (SSE) and variational RRKM.

090. THE PHOTODISSOCIATION OF $Co(CO)_3NO$ AT 290 NM: A QUESTION OF ENERGY DISPOSAL. Aaron P. Rybar, Yu-Liang Hsiao, Emily L. Reichert, Jeb L. Adams, Eric E. Niederkofler, Stacy M. Coppin, and Jeffrey A. Bartz, Department of Chemistry, University of Redlands, Redlands, CA 92373-0999

It is well-established that multiphoton photodissociation of metal carbonyl compounds results in the ejection of the carbonyl ligands while the metal atom emerges in an excited state. The excited metal atom fluoresces and the specific frequencies emitted suggest photodissociation pathways as well as energy disposal into the photoproducts. In the photolysis of tricarbonylnitrosylcobalt, $Co(CO)_3NO$, at 290 nm, the frequencies of dispersed metal atom fluorescence reveal that a photoproduct other than the metal atom carries away a significant amount of energy. These results will be discussed as well as an analysis of the metal atom fluorescence for evidence of statistical energy redistribution.

091. PHOTOGENERATION AND SPECTROSCOPY OF OZONE MONOMERS AND DIMERS IN OXYGEN MATRICES. **C.G. BRESSLER,** M.J. Dyer, and R.A. Copeland, Molecular Physics Laboratory, SRI International, Menlo Park, CA 94025

Irradiation of solid oxygen (12 K) with ultraviolet laser light (210-250 nm) produces ozone monomers and dimers in specific sites in the matrix analyzed by means of Fourier-transform infrared spectroscopy. Ozone species are produced with greater concentration as the wavelength is decreased in this range. Interestingly, ozone is even generated when the photon energy is insufficient to dissociate an isolated oxygen molecule. Dose dependant growth characteristics give initial clues about the nature of these species. By varying the temperature and isotopic composition we examine the formation mechanics, structure and stability of the photoenergized matrix. Additional studies on ozone doped oxygen matrices add to our understanding of the ozone monomer and dimer generation in oxygen matrices.

092.

RATES AND PATHWAYS FOR COLLISIONAL PROCESSES INVOLVING THE HERZBERG STATES OF OXYGEN. **E. S. Hwang** T.P. Shiau, B. Buijsse, and R. A. Copeland, Molecular Physics Laboratory, SRI International, Menlo Park, California 94025

The emissions from the oxygen Herzberg states (A, A', c) and lower-lying electronic states (b, a) are important observables in the earth's nightglow. These electronic states are generated in the upper atmosphere by the three body recombination of two oxygen atoms. Understanding the details of the collisional processes involving these states is important in modeling this emission. At SRI, we have performed several experiments using direct excitation with an ultraviolet laser pulse. We have measured the collisional removal rate constants of the oxygen A state for the atmosphericaly important colliders at mesospheric temperatures (150 - 225 K). In addition, the vibrational-level-dependence of b-state production following A-state excitation is characterized and possible reactions producing ozone and nitrous oxide are studied.

093. ROTATIONAL STATE DEPENDENCE OF ELECTRONIC QUENCHING OF A $^2\Sigma^+$ OH. Brooke L. Hemming, Joel E. Harrington, Volker Sick* and David R. Crosley, Molecular Physics Laboratory, SRI International, Menlo Park, CA 94025 (*University of Heidelberg, Germany)

Using laser induced fluorescence, we studied the role of rotational energy in the quenching of electronically excited OH by several collider molecules, with particular emphasis on high rotational energies. The colliders studied include O_2, N_2, H_2, Xe and Kr. Photolysis of HNO_3 by an ArF excimer laser (193 nm) produces rotationally-excited X $^2\Pi$ OH, which made possible state-specific measurement of the quenching rates of rotational levels as high as N' = 20. The same trend was observed for all colliders: a quenching cross-section that declines until approximately N'=10, after which it plateaus. For example, O_2 quenches N' = 1 OH with a cross-section of 21 Å^2, while at N' = 10 and thereafter, its quenching cross-section is about 7 Å^2. The decline in quenching cross-section can be attributed to an anisotropic-attractive potential surface.

094.

SEMICLASSICAL INITIAL VALUE REPRESENTATION FOR ELECTRONICALLY NON-ADIABATIC MOLECULAR DYNAMICS. **XIONG SUN,** W.H. MILLER, DEPARTMENT OF CHEMISTRY, UNIVERSITY OF CALIFORNIA, BERKELEY, CA, 94720.

The semiclassical initial value representation (SC-IVR), which has recently seen a great deal of interest for treating nuclear dynamics on a single potential energy surface, is generalized to be able to describe electronically non-adiabatic (i.e. multi-surface) processes. The essential idea is a quantization of the classical electron-nuclear Hamiltonian of Meyer and Miller [J. Chem. Phys. 70, 3214 (1979)] within the SC-IVR methodology. Application of the approach to a series of test problems suggested by Tully shows it to provide a good description of electronically non-adiabatic dynamics for a variety of situations.

095. SUBSTITUENT EFFECTS IN THE RING-OPENING REACTION OF TRIAZIRIDINE. Carl Salter, Shirin Arastu, Chemistry Department, Moravian College, Bethlehem, PA 18018.

Triaziridine, though it appears to be a saturated nitrogen ring compound isoelectronic with cyclopropane, actually undergoes conrotatory electrocyclic ring-opening reactions like those of cyclobutene. These reactions are the first examples of Woodward-Hoffmann electrocyclic reactions in completely inorganic rings. Like the substituted cyclobutenes which have been studied by Houk, we expect substituted triaziridines to tell us about electron demand in electrocyclic ring-openings. We present the results of preliminary investigations of the ring-opening reaction of hydroxy triaziridine and methyl triaziridine, including transition states found at SCF/6-31G** using GAUSSIAN.

096.

ULTRAFAST PHOTO-INDUCED DYNAMICS IN TRANSITION METAL COMPLEXES: EFFECTS OF ELECTRONIC STRUCTURE AND SOLVATION ON FRANCK-CONDON STATE EVOLUTION. **James K. McCusker,*** Alvin Yeh, Niels H. Damrauer, and Charles V. Shank, Department of Chemistry, University of California at Berkeley, California, 94720-1460, and the Material Sciences Division, Lawrence Berkeley National Laboratory, Berkeley, California 94720

Ruthenium polypyridyl complexes, for which [Ru(bpy)3]2+ is the prototype, represents one of the most widely studied classes of molecules in the inorganic photophysical community. We have used this compound and related derivatives as a template for studying Franck-Condon state evolution in transition metal complexes via femtosecond time-resolved absorption spectroscopy. Data that reveal dynamics associated with wavepacket motion from the initially formed excited state in these molecules as a function of both intramolecular and extramolecular effects will be described. Results from variable wavelength excitation studies directed at understanding the effect of excess vibrational energy on excited-state evolution will also be discussed.

097.

QUANTUM NUMBER DEPENDENCE OF TUNNELING PROBABILITIES IN HIGHLY EXCITED STATES OF SEVERAL SIMPLE QUANTUM MECHANICAL SYSTEMS. **James Diamond,** Darsi Adams, and Amy Germaine, Chemistry Department, Linfield College, 900 S.E. Baker S., McMinnville, OR 97128-6894

We are interested in the examining the Bohr Correspondence Principle. One formulation of this principle is that the classical result for a mechanical system will be obtained from the quantum mechanical result in the limit of large quantum numbers. We have investigated several simple quantum mechanical model systems in order to compare the behavior of the quantum systems with their classical counterparts. In particular, we have studied the probability of locating the system within the classically forbidden region. We have attempted to establish the asymptotic manner in which these tunneling probabilities approach the classical limit as the quantum number n becomes large and to determine the influence of other physical parameters (such as angular momentum or barrier strength). We will present results for the simple square well, the harmonic oscillator, and the hydrogen atom. This provides an opportunity for a deeper examination of the Bohr correspondence principle.

097A PROMOTING MULTISCALE COMPUTATIONAL SCIENCE AND ENGINEERING PROJECTS ON THE CORNELL THEORY CENTER'S IBM SP. **Marcy E. Rosenkrantz,** Cornell Theory Center, Cornell University, Ithaca, NY

We have implemented a new "Strategic Applications" program designed to foster solutions of complex computational problems and to help us develop and integrate a coherent infrastructure for scientific computing. Our goal is to enable new breakthroughs in important areas of science which require integration of widely diverse computational methods and models to address phenomena over a wide range of length and time scales; manipulation, storage, retrieval, display, and analysis of massive data sets; development of software useable and reuseable within and across problem domains; integration of computer simulation into the larger process of scientific research. Examples of areas that could benefit include: fracture mechanics, structural biology, bio- and earthquake engineering, thin-film growth, computational fluid dynamics, geographical information systems, pulsar astronomy, epidemiolgy, and urban studies among many others.

098.

DETERMINATION OF CONFORMATIONAL DISORDER IN THE FLUID PHASE LIPID BILAYERS BY MOLECULAR DYNAMICS SIMULATION. **W.-J. Sun#**, K.-C. Tu*, D. J. Tobias*, M. L. Klein*, H. L. Strauss#, and R. G. Snyder#, #Department of Chemistry, University of California, Berkeley, CA 94720, *Department of Chemistry, University of Pennsylvania, Philadelphia, PA 19104

The influence of packing constraints on the conformation of the chains in fluid phase DPPC bilayers has been estimated from molecular dynamics (MD) simulation (Primary results published in Tu, Tobias, and Klein, Biophys. J. 69, 2558(1995)). The estimate is based on a comparison between the MD-derived conformational statistics of the fluid bilayer and similarly derived statistics for unconstrained n-hexadecane. The gauche concentration in the bilayer chains was found to be about 30% lower than in the n-alkane. In the DPPC bilayers, the gauche concentration increases from about 20% at the middle of the chains to about 30% at the methyl end. As expected, the two chains of the lipid are conformationally indistinguishable. The calculated gauche distribution will be described and related to previous experimental values.

099. FINDING REACTION PATHWAYS OF COMPLEX SYSTEMS. Félix S. Csajka, Jordi Marti and David Chandler. Department of Chemistry, University of California at Berkeley, Berkeley, CA 94720

We introduce a stochastic method to sample trajectories in complex systems. The method connects two boundary regions (for example a reactant and a product region) by a set of directed paths in space-time. We construct an action for directed paths and sample reaction pathways using Monte Carlo algorithms for polymer systems. We apply the directed path method to elementary dynamical processes in water. We study hydrogen-bond breaking in ambient water; we obtain the profiles of structure and energy averages along the trajectory. The role of the surrounding water molecules is traced through a set of specific pair distribution functions. We also present results for the dissociation of an ion pair: we study the evolution of a sodium chloride ion pair from contact to solvent-separated configurations.

100. INFORMATION THEORY AND THE HARD SPHERE FLUID.

Gavin E. Crooks and David Chandler
Department of Chemistry,
University of California, Berkeley, California 94720

A computer simulation of the hard sphere fluid was used to measure the probabilities of observing N molecular centers within molecular sized volumes of the fluid. These probability distributions are found to be almost exactly gaussian at medium densities. A maximum entropy prediction of the overall occupation distribution was constructed from knowledge of the first two moments of the distribution. These moments can, in principle, be experimentally measured for physical systems. After correcting for deviations from gaussian behavior at low densities we can quantitatively predict the occupation distribution for low and medium densities.

101. KINETICS & MECHANISM OF OXIDATION OF ORGANIC COMPOUNDS BY $Os\ O_4$ IN DIFFERENT MEDIA. STRUCTURE REACTIVITY CORRELATION. P.VEERA SOMAIAH, POST-GRADUATE SCHOOL IN APPLIED CHEMISTRY, OSMANIA UNIVERSITY, MIRZAPUR - 502 249, A.P., INDIA.

The use of transition metal ions either alone or as binary mitures as homogeneous Catalysts in the oxidation of several organic compounds by various oxidants is of recent interest. $Os\ O_4$ alone act as a better oxidant in the oxidation of various organic compounds in different media. To see this, oxidation of various organic substances with different functional groups were studied both in alkaline and acid medium. The nature of reactive species of Os (VIII) in both medium are proposed from the experimental results and also confirmed with the spectral evidences. Possible mechanisms are proposed for the oxidation process by calculating reaction constants (ρ^*) from Taft's Plot.

102. MEASUREMENTS OF DIFFUSION COEFFICIENTS IN SULFURIC ACID SOLUTIONS. J. K. Klassen, Z. Hu, L. R. Williams, Molecular Physics Laboratory, SRI International, Menlo Park, CA 94025

Diffusion coefficients in sulfuric acid play a critical role in many stratospheric processes including calculating nucleation rates in sulfate aerosols and determining whether heterogeneous reactions occur in the bulk or on the surface of aerosols. In order to provide quantitatively accurate diffusion coefficients, the experimental challenges of measuring solute diffusion in strong electrolyte environments must be overcome. We use a modified diaphragm cell technique and report diffusion coefficients of 1.1×10^{-5} cm^2/s for HCl in 30 wt % sulfuric acid and 4.2×10^{-6} cm^2/s for HCl in 60 wt % sulfuric acid at 293 K. These values are 20% lower than diffusion coefficients estimated from the diffusion coefficient of HCl in water adjusted for sulfuric acid's viscosity dependence. However, we find that the temperature and viscosity dependence of these diffusion coefficients agree quite well with the Stokes-Einstein equation (D=cT/η) yielding c = 7.8×10^{-8} cm^2 cP/s K over the 220-293K temperature range.

103. MOBILITY AND SOLVATION OF IONS IN CHANNELS: Ruth Lynden-Bell, Atomistic Simulation Group, School of Mathematics and Physics, The Queen's University, Belfast BT7 INN, U.K. and Jayendran C. Rasaiah, Department of Chemistry, University of Maine, Orono, ME 04469.

We present results from a simulation of the mobility and solvation of ions and uncharged molecules in aqueous solution in smooth cylindrical channels of radii between 2.5 and 5.5 Å at room temperature. The water molecules in channels form a cylindrical solvation shell inside the channel walls with some evidence of a second shell in the center of the largest channel. When a sodium ion is added it tends to lie in the center of the channel where it can form the most complete solvation shell. Its diffusion rate decreases in smaller channels until it moves too slowly to be detected in a channel of 2Å radius. A range of ions of different sizes were studied in a channel with radius 3Å. While the smaller of these ions (F$^-$, Na$^+$ and Ca^{++}) lie preferentially in the center of the channel, larger ions (Cl$^-$ and Cs$^+$) penetrate some way into the sheath of water inside the wall and methane and ions with the charges turned off move next to the wall. A Landau free energy analysis shows that this change is due to a change in the balance between entropy and energy.

104. NONLINEAR CHEMICAL DYNAMICS IN THE HOMOGENEOUS SYSTEM OF IODATE OR BROMATE WITH SULFITE-CARBONATE-SULFURIC ACID IN A CSTR. Glen A. Frerichs, Department of Chemistry, Westminster College, Fulton, MO 65251; Edward W. Richmond and Richard C. Thompson, Department of Chemistry, University of Missouri-Columbia, Columbia, MO 65211

Nonlinear kinetic behavior has been observed in the KIO$_3$-Na$_2$SO$_3$-Na$_2$CO$_3$-H$_2$SO$_4$ and NaBrO$_3$-Na$_2$SO$_3$-Na$_2$CO$_3$-H$_2$SO$_4$ systems using a continuous-flow stirred tank reactor. Regular, small-amplitude oscillations in the potential of a platinum electrode were obtained for the iodate system. Large-amplitude oscillations in both the potential and pH were found for the bromate system. Proposed models for the systems will be discussed. The observed bistability between oscillations and the low pH steady state can be compared to that predicted (but not observed) by others for iodate or bromate with sulfite and ferrocyanide in sulfuric acid.

105. RELAXATION CONSTANTS AND MECHANISMS OF HYDROGEN BOND DYNAMICS IN WATER AT LOW TEMPERATURES, Alenka Luzar, Hilgard Hall -3110, University of California at Berkeley, CA 94720-3110

We study the statistical evolution of the formation and rupture of hydrogen bonds in water below ambient temperature including supercooled regime. In particular, we use molecular dynamics simulations to calculate hydrogen bond reactive flux correlation functions for a variety of specific conditions, and identify rate constants for breaking and making hydrogen bonds. By studying temperature dependence of the reactive flux functions we estimate activation energy and activation entropy at the apparent transition state. Our calculations reproduce Arrhenius behavior of hydrogen bond lifetimes observed in depolarized Rayleigh scattering experiment. This behavior suggests that there is no significant cooperativity between different hydrogen bonds. We determine the extent of correlations with the fluctuations of neighboring bonds. We show that different (temperature dependent) environment affects statistical factors while it does not influence the nature of the relaxation that is governed by the interplay between hydrogen bond dynamics and diffusion.

106. SOLVATION DYNAMICS AND ION MOBILITIES IN AQUEOUS SOLUTIONS
S. Koneshan and Jayendran C Rasaiah, Department of Chemistry, University of Maine, Orono, ME.04469 and Song Hi Lee, Department of Chemistry, Kyung Sung University, Pusan, Korea.

We describe molecular dynamics simulation of the metal cations (Li^+, Na^+, K^+, Rb^+, Cs^+, Ca^{2+}) and the halides (F^-, Cl^-, Br^-, I^-) at infinite dilution using the SPC/E model for water at 25^oC. The same simulation is also described for CH_4 and the uncharged species (Li, Na, K, Rb, Cs, Ca, Br) obtained by discharging the corresponding ions. Mobilities calculated from the mean square displacement and the Einstein relation show the same trends with ion size and charge as the experimental results which lie on separate curves as a function of the radii of cations and anions. When the charges are turned off the diffusion coefficients of Li, Na, K, Rb and F increase above the corresponding results for the charged ions, whereas for Cl, Br and I, it decreases below the values of the anions. For Cs^-, the diffusion coefficient remains unchanged on discharging the ion. These results can be explained by assuming, following Frank and Gurney, that the anions (except F^-) break water structure and the cations have the opposite effect. The solvation dynamics of cations and anions are different.

107. SOLVENT EFFECTS ON THE SOLVATION SHELL EXCHANGE KINETICS Tsun-Mei Chang and Liem X. Dang, Environmental Molecular Sciences Laboratory, Pacific Northwest National Laboratory,* Richland, WA 99352

Classical molecular dynamics simulations are carried out to investigate the solvation shell exchange kinetics of potassium ion in bulk water, chloroform, and carbon tetrachloride. The reactive flux method is used to compute the dissociation rates. The influence of the solvent on the reaction rates and kinetic properties of this exchange process will be discussed. Comparison will be also be made to the prediction of the Grote-Hynes theory.

*This work was performed under the auspices of the Division of Chemical Sciences, Office of Basic Energy Sciences, US Department of Energy under Contract DE-AC06-76RLO 1830 with Battelle Memorial Institute, which operates the Pacific Northwest National Laboratory, a multiprogram national laboratory.

108. STUDIES ON ION PAIR EQUILIBRIA : CONDUCTANCES OF AQUEOUS METHANOLIC SOLUTIONS OF SILVER CHLORATE.
D. RAJESHWAR REDDY AND V.MANAIAH*
DEPARTMENT OF CHEMISTRY, NIZAM COLLEGE, OSMANIA UNIVERSITY, HYDERABAD - 500 001, (A.P.), INDIA.

Molar conductances of silver chlorate measured at 25^oC in water, 10,20,40 & 60% (w/w) methanol-water mixtures were analysed using Fuoss-Hsia equation for thermodynamic association constants (K_A) and conductances at infinite dilution (Λo). From a consideration of the functional dependance of log K_A on $1/D$, to test the validity of various theories of Ionpair formation, and ionsize parameter (ISP) values, the modified Dension-Ramsey theory in the form $K_A = K_A^o \exp(b)$ was found to be applicable to describe the ionpair equilibria. In view of the observed association constants and ISP values, a dynamic equilibrium between solvent separated and contact ionpair was proposed and corresponding conversion constants are also presented.

109. THEORY OF TURING PATTERNS IN A BELOUSOV-ZHABOTINSKI REACTION WITH IMPOSED CHEMICAL GRADIENTS
Zhang, Y. H., Burningham, A; Weise, W; and Neshyba, S;
Department of Chemistry, University of Puget Sound,
Tacoma, WA 98416

This is a theoretical study of the effects of chemical gradients on Turing patterns in a reaction-diffusion system. Specifically, we numerically solve equations of motion for a Belousov-Zhabotinski system with boundary conditions corresponding to a diffusive medium situated between two continuously stirred tank reactors (CSTRs). Our investigations are guided by the following concerns: 1) why do Turing patterns in analogous experimental systems occur in narrow bands perpendicular to the imposed gradient, and 2) how must the theoretical apparatus for gradient-free Turing instability be modified in order to accomodate imposed chemical gradients.

110.

CHIRAL DISCRIMINATION IN INTERMOLECULAR ELECTRONIC ENERGY TRANSFER PROCESSES IN SOLUTION Sarah J. Chisdes F. S. Richardson, Department of Chemistry, University of Virginia, Charlottesville, VA 22901

Time-resolved chiroptical luminescence (TR-CL) measurements are used to study the kinetics of chirality dependent excited state quenching processes in solution. Solution samples typically contain a racemic mixture of chiral luminophore molecules (L) and a small, optically resolved concentration of chiral quencher molecules (Q). The luminophores are excited with a pulse of unpolarized light to create an initially racemic excited state population of ΔL^* and ΛL^* enantiomers. TR-CL measurements are then used to monitor the differential decay kinetics of the ΔL^* and ΛL^* subpopulations. Observed differences between the ΔL^* and ΛL^* decay kinetics reflect differential rate processes and efficiencies for each ΔL^*-Q and ΛL^*-Q quenching actions, and are diagnostic of chiral discriminatory interactions between luminophore and quencher molecules. In this study, lanthanide coordination compounds with three bladed propeller structures were used as luminophores and cobalt coordination compounds, also with three bladed propeller structures, were used as quenchers.

111.

DYNAMICS OF ELECTRONS IN INTERFACIAL QUANTUM WELLS. **Jason D. McNeill**, N.-H. Ge, R. L. Lingle, Jr., C. M. Wong, and C. B. Harris, Department of Chemistry, University of California, Berkeley, California 94720 and Chemical Sciences Division, Lawrence Berkeley National Laboratory, Berkeley, California 94720

Time-resolved two-photon photoemission with sub-100 fs time resolution is used to probe the dynamics of excited electrons in ultrathin layers of Xe on a Ag(111) surface. The Xe layer presents a potential well, bounded on one side by the Ag substrate and on the other side by the vacuum, which supports quantum well states derived from the Xe conduction band. The dynamics of the electrons in excited conduction band-derived quantum well states are found to possess a complex dependence on layer thickness and quantum number. A simple quantum-mechanical model is proposed which accounts for the energies and lifetimes of the lowest members of the quantum well series for 1-6 layers of Xe. These results contribute to a fundamental understanding of carrier dynamics in quantum-confined systems and interfaces and the relationship between quantum well electronic structure and bulk band structure.

112. ENHANCED FLUORESCENCE OF 2,6-DIPHENYPYRIDINE BY HYDROGEN BOND TRANSFER. A. C. TESTA and F. Deng, Department of Chemistry, St. John's University, Jamaica, New York 11439

Phenylpyridines fluoresce in water as a result of an excited state proton transfer, (ESPT), which is well understood in terms of the increased basicity of aromatic monoazines in

S_1. We have recently initiated a fluorescence study of 2,6-diphenypyridine, which has an unexpectedly low pK_a and fluoresces in hydrocarbon, polar and aqueous solvents. This molecule does not undergo an ESPT, however, its fluorescence is larger in acidic aqeous media. Experiments in a hydrocarbon solvent indicate that the formation of a 1:1 EDA complex can lead to a significant increase in fluorescence yield. Since a proton is unlikely in a hydrocarbon solvent, the result is attributed a hydrogen bond transfer in S_1. Quantum yield data and the properties of the EDA complex will presented, and the hydrogen bond transfer in the excited state will be discussed.

113.

EXCIPLEX DIPOLE MOMENTS: EXCITED CYANOANTHRACENE ACCEPTORS AND METHYLBENZENE DONORS. **Steven E. Mylon,** Karyn N. Grzeskowiak, Sergei N. Smirnov, Charles L. Braun, Department of Chemistry, Dartmouth College, Hanover, NH 03755

Dipole moments have been measured for the exciplexes of 9,10-dicyanoanthracene (DCA) and 2,6,9,10-tetracyanoanthracene (TCA) as photoexcited electron acceptors and methylbenzenes as electron donors. Measurements were made in both neat donor solvents and in benzene solutions with varying donor concentrations. For the same donor large dipole moments are seen in neat donor solvents in comparison to smaller moments measured in benzene solutions. This is evidence for the formation of 2:1 complexes (triplexes) at high donor concentrations. Dipole moments reach a constant value as the donor concentration is reduced. Results are compared with other studies which attempted to quantify the fraction of charge transfer in 1:1 exciplexes. Assuming a donor/acceptor separation of 3.5Å, the maximum charge transfer never exceeds 75%. Reasons for the observed failure of the fractional charge transfer to reach 100% are addressed.

114. FEMTOSECOND STUDY OF EXCITON DYNAMICS IN STATISTICAL COPOLYMERS IN SOLUTION AND THIN FILMS. M. A. Kreger and J. Z. Zhang, Department of Chemistry, University of California, Santa Cruz, CA 95064, J. C. Scott and R. D. Miller, IBM Almaden Research Center, 650 Harry Road, San Jose, CA 95120

The exciton dynamics in newly developed electroluminescent statistical copolymers have been characterized using femtosecond transient absorption spectroscopy. A comparison of the exciton formation and decay dynamics in these copolymers in solutions and in films show that the exciton is strongly affected by inter- and intra-molecular interactions. The exciton in solution is long-lived (~270 ps), which correlates with strong fluorescence. The lifetime is drastically reduced in films (~2 ps) and this is tentatively attributed to predominant non-radiative relaxation due to aggregation of the polymer. The dependence of the exciton behavior on the environment is significant to the understanding and development of LEDs and other electro-optical devices based on conjugated polymers.

115. Femtosecond Time Resolved Infrared Studies of Si-H Bond Activation
H. Yang, K.T. Kotz, M. Asplund, S.E. Bromberg, H. Frei[†] and C.B. Harris[†*]
[*]Department of Chemistry, University of California, Berkeley, CA 94720, and
[†]Chemical Science Division, Lawrence Berkeley National Laboratory, Berkeley, CA 94720

We report the first time-resolved IR studies of the silane Si-H bond activation mechansim on timescales from sub-picosecond to nanosecond. It has been accepted that the primary photochemical process following UV irradiation of η^5-CpMn(CO)$_3$ [**A**] in room temperature solution is the loss of a CO ligand. By monitoring the time evolution of the CO stretching frequencies, we have observed three distinct pathways for the formation of the activated hydridosilyl complex η^5-CpMn(CO)$_2$(H)(SiEt$_3$) [**B**]: (1) direct formation after CO dissociation; (2) formation through an intermediate [**C**] whose lifetime is about 90 ps; and (3) formation through another intermediate [**D**] which can be assigned as a solvated dicarbonyl species that eventually rearranges to form **B** on a time-scale of > 1 ns. Considering possible chemical processes, we have tentatively assigned **C** as a ring-slipped η^3 species which changes its hapticity from η^3 to η^5 in ~90 ps. Detailed structures of the intermediates are being studied using *ab initio* electronic structure calculations. Experiments that extend beyond 1 ns and those on Si-H bond activation by other metal complexes such as η^5-CpRe(CO)$_3$ are underway.

115A

PHOTOPHYSICS'OF (O2(1DG))2 AND O2(1S+G) IN SOLUTION PHASE. Pi-Tai Chou Youn-Chan Chen and Ching-Yen Wei, Department of Chemistry, The National Chung-Cheng University, Chia-Yi, Taiwan, R.O.C.

Using the C60 fluorescence as a reference in combination with the direct spectroscopic observation of 1O2 visible emission, photophysics of 1O2 dimol and 1S+g states have been studied. The quantum yield of O2 1S+g * 3S-g (765 nm) emission has been measured to be 1.97 x 10-7, and consequently, the radiative decay rate of 1S+g * 3S-g transition was determined to be 1.1 s-1 which, within the experimental error, is consistent with the previous report of 0.46 s^-1. Further determination of the radiative decay rate of the dimol (1Dg)2 * (3S-g)2 transition and the dissociation rate of dimol has been achieved by the measurement of tetra-tert-butylphthalocyanine delayed fluorescence induced by the energy transfer from the 1O2 dimol. Due to the lack of observing O2 (S+g) 765 nm emission upon C60 sensitizing 1O2 emission, an upper limit of the triplet state energy of 37 kcal/mol in CCl4 was concluded for C60.

116. SPIN DYNAMICS IN MANY-BODY DIPOLAR SYSTEM

Yung-Ya Lin and Alexander Pines
Department of Chemistry, University of California, Berkeley, CA 94720, and
Material Science Division, Lawrence Berkeley National Laboratory, Berkeley, CA 94720

Our studies of the many-body character of the spin dipolar interaction in solid-state NMR reveal some unusual and interesting physical chemistry. (i) Rigid Lattice. The transverse correlation functions are calculated by the Kubo-Tomita theory and compared with the measurements. The results suggest that the damped oscillating long-time tail of the dipolar dephasing may be ascribed to the long-range cross correlation governed by the Ising interaction. This gives satisfactory explanation to the recent surprising observation of substantial line-narrowing with long acquisition delays. (ii) Slow Motion. The possibility of effective dipolar decoupling induced by slow lattice motion is demonstrated based on a generalized relaxation formulation and real experiments. (iii) Motional Narrowing. Comparison of some diffusion jump rates determined from NMR motional-narrowing studies and other more direct techniques yields order of magnitude differences in the prefactor of the Arrhenius equation and uncertainties of the activation energies. Numerical experiments illustrate that these"prefactor anomalie" may arise from the highly nonexponential time correlation of the fluctuating local fields due to correlated vacancy migration and reduced dimensionality.

117.

TEMPERATURE-DEPENDENT VIBRATIONAL AND ORIENTATIONAL RELAXATION OF MONOCARBONYLS IN POLYATOMIC SOLVENTS. **Paul H. Davis** Christopher M. Aubuchon, Michael D. Fayer; Department of Chemistry, Stanford University, Stanford, CA 94305-5080

Temperature-dependent vibrational lifetimes and orientational relaxation times of the asymmetric carbonyl stretch of carbonylchlorobis(triphenylphosphine)rhodium(I) in various polyatomic solvents have been measured. These studies employed a novel laser system consisting of a regeneratively amplified Ti:sapphire laser used to pump a two-stage OPA (optical parametric amplifier). The resultant signal and idler beams were difference mixed to produce ~800 fs pulses centered at the maximum of the carbonyl absorption (~5 microns). This unique laser system allowed for both wavelength tunability and bandwidth selection. The experimental results obtained serve both to highlight the utility of this laser system in picosecond and femtosecond IR spectroscopic studies and to elucidate the solvation dynamics and molecular motions of complex molecules in polyatomic solvents.

118.

THE EMISSION OF C2 FROM WATER/ORGANICS MIXTURES UNDER ULTRASONIC IRRADIATION. **Y.T. Didenko,** W.B.McNamara, III, and K.S.Suslick, School of Chemical Sciences, University of Illinois at Urbana-Champaign, Urbana, IL, 61801

High intensity ultrasonic waves in liquids produce cavitation, the formation of small nonlinearly-pulsating bubbles that are responsible for a number of physico-chemical effects of ultrasound. One of the cavitational effects is the emission of light - sonoluminescence, which occurs during the violent collapse of a cavitation bubble. The purpose of this work was to estimate the temperature inside the cavitation bubbles formed in water. Several organic substrates were tested as possible sources of C2 -emission from water/organics mixtures. The comparison of the relative intensities of two C2 bands observed under ultrasonic irradiation of water/ benzene mixture gave a temperature inside the cavitation bubble of 4500 K. To our knowledge, this is the first experimental estimate of the temperature of cavitation event in water based on the spectra of sonoluminescence.

119.

TWO-DIMENSIONAL FIFTH ORDER RAMAN SPECTROSCOPY OF THE INTERMOLECULAR MODES OF LIQUIDS. **A. Tokmakoff** and G. R. Fleming, Department of Chemistry and the James Franck Institute, University of Chicago, Chicago, IL 60637

Two-dimensional vibrational spectroscopy has been used to study the intermolecular modes of liquid CS_2. While linear third order Raman spectroscopies observe the spatially averaged spectrum of solvent motions, time-domain fifth order Raman spectroscopy allows the microscopic composition of this spectral density to be probed. The power of this technique is the sensitivity of the two-dimensional response to spatial heterogeneity in the liquid dynamics. Simulations of the response confirm a picture of the liquid CS_2 in which dynamics are dictated by a distribution of fixed structures at short times (<300 fs), and by spatially uniform structural relaxation at long times (>700 fs). The data suggest that the microscopic dynamics show highly damped oscillatory behavior at intermediate times.

120. ULTRAFAST DYNAMICS OF ELECTRON LOCALIZATION AT INTERFACES
N.-H. Ge, C. M. Wong, J. D. McNeill, R. L. Lingle, Jr., K. Gaffney, and C. B. Harris
Department of Chemistry, University of California, Berkeley, CA 94720, and
Chemical Science Division, Lawrence Berkeley National Laboratory, Berkeley, CA 94720

The dynamics of image potential electrons at alkane/Ag(111) interfaces were studied by angle-resolved two-photon photoemission spectroscopy (TPPE) with femtosecond time resolution. Optical excitation creates interfacial electrons in extended states which are characterized by a nearly free-electron mass for electron motion parallel to the interface. The initially delocalized state decays into a high-mass localized state within a few hundred femtoseconds via a self-trapping mechanism. The energy dependence of the self-trapping rate can be modeled as an activated barrier crossing between the delocalized and localized states. The lattice relaxation energy and the characteristic phonon energy that results in 2–D small polaron localization are determined. The subsequent decay of the localized electron back to the metal is on a longer time scale. This is due to the tunneling barrier created by the repulsive electron affinity of the adlayer. Our results show that angle and time resolved femtosecond TPPE serves as a powerful and general probe of electronic structure and carrier dynamics at interfaces.

121.

VIBRATIONS OF A SOLVATED DIATOMIC MOLECULE USING INSTANTANEOUS NORMAL MODES. **Stephen J. Schvaneveldt,** Department of Chemistry, Whitman College, Walla Walla, Washington 99362

An analytic theory for the vibrational absorption lineshape of a solvated molecule using instantaneous normal mode (INM) dynamics is presented. Within the INM approximation, the potential surface on which a fluid evolves is replaced by a harmonic potential surface whose curvature matches the curvature of the true potential surface at the initial configuration of the fluid. This approximation captures the short time dynamics of the fluid, and describes these motions in terms of phonons. The lineshape is related to a configurationally averaged phonon Green's function, and a diagrammatic analysis of this Green's function leads to a self-consistent approximation to the lineshape. This theory is applied to a harmonic diatomic molecule dissolved in an atomic Lennard-Jones solvent and compared to lineshapes based on simulation.

122.

A COMPUTATIONAL STUDY OF THE GAS PHASE ACIDITIES AND BASICITIES OF DOUBLY-CHARGED IONS -- AN ESTIMATE OF THE EFFECTIVE DIELECTRIC CONSTANTS OF SPACER GROUPS. **Scott Gronert** Department of Chemistry and Biochemistry, San Francisco State University, San Francisco, CA 94132

Multiply-charged ions are very useful in the mass spectrometric analysis of complex substrates such as peptides, but to interpret the physical properties of these ions (i.e., acidity or basicity),

Coulombic repulsion must be taken into account. In the present study, ab initio calculations at the MP2/6-31+G** level have been used to determine the gas phase acidities of a series of diammonium ions ($^+H_3NCH_2(Z)_nCH_2NH_3^+$) where the Z group = CH_2 or CH_2OCH_2. In addition, basicities have been calculated for an analogous series of dialkoxides ($^-OCH_2(Z)_nCH_2O^-$). By varying the length of the spacer group (n), a correlation between the acidity (or basicity) of the doubly-charged ion and the distance between the charges is established. This correlation leads to an effective dielectric constant for the spacer group.

123. A dissociative model of water

By L. René Corrales
Envirnmental Molecular Sciences Laboratory
Pacific Northwest National Laboratory [a]
Richland, WA 99352

We present a dissociating model of water based on a semi-empirical methodology initially developed to model covalently bonded systems and more recently shown to be applicable in modelling heteroatomic systems. The methodology, as used for water, is consistent with the original methodology for networked materials in which nearest bonding neighbors are defined by a semiempirical interaction and all other interactions are defined by classical-like interactions. In this method, the hydrogen-bond is formed by a hydrogen overlapping with a lone pair orbital of a neighboring oxygen and is highly directed. We present the results and discussion of the dissociation energies and geometries of a single water molecule and the water dimer. The geometric structure of several clusters and of ice demonstrate the effect of charge transfer in the energetics and the oxygen-oxygen bond distance.

[a] *Pacific Northwest Laboratory is operated by Battelle Memorial Institute for the U.S. Department of Energy under contract DE-AC06-76RLO 1830*

124.

A THEORETICAL STUDY OF UREA MONOMERS AND DIMERS. **Artem E. Masunov,** and J.J. Dannenberg, Department of Chemistry, City University of New York, Hunter College and the Graduate School, 695 Park Avenue, New York NY 10021

Small changes in molecular structures can have large influences upon crystal structures. Understanding the basis for these effects can be of singular importance for understanding and designing the intermolecular interactions that dictate crystal packing and will eventually play important roles in crystal engineering and material science in general. Urea and thiourea provide interesting and contrasting examples of different crystal packing for molecules of similar structures. Semiempirical and ab initio geometry optimizations and vibrational frequencies calculations at different levels of theory were performed. Eight stationary points (nonplanar minima and transition states for inversion and internal rotation) were characterized, based on which dynamical structure of the molecule will be discussed. Four different classes of dimers, related to the known crystal structures were found.

125. BOND ENERGIES, BOND LENGTHS, AND POTENTIAL CURVES OF SEVERAL EXCITED ELECTRONIC STATES OF M·RG VAN DER WAALS COMPLEXES (M = Ca, Mg; RG = Ar, Kr, Xe). John G. Kaup, Allen W.K. Leung, Steven Massick, Brian Gau, and W.H. Breckenridge. Department of Chemistry, University of Utah, Salt Lake City, Utah 84112.

We present very recent spectroscopic data on several excited states of M·RG van der Waals complexes, where M = Mg, Ca and RG = Ar, Kr, Xe. From these data, we have determined bond energies, bond lengths and potential curves for many of the states. This information, along with *ab initio* calculations of some of the states, will be presented and analyzed with the goal of understanding the bonding and spin-orbit effects in such van der Waals molecules. Finally, dissociation energies of the ground states of the analogous M^+·RG complexes, determined by two-color photoionization threshold measurements, will also be presented and compared with values estimated from ion-dissociation measurements of Duncan and co-workers.

126. DENSITY FUNCTIONAL STUDIES OF THE INTERACTION BETWEEN SINGLET METHYLENE AND WATER IN THE GAS PHASE AND IN SOLUTION. Jean M. Standard and Jennifer M. Tucker, Department of Chemistry, Illinois State University, Normal, IL 61790-4160

Results will be presented from density functional calculations exploring the singlet methylene-water interaction in the gas phase and in solution. The solution phase studies employed the Onsager reaction field model with a range of dielectric constants. All calculations were carried out using density functional theory with the B3LYP functional and 6-31G* or 6-311++G** basis sets. Equilibrium geometries, energies, and vibrational frequencies were determined for a number of stationary points on the methylene-water potential energy surface, including the transition state for a 1,2-hydrogen shift rearrangement of the methylene-water ylide to form methanol. The effect of solvation on the methylene-water interaction will be discussed. Results will be compared with high level ab initio MP2 and QCISD studies.

127. ELECTRIC FIELD GRADIENT PERTURBATIONS IN SOME N_2O-CONTAINING COMPLEXES. Helen O. Leung, Department of Chemistry, Mount Holyoke College, South Hadley, MA 01075

The rotational spectra of some N_2O-containing complexes have been recorded via Fourier transform microwave spectroscopy. Due to the presence of two quadrupolar ^{14}N nuclei, each rotational transition exhibits complicated hyperfine structure, which, upon analysis, has yielded nuclear quadrupole coupling constants for each ^{14}N nucleus. These constants are used to furnish angular information for the complex and are sensitive to the electric field gradient at each ^{14}N nucleus. If van der Waals interactions do not significantly disturb the electrical environment of the subunits, as is commonly assumed, then properties derived from the coupling constants for one ^{14}N nucleus should be the same as those from the other nucleus. We have found that while some complexes support this assumption, there exist others in which the electrical field gradient at the central nitrogen nucleus is greatly distorted, prompting us to carefully examine the effect of one molecule on its partner.

128.
Gas Phase NMR Studies of N,N-Dimethylthioamides. Influence of the Thiocarbonyl Substituent Upon the Internal Rotation Activation Energies.
S.M. Neugebauer Crawford, A.N. Taha, C.B. LeMaster, and N.S. True
Chemistry Department, University of California, Davis, California 95616.

Temperature-dependent gas phase ^1H NMR spectra of seven thiocarbonyl substituted N,N-dimethylthioamides ($XCSN(CH_3)_2$) obtained at 300 MHz are consistent with the following free activation energies $\Delta G^{\neq}$ (kcal /mol) : X= H, 22.5 (0.1); CH_3, 18.0 (0.1); F 18.3 (0.1); Cl, 16.9 (0.2); CF_3, 17.2 (0.1); CH_2CH_3, 17.6 (0.1); $CH(CH_3)_2$, 16.3 (0.1). The results are compared to condensed phase values and to the corresponding gas phase oxo amides.

129. GAS PHASE PROTON NMR STUDIES OF ^{15}N-FORMAMIDE
S. M. Neugebauer-Crawford, A.N. Taha; and N.S. True
Department of Chemistry, University of California, Davis, California 95616

The activation parameters characterizing the internal rotation about the C-N partial double bond in gaseous ^{15}N-formamide have been determined from variable-temperature exchange broadened proton magnetic resonance spectra. The obtained Gibbs energy of activation compares favorably with previously obtained rotational barrier predictions based on *ab initio* molecular orbital calculations. RRKM calculations indicate that the rate constants will be pressure dependent up to pressures of several atmospheres. Experimental pressure dependent studies will be compared to the theoretical predictions.

PHYS

130. MOLECULAR DYNAMICS STUDY OF WATER AND WATER/CHLORINATED HYDROCARBON MIXTURES WITH POLARIZABLE POTENTIAL MODELS Liem X. Dang*, Environmental Molecular Sciences Laboratory, Pacific Northwest National Laboratory, Richland, WA 99352

A series of molecular dynamics simulations were carried out to study water and water/chlorinated hydrocarbon mixtures. The properties of water clusters containing up to six water molecules were evaluated. A prism-like structure is predicted to be lowest in energy for the $(H_2O)_6$ cluster and a cage-like structure is the second lowest in energy with the energy about 0.2 kcal/mol higher than the prism-like structure. The computed dipole moments of water molecules in clusters indicated that there is a transition from cyclic planar configurations to three dimensional structure networks. The computed thermodynamic properties for the model including the liquid density, the enthalpy of vaporization, as well as the diffusion coefficient at room temperature, are in excellent agreement with experimental values. The computed density profile of the water of liquid/valor interface shows that the interface is not sharp at a microscopic level and has a thickness of 3.2 Å at 298 K. The calculated surface tension at room temperature is in reasonably agreement with the corresponding experimental data. The computed average dipole moments of water molecules near the interface are close to their gas phase values. The thermodynamic and structural properties of water/chlorinated hydrocarbon mixtures as a function of mole fraction were evaluated.

*This work was performed under the auspices of the Division of Chemical Sciences, Office of Basic Energy Sciences, U.S. Department of Energy under Contract DE-AC06-76RLO 1830 with Battelle Memorial Institute, which operates the Pacific Northwest Laboratory, a multiprogram national laboratory.

131.

NATURAL RESONANCE THEORY: THE ORIGIN OF BARRIERS TO AMIDE ROTATION. **Eric D. Glendening** and John A. Hrabal II, Department of Chemistry, Indiana State University, Terre Haute, IN 47809

Conventional resonance theoretic concepts have been frequently criticized during the last decade. Amide resonance and its influence on the rotation barrier of the peptide bond have been the focus of much of this criticism. Whereas the traditional view attributes these features to a resonance hybrid consisting largely of the parent Lewis and secondary dipolar forms, structural evidence and recent computational analyses of the charge density suggest instead that the resonance picture fails. We present here natural resonance theory (NRT), a novel computational approach to resonance theory based on Weinhold's natural bond orbital (NBO) methods. Application of NRT to formamide and its S, Se, and Te replacement analogs provides a remarkably simple resonance description in which both the calculated rotation barriers and geometrical features correlate remarkably well with the natural resonance weights and bond orders, respectively.

132.

Si3OY (Y = 1-6) CLUSTERS: MODELS FOR OXIDATION OF SILICON SURFACES AND DEFECT SITES IN BULK OXIDE MATERIALS. **Lai-Sheng Wang(1,2)**, J. B. Nicholas(2), M. Dupuis(2), H. Wu(1), and S. D. Colson(2), (1)Department of Physics, Washington State University, Richland, WA 99352; (2)Environmental Molecular Sciences Laboratory, Pacific Northwest National Laboratory*, Richland, Washington 99352.

We studied the structure and bonding of a series of silicon oxide clusters, Si3Oy (y=1-6), using anion photoelectron spectroscopy and ab initio calculations. For y=1-3 the clusters represent the sequential oxidation of Si3, and provide structural models for the oxidation of silicon surfaces. For y=4-6, the clusters contain a central Si in a tetrahedral bonding environment, suggesting the onset of the bulk-like structure. Evidence is presented that suggests that the Si3O4 cluster (D2d) may provide a structural model for oxygen-deficient defect sites in bulk SiO_2 materials.
*Pacific Northwest National Laboratory is operated for the U.S. Department of Energy by Battelle under contract DE-AC06-76RLO 1830.

133.

SPECTROSCOPIC DETERMINATION OF DISSOCIATION CONSTANTS OF ALPHA-CYCLODEXTRIN INCLUSION COMPOUNDS; Holly C. Gaede, Stephen B. Randall, and Sruthi Tallapragada; Ursinus College, Department of Chemistry; P.O. Box 1000, Collegeville, PA, 19426.

Cyclodextrins (CDs) are cyclic oligosaccharides composed of six or more glucose monomers linked by alpha - 1, 4 bonds. CD molecules have a toroidal shape with a hydrophobic cavity and a hydrophilic exterior, rendering the molecule soluble in water. CDs are capable of hosting various guest molecules in the cavity, with practically the only requirement of the guest is that it fits at least partially into the cavity. Many forces are known to contribute to the formation of CD inclusion compounds, but the relative importance of each remains unclear. In this study the role of hydrogen bonding was investigated by forming alpha- CD inclusion compounds with guests of varying abilities to hydrogen bond. The stability of each complex was elucidated by spectroscopic determination of its dissociation constant in solution.

134.

STRUCTURES AND THERMOCHEMISTRY OF MEDIUM SIZED BORON OXIDE CLUSTER IONS AND NEUTRALS. **Douglas P. Linder** and Michael Page, Department of Chemistry, North Dakota State University, Fargo, ND 58105

Boron has long been of interest in high energy density fuels and propellants, due to its large heat of combustion, and the chemistry/thermochemistry of boron oxide compounds plays a very important role in this process. However, the thermochemistry of even small B/O-containing molecules is not well established, and recent experimental data point to several significant uncertainties in the literature for B/O/F/H species. Therefore, in an effort to gain a better understanding of B/O thermochemistry and B/O bonding in general, a comprehensive study of the structures and thermochemistries of several medium sized boron oxide cluster ions and neutrals has been undertaken. This poster will focus on neutral and cationic isomers of $(BO)_n$ and $O(BO)_n$, where n=3 and possibly 4. The theoretical methods employed are the density functional theory as well as more traditional ab initio techniques.

135.

X-RAY SCATTERING FROM DISCOTIC LIQUID CRYSTALLINE COMPOUNDS AT THE AIR-WATER INTERFACE. **D. Gidalevitz**, O. Mindyuk, J. Strzalka (University of Pennsylvania), B. Ocko (Brookhaven National Laboratory) and P. A. Heiney, Department of Physics and Astronomy, University of Pennsylvania, Philadelphia, PA 19104

We present the first X-ray diffraction study of Langmuir films of discotic compounds at the air-water interface. The subjects of our study were discotic liquid crystalline compounds of two types: hexasubstituted azacrown derivatives and triphenylene derivatives with various lengths of side chains. The X-ray grazing incidence and reflection measurements were performed on Beamline X22 at the NSLS, Brookhaven. The azacrown derivative was seen to adopt a "face-on" arrangement, consistent with pressure-area isotherm measurements. All triphenylene derivativesshowed one peak in grazing incidence, consistent with intercolumnar distances of 13-20Å. Mixtures of triphenylene isomers that were selectively disubstituted at the 2,3 and 3,6 positions were seem to have smaller intercolumnar spacings than each individual pure compound.

136.

AN EXPERIMENTAL STUDY OF THE ELECTRONIC STRUCTURES OF THE FIRST ROW TRANSITION METAL MONOXIDES AND DIOXIDES BY ANION PHOTOELECTRON SPECTROSCOPY . **Hongbin Wu(1)** and Lai-Sheng Wang(1,2), (1) Department of Physics, Washington State University, Richland, WA 99352, (2) Environmental Molecular Sciences Laboratory, Pacific Northwest National Laboratory *, Richland WA 99352

The electronic structures of the monoxides and dioxides of the first row transition metals from Ti to Zn are investigated systematically by anion photoelectron spectroscopy. The electron affinities of these novel molecules are measured. The electronic states of the monoxides are assigned according to existing theoretical calculations. We also observed an excited state for the CuO- anion at 0.5ev above the anion ground state. The vibrational frequencies of the v1 mode for the ground states of VO_2, MnO_2, CoO_2, and NiO_2 are observed, which are 906cm-1, 828cm-1, 809cm-1, and 751cm-1 respectively.   * Pacific Northwest National Laboratory is operated for U.S. Department of Energy by Battelle under Contract DE-AC06-76RLO 1830

137.
ELECTRON SPIN RESONANCE OF THE H-NH2 RADICAL PAIR. **A.P. Williams,** R.J. Van Zee, and W. Weltner, Jr., Department of Chemistry, University of Florida, Gainesville, Florida 32611

Exchange within the magnetic-dipole coupled radical pairs H-NH2 and D-ND2 trapped in solid argon and krypton at 4 Kelvin was observed via Electron Spin Resonance (ESR). They were produced from ammonia by radiation from a corona discharge through a rare gas. From the zero field splittings of the triplet species it is concluded that the H atom and NH2 are separated by one rare-gas atom in each of these matrices.

138. EMISSION AND TRIPLET-STATE ODMR SPECTRA OF THE FULLERENE OXIDES $C_{60}O$, $C_{60}O_2$ AND $C_{120}O$ IN GLASSY SOLVENTS AT LOW TEMPERATURES. M.P. May, D.A. Costa, A.L. Balch, and D.S. Tinti, Department of Chemistry, University of California, Davis, CA 95616

Fluorescence, phosphorescence and triplet-state ODMR spectra of the fullerene oxides $C_{60}O$, $C_{60}O_2$ and $C_{120}O$ are compared in glassy solvents at ≤ 4.2 K. The fluorescence spectra are very similar and show strong electronic origins with only weak to moderate vibronic activity. The fluorescence of the C_{60} parent is very different with a weak origin and strong vibronic activity. The phosphorescence spectra, detected only in heavy-atom glasses, are strikingly similar for the oxide derivatives and the parent. Absorption detected ODMR spectra at zero field were obtained in a light-atom glass and analyzed to yield the splitting parameters D and E of the lowest triplet state. Certain kinetic parameters of the triplet state were also deduced from ODMR studies.

139. FAR-INFRARED AND FLUORESCENCE SPECTRA AND POTENTIAL ENERGY SURFACES FOR TETRAHYDROFURAN-3-ONE IN ITS S_0 AND $S_1(n,\pi^*)$ ELECTRONIC STATES. Soo-No Lee, Paul Sagear, Niklas Meinander, and Jaan Laane, Department of Chemistry, Texas A&M University, College Station, Texas 77843-3255.

The far-infrared spectrum of tetrahydrofuran-3-one, $O=\overline{CCH_2OCH_2CH_2}$, shows two series of bands in the 100-120 and 220-240 cm^{-1} regions corresponding to the ring-bending and ring-twisting vibrations, respectively. A two-dimensional potential energy surface with minima corresponding to twisted structures and a barrier to planarity of 1550 cm^{-1} fits the data nicely. For the $S_1(n,\pi^*)$ excited state the fluorescence excitation spectra of the jet-cooled sample were used to determine the energy levels for the carbonyl wagging as well as the two ring modes. The carbonyl inversion potential function has a barrier to planarity of 1152 cm^{-1} (13.8 kJ/mole) and a bending-twisting surface similar to that in the ground state. However, a saddle point corresponding to the bent ring conformation gives rise to a hindered pseudorotational pathway with a barrier of approximately 1000 cm^{-1}.

140. INFORMATION CAPACITY OF A CHEMICAL BOND FOR VIBRATIONAL INFORMATION QUANTUM SPECTROSCOPY

Dr SYED AMEEN (PH.D)
6622 MONTEZUMA AVE SUITE 64
SAN DIEGO, CA. 92115
Tel:(919)-582o633

ABSTRACT:
An approach to VIBRATIONAL INFORMATION QUANTUM SPECTROSCOPY(VIQS) is developed where the sequence of well-tailored Laser Pulses corresponding to the vibronic frequencies of CHEMICAL BONDS is used to calculate the INFORMATION CAPACITY of a Chemical Bond in molecules. A relationship between the THERMODYNAMICS and Quantum Vibronic Spectroscopy is established by introducing a new concept of "THE INFORMATION CAPACITY OF A CHEMICAL BOND" which is used to investigate the TRANSFER AND STORAGE OF INFORMATION AND ENERGY in molecules. The application of sequence of ULTRASHORT INFRA RED FREQUENCIES, as used in COMMUNICATION THEORY, shows that the Information Capacity of a bond increases when shorter pulse-widths are used, thus indicating a newer probe to low energy quantum spectroscopy is on its way through the incorporation of INFORMATION THEORY, THERMODYNAMICS AND QUANTUM THEORY. A relationship between the ENTROPY, ENTHALPY, TEMPERATURE and INFORMATION CAPACITY OF A CHEMICAL BOND is established which becomes applicable to low energy spectroscopy.

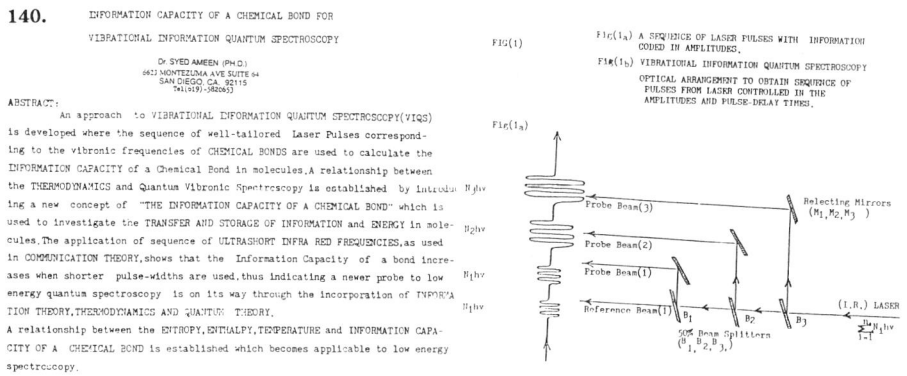

FIG(1)
FIG(1a)
FIG(1b) A SEQUENCE OF LASER PULSES WITH INFORMATION CODED IN AMPLITUDES.
FIG(1b) VIBRATIONAL INFORMATION QUANTUM SPECTROSCOPY OPTICAL ARRANGEMENT TO OBTAIN SEQUENCE OF PULSES FROM LASER CONTROLLED IN THE AMPLITUDES AND PULSE-DELAY TIMES.

141. LOW TEMPERATURE SPECTROSCOPY FOR CRYSTALLIZATION DIAGNOSTICS
Robert F. Ferrante, Chemistry Department, U.S. Naval Academy, Annapolis, Maryland, 21402

The crystalline or amorphous nature of ices in both cometary and interstellar environments has important ramifications for describing the history and behavior of the object. Vapor-phase deposition onto very cold surfaces is usually expected to produce amorphous ices. However, laboratory simulations in cometary model systems containing silicate grain analogs show that predominantly crystalline phase deposits can be formed at temperatures well below the crystallization point. For many ices, the crystalline or amorphous nature of the deposit can be determined spectroscopically; significant changes in the IR spectrum accompany crystallization of H_2O, CH_3OH, etc. In an attempt to elucidate the mechanism for the observed low-temperature crystallization, we have begun to explore Raman spectroscopy as a possible diagnostic tool. Results of such studies will be presented.

142.

MANY-ELECTRON CRYSTAL FIELD MODEL OF THE MAGNETIC PROPERTIES OF TRANSITION-METAL COMPLEXES AND ESEEM EXPERIMENTAL RESULTS. **Nikolai V. Shokhirev**, Arnold M. Raitsimring and F. Ann Walker, Department of Chemistry, University of Arizona, Tucson, AZ 85721

A many-electron multi-configuration quantum-chemical approach that takes into account the spin-orbit interaction for d-electrons was developed for the case of arbitrary crystal field. The model allows quantitative description of NMR paramagnetic, g- and HFI-tensors and zero-field splitting, and the results can also be applied to magnetic Mössbauer and MCD spectroscopies. The method was implemented by preparation of the program for PC. Our initial applications have been to low-spin d^5 systems, where inclusion of all electrons modifies somewhat the range of possible g-values predicted from the d^1-hole model of Griffith. Counter-rotation of the in-plane g-tensor and planar axial ligands bound to ferriheme centers is described theoretically. ESEEM experiments on selected ferriheme models confirm the counter-rotation.

143. NONLINEAR OPTICAL PROPERTIES OF COMPOUNDS CONTAINING BENZIMIDAZOLE AND BENZITHIAZOLE UNITS USING THE Z-SCAN TECHNIQUE. *Paul A. Fleitz and *Richard L. Sutherland, Materials Directorate, Wright Laboratory, WL/MLPJ, Wright-Patterson AFB, OH 45433-7702. FLEITZPA@ml.wpafb.af.mil

The nonlinear optical properties of a series of organic compounds containing benzimidazole and benzithiazole units were investigated using the Z-scan technique. The experiments were performed using a Nd:YAG laser frequency doubled to 532nm with 25ps laser pulses. The samples were prepared as solutions in THF or heated above their melting point into the isotropic liquid phase. All samples show strong evidence of nonlinear absorption. Because the samples absorb at the laser wavelength, this nonlinearity is most likely the result of an excited state absorption process. The nonlinear refraction varies greatly from one compound to another and depends on whether the measurement is made on molten or solution samples. One compound has a negative nonlinear refractive index in the melt, but shows no evidence of nonlinear refraction in solution.

* Science Applications International Corporation, 101 Woodman Drive, Dayton, OH 45431.

144.

OBSERVATION OF NEW VIBRONIC TRANSITIONS IN THE $\widetilde{B}\,^2A"$ - $\widetilde{X}\,^2A"$ MANIFOLD OF THE CH_2CHO RADICAL AND THE MODE-TO-MODE ENERGY FLOW WITHIN $CH_2CHO(\widetilde{B}\,^2A")$
Ruolian Wan, Katherine I. Barnhard and Brad R. Weiner
Department of Chemistry, University of Puerto Rico, Rio Piedras, P.O.BOX23346, San Juan, PR00931

New laser induced fluorescence spectroscopic transitions of the vinoxy (CH_2CHO) radical have been observed and assigned in the 310-330 nm wavelength region fluorescence. Both excitation and emission spectra have been recorded. In the excitation spectrum, the peaks at 30,211, 30,381 and 30,637 cm^{-1} have been assigned to the 6_0^1, 5_0^1 and 4_0^1 vibronic transitions of the $\widetilde{B}\,^2A"$ - $\widetilde{X}\,^2A"$ manifold.

State specific radiative lifetimes and quenching cross-sections are reported for different vibrational modes, and the lifetime shortening with increasing energy is found. Mode-to-mode vibrational energy flow within the ($\tilde{B}^2A$") state has been observed and measured. Vibrational relaxation channels and non-radiative decay routes will be discussed.

145.

PENDULAR STATE SPECTROSCOPY OF THE INTERMOLECULAR VIBRATIONS IN BENZENE-AR COMPLEX. **W. Kim**, P. M. Felker, Department of Chemistry and Biochemistry, University of California, Los Angeles, California 90095-1569

Mass-selective, ionization-detected stimulated Raman spectroscopies (IDSRS) of the ground state intermolecular vibrational transitions in van der Waals complexes of benzene isotopomers with Ar at rotational resolution (0.03 cm-1) are reported. From the dependency of the observed spectra on the relative polarization of the stimulated Raman fields and the agreement with the simulated spectra, it is concluded that the interaction of the permanent polarizability anisotropy of the complex with a strong oscillating field results in a pendular state and transitions between pendular states can be spectroscopically observed. With the aid of simulations, we can assign each of the five bands for benzene-Ar isotopomers by using the fact that the rotational structure of the transition between pendular states changes in different patterns as the relative polarization alters depending upon the symmetry of the state involved in the transition.

146.

POLYSELENIDES AND THEIR RADICAL IONS. A. J. Goldbach[a], J. A. Johnson[a], M. L. Saboungi[a], L. A. Curtiss[b], A. R. Cook[c], and D. Meisel[c], [a]Material Sciences, [b]Chemical Technology, and [c]Chemistry Divisions, Argonne National Laboratory, Argonne, IL 60439

Several polyselenides, Se_n^{2-} (n=2-4), were synthesized in aqueous solutions by oxidation of hydrogen selenide with hydrogen peroxide. Both absorption and Raman spectra were recorded and the observed bands were assigned to the various species and their vibrational modes. Additionally, single-electron reduction and oxidation radical ion products were generated in solution by reaction with e^-_{aq} and OH radicals, respectively, using the pulse radiolysis technique. Spectra of these products, as well as their formation and decay dynamics, were recorded and will be discussed. Ab initio electronic structure and vibrational mode calculations were conducted and the results were compared with experimental observations.

147.

SPECTROSCOPIC CHARACTERIZATION OF DEUTERATED DIMETHYLZINC CATION. **Timothy R. Brewer** Stephen Cotton and Fred J. Grieman, Pomona College, Seaver Chemistry Laboratory, 645 N. College Avenue, Claremont, CA, 91711

Abstract: The perdeutero compound of dimethylzinc was synthesized by reacting a zinc-copper couple complex with deuterated methyl iodide under vacuum. A rotationally-cold electronic spectrum of perdeutero dimethylzinc cation was obtained by the use of a crossed free-jet/electron beam technique. The spectrum is compared to earlier studies of the non-deuterated dimethylzinc cation where a weak progression signal was assigned to the symmetric CH3 deformation mode of the ground electronic state. The wavenumber of the deuterated vibrational energy was found to be 857 cm-1 in our spectrum. The close agreement between this value and that obtained by other researchers confirms the vibrational assignment of the dimethylzinc cation spectrum. Quantum calculations using Gaussian '94 will also be discussed on the neutral and ionic species studied.

148.

SPECTROSCOPIC STUDIES OF THE JET-COOLED VINOXY RADICAL. **Lori R. Brock** and Eric A. Rohlfing, Combustion Research Facility, Sandia National Laboratories, MS9055, Livermore, CA 94551

We use LIF excitation, hole-burning two-color resonant four-wave mixing (TC-RFWM), and dispersed fluorescence (DF) to characterize the $\tilde{B}\ ^2A$" - $\tilde{X}\ ^2A$" system of jet-cooled CH_2CHO. The LIF and TC-

RFWM excitation spectra reveal a large number of previously unobserved, weakly fluorescent bands. The collision-free lifetimes of 18 levels in the B state show a rapid increase in predissociation rate at ~1100 cm^{-1} above the origin, possibly due to vibronic interaction with the $\tilde{A}$ $^2A'$ or $\tilde{C}$ $^2A'$ states. High-resolution LIF spectra provide rotational constants for 6 levels in the B state that help confirm our vibrational assignments, particularly those involving out-of-plane (a") overtones. The DF spectra, obtained from excitation of 7 levels in the B state, show evidence of extensive IVR that mixes optically bright and dark ground-state levels. In total, we assign 54 levels in the B state and 56 levels in the first 3100 cm^{-1} of the X state, including the 6 a' fundamentals and 3 a" overtones in both states.

This work supported by the U. S. Department of Energy, Office of Basic Energy Sciences, Chemical Sciences Division.

149.

SPECTROSCOPY AND DYNAMICS OF THE LOWEST ELECTRONICALLY EXCITED STATE OF NO. **Hannelore Bloemink, Richard A. Copeland, and Tom G. Slanger, Molecular Physics Laboratory, SRI International, Menlo Park, California 94025**

The NO(a) state, lying 4.74 eV above the ground state, has been little studied, primarily because there have been no adequate production methods. Our recent investigations have provided the first spectroscopic gas-phase data on the a-X transition, and these studies have been extended to include collisional energy transfer measurements on the v = 11 vibrational level. Electronic energy transfer between the NO(a) and NO(B) states is the basis of the NO(a) detection method. In this work we examine the v = 8 level of NO(a), reporting excitation spectra of the a-X 8-0 band. We measure loss rate coefficients with various colliders, and we investigate the interaction between the a(v = 8) level and the neighboring B(v = 0) level, which is a determining factor in previous studies of apparent NO(B) reactivity.

150.

SPIN DYNAMICS METHOD OF COMPUTING ELECTRON AND NUCLEAR SPIN RELAXATION TIMES IN PARAMAGNETIC SOLUTIONS. **Robert R. Sharp and Shawn M. Abernathy,** Department of Chemistry, The University of Michigan, Ann Arbor, MI 48109

A spin dynamics (SD) simulation algorithm has been developed to compute nmr paramagnetic relaxation enhancements (nmr-PRE) produced by S≥1 metal ions in solution. This study is part of an experimental and computational effort to develop new methodologies for extracting spin Hamiltonian and molecular structure information from nmr-PRE experiments. The SD approach is analogous to molecular dynamics calculations, except that the quantum mechanical equations of motion of the spin operators are used to compute the time correlation functions of spin relaxation theory. SD simulations provide flexible, accurate and rapid calculations of nmr-PRE's under very general physical conditions, including spin Hamiltonians which contain non-commuting Zeeman, ZFS, and hyperfine interactions of arbitary magnitude. The program also simulates the effects of rapid Brownian reorientational motions in solution.

151. A THEORETICAL STUDY OF THE DIPOLAR AND SCALAR CONTRIBUTIONS FROM THE UNPAIRED ELECTRON TO ^{1}H NMR RELAXATION RATES IN LOW-SPIN METALLOPORPHYRIN COMPLEXES. **Konstantin I. Momot,** and F. Ann Walker, Department of Chemistry, University of Arizona, Tucson, AZ 85721

MO Huckel calculations and HF *ab initio* calculations have been used to determine the dipole-dipole and scalar contributions from the unpaired electron to the longitudinal and transverse ^{1}H NMR relaxation rates in several Fe^{3+} porphyrinates. The studied complexes included [TMPFe(III)(2-MeImH)$_2$]Cl (1), [(p-Cl)$_3$(p-NEt$_2$)TPPFe(III)(N-MeIm)$_2$]Cl (2), and [(p-Cl)(p-NEt$_2$)$_3$TPPFe(III)(N-MeIm)$_2$]Cl (3), where TMP=tetramesitylporphyrin, TPP=tetraphenylporphyrin. For the *ab initio* calculations, the molecules have been "cut" by removing some of the peripheral groups. The appropriate extent of cutting has been determined by observing the changes in the calculated molecular orbitals. The effects of peripheral substituents, axial ligand orientation, distortion of the metalloporphyrin core from planarity, and "pure" electronic effects (the effect of MO coefficients) are discussed. The computational results were interpreted in light of our previous experimental measurements of T_1s and T_2s of complexes (1) through (3) at a series of temperatures.

152. STUDY OF THE TEMPERATURE DEPENDENCE OF RELAXATION TIMES FOR SEVERAL PARAMAGNETIC METALLOPORPHYRIN COMPLEXES. **Konstantin I. Momot**, and F. Ann Walker, Department of Chemistry, University of Arizona, Tucson, AZ 85721.

Relaxation times (T_1 and T_2) temperature dependence has been studied for pyrrole protons of [(p-Cl)$_3$(p-NEt$_2$)TPPFe(III)(N-MeIm)$_2$]Cl, [(p-Cl)(p-NEt$_2$)$_3$TPPFe(III)(N-MeIm)$_2$]Cl, and [TMPFe(III)(2-MeIm)$_2$]Cl, where TPP=tetraphenylporphyrin, TMP=tetramesitylporphyrin. Similar trends have been observed in all complexes for variation of relaxation times with temperature. In the temperature range 200-265K, relaxation times increase as temperature increases. When temperature is increased further (up to 300K for the TPP complexes, and up to 273K for the TMP complex), relaxation times decrease. This latter phenomenon has been attributed to chemical exchange processes occuring due to rotation of imidazole ligands in the TMP complex and dissociation of imidazole ligands in the TPP complexes. In all complexes, T_2s are considerably shorter than T_1s. Relaxation times in the TMP complex are generally larger than corresponding values for the TPP complexes. Chemical shift dependence vs temperature has also been measured, and is close to the so called Curie behavior. Application of obtained T_1 data for measurements of NOE buildup in NOESY spectra will also be discussed.

153.

SYSTEMATIC SPECTROSCOPIC STUDIES OF 1-SUBSTITUTED-2-METHYLBENZIMIDAZOLES AND CORRELATION WITH THEORETICAL CALCULATIONS **R. Infante**, S. P. Hernandez and L. A. Rivera. Department of Chemistry, University of Puerto Rico, Mayagüez Campus, P. R. 00681

Benzimidazoles are heterocyclic compounds which show high biological activity. We have synthesized a family of 1-substituted-2-methylbenzimidazoles from base catalyzed susbtitution on 2-methylbenzimidazole using primary and secundary halides. Five 1-susbtituted-2-methylbenzimidazoles were prepared, purified and characterized by spectroscopic methods (GC-MS, FT-Raman, FT-IR, C-13 NMR, H-1 NMR and UV-Vis). These results were compared with theoretical calculations of vibrational spectroscopy and C-13 NMR chemical shifts predicted by Gauge Independent Atomic Orbital (GIAO) methods. Results of the calculations will also be presented.

154.

TEMPERATURE AND VIBRATIONAL LEVEL DEPENDANCE OF OH COLLISIONAL ENERGY TRANSFER. **Jean Lacoursiere**, Mark J. Dyer, and Richard A. Copeland, Molecular Physics Laboratory, SRI International, Menlo Park, CA 94025.

Emission from vibrationally excited OH(v = 1-9) in the ground electronic state is used to measure temperature, wind velocities, and periodic changes in earth's upper atmosphere. To model this emission, the collisional energy transfer rate constants and mechanisms for these excited states must be understood. At SRI we are studying these processes using a pump-probe technique. An infrared laser pulse excites OH radicals to a specific vibrational level, and a time-delayed ultraviolet laser pulse probes the decay of the excited state as a function of the collider gas partial pressure. Measurements of the vibrational removal rate constants for v = 8, 10, and 11 at room temperature will be presented. Interesting resonant behavior is observed with the colliders CO2 and N2O. The removal rate constant of v = 10 with the atmospherically important O2 collider are measured between 200 and 300 K.

155. THE INFLUENCE OF HIGH-FREQUENCY MODES IN ULTRASHORT PULSE SPECTROSCOPY. **Mark N. Kobrak**, Eric M. Hiller and Stuart A. Rice, The James Franck Institute, The University of Chicago, Chicago, IL 60637

Many molecules studied by ultrashort pulse spectroscopic techniques contain modes which have both short and long periods relative to the timescale of excitation. The influence of the high frequency modes on the ultrashort pulse absorption process is usually neglected. In this paper we examine the effect of the high frequency modes on the energetics of one- and two-pulse processes, and construct a method for calculating observables for polyatomic systems containing both high and low frequency modes. We compare our results with experimental data for the photodissociation of iodobenzene and the time-resolved fluorescence data in bacteriorhodopsin.

156.

ALN THIN FILMS GROWN BY MOCVD ON MAGNESIUM OXIDE AND THEIR CHARACTERIZATION. **Hugh H. Richardson** Noel P. Magtoto, Department of Chemistry and Condensed Matter and Surface Science Program, Ohio University, Athens, Ohio 45701

AlN is a semiconductor well-known for its excellent thermal conductivity, high electrical resistivity and superior piezoelectric properties. Considerable effort has been expended in growing AlN due to its possible application for blue and ultraviolet semiconductor optical devices. Currently the vast majority of research on AlN is focused on the wurtzite crystals grown largely on sapphire substrates. The difficulty arising from the thermal and lattice mismatched between AlN and the currently used substrates bars the way towards establishing this semiconductor as an excellent candidate for blue light emitting devices. Because the lattice parameters and thermal properties of MgO are comparable to those of AlN, we have explored the possibility of growing thin films of AlN on MgO (100). In this presentation, we shall present the results of FTIR, XPS and RBS measurements used to characterize these films..

157.

FTIR INVESTIGATION OF GLASS IONOMER DENTAL CEMENTS. S. Sneckenberger*, Z. Ouyang*, and E. Kao**, P. W. Jagodzinski*, Department of Chemistry* and School of Dentistry**, West Virginia University, Morgantown, West Virginia 26505

Glass ionomer dental cements are produced by the hardening reaction between aqueous polyacrylic acid (PA) and metal ions released during the decomposition of the glass ionomer particles. The need to improve the physical properties of these cements, while maintaining a reasonable working time, has led to the development of improved ionomer formulations and variations in the polyacid sidechains. FTIR has been used to monitor the hardening reaction involving a variety of polyacids with varying degrees of accessibility of the carboxyl groups to the aqueous metal ions. Decrease in the intensity of the COOH band of the free polyacids at 1710 cm^{-1}, and increases in the 1634 cm^{-1} band due to the Ca-PA complex and in the 1570 cm^{-1} band due to the Al-PA complex allow for convenient and unique reaction monitoring. Using this data, the cross-linking degree for each polyacid is calculated and the differences among glass ionomer cements are compared. Recommendations for optimum sidechain polyacid configurations will be presented.

158.

INFRARED SPECTRAL HOLEBURNING OF POLYMERS: A PROBE OF LOCAL STRUCTURE. **Zhan Chen,** Suli Fei and Herbert L. Strauss, Department of Chemistry, University of California, Berkeley, California 94720

Infrared holeburning studies of the ionic polymers poly(ammonium styrene sulfonate) and ammonium Nafion are reported. The ammonium groups are doped with a few percent of deuterium and the irradiation is in the N-D stretching bands. Both polymers show only broad spectral absorption bands, but the narrow spectral holes reveal details of the ammonium group interactions with their surroundings. The results are interpreted by comparison with the holeburning of the model compounds ammonium tosylate and ammonium triflate (J. Phys. Chem. 1996). Characteristic behavior of the holes shows that some of the local conformations of the models persist in the polymers. Particularly unusual are the changes of hole depth with time: for the Nafion, some holes decay for a period of time, then grow and then continue to decay -- a behavior also observed in the model triflate.

PHYS

159.

IR DETERMINATION OF MICRO-AGGREGATION IN MIXED PHOSPHOLIPID BILAYERS IN THE GEL STATE. **Wenhong Yan,** Herbert L. Strauss, Robert G. Snyder, Department of Chemistry, University of California, Berkeley, CA 94720

Lateral micro-aggregation in lipid bilayers of gel-state binary phospholipid mixtures has been measured over temperatures ranging from -19°C up to the main transition. The measurements were made using an isotope infrared method in which domain size (1-100 lipid molecules) is determined from the splitting (or width) of the methylene scissors band associated with the acyl chains. The method requires the chains of one component be deuterated. Micro-aggregation was studied in fully hydrated mixtures of the type $diC_n^D X/diC_{n'}^H Y$, where typically n=18, and n' =18, 20, 22, or 24, and X,Y = PC, PA, or PE. We report here primarily on mixtures whose components have like head groups but different chain lengths ($diC_{18}^D PC/diC_{n'}^H PC$, n'=20, 22, 24), and on mixtures with unlike head groups and equal or nearly equal ($diC_{18}^D X/diC_{20}^H Y$) chain lengths.

160.

MOLECULAR DYNAMICS, CIRCULAR DICHROISM AND ICE CRYSTAL GROWTH STUDIES OF SYNTHETIC ANTIFREEZE COMPOUNDS. A. Wierzbicki* C. A. Knight#, D. D. Muccio†, J. D. Madura*, and J. P. Harrington*,*Department of Chemistry, University of South Alabama, Mobile, Alabama 36688, #National Center for Atmospheric Research, Boulder, Colorado 80307, †Department of Chemistry, University of Alabama, Birmingham, Alabama, 35294

Results of our earlier investigations of shorthorn sculpin antifreeze protein structure-function relationship in the antifreeze activity of this protein suggested a possibility of designing synthetic alanine/lysine based antifreeze polypeptides. We will present the results of our recent investigations in which we designed, synthesized and analyzed for antifreeze activity, using nanoliter osmometer, several alainie/lysine rich macromolecules. Molecular dynamics, circular dichroism and ice crystal growth studies of these compounds will be presented and the design of synthetic noncolligative antifreeze polypeptides will be discussed.

161.

MULTISITE INCLUSION COMPLEXATION OF REDOX ACTIVE DENDRIMERIC FERROCENE GUESTS. René Castro; Isabel Cuadrado; Carmen M. Casado; Moisés Morán; and Angel E. Kaifer, Department of Chemistry, University of Miami, Coral Gables, Florida, 33124, and Departamento de Química Inórganica, Facultad de Ciencias, Universidad Autónoma de Madrid, Cantoblanco 28049-Madrid, Spain.

Three generations of polyimine dendrimers were synthesized having 4, 8, and 16 peripheral ferrocene units and their complexation with B-cyclodextrin (B-CD) probed using electrochemical and spectroscopic techniques. The water insoluble dendrimers are solubilized upon complexation of the exterior ferrocene units by B-CD. Competitive binding experiments show that the complexation is of the inclusion type. Upon progressing to the larger generations, the steric environment around the ferrocene units becomes highly congested and effects due to this are seen in the solubility behavior and in the electrochemical experiments. An additional wave in the voltammetric experiments involving the 16 unit dendrimer (not seen in the 4 or 8) suggests that not all of the ferrocene units are available for complexation. This is the first example of an inclusion complexed dendrimer.

162.

POLYMERIZATION OF ACRYLONITRILE COUPLED TO AN OSCILLATING REACTION. **Randy P. Washington** John A. Pojman, Dept. of Chemistry and Biochemistry, S. S. Box 5043, University of Southern Mississippi, Hattiesburg, MS, 39406

Polymerization of acrylonitrile has been shown to occur periodically with oscillations in a batch reactor. Two radicals that are formed in the Belousov-Zhabotinsky system play a major role. The malonyl radical initiates polymerization, while the bromine dioxide radical is suspected of terminating the polymer chain. We discuss here the study of acrylonitrile polymerizing in the Belousov-Zhabotinsky and Racz oscillating reaction in a batch, semibatch and continuous-flow stirred tank reactor. The molecular weight distribution of polyacrylonitrile as a function of the oscillating reaction's dynamics for the various reactor's designs. Temperature profiles, light transmittance (monitoring polymer precipitation), and platinum potential data indicate that the dynamics of the reaction affects the molecular weight distribution of the resulting polymer. Structural analysis using NMR and IR of the polymer is also presented.

163. SEGMENTAL MOTIONS OF POLY(ETHYLENE TEREPHTHALATE) COPOLYMERS BY SOLID-STATE NMR AND THERMAL ANALYSIS. D. M. Connor, H. W. Beckham, and D. M. Collard, School of Chemistry and Biochemistry, Georgia Institute of Technology, Atlanta, Georgia 30332-0400

The effects of comonomers added to poly(ethylene terephthalate) (PET) as crystallization modifiers are not well understood on the molecular scale. We are using solid-state NMR to examine the effect of these comonomers on segmental motions. PET was synthesized with a series of difunctional comonomers (1,4-cyclohexanedimethanol, isophthalic acid, 2-methyl-1,3-propane diol, and diethylene glycol) under identical conditions to provide well-defined materials. Thermal analysis by DSC indicates a large influence of comonomers on the rate of crystallization, which can be assessed by Avrami plots. Broad-line and CP-MAS spectra as well as T_1 and $T_{1\rho}$ relaxation times are being measured. This information will provide a foundation for the design of new additives or comonomers for controlling the degree of crystallinity, crystallite size, and crystallization rate in semicrystalline PET.

164. SMALL ANGLE NEUTRON SCATTERING AND POSITRON ANNIHILATION INVESTIGATIONS OF METAL ALLOYS FOR HYDROGREN STORAGE. Peter J. Hall, L. E.A.B. Berlouis, A. J. Mackinnon, Department of Chemistry, University of Strathclyde, Glasgow G1 1X1, UK

A series of FeTi based matel alloys have been produced by vapour deposition. The alloys were loaded with hydrogen both electrochemically and under pressure. The fresh, hydrogen loaded and deloaded samples were characterized by SANS and positron annihilation. Thermal analytical techniques were used to determine the degree of hydrogen loading. SANS showed an increase in the voidage of the alloys following repeated loading and deloading. Positron annihilation of the loaded samples revealed three distinct sites in the structure: solid hydride, empty voids and high pressure gas bubbles. The ratio of these sites varied with the alloy composition and method of loading. The possibility of producing an alloy to maximize the degree of high pessure hydrogen storage is discussed.

165. SOLUTION AND SOLID STATE PHOTOPHYSICAL CHARACTERIZATION OF POLYFLUORENE DERIVATIVES USED IN ORGANIC LIGHT-EMITTING DIODES. Julie A. Teetsov and Marye Anne Fox, Department of Chemistry and Biochemistry, University of Texas at Austin, Austin, Texas, 78712

Organic LEDs are of interest because of the significant processing, mechanical and size advantages they offer as compared with conventional inorganic semiconductors. Because of high fluorescence quantum yields, rigid backbone π-conjugated alkyl substituted polyfluorene derivatives are under investigation as the semiconducting and radiative component in LED devices. A photophysical study of fluorescence quantum yields and fluorescence lifetimes of a series of polyfluorenes disubstituted at the 9 position with alkyl chain lengths increasing from C_6 to C_{18} as solution and thin films has been conducted. Phase transition behavior and the effects of introducing liquid crystalline order to thin films is being investigated. Solution and thin film response to photoexcitation will be compared to LED electroluminescence in order to determine more precisely the conditions that provide optimal light emission in LEDs.

166.

SYNTHESIS AND EVALUATION OF NAPHTHALIMIDE DERIVATIVES AS ORGANIC LIGHT EMITTING DIODE MATERIALS. **D.L. Orth** and B.C. Anderson, Department of Chemistry, Wisconsin Lutheran College, 8800 W. Bluemound Rd., Milwuakee, WI 53226 and M.E. Thompson, Department of Chemistry, University of Southern California. Los Angeles 90089-0744

Organic Light Emitting Diodes (OLEDs) hold great promise in color display technology. Important to their development is the identification of good emitters in the red, green, and blue regions. N-propyl-4-bromo-1,8-naphthalimide is known to fluoresce green (530 nm) in ethanol. However, the emission depends greatly upon solvent, and the material does not emit green under the conditions of an OLED. We present emission data for several other naphthalimides as possible molecules for incorporation into OLEDs.

167. THEORY FOR LONG TIME DYNAMICS OF CHAIN MOLECULES: TESTS FOR THE INTERNAL DYNAMICS OF BIOPOLYMERS. K. S. Kostov and K. F. Freed, James Franck Institute, University of Chicago, Chicago, IL 60637

A recent theory, which extends the optimized Rouse-Zimm model for polymer dynamics, is tested for the long time dynamics of polyglycines, the neurotransmitter met-enkephalin, and the vasoconstrictor endothelin. The theory provides a systematic procedure for including the contributions of the neglected internal friction and memory functions. The peptide dynamics are described by time correlation functions expressed in terms of the eigenvalues and eigenfunctions of the diffusion operator. The eigenvalues are found by expanding the eigenfunctions in a basis set. The basis functions are chosen using a sorting procedure which greatly reduces the number of functions required to account for the memory functions. Side chain conformational transitions (including aromatic ring flips) in met-enkephalin and endothelin present a new level of complexity as compared to the previously studied alkanes and glycines. The theory is compared with Brownian dynamics simulations so that both simulations and theory use identical and realistic potentials and the same solvent model. Inclusion of the memory functions results in excellent agreement with the simulations. The new method describes motion on time scales more than an order of magnitude longer than those accessible to molecular dynamics simulations. The results indicate that the theory can be successfully applied to biologically relevant peptides and proteins.

168. ACCURATE CALCULATION OF MOLECULAR ENERGETICS. T. H. Dunning, Jr., K. A. Peterson, T. van Mourik, and A. K. Wilson, Environmental Molecular Sciences Laboratory, Pacific Northwest Laboratory, Richland, WA 99352 and D. E. Woon, Molecular Research Institute, 845 Page Mill Road, Palo Alto, CA 94304.

Using advanced methods for treating electron correlation and correlation consistent basis sets, it is now possible to compute accurate energetics for small molecules. We illustrate the accuracy achievable with both perturbation theory and coupled cluster techniques and basis sets up to sextuple zeta. A number of different molecules will be considered, including ArHX, the CH_n/C_2H_n series, HCO^x (x=+1,0,-1), and H_nSO_m.

The Pacific Northwest Laboratory is operated for the U.S. Department of Energy by Battelle Memorial Institute under Contract No. DE-AC06-76RLO 1830.

169.

APPLICATION OF B3LYP TO TRANSITION METAL SYSTEMS. **Charles W. Bauschilcher, Jr.** and Harry Partridge, STC-230-3, NASA Ames Research Center, Moffett Field, CA 94035.

The application of B3LYP to several systems will be presented. Examples include the successive metal-ligand bond energies for Sc(OH)+n and the vibrational frequencies of metal oxide systems. While the B3LYP is shown to work well for many systems, provided that errors in the atomic splitting are accounted for, we illustrate some cases where the B3LYP results are not as accurate as required. In these cases, we find that CCSD(T) energetics are in excellent agreement with experiment. The B3LYP frequencies are compared with CASSCF results for some cases where near degeneracy effects are large.

170. ACCURATE THERMOCHEMISTRY FOR MEDIUM-SIZED AND LARGE MOLECULES
Krishnan Raghavachari and Boris B. Stefanov, Bell Laboratories, Lucent Technologies, Murray Hill, NJ; Larry A. Curtiss, Argonne National Laboratory, Argonne, IL;

Accurate techniques such as Gaussian-2 (G2) theory have been proposed in recent years to evaluate the thermochemistry of small molecules from first-principles. However, as the molecules get larger, the errors in G2 theory and similar approaches tend to accumulate. For example, the computed heats of formation of benzene and naphthalene with G2 and G2(MP2) theories, respectively, have errors of 3.9 and 7.2 kcal/mol.

In this work, we explore strategies for computing accurate heats of formation for medium-sized and large molecules. In our first scheme, G2 theory is combined with isodesmic bond separation reaction energies to yield accurate thermochemistry for larger molecules. For a test set of 40 molecules composed of H, C, O, and N, our method yields enthalpies of formation, ΔH_f^0(298 K), with a mean absolute deviation from experiment of only 0.5 kcal/mol. This is an improvement of a factor of three over the deviation of 1.5 kcal/mol seen in standard G2 theory. Similar techniques can also be applied to larger molecules using gradient corrected density functional techniques. We present accurate thermochemical information on the isomers of C_{20}, a notoriously difficult molecule to describe by means of conventional calculations.

171.

QUANTUM MONTE CARLO AS A HIGH-ACCURACY METHOD FOR TREATING CHEMICAL REACTIONS; Jeffrey C. Grossman, *University of California, Berkeley, 94720*, Lubos Mitas, *University of Illinois, Urbana-Champaign, 61801*

Barrier heights (BH) and heats of formation (ΔH) have been calculated for a variety of reactions involving sp atoms and molecules by the quantum Monte Carlo (QMC) methods as well as seven other standard first principles methods. QMC calculations of BH and ΔH are within 1 Kcal/mol (0.04 eV) of experimental values for all reactions considered. By contrast, we have found significant deficiencies of density functional and related methods for these systems. The transition state of the largest reaction studied, involving C_8H_8, has recently been observed experimentally to be an open-shell singlet. While this challenging case cannot be correctly described by the traditional approaches, we demonstrate that QMC is the *only* method which agrees well with experiment.

172.

FIRST EVIDENCE FOR ANCHIMERIC SPIN DELOCALIZATION IN TRYPTOPHAN RADICAL CATION Susan E. Walden and Ralph A. Wheeler Department of Chemistry and Biochemistry, University of Oklahoma, 620 Parrington Oval, Room 208, Norman, Oklahoma, 73019

From redox enzymes and photosynthesis to protein damage due to chemical and radiation exposure, the biological importance of amino acid radicals is only recently emerging. For example, the tryptophan radical cation, TrpH•+, is now believed to serve as an electron transfer intermediate in CcP compound I and DNA photolyase. To aid the experimental detection and characterization of TrpH•+, a hybrid Hartree-Fock/density functional method, B3LYP, was used to predict its structure and spin density distribution. We have the first evidence that in certain minimum energy conformations the unpaired spin from the positively charged indole side chain is delocalized onto the heteroatoms of the alanyl group. As this is similar to the neighboring group stabilization of a vacant p-orbital of a carbocation, we have named this effect anchimeric spin delocalization (ASD). We present our evidence for ASD including conformational dependence and discuss its implications for electron transfer in proteins.

173.
QUANTUM CHEMICAL STUDIES OF FUNCTIONALIZED FULLERENES.
Celeste M. Rohlfing, Combustion Research Facility, Sandia National Laboratories, Livermore, CA 94551–0969

Quantum chemical methodologies can assist the tailoring of the physical and chemical properties of molecular materials. The targeted functionalization of fullerenes is an important goal in realizing the eventual applications of these novel materials. Several examples will be presented in which ab initio predictions of relative isomer stabilities for $C_{60}H_2$, $C_{60}H_4$, $C_{70}H_2$, and $C_{60}(CH_2)_2$ were critical in the synthesis and characterization of these fullerenes.

174. DETECTION AND MANIPULATION OF SINGLE MOLECULES AND NANOENVIRONMENTS IN ROOM-TEMPERATURE SOLUTIONS
Richard N. Zare and Daniel Chiu. Department of Chemistry, Stanford University, Stanford, CA 94305-5080

Advances in optical microscopy have permitted not only the detection of a single molecule, but also its manipulation in a room-temperature solution. It is also important to exert control on the nanoenvironment surrounding a single molecules to investigate and interpret its behavior. Various approaches are being pursued to achieve these feats. Uses of optical trapping, vesicles, and capillary electrophoresis are described in which singe molecules are trapped, manipulated, and transferred. The use of liposomes to create suitable nanoenvironments for single-molecule experiments is also presented.

175, DETECTION OF SINGLE FLUORESCENT MOLECULES IN MICRODROPLETS: EXPERIMENT AND PHOTOPHYSICS. J. Michael Ramsey, Chemical and Analytical Sciences Division, Oak Ridge National Laboratory, P. O. Box 2008, Oak Ridge, TN 37831-6142

Several laboratories around the world are now working on experiments to detect single molecules. Many elegant experiments have been performed on single molecules in host crystals at liquid helium temperatures, and more recently near field optical microscopy experiments have probed single molecules on surfaces. Single molecules in liquids have been probed directly or indirectly through an amplification process. All of these studies have used fluorescence as the technique for observing the molecule. The most challenging of all these experiments is the direct detection of a single molecule in solution. Finite photochemical lifetimes limit the fluorescence signal from a single molecule. This limitation necessitates extremely small probe volumes ($\approx$ 1 pL) for detection. Our laboratory has chosen to use microdroplets with diameters of approximately 10 μm or less. Early experiments on levitated droplets have allowed detection of single R6G molecules in 10-μm diameter droplets of glycerol with a signal-to-noise ratio of 50. More recent experiments in droplet streams have yielded single molecule detection results with a signal-to-noise ratio near 10. These results were obtained with analysis rates of $\approx$ 2 Hz but higher rates are feasible. The microdroplet format also allows the observation of cavity QED enhanced spontaneous emission rates and through solvent mediated effects, enhancements in fluorescence yields. Thus the possibility of increased detection sensitivity in the microdroplet format is promising.

176. SINGLE MOLECULE FLUORESCENCE BURST COUNTING OF DNA IN CAPILLARY ELECTROPHORESIS: OPTIMIZATION AND APPLICATION TO DETECTION OF 100-1000 BP FRAGMENTS. Brian B. Haab and Richard A. Mathies, Department of Chemistry, University of California, Berkeley, CA 94720

Single molecule fluorescence burst counting is a highly sensitive and effective method for detecting capillary electrophoresis separations of ds-DNA fragments (Haab and Mathies, *Anal. Chem.*, **34**, 3253). Confocal microscopy is used to excite and observe fluorescence bursts from individual DNA fragments labeled with the intercalation dye TO6 as they electrophorese through the ~2 μm diameter focused laser beam. In previous work, separations of 4-7 kbp DNA fragments were easily detected when only 50-100 molecules passed through the probe volume. In current work, the concentration and laser power were optimized by studying fluorescence burst intensities from pBluescript DNA as these parameters were varied. The optimal dye concentration was $\leq$ 100 nM, and the optimal laser power was $\leq$ 1 mW. To explore the fragment size limit of detection under the optimized conditions, single molecule counting was applied to separations of a 100-1000 bp DNA sizing ladder in 3% linear polyacrylamide. Discrete and baseline resolved fluorescence bursts were observed in each of the bands, and the average burst size within each band increased linearly with fragment size. The mean occupancy in the probe volume ranged from 0.002 to 0.02, indicating that the fluorescence bursts arise from individual DNA fragments. This method should be valuable for the detection of trace bacterial or viral contamination and in DNA-based health care diagnostics.

177. SINGLE-MOLECULE DETECTION AND SPECTROSCOPY BY SURFACE-ENHANCED RAMAN SCATTERING. Shuming Nie and Steven R. Emory, Department of Chemistry, Indiana University, Bloomington, IN 47405

Recent advances in ultrasensitive instrumentation have allowed the detection, identification, and dynamic studies of single molecules in low-temperature solids, in room-temperature liquids, and on dielectric surfaces. This capability opens many opportunities for scientists to address current problems and to explore new frontiers in various disciplines such as analytical chemistry, molecular biology, and nanostructured materials. Current methods for probing single molecules, however, are restricted to a limited set of fundamental principles, including laser-induced fluorescence, frequency-modulated optical absorption, and microelectrodes. We report a fundamentally new methodology based on surface-enhanced Raman scattering (SERS) to study single molecules and single nanoparticles at room temperature. Individual silver colloidal nanoparticles were screened from a large heterogeneous population for special size-dependent properties and were used to amplify the spectroscopic signatures of adsorbed molecules. For single rhodamine 6G molecules adsorbed on the selected nanoparticles, the intrinsic Raman enhancement factors are on the order of 10^{14}-10^{15}, much larger than the ensemble-averaged values derived from conventional measurements. This enormous enhancement leads to vibrational Raman signals that are more intense and more stable than single-molecule fluorescence.

178. SINGLE MOLECULE ELECTROCHEMISTRY, Allen J. Bard and Fu-Ren F. Fan. Department of Chemistry and Biochemistry, The University of Texas at Austin, Austin, TX 78712

The scanning electrochemical microscope (SECM) can be used to observe the electrochemistry of single molecules. Specially constructed ("trapping") tips of sharpened Pt-Ir wire in a wax sheath are used to trap a small volume of solution containing a few molecules between the conductive tip electrode and a substrate electrode. Because the trapped molecules shuttle rapidly between tip and substrate a single molecule will undergo repeated electron transfer reactions and produce a current of the order of a pA.

179.

DETECTION OF INDIVIDUAL CHEMICAL REACTIONS BY ELECTROGENERATED CHEMILUMINESCENCE. **R. Mark Wightman**; Ernest Ritchie, Department of Chemistry, University of North Carolina at Chapel Hill, Chapel Hill, NC 27599-3290

Individual chemical reactions that result in photon emission have been observed following electrochemical generation of radical ions with a microelectrode. Potential pulses of submillisecond duration limit the volume of solution where reactions occur to the femtoliter region adjacent to the electrode. In dilute solution, the reactions occur at a sufficiently low rate that each can be counted. The reaction events were found to follow Poison statistics.

180.

LAYER-BY-LAYER GROWTH OF SIO2 FILMS ON SILICON SURFACES
H. Hoffmann, U. Mayer, H. Brunner, T. Vallant,
Department of Inorganic Chemistry, Technical University of Vienna, A-1060 Vienna, Austria

A novel procedure for a controlled monolayer growth of silicon oxide films on silicon substrates is presented. It is based on a binary A-B reaction sequence involving the formation of an alkylsiloxane monolayer through self-assembling of alkyltrichlorosilane precursor molecules from solution (Step A) followed by UV/ozone oxidation of the hydrocarbon groups (Step B), which leaves behind a monolayer of silicon oxide (d = 0.27 nm) on the surface. Repeated application of this A-B cycle results in a layer-by-layer growth of the oxide film with a strictly linear thickness increase of 0.27 nm per cycle, as evidenced by ellipsometry and infrared reflection spectroscopy. The mechanisms of the film formation and oxidation processes will be discussed and some technological applications of this method will be considered.

181.

ATOMIC LAYER CONTROLLED GROWTH OF SIO2 FILMS USING SELF-LIMITING SURFACE CHEMISTRY. **Jason W. Klaus** A. W. Ott, J. M. Johnson and S. M. George, Department of Chemistry and Biochemistry, University of Colorado, Boulder, CO 80309-0215

SiO2 thin films were deposited with atomic layer control using self-limiting surface chemistry. The SiO2 growth was accomplished by separating the binary reaction SiCl4 + 2H2O --> SiO2 + 4HCl into two half-reactions. Successive application of the half-reactions in an ABAB... binary reaction sequence produced SiO2 deposition at 600-800 K and reactant pressures of 1-10 Torr. The self-limited growth of the SiO2 thin films was monitored with an in-situ spectroscopic ellipsometer. SiO2 growth rates were measured versus temperature and reactant exposure time. The deposition rate per AB cycle was 1.17 A/AB cycle at 600 K, 0.87 A/AB cycle at 700 K and 0.74 A/AB cycle at 800 K. The deposition rate per AB cycle was shown to be determined by the coverage of Si-OH* surface species. The surface topography of the SiO2 films measured using atomic force microscopy was extremely smooth with a roughness nearly identical to the initial substrate.

182. GROWTH OF METAL PARTICLES AND THIN FILMS BY MOCVD. B. Fraser,[a] A. Hampp,[a] Tue Ngo,[a] R. Stanley Williams,[a,b] H. D. Kaesz,[a] N. Viguier,[c] and F. Maury,[c] [a]Dept. of Chem. & Biochem., Univ. of California, Los Angeles, CA 90095-1569, USA; [b] Hewlett Packard Labs., Palo Alto, CA 94304, and [c] CNRS-INPT, Ecole Nationale Supérieure de Chimie, 31077 Toulouse cedex 4, FRANCE.

Monodisperse nickel particles were grown on Si(100) by MOCVD from $Ni(\eta^5-C_5H_4Me)$ under hydrogen at 230 °C. Precursor temp. (35-65 °C) and deposition time were used to control particle size (100 to 500 nm) as measured by SEM; relative standard deviation of particle size is 15%. The observed particle distribution is controlled by island nucle-ation and growth as modelled by kinetic theory. Growth patterns can be altered by the nature of additives in the gas stream.

Thin films of β-CoGa have been successfully grown on (100)GaAs substrates by metalorganic molecular beam epitaxy from the separate precursors $(\eta^5-C_5H_5)Co(CO)_2$, **1**, and Et_3Ga, **2**. The composition of the intermetallic films is adjustable by control of the composition of the gas phase to achieve lattice match between a Ga-rich CoGa film and the GaAs substrate. CoGa films of the desired cubic β (CsCl) structure are grown epitaxially on (100)GaAs in the temperature range 260 - 300°C from gas phase compositions **2**: **1** ≥ 5:1. These films are thermodynamically stable on GaAs up to 500 °C.

183. MODELLING CHEMICAL VAPOR DEPOSITION OF EARLY TRANSITION METALS, F. A. Houle and W. D. Hinsberg, IBM Almaden Research Center, 650 Harry Road, San Jose, CA 95120 and K. A. Singmaster, Department of Chemistry, San Jose State University, San Jose, CA 95193.

Deposition of clean films from the early transition metals by chemical vapor deposition (CVD) is seriously complicated by the reactivity of these elements. Through a combination of experiment and simulations on a prototypical system, the Cr, Mo and W hexacarbonyls, we have developed a model that provides new insight to the links between film composition and the macroscopic deposition process. The model includes full chemical detail such as elementary reactions and gas diffusion, adsorption and desorption, allowing identification of processes that control the incorporation of C and O contaminants. Of particular importance are kinetic factors introduced by the coupling between gas transport and surface chemistry, an aspect of the reaction that is not accessible in standard surface science experiments. These factors influence the effective residence time of surface adsorbates, and hence branching ratios between competing reactions. A model of this complexity is not complete, and areas where improvement is needed will be noted. Implications of the kinetics for optimal design of CVD processes will be discussed.

PHYS

184. ATOMIC LAYER EPITAXY (ALE) FOR II-VI MATERIAL GROWTH: A DETAIL STUDY OF THE SURFACE CHEMISTRY
Ming Han, Yi Luo, Dave A. Slater, Joseph E. Moryl, Richard M. Osgood, Columbia Radiation Laboratory, Columbia University, New York, NY 10027

Atomic layer controlled growth of CdS on a ZnSe(100) surface has been studied. Growth is achieved through the operation of a binary chemical reaction sequence employing the sequential supply of the precursors $(Cd(CH_3)_2)$ and H_2S to a ZnSe substrate at room temperature. This approach improves the interface abruptness by reducing material interdiffusion frequently observed in a high temperature ALE process. We present a study of the surface chemistry and kinetics underlying this growth process. The chemical composition of the each grown layer was characterized by AES, XPS, and LEIS, and LEED was employed to study the surface order. These techniques show this binary reaction sequence results in self-limiting growth at each half step of the reaction sequence, in turn, leading to the growth of a single layer of CdS during each full reaction cycle with a high quality epitaxial film obtained. TPD and NEXAFS are used to further probe the reaction kinetics, the surface reaction species and their geometry in each half reaction.

185. REACTIONS OF TRIETHYLGALLIUM ON GAAS(100). **K. C. Wong,** M. T. McEllistrem, M. S. Jackson and J. G. Ekerdt, Science and Technology Center, University of Texas Austin, Texas 78712

The thermal decomposition reaction mechanisms of triethylgallium (TEGa), d_0 and d_{15}, were studied on the GaAs(100) surface, Ga- and As-rich, using temperature programmed desorption (TPD) and static secondary ion mass spectroscopy (SSIMS). TPD studies reveal the ethyl ligands undergo homolysis and hydride elimination with a peak temperature near 600 K. Homolysis occurs at a slightly lower temperature than hydride elimination. There is no evidence for alkane formation on the surface. A kinetic isotope effect of 0.083 eV is observed that is consistent with a transition state for hydride elimination involving C-H bond scission. SSIMS studies reveal TEGa dissociatively adsorbs at temperatures as low as 150 K; the dissociated ethyl groups migrate to Ga lattice sites. TPD and temperature programmed SSIMS results are consistent with a model in which it is the lattice-site ethyl groups that undergo hydride elimination. The studies further suggest that ligand removal, and in turn precursor deposition, will be limited by the availability of Ga lattice sites during MOCVD.

186. ROLE OF SURFACE ELECTRONIC STATES IN DETERMINING CVD GROWTH RATES ON SI(100). **D.J. Doren** and R. Konecny, Department of Chemistry and Biochemistry, University of Delaware, Newark, Delaware 19716

Low temperature film growth on Si from hydride precursors is limited by the rate of desorption of hydrogen to create dangling bond sites on the surface. A great deal of effort has gone into studying this desorption reaction with no clear consensus on the mechanism. New theoretical work demonstrates a desorption mechanism involving an excited surface electronic state that rationalizes several paradoxes in the experimental data on desorption. The mechanism implies that the desorption rate can be altered by chemical effects that change the energy of the surface electronic excited state. This is proposed as the explanation for the known effects of dopants on silicon film growth rates (growth is typically faster while doping with p-type dopants, and slower while doping with n-type dopants). Implications for growth of Si-Ge alloys will also be discussed.

187. "ULTRAFAST STUDIES ON THE HYDRATED ELECTRON AND ELECTRON TRANSFER IN DNA" P.F. Barbara, C. Silva, P.K. Walhout, K. Yokoyama, E.J.C. Olson and D. Hu, University of Minnesota, Department of Chemistry, Minneapolis, MN 55455

Femtosecond spectroscopy has been used to study the excited-state solvation dynamics of the hydrated electron. The data clearly reveal dynamics on the ~30fs time scale which we assign to the inertial molecular dynamics of the solvent cage. The implications of these data on validity of theoretical models for the solvated electron will be described in detail. On a separate subject, the distance dependence of electron transfer (ET) rates between donors and acceptors in a DNA environment has been an area of intense research and heated discussion. We will describe a quantitative model for the photodynamics and gel electrophoresis/photocleavage experiments of the untethered DNA/metallointercalator system. Comparison of the simulation results with a broad range of experimental data strongly indicate that a fairly typical distance dependence for ET ($\beta \approx 1$ Å^{-1}) coupled to donor/acceptor clustering is able to account for all features in the data.

188. DIFFUSION IN QUASI-TWO DIMENSIONAL COLLOID SUSPENSIONS. S. A. Rice, The James Franck Institute, The University of Chicago, Chicago, IL 60637

Recent experimental and theoretical studies of diffusion in quasi-two dimensional colloid suspensions will be reviewed. Attention will be focused on the functional form of the crossover from short time to long time dynamical behavior, and how that behavior depends on the colloid particle concentration.

189. ELECTRICAL MANIPULATION OF MOLECULES IN PLANAR SUPPORTED BILAYERS. Steven G. Boxer, Jay T. Groves, Nick Ulman, Department of Chemistry, Stanford University, Stanford, CA 94305-5080

Unilamellar phospholipid vesicles fuse spontaneously with a hydrophilic surface, such as that of appropriately treated glass, to form an extended, planar, bilayer membrane. This structure is generally separated from the solid support by 10-20 Å of water which serves to preserve the fluidity of both surfaces of the membrane. These supported bilayers have many properties in common with native biological membranes and are a useful tool for studying membrane properties. Experiments demonstrating the fluidity will be presented. Electric fields can be used to generate concentration gradients of charged components in confined regions of the supported membrane. A quantitative model describing the concentration profiles will be presented.[1] This technique is used to manipulate and concentrate charged lipids and GPI-linked forms of MHC complexes.[2] When components are not dilute, the behavior of these concentration profiles provides information on the effective molecular size. Photosynthetic reaction centers are used to demonstrate the orientation and function of membrane proteins in supported bilayers.[3] Fabrication of barriers to lateral diffusion using photolithographic techniques will be described.[4] This and more sophisticated field geometries open new opportunities to study complex dynamics in two-dimensional systems and to manipulate interactions with living cells.
[1] J.T. Groves and S. G. Boxer, *Biophys. J.* **69**, 1972-5 (1995); .[2] J.T. Groves, C. Wolfing, and S.G. Boxer, *Biophys. J.* **71**, 2716- (1996).[3] J. Salafsky, J.T. Groves and S.G. Boxer, *Biochemistry* **35**, 14773-14781 (1996); J.T. Groves, N. Ulman and S.G. Boxer, *Science*, in press.

190. AB INITIO STUDIES OF ACID SOLUTIONS, M. L. Klein, Department of Chemistry, University of Pennsylvania, Philadelphia, PA 19104-6202

The structure and dynamics of liquid HF with dissolved KF has been investigated. Specifically, a system with stoichiometry KF-2HF was simulated using Car-Parrinello ab initio molecular dynamics at both 400 K and 1000K. The system, which was started from a phase separated mixture, rapidly formed into solvated potassium ions and $H_nF_{n+1}^-$ polyfluoride anions, with n=1, 2, 3, and 4. These polyfluoride anions were classified and their structures and dynamical behavior were compared with the known structures and spectra of crystalline compounds KF-xHF and with theoretical predictions of isolated gas phase species. The present study reveals dramatic frequency shifts in the HF stretching mode with variation in the HF coordination number of the polyfluoride anion. Previous infrared modes assigned as combination bands have been reinterpreted as fundamental modes of vibration of the polyfluoride anions.

191. QUANTUM MONTE CARLO BINDING ENERGIES FOR SILICON HYDRIDES. W. A. Lester, Jr. and C. W. Greeff, Chemical Sciences Division, Lawrence Berkeley National Laboratory and Department of Chemistry, University of California at Berkeley, Berkeley, CA 94720-1460

Variational and diffusion Monte Carlo calculations using effective core potentials have been carried out on SiH_n (n = 1-4), Si_2, Si_2H_6, and Si_3H_8. The atomization energies of SiH - SiH_4 are in good agreement with experiement and with the best other calculations. The cumputed electron affinity of SiH_3 is within 0.02 eV of the measured value. The calculated atomization energies of SiH_4, Si_2H_6 and Si_3H_8 are found to be more consistent with the original heats of formation of Gunn and Green, without subsequent corrections made on the assumption that the final state of Si is amorphous. (This work was supported by the Director, Office of Energy Research, Office of Basic Energy Sciences, Chemical Sciences Division of the U.S. Department of Energy under Contract No. DE-AC03-76SF00098.)

192. GENERALIZED MOLECULAR ORBITAL THEORY II (GMO2). Michael B. Hall Department of Chemistry, Texas A&M University, College Station, Texas 77843-3255; Marc Couty, Laboratoire de Chimie Théorique, Université de Marne-la Vallée, M2, Bois de l'Etang, rue Galilee, 77420 Champs sur Marne, FRANCE

The generalized molecular orbital (GMO) concept is extended to a higher order method, which begins with a pair-excited multi-configuration self-consistent field (PEMCSCF) for the orbital optimization, and is followed by a multi-reference configuration interaction calculation. Here, this method is referred to as GMO2. The method has the advantage of being variational, of handling large numbers of active electrons, and of only needing the user to specify the number of active electrons and orbitals without specifying a dominant MO or VB configuration. We will briefly review the PEMCSCF theory, describe in more detail a new and more efficient optimization procedure, and propose determining the energy with configuration interaction (CI) at the single, double, triple and quadruple excitations level (SDTQ) as a replacement for the full CI, which is needed in a complete active space (CAS) method. Several examples of the application of the method are investigated: methane, tetrahydrogen, benzene, dinitrogen dissociation, acetylene dissociation. For the five systems studied, the PEMCSCF orbital optimization produces orbitals that only differ from those of a CASSCF by an average of 4 kcal/mol when both localized bond type or symmetry adapted orbitals are used in a full CI GMO2(FCI). The additional error of replacing full CI with a SDTQ CI, GMO2(SDTQ) is usually less than 1 kcal/mol.

193. INTERACTIONS OF WATER WITH NUCLEIC ACID BASES AND THEIR H-BONDED COMPLEXES: A RIGOROUS AB INITIO STUDY.
Jerzy Leszczynski, Dept of Chemistry, Jackson State University, Jackson, MS 39217

Reliable modeling of the DNA molecule and its interactions is very difficult. The results of accurate ab initio calculations with inclusion of electron correlation contributions have significantly modified our view on nucleic acid bases (NABs) and the mechanisms of their interactions. This talk will discuss recent developments in our laboratory concerning: (i) Planarity of NABs. The amino groups of NABs are now believed to be very flexible and intrinsically nonplanar, allowing for hydrogen-bond-like interactions oriented out of the NAB planes. Many H-bonded DNA base pairs have been shown to be intrinsically nonplanar. (ii) Proton transfer in NAB and their complexes. Despite small energy differences between oxo-hydroxy tautomers of NABs, the energy barrier for such processess is significant. However, explicit inclusion of the water molecules of the first solvation shell results in a notable decrease in such barriers. (iii) Interaction of hydrogen-bonded DNA base pairs with water molecules. The first solvatation shell of 8 (GC) or 7 (AT) water molecules stabilizes these complexes and changes the geometric arrangements.
All presented calculations were carried out at the HF and MP2 levels of theory with basis sets of at least 6-31G(d) quality. Molecular parameters of the considered systems were fully optimized, and harmonic vibrational frequencies were used to determine if the considered structures are minimum-energy or transition structures.

194.

QUANTUMCHEMISTRY OF ZEOLITE ACIDITY Rutger A. van Santen Laboratory of Inorganic Chemistry and Catalysis, Eindhoven University of Technology, PO Box 513, 5600 MB Eindhoven, The Netherlands

Using cluster models of zeolitic protons, the chemistry of methanol activation will be discussed. Results of Density Functional Theory calculations are presented on transition states corresponding to elementary reaction steps in the formation of dimethylether as well as ethylene. Alternative mechanisms are compared. The formation of dimethylether proceeds via an associative mechanism of two methanol molecules. Whereas methoxy formation does not play a role in dimethylether formation, methoxy species are essential intermediates in the reaction of methanol to ethanol. Whereas oxonium ion formation can occur at reactionconditions, they are spectator ions with respect to the process that leads to carbon-carbon bond formation.

195.

RECENT DEVELOPMENTS IN OPEN-SHELL COUPLED-CLUSTER THEORY. T. Daniel Crawford and Henry F. Schaefer III, Center for Computational Quantum Chemistry, University of Georgia, Athens, Georgia 30602-2556

Recent work in our group has focused on the efficiency of high-spin open-shell coupled-cluster calculations, particularly when based on a spin-restricted reference determinant. We report here the development of a new perturbational triple-excitation correction, referred to as (zT), which offers the same impressive accuracy as the commonly applied (T) correction, but which potentially requires storage of only one-third the number of cluster amplitudes and two-electron integrals. In addition, we are currently working on an approximate Brueckner-orbital-based coupled-cluster wavefunction of open-shell systems which maintains the spin restriction on the component molecular orbitals. As a result, the disk storage requirements are significantly reduced, but without compromising the method's accuracy.

196. **The Chemistry of Single Enzyme Molecules** N. Dovichi, D. Craig, J. Wong, and E. Arriaga, Chemistry Department, University of Alberta, Edmonton, Alberta T6G 2G2 CANADA.

Single molecules of calf intestinal alkaline phosphatase are captured in a capillary filled with a fluorogenic substrate. During incubation, each enzyme molecule creates a pool of fluorescent product. After incubation, the product is swept through a high sensitivity laser-induced fluorescence detector. Replicate incubations are used to construct a kinetic curve for a single molecule. While the activity of a single molecule is constant and independent of incubation time, the activity of different molecules is heterogeneous. We measure the activation energy for the reaction catalyzed by a single enzyme molecule. The activation energy is also heterogeneous for different molecules, but the average activation energy is identical to the activity of a bulk solution. This result confirms the first postulate of statistical mechanics: a thermodynamic property of a single molecule, averaged over time, is identical to the value determined from an ensemble of molecules. Finally, we monitor the effect of thermal denaturation of individual molecules; thermal denaturation is catastrophic, wherein enzyme molecules undergo an irreversible conversion to an inactive form.

197.

NOVEL IMAGING MICROSCOPIES FOR SINGLE BIOMOLECULES. **Prof. H. Schindler** Institute of Biophysics, University of Linz, A-4040 Linz-Auhof, Austria.

198. DUAL MOLECULE SPECTROSCOPY: MULTI-COLOR RULERS FOR MACROMOLECULES. S. Weiss, T. Ha, T. Enderle, P.R. Selvin and D.S. Chemla, Molecular Design Institute, Materials Sciences Division, MS 2-300, Lawrence Berkeley National Laboratory, Berkeley, CA 94720

The fluorescence from two nearby fluorophores can be used to measure distances, relative orientations, and changes in distances/orientations of biological macromolecules with very good spatial, angular and temporal resolutions (a few nm, a few degrees, a few ms). Two techniques, based on the labeling of single macromolecules with a pair of fluorophores which differ in color have been developed. In the first technique, energy is allowed to transfer from one dye molecule to the other (by a proper choice of dyes). The amount of transferred energy, which is measured spectroscopically, gives the required distance (or distance changes) information on a 10 -100 Å scale. In the second method, the two dyes are chosen such that energy transfer is not allowed. The fluorescent light coming from the pair is spectrally separated and sent to two detectors. The distance (on a scale of 50 -2500 Å) is determined from the difference in the centers of the point-spread-function (PSF) of the two molecules. We will present experimental studies based on the two techniques and discuss possible applications in life sciences.

* Supported by the Laboratory Directed Research and Development Program of Lawrence Berkeley National Laboratory under U.S. Department of Energy, Contract No. DE-AC03-76SF00098 and by the Office of Naval Research, Contract No. N00014-95-F-0099.

199. DIRECT MEASUREMENT OF SINGLE-MOLECULE DYNAMICS IN FREE SOLUTION. Edward S. Yeung and Xiao-Hong Xu, Ames Laboratory-USDOE and Department of Chemistry, Iowa State University, Ames, IA 50011

Continuous monitoring of the solution dynamics of individual rhodamine-6G molecules and 30-base-ss-DNA tagged with rhodamine is achieved by total internal reflection fluorescence microscopy. A small observation depth is defined by exciting the molecules either through the evanescent wave at the quartz-liquid interface, or by using micron-size wires to form a thin layer of solution. A microscope thus affords diffraction-limited resolution of interconnected volume elements that are 13 aL to 360 aL, respectively. An intensified CCD camera repeatedly records fluorescence from the same set of molecules to provide rate information on each as they diffuse and photobleach. With a special detection arrangement, time resolution down to 0.37 ms was achieved. The present technical limit is 10 µs for direct clocking of events. Statistical variations in molecular diffusion coefficients and in photobleaching rates are found. The average diffusion coefficients are smaller and the average photobleaching lifetimes are longer for the dye-DNA covalent complex compared to the dye molecule by itself.

200. IN-SITU STM STUDIES OF THE KINETICS OF STRUCTURAL DEFECTS IN SURFACE LAYERS.
A.K. Schmid, J.C. Hamilton, N.C. Bartelt, C.B. Carter*, R.Q. Hwang, Sandia National Laboratory, Livermore, CA94550. *University of Minnesota, Minneapolis, MN 55455

It is well known that diffusive mass transport is often immensely faster on surfaces than in the bulk. In this talk, we show how in surface layers the kinetics of other fundamental processes are greatly enhanced, due to similar atomic mechanisms as those responsible for fast surface diffusion. The mobility of dislocations present in a monolayer of Cu on Ru(0001) was found to be so high that Brownian motion can be observed at room temperature. The motion is interpreted as being caused by a mechanism involving atomic exchange between the adatom lattice gas covering the surface with the solid film. In quenched oxygen films on Ru(0001), grain boundaries are mobile at room temperature, leading to coarsening of the film. Based on an understanding of the atomic mechanisms driving this coarsening, we have developed a microscopic model which accurately predicts the mesoscopic time evolution of the system.
This work was supported by DOE under Contract No. DE-AC04-94AL85000

201. SIMULATIONS OF METAL(100) HOMOEPITAXY: SUBMONOLAYER ISLAND FORMATION, "MOUND" FORMATION DURING MULTILAYER GROWTH, AND KINETIC PHASE TRANSITIONS TO RE-ENTRANT SMOOTH GROWTH. M.C. Bartelt, Sandia National Laboratories, Livermore, CA 94550, and J.W. Evans, Ames Laboratory, Iowa State University, Ames, IA 50011

We describe recent developments in the modeling of metal(100) homoepitaxy, and in the interpretation of STM and high-resolution diffraction experiments. Our model includes (i) a detailed description of nucleation and growth of near-square islands in each layer; (ii) an additional barrier to downward transport at island or step edges; and (iii) downward "funneling" of atoms deposited at step-edges to lower fourfold-hollow adsorption sites [Bartelt & Evans, Phys. Rev. Lett. **75** (1995) 4250; MRS Proc. **399** (1996) 89]. For Ag(100), Cu(100), and Fe(100) homoepitaxy, we compare model predictions with observed behavior of the submonolayer island densities and size distributions, as well as layer populations at the onset of multilayer growth. At "low" temperature, T, where island formation is irreversible, we thus determine effective barriers for adatom diffusion on terraces and across step-edges [Evans & Bartelt, Proc. NATO ASI on *Surface Diffusion* (Plenum, New York, 1997)]. Adatom-adatom bonding is assessed from the transition to reversible island formation at higher T [Zhang, Bartelt *et al.*, Surf. Sci. (in press)]. Results for multilayer growth show strong dependence of film roughness and morphology on T. At typical T, the presence of step-edge barriers produces "mounds" with fixed average slope. At "low" T, downward funneling dominates, due to higher island and step densities, driving re-entrant smooth growth, and eventually a transition from mounded to smooth films [Bartelt & Evans, Surf. Sci. Lett. (submitted)]. Possible re-emergence of rough growth at "very low" T, due to void formation, is discussed.

202.

FILM GROWTH OF METALS ON CLOSE PACKED METAL SURFACES. **J. C. Hamilton** Sandia National Laboratories, Livermore, CA 94550

Heteroepitaxial metal films grown on close packed metal surfaces exhibit a wide range of structures due to the different bond lengths of the substrate and overlayer metals. The strain associated with this mismatch of lattice constants can be accomodated by periodic dislocation structures in the metal film. These structures have been modeled using a Frenkel-Kontorova model with simple pair potentials for the interactions between metal atoms. These models predict rapid diffusion of monomers, dimers, trimers, and quadrimers on close packed surfaces. They also predict rapid diffusion of larger islands having "magic sizes". Finally they provide excellent models of scanning tunneling microscope images of copper films on the Ru(0001) surface.

203. MONOLAYER POLYMERIZATION MEDIATED BY SUBSTRATE TOPOGRAPHY, Christine Evans, Mark Mowery; Chemistry Department; University of Michigan; Ann Arbor, MI 48109-1055

A limiting factor in the development of self-assembled monolayer systems for use in applications such as sensor design and adhesion studies has been the inadequate ruggedness of monolayers formed from n-alkane and ω-terminated thiols and disulfides. By incorporating conjugated diacetylene groups within the monolayer structure and exposing these monolayers to UV radiation, a polymer has been successfully formed within a single molecular layer. Molecules of the form, $[CH_3(CH_2)_n C \equiv C - C \equiv C(CH_2)_m S-]_2$, will spontaneously self-assemble into monolayer films on gold surfaces and are subsequently photo-polymerized to yield conjugated monolayer polymers that are very robust. Not only are these polymerized monolayers able to withstand temperature, solvent, and electrical extremes not previously possible with many monolayer systems, the conjugated backbone inherent in these polymerized monolayers provides unique optical and electronic properties which can be utilized in sensor applications. While these systems are very rugged, the polymerization process is highly dependent on the distance between diacetylene groups on neighboring molecules within a monolayer. Based on initial molecular structures using m=9 and n=7, 11, or 15, we present evidence that the topography of the gold substrate has a direct impact on the polymerization process. This substrate mediated polymerization may be used to advantage in the fabrication of nanostructured interfaces.

204.
SUBSTRATE EFFECTS ON THE GROWTH OF MGCL2 THIN FILMS. Joel G. Roberts, D. Howard Fairbrother and Gabor A. Somorjai, Department of Chemistry, University of California at Berkeley and Materials Science Division, E. O. Lawrence Berkeley National Laboratory, Berkeley, CA 94702

The dependence of the overlayer growth on the underlying substrate is illustrated in this study of $MgCl_2$ thin films on the following substrates: Pd(111), Pt(111), Pd(100), Pt(100) and Rh(111). On Pd(111) and Pt(111), the TPD of the deposited $MgCl_2$ showed a significant substrate-adsorbate interaction as evidenced by a monolayer desorption feature. The interaction was further attested by the formation of two monolayer LEED patterns -- Pd(111)-(4 x 4)-$MgCl_2$ and Pd(111)-($\sqrt{13}$ x $\sqrt{13}$)-R 13.9°-$MgCl_2$. Also, on Pd(111) and Pt(111), a multilayer coverage pattern was grown, $MgCl_2$ (1 x 1). When Pd(100) was used as the substrate, the monolayer desorption feature disappeared from the TPD as well as the two monolayer patterns seen on Pd(111), but a $MgCl_2$ (1 x 1) pattern with multiple rotated domains was created at the multilayer coverage. This difference resulted from the fact that the Pd(100) does not possess the correct angle for the (0001) face of the $MgCl_2$. To preserve this angle, the deposition of $MgCl_2$ was performed on Rh(111) and the reconstructed face of Pt(100). Again, evidence of the strong substrate-adsorbate interaction was gone. The buckling of Pt(100)'s surface layer caused this result. For the Rh(111), the lattice match was not preserved with the angle.

205.
CF_4 ADSORBATE SPECTROSCOPY AS A PROBE OF SURFACE DEFECT STRUCTURE AND OF ADSORBATE ISLAND GROWTH. O. Alpert, V. Buch, Department of Physical Chemistry, The Hebrew University, Jerusalem 91904, Israel, and J. P. Devlin, Department of Chemistry, Oklahoma State University, Stillwater, Oklahoma 74078, US

The focus of this joint computational - experimental effort is spectroscopy of CF_4 adsorbate on molecular surfaces. The observable of the present interest is the vibrational fundamental of the triply degenerate v_3 asymmetric CF stretch of CF_4. This vibration is characterized by a very large direction independent dipole derivative. Therefore, oscillations of neighboring CF_4 molecules are strongly coupled to each other by the dipole - dipole interaction. Infrared spectra of CF_4 adsorbate are dominated by collective in-phase vibrations, and highly sensitive to the structure of the adsorbate layer. At low coverages, CF_4 adsorbate lineshapes are suggested as a probe of surface defect structures. At higher coverages, CF_4 spectroscopy is used to follow adsorbate island formation on the surface. Theoretical analysis is presented of the experimentally measured CF_4 adsorbate spectra on a variety of molecular nanoparticles (CO_2, crystalline and amorphous mesitylene etc).

206. MOLECULAR AND PROTEIN DYNAMICS AS SEEN THROUGH THE VIBRATIONS. Robin M. Hochstrasser, Department of Chemistry, University of Pennsylvania. Philadelphia, PA 19104-6323

Ultrafast spectroscopies, particularly in the infrared, are providing new insights into molecular and biological processes that are initiated or triggered by light pulses. Such diverse processes as isomerization in bacteriorhodopsin, coherence and energy delocalization in the light harvesting pigments, energy fluctuations within the interior of proteins and protein folding all contain fundamental features that were more difficult to examine in model systems. This talk will include examples of these and other biological systems that are generating novel theoretical and experimental challenges to chemical physics.

207. ULTRAFAST PROCESSES INVOLVED IN ACTIVATING CHEMICAL BONDS. Charles B. Harris, S. E. Bromberg, H. Yang, T. Lian, M. C. Asplund, K. T. Kotz Department of Chemistry, University of California, Berkeley, CA 94720.

Certain organometallic complexes in solution can break or "activate" the strong bonds of alkane or silane solvents after UV excitation. Elucidating the elementary steps of the reaction mechanism is critical to understanding this activation process. Ultrafast spectroscopic experiments are necessary to observe the intermediates because the reaction proceeds too quickly for conventional microsecond time-resolved systems to detect them. In light of this, we have carried out the first ultrafast IR study of C-H bond activation using Tp*Rh(CO)2 in room temperature alkane solution, taking advantage of the structural sensitivity of the CO stretching frequencies. Two subsequently formed intermediates are involved in the reaction, one of which may be a s-complex, postulated to be a key intermediate preceding activation. In the related Si-H bond activation reaction of CpM(CO)3 (M=Mn, Re), we have observed a new intermediate and corresponding pathway for silyl hydride product formation. From these observations we propose mechanisms for the overall reactions. Experiments are currently underway to better characterize the structures of the intermediates.

208. EARLY EVENTS IN PROTEIN FOLDING. W. A. Eaton, Laboratory of Chemical Physics, National Institutes of Health, Bethesda, MD 20892-0520.

Understanding *how* proteins fold is one of the central problems in biochemistry. To investigate the mechanisms of protein folding on the previously inaccessible submillisecond time scale, we have used photochemical triggering, laser temperature jump, and ultrarapid mixing methods. These experiments provide the first glimpse of processes such as secondary structure formation, global collapse to compact denatured states, and fast barrier crossings to the native state. The results will be discussed in terms of the statistical energy landscape theory of protein folding.

209.
FEMTOSECOND X-RAY PULSES: A NEW TOOL FOR DYNAMIC STRUCTURAL DETERMINATION. **Charles V. Shank,** Chemistry Department, University of California and Lawrence Berkeley National Laboratory, Berkeley, California

Over the last two decades femtosecond optical measurement techniques have been developed which have permitted the investigation of ultrafast events with unprecedented time resolution across a broad range of scientific activities. At the same time, synchrotron light sources have revolutionized static structural determinations. In this talk I will describe progress in developing a femtosecond X-ray pulse source suitable for dynamic structural determination.

210.

FINDING TRANSITION STATES IN COMPLEX SYSTEMS -- THROWING ROPES OVER ROUGH MOUNTAIN PASSES, IN THE DARK

The concept of a transition state is central to theoretical studies of rare but important events. To exploit the concept, reaction coordinates must be identified. In a complex system, however, the potential energy surface is intrinsically rough. There is not a small number of singly important saddle points dividing stable states. As such, explicit specification of reaction coordinates is generally difficult if not impossible. Nevertheless, statistical characterizations are generally possible and sufficient. This lecture describes a physical picture and applications of a class of stochastic algorithms that can be used to locate dynamical pathways and transition states in complex systems. The approach builds on a nascent idea due to Pratt [L.R. Pratt, J.Chem Phys. 85, 5045 (1986)].

211. NONLINEAR SPECTROSCOPY OF SPECIES WITHIN MICROCAVITIES
R. K. Chang, M. H. Fields, J. Popp, and J. M Hartings, Applied Physics Department, Yale University, New Haven, Connecticut 06520-8284

Micrometer-sized liquid droplets form extremely high quality factor ($Q > 10^7$-10^8) microcavities in the UV to the near-infrared region. The incident radiation can be stored within the droplet rim, resulting in greatly enhanced internal intensity. Nonlinearly generated radiation, such as from species-specific stimulated Raman scattering (SRS), can experience high-Q optical feedback. Thus, the pump-intensity threshold for SRS is greatly reduced. We will review the spatial and spectral characteristics of the microcavity resonant modes. Our efforts to detect SRS from minority species within the droplet have included: (1) external seeding; (2) internal seeding with fluorescence; (3) resonance Raman effect; (4) cascade pumping with the SRS from majority species; and (5) enhanced input coupling with laser-induced droplet-shape deformations. Speculations will be made regarding the potential of using microcavities to enhance the interaction of optical radiation with nanostructured materials placed inside such liquid or dielectric microcavities.

212. FRACTAL-SURFACE-ENHANCED OPTICAL NONLINEARITIES. Vladimir M. Shalaev, Department of Physics, New Mexico State University, Las Cruces, New Mexico 88003 U.S.A.

Optical phenomena experience giant enhancements in fractal nanostructured materials consisting of small ~10 nanometer-sized features. This talk will discuss recent developments in nonlinear optics of fractal nanocomposites and self-affine thin films. Optical excitations of nanomaterials with fractal morphology result in highly localized areas of large fields, with the spatial location of the "hot" spots depending on the light frequency. Because of the localization and large varieties of local structures, fractal nanomaterials resonate in a broad spectral range, providing strong enhancements for optical nonlinearities in the visible and infra-red parts of the spectrum.

213. QUANTUM WELL OPTOELECTRONICS: PHYSICS TO APPLICATIONS. D. A. B. Miller, Department of Electrical Engineering, Stanford University, Stanford, CA 94305-4085

Quantum confinement leads to interesting physics and practical devices in optoelectronics. The talk will review some of the physics and applications of quantum wells in particular. Quantum wells, typically made of thin (e.g., 10 nm) layers of semiconductors, offer both an intriguing "laboratory" to do quantum mechanics, and also allow real, practical, quantum mechanical engineering of novel and useful devices. They can be used, for example, to investigate "atomic" physics under extreme conditions, investigate the transition between nonlinear optics and coherent electronic transport, and study basic quantum optics. In applications, they are key to many of the new optoelectronic devices that will enable the emerging information age, from low power semiconductor lasers through femtosecond optical pulse generation to massively parallel optical interconnections from silicon integrated circuits.

214. THE IMPACT OF DIELECTRIC CONFINEMENT ON THE OPTICAL NONLINEARITIES AND ELECTRON DYNAMICS IN METAL AND SEMICONDUCTOR NANO-CRYSTALS IN GLASS MATRICES. Ch. Flytzanis, Laboratoire d'Optique Quantique, Ecole Polytechnique, 91128 Palaiseau cédex, France

The main aspects of the dielectric confinement in nanocrystals with delocalized electrons uniformly dispersed in a transparent dielectric are analysed and their impact on the nonlinear optical response of such materials is assessed.

The case of metal and semiconductor nanocrystals in glass matrices is explicitly considered and the recent studies of the electron wave packets in these nanostructures using femtosecond pump and probe femtosecond techniques are reviewed to obtain information about the dynamics of the collective electronic excitations and the optical nonlinearities in these systems.

The roles of the form and spatial configuration will be discussed and extensions to other classes of composite materials will also be presented.

215. **THIRD-ORDER NONLINEAR OPTICAL PROPERTIES OF SELECTED COMPOSITES.**
George L. Fischer and Robert W. Boyd,[*] *Rome Laboratory, RL/ERXO, Hanscom AFB, MA 01731*

Several examples of the authors' current research in the nonlinear optics of composite materials are presented. Each composite possesses qualities not found in either of the starting components. Layered composites show enhanced optical nonlinearities when the electric field amplitude of an incident laser beam is concentrated in the layers containing the nonlinear material. This nonuniform distribution of the field results from differences in linear refractive indices between layers. Nonlinear absorption is eliminated in composites containing a saturable absorber as well as a reverse saturable absorber. $\chi^{(3)}$ is doubled in blending two rodlike polymers. In the nonlinear Christiansen filter differences in the nonlinear index of refraction between glass and a linear index matched solution of carbon disulfide and acetone result in optical limiting. The size scale in the composites ranges from molecular in the blended polymers, to nanometer in the layered and colloidal composites, to sub millimeter in the Christiansen filter.
[*]*University of Rochester, Institute of Optics.*

216.

STUDYING THE DIRECT PROCESS CHEMISTRY IN VACUUM: REACTION OF METHYL + CHLORINE MONOLAYERS ON COPPER SILICIDE SURFACES. Dong-Hong Sun, Alejandra B. Gurevich, Laura J. Kaufman and Brian E. Bent, Dept of Chem, Columbia Univ, NY, NY 10027; Antony P. Wright and Brian M. Naasz, Dow Corning Corp, Midland, MI 48686.

The industrial Direct Process for synthesizing methylchlorosilanes, in particular dimethyldichlorosilane, reacts gaseous methyl chloride under 1-7 atms pressure with fluidized silicon powder at 280-320°C, and is catalyzed by copper powder and trace promoter elements. We modeled this reaction by sequentially adsorbing methyl radicals and chlorine on polycrystalline Cu_3Si alloy under ultra-high vacuum conditions. Our results have shown that mixed monolayers of methyl groups with chlorine atoms produce methylchlorosilanes, preferentially dimethyldichlorosilane. Methyl monolayers in the absence of chlorine produce primarily trimethylsilane, and chlorine monolayers in the absence of methyl groups produce $SiCl_4$. We have extended our studies to show the effect of surface compsition/structure and the effect of promoters (zinc, tin and aluminum) on reaction kinetics and product distribution. The chemistry has also been studied, using as a substrate, a copper silicide overlayer on a copper single crystal. The silicide was formed by decomposition of adsorbed SiH_4.

217.

LOCAL SECOND-ORDER CALCULATION OF CORRELATION ENERGY IN LARGE MOLECULES AND CLUSTERS. Peter Pulay and G. Rauhut, Department of Chemistry and Biochemistry, University of Arkansas, Fayetteville, Arkansas 72701

We have recently developed a program for calculating correlation energies for large molecules and clusters by local second-order many-body perturbation theory [1,2]. Unlike other methods [3,4] which use a series of one-index transformations, we transform two indices at a time. This had been considered inefficient because the computational work scales with the sixth, instead of the fifth power of the molecular size. However, it diminishes the demand for fast memory, the main bottleneck in MP2 calculations for large molecules. The program consists of an in-core direct integral transformation, followed by a local second-order Moller-Plesset procedure. Using thresholds on the products of density matrix elements and integrals, the actual scaling reduces from the formal sixth-power to between second and third power for large molecules. The method will be described and compared with alternative techniques. Possible further improvements will be discussed. Test calculations, e.g. for clusters of glycine molecules, will be presented.

[1] G. Rauhut and P. Pulay, to be published.
[2] G. Rauhut and P. Pulay, presented at "Molecular Quantum Mechanics: Methods and Applications", Cambidge, England, Sept. 1995.
[3] S. Saebo and J. Almlof, Chem. Phys. Lett. 1989, 154, 83.
[4] M. Head-Gordon, J. A. Pople, and M. J. Frisch, Chem. Phys. Lett. 1988, 253, 503.

218.

CORRELATED WAVE FUNCTIONS FOR SINGLET BIRADICALS. E. R. Davidson, Department of Chemistry, Indiana University, Bloomington, Indiana 47405

Singlet biradical wave functions are inherently multi-determinantal. CAS-SCF approaches, however, often fail to give qualitatively correct potential energy surfaces because they underestimate the effects of ionic configurations. Correlated multi-reference methods that allow for redefinition of the reference space after correlation are needed. Some examples will be given contrasting the results of various approaches to some typical organic biradicals.

219.

NEW DEVELOPMENTS IN OPEN-SHELL COUPLED-CLUSTER THEORY. Marcel Nooijen and Rodney J. Bartlett, Quantum Theory Project, University of Florida, Gainesville, Florida 32611-8435.

The crucial concept in various recent developments in open-shell coupled-cluster theory is the many-body similarity transformation. The amplitudes that define this transformation are chosen such that relevant second quantized matrix elements of the transformed Hamiltonian are set to zero, decoupling the electronic eigenvalue problem to a large extent. In similarity transformed equation-of-motion coupled-cluster theory (STEOM-CC) the transformed Hamiltonian is subsequently diagonalized over a small sub-space to yield a method for (generalized) excitation energies of exceptional efficiency. We will discuss additional approximations including a mix of perturbation and coupled-cluster theory that furthers the efficiency of the approach with little loss of accuracy. STEOM is easily applicable to systems the size of porphyrin. On the other side of the computational spectrum the concept of many-body similarity transformations is used to derive general schemes for spin-adapted open-shell and multi-reference coupled-cluster methods in the spirit of contracted multi-reference CI. Preliminary examples will be presented.

220.

ANALYTIC ENERGY GRADIENTS FOR CI WAVE FUNCTIONS. **Ron Shepard** and G. S. Kedziora, Chemistry Division, Argonne National Laboratory, Argonne, IL, 6043

Analytic energy gradients have been developed for several ab initio wave functions and energy expressions, but the configuration interaction (CI) case is particularly appealing for several reasons. Large-scale multireference CI potential energy surfaces are among the most accurate and balanced for describing chemical reactions. The method can be applied to both ground and excited electronic states, to closed-and open-shell electronic states, and to geometrical conformations that may be highly multiconfigurational in nature. The flexible nature of the CI method, in addition to its high computational efficiency, such as displayed by the COLUMBUS Program System, poses additional challenges in the development of correspondingly flexible and efficient analytic energy gradient procedures. The current state of development of analytic energy gradients for large-scale CI wave functions will be discussed from this point of view.

221. TAMING THE CONVERGENCE OF MULTIREFERENCE PERTURBATION THEORY. R. K. Chaudhuri, K. F. Freed, and J. P. Finley, The University of Chicago, Chicago, Illinois 60637

Following the computational methods developed by Bartlett and coworkers, we study the high order multireference perturbation (MRPT) expansions of several standard systems for which the traditional Moller-Plesset and Epstein-Nesbet MRPT expansions diverge badly. These systems are the rectangular H_4 molecule, the ^1S states of the Be atom, the excited $^1A_1$ surfaces for the insertion of Be into H_2, and the $^1A_1$ states of CH_2. Simple two-state models are shown to explain the perturbative divergences in all cases, while the forced degeneracy (FD) effective valence shell Hamiltonian and optimized perturbation (OPT) partitioning methods are found to produce either convergent MRPT expansions or usefully asymptotic series that are accurate in low orders and that may be resummed with Pade approximants. The molecular systems are analyzed for the convergence of MRPT methods with increasing size of the reference space to treat the large reference spaces used in practical calculations.

222. DIRECT CALCULATION OF CORRECTIONS TO TRIAL WAVEFUNCTIONS AND ENERGIES. James B. Anderson, Department of Chemistry, Pennsylvania State University, University Park, PA 16802

The difference δ between the exact wavefunction ψ and a trial wavefunction ψ_0 may be determined directly in quantum Monte Carlo calculations. For an analytic trial function from any source the difference δ may be calculated and used to correct a trial function to obtain a wavefunction of higher accuracy and a more accurate eigenvalue. The difference between the exact energy E and the expectation value of the energy <E> for a trial wavefunction may be similarly determined. Applications to date include those to H_2 and H_2O with improvements in accuracy in energy by factors of 10 to 100. Prospects for applications to larger systems will be discussed.

223. SINGLE MOLECULE TRAJECTORIES REFLECTING NON-LINEAR BIOMOLECULAR AND PHOTOPHYSICAL DYNAMICS IN CELLS AND SOLUTIONS:

Watt W. Webb, Applied Physics, Cornell University, Ithaca, N.Y. 14853

Trajectories of individual cell surface receptor macro molecules in sparse populations on living cells reveal anomalous subdiffusion with non-linear power laws for mean square displacements growing with less than linear exponents. The responsible mobility restraints are attributed to the random potential field created by transient interactions amongst mobile and some immobile molecules. Trajectories of individual fluorophore molecules diffusing in very dilute solution can transit the sub-femtoliter focal volumes of focused femtosecond modelocked lasers where transient multiphoton molecular excitation of fluorescence occurs. Analysis of the fluorescence signals can yield diffusion coefficients and features of intramolecular dynamics and of the photophysics of non-linear molecular excitation. Optical force microscopy measures the fracture forces of individual weak (~10 picoNewton) intermolecular bonds between biomolecular pairs. Appropriate techniques and results of some of these quantitative measurements on individual biomolecules are illustrated.

224.

POLYACRYLAMIDE GELS FOR SINGLE MOLECULE BIOPHYSICS. **Robert M. Dickson** and W. E. Moerner, Department of Chemistry and Biochemistry, University of California, San Diego, 9500 Gilman Drive, La Jolla, CA 92093-0340

Individual fluorescent molecules and individual singly-labeled proteins have been observed in the water-filled pores of poly(acrylamide) gels with far-field microscopy. Brownian motion of small molecules was dramatically reduced by the gel framework, thus enabling extended study of single fluorophores in aqueous environments. A highly axially dependent laser field was used to both excite the fluorophores and image the molecules in three dimensions. Single molecules have been followed as they move within and through the porous gel structure. In contrast to dry polymeric hosts, these water-based gels form a useful medium for single-molecule studies of biological systems in vitro. Several new observations of single protein behavior have provided insight into biochemical function and will be discussed.

225. SINGLE MOLECULE IMAGING SPECTROSCOPY AND NANOMANIPULATION OF BIOMOLECULES IN AQUEOUS SOLUTION. T. Yanagida, Department of Physiol., Medical School, Osaka University, Osaka, Japan

We have refined total internal reflection fluorescence microscopy (TIRFM) to visualize single fluorescent dye molecules in aqueous solution at a full video rate. This new fluorescence microscopy allowed us to directly image individual ATPase reactions and motions by a single motor protein (Nature, 374, 555- '95; Nature, 380, 451- '96). Furthermore, we have extended this method to single molecule spectroscopy and FRET. Using the single molecule spectroscopy and FRET, we could directly monitor subunit dynamics in a single Ca-binding protein and protein-protein communication at a single molecule level (BS (US), Abs. '96, '97). Combining this method for single molecule imaging with manipulation of a single protein molecule by laser trapping nanometry, we have succeeded in simultaneous measurements of single molecule mechanics and the ATPase reaction by a motor protein (BS (US), Abs. '96, '97). We are applying these new methods to studies on mechnaochemical energy transduction in molecular motors, DNA - protein interaction and signal transduction in cell.

226. INTERMITTENCY OF SINGLE MOLECULE REACTION DYNAMICS IN FLUCTUATING ENVIRONMENTS. J. Wang, Department of Chemistry, University of Illinois, Urbana, IL 61801

Individual activated events in slowly fluctuating environments are now accessible to study by single molecule spectroscopies. The statistics of such events should exhibit intermittency and will not always obey the Poisson Law. Examples illustrate the ideas when the environments relax exponentially and non-exponentially as in glasses or biomolecules.

227. PROBING PROTEIN CHANNEL DYNAMICS AT THE SINGLE MOLECULE LEVEL. Robert C. Dunn and M. Ann Lee, Department of Chemistry, University of Kansas, Lawrence, Kansas 66045

It would be difficult to overstate the importance played by protein ion channels in cellular function. These macromolecular pores allow the passage of ions across the cellular membrane and play indispensable roles in all aspects of neurophysiology. While the patch-clamp technique continues to provide elegant descriptions of the kinetic processes involved in ion channel gating, the associated conformational changes remain a mystery. We are using the spectroscopic capabilities and single molecule fluorescence sensitivity of near-field scanning optical microscopy (NSOM) to probe these dynamics at the single channel level.

Using a newly developed cantilevered NSOM probe capable of probing soft biological samples with single molecule fluorescence sensitivity, we have begun mapping the location of single NMDA receptors in intact rat cortical neurons with <100 nm spatial resolution. We will also present recent results exploring the conformational changes accompanying activation of nuclear pore channels located in the nuclear membrane of *Xenopus* oocytes. Our recent NSOM and AFM measurements on single nuclear pore complexes reveal large conformational changes taking place upon activation, providing rich, new molecular level details of channel function.

228. HOW DO METAL ORBITALS AFFECT THE STM IMAGES OF ORGANOMETALLIC COMPOUNDS AND HOW CAN THESE EFFECTS BE USED? K. W. Hipps, X. Lu, and Ursula Mazur, Chemistry and Materials Science, Washington State University, Pullman, WA 99164-4630

In recent years there has been intense interest in two seemingly different aspects of nano-technology and nano science: a) the creation of electronic devices based on single molecular units; and b) incorporation of chemical analysis capability into the scanning probe microscope. The work we will present indicates that these may in fact be closely related goals and points the way to some special solutions of these problems. We will show that STM images obtained from a series of transition metal phthalocyanines adsorbed on Au(111) are intensely characteristic of the orbital occupation of the central metal ion. Further, the symmetry and energetics of the valence d orbitals are critical factors in determining the possibility of analytical specificity in STM based single molecule analysis. An equally exciting potentiality is that macrocyclic organometallics may be used as molecular electronic building blocks--atomic wires (certain metal ions) contained in a planar insulating sheathes, or insulators sheathed in semiconductors.

229. ORIENTATION AND ALIGNMENT IN ADSORBATE PHOTOCHEMISTRY. J.C. Polanyi, Department of Chemistry, University of Toronto, Toronto, Ontario M5S 3H6.

This paper will review some recent work in the area of 'surface aligned photochemistry'. Photolysis of halide adsorbates (HX and RX) on insulator and semiconductor surfaces has been shown to give aligned and oriented adlayers. Irradiation with UV then leads to the formation of photofragments directed (a) *downward* at the surface, and/or (b) *upward* away from the surface and/or (c) *along* the surface. Unpublished examples of each category of behaviour will be given. In cases (a) and (c) the scattering event constitutes a 'localised' encounter, (a) with a surface atom or (c) with a neighbouring adsorbate molecule. Collisions with a surface atom (case(a)) can lead either to 'Localised Atomic Scattering, LAS', or to 'Localised Atomic Reaction, LAR'. The use of LAR will be exemplified for the adsorbate photoreaction of chlorobenzene on Si(111) 7×7, studied by STM. Collisions with a neighbouring adsorbate (case (c)) can lead to 'Surface Aligned Reaction, SAR', e.g., CH_3 + $BrCH_3' \rightarrow CH_3Br + CH_3'$. Reference will also be made to oriented and aligned harpooning reaction in Na..XR complexes, studied in the same laboratory.

230 ALIGNMENT OF RECOMBINATIVELY DESORBED HYDROGEN FROM Cu(111): QUANTUM STATE AND VELOCITY DEPENDENCE, Alec M. Wodtke, Stephen Gulding, Hong-Tao Hou, Department of Chemistry, UCSB, Santa Barbara, CA 93106 and Charles Rettner, Daniel J. Auerbach, I.B.M. Research Division, Almaden Research Center, San Jose, CA 95120-6099

Recombinative desorption of atomic Hydrogen from Copper produces aligned molecular Hydrogen which is probed state-specifically by polarized 1+1' Resonance Enhanced Multi-photon Ionization. Field-free ion time-of-flight reveals that the alignment, which resembles molecular helicopters lifting off of the Copper surface, has a strong inverse dependence on recoil velocity, a direct dependence on rotational quantum number and is nearly independent of vibrational quantum number. The application of the principle of detailed balance allows the desorption data to be inverted, providing the kinetic energy dependence of the sticking probability to be obtained as a function of molecular Hydrogen's: rotational excitation, vibrational excitation and alignment. These results provide an excellent test of theoretical models of surface reactivity.

231.
CLASSICAL TRAJECTORY STUDIES OF A PHOTO-INITIATED SURFACE-ALIGNED REACTION: H+CO_2 ON LiF(001). J. V. Setzler, H. Guo, Department of Chemistry, University of Toledo, Toledo, Ohio 43606, and G. C. Schatz, Department of Chemistry, Northwestern University, Evanston, Illinois 60208

We report preliminary results of the classical trajectory study of a chemical reaction between H and CO_2 on LiF(001). The reaction is initiated by photodissociation of well-aligned HBr(ad) at 193 nm, which produces a 'hot' H atom directed towards a nearby CO_2(ad). Single molecules of each reactant are placed on a static surface and several energetically-favorable configurations representative of low coverage conditions are determined by a Monte Carlo method. Quasiclassical trajectories of the system at 80 K are calculated for four different adsorption configurations. A full dimensional HCO_2 potential derived from ab initio calculations by Schatz and coworkers is used. For several adsorption configurations we find that the reactivity can be enhanced by a factor of 3-4 compared with the gas phase reaction. The calculated impact parameters and incident angles of the surface-aligned collisions indicate that the enhanced reactivity can be attributed to the alignment of the co-adsorbates on the surface. Most product state-distributions show little memory with regard to the configuration from which they are derived and they are similar to those obtained in the gas phase.

232. TUNING THE ORIENTATION OF MOLECULES AT SOLID-LIQUID INTERFACE USING ELECTROCHEMICAL POTENTIAL. Zhi Xu and Gregory He, Department of Chemistry, University of Missouri-St. Louis, St. Louis, Missouri 63121

Sum-frequency generation (SFG) has been applied to study the effect of electrochemical potential on the orientation of 4-(4-(diethylamino)styryl)-N-methylpyridinium iodide (**D289**) molecules in the electrical double layer region. Our results have shown that that the average molecular orientation of D289 molecule changes from 90° to about 30°, corresponding to the surface normal, when the electrochemical potential is changed from PZC to -0.9 V. A statistical thermodynamics model employing Boltzmann distribution has been developed which is in excellent agreement with the experimental results. From the orientation data and the statistical model, the electrical field experienced by molecules in the electrical double layer region and the effective thickness of the electrical double layer have been extracted as $E \approx 5 \times 10^6$ V/cm and $d \approx 20$ Å, respectively.

233.

LASER-INDUCED MOLECULAR OPTICS: FROM STEREODYNAMICS TO NANOSCALE MACHINING. Tamar Seideman, Steacie Institute, National Research Council, Ottawa, Ontario K1A 0R6

Moderately intense laser fields are capable of aligning and orienting molecules, providing a route to stereochemical studies and to time-resolved molecular dynamics. Laser photons can also focus molecular beams to a fraction ($< 1\%$) of their initial size, hence a route to nanoscale lithography. The combined focusing and alignment effect induced by the coherent light provides potentially a route to new materials, possessing predesigned electric or magnetic properties. Molecular focusing extends to a general field of molecular optics. The possibilities of collimating, steering, reflecting and waveguiding molecular trajectories suggest a photon-controlled molecular-beam technology.

234.
ROTATIONAL STATE-RESOLVED STICKING FOR H_2 ON PD(111): TESTING DYNAMICAL STEERING IN DISSOCIATIVE CHEMISORPTION. Micheal Gostein and Greg O.Sitz, Physics Dept., The University of Texas at Austin, Austin TX 78712.

We report measurements of the sticking coefficient of H_2 on Pd(111) as a function of incident translational energy and initial rotational state. We find the sticking coefficient is sensitive to the rotational quantum number, first decreasing as J is raised from 0 to 3, then increasing for J equals to 4 and 5. We also find significant rotational excitation (from J=0 to 2, and J=1 to 3) during scattering, with a probability that depends strongly on surface temperature but only weakly on incident translational energy. This excitation occurs in direct scattering and does not result from associative recombination. These measurements test recent theoretical predictions that low rotational states of H_2 should stick better on a Pd surface because they are more easily steered to high dissociation probability geometries.

Work supported by the Robert A. Welch Foundation, grant F-1198 and the National Science Foundation CHE-9512232.

235. THE SELECTIVE ABSTRACTION OF IODINE FROM I_2Cl_6 BY Si(111)-7×7. Yong Liu and Andrew C. Kummel, Department of Chemistry, University of California at San Diego, 9500 Gilman Drive, La Jolla, California 92093-0358.

WITHDRAWN

The Si(111)-7×7 surface selectively abstracts iodine from planar I_2Cl_6 and preferentially ejects chlorine into the gas phase during adsorption. This results in an experimentally measured absolute ratio of Cl:I=1:1 on the Si(111)-7×7 surface dosed with 0.5-2.5 eV I_2Cl_6 at low coverages. Scanning tunnelling microscopic images show that the adsorbed halogen atoms form isolated single sites (45%), adjacent double sites (20%), and adjacent triple or quadruple sites (35%). The iodine-selective abstraction and the relatively small adsorbate sizes suggest that I_2Cl_6 molecules collide with the surface with a face-on orientation to form an adsorption transition state containing a central iodine atom partially bonded to a Si adatom and four Cl atoms. The observed chemical selectivity is a direct result of the stereoselectivity of the gas-surface reaction.

236. SYNTHESIS AND HANDLING OF SINGLE SHEETS OF A COVALENT MONOLAYER SQUARE GRID POLYMER. Thomas F. Magnera, Jaroslav Pecka, and Josef Michl, Department of Chemistry and Biochemistry, University of Colorado, Boulder, CO 80309-0215

The molecular construction kit concept ("Tinkertoy rods and connectors") has been extended to the synthesis of 0.7 nm thick and over 100×100 nm wide planar covalent polymer grid sheets comprising contiguous 2.5 nm squares with 1.5 nm square hollow centers. The vertices of the squares contain porphyrin rings and their edges consist of aromatic rings. The synthesis was performed by two-dimensional cross-linking of monomers confined to two dimensions by strong adsorption on a liquid surface. The grid sheets survive boiling with ethanolic HCl and transfer from one surface to another. Their structure has been proven by a combination of grazing incidence IR spectroscopy and STM imaging.

237. COVALENT ATTACHMENT OF POLYMER MICELLES TO SiO_2 SURFACES
Jungsoek Hahn and Stephen E. Webber, Department of Chemistry and Biochemistry and Center for Polymer Research, The University of Texas at Austin, Austin, Texas 78712

Polystyrene-*block*-poly(methacrylic acid) polymers will self-assemble into polymer micelles in the appropriate mixed solvent (*e. g.* dioxane-water). If the polyacid micelle corona is reacted with 1-(3-dimethylaminopropyl)-3-ethylcarbodiimide and then exposed to a SiO_2 surface that has been aminated with γ-aminopropyltrimethoxysilane, the surface is densely covered with polymer micelles. This permits one to chemically modify the SiO_2 surface with these "nanoclusters" of polymers. These *ca.* 50 nm spherical clusters each contain several hundred polymers. Their dimensions are consistent with their solution dimensions obtained from light scattering. We will discuss our characterization of these surfaces, the modification of the polymer micelles after surface attachment, and the use of these surfaces for subsequent chemical modification.

238. HETEROSUPRAMOLECULAR CHEMISTRY : THE SELF-ASSEMBLY AND SELF-ORGANISATION OF NANOSTRUCTURED MATERIALS. L. Cusack, A. Gorelov, X. Marguerettaz, N. Rao, R. Rizza and D. Fitzmaurice, Department of Chemistry, University College Dublin, Belfield, Dublin 4, Ireland.

Recent years have seen the development of powerful techniques and methodologies for the self-assembly and self-organisation of supermolecules and for the preparation of metal and semiconductor nanocrystallites. Together, these developments allow us explore the possibility of self-assembling and self-organising complex nanostructures possessing new properties and diverse function. The term heterosupramolecular chemistry has been used to describe these and related studies. This talk will describe the self-assembly in solution of two appropriately modified semiconductor nanocrystallites by complementary hydrogen bonding to form a heterosupermolecule. Also described will be the subsequent self-organisation of these heterosupermolecule to form an ordered nanocrystallite array, or superlattice. Finally, our most recent studies directed toward the self-assembly and self-organisation of mixed metal and semiconductor nanocrystallite superlattices will be described.

239. COMPOSITE MATERIALS WITH FIELD-INDUCED ANISOTROPY. J. E. Martin and R. A. Anderson, Sandia National Laboratories, Albuquerque NM 87185-1421

When a colloidal suspension is subjected to a sufficiently large electric field, the induced dipolar particle interactions will cause chaining along field lines. If the continuous phase is polymerizable, e.g. a thermosetting resin, the anisotropic structures produced by the field can be pinned by the sol-gel transition of the resin. The solid materials so produced have marked anisotropies in permittivity and other properties, that are strongly dependent on the particle concentration, pinning time, and field strength. In addition to synthesizing these materials, we have written a Langevin dynamics simulation of structure formation in an applied field, using the simple dipole interaction potential, Stokes friction for the spheres, and a correlated fluctuating force. We have simulated particle 'column' formation in a uniaxial electric field and particle 'sheet' formation in a rotating field and developed a number of ways of operationally defining anisotropy, including the dipolar interaction energy, dielectric constants and optical attenuation along different axes, and scattering functions. We will describe the concentration dependence of these properties, and will compare these to experimental light scattering and permittivity measurements.

This work was supported by the U.S. Department of Energy under Contract No. DE-AC04-94AL85000. Sandia is a multiprogram laboratory operated by Sandia Corporation, a Lockheed Martin Company, for the U.S. Department of Energy.

240. SYNTHESIS OF DOPED FULLERENE CLUSTERS AND BORON-NITROGEN TUBULES USING LASER ABLATION. **Z. Charles Ying,** Jane G. Zhu, Department of Physics, New Mexico State University, Las Cruces, NM 88003; R. N. Compton, Departments of Chemistry and Physics, University of Tennessee, Knoxville, TN 37996; R. L. Hettich, L. F. Allard, Jr., Oak Ridge National Laboratory, Oak Ridge, TN 37831; R. E. Haufler, Comstock Inc., Oak Ridge, TN 37830.

A variety of nanostructured materials, including clusters, nanoparticles, nanotubules, and thin films, can be produced by using laser-ablation technique. This paper reports our production of (i) nitrogen doped fullerene clusters by ablation of graphite under a nitrogen environment and (ii) boron-nitrogen nanoparticles and nanotubules by ablation of boron nitride. These novel materials have been extensively characterized by mass spectrometry, electron microscopy, and x-ray photoemission spectroscopy. The boron-nitrogen nanoparticles and nanotubules were separated from other laser-ablation products, such as boron particles, by a hydrogen-peroxide treatment.

241.
ALIGNMENT OF ORDERED MESOPOROUS SILICA USING MAGNETIC FIELDS AND LIQUID CRYSTAL PROCESSING. **Sarah H. Tolbert,** Ali Firouzi, D.J. Schaefer, Galen D. Stucky, and Bradley F. Chmelka. Departments of Chemistry and Chemical Engineering, University of California, Santa Barbara, CA 93106.

The liquid crystalline properties of unpolymerized silicate/surfactant composites at high pH can be used to produce ordered mesoporous hexagonal silicates (MCM-41) with macroscopic pore orientation. Silicate/surfactant liquid crystals are oriented by cooling through their isotropic-to-anisotropic phase transitions in the presence of a magnetic field. The extent of alignment can be quantified using deuterium NMR spectroscopy. X-Ray diffraction shows that this alignment is preserved when the silicate species are polymerized by reduction of pH and thermal treatment. These samples can further be calcined to remove the surfactant without substantial loss of macroscopic orientational order (although with some loss of periodicity). The net result is a mesoporous hexagonal silicate with long range orientational order of the pores.

242. DIPOLAR AND ELECTRORHEOLOGICAL FLUIDS. Thomas C. Halsey, Exxon Research and Engineering Company, Annandale, NJ 08801

Electrorheological fluids are colloidal suspensions of highly polarizable particles in a non-polarizable solvent, and are a special case of "dipolar fluids." In a strong electric field, the behavior of the fluid is dominated by the dipolar interactions between the colloidal particles. When a high electric field is applied to a fluid, dielectric columns oriented parallel to the field rapidly form. The relaxation of these columns to a phase separation is a much slower process. I calculate the shape of the columns formed and the time scale for the relaxation of the columnar structure. In a shear flow, the fluid exhibits shear-thinning behavior. This can be accounted for by an "independent droplet" model, in which the shape and size of structures in the fluid is set by a competition between dipolar and hydrodynamic forces. This model also accounts for light-scattering results in steady shear, as well as some aspects of viscoelastic measurements.

243. ULTRAFAST LASER SPECTROSCOPY. A. H. Zewail, Department of Chemistry, California Institute of Technology, Pasadena, California 91125

In this talk, we will highlight the progress made over the past two decades in studying structure and dynamics of molecular systems with ultrashort time resolution.

244. Interaction of Theory and Experiment in Reaction Rates and Intramolecular Dynamics

There are many illustrations of the stimulating effect of experiments on theory and the reverse. Several examples drawn from our own experience include the areas of electron transfer, unimolecular reactions in gas phase and in clusters, bimolecular associations, solvent dynamics, and intramolecular dynamics. Examples will be given where particular experiments suggested the development of new theory, and where the theory in turn served to stimulate further experiments. One challenge has been to try to introduce suitable physical or mathematical approximations to simplify a complicated problem, yield simple expressions for the reaction rate, but remain physically credible.

245. LIKE-SIGN ION PAIRS. Harold L. Friedman and Fernando O. Raineri, Department of Chemistry, State University of New York at Stony Brook, Stony Brook, New York 11794-3400, USA.

The term "ion pair" usually refers to a positive ion and a negative ion in solution forming a complex or at least having a close encounter. The concept was developed by Bjerrum (1925) and others to account for the effect of ion association on the properties of strong electrolyte solutions. We point out that like-sign ion pairs also may be invoked to explain certain effects in the kinetics, spectroscopy, and neutron diffraction of certain ionic solutions. Candidates for like-sign ion pairs are the dimer of hexaaquo nickel ion, the hexaaquo ferrous-hexaaquo ferric pair, and the ferrocyanide-ferricyanide pair.

246.

SYNTHESIS AND CHARACTERIZATION OF II-VI QUANTUM DOTS AND THEIR ASSEMBLY INTO 3D QUANTUM DOT SUPERLATTICES. **Christopher B. Murray**, IBM Corporation, T. J. Watson Research Center, Yorktown Heights NY 10598.

Lyothermal synthesis procedures and size selective precipitation methods are developed to prepare and isolate nanocrystal "quantum dots" (QDs) of II-VI semiconductors which are monodisperse to the limit of atomic roughness (st.dev.<2A). Methods are then established to "self-organize" the QDs into three dimensional superlattices; i.e., colloidal crystals in which a single QD nanocrystal sits on each lattice site. The size, spacing and orientation of the QDs can be rigorously controlled. These superlattices were assembled epitaxially on the surface of silicon wafers and sapphire flats, thereby creating close packed QD arrays which can be probed and addressed optically and electronically. The fabrication techniques were optimized to permit rapid production of an entire series of 3D nanocrystal superlattices from a single reaction. Each superlattice was assembled with near-atomic precision using only bulk chemicals, solvents,

247. NOVEL MAGNETIC BEHAVIOR IN SELECTED PURE, DILUTE AND MIXED MAGNETIC SYSTEMS.*
Gary C. DeFotis, Department of Chemistry, College of William and Mary, Williamsburg, VA 23187

Recent work in this laboratory on a variety of pure, dilute and mixed magnetic systems will be reviewed. The materials examined are insulating in nature, and any exchange interactions present are short-range and of such strength as to lead to magnetic phase transitions in or near the liquid helium temperature range. Important single-ion anisotropy effects also contribute to the magnetic behavior in certain of the materials, and sometimes a low magnetic lattice dimensionality is also present. Examples discussed will include systems from the recently developed series $MCl_2 \cdot H_2O$ (M = 3d transition metal), in some of which novel nonequilibrium effects appear; an iron(III) molecular ferromagnet exhibiting unusual critical behavior, and an interesting variation of critical properties with diamagnetic dilution; and certain mixed magnetic systems, where interest centers on the structure of the T-composition phase diagram and the types of phases present, including spin glass regimes. *Supported by National Science Foundation-Solid State Chemistry-Grant No. DMR-9527357 and by a grant from the Petroleum Research Fund of the American Chemical Society.

248.

DENSITY-FUNCTIONAL THEORY: WHERE DO WE GO FROM HERE? Axel D. Becke, Department of Chemistry, Queen's University, Kingston, Ontario, Canada K7L 3N6

Despite the encouraging recent advances in Kohn-Sham DFT, systematic improvement of the exchange-correlation approximations on which it depends remains elusive. As a result, the literature contains a confusing choice of functionals with various advantages and disadvantages, and derived from various points of view. Our current strategy for exchange-correlation functional development will be discussed, and some outstanding problems (and hopefully their solutions) also presented.

249. 'DETERMINING NEW FUNCTIONALS FROM AB INITIO DATA', by David J. Tozer and Nicholas C. Handy, Department of Chemistry, Lensfield Road, Cambridge CB2 1EW, United Kingdom.

We have been determining high quality numerical exchange-correlation potentials from high quality ab initio analytic electron densities. Some details of the approach will be given. We are now trying to improve analytic exchange-correlation functionals using this numerical data. Our progress to date will be described. We believe that the use of high quality ab initio quantum chemistry is a good way to refine exchange-correlation functions. The question of basis sets for DFT calculations will also be addressed.

250. EMPIRICAL DENSITY FUNCTIONALS. R. D. Adamson, P.M.W. Gill and J. A. Pople, Department of Chemistry, Northwestern University, Evanston, Illinois 60208.

A number of commonly used density functionals have been reparameterized to give optimal fits to an extensive database of experimental energy data. The resulting empirical functionals are potentially useful in DFT applications. They show a number of significant differences with those using existing parameters; the importance of these variations is discussed.

251. AB INITIO THEORY OF NMR CHEMICAL SHIFTS IN SOLIDS. Steven G. Louie, Department of Physics, University of California, Berkeley, CA 94720, and Lawrence Berkeley National Laboratory, Berkeley, CA 94720

A new formalism for *ab initio* calculation of the orbital magnetic susceptibility and the NMR chemical shifts in solids and liquids is presented. The approach can be applied to periodic systems such as crystals, surfaces or polymers, and with a supercell technique, to nonperiodic systems such as amorphous materials, liquids, or solids with defects. The formalism is based on the density functional theory in the local density approximation and makes use of a generalized f-sum rule to eliminate the divergent terms that plagued previous theories. Calculations have been successfully carried out for the diamagnetic susceptibility of a number of insulators and for the NMR chemical shifts of a variety of systems including free molecules, ionic crystals, hydrogen-bonded materials and amorphous carbon.

252. AB INITIO STUDIES OF SOLVENT EFFECTS ON THE ELECTRON TRANSFER MATRIX ELEMENT. Robert J. Cave, Thomas Henderson, and Matthew Wander, Department of Chemistry, Harvey Mudd College, Claremont, CA 91711

Using the recently developed Generalized Mulliken-Hush method (CPL 249, 15, (1996)) we present theoretical electronic coupling matrix elements for a model system (Zn_2X^+, where X is one or more solvent molecules). The results are used to discuss specific orientation-dependent and energetic superexchange effects on electron transfer matrix elements for both ground and excited states. Specific solvent effects on the electronic coupling are also examined in the linked donor-acceptor systems of Zimmt and coworkers (JACS 118, 243 (1996)). Finally, results are also presented for solvent reorganization energies and continuum dielectric effects on the electronic coupling in the contact ion-pair benzene+Cl.

253.

THEORETICAL AND COMPUTATIONAL RESEARCH DEVELOPMENT FOR ELECTRO-OPTICAL (EO) PROPERTIES OF LARGE MOLECULES. **H Sekino** Analatom, Sunyvale, CA 94089

The canonical Molecular Orbitals (MO) of large molecules are immense in its number and similar to each others in its orbital energies. Truncation schemes often taken in semiempirical methods make little sense. We developed direct methodologies for the EO properties of large molecules consistently in semi-empirical, Density Functional Theory (DFT) as well as ab-initio methods. The Quasi Particle (QP) formalism where alternative truncation schemes may be applied on transition densities rather than MO, is used for the excitation energy calculations. Hyperpolarizabilities as well as excitation energies of organic molecules representing a unit of NLO enhancing materials are computed in the computational scheme described above.

254. THE ROLE OF ADSORBATE DIFFUSION IN SURFACE REACTIONS. G. Ertl, Fritz-Haber-Institut der Max-Planck-Gesellschaft, D-14195 Berlin, Germany

Thermally activated surface diffusion of N atoms adsorbed on a Ru(0001) surface was followed by STM, and the analysis revealed good agreement between the microscopic (random walk) and the macroscopic (Fick's) diffusion coefficients. Complications may arise by interactions between adsorbed particles at higher coverages, by the presence of surface defects and by the existence of transient 'hot' precursors. Coupling between reaction and diffusion may cause the formation of spatio-temporal concentration patterns on mesocopic (i.e. μm) length scales, as will be exemplified with the oxidation of CO on Pt(110).

255. H_2O DIFFUSION AND DESORPTION KINETICS ON SINGLE-CRYSTAL ICE: EFFECT OF HCl AND HNO_3. **Frank E. Livingston** and Steven M. George, Dept. of Chemistry and Biochemistry, University of Colorado, Boulder, CO 80309-0215.

Heterogeneous reactions play a key role in stratospheric ozone depletion. The presence of stratospheric species, such as HCl and HNO_3, may affect the dynamical properties of ice. HDO diffusion into ultrathin (20-125 BL) neat, HCl-, and HNO_3-dosed single-crystal $H_2^{16}O$ ice grown epitaxially on Ru(001) was investigated using a novel combination of laser-induced thermal desorption (LITD) probing and isothermal desorption depth-profiling. The measured HDO diffusion coefficients yielded surface lifetimes of ~40 seconds to ~50 milliseconds before diffusion into the bulk for T=150 K to 170 K, respectively. The HDO diffusion rate increased by ~10 for single-crystal ice dosed with HCl. Conversely, the HDO diffusion rate decreased by ~30-100 for ice multilayers exposed to HNO_3. LITD was also used to study the effect of HCl and HNO_3 on H_2O isothermal desorption kinetics from crystalline ice multilayers. H_2O desorption rates from HCl- and HNO_3-dosed single-crystal ice were respectively ~2 times larger and ~5 times smaller compared with H_2O desorption rates from neat crystalline ice films. These LITD measurements indicate that the diffusion and desorption kinetics of H_2O on ice are significantly altered by stratospheric species. These changes may affect heterogeneous chemistry on ice.

256. ELEMENTARY PROCESSES OF SURFACE DIFFUSION STUDIED BY HELIUM ATOM SCATTERING. J. Peter Toennies, Max-Planck-Institut für Strömungsforschung, 37073 Göttingen, Germany

As a result of further refinements in technique it has become possible to use high resolution helium atom scattering to study (1) the angle resolved elastic scattering cross section from single adsorbates [1], (2) their low frequency (< 10 meV) frustrated translational vibrations and (3) their microscopic diffusion mechanisms (length scale 2 -30 Å). From an analysis of experiments (2) and (3) for 3 % coverages of Na on the Cu (001) surface it has been possible to accurately determine the potential surface governing the lateral motion of the Na atoms [2]. Similar experiments have been carried out for CO molecules on Cu (001), Ni (001), Ni(110) and Pt(111) and potential surfaces for the motion of the molecule, in which the tilt angle is accounted for, have been obtained [3].

[1] A.P. Graham, F. Hofmann, J.P. Toennies and J.R. Manson; J. Chem. Phys. 105, 2093, (1996)
[2] A.P. Graham, F. Hofmann, J.P. Toennies, L.Y. Chen and S.C. Ying; submitted
[3] A.P. Graham, F. Hofmann, J.P. Toennies; Chem. Reviews 96, 1307, (1996), article in preparation.

257. ORIENTING REACTANTS IN ENTRANCE CHANNEL COMPLEXES. Marsha I. Lester, Department of Chemistry, University of Pennsylvania, Philadelphia, PA 19104-6323

Recently, this laboratory has shown that a hydroxyl radical and a reactive partner such as molecular hydrogen can be stabilized within a shallow well in the entrance channel to chemical reaction. The entrance channel complex serves to orient the OH and H_2/D_2 reactants in a T-shaped O-H--H_2/D_2 configuration. The characteristics of the binary OH-H_2/D_2 complexes are probed by electronic spectroscopy on the OH $A\ ^2\Sigma^+ - X\ ^2\Pi$ transition as well as by infrared overtone spectroscopy on the OH $X\ ^2\Pi\ v''=0 \rightarrow 2$ transition which is detected using a fluorescence depletion technique. Both excitation processes have the possibility of inducing a hydrogen abstraction reaction between the pre-oriented reactants.

258. NONADIABATIC TRANSITIONS AND ORBITAL REORIENTATION EFFECTS IN CLUSTERS AND SOLIDS. R. B. Gerber, A. I. Krylov and M. Niv, Department of Physical Chemistry, Hebrew University, Jerusalem 91904, Israel and Department of Chemistry, University of California, Irvine 92697-2025, USA

The photodissociation of HCl in solid Ar, and in (HCl)Ar_n clusters is studied by nonadiabatic Molecular Dynamics simulations, based on a surface-hopping treatment of transitions between different electronic states. The relevant 12 potential energy surfaces and the nonadiabatic interactions between them were generated by a Diatomics-in-Molecules (DIM) approach, which incorporates also spin-orbit coupling. The focus of the study is on the nonadiabatic transitions and on their role, both in the cage exit of the H-atom and in the recombination process. Reorientation of the unpaired potential of the Cl atom plays a key role in the process and in the competition between recombination and cage exit. It is found that nonadiabatic transitions occur very frequently in the first picosecond of the process. In some of the trajectories all the 12 electronic states are visited during the timescale studied. At least one nonadiabatic transition was found to occur even in many of the fastest cage-exit events.

259.

MODE SPECIFIC VIBRATIONAL PREDISSOCIATION OF ORIENTED HCN-HF.
Lukas Oudejans, Roger E. Miller; Dep. of Chemistry, University of North Carolina at Chapel Hill, NC 27599

Photofragment angular distributions have been obtained for the dissociation of oriented HCN-HF. The complex is oriented in a large dc electric field, prior to its dissociation, using the "brute force" method. Therefore, the two fragments recoil in opposite directions in the laboratory frame of reference, resulting in a separate detection by a bolometer. For comparison, in the non-oriented case, the bolometer detects both fragments resulting in less structured angular distributions as a result of the difference in mass and internal energy. The bolometer is also used as a microcalorimeter yielding the internal energy ratios of the two fragments. A second laser has been used to probe the rotational state of the fragment resulting in a complete assignment of the rotational and vibrational states of the fragments. These results confirm that the photodissociation process leads to the production of a vibrationally excited HCN fragment.

260. **STATE-RESOLVED ANGLE-VELOCITY MEASUREMENTS OF REACTIVE AND UNREACTIVE SCATTERING IN VAN DER WAALS COMPLEXES**
Jack A. Syage, The Aerospace Corporation, P. O. Box 92957, Los Angeles, CA 90009.

Using a photofragment imaging by sections method, we can record state-resolved differential cross sections in angle and velocity without resorting to an inversion routine. Our current work focuses on reactive and unreactive scattering initiated by photodissociation in aligned van der Waals complexes. Previous work focused on unreactive scattering of $I(^2P_{3/2})$ and $I(^2P_{1/2})$ from photodissociation of $(CH_3I)_n$ (dominated by n=2) as well as reaction leading to I_2. In this talk we will present our latest results on O_3-H_2O. The reactive and unreactive branching under study is

$$O_3 \cdot H_2{}^{18}O \rightarrow OH + {}^{18}OH + O_2 \quad \text{(reactive scattering)}$$
$$O_3 \cdot H_2O \rightarrow O + O_2 + H_2O \quad \text{(unreactive scattering)}$$

The reactive channel is detected by 2+1 REMPI of OH using transitions $D^2\Sigma^- \leftarrow X^2\Pi$ (242-247 nm region) and $3^2\Sigma^- \leftarrow X^2\Pi$ (227-229 nm) region. The unreactive $O(^3P_j)$ channel is detected by 2+1 REMPI at around 226 nm for the different spin-orbit states, j. We hope to have results for the $O(^1D)$ channel.

261.

DISSOCIATION OF H_2, D_2 AND HD IN INTENSE LASER FIELDS. D. W. Chandler[a], D. W. Neyer[a] and A. J. Heck[b], a) Sandia National Laboratories, PO BOX 969, MS 9055, Livermore CA, 94550 b) Department of Chemistry, University of Warwick, Coventry CV4 7AL, UK

Under the influence of intense laser pulses protons are observed from the irradiation of hydrogen molecules. The mechanism of their origin changes with laser power and frequency. Photofragment imaging techniques and photoelectron imaging are used to unravel the several pathways for the production of protons at both intermediate and high laser intensity. They include molecular ionization followed by dissociation, neutral dissociation followed by ionization of the atomic fragments and direct three-body dissociative ionization. High laser intensity effects such as alignment of the molecular axis, pondermotive forces, saturation and AC Stark shifting of states lead to unusual angular and velocity distributions. Data for H_2, D_2 and HD will be presented and discussed. In addition a new advance in the 'Ion Imaging' technique allowing an order of magnitude better resolution than previously reported will be presented.

262. ALIGNMENT OF OPEN-SHELL P-ORBITALS AND D-ORBITALS: BONDING IN ELECTRONIC STATES OF DIATOMIC METAL-ATOM/RARE-GAS VAN DER WAALS COMPLEXES. Steven Massick, John G. Kaup, Allen W.K. Leung, Brian Gau, and W.H. Breckenridge, Department of Chemistry, University of Utah, Salt Lake City, Utah 84112.

Recent results from our laboratories illustrate the dramatic effects of alignment of outer-shell p or d orbitals on the bonding in electronic states of diatomic metal-atom/rare-gas van der Waals complexes. For a state with an outer-shell valence p-orbital, π-alignment leads to much stronger bonding than σ-alignment. Similarly, for states with outer-shell valence d-orbitals, δ-alignment is more favorable than π-alignment, with σ-alignment being much less favorable. For the recently-discovered doubly-excited npnp valence states, π-alignment of both p electrons leads to unusually strong van der Waals bonding. For outer-shell Rydberg p and d electrons, alignment becomes less and less important, because as the principal quantum number n increases the RG atom can more readily penetrate the increasingly diffuse Rydberg electron clouds and approach the positively charged "cores." The bond energies thus approach those of the M$^+$·RG ions. New data on M·RG and M$^+$·RG complexes, where M = Mg, Ca and RG = Ar, Kr, Xe will be discussed within the framework of our ideas on the maximization of attractive forces and (especially) the minimization of repulsive forces in the bonding for such complexes.

263.
ORIENTATIONAL ORDER - DISORDER TRANSITION IN ICE. V. Buch, Department of Physical Chemistry, The Hebrew University, Jerusalem 91904, Israel

Computational study is presented of energetics and structure of proton disordered ice Ih, and different proton ordered forms of hexagonal ice. The study includes optimization of crystalline structures, and molecular dynamics simulations. Special attention was devoted to treatment of long range forces. Correlation is demonstrated between the energy of the ice configuration and the fraction of nearest neighbor "cis" configurations; the correlation is relaxed for some proton ordered forms of ice due to structural relaxation. The important role of the very long range dipole-dipole interaction is elucidated. Sensitivity of ice properties to different features of the potential (polarizability, distribution of charges, anisotropy of the repulsive core) is probed by simulations. It is finally argued that the well known orientational order - disorder transition in ice doped by KOH is most likely induced by the coulombic interactions between the water molecules and the impurity ions.

264.

FEMTOSECOND STUDIES OF ELECTRON DYNAMICS IN 2-D: SURFACES, INTERFACES, AND DEVICES. **C.B. Harris,** N.-H. Ge, R. E. Jordan, R. L. Lingle, J. D. McNeill, and C. M. Wong, Department of Chemistry, University of California, Berkeley, California 94720 and Chemical Sciences Division, Lawrence Berkeley National Laboratory, Berkeley, California 94720

Angle and time resolved femtosecond two-photon photoemission is utilized to probe the dynamics of electrons at surfaces and interfaces. A femtosecond laser is used to create an excited electronic distribution at an interface which is photoejected some time later by a probe pulse after which the kinetic energy is measured. Studies of various insulator/metal interfaces show that the dynamics are largely determined by the electron affinity of the adsorbate. In the case of a repulsive electron affinity, the dynamics can best be described as due to tunneling through the repulsive barrier presented by the adlayer. For two or more layers, 2-D small-polaron localization of the initially delocalized electron occurs and can be directly measured on the fs timescale. Adsorbates with a positive electron affinity form an attractive well, giving rise to quantum well states which contain important information about band structure and carrier dynamics at the interface.

265. ULTRAFAST PHOTO-INDUCED ELECTRON-TRANSFER FROM COUMARIN DYES ADSORBED ON SEMICONDUCTOR NANOCLUSTER SURFACES.
Edward W. Castner, Jr., Chemistry Department, Brookhaven National Laboratory, Upton, NY 11973-5000 and Kei Murakoshi, and Shozo Yanagida, Material and Life Sciences, Graduate School of Engineering, Osaka University, Yamadaoka, Suita, Osaka 565, Japan

Photosensitizers adsorbed directly on semiconductor surfaces provide the basis for solar photoelectrochemical devices. Semiconductor nanoclusters in solution allow for simple preparation of the photosensitized surface, with very large surface area. To optimize the efficiency of the photoelectrochemical cycle, one must know the rates of forward and reverse electron-transfer from the excited state dye to the semiconductor conduction band. We present measurements of the ultrafast forward electron-transfer rates for three coumarin dyes (Coumarins 343, D-1421, and D-126) bound by the carboxylic acid group to the metal cationic sites of TiO_2 and ZnO semiconductor nanocluster surfaces. These rates are obtained directly from the ultrafast fluorescence dynamics, measured using the femtosecond upconversion technique. Coumarins 343 and D-1421 on TiO_2 nanoclusters in aqueous solution display very rapid forward electron-transfer with rate constants exceeding 10^{13} s^{-1}. Substantially slower rates are observed for the same coumarin TiO_2 systems in methanol solution.

266.

SIZE-DEPENDENT ELECTRON DYNAMICS OF GOLD NANOPARTICLES PROBED BY SUBPICOSECOND TRANSIENT SPECTROSCOPY
Temer S. Ahmadi, Stephan L. Logunov, and <u>Mostafa A. El-Sayed</u>, School of Chemistry and Biochemistry, Georgia Institute of Technology, Atlanta, Georgia 30332-0400

The dynamics of gold nanoparticles with average particle sizes of 1.5, 2.5, 3.3, and 30 nm were studied by transient spectroscopy after excitation with subpicosecond laser pulses. The photoexcitation leads to the heating of the electron gas and its subsequent thermalization through electron-electron and electron-phonon interaction. Spectral responses for different sizes of gold nanoparticles were different. The size dependent spectral changes are attributed to the reduction of the density of states for small nanoparticles. The variation of the laser intensity changed kinetics of the cooling process in gold nanoparticles due to temperature dependence of the heat capacity of electron gas.

267.

TIME RESOLVED CARRIER DYNAMICS NEAR THE INSULATOR-METAL TRANSITION. **M.J. Feldstein** C.D. Keating, W. Zheng, Y.H. Liau, A.G. MacDiarmid, M.J. Natan, N.F. Scherer, Department of Chemistry, Univerisity of Pennsylvania, Philadelphia, PA 19104-6323

The study of insulator-metal transitions can provide insight into the dynamical properties of complex, disordered systems. Specifically, the subtleties of phenomena like carrier localization and transport in systems such as conducting polymers and colloidal metal films can be elucidated. This paper reports the results of optically excited carrier dynamics near the insulator-metal transition in a series of thin films composed of 12 nm colloidal gold particles and a series of thin polyaniline films of varying conductivity. Carrier lifetimes and dynamics have been probed directly by femtosecond time-resolved laser spectroscopy measurements. Additionally, the films' mesoscopic structure has been determined with atomic force microscopy. Correlations are made between film structure, carrier dynamics, and conductivity. The issues of carrier localization, transport, and scattering are addressed.

268. MECHANISITIC ASPECTS OF ENERGY TRANSFER ACROSS NANOMETRIC STRUCTURES, BOUNDARIES, AND INTERFACES: A NEAR-FIELD SCANNING OPTICAL MICROSCOPY STUDY OF NANOCRYSTALLINE MOLECULAR SEMICONDUCTORS. David M. Adams, Josef Kerimo, Arie Zaban, Brian A. Gregg, Paul F. Barbara, University of Minnesota, Department of Chemistry, 207 Pleasant St., Minneapolis, MN 55455

The mechanism of energy transfer across nanostructured assemblies is investigated with near-field scanning optical microscopy(NSOM). NSOM is a high resolution imaging technique which provides simultaneous high resolution topographic images and subdiffraction limited visible light images. High resolution NSOM with ~30nm optical resolution is used to directly image small spatial regions of energy transfer between nanocrystals of donor perlyene-*bis*-phenethyl imide(PPEI) and acceptor titanyl phthalocyanine (TiOPc) molecular semiconductors. Correlated local variations in the polarized emission and transmission intensities on the nanometric scale indicate regions of increased coupling between donor and acceptor. These variations are interpreted as arising from variations at the interface in local orientation, local populations, and effective contact between donor and acceptor. Quantitative estimates of the exciton diffusion length in the donor PPEI are obtained from measurements of quenched and unquenched emission intensity as a function of nanocrystalline height.

269.
ORDERED 2-D PHASES OF NANOCRYSTALS: FORMATION AND COOPERATIVE PHYSICAL PHENOMENA. **J. R. Heath,** Dept. of Chemistry and Biochemistry, UCLA 405 Hilgard Ave., Los Angeles, CA 90095-1569

Metal and semiconductor nanocrystals can be assembled into various two-dimensional phases, including annular-ring structures, lamellae, closest-packed phases, and foams. The non-linear optical response of these phases as a function of interparticle separation reveals strong coupling between the particles that is not well described by a classical multipole model, and may be indicative of particle-particle wavefunction overlap. These phases have been incorporated (as thin films) into simple capacitive devices. Coulomb blockade and coulomb staircase measurements have been carried out and provide insight into how cooperative particle interactions can manifest themselves in electron transport.

270.
NANOSTRUCTURED PARTICLES FOR ELECTRORHEOLOGICAL APPLICATIONS, Ping Sheng, W. Y. Tam, G. H. Yi, W. Wen, H. Ma, and M. M. T. Loy, Department of Physics, The Hong Kong University of Science and Technology, Clear Water Bay, Kowloon, Hong Kong

The electrorheological (ER) fluids constitute a class of materials whose rheological properties are controllable through the application of an electric field. Their potential applications include improved shock absorbers, engine mounts, clutches, valves, and advanced automatic transmissions. With the guidance of first principles understanding of the ER mechanism [1], we have predicted and experimentally verified a 300-fold increase in the ER effect through the use of nanostructured particles [2], consisting of 1.5 μm to 50 μm sized glass spheres coated with a 10nm Ni layer and a 25 nm to 50 nm outer layer of TiO_2

271. FACILITATED TARGET LOCATION AND TRANSCRIPTION BY E. COLI RNA POLYMERASE IMAGED IN BUFFER BY SCANNING FORCE MICROSCOPY
Carlos Bustamante, Howard Hughes Medical Institute and Institute of Molecular Biology, University of Oregon, Eugene, OR 97403

Non-specific and specific interactions of E. coli RNA polymerase (RNAP) with DNA have been visualized using tapping-mode SFM operating in buffer. The first direct evidence that RNAP can use various mechanisms of facilitated promoter location has been obtained. Utilizing tapping mode SFM in fluid, both RNAP holoenzyme and the core enzyme can be seen diffusing along the DNA. These observations indicate that the polymerase can use this form of "sliding" along the DNA to increase its rate of promoter location. Other mechanisms of facilitated target location such as intrasegment hopping and intersegment transfer have also been observed.
In related experiments, transcription by single RNAP molecules was followed in real-time. Upon addition of nucleoside triphosphates (NTPs) to mica-bound stalled elongation complexes, the polymerase molecules threaded their template toward the terminator. Unlike the random motion displayed by non-specifically bound RNAP, transcribing polymerases moved in a processive and unidirectional manner along the DNA at an average rate of 1.5 ± 0.8 nt·s^{-1}. Biochemical assays performed under identical conditions confirmed the viability of mica-bound complexes and the transcription rate determined by SFM, which is about five-fold slower than that in solution. Future developments in SFM design are likely to improve the temporal and spatial resolution of the images. This will enable the SFM to gain new insights into complex biological processes at the nanometer scale.

272. PROBING BIOMOLECULES WITH THE ATOMIC FORCE MICROSCOPE. Helen G. Hansma, Department of Physics, University of California, Santa Barbara, CA 931106

Biomolecules in which substructure has been seen by AFM include single-stranded, double-stranded and triple-stranded nucleic acids, proteins, and membrane protein channels. Proteins, including some active enzymes, DNA molecules and DNA-protein complexes have also been probed by AFM while moving in aqueous buffer on a mica surface. Thus AFM is useful for characterizing molecular structures and for observing and tracking the motion of biomolecules.

273. ATOMIC AND NANOMETER SCALE MODIFICATION OF MATERIALS WITH PROXIMAL PROBES, Phaedon Avouris, IBM Research Division, T.J. Watson Research Center Yorktown Heights NY, 10598.

I will first consider the selective breaking of chemical bonds at surfaces using STM tip-induced excitations. Hydrogen desorption from silicon will be used as an example. Two types of desorption mechanisms will be demonstrated: (a) desorption as a result of an electronic transition with the STM operating in the field emission mode, and (b) desorption at energies below the onset of electronic transitions in the Si-H unit via a novel mechanism involving the multiple-vibrational excitation of the Si-H bond. The characteristics of the two mechanisms, i.e. energy thresholds, cross-sections, current dependencies, resolution, etc. will be analyzed. As an example of modification induced by the transient capture of tip-emitted electrons, I will discuss electron attachment to the $3\sigma^*$ shape-resonance of O_2 chemisorbed on Si(111). This process leads to Si-O and O-O bond breaking. Finally, as an example of a process with immediate technological importance, I will discuss the kinetics, mechanism and applications of an AFM tip-induced local oxidation of Si.

274. OBSERVATION AND PREDICTION OF SCANNING TUNNELING MICROSCOPY IMAGES OF INDIVIDUAL ADSORBED MOLECULES ON METAL SURFACES, S. Chiang, Department of Physics, University of California, Davis, CA 95616.

An overview of earlier work on imaging individual molecules, including naphthalene, azulene, and several methylazulenes, on Pt(111) will be presented. Naphthalene and its isomer azulene have been imaged with submolecular resolution in both pure and mixed monolayers using the scanning tunneling microscopy (STM). These molecules can be distinguished in mixed monolayers by their differing asymmetry. Several isomers of methylazulene can also be distinguished by their observed shapes on the surface: 1-methylazulene (kidney bean), 2-methylazulene (pear), and 6-methylazulene (diamond). Diffusion rates, sticking coefficients, and molecular orientations relative to the substrate have also been compared. These images have also been predicted using a computational method, based on extended Hückel theory, for calculating the local density of states for the molecule adsorbed on a cluster of metal atoms. The method has recently been used to predict STM images for benzene on Pt(111) and furan and pyrrole on Pd(111). New molecular images for aromatic molecules on Pd(111) will also be presented.

275.
ATOMIC-SCALE VIEWS OF THE DYNAMICS, INTERACTIONS, AND PROPERTIES OF SINGLE ADSORBED MOLECULES. P. S. Weiss, J. J. Arnold, L. A. Bumm, M. T. Cygan, J. H. Ferris, J. A. Johnson, M. M. Kamna, and J. G. Kushmerick, Department of Chemistry, The Pennsylvania State University, University Park, PA 16802-6300.

Scanning probe microscopes allow unprecedented views of the dynamics of single adsorbates and the effects that they have on their surface environment. We are able to deduce the trajectories followed by individual adsorbates in the adsorption process and to infer how quickly they accommodated their incident kinetic energy to the surface. We discuss the motion of individual adsorbates at reduced temperature and the motion and electronic properties of constrained "marker" molecules in densely packed films.

276.

AB INITIO MOLECULAR DYNAMICS SIMULATION OF CHEMICAL PROCESSES. **M Parrinello**, University of Pennsylvania, 3231 Walnut Street, Philadelphia, PA 19104-6202

277. NEW METHODS FOR DIRECT DYNAMICS CALCULATIONS. Donald G. Truhlar, Yao-Yuan (John) Chuang, Patton Fast, Elena Laura Coitiño, Wei-Ping Hu, and Jordi Villa, Department of Chemistry and Supercomputer Institute, University of Minnesota, Minneapolis, Minnesota 55455-0431

This talk will overview several recent improvements in and applications of direct dynamics methods to the calculation of chemical reaction rates. The goal of this research is to enable the convenient use of high-level electronic structure calculations in conjunction with dynamics calculations. The work to be discussed in this talk is particularly focused on bimolecular gas-phase reactions, variational transition state theory, and multidimensional semiclassical tunneling calculations. Enhancements under current study include minimizing the amount of electronic structure data required for a given level of dynamical theory, improvements in dual-level interpolation methods, and implementation of curvilinear coordinates. Applications under curent study include the reactions of H, OH, and Cl with hydrocarbons.

This talk will overview a selection of this work. This work was supported in part by the U. S. Department of Energy, Office of Basic Energy Sciences.

278. AB INITIO DIRECT DYNAMICS CALCULATIONS OF ATMOSPHERIC REACTIONS. R. Steckler and G.M. Thurman, San Diego Supercomputer Center and SDSU, P.O. Box 85608, San Diego, CA 92186 and J.D. Watts and R.J. Bartlett, Quantum Theory Project, University of Florida, Gainesville, FL 32611

Few quantities are more fundamental to chemistry than rate constants; but often, particularly for reactions involving transients as occur in flames and in atmospheric chemistry, accurate experimental values are not known, or are only known over a limited temperature range. One of the long term goals for modern theory has been to provide reliable rate constants solely from first principles. However, their accurate determination requires both state-of-the-art dynamics and electronic structure for the determination of potential energy surfaces. Until recently, it has not been practical to combine the two components. Recently a new method has been developed called direct dynamics. This talk will discuss new developments in direct dynamics and its application to the first-principles computation of rate constants for atmospheric reactions. Included in the discussion will be a detailed discussion of the rate constants for OH+HCl which were computed directly using ab inito electronic structure theory at the MBPT(2) and CCSD(T) levels and VTST.

279. QUANTUM SIMULATIONS OF ION-WATER CLUSTERS, Hai-Ping Cheng, Department of Physics and Quantum Theory Project, University of Florida, Gainesville, Florida 32611

High-level quantum molecular dynamics simulations are performed to study the proton motion in $(NH_3)_m(H_2O)_nH^+$ (m=0, 1 and n=1,2,3,4), clusters. These small clusters are chosen as prototype systems for studying the mechanisms of proton transfer at an atomistic level. We focus on the fundamental steps of proton motion in molecular clusters, the effects of finite temperature, the dynamical consequences of proton affinities, and the interplay between proton motion and proton affinity in these systems. Vibrational spectra of $H_5O_2^+$ and $NH_3(H_2O)_4H^+$ are obtained and analyzed in detail. The features of the spectrum can be used, in principle, to probe proton motion in the transition state region reactions. Other ion-water clusters such as $NO^+(H_2O)_n$ (n=1,2,3...) are also studied. In the calculations, the electronic charge distribution is calculated concurrently with the nuclear dynamics. An analysis of iso-charge density surfaces gives qualitative and quantitative descriptions of the dynamics of electronic redistribution. The BOMD is performed in the framework of density functional theory (DFT) with generalized gradient approximations (GGA).

280.

FIRST PRINCIPLES SIMULATION OF THE PHOTODISSOCIATION ON THE COUPLED $1^1A''$ AND $2^1A''$ POTENTIAL SURFACES OF OZONE - ELECTRONIC STRUCTURE CALCULATIONS. Clemens Woywod, Department of Chemistry, University of California, Berkeley, CA 94720, Martin Stengle and Wolfgang Domcke, Institute of Theoretical Chemistry, Heinrich-Heine-University, D-40225 Düsseldorf

The electronic wavefunctions of the two lowest excited singlet states of ozone have been determined for the photodissociation channels. In this 2D electronic subspace the interacting states have been directly obtained in the diabatic representation at the CASSCF level of theory employing a two-step procedure [Domcke and Woywod, CPL **216**, 362 (1993); Domcke, Woywod, and Stengle, CPL **226**, 257 (1994)]. In step 1 a smoothly varying one-electron basis is constructed for the complete set of *ab initio* grid points, in step 2 block-diagonalization of the Hamiltonian matrix is used to generate diabatic wavefunctions. It is shown how the diabatic-to-adiabatic mixing angle thus defined at the CASSCF level can be used to subsequently transform the transition-dipole-moments and the CASPT2 potential functions from the adiabatic into the diabatic picture. The matrix elements of the diabatic electronic wavefunctions vary slowly with the nuclear coordinates. This property has been exploited to reduce the number of grid points for the *ab initio* calculations and to eliminate degeneracy artifacts obtained with the CASPT2 electronic structure model.

281.

ULTRAFAST NON-ADIABATIC DYNAMICS IN ELECTRONICALLY EXCITED POLYATOMIC SYSTEMS WITH "ON THE FLY" GENERATED PES. A.I.Krylov C.D.Sherrill, M.Head-Gordon, Dept. of Chemistry, University of California, Berkeley, CA, 94720

The purpose of this work is to understand the mechanisms of photodissociation, electronic relaxation, and photoinduced isomerization in polyatomic molecules. The theoretical treatment of electronically excited states in polyatomic systems poses significant problems for theoreticans due to unique features of these states, such as delocalized character of excited states, many-body nature of interactions, and electronic degeneracy. To solve this problem, we develop new approach for modeling non-adiabatic dynamics in polyatomic molecules by means of semiclassical Molecular Dynamics with "on the fly" generated ab initio PES. This requires reformulation of standard quantum chemical methods to be feasible to dynamical calculations. We study ultrafast internal conversion in hexatriene and photodissociation of nitric acid.

282. SURFACE DYNAMICS IN THE PRESENCE OF LATTICE STRESS: SiGe GROWTH ON Si(001),*
Max G. Lagally, University of Wisconsin-Madison, Madison, WI 53706

The epitaxial growth of one material on itself has presented the ideal opportunity for exploring the broad range of dynamic properties on surfaces connected with atomic transport: surface adsorption/ desorption/decomposition kinetics, surface diffusion, desorption from and sticking to steps, and transport over steps, along steps, and around island corners and kinks. Non-homoepitaxial systems, the vast majority, of necessity involve a lattice mismatch, which produces stress that contributes a new term to the total system free energy and produces possible new kinetic effects associated with the existence of (at minimum) two interacting components. In this talk, we consider the growth (by both MBE and CVD) of Ge on Si(001), a model system for the influence of stress on surface kinetic and thermodynamic processes. We demonstrate the kinetics of stress-driven place exchange of Si and Ge between surface and subsurface layers using H titration, RHEED, and infrared spectroscopy. We use STM to explore surface diffusion of Ge and Si with and without the presence of H. We explore with AFM and x-ray diffraction the stress-induced bunching of substrate steps when a thin film of Ge is deposited. We present a model that demonstrates the interplay between the effects of step kinetics and thermodynamics. As time permits, we will describe areas of surface transport that stress may additionally affect, but about which nothing quantitative is as yet known.
*Supported by NSF and AFOSR

283. THEORETICAL STUDIES OF MICROSCOPIC MECHANISMS OF DIFFUSION AND GROWTH ON SEMICONDUCTORS. E. Kaxiras, Department of Physics and Division of Applied Sciences, Harvard University, Cambridge MA 02138

The dynamics of atoms on semiconductor surfaces are complicated due to surface reconstruction and the nature of covalent bonding. The issues of surface diffusion and island nucleation are central to understanding growth, both in homo-epitaxy and hetero-epitaxy. We investigate these phenomena in systems involving Si and Ge substrates, with attention to diffusion and nucleation events at the initial stages of growth. We use ab-initio Local Density Functional theory to analyze the energetics, and solid-on-solid kinetic Monte Carlo simulations to make contact with realistic systems. The results provide fresh insight to a number of recent growth experiments.

284. ACCELERATING THE DYNAMICS OF INFREQUENT EVENTS.
Arthur F. Voter, Theoretical Division, Los Alamos National Laboratory, Los Alamos, NM 87545

Molecular dynamics (MD) is a powerful tool for investigating detailed atomic-scale behavior on time scales of a nanosecond or less. For slower, infrequent-event processes, transition state theory (TST) can be employed, provided the the nature of the transition states are known – i.e., if the relevant saddle points can be found. However, in many cases, the reactive events that will occur are not known in advance, or the transition states are very complicated. I discuss a new method for treating this type of case for solid-state systems. A bias potential, constructed from the gradient and Hessian, raises the energy of the system without affecting the TST dividing surfaces. Performing MD on the biased potential leads to accelerated transitions from state to state. In this "hyper-MD" approach, time is no longer an independent variable; the elapsed time is estimated as the simulation proceeds, converging on the correct time in the long-time limit. Hyper-MD simulations of metallic surface diffusion on the microsecond time scale will be presented.

285. PHASE CONTROL OF ORIENTATION AND ALIGNMENT AND THE REAL TIME DEPENDENCE OF THE β ANISOTROPY PARAMETER. Moshe Shapiro, Department of Chemical Physics, The Weizmann Institute of Science, Rehovot, 76100 Israel

In the first part of the talk we discuss various laser control scenarios for inducing orientation in molecules and electrons. Amongst such scenarios are: laser control of the directionality of photo-ejected electrons in semiconductors, a prediction recently verified experimentally; elliptic polarization control of orientation in photodissociation reactions, an effect postulated in the strong field regime to enable the generation of "atto"-second lasers; strong field phase control of spatial orientation of photodissociation fragments and intensity control over the direction of above threshold electrons. In the second part of the talk we discuss how vector properties of photodissociation with a coherent laser pulse evolve in time during and after the excitation process. In particular, we show that the *transient* anisotropy parameter, $\beta(t)$, displays a *beating* phenomenon, in which periodic oscillations between a parallel-type and a perpendicular-type angular distribution are observed. Depending on the excitation pulse, this "vectorial" beating may be accompanied by a faster beating (similar to the "scalar" vibrational beating observed in the past in predissociating systems, such as NaI) if more than one resonance is excited. The transient anisotropy beating may be used to obtain experimentally the time-dependent average interfragment separation.

286. ALIGNMENT AND CONTROL OF ROVIBRONIC WAVE PACKETS. Stephen R. Leone, JILA and Department of Chemistry and Biochemistry, NIST and University of Colorado, Boulder, Colorado 80309-0440

Rovibronic wave packets are prepared and probed with ultrafast laser pulses in lithium dimers. Controlled composition of the state amplitudes in the wave packets is facilitated by intermediate state selection with a single frequency cw laser. A new form of two-level rotational coherence spectroscopy is demonstrated. Wave packet amplitudes are controlled by both intermediate state selection and laser pulse shaping. The rotational alignment can be controlled independently of the vibrational motion. The results will be discussed in terms of alignment control dynamics.

287. SPATIAL TAMING AND TRAPPING OF MOLECULES. Dudley R. Herschbach, Department of Chemistry and Chemical Biology, Harvard University, Cambridge, MA 02138

The translational wildness of molecules has long been subdued by herding them into supersonic beams. Rotational tumbling in such beams has recently also been suppressed by use of strong static electric or magnetic fields. These act on permanent dipole moments to form spatially oriented or aligned pendular states, superpositions of the field-free rotational states in which the molecular axis librates about the field direction. A more general method is now being developed, not restricted to polar or paramagnetic molecules, and applicable both to alignment and to spatial trapping. This creates pendular states by means of the anisotropic interaction of the electric vector of intense laser radiation with the dipole moment induced in a polarizable molecule. This talk will discuss current efforts and prospective applications, such as reacting Bose-Einstein condensates.

288. NEW DIRECTIONS OF ORIENTATIONAL AND STATE SELECTIVE EFFECTS IN MOLECULAR REACTION DYNAMICS. S. Stolte, Laser Centre, Department of Chemistry, Vrije Universiteit, De Boelelaan 1083, 1081 HV Amsterdam, The Netherlands

Pro's and con's of the two current methods of orienting polar molecules will be discussed. Brute-Force orienting the CH_3X (X=Br,I) reactant molecule made it possible to observe the steric effect of the $K+CH_3X \rightarrow KBr+CH_3$ reaction at a subthermal collision energy (~1.3 eV) as a function of the scattering angle. Elucidating (unexpected) features of the dynamics of this benchmark "harpooning" reaction have been found.
Recent studies show that the steric effect (head vs. tail approach) plays an important role in rotationally inelastic (NO-Ar) collisions. This type of measurement promises a much more detailed determination of the anisotropy of the intermolecular potential as currently possible.
Orienting the parent molecule is becoming also a powerful method to investigate the detailed dynamics of photolysis. With respect to the molecular fixed frame of the (oriented) parent it allows the state specific probing of the recoiling fragments. The angular recoil distributions of the fragments resulting from oriented CH_3I+266 nm $\rightarrow CH_3+I$ and oriented N_2O+205 nm $\rightarrow N_2+O(^1D)$ observed via two-dimensional ion-imaging will be presented.
Finally the novel feasibility of orienting molecules by applying coherent phase-control will be explored theoretically.

289. DEGENERATE FOUR-WAVE MIXING AS A PROBE OF SPATIAL ANISOTROPY IN MOLECULAR DYNAMICS. P. H. Vaccaro, T. A. W. Wasserman, P. Staats, and B. R. Johnson, Department of Chemistry, Yale University, New Haven, Connecticut 06511.

Ongoing studies have demonstrated the feasibility of exploiting fully-resonant variants of four-wave mixing spectroscopy for the elucidation of molecular structure and dynamics. As spectroscopic tools, the primary advantage of such nonlinear optical techniques stems from their absorption-based nature and background-free response which makes possible the interrogation of trace species that are not amenable to conventional fluorescence detection schemes. While these characteristics provide a facile means for investigating the scalar properties (e.g., internal state distributions) of nascent reaction products, the inherent polarization specificity and velocity discrimination afforded by sub-Doppler implementations of Degenerate Four-Wave Mixing (DFWM) spectroscopy can also be employed for vector correlation measurements which provide a quantitative signature for the anisotropic forces mediating a chemical transformation. Recent developments in the theory and application of DFWM to problems of molecular photodissociation dynamics will be discussed, with particular emphasis on the extraction and interpretation of spatial information (i.e., rotational and translation anisotropy) from unrelaxed photofragments.

290. NANOSPHERE LITHOGRAPHY: STRUCTURAL AND OPTICAL PROPERTIES OF TUNABLE PERIODIC PARTICLE ARRAYS. R.P. Van Duyne Department of Chemistry, Northwestern University, Evanston, Illinois 60208-3113

Nanosphere lithography (NSL) is an inherently parallel, "benchtop" nanofabrication technique based on the self-assembly of polymer or silica nanospheres to form highly ordered 2D colloidal crystals. These crystals act as material, M, deposition masks. Both single layer, SL, and double layer, DL, nanosphere masks have been made. SL- and DL-periodic particle arrays (PPAs) result from the "lift-off" of the nanosheres and excess M. NSL is materials general and has been implemented using both physical vapor deposition (PVD) and pulsed laser deposition (PLD) methods. PPAs with tunable in-plane diameter, a, out-of-plane height, b, and interparticle spacing, d_{ip}, are made by systematic variation of the: (I) nanosphere diameter, D; (ii) mass thickness of M, d_m; and (iii) angle between collimated vapor deposition beam and the surface normal, Θ. The PPAs are structurally characterized by atomic force microscopy (AFM) and optically characterized by UV-VIS absorption spectroscopy and surface-enhanced Raman excitation spectroscopy (SERES).

291.

SELF ORGANIZATION INTO 2D AND 3D SUPERLATTICES OF NANOSIZED PARTICLES DIFFERING BY THEIR SIZE, **M. P. Pileni**, F. Billoudet[1], L. Motte[1,2], C. Petit[1,2], A. Talef[1], (1)Laboratoire S. R. S. I., U. R. A. C. N. R. S. 1662, Universite P. et M. Curie (Paris VI), B. P. 52, 4 Place Jussieu, - 75231 Paris Cedex 05, FRANCE (2)C. E. A.- C. E. N. Saclay, DRECAM, S. C. M., 91191 Gif-sur-Yvette, Cedex, FRANCE

Self assemblies made of silver sulfide and silver metallic nanoparticles are observed. The domain of monolayer and the number of large aggregate formed strongly depend on the preparation of the sample. Monolayers organized in a hexagonal network are observed in a very large domain. This is observed for various particle sizes. Large and small aggregates made of nanoparticles differing by their size have been obtained. The particles are highly oriented and form a face centered cubic structure. The size and the shape of the 3D crystals are not yet controlled. This could be due to the preparation mode but also to the quality of the support used.

292.

SEMICONDUCTOR NANOCRYSTALLITES AS ARTIFICIAL ATOMS: A SPECTROSCOPIC APPROACH. **Moungi G. Bawendi** Department of Chemistry, The Massachusetts Institute of Technology, Cambridge, MA 02139

Nanometer size crystallites of semiconductors are often called "quantum dots" or "artificial atoms" because of their predicted discrete electronic states and delocalized wavefunctions. The emerging spectroscopic picture of cadmium selenide nanocrystallites seems consistent with this description, but also shows some interesting complexities. In this talk we summarize a number of optical studies, including photoluminescence excitation, Stark and Zeeman experiments, and single dot fluorescence spectroscopy to build a quantum mechanical description of this system.

293.

INORGANIC SELF-ASSEMBLY: A DIRECT ROUTE TO ORDERED METAL OXIDE ARRAYS. **C. G. Wall**, B. K. Niece, L. Ge, A. A. Gewirth, W. G. Klemperer, Department of Chemistry and The Beckman Institute for Advanced Science and Technology, University of Illinois at Urbana-Champaign, Urbana, IL, 61801

Self-assembly of some heteropolytungstates on Ag and Au single crystal surfaces in redox-active solution was studied using in situ Scanning Tunneling Microscopy, electrochemical, and ex situ reflectance infrared measurements. Heteropolyoxoanions spontaneously form ordered, adherent monolayers on Ag(111) surfaces, the structure of which is controlled by the shape of the anion. These monolayers exhibit domains, spanning tens of nanometers, which arise as a consequence of the oxophillic interactions between the polyoxoanion and the Ag surface. On Au substrates monolayer structures are also formed, but only become well order under the influence of an applied external potential. These monolayers exhibit fluxional structures due to their relatively weaker interaction with the Au surface as compared to Ag.

294. MOLECULAR METAL NANOCRYSTALS & THEIR ARRAYS. Robert L. Whetten, *Schools of Physics & Chemistry, Georgia Institute of Technology, Atlanta, GA 30332-0430*

A longstanding ideal at the intersection of the molecular and solid-state sciences has begun to be realized. It has proved possible to generate metal nanocrystals with commensurately passivated surfaces, replicated so identically that they constitute molecular matter: they possess molecular vapor and solution phases, can be separated from mixtures to yield pure substances, and condense reversibly into crystalline solids with long-range translational order.[1-8] The archetypal system of *Au* nanocrystals, passivated by a weakly bound compact monolayer of *n*-alkylthiol(ate) groups, has distinct cores ranging from a critical minimum of 1.4 nm to ~ 3 nm (70 to ~ 1000 atoms, determined mass-spectrometrically in the 15 to 300 k mass range), and thus shows quantum effects in optical and tunneling spectra at ordinary to below liquid-nitrogen temperatures. The faceted core structure changes subtly from fcc to decahedral packing just below 2.0 nm; the surfactants can be desorbed to give pure metal clusters in macroscopic quantities. Solids formed by reversible condensation include $hcp, fcc, bcc,$ and bct (super)lattice arrays, depending on chain-length; these have a tunable inter-metal spacing, giving rise to novel structural, dielectric, optical, and transport properties.

1. R. L. Whetten, et al., *Adv. Mater.* **8**, 428-433 (1996). 2. S. A. Harfenist, et al., *J. Phys. Chem.* **100**, 1390 (1996). 3. D. Luedtke, et al., *ibid* **100**, 13328 (1996). 4. I. Vezmar, et al., *Z. Phys.* **D**, in press. 5. C. L. Cleveland, et al., *ibid*, in press. 6. M. M. Alvarez, et al., *Chem. Phys. Lett.* in press. 7. M. M. Alvarez, et al., *J. Phys. Chem.* in press; S. L. Logunov, et al.; *ibid* in press. 8. M. N. Shafigullin, et al., submitted.

295.

SELF-ASSEMBLY OF AU COLLOID MONOLAYERS AND MULTILAYERS Michael J. Natan, Department of Chemistry, The Pennsylvania State University, University Park, PA 16802.

Two- and three-dimensional arrays of colloidal Au nanoparticles can be organized on a variety of substrates using organosilanes and/or bifunctional organic crosslinkers. Moreover, the diameters of surface-confined particles can be enlarged by selective deposition of metal onto the particle surface. The kinetics and thermodynamics of these processes are understood, and will be described. Moreover, these processes can be exploited to control interparticle spacing in two dimensions and overall particle coverage in three dimensions, the key parameters dictating bulk optical, electrochemical, and electronic behavior. For example, 10 layers of 11-nm diameter colloidal Au particles crosslinked by the short spacers 2-mercaptoethanol or 2-mercaptoethylamine yield extraordinarily conductive films, while those crosslinked by longer alkyl chains are far less conductive. Examples from SERS and electrochemistry will also be discussed.

296. **Ab Initio CBS-QCI Calculations of the Ammonia Inversion Mode.**
Daniel J. Rush* and Kenneth B. Wiberg*
Department of Chemistry, Yale University, New Haven, CT 06520-8107

The Complete Basis Set (CBS) extrapolation model chemistry of Petersson and co-workers was used to explore the potential energy surface of the ammonia inversion mode. The CBS-QCI theoretical energies were calculated using 41 points along the inversion surface at the MP2/6-311++G** geometries. A variety of techniques were explored to model the potential surface. Subsequent numerical solution of the one-dimensional Schrödinger equation produced energy levels for ammonia isotopomers in good agreement with experimental transitions. Accounting for the variable nature of the reduced mass with inversion coordinate is shown to be of significance. This study is an important first step in producing reliable methods for making ab initio thermodynamic corrections from $\Delta E(0~K)$ to $\Delta G(298~K)$ in other nitrogen containing systems. Because no experimental methods generate data at 0 K, these corrections provide a crucial link between experimental thermochemical energies and ab initio theory.

297. ABINITIO SCF AND DFT STUDY OF N-ARYL-N-NITROSOHYDROXYLAMINE DERIVATIVE SPECIES AND PARA-SUBSTITUTED ANALOGUES. **L. L. Griffin**, D. Bonchev (a,b), A. T. Balaban (a,c), W. A. Seitz(a), R. E. Garfield(d). (a) Texas A&M University at Galveston, Galveston, TX 77553-1675, USA; (b) Burgas University of Technology, Burgas 8010, Bulgaria; (c) Romanian Academy of Sciences, Calea Victoriei, Bucharest, Romania, (d) The University of Texas Medical Branch, Galveston, Texas 77555-1062

Salts of Aryl-substituted phenylnitrosohydroxylamines act as nitric oxide donors in physiological solutions. The anions may undergo transformations to other species prior to loss of NO. We have applied ab initio HF and DFT optimizations to the anions and four derivative species/conformers of ortho and para substituted cupferrate (Ph-N(O)NO anion). Trends in relative species energy, electronic and structural properties, are observed relative to the classical electron-releasing or withdrawing nature of the group at the para position. The structure of cupferrate optimized at B3LYP//6-31G** agrees excellently with the X-Ray diffraction structure.

298.
AB INITIO CALCULATIONS OF FIVE- AND SIX-COORDINATE SILICON COMPLEXES.
William D. Thweatt, Steven D, Kloos, Beon-Kyu Kim, Michael Page, and Philip Boudjouk, Department of Chemistry, North Dakota State University, Fargo, ND 58105

Five- and six-coordinate silicon systems are important for understanding the mechanisms of silicon reactions and the role of silicon intermediates in these reactions. Some novel six coordinate silicon complexes have been studied by inorganic chemists in recent years. One of the surprising results has been that hydrogens rather than chlorines occupy the axial positions of some six-coordinate compounds. Another result has been remarkably high Si(29)-H coupling constants in the Si(29) NMR data. Ab Initio calculations including MP2 geometry optimizations, frequency calculations, studies of the Mullikan populations of overlaps of the atomic orbitals within the Hartree-Fock molecular approximation, and mapping of features of the potential energy surfaces were performed in order to shed some light on experimental results.

299. *AB INITIO* CALCULATIONS OF THE CHARGE-DENSITY SUSCEPTIBILITY OF THE WATER MOLECULE. Xiaoping Li, Chetan Ahuja, James F. Harrison, and Katharine L. C. Hunt, Department of Chemistry, Michigan State University, East Lansing, Michigan 48824

The nonlocal charge-density susceptibility $\chi(\mathbf{r}, \mathbf{r}')$ gives the change in electronic charge density $\Delta\rho(\mathbf{r})$ due to a perturbing, delta-function potential acting at $\mathbf{r}'$. We have evaluated $\chi(\mathbf{r}, \mathbf{r}')$ *ab initio* for the water molecule, including correlation effects. In each individual calculation, the point of perturbation $\mathbf{r}'$ is fixed, and $\chi(\mathbf{r}, \mathbf{r}')$ is generated as a continuous, analytic function of $\mathbf{r}$. Results for $\chi(\mathbf{r}, \mathbf{r}')$ permit the analysis of electrical properties, characteristics of molecular potential surfaces, and intermolecular forces--on the *intramolecular* scale. Properties that depend on $\chi(\mathbf{r}, \mathbf{r}')$ include infrared absorption intensities, harmonic force constants, Sternheimer electric-field shielding tensors, polarizabilities (including quadrupole and higher-multipole response tensors), and induction energies. Dispersion energies depend on the susceptibility densities $\chi(\mathbf{r}, \mathbf{r}'; i\omega)$ of the interacting molecules, taken at imaginary frequencies.

300.
AB INITIO STUDIES OF STRUCTURES, THERMODYNAMIC STABILITIES, AND PROTON AFFINITIES OF H2SI=X AND HSI-XH (X = O, S, NH, PH, CH2, BH, ALH). **Matthew J. Alm**, Michael Page, and Denley B. Jacobson, Department of Chemistry, North Dakota State University, Fargo, ND 58105

The chemical properties of unsaturated heavy main group element compounds (period>2)-- electronic structure, reactivity, and relative thermodynamic stability-- are fundamentally different from compounds of period 2 elements. Silicon has been extensively studied due to its similarity to carbon and its use in chemical vapor deposition. Gas phase ion studies have been used to characterize the stability of heavy main group element compounds by measuring the proton affinities of relevant isomers. The structures and thermodynamic stabilites of a range of unsaturated silicon species have been studied. G2 calculations were performed to compare the stability of of SiH2=X and SiHXH (X=O, S, NH, PH, CH2, BH, AlH). In addition, the proton affinity of each species was determined, yielding the stability of the corresponding silylenium ion.

301.
AB INITIO STUDIES OF THE GEOMETRICAL AND ELECTRONIC STRUCTURE OF NICKEL PORPHYRINS. M. C. Piqueras and C. M. Rohlfing, Combustion Chemistry Department, Sandia National Laboratories, Livermore, CA 94551-0969

Ab initio calculations on nickel porphyrin (NiP) and nickel tetra(t-butyl) porphyrin (NiTtBuP) have been performed to analyze the conformationally dependent photophysical properties of these systems. Geometry optimizations and harmonic frequency analyses have been performed at the Hartree-Fock (HF) and second-order perturbation (MP2) theory level on NiP. For NiTtBuP, HF calculations have been carried out on domed, ruffled and waved conformers.

302. ABINITIO STUDY OF THE CONVERGENCE OF CLUSTER EXPANSION PARAMETERS FOR ACYCLIC CONJUGATED HYDROCARBONS. L. L. Griffin, T. G. Schmalz, D. J. Klein, Texas A&M University at Galveston, Galveston, Texas 77553-1675

The pi-electron energies of acyclic conjugated hydrocarbon compounds can be expressed as expansions over the energies of a small number of bond types in the Huckel model. These expansions are contained within a Hubbard model expansion based on energies of clusters of hydrogen-deleted graphs and the number of occurrences of each cluster within the molecular graph. The Hubbard model expansion is rapidly convergent so that pi energies are expressed in terms of seven or fewer clusters in both the HMO and VB limits. Abinitio SCF (HF//3-21G and HF//6-31G**) energies of a series of linear and branched conjugated hydrocarbons are presented to show that Hubbard expansion parameters based on these energies also converge within a small number of local clusters.

303.
AB INITO INVESTIGATION OF KRYPTON MONOFLUORIDE AND KRYPTON DIFLUORIDE. **G. J. Hoffman** Laura A. Swafford, and R. J. Cave, Department of Chemistry, Pomona College, Claremont, CA 91711

Results from ab initio calculations on krypton monofluoride and krypton difluoride are presented. Calculations are performed at the level of CCSD(T) using a variety of basis sets for both Kr and F. The Gaussian 94 package was used on the Cray 90 at the San Diego Supercomputer Center. This is the first quantum chemical study showing the ground state of KrF to be bound, in agreement with experiment. Basis set superposition error (BSSE) must be taken into account to obtain accurate energies. Calculated results for krypton monofluoride and krypton difluoride in their ground states are compared with experiment and previous calculation. Results from calculations on the first triplet state of krypton difluoride are also presented.

304. COMPLEXES OF $(H_3B,C,F)^+$: A COMPUTATIONAL STUDY. Carol A. Deakyne, Department of Chemistry, Eastern Illinois University, Charleston, IL 61920; Joel F. Liebman, Department of Chemistry and Biochemistry, University of Maryland Baltimore County, Baltimore, MD 21228

$(H_3B,C,F)^+$ has been investigated via conventional ab initio calculations and via density functional methods including exact exchange contributions. Cyclic and acyclic systems with both B-C and B-F connectivities have been examined for both singlet and triplet states. Equilibrium structures were first identified at the B3LYP/6-31+G(d,p) level of calculation and then reoptimized at the QCISD(T) level with several more flexible basis sets. A variety of isomeric forms have been found in this work, including some forms involving hydrogen transfer.

305. DENSITY FUNCTIONAL CALCULATIONS OF THE RAMAN SPECTRA of 3-METHYLINDOLE AND Ti_2Cl_6 K.J. Jalkanen[a], R. DeKock[a], S.W. Bunte[b], K.L. McNesby[b], D.B. Goodin[c], & G.M. Jensen[c] [a]Calvin College, 3201 Burton St. SE, Grand Rapids, MI 49546; [b]US Army Research Laboratory, AMSRL-WM-PC, Aberdeen Proving Ground, MD 21005-5066; [c]The Scripps Research Institute, Department of Molecular Biology, MB8, 10666 N. Torrey Pines Rd., La Jolla, CA 92037.

We have calculated at the RHF/6-311++G(2d,2p) level of theory the polarizability derivatives of the tryptophan analog 3-methylindole, the neutral radical of 3-methylindole, and the cation radical of 3-methylindole. These calculations have been combined with Density Functional calculations of the Hessian matrix completed at the B3LYP/TZ2P level to predict the Raman intensities and depolarization ratios for each of these molecules. Our calculated Raman intensities will be compared to experiment for 3-methylindole. In addition, we have calculated conformational energies, vibrational frequencies and intensities, and Raman intensities and the depolarization ratios of Ti_2Cl_6. The conformational energies, vibrational frequencies, and IR intensities are at the B3LYP level and the Raman intensities are at the RHF level (polarizability derivatives) combined with the DFT Hessians.

306. COMBINED *AB INITIO* AND ELECTROSTATIC CALCULATIONS OF TRYPTOPHAN-191 OXIDATION IN COMPOUND ES OF CYTOCHROME *C* PEROXIDASE S.W. Bunte[a], G.M. Jensen[b], A. Warshel[c], & D.B. Goodin[b], [a]US Army Research Laboratory, AMSRL-WT-PC, Aberdeen Proving Ground, MD 21005-5066; [b]The Scripps Research Institute, Department of Molecular Biology, MB8, 10666 N. Torrey Pines Rd., La Jolla, CA 92037; [c]University of Southern California, Dept. of Chemistry, Los Angeles, CA 90089.

The ferric Cytochrome *c* peroxidase (CCP) reacts with hydrogen peroxide to form a two electron oxidized intermediate: compound ES. One oxidizing equivalent is stored as an oxyferryl heme, and the other is localized on the indole ring of tryptophan-191. ENDOR experiments and Becke3LYP density functional *ab initio* calculations of the spin densities of the oxidized forms of 3-methyl indole are consistent with the conclusion that the oxidized indole in compound ES is a cation radical. The energetics of the preferential oxidation of Trp-191 to a cation radical has also been modeled using the Protein Dipoles Langevin Dipoles (PDLD) method of Warshel and co-workers in combination with the (ostensibly gas phase) *ab initio* calculations. These latter include Mulliken and ESP partial charge sets and the gas phase energies for the 3-methyl indole and the neutral and cation radicals of 3-methyl indole. These methods in combination lead to predictions for the oxidation potential of Trp-191 and for the pK_a of the indole radical (for the neutral/cation radical equilibrium).

307.

"Direct" Computations in Time-Dependent Hartree-Fock Calculations of Nonlinear Optical Properties. Shashi P. Karna, *US Air Force Phillips Laboratory, Space Mission Technologies Division, 3550 Aberdeen Ave, SE, Kirtland AFB, NM 87117-5776*

The "direct" method of computing Fock matrix in *ab initio* self-consistent-field theory pioneered by Almlöf et al[1] marked the beginning of a new era in computational quantum chemistry. Unlike the conventional compute-store-and-retrieve algorithm, the direct method relies on computing the two-electron integral as and when needed, thus obviating the integral storage requirement This unique and desirable feature has made the "direct" method one of the most popular techniques in modern *ab initio* quantum chemistry. Recently, the concept of "direct" SCF method has been used for the development of a "direct" time-dependent Hartree-Fock (TDHF) theory for the calculation of linear and nonlinear optical (NLO) properties.[2] In this talk, I shall discuss the implementation of the "direct" TDHF method and its application to calculate the NLO properties of oligo-thiophenes and endohedral C_{60} complexes.

[1]J. Almlöf, K. Fægri, Jr, and K. Kossel, *J. Comp. Chem.* **3**, 385 (1992).
[2]S. P. Karna, *Chem. Phys. Lett.* **214**, 186 (1993).

PHYS

308.

EXAMINATION OF THROUGH-SPACE AND THROUGH-BOND INTERRING INTERACTIONS IN [2.2]PARACYCLOPHANES WITH DENSITY FUNCTIONAL THEORY Susan E. Walden Daniel T. Glatzhofer Department of Chemistry and Biochemistry, University of Oklahoma, 620 Parrington Oval, Norman, Oklahoma 73019

Since first synthesized nearly 50 years ago, [2.2]paracyclophanes have elicited intensive curiosity, speculation, and investigation. The distorted benzene rings held in close proximity result in intriguing electronic and spectroscopic properties. One question regarding these properties is whether interring communication arises from through-space or through-bond interactions. We present hybrid Hartree-Fock/density functional theory calculations designed to explore interring interactions in [2.2]paracyclophanes, including radical cations. Our analysis includes optimized geometries, spin density distributions, and frontier molecular orbitals.

309.

STEREOSELECTIVE MODULATION OF RING OPENING OF 3-SUBSTITUTED CYCLOBUTENES BY METAL CATIONS. Ramkumar Rajamani, Carlos F. Lopez and Jeffrey D. Evanseck, Department of Chemistry, University of Miami, 1301 Memorial Drive, Coral Gables, Florida 33124–0431.

The origin of metal assisted modulation of stereoselectivity in the ring–opening of 3-substituted cyclobutenes has been identified by using a combination of high–level ab initio molecular orbital theory, self–consistent variational energy decomposition, natural energy decomposition analysis and continuum electrostatics. Hartree–Fock and density–functional calculations with large basis sets uniformly compute a stabilization of the transition structure favoring inward rotation in the presence of lithium cation. The polarizable continuum model with the dielectric constant of water was used to approximate solvent effects. Calculations show an enhancement over the gas-phase estimations when the substituent is a strong donor while de-emphasizing the effect for strong acceptors. Both energy decomposition methods indicate that an electrostatic effect is the origin of stereoselectivity modulation in the ring opening of 3-substituted cyclobutenes.

310. THEORETICAL STUDY OF ISOMERS OF H_NGaP. W.W. Brown and M.R. Hoffmann, Department of Chemistry, University of North Dakota, Grand Forks, ND 58202

Ab initio electronic structure calculations of the equilibrium structures and relative energies of various isomers of H_2GaP, H_3GaP, and their anions, were performed. It was found that several of the structures exhibit significant pi-bonding. Calculations were performed at the SCF, MP2 and B3LYP levels of electron correlation, and the 6-31++G** single particle basis set, using the Gaussian94 package. Solvent effects were included in some of the calculations at the SCI-PCM reaction field level.

311. TRANSITION METAL-SILICON PI BOND STRENGTHS
Brett M. Bode and Mark S. Gordon, Department of Chemistry, Iowa State University, Ames IA 50011

Pi bond strengths for the transition metal-silicon double bonds in Silylene complexes of the form $MSiH_2^+$ are studied for the first row transition metals using ab initio wavefunctions. Electron correlation is included using FORS-MCSCF (full optimized reaction space multiconfiguration self-consistent field) and MCQDPT2 (multiconfiguration quasi-degenerate 2nd order perturbation theory) methods. A comparison between all-electron and effective core potential (ECP) methods is made.

312.

ACTIVE NICKEL NANOPARTICLES GENERATED BY UV LASER-ASSISTED GAS PHASE PHOTONUCLEATION. Donald H. He[*†], Richard H. Heist[†], Brian L. McIntyre[‡] and Thomas N. Blanton[§], [†]Department of Chemical Engineering and [‡]School of Engineering and Applied Sciences, University of Rochester, Rochester, New York 14627; [§]Analytical Technology Division, Eastman Kodak Company, Rochester, New York 14652-3712.

Vapor phase synthesis is a versatile technique for creating nanostructured materials with potentially useful properties. Comparatively less studies have been done involving the chemical reactivity of transition metal nanoparticles (unsupported), especially in gas-solid catalysis. In this research, a UV laser-assisted gas phase photonucleation process is used to generate Ni nanoparticles from room temperature $Ni(CO)_4$ vapor. The structure and morphology of the nano-aggregates of ca. $1 \sim 2$ nm particles are characterized by transmission electron microscopy (TEM) and X-ray diffractometry (XRD). The ethane hydrogenolysis reaction is used to probe the reactivity of the vapor nucleated Ni nanoparticles. Observations of high reaction rates and extraordinary resistance to deactivation are described and suggest novel behavior of these nickel nanoparticles.

313.

ELECTRONIC STRUCTURE OF SILICON NANOCRYSTALS AS A FUNCTION OF PARTICLE SIZE. **T. W. H. van Buuren,** L. N. Dinh, I. Jimenez, L. L. Chase, L. J. Terminello, J. A. Carlisle* LLNL, Livermore, CA 94551; T. A Callcott, M. Grush, and J. J. Jia, University of Tennessee, Knoxville, TN; D. L. Ederer,Tulane University, New Orleans, LA; F. J. Himpsel, University of Wisconsin, Madison, Madison, WI 53706

X-ray absorption, photoemission, and soft x-ray fluorescence were measured with synchrotron radiation on thin films of Si nanoclusters.. These clusters were synthesized in situ by thermal evaporation of Si in an Ar buffer gas followed by subsequent exposure to atomic hydrogen to passivate the surface. The size distribution for these clusters were measured using STM and AFM and ranged from 1 - 10's of nanometers in mean diameter. Blue shifts of the conduction band as large as 0.38 eV were observed for clusters with a 1.5 nM mean diameter. Valence band shifts were also observed in the nanocrystals and are consistent with the conduction band shifts. These results were compared to recent theories predicting Si nanocluster quantum effects.

314.

FACETING OF SEMICONDUCTOR NANOCRYSTALS STUDIED BY HRTEM . **Andreas V. Kadavanich** A. P. Alivisatos, University of California, Department of Chemistry, Box 101, Latimer Hall, Berkeley, CA 94720-1460

High Resolution Transmission Electron Microscopy (HRTEM) was used to determine the morphology of various colloidal semiconductor nanocrystal systems. Nanocrystals prepared under quasi-equilibrium conditions have well-defined, faceted shapes. The observed shapes can be explained by the Wulff theorem and consideration of the effects of the passivating ligands on the surface energies. Facetting is most pronounced in II-VI systems, but recently prepared III-V nanocrystals are shown to be faceted as well.

315.

FLUORESCENCE SPECTRA AND DYNAMICS OF EU2O3 AND EU:Y2O3 NANOCRYSTALS. **Brian M. Tissue** Bipin Bihari, Chemistry Department, Virginia Tech, Blacksburg, VA 24061-0212, tissue@vt.edu

This paper describes the preparation and optical properties of nanocrystalline Eu2O3 and Eu3+:Y2O3. The nanocrystals form by gas-phase condensation after vaporizing oxide starting material with a cw-CO2 laser. The resulting particle size is controllable by adjusting the pressure of the inert gas atmosphere. Particles larger than 10 nm show sharp-line excitation and fluorescence spectra similar to bulk material. Particles smaller than 10 nm have broad shifted lines, that are tentatively attributed to a new disordered Eu-oxide phase. Preliminary intensity-dependence experiments show non-linear effects that appear to arise from the nanocrystals acting as a resonance cavity.

316.

KINETICS BARRIERS TO STRUCTURAL PHASE TRANSITIONS IN SEMICONDUCTOR NANOCRYSTALS. **Amy B. Herhold,** Chia-Chun Chen, and A.P. Alivisatos, Department of Chemistry, University of California, Berkeley and Materials Sciences Division, Lawrence Berkeley National Laboratory, Berkeley, CA 94720

Pressure-induced structural phase transitions in semiconductors typically exhibit a large pressure hysteresis, indicating a significant kinetic barrier to transformation. In bulk solids, the kinetics are complex, involving multiple nucleation events at defects and domain fracture. Nanocrystals, however, transform via a single nucleation event per nanocrystal. We have studied the kinetics of pressure-induced phase transitions in CdSe nanocrystals with high pressure, high temperature X-ray diffraction and optical absorption measurements in a diamond anvil cell. Unimolecular kinetics are observed. The variation of the barrier height with nanocrystal size will be discussed.

317.

LARGE, ELECTRONIC GROUND-STATE, PERMANENT DIPOLES IN COLLOIDAL CDSE NANOCRYSTALS. **S. A. Blanton,** R. L. Leheny, M. A. Hines, S. R. Nagel, & P. Guyot-Sionnest, The James Franck Institute, The University of Chicago, 5640 S Ellis Ave, Chicago, IL, 60637

We observe 40-80 Debye permanent dipoles in CdSe colloidal nanocrystals through dielectric dispersion measurements. The existence of such a dipole suggests that current theories of nanocrystal electronic structrure be modified to take into account mixing of the exciton states by this non-negligible perturbation. We also observe an anomalous dependence of colloid conductivity on nanocrystal concentration, possibly due to stirred percolation, and demonstrate that the hydrodynamic properties of the nanocrystals such as the rotational diffusion time scale with nanocrystal size as expected from simple Debye theory.

318. OPTICAL PROPERTIES, ELECTRON TRANSFER, AND ENERGY TRANSFER IN COLLOIDAL InP QDs AND CLOSED-PACKED QD ARRAYS. O.I. Mićić, K.M. Jones, A.G. Cahill, J.R. Sprague, A. Zaban, Zhenghao Lu, and A.J. Nozik, NREL, Golden, CO 80401

Quantum dots of InP with diameters ranging from 20 to 65 Å were synthesized as well-crystallized nanoparticles with the bulk zinc blende structure. Colloidal solutions of InP quantum dots exhibit excitonic features in their absorption spectra. The InP QD preparation showed highly efficient band-edge photoluminescence after etching the particles with hydrogen fluoride; the quantum yield is about 60% at 10 K. Fluorescence line narrowing experiments and photoluminescences excitation spectra can be used to derive the Stokes shifts as a function of particles size (45 to 26 Å). The measured PL lifetime of the band edge emission was found to be ~200 ns at 10 K. InP QD close packed films can be prepared with short range hexagonal ordering. A close packed QD array shows long range energy transfer from smaller to larger particles. Photoinduced electron transfer from InP QDs to SnO_2, TiO_2 and ZnO and p-GaN has been studied. We will also describe efforts to measure transport properties of close packed InP QD arrays.

319.

PHOTOCONDUCTIVITY IN CDSE QUANTUM DOT SOLIDS **C. A. Leatherdale** C. R. Kagan, M. A. Kastner, and M. G. Bawendi, MIT, 77 Massachusetts Avenue, Cambridge, MA, 02139

Solids formed from assemblies of quantum dots are a novel, tunable solid state material. The conductive properties can be modified by changing the distance between the dots, the nature of the spacer material, and the size of the quantum dots themselves. We present measurements of photoconductivity in close packed, glassy solids of CdSe quantum dots. We study the mechanisms of carrier generation and transport as a function of quantum dot size and spacer material.

320. PHOTOCONDUCTIVITY IN CLOSE PACKED CdSe QUANTUM DOT SOLIDS. C. R. Kagan[*], C. A. Leatherdale, M. A. Kastner, and M. G. Bawendi, MIT, 77 Massachusetts Ave., Cambridge, MA 02139

We demonstrate and analyze size dependent photoconductivity in close packed quantum dot (QD) solids prepared from CdSe quantum dot (QD) samples tunable in size from 17 to 150 Å in diameter. The individual QDs and the close packed QD solids are both structurally and optically well-characterized. Organic ligands coordinate the surfaces of the QDs maintaining 11±1 Å interdot separations. In the dark, the QD solids are highly insulating (>500 MΩ•cm). Photoexcitation of the QD solids generates carriers in the QDs which are separated and transported through the QD solids. We measure the wavelength, voltage, temperature, and intensity dependence of the photocurrent to elucidate the carrier generation, separation, and transport mechanisms.

[*]Currently at Lucent Technologies

321. QUANTUM CONFINEMENT EFFECTS ENABLE PHOTOCATALYZED NITRATE REDUCTION AT NEUTRAL PH USING CDS NANOCRYSTALS. Brian A. Korgel and Harold G. Monbouquette, Department of Chemical Engineering, University of California, Los Angeles, CA 90095-1592.

Size-quantized CdS nanocrystals serve as photocatalysts for nitrate reduction at pH 7.0 under conditions that mimic illumination by sunlight with overall product quantum yields of up to 4% for ~20 Å, amine-terminated particles. Due to the effects of quantum confinement on electron and hole redox potentials, photocatalyzed nitrate reduction rates depend strongly on the particle size, and the fastest reduction rates were observed with the smallest amine-capped nanocrystals studied. Nitrate reduction rates are reduced two-fold or more on negatively charged, carboxy-terminated nanocrystals that electrostatically repel nitrate. The effect of nitrate adsorption on photoreduction rates is described well by a Langmuir-Hinschelwood expression. Chloride competes with nitrate for access to particle surfaces and reduced photoreduction rates are observed for both amine- and carboxy-terminated particles with increased NaCl concentration. The rate of photocatalyzed nitrate reduction on the amine-capped particles goes through a minimum at about pH 6.5, whereas the efficiency of nitrate reduction for the carboxy-terminated system decreases monotonically with increasing pH.

322. PREPARATION OF ACICULAR MAGNETIC NANOPARTICLES BY REDUCTION OF FERROUS ION IN THE PRESENCE OF ASSOCIATION COLLOIDS. David E. Nikles, Jason L. Cain, Jiaping Chen, and Min Chen. Department of Chemistry and the Center for Materials for Information Technology, The University of Alabama, Tuscaloosa, Alabama 35487-0338, DNIKLES@UA1VM.UA.EDU.

Acicular iron particles were prepared by the sodium borohydride reduction of ferrous ion trapped in the discrete aqueous phase of an association colloid. The colloid and the application of a magnetic field gave cylindrical (acicular) iron particles with moderate coercivities. Two different association colloid phases have been investigated. Reduction in the presence of tubular lecithin assemblies gave single crystal, acicular iron particles with lengths in the range of 30 nm to many microns, depending on the concentration of ferrous. The aspect ratio exceeded 10:1. Reduction in the presence of a lamellar liquid crystal phase of $C_{12}H_{25}$-$(OCH_2CH_2)_4$-OH gave iron needles, 50 nm long and 5 nm in diameter. This new method provided acicular iron particles of a size range that the magnetic tape industry has trouble achieving using conventional iron particle syntheses.

323. PROBING ELECTRONIC RELAXATION IN SINGLE-SIZED METAL NANO-PARTICLES. B. A. Smith and J. Z. Zhang, Department of Chemistry, University of California, Santa Cruz, CA 95064, U. Giebel and G. Schmid, Institut für Anorganische Chemie, der Universität GH Essen, Universittatsstr. 5-7, D-45117 Essen, Germany.

Direct measurements of electronic energy relaxation in single-sized metal nano-particles have been performed using femtosecond laser spectroscopy. The relaxation time in large (10-40 nm) Au particles is 7 ps, which is slower than that in bulk (1 ps), due to weaker electron-phonon coupling. The relaxation becomes faster (1.2 ps) in smaller single-sized particles, Au55 (2 nm), due to increased surface collisions. The weakening of electron-phonon coupling and increase in surface collisions therefore have competing effects on the electronic relaxation rate, with the former dominating at large sizes and the latter dominating at smaller sizes. Similar observations have been made in Pt nano-particles. However, the relaxation time in very small clusters such as Au13 becomes much slower (300 ps), indicating the emergence of a molecular behavior. The results suggest that the bulk to molecule transition for Au occurs between 55 and 13 atoms.

324.

SIZE DEPENDENT ELECTRONIC LEVEL STRUCTURE OF INDIUM ARSENIDE NANOCRYSTAL QUANTUM DOTS. **U. Banin,** J. C. Lee, A.A. Guzelian, and A.P. Alivisatos, Department of Chemistry, University of California, Berkeley, California 94720

Semiconductor nanocrystals serve as model systems for evolution of bulk properties from the solid state to the molecular regime. We use size selective optical spectroscopies to examine the dependence of the electronic level structure on size in highly quantum confined nanocrystals of the III-V semiconductor InAs. The colloidal InAs quantum dots are crystalline and monodisperse with diameters ranging between 70 to 25 Å. Up to ten discrete features in the absorption are observed by photoluminescence excitation spectroscopy. The size dependence of these transitions is studied and compared with quantum confinement theories. Line narrowed fluorescence measurements reveal a size dependent Stokes shift of the emission. The observed shift is assigned to a splitting of the band gap transition induced by enhanced electron-hole exchange interaction in quantum dots.

325.

SYNTHESIS AND CHARACTERIZATION OF MANGANESE-DOPED CDSE QUANTUM DOTS. **F.V. Mikulec,** M.K. Kuno, M. Bennati, R.G. Griffin, M.G. Bawendi, Department of Chemistry, Massachusetts Institute of Technology, Cambridge, MA 02139

Doping a semicondutor with a transition metal can influence the optical and magnetic properties of the host material due to a strong exchange interaction between the carriers and the localized spins of the dopant. We employ a method based on the high temperature decomposition of organometallic compounds in a coordinating solvent to prepare nearly monodisperse, manganese-doped CdSe quantum dots. This material is characterized by EPR, optical spectroscopy, and NMR.

326. ULTRAFAST DYNAMICS OF DYE MOLECULES ADSORBED ONTO NANOMETER SIZED SEMICONDUCTOR PARTICLES. Gregory V. Hartland, Ignacio Martini, and Prashant Kamat, Department of Chemistry and Biochemistry and the Radiation Laboratory, University of Notre Dame, Notre Dame, IN 46556

Dye molecule sensitization of wide bandgap semiconductor particles is important in a variety of applications, such as photography and solar energy conversion. In order to fully understand these process, ultrafast measurements of the timescale for electron-transfer from the dye to the semiconductor particle are required. We have examined the relaxation dynamics of dye molecules adsorbed onto SnO_2 or SiO_2 colloidal particles with ca. 200 fs time resolution. In these systems excitation of the adsorbed dye molecules can lead to electron transfer to SnO_2 but not to SiO_2, because SiO_2 is a nonconductor. Thus, comparing the results obtained for these two particles allows us to separate the electron transfer dynamics from other photophysical processes, such as internal conversion and vibrational relaxation. For example, our measurements show that for cresyl violet adsorbed onto SnO_2 the majority of the excited dye molecules (80%) are deactivated by internal conversion/vibrational relaxation (2.9 ps time constant). The remaining molecules undergo electron transfer to SnO_2, with the ground state of the dye being rapidly repopulated by the back electron transfer reaction (11.3 ps time constant).

327. DYNAMICS OF CRYSTALLINE COLLOIDAL ARRAYS DURING PHOTOPOLYMERIZATION OF ACRYLIC MONOMER MATRIX. H. B. Sunkara[*], B. G. Penn, NASA Marshall Space Flight Center, Space Sciences Laboratory, Huntsville, AL 35812. [*] National Research Council Fellow

The photoinitiated bulk polymerization process, which has been used recently to manufacture nanocomposite laser filters, is examined to understand the dynamics of both the crystalline colloidal arrays (CCA) and the host monomer matrix. The observed discrepancies in optical diffraction properties of CCA in poly(methyl methacrylate) composite films compared to the liquid state and heterogeneities within a film are ascribed to the dielectric constant dependent electrostatic interparticle repulsive interaction which couples with gravity driven convection and sedimentation. The origins of thermal, solutal, double-diffusive and Marangoni convections and factors which influence the convective motion during the photopolymerization process are identified. For a clear understanding of the gravitational effects on these multiphase systems, processing of these materials in a low gravity space environment is recommended.

328.
DYNAMICS OF POLYMERS IN NOVEL SILICATE-POLYMER NANOCOMPOSITE MATERIALS. E. Manias E.P. Giannelis, Materials Science & Engineering, Cornell University, Ithaca, NY14853

Layered-silicate based polymer nanocomposites is a novel class of materials that has recently attracted considerable scientific interest, mainly due to the potential technological applications. Moreover, in the form of intercalated hybrics represent new model systems to study the dynamics of polymer chains in extreme --nanoscopic-- confinements or at the solid-polymer interface. The structure of the nanocomposite materials is a multilayer with alternating polymer and mica-type silicate layers with a periodicity of a few nanometers. The effect of the wall-polymer interactions on the intercalation kinetics of the polymer chains the layers is explored by X-ray Diffraction and Rutherford Backscattering analysis. In addition, for these highly adsorbed coils in nanometer geometries, the chain and segmental dynamics of the polymers are further investigated by Differential Scanning Calorimetry (DSC) and Nuclear Magnetic Resonance (NMR).

329.
EVOLUTION OF NANOSTRUCTURES IN POLYMERIC BLENDS Abdelsamie Moet Dept of Chemical Engineering, College of Engineering, United Arab Emirates University, P.O. Box 17555 Al-Ain, UAE and Varaporn Tanrattanakul, Polymer Science Program, Prince of Sngkla University, Hat-Yai, Thailand

Conditions under which nanostructured phase development in two sets of polymeric blends are described. In both, a styrene-ethylene block copolymer grafted with maleic anhydride functionality (MA-g-SEBS) was employed to enhance mixing by reducing the interfacial tension at the interface of immiscible polymer pairs. The experiments were conducted in a melt extruder, followed by microscopic examinations and measurements of the mechanical properties. In a blend of nylon 6 and polyethylene the block copolymer was found to form nanoscale bridges which have been shown to exhibit significant load transfer functions. In the second pair, polyethylene and poly(ethyleneterephthalate), the block copolymer is believed to have effected the formation of a complex phase morphology mainly involved nano scale dispersion. The later blend was found to exhibit superior plasticity. Effects of shear rate and composition on the foregoing features will be also discussed.

330. **Kinetic Modeling of Switchable Hologram Formation in Liquid Crystalline Photopolymers.** Michael B. Sponsler, Department of Chemistry and W. M. Keck Center for Molecular Electronics, Syracuse University, Syracuse, NY 13244-4100

A kinetic simulation program has been written that models the chemical and physical processes involved in the formation of electrically switchable holographic gratings. The gratings are produced by photopolymerization of liquid crystalline monomers and electric-field reorientation of the non-polymerized, dark regions. The program, written with the aid of the commercial program Stella II, has allowed independent exploration of many variables, including exposure, initiation efficiency, inhibitor concentration, exposure pattern (with or without a uniform preexposure), percentage of difunctional monomer (the balance being monofunctional), grating spacing, electric field, and temperature. The model has proven useful in the rationalization and prediction of trends concerning the effects of these variables upon the diffraction and switching properties of the gratings, and trends concerning the interactions between some of the variables.

331.

LIGHT EMITTING HETEROSTRUCTURES WITH INORGANIC NANOCRYSTALLITES AND CONJUGATED POLYMERS. **Hedi Mattoussi,** Bashir O. Dabbousi, Deryn Fogg, Michael F. Rubner, Moungi G. Bawendi, Center for Materials Science and Engineering, MIT, Cambridge, MA

We are investigating light emitting devices made of thin films of organic/inorganic heterostructures. Our approach has been to use molecular level processing to control device architecture. An organic hole transport layer is made of self assembled poly (phenylene vinylene), built using a layer-by-layer adsorption approach from its charged precursor form. Nanocrystallites of CdSe form the emitter layer. As a result of quantum confinement, the nanocrystallites offer the unique opportunity of tuning the emission wavelength over the visible spectrum by varying the particle size. We find that the electroluminescent signal is dominated by the emission from the nanocrystallites. Furthermore, we find that overcoating the nanocrystallites with few monolayers of ZnS substantially improves the device stability and its performance. Effects of subtle variation in the organic layer on the electroluminescence will be discussed.

332.

NANOPHASE SEPARATION OF POLYMER BLENDS: MOLECULAR STRUCTURE AND CRITICAL BEHAVIOR. **D. M. Hussey,** L. Keller and M. D. Fayer, Dept. of Chemistry, Stanford University, Stanford, CA 94305-5080

We use electronic excitation transfer (EET) to study the process of phase separation of molecularly mixed polymer blends. Chromophores are covalently incorporated into one of the blend polymers, and EET between these chromophores contributes to the rate of fluorescence anisotropy decay. EET is very sensitive to chromophore distribution, revealing the single-chain structure and spatial distribution of molecules in the blend on the Angstrom scale. We find that not only the annealing temperature, but also the rate of heating can also have a dramatic effect on the nanoscopic structure of the blend. The nanodomains formed in the initial stages of phase separation can be trapped in the material by allowing the sample to undergo a glass transition. Thus we can learn how to influence the molecular structure and physical properties of the blend while developing an understanding of the kinetics and mechanisms of phase separation.

333.

NEAR-FIELD SCANNING OPTICAL MICROSCOPY OF LIGHT EMITTING OPTICAL POLYMER THIN FILMS. **Jessie A. DeAro** Kenneth D. Weston, Grace A. Credo, Steven K. Buratto Department of Chemistry, University of California at Santa Barbara, Santa Barbara, CA 93106

We will discuss our application of Near-Field Scanning Optical Microscopy (NSOM) and Near-Field Scanning Optical Spectroscopy (NFOS) to the study of nanoscale optical properties of thin films of poly(p-phenylene vinylene) (PPV) and derivatives. These techniques provide the means for directly probing the nanoscale region of polymer thin films. Photoluminescence and polarized-light absorption NSOM images, as well as photoluminescence NFOS spectra, show a direct correlation to the polymer film morphology. These NSOM and NFOS results on non-oriented and stretch-oriented thin films of PPV will be presented and discussed. The photoluminescence NSOM images of non-oriented PPV thin films show 100 nm aggregates which quench the polymer luminescence. The stretch-oriented thin films have 100 nm tall fibril topographical features which alter the molecular chain alignment revealed by polarized-light absorption NSOM experiments. Polymer blends with TiO_2 crystallites (used to form microcavities) are also currently being investigated using the NSOM technique and the progress on these films will also be presented.

334.

QUANTUM CONDUCTION MECHANISMS IN POLYENE MOLECULAR WIRES. **Victor V. Belyaev** and C. H. Mak, Department of Chemistry, University of Southern California, Los Angeles, CA 90089-0482

The dominant conduction mechanism in long undoped polyacetylene chains is due to soliton motions. Solitons have very light effective masses and are therefore heavily quantum mechanical in character. In long molecules where the length of the wire is large compared to the soliton dispersion width, soliton conduction mechanism have been thoroughly studied in the past. In the new era of nanotechnology, polyene molecules have found new applications as potential candidate for molecular wires. When the length of these wires becomes short enough compared to the soliton width, the conduction mechanism transitions into a fully quantum mechanical regime. We will examine this transition by studying the dispersion width of solitons in polyene molecular wires with quantum Monte Carlo simulations and discuss its implications on the conduction mechanism.

335. A MODULAR SYNTHON FOR BUILDING HYDROGEN BONDED LAYERS IN CRYSTALLINE SOLIDS
G.Tayhas R. Palmore and Mary T. McBride, Department of Chemistry, University of California - Davis, Davis, CA 95616

The key to the development of new solid state materials with specifically engineered properties lies in the precise positioning of components within the crystal lattice. Judicious selection of a molecule or, more specifically, a functional group can lead to controlled, well-ordered and predictable solids. We have initiated a program of research aimed at controlling the arrangement of molecules in crystalline solids by exploiting the strength, directionality and selectivity of hydrogen bonds. We have focused on the cyclic dipeptide of S-aspartic acid, a molecule containing four functional groups (two cis-amides and two carboxylic acids). These groups are positioned to encourage the self-assembly of two dimensional aggregates, that is, hydrogen-bonded layers. We will discuss how the alignment of these groups promotes the formation of layers and describe studies of co-crystallization that probe the relative selectivity of hydrogen bonding functional groups and introduce function into these layers.

336.

CF_4 ADSORBATE SPECTROSCOPY AS A PROBE OF SURFACE DEFECT STRUCTURE AND OF ADSORBATE ISLAND GROWTH. O. Alpert, V. Buch, Department of Physical Chemistry, The Hebrew University, Jerusalem 91904, Israel, and J. P. Devlin, Department of Chemistry, Oklahoma State University, Stillwater, Oklahoma 74078, US

The focus of this joint computational - experimental effort is spectroscopy of CF_4 adsorbate on molecular surfaces. The observable of the present interest is the vibrational fundamental of the triply degenerate ν_3 asymmetric CF stretch of CF_4. This vibration is characterized by a very large direction independent dipole derivative. Therefore, oscillations of neighboring CF_4 molecules are strongly coupled to each other by the dipole - dipole interaction. Infrared spectra of CF_4 adsorbate are dominated by collective in-phase vibrations, and highly sensitive to the structure of the adsorbate layer. At low coverages, CF_4 adsorbate lineshapes are suggested as a probe of surface defect structures. At higher coverages, CF_4 spectroscopy is used to follow adsorbate island formation on the surface. Theoretical analysis is presented of the experimentally measured CF_4 adsorbate spectra on a variety of molecular nanoparticles (CO_2, crystalline and amorphous mesitylene etc).

337.

CF_4 ADSORBATE SPECTROSCOPY AS A SURFACE STRUCTURE PROBE FOR MOLECULAR NANOPARTICLES. V. Buch, Department of Physical Chemistry, The Hebrew University, Jerusalem 91904, Israel, and J. P. Devlin, Department of Chemistry, Oklahoma State University, Stillwater, Oklahoma 74078, US

This work represents a joint computational - experimental effort to investigate spectroscopy of CF_4 adsorbate on molecular surfaces. The observable of the present interest is the vibrational fundamental of the triply degenerate v_3 asymmetric CF stretch of CF_4. This vibration is characterized by a very large direction independent dipole derivative, and therefore oscillations of neighboring CF_4 molecules are strongly coupled to each other by the dipole - dipole interaction. Infrared spectra of CF_4 adsorbate are dominated by collective in-phase vibrations, and are shown here to be highly sensitive to the structure of the adsorbate layer, and of the underlying surface. Particularly remarkable is the sensitivity to the extent of surface (dis)order and roughness. Thus, the CF_4 adsorbate spectroscopy is suggested here as a useful tool for characterization of surface properties of molecular nanocrystals and solids. The most striking application is a demonstration of disordered surface structure of annealed ice nanocrystals.

338. CHARACTERIZATION, ADSORPTION, AND IRRADIATION OF SELF ASSEMBLED MONOLAYERS OF 2-(10-THIODECOXY)ANTHRACENE AND 9-(10-THIODECYLMETHOXY)ANTHRACENE. Marilyn D. Wooten and Marye Anne Fox, Department of Chemistry and Biochemistry, University of Texas at Austin, Austin, Texas, 78712

The photochemical reactivity of anthracene, its capacity to photodimerize, has been studied extensively in solution and the solid state. We report on the synthesis, characterization, and the excited state reactivity of 2-(10-thiodecoxy)anthracene (1) and 9-(10-thiodecylmethoxy)anthracene (2) as self assembled monolayers on a crystalline gold surface via a thiol linkage. Irradiation at λ=350nm of 1 and 2 in solution and on the gold surface resulted in the alteration of various chemical and physical properties of the adsorbed and non adsorbed species

339.

Comparison of Surface Adsorption Structures in 1-Octanethiol Self-Assembled on Cu(111) and Ag(111)

Hugh Rieley, Gary K. Kendall

Department of Chemistry, University of Liverpool, Liverpool, L69 3BX, UK

Surface adsorption structures of 1-octanethiol self-assembled monolayers (SAMs) adsorbed on Cu(111) and Ag(111) have been determined through X-ray absorption by thiolate sulphur atoms. Angle-resolved Near-Edge X-ray Absorption Fine Structure (NEXAFS) above the S K-edge revealed that the molecular axis through the alkane chain is canted at an angle of 9±5° to the surface normal. Normal Incidence X-Ray Standing Wave (NIXSW) profiles of the S absorber close to the (111) and ($\bar{1}$11) bulk reflections were used to locate the position of the S atoms with respect to the substrate. In each case, the S atoms penetrate the surface in an equal occupation of FCC and HCP three-fold hollow sites. The incoherence of the distribution with respect to the ($\bar{1}$11) planes of the bulk indicates the existence of more than one adsorption geometry, displaced from the true three-fold positions.

340. ELECTROCHEMICAL STUDIES OF COMPLEX SELF-ASSEMBLED MONOLAYERS OF THIOLS ON GOLD. R. S. Reese, Andreas Knorr, M. D. Wooten, and M. A. Fox, The Department of Chemistry and Biochemistry, The University of Texas at Austin, Austin, Texas, 78712.

Self-assembled monolayers (SAMs) of thiols on gold present a promising methodology for the construction of complex devices while maintaining control of the architecture on a molecular level. Our interest in SAMs stems from their possible uses in composite systems for directional photoinduced energy and electron transfer. An understanding of the effect of chemical composition on the morphological, electronic, and photonic properties is required for the rational design of such devices. To this end, we have designed and constructed several SAM systems incorporating conjugated, redox-active, and photoactive substructures. Grazing angle FT-IR spectroscopy was used as a means of characterizing the SAMs and for detecting structural changes in the monolayer induced by electrochemical and photochemical events. The electrochemical behavior of the SAMs were investigated using a variety of electrochemical methods including potential step, linear sweep, and time resolved techniques.

341.

ELIMINATION OF CROSS-TALK AND MODULATION OF FUNCTION IN ORGANISED HETEROSUPRAMOLECULAR ASSEMBLIES. Xavier Marguerettaz, S.N. Rao, Steven Connolly, Steven Doherty, Alan Merrins, Donald Fitzmaurice, University College Dublin, Deptartment of Chemistry, DUBLIN 4, IRELAND.

In this poster we report the use of LB for the preparation of an organised heterosupramolecular assembly. A monolayer of TiO2 nanocrystallites is deposited on conducting glass using LB techniques and fired. A mixed monolayer of icosylphosphonic acid (I) and 1-icosyl-1'-[2-phosphonoethyl]-4,4'-bipyridinium dichloride (II) molecules is subsequently deposited, using LB techniques, on the above TiO2. The resulting organised assembly is incorporated as working electrode in an electrochemical cell. Bandgap excitation of TiO2 results in electron transfer to the viologen component of II. Co-deposition of I and II ensures that at sufficiently high dilutions, electrons are transferred only to the viologen adsorbed at a given nanocrystallite, i.e. cross-talk is eliminated. Potentiostatic modulation of the electron transfer function has also been demonstrated for the above assembly. Some implications of the elimination of cross-talk and the modulation of function in molecular devices are considered.

342. FORMATION AND PROPERTIES OF Au AND Ag COLLOID MONO- AND MULTILAYERS Keating, C. D.; Musick. M. D.; Keefe, M. H.; Natan, M. J., Department of Chemistry, The Pennsylvania State University, University Park, PA 16802.

The thermodynamics of nanoparticle adsorption to organosilane-modified surfaces has been investigated; fitting the data to a Frumkin isotherm, an equilibrium constant (K_{eq}) for colloid adsorption has been calculated. The effect of nanoparticle identity, pH and surface functional group on K_{eq} are discussed. Knowledge of K_{eq} has been used to prepare surfaces with varied interparticle spacings. From these submonolayers, colloid multilayers have been built up, using bifunctional crosslinkers to bind surface nanoparticles and allow additional colloids to adsorb from solution. Multilayer properties have been investigated as a function of interparticle spacing in the initial submonolayer. Mono- and multilayer films have been imaged by atomic force microscopy (AFM) and field-emission scanning electron microscopy (FE-SEM). With increasing film thickness, colloid multilayer films begin to behave as bulk metal, taking on wavelength-independent near infrared transmittance, high conductivity, and an improved heterogeneous rate constant for electron transfer characteristic of bulk metal.

343. HETEROGENEOUS PATCHING OF SELF-ASSEMBLED RIGID ROD ARYLTHIOLS. Robert W. Zehner, Lawrence R. Sita, Department of Chemistry, The University of Chicago, Chicago, IL, 60637

Electrochemical characterization of self-assembled monolayers (SAMs) of arylthiol **1** on Au(111) reveals that these SAMs have a high density of defect sites, potentially as a result of the deposition solvent. However, arylthiol SAMs subjected to a heterogeneous "patching" process involving addition/exchange of HDT into the original monolayer exhibit superior electrochemical blocking/passivating behavior, while retaining most of the original material, as determined by XPS. This reproducible technique provides a pathway to highly blocking arylthiol SAMs. The results of optical and electronic studies with terminally functionalized arylthiols will also be presented.

1, R = H

344.

HIGH CONTRAST NEAR-FIELD OPTICAL MICROSCOPY OF METAL PHOSPHONATE SELF ASSEMBLED MULTILAYERS. **Kenneth D. Weston**, Jessie A. DeAro, Grace A. Credo, Steven K. Buratto Department of Chemistry, University of California at Santa Barbara, Santa Barbara, CA 93106

A dual mode near-field scanning optical microscopy (NSOM) technique, in which light is both emitted from and collected by a sub-wavelength aperture is described and its utility in probing the nanoscale optical properties of metal phosphonate self-assembled multilayer (SAM) films is demonstrated. Contrast in dual mode images is easier to interpret than in other NSOM configurations. This simplification allows true optical contrast to be separated from topography induced contrast, a task that is nearly impossible for other reflection NSOM geometries. High contrast (via the refractive index) and high resolution (~100 nm) dual mode NSOM images reveal clusters and voids in the films, local variations in the film thickness, and local variations in the index of refraction.

345. LARGE-SCALE MOLECULAR DYNAMICS STUDIES OF THE ADSORPTION OF POLYELECTROLYTES ONTO CHARGED SURFACES. Jessie C. Chu and C. H. Mak, Department of Chemistry, University of Southern California, Los Angeles, CA 90089-0482

The fabrication of nanostructures often relies on the adsorption of self-assembly systems onto solid substrates. While this method is simple, it lacks specificity because the resulting structures are largely determined by the adsorbate interactions but not by the adsorbate-surface interactions. Using large-scale molecular dynamics simulations, we have investigated a potentially useful alternative method of defining nanostructures by the adsorption of polyelectrolytes from an ionic solution onto charged solid substrates. We will address the fundamental interactions of flexible or stiff polyelectrolyte chains with the surface as well as between multiple charged chains themselves, and discuss their implications for substrate pattern recognition applications.

346.

PREPARATION AND CHARACTERIZATION OF Ag COLLOID MONOLAYERS
Robin M. Bright, Michael D. Musick, and Michael J. Natan, Department of Chemistry, The Pennsylvania State University, University Park, PA 16802

The process of Ag colloid self assembly onto organosilane-functionalized glass, carbon, and Au surfaces has been studied using uv-vis, quartz crystal microgravimetry (QCM), atomic force microscopy (AFM), and field-emission scanning electron microscopy (FE-SEM). Colloids prepared by citrate reduction and by EDTA reduction of Ag^+ yield different deposition kinetics, primarily a result of differences in particle concentration and monodispersity. The sticking probabilities and equilibrium constants for each type of colloid were measured. Surface enhanced Raman scattering (SERS) and electrochemical properties of Ag colloid monolayers were compared to bulk Ag. Deposition of Ag colloid onto carbon electrodes improves the heterogeneous rate constant for electron transfer to $Ru(NH_3)_6^{3+}$, but the modified electrodes still fall short of values obtained for bulk Ag. Likewise, Ag colloid monolayers are less SERS-active than roughened bulk Ag. The effects of particle polydispersity on these phenomena are discussed.

347.

SELF-ASSEMBLY AND SELF-ORGANISATION OF A SEMICONDUCTOR NANOCRYSTALLITE SUPERLATTICE. Lucy Cusack, Rosalba Rizza, Alexander Gorelov and Donald Fitzmaurice, Department of Chemistry, University College Dublin, Dublin 4, Ireland.

Abstract The self-assembly and self-organisation of nanocrystallites and molecules to form complex nanostructures in solution is an important objective of materials chemistry and physics. The importance of this objective is as a result of a desire to be able to 'programme' the 'bottom-up' assembly of nanometer-scale devices in solution. In this context, an important interim objective is the self-assembly and self-organisation of nanocrystallite arrays, or superlattices, in solution. A general strategy that permits the self-assembly of both nanocrystallites and molecules, as opposed to only nanocrystallites or molecules is favoured, as ultimately, this will permit greater functional diversity. We note, the ability to self assemble and self-organise molecules in solution is far advanced by comparison. We will describe the recognition directed self-assembly and self-organisation of a superlattice of TiO_2 nanocrystallites in solution using an approach that offers the prospect of wide applicability.

348.

STRUCTURE OF ADLAYERS ON UNDERPOTENTIALLY DEPOSITED MONOLAYERS OF SILVER ON GOLD. **Paul E. Laibinis,** G. Kane Jennings, and Bentley J. Palmer, Department of Chemical Engineering, Massachusetts Institute of Technology, Cambridge, MA 02139

Underpotential deposition of silver onto gold forms regular two-dimensional structures that are influenced in their packing density and coverage by the electrochemical conditions governing their formation. We have used these well-defined structures as platforms for further assembly. Alkanethiols adsorb onto these subtrates to entrap the silver atoms between the gold and organic films. The alkanethiols adopt well-defined adsorbed structures that have been characterized by infrared and x-ray photoelectron spectroscopies. The silver atoms -- although present at less than a monolayer -- afford greater stability to the adsorbed organic layer. Copper atoms have been incorporated within the silver on gold platform to manipulate the two-dimensional structure of the silver atoms. The structure of this heterolayer will be discussed based on results from x-ray photoelectron spectroscopy, electrochemistry, and scanning probe microscopy.

349.

BINARY LIQUID MIXTURES IN NANOPOROUS MATERIALS: PHASE SEPARATION KINETICS, WETTING AND EQUILIBRIUM PROPERTIES FROM NUMERICAL SIMULATION. **Lev D. Gelb,** K. E. Gubbins, School of Chemical Engineering, Olin Hall, Cornell University, Ithaca, NY 14853-5201

Liquid mixtures in narrow pores can have very different properties than bulk mixtures. We have performed Molecular Dynamics and Monte Carlo simulations of the behavior of simple liquid mixtures in micropores of several different geometries. The effects of confinement on the kinetics of spinodal phase separation are characterized, and the equilibrium phase diagrams for these adsorbed mixtures are determined. We have also investigated the wetting transitions that occur in these pores, and combine these results into a general picture of the phase behavior of these systems.

350. CLUSTER FORMATION AND CHARGE TRANSFER REACTIONS IN MIXED ALKALI METAL BROMO SODALITE. Lj. Damjanovic, G. D. Stucky and V. I. Srdanov, Department of Chemistry, University of California at Santa Barbara, Santa Barbara CA 93106

Sodium bromo sodalite in which 10% of the cages have a missing NaBr unit was doped with different alkali metals. The "defect" cages, which contain just three charge compensating Na^+ anions, are capable of absorbing and ionizing the excess alkali atom. Doping with sodium atoms yields the familiar 13 peak EPR pattern of the Na_4^{3+} paramagnetic cluster: $Na + \{3Na^+\}_{cage} \longrightarrow Na_4^{3+}$. The reaction $K + \{3Na^+\}_{cage}$, however, leads to a transient $(KNa_3)^{3+}$ paramagnetic species that dissociates to $\{K^+2Na^+\}_{cage}$ and atomic sodium. The later reacts with another $\{3Na^+\}_{cage}$ to form Na_4^{3+} cluster. After all accessible $\{3Na^+\}_{cage}$ convert to $\{3K^+\}_{cage}$, the paramagnetic K_4^{3+} clusters begin to form and eventually replace existing Na_4^{3+} clusters. The time evolution of $Na_4^{3+} \longrightarrow K_4^{3+}$ reaction will be presented.

351.

COLUMNAR AND MICELLAR CUBIC LIQUID CRYSTAL STRUCTURES FROM TAPERED DENDRON MOLECULES   Goran Ungar V.S.K. Balagurusamy, University of Sheffield, Sheffield S1 3JD, U.K.; Virgil Percec , Gary Johannson and Jim Heck, Case Western Reserve University, Cleveland OH 44106

Self-assembly of fan-shaped or cone-shaped molecules can lead to the formation of a number of liquid crystal phases with 2- or 3-dimensional long-range correlation in density fluctuations. Cylindrical and spherical supramolecular aggregates of such molecules were created by controlling their self-assembly via molecular shape and specific interaction. X-ray studies, using quantitative low-resolution crystallography, showed hexagonal columnar structures from fan-shaped tricatenar compounds and revealed the internal structure of the columns. The ionic channel model was confirmed in the case of salt complexes. Cone-shaped dendrons, of generation number 1 to 4 form cubic phases. Detailed structure of one of these new thermotropic phases is presented, consisting of spherical and nearly spherical aromatic clusters dispersed in an aliphatic continuum.

352.
CORE LEVEL PHOTOABSORPTION STUDY OF DEFECTS AND METASTABLE BONDING IN BORON NITRIDE. **L. J. Terminello** I. Jimenez, A. F. Jankowski, D. G. J. Sutherland, J. A. Carlisle, LLNL, Livermore, CA 94551; G. L. Doll, J. V. Mantse, GM Research Laboratory, Warren MI 48090; W. M. Tong, D. K. Shuh, LBNL, Berkeley, CA 94720; F. J. Himpsel, University of Wisconsin, Madison, Madison, WI 53706

A comprehensive study of different bonding environments and defect structure in boron nitride films and materials has been performed using core level photoabsorption. We have identified nitrogen vacancies in the hexagonal bonding of BN, nitrogen interstitials, boron clustering, sp3 like metastable phases and sp3 phases in BN films grown by various methods. Quantitative information on the concentration and distribution of point defects can be determined and is directly related to the formation of new phases. The characterization of thin films using the chemically selective technique of synchrotron radiation based core level photoabsorption can overcome the limitations of other structural probes.

353.

Electrochemiluminescence from calixarene-coated porous silicon liquid junction cells. Libing Zhang, Jeffery L. Coffer, Texas Christian University, Fort Worth, TX 76129

The electrochemiluminescence (ECL) behavior of porous Si (PS) coated with macrocyclic calixarene carboxylic acid derivatives has been studied. Efficient ECL is observed from the anodic oxidation of calixarene-coated PS in an electrolyte containing 1.0 M H_2SO_4. The efficiency and stability of ECL is dependent on the cavity size of the calixarene macrocyclic molecules comprising the film. Compared to fast degradation behavior of untreated porous Si (in minutes), longer performance of ECL is observed with calixarene carboxylic acid derivativation (ca. 4 hrs for calix[4]arene-COOH coated PS; ca > 50 min for calix[n]arene (n = 5, 7, 8) analogs). With increasing time at constant bias, the ECL peak maximum from these coated porous Si is gradually red-shifted beyond 700 nm. The measured quantum efficiency of ECL is also observed to be dependent upon the size of the calixarene carboxylic acid used. The above phenomena are attributed to packing differences between the different types of calixarene molecules and size-selective excitation of Si nanoparticles at the solid/liquid interface.

354.

FRICTION BETWEEN DIAMOND SURFACES IN THE PRESENCE OF SMALL, THIRD-BODY MOLECULES. **Martin D. Perry** and Judith A. Harrison, Department of Chemistry, United States Naval Academy, Annapolis, MD 21402

Molecular dynamics simulations have been used to examine the friction between the hydrogen-terminated (111) faces of diamond with small hydrocarbon (third-body) molecules trapped between them. In general, the presence of the trapped, third-body molecules reduced the friction between the diamond surfaces, particularly at high loads. The size and shape of the third-body molecule, as well as the alignment of atoms on opposing diamond surfaces, was found to be paramount in determining the magnitude of the friction reduction. These results are compared to results from previous simulations which examined the effects of chemically-bound hydrocarbons on the friction between diamond surfaces and to available experimental data.

355. NANOCRYSTALLINE SEMICONDUCTORS IN NANOCHANNEL GLASS ARRAYS. A.A. Guzelian, A.D. Berry, and R.J. Tonucci, Naval Research Laboratory, Washington, DC 20375.

We report on two methods for depositing nanocrystalline semiconductor material into nanochannel glass (NCG) arrays. Both methods are general and should be applicable to a variety of semiconductors. The first method employs CVD techniques to deposit GaN into the nanochannel glass resulting in a periodic array of polycrystalline wires with high aspect ratios. The second method involves the deposition of soluble semiconductor nanocrystals such as CdSe and InP from solution into the NCG arrays. This method can allow for the control of the amount of material deposited into the arrays and we have made progress in obtaining a high fill fraction of nanocrystal material in the channels. The structures are characterized using optical and electron microscopies. These composite structures have the potential to exhibit interesting photonic band gap effects while the NCG also provides an excellent template for ordering quantum confined materials into a macroscopic array.

356. NEAR-FIELD AND CONFOCAL SCANNING SPECTROSCOPY/MICROSCOPY OF PORPHYRIN WHEELS
F.C. De Schryver*, S. De Feyter*, K. Grim*, H. Faes*, J. Hofkens*, K. Jeuris*, L. Latterini*, R.J. Nolte**, A. Rowan**, M. Ruiter*, A. Schenning**, P. Vanoppen*
* K.U.Leuven, Dept. of Chemistry, Celestijnenlaan 200F, B-3001 Heverlee, Belgium
** K.U.Nijmegen, Dept. of Chemistry, Toernooiveld 1, NL-6525 ED Nijmegen, The Netherlan
Schenning, et.al. recently reported the preparation of a unique type of self-assembly, comprised of certain porphyrin compounds, and in the shape of a ring with a diameter in the micron range and a ring height and width on the order of tens of nanometers. These wheel-like molecular aggregates are related to ring-shaped molecular aggregates known to occur in nature, e.g. in the bacterial light-harvesting complex, LH2. This paper is concerned with a detailed microscopic investigation of porphyrin wheels by confocal visible-near-IR fluorescence microscopy and fluorescence scanning near-field optical microscopy and scanning force microscopy. These methods are used to study the spatially resolved fluorescence and absorption spectra, including measurements with polarized light and polarized detection. The results offer insight into several aspects of the structure and photophysical dynamics of the rings, the degree of molecular order and orientations.

357. PERCOLATION TRESHOLD FOR EXCHANGE NARROWING IN SODIUM-DOPED SODIUM BROMO SODALITE. V. I. Srdanov, P. Sieger and G. D. Stucky. Department of Chemistry, University of California at Santa Barbara, Santa Barbara CA 93106

A solid solution of paramagnetic $[(Na^+)_4(e^-)]_x$ and diamagnetic $[(Na^+)_4Br^-]_{1-x}$ ($0<x<1$) clusters was synthesized in order to study percolation treshold for exchange interaction in sodalites. The x=0.05 stoichiometry produces a 13 peak EPR signal while the x=0.15 shows an additional narrow resonance (FWHM = 2G) that was previously atributed to metallic sodium clusters. Our studies show that the narrow resonance is due to exchange interaction among electrons associated with Na_4^{3+} clusters in neighboring sodalite cages. Highlights of power-dependent and temperature-dependent EPR studies will be presented.

358.
PHOTOELECTRON SPECTROSCOPY OF METALLOCARBOHEDRENES M8C12- (M=TI, V, CR, ZR, NB)*. **San Li**, Hongbin Wu, Xi Li, Chuanfan Ding, and Lai-Sheng Wang, Department of Physcis, Washington State University, Richland, WA 99352 and Evironmental Molecular Sciences Laboratory, MS K2-14, Pacific Northwest National Laboratory, Richland, WA 99352

Photoelectron spectroscopy experiments have been performed on five metallocarbohedrene (met-car) anions, Ti8C12-, V8C12-, Cr8C12-, Zr8C12-, Nb8C12-. These met-car anions were produced by two different methods. We found that the Ti and Zr met-cars show unusually low electron affinities (EAs) and that the EAs increase from Ti to Cr met-cars. The observed photoelectron spectra and the electronic structure of the met-cars are interpreted using existing theoretical calculations.
*This work is conducted at Pacific Northwest National Laboratory, operated by Battelle Memorial Institute for the DOE under Contract DE-AC06-76RLO1830.

359. PORPHYRIN / TiO2 ADSORPTION AND PHOTOSENSITIZATION.
Carl C. Wamser, Suman Cherian, and Margaret Fyfield, Departments of Chemistry and Physics, Portland State University, Portland, Oregon 97207-0751

We have studied the adsorption of tetra(4-carboxyphenyl)porphyrin (TCPP) on colloidal TiO2 (Degussa P25 - 70% anatase, 30% rutile, 30 nm particles). TCPP was adsorbed from ethanol solution either onto colloidal dispersed particles or onto a thin film of particles that had been sintered at 450° in the standard manner for making a Grätzel cell. For TCPP adsorbed onto dispersed colloidal particles, the Langmuir adsorption isotherm indicates that each TCPP molecule occupies an average site area of 1.2 nm^2. Since this is smaller than its flat cross-sectional area (2 nm^2), the porphyrin macrocycle must be upright rather than flat on the surface, probably attached by one or two carboxyl groups to Lewis acidic Ti sites. Adsorption of TCPP on the sintered particles gave a saturation adsorption that was about 60% of that for the dispersed particles, suggesting a correspondingly reduced surface area available for TCPP binding. The effect of fractal dimensionality on the adsorption data and the effectiveness of TCPP as a photosensitizer in a Grätzel cell will also be discussed.

360.
PROBING THE ELECTRONIC STRUCTURE EVOLUTION OF TRANSITION METAL CLUSTERS FROM ATOM TO BULK. **Lai-Sheng Wang** and Hongbin Wu, Department of Physics, Washington State University, Richland, WA 99352 and Environmental Molecular Sciences Laboratory, Pacific Northwest National Laboratory, Richland, WA 99352

The electronic structure of transition metal clusters is important but has been difficult to study both experimentally and theoretically. Using photoelectron spectroscopy of size-selected anions, we have been able to successfully probe the electronic structure evolution of the first role transition metal clusters from Ti to Ni, in the size range from the atoms to about 100 atoms. Transition from molecular to bulk-like electronic structure is observed clearly for the early transition metal clusters. Direct correlations between the electronic structure and the chemical and magnetic properties are observed for the late transition metal clusters. Using density functional calculations, we also discovered a dimer growth route for small Cr clusters and a distinct transition from the dimer growth to bulk-like structures.

361.
SPREADING OF SOLID NANO- AND MICROPARTICLES AT WATER-AIR INTERFACES: A FILM BALANCE STUDY. **Z. Hórvölgyi(1)**, M. Máté(1), Ramsden(2), Szalma(3) and J. H. Fenderl(4), (1)Dept. of Phys Chem, Technical Univ. of Budapest, H-1521 Budapest, Hungary, (2)Biozentru, Univ. of Basel, Switzerland, (3)Loránd Eötvös Univ., H-1117 Budapest, Hungary, (4)Syracuse University, Syracuse, NY

The spreading of nano- and microparticles at water-air interfaces can lead to mono- or multiparticulate layer formation. The compressional studies of such layers in film balances can provide useful information about the layer-structure as well as the particle-particle and particle-subphase interactions. The surface presure vs. surface area isotherms of surface modified silica nano- and microparticles have been determined and the role of colloidal interactions in the "two-dimensional" structure formation is discussed in this lecture. The cohesive and non-cohesive layer formation is interpreted in terms of the wettability of particles and their colloid stability in the spreading liquid.

362. TIME-RESOLVED SPECTROSCOPIC STUDIES OF QUENCHING OF POROUS SILICON PHOTOLUMINESCENCE BY ALKYL AMINES.
Beata Sweryda-Krawiec, Jeffery L. Coffer, Texas Christian University, Fort Worth TX 76129

In an effort to determine the mechanism of quenching of porous Si (PS) photoluminescence (PL) by Lewis bases, the interaction of alkyl amines with porous Si was examined by steady-state and time-resolved photoluminescence spectroscopy.

It is well known that dilute solutions of strong Lewis bases such as alkyl amines results in the diminution of integrated PS PL intensity; such PL quenching is a function of the quencher concentration exposed to the PS substrate. In principle, this quenching can result from dynamic modification of the lifetime of the PS lumophores, or alternatively, a static process between quencher and fluorophore.

We have measured the effects of exposure of dilute solutions of simple alkyl amines on the decay curves of porous Si photoluminescence. These curves were fit to a stretched exponential function in order to obtain average lifetime data for the Si lumophores. These calculated average lifetimes, as well as changes in integrated photoluminescence intensity, were fitted and analyzed according to a simple Stern-Volmer equation.

363. VIBRATIONAL MODES OF THE SURFACE AND SUBSURFACE MOLECULES OF ICE. J. P. Devlin, and V. Buch, Departments of Chemistry, Oklahoma State University, Stillwater, OK 74078 and The Hebrew University, Jerusalem 91904, Israel

Assemblies of nanocrystals, on ZnS infrared windows, have been prepared from aerosols supported in N_2(g) at 70 K. These stable assemblies, with nanocrystals of average diameter of ~20 nm, retain the surface area of the aerosol phase, and allow versatile sampling of the surfaces of solid particles of many substances. Particular emphasis has been placed on the FT-IR spectroscopy of the surface and subsurface of crystalline ice. The differences in spectra, obtained before and after Ostwald ripening (or as bare and adsorbate-coated nanocrystals), when combined with simulated spectra of the disordered reconstructed surface of cubic ice, have permitted a detailed assignment of the vibrational modes of the three categories of water molecules at the ice surface. More recently, this study has been extended to the ice subsurface, 2 - 3 bilayers that separate the disordered ice surface from the crystalline interior. A detailed spectrum of the subsurface has been obtained through the use of strong adsorbates that restore much of the order of the ice surface.

364.

A NEW APPROACH TO THE STUDY OF MOLECULAR REACTIVITY Hugh Doyle Geoffery Will, Donald Fitzmaurice, Department of Chemistry,University College Dublin, Dublin 4, IRELAND

The energetics and kinetics of chemical reactions are dependent on molecular approach and reactivity. In this context, it is important to study the effects of orientation and spatial separation on molecular reactivity. In a new approach, the interaction between molecules possessing precisely defined orientations has been directly studied as a function of spatial separation. Using Langmuir-Blodgett techniques, orientated monolayer films of octadecyl-3,4-dihydroxybenzoate and TiO2 were prepared on muscovite mica substrates and mounted in a modified surface forces apparatus (SFA). The charge transfer absorption band, assigned to adsorption of the catechol moiety at Ti4+ states on the surface of a TiO2 nanocrystallite, was studied as a function of spatial separation, at angstrom resolution, by UV-Visible spectroscopy. In effect, this allows wavefunction overlap to be monitored in real space. It is expected this approach will yield an improved understanding of the interaction between orientated molecules.

365.
AB INITIO CALCULATIONS OF THE INTERACTIONS OF HE WITH $CL_2(B^3\Pi_u)$ AND $CL(X^2P)$. R. Burcl, S. M. Cybulski, M. M. Szczęśniak, G. Chałasiński, Department of Chemistry, Oakland University, Rochester, Michigan 48309

Potential Energy Surfaces (PESs) of He+$Cl_2(B^3\Pi_u)$ complex were evaluated using 4th order Møller-Plesset perturbation theory (UMP4) and coupled cluster theory (CCSD(T)) and analyzed in terms of fundamental contributions: electrostatic, exchange, induction, and dispersion. The lowest triplet state of Cl_2 is split by the presence of He atom into A' and A" states. Although this splitting is relatively small, the resulting surfaces have different topology: A' has a single T-shaped minimum; A" possesses global T-minimum and local linear minimum. The changes in PESs upon increase of intramolecular Cl-Cl distance were also investigated. It is found that CC level of theory is needed to properly describe these effects; the discrepancies between MP4 and CCSD(T) become very important for stretched geometries. In addition the interaction of He with Cl atoms was investigated by the same approach. Two calculated states, Σ and Π differ from empirical potentials. The calculated well depths are 29.0 cm^{-1} at 3.35Å (Σ) and 15.1 cm^{-1} at 3.80Å (Π).

366.

ATOMIC ORBITAL ALIGNMENT IN PHOTODISSOCIATION. **Allan S. Bracker,** Holly M. Bevsek, David A. Blunt, F. Cortney Sailes, Arthur G. Suits, Yuan T. Lee, Department of Chemistry, University of California and Chemical Sciences Division, Lawrence Berkeley Laboratory, Berkeley, CA 94720.

The polarization of the electronic angular momentum of atomic photofragments is governed by the details of the electronic rearrangment which occurs during dissociation. Measurements of this property can yield information about the symmetry of the dissociative states, coherence effects, non-adiabatic effects and the long-distance interactions between the departing fragments. We discuss ion imaging measurements of atomic orbital aligment for the dissociation of ozone and diatomic halogens. These results are analyzed using the bipolar moments method and the 'dynamical functions' of Siebbeles and coworkers (J.Chem.Phys 100, 3610 (1994)).

367. CLASSICAL VARIATIONAL PORTRAITS OF THE DYNAMICAL STEREOCHEMISTRY OF ATOM–DIATOM REACTIONS. G. W. Koeppl, Department of Chemistry and Biochemistry, Queens College of CUNY, Flushing, NY 11367

A classical variational theory of chemical reaction rates [J. Chem. Phys. 87, 5746 (1987)] is applied to obtain portraits of the dynamical stereochemistry of several atom–diatom reactions. The variational theory gives the equilibrium flux through a trial surface; the surface is varied to obtain a least upper bound for the rate. The dividing surface (DS) for an atom–diatom reaction, A + BC, is defined by a power–series expansion of the B–C internuclear separation (r) in internal coordinates R and θ where R is the distance between A and the center-of-mass of BC and θ is the angle which the BC internuclear axis makes with a line from the center-of-mass of BC to atom A. The downhill simplex algorithm is used to search the space of three, six, and ten variational parameters of first, second, and third–order expansions of the DS and obtain minimum values for the canonical rate constant or energy–dependent mean reaction cross section. The presence of angle–dependent terms in the DS makes it possible to describe the dynamical stereochemistry of atom–diatom reactions in a new and useful manner. Portraits of the dynamical stereochemistry are obtained by plotting contours of the flux through the DS using polar coordinates R and θ; such plots are reactivity relief maps of the DS. The portraits can be used to show how the field of reactivity which surrounds a diatomic reactant expands with increasing temperature and energy.

368. CONFORMATIONAL AND STRUCTURAL INFLUENCES ON STEREODYNAMICS: PHOTOFRAGMENTATION OF LARGE CHLORONITROSOALKANES. Radoslaw Uberna, Kazuhiko Fukui, John H. Frederick, and Joseph I. Cline, Department of Chemistry and Program in Chemical Physics/216, University of Nevada, Reno, NV 89557-0020

The mechanism of the 650 nm, S_1 $\tilde{A}^1A'' \leftarrow S_0$ $\tilde{X}^1A'$ photodissociation of 2-chloro-2-nitrosopropane, 1-chloro-1-nitrosocyclohexane (CNCH), and 2-chloro-2-nitroso-6,6-dimethylbicyclo[3.1.1]heptane (CNMH) is studied by polarized $1+1'$ resonance-enhanced multiphoton ionization probing of the NO $X^2\Pi_{\Omega=1/2,3/2}(v=0)$ product with time-of-flight mass spectrometry detection to measure NO fragment μ-v-j correlations. The average $\beta_0^0(22)$ bipolar moments for NO produced in the T_1 channel of CNP and CNMH are approximately twice that of CNCH. The $\beta_0^2(20)$ moment is small and approximately the same for the three molecules. These results show that (1) significant v-j correlations can exist in dissociation mechanisms with essentially statistical energy distributions and (2) the v-j correlation is relatively insensitive to molecular size but is diminished by structural nonrigidity. Classical trajectory calculations show the measured preference for $\mathbf{v} \perp \mathbf{j}$ NO alignment arises from interplay of a barrier in the exit channel with the C-N-O bending potential.

369.

DISSOCIATION DYNAMICS OF QUASIBOUND LEVELS IN THE $g^3\Sigma_g^+$ STATE OF H2: CHARACTERIZATION OF THE CONTINUUM. **Eloy R. Wouters[+]**, B.Buijsse, J. Los and W. J. van der Zande, FOM Institute for Atomic and Molecular Physics, Kruislaan 407, 1098 SJ Amsterdam, The Netherlands; [+]Present Address: Dept. of Chemistry, Univ. of California, Berkeley, CA 94720.

Decay pathways and photofragment anistropy parameters have been determined for a series of rovibrational levels of $g^3\Sigma_g^+$ state in molecular hydrogen using fast-beam photofragment spectroscopy. The $g^3\Sigma_g^+$ (v=4) levels are found to be weakly predissociated by homogeneous interaction with the lower lying n=2 $a^3\Sigma_g^+$ state and by rotational coupling with the $^3\Pi_g$ state. The observed photofragment anisotropy reveals fine and hyperfine structure depolarization. The decay of the $g^3\Sigma_g^+$ (v=5) levels is dominated by barrier tunneling. The anisotropy parameters identify the symmetry of the continuum to be of mixed Σ_g and Π_g character, which implies that hydrogen molecules can tunnel through two different barriers. The determination of the kinetic energy and the anisotropy parameter of the photofragments have resulted in a number of spectral reassignments.

370.
ENERGETICS AND ELECTRONIC PROCESSES IN THE UV PHOTOLYSIS OF NICKEL TETRACARBONYL. **Tycho J. Speaker and Roger W. Anderson** Department of Chemistry,. University of California, Santa Cruz, CA 95064

The photolysis of nickel tetracarbonyl is studied at 193 and 248 nm. The photolyis results in two photon dissociation of the carbonyl ligands, and the excitation of the nickel atom with a substantial amount of energy. The excitation of the nickel atom is explained by a sequence of two photon absorptions by charge transfer transfer transitions where each transition is followed by an electron jump back to the nickel core. The energy of the highest observed nickel state is used to establish a maximum total bond for the nickel tetracarbonyl molecule. The highest energy nickel atom state that is not observed is used to establish a lower bond to the total dissociation energy. The photolyses at 193 and 248 nm yield a total dissociation energy of 6.20 ± 0.02 and 6.17 ± 0.02 eV respectively.

371.

FEMTOSECOND CONTROL OF BIMOLECULAR REACTIONS: EXPERIMENTS AND THEORY. **M. Dantus,** P. Gross and U. Marvet, Department of Chemistry, Michigan State University, East Lansing MI, 48824-1322

Bimolecular reactions are characterized by the energy of collision, orientation between reactants and impact parameter. We have developed a novel technique with which these reactive parameters can be controlled as well as the time at which the reaction occurs. The method used is femtosecond photoassociation spectroscopy (FPAS), whereby an ultrafast laser pulse causes binding between free atoms or molecules, U. Marvet and M. Dantus, Chem. Phys. Letters 245, 393 (1995). The process may be thought of as the reverse of photodissociation. This technique provides a 'time zero' for the bond formation process and restricts the alignment and orientation between reactants, and therefore one is able to obtain information such as the bond formation time and transition state dynamics. The requirements for achieving control of the impact parameter using FPAS will be discussed.

372.
FEMTOSECOND DYNAMICS OF CONCERTED ELIMINATION PROCESSES. **M. Dantus,** U. Marvet, Q. Zhang and E. J. Snyder, Department of Chemistry, Michigan State University, East Lansing MI, 48824-1322

Excitation at high energies opens several reaction channels of photodissociation to polyatomic molecules. We are investigating in our laboratory these processes which present intriguing dynamics caused by surface crossings. We will present our results on the concerted elimination of diatomic molecules following the photodissociation of alkyl halides. Discrimination between the different reaction channels is achieved by selecting the LIF signal arising from these fragments only. Our experiments focus on the electronic states involved in the reaction, the dynamics of the process as a function of the mass and symmetry of the parent molecule and vibrational coherence in the products.

373. ISOMERIZATION AND DISSOCIATION IN COMPETITION - UNRAVELING COMPLEX DISSOCIATION PATHS. Tomas Baer, Oleg Mazyar, Jeffrey W. Keister, and Paul M. Mayer, Department of Chemistry, University of North Carolina, Chapel Hill, NC 27599-3290

The complex dissociation dynamics of energy selected ions are investigated by photoelectron photoion coincidence (PEPICO) spectrometry. Ions are formed by continuous vuv photoionization of a seeded molecular beam. Electrons and ions are extracted in opposite directions in a low electric field. Dissociation rates are measured by modeling the asymmetric time of flight (TOF) fragment ion distribution. A number of ions have been found which dissociate via two-component decay rates. This is clear evidence for the participation of lower energy isomers. Some complex dissociation paths involve as many as six isomerization or dissociation steps. In favorable cases, e.g. cyclopentane and tri-methyl borate ions, all six rate constants can be extracted from the data. The results are analyzed by *ab initio* MO calculations and the statistical RRKM theory.

374. ORIENTATION DEPENDENCE OF H ABSTRACTION REACTIONS. Edgar E. Arcia and Trina Valencich, Department of Chemistry, The University of Montana, Missoula, MT 59812

We compare the characteristics of a new empirical potential energy surface form for the CH_5 system with *ab initio* calculations and recent experimental measurements. We will report the results of quasiclassical trajectory simulations of collisions of H atoms with CD_4, and CH_3D, at relative collisional energies of 138 and 223 kJ/mol. The internal energy states of the reactants were prepared by normal mode sampling. The cross-section, on V4, for abstraction by H from CD_4 is 0.151×10^{-20} m^2, and for substitution is 0.099×10^{-20} m^2, at 138 kJ/mol and 223 kJ/mole respectively. The barrier height for abstraction must increase as the angle between the forming and breaking bonds in the D_3C-D-H transition state deviates from a collinear arrangement to agree with Gaussian92 predictions. Surfaces without this feature calculate a larger cross-section for abstraction and favor different microscopic mechanisms.

375. PHOTOCHEMISTRY AND DYNAMICS OF CHARGE-TRANSFER COMPLEXES. Gary DeBoer and Mark A. Young, Department of Chemistry, University of Iowa, Iowa City, Iowa 52242

We have examined a number of weakly bound complexes formed in a supersonic expansion which can undergo a charge-transfer (CT) process upon electronic excitation. The CT mechanism is reflective of the entire supramolecule and any subsequent chemistry may be unique to the complex environment. We will report our findings for several interesting CT species. Studies of the prototypical C_6H_6-I_2 complex indicate that a fast charge-recombination step initiates chemistry and that the complex geometry is of critical importance. We are now attempting to extend these insights to other systems. We have studied a series of O_2 containing complexes with benzene or alkene donors. These complexes are found to produce O atom fragments with a greatly enhanced yield relative to excitation of the bare oxygen molecule. Results for other CT systems will also be presented.

376. PHOTOFRAGMENTATION OF ClNO AND REACTIONS OF ClNO WITH CHLORINE ATOMS IN CLUSTERS
Carlos Conde, Christof Maul, and Edwin Quiñones
Department of Chemistry, University of Puerto Rico, PO Box 23346, Río Piedras, PR 00931-3346

The excitation of jet-cooled ClNO at 355 nm produces an ensemble of NO fragments translationally hot and with a Gaussian rotational state distribution. However, upon increasing the backing pressure of the ClNO/He mixtures above 1 atm, we observed a second NO ensemble described by a Boltzmann rotational distribution emanating from the Cl + ClNO reaction initiated within the $(NOCl)_2$ cluster. This was confirmed studying the Cl + ClNO reaction in the bulk. The same reaction was studied forming the Cl_2–ClNO van der Waals precursor and lower NO rotational states were detected indicating that there is a much narrower range of impact parameters available to the reaction than in the bulk. Increasing the backing pressure to 6 atm favors the formation of larger NOCl clusters from which only rotationally relaxed but translationally excited NO molecules are detected due to the cluster 'cage'.

377. PHOTOFRAGMENTATION STUDY OF Cl_2 USING ION IMAGING. P.C. SAMARTZIS, I. Sakellariou, T.N. Kitsopoulos, Department of Chemistry, University of Crete and Institute of Electronic Structure and Laser, Foundation for Research and Technology-Hellas, 711 10 Heraklion-Crete, Greece

The velocity distributions of the $Cl(^2P_{3/2})$ and $Cl(^2P_{1/2})$ photofragments following the photolysis of Cl_2 at 3.477 and 3.440 eV respectively, are measured using ion imaging. Chlorine atoms are detected using (3+1) resonance enhanced multiphoton ionization through the $4s'(^2D_{3/2})$ excited electronic state. The measured photofragment translational energy distributions yield a bond dissociation energy of $D_0 = 2.474 \pm 0.008$ eV for ground electronic state $Cl_2(X^1\Sigma g)$, while the angular distributions determine the anisotropy parameters to be $\beta(^2P_{3/2}) = -0.70 \pm 0.03$ and $\beta(^2P_{1/2}) = 1.95 \pm 0.05$. The branching ratio $(Cl(^2P_{1/2}))/(Cl(^2P_{3/2}))$ at the specified energies is estimated at 0.19 ± 0.05, approximately one order of magnitude higher than previously determined.

378.

QUANTUM REACTION RATE CALCULATIONS. **D.E. Skinner** W.H. Miller, Department of Chemistry, U.C. Berkeley, Berkeley, California, 94720

The calculation of quantum mechanical reaction rates provides an important link between the highly detailed theory of reactive scattering and experimentally observable quantities. Recently Miller proposed a method for treating the deactivating effects important in recombination reactions within the theoretical framework of flux correlation functions. The goal of our present research is to calculate quantum reaction rates from exact quantum dynamics, but taking into account dissipative effects of the reaction environment in an approximate way. Applications of this theory include unimolecular dissociation and collisional and radiative recombination.

379.

STEREODYNAMICS IN PHOTODISSOCIATION: MEASUREMENTS OF MU-V-J CORRELATIONS USING 2+N REMPI WITH TIME OF FLIGHT MASS SPECTROSCOPY Pisano, Patrick J. and J. I. Cline

A practical method is described for 2+n polarized REMPI probing of photofragment mu-v-j correlations arising in the fragmentation of molecules excited on an electric dipole transition. The technique combines the two photon alignment and orientation theory of A. Kummel, G.O. Sitz and R.N. Zare (J.Chem.Phys. 85, 6874 and J.Chem.Phys. 88, 6707) with R. Dixon's bipolar moment description of vector correlations (J.Chem.Phys. 85, 1866). This approach allows one to express the intensity profile of a time-of-flight transient in terms of the bipolar moments and the polarization geometry. Careful selection of the polarization geometry allows the measurement of a small subset of the many bipolar moments contributing to the intensity profile. The approach presented is applicable to the detection of spherical and symmetric tops. Previous work by Mons and Dimicoli will be shown to be related to a restrictive application of our technique. Several photodissociation experiments (Methyl Iodide, Nitromethane, etc.) will be discussed as examples of the application of this technique.

380. **THE PHOTODISSOCIATION OF NITRIC ACID AT 193 NM.** T. L. Myers, N. R. Broughton and L. J. Butler, The James Franck Institute and Department of Chemistry, The University of Chicago, Chicago, Illinois 60637

The experiments presented here investigate the competing photodissocation pathways of nitric acid upon excitation of the $\pi(nb,O) \rightarrow \pi^*(NO_2)$ transition at 193 nm. We measure the branching ratio between the production of OH + NO$_2$ and O + HONO in a crossed laser-molecular beam apparatus. To help assign the product electronic states, we constuct adiabatic correlation diagrams. We present a modified correlation diagram for the OH + NO$_2$ reaction coordinate, which restricts the orientation of the radical OH p electron to remain in the parent molecular plane. Our kinetic energy distributions evidence two pathways for the production of OH + NO$_2$. One channel produces stable NO$_2$ fragments in the 1^2B_2 electronic state whereas the other channel yields fragments which undergo secondary dissociation to NO + O. We examine the influence of nonadiabaticity along the reaction coordinate to explain the significant branching to this other mechanism. We also evidence two pathways for the production of O + HONO: the dominant channel correlates to O(^{1}D) + HONO (1^1A'), and the minor channel yields O(^{3}P) + HONO (1^3A").

381.

VIBRATIONALLY MEDIATED PHOTODISSOCIATION OF ISOCYANIC ACID(HNCO) IN A MOLECULAR BEAM. **Martin J Coffey** Steven S. Brown, H. Laine Berghout, Rex K. Frost, and F. Fleming Crim. Department of Chemistry, University of Wisconsin-Madison, 1101 University Ave., Madison, WI 53706.

Isocyanic acid photodissociates to form NH* + CO, NH + CO, or H + NCO, where NH* is the metastable singlet state of NH as opposed to its ground triplet state. The NH* + CO channel is a spin-foribben dissociation pathway, but effectively competes with the two spin allowed pathways. We have studied the photolysis in a molecular beam at various photolysis energies by detecting the H atom and NH radical by laser-induced fluorescence. We have also performed the isoenergetic photodissociation out of the 3nu1 excited state. By calibrating our detection of H atoms to H2S, we are able to determine the relative importance of all three decomposition pathways, and determine on a more absolute scale the effectiveness of our ability to vibrationally control this reaction.

382.

A NEW METHOD FOR THE STUDY OF THE COMPOSITION AND ORIENTATION OF MOLECULAR LIQUID SURFACES. **Philip R. Watson**, Thomas A. Gannon and Michael Tassotto Department of Chemistry, Oregon State University, Corvallis, Oregon 97331-4003

We describe the first results from a new experiment that probes the composition and molecular orientation at the surface of molecular liquids. The experiment involves directing rare gas ions in the 1-5 keV energy range onto freshly generated liquid surfaces in vacuo and measuring the time-of-flight of scattered rare gas ions and recoils from the liquid surface. The simple physics of the interaction and the extremely low penetration depth of the incident ions (about 1 atomic layer) results in a simple and direct interpretation without any extra information being necessary. The experiment yields surface atomic concentrations, and by inference, surface orientation. We will present data for glycerol that can be compared with molecular dynamics simulations of surface orientation.

383.

DIRECT OBSERVATION OF METASTABLE O_4^* Holly M. Bevsek, F. Cortney Sailes, Allan S. Bracker, and Arthur G. Suits
Department of Chemistry, University of California and Chemical Sciences Division, Lawrence Berkeley National Laboratory, Berkeley, CA USA 94720

We report the first direct detection of a metastable state of tetraoxygen, O_4^*. A covalently bound form of O$_4$ has long been theorized to exist[1-6] with two proposed structures: a cyclic form (D_{2d}) predicted to lie 100 kcal/mol above two isolated O$_2$ molecules[3], and a D_{3h} structure, most recently predicted to be 54 kcal above this asymptote. However, the only experimental observations of this system have been through producing dissociative states of O_4^* using charge transfer neutralization[7] of O_4^+ or photodetachment[8] of O_4^-, and detecting the O$_2$ products. In the present study, our excitation source produces a molecular beam of O_4^* with significant intensity and a lifetime exceeding 80 µs, while detection is accomplished via (1+1) REMPI. The observed spectrum consists of a series of vibrational bands which show rotational structure. Our discussion will focus on both the energetics and structure of this unique species and its possible role in the reaction of oxygen atoms with ozone.

[1] V. Adamantides, D. Neisius, and G. Verhaegen, Chem. Phys., **48** 215 (1980).
[2] V. Adamantides, Chem. Phys., **48** 221 (1980).
[3] E.T. Seidl and H.F. Schaefer, J. Chem. Phys., **88** 7043 (1988).
[4] K.M. Dunn, G.E. Scuseria, and H.F. Schaefer, J. Chem. Phys., **92** 6078 (1990).
[5] E.T. Seidl and H.F. Schaefer, J. Chem. Phys., **96** 1176 (1992).
[6] E. Ferreira, P. Gardiol, R.M. Sosa, and O.N. Ventura, J. Molec. Struct., **335** 63 (1995).
[7] H. Helm and C.W. Walter, J. Chem. Phys., **98** 5444 (1993).
[8] K.A. Hanold, M.C. Garner, and R.E. Continetti, Phys. Rev. Lett., **77** 3335 (1996).

384.

EXCITATION; ORIENTATION AND ALIGNMENT OF MOLECULES BY STIMULATED RAMAN PUMPING: $C_2H_2\ 2_0^1$. Joshua B. Halpern, Department of Chemistry, Howard University, Washington, DC 20059, Helmut Zacharias, Armin Rudert, Jose Martin and Ralf Dopheide, Physikalisches Institut, Universität Munster, Munster, Germany

We have previously used stimulated Raman pumping to produce ensembles of vibrationally excited molecules with spatially aligned angular momentum vectors. This results in a negative alignment, $A_0^{(2)}$, where the angular momentum vectors lie preferentially in a plane perpendicular to the exciting lasers' electric fields. Raman puming with circularly polarized light yields a significant positive orientation, $A_0^{(1)}$, relative to the lasers' propagation direction. For linear or planar molecules, the „propeller" like molecular orientation makes it possible to study collision dynamics and reactions both parallel and perpendicular to the rotational axis. As a first example stimulated Raman pumping the actylene 2_0^1 transition has produced a strongly oriented excited state population. Laser induced fluorescence has been used to verify the initial degree of orientation and follow collisional decays. First results will be presented.

385.

FLUORESCENCE, QUENCHING AND PHOTOLYSIS OF THE C_2N_2 ($A^1\Sigma_u^-$) AND ($B^1\Delta_u$) STATES: EVIDENCE FOR STERICALLY HINDERED, CROSSINGS TO THE ($X^1\Sigma_g^+$) GROUND STATE. Joshua B. Halpern and Yuhui Huang. Department of Chemistry, Howard University, Washington, DC 20059.

$C_2N_2\ A^1\Sigma_u^- \leftarrow X^1\Sigma_g^+$ and $B^1\Delta_u \leftarrow X^1\Sigma_g^+$ absorption, LIF and CN photodissociation fragment spectra have been measured between 230 and 206 nm. Photophysical properties of individual vibrational levels of both electronically excited states are reported. Lifetimes, quenching rates and fluorescence and dissociation quantum yields have been measured for the $A^1\Sigma_u^-$ state. All bands of the $B^1\Delta_u$ state are pre-dissociative with a 100% yield to CN fragments. The results are interpreted using a model originally proposed for acetylene by Fields and Dupre, where crossings from the ungerade excited singlet to triplet levels can only occur to vibrational levels with overall ungerade character, and further crossings to the gerade singlet ground state from these triplet levels do not happen. Only above a small barrier do the triplet levels acquire enough gerade character to allow crossings to the dense manifold of equienergetic ground state levels.

386.

HIGH PURITY, HIGH INTENSITY BEAMS OF FLUOROCARBON RADICALS PRODUCED VIA HEXAPOLE FOCUSING: APPLICATIONS IN SILICON AND SiO_2 SURFACE CHEMISTRY. M. A. Weibel, T. D. Hain, T. J. Curtiss, Department of Chemistry, University of Utah, Salt Lake City, UT 84112

Intense molecular beams of fluorocarbon radicals have been generated with a corona discharge passing through a free jet expansion of C_2F_6/Ar. An electrostatic hexapole selectively focused polar species through an aperture enabling us to isolate (at least partially) specific radicals (e.g., CF, CF_2, and CF_3) in the transmitted beam. Pure beams of CF radicals populating a single rotational state ($|J\ \Omega\ M_J> = |1/2\ 1/2\ 1/2>$) were generated. Preliminary results from work in progress on reactive scattering from Si and SiO_2 surfaces will be reported.

387. MOLECULAR ALIGNMENT BETWEEN NANOCRYSTALLINE DOMAINS TRAPPED IN MOLECULE CORRALS AT A LIQUID-CRYSTAL INTERFACE. D. L. Patrick, M. D. Morse, V. J. Cee and T. P. Beebe, Jr., Department of Chemistry, University of Utah, Salt Lake City, UT 84112.

The existence of a previously unknown long-range effective interaction potential is inferred from a scanning tunneling microscopy-based analysis of the statistics of mutual alignment between nearby molecular domains in an adsorbed crystalline monolayer in contact with a bulk liquid-crystal film. These domains are formed and trapped for observation in 'molecule corrals', which are flat-bottomed, nanometer-sized pits produced on the graphite surface by an etching reaction. The aligning interaction appears to be propagated by molecules in the interfacial liquid-crystal layer interacting with those on the surface. The different orientational correlations observed between domains of the same versus opposite chirality implies chiral ordering of achiral liquid-crystal molecules in the interfacial layer. For the alkylcyanobiphenyl liquid crystal used in the study, alignment is found to be propagated by, and depend almost exclusively on the relative orientation of the alkyl tails, not the more polar cyanobiphenyl head group, as one might have expected.

388. ORIENTATION OF NONSYMMETRIC LINEAR TRIATOMIC USING ω+2ω LASER EXCITATION SCHEMES - NUMERICAL SIMULATIONS. A.D. Bandrauk, C. Dion, O. Atabek, Laboratoire de Chimie Théorique, Université de Sherbrooke, Que, Canada and Laboratoire de Photophysique Moléculaire, Orsay, France.

Nonsymmetric linear molecules such as HCN possess dipole moments which are both permanent and induced during vibrations. In intense low frequency laser fields, nonlinear molecule-field effects will arise from all interactions resonant and nonresonant, permanent and induced. These will combine to enhance or reduce laser-induced alignment. In particular, a double laser excitation scheme, ω+2ω, produces nonsymmetric temporal electric fields which are necessary for orientation of molecules as opposed to simple alignment. Solutions of the time-dependent Schroedinger equation for HCN in a combination of ω and 2ω laser fields will be presented in order to establish the virtues and limits of such excitation schemes in order to orient and focus polyatomics at high laser intensities, beyond the perturbative regime.

389.

RESONANCE AND FARADAY DETECTION OF POLARIZED GROUND STATE ATOMIC PHOTOFRAGMENTS, **K. O. Korovin**, B. V. Picheyev, and O. S. Vasyutinskii; Ioffe Physico-Technical Institute, Russian Academy of Sciences, 194021 St. Petersburg, Russia

Polarization spectroscopy of atomic photofragments found wide applications as allows to obtain detailed information on photodissociation dynamics [1,2]. The contribution presents new experimental methodic that can be used to detect and study nonflourescing polarized photofragments: Faraday detection of polarized photofragments. The idea of this detection technic came from optical pumping experiments [3]. The presentation contains general theoretical treatment and possible experimental geometries that can be used to measure oriented and aligned photofragments as well as our experimental results for oriented Rb atoms produced in photodissociation of RbI by ciucularly polarized light at 266 nm. Our studies show that particularly in photodissociation experiments the Faraday detection of polarized photofragments has important advantage compare to usually used resonance technics allowing to decrease greatly absorption, saturation, and thick optical layer effects and increase noticeably obtained experimental signals.

390.

ROVIBRATIONAL ENERGY TRANSFER IN CROSSED BEAMS FROM ALIGNED S1 GLYOXAL. **Miles J. Weida**, Samuel M. Clegg, Andrew S. Burrill, and Charles S. Parmenter, Department of Chemistry, Indiana University, Bloomington, IN 47405

The role of precursor alignment in rovibrational energy transfer is investigated for inelastic collisions involving S1 glyoxal. Glyoxal is aligned via S1 - S0 excitation with a polarized pump laser and then collided with a second gas in a well-defined crossed beam geometry. By varying the polarization of the pump laser with respect to the relative collision velocity, the planar glyoxal molecule is placed in an aligned state such that the collisions are predominately "edge-on" or "broadside." The cross sections for rovibrational energy transfer are then determined via dispersed fluorescence spectra of the states populated after a single collision. The study addresses the importance of collisional steric effects in the competition between vibrational and rotational energy transfer, and also considers the propensity to excite or de-excite in-plane vs. torsional out-of-plane vibrations of glyoxal depending on the collision geometry.

391. STEREOCHEMICAL ASPECTS OF ANCHIMERIC ASSISTANCE IN SOLVOLYSES OF 3-AZETIDINYL HALIDES. Robert. H. Higgins. Department of Natural Sciences, Fayetteville State University, Fayetteville, North Carolina 28301

Hydrolyses of 3-azetidinyl chlorides, tosylates, and mesylates occur by 1-azabicyclo[1.1.0]butyl cations as intermediates formed directly by anchimeric assistance or by collapse of the resulting 3-azetidinyl carbocations. High-level Hartree-Foch *ab initio* calculations indicate that the azabicyclobutyl cations are more stable than the corresponding azetidinyl carbocations and that transition states for the anchimeric assistance pathway possess lower energies than those for ionization to 3-azetidinyl carbocations. Furthermore, these calculations indicate that the N-alkyl substituent and the leaving group must be *trans* to each other (for orbital symmetry reasons). The latter observation requires that solvolyses of 2,3-disubstituted azetidinyl compounds occur such that *trans* compounds yield 2-substituents at the equatorial position in the bicyclic ion and in the axial position in the case of the *cis*-2,3-disubstituted azetidinyl compounds. When bulky N-alkyl substituents are present, severe eclipsing interactions occur between the N-alkyl and 2-substituents in the anchimeric assistance pathway allowing competitive direct ionization to the 3-azetidinyl carbocations. Supported by NIH (S06GM08206).

392. THE DYNAMICS OF INVERTING MOLECULES IN INHOMGENEOUS ELECTROSTATIC FIELDS Elva A. Torres, Leonard C. Pipes, and Delroy A. Baugh, Dept. of Chemistry and Biochemistry, University of California, Los Angeles, CA 90095-1569

The ability to state-selectively focus molecules provides a powerful means of elucidating the nature of molecular processes in great detail. The state-selection of noninverting symmetric tops in an electrostatic hexapole field (EHF) via the first-order Stark effect is now well understood. A distinct case, however, occurs when two nearly degenerate levels are present as in the case of inverting molecules. Here we examine the state-selection and focusing dynamics of a wide class of inverting molecules such as non-rigid symmetric tops, asymmetric tops, and clusters. We provide an analytical solution which shows that a molecule which undergoes quantum mechanical tunneling while traversing an EHF can be described by a two-dimensional harmonic oscillator with angular momentum L_υ. The explicit dependence of this "virtual" angular momentum L_υ on the molecular inversion frequency shows how inversion tunneling influences the classical dynamics of inverting molecules in an inhomogeneous EHF. A comparison between simulated and experimental focusing curves is also presented.

393. THE FIRST EVIDENCE FOR TORSIONAL TUNING OF ELECTRON TRANSFER ACTIVATION ENERGIES: DENSITY FUNCTIONAL STUDIES OF UBIQUINONE. Scott E. Boesch and Ralph A. Wheeler; Dept. of Chemistry & Biochemistry; Univ. of Oklahoma; Norman, OK 73019

One of a ubiquinone molecule's methoxy substituents (see **1**) is proposed to provide a handle for tuning electron transfer activation energies and rates. Hartree-Fock/density functional calculations for 5-allyl-2,3-dimethoxy-1,4-benzoquinone (UQ) and UQ$^{\bullet-}$ (models for ubiquinone-1 and its radical anion with isoprenyl methyls replaced by hydrogens) show that 50-60% of UQ's inner-sphere electron self-exchange reorganization energy originates from a large change in a single methoxy torsional angle. Implications of torsional tuning of activation energies in photosynthetic, respiratory, and synthetic electron transfer will be discussed. Torsional tuning of activation energies may also be observed in other reactions such as hydrogen atom or group transfers.

394.

DISCRETE STEREODIRECTED REPESENTATIONS IN QUANTUM MECHANICS. **Roger W. Anderson and Vincenzo Aquilanti,** Department of Chemistry,. University of California, Santa Cruz, CA 95064 and Departimento di Chemica, University of Perugia, Perugia, Italy

Discrete representations are introduced for the interaction of an atom with a diatomic molecule. These discrete representations are are characterized by a steric quantum number and they are particularly useful to describe the atom-diatom system when the interaction between the two species is greater than the rotational spacing of the diatomic molecule. Calculations are described for rotational excitation in this representation, and the correspondence of the discrete representation with discrete orthogonal polynomials is shown. Several applications of discrete representations and discrete orthogonal polynomials are presented. One important application is the summation of partial waves that requires calculation at only a few such waves.

395.

TRACKS OF A SYMMETRIC TOP MOLECULE IN HEXAPOLE ELECTRIC FIELDS. **Roger W. Anderson** Department of Chemistry,. University of California, Santa Cruz, CA 95064

In recent years high resolution experimental focusing curves have been determined for symmetric top molecules that have traversed hexapole fields. However these focusing curves are often inconsistent with linear stark effect focusing. This work compares the first order stark focusing with the focusing expected for an exact treatment of the interaction of the electric field and the polar molecule. Some of the symmetric top states have less focusing with the exact interaction and others less. This work further presents the effect of nuclear quadropole-electric field gradient interaction on the focusing. The result is that one of the MF levels will focus as if it were a pure |JKM> state, and the others have anharmonic focusing. The final part of the work is the effect of approximating a hexapole field with a device constructed of cylindrical rods.

396. A SPECTROSCOPIC STUDY OF EPIDERMAL TRANS- AND CIS-UROCANIC ACID
K.M. Hanson and J.D. Simon, Department of Chemistry and Biochemistry, University of California, San Diego, La Jolla, CA 92093-0341

Because accumulation of cis-urocanic acid (c-UA) in the skin has been implicated as a mediator for skin cancer, a detailed spectroscopic investigation of both UA isomers is underway to elucidate the nature of these chromophores' wavelength-dependent photochemistry under physiological conditions. Specifically, we have recorded and analyzed data from fluorescence, photoacoustic calorimetry, singlet-oxygen emission, and time-resolved femtosecond laser measurements as a function of excitation wavelength, pH, and isomer. The data confirm that the isomerization of t-UA to c-UA dominates at 308 nm, yet t-UA intersystem crosses to a long-lived electronically excited triplet state at 266 nm (the absorption maximum) and directly absorbs into a triplet state at 355 nm (the tail of the absorption spectrum) where singlet oxygen is found to be produced. The wavelength-dependent photochemistry of both isomers will be compared and explained to result from the presence of multiple electronic states under the structureless absorption spectra.

397.

APPLICATION OF PERSISTENT INFRARED HOLE BURNING TO BIOLOGICALLY RELEVANT MATERIALS: AMINO ACID SALTS. **Hung-Wen Li,** Gu-Sheng Yu, Herbert L. Strauss, Department of Chemistry, University of California, Berkeley, CA 94720-1460

Persistent Infrared hole burning (PIRHB) has been successfully used to investigate NH_3D^+ ions doped in inorganic salts, organic salts and polymer matrix at low temperatures. We have now extended the application of PIRHB to biological-related systems such as fatty acid ammonium salts and amino acid hydrochlorides. Here, we present PIRHB results on glycine-N^{14} hydrochloride, glycine-N^{15} hydrochloride and L-alanine hydrochloride, each with one deuterium on the amino group. At low temperatures, infrared radiation drives the rotation of the covalent bonded -NH_2D^+ group and produces persistent holes, satellite holes and antiholes. The hole burning spectra, the difference between the absorption spectra before and after IR radiation, provide information on the local environment and the vibrational mode coupling of the -NH_2D^+ group. The assignments of the bands and the specific effects of the local environment will be discussed.

398. CHANGES IN THE SECONDARY STRUCTURE OF THE NADP-LINKED OXIDOREDUCTASE OF *THERMOANAEROBIUM BROCKII*: RESULTS FROM A RAMAN INVESTIGATION OF THE EFFECTS OF ELEVATED TEMPERATURE AND THIOL-BLOCKING AGENTS. Z. Ouyang and P.W. Jagodzinski, Department of Chemistry, West Virginia University, Morgantown, West Virginia, 26506

The activity of NADP-linked alcohol-aldehyde/ketone oxidoreductase from the thermophilic bacterium *Thermoanaerobium brockii* (TADH) is known to be remarkably resistant to the effects of heat. It also possesses some properties that are distinctly different from other dehydrogenases. We have collected Raman spectra in order to monitor the secondary structure changes that occur as a function of temperature with particular attention to the changes that might occur near 50°C. These experiments were conducted in the presence and absence of the thiol-protecting agent dithiothreitol. We have also monitored the conformational changes that occur as the result of addition of the thiol-blocking agent p-chloromercuribenzoate. These results will be compared to the heat denaturation results and all results obtained will be compared to that of other thermophilic enzymes.

399.
CHARGE RESONANCE EFFECT ON HETERODIMER ABSORPTION IN BACTERIA PHOTOSYNTHETIC REACTION CENTER. **Huilin Zhou** Steven G. Boxer, Dept. of Chemistry, Stanford University, Stanford, CA 94305-5080

Low temperature absorption spectra of the heterodimer special pair in heterodimer/hydrogen-bond mutants of Rb. sphaeroides were studied, which show two broad bands with interesting lineshape variations among different mutants. It is postulated that this results from a charge resonance interaction between the lower exciton and an intradimer charge transfer (CT) state. A theory was developed. In the weak coupling limit, a Lorenzian lineshape is expected whose linewidth corresponds to the lifetime of exciton state "decaying" into the CT state. In the strong coupling limit, the lineshape is predicted to consist of two narrow bands. In the intermediate coupling limit, two broad bands are expected. Comparison between theory and experiment is given, and we conclude that charge resonance interaction of heterodimer belongs to the intermediate coupling limit and only one intradimer CT state, Bchl+Bphe-, is primarily involved.

400.
CO PHOTOLYSIS AND REBINDING DYNAMICS IN WILD TYPE AND MUTANT CYTOCHROME OXIDASE FROM RHODOBACTER SPHAEROIDES. **J.P.M. Schelvis,** Y. Hilmi, S. Ferguson-Miller, and G.T. Babcock, Departments of Chemistry and Biochemistry, Michigan State University, East Lansing, Michigan 48824

Picosecond and cw resonance Raman, and transient absorption spectroscopy are used to study heme a3 and its ligands before and after CO photolysis in cytochrome oxidase from Rb. sphaeroides WT and its H333N mutant. H333 is one of three histidines that ligates to CuB. This histidine has been proposed to be involved in proton translocation and in regulating the binding of small exogenous ligands to heme a3. We will discuss the effect of the mutation on CO photolysis and (re)binding of CO to heme a3, and the role of H333 and CuB in regulating the accessibility to heme a3 for small molecules.

401.

COMPUTER SIMULATION STUDIES OF CTG TRIPLET REPEAT SEQUENCES: Joshua Lynch, Department of Physics, University of Maine Orono, ME 04469 and Jayendran. C Rasaiah, Department of Chemistry, University of Maine, Orono, ME, 04469.

Long segments of CTG trinucleotide repeats in human DNA have recently been correlated with a class of neurological diseases which includes myotonic dystrophy, fragile-X syndrome, and Kenndy's disease. These diseases are characterized by genetic anticipation, or the tendency for symptoms to increase in severity and appear at earlier ages in consecutive generations. This is thought to arise from replication errors caused by unusual conformations of CTG repeat segments. We have studied the conformation, elasticity, and mobility of short segments of CTG repeats in 0.5 M sodium chloride solution with molecular dynamics simulations. The simulations are carried out in the microcanonical ensemble using an all-atom force field with CHARMM parameters. The TIPS3 water model is used to simulate a molecular solvent. Electrostatic interactions are calculated by Ewald summation and the equations of motion integrated using a Verlet algorithm in conjunction with SHAKE constrained dynamics to maintain bond lengths. The simulation of CTG repeat sequence is compared with a control system containing CAG triplet repeats to determine possible differences in the conformation and elasticity of the two sequences.

402. DETERMINATION OF MICRO ACIDITY CONSTANTS AND RESOLUTION OF TAUTOMERIC SPECIES SPECTRA VIA TEMPERATURE VARIANT SPECTROTITRIMETRY Joseph C. D'Angelo and Timothy W. Collette, U.S. Environmental Protection Agnecy, 960 College Station Road, Athens, Georgia 30605

The chemical and biological reactivity of proteins and other compounds vary according to the precise location of monopolar charges residing upon them. We have recently devised an accurate and generally applicable method for the measurement of micro equilibrium constants as in prototropic tautomerism and zwitterionic species ratios. Such information should be useful for, e.g., elucidating the energetics of enzyme-substrate interactions. Concomitant with this development is the ability to resolve the spectra of such tautomers, which enables their qualitative identification. We will provide mathematical and experimental proof of method efficacy and resolved tautomeric species spectra using various spectroscopies. We will compare our results with those obtained using previous methods, e.g., isothermal nmr, and will rationalize any observed differences.

403.

DNMR, X-RAY, AND AB INITIO STUDIES OF THIOSEMICARBAZIDE AND 4,4-DIMETHYLTHIOSEMICARBAZIDE **Sam M. Mazhari** Ali Jabalameli, Andrew Nowek and Richard H. Sullivan Department of Chemistry, Jackson State University, City Jackson, State MS, Zip 39217

Abstract: Thiosemicarbazide and its derivatives (backbone = N-N-CS-N) have long been the subject of much study. The interest in these compounds stems, mainly, from the biological activity that they exhibit. Their applications as anti-microbial, anti-fungal, anti-cancer agents, and other pharmacological uses as intermediates for synthesis of valuable drugs has been effective. A specific property associated with these compounds is their ostheolathrismic effect, in which activity is greatly enhanced by alkylation at the 4-position. Thiosemicarbazide (TSC) and 4,4-Dimethylthiosemicarbazide (DMTSC) were investigated using dynamic proton NMR spectroscopy. The study illustrated only one stable conformation for each compound over the temperature range of study. Furthermore, the barriers to internal rotation about the amide bonds

404.

HELIX-FORMING PEPTIDES: A MODE-SPECIFIC ANALYSIS OF CHAIN LENGTH EFFECTS AND THERMAL DENATURATION USING 2D FT-IR. **Darla K. Graff,** Belinda Pastrana-Rios, Sergei Yu. Venyaminov, Franklyn G. Prendergast, Mayo Clinic and Foundation, 200 First Street SW, Rochester, MN 55905.

Several helix-forming peptides of the form, Ac-W(EAAAR)nA-NH2, where n = 1, 3, 5 and 7, were synthesized and their thermal denaturation (-4 to 94 °C) monitored in D2O using FT-IR spectroscopy. Because the amide-I' region of the IR spectra is congested, composed of features derived from conformation of the peptide backbone and from amino acid side chains, unambiguous resolution of peaks is difficult. Two-dimensional (2D) correlation analysis was coupled with a peak-fitting procedure to deconvolute and faciliate a physical interpretation of the spectra. The results of the fitting and 2D analysis, which show normal-mode frequency shifts as well as bandwidth and intensity changes, are discussed in terms of the effect of both chain length and thermal variation on peptide secondary structure. (Supported by GM34847 and DK-07198.)

405.

HYPERICIN, HYPOCRELLIN AND MODEL COMPOUNDS: PRIMARY PHOTOPROCESSES OF LIGHT-INDUCED ANTIVIRAL AGENTS. **D. S. English,** K. Das, J. W. Petrich, Department of Chemistry, Iowa State University, Ames, Iowa 50011

The excited-state photophysics of the natural products hypericin and hypocrellin are compared with their corresponding synthetically prepared methoxy analogs through the use of ultrafast transient absorbance measurements. The results are consistent with both the interpretation of the primary photoprocess in hypericin and hypocrellin as that of excited-state intramolecular proton or atom transfer and with the presence of a ground state heterogeneity in hypericin. A unifying picture of the hypericin and hypocrellin photophysics is presented.

406. MYOGLOBIN AND MYOHEMERYTHRIN SOLVOLYTIC AUTOXIDATION MECHANISM DIFFERENCES. Christopher R. Lloyd, Alicia Denison, Edward M. Eyring, Department of Chemistry, University of Utah, Salt Lake City, UT 84112

The normal physiological function of both myoglobin and myohemerythrin is the reversible binding of O_2 in the muscle cells; this can occur only with the reduced (deoxy) forms of these proteins. O_2 binding is accompanied by a concurrent electron transfer; oxygen is bound as superoxide in myoglobin and peroxide in myohemerythrin. These oxygenated proteins also undergo a much slower autoxidation reaction in which the bound oxygen is lost as superoxide or peroxide. Autoxidation leaves the protein oxidized and inactive. In both cases, autoxidation is promoted by the solvent (H_2O). The solvolytic autoxidation mechanisms of myoglobin and myohemerythrin have been elucidated with high-pressure kinetic techniques and the consequences of site-directed mutagenesis and synthesis of active site model compounds will be discussed.

407. PHOTOINDUCED ELECTRON TRANSFER IN SYNTHETIC SELF-ASSEMBLING POLYPEPTIDE SYSTEMS. Guilford Jones, II,* Valentine I. Vullev, Department of Chemistry, Boston University, 590 Commonwealth Avenue, Boston, MA 02215

Investigation of electron transfer in synthetic protein environments is relevant to achieving a better understanding of the mechanism of natural photosynthesis. A significant part of the effort in *de novo* protein synthesis has been devoted to the design and the creation of helix "bundles" - complexes of amphipathic peptides that aggregate as dimer, trimer or tetramer arrays. Here we describe the design, preparation and structural characterization of a photoactive helical peptide dimer (TT1) that has an anionic dye, eosin Y, covalently attached to its N terminal and a tryptophan, serving as an electron donor located in the middle of the chain. In aqueous media aggregation of TT1 is observed; upon addition of a polycationic electrolyte, poly-L-lysine (PLL), the eosin moieties dimerize and on photolysis participate in intrapeptide electron transfer. Auxiliary electron acceptors, dinitrobenzamide groups attached to PLL were also shown to engage in photoinduced electron transfer. The photophysical properties of the aggregates were investigated with laser flash photolysis. The dye dimers were excited selectively at 532 nm with 7 ns laser pulses and the transient kinetics monitored in the nanosecond and microsecond time periods.

408. PYRIDINYL AND KETYL RADICALS OF VITAMIN B6 ON MICELLAR AND LIPOSOMAL SURFACES. John W. Ledbetter and Stephanie Schaertel, Department of Biochemistry and Molecular Biology, Medical University of South Carolina, Charleston, SC 29425 and Department of Chemistry, Grand Valley State University, Allendale, MI 49401.

Excitation by 337-nm laser light populates both the $n\pi^*$ and $\pi\pi^*$ triplet states of the aldehyde form of vitamin B6, pyridoxal-5'-phosphate, in neutral aqueous solutions. Both of these states can abstract a hydrogen from a suitable reductant to form ketyl and 1-hydropyridinyl radicals, respectively. The radicals were observed in absorption in the 400-nm region. The reactivities are significantly different, with the $\pi\pi^*$ state and pyridinyl radical being the least.

Quenching of the ketyl radical lifetime was determined for the reaction with several antioxidants and in the presence of positive and negative micelles and liposomes. Second order rate constants increased in the order ascorbate<melatonin<catechin<NADH<CoQo<Trolox, the last of the which yielded 2.4×10^9 $M^{-1}s^{-1}$ in buffer at pH 7. The micellar charge exhibited a strong effect on the radical reactivity. Negative micelles retarded the reaction with antioxidants and positive micelles accelerated it. Liposomes of dipalmitolylphosphatidylcholine further enhanced the reaction.

409. SPECTROSCOPIC STUDIES OF MELANIN. Susan E. Forest and John D. Simon, Department of Chemistry and Biochemistry, University of California San Diego, La Jolla, CA 92093-0341

It is generally accepted that melanin is a photoprotective molecule that shields the skin against ultraviolet and visible radiation. But melanin also generates active oxygen species, such as superoxide, hydroxyl radicals, and hydrogen peroxide, which are known to damage DNA. Published electron spin resonance experiments suggest the involvement of an excited triplet state in melanin upon irradiation. In constrast, non-radiative studies report that all the energy deposited into melanin upon excitation is released in the form of heat. To understand these results, we have carried out spectroscopic studies of melanin under physiological conditions in order to develop a comprehensive model for melanin photoreactivity. Specifically, we have found that the photoacoustic calorimetry of melanin is wavelength-dependent. These results can be important in understanding the photoprotective role and damaging effects of this ubiquitous biopolymer pigment.

410. THERMODYNAMIC APPROACH TO THE UNFOLDING OF SEVEN MUTANT FORMS OF PHAGE T4 LYSOZYME, P.W. Chun, Dept. of Biochem. and Mol. Biology, University of Florida, Gainesville, FL 32610-0245

The Planck-Benzinger thermal work function $\Delta W(T)$ represents the strictly thermal components of any intra- or intermolecular bonding term in a system, that is, energy other than the inherent difference of the 0 K portion of the interaction energy. The latter, the temperature-invariant enthalpy, is the only energy term at absolute zero Kelvin. The magnitude of $\Delta H(T_o)$ or $\Delta H°(T_o)$, the temperature-invariant enthalpy, is determined by the type of macromolecular interaction taking place under experimental conditions, and thus, this thermodynamic function should be particularly applicable to studies involving a site-directed mutagenic approach to the examination of structure-function problems in protein. The temperature-invariant enthalpy has been evaluated for the unfolding of T4 lysozyme wild-type T157, and various mutants in which the threonine residue at position 157 has been replaced by one of seven different residues—R, A, N, I, V, E or L. In wild-type (WT) and T157R, $\Delta H(T_o)$ is about 15 kcal mol^{-1}. In T157A and T157N, this value is 14.5 kcal mol^{-1}; in T157I and T157V, it is 13.6 kcal mol^{-1}. The mutants T157E and T157L have a temperature-invariant enthalpy of about 12.5 kcal mol^{1}. Only one mutant, T157R(Thr → Arg) is as stable as the wild-type. In this one mutant, $<T_m>$ is raised to 345 K, while the remaining seven, $<T_m>$ is 330 K. As we have previously reported,[2] in the temperature-sensitive mutant R96H (Arg → His), $<T_m>$ decreases from 315 to 295 K when pH is lowered from 3.0 to 2.0. This suggests that arginine plays an important role in stabilizing the mutant protein. (Supported by a Faculty Development Award, University of Florida)

411. THERMODYNAMIC APPROACH TO THE UNFOLDING OF THE LYSOZYME PHAGE T4 WILD-TYPE R96 AND TEMPERATURE-SENSITIVE MUTANT R96H (ARG → HIS), P.W. Chun, Dept. of Biochem. and Mol. Biology, University of Florida, Gainesville, FL 32610-0245

The Planck-Benzinger thermal work function $\Delta W(T)$ represents the strictly thermal components of any intra- or intermolecular bonding term in a system, that is, energy other than the inherent difference of the 0 K portion of the interaction energy. The latter, the temperature-invariant enthalpy, is the only energy term at absolute zero Kelvin. The magnitude of $\Delta H(T_o)$ or ΔH_o, the temperature-invariant enthalpy, is determined by the type of macromolecular interaction taking place under experimental conditions, and thus, this thermodynamic function should be particularly applicable to studies involving a site-directed mutagenic approach to the examination of structure-function problems in protein. The temperature-invariant enthalpy, $\Delta H(T_o)$, for wild-type R96 and temperature-sensitive mutant R96H of the phage T4 lysozyme unfolding has been evaluated at two different pH values. The $\Delta H(T_o)$ value for R96 was 12.32 kcal mol^{-1} at pH 3.0, and 7.35 kcal mol^{-1} at pH 2.0. For the temperature-sensitive mutant R96H at pH 3.0, $\Delta H(T_o) = 7.61$ kcal mol^{-1}. At pH 2.0, $\Delta H(T_o) = 2.4$ kcal mol^{-1} for R96H. This difference in $\Delta H(T_o)$ implies that a structural alteration in the temperature-sensitive mutant R96H makes this form more accessible to solvent or solvent additives. With decreasing pH, the melting temperature $<T_m>$ decreases, as does the stability of the molecule, in both the wild-type R96 and mutant R96H. At both pH 2.0 and 3.0 for the mutant R96H in which arginine is replaced by histidine (a single amino acid mutation), there is a reduction in the temperature-invariant enthalpy of 5 kcal mol^{-1} from that of the wild-type R96. (Supported by a Faculty Development Award, University of Florida)

412. Adsorption of Cyanopyridines on Copper Colloids: Effects on the C≡N Stretching Normal Mode, Candace M. Coyle and Paul W. Jagodzinski, Department of Chemistry, West Virginia University, Morgantown, West Virginia 26505

Surface-Enhanced Raman (SERS) and Ultraviolet-Visible (UV-VIS) spectra of 2-, 3-, 4-cyanopyridines (2-, 3-, 4-CP) adsorbed on copper colloids have been obtained and compared with previous studies of these molecules on silver and copper electrodes. The copper colloid absorption band at 566 nm undergoes a shift to 571 nm upon adsorption of the 4-CP in the observed UV-VIS spectrum. Excitation at 647.1 nm leads to surface-enhanced Raman spectra of the adsorbed isomeric CP's. Prior investigations of cyanopyridines on silver and copper electrodes indicate a surface enhancement of the -C≡N stretching normal mode that is dependent upon the position of the -C≡N substitutent on the aromatic ring. The SERS collected for the three CP's adsorbed onto the surface of a copper colloid show a complete disappearance of the -C≡N stretching normal mode. Isotopic species of the CP's have been prepared in order to confirm the mechanics of binding of the CP's to the copper surface. In addition, these CP's will be adsorbed on copper electrodes and silver colloids in order to extend the comparison with silver surfaces.

413.

CHARGE TRANSFER EXCITATIONS AND SURFACE ENHANCED RAMAN SCATTERING: DEPENDENCE UPON CRYSTAL FACE. Patanjali Kambhampati C.M. Child, Michelle C. Foster, Alan Campion, Department of Chemistry and Biochemistry, The University of Texas at Austin, Austin, Tx 78712

The surface-enhanced Raman spectra and electron energy loss spectra (EELS) of pyromellitic dianhydride (PMDA) on Cu(100) and Cu(111) are reported. The Raman spectra are sensitive to surface structure and to the incident laser polarization but in different ways. These observations are shown to be inconsistent with a simple classical picture which considers only the interaction of the surface electromagnetic fields with the anisotropic molecular polarizability. The EEL spectra are also sensitive to surface structure. These observations are shown to be the result of variations in the electronic structure of the excited state of the adsorbate - substrate complex.

414. ELEMENTAL COMPOSITION OF METALLIC FILMS DEPOSITED BY MOCVD FROM Mo(CO)$_6$. J. Ure, T. Clemes, J. Bond, P. Sharma, M. Coughlan and K. A. Singmaster, Department of Chemistry, San Jose State University, San Jose, California 95192-0101, and F. A. Houle, IBM Research Division, Almaden Research Center, 650 Harry Road, San Jose, California, 95120-6099

The composition of films deposited by large area resistive heating of Mo(CO)$_6$ has been investigated using Scanning Auger Electron Spectroscopy. Pure metal deposits were obtained for temperatures above 450 C. Below 450 C the films contained significant amounts of oxygen and carbon.

This differs significantly from what had previously been observed for films deposited by localized laser heating of the substrate, where pure metallic films were obtained at temperatures below 300 C. The data indicate that the coupling between mass transport and surface chemistry plays an important role even under mass transport limited kinetics. Mechanistic aspects of the chemistry will be discussed. This work is supported by NSF and NIH-MARC Program.

415. **ELLIPSOMETRIC PROBES OF MONOLAYER FILMS.** Susan E. Ross and Joyce A. Guest, Department of Chemistry, University of Cincinnati, Cincinnati, OH 45221-0172.

We have constructed a sensitive photoelastic modulator-based ellipsometer for use in examining molecular adsorption at liquid and solid surfaces. The ultraviolet laser source for this system increases the refractive index contrast for common chromophores adsorbed onto aqueous and silica surfaces significantly over the contrast obtainable using wavelengths far from electronic resonance of either the adsorbate or the substrate. The detection sensitivity for extremely thin films is therefore enhanced. Demonstration of this method's effectiveness will be presented.

416.

LEED STUDY OF SMALL MOLECULES ADSORBED ON THIN EPITAXIAL IRON OXIDE LAYERS. **F. Reniers,** M. Van Hove, G. Somorjai, Lawrence Berkeley National Laboratory, Materials Science Division, MS 66-200, 1 Cyclotron Road, Berkeley, California, 94720

1 monolayer FeO and 8 monolayers Fe_3O_4 films were formed by UHV deposition of iron on a Pt (111) single crystal followed by oxidation at 800 K. The composition of the films was determined by AES and their structure by LEED. Intensity-Voltage curves were recorded. CO, H2O, Methanol, ethylene and benzene were then adsorbed on the oxide films at temperatures between 130 K and 300 K. The adsorbed layers were characterized by AES, TPD and LEED.

417.

SCANNING TUNNELING MICROSCOPY STUDIES OF THE ORIENTATION AND CONFORMATION OF DOCOSANE DISULFIDE PHYSISORBED ON GRAPHITE AND MOLYBDENUM DISULFIDE. **Karen J. C. Muyskens(1),** Donna M. Cyr(2), George W. Flynn(2), Andrew Black(3), and George M. Whitesides(3), (1)Chemistry Department, Calvin College, Grand Rapids, MI 49546, (2)Chemistry Department, Columbia University, New York, NY 10027, and (3)Chemistry Department, Harvard University, Cambridge, MA 02138.

Films of docosane disulfide physisorbed on graphite and molybdenum disulfide have been studied at the liquid-solid interface using scanning tunneling microscopy. The images show that docosane disulfide adopts a linear conformation on molybdenum disulfide (CCSS dihedral angle of 60°) and a V-shaped conformation on graphite (180° dihedral angle). In addition, STM images of docosane disulfide on graphite reveal submolecular features. The spot pattern suggests that one of the hydrocarbon side chains is oriented with its hydrocarbon backbone parallel to the graphite surface while the other is tilted with respect to the surface.

418.

SURFACE 29SI NMR BY SPINOE-CP. **T. Pietrass** R. Seydoux, A. Pines, Dept. of Chemistry, University of California, Berkeley, CA 94720, and New Mexico Tech, Socorro, NM 87801.

Optically polarized xenon has been used for the study of surfaces with direct 129Xe NMR detection, cross polarization to surface protons and nuclear Overhauser enhancement of surface nuclei (1H and 13C). Cross polarization and the nuclear Overhauser effect (NOE) both rely on dipolar couplings between the xenon and surface spins. Favorable polarization transfer conditions prevail when studying proton spins. In this work, the 129Xe/1H NOE was exploited to observe other nuclei via Hartmann-Hahn cross polarization. Highly spin polarized xenon was adsorbed onto the precooled sample (Aerosil 300). Surface proton spins became spin polarized during contact with the adsorbed xenon via the nuclear Overhauser effect. The 29Si NMR was then acquired in single scans using a 1H/29Si cross polarization pulse sequence, producing enhanced signals. The enhancement factor of the 29Si signal was comparable to that achieved for proton spins via the NOE, indicating negligible losses in the second polarization transfer step.

419.
The Chemisorption of Polyimide Precursors and Related Molecules on Si(100)-2x1

Talal Alkunshalie, T. Bitzer, N. V. Richardson

IRC in Surface Science, Univ. of Liverpool, Liverpool, Merseyside, L69 3BX

There is considerable interest in developing the reactive growth of polyimide films on metal and semiconductor surfaces. In this study, we report the adsorption of the polyimide precursors pyromellitic dianhydride(PMDA) and phenylene diamine (PDA) on Si(100)-2x1. For the dosage of PMDA on Si, a doser was specifically designed to minimize water contamination.
LEED images of Si(100)-2x1 exposed to the precursors indicate that the Si substrate remains reconstructed upon adsorption. The vibrational analysis of the conditioned surfaces by HREELS (High Resolution Electron Energy Loss Spectroscopy), shows that Si-O-C and Si-N-C linkages are formed upon the chemisorption of PMDA and PDA respectively. The chemisorption process and the coordination of the chemisorbed species will be discussed. We also present first results on the reactive growth of a obigimide film on Si(100)-2x1.

420. A COMPUTATIONAL INVESTIGATION OF ALCOHOLS AND ALDEHYDES ON Rh(111) AND Pd(111). Hiram W. Blanken and Kimberly A. Lawler, Department of Chemistry, California State University, Fresno, Fresno, CA 93740.

There is experimental evidence that alcohols undergo different decarbonylation pathways on Rh(111) and Pd(111) surfaces. Aldehydes, however, undergo identical degradation pathways on Pd and Rh. We have undertaken an approximate molecular orbital computational investigation of the decarbonylation reaction. The interaction of a methoxy radical with the Pd(111) and Rh(111) surfaces was explored as a model for an early step in alcohol decarbonylation, where the differences between the two surfaces begin to become apparent. On the Rh(111) surface the methoxy radical showed a strong preference for the on-top site over a range of distances, however the interaction of methoxy radical with the Pd(111) surface showed no such preference. Formaldehyde showed weak stabilization with the Pd(111) surface, however the interaction of formaldehyde with Rh(111) resulted in no net stabilization. The trends are consistent with some of the experimentally observed differences in the decarbonylation mechanisms on the two surfaces.

421. ACETYLENE CYCLIZATION TO BENZENE ON Pd/W(211): RELEVANCE TO NANOSCALE PYRAMIDAL FACETS, I. M. Abdelrehim, K. Pelhos, and T. E. Madey, Laboratory for Surface Modification, Dept. of Physics and Astronomy, Rutgers University, Piscataway, New Jersey, 08855-0849.

Pyramidal structures with facets having {211} orientation and nanoscale dimensions are formed when a Pd-covered W(111) surface is heated above ~750K; the structures are identified by scanning tunneling microscopy (STM). As part of a program to study structure-reactivity relationships in the morphologically-unstable Pd/W system, we are examining a known structure sensitive reaction, acetylene tricyclization to benzene. We find using temperature programmed desorption (TPD) that benzene forms reactively from acetylene on Pd/W(211). Whereas clean W(211) reveals little or no benzene formation, various Pd coverages ranging from submonolayer to multilayers catalyze benzene formation. A single TPD peak at 460K is observed for the submonolayer and monolayer Pd coverages. In the case of an unannealed, as-deposited multilayer (~5 ML) of Pd, benzene evolution is not observed. After annealing a 5 ML Pd film to 700K, acetylene adsorption leads to benzene desorption peaks at 190K and 330K, as well as 460K. The appearance of three desorption states is consistent with the coexistence of monolayer Pd/W and 3 dimensional crystalline Pd nanoclusters; the low temperature peaks are associated with the Pd clusters. In addition, self-poisoning by residual carbon is observed to decrease the benzene yield as well as to shift the 460K peak to lower temperatures. Work in progress focuses on the reactivity of planar and faceted Pd/W(111).

422. COLLISIONS AND REACTIONS OF ORGANIC MOLECULES WITH SULFURIC ACID. K. M. Fiehrer, M. D. Antman, J. K. Klassen, and G. M. Nathanson, Department of Chemistry, University of Wisconsin, Madison, WI, 53706.

We will report on the ways in which organic molecules scatter from, accommodate on, and react with sulfuric acid. These molecules include olefins, alcohols, ethers, aldehydes, and carboxylic acids. Using molecular beam time-of-flight and absolute uptake measurements, we are investigating how scattering and trapping compete at high and low impact energies and at grazing and perpendicular approach directions. In particular, we hope to describe experiments to determine the roles of scattering, transient solvation, and protonation in collisions of these gases with 98.8 wt% sulfuric acid and with the dilute, supercooled acid near 70 wt% and -60 °C.

423. DEHYDROGENATION OF C6 HYDROCARBONS TO BENZENE ON PT(111) AT HIGH PRESSURES INVESTIGATED WITH INFRARED-VISIBLE SUM FREQUENCY GENERATION. **Xingcai Su** Ron Y. Shen, G. A. Somorjai

Abstract The dehydrogenation of C6 hydrocarbons (C_6H_{14}, C_6H_{12}, C_6H_{10}, C_6H_8) to benzene over platinum single crystal model catalyst was carried out under both ultra high vacuum (UHV) and high pressure conditions. The surface species were monitored with infrared-visible sum frequency generation (SFG). Being a surface specific vibrational spectroscopy, SFG can be used to identify surface intermediate during high pressure catalysis. SFG spectra revealed that 1, 3-cyclohexadiene converted to 1, 4-cyclohexadiene on Pt(111) between 200 K~270 K. Both 1, 3- and 1, 4-dienes dehydrogenated to benzene above 300 K. UHV studies indicated that cyclohexane and cyclohexene dehydrogenated to benzene through a C_6H_9 intermediate. However, at high pressure, this conversion happened via a 1, 4-diene intermediate. Primary results indicate that 1, 4-diene is an important intermediate in aromatization reaction.

424.

ENERGY TRANSFER FROM SELF-TRAPPED EXCITONS OF NACL TO PHYSISORBED CO. **S.K. Dunn** T.L. Swiger and C.L. Shofner, Department of Chemistry, Centre College, Danville, Kentucky 40422

The photophysical and photochemical behavior of CO on NaCl crystallites has been studied. Absorption of far-UV radiation results in the production of self-trapped excitons in the salt substrate. Subsequent electronic to vibrational energy transfer produces vibrationally hot carbon monoxide molecules that react to form carbon dioxide. The photochemical rate of production of carbon dioxide shows neither coverage nor temperature dependence in the range from 80-120K.

425.

IN SITU INFRARED REFLECTION-ABSORPTION SPECTROSCOPY (IRAS) STUDY OF SELF-SUSTAINED OSCILLATIONS IN THE PT(100)/NO + CO SYSTEM. **Hugh H. Richardson** Noel P. Magtoto, Department of Chemistry and Condensed Matter and Surface Science Program, Ohio University, Athens, Ohio 45701

We have observed and characterized self-sustained oscillations during the reaction of NO and CO on Pt(100). These oscillations behave quite differently from the ones observed in earlier investigations in terms of the temperature range, pressure range, extreme sensitivity to variations in the pCO/pNO ratio. IRAS measurements show that NO initially adsorbs molecularly in the upper reaction rate branch, but dissociates completely before the reaction evolves into the lower rate branch. These measurements also show unequivocally that only the CO adlayer and not the NO adlayer nor the coadsorbed state of NO and CO is forming on the surface as the reaction slides back to the lower rate branch. We shall present a model that will rationalize the different events observed during oscillations.

426.

INTERMOLECULAR ELECTRON TRANSFER ON MICELLE SURFACES: THE ROLE OF DIFFUSION AND IMPORTANCE OF SOLVENT EFFECTS. **Kristin Weidemaier** H. L. Tavernier, K. T. Chu, and M. D. Fayer, Department of Chemistry, Stanford University, Stanford, CA 94305

Intermolecular electron transfer on micelle surfaces is strongly influenced by diffusion of the donor and acceptor chromophores. Tiime correlated single photon counting has been used to study forward electron transfer between dimethylaniline and octadecylrhodamine B on the surfaces of two micelles (DTAB and Triton X-100). Analysis of the data requires a detailed theory capable of describing through-solvent electron transfer occurring over a variety of distances. The importance of diffusion is clearly demonstrated. In addition, solvent structure and hydrodynamic effects play a critical role in the dynamics.

427.   **Is the uptake of alcohols by H2O droplets governed by equilibrium solvation dynamics?**
Ramona S. Taylor, Liem X. Dang, and Bruce C. Garret
Environmental Molecular Sciences Laboratory
Pacific Northwest National Laboratory*
Richland, WA 99352

Heterogeneous processes are important components of the earth's atmospheric system. Experiments suggest that the interaction of small gas phase molecules with the liquid /vapor interface of aqueous droplets and their subsequent accommodation into the bulk of droplet is an integral part of these processes. A recently devised, molecular-level model predicts that this mass accommodation process is solely dependent on the rates of solvation and desorbtion of the impinging molecule and that the molecule must cross a large energy barrier before becoming solvated in the bulk water. In this work, we are using molecular dynamics computer simulations to examine this model. Due to the availability of experimental data, the ethanol-water and ethylene glycol-water systems have been chosen as prototypes for the mass accommodation process of non-reacting molecules. The potential of mean force technique is used to explore the equilibrium free energy surface for inserting these alcohols into bulk water from its vapor. These free energy profiles do not correspond to those predicted by the above mentioned model. The non-equilibrium aspects of this process are also explored.

**Pacific Northwest Laboratory is a multiprogram national laboratory operated for the U.S. Department of Energy by Battelle Memorial Institute under Contract DE-AC06-76RLO 1830.*

428. KINETIC STUDY OF COPPER CATALYZED DECOMPOSITION OF TRIPHENYLBORON, DIPHENYLBORONIC ACID AND PHENYLBORONIC ACID IN AQUEOUS ALKALINE SOLUTIONS AT 40 TO 70 °C. C. L. Crawford and R. A. Peterson, Savannah River Technology Center, Westinghouse Savannah River Co., Aiken, SC, 29808.

Kinetics of alkaline solution-phase decomposition reactions of triphenylboron, diphenylboronic acid and phenylboronic acid in the presence of copper have been analyzed by using High Performance Liquid Chromatography (HPLC). The reactivity of each organic component was determined over the statistically designed experimental test condition ranges of 0.5 - 2.5 M OH⁻, 0.1 - 10 ppm added soluble Cu(II) and 40 - 70 °C in air-sealed carbon-steel reaction vessels. Results are discussed in terms of the relative reaction rates of the starting reagents. Activation energies and end-product distributions of phenol and benzene will be presented.

429. SURFACE PHOTOCHEMISTRY AND DYNAMICS OF ETHYL BROMIDE ADSORBED ON GALLIUM ARSENIDE (110). J.E. Moryl, K.A. Khan, D.A. Slater, P.J. Lasky and R.M. Osgood Jr., Columbia Radiation Laboratory, Columbia University, New York, NY 10027

Ethyl bromide adsorbs molecularly on single crystal GaAs surfaces at low temperatures. We have carried out experiments probing the UV photodissociation dynamics of this adsorbate under ultra-high vacuum conditions. Experimental techniques employed include temperature programed desorption (TPD) and angle resolved time-of-flight mass spectrometry. Our results indicate that the photodissociation does not occur by direct photon absorption by the ethyl bromide but through dissociative electron attachment (DEA) of sub-vacuum level electrons produced in the substrate. The time-of-flight measurements reveal that ethyl fragments are ejected into the gas phase via two dynamically distinct mechanisms. One mechanism produces fragments with higher average velocities and an angular distribution that reflects the adsorbate's initial surface orientation, while the other channel appears to reflect partial energy accommodation on the surface. A third channel results in surface bound ethyl fragments. These results will be compared to similar studies of methyl halides on the same surface.

430.

THE ADSORPTION AND OXIDATION OF SULFUR DIOXIDE GAS ON ALKALI HALIDE SURFACES. Gregory A. Rechtsteiner, Jane A. Ganske, Department of Chemistry, Pepperdine University, Malibu, CA 90263.

The oxidation of sulfur dioxide gas to sulfates and sulfuric acid aerosols is a reaction known to be accelerated on the surfaces of condensed phases in the atmosphere. We have examined the oxidation of SO_2 in air on the surfaces of alkali halides (KBr, NaCl and KCl) using diffuse reflectance infrared Fourier transform spectroscopy (DRIFTS), which has allowed interrogation of both physisorbed and chemisorbed species. Results indicate that dry alkali halide particles (<30 μm diameter) reversibly adsorb SO_2 gas at room temperature. Physisorbed SO_2 is effectively removed from the salt surface by gentle heating under vacuum. The kinetics of this desorption process has been followed over the temperature range 25°C-200 °C in an Arrhenius-type fashion. The activation energy for desorption has been measured to be $E_d = 28 \pm 3$ kJ/mol. Physisorbed sulfur dioxide gas reacts with oxygen on the salt surfaces beginning near 175 °C, forming the alkali sulfate. Infrared spectral features identify the sulfate products as bulk bidentate-bound alkali metal sulfates.

431. TWO APPROACHES TO STUDY SURFACE LIGHT-INDUCED DRIFT ARISING FROM ANGULAR-DEPENDENT SCATTERING. Michael A. Vaksman and Fatima Burovich, Department of Chemistry, University of Detroit Mercy, PO Box 19900, Detroit, MI 48219-0900

Surface light-induced drift (SLID) which occurs under velocity-selective excitation due to the state-dependence of the probability of trapping or chemisorption has demonstrated its extreme sensitivity in studying the state-specificity of gas-surface interactions. We show that SLID can be used to obtain the information on the *angular* dependence of the gas-surface scattering. For this purpose one needs to deconvolute the data from the shape of the $\Delta P(\Omega)$ curve, where ΔP is the pressure drop along the cell and Ω is the detuning of the laser frequency from resonance. We present the initial numerical results obtained using two different approaches. In the first approach, gas-surface scattering is described by specific terms inside the kinetic equations, while in the second approach it is accounted for by the boundary conditions. We compare those results and discuss the ways to study the scaling laws for the angular dependence of surface scattering by the SLID technique.

432.

HOST-GUEST EQUILIBRIUM BETWEEN 4-(N,N-DIMETHYLAMINO)BENZONITRILE AND β-CYCLODEXTRIN IN DMSO AND DMF. Caleb E. Feliciano and Edwin Quiñones, University of Puerto Rico, Department of Chemistry, P.O. Box 23346, Río Piedras, PR 00931-3346

The complexity of water as a solvent has prompted studies of molecular recognition in non-aqueous media. In this context, the host-guest equilibrium between DMABN and CD is an ideal system because both species are soluble in dimethyl sulfoxide (DMSO) and in dimethyl formamide (DMF), and studies of these systems in aqueous solution were reported from our laboratory. Fluorimetric and absorption studies indicate the formation of host-guest complexes between CD and DMABN in DMSO and DMF. In DMSO, the values of the equilibrium constant and the $\Delta G°$ at 298 K are 6.0 ± 0.6 M^{-1} and -4.4 ± 0.4 kJ/mol, respectively, and temperature variation studies show that the process is temperature independent in DMSO ($\Delta S \approx 0$). On the other hand, the equilibrium constant in DMF at 298 K equals 7.3 ± 0.7 M^{-1}. ΔH in both media is negative. There is a strong tendency for the formation of a DMABN excimer in DMSO as evidenced by the appearance of a band around 500 nm at concentrations above 5×10^{-7} M, which is decreased upon the addition of CD because the host-guest equilibrium is also established.

433. ON THE PHASE TRANSFER OF SOLID CHEMICAL ELEMENTS (MELTING (T_m) AND CRITICAL (T_c) TEMPERATURES). Sándor Kristyán, Department of Chemistry, Emory University, Atlanta, GA 30322, U.S.A.

In the author's previous studies [J.Phys.Chem. Vol.95 No.2 (1991) 921-932, Surf.Sci.Letters 255 (1991) L562 - L570, Il Nuovo Cimento Vol.15D, N.6 (1993) 815 - 827, Langmuir 1994 Vol.10, 1987-2005] the average number of neighbors of an atom in the surface phase was found to increase by about 5 - 20 % between 0 K and the melting temperature for all solid chemical elements. This increment makes the surface phase of solids "geometrically impossible" to exist at from a particular temperature (T_m), resulting the formation of vacancies in the surface layers and diffusing into the bulk. The latter results in the collapse of crystal structure, beginning with the formation of liquid in the surface layers a few atoms thick, in agreement with recently published experimental studies. As that process continues, the critical point is characterized as the number of neighbors in the liquid bulk is equal to the number of neighbors in the surface phase ("disappearance" of surface). Many simple expressions ($\gamma = \alpha \Delta H'$; $\partial \alpha/\partial T = - R \ln m/\Delta H'$; $\Delta \alpha_{cr} \equiv \alpha(T_m) - \alpha(T = 0 K) = -R T_m \ln m /\Delta H'$; $T_m = |\Delta \alpha_{cr}| \Delta H'/R \ln m = \Delta n_{cr} \Delta H'/z R \ln m$; $\Delta S_{m,calc} = \Delta H_{m,calc}/T_m = R \ln m \approx - (\partial \gamma/\partial T)_{avrg}$; $T_c/T_m = \alpha(T = 0 K)/|\Delta \alpha_{cr}| \approx 3.6$; $T_c/T_m \approx \gamma/\Delta H_m$; $\lambda \approx 2a/b$) are compared to experiments.

434. ORIENTATIONAL ORDER AND CHARGE FRUSTRATION IN MODELS OF THE GLASS TRANSITION. Carlo Carraro, Hyung-June Woo, and David Chandler, Department of Chemistry, University of California, Berkeley, CA 94720.

Geometric frustration has long been understood as underlying the formation of some types of structural glasses. Recently, the Kivelsons and their coworkers have proposed a model based on charge frustration. It successfully predicts many features of the glass transition in quantitative detail. But its connection to geometric frustration is unclear. To address this uncertainty, we present a field theory of frustrated orientational order coupled to local density fluctuations and discuss the way in which the field theory can be mapped onto charge frustrated models. Implications of the analysis to collective effects and slow kinetics will be described.

435. PHASE TRANSITIONS OF POLAR FLUIDS: Seamus F. O'Shea, Department of Chemistry, University of Lethbridge, Alberta, Canada, T1K 3M4, Girija S. Dubey, Department of Physics and Astronomy, Hunter College of the City University of New York, New York, NY 10021 and Jayendran. C Rasaiah, Department of Chemistry, University of Maine, Orono, ME, 04469.

Gibbs ensemble simulations are reported for Lennard-Jones particles with embedded quadrupoles of strength $Q^* = Q/(\epsilon\sigma^5)^{1/2} = 2.0$ where ϵ and σ are the Lennard-Jones parameters. Calculations revealing the effect of the dispersive forces on the liquid-vapor coexistence were carried out by scaling the attractive r^{-6} term in the Lennard-Jones pair potential by a factor λ ranging from 0 to 1. Liquid-vapor coexistence is observed for all values of λ including $\lambda = 0$ for $Q^* = 2.0$, unlike the corresponding dipolar fluid studied by van Leeuwen and B. Smit, et al (Phys. Rev. Lett., 71, 3991 (1993)) which showed no phase transition below $\lambda = 0.35$ when the reduced dipole moment $\mu^* = 2.0$. The critical temperature and pressure show a clear quadratic dependence on λ.

436. SURFACE-INDUCED PHASE TRANSITION OF A CONFINED FLUID. Ka Lum, and David Chandler, Department of Chemistry, University of California at Berkeley, CA 94720

The phase transition between the gas and liquid phases can be strongly affected by the presence of walls. When confined between two drying walls, a liquid close to liquid-gas coexistence becomes metastable and evaporates at small separations. This phenomenon of surface-induced phase transition is investigated with a lattice-gas model in a L×L×D geometry. The effect of the drying walls is introduced by coupling the sites adjacent to the surfaces to a repulsive surface field. Monte Carlo simulations on the system have revealed that the dynamics that leads to the phase transition is very different from what might be imagined from classical nucleation theory. The simulation results can be understood by employing a Landau-Ginzburg form of free energy for the confined system. Within the mean-field approximation, the dependence of the density profiles and their relative stabilities on the phenomenological parameters are analyzed. The fluctuations about the stationary points and the two-point correlation functions are characterized by going beyond the mean-field theory. The path of evaporation in the configurational space are also studied.

437. THERMODYNAMIC PROPERTIES OF LITHIUM VAPOR AT HIGH TEMPERATURES INCLUDING CONTRIBUTIONS FROM IONS AND FIRST EXCITED STATE ATOMS.
Paul M. Holland, RVM Scientific, 722 Camino Cascada, Santa Barbara, CA 93111 and Louis Biolsi, Chemistry Department, Univ. of Missouri-Rolla, Rolla, MO 65401.

High temperature thermodynamic properties of lithium vapor are calculated using statistical mechanics and a method relating thermodynamic properties to the second virial coefficient. In these calculations, contributions are included from atoms in both the ground and first excited states, and ground state ions. Unlike the usual methods employing partition functions, the virial coefficient method depends on interaction potentials which provide a useful description of energies even near the top of the potential well, where the vibrational-rotational energy levels needed for the partition function approach are not readily accessible. The interaction potentials between (^{2}S) lithium ground state atoms, first excited state (^{2}P) atoms, and ground state (^{1}S) lithium ions are obtained by fits of the Hulburt-Hirschfelder potential to *ab initio* results. Standard state heat capacities, entropies, free energies, and enthalpies are then calculated from virial coefficients and their derivatives, using a degeneracy weighted average over the contributing states.

438. THERMODYNAMICS OF REACTIONS INVOLVING PHASES AT DIFFERENT ELECTRIC POTENTIALS,
Robert A. Alberty and Irwin Oppenheim, Department of Chemistry, Massachusetts Institute of Technology, Cambridge, MA 02139

Since the electric work is proportional to the amounts of species, the involvement of electric work terms in the fundamental equation does not change the natural variables of the Gibbs energy G, which are T, P, and amounts of species when there are no chemical or phase equilibria. The electric potentials of phases can be introduced in calculations with the Gibbs energy by including the electric work term $Fz_i\phi_i$ in the definition of the activity a_i of an ion in terms of its chemical potential, but this does not make the electric potential ϕ_i of the phase that the ion is in a natural variable. When it is desirable to make calculations with the electric potentials of phases as natural variables, it is necessary to define a transformed Gibbs energy G' by using a Legendre transform. It is shown that the chemical potential μ_i of ion i in a phase is uniform throughout a system at equilibrium and is equal to its transformed chemical potential μ_i' plus the electric work term $Fz_i\phi_i$. These concepts are applied to a system involving a chemical reaction in which the product is an ion in a phase with a different electric potential from the reactant phase.

439. INCLUSION OF TRIPLE EXCITATIONS IN COUPLED-CLUSTER TREATMENTS OF ELECTRONICALLY EXCITED STATES OF MOLECULES: METHODOLOGY AND APPLICATIONS.
John D.Watts and Rodney J. Bartlett, University of Florida, Quantum Theory Project, P.O. Box 118435, 362 Williamson Hall, Gainesville, Florida 32611

Equation-of-motion coupled-cluster methods are a useful tool for the study of electronically excited states of molecules. We have proposed several approaches for including triple excitations in these methods. The basic methodology will be reviewed, and the capabilities of the methods will be illustrated by several examples, including the study of triplet electronic states.

440. EXPERIMENTAL ASSIGNMENTS AND EVALUATIONS OF RECENT ACETONE AND FORMALDEHYDE EXCITED STATE CALCULATIONS, Ruth McDiarmid and Xing Xing, National Institutes of Health, Bethesda, MD 20892-0510

The experimental and theoretical determinations of the energies of the valence excited states have been, and remain, significant problems in acetone and formaldehyde spectroscopy. In recent resonantly enhanced multiphoton ionization and photoacoustic spectroscopic investigations of acetone frequency, intensity, lifetime, and technique-related anomalies in the experimental spectra of the 3p and 3d Rydberg states have been interpreted as arising from perturbations of these states by the A_1 ($\pi\pi^*$) valence state. The relations between these results and the results of recent calculations on acetone and formaldehyde will be discussed.

441.
COUPLED-CLUSTER CALCULATIONS OF THE EXCITATION ENERGIES OF BENZENE AND THE AZABENZENES. Janet E. Del Bene, Department of Chemistry, Youngstown State University, Youngstown, Ohio 44555 USA

The equation of motion coupled-cluster method with all single and double excitations (EOM-CCSD), and EOM-CCSD with noniterative triple excitations [EOM-CCSD($\widetilde{T}$)], have been employed with the POL1 basis set to determined the electronic excitation energies of benzene and the azabenzenes. Vertical n→π* and π→π* transition energies of benzene, pyridine, 1,4-, 1,3-, and 1,2-diazine, symmetrical triazine, and symmetrical tetrazine are reported, and compared with experimental data. While EOM-CCSD excitation energies reflect the trends in relative excitation energies in this series of molecules, the computed energies are too high, with an average error of 0.35 eV. The inclusion of triples lowers excitation energies and brings them into better agreement with experiment, with an average difference of 0.10 eV. The computed results provide new insights into these excited states, and their experimental assignments.

442.

DEVELOPMENT AND APPLICATION OF SINGLE-REFERENCE PERTURBATION AND COUPLED-CLUSTER THEORIES FOR EXCITED ELECTRONIC STATES. Timothy J. Lee, MS230-3, NASA Ames Research Center, Moffett Field, California 94035-1000

Recent work on the development of single-reference perturbation theories for the study of excited electronic states will be discussed. The utility of these methods will be demonstrated by comparison to linear-response coupled-cluster excitation energies. Results for some halogen molecules of interest in stratospheric chemistry will be presented.

443.
A HIERARCHY OF COUPLED CLUSTER APPROXIMATIONS FOR LARGE SCALE CALCULATIONS OF EXCITATION ENERGIES, TRANSITION MOMENTS AND FREQUENCY DEPENDENT MOLECULAR PROPERTIES. Ove Christiansen, Asger Halkier, Christof Hättig, Trygve Helgaker,* Henrik Koch, Poul Jørgensen and Jeppe Olsen, ** Department of Chemistry, University of Aarhus, Denmark. *Dept. of Chemistry, University of Oslo, Norway, ** Dept. of Chemistry, University of Lund, Sweden

In a recent series of papers we have advanced a hierarchy of coupled cluster models for evaluation of excitation energies, transition moments and frequency dependent molecular properties. The standard CCS and CCSD models have been augmented with iterative CC2 and CC3 models that are approximations to CCSD and CCSDT respectively. From the response functions for these models we have identified expressions for excitation energies, transition moments and frequency dependent molecular properties. Higher levels in this hierarchy is found to give molecular properties of increased accuracy at increased computational cost. The computational scaling of CCS, CC2, CCSD and CC3 is N^4 N^5 N^6 and N^7 where N is the number of orbitals. Benchmark calculations of excitation energies have shown that the error decrease about a factor three at each level in the hierarchy. Excited states of benzene and other molecules have been investigated performing large scale integral direct calculations.

444. ELEMENTARY ATOMIC DIFFUSION PROCESSES AT METAL SURFACES, Karsten W. Jacobsen, Department of Physics, Technical University of Denmark, DK-2800 Lyngby, Denmark

The understanding of mobility at metal surfaces is an important ingredient in a more general understanding of metal growth and of chemical reactions at surfaces. This talk will focus on the calculation of rates of elementary atomic diffusion processes. For simple situations like an atom diffusing over a flat metal surface the location of the transition state is largely determined by symmetry and the barrier for the diffusion process can be calculated within the density functional approach. This will be illustrated with a study of the anisotropy of adatom diffusion on the Cu(110)-surface. For more complicated processes the location of the transition state can be non-trivial and special techniques for finding it are required. A configuration-space-path technique for doing this will be discussed and applied to contact formation between an STM tip and a metal surface. Finally the possibility of correlated "long jumps" at surfaces will be discussed on the basis of molecular dynamics simulations and rate theory.

445. **Bond Breaking and Bond Making in Self-Diffusion and Crystal Growth** — <u>Matthias Scheffler</u>, Byung-Deok Yu, Christian Ratsch, and Paolo Ruggerone
Fritz-Haber-Institut der Max-Planck-Gesellschaft, D-14195 Berlin-Dahlem, Germany

For Pt(100) and Ir(100) experiments revealed that self-diffusion proceeds by atomic exchange. For Al(100) Feibelman had performed a thorough theoretical analysis, showing that exchange diffusion should also occur for this system, actuated by the formation of covalent bonds of the sp valence electrons at the transition state. Why Pt and Ir should behave similar to Al, but the other transition metals should not, remains an open question.

Using density-functional theory we find that the origin of exchange diffusion at the late $5d$ transition metals is different to that at Al. The calculations predict that exchange diffusion should also occur on Au(100) and that two mechanisms play a crucial role: $i)$ the tensile surface stress and $ii)$ the correlation of bond strength and local coordination.

We also discuss the adsorption and diffusion at steps, and in order to determine the dependence of growth phenomena on temperature, deposition rate, and different diffusion processes the ab initio total-energy results are complemented by a kinetic Monte-Carlo simulation.

446. DIFFUSION ON STRAINED AND PATTERNED SURFACES.
Klaus Kern, Institut de Physique Expérimentale, Ecole Polytechnique Fédérale de Lausanne, CH-1015 Lausanne, Switzerland

The diffusion of a single adatom on terraces and across or along step edges are fundamental processes that determine the shape and morphology during epitaxial growth. Recent experiments have revealed that isotropic two-dimensional strain as well as its relief via dislocations have a drastic effect on surface diffusion. For metal-on-metal systems compressive strain was found to decrease and tensile strain to increase activation energies for diffusion. Misfit dislocations, on the other hand, were found to constitute very efficient repulsive barriers for migrating atoms. Both effects can be used to manipulate the growth morphology and to tailor specific surface morphologies on the nanometer-scale.

447. PRODUCT STATE-RESOLVED POLARIZATION OF ELEMENTARY BIMOLECULAR REACTIONS, M. Brouard, The Physical and Theoretical Chemistry Laboratory, South Parks Road, Oxford, OX1 3QZ, U.K.

Product state-resolved differential cross-sections, correlated product pair internal energy releases, and centre-of-mass rotational polarizations have been determined for a number of elementary, photon-initiated reactions, including the chemical activation reactions of O(1D) with

H2 and CH4 and the reactions of hot H atoms with H2O and CO. The studies employ Doppler-resolved Laser induced fluorescence to probe the nascent reaction products under thermal 'bulb' conditions. The results highlight the new stereodynamical insights obtained in studies which employ product state resolution. The data will be compared with the predictions of quasi-classical trajectory calculations, with particular emphasis on the interpretation of rotational polarization effects.

448.
VECTOR CORRELATIONS IN REACTIVE COLLISIONS WITH ORIENTED MOLECULES. H.J. Loesch, Fakultat fur Physik, Universitat Bielefeld, D-33501 Bielefeld, Germany

The introduction of the brute force technique has initiated considerable progress in dynamical stereochemistry. In contrast to the well known hexapole field method it can be used to create axis orientation also in beams of linear and asymmetric top molecules [H.J. Loesch. Ann. Rev. Phys. Chem. 46, 555-594 (1995)]. After a brief description of the method we report results of our recent crossed beam experiments on the reactive scattering of K atoms from oriented Iodobenzene, 2,3 and 4 Iodotoluene and other asymmetric top molecules. They are compared to findings of earlier investigations on K+ICl and K+CH$_3$I. In all cases an extremely strong correlation between the direction of the molecular electric dipole moment and the most probable direction of product velocity has been found. The results can be rationalized to some extent by a simple impulsive reaction model (direct interaction with product repulsion). The remaining discrepancies are used to improve the model.

449.
STERICAL EFFECTS IN BIMOLECULAR REACTIONS OF ATOMIC CARBON WITH UNSATURATED HYDROCARBONS AS STUDIED VIA CROSSED BEAM EXPERIMENTS R.I. Kaiser, D. Stranges, Y.T. Lee, and A.G. Suits, Department of Chemistry, University of California, Berkeley, CA 94720

The reaction of carbon atoms C(3Pj) with C2H2 (acetylene), CH3CCH (methylacetylene), C2H4 (ethene), and CH3CHCH2 (propylene) were investigated under single collision conditions to elucidate sterical effects of the methyl (CH3) group on the chemical dynamics of atomic carbon reactions with unsaturated hydrocarbons. Reaction with C2H2 proceeds via two microchannels thru c-C3H2 and CCHCH intermediates within orbiting limits decomposing to c-C3H + H and after H migration to l-C3H + H. The CH3 group in methylacetylene reduces the cone of acceptance and prevents formation of any cyclic C4H4 intermediate. Here, reaction dynamics are governed via addition to form CCHCCH3 which fragments after rearrangement to methylpropargylene to l-C4H3 + H. In a similar manner, the steric hindrance of the propylenic CH3 group quenches a second, direct microchannel as found in the reaction with ethylene, leading to long lived reaction intermediates decomposing to methylpropargyl (C + C3H6) and propargyl radicals (C + C2H4), respectively.

450. Effect of the Atomic Orbital Alignment on the Chemiluminescent Reaction $Ca(^1P_1)+CH_3X(X=Cl,I)$. Guo-wen Ding, Da-li Xu, and Guo-zhong He, State Key Laboratory of Molecular Reaction Dynamics, Dalian Institute of Chemical Physics, Chinese Academy of Sciences, Dalian, 116023, P.R.China

We investigated the influence of the alignment of laser-excited $Ca(^1P_1)$ reagent on the outcomes of the chemiluminescent reactions $Ca(^1P_1) + CH_3X \rightarrow CaX(A^2\Pi, B^2\Sigma) + CH_3$, where X=Cl,I in a beam-gas apparatus. The results show that:

1. The branching ratio σ_A/σ_B becomes larger for $E \parallel V_{rel}$ than for $E \perp V_{rel}$.

2. The $CaX(A^2\Pi-X^2\Sigma^+)$ transition displays a remarkable alignment dependence while $CaX(B^2\Sigma^+-X^2\Sigma^+)$ shows some alignment preference but with a much smaller value.

3. The alignment dependence for the $Ca(^1P_1) + CH_3Cl$ chemiluminescent reaction is stronger than that for the $Ca(^1P_1) + CH_3I$ reaction.

451.
REAGENT AND PRODUCT POLARIZATION IN THE REACTIONS OF ATOMIC CHLORINE WITH METHANE AND ETHANE, P. Rakitzis, Dept. of Chemistry, Stanford University

We use the core-extraction technique to measure product polarization-dependent differential cross sections and scattering-angle-dependent reagent steric effects.

a) We present polarization-dependent differential cross sections for the DCl (v'=0,J'=1) and HCl (v'=1,J'=1) products from the reactions of Cl with CD_4, C_2D_6 and CH_4 (v3=1). We measure all five of the polarization moments of the differential cross section of the DCl (v'=0,J'=1) product from the Cl + C_2D_6 reaction. We find that the product rotation from the Cl + C_2D_6 and CH_4 (v3 = 1) reactions is aligned in the plane perpendicular to the product separation direction. We explain the measured product polarization in terms of the line-of-centers model with hard sphere scattering and the location of the atom transfer along the reaction coordinate.

b) We present scatterin-angle-dependent steric effects for the HCl (v'=1,J'=1) and HCl (v'=0,J'=5) product from the reaction of Cl with CH_4 (v3=1).

452. A THEORETICAL STUDY OF OH ANGULAR MOMENTUM ALIGNMENT IN THE REACTION $H + CO_2 \rightarrow OH + CO$. Kimberly S. Bradley and <u>George C. Schatz</u>, Department of Chemistry, Northwestern University, Evanston IL 60208-3113

We present the results of a quasiclassical trajectory study of the reaction $H + CO_2 \rightarrow OH + CO$, with emphasis on modelling recent experimental measurements of the product OH rotational angular momentum alignment due to Brouard, Lambert, Rayner and Simons. A new potential surface has been developed for this calculation that describes the entrance and exit channel bottlenecks to reaction in accord with the best estimates of *ab initio* calculations. Our product state distributions, angular distributions and translation distributions are in generally good agreement with experiment. Very little alignment is seen in the OH angular momentum distribution. An analysis of the moments of this distribution indicates that a large number of moments (around 7) are needed to map out the polarization distribution accurately.

453. ORIENTATION AND STATE-SELECTION OF HYDROXYL RADICALS BY ELECTRIC HEXAPOLE FOCUSING: APPLICATIONS IN DYNAMICAL STEREOCHEMISTRY. <u>T. J. Curtiss</u>, T. D. Hain, M. A. Weibel, Department of Chemistry, University of Utah, Salt Lake City, UT 84112

Hydroxyl radicals generated in a corona-discharge free jet expansion have been focused and rotationally state-selected with an electrostatic hexapole. Tuning the hexapole resonance voltage selectively focused into a reactive scattering volume either of two populated $|J \Omega M_J \rangle$ states, the $|3/2\ 3/2\ 3/2\rangle$ state or the $|3/2\ 3/2\ 1/2\rangle$ state. Classical trajectory simulations incorporating effects due to first and second order Stark interactions, spin-orbit mixing, -doubling, and hyperfine interactions successfully reproduced experimental focusing spectra for OH and OD seeded in He and Ar. Theoretical calculations of the field dependent orientational probability distribution functions for the two focused states demonstrate extraordinary experimental control over the collision geometry in bimolecular reactions. Preliminary results of work in progress on the stereochemistry of the $OH + D_2O$ and $OD + H_2$ reaction systems will be reported.

454.
SIZE DEPENDENCE OF STRUCTURAL METASTABILITY IN SEMICONDUCTOR NANOCRYSTALS. **A.P. Alivisatos,** University of California, Berkeley, Department of Chemistry, Berkeley, California, 94720-1460

Nanometer size crystals are investigated extensively because of their size dependent optical and electrical properties. However, such crystals are also of great interest in the study of phase transitions. The conversion of an extended crystal from one structure to another is usually nucleated at a defect, and proceeds by complex kinetics involving multiple nucleation and fragmentation into domains. Defect free nanometer size crystals undergo pressure and temperature induced first order solid phase transitions via single nucleation events, at much higher pressures than the corresponding extended crystal. In this talk the various factors which influence the structural stability of nanometer size crystals will be discussed (size, surface composition), as well as the implications of these results for the preparation of high energy density metastable materials.

455. ELECTRON-HOLE INTERACTIONS IN SILICON NANOCRYSTALS   K. Leung and K. B. Whaley, Department of Chemistry, University of California at Berkeley, Berkeley, CA 94720

We investigate the electron-hole interactions in spherical silicon nanocrystals by incorporating Coulomb, exchange, and spin-orbit couplings into tight-binding models. We study the effect of the electron-hole attraction no the absorption spectra and on the dielectric constant using a real-time propgation technique. Diagonalizing the two particle states colse to the band gap gives exchange splittings in agreement with the effective mass approximation (EMA). The splittings persist in the presence of spin-orbit coupling for nanocrystals of diameter up to 30 Angstroms, suggesting that dark triplet states below the absorption threshold can be the origin of the Stokes shifts and temperature-dependent lifetimes observed in luminescence experiments.

456.   UNIVERSALITY IN FINITE-SIZE EFFECTS IN PERCOLATION. R. M. Ziff, Dept. of Chemical Engineering, Univ. Michigan, Ann Arbor MI 48109-2136

The percolation model relates to clustering, conductivity, and porosity in nanostructured materials. The model itself is also of a very fundamental nature and has been studied quite extensively in the four decades of its existence -- yet it remains to hold many new and surprising results. For small systems, finite-size effects can be very important. We show that, for two-dimensional systems, the number of clusters in a square system of size LxL differs from its bulk value ($L^2*(3*sqrt(3)-5)/2$ for example for a square-bond system) by a universal number ≈ 0.89, which is independent of the lattice and percolation type (periodic boundary conditions). Likewise for an infinite strip of width L with periodic boundary conditions, the number per unit length is larger by a universal 0.361. The number of percolating clusters is also shown to obey universality.

457.

THE BAND EDGE LUMINESCENCE OF CDSE NANOCRYSTALLITES. **M. Kuno** J. K. Lee, B. Dabbousi, F. V. Mikulec, M. G. Bawendi, MIT Department of Chemistry, 77 Massachusetts Avenue, Cambridge, MA, 02139

We study the band edge luminescence of CdSe nanocrystallites (NCs). The origin of this emission is controversial as certain observed properties have been unexpected. For instance, the emission has an unusually long radiative lifetime (~1 microsecond at 10K). This is in contrast to the nanosecond lifetimes seen in bulk CdSe. In addition, a prominent redshift of the luminescence is observed which cannot be explained by simple exciton-phonon coupling. The band edge luminescence has often been attributed to recombination from surface related states. However, recent theories which take into account the electron-hole exchange interaction predict the presence of a "dark" or optically inactive exciton ground state. We perform photoluminescence and fluorescence line narrowing experiments on a series of surface modified CdSe NCs and explain the band edge luminescence in the context of a dark exciton theory.

458. STRUCTURAL FORMS AND THERMAL PROPERTIES OF NICKEL-ALUMINUM ALLOY CLUSTERS. J. Jellinek and E. B. Krissinel, Chemistry Division, Argonne National Laboratory, Argonne, Illinois 60439, USA

A detailed analysis of the structural forms - *isomers* (different geometries) and *homotops* (different distributions of the component materials within a given geometry) - of mixed nickel-aluminum clusters is presented. The results are obtained using numerical simulations with a many-body potential. New definitions of the mixing energy and mixing coefficient, valid not only for pairwise-additive but also many-body potentials, are introduced. It is shown that for each fixed stoichiometric composition the hierarchy of the structural forms can be understood by grouping the clusters into classes defined by the isomeric form and the type of the central atom. Within each such class, the energy ordering of the homotops is governed by the mixing coefficient. The simulations show that the clusters undergo a solid-to-liquid-like transition as their internal energy is increased. The possible individual stages in this transition include isomerizations involving only surface atoms, isomerizations involving all atoms, surface melting and complete melting. The stages actually present (or absent) in the transition experienced by a given cluster depend on its size and stoichiometric composition. A close correlation between the dynamical features and the patterns of the energy spectra of the structural forms is established and described.

This work supported by the Office of Basic Energy Sciences, Division of Chemical Sciences, US-DOE under contract No. W-31-109-ENG-38. EBK is also supported by the NIS-IPP Program.

459. NEAR-FIELD AND CONFOCAL OPTICAL SPECTROSCOPY OF SINGLE NANOPARTICLES. Steven R. Emory and Shuming Nie, Department of Chemistry, Indiana University, Bloomington, IN 47405

Metal and semiconductor particles on the nanometer scale have "size-tunable" optical, electronic, and structural properties that are not available in either isolated molecules or bulk solids. These properties are currently under intense study for potential use in microelectronics, quantum dot lasers, nanoscale smart sensors, and high-density data storage. However, the intrinsic size-dependent properties of these nanoparticles remain largely undetermined because of the multiple sources of sample inhomogeneities such as particle size, shape, local fields, and surface defects. Most current studies are based on measurements of ensemble averages in which a large population of nanoparticles are probed and average responses recorded. These experiments yield only population-averaged results and not the intrinsic properties of individual particles. We report the use of near-field and confocal Raman spectroscopy to examine the intrinsic optical properties of single silver colloidal nanoparticles. A novel finding is that a few rare nanoparticles exhibit unusually high efficiencies for surface-enhanced Raman scattering (SERS). We have developed a screening procedure to identify these optically "hot" particles in a heterogeneous population of nanoparticles. We have also observed an irregular on and off (intermittent) photon emission behavior from single silver nanoparticles, similar to intermittent luminescence in single CdSe quantum dots at room temperature.

460. MODE LOCALIZATION IN SELF-AFFINE FRACTAL INTERFACES OBSERVED BY NEAR-FIELD MICROSCOPY. Peng Zhang, Tom L. Haslett, Constantine Douketis and Martin Moskovits, Department of Chemistry, University of Toronto, Toronto M5S 1A1, Canada

Near field microscopy with simultaneous topographic imaging shows that the electromagnetic fields excited in a self-affine film are localized in very small portions of the film. The pattern of high-field areas varies greatly with excitation wavelength and does not correlate simply with the topography. This observation validates a recent theory of the optical response of self-affine fractal objects and also indicates that the locus of optical effects such as surface enhanced Raman does not reside in special topographic elements such as interstices. Additionally, we show that the conclusions of the theory which was based on a quasi-local approximation are qualitatively valid for self-affine films considerably larger than the exciting optical wavelength.

461. HOW ACCURATELY CAN WE COMPUTE MOLECULAR PROPERTIES? LIES, DAMNED LIES AND STATISTICS. David F. Feller, Environmental Molecular Sciences Laboratory, Pacific Northwest National Laboratory, Richland, Washington 99352

As the power of computer hardware and software has continued to double every 12-18 months, computational chemists have grown increasingly ambitious in applying their tools to a ever wider range of molecular properties. As a result, it has become increasingly difficult for an individual scientist to stay informed of the typical accuracy to be expected from popular basis sets and levels of theory in areas of chemistry outside of their own area of expertise. The impact of an on-line database of electronic structure properties on experts and novice users of these applications is discussed in light of a prototype.

462.

THE FIRST PRINCIPLE PREDICTION OF COUPLING CONSTANTS FOR CARBOCATIONS AND OTHER MOLECULES. S. Ajith Perera, Marcel Nooijen and Rodney J. Bartlett, Quantum Theory Project, Department of Chemistry and Physics, University of Florida, Gainesville, Florida 32611-8435.

The equation-of-motion coupled-cluster (EOM-CC) method offers a unified approach to molecular ground and excited state and their properties, particularly, NMR spin-spin coupling constants. Prior *ab initio* or semi-emperical attempts for the theoretical predictions of coupling constants have not offered a reliable and at the same time widely applicable tool. However, EOM-CC offers a highly correlated *ab initio* approach which is of predictive accuracy. We illustrate the approach with applications to transient systems like the ethyl carbocation, where such predictions fill a void in the absence of experiment; and to the 2–norbornyl cation, a principal in the long term controversy as to whether it is a "classical" or "non-classical" carbonium ion, where theory augments the experimental data. Other application on conformational analysis will also be addressed. In addition, we will present some recent work on chemical interpretation of coupling constants.

463.

ELECTRIC AND MAGNETIC FIELD EFFECTS IN MAGNETIC FIELD DENSITY FUNCTIONAL THEORY. Robert A. Harris, Department of Chemistry, University of California at Berkeley, Berkeley, California 94720

The magnetic field density functional theory of Grayce and Harris [Phys. Rev. A,50,3089(1994) & J.Phys. Chem.99,2724(1995)] provides a natural description of electric field effects in all second order magnetic responses. Magnetic field effects in spin-spin coupling and chemical shielding are also naturally manifested in density functional theory. A "hybrid" density functional theory is also presented. A modified form of the electron gas theory developed by the above mentioned authors is used to obtain the magnetic responses of rare gas atoms as probes of complex molecules. {Research done in collaboration with F.Salsbury and A. Pines}

464. LINEAR RESPONSE COUPLED-CLUSTER APPROACH TO STATIC MOLECULAR PROPERTIES. P. Piecuch, A. E. Kondo, J. Paldus, and V. Spirko, Department of Chemistry, University of Toronto, Toronto, Ontario, Canada M5S 3H6

The single-reference (SR) coupled-cluster (CC) theory is nowadays widely used in electronic structure calculations. However, the CC calculations of electrostatic properties [multipole moments, (hyper)-polarizabilities] and of the corresponding property functions are less routine and usually restricted to geometries near the equilibrium, where the finite-field approaches provide a reasonable alternative. In this talk, we shall demonstrate that the linear-response (LR) CC method, first suggested by Monkhorst and further developed and implemented by ourselves using the orthogonally spin-adapted SRCC theory, can be successfully employed to generate various property functions that are in turn needed in order to calculate rovibrational corrections to molecular properties and the corresponding transition moments. As in other analytic approaches, in LRCC calculations no numerical differentiation of energies with respect to field components is required and cumbersome sum-over-states formulas are never used. Formal developments will be illustrated by LRCC results for diatomics and small polyatomics, rovibrational corrections to properties, and rovibrational transition moments. The usefulness of LRCC approach in determination of radii of convergence for energy expansions describing molecules in external fields will also be discussed. The usefulness of extending the LR-SRCC theory to quasidegenerate cases and properties in excited states will be illustrated by showing the results obtained with finite-field multi-reference CC methods.

465. AB INITIO CALCULATIONS OF VIBRATIONAL CIRCULAR DICHROISM AND INFRARED SPECTRA USING SCF, MP2 AND DENSITY FUNCTIONAL THEORIES FOR A SERIES OF MOLECULES. Cary F. Chabalowski, Us Army Ballistic Research Laboratory, Aberdeen Proving Ground, Maryland 21005-5066 and James O. Jensen, Philip J. Stephens, Frank J. Devlin, Michael J. Frisch, and James R. Cheeseman.

In this study we compare SCF, Second-order Moeller-Plesset (MP2) and Density Functional Theory (DFT) results for Vibrational Circular Cichroism (VCD) and Infrared (IR) spectra using atomic orbital basis sets ranging from double-zeta to triple-zeta plus polarization. Structures, force constant matrices and atomic polar tensors have been calculated at the, SCF, MP2 and DFT levels; and atomic axial tensors at the SCF and DFT levels. DFT calculations were carried out using various exchange-correlation functionals, and Gauge-Invariant Atomic Orbitals. The results from these methods are compared amongst themselves and to experiment. This study documents; 1) the similarity in accuracy between the MP2 and DFT methods in predicting the VCD and IR absorption spectra in the mid-IR region; 2) the variation in accuracy for the DFT and MP2 results over a range of basis set sizes; 3) the substantial increase in accuracy from SCF to MP2 and DFT levels for all but the smallest basis set; and 4) the dependence of the VCD and IR absorption spectra obtained from DFT force fields upon the choice of exchange-correlation functionals.

466. MOLECULAR DYNAMICS SIMULATION OF HYDROGEN ADSORPTION ON Si(100). M. Pullon, R. Maboudian and C. Carraro, Departments of Chemical Engineering and Chemistry, University of California, Berkeley, CA 94720

At surface temperatures as low as 150 K and for coverages as low as 0.2 monolayers, atomic hydrogen is observed to adsorb in pairs on the Si(100)(2x1) surface, occupying both sites on the Si dimer [1]. To understand this phenomenon a classical molecular dynamics simulation has been used to investigate the dynamics of hydrogen adsorption on Si(100). The results of these simulations are used to explore the mechanism responsible for hydrogen pairing.

[1] W. Widra, S.I. Yi, R. Maboudian, G.A.D. Briggs, and W.H. Weinberg, Phys. Rev. Lett. 74, 2074 (1995).

467. SURFACE DIFFUSION OF Ge ON Si(111): EXPERIMENT AND SIMULATION. C. E. Allen, R. Ditchfield, and E. G. Seebauer, Department of Chemical Engineering, University of Illinois, Urbana, IL 61801

Surface diffusion of Ge on Si(111) at high temperatures has been examined experimentally by second harmonic microscopy and computationally by molecular dynamics simulations with a Stillinger-Weber potential. Experimentally, the activation energy and pre-exponential factor for mass transfer diffusion equalled 2.48 ± 0.09 eV and $6 \times 10^{2 \pm 0.5}$ cm^2/s, respectively. Simulational results yielded essentially the same numbers, confirming the utility of the Stillinger-Weber potential for diffusional studies. A previously developed semiempirical correlation also did fairly well. The simulations also provided estimates for the corresponding parameters for intrinsic diffusion and for the enthalpy and entropy of Ge adatom-vacancy pair formation on Si. The simulations further yielded evidence for minor contributions of atom exchange to intrinsic diffusion, as well as the complex high-temperature islanding phenomena on picosecond timescales.

468. NUMERICAL AND ANALYTICAL MODELING OF LOW-FREQUENCY ADSORBATE DYNAMICS, Steven P. Lewis and Andrew M. Rappe, Department of Chemistry, University of Pennsylvania, Philadelphia, PA 19104.

In order to explore the low-frequency dynamics of adsorbed surfaces, we have performed detailed, density functional theory calculations of the normal modes of vibration for a carbon monoxide adlayer on copper (100), treating all atoms of the system on an equal footing. We find that the frustrated translation of CO is not a normal mode of the system, contrary to the usual interpretation. Rather it couples strongly to the bulk copper phonons and broadens into a resonance. From the width of the resonance peak, we predict a characteristic decay time of 3 ps, in excellent agreement with the recently measured value of 2.3 ± 0.4 ps.[1]

Based on this physical picture, we have constructed an analytical model of resonant damping that gives a simple rate law depending only on fundamental properties of the system, such as resonant frequency and surface coverage. When applied to the above system, the model gives a remarkably accurate decay time of 2.0 ps, and thus can be relied on to accurately describe the frustrated translational damping in a broad class of systems.

[1] T. A. Germer, J. C. Cavanagh, E. J. Heilweil, and R. R. Cavanagh, J. Chem. Phys. **101**, 1704 (1994).

469. DIFFUSION OF WATER MOLECULES ON ICE Ih SURFACE.
E.R. Batista and H. Jonsson, University of Washington, Department of Chemistry, Seattle, WA 98195.

We present results of calculations of binding and diffusion of water admolecules on the basal face of ice Ih. Relaxation of several substrate molecules near the admolecule greatly affects the binding energy and diffusion barriers. At low coverage, the admolecules prefer to sit at non-epitaxial sites with a surprisingly large binding energy, ranging from 0.53 to 0.63 eV depending on the proton ordering. The calculated activation barriers for diffusion are between 0.16 eV and 0.28 eV which agrees well with recent experimental upper bound on the surface diffusion coefficient and measurements of admolecule diffusion length during ice growth.

470. THE DIFFUSION OF LARGE, ADSORBED MOLECULES ON SOLID SURFACES, Kristen A. Fichthorn and Janhavi S. Raut, Departments of Chemical Engineering and Physics, The Pennsylvania State University, University Park, PA 16802.

Although the diffusion dynamics of adsorbed atoms and small molecules have been the subject of many investigations, relatively few studies have focused on the diffusion of large, adsorbed molecules. To probe the unique features of molecular diffusion, we have conducted molecular-dynamics-simulation studies of diffusion in a series of physically adsorbed n-alkanes (C2-C50) on a model Pt(111) surface. In addition, we have used transition-state theory to estimate theoretical diffusion coefficients for these molecules and to determine the characteristic mechanisms by which they diffuse. These combined studies have yielded much insight into the unique mechanisms of molecular diffusion and their ramifications for molecular diffusion coefficients. Our studies also highlight differences between relatively small molecules, which diffuse in a concerted manner and have unique (but related) diffusion barriers, and large molecules, which share a common diffusion barrier that is independent of chain length. We discuss the similarities between the diffusion of large adsorbed molecules and that of atomic clusters.

471. DECAY OF ISOLATED SURFACE FEATURES DRIVEN BY THE GIBBS-THOMSON EFFECT.
James G. McLean, B. Krishnamachari, D.R. Peale, E. Chason, James P. Sethna and B.H. Cooper, LASSP, Cornell Univ., Ithaca, NY 14853

Isolated, well-ordered nanoscale features on surfaces are expected to decay in size over time due to the Gibbs-Thomson effect (the same effect responsible for Ostwald ripening). The classic analytic theory of Ostwald ripening can be extended to the case of isolated features. A new limiting case arises, wherein the decay rate is limited by escape of the atoms through an outer boundary, such as a surrounding terrace edge. We have performed computer simulations for a solid-on-solid model with nearest-neighbor interactions of a cluster of atoms within a pit, which is seen to decay as expected. By making appropriate measurements from the simulated system, we have determined the macroscopic parameters which enter into the analytic theory. This yields an analytic prediction for the decay, which is found to agree well with the decays actually simulated. These detailed measurements yield insights into the microscopic nature of the Gibbs-Thomson effect.

472. ANISOTROPY IN THE PHOTODISSOCIATION OF OZONE, P. L. Houston, Ruth Wilson, and Julie Mueller, Department of Chemistry, Cornell University, Ithaca, NY 14853

The $O(^3P) + O_2(3\Sigma)$ channel in the photodissociation of ozone has been examined for several wavelengths between 226 and 266 nm by the product imaging technique. The speed distribution of the products shows two distinct peaks; for 226 nm dissociaiton, the maxima correspond to O_2 vibrational levels of 14 and 27. The ratio of the two peaks varies with wavelength. The anisotropy of fragment recoil vectors is a function of speed: those fragments corresponding to channels with the highest vibrational internal energy in the O_2 recoil with an angular distribution corresponding to $\beta=0.70$-0.81, while those corresponding to intermediate vibrational energy recoil with $\beta=1.33$-1.37. These values are nearly independent of dissociation wavelength. The implications of these results for the mechanism of ozone dissociation to the triplet product channel will be discussed.

473.

Intermolecular Orientation/Alignment Effects in Unimolecular Photolysis of Clusters

David J. Nesbitt
JILA/Univ. of Colorado, Boulder, CO 80309-0440

The use of weak intermolecular forces as a means of orienting/aligning reactants prior to photolysis has proven to be a very powerful tool for studying photophysical dynamics in clusters. The control of such methods can be considerably enhanced by combining UV photolysis with high resolution overtone excitation from single mode injection seeded OPO to photolyze a specific bond in a cluster with quantum state and cluster size selectivity. This talk will discuss recent applications of these and related methods in our group:1) IR+UV studies of Ar-HOH and Ar-HOD clusters, in which the internal rotor quantum state significantly influences the single atom cage effect. 2) "gentle recoil" UV photolysis of Ar_n-H_2S clusters (n=1,2), which indicate a surprising efficiency (> 5%) for forming internally *hot but bound* free radical Ar-HS and Ar_2-HS complexes despite excess photolysis energies 50-200 fold above the cluster dissociation limit.

474. PHOTOELECTRON DYNAMICS IN ORIENTED PERFLUORO MOLECULES, P. Downie & I. Powis, Department of Chemistry, The University of Nottingham, Nottingham NG7 2RD, UK.

Orientation of free gaseous molecules can be inferred by imaging the angular distribution of molecular/atomic photofragments. In the case of a dissociative molecular photoionization the ARPEPICO technique (Angle Resolving Photoelectron-Photoion Coincidence) can simultaneously image the distribution of photoelectrons. A recoil vector correlation can be established for individual photoelectron-photofragment ion pairs, yielding molecule frame photoelectron angular distributions. Both experimental and theoretical investigation of the fixed-molecule and lab-frame distributions are presented. The orientation dependence generates considerable insight into the overall photofragmentation dynamics, with fluorinated species offering particularly instructive examples. Results obtained for oriented adsorbates are compared with the free molecule data, prompting consideration of the scope for either inducing or probing molecular orientation.

475.

K QUANTUM NUMBERS IN LONG LIVED INTERMEDIATES Ilya Bezel, **Curt Wittig**, Department of Chemistry, USC, Los Angeles, California 90089

A molecule's approximate constants of motion can be examined in the bound region by using spectroscopy to identify the effects of couplings that lessen the goodness of its quantum numbers. This is also possible in the region of the continuum as long as the resonances are isolated. However, when the resonances overlap it is best to probe the dynamical process directly. This has been done to examine conservation of the K quantum number in the unimolecular decomposition of NO_2, an almost prolate rotor. As J increases near reaction threshold with other parameters held constant, K mixing causes the reaction rate to decrease. In the experiment, molecules having well-defined rotational excitations were selected by pumping rotational components of the (101) ← (000) combination band. Photoinitiated reaction rates were recorded by applying the picosecond-resolution pump-probe technique. The rotational levels accessed in the first step cover a significant range (odd N, $1 \leq N \leq 15$, K = 0). With the photolysis photon energy fixed, the observed rate is independent of rotational energy. This is consistent with minimal K-mixing in the decomposition process.

476. FACILE ORIENTATION/ALIGNMENT: PHOTOFRAGMENTATION OF MAGNETIC STATE POLARIZED PARENT MOLECULES. Delroy A. Baugh, Dept. of Chemistry and Biochemistry, University of California, Los Angeles, CA 90095-1569

A major objective in state-to-state photofragmentation spectroscopy is to determine structural attributes, such as the transition dipole vector field(TDVF), and dynamical information, such as product rotational energy-level populations from studies of elementary unimolecular reactions. While final energy-level specific experiments are now routinely performed in a number a laboratories, the more detailed experiments that are required for the determination of structural attributes such as the TDVF have only been recently reported by our group. Our method uses laser-polarization-analysis of magnetic state-selected parent molecules and magnetic-state polarized products to directly determine the TDVF. Our theoretical formalism for extracting the TDVF and the results from several recent experiments on model molecular systems will be presented.

477.

SURFACE AND INTERFACES OF CLUSTERS. Uwe A. Kreibig, I. Physikalisches Institut der R W T H Aachen, D-52056 Aachen, Germany.

Uncommon *physical* behavior of solid state clusters usually is ascribed to cluster size effects. Whereas the importance of cluster surfaces and, in particular, of cluster interfaces is well known, since long, by chemists who apply them on large scale e.g. for heterogeneous catalysis, only recently cluster surfaces and interfaces were recognized to be important for various *physical* properties, as well.
Results of optical experiments on 2 nm silver clusters, produced and characterized in a supersonic beam, and, afterwards, embedded in various surroundings, are presented which prove that interface effects exceed size effects by far, though the clusters consist of 2.10^2 atoms, only. These effects are explained by modelling static and dynamic charge transfer processes at the interfaces in view of their implications on the excitation of cluster plasmon polaritons.
As well, the non-linear optical response of clusters can be strongly modified via surface manipulation, as is demonstrated by experimental results concerning optical bistability.

478.

CHANGES IN THIN METAL FILM NANOSTRUCTURE AT NEAR-AMBIENT TEMPERATURES. **Kathy L. Rowlen, Department of Chemistry and Biochemistry, University of Colorado, Boulder, CO 80309**

Atomic force microscopy has been used to quantitatively characterize the nanostructure of thin films of silver. The influence of substrate, deposition conditions and temperature on both the nanostructure and optical properties has been examined. It was found that 5 nm silver films can be annealed at temperatures only slightly above ambient. The kinetics and mechanism of the annealing process will be discussed along with the implications for surface enhanced Raman scattering.

479. NANOSTRUCTURES AND SURFACE EVOLUTION IN EPITAXIAL GROWTH
Fereydoon Family, Department of Physics, Emory University, Atlanta, GA 30322

The results of detailed theoretical and simulational studies of the formation, evolution, and coalescence of clusters in the submonolayer and multilayer growth regimes in molecular-beam epitaxy will be presented and compared with experiments. In particular, the dependence of island morphology and density on coverage, the ratio of the deposition rate to diffusion rate, critical island size, as well as edge relaxation mechanisms will be discussed. It will be shown that the dynamic scaling of the cluster size distribution provides an effective approach for understanding the growth and evolution of islands in the submonolayer regime. In the multilayer growth regime the role of instabilities such as the step-barrier and the adatom-cluster attraction on surface evolution and morphology will be discussed. Theoretical and simulational results for the formation, growth and coarsening of mounds, will be presented and compared with a number of recent experiments.

480.

ION BEAM SYNTHESIS AND OPTICAL PROPERTIES OF SEMICONDUCTOR NANOCRYSTALS AND QUANTUM DOTS. **Jane G. Zhu,** Department of Physics, New Mexico State University, Las Cruces, NM 88003; C. W. White, S. P. Withrow, and J. D. Budai, Oak Ridge National Laboratory, Solid State Division, Oak Ridge, TN 37831, D. O. Henderson, Fisk University, Physics Department, Nashville, TN 37208

Nanocrystals of elemental and compound semiconductors have been fabricated in SiO_2 and Al_2O_3 matrices by the ion implantation technique. The nanocrystal size distributions can be controlled by the ion implantation energy, dose and annealing temperatures. The implantation sequence of the constituent species is found to significantly influence the formation of compound semiconductor nanocrystals. Quantum confinement effects occur for the nanocrystals with sizes smaller than the exciton diameters. The optical characterizations of these semiconductor nanocrystals include photoluminescence, optical absorption and infrared reflectance measurements.

481.

LASER ASSISTED MOLECULAR BEAM DEPOSITION Robert L. DeLeon, W. M. K. P. Wijekoon, Paras N. Prasad & James F. Garvey* Department of Chemistry, SUNY/Buffalo, Buffalo, NY 14260-3000

A Smalley-type cluster source has been employed to generate new bulk materials, in the form of thin films. By controlling the molecular beam expansion, we are capable of "spraying" clusters onto a particular substrate which then coalesce to generate a uniform coating. Subtle modifications of laser fluence or expansion conditions result in dramatic changes in the chemical properties and morphology of the film generated. Indeed by employing reagent gasses, we are capable of generating novel molecular species within the beam which are then deposited as a bulk film. Through this technique of laser assisted molecular beam deposition (LAMBD) we have been able to tailor the experimental conditions to produce a variety of films with industrial and electronic applications (superconducting thin films, polymeric thin films, diamond-like carbon thin films, patterned or multi-layered thin films, wide bandgap semiconductors, etc....). We will focus our discussion on our recent efforts to fabricate inorganic host:organic dopant films (i.e., metal oxide:organic dye) suitable as photonic media.

482. STRUCTURES OF THIN FILM SURFACE ALLOYS, Robert Q. Hwang and Andreas K. Schmid, Surface Chemistry Department, Sandia National Laboratories, Livermore, CA 94550

In recent years, thin metal film growth has attracted much attention both experimentally and theoretically. Due to the complex combination of thermodynamic and kinetic phenomena, a variety of film morphology and atomic structures have been observed. In the case of surface alloys, in which there are more than one film component, the situation is further complicated due to interactions both directly and mediated through the substrate. This can lead to new growth mechanisms and alloy formation that produce a variety of morphologies and stoichiometries. These effects will be discussed in the context of the two component films of Ag/Co, Ag/Cu and Cu/Co on Ru(0001). In their bulk states, all four metals exhibit considerable miscibility gaps up to their melting points. In the thin film systems, however, the effect of substrate induced strain plays a crucial role in determining the final film configuration. This leads to two dimensional alloying and greatly influences the morphology of the films. The mechanisms driving the alloying and the observed pattern formation will be discussed. This work supported by the Office of Basic Energy Sciences of DOE under contract number DE-AC04-94AL85000.

American Chemical Society
DIVISION OF POLYMER CHEMISTRY

213th ACS National Meeting

San Francisco, CA
April 13-17, 1997   **T.E. Long, Program Chairperson**

SUNDAY MORNING

- **Special Topics in Polymer Chemistry**
K.L. Wooley, Presiding   Papers 1-9

- **Organic Materials and Devices in Display Technology: Tutorial on Organic Electroluminescence Technology**
B.R. Hsieh, Presiding   Papers 10-13

- **Acrylate Polymerization: New Mechanisms and Polymers Ionic and Coordination Polymerization**
W.J. Brittain, Presiding   Papers 14-20

- **Fundamental Properties of PEG and New Copolymers**
D.W. Grainger, Presiding   Papers 21-29

SUNDAY AFTERNOON

- **Special Topics in Polymer Chemistry**
R.B. Moore, Presiding   Papers 30-38

- **Special Topics in Polymer Chemistry**
K.E. Ulrich, Presiding   Papers 39-47

- **Actrylate Polymerization II: New Mechanisms and Polymers Free Radical Polymerization**
R. Quirk, Presiding   Papers 48-54

- **PEG Liposomes, Particulates, and Surfaces**
S. Zalipsky, Presiding   Papers 55-63

SUNDAY EVENING

- **Posters**
R.B. Moore, Presiding   Papers 64-156

MONDAY MORNING

- **Industrial Milestones in Polymer Science**
  T.E. Long, Presiding — Papers 157-162

- **New Light Emitting Device Concepts**
  C.W. Tang, F. Papadimitrakopoulos, Presiding — Papers 163-170

- **Acrylate Polymerization III: New Mechanisms and Polymers Ionic Polymerization**
  D. Haddelton, Presiding — Papers 171-177

- **PEG Surfaces and PEG-Protein Conjugates**
  O. Kinstler, Presiding — Papers 178-187

MONDAY AFTERNOON

- **Industrial Milestones in Polymer Science**
  S.R. Turner, Presiding — Papers 188-194

- **Polymeric LED Systems**
  A.B. Holmes, C.Y.C. Lee, Presiding — Papers 195-203

- **Acrylate Polymerization IV: New Mechanisms and Polymers Architecture and Properties**
  A.H.E. Muller, Presiding — Papers 204-211

- **PEGylated Proteins and Small Molecules**
  J.M. Harris, Presiding — Papers 212-220

MONDAY EVENING

- **Posters**
  J.M. Harris, D.W. Grainger, Presiding — Papers 221-245

TUESDAY MORNING

- **Carl S. Marvel Creative Polymer Chemistry Award In Honor of Bruce M. Novak**
  D.A. Tirrell, Presiding — Papers 246-250

- **Organic Materials for Liquid Crystal Display Applications**
  H.W. Schmidt, C.S. Hsu, Presiding — Papers 251-259

- **Advances in Free Radical Polymerization: Fundamentals of Free Radical Reactions**
  M. Buback, Presiding — Papers 260-264

- **NMR Spectroscopy of Synthetic Macromolecules: Tutorials**
  H.N. Cheng, Presiding — Papers 265-267

TUESDAY AFTERNOON

- **Carl S. Marvel Creative Polymer Chemistry Award**
  In Honor of Bruce M. Novak
  T.J. Deming, Presiding Papers 268-271

- **Stability of Organic Light Emitting Diodes**
  Y. Yang, K.J. Wynne, Presiding Papers 272-280

- **Advances in Free Radical Polymerization: Advances in Conventional Radical Polymerization**
  B. Yamada, Presiding Papers 281-285

- **NMR Spectroscopy of Synthetic Macromolecules: Polymer Microstructure in Solution**
  P.A. Mirau, Presiding Papers 286-292

- **Special Topics in Polymer Chemistry**
  C.N. Bowman, Presiding Papers 293-301

TUESDAY EVENING

- **Posters**
  T.E. Long, Presiding Papers 302-442

WEDNESDAY MORNING

- **Organic Thin Films: Monolayers**
  C.W. Frank, Presiding Papers 443-451

- **Plasma and Polymers**
  R. d'Agostino, Presiding Papers 452-460

- **Advances in Free Radical Polymerization: Nitroxide Mediated Controlled Radical Polymerization**
  D. Priddy, Presiding Papers 461-466

- **NMR Spectroscopy of Synthetic Macromolecules: Solid State NMR of Non-Crystalline Polymers**
  H.W. Spiess, Presiding Papers 467-473

- **ACS Polymer Chemistry Award: Polymer Structure and Thermodynamics**
  In Honor of William J. MacKnight
  P. Norling, Presiding Papers 474-479

WEDNESDAY AFTERNOON

- **Organic Thin Films: Grafting and Surface Segregation**
  W. Knoll, Presiding — Papers 480-489

- **Plasma and Polymers**
  H.J. Griesser, Presiding — Papers 490-499

- **Advances in Free Radical Polymerization: Metal Catalyzed Controlled Radical Polymerization**
  J. Harwood, Presiding — Papers 500-504

- **NMR Spectroscopy of Synthetic Macromolecules: Industrial Applications of Polymer NMR**
  A.O. English, Presiding — Papers 505-510

- **ACS Polymer Chemistry Award: Multiphase Polymers In Honor of William J. MacKnight**
  R.A. Weiss, Presiding — Papers 511-516

THURSDAY MORNING

- **Organic Thin Films: Supramolecular Assemblies**
  N.L. Abbott, Presiding — Papers 517-526

- **Plasma and Polymers**
  D.G. Castner, Presiding — Papers 527-537

- **Advances in Free Radical Polymerization: Other Methods for Controlled Radical Polymerization**
  B. Gilbert, Presiding — Papers 538-543

- **NMR Spectroscopy of Synthetic Macromolecules: Blends and Interphases**
  R.A. Byrd, Presiding — Papers 544-549

THURSDAY AFTERNOON

- **Organic Thin Films: Optoelectronics and Other Applications**
  A. Knoesen, Presiding — Papers 550-559

- **Plasma and Polymers**
  E.R. Fisher, Presiding — Papers 560-569

- **Advances in Free Radical Polymerization: Controlled Architectures by Radical Polymerization**
  C.J. Hawker, Presiding — Papers 570-574

- **NMR Spectroscopy of Synthetic Macromolecules: NMR in the Solid State**
  D.G. Cory, Presiding — Papers 575-580

DIVISION OF POLYMER CHEMISTRY, Inc

001.

STEP GROWTH OF AN AB_2 MONOMER, WITH CYCLE FORMATION

Colin Cameron, Courtaulds Coatings, Stoney Gate Lane, Felling, Gateshead, England, UK.
Allan H. Fawcett*, Cecil R. Hetherington, Richard A.W Mee,
School of Chemistry, The Queen's University of Belfast, Northern Ireland, UK, and
Frederic C. McBride, The Computer Centre, The Queen's University of Belfast, Northern Ireland, UK.

A lattice bound model of the polymerization of and AB_2 monomer allows the formation of cycles within the molecules in competition with the conventional growth process, and shows that this factor eventualy limits the kinetic growth. We show that throughout the reaction the number of rings of size **m** in the system is given by $N_{r,m} = C\, p_a^m\, m^{-2.71}$, the constant being related to the final molecular weight achieved and p_a being the extent of reaction of the **A** groups. The model provides adjacency or Kirchof matrices to describe the mean structures of the oligomers of each size.

002.

HYPERBRANCHED PERFLUORINATED POLYMERS. **Anja Mueller,** Karen L. Wooley*, Department of Chemistry, Washington University, One Brookings Drive, St. Louis, MO, 63130

Hyperbranched perfluorinated poly(benzyl ether)s (HBPF polymers) were prepared from 3,5-di(pentafluorobenzyloxy)benzyl alcohol (F10-[G-1]-OH). The polymerization is based upon nucleophilic substitution of the p-fluorines of the two pentafluorophenyl groups by the benzylic alcohol (AB2 monomer). Optimized reaction conditions for the polymerization to an HBPF polymer involve the addition of sodium metal to a solution of monomer in THF at refluxing temperature. The molecular weight and molecular weight distribution of the resulting polymer are affected by the surface area of the sodium particles. The HBPF polymer has a high number of accessible pentafluorophenyl chain ends which can be used to change the physical and chemical properties of the polymer. Polymers which have the chain ends substituted with perfluorinated alkyl chains were also synthesized. Contact angle measurements show that all HBPF polymers prepared are highly hydrophobic.

003. STEREOCONTROL OF VINYL POLYMERS VIA CYCLOPOLYMERIZATION. Shiying Zheng and Dotsevi Y. Sogah, Chemistry Department, Cornell University, Ithaca, NY 14853

A new family of bismethacrylate monomers **1 - 6** templated by tartrate derivatives were synthesized and polymerized by free radical and group transfer polymerization (GTP). The resulting polymers contained no unreacted pendant vinyl groups and were soluble in common organic solvents, suggesting that the polymer main chain consisted of cyclic repeat units. The stereochemical results showed that by increasing the bulkiness of the monomer, the isotactic content of the polymer could be increased significantly. For example, the triad distribution mm/mr/rr increased from 15/51/34 for poly-**1** to 87/10/3 for poly-**6** by GTP; this is the highest isotacticity ever reported for GTP. The chiroptical properties, stereochemistry, and copolymerization results showed that poly-**3**, **-4**, and **-6** may assume a helical conformation.

004.

THE STUDY OF SHELL CROSS-LINKED KNEDELS (SCK): FORMATION AND APPLICATION. **K. Bruce Thurmond II,** Tomasz Kowalewski, and Karen L. Wooley*, Department of Chemistry, Washington University, One Brookings Drive, St. Louis, MO, 63130

Water soluble, amphiphilic, organic nanometer-sized spheres composed of a hydrophilic cross-linked shell surrounding a hydrophobic mobile core have been prepared. Using anionic "living" polymerization, amphiphilic diblock copolymers were prepared consisting of a polystyrene block and a poly(4-vinylpyridine) block. Quaternization of the PVP block with 4-chloromethyl styrene allowed for micellar aggregates to be formed. These micellar aggregates are spherical in shape with the hydrophobic polystyrene making up the core. Cross-linking of the styrenyl groups by a radical process locked in the spherical shape. Techniques such as NMR, flourescence, and atomic force microscopy (AFM) provided characterization of the SCK's. The SCK's display no critical micelle concentration (cmc) and are able to encapsulate hydrophobic organic molecules (e.g. pyrene, THF, BDPA) into their core.

005. **POLYROTAXANES BY SELF THREADING OF FUNCTIONALIZED MACROCYCLES: POLYAMIDES,** H. W. Gibson, D. Nagvekar, N. Yamaguchi, W. S. Bryant and S. Bhattacharjee, Department of Chemistry, Virginia Polytechnic Institute & State University, Blacksburg, VA 24061-0344

Bis(5-carboxy-1,3-phenylene)-26-crown-8 was utilized to investigate the effect of the ring size on polyrotaxane formation by *in situ* threading during polymerization. Direct polycondensation of the diacid with 4,4'-oxydianiline and bis[4-(m-aminophenoxy)phenyl]phenylphosphine oxide gave two new poly(amide crown ether)s, **4a** and **4b**, respectively. Both **4a** and **4b** were soluble in dipolar aprotic solvents, unlike the analogous polyaramides **5a** and **5b** derived from reaction of homologous 32-membered crown ether diacid, which were insoluble in all solvents examined, including H_2SO_4. However, the GPC traces of polyaramides **4a** and **4b** exhibited bimodal behaviour indicative of two distinct molecular weight fractions. The fraction at lower retention volume is attributed to the high molecular weight (M_n = 597 and 800 kg/mol for **4a** and **4b**, respectively, by universal calibration with a viscosity detector), branched poly(amide crown ether) formed by *in situ* threading of the macrocyclic moieties by other backbone elements, i.e., rotaxane formation, the same process which produces insoluble networks in **5a** and **5b**. The driving force for threading is believed to be hydrogen bonding. These results constitute a method for formation of physically crosslinked networks of potential utility as membranes and elastomers.

006. SYNTHESIS AND MODIFICATION OF FUNCTIONALIZED POLY(ARYLENE ETHER KETONE)S. M. Klapper, A. LeGuen, U. Hoffmann, T. Wehrmeister, S. Mullins, K. Müllen, Max-Planck-Institut für Polymerforschung, Ackermannweg 10, D-55128 Mainz, Germany

Poly(arylene ether ketone)s (PEEKK) are high performance engineering thermoplastics, widely used for electronic, medical, aircraft and aerospace application. Initially, considerable effort was made to modify their chemical structure, essentially to improve their thermal and mechanical properties. Our work, however, focusses more on tuning polymer properties by functionalization. This can be achieved either by chemical modification of the polymer or by direct synthesis using functionalized monomers. New poly(arylene ether ketone)s with different π-moieties, such as anthracene or pyridine or functionalities such as the amino group in the polymer backbone. These new polymers are easily be modified by a broad variety of polymer-analogous reactions, e.g. quaternisation, Diels-Alder reaction, acylation or grafting processes. The influence of hydrogen bonding of oligoamide side chains attached to a PEEKK backbone is demonstrated. Furthermore, the synthesis and properties of comb-type polymers of PEEKK with polystyrene or polyisoprene backbones is presented.

007. DEVELOPMENT OF A LIVING POLYMERIZATION APPROACH TO DENDRITIC POLYMERS; Hasan A. Al-Muallem and Daniel M. Knauss, Department of Chemistry and Geochemistry Colorado School of Mines, Golden, CO 80401

Synthetic approaches to dendrimers and dentritic polymers have developed tremendously over the past few years. This presentation will present a new approach to dendritic polymers utilizing anionic polymerization techniques. The resulting polystyrene-based polymers are found to have narrow polydispersities and to be of generational growth between 3 and 6. The low intrinsic viscosities found for these high molecular weight materials is one representative property of their novel structure.

008. SYNTHESIS OF CYCLIC OLiGO-ETHERIMIDES: T. Takekoshi and J.M. Terry, General Electric Co., Corporate Research and Development, Schenectady, NY 12301

Applications of high temperature polymers are often hampered by the process difficulty associated with their high Tg and Tm, and high melt viscosity. One of the solutions to the problem is to employ low melting cyclic oligomers and polymerize them *in situ*. Recently such macrocyclic oligomers have been synthesized for polycarbonate, polyester, polyaramide, and polysulfone systems. Effective catalytic ring-opening polymerizations have been also developed for some of these oligocyclics. This paper describes an efficient method of synthesizing macrocyclic oligoetherimides in high yield. The process is based on Kricheldorf polycondensation in which a bistrimethylsilyl ether of bispenols is reacted with an arylenebis(fluorophthalimide) derived from fluorophthalic anhydrides and aromatic diamines. The oligomerization was performed under a *pseudo* high dilution condition in NMP in the presence of alkali fluoride as a catalyst. Cyclic oligoetherimides were greatly more soluble than their linear counter parts in conventional solvents such as toluene.

009. SHAPE PERSISTENT MACROCYCLES: VARIOUS ISOMERS AND SIZES BY THE REPETITIVE HAGIHARA-COUPLING/DESILYLATING STRATEGY. S. Höger, S. Müller, L. Karcher, Max-Planck-Institut für Polymerforschung, Ackermannweg 10, 55128 Mainz, Germany.

The shape persistent macrocyclic amphiphile **1**, as well as an isomer of **1** and amphiphiles with extended switchable portions are prepared via Glaser coupling of the corresponding bisacetylenes. These compounds are available in multi-gram amounts by a repetitive Hagihara/desilylating strategy.

010. OVERVIEW OF ORGANIC ELECTROLUMINESCENT DEVICES FOR DISPLAY APPLICATIONS, Ching W. Tang, Imaging Research and Advanced Development, Eastman Kodak Company, Rochester, New York 14650-2110

In this tutorial, the fundamental performance characteristics and limitations of organic electroluminescent (EL) devices will be described and issues critical to the development of this EL technology for display applications ranging from efficent backlights to high-resolution dot-matrix color panels will be reviewed.

011. CHARACTERIZATION OF INJECTING CONTACTS IN OLED. M. STOLKA and M. Abkowitz, Xerox Corporation, Webster Center for Research and Technology, Webster NY14580

Organic charge transport molecules and polymers are used as active components in optoelectronic devices such as Organic Light Emitting Diodes (OLED), electrophotographic receptors, photovoltaic cells, field-effect transistors etc. and also as electrodes in some of these devices. For example, the light output and useful life of OLEDs are determined by the nature and stability of contacts between anode and the hole transport material.

Ideally, the contacts should be ohmic, i.e. should behave as infinite reservoirs of charges to sustain Space Charge Limited Currents in the organic materials. A test for distinguishing ohmic from partially blocking or even fully blocking contact is described. It is shown that TPD, the model molecule often used in OLED devices, and polymers derived from TPD, form ohmic contacts with some electrodes such as Au and Ag. The quality of contact with ITO strongly depends on the treatment and purification of the electrode. Space charge limited currents, a clear manifestation of ohmic contacts, have been observed in these materials. In this context, we will also discuss the effect of partial oxidation of the hole transport molecules and provide our perspective on electrode contacts in ECL (Electrogenerated ChemiLuminescence) cells.

012.

CHARGE TRANSPORT IN MOLECULAR DEVICES. A. Dodabalapur, Bell Laboratories, Lucent Technologies, Murray Hill, NJ 07974

This tutorial will review the important charge injection mechanisms and transport phenomena in molecular devices. Such materials possess an astonishing range of mobilities. In single-crystals and ordered polycrystalline films, polaronic effects are very important in determining the principal transport characteristics. In less ordered and amorphous materials, disorder plays a dominant role. We describe transport in materials such as diamines and 8-hydroxyquinolinato aluminum which are employed in light emitting diode (LED) fabrication. The mobility in such materials is strongly electric field dependent over a range of temperatures and fields. Charge transport in heterostructures of such materials, commonly used in high performance LEDs, is influenced by numerous factors including contacts, space charge effects, trapping, and volume effects. In organic field-effect transistors, a high carrier mobility is desired, and for this reason ordered materials with strong π or π^* orbital overlap are preferred. The carrier mobility depends on the electronic overlap, polaron binding energies, phonon frequencies, etc. The tutorial will conclude with a discussion on how molecular orbital energy levels influence device performance.

013. ORGANIC ELECTROLUMINESCENT MATERIALS. **B. R. Hsieh,** Xerox Corporation, 800 Phillips Road, 114-39D, Webster, NY 14580

Both organic polymeric and molecular electroluminescent (EL) materials will be reviewed. Numerous organic photoconductor (OPC) materials have been disclosed and tested for EL applications in the past 20 years. But only very few has the real potential for EL applications. One of the current benchmark EL polymer is MEH-PPV; naphthalene-TPD and Alq are the current benchmarks for hole transport and electron transport molecules respectively. The only common feature of these three vastly different materials is simple synthesis. The challenge for organic material chemists is to design **easily accessible** new materials that show improved EL properties over the current benchmarks.

014. STEREOSPECIFIC LIVING POLYMERIZATION OF FUNCTIONAL METHACRYLATES WITH t-BuLi /BULKY ALUMINUM PHENOXIDE.   T. Kitayama, T.Hirano, S.He, S.Urbanek, T.Yabuta, and K.Hatada, *Department of Chemistry, Faculty of Engineering Science, Osaka University, Toyonaka, Osaka 560, Japan*

A combination of t-BuLi and bis(2,6-di-t-butylphenoxy)methylaluminum [MeAl(ODBP)$_2$] has been proved to exhibit unique stereoregulating power in methacrylate polymerization. In this communication is described the utility of the initiator for the control of livingness, stereospecificity, and monomer selectivity in the polymerization and copolymerization of methacrylates.
The polymerization of trimethylsilyl methacrylate (TMSMA) with t-BuLi/MeAl(ODBP)$_2$] gives highly syndiotactic (st-) polymer with narrow molecular weight distribution (MWD). Copolymerization of ethyl (EMA) and TMSMA with t-BuLi/MeAl(ODBP)$_2$ in toluene at $-78°C$ proceeded in a monomer-selective manner without loosing stereospecificity for both monomers, giving a stereoregular block copolymer. Similarly, terpolymerization of EMA, TMSMA and t-butyl methacrylate ($-78°C \longrightarrow -40°C$) gave a triblock copolymer. Polymerization of a dimethacrylate having primary and tertiary ester groups proceeded preferentially through the primary ester side, giving a soluble polymer with pendant methacrylate groups. Polymerization of a monomer having a phenolic OH with two adjacent t-butyl groups successfully gave a polymer with narrow molecular weight distribution (M_w/M_n=1.08), though the initiator efficiency was low (34%). Polymerization of this monomer with oligo(EMA) anion formed with t-BuLi/MeAl(ODBP)$_2$ gave a block copolymer with quantitative efficiency.

015. NEW PLURIFUNCTIONAL INITIATORS FOR ANIONIC POLYMERIZATION OF ALKYL METHACRYLATES. C.Hubert, Y.Gnanou, Laboratoire de Chimie des Polymères Organiques, U.M.R.CNRS-ENSCPB-Av.Pey Berland BP108, 33402 Talence, FRANCE

A novel hexafunctional initiator ($C_6[CH_2OC_6H_4C(CH_2)C_6H_5]_6$) has been purposely synthesized to subsequently serve in the anionic polymerization of methyl methacrylate. This initiator is obtained in four steps:
1. Addition of 4-hydroxybenzophenone to hexakis(bromomethyl)benzene
2. Grignard reaction using methylmagnesiumiodide
3. Dehydration reaction
4. Addition of stochiometric amounts of trimethylsilylmethyllithium

The polymethacrylate stars that are obtained using this hexacarbanionic initiator exhibit reasonably narrow molar mass distribution. Their actual functionality is currently investigated.

016. GROUP-TRANSFER POLYMERIZATION OF METHYL METHACRYLATE USING ZIRCONOCENE INITIATORS: POLYMERIZATION MECHANISMS AND APPLICATIONS. Yufang Li, David G. Ward, Srinivasa S. Reddy, and Scott Collins*, Department of Chemistry, University of Waterloo, Waterloo, Ontario, Canada N2L 3G1

The polymerization of methyl methacrylate (MMA), initiated by zirconocene complexes Cp_2ZrMe_2 (**1**) and $[Cp_2ZrMe(THF)][BPh_4]$ (**2**), provides partially syndiotactic poly(methyl methacrylate) (PMMA) in high yield and with a narrow molecular weight distribution (MWD). The kinetics of this process were studied and reveal that in this system, the rate of initiation is much slower than that of propagation. Initiation appears to involve the rate-limiting reaction of complex **2** with monomer to generate a cationic enolate complex (**3**). The latter compounds also initiate polymerization of MMA, albeit at a much slower rate than that observed using initiators **1** and **2**. The mechanism for propagation in polymerizations initiated by complexes **1** and **2** has a rate-limiting step involving the reaction of neutral zirconium enolate species (**4**), produced in situ from **3** and **1**, with monomer, activated by coordination to complex **2**.

Neutral enolate complexes **4** and complex **2** function as effective initiators of MMA polymerization in which the rate of initiation is greater than or equal to the rate of propagation. Under these conditions, PMMA can be produced with very narrow MWD and the polymerization process is living at ≤ 0 °C.

017.

BIMETALLIC SAMARIUM(III) CATALYSTS VIA ELECTRON TRANSFER INITIATION: THE FACILE SYNTHESIS OF WELL-DEFINED (METH)ACRYLATE TRIBLOCK COPOLYMERS, B. M. Novak, and L. S. Boffa, Department of Polymer Science and Engineering, and the Materials Research Science and Engineering Center, University of Massachusetts, Amherst, Massachusetts 01003

A method for generating a bifunctional organolanthanide(III) initiator *in situ* from a (meth)acrylate monomer and a divalent samarium precursor, $(C_5Me_5)_2Sm$ (**1**) or $(C_5Me_5)_2Sm(THF)_2$ (**2**), is described. This process involves one-electron transfer from the Sm(II) species to monomer, forming radical anions which couple to give a bimetallic samarium(III) enolate that acts as a bisinitiator for "living" polymerization. Well-defined, highly syndiotactic ABA triblock copolymers containing both methacrylate and acrylate segments were prepared in two monomer addition steps with this methodology. Selective side chain deprotection may be carried out on copolymers containing poly(*tert*-butyl acrylate) or poly(benzyl methacrylate) blocks to give syndiotactic ester-acid copolymers.

018. RARE EARTH METAL INITIATED POLYMERIZATIONS AND COPOLYMERIZATIONS OF ALKYL (METH)ACRYLATES H. Yasuda,* E. Ihara, M. Morimoto, and T. Kakehi, Department of Applied Chemistry, Faculty of Engineering, Hiroshima University, Higashi-Hiroshima 739, Japan

Rare earth metal complexes such as $[SmH(C_5Me_5)_2]_2$ and $SmMe(C_5Me_5)_2(THF)$ were found to proceed the living polymerization of alkyl methacrylate. The polymerization satisfy the following 4 requirements, i.e. 1)high molecuylar weight polymers($M_n > 400,000$), 2)extremely narrow molecular weight distribution (M_W/M_n < 1.05), 3) high conversion of polymers in a short period(> 95 %), and 4) high syndiotacticity (> -95%). Alkyl acrylate also performed the living polymerization by the efficient catalysis of rare arth metal complexes. However, in this case, stereoregularity of the resulting polymers was very low. By taking advantages of the living nature observed for the polymeriztion of alkyl methacrtylate and alkyl acrylates, we have performed the ABA type block copolymerization of methyl methacrylate/butyl acrylate/methyl methacry-late to find the rubber like elastic properties. As a result, the 8:72:20 block copolymer showed the good elastic properties. As an extension of this study, we have prepared the block copolymer of methyl methacrylate with methacrylic acid by the polymerization of methyl methacrylate and trimethylsilyl methacrylate followed by the hydrolysis.

019. LIGATED SYNTHESIS OF STYRENE-(METH)ACRYLATE BLOCK COPOLYMERS AND FUNCTIONALIZED POLYMETHACRYLATES. P. Vlček, J. Otoupalová, J. Kříž, Institute of Macromolecular Chemistry, Academy of Sciences of the Czech Republic, 162 06 Prague 6, Czech Republic; R.J. Composto, R. Oslanec, Department of Materials Science & Engineering, University of Pennsylvania, Philadephia PA 19104-6272, U.S.A.

Block copolymerization of non-polar and polar vinyl monomers might be complicated by condensation reactions of carbanion and carbonyl group, e. g. in methacrylate. The present method using organometalics/*tert*.alkoxide complexes for block copolymerization of styrene and (meth)-acrylates, does not need any modification of highly nucleophilic polystyrene carbanion and produces block copolymers with narrow MWDs containing no homopolymer. The initiators based on an ester enolate/*tert.* butoxide complex or *tert.* butoxide alone give poly(meth)acrylates with masked COOH or OH groups, which can then be simply released. The method provides polymers quantitatively functionalized with a reactive group.

020. **Lanthanoid Alkoxide - Ketene System as A Simple and Novel Lanthanoid Initiator for Polymerization of Methacrylic Ester**

H. Sugimoto and S. Inoue

Faculty of Engineering, Science University of Tokyo, Kagurazaka, Shinjuku-ku, Tokyo 162, Japan
TEL: 81-3-3260-4271 (ext. 3493) FAX : 81-3-5261-4631 e-mail: hrssgmt@ci. kagu. sut. ac. jp

We have recently discovered that La(O^{i}Pr)$_3$ brings about the polymerization of diethylketene (Et$_2$CCO) via a La enolate as the growing species to afford a polyester. Taking into account this finding, in the present study, by using as initiator a La enolate obtained by the reaction of La(O^{i}Pr)$_3$ and ketene, we succeeded in the polymerization of MMA which cannot be achieved with La(O^{i}Pr)$_3$ alone as initiator. For example, the reaction mixture of La(O^{i}Pr)$_3$ with 10 equiv of Et$_2$CCO in THF for 4 min brought about the polymerization of MMA (200 eq) at room temperature to give a polymer in 69 % in 24 h (Mn = 15,000). Under similar conditions, the polymerization of MMA (100 eq) with the La(O^{i}Pr)$_3$ - diphenylketene system (1 : 10) gave a polymer with MMA and diphenylketene units as revealed by GPC with UV detection in 17 % (Mn = 12,000).

021. INTRODUCTION TO BIOMEDICAL AND BIOTECHNICAL APPLICATIONS OF POLYETHYLENE GLYCOL. J. Milton Harris, Department of Chemistry, University of Alabama in Huntsville, Huntsville, AL 35899

The goal of this presentation is to provide an introduction to the current symposium by giving a brief overview of the following applications: 1. PEG-protein conjugates for pharmaceutical applications; 2. PEG-enzyme conjugates for industrial processing; 3.

surface modification with PEG to provide protein- and cell-rejecting surfaces; 4. surface modification with PEG to provide control of electroosmosis; 5. two phase partitioning for protein and cell purification; 6. PEG hydrogels for cell encapsulation, drug delivery and wound covering; 7. PEG-modification of small-molecule pharmaceuticals; 8. PEG tethers for synthesis of biomolecules; 9. PEG tethering of molecules for biological targeting and signaling; 10. PEG-liposomes and micelles for drug delivery. There have also been important advances in the key ancillary areas of synthesis of PEG derivatives and new PEG backbones or architectures, and analysis of PEG conjugates.

022. PEG-GRAFTED POLYMERS AS DRUG CARRIERS. E.H. Schacht, K. Hoste, Polymer Materials Research Group, Institute for Biomedical Technologies (IBITECH), University of Ghent, Krijgslaan 281, B-9000 Ghent, Belgium

Poly(ethylene glycol) (PEG) grafted derivatives of dextran and poly[N(2-hydroxyethyl)glutamine] (PHEG) were prepared by reaction of α-methyl ω-amino PEG with 4-nitrophenyl chloroformate activated polymer. The PEG-grafted polymers behave as amphiphatic compounds. In solid state the PEG side groups can form a crystalline phase. PEG-grafter PHEG can still be degraded by lysosomal enzymes. The PEG-grafted polymers are suitable as carriers for the preparation of macromolecular prodrugs.

023. **GRAFT COPOLYMERS OF PEO-PPO-PEO TRIBLOCK POLYETHERS ON BIOADHESIVE POLYMER BACKBONES: SYNTHESIS AND PROPERTIES**

Allan S. Hoffman, Guohua Chen, Xiangdong Wu, Zhongli Ding
Center for Bioengineering, University of Washington, Seattle, WA;
Baghwati Kabra and Kiran Randeri, Alcon Corp., Ft. Worth, TX*;
Matt Schiller and Eyal Ron, GelMed Corp., Bedford, MA*;
Nikolaos Peppas and Chris Brazel, Purdue University, Lafayette, IN**

We have synthesized novel graft copolymer compositions by grafting temperature-sensitive Pluronic® side chains onto PAAc and chitosan backbones. These interesting graft copolymers form low viscosity solutions at room temperature that gel when the temperature is raised above the cloud point temperatures of the grafted side chains. Crosslinked hydrogels of the PAAc graft copolymers are shown to be bioadhesive. These unique graft copolymers should be suitable as drug carriers for prolonged drug release to mucosal surfaces such as the eye, nose, intestines and vagina. (These collaborators carried out viscosity vs. temperature*, or bioadhesion** tests).

024. NOVEL DEGRADABLE PEG ESTERS FOR DRUG DELIVERY: SYNTHESIS AND CHARACTERIZATION. Xuan Zhao and J. Milton Harris, Materials Science Program, University of Alabama in Huntsville, Huntsville, AL 35899

In this presentation we describe two applications of new hydrolytically-degradable PEG derivatives. The first involves preparation and characterization of degradable PEG hydrogels. These PEGs contain ester linkages and can be prepared by reaction of PEG carboxylic acids with PEG. The degradation rates of these gels can be fine-tuned by variation in molecular structure of the esters and in the degree of branching of the PEGs. The second application involves preparation of soluble PEG-protein conjugates in which the PEG is attached to the protein via a hydrolytically degradable ester linkage. The ester-containing PEGs are prepared by condensation of PEG carboxylic acids with small hydroxy-acids. Conversion to succinimidyl active esters permits coupling to amino groups of proteins. Varying the types of PEG acid and hydroxy-acid gives control over the rates of hydrolysis. Applications to drug delivery will be presented.

POLY

025. PEG-BASED HYDROGELS FOR BLOCKING, PRESENTING AND RECEIVING INFORMATION IN THE CONTROL OF WOUND HEALING. J. A. Hubbell, Division of Chemistry and Chemical Engineering, California Institute of Technology, Pasadena, CA 91125

Biocompatibility can be thought of as relating to information transfer, communication between a material and nearby cells in a biological environment. Many of communication signals are biochemical in nature, based on glycoprotein or glycosaminoglycan structures that are either immobilized or diffusible. A precise knowledge of these biochemical communication signals can be exploited in the development of materials that either mimic or block these signals. Examples in material design for wound healing and tissue engineering will be addressed to illustrate the following basic operations in biomaterials communication: (1) blocking untoward information provided by a tissue to cells in a healing response; (2) providing new immobilized or diffusible information in the aforementioned situation; and (3) providing in a material the ability to respond to a healing environment, i.e. building into a material the ability to listen, not just to speak.

026. Molecular Design of Heterotelechelic Poly(ethylene glycol) for Superbiomolecule Yukio Nagasaki, Masao Kato, Kazunori Kataoka Department of Materials Science and Technology, Science University of Tokyo, Noda 278, Japan (nagasaki@rs.noda.sut.ac.jp)

This paper summarizes 1) synthetic methods of heterotelechelic poly(ethylene glycol) (heteroPEG); 2) preparation of PEG surface brushes with functional free end; and 3) preparation of reactive polymeric micelles. Potassium alkoxide having functional group such as acetal and cyano groups initiated the polymerization of ethylene oxide to form PEG with functional group at α-end. By the introduction of different functional group at w-chain end, new heterotelechelic PEGs were obtained with high yield and high purity. The heteroPEG having a polymerizable end group can be copolymerized with common hydrophobic monomers to form a graft copolymer. On a surface, the graft copolymer formed polymer brushes having functional free end. Block copolymerization of lactide from the end group of heteroPEG gave PEG-PLA block copolymer having functional group at PEG chain end. In aqueous media, these block copolymers formed polymeric micelle which possesses functional groups on their surface and regarded as a reactive polymeric micelle. Such heteroPEG derivatives are promising as high performance biomaterials.

027. SYNTHESIS OF PEO STAR MOLECULES BASED ON PAMAM DENDRIMER CORES D. R. Yen and E. W. Merrill, Department of Chemical Engineering, Massachusetts Institute of Technology, Cambridge, MA 02139

A new method of synthesizing a monodisperse sample of poly(ethylene oxide) star molecules is described. It is based on reacting preformed arms consisting of functionalized derivatives of linear PEO with polyamidoamine dendrimer cores. The molecular weight and polydispersity of the newly formed star molecules were determined using gel permeation chromatography in series with light scattering. Stars with up to 144 arms have been prepared using this method.

028. POLY(ETHYLENE GLYCOL) AND CYCLODEXTRIN AS NOVEL BUILDING UNITS FOR MOLECULAR ASSEMBLY AND POLYROTAXANES. Akira Harada*, Jun Li*, Mikiharu Kamachi*, Yasuyuki Kitagawa[+], and Yukiteru Katsube[+], *Department of Macromolecular Science, Graduate School of Science, Osaka University, Toyonaka, Osaka 560, Japan. [+]Institute of Protein Research, Osaka University, Suita, Osaka 565, Japan

Complex formation between α-cyclodextrin (α-CD) and oligo(ethylene glycol)s (OEG) which have been prepared by stepwise synthesis coupled with preparative SEC has been studied. α-CD was found to form complexes with tetra(ethylene glycol) and larger OEGs.

The yields of the complexes increase with an increase in the degree of polymerization. Eicosakis(ethylene glycol) and larger OEGs formed complexes with α-CD almost quantitatively. The complexes are 2:1(two ethylene glycol units and one α-CD). The X-ray patterns and ^{13}C CP/MAS NMR spectra of the complexes suggest that an OEG chain is included in the channel formed by α-CDs. X-ray studies on a single crystal of a complex of α-CD with hexaethylene glycol showed that CDs form an column in head-to-head and tail-to-tail fashion and a hexaethylene glycol chain is included in a channel formed by CDs.

029. APPLICATION OF ION MOBILITY/ION CHROMATOGRAPHY METHODS TO DETERMINE GAS PHASE CONFORMATIONS AND SIZE DISTRIBUTIONS OF PEG AND OTHER POLYMERS. M. T. Bowers, T. Wyttenbach, J.J. Batka, P. Weis and J. Gidden, Department of Chemistry, University of California, Santa Barbara, CA 93106

A matrix assisted laser desorption ionization (MALDI) source has been attached to our ion mobility/ion chromatography instrument. This configuration allows us to deliver intact polymer units into the gas phase. These are cationized by various alkali ions. In this talk I'll discuss methods for determining conformations of these species in the gas phase and how these conformations change with repeat number. The talk will primarily deal with PEG 600 polymers containing 9 to 20 monomer units. If there is time I will also discuss our efforts at determining accurate polymer size distributions using MALDI and extensive theoretical data analysis. Hopefully a general method will result suitable for application to most polymer systems. Initial applications have been on a number of poly (ethylene terephthalate) (PET) polymer systems but should work equally well on PEG.

030. CONTROLLING THE RESPONSE OF PDMS TO DYNAMIC FATIGUE BY FILLER CHOICE. Susan A. Visser, Eastman Kodak Company, Rochester, NY 14650-2129.

Filled polysiloxane elastomers are useful for high temperature applications because of their superior thermal stability; however, they can exhibit great instability when a cyclic load is added to the high temperature. A variety of inorganic fillers are investigated to determine their impact on the response of filled polysiloxane elastomers to cyclic stress at elevated temperature. For the first time, three fillers that give stable networks are identified: zinc oxide, tin oxide, and copper oxide. Four fillers are examined that enhance network degradation: aluminum oxide, calcium oxide, titanium oxide, and tungsten oxide. Silane-treated fillers are used to demonstrate that differences based on filler type do not arise because of changes in the strength of polymer-filler interactions. No correlation between stability of the filled elastomer to cyclic stress in hot air and filler particle size distribution, filler surface acid/base character, or the presence of specific inorganic contaminants in the filler is observed. It is proposed that the intrinsic ability of certain inorganic oxides to catalyze and/or inhibit polysiloxane degradation reactions determines the response of the filled elastomers to cyclic stress at elevated temperature.

031. PREPARATION OF PTFE AND PTFE / PMMA CORE-SHELL COMPOSITE PARTICLES IN CO_2 / AQUEOUS SYSTEMS. B. E. Kipp and J. M. DeSimone, Dept. of Chemistry, University of North Carolina at Chapel Hill, Chapel Hill, NC 27599. D. J. Pochan and S. P. Gido, Dept. of Polymer Science and Engineering, University of Massachusetts - Amherst, Amherst, MA 01003.

A new polymerization methodology for the synthesis of water insoluble polymeric materials in a bi-phasic CO_2 / aqueous system has been developed. CO_2 forms a second, unique condensed phase in which the density, viscosity, and dielectric constant can be manipulated by simple changes in pressure and temperature. In addition, CO_2 has the ability to plasticize glassy polymer particles, and in combination with an aqueous phase, can depress the pH. This process was first applied to the polymerization of tetrafluoroethylene initiated by ammonium persulfate in an aqueous phase both in the presence and absence of a dispersing agent, sodium perfluorooctanoate. Polytetrafluoroethylene (PTFE) was formed in high yields and reasonable molecular weight under these conditions. The latex produced in the dispersion polymerization was employed as a first stage latex in the formation of composite core-shell materials with methyl methacrylate as a second stage monomer. These spherical particles ranged from 50-70 nm in diameter, and contained a PTFE core engulfed by a poly(methyl methacrylate) shell.

032. ELECTROSTATIC INTERACTIONS IN IONOMER SOLUTIONS. Srinivas Nomula and Stuart L. Cooper, Department of Chemical Engineering, University of Delaware, Newark, Delaware 19716

Model polyurethane ionomers form a simple model system with regularly spaced ionic groups. Dynamic Light Scattering applied to the ionomer solutions reveals two diffusive modes in non-polar as well as polar solvents. Only one diffusive mode is obtained for the underivatized polyurethane. The two modes, a fast mode ($\sim 10^{-6}$ cm^2/s) and a slow mode ($\sim 10^{-8}$ cm^2/s), differ by two orders of magnitude. In a non-polar solvent, toluene, the fast mode and slow mode correspond to the diffusion of single chains and aggregated ionomer chains in solution respectively. In polar solutions, the fast mode diffusion coefficient is seen to increase with the polarity of the solvent showing that diffusion coefficient of single polyions is enhanced by electrostatic interactions. The slow mode is found to be due to structures which are easily disrupted. This talk will discuss the effect of polarity of the solvent on the structures in ionomer solutions. In a non-polar solvent, toluene, the aggregates are made of physically crosslinked chains where as in a polar solvent, DMAc, the aggregates consist of interaction domains.

033. A REEVALUATION OF NAFION® MORPHOLOGY. Morton Litt. Macromolecular Science Dept., Case Western Reserve University. Cleveland, OH 44106-7202

While the cluster-network morphology proposed by Gierke for Nafion® is generally accepted, there are major conceptual difficulties in the processes implied by the ionic cluster reorganization. An irregular lamellar structure agrees better with normal polymer physics concepts and has experimental confirmation. Planar micelles form when the polymer is "solubilized". Since films from "solubilized" polymer have identical properties as hydrolyzed films, this should be true for all Nafion-like polymers. A micellar aggregate held together by tie molecules absorbs water and expands in one dimension as the water pushes the micelles apart. This is seen in the long spacing (3.5 to 5 nm) that varies with the water content. This and other data have been used to test the hypothesis of a micellar aggregate morphology for Nafion.

034. CHROMOPHORE AGGREGATION AND THE PHOTOCHEMISTRY OF POLYMERS AND MODEL COMPOUNDS WITH ARYL CINNAMATE AND STILBENE BIS-CARBOXYLATE CHROMOPHORES. David Creed, Richard A. Cozad, Charles E. Hoyle, Lixin Jin, Venkataram Krishnan, Sangya S: Varma, Center for Macromolecular Photochemistry and Photophysics, Departments of Chemistry and Biochemistry and Polymer Science, University of Southern Mississippi, Hattiesburg, MS 39406

LC polymers in which aryl cinnamate and stilbene bis-carboxylate chromophores are also the rigid mesogenic groups exhibit dramatic chromophore aggregation effects that can perturb their photochemistry and photophysics in thin films. The effects observed include wavelength dependent photochemistry, enhanced triplet energy migration, and UV-irradiation induced hyperchromism. We now report that these polymers in poor solvents exhibit similar photochemistry and photophysics to that observed in thin films. Some of the spectroscopic and photochemical effects can be duplicated using small molecule model compounds dispersed in poly(methyl methacrylate) and/or poly(butyl acrylate). In the case of stilbene polymers and model compounds three types of fluorescence spectra can be observed: 'isolated chromophore' fluorescence, and 'excimer like' and 'structured' fluorescence. Both of the latter arise from excitation of ground state aggregates.

035. SYNTHESIS AND DYNAMIC MECHANICAL PROPERTIES OF ALIPHATIC POLYURETHANE ROTAXANES. Jason Hunt, Karthik Nagapudi, Haskell W. Beckham *Polymer Education and Research Center, School of Textile and Fiber Engineering, Georgia Institute of Technology, Atlanta, GA 30332-0295.*

Aliphatic polyurethane rotaxanes have been synthesized to study how phase structures and molecular dynamics influence bulk material properties in polyrotaxanes. Hexamethylene diisocyanate and ethylene glycol were polymerized in melted 30-crown-10 to yield poly(ethylene hexamethylenediurethane)-

rotaxa-30-crown-10 in multigram quantities. Repeated dissolution in hexafluoroisopropanol followed by precipitation in ethyl acetate (a good solvent for the crown) removed the unthreaded crowns to give pure polyrotaxane with 2% threaded cyclics (molar basis). Evidence of polyrotaxane formation includes the persistent crown ether signals (3.5 ppm in ^{1}H spectrum, 70 ppm in ^{13}C spectrum) after multiple reprecipitations. Dynamic mechanical spectra showed two transitions: a T_g transition at 45 °C, and a sub-T_g transition at -85 °C. The DMS spectrum of the polyurethane backbone was essentially identical. Solid-state ^{13}C NMR spectra revealed the crown ethers to be highly mobile compared to the polyurethane backbone.

036. THERMOSET SECOND-ORDER NLO MATERIALS FROM A TRIFUNCTIONALIZED CHROMOPHORE. Youn Soo Ra, Shane S. H. Mao, Bo Wu, Lan Guo and Larry R. Dalton*, Loker Hydrocarbon Research Institute, Department of Chemistry and Department of Materials Science and Engineering, University of Southern California, Los Angeles, CA 90089-1661; Antao Chen and William H. Steier, Center for Photonic Technology, Department of Electrical Engineering, University of Southern California, Los Angeles, California 90089-0483

A new trifunctionalized Disperse-Red type chromophore, Disperse-Red Triol (DRTO), was synthesized. Reaction of this chromophore with tolylenediisocyanate (TDI) at 80 oC afforded soluble oligomers, which were spin cast to form high optical quality films. Concurrent corona poling and thermosetting of these films resulted in a polyurethane network with an electro-optic coefficient (r33 at 1.06 mm) of 14.5 pm/V and excellent dynamic thermal stability. This system (PU-DRTO) offers some prominent advantages over the existing PU-DR19 system, in terms of simpler material processing, wider processing window and higher temporal stability of dipole

037. CHARACTERIZATION OF BLENDS OF MALEIC ANHYDRIDE GRAFTED EPDM AND NYLON 12. David P. Ching and Frank W. Mercer, Raychem Corporation, Menlo Park, CA, 94025 and Art F. Diaz, Material Science Department, San Jose State University, CA, 95192

There is need for soft but tough thermoplastic elastomers (TPEs) that could be used from room temperature up to 150 °C for mechanical sealing in automotive applications. A survey of low cost, commercial TPEs revealed that those materials do not meet the mechanical performance requirements. This presented an opportunity to design some new TPEs by graft copolymerization of commercial Nylon 12 with maleic anhydride functionalized EPDM rubbers. The TPEs were prepared by melt mixing using a Brabender Plasticorder® system. Preliminary results showed that compatibilized blends of Nylon 12 (10 - 40 wt%) and EPDM rubber were produced but evidence of grafting could not be determined by transmission FTIR. The grafting (less than 2 wt%) is probably below the lower limit of detection. SEM micrographs showed smaller Nylon 12 domains in the EPDM matrix for compatibilized blends. Melt rheology showed Nylon 12 with a suppressed recrystallization temperature (90 °C - 110 °C) in compatibilized blends.

038. FTIR SPECTROSCOPIC STUDIES OF ORIENTED SYNDIOTACTIC POLYPROPYLENE. K. Song and J. Ok, Department of Chemical Engineering, Kyung Hee University, Kyunggi-do 449-701, Korea

In a semicrystalline polymer such as s-PP, molecular orientations in crystalline and amorphous phases exhibit different behaviors during orientation of a sample. We studied such behaviors using a polarized FTIR spectroscopic technique. From the dichroic ratios of the characteristic bands of crystalline and amorphous phases, the average orientation of polymer chains relative to the drawing direction for each phases were expressed in terms of the Hermans orientation function. The crystalline orientation function rises rapidly and reaches a maximum value at low draw ratios while the chains in the amorphous phase are found to be oriented along the drawing direction much slowly to the lower orientation value. The dichroism of annealed sample was investigated to find out if the conformation change occurs without losing chain orientation along stretch direction.

039. CYCLOPOLYMERIZATION OF NOVEL DIALLYL QUATERNARY AMMONIUM MONOMERS TO YIELD ORDERED STRUCTURES IN AQUEOUS MEDIA
Charles L. McCormick and Yuxin Hu, Department of Polymer Science, The University of Southern Mississippi, Hattiesburg, Mississippi 39406-0076.

Series of amphiphilic homo- and copolymers of substituted N,N-diallyl ammonium salts were prepared utilizing photo-cyclopolymerization. These substituted diallyl ammonium salt monomers with long single and twin-tailed alkyl moieties form stable organized architectures in aqueous media. The photo-cyclopolymerization of these monomers and hydrophobe spacer comonomers can produce polymerized vesicles. Atomic force microscopy has been used to characterize the size and size distribution found in dilute aqueous media. In cases thus far examined, mixed comonomer vesicles and polymerized vesicles range in size from 200-300 nm. Size distribution of both appear to increase with increasing incorporation of the hydrophobic spacer comonomer, N, N-diallyl dimethyl ammonium chloride.

040. KINETIC MODELING OF CROSSLINKING REACTIONS OF CYCLOALIPHATIC EPOXIDES WITH HYDROXYL AND CARBOXYL FUNCTIONALIZED ACRYLIC COPOLYMERS: II. CATALYST EFFECTS
Shaobing Wu, Mark D. Soucek*, Department of Polymers & Coatings, North Dakota State University, Fargo, ND 58105.

The crosslinking reaction kinetics of cycloaliphatic diepoxides with both hydroxyl and carboxyl functionalized acrylic copolymers were investigated using model compounds. Cyclohexene oxide, methanol, and acetic acid were used to model cycloaliphatic diepoxides, hydroxy ethylmethacrylate, and acrylic acid, respectively. The reactions of cyclohexene oxide with methanol and acetic acid were performed as a function of both acid and base catalysts. It was observed that esterification and epoxy hydrolysis were significantly increased when p-toluene sulfonic acid was used as the catalyst. Tertiary amines were effective catalysts only for the formation of ester products at low concentrations. As the amine concentrations were increased, the amines were shown to generally inhibit the formation of both the ether and ester products.

041. N,N'-DIALKYL SUBSTITUTED POLYOXYALKYLENE AMINES AS CURING AGENTS FOR BLOCKED ISOCYANATES. J. J. Lin, Department of Chemical Engineering, National Chung-Hsing University, Taichung, Taiwan, and G.P. Speranza. M. Cuscurida, Huntsman Corporation, Austin, TX 78752.

Commercially available polyoxyalkylene polyamines are produced by ammonia amination of the corresponding polyether polyols. It is found that those polyether polyamines can be converted into secondary amines by an one-step reductive amination process of amines and ketones. More specifically, a series of secondary isopropylamines terminated polyoxyethylene and polyoxypropylene derivatives are produced by hydrogenation of the mixtures of polyether primary amines /acetone in the presence of catalyst. These new products, including polyether diamines and triamine with molecular weights ranged from ca. 200 to 2000, have been evaluated as curing agents for epoxy resins and blocked isocyantes. As a curing agent for the blocked isocyanate prepolymer, N,N'-dialky polyetherdiamine shows the improved properties of a heated cured coating, as well as the improved storage stability of uncured one-component coating, as compared with those made from its premiary amine precursor.

042. NITROXIDE INITIATED/MEDIATED POLYMERIZATION OF STYRENE: ANALYSIS OF END-GROUPS. Yucheng Zhu and B. A. Howell, Center for Applications in Polymer Science, Central Michigan University, Mount Pleasant, MI 48859 and D. B. Priddy, Dow Polystyrene R&D, 438 Bldg, Midland, MI 48667

Recently much attention has been devoted to developing an understanding of nitroxide mediated polymerization (NMP). It is believed that NMP virtually eliminates chain transfer and termination resulting in excellent control (>95%) of chain-end architecture. To further study the mechanism of NMP, we synthesized NMP initiator/mediators having a UV/vis chromophore attached to either the initiating or terminating fragments. This allowed us to both qualitatively and quantitatively analyze the chain-ends using a GPC-UV/vis technique. This technique showed that: 1) the chromophores are attached to the chain-ends, 2) initiating fragments give higher chain-end functionalization than the terminating nitroxyl fragment, 3) chain-end purity diminishes with both conversion and MW, 4) suppression of styrene autopolymerization by the addition of camphorsulfonic acid increases end-group purity, and 5) chain-end purity of >90% is only achieved when making very low MW (i.e., <10,000) polystyrene.

043.

UV CURABLE POLYIMIDES FROM PHOTOCHEMICALLY GENERATED BISDIENES AND BISMALEIMIDES. **M.A.Meador*** J.Jones**, M.A.B. Meador*, D.A. Scheiman***, and L.L. Williams****, * Materials Division, NASA Lewis Research Center, Cleveland, OH 44135; ** Department of Chemistry, Clark Atlanta University, Atlanta, GA 30314; *** NYMA, Inc., 2001 Aerospace Pkwy, Brook Park, OH 44142; **** Department of Chemistry, Spellman College, Atlanta, GA 30314

A new route to polyimides has been developed based upon the Diels-Alder cycloaddition between various bismaleimides and bis(o-quinodimethanes) generated from the photolysis of 2,5-diaroyl-p-xylenes. The resulting polyimides have onsets of decomposition (measured by TGA) in air above 310C and values above 360C in nitrogen. Glass transition temperatures for some of these polyimides were as high as 300C. Polyimides prepared via this route from 2,5-(bis(4-dodecyloxybenzoyl))-p-xylene were soluble in NMP, 1,2-dichloroethane, and benzene.

044.

SYNTHETIC ANALOGS OF MARINE MUSSEL CEMENT PROTEINS. Miaoer Yu, Tiziana DeSimone, and Timothy J. Deming*, Materials and Chemistry Departments, University of California, Santa Barbara, California 93106

The design, synthesis, and characterization of chemically prepared polymeric analogs of naturally occurring marine adhesive proteins will be described. Using α-amino acid-N-carboxyanhydride monomers we have prepared copolypeptides containing the catechol and amino functional side chains of L-dihydroxyphenylalanine (L-DOPA) and L-lysine. Through variation of copolymer composition and reaction conditions, we have been able to develop polymers which are able to form crosslinked networks similar in character to those formed with the natural adhesive proteins. The relationship of copolymer composition to adhesive forming capability will also be discussed. In addition to studies on polymeric materials, we have also examined small molecule reactions involving the putative reactive functional groups contributing to crosslink and adhesive bond formation. UV/vis, ^{1}H NMR, and HPLC analyses of these reactions will be reported and the results compared to the behavior of the polymer chains.

045. DNA AS A REACTION MEDIATOR ? A COMPARISON OF 9-ANILINOACRIDINE THIOLYSIS RATES IN THE PRESENCE AND ABSENCE OF DOUBLE-STRANDED DNAs. Gary D. Jaycox, Department of Chemistry, Dartmouth College, Hanover, NH 03755.

The thiolytic cleavage of a series of 9-anilinoacridines was attempted both in the presence and absence of double-stranded DNA polymers. In the absence of DNA, all of the compounds evaluated were aggressively cleaved by ethanethiol and glutathione under the reaction conditions employed. Thiolysis reactions were pseudo first-order with rate constants falling near 5.7×10^{-3} min^{-1} at the 25 °C isotherm. Striking differences in reaction rates were noted when the thiolysis assays were carried out in the presence of DNA polymers. Anilinoacridines that were capable of fully intercalating into the DNA base stack were totally protected from the cleavage reaction. The same level of protection was not afforded to derivatives that lacked proper geometries for DNA intercalation. This shape/size selectivity shown by the DNA template is reminiscent of other host systems that have been successfully utilized to mediate chemical transformations in geometrically constrained environments (e.g. cyclodextrin cavities, zeolite channels). The implications of these results will be discussed along with opportunities for further work in this area.

Current Address: DuPont Central Research and Development, Exp. Station, Wilmington, DE 19880-0328

046. ROLE OF THE NONIONIC SURFACTANT TRITON X-405 IN EMULSION POLYMERIZATION OF STYRENE. E. Özdeğer, E.D. Sudol, M.S. El-Aasser, and A. Klein, Emulsion Polymers Institute and Department of Chemical Engineering, Lehigh University, Bethlehem, PA 18015

The emulsion polymerization of styrene was studied using the nonionic surfactant Triton X-405 (octylphenoxy polyethoxy ethanol). Two separate nucleation periods were noted in these polymerizations resulting in bimodal final latex particle size distributions. The partitioning of the surfactant between the phases was found to play the major role in the nucleation mechanism(s) in these polymerizations. Although the total concentration of the emulsifier was always added at a level above its critical micelle concentration (CMC) based on the water phase in the recipe, it was found that the portion of the surfactant initially present in the aqueous phase was below its CMC due to the partitioning. This CMC was also found to increase with increasing total surfactant because of the distribution of the surfactant (varying ethylene oxide chain length) between the phases. Under these conditions, the first of the two nucleation periods was attributed to homogeneous nucleation, followed by micellar nucleation.

047. MODIFICATION OF EPOXY RESINS BY SECONDARY AMINE TERMINATED POLY(DIMETHYL SILOXANE). Q. Ji, O. Kwon, H. Zhuang, Y. N. Liu, M. Muggli, K. Kwan, T. C. Ward and J. E. McGrath, Virginia Polytechnic Institute and State University, Department of Chemistry and NSF Science and Technology Center, High Performance Polymeric Adhesives and Composites, Blacksburg, VA 24061-0344.

Secondary amine terminated poly(dimethyl siloxane) (PDMS) has been used to chemically modify bisphenol A based epoxy resins. The compatibility of epoxy resin and PDMS were controlled by PDMS concentration and curing condition. It was found that the siloxane and epoxy resins are miscible with higher siloxane concentration and lower temperature. The fracture toughness was increased by introducing PDMS

048. (METH)ACRYLATES POLYMERIZATION IN THE PRESENCE OF TRANSITION METAL COMPLEXES : THE KHARASCH REACTION REVISITED, Ph. Teyssié, Ph. Dubois, C. Granel, R. Jérôme, Ph. Lecomte, G. Moineau, CERM, University of Liège, Sart-Tilman, B6, 4000 Liège, Belgium

Controlled ("pseudo-living") alkyl(meth)acrylate radical-type polymerization has been performed using combinations of transition metal complexes, and promoters bearing an easily leaving group (halogen atoms, carboxylic groups,...). The M_T complexes include the van Kooten chelated phenylnickel salts, and different palladium (II) and rhodium (I) ligated salts. Several aspects of these reactions will be illustrated, such as : complex kinetic behaviour, catalytic nature versus M_T, relative importance of different oxidation states, tolerance versus water and emulsifiers, apparent sensitivity towards monomer nature and final polymer structure. A tentative mechanistic discussion will be based on these data. The synthesis of narrow MWD, high MW (up to 150 KD) and highly regioregular polymers will be reported, as well as that of different block copolymers.Finally, the apparent lack of stereoselectivity of this type of reactions will be discussed.

049. AN INVESTIGATION INTO THE MECHANISM OF CATALYTIC CHAIN TRANSFER POLYMERISATION (CCTP) BY FOURIER TRANSFORM MASS SPECTROMETRY, David M Haddleton*, Peter J Derrick, Jan Axelsson, Albert J R Heck, Kevin G Suddaby and Darren R Maloney, Department of Chemistry, and Institute of Mass Spectrometry, University of Warwick, Coventry, CV4 7AL, U.K.

Catalytic chain transfer by certain low spin cobalt(II) compounds in methacrylate polymerisation has been demonstrated to be more than four orders of magnitude greater than conventional chain transfer agents such as mercaptans. As such it is attracting widespread industrial interest so as to reduce cost, toxicity and odours arising from conventional chain transfer. The mechanism of CCTP is believed to involve the perturbation of propagation by abstraction of a hydrogen atom from propagating polymer; yielding an unsaturated macromonomer product. We have been utilising high resolution ESI-FTMS to investigate this problem. ESI-FTMS has enabled us to not only tell the empirical formula unequivocally but by the combination of polymerisation experiments with FTMS we have been able to determine the position of structural units within an acrylic polymer chain. Both ESI-MS and FTMS are proving extremely powerful techniques for polymer analysis.

050. **The "Living" Stable Free Radical Polymerization - The Acrylate Conundrum** Michael K. Georges, Nancy A. Listigovers, Peter G. Odell, Gordon K. Hamer, Marion Quinlan, and Richard P.N.Veregin *Xerox Research Centre of Canada, 2660 Speakman Drive., Mississauga, Ontario, Canada L5K 2L1*

Initial attempts to prepare homopolymers of acrylates under a variety of conditions by the nitroxide mediated stable free radical polymerization (SFRP) process proved unsuccessful. These result were baffling in light of the fact that synthesizing random copolymers of acrylates with styrene was possible. In addition, chain extensions of TEMPO-terminated polystyrene was possible with dienes and substituted styrenes but not with acrylates. In this presentation, these results will be discussed along with experiments that were performed to gain further insight into the acrylate problem. The knowledge gained from these studies has enabled the successful synthesis of acrylates.

051. **CONTROLLED/"LIVING" RADICAL POLYMERIZATION OF METHYL METHACRYLATE AND VARIOUS FUNCTIONAL ACRYLATES BY ATOM TRANSFER RADICAL POLYMERIZATION.** Krzysztof Matyjaszewski and Thomas Grimaud, Department of Chemistry, Carnegie Mellon University, 4400 Fifth Avenue, Pittsburgh PA 15213.

The polymerization of methyl methacrylate (MMA) and various substituted (meth)acrylates using Atom Transfer Radical Polymerization (ATRP) is reported. By using a homogeneous catalytic system, copper (I) bromide (CuBr) and 4,4'-di-(5-nonyl)-2,2'-bipyridine (dNbipy) as the ligands, in conjunction with various sulfonyl chlorides or reactive alkyl halides as initiators, the polymerization of methyl methacrylate proceeded in a controlled way. Poly(methyl methacrylates) with polydispersities as low as 1.1 were synthesized, while molecular weights were controlled until $M_n \approx 180,000$.

052. **END FUNCTIONAL COPOLYMERS BY FREE RADICAL ADDITION-FRAGMENTATION CHAIN TRANSFER**
Brian P. Devlin, Thomas R. Darling, Charles T. Berge, Michael J. Darmon, Michael C. Grady, Joan E. Hansen, William J. Simonsick, Robert R. Matheson, Lance L. Litty, Donald A. Paquet[1], Lech Wilczek[1] and Alexei A. Gridnev[1]

DuPont Automotive Finishes, Marshall R&D Laboratory, 3401 Grays Ferry Ave., Philadelphia, PA 19146
[1]DuPont Central Science and Engineering, Experimental Station, Wilmington DE 19898

A method for producing acrylic polymers of controlled m. wt. and end group functionality is described. The procedure utilizes pure dimeric and oligomeric ω-vinyl terminated polymethacrylates as addition-fragmentation chain transfer agents. Functional chain transfer agents, for example, those made from hydroxyl, epoxy, etc. substituted methacrylate esters, provide a convenient route to end group functional polymers under high yield, conventional free radical polymerization conditions. The efficient use of these transfer agents in the preparation of copolymers with a minimum of two functional groups per chain is discussed.

POLY

053. STABLE FREE RADICAL POLYMERIZATION OF ACRYLATES USING MACROINITIATORS. Paula J. MacLeod, Michael K. Georges, Marion Quinlan, Karen A. Moffat, and Nancy A. Listigovers, *Xerox Research Centre of Canada, 2660 Speakman Drive, Mississauga, Ontario, Canada, L5K 2L1*

A TEMPO-terminated polystyrene oligomer (PS-T) can be used as a macroinitiator for acrylate monomer. The resulting polymers are living in that the molecular weight distribution shifts to higher molecular weight as a function of time. Over time, the propagation rate tends to slow down presumably due to increased levels of free nitroxide. If the free nitroxide level is reduced, for example by adding 3-indole butyric acid, a reasonable propagation rate can be maintained. Initiator adducts, such as MB-TMP and MB-DBN, can also be used to initiate controlled acrylate polymerizations and results using adducts will also be presented.

MB-TMP   MB-DBN

054.

MINIEMULSION COPOLYMERIZATION OF ACRYLATES, J. L. Reimers and F. J. Schork, School of Chemical Engineering, Georgia Institute of Technology, Atlanta, GA 30332-0100

The diffusional degradation of emulsions can be eliminated by the addition of a water-insoluble reagent referred to as a cosurfactant. The resulting emulsions exhibit long-term stability, and produce droplets in the range of 100 to 500 nm diameter. Under these conditions, and unlike conventional emulsion polymerization, the droplets become the locus of particle nucleation, and the polymerization process is referred to as *miniemulsion polymerization*.

Since there is no dependence on the transport of monomer across the aqueous phase, miniemulsion polymerization is expected to yield copolymers that more nearly follow the copolymer equation than those resulting from conventional emulsion polymerization. This paper deals with the miniemulsion and convectional emulsion copolymerization of highly water-insoluble comonomers with methyl methacrylate. The effect of the polymerization process (mini- or macroemulsion) on the copolymer composition will be examined.

055. POLY(ETHYLENE OXIDE)-BEARING LIPIDS RECONSTITUTED IN LIPOSOMES AND THEIR BIOLOGICAL FUNCTIONS. J. Sunamoto, Department of Synthtic Chemistry & Biological Chemistry, Graduate School of Engineering, Kyoto University, Sakyo-ku, Yoshida Hommachi, Kyoto 606, Japan

Liposome is composed of a closed lipid bilayer membrane, and essentially the same membrane structure constitutes fundamentals of cytoplasm membrane. Therefore, cell-liposome fusion has a potential application in medicinal fields or in cellular engineering. On the other hand, poly(ethylene oxide) (PEO) has been commonly used as an artificial fusogen for fusion of protoplasts or of mammalian cells. However, for the fusion with PEO, cytotoxicity of PEO itself or undesired fusion homophilic between the same cells or the same liposomes are accompanied. For a quest of simple and well-regulated cell-liposome fusion under mild conditions, we have been studying an artificial lipid that bears PEO moiety at the head group (PEO-lipid). The PEO moiety is to be localized near the outer surface of liposome, where the fusion is supposed to take place. In this article, we would like to describe function of the PEO-lipid reconstituted liposome paying attention also to be physicochemical characteristics of the PEO-lipid.

056. RECENT DEVELOPMENTS IN LIGAND-BEARING POLYMER-GRAFTED LIPOSOMES Samuel Zalipsky, SEQUUS Pharmaceuticals, Inc., 960 Hamilton Court, Menlo Park, CA 94025

Various synthetic methods for preparation of end-group functionalized polyethylene glycol-distearoyl-phosphatidylethanolamine (PEG-DSPE) derivatives were established using heterobifunctional PEGs. These new materials are used for preparation of PEG-grafted liposomes, which contain biologically relevant ligands covalently fixed at the periphery of the polymeric "brush" covering the vesicles. A balance between preservation of biological activity of the ligand and reasonably long circulating plasma lifetime could be found in almost every case. This suggests that this technology has an array of applications in targeted liposomal drug delivery as well as in platform presentation of useful ligands, which in their free form are cleared from systemic circulation too fast to be useful. Recently developed approaches to the systems containing ligands bound to the termini of liposomal surface grafted PEG chains, as well as their properties and applications will be reviewed.

057. TARGETABILITY OF PEG-IMMUNOLIPOSOMES CONJUGATING ANTIBODIES AT THE ENDS OF PEG CHAINS K. Maruyama[1], T. Takizawa[1], T. Tagawa[2], K. Nagaike[2] and M. Iwatsuru[1]. [1]Faculty of Pharmaceutical Sciences, Teikyo University. Sagamiko, Kanagawa 199-01 Japan. [2]Mitsubishi Chemical Co., Bioscience Laboratory, Aoba-ku, Yokohama 227 Japan.

Drug delivery to specific site by immunoliposomes represents a potentially attractive mode of therapy. For a much less accessible target sites, such as tumor cell surface antigens in a solid tumor, only PEG-immunoliposome would be expected to circulate long enough to reach the target. Dipalmitoyl-N-(6-maleimide caproyloxy polyethyleneglycol succinyl) PE (DPPE-PEG-Mal), distearoyl-N-(3-carboxypropionoyl poly ethyleneglycol succinyl) PE (DSPE-PEG-COOH) were newly synthesized and used to prepare immuno-liposomes carrying anti-human CEA antibodies (21B2) or their Fab' fragments at the ends of the PEG chains. CEA-positive human gastric cancer strain MKN-45 cells were inoculated in BALB/c nu/nu mice. The increasing amount of 21B2 showed a high liver uptake of PEG-mrnunoliposomes by Fc receptor-mediated mechanism. The conjugation of Fab' fragments of 21B2 showed a prolonged circulation time due to low liver uptake. Fab'-PEG-immunoliposomes sized to ~100nm in mean diameter showed significantly greater accumulation in solid tumor. Small Fab'-PEG-immunoliposomes could predominantly pass through the leaky tumor endothelium by passive convective transport and bind tumor cells.

058. POLY(ETHYLENE OXIDE) COATED LIPOSOMES: THEORY AND PRACTICE
D.D. Lasic, Consultant, 7512 Birkdale Dr., Newark, CA 94560

Nonspecific reactivity with the milieu was the major reason why liposomes did not fulfill expectations as a drug carrier for systemic delivery, with short blood circulation times being its most important manifestation. Biological stability of liposomes, however, increased up to two orders of magnitude upon coating their surface with inert polymers, such as PEO.

Qualitative explanation that the prolonged blood circulation is due to steric stabilization was confirmed by force-distance measurements using osmotic stress technique and surface force apparatus. Simple scaling laws could fit measured profiles and this analysis showed that for effective stabilization, polymer thickness should be around 5-10% of the vesicle diameter. In practice, such liposomes laden with anti-tumor agent doxorubicin exhibited prolonged blood circulation and accumulation in tumors where vasculature is leaky, resulting in first commercially available liposomal formulation.

059. AMPHIPHILIC POLYETHYLENE GLYCOL DERIVATIVES: LONG-CIRCULATING MICELLAR CARRIERS FOR THERAPEUTIC AND DIAGNOSTIC AGENTS. V.P.Torchilin and V.S.Trubetskoy, Center for Imaging and Pharmaceutical Research, Massachusetts General Hospital and Harvard Medical School, Charlestown, MA 02129.

Micellar pharmaceutical carriers have been prepared using amphiphilic polyethylene glycol (PEG) derivatives. Micelles obtained are very stable (even at concentrations below 1 µg/ml), have small size of 10 to 50 nm, narrow size distribution, and long circulation time *in vivo*. Micelles can be loaded with both, therapeutic and diagnostic agents. The loading can be performed chemically - by modification of hydrophobic block in PEG derivative with the molecule of interest, or physically - by entrapment of poorly soluble or artificially insolubilized drug or diagnostic label into the hydrophobic core of the micelle. PEG-based micelles were prepared suitable for magnetic resonance imaging (contained Gd), gamma-imaging (contained radiometal, such as 111-In), and for computed tomography (contained organic iodine). Alternatively, micelles were loaded with poorly soluble biologically active compounds, such as ellipticine. Typical example of preparation of PEG-based diagnostic micelles (for computed tomography) includes PEG derivatization with polylysine, modification of polylysine block with organic iodine, and spontaneous micellization of heavily iodinated amphiphilic PEG.

POLY

060. WATER-SOLUBLE POLYION COMPLEX ASSOCIATES OF DNA AND PEG-P(L-LYSINE) BLOCK COPOLYMER. Kazunori Kataoka and Satoshi Katayose, Department of Materials Science and Technology, Science University of Tokyo, Noda 278, Japan (kataoka@rs.noda.sut.ac.jp)

Complex formation of poly(ethylene glycol)-poly(L-lysine) (PEG-PLL) AB type block copolymer with salmon testes DNA or Col E1 plasmid DNA in aqueous milieu were studied from a standpoint of designing novel gene vector system used *in vivo*. The PLL segment of PEG-PLL conjugates with DNA through an electrostatic interaction to form a water-soluble and electrically stoichiometric complex. PEG segments surrounding the core of the polyion complex prevented the complex from precipitation even under electrically neutralized condition. The profile of the thermal melting curve revealed a higher stabilization of DNA structure in PEG-PLL/DNA complexes compared to that in the complex made from DNA and PLL homopolymer with the same molecular weight as the PLL segment in PEG-PLL. This stabilizing effect on the DNA structure may be due to the compartmentalization of DNA into the microenvironment of PEG with low permittivity. The reversible nature of PEG-PLL/DNA complex was further verified through the addition of polyanion (Poly-L-aspartic acid): Poly-L-aspartic acid replaced DNA in the complex with PEG-PLL, resulting in the release of free DNA in the medium. Further, PEG-PLL/DNA complex showed high resistance against DNase I attack, suggesting DNA protection through the segregation into the core of the associate having PEG palisade.

061. BARRIER PROPERTIES OF SURFACE-GRAFTED PEG FOR MACROMOLECULE-SURFACE INTERACTIONS. David Needham and Doris Noppl-Simson Department of Mechanical Engineering and Materials Science, Duke University, Durham, NC 27708-0300.

When grafted to a lipid bilayer membrane (by incorporating PEG-lipids), Polyethylene Glycol can act as an effective barrier that inhibits or prevents the close approach of other surfaces, macromolecules, and other colloidal particles. In order to explore the barrier properties of PEG-lipids for macromolecular interactions we have used micropipet techniques on single vesicles to investigate the extent to which grafted-PEG can inhibit the adsorption and binding of a test "ligand" (avidin) for its surface "receptor" (biotin-lipid). The incorporation of 2 mol% PEG^{750}-lipid slowed the diffusion of fluorescent avidin to the vesicle surface by a factor of 5. By adding increasing amounts of surface-grafted PEG^{750}-lipid, the binding of avidin to a surface containing 5 mol% biotin was decreased to background levels when the surface concentration reached 10 mol% PEG-lipid. Thus, PEG "mushrooms" provide a surface pressure that resists the creation of excess area at the surface equivalent to the size of the avidin, and this represents an additional energy barrier against diffusion of avidin to the surface. Thus, a relatively small 750 molecular weight polymer, at relatively low surface density (up to 10 mol%), extending only 2.5 nm from the lipid bilayer surface, inhibits the surface-access of a macromolecule of only a few nanometers in dimension.

062. **THE FORMATION AND SURFACE BIOMOLECULAR INTERACTIONS OF PEG GRAFTED MONOLAYERS USING SPR, XPS AND AFM** Martyn C Davies, Richard Frazier, Saul J B Tendler, Clive J Roberts, Phil M Williams, Gert Matthijs+ and E Schacht+. Laboratory of Biophysics and Surface Analysis, Department of Pharmaceutical Sciences, University of Nottingham, Nottingham NG7 2RD. +University of Gent.

PEG, or poly(ethylene oxide), is a hydrophillic polymer that is widely recognised as one of the most effective protein resistant polymers for use in biomaterial design. The effectiveness of protein resistance by PEG is highly dependent upon its conformation at the surface. In this paper, we show the role of surface analytical tools in characterising the formation and biointeractions of PEG molecules adsorbed or tethered to surfaces. In particular, we describe the power of surface plasmon resonance (SPR) together with XPS and AFM, as a tool for the *in-situ* analysis of the formation of PEG monolayers and their biomolecular interactions with selected proteins. A number of specific systems will be described including PEG-containing block copolymer surfactant films which have been widely demonstrated as effective protein resistant coatings, thiol terminated-PEG molecules and finally PEG chains grafted onto a chemisorbing dextran.

063. PEG-GRAFTED POLYSILOXANES AS CHEMICALLY IMMOBILIZED MONOLAYERS: INTERFACIAL PROPERTIES. Guoqiang Mao, David W. Grainger, David G. Castner[*], Dept. of Chemistry, Colorado State University, Fort Collins, CO 80523-1872, [*]Dept. of Chemical Engineering, NESAC/Bio, University of Washington, Seattle, WA 98195-1750

Polyethylene glycol's unique ability to passivate the surface adsorption of biologicals has generated much activity in PEG-surface immobilization. The quandary in understanding this effect is the current experimental problem of simultaneously controlling PEG immobilization density and PEG brush length. We have immobilized PEG to hydrophobic polymer backbones that chemically react with solid interfaces to yield polymer-immobilized brush systems. PEG of various molecular weights is chemically attached to polysiloxane polymers already bearing alkylsilane anchor chains. The terpolymers are chemisorbed to SiO_2 surfaces, yielding bound polymer monolayers. Surface analysis using XPS, SIMS and contact angle methods shows that PEG yields are limited, and surface "expression" of PEG properties is also limited. Sequential approaches to sorting out these PEG immobilization limitations have also been studied.

064. SYNTHESIS AND CHARACTERIZATION OF POLYPHENYLENE SULFIDE SULFONES: PART I. Y. N. Liu, A. Bhatnagar, Q. Ji, H. Zhuang, J. F. Geibel[1] and J. E. McGrath, Virginia Polytechnic Institute and State University, Department of Chemistry and NSF Science and Technology Center, High Performance Polymeric Adhesives and Composites, Blacksburg, VA 24061-0344; [1]Phillips Petroleum Company, Batlesville, OK, USA.

Poly(phenylene sulfide sulfone)s have been successfully prepared in high molecular weight by nucleophilic aromatic substitution step- or polycondensation polymerization. These materials are closely related to the well-known arylene ethers, but have been relatively unexplored. The reactions for formation of poly(phenylene sulfide sulfone) were conducted in a pressure reactor at 100~150 Psi. Dichlorodiphenyl sulfone as the activated aromatic dihalide was reacted with either sodium sulfide nonahydrate or sodium hydrosulfide in NMP at elevated temperature. The indications to data are that the reaction is essentially complete after about 3 hours at 200°C for the stiochiometry utilized. Commercially useful molecular weights have been achieved that range from 14,000 to about 44,000 number average molecular weight. The molecular weight distribution appears to be in the expected range of about 2.0. The resulting polymers have T_g's from 207°C to 222°C and exhibit high thermooxidative stability in air as well as relatively high refractive index number.

065. SYNTHESIS AND CHARACTERIZATION OF POLYPHENYLENE SULFIDE SULFONES: PART II. A. Bhatnagar, Y. N. Liu, M. Muggli, T. C. Ward, D. A. Dillard, H. Parvatareddy and J. E. McGrath, Virginia Polytechnic Institute and State University, Department of Chemistry and NSF Science and Technology Center, High Performance Polymeric Adhesives and Composites, Blacksburg, VA 24061-0344.

Polyphenylene sulfide sulfone (PPSS) and copolymers with triarylphosphine oxide (PPSS-*co*-PO) were prepared in high molecular weight *via* nucleophilic aromatic substitution reaction of the appropriate bishalide with sodium sulfide in NMP at elevated temperatures and pressure. The sulfide-sulfone homo and copolymers exhibited very good stress-crack solvent resistance unlike analogous polyarylene ether sulfones. The behavior may be related to the fact that these materials undergo solvent induced crystallization. Polyphenylene sulfide sulfone homopolymer display a very high refractive index, high tensile modulus and high glass transition temperature. Mechanical properties of PPSS films are in the range of polyarylene ether sulfones and would be considered to represent ductile behavior. Incorporation of triarylphosphine oxide into the backbone of PPSS is also being investigated to further enhance the already good flame-resistance.

066. FLAME RESISTANT EPOXY NETWORKS BASED ON ARYL PHOSPHINE OXIDE CONTAINING DIAMINES FLAME RESISTANT EPOXY NETWORKS BASED ON ARYL PHOSPHINE OXIDE CONTAINING DIAMINES. C. Tchatchoua, Qing Ji, S. A. Srinivasan, H. Ghassemi, T. H. Yoon, M. Martinez-Nunez, [1]T. Kashiwagi and J. E. McGrath, Virginia Polytechnic Institute and State University, Department of Chemistry and NSF Science and Technology Center, High Performance Polymeric Adhesives and Composites, Blacksburg, VA 24061-0344; [1]National Institute of Standards and Technology, Materials and Fire Research Group, Gaithersburg, MD 20899.

Bis(3-aminophenyl)methyl phosphine oxide (DAMPO) and bis(3-aminophenyl)phenyl phosphine oxide (DAPPO) have been synthesized and isolated in highly pure form and high yield. These phosphine oxide diamines were successfully used to cure a commercial epoxy resins. Epoxy resins cured with phosphine oxide diamines exhibit higher char yield compared with those cured by 3,3'-diaminophenyl sulfone (DDS) as a commercial curing agent. These phosphorus containing epoxy resins showed self-extinguishing characteristics in comparison to the DDS cured sample which continued to burn. Preliminary cone calorimetry test showed the heat release rate decreased as the phosphorus concentration increased.

067. **POLYROTAXANES BY SELF THREADING OF FUNCTIONALIZED MACROCYCLES: POLYESTERS.** H. W. Gibson, D. Nagvekar, W. S. Bryant, J. Powell and S. Bhattacharjee, Department of Chemistry, Virginia Polytechnic Institute & State University, Blacksburg, VA 24061-0344

Poly(ester crown ether)s were synthesized by direct condensation of bis(5-carboxy-m-phenylene)- (3x2)-crown-(x), x = 8 and 10 (**1a** and **1b**) with bisphenols via the Higashi method. The formation of higher molecular weight poly(ester crown ether) **5** from the 32-membered crown ether **1b** and bisphenol-A (**2a**) as compared to polyester **4** made from the 26-membered crown ether **1a** and bisphenol-A (**2a**) and polyester **6** derived from 32-membered **1b** and bis(p-t-butylphenyl)-bis(p-hydroxyphenyl)methane (**2b**) under identical conditions is believed to be the result of threading (rotaxane formation) in **5** but not in **4** or **6**. However, threading in **5** was undetectable through NOESY experiments. However, melt polymerization of bis(5-hydroxymethyl-m-phenylene)-32-crown-10 (**7**) with sebacoyl chloride (**8**) gave an insoluble, but swellable elastomeric material **9**, a polyrotaxane network, as a result of more extensive *in situ* threading brought about by hydrogen bonding of the crown ether with the alcohol moieties in the more concentrated medium.

068. HOMOGENEOUS SYNTHESIS OF POLYANILINE USING DDQ AS THE OXIDANT

Phelesia Y. Cooper,[1] Floyd L. Klavetter,[2*] William H. Starnes, Jr.[3]
[1]Department of Chemistry and The Center for Materials Research, Norfolk State University, Norfolk, VA 23504
[2]Department of Applied Science, The College of William and Mary, Williamsburg, VA 23187-8795
[3]Departments of Chemistry and Applied Science, The College of William and Mary, Williamsburg, VA 23187-8795

A homogeneous polymerization of aniline is accomplished with the organic oxidant 2,3-dichloro-5,6-dicyano-1,4-benzoquinone (DDQ). Under homogeneous conditions, through counter-ion effects, polyaniline remains dissolved in solution during the polymerization. Stoichiometric analysis confirms that DDQ is acting as a two-electron oxidant.

069.
WATER-SOLUBLE SHELL CROSS-LINKED KNEDELS (SCK): AMPHIPHILIC NANOSPHERES COMPOSED OF POLYSTYRENE AND POLYACRYLIC ACID DOMAINS. **Haiyong Huang**, Rolf Gertzmann, Tomasz Kowalewski and Karen L. Wooley*, Department of Chemistry, Washington University, One Brookings Drive, St. Louis, MO 63130

The preparation of shell cross-linked knedel (SCK) nanospheres with a hydrophilic negatively-charged shell and a mobile hydrophobic core is described. The synthetic steps which lead to the final nanometer-sized spherical macromolecules (25 nm diameter) include preparation of an amphiphilic block copolymer (PS-b-PAA), followed by micellar self assembly and then crosslinking of the shell (PAA) block by reaction with diamino linking reagents in the presence of a water soluble carbodiimide, 1-(3-dimethylaminopropyl)-3-ethyl carbodiimide. Atomic force microscopy (AFM) was used for characterization of the particle size and size distribution, which revealed the SCK's to possess narrow size distribution and a globular shape.

070. CATIONIC COPOLYMERIZATION BETWEEN CYCLIC KETENE ACETALS, Zhihong Wu, Liwei Cao, and C. U. Pittman, Jr., Department of Chemistry, Mississippi State University, MS 39762

Cationic copolymerizations of 4-methyl-2-methylene-1,3-dioxane, **1** (M_1), with 2-methylene-1,3-dioxane, **2** (M_2); and of 4,4,6-trimethyl-2-methylene-1,3-dioxane, **3** (M_1), with **2** (M_2) were conducted in THF at R.T. Copolymerization at different $M_1°/M_2°$ ratios (10 and 12 experiment points respectively) were carried out to low conversion and the composition data was fitted by a computer program nonlinear minimization algorithm to obtain best values of the reactivity ratios. It was found that $r_1 = 1.727$ and $r_2 = 0.846$ for the copolymerization of **1** with **2**; and $r_1 = 2.262$ and $r_2 = 0.310$ for the copolymerization of **3** with **2**. Substituents at the 4- and 6- positions of six membered ring monomers increase their reactivities in the copolymerization.

071. BIPYRIDINE-CONTAINING AROMATIC POLYAMIDES: SYNTHESIS AND PROPERTIES. S. C. Yu, W. K. Chan, Department of Chemistry, University of Hong Kong, Pokfulam Road, Hong Kong.

A series of aromatic polyamides containing 2,2'-bipyridine moiety were synthesized by polycondensation of the 2,2'-bipyridine-5,5'-dicarboxylic acid with various aromatic diamines in hexamethylphosphoramide and LiCl using triphenylphosphite as the condensing agent. For those polyamides containing rigid mainchain, they had high thermal stability and their solutions in HMPA-LiCl exhibited lyotropic liquid crystal phases. The phase transition behaviour were studied by polarizing microscopy and X-ray diffraction. It was found that the clearing temperature depended on the solution concentration and the mainchain flexibilities. The polymers also formed metal complexes with $Ru(bpy)_2Cl_2$, which was demonstrated by the changes in electronic spectrum.

072. SYNTHESIS AND RING-OPENING POLYMERIZATION OF THE CYCLIC OLIGOMERS OF POLY(OXY-1,3-PHENYLENECARBONYL-1,4-PHENYLENE). M. F. Teasley, R. L. Harlow, and D. Q. Wu, DuPont Central Research & Development, Experimental Station, Wilmington, Delaware 19880-0328

The cyclic oligomers (cyclomers) were prepared from 4-fluoro-3´-hydroxybenzophenone by nucleophilic aromatic substitution. The cyclic dimer was enriched in the crude mixtures from DMSO/toluene using pseudo-high dilution and isolated by sublimation. Clean cyclomers were prepared in o-dichlorobenzene using 18-crown-6. The ring-opening polymerization of the cyclic dimer was studied using nucleophilic alkali metal salts and 18-crown-6, then extended to the cyclomers. The polymers were analyzed by GPC, and absolute molecular weights were estimated from light-scattering studies. Catalysts, such as cesium fluoride, gave incomplete conversion of the cyclomers to high molecular weight polymer. Initiators, such as potassium and cesium 4-benzoylphenolate, gave essentially complete conversion and stoichiometrically controlled the molecular weights. The reactivity differences in the nucleophiles were apparent due to the structure of the cyclic dimer, as shown by X-ray diffraction, which reduces its reactivity relative to the higher cyclomers.

073. HYBRID, NORBORNENYL-BASED POLYHEDRAL OLIGOSILSESQUIOXANE (POSS) POLYMERS. T. S. Haddad,[†][*] A. R. Farris[†] and J. D. Lichtenhan[‡], [†]Hughes STX Corporation, Phillips Laboratory, [‡]Propulsion Directorate, OLAC PL/RKS Bldg. 8451, Edwards Air Force Base, CA, 93524-7680

We have begun a program to design and synthesize well-defined, linear polyhedral oligosilsesquioxane (POSS) polymers, $[RSiO_{1.5}]_n$, in order to further develop the structure/property relationships of this important class of technologically useful

POLY

compounds. Previously, we have reported the synthesis and characterization of a variety of "bead-type" POSS copolymers wherein the POSS unit is an integral part of the polymer backbone. This paper presents our initial attempts at synthesizing "pendant-type" POSS polymers and copolymers using a POSS-macromer onto which a norbornenylethyl moiety has been appended. Homopolymers and copolymers with varying proportions of norbornene have been synthesized by carrying out Molybdenum catalysed ring opening metathesis polymerizations.

074. TWO-COMPONENT INITIATOR SYSTEMS for RING-OPENING POLYMERIZATION of CYCLIC BPA-CARBONATE OLIGOMERS. The *IN SITU* CLEAVAGE of DISULFIDES by TRIARYLPHOSPHINES H.O.Krabbenhoft, D.J. Brunelle, E.J. Pearce, GE Corporate Research & Development, Schenectady, NY 12301

A two-component initiator system based on the *in situ* cleavage of phenyl disulfide by a triarylphosphine has been developed for the ring-opening polymerization (ROP) of cyclic BPA-carbonate oligomers for potential use in composite applications so that the pre-polymer can suitably wet the composite material *before* being converted to high molecular weight polymer. The initiator precursors (phenyl disulfide and triphenylphosphine) did not independently initiate significant ROP of the oligomeric cyclic BPA carbonate mixture. However, when a mixture of cyclic BPA carbonate oligomers and one of the initiator components was combined with a mixture of cyclic BPA carbonate oligomers and the other initiator component, a high molecular weight polymer (M_w = ~70,000) was produced. The polymerization-initiating species is thought to be thiophenyltriphenylphosphonium thiophenoxide. The effects of concentration of the initiator components, reaction temperature, reaction time, etc. on the ROP were studied; in general, the degree of polymerization ranged from about 65 to 75%.

075. PSEUDO-LADDER RIGID-ROD POLYMERS WITH 2,3-DIHYDROXY-1,4-PHENYLENE UNITS. J. L. Burkett[a], T. D. Dang[a], L. S. Tan[b]*, and F. E. Arnold[b]. a. U. of Dayton Research Institute, 300 College Park, Dayton, OH. 45469-0110. b. Polymer Branch, WL/MLBP, Wright Laboratory, 2941 P Street Suite 1, Wright-Patterson Air Force Base, OH 45433-7750
* Corresponding author.

Pseudo-ladder rigid-rod polybenzobisthiazole (2,3-DiOH-PBZT) and polybenzobisoxazole (2,3-DiOH-PBO) were successfully synthesized from 2,3-dihydroxyterephthalic acid and 2,5-diamino-1,4-benzenedithiol dihydrochloride (**DABDT.2HCl**) or 2,4-diamino-1,5-benzenediol dihydrochloride (**DABDO.2HCl**) in polyphosphoric acid (PPA). However, several attempts to prepare the 2,3-DiOH-PBI failed, despite the fact that its model compound was isolated in 96% yield from 2,3-$(OH)_2$TA and o-phenylenediamine under similar reaction conditions. Intrinsic viscosity values for 2,3-DiOH-PBZT were 11.54-22.06 dL/g in MSA at 30°C but those for PBO analog were much lower, about 2.28 dL/g. The results seemingly paralleled the relative chemical reactivities of the tetrafunctional monomers in PPA: DABDT>DABDO>TBA (tetraaminobenzene). In addition, they also suggested 2,3-$(OH)_2$TA is less reactive than its 2,5-isomer.

076. POLYMERIZATION AND 1D AND 2D NMR ANALYSIS OF ALPHA-OLEFINS FROM LATE TRANSITION METAL CATALYSTS L. T. J. Nelson, E. F. McCord, L. K. Johnson, S. J. McLain, and S. D. Ittel, *DuPont, CR&D, Experimental Station, Wilmington, DE 19880-032;* C. M. Killian and M. Brookhart *The University of North Carolina at Chapel Hill, Chapel Hill, NC 27599-3290*

A new class of polymerization catalysts based on α-diimine nickel and palladium systems has recently been reported. (Johnson, L. K.; Killian, C. M.; Arthur, S. D.; Feldman, J.; McCord, E. F.; McLain, S. J.; Kreutzer, K. A.; Bennett, M. A.; Coughlin, E. B.; Ittel, S. D.; Parthasarathy, A.; Tempel, D. J.; Brookhart, M. S., *PCT Int. Appl.*, WO 9623010, **1996**) These new catalysts polymerize various olefins to give unprecedented high molecular weight polymers with unique microstructures. Here we report the polymers of a-olefins, from propylene (C3) to 1-eicosene (C20), that were analyzed by various 1D and 2D NMR techniques. The combination of NMR spectra and calculated chemical shifts shows that the substituted a-diimine palladium catalyst gives monomer "straightened" poly(a-olefins) with unusual branching distributions. By using these structures and branching distributions, a preliminary set of predictive "polymerization rules" were developed. The amounts and type of structures observed are influenced by such factors as the transition metal used, the temperature, and the pressure. However, the results here show the power and usefulness of NMR spectroscopy in the field of polymer chemistry.

077. SYNTHESIS OF POLY(ALKYLESTER) DENDRIMERS. A. A. Coleshill,[a] D. M. Haddleton,[a] P. C. Taylor*[a] and S. G. Yeates[b], [a]Dept. of Chemistry, University of Warwick, Coventry, UK CV4 7AL, [b]Zeneca Resins, The Heath, Runcorn, Cheshire, UK WA7 4QD

Dendrimers have captured the imagination of a growing number of chemists. For example dendrimers have lower viscosities than their linear polymer counterparts of the same molecular weight, which should lead to significant improvements in handling properties of, in particular, surface coatings. Our group has previously prepared poly(arylester) dendrimers for this reason. In this paper we describe the extension of our divergent polyester dendrimer synthesis to alkyl analogues, which we expect to possess better processing properties.

Preparation of suitably protected monomers is the greatest obstacle to divergent synthesis of poly(alkylester) dendrimers; metal containing bases must be avoided in any case where the "monomer" can behave as a chelating ligand. Silyl protection obviates this problem and permits the preparation of a range of alkylester dendrimers. The full range of structures and derivatives prepared, together with their analytical data, will be reported.

078. CLEAVEGE OF t-AMYL ESTER AND APPLICATION FOR IMAGING SYSTEM CATALYZED WITH PHOTOGENERATED ACID. M. Hiro, T. Akahori, S. Tachiki, Ibaraki Research Laboratory, Hitachi Chemical Co., Ltd., 4-13-1, Higashi-cho, Hitachi, Ibaraki, 317 JAPAN

A novel polymeric imaging material containing *tert*-amyl ester group and photoacid generator has been designed. The polymer having *tert*-amyl ester group in its side chain was prepared by free radical polymerization. The polymer was converted to the polymer having carboxylic acid group through a side chain cleavage reaction by thermolysis catalyzed with photogenerated acid. Thermal Gravimetry Analysis (TGA) study shows the side chain scission occurred below 100 °C with the presence of photogenerated acid-catalyst comparing to 230 °C through thermolysis without acid-catalyst. Gas Chromatograph-Mass Spectrometry (GC-MS) analysis study that detected mass numbers of the cleaved fragments suggests *iso*-amylene was released through the side chain cleavege reaction.

079. THERMOSETS DERIVED FROM PHENETIDINE-BASED TELECHELICS
Tito Viswanathan, Janna Helmich, Stephen Rodriguez, Qiuwei Feng
Department of Chemistry, University of Arkansas at Little Rock, Little Rock, AR 72204

The development of soluble electroactive polyconjugated telechelics based on aniline are highly desirable due to numerous applications that are possible especially in the coatings and adhesive industry. Combined with the higher solubility and reactivity of phenetidine-based telechelics, curing of these functionally reactive oligomers has been accomplished using aqueous formaldehyde and water-dispersed epoxy resins. Spectroscopic studies have been done to characterize the prepolymers as well as the cured materials. The focus of this paper is to report work done on the reaction of o-phenetidine with phenylenediamine in 2:1 and 4:1 ratios. A study of the mass spectrum data of the 1:2 ratio product shows that a branched oligomer exists in the final product, along with the expected trimer.

080. Synthesis of Dendrimer Based on s-triazine/aniline. Tito Viswanathan, Qiuwei Feng and Alan Toland, Department of Chemistry, University of Arkansas at Little Rock, Little Rock, Arkansas 72204

Dendrimers are regularly branched molecules which cannot form any entanglements leading to the interaction with the surrounding media primarily through a surface effect. This can result in specific potential uses. The novelty in structure and use has generated an intense interest in these type of compounds. This paper reports the synthesis of a s-triazine/aniline based dendrimer formed by reacting one mole of cyanuric chloride with three moles of a diamine telechelic polymer which has been synthesized earlier in our lab. These new class of oligomers may be classified as conducting dendrimers. Characterization of the product has been done using IR, NMR, tlc, and UV-Vis.

081. COPOLYMERIZATION OF ZINC METHACRYLATE AND PERFLUOROALKYL ACRYLATE. T. Ikeda, Research & Development Department, Machinery Parts Division, Nippon Valqua Industries, Ltd., Yasunaka-cho, Yao 581, Japan and B. Yamada, Material Chemistry Laboratory, Faculty of Engineering, Osaka City University, Sugimoto-cho, Sumiyoshi-ku, Osaka 558, Japan

The copolymerization of zinc dimethacrylate (ZMA) and 2-(N-ethylperfluorooctanesulfonamide)ethyl acrylate (RfsA) was carried out at various comonomer compositions in relation to the copolymerization of ZMA and RfsA in synthetic rubber during the peroxide curing process ("*in situ*" copolymerization). The copolymerizations of fluorine-containing acrylate and methacrylate without sulfonamido group with ZMA in different solvents were also studied as models for "*in situ*" copolymerization. The following characteristics were found. (1) In the copolymerization of ZMA and RfsA, the composition curve was shifted with increasing the conversion except at an equimolar mixtures of the monomers, and the copolymerization behavior seemed to be changed at an equimolar amount of the monomers as a boundary. (2) The compositions of the ZMA-RfsA copolymer at an equimolar mixture of the monomers yielded the copolymer of which the compositions was almost the same with that of the comonomer regardless of the conversion and polarity of solvent. The copolymerizations of ZMA with the fluorine-containing monomers seemed to be influenced by the monomer reactivity and the polarity of the polymerization medium. It was considered that the sulfonamido group in RfsA and copolymer of RfsA could specifically improve the miscibility of RfsA with ZMA.

082. NEW TYPE OF CROSSLINKING AGENTS FOR VINYL POLYMERS. A. Zada, Y. Avny and A. Zilkha, Department of Organic Chemistry, The Hebrew University of Jerusalem, Jerusalem 91904, Israel

Cyclic monomers having on the one hand one polymerizable double bond and on the other hand a macrocyclic ring (ring size $\geq$ 28) which can be threaded by a growing chain of a vinyl polymer are shown to function as a new type of crosslinking agents for vinyl polymers. The crosslinking is through threading, i.e. rotaxane formation, and differs from the classical crosslinking through covalent bonds formed with difunctional vinyl monomers such as divinyl benzene. Thus, a cyclic ester of fumaric acid with octaethylene glycol, a 29-membered ring was prepared using high dilution techniques by the reaction of fumaryl chloride with octaethylene glycol in tolune. It was homopolymerized by benzoyl peroxide at $60°$ to a soluble polymer, and it was copolymerized with styrene and methyl methacrylate to yield crosslinked, insoluble polymers that swelled in various solvents.

083.

SYNTHESIS AND PHOTOPOLYMERIZATION OF LIQUID CRYSTALLINE DIACRYLATE RESIN

Jay C. Bhatt, Minhhoa H. Dotrong, Robert T. Pogue, Jill S. Ullett and Richard P. Chartoff
Center for Basic and Applied Polymer Research, University of Dayton,
300 College Park Avenue, Dayton, OH 45469

Morton H. Litt and Xian-Tong Bi
Macromolecular Science Department, Case Western Reserve University, Cleveland, OH 44106-7202

Abstract

Photo-polymerization of liquid crystalline monomers in a mesophase should lead to the formation of highly ordered polymeric networks. The liquid crystalline orientation of the monomer can be frozen in by crosslinking and will be retained in the resulting polymer. The highly crosslinked networks thus formed are expected to have unusually high glass transition temperatures (T_g). In this paper we report on the design and synthesis of a mesogenic diacrylate monomer with liquid crystalline phases at ambient temperature. Formulations containing the monomer and photo-initiators were developed to facilitate photopolymerization of the monomer in the mesophase and in the isotropic phase. Conditions for obtaining rapid cure and a high T_g will be reported.

084.

SYNTHESIS AND PHOTOCURING OF ACETYLENE-CONTAINING DIACRYLATE MONOMERS

Jay C. Bhatt, Minhhoa H. Dotrong, Robert T. Pogue, Jill S. Ullett and Richard P. Chartoff
Center for Basic and Applied Polymer Research, University of Dayton,
300 College Park Avenue, Dayton, OH 45469-0130
polymers@udri.udaytonedu

Abstract

In our efforts to develop new liquid crystalline monomers, we have shown that a stilbene-containing diacrylate monomer, when photocured in both liquid crystalline and isotropic states, results in the formation of a highly crosslinked polymer with a high glass transition temperature (T_g). In this paper we report on the design and synthesis of two new diacrylate monomers incorporating a diphenyl acetylene unit. Thermal and spectroscopic characterization as well as photocuring conditions of these monomers will be presented.

085. SYNTHESIS OF CONJUGATED MONO-ANIL DERIVATIVES AND THEIR POLYMERS FORMED BY THERMOLYSIS, S. E.Reybuck, P. G. Rasmussen*, T. Jang, R.G. Lawton, Dept. of Chem., Univ. of Mich., Ann Arbor MI 48109-1055

In this report we describe the synthesis, characterization and polymerization of a new family of conjugated monomers derived by Schiff base reactions of diaminomalonitrile (DAMN) with simple unsaturated aldehydes. These mono-anils, such as the one formed from DAMN and acrolein, can polymerize by thermal initiation to afford soluble polymers. Since the polymerization proceeds by Michael addition, no gases are evolved and the product is a hyperbranched polymer of high thermal stability. Anaerobic pyrolysis gradually converts these materials, in high yield, to carbon nitrogen materials.

086. FUNCTIONALIZATION AND POLYMERIZATION OF [5]HELICENE AND ITS PRECURSORS. Z. Y. Wang*, J. P. Gao, T. P. Bender, Y. Qi, S. MacKinnon, L. Kuang, J. P. Chen, and X. S. Meng. Ottawa-Carleton Chemistry Institute, Department of Chemistry, Carleton University, 1125 Colonel By Drive, Ottawa, Canada K1S 5B6

Carbohelicenes such as [5]helicene are polycyclic aromatic compounds that display a level of handedness in their molecular conformations. We have shown several approaches to the introduction of [5]helicene into polymers. All monomers were derived from a single tetrahydro[5]helicene building block **2** which can be prepared on a large scale from readily available starting materials. Several high molecular weight polymers were made from these monomers, including polyimides and polyisocyanides. Chemical or thermal transformation of tetrahydro[5]helicene to [5]helicene, such as **2** to **1**, was also established. Thus, the polyimide derived from **2** could be converted to the [5]helicene-based polyimide simply upon heating at 400 °C in air or treatment with elemental bromine at elevated temperatures.

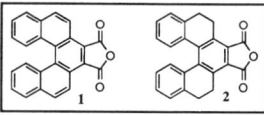

087. **PHENYL SUBSTITUTED POLY(3,4-ETHYLENEDIOXYTHIOPHENE).** Anil Kumar, Mark C. Morvant and John R. Reynolds, Department of Chemistry, Center for Macromolecular Science and Engineering, University of Florida, Gainesville, Florida 32611

A new monomer, phenyl substituted 3,4-ethylenedioxythiophene (EDOT-Ph), was synthesized starting from 3,4-dihydroxy-2,5-carboethoxythiophene and electropolymerized to form thin films. Cyclic voltammetry of the monomer in 0.1 M TBAP/CH$_3$CN showed the monomer oxidation at 1.1 V and a reversible redox process ($E_{1/2}$) at -0.1 V vs. Ag/Ag$^+$ for the polymer which quickly grew in on repeated scanning. A linear relationship was found between the peak current and scan rate indicating that the electroactive polymer film is well-adhered. The polymer showed a band gap onset of 1.6 eV and a peak at 2.0 eV in the neutral form. These films were switched electrochemically between a highly absobing dark blue reduced state and a highly transmissive oxidized state. This paper will present the synthesis and electropolymerization of the monomer and discuss the optoelectrochemical behavior of the resultant polymer.

088. IN-SITU STUDIES OF THE ELIMINATION KINETICS FOR ANIONIC SYNTHESIZED POLY(PHENYL VINYL SULFOXIDE) DIBLOCK COPOLYMERS - SOLUBLE PRECURSORS TO POLYACETYLENE DIBLOCK COPOLYMERS. Louis M. Leung and T.S. Lam, Department of Chemistry, Hong Kong Baptist University, Kowloon, Hong Kong

A series of diblock copolymer consists of phenyl vinyl sulfoxide and a styrenic monomer (including styrene, vinyle toluene, α-methyl styrene) was prepared by the anionic synthesis method. Upon thermolysis, the thermally labile phenyl vinyl sulfoxide converts to acetylene via a concerted cyclic sigmatropic reaction. The elimination kinetics of the copolymer either as solid thin films or solution form were monitored by a UV-vis spectroscopy in the temperature range from 50 to 90 °C. Due to the high styrenic composition, the acetylene copolymers remained soluble in solution throughout the kinetic studies. Using a first-order kinetic model, the elimination reaction was divided into three linear regimes : an initiation, a zipper-like elimination, and an oxidative degradation reaction. The copolymers were further characterized by FTIR, GPC and TGA analyses.

089. **Deprotonation-Substitution Reactions of Poly(dialkylphosphazenes) and Their Phosphoranimine Precursors** Patty Wisian-Neilson, Karl A. Koch, and Cuiping Zhang, *Department of Chemistry, Southern Methodist University, Dallas, TX 75275*

Diethyl ether solutions of the phosphoranimines, Me$_3$SiN=P(OPh)(Me)(R) {R = *n*-Bu, *n*-Hex} and THF solutions of the polymers, [Et(Ph)PN]$_n$, [Me(Bu)PN]$_n$, and [Me(Hex)PN]$_n$ were sequentially treated with *n*-BuLi and electrophiles to afford new phosphoranimines, Me$_3$SiN=P(OPh)(CH$_2$SiMe$_3$)(R) and Me$_3$SiN=P(OPh)(CH$_2$PPh$_2$)(*n*-Bu), and new copolymers {Et(Ph)PN}$_x${(Me$_3$Si)(CH$_3$)HC](Ph)PN}$_y$, {Me(R)PN}$_x${[(Me$_3$Si)H$_2$C](R)PN}$_y$, and {Me(R)PN}$_x${[(η^5-C$_5$H$_5$)Fe(η^5-C$_5$H$_4$)CH(OH)CH$_2$](R)PN}$_y$ where R = *n*-Bu, *n*-Hex.

The compounds were characterized by elemental analysis, NMR spectroscopy, gel permeation chromatography (GPC), differential scanning calorimetry (DSC), and water contact angle measurements. The glass transition' temperatures of the new polymers and several related poly(phosphazenes) were correlated to their free volumes, which were calculated using the densities that were measured for each polymer system.

090. TACTICITY OF POLY(METHYL METHACRYLATE) FORMED IN MICROEMULSIONS. Spence C. Pilcher and Warren T. Ford, Department of Chemistry, Oklahoma State University, Stillwater, OK 74078

Polymerizations of methyl methacrylate (MMA) were carried out in microemulsions using stearyltrimethylammonium chloride (STAC), cetyltrimethylammonium bromide (CTAB), dodecyltrimethylammonium bromide (DTAB), or a mixture of nonionic Triton surfactants (nonylphenols with varying amounts of ethylene oxide units) already known to produce particles having radii 5-25 nm that contain only a few macromolecules. Particle sizes were measured by dynamic light scattering (DLS) and transmission electron microscopy (TEM). Ultra-high molecular weights (~4.0 x 10^6 g/mole) were obtained by gel permeation chromatography (GPC) based upon polystyrene standards. Polymerizations were also performed using STAC and CTAB at varying surfactant to monomer ratios using both water-soluble and oil-soluble initiators. Tacticities of the polymers at the different surfactant to monomer ratios were measured to the triad level by ^{1}H NMR spectroscopy at 400 MHz. The tacticities were then correlated with the sizes and the compositions of the starting microemulsions.

091. REACTION OF C_{60} WITH OXYGEN INITIATED BY RADICALS FROM AZO(BIS-ISOBUTYRONITRILE). Amanda G. Camp, Warren T. Ford,* Alanta Lary, and Dilip K. Sensharma, Department of Chemistry, Oklahoma State University, Stillwater, Oklahoma 74078

In an o-dichlorobenzene solution exposed to an atmosphere of air or pure oxygen, C_{60} reacts with radicals generated by thermal decomposition of azo(bisisobutyronitrile) and with oxygen to form mixtures of highly substituted and oxygenated monomeric, dimeric, and trimeric derivatives. LSIMS, LDMS, and SIMS (Liquid Secondary Ion, Laser Desorption, and Secondary Ion Mass Spectrometry) of these mixtures give clusters of peaks for C_{2n} from C_{40} to greater than C_{200}. Elemental analyses show large H, N, and O contents that increase with the amount of AIBN and the partial pressure of oxygen during the reaction. ^{1}H and ^{13}C NMR spectra consist of broad bands assigned to complex mixtures of substituted fullerenes containing many 2-cyano-2-propyl groups derived from AIBN. Diffuse reflectance infrared spectra show hydroxyl and carbonyl functional groups. Thermal gravimetric analysis (TGA) of the mixture prepared under oxygen retained 40% of the initial weight at 640 °C, while <19% of the initial weight remained from the mixture prepared under argon.

092. ANIONIC SYNTHESIS OF HIGHLY ISOTACTIC POLYSTYRENE IN THE PRESENCE OF LITHIUM HYDROXIDE OR LITHIUM ALKOXIDES. T. Makino, and T. E. Hogen-Esch., Loker Hydrocarbon Research Institute and Department of Chemistry, University of Southern California, Los Angeles, CA 90089-1661.

Styrene is polymerized by t-butyllithium in the presence of lithium hydroxide or lithium alkoxides in hexane at -30 °C producing a mixture of highly isotactic (>90%) and atactic polystyrene. Although these fractions of isotactic polystyrene for the case of t-BuOLi are small ($\leq$ 5%) more isotactic polystyrene is formed in the presence of lithium 2-adamantoxide. (15-20%) Surprisingly in the presence of LiOH (~1 equiv.) the overall isotactic content is greatly enhanced. (mm $\geq$ 60%) A rational for some of these finding is discussed.

093. **STERIC EFFECT OF THE TEMPLATE ON THE ASYMMETRIC CYCYLOCOPOLYMERIZATION OF BISVINYLBENZOATE WITH STYRENE**
Takahiro Uesaka, Kazuaki Yokota, Toyoji Kakuchi, and Osamu Haba*
Hokkaido University, Sapporo 060, Japan, and *Yamagata University, Yamagata 992, Japan

Three 1,2-diols, (2S,3S)-2,3-butanediol, (R,R)-1,2-diphenyl-1,2-ethanediol, and 1,2:5,6-di-O-isopropylidene-D-mannitol, were used as templates to investigate the influence of the bulkiness of the template on the asymmetric cyclocopolymerization. The copolymerizations were carried out with AIBN in toluene at 60 °C under N_2 atmosphere. The resulting copolymers were hydrolyzed and methylated by treatment with CH_2N_2. The methylated copolymer showed optical activity even after complete removal of the templates, and thus the new chirality was induced in the main chain. The intensities in the CD spectra of the methylated copolymers increased as follows: 1,2:5,6-di-O-isopropylidene-D-mannitol>(R,R)-1,2-diphenyl-1,2-ethanediol>(2S,3S)-2,3-butanediol. As a result, the bulkiness of template is one of important factors for the asymmetric induction.

094. CHAIN EXTENSION OF POLY(ETHYLENE TEREPHTHALATE) WITH DIEPOXIDES BY A REACTIVE EXTRUSION-SIMULATING PROCESS. A. A. Haralabakopoulos, D. Tsiourvas and C. M. Paleos*, Institute of Physical Chemistry, NCSR "Demokritos", 15310 Aghia Paraskevi, Attiki, Greece.

Bifunctional chain extenders bearing terminal epoxide groups were interacted with the terminal carboxylic moieties of PET in order to maintain or increase the molecular weight of the melt processed material. The reactions were carried in a miniature reactor operating under reactive extrusion conditions. The chain extenders used, were commercially available diglycidyl ethers or cycloaliphatic diepoxides bearing various spacers. Intrinsic viscosity maxima were dependent, as expected, on the nature and amount of the extender that was employed. Thus effective chain extension, as indicated by relatively high intrinsic viscosity values, was obtained mostly with the cycloaliphatic diepoxy extenders, while the diglycidyl ethers showed inferior behavior. It seems that the extenders with short spacers favor the chain extension of PET whereas those spacers bearing aromatic rings had the opposite effect. In addition the decreasing Tm values obtained with increasing amount of extender support, in some cases, the above mentioned findings.

095.

POLY(P-PHENYLENE) DERIVATIVES VIA NI-CATALYZED COUPLING OF AROMATIC DICHLORIDES. A.J. Pasquale, T.K. Vonhof, and V.V. Sheares. Department of Chemistry, Iowa State University, Ames, IA 50011

Poly(p-phenylene)s (PPP)s are currently receiving considerable attention because of their high thermal stability and potential in numerous thermally robust organic materials including composites, lubricant additives, and thermoset precursors for high-performance aerospace material applications. However, because the material is quite rigid, it is difficult to synthesize and to form into useful products. Appropriately substituted materials are soluble and yet exhibit thermal and mechanical properties comparable to the unsubstituted polymer. Previous work in our laboratory focused on the synthesis of poly(2,5-benzophenone), a benzoylated analogue of poly(p-phenylene). The resulting polymer was a soluble, amorphous polymer of high molecular weight which was thermally processable. The synthesis of new alkylated and fluorinated monomers based on 2,5-dichlorobenzophenone and the subsequent polymerizations will be discussed in addition to the monomer and polymer analyses.

096. THE SYNTHESIS, CHARACTERIZATION AND MOLECULAR MODELING OF CYCLIC ARYLENE ETHER OLIGOMERS.
S. Ahmed and R. A. Orwoll, Department of Applied Science, College of William and Mary, Williamsburg, VA 23187
T. V. Olah, Department of Drug Metabolism, Merck Research Laboratories, West Point, PA 19486

The synthesis of several novel cyclic arylene ether oligomers, based on 1,3-bis(4-fluorobenzoyl) benzene with 3,3'-methylene diphenol and 3,3'-(ethylene dioxy) diphenol, and their characterization by LC-MS/MS and 1-D ^{1}H-NMR are presented. These cyclic oligomers were synthesized in high yield by regulating the reaction time, concentration, temperature and solvent systems. The relevance of reaction conditions is discussed. Electrospray ionization (ES) along with MS/MS was used to characterize cyclic oligomers of molecular weight up to 1600 Da. Molecular dynamics and potential energy calculations were also carried out to understand the molecular characteristics of the diphenols that lead to the formation of these cyclic species.

097. NOVEL SYNTHETIC ROUTES TO END FUNCTIONALIZED POLYISOBUTYLENES, Robson F. Storey and Andrew B. Donnalley, Department of Polymer Science, University of Southern Mississippi, Southern Station Box 10076, Hattiesburg, MS 39406-0076

In recent years there have been a number of advances made in the synthesis of polymers of controlled architecture via the technique of living cationic polymerization. Telechelic polyisobutylenes are of commercial interest and utility for the formation of ionomers, thermoplastic elastomers, and crosslinking agents. This paper will investigate novel synthetic pathways to hydroxyl and carboxylic-acid terminated PIBs.

098. PHOTOINDUCED POLYMERIZATION OF FUMARATES WITH ELECTRON DONOR MONOMERS. Shigeki Katogi, Chris W. Miller, Charles E. Hoyle and Sonny Jönsson Department of Polymer Science, The University of Southern Mississippi Box 10076, Hattiesburg, MS 39406-0076

In UV curing processes, acrylate and methacrylate monomers have long been used as prepolymers and reactive dilluents. We report here in on photoinduced polymerization via charge transfer complex between electron donor and acceptor monomers which proceed at rapid rates. This system has the usual advantages of UV curables such as the absence of solvents. We present results for numerous acceptor monomers and fumarate esters mixed with typical electron donor monomer systems which show that photopolymerization proceeds even in the absence of photoinitiator. In the presence of photoinitiator, certain of these systems exhibit photopolymerization rates much faster than acrylates in air atmosphere.

099.
PHOTOINDUCED POLYMERIZATION OF MONOFUNCTIONAL MALEIMIDES AND DONOR MONOMERS. **S.C. Clark**, G.J. Doucet, G.A. Mattson and C.E. Hoyle. Department of Polymer Science, University of Southern Mississippi, Hattiesburg, MS 39406-0076. S. Jonsson, Fusion UV-Curing Systems, Gaithersburg, MD 20878.

The inherent reactivity of a maleimide/donor formulation was established by polymerizing in the presence of a photoinitiator. Donors were polymerized with maleimides where the substituent on the maleimide was either t-butyl or a carbonate group. The carbonate containing maleimide had a greater reactivity with corresponding donor monomers than t-butyl maleimide.

100. **SYNTHESIS AND CHARACTERIZATION OF DIPHENYLETHYNYL TERMINATED BENZOXAZOLE OLIGOMERS** Jennifer J. Jensen, Lon J. Mathias Department of Polymer Science, University of Southern Mississippi, Hattiesburg, Mississippi, 39406-0076

3,5-Diphenylethynylbenzoyl chloride was synthesized by palladium catalyzed coupling of phenylacetylene with 3,5-dibromobenzoic acid followed by conversion to the acid chloride. 3,5-Diphenylethynylbenzoyl chloride was used as an endcapper in bisbenzoxazoles and benzoxazole oligomers. The bisbenzoxazole model compound exhibited a curing exotherm from 330 - 440 °C followed by a 5% weight loss at 545 °C in nitrogen. The model compound was soluble in chloroform and NMP and insoluble after cure. Preliminary studies using Raman Spectroscopy suggest efficient reaction of the ethynyl groups by monitoring the disappearance of the ethynyl stretch throughout the cure cycle. A series of diphenylethynyl endcapped oligomers based on 6F diamino phenol and oxydibenzoyl chloride were synthesized with molecular weights of 5,000 - 15,000. The hydroxyamide oligomers were isolated and characterized for molecular weight using GPC, ^{1}H NMR and ^{13}C NMR. These oligomers were chemically cyclized to form benzoxazole oligomers which had T_g's of 270 °C. They were cured between 320 - 440 °C and then showed a 5% weight loss at 545 °C.

101. **HIGHLY SOLUBLE POLYIMIDES FROM STERICALLY HINDERED DIAMINES**
Victoria L. Ulery, Tara J. Smith, Tina L. Grubb, Gordon L. Tullos, and Lon J. Mathias, University of Southern Mississippi, Hattiesburg, Mississippi 39406-0076, Michael Langsam, Air Products & Chemical Company, 7201 Hamilton Boulevard, Allentown, PA 18195-1501

Polyimides were synthesized from thianthrene-2,3,7,8-tetracarboxylic dianhydride, 4,4'-(hexafluoroisopropylidene)diphthalic anhydride, 3,3',4,4'-benzophenonetetracarboxylic dianhydride, or 3,4,3',4'-biphenyl tetracarboxylic dianhydride and sterically hindered diamines including diethyl toluene diamine, 4,4'-methylenebis(3-chloro-2,6,-diethylaniline), 4,4'-thiodianiline, or 2,4-diaminomesitylene. Poly(amic acids) were synthesized through the conventional low temperature technique in NMP, followed by thermal cyclization in solution to polyimides. Incorporation of bulky alkyl groups gave polyimides which were soluble in common organic solvents such as CH_2Cl_2, $CHCl_3$, and THF. Monomer and polymer synthesis, as well as polymer thermal and physical properties will be discussed.

102. **Ru Catalyzed Co-Polymerization of 4-Acetylstilbenes and 1,3-Divinyltetramethyldisiloxane**
T. M. Londergan and W. P. Weber,* K.B. and D. P. Loker Hydrocarbon Research Institute, Dept. of Chemistry, University of Southern California, Los Angeles, California 90089-1661

Successful dihydridocarbonyl*tris*(triphenylphosphine)ruthenium (Ru) catalyzed polymerization of 4-acetylstilbene with 1,3-divinyltetramethyldisiloxane (I) yields copoly(3,3,5,5-tetramethyl-4-oxa-3,5-disila-1,7-heptanylene/ 4-acetyl-3,5-stilbenylene). Polymerization results from Ru catalyzed activation of aromatic C-H bonds which are *ortho* to the acetyl for addition across the terminal C-C double bonds of I. The internal C-C double bond of 4-acetylstilbene does not react.

103. **PREPARATION OF POLY(DIPROPYLENECARBONATE-CO-DISILOXANE) BY A CONDENSATION POLYMERIZATION. THERMALLY DEGRADABLE POLYMERS**

Douglas A. Loy* and Brigitta M. Baugher

Porous Materials and Encapsulants Department, Sandia National Laboratories, Albuquerque, NM 87185-1407

Poly(dipropylenecarbonate-co-disiloxane) was prepared through a simple acid-catalyzed hydrolysis and condensation reaction of bis(3-dimethylethoxysilylpropylene) carbonate **1**. The polymer was thermally stable to 300 °C. Higher temperatures resulted in the complete degradation of the polymer to monomeric products including 1,3-bis(γ-hydroxypropyl)-1,1,3,3-tetramethyldisiloxane **2** through acidolysis of the polymer, or a mixture of **2** with 1,3-diallyl-1,1,3,3-tetramethyldisiloxane **3** and 1-allyl-3-(γ-hydroxypropyl)-1,1,3,3-tetramethyldisiloxane **4** through decarboxylation. The nature of degradation products was markedly dependent on the manner in which the polymer had been processed and was particularly sensitive to traces of acid.

104. **SYNTHESIS AND CHARACTERIZATION OF STYRENIC POLYMER WITH PENDANT BIIMIDAZOLE MOIETY**

Geng Lin and Harvest L. Collier, Department of Chemistry University of Missouri-Rolla, Rolla, MO 65401

The monomer, 1-hydroxyethyl-1'-vinylbenzyl-2,2'-biimidazole, was synthesized from 1-hydroxyethyl-2,2'-biimidazole and chloromethyl styrene. The model monomer, 1-vinylbenzylimidazole was also synthesized. The synthesized HVBI and VBI were identified by IR and ^{1}H-NMR. The homopolymer of HVBI was obtained using AIBN as an initiator. The properties and applications of polymer-metal complex pf poly(HVBI) are still under investigation.

105. **SYNTHESIS AND STUDY OF CONJUGATED POLYMERS CONTAINING SILOLE UNITS**

Wu Chen, Sina Ijadi-Maghsoodi, Thomas J. Barton*
Chemistry Department, Ames Laboratory (US DOE)
Iowa State University, Ames, IA 50010

A series of silole (silacyclopentadiene)-containing polymers were synthesized and studied. The UV/Vis absorption were unusually longer than those of analogues conjugated polymers, indicating that silicon is lowering the LUMO energies. The emission spectra and electrical conductivity of these polymers were measured. Ethynylene-linked silole polymers showed very low conductivities, but a silicon-bridged polythiophene had very large, doped conductivity of 400 S/cm.

106 **POLYMERS OF FLUOROALKYL- AND SEMIFLUOROALKYL-SUBSTITUTED THIOPHENES.** J. Cameron Tyson, Jennifer S. Middlecoff, and David M. Collard*. School of Chemistry and Biochemistry, and The Polymer Education and Research Center, Georgia Institute of Technology, Atlanta, GA 30332-0400.

Fluorinated polymers have attracted attention because of their unusual properties which are a consequence of the hydrophobicity, rigidity, thermal stability, chemical and oxidative resistance, and self-organization of perfluoroalkyl chains. With the aim of controlling molecular organization and electronic structure of polythiophenes, we studied 3-perfluoroalkylthiophenes. Studies were performed on a series of conjugated polymers derived from 2-(3-thienyl)ethyl)perfluoroalkanoates, semifluorinated alkylthiophenes, and fluoroalkyl/alkyl substituted bithiophenes. Chemical (ferric chloride and copper-mediated Ullmann) and electrochemical polymerizations were performed, and properties of these perfluoro polymers are compared to hydrocarbon analogs.

107. **METAL-CENTERED STAR-SHAPED POLYMERS: LIVING POLYMERIZATION USING MULTIFUNCTIONAL IRON TRIS BIPYRIDYL COMPLEX INITIATORS** Cassandra L. Fraser, Jaydeep J. S. Lamba, John E. McAlvin, Benjamin P. Peters, *Department of Chemistry, University of Virginia, Charlottesville, VA 22901*

Since transition metals are structurally diverse and possess a variety of physical properties and reactivities, they serve as valuable templates for the construction of materials. These features have been exploited in the synthesis of dendrimers by convergent, divergent and self assembly routes. In comparison, relatively little has been reported about the compatibility of metal complexes with living polymerization techniques, which also allow entry into a variety of interesting supramolecular architectures. Recently we reported that hexafunctional Fe(II) *tris* -(4,4'-halomethyl-2,2'-bipyridine) complexes were effective initiators for oxazoline polymerizations, producing monodisperse six-arm star-shaped Fe core polymers. Further experiments were conducted to determine whether these polymerizations are "living." It was demonstrated that a linear relationship exists between M_n and monomer/initiator ratio, and that diblock copolymers could be produced from 2-ethyl and 2-phenyl oxazolines. Fragmentation of star block copolymers yields bpy-centered hydroxytelechelic ABA triblocks.

108. **PHOTOCATALYTIC AND PHOTOINITIATING PROPERTIES OF IRON ORGANO-METALLIC COMPLEXES IN SOLUTION AND AROMATIC DICYANATE ESTERS.** Vladimír Jakúbek and Alistair J. Lees, *Department of Chemistry, State University of New York at Binghamton, Binghamton, NY 13902-6016*, Stephen J. Fuerniss and Kostas I. Papathomas, *IBM Microelectronics, Endicott, NY 13760*

The photochemistry of a series of iron-arene complexes with the general formula $[CpFe(\eta^6\text{-arene})]X$ ($Cp = \eta^5\text{-}C_5H_5$; arene = benzene, toluene, naphthalene, pyrene; $X = PF_6$) has been investigated in deoxygenated methylene chloride solutions containing phen (phen = 1,10-phenanthroline). The solution photoreactions, which form the $[Fe(phen)_3]^{2+}$ product, have been monitored by diode-array UV-visible spectroscopy following excitation at 366 nm. Absolute photochemical quantum efficiencies for arene dissociation (Φ_{cr}) have been determined by a photokinetic procedure and are in the range 1.80-3.47, indicating that a photocatalytic reaction takes place on 366-nm excitation. DSC measurements of $[CpFe(\eta^6\text{-arene})]X$ in Bisphenol E dicyanate (AroCy L-10) solution suggest that three different photoproducts, $[CpFe(\eta^4\text{-arene})L]X$, $[CpFe(\eta^2\text{-arene})L_2]X$, $[CpFeL_3]X$ (L = AroCy L-10), are formed following UV irradiation.

109. SYNTHESIS, PREPARATION, AND CHARACTERIZATION OF NOVEL ORDERED POLYDIACETYLNE THIN FILMS. D. B. Wolfe, Department of Chemistry, Rice University, Houston, Texas 77005. M.S. Paley and D. O. Frazier, Chemistry and Polymeric Materials Group, Space Science Laboratory - ES 76, National Aeronautics and Space Administration, Marshall Space Flight Center, Huntsville, Alabama 35812.

The development of second and third order nonlinear optical devices has lagged because of the lack of efficient materials. For many applications orientation of these materials in one direction is necessary to optimize the nonlinear optical response. We present derivatives of a polydiacetylene, PDAMNA, which possess side groups that encourage the orientation of the polymer chains. Photodeposition of these compounds yield good optical quality thin films which display orientation normal to the substrate as determined by polarized UV-Vis spectroscopy. Furthermore, SEM micrographs show the chains normal to the surface and the subsequent decrease in order as the thickness increases. Energy calculations of the different conformers of the dimers shows a 24 kcal/mole difference between the alternating and non-alternating structures. These data show that partially ordered polydiacetylene thin films can be obtained through photodeposition.

110. PHOTOINITIATED CATIONIC POLYMERIZATION OF MONO- AND DIFUNCTIONAL CYCLIC DIOXOLANES. Kevin D. Belfield* and Feras B. Abdelrazzaq, Department of Chemistry, University of Detroit Mercy, P.O. Box 19900, Detroit, MI 48219-0900.

The ring-opening polymerization of heterocyclic monomers has been reported to occur with small volume decreases (shrinkage) or, in a limited number of examples, volume increases (expansion). Of these, 1,3-dioxolane derivatives have been the object of considerable attention. A difunctional dioxolane momomer, 2,2'-(1,4-phenylene)bis-4-methylene-1,3-dioxolane, was synthesized and characterized by ^{1}H NMR, MS, and UV-visible spectrophotometry. This difunctional dioxolane momomer was found to undergo photoinitiated cationic ring-opening polymerization with 4-methylene-2-phenyl-1,3-dioxolane, resulting in crosslinked materials with Tgs up to 30 °C higher than the corresponding homopolymer of the monofunctional dioxolane. Polymerizations were more efficient using an iodonium salt photoinitiator relative to a sulfonium salt.

111. A NEW THERMALLY-REVERSIBLE PHOTOCROSSLINKABLE SILOXANE. Kevin D. Belfield* and Peng Geng, Department of Chemistry, University of Detroit Mercy, P.O. Box 19900, Detroit, MI 48219-0900.

Photocrosslinking, the utilization of electromagnetic radiation (or light) as the energy source for polymerization of functional monomers, oligomers, and polymers, is the basis of important commercial processes. Photocrosslinking may be achieved by photoinitiated radical or cationic polymerization, the use of photocrosslinking agents, and photocyclization reactions, such as the photochemical [π2s+π2s] cycloaddition. We wish to report a novel approach to make thermally-reversible polymeric siloxane systems based on the [π2s+π2s] photocycloaddition of conjugated olefins, and subsequent thermal reversion of the corresponding cyclobutane adducts.

112. SYNTHESIS OF NEW LOW Tg PHOTOREFRACTIVE POLYMERS. Kevin D. Belfield*, Chandra Chinna, Suresh Sriram, and Ousama Najjar, Department of Chemistry, University of Detroit Mercy, P.O. Box 19900, Detroit, MI 48219-0900.

The photorefractive effect, found in materials that are both photoconductive and have nonlinear optical properties, is avidly being pursued for optical processing applications. In principle, information can be stored optically in a photorefractive material then erased by illuminating the material with light of uniform intensity (to uniformly redistribute charges). Thus, photorefractivity holds great potential in holographic optical data storage and processing applications, as well as frequency doubling of laser light. The development of new photorefractive systems is deemed important to advancement in these fields. Herein, we report the synthesis of fully functionalized siloxane and acrylate based polymers having CT and NLO functionality attached to each repeat unit of the polymer. These polymers, when doped with CG dopants, should exhibit photorefractive behavior. This is the first report of an attempt to synthesize fully functionalized low T_g photorefractive polymers with phosphonate ester groups.

113. MIXED METAL OXIDE INORGANIC/ORGANIC CERAMER COATINGS. Chad R. Wold and **Mark D. Soucek***, **Department of Polymers and Coatings, North Dakota State University, Fargo, North Dakota 58105**

Mixed metal oxide hybrid ceramer coatings were developed using linseed oil as the organic phase with titanium *i*-propoxide, titanium (di-*i*-propoxide) bis(acetylacetonate), and Zirconium *n*-propoxide as the inorganic sol-gel precursors. The overall goal of this study is to develop a primer that will provide better adhesion and corrosion protection for metal substrates with minimal environmental impact. Zinc acetate dihydrate was introduced to the system as an anticorrosive agent. Various coatings properties such as adhesion, hardness, impact resistance and flexibility were observed as a function of sol-gel precursor type and content. In addition, electrochemical impedance spectroscopy (EIS) was used to evaluate the corrosion properties in which low concentrations of sol-gel precursor content demonstrated excellent electrical barrier properties.

114.
SOLID ACID SUPPORTED METALLOCENE CATALYSTS FOR OLEFIN POLYMERIZATION.
T. Suzuki, Y. Suga, Yokohama Research Center, Research and Development Division, Mitsubishi Chemical Corporation, Yokohama, Kanagawa 227 Japan

Recently, much effort has been paid to find out the MAO substitute in metallocene-MAO catalyst system with the aim of lowering the catalyst cost. It has been known that metallocene compound supported on inorganic carriers having Lewis acidity such as Al2O3 and MgCl2 is activated by ordinary alkylaluminium. However, the catalytic activity is not so high for the practical use. Among our study on Lewis acid, we found out that the metallocene compound in combination with certain clay minerals or zeolite beta and ordinary alkylaluminium such as AlR3 showed high activity in olefin polymerization. When solid acid such as phosphowolframic acid, silicomolybdic acid and zeolite A3 were used, the catalysts did not show sufficient polymerization activity. The role of solid acid for ethylene polymerization will be discussed.

115. DESIGN AND SYNTHESIS OF NOVEL FLUORESCENT CHEMOSENSORS FOR BIOLOGICALLY ACTIVE MOLECULES. M. E. Cooper, B. P. Hoag, and D. L. Gin,* Department of Chemistry, University of California, Berkeley, CA 94720

The development of strategies for the specific detection of biologically active molecules is described. In the first, employing molecular imprinting illustrated in Scheme I, cholesterol is linked to a fluorescent "reporter" molecule and imprinted into an ethyleneglycol dimethacrylate polymer network. Hydrolysis of the carbonate linkage creates specific recognition sites for cholesterol to reversibly re-bind through hydrogen bonding to the hydroxyl functionality on the fluorescent molecule. It was our intention to be able to detect the cholesterol re-binding event by attenuation of the fluorescent signal. Our observations show, however, that preferential re-binding of cholesterol was not observed, possibly due to the nature of the weak imprinting interaction. Furthermore, no fluorescence attenuation was detected through hydrogen bonding to the fluorescent reporter molecule. A new, more sensitive fluorescent monomer was prepared and polymerized into a network which is highly sensitive to proton donor molecules. Its fluorescence is quenched by cyclohexane solutions of aniline and phenol but not by anisole, for example. We intend to integrate this functional fluorescent molecule into an imprinted polymer format once the intricacies of the imprinting technique are elucidated and reproduced.

116. ORDERED PIEZOELECTRIC NETWORKS VIA A LIQUID-CRYSTALLINE MONOMER STRATEGY: SYNTHESIS OF THE LC MONOMER. Brian C. Baxter and Douglas L. Gin, Department of Chemistry, University of California, Berkeley, Berkeley, CA 94720

Piezoelectric materials are used in many technical applications. Single crystal ceramics are the most extensively used due to their long term stability and order; however, they are brittle and difficult to fabricate and process. More recently, poled piezoelectric polymers have been introduced which offer ease of fabrication and processability, but lack long term stability. Highly crosslinked, highly ordered piezoelectric networks will offer the order and stability or the ceramics while maintaining the ease of fabrication and processability of polymers. A highly ordered network can be achieved through the use of crosslinkable liquid crystalline monomers. Molecules can be designed which exhibit a chiral smectic C LC phase, which can then be oriented by an external electric field to form a noncentrosymmetric material with a macroscopic dipole moment. In situ photopolymerization to form the highly ordered, network which will then be piezoelectric. This presentation will discuss the design and synthesis of the chiral smectic C monomer.

117. HIGHLY ORDERED POLYMER-INORGANIC NANOCOMPOSITES VIA A MONOMER SELF–ASSEMBLY – IN SITU CONDENSATION APPROACH. D. H. Gray, D. L. Gin; Department of Chemistry, University of California, Berkeley, CA 94720

A new method for synthesizing organic–inorganic composites with architectural control on the nanometer scale has been established. A polymerizable lyotropic liquid crystal, sodium p-styryloctadecanoate, is used to form an inverse hexagonal phase at room temperature in the presence of a sol-gel silica precursor solution. The monomer system acts as an organic template, providing the underlying order of the composite system. Subsequent photopolymerization of the template in the presence of a cross-linking agent with retention of the liquid-crystalline order was confirmed using low-angle x-ray diffraction spectroscopy and polarized light microscopy. Simultaneous photoacid-catalyzed condensation of the silica precursor within the channels of the polymer template yields an ordered nanocomposite material. Infrared, UV-visible, and ^{29}Si solid-state NMR spectroscopic methods were used to confirm polymerization and silica condensation within the composite.

118. APPLICATION OF THE STOPPED-FLOW METHOD FOR THE STUDY OF THE EFFECT OF HYDROGEN ON THE OLEFIN POLYMERIZATION AS WELL AS THE SYNTHESIS OF NOVEL OLEFIN BLOCK COPOLYMER
Hideharu Mori, Kunihiko Tashino, Mikio Yamahiro, <u>Minoru Terano</u>*, School of Materials Science, Japan Advanced Institute of Science and Technology, Asahidai 1-1, Tatsunokuchi, Ishikawa 923-12, JAPAN

The stopped-flow method was successfully applied for the study of the mechanism of the chain transfer reaction and the activity enhancement by hydrogen in the propylene polymerization with Ziegler catalyst. The synthesis of novel olefin block copolymer, polypropylene-*block*-poly(ethylene-*co*-propylene), is also conducted by using the method. One of the most attractive features of the method is the capability for the observation of the polymer produced in the initial stage of polymerization, which reflects the nature of active sites directly. Therefore, the behavior of hydrogen for the chain transfer reaction can be observed clearly through the analysis of the polymer produced by the method.The combination of propylene polymerization and H_2-D_2 exchange reaction using stopped-flow method indicates that the chain transfer reaction occurs in the system in which HD is produced by H_2-D_2 exchange reaction. In other words, atomic hydrogen produced by the dissociation of molecular hydrogen acts as a transfer agent. The mechanism of the activity enhancement by hydrogen will be discussed as well. The stopped-flow polymerization method was successfully improved to apply for the synthesis of novel and neat olefin block copolymer, polypropylene-*block*-poly(ethylene-*co*-propylene) with $MgCl_2$-supported Ziegler catalyst. Several experiments were conducted to demonstrate the formation of a covalent linkage between PP and EPR in the copolymer obtained.

119. MULTI-METHACRYLATE DERIVATIVES OF STYRENE-ALLYL ALCOHOL COPOLYMERS IN COMPOSITES. B. M. Culbertson, Y. Tong and Q. Wan, The Ohio State University, Columbus, Ohio 43210-1241.

Seeking to improve dental composites, we have recently prepared and evaluated multi-methacrylate systems (MSSA) derived from low molecular weight styrene-allyl alcohol(SAA) copolymers. The hydroxyl groups on SAA are readily converted to methacrylate residues, via transesterification with MMA. In one test series, 35.5 % of the OH groups on SAA were converted to a MSAA oligomer. A control system, consisting of BisGMA/TEGDMA(50/50,wt/wt), having 0.5 wt % camphoroquinone and 1.0 wt % N,N-dimethylamino-ethyl methacrylate, was modified with 5 wt % of the MSAA monomer. The systems were cured under N_2 with visible light for 5 min at ambient temperatures. Test specimens (n=5) for compressive strength (CS) and glass transition tempeature(T_g) determinations were conditioned in water for 7 days at 37°C. T_g (wet samples) were measured by thermomechanical analysis (TMA). The remaining (unpolymerized) double bonds (RDB) were estimated by FT-IR techniques. In this example, the experimental had CS = 511 MPa, T_g = 83 °C, and RDB =20.9 %. In contrast, the control had CS = 375 MPa, T_g (wet)= 50 °C, and RDB = 18.7 %. Preliminary data suggests these MSAA oligomers could be useful in a variety of composites.

120. SYNTHESIS AND CHARACTERIZATION OF MODIFIED SEGMENTED POLYURETHANES DISPLAYING IMPROVED FIRE RESISTANCE. <u>Q. Ji</u>, M. Muggli, F. Wang, T. C. Ward, G. Burnsa[a], U. Sorathiab[b] and J. E. McGrath, Virginia Polytechnic Institute and State University, Department of Chemistry and NSF Science and Technology Center, High Performance Polymeric Adhesives and Composites, Blacksburg, VA 24061-0344; [a]Dow Corning Corporation, Midland, MI 48486-0994; [b])Naval Surface Warfare Center, Annapolis, MD 21402.

Several series of modified segmented polyurethanes have been synthesized by the incorporation of bis(hydroxy propyl) isobutyl phosphine oxide (BHBPO) and oligomeric secondary amine functional poly(dimethyl siloxane) (PDMS) as a chain-extender and soft segment or co-chain extender and co-soft segment, respectively. The physical properties were characterized by various techniques. The results showed that it is possible to synthesize modified segmented polyurethanes with good physical properties. The self-extinguishing characteristics can be demonstrated for phosphorus containing polyurethane systems. Preliminary cone calorimetry measurements showed a heat release rate drops 67% for 20% siloxane soft segment containing polyurethane system.

121. DIELECTRIC RELAXATION BEHAVIOR OF MODIFIED SEGMENTED POLYURETHANES. Q. Ji, Y. N. Liu and J. E. McGrath, Virginia Polytechnic Institute and State University, Department of Chemistry and NSF Science and Technology Center, High Performance Polymeric Adhesives and Composites, Blacksburg, VA 24061-0344; D. Hayward and R. A. Pethrick, Department of Pure and Applied Chemistry, University of Strathclyde, 295 Cathedral Street, Glasgow G1 1XL, Scotland.

Siloxane containing segmented polyurethanes have been synthesized. Their extensive dielectric studies have been performed over wide temperature and frequency ranges. Four dielectric relaxation process were observed, labeled γ, β, α and α^+. The and relaxation are simple relaxations and do not change significantly with the polymer composition. The α relaxation, glass transition relaxation of soft phase can only be observed in richer soft phase samples. As the hard segmented concentration increased, there is the fourth relaxation process, α^+ relaxation, appeared which is part masked by d.c. conduction. A subtraction treatment was successfully used and a clear relaxation process was obtained. Unlike the dipolar relaxation process, this α^+ relaxation is a simple interfacial polarization relaxation process with a lower activation energy.

122. NEW FIRE RESISTANT EPOXY RESINS WITH IMPROVED TOUGHNESS. Y. N. Liu, Q. Ji, and J. E. McGrath, Virginia Polytechnic Institute and State University, Department of Chemistry and NSF Science and Technology Center, High Performance Polymeric Adhesives and Composites, Blacksburg, VA 24061-0344.

A new type of phosphine oxide containing secondary diamine, bis(4-cyclohexyl aminophenyl) phenyl phosphine oxide (BCAPPO), was successfully synthesized by nucleophilic aromatic substitution reaction. The structure of the new type of the secondary diamine was confirmed by elemental analysis, ^{1}H-NMR, ^{13}C-NMR and ^{31}P-NMR spectroscopies. The novel epoxy materials were prepared via generation of linear epoxy oligomers by reacting the newly synthesized secondary amine functional BCAPPO with EPON 828 followed by crosslinking the oligomers with 4,4'-diaminodiphenylsulfone (DDS). The modified epoxy networks exhibited superior char yields and comparable thermal properties, in relation to the common epoxy networks. It is proposed that the incorporation of phosphine oxide based secondary diamine into the epoxy backbone would enhance the flame-resistance and improved the toughness. Furthermore, the use of this new diamine in the formation of composites is of great interest since the polar phosphine oxide groups may improve the interactions between the epoxy matrix and carbon fibers..

123. CONTROLLING MATERIALS ARCHITECTURE ON THE NANOMETER SCALE: COMPOSITE SYNTHESIS USING LYOTROPIC LIQUID-CRYSTALLINE MONOMERS. W. M. Fischer, D. H. Gray, R. C. Smith, and D. L. Gin,* Dept. of Chemistry, University of California, Berkeley, CA 94720, and the Center for Advanced Materials, Lawrence Berkeley National Laboratory

Nanometer-scale engineering is a characteristic of biological materials, which is primarily responsible of their impressive properties. Due to the limitations of conventional processing techniques, control over this realm has been inaccessible for the construction of synthetic polymers and composite materials. We have successfully developed a novel synthesis strategy for the construction of ordered nanocomposite materials, using monomers based on lyotropic (i.e., amphiphilic) liquid crystals (LCs). In this strategy, self-organizing lyotropic LC monomers are used to form an ordered template matrix around reactive hydrophilic solutions. Subsequent in situ photopolymerization to lock-in the matrix architecture, followed by initiation of chemistry within the resulting ordered hydrophilic domains, yields the anisotropic nanocomposites. This technique offers unparalleled control on the small-scale for the synthesis of man-made materials. Using this technique, composites have been synthesized that have a regular hexagonal arrangement of extended "filler" domains 4 nm in diameter. Anisotropic polymer–silica and polymer–conjugated polymer nanocomposites have been successfully made, thereby demonstrating the versatility of this method.

124. SPECTROSCOPIC AND MODEL COMPOUND STUDIES IN THE CURE CHARACTERIZATION OF A BISMALEIMIDE/DIALLYLBISPHERAL-A THERMOSET RESIN; J.C. Phelan and C.S.P. Sung, Institute of Materials Science, University of Connecticut, Storrs, CT 06269-3136

Fluorescence, FT-IR and UV-reflectance changes which accompany bismaleimide/diallylbisphenol-A resin cure are reported for three cure schedules. The formation of succinimide moieties is attributed to Ene, Diels-Alder, and alternating copolymerization reactions involving phenylmaleimide and allylphenol groups. The formation of 2-propenylphenol vinyl groups is attributed to Ene reaction between allylphenol groups and maleimide groups as well as by direct isomerization of allylphenol. The destruction of 2-propenylphenol vinyl groups is assigned to crosslinking reactions. Corroberating evidence for the formation and destruction of 2-propenylphenol vinyl groups is provided by UV-reflection spectroscopy. Evidence for Diels-Alder reactions followed by re-aromatization processes is provided by UV-reflectance spectroscopy. Fluorescence signals are initially quenched, but increase and then level off as the resin cures. Model compound studies indicate that emissions that occur at 356 nm when the resin is excited at 280 nm are from the phenolic portion of the resin, while emissions that occur at 440 nm when the resin is excited at 380 nm are from phenyl-succinimide. Structure arising from Diels-Alder-Ene and alternating copolymerization reaction sequences have also been confirmed by model studies. Support that the final stages of cure involve re-aromatization of phenolic groups and crosslinking reactions of 2-propenylphenol vinyl groups is presented.

125. COEXISTENCE CURVE OF POLYSTYRENE IN METHYLCYCLOHEXANE AT THE TRICRITICAL POINT

T. Dobashi
Faculty of Engineering, Gunma University, Kiryu 376, Japan
M. Nakata
Faculty of Science, Hokkaido University, Sapporo 060, Japan

The coexistence curves for the system polystyrene + polystyrene + methylcyclohexane both at the tricritical point and off-tricritical point are studied by experiments and calculation. The coexistence curve for the tricritical concentration is extremely asymmetrical near the tricritical temperature on the diagram total polystyrene vs temperature. The diameter moves toward dilute side just below the tricritical temperature. The coexistence curve for the off-tricritical concentration is in distorted sand-clock shape. These characteristic features are explained with an empirically determined free energy.

126. CRITICAL BEHAVIOR OF POLYDISPERSE POLYSTYRENE SOLUTION
R. Kita, K. Kubota, M. Nakata[a], and T. Dobashi
Faculty of Engineering, Gunma University, Kiryu 376, Japan
[a]Faculty of Science, Hokkaido University, Sapporo 060, Japan

The critical exponents for the shape of coexistence curve (β), the osmotic compressibility (γ), and the long-range correlation length (ν) were determined for polydisperse polystyrene (M_w/M_n=2.8, M_w=23.9 $\times$ 10^4) in cyclohexane near its critical mixing point. Here M_w and M_n are weight- and number-averaged molecular weight, respectively, and the distribution function of molecular weight is Schulz-Zimm type. The obtained critical exponents were β =0.363 $\pm$ 0.005, γ =1.33$\pm$0.06, and ν =0.70$\pm$0.02. They show distinct deviations from the three-dimensional Ising values and are in good agreement with Fisher's renormalized exponents concerning the hidden variables.

127. SMECTIC MESOMORPHISM OF LONG-CHAIN n-ALKYLAMMONIUM POLYACRYLATES, D. Tsiourvas, M. Arkas, C. M. Paleos and A. Skoulios, a. Institute of Physical Chemistry, NCSR "Demokritos", Aghia   Paraskevi, Attiki, Greece and b. Groupes Des Matériaux Organiques, Institut de Physique et Chimie des Matériaux, 23, rue du Loess, 67037 Strasbourg, France

The liquid crystalline character of monomeric or polymeric amphiphilic compounds was attributed to the formation of supramolecular structures consisting of isolated lipophilic and hydrophilic segments. Concerning protonated ammonium salts, the investigations primarily concerned   salts bearing monomeric counter ions. We have been the first to report the preparation of liquid crystalline ammonium salts by the interaction of long-chain amines with polyacids including polyacrylic and polymaleic acids. In this case liquid crystalline phases were characterized as "template induced" since the ammonium salts were formed by neutralization of the amines by the polyacids following a mechanism analogous to template polymerization. Within this framework we extent our preliminary work on the preparation and characterization of n-alkylammonium polyacrylates highlighting on thermal and X-ray diffraction studies.

128. THERMAL PROPERTIES OF POLYETHYLENE MODELS

Messaoud El Hadad, Mohamed Maïssara, Laboratoire de Chimie Quantique et Spectroscopie Moléculaire, Université Hassan II-Mohammedia, BP 7955, Casablanca, Marocco ; <u>François J. Carrière</u>, laboratoire de Chimie Macromoléculaire, case 185, Pierre & Marie Curie University, 75005 Paris, France.

Considered as models, the linear n-alkanes and their binary mixtures were studied by DSC and infrared to elucidate the mechanisms which govern the solid-solid phase transitions. Two series of n-alkanes binary mixtures were prepared : $(1-x)C_n + xC_{n+1}$ with n = 19 to 22 and $(1-x)C_n + C_{n+2}$ with n = 19 to 21. These mixtures showed always one (or more) solid-solid phase transition, whereas in the case of pure n-alkane, the presence of these transitions depended essentially on the parity of n : odd n, from n = 9 ; even n, n ≥ 24. The curves, transition temperature (T_t) vs x, presented always a minimum at about x = 0.30, whereas the melting temperature, (T_m) vs x, increased almost linearly. At this minimum, the difference $T_m - T_t$ varied from 5 to 25 K according to the parity, the length of the chains and the difference of one or two carbon atoms. The enthalpies of the transitions are about six times lower than the corresponding melting enthalpies. Infrared study showed the influence of conformational defects : end gauche, kink and translation along the chain axis, on the mechanism of the solid-solid phase transitions.

129. GRAFTING OF MESOGENIC GROUPS ONTO POLYMERS HAVING A POLYETHER MAIN CHAIN : DSC STUDY OF THE POLYMERS.

H. Pham, A. Le Borgne, <u>François Carrière</u>, N. Spassky, Laboratoire de Chimie Macromoléculaire, Université Pierre & Marie Curie, 75005 Paris, France

An atactic sample of polyepichlorohydrin (PECH) of molecular weight $\overline{M}n$ = 5700 prepared by cationic initiation was chemically modified with introduction of -O-ΦΦ-OCH$_3$ mesogenic group. The degree of substitution varied from 19 to 78 %. The studies of thermal properties by DSC indicate that T_g's increase linearly with the degree of substitution. The Cp vs T curves show that the values of ΔCp are much lower for a second heating. Melting temperatures appeared starting from 30 % substitution and increase linearly with degree of substitution as well as the corresponding ΔH values. A comparison with PECH's substituted with cyano-biphenyl group is established.

130. CONDUCTING COMPOSITES DERIVED FROM POLYANILINE AND BENZOBISTHIAZOLE RIGID ROD POLYMER.

Loon-Seng Tan*[a], S. R. Simko[b], S. J. Bai[b], R. A. Vaia [a] and R. J. Spry [a]   a. Polymer Branch, WL/MLBP, Wright Laboratory, 2941 P Street Suite 1, Wright-Patterson Air Force Base, OH 45433-7750 b. University of Dayton Research Institute, 300 College Park, Dayton, OH. 45469-0001.   * Corresponding author.

As an alternate method for processing polyaniline (PANI) from its conducting (protonated) state, vacuum-casting of PANI from its methanesulfonic acid (MSA) solution can provide films with electrical conductivity values of 128-152 S/cm. In addition, we have also shown that PANI.MSA/PBZT* composite films with PANI content ranging from 2 - 90 wt% can also be prepared similarly. Conductivity values spans from less than 10^{-10} S/cm to 124 S/cm and nicely followed a percolation pattern. Although grossly phase-separated with PBZT domain sizes of 1-10 mm by 0.5-2 mm, these cast films showed relatively low percolation threshold ($f_c \sim 3.5\%$). This compared well with the conductivity values for the PANI.H$_2$SO$_4$/PPTA* composite fibers (A. Andreatta, et al., *Polym. Commun.* **1990**, *31*, 275), which were also phase-separated. Efforts are currently underway to delineate the relationship of degree of phase separation (PBZT domain size) and the bulk conductivity of the composite films. *PBZT=poly(p-phenylene benzobisthiazole); PPTA=poly(p-phenylene terephthalamide)

131. PREPARATION AND CHARACTERIZATION OF GRISEOFULVIN/PVP SOLID DISPERSIONS, H. Akin, J. Heller* and F.W. Harris, Maurice Morton Institute and Department of Polymer Science, The University of Akron, Akron, OH 44325-3909
* Advanced Polymer Systems, 3696 Haven Avenue, Redwood City, CA 94063

Solid dispersion technology has been widely investigated over the past 25 years as a means of increasing the dissolution rates and absorption of many poorly soluble drugs. In this technology, drugs are dispersed in an inert carriers so that they are present in an amorphous form or as an amorphous drug-polymer complex. Griseofulvin (gris) and polyvinylpyrrolidone (PVP) were dissolved in a common solvent, then isolated as an intimate mixture or complex. The solvent, methylene chloride appeared to enhance the interactions between gris and PVP and, thus, promote the formation of amorphous solid dispersions. The ratio of gris-to-PVP had to be less than 1:1 (w/w) in order to prevent the gris crystallizing. The amorphous phase of the drug in the solid dispersions was stable for one year. Higher molecular weight PVP samples were more efficient as complexing agents than those with lower degrees of polymerization. The dissolution of gris from dispersions was greater than pure gris.

132. USE OF NEURAL NETWORKS FOR THE STUDY OF CORRELATIONS BETWEEN MOLECULAR ORBITALS AND ELECTRONIC PROPERTIES OF POLYMERS. Qing Luo[1], Jerry A. Darsey[1], Cesar M. Compadre[2], (1) Department of Chemistry, University of Arkansas at Little Rock, AR 72204, (2) Department of Biopharmaceutical Science, University of Arkansas for Medical Sciences, Little Rock, AR 72205

The ability to predict electronic properties of molecules is of great value in optimizing their design, modification and processing. Such capabilities represents a great challenge to present computational methods since there are a large number of possible conformations which influence the physical properties for any given molecule. This presentation will discuss one approach which is to employ SCF-MO quantum mechanical calculations in conjunction with artificial neural networks (ANN). We have calculated the eigenenergies of polymers using *ab initio* SCF-MO program and using ANN to predict the dielectric constants of a series of 14 polymers.

133. **SURLYN®/[SILICON OXIDE] HYBRID MATERIALS VIA POLYMER IN SITU SOL-GEL CHEMISTRY.** David A. Siuzdak, Kenneth A. Mauritz, Department of Polymer Science, University of Southern Mississippi, Hattiesburg, MS 39406-0076.

Ethylene-methacrylic acid copolymers and their ionomeric derivatives (Surlyn®) were used as interactive matrices for the in situ polymerization of tetraethoxysilane. It was assumed that a limited sol→gel process initiates within ionic clusters or around acid groups given the presence of water, and that a highly-dispersed silicon oxide phase results upon drying. FTIR spectra and environmental scanning electron microscopy-EDAX verified that a silicon oxide phase can indeed be incorporated in Surlyn® via this route. DSC first-scans show crystallite melting and a sub-T_m feature which disappears upon reheating after quenching from the melt; the relative magnitudes of these transitions vary with filler content. Stress-strain tests show progressive increase in modulus with reduction of elongation-at-break for increasing silicon oxide content, which is attributed to inorganic structures that become increasingly more invasive throughout Surlyn®. ^{29}Si solid state NMR spectroscopy revealed that tri- and tetrasubstitution of Si around SiO_4 units predominates, regardless of silicon oxide content.

134. MODIFICATION OF THE CLUSTER DOMAINS OF PERFLUOROSULFONATE IONOMERS VIA IN SITU BASE-CATALYZED INORGANIC SOL-GEL REACTIONS.
John T. Payne, David A. Reuschle, and Kenneth A. Mauritz, Department of Polymer Science, University of Southern Mississippi, Southern Station Box 10076, Hattiesburg, MS 39406-0076

Nafion®/(silicon oxide) hybrids were produced via in situ, base-catalyzed sol-gel reactions of tetraethylorthosilicate in prehydrated, ethanol-swollen Group 1 metal counterion form membranes. Uptake vs. pH was investigated for base pre-soaked and non-base pre-soaked systems. As pH increased, wt.% uptake decreased. Maximum uptake in both base systems occurred at ~ pH 8 and decreased as pH increased, the Li^+ form having highest wt.% gain. As counterion radius increased, wt.% uptake decreased over the pH range. Cs^+ form membranes maintained constant uptake from pH 8 to pH 11. ^{29}Si solid state NMR spectra showed that for low uptakes, Q_3 and Q_4 coordination about SiO_4 units predominates, with base pre-soaked samples showing a larger $Q_4:Q_3$ peak ratio. High uptake yielded predominance of Q_3 and a slight Q_2 peak associated with un-reacted SiOH groups. The results of FTIR studies of these systems are in harmony with the NMR results.

135. INORGANIC MODIFICATION OF ELASTOMERIC POLY(STYRENE-b-ISOBUTYLENE-b-STYRENE) BLOCK COPOLYMER IONOMERS VIA IN SITU SOL-GEL REACTIONS OF TETRAETHOXYSILANE.
John T. Payne, David A. Reuschle, L. Brian Brister, Christopher L. Curry, Kelly A. Shoemake, Robson F. Storey, and Kenneth A. Mauritz, Department of Polymer Science, University of Southern Mississippi, P.O. Box 10076, Hattiesburg, MS 39406-0076

Unique poly(styrene-b-isobutylene-b-styrene) block copolymers have been created via living cationic polymerizations. These block copolymers vary in molecular weight, block lengths of polyisobutylene and polystyrene segments as well as architecture, to include linear, three arm and multi-arm star polymers, and telechelic ionomers. The telechelic ionomer block copolymers have end group functionalization to include silicon-alkoxide-functional polyisobutylene and carboxylic acid groups, which can easily lead to post-modification by a variety of means. We are formulating unique organic/inorganic hybrids by both the acid- and base-catalyzed growth of nanometers-in-size silicon oxide clusters and will attempt to incorporate these quasi-networks via *in situ*, sol-gel reactions of tetraethylorthosilicate around the hard blocks of the copolymer morphology. This type of inorganic modification has the potential to improve inherent polymer physical properties, such as mechanical, thermal, gas permeation, and chemical resistance. Depending on the catalyst used, different in situ-grown silicon oxide cluster structures and organic/inorganic interfacial interactions, are being tailored.

136. Evaluation of Apolipophorin III and a Site-Specific Mutant of Apolipophorin III as a Primary Emulsifier of Tetradecane
Monica Tisack, Johanna Kahalley, Gordon Cannon, Charles McCormick, and Robert Lochhead
The University of Southern Mississippi, Department of Polymer Science, Hattiesburg, Mississippi 39406-0076

Apolipophorin III is an amphipathic protein found in the *Manduca sexta* insect. In nature it associates with small molecule surfactants and diglycerides in a micellar fashion which serves to stabilize and transport fats within the water-soluble hemolymph. In order to address this interaction, a site-specific mutation has been made using the inverse polymerase chain reaction to convert the genetic codon for a specific residue, leucine, to that of a cysteine. This conversion imparts a functionality which will be chemically modified with fluorescent, hydrophobic, and hydrophilic groups. Comparisons are made between the native protein and the cysteine mutant in terms of interfacial and surface tension, structural stability, molecular weights, and isoelectric focusing. Results show that the native and mutant protein yield similar molecular weights and isoelectric points. In addition, preliminary surface studies indicate that the conversion of an isopropyl group to a sulfydryl group imparts increased interfacial activity at pH 8 for lower protein concentrations, yet the extent of interfacial tension depression remains constant.

137. MOLECULAR CHARACTERIZATION OF SELF-ASSEMBLY OF FUNGAL HYDROPHOBINS AND AN ASSOCIATED POLYSACCHARIDE FROM AQUEOUS MEDIA. G. G. Martin, M. F. Richardson, G. C. Cannon, and C. L. McCormick, Department of Chemistry and Biochemistry and Department of Polymer Science, The University of Southern Mississippi, Hattiesburg, MS 39406-0076

During growth of certain fungal organisms including the plant pathogen, *Schizophyllum commune*, a family of hydrophobic polypeptides, termed "hydrophobins," and an associated polysaccharide, schizophyllan, are secreted into the aqueous medium. These molecules are the basis of ordered macromolecular structures that play a significant role in stabilization, attachment, and protection during the biological life cycle of the organism. Using various techniques including high-speed centrifugation, size exclusion chromatography, light scattering, and atomic force microscopy, we have begun an investigation of the molecular components involved in this assembly process. These studies have revealed at least two populations of macromolecular structures that may be involved in self-assembly: small aggregates that are 50-100 nm in size and larger structures about one micron in size consisting of 50-100 nm-size aggregates. Preliminary experiments in our laboratory indicate that these biopolymers may have immense technological potential in areas such as surface modification, fouling resistance, controlled delivery, protective encapsulation, and water remediation.

138. THE MACROGALLERIA: USE OF THE WORLD WIDE WEB IN POLYMER SCIENCE EDUCATION; Mark Michalovic, Kelly Anderson, Gregory Brust, and L.J. Mathias*, Department of Polymer Science, University of Southern Mississippi, Hattiesburg, Mississippi 39406-0076

A new multimedia internet offering is described that provides both introductory and advanced material in polymer science and science in general. Target audiences include the general public, K-12 students and college undergraduates. The novel presentation method (based partly on a shopping mall) combines with extensive and varied examples, real-life stories of discovery, and real-world applications to make this an exciting and relevant educational tool. Use on-campus (college freshman) and off-campus have already generated excitement and positive feedback that supports extension and expansion of our approach. The website may be visited at http://www.psrc.usm.edu/polyclass/macrogalleria.html.

139.
GPC FOR WINDOWS - THE FIRST SHAREWARE GPC SOFTWARE M.J. Ballard PO Box 2197, Mount Waverley, Vic AUSTRALIA, 3149.

GPC for Windows is a new, powerful, inexpensive and yet easy-to-use alternative to commercial GPC software. It is capable of acquiring data from a variety of instruments, importing data from existing acquisition software, and analysing that data. It also offers a number of technical improvements over existing GPC software, including: easy calibration with a variety of fitted functions (and not just polynomials); Universal Calibration, using both Mark-Houwink-Sakurada and Stockmayer-Fixman equations; a variety of scientifically interesting manipulations (integration, differentiation, smoothing, lines of best fit) of the raw and transformed data; "one click" calculation of number, weight and log weight MWDs; and a rapid "De-Broadening" algorithm that removes the effects of column broadening. Disks of the shareware version will be available at the lecture or can be downloaded from http://www.chem.csiro.au/ballard/gpc10.htm or from ftp.winsite.com/pub/pc/win3/demo/gpc10.zip .

140. OXYGEN EFFECTS IN PHOTO-COPOLYMERIZATION OF MONOFUNCTIONAL MONOMERS WITH DIFUNCTIONAL ACRYLATES AND METHACRYLATES.
C. W. Miller, R. Kess, T. Iijima, and C.E. Hoyle, Department of Polymer Science, University of Southern Mississippi, Hattiesburg, MS 39406; Sönny Jonsson, Fusion UV-Curing Systems, Gaithersburg, MD.

Molecular oxygen exists in its ground electronic state as a triplet biradical. Therefore it can easily react with radical species produced during free-radical polymerization, decreasing polymerization efficiency. Despite the fact that most industrial free-radical UV-curing processes are conducted in the presence of oxygen (air), little data exists describing the kinetics of photopolymerization in the presence of air or the kinetics of the photopolymerization of mixtures of monomers with differing functionality. Herein, we present data describing the effect of adding various percentages of hexylacrylate (HA), N-vinylpyrrolidinone (NVP), and N-vinylformamide (NVF) on the efficiency of photoinitiated polymerization of 1,6-hexanedioldiacrylate (HDDA) in nitrogen and air. Data is also presented describing the effect of adding hexylmethacrylate (HM) to 1,6-hexanedioldimethacrylate (HDDM) in nitrogen and air. Results indicate that HA decreases the efficiency of polymerization of HDDA in both nitrogen and air. Similar trends are found in the HDDM/HM mixtures, though not as pronounced. In contrast to results for HA in HDDA and HM in HDDM, certain concentrations of NVP and NVF increase the efficiency of polymerization of HDDA in nitrogen and air.

141. **CHIRAL RECOGNITION PROPERTIES OF (1→6)-2,5-ANHYDRO-3,4-DI-*O*-ALKYL-D-GLUCITOL TOWARD RACEMIC AMINO ACIDS AND AMINES**
Toshifumi Satoh, Kazuaki Yokota, and Toyoji Kakuchi
Hokkaido University, Sapporo 060, Japan

The formation of complex between (1→6)-2,5-Anhydro-3,4-di-*O*-alkyl-D-glucitol (**1**) and DL-amino acid are discussed using NMR spectral measurements. The optical resolution abilities of amino acids and amines by liquid chromatography using polymer **2** bond on silica gel are also studied. For the ^{1}H NMR measurement of the complex, all the protons in the repeating units for polymer **1** changed before and after the formation of a complex between **1** and DL-amino acid. The coordination environment of polymer **1** with L-amino acid differs from that with D-amino acid. In the optical resolutions of racemic amino acids using polymer **2**, D-isomers were eluted first. The separation factors of many racemates were 1.1-1.2, and the resolution factors were 0.60-3.42.

1a: R= CH$_3$
1b: R= C$_2$H$_5$

2a: R= CH$_3$
2b: R= C$_2$H$_5$

142. DYNAMICS OF PHYSICAL NETWORK IN ABA TRIBLOCK COPOLYMER GELS: SOLVENT PENETRATION INTO JUNCTION ZONES. N. Mischenko, R. Kleppinger, K. Reynders, M. H .J. Koch[1], H.Reynaers. Department of Chemistry K.U.Leuven, B-3001 Heverlee, Belgium; [1]EMBL-Outstation, DESY, D-22603 Hamburg, Germany

ABA triblock copolymers (Mn=1-2*10^5, 30 to 50 wt % of A-blocks, A=PS) being dissolved at elevated temperatures in an oligomeric hydrocarbon solvent (Mn=400) selective for the B-block (B=rubber) result on cooling in a physical gel where microphase separated A-blocks form the junctions of the spatial network. Time-resolved SAXS on cooling from the disordered state (monotonously decreasing SAXS) shows an appearance of an interference between A-domains and a steep increase of the interdomain distance D followed (at t<t**) by a region of relatively slow increase of or even constant D. The value of t** depends on the size and concentration of the PS end-blocks and ranges from 70-85°C (M_{PS}=15,000) to 190-200°C (M_{PS}=50,000). At even lower temperature (t*) the interdomain distance can further increase or remain constant. We suppose that the PS domains existing in t*<t<t** range consist of already phase separated solvated PS blocks. Values of t** and t* correspond to the beginning and to the end of the solvent expelling which correlates with the change from a visco-elastic to an elastic behavior (oscillating shear at ω=1rad/s).

143. EFFECTS OF ELECTRON IRRADIATION AND PERCENT COMPOSITION ON THE T_g AND DAMPING OF BORON-EPOXY COMPOSITES. I. I. Nieves, R. L. Kiefer, M. B. Glasgow, and R. A. Orwoll, Departments of Chemistry and Applied Science, College of William and Mary, Williamsburg, VA 23187, and S. A. Thibeault, Materials Division, NASA Langley Research Center, Hampton, VA 23665

Samples of a space-certified epoxy containing varying quantities of crystalline and amorphous boron additives were evaluated for their ability to withstand electron radiation. Samples of these materials were divided into two groups, one serving as a control while the other was irradiated with 1 MeV electrons. The samples were subsequently examined with dynamic mechanical analysis in order to obtain frequency and damping data as a function of temperature. The frequency information then was used to determine the glass transition temperature (T_g) for each sample. Comparisons of the T_g and the damping maximum for different samples were utilized to qualitatively determine how material properties vary according to sample composition and how these properties are altered by irradiation. The results showed that both the T_gs and damping maxima were increased through the addition of boron additives. By contrast, irradiation produced virtually no change in the T_gs, yet produced a significant lowering of the damping maxima.

144. CHARACTERIZATION OF CURE CHEMISTRY OF 4-PHENOXY-4'-PHENYLETHYNYLBENZOPHENONE

K.H. Wood, R. A. Orwoll, College of William and Mary, Williamsburg, VA 23187-8795
B. J. Jensen, NASA Langley Research Center, Hampton, VA 23681-0001
P. R. Young, Emory & Henry College, Emory, VA 24327
H. M. McNair, Virginia Polytechnic and State University, Blacksburg, VA 24061-0344

The ability to process high performance oligomers into useful adhesives and high quality composites has been significantly advanced by synthetic techniques in which oligomers terminated with reactive groups cure or crosslink at elevated temperature after the article has been fabricated. Utilizing phenylethynyl endcaps is one means of achieving this advantage The present paper addresses the cure chemistry a phenylethynyl-terminated model compound examined in an attempt to understand reaction mechanisms and cure products of more complex oligomer systems. The thermal cure of 4-phenoxy-4'-phenylthynylbenzophone (4-PPEB) was studied using a variety of analytical techniques including differential scanning calorimetry, Fourier transform infrared, nuclear magnetic resonance spectroscopies and liquid chromatography-mass spectroscopy. The intent of this research is to provide fundamental insight into the molecular structure of these new engineering materials so that their performance and durability can be more adequately appreciated.

145. PHOTODEGRADABLE VINYL POLYMERS CONTAINING METAL–METAL BOND CROSSLINKS. Brandi L. Langsdorf, Jeffrey J. Wolcott and David R. Tyler, Department of Chemistry, University of Oregon, Eugene, OR 97403

Vinyl polymers containing Fe-Fe bond crosslinks were synthesized using the di-vinyl derivative of $Cp_2Fe_2(CO)_4$ shown in the reaction. The synthesis of the di-vinyl derivative will be described, as will the preparation of its copolymers with styrene, methyl methacrylate, acryonitrile, and N-vinyl pyrrolidinone. The polymers are photoreactive because metal-metal bonds can be cleaved with visible light. Irradiation of the copolymer films with visible light cleaved the Fe–Fe bonds, as indicated by the disappearance of the characteristic absorbance at 568 nm attributed to the Fe–Fe chromophore. Potential uses for these new materials will also be discussed.

146. TEMPLATE SYNTHESIS OF CONDUCTIVE POLYANILINE/SULFONATED POLYSTYRENE COMPOSITES

Yueping Fu, R. A. Weiss
Polymer Science Program and Department of Chemical Engineering
University of Connecticut, Storrs, CT 06269-3136

Highly conductive polyaniline/sulfonated polystyrene (PANI/HSPS) composites were prepared by template polymerization of aniline within the HSPS matrix. The amount of polyaniline, the conductivity and the morphology of the composite films depend on the sulfonation level of HSPS and the content of PANI. A maximum conductivity of 0.1 S/cm has been achieved.

POLY

147.

THERMAL TRANSITIONS AND MECHANICAL RELAXATIONS OF PERFLUOROSULFONATE IONOMERS. **Robert B. Moore**, Kevin M. Cable, Department of Polymer Science, University of Southern Mississippi, P.O. Box 10076, Hattiesburg, Mississippi 39406-0076

Differential scanning calorimetry (DSC) was used to evaluate the effect of counterion type and thermal history on the thermal transitions of perfluorosulfonate ionomers (PFSIs). Endothermic peaks similar to those observed for the as-received samples were found to reappear, for annealed samples, and grow in intensity with annealing time. Dynamic mechanical (DMA) studies were conducted in order to evaluate the effect of counterion size and strength of electrostatic interactions on the ionic network within PFSIs. As the counterion size increased, relaxations associated with the ionic domains and perfluorocarbon matrix were observed to shift to lower temperatures. Based on these studies, the thermal transitions observed in DSC thermograms of PFSIs are proposed to be more indicative of crystalline melting while the mechanical relaxations are more indicative of chain motions within the electrostatic network.

148. ^{13}C AND ^{1}H RELAXATION TIMES OF COLUMNAR TYPE RIGID-ROD POLYESTERS WITH FLEXIBLE SIDE CHAINS OBTAINED BY HIGHRE SOLUTION AND SOLID STATE ^{13}C NMR

Mingming Guo*, Institute of Polymer Science, The University of Akron, Akron, OH 44325-3909
H. R. Kricheldorf, Institut fur TMC, Universitat Hamburg, Bundesstr. 45, D-20146, Hamburg, Germany

Rigid-rod poly(p-biphylene terephthalate) with flexible side chains forms the unique hexagonal columnar phase, but not the well known layered mesophase. The conformation change, molecular motion and morphology of these new samples, with side chain length 8,12,16, were studied by solid state ^{1}H and ^{13}C NMR relaxation times and variable temperature spectra. At room temperature the side chains in all samples are in the melt or disorder state. This is quite different with similar samples that have been reported, in which crystallinities are 68, 31, 29 respectively for side chain length 8, 12, 16. There are almost neither linewidth nor chemical shift difference between Block decay and CP/MAS spectra in these samples. These results indicate that these systems are pretty uniform from the molecular motion and morphology point of view. While $^HT_1$ and $^HT_1\rho$ were identical, different $^HT_2$, $^CT_1$ and $^HT_1\rho$ at low spin lock field were observed. This indicates that systems are homogenous at scale above 50 Å and become heterogeneous when the observation scale goes down to 20 Å. Low temperature spectra recorded at -20, -40, -60 °C indicate that the side chain in this columnar phase can crystallize although it is much more difficult.

149. SIDE CHAIN CONFORMATIONS AND ITS FORMATION KINETICS IN SANIDIC LC POLYMERS OBTAINED BY VARIABLE TEMPERATURE SOLID STATE ^{13}C NMR

Mingming Guo*, Institute of Polymer Science, The University of Akron, Akron, OH 44325-3909
H. R. Kricheldorf, Institut fur TMC, Universitat Hamburg, Bundesstr. 45, D-20146, Hamburg, Germany

The conformation, mobility, chain packing, and crystalline kinetics of the flexible side chains of the rigid-rod poly(phenylene-terephthalate) with both flexible mono- and disubstitued alkylthio side chains were studied using variable temperature ^{13}C solid state NMR. Information on the conformation probability and motion of alkyl side chain can be obtained not only from the temperature dependence of the integral of the ordered peak area but also from the ^{13}C chemical shifts. While it is generally agreed in the literature that the ordered region of the side chain in sanidic samples is composed of all trans conformation, we find that the conformation in both ordered region and disordered region changes with temperature, the length, and the number of the side chains. About a 6.5 % gauche conformation increase was detected for the carbons in the middle of the ordered region when it passes the solid-solid transition temperature. While the gauche conformation probability in the disordered region does not depend on temperature change, it decreases when the side chain length decreases. The side chain isothermal crystallization rate of mono substitute sample is slower than the disubstituted sample.

150. CONSTITUTION, CONFIGURATION, AND THE OPTICAL ACTIVITY OF CHIRAL DENDRIMERS
James R. McElhanon and Dominic V. McGrath*
Department of Chemistry, University of Connecticut, Storrs, CT 06269-4060

Dendrimers with chiral surfaces or interiors possess the potential to adopt chiral conformations and may exhibit macromolecular asymmetry. To study such chiroptical anomalies, it is necessary to investigate the role of constitutional changes as dendrimer generation increases. We present here a chiroptical study of two series of cinnamate-based chiral dendrimers obtained by convergent synthesis using chiral, non-racemic AB_2 monomers. We have prepared appropriate model compounds to investigate the role of effect of molecular constitution on optical activity. Although we observe dramatic changes in optical activity with increasing generation, we have been able to determine that conformation does not contribute to this enhancement.

151.
PERMEABILITY, SOLUBILITY AND DIFFUSIVITY OF ACETIC ACID AND WATER IN POLYANILINE MEMBRANES.
Ian J. Ball, Shu-Chuan Huang, Annise P. Berger and Richard B. Kaner*
Dept. of Chemistry and Biochemistry & the Solid State Science Center
University of California, Los Angeles, CA 90095-1569

Doped polyaniline membranes have the ability to separate water from carboxylic acids. The mechanism of permselectivity of water over acetic acid in polyaniline membranes is under investigation. It appears that the high permselectivity of water over acetic acid in doped polyaniline membranes is due to both favorable solubility and diffusivity for water. This is in contrast to undoped polyaniline where only diffusivity favors water, while solubility favors acetic acid. Temperature-dependent pervaporation experiments with both doped and undoped polyaniline membranes have been performed with acetic acid and water mixtures, and the activation energies of permeation (E_P) have been calculated. The highest activation enegry was found to be 7.8 kcal/mol for 50% acetic acid/50% water through undoped polyaniline. Conversely, concentrated acetic acid does not permeate through doped polyaniline at all. By measuring the accumulation of permeamt over time, the time lag and the diffusion coefficient can be calculated. The time lag of water through doped polyaniline is on the order of minutes, which yields a diffusion coefficient of $\sim 10^{-8}$ cm^2/s. The time lag of acetic acid through undoped polyaniline is on the order of hours.

152. SUPERCRITICAL FLUID INFUSION OF SILVER INTO POLYIMIDE FILMS OF VARYING CHEMICAL COMPOSITION
R.K. Boggess,[a] L.T. Taylor, J. Rosolovsky,[b] A.F. Rubira,[c] D.M. Stoakley,[d] A.K. St.Clair,[d] Department of Chemistry, Virginia Tech, Blacksburg, VA 24061-0212.

Highly reflective polyimide films were formed by infusing (1,5-cyclooctadiene-1,1,1,5,5,5-hexafluoroacetylacetonato)silver(I) [Ag(COD)(HFA)] into a number of polyimides using supercritical CO_2 and then thermally curing those films at 300°C for time intervals between 30 minutes and 3 hours. Reflectivities of the films exhibited strong dependence on the infusion and cure conditions as well as on the type of polyimide used. Highest reflectivity was achieved with a silvered film prepared from 3,3',4,4'-benzophenonetetracarboxylic acid dianhydride (BTDA) and 4,4'-oxydianiline (ODA) infused at 5000 psi, 100°C, for 30 minutes and cured for 1 hour at 300°C. Reflectivities of silvered surfaces varied from 39% to 67%. A strong correlation between the presence of a ketonic group in the polyimide structure and the formation of mirror surfaces was detected.

[a]Department of Chemistry and Physics, Radford University, Radford, VA 24142; [b]Department of Chemical Engineering, MIT, Cambridge, MA 02139; [c]Universidade Estadual de Maringa, Departamento de Quimica, Av. Columbo 5790, 87020-900, Maringa-PR, Brazil; [d]NASA., Langley Research Center, Hampton, VA 23665

153. HYDROGEN BOND ANALYSIS OF POLYMER BLENDS BY MOLECULAR ORBITAL CALCULATIONS.
Sukmin Lee, Seung Han Chun, Jung Goo Lee, Soonja Choe*, and Hosull Lee
Kumho Chemical Laboaroty, Hwaam-dong, 57-1, Yuseong-ku, 305-348, Korea and *Department of Chemical Engineering, Inha University, Inchon 402-751, Korea.

In the miscible blends containing aromatic polybenzimidazole (PBI) and polyimides, the hydrogen bonding between the N-H group of PBI and C=O group of polyimides has been reported to be an important factor responsible for the miscibility. We attempted to anlayze the hydrogen bonding using model compounds of PBI and polyimides by the quantum mechanical calculation. Using the model compounds from PBI and polyimides derived from various dianhydrides and diamines, the optimum geometries of model compounds and their coupled pairs were obtained to calculate the energetics, and vibrational frequencies. The hydrogen bonding strength calculated was compared qualitatively with the experiments through thermal analysis and FT-IR. The hydrogen bond distance was not relevant to the bond strength presumably due to the different chemical nature of polyimides. We found that the hydrogen bonding originating from the dianhydrides in polyimides should be more responsible for the miscibility behaviors of PBI and polyimides.

154. THERMOTROPIC LIQUID CRYSTAL POLYMER AND PBT MOLECULAR COMPOSTIE

Seong Hun KIM, Jung Gyu LEE, Han Sang KIM
Department of Textile Engineering, College of Engineering, Hanyang University, Seoul, 133-791, KOREA.

Thermotropic liquid crystal polymer based on p-hydroxybenzoic acid and poly(ethylene terephthalate) (60/40) with PBT blends were prepared for various compositions. Thermal, mechanical, rheological and morphological properties were investigated with change of the composition of the blends. The crystallization behavior of the PBT was affected by the adding of LCP, and a synergistic effect of thermal stability and tensile strength was observed in the blends containing 5 or 7% LCP content. While the blends with more than 20 wt.% of LCP content revealed the depression of shear viscosity induced by phase separation between the two phases, however shear viscosity was increased in the blends at low LCP content less than 10 wt.%. Morphology of the system showed excellent dispersity of the LCP in the PBT matrix and good compatibility at low content of LCP.

155. LYOTROPIC LIQUID CRYSTALLINE TEMPLATE FORMATION OF COMPOSITE MATERIALS. R.C. Smith, W.F. Fischer, D.L. Gin; Department of Chemistry, University of California, Berkeley, CA 94720

A highly ordered polymer nanocomposite containing poly(*p*-phenylenevinylene) (PPV) in hexagonally-packed tubular channels was synthesized by a monomer self-assembly approach. A lyotropic liquid crystal monomer was combined with an aqueous solution of poly(*p*-xylylenedimethylsulfonium chloride) and an organic radical photoinitiator. The LC monomer self-assembles in the presence of the aqueous solution to afford a stable inverse hexagonal mesophase in which the ionic PPV precursor polymer resides in the extended aqueous channels. Subsequent photopolymerization followed by thermal conversion yields the PPV-polymer nanocomposite. The resulting material has a regular hexagonal nanoarchitecture with channels approximately 40 Å in diameter. Highly oriented films and fibers of the nanocompostie can be easily formed by applying conventional processing techniques on the initial liquid crystal monomer mixture.

156. TAILOR-MADE MULTIPOLYMERS AND POLYMER ALLOYS AT WILL FOR ADVANCED TECHNOLOGIES. SYNTHESIS, ANALYSIS, CHARACTERIZATION, AND STATISTICAL THERMODYNAMICS.
H. Craubner, Institut für Physikalische Chemie der Universität Düsseldorf, Gauss-Str. 62 D-70193 Stuttgart, Germany.

Initiated mechano-chemical polyreactions for the synthesis of tailor-made multi-polymers and polymer alloys at will for advanced technologies are reported and theoretically analysed. Block-graft copolymers of, e.g., polyethylenes and polyamides, as well as a series of different other polymers are synthesized mechano-chemically. Batches of block-graft copolymers are used to produce polymer alloys. The rather complex highpolymeric materials are analysed by corresponding techniques and characterised by their physico-chemical properties. The results are corroborated by a "Reliability Theory "(1). The polyreactions may be analysed by making use of the "Statistical Thermodynamics of Multicomponent Polymer Systems" (2), combined with adequate measuring techniques (3).

References: 1) H. Craubner in: Polymers for Adv. Technol.(IUPAC Internat. Sympos., Jerusalem, Israel, 1987); M. Lewin, Ed.; VCH Publ., New York, 1988, p 905.- 2) H. Craubner, Pol. Prepr. **33**(2),314(1992); **35**(1),150(1994).-3) H. Craubner, Rev. Sci. Instrum. **57**,2817(1986).

157. DEVELOPMENT AND COMMERCIALIZATION OF IMAGING POLYMERS AT KODAK. Richard W. Blevins, Manufacturing Research and Engineering, Eastman Kodak Company, Rochester, NY 14650

The first permanent photograph, made by Niepce around 1825, used "bitumen of Judea" coated on a pewter plate and crosslinked by exposure to the sun. Less than century later, George Eastman popularized silver halide photography with naturally derived polymers including cellulose nitrate for the film base, gelatin for the emulsion, and cellulose paper for the final print. Throughout the following decades, Kodak chemists struggled to find synthetic replacements for natural polymers as well as for silver halide. Research in light-sensitive polymers lead to the development of two of the first synthetic lithographic photopolymers, poly(vinylcinnamate) and cyclized polyisoprene, both of which are still in use. Kodak's extensive investment in polymer research has spawned a variety of new products and improvements in traditional photography, printing, x-ray imaging, microlithography, xerography, and digital photography. This talk will present the history and development of a small selection from the broad range of imaging systems and polymers that have been developed and commercialized at Kodak.

158. STYRENIC THERMOPLASTIC ELASTOMERS. Geoffrey Holden, Shell Development Co. (Retired), Holden Polymer Consulting Incorporated, Prescott, AZ 86301

This paper describes the invention and development of styrenic thermoplastic elastomers. The original discovery in 1963 was clearly serendipitous. It was based on the observation that the melt viscosity of styrenic block copolymers is anomalously high. At first, applications of these polymers were foreseen in reducing the cold flow and improving the green strength of diene rubbers. However, it soon became apparent the most important ones would be as thermoplastic elastomers – that is, materials that have many of the physical properties of vulcanized rubbers but can be processed as thermoplastics.

This paper also reviews the synthesis, morphology, structure – property relationships and commercial applications of these polymers. These include such diverse end uses as adhesives, footwear, medical devices, automobile parts and asphalt modification.

159. THE VECTRA® STORY. G. Calundann, Hoechst Celanese Corporation, Summit, New Jersey 07901

Development of the initial members of a unique family of thermotropic liquid crystalline polymers (LCPs), now commercialized worldwide under the trade names Vectra® and Vectran®, began at Hoechst Celanese Corporation (Summit, NJ) in 1974-75. Today, Vectra liquid crystalline engineering resins are key materials for a broad range of electronic, medical and industrial applications. This talk describes research at Hoechst over the last twenty years that has defined the molecular architecture necessary to prepare high performance polyarylates with a unique melt structure; architecture that allows these polymers to form ordered, that is, liquid crystalline melts within preselected temperature boundaries. It is this melt structure that enables these materials to be quickly and easily processed into a variety of articles - from fiber and film to finely detailed, precision molded parts.

160. KEVLAR(R): FROM DISCOVERY TO APPLICATIONS. Gabara, V., E. I. DuPont de Nemours & Co., Inc., P. O. Box 27001, Richmond, VA 23261.

Twenty-five years since the commercialization of Kevlar(R) marks a new period in the development of advanced fibers. In this paper, we will re-trace major technological developments which formed the basis for this major advance in our industrial capability. We will attempt to establish a linkage between Kevlar(R) properties and development of its major applications. We expect that this look will help us to identify patterns which are critical in achieving a combination of scientific and industrial success.

POLY

161. THE DEVELOPMENT AND COMMERCIALIZATION OF WELL DEFINED OLEFIN POLYMERIZATION CATALYSTS. H. W. Turner, Symyx Technologies, 420 Oakmead Parkway, Sunnyvale, CA 94086

The field of single-sited α-olefin polymerization has developed rapidly over the past 15 years. Single-sited catalysts are derived from ancillary ligand-stabilized transition metal complexes having a reactive metal-carbon (or hydrogen) σ-bond cis to a vacant coordination site. The commercial driving force for this technology is to provide cost competitive catalyst systems which offer higher levels of synthetic control over polymer microstructure, molecular weight, molecular weight distribution, topology, and choice of comonomer compared to conventional Ziegler-Natta catalysts. Technical achievements in catalyst and activator design will be reviewed and related to improvements in the ability to tailor polymer structure and performance.

162. MILESTONES IN POLYMER BLENDS. L. M. Robeson, Air Products and Chemicals, Inc., Allentown, PA 18195

Polymer blend technology has been a major area of academic and industrial research in polymer science since the 1970's. Many important commercial products have resulted from this increased attention. The ability to rapidly develop new materials to meet emerging application requirements often favors polymer blend approaches over new polymer synthesis. The key commercial milestones involving miscible blends (e.g. PVC permanent plasticizers; PPO/polystyrene) and the multitude of phase separated blends (involving both commodity polymers and engineering polymer blends) will be reviewed. Advances in science/technology of polymer blends including equation of state approaches applied to polymer blends, lcst (lower critical solution temperature) behavior, importance of specific interactions, analog heat of mixing, reactive extrusion compatibilization, concept of specific rejection, and important advances in polymer blend analysis will be reviewed. The emerging areas of interest for future polymer blend technology (including liquid crystalline polymer blends, molecular composites, conducting polymer blends) will be described. As we go into the 21st century, polymer blend technology will continue to be a key area in polymer science.

163. SEMICONDUCTING POLYMERS AS A NEW CLASS OF SOLID-STATE LASER MATERIALS. Benjamin J. Schwartz,[a] Maria A. Díaz-García, Fumitomo Hide, Mats R. Andersson,[b] Qibing Pei[c] and Alan J. Heeger, Institute for Polymers and Organic Solids, University of California, Santa Barbara, Santa Barbara, CA 93106-5090
 [a] Permanent Address: Department of Chemistry and Biochemistry, UCLA, Los Angeles, CA 90095-1569
 [b] Present Address: Dept. of Polymer Tech., Chalmers University of Technology, S-412 96, Göteborg, Sweden
 [c] Uniax Corporation, 6780 Cortona Road, Santa Barbara, CA 93117

Optically pumped lasing is demonstrated in undiluted films of over a dozen different photoluminescent conjugated polymers whose emission spans the visible spectrum. Lasing is evidenced by a dramatic collapse of the emission line width at very low pump energy thresholds. The short gain lengths in these materials are attributed to the high density of chromophores, the large density of states associated with the interband (π-π^*) transition in quasi-one-dimensional systems, and the Stokes shift which minimizes self-absorption and enhances stimulated emission in the absence of excited state absorption. The observation of lasing in this new class of solid-state laser materials is explained in terms of photon confinement by planar waveguiding. The dependence of the threshold and the gain narrowed line width on the solvent from which the polymer films are cast suggests that chain packing can be used to control lasing in these materials.

164. MICROCAVITY DESIGN FOR ORGANIC ELECTROLUMINESCENCE: OPPORTUNITIES AND CHALLENGES
L. J. Rothberg [a], R. H. Jordan [b], A. Dodabalapur, and R. E. Slusher, Lucent Technologies Bell Laboratories, 700 Mountain Ave., Murray Hill, NJ 07974

Several potential applications of organic electroluminescent technology to display applications are considered. We show how fabricating these devices in a resonant cavity configuration can address some of their shortcomings. The modifications relative to standard organic electroluminescent devices are illustrated by construction of microcavity light-emitting diodes, and include gross changes in the device emission color and spatial pattern. This enables one to improve the control of color, the overall usable brightness and the ability to integrate red, green and blue into a display. The physical basis for these changes and some of the attendant difficulties in using microcavity emitters will be discussed.

(a) present address, University of Rochester, Dept. of Chemistry, Rochester, NY 14627
(b) present address, Eastman Kodak Research Labs, 1999 Lake Ave., Rochester, NY 14650.

165.
PHOTONIC POLYMERS FOR THE DEVICES OF THE 21ST CENTURY G.Hadziioannou Department of Polymer Chemistry and Materials Science Center, University of Groningen, Nijenborgh 4, Groningen 9747 AG, The Netherlands

This paper presents research activities towards the development of polymer materials and devices for optoelectronics. An approach to controlling the conjugation length and transferring the luminescence properties of organic molecules to polymers through block copolymers containing well-defined conjugated sequences is discussed. Electroluminescent and photovoltaic devices from polymers or oligomers are described, with an emphasis on the color tunability and the morphology-performance relationship. Lasing from optically pumped diluted solution and neat polymers with low energy threshold of generation is demonstrated.

166. NEARLY MONOCHROMATIC LIGHT EMITTING DIODES BASED ON ORGANIC THIN FILMS

S. Dirr[1], H.-H. Johannes[1], S. Wiese[1], D. Ammermann[1], A. Böhler[1], W. Grahn[2], W. Kowalsky[1]
[1]Institut für Hochfrequenztechnik and [2]Institut für Organische Chemie, Technische Universität Braunschweig Postfach 3329, D-38023 Braunschweig, Germany, Phone: +49-531-391-2467, Fax: +49-531-391-5841, e-mail: s.dirr@tu-bs.de

Electroluminescent devices with narrow emission bandwidths in the red, green, and blue spectral region based on thin films of lanthanide complexes or microcavity structures are investigated in terms of color purity for multicolor display applications. Multilayer organic light emitting diodes (OLEDs) with the β-diketonates tris-(2,4-pentanediono)-1,10-phenathroline terbium (III) (Tb(ACAC)$_3$Phen) and tris(4,4,4-trifluoro-1-(2-thienyl)-1,3-butanediono)-1,10-phenanthroline europium (III) (Eu(TTFA)$_3$Phen) show nearly monochromatic electroluminescence in the green and red spectral region, respectively. We have also demonstrated, that vacuum-deposited thin films of the novel complex tris(4,4,4-trifluoro-1-phenyl-1,3-butanediono)-1,10-phenanthroline dysprosium (III) show luminescent lines for use as active layer in white-light emitting diodes. In case of microcavity electroluminescent cells with the fluorescent dye aluminum-tris(8-hydroxychinoline) (Alq$_3$), spectral bandwidths of about 23 nm have been achieved at a resonance wavelength of 475 nm. This corresponds to a spectral narrowing by a factor of about 3.5 in comparison with noncavity OLEDs.

167. THIN FILMS OF CONJUGATED POLYMERS: APPLICATION IN SENSORS FOR HYDROCARBON VAPORS, MICROCONTACT-PRINTED LIQUID CRYSTAL DISPLAYS AND LIGHT-EMITTING DEVICES, A. G. MacDiarmid, W. Zhang, Z. Huang, P-C. Wang, F. Huang and S. Xie, Department of Chemistry, University of Pennsylvania, Philadelphia, PA 19104-6323.

Thin films of doped polypyrrole, polyaniline and polyaniline tetramer serve as excellent sensors for hydrocarbon vapors. Rapid and reversible changes in resistance are observed on alternate exposure to, e.g., toluene vapor and nitrogen respectively. Computer derived patterned hydrophobic films are produced on hydrophilic glass substrates by a microcontact "stamp" printing technique. Thin films of polypyrrole and polyaniline are deposited preferentially on the hydrophobic pattern either electrochemically or from dilute aqueous solutions of the polymerizing monomer. When the patterned polypyrrole on a glass microscope slide is used as one electrode and ITO conducting glass is used as the other electrode, patterned polymer dispersed liquid crystal displays are readily produced. Characteristics of LED's markedly improve from ITO/MEH-PPV/Al to ITO/EB/MEH-PPV/Al to ITO/doped EB/MEH-PPV/Al where EB=polyaniline emeraldine base.

168. POLYMER LIGHT-EMITTING ELECTROCHEMICAL CELLS. Yang Yang and Qibing Pei, UNIAX Corporation, Santa Barbara, CA 93117, U.S.A.

A polymer light-emitting electrochemical cell (LEC) is a bipolar p-n junction device; when biased, p- and n-doping regions, in the adjacent to the anode and cathode, are electrochemically induced respectively. A dynamic p-n junction is therefore created between the p- and n-doped regions. The active layer of an LECs is a blends of luminescent conjugated polymers and solid electrolytes. The device operating mechanism is completely different from traditional polymer light-emitting diodes, in which charge injection is limited by the barriers at the polymer/electrode interfaces. In this presentation, fundamental mechanism of the LEC will be discussed, and several novel devices based on the unique characterristics of the LECs will also be introduced.

169.

"Organic Polymer LEDs with mobile and immobile ions", Herman F.M. Schoo, Rob C.J.E. Demandt, Jeroen J.M. Vleggaar and Coen T.H. Liedenbaum, Polymer & Organic Chemistry Department, Philips Research Laboratories Eindhoven, The Netherlands

Partially conjugated alkoxy-PPV materials with tertiary amines and cationic species covalently bound to the polymer backbone are described. The amine/cationic species are introduced via a polymer analogous reaction of a functional alcohol and the chloro-precursor polymer. Polymer LEDs prepared with the modified polymer materials show a strong 'burn-in' effect, leading to very high EL quantum efficiencies especially with higher workfunction cathodes.

170. ELECTROCHEMILUMINESCENCE AND EFFICIENT LIGHT EMITTING DIODES BASED ON MULTIFUNCTIONAL DYES: DONOR/ACCEPTOR SUBSTITUTED STILBENOIDS AND DIFLUOROBORADIAZA-S-INDACENES. J. Daub, Institut für Organische Chemie, Universität Regensburg, D-93040 Regensburg (Germany)

The fabrication of organic light-emitting diodes (OLEDs) by spin coating or vapor deposition requires the skillful design of polymeric, oligomeric or monomeric dye systems. The aim of our work is to use multifunctional and low molecular weight dyes which can be easily reduced and oxidized and additionally are luminescent. On the basis of this class of compounds one and two layer OLEDs are fabricated. In order to optimize the electroluminescent properties of the dye components the OLED fabrication was preceded by electrochemical, spectroelectrochemical and ECL (electrochemically induced luminescence) studies. Energy level structures for the LED assemblies are developed in order to predict the transport rates of the charge carriers qualitatively. The dye compounds involved in the present report are derivatives of donor/acceptor substituted stilbenoids and difluoroboradiaza-s-indacenes. These investigations were undertaken in cooperation with Prof. H. Bässler (Marburg University) and his group.

171.

THE ANIONIC SYNTHESIS OF FUNCTIONALIZED LINEAR AND STAR-BRANCHED POLY(METHYL METHACRYLATES) USING PROTECTED HYDROXY-FUNCTIONALIZED ALKYLLITHIUM INITIATORS. Roderic P. Quirk and R. Shane Porzio, The Maurice Morton Institute of Polymer Science, University of Akron, Akron, Ohio, 44325, USA.

Living anionic polymerization using suitably protected, functionalized initiators is one of the best methods for preparing end-functionalized polymers. The protected hydroxy-functionalized initiator, 3-(t-butyldimethylsilyloxy)-1-propyl lithium, was used to prepare linear and star-branched poly(methyl methacrylates). The linear polymer was synthesized in THF at -78° C in the presence of LiCl and exhibited a narrow molecular weight distribution (M_w/M_n = 1.09) as well as excellent agreement between calculated (M_n = 8,000 g/mole) and observed (M_n = 7,800 g/mole) molecular weights. The star-branched sample was prepared in a similar fashion employing ethylene glycol dimethacrylate as a linking agent. Absolute molecular weight characterization by the Universal Calibration Method showed that the star had an average of 10.4 arms. Deprotection of the silyloxy protecting group was achieved by acid hydrolysis. Titration of the resulting hydroxyl groups gave a functionality of 0.99 for the linear polymer and 10.7 for the star-branched polymer.

172. SYNTHESIS AND CHARACTERIZATION OF NARROW MOLECULAR WEIGHT DISTRIBUTION POLY(SULFOPROPYLBETAINE) BLOCK COPOLYMERS.
Andrew B. Lowe, Norman C. Billingham and Steven P. Armes, School of Chemistry, Physics and Environmental Sciences, University of Sussex, Falmer, Brighton, U.K.

Poly(sulfopropylbetaine) block copolymers of narrow molecular weight distribution have been synthesised directly for the first time. Group transfer polymerisation of 2-(dimethylamino)ethyl methacrylate (DMAEMA) with various alkyl methacrylates, e.g. methyl methacrylate, yields well-defined, near-monodisperse precursor block copolymers. Subsequent quantitative betainization of the tertiary amine groups on the DMAEMA residues with 1,3-propanesultone under mild conditions leads to hydrophilic-hydrphobic sulfopropylbetaine block copolymers. Aqueous solution studies indicate that these betainized block copolymers exhibit little surface activity, while the corresponding precursor copolymer shows considerable surface activity. Dynamic light scattering and aqueous gel permeation studies indicate the sulfopropylbetaine block copolymers form micelles in both aqueous and salt solutions. The hydrodynamic diameter of the micelles is dependent on copolymer composition, hydrophobicity of the comonomer, salt concentration, and whether 2,2,2-trifluoroethanol is used as a co-solvent to promote molecular dissolution.

173. KINETICS AND MECHANISMS OF ANIONIC POLYMERIZATION OF (METH)ACRYLATES IN THE PRESENCE OF LITHIUM 2-METHOXYETHOXIDE.
A. Maurer , X. Marcarian, A.H.E. Müller, Institut für Physikalische Chemie, Universität Mainz, D-55099 Mainz, Germany; *B. Vuillemin, C. Navarro,* Elf-Atochem S.A., GRL, F-64170 Artix, France

The anionic polymerization of methyl methacrylate (MMA) using 1,1-diphenylhexyllithium as initiator and lithium-2-methoxyethoxide as a ligand has been studied in the temperature range of -40 °C to +90 °C in toluene using a flow tube reactor. For ligand to initiator ratios ≥ 5 linear first-order time-conversion and unimodal polymers with a narrow MWD ($M_w/M_n < 1.1$) are obtained. Up to 5% of added THF does not affect the results. The degree of polymerization is a linear function of conversion and initiator efficiencies are high. For temperatures above 0 °C, termination is observed which is attributed to "back-biting" according to MALDI-TOF-MS measurements. The determined rate constants of propagation are three orders of magnitude higher than in the absence of ligand and correspond to those obtained with cryptated counterions in THF. This may indicate an analogous formation of "ligand-separated" ion pairs in this system. In addition, block copolymers of MMA and n-butyl acrylate were synthezised at -10 °C. SEC shows more than 95% coupling efficiency and low polydispersity.

174. TERMINATION REACTIONS IN THE TETRAPHENYLPHOSPHONIUM ION MEDIATED ANIONIC POLYMERIZATION OF ALKYL ACRYLATES
D.K.Dimov, T.E.Hogen-Esch, Loker Hydrocarbon Research Institute and Department of Chemistry, University of Southern California, Los Angeles, CA 90089-1661

In contrast to the essentially "living" process with MMA, the tetraphenylphosphonium ion mediated anionic polymerization of alkyl acrylates is plagued by side reactions preventing complete monomer conversions and MW control. Studies with model compounds, i.e. methylisobutyrate and ethyl-2-methyl-3-oxobutanoate, and ^{1}H NMR data on the end groups of poly(t-butylacrylate) indicate that the main termination pathway is an intramolecular Claisen reaction followed by immediate deprotonation of the β-ketoester by a second chain-end anion that is significantly (ΔpKa = 14-16) more basic than the β-ketoester enolate formed. This scheme may account for the significant differences between anionic polymerizations of acrylates and methacrylates.

POLY

175. METAL-FREE ANIONIC POLYMERIZATION OF METHYL METHACRYLATE.
D.Broska, A.Fieberg, C.Heibel, F.Bandermann, Institut für Technische Chemie, Universität GH Essen, D-45117 Essen, FRG

In the metal-free anionic polymerization of MMA with tetrabutylammonium diethyl 2-ethylmalonate or tetrabutylammonium 2-nitropropanate as initiators an inhibition period is caused by traces of initiator precursor molecules, undergoing a transfer reaction with the active species. Induction periods in time conversion curves point to an initiation reaction, being slower than the propagation reaction. All polymerizations of MMA are usually going to complete conversion. Surprisingly, the well-known intramolecular cyclization reaction was not observed in the MMA system. However, a Hofmann elimination reaction terminates growing chains. This is due to the butyl ligands of the ammonium cation. A ß-hydrogen atom can be transferred to the growing chain under the formation of tri-n-butyl amine and 1-butene. To eliminate this termination step, tetramethylammonium diethyl 2-ethylmalonate has to be used as initiator. But now a transalkylation under the formation of a methyl group terminated saturated chain end and trimethyl amine arises as further termination reaction. However, since the termination reactions are not very fast, the metal-free anionic polymerization of MMA exhibits some kind of living character.

176. NEW INITIATOR SYSTEMS FOR THE LIVING ISOTACTIC LIGATED ANIONIC POLYMERIZATION OF METHYL METHACRYLATE IN TOLUENE AT 0°C. T. Zundel, Ph. Teyssié, R. Jérôme, Center for Education and Research on Macromolecules, University of Liège, Sart-Tilman, B6, 4000 Liège, Belgium

Sec-butyllithium combined with a new family of ligands were used to initiate the ligated anionic polymerization (LAP) of MMA in toluene at 0°C. Compared with other µ-ligands, the new species are more efficient stabilizers of active centers. Under selectively chosen reaction conditions, the polymerization proceeds quantitatively without self-termination at 0°C producing a high molecular weight polymer with very narrow molecular weight distribution ($\overline{Mw}/\overline{Mn}$ < 1.2). The microstructure of the PMMA is highly isotactic (90%). The ^{7}Li NMR spectra of sBuLi and ligand mixtures in toluene display that only one single type of mixed species prevails, for high [ligand]/[sBuLi] molar ratio. As also evidenced by ^{7}Li NMR, the quantitative formation of {PMMALi/aggregated ligand} species on adding MMA to that initiator solution, confirms that only one type of ionic species is active during the propagation of MMA.

177.

CONTROL OF ANIONIC POLYMERIZATION OF METHYLMETHACRYLATE *VIA* LITHIUM AMINOALKOXIDES

J. MARCHAL, Y. GNANOU, M. FONTANILLE

Laboratoire de Chimie des Polymères Organiques
UMR CNRS-ENSCPB- Université Bordeaux-1 BP 108, 33402 TALENCE Cedex (FRANCE)

The use of tertiary amine alkoxides in the anionic polymerization of alkyl methacrylates is found greatly beneficial. Controlled polymerization of this class of monomers can indeed be achieved at elevated temperatures in THF and also in toluene, in the presence of these novel σ/µ ligands. In the latter solvent and with lithium 2-(dimethylamino)ethoxide as additive, the anionic polymerization of MMA exhibits a controlled character at a temperature as high as 70°C.

178. SURFACE CHARACTERIZATION OF POLY(ETHYLENE GLYCOL) COATINGS VIA ANALYTICAL ELECTROPHORESIS
Kazunori Emoto, J. Milton Harris, and James M. Van Alstine
Department of Chemistry and Materials Science Program,
University of Alabama in Huntsville, Huntsville, Alabama 35899

We are exploring two methods for evaluating surfaces coated with poly(ethylene glycol) (PEG) or other neutral polymers. They are applicable to a variety of stationary surfaces or colloids (*e. g.*, proteins, liposomes, microspheres). One method involves determination of contact angle or partition coefficient related to a surface or colloid interacting with an aqueous polymer two-phase system (J. Chromatog. B, 1996, 680, 145). The other involves determining electroosmosis or particle electrophoretic mobility (Anal. Chem., 1996, 70, in press). The methods provide semi-quantitative data related to surface and grafting chemistry, as well as polymer coating density, thickness, stability, and surface alteration or "masking". Recent studies are related to control of coating oxidation, surface grafting density, and ability to tether proteins. In collaboration with N. Burns, M. Malmsten and K. Holmberg (Inst. for Surface Chemistry, Stockholm) these and complementary methods such as ellipsometry are being used to evaluate materials of biomedical significance (J. Biomed. Materials Res., 1995, 29, 741 and J. Colloid Interf. Sci., 1996, 177, 502 plus 178, 123).

179.

POLY (ETHYLENE OXIDE) GRAFTED TO MICA SURFACES - A MODEL SYSTEM FOR POLYMER BRUSHES. M.O. Geke R.A. Shelden, W.R. Caseri, and U.W. Suter, Department of Materials, Institute of Polymers, ETH Zürich, 8092 Zürich, Switzerland

Polymer brushes consist of macromolecules grafted at one end to a surface at a density high enough so that the chains are stretched away from the surface. Depending on the grafting density, the stretching may be even larger than the unpertubed coil dimensions of the polymer. We have developed an experimental model system, based on mica and poly(ethylene oxide) (PEO) chains bearing cationic trimethylammonium bromide endgroup. These polymers attach to the surface via ion-exchange replacing the inorganic ions, in our case lithium. This model system allows a variety of experimentally controllable parameters such as solvent quality temperature, or the type of inorganic ions on the surface. We have investigated the equilibrium ion-exchange adsorption reaction for these modified PEOs with various polymer chain length. The experimental ion-exchange isotherms of PEO on the mica surface were interpreted using expressions based on the analytical SCF theory. This treatment yields the entropic contribution to the polymer exchange reaction.

180. THE ADSORPTION AND FUNCTIONALITY OF FIBRINOGEN ON HYDROPHOBIC SURFACES MODIFIED WITH POLY(ETHYLENE OXIDE)-CONTAINING COPOLYMER FILMS. S. M. O'Connor*, S. J. Patuto#, S. H. Gehrke*, and G. S. Retzinger#, *Department of Chemical Engineering and #Department of Pathology and Laboratory Medicine, University of Cincinnati, Cincinnati, OH 45221

To study the interactions of fibrinogen with surfaces, triblock copolymers of the form $PEO_\alpha/PPO_\beta/PEO_\alpha$ (where PEO is polyethylene oxide and PPO is polypropylene oxide) are ideally suited since differences in interactions can be attributed solely to differences in physicochemical properties of the coatings. We have developed a model system in which well-defined monolayers of these copolymers supported by solid, hydrophobic beads, are used to assess the influence of the surface microenvironment on the adsorption and proteolytic degradation of fibrin(ogen). The data demonstrate that copolymer identity and/or packing density influence protein adsorption. They also suggest that copolymer films that adsorb fibrinogen may function as clot nucleation sites and, thus, influence a host of fibrin(ogen)-dependent phenomena. While beads coated with copolymers with long PEO segments bind little protein, in fibrinogen-dependent fashion, beads coated with copolymers with short PEO segments adhere readily to macrophages; such beads also aggregate when stirred in the presence of thrombin, a consequence of interbead fibrin formation.

181. STRUCTURAL INTEGRITY IN PROTEIN IMMOBILIZATION. S-C. Huang, P.A. Tresco, and K.D. Caldwell, Center for Biopolymers at Interfaces, University of Utah, Salt Lake City, UT 84112

The immobilization of fragile macromolecules with a pre-determined surface concentration and with retained biological function is an often desired goal. Due to the general process simplicity, adsorption is a method frequently used to attach proteins to surfaces. However, recent calorimetric analyses have indicated that such procedures often lead to significant structural losses and, by implication, losses of activity. The well known protein repelling qualities of PEO suggest that immobilization via a PEO tether might be a constructive route to reducing protein denaturation while protecting the surface from unwanted non-specific fouling. Certain PEO-PPO-PEO triblock copolymers have been shown to adsorb rapidly and for practical purposes irreversibly to many polymeric materials with desirable mechanical properties. By activating the endgroups of such block-copolymers, and subsequently mixing derivatized and underivatized surfactant, it is possible to provide protein attachment with a desirable spacing. This strategy has been used to link a variety of proteins to particulate surfaces. While the adsorption complexes show massive loss of structure, the tethered protein appear structurally nearly intact. The implications for tissue culture, immunodiagnostics, and enzyme immobilization will be discussed.

182. INCORPORATION OF PEG-PROTEINS INTO POLYMERS. K.E. LeJeune[1], J. Panza[2], A.J. Russell[2], Dept. of Chemical Engineering, [1]Carnegie Mellon University, [2]University of Pittsburgh, Pittsburgh, PA 15219.

While the attachment of proteins to polymers is straightforward, their incorporation during polymer synthesis holds several advantages. Unfortunately, the vast majority of polymerizations occur in harsh organic solvents which have little to no ability to solublize protein. The attachment of polyethylene glycol to a protein molecule can greatly enhance organic solvent solubility and facilitates protein polymer synthesis. Since a PEG-protein can be dissolved in an organic solvent in close proximity to its native structure and contains functionalities capable of reacting with a growing polymer chain, the enzyme could become intrinsically coupled to a polymeric material during polymer synthesis. In order to react PEGylated proteins with monomers in organic solution without significant deleterious effects, a heterofunctional PEG must be employed with one end of the PEG designed to couple to a protein and the other with a growing polymer chain. We have synthesized subtilisin polymers through using various heterofunctional PEG acrylates. Resultant PEG-subtilisin macromonomers and biopolymers have significant activity retention in both aqueous and organic media. Significant enzyme stabilization upon PEG modification and immobilization have also been observed.

183. STRATEGIES FOR THE PREPARATION AND CHARACTERIZATION OF POLYETHYLENE GLYCOL (PEG) CONJUGATED PHARMACEUTICAL PROTEINS. P. Bailon, W.-J. Fung and J.E. Porter, Department of Biopharmaceuticals, Hoffmann-La Roche, Inc., Nutley, NJ 07110

Polyethylene glycol conjugated proteins belong to a new class of biomolecules which are neither a protein nor a polymer, but are hybrids between the two. Though PEG conjugation of biotherapeutics is now a common practice to attain sustained clinical response, yet very little is known about the strategies and criteria used to produce a well-characterized pegylated biomolecule. In this paper, we address these issues which include, selection of pegylation reagents, reaction conditions, process considerations, purification and biochemical characterization. The latter involves peptide mapping, amino acid sequencing and mass spectrometric analyses to determine pegylation sites and identification of positional isomers. In most cases, a direct or inverse relationship exists between the attached PEG mass and the *in vivo* and *in vitro* bioactivities, respectively.

184. BIOCHEMISTRY AND IMMUNOLOGY OF POLYETHYLENE GLYCOL-MODIFIED ADENOSINE DEAMINASE (PEG-ADA). M.S. Hershfield, Departments of Medicine and Biochemistry, Duke University Medical Center, Durham, NC 27710

The first therapeutic trial of a PEGylated enzyme began in 1986, when we used intramuscular bovine adenosine deaminase (ADA) conjugated with 5 kDa PEG, manufactured by Enzon, Inc., to treat severe combined immunodeficiency due to inherited ADA deficiency (Hershfield et al, N Engl J Med 316:589, 1987). During the past decade PEG-ADA, injected once or twice weekly, has been used in over 60 patients with this rare, fatal disorder; 40 have been treated for >2 years (mean 5.5 years). We have monitored biochemical parameters in 50 of these patients. PEG-ADA therapy maintains high plasma ADA activity, effectively eliminating toxic ADA substrates as indicated by correction of metabolic abnormalities in erythrocytes. Treatment restores immune function to a level sufficient to sustain good health. The majority of patients develop antibody to PEG-ADA, but there have been no allergic or other adverse reactions, and clearing or blocking antibody has developed in only a few cases. In 1990, PEG-ADA (ADAGEN™) became the first parenteral enzyme to be approved by the US Food and Drug Administration for replacement therapy in an inherited metabolic disease. These results suggest that PEG-modified enzymes can be used safely and effectively for long-term therapy.

185. PREPARATION AND CHARACTERIZATION OF PEGYLATED HGH AND HGH MUTANTS. Kenneth Olson§, Richard Gehant§, Venkat Mukku**, Kathy O'Connell⁜, Brandon Tomlinson¶, Klara Totpal** and Marjorie Winkler§. Departments of §Recovery Sciences, **QC Biochemistry, ⁜Protein Chemistry & ¶Analytical Chemistry. Genentech, Inc. 460 Point San Bruno Blvd., South San Francisco, CA 94080

In order to prepare long-lasting human growth hormone (hGH) and human growth hormone antagonists (GHA), the hGH and GHA proteins were conjugated with polyethylene glycol (PEG) to increase their molecular size. Following conjugation, each reaction mixture was fractionated by cation exchange chromatography fc: biological assessment and chemical characterization. Because the PEGylation reagents chosen for protein conjugation react with lysine residues, tryptic peptide analysis was found to be a useful tool to identify the PEGylation sites. A reduction in the amount of a peptide was correlated with each site of PEGylation. The results obtained by tryptic analysis were confirmed by MALDI-TOF mass spectrometry. Analysis of the data showed the pattern of PEGylation as well as the extent of PEGylation were consistent from lot to lot. In all cases, the PEGylated proteins were active with a sustained half life in vivo. The optimal in vivo activity was found with PEGylated molecules between 40-50 kD in mass.

186. MANUFACTURING, CHARACTERIZATION, AND PHARMACOKINETICS OF A MONODISPERSE PEG-HIRUDIN. Brian Fenton, Kevin Lauziere, Jeffrey Welch, Richard Crowley, Peter Licari, Klaus Ruebsamen and Brian Turner, BASF Bioresearch Corp., Worcester, MA 01605

The 65 amino acid polypeptide hirudin is a potent thrombin inhibitor, but its relatively short circulatory half-life limits its usefulness in certain clinical indications. Coupling of PEG to native hirudin is one approach to increasing half-life, but the result is a polydisperse population of molecules, some of which are biologically inactive or have excessively long half-lives. We instead elected to manufacture a well-defined, monodisperse PEG-hirudin for clinical development. Our approach was to couple PEG5000 to a recombinant hirudin containing fewer lysine residues than native hirudin to minimize heterogeneity, then purifying the clinically useful form of PEG-hirudin. Recombinant hirudin was secreted from cultures of *Hansenula polymorpha* and purified to homogeneity. Following coupling with activated PEG, the desired form of PEG-hirudin was purified using anion exchange and hydrophobic interaction chromatography. The result is a fully active, monodisperse product containing two moles of PEG per mole of hirudin. We will present an overview of the manufacturing process, physicochemical characterization results, and *in vivo* data comparing the pharmacokinetics of hirudin and PEG-hirudin.

187. **Issues Encountered in the Production of Site-specific Mono-PEGylated Therapeutic Proteins.**
J. Seely*, C. Richey*, T. Grasel** and J. Wilson**
Amgen Process Development
Boulder, CO 80301* and Thousand Oaks, CA 91320**

Adding a single PEG molecule to a protein can markedly extend its serum half-life. By directing where the mono-PEGylation occurs, we can reduce the clearance rate without adversely affecting the biological activity of the protein.   We have used PEG vinylsulfone for site-specific modification at free cysteine residues and PEG aldehyde for selective modification of the N-terminal alpha amino group. Several issues have been encountered that effect both the yield and quality of monoPEGylated proteins. These issues include PEG linker purity (both in terms of the degree of activation and size homogeneity), PEGylation conditions, PEG linker stability and, in the case of PEG aldehyde, sodium cyanoborohydride quality.   Examples of each of these will be presented, as will some of the ways in which these problems have been addressed.

188. POLYETHERIMIDES: T. Takekoshi, General Electric Co.,Corporate Reserch and Development, Schenectady, NY 12301

In 1967 rediscovery of usefulness of activated aromatic nitro compounds in nucleophilic substitution reactions (NSR) (J.H. Gorvin, *Chem. Ind.*, **36**, 1525 (1967)) prompted us to investigate its potential as a basic tool of synthesizing high temperature polymers. Nitro-phthalimides were found to be some of the most reactive substrates toward NSR. In contrast to their halogenated counterparts the nitro derivatives can be produced more economically because nitration is conducted in more controlled manner at low cost. Subsequently, a composition of thermally processible polyetherimide was conceived for commercialization. After exhaustive study on numerous options, General Electric Company began commercial production of its ULTEM® polyetherimide resin in 1982. ULTEM® resin is composed of phthalic anhydride and bisphenol A as dianhydride components and m-phenylenediamine. It has high modulus (3.0 GPa), strength (105 MPa) and excellent electrical properties, as well as unique fire-resistant property. It is widely used in electrical, electronic, automotive and other industrial applications.

189. HISTORY OF PET FOR CONTAINER APPLICATIONS. G. D. Millington, Eastman Chemical Co., Container Plastics Business Organization, Kingsport, TN, 37662.

This presentation covers the history of the development of PET (poly(ethylene terephthalate)) for packaging applications with a focus on beverage containers. PET was commercialized in 1978 for beverage containers and has enjoyed double-digit growth rates. The presentation will cover materials and chemistry, applications, major growth areas, and emerging products such as poly(ethylene naphthalate) (PEN).

190. RECENT ADVANCES IN STRUCTURAL ADHESIVES. Kurt C. Frisch Jr. 3M, Adhesive Technologies Center, 209-1C-23, St. Paul, MN 55144

The global adhesive sealant industry based on polymeric materials is valued at about $18 billion. Approximately one-third of this total can be described as specialty materials, products formulated to meet demanding performance requirements. This talk will focus on structural and semi-structural specialty materials and describe some recent technological developments which has helped shape this industry. Topics selected will include: urethanes (including moisture-curing hot melts), third generation epoxies, acrylics and high temperature adhesives. The most significant trends affecting the adhesive and sealant industry are related to concerns about health, safety and environmental impact and will be discussed.

191.
THE IMPACT OF METALLOCENE AND CONSTRAINED GEOMETRY CATALYST TECHNOLOGY ON THE PLASTICS INDUSTRY. Kurt W. Swogger, The Dow Chemical Company, Polyethylene & INSITE™ Technology R&D, 2301 Brazosport Blvd., B-1607 Bldg, Freeport, TX 77541

The recent development of metallocene single site catalyst technology has had a profound impact on the polyolefin industry. The Dow INSITE* Technology catalyst, a latest version of constrained geometry single site catalyst, exhibits high reactivity to high alpha-olefins and other vinyl monomers such as styrene. This unique chemistry allows Dow to design and produce, via molecular architecture, many new families of polymers, included AFFINITY* polyolefin plastomers, EPDM's, enhanced polyethylene's, ethylene/styrene interpolymers and a new family of adhesive polymers. This broad portfolio of polymer families enables the plastic industries to readily develop new applications, from packaging to durable to automotive, with increased speed and profit margin. This paper will address the chemistry of the Dow INSITE Technology catalyst and the structure and applications of the polymers produced by this technology.

192. PHOTORESISTS FOR MICROLITHOGRAPHY: FROM KILOBIT TO GIGABIT DEVICES
Ralph R. Dammel, AZ Photoresist Products, Hoechst Celanese Corporation, Somerville, NJ 08876.

The meteoric rise of the semiconductor industry is undoubtedly one of the dominant economic influences of the last quarter of this century. Often the contribution of photochemistry to this success story is forgotten: microlithography, the technique by which all semiconductor devices are made, depends critically on the availability of suitable photoresist materials. This presentation will provide a historical perspective of the origins of present-day near-UV photoresists, discuss recent advances in the understanding of their mechanism, and show examples of their present performance. The manufacture of the 256 Mbit DRAM and more advanced device generations requires a switch to shorter wavelength, which in turn will necessitate a major change in photoresist design. The second part of the presentation will deal with the requirement profile, chemistry and performance status of photoresists for 248 and 193 nm DUV lithography.

193. COMMERCIALIZATION AND PERFORMANCE OF AMODEL POLYPHTHALAMIDES. Dean, B. D., Amoco Polymers Inc., 4500 McGinnis Ferry Rd., Alpharetta, GA 30202.

Polyphthalamides, or PPAs, are nylons formed from the reaction of aromatic diacids with aliphatic diamines. The aromatic diacid(s) comonomers allow for tailoring of mechanical and thermal performance, through the higher melting points, higher glass transition temperatures, lower moisture absorption, and enhanced dimensional stability versus aliphatic nylons. In 1991, Amoco Polymers, Inc. announced the commercialization of Amodel polyphthalamide for injection molding applications in the automotive, electronic, and consumer markets.

194.
THE WORLD'S STRONGEST FIBER: DSM'S GEL-SPINNING PROCESS OF ULTRA-HIGH MOLECULAR WEIGHT POLYETHYLENE, Paul Smith, ETH Zürich, Department of Materials, CH 8092 Zürich, Switzerland and Piet J. Lemstra, EPL, Eindhoven University of Technology, 5600 MB Eindhoven, The Netherlands

An historical account will be presented that details the development in the period 1978-1980 of the world's strongest fiber, which exhibits specific mechanical properties that surpass those of steel by a factor 10. The overview will include significant preceding inventions that ultimately led to this development as well as the partly fortuitous findings that culminated in the associated theoretical framework for the deformation of semi-crystalline polymers. Finally, present applications and currently untapped opportunities will be discussed.

195. SPONTANEOUS AND STIMULATED LUMINESCENCE IN CONJUGATED POLYMERS
Graham J. Denton, Nick T. Harrison, Nir Tessler and Richard H. Friend, Cavendish Laboratory, Madingley Road, Cambridge CB3 0HE, UK.

Conjugated polymers such poly(p-phenylenevinylene), PPV, are now known to show efficient luminescence, either optically or electically stimulated (the latter in electroluminescent diodes). However, the nature of the emitting state in these materials is poorly understood. We present here a range of measurements on PPV, which probe both spontaneous and stimulated emission processes. These include measurements of photoluminescence excitation spectra, sub-picosecond transient induced absorption, and emission measurements on a range of microcavity structures. These indicate that the primary photoexcitation in PPV is an singlet exciton which shows efficient luminescence. Generation of other excitations, including separated charges, is facilitated by extrinsic factors, such as photo-oxidation. Conditions for lasing in microcavity structures will be discussed.

POLY

196. NEW COPOLYMER BLENDS FOR LEDs, F.E. Karasz, B. Hu and L. Wang, University of Massachusetts, Amherst, MA 01003

The use of multicomponent systems in polymer light emitting diodes (PLEDs) can confer substantial advantages in terms of brightness and efficiency. In blends of high and low band gap electroluminescent chromophores energy can in certain cases migrate from the former to the lower thereby providing an enhanced output; such blends can offer a supplementary route for fine-tuning the color emission of the device. Blends of copolymers can be used to control overall morphology of the light emitting layer in a PLED and thereby provide luminescing domains in which excitons are more efficiency confined by the presence of a non-electro-active matrix. This principle has been further developed by incorporation of additional components in the chromophore layer which serve as carrier transport and blocking species, eliminating the necessity for sequential layer deposition in the device. We show that highly efficient diodes can be obtained in a single layer multifunctional architecture by blending chromophores with appropriate hole and electron transporting components. The operation of this device depends *inter alia* on the multi-domain morphology of the polymer blend and a judicious choice of the carrier transport components. This concept offers a systematic, rational approach to designing simplified efficient PLEDs with emission in any designated portion of the visible spectrum.

197. ELECTROLUMINESCENCE From Polymers With Isolated Chromophors. Y. Chen, R. Dujardin, A. Elschner, F. Jonas, A. Karbach, D. Quintens, **R. Wehrman**. Bayer AG, Central Research and Development, Rheinuferstr. 7-9, D-47829 Krefeld, Germany

Multiblock polycondensates consisting of distyryl-blocks and various non-conjugated spacer groups, polymerisates with linked chromophors, and blend systems based on molecular dispersed fluorescent dyes were synthesized and tested as electroluminescent materials in one-layer devices. Their electrooptical characteristics will be discussed with regard to other important physical properties..

198. ELECTROLUMINESCENT PROTON TRANSFER POLYMERS. S. A. Jenekhe*, X. Zhang, and R. M. Tarkka, Department of Chemical Engineering and Center for Photoinduced Charge Transfer, University of Rochester, Rochester, New York 14627-0166.

Intramolecularly hydrogen-bonded molecular materials which exhibit excited state intramolecular proton transfer are of interest as polymer photostabilizers, laser dyes, triplet quenchers, and as photochromic materials. Recent studies in our laboratory have shown that such materials can also give rise to *proton transfer electroluminescence* which is a promising approach to electrically pumped organic diode lasers (1). In this presentation, we report on the synthesis, photophysics, and photochemistry of a series of new proton transfer polymers and their application in electroluminescent devices. It is shown that a large extent of electronic delocalization inhibits the excited state intramolecular proton transfer process in these polymers whereas molecular size does not. Emission from electroluminescent devices based on these polymers is shown to originate from electrically generated intramolecular proton transfer.

199. SPIRO LINKED COMPOUNDS AS ACTIVE MATERIALS IN ORGANIC LIGHT EMITTING DIODES. J. Salbeck[a,b], J. Bauer[a], F. Weissörtel[b], [a]Hoechst AG, Corporate Research, D-65926 Frankfurt, Germany, [b]Max-Planck-Institute for Polymer Research, D-55128 Mainz, Germany

Recent studies have shown that the degradation of organic electroluminescence (EL) devices is attributable to morphological change resulting from the thermal instability of the amorphous organic layers used in the EL devices. This talk will discuss a variable concept for the design of amorphous low molecular weight materials with high glass transition temperatures (T_g), for the use as luminescent or charge transport materials in EL devices. The concept is based on the specific introduction of spiro centers into defined structures. These spiro-linked low molecular weight compounds form stable glasses with high T_g, usually associated with amorphous polymers. High quality amorphous films of these luminescent or charge transport materials can be prepared by conventional spin-coating as well as by vapor deposition.

200. THIN FILM LIGHT EMITTING HETEROSTRUCTURES: FROM CONJUGATED POLYMERS TO RUTHENIUM COMPLEXES TO INORGANIC NANOCRYSTALLITES. J-K. Lee, H. Mattoussi, D. Yoo, A. Wu and M. F. Rubner, Department of Materials Science & Engineering, MIT, Cambridge, MA 02139

Molecular level processing techniques have been utilized to fabricate light emitting heterostructures based on conjugated polymers, new trisbipyridyl ruthenium (II) polyesters and nanocrystallites of cadmium selenide (CdSe). In all cases, thin film multilayers were fabricated in an attempt to provide balanced charge transport and efficient carrier recombination. For the nanocrystallites, sequentially adsorbed layers of poly(p-phenylene vinylene) (PPV) and poly(methacrylic acid) (PMA) were used as hole transport layers in devices with the following architecture: ITO/(PPV/PMA)$_n$/CdSe nanocrystallites/Al. The color emitted from such devices can be tuned from red to blue by changing the diameter of the nanocrystallites. Multilayer thin films fabricated from sequentially adsorbed layers of a new trisbipyridyl ruthenium (II) polyester and PMA were found to exhibit external quantum efficiencies as high as 1%. Details concerning the fabrication and properties of thin film devices made from these various materials will be discussed.

201. POLYMERIC 8-HYDROXYQUINOLINE METAL CHELATE LIGHT EMITTING DIODES. D. L. Thomsen III, K. A. Higginson, F. Papadimitrakopoulos*, Department of Chemistry, Institute of Materials Science, University of Connecticut, Storrs, CT 06269-3136

We are presently reporting a novel light emitting diode (LED) made from polymeric metal chelate derivatives of 8-hydroxyquinoline. Reactive self-assembly of diethyl zinc and 8-8'-dihydroxy-5-5'-biquinoline (bisquinoline) onto hydroxy-functionalized ITO resulted in ultra-thin, homogeneous films of insoluble and intractable metal chelates. Single layer electroluminescent (EL) devices, with films as thin as 400 Angstroms exhibited a strong orange emission, similar to their photoluminescence (PL) response. The growth of a 1,200 Angstrom film in about 2.7 hours, with the potential for multilayer fabrication at high purity levels, offers an alternative to vacuum deposition and spin coating.

202. MOLECULAR WEIGHT DISTRIBUTION AND CONFIGURATION OF MEH-PPV IN DILUTE SOLUTION, Patricia M. Cotts, DuPont CR&D, Experimental Station, Wilmington, DE 19880

Despite the widespread use of poly(phenylenevinylene) and its analogs as an electroluminescent and conducting polymer, there has been very little physical characterization of this polymer reported. In this study we report measurements of the molecular weight distribution and configuration (root-mean-square radius of gyration (R_g) as a function of M) of poly(2-methoxy,5-(2'-ethylhexoxy)-p-phenylenevinylene), MEH-PPV, in dilute solution in chloroform. Determination of M and R_g for monodisperse fractions eluting from a size exclusion column was accomplished by light scattering using a Wyatt Dawn DSP multi-angle light scattering detector. The SEC comprised a Waters WISP sample injector, 590 solvent delivery system and R410 refractive index detector. The SEC columns were Polymer Laboratories PLGel, 20μm Mixed A. Measurement of two different molecular weight polymers extended the range of M and R_g so that $30,000 < M < 3,000,000$ and $10 < R_g < 80$ nm. At these high M, the dependence of R_g on M is consistent with a random coil despite the highly conjugated backbone. The molecular weight distributions are broad, with $M_w/M_n = 1.7- 2.3$.

203. Synthesis and Characterization of a Soluble Blue Light Emitting Alternating Copolymer. H.N. Cho, J.K. Kim, D.Y. Kim and C.Y. Kim, Polymer Materials Lab., KIST, Seoul 136-791, Korea

An alternating copolymer with monomer units of fluorene and phenylenevinylene is synthesized by reaction between n-hexyl fluorene phosphonium salt and isophthalic aldehyde based on the Wittig reaction. The polymer solution in chloroform is made into a film with a very fine surface by spin-casting onto indium-tin oxide glass to make an light emitting diode with an aluminum negative electrode. The optical absorption spectrum shows the peak at 370nm while both the PL and EL spectra have the peak at 560nm showing the Stokes shift of 100nm. The peak in the PL spectrum shifts to 420nm with a fine structure at 440nm on either dilution by chloroform or blending with polyvinylcarbazole (PVK). The peak in the EL spectrum also shifts to 440nm on blending of 20% of the copolymer with PVK. Further dilution for 10% of the copolymer shifts the EL peak to 420nm.

204. THERMOPLASTIC ELASTOMERS BASED ON ABC TRIBLOCK COPOLYMERS.
R. Stadler[§], S. Brinkmann[§], V. Abetz[§], E. L. Thomas[#];
[§]Institut für Organische Chemie, Universität Mainz, Postfach 3980, 55099 Mainz, Germany,
[#]Massachusetts Institute of Technology, 77 Mass. Ave., Cambridge, MA 02139, USA

Thermoplastic elastomers have been of great interest due to the unique combination of properties. Comparing the mechanical properties of ABA-type polymers to novel ABC type copolymers should proove interesting: In microphase separated block copolymers of the ABA type B-chains can arrange as elastic ineffective loops or as elastic effective bridges. In ABC triblock copolymers however, where all components are phase segregated, only bridges should be formed. We report the preparation of a variety of polystyrene-b-1,4-polybutadiene-b-polymethylmethacrylate (SBM) block copolymers and their polystyrene-b-1,4-polybutadiene-b-polystyrene (SBS) analogues via sequential anionic polymerization. The polymers are characterized using ^{1}H-NMR and GPC. The microstructure is investigated by transmission electron microscopy (TEM) and small angle X-ray scattering (SAXS). The mechanical poperties are compared in stress-strain experiments.

205. NEW METHACRYLATE BLOCK AND RANDOM COPOLYMERS FOR SUBMICRON LITHOGRAPHIC IMAGING. C. K. Ober, M. A. Z. Hupcey and A. H. Gabor, Materials Science and Engineering, Cornell University, Ithaca, NY 14853.

Increasing circuit density in semiconductors is leading to the adoption of new imaging technologies and the design of resist compositions suited to them. Polymer resists based on acrylates and methacrylates provide the needed versatility and transparency. Cleavage of the appropriate ester in a poly(methacrylate) using a photoacid generator produces a carboxylic acid which provides aqueous base solubility. In this paper are described the syntheses of block and random copolymer resists based on t-butyl methacrylate combined with isobornyl methacrylate, the latter monomer having etch resistance due to the presence of a ring structure rich in tertiary carbon. Polymers of identical molecular weights and composition in both block and random architecture were produced using group transfer polymerization and examined for their lithographic performance. The superior processing behavior of block copolyemrs will be described.

206. POLYMER NANOPARTICLES WITH NEW ARCHITECTURES PREPARED VIA NON-AQUEOUS DISPERSION COPOLYMERIZATION OF OXAZOLINE-FUNCTIONAL METHACRYLATES, Matthias Hölderle, Matthias Bruch, Rüdiger Schäfer, Rolf Mülhaupt, Freiburger Materialforschungszentrum und Institut für Makromolekulare Chemie der Albert-Ludwigs Universität, Stefan-Meier-Str.21, D-79104 Freiburg i.Br., Germany

Oxazoline-functional acrylic nanoparticles with average sizes varying between 100 and 500 nm were prepared by non-aqueous dispersion copolymerization of MMA, EGDMA and a novel oxazoline-functional methacrylate (OMA), derived from ε-caprolactone and 2-

aminoethanol, in n-heptane medium. Particle sizes were controlled by the content of the polystyrene-*block*-poly(ethene-*alt*-propene) dispersing agent. OMA was varied between 0 and 95 mol-%. Stable dispersions were obtained after removing dispersing agent. In mesitylene dispersion at 163 °C monocarboxylate-terminated polystyrene of M_n=15900 g/mol was grafted onto the acrylic particles via esteramide formation, resulting from nucleophilic attack of the carboxylate end group at the oxazoline ring. Core/shell nanoparticles were characterized by transmission electron microscopy and light scattering.

207. **POLYROTAXANES BY SELF THREADING OF FUNCTIONALIZED MACROCYCLES: POLYMETHACRYLATES.** H. W. Gibson, D. Nagvekar and S. Bhattacharjee, Department of Chemistry, Virginia Polytechnic Institute & State University, Blacksburg, VA 24061-0344

Novel poly(methacrylate crown ether)s were synthesized to investigate polyrotaxane formation by *in situ* self threading. Thermal polymerization of the methacrylate derived from 5-hydroxymethyl-m-phenylene-m-phenylene-32-crown-10 gave soluble, very high molecular weight (M_n 9.4 x 10^6 g/mol, PS standards) polymethacrylate. NMR analysis by NOESY demonstrated through-space coupling of the backbone protons with the intra-annular aromatic protons of the macrocycle, a direct proof of rotaxane formation. In a control experiment, polymerization of the methacrylate ester of 5-hydroxymethyl-m-phenylene-16-crown-5 gave a low molecular weight material and NOESY did not reveal any through-space correlation, indicating the absence of threading, as expected because the ring is too small to undergo threading.

208. GEL PERMEATION CHROMATOGRAPHY COUPLED TO FOURIER TRANSFORM MASS SPECTROMETRY FOR THE CHARACTERIZATION OF THE PRODUCTS OF METHACRYLATE POLYMERIZATIONS. William J. Simonsick, Jr., David J, Aaserud, Michael C. Grady, DuPont Automotive Finishes, Marshall R&D Laboratory, 3401 Grays Ferry Ave., Philadelphia, PA 19146. Laszlo Prokai, Center for Drug Discovery, College of Pharmacy, University of Florida, Gainesville, FL 32610.

We have coupled a gel permeation chromatograph to a Fourier transform mass spectrometer (FTMS) through an electrospray ionization interface. The GPC/ESI/FTMS was used to characterize the products of free-radical methacrylate polymerizations. Macromonomers containing glycidyl methacrylate and n-butyl methacrylate monomers were characterized by GPC/ESI/FTMS. Sodiated GMA/BMA oligomers in excess of 4 kDa were detected. End-group information was furnished. ^{13}C isotopic resolution was observed to about 3 kDa and GMA/BMA isobaric tetramers were resolved in the on-line GPC/ESI/FTMS studies. ESI/FTMS afforded resolutions in excess of 80,000 which allowed the separation of isobaric GMA/BMA macromonomers under 1000 Da.

209. COMPUTER SIMULATIONS OF LIQUID-CRYSTALLINE DIACRYLATE MONOMER POLYMERIZATIONS. Henk M.J. Boots, Philips Research Laboratories, Prof. Holstlaan 4, 5656AA Eindhoven, The Netherlands; Nikolaos A. Peppas, School of Chemical Engineering, Purdue University, West Lafayette, IN 47907-1283, USA; Kristi S. Anseth, Department of Chemical Engineering, ECCH 128, University of Colorado, Boulder, CO 80309-0424, USA.

An off-lattice, kinetic-gelation model was developed to simulate the chain crosslinking polymerization of liquid-crystalline (LC) diacrylate monomers. A detailed description of the simulation rules for propagation and particle relaxation is provided. LC molecules were pre-equilibrated in the isotropic, nematic, and smectic A states. The influence of the polymerization conditions on the ordering of the LC states was examined. In particular, the focus of this work was to examine the conditions under which increased or decreased ordering occurs as a result of polymerization.

210. HIGH-SPEED POLYMERIZATION OF ACRYLATE MONOMERS BY UV IRRADIATION
C. Decker, T. Bendaikha, D. Decker, K. Zahouily Laboratoire de Photochimie Générale (URA-CNRS n°431) - Ecole Nationale Supérieure de Chimie - 3, rue Werner - 68200 Mulhouse (France)

The photoinitiated polymerization of multifunctional acrylate monomers has been studied by real-time infrared spectroscopy in a millisecond timescale. The important kinetic parameters have been evaluated from the conversion *versus* time curve recorded during and after the UV exposure. The cross-linking polymerization proceeds with long kinetic chains (30,000 mol per radical) because of both an effective propagation step and an inefficient termination step. By using a fast photobleachable initiator (acylphosphine oxide), frontal polymerization has been performed to produce up to 3 cm thick polymer specimens within a few seconds. The same type of acrylate formulation was found to undergo rapid polymerization upon exposure to sunlight with formation of a highly cross-linked polymer well suited for a fast production of large dimension glass laminates.

211. CURE REACTIONS OF VINYL ESTER/STYRENE COMPOSITE MATRIX RESINS. H. Li, A.C. Rosario, S. V. Davis, T. E. Glass, J. J. Lesko, and J. S. Riffle, Department of Chemistry and NSF Science and Technology Center: High Performance Polymeric Adhesives and Composites, Virginia Polytechnic Institute & State University, Blacksburg, VA 24061-0344 and J. J. Florio, Hexcel, Inc., Kent, WA

"Vinyl ester" oligomers diluted with styrene are important resins for thermosetting polymer matrix composites. One major objective of this research is to investigate the chemistry and mechanism of the cure reactions of vinyl ester resins in rapid reactions at elevated curing temperatures, which would be typical of composite processing conditions. FTIR was utilized to study the cure reaction of vinyl ester resins. At 140°C almost 100% conversion of vinyl ester and ≈90% conversion of styrene was reached only 3 minutes after the induction period under the conditions investigated. This fast reaction rate is important for continuous, on-line, processing. Solidstate NMR was utilized to determine the final conversion of methacrylate endgroups of vinyl ester oligomers. Final conversion approached 100% for vinyl ester oligomers with M_n ~680g/mol and 30wt % styrene at 120°C. The important factors affecting the conversion of methacrylate groups are temperature and styrene contents. Lower cure temperatures (below Tg) result in residual unsaturation.

212. SITE PROTECTED PEGYLATION OF RECOMBINANT (p75) TUMOR NECROSIS FACTOR RECEPTOR. D.K. Pettit, N. Nightlinger, A. Goetze, and W.R. Gombotz, Department of Analytical Chemistry and Formulation, Immunex Corp., Seattle, WA 98112

The process of protein PEGylation is known to improve circulation half-lives, enhance solubility, and reduce immunogenicity; however, for many proteins this process also results in significant reduction in biological activity. Several approaches have been developed to direct PEGylation to certain regions of protein molecules in order to reduce the loss in bioactivity. We have developed a novel process to PEGylate proteins in a manner which protects the binding site or active region of the protein and allows PEGylation of amino acids outside the ligand binding region by conventional lysine conjugation methods. As a demonstration of this process monoclonal antibodies which neutralize the activity of TNFR:Fc (a p75 tumor necrosis factor receptor - IgG fusion protein) were covalently immobilized to biosupport beads. TNFR:Fc was allowed to bind to the antibodies and lysine reactive PEG was conjugated to the exposed regions of the TNFR:Fc protein. Following elution from the column the PEGylated TNFR:Fc was found to be physically altered (as assessed by size exclusion chromatography, fluorescamine assay, and tryptic peptide mapping), yet retained complete biological activity. This result was contrasted by the significant loss of biological activity when TNFR:Fc was PEGylated without the aid of the monoclonal antibody affinity column.

213. IMPROVED CONJUGATION OF CYTOKINES USING HIGH MOLECULAR WEIGHT POLY(ETHYLENE GLYCOL): PEG-GM-CSF AS A PROTOTYPE. Mark G. P. Saifer,[1] L. David Williams,[1] Merry R. Sherman,[1] John A. French,[1] Larry W. Kwak[2] and Joost J. Oppenheim.[2] [1]Mountain View Pharmaceuticals, Inc., San Carlos, CA 94070 and [2]National Cancer Institute, Frederick, MD 21702

Covalent attachment of poly(ethylene glycol) (PEG) to enzymes can increase their stability and half-life in plasma and decrease their immunogenicity and antigenicity, often with partial retention of enzymatic activity. Cytokines, *e.g.* granulocyte-macrophage colony-stimulating factor (GM-CSF), are small receptor-binding proteins that function as intercellular signals between white blood cells (WBC). Previous attempts to improve the efficacy of cytokines by attaching several strands of 5-kDa PEG resulted in decreased receptor-binding activity and/or insufficient enhancement of plasma half-life. We report here the efficient coupling of a few strands of 36- or 42-kDa PEG to recombinant murine or human GM-CSFs and the optimized production of biologically active conjugates. Injecting mice with adducts of rmuGM-CSF containing one or two strands of 36-kDa PEG produced dose-dependent increases in the numbers of most types of WBC. At the highest dose, total WBC counts were 15 times higher than those observed after injecting GM-CSF.

214. POLY(N-ACRYLOYLMORPHOLINE) CAN BE USED IN ALTERNATIVE TO POLY(ETHYLENE GLYCOL) FOR PROTEIN MODIFICATION. F. M. Veronese[1], O. Schiavon[1], P. Caliceti[1] and P. Ferruti[2], (1) Department of Pharmaceutical Sciences, University of Padua, Via Marzolo 5, 35131 Padua, Italy, (2) Department of Organic and Industrial Chemistry, University of Milan, Milan, Italy.

The properties of poly(N-acryloylmorpholine) (PAcM) enzyme conjugates were investigated in relation to the most known linear PEG and the more recently proposed branched PEG (PEG-2) in order to understand the potentials and limits of this new oligomer. The proposed synthesis yields hydroxylated or carboxylated end terminal groups suitable for activation and binding to protein amino group, and the polymerization conditions may be designed to yield fractions of different molecular weight that can be further fractionated to narrow molecular weight products. The conjugation of rapresentative enzymes showed that the products maintain enzymatic activity with increased resistance to proteolytic digestion. The conjugation of uricase or superoxide dismutase, enzymes of therapeutic interest, with PAcM increases also the residence time in rats. Immunological investigations in mice showed decrease in antibodies production. All of these effects, relevant for applications in therapy, are similar to those found using PEG as polymer. A comparative study carried out among various conjugates demonstrated that, generally, the order of efficacy of the polymers is PEG2>PEG≥PAcM.

215. PREPARATION, PURIFICATION, AND CHARACTERIZATION OF POLY(ETHYLENE GLYCOL)-MODIFIED MICROPEROXIDASE-11 Kwai Dzy Mak, Jing Liu Nuan, LiCi Liang, Xing Qi Zhou, and Patricia Ann Mabrouk, Department of Chemistry, Northeastern University, Boston, MA 02115

Poly(ethylene glycol)-modified horse microperoxidase-11(MP-11-PEG) has been prepared, purified by SEC HPLC, and characterized by UV-vis spectroscopy, enzyme assay, and cyclic voltammetry for use as a peroxidase model in our non-aqueous spectroelectrochemical investigations of heme enzyme catalysis. The modified heme peptide is spectroscopically indistinguishable from microperoxidase-11. The peroxidase activity of the MP-11-PEG has been assessed for the oxidation of mesidine at 10^0C and pH 4.9 and is comparable to that of microperoxidase-11. Cyclic voltammetric data for the imidazole complex of MP-11-PEG at glassy carbon in aqueous solution, pH 7.0 indicates that the modified heme peptide undergoes quasi-reversible electron transfer at GC ($k_s' = 1.4 \pm 0.1 \times 10^{-3}$ cm/s). Together the data suggest that pegylation does not affect the heme active site structure, catalytic activity, or electron transfer characteristics of the heme peptide in MP-11-PEG.

216. DESIGN OF ANTITUMOR AGENT-TERMINATED POLY(ETHYLENE GLYCOL) CONJUGATE AS MACROMOLECULAR PRODRUG. T. Ouchi, H. Kuroda, Y. Ohya, Department of Applied Chemistry, Faculty of Engineering, Kansai University, Suita, Osaka 564, Japan

Since poly(ethylene glycol) (PEG) is a water-soluble and biocompatible polymer, PEG is very interesting material as a carrier of macromolecular prodrug. Although doxorubicin (DXR) is one of the most prominent antitumor agents in cancer chemotherapy, its very strong side-effects, its no good water-solubility and no good stability have also be cited as the unsolved problems. In order to solve their defects of DXR, we investigated the design of DXR-terminated PEG conjugate as a macromolecular prodrug of DXR. MeO-PEG/Schiff's base/DXR conjugate showed a lysosomotropic release behavior of free DXR, very good stability in PBS for a long period, and exhibited a significant cytotoxic activity against p388 lymphocytic leukemia cells in vitro. Moreover, this conjugate exhibited more strong cytotoxicity than free DXR against DXR resistance p388 leukemia cells. Furthermore, Lactose/PEG/amide/DXR conjugate showed stronger cytotoxic activity against HLE human hepatoma cells than MeO-PEG/amide/DXR conjugate. The cytotoxic activity of Lactose/PEG/amide/DXR conjugate against HLE cells was inhibited by addition of galactose.

217. PEG-LINKED ARTEMESININ ANTIMALARIALS. Michael D. Bentley and Bong-Youl Chung, Department of Chemistry, University of Maine, Orono, Maine 04469; William Ellis and Patrick McGreevy, Walter Reed Army Institute of Research, Washington, D.C.20307-5100

On a global basis, malaria remains one of our most devastating diseases. With increasing resistance of the disease-causing plasmodia to current drugs, there is an urgent need for new antimalarials. Of considerable current interest is artemisinin (qinghaosu) (I), an antimalarial from the Chinese traditional medicinal plant, *Artemisia annua* (qinghao). With the goal of improving the activity and pharmacokinetics of (I), we have prepared a series of polyethylene glycol (PEG) derivatives of this compound and here present the preparation and assay results against two strains of *Plasmodium falciparum* as well as some preliminary tests of one of these derivatives in mice infected with *P. berghei*. Ester linkages to PEG 400, 1500, and 3400 were formed by reacting dihyroartemisinin with PEG disuccinic acids in the presence of DCC/DMAP. MPEG 550 was reacted with dihydroartemisini/BF_3 to form the acetal. All of the PEG derivatives displayed higher activity against both strains of *P. falciparum* than did (I). Activity increased with decreasing molecular weight of the PEG esters. Oral and subcutaneous treatment of *P. berghei* with the PEG 1500 ester derivative improved mouse longevity and reduced parasite levels.

218. A NOVEL BIOCIDAL PEG COMPOUND. S.D. Worley and M. Eknoian, Department of Chemistry, Auburn University, Auburn, Alabama 36849 and J.M. Harris, Department of Chemistry, University of Alabama, Huntsville, AL 35899.

The monomer p-methylacetophenone was converted into 5-methyl-5-(4'-methylphenyl) hydantoin by treatment with ammonium carbonate and potassium cyanide in an ethanol/water mixture. Then the hydantoin monomer was brominated at the aromatic methyl group by use of NBS in carbon tetrachloride in the presence of AIBN initiator. Reaction of the 5-methyl-5-(4'-bromomethylphenyl)hydantoin with PEG-amine in methylene chloride in the presence of sodium hydroxide produced a PEG polymer with a terminal 5-methyl-5-(4'-methylphenyl)hydantoin group. This polymer was made biocidal by chlorination of the hydantoin ring using 10% sodium hypochlorite. The final biocidal polymer was soluble in water, and a solution containing 1000 ppm of it inactivated the bacterium Staphylococcus aureus within a 10 min contact time. The solution only declined in total chlorine content from 3.5 to 1.8 ppm over a period of 110 days. The material may be useful as a surface-anchored biocidal film.

219. POLYETHYLENE GLYCOL-CONTAINING SUPPORTS FOR SOLID-PHASE SYNTHESIS OF PEPTIDES AND COMBINATORIAL ORGANIC LIBRARIES. George Barany,[1] Fernando Albericio,[1,2,3] Steven A. Kates,[3] and Maria Kempe.[1,4] from [1]Department of Chemistry, University of Minnesota, Minneapolis, MN 55455, USA; [2]Department of Organic Chemistry, University of Barcelona, E-08028 Barcelona, Spain; [3]PerSeptive Biosystems Inc., 500 Old Connecticut Path, Framingham, MA 01701; and [4]Department of Pure and Applied Biochemistry, University of Lund, S-221 00 Lund, Sweden.

The choice of polymeric support is a key factor for the success of solid-phase methods for syntheses of organic compounds and biomolecules such as peptides and oligonucleotides. Classical Merrifield solid-phase peptide synthesis, performed on hydrophobic polystyrene beads, sometimes suffers from sequence-dependent coupling difficulties. This lecture reviews the preparation of families of polyethylene glycol-polystyrene (PEG-PS) graft and Cross-Linked Ethoxylate Acrylate Resin (CLEAR) supports, and demonstrates their applications to synthesis. PEG-PS supports can be prepared conveniently starting from PEG's of different molecular weights which are coupled covalently to PS, whereas CLEAR supports are prepared from a branched cross-linker containing short PEG chains. Both of these supports swell in a broad range of hydrophobic and hydrophilic solvents, and have excellent physical and mechanical properties for both batchwise and continuous-flow systems. Final loadings (0.1 to 0.6 mmol/g) can be readily tailored depending on the synthetic goal (supported in part by NIH GM 42722 and 51628 to G.B.).

220. POLY(ALKYLENE OXIDES) IN PROCESSES OF MOLECULAR SELF-ASSEMBLY. I.N.Topchieva, I.G.Panova, V.I.Gerasimov, K.I.Karezin, and V.A.Kabanov, Department of Chemistry, Lomonosov State University, 119899 Moscow, Russia

Complexation between linear poly(alkylene oxides) and cyclodextrins (CD) resulting in the formation of new types of inclusion compounds, "molecular necklaces", presents an approach for the design of molecular assemblies with diverse architecture. A step forward in this direction was made by using di- and triblock copolymers of ethylene and propylene oxides (Proxanols, Pluronics) and α-, β- and γ-Cds. We demonstrated the possibility of formation of inclusion complexes by means of selective interactions between copolymers and CDs. Poly(ethylene oxide) block forms one-stranded complexes with α-CD and double-stranded with γ-CD, whereas poly(propylene oxide) block - one-stranded complexes with β- and γ-CD. Crystalline complexes based on block copolymers and Cds are characterized by the type of crystalline lattice differing from each other, and that of the initial components. They have the general composition, corresponding to one molecule of α- and β-CD per two monomer units of polymer. The complexation process is a brilliant example of self-assembly.

221. **EFFECT OF MACROMOLECULAR ARCHITECTURE ON THE SELF-ASSEMBLY OF POLY(ETHYLENE OXIDE)/POLY(PROPYLENE OXIDE) COPOLYMERS IN SELECTIVE SOLVENTS: A–B–A VS. B–A–B BLOCK SEQUENCE**
Paschalis Alexandridis, Ulf Olsson, and Björn Lindman, Physical Chemistry 1, Center for Chemistry and Chemical Engineering, University of Lund, P.O.B. 124, Lund, S-22100, Sweden.

The phase behavior of the $(EO)_{13}(PO)_{30}(EO)_{13}$ (Pluronic L64) and $(PO)_{19}(EO)_{33}(PO)_{19}$ (Pluronic 25R4) copolymers in the presence of water [selective solvent for poly(ethylene oxide), PEO] and p-xylene [selective solvent for poly(propylene oxide), PPO] has been explored at 25°C [*Macromolecules* **1995**, *28*, 7700; *J. Phys. Chem.* **1996**, *100*, 280]. L64 and 25R4 have similar molecular weights and chemical composition, but opposite block sequence; the effect of the macromolecular architecture on the block copolymer self-assembly can thus be elucidated. The self-assembly behavior of the two copolymers in the selective solvents is strikingly similar. This suggests that the block architecture (A–B–A vs. B–A–B) does not influence significantly the global phase behavior of the triblock copolymers, and that the volume fraction of the apolar (PPO+xylene) components is the major parameter governing the copolymer phase behavior. The ability of both L64 and 25R4 to attain the same ordered structures (lamellar, oil-in-water and water-in-oil hexagonal, and bicontinuous cubic) at roughly the same compositions is consistent with very little angular correlation in the location of the two end-blocks of a copolymer molecule in an assembly.

222. **"GELS" FORMED BY A POLY(ETHYLENE OXIDE)/POLY(BUTYLENE OXIDE) BLOCK COPOLYMER IN WATER: THERMODYNAMICS, STRUCTURE, AND DYNAMICS**
Paschalis Alexandridis and Marina Tsianou, Physical Chemistry 1, Center for Chemistry and Chemical Engineering, University of Lund, P. O. Box 124, Lund, S-22100, Sweden.

The concentration-temperature stability regions, crystallographic structure, and rheological properties of the "gels" formed by an amphiphilic poly(ethylene oxide)-*b*-poly(butylene oxide)-*b*-poly(ethylene oxide) block copolymer, $(EO)_{17}(BO)_{14}(EO)_{17}$, in water have been resolved. Three different lyotropic liquid crystalline "gel" structures can be formed with increasing copolymer concentration: a cubic arrangement of spherical micelles (I_1), a hexagonal array of cylindrical micelles (H_1), and a lamellar arrangement of planar copolymer assemblies (L_α), in the 30-40, 45-55, and 65-85 copolymer wt. % (20°C), respectively. The thermal stability of the "gel" regions increases in the order cubic < hexagonal < lamellar. Both the lamellar and cylinder spacing (70-80 Å) and the interfacial area per copolymer molecule (70-80 Å^2) decrease with increasing copolymer content, but are not affected much by temperature. The "gels" exhibit a notable elastic behavior (and high viscosity) which becomes more pronounced as their structural order increases from lamellar to hexagonal to cubic. The self-assembly behavior in water of the $(EO)_{17}(BO)_{14}(EO)_{17}$ triblock copolymer is similar to that of the diblock copolymer $(BO)_{10}(EO)_{17}$ [*Langmuir* **1997**, *13*, in press] and that of triblocks PEO-PPO-PEO with double its molecular weight [*Langmuir* **1996**, *12*, 2690].

223. **BIOACTIVE POLYETHYLENE GLYCOL-INSULIN CONJUGATES WITH ENHANCED STABILITY.** F. Liu, M. Baudys, D. Mix, K. Hinds, and S. W. Kim, Department of Pharmaceutics and Pharmaceutical Chemistry, Center for Controlled Chemical Delivery, University of Utah, UT 84112

Polyethylene glycol (PEG) has been widely used for modification of protein drugs to decrease immunogenicity and improve pharmacokinetic parameters of parent proteins. Monocarboxyl derivatives of monomethoxyPEG (mPEG) of different M. Wt. (600 - 2000) were synthesized from mPEG by either oxidation with $KMnO_4$ or nucleophilic substitution with ethyl bromoacetate. The oxidative method gave higher yields but also led to unwanted fragmentation of the mPEG chain. Preferred alkylation method provided only medium yields but avoided any cleavage of the mPEG chain. The carboxyl group of mPEG-O-CH$_2$COOH was activated by the mixed anhydride method and then reacted with the insulin Phe B1 amino group. The kinetics of pegylation reaction was studied and found to be mPEG M. Wt. and insulin concentration dependent. The physical stability of mPEG(600) insulin derivative increased over 40 times compared to insulin. Finally, we studied the effect of mPEG M. Wt. on the bioactivity and stability of insulin. The bioactivity of insulin derivatives gradually decreased as the M. Wt. of the attached mPEG chain increased, while the stability of insulin derivatives slightly increased. The bioactivity of mPEG600 and mPEG800 insulin derivatives was completely preserved.

224. PEO MACROMONOMERS WITH HYDROPHOBICALLY ENHANCED METHACRYLATE END GROUPS. Hironao Furuhashi, Seigou Kawaguchi, Shinichi Itsuno, and Koichi Ito, Department of Materials Science, Toyohashi University of Technology, Tempaku-cho, Toyohashi, 441 Japan

Poly(ethylene oxide) macromonomers carrying methoxy group on the one (α-) end and methacryloyloxyhexyl or methacryloyloxydecyl group as the other (ω-) end were prepared, homopolymerized in water, and dispersion-copolymerized with styrene or methyl methacrylate in a methanol-water mixture. They were found to polymerize more rapidly and to produce stable polystyrene dispersions more effectively, as compared to the corresponding macromonomers carrying either α-methoxy or α-dodecyloxy and ω-methacryloyloxy end groups. Thus the amphiphilic constitution of the macromonomers, such that favors the polymerizing methacrylate end groups to locally concentrate into the micelle core or to the particle surface while the poly(ethylene oxide) chains extending to the medium, appears to be most important in enhancing their polymerizability and effectiveness as reactive steric stabilizers. On the other hand, stable poly(methyl methacrylate) particles with a number of craters or pleats on the surface were produced with a simple PEO macromonomer with α-methoxy and ω-methacryloyloxy end groups.

225. RECENT DEVELOPMENTS IN THE SYNTHESIS OF BRANCHED POLY(ETHYLENE OXIDES) VIA ANIONIC POLYMERIZATION, K. Naraghi, Y. Ederlé, D. Haristoy, P.J. Lutz, Institut Charles Sadron, CNRS,67 083 Strasbourg Cedex, France.

The synthesis via anionic polymerization of branched molecules consisting of an hydrophobic core or backbone carrying a number of hydrophilic poly(ethylene oxide) branches has been investigated. Different approaches are discussed. The "core-first" method in which living poly(DVB) cores are used to initiate the polymerization of oxirane, is easy to perform but the samples obtained exhibit broad molar mass distributions arising from large distributions of the core functionalities. Another limitation of that method may be the reproducibility. The "grafting from" method involves a poly(diisopropenylbenzene) (DIB)backbone. The active sites subsequently created on the poly(DIB)backbone can serve as efficient initiators for the anionic polymerization of oxirane. That method allows a far better control in advance of the functionality than the "core-first" method. Two other approaches were attempted: the anionic homopolymerization of PEO macromonomers whereby no functionalization of the grafts is yet possible and "grafting from" methods based on a polynorbornene backbone exhibiting hydroxyl groups. After reaction with diphenylmethyl-potassium these sites can also initiate the anionic polymerization of oxirane.

226. MOLECULAR DYNAMICS SIMULATIONS OF POLY(OXYETHYLENE) IN SOLUTIONS AND IN THE PRESENCE OF CATIONS. Kenzabu Tasaki, Mitsubishi Chemical America, 99 W Tasman Dr. San Jose, CA 95134.

We have performed molecular dynamics simulations of a poly(oxyethylene) (POE) chain in benzene solution and an aqueous solution with and without potassium ions. Our interest is the conformation and the hydration of POE, the solvent effect, and the effect of cations on POE in order to understand its unique solution characteristics in water and in the presence of salts. The POE chain assumed a helical conformation in an aqueous solution and transformed to a random coil in benzene solution. The pair distribution functions for the water oxygen atoms indicated the enhanced water structures near POE. In the presence of potassium ions, the helix was partially broke down and the terminal groups formed a complex with a potassium ion. The POE oxygen–water oxygen pair distribution function demonstrated strong perturbations of hydration by potassium ions . We discuss the solubility and phase behavior of POE in aqueous solution with and without potassium ions.

227. NONCOVALENT COMPLEXES OF α–CHYMOTRYPSIN AND BLOCK COPOLYMER OF ETHYLENE AND PROPYLENE OXIDES.
I.N. Topchieva, E.M. Sorokina, N.V. Efremova, School of Chemistry, Lomonosov State University, Moscow, 119899 Moscow, Russia, Bakh institute of Biochemistry, Russian Academy of Sciences, 117071 Moscow, Russia.

Two principal methods of the formation of complexes between proteins (α-chymotrypsin (CHT) and bovine serum albumin (BSA)) and amphiphilic block copolymer of ethylene and propylene oxides (Proxanol) are elaborated. The first is based on the heating of the components at temperatures of 44 to 60°C. The composition of the complexes was estimated after their isolation from solution by determination of their protein content and corresponds to 4–11 polymer chains per CHT molecule. The increase of the temperature of complexation brings to the higher content of polymer in complex, and to decrease of its enzymatic activity. The second method of receiving polymer–protein complexes is based on the action of high pressures (1–400 MPa). This method allows to obtain complexes based both on Proxanol and PEG. The composition of high-pressure-induced complexes does not depend on the value of pressure and corresponds approximately to seven polymer chains per CHT molecule. The complexes fully retain the catalytic activity of CHT. Both types of complexes are characterized by higher thermostability as regards to the native CHT.

228. CONJUGATES OF PROTEINS WITH DIBLOCK COPOLYMERS OF ETHYLENE AND PROPYLENE OXIDES. N.V. Efremova[1], I.N.Topchieva[2], B.I.Kurganov[1], (1) A.N.Bakh Institute of Biochemistry, 117071 Moscow, Russia, (2) Chemistry Department, M.V.Lomonosov Moscow State University, 117234 Moscow, Russia

Conjugates of proteins (bovine serum albumin and α-chymotrypsin (CT)) with poly(ethylene glycol) and amphiphilic diblock copolymers of ethylene oxide and propylene oxide (BC) were synthesized, using monoaldehyde polymer derivatives as the amino group modifying reagents. Four types of conjugates varying in the placement of hydrophobic block and type of polymer chain distribution were obtained. All of them were characterised by higher thermostability comparing to initial protein, the stabilizing effect increasing in parallel with the degree of modification. The amphiphilic character of the attached polymer chains provided new functions of modified hydrophilic proteins, such as the capability to interact with biological and model membranes, maintenance of effective solubilization and transport of insoluble biologically active compounds, and the ability to form molecular assemblies due to the interaction with various amphiphilic compounds (BC, lipids, polysacchrides).

229. PREPARATION AND CHARACTERIZATION OF OLIGOSACCHARIDE- AND OLIGOPEPTIDE-BEARING STEALTH® LIPOSOMES Joshua Gittelman, Jennifer Harding, Nasreen Mullah, Luke Guo, Shawn DeFrees[§], and Samuel Zalipsky, SEQUUS Pharmaceuticals, Inc., 960 Hamilton Court, Menlo Park, California 94025, and [§]Cytel Corporation, 3525 Hopkins Court, San Diego, California 92121

Two conjugates designed for presentation of cell adhesion ligands on the surface of Stealth® liposomes were synthesized. Sialyl Lewis^X (SLX) containing pentasaccharide or laminin pentapeptide YIGSR was linked to a heterobifunctional polyethylene glycol (PEG), which was bound to distearoylphosphatidylethanolamine (DSPE) lipid anchor. The conjugates were characterized by NMR, HPLC, and MALDI-TOFMS. Ligand-bearing liposomes (PHPC / cholesterol / mPEG-DSPE / ligand-PEG-DSPE, 55:40:3:2, 100 nm) were formulated by two methods: thin film hydration / extrusion (THE), and a new method of ligand-PEG-DSPE insertion into preformed vesicles. THE resulted in 55-65 % of the ligand residues presented on the external surface, whereas insertion produced liposomes with the same ligand densities, all incorporated on the outer monolayer. Since only the external ligand moieties can be available for interaction with the target receptors, the insertion approach is the more efficient for incorporation of ligand-PEG-DSPE conjugates into mPEG-grafted liposomes.

230.

EVAPORATIVE LIGHT SCATTERING DETECTION WITH REVERSED PHASE HPLC AS A TOOL TO ANALYZE AND CHARACTERIZE PEG LINKERS. Pat D. Green, Hsi Meng and James E. Seely, Boulder Analytical Resources, Amgen Boulder Inc., 3200 Walnut St., Boulder, CO 80301

Evaporative Light Scattering (ELS) detection has proven to be a powerful tool to detect PEG linkers. Unlike Refractive Index (RI) detection, this detector is gradient compatible for improved resolution and faster separation of PEG. In conjuction with reversed phase HPLC, PEG linkers can be separated by size and end group functionality.
This method can also be used to directly quantitate the amount of aldehyde linker in a PEG mixture. Using p-aminobenzoic acid (PABA) to derivatize the aldehyde functional group, we have been able to separate the PEG aldehyde species from unreactive PEG. PABA reacted PEG aldehyde has increased retention time such that the derivatized PEG is readily separated. This reaction of PABA allows for UV detection of PEG aldehyde content prior to ELS detection via the PABA chromophore. The percent PEG aldehyde content in the sample is measured from the area percent of the shifted derivatized peak in the ELS trace.

231.

Synthesis and Characterization of PEO-based Networks from Azide Precursors. M. Pinteala[1], V. Harabagiu[1], B. Simionescu[2], P. Guégan[3], H. Cheradame[3].
Aziridine modified polyethylene oxides (PEO) were synthesized by condensation of 3,3'-diazido dephenyl sulfone with a,w-diallyl PEO. The true structure of the addition product was established by IR spectroscopy, ^{1}H NMR and elemental analysis. Some residual triazoline functions were oberved but no enamine were detected. The aziridine containing polymers were crosslinked by cationic polymerization leading to networks with low amount of extractible products. Thermal properties measurements were conducted on both networks precursors and crosslinked polymers. The introduction of imine bridges is strongly decreasing the chains mobility.

232. SYNTHESIS AND PROPERTIES OF NEW INTERPENETRATING NETWORKS BASED ON PEO. M. Grosz[a], S. Boileau[a], P. Guégan[b], H. Cheradame[b], A. Deshayes[c]; [a] Collège de France, 11 place M. Berthelot, 75 231 Paris Cédex 05, France; [b] L.P.C.B. CNRS, 2 rue Henri Dunant, BP 28, 94 320 Thiais, France; [c] F. T. CNET, 38 av. du Général Leclerc, 92 131 Issy-les-Moulineaux Cédex, France.

New polymeric systems based on poly(ethylene oxide) (PEO) polysiloxane semi-interpenetrating and interpenetrating networks (IPNs) were prepared according to the sequential in-situ method. Polysiloxane bearing PEO grafts and room temperature crosslinkable -Si(OEt)$_3$ side groups were mixed with poly(ethylene glycol) dimethacrylate (PEGDM) and AIBN. After the formation of the first network at r. t., the crosslinking of PEGDM was performed by raising the temperature up to 60°C. Semi IPNs were obtained by forming PEO networks in the presence of polysiloxanes bearing only PEO grafts. Thermal properties and conductivity of those systems containing lithium bistrifluoromethanesulfonimide were studied and compared to those of PEO networks with siloxane crosslinks.

233. AN AMPHIPHILIC APPROACH FOR PREPARING HOMOPOLYROTAXANES OF POLY(ETHYLENE OXIDE). Coleen Pugh, Frank A. Brandys and Jin-Young Bae, Department of Chemistry, Macromolecular Science and Engineering Center, University of Michigan, Ann Arbor, Michigan 48109-1055

A new and potentially highly efficient statistical method for preparing polyrotaxanes which lack an enthalpic driving force for threading is described. The approach is based on the ability of amphiphilic macrocrown ethers, in which a hydrophobic alkyl tail is attached to a hydrophilic macrocrown ether, to form reversed micelles/aggregates in hydrocarbon solvents which selectively solvate the hydrophobic tails. Light scattering and NMR demonstrate that micellar aggregation occurs when small amounts of water are added to toluene and benzene solutions of the amphiphile; many of the systems form lamellar mesophases. When a hydrophilic poly(ethylene oxide) derivative is then added to such an organized solution, it is forced to the interior of the micelle where it threads the crown ethers.

234. INVESTIGATING THE MOLECULAR MECHANISM OF PROTEIN INTERACTIONS WITH POLYETHYLENE GLYCOL (PEG). Shailesh R. Sheth and Deborah E. Leckband, Department of Chemical Engineering, University of Illinois at Urbana-Champaign, Urbana, IL 61801.

Performance of biomedical devices incorporating end-grafted PEG is presently handicapped by a lack of fundamental insight into the actual molecular mechanism governing PEG-protein interactions. We have used the Surface Forces Apparatus (SFA) to directly measure the range and magnitude of the molecular forces between end-grafted PEG and tethered streptavidin monolayers in aqueous electrolyte solutions. The effect of the surface density of PEG, its molecular weight, and temperature on the range and magnitude of forces were systematically investigated. Our results reveal for the first time that structural attributes of PEG in solution may play a crucial role in determining these molecular interactions. Indeed, repeated PEG (MW 2000, 25 °C) -protein contacts in these studies transformed the initially all-repulsive interaction profile into a weakly attractive one at ca. 105 Å. This was accompanied by a thickening of the polymer steric barrier consistent with the new more extended state of PEG. These changes are attributed to conformational changes in the polymer induced by close interactions with the protein layer.

235. Improved Solubility and Dissolution Characteristics of Poorly-Water Soluble Drugs with High Molecular Weight Polyethylene Oxides. D. H. Donabedian, E. M. Clark and I. M. Braddock, Research and Development Center Specialty Polymers and Products, Union Carbide Corporation, Bound Brook, NJ 08805.

Many hydrophobic drugs have limited solubility and dissolution rates due to their low water solubility. Solubility, and thus bioavailability becomes less due to this slow absorption. The use of higher molecular weight polyethylene oxides on improving the solubility and dissolution characteristics of poorly water soluble drugs such as ibuprofen and tolbutamide has been investigated. For ibuprofen based systems, a ten fold increase in equilibrium solubility was determined for polyethylene oxides with an average MW of 100,000 g/mol at 5.0 % (w/v) concentrations. For tolbutamide, a five fold increase in solubility was determined for polyethylene oxides with an average MW of 300,000 g/mol at 2.0 % (w/v) concentrations. The dissolution profile for both of these drugs from polyethylene oxide matrix tablets showed a sustained release profile and enhanced solubility as a function of molecular weight. Other commonly used excipients did not enhance ibuprofen solubility.

236. SYNTHESIS AND CHARACTERIZATION OF ONE-ENDED FLUOROCARBON DERIVATIVE OF POLY(ETHYLENE GLYCOL)S
Huashi Zhang, Jing Pan and Thieo E. Hogen-Esch, Loker hydrocarbon Research Institute, University of Southern California, Los Angeles, California 90089-1661,

One-ended fluorocarbon functionalized poly(ethylene glycol) derivatives with different PEG MW and different length of R_F end group were synthesized, and narrow MW distribution and quantitatively degree of functionalization ($\supseteq$ 90%) were achieved as shown by GPC and ^{19}F NMR. At 10^{-5} to 10^{-3} M concentration region, the equilibrium between the dissociated form and the associated form was observed by ^{19}F NMR for 5K and 10K MW of the same C_6F_{13} end group of PEG derivatives (4a and 4a'). The association number (n) is close to 2 for both 4a and 4a', which indicates the dimeric association structure for the C_6F_{13} end group of PEG derivatives at low concentration region. The association constant (K) of 4a and 4a' are 1.8×10^4 M^{-1} and 780 M^{-1}, respectively, indicating the hydrophobic association formation is favor by smaller MW of PEG derivative. The addition NaCl salt also shifts the equilibrium towards the association form as shown by ^{19}F NMR.

237. **NOVEL NETWORKS AND GELS CONTAINING INCREASED AMOUNTS OF GRAFTED AND CROSSLINKED POLY(ETHYLENE GLYCOL)**
A.M. Lowman, T.D. Dziubla, and N.A. Peppas,
School of Chemical Engineering, Purdue University, West Lafayette, IN 47907-1283

Poly(ethylene glycol) (PEG)-rich hydrogels containing PEG grafts and bridges were prepared by free-radical solution polymerization of mono and difunctional PEG macromonomers. The difunctional PEG macromonomer served as the crosslinking agent as well as providing the network with a PEG backbone. The addition of monofunctional PEG macromonomers reduced the degree of crosslinking in the system and provided grafted chains of free PEG. An analysis of the swelling behavior of these gels was performed to examine the effect of comonomer feed ratio on the properties of the network. Additionally, solute transport experiments were performed. The materials containing the smallest fraction of difunctional PEG macromonomer swelled the most rapidly and to the greatest extent. Similarly, solute diffusion coefficients in the gels increased as the amount of monofunctional monomer increased.

238. USING CAPILLARY ELECTROPHORESIS AND MALDI-TOF MS TO CHARACTERIZE POLY(ETHYLENE GLYCOL)-PROTEIN CONJUGATES. Michael J. Roberts and J. Milton Harris, Materials Science Program, University of Alabama in Huntsville, Huntsville, AL 35899

Proteins of therapeutic interest are often immunogenic, susceptible to proteolysis and rapidly cleared through the kidney. These problems can be overcome by conjugating the protein with PEG. Determining the extent of covalent modification of therapeutic proteins with PEG is important in understanding the therapeutic efficacy of the conjugates, and it is important in demonstrating the reproducibility of conjugate preparations. We present the development of a PEG capillary coating for improved separation of individual "PEGmers" by capillary electrophoresis (CE). These results are compared with those from MALDI. The ability of CE and MALDI to measure polydispersity of PEG and PEG-protein conjugates is also addressed.

239. HYDROPHILIC MODIFICATION OF PET BY PET/PEG BLOCK COPOLYMER. C. G. Cho, K. L. Choi, S. S. Hwang, and S. W. Woo, Department of Textile Engineering, Hanyang University, Seoul 133-791, Korea.

Poly(ethylene terephthalate) (PET) is one of the most important industrial polymers with good mechanical and thermal properties. But due to its high crystallinity and poor hydrophilicity, it has difficulties in controlling bulk and surface properties such as dyeability, antistatics, and water absorptivity. In this research we prepared block copolymers of PET and poly(ethylene glycol) (PEG) with various molecular weights and compositions. Efficiency of block copolymer formation was estimated with increasing PEG molecular weight. When the molecular weight of the PEG was between 200 and 4,000, the efficiency of the block copolymer formation was stayed almost same, but decreased as that of PEG increased to 10,000. Their thermal properties and water contact angle were measured, and tried to be correlated with PEG molecular weight, in order to find out some key factors in controlling hydrophilicity of PET.

240. POLYMER-SALT COMPLEXES AS SOLID ELECTROLYTES
V.Saketharaghavan and Kathryn Uhrich*
Department of Chemistry, Rutgers University,
Piscataway, NJ 08855

Hyperbranched polymeric systems are proposed for use as solid electrolytes in electrochemical devices such as high density batteries. The hyperbranched structures consist of mucic acid, alkyl chains, 1,1,1-tris(hydroxyphenyl)ethane as the multifunctional core molecule and triethylene glycol and polyethylene glycol amine chains . The synthesized polymers are treated with $LiClO_4$ in a mutual solvent to form a polymer-salt complex. The polymer-salt material is cast as a film and the ionic conductivity measured as a function of temperature and $LiClO_4$ concentration. The polymers are thoroughly characterized by a variety of techniques such as nuclear magnetic resonance (NMR) spectroscopy, infrared (IR) spectroscopy, gel permeation chromatography (GPC), differential scanning calorimetry (DSC) and thermogravimetric analysis (TGA).

241. **HYDROPHOBICALLY TERMINATED POLYOXYETHYLENES FOR THE DECORTION OF LIPID MEMBRANES**
Werner A. Goedel, Max-Planck Institut f. Colloids & Surfaces, Hs.9.9, Rudower Chaussee 5, 12489 Berlin, Germany, e.mail:goedel@mpikg.fta-berlin.de

Polyoxyethylenes hydrophobically modified at with a cholesteryl end group have been synthesized and characterized via HPLC and MALDI-ToF mass spectroscopy. The hydrophobic end group is attatched to the polymer via a hydrolyzable link. We study the interactions of the polymers with lipid membranes, especially the effect of hydrolyzing the link between polymer and hydrophobic head group.

242. $LiClO_4$ SALT COMPLEXES OF UNSATURATED ETHYLENE OXIDE-SEGMENTED POLYMERS. Jun Qiao and Gregory L. Baker*, Department of Chemistry and Center for Fundamental Materials Research, Michigan State University, East Lansing, MI 48824

Polymer electrolytes were prepared from unsaturated ethylene oxide-segmented polymers and lithium perchlorate. The thermal behavior of the electrolytes was studied by differential scanning calorimetry (DSC) measurements. It was found that the glass transition temperature (T_g) of the electrolytes increases with increasing salt content for each polymer, and that the polymer crystallinity decreases with increasing salt content. Typical ionic conductivities for these complexes were $1-2 \times 10^{-5}$ S/cm, with polymers with higher ethylene oxide fraction having higher conductivity. It was shown that the temperature-dependent conductivities can be fit by the Vogel-Tammann-Fulcher (VTF) equation.

243. UNSATURATED ETHYLENE OXDIE-SEGMENTED POLYMERS: PREPARATION AND THEIR PROPERTIES. Jun Qiao and Gregory L. Baker*, Department of Chemistry and Center for Fundamental Materials Research, Michigan State University, East Lansing, MI 48824

A novel series of unsaturated ethylene oxide-segmented polymers were prepared by acyclic diene metathesis (ADMET) chemistry. The polymers have high molecular weights (50-100,000) and are thermally stable to 300 °C. DSC studies indicate that the T_gs range from -70 to -80 °C, and increase with ethylene oxide content. The polymers form several crystalline phases, with the highest melting phase showing an unusual odd-even effect.

244. SYNTHESIS AND CHARACTERIZATION OF ETHYLENE OXIDE SUBSTITUTED POLY(*p*-PHENYLENE)S. Cory J. Ruud and Gregory L. Baker, Department of Chemistry and the Center for Fundamental Materials Research, Michigan State University, East Lansing, Michigan 48824.

Polyphenylenes are among the most thermally stable and intractable polymers known. Typically, polyphenylenes are obtained as yellow powders which are completely insoluble. Despite these limitations, polyphenylenes are attractive for applications as electrical conductors, dielectric layers and luminescent materials. Recent work has shown that adding substituents to the polyphenylene backbone increases solubility and processibility. We have prepared a series of soluble ethylene oxide substituted poly(*p*-phenylene)s using transition metal catalyzed reactions and have studied the physical properties of these polymers. Polyphenylenes with short side chains were pale yellow powders, but high molecular weight samples with longer side chains were tacky solids. These polymers show high thermal stability and some exhibit liquid crystalline behavior.

245.
CRITICAL FLUCTUATIONS IN A BINARY MIXTURE OF POLY(ETHYLENE GLYCOL) AND POLY(PROPYLENE GLYCOL) STUDIED BY ULTRASONIC AND LIGHT SCATTERING EXPERIMENTS. I. Alig, S. Hoffmann, W. Mayer, Deutsches Kunststoff-Institut, Schlossgartenstraße 6, D-64289 Darmstadt, Germany; G. Meier, Max-Panck-Institut für Polymerforschung and IFF Jülich, D-5245 Jülich, Germany

The critical mixture of polyethylene glycol (M=400 g/mol^{-1}) and polypropylene glycol (M=1000 g/mol^{-1}) was investigated by light scattering and ultrasonic experiments in the homogeneous one-phase region. The ultrasonic measurements are interpreted by the dynamic scaling theory for binary critical mixtures. The low mutual diffusion coefficient of the mixture allows to study the high frequency behavior of critical ultrasound attenuation. The mutual diffusion coefficient and the correlation length of the composition fluctuation are estimated from light scattering experiments and compared to the ultrasonic results. To get further insight into the physical mechanism of critical ultrasound behavior of a polymer blend, measurements of the pressure dependence of the critical temperature dT_c/dP were performed.

246. Controlled Chain Conformation and Rigidity of Polymer Backbones Jacketed with Dendritic Coats. Virgil Percec, The W.M. Keck Laboratories for Organic Synthesis, Department of Macromolecular Science, Case Western Reserve University, Cleveland, OH 44106-7202

The design of conformationally flexible flat tapered and conical dendritic building blocks which self-assemble into cylindrical and respectively spherical supramolecular dendrimers will be discussed. When polymer backbones are jacketed with dendritic coats, their conformation and rigidity is manipulated by the supramolecular structure generated by the assembly of their dendritic side groups. Examples of systems inducing random-coil, helical and extended chain conformations and their interchange via different mechanisms will be presented.

247.
WELL DEFINED POLYPEPTIDES VIA COORDINATION POLYMERIZATION. Timothy J. Deming, Materials and Chemistry Departments, University of California, Santa Barbara, California 93106

We have designed transition metal based initiators for the controlled polymerization of α-amino acid-N-carboxyanhydrides (NCAs). Coordination complexes which contain primary amine initiating groups were synthesized and screened for polymerization activity using γ-benzyl-L-glutamateNCA and ε-carbobenzyloxy-L-lysineNCA monomers. Through variation of ligand substituents, polymerization solvent, and transition metal, polymerization systems were developed which were able to prepare poly(γ-benzyl-L-glutamate) in quantitative yield, with narrow molecular weight distribution, and with control over molecular weight. Requirements for successful polymerization will be discussed, as well as potential for polypeptide end-group functionalization and block copolymer formation.

248. SYNTHESIS AND ASSEMBLY OF MONODISPERSE MOLECULAR RODS
Michael S. Yu and David A. Tirrell, Department of Polymer Science and Engineering, University of Massachusetts, Amherst, MA 01003 USA

This lecture will describe the synthesis of monodisperse derivatives of poly(γ-benzyl α,L-glutamate) (PBLG), and the assembly of such polymers in solution and in the solid state. Of special interest is the possibility that uniform PBLGs can be induced to form smectic liquid crystal phases in solution, and that those phases can be processed into solid films that preserve smectic-like order. Textural and x-ray scattering evidence for the anticipated structures will be presented.

249. **ORGANOMETALLIC CATALYSTS IN THE PREPARATION OF NANOSTRUCTURED MATERIALS.** Thomas A.P. Seery, Dale L. Huber, Polymer Program and Department of Chemistry, University of Connecticut, Storrs, CT 06269;

Recent advances in the field of homogeneous polymerizations using organometallic catalysts have set the stage for advances in the preparation of functional surfaces. Linear polymers can be produced via surface initiated polymerizations (SIP) using organometallic catalysts. Initiator ligands to the catalyst provide covalent bonds from the substrate to the incipient polymer layer. New structures formed with this approach will provide great variation in bond density and molecular orientation at the interface. Organometallic catalysts provide synthetic access to novel materials and are currently the focus of much attention due to their precise control of polymer stereochemistry. Both flexible and rigid polymers have been prepared with unique properties of optical rotation, solubility, conductivity, electroluminescence, and liquid crystallinity. This work features both flat surfaces and particle models where the particle surfaces provide a testing ground for the surface chemistry.

250. MECHANISTIC STUDIES OF OLIGOESTER CYCLIZATION. Patricia Hubbard and William J. Brittain*, Department of Polymer Science, The University of Akron, Akron, OH 44325-3909.

We have studied the formation of cyclic oligoesters under conditions of kinetic and thermodynamic control. The reaction of acid chlorides with alcohols in the presence of tertiary amines provides high cyclic yields under kinetic control. The role of the amine has been probed using stopped-flow spectroscopy. We have used rate measurements to help clarify whether the amine serves as a basic or nucleophilic catalyst. The depolymerization of polymer has been studied as a method for cyclic formation under thermodynamic control. Of particular interest is the influence of the depolymerization conditions on the ring size distribution. Experimental data has been compared to modeling results where rotational isomeric state theory was used to predict ring size distribution.

251. Reflective Liquid Crystal Displays and Materials. **Haiji Yuan**, dpiX, A Xerox Company 3406 Hillview Ave. Palo Alto, CA 94304

State of art reflective liquid crystal displays are reviewed. The operating mechanisms for different types of reflective LCDs are described. The challenges for the key organic materials, such as liquid crystals, polymers, photoinitiators, dyes, chiral agents, and polyimides that are used for these displays are outlined.

252. **POLYMER STABILIZED CHOLESTERIC LIQUID CRYSTALS AND DISPLAY APPLICATIONS.** Liang-Chy Chien,* Frédéric Vicentini, Yuhui Lin and Uwe Müller, *Chemical Physics, Liquid Crystal Institute and NSF ALCOM Center, Kent State University, Kent, Ohio, 44242, USA.*

We prepared a polymerizable tunable chiral material (TCM), whose chirality can be rendered by light, for the color patterning of a reflective cholesteric liquid crystal display (RC-LCD). The color patterning is achieved by varying the dosage of UV light irradiation, the pitch length of various regions can be tuned to reflect different colors, "thereby creating different color pixels in liquid crystal materials itself." Pixels can be tuned over the entire visible range. In a RC-LCD, the patterned colors were stabilized with a polymer network. In case of using a non-polymerizable TCM, we formed polymer barrier walls, resulting from the phase separation upon the masked photopolymerization, to provide a better defined color pixels.

POLY

253. INTERPLAY OF MORPHOLOGY AND ANCHORING ON PDLC LIGHT SCATTERING
Paul S. Drzaic, Raychem Corporation, Menlo Park CA 94025

Emulsion-based polymer dispersed liquid crystal PDLC films typically show strong light scattering at zero fields, the basis of a number of useful electro-optical effects. There is one exception to this situation: with strong perpendicular anchoring, emulsion-based films become substantially transparent. This effect arises from the flattened nature of the cavities in emulsion-type PDLC films, which tend to align the liquid crystal along the film normal when the liquid crystal anchoring is perpendicular. In phase separation-type films at medium droplet densities, in contrast, perpendicular alignment can still produce strong light scattering, as the alignment is more randomized. Some subsistent effects in the anchoring properties of acid-doped PDLC films with polyvinylalcohol matrices are also discussed.

254.

SYNTHESIS OF CHIRAL NETWORKS FOR POLYMER STABILIZED CHOLESTERIC TEXTURE (PSCT) DISPLAYS. **James A. Rego**, Paul A. Cahill, Electronic and Optical Materials Department, Sandia National Laboratories, Albuquerque, NM 87185-1405.

Mono- and di-functional acrylate twist agents based on a chiral phenyl ethanediol core have been synthesized and polymerized to form PSCT devices in which the sole source of the chirality is the polymer network. The helical twisting power of the networks retains that of the monomers which range from 14.0 to 15.1 µm-1. With increasing crosslink density the reflection band blue-shifts and broadens, and with difunctional monomer, a broadband (white) reflection is observed, but bistability is lost. This work was suported by the United States Department of Energy under Contract DE-AC04-94AL85000. Sandia is a multiprogram laboratory operated by Sandia Corporation, a Lockheed Martin Company, for the United States Department of Energy.

255. POLYMERIZATION OF POLYMER STABILIZED FERROELECTRIC LIQUID CRYSTALS FORMED FROM BRANCHED LIQUID CRYSTALLINE DIMETHACRYLATES. C. A. Guymon and C. N. Bowman, Department of Chemical Engineering, University of Colorado, Boulder, CO 80309-0424; R. Shao, Department of Physics, University of Colorado, Boulder, CO 80309-0390; D. Hoelter and H. Frey, Institut fuer Makromolekulare Chemie und Freiburger Materialsforschungszentrum der Albert-Ludwigs-Universitaet, Stefan-Meier-Str. 21/31, 79104 Freiburg (FRG)

To produce novel phase behavior and and properties, a variety of polymer/liquid crystal have recently been developed. Systems of particular interest include ferroelectric liquid crystals (FLCs) mechanically stabilized by a small amount of polymer. This study examines the effects of adding a branched liquid crystalline dimethacrylate to an FLC both before and after polymerization. The monomer acts to increase the temperature range of the smectic A phase and at relatively low concentrations this range is over ten times that observed in the FLC. After polymerization, the phase transition temperatures return to values close to those of the FLC, but considerable birefringence is observed at temperatures well above the isotropic transition. Polymerization behavior is significantly different than found in other systems and at appropriate temperatures, polymerization driven endothermic phase transitions may be observed.

256. LINEAR OPTICAL ANISOTROPY IN AROMATIC POLYIMIDE FILMS AND ITS APPLICATIONS AS NEGATIVE BIREFRINGENT COMPENSATORS IN LIQUID CRYSTAL DISPLAYS, Fuming Li, Edward P. Savitski, Jyh-Chien Chen, Yeocheol Yoon, Frank W. Harris and Stephen Z. D. Cheng, Maurice Morton Institute and Department of Polymer Science, The University of Akron, Akron, Ohio 44325-3909

Samples of soluble aromatic polyimides of varying chemical structure and molecular weight were synthesized in refluxing *m*-cresol at elevated temperatures through a one-step polymerization route.

Modifications of the dianhydride and diamine monomers were designed to prepare aromatic polyimides having new architectures based on the requirements of liquid crystal display applications. The solution-cast films exhibit linear optical anisotropy, which is called *uniaxial negative birefringence* (*UNB*) and is characterized by the presence of a larger refractive index along the in-plane direction than in the out-of-plane direction. It is found that the *UNB* is critically associated with the backbone linearity, rigidity, and intrinsic polarizability as well as film thickness and molecular weight of the polyimides.

257. SYNTHESIS OF ANISOTROPIC POLYMER FILMS FOR DISPLAY APPLICATIONS Chain-Shu Hsu and Hwai-Len Chen, Department of Applied Chemistry, National Chiao Tung University, Hsinchu, Taiwan 30050

Two liquid crystalline diacrylate and diepoxide with a very wide temperature range of mesophase were synthesized and characterized. The liquid crystal (LC) mixtures of diacrylates or diepoxides were photo-polymerized after macroscopic orientation of monomers via interaction with a unidirectionally rubbed polyimide substrate. The obtained LC networks show optical anisotropic property. The birefringence of these networks depends on polymerization temperature. Their birefringences decrease as the polymerization temperature increases. Addition of the non-mesogenic diluent into the LC monomer also lead to the decrease of the birefringence of the LC networks. As the concentration of the non-mesogenic diluent increases, it not only leads to a lower birefringence but also exhibits a weaker dispersion in wavelength for the LC networks obtained. The prepared LC networks can be used as retardation films for liquid crystal display.

258. LIQUID CRYSTAL ALIGNMENT LAYER POLYIMIDES AND NEMATIC LC PRETILT ANGLES. K.-W. Lee, S.-H. Paek[#], A. Lien, C. Durning[#] and H. Fukuro[##]. IBM T.J. Watson Research Center, PO Box 218, Yorktown Hts., NY 10598. [#]Dept of Chemical Engineering, Columbia University, NY, NY 10027. [##]Nissan Chemical Central Research Lab., Funabashi, Ciba 274, Japan

Eight polyimides containing alicyclic dianhydride groups and (or) long alkyl side chains were designed and synthesized to study control of LC pretilt angles which are critical to electro-optical properties of modern thin film transistor-liquid crystal displays. Several commercially available polyimides such as Nissan SE5291, JSR AL1054, DuPont PMDA-ODA) were also employed for this study. An LC pretilt angle is affected by both electronic interaction and steric repulsion between LC molecules and the alignment layer polymer surface. The stronger the electronic interaction, the smaller the LC pretilt angle. The greater the steric repulsion, the larger the LC pretilt angle. The degree of poly(amic acid) imidization and the rubbing force also affect the LC pretilt angles while the polyimide crystallinity does not.

259. MICROPHASE STABILIZED FERROELECTRIC LIQULID CRYSTAL (MSFLC): BISTABLE SWITCHING OF FLC-COIL DIBLOCK COPOLYMER
G. Mao[1], J. Wang[1], C. K. Ober[1]*, M. J. O'Rourke[2], E. L. Thomas[2], M. Brehmer[3], and R. Zentel[3]
1. Department of Materials Science and Engineering, Cornell University, Ithaca, NY 14853;
2. Department of Materials Science and Engineering, Massachusetts Institute of Technology, Cambridge, MA 02139;
3. Institut für Organische Chemie, Universität Mainz, Becherweg 18-20, 55099 Mainz, Germany

Ferroelectric liquid crystals (FLC) have drawn great attention for displays due to their microsecond bistable switching times. Poor mechanical properties, lack of shock resistance due to thin cell thicknesses, and stabilization of the bookshelf arrangement are major challenges. We proposed and confirmed that block copolymers with chiral S_c^* mesophases offer excellent properties for a microphase stabilized FLC (MSFLC) cell to produce untwisting of the supermolecular pitch of the S_c^* structure. Orientation of the mesogens can be stabilized via the alignment of the block microdomains. By employing polymer analogous reactions, well defined block copolymers with S_c^* mesophases were synthesized. After shearing the block copolymer, bistable switching behavior was observed. TEM was used to confirm lamellar microdomain morphology. Currently, we are extending our studies of the effect of microdomain morphology and molecular weight on switching properties.

260. **THEORETICAL ASPECTS OF FREE RADICAL REACTIONS.** Leo Radom, Ming Wah Wong, Addy Pross, Johan P. A. Heuts and Robert G. Gilbert, Research School of Chemistry, Australian National University, Canberra, ACT 0200, Australia, Department of Chemistry, University of Queensland, Brisbane, Queensland 4072, Australia, School of Chemistry, University of Sydney, Sydney, NSW 2006, Australia, and Department of Polymer Science, School of Chemical Engineering and Industrial Chemistry, University of New South Wales, Sydney, NSW 2052, Australia

The use of *ab initio* molecular orbital calculations to study the reactions of free radicals will be reviewed. Particular emphasis will be placed on our recent studies of radical addition to alkenes, a reaction that is of basic interest as a fundamental bond-forming process and of practical importance as the central propagation reaction in many polymer processes. The studies involve three stages: (a) *assessment* of the various possible theoretical procedures, (b) *application* of a selected procedure to specific chemical problems, and (c) *qualitative rationalization* of the results. Both published and unpublished work will be discussed.

261. **ABSOLUTE KINETICS OF FREE RADICAL REACTIONS.** Janusz Lusztyk, Steacie Institute for Molecular Sciences, National Research Council of Canada, 100 Sussex Drive, Ottawa, Ontario K1A 0R6, Canada.

Applications of nanosecond laser flash photolysis with either UV-visible or infrared detection to measurements of kinetics of organic free radicals will be presented. The kinetics of radical species derived from such radical initiators as dialkyl, diacyl peroxides and peroxydicarbonates will be discussed including their unimolecular fragmentation processes and radical-molecule reactions. Solvent effects on the kinetics of these reactive intermediates will be quantitatively described. Kinetic studies of fluorinated alkyl radicals, particularly of their addition reactions will also be presented providing the first relatively comprehensive set of data for these strongly electrophilic carbon-centered species.

262. STRUCTURE-REACTIVITY CORRELATIONS FOR RADICAL REACTIONS WITH ALKENES. B. Giese, Department of Chemistry, University of Basel, St. Johanns Ring 19, CH-4056 Basel, Switzerland.

In recent years the factors that govern the regioselectivity, chemoselectivity, and stereoselectivity of radical reactions have been elucidated in detail using product studies, kinetic and thermodynamic data, spectroscopic measurements as well as quantum chemical calculations. It has turned out that polar and steric effects play a major role, whereas the stabilities of the radicals formed during the addition reactions are often of minor importance. Although these data are derived from reactions of small molecules, their correlation with parameters and effects of radical polymerization is possible in many instances. Thus, many effects of radical polymerization (penultimate unit effect, copolymerization parameters) can be, at least in part, explained by the rules of the radical addition reactions to alkenes.

263. CASCADE RADICAL REACTIONS IN ORGANIC SYNTHESIS: AN OVERVIEW. Dennis P. Curran, Department of Chemistry, University of Pittsburgh, Pittsburgh, PA 15260, USA

This lecture will provide an overview of the past decade of developments of cascade radical reactions in organic synthesis that is targeted towards a polymer audience. Many of the principles in play in today's small molecule cascades emerged from polymer chemistry and were modified accordingly. Indeed, the lecture might instead be titled "How to Start a Radical Polymerization and Then Stop It Before It Really Gets Going". The similarities between polymer chemistry and small molecule synthesis will be apparent. The key difference is that the goal of a radical polymerization is to get every radical to do the same thing while the goal of a cascade radical reaction is to get every radical to do a different thing. Prototypical sequences of inter- and intramolecular radical reactions will be discussed along with the methods that are used to conduct them.

264. SUPRAMOLECULAR ARYL TRANSITION METAL CATALYSTS; SELECTIVE PRODUCT FORMATION WITH ORGANOMETALLIC RADICALS, Gerard van Koten, David M. Grove, Johann T.B.H. Jastrzebski, Department of Metal-Mediated Synthesis, Utrecht University, 3584 CH Utrecht, The Netherlands.

The chemistry of organometallic radicals is a current research topic of the Utrecht group. The first topic concerns aryldiamine d^8-metal compounds $[MX_n(C_6H_2(CH_2NMe_2)_2-2,6-R-4]$ of which the species with $M=Ni^{II}$ and n=1 is an active homogeneous catalyst for the Kharasch addition reaction of halocarbons (e.g. CCl_4) to alkenes. Instead of polymerization selective 1:1 addition occurs. Some mechanistic aspects involving one electron transfer reaction steps as well as the results of a dendritic version of this catalyst (Mol.w.t. 7.032, 12 Ni-sites, anchoring via R) in a membrane reactor will be discussed. A second topic is the chemistry of the persistent alkylzinc-α-diimine radicals derived from reactions of dialkylzincs with RN=CHCH=NR-derivatives. Selective formation of radical and radical anionic species and their role in the selective group transfer reactions will be discussed.

265. MULTIDIMENSIONAL NMR: PAST, PRESENT, AND FUTURE.
R. Andrew Byrd, Macromolecular NMR Section, ABL-Basic Research Program, NCI-FCRDC, Frederick, MD, 21702-1201

NMR has provided one of the most fruitful spectroscopies for the characterization of chemical, conformational, and dynamical structure of molecules. It's applications have circumscribed chemistry and biology and range from liquid to solid state. A large part of this breadth and impact has derived from the development of techniques to both separate and encode spectral information in more than one dimension, leading to both greater resolution and selectivity. The extension of spectral information into two-, three-, or four-dimensions relies on Fourier methods combined with the design of pulse sequences which isolate and correlate selected spin-spin interactions. This lecture will review some of the basic concepts for information encoding, the separation of interactions, and the design of correlations among interactions which lead to the extremely powerful analysis of the chemical and conformational structure of large molecules. The growth of NMR will clearly continue with further development of these multidimensional methods and instrumentation.
Research sponsored by the National Cancer Institute, DHHS, under contract with ABL.

266. NMR IMAGING OF POLYMERS, D. G. Cory, NW14-4111, 150 Albany St., Department of Nuclear Engineering, Massachusetts Institute of Technology, Cambridge, MA 02139

This tutorial will provide a brief introduction to the methods, limitations and information content of NMR images of synthetic polymers. The background will include the length scales over which spatial information may be determined, how magnetization grating are created and used to record an image, and a linear model of the imaging process. These provide a natural setting to discuss both resolution limits and distortions. One of the most important constraints is the limited time available to form a magnetization grating and this has lead to the development of coherent averaging methods. In addition, the information content of an image will be explored including how to develop contrast, and how to extract statistical information. Examples will be provided from high resolution studies of elastomers and engineering plastics.

267. **MULTIDIMENSIONAL SOLID STATE NMR OF POLYMERS**
H.W. Spiess,
Max-Planck Institut für Polymerforschung, Postfach 3148, D-55021 Mainz, Germany

Two-dimensional solid-state NMR [1] emerges as an extremely selective and highly versatile tool for studying structure and dynamics of synthetic macromolecules. A general introduction to solid state NMR with various experimental examples including the characterization of chain conformation and molecular dynamics of amorphous polymers, chain alignment in oriented polymers and interface structures in core-shell systems will be given. In addition, new developments in high resolution multiple quantum NMR of solids will be described. This new technique allows structural elucidation with atomistic resolution including hydrogen positions and the determination of the anisotropy of chain motions by probing residual dipolar couplings between adjacent groups along a macromolecule. Experimental results on polymer melts and crosslinked elastomers will be presented.

[1] K. Schmidt-Rohr and H.W.Spiess "Multidimensional Solid-State NMR and Polymers", Academic Press, London (1994).

268. **Creating New Well-Defined Polymer Structures using Controlled / Living Radical Polymerization.** Timothy E. Patten* and Raj Virk, Department of Chemistry, University of California at Davis, Davis, CA 95616.

The recent development of controlled / living radical polymerizations has important implications for the preparation of new polymers and materials. For the first time radical intermediates, with their known reactivity patterns, can be used in the synthesis of (co)polymers with well-defined endgroups, molecular weights, and molecular weight distributions. The tolerance of radicals towards many reaction conditions (polar functional groups, water, etc.) normally detrimental to anionic, cationic and many transition-metal catalyzed polymerizations and the range of monomers that can be homo- and copolymerized using such a technique are advantageous properties that can now be exploited in controlled polymer synthesis. We have incorporated these advantages into strategies for the synthesis of new types of well-defined polymers and copolymers using controlled / living radical polymerizations. Aspects of these strategies will be discussed, as well as experiments elucidating the roles of copper (I/II) complexes in regulatory mechanisms and side reactions in atom transfer radical polymerization (ATRP).

269. **Converting Non-living Systems to "Living"/Controlled Polymerization Systems or Learning "Tricks" which Improve Control of Molecular Weights, Polydispersities and Functionalities.**
Krzysztof Matyjaszewski, Department of Chemistry, Carnegie Mellon University, 4400 Fifth Avenue, Pittsburgh, PA 15213.

Synthesis of well-defined polymers requires low contribution of chain breaking reactions (transfer and termination), fast initiation and fast exchange between propagating species of different lifetimes and reactivities. If these requirements are not met, then polymers with poor control of end-functionalities, high polydispersities, and degrees of polymerization either higher or lower than predetermined by $DP_n = \Delta[M]/[I]_0$ are obtained. Several "tricks" how to convert such non-living systems to well-defined systems will be presented together with a rational explanation of the chemistry involved. This will include fixing various imperfections in initiation, reducing contributions of transfer and termination and accelerating exchange in carbocationic and radical polymerization.

270. **Synthesis of Organic Molecules and Materials Using Ruthenium Carbene Complexes.** R.H. Grubbs, Helen Blackwell, Bobby Maughon, Eric Dias, Division of Chemistry and Chemical Engineering, California Institute of Technology, Pasadena, California 91125 USA

Over the past eight years, early transition metal catalysts for the ring opening metathesis polymerization of cyclic olefins have been developed. These catalysts are simple organometallic complexes containing metal carbon multiple bonds that in most cases polymerize olefins by a living process. These catalysts have been used to prepare a family of near monodispersed and structurally homogeneous polymers. A series of group VIII ROMP catalysts that allow a wide range of functionality to be incorporated into the polymer side chains have been prepared. The most important members of this family of complexes are the bisphosphine-dihalorothenium carbene complexes. These same catalysts can also be used in the synthesis of fine chemicals as well as block copolymers. The ability of these catalysts to function in the presence of water allows the emulsion polymerization of cyclic olefins and the polymerization of highly functionalized monomers. A new class of water soluble phosphines have been prepared which yield active water soluble catalysts. These catalysts have also been used in the modification of polypeptides and the synthesis of polycyclic olefins.

271.

FROM MONOMERS TO POLYMERS THROUGH CATALYSIS, Auerbach, D.; Baldwin, K.; Boffa, L.; Brainerd, A.; Cafmeyer, J.; Cederstav, A.; Charles, S.; Cramer, S.; Davies, C.; David, J.; Deming, T.; Dunbar, M.; Ellsworth, M.; Ellzey, K.; Goh, C.; Goodson, F.; Goodwin, A.; Grubbs, B.; Hagen, E.; Hoff, S.; Jiang, R.; Jin, X.; Jones, N.; Kepley, L.; Khatri, C.; Kwark, Y.-G.; Landenberger, P.; Larson, R.; Lim, A. R.; Loos, K.; Mann, G.; Matzger, A.; Maxin, Meyer, T.; G.; Nakano, M.; Novak, B.; Perrott, M.; Patten, T.; Rosenau, B.; Safir, A.; Schlitzer, D.; Schueneman G.; Seery, T.; Shibayama, K.; Seidel, S.; Stewart, J.; "Tom, the Postdoc"; Yu, J., Verrier, C.; Wallow, T. Department of Polymer Science & Engineering, and the Materials Research Science & Engineering Center, University of Massachusetts, Amherst, MA 01002.
There is not a single chemical reaction of consequence that is not catalyzed by some means. The catalyst can be as simple as a proton or as complex as an biological enzyme. We have had some interest in applying new catalytic approaches to the synthesis of new and old polymers. Selected examples will be discussed.

272. **IMPROVED STABILITY OF MOLECULAR ORGANIC EL DEVICES. H. Fujii**, New Materials Research Center, SANYO Electric Co.,Ltd., Osaka 573, Japan

This talk will discuss the influence of the emission site on the running durability of molecular organic electroluminescent devices. Fundamental device structure of MgIn / $BeBq_2$ / TPD / MTDATA / ITO was employed. When rubrene was doped into the TPD layer, the device emitted bright yellow light originated from rubrene with a maximum luminance of 60 600 cd/m^2. Half-decay time of the device from the initial luminance of ca. 500 cd/m^2 under constant direct current was 3554 h. On the other hand, when rubrene was doped into the $BeBq_2$ layer, the half-decay time was 110 h. The rubrene-TPD emitter was found to have high efficiency and improved stability. It was thus found that the emission site exerts a significant influence on the running durability.

273. TEMPERATURE AND ENVIRONMENTAL STABILITY STUDIES OF OLED's BASED ON SMALL MOLECULES. Z. D. Popovic, S. Xie, N. Hu, A. Hor, B. Ong, P. F. Smith, and M. L. Hair, Xerox Research Center of Canada, Mississauga, Ontario L5K 2L1

We investigated environmental and temperature stability of organic electroluminescent devices based on small molecules deposited by the vacuum evaporation method. As model systems OLED's with the following structures were used: ITO/TPD/AlQ/MgAg and ITO/CuPc/NPB/ AlQ/MgAg. The presence of water is well known to cause a rapid degradation of OLED's by formation of black spots. In TPD based cells we found evidence for TPD crystallization in the black spot areas. The presence of oxygen, on the other hand, is not found to significantly impact device life. In TPD devices rapid degradation is observed already at 40°C which also effects device shelf life. NPB based devices are much more stable. Accelerated aging at high currents (> 30 mA/cm^2) showed a superlinear dependence of cell life on the reciprocal driving current. To separate thermal and current induced degradation, cells were cooled during high current operation and maintained at room temperature. Accelerated aging results for different cell temperatures will also be discussed.

274. STUDY OF DEGRADATION MECHANISM OF ORGANIC ELECTROLUMINESCENCE (EL) DEVICES BY SCANNING FORCE MICROSCOPIES. M.Fujihira,* L.-M.Do, C.Ganzorig, A.Koike, and T. Kato, Department of Biomolecular Engineering, Tokyo Institute of Technology, 4259 Nagatsuta, Midori-ku, Yokohama 226, Japan

Degradation mechanism of layered organic molecular EL devices, i.e. ITO / TPD / Alq$_3$ / Al, has been studied with scanning force microscopies, such as AFM and friction force microscopy (FFM) together with other methods, such as Auger electron spectroscopy, fluorescence spectroscopy and microscopy, and ionization potential measurements. From observed results, we concluded that by Joule heating a small amount of Alq$_3$ diffused into the TPD layer which results in the drastic change in electronic properties of the hole transport layer (HTL), and thus the change in the EL characteristics. The effect of diffusion was confirmed by fabrication of organic EL cells with a mixed HTL of TPD and Alq$_3$. We found also that the organic EL cells with the mixed HTL has a longer lifetime than the conventional layered EL cells. The increase in the lifetime is attributed to the increase in the thermal stability of the mixed organic films. Much less morphological change of the mixed HTL than the pure TPD layer by heating was observed.

275. FAILURE MODES IN ORGANIC LIGHT-EMITTING DIODES, H. Antoniadis, M. R. Hueschen, J. McElvain, J. N. Miller, R. L. Moon, D. B. Roitman, and J. R. Sheats, Hewlett-Packard Labs, 3500 Deer Creek Rd, Palo Alto, CA 94304

A comprehensive study of the main failure mechanisms in vacuum vapor-deposited organic light-emitting diodes (LEDs) is presented. Three degradation modes were identified for a prototype bilayer ITO/TPD/Alq$_3$/Mg/Ag device: a) formation and growth of black non-emissive spots, b) abrupt ceasing of light emission associated with catastrophic failures caused by electrical shorts, and c) long term wearout associated with the decrease of quantum efficiency and luminance along with an increase in voltage, while the device is stressed under constant current. Characterization of post-stress devices by luminescent microscopy revealed that black spots result from delamination of the metal at the Alq$_3$/Mg interface which is initiated by pinholes on the cathode. Moreover the addition of a thin conductive polymer layer between the TPD and the ITO/glass substrate was found to suppress the leakage currents and eliminate the associated catastrophic failures.

276. POLYMERIC ANODES FOR REDUCED OXIDATION RATE IN ORGANIC LIGHT-EMITTING DIODES. J. Campbell Scott, IBM Research Division, IBM Almaden Research Center, 650 Harry Rd., San Jose, CA 95120-6099, Sue A. Carter, Physics Department, University of California, Santa Cruz, CA 95064, Siegfried Karg, Physikalisches Inst., Universitat Bayreuth, D-8580 Bayreuth, Germany, and Marie Angelopoulos, IBM Research Division, T.J. Watson Research Center, Yorktown Heights, NY 10598.

Polymer light-emitting diodes, based for example on MEH-PPV, have been shown to be susceptible to oxidative degradation, leading to loss of conjugation, reduced carrier mobility and higher the operating voltage. The formation of carbonyl species quenches luminescence. *In situ* FTIR reveals that indium-tin oxide can act as the source of oxygen. Confirmation of this mechanism of degradation and guidance for its elimination is provided by studies of MEH-PPV LEDs prepared with a variety of conducting polymer anodes, including structures where the polymer layer, for example polyaniline or polythiophene, is present between the ITO and the MEH-PPV. In addition to improvement in initial device characteristics, the rate of decay of the luminance decreases by two orders of magnitude. These data not only confirm that ITO is the source of oxygen, but also imply that oxidation is the result of an interfacial reaction and not merely the result of oxygen diffusion out of the ITO.

277. **AMORPHOUS MOLECULAR MATERIALS FOR ORGANIC ELECTROLUMINESCENT DEVICES.** Yasuhiko Shirota, *Department of Applied Chemistry, Faculty of Engineering, Osaka University, Yamadaoka, Suita, Osaka 565, Japan*

A novel amorphous molecular material, 5,5"-bis{4-[bis(4-methylphenyl)-amino]phenyl}-2,2':5',2"-terthiophene (BMA-3T), has been found to function as a yellow-emitting material in organic electroluminescent (EL) devices. Both the single layer EL device using BMA-3T alone and the double layer of tris(8-quinolinolato)-aluminum sandwiched between indium-tin-oxide (ITO) and an alloy of magnesium and silver (ca. 10:1) electrodes emitted a bright yellow light resulting from BMA-3T. The double layer EL device showed much better performances than the single layer EL device, exhibiting a maximum luminance of ca. 13000 cd m^{-2} at a driving voltage of 18 V, and a luminous efficiency of 1.1 lm W^{-1} at a luminance of 300 cd m^{-2}.

278. THERMAL STABILITY OF ORGANIC ELECTROLUMINESCENT DEVICES FABRICATED USING NOVEL CHARGE-TRANSPORTING MATERIALS
S. Tokito, H. Tanaka, K. Noda, A. Okada and Y. Taga
Toyota Central R & D Labs., Inc., Nagakute, Aichi-gun, Aichi 480-11, Japan

Novel hole and electron-transporting materials have been synthesized to improve the thermal stability of organic electroluminescent (EL) device. Molecular structures of the hole and the electron-transporting materials have been designed based on triphenylamine (TPA) and oxadiazole (OXD) moieties, respectively. The resulting materials have a high glass transition temperature. For the vacuum-deposited EL devices using the TPA derivatives and a typical emitting material, tris(8-quinolinolato)aluminum, the thermal stability was clearly seen to depend on the Tg of the hole-transporting material. Lowest turn-on-voltage for light emission and highest luminous efficiency were observed at around the Tg. Excellent durability of continuous operation was also achieved at a high temperature. For the EL device using the novel OXD derivative as an electron-transporting layer, good emission efficiency and stability were achieved. The OXD derivatives have been also applied to the polymeric system with poly(N-vinylcarbazole) matrix.

279.
DOPING OF THE CHARGE TRANSPORT LAYER WITH HIGHLY LUMINESCENT MOLECULES. Zakya H. Kafafi, Darius J. Fatemi, Hideyuki Murata, and Charles D. Merritt, U. S. Naval Research Laboratory, Washington D.C. 20375

Highly luminescent molecular composites were prepared by high vacuum deposition and used as the emitter layers in molecular organic light-emitting diodes (MOLEDs). These composites consist of a highly fluorescent molecule as the dopant and a charge transporter as the host. Examples of emitters are derivatives of perylene, anthracene, naphthacene, quinacridone and pyrromethene difluoroborate complexes. In addition, some dopants were selected to act as an electron/hole trap in the charge transport layer. The paper discusses the observed enhancement in photoluminescence quantum yield (a factor of 4) and electroluminescence quantum efficiency (a factor of 30 for single-layered MOLEDs) in terms of efficient energy transfer from host to guest molecules and efficient radiative recombination of trapped carriers.

280. EFFECTS OF METALS ON LUMINESCENCE OF ORGANIC LIGHT EMITTING MATERIALS*. V.-E. Choong, Y. Park, and Y. Gao, Department of Physics and Astronomy, University of Rochester, Rochester, NY 14627. B. R. Hsieh, Xerox Corporation, Webster, NY 14580. C. W. Tang, Imaging Research and Advanced Development, Eastman Kodak Co., Rochester, NY 14650.
Metal/polymer interfaces play an important role in polymeric light emitting diodes (LEDs). In typical organic light-emitting devices, metallic electrodes are used to inject charged carriers into the organic electroluminescent (EL) medium. However, what other effects the metals have on the

organic medium is not well known. In this work, we report severe photoluminescence (PL) quenching of organic thin films comprising of the most useful materials, namely 1,4-bis[4-(3,5-di-tert-butylstyryl)styryl]benzene (4PV), upon sub-monolayer deposition of Al, Ag, and Ca in an ultra high vacuum environment. The severity of the luminescence quenching, which depends on the type of metal used, can greatly affect the EL device performance. Our results provide us with a better understanding of how the interfaces affect the device performance, and its potential impact on the LED industry is evident.

* Supported in part by the National Science Foundation under Grant No. DMR-9612370 and by DARPA DAAL 0196R9133.

281. PULSED LASER EXPERIMENTS DIRECTED TOWARD THE DETAILED STUDY OF FREE-RADICAL POLYMERIZATION. M. J. Buback, Institute for Physical Chemistry, University of Goettingen, Tammannstrasse 6, D-37077 Goettingen

Over the past ten years, introduction of several pulsed laser techniques enabled a significant improvement in the detailed study of free-radical polymerization kinetics. Combining pulse initiation with an SEC analysis of polymer molecular weight and with time-resolved vibrational spectroscopic measurement of monomer conversion induced by a single laser pulse (or by a pulse sequence), allows to accurately measure individual rate coefficients of propagation and of termination within extended ranges of pressure and temperature. During polymerization to high conversion, termination rate k_t may change by orders of magnitude. The various types of diffusion control: segmental diffusion, translational diffusion, and reaction diffusion, are associated with different kinds of temperature, pressure, and chain-length dependence of k_t.

282. RATE COEFFICIENTS CONTROLLING MOLECULAR WEIGHT DISTRIBUTIONS IN EMULSION POLYMERIZATIONS. Paul A Clay, David I Christie and Robert G Gilbert Chemistry School, University of Sydney, Sydney, NSW 2006, Australia

A novel means is deduced to obtain rate and mechanistic information from experimental molecular weight distributions (MWDs), through using both the full pseudo-instantaneous number MWD and rate data. These are applied in an extensive series of experiments on styrene seeded emulsion polymerizations. The data show unambiguously, through the linearity of ln(number MWD) at higher molecular weights, that termination is chain-length dependent, being dominated by short-long events. Average termination rate coefficients as a function of conversion show acceptable accord with an *a priori* theory, with relatively minor adjustment of parameters; however, the data for entry rate coefficients (initiator efficiency) for large particles (130 nm unswollen diameter) do not agree with a model found to work for smaller particles. An excess of low molecular weight species seen at high conversion is ascribed to surface anchoring effects causing increased termination as entering radicals are confined to a shell.

283.
COPOLYMERIZATION PROPAGATION KINETICS OF STYRENE AND
METHYL METHACRYLATE - REVISITED

Michelle L. Coote [1], Michael D. Zammit [1], Thomas P.Davis [1], Gary D.Willett [2]

[1] School of Chemical Engineering and Industrial Chemistry, [2] School of Chemistry, UNSW, Sydney, NSW 2052 Australia.

Pulsed - laser polymerization was used to measure the propagtion rate constants in the copolymerization of styrene and methyl methacrylate over a temperature range 20-60 °C. Molecular weight analyses were performed using size-exclusion-chromatography with LALS and viscometry detectors. The implicit penultimate unit model was fitted to the data allowing the evaluation of s_1 and s_2. Initial results show that temperature effects on s_1 and s_2 are minimal, as expected for reactivity ratio parameters. A study of the variation of activation energy with comonomer feed composition suggests that the origin of the penultimate unit effect is mainly enthalpic.

284. COBALOXIME BORON FLUORIDE - MEDIATED FREE-RADICAL COPOLYMERIZATION. J.P.A. Heuts, D. Kukulj, D.J. Forster, T.P. Davis, School of Chemical Engineering and Industrial Chemistry, University of New South Wales, Sydney, NSW 2052, Australia

Within the past decade, catalytic chain transfer with low-spin Co(II) complexes has received an increasing attention as an effective means of reducing the molecular weight. Although the behaviour of these catalysts, and in particular cobaloxime and colbaloxime boron fluoride (COBF), has been studied to relatively large extent in homopolymerization, hardly anything has been reported on their behaviour in copolymerization. In the present paper, we report experimental and modeling results of the copolymerization of styrene and methyl methacrylate in the presence of COBF at 40°C. It is shown that a significant fraction of the polymers originating from transfer with COBF has a styrene end-group. The reason for this result lies in the fact that the fraction of propagating styrene radicals is large over the whole range of monomer feed compositions. Furthermore, the use of catalytic chain transfer experiments for the determination of s-values in the penultimate model is suggested.

285. RECENT ADVANCE IN ESR STUDY ON THE RADICAL POLYMERIZATION

Mikiharu Kamachi

Department of Macromolecular Science, Graduate School of Science, Osaka University, Toyonaka Osaka 560 Japan

Recently, we have found that the well-resolved ESR spectra for several vinyl and diene monomers could be observed even in a single scan without the aid of the computer accumulation and the ESR measurements are possible in the steady state of the concentration of the propagating radicals if the optimum conditions in the initiator concentration and irradiation light intensity.are chosen. Accordingly, propagation rate constants for vinyl and diene compounds were determined by the ESR methods. In the course of the determination of k_p for styrene, we found a difference in k_p between photo- and thermal-initiated polymerizations of styrene. Furthermore, ESR measurements provided direct information on the structure, properties, and concentration of propagating radicals. Moreover, application of the time-resolved ESR spectrcopy to the elucidation of the mechanism of initiation process in the radical polymerization will be mentioned. In this review, I will mention recent advance in ESR study on radical polymerization.

286. CHARACTERIZATION OF POLY(1-FLUORO-1-CHLOROETHYLENE) AND POLY(1-FLUORO-1-CHLOROETHYLENE-CO-ISOBUTYLENE) MONOMER-AND STEREOSEQUENCE DISTRIBUTION USING $^1H/^{19}F/^{13}C$ TRIPLE RESONANCE 3D-NMR. Peter L. Rinaldi[a], Lan Li[a], Hsin-Ta Wang[b], and H. James Harwood[b], [a]Knight Chemistry Laboratory, Department of Chemistry, The University of Akron, Akron, Ohio 44325-3601, [b]The Maurice Morton Institute of Polymer Science, Department of Polymer Science, The University of Akron, Akron, Ohio 44325-3909

In an earlier communication,[1] a 3D-$^1H/^{13}C/^{19}F$ triple resonance NMR experiment was used to unambiguously determine the resonance assignments for mm, mr/re, and rr triad stereosequences in poly(1-chloro-1-fluoroethylene) (PCFE) without resorting to the preparation of a stereoregular polymer with known relative configuration. In this paper, we show that a better resolution 3D-NMR spectrum provides assignments for tetrad and pentad resonances in PCFE. We also use 3D-$^1H/^{13}C/^{19}F$ triple resonance NMR combined with HMQC and HMBC 2D-NMR to obtain complete resonance assignments for poly(1-fluoro-1-chloroethylene-co-isbutylene)(PCFE) monomer-and stereosequences.

287. 13C AND 2D NMR OF NOVEL ETHYLENE AND OLEFIN POLYMERS MADE WITH NEW LATE METAL CATALYSTS. S. J. McLain, E. F. McCord*, L. K. Johnson, S. D. Arthur, S. D. Ittel, L. T. J. Nelson, D. J. Tempel, C. Killian and M. S. Brookhart, E. I. du Pont de Nemours & Company, Inc., Experimental Station, P. O. Box 80328, Wilmington, DE 19880-0328

α-Diimine complexes of Ni and Pd catalyze the addition polymerization of ethylene and α-olefins to give polymers with unique microstructures. The unusual polymerization features of these catalysts are evidenced in the polymer microstructures which can be discerned by NMR: 1) they give highly branched and probably hyperbranched polyethylenes with unusual branching distributions; 2) they polymerize α-olefins in a "straightened" fashion, but with unusual branches as well; 3) they copolymerize polar monomers such that most of the comonomer is located at the ends of branches; and 4) they form polymers in which the terminal olefin bond can become scrambled among various positions. 2D NMR spectra which elucidate these microstructures, and mechanisms which explain their formation, will be presented.

288. 1D AND 2D STUDIES OF POLYMER END GROUPS AND POLYMERIZATION MECHANISMS. John R. Ebdon, Thomas N. Huckerby and Melanie J. Shepherd, The Polymer Centre, School of Physics and Chemistry, Lancaster University, Lancaster, LA1 4YA, UK.

Identification and quantification of end groups in polymers made by chain-reaction processes can provide insight into mechanisms of initiation and termination. In this paper we describe the application of 1D and 2D ^{1}H and ^{13}C NMR to the study of end groups in polyacrylonitriles initiated with ammonium persulfate and sodium metabisulfite and in poly(methyl methacrylate)s initiated with phenylazotriphenylmethane. In the former case it is shown that the polymerization process is dominated by transfer to the bisulfite ion; in the latter case it is found that there is significant termination by disproportionation despite the polymerization being widely regarded as having "quasi-living" character.

289. **SOLUTION NMR STUDIES OF END GROUP FORMATION IN PROPYLENE POLYMERIZATION WITH *ANSA*-ZIRCONOCENE CATALYSTS**
Luigi Resconi*[†], Fabrizio Piemontesi[†], Olof Sudmeijer[‡]
[†]*Montell Polyolefins, P.le Donegani 12, I-44100 Ferrara.* [‡]*SRTCA, 1031 CM Amsterdam*

The kinetics and mechanisms of chain transfer reactions in propylene polymerization with several C_2-symmetric *ansa*-zirconocene catalysts have been analyzed. For example, MAO-activated *rac*-ethylenebis(1-indenyl)ZrCl$_2$ (**1**/MAO) and *rac*-ethylenebis(4,7-dimethyl-1-indenyl)ZrCl$_2$ (**2**/MAO) produce low molecular weight iPP terminated with both the *cis*-2-butenyl end group (chain transfer to the monomer after a secondary propene insertion) and vinylidene end groups (chain transfer after a primary propene insertion). Allyl end groups (from unimolecular β-methyl transfer) are observed in iPP from **2**/MAO at lower propene concentration. The novel *rac*-Me$_2$C(3-*t*-butyl-1-indenyl)$_2$ZrCl$_2$ (**3**/MAO) produces allyl and vinylidene terminated iPP. **3** shows the highest selectivity for β-methyl transfer so far observed in an isospecific zirconocene, providing an entry to isotactic propylene macromers.

290.
Full Assignment of ^{13}C NMR Spectra of Polypropene: Mechanistic implications. V. Busico, R. Cipullo, M. Vacatello, A.L. Segre* Chem.Dept, Università di Napoli "Federico II" via Mezzocannone 4, 80134 Napoli *Inst.of Structural Chemistry and NMR Service of CNR C.P.10 Monterotondo Staz. ,00016 Roma, Italy

The assignment of ^{13}C spectrum of polypropylenes has been the subject of studies and research from many different groups. In this paper, we will show that, combining high field spectra (^{13}C at 150M Hz), the proper use of solvents (CDCl$_2$-CDCl$_2$ and C$_6$D$_{12}$), and a rather low temperature (70°C), leads to a noticeable improvement of resolution. In the high-resolution ^{13}C NMR spectra of propene polymers a full assignment was obtained for methyl, methine and methylene resonances up to the undecad, heptad and decad levels, respectively. The assignment has been made by analyzing model polymers with known configurational statistics, prepared *via* suitable homogeneous Ziegler-Natta catalysts, than cross-checking chemical shift calculations based on the γ-gauche effect with best-fitting procedures of integrals in different spectra.
Mechanistic implications derived from this assignment will also be discussed.

291.

CHARACTERIZATION OF COPOLYMERS BY MEANS OF HIGH RESOLUTION NMR SPECTROSCOPY.
A.M. Aerdts and A.L. German, Department of Polymer Chemistry, Eindhoven University of Technology, PO Box 513, 5600 MB Eindhoven, The Netherlands.

High resolution Nuclear Magnetic Resonance (NMR) spectroscopy has been particularly effective in the determination of the intramolecular chain structure of polymers. The intramolecular (sequence distribution and tacticity) and intermolecular (chemical composition molar mass distribution) copolymer microstructure is important, because it may supply information about the monomer addition process, e.g. about the preference of monomers to add in (co)iso- or cosyndiotactic configuration. Moreover, knowledge about the inter-and intramolecular structure is of paramount importance for the understanding of relations between molecular structure and polymer properties. Not only the microstructure of copolymers prepared in solution and emulsion polymerization have been investigated, but also the microstructure of copolymers formed during melt-mixing.

292. TACTICITY CONTROL IN A FREE-RADICAL COPOLYMERISATION: (-)-POLY(R-1-OCTEN-3-OL SULFONE)

Allan H. Fawcett, R. Karl Malcolm
Department of Pure and Applied Chemistry
The Queen's University of Belfast,
Belfast, Northern Ireland.

ABSTRACT: A predominantly isotactic poly(olefin sulfone) has been prepared by a free-radical reaction with sulfur dioxide and the olefin **R**-1-octen-3-ol, a monomer in which the chiral centre is adjacent to the double bond that becomes part of the backbone. Large dispersions of chemical shift and well-resolved fine structure in the ^{13}C-nmr spectra have allowed assignments to be made of sequence and intra-residue effects in isotactic and atactic polymers at the main chain tetrad level. The asymmetric induction value, $P_{m'}$, is barely significantly higher for the **RS**-polymer, (0.66 ± 0.01) than for the **R**-polymer, (0.61 ± 0.02). The controlling effect over one bond within the residue is much greater than any over four bonds along the main chain.

293. MECHANISMS FOR ACCELERATED POLYMERIZATION RATE BEHAVIOR IN POLYMER/SMECTIC LIQUID CRYSTAL COMPOSITES. C. A. Guymon and C. N. Bowman, Department of Chemical Engineering, University of Colorado, Boulder, CO 80309-0424

Recently, considerable attention has been devoted to polymerizations performed in a liquid crystal (LC) matrix. Systems of particular interest include ferroelectric liquid crystals (FLCs) mechanically stabilized by a small amount of polymer. The polymerization of these systems often results in accelerated polymerization rates when performed in the ordered LC phases of the mixture. This study examines the mechanisms behind this behavior in polymer/smectic liquid crystal composites. Two distinct mechanisms arise in these polymerizations to drive increases in polymerization rate; one in which the termination rate is depressed and another in which both termination and propagation rates increase. The second type of behavior can be well modeled by assuming that monomers are segregated in a reduced volume and that diffusional limitations are negligible. If diffusional limitations are a factor, then the first type of behavior is observed. Although segregation of monomer and double bonds plays an important role in the accelerated polymerization rates, the specific mode of segregation does not appear to significantly alter the mechanism behind this behavior.

294. **A NEW APPROACH TO MEASUREMENT OF PROPAGATION RATE CONSTANTS OF SMALL RADICALS** A. A. Gridnev and S. D. Ittel, DuPont, Central Research and Development, Experimental Station, Wilmington, DE 19880-0328

Cobalt porphyrins, known for their ability to react with free radicals, can be used in the measurements of propagation rate constant of small radicals. One approach involves cobalt-catalyzed chain transfer in methyl methacrylate (MMA) or methacrylonitrile (MAN) polymerizations employing high concentrations of catalyst. These conditions assure that most of the free radicals are monomeric species (H-MMA• and H-MAN•, respectively). These monomeric radicals react with additional monomer yielding the dimer radicals, H-MMA$_2$• and H-MAN$_2$• respectively, or with the cobalt catalyst to yield the starting monomer. These reactions generate the respective dimers relatively free of higher oligomers. Simple equations can be derived to calculate rate constant values for each specific radical under these conditions. It was found, that H-MAN• reacts with methacrylonitrile at a rate (k_{p1} = 340±40 $M^{-1}s^{-1}$ at 60°C) which is considerably faster than does a high polymer radical (literature: $k_{p\infty}$ = 55 $M^{-1}s^{-1}$ for MAN at 60°C). For H-MMA• at 60 °C, k_{p1} = 14,000 ± 1500 $M^{-1}s^{-1}$ and for H-MMA$_2$•, k_{p2} = 3600 $M^{-1}s^{-1}$, relative to the literature value for H-MMA$_\infty$•, $k_{p\infty}$ = 843 $M^{-1}s^{-1}$.

295. CHIRAL INDUCTION OF ACHIRAL POLYCARBODIIMIDES. D. S. Schlitzer, and B. M. Novak*, Department of Polymer Science and Engineering, Materials Research Science and Engineering Center, University of Massachusetts, Amherst, Massachusetts 10003

The polymerization of 1-(R)-2,6-dimethylheptyl)-3-hexyl carbodiimide is under kinetic not thermodynamic contol as evidenced by a four-fold increase in the optical rotation of the polymer after annealing at 65 °C for 24 h. The circular dichroic spectrum shows a four-fold increase in the molar absorptivity upon annealing as well. The guanidine repeat units in polycarbodiimides are quite basic and susceptible to protonation by an acid in organic media. The addition of a chiral acid, (S)-camphorsulfonic acid, to a racemic mixture of right and left handed helices of poly(di-n-hexyl-carbodiimide) ([α] = 0, 365 nm, 20 °C, chloroform) produced ion pairing between the polymer and the chiral anion to force the racemic mixture of helices to adopt one handedness resulting in a material which displayed large optical rotations ([α] = 257, 365 nm, 20 °C, chloroform). The optical rotation at an 8:1 molar ratio of polymeric repeats:chiral acid is nonlinear. Isolation of the polymeric chiral salt complex has been obtained via precipitation; dissolution of the chiral polymeric salt leads to a polymer with a non-zero optical rotation.

296. TOTAL SYNTHESIS OF POLY(TETRAMETHYL-M-SILPHENYLENE-SILOXANE), AN ELASTOMER OF ENHANCED HIGH-TEMPERATURE STABILITY. Ruzhi Zhang, Allan R. Pinhas, and James E. Mark*, Department of Chemistry and the Polymer Research Center, The University of Cincinnati, Cincinnati, Ohio 45221-0172

Poly(tetramethyl-m-silphenylene-siloxane) has been successfully synthesized in a step-growth polymerization. It was found that a high-temperature, high-vacuum drying process is crucial for obtaining high molecular weight polymer, and that cyclic dimer was one of the major by-products obtained. The molecular weight of the polymer was M_n=126,000 g/mol and M_w/M_n=2.06, and its glass transition temperature was -52 °C. No melting temperature was detected by DSC. TGA measurements revealed very good high temperature properties, with the onset temperatures for degradation being 415°C under nitrogen, and 495°C in air.

297. POLYSTYRENE-block-POLY(2-CINNAMOYLETHYL METHACRYLATE) MICELLES : THEIR RELAXATION KINETICS. Royale S. Underhill, Jianfu Ding, and Guojun Liu, Department of Chemistry, University of Calgary, 2500 University Dr. N.W., Calgary, Alberta, Canada, T2N 1N4.

Polystyrene-block-poly(2-cinnamoylethyl methacrylate) (PS-b-PCEMA) forms micelles with a PCEMA core in high cyclopentane (CP) content CP/THF solvent mixtures. After mixing a micellar solution of PS-b-PCEMA with a micellar solution of PS-b-PCEMA-Py, where PCEMA is labeled with

2-3 pyrene groups per chain, the pyrene monomer emission intensity increased with time. This increase is caused by the mixing of chains of different micelles and the formation of new micelles containing fewer pyrene groups. The decrease in pyrene concentration causes pyrene monomer emission intensity to increase and excimer emission intensity to decrease. To be reported are the change in this micelle relaxation rate as a function of the block lengths, solvent composition, and temperature.

298. VARIABLE TEMPERATURE SOLID STATE ^{13}C NMR STUDIES OF THE SELF-ASSEMBLY OF 5-OCTADECYLOXY ISOPHTHALIC ACID

Mingming Guo, Institute of Polymer Science, The University of Akron, Akron, OH 44325-3909

The supermolecular organization, molecular motion of model high temperature membrane model compound, 5-octadecyloxyisophthalic acid, ODOIA, has been studied by DSC and variable temperature ^{13}C solid state NMR. The molecular level origin of the assembling at crystalline, liquid crystal and isotropic melt were obtained based on the high resolution solid state NMR results. At temperatures lower than 385 K the side chains are accommodated in the ordered crystal region which resonate at 33.8 ppm and is composed of extended all trans conformation. At temperature 385 to 432 K, ODOIA is in the liquid crystalline state. The alkyl side chain carbons are mobile and resonate at 32.0 ppm, while the rigid rod assembled by hydrogen bonding and aromatic ribbon maintain the layered liquid crystal. At temperature higher than 432 K, ODOIA is in the isotropic state and the side chains resonate at 30.0 ppm. The quenched random distribution of the side chains around the aromatic ribbon are packed in the crystal region with dominant in plane orientation when annealing the sample at temperature higher than 330 K. The $^{H}T_1$ and $^{C}T_1$, as well as their relation to the chain packing, of as prepared sample and the quenched sample are also presented.

299. INFLUENCE OF DIPHENYL METHYL PHOSPHINE OXIDE DIAMINE INCORPORATION ON THE SYNTHESIS, PHYSICAL AND FIRE RESISTANT BEHAVIOR OF THERMOPLASTIC POLYETHER IMIDES. H. Zhuang, B. Tan, C. Tchatchoua, Q. Ji and J.E. McGrath, Virginia Polytechnic Institute and State University, Department of Chemistry and NSF Science and Technology Center, High Performance Polymeric Adhesives and Composites, Blacksburg, VA 24061-0344.

DAMPO) was synthesized from commercial available methyl triphenyl phosphonium bromide and incorporated into thermoplastic poly(arylene ether imide) in controlled concentration of 25, 50, 75 and 100 mol percent. The resulting material were prepared in controlled high molecular weight, about 20,000g/mol in a very predictable way with controlled phthalimide endblocks. The materials have demonstrated significantly increased residual char as a function of the phoshine oxide content and thus shown expectation of increased fire resistance. High glass transition temperature, similar to the control, were determined by DSC analysis. The mechanical behavior was investigated at room temperature and the modulus, tensile strength and elongation at failure values appear quite comparable to the control system.

300. EFFECT OF END GROUPS ON THE THERMAL STABILITY OF A SEMICRYSTALLINE POLYIMIDE. M. J. Graham, S. Srinivas, S. Gardner, G. L. Wilkes and J. E. McGrath, Virginia Polytechnic Institute and State University, Department of Chemistry and Chemical Engineering, and NSF Science and Technology Center, High Performance Polymeric Adhesives and Composites, Blacksburg, VA 24061-0344.

Presently, the interest in semicrystalline thermoplastic polymers for applications as high performance materials is increasing. Polyimides are being investigated for such a purpose. Semicrystalline polyimides offer advantages such as excellent solvent resistance and good thermooxidative stability. An important requirement of these materials is that they be able to undergo melt processing. This is only achievable for reasonable melting materials (i.e. melting points not exceeding 400°C). In doing so the material must have the capacity to recrystallize in order to retain physical properties. Recently, a recrystallizable polyimide was developed as a possible usage as a melt processable semicrystalline thermoplastic polymer. The effect of phthalimide endgroups on the thermal stability of this polyimide is discussed herein..

301. SYNTHESIS AND CHARACTERIZATION OF POLY(DIMETHYL SILOXANE) MODIFIED POLYURETHANE. F. Wang, Q. Ji and J. E. McGrath, Virginia Polytechnic Institute and State University, Department of Chemistry and NSF Science and Technology Center, High Performance Polymeric Adhesives and Composites, Blacksburg, VA 24061-0344.

Polyurethanes and poly(dimethyl siloxane) (PDMS) containing polyurethane ureas were synthesized via one step solution polymerization, using poly(tetramethylene oxide) (PTMO) and/or secondary amine terminated PDMS as soft segment, 4,4'-methylene diphenyl diisocyanate, and 1,4-butanediol as chain extender. The absolute molecular weight was measured by GPC. The structure of the polymers were characterized by 1H NMR, FTIR, and elemental analysis. The glass transition temperature of the soft segments, PTMO and PDMS, were observed at -40°C and -117°C in polyurethane and PDMS containing polyurethane urea system, respectively. The 5% weight loss temperature was around 310°C for both system. However, the char yields at 700°C in air of the PDMS containing polyurethane urea system was higher than control system and increased with silicon content.

302. SYNTHESIS OF A NEW CLASS OF NITROXYL RADICALS AND APPLICATION TO " LIVING " FREE RADICAL POLYMERIZATION, S. Grimaldi, J.-P. Finet, A. Zeghdaoui, P. Tordo, Université Aix-Marseille I et III, 13397 Marseille Cedex 20 FRANCE, D. Benoit, Y. Gnanou, M. Fontanille, ENSCPB, 33402 Talence Cedex FRANCE, P. Nicol, Elf-Aquitaine, 64170 Lacq FRANCE, J.-F. Pierson, Elf-Atochem, 92091 Paris La Défense 10 FRANCE

The synthesis and the ESR study of a new class of nitroxyl radicals have been developed. These nitroxides exhibit substituents which can influence the bond energy of the NO-C bond in the corresponding N-alkoxyamines. They have been tested in " living " free radical polymerization and some of them allow a very efficient control of the free radical (aqueous or non-aqueous) polymerization of styrene and other monomers. In comparison with TEMPO (2,2,6,6-tetramethyl-1-piperidinyloxy), the polymerization of styrene proceeds at a significantly faster rate, with a narrow polydispersity range 1.1 to 1.6, the number-average molecular weight M_n reaching around 130.000.

303. SUBSTITUENT EFFECT ON STYRENE POLYMERIZATION IN THE NITROXIDE MEDIATED STABLE FREE RADICAL POLYMERIZATION. Katsumi Daimon, Peter M. Kazmaier, Michael K. Georges, Gordon K. Hamer and Richard P. N. Veregin, Xerox Research Centre of Canada, 2660 Speakman Drive, Mississauga, Ontario, Canada, L5K 2L1

In recent years, the living Stable Free Radical Polymerization (SFRP) process has been extensively studied to gain a fundamental understanding of the reaction and its mechanism. As part of this study, kinetics of a series of substituted styrenes in this polymerization were investigated. Electron withdrawing substituents on the styrene ring were found to increase kp, and homolytic bond dissociation is relatively independent of substituents. Trapping rate must be different for various monomers and it plays a significant role in determining substituent effects.

304.

KINETICS AND MECHANISM OF "LIVING" FREE RADICAL POLYMERIZATION OF STYRENE MEDIATED BY STABLE NITROXIDE RADICALS

S. Oulad Hammouch, J.-M. Catala*

Institut Charles Sadron (CNRS-ULP) 6 rue Boussingault, 67083 Strasbourg Cedex- France

The free radical polymerization of styrene mediated by di-*ter*-butyl nitroxide adduct (A-T or T-A-T) has been studied. The influence of parameters such as nitroxide adduct concentration and active sites concentration on styrene polymerization was examined. It was shown that the polymerization rate is regulated by the concentration of radicals (thermally or using an azo initiator) generated in the medium but that the molecular weight of the polymers formed is controlled by the nitroxide adduct, giving yield to the typical feature of a so-called "living" system.

305. COMPUTER SIMULATION ON THE KINETICS OF NITROXIDE-CONTROLLED FREE RADICAL POLYMERIZATION, Y. Tsujii, T. Fukuda, and T. Miyamoto, Institute for Chemical Research, Kyoto University, Uji, Kyoto 611, Japan

The 2,2,6,6-tetramethylpiperidinyl-1-oxy(TEMPO)-controlled free radical polymerization of styrene was studied by means of the computer simulation on the basis of the proposed kinetic scheme taking account of the reversible dissociation of a polystyryl adduct with TEMPO, thermal initiation, bialkyl termination and the decomposition of the active chain end with their experimentally determined rate constants. The polydispersity index M_w/M_n as well as the concentration of each species could be quantitatively predicted by calculating statistical moments of total products without introducing any adjustable parameters. The results of the simulation with the non-stationary state approach revealed that the stationary state with respect to both [P*] and [TEMPO] was quickly reached, typically within 5 min for systems with no excess TEMPO. Requisites for a narrow polydispersity were shown to be a relatively low concentration of thermally produced polymer chains and frequent dissociations of PS-TEMPO adduct.

306. CONTROLLED RADICAL POLYMERIZATION OF VINYLIC MONOMERS IN THE PRESENCE OF UNUSUAL STABLE RADICALS
Daniel Colombani [a,b], Marco Steenbock [a], Markus Klapper [a], Klaus Müllen [a,*]
[a] Max-Planck-Institut für Polymerforschung, Ackermannweg 10, D-55128 Mainz, Germany
[b] Institut Charles Sadron, UPR 22 CNRS, 6 rue Boussingault, F-67083 Strasbourg, France

The controlled radical polymerization of vinylic monomers was investigated with azo or peroxy derivatives as initiators, in the presence of unusual stable radicals to afford narrowly distributed polymers. Polymerizations were performed in bulk, at various temperatures depending from the monomer involved. It was confirmed that the polymerization proceeds in accordance with a "living" mechanism. The variations of the conversion and of the molar mass were monitored with time and the thermal degradation of the obtained polymers was studied versus temperature. Re-initiation of the growth of the chains in the presence of a second monomer was investigated to give block copolymers. A study of the structural effect of diverse counter-radicals (*e.g.* in relation with the presence of substituents close to the radical center) on the control of the polymerization has also be done.

307. FREE RADICAL RING-OPENING POLYMERIZATION OF CYCLIC MONOMERS APPLICATIONS AND KINETICS
Daniel Colombani [a,*], Anne Baudouin [a,b], Firouz Asgarzadeh [a,b], Philippe Chaumont [b,*]
[a] Institut Charles Sadron, UPR 22 CNRS, 6 rue Boussingault, F-67083 Strasbourg, France
[b] Laboratoire d'Etude des Matériaux Plastiques et des Biomatériaux, UMR 5627 CNRS, 43, Bd du 11 Novembre 1918, F-69622 Villeurbanne, France

The addition-fragmentation polymerization (AFP) of cyclic monomers has been investigated, with a special emphasis given to the potential of this procedure for access degradable polymers. The term AFP means that an intramolecular process (*e.g.* a ring-opening) occurs, following the addition step on a monomer. Numerous ring-opening monomers (without addition step) are involved by ionic polymerizations whereas only few of them involving a radical addition processes have been reported reported. The free radical homo- and copolymerizations involving such monomers are of great interest because of their specific properties of incorporation of functional groups in the main chain.

POLY

308. SYNTHESIS AND CHARACTERISATION OF DIENE-FUNCTIONAL MACROMONOMERS
OBTAINED THROUGH A FREE RADICAL ADDITION-FRAGMENTATION PROCESS.
Marie-Odile Zink [a], Daniel Colombani [a], Philippe Chaumont [b]
[a] Institut Charles Sadron, UPR 22 CNRS, 6 rue Boussingault, F-67083 Strasbourg Cedex, France. [b] Université Claude Bernard Lyon 1, LEMBP Bâtiment 305, 43 Bvd du 11 Novembre 1918, F-69622 Villeurbanne Cedex, France.

Pentadienyl-type chain transfer agent, 5-tert.-butylthio-1,3-pentadiene (TBPD) was used in free radical polymerization of methyl methacrylate to produce conjugated diene-end capped macromonomers by a free radical addition-fragmentation mechanism. The chain transfer constant was calculated using a modified Mayo's equation and revealed good transfer proprieties for MMA. The macromonomers were synthesized at low and up to high conversions. Characterization of the macromonomers revealed that the functionality is close to 0.7 and is not affected by copolymerization. Copolymerization of these macromonomers with vinylic monomers were carried out leading to the formation of well defined graft copolymers.

309. RADICAL POLYMERIZATION OF METHYL METHACRYLATE
USING ARENEDIAZONIUM SALTS AND SODIUM ACETATE AS INITIATING SYSTEM

Daniel Grande[1,2,3], Yves Gnanou[2], Ramiro Guerrero[1]

[1] Centro de Investigación en Química Aplicada, Blvd. Reyna Hermosillo, 140, Saltillo, Coah. 25100 (México)
[2] Laboratoire de Chimie des Polymères Organiques, UMR CNRS 5629, ENSCPB - Université Bordeaux I, 33402 Talence (France)
[1,2,3] Also belonging to the UAC - CIQA - UANL Ph.D. Program

The free radical polymerization of methyl methacrylate (MMA), mediated by a couple of stable and reactive radicals that are produced by the Gomberg-Bachmann reaction of arenediazonium salts with sodium acetate, has been investigated. The system exhibits two distinct stages: a first induction period during which extensive primary radical termination occurs and a second chain growth step. Although the efficiency of such initiating system is rather poor, a certain degree of control of the polymerization can, however, be achieved as long as the window of conversion remains low.

310. CONTROL OF FREE-RADICAL POLYMERIZATION OF
METHYL METHACRYLATE BY A DIPHENYLETHANE-TYPE INITER

M. E. De León[1,2,3], Y. Gnanou[1], R. Guerrero[2]
[1] Laboratoire de Chimie des Polymères Organiques, UNMR CNRS 5629,
ENSCPB - Université Bordeaux I, 33402 Talence (France)
[2] Centro de Investigación en Química Aplicada, Blvd. E. Reyna H. 140, Saltillo, Coah. 25100 (México)
[3] Also belonging to the UAC-CIQA-UANL Ph. D. Program

1,1,2,2-tetraphenyl-1,2-bis(trimethylsilyloxy)ethane gives upon decomposition two diphenylmethyl-type radicals that are slow to initiate the polymerization of methyl methacrylate. These radicals also contribute to the formation of a reservoir of dormant species through reversible termination with the growing radicals of poly(methyl methacrylate) (PMMA). As the conversion increases, this reservoir of dormant species depletes because of the concomitant occurrence -through in small proportion at the beginning of the polymerization - of bimolecular termination reactions. However, this initer system gives access to PMMA-based block copolymers with good blocking efficiency, provided polymerization is discontinued at low conversion.

311. KINETIC STUDIES ON THE COPOLYMERISATION OF
N-PHENYLMALEIMIDE AND MMA BY AN ON LINE NEAR-IR METHOD.
D.J.T.Hill, P.J.Pomery, L.Y. Shao, A.K. Whittaker, Polymer and Radiation Group,
Department of Chemistry, University of Queensland, Brisbane, Australia 4072

The FT-NIR absorption frequency of the maleimide double bond is well resolved from those of other vinyl monomers, such as methyl methacrylate (MMA) and styrene. Thus, FT-NIR spectroscopy offers a convenient tool with which to investigate the kinetics of these copolymerisations. Kinetics of thermally initiated solution (tetrahydrofuran) copolymerisation of N-phenylmaleimide (PHMI) with MMA have been studied by an on line NIR method at 60 °C. Sample were sealed in glass tubes under vacuum after O_2 was removed. The real time concentrations of both monomers were monitored throughout the copolymerisation, thus the corresponding instantaneous copolymer composition were derived by appropriate data treatment. After fitting instantaneous f_1 and F_1 data to the different models, it was found that the copolymerisation of PHMI with MMA is best described by the penultimate model with reactivity ratios: r_{11}=0.118, r_{01}=0.118, r_{10}=1.338, r_{00}=0.438. The standard error of S_y is 0.011.

312. **EFFECT OF METAL COMPLEXATION ON THE SPONTANEOUS COPOLYMERIZATION OF 4-VINYLPYRIDINE,** Michael G. Mikhael, Takaki Kanbara, Anne Buyle Padias and H.K. Hall, Jr.*, Chemistry Department, The University of Arizona, Tucson, AZ 85721.

The spontaneous copolymerization of 4-vinylpyridine (4-VP) complexed with two different metal ions (zinc and lithium salts) with various electron-rich vinyl monomers (p-methoxystyrene **MeOSt**, p-methylstyrene **MeSt**, and styrene **St**) was investigated at 75°C. Increasing the metal salt concentration or the nucleophilicity of the electron-rich monomer increased the copolymer yields. All obtained copolymers are characterized by high molecular weight (10^5). Those prepared by using zinc complexes are characterized by a very broad molecular weight distribution. Complexation with lithium perchlorate resulted in a narrower molecular weight distribution. Both ^{1}H NMR and elemental analyses confirmed the almost 1:1 copolymer structure. Changing the anion of the zinc salt (chloride, acetate, triflate) does not show any considerable effect either on the copolymerization rate or on the molecular weight. The proposed mechanism exhibits the formation of a σ-bond between the β-carbons of the two donor-acceptor monomers. This creates the 1,4-tetramethylene biradical intermediate which can initiate the copolymerization reaction.

313. NOVEL RADICAL GRAFT POLYMERIZATION ONTO SILYLMETHYL GROUP IN POLYDIMETHYLSILOXANE. Motoki Okaniwa, Yokkaichi Research Laboratory, Japan Synthetic Rubber Co., Ltd., 100, Kawajiri-cho, Yokkaichi, Mie, 510, Japan

Graft polymerization of styrene and acrylonitrile onto silylmethyl group in polydimethylsiloxane were carried out with various initiators in emulsion system. When customary graft polymerization initiators such as benzoyl peroxide were used, monomer was scarcely grafted. By selection of adequate initiator; the combination in tert-butyl perlaurate with redox system, highly grafted polymer was obtained. The relationship between the kinds of initiators and the grafting reactivity were analyzed from seeing some properties of initiator, and the essential conditions for initiator to get high graft ratio are clarified as follows; (1)Only tert-butoxy radical which is high abstracting reactivity of hydrogen atom, is produced by redox system. (2)Oil soluble initiator. (3)Initiator does not have tert-hydrogen atom abstracted easily. The obtained graft polymer with high graft ratio could be used as modifying agents of poly(styrene-co-acrylonitrile)-g-polybutadiene (ABS) for lubricity, heat resistance and water-repellencey.

314. THE EFFECT OF MOLECULAR ARCHITECTURE ON THE THERMOTROPIC BEHAVIOR OF SIDE-CHAIN LIQUID CRYSTALLINE POLYMERS AND ITS RELATION TO POLYDISPERSITY. A. M. Kasko, A. M. Heintz, Y. Wang and C. Pugh, Department of Chemistry, Macromolecular Science and Engineering Center, University of Michigan, Ann Arbor, MI 48105-1055

The broad phase transitions of side-chain liquid crystalline polymers (SCLCPs) have always been attributed to their polydispersity. However, polydisperse mixtures of linear SCLCPs exhibit narrow transitions. Since SCLCPs produced by radical polymerization contain branched structures at high monomer conversion, the broad transitions are likely due to non-uniform mixtures of molecular architectures. We will therefore present a systematic study of the effect of molecular architecture (linear, branched, star and comb) and their polydisperse blends on the thermotropic behavior of SCLCPs based on poly {4-[ω-(propenyloxy)undecyloxy]-4'cyanobiphenyl}.

315. REDOX TELOMERIZATION OF STYRENE WITH RCCl$_3$-TYPE TELOGENS CATALYZED BY COPPER (I) CHLORIDE : CONTRIBUTION OF LIGAND 2,2'-BIPYRIDINE TO THE ACTIVATION OF THE CATALYTIC SYSTEM. M.DESTARAC, J.M.BESSIERE, B.BOUTEVIN, Laboratoire de Chimie Appliquée UPRES A 50760, Ecole Nationale Supérieure de Chimie de Montpellier, 8 rue Ecole Normale, 34296 Montpellier Cedex 5, France.

Redox telomerization of styrene catalyzed by copper (I) chloride has been widely studied. Matyjaszewski and his co-workers recently developed the transition metal catalyzed atom transfer radical polymerization (ATRP) process, based on a RX/CuX/2,2'-bipyridine system, leading to polymers with well-controlled architectures. In this system, the ligand plays a major role in redox properties of the copper complex and gives an "immortal" nature to redox telomerization. Considering the important effects of the R group on the activity of RCCl$_3$-type telogens in conventional redox telomerization, this study has been extended to "immortal redox telomerization". Differences of reactivities in both conventional and immortal redox telomerizations are discussed.

316. **POLYMERISATION OF STYRENE USING A SOLUBLE COPPER(I) COMPOUND IN CONJUNCTION WITH AN ALKYL BROMIDE,** David M Haddleton*, Andrew J Shooter, Michael J Hannon and John A Barker. Department of Chemistry, University of Warwick, Coventry, CV4 7AL, U.K.

Atom transfer radical polymerisation (ATRP) has emerged from two laboratories independently whereby low valent metal halides are used in conjunction with an alkyl halide. Sawamoto reports the use of Ru(II)(PPh$_3$)Br$_2$ for the living polymerisation of MMA. Matyjaszewski has utilised copper(I) halides for styrene, methyl acrylate and methacrylates. Perhaps the most widely used method to solubilise copper(I) is the formation of acetonitrile complexes with large counter-ions e.g. [Cu(CH$_3$CN)$_4$][PF$_6$]. This present study was carried out in order to determine the usefulness of soluble copper (I) and to investigate whether the role of bpy is to solubilise copper(I) or to accept electron density and stabilise copper(I) towards oxidation to Cu(II). Polymerisation of styrene in the presence of an alkyl halide and [Cu(CH$_3$CN)$_4$]$^+$ salts proceeds via heterolytic cleavage of the carbon halogen bond. Thus this system is unsuitable for atom transfer radical polymerisation. Nevertheless the use of [Cu(CH$_3$CN)$_4$]$^+$ salts has been demonstrated to lead to extremely rapid, but uncontrollable, cationic polymerisation of both styrene and isobutyl vinyl ether.

317. **QUASI-LIVING RADICAL POLYMERIZATION OF ACRYLATES BY AN ORGANOCOBALT PORPHYRIN COMPLEX AT ROOM TEMPERATURE.** Mingli Wei and Bradford. B. Wayland,* Department of Chemistry, University of Pennsylvania, Philadelphia, PA 19104, Michael Fryd, Marshall Laboratories, E.I. du Pont de Nemours and Company, Philadelphia, PA 19146

A new porphyrin cobalt complex, (Br$_8$TMP)Co-R ((Br$_8$TMP)H$_2$ =*meso*-tetramesityl-β-octabromoporphyrin) has been synthesized by introducing eight bromine substituents to the β-pyrrole positions. The cobalt complex (Br$_8$TMP)Co-R is found to have weaker Co-C bond than of (TMP)Co-R which permits reducing the polymerization temperature for acrylates from 60°C to 25°C. The polymerization of methyl acrylate initiated and controlled by (Br$_8$TMP)Co-R at room temperature (25°C) has yielded linear increase of molecular weight (M$_n$) with monomer conversion and relatively narrow molecular weight distribution (~1.1) which are indicative of an effective living radical polymerization. Temperature effect on the polymerization is also discussed.

318. **MODEL STUDIES OF THE SLOW TERMINATION PROCESS IN THE POLYMERIZATION OF STYRENE BY ATOM TRANSFER RADICAL POLYMERIZATION.** Mingli Wei and Krzysztof Matyjaszewski,* Department of Chemistry, Carnegie Mellon University 4400 Fifth Avenue, Pittsburgh, PA 15213. Timothy E. Patten, Department of Chemistry, University of California, Davis, CA 95616

In the Atom Transfer Radical Polymerization (ATRP) of styrene, the polymerization is well-controlled up to a molecular weight of 30,000, with relatively narrow molecular weight distribution (1.05 to <1.5). Above this molecular weight range a chain termination process can be observed as evidenced by broader molecular weight distributions and deceleration in the polymerization kinetics. Various factores affecting the polymerization process have been investigated by a model reaction of a polymer endgroup analogue (1-phenylethyl bromide) and a copper(I) catalyst (CuBr) in benzene. The model reaction suggests that the major contribution to the chain termination process is from the reactions of Cu(II) species with propagating radicals and halogen-containing chain-ends.

319. **ATOM TRANSFER RADICAL POLYMERIZATION OF STYRENE IN THE PRESENCE OF COPPER CARBOXYLATES.** Mingli Wei, Jianhui Xia, Scott G. Gaynor and Krzysztof Matyjaszewski* Department of Chemistry, Carnegie Mellon University, 4400 Fifth Avenue, Pittsburgh, PA 15213.

The atom transfer radical polymerization (ATRP) of styrene by copper acetate (CuOAc) and copper 2-thiophenecarboxylate (CuTC) has been studied. Some new features have been observed compared with the ATRP by Cu(I) halides. The polymerization rate by CuOAc is faster than that by CuBr at similar conditions. The polymerization rate of styrene by a 1:1 ratio of Cu(I) halide to dNbipy gives about half of the polymerization rate by a 1:2 ratio of Cu(I) halide to dNbipy, but the polymerization rates are similar when 1:1 and 1:2 ratios of Cu(I) carboxylate to dNbipy are employed. The molecular weights are higher than the theoretical values and the molecular weight distribution is broader compared with the ATRP by Cu(I) halides, which are indicative of insufficient initiation and termination process. The molecular weight and molecular weight distribution can be better controlled by regulating the polymerization with added $CuBr_2$ or CuBr complexes.

320. **EFFECT OF THE COUNTERANION IN ATOM TRANSFER RADICAL POLYMERIZATION USING ALKYL (PSEUDO)HALIDE INITIATORS,** Kelly Davis, Jennifer O'Malley, Hyun-Jong Paik, and Krzysztof Matyjaszewski*, Carnegie Mellon University, Department of Chemistry, 4400 Fifth Avenue, Pittsburgh, PA 15213

Control in Atom Transfer Radical Polymerization is achieved through the establishment of a rapid equilibrium between dormant and active species in the growing radical chain. Cu(I) complexed with substituted bipyridines has proven to be among the most successful catalysts for this process, promoting the reversible transformation between the dormant alkyl halide and the active radicals. A non-coordinating Cu(I) species, $CuPF_6$, is used to obtain more precise information of the effect of the (pseudo)halogen atom as well as the counteranion on ATRP. Results show ATRP of styrene initiated by 1-phenylethyl chloride, using $CuPF_6$ complexed by two bipyridine ligands as the catalytic system, produces the best controlled system with linear kinetic behavior and good molecular weight control and rates of polymerization double that of reactions catalyzed by CuCl. Using the $CuPF_6$ catalytic system for ATRP of methyl acrylate initiated by either methyl 2-bromopropionate or methyl 2-chloropropionate shows an increase in the rate of polymerization to more than 10 times that of the Cu(I)halide catalyzed reactions and producing molecular weights in agreement with the theoretical values.

321. **Polymerization of Acrylates by Atom Transfer Radical Polymerization. 2. Homopolymerization of *tert*-Butyl- and Isobornyl Acrylate.** Simion Coca, Kelly Davis, Peter Miller, and Krzysztof Matyjaszewski*. Department of Chemistry, Carnegie Mellon University, 4400 Fifth Avenue, Pittsburgh, PA 15213.

Atom transfer radical polymerization was applied to the bulk homopolymerization of isobornyl acrylate and tert-butyl acrylate. The reactions, performed in bulk, exhibited the characteristics of a "living" radical polymerization: no transfer or termination was observed in kinetic analyses and molecular weight was controlled according to the initial concentrations of monomer and initiator. The difunctional initiator, 1,2-bis(bromopropionyloxy)ethane, was synthesized and used in the polymerization of tert-butyl acrylate. Activation of this telechelic initiator was efficient leading to addition of monomer on both sides of the polymeric chain as evidenced by GPC data.

322.
Polymerization of Acrylates by Atom Transfer Radical Polymerization. 1. Homopolymerization of 2-Hydroxyethyl-, Glycidyl-, Vinyl- and Allyl Acrylate. Simion Coca and Krzysztof Matyjaszewski*, Department of Chemistry, Carnegie Mellon University, 4400 Fifth Avenue, Pittsburgh, PA 15213

"Living"/controlled Atom Transfer Radical Polymerization (ATRP) has been successfully used for polymerization of 2-hydroxyethyl acrylate, glycidyl acrylate and vinyl acrylate leading to well defined materials. Thus, polymerization of these monomers in the presence of methyl 2-bromopropionate as the initiator and CuBr/4,4'-di-(5-nonyl)-2,2'-bipyridine as the catalyst yields products, exhibiting molar mass controlled by ratio $[monomer]_0/[initiator]_0$ and narrow molecular weight distribution ($M_w/M_n \approx 1.1$-1.3). NMR studies have confirmed that hydroxyl group, oxirane ring, and vinyl groups remain unaffected, and the polymerization involves exclusively acrylic unsaturation of monomer.

323.
Block Copolymers by Transformation of "Living" Carbocationic into "Living" Radical Polymerization. Simion Coca and Krzysztof Matyjaszewski*, Department of Chemistry, Carnegie Mellon University, 4400 Fifth Avenue, Pittsburgh, PA 15213

We report that a general method of transformation of "living" carbocationic into "living" radical polymerization, without any modification of initiating sites, is now applicable for preparation of AB and ABA-block copolymers. For example, a polystyrene with chlorine chain end, polystyrene-Cl (PSt-Cl) prepared cationically, was used as an efficient macroinitiator for homogeneous "living" Atom Transfer Radical Polymerization (ATRP) of styrene and (meth)acrylates. Futhermore, ABA-type block copolymers comprising rubbery polyisobutene (PIB) was obtained using a Cl-St-PIB-St-Cl prepared cationically, as an efficient difunctional macroinitiator for ATRP of styrene and (meth)acrylates. This procedure is now being tested for other monomers (e.g. vinyl ethers) polymerizable only by cationic mechanism.

324. GRAFT / COMB COPOLYMERS BY ATOM TRANSFER RADICAL POLYMERIZATION: HYDROGELS, Kathryn L. Beers, Alison Kern and Krzysztof Matyjaszewski*, Department of Chemistry, Carnegie Mellon University, 4400 Fifth Avenue, Pittsburgh, PA 15213

Atom Transfer Radical Polymerization has been used to make polystyrene macromonomers with low polydispersity and molecular weights close to theoretical values for use in synthesizing graft copolymers with a water soluble poly(N-vinyl pyrrolidone) backbone. Relative amounts of each polymer in the sample, final molecular weights of the copolymers and, therefore, the average number of macromonomer units incorporated into each chain depend on the size and initial weight percent of the macromonomer in solution. The resultant materials possess amphiphilic segments which exhibit microphase separation as shown by differential scanning calorimetry. This phase separation leads to the formation of physical crosslinks which prevent dissolution of the polymer in water, causing the materials to swell. Water content of the swollen networks was determined and values were confirmed using thermal gravimetric analysis.

325. Polyacrylonitrile with Low Polydispersities by Atom Transfer Radical Polymerization. Seong Mu Jo, Hyun-jong Paik, and Krzysztof Matyjaszewski* Department of Chemistry, Carnegie Mellon University, 4400 Fifth Avenue, Pittsburgh, PA 15213

Kinetic plots for homopolymerizations of AN using 2-bromopropionitrile initiator show curvature after the initial linearity of $\ln[M]_0/[M]$ vs. time. The decrease of reaction rates may be due to complex formation between the nitrile group of PAN and copper ions. All reactions were run at $[CuBr]_0/[bipy]_0 = 1/3$. Increase of the $[I]_0/$

[CuBr]$_o$ ratio resulted in an increase of reaction rate, but 1/1 ratio showed a decrease of reaction rate, higher molecular weights and polydispersities than those at 1/0.1 and 1/0.5 ratios. At 1/0.1 ratio, polyacrylonitrile with very low polydispersity [M_w/M_n(GPC) =1.04] at 38.3% conversion was produced. Polyacrylonitriles with polydispersities of 1.05 (32.4%) and 1.12 (94.3%) were obtained at 1/0.5/1.5 ratio.

Polyacrylonitrile prepared by using 2-bromopropionitrile have end groups which can be used as macroinitiators for block copolymerization. The molecular weights determined by GPC (polystyrene standards) were much higher than predicted molecular weight. The molecular weights by ^{1}H-NMR and by mass spectroscopy (MALDI) are very close to predicted molecular weights. MALDI MS of polyacrylonitriles having M_n(NMR) / M_w/M_n(GPC) 1900 / 1.05 and 3160 / 1.04 show 1716 / 1.02 and 2656 / 1.01, respectively.

326. **Effects of Various Copper Salts and Additives on Polymerization of Acrylonitrile by Atom Transfer Radical Polymerization.** Seong Mu Jo, Hyun - Jong Paik, and Krzysztof Matyjaszewski * Department of Chemistry, Carnegie Mellon University, 4400 Fifth Avenue, Pittsburgh, PA 15213

In acrylonitrile polymerization using benzyl bromide as an initiator, addition of styrene (5 mol% to AN) produced polymer with molecular weight close to theoritical value and lower polydispersity than those with no additive, which indicated some improvement of benzyl bromide initiation efficiency. However, use of copolymerizable monomers such as methyl methacrylate, methyl acrylate, fumaronitrile, vinylidene chloride, and 4-methylstyrene did not show significant improvement in benzylbromide initiation efficiency.
Lowering the reaction temperature resulted in reduced polydispersity because some side are supressed.
Benzyl bromide / CuCl or copper 2-thiophenecarboxylate (CuTc) initiating systems showed slower reaction rate, higher M_n, and higher M_w/M_n than the benzyl bromide / CuBr system. In the case of CuTc, M_n at initial time was similar to that of the CuBr system, but M_n increased with conversion and polydispersities were higher than those at CuBr and CuCl.

327. **Synthesis of Block and Graft Copolymers from Poly(dimethylsiloxane) Macroinitiators by Atom Transfer Radical Polymerization.** Yoshiki Nakagawa, Peter Miller, Cristina Pacis and Krzysztof Matyjaszewski*. Department of Chemistry, Carnegie Mellon University, 4400 Fifth Avenue, Pittsburgh, PA 15213

Poly(dimethylsiloxane) (PDMS) macroinitiators were prepared from vinyl and hydrosilyl terminal and pendant functionalized PDMS by hydrosilation with an attachable initiator. Two attachable initiators were used: one was vinyl benzyl chloride and the other contained a benzyl chloride and a hydrosilyl group. With these macroinitiators styrene and acrylates were polymerized by ATRP to give well-controlled PDMS block and graft copolymers. Some of these copolymers showed typical properties of thermoplastic elastomers; Tg values for both polymer blocks were elucidated by differential scanning calorimetry.

328. **From Step Growth to "Living" Radical Polymerization: The Synthesis of ABA Block Copolymers of Vinyl and Step Growth Polymers by ATRP.** Scott G. Gaynor, Shane Z. Edelman, and Krzysztof Matyjaszewski,* Department of Chemistry, Carnegie Mellon University, 4400 Fifth Avenue, Pittsburgh, PA 15213

Atom Transfer Radical Polymerization, ATRP, has been developed as a method to control radical polymerizations. This system has been demonstrated to allow for the control of polymer molecular weight, in terms of size and distribution (M_w / M_n), functionality, composition (homo-, random, and block copolymers), and architecture (graft, branched or linear). As initiators, many small organic molecules have been used that have activated halogens. We report here the use of macroinitiators comprised of step growth polymers to prepare ABA block copolymers where the A blocks are comprised of vinyl monomers and the B block is a step growth polymer. The macroinitiators were obtained by first preparing a step growth polymer with a slight excess of one monomer, so as to prepare a step growth polymer with reactive end groups, such as -OH. These polymers were then treated with 2 - bromopropionyl bromide to yield step growth polymers with 2 - bromopropionyloxy end groups that can initiate ATRP in the presence of a vinyl monomer and copper (I) catalyst. The step growth polymers that were prepared are polysulfone, and a polyester from 1,6-hexanediol and adipic acid. The vinyl monomers used included styrene and butyl acrylate. The preparation of ABA block copolymers was confirmed by ^{1}H NMR and observation of an increase in the molecular weight of the polymers from macroinitiator to block copolymer.

329. GRADIENT COPOLYMERS OF STYRENE AND *n*-BUTYL ACRYLATE THROUGH ATOM TRANSFER RADICAL POLYMERIZATION, Stephen V. Arehart, Dorota Greszta, and Krzysztof Matyjaszewski*, Department of Chemistry, Carnegie Mellon University, 4400 Fifth Avenue, Pittsburgh, PA 15213.

ABSTRACT: Gradient copolymers of styrene and *n*-butyl acrylate can be prepared through atom transfer radical polymerization and the sequence distribution can be controlled through the rate of addition of *n*-butyl acrylate to the reaction mixture. Due to the similar reactivity ratios of styrene and *n*-butyl acrylate, simultaneous copolymerization of these two monomers results in copolymers with little variation of composition along the chain and the overall composition remains constant to approximately 70% total monomer conversion. Preliminary thermal analysis shows that gradient copolymers have unique thermal behavior relative to random and diblock copolymers.

330. THERMAL PROPERTIES OF GRADIENT COPOLYMERS AND THEIR COMPATIBILIZING ABILITY. Krzysztof Matyjaszewski*, Dorota Greszta, Department of Chemistry, Carnegie Mellon University, 4400 Fifth Ave., Pittsburgh, PA 15213, USA, Tadeusz Pakula*, Max-Planck-Institute für Polymerforschung, Postfach 3148, 55021 Mainz, Germany

Gradient copolymers (i.e. copolymers with a gradient of composition along the chain) of styrene and methyl acrylate were synthesized by Atom Transfer Radical Polymerization. As determined experimentally by Differential Scanning Calorimetry and Dynamic Mechanical Analysis, thermal and mechanical properties of gradient copolymers differ significantly from those of random or block copolymers with similar molecular weights and compositions. Additionally, Cooperative Motion Algorithm simulations of the interface behavior of incompatible blends modified by gradient copolymers indicate that gradient copolymers are very efficient compatibilizers as compared to corresponding random and block copolymers.

331. GRADIENT COPOLYMERS OF STYRENE AND ACRYLONITRILE VIA ATOM TRANSFER RADICAL POLYMERIZATION. Dorota Greszta and Krzysztof Matyjaszewski*, Department of Chemistry, Carnegie Mellon University, 4400 Fifth Avenue, Pittsburgh, PA 15213, Tadeusz Pakula, Max-Planck-Institute für Polymerforschung, Postfach 3148, 55021 Mainz, Germany

Gradient copolymers of styrene and acrylonitrile have been prepared via Atom Transfer Radical Polymerization. Two synthetic methods have been successfully employed to obtain the compositional gradient during the copolymerization. First, a possibility of spontaneous gradient formation by copolymerizing the two monomers mixed at appropriate ratios was investigated. In this case the gradient formation was governed by monomers feed composition and the reactivity ratios. However, the interpretation of the results was complicated by the penultimate effect observed in the copolymerization of styrene and acrylonitrile. Second method, a copolymerization with addition of acrylonitrile to styrene has proven to be more feasible yielding the desired gradient copolymers with controlled molecular weights and narrow polydispersities. The shape of the gradient was varied by changing the rate of the addition of acrylonitrile.

332. Polymerization of Substituted Styrenes by Atom Transfer Radical Polymerization Jian Qiu and Krzysztof Matyjaszewski *, Department of Chemistry, Carnegie Mellon University, 4400 Fifth Avenue, Pittsburgh, PA 15213

A series of substituted styrenes were polymerized by Atom Transfer Radical Polymerization (ATRP), and the correlation between monomer structure and polymerization rate was studied. The effect of substituents is discussed with regard to the Hammett equation. The results show that most of the monomers can polymerize in a controlled way by ATRP, i.e. the apparent polymerization rate is first order with respect to monomer concentration and molecular weights increase linearly with monomer conversion. The molecular weights obtained fit the theoretical values and polydispersities are relatively low (<1.5). Monomers having electron withdrawing (EW) substituents result in better polymerization control and polymerize faster than those with electron donating (ED) substituents. The apparent polymerization rate constants follow the Hammett equation with ($\rho=1.5$). Further study indicates that the difference of polymerization rate can be attributed to both k_p and the equilibrium constant, K_{eq}. Monomers with EW substituents have larger k_p and K_{eq} than those with ED substituents, therefore EW substituents increase the monomer reactivity and decrease the stability of dormant species, while ED substituents have the opposite effects.

333. CHAIN LINEAR TERMINATION INFLUENCE ON MOLECULAR WEIGHT DISTRIBUTIONS FORMED UNDER RADICAL POLYMERIZATION INITIATED BY LASER PULSES. A.N. NIKITIN and A.V. EVSEEV, NICTL - Laser Research Center, Academy of Sciences of Russia, Shatura, Moscow Region, 140700, Russia

The scheme of radical polymerization which takes into account the reactions of polymer chains initiation, propagation, termination by disproportionation and combination and chain linear termination processes (inhibition, retardation and chain transfer to monomer) is investigated theoretically. For chain-length-independent termination (by disproportionation and/or combination) case the analytical expressions are derived to calculate the molecular weight distributions (MWDs) formed under the initiation by arbitrary sequence of short laser pulses. The new relatively simple numerical technique for MWDs calculation is proposed for the chain-length-dependent termination case. The new method for determination of the rate constant of linear termination (as well as the constant of chain transfer to monomer) is proposed. The methods of the termination rate constants determination (measurement of monomer-to-polymer conversion for different initiation and polymerization by packets of pulses) are considered and the limits of application of these methods are determined. The MWDs calculated for methyl methacrylate and styrene bulk polymerization are presented.

334. POLYSTYRENE-B-POLYISOBUTYLENE-B-POLYSTYRENE BLOCK COPOLYMERS BY COMBINING LIVING CATIONIC AND "LIVING" FREE RADICAL POLYMERIZATION. X. Chen, B. Iván[a], J. Kops, and W. Batsberg[b], Department of Chemical Engineering, Technical University of Denmark, DK-2800 Lyngby, Denmark
[a]Permanent address:CRIC, Hung. Acad. Sci., H-1525 Budapest, P.O. Box 17, Hungary
[b]Department of Solid State Physics, Risø National Laboratory, DK-4000 Roskilde, Denmark

Well-defined triblock copolymers polystyrene-b-polyisobutylene-b-polystyrene have been prepared by first synthezing a PIB macroinitiator having narrow MWD capped with several units of styrene by living cationic polymerization. Then, styrene was polymerized in bulk and in xylene, initiated by the chloro endgroups of the macroinitiator in the presence of CuCl/2,2'-bipyridine. ^{1}H NMR spectra of the triblocks and PIB show that different PS content in the block copolymers can be obtained at different reaction conditions. The GPC curves show parallel shifts towards higher molecular weights compared to the initial PIB, and the MWD remained narrow and monomodal. High initiation efficiency and no autopolymerization of styrene were observed.

335. FRONTAL POLYMERIZATION IN MICROGRAVITY. **John A. Pojman**, Akhtar M. Khan, Department of Chemistry and Biochemistry and Lon Mathias, Department of Polymer Science, The University of Southern Mississippi, Hattiesburg, MS 39406-5043

Frontal polymerization is a mode of converting monomer into polymer via a localized reaction zone that propagates, most often through the coupling of thermal diffusion and Arrhenius reaction kinetics. Because of convective instabilities, it is not possible to perform frontal polymerization with many monomers that produce thermoplastics, such as n-butyl acrylate, without the addition of a viscosity enhancing agent. Performing propagating fronts of n-butyl acrylate polymerization on a sounding rocket allowed us to determine that the ultrafine silica gel (CAB-O-SIL) used in ground based research may affect the molecular weight of the polymer produced. Samples prepared with CAB-O-SIL did have slightly broader molecular weight distributions that could reflect the decrease in termination because of the higher viscosity. However, the difference could also be caused by differences in front temperatures because of the lack of convective heat losses under weightlessness.

336. SYNTHESIS OF WATER SOLUBLE HOMOPOLYMERS AND BLOCK COPOLYMERS BY LIVING FREE-RADICAL POLYMERISATION, Laurence I. Gabaston, Steven P. Armes, Richard A. Jackson, School of Chemistry, Physics and Environmental Science, University of Sussex, Falmer, Brighton, BN1 9QJ, U.K.

Living free-radical polymerisation has been used to synthesis novel water-soluble block copolymers. Homopolymers of sodium poly(4-styrenesulfonate) (PSSNa) and sodium poly(4-styrenecarboxylate) (PSCOONa) have been synthesised with narrow molecular weight distributions (M_w/M_n = 1.21 and 1.14, respectively). Block copolymers of PSSNa-PSCOONa have been synthesised with molecular weights of 10^3 to 10^4 g mol^{-1} and polydispersities less than 1.30, as measured by aqueous GPC. These block copolymers were synthesised by chain extension of pre-formed PSSNa homopolymer with sodium 4-styrenecarboxylate, which allowed the synthesis of blocks of different copolymer compositions and chain lengths. Copolymer compositions were assessed using both infra-red and NMR spectroscopy, and also sulfur microanalyses. Preliminary dynamic light scattering studies indicate that these novel polyacid block copolymers form micellar aggregates of ca. 18 nm in dilute acidic solution.

337. *tert.* BUTOXIDE ASSISTED ANIONIC POLYMERIZATION OF ETHYL ACRYLATE. P. Vlček, J. Otoupalová, J. Jakeš, Institute of Macromolecular Chemistry, Academy of Sciences of the Czech Republic, 162 06 Prague 6, Czech Republic

As acrylates with non-branched alkyls exhibit high tendency to side reactions, they cannot be polymerized in a controlled way by anionic polymerization with alkali metal initiators. The ligated polymerization in the presence of stabilizing additives, lowering selectively the rate of self-termination, has been applied to polymerization of butyl and 2-ethylhexyl esters. This communication shows preliminary results of the ligated polymerization of ethyl acrylate initiated with a Li-ester-enolate/*tert.* butoxide complex. The process must be performed at low temperatures in a medium of low polarity and with a high excess of the alkoxide over the initiator. The rate of self-termination is high even under these conditions and increases with reaction temperature and with decreasing alkoxide concentration.

338. CYCLOALIPHATIC EPOXIDE CROSSLINKABLE CARBOXYL FUNCTIONALIZED ACRYLIC LATEXES Shaobing Wu, Mark D. Soucek*, Department of Polymers & Coatings, North Dakota State University, Fargo, ND 58105

Thermoset acrylic latexes were synthesized using methyl methacrylate (MMA), butyl acrylate (BA), acrylic acid (AA), and 2-sulfoethyl methacrylate (SEM). The resulting latex was crosslinked with a cycloaliphatic diepoxide, 3,4-epoxycyclohexylmethyl-3,4-epoxycyclohexane carboxylate. The crosslinking reaction of the coating films was investigated by gel content, IR, TGA, and DSC. The differential scanning calorimeter (DSC) revealed an exothermic peak indicative of a crosslinking reaction. An increase in the temperature of the glass transition (Tg) and the thermal stability of the crosslinked coatings were observed. Furthermore, the pencil hardness, solvent resistance, and the reverse impact resistance were shown to increase significantly compared with the un-crosslinked acrylic coating films. From these results, it was postulated that the cycloaliphatic diepoxide reacted with the carboxylic group (AA) of the latex copolymer to form the crosslinked latex films.

339. TETRAPHENYLPHOSPHONIUM MEDIATED POLYMERIZATION OF METHYLMETHACRYLATE. ASPECTS OF INITIATION AND TERMINATION REACTIONS. D. K. Dimov, T. E. Hogen-Esch, Loker Hydrocarbon Research Institute Deptartment of Chemistry, University of Southern California, Los Angeles, CA 90089

The polymerization of MMA in THF may be initiated by tetraphenylphosphonium (TPP) enolates. The TPP adduct of methylisobutyrate produces quantitative yields of PMMA with narrow polydispersity, similar to the recently reported triphenylmethyl TPP initiators, and is less prone to thermal degradation. However, the MW control and initiator efficiencies of TPP enolates decrease with increasing acidity of the parent esters. In certain cases, there is evidence for involvement of K enolates in the MMA polymerization, e.g. incomplete conversions, broad MW distributions, polymer tacticities. This is attributed to the presence of KOH generated from residual water in the TPPCl used as TPP source in the initiator synthesis. The problem can be corrected by TPPCl titration with TPMK followed by extensive washing with solvent to remove the traces of KOH.

340. Architectural Control in Acrylic Polymers by Atom Transfer Radical Polymerization. Scott G. Gaynor, Priya Balchandani, Anthony Kulfan, Matthew Podwika, and Krzysztof Matyjaszewski,* Department of Chemistry, Carnegie Mellon University, 4400 Fifth Avenue, Pittsburgh, PA 15213.

Atom Transfer Radical Polymerization, ATRP, allows for the controlled polymerization of vinyl monomers by establishment of an equilibrium between active and dormant radical species. The dormant species are organic halides where the halogen is adjacent to a group that can stabilize the resulting radical such as a carbonyl or benzylic group. Incorporation of an initiating site, i.e., an "activated" halogen, with a vinyl monomer results in a monomer that can not only be polymerized but also initiates the polymerization. We describe these monomers as A=B-C*, where the double bond is described by A=B, and the activated halogen is C*. Such A=B-C* monomers include p - chloromethylstyrene and 2-(2-bromopropionyloxy) ethyl acrylate. When these A=B-C* monomers were homopolymerized by using ATRP, hyperbranched polymers were obtained. Copolymerization of the A=B-C* monomer with a conventional vinyl monomer resulted in a branched polymer, whose density of branch points could be controlled by varying the concentration of the A=B-C* monomer. Macroinitiators were prepared by copolymerization, initiated by AIBN, of an A=B-C* monomer with a conventional monomer. The macroinitiator that is obtained has pendent activated halogen groups that can initiate ATRP. This macroinitiator was then used to prepare graft copolymers using ATRP. Some of these graft copolymers behaved as thermoplastic elastomers.

341. HYPERBRANCHED METHACRYLATES BY SELF-CONDENSING GROUP TRANSFER POLYMERIZATION. P. F. W. Simon, W. Radke, A. H. E. Müller, Institut für Physikalische Chemie, Universität Mainz, D-55099 Mainz, Germany.

The principle of self-condensing vinyl polymerization (SCVP) was applied to group transfer polymerization (GTP) of a methacrylic monomer. The polymerization of 2-[(2-methyl-1-triethoxysiloxy-1-propenyl)oxy]ethyl methacrylate (MTSHEMA) catalyzed by tetrabutylammonium bibenzoate (TBABB) provides a hyperbranched polymer. The monomer was synthesized by hydrosilylation of ethyleneglycol dimethacrylate in the presence of tris(triphenylphosphine)rhodium(I) chloride in 60 % yield. In a typical polymerization process, 0.1 % of TBABB was added to a 0.5 molar solution of MTSHEMA in THF. At room temperature, the obtained molecular weights are rather low due to "back-biting" reactions which terminate living centers, thus cleaving parts of the molecule. The resulting β-ketoesters were identified by multidetector SEC. The extent of back-biting could be decreased by lowering the temperature to -50 °C. The linear analog of PMTSHEMA, poly[2-(isobutyryl)ethyl methacrylate] (PIBHEMA), was polymerized radically in toluene. SEC with a both viscosity and light scattering detectors shows the hyperbranched topology of the PMTSHEMA obtained.

POLY

342. SYNTHESIS OF METHACRYLATE-BASED COPOLYMERS VIA GROUP TRANSFER POLYMERISATION. Maria Vamvakaki, S. P. Armes and N. C. Billingham, School of Chemistry Physics and Environmental Sciences, University of Sussex, Falmer, Brighton, BN1 9QJ, U.K.

Group-transfer polymerization (GTP) has been used to synthesise a range of statistical and block copolymers of methacrylates with a small fraction of acidic and basic units. More specifically, n-butyl methacrylate (nBuMA) and benzyl methacrylate (BzMA) were copolymerised using Group Transfer Polymerization. The precursor copolymers were then deprotected via catalytic hydrogenolysis to selectively remove the benzyl groups, giving statistical and block copolymers containing carboxylic acid groups. Similarly, copolymerisation of nBuMA with 2-(dimethylamino)ethyl methacrylate (DMAEMA) yielded copolymers containing basic groups directly. The zwitterionic derivatives of these copolymers were obtained by betainisation of the amine groups. In both cases, the copolymers produced were of low polydispersity ($\overline{Mn}/\overline{Mw} \leq 1.15$) and good molecular weight control was achieved. Gel permeation chromatography (GPC) was used to determine copolymer molecular weights, while proton NMR spectroscopy was used to assess their copolymer composition.

343. ABA TRIBLOCK COPOLYMERS *VIA* GROUP TRANSFER POLYMERISATION
J. Purcell, S. P. Armes and N. C. Billingham, School of Chemistry, Physics and Environmental Science, University of Sussex, Falmer, Brighton, Sussex, BN1 9QJ, U.K

The production of acrylic thermoplastic elastomers requires the synthesis of ABA triblock copolymers in which the B blocks are "hard" (high T_g) and the A block is rubbery. We have used Group Transfer Polymerisation (GTP) to produce such polymers in which A is Lauryl Methacrylate (LMA) and B is Benzyl Methacrylate (BzMA). The bifunctional initiators 1,5-bis(trimethylsiloxy)1,5-dimethoxy-2,4-dimethyl-1,4-pentadiene (BDDP) and 1,4-bis(methoxytrimethylsiloxymethylene) cyclohexane (BMMC) were synthesised and used with bibenzoate catalyst in THF to produce a range of polymers with molecular weights up to 135,000 and polydispersities as narrow as 1.12. The trifunctional and tetrafunctional initiators mesitylene trimethylsilyl dimethylketene acetal, Mes(TS)$_3$, and pentaerythritol (trimethylsilyl ketene acetal)$_4$, PET(TMS)$_4$ are also being investigated.

344. **HETEROTACTIC LIVING POLYMERIZATION OF METHACRYLATE — INSIGHT INTO MECHANISM FROM CHAIN-END STEREOSTRUCTURE.** T. Kitayama, T. Hirano, and K. Hatada, *Department of Chemistry, Faculty of Engineering Science, Osaka University, Toyonaka, Osaka 560, Japan*

Polymerizations of methacrylates with *t*-BuLi / bis(2,6-di-*t*-butylphenoxy)methylaluminum [MeAl(ODBP)$_2$] in toluene at low temperature give highly heterotactic (*ht*-)polymers, which comprise an alternating sequence of meso (*m*) and racemo (*r*) diads. Stereochemical structure in the main chain and near the chain ends of *ht*-poly(ethyl methacrylate) [poly(EMA)] was studied by ^{13}C NMR to examine the characteristics of this peculiar stereoregulation. Sequence distribution in the main chain indicates that propagating anion having *r* diad at the chain end ($\sim rM^-$) exhibits higher stereospecificity than that with *m* diad ($\sim mM^-$): The analysis for the initiating chain-end revealed that (1) the dimer anion preferentially undergoes racemo addition to form a trimer anion with racemo (*r*) diad (*t*-Bu-M-$M^- \longrightarrow$ *t*-Bu-*r*-M^-), (2) the trimer anion with *r* diad favors meso(*m*) addition (*t*-Bu-*r*-$M^- \longrightarrow$ *t*-Bu-*r*-*m*-M^-), and (3) the *m*-selectivity is much enhanced by lowering the polymerization temperature. The analysis of the terminating chain-end showed that $\sim rM^-$ anions exist more abundantly than $\sim mM^-$ anions, suggesting the higher stability of the *r*-ended anion. Based on these structural analyses the mechanism of stereospecificity in the heterotactic polymerization is discussed.

345. SYNTHESIS OF CORE-SHELL COMPOSITE MATERIALS BY DISPERSION POLYMERIZATION METHODS IN LIQUID CARBON DIOXIDE. J. L. Young, Y.-L. Hsiao and J. M. DeSimone, Department of Chemistry, CB #3290, Venable and Kenan Laboratories, University of North Carolina, Chapel Hill, North Carolina 27599.

Composite materials of poly(methyl methacrylate) (PMMA) and poly(acrylic acid) (PAA) exhibiting a variety of morphologies were prepared in CO_2 and freon-113 by dispersion polymerization. In CO_2, batch addition of AA resulted in PMMA particles containing PAA-rich microdomains while continuous addition of AA resulted in materials with core-shell, inverted core-shell and concentric shell morphologies. In freon-113, a core-shell morphology was exhibited for both batch and continuous addition of the second stage monomer, and particle sizes were smaller. It is suggested that CO_2 plasticization of the particles has a dramatic effect on the composite material morphology by affecting the rates of monomer diffusion as well as those of polymerization. By adjusting the mode of addition of second stage monomer or the CO_2 pressure, it may be possible to control the morphology of the core-shell materials.

346. **POLYMERISATION OF METHYL METHACRYLATE USING A NOVEL CHROMIUM COMPOUND.** Mark A Stump, David M Haddleton*, Andrew McCamley, David Duncalf, J A Segal[†] and Derek J Irvine[†], Department of Chemistry, University of Warwick, Coventry, CV4 7AL, U.K; [†] ICI Acrylics Limited, PO Box 90, Wilton Centre, Middlesbrough, Cleveland, TS90 8JE, U.K

The use of organometallics in the polymerisation of acrylics has been widespread for many years [1]. Transition metals may be involved as accelarators for free radical polymerisation, co-ordination polymerisation or in ionic polymerisation. Chromium has been used previously to promote "anionic polymerisation" and as d^3 in the +3 oxidation state shows potential for atom transfer chemistry in a Cr(II)/Cr(II) oxidation couple as well as being suited to co-ordination polymerisation. A novel Cr/Zn compound has been synthesised which has been shown to have a very unusual structure. In the presence of an alkyl bromide polymerisation of methyl methacrylate occurs presumaly via abstraction of bromide and production of an alkyl radical. This system shows promise for ATRP via a Cr(II)/Cr(III) redox couple.

1) Davis, T. P.; Haddleton, D. M.; Richards, S. N. *J. Macromol. Sci.- Rev. Chem. Phys.* **1994**, *C34*, 243.

347. **PHOTOSENSITIVE POLYMERS: SYNTHESIS AND PROPERTIES.** Gaddam N. Babu,[1] Richard A. Newmark,[2] Steven M. Heilmann,[2] Pamela A. Percha[2] and David A. Weil[2], [1]Adhesive Technologies Center, and [2]Corporate Research Laboratories; 3M Company, St. Paul, MN 55144.

Polymeric systems containing photoreactive pendant groups have been extensively investigated in a number of commercial applications such as coatings, inks, photoresists, photoimaging, photolithography and adhesives. Recently, polymers containing photodissociable side chains have gained much interest in the area of UV curable coatings. We report the monomer synthesis of 4-(2-hydroxy-2-methylpropanoyl)phenoxy ethyl 2-(2-propenylamino)-2-methyl propanoate 1 and 4-(2-hydroxy-2-methyl propanoyl)phenoxy ethyl 2-(2-methyl-2-propenylamino)-2-methyl propanoate 2 and radical polymerization to homopolymers and copolymers with methyl methacrylate and styrene. Reactivity ratios were determined by the Kelen-Tüdŏs equations. Thermal stabilities of these copolymers have been determined by thermogravimetric analysis. Their glass transitions temperatures have been measured by differential scanning calorimetry.

348. **SYNTHESIS OF PERFLUORINATED ACRYLATE AND METHACRYLATE FOULING-RELEASE POLYMERS.** E. M. Arias, M. D. Putnam, P. A. Boss, R. D. Boss, A. A. Anderson, and R. D. George*, Naval Command, Control, and Ocean Surveillance Center, RDT&E Division, Enviromental Chemistry/Biotechnology, Code D361, 53475 Strothe Road, San Diego, CA 92152-6325

Historically, toxic coating systems have been an efficient solution to the centuries-old problem of fouling by marine organisms, but contribute to a growing need for coatings in which there is no risk of environmental pollution through release of antifouling agent (biocide). Our work is motivated by this need for new, non-toxic, non-polluting fouling-resistant coatings. The challenge is to develop materials which work well in all types of fouling environments, in the presence of both physical and chemical mechanisms of fouling organism attachment. Our approach minimizes the adhesive properties of the surface, leading to fouling organism attachment with low adhesion strength or high fouling-release (FR), a parameter directly related to coating cleanability. Consequently, the physico-chemical nature of the FR material and coating surface becomes increasingly important with respect to strength of organism attachment. In our work, we have taken advantage of the weak adhesion characteristics of low surface free energy polymers which have perfluorinated alkyl sidechains incorporated onto a comb-type polymer with backbones prepared from acrylate and methacrylate precursors. We are using these model polymers to assess the effect(s) of physical properties on coating preparation, fouling performance, and cleaning behavior. Polymer series have been prepared with systematic variation of copolymer type, molecular composition, and polymer structure. Synthetic reproducibility and consistency issues are critical for correlating physical properties to FR performance and have been addressed through optimization and automation of key synthetic processes. This presentation will focus on syntheses, characterization, and physical properties of perfluorinated copolymers. Subsequent FR coating preparation and properties will also be described.

POLY

349. **An anionic polymerization of 2-(trialkylsiloxyethyl) methacrylate - A new route to semitelechelic poly(HEMA) synthesis** -<u>Yukio Nagasaki</u>, Michihiro Iijima, Hotaka Ito, Kazunori Kataoka, Masao Kato, Department of Materials Science and Technology, Science University of Tokyo, Noda 278, Japan

Anionic polymerization of 2-(trialkylsiloxyethyl) methacrylates (ProHEMA) was investigated by butyllithium as an initiator in THF. In the early stage of the polymerization, unreacted monomer molecules interacted to the counter cation to stabilize the growing species. Once surrounding monomer molecules consumed completely, the growing species induced some side reactions rapidly. We proposed the stabilization mechanism of the growing species by the chelation of monomer molecule to Li cation. The chelation of ProHEMA increased the nucleophilicity of oxonium anion. As a result, alkalimetal alkolate such as potassium ethoxide had initiation ability of the ProHEMA polymerization. This polymerization system will open new synthetic route for semitelechelic poly(HEMA) for biomedical applications.

350. **PREPARATION AND CHARACTERIZATION OF RANDOM COPOLYMERS OF ACRYLIC ACID AND MULTIFUNCTIONAL ACRYLATES.** <u>Robert A. Scott</u> and Nicholas A. Peppas, School of Chemical Engineering, Purdue University, West Lafayette, IN 47907-1283

Incorporation of an ionizable comonomer, such as (meth)acrylic acid, into highly crosslinked polymer networks gives materials of great interest both from a theoretical viewpoint and from an applications perspective. The present work represents a preliminary investigation into the properties and structure of materials obtained by copolymerization of highly functional acrylates with acrylic acid. The effects of acid content, crosslinking monomer functionality, and network chain composition on the dynamic swelling behavior of polymer networks prepared by photopolymerization were examined. The information obtained from the swelling data was used to elucidate the network structure of these materials. A strong effect of ionizable comonomer content on the network structure of highly crosslinked copolymers was demonstrated. This effect manifests itself in retarded swelling kinetics for systems containing large amounts of the ionizable monomer, despite the anticipated reduction in covalent crosslinking density. The polymer systems examined exhibited non-Fickian swelling behavior at all ion contents.

351. **INVESTIGATIONS INTO STRUCTURE/PROPERTY RELATIONSHIPS FOR POLYHEDRAL OLIGOMERIC SILSESQUIOXANE (POSS) BASED METHACRYLATE POLYMERS**

<u>Joseph J. Schwab</u>*[†], Joseph D. Lichtenhan[‡], Kevin P. Chaffee[‡], John C. Gordon[†], Yoshiko A. Otonari[‡], Micheal J. Carr[‡], Alan G. Bolf[†]

[†]Hughes STX, Phillips Laboratory, Edwards AFB, CA 93524. [‡]Phillips Laboratory, Propulsion Directorate, Edwards AFB, CA 93524.

A catalogue of hybrid plastic building blocks based on Polyhedral Oligomeric Silsesquioxanes (POSS) reagents has been developed. POSS reagents are unique in that they are physically large (approx. 15 Å diameter and 1000 amu) and are composed of a robust silicon-oxygen framework that can be easily functionalized with a variety of organic substituents. Appropriate functionalization of POSS cages allows for their incorporation into traditional thermoplastic resins without modification of existing manufacturing processes. In the case of methacrylate polymers, incorporation of POSS segments leads to dramatic increases in glass transition temperatures and oxygen permeability. A conceptual model for the effects that POSS units exert on polymer systems as well as insights into the origins of the above property enhancements will be discussed.

352. **ELECTROCHEMICAL CHARACTERIZATION OF HOLE TRANSPORT AND ELECTRON TRANSPORT MATERIALS FOR ORGANIC LEDs**
Mukundan Thelakkat, Ralf Fink, Peter Pösch, Johann Ring and Hans-Werner Schmidt, Makromolekulare Chemie I, Universitätstr. 30, 95440 Bayreuth, Germany

A variety of hole transport materials (HTL) such as triphenyldiamines (TPDs), 4,4´,4´´-tris-(N,N-diphenyl amino) triphenylamine (MTDATA) and 1,3,5-tris-(diphenyl amino) benzenes (TDABs) are examined and compared for their electrochemical stability and reversibility of redox processes. In all cases, at least the first oxidation process is reversible. Using an internal redox standard, ferrocene/ ferrocenium (FOC) for each measurement, the HOMO and LUMO energy levels are determined from redox potentials. These HOMO values are 0.2-0.3 eV lower than the ionisation potentials (Ip) measured in solid state. In general, the electron rich derivatives show the lowest oxidation potential values. Similarly, a series of variously substituted Π-electron deficient hetrocyclics belonging to the classes of oxadiazoles (OXD), 1,2,4-triazoles (TAZ), 1,3,5-triazines (TRZ) and 1,4-diazines (DAZ) are examined. The results reveal that all these heterocyclics possess one reversible reduction per heterocyclic ring. The electron affinity and the hole blocking property of these heterocyclics increase in the order: oxadiazoles < triazoles < triazines < diazines.

353.
NEW HOLE TRANSPORT MATERIALS FOR ORGANIC LIGHT EMITTING DEVICES
Mukundan. Thelakkat, Andreas Bacher, Ralf Fink, Frank Haubner and Hans- Werner Schmidt
Makromolekulare Chemie I, Universität Bayreuth, Universitätsstr. 30, 95440 Bayreuth, Germany

A variety of triphenylamine derivatives having low ionisation potentials and high glass transition temperatures (T_g) as hole transport materials (HTL) and as emitters for organic light emitting devices (LEDs) are reported. We synthesized low molecular weight triphenyldiamines (TPDs), new 1,3,5-tris(diaryl amino) benzenes (TDABs), polymeric triphenyldiamines (poly-TPDs) and insoluble triphenylamine networks based on tris(4-ethynyl phenyl)amine (TEPA). Their thermal (DSC) and electrochemical (Cyclic Voltammetry) behaviour are examined in order to determine the solid state properties and HOMO energy levels respectively. The TDABs carrying naphthyl and anthracyl substituents have very low HOMO values and act themselves as emitters also. These materials are tested as HTLs and as emitters in multi-layer LEDs.

354. COORDINATION POLYMER FILMS BY REACTIVE SELF-ASSEMBLY. D. L. Thomsen III and F. Papadimitrakopoulos*, Department of Chemistry, Institute of Materials Science, University of Connecticut, Storrs CT 06269-3136

We hereby introduce a novel self-assembly technique to grow thin films made of polymeric metal chelate derivatives of 8-hydroxyquinoline, which are insoluble and intractable, suitable for semiconducting applications. Polymeric 8,8'-dihydroxy-5,5'-biquinoline (bisquinoline)-zinc chelates were self-assembled on hydroxy-functionalized surfaces to yield uniform films by alternately dipping in THF solutions of diethyl zinc and bisquinoline, separated each time by a THF wash. NMR, FTIR, and UV/VIS spectroscopy were used to characterize these new assemblies.

355.
PREPARATION AND CHARACTERIZATION OF POLYMER DISPERSED LIQUID CRYSTAL FILMS USING POLY(BORNYL METHACRYLATE) J. H. Liu, H. T. Liu and F. R. Tsai, Department of Chemical Engineering, National Cheng Kung University, Tainan 70101, Taiwan, R.O.C.

Chiral (+)-bornyl methacrylate was synthesized from (+)-camphor. To investigate the influence of steric environment of chiral polymers on the electro-optical characteristics of polymer dispersed liquid crystal (PDLC) films, a commercial available positive nematic liquid crystal E7 was dispersed in chiral and achiral polymer matrixes. The electro-optical characteristics and the micro-structures of the PDLC films with chiral and achiral racemized polymers were investigated. It was found that transient damping response while a square pulse of 20 ms and 60 voltage was applied. The reversible turbid and transparent changes corresponding to a.c. electric field were also investigated.

356.

PREPARATION OF GRADIENT REFRACTIVE INDEX ROD BY SWOLLEN-GEL POLYMERIZATION
H. T. Liu, J. H. Liu, Department of Chemical Engineering, National Cheng Kung University, Tainan 70101, Taiwan, R.O.C.

A new method for preparing Gradient Refractive Index (GRIN) rod by using the swollen-gel polymerization technique was fabricated. Monomer pair systems of methyl methacrylate (MMA) with reactive benzyl methacrylate (BzMA) and unreactive bromonaphthalene (BN) and benzyl phenyl acetate (BPAc) were investigated. Two-dimentional refractive index profile of rods were estimated. A high numerical aperture (NA) value of 0.336 in system of MMA/BN= 2/1 was obtained. A good symmetric GRIN rod can be fabricated by using the swollen-gel polymerization. The image with no distortion can be observed through the GRIN rods fabricated by using the swollen-gel polymerization method.

357. POLARIZED PHOTOLUMINESCENCE FROM ORIENTED FILMS BASED ON BLENDS OF POLYETHYLENE AND CONJUGATED POLYMERS
Ch. Weder*, C. Sarwa, C. Bastiaansen, P. Smith; Department of Materials, Institute of Polymers, ETH Zürich, CH-8092 Zürich, Switzerland

With the objective to design devices, which efficiently emit light in a highly anisotropic, i.e. polarized, fashion we have prepared blends of poly(2,5-dioctyloxy-*p*-phenyleneethynylene) (O-OPPE) and ultra-high-molecular-weight polyethylene (UHMW-PE). Solution-casting followed by tensile deformation was employed in order to obtain oriented films. Highly polarized absorption and emission were observed for the drawn films. Based on the emission spectra, a dichroic ratio of about 13 was determined. Thus, already at modest draw ratios as they were applied in this work, highly polarized emission can be obtained with the processing scheme employed. The PL spectrum of undrawn films was found to be comparable to the one of pure films of the conjugated polymer. At the other hand, the emission spectrum for drawn films matches the one of the pure O-OPPE in solution. This behaviour suggests different morphologies for the pristine and the drawn blends. In undrawn blends, a phase separation between O-OPPE and UHMW-PE seems to occur, and consequently the photophysical properties of the blend match essentially those of pure O-OPPE. In drawn films, O-OPPE behaves as if it was molecularly dispersed or dissolved in a hydrocarbon matrix and therefore the photophysical properties are comparable to a dilute solution of the conjugated polymer.

358.

SYNTHESIS OF HIGHLY PHENYLATED POLY(P-PHENYLENE VINYLENES) VIA HALOGEN PRECURSOR ROUTE (HPR) T. E. Goodwin,1 S. S. Gonzales,1 R. Bohra,1 W. E. Feld,2 B. R. Hsieh,3 1Dept. of Chemistry, Hendrix College, Conway, AR 72032; 2Dept. of Chemistry, Wright State University, Dayton, OH 45435, 3Xerox Corp. 800 Phillips Road, 114-39D, Webster, NY 14580.

We have found that halogen precursor route (HPR), which involves the polymerization of a bis(halomethyl)benzene monomer with about one equivalent amount of base to give a soluble halogen precursor polymer followed by thin film deposition and thermal conversion to give PPV derivative thin film,. is much superior than the well known sulfonium precursor routes for the fabrication of PPV thin films. As a result, we have been able to prepare a wide range of PPV derivatives via HPR. Here we report our recent synthetic effort on several new highly phenylated PPVs. The objective of this work is to define the scope of HPR.

359. TOWARD THE SOLUBLE DERIVATIVES OF POLY(2,3-DIPHENYLPHENYLENE VINYLENE) (DP-PPV). G. M. Schaaf,[1] Z. A. Pfeiffer[1], **W. A. Feld**[1], B. R. Hsieh,[2] [1]Dept. of Chemistry, Wright State Univ., Dayton, OH 45435; [2]Xerox Corp. 114-39D, Webster, NY 14580

In order to avoid thermal conversion and to ease device fabrication, we are working toward the synthsis of soluble DP-PPV derivatives. Here we report the synthesis of a DP-PPV derivative, namely **DP6-PPV**. Polymerization of the monomer in the presence of large excess t-BuOK gave green-yellow soluble DP6-PPV. The properties of DP6-PPV are being characterized and will be reported along with those of other new DP-PPV derivatives.

360.
MICROCAPSULED PIGMENT AND ITS APPLICATION TO LIQUID TONER. - CONTROL OF NEGATIVE CHARGE.
N.Yabuuchi, A.Magosi, Y.Shibai, K.Ishii, Nippon Paint Co., Ltd. 19-17,Ikedanakamachi, Neyagawa-city, Osaka, 572, Japan.

Charge control technologies for microcapsuled liquid toner have been developed by using electrodeposition method. Pigment dispersant containing long alkyl chain(C_{12}) was used in place of the conventional polar agents (acid or basic) to disperse phthalocyanine blue, and had little influence to toner charge. For generating negative charge, acid polymer was added to toner particle. The quantity of deposited toner was increased in proportion to acid content and acidity and was further increased by adding a charge director. The charge director is an acrylic polymer consisting of basic monomer and hydrophobic monomer of which resistivity is more than $10^{13} \Omega \cdot cm$. The resistivity of toner solution was kept more than $10^{12} \Omega \cdot cm$ even in the presence of charge director. As a result, deposited toner and the value of resistivity were increased over the standard toner used for SAVIN-9040 commercial copy machine.

361.
SYNTHESIS AND CHARACTERIZATION OF SOLUBLE ALKOXY-SUBSTITUTED POLY(PARA-PHENYLENES) Yueping Fu, Polymer Science Program, University of Connecticut, Storrs, CT 06269-3136

Soluble poly(para-phenylenes) with alternating phenyl and didecyloxy-substituted phenyl rings were synthesized via Grignard coupling and characterized. The degree of polymerization is relatively low. The optical properties of the poly(para-phenylenes) are dependent on the steric and electronic effects of the alkoxy substitutents.

362.   SYNTHESIS, POLYMORPHISM AND ELECTRO-OPTIC PROPERTIES OF A NEW CLASS OF LIGATED TWIN Sc* LIQUID CRYSTALS. David R. Medeiros, Michael A. Hale, Jeffrey K. Leitko, C. Grant Willson, U. Paul Schroder, Departments of Chemistry and Chemical Engineering, University of Texas, Austin, TX 78712. IBM Research Division, Almaden Research Center, 650 Harry Road, San Jose, CA 95120.

A new class of calamitic laterally-linked dimeric liquid crystals possessing two asymmetric carbons have been designed and synthesized. These materials have been shown to possess a rich polymorphism including ferro-, ferri- and antiferroelectric phases. Analysis of these compounds has been performed using x-ray diffraction, thermal analysis and electro-optic measurements. These materials show dramatically different properties from their corresponding monomers. Dimeric molecules of this type which exhibit ferroelectric and antiferroelectric phases constitute an interesting class of materials with potential applications in high speed electro-optic displays.

POLY

363. THERMAL TRANSITIONS IN LIQUID CRYSTAL ALIGNMENT LAYER FILMS
David C. Rich, Enid K. Sichel, Peggy Cebe, Tufts University, Dept. Physics and Astronomy, Medford, MA 02155; Anne K. St. Clair, Materials Division, NASA Langley Research Center, Hampton, VA 23681

A new method for investigating thermophysical transitions in thin liquid crystal alignment layers is demonstrated using polyamide-imide and PMDA-APB polyimide alignment layer films. Thin films, 450Å-1500Å thick, are spin coated onto glass slides and aligned by mechanical brushing. The technique involves annealing a set of alignment films at an array of temperatures after the films have been brushed-aligned with a cloth, but before liquid crystal cells are constructed from the films. At annealing temperatures below the glass transition, the alignment layers maintain the brush-induced orientation. When the glass transition temperature in the polymer is exceeded by post-brush annealing, the aligning ability of the brushed films is destroyed. This method relies on optical detection of liquid crystal alignment, and is therefore an indirect measure of the thermo-physical properties of the alignment film. The method is useful for brushed films where, because of surface roughness, ellipsometric techniques can not be used.

364.

DIBLOCK AND MULTIBLOCK LIQUID CRYSTALLINE POLYMERIC SYSTEMS. **Bindu Nair** Wen-Yue Zheng, Aaron Moment, Paula Hammond, MIT, 25 Ames Street, Cambridge, MA 02139

Multiphase copolymers with side chain liquid crystal mesogenic groups have been synthesized which exhibit interesting liquid crystalline properties at room temperature. We believe these organic materials constitute a new and exciting class which will exhibit both electro-optic and mechano-optic behaviors. New in this work is the emphasis on morphology and processing as a tool for design of new materials with unique liquid crystalline properties. Diblock studies underline the interaction between the phase behavior within a liquid crystalline domain and the microphase separation of the system. Design, synthesis, and preliminary results on a siloxane-based segmented polyurethane will also be presented; these thermoplastic elastomers present the possibility of coupling mechanical fields to the liquid crystalline response. Phase behavior of both nematic and smectic C* mesogens in these systems will be discussed using optical microscopy, DSC, and WAXD data.

365. Characteristics of an Emissive Polymer Blend on LED, D.Y. Kim, J.K. Kim, H.N. Cho, and C.Y. Kim, Polymer Materials Lab., KIST, Seoul 136-791, Korea

Two alternating copolymers with monomer units of dialkyl fluorene and phenylene vinylene is blended with each other by solution blending. A polymer with a meta linkage on the phenylenevinylene group shows the PL spectrum peak at 430nm while the other polymer with a para linkage has the peak at 480nm. The blend shows the PL peak at 480nm indicating the polymer with the para linkage only gives a radiative decay. The EL spectra of the polymers and the blend are almost the same as the PL spectra of the respective ones. The EL spectrum of the blend is much sharper to give clear color and the quantum efficiency of an LED fabricated with the blend as an emissive layer is 0.026%, almost 10 times better than that of each copolymer. The onset potential for current is about 12V.

366.

NOVEL POLY(p-PHENYLENEVINYLENE)-RELATED POLYMERS CONTAINING ORGANOSILYL AND CARBAZOLE UNITS FOR BLUE LIGHT-EMITTING DIODE. Mi-Kyung Ryu*, Ki-Dong Kim, Ji-Hoon Lee[+], Soo-Min Lee*, and Hwan Kyu Kim; Department of Macromolecular Science, Hannam University, Taejon, 300-791 Korea; *Department of Chemistry, Hannam University, Taejon, Korea; [+]Department of Chemistry, KAIST, Taejon, 305-701, Korea.

The silicon-containing poly(p-phenylenevinylene)-related polymers were synthesized using well-known Wittig reaction between the N-(2-ethylhexyl)-3,6-diformylcarbazole monomer and the appropriate diphosphonium salts. The resulting polymers were highly soluble in common organic solvents and they could be spin cast onto glass plate to give highly transparent homogeneous thin films. The number-average molecular weight of the resulting polymers is in the range of 1500 to 2000, associate with the polydispersity range of 2.6 to 3.8. The present polymers have strong UV absorption bands around 351-353 nm. The photoluminescence and electroluminescence spectra appeared around 420-480 nm in blue emission region, due to the interruption of the regular π-conjugated system by organosilicon unit.

367. THERMALLY STABLE, BLUE-LIGHT EMITTING COPOLYMERS DERIVED FROM 2,7-BROMO-9,9-DIALKYLFLUORENES. R. D. Miller, J. C. Scott, M. Kreyenschmidt, J. Kwak, V. Lee, T. Fuhrer, S. Karg, IBM Almaden Research Center, 650 Harry Road, San Jose, CA 95120-6099

Thermally stable, poly-p-phenylenes, including poly(fluorene) derivatives have been shown to be efficient electroluminiscent materials with emission spectra predominately in the blue. We will present results on a series of novel, soluble random copolymers of this general class produced by the nickel (0) mediated polymerization of the respective aryl dibromides. The conjugation length, solubility and glass temperature can be controlled by the copolymer composition. The comonomer is chosen to control the solubility and electronic properties, the latter either by twisting the rings of the polymer main chain through steric interactions or by the controlled introduction of tetrahedral conjugation interrupts. Light-emitting diodes with blue to blue-green emission have been fabricated using either ITO or polyaniline as the anode and calcium for the cathode and turn-on voltages below 10V have been achieved.

368. ORGANIC n-CHANNEL MATERIALS FOR FIELD-EFFECT TRANSISTORS. Joyce G. Laquindanum, Howard E. Katz, Ananth Dodabalapur, and Andrew J. Lovinger, Bell Laboratories- Lucent Technologies, 600 Mountain Ave., Murray Hill, NJ 07974

Compounds containing the naphthalene framework were investigated as semiconducting layers in field-effect transistors (FETs). These compounds, 1,4,5,8-naphthalene tetracarboxylic dianhydride (NTCDA), 1,4,5,8-naphthalene tetracarboxylic diimide (NTCDI), and 11,11,12,12-tetracyanonaphtho-2,6-quinodimethane (TCNNQ), were shown to behave as n-type semiconductors displaying encouraging field-effect mobilities (μ_{FET}) ranging from 3×10^{-4} cm^2/Vs to 3×10^{-3} cm^2/Vs. Since high mobility p-channel organics are already known, this enables the fabrication of all-organic complementary circuits, a class of logic circuits that require both both n- and p-channel transistors. The first all-organic complementary inverter circuit will be described.

369. Polythiophene with Urethane Substitute for Light Emitting Devices. Sang-Don Jung, Taehyoung Zyung, Woohong Kim*, K. G. Chittibabu** and S. K. Tripathy**, Electronics and Telecommunications Research Institute, P.O.Box 106, Yusong, Taejon, 305-600, Korea, *Samsung Advanced Institute of Technology, Central Research Institute of Chemical Tech., 103-12, Moonji-dong, Yusong-gu, Taejon, Korea, **Department of Chemistry, University of Massachusetts Lowell, One University Ave., Lowell, MA 01854 USA

We have fabricated the polymer light emitting devices based on the highly soluble polythiophene containing a urethane side chain, poly[2-(3-thienyl) ethanol *n*-butoxy carbonylmethyl urethane] (PURET), and found that the devices emit orange-red light. The single-layered devices with indium-tin oxide (ITO) as a hole injection electrode and aluminum as an electron injection electrode emit the light under both forward and reverse bias. The quantum efficiency of the electroluminescence under the reverse bias is always higher than that under the forward bias, indicating that the injection of the electron from the ITO is more efficient.

370. Role of Dopant Counter-Anion Functionality in Polyaniline Salts / Blends and Implications to Morphology

Alan R. Hopkins and Paul G. Rasmussen, Center for Macromolecular Science and Engineering, The University of Michigan, Ann Arbor, MI 48109-1055; Rafil A. Basheer, Polymers Department, General Motors Research and Development Center, Warren, MI 48090-9055; B. K. Annis and G. D. Wignall, Oak Ridge National Laboratory, Oak Ridge, TN 37831-6393

Wide angle X-ray scattering (WAXS), small angle X-ray scattering (SAXS) techniques and light microscopy have been used to characterize the influence of dopant structure on the morphology of the polyaniline emeraldine salt (PANI-ES) solutions blended with commercial nylons, especially nylon 6, using hexafluoroisopropanol (HFIP) as a solvent. Larger and more bulky sulfonic acids are found to cause greater disruption of the PANI-ES chain compared to the smaller ones. The changes in ordering of the chains are reflected in the lower electrical conductivity observed with the bulkier dopants. The lower conductivities parallel a courser, more aggregated morphology whereas the higher conductivities accompany a finer, more dispersed network.

371. CORRELATION OF STRUCTURE AND DYNAMICS OF LIQUID-CRYSTALLINE POLYMETHACRYLATES HAVING NOVEL AZOBENZENE GROUPS IN THE SIDE CHAIN. A. Schönhals[1], R Ruhmann[1], H.-E. Carius[1] D. Wolff[2]; [1]Institut für Angewandte Chemie, Rudower Chaussee 5, D-12489 Berlin; [2]Technische Universität Berlin, Straße des 17. Juni 135, D-10623 Berlin, Germany

Dielectric spectroscopy from 10^{-2} Hz to 10^9 Hz and in a temperature range from 170 K to 430 K is employed to study the dynamic behavior of liquid-crystalline polymethacrylates with 4'-trifluoromethoxy-azobenzene units as mesogenic groups. Different mesophase structures were achieved by the variation of the lengths of the number n of the alkyl spacer groups from 2 to 8. Also the behavior of corresponding non-fluorinated polymer (methoxy-tail) for n=6 is also investigated. The dynamic behavior is directly correlated to both the chemical and mesomorphic structure which is estimated from DSC and X-ray investigations.

372. DYNAMICAL BEHAVIOR OF AMORPHOUS COMB-LIKE COPOLYMERS HAVING PHOTOCHROMIC AZOBENZENE IN THE SIDE CHAIN. A.Fritz, A. Schönhals, R. Ruhmann, Institut für Angewandte Chemie, Rudower Chaussee 5, D-12489 Berlin, Germany

Dielectric spectroscopy in the frequency range from 10^{-4} Hz to 10^6 Hz is employed to study the dielectric behavior of polymeathacrylates having the photochromic azobenze group in the side chain copolymerized with methacrylate units having no mesophase forming unit in the side chain. So amorphous materials are obtained. Also the related homopolymers are investigated. Compared to the corresponding liquid crystalline materials a broad variety of relaxation processes is observed. The different relaxation processes are assigned to molecular motional processes and discussed in the frame work of the structure of the polymers under consideration.

373.
POLARIZED ELECTROLUMINESCENCE OF LIQUID CRYSTALLINE POLYMERS WITH ISOLATED ARYLENEVINYLENE SEGMENTS IN THE MAIN CHAIN; Georg Lüssem, F. Geffarth, A. Greiner, W. Heitz, M. Hopmeier, M. Oberski, C. Unterlechner, J. H. Wendorff, Department of Physical Chemistry, University of Marburg, Hans–Meerwein–Str., 35032 Marburg, Germany

Soluble, liquid crystalline main chain polymers (LCPs) for applications in organic LEDs emitting polarized light were obtained by attaching lateral substituents to the polymer backbone. The mesophase structure of these substances was characterized by X-ray scattering as being of a smectic A phase type. Thin films of the substances were oriented by using a rubbed polyimide film as an aligning substrate, tempering and cooling below T_g to freeze the orientation. Measurements of UV-dichroism, photo- and electroluminescence reveal that the transition dipole moments of the chromophores are oriented along the rubbing direction (S=0.67). The spectra exhibit a substantial difference in the intensity measured in the direction of rubbing and perpendicular to it, respectively. The LEDs emit partially polarized light.

374.
CONTROLLED BATHOCHROMIC SHIFT OF THE EMISSION MAXIMUM OF TRI-(STILBENE)-AMINE DYES FOR THE APPLICATION IN ORGANIC LED DEVICES, C. Unterlechner, T. Christ, R. Sander, V. Stuempflen, W. Heitz, J. H. Wendorff, Department of Physical Chemistry, University of Marburg, Hans-Meerwein-Str., 35043 Marburg, Germany

Tri-(stilbene)-amine dyes are of special interest for the application in organic LED devices because they show a large shift of their fluorescence maximum in different solvents as well as in different matrix polymers. Therefore it is possible to produce LEDs with different luminescence maximums with just one chromophore. We found out that the fluorescence shift results from an overlap of two emission bands that are not always sharply separated. The intensity ratio of these bands is actually responsible for the wavelength of the fluorescence maximum. We suppose that the forming of excimeres is the reason for the existence of two emission bands.

375.
Preparation of Micronsized Polystyrene Beads by Seeded Emulsifier-free Emulsion Polymerization without Swelling Process

Soonja Choe, Sang-Eun Shim, Yoon-Jong Cha*, Tae-Kwang Ahn
Department of Chemical Engineering, Inha university, Namgu, Inchon, 402-751, South Korea
*Cemkorea co., ltd., Gojandong, Namdong-gu, Inchon, 405-310, South Korea

Micron sized monodisperse crosslinked polystyrene(PS) beads were prepared by a multi-step emulsion polymerization using styrene monomer, divinylbenzene(DVB) crosslinking agent and water-soluble initiator in the absence of emulsifier. The three-step emulsion polymerization using the pre-existing polymer seeds at relatively low temperature was required to bring monodisperse crosslinked PS beads greater than 2 microns. In the result, the monodisperse beads above 1 micron of the diameter were produced at the second step and continued to grow without the generation of small secondary particles caused by the nucleation in the aqueous phase.

376.
EXAMINATION OF LANGMUIR FILM MODEL FOR SELF-ASSEMBLY FORMATION OF ALKYLTRICHLOSILANE ON SILICON OXIDE. S.-Y. Kim, C. W. Frank. Department of Chemical Engineering, Stanford University, Stanford, CA 94305-5025

Octadecyltrichlorosilane (OTS) and dodecyltrichlorosilane (DTS) are adsorbed on a silicon oxide surface using a self-assembly technique. Contact angle measurements on these films support the existence of a physisorbed precursor state, analogous to a Langmuir film, where the adsorbents have sufficient lateral mobility to reach thermodynamic equilibrium before final chemical anchoring to the substrate. However, the FTIR spectra and atomic force microscopy (AFM) images of these films suggest a very disrupted structure that is far more defective than a Langmuir film. Structural disruption increased with an increase in the amount of water in the solvent. These results suggest that there exists a precursor, physisorbed state, with the adsorbents undergoing surface migration as well as an adsorption-desorption with the bulk solution.

POLY

377. The *"Living"* Free Radical Synthesis of Random and Block Copolymers Containing Hydrophobic and Hydrophilic Moieties: Novel Thin Film Dissolution Behavior. G.G. Barclay*, M. King and R. Sinta, Shipley Company, 455 Forest Street, Marlborough, MA 01752-3092. E. Malmstron, H. Ito and C.J. Hawker* IBM Almaden Research Center, Center for Polymeric Interfaces and Macromolecular Assemblies, 650 Harry Road, San Jose, CA 95120-6099

The preparation of copolymers is the most important synthetic strategy available to the polymer chemist to tailor the physical, solution and mechanical properties of macromolecules. In this paper the synthesis of well defined random and block copolymers containing monomer units of very different polarity is described. Well defined random and block copolymers incorporating 4-hydroxystyrene and styrene have been prepared to investigate the combined effects of macromolecular architecture and hydrophobicity on thin film aqueous base dissolution.

378. FUNCTIONALIZATION OF ALKYL MONOLAYERS ON SURFACES WITH DIVERSE AMINES: PHOTOCHEMICAL CHLOROSULFONATION FOLLOWED BY SULFONAMIDE FORMATION.
Ronald L. Cicero*, Peter Wagner°, Matthew R. Linford*', Craig J. Hawker⁺, Robert M. Waymouth*, Christopher E. D. Chidsey*. * Department of Chemistry, Stanford University, Stanford, CA 94305-5080, ° Department of Biochemistry, Stanford University, Stanford, CA 94305-5307, ⁺ IBM Almaden Research Center, San Jose, CA 95120-6099, ' Present address: Max Planck Institut, Berlin, Germany 12489.

We report a simple method for the photoinduced chlorosulfonation of methyl-terminated alkyl monolayers on Si(111). The resulting sulfonyl chloride groups react quantitatively with a wide variety of primary amines to form sulfonamides. These functionalized surfaces are stable to dioxane, aqueous sulfuric acid and aqueous ammonia at temperatures near 100°C.

379. SYNTHESIS AND POLYMERIZATION OF LIPID SUPRAMOLECULAR *"ORGANIC ZEOLITE"* ASSEMBLIES. W. Srisiri,[a] D.F. O'Brien,[a] G. Orädd,[b] S. Persson,[b] G. Lindblom[b] a) Department of Chemistry, C.S. Marvel Laboratories, The University of Arizona, Tucson, AZ 85721, b) Department of Physical Chemistry, Umeå University, S-90187, Umeå, Sweden

Hydrated lipids form nonlamellar, as well as lamellar assemblies at certain conditions of concentration, temperature and pressure. Nonlamellar assemblies can be utilized for both biological and material sciences due to their well-defined porous nanoscale architecture and biocompatible nature of lipids Potential use of these lipids however depends at least in part on the prospects for stabilization of a nonlamellar architectures. Here we reported the stabilization of a nonlamellar bicontinuous cubic (Q_{II}) phase with Ia3d symmetry by polymerization. Monoacylglycerol (MAG) and 1,2-diacylglycerol (DAG) lipids, containing polymerizable dienoyl groups in the hydrocarbon chains were synthesized. Upon hydration, a 9:1 molar mixture of lipids MAG and DAG formed a well-defined Q_{II} phase at temperatures from 19°C to 45°C, as determined by ^{2}H-NMR spectroscopy and x-ray diffraction. The polymerization of these Q_{II} phases was accomplished using thermally decomposition of H_2O_2. The polymers obtained were linear in nature, and the Q_{II} architecture was maintained after polymerization. Polymerized Q_{II} phases can be considered as organic zeolites, providing sites for sequestering drugs or biomolecules, and are also expected to be useful models for the investigation of bicontinuous cubic phases.

380.
PHYSICAL PROPERTIES OF POLY(BENZYLETHER) DENDRIMERS AT THE AIR-WATER INTERFACE. **J.P. Kampf** and C. W. Frank, Department of Chemical Engineering, Stanford University, Stanford, CA 94305-5025. C. J. Hawker, IBM Almaden Research Center, 650 Harry Road, San Jose, CA 91520-6099.

Surface pressure versus area isotherms were taken for a series of fourth generation poly(benzylether) dendrimers with hydrophilic oligo(ethyleneglycol) chains of varying length attached to the focal point of the molecules. We found that those molecules which formed stable monolayers underwent a transition from a monolayer to a bilayer, represented by a peak in the isotherm. Hysteresis experiments showed that, although the monolayer was stable, the bilayer was unstable under the experimental conditions. The length of the hydrophilic tail directly affected the surface pressure at which the bilayer transition occurred but had a negligible effect on the projected area and compressibility of the monolayer. These results demonstrate that it is possible to control the interactions between these dendritic molecules and the water surface

381. **MONOLAYERS OF A POLYMER MELT, TETHERED TO THE WATER SURFACE**
Werner A. Goedel, Max-Planck Institut f. Colloids & Surfaces, Hs.9.9, Rudower Chaussee 5, 12489 Berlin, Germany, e.mail:goedel@mpikg.fta-berlin.de

Hydrophobic polymers with low glass transition temperature (polyisoprenes) and a single head group (sulfonate) have been synthesised and characterised as insoluble monolayers on a water surface. A single ionic head group is sufficient to make these hydrophobic polymers surface active and isotherms of an expanded type have been recorded. These isotherms can be described quantitatively using a theory based on constant density of the hydrophobic region, Gaussian chain statistics and affine deformation: At a fixed area per head group, the surface pressure is proportional to the length of the hydrophobic tail. The slope of this relation is proportional to the third power of the area per head group.
Chains shorter than 300 repeat units systematically deviate from the theory.
(*Langmuir* 1994, 10, 4209 and 1993, 9, 1184, *macromolecules* submitted 1996, *progr. polymer & colloid sci.* submitted 1996).

382.
LANGMUIR AND LBK FILMS OF POLY(VINYLIDENEFLUORIDE) COPOLYMERS AND ORDERED LAYERS WITH PMMA AND POMA R.C. Advincula, C.W. Frank, Stanford University, Department of Chemical Engineering / CPIMA, Stanford, CA 94305-5025, W. Knoll, MPI for Polymer Research, D-55128 Mainz, Germany

Ultrathin LBK films Poly(vinylidine fluoride)and its copolymers have been recently investigated for their ferroelectric switching behavior which is characterized by "conductance switching". Here we report our preliminary investigations on Langmuir and LBK films of the polymers with the view of understanding and optimizing the multilayer film-forming properties at the air-water interface. Only the copolymers are capable of exhibiting monolayer behavior. The multilayers were investigated by ellipsometry, surface plasmon spectroscopy, and X-ray diffractometry. We have investigated the miscibility behavior of these polymers with poly(methylmethacrylate)(PMMA) and poly(octadecylmethacrylate) (POMA) as ultrathin films. Atomic force microscopy showed the formation of fractured domains which could be a consequence of film transfer between the two different polymers.

383.
INTERFACIAL PROPERTIES OF POLYIMIDE LANGMUIR-BLODGETT FILM **Dong-Myung Shin*,** Curtis W. Frank**,Robert D. Miller*** , *Department of Chemical Engineering, Hong-Ik University, 72-1, Sangsu-Dong, Mapo-Ku, Seoul, 121-791, Korea. **Department of Chemical Engineering, Stanford University, Stanford, CA, 94305. *** IBM Almaden Research Center, 650 Harry Road, San Jose, CA 95120-6099

Polyamic acid film was fabricated using the Langmuir-Blodgett technique. Imidization of polyamic acid was carried out using chemical imidization followed by thermal imidization at 220C. The influence of interfacial solvent polarity and specific solvent-solute interaction at the interface of the polyimide film and organic solvents were investigated using UV-visible spectroscopy. A hypsochromic shift of the push-pull azobenzene was observed in polar solvents.
The physical characteristics of the polyimide film formed from the Langmuir-Blodgett approach were characterized by many spectroscopic techniques, including FT-IR, surface plasmon

384. **FILM FORMING POLY(P-XYLYLENE)S BY CHEMICAL VAPOR DEPOSITION: SYNTHESIS, PROPERTIES, AND POLYMER ANALOGOUS REACTIONS** A. Greiner, S. MANG, O. Schäfer, Philipps-Universität Marburg, Fb Physikalische Chemie, 35032 Marburg / Germany

High molecular weight, film forming poly(p-xylylene)s (PPX)s for sensor applications were obtained by vapor phase pyrolysis of halogeno-functionalized p-xylenes. Quinodimethanes were formed from halogeno-functionalized p-xylenes at pyrolysis temperatures between 700-900°C. The quinodithanes formed the corresponding PPXs upon physical condensation on substrates at temperatures below 100°C.
Solubility, fusability and glass transitions of PPXs were controlled by structural modification on the phenylene moiety or on the ethylene bridge.

385.
GLASS TRANSITION IN ULTRATHIN FILMS OF GRAFTED PMMA O. Prucker C.W. Frank, Stanford University, Department of Chemical Engineering / CPIMA, Stanford, CA 94305-5025, H. Bock, S. Christian, W. Knoll, MPI for Polymer Research, D-55128 Mainz, Germany

Thin films of polymers are subject to high thermal and mechanical stress when used as protective coatings or lubricants. However, up to now only little is known about how the restricted geometry of those arrangements influence their thermal and dynamical properties. In this study we investigated the glass transition of grafted PMMA films with thicknesses of 4 - 600 nm on silicon oxide surfaces. Tg was detected by surface plasmon and waveguide spectroscopy showing that a very pronounced Tg depression can be found as the films get thinner than app. 100 nm. Similar experiments on PMMA films deposited by spincoating and by the Langmuir-Blodgett-Kuhn technique revealed a comparable thermal behavior of those films. Thus, we did not find any influence of the varying strength of adhesion and the different (internal) organization of the three types of films.

386.
SYNTHESIS AND PREPARATION OF GRAFTED POLY(CAPROLACTONE) ULTRATHIN FILMS S. Gullerud*C.W. Frank, Stanford University, Department of Chemical Engineering / CPIMA, Stanford, CA 94305-5025, D. Mecerreyes, P. Dubois, R. Jerome, Center for Education and Research on Macromolecules (CERM), University of Liege, Sart-Tilman, B5, 4000 Liege, Belgium, M. Trollsas, J. L. Hedrick, IBM Research Division, Almaden Research Center / CPIMA, 650 Harry Road, San Jose, CA 95120-6099

The synthesis of poly(caprolactone)s with triethoxysilane functional groups as well as pyrene tagged side chains is described. These poly(caprolactone) materials were used to create ultrathin films on a silicon oxide surface by a grafting to approach. Film of monofunctional poly(caprolactone) were limited to 3 nm thickness, while randomly functional poly(caprolactones) in water saturated solvent yielded films as thick as 10 nm. Control of the film thickness was achieved by varying experimental parameters for the grafting to process.

387. Atomic Force Microscopy and Scanning Tunneling Microscopy of Rubbed Polyimide Surfaces. C. Devlin and S. Chiang, Department of Physics, University of California, Davis, CA 95616

Rubbed polyimide films are important in the alignment of liquid crystals in liquid crystal displays. We have used atomic force microscopy (AFM) and scanning tunneling microscopy (STM) to investigate the effects of the rubbing process on the polyimide film surface and the properties of this surface, both bare and covered by a liquid crystal. In

AFM studies of very thin films, we have found the polyimide can be moved around and even scraped away by rubbing. We have also seen nanometer-scale alignment in the polyimide caused by rubbing in STM. The effect of applying liquid crystal to the polyimide has been viewed using STM, where we see "holes" or areas of lesser tunneling current appear on deposition of liquid crystal which remain after the liquid crystal is removed.

388. FORMATION AND PROPERTIES OF POLYSTYRENE-*BLOCK*-POLY(2-CINNAMOYLETHYL METHACRYLATE) BRUSHES. Guojun Liu, Department of Chemistry, The University of Calgary, 2500 University Dr., NW, Alberta, Canada, T2N 1N4

In cyclopentane/THF or cyclohexane/THF with 10 to 40% THF by volume, polystyrene-*block*-poly(2-cinnamoylethyl methacrylate) (PS-*b*-PCEMA) formed micelles with PCEMA as the core and PS as the corona. The micelles were adsorbed by silica and disintegrated on silica surfaces to form polymer brushes in which the PCEMA block formed the anchoring layer and the PS block stretched like bristles of a brush into the solution phase. The brush structure was confirmed by our transmission electron microscopic studies. Scaling relations describing the dependence of surface coverages on the block lengths were established. These relations are in agreement with the theoretical predictions of Marques et al. and are different from those established for polymer adsorption below the critical micelle concentration in a block-selective solvent.

389.
IN-SITU IMAGING OF SWELLING OF END-GRAFTED POLYMER LAYERS UNDER VARIABLE SOLVENT QUALITY. **1.K. Sinniah** 2. A. Karim, J.F. Douglas, 3. J.E. Reutt-Robey, (1). Calvin College, Dept. of Chemistry & Biochemistry, Grand Rapids, MI 49546, (2). NIST, Gaithersburg, MD 20899, (3). Univ. of Maryland, Dept. of Chemistry & Biochemistry, College Park, MD 20742

We have examined the in-situ swelling and lateral distribution of a model end-tethered polystyrene system in good (toluene), marginal (cyclohexane), poor (water), and non-solvent (air) conditions using atomic force microscopy (AFM). Results from grafted polystyrene prepared from its solution in a good solvent are compared to those grafted under marginal solvent conditions. In instances where coverage was incomplete, the height and the average width of grafted polymer clusters on the surface were measurable by AFM in air. The influence of solvent quality on their swelling behavior was investigated by imaging the layers under toluene and cyclohexane solvents. The kinetics of evolution of the grafted layer will also be presented. Finally, we observe the formation of polystyrene micelles in solvents which are inadequately dried.

390.
THERMOSET SECOND-ORDER NLO MATERIALS FROM A TRIFUNCTIONALIZED CHROMOPHORE. Youn Soo Ra, Shane S. H. Mao, Bo Wu, Lan Guo and Larry R. Dalton*, Loker Hydrocarbon Research Institute, Department of Chemistry and Department of Materials Science and Engineering, University of Southern California, Los Angeles, CA 90089-1661; Antao Chen and William H. Steier, Center for Photonic Technology, Department of Electrical Engineering, University of Southern California, Los Angeles, California 90089-0483

A new trifunctionalized Disperse-Red type chromophore, Disperse-Red Triol (DRTO), was synthesized. Reaction of this chromophore with tolylenediisocyanate (TDI) at 80 oC afforded soluble oligomers, which were spin cast to form high optical quality films. Concurrent corona poling and thermosetting of these films resulted in a polyurethane network with an electro-optic coefficient (r_{33} at 1.06 mm) of 14.5 pm/V and excellent dynamic thermal stability. This system (PU-DRTO) offers some prominent advantages over the existing PU-DR19 system, in terms of simpler material processing, wider processing window and higher temporal stability of dipole

391. PREPARATION AND NLO PROPERTIES OF 1,3-BIS(DICYANOMETHYLIDENE)INDANE (BDMI) BASED CHROMOPHORES IN PMMA THIN FILMS. Sam-Shajing Sun and Larry R. Dalton, Loker Hydrocarbon Research Institute, Departments of Chemistry and of Materials Sciences and Engineering, University of Southern California, Los Angeles, CA 90089-1661. Sean M. Garner and William H. Steier, Center for Photonic Technology, Department of Electrical Engineering, University of Southern California, Los Angeles, CA 90089-0271

Amino-phenylenethienylidene (APT) donor-bridge systems coupled with strong electron acceptors such as 1,3-bis(dicyanomethylidene)indane (BDMI) have recently been demonstrated to exhibit exceptionally large second order optical nonlinearities as well as excellent chemical stability in composite polymer thin films. In this paper, "centric alignment" mechanism was proposed to explain both the observed APT-BDMI nonlinearity reduction from anticipated results and the poling induced absorbance drop non-recovery phenomena.

392. SURFACE PATTERNING OF PLASMA DEPOSITED FLUOROPOLYMER FILMS BY ATOMIC FORCE MICROSCOPY
Michael D. Garrison and Buddy D. Ratner
Center for Bioengineering, University of Washington, Seattle, WA 98195

Deposition of fluoropolymers via radio-frequency glow discharge (RFGD) allows for control of film thickness, roughness, and surface chemistry. The resulting films provide a smooth, planar, substrate suitable for atomic force microscopy (AFM). In certain applications, design of surface patterns in RFGD fluoropolymer films might be advantageous. The AFM can be used to rupture, etch, or erode surface films at the nanometer scale. However, variations in the film chemistry may affect the surface modulus under the loading and shear forces of the AFM tip. This work describes the preparation of fluoropolymer films by RFGD. Surface chemical composition was determined by electron spectroscopy for chemical analysis (ESCA). Surface topography and morphology was assessed by AFM. Using the AFM as a lithographic tool, patterns were introduced onto the surface film. The types of surface patterns observed, and the relevant control parameters are discussed.

393. PLASMA-DEPOSITED HYDROCARBON FILMS FOR ORGANIC SURFACE CHEMISTRY ON METALLIC DEVICES. C. Cassinelli, M. Morra, Nobil Bio Ricerche, 14018 Villafranca d'Asti, ITALY, and F. Garbassi, L. Meda, Istituto G. Donegani, 28100 Novara, ITALY

Surface modification is often required to tune surface properties of metallic devices. Our goal was to use the powerful and well developed methods of surface modifcation of organic materials to direct surface properties of titanium devices. To this end, a 20 nm thick hydrocarbon layer was deposited from ethylene plasma on Ti foils and implant parts, and the surface organic layer was further modified using several organic surface chemistry recipes. Cell adhesion was promoted by air plasma treatment of the surface hydrocarbon layer. Primary amino groups on a 6 atoms spacer were introduced by carbodiimide promoted condensation of hexamethylenediamine and carboxyl groups of poly acrylic acid. The latter was grafted (*via* Ce(IV) promoted polymerization) on the air plasma treated hyrocarbon coating. A polysaccharide was covalently linked to the surface amino groups. X-ray Photoelectron Spectroscopy and cell adhesion data show that the plasma deposited hydrocarbon layer can effectively act as the first building block of more specific surface chemistries and allows to apply the established body of knowledge on surface modification of plastics to metallic devices.

394. PLASMA POLYMERIZATION UNDER D. C. GLOW. ESTIMATION OF FORMATION OF POLYMERIZED FILMS USING AFM TECHNIQUE, Daisuke Sato, Toshihiro Suwa, and Masa-aki Kakimoto, Department of Organic and Polymeric Materials, Tokyo Institute of Technology, Meguro-ku, Tokyo 152, Japan

Plasma polymerized films of octamethyltetrasiloxane (OMCTS) was prepared under D.C. glow discharge. Surface of the polymerized films were observed by the atomic force micro scope (AFM). The particles were seen as the white patches in the AFM picture. It is observed that diameter of the particles increased and contour of them became dim with increasing the monomer gas pressure. When the applied voltage changed from 0.8 to 1.6 kV under the constant pressure of 10 Pa, the contour of the particles became clear with increasing the voltage. It is assumed that formation of the thin films by the plasma polymerization consists of two kinds of reactions. 1) The first is formation of small particles (clusters) by agglomeration of active species under plasma state. 2) The second is polymerization at the surface of the substrate or films formed.

395. SURFACE ANALYSIS AND BIOREACTIONS ON SILVER-CONTAINING AMORPHOUS HYDROGENATED CARBON FILMS.
R. Hauert[1], R. Gampp[1], U. Müller[1], A. Schroeder[2], J. Blum,[2] J. Mayer[2], F. Birchler[2], E. Wintermantel[2]
[1] Swiss Federal Laboratories for Materials Testing and Research (EMPA), CH-8600 Dübendorf, Switzerland.
[2] Chair of Biocompatible Materials Science and Engineering, Department of Materials, ETH Zürich, CH-8952 Schlieren, Switzerland.

For applications like catheters, sensors etc. biologically inert surfaces with as little interaction with biological media as possible are needed. Silver-containing films of hard amorphous hydrogenated carbon (Ag - a-C:H) have been prepared by simultaneous dc magnetron sputtering of a silver target and rf plasma assisted chemical vapor deposition (PACVD) of methane. XPS analysis showed the Ag $3d_{5/2}$ line at the position for pure silver, so silver is expected to be incorporated in the films in a metallic state. Compared to pure a-C:H, Ag - a-C:H films containing 8 at% (atomic percent) silver did not have a significant influence on the growth rate of osteoblasts, whereas Ag - a-C:H films containing 16 at% decreased the cell growth rate by more than a factor of 2.

396.

A novel technique suitable for generation of large volume plasmas
Kristian Glejbøl, Bjørn Winther-Jensen, NKT Research Center, Sognevej 11, DK - 2605 Broendby

For many parts made from polymers, it is cost efficient to substitute the used polymer with a less expensive grade and then paint the part, to ensure a high surface quality. To guarantee good adhesion of the paint, it is often necessary to treat the surface prior to painting. The treatment can preferably be carried out using the energetic particles and radiation from a plasma.
Here a new approach to generation of plasmas, the 3-phase plasma is discussed. The equipment is made solely from inexpensive standard components, readily available. The 3-phase plasma technique has been used for surface modifications for more than two years and has proven to be a simple, highly reliable tool. The 3-phase plasma offers the same advantages as a conventional DC-plasma, without the disadvantages normally associated with this type of plasma (arching, expensive power supply, inhomogenious plasma).

397. INTERPHASE CHARACTERIZATION OF SILICON-BASED BARRIER LAYERS ON PET. N. Schühler, A. S. da Silva Sobrinho, J. E. Klemberg-Sapieha, M. Andrews* and M. R. Wertheimer, Groupe des Couches Minces (GCM) and Dept. of Engineering Physics, École Polytechnique, Montréal, QC H3C 3A7. *Dept. of Chemistry, McGill University, Montréal, QC H3A 2K6.

Thin films of SiO_2 or Si_3N_4 on flexible or rigid polymeric substrates are receiving much attention in numerous application areas, since they are excellent barriers against the permeation of gases (e.g. O_2) or vapors (e.g. H_2O), they permit recycling are optically transparent and inert, etc. In this laboratory, SiO_2 and Si_3N_4 films are deposited onto commercial and spin-coated polyethylene terephthalate (PET) using "dual"-frequency (microwave/radiofrequency) plasma-enhanced CVD (PECVD). These films adhere extremely well to the polymer substrates, the reason for this behavior being an extended "interphase" region. In this paper we report results obtained by probing the interphase by various destructive (ERD, XPS, EDX) and non-destructive (ATR-FTR, IRRAS) techniques. For PECVD-deposited SiO_2 films, the thickness of the interphase was found to range from about 30 to about 100 nm, while the thickness obtained for PVD-SiO_2 films by the same techniques was considerably smaller. Infrared spectroscopy indicates, for both SiO_2 and Si_3N_4 films, a predominant organosilicon phase close to the substrate. We propose an ablation/redeposition mechanism for the early stage of the coating process, during which the plasma contact leads to a mobilization of fragments from the polymer surface into the plasma: Reaction and redeposition with activated monomer yields the organosilicon-rich interphase.

398. ORGANOSILICON PLASMA POLYMERS FOR PERVAPORATION MEMBRANES. M. S. Silverstein, L. Zuri and M. Narkis, Departments of Materials Engineering and Chemical Engineering, Technion - Israel Institute of Technology, Haifa 32000, Israel

Water often contains trace amounts of toxic chlorinated volatile organic compounds. In pervaporation, a membrane separation process, the process efficiency is proprotional to membrane thickness and composite membranes with an ultra-thin selective layer and a porous support are preferred. Plasma polymerization can be used to deposit ultra-thin, selective films in a process that avoids environmentally unfriendly etchants and solvents. The molecular structure and transport properties of plasma polymerized hexamethyldisiloxane films were investigated. The maximum rate of deposition, at the lowest plasma power and highest monomer flow rate used, yielded a more organic polydimethylsiloxane-like structure. The highest plasma power and lowest monomer flow rate used yielded a more inorganic silica-like structure. The flux through the plasma polymer membranes was exceptionally high and the permeate was enriched in water from a trichloroethane in water feed.

399. LASER SENSITIVE PLASMA POLYMER FILMS. M. S. Silverstein and I. Visoly, Department of Materials Engineering, Technion - Israel Institute of Technology, Haifa 32000, Israel and M. Janai and Y. Cassuto, Chip Express, P.O. Box 2401, Advanced Technology Center, Haifa 31000, Israel

Rapid production of new integrated circuit (IC) designs involves using a frequency doubled NdYAG laser to cut metal lines. This rapid production technique could be extended if the same laser could be used to micromachine ultra-thin dielectric and protective polymer films for multilayer IC lithography. Plasma polymerization deposits ultra-thin, pinhole-free films in a dry process that avoids the etchants, solvents and temperatures that limit the applicability of other deposition techniques and uses standard microelectronics processing equipment. The molecular structure and laser sensitivity of plasma polymerized ethylene films were investigated. The molecular structure, optical absorption and adhesion were affected by the addition of hydrogen or nitrogen to the plasma. Square laser pulses on the order of microns were reproduced quite accurately with square corners and straight sidewalls.

400. IMPORTANCE OF SURFACE RESTRUCTURING IN THE EVOLUTION OF THE SURFACE PROPERTIES OF PLASMA POLYMERS. T.R. Gengenbach, R.C. Chatelier and H.J. Griesser, Division of Chemicals and Polymers, CSIRO, Private Bag 10, Clayton South MDC, Clayton 3169, Australia.

It is well known that the surface layers of conventional polymers can adapt their compositions in response to interfacial forces. Plasma polymers, on the other hand, are generally viewed as highly crosslinked, imperturbable systems. We have investigated the surface mobility of a number of plasma polymers and found a degree of mobility, although substantially less extensive and slower than that of conventional polymers. The characteristic time constant is of the order of 50 days. Angle dependent XPS analyses show changes in the depth distribution of polar groups with time; together with time-dependent contact angle measurements they enable interpretation of the changes in surface compositions with time in terms of concurrent oxidation and restructuring. In general, the mobility of plasma polymer segments cannot be neglected.

401. PLASMA POLYMER COATINGS ON TRACK-ETCHED MEMBRANES. G.J. Beumer, A. Fuhrer, T. Vaithianathan and H.J. Griesser, Division of Chemicals and Polymers, CSIRO, Private Bag 10, Clayton South MDC, Clayton 3169, Australia.

Thin plasma polymer coatings are useful for achieving modification of the interfacial properties of track-etched membranes, and to provide reactive surface groups that can be used for further covalent interfacial reactions such as the immobilization of proteins. However, such coatings reduce the effective pore size of the membrane. In this study we have investigated the effects of coating Poretics membranes of 100 nm and 50 nm pore size with n-heptylamine and acetaldehyde plasma polymers of various thicknesses. Statistical evaluation of scanning electron micrographs indicates that the narrow pore size distribution of the original membrane was retained on plasma coating. By controlling the deposition time, any arbitrary pore size is achieved, thus enabling fabrication of membranes with controlled interfacial properties and narrow pores.

402. MALDI-TOF-MS STUDY OF PROTEIN ADSORPTION TO HEPTYLAMINE PLASMA POLYMER. Peter Kingshott, Heather A.W. St John, Ronald C. Chatelier, Frank Caruso and Hans J. Griesser. Division of Chemicals And Polymers, CSIRO, Private Bag 10, Clayton 3168, Australia.

We introduce matrix-assisted laser desorption ionisation time-of-flight mass spectrometry (MALDI-TOF-MS) as a direct method for the investigation of protein adsorption to plasma polymer surfaces. Lysozyme was adsorbed to freshly deposited n-heptylamine plasma polymer (HA pp), and after application of a photoadsorbing matrix (sinapinic acid) and bombardment of the surface with a pulsed nitrogen laser (337 nm) the resultant spectrum contained a peak at m/z 14106.6 corresponding to the $[M + H]^+$ parent ion of lysozyme. Confirmation that this peak is in fact lysozyme comes from the presence of both a charged dimer peak at m/z 28482.4 and a doubly charged monomer peak m/z 7060.4. XPS and QCM analysis confirm that a side-on hexagonal monolayer of lysozyme exists on the HA pp surface.

403. STUDY OF THE MECHANICAL STABILITY OF HMDSN PLASMA POLYMERIZED THIN FILMS. C. Z. Ramalho, M. O. H. Cioffi, B. Klaffke Dupoy, R. P. Mota, R. Y. Honda, M. A. Algatti, H. J. C. Voorward and M. E. Kayama, UNESP-Universidade Estadual Paulista, Campus de Guaratinguetá, 12500-000 Guaratinguetá, SP, Brazil

The adhesion of thin films to different substrates is a subject that has received considerable attention recently. The high degree of adhesion of a film to a surface is fundamental for most of the applications. In this paper is realised the study of adhesion and mechanical stress in HMDSN thin films deposited over glass substrates. All the films resisted to the Scotch-test without peeling out. A further test using boiling distillate water and seawater demonstrated that the films were stable presenting only small spots due, in the case of distillate water, imperfections in film structure. The boiling seawater removed the HMDSN film producing a great number of spots, mainly due the corrosive processes caused by the presence of Chlorine and Sodium ions in aqueous media. The films presented a compressive stress that relaxed to values six times lower the initial one after 60 hours, due to incorporation of water vapour and oxygen present in atmosphere.

404. PLASMA POLYMERIZATION/CO-POLYMERIZATION TO ENGINEER SURFACES AT A MOLECULAR LEVEL FOR PROBING CELLULAR RESPONSE - R. DAW ET AL, Univ. of Sheffield, UK.

Cellular responses are specific to a material's surface features. These features may be physical e.g. topography or chemical e.g. acidity. The surface chemistry most importantly infulences which proteins initially attach to an implant material. To probe the effect of chemistry on cellular response, there is a need to design surfaces at a molecular level.

Many researchers have utilised SAM to exert a high level of control over surface chemistry. However, plasma polymerisation/co-polymerisation provides workers with an alternative and versatile technique.

This talk will discuss the use of low power plasma polymerisation and co-polymerisation to engineer surfaces at a molecular level. 'Stable' films rich in specific functionalities (e.g.**acid,** alcohol or amine) can be produced and their concentration tailored by the addition of an inert co-reactant hydrocarbon to the 'functionalised' monomer.

These surfaces will be used to study the response of osteoblast cells to various chemistries.

405. PLASMA-TREATED POLYESTERS: POWER-TIME RECIPROCITY. J. M. Grace and L. J. Gerenser, Eastman Kodak Company, Rochester, New York, 14650.

Investigations of the effects of plasmas on polymer surfaces are of increasing interest in industry. The effects of plasmas are often studied by fixing the plasma power and varying the treatment time or by fixing treatment time and varying the plasma power. These approaches rely implicitly on the assumption that the effects of plasma power and treatment time are reciprocal (i.e., that the effect of treatment can be directly related to the dose, defined as power x time x A, where A is a geometrical constant). While power-time reciprocity is sometimes observed in plasma treatments, this reciprocity is by no means universal. We will discuss examples of reciprocity and nonreciprocity, as observed by probing the surface chemistry of polyesters treated in nitrogen and oxygen plasmas. Specifically, for nitrogen plasma treatments of a polyester, we have found that surface chemistry, as probed by X-ray photoelectron spectroscopy, can exhibit power-time reciprocity or non-reciprocity, depending on pressure. In comparison, oxygen plasma treatments of the same material exhibit power-time reciprocity over the same pressure range. Possible reasons for the observed behavior will be discussed.

406. INTERACTIONS BETWEEN *PSEUDOMONAS AERUGINOSA* AND PLASMA-DEPOSITED PEO-LIKE THIN FILMS DURING INITIAL ATTACHMENT AND GROWTH. Erika E. Johnston*, James D. Bryers**, Buddy D. Ratner, Depts. of Bioengineering and *Chemical Engineering, University of Washington, Seattle, WA 98195-1750 and **Center for Biofilm Engineering, Montana State University, Bozeman, MT 59717

Bacterial biofilms are morphologically complex layers of bacteria and polysaccharides that accumulate at solid-liquid interfaces in aqueous environments. Biofilms contribute to the fouling of marine surfaces and heat transfer equipment and to the persistence of implant infections. Therefore there exists an acute need for surface treatments that can slow or prevent the formation of bacterial biofilms. It has been previously shown that PEO-like plasma deposited films (PDFs) resist adsorption of blood proteins and platelets and slow the initial accumulation of bacteria. The current investigation was undertaken to improve our understanding of how bacteria interact with plasma-deposited PEO-like surfaces. The individual processes (adsorption, desorption, cell erosion from colonies, and growth) that contribute to accumulation were examined under two sets of conditions: initial attachment, and growth during exposure to glucose rich media.

407. MORPHOLOGICAL STUDIES OF SIO$_x$ COATINGS DEPOSITED ON POLYMER SUBSTRATE TO ENHANCE BARRIER EFFICACY: EFFECT OF DEPOSITION POWER. B.-C. Wang, A. G. Erlat, R. J. Spontak, Department of Materials Science & Engineering, North Carolina State University, Raleigh, NC 27695; Y. G. Tropsha, E. A. Vogler, M. Dalvie, K. D. Mar, Polymer Science & Technology Department, Becton Dickinson Research Center, RTP, NC 27709

Transparent coatings capable of enhancing gas impermeability through polymers have recently received considerable attention in packaging and medical applications. Plasma-based processing affords the advantage of depositing multi-directional coatings. In this work, analytical methods, such as energy-filtered transmission electron microscopy (EFTEM) and atomic force microscopy (AFM), have been successfully employed to help elucidate key relationships between the morphology of thin SiOx films and the barrier properties of coated polymers. It has been found that discrete particles constitute the principal morphological characteristics of plasma-deposited SiOx coatings on polycarbonate. Under controlled deposition conditions (e.g., power, pressure, flow rate and oxygen: hexamethyldisiloxane ratio), the particle characteristics remain consistent in both size and porosity. However, from AFM and ellipsometry, the porosity is power-dependent. A critical high power level is identified for producing films with consistently high barrier quality. Depositions conducted at low power, on the other hand, result in SiOx films of higher porosity and poorer barrier properties. Along with the data presented, a conceptual model will be introduced to interpret the observed relationships.

408. SURFACE AND CORROSION CHARACTERISTICS OF FLUORINE-RICH CARBON FILMS. C. Srividya & S. V. Babu, Center For Advanced Materials Processing & Department of Chemical Engineering, Clarkson University, Potsdam, NY 13699, S. A. Visser, Eastman Kodak Company, Rochester, NY 14650-2129

Fluorine-rich carbon films with excellent corrosion resistance and surface energies as low as 20 mN/m were deposited on type 301 stainless steel substrates by radio frequency (13.56 MHz) plasma deposition from a mixture of C_2F_6 or CF_3COCF_3 and C_2H_2 or C_4H_6. Surface energies of the films were estimated using contact angle values obtained with various solvents. The composition of the surface layer of these films was analyzed using x-ray photoelectron spectroscopy (XPS). The film surface energy decreased with increasing flourine content and CF_3 and CF_2 groups at the film surface. The corrosion characteristics of the fluorinated carbon films were investigated using standard potentiometric methods in electrolytic solutions containing 0.1 M NaCl & 0.1 M Na_2SO_4 and 0.1 M HCl and 0.1 M Na_2SO_4, respectively, in deionized water. The measured corrosion current was 3 orders of magnitude smaller than that for bare steel and decreased with increasing fluorine content in the film.

409. ^{17}O NMR INVESTGATIONS OF OXIDATIVE DEGRADATION IN POLYMERS T. M. Alam, M. Celina, R. A. Assink, K. T. Gillen, and R. Clough, Aging and Reliability, Bulk Properties Department, Sandia National Laboratories, Albuquerque, New Mexico 87185-1407.

An improved understanding of mechanisms for oxidative degradation in polymers may be realized by identification of the degradation species. Both ^{1}H and ^{13}C NMR have been used to study polymer degradation, but the interpretation is complicated by the relatively small concentration of degradation products in comparison to the native polymer signal. In this study we report both solution and solid state magic angle spinning (MAS) ^{17}O NMR spectra for a series of oxidatively aged polymers. By utilizing enriched O_2 during the aging process, the ^{17}O NMR spectra provide a unique probe in that all observed resonances result directly from degradation. For the polymers investigated a wide range of degradation products were observed including aldehydes, ketones, carboxylic acid, alcohols and ethers. The resulting decomposition products can be identified by utilizing polarization transfer techniques and model compound studies. Work supported by United States Department of Energy under Contract DE-AC04-94AL85000.

POLY

410. SOLID STATE ^{13}C NMR INVESTIGATIONS OF PERFLUOROOCTYL ACRYLATE / METHYL ACRYLATE COPOLYMER BLENDS. K. L. Altmann and L. H. Merwin, Naval Air Warfare Center Weapons Division, Code 474230D, China Lake, CA 93555 and R. D. George, Naval Command, Control and Ocean Surveillance Center, RDT&E Division, Environmental Chemistry & Biotechnology, Code D361, 53475 Strothe Road, San Diego, CA 92152-6325

Coatings used to protect ship hulls from marine fouling organisms are essential for efficient and economic Fleet performance. Current antibiofouling strategies involve the use of copper-containing compounds as toxic agents that poison the organisms. These materials are highly regulated and pose significant environmental hazards. The present effort uses a different mechanism for fouling control, in which the physical properties of the surface are utilized to allow easy release (and removal) of the organisms. A series of environmentally-friendly fluorinated polymer blends are under investigation as candidate replacements for copper-containing coatings. In the process of characterizing these novel materials, solid state NMR experiments, including ^{13}C CP/MAS and ^{1}H $T_{1\rho}$ measurements, have been performed to investigate the homogeneity of the blends.

411. END GROUP CHARACTERIZATION OF PET AND ITS COPOLYMERS. T. Amari, Mitsubishi Chemical Corporation, Yokkaichi Research Center

In production of aromatic polyesters such as poly(ethylene terephthalate) and its copolymers, the understanding of degradation mechanism of polymer chains is very important. We have investigated degradation via end groups by NMR and titration techniques. We have determined various end groups with NMR, especially unsaturated groups which are believed to have formed as a result of thermal breakage of ester bonds. We have also measured the total concentration of unsaturated groups by bromination titration, and compared the results with the NMR method. Together with data of other end groups, we have attmpted to describe the mechanism for degradation.

412. THE MECHANICAL α RELAXATION IN POLY(ε-CAPROLACTONE) INVESTIGATED BY SOLID-STATE NMR SPECTROSCOPY. Karthik Nagapudi, Haskell W. Beckham *Polymer Education and Research Center, Georgia Institute of Technology, Atlanta, GA 30332-0295.*

Molecular motions underlying the mechanical α transition in poly(ε-caprolactone) (PCL) are investigated with solid-state ^{13}C NMR spectroscopy. Dynamic mechanical spectroscopy (DMS) data indicate the presence of a glass transition at -45 °C and an α transition around 10 °C in solvent-cast films of PCL. Melt-cast films exhibit the glass transition, but not the α transition. X-ray diffraction indicated both samples were highly crystalline with diffraction peaks at the same 2θ values. ^{13}C CP/MAS spectra of melt-cast and solvent-cast PCL revealed the absence of motions on the timescale of the carbonyl chemical shift anisotropy (11.5 kHz) at room temperature. 2D exchange experiments are underway that utilize the orientation dependence of the carbonyl carbon chemical shift.

413.
Relaxometric Study of Oxygen Absorption on Aromatic Polymers. D. Capitani, A.L. Segre
Ist. Strutturistica Chimica del CNR M.B. 10 Monterotondo Staz.   00016 ROMA (Italy)

Oxygen can be selectively absorbed on aromatic systems. The selective absorption of paramagnetic oxygen on aromatic polymers causes effects observable by ^{1}H NMR relaxometric methods since it acts as a relaxation contrast agent. By lowering the temperature a marked shortening of the spin lattice relaxation (T_1) is observed. In some aromatic polymer, at temperatures adequately low, a full interruption of the spin diffusion process is observable. As a consequence protons belonging to the aromatic rings relax much faster

than backbone or aliphatic protons. The amount of absorbed oxygen is different in different polymorphs and much higher in the amorphous fractions on respect to their crystalline counterparts. Different polymorphs exhibit very different T_1 relaxation times and can be easily measured. In semicrystalline aromatic polymers, NMR relaxometry can be used to obtain the crystalline/amorphous ratio. The extreme spread of T_1 values, induced by oxygen absorption, can be used to measure second order transitions. Examples of the usefulness of the method will be given shoving:
i) interaction of a guest polymer, syndiotactic polystyrene, with a host small molecule. ii) evaluation of a polymer as a possible oxygen scavenger.   iii) presence of secondary transition in copolymers.

414. SOLID STATE ^{13}C NMR CHARACTERIZATION OF SPACER CONFORMATION IN THE THERMOTROPIC LIQUID CRYSTALLINE POLYETHERS TPP-*n*, Jinlong Cheng, Raychem Corporation, 300 Constitution Drive, Menlo Park, CA 94025; Stephen Z. D. Cheng, Department of Polymer Science, The University of Akron, Akron, OH 44325

The mesophase behavior of a couple of polyethers based on 1-(4-Hydroxy-4'-biphenyl)-2-(4-hydroxyphenyl)propane and α, ω-Dibromoalkanes (TPP-*n*), where *n* = 12 and 15, the number of CH_2 groups in the spacer, have been characterized by solid state ^{13}C NMR. The chain conformations of the spacer are studied as a function of temperature by the γ-*gauche* effect of the ^{13}C NMR chemical shift which depends largely on the C-C conformation under the conditions of magic angle spinning. Empirical expressions based on the ^{13}C chemical shift and signal intensity are proposed to quantitatively reflect the *trans* content in the spacer.

415.
A NEW CPMAS 13C NMR METHOD FOR THE STUDY OF BIOGEOPOLYMERS Robert L. Cook Cooper H. Langford, Department of Chemistry, University of Calgary, 2500 University Dr. N.W. , Calgary, Alberta, T2N 1N4

In the study of biogeoploymers, such as humic materials, solid state cross polarization magic angle spinning (CPMAS) 13C NMR has been a very useful tool for interrogation. It has been a long held belief that the best spectra are obtained on low field instruments using slow to moderate sample spinning speeds. A new method of CPMAS 13C NMR for the study of biogeoploymers is presented. This new method uses high fields, high sample spinning speeds, and the novel Ramp-CP method. The way in which this method overcomes the limitations of the previous CPMAS methods will be discussed. The way in which this new method is superior in terms of qualitative and quantitative data will be discussed and results on a well characterized fulvic acid will be presented

416. DETERMINATION OF REACTIVITY RATIOS FROM ^{13}C NMR ANALYSIS OF TRIAD-FRACTIONS FOR STYRENE-ACRYLONITRILE COPOLYMERISATION AS A FUNCTION OF TEMPERATURE. Angelo Ferrando, Aldo Longo, EniChem Research Center, Mantova, Italy

The copolymerisation of styrene and acrylonitrile has been investigated with the purpose of predicting the changes in copolymer composition and comonomer distribution with the copolymerisation temperature. Five different temperatures were studied, covering the range between 25° and 130°C. For all the samples the composition were determinated by elementary analysis and ^{1}H NMR, and by ^{13}C NMR the triad-fractions; from these data the four different reactivity ratios expected for the penultimate model were then calculated: raa = 0.520 exp[-836/T(K)], rsa = 0.652 exp[-623/T(K)], ras = 0.742 exp[-90/T(K)] and rss = 1.105 exp[-492/T(K)]. As already seen for other aspects of this copolymerisation, also in this case the only compositions data were more or less insensitive or at least not conclusive respect the effect of temperature of copolymerisation. On the contrary, the data collected from the triad-fractions were found really able to show selectively the variations induced by the different conditions of reaction on the microstructure of the copolymer.

POLY

417. THE EFFECT OF COPOLYMER COMPOSITION ON THE DIFFUSION OF WATER INTO POLY(HYDROXY-ETHYL METHACRYLATE-CO-TETRAHYDROFUFURYL METHACRYLATE). P.Y. Ghi, D.J.T. Hill, P.J. Pomery & A.K. Whittaker#, Department of Chemistry, #Centre for Magnetic Resonance, University of Queensland, QLD 4072, Australia.

The study of the diffusion of water into swellable polymer matrices such as hydrogels is significant to the pharmaceutical industry, since the controlled release of active agents incorporated within the polymer matrix is related to the kinetics of the diffusion of water into these matrices. The aim of this paper is to investigate the effect of copolymer composition of a hydrogel system, poly(hydroxyethyl methacrylate - co-tetrahydrofurfuryl methacrylate), on the diffusion of water by mass uptake measurements and by NMR imaging of the water concentration within the polymer matrix.

418. **Characterization of Flame Retarded Polymer Combustion Chars by Solid-State ^{13}C and ^{29}Si NMR and EPR.** Jeffrey W. Gilman*, Serge Lomakin, Takashi Kashiwagi, David L. VanderHart, and Vitaly Nagy; NIST, Gaithersburg, MD 20988

Abstract: Additives that increase the amount of charcoal-like residue or char that forms during polymer combustion are effective fire retardants. However, very little is understood about the structure of char or how it forms. We have reported that silica gel combined with potassium carbonate is an effective fire retardant for a wide variety of polymers. One of the goals of this work is to determine the role of non-protonated carbon in controlling char performance, and thereby in controlling flammability. Using a combination of CP/MAS and SP/MAS ^{13}C NMR techniques we report here on a method for estimating the ratio of non-protonated carbon to protonated carbon in hydrogen-depleted chars containing high concentrations (10^{19} spins/g) of paramagnetic centers.

419. **2D ROTATIONAL FRAME NOE SPECTROSCOPY (ROESY) STUDIES ON THE CONFORMATION OF MISMATCHED NUCLEOTIDE**

Mingming Guo*, Institute of Polymer Science, The University of Akron, Akron, OH 44325-3909
Shiping Zhang, Chemistry Department, New York University, New York, NY 10003

The structure and conformation of a partially self complementary pentanucleotide, dGTCAC, designed to form a C-C internal mismatch in the duplex state, were investigated using ^{1}H 1D Jump-and-return, presaturation and the 2D TOCSY, NOESY and ROESY NMR at different temperature. In the HOHAHA spectra, relayed intranucleotide coherence from H1'-to-H3', from H1'-to-H4' and from H3'-to-H5', H5" facilitated the assignment procedure. The crosspeaks of 2D NOESY spectra acquired at different temperature and mixing time are very weak, even for some short distance proton pairs. This indicated that the dynamic of the DNA system is in the motion unfavorable region. 2D ROESY spectra offer necessary though space information of the DNA. The pentanucleotide dGTCAC forms a weak duplex at NMR concertration and at low temperature. The presence of the C-C mismatch interrupts the duplex at C3. The base involved in the C-C mismatch lie at the periphery of the double helix, in proximity with the intrareseduial sugar proton H2' and H2", but not the preceding H2' and H2" sugar protons.

420.

SOME APPLICATIONS OF GRADIENT STIMULATED ECHO NMR METHOD IN BIOPOLYMERS

Mingming Guo*, Institute of Polymer Science, The University of Akron, Akron, OH 44325
Shiping Zhang, Chemistry Department, New York University, New York, NY 10003

PFG stimulated echo method was used to quantitatively measure the diffusion constants for both water and solute for the aqueous DNA junction systems. As a water suppression method, the advantages of this technique are absence of phase distortion, large suppression factor for solutions of macromolecules, and no needs for phase cycling. With short (3 ms) and strong PFG (32 G/cm), the sensitivity increases significantly and the exchangeable imino resonances appear. In the FT mode the diffusion constant D of water and different DNA were obtained simultaneously by arraying PFG strength. For water, two component diffusion behavior appears. The fast one is related to the free water, while the slow one related to the hydrated water. Diffusion constant D, as a function of shape, size, binding etc., of four arm junction DNA is larger than that of the related two hairpins. Because the junction have high molecular weight, more double helix and slow migration rate. As a sensitive probe of molecular shape, the D of the two single strand which have similar numbers of base pairs are different. The one with large D has stronger hydrogen bonding, which matches the electrophoresis results.

421. 2D ROESY AND NOESY STUDIES ON THE INTERACTION OF ETHIDIUM WITH A DNA CONTAINING A BASE-PAIR MISMATCHED

Mingming Guo,* Institute of Polymer Science, The University of Akron, Akron, OH 44325-3909
Shiping Zhang, Chemistry Department, New York University, New York, NY 10003

^{1}H 1D 1-1 echo and presaturation spectra as well as 2D TOCSY, NOESY, and ROESY NMR at different temerature and concentration were used to investigate the structure and dynamics of the binding of ethidium bromide (EB) to a partially self complementary pentanucleotide, dGTCAC, designed to form a C-C internal mismatch in the duplex state. In the absence of EB, the pentamer forms a weak duplex at NMR concentration and at low temperature. The presence of the C-C mismatch interrupts the duplex at C3. After combining with EB the DNA become rigid, the correlation time become short, and the NOE's become much stronger. EB prefers to intercalate to the mismatch position between the T2 and C3, to stabilize the conformation of the DNA based on several intermolecular NOE's and ROE's. The NOE and ROE cross peaks allow construction of a model placing the phenanthridine ring system of EB in the duplex, with the phenyl and ethyl moieties lying out side. The symmetric conformation of the DNA become unsymmetric after combining with EB, which returns to symmetric above 10 °C.

422. SOLID STATE NMR INVESTIGATION OF THE MISCIBILITY OF MOLECULAR COMPOSITES CONTAINING AROMATIC POLY(PYRIDINIUM SALTS)

Mingming Guo,* Feiyue Lin, and Frank. W. Harris*
Mourice Morton Institute of Polymer Science, The University of Akron, Akron, OH 44325-3909

The miscibility and morphology of ionomeric molecular composites, comprised a rigid-rod aromatic poly(pyridinium triflate) (PPT) and poly(N,N-dimethylacrylamide-co-(2-acryamido-2-methyl-2-propanesulfonic acid))(PDMAS), were investigated using solid state high-resolution NMR. Composites contain 5, 10 and 15 wt% PPT were prepared by mixing a 1 wt% DMF solution of PPT with a 5 wt% DMF solution of PDMAS and isolated by precipitation in acetone. Changes in solubility and DSC measurements indicate that there was considerable intermolecular interaction between the two polymers. The $^H T_1$ relaxation time determinations indicated that PPT and PDMAS are intimately mixed on a scale of 300-600 Å due to intermolecular ionic interaction in the composite. However, The $^H T_{1\rho}$ relaxation time determinations indicated that: the molecular composites were heterogeneous on a scale of 30-50 Å. The difference in the $^H T_{1\rho}$ of the two components in the composites decreased and approached 0 when the ratio of cations in PPT and anions in PDMAS approached 1.

423.
DETERMINATION OF SPECIFIC RUBBER-FILLER INTERACTIONS BY SOLID STATE NMR TOOL B. Haidar, R. Mansancal, A. Vidal, L. Delmotte, Institut de Chimie des Surfaces et Interfaces, CNRS, P.O.Box 2488, F68057 MULHOUSE-France

Solid state NMR has been widely used as an excellent tool in the investigation of rubber reinforcement. Attention is often limited to the determination of T2, the relaxation time of the polymer protons as a whole. We tried in the present work to examine the effect of specific interactions between the filler surface and each one of the different groups constitutive of a polymer chain. Spin-echo technique was used in which pulse sequences were 90° - t - 180°- t. High resolution was achieved by rapidly spinning (up to 18 kHz) the sample at the magic angle. The actual values of the signal intensity as function of the delay time t of each peak was found to be reasonably expressed by two exponential equations. Results show that some groups are more immobilized than others under the effect of the filler (carbon black)/polymer (SBR and different BR polymers) interactions.

424.

CHARACTERIZATION OF AUTOMOTIVE CLEAR COATS CONTAINING SILANE REACTIVE DILUENTS. J. E. Hansen[1], L. G. Galya[2], F. Davidson[2], G. C. Campbell[2], and A. D. Calibeo[2]; [1]DuPont Automotive Products, 3401 Grays Ferry Avenue, Philadelphia, PA 19146; [2]DuPont Central Research and Development, Experimental Station, Wilmington, DE 19880

In response to new environmental legislation, reactive diluents are being used to produce low volatile organic content (VOC) automotive coatings. The use of hydrosilylated terpenes to produce high solids clear coats has proved to be an attractive response to this drive. Hydrosilylation produces several different isomers which have been characterized by ^{29}Si, ^{13}C and ^{1}H NMR. As the hydrosilylated terpenes are incorporated into cured films, a range of isomers is produced. The extent of hydrolysis and subsequent condensation has been determined by ^{29}Si CP/MAS NMR.

425. Ferroelectric Phase Transition of P(VDF/TrFE) Studied from CP/MAS NMR Method. F. ISHII and A. TSUTSUMI. Department of Applied Physics, Hokkaido University, Sapporo 060, Japan.

The ^{1}H-^{13}C cross nuclear magnetic relaxation of CH_2, CHF and CF_2 groups in P(VDF$_{73}$/TrFE$_{27}$) was investigated by a CP/MASS NMR technique. In rising(↑)- and falling(↓)-temperature processes between 30 and 130°C, the ^{1}H nuclear spin-lattice relaxation time, $T_1\rho_H$ in the rotate frame for each group were minimal at 110°C(↑) and 80°C(↓) in the ferroelectric transition regions, respectively.

The $T_1\rho_H$ vs. T curve for each group was analyzed in terms of the rotational motion about chain axis accompanied to the local defect diffusion motion of trans to gauche conformation. The local defect diffusion and axial rotation motions took place predominantly in the lower and higher temperature sides of the transition region, respectively. The respective activation energy values were estimated to be 10 and 3 kcal/mol.

426. A METHOD FOR DETERMINING TRIAD DISTRIBUTION IN ETHYLENE-1-BUTENE COPOLYMERS BY C-13 NMR. Do H. Lee and Jae Y. Jho, Department of Chemical Technology, Seoul National University, Seoul 151-742 Korea

Triad distributions of the 1-butene comonomers in ethylene-1-butene copolymers were determined by analyzing the carbon-13 NMR spectra using an optimization calculation technique. Problems were encountered in using the conventional methods especially when the comonomer contents were low. We propose an optimization method for calculation that takes the possible errors in NMR measurements into account. Applying the method to the spectra of LLDPE with 1-butene as the comonomer gave physically reasonable values of comonomer distribution. The method was thought to be useful by giving the possible maximum and minimum triad concentrations reflecting the chain structures. A comparison of the triad concentrations in LLDPEs prepared by Ziegler-Natta catalysts with those by a metallocene catalyst shows that metallocene catalyst gave a more uniform comonomer distribution when comonomer content was the same.

427.

MECHANOCHEMICALLY GENERATED OLEFINIC CHAIN ENDS IN EPDM : A HIGH RESOLUTION NMR AND FTIR STUDY Andrew C. Kolbert, Lisheng Xu, Joseph G. Didier, DSM Copolymer Inc., P.O. Box 2591, Baton Rouge, LA 70821-2591

We have observed chain ends in ethylene-propylene copolymers created by subjecting the polymer to high temperature extrusion. New chemical structures, not present prior to shearing, corresponding to vinyl, vinylidene, and t-vinylene chain ends were observed via NMR and FTIR spectroscopy. From the relative intensities of the NMR resonances, we can draw some conclusions regarding the mechanism of polymer shearing which appears to be hydrogen abstraction followed by disproportionation via beta-scission of the main chain radicals to form olefins. A strong preference was shown for C-H scission of methine protons, presumably due to the stability of the tertiary radical intermediate. The presence of olefinic groups leads to a particular sensitivity to oxidation, with vinylidene groups being preferentially attacked.

428. MAIN CHAIN DYNAMICS IN PMMA AND PEMA AS OBSERVED BY 2D SOLID-STATE NMR

Sigrid C. Kuebler, Dieter J. Schaefer*, Christine Boeffel, Ute Pawelzik and Hans W. Spiess
Max-Planck-Institut für Polymerforschung, Postfach 3148, D-55021 Mainz, Germany
Dept. of Chemical Engineering, University of California, Santa Barbara, California 93106

The geometry of main chain motions in PMMA and PEMA are investigated focusing on ^{2}H 2D solid-state exchange NMR. Unlike in other polymeric systems, a restricted rotation of the main chain, which is coupled to a 180° flip of the asymmetric ester side group (β-process), is already present below the glass transition temperature. Above T_g, both systems show different behavior. While in PMMA, the fast side group motion implies a high mobility of the main chain itself, the β-process predominantly influences the geometry of motion in PEMA. In the latter, the dynamics of the larger ethylester side group retains its anisotropy above the glass transition and induces an anisotropic backbone motion parallel and perpendicular to the local chain axis, which leads to slower mean correlation times of the α-relaxation than in other amorphous polymers.

429.

DETERMINATION OF THE ORIENTATION DISTRIBUTION IN AROMATIC POLYESTERS BY SLOW-MAGIC-ANGLE-SPINNING DECODER NMR. **M.-Y. Liao** and **G. C. Rutledge** Department of Chemical Engineering, Massachusetts Institute of Technology, Cambridge, MA 02139

Slow-Magic-Angle-Spinning DECODER NMR is used to determine the orientation distribution in solution-spun and melt-spun fibers of an aromatic terpolyester comprised of p-hydroxybenzoic acid, isophthalic acid, and hydroquinone. This techniques reorients the sample by continuous rotation at slow speeds about the magic angle, which is simpler to implement than the static mechanical flip employed in the original DECODER study. The experimental results show a conventional powder spectrum for a sample with the fibers packed randomly. For a sample with the HIQ fibers packed parallel to the rotor axis a high degree of orientation is measured. Due to the overlap of the anisotropic chemical shift from different sites, a three dimensional technique combining DECODER with isotropic chemical shift separation is used.

430. Investigation of *Cis-Trans* Isomerization for Poly(phenylacetylene-d$_1$) by Deuterium Nuclear Quadrupole Coupling and Semiempirical MO Calculations.

S. Matsunami, T. Kakuchi, and F. Ishii
Graduate School of Environmental Earth Science, Hokkaido University, Sapporo 060, Japan

Deuterium nuclear quadrupole coupling constants of poly(phenyl-acetylene-d$_1$) (PPA-d$_1$) and its annealing sample, An-PPA-d$_1$ were the values of 164 and 168 kHz, respectively. A quadrupole coupling constant and the formation energy of various chain configurations were calculated using semiempirical MO methods (AM1 and PM3). Agreement between theory and experiment confirmed that the 4 kHz difference between PPA-d$_1$ and An-PPA-d$_1$ NQCC values is attributed to the difference of main chain structure between *cis*-transoidal for PPA-d$_1$ and 80°-deflected *trans*-transoidal for An-PPA-d$_1$.

431.

USE OF C-13 LABELING TO PROBE REACTIONS OCCURING ON CURING AND DEGRADATION OF HIGH TEMPERATURE POLYMERS. **Mary Ann B. Meador**, J. Christopher Johnston, NASA Lewis Research Center, Cleveland, OH 44136; Aryeh A. Frimer, Bar-Ilan University, Ramat Gan 52900, Israel; Paul J. Cavano, Case-Western Reserve University, Cleveland, OH 44106

High temperature polymers, such as crosslinked polyimides, are typically insoluble, intractible materials. Consequently, it has been difficult in these systems to follow high temperature curing or long term degradation reactions. Selective labeling of the polymers with C-13, coupled with solid NMR, allow the reactions occuring at high temperatures to be followed one carbon at a time. This technique has been successfully employed to provide insight into reactions occuring on both curing and degradation of PMR-15, a polymer matrix resin used extensively in aircraft engine applications.

432. STRUCTURE AND DYNAMICS OF CARBON TYPES IN ASPHALTS. Daniel A. Netzel, Thomas F. Turner, Gerald E. Forney, Jr., Western Research Institute, 365 North 9th St., Laramie, WY 82070-3380; Michelle L. Serres, Chemical Engineering Department, University of Wyoming, Laramie, WY 82071.

An asphalt is a highly complex mixture of ~30% polynuclear aromatics, heteroaromatics carbon types and ~70% branch, cyclic and normal alkane carbon types. It is a viscoelastic material having a reasonably well defined, but broad, glass transition temperature that marks the change from a viscous liquid to a brittle solid. Carbon-13 solid-state NMR techniques were used to characterize the static and time-dependent structure and the molecular dynamics of the carbon types in asphalts at temperatures above and below the glass-transition temperature. NMR studies have shown that the amorphous and semicrystalline phases of the aliphatic carbons exist in asphalts over temperature range from 20 to -45°C. Based on the differences in the chemical shifts of the amorphous and crystalline phases of the methylene carbons, the crystal structure is orthorhombic. In addition, the crystallization of the paraffinic-like materials in asphalts occurs rapidly during initial cool-down (DSC measurements) and continue more slowly for several years (NMR measurements). The crystallization process is probably the result of the rearrangement of the random coil gauche conformer of the n-alkanes to the all-trans conformer.

433. PHOTOSENSITIVE POLYMERS: MICROSTRUCTURE AND SEQUENCE OF COPOLYMERS Richard A. Newmark[1], Steven M. Heilmann[1], and Gaddam N. Babu[2]
[1]Corporate Research Laboratories and [2]Adhesive Technologies Center, 3M Company, St. Paul, MN 55144.

The NMR spectra of 4-(2-hydroxy-2-methylpropanoyl)phenoxy ethyl 2-(2-propenylamino)-2-methyl propanoate **1** and 4-(2-hydroxy-2-methyl propanoyl)phenoxy ethyl 2-(2-methyl-2-propenylamino)-2-methyl propanoate **2** have been completely assigned. Sequence analyses of the copolymers of **1** and **2** with methyl methacrylate and styrene have been studied by ^{13}C NMR. The results indicate the copolymers are predominantly random.

434. CORRELATION BETWEEN LIGAND STRUCTURE AND CATALYST REGIO-SPECIFICITY IN PROPYLENE POLYMERIZATION WITH C$_2$-SYMMETRIC *ANSA*-ZIRCONOCENE CATALYSTS Luigi Resconi and Gilberto Moscardi
Montell Polyolefins, P.le G. Donegani 12, 44100 Ferrara, Italy

The regiospecificity of C$_2$-symmetric *ansa*-zirconocene catalysts in propylene polymerization, as well as the mechanism of isomerization of the secondary growing chain-end, have been analyzed and correlated to the *ansa*-ligand structure. In liquid monomer at 50 °C the amount of secondary insertions increases on going from the more open to the more hindered system, that is from ca. 0.4 % for *rac*-Me$_2$C(Ind)$_2$ZrCl$_2$ up to 20 % for *rac*-C$_2$H$_4$(4,7-Me$_2$-H$_4$Ind)$_2$ZrCl$_2$. At the same time, there is an increase of the fraction of secondary units undergoing isomerization from the normal 2,1 unit to the 3,1 unit (tetramethylene sequence), from none to complete. These observations suggest that the mechanism of the 2,1 → 3,1 isomerization through a direct, concerted 1,2-Zr / 2,1-H shift is more likely than the two-step one involving terminal β-H elimination after a secondary insertion, to produce a metal-hydride-olefin complex, followed by olefin rotation and reinsertion into the metal-hydride bond.

435. SOLID STATE NMR STUDIES OF HYPERCROSSLINKED RESINS
W.P. Steckle, Jr. and J.R. Duke, Jr., Los Alamos National Laboratory, Polymer and Coatings Group, MST-7, MS E549 Los Alamos, NM 87545

Polycondensation of polyaromatic hydrocarbons using rigid aromatic crosslinking agents provides a simple, yet versatile, method for preparing rigid hypercrosslinked foams. This straightforward polycondensation, known as Friedel-Crafts polycondensation typically produces insoluble infusible polymers with very high surface areas resulting from the microporosity of the resins. In ther past numerous attempts have been made to elucidate the structure of these materials by techniques such as FTIR, solid state NMR, and swelling experiments. ^{13}C CPMAS and deuterium NMR will be used to investigate the level of crosslinking in deuterated polystyrene, hypercrosslinked d_8-polystyrene, and hypercrosslinked d_6-benzene resins. This is observed by the reduction in mobility in the phenyl rings as the level of crosslinking increases.

436. SYNTHESIS AND CHARACTERIZATION OF POLYPHOSPHAZENE COPOLYMERS USING PHOSPHORUS-31 NMR SPECTROSCOPY
Frederick F. Stewart[*], E.S. Peterson[*], M.L. Stone[*], and R.E. Singler[**]
[*]Idaho National Engineering Laboratory, Lockheed Martin Idaho Technologies Co., P.O. Box 1625, Idaho Falls, ID 83414-2208; [**]Department of Chemistry, U.S. Military Academy, West Point, NY 10996

In the synthesis of organo-substituted phosphazenes, it is often helpful to follow the process using phosphorus-31 NMR spectroscopy. Linear phosphazenes are synthesized by the ring-opening polymerization of phosphonitrilic chloride to yield linear poly(dichlorophosphazene). Attachment of organic pendant groups serve two functions. The first is stabilization of the polymer, poly(dichlorophosphazene) is hydrolytically unstable. The second is to give the polymer predictable physical characteristics. The attachment of a mix of substituents has been determined in our lab to give a higher degree of tailorability to the polymer. In this paper, the synthesis of mixed aryloxy/polyether phosphazenes will be discussed. Differing synthetic approaches will be discussed and their reaction course observed using phosphorus-31 NMR spectroscopy

437. NMR SPECTROSCOPY OF NOVEL SULFONATED POLY (p-PHENYLENE TEREPHTHALAMIDE) AND ITS DERIVATIVES
R. Tirasirichai, A.C. Watterson, K. Power ; Polymer Science Program, Department of Chemistry, University of Massachusetts-Lowell, Lowell, MA 01854

The synthesis and characterization of a novel sulfonated poly(p-phenylene terephthalamide) and its derivatives are described. A model compound was also synthesized to investigate the microstructure of these polymers by NMR techniques. In order to improve the solubility of PPT-S, the proton in the sulfonic acid group was replaced by a lithium cation. The unambiguous assignment of the peaks in the spectra has been successfully made for the first time by comparison of the model compound spectra with the polymer spectra.

438.

NMR SPECTROSCOPICAL CHARACTERIZATION OF DECOUPLING POLYMERS. Frans S.B. Tjan, J.C.S. Niël and J.C. van der Pas, Unilever Research Laboratory, P.O. Box 114 3130 AC Vlaardingen The Netherlands

Physically stable, concentrated (in surfactant and electrolyte) and pourable lamellar dispersions can be made using *decoupling polymers* (DCPs). It has been demonstrated that poly(sodium acrylate-co-lauryl methacrylate) is capable to *decouple* the intra- from the interlamellar droplet interactions. Dependent on the isopropyl alcohol (IPA)/water ratio in the polymerization process different forms can be made e.g. called *collapsed* and *extended*. From combined physical and NMR spectroscopical characterization results, it can be derived that in the collapsed form the lauryl anchors are not uniformly distributed over the polymer backbone (occurrence of *LMA-blocks*). This leads to strong *intra*molecular interactions between lauryl anchors which are close or next neighbours. In the extended forms the lauryl anchors are much more randomly distributed over the polymer backbone.

439. NUCLEAR MAGNETIC RESONANCE STUDIES OF X-RAY DEGRADED POLY(METHYL METHACRYLATE). E. E. Waali,[a] J. D. Scott,[b] J. M. Klopf,[b] Y. Vladimirsky[b] and O. Vladimirsky.[b] [a]Department of Chemistry, University of Montana, Missoula, MT 59812 and [b]The J. Bennett Johnston, Sr. Center for Advanced Microstructures and Devices, Louisiana State University, Baton Rouge, LA 50803.

The 1-D hydrogen and 2-D COSY, HMQC and HMBC NMR spectra of X-ray degraded poly(methyl methacrylate) have been obtained. These allow the assignment of 34 hydrogen and carbon NMR signals for chain-scission products (**1** and **2**) and chain-intact products (**3** and **4**). Methyl formate and acetaldehyde were also observed. The relevance to x-ray micromachining and lithography will be discussed.

P = polymer chain

440. SPIN DIFFUSION AND RELAXATION IN HETEROGENEOUS POLYMERS. Jiahu Wang, Kevin S. Jack and Almeria L. Natansohn, Department of Chemistry, Queen's University, Kingston, Ontario, Canada K7L 3N6

Spin diffusion and spin-lattice relaxation in heterogeneous polymers have been modeled on the same footing. This new approach extends the applicability of spin diffusion and spin-lattice relaxation measurements for the determination of domain sizes and interfacial thickness in heterogeneous polymers. It has been shown that the agreement between NMR measurements and the simulated results can be significantly improved with the incorporation of spin relaxation processes into the spin diffusion experiments. Changes of the spin-lattice relaxation times (T_1) in different domains can be predicted satisfactorily based on the T_1s of consisting homopolymers and the microdomain structures in the present model.

441. **Structural Characterization of Unique Polyisobutylene Olefin End-Groups by 2D Pulsed-Field-Gradient HMBC NMR.** Kurt F. Wollenberg and Christopher J. Kolp, Physical and Analytical Sciences Department, The Lubrizol Corporation, Wickliffe, Ohio, 44092

Unique and typical olefin end-groups of $AlCl_3$ initiated polyisobutylene (PIB) have been studied by 2D PFG HMBC (heteronuclear multiple-bond correlation) NMR. The structures of the unique olefin terminals have been confirmed with carbon and proton peak assignments that have extended beyond the olefin terminal by two to three repeating units along the PIB backbone. The HMBC technique has shown tremendous sensitivity by revealing numerous long-range C,H connectivities of these dilute structural units. The C, H through-bond correlations observed in the HMBC experiment coupled with one-bond correlations afforded by the PFG HSQC (heteronuclear single quantum correlation) experiment provides a significant gain, in terms of facilitating olefin terminal assignments for low molecular weight polymers, over the more traditional phase cycled HMBC and direct heteronuclear analogues of the HMBC and HSQC techniques. Schematic representations of the through-bond correlations and cross-peak assignments of these olefin terminals will be presented. Sensitivity and dynamic range have been increased in some instances by using selective off-resonance excitation pulses on the repeating methylene and methyl protons of the polyisobutylene backbone.

442. COPOLYMER MICROSTRUCTURES AS FINGERPRINTS OF ZIRCONOCENE CATALYSTS. R.C. Zeigler and H. Rychlicki, Montell USA Inc., R&D Center, 912 Appleton Rd., Elkton, MD 21921.
L. Resconi, F. Piemontesi and G. Baruzzi, Montell Polyolefins. G. Natta Research Center, P.le G.Donegani 12, I-44100, Ferrara, Italy.

The *rac*-[ethylenebis(4,7-dimethyl-η^5-1-indenyl)]zirconium dichloride catalyzed copolymerization of propene with a low concentration of ethylene produced higher molecular weight polymer than the analogous propene homopolymerization with the same catalyst. NMR analysis of the copolymer revealed that the mechanism of this molecular weight increase was the insertion of an ethylene unit after a regioirregular (2,1) propene insertion. This insertion of the relatively small ethylene molecule into the sterically restricted catalyst-2,1 propene-polymer species competed with chain termination processes resulting in longer polymer chains. The effect of NOE on the quantitation of ethylene incorporation in ethylene-propene copolymers is also described.

443. A REVIEW OF ORGANIC THIN FILMS: THEIR CHARACTERIZATION AND APPLICATIONS, J. D. Swalen, University of California Santa Cruz, Santa Cruz, CA.

Ordered thin organic films in the thickness range of a few to several hundred nanometers have considerable technological promise. Current electronic and optical devices incorporate structures that are in this thickness range. Organic thin films have been proposed, and in some cases applied, as passive or active components traditionally fabricated with other materials. Scientific studies of molecular interactions in thin-film structures leading to an understanding of the collective properties of ordered arrays have only been possible by characterizations with a number of new surface science techniques. These will be reviewed and some recent significant results presented. A number of applications will also be discussed.

444. SELF-ASSEMBLED MONOLAYER GROWTH OF 16-MERCAPATO-1-HEXANOL ON GOLD. Sheng Pan,[1] Anna Belu,[1] Mimi Mar,[2] Buddy D. Ratner,[1] [1]Department of Chemical Engineering and Center for Bioengineering, [2]Department of Electrical Engineering, The University of Washington, Seattle, WA 98195

An understanding of the kinetics and mechanism of self-assembled monolayer formation and ordering is important for the progress in their applications. The chemical composition, film structure and surface properties at different self-assembly formation stages are investigated systematically using various surface analytical techniques (IR, TOF-SIMS, ESCA etc.). Monolayer formation involves a two-step process. The first rapid stage leads to an adsorption of molecules to the surface in an island type of growth. It is followed by a slow process where the monolayer undergoes a phase transition by reconstruction of structural conformations and realignment of molecular axes with the surface normal. The surface wettability approaches the equilibrium value at relatively long times.

445. DESIGN AND CHARACTERIZATION OF MINIMALLY ADHESIVE SURFACES: THE ROLE OF CHEMISTRY, MOBILITY AND TOPOGRAPHY, John F. Rabolt, Materials Science Program, University of Delaware, Newark, DE 19716

A detailed study of the semifluorinated alkylamidethiol, $CF_3(CF_2)_7N(H)C(O)(CH_2)_2SH$ (F8AT), was carried out using FTIR, Raman, contact angle, ellipsometry, angular dependent XPS and NEXAFS. Recently, related molecules with short and long alkyl segments between the fluorinated blocks and thiol groups, namely, $CF_3(CF_2)_7(CH_2)_2SH$ (F8H2SH) and $CF_3(CF_2)_7(CH_2)_{11}SH$ (F8H11SH), together with their mixtures with F8AT, were investigated as self-assembled films. The helical fluorinated part of the these molecules is rigid due to intramolecular stablization while the alkyl part is, on the other hand, quite flexible and must be stablized intermolecularly by other chains in order to maintain its planar zigzag conformation. Our results on F8H11SH show that the inclusion of a long hydrocarbon sequence into the backbone of a semifluorinated n-alkyl thiol molecule has a marked effect on its orientation on a gold surface. Furthermore, the orientation of the fluorocarbon helix appears to be more tilted relative to the surface normal in marked contrast to other semifluorinated n-alkyl thiols with short hydrocarbon sequences which were found to orient normal to the surface.

446. **USING SELF-ASSEMBLED MONOLAYERS TO MODIFY ELECTRODE INTERFACES IN POLYMERIC LIGHT-EMITTING DIODES.** Weidong Chen[1], Sandra Burnham[2], Christopher E.D. Chidsey[2] and J. Campbell Scott[1], [1]IBM Research Division, IBM Almaden Research Center, 650 Harry Rd., San Jose, CA 95120-6099, [2]Department of Chemistry, Stanford University, Stanford, CA 94305-5080.

Charge injection from the electrodes of polymer light-emitting diodes into the emissive layer plays a critical role in the overall efficiency of such devices. In order to understand in more detail the mechanism of the injection process and to provide guidance for optimizing the charge injection rate, we have studied chemically well controlled and well characterized electrodes, consisting of self-assembled monolayers (SAMs) at the interface between gold anodes and thin films of MEH-PPV. The SAMs are prepared from alkane-thiols of length up to C12, which are used to tether various electro-active species, such as ferrocene or phenyleneethynylene oligomers, at a well-defined distance from the gold surface. Current-voltage and radiance measurement are employed to determine the effect on change injection properties. The efficiency of charge transfer is found to depend on the length of the alkane spacer and on the oxidation potential of the active group.

447. TUNING THE ELECTRONIC PROPERTIES OF SILICON VIA MOLECULAR SELF-ASSEMBLY Noemi Zenou, Alexander Zelichenok, and Shlomo Yitzchaik* The Hebrew University of Jerusalem, Department of Inorganic Chemistry, Jerusalem 91904, ISRAEL Rami Cohen and David Cahen Weizmann Institute of Science, Department of Materials and Interfaces, Rehovot 76100, ISRAEL

Control over the surface chemistry and physics of electronic and optical materials is essential for constructing devices and fine-tuning their performance. In the past few years we have started to explore the use of organic molecules for systematic modification of semiconductor surface electronic properties. In this paper, controlled modification of silicon surfaces by self-assembling of a series of quinolinium-based chromophores is reported. The progress of the assembly process is monitored by XPS, UV-Vis, and FTIR spectroscopies as well as with AFM. The effect of monolayer dipole-moment on Silicon's surface potential and the interaction with surface states is monitored by CPD measurements. A pronounced effect of a sub-nanometer coupling-agent layer alone on silicon's electron affinity and band-banding was observed. We also show a way to modulate silicon's work-function by tuning the dipole strength of an organic self-assembled monolayer, that is pointing toward the silicon surface.

448. **SURFACES AND INTERFACES OF LIQUID CRYSTAL ALIGNMENT LAYER POLYIMIDES.** K.-W. Lee, S.-H. Paek[#], A. Lien and C. Durning[#]. IBM T.J. Watson Research Center, PO Box 218, Yorktown Hts., NY 10598. [#]Dept of Chemical Engineering, Columbia University, NY, NY 10027.

A thin layer ($\simeq$60 nm) of polyimide (PI) film is employed to align nematic liquid crystals in modern thin film transistor-liquid crystal displays (TFT-LCDs). The PI film is usually rubbed with a cloth to induce LC molecules uniformly aligned as well as to obtain optimum LC pretilt angles for the desired electro-optical properties. Polar functional groups of PI surfaces are reoriented out-of-the-plane of the surface by rubbing with a cotton cloth (used to minimize the static charge) while non-polar groups fold in ward, toward the bulk of the PI film. The driving force is an electronic attraction and (or) repulsion between LC and PI molecules. This molecular orientation increases in both depth and area as the rubbing force or density increases, but a great rubbing force provides a thermally more stable LC pretilt angle than a high rubbing density.

449.
SIMULATION OF 4'-N-PENTYL-4-CYANOBIPHENYL MONOLAYER ANCHORING ON POLYALKANPYROMELLITIMIDE SURFACES. **Vladimir A. Zubkov,** Curtis W. Frank, Department of Chemical Engineering, Stanford University, Stanford, CA 94305-5025

Molecular dynamics smulation was performed on 2D interfaces comprising the 4'-n-pentyl-4-cyanobiphenyl (5CB) monolayer and an polyalkanpyromellitimide (polyAPM) surface having three (m=3) or four (m=4) methylene spacers. The calculated orientation distribution function revealed a sharp difference in the monolayer anchoring for APM surfaces with m=3 and m=4, which is not that evident from the experimental tilt angle values. Simulation showed graphically how the monolayer anchoring follows the surface morphology. In agreement with experiment, the calculated tilt angle for the monolayer above the fairly smooth polyAPM surface (m=4) is less than that above the more corrugated polyAPM surface (m=3) although calculated tilt values are lower than experimental ones. Simulation also revealed that the 5CB tilt is influenced not only by coulomb interactions of 5CB with the surface but also by interactions within the 5CB monolayer.

450. ORIENTATIONS OF LIQUID CRYSTALS ON SELF-ASSEMBLED MONOLAYERS FORMED FROM ALKANETHIOLS ON GOLD. Nicholas L. Abbott, Vinay K. Gupta, William J. Miller, Rahul R. Shah, Department of Chemical Engineering and Materials Science, University of California at Davis, Davis CA 95616

We report an experimental study of the orientations assumed by nematic and smectic liquid crystals on self-assembled monolayers formed from alkanethiols on gold and silver. In contrast to the homeotropic orientation of nematic phases of 4-n-pentyl-4'-cyanobiphenyl (5CB) observed on SAMs formed from octadecyltrichlorosilanes on silica, we find planar orientations of 5CB on alkanethiols on gold. When supported on obliquely deposited films on gold, a preferred azimuthal orientation within the liquid crystal was found: the director was found to be parallel to the direction of incidence of the gold when the SAMs were formed from $CH_3(CH_2)_{n-1}SH$ with n an even integer, and perpendicular to the direction of incidence for odd values of n. These observations and other will be discussed in this presentation.

451. LIQUID CRYSTAL ANCHORING ON BINARY SILANE SELF-ASSEMBLED MONOLAYERS, J.M.Y. Yang, C.W. Frank, Stanford University, Stanford, CA.

Liquid crystal anchoring phenomena are widely applied in displays, but the molecular basis for the anchoring is not well understood. We have used binary self-assembled monolayers as model surfaces for the study of anchoring of 5CB, a typical nematic liquid crystal. We examine several binary systems including hydroxyl/methyl, bromo/methyl, naphthyl/methyl, hydroxyl/methyl and phthalimido/methyl. A surface composition-induced change in liquid crystal anchoring from homeotropic to near planar occurs as the surface concentration of the polar end group increases. The transition behavior may be explained in terms of a modified Creagh-Kmetz correlation in which we account for dispersive, polar and acid-base type interactions separately.

452. TUTORIAL: AN INTRODUCTION TO THE DEPOSITION OF POLYMERIC THIN FILMS USING PLASMAS. Buddy D. Ratner, Center for Bioengineering and Dept. of Chemical Engineering, Box 351750, University of Washington, Seattle, WA 98195

Thin, polymeric films are readily deposited from gas-phase, glow discharge plasma environments for surface modification, barrier coatings or creating novel polymeric structures. Plasmas are also valuable for surface modification of polymers by etching, directly introducing new functional groups or activating surfaces to permit further grafting or reaction. They offer a number of special properties including: solvent-free processing; pin-hole free, conformal coating; rapid treatment; tenacious adhesion to the substrate; a wide range of possible substrates; excellent control of chemistry over a wide range of compositions; negligible monomer costs; and continuous (reel-to-reel) processing options. This short review outlines some of the key ideas important to plasma deposition of polymeric (organic) thin films.

453. STUDY OF PLASMA-TREATED POLYMERS AND THE STABILITY OF THE SURFACE PROPERTIES. F. Arefi-Khonsari, M. Tatoulian, G. Placinta, J. Kurdi, J. Amouroux, Laboratoire de Genie des Procedes Plasma, Universite Paris VI, ENSCP, 11 rue Pierre et Marie Curie,75231 Paris Cedex 05, France

The surface treatment of polymers in the aim to improve the macroscopic properties such as adhesion requires a thorough understanding of the interfacial phenomena. That is why in this paper the surface properties of plasma-treated polymers will be discussed in terms of the modern theory of acid-base interactions. An attempt has been made in order to explain the adhesion improvement of aluminium to two differently treated Polypropylene films (1) in an oxidizing atmosphere such as Oxygen and (2) in a reducing one such as ammonia in terms of acid-base interactions.
On the other hand, the stability of the surface properties being an important issue for the industrial development of plasma processes, the ageing effect of such plasma-treated surfaces on the wettability and adhesion properties were studied. In order to slow down the ageing effect, the surface layer can be stabilized via crosslinking using the noble helium plasma gas.Therefore this can be accomplished either by a two step process (before ammonia or oxygen plasma) or a one step process in mixtures of helium and ammonia or helium and oxygen discharges.These mixtures were optimized in order to obtain crosslinking and functionalization in a one step process.

454. EFFECT OF MONOMER COMPOSITION ON THE STRUCTURE AND PROPERTIES OF PLASMA FLUOROPOLYMERS. M. S. Silverstein and R. Chen, Department of Materials Engineering, Technion - Israel Institute of Technology, Haifa 32000, Israel

Ultra-thin, high-performance and high-temperature fluoropolymer films are of interest for advanced microelectronic, biomedical and protective coating applications. Plasma polymerization can be used to deposit ultra-thin, pinhole-free films in a process that avoids the etchants, solvents and temperatures that limit the applicability of other deposition techniques. The molecular structure and properties of plasma polymerized hexafluoropropylene (PPHFP) films were investigated. The rate of deposition was found to be more strongly influenced by monomer flow rate than by plasma power. The variation of the deposition rate with plasma conditions, specifically the ratio of power to flow rate cubed, was described using exponential master curves. Adding hydrogen or nitrogen to the plasma yielded significant changes in the deposition rate, molecular structure, surface topography, surface tension and electrical breakdown strength.

455. PREPARATION AND CHARACTERIZATION OF AN IONOMERIC PLASMA POLYMERIZED FILM FOR BIOMEDICAL APPLICATIONS.
Robert C. Tucker[*], Michael J. Danilich[‡], and Roger E. Marchant[*], [*]Dept. of Biomedical Engineering, Case Western Reserve University, Cleveland, OH 44106, [‡]Bethlehem Steel, Bethlehem, PA 18017

Thin novel ionomeric plasma polymerized perfluoroallyl phosphonic acid (PPPAPA) films have been developed to improve biomedical sensor ionic selectivity. This film's composition, morphology, electrochemistry, and adhesion properties were studied. ATR-FTIR indicates retention of phosphonic acid groups in these films after plasma polymerization. Ellipsometry and AFM indicate complete surface coverage on these substrates. Cyclic voltammetry and chronocoulometry show that the PPPAPA films significantly reduce transport of a negatively charged organic interferant, ascorbate, compared with a positively charged analyte, dopamine, even though both analytes have similar molecular weights, sizes, and diffusion coefficients in water. The experimental reproducibility of the electrochemistry provide evidence that the deposited PPPAPA films remain adhered to gold in buffer solution. This indicates PPPAPA films have promise as biomedical sensor coating with ionic selectivity.

456. DEPOSITION AND ETCHING OF HMDSO/SF$_6$ PLASMA POLYMERIZED THIN FILMS. R. G. Santana, R. Y. Honda, M. A. Algatti, M. E. Kayama and R. P. Mota, UNESP-Universidade Estadual Paulista, Campus de Guaratinguetá, 12500-000 Guaratinguetá, SP, Brazil

In this paper is presented the results of deposition and etching of mixture of Hexamethyldisiloxane (HMDSO) and Sulphur Hexafluorine (SF$_6$) in different proportions in order to study the changes in molecular structure of the polymer as a function of SF$_6$ concentration. It is shown that the increasing of SF$_6$ concentration produces the decreasing of all peaks in infrared absorption spectra. For concentrations about 25% is observed the enhancement of the incorporation of Fluorine in polymer structure. For higher concentrations the etching processes is dominant. The exposition of plasma polymerized HMDSO to DC excited SF$_6$ plasmas demonstrated a decreasing in the etching rate with the residence time under the action of the corrosive plasma, due to competition between deposition and etching processes

457. DC PLASMA POLYMERIZATION OF PYRROLE AND HEXAMETHYLDISILOXANE: MECHANISMS, CHARACTERIZATION AND APPLICATIONS. W.J. van Ooij and S. Guo, Department of Materials Science and Engineering, University of Cincinnati, Cincinnati, OH 45221

We have modified a simple, parallel-plate DC glow discharge reactor to include pulsing and a hollow cathode. This enables us to deposit plasma-polymerized films on any substrate without sparking or arching. Polymers that we have extensively characterized using a multitude of analytical techniques, include films of pyrrole and hexamethyldisiloxane. The power density in our process can be as low as 5×10^{-4} W/cm^2. The substrates we have coated include metal sheets, plastic films, fibers, cords and powders. The mechanism of film growth has been studied and appears to be markedly different for the two monomers.

458. A DEPOSITION MODEL OF PLASMA POLYMERIZED ACETYLENE FILMS BY PULSED RF DISCHARGE. T. Uchida and S. Morita, CCRAST, Nagoya University, Nagoya 464-8603, Japan

Plasma polymerization of acetylene by a pulsed RF discharge was performed in a box-type reactor. The mixture gas with acetylene and argon was used here. The growth rate of polymer was calcurate theoretically, where a disociation of monomer as an initiation, a reaction between the radical and a monomer as a propagation, an adsorption of radical as a growth of polymer were used in the calcuration. The film thickness increased nonlinearly with increasing the distance from the monomer inlet. The deposition rate saturated with increasing the discharge period. The experimental results were fitted to the theoretical equations and the reaction constants were evaluated from the fittings. The values are not unusual from the other experimental results and the initial stages of reaction were discussed.

459. RELATIVE RATES FOR PLASMA HOMO- AND COPOLYMERIZATIONS OF OLEFINS IN A HOMOLOGOUS SERIES OF FLUORINATED ETHYLENES. M. A. Golub, NASA Ames Research Center, Moffett Field, CA 94035-1000; T. Wydeven, Lockheed Martin Engineering and Sciences, NASA Ames Research Center, Moffett Field, CA 94035-1000.

The relative rates of plasma (co)polymerizations of ethylene, vinyl fluoride, vinylidene fluoride, trifluoroethylene and tetrafluoroethylene (VF$_x$; x = 0–4, respectively) were determined in an rf, capacitively coupled, tubular reactor with external electrodes using identical plasma parameters. The averages of deposition rates obtained by both microgravimetry and ellipsometry were plotted versus the F/C ratios of the monomers or monomer blends. The deposition rates for VF_x (x = 1-3) and 20 monomer blends were all located above a straight line joining the rates for VF_0 and VF_4, and yielded a concave-downward plot of deposition rate versus F/C ratio similar to that reported previously for VF_0/VF_4 blends. The deposition rates for VF_m/VF_n blends (m = 3 or 4; n = 0-2) were all greater than expected for non-interacting monomers; those for VF_0/VF_2 and VF_1/VF_2 blends were all lower than expected; while those for VF_0/VF_1 and VF_3/VF_4 blends fell on a straight-line plot versus F/C ratio, indicative of apparent non-interaction between monomers. The mechanisms for plasma (co)polymerizations of VF_x monomers responsible for the wide range of relative deposition rates remain to be elucidated.

460. ALDEHYDE PLASMA POLYMERS. G. J. Beumer, X. Gong, L. Dai, H.A.W. St John and H.J. Griesser, Division of Chemicals and Polymers, CSIRO, Private Bag 10, Clayton South MDC, Vic, Australia 3169.

Plasma polymer coatings have been produced from a range of aldehyde monomers, and characterised by XPS, FTIR, contact angles and derivatization. One application of interest for these thin-film coatings is that they provide a means to covalently attach proteins to surfaces. For this reason it is of interest to characterise the composition of the outermost few nm which will determine interfacial interactions, and in particular to confirm the presence of aldehyde species required for the immobilisation scheme. Both XPS and FTIR revealed contributions of C=O species, as well as other functionalities, and showed that aldehyde abstraction was quite sensitive to small changes in plasma parameters. Due to the difficulty in distinguishing between ketone and aldehyde groups in IR and XPS, several derivatization schemes were also employed to probe for aldehyde species. Contact angles were used to monitor changes in surface properties on ageing.

461. **The Nitroxide-Mediated Stable Free Radical Polymerization Process-Mechanistic Considerations.** Michael K. Georges, Peter G. Odell, Richard P. N. Veregin and Barkev Keoshkerian *Xerox Research Centre of Canada, 2660 Speakman Drive., Mississauga, Ontario, Canada L5K 2L1*

The reversible capping mechanism of the nitroxide-mediated stable free radical polymerization (SFRP) process is reviewed. Experimental proof to support the proposed reversible capping mechanism is provided, contradicting the statement that reversible combination has not been demonstrated experimentally. Suggestions that the initiation is dominated by thermal initiation is also discussed in detail. The SFRP process has been performed without thermal polymerization, enabling the synthesis of high molecular weight, narrow polydispersity polystyrene. The details of this work will be presented.

462. MECHANISM AND KINETICS NITROXIDE-CONTROLLED FREE RADICAL POLYMERIZATION, T.Fukuda,* Y.Tsujii,and T.Miyamoto, Institute for Chemical Research, Kyoto University, Uji, Kyoto 611, Japan

The mechanism and kinetics of nitroxide-mediated free radical polymerization were discussed on the basis of a set of experimental data collected for the polymerization of styrene at high temperatures in the presence of a polystyryl adduct with 2,2,6,6-tetramethylpiperidinyl-1-oxy (PS-TEMPO) or its unimer model (BS-TEMP). The existence of the reversible dissociation of PS-TEMPO or BS-TEMPO, thermal initiation, bialkyl termination, the decomposition of the active chain end, and the stationary state with respect to the radical concentrations were all evidenced and incorporated by means of a computer simulation into a quantitative description of this branch of controlled/"living" radical polymerization.

463. STYRENE POLYMERIZATION MEDIATED BY FIVE-MEMBERED CYCLIC NITROXIDES. B. Yamada, Y. Miura, Y. Nobukane, and M. Aota, Material Chemistry Laboratory, Faculty of Engineering, Osaka City University, Sugimoto,Sumiyoshi-ku, Osaka 558, Japan

In order to attain the living radical polymerization of styrene at lower temperatures, pyrrolidinyloxyls bearing substituents on the 2-, 3-, and 5-carbons were chosen as the mediators. The polymerization with 2,2,5,5-tetramethylpyrrolidinyloxyl/BPO (1.3 molar ratio) at 110 °C gave the polymer of M_n = 38300 and M_w/M_n = 1.66, and an increase in conversion to 64% from 16% brought about increase and decrease in M_n and M_w/M_n to 74900 and 1.41, respectively. Similar tendencies were also observed in the polymerization in the presence of the nitroxides bearing different substituents; the controlled and conventional polymerizations yielded the polymer. The polymerization in the presence of 2,2,3,3,5-pentamethyl-5-phenylpyrrolidinyloxyl yielded the polymer of a higher M_n and a narrower M_w/M_n, and the significantly weakened C-O bond of the dormant species by the steric hindrance is expected. The optimum conditions for the living polymerization of styrene depending on the structure of the cyclic nitroxide will be discussed based on the experimental findings.

464. 'LIVING' FREE RADICAL POLYMERIZATIONS: ROLE OF AUTOPOLYMERIZATION AND NITROXIDE MOBILITY. L. Michalak, E. Malmström, W. Devonport, M. Mate, C. Hawker, IBM Almaden Research Center, 650 Harry Road, San Jose, CA, 95120-6099, G. Barclay and R. Sinta, Shipley Company, 455 Forest Street, Marlborough, MA 01752

The role of autopolymerization and mobility of the mediating nitroxide free radicals is probed in a series of experiments using unimolecular initiators. It was demonstrated that the nitroxide mediated autopolymerization of styrenic derivatives in the absence of initiators is a controllable, 'living' process. Evidence is presented which supports the insitu generation of unimolecular initiators which have structures corresponding to those expected from a Mayo mechanism. For a typical 'living' free radical polymerization using either unimolecular or bimolecular initiating systems these autopolymerization reactions are controlled by exchange between the mediating nitroxide moieties at the termini of the growing polymer chains and the radical produced by autopolymerization. A series of radical crossover experiments using functionalized unimolecular initiators demonstrated that this exchange reaction is extremely efficient and leads to scrambling of the nitroxide chain ends at very low conversions.

465. Controlled Free-Radical Polymerization In The Presence Of A Novel Asymmetric Nitroxyl Radical
D. Benoit[1], S. Grimaldi[2], J.P. Finet[2], P. Tordo[2], M. Fontanille[1], Y. Gnanou[1]
[1] Laboratoire de Chimie des Polymères Organiques - UMR CNRS-ENSCPB, Université Bordeaux-1, BP 108, 33402 Talence Cedex (France) ; [2] Laboratoire de Structure et Réactivité des Espèces Paramagnétiques UMR CNRS université de Provence, Case 521, av. Normandie-Niemen, 13397 Marseille Cedex 20 (France)

An analogue of ditertiobutyl nitroxyl (DTBN) radical, that carries in α-position a substituent purposely introduced to induce both electronic and steric effects, is found to bring about a controlled process when used in the free-radical polymerization of miscellaneous monomers. N-tertiobutyl-1-diethylphosphono-2,2 dimethylpropyl nitroxyl (DEPN) is a novel stable radical that affords a faster polymerization of styrene than that mediated by DTBN and yet an excellent control of the molar masses, along with samples of low polydispersity index. Styrene apart, DEPN also functions satisfactorily for alkyl acrylates. The polymerization of the latter monomer, when carried out in the presence of DEPN, fulfills all the features that are exhibited by genuinely controlled processes. Di- and triblock copolymers based on polystyrene and polyalkylacrylate have been obtained by sequential polymerization of the corresponding monomers.

466. **Design and Synthesis of New Nitroxides and their Application in Living Polymerization.** Dotsevi Y. Sogah, Rutger D. Puts, Alex Trimble and Oren Sherman *Department of Chemistry, Baker Laboratory, Cornell University, Ithaca, NY 14853*

This presentation will describe our current efforts in the design of new chiral nitroxide radicals suitable for lower temperature polymerization. The kinetics and mechanistic implications of using chiral nitroxides will be described. Additionally, the electronic and steric influence on living free radical polymerization will be described. We have designed and prepared a universal multifunctional initiator in which the reactive sites are all orthogonal which offers a revolutionary opportunity for development of one-pot syntheses of complex structures. For example, anionic, free radical and cationic polymerizations could be carried out either sequentially or simultaneously. The versatility of the structure is such that different types of macromonomers suitable for any of the major methods of polymer synthesis can be made. The details of the above will be provided.

467. POLYMER SEGMENTAL DYNAMICS AND ENTANGLEMENT CONSTRAINTS A. D. English, Central Science and Engineering, DuPont Central Research and Development, Experimental Station, Wilmington, DE 19880-0356 P. T. Inglefield, A. A. Jones and Y. Zhu, Department of Chemistry, Clark University, Worcester, MA 01610

The relaxation time (T_e) associated with the transition from liquid like behavior to pseudo-solid like behavior for polybutadiene and polyisoprene, as observed in simple Bloch decay NMR experiments, has been shown to be linearly dependent upon the molecular weight of the polymer. This transition begins as the molecular entanglement coupling regime is entered at the critical molecular weight, M_c, as classically determined from a variety of rheological experiments. Because the length scale that is probed is of the order of M_c, these types of NMR experiments do not afford the opportunity to experimentally test different models of large scale motion such as Rouse dynamics or reptation.

468. **Nuclear spin correlations in entangled polymeric liquids.** Robin C. Ball, Paul T. Callaghan and Edward T. Samulski
Cavendish Laboratory, Cambridge University, Madingley Road, Cambridge CB3 0HE, Department of Physics, Massey University, Palmerston North, New Zealand, Department of Chemistry, University of North Carolina, CB #3290, Venable and Kenan Laboratories, Chapel Hill, NC 27599-3290

We derive closed-form expressions for a sine correlation function $\beta(2t,t)$ which is particularly sensitive to modulations of nuclear dipolar (and quadrupolar) interactions caused by slow molecular reorientations. This function is placed in the context of traditional NMR experiments—the Free Induction Decay and the solid echo amplitude—and a closed-form expression for β is evaluated for macromolecular motion dominated by reptation. We introduce a simple correlation function for reptation that depends on only two parameters, the pre-averaged dipolar interaction strength and t_d, the tube disengagement time. From proton NMR measuremnents of β in poly(dimethyl siloxane) melts we estimate a (preaveraged) second moment value of 1.3×10^6 s^{-2} and $t_d = 0.019$ s at 300K, in reasonable agreement with a calculated value for the second moment and an independent NMR estimate of the terminal relaxation time.

469. CONFORMATION AND MOBILITY IN SPACERS OF SMECTIC POLY(ESTER IMIDE)S
C. Wutz, H.R. Kricheldorf, University of Hamburg, Bundesstr. 45, D-20146 Hamburg, Germany

Poly(ester imide)s based on trimellitimide mesogens and long aliphatic spacers form different smectic phases depending on the thermal treatment. ^{13}C NMR CP/MAS spectra reveal that the spacers with ≥12 methylene groups have a high amount of trans-trans-conformations. Conformation and mobility of spacers with four central deuterated methylene groups were investigated by ^{2}H NMR and ^{13}C NMR CP/MAS with dipolar dephasing and ^{13}C/^{1}H-WISE. The non-quaternary suppression indicates, that the tt-conformations are situated rather at the border than in the center of the spacer layer. Hence, the conformational order is not due to lateral enthalpic interactions but is induced by the rigidity of the mesogens. This conception of a highly extended but mobil spacer is confirmed by WISE-spectra which reveal a comparable mobility of tg-, gt- and tt-conformations. Comparison of measured and calculated ^{2}H NMR spectra indicate, that the spacer segments preferable do not undergo trans-gauche-jumps but librations in the rapid exchange limit with a distribution of angles on the tetrahedral cone. The -$(CH_2)_{22}$-spacers are capable of crank-shaft-motion at elevated temperatures. The packing of the mesogens has only very little influence on conformation and mobility of the spacer.

470. NMR DETERMINATION OF CURE LEVEL AND SELECTIVITY IN EPOXY AND VINYLSILOXANE RESINS. Frederic C. Schilling and Howard E. Katz, Bell Laboratories, Lucent Technologies, Murray Hill, NJ 07974-0636

The thermal cure of epoxides or vinyl siloxanes in the presence of vinyl monomer that is subsequently photocured should lead to sequential semi-interpenetrating network (SIPN) films that are optically clear and possess good mechanical properties. In this report we describe the use of NMR methods to monitor the level and selectivity of cure in two sequential SIPNs. The Pt-catalyzed thermal crosslinking of vinyl-substituted siloxanes with silyl hydrides in the presence of isobornyl acrylate produces a network structure. Subsequently, the acrylate is photocured to complete the SIPN. In a second system, diepoxide oligomers are thermally cured in the presence of phenoxyethylacrylate that is sequentially photocured. Solid-state ^{13}C and ^{29}Si NMR data identify an undesired reaction of acrylates with the hydride crosslinking agent that limits the utility of vinylsiloxane/acrylates in the formation of SIPNs. In contrast, the cure chemistries of diepoxide/acrylates are found to be very selective and permit formation of optically clear SIPN materials that possess good mechanical properties.

471. CHARACTERIZATION OF PHENOLIC RESINS FOR CARBON COMPOSITES. Sean A. Curran, Terry B. Walker, Raymond Brambilla, AlliedSignal Inc., Morristown, NJ 07962

Phenolic resoles are used as matrix resins in carbon composites. These resins have complex structures based on phenol formaldehyde condensation. They include multiple types of substitution, branching and crosslinks. NMR has proven to be a valuable approach to measure the initial resin chemistry. In addition it has been possible to monitor changes which occur in the resin as the composite proceeds through the stages of pre preg, molding and post baking. The matrix resins undergo multiple stages of reaction, with the final structure dependent on initial structure and processing. This talk will present evidence for processing dependent phase separation, and show how the chemistry of these resoles changes with thermal history.

472. CHARACTERIZATION OF A THERMOSTABLE POLYMER BY ^{13}C SOLID AND LIQUID STATE NMR. RELATIONSHIP BETWEEN CHEMICAL STRUCTURE AND THERMAL PROPERTIES
M.F. GRENIER-LOUSTALOT, Service Central d'Analyse, C.N.R.S., USR B0059 Echangeur de Solaize - B.P. 22 - 69390 - VERNAISON - FRANCE

New thermosetting materials with thermal propargylic groups have been studied (synthesis, mechanism and kinetics of polymerization) as high performance prepolymers. The results obtained in molten state by ^{13}C, ^{1}H, ^{19}F RMN solid and liquid state show that beyond a certain degree of advancement towards the chromene group whereas residual propargylic functions remains in the reaction medium. The chromene group homopolymerizes by opening the double bond in α position of the methylene group. These two reactions occur simultaneously in the temperature range of 200-400°C.

473. **NMR RELAXATION AND DIFFUSION STUDY OF ULTRASOUND DEVULCANIZATION OF SBR NETWORKS.** E. von Meerwall[a,b,c], J. Massey[a,b], S. T. Johnston[a], Departments of (a) Physics, (b) Chemistry, and (c) Maurice Morton Institute of Polymer Science; S. H. Kim, V. Yu. Levin, and A. I. Isayev, Institute of Polymer Engineering, University of Akron, Ohio 44325

To aid in recycling of rubbery polymers and composites, ^{1}H NMR relaxation and pulsed-gradient spin echo diffusion measurements have been performed on virgin and unfilled vulcanized SBR, and networks after various extents of devulcanization by an ultrasound technique. The NMR methods recognize light sol, but do not distinguish between unattached entangled large molecules and chemical network. Devulcanization adds sol with a wide mass distribution, increasing both sol and gel mobility, but lowering sol mean diffusivity. The latter effect is accounted for by the larger mean sol molecular mass but mainly by the increasing glass transition temperature, which is related to main-chain stiffening seen in our earlier ^{13}C NMR experiments. In addition to improving the molecular-level understanding of devulcanization, this study provides guidance for the optimization of the devulcanization parameters.

474. **MOLECULAR PROFILES OF HYPERBRANCHED POLYARYLATES**, B. Ma and F.E. Karasz, Department of Polymer Science and Engineering, University of Massachusetts, Amherst, MA 01003

High molecular weight hyperbranched polyarylates based on the condensation of 5-acetoxyisophthalic acid have been synthesized and fractionated. Their structure and scaling behavior were characterized by ^{1}H NMR, light scattering, molecular simulation and rheology. The degree of branching was found to be invariant with the size of the clusters. Light scattering studies of two modified hyperbranched polyarylates resulted in an exponent $v=0.41\pm0.03$ in the relation $R_g \sim M^v$ in accordance with the prediction of percolation theory. Measurements of A_2 show an exponent $c = 0.81\pm0.04$ in the relation $A_2 = KM^c$. These results point to a stable and relatively compact structure, and this is supported by a proton NMR-shift reagent study. Finally an absence of the plateau zone was the principal characteristic of the G', G"~ω rheology even for high molecular weight samples; also η_0 of the hyperbranched polymer is significantly lower than that of a linear analog.

475. **SURFACE MECHANICAL PROPERTIES OF POLYMERIC SOLIDS** Tisato KAJIYAMA, Keiji TANAKA, Atsushi TAKAHARA, Department of Chemical Science & Technology, Kyushu University, Higashi-ku, Fukuoka 812-81, JAPAN

Surface dynamic storage modulus, E' and surface loss tangent, tan δ of monodisperse polystyrene (PS) films with various molecular weights were evaluated at 293 K on the basis of scanning viscoelasticity microscope (SVM). In the case of the PS film with the number-average molecular weight, Mn lower than ca. 30k, the magnitudes of surface E' and surface tan δ were smaller and larger than the magnitude showing a glassy state, respectively. Thus, it seems reasonable to conclude that the PS film surface with Mn less than ca. 30k is in a glass-rubber transition state or a rubbery state even at 293 K. The scanning rate dependence of lateral force for the monodisperse PS films was investigated at 293 K by lateral force microscopy (LFM). Since the magnitude of lateral force was apparently dependent on the scanning rate, especially in the case of Mn lower than ca. 40k, it seems reasonable to conclude that the PS film surface with Mn less than ca. 40k is in a glass-rubber transition state or rubbery state at 293 K. The LFM results agreed well with the SVM results if the scanning rate of cantilever tip for LFM measurements corresponded to the measuring frequency for SVM measurements. The remarkable depression of Tg at the polymeric surface was explained by the excess free volume induced due to the surface localization of chain end groups. The surface enrichment of chain end groups was confirmed by dynamic secondary ion mass spectroscopic measurement.

476. POLYMER INTERFACES STUDIED BY NEUTRON REFLECTION. J. Higgins, S. Butler, H. Hermes, Department of Chemical Engineering, Imperial College, Prince Consort Road, London, SW7 2BY, U.K. and D. Bucknall, Rutherford Appleton Laboratory, Chilton, Didcot, OX11, U.K.

Interfaces in heterogeneous polymer blends are crucial in determining the properties of the material. The neutron reflection technique has proved unique in providing detailed information on the development of the composition profile of interfaces between both miscible and immiscible polymers with and without the addition of compatibilising di–block copolymers. Data will be presented showing how neutron reflection measurements give clear information on interfaces formed in model systems, how these results can be correlated with mechanical tests on the same interfaces, but also how these results may not translate into clear predictions of the properties of heterogeneous blends after processing.

477. FROM SELF-ORGANED LIQUID CRYSTAL POLYURETHANES TO SELF-ASSEMBLED LIGHT EMMITING DIODES: F. Papadimitrakopoulos, Polymer Science Program, Department of Chemistry, Institute of Materials Science, University of Connecticut, Storrs, CT 06269-3136.

Self-organization has been a topic where significant scientific advances and understanding were attained during the past decades. Concepts controlling the ordering of liquid-crystalline (LC) state, find repeated use in ordering of rigid semiconductive polymers, that recently have attracted the attention of the scientific community. We hereby, couple these concepts and lessons learned from self-organized liquid crystalline polyurethanes to poly(p-phenylenevinylene) (PPV) and its light-emitting diodes (LEDs). In addition, we introduce a novel self-assembly technique to grow thin films made of polymeric metal-chelate derivatives of 8-hydroxyquinoline, which are insoluble and intractable, suitable for semiconducting applications. Considerable attention is given to newly surfaced issues of chemical and structural impurities as well as the effect of structure in the overall performance of polymeric vs. oligomeric LEDs.

478. CRYSTALLINITIES AND TOPOLOGIES OF POLYESTERS FROM RING EXPANSION POLYMERIZATION OF CYCLIC OLIGOMERS. R.P. Kambour[1,2], A. Boulares[1,2,3], D.J. Brunelle[2], W.J. MacKnight[1], L.L. McCracken[2], and T. Takekoshi[2], [1]Polymer Science & Engineering Dept., University of Massachusetts, Amherst, MA, [2]General Electric Research & Development Center, Schenectady, NY 12301, [3]Currently University of Genoa, Genoa, Italy.

With certain organo-metallic initiators the cyclic oligomers (BTCs and ETCs) of polybutylene terephthalate (PBT) and polyethylene terephthalate (PET) can be polymerized rapidly to their respective high polymers (c-PBT and c-PET). Nascent c-PBTs (n-c-PBTs) made from the BTC melt at, e.g., 190¡C with one initiator, a cyclic diorgano- distannoxane, yield heats of fusion 40% greater than those of the conventional linear condensation polymer (l-PBT) crystallized at, e.g.,190¡. l- and c-PBTs are spherulitic, have the same x-ray unit cells, and have low angle x-ray long periods that differ only moderately. Annealing n-c-PBT below the melting range raises its upper end from 232¡, ca., to [3] 250¡C.---15 or 20¡ higher than the highest literature value for l-PBT. Held above Tm c-PBTs show a temperature-activated loss in the ability on subsequent cooling to recrystallize to the original extent, reverting slowly to the "crystallizability" of l-PBTs. Several concurrent chemical changes, e.g., breakdown in the initiator residue, suggest that n-c-PBT consists of macro-rings each held together by the initiator residue and that thermal breakdown of this residue opens the macro-ring irreversibly. The "excess" crystallinity of n-c-PBTs may arise from a relative lack of chain entanglements, a lack arising from the cyclic topology of the n-c-PBT chain. It is also speculated that the unusually high Tm may reflect a topology-related lower entropy of the macro-ring melt.

479.

PSEUDO IDEAL BEHAVIOR IN HARD MOLECULE MIxTURES I. C. Sanchez, Chemical Engineering Department, University of Texas at Austin, 78746

We have investigated by Monte Carlo simulation the thermodynamics of a broad class of hard molecule mixtures that differ in size. Surprisingly, these systems, which include mixtures of spheres, spherocylinders, and tangent sphere chains, closely approximate a generalized ideal mixing law. The ideal mixing law is not the well-known one associated with Raoult's law, but a more general law based on the change in translational degrees of freedom that occur on mixing. It is shown that the pseudo ideal mixing results from a cancellation of favorable excluded volume interactions and unfavorable equation of state work terms. At constant temperature and pressure, the constituent chemical potentials follow a functional form that is closely related to the classical Flory-Huggins chemical potential.

480. BROWNIAN DYNAMICS SIMULATIONS OF SHEARED "WET" POLYMER BRUSHES
Eric S.G. Shaqfeh, Patrick S. Doyle, Thomas Kwan, Alice P. Gast, Department of Chemical Engineering, Stanford University, Stanford, California 94305-5025

We present novel Brownian dynamics simulations of non-dilute polymer brushes under both steady and oscillatory shear. Shear is created by the motion of a single plate above the tethered brush, the gap in the shear simulation is adjusted to span brushes which are uncompressed and high compressed. Our brush is composed of tethered, many bead Kramer's chains in a periodically replicated cell.
Our formulation of the dynamics is novel and includes self-consistent calculations of the solvent flow through the brush. We demonstrate that under steady shear large normal forces exerted by the brush on the top plate are observed, and these increase monotonically with decreasing gap width. Moreover, these normal forces are substantially greater than the concomitant shear forces. Finally, the shear and normal forces shear thin monotonically beyond a certain flow strength or Weissenberg number. Under oscillatory shear, we find that especially for compressed brushes, large normal stress thickening can occur under high Weissenberg number flows and at large enough frequency. This behavior is shown to depend on intermolecular collisions between polymers and occurs at a critical value of the strain amplitude. These simulations served to explain the recent experiments by Klein *et. al.* 1994

481. THIN HYPERBRANCHED FILMS GRAFTED TO GOLD, SILICON AND ALUMINUM.
David E. Bergbreiter, Richard M. Crooks, Merlin L. Bruening, Yuefen Zhou, Yuelong Liu, Gaston Aguilar and Mingqi Zhao, Chemistry Department, Texas A&M University, College Station, TX 77843.

A new approach to grafting thin organic films onto inorganic supports that uses hyperbranched grafts will be discussed. This work is new way of grafting that uses a forgiving synthetic strategy to take into account problems associated with multistep organic synthesis at interfaces. The films prepared in this chemistry are poly(acrylic acid) grafts and poly(acrylic acid) derivatives. Films are prepared as functionally dense thin films with covalent binding to the support and with thicknesses that can be controlled through synthesis. Inorganic supports that have been successfully used include gold, glass, quartz, silica gel, silicon and aluminum. These films have potential for use as sensors and in development of corrosion resistant coatings. A second sort of hyperbranched grafting in which multilayers of dendrimers are simultaneously attached to a surface via a reactive polymer brush will also be discussed.

482. THE AUTOGENOUS POLYMERIZATION OF FORMALDEHYDE ON SURFACES. John G. Van Alsten, Central Research & Development, E. I. duPont de Nemours & Co., Wilmington, DE 19880-0356

We discuss the mechanism of polymerization of formaldehyde on surfaces of derivitized gold and aluminum. Polymerization occurs by an addition mechanism with either adsorbed or condensed formaldehyde. In the absence of added initiator, polymerization is self-initiated by adsorbed or condensed formic acid or formate ion, presumably produced by a self-oxidation and reduction of formaldehyde. The rate of polymerization and chain length is dictated by the concentration of initiator and chain transfer agent (generally water) at the surface. As such, the rate and DP show a remarkable dependence on the chemical nature of the surface. The morphology of the material deposited also changes dramatically with time, and these changes also help to elucidate the process mechanism.

483.

GRAFTING POLYGLUTAMATES FROM FUNCTIONALIZED SELF ASSEMBLED MONOLAYERS
H. Menzel A. Heise Institut für Makromolekulare Chemie, Universität Hannover, Am Kleinen Felde 30, 30167 Hannover, Germany; Hyun Yim, M. D. Foster Institute of Polymer Science, The University of Akron, Akron, OH 44325-3909, USA; R. H. Wieringa, A. J. Schouten Department of Polymer Chemistry, University of Groningen, Nijenborgh 4, 9747 AG Groningen, The Netherlands

Due to their helical structure rigid rodlike poly(glutamate)s show a large dipole moment along the molecular axis, as well as a high hyperpolarization. Therefore, polyglutamates are promising materials for the creation of films for NLO-applications, provided that an unidirectional orientation of the backbones can be achieved. An approach to such an orientation is the covalent attachment of one of the endgroups to the substrate in combination with a high grafting density. Grafting from the surface gives high grafting densities. By using mixed self assembled monolayers and subsequent in-situ modification we were able to install amino groups as initiator sites in controlled density at the surface. We investigated the effect of initiator density on the grafted polymer layer with regard to the layer thickness, layer roughness, polymer conformation, and orientation.

484. THERMAL POLYMERIZATION OF N-CARBOXY ANHYDRIDES OF POLYPEPTIDES ON SOLID SUBSTRATES. Ying-Chih Chang and Curtis W. Frank, Department of Chemical Engineering, Stanford University, Stanford, California 94305-5025

A dry process involving surface polymerization of α-aminoacid-N-carboxy anhydride (NCA) monomers is demonstrated here. Both poly(γ-methyl-L-glutamate) (PMLG) and poly(γ-benzyl-L-glutamate) (PBLG) are initiated from 1-amino propyl triethoxysilane (APS) modified silicon (100) native oxide substrates at an elevated temperature from 80-120°C in *vacuo*. Ellipsometry and transmission Fourier transform infrared spectroscopy (t-FTIR) are applied to characterize the thickness and the secondary conformation of the resulting polypeptide films. Under an identical reaction condition, both PMLG and PBLG films have an average film thickness ca. 20 nm. However, from a t-FTIR study, the spectra indicate that the PBLG film contains pure α-helical structure while the PMLG film contains both α-helical and β-sheet structures; this result is consistent with the previous observation of their molecular helicity in bulk.

485. THE BULGE-BLISTER TEST AS A PROBE OF ADHESION AT THE POLYMER/SOLID INTERFACE. C. E. D. Chidsey, Huihong Luo, W. H. Limburg, W. D. Nix and R. J. Hohlfelder, Departments of Chemistry and Materials Science and Engineering, Stanford University, Stanford, CA 94305

Interfacial chemistry plays a key role in determining the strength of adhesion at a polymer-solid interface despite the fact that most of the energy of debonding is expended in the deformation of the polymer rather than the breaking of the interfacial bonds. To achieve a microscopic understanding of the interrelation of bond breaking and polymer deformation, we need a reliable measure of the debonding energy and methods to deconvolute the macroscopic and microscopic energy dissipation. We have developed a bulge-blister testing apparatus and procedures for extracting both debonding energies, and the elastic, anelastic and plastic responses of a polymer film on a solid substrate. This approach is illustrated here with a crosslinked acrylate polymer on hydrated silicon nitride with and without the conventional adhesion promoter, 3-mercaptopropyltriethoxysilane.

486. PERFLUORO-GRAFTED POLYSILOXANES AS IMMOBILIZED MONOLAYERS: INTERFACIAL PROPERTIES FOR PERFLUOROETHERS VERSUS PERFLUOROALKYLS. David W. Grainger, Wei Wang, David G. Castner[*], Dept. of Chemistry, Colorado State University, Ft. Collins, CO 80523-1872, [*]Dept. of Chemical Engineering, NESAC/Bio, University of Washington, Seattle, WA 98195-1750

We have previously shown using XPS work that perfluoroalkyl-derivatized polymer monolayers on gold have a highly perfluoro-enriched outer surface. We now have good evidence from recent NEXAFS work that the perfluoroalkyl grafted side chains are able to organize as a function of side chain density to "stand up" at the outer surface, yielding an anisotropic brush orientation 15Å thick. In contrast, the same polymer systems bearing perfluoropolyether side chains on gold show reduced surface enrichment and virtually no structural anisotropy. The interfacial properties of these two systems from the standpoints of angle-resolved XPS, polarized reflection FTIR, ToF-SIMS, and NEXAFS will be compared. In addition, new results for these perfluoro-grafted polysiloxane monolayer systems on aluminum oxide using alkanoic acid side chain anchors will be presented.

487. **SURFACE-ACTIVE POLY(ETHYLENE OXIDE)S** Zhaohui Su, Dacheng Wu, Shaw Ling Hsu and Thomas J. McCarthy, Polymer Science and Engineering Department, University of Massachusetts, Amherst, Massachusetts 01003

Samples with one or both ends of monodisperse poly(ethylene oxide) (PEO) functionalized with perfluorodecanoyl groups (PEO^F or PEO^{2F}) were synthesized and blends of these end-capped PEOs with PEOs of the same molecular weight (M_n ~2000 - ~16000 K) were prepared as cast films. Samples were analyzed by X-ray photoelectron spectroscopy (XPS) and contact angle analysis. XPS analysis was also performed on samples at elevated temperatures (melts). Due to the lower surface energy of the fluorocarbon end groups, the modified PEOs preferentially adsorb to the free polymer surface. The surface concentration of the perfluoroalkyl end groups was measured by XPS which indicates that perfluoroalkyl chain ends adsorb to the polymer surface in a reasonably close - packed fashion (at all molecular weights) and leave a zone depleted of fluorine immediately beneath the highly fluorinated surface region. PEO^{F2} samples are found to have a slightly higher surface affinity than PEO^F of the same molecular weight. There is only a slight effect of molecular weight on surface fluorine content indicating a "stretched brush" conformation for the higher molecular weight samples.

488. SELF-ASSEMBLY OF FLUORINATED LC BLOCK COPOLYMERS WITH A STABLE LOW ENERGY SURFACE. Jianguo Wang, Guoping Mao, C. K. Ober and E. J. Kramer, Department of Materials Science and Engineering, Cornell University, Ithaca, NY 14853-1501

Monodisperse polystyrene/liquid crystalline semifluorinated block copolymers have been synthesized by anionic polymerization of polystyrene-b-1,2/3,4 polyisoprene and corresponding polymer analogous reactions. The self-assembly of these side chain block copolymers occurs on two different size scales including microphase separation at the macromolecular level and on the molecular scale of mesogen. By controlling the block copolymer compositions and the length of the fluorocarbon and hydrocarbon unit in the side chain, the effect of the structure on the surface properties, as well as the influence of liquid crystalline structure on the critical surface tension and surface rearrangement have been evaluated in detail.

489. HYBRID ORGANIC/INORGANIC COPOLYMER FILMS WITH STRONGLY HYDROGEN-BOND ACIDIC PROPERTIES FOR VAPOR SENSING
Jay W. Grate and Steven N. Kaganove, *Environmental Molecular Sciences Laboratory, Pacific Northwest National Laboratory, P.O. Box 999. Richland, WA 99352;*

Hybrid organic/inorganic polymers have been prepared incorporating fluorinated bisphenol groups linked using oligosiloxane spacers. These hydrogen-bond acidic materials have glass-to-rubber transition temperatures below room temperature and are excellent sorbents for basic vapors. The physical properties are tunable by varying the length of the oligosiloxane spacers and the molecular weight, and the materials are easily crosslinked to solid elastomers. The potential use of these materials for chemical sensing has been demonstrated by applying them to surface acoustic wave vapor detectors as thin films for detecting hydrogen-bond basic vapors, and it has also been demonstrated that one of these materials with suitable viscosity and refractive index can be used to clad silica optical fibers using an automated fiber drawing tower. These fibers have potential as evanescent wave optical fiber sensors.

490. GRAFTING OF FUNCTIONAL GROUPS ONTO POLYETHYLENE BY MEANS OF RF GLOW DISCHARGES AS FIRST STEP TO THE IMMOBILISATION OF BIOMOLECULES. P. Favia, F. Palumbo, R. d'Agostino; Dipartimento di Chimica, Università di Bari, Centro di Studio per la Chimica dei Plasmi, CNR. Via E. Orabona 4, 70126, Bari, ITALIA

The strategy of grafting chemical groups via Radio Frequency Glow Discharges (RFGD) can be utilised for providing the surface of polymeric biomaterials with selected 'anchor sites' in order to harpoon biomolecules. In this contribution we show the last results of our research devoted to obtain blood-compatible surfaces through the immobilisation of anti-thrombotic molecules on plasma-treated polyethylene (PE). Heparin and highly sulphated Hyaluronic Acid have successfully been immobilised onto PE RFGD-grafted with -COOH groups by using diamines as spacer molecules. Plasma-process control has been achieved with Actinometric Optical Emission Spectroscopy, X-ray Photoelectron Spectroscopy has been utilised for monitoring each step of the whole surface modification process.

491. HYDRATION INDUCED CHANGES IN PLASMA DEPOSITED FLUOROCARBON FILMS. S. Sarma, D. G. Castner, and B. D. Ratner, NESAC/BIO, Dept. Chemical Engineering and Center for Bioengineering, Box 351750, Univ of Washington, Seattle, WA 98195.

Fluorocarbon (FC) plasma deposition typically produces hydrophobic films considered to be highly crosslinked, pin hole free networks. To investigate the stability of these films in hydrophilic biological environments, FC coatings deposited onto poly(methyl methacrylate) (PMMA) substrates were examined with XPS, static ToF-SIMS, and AFM after soaking in buffer solutions. XPS analysis of unsoaked films showed surface concentrations of ~60 at. % F and ~40 at. % C. Soaking in buffer decreased the F surface concentration up to 20 at. % after 1 day and up to 30 at. % after 3 weeks. As the F decreased, both O and C concentrations increased. AFM images indicated this change was not due to delamination of the FC film. Buffer soaking of bulk FC films showed contaminant deposition from the buffer only accounts for a small part of these changes. It appears that low molecular weight PMMA fragments, formed during the RFGD deposition, were diffusing to the surface of the FC film. PMMA samples that were Ar etched for 2 min. just prior to the FC deposition had the poorest stability in buffer, probably due to a higher degree of PMMA fragmentation caused by the Ar etch. ToF-SIMS experiments were consistent with fragments of PMMA, not intact PMMA chains, diffusing through the FC film.

492. SURFACE MODIFICATION OF TISSUE CULTURE POLYSTYRENE WITH LOW-TEMPERATURE PLASMAS

Dorit Keller, Karsten Schröder, Bettina Husen, Andreas Ohl
Institut für Niedertemperatur-Plasmaphysik e. V., 17489 Greifswald, Germany

A clean downstream microwave excited Ar/H_2 plasma treatment of tissue culture polystyrene (PRIMARIA®) is used to vary the type and density of oxygen and nitrogen containing groups. XPS results indicate a combination of modification and etching processes leading to a decrease of oxygen and nitrogen density, less amount of carbon in highly oxidized functional chemical groups and an increase of the $-NH_2/N$-ratio at the surface. Consequently, maximum water contact angles and amine group densities are found at moderate treatment times. Long H_2 plasma treatment (1000 s) reestablishes a hydrophobic surface with very small amounts of oxygen and nitrogen and a very high $-NH_2/N$-ratio. The increase of amine groups did not influence cell response positively. Culturing of cells on the long times plasma treated material shows a further reduced cell adhesion.

POLY

493. LIFE AFTER PLASMAS : OXIDATIVE REACTIONS IN PLASMA POLYMERS AND PLASMA TREATED CONVENTIONAL POLYMERS. T.R. Gengenbach, R.C. Chatelier and H.J. Griesser, Division of Chemicals and Polymers, CSIRO, Private Bag 10, Clayton South MDC, Clayton 3169, Australia.

While the notion that many plasma polymers take up oxygen after deposition has been known for years, few data exist on the rates, extents, chemical groups produced, and reaction mechanisms. Likewise, the post-treatment oxidation of plasma-treated conventional polymers has not been well characterized. We have studied the compositional changes that occur as a result of post-plasma oxidation for periods of up to 4 years for a range of plasma-prepared materials, by a combination of XPS and grazing angle FTIR methods. Here we will present a broad overview of similarities and differences in the post-plasma ageing processes of various systems. For plasma polymers deposited from amine monomers for instance, XPS analyses can distinguish between various possible oxidative products.

494. BARRIER COATINGS FOR FOOD PACKAGING. Eugene S. Lopata, BOC Coating Technology, 4020 Pike Lane, Concord, CA 94520-1297

Dense silicon oxide barrier coatings are deposited at high rate onto polymer webs using PECVD technology. These barriers decrease O_2 and H_2O permeation by up to two orders of magnitude on 12 micron PET with the advantages of clarity, microwaveability, and environmental friendliness. This paper will detail properties of silicon oxide coatings which affect barrier performance levels and packaging conversion issues. Barrier performance is shown to be a function of coating thickness and density. PECVD coatings are demonstrated to show 2-4% strain before failure, and although surface energy is low as deposited (31 dynes/cm), appropriate oxidative post-glows are capable of raising the surface energy to 66 dyne/cm.

495. PLASMA-INDUCED SURFACE RADICALS OF LOW DENSITY POLYETHYLENE STUDIED BY ELECTRON SPIN RESONANCE Masayuki Kuzuya,[*] Shin-ichi Kondo, Masami Sugito Laboratory of Pharmaceutical Physical Chemistry, Gifu Pharmaceutical University, 5-6-1, Mitahora-Higashi, Gifu 502, Japan.

Plasma-induced low density polyethylene (LDPE) radicals were studied in detail by electron spin resonance (ESR) on its comparison with those of high density polyethylene (HDPE). The observed ESR spectra of plasma-irradiated LDPE are largely different from those of HDPE. The systematic computer simulation disclosed that such observed spectra of LDPE consist of three kinds of radicals, mid-chain alkyl radical, allylic radical and dangling bond sites (DBS) at the surface cross-linked region. All these component radicals are essentially identical to those of HDPE. One of the most special features in LDPE, however, is the fact that DBS is a major component unlike a mid-chain alkyl radical in HDPE. Thus, the nature of radical formation of PE was found to be affected by the polymer morphology in a very sensitive manner.

496. PLASMA VACUUM-ULTRAVIOLET INDUCED REACTIONS OF POLYOLEFINS. A. Holländer, J. Behnisch, Fraunhofer-Institut für Angewandte Polymerforschung, Kantstr. 55, D-14513 Teltow, Germany

The interaction between low pressure plasmas and organic polymers is still not understood completely. The ultraviolet, in particular the vacuum-ultraviolet (VUV, 100 nm $< \lambda <$ 200 nm) radiation from a low-pressure plasma is known to initiate chemical reaction in organic polymers which contribute to the structural alterations observed after a plasma treatment. In order to understand the mechanisms of the VUV-induced alterations, we investigated the reactions at the surface of polyethylene and polypropylene during the exposure to VUV radiation in high vacuum. This talk will present results of a mass spectrometric study of low molecular weight fragments evolved from the polymer during VUV irradiation and the ESCA analysis of the irradiated surfaces. Based on these data, reactions paths and their importance for plasma treatments will be discussed.

497. PROCESS CONTROL FOR PLASMA TREATMENT OF POLYMERS

Riccardo d'Agostino
Centro di Studio per la Chimica dei Plasmi- Department of Chemistry, University of Bari- via Orabona 4 - 70126 Bari - Italy tel. +(39)-80-544 2080 fax 544 3405 e-mail: dagostino@area.ba.cnr.it

It is shown that a systematic combined use of gas phase (actinometry) and surface diagnostics (XPS) can effectively lead to an accurate insight of the microscopic process of deposition and plasma treatment, provided the proper reactor architecture is adopted for the experiments. The combined diagnostics also enable the identification of the proper tools for *in situ* checking process performances. Examples to be dealt in the oral contribution include fluoropolymers, silicones, metal-filled polymers, diamond-like films.

498.
PLASMA TREATMENT OF POLYOLEFINS: INFLUENCE OF PLASMA PARAMETERS AND MATERIAL COMPOSITION ON PAINT ADHESION. A. Nihlstrand and T. Hjertberg, Department of Polymer Technology, Chalmers University of Technology, SE-412 96 Gothenburg, Sweden and K.S. Johansson, Institute for Surface Chemistry, P.O. Box 5607, SE-114 86 Stockholm, Sweden

The adhesion properties obtained after oxygen plasma treatments of four commercial TPOs (thermoplastic polyolefins) and also a number of model (PP) and (TPO) materials of different well defined compositions have been studied. The adhesion between a polyurethane (PUR) lacquer and the plasma-treated materials was strongly affected by the plasma treatment conditions and the chemical composition of the materials. It was found that the degree of surface modification was not the determining factor for obtaining successful adhesion. Instead, the cohesive strength of the near-surface region of the substrate is the critical factor. All failures, independent of the absolute peel strength, are located below the oxidized surface layer, the only difference being the depth at which they occur. The depth is influenced in turn by both the plasma processing parameters and the material composition. In particular, double bonds and ethylene in the form of blocks were shown to be favourable.

499. R.F. PLASMA DEPOSITION OF BARRIER THIN FILMS FROM HEXAMETHYLDISILAZANE AND HEXAMETHYLDISILOXANE CONTAINING FEEDS. R. Lamendola and R. d'Agostino, Department of Chemistry, University of Bari, Via Orabona 4, 70126, Bari, Italy.

A comparative study has been performed on hexamethyldisiloxane-oxygen and hexamethyldisilazane-oxygen fed radiofrequency glow discharges at low substrate temperature. Actinometric Optical Emission Spectroscopy and X-Ray Photoelectron Spectroscopy have been utilised as diagnostic tools of plasma phase and of thin films composition, respectively. Gas Transmission Rate measurements have been performed on deposited films at different working conditions. Both in hexamethyldisiloxane and hexamethyldisilazane containing feeds, CH-to-Ar emission ratio has been utilised as a *in situ* probe of carbon content and of the barrier properties of the films.

500. METAL CATALYZED "LIVING" RADICAL POLYMERIZATION INITIATED WITH SULFONYL HALIDES.
V. Percec and B. Barboiu, Department of Macromolecular Science, Case Western Reserve University, Cleveland, Ohio 44106 - 7202

Substituted and unsubstituted sulfonyl halides (chlorides, bromides and iodides) are known to add without and with extrusion of SO_2 group to substituted and unsubstituted olefins and acetylenes. This reaction ('initiation") is catalyzed by UV, various sources of radicals and by transition metals. Depending on the nature of the substituent(s) present on the parent olefin and acetylene, the resulting alkyl halide and, respectively, vinyl halide are known to undergo addition to the same olefin ("propagation") under similar reaction conditions. This sequence of reactions was recently exploited in our laboratory to generate a metal catalyzed "living" radical polymerization. Determination of the apparent rate constants of initiation and propagation showed that arylsulfonyl chlorides are universal initiators for the "living" radical polymerization of styrenes, methacrylates and acrylates.

501. Atom Transfer Radical Polymerization. Role of Various Components and Reaction Conditions Krzysztof Matyjaszewski, Department of Chemistry, Carnegie Mellon University, 4400 Fifth Avenue, Pittsburgh, PA 15213

Analogous to atom transfer radical addition reactions, atom transfer radical polymerization (ATRP) occurs through the repetitive addition of a monomer to the growing radicals generated from dormant alkyl (pseudo)halides by a reversible redox process. This process is catalyzed by a transition metal compound, such as a cuprous halide complexed by two 2,2'-bipyridine (bipy) molecules:

$$P_n\text{-}X + [Cu(I)/2bipy]^{\oplus} \underset{k_{deact}}{\overset{k_{act}}{\rightleftharpoons}} P_n^{\bullet} + [X\text{-}Cu(II)/2bipy]^{\oplus} \quad (1)$$

$$\downarrow k_p$$

Monomer

ATRP is a multicomponent system usually consisting of an alkyl halide, a redox active lower oxidation state transition metal compound with the corresponding ligands (M_t^n), deactivator (XM_t^{n+1} species), either formed spontaneously or present at the very beginning of the reaction, monomer, growing chains, and some additive. The reaction can be performed in bulk or in solution at various temperatures and for certain periods of time. The effect of structure and concentration of these components on the polymerization rate and properties of the obtained polymers as well as the factors that must be considered for optimal reaction conditions is explained.

502. THE EFFECT OF PHENOLS ON ATOM TRANSFER RADICAL POLYMERISATION (ATRP) OF METHYL METHACRYLATE AND STYRENE, David M Haddleton* and Andrew J Shooter, Department of Chemistry, University of Warwick, Coventry, CV4 7AL, U.K.

Controlled polymerisation of vinyl monomers to give living, or 'pseudo-living', polymerisation with predictable Mn and narrow polydispersity index (PDI) is of continuing interest. Traditionally this has been confined to ionic polymerisation. We have conducted a study of the use of Cu(I)Br/alkyl bromide/bipyridine in ATRP of styrene and MMA in the presence of a range of phenols. This was designed with two purposes in mind (1) phenols are relatively acidic and are certainly not compatible with anionic polymerisation and (2) phenols are widely utilised as free radical inhibitors. It is noted that the mechanism of such ATRP reactions is far from clear and current postulations are undoubtedly simplifications. ATRP has been demonstrated to be effective in the presence of large amounts of free radical inhibitor. These results suggest that propagation is not via a free radical and propagation probably takes place via an in cage process where the copper(I) species is in close proximity to the propagating polymer.

503. **TRANSITION METAL-MEDIATED LIVING RADICAL POLYMERIZATION: RECENT ADVANCES.** **Mitsuo Sawamoto** and Masami Kamigaito, *Department of Polymer Chemistry, Kyoto University, Kyoto 606-01, Japan*

This paper discusses *living* radical polymerization of vinyl monomers mediated by transition metal complexes. The table shows their typical initiating systems where initiators are alkyl halide derivatives and complexes involve groups 8–10 transition metals. The subjects herein include:

- Propagation mechanism: effects of possible quenchers and polymer terminal structures;
- Living polymerization in polar protic solvents such as methanol and water;
- The scope of the metal complexes to include Ru(II), Fe(II), and Ni(II) halides with PR_3.

$$R-X \underset{(X = Cl, Br)}{\overset{Ru^{II}}{\rightleftarrows}} R^\bullet - Ru^{III}-X \xrightarrow[Al(OR)_3]{MMA} \sim\sim\sim C^\bullet - Ru^{III} - X \quad \text{Living Polymer}$$

Transition Metal Complexes	Monomers		
$RuCl_2(PPh_3)_3$ $FeCl_2(PPh_3)_2$ $NiBr_2(PPh_3)_2$	$CH_2=CH$ $\mid$ CO_2R' (R' = Me, Et, nBu)	$CH_2=C$ $\mid$ Me $\mid$ CO_2R'	$CH_2=CH$ $\mid$ R'' (R'' = H, Me, Cl)
Initiators (R–X)			
CCl_4 CCl_3 $\mid$ CO_2Me CCl_3 $\mid$ $COMe$ $CHCl_2$ $\mid$ CO_2Me $CHCl_2$ $\mid$ $CO-\phi$	CH_3-CH-X $\mid$ CO_2R' CH_3-C-Br $\mid$ CO_2R'	CH_3-CH-X $\mid$ Ph CH_3-C-Br $\mid$ CO_2Et	(X = Cl, Br) $Y-\phi-SO_2Cl$ (Y = H, Me, Cl, OMe, NO_2)

504. CONTROL OF RADICAL POLYMERIZATIONS BY METALLORADICALS. **Bradford B. Wayland**,[a] and Michael Fryd,[b] a) Department of Chemistry, University of Pennsylvania, Philadelphia, PA 19104-6323, b) DuPont Marshall Laboratory, Philadelphia, Pennsylvania 19146.

Bond homolysis of an organometallic complex $(M-C(CH_3)(R)X)$ in solution proceeds through the intermediacy of a caged radical pair $(M\bullet \; \bullet C(CH_3)(R)X)$ that can recombine, separate into freely diffusing radicals and react by $M\bullet$ abstracting a β-H from the organic radical to form a metal hydride (M-H) and an olefin. In the absence of events that irreversibly terminate radicals and metal hydride the homolytic dissociation of an organo-metal complex can potentially provide a constant equilibrium source of both an organic radical and a metal hydride. Variation of the steric demands and electronic factors of the $M\bullet$ and $R\bullet$ units can be used to promote or inhibit alternate pathways that produce interrelated chain transfer catalysis and quasi-living radical polymerizations. A series of metalloradical complexes and olefin monomers with varying steric and electronic factors will be used in illustrating the factors that optimize the living character of radical polymerizations. Organocobalt complexes that initiate and control radical polymerizations of acrylates that produce linear increase in M_n with conversion, relatively low polydispersity homo polymers and block copolymers will be used in illustrating the efficacy of this approach. The scope and limitations of the metalloradical method in achieving quasi-living radical polymerization of olefin monomers will be discussed.

505. NMR STUDIES AND SIMULATION OF POLYMERIZATION. **H. N. Cheng** and L. J. Kasehagen, Hercules Incorporated, Research Center, 500 Hercules Road, Wilmington, DE 19808-1599

^{13}C NMR is commonly used to study polymer microstructure which often provides information on the polymerization mechanism. In simple cases, NMR studies can be assisted through the use of reaction probability models. However, for more complex polymerizations, the reaction probability models are often inadequate. One possible approach in such cases is to consider the reactions involved in the polymerization and employ a realistic kinetic scheme for NMR analysis. As an example, we examined high-pressure, free-radical polymerization of ethylene. We used a large set of free-radical reactions, including initiation, propagation, chain transfer, backbiting, β-scission, and termination. Through this reaction scheme, we obtained detailed information on polymer microstructure and molecular weight distribution of low density polyethylene (LDPE). Furthermore, knowing the structure/shift dependence, we simulated the ^{13}C NMR spectra of LDPE prepared under different polymerization conditions (e.g., temperature, pressure, and initiator concentrations). Thus, this approach permits the correlation of polymerization process, reaction kinetics, polymer microstructure, and NMR spectra.

506. NMR STUDIES OF MODIFIED EPDM ELASTOMERS. <u>Geert van der Velden</u> and Jo Beulen, DSM Research, PAC-MC, P. O. Box 18, 6160 MD Geleen, The Netherlands

Ethylene-propylene-diene rubbers (EPDM) and related elastomers are important industrial materials. In this work, systematic studies have been made of modified and unmodified EPDM rubbers by ^{1}H and ^{13}C solution NMR. These studies include: (1) Effect of solvents (e.g., 1,1,2,2-tetrachloroethane) on the quality of ^{1}H and ^{13}C NMR spectra of unmodified rubbers; (2) ^{1}H NMR analysis of EPDM rubbers containing a chlorinated termonomer; (3) ^{1}H NMR analysis of EPDM rubbers containing hydroxy modified termonomers. The photosensitive property of the rubbers has been used in the modification step. (4) ^{13}C NMR analysis of EPDM rubbers containing high amounts (above 10 weight %) of ethylidene norbornene (ENB); (5) Preliminary results on the ^{13}C NMR spectra of EPDM rubbers containing doubly ^{13}C enriched ENB termonomer. It has been shown that the use of solution NMR in a systematic fashion permits detailed structural information to be obtained on the EPDM and related elastomers.

507. ^{1}H NMR CHARACTERIZATION OF SWELLING IN CROSSLINKED POLYMER SYSTEMS. P. J. O'Connor, S. S. Cutié, <u>P. B. Smith</u>, S. J. Martin, R. L. Sammler, W. I. Harris, M. J. Marks, L. Wilson, Analytical Sciences Laboratory, The Dow Chemical Company, Midland, MI 48667

A ^{1}H NMR method capable of determining the level of swelling of microscopic volume elements (about 20 μm in diameter) within crosslinked materials is described. This is extremely useful for the characterization of the swelling heterogeneities which may exist within common network systems, such as core/shell or other morphologies. The method utilizes the differences in chemical shift between solvent absorbed into the crosslinked polymer and that of solvent outside the polymer. This chemical shift difference is then correlated to macroscopic swelling (rather than crosslinking) through a simple model which encompasses both the effective chemical crosslinks and the entanglement crosslinks in the manner of classical swelling experiments. The analysis is demonstrated for styrene, divinyl benzene copolymer beads, crosslinked polycarbonates, ion-exchange cation resins and crosslinked poly(acrylic acid). A calibration is, in each case, developed with a series of standard materials whose bulk swelling characteristics were determined. An example of the analysis of the crosslinking morphology within a single cation exchange bead is also presented. This method is useful for the swelling studies of any network system with aromatic or acid functionality and can enable the identification of new structure/property relationships critical for developing advanced materials.

508. NMR DIFFUSION STUDY OF SWELLED AMPHIPHILIC NETWORKS FOR USE AS PERMSELECTIVE BIOMEMBRANES. <u>E. von Meerwall</u>[a,b,c], T. Cregger[a], S. Shamlou[c], and J. P. Kennedy[c], Departments of (a) Physics, (b) Chemistry, (c) Maurice Morton Institute of Polymer Science, University of Akron, Ohio 44325.

Using pulsed-gradient spin-echo NMR we have studied the permselectivity of certain biocompatible membranes made of poly(N,N-dimethylacrylamide) or poly(N,N-dimethylaminoethylmethacrylate) crosslinked with polyisobutylene. The molecular weight of this crosslinker determines the permeability to moderately sized molecules. The networks were sol-extracted and then swelled to equilibrium in 99.9% D_2O; diffusants were then added at a 2% concentration. Diffusion measurements were performed for glucose (MW near 200 g/mol) and dextran (MW near 11 and 19 kg/mol). It is found that the diffusion of glucose in not slowed substantially in the swelled networks but depends weakly on network properties, and that dextran is unable, or very slow, to enter the networks. To obtain usable insulin concentrations the pH requires careful adjustment, performed by addition of a deuterated buffer solution.

509. Oxazoline Hydrolysis, Synthesis and Polymerization of a New Oxazoline Methacrylate   H.A. Rasoul, D.B. Obuchowski, W.M. Culbertson, <u>D.L. Trumbo</u>*, SC Johnson Polymer, 1525 Howe St., Racine, WI 53404

The hydrolytic stability of several oxazoline model compounds were evaluated by ^{1}H-NMR spectroscopy. It was found that aromatic substitution of the oxazoline ring yielded hydrolytically stable materials whereas aliphatic substitution did not. These results were used to guide the synthesis of a methacryloxazoline monomer. The monomer's homo- and copolymerization behavior was then evaluated.

510. PORTABLE SPECTRAL INTERPRETERS FOR FIELD APPLICATIONS: RAPID SPECTRAL ASSIGNMENTS OF MOLECULAR FUNCTIONALITY OR SYSTEM COMPOSITION
Karlis Adamsons, DuPont Company, Marshall R&D Laboratory
3401 Grays Ferry Ave., Philadelphia, PA. 19146

PC-based tools have been developed to assist in providing rapid spectral assignments for NMR, IR, UV-VIS and fluorescence spectra. The tools have been integrated into a system toolbox that can be made resident on most of today's 386/486/Pentium-based laptops, as well as office/lab workstations. The benefit of using laptops is inherent in their portability giving users the ability to perform spectral interpretations in field, office or laboratory settings. The system has a simple frontend menu from which all of the tools can be accessed. The spectral databases incorporated derive from materials that are typically encountered in the paint, coatings, and plastics industry. Some of the tools (i.e., NMR, IR) offer two features: assignment of functional groups and/or bonds; and identification of compounds or components in a blend or mixture. The system includes detailed relational databases on the following: (1) Proton NMR; (2) ^{13}C NMR; (3) ^{19}F NMR; (4) ^{29}Si NMR; (5) IR of organic materials; (6) IR of inorganic materials (primarily pigments); (7) UV-VIS of organic materials; and (8) Fluorescence (UV/VIS) of organic materials. The system uses dBase type relational databases, and VP-INFO macros for the interactive frontend, interpreter, and output modules. Although the system was originally designed for specific needs in the Automotive Products (and related finishes) industry, the approach can be readily adapted to other areas (i.e., pharmaceutical, gas/oil, wood/paper, environmental, biotechnology, and agricultural product industries; and at Universities for training and R&D purposes).

511. MODEL STUDIES OF MULTICOMPONENT AND MICROPHASE SEPARATED POLYMER SYSTEMS. C.D. Eisenbach, Institut für Technische Chemie II, Universität Stuttgart, D-70569 Stuttgart, Germany

One successful approach to enlarge the knowledge about structure-properties relationships and also to explore the innovative potential of polymer materials is the design and study of model systems. This will be illustrated with the example of segmented block and graft copolymers of various chain architecture, which could be precisely controlled by appropriate synthetic routes; examples include thermoplastic elastomers and polymer-ion complexes. Both the molecular design of the individual macromolecule and the composition of the macromolecular system affected the phase morphology and the physical as well as material properties. These studies have also demonstrated that even small changes in the macromolecular constitution could result in distinct morphologies and novel properties.

512. MULTIPLE MORPHOLOGIES IN CREW-CUT AGGREGATES OF ASYMMETRIC AMPHIPHILIC DIBLOCK COPOLYMERS. A. Eisenberg, Department of Chemistry, McGill University, 801 Sherbrooke Street West, Montreal, Quebec, Canada H3A 2K6.

Asymmetric diblocks consisting of long polystyrene chains attached to short acrylic acid or ethylene oxide chains can be made to self assemble in solution into crew-cut aggregates with large hydrophobic cores and short hydrophilic coronae. Depending on the detailed method of preparation (the relevant parameters being, among others, block lengths, polymer concentration, micro-ion content, and rate of addition of precipitant to the solution) one can obtain a very wide range of morphologies. These include not only simple spherical aggregates, rods vesicles, lamellae and tubules, but also a wide range of topologically more complex structures such as branching and bicontinuous rods, branching and bicontinuous tubes, hollow "doughnuts," and a range of compound aggregates, including hollow spheres with partitions as well as porous spheres. The preparation involves the addition f a non-solvent, e.g. water, to a molecular solution of the block copolymer in a solvent (e.g. DMF) or to a solvent/non-solvent mixture. In the presentation, the thermodynamic and kinetic aspects of the self-assembly are discussed in addition to a description of the new morphologies.

513. EFFECT OF CONCENTRATION ON NETWORK TOPOLOGY AND DYNAMICS OF TELECHELIC IONOMERS IN TOLUENE. Sam Bhargava and Stuart L. Cooper, Department of Chemical Engineering, University of Delaware, Newark, DE 19716.

Telechelic ionomers in non-polar solvents form a physically crosslinked network in semi-dilute solutions. These solutions often exhibit shear-thickening and shear-thinning thus making these materials useful as viscosifiers and rheology modifiers. Transient network models describe the dynamics of such associating polymer solutions under shear. In the framework of Tanaka and Edwards transient network model (1991), a single relaxation time relating directly to the strength of association describes the longest relaxation time. Marrucci et al. (1993) challenged this concept and proposed that this relaxation time should increase with concentration because neighboring aggregates influence chain relaxation. Further, several theories predict concentration dependence of shear-thickening behavior. In this study, we report the effect of concentration on the number of effective chains, the network relaxation time, the critical shear-rate for shear-thickening and viscosity enhancement. Our results support the theory of Marrucci et al.

514. MISCIBILITY AND MORPHOLOGY OF POLYOLEFIN BLENDS AND BLOCK COPOLYMERS. K. Sakurai, W. J. MacKnight, University of Massachusetts, Amherst, Massachusetts 01003; D. J. Lohse, R. T. Garner, D. N. Schulz, and J. A. Sissano, Exxon Research and Engineering Co., Annandale, New Jersey 08801.

The work described here focuses on blends of polypropylene (PP) with ethylene-propylene copolymers (EP) of varying sequence distributions. By light scattering we have seen that these blends have an LCST type of phase diagram; this confirms neutron scattering measurements of the Flory χ parameter. Moreover, we show that the degree of miscibility increases as the EP becomes more alternating (at an overall 50 mole % composition). The morphologies of blends of an EP diblock copolymer with the PP homopolymer have also been examined, particularly with regard to the ability of short homopolymers to swell the PP block, and the effect of blending on the structure of the PE crystalline lamellae.

515. PHASE BEHAVIOR AND STRUCTURE PINNING DURING PHASE-SEPARATION OF POLY-AMIDE/IONOMER BLENDS. R. A. Weiss[1], Y. Feng[1], R. Xie[1], R. Tucker[1], C. C. Han[2], [1]Dept. of Chemical Engineering, University of Connecticut, Storrs, CT 06269; [2]Polymer Division, National Institute of Standards & Technology, Gaithersburg, MD 20899

Blends of the lithium salt of lightly sulfonated polystyrene and poly(N,N'- dimethylethylene sebacamide) (Li-SPS/mPA) are miscible as a result of strong ion-amide complexation. The blends exhibit LCST phase behavior and an increase of the sulfonation level from 4 to 9.5 mol% raises the critical temperature ca. 150°C. Phase separation may be thermally induced and is thermodynamically reversible. However, the kinetics of phase separation and the resulting blend morphology are unusual. The phase separation kinetics as determined by light scattering following a temperature-jump deep into the spinodal region of the phase diagram deviate from conventional Cahn-Hilliard theory and the phase separation process stalls after a couple of hours, essentially *pinning* the structure. Small angle neutron scattering experiments indicate that a microstructure with a characteristic dimension of ca. 20 nm also grows following a temperature-jump into the spinodal, and this structure also becomes pinned after about 30 min. The phase separated blend, therefore possesses a two-phase macrostructure with domains of the O(mm), plus microphase separation of the O(nm).

516. STOICHIOMETRIC COMPLEXES OF SYNTHETIC POLYPEPTIDES AND OPPOSITELY CHARGED SURFACTANTS. E.A. Ponomarenko, A.J. Waddon, K.N. Bakeev, D. Tirrell and W.J. MacKnight, Polymer Science and Engineering Department, University of Massachusetts, Amherst, MA 01003.

We have studied the solution and solid state properties of complexes formed by the interaction of poly(α,L-glutamic acid)PG or poly(l-lysine) PL with alkyltrimethyl ammonium bromides and alkyl sulfates, respectively. The complexes are soluble in various organic solvents. The PG complexes in the solid state consist of alternating layers of PG chains in the α helical conformation separated by bimolecular layers of the surfactants. The surfactant alkyl chains are interdigitated and perpendicular to the lamellar surfaces. In PG complexes of related structure with covalent bonds formed by esterification, the alkyl side chains crystallize when they consist of 10 or more carbon atoms. In the case of complexes formed by coulombic interactions more than 16 carbon atoms in the side chains are necessary for crystallization. The PL complexes are similarly organized into lamellar structures in the solid state, but the PL chains can adopt either α helical or β sheet conformations, depending upon the method of preparation. In dilute solution in the mixed solvent chloroform and trifluoracetic acid (TFA) the PL complexes neither form interchain aggregates nor dissociate up to 10 volume per cent TFA. The PL chains adopt the α helical conformation below 4 volume percent TFA and a transition to a disordered, coil form occurs at high TFA concentrations.

517.
TETHERED MEMBRANES ON SOLID SUPPORTS Knoll, W., Heibel, C., Rühe, J., Max-Planck-Institut für Polymerforschung, Ackermannweg 10, 55128 Mainz, FRG and Frontier Research Program, The Institute of Physical and Chemical Research (RIKEN), Wako, Saitama, Japan

We report on synthetic supramolecular strategies for the build-up of polymer-supported lipid bilayer membranes on solid substrates. We will concentrate on the sequential chemisorption concept with an anchor layer, a soft polymer buffer layer, and a partially covalently linked lipid monolayer which after transfer of a second lipid layer constitutes the supported, tethered membrane. Aiming at functionalizing SiOx surfaces we choose silane derivatives as anchor systems. Their N-hydroxysuccineimide (NHS) endgroups allowed us to covalently attach linear, poly(ethyloxazoline-co-ethyleneimine) as the polymeric buffer layer. The structural features of these interfacial architectures were characterized by IR-, XPS, surface plasmon, and 29Si-CP-MAS-NMR-spectroscopies. The final lipid monolayer was coupled to the polymer support through NHS-modified fatty acids doped into a phospholipid Langmuir layer prepared at the water/ air interface.

518. CROSSLINKING POLYMERIZATION IN SUPRAMOLECULAR ASSEMBLIES: LAMELLAR AND NONLAMELLAR PHASES.
Thomas M. Sisson, Henry G. Lamparski, Silvia Kölchens, Tina Peterson, Anissa Elayadi, and David F. O'Brien Department of Chemistry, C.S. Marvel Laboratories, The University of Arizona, Tucson, AZ 85721

Polymerization of monomeric lipids in an assembly proceeds in a linear or crosslinking manner depending on the number of polymerizable groups per monomeric lipid. This report describes three approaches to the characterization of the gel point for polymerizations constrained by the two-dimensional nature of lipid bilayers. The gel point for two-dimensional lipid assemblies was determined by correlation of the onset of significant changes in the physical properties of the polymerized bilayers with the bilayer composition. The properties examined in this study were the lateral diffusion of a small molecule probe of the bilayer, the stability of polymerized bilayer vesicles in the presence of surfactants, and the solubility of lipid polymers isolated from the bilayers after removal of water. Each method used indicated a substantial mole fraction (0.25 to 0.35) of the bis-substituted lipid was necessary to cause crosslinking of the bilayer. Symmetrical crosslinking agents were synthesized to investigate the inefficient crosslinking in organized assemblies. Possible explanations for the inefficient nature of the lipid bilayer crosslinking will be discussed.

519. RECOGNITION AND DISCRIMINATION OF SURFACE-BOUND DNA. Laura T. Mazzola and Curtis W. Frank, Stanford University, Stanford, CA 94305; Stephen P. A. Fodor, Affymetrix, Santa Clara, CA 95051

We describe here a method for detecting biomolecular recognition using atomic force microscopy (AFM). AFM detects surface morphology by rastering a sharpened tip across the sample surface. The probe tip interacts with the substrate at molecular (if not atomic) dimensions, which suggests that this technique can also be used to probe the forces involved in molecular recognition. We have used photolithography to direct the synthesis of a two-dimensional matrix of oligonucleotides containing the sequence 3'-CAGTTCTACGATGGCAAGTC, alternating with regions containing the reverse complement. These arrays were then probed with AFM tips obtained from BioForce Laboratory (Santa Barbara, CA 93103), to which the aforementioned oligonucleotide had been chemically bound. When used as an imaging probe, these modified tips display chemical recognition and sequence-specific discrimination of the substrate DNA.

520. **INTRODUCING RUBBER INTO THE LANGMUIR-BLODGETT TECHNIQUE**, **Werner A. Goedel,** Max-Planck Institut f. Colloids & Surfaces, Hs.9.9, Rudower Chaussee 5, 12489 Berlin, Germany, e.mail:goedel@mpikg.fta-berlin.de

Hydrophobic polymers with low glass transition temperature (polyisoprenes) and a single head group (sulfonate) bearing photoreactive side groups (anthracene) have been synthesised and characterised as insoluble monolayers on a water surface. The isotherms are similar to the parent polymers without anthracene side groups and the films can be transferred to solid substrates. The films on solid substrates as well as on the water surface can be crosslinked via irradiation with UV-light. Irradiation of the monolayers through a mask followed by solvent treatment gives rise to laterally structured monolayers.
40 nm thick films crosslinked on the water surface can be transferred to solid substrates grids spanning openings of up to 0.3 mm diameter. The films are rubber elastic and can be elastically deformed by applying an overpressure from one side of the freely suspended membrane.
(relevant publications: *Langmuir* 1993, 9, 1184 and 1994, 10, 4209, *macromolecules* submitted 1996, *progr. polymer & colloid sci.* submitted 1996).

521. ORIENTATIONAL DYNAMICS OF A TWO-DIMENSIONAL POLYMER NEMATIC. G. G. Fuller, T. Maruyama, C. W. Frank, C. R. Robertson, Department of Chemical Engineering, Stanford University, Stanford, CA 94305-5025

In situ measurements of dichroism have been used to measure the time dependent orientation of "hairy rod" amphiphiles residing within a two dimensional solution as a Langmuir films. The system is shown to behave as a two-dimensional nematic. Application of extensional flows is shown to reorient and improve the alignment of the system. Using a simple molecular model developed by Marrucci and Maffettone, it is demonstrated that the rotational diffusivity and nematic potential can be measured. Furthermore, the prediction of the model of the existence of a strong flow/weak flow criteria has been verified. This criteria states that a critical velocity gradient must be applied to the nematic before the flow can overcome the nematic potential. Below this flow strength, the flow can actually diminish the degree of alignment, depending on the initial orientation of the director relative to the principal axis of strain.

522. DENDRIMER SELF-ASSEMBLED FILMS; V. N. Bliznyuk, Institute of Semiconductor Physics, Ukrain. Acad. Sci., Kiev, Ukraine 252650; F. Rinderspacher, Dept. Plastics Engineering, University of Paderborn, Paderborn, Germany 33098; V. V. Tsukruk, College of Engineering & Applied Sciences, Western Michigan University, Kalamazoo, MI, USA 49008

Composite molecular films of $(AB)_x$ type are fabricated by self-assembly of dendritic macromolecules of two adjacent generations using the electrostatic layer-by-layer deposition. Polyamidoamine dendrimers with surface amine groups for generations 4, 6, and 10 (G4, G6, G10) and carboxylic groups for generations G3.5, G5.5 and G9.5 have been used as building units. The molecular films are explored by scanning probe microscopy and X-ray reflectivity. All even generations are observed to form homogeneous, stable monolayers on a silicon surface. The thickness of a single monolayer varies with generation (molecular weight) from 1.8 nm (G4) to 2.8 nm (G6) and 5.6 nm (G10). The average thickness of a molecular layer in multilayer films is much smaller than the diameter of an ideal spherical dendritic macromolecule. This difference indicates compression of dendrimers along the surface normal caused by exceeding electrostatic interactions between adjacent oppositely charged dendrimer layers. The model of molecular ordering of dendrimer films fabricated at given conditions assumes highly compressed dendritic macromolecules of oblate shape with a degree of anisometry more than 1 : 3.

523. **STEPWISE POLYMER SURFACE MODIFICATION: CHEMISTRY - LAYER-BY-LAYER DEPOSITION.** Vipavee Phuvanartnuruks and Thomas J. McCarthy, Polymer Science and Engineering Department, University of Massachusetts, Amherst, Massachusetts 01003.

Layer-by-layer deposition of polyelectrolytes onto a chemically derivatized fluoropolymer film is reported as a method for step-wise polymer surface modification. Alcohol functionality is introduced onto the surface of poly(chlorotrifluoroethylene) (PCTFE) by reaction with a protected alcohol - containing lithium reagent and subsequent hydrolysis (PCTFE-OH). Poly(allylamine hydrochloride) (PAH) and poly(sodium styrene sulfonate) (PSS) are then sequentially adsorbed to PCTFE-OH using a variety of conditions. X-ray photoelectron spectroscopy and contact angle analyses indicate that the layers are stratified and that the thickness of individual layers and multilayer assemblies can be controlled by experimental variables. The structure of the first adsorbed layer (PAH) is a strong function of the pH of the polyelectrolyte adsorption solution. Subsequent layer thicknesses depend on the first layer thickness (surface charge density) as well as the ionic strength of the polyelectrolyte solutions. The wettability of the multilayer film is controlled by the outermost polyelectrolyte layer and the thickness of the individual layers (sublayers have an influence in multilayers with thinner layers).

524. COMPOSITE SELF-ASSEMBLED FILMS FROM CHARGED LATEX NANOPARTICLES V. N. Bliznyuk, Institute of Semiconductor Physics, Ukrain. Acad. Sci., Kiev, Ukraine 252650; V. V. Tsukruk, College of Engineering & Applied Sciences, Western Michigan University, Kalamazoo, MI, USA 49008

We study several multilayer latex-latex films fabricated by electrostatic self-assembly and composed of charged PS latexes by combined scanning probe microscopy and X-ray refelectivity. PS nanoparticles of 20 - 200 nm in diameter possess carboxy, sulfate, and amidine surface groups. Most of latexes used are able to form monolayers with short-range local ordering of nanoparticles within monolayer. Layer-by-layer deposition of latexes with alternating charges results in steady increase of film thickness that is consistent with dense centered cubic packing of nanospheres up to the first five layers. Virtually linear growth of film thickness is observed for further deposition (up to 50 layers) with average increment of 7.3 nm per added layer for 20 nm nanoparticles due to the incomplete monolayer formation. During thermal treatment in a wide range of temperatures, we observe that strong tethering of charged nanoparticles to surfaces prevents their surface diffusion and rearrangement required for formation of perfect lateral ordering.

525. GROWTH OF IRON OXIDE NANOPARTICLES ON SELF-ASSEMBLED MONOLAYERS. P. Stroeve, Department of Chemical Engineering and Materials Science, University of California at Davis, Davis, CA 95616, M. Nagtegaal, W. Tremel, Department of Inorganic Chemistry, University of Mainz, Mainz, Germany, and W. Knoll, Max-Planck Institute fur Polymerforschung, Mainz, Germany

Substrate surfaces, derivatized with self-assembled monolayers (SAM), can be used to control the growth of inorganic particles from aqueous solutions. Our work involves self-assembled thiol monolayers on gold to study the influence of functional groups on the nucleation and growth of iron oxide nanoparticles. In particular, we examine the growth of iron oxide nanoparticles on thiol monolayers containing different acidic functional groups: carboxylate or sulfonate terminated alkyl thiols. By exposing the thiol monolayers to the appropriate reactants, the growth of the iron oxide nanoparticles is monitored by surface plasmon spectroscopy (SPS), atomic force spectroscopy (AFM) and grazing angle FTIR. The growth of iron oxide particles is influenced by the pH of the aqueous solution, the charge of iron ions in the solution and the type of acidic functional group on the thiols. Nanoparticles from 1 nm to 40 nm can be grown depending on the above variables and the number of reaction cycles.

526.

THE EFFECT OF ION TYPE AND IONIC CONTENT ON TEMPLATING PATTERNED IONIC MULTILAYERS. **Sarah L. Clark** Martha F. Montague, Paula T. Hammond, Department of Chemical Engineering, Massachusetts Institute of Technology, 25 Ames Street, Cambridge, Massachusetts 02139

We utilize cations (Na^+, Li^+) and anions (Cl^-, sulfate) to enhance or destroy the adsorption resistance of an oligo-ethylene glycol (EG)-terminated self-assembled monolayer (SAM). Dipping a micro-patterned surface of EG and acid (COOH)-terminated SAMs on gold alternately in dilute solutions of polydiallyldimethylammonium chloride (PDAC) and sulfonated polystyrene (SPS), the polyelectrolyte selectively adsorbs on the acid regions. Adding NaCl to the aqueous polyelectrolyte solution alters the polyelectrolyte conformation and the adsorption characteristics of the patterned surface. Under especially high NaCl concentrations (~1 M), the selective adsorption of the two regions can be reversed. Combined with the supramolecular control of the ionic multilayer technique, this control over patterning the multilayers has potentially promising

527. AN INVESTIGATION OF AROMATIC AND ALIPHATIC FLUOROCARBON POLYMER FILMS: IS CONTROL OF FILM PROPERTIES POSSIBLE VIA PULSED PLASMA POLYMERIZATION? N. M. Mackie and <u>Ellen R. Fisher</u>, Department of Chemistry, Colorado State University, Fort Collins, CO 80523-1872

We have examined pulsed plasmas as a means to control the composition of plasma polymerized films. Pulsed plasmas have an advantage over traditional continuous wave (CW) plasmas by retaining the starting monomer intact while depositing polymer films. This tailoring of polymers allows us to deposit new materials which would be difficult by other plasma polymerization methods. We have studied the effect of pulse parameters on the bulk and surface properties of plasma polymerized films for C_2F_6/H_2, C_6H_6, $C_6F_3H_3$, and C_6F_6 plasmas. Bulk structure of the films is characterized using FTIR spectroscopy and surface structure is analyzed using SEM, XPS, and contact angle measurements. For plasma polymers deposited from a 90%C_2F_6 (10% H_2) plasma, we see a dramatic increase in the CF_2 content in the film. With aromatic precursors, the aromatic structure is retained in the deposited films. Preliminary results on the surface reactivity of radicals (CF and CF_2) during film deposition will also be presented.

528. CONTACT VERSUS NON-CONTACT MODE IN THE SCANNING FORCE MICROSCOPY OF PLASMA-TREATED POLY(ETHYLENE TEREPHTHALATE) FILMS. John S. G. Ling[1], Andrew J. Murray[2] and Graham J. Leggett[1]*, Department of Materials Engineering and Materials Design, The University of Nottingham, University Park, Nottingham NG7 2RD, United Kingdom[1] and TopoMetrix Corporation, 18 Hill Street, Saffron Walden, Essex CB10 1JD, United Kingdom[2]

Argon plasma treated poly(ethylene terephthalate) (PET) films have been characterised using a range of surface analytical techniques, including contact angle goniometry, X-ray photoelectron spectroscopy (XPS) and scanning force microscopy (SEM). Treatment of the polyester for 20 min with a discharge power of 100 W yields a wetting surface. XPS spectra of the treated material reveal an apparent broadening of the main features in the C1s envelope, in accordance with the observations of other workers. Scanning force micrographs reveal a complex and pronounced morphology in the treated material, compared to the relatively flat virgin material. This morphology is observed in both contact and non-contact mode; however, the quality of resolution obtained in non-contact mode is superior to that achieved in contact mode. It is concluded that the surface structures formed by plasma-treatment of PET are more delicate than the untreated material.

529. **A COMPARISON OF OXIDIZED POLYSTYRENE SURFACES PRODUCED BY PLASMA AND NEUTRAL BEAM METHODS.** Terry L. Thompson,[1,2] Bonnie J. Tyler,[1,3] and Timothy K. Minton,[1,2] [1]Center for Biofilm Engineering, [2]Department of Chemistry and Biochemistry, [3]Department of Chemical Engineering, Montana State University, Bozeman, MT 59717 USA.

We are investigating a new method for surface chemical modification of polymers based on controlled exposure to a directed beam. The objectives of this study are to identify and characterize differences between surfaces modified by a neutral beam and by more conventional plasma treatments. We have exposed polystyrene surfaces to beams of oxygen atoms and to oxygen plasmas, and we have characterized the modified surfaces with X-ray photoelectron spectroscopy. Both plasma and beam-exposed samples show substantial and similar total oxygen atomic concentrations; however, the resulting distribution of carbon-oxygen bonding environments differs greatly between the two exposure methods. In order to facilitate identification and quantification of functional groups, we have initiated chemical derivatization studies of the modified surfaces.

530. SURFACE MODIFICATION OF POLY(TETRAFLUOROETHYLENE) FILM BY PLASMA GRAFT POLYMERIZATION OF SODIUM VINYLSULFONATE. Norihiro Inagaki, Shigeru Tasaka and Yoh-ichiro Goto, Laboratory of Polymer Chemistry, Faculty of Engineering, Shizuoka University, 3-5-1 Johoku, Hamamatsu, 432 JAPAN

PTFE surface was modified by the graft polymerization of sodium vinysulfonate, and the chemical composition of the graft-polymerized PTFE surface was analyzed by XPS. Peroxides acted as an initiator for the graft polymerization were formed on the PTFE surface by a combination procedure of the argon plasma irradiation and the air exposure, and the graft polymerization of sodium vinylsulfonate was initiated at 65 - 80°C from the peroxide groups. The peroxide concentration is $3 - 5 \times 10^{-13}$ numbers/cm^2. The graft polymerization was ascertained by the weight increase and appearance of C_{1s} spectrum at 285.0 eV (CH$_2$ and CH groups) and S_{2p} spectrum at 168.4 and 169.8 eV (SO$_3^-$ groups). An average degree of polymerization of the graft polymers was 3.4×10^3. The graft polymer does not distribute over all PTFE surface, but a part of the PTFE surface comes out the surface. The coverage of the PTFE surface with the graft polymer is 43 %.

531. C_2F_4/PROPANOL COPOLYMERIZED FILMS PRODUCED USING A MONOMER EVAPORATOR. K. Bernhardt, G. Urban, H.Wanzenboeck, Department of Electronics, University of Technology, Vienna, Austria

In order to control precisely the amount of a vapor component fed into a plasma polymerization process an evaporation device has been constructed. It enables to set values of a given vapor pressure of liquid monomers and to prepare defined mixtures with a carrier and/or other reactive gas components. Construction and control strategy of the source will be briefly described. Major attention will be focused on the effectiveness of the device which will be demonstrated at an example of films copolymerized out of C_2F_4 and propanol vapor mixtures using argon as carrier gas. Film properties determined by means of contact angle measurement, infrared spectroscopy and their dependence on the vapor to carrier as well as to reactive gas ratio will be presented.

532. SURFACE CHEMISTRY OF A NITROGEN-PLASMA-TREATED POLYESTER. L. J. Gerenser and J. M. Grace, Eastman Kodak Company, Rochester, NY 14650-2132

The surface chemistry of nitrogen-plasma-treated poly(ethylene-2,6-naphthalate) (PEN) was characterized with x-ray photoelectron spectroscopy (XPS), high resolution electron energy loss spectroscopy (HREELS), and static secondary ion mass spectrometry (SSIMS). Plasma pressure and dose were found to significantly influence the surface chemistry. Nitrogen-plasma treatment incorporates nitrogen in the upper 5 nm of the PEN surface and disrupts the naphthalene rings, producing an aliphatic-like species. The incorporated nitrogen is primarily in the form of amine, imine, and amide groups. Nitrogen-plasma treatment also induces rearrangement of the PEN ester groups into carbonyl and carboxyl groups. The distribution of nitrogen functionalities and the ester group rearrangement are both pressure and dose dependent. By mixing other gases with nitrogen, the distribution of plasma-induced nitrogen functionalities can be altered to a greater extent.

533. SURFACE MODIFICATION OF CHLOROBUTYL AND NITRILE RUBBER VIA PLASMA POLYMERIZATION OF VINYLIDENE FLUORIDE. R. J. Ratway and C. M. Balik, Dept. of Materials Science and Engineering, Box 7907, North Carolina State University, Raleigh, NC 27695-7907.

Radio frequency plasma polymerization of vinylidene fluoride (VDF) has been used to modify the surface properties of chlorobutyl and nitrile rubber. The structure and properties of the plasma layer were studied as a function of plasma power and VDF flow rate, and were characterized with FTIR, SIMS, SEM, and sliding friction measurements. As plasma power increased, the degree of fluorination decreased in the plasma polymer. Both types of rubber exhibited a lower coefficient of friction after plasma treatment, but the reduction was more significant for chlorobutyl than for nitrile rubber. The friction coefficient exhibited no apparent trend with plasma power or flow rate within the ranges studied. Repetitive sliding friction tests were conducted to assess the wear behavior of the plasma layer. It was found that plasma treatment was as effective as a commercially used silicone oil treatment in reducing friction.

534. PULSED PLASMA POLYMERIZATION OF ACRYLOYL CHLORIDE: FUNCTIONALIZATION OF SURFACES WITH REACTIVE ACID CHLORIDE GROUPS. Jose G. Calderon and Richard B. Timmons, Department of Chemistry and Biochemistry, Box 19065, The University of Texas at Arlington, Arlington, Texas 76019-0065.

A variable duty cycle RF pulsed plasma technique was employed to generate thin polymeric films from acryloyl chloride monomer. Significant retention of the acid chloride functional group of the starting monomer in these films was achieved by operating the pulsed plasma at exceptionally low duty cycles. XPS, FT-IR and AFM analyses of the plasma generated polymers revealed large scale progressive changes in both chemical and physical properties of these films with sequential changes in the plasma duty cycle employed during deposition. The -C(O)Cl containing films can be successfully employed for subsequent molecular surface structuring via covalent attachment of various molecules, as illustrated in this work.

535. Possibilities Offered by a Mixed Plasma Treatment to Promote Endothelialization on Synthetic Materials. A. Hamzaoui [a,b], A. Mas [a], F. Schué [a]
[a] Laboratoire de Chimie Macromoléculaire, Université de Montpellier II, Place Eugene Bataillon, 34095 Montpellier II - Cedex 05 (France)
[b] Department of Mechanical Engineering, Florida International University, Miami, Florida, 33199

A mixed plasma treatment related to vascular graft improvement was performed on Silastic®disks (Dow Corning Corporation). The concept of our work, presented here, is to modify the surface chemical composition of synthetic materials in order to promote the growth of endothelial cells (EC). Porcine aortic endothelial cell (PAEC) adhesion and growth were both studied. It is important to study not only the short-term adhesion characteristics of EC on the modified surface, but also the capacity of EC seeded at low density to grow and finally cover the treated surface with a non thrombogenic confluent layer of cells. The results of our study indicate that the mixed plasma treatment allows cell adhesion and promotes cell growth on the treated Silastic®. The adhesion and doubling time as well as the final cell density at post-confluence of PAEC seeded on treated Silastic® were similar to those of PAEC seeded on Tissue culture polystyrene (TCP) and on TCP coated with purified human fibronectin. There was no adhesion (and therefore no growth) on untreated Silastic®. The treated materials were characterized by contact angle measurements, as well as Infra Red Spectroscopy (ATR) and scanning electron microscopy (SEM).

536. INFECTION-RESISTANT BIOMATERIALS RELEASING ANTIBIOTICS THROUGH A PLASMA-DEPOSITED THIN FILM. Connie S.K. Kwok[1], Buddy D. Ratner[1,2], Thomas A. Horbett[1,2], [1]Department of Chemical Engineering and [2]Center for Bioengineering, Box 351750, University of Washington, Seattle, Washington, 98195

Bacterial infection on implanted endoprosthetic devices is a major clinical problem, caused by adhesion of bacterial cells and subsequent biofilm formation on the implant surface. In order to control this biofilm formation, we are developing a new antibiotic-loaded polyurethane polymer that itself releases an antibiotic (ciprofloxacin) at the implant interface, thus preventing initial bacterial adhesion and growth. The amount of the ciprofloxacin released is carefully controlled in such a manner that its local concentration is just high enough to kill bacteria in the near wall zone. The desired constant release rate is achieved by depositing a rate-limiting barrier onto the biomaterial surface using radio-frequency glow discharge plasma deposition (RF-GDPD) technique. Various plasma conditions (power, system pressure, deposition time) were examined leading to surface coatings with different cross-linking densities and thicknesses. The coating surface and its corresponding release behavior are investigated and correlated by ESCA and elution studies respectively.

537. IMPROVED ADHESION, GROWTH AND PROLIFERATION OF ENDOTHELIAL CELLS ON PLASMA MODIFIED SURFACES OF EXPANDED POLYTETRAFLUOROETHYLENE U. König, A. Augsburg, F. Simon, K. Lunkwitz, G. Hermel, C. Werner, M. Müller, H.-J. Jacobasch, Institute of Polymer Research Dresden, Hohe Str. 6, 01069 Dresden, Germany

Physical modification of polymer surfaces by means of low pressure plasma treatment has shown to enhance adhesion, growth and proliferation of human umbilical veins endothelial cells (HUVEC`s).
Expanded polytetrafluoroethylene (ePTFE), a frequently applied medical polymer in cardiovascular surgery, was treated with NH_3- and H_2/H_2O-plasma. The purpose of the plasma modification was the introduction of amino groups with ammonia and oxygen-containing groups with H_2/H_2O into the PTFE surfaces. The results of the treatment were evaluated by XPS, contact angle and zeta-potential measurements. Furthermore, adhesion, growth and proliferation of HUVEC`s onto modified samples were observed in vitro in culture media. The HUVEC`s adhesion correlated with the different plasma treatment parameters.

538. **CONTROLLED GROWTH FREE RADICAL POLYMERIZATION OF METHACRYLATE ESTERS - REVERSIBLE CHAIN TRANSFER vs. REVERSIBLE TERMINATION.** Graeme Moad,* Frances Ercole, Julia Krstina, Catherine L. Moad, Ezio Rizzardo, and San H. Thang, CSIRO, Division of Chemicals and Polymers, Clayton, Victoria 3169, Australia.

There has recently been a marked growth in interest in methods for producing narrow polydispersity polymers by free radical polymerization. This paper will contrast two processes that have been developed for the synthesis of these materials and consider their application to the synthesis of narrow polydispersity polymers of methacrylate esters pointing out some of the advantages and limitations of each. The use of kinetic modeling to choose reaction conditions, minimize side reactions, and gain further understanding of the mechanism will be described.

539. **FREE RADICAL SYNTHESES OF FUNCTIONAL POLYMERS INVOLVING ADDITION-FRAGMENTATION REACTIONS,** P. Chaumont, Université C. Bernard-Lyon1, Villeurbanne, France 69600, D. Colombani, Institut C. Sadron, Strasbourg, France 67000, et al.

End-functional polymers have attracted growing attention as precursors for synthesizing copolymers, both block and graft structures. Classical syntheses of such species involve anionic or cationic polymerizations, and chemical modifications. Many recent studies are devoted to the functionalization of free radical polymers by way of the addition-fragmentation reactions. The addition-fragmentation agents are known to act as chain transfer agents, i.e. these compounds are capable of controlling the molar mass of the formed polymers. Furthermore, this type of transfer reactions induces the formation of functional groups at both polymer-ends. The present work is dealing with the use of such chain transfer agents to prepare functional polymers, specially macromonomers, and it is focused on the study of peroxyketal homologues and pentadiene derivatives.

540. **CONTROLLED RADICAL POLYMERIZATION: CONTROL OF MOLECULAR WEIGHT, COMPOSITION AND STEREOREGULARITY IN HOMOGENEOUS AND HETEROGENEOUS SYSTEMS.** Stefan A.F. Bon, Gregory Chambard, Frank A.C. Bergman, Erwin H.H. Snellen, Bert Klumperman and Anton L. German. Eindhoven University of Technology, Laboratory of Polymer Chemistry, P.O. Box 513, 5600 MB Eindhoven, The Netherlands.

The kinetics of CRP based upon alkoxyamines/nitroxides mediated systems, are described. The possibility to control the stereoregularity via this technique is shown. Trans-2,2,5,5-substituted pyrrolidine based nitroxides, which preferred a half-chair (C_2) conformation at room temperature, were synthesized. The preparation of alkoxyamines with a racemic mixture of trans-2,5-bis(methoxycarbonyl)-2,5-dimethyl-1-oxy-pyrrolidine and styrene, acrylonitrile, methyl acrylate and methyl methacrylate all showed exclusively one pair of diastereoisomeric enantiomers. The parameters for the Arrhenius equation for the homolytic dissociation of 1-t-butoxy-2-phenyl-2-(1-oxy-2,2,6,6-tetramethylpiperidinyl)-ethane were determined. E_a = 138.8 kJ/mol $\pm$ 18 kJ/mol and ln(A) = 34.5 $\pm$ 0.6. A method to determine both the chain-length and viscosity dependence of the rate parameters of radical trapping by the nitroxide and bimolecular termination based upon the kinetics in a CRP system is given.

541. **LIVING FREE-RADICAL DISPERSION POLYMERISATION OF STYRENE,** Laurence I. Gabaston, Steven P. Armes, Richard A. Jackson, School of Chemistry, Physics and Environmental Science, University of Sussex, Falmer, Brighton, BN1 9QJ, U.K.

Micrometer-sized polystyrene particles were synthesised by living free-radical dispersion polymerisation in alcoholic and mixed aqueous media. Potassium persulfate and benzoyl peroxide were used as free-radical initiators and poly(N-vinylpyrrolidone) (molecular weight 360,000 g/mol and 40,000 g/mol) was used as the steric stabiliser. In order to produce pseudo-living free-radical polymerisation, 2,2,6,6-tetramethyl-1-piperidinyloxyl (TEMPO) was used as the reversible capping agent. Since the reaction temperature required for the TEMPO-based syntheses is relatively high (120-145°C), high boiling point alcohols such as n-butanol, n-pentanol and n-octanol were used, as well as solvent mixtures such as ethylene glycol / water. Gel permeation chromatography studies confirmed that the polystyrene had narrow molecular weight distributions (M_w/M_n's were in the range 1.06-1.23). Particle size distributions were assessed by electron microscopy and disc centrifuge photosedimentometry.

542. RADICAL POLYMERIZATIONS AND COPOLYMERIZATIONS MEDIATED BY ORGANOCOBALT COMPOUNDS.
Labros D. Arvanitopoulos, Brian M. King, Chung-Yu Huang and H. James Harwood, Maurice Morton Institute of Polymer Science, The University of Akron, Akron, Ohio 44325-3909.

A variety of mono-, di-, and poly-functional alkyl cobaloximes have been synthesized and used as photoinitiators for the polymerization of acrylates. The polymerizations yield polymers with useful end group functionality (-OH, -COOH, -COOR, -halogen, -CN, etc.) and polymers with block, star, and radical-block architectures. Molecular weights increase with conversion as expected for pseudo-living polymerization and products with narrow-molecular weight distributions are found.

543. Applications of Barton Esters in Polymer Synthesis and Modification via Free Radical Processes

T.S. Evenson and W.H. Daly

Louisiana State University, Department of Chemistry and Macromolecular Studies Group, Baton Rouge, LA, 70803

Esters of N-hydroxypyridine-2-thione (Barton esters) were used to initiate the free radical polymerization of vinyl monomers. Decomposition of the Barton esters by heat or visible light generates initiating alkyl and benzoyloxy radicals and non-initiating 2-pyridinesulfide radicals. A study of the free radical polymerization of styrene initiated by phenyl Barton ester (1) has been performed. Heat and visible light have each been used to effect initiation. Factors such as chain transfer to initiator and rate of polymerization have been measured. Incorporation of Barton esters into the side groups of polymer chains has led to the formation of graft copolymers with minimal concommitant homopolymerization.

544. SOLID-STATE NMR CHARACTERIZATION OF POLYMER BLENDS FOR PHOTOLITHOGRAPHY. Peter A. Mirau and Sharon A. Heffner. Bell Laboratories, Lucent Technologies, 700 Mountain Ave., Murray Hill, NJ 07974

Solid-state multipulse proton NMR, wideline deuterium NMR and 2D NMR in solution have been used to study blend formation is resist formulations of novolacs and the radiation-sensitive dissolution inhibitor poly(2-methyl-1-pentene sulfone). Molecular level mixing was demonstrated using combined rotation and multipulse NMR and proton spin diffusion, while 2D solution NMR experiments were used to demonstrate that the intermolecular interactions that lead to blend formation are extremely weak. Wideline deuterium NMR studies of novolacs and poly(vinyl phenol) with deuterated hydroxy groups showed that the differences in blend formation could be related to differences in self-association of the hydroxy groups. These results suggest a possible method to enhance the stability of resist formulations containing dissolution inhibitors.

POLY

545. SPIN DIFFUSION METHODS FOR THE DETERMINATION OF STRUCTURE IN POLYMER BLENDS: ZERO QUANTUM CORRELATION SPECTROSCOPY. Paul T. Inglefield*, Alan A. Jones, Jeff M. Koons and Galina E. Pavlovskaya, Carlson School of Chemistry, Clark University, Worcester, Ma 01610.

The detection of zero quantum coherences in carbon -13 NMR solid state spectra is demonstrated. The use of these coherences to determine distance geometries in crystalline and amorphous materials is shown. The application of this technique to the determination of local structure in polymer blends is presented. Local structure is evidenced in the case of a compatible blend of Polystyrene and Poly(phenylene oxide). The potential advantages of this technique with respect to other methods based on spin diffusion are discussed.

546. NMR Spin-Lattice Relaxation Time Distribution: A Novel Method for the Study of Polymer Blends. Peter. P. Chu*, Shan-Hong Liu, Han-Wen Liu. Department of Chemistry, National Central University, Chung-Li, Taiwan, R.O.C.

Applying proton NMR spin-lattice relaxation time distribution in the studies of molecular entanglement, miscibility and morphology of polymer blend systems are demonstrated. Polymer blend systems with varying degree of miscibility and molecular interactions are present. These systems include: (1) Immiscible blends such as Polystyrene- polyethylene belnd, (2) Semicrystalline blend such as Polystyrene- Polyethylene-terephthalate, and (3) Highly miscible blends such as Phenolic-Phenoxy with hydrogen bonding. Since the method is most sensitive to domains in the range of few hundred A, the result provide direct information which are inaccessible from conventional mechanical tests (few cm); the conventional morphological tools such as DSC and SEM (few um), and high resolution ^{13}C NMR relaxation studies such as T_1^C, $T_{1\rho}^H$, T_{CH} (<100 A) measurements. The minimal measurement time, little sample pretreatment, nondestructive, direct information and simple interpretation makes low field NMR a convenient tool to the study of polymer blends.

547.
NMR RELAXATION OBSERVATIONS OF THE CONSTRAINED SEGMENTAL MOTIONS OF BLOCK COPOLYMERS IN THE NARROW CHANNELS OF THEIR INCLUSION COMPOUND CRYSTALS FORMED WITH UREA. I. D. Shin[a], N. Vasanthan[b], S. Nojima[c] and A. E. Tonelli, Fiber and Polymer Science Program, College of Textiles, North Carolina State University, Raleigh, NC 27695-8301

Two block copolymer-urea inclusion compounds (ICs) have been formed, and segmental motions in each of their blocks observed by T_1 and $T_{1\rho}$ ^{13}C NMR relaxation measurements. The urea (U)-ICs formed with a poly(ε-caprolactone) (PCL)-polybutadiene (PBD) diblock and a PCL-poly(ethylene oxide) (PEO)-PCL triblock copolymer differ in at least two major respects. First the structures of their urea host lattices are distinct, and second both blocks of PCL-PBD are included in the host urea channels, while only the terminal PCL blocks of PCL-PEO-PCL reside in the U-IC channels. As a consequence, we may observe PCL blocks in two distinct U-IC environments, and may compare their behaviors to those of PCL chains in the homopolymer PCL-U-IC and homo- and block copolymer bulk crystals. In addition, $T_{1\rho}$ measurements of ^{1}H spin diffusion reveal structural aspects of the block copolymer-U-ICs, and the isolation of U-IC included polymer chains from their neighbors may permit the probing of 1-dimensional ^{1}H spin diffusion by observing the $T_{1\rho}$ (^{1}H) relaxation in these block copolymer-U-ICs.
A) Department of Chemistry, North Carolina State University, Raleigh, NC 27695-8204
b) Allied-Signal Corp.,Bldg. CRL-G15, Morristown, NJ 07962-1057
c) School of Material Science,Japan Advanced Instit. of Sci. and Technology(JAIST),Tatsunokuchi, Ishikawa 923-12, Japan

548. NMR STUDIES OF TRANSESTERIFICATION AND MISCIBILITY OF PEN/PET BLENDS

Mingming Guo, Institute of Polymer Science, The University of Akron, Akron, OH 44325-3909

Both solution and solid state NMR have been used to study the transesterification and miscibility of poly(ethylene naphthalene) (PEN) and PET blends. There are difficulties to directly use the two NMR methods for this blend system due to the chemical shift and proton relaxation time similarity of the two homopolymers. The deuterated PEN was solution blended with protonated PET and melt-pressed at 280 °C for 0, 0.5, 2, and 10 minutes respectively. The intermolecular cross polarization has been used to extract the miscibility information of the blends. The result indicates that the blend system become miscible when melt-pressed time longer than 0.5 minutes, while for the 30s and 2 min. melt-pressed blends there is no transesterification take place based on solution NMR results. While there is general agreement in the literature that transesterification leads to or enhance miscibility, we present here for the first time the direct experiment evidences, which indicate that transesterification is not a necessary condition for miscibility in the polyester blends.

549. **Deuterium NMR of Adsorbed Polymers. Frank D. Blum***, Wuu-Yung Lin*, Gu Xu*, Minghua Liang* and Charles G. Wade§. *Department of Chemistry and Materials Research Center, University of Missouri-Rolla, Rolla, MO, 65409-0010. §IBM Research Division, IBM Almaden Research Center, 650 Harry Rd., San Jose CA 95120-6099.

Deuterium NMR spectra were obtained for poly(vinyl acetate)-d_3 (PVAc-d_3) and poly(methyl acrylate)-d_3 (PMA-d_3) in bulk and adsorbed on silica. At low temperatures a Pake pattern was found. The collapsed spectra was due to the onset of backbone motion on the NMR time scale. This collapse occurred at a temperature about 40 °C higher than that found for the glass transition from differential scanning calorimetry. For the surface species, components more and less mobile than those found in bulk were observed. These components were assigned to the segments near the air and solid interface, respectively.

550. **OPTICAL DISPERSION PROPERTIES OF TRICYANOVINYLANILINE POLYMER THIN FILMS FOR ULTRASHORT OPTICAL PULSE DIAGNOSTICS**

Ph. Prêtre[1], E. Sidick[1], A. Knoesen[1], D. J. Dyer[2], R. J. Twieg[2]

NSF Center on Polymer Interfaces and Macromolecular Assemblies
[1]University of California, Davis, CA 95616
[2]IBM Research Division, Almaden Research Center, San Jose, CA 95120-6099

We have investigated a series of tricyanovinylaniline (TCV) polymer thin films for their use in ultrashort optical pulse diagnostics of femtosecond Ti:Sapphire lasers. These thin films are ideally suited for ultrashort pulse diagnostic since they eliminate the angle tuning associated with birefringent phase-matched crystals, minimize pulse distortion introduced by group velocity dispersion, and exhibit excellent photochemical stability. The linear optical dispersion of these polymers can be tailored over a wide range for efficient and distortionless frequency conversion. Coherence lengths between 400 nm and 27 microns at a wavelength λ = 800 nm have been found for the two extreme cases of dispersion in these materials. Film thicknesses of at least two microns are tolerable without introducing any significant pulse distortion at the same wavelength.

551.
DESIGN AND SYNTHESIS OF A PERFLUOROALKYLDICYANYVINYL-BASED NLO MATERIAL FOR ELECTRO-OPTIC APPLICATIONS. **Fang Wang,** Aaron W. Harper, Minqian He, Larry R. Dalton, Loker Hydrocarbon Research Institute, University of Southern California, Los Angeles, CA 90089-1661

We report our preliminary results of a new polymer system for electro-optic applications. A new chromophore based on a perfluoroalkyldicyanovinyl acceptor was synthesized. Absorption and nonlinear optical properties of the chromophore were intermediate of the dicyanovinyl and tricyanovinyl analogs. A thin film of 32 wt% of this chromophore doped into PMMA gave an electro-optic coefficient r_{33} = 25 pm/V at 1.06 microns, as determined by ATR spectroscopy. For comparison, the corresponding value for Disperse Red-type chromophore at the same number density is 6 pm/V.

552. PROGRESS TOWARDS THE TRANSLATION OF LARGE MICROSCOPIC NONLINEARITIES TO LARGE MACROSCOPIC NONLINEARITIES IN HIGH- μβ MATERIALS. **Jingsong Zhu,** Mingqian He, Aaron W. Harper, Sam-Shajing Sun, Larry R. Dalton, Department of Chemistry, Loker Hydrocarbon Research Institute, University of Southern California, Los Angeles, CA, 90089-1661, Department of Materials Science and Engineering, University of Southern California, Los Angeles, CA, 90089-0483; Sean M. Garner, William H. Steier, Center for Photonic Technology, Department of Electrical Engineering, University of Southern California, Los Angeles, CA, 90089-0483

Rational structural modification of a high-nonlinearity chromophore effectively prevented intermolecular electrostatic interactions from interfering with electric field poling of the chromophores. Guest-host polymer systems incorporating this chromophore has yielded extremely large electro-optic coefficients (up to 42 pm/V), which are twice that of the unmodified chromophore at the same loading density.

553. NONLINEAR OPTICAL FILMS FROM PAIRWISE-DEPOSITED SEMI-IONOMERIC SYNDIOREGIC POLYMERS. M. J. Roberts, G. A. Lindsay, K. J. Wynne,[1] R. A. Hollins, P. Zarras, J. D. Stenger-Smith, M. Nadler, A. P. Chafin, Naval Air Warfare Center Weapons Division, Code 4B2200D, China Lake, CA 93555-6100; [1]Office of Naval Research, Arlington, VA 22217-5000

We are reporting recent progress toward the goal of processing polymers at low temperature to produce all-polymeric nonlinear optical films which may ultimately find application as active waveguides. Our previously reported work has utilized the Langmuir-Blodgett (LB) technique to produce such films but the noncentrosymmetric order in those films was lost when heated above 80°C. New polymers were designed to have higher temperature stability by eliminating the long alkyl side chains. Also, the new polymers were designed for a new LB film deposition scheme which results in an electro-optic (EO) film which has never undergone electric field poling nor high temperature treatment. Two complimentary polymers, a polycation and a polyanion, are used to build up a polar film. A quadratic enhancement of the SHG intensity with increasing film thickness was observed.

554. DEPENDENCE OF PHOTOISOMERIZATION OF AZO DYE ON THE MOLECULAR STRUCTURE OF VERY HIGH (UP TO 350 °C) GLASS TRANSITION TEMPERATURE AZO-POLYIMIDE POLYMERS: ROOM TEMPERATURE LIGHT INDUCED ORIENTATION
Zouheir Sekkat[†,‡], Jonathan Wood[‡], Wolfgang Knoll[‡,*], Willi Volksen[§], Victor Y. Lee[§], Robert D. Miller[§], André Knoesen[†]; [†]Department of Electrical and Computer Engineering, University of California, Davis, California 95616; [‡]Max-Planck-Institut für Polymerforschung, Ackermannweg 10, 55128 Mainz, Germany. [§]IBM Research Division, Science and Technology, 650 Harry Road, K13/801, San Jose, California 95120-6099; [*]Frontier Research Program, The Institute of Physical and Chemical Research (RIKEN), Wako, Saitama 351-01, Japan.

This talk will report on light-induced molecular motion in high glass transition temperature polyimides containing a photochromic azo dye for optical data storage. We will show that the fundamental aspects of this process are of importance, as they open a new way of viewing sub-Tg (sub-glass transition temperature) polymer dynamics in photochromic polymers: Molecular motion is shown to depend on the molecular structure of the unit building blocks of the polymer; the process of photoisomerization itself depends on the polymer molecular structure. We will also present evidence of light-induced nonpolar orientation in very high Tg polyimides (Tg up to 350 °C) containing no flexible connectors or tethers. The azo dye chromophore is incorporated through the donor substituent as a part of the polymer backbone. This impressive photoinduced orientation occurs at room temperature, i.e. at least 325 °C below the Tg for one of the polymers we investigated. After photo-induced ordering, this polymer must be heated to 350 °C, to induce main-chain movement and erase the orientation; while the direction of the orientation of the chromophores can be easily controlled at room temperature simply by choosing the appropriate polarization of the irradiating light. This light-induced orientation process will be shown to be useful for encoding images in waveguides.

555. PREPARATION OF NOVEL CROSS-LINKED POLYMERIC THIN FILMS FOR SECOND-ORDER NON-LINEAR OPTICS. M. Trollsas[a], F. Sahlen[a], U. W. Gedde[a], C. Orrenius[b], A. Hult[a] and S. T. Lagerwall[c], [a]*Department of Polymer Technology,* [b]*Organic Chemistry, Department of Chemistry, Royal Institute of Technolgy, S-100 44 Stockholm, Sweden,* [c]*Physics Department, Chalmers University of Technology, S-412 96 Goteborg,, Sweden.*

Starting with a ferroelectric liquid crystal (which has polar order perpendicular to the director) we have succeeded in making materials with true polar order. The materials are ferroelectric liquid crystalline monomers or ferroelectric liquid crystalline monomer mixtures which have been photo-cross-linked into pyroelectric polymers. The polar order of the pyroelectric materials have been proven by Pockels measurements and second harmonic generation experiments. The novel materials which after cross-linking not anymore are ferroelectric have been prepared by multistep syntheses

556. SELF-ASSEMBLED ORGANIC SEMICONDUCTOR QUANTUM BOXES AND WIRES FROM BLOCK COPOLYMERS. S. A. Jenekhe * and X. L. Chen, Department of Chemical Engineering and Center for Photoinduced Charge Transfer, University of Rochester, Rochester, New York 14627-0166.

Low dimensional organic semiconductors such as quantum boxes, quantum wires, or superlattices are predicted to exhibit enhanced or novel electronic and optical properties for applications in solid state electronic and optoelectronic devices in ways similar to low dimensional inorganic semiconductors. However, the relatively small charge carrier and exciton Bohr radii of organic semiconductors (ca. 1.0-1.5 nm) has heretofore made it difficult to prepare organic semiconductor heterostructures with quantum confinement effects. Ultrasmall semiconducting polymer quantum boxes and wires with confinement volumes of 1-2 nm^3 have been prepared in our laboratory by self-organization of blends of ABA block conjugated copolymers with a parent homopolymer. Observation of new optical transitions, enhanced luminescence, and stability of luminescence at high electric fields confirmed the confinement of excitons in the organic quantum boxes and wires.

557. MECHANISTIC STUDIES OF CHEMICALLY AMPLIFIED PHOTORESISTS. W. Hinsberg, G. Wallraff, F. Houle, J. Frommer and R. Beyers, IBM Almaden Research Center, San Jose, CA 95120

The detailed structure of the relief image formed upon processing a chemically amplified (CA) photoresist is the result of a complex interplay of the projected optical image, the resist optical and photochemical properties, and the chemistry and the physics of post-exposure thermal processing. Rational design of new CA resist materials is greatly facilitated by a quantitative understanding of how structure and properties of the film components determine resist imaging properties. We describe here results of our studies aimed at quantifying the chemical deprotection kinetics and diffusion of photogenerated acid within the latent image in thin polymer resist films, as a function of the polymer and photosensitizer properties. To support these studies, we are developing spectroscopic and microscopic methods for the characterization of film composition at high spatial resolution.

558. NANOSCOPICALLY ENGINEERED ORGANIC-INORGANIC HYBRIDS AS LOW DIELECTRIC CONSTANT, HIGH MODULUS INSULATING MATERIALS FOR MICROELECTRONIC APPLICATIONS. J. L. Hedrick, R. D. Miller, D. Yoon, H. J. Cha, H. R. Brown, S. A. Srinivasan, R. Di Pietro and V. Flores, IBM Almaden Research Center, 650 Harry Road, San Jose, CA 95120

As device dimensions and functionality density increases, new materials and processes will be required to optimize performance. This is particularly evident in the back-end-of-the-line where functional transistors on the chip are wired and interconnected with feature sizes decreasing to 0.18 μm and densities increasing to many metal levels by the turn of the century. Without design and insulator change, the cycle time will not decrease significantly due to RC delays which comprise nearly half the total cycle time in current microprocessors. To minimize these delays, the dielectric constant of the insulator must decrease. Network silicate derivatives (e.g., spin-on-glasses, silsesquioxanes, etc.) possess many of the desirable characteristics required for ILD materials. These highly crosslinked, network polymers, however, often show poor mechanical properties which tend to crack when film thicknesses exceed one micron. We will describe our recent work on toughened nanoscopically engineered hybrid materials with dielectric constants below 3.0 which form thermally stable, hard, tough films for spin-on applications.

559. FUNCTIONAL OLIGOMERS AND REACTIVE COPOLYMERS FOR THE GENERATION OF THIN FILM NANOFOAMS. Kenneth R. Carter, Richard A. DiPietro, James L. Hedrick, Robert D. Miller and Paul T. Furuta. IBM Research Division, Almaden Research Center, 650 Harry Rd., San Jose, CA 95120-6099

A means of generating foams of high temperature polymers, polyimides, has been developed for use in dielectric layers in microelectronics. In these systems, the pore sizes generated are in the tens of nanometers range, thus the term "nanofoams". The foams are generated by preparing phase separated block copolymers with the majority phase comprised of polyimide and the minor phase consisting of a thermally labile block. Films are cast, solvent removed and the copolymers are cured, causing phase separation of the two blocks. The labile blocks are subsequently removed via thermal treatments leaving pores the size and shape of the original copolymer morphology. New poly(propylene oxide) (PO) oligomers have been synthesized and used as labile blocks for the formation of polyimide nanofoams. The necessary amine-functionalization of the PO oligomers was achieved by functionalization of hydroxy-terminated PO as well as the direct synthesis of functional PO using a new series of organometallic catalysts. The synthesis of PO oligomers, PO-containing triblock copolymers and the resulting thin film nanofoams was explored.

560. THE ROLE OF ENTROPY IN THE RESTRUCTURING OF MODIFIED POLYMER SURFACES. H.J. Griesser, L. Dai, T.R. Gengenbach and R.C. Chatelier, Division of Chemicals and Polymers, CSIRO, Private Bag 10, Clayton South MDC, Clayton 3169, Australia.

The restructuring ("hydrophobic recovery") of polymer surfaces modified by plasma or other surface treatment techniques has traditionally been interpreted in terms of interfacial enthalpy. We have measured air/water contact angles of ammonia-plasma-modified Teflon FEP samples as they were stored in various media after treatment. In all cases, including storage in polar solvents such as water and methanol, the contact angles increased with time. Angle dependent XPS analyses showed the modified region to be very shallow, and that the depth distribution of N was altered on storage, by diffusion towards deeper regions. This boundary layer spreading was interpreted as resulting from translational entropy which together with interfacial enthalpy defined the redistribution of polar groups as the materials aged.

561. ACID-BASE PROPERTIES OF PLASMA DEPOSITED Vs. CONVENTIONAL HYDROPHILIC POLYMER FILMS. M. Morra, C. Cassinelli, Nobil Bio Ricerche, 14018 Villafranca d'Asti, ITALY

Films deposited from hydroxyethylmethacrylate (HEMA) and oligoglyme plasma have been investigated in several published studies. It is generally observed that the molecular structure of these plasma deposited films is affected by the deposition conditions. The details of the surface structure, in turn, affect the Lewis acid-base or electron acceptor-electron donor characteristics of these films and the way they communicate with the surrounding environment. We have used the two suggested methods for the calculation of acid-base surface properties from wetting measurements (i.e. electron donor-electron acceptor surface free energy components and the calculation of the acid-base work of adhesion) to try to gain insights on the molecular structure and the surface field of forces of films deposited from HEMA and tetraethylene glycol dimethyl ether plasma, and compared the results to analogous measurements performed on conventional polymers and to ESCA data. Results show that both quoted methods bring information on the effect of deposition conditions and details of the film structure, but some uncertainty on quantitative aspects of the quoted methods does not allow to completely exploit the obtained data.

562. XPS ANALYSIS OF O_2-PLASMA TREATED POLYETHERETHERKETONE (PEEK). S.-W. Ha[1], R. Hauert[2], K.-H. Ernst[2], E. Wintermantel[1], [1] Chair of Biocompatible Materials Science and Engineering, Department of Materials, ETH Zürich, CH-8952 Schlieren, [2] Swiss Federal Laboratories for Materials Testing and Research (EMPA), CH-8600 Dübendorf

PEEK is a high-molecular-weight aromatic thermoplastic polymer, which is characterized by high chemical and thermal stability combined with good mechanical properties. Due to these properties PEEK is now used in a wide range of applications in the automotive, aerospace, electronics, medical and chemical processing industries. Like most polymers, PEEK has a relative low surface energy which is disadvantageous with regards to subsequent coating, glueing or painting processes. In this study, plasma treatment of PEEK surfaces was performed in an oxygen plasma and the chemical modifications occurring on the PEEK surface due to this surface treatment were analysed by X-ray photoelectron spectroscopy (XPS). Additionally, angle resolved XPS analysis was performed to investigate the penetration depth of plasma activation.

563. PLASMA TREATMENT OF POLYDIMETHYLSILOXANE. Michael J. Owen and Jennifer L. Stasser, Dow Corning Corporation, Midland, Michigan 48686-0994.

Plasma treatment is a useful way of enhancing wettability to improve adhesion to silicone substrates such as polydimethylsiloxane (PDMS). The main problems associated with such treatment are the subsequent recovery of hydrophobicity that ensues once treatment is discontinued and the appearance of microcracking if the treatment is too harsh or prolonged. Diffusion of low molecular weight species is a likely cause of this hydrophobic recovery. We have used contact angle studies and X-ray photoelectron spectroscopy to monitor the progress of this recovery and gel permeation chromatography to characterize unbound solvent-extractable material. This is typically a main peak centered near 2000 MW with lower MW cyclic peaks. Contact angle studies at elevated temperatures are consistent with a diffusion controlled recovery of such species. Optical and scanning electron microscopy have been used to detect microcracking. Conditions have been found for complete water wettability of PDMS without microcrack formation.

564. REACTIVE ATOM ETCHING OF POLYMERS. James W. Seale,[1] Timothy K. Minton,[2] and Joseph W. Perry.[3] [1]Department of Chemical Engineering, [2]Department of Chemistry and Biochemistry, Montana State University, Bozeman, MT 59717; [3]Jet Propulsion Laboratory, California Institute of Technology, Pasadena, CA 91109.

A novel approach is under investigation for the etching of polymers based on a hyperthermal atomic beam source. Neutral beam etching may result in significant improvements in the processing and fabrication of devices relative to the state-of-the-art reactive ion etching (RIE). RIE is currently used extensively for the fabrication of polymer waveguide devices, but it has limitations associated with charge- and bombardment-induced damage. In neutral beams with relatively low kinetic energies, charging and bombardment damage would be absent. We have used a directed neutral beam of oxygen atoms with 5 eV of kinetic energy to etch a polyimide sample and have achieved etch profiles with very straight and smooth sidewalls. These sidewalls meet the design criteria for low-extrinsic-loss optical waveguides. Fundamental beam-surface scattering experiments suggest a close relationship between atom-surface scattering dynamics and etch profile evolution.

565.

Radical trapping analysis on vacuum-ultraviolet treated polymers, R. Wilken, A. Hollander, J. Behnisch, Fraunhofer-Institut für Angewandte Polymerforschung, 15413 Teltow, Germany

During a low pressure plasma treatment, a polymer is subjected to vacuum-ultraviolet (VUV, $100nm<\lambda<200nm$) radiation which produces radicals in the surface as a main initiation reaction. We investigated the formation of radicals on PE, PP and PS under ultrahigh vacuum conditions by trapping them with nitrous oxide. The rate of N- and O-incorporation were analyzed by XPS. The structure of the N-O-containing functional groups were characterized by the chemical shifts in the N1s region.

566. MOLECULAR REACTION PATHWAYS DURING PLASMA MODIFICATION OF POLYMER SURFACES. J. Hopkins, R. D. Boyd, J. P. S. Badyal, Department of Chemistry, Science Laboratories, Durham University, Durham DH1 3LE, England, UK.

CF_4 and O_2 glow discharge treatment of biaxially oriented polypropylene film results in the surface incorporation of fluorine and oxygen atoms respectively. The stability and extent of substrate modification has been investigated by X-ray photoelectron spectroscopy (XPS). The observed differences in physicochemical behaviour can be accounted for on the basis of extended Huckel molecular orbital theory.

567.
INFLUENCE OF O_2^+-IONS IN OXYGEN PLASMA MODIFICATION OF POLYMERS: J. Meichsner, Institut für Physik, TU Chemnitz-Zwickau D-09107 Chemnitz, Germany

Polystyrene and Polyethylene films (20-50 nm) prepared by a dip coating technique were modified in a capacitively coupled 13.56 MHz oxygen discharge at defined plasma conditions. Ion energy distribution function at the powered and grounded electrode and plasma ion density were measured by means of energy selective mass spectrometry and Langmuir probe. Polymer modification characterized by FTIR and ellipsometry was investigated on the powered and grounded electrode and in the remote plasma. Plasma surface treatment results in polymer degradation and formation of new functional groups. Degradation was influenced mainly by the intensity of the ion flux whereas ion energy was not important in the studied range (10 to 300 eV). Formation of carbonyl groups showed no significant correlation to the ion bombardment.

568. OXIDATION AND ABLATION OF POLYMERS BY VACUUM-UV RADIATION FROM LOW PRESSURE PLASMAS. A. Fozza, J. Roch, J.E. Klemberg-Sapieha, A. Kruse[1], A. Holländer[1] and M.R. Wertheimer, Groupe des Couches Minces and Dept. of Engineering Physics, Ecole Polytechnique, Box 6079, Station Centre-Ville, Montreal., QC H3C 3A7, Canada; [1]Fraunhofer - Institute, Germany

Low pressure glow-discharge plasmas are increasingly used as an effective method for the surface modification of polymers; they can also serve in the laboratory to simulate low Earth orbital environment (LEO). Although vacuum-ultraviolet (VUV, $\lambda < 200$ nm) is an important component of plasma environment, only few studies have focused on its effects so far. The emission from low-pressure microwave plasmas of various gases in the VUV-UV regions was investigated in order to use these plasmas as light sources for the study of the VUV/UV effects on various polymers (polyethylene, polymethylmethacrylate, etc.) or high molecular weight oligomers (hexatriacontane). We have employed a quartz crystal microbalance (QCM) in order to measure in-situ the net mass change of the polymeric films exposed to VUV/UV radiation originating from hydrogen plasmas. Measurements were made with the specimens either in vacuum, or immersed in VUV-generated atomic oxygen. Following irradiation, samples were analyzed by X-ray photoelectron spectroscopy (XPS) in order to study the evolution of the oxygen content and of the various functional groups.

569. ON THE EFFECTS OF ENERGY TRANSFER TO PS, PP, PE, AND PET FORM AN AR PLASMA. Richard M France, Robert D Short, Laboratory for Surface and Interface Analysis, Department of Engineering Materials, University of Sheffield, Sir Robert Hadfield Building, Mappin Street, Sheffield, S1 3JD.

Argon plasma treatment and subsequent exposure to atmospheric oxygen has been used to introduce new carbon-oxygen functionalities to polymer surfaces. We report on polystyrene (PS), low density polyethylene (LDPE), polypropylene (PP) and polyethylene terephtalate (PET). These polymers exhibit 'saturated' and 'stable' levels of oxygen incorporation. 'Saturated' is the maximum amount of oxygen that can be introduced into the surface; 'stable', the maximum level of oxygen incorporation which is stable to both washing with a polymer non-solvent, and ageing. The vertical homogeneity of oxygen uptake is probed through the O1s/O2s ratio and by high energy-resolution XPS, at various take off-angles. The 'stable' level of treatment is characterised by a high selectivity towards C-O functionalities. Attention is drawn to the similarity between plasma and other energy transfer phenomena (e.g. photo-degradation and high-energy irradiation). Our results are rationalised on the basis of the accepted mechanisms of degradation (cross-linking vs. chain scission, production of volatiles) for these polymers.

570. **LIVING FREE RADICAL POLYMERIZATION AND DENDRITIC POLYMERS.**
J. M.J. Fréchet, M. Leduc, M. Weimer, R. B. Grubbs, M. Liu, and C.J. Hawker
Department of Chemistry, University of California-Berkeley, Berkeley, CA. 94720-1460

The preparation of dendritic polymers by "living" free-radical polymerization is explored with four different approaches. (1) Linear-dendritic block copolymers in which a regular dendron is used to initiate the polymerization of a linear chain from its focal point to afford a "kite-like" structure. (2) Linear dendritic star copolymers prepared by growth from the reactive chain-ends of a dendrimer. (3) Hyperbranched polymer obtained by Self-Condensing Vinyl Polymerization[1] of a bifunctional monomer. And, (4) Dendritic-graft or "dendrigraft" copolymer prepared by a graft-on-graft multistep process. In all of these cases, the "living" free-radical polymerization process is started using either TEMPO mediated systems[2] or Matyjaszewski's ATRP technique[3]. A comparison of the methods and their applicability to different target structures reveals the advantages and drawbacks of each.
[1]*Fréchet, J.M.J.; Henmi, M.; Gitsov, I.; Aoshima, S.; Leduc, M.; Grubbs, R.B. Science, 1995, 269, 1080*
[2]*Hawker, C.J.; Fréchet, J.M.J., Grubbs, R.B.; Dao, J.; J. Am. Chem. Soc. 1995, 117, 10763*
[3]*Patten, T.E.; Xia, J.; Abernathy, T.; Matyjaszewski, K; Science 1996, 272, 866-8.*

POLY

571. **Atom Transfer Radical Polymerization: How to Make Polymer Chains of Various Shapes, Compositions, and Functionalities.** Scott G. Gaynor and Krzysztof Matyjaszewski,* Department of Chemistry, Carnegie Mellon University, 4400 Fifth Avenue, Pittsburgh, PA 15213

Atom Transfer Radical Polymerization, ATRP, is a "living" radical polymerization. ATRP involves the establishment of a equilibrium between active and dormant species by reaction of simple organic halides with copper (I) / dipyridyl complexes. This process has been demonstrated to produce linear polymers with narrow polydispersity, $1.03 < M_w / M_n < 1.5$, and with DP = $\Delta[M] / [I]_0$ up to $M_n \approx 150,000$. This report is to describe the materials that have been prepared using ATRP, both alone and in conjunction with other polymerization methods. Homopolymers have been prepared from styrene, acrylics, methacrylics, and acrylonitrile. These monomers can be copolymerized with each other to form random copolymers. Also, alternating copolymers have been prepared using isobutene and methyl acrylate. By controlling the rate of addition of a second monomer to a polymerization, copolymers with a compositional gradient along the polymer chain have been prepared. Block copolymers have been prepared by sequential addition of a second monomer to a completed polymerization. Also, macroinitiators have been prepared using other polymerization techniques such as step growth and cationic polymerizations. Telechelic styrenes have been prepared by conversion of halogen end groups to amine end groups using simple organic synthesis. By development and use of vinyl monomers that are functionalized to contain a halogen that can be used to initiate ATRP (described as A=B-C* monomers), branched and hyperbranched polymers have been prepared by radical polymerization. Also, these A=B-C* monomers have been used to prepare graft copolymers by grafting from and grafting through methods.

572. CONTROLLED RADICAL POLYMERIZATION METHODS FOR THE SYNTHESIS OF NON-IONIC SURFACTANTS FOR CO_2. D. E. Betts, T. Johnson, C. Anderson, and J. M. DeSimone, Department of Chemistry, CB #3290, Venable and Kenan Laboratories, University of North Carolina, Chapel Hill, North Carolina 27599.

Currently we are considering the design and synthesis of several amphiphilic diblock copolymers composed of a CO_2-philic fluorocarbon block and a CO_2-phobic lipophilic or hydrophilic block. We have found the synthesis of these materials to be ideally suited to controlled free radical polymerization techniques. Specifically these block copolymers are being synthesized using either the iniferter technique as developed by T. Otsu or atom transfer radical polymerization (ATRP) as developed by K. Matyjaszewski. These block copolymers will find use as surfactants in CO_2 for the dispersion polymerization of CO_2 insoluble polymers and in cleaning and surface treatment technologies.

573. **BLOCK COPOLYMER PREPARATION USING NORMAL/LIVING TANDEM POLYMERIZATION.** I. Q. Li and B. A. Howell, Center for Applications in Polymer Science, Central Michigan University, Mount Pleasant, MI 48859 and M. T. Dineen and D. B. Priddy, Dow Polystyrene R&D, 438 Bldg, Midland, MI 48667

Recently much attention has been devoted to developing an understanding of nitroxide mediated free radical polymerization. Most of the work has focused on obtaining narrow polydispersity polystyrene. We believe that the real value of this technology is not for the preparation of narrow polydispersity polymers, but for the preparation of block copolymers without having to use anionic polymerization.

Some monomers do not polymerize well using living free radical polymerization (e.g., n-butyl acrylate). We have overcome this problem and have demonstrated the preparation of several block copolymers (e.g., styrene-n-butylacylate) using a novel normal/living tandem polymerization scheme.

574. STEREOCHEMICAL CONTROL OF FREE-RADICAL POLYMERIZATION OF VINYL MONOMERS. T. Nakano, Y. Ishigaki, A. Matsuda, and Y. Okamoto, Department of Applied Chemistry, School of Engineering, Nagoya University, Chikusa-ku, Nagoya 464-01, Japan

Three methods of stereospecific free radical polymerization are presented. The first method aims at stereochemical induction based on the interaction between monomer and the added "template" molecule through H-bonding or ionic interaction. For this purpose, polymerization of methacrylic acid was carried out in the presence of amine compounds including 1,2-diaminocyclohexane in chloroform. The presence of amine compounds led to increased isotacticity (mm) and heterotacticity (mr) at the expense of syndiotacticity (rr) up to ca. 16% mm and 49% mr compared with 8% mm and 41% mr in the control polymerization. The second method is based on the effects of monomer concentration, temperature, and solvent on polymerization stereochemistry. The three factors of reaction conditions have remarkable influences on the polymerization of trityl and diphenyl-2-pyridylmethyl methacrylates and isotacticity range of >99% to 63% was attained in the polymerization of trityl ester. The third method involves chiral cobaltus complex. Polymerization of methacrylates, acrylates, vinyl acetate, and styrene in the presence of cobaltus complex exhibited a living-like nature. Stereochemistry of polymerization was affected by the complex.

575. SPIN DIFFUSION IN POLYSTYRENE-POLYISOPRENE DIBLOCK COPOLYMERS. Kevin S. Jack[1], Almeria Natansohn[1], Richard P. Register[2] and Jiahu Wang[1], [1]Department of Chemistry, Queen's University, Kingston, Ontario, Canada K7L 3N6 and [2]Department of Chemical Engineering, Princeton University, Princeton, NJ, USA 0854-5263.

The size of domains in a range of polystyrene-polyisoprene block copolymers were determined from measurements of ^{1}H spin diffusion. It was found that the domain sizes determined from the NMR method, using a two-phase model for the copolymers phase structure, were in very good agreement with those determined from SAXS and theoretical predictions using similar models. Accurate measurements of the domain sizes in these copolymers were obtained using a model for the copolymer phase structure which included an interphase and the effects of spin-lattice relaxation on the spin diffusion experiments. Furthermore, the average distances between PS domains were shown to be in good agreement with those determined by the more established SAXS method.

576. **LOCAL ORDER AND DYNAMICS IN POLYISOBUTYLENE-BASED COPOLYMERS.** J. L. White, Exxon Chemical Co., 5200 Bayway Drive, Baytown, Tx. 77522-5200

Polyisobutylene (PIB) copolymers with isoprene represent commercially valuable materials with primary use in tire-related applications. We have investigated dynamics in a different class of PIB copolymers, containing p-methylstyrene and bromo-p-methylstyrene as comonomer. Solid state ^{2}H and 2D correlation results indicate different motional regimes for the methylstyrene comonomer vs. the PIB comonomer. Solution NMR data indicates that the methylstyrene comonomer does not incorporate randomly, resulting in reduced segmental motion as the clustering increases. These copolymers are structurally intermediate between random copolymers and true block copolymers prepared by living polymerizations. 2D WISE and HETCOR data provides information regarding domain sizes for high methylstyrene contents. We have also investigated the T_2 dependence on curing in PIB copolymers with bromo-p-methylstyrene as the comonomer, as well as the relationship between spin-spin relaxation constants and rheometer torque values.

577. PHASE STRUCTURE OF POLYMERS VIA GOLDMAN-SHEN PULSE SEQUENCE: MODEL SPIN DIFFUSION CALCULATIONS. T. T. P. Cheung, Phillips Petroleum Company, Bartlesville, OK 74004.

The time evolution of the magnetization in the Goldman-Shen type of experiments is investigated theoretically for different models of a multi-phase system. It is shown that the decay of the total magnetization associated with the domains where the magnetization initially resides is affected by the size distribution of the domains and the distribution in the domain separations. The models examined include the two-phase-multi-domain (2P-MD) case, in which the system consists of two phases with infinite number of domains in each phase, and the two-phase-four-domain (2P-4D) model, in which the system has two phases with two domains in each. The spin diffusion equation is solved for these models using a simple two-value distribution in the domain width or the domain separation. The solutions for the magnetization of the 2P-MD model show a slower decay in the long time regime in comparison to those of the 2P-4D model. The solutions of the 2P-4D model give a reasonable approximation to the 2P-MD model provided that the spread in the distribution is sufficiently smaller than the average domain width or domain separation. The simple two-phase-two domain (2P-2D) model, in which the two phases are represented by a single domain each, is also investigated. In this case, the effect of the distribution is incorporated by taking an ensemble average. It is shown that there is no correct way of taking the average and the results always yield erroneous long time behavior of the magnetization.

578. POLY(ETHYLENE TEREPHTHALATE) REVISITED. Nicholas Zumbulyadis, Imaging Research and Development Laboratories, Eastman Kodak Company, Rochester, NY 14613-2132 and Klaus Schmidt-Rohr, Polymer Science and Engineering, University of Massachusetts at Amherst, Amherst, MA 01003-4530.

The conformation and morphology of poly(ethylene terephthalate) (PET) has been the subject of numerous investigations using spectroscopic, thermal, scattering and diffraction techniques. Nevertheless, several fundamental questions are still outstanding, particularly the torsion angle around the ethylene C-C bond and the unambiguous determination of crystalline content. We have applied the recently published method by Schmidt-Rohr (*J. Amer. Chem. Soc.*, 1996, *118*, 7601-7613) for the determination of dihedral angles in solids to the conformational analysis of doubly ^{13}C-labeled PET. We have studied both amorphous and thermally crystallized samples. We will discuss our results in the context of previously published IR measurements and molecular dynamics simulations. The implications of our results for the determination of crystallinity from NMR line shape analysis and IR measurements will also be discussed.

579. ESTIMATION OF MECHANICAL PROPERTIES OF AGED SILK WITH THE AID OF NMR RELAXATION TIME AND FLUCTUATION-DISSIPATION THEOREM. Riichirô Chûjô, Department of Materials Engineering, Teikyo University of Science and Technology, Uenoharaæmachi, Yamanashi 409-01, Japan

As an extension of a previous work [*ACS Polym. Prepr.*, **37**(2), 184 (1996)] on archeological silk, a method is proposed for the estimation of mechanical properties of aged silk. Active measurements of the mechanical properties are impossible for aged samples because the applied stress tends to concentrate on the degraded regions. NMR relaxation, which provides passive measurements of the mechanical behavior, does not have this limitation. The strategy therefore is to carry out both active measurements and NMR relaxation measurements on fresh samples and to establish correlations. Fluctuation-dissipation theorem guarantees the equivalency of these two types of measurements. NMR relaxation measurements are then done on aged samples. From the relationship between the active and the passive measurements, the mechanical properties of the aged silk samples are estimated.

580. The Application of NMR Techniques to the Development of Polymer Derived Ceramic Materials.

J. Lipowitz, K. Su, A. Szweda, R. B. Taylor, H. Yue* and G. A. Zank.
Dow Corning Corporation, Midland, MI 48686

Since the early pioneering work of Yajima, where polycarbosilane was used as a precursor to a silicon carbide ceramic material, the use of various silicon-based polymers as ceramic precursors has extensively been studied. Dow Corning Corporation has been actively developing polymer derived ceramics, fibers, and ceramic matrix composites (CMC) for more than 15 years. These material are now becoming commercially available as SYLRAMIC™ SiC fiber and SYLRAMIC™ Composites. These Sylramic materials possess high strength, lightweight, toughness and can be used at temperatures in excess of 1000°C and have potential applications in defense and aerospace industries. NMR techniques play an important role in the development of these materials. ^{29}Si solution NMR has been used in the characterization of the preceramic polymer microstructure in order to improve the polymer synthesis. ^{29}Si solid state NMR has also been used for monitoring the polymer cure process as well as the decomposition of a siloxane polymer to ceramic materials. A newly developed ^{1}H NMR microimaging application has also been uniquely applied for the determination of pore size and its distribution within ceramic matrix composites. The application of NMR techniques in the development of polymer derived ceramic materials will be presented.

American Chemical Society
DIVISION OF POLYMERIC MATERIALS: SCIENCE AND ENGINEERING

213th ACS National Meeting

San Francisco, CA
April 13-17, 1997   D.J. Lohse, Program Chairperson

SUNDAY MORNING

- **Functional Polymers**
Anionic/Cationic/Ring Opening Methods
R.P. Quirk, D.N. Schulz, Presiding   Papers 1-8

- **Engineering Polyolefins**
Cyclic Olefin Copolymers
B.A. Harrington, Presiding   Papers 9-13

- **Low VOC Coating Systems**
Analytical and Application Techniques in Low VOC Coatings
P.R. Sperry, Presiding   Papers 14-20

- **Polymers for Ophthalmic Applications**
J.F. Kunzler, Presiding   Papers 21-26

SUNDAY AFTERNOON

- **Functional Polymers**
New Concepts/Heteroatom Polymers
S.I. Stupp, H.R. Allcock, Presiding   Papers 27-33

- **Engineering Polyolefins**
Cyclic Olefin Copolymers
T. Wehrmeister, Presiding   Papers 34-37

- **Low VOC Coatings Systems**
Radiation-Cured Low VOC Coatings
P.M. Lesko, Presiding   Papers 38-44

- **Polymers for Ophthalmic Applications**
P.L. Valint, Jr., Presiding   Papers 45-49

MONDAY MORNING

- **Functional Polymers**
Post Functionalization / Applications
J. Bock, J. Dias, Presiding   Papers 50-57

- **Engineering Polyolefins**
  Ethylene / Carbon Monoxide Copolymers
  T. Burkhardt, Presiding — Papers 58-64

- **Low VOC Coatings Systems**
  High Solids and Water-Borne Low VOC Coatings
  B. Lestarquit, Presiding — Papers 65-71

- **Polymers for Ophthalmic Applications**
  Y.C. Lai, Presiding — Papers 72-76

- **Applied Polymer Science Award in Honor of J.L. Koenig**
  H. Ishida, Presiding — Papers 77-82

MONDAY AFTERNOON

- **Functional Polymers**
  Living Radical Methods / Designed Materials
  S.R. Turner, C.A. Costello, Presiding — Papers 83-89

- **Engineering Polyolefins**
  Blends and Composites
  P. Arjunan, Presiding — Papers 90-95

- **Low VOC Coatings Systems**
  Water-Borne Low VOC Coatings
  J.M. Friel, Presiding — Papers 96-102

- **General Papers / New Concepts in Polymeric Materials**
  High Performance Polymers
  D.W. Thompson, Presiding — Papers 103-109

- **Applied Polymer Science Award in Honor of J.L. Koenig**
  P.C. Painter, Presiding — Papers 110-115

TUESDAY MORNING

- **Functional Polymers**
  Advanced Optical Materials
  A.O. Patil, E. Reichmanis, Presiding — Papers 116-123

- **Structure and Properties of Glassy Polymers**
  B.D. Freeman, Presiding — Papers 124-132

- **Intelligent Materials and Novel Concepts for Controlled Release Technology**
  Responsive Polymers, Part I
  E.S. Ron, D.G. Rethwisch, Presiding — Papers 133-139

- **Cooperative Research Award in Honor of H.K. Hall, Jr.**
  D.N. Schulz, A.B. Padias, Presiding — Papers 140-144

- **Creative Invention Award Symposium on Laser Ablation in Honor of Rangaswamy Srinivasan**
  B.J. Garrison, Presiding   Papers 145-150

TUESDAY AFTERNOON

- **Functional Polymers**
  Organometallic Polymerization / Materials
  B.M. Novak, J.R. Reynolds, Presiding   Papers 151-157

- **Structure and Properties of Glassy Polymers**
  G.M. Wilkes, Presiding   Papers 158-166

- **Intelligent Materials and Novel Concepts for Controlled Release Technology**
  Responsive Polymers, Part II
  E.S. Ron, D.G. Rethwisch, Presiding   Papers 167-175

- **General Papers / New Concepts in Polymeric Materials**
  Synthesis and Blends
  P. Arjunan, Presiding   Papers 176-180

TUESDAY EVENING

- **Posters**
  D.J. Lohse, Presiding   Papers 181-251

WEDNESDAY MORNING

- **Polymer Ecology: Biodegradation and Recycling**
  S.P. McCarthy, Presiding   Papers 252-257

- **Structure and Properties of Glassy Polymers**
  M. Forsyth, Presiding   Papers 258-266

- **Intelligent Materials and Novel Concepts for Controlled Release Technology**
  Novel Delivery Concepts and Technologies
  A. Comfort, B. Berner, Presiding   Papers 267-274

- **Polymers for Biosensors and Biosensing**
  Optical, Nucleic Acid and Immuno Sensing
  P.G. Edelman, Presiding   Papers 275-282

WEDNESDAY AFTERNOON

- **Polymer Ecology: Biodegradation and Recycling**
  R.A. Gross, Presiding   Papers 283-289

- **Structure and Properties of Glassy Polymers**
  C.A. Angell, Presiding   Papers 290-298

- **Intelligent Materials and Novel Concepts for Controlled Release Technology**
  Lipsome Technology
  P.W. Taylor, D.H. Thompson, Presiding — Papers 299-306

- **Polymers for Biosensors and Biosensing**
  Electrochemical Sensors
  D. Pfeiffer, Presiding — Papers 307-314

THURSDAY MORNING

- **Polymer Ecology: Biodegradation and Recycling**
  G. Swift, Presiding — Papers 315-320

- **Structure and Properties of Glassy Polymers**
  H. Marand, Presiding — Papers 321-329

- **Intelligent Materials and Novel Concepts for Controlled Release Technology**
  Microanalysis
  M.A. Burns, C. Knors, J. DeNuzzio, Presiding — Papers 330-337

- **Polymers for Biosensors and Biosensing**
  In Vivo and Non-Invasive Sensing
  J.R. Woodward, Presiding — Papers 338-345

THURSDAY AFTERNOON

- **Polymer Ecology: Biodegradation and Recycling**
  S.J. Huang, Presiding — Papers 346-351

- **General**
  P. Zarras, Presiding — Papers 352-357

- **General**
  S.A. Carter, Presiding — Papers 358-363

- **Polymers for Biosensors and Biosensing**
  Bioengineering for Enhanced Performance
  D.A. Butterfield, Presiding — Papers 365-371

DIVISION OF POLYMERIC MATERIALS: SCIENCE & ENGINEERING

001. NOVEL FUNCTIONAL CO-POLYMERS BY THE COMBINATION OF LIVING CARBOCATIONIC AND ANIONIC POLYMERIZATIONS. Jesper Feldthusen, Béla Iván*, and Axel H. E. Müller, University of Mainz, Institute of Physical Chemistry, Welderweg 11-15, D-55099 Mainz, Germany; *Permanent address: Central Research Institute for Chemistry of the Hungarian Academy of Sciences, H-1525 Budapest, Pusztaszeri u. 59-67, P. O. Box 17, Hungary

New low molecular weight compounds and polymers, e. g. polyisobutylene (PIB), containing 1-methoxy-1,1-diphenylethyl (MDPE) group have been found as efficient initiator for (meth)acrylic monomers after quantitative metallation with K/Na alloy. Homopolymers and PIB-based AB and ABA block copolymers were synthesized by living anionic polymerization of methyl methacrylate (MMA) and *tert*-butyl methacrylate (tBMA) with nearly quantitative initiating efficiencies. The resulting block copolymers are potential new thermoplastic elastomers, blending and dispersing agents, emulsifiers, nonionic surfactants, biomaterials etc.

002. USE OF PROTECTING GROUPS IN POLYMERIZATION. D. N. Schulz, Exxon Research and Engineering Co., Rt. 22 East, Annandale, NJ 08801; S. Datta, Exxon Chemical Co., 5200 Bayway Drive, Baytown, TX 77520; R. M. Waymouth, Department of Chemistry, Stanford University, Stanford, CA 94305

Functional polymers are macromolecules containing functional groups that have polarity or reactivity different from backbone chains. Such polymers often show improved properties by virtue of enhancements in phase separation, reactivity, or interpolymer associations. Unfortunately, functional polymers are often difficult to synthesize because of the antagonism or competition of functional groups with the active sites of catalysts or initiators. In addition, functional groups can sometimes act as sites for chain transfer or termination. Following the lead of peptide chemists, polymer chemists have increasingly used protecting groups during polymerization to mask reactive functionality. This paper reviews recent advances in the application of protecting groups in anionic, cationic, Ziegler-Natta, metallocene and metathesis polymerization.

003. NOVEL, PROTECTED FUNCTIONALIZED INITIATORS FOR ANIONIC POLYMERIZATIONS.
Douglas E. Sutton and James A. Schwindeman*
FMC Lithium Division, P.O. Box 795, Bessemer City, N.C. 28016

A new class of protected hydroxyl containing functionalized initiators were recently disclosed by the Defense Evaluation and Research Agency (DERA). These novel initiators have the general structure: TBS–O–$(CH_2)_n$–Li. Excellent solubility in hydrocarbon solvents was exhibited by these materials, which allowed the preparation of telechelic, high 1,4–microstructure polybutadienes. The two-step synthesis of these functionalized initiators from commercially available raw materials will be presented in detail. The first step involved reaction of an omega–haloalcohol with t–butyldimethylchlorosilane, in the presence of an acid acceptor, to form the precursor. This precursor was then reacted with lithium metal in a hydrocarbon solvent, to afford a solution of the functionalized initiator. The thermal stability of these initiators in hydrocarbon solution will also be presented. The application of the precursors and functionalized initiators in anionic polymerization of dienes will be briefly discussed.

PMSE

004. ANIONIC SYNTHESIS OF HYDROXY-FUNCTIONALIZED POLYBUTADIENES USING PROTECTED, FUNCTIONALIZED ORGANOLITHIUM INITIATORS. Roderic P. Quirk and Sung H. Jang, Maurice Morton Institute of Polymer Science, University of Akron, Akron, Ohio 44325-3909

Anionic synthesis of hydroxy-functionalized polybutadienes (M_n = 2.0 x 10^3- 2.5 x 10^4) was carried out using functionalized initiators which contain *tert*-butyldimethylsiloxy protecting groups. The removal of protecting groups was effected by hydrolysis using hydrochloric acid. Polybutadienes synthesized with functionalized initiators exhibit narrow molecular weight distributions (M_w / M_n =1.07-1.09) and high 1,4-microstructures (81-93 %). Telechelic dihydroxypolybutadienes were prepared either via ω-hydroxyethylation with ethylene oxide or by coupling reactions followed by removal of protecting groups. Functionalization yields were in the range of 0.97-1.05 for monohydroxypolybutadienes and 1.95-2.08 for telechelic polybutadienes. These synthetic methods avoid the gelation problems which are associated with high degrees of aggregation when dilithium initiators are employed to make telechelic polymers. The linking reaction of functionalized poly(butadienyl)lithium with a tetrafunctional linking agent (i.e., $SiCl_4$) resulted in 4-arm, star-branched polybutadienes with hydroxy groups at each arm end. Multi-hydroxy polydienes can be used as precursors for polyurethane synthesis and as useful components of coatings, adhesives, sealants, and paints.

005. MODEL END-FUNCTIONALIZED LINEAR AND 3-ARM STAR POLYMERS. ORGANIZATION IN SOLUTION AND IN BULK. S. Pispas, M. Pitsikalis, N. Hadjichristidis, Department of Chemistry, University of Athens, Greece

Model block copolymers of styrene/isoprene, linear and 3-arm star polybutadienes with dimethylamine end groups were prepared by anionic polymerization using [3-(dimethylamino) propyl] lithium and sec-BuLi as initiators and methyltrichlorosilane as linking agent. The dimethylamine end groups were transformed to the highly polar sulfozwitterionic ones by reaction with cyclopropansultone. Association of the different ω-functionalized species was studied either in dilute solution by LALLS, MO and dynamic light scattering or in bulk by rheology, SAXS and dielectric spectroscopy. The influence of the environment, the macromolecular architecture and the number of sulfozwitterionic groups on the properties of functionalized and the corresponding neutral compounds will be presented.

006. Stereospecific Anionic Polymerization of *N,N*-Diethylacrylamide
S. Nakahama, M. Kobayashi, T. Ishizone, A. Hirao, and, #M. Kobayashi
(Dept. of Polymer Chemistry, Tokyo Institute of Technology, Meguro-ku, Tokyo 152, Japan
#Lintec Corp., Warabi, Saitama 335, Japan)

Anionic polymerizations of *N,N*-diethylacrylamide (DEAA) were carried out with *sec*-butyllithium / diphenylethylene and potassium diphenylmethanide in the absence and presence of diethylzinc in THF at -78℃ for 60 min. Triad tacticities of the resulting polymers were determined by the ^1H and ^{13}C NMR spectroscopies. Poly (DEAA)s produced in the absence of Et_2Zn have broad molecular weight distributions, while the addition of Et_2Zn as a weak Lewis acid to the polymerization system gave the polymers of narrower MWD. The polymer produced with Li^+ coutercation in the absence of Et_2Zn is highly isotactic with broad MWD, while the poly(DEAA) generated in the presence of Et_2Zn has high syndiotactic configuration. The rate constant of polymerization (kp) with Li^+ countercation was estimated to be more than 5.66x10^2 l/mol/min. In the presence of Et_2Zn, the value of kp was lowered to be 2.60x10^2 l/mol/min. These results suggest that Et_2Zn coordinates with the anionic propagating species to lower its activity affording the narrow MWD. Anionic polymerizations of DEAA with K^+/Et_2Zn at higher temperature results in the polymers of very narrow MWD (Mw/Mn=1.08) and highly heterotactic configurations (mr>90%).

007. SYNTHESIS OF α-HYDROXYL, ω-METHOXYCARBONYL ASYMMETRIC TELECHELIC POLYISOBUTYLENE. B. Koroskenyi and R. Faust. Polymer Science Program, Chemistry Department, University of Massachusetts-Lowell, One University Avenue, Lowell, MA 01854.

A facile route has been developed for the synthesis of α-hydroxyl, ω-methoxycarbonyl asymmetric telechelic polyisobutylene (PIB). It involves living polymerization of isobutylene by direct initiation with BCl_3, followed by chain end capping with 1,1-diphenyl ethylene. The diphenyl carbenium ion is then quenched with 1-methoxy-1-(trimethyl siloxy)-2-methyl propene to yield the carboxylic acid methyl ester. The last step of the synthesis is oxidation of the alkylboron head group by alkaline H_2O_2, yielding the primary hydroxyl functionality. According to ^1H NMR and FT-IR spectroscopic studies, α-hydroxyl, ω-methoxycarbonyl asymmetric telechelic PIB is obtained in quantitative yields.

008.
Synthesis of Biodegradable Copolymers with Hydrophilic Functional Groups
S. Jin and K. E. Gonsalves*, Department of Chemistry and Polymer Program at the Institute of Materials Science, University of Connecticut, Storrs, CT 06269

By ring opening polymerizations, we synthesized a series of novel biodegradable copolymers which include polylactide and poly(ϵ-caprolactone) backbones and pendant hydrophilic functional groups, such as hydroxy, carboxylic acid, phosphoric acid and dimethylphosphonate groups. These copolymers have both hydrophobic polyester backbones and hydrophilic side groups. The polymers were characterized by elemental analysis, FTIR and NMR spectroscopy, DSC and GPC techniques. The properties of the polymers, such as the formation of hydrogels and solubility behavior, were discussed.

009. CYCLIC OLEFIN POLYMERS: A HISTORICAL OVERVIEW. Brian L. Goodall, The BFGoodrich Company, 9921 Brecksville Rd., Brecksville, OH 44141

Interest in the polymerization of strained cyclic olefin monomers, such as the norbornene family, can be traced back to the 1950's. Since those early days both academic and industrial research as well as commercial interest have continued at a steady level over the last four decades. In the 1960's and 1970's the major developments were in the area of ring-opening metathesis polymerization (ROMP) and the only commercial product was Norsorex (ROMP homopolymer of norbornene). In the 1980's hydrogenated ROMP polymers were developed and commercialized, reaction injection molding (RIM) of dicyclopentadiene was introduced commercially and well-defined single component ROMP catalysts emerged. Also in the 1980's the addition polymerization of norbornenes for the manufacture of heat resistant thermoplastics became an area of increasing interest. The 1990's heralds an era of unprecedented interest in this area with metallocene-derived norbornene/ethylene copolymers being commercialized and breakthroughs in Group VIII transition metal catalysts for both ROMP and addition polymerization.

010. NEW MATERIALS AND KINETIC ASPECTS BY COPOLYMERIZATION OF CYCLIC OLEFINS WITH METALLOCENE CATALYSTS. W. Kaminsky, M. Arndt, I. Beulich, Institute for Technical and Macromolecular Chemistry, University of Hamburg, D-20146 Hamburg

The copolymerization of cycloolefins with ethene was investigated using C_1-, C_2- and C_s-symmetric metallocenes and methylalumoxane as cocatalyst. Different types of copolymers from random distributed to alternating structures can be obtained

Substitution of the Cp-ring in $[Me_2C(t-BuCp)(Flu)]ZrCl_2$ yields in an alternating structure of ethene/norbornene copolymers, because the rigid norbornene can be inserted only from the open side of the metallocene. These copolymers are crystalline and have melting points up to 320 °C. By variation of the polymerization parameters, copolymers with glass transition temperatures above 180 °C and molecular weights > 100 000 are synthesized.

011. POLY(CYCLOPENTENE): A NEW PROCESSIBLE HIGH-MELTING POLYOLEFIN MADE FROM NICKEL AND PALLADIUM CATALYSTS. S. J. McLain, J. Feldman, E. F. McCord, K. H. Gardner, M. F. Teasley, E. B. Coughlin, and K. J. Sweetman, DuPont Central Research & Development, Experimental Station, P.O. Box 80328, Wilmington, DE 19880-0328. L. K. Johnson and M. Brookhart, Department of Chemistry, University of North Carolina at Chapel Hill, Chapel Hill, NC 27599-3290.

Alpha-diimine complexes of Ni and Pd polymerize cyclopentene to give high molecular weight semi-crystalline polymers with melting points ranging from 241-330 °C. The polymers have exclusively cis-1,3-enchainment and are partially tactic. X-ray powder diffraction shows that they have a new crystalline form that is different from the highly isotactic polymer reported by Kaminsky. The relationship between catalyst structure, polymer tacticity, and melting behavior will be described.

012. TOPAS® - NEW CYCLOOLEFIN COPOLYMERS APPLYING METALLOCENE TECHNOLOGY. H.T. Land, F. Osan, T. Wehrmeister, Hoechst AG, CRT-TP, 65926 Frankfurt, Germany

TOPAS® (*T*hermoplastic *O*lefin *P*olymer of *A*morphous *S*tructure) is the trade name for Hoechst´s Cycloolefin Copolymers (COC). Cycloolefin copolymers are a new class of polymers, which are based on cycloolefins and ethylene. Manufactured via metallocene technology, TOPAS® grades exhibit a unique combination of attractive properties. Features include high transparency, high heat deflection temperature, and very good electrical insulating properties as well as excellent water vapour barrier properties. Therefore, TOPAS® is suitable for a broad range of applications. The spectrum of applications covers optical storage media as well as films for packaging or capacitors. In addition, medical equipment markets are accessible for TOPAS®, since highly efficient metallocene catalysts account for the high purity of TOPAS® products.

013. PROPERTIES AND APPLICATIONS OF CYCLOOLEFIN COPOLYMER (COC) A.Toyota and M.Yamaguchi, Mitsui Petrochemical Industries, Yamaguchi-ken, 740, Japan

Cycloolefin copolymer (COC) was a new type of polyolefin which Mitsui Petrochemical Ind. first sampled in 1992. COC comprises of α-olefin and cycloolefin. COC is an amorphous polyolefin and has high glass-transition-temperature because it has bulky cycloolefin unit randomly attached to the polymer backbone. COC exhibits many unique properties ; high transparency, high heat resistance, extremely low birefringence, excellent moisture barrier, high chemical resistance, etc. Due to these features, COC generates new application for polyolefins, for example, optical discs, lenses, medical package, food film, capacitor film etc.

014. ENVIRONMENTAL LEGISLATION: A EUROPEAN PERSPECTIVE. J. Bernie, The Paint Research Association, Teddington, England.

This paper will review the development of legislation in Europe and show the U.S. influence on current regulations. The role of the European Government will be explained. Local variations and procedures to protect national interests will be highlighted.

015.

US-EPA's REFERENCE METHOD 24, IS IT APPLICABLE FOR THE MEASUREMENT OF ALL REGULATORY VOCs? K. HIROSHI FUJIMOTO, ADV. TECHNOLOGIES OF MICHIGAN, LIVONIA, MI 48151 AND S. RAMESH, BASF, SOUTHFIELD MI 48086.

The ASTM test methods used in US-EPA's RM 24 for the determination of volatile organic compounds (VOCs), the latest revisions, and the precision statements developed for each test method will be discussed, as well as, the ubiquitious problem in the calculation of some of the low VOC containing products. Initially, these test methods were developed for the measurement of paint and coating VOCs. However, due to the amended Clean Air Act of 1990 and a lack of another test method, RM 24 is used to measure the VOCs of treatment and process chemical compounds, adhesives, paper processing chemicals, personal products to name a few. This information is needed to comply with Title V and Title III of the Clean Air Act. The problems inherent in this transition will be discussed, as well as, some techniques on how to solve them.

016. FILM FORMATION STUDIES OF LATEX BLENDS AND CORE-SHELL LATICES BY ATOMIC FORCE MICROSCOPY. Yaqiang Ming, Dale J. Meier, Michigan Molecular Institute, Midland, MI 48640.

Atomic force microscopy is employed to study film formation of latex blends of compatible and incompatible polymers, as well as "core-shell" latices. The films dried at different temperature have different surface morphologies. A maximum in the surface roughness occurs during annealing when the annealing temperature is above that of one latex and below that of the other. With annealing temperature above both T_g's, the surface roughness decreases with annealing time. Blends of latices of compatible polymers form smooth, homogenious films, but incompatible blends form rough, heterogenious films. The surface roughness of annealed films first increase with increasing concentration of the higher T_g latex, but then decrease with further addition. Particles of different sizes are randomly distributed in thick layer films, and show no tendency to phase separate or aggregate. The film roughness increases with increasing size of the higher T_g latex. Various morphologies are found with incompatible "core-shell" latex particles, and they form heterogenious films upon annealing.

017.

DARK CURE: AN IN SITU SPECTROSCOPE EXAMINATION OF UV CURED, ZERO-V.O.C. COATINGS

A. Sokol, President, UV Coatings, L.L.P., Cleveland, Ohio 44122

Ultraviolet (UV) cured coatings are known to undergo a post processing mechanism known as the "Dark Cure". FTIR/ATR spectroscopic techinques were incorporated to profile and track in situ changes (from liquid state to dark cure) in molecular structure as well as examining moiety motility within the polymer matrix of several UV cured, zero V.O.C. coatings. Understanding and controlling this dynamic, results in the ability to design polymeric matrices which exhibit unique and enhanced physical properties.

018. NMR DIFFUSION STUDIES OF ASSOCIATING POLYMER INTERACTIONS WITH SURFACTANT AND LATEX. P.M. Macdonald, Department of Chemistry, University of Toronto, Toronto, Ontario, Canada M5S 3H6

Associating polymers (APs) employed as rheology modifiers in water-borne coatings applications. They consist of a water-soluble polymer backbone to which hydrophobic modifiers are attached. The use of APs in coatings formulations is complicated by their tendency to interact with many of the typical coatings constituents. NMR diffusion measurements are capable of revealing the molecular details of such interactions, and studies of AP interactions with latex and with surfactant will be described here. In combination with parallel rheology measurements, NMR diffusion studies reveal the impact such interactions have on the network forming capabilities of APs and the resulting rheological properties.

019. SPRAY APPLICATIONS OF LOW-VOC AND ZERO-VOC COATINGS

Lin-Lin Xing[1], J. Edward Glass,[1] and Raymond H. Fernando[2]
[1]Polymers and Coatings Department, North Dakota State University, Fargo, ND 58105
[2]Armstrong World Industries, Inc., Research Development Center, Lancaster, PA 17604

The spray behaviors of latex and high pigment coatings (PVC = 86 %) were examined respectively by high-speed photography. The rheology, dynamic and static surface tension of the coatings were measured in formulations containing different thickener and binder types. The dynamic and static surface tension have little effect on the spray behaviors of the coatings, relative to the viscoelastic response of these coatings, and the latter is influenced by the components in the formulation. Large-hydrophobe HEUR thickeners have been found to be effective viscosifier of architectural coatings without the imparting poor sprayability.

020. ATOMIZATION OF NARROW-DISPERSED EPOXY MICROSPHERES FOR ELECTROSTATIC POWDER SPRAY. Y. Zweifel, P. Fransson, Y. Senuma, J. Hilborn, Polymer Laboratory, Ecole Polytechnique Federale de Lausanne, Switzerland, and J-A. Cotting, Ch. Rickert, CIBA, Polymer Division, Basel, Switzerland.

Epoxy resins suitable for powder coatings have been atomized using a new centrifugation technique. A specially designed rotating disc apparatus makes it possible to produce micron-sized spherical particles. with a narrow size distribution, directly from melt. The particle mean diameter can be accurately tuned by varying the rotation speed of the disc, higher speeds producing smaller particles. As a result the particles are perfectly spherical, which gives interesting handling properties. Current work involves powder charging and coating of metallic probes with conventional electrostatic spraying, both tribo and corona. Charging, coalescence, surface thickness, and homogeneity will be studied as a function of the size, the distribution and the shape of the powders. Relatively large scale production is envisaged, > 100 kg./hour with multi-disc atomizers using a compact installation.

021. EXPERIMENTS AND MODELING OF THE PHOTOPOLYMERIZATION KINETICS OF 2-HYDROXYETHYL METHACRYLATE Michael D. Goodner and Christopher N. Bowman, Department of Chemical Engineering, University of Colorado, Boulder, CO 80309-0424.

Polymerizations of 2-hydroxyethyl methacrylate (HEMA) are often used in the formation of contact lenses. In this work we present experimental results for the polymerization kinetics and kinetic constants for photopolymerizations of HEMA. In addition to the kinetic results a detailed kinetic model has been developed to account for various amounts of crosslinking agent and/or plasticizers in the polymerizing mixture. Results are presented from unsteady state polymerizations which provide information on the kinetic constants and the reaction diffusion parameter along with model predictions and experiments for steady state polymerizations. Finally, model results show that as the amount of crosslinking agent is increased, the polymerization rate increases but the maximum attainable conversion actually decreases.

022.
THE ROLE OF IONIC HYDROPHILIC MONOMERS IN SILICONE HYDROGELS FOR CONTACT LENS APPLICATION. Yu-Chin Lai, Paul L Valint, Jr. and Gary D. Friends, Vision Care Global Scientific Affairs, Bausch & Lomb Inc. Rochester, NY 14692

In the quest for hydrogel contact lenses with improved extended wear capability, the use of siloxane moieties in the lens materials was investigated. However, the introduction of the hydrophobic siloxane groups gave rise to wettability and lipid-like deposit problems. It was found that, while these problems could be reduced through the addition of ionic monomers to the formulations, other key properties were also affected. In this paper, the advantages and disadvantages of using ionic monomers will be discussed.

023. **SILOXANE FUNCTIONALIZED MALONATE BASED POLYESTERS: SYNTHESIS, CHARACTERIZATION, AND THEIR USE IN THE DESIGN OF OXYGEN PERMEABLE HYDROGELS.** R. Bambury and J. Künzler, Department of Polymer Development, Bausch and Lomb, Rochester, NY 14692

This report presents preliminary results on the design of a new oxygen permeable hydrogel for contact lens application based on methacrylate end-capped siloxane functionalized malonate based polyesters. The polyesters are prepared in two relatively simple steps. The first step consists of the polyesterification reaction of Tris(trimethylsiloxy)silyl propyl diethyl malonate with a variety of commercially available aliphatic diols. The second step consists of reacting the resultant hydroxy terminated malonate polyester with isocyanoethylmethacrylate. The methacrylate capped malonate based polyesters, when copolymerized with hydrophilic monomers such as dimethylacrylamide, results in transparent, oxygen permeable hydrogels.

024.

NOVEL SILICONE HYDROGELS BASED ON FUMARATE-CAPPED PREPOLYMERS OF POLYDIMETHYLSILOXANE Yu-Chin Lai, Edmund T. Quinn and Ronald E. Bambury, Vision Care Global Scientific Affairs, Bausch & Lomb Inc. Rochester, NY 14692-0450

A series of novel hydrogels derived from compositions containing a fumarate-capped siloxane prepolymer, N, N-dimethyl acrylamide and 3-Methacryloxypropyl tris(trimethylsiloxy)silane (TRIS) were prepared. The fumarate prepolymers were derived from the reaction of fumaryl chloride and α, ω-bis-(4-hydroxy-butyl) polydimethylsiloxane (HB-PDMS) of M_n 1600-4000, and end-capped with t-butylamine, diethanolamine or 1,1,3,3-tetramethyl butylamine. The effect of molecular weight of polysiloxane, end-capping amine, TRIS and organic diluent used in polymerization, on hydrogel properties is discussed. These hydrogels had extremely high oxygen permeability and other bulk properties useful for contact lens application.

025.

THE INFLUENCE OF SILOXANE HYDROGELS PHASE SEGREGATION ON EXTENDED WEAR CONTACT LENS USE. R. Baron, E. Ajello, Q. Liu, L. Terry, L. Winterton, CIBA Vision, Duluth, GA, 30155, J. Riffle, Virginia Polytechnic Institute, Blacksburg, VA, 24061

For over 40 years, various groups have tried to make soft contact lenses with high oxygen permeability and good mobility on-eyes. To the best of our knowledge, all these attempts have failed. This talk will examine the first such lens and it's characterization. From this characterization, principles were devised which lead to the creation of a second series of materials. This in turn led to the expansion of the original principles into a new set of standards. These standards were used to the develop many other families of materials which also met the performance criteria.

PMSE

026. MORPHOLOGY REQUIREMENTS FOR ON-EYE MOBILITY OF SOFT OXYGEN PERMEABLE CONTACT LENSES.
A. Domschke, L. Winterton, Business and Technology Innovation, Ciba Vision, Duluth, GA 30155, USA; D. Lohmann, Corporate Research Unit, Ciba Geigy Limited, Basel, Switzerland,

Soft contact lenses with high oxygen permeability are the focus of R&D in contact lens industry. Thus high comfort is combined with biocompatibility which allows the extended wear of contact lenses for several weeks including over night wear. Distinct from conventional soft hydrogel lenses or rigid oxygen permeable lenses, the new, soft and oxygen permeable lenses require a specific phase morphology in order to maintain the lens on-eye mobility. This presentation will discuss the phase morphology and its role in soft oxygen permeable contact lens materials. Permeation measurements and small angle neutron scattering experiments gave evidence for a bicontinuous phase morphology which turned out to be the key issue for on-eye mobility.

027. SUPRAMOLECULAR POLYMERS AS MULTIFUNCTIONAL MATERIALS. S. I. Stupp, Departments of Materials Science and Engineering and Chemistry, Beckman Institute for Advanced Science and Technology, Materials Research Laboratory, University of Illinois at Urbana-Champaign, Urbana, IL 61801

An interesting target in polymer science is to find pathways to highly regular supramolecular units with dimensions similar to those of high molar mass linear or hyperbranched polymers. These units lack a polymeric backbone but could serve as precursors to shape invariant covalent polymers analogous to folded proteins. We are pursuing access to these polymers with designed molecules programmed to assemble into nanostructures of regular shape and dimension. Space filling requirements for shape invariant nanostructures, and the great structural diversity that can be achieved by molecular synthesis, can lead to polymeric materials with unique potential to be multifunctional and easily processable. Another interesting aspect of *shape invariant* polymers is the possibility of forming 2D or 3D networks by *chemical coordination of nanostructures*. This lecture will describe the discovery of systems that yield supramolecular polymers and discuss the integration of properties in these systems. Because of space filling behavior these supramolecular polymers can integrate properties such as adhesion, nonlinear optics, luminescence, nano-patterning, and control of surface chemistry.

028. SUPRAMOLECULAR ASSOCIATIONS OF POLYSILOXANES WITH HYDROGEN BOND DONORS OR ACCEPTORS. S. Abed, S. Boileau, L. Bouteiller, J.R. Caille, N. Lacoudre, D. Teyssié, J.M. Yu. Collège de France, 11 place M. Berthelot, 75231 Paris Cédex 05, France.

Polysiloxanes were modified with different aminoacids in order to induce potential interactions with polar surfaces. N-acetylcysteine was successfully grafted onto polymethylvinylsiloxane (PMVS) via a free-radical addition of the thiol function onto the vinyl groups. PMVS and polymethylmercapto-propylsiloxane (PMMPS) were also quantitatively modified with other hydrogen bond donors or acceptors among which mercaptosuccinic acid or allyloxybenzoic acid (ABA). Those polymers show a strong tendency to associate through hydrogen bonding. Moreover the ABA modified PMPS exhibits mesomorphous properties both on its own and as a complex with bipyridine. The synthesis and properties of telechelic polysiloxanes with ABA moieties will be discussed. Finally, the modification of polysiloxanes with allyloxyuracil and allyldiaminopyridine derivatives was investigated as a design for the formation of "double stranded polysiloxane complexes" by molecular recognition.

029. SYNTHESIS OF NOVEL POLYPHOSPHATES AND PREPARATION OF FUNCTIONAL MATERIALS BY BLENDING WITH POLY(VINYL ALCOHOL). S. Nakamura, C. Wang, D. Iwami, and T. Takayama, Faculty of Engineering, Kanagawa University, Kanagawa-ku, Yokohama 221, Japan

Phosphorous-containing polymers and their polymer blends are expected to find various applications. We have synthesized novel polyphosphates having pendant tertiary amine salt groups (PPASs) from a cyclic phosphate (H_3PO_4-DMPA) and polymer blends of PPAS and poly(vinyl alcohol) (PVA) have been also prepared.

H_3PO_4-DMPA was obtained by reacting phosphoric acid with N,N-dimethyl-1,3-propanediamine. Polyphosphates, PPASs were obtained by heating H_3PO_4-DMPA to 120 or 160°C for 3h under nitrogen. The reaction mechanism of H_3PO_4-DMPA to PPAS is discussed. Blends of PPAS and PVA were prepared by adding PPAS to a PVA solution in formamide and coagulating the solution with acetone. No macroscopic phase separation nor precipitation is observed in the blend solution. Miscibility of the blend was examined by DSC, CP/MAS NMR. Glass transition temperature of the blends are higher than that of PVA due to the interaction between the tertiary amine salt groups of PPAS and hydroxyl groups of PVA. The resulting blends are soluble in methanol and polymer reaction may occur partly between the end groups of PPAS and the hydroxyl groups of PVA.

030. **FUNCTIONAL POLYPHOSPHAZENES.** Harry R. Allcock, Department of Chemistry, The Pennsylvania State University, University Park, Pennsylvania 16802

Perhaps the ultimate functional polymer is poly(dichlorophosphazene), $(NPCl_2)_n$, in which ~30,000 chlorine atoms per molecule can be replaced by other side groups, especially by a wide variety of organic units. Some poly(organophosphazenes) undergo additional reactions that modify the organic side groups without cleavage of the main chain. By such reaction sequences it has been possible to prepare polyphosphazenes with a wide variety of functional groups that are useful as solid polymer electrolytes, electro-optical and optical materials, metal-coordination substrates, biomedical polymers, and surface active materials. This talk will deal with several of these aspects.

031. ORGANOSILICON 1,3-OXAZOLINES: REACTIVE SILICONE LIQUID RUBBERS AND SILANE COUPLING AGENTS TAILORED FOR REACTIVE PROCESSING APPLICATION AND NANOCOMPOSITE FORMATION, P. Müller, C. Wörner, R. Schäfer, R. Mülhaupt, Freiburger Materialforschungszentrum und Institut für Makromolekulare Chemie der Albert-Ludwigs Universität, Stefan-Meier-Str.21, D-79104 Freiburg i.Br., Germany

Novel families of oxazoline-functional silanes, silsesquioxanes, silicones, and silica particles were prepared and applied in reactive processing and network formation. Oxazoline silanes proved to be very effective coupling agents to improve filler adhesion in composites. Oxazoline-terminated silicones produced novel silicone rubbers upon cationic cure. When applied as chain extenders in PBT processing, oxazoline silicone liquid rubbers produced segmented PBT containing silicone segments. Interfacial coupling reaction of oxazoline-functional pyrogenic silica and grafting of oxazoline-functional silsesquioxane gave novel core/shell nanoparticles.

032. **Functionalized polymers: Polymer chemistry with 6-membered mesoions**
Helmut Ritter
Bergische Universität GH Wuppertal, FB 9, Organic -and Macromolecular Chemistry, Gauß-Str.20, D-42097 Wuppertal, Germany

The presentation deals with synthesis and behaviour of polymeric mesoions. We recently described the preparation of functionalized polystyrenes containing six membered mesoionic 4,6-dioxo-1,3-diazines as pendant groups. It was demonstrated that these attached **highly polar mesoionic heterocycles** are photoactive yielding less polar polymeric bis(ß-lactames) during UV-irradiation. A styrene modified mesoion can be polymerized radically with AIBN yielding a polyvinyl system, as expected. Surprisingly, a polyaddition via regioselective Diels-Alder type mechanism was also observed in solution. The obtained product was proved by mass spectrometry to be an oligomer.
Recently, N-phenylmethacrylamide was successfully modified with chloro-carbonyl-ketene yielding mesoionic oxazinium system. From preliminary studies it was demonstrated that the mesoionic monomer can be copolymerized radically with methymethacrylate.

033. HYDROPHILIC BLOCK COPOLYMERS: INTERACTION WITH METAL SALTS AND METAL COLLOID FORMATION IN AQUEOUS MEDIUM.
L. Bronstein, Nesmeyanov Institute of Organoelement Compounds, 28 Vavilov St., Moscow 117813, Russia. Milos Sedlak, Faculty of Chemical Technology, Organic Chemistry, University of Pardubice, 53210 Pardubice, Czech Republic. J. Hartmann, M. Breulmann, H. Cölfen, M. Antonietti, Max Plank Institut für Kolloid- & Grenzflächenforschung, Kantstrasse 55, D-14513 Teltow-Seehof, Germany

The micellization of polyethylene-b-polyethyleneimine (PEO-b-PEI) due to interaction of Au, Pd, and Pt compounds with the imine groups of the polymer was studied. The characteristics of these micelles were found to depend on metal type, the molar ratio of polymer/metal salts, and other parameters. For gold species both micellization of polymer and self-reduction of gold ions by imine groups proceeds; this process seemed to be extremely sensitive to gold salt loading. The influence of type of reducing agent (hydrazine, $NaBH_4$, H_2) on micellar parameters and the Pd and Pt colloid formation was investigated. It was shown that size and shape of Pt colloids can be controlled by changing the molar ratio of PEI/metal salt. Spheres, cubes and triangles were observed in this case.

034. NOVEL HEAT-RESISTANT CYCLIC OLEFIN POLYMERS MADE USING SINGLE COMPONENT NICKEL AND PALLADIUM CATALYSTS. Brian L. Goodall, Dennis A. Barnes, George H. Benedikt, Lester H. McIntosh and Larry F. Rhodes, The BFGoodrich Company, 9921 Brecksville Rd., Brecksville, OH 44141

Using a new family of single component cationic nickel and palladium catalysts we are able to polymerize norbornenes at unprecedented rates to afford a family of amorphous, transparent polymers with excellent heat resistance. The glass transition temperature (T_g) of the polymers can be conveniently controlled in the range of about 150°C to about 400°C by selecting the appropriate norbornene monomer or mixture of monomers. The molecular weights of the polymers can be tailored from thousands to millions using a new chain transfer process involving the use of olefins such as ethylene and isobutylene. This chain transfer process also serves to append an olefinic double bond at the end of each polymer or oligomer chain. Catalysts, polymerization mechanism and polymer properties are all presented in some detail.

035. SYNTHESIS OF CONFORMATIONALLY RESTRICTED POLYMERS. W. Heitz, Philipps-Universität Marburg, Fb Physikalische Chemie / Polymere und Wissenschaftliches Zentrum für Materialwissenschaften, 35032 Marburg / Germany

To the extreme conformational restriction can result in a polymer with a conformation defined during synthesis. Vinylpolymerisation of norbornene is not an ideal representative but comes close to that idea. It can be accomplished by zirconocene catalyst. The polymer thus formed is semicrystalline and has a low solubility which limits the molecular weight for homopolymerization. Using Pd-catalysts amorphous polymers are formed soluble in chloroform, toluene, or cyclohexane with $M>10^6$. The variation of the Pd-catalyst as well as initiators like CoX_2 / MAO, Cp^*CrX_2 / MAO and the influence on the properties of the resulting polynorbornenes are presented. The polymerization of norbornene carboxylic esters is discussed as well. Oxidative addition and polymerization can be combined to introduce functional end groups. This is demonstrated by oxidative addition of substituted allyltriflates to $Ni(COD)_2$ and subsequent polymerization of butadiene. The [2+2] polycycloaddition of norbornene with Ni(0) results in a ribbon type polymer. The polymer should have a two dimensional coil structure.

036. ADMET POLYMERIZATION USING CLASSICAL CATALYTIC SYSTEMS. Fernando J. Gómez and Kenneth B. Wagener,* The George and Josephine Butler Polymer Research Laboratory, Department of Chemistry and Center for Macromolecular Science and Engineering, University of Florida, Gainesville, FL 32611-7200.

The synthesis of high molecular weight, linear polyoctenylene via ADMET polymerization has been achieved by means of four tungsten-based classical catalytic systems. The tungsten aryloxo complexes $W(O)Cl_2(O-2,6-C_6H_3Br_2)_2$ and $WCl_4(O-2,6-C_6H_3Ph_2)$ were used as the pre-catalytic species while tributyltin hydride and tetrabutyltin were the cocatalytic complexes of choice. The reaction was carried out under solvent-free conditions which ensures high molecular weights. Soluble, linear products were obtained, indicating the absence of side addition-type reactions that could yield cross-linked polymers.

037. SYNTHESIS AND PROPERTIES OF HYDROGENATED RING OPENING METATHESIS POLYMER M. Hashimoto, Research & Development Center, Nippon Zeon Co., Ltd. 1-2-1 Yako, Kawasaki-ku, Kawasaki 210 Japan

Novel cyclic olefin polymers have been synthesized by ring opening metathesis polymerization (ROMP) of norbornene derivatives, followed by hydrogenation of double bonds on the main chain. Following Hydrogenation give outstanding features to the unsaturated polymers and the physical properties of the hydrogenated Ring-Opening polymers (HROP) has been systematically investigated. They are generally amorphous and have improved heat stability. In addition, they show excellent properties as optical materials; high transparency; low birefringence; low water absorption; and others. ZEONEX®, commercialized by Nippon Zeon Co., Ltd., is a typical example of HROP and has been used in varied fields, especially for optical use.

038. OVERVIEW OF SAFETY AND HANDLING OF UV/EB CURING MATERIALS, Ronald Golden, UCB Chemicals Corporation, Smyrna, GA, 30080

Ultraviolet (UV) and electron beam (EB) curing has been recognized as a successful low to zero VOC coating and adhesives technology for over 25 years. However, many misconceptions still exist concerning the safety of the materials and equipment used in UV/EB curing. Such misconceptions have overshadowed the fact that UV/EB materials are, in general, less hazardous and easier to control than most solvents. In some cases, a misunderstanding of safety issues has blocked adoption of this powerful compliance technology.

This paper discusses the physical hazards of UV/EB technology, provides data on raw material toxicity, and describes the engineering controls and industrial hygiene practices which have been developed to assure workplace safety. The practical evaluation of workplace hazards and safe handling practices will enable potential end users of UV/EB technology to make a rational decision when choosing options for compliance.

039. ADVANCES IN THE PHOTOINITIATED POLYMERIZATION OF ACRYLATES AND FUNCTIONAL MALEIMIDES. Shan C. Clark[1], Sonny Jönsson[2] and Charles E. Hoyle[1]
[1]Department of Polymer Science, University of Southern Mississippi, Hattiesburg, MS 39406
[2]Fusion UV-Curing Systems Gaithersburg, MD 20878

Acrylates containing functional maleimides photopolymerized in the absence of conventional photoinitiators attain acceptable rates of polymerization. An acrylate formulation containing a maleimide with a pendent carbonate functionality increased the initiation efficiency of the system and the corresponding rate, compared to an aliphatic maleimide. FTIR was used to monitor the photopolymerization of a maleimide/diacrylate system utilizing a medium pressure mercury lamp and a XeCl lamp. Higher conversions were obtained for both the maleimide and the acrylate using the XeCl lamp.

040. EFFECT OF EXCIMER LIGHT SOURCES ON THE COLOR STABILITY AND MECHANICAL PROPERTIES OF UV CURABLE INKS DESIGNED FOR FIBER OPTIC APPLICATIONS. Dave M. Szum, Lindsay S. Coons, DSM Desotech Inc., Elgin, IL 60120

Recent developments in UV technology have created significant opportunities in the telecommunications industry. In particular, greater emphasis is being placed on the development of reliable coloring systems as these networks face increasingly harsh environments. Development of solventless, no VOC, UV-curable inks have resulted in improved application and performance properties in comparison to the current inks that are widely used in the fabrication of optical fiber cables and ribbon. In addition, improvements in UV-curing technology such as the recent development of high-intensity light sources based on excimers have made UV inks a viable alternative to the current ink technology. This talk will compare the color stability of several ink formulations under various aging conditions including heat, fluorescent light and extreme humidity. The impact of the excimer technology on the mechanical properties of these ink formulations including the cure speed, tensile strength and modulus and percent elongation will also be discussed.

041. Novel Reactive Diluents for Cationic UV Coatings, Shaobing Wu, Mark D. Soucek*, Department of Polymers & Coatings, North Dakota State University, Fargo, ND 58105

Reactive diluents for cationic UV cure coatings were synthesized using di and tri functional ε-caprolactone polyols with tetraethyl orthosilicate (TEOS). The chemical structures of the diluents were characterized using IR and ^{1}H-NMR spectroscopy. The resulting siloxane functionalized products were used to reformulate cationic UV initiated cycloaliphatic epoxide coatings. The film properties were evaluated in terms of pencil hardness, crosshatch adhesion, reverse impact resistance, and MEK double rubs. Also, the viscosity of the coatings were measured and compared with the parent polyols. It was observed that the addition of the TEOS functionalized polyols can lower the viscosity by a factor of 2. The addition of the TEOS functionalized polyols also improved the hardness and solvent resistance of the coatings. Based on the coating properties, a crosslinking reaction mechanism was proposed.

042.

TUNED PHOTOINITIATORS. Rajamani Nagarajan, Joseph S, Bowers, Jr., First Chemical Corporation, Pascagoula, Mississippi 39568 and Sonny Jonsson, Fusion UV Curing Systems, Gaithersburg, Maryland 20878

It is well known in the field of ultraviolet (UV) curing that the efficiency of a photoinitiator will be maximum if its UV absorption matches the spectral emission of the lamp used for curing. Research in our laboratories have led to the development of photoinitiators that have excellent absorption around 310 nm, the region where the UV emission of the commercial "excimer" lamp is concentrated. These initiators show exceptionally high efficiencies compared to the conventional initiators, when used in conjunction with this lamp for curing formulations. This paper will present basic as well as practical results obtained from the studies on these initiators.

043. CATIONIC UV-CURING OF VINYL ETHER-BASED COATINGS. C. Decker, F. Morel. Laboratoire de Photochimie Générale (URA-CNRS n°431) - Ecole Nationale Supérieure de Chimie 3, rue Werner - 68200 Mulhouse (France)

Solvent-free formulations of vinyl ether monomers have been cured by UV-irradiation in the presence of a cationic photoinitiator. The polymerization kinetics was monitored *in situ* by real-time infrared spectroscopy and shown to proceed extensively within a few seconds upon intense illumination. The rate of polymerization increases linearly with the light intensity, as expected for a photoinitiated chain reaction involving monomolecular termination. An increase of the formulation viscosity through the addition of vinyl ether functionalized oligomers leads to a drastic drop of the cure speed, because of a sharp reduction of the propagation rate constant. Temperature was found to have a minor effect on the polymerization rate of divinyl ether of triethyleneglycol, in the 20-75°C range. The main interest of UV-curable coatings based on vinyl ethers lies in their low VOC emission and their high cure speed at ambient temperature.

044. COMPARATIVE STUDY OF PHOTOINITIATOR CURING EFFICIENCY USING UNIVERATE REGRESSION ANALYSIS

Chia - Hu Chang, Thai Nguyen and Keith Cooper
Ciba Specialty Chemical Company, Additives Division
540 White Plains Road, PO Box 2005
Tarrytown NY 10591-9005

Abstract

Two alpha hydroxy ketone and two phosphine oxide photoinitiators are compared in a clear acrylate based formulation. Regression technique is employed for data analysis. The technique allows us to use a simple industrial type of curing unit and conditions for experiments. Consequently, results from the study will be more apparent and meaningful to formulators. Our data suggests that on the weight basis, hydroxy ketone appears to be less vulnerable to oxygen inhibition. On the contrast, phosphine oxide is more reactive in nature. Furthermore, Using a nitrogen blanket is a good method to increase productivity and reduce formulation cost in a very thin coating applications.

045. ULTRATHIN PMMA-LIKE CONTACT LENS COATINGS DEPOSITED FROM A RADIOFREQUENCY GLOW DISCHARGE. T.R. Gengenbach, P. Pasic, P. Zientek, H.J. Griesser, T. Davis, E. Papas and R. Austen Division of Chemicals and Polymers, CSIRO, Private Bag 10, Clayton South MDC, Clayton 3169, Australia.

Thin coatings were deposited from a glow discharge plasma established in methyl methacrylate vapor, onto hydrophobic contact lens materials with high oxygen permeability. The composition of the coatings depended markedly on the deposition parameters. Coatings deposited at low rf power input possessed FTIR spectra similar to that of conventional PMMA, whereas coatings deposited at high power input showed considerable scrambling of the ester group, resulting in diverse polar functionalities in the deposited film. In addition, some coatings were modified by alkaline hydrolysis of the surface ester groups. The coatings lowered the oxygen permeability. On-eye testing showed performance as for hydrogels.

046.
EFFECTS OF SURFACE CHARGE AND COATING STRUCTURE ON LYSOZYME ADSORPTION. P.Kingshott, R.C.Chatelier, H.A.W. St John and H.J.Griesser, Division of Chemicals and Polymers, CSIRO, Private Bag 10, Clayton South MDC, Clayton, 3169, Australia.

The effect of electrostatics upon lysozyme adsorption to the protonated amine *n*-heptylamine plasma polymer (*n*-ha) surface is investiagted using x-ray photoelectron spectroscopy (XPS). At physiological pH (7.4) and pH 5.5, where the *n*-ha surface and the lysozyme molecules carry a net position charge, no repulsion of lysozyme occurs regardless of the ionic strength of the buffer solution suggesting that other forces are more dominant in the adsorption of lysozyme. The rate of protein adsorption to the n-ha surface dramatically increases when an artificial tear fluid formulation (ATF) is used suggesting that even if lysozyme was capable of being electrostatically repelled other proteins have a high affinity for that surface. The rate of protein adsorption is however, reduced when the n-ha surface is coated with a covalently coupled layer of oxidised dextran, but after a significant adsorption time (overnight) the final coverage for the two surfaces reaches that expected for a monolayer of protein.

047. POLYSACCHARIDE COATINGS FOR CONTACT LENSES. H.J. Griesser, R.C. Chatelier, L. Dai, H.A.W. StJohn, T. Davis and R. Austen, Division of Chemicals and Polymers, CSIRO, Private Bag 10, Clayton South MDC, Clayton 3169, Australia.

We have developed highly wettable coatings for contact lens materials that are hydrophobic. The coatings are applied in a two-step process. The first step comprises amination of the bulk polymer material using a gas plasma (glow discharge) procedure. In the second step, a polysaccharide is covalently coupled onto the surface amine groups. Surface analytical methods were used to verify the successful deposition of the two thin layers. The coatings were resistant to delamination on autoclaving. Dk measurements showed a measurable but tolerable decrease in the oxygen permeability upon coating. On-eye evaluation of coated contact lenses showed performances equivalent to that of current hydrogel lenses while offering higher oxygen fluxes.

048.

STUDY OF PROTEIN ADSORPTION ONTO POLYSACCHARIDE CONTACT LENS COATINGS BY MALDI-TOF-MS AND XPS. P.Kingshott, H.A.W. St John, R.C.Chatelier and H.J.Griesser, Division of Chemicals and Polymers, CSIRO, Private Bag 10, Clayton South MDC, Clayton, 3169, Australia.

We utilise matrix-assisted laser desorption ionisation time-of-flight mass spectrometry (MALDI-TOF-MS) and x-ray photoelectron spectroscopy (XPS) to study the interfacial interactions between tear proteins, from an artificial tear fluid formulation (ATF), and polysaccharide (PS) contact lens coatings. All PS coatings exhibited excellent water wettability and their and XPS confirmed that a uniform PS coating is successfully attached to the aminated n-heptylamine plasma polymer film using our attachment chemistry. XPS also shows that all PS surfaces have a strong affinity for the proteins of ATF but the rate of protein adsorption varies between PS surfaces. MALDI-TOF-MS directly identifies which species of protein exists on each surface by the generation of intact protein molecular ions. For example, our oxidised dextran (OD) surface contains both adsorbed lysozyme and lactoferrin after immersion in ATF for 1 hour.

049.

SURFACE CHARACTERIZATION OF WORN HYDROGEL CONTACT LENSES. H.A.W. St John, P. Kingshott, H.J. Griesser, C. Morris and S. Bolis, Division of Chemicals and Polymers, CSIRO, Private Bag 10, Clayton South MDC, Clayton 3169 Australia.

Surface sensitive techniques such as XPS, SSIMS and MALDI-MS have been used to study the initial stages of interface conversion for hydrogel contact lenses worn by human patients. Proteinaceous material was found to be irreversibly bound to lens surfaces within 5 seconds of wear. The rate of further deposition was shown to dependent on the patient and composition of the lens polymer. The spatial distribution of deposits on the front and back surfaces of the lens was also dependent on the polymer composition. Lipid deposition was highly patient dependent and independent of protein adsorption. SIMS analysis confirmed the conversion of the outermost surface of Acuvue lenses to mostly proteinaceous material within 4 hours of wear. MALDI-MS identified the deposition of lysozyme on all worn hydrogel lenses, and also lower molecular weight species such as protein fragments.

050. POLYANILINE AND FULLERENE GRAFTED FUNCTIONAL POLYMERS. A. O. Patil, Corporate Research Laboratory, Exxon Research and Engineering Company, Route 22 East, Clinton Township, Annandale, NJ 08801

We have grafted polyaniline and fullerenes to amine containing flexible hydrocarbon polymers such as amino-ethylene propylene terpolymer (EPDM-amine) to obtain novel functionalized polymers. These materials are soluble in solvents such as heptane or THF, in which the polyaniline or fullerene is essentially insoluble. The reactions are followed by various spectroscopic techniques. Surprisingly, one doesn't need the amine functionality to make fullerene grafted polymers. Recently, we have grafted fullerene onto nonfunctional saturated hydrocarbon backbones, such as ethylene-propylene copolymers by free radical means. This paper will discuss the synthesis characterization and applications of these functional polymers.

051. NOVEL FUNCTIONALIZATION OF POLY(ISOBUTYLENE-CO-p-METHYLSTYRENE) RUBBER. S. A. Haque, K. W. Powers, and H. C. Wang, Exxon Chemical Co., Baytown Polymers Center, Baytown, TX 77520, J. M. J. Fréchet and J. H. G. Steinke, Department of Chemistry, Baker Laboratory, Cornell University, Ithaca, NY 14853

Poly(isobutylene-co-p-methylstyrene [poly(IB-PMS]) containing 2-5 mol% of p-methylstyrene is an elastomer requiring a preferred selective functionalization at the pendant aromatic group for its versatile use. Methods have been developed for efficient and selective functionalization in solid state as well as in hydrocarbon solvents. For example, one can selectively brominate at the p-methyl position of the aromatic unit of PMS. It may further be quantitatively derivatized to form esters of acrylic, methacrylic, 2,4-hexadienoic acids, and ethers of hydroxybenzophenone. Oxidation is an alternative method for selective functionalization to obtain benzoic acid derivative. Catalytic CrO_3 in large excess of *tert*-butylhydroperoxide is effective for selective oxidation in the temperature range between 20°C-50°C in hydrocarbon solvents. Metalation by superbase (SB) and then quenching by electrophiles yields other new groups of functional polymers. SB derived from *sec*-BuLi and Cs -1-(-)-menthoxide allowed quantitative metalation at the primary benzylic position (99%) and least aromatic ring metalation (<2%).

052. SELECTIVE MODIFICATION OF 4-METHYLSTYRENE POLYMERS. Q. Sheng, H.D.H. Stöver*, Dept. of Chemistry, McMaster University, Hamilton, ON, Canada L9H 6P8

We report on the selective oxidative functionalization of polymers incorporating 4-methylstyrene. Cobalt(II) acetate is used as oxidation catalyst in the presence of air, to convert 4-methyl groups to aldehyde and carboxylic acid. Cerium(IV) ammonium nitrate and cobalt(III) acetate are used as stoichiometric oxidants under nitrogen, to convert 4-methyl groups into acetoxymethyl, trichloroacetoxymethyl, and chloromethyl groups. In the 4-methylstyrene homopolymers, and particularly in its copolymers with α-substituted comonomers such as α-methylstyrene and alkylmethacrylates, we exploit the steric hindrance along the backbone to protect the backbone benzylic methine sites, and direct oxidation towards the benzylic 4-methyl group. The resulting functional polymers have interesting properties, and may help clarify the mechanism of metal-mediated oxidation of alkylbenzenes.

053. TOWARD MINIMALLY ADHESIVE POLYMER SURFACES. M. B. Goodwin[1a,2], A. D. Vu[1a,2], S. Bullock[1a,2], A. Ervin[1b], P. Gatenholm[3], K. J. Wynne[1a,4], [1a]Materials Chemistry Branch and [1b]Chemical Dynamics and Diagnostic Branches, Naval Research Laboratory, Washington, D,C. 20375, [2]Department of Chemistry, George Mason University, Fairfax, Virginia 22030, [3]Department of Polymer Technology, Chalmers University of Technology, S-412 96, Göteborg, Sweden, and [4]Physical Sciences S&T Division-331, Office of Naval Research, Arlington, VA.

This work addresses the development of minimally adhesive polymer surfaces with the goal of discerning compositional and morphological features which create a surface unattractive to the settlement and adhesion of marine organisms. In continuing this effort, the evolution of surface properties as a function of cure time for filled, sol-gel polydimethylsiloxane (RTV11®) samples has been determined via dynamic contact angle analysis. The effect of humidity on cure and time of storage in aqueous solution after air cure have been evaluated. The results show that, depending on cure conditions, surface wettability changes rapidly in the first few days of cure and may continue to change up to one or two months after preparation. Tapping mode atomic force microscopy (AFM) has been used to elucidate changes which occur on immersion of RTV11® in water. Samples cured under standard conditions erode in water, with surface topography changing continuously over a 12 day's immersion. These results agree with weight loss, contact angle, and water analysis data.

054. SURFACE-ACTIVE MACROMONOMERS FOR COATING OF CONTACT LENS POLYMERS. P. L. Valint, Jr., G. L. Grobe, III, J. A. McGee, D. M. Ammon and E. J. Leibenguth, Global Vision Care, Bausch & Lomb, Rochester, NY 14692

Soft contact lens materials have bulk physical properties that dictate the type of polymer to be used, such as water content, refractive index, elasticity, light transmittance and tear strength. These properties are sometimes in conflict with surface properties, i.e. wettability, lubricity and tear component deposition. Since the biocompatibility of a contact lens polymer is almost entirely the result of the interactions of the polymer surface with ocular tissue and tear fluid, it may be necessary to modify the surface chemistry to achieve both the desired bulk and surface properties. This paper describes the use of polymerizable surface-active materials to modify surface properties. Synthesis of the surface-active macromonomers, SAM, will be presented. Characterization of the resultant surfaces carried out by a multi-technique approach utilizing critical surface tension analysis, x-ray photoelectron spectroscopy (XPS), Time of flight-secondary ion mass spectroscopy (ToF-SIMS), atomic force (AFM) and lateral force (LFM) microscopy will be presented.

055.

AMINE CONTAINING EPDM TO CONTROL THE CARBON BLACK DISTRIBUTION AND COVULCANIZATION OF EPDM AND NITRILE RUBBERS: S. Datta*, R. C. Keller and F. T. Morrar, Exxon Chemical Company, 5200 Bayway Drive, Baytown, TX 77520

Blends of amine containing EPDM (≈ 0.8 mole % -NH_2) with nitrile rubber (NBR) are immiscible but on vulcanization show excellent combination of resistance to thermal aging and solvents. They thus combine the beneficial properties of polyolefin and acrylonitrile containing elastomers. These properties arise not only from the compatibilization of the blend due to specific chemical interaction between the amine group and the nitrile group but also due to interaction between the carbon black and the amine containing EPDM which leads to even interphase distribution of curatives and carbon black.

056. **Multi-functional Coordination Crosslinks in Poly(vinylamine) Complexes with Cobalt Chloride.** Laurence A. Belfiore*, Erik Indra, & Pronab Das, Polymer Physics & Engineering Laboratory, Department of Chemical & Bioresource Engineering, Colorado State University, Fort Collins, Colorado, USA 80523

Poly(vinylamine) complexes with cobalt chloride hexahydrate exhibit a 45^0C enhancement in the glass transition temperature per mol% cobalt relative to the undiluted polymer. Multiple amino ligands in the coordination sphere of pseudo-octahedral cobalt(II) have been proposed to rationalize this increase in T_g. Estimates of the pseudo-octahedral ligand field splitting and the ligand field stabilization energy for high-spin d^7 Co(II) complexes provide support for the proposed amorphous structures. The concept of multi-functional coordination crosslinking in polymeric materials via transition metal chemistry is quite novel, but consistent with several small-molecule examples in the inorganic literature. Thermal energy is required to remove a sufficient number of amino ligands from the coordination sphere of cobalt and induce the glass transition. This endothermic process is correlated with the energetic stabilization of cobalt's d^7 electronic configuration in the presence of multiple amino sidegroup ligands.

057. EFFICIENT FUNCTIONALIZATION AND DERIVATIZATION OF POLYMERIC OLEFINS. J. I. Emert, Exxon Chemical Company, P.O. Box 536, Linden, N.J. 07036

Homo and copolymers of α-olefins and ethylene/α-olefins prepared via metallocene catalysts could be readily functionalized to carboxylic acid derivatives and condensed with polyethyleneamines. Some of the chemistries recently applied to those classes of polymers will be discussed.

058. EFFICIENT SYNTHESIS OF ALIPHATIC POLYKETONES BY COPOLYMERIZATION OF OLEFINS AND CARBON MONOXIDE CATALYZED BY CATIONIC PALLADIUM-CHELATE LIGAND COMPLEXES, E. Drent, W.W. Jager, *Shell Research and Technology Centre, Amsterdam, P.O. Box 38000, 1030 BN Amsterdam, The Netherlands.*

In the late 1940's, Reppe discovered a nickel catalyst for the co-oligomerization of ethene and carbon monoxide. Since then, various groups have made attempts to develop more efficient catalysts for this reaction. The recent discovery of a new class of very active palladium catalysts now allows, for the first time, ready access to high molecular weight aliphatic polyketone polymers. The catalysts comprise a cis-ligated palladium (II) species associated with weakly or non-coordinating anions. A unique feature of the catalysts is their ability to also catalyze the copolymerization of higher olefins than ethene with carbon monoxide. Some key-points of "polyketone catalysis" are highlighted. Similarities and dissimalarities between polyketone catalysis by cationic palladium complexes and polyolefin catalysis by cationic metallocene complexes will be outlined.

059. LIVING COPOLYMERIZATION OF PROPYNE WITH ALLENE: ALTERNATIVE ROUTES INTO FUNCTIONAL POLYOLEFINS, Bruce M. Novak* and Mitsuru Nakano, Department of Polymer Science and Engineering and the Materials Research Science and Engineering Center, University of Massachusetts, Amherst, MA 01002

Although the incorporation of polar monomers into coordination polymerization systems has been an active area of research for over forty years, no satisfactory solution to this problem has emerged. Given this reality, we are interested in reaching the same goal by another route, namely, the reliable (co)polymerization of allene to form unsaturated polymers that can be modified in specific ways. We have found that allyl nickel complexes will homo and copolymerize allene in a living fashion to form non-crosslinked unsaturated polymers. Plots of molecular weight as a function of monomer to initiator ratio are linear indicating that chain transfer does not occur and chain extension experiments show that termination is not a dominant factor. Economic viability is introduced into this system by demonstrating that allene can be copolymerized with propyne, its ubiquitous, isomeric form.

060. CATALYTIC ALTERNATING ALKENE-CARBON MONOXIDE COPOLYMERIZATION USING PALLADIUM COMPOUNDS. Ayusman Sen, Department of Chemistry, The Pennsylvania State University, University Park, PA 16802

The presentation will focus on aspects of the palladium(II) catalyzed synthesis of alternating alkene-carbon monoxide copolymers which are of great current interest. The synthesis of chiral, isotactic, alternating α-alkene-carbon monoxide copolymers, as well as alternating copolymers of functional alkenes with carbon monoxide will be discussed. The use of cationic Pd(II) complexes with chiral chelating bisphosphine ligands led to the synthesis of chiral, isotactic, polymers with the degree of enantioselectivity exceeding 90%. The enantiomerically pure, isotactic, alternating α-alkene-carbon monoxide copolymers formed stereocomplexes with α-alkene-carbon monoxide copolymers with opposite chiral sense for the tertiary carbons in the main chain. The alternating copolymerization of functional alkenes, $CH_2=CH(CH_2)_xOH$ (x = 2, 3, 4, 9) and $CH_2=CH(CH_2)_xCOOH$ (x = 1, 2, 4, 8), with carbon monoxide was also achieved.

061. OLEFIN-CARBON MONOXIDE ALTERNATING OPTICALLY ACTIVE COPOLYMERS AND TERPOLYMERS. Simona Bronco, Giambattista Consiglio and Enrico L. Gindro, Eidgenössische Technische Hochschule, Laboratorium für Technische Chemie, ETH Zentrum, CH-8092 Zürich, Switzerland.

Cationic palladium complexes modified with ferrocenyl diphosphine ligands such as [{(R)(S_p)-1-[2-(diphenylphosphino)ferrocenyl]ethyl-dicyclohexylphosphine}palladium(II)bisaquo]triflate give very active catalyst precursors for the regioregular alternating isotactic enantioselective copolymerisation of propene (and other aliphatic olefins) with carbon monoxide. With the mentioned catalyst precursor a reactivity ratio of ~14 was found between ethene and propene for the terpolymerization reaction. The distribution of the two olefins in the terpolymers seems to be statistic. The poly(1-oxo-2-methyltrimethylene) produced has a *llll*-pentad content of about 93% that arises from an enantioface discrimination close to 99%. The high discriminating ability of this catalyst (and previously investigated catalysts) is interpreted on the basis of a two-parameter model based on the enantiomorphic catalyst and the control of the chain end. In fact, the optical activity of the various terpolymers with different propene content (when extrapolated to 100% propene) is, in general, lower than the optical activity of the copolymer.

062. KETONEX™ ALIPHATIC POLYKETONE/COMMODITY POLYMER BLENDS
J.G. Bonner and A.K. Powell, BP Chemicals Ltd., Grangemouth, FK3 9XH, UK.

Ketonex™ aliphatic polyketones are a family of polymers which are prepared by polymerisation of olefins and carbon monoxide in a strictly alternating sequence. They are semicrystalline thermoplastics which have a unique combination of mechanical, high temperature, chemical resistance, wear and barrier properties and are being developed for a range of engineering, barrier packaging and fibre applications. In addition these aliphatic polyketones have been found to be highly compatible with a variety of other polymers, including polyolefins and PVC. Mechanical, thermal and barrier properties of commodity resins can be significantly improved by the addition of aliphatic polyketones.

063.

ALIPHATIC POLYKETONE POLYMERS: PROPERTIES AND APPLICATIONS.
Carlton E. Ash and John E. Flood, Westhollow Technology Center, Shell Chemical Company, Houston, Texas 77251

Aliphatic polyketone polymers (PK) are a new family of engineering polymers commercially available from Shell. Polyketone polymers comprised of alternating C_2H_2 and CO monomers exhibit an excellent range of performance properties combining strength, stiffness, and toughness, with thermal, chemical, wear, and permeability resistance. This breadth of properties will permit their use in a wide variety of applications. The structure and property attributes of PK polymers are briefly discussed along with their utility in a number of potential applications.

064. HIGH COOLING RATE NON-ISOTHERMAL CRYSTALLIZATION KINETICS OF ALIPHATIC POLYKETONES. G. A. Holt, Jr. and J. E. Spruiell, Center for Materials Processing and Department of Materials Science and Engineering, The University of Tennessee, Knoxville, TN 37996

The non-isothermal crystallization kinetics of three aliphatic polyketones were studied using a newly developed light depolarization microscopy technique under cooling rates ranging between 10°C and 3,500°C/min. Both overall crystallization rates and spherulite growth rates were measured. The overall crystallization rates were interpreted in terms of the classical Avrami type theory. The Avrami exponent did not display a cooling rate dependence for the materials.

065. ACRYLATE FORMULATIONS FOR A SOLVENTLESS MAGNETIC TAPE MANUFACTURING PROCESS Matthew M. Ellison, Jin-Yong Huh, Adam M. Power, Jonna B. Purse and David E. Nikles Department of Chemistry and Center for Materials for Information Technology, The University of Alabama, Tuscaloosa, Alabama 35487-0336 dnikles@ualvm.ua.edu

Magnetic tape is manufactured by a continuous web coating process using organic solvents, some of which are on the EPA's list of hazardous air pollutants. Our objective is to replace the organic solvents with a mixture of acrylate monomers, which would be the solvent for the magnetic dispersions, containing commercial iron particles. A combination of particle surface modification, with acrylate coupling agents, and the use of dispersing agents allowed us to prepare solventless dispersions with rheological properties suitable for coating by hand draw-down techniques. The coatings were irradiated with an electron beam to give solid magnetic films. However, the magnetic hysteresis loops showed the coating to have porr magnetic dispersion and orientation, a focus of further research.

066. SAG-RESISTANT HIGHER-SOLIDS BAKED COATINGS WITH MIXED POLYISOCYANATE AND MELAMINE-FORMALDEHYDE CROSSLINKERS. G. Teng, R.P. Subrayan, and F. N. Jones, Coatings Research Institute, Eastern Michigan University 430 W. Forest Avenue, Ypsilanti, MI 48197 and V. Swarup, Exxon Chemical Company, Intermediates Technology, 5200 Bayway Drive, Baytown, TX 77520

Coatings containing two polyols (an oligoester diol and a phenolic ester alcohol) and two crosslinkers (a triisocyanate and a melamine-formaldehyde resin) were studied. Proportions of the four reactants and of catalysts were adjusted with the intent of effecting sequential crosslinking reactions in which all aliphatic hydroxyl groups react with the triisocyanate to form aliphatic carbamate crosslinks and the phenolic groups subsequently react with the melamine-formaldehyde resin to form benzoxazine crosslinks. While actual crosslinking mechanisms can only be inferred from model studies, coatings formulated with this intent had excellent film mechanical properties. Pot life and sag resistance can be adjusted by changing the proportion of soluble tin catalyst, which governs the rate of carbamate formation. Formulations with relatively high (0.1 parts per hundred) levels of soluble tin catalyst had excellent sag resistance by ASTM 4400 - 89a.

067. SCRATCH RESISTANCE OF A SERIES OF CATIONICALLY CURED EPOXY RESINS
Jakob Lange, Alexandre Luisier and Anders Hult*
Department of Polymer Technology, Royal Institute of Technology, S-100 44 Stockholm
*To whom correspondence should be addressed

The scratch resistance of a series of cationically cured epoxy coatings with different crosslink densities is investigated. Three scratch tests, simulating different scratch modes, are used. The effect of adding pigment and wax to the coating on the scratch resistance is also examined. The results show the scratch resistance to increase with increasing crosslink density in all tests. Adding pigment decreased the scratch resistance in tests where the material removed in the scratch process remains between the substrate and the scratching device. In tests where there is no interference from removed material, adding pigment increased the scratch resistance. Wax was found to increase the scratch resistance in tests where a scratch is created in a single draw, but to decrease the scratch resistance when scratching occurs as the result of repeated movement.

068. INFLUENCE OF SPACER LENGTH ON THE RHEOLOGY OF HMHEC COATINGS RHEOLOGY, Melissa A. Grunlan and J. Edward Glass, Department of Polymers and Coatings North Dakota State University, Fargo, ND 58105

Hydrophobe-modified hydroxyethyl celluloses (HMHECs) are prepared with an emphasis on distancing the hydrophobe from the segmentally rigid anhydroglucose repeating units of cellulose. This in concept would serve two purposes, it could facilitate thickening by nonionic surfactants, the primary surfactant in an architectural latex coating, and it could modify the shear thinning behavior of hydroxyethyl cellulose thickened coatings containing small particle latices, arising from depletion floc-culation phenomenon. These observations are realized in aqoueus solutions and are now being examined in fully formulated architectural coatings.

069. INFLUENCE OF LATEX MEDIAN PARTICLE SIZE ON THE RHEOLOGY OF HEUR THICKNED FORMULATIONS, Peter T. Elliott, Wylie H. Wetzel, and J. Edward Glass, North Dakota State University, Polymers and Coatings Dept. Fargo N.D. 58105.

The rheology of model HEUR thickeners were studied in the presence of model latexes. High and low T_g methyl methacrylate based latexes were synthesized with monodisperse particle sizes and varying amounts of oligomeric surface acid. The model HEUR thickeners are based on 8,000, 12,000, and 35,000 M_n POE capped with $H_{12}MDI$ and a $C_{12}H_{25}$ hydrophobe. The latices examined in this study were 198 and 423 nm, both stabilized with methacrylic acid surface segments. The influence of their interaction on the dispersion's rheology and the coatings T_g are examined in this study.

070. *UNDERSTANDING HEUR ASSOCIATIVE THICKENER INFLUENCES IN WATER-BORNE COATINGS*, Wylie Wetzel, Mao Chen, and J. Edward Glass, North Dakota State University, Polymers and Coatings Dept, Dunbar Hall, Fargo, North Dakota 58105

The structural features of associative thickeners influence their viscosifying properties in neat and surfactant containing aqueous solutions and in architectural coating formulations. Proprietary commercial materials with unknown structural features provide an array of performance variations. The study of model or commercial thickeners in aqueous solutions provide general concepts, but they do not provide a quantitative model for understanding the influence of associative thickeners in coating formulations or on applied film properties. Our understanding of the mechanism of thickening in a coatings, based on model associative thickener of the Hydrophobically-modified Ethoxylate URethane (HEUR) type are presented in this study. To provide a more quantitative description, the interactions of associative thickeners with the disperse phases of coatings, i.e., the film former and TiO_2, in the presence of varying surfactant types and concentrations, are considered in this study.

071. EFFECTS OF HYDROPHOBIC INTERACTION ON SWELLING OF CARBOXYLATED CORE-SHELL STRUCTURED LATEX PARTICLES H. Nakamura,and K. Tachi, Toyota Central R& D Labs,Inc.,Aichi, Japan

Carboxylated core-shell particles which contain carboxyl group in shell swell in aqueous dispersion by electrostatic repulsion between carboxylate ions. Effects of hydrophobic interaction on the swelling have been examined by changing hydrophobicities of shell and amine. As hydrophobicity of shell increases, the swelling is suppressed. This means the swelling depends not only on electrostatic repulsion but also on hydrophobic attraction between hydrophobic group of shell and that of shell or core. When the dispersion is neutralized by hydrophobic amine which does not contain hydroxyl group, the swelling is promoted in low degree of dissociation and suppressed in high degree with increasing of hydrophobicity of the amine. This suggests the amine is concentrated near the particles by hydrophobic attraction. This concentrated amine promotes dissociation of carboxyl group, but combines with carboxylate ion electrostatically just like a crosslinker. As hydrophobicity of the amine increases, the dissociation and the crosslink are promoted.

072. ATOMIC FORCE MICROSCOPY STUDY OF INTRAOCULAR LENS IMPLANTS - TOPOGRAPHY AND SURFACE PROPERTIES T. Kowalewski[*] and V.Nathan Ravi[#], [*]Department of Chemistry, [#] Departments of Ophthalmology and Chemical Engineering, Washington University, MO, 63130 and [#]Department of Surgery, V.A. Medical Center, St.Louis.

We used Tapping Mode Atomic Force Microscopy (AFM) to study the surfaces of silicone and PMMA intraocular lens implants. Non-scanning experiments clearly revealed high deformability of silicone as compared to PMMA. By varying the fraction of kinetic energy of the probe transferred to the surface of silicone, we have selectively imaged the highly compressible surface of polysiloxane and less compressible subsurface domains. In the maps of phase shift of oscillating cantilever the polysiloxane-rich, highly viscous, regions exhibited phase shifts larger than the areas located directly above rigid domains. Surfaces of lenses exposed to a model intraocular fluid revealed the presence of aggregated adsorbates and exhibited decrease of compressibility (silicone). Obtained results point to the utility of AFM for the studies of the mechanisms of formation of biofilms on lens implants as a function of surface properties, including surface viscoelasticity.

073. EVALUATION OF AN OPHTHALMIC DRUG DELIVERY SYSTEM: OCCUMER™ POLYMER AND ANIONIC THERAPEUTICS BY ESCA AND AFM. A. D. Westwood, D. J. Leder, and D. H. Donabedian, Union Carbide Corporation, Specialty Polymers and Products, P.O. BOX 670, Bound Brook, NJ 08805.

An innovative approach to obtain a sustained-release drug delivery system would be to incorporate a cationic, mucoadhesive, polymer as a carrier that could deliver anionic therapeutics to the eye surface and release it in a controlled fashion. It has been shown using Atomic Force Microscopy (AFM) and Electron Spectroscopy for Chemical Analysis (ESCA) that anionic therapeutic molecules, such as Hyaluronic acid (HA) and Penicillin G, can complex with the Occumer™ Polymer, which can then deliver these anionic therapeutics to an anionic surface, such as mica. The substantivity of the complex on the anionic surface has been demonstrated, and that after subsequent washing with water the therapeutic is still adhered to the anionic surface via the cationic polymer. This result suggests that drug loss due to natural tear turnover may be minimized allowing for lower drug loading and reduced side effects.

074. SOLUTION AND RHEOLOGICAL PROPERTIES OF THE OCCUMER™ POLYMER FOR USE IN DRY EYE APPLICATIONS. D. H. Donabedian and I. M. Braddock, Union Carbide Corporation, Specialty Polymers and Products, P.O. BOX 670, Bound Brook, NJ 08805.

For topical therapeutics, such as dry eye solutions, the lubricating and wetting agents are quickly washed out to ineffective levels due to natural tear turn over. Union Carbide, has developed a family of cationic, water-soluble polysaccharides (the Occumer™ Polymer series), that are substantive to the surface of the eye and may be useful as drug delivery vehicles for actives such as hyaluronic acid. Clear, colorless, particulate free solutions with viscosities ranging from 5-50 dL/g have successfully been filtered through 0.22μ filters. The rheological properties of these solutions have been evaluated and are similar to natural tears. Compatibility and efficacy of the Occumer™ Polymer:HA solutions with preservatives such as BAC, EDTA, and thimerosol have been completed and all solutions are microbial free. Accelerated aging studies at 40 °C for three months have been performed on these solutions and no appreciable loss in viscosity over this time period has been observed for solutions containing less than 0.10% HA.

075. **DEVELOPMENT OF VISCOELASTIC PROBES FOR UNDERSTANDING THE MECHANISM OF PRESBYOPIA.** V Nathan Ravi, Departments of Ophthalmology and Chemical Engineering, Washington University, and Department of Surgery, Veterans Affairs Medical Center, St. Louis, Missouri, 63110

Presbyopia is the inability of the normal eye to read clearly at close distances. Recurring observations implicate changes in the viscoelastic properties of the lens matrix as one the important facets in the mechanism of this condition. The extent to which changes in lens viscoelasticity contribute to presbyopia is not clear. Our ultimate goal is to understand the molecular mechanism of presbyopia and develop surgical strategies for treatment of this condition. In this paper our preliminary research on the synthesis of hydrogels derived from polyethyleneglycol that could have viscoelastic properties in the physiological range is discussed. We plan to replace the lens matrix with these well-characterized hydrogels or "viscoelastic probes" to provide insight into the mechanism of presbyopia and cataract, and thereby lead to alternative treatments of this condition.

076. **IMMUNOHISTOCHEMICAL ANALYSIS OF COMPLEMENT AND IMMUNE MEDIATORS ON SOFT CONTACT LENSES IN NORMAL AND CONTACT LENS INTOLERANT EYES.** Jean T. Jacob and Jonah J. Lin, LSU Eye Center 2020 Gravier St, Ste B, New Orleans, LA 70112

To determine if the ionicity of the soft lens polymer plays a role in inducing red eyes in contact lens intolerant patients, we looked for differences in Type 1 (allergic) hypersensitivity mediators and polymer-induced complement activation mediators eluted from Type IV (ionic) and Type II (non-ionic) soft contact lenses worn by 5 normal and 5 intolerant contact lens patients in both eyes for 8 hours. Mediators were identified by Western blot using monoclonal antibodies to IgG, IgE, and iC3b. In general, non-ionic lenses sequestered significantly higher amounts of immune and complement mediators than ionic lenses. However, high amounts of IgE eluted from both types of lenses worn by contact lens intolerant patients suggest that intolerant lens wearers may develop red eyes as a result of an allergic response that is independent of the ionicity of the lens material. The finding that ionic lenses apparently sequester a significantly increased amount of iC3b (complement) in intolerant wearers, compared with normals, suggests the possibility of increased sensitivity that could be related to the material itself.

077. **STUDY OF CHEMICAL REACTIONS BY IN-SITU NUCLEAR MAGNETIC SPECTROSCOPY** K. Shridhara Alva, Ashok L. Cholli, Kenneth A. Marx, Jayant Kumar and Sukant Tripathy, Departments of Chemistry and Physics, Center for Advanced Materials, University of Massachusetts Lowell, Lowell MA 01854, USA

Spectroscopic techniques have been extensively used to elucidate the kinetics and mechanism of chemical reactions. In-situ study of structural changes during the chemical reaction will provide vital information in following the mechanism of such a reaction. In this report, we discuss the use of nuclear magnetic resonance spectroscopy to study the biochemical synthesis of polyphenols and polyanilines. Horseradish peroxidase catalyzed oxidative free radical coupling of phenols and anilines in the presence of hydrogen peroxide has been studied under various experimental conditions but little is known about the structural changes during the polymerization. We utilize *in-situ* proton NMR spectroscopy to study the polymerization of some of the water soluble phenol and aniline derivatives. Based on the structural changes, we establish the involvement of ortho and para positions of the hydroxyl group with their relative preferences in the oxidative free radical coupling. For example, in 8-hydroxyquinoline-5-sulfonate, we establish that the positions 2, 4 and 7 are involved in the chemical bonding with the order of preference being $7 \geq 2 > 4$. Similar observation was also made with some of the anilinic derivatives studied. We discuss the novel features of these findings and their importance in the design of polymeric materials will be discussed.

078. **TERNARY HYDROGEN BONDED POLYMER SOLUTIONS: THE SYSTEM POLY(ETHYLENE OXIDE), POLY(n-ALKYL METHACRYLATE) AND 4-ETHYL PHENOL** Michael M. Coleman, Yuhong Hu and Paul C. Painter, Department of Materials Science and Engineering, The Pennsylvania State University, University Park, PA 16802.

Ternary phase diagrams have been experimentally determined at 100°C for systems containing a series of poly(n-alkyl methacrylates), poly(ethylene oxide), and a solvent(4-ethyl phenol) (EPh). A totally miscible phase diagram is experimentally determined for the PMMA/PEO/EPh system, while closed loop diagrams are typically observed for PEMA/PEO/EPh and similar systems. The association model developed in our laboratory has been applied to these ternary polymer solutions. The phase behavior of these ternary systems are predicted to be sensitive not only to the competing physical dispersive forces between two polymer-solvent pairs, but also the competing chemical forces (hydrogen bonds). The balance of these two competing factors controls the miscibility of these systems.

079. CORRELATION AND CONNECTIVITY EFFECTS IN POLYMER SOLUTIONS AND BLENDS
Paul Painter, Lennart Berg, Boris Veytsman and Michael Coleman, Polymer Science Program, The Pennsylvania State University, University Park, PA 16802

Flory-Huggins expression for free energy of mixing is modified to include effects of local screening due to chain connectivity and folding. Assuming that the fraction of screened contacts remains roughly the same in the wide interval of concentrations, we obtained phase diagrams and thermodynamics in a very good agreement with the experimental data. The effect of molecular weight on the fraction of screened contacts as well as the ramifications for thermodynamics of polymer solution and melts are discussed.

080. SPECTROSCOPIES I HAVE KNOWN. J. L. Koenig, Department of Macromolecular Science, Case Western Reserve University, Cleveland, OH 44106-7202

This award address will cover the evolution of vibrational and NMR techniques for the characterization of polymers. In particular, computerized dispersive infrared will be evaluated in the time frame of its utility. Laser excited Raman spectroscopy will be discussed in terms of its evolution and impact on the determination of polymer structure. Solid state NMR will be discussed relative to its impact on our knowledge of the solid state structure of polymers. NMR imaging will be evaluated in terms of the determination of spatially distributed structures. A perspective will be offered of the future in these evolving spectroscopies.

081. THE MOLECULAR ORIGIN OF UNUSUAL PHYSICAL AND MECHANICAL PROPERTIES IN NOVEL PHENOLIC MATERIALS BASED ON BENZOXAZINE CHEMISTRY

Supa Wirasate[1], Sujitra Dhumrongvaraporn[1], Douglas J. Allen[2], and Hatsuo Ishida[2]*,
[1] The Petroleum and Petrochemical College, Chulalongkorn University, Bangkok 10330, Thailand
[2] The NSF Center for Molecular and Microstructure of Composites (CMMC), Department of Macromolecular Science, Case Western Reserve University, Cleveland, Ohio 44106-7202

Abstract
The effect of temperature on hydrogen bonding of polybenzoxazines and novolac phenolic resin is investigated by measuring the integrated infrared absorbance of various hydrogen bonding modes as a function of temperature. It is found that conformationally preferred hydrogen bonding modes maintain constant intensities over the wide temperature ranges studied. In particular, the O---H---N hydrogen bond shows strong bonding that does not change over the temperature range. On the other hand, statistically distributed hydrogen bonding is more sensitive to the temperature change and its infrared intensities start decreasing around the β-transition of polybenzoxazines. The unusual physical and mechanical properties of polybenzoxazines, such as volumetric expansion upon polymerization, high modulus, and high glass transition temperature of polybenzoxazines, are explained based on the complex hydrogen bonding.

082. MOLECULAR STRUCTURE OF INTERFACES BETWEEN PLASMA-POLYMERIZED ACETYLENE FILMS AND STEEL SUBSTRATES. F. J. Boerio and Y. M. Tsai, Department of Materials Science and Engineering, University of Cincinnati, Cincinnati, OH 45221-0012

Thin films of plasma polymerized acetylene deposited onto polished steel substrates are excellent primers for rubber-to-metal bonding. Although reactions occurring at the interface between rubber and plasma polymerized films of acetylene have been characterized, little is known about the molecular structure of interfaces between the films and the steel substrates. The goal of this research was to determine the molecular structure of the primer/substrate interface by using surface analysis techniques such as reflection-absorption infrared spectroscopy (RAIR) and X-ray photoelectron spectroscopy (XPS) to examine films having thicknesses between 20 Å and 900 Å. RAIR spectra of the thinnest films were characterized by a band near 3250 cm^{-1} that was assigned to the CH stretching mode in acetylide species [H-C≡C$^-$], suggesting that the early stages of deposition involved dissociative chemisorption of acetylene. As deposition continued, the band near 3250 cm^{-1} disappeared and a band appeared near 750 cm^{-1} which was possibly related to aromatic species formed in the latter stages of the reaction through recombination of unsaturated groups. XPS indicated that plasma polymerization of acetylene films resulted in reduction of the oxide/hydroxide layer on steel substrates.

083. **Functional Polymers by Atom Transfer Radical Polymerization.** Krzysztof Matyjaszewski*, Simion Coca, Yoshiki Nakagawa, Jianhui Xia. Department of Chemistry, Carnegie Mellon University, 4400 Fifth Avenue, Pittsburgh, PA 15213

Atom transfer radical polymerization (ATRP) has been used successfully for polymerization of various functional monomers such as glycidyl acrylate and hydroxyethyl acrylate, and for the corresponding functional initiators which contain cyano, allyl, ester, hydroxy, epoxide, lactone and vinyl ester groups. Polymers prepared by ATRP with the epoxy and vinyl ester functionalized initiators are macromonomers and can be used for the preparation of the corresponding graft copolymers. The macroinitiators for ATRP yield block copolymers. The terminal halogen atoms of the produced polymers have been replaced quantitatively by nucleophilic and electrophilic reagents.

084. **The Nitroxide Mediated Free Radical Polymerization Process - Block Copolymer Synthesis in the Presence of DMSO, a Mechanistic Understanding** Michael K. Georges, and Nancy A.Listigovers *Xerox Research Centre of Canada, 2660 Speakman Drive., Mississauga, Ontario, Canada L5K 2L1*

The nitroxide-mediated stable free radical polymerization (SFRP) process provides the opportunity to prepared well-defined blocks by a free radical polymerization process. Initial attempts to prepare styrene-based blocks with acrylates and methacrylate proved unsuccessful. It was, however, possible to prepare these blocks when the polymerizations were performed in the presence of dimethyl sulfoxide (DMSO). A possible mechanism for the DMSO based polymerizations is presented along with a discussion of whether this mechanism is still valid in light of our current understanding of acrylate polymerization by the SFRP process.

085. **FUNCTIONAL POLYMERS BY STABLE FREE RADICAL POLYMERIZATION**
Paula J. MacLeod, Michael K. Georges, Gordon K. Hamer, *Xerox Research Centre of Canada, Mississauga, Ontario, Canada, L5K 2L1*

Stable free radical polymerization (SFRP) can be used to prepare existing functional polymers and novel functional polymers. SFRP was used to prepare thermoplastic elastomers of the general structure, polystyrene-*b*-polydiene-*b*-polystyrene. When the diene used was butadiene, the polybutadiene block was found to have essentially the same microstructure as polybutadiene prepared by a conventional radical polymerization. Branching and functionalization of these materials will be discussed. As well, the stable free radical polymerization and copolymerization of monomers such as N-vinylcarbazole will be presented.

086. CONTROLLED SYNTHESIS OF REACTIVE LINEAR POLYMERS AND HYBRID DENDRITIC MACROMOLECULES BY 'LIVING' FREE RADICAL PROCEDURES. A. Desai, E. Malmström, W. Devonport, M. Mate, C. Hawker, IBM Almaden Research Center, 650 Harry Road, San Jose, CA, 95120-6099

The compatibility of nitroxide mediated 'living' free radical polymerizations with synthetically useful functional groups is demonstrated by the preparation of a wide variety of reactive linear polymers. Functionalized monomer units containing substituents such as chloromethyl, hydroxymethyl, amino, active ester, aldehyde and acid chloride can be used while still maintaining control over molecular weight and polydispersity. These well defined reactive linear polymers are then used to coupled with the corresponding monofunctionalized dendritic macromolecules to give hybrid dendritic-linear macromolecules with extremely high molecular weights. This methodology allows the synthesis of large 3-dimensional macromolecules in only a limited number of steps.

087. FUNCTIONAL POLYMERS BY TRANSITION METAL-MEDIATED LIVING RADICAL POLYMERIZATION. Mitsuo Sawamoto and Masami Kamigaito, *Department of Polymer Chemistry, Kyoto University, Kyoto 606-01, Japan*

This paper discusses the controlled synthesis of functional polymers by living radical polymerization of methacrylates and styrene derivatives initiated with a ruthenium(II) complex [RuCl$_2$(OPh$_3$)$_3$] coupled with an alkyl halide as an initiator (see Chart below). Specific targets include:
  (a) *AB- and ABA-Block Copolymers,* by sequential living polymerization of methyl methacrylate (MMA) and butyl methacrylate, ethyl acrylate, or styrene;
  (b) *Living Random Copolymers,* by living polymerization of mixtures of these monomers;
  (c) *Triarmed Star Polymers,* with use of trifunctional initiators carrying dichloroacetate groups for initiation;
  (d) *End-Functionalized Polymers,* with use of functionalized initiators such as Cl$_2$CHCO$_2$CH$_2$CH$_2$OH.

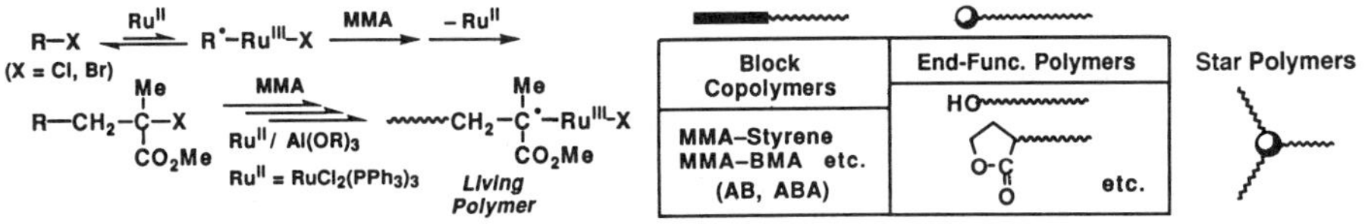

088.
A MODULAR APPROACH TO POLYMER ARCHITECTURE CONTROL VIA CATENATION OF BIOMOLECULAR *LEGO*® SETS. POLYMERS CONTAINING TEMPLATED β-SHEETS. Michael J. Winningham and Dotsevi Y. Sogah. Department of Chemistry, Baker Laboratory, Cornell University, Ithaca, New York 14853-1301

A biomolecular *Lego*® set modular method whereby prefabricated building blocks are linked block by block has been developed and applied to the synthesis of peptide-based polymers containing parallel β-sheets induced by phenoxathiin derivatives acting as reverse turn mimics. Solid state FTIR of the polymers revealed that the expected parallel β-sheets were retained in the polymer. Conformationally unrestricted units incapable of inducing sheet formation provide mostly random coils and contribute to interchain and/or intersheet antiparallel hydrogen bonding. DSC and TGA studies reveal that as the flexibility of the linkers decreases T$_g$ and onset decomposition temperature also decrease. The properties of the polymers are compared with those of silk and nylon. Powder x-ray diffraction of the unoriented polymers show that they are semi-crystalline. Details will be discussed.

PMSE

089. ENGINEERING OF MACROMOLECULAR CRYSTALS: SYNTHESIS AND ANALYSIS OF A LINEAR POLYETHYLENE INCORPORATING FUNCTIONAL "IMPURITIES" AT LONG REGULAR DISTANCES. C. Le Fevere de Ten Hove, A. Jonas, and J. Penelle, Departement de Chimie et Departement de Science des Materiaux et des Procedes, B-1348 Louvain-la-Neuve, Belgium.

This paper will present our preliminary results on a synthetic approach allowing to obtain polyethylene-like macromolecular crystals incorporating chemical defects on every 22nd carbon atom. Thermal properties and structure analysis by WAXS and SAXS will be presented.

090. Functional Polyolefins Ranging from Semicrystalline Plastics to Amorphous Elastomers; Their Applications in Polyolefin Blends and Composites
T. C. Chung, H. L. Lu and Bing Lu, Department of Materials Science and Engineering, The Pennsylvania State University, University Park, PA 16802

This presentation will focus on new functional polyolefins prepared by the reactive polyolefin process, involving reactive p-methylstyrene and borane comonomers and metallocene catalysts with constrained ligand geometry. Both comonomer units in polyolefin are very versatile, not only transformed to a broad range of functional groups, such as -COOH, anhydride, -OH, -NH_2 and expoxyl groups, but also converted to free radical and anionic initiators for graft-from polymerization to produce the graft copolymers with polyolefin backbone and functional polymer side chains. The combination of two reactive comonomer approach, a broad range of functionalized copolymers and terpolymers, with the property ranging from semicrystalline plastic, thermoplastic elastic to completely amorphous elastic, have been prepared. In turn, the functionalized PE and PP copolymers are very useful interfacial reagents which promote the adhesion and compatibility between polyolefins and other materials, such as glass, metals and other polymers.

091. PRODUCING A STABLE ALLOY OF POLYPROPYLENE WITH AN IMMISCIBLE POLYMER USING IN-REACTOR TECHNOLOGY.
A. J. DeNicola and R. A. Fava, Montell Polyolefins, Research and Development Center, 912 Appleton Road, Elkton, MD 21921.

Polypropylene and polystyrene are immiscible polymers and can only be made to coexist as a stable alloy if a compatibilizing agent is present at the interfaces. A polystyrene-grafted polypropylene compatibilizing agent can be created *in situ* by allowing styrene with initiator to diffuse into polypropylene granules and polymerize. The factors affecting this process are discussed. It is shown that the uniformity of the product is sensitive to the morphology of the polypropylene. There is a balance between diffusion and polymerization rates and it is also desirable for the polypropylene to be grossly porous. It is shown that, in this case, a uniform composition distribution through mm-sized, spherical polypropylene particles can be achieved. Properties such as grafting efficiency and molecular weight are shown to be sensitive to reaction conditions.

092. PLASTIC BEHAVIOR OF HOMOGENEOUS ETHYLENE COPOLYMERS FROM THE METALLOCENE TECHNOLOGY, Saïd Elkoun, Valérie Gaucher-Miri, and Roland Séguéla, Laboratoire "Structure et Propriétés de l'Etat Solide", Université des Sciences et Technologies de Lille, 59655 Villeneuve d'Ascq, France.

The tensile drawing of a homogeneous ethylene copolymer from the metallocene technology is studied in comparison with that of a conventional Ziegler-Natta heterogeneous copolymer. The former displays greater propensity for initiation of homogeneous plastic deformation. Two models for homogeneous and heterogenous crystal slip are discussed to account for this phenomenon. The lower value of the most probable crystal thickness of the homogeneous copolymer compared to the heterogenous copolymer, at equivalent crystallinity, is shown to play a major role in the activation of crystal slip. The homogeneous copolymer displays a stronger strain-hardening that is ascribed to a more disordered chain topology. It is suggested that both phenomena of greater propensity for homogeneous crystal slip and higher strain-hardening are responsible for the improved use properties of homogeneous copolymers.

093. TRANSMISSION ELECTRON MICROSCOPY AND SYNCHROTRON SMALL-ANGLE X-RAY SCATTERING STUDY OF DEFORMATION MECHANISM OF LINEAR LOW DENSITY POLYETHYLENE. Andrew J. Peacock, Sharon X. Lu and Gary M. Brown, Baytown Polymers Center, Exxon Chemical Co., 5200 Bayway Dr., Baytown, TX 77522

Despite the fact that linear low density polyethylene (LLDPE) has been known since the 1970s, we still lack fundamental understanding of its deformation mechanism. This may come from its complicated semi-crystalline nature or may be impeded by our pre-conceived ideas based on high density PE. In this work, we have explored the deformation mechanism of LLDPE film using transmission electron microscopy (TEM) and synchrotron small angle X-ray scattering (SAXS). The resin is synthesized using Ziegler-Natta technology and blown into film under typical film blown conditions. Progressive changes in polymer morphology were observed as a function of elongation. Herman's orientation functions calculated from SAXS data analysis suggests crystal lamellae rotate towards the stretching direction above the yield point. TEM pictures reveal the formation of a new crystal morphology in the strain hardening region. A novel deformation mechanism is proposed for LLDPE films.

094. TOUGHENING BEHAVIOR OF RUBBER MODIFIED POLYPROPYLENE.
Reinoud J. Gaymans and Allard van der Wal
University of Twente, PO Box 217, 7500 AE Enschede, The Netherlands

Polypropylene (PP) is a semi ductile material but with a notch it fractures in a brittle manner. PP can be toughened by dispersed rubber particles to very high values 50 kJ/m^2. On extrusion blends are studied the influence of materials, structural and test parameters. The blends are characterized by their fracture toughness and by their brittle-tough transition. Also studied is the development of the deformation zone. The brittle-tough transition in PP is in all cases well above its glass transition temperature (-5°C). With rubber modification the brittle-tough transition temperature is strongly lowered. The brittle-tough transition temperature of the blends is a function of the rubber concentration, the rubber particle size, matrix properties and test speed. Due to the plastic deformation heat is generated and at high speed testing in the fracture plane melting is taking place. A model for the deformation of the rubber toughened PP is proposed.

095. MORPHOLOGY OF THE TRANSCRYSTALLINE INTERPHASE IN POLYPROPYLENE-GLASS COMPOSITES. A. Lustiger, C. N. Marzinsky, R. R. Mueller, H. D. Wagner, Exxon Research and Engineering Co., Annandale, New Jersey 08801.

By coating glass fibers with the appropriate nucleating agent, transcrystallinity can be generated in polypropylene/ glass composites. Transcrystallinity can consist either of the alpha (monoclinic) or beta (hexagonal) crystal structure. Through the use of directional solidification, the transcrystalline morphology can be duplicated in polypropylene films on a level large enough for mechanical and morphological study. Permanganic etching and subsequent electron microscopy reveals that lamellar orientation in alpha transcrystallinity differs significantly from the beta form. Alpha transcrystallinity consists of lamellae which are edge-on relative to the polypropylene film thickness, while beta transcrystallinity consists of lamellae which are primarily flat-on. This difference in morphology results in significant variations in mechanical properties and damage mechanisms.

096. *ZERO AND NEAR-ZERO V. O. C. INTERIOR GLOSSY ARCHITECTURAL COATINGS*

M. J. Collins, J. W. Taylor and R. A. Martin, Eastman Chemical Company - Waterborne Technology Research Laboratory, P. O. Box 1722, Kingsport, TN 37662

Zero and near-zero volatile organic compound (V. O. C.) gloss interior architectural coatings have been studied. Using an experimental emulsion developed at Eastman Chemical Company for zero and near-zero V. O. C. coatings, formulation variables in gloss architectural formulations are explored. The effects of thickener and surfactant on performance characteristics of the coating are discussed. A balance between block and print resistance versus low temperature coalescence of glossy zero V. O. C. coatings is shown. The effect of additives on wet edge at controlled temperature and humidity is investigated and coating film formation at various relative humidities is demonstrated.

097. **WATERBORNE EPOXY RESIN EMULSIONS FOR PRINTED WIRING CIRCUIT BOARD APPLICATIONS**, Jeffrey C. Hedrick, Christy Sensenich, Alfred Viehbeck and Kostantinos Papathomas, IBM Corporation, Thomas J. Watson Research Center, Yorktown Heights, New York, 10598-0218.

Printed wiring circuit boards (PCB) are manufactured by impregnating glass cloth with a laminating resin using a treater tower. During the impregnation process large volumes of organic solvents such as acetone, methyl ethyl ketone, toluene, methoxyethanol, dimethylformamide, etc. are utilized to solvate the resin and control viscosity. A single manufacturing treater tower can utilize over 3000 lbs of solvent per day which must be collected and incinerated. The possibility of utilizing water-based epoxy resin emulsions to fabrication PCB's was investigated. A bromine-containing, flame retardant water-based epoxy resin emulsion was formulated with a water soluble hardening agent and catalyst and the chemistry and cure kinetics were investigated. Prepreg and laminates were fabricated from the waterborne epoxy resin and the thermal, adhesive, and moisture absorption properties were evaluated and compared to a solvent-based epoxy resin control.

098. LATEX BLENDS AS AN APPROACH TO ZERO VOC COATINGS. J. Feng, M. A. Winnik, Department of Chemistry, University of Toronto, Toronto Canada M5S 3H6

Blends of hard latex and soft latex in a common dispersion form void-free films at room temperature in a process which appears to be controlled by the modulus of the soft component. Nevertheless, the film itself is mechanically reinforced by the presence of the hard component, and exhibits an elastic modulus sufficiently high that a homopolymer latex with that modulus could not deform upon drying to form a continuous film. Here we look at aspects of the film formation process. On flat substrates, these films dry initially at the edges, and a drying front propagates from the edges toward the center of the coating. We show that drying of these dispersions occurs much more slowly for the blends than for either pure component. This requires that water evaporation takes more rapidly at the wet-dry interface of the film than from the surface of the water droplet. In some systems, we label the hard and soft components with fluorescent dyes. This allows us to examine the interface between the hard and soft components by energy transfer measurements to assess the intimacy of contact of the two different polymer phases.

099. CROSSLINKERS FOR AQUEOUS POLYMERS, John L. Gardon, Consultant, Bloomfield Hills, Michigan 48301

Most latexes in architectural and related applications are used in an uncrosslinked thermoplastic form. Latexes and water dispersible polymers for industrial baking finishes employ traditional amino resin or blocked isocyanate crosslinkers both active towards hydroxyl groups. New needs for crosslinkers arise for high performance low bake coatings or for markets where the traditional crosslinkers represent toxic hazards. One class of increasingly important crosslinkers is carboxyl reactive and includes aziridines, epoxies, carbodiimides, and zinc amine complexes. Double bonds attached either to resin backbones or to crosslinkers react by the "real" or "pseudo" Michael reaction with carbonyl activated methylene groups or with amine groups. Double bonds can also provide crosslinking sites for U.V. cure or for oxidative reactions. Carbonyl groups are the bases for crosslinking by carbohyrazides or amines. The latest advances in high performance coatings for low temperature applications involve amine cure of aqueous epoxy dispersions and isocyanate cure of aqueous polyols.

100. REACTIVE LATICES: DIFFERENT APPROACHES TOWARDS CROSSLINKABLE WATERBORNE COATINGS.
Bert Klumperman, John Geurts, John Verstegen, Steven van Es, and Anton German
Department of Polymer Chemistry, Eindhoven University of Technology
P.O. Box 513, 5600 MB, Eindhoven, The Netherlands

Two different approaches are presented to develop a crosslinkable waterborne coating. The first system is a so-called "Two Pack in One Pot" system. This system consists of a latex blend of two latices carrying complimentary reactive groups. Due to the stabilization of the latex particles, the crosslinking reaction is inhibited until the film formation stage. Upon film formation, the copolymers carrying the reactive groups get into close proximity and react. Various experimental approaches indicate that the ratio between diffusion rate and chemical reaction rate is extremely important in the control of the homogeneity of the crosslinked material.
The second system is based on photo-acid generators (PAGs) as a method to catalyze the crosslinking reaction by UV light. The feasibility of the basic ingredients of this system was proven. The incorporation in a UV crosslinkable waterborne coating is still under investigation.

101. AMBIENT TEMPERATURE CROSSLINKING ACRYLIC LATEXES. G. Monaghan, Rohm and Haas company, 8901 Research Drive, Charlotte, North Carolina 28262.

Acrylic latexes which are capable of internal crosslinking at room temperature have been introduced. These latexes utilize an oxidative crosslinking mechanism to produce latex paints that can be formulated to have good block, scrub, low temperature coalescence, and household chemical resistance without using external coalescents. This technology offers advantages over conventional latex binders, but paints must be formulated carefully to take full advantage of the crosslinking mechanism. The formulation of the crosslinking latex into solvent-free high quality stain-resistant flat and semi-gloss paints is explored.

102. IN-SITU PHOSPHATIZING COATINGS: WATER-BASED PAINTS. Tao Yu and Chhiu-Tsu Lin, Department of Chemistry, Northern Illinois University, DeKalb, IL 60115

A single-step *in-situ* phosphatizing coating (ISPC) can be developed by pre-dispersing an optimum amount of *in-situ* phosphatizing reagents into the coating formulation. The technique is applied respectively to a low VOC water-reducible alkyd baking enamel and a solventless thermoset acrylic latex system to form compatible and stable coatings. Phosphate chemistry proceeds *in-situ* on steel substrate via an acid-base type of interaction, P-O$^-$---M^{n+}, as evidenced by the observed FTIR spectral peaks at 1066 cm^{-1} and 541 cm^{-1}. The coating properties and protective performances of ISPCs are compared respectively to those of the multi-step control coatings on bare cold-rolled steel, and on pre-phosphated B-1000 testing panels with and without chromate passivating. Comparable curing condition of the ISPC paint films to those of control sample is assured via DSC measurements. Enhanced electrochemical impedance obtained for ISPCs suggest a better corrosion resistance of the coatings. This result is confirmed by the coatings' improved disbonding resistance from the "X" scribe after 144 hours in a 3% salt water immersion test, and in standard salt spray testings. Slower rates of cathodic delamination are indicative for the enhanced adhesion of the ISPCs. The coatings' storage stability is verified by monitoring the rheology behavior. The low VOC ISPCs are shown to provide superior protection of metal surfaces, and can eliminate several process steps, avoid waste generation at the source, and more importantly, be a selfheal of chrome (Cr^{6+}) alternative.

103. SYNTHESIS OF THREE-DIMENSIONAL POLYIMIDES BASED ON THE FLUORENE UNIT FOR COMPRESSIVE STRENGTH STUDIES. Odessa N. Petzold[1], Issifu I. Harruna[1], Kofi B. Bota[1], Derrick R. Dean[2], [1]Department of Chemistry and the High Performance Polymers and Composites Center, Clark Atlanta University, Atlanta, GA 30314. [2]Air Force laboratory, Polymer Branch, Wright-Patterson Air Force Base, OH 45433.

The star-like molecule, 2,7-diamino-9,9-bis(4-aminophenyl)fluorene, was used to effectively control the orientation of linear polyimide chains so as to prepare star-like polyimide systems that are excellent candidates of improved compressive properties compared to the linear analogues. The incorporation of this favorable geometric molecule by condensation with the C-terminus of the linear polyimides resulted in the synthesis of three-dimensional polyimides. Polyimide systems exhibiting improved solubility in organic solvents and strong acids, and transitions at lower temperatures were prepared by introducing hexafluoro groups (bis-4,4'-aminophenylhexafluoropropane and 4,4'-(hexafluoroisopropylidene)diphthalic anhydride), and by using a dianhydride (bicyclo[2.2.2]oct-7-ene-2,3,5,6-tetracarboxylic dianhydride) with reduced symmetry and decreased π–electron density relative to benzene tetracarboxylic dianhydride. An improvement in the solubility of the thermally stable three-dimensional polyimides was observed as a result of a decrease in linearity of the star-like polyimides.

104. MORPHOLOGY AND PHYSICAL PROPERTIES OF POLYIMIDE-INORGANIC HYBRIDS P. R. McDaniel T. L. St Clair

Improvements in high performance polymer systems are motivated by the search for advanced materials with improved or unique properties. Modification by incorporation of a second component of differing chemical structure or composition utilizing the morphology of multiphase systems with phases of differing chemical structure allows for a remarkable balance of diverse properties. Polyimide-titanates have been evaluated and polyimide-silicate hybrids have been produced by four different imidization methods. Each morphology was examined by SEM; thermal properties determined by Thermal Gravimetric Analysis, and Differential Scanning Calorimetry Thermal Mechanical Analysis; along with determining bulk mechanical properties.

105. SYNTHESIS OF HIGHLY REFLECTIVE COMPOSITE POLYIMIDE FILMS VIA A SELF-METALLIZING PROCEDURE UTILIZING (HEXAFLUOROACETYLACETON-ATO) SILVER(I) Robin E. Southward,[1] D. S. Thompson,[1] D. W. Thompson,[1] and A. K. St. Clair[2], Department of Chemistry and Department of Applied Science, College of William and Mary,[1] Williamsburg, VA 23187 and Langley Research Center, NASA,[2] Hampton, VA 23665-5225

Optically reflective composite polyimide films have been prepared by casting a dimethylacetamide solution of silver(I) acetate, hexafluoroacetylacetone, and the poly(amic acid) formed from 3,3',4,4'-benzophenone tetracarboxylic acid dianhydride (BTDA) and 4,4'-oxydianiline (4,4'-ODA) onto an fully imidized parent BTDA/4,4'-ODA base. Thermal curing of the silver(I)-containing poly(amic acid) topcoat leads to imidization with concomitant silver(I) reduction followed by silver(0) migration yielding a reflective (70-80 %), but not conductive, silver surface. This "film-on-film" composite approach minimizes the silver required for the formation of a reflective surface and preserves the essential mechanical and thermal properties of the parent polymer. The metallized films exhibit outstanding metal-polymer and polymer-polymer adhesion, with the strong metal-polymer adhesion attributable to mechanical interlocking. Films were characterized by X-ray, DSC, TGA, XPS, TEM, and AFM.

106. LOWERING COEFFICIENTS OF THERMAL EXPANSION IN SOLUBLE FLUORINATED POLYIMIDES VIA THE *IN SITU* FORMATION OF LANTHANUM-OXO CLUSTERS Robin E. Southward,[1] D. S. Thompson,[1] D. W. Thompson,[1] T. A. Thornton,[1] and A. K. St. Clair[2], Department of Chemistry and Department of Applied Science, College of William and Mary,[1] Williamsburg, VA 23187 and Langley Research Center, NASA,[2] Hampton, VA 23665

The synthesis and characterization of composite inorganic-polyimide films is described using tris(2,4-pentanedionato)-lanthanum(III) dihydrate, La(acac)$_3$(H$_2$O)$_2$, as the inorganic sol-gel precursor and two soluble polyimides formed from 2,2-bis(3,4-dicarboxyphenyl)hexafluoropropane (6FDA) and 1,3-bis(3-aminophenoxy)benzene (3,3'-APB) or 2,2-bis[4-(4-aminophenoxy)phenyl]hexafluoro-propane (4-BDAF). The goal is to lower CTE's without compromising mechanical or thermal properties while maintaining the visual clarity of the polymer films. Hydrolytically and thermally labile eight coordinate La(acac)$_3$(H$_2$O)$_2$ was added to a DMAc solution of polyimide. Cast films were thermally cured during which the La complex underwent hydrolysis to form a nanometer-sized La-oxo phase The CTE's were lowered up to 35 percent.

107. FROM DECACYANOOCTATETRAENE TO POLY(DICYANOACETYLENE). Huinan Yu, Fred Wudl, Institute for Polymers and Organic Solids, University of California, Santa Barbara, CA 93106

Decacyanooctatetraene (DCOT) **1** has been prepared as its dianion and studied as a model compound for the prototypical n-dopable polymer poly(dicyanoacetylene) **2**. Single crystal X-Ray diffraction studies revealed that the DCOT dianion adopts an all *cis* configuration. The backbone forms a helical structure with all the cyano groups pointing away from the center of the helix. Electrochemical studies showed that the *cis* DCOT dianion oxidizes in a single two-electron step to give the neutral DCOT, indicating poor conjugation in the backbone. Also, the neutral DCOT is a very strong oxidizing reagent, whereas its dianion, which could be considered as n-doped poly(dicyanoacetylene), is chemically and thermally very stable.

108.

PREPARATION AND EVALUATION OF POLY (3-PHENYLTHIOPHENES) DERIVATIVES FOR ELECTROCHEMICAL ULTRACAPACITOR APPLICATIONS   Mona M. Eissa, Andrew A. Moxey and John P. Ferraris, University of Texas at Dallas, Box 830688, Richardson, TX 75083.

The growing interest in electrochemical ultracapacitors is stimulated by their potential application for secondary power-storage in electric vehicles. We have been investigating polythiophene derivatives which n and p doped reversibly over a more than two volt range. In this paper, we report on the preparation and the electrochemical characterization of a series of 3-(monofluorophenyl)thiophene, 3-(difluorophenyl) thiophene and 3-(Cyanophenyl)thiophene monomers. Monomer and polymer oxidation and reduction potentials are correlated with the number and position of the fluorine substituents on the phenyl ring. The electrochemical polymerization and charge-discharge properties of the resulting polymers were studied in various organic electrolytes. A strong substituent effect on the polymerization efficiency and on the charge -discharge behavior of the polymer, particularly in the cathodic region, was observed. Performance data of a symmetric supercapacitor based on 3-(monofluorophenyl)thiophene and 3-(difluorophenyl)thiophene is presented.

109. PERFORMANCE OF POLYPYRROLE-BASED FILLER MATERIALS IN GENERATOR STATOR COILS. K. F. Schoch, Jr. and J. D. B. Smith, Westinghouse Science and Technology Center, 1310 Beulah Rd., Pittsburgh, PA 15235 and F. T. Emery, Westinghouse Power Generation Business Unit, 4400 Alafaya Trail, Orlando, FL 32816

Partial discharge activity in voids in the insulation of generator stator coils is a significant cause of premature failure of electrical insulation. Insulating and partially conductive fillers are presently used to minimize void content and eliminate the effects of any remaining voids. When partial discharge activity is present in a coil, the loss tangent of the coil increases over a range of applied voltages from 2 kV to 28 kV. Polypyrrole-coated non-woven materials developed at Milliken Research represent an excellent means of putting a conductive filler in a coil in a controlled manner. This paper describes the results of electrical testing of test bars and coil sections fabricated using this material. Coils made using the partially conducting fillers show reduced partial discharge activity compared to coils made using insulating fillers, with the reduction in partial discharge activity correlating with the conductivity of filler material.

110. RAMAN SCATTERING IN POLYMERIC MATERIALS: A TECHNIQUE COMING OF AGE, Bruce D. Chase; Alan D. Kennedy; DuPont Company, Experimental Station, P.O. Box 30328, Wilmington, Delaware 19880-0328

It has been recognized since the early 1970's that Raman scattering can potentially offer a great deal of information to the polymer scientist. The inherent sensitivity to backbone vibrational modes and the ease of sampling should result in widespread use of Raman measurements on polymeric materials. However, the perpetual problem of background fluorescence inhibited the use of the technique. Within the last decade, the development of both FT-Raman and dispersive CCD based systems with red excitation have dramatically turned this situation around. It is now quite easy to generate high quality Raman spectra on most polymer materials. The development of these new instrumental approaches will be reviewed. Current work on both PET fibers and nylon fibers will be used to illustrate the applicability of Raman scattering to determination of structure and orientation in fiber systems.

PMSE

111. DESIGNING ORGANIC MOLECULES FOR NONLINEAR OPTICS: THE SPECIFIC AND UNIQUE CONTRIBUTION BY SPECTROSCOPY. G. Zerbi, C. Castiglioni, M. Delzoppo and P. Zuliani, Department of Industrial Chemistry, Politecnico. 20133, Milano, Italy.

NLO responses are known to be large for organic molecules which contain pi electrons delocalized over large delocalisation paths. Large delocalisation generates small band gaps and large electron-phonon coupling. We have recently shown the relation between vibrational infrared, Raman and hyper-Raman spectra and NLO properties and have shown and the useful use of such spectra for the direct measure of molecular nonlinearities of polyconjugated low band-gap materials. We will show that by focussing at specific nuclear displacements it becomes possible by experimental and/or theoretical spectra to extract which vibrations are most effective in determining NLO properties. Molecular engineering and synthetic procedures follow aiming at materials with optimal properties.

112. MULTI-DIMENSIONAL SPECTROMETRY AND MAPPING. W. G. Fateley, R.A. DeVerse and R. M. Hammaker, Department of Chemistry, Kansas State University, Manhattan, KS 66506-3701

Improvements in infrared and Raman spectrometers have provided the pathways to new developments in elucidation of structure in molecules in general and specifically in polymeric materials. The use of two-dimensional detectors adds important new information to structural studies. This presentation will include the combination of interferometers and Hadamard transform techniques to perform multi-dimensional spectrometry and mapping.

113. SURFACE AND INTERFACIAL STUDIES OF WATER-BORNE POLYMERS AND COATINGS; ATR AND RPA STEP-SCAN FT-IR SPECTROSCOPIC ANALYSIS. M. W. Urban, Department of Polymers and Coatings, North Dakota State University, Fargo, ND 58105.

Our recent efforts focus on the understanding of molecular level processes and interactions near film-air (F-A) and film-substrate (F-S) interfaces of water-borne polymeric films and coatings. It appears that attenuated total reflectance (ATR) and rheo-photoacoustic (RPA) step-scan (SS) Fourier transform infrared (FT-IR) spectroscopy can be utilized in quantitative analysis of stratification processes occurring near the F-A and F-S interfaces of latex homo and copolymers, water-reducible urethanes, and other environmentally compliant thermosetting and thermoplastic systems. In this paper, fundamental aspects of these surface selective and sensitive techniques will be addressed, along with a molecular level analysis of adhesion, kinetics and diffusion, and the effect of the environment on crosslinking reactions. A particular emphasis will be given to quantitative aspects of analysis in a context of surface depth profiling experiments.

114. TOWARDS A BETTER UNDERSTANDING OF SEGMENTAL MOBILITY IN POLYMERS UNDER EXTERNAL PERTURBATIONS: TIME-RESOLVED FTIR-STUDIES. A. Ameri, S. Ekgasit, C. Hendann, S. Michel, S. Okretic, F. Pfeifer, P. Wu, I. Zebger and H. W. Siesler, Department of Physical Chemistry, University of Essen, D 45117 Essen, Germany

The elucidation of the mechanisms and dynamics of segmental mobility in polymers under the influence of external perturbations (electric, electromagnetic or mechanical fields) can tremendously contribute to the improvement of certain technical processes or to the optimisation of the end-use properties of technically important polymeric materials. Thus, in the case of liquid-crystalline polymers, for example, the influence of an external electric or electromagnetic field is the basis of liquid-crystal displays and information storage, respectively. Application of a mechanical field, on the other hand, is a wide-spread pretreatment of a polymeric material (e.g. a film) to improve its mechanical end-use properties. Selected examples will demonstrate the potential of time-resolved FTIR-spectroscopy to investigate the addressed topics.

115. SOLID-STATE ^{13}C NMR STUDIES OF CHANGES IN CROSS LINKED CARBON STRUCTURE DURING HEATING UNDER AIR AND NITROGEN ENVIRONMENTS Dallas D. Parker and Jack L. Koenig, Department of Macromolecular Science, Case Western Reserve University, Cleveland, Ohio 44106-7202

In the breakdown of natural rubber networks at higher temperatures, both thermal and oxidative factors interact together resulting in "thermo-oxidative" degradation. The changes that occur in sulfur crosslinks during thermo-oxidative degradation of sulfur cured unfilled natural rubber were investigated using ^{13}C NMR analysis. In the NMR spectra, the crosslinked carbons formed during the vulcanization process appear as peaks whose chemical shift corresponds to the various possible crosslink structures formed. Samples of unfilled vulcanized rubber were heated at 100 °C and 150 °C under both air (oxygen) and nitrogen environments. Changes were observed by measuring peak areas under the small crosslinked carbon peaks using the main chain carbon peaks as reference. Changes in the rate of decomposition of the different peaks were found under the varying conditions as well as the rate of *cis-trans* isomerization in the main carbon chain.

116. N-DOPABLE POLYMERS, F. Wudl, Y. Greenwald, M. Fourmigue, H. Yu Departments of Chemistry and Materials, University of California, Santa Barbara, CA 93106

Recent developments in the synthesis and property evaluation of n-dopable conducting polymers at Santa Barbara will be presented. Investigations into poly(3,4-dicyanothiophene), poly(cyanoacetylene-co-acetylene) and a poly(dicyanoacetylene) oligomer will be presented. An all organic polymeric diode fabricated from poly(3,4-dicyanothiophene and MEH-PPV will be described.

117. **POLY(P-ARYLENEVINYLENE)S VIA FUNCTIONALIZED POLY(P-XYLYLENE)S: SYNTHESIS, PROPERTIES, AND POLYMER ANALOGOUS REACTIONS. A. Greiner, O. Schäfer, Philipps-Universität Marburg, Fb Physikalische Chemie, 35032 Marburg / Germany**

High molecular weight, film forming, Br- and Cl-functionalized PPXs were obtained by vapor phase pyrolysis of dihalogeno-functionalized p-xylenes. PPX was obtained as side product with α,α´-dihalogeno-p-xylenes but not with α,α-dihalogeno-p-xylenes. The Cl-functionalized PPXs were transformed to poly(arylenevinylene)s by annealing between 150°-300°C. PPV prepared below 200°C represents a glass transition temperature at 220°C followed by a fast crystallization regime. The deposition of functionalized PPXs onto ITO layers for LED cell production is limited by corrosive HHal liberated during pyrolysis. The problem was solved by replacing ITO by gold.

118.
Poly(*p*-phenylenevinylene) in Light Emitting Diodes. S. Son, F. Papadimitrakopoulos and M. Galvin, Bell Laboratories, Lucent Technologies, Murray Hill, NJ 07974.

Poly(*p*-phenylenevinylene) (PPV) is frequently used as the emissive layer in polymeric light emitting diodes (LED's). It's performance in these devices is very dependent on what synthetic route is used to produce it. When the Wessling route is used, hydroxyl substituents in the precursor polymer can result in carbonyl formation in PPV. These carbonyl moieties reduce the photoluminescence and the electroluminescence of PPV. An alternative route which utilizes a xanthate precursor to PPV has resulted in better performance of LED's. This talk will describe this research.

119. DESIGN CONCEPTS FOR SOLUTION DEVELOPED 193 NM LITHOGRAPHIC MATERIALS. E. Reichmanis, O. Nalamasu, T. I. Wallow, F. M. Houlihan, A. E. Novembre, Bell Laboratories, Lucent Technologies, 600 Mountain Avenue, Murray Hill, NJ 07974

The intense absorption of aromatic molecules at 193 nm severely limits the use of conventional matrix resins such as novolacs and polyvinylphenols, for 193 nm lithography. This has both necessitated a paradigm shift in the approach to lithographic materials and process design and spawned the design of revolutionary resist schemes. In recognition of the strong motivation to retain as much of the acquired knowledge base regarding solution developed resists as possible in the design of materials for advanced lithographic applications, many of the current research efforts related to 193 nm lithographic materials involve the design of new chemistries that provide for aqueous base solubility, etching resistance, resolution, photospeed and process latitude. In this paper, we will identify novel functionalities that can lead to robust, aqueous base developed 193 nm resist materials. For instance, a new class of polymers based on cycloolefin-maleic anhydride copolymers and terpolymers were synthesized, their aqueous base dissolution characteristics evaluated, and imaging performance determined. The effect of various polymer functionalities on system performance will be discussed.

120. DESIGN, PREPARATION, CHARACTERIZATION, AND LITHOGRAPHIC IMAGING OF HIGH PERFORMANCE CHEMICALLY AMPLIFIED DEEP UV RESISTS, H. Ito, G. Breyta, D. Fenzel-Alexander, D. Hofer (IBM Almaden Research Center, 650 Harry Road, San Jose, CA 95120), W. Conley, B. Brunsvold, R. Nunes (IBM Microelectronics Division, Hopewell Junction, NY 12533), P. Hagerty, J. Thackeray (Shipley Company, 455 Forest Street, Marlboro, MA 01752)

This paper describes the design concept, preparation, and lithographic characterization of our environmentally stable chemical amplification positive resist called ESCAP, upon which the IBM/Shipley deep UV resist alliance has built commercial products UVIIHS and UVIII. The environmental stabilization is achieved by reduction of the free volume in the resist film by annealing, which demands a robust resist resin and acid generator. To minimize evaporation and diffusion of an acid generator and photochemically generated acid during the high temperature bake processes, a bulky acid generator which produces a bulky camphorsulfonic acid is employed. The good annealing and the use of a bulky acid minimize acid diffusion and provide excellent lithographic performance. As the aqueous base development step is crucial in converting the sinusoidal aerial image to a step function, we have extensively studied the dissolution behavior of the exposed resist film using a quartz crystal microbalance, which has revealed an extremely high developer selectivity of fifteen.

121. SYNTHESIS AND APPLICATION OF POLYMERIC QUENCHERS FOR LITHOGRAPHY. D. Bi, H.F. Evans, M.J. Fitzgerald, R.C. Liang*, S.S. Pang, W.C. Schwarzel, X. Wang and T.F. Yeh, POLAROID CORPORATION, Cambridge, Massachusetts 02139

Poly-(ethylene)-co-(maleic anhydride) [EMA] was reacted with the nitroxyl radical, 4-amino-TEMPO and 2-aminoethyl sulfonic acid (taurine) in water to give amide and ammonium ion substitution. ^{13}C NMR analyses of this polymer as prepared, after dialysis, and after reduction and dialysis confirmed amide formation, which was quantified by HPLC and elemental analysis. Significant improvement in rate of development, background cleaning, ink receptability, contrast and resolution of the image has been demonstrated when this polymer was used in the aqueous overcoat of a lithographic printing plate. This polymeric quencher appears to inhibit free radical polymerization at the image coat/overcoat interface, efficiently suppressing the detrimental effects of intermixing. It also appears to be a good contrast enhancing material which suppresses the undesirable photoreaction induced by scattered light.

122. PHOTOINITIATED CROSS-LINKING OF EPOXY-FUNCTIONALISED POLYISOPRENE C. Decker, T. Hoang Ngoc, Laboratoire de Photochimie Générale (URA-CNRS n°431) - Ecole Nationale Supérieure de Chimie - 3, rue Werner - 68200 Mulhouse (France)

Epoxy-functionalized polyisoprene has been cross-linked by UV-irradiation in the presence of a cationic photoinitiator. The reaction kinetics has been monitored by IR spectroscopy and shown to proceed extensively within less than 1 s upon intense illumination. From the conversion dependence of the insolubilisation process, it was concluded that both inter- and intramolecular propagation reactions occur during the polymerization of the epoxy ring. Blends of epoxidized polyisoprene and difunctional vinyl ether or acrylate monomers were shown to undergo a very fast and extensive cross-linking polymerization, with formation of interpenetrating polymer networks. The UV-cured films are both hard and flexible, and show therefore a good resistance to abrasion and scratching.

123. **POLYMERS WITH ARYLAZOPHOSPHONATE UNITS APPLICATION FOR LASER ABLATION**
O. Nuyken*, C. Scherer, M. N. Nobis, Technische Universität München,
Lehrstuhl für Makromolekulare Stoffe, Lichtenbergstr. 4, D-85747 Garching, Germany
J. Ihlemann, Laser-Laboratorium Göttingen e.V.,
Hans-Adolf-Krebs-Weg 1, D-37077 Göttingen, Germany

The laser ablation of polymer films is a promising lithographic technique in opposite to the conventional ablation procedures. Polymers are suitable for laser ablation if they absorb laser light and disintegrate into small gaseous products after irradiation. Here we report about a new class of photosensitive polymers with azophosphonate units in the main chain, which are applied to laser ablation for the first time. The poly(arylazophosphonate)s were prepared by interfacial polycondenzation of bifunctional phosphoric diesters and diazoniumsalts. The resulting polymers show excellent film forming properties and were exposed to a pulsed XeCl excimer laser. The optical resolution can be demonstrated with REM images of the microstructured polymer films. The quality of the microstructures makes the poly(arylazo-phosphonate)s suitable fore further complex pattering experiments.

124. **ENTROPY AND FRAGILITY IN CHAIN POLYMERS AND THE ΔC_p PROBLEM.** C. A. Angell, Department of Chemistry, Arizona State University, Tempe, Arizona 85287-1604

The fragility of glassforming systems, which is correlated strongly to the magnitude of ageing affects below T_g, is related to the rate per K at which the liquid (or rubber) is approaching its configurational ground state, *i.e.* the rate at which its excess (non-vibrational) entropy is "running out." For simple liquids, it is usually correlated to the increase in C_p normalized to the glass value at T_g. The most fragile glassformers identified to date have been chain polymers, PVC being the extreme case (for structural reasons which are far from clear). For chain polymers, however, the change of heat capacity at T_g relative to the glass value of T_g is small, ~10% compared to ~80% for fragile molecular liquids. We examine the origin of this difference and discuss its relevance to the vibrational dynamics of amorphous systems.

125. DYNAMIC PROPERTIES OF POLYMER MELTS ABOVE THE GLASS TRANSITION: MONTE CARLO SIMULATION RESULTS. J. BASCHNAGEL, K. OKUN, M. WOLF-GARDT, E. ANDREJEW, AND K. BINDER, *Institut für Physik, Universität Mainz, 55099 Mainz, Germany.*

Dynamic properties of a glassy polymer melt are studied by a Monte Carlo simulation of the bond-fluctuation (lattice) model. The chains interact by a hard-core potential and possess an internal energy which makes them to become stiffer during supercooling. Therefore a competition between energetic and packing constraints develops at low temperatures, which causes the glassy behavior of the model. The structural relaxation of the melt is studied over the whole temperature range from the high to the (strongly) supercooled state by means of various mean-square displacements, the self-diffusion constant of the chains, the end-to-end vector correlation function, the intermediate incoherent scattering function, the correlation functions of the Rouse-modes, etc. This analysis provides evidence for the time-temperature superposition principle and for mode-coupling theory. Additionally, it shows that the structural relaxation time increases in a Vogel-Fulcher-like fashion, yielding a Vogel-Fulcher temperature of about $T_0 \approx 0.12 - 0.13$, whereas the critical temperature of mode-coupling theory is $T_c \approx 0.15 > T_0$.

126. THE USE OF CONFIGURATIONAL ENTROPY TO DERIVE THE KINETIC PROPERTIES OF POLYMER GLASSES. Edmund A. Di Marzio, National Institute of Standards and Technology, Gaithersburg, MD 20899

We derive from first principles the viscosity $\eta(\omega,T)$, the diffusion coefficient $D(\omega,T)$ and dielectric response $\epsilon(\omega,T)$. The approach examines the topology of the potential energy surface in configuration space and the motion of the phase point in this space. This configuration space is a mix of deep and shallow wells separated by energy barriers. We use a trapping model for escape from the deep wells and use the principle of detailed balance to relate kinetic and equilibrium quantities. We obtain $\text{Log}\eta(0,T) = B - AG_c/kT$. **Thus, although the vanishing of the configurational entropy S_c locates the glass transition in temperature-pressure space the configurational free energy G_c determines the zero frequency viscosity**. Our viscosity equation is to be compared with the Vogel-Fulcher equation $\log\eta(0,T) = B + A/(T-T_c)$. We correctly predict the main features of the relaxation response in a homologous series of Polydimethysiloxane.

127. FAILURE AND DEFORMATION STUDIES OF SYNDIOTACTIC POLYSTYRENE. M. A. Jones and C. J. Carriere, Central Research and Development, The Dow Chemical Company, Midland, MI 48674

Syndiotactic polystyrene (sPS) is a new chemically resistant, semi-crystalline polymer. The research reported in this presentation was aimed at developing a basic understanding of the failure mechanism and toughness of sPS. This work included investigations of the effect of molecular weight, as well as flow-induced anisotropy. Scanning electron microscopy (SEM) was used to aid in the determination of the failure mechanism. Syndiotactic polystyrene was observed to fail with a slow, controlled crack growth and ruptured with an almost non-detectable amount of yielding. The proposed failure mechanism, based on SEM micrographs, is one of constrained crazing, followed by void coalescence with the spherulite nucleators acting as stress concentrators in the system. The damage appears to be greatly confined, with little initial cold-drawing of the spherulites. Addition of nucleator reduces the fracture toughness somewhat, as added nucleation sites proliferate the sites for stress concentration across the sample.

128. AFM STUDY OF SURFACE TOPOGRAPHY OF GLASSY POLYMERS. T. Kowalewski and J. Schaefer, Department of Chemistry, Washington University, St. Louis, MO 63130.

Tapping Mode Atomic Force Microscopy (AFM) has been used to map the nanoscale topography of surfaces of a variety of glassy polymers solidified by different methods (spin-coating, solvent casting, melt solidification). Quantitative characterization of AFM topographs was performed with the aid of Fourier analysis. We have determined that over the scale range from ~ 5 nm to several micrometers the surfaces of glassy polymers are not fully random. They are characterized by the presence of "grains", with lateral sizes in the range of 5-10 nm and height of about 0.25 nm (based on root mean square surface roughness). The "grains" tend to form "clusters of clusters of clusters". Based on the analysis of power spectra of 2-dimensional Fourier transforms of AFM topographs the lateral sizes of those aggregates follow a simple $1/f^\beta$ scaling law where f denotes the spatial frequency and β ranges from 0.7 to 1.0. We will discuss the possible relation of such surface morphology to bulk organization of glassy polymers and its implications for our understanding of the nature of glass transition.

129. TEMPERATURE-MODULATED CALORIMETRY OF THE FREQUENCY DEPENDENCE OF THE GLASS TRANSITION OF POLY(ETHYLENE TEREPHTHALATE) AND POLYSTYRENE. B. Wunderlich and I. Okazaki, Department of Chemistry, The University of Tennessee, Knoxville,Tennessee 37996-1600 and Oak Ridge National Laboratory, Oak Ridge, Tennessee 37831-6197

Temperature-modulated differential scanning calorimetry, TMDSC, is a new technique that permits to measure the apparent heat capacity as a function of modulation frequency. The method is briefly described and a quasi-isothermal measurement method is used to derive the kinetic parameters for poly(ethylene terephthalate) and polystyrene. A first-order kinetics expression was used to describe the approach to equilibrium and point out the limits caused by asymmetry and cooperativity of the kinetics. The use of a complex description of heat capacity and entropy is discussed. Activation energies vary from 75 to 350 kJ/mol, dependent on thermal pretreatment of the samples and the preexponential factor is correlated with the activation energy.

130. POLYMER-POLYMER MISCIBILITY INVESTIGATED BY TEMPERATURE MODULATED DIFFERENTIAL SCANNING CALORIMETRY. G.O.R. Alberda van Ekenstein, G. ten Brinke, Laboratory of Polymer Chemistry, University of Groningen, The Netherlands; T.S. Ellis, Generals Motors Research and Development Center, Polymers Department, Warren, MI 48090-9055

Enthalpy relaxation in the amorphous state of glassy ans semi-crystalline polymers has been shown be a valuable tool for elucidating phase phenomena in polymer blends. When characterized by lorimetric methods, as an enthalpy recovery peak, mixtures of polymers possessing indistinguishable or nilar thermal properties can be analyzed quite effectively. Recent instrumentation developments, in the m of modulated or oscillating heating profiles (MDSC) have added considerably to this approach. In this per we will review the main features of the technique, using both modulated and conventional DSC, and amine the additional information available from MDSC.

131. VOLUME RECOVERY AND PHYSICAL AGING IN GLASSY POLYCARBONATE FOLLOWING TEMPERATURE JUMPS. Carl R. Schultheisz and Gregory B. McKenna, Polymers Division, National Institute of Standards and Technology, Gaithersburg, Maryland 20899 USA

When subjected to temperature changes in the neighborhood of the glass transition (T_g), amorphous polymers exhibit evolution of both structure (as characterized by the volume or enthalpy) and mechanical response (physical aging). Understanding and modeling these changing properties is important for predicting the dimensional stability and mechanical behavior of injection-molded thermoplastics (as well as other polymer systems). We have performed dilatometric experiments on polycarbonate using temperature-jump histories, and modeled the results using a nonlinear volume response incorporating an internal or material time clock that depends on the current volume. In down-jumps close to T_g, a direct comparison between the evolution of the volume and the mechanical responses (measured in separate torsion experiments) indicates that the mechanical properties cease evolving (reach equilibrium) before the volume does. These results are similar to behavior seen previously in this laboratory with epoxy glasses near to the glass transition. We interpret the results to imply that each material process has a different fictive temperature.

132. IONIC CONDUCTIVITY IN GLASSY PVA/LITHIUM SALT SYSTEMS. M. Forsyth, H.A. Every, F. Zhou and D.R. MacFarlane, Department of Materials Engineering and Department of Chemistry, Monash University, Wellington Road, Clayton, Victoria, Australia 3168

Ionic conductivity has been observed in glassy PVA/lithium salt complexes. ^{7}Li solid state NMR spectroscopy has shown that lithium ion motion is present even at temperatures well below the DSC glass transition temperature suggesting that at least in part, lithium ion motion is responsible for the ionic conductivity and that this motion is decoupled from the main chain polymer segmental motions. The degree of hydrolysis of the PVA polymer appears to influence the level of lithium ion mobility and consequently the conductivity, with the 88% hydrolysed sample giving almost and order of magnitude higher conductivity than the 99% PVA/salt systems. FTIR spectroscopy indicates only weak interactions between the salt and the hydroxyl groups.

133.

RESPONSIVE GELS AND PHASE TRANSITION
Toyoichi Tanaka, Physics Department, MIT, Cambridge, MA 02139

Marvelous functions of proteins such as memory of conformation, molecular recognition, catalysis, and motions are a result of rich variety of molecular interactions between monomers. Recent theoretical studies indicate that heteropolymers, made of monomers capable of different interactions, can generally take four distinct phases: (1) swollen and fluctuating (gas-like phase), (2) collapsed and fluctuating (liquid-like phase), (3) frozen in a degenerate conformation (glassy phase), and (4) frozen in a unique conformation (crystal-like phase). Experimental studies of phase behavior of heteropolymer gels and possible usage of them for molecular recovery and other applications will be discussed. A suggestion will be given for synthesis of polymers that can recognize a given target molecule.

134. MIMICKING THE SMART POLYMER MATRIX OF THE SECRETORY GRANULE: A NEW CONCEPT IN CONTROLLED RELEASE DRUG DELIVERY. Patrick F. Kiser[‡]*, David Needham* and Glynn Wilson[‡], [‡]Access Pharmaceuticals Inc., Dallas, TX 75207; *Department of Mechanical Engineering and Materials Science, Duke University, Durham, NC 27708

Regulated secretion of intercellular signaling molecules is controlled by specialized secretory cells. These cells synthesize secretory granules composed of a poly-anionic condensed polymer matrix encapsulated within a lipid membrane. The condensed matrix functions as a storage vehicle for the stable encapsulation of small molecules and proteins and as a triggered release vehicle through an exocytosis induced polymer phase transition. We have begun to systematically engineer a multi-component smart polymer drug delivery system which can closely mimic the properties of the secretory granule and can be used for parenteral and non-parenteral triggered burst release of drug. The system is composed of a drug loaded condensed phase poly-electrolyte particle which has been coated with an adhered ion-impermeable lipid membrane. This work describes the assembly of an environmentally responsive hydrogel microparticle encapsulated with a supported membrane which mimics many properties of the secretory granule such as a high drug loading capacity, swelling, and burst release of drug.

135. NOVEL LYOPHILIC COLLOIDS AND SOLUBLE POLYMER COMPLEXES IN DRUG DELIVERY: AN OVERVIEW. A.V. Kabanov, University of Nebraska Medical Center, Omaha, NE 68198-6025.

One major trend in pharmaceutics that has emerged within the past several years is the use of lyophilic colloids and soluble polymer complexes for drug delivery. The systems are designed using self assembly principles to (*i*) carry the drug to the target in the body, (*ii*) transport the drug into the cell, (*iii*) and direct the drug into the intracellular compartment. An overview of these systems includes: (A) Micelles of amphiphilic block copolymers designed to target drug across blood-brain and intestinal barriers as well as treat multiple drug resistant tumors. (B) Block ionomer complexes formed between polynucleotides and cationic block copolymers for gene delivery. (C) Water soluble stoichiometric complexes formed between block polyelectrolytes and oppositely charged surfactants. (D) Biospecific surfactants that alter morphology through interactions with specific ligands. The formation, solution behavior and transitions as well as the relationship between structure and biological activity with these systems *in vitro* and *in vivo* are discussed.

136. SYNTHESIS OF SUCCINIC ACID/PEO HYDROGELS AND THEIR USE AS CONTROLLED RELEASE VEHICLES. J. Elisseeff, W. McIntosh and R. Langer, Harvard/MIT Div. of Health Sciences and Technology and Dept. of Chemical Engineering, Massachusetts Institute of Technology, Cambridge, MA 02139

Hydrogels based on succinic acid and poly(ethylene oxide) were synthesized. Succinic acid was substituted using methacrylic anhydride. Substitution was confirmed by loss of the wide acid peak at 3100cm^{-1}. The dimethacrylated succinic acid (SAD) was subsequently photopolymerized with dimethacrylate PEO (PEOD) in varying proportions to form hydrogels. The equilibrium swelling volume (Q) of the hydrogels decreased as SAD concentration increased. To test for the ability of these hydrogels to release drugs such as proteins, a model protein drug, albumin, and a model low molecular weight drug, rhodamine, were used. The hydrogels showed a fast initial release of bovine albumin for 10 days after which release slowed. As SAD concentration in the hydrogels increased the percent of albumin released at day 10 decreased. After the initial 10 day release, the hydrogels with higher SAD concentrations continued to release albumin at an average rate of 1% per day for up to 40 days. Hydrogels of 17% and 21% SAD released 70-90% of encapsulated rhodamine during the initial 10 day fast release as measured by flourimetry.

137. **TEMPERATURE-SENSITIVE DRUG DELIVERY SYSTEM COMPOSED OF POLY (N,N-DIMETHYLAMINOETHYL METHACRYLATE-CO-ACRYLAMIDE).** S. H. Yuk, S. H. Cho, and H. B. Lee, Biomaterials Laboratory, Korea Research Institute of Chemical Technology, P. O. Box 107, Yusung, Taejeon 305-600, Korea

Copolymers of N,N-dimethylaminoethyl methacrylate (DMAEMA) and acrylamide (AAm) were prepared to demonstrate a temperature-sensitive drug delivery system. Poly DMAEMA has a lower critical solution temperature (LCST) around 50 °C in water. With copolymerization of AAm with DMAEMA, the LCST shifts to the lower temperature were observed, probably due to the formation of hydrogen bonds between amide and N,N-dimethylamino groups. FT-IR studies clearly show the formation of hydrogen bonds which protect N,N-dimethylamino groups from exposure to water and result in a hydrophobic contribution to the LCST. Using this copolymer system, the swelling and drug release pattern in response to temperature were examined.

138. **A COMPLEXATION/DECOMPLEXATION MECHANISM IN pH-RESPONSIVE COPOLYMER NETWORKS** A.M. Lowman and N.A. Peppas, School of Chemical Engineering, Purdue University, West Lafayette, IN 47907-1283.

Copolymer networks which exhibit pH-dependent swelling behavior due to the formation of interpolymer complexes have been synthesized. At low pH, these networks remain relatively unswollen due to the formation of interpolymer complexes. In aqueous media of pH greater than the pK_a of the network, the complexes are broken, and the network swells to a high degree. These materials have the ability to respond rapidly to changes in environmental pH. Additionally, a model has been developed to describe the mechanism of complex formation in copolymer networks. This analysis accounts for the effects of the copolymer structure and composition as well as the pH, ionic strength and temperature of the swelling agent.

139. GEL-COATED CATHETERS AS DRUG DELIVERY SYSTEMS. S. H. Gehrke, J. F. McBride, S. M. O'Connor, H. Zhu, J. P. Fisher, Department of Chemical Engineering, University of Cincinnati, Cincinnati, OH 45221

Medical catheters are often coated with hydrogels to increase lubricity to aid insertion. This coating can absorb therapeutic agents which then will be released gradually during use. We have developed techniques for increasing sorption of large molecules such as proteins above that which can be achieved by simple soaking the catheter in a solution of the macromolecule. The technique is based on the thermodynamics of protein partitioning in two-phase aqueous polymer solution extraction. Furthermore, we have shown that release of drugs from currently marketed catheters occurs by a diffusion-controlled mechanism. However, using catheter models coated with environmentally responsive gels it is possible to significantly modify the release kinetics such that drug release can be triggered at a desired location upon application of a triggering stimulus.

140.

INDUSTRY-UNIVERSITY INTERACTIONS. James M Pearson, Eastman Kodak Company, Research Laboratories 6/83, Rochester, NY 14650-2211

This presentation will explore the role of the Research Universities in providing science and engineering talent and knowledge to US Industry. These research and education functions were largely made possible by generous Federal funding to the universities over the past half century. Recent budget pressures at both the Federal and State level, however, are forcing a re-examination of this investment strategy. Significant changes have also taken place in US Industry as a result of intense global economic and technological competition. Industrial R&D has become highly focused and strategically driven. In both cases the sponsors of the research are seeking value from their financial investment.

I will examine how these recent changes are impacting University-Industry interactions. A number of interaction models, e.g., consortia, centers, one-on-one, will be evaluated, and research strategies will be discussed in light of today's industrial needs and constraints.

141. SUCCESSFUL INDUSTRY-UNIVERSITY COLLABORATION. M. Jaffe, Hoechst Celanese Corp., 86 Morris Avenue, Summit, NJ 07901

While measurement of success in industry-university collaborations has proven elusive, essential elements of an effective collaboration can be identified. These include accurate problem definition, willingness to experiment and risk failure, a shared mission and vision, the retention of each organization's cultural strengths, responsive and sustained support by management, trust between the participants and mutually beneficial outcomes. It is the different perception of "benefit" by industry and academe ("business value" versus scholarship) that confounds the definition of success. Other confounding issues include ownership of intellectual property, "unintentional" technology transfer, and the ever present difficulty of effective communication of results (internally and externally). In an environment of shrinking technical assets in industry, changing funding patterns in the university, and a shifting of government science goals from defense to commercial, it is critical to all sectors that these issues are understood and effective strategies for collaboration be established.

142. CREATION OF VALUE ACROSS THE ACADEMIC INDUSTRIAL INTERFACE.
J. Spanswick, BRIDGES Technology Innovation Capitalization, Wheaton, IL 60187

Technology Vision 2020: Report of the US Chemical industry indicated that the current level of industrial R&D may be insufficient to sustain a competitive edge but that partnership can help leverage R&D resources. The industrial/Academic partnership is a long standing relationship that has changed over the past 25 years and will continue to evolve in the future. I would like to draw on 30 years experience in industry working with academic institutions in the US and Europe and my recent interaction from the academic side assisting in setting up the ATRP Research Consortium at CMU to identify what works and what does not. Some questions addressed are:

what are the critical requirements for successful interaction/collaboration? How do you link collaboration/funding and results? Funding alone does not work. What type of partnerships are required in the future as constraints are placed on academic funding just as it has been placed on industrial research? What types of consulting relationships are best suited to industrial project needs? What type of collaboration will work in the future?

143. MONITORING MACROMOLECULAR REACTIVITY BY SMALL MOLECULE SIMULATION.

J.I. Emert, D.C. Dankworth, A. Gutierrez, D.J. Martella, S.P. Rucker, J.E. Stanat, and J.P. Stokes, Exxon Chemical Co., P.O. Box 536, Linden, NJ 07036

H.K. Hall, Jr. and A. Worku, C.S. Marvel Laboratories, Department of Chemistry, The University of Arizona, Tucson, AZ 85721

One of the key elements of successful cooperative industry/university research is to translate practical problems of commercial significance into simple science questions that could be answered in a less complex system. This is especially critical in studying macromolecules with multiple environments with a potentially broad range of reactivities. Construction of specifically designed models which mimic individual environments greatly simplifies analysis of the relevant mechanisms and kinetics. Results could then be extrapolated to the polymer to direct the course of the reaction to the desired products. An elegant example of this strategy will be described whereby the reactivity of hindered polymeric hydrocarbon esters to condensation with amines is monitored by kinetic and spectroscopic analysis of suitable models for the reactive centers.

144. POLYMERIZATION MECHANISM AND KINETICS FOR LIQUID CRYSTAL POLYMERIZATION

H. K. Hall, Jr. Chemistry Department, The University of Arizona, Tucson, AZ 85721.

The kinetics of the acidolysis reaction has been investigated in the bulk homo- and copolymerization of 4-acetoxybenzoic acid (ABA) and 6-acetoxy-2-naphthoic acid (ANA) by the measurement of the rate of the acetic acid evolution. A mechanism change from acidolysis in the early stages to phenolysis in the later stages of the copolymerization of ABA and ANA has been proposed based on the end groups of the polymer chains. This hypothesis is substantiated by the nature of the byproducts and the measurement of polymer molecular weight, and the study of model reactions. The polymerization of diacids with diacetates was also examined because it would lead to low cost LCPs. Their kinetics were investigated using a newly developed GC method.

145. LASER POLYMER ABLATION IN THE NEAR-INFRARED: APPLICATIONS IN INDUSTRY AND ACADEMIA.

Dana D. Dlott, David E. Hare, Richard J. D'Amato and Thomas E. Lewis, School of Chemical Sciences, University of Illinois at Urbana Champaign, Champaign, IL 61801.

By careful design and engineering of polymeric materials, it is possible to have extremely efficient laser ablation processes using pulses from near-infrared solid state lasers. In this talk, I will describe studies of the fundamental mechanisms of near-infrared polymer ablation processes, and the applications of these processes in industry, primarily in the area of graphic arts and microlithography, and in academia, where they are used to study the physical properties of materials under extreme conditions of shock compression.

146. NANOSECOND PHOTOGRAPHIC, INTERFEROMETRIC, AND SPECTROSCOPIC STUDIES ON LASER ABLATION AND RELATED PHENOMENA OF POLYMER FILMS.

H. Furutani, H. Fukumura, and H. Masuhara, Dept. of Appl. Phys., Osaka University, Suita, Osaka 565, Japan.

Laser-induced expansion and contraction dynamics of polymer films below ablation threshold was investigated with a time-resolved interferometry. Dynamic analysis of the morphological changes with a few tens nm and 10 ns resolution provides a new insight on photothermal mechanism. For a reactive polymer films, laser-induced decomposition and following plume ejection processes was examined by applying further ns photographic method. Time-resolved spectroscopic measurement during excimer laser irradiation was very useful to identify transient species responsible to morphological changes. Recent advances and some future perspective will be presented.

147. MECHANISTIC ASPECTS OF 'LIVING' FREE RADICAL POLYMERIZATION AND INITIATION MEDIATED BY NITROXIDES. J.C. Scaiano, T.J. Connolly, N. Mohtat, C.N. Pliva, W. Skene, S. Belt and M.V. Baldovi, Department of Chemistry, University of Ottawa, Ottawa, Canada K1N 6N5.

Polymers such as polystyrene, prepared in the presence of nitroxides such as TEMPO, behave as *living* or *dormant* polymers, capable of undergoing further polymer growth at temperatures (e.g. 120 °C) under which the nitroxide cap can readily come off. This methodology is the subject of much current attention since it allows low polydispersity polymers to be prepared under conditions of free radical polymerization. Our studies have centered on the synthesis of novel initiators; mechanism of *living* free radical polymerization; possible role of acids; initiation by nitroxides; thermodynamics of bond cleavage; and possible photoinduced routes to *living* free radical polymerization. Laser techniques have proven very useful in the study of some of these processes.

148. MOLECULAR DYNAMICS MODEL FOR LASER ABLATION OF ORGANIC SOLIDS. L. V. Zhigilei, P. B. S. Kodali, and B. J. Garrison, Department of Chemistry, The Pennsylvania State University, University Park, PA 16802

A breathing sphere model is presented for molecular dynamics simulations of laser ablation of organic solids. An approximate representation of the internal molecular motion permits a significant expansion of the time- and length-scales of the model and still allows one to reproduce a realistic rate of the vibrational relaxation of excited molecules. The model provides a realistic description of the ablation of molecular films and matrix-assisted laser desorption (MALDI). A well-defined threshold fluence has been found to separate two mechanisms for the ejection of molecules - surface vaporization for low laser fluences and collective ejection or ablation for high fluences. Above the threshold the laser induced high pressure and the explosive homogeneous phase transition leads to the strongly forwarded emission of ablated material and high, from 500 up to 1500 m/s, maximum velocities of the ejected plume expansion. Large analyte molecules in MALDI get axial acceleration from an expanding plume and move along with the matrix molecules at nearly the same velocities. Big molecular clusters are found to constitute a significant part of the ejected plume at fluences right above the ablation threshold. The processes in the plume are found to have a strong influence on the final velocities and angle distributions of ejected molecules and molecular clusters.

149. SHORT PULSE UV LASER ABLATION. M. Stuke, Max-Planck-Institut für biophysikalische Chemie, P.O. Box 2841, D-37018 Göttingen, Germany

The pioneering work of Srinivasan and coworkers on the UV laser ablation of polymers - more than 15 years ago - has opened a whole new field ranging now from precise pulsed removal of material and surface modification to pulsed laser deposition of thin films. The role of photochemical effects and incubation was pointed out very early. Applications now include patterning of polymers in microelectronics, ink jet printing technology and in the medical field the shaping of the cornea.
In this talk the influence of the laser pulse duration will be discussed. Short pulses down to the femtosecond regime increase efficiency and precision and make more materials accessible for precise materials processing.

150. THE TECHNOLOGY OF ULTRAVIOLET LASER ABLATION OF ORGANIC POLYMERS AND TISSUE. R. Srinivasan, UVTech Associates, 98 Cedar Lane, Ossining, NY 10562.

The interaction of nanosecond, ultraviolet laser pulses with condensed organic material such as polymers and tissue has been the subject of numerous scientific investigations over the past 15 years. Considerably less attention has been paid to the nature of the process that occurs between UV laser pulses (usually 248nm or 308nm) and a polymer film in an actual manufacturing set-up. A train of UV pulses when used to drill a single hole or a matrix of holes in a polymer film introduces complications which show that our understanding of the interaction that is involved is far from complete. The use of pulsed, 193nm laser radiation to pattern tissue has also brought out the complexities that exist when the etching involves extremely small (10^{-3}) aspect ratios. The growth of this technology will depend on increased understanding of these practical problems.

151. RARE EARTH METAL INITIATED COPOLYMERIZATIONS TO PRODUCE NOVEL FUNCTIONAL POLYMERS, H. Yasuda,* E. Ihara, M. Morimoto, and T. Tokimitsu Department of Applied Chemistry, Faculty of Engineering, Hiroshima University, Higashi-Hiroshima 739, Japan

By taking advantages of the living polymerization ability of rare earth metal complexes, we have performed the copolymerizations of alkyl acrylates with caprolactone to give high molecular weight polymers with narrow molecular weight distribution, block copolymerization of MMA/butyl acrylate/MMA to give rubber like elastic polymers, block copolymerization various type of lactones with MMA, and bock copolymerization of ethylene with MMA, alkyl acrylates or lactones. Especially noteworthy is the formation of copolymer of ethylene/polar monomers such as MMA, alkyl acrylates and lactone. Resulting ethylene copolymer exhibit very high chemical reactivity as revealed by the excellent dyeing with dispersive dyes. Although block copolymerization of alkyl8meth)acrylates with lactone is possible when the polymerization of alkyl (meth)acrylate was peformed prior to the polymerizationnof lactones, the polymerization of lactones followed by the addition of MMA does not produce the block copolymer. In this case, only the lactone homopolymer was obtained.

152. SIMPLE ISOMERIZATION ROUTES TO POLY(VINYL AMINE) PRECURSOR MONOMERS, Bruce M. Novak* and Jeffrey T. Cafmeyer, Department of Polymer Science and Engineering, and the Materials Research Science & Engineering Center, University of Massachusetts, Amherst, MA 01002

Amine functionalized polymers are of interest for use in many applications. One of the more elusive simple vinyl polymers is poly(vinyl amine) (PVAm) due to the lack of readily available (read inexpensive) monomers. We have therefore investigated both the homo- and copolymerization of vinyl vinylamine (enamine) derivatives. Enamines like enols, are unstable relative to their imine tautomer. Enamines in this study were formed by isomerization of double bond in allyl amine and its derivatives using Rh(I) and Ru(II) complexes. We found that this meta-stable species could be copolymerized (AIBN, hʋ, or heat) with maleic anhydride or maleimides. Likewise, we have found that allyl formamides and allyl acetamides can be isomerized to their N-vinyl isomers. These vinyl monomers are also active toward both homo- and copolymerization under radical or cationic conditions.

153.

STEREOSPECIFIC POLYMERIZATION OF FUNCTIONALIZED OLEFINS WITH METALLOCENE CATALYSTS. Udo M. Stehling, Andrew L. McKnight, Kevin M. Stein, Robert M. Waymouth, Department of Chemistry, Stanford University, Stanford, CA 94305

The direct synthesis of cationic, group 4 metallocene complexes has led to catalysts which show some tolerance for Lewis basic solvents (e.g.: N,N-dimethylaniline, tetrahydrofuran, chlorobenzene, anisole, etc.) during the polymerization of olefins. This provided the precedent for our efforts of the direct polymerization of functionalized monomers. One goal was to study the influence of metallocene ligand structures on the stereospecifity of polymerizations with functionalized monomers. We used a set of different zirconocenes with C_2-, C_1- or C_s-symmetric ligands but also asymmetric catalysts. The results indicate that utilization of 5-N,N-diisopropyl-amino-1-pentene as the monomer has no negativ effect on the expected stereospecifity of the polymerization. Copolymerizations of 5-N,N-diisopropylamino-1-pentene and 1-hexene were carried out and led to copolymers with a statistical distribution of both monomers. That means ideal copolymerization behavior was observed. This work provided us with new insights in the direct synthesis of functionalized poly(α-olefins).

154. NEW ROUTES TO ESTER AND ACID FUNCTIONALIZED POLYETHYLENE
S. J. McLain, E. F. McCord, S. D. Arthur, E. Hauptman, J. Feldman, and W. A. Nugent, DuPont Central Research & Development, Experimental Station, P.O. Box 80328, Wilmington, DE 19880-0328.
L. K. Johnson, S. Mecking, and M. Brookhart, Department of Chemistry, University of North Carolina at Chapel Hill, Chapel Hill, NC 27599-3290

Three low pressure routes to ester and acid functionalized polyethylene are detailed:

1. Hydroformylation/oxidation of the Ring Opened Metathesis Polymer (ROMP) of cyclododecene.

2. ROMP / hydrogenation of 5-carbomethoxycyclooctene and 4-carbomethoxycyclopentene.

3. Direct copolymerization of ethylene and methyl acrylate using an alpha-diimine Pd catalyst.

155. Functional Polyolefins with Graft Structure Prepared by Metallocene Catalyst and p-Methylstyrene Comonomer
T. C. Chung, H. L. Lu, W. Janvikul and R. D. Ding Department of Materials Science and Engineering, The Pennsylvania State University, University Park, PA 16802

This presentation updates new experimental results in the preparation of functional polyolefins with graft copolymer stucture, containing polyolefin backbone and functional polymer side chains. The chemistry involves two reactive comonomers, borane-containing α-olefin and para-methylstyrene (p-ms), which can be effectively incorporated into polyolefins, especially using metallocene catalysts with constained ligand geometry. The borane groups in polyolefin are very versatile intermediates and can be spontaneously oxidized by oxygen to form stable polymeric radicals for living "radical" graft-from polymerization. In turn, the p-methylstyrene groups in polyolefins can be easily metallated to produce polymeric anions for living "anionic" graft-from polymerization. Many graft copolymers, such as PE-g-PS, PE-g-PMMA, PE-g-PAN, PP-g-PS, PP-g-PB, PP-g-PI and PP-g-PMMA, have been prepared with relatively well-defined molecular structure. These graft polymers are very effectively interfacial reagents, which are not only dramatically increase the interaction between polyolefin and other materials but also preserve most of physical properties, such as crystallinity and melting point, of polyolefin.

156. POLYCARBOSTANNANES VIA ACYCLIC DIENE METATHESIS (ADMET) POLYMERIZATION. Patrick S. Wolfe, Fernando J. Gómez and Kenneth B. Wagener,* The George and Josephine Butler Polymer Research Laboratory, Department of Chemistry and Center for Macromolecular Science and Engineering, University of Florida, Gainesville, FL 32611-7200.

The synthesis of a polycarbostannane via Acyclic Diene Metathesis (ADMET) polymerization has been performed utilizing either the well-defined molybdenum alkylidene $Mo(CHCMe_2Ph)(N-2,6-C_6H_3-^iPr_2)(OCMe_2CF_3)_2$ (**1**) as the catalyst or the tungsten aryloxo complexes $W(O)Cl_2(O-2,6-C_6H_3Br_2)_2$ (**2**) or $WCl_4(O-2,6-C_6H_3Ph_2)_2$ (**3**) as precatalysts in a classical-type metathesis catalytic system. In the latter cases, the monomer itself performs as the cocatalyst in the first monomer-cocatalyzed metathesis polymerization. Linear, unsaturated polymers were obtained in all cases.

157. FUNCTIONALIZED POLY(P-PHENYLENE)S VIA SUZUKI CROSS-COUPLING. Peter B. Balanda and John R. Reynolds, Center for Macromolecular Science and Engineering, Department of Chemistry, University of Florida, Gainesville FL 32611-7200

We report the synthesis of poly(*p*-phenylenes) via Suzuki cross coupling conditions with Pd(OAc)$_2$ as the catalyst species wherein attempts were made to limit the molecular weight of the growing chain via stoichiometric imbalance and choice of solvent. PPPs with cation coordinating methoxyethoxy- and triethoxy- sidechains, tertiary and quaternary ammonium sidechains, and a material with a pendant *p*-methoxyphenyl group were produced. Polymers with alkyl attachment to the ring displayed absorption maxima of ca. 295 nm, while those with alkoxy attachment had absorption maxima in the range of ca. 335 to 360 nm and were blue light emitters. Polymers reported are poly[{2,5-bis(1,4-dioxapentyl)-1,4-phenylene}-*alt*-1,4-phenylene] (**1**), poly[{2,5-bis(1,4,7-trioxanonyl)-1,4-phenylene}-*alt*-1,4-phenylene] (**2**), poly[{2,5-bis(2,5-dioxahexyl)-1,4-phenylene}-*alt*-1,4-phenylene] (**3**), poly[{2,5-bis(2,5,8-trioxadecyl)-1,4-phenylene}-*alt*-1,4-phenylene] (**4**), poly[2,5-bis(1,4,7-trioxanonyl)-1,4-phenylene] (**5**), poly[{2,5-(5-[4-methoxyphenyl]-1,5-dioxapentyl)-1,4-phenylene}-*alt*-1,4-phenylene] (**6**), poly[{2,5-bis(3-*N*,*N*-diethylamino-1-oxapropyl)-1,4-phenylene}-*alt*-1,4-phenylene] (**7**), poly[{2,5-bis(3-*N*,*N*,*N*-triethylamino-1-oxapropyl)-1,4-phenylene}-*alt*-1,4-phenylene] dibromide (**8**), and a copolymer of **2** and **4**.

158. MOLECULAR MOBILITY AND MECHANICAL PROPERTIES OF GLASSY POLYMERS. L. Monnerie, F. Laupretre, J. L. Halary, P. Tordjeman, L. Teze, F. Beaume, Laboratoire de Physicochimie Structurale et Macromoléculaire, URA-CNRS 278, ESPCI, 10, rue Vauqueline, 75231 Paris Cedex 05

This lecture deals with the molecular motions occurring in the glassy state and their relationship with mechanical properties such as plasticity, shear banding, crazing and fracture.
As regard the local motions, a detailed molecular description implies the combination of viscoelastic measurements and spectroscopic investigations through dielectric relaxation, ^{13}C and ^{2}H NMR. This is illustrated on a series of semi-aromatic polyamides.
The effect of molecular mobility on the mechanical properties is put forward on a series of copolymers of methylmethacrylate with either cyclohexyl maleimide or methyl glutarimide. The important feature is the cooperativity of the molecular motions involved in the β transition as well as their coupling with the α transition molecular processes.

159. THE THERMAL GLASS TRANSITION BEYOND THE TIME TRAP. K.-P. Bohn, J. K. Krüger, FB 10.2 Experimentalphysik, Universität es Saarlandes, 66041 Saarbrücken, Germany

Time Domain Brillouin Spectroscopy (TDBS) and modulated Differential Scanning Calorimetry (MDSC) were used as a crucial experimental approach to demonstrate the intrinsic nature of the thermal glass transition in Polyvinylacetate (PVAC). The transition from the equilibrium liquid to the ideal glassy state appears at an ultimate temperature T_{gs}.
It was found that the fully relaxed sound frequency shows in the limit of infinitely small temperature jumps a thermoreversible kink at T_{gs}, which characterizes the transition into the ideal glassy state. At the intrinsic glass transition temperature T_{gs} the instantaneous frequency response per Kelvin shows a discontinuity, which disappears if $T_g = T_{gs}$. Close to but still above the glass transition the average relaxation times τ_α of the α-process behaves Arrhenius-like, becoming as large as $7 \cdot 10^6$ s. At T_{gs}, τ_α displays a cutoff.
Combining the Brillouin scattering data with those obtained by MDSC, the mode and the thermal Grüneisen parameter araound the glass transition of PVAC were analyzed. It turns out that both types of Grüneisen parameters behave anomalously at T_{gs}.

160. A MOLECULAR MODEL FOR CRAZING MECHANISMS IN GLASSY POLYMERS. HZY Han[+], RA Duckett[+], TCB McLeish[+], NJ Ward[#] & AF Johnson[#], Polymer IRC, [+]Dept of Phys & Astron, [#]Sch of Chem, Univ of Leeds, Leeds LS2 9JT, UK

The strains to craze for monodisperse linear polystyrenes at different temperatures and strain rates have been investigated using a thin film technique with optical microscope. The surface energies of crazing have been deduced from the experimental data. Using the "tube" model of crazing of McLeish, Plummer and

Donald, and Kramer's geometrically necessary entanglement loss concept, we calculated the frictional force acting on a single chain and related it to craze growth mechanisms and craze void advance surface energy. The entanglement loss in the chain retraction process and its effect on craze growth surface energy were calculated and compared to the previous theories. We argue that pure scission starts at critical chain length equals M_e (not $2M_e$ as previously used). The calculated results were supported by experimental data. We demonstrated that the discontinuity in the craze growth surface energy at transition temperature from pure disentanglement to scission and disentanglement mixed crazing is a general phenomenon in glassy polymers. The sign and strength of the discontinuity depends on the molecular weight.

161. ON THE PHYSICAL AGING OF SEMICRYSTALLINE POLYMERS BELOW AND "ABOVE" T_G. H. Marand[1,*], V. Velikov[1], V. Prabhu[2], S. Srinivas[1], S. Christian[1], (1) Virginia Polytechnic Institute and State University, NSF Science and Technology Center for High Performance Polymeric Adhesives and Composites, Chemistry Department, Blacksburg, VA 24061-0212, (2) University of Massachusetts, Polymer Science and Engineering Department, Amherst, MA 01003.

The momentary small strain creep behavior of semicrystalline PEEK was investigated as a function of annealing time in a temperature range encompassing the glass transition (120C to 260C). DSC was used to follow the enthalpic recovery after annealing below Tg and the development of the "low endotherm" after annealing above Tg. Using DSC we also followed the evolution of the glass transition temperature with annealing above Tg. Although the shift of the creep compliance curves to longer time with increasing annealing time above Tg has very similar characteristics to that observed below Tg, we explain the time dependence of the properties above Tg by a model that differs significantly from that proposed by Struik. We propose that the "physical aging" like behavior observed above Tg is a direct consequence of the effect of secondary crystallization on the number of constraints in the amorphous fraction. Our studies further suggest that the existence of a rigid amorphous fraction above Tg may need to be reconsidered.

162. DIELECTRIC, DYNAMIC MECHANICAL AND ENTHALPIC RELAXATION OF FLUORINATED AROMATIC POLY(ETHERS) AND POLY(ETHER KETONES). A. A. Goodwin, Materials Engineering, Monash University, Australia. F.W. Mercer and M.T. McKenzie, Raychem Corporation, CA, USA. J.N. Hay and J.R. Atkinson, Metallurgy and Materials, Birmingham University, UK.

Linear aromatic thermoplastics have long been known for meeting the property requirements for structural resins, coatings and films required in high performance applications. In the area of microelectronics considerable attention has been devoted to fluorinated polymers since this results in materials with low dielectric constant and low moisture absorption, which makes them good candidates for use as insulators. This talk will discuss the relaxation behaviour of aromatic fluorinated poly(ethers) and poly(ether ketones) measured above and below the calorimetric glass transition temperature using different probes of molecular mobility. Correlations between the temperature dependence of the relaxation time, the relaxation broadness and intermolecular cooperativity and chemical structure will be included. The sensitivity of the dielectric and dynamic mechanical probes to different processes will be compared, and the relationship of these techniques to enthalpic relaxation will be discussed.

163. INVESTIGATION OF PHYSICAL AGEING IN POLYMETHYL METHACRYLATE USING POSITRON ANNIHILATION, DIELECTRIC RELAXATION AND DYNAMIC MECHANICAL THERMAL ANALYSIS W. J. Davis and R. A. Pethrick, Department of Pure and Applied Chemistry, University of Strathclyde, Thomas Graham Building, 295 Cathedral Street, Glasgow G1 1XL, U.K.

The spontaneous change in polymer properties with time is a well documented phenomena and is known as physical ageing or entropy - structural relaxation. Free volume and entropy as characteristic variables of the system undergoing ageing process. Positron annihilation spectroscopy [PAS] has the capability of probing the void structure in amorphous materials and providing information on changes which occur with time and their effects on molecular mobility. Dielectric relaxation spectroscopy [DRS] characterises the microscopic mobility, whereas dynamic mechanical thermal analysis [DMTA] probe longer range more macroscopic properties. Comparison of ageing 'rates' from temperature plots showed an excellent correlation with previously published results and provides insight into the molecular mechanisms associated with physical ageing.

164. DYNAMIC-MECHANICAL AND CALORIMETRIC SIGNATURES OF PHYSICAL AGING IN PET FILMS. J. Greener, J. M. O'Reilly and B. A. Contestable, Eastman Kodak Company, Rochester, NY 14650-2158

Physical aging has been widely studied in polymers using a variety of techniques. In this study DMTA and DSC techniques are combined to characterize the physical aging of biaxially oriented semicrystalline PET films. Annealing at 20-50°C below Tg for 2-12 days leads to a reduction in tan δ in the vicinity of the α transition and to a corresponding drop in the enthalpy of the sample. An enthalpy peak in the specific heat curve emerges and the energy of the peak (ΔH) and its position (Tp) increase with increasing annealing temperature and time. Similarly, the change in tan δ exhibits a characteristic peak whose intensity and position increase with annealing temperature and time in a manner that parallels the changes in Tp and ΔH. The differences and similarities between these techniques will be discussed in terms of the mechanisms of physical aging.

165.

EFFECTS OF THE APPLICATION AND RELEASE OF STATIC DEFORMATION ON THE DYNAMIC MODULUS OF AMORPHOUS GLASSY POLYMERS B. Haidar, A. Vidal, Institut de Chimie des Surfaces et Interfaces, CNRS, P.O.Box 2488, F68057 MULHOUSE-France

Macroscopic deformation, like temperature jumps, allows the observer to detect and to quantify physical aging phenomenon. In the present work and in order to identify the origin of aging, we proposed to study, under isothermal conditions, the effect of the application and the (partial or complete) release of a static deformation (in tension, compression or sheer) on the dynamic modulus, E', of amorphous glassy polymers. In a typical experiment, a well aged specimen was first subjected to constant deformation (up to 3%), then static strain was modulated intermittently with a sinusoidal strain (amplitude <0.1% at 0.1 to 25 Hz). It has been found that application and release of any static deformation provoke, first, E' decrease "deaging", followed by E' increase "aging" effects. Aging rate depends primarily on the starting pseudo-equilibrium state and equally on the application or the release of a given deformation jump.

166. THE EFFECTS OF MOLECULAR ORIENTATION ON THE PHYSICAL AGING BEHAVIOR OF GLASSY AMORPHOUS POLYMERS, M. D. Shelby and G. L. Wilkes, Polymer Materials and Interfaces Laboratory, Virginia Tech, Blacksburg, VA 24061-0211

Physical aging rates were monitored as a function of molecular orientation in samples of atactic polystyrene using dilatometry and creep measurements. Samples were prepared by drawing to varying stretch ratios at $T_g+20°C$ using a T. M. Long film stretcher. Orientation was quantified using optical birefringence and the Hermans' orientation function, f, which attained a maximum value of 0.11. Density in the unaged samples increased linearly with f due to better chain packing. Correspondingly, this implies a decrease in the free volume and should result in a decrease in molecular mobility based on the free volume theory. In contrast, the rate of volume relaxation, β, was about 40% higher for the oriented samples as compared with the isotropic controls and this value did not change with further increases in orientation. The mechanical shift rate, μ, was found to decrease slightly with orientation although data interpretation is complicated by the superimposed sample shrinkage during the creep test. As a check of consistency, the same tests were also performed on oriented bisphenol-A polycarbonate (aged at 90°C) yielding almost identical relative changes in β and μ.

167. **GRAFT COPOLYMERS OF PEO-PPO-PEO TRIBLOCK POLYETHERS ON BIOADHESIVE POLYMER BACKBONES FOR USE AS DRUG DELIVERY CARRIERS**

Allan S. Hoffman, Guohua Chen, Xiangdong Wu, Zhongli Ding
Center for Bioengineering, University of Washington, Seattle, WA
Baghwati Kabra and Kiran Randeri, Alcon Corp., Ft. Worth, TX*; Matt Schiller and Eyal Ron, GelMed Corp., Bedford, MA*; Nikolaos A. Peppas and Christopher Brazel, Purdue University, Lafayette, IN**

Bioadhesive polymers are often added to drug formulations in order to prolong the residence time of the formulation on mucosal surfaces such as the eye, nose, vagina or intestines. However, the drug may still be released too rapidly. We have recently synthesized novel graft copolymer compositions by grafting temperature-responsive Pluronic® side chains onto bioadhesive PAAc and chitosan backbones. These interesting graft copolymers form low viscosity solutions at 4°C that gel when the temperature is raised to 34°C. Using the Pluronic®-g-PAAc graft copolymers as carriers for timolol maleate, the gelation of the drug-polymer solution causes significant prolongation of both the drug release rate from the graft copolymers and the erosion (dissolution) rate of these copolymers. Hydrogels of these graft copolymers are bioadhesive. Insulin has been similarly released from the gel formed by the Pluronic® L-122-g-chitosan graft copolymers at 37°C. These interesting bioadhesive and thermally-gelling graft copolymers may be especially suitable as carriers for drugs of opposite charge to that of the backbone polymer, for topical delivery to mucosal surfaces.
(These collaborators carried out viscosity vs. temperature*, or bioadhesion** tests).

168. APPLICATIONS OF POLY (OXYETHYLENE-*b*-OXYPROPYLENE-*b*-OXYETHYLENE)-*g*-POLY(ACRYLIC ACID) POLYMERS (SMART HYDROGEL™) IN DRUG DELIVERY.
L. Bromberg, T.H.E. Mendum, M.J. Orkisz, E.S. Ron, E.C.Lupton. GelMed, Inc., 213 Burlington Road, Bedford, Massachusetts 01730

Poly(oxyethylene-*b*-oxypropylene-*b*-oxyethylene) triblock copolymers (trademarks Poloxamer or Pluronic) have a potential in drug delivery due to their inherent non-toxicity, transparency, and ability to gel at body temperatures. Grafting of poly(acrylic acid) to Poloxamers imparts novel properties, such as pH-responsiveness, bioadhesiveness and gelation at low polymer concentration (1-3 w%) and therefore leads to a new family of polymeric drug vehicles (Smart Hydrogel™). Solubilization of steroid hormones (estradiol, progesterone) in Smart Hydrogel™ is temperature-dependent and is related to structural changes in the polymer solution upon aggregation of poly(oxypropylene) segments. Once solubilized, the drug is held in the Smart Hydrogel™ and released with the rate depending upon temperature and polymer concentration.

169. RHEOLOGICAL PROPERTIES OF REVERSE THERMOGELLING POLY(ACRYLIC ACID)-g-(OXYETHYLENE-b-OXYPROPYLENE-b-OXYETHYLENE) POLYMERS (SMART HYDROGEL™). Michal J. Orkisz, Lev Bromberg, Rebecca Pike, E.C. Lupton, Eyal S. Ron; GelMed Inc., 213 Burlington Rd, Bedford, MA 01730

The recently developed Smart Hydrogel™ materials (poly(acrylic acid)-*g*-(oxyethylene-*b*-oxypropylene-*b*-oxyethylene) polymers) offer a significant potential for applications in drug delivery systems. The combined properties of solubilization of hydrophobic drugs, bioadhesion and thermo-viscosification, all at low Smart Hydrogel™ concentrations, make this polymer system suitable for use in many applications involving drug delivery via ocular, nasal, dermal, vaginal etc. routes. The mode of application of a Smart Hydrogel™ based material, such as spraying, spreading, injecting, as well as its on-site residence is dependent on the viscoelastic properties of Smart Hydrogel™ polymers. These properties are investigated here. Temperature, concentration, shear and shear history dependence of Smart Hydrogel™ polymer viscosity is presented, as well as the temperature dependence of its elastic moduli.

170. USING MICROBICIDAL GELS TO PREVENT SEXUAL TRANSMISSION OF HIV. M.G. Bergeron, N. Gagné, P. Gourde, S. Perron, M. Tremblay, J. Juhasz, D. Beauchamp, A. Désormeaux. Centre de Recherche en Infectiologie, CHUL, Québec, QC, Canada G1V 4G2.

As the number of individuals infected with HIV is growing dramatically throughout the world, there is an urgent need to develop innovative preventive measures that can reduce the transmission of HIV. We have recently developed microbicidal gels that could be highly effective to achieve such a goal. The advantage of our gel preparations is that it may prevent the sexual transmission of HIV by blocking its diffusion through vaginal, cervical and ano-rectal mucosa either physically (gel alone) or chemically by destroying the virus (spermicides). The efficacy of the gel has been evaluated *in vitro* using a p24 assay. Our studies showed that the gel formulation acts as an effective barrier against HIV. Experiments performed on rabbits revealed that the gel is non toxic when applied to the vaginal and cervical mucosa of animals. Moreover, we have clearly demonstrated that nonoxynol-9 incorporated into the gel was less toxic to the mucosa of animals when compared to free nonoxynol-9. Furthermore, entrapment of antivirals into liposomes could be effective to prevent virus infection by delivering high intracellular concentration of drugs in cells susceptible to infection. Studies are in progress to determine the cellular distribution of liposomes incorporated into our gel formulation. The use of microbicidal gels could thus represent a convenient strategy to reduce the transmission of HIV and other sexually transmitted diseases.

171. PHOTORESPONSIVE CONTROL OF ION-EXCHANGE IN LEUCOHYDROXIDE CONTAINING HYDROGEL MEMBRANES. M.G. Kodzwa and D.G. Rethwisch, Department of Chemical and Biochemical Engineering, University of Iowa, Iowa City, Iowa 52242

Polyacrylamide hydrogels containing bis-[4-(dimethylamino)phenyl] (4-vinyl-phenyl)methyl leucohydroxide (leucohydroxide) as a pendant photochrome have been studied as photoresponsive membranes. Leucohydroxide is reversibly ionized upon irradiation with UV light to form fixed cationic charges on the membrane. These charged membranes act as anion exchange membranes. This provides a mechanism to photocontrol the permeability of these membranes. Indeed, the permeability of methyl orange (an anionic permeant) was nearly doubled by irradiation of the membrane. The permeability of a neutral species was unaffected by irradiation. A simple transport model with no adjustable parameters was developed to explain these results. These membranes provide a potential mechanism for photocontrolled drug delivery.

172. PREPARATION AND CHARACTERIZATION OF GLUCOSE-SENSITIVE P(MAA-g-EG) HYDROGELS. Christie M. Dorski, Francis J. Doyle III, and Nikolaos A. Peppas, Department of Chemical Engineering, Purdue University, West Lafayette, Indiana, 47907-1283

Graft copolymers of pH-sensitive poly(methacrylic acid-g-ethylene glycol) (P(MAA-g-EG)) were investigated for use in a novel self-regulated device that delivers appropriate amounts of insulin in response to changing glucose levels. This approach involves the incorporation of pH-sensitive P(MAA-g-EG) with immobilized glucose oxidase to yield a glucose-sensitive polymer. Glucose-sensitive gels were synthesized by first activating glucose oxidase and then polymerizing P(MAA-g-EG) in the presence of the activated enzyme. The equilibrium swelling behavior of the gels was examined as a function of pH. At low pH values the gels were in a collapsed state due to complexation. At high pH values, the gels swelled to approximately 20 times their dry weights. Glucose-sensitive P(MAA-g-EG) gels showed lower initial degrees of swelling for higher glucose concentrations. The swelling/syneresis behavior of the glucose oxidase-containing gels was investigated under varying pH conditions to characterize their dynamic swelling behavior. Glucose-sensitive gels were capable of releasing 0.5 mg of insulin in 5 minutes and an additional 0.5 mg over 3.5 hours.

173. GLUCOSE INDUCED SWELLING KINETICS OF POLY(HEMA/DMA) HYDROGEL CONTAINING BOUND GLUCOSE OXIDASE. Ying Chu[1], Vivek Agarwal[1], Padma Prabodh Varanasi[2], and Sasidhar Varanasi[1]
[1]Department of Chemical Engineering, University of Toledo, Toledo, Ohio 43606.
[2]S.C. Johnson & Son, Inc., Racine, WI 53403

The kinetics of gel swelling plays a crucial role in the design of devices employing stimuli-sensitive gels. We have recently reported on the pH sensitive swelling kinetics of poly(HEMA/DMA) hydrogels (*J. Appl. Polym. Sci.*, **58**, 2161 (1995)). Herein, we extend these kinetic studies to the case where the hydrogel is made sensitive to glucose concentration via a reaction catalyzed by glucose oxidase immobilized in the gel. We present both experimental data and theoretical predictions. The system studied here has been proposed as a promising candidate for *self-regulated release* of insulin. The effect of important design variables, viz., the gel charge density, enzyme loading, and glucose and buffer concentration, on the gel performance were investigated. The model explicitly accounts for (1) the continuous variation of gel dimensions during swelling and its effect on the rection-diffusion processes in the gel, and (2) in the dynamic role played by the buffer in augmenting proton transport through the gel. The model is able to capture the experimentally observed trends in swelling kinetics.

174. REVERSIBLE BLOCK/GRAFT COPOLYMERIC EMULSIFIERS BASED UPON INTRAMOLECULAR COMPLEXATION. A. M. Mathur[*], A. B. Scranton[*] and J. Klier[\#]
[*]Dept. of Chemical Engineering, Michigan State University, East Lansing, MI 48824
[\#]Central Research, Dow Chemical Company, Midland, MI 48674

Graft copolymers comprising a backbo..e of poly(methacrylic acid) and oligomeric ethylene glycol grafts were synthesized and evaluated for their reversible oil-emulsification properties and surface activity. The backbone Lewis acid repeat units are capable of forming reversible hydrogen bonded complexes with the Lewis base ether oxygens of the oligomeric grafts. These complexes are formed under complex promoting (acidic) conditions, and are considerably more hydrophobic than either of the constituent polymers. The copolymers would assume a multiblock architecture with alternating hydroph..ic and hydrophobic blocks under acidic conditions, and revert back to a hydrophilic gra[ft c]opolymer when the complex is broken. The copolymers formed oil-in-water emulsions under acidic conditions, while under basic conditions, no emulsification was observed. The oil-emulsification ability was found to be reversible and could be turned "on" or "off" by manipulating the solution pH. The polarity sensitive fluorescence of solubilized pyrene was used to study the formation of hydrophobic aggregates as a function of pH and aqueous copolymer concentration.

175.

DESIGN OF HIGHER PRECISE RELEASE CONTROL FORMULATION FOR PROTEIN DRUG.
A. Sano M.Maeda,M.Kjihara,S.Tani and K.Fujioka Formulation Research Laboratories,Sumitomo Pharmaceuticals Co.,Ltd. 3-45 Kurakakiuchi 1-Chome Ibaraki-Shi,Osaka,567

We have been developing practical controlled-release system suitable for protein drug. For this purpose we selected the strategy that using carrier material from already approved biocompatible materials as implant for human , and control the release by structural contrivance of formulation. One of developed technology is "Minipellet formulation" which is cylindrical dosage forms using collagen as a carrier material. Minipellet formulation of interferon was already filed NDA in Japan. We enabled to control release of drug without crosslinking of collagen. We present the analysis of structure and release mechanism of our formulations and design of higher precise release control formulation.

176. COMPARATIVE ESTERIFICATION RATES OF FOUR DIOLS USED FOR POLYESTER SYNTHESES Dale E. Van Sickle, Mary A. Taylor and Larry A. Bass, Research Laboratories, Eastman Chemical Company, P.O. Box 1972, Kingsport TN 37662

While a large amount of data has accumulated on rates of esterification and polyesterification, direct comparison of the esterification reactivities of a series of diols is difficult because, usually, different experimental conditions were used to collect the data to be compared. In this work we have compared the rates of self catalyzed esterification of four diols used in polyester coatings technology with a single "model" acid, cyclohexanecarboxylic acid (CHCA). Measurements have been made using both a large excess of diol and a large excess of CHCA. The diols investigated were 2,2,4-trimethylpentane-1,3-diol (TMPD), hydroxypivalyl hydroxypivalate (HPHP), neopentyl glycol (NG) and 1,4-cyclohexanedimethanol (CHDM). The second-order rate constants for the disappearance of CHCA in a large excess of the diols at $210°C$ are: TMPD, 0.011; HPHP, 0.017; NPG, 0.018 and CHDM, 0.032, all in units of kg/mol/min. When a small amount of diol is reacted with an excess of CHCA, the disappearance of the diol is well modeled by first-order kinetics. The $150°C$ first-order rate constants correlate with the second-order constants, possibly in a linear fashion.

177. FRONTAL CURING OF EPOXY RESINS: COMPARISON OF MECHANICAL AND THERMAL PROPERTIES TO BATCH CURED MATERIALS. Yuri Chekanov, Institute of Chemical Physics, Russian Academy of Sciences, Russia, Chernogolovka, 142432 and **John A. Pojman**, Department of Chemistry and Biochemistry, University of Southern Mississippi, Hattiesburg, MS 39406-5043

The epoxy resin diglycidyl ether of bisphenol F (EPON 862) was cured by the aliphatic amine curing agent Epicure 3371 in stoichiometric ratio both frontally and in batch cure schedule. Glass transition temperatures (Tg) were determined for cured epoxy system using differential scanning calorimetry (DSC) and dynamic mechanical analysis (DMA). Tensile properties of epoxy samples were tested according to ASTM D638M - 93. The properties of frontally cured epoxy resin were shown to be very close to that of batch cured epoxy resin. The velocity of cure front propagation was measured for both neat and filled epoxy. Rubber particles (ground tires) were used as a filler. The maximum percentage of filler that would sustain propagation was 30%. Advantages and disadvantages of frontal curing of epoxy resins will be discussed.

178. **Simultaneous Interpenetrating Networks for the System Epoxy-Acrylate.** Alexandre A. Baidak[3], Jean-Marie Liegeois[6] and Leslie H. Sperling[1-5], [1]Department of Chemical Engineering, [2]Department of Materials Science and Engineering, [3]Materials Research Center, [4]Center for Polymer Science and Engineering, [5]Polymer Interfaces Center, Lehigh University, Bethlehem, PA 18015-3194, [6]Université de Liège, L.M.P.C. Building B6, Institute de Chimie, B-4000 Liège, Belgium

Simultaneous Interpenetrating Polymer Networks (SINs) based on an epoxy / poly(n-butyl acrylate) system were synthesized at 120 °C. The polymerization kinetics were studied in situ by Fourier Transform Infrared Spectroscopy (FTIR). Three key events occur during the polymerization, namely the gelation of the network I, gelation of the network II, phase separation of one polymer from the other. Thus, metastable phase diagrams describing the relations of the three events were constructed. Three dimensional tetrahedrons characterizing the four component system (the two monomers and the two polymers) allow the visualization of these three key events and also to define some critical points, e.g. the loci of the points where simultaneous gelation of the two networks occurs. The inside of the tetrahedron was also investigated using partially reacted model compounds. Those tetrahedrons can be used as guidelines for setting up a synthesis strategy leading to desired morphologies.

179. POLYETHERESTER THERMOPLASTIC ELASTOMER AS IMPACT MODIFIER FOR RIGID POLYVINYL CHLORIDE. Rudolph D. Deanin, Neeraj Munot, Carol M. Barry, and Candace L. Ellis, Plastics Engineering Department, University of Massachusetts Lowell, Lowell, MA 01854.

Addition of soft-grade polyetherester engineering thermoplastic elaster into rigid polyvinyl chloride produced dramatic synergistic improvement in impact strength up to 44X. It also increased melt flow rate synergistically, and ultimate elongation exponentially, without lowering heat deflection temperature.

180. IONOMER POLYBLENDS WITH LOW-DENSITY POLYETHYLENE, POLYBUTYLENE TEREPHTHALATE, AND NYLON 66. Rudolph D. Deanin and Wei Chu, Plastics Engineering Department, University of Massachusetts Lowell, Lowell, MA 01854.

Adding an ethylene/sodium methacrylate ionomer to low-density polyethylene increased melt flow, tensile modulus and strength, and gave synergistic increase in ultimate elongation. Adding it to polybutylene terephthalate increased ultimate tensile strength synergistically. And adding it to nylon 66 increased impact strength synergistically.

181.
NEW STARS: EIGHT POLYISOBUTYLENE ARMS EMANATING FROM A CALIXARENE CORE
Sunny Jacob, Istvan Majoros, and Joseph P. Kennedy
The Maurice Morton Institute of Polymer Science
The University of Akron, Akron, Ohio 44325-3909

The first synthesis of well-defined star polymers consisting of eight polyisobutylene (PIB) arms emanating from a calix[8]arene core is described. The synthesis was accomplished by the use of octafunctional calixarene-derivative initiators which, in conjunction with $BCl_3/TiCl_4$ coinitiators, induce the living polymerization of isobutylene (IB). Specifically, the initiators were tert-hydroxy- and tert-methoxy calixarene derivatives. The tert-methoxy derivative is soluble in polymerization charges (CH_3Cl/hexanes) at -80°C and was used preferentially. To gain insight into the mechanism of polymerization, model experiments using a monofunctional analogue, 2-(p-methoxyphenyl)-2-methoxypropane of the octafunctional initiators, were also conducted. Well-defined stars with close to theoretical composition and molecular weights were obtained. Gel permeation chromatography (GPC), by the use of on-line RI, UV and laser light scattering (LLS) detectors, was used to determine the molecular weight and composition of the linear and octa-arm star polymers. Determination of number of arms of the star by core-destruction indicated presence of close to theoretical number of arms.

182.
FUNCTIONAL POLYISOBUTYLENES BY END QUENCHING. S. Hadjikyriacou and R. Faust. Polymer Science Program, Chemistry Department, University of Massachusetts Lowell, One University Avenue, Lowell, MA 01854.

In-situ functionalization of living polyisobutylene (PIB) ends was accomplished by capping the living PIB chain end with 1,1 diphenylethylene, in Hexane/CH_3Cl 60/40 v/v in the presence of $TiCl_4$ at -80 °C, followed by end-quenching using a variety of nucleophiles.
Quantitative transfer of hydride ion was obtained when the chain ends were quenched with tributyltin hydride. Quenching with Dimethylaminotri-n-butyltin lead to the formation of dimethylamino functional PIB in close to quantitative yields. Quenching with excess NH_3 solution in dichloromethane resulted in quantitative amino functionality. The structures of the functional PIBs were verified by ^1HNMR spectroscopy and by the synthesis of model compounds.

183. SYNTHESIS OF CYCLIC CARBONATE FUNCTIONAL POLYMERS. Dean C. Webster and Allen L. Crain. Research Laboratories, Eastman Chemical Company, P.O. Box 1972, Kingsport, TN 37662.

The copolymerization behavior of vinyl ethylene carbonate (4-ethenyl-1,3-dioxolane-2-one, VEC) with other unsaturated monomers in solution free radical copolymerizations was studied. VEC copolymerizes well with vinyl ester monomers. Conversions of monomer to polymer are high with complete incorporation of VEC into the copolymers. When copolymerized with acrylic monomers, however, VEC is not completely incorporated. Sufficient levels can be incorporated to provide sufficient cyclic carbonate functionality for subsequent reaction and crosslinking. The unincorporated VEC can be removed using a thin film evaporator.

184. Polymerization of *N,N*-Dialkylacrylamides with Anionic Initiators Modified by Diethylzinc
M. Kobayashi, T. Ishizone, A. Hirao, and S. Nakahama, #M. Kobayashi
(Dept. of Polymer Chemistry, Tokyo Institute of Technology, Meguro-ku, Tokyo 152, Japan
#Lintec Corp., Warabi, Saitama 335, Japan)

Anionic polymerization of *N,N*-dimethyl-(DMAA) and dipropyl(DPAA)acrylamide were carried out with sec-butyllithium / diphenylethylene and potassium diphenylmethanide in the absence and presence of diethylzinc in THF at -78°C for 60 min. Tacticities of the resulting polymers were observed by the ^1H and ^{13}C NMR spectroscopies. Poly(DMAA) and poly(DPAA) produced in the absence of Et_2Zn have broad molecular distribution in the both cases of Li$^+$ and K$^+$ countercation, while the addition of Et$_2$Zn as a weak Lewis acid to polymerization system gave the polymers of narrower MWD. Poly(DMAA) and poly(DPAA) produced with Li$^+$ coutercation is highly isotactic, while these polymers obtained with K$^+$ countercation were atactic. Tacticity of poly(DMAA) was changed from isotactic to syndiotactic by addition of Et$_2$Zn to the initiator system of *sec*-butyllithium / diphenylethylene. However, tacticity of poly(DPAA) was not change remarkably regardless of the addition of Et$_2$Zn. It is found that the polymerization of *N,N*-dialkylacrylamides are significantly affected by the alkyl substituents (R : methyl, ethyl, propyl). Anionic polymerizations of *N,N*-diethylacrylamide (DEAA) were carried out with K$^+$/Et$_2$Zn at higher temperatures condition result in narrow MWD (Mw/Mn=1.08, at 0℃) and highly heterotactic polymer (mr>90%, at 18℃). The cloud point (Tc) of poly(DEAA) in water was found to strongly depend on the molecular weight and stereospecificity.

185. **Synthesis of Side-Chain Liquid Crystalline Homopolymers and Block Copolymers with Well-Defined Structures by Living Anionic Polymerization and Their Thermotropic Phase Behavior**
M. Yamada, T. Iguchi, A. Hirao, S. Nakahama, and J. Watanabe
(Dept. of Polymer Chemistry, Tokyo Institute of Technology, Meguro-ku, Tokyo 152, Japan)

Side-chain liquid crystalline polymers of 6-[4-(4'-methoxyphenyl)phenoxy]hexyl methacrylate (**1**) and 6-[4-(4'-cyanophenyl)phenoxy]hexyl methacrylate (**2**) with well-regulated molecular weights and narrow distributions were successfully obtained by the anionic loving polymerization. On the basis of on DSC and X-ray analyses, it is clarified that poly(**1**) exhibits smectic mesophase while nematic mesophase is observed in poly(**2**). It is also demonstrated that their phase transition temperatures are depended on molecular weight.
Novel well-defined block copolymers of styrene with **1** or **2** were successfully prepared by sequential anionic living polymerizations. Interestingly, all the block copolymers that have equal weight fraction exhibit smectic mesophase. For PSt-*b*-poly(**1**), a lamellar microphase-separated structure was observed by TEM observation. The SAXS measurement at various temperatures shows that the lamelar spacing decreases significantly through the liquid crystalline state from crystalline to isotropic state. For block copolymers of styrene with **2** with different molecular weights, lamellar and cylinder types of microphase separations were observed.

186. FUNCTIONALIZED ALUMOXANES: ORGANIC/INORGANIC HYBRID POLYMER MATERIALS. D. Brent MacQueen, Georgette Siparsky and Ronald L. Cook, TDA Research, Inc., 12345 West 52nd Avenue, Wheat Ridge, CO 80033

The Clean Air Act of 1990 and the enactment of more stringent state and local environmental regulations are forcing coatings and paint producers to reduce emissions of volatile organic compounds (VOC's) and hazardous air pollutants (HAP's) from their products. In addition environmental regulations are also forcing reduction in the use of chromate conversion coatings in recognition of the hazardous nature of Cr(VI) compounds. At the same time customers are demanding that the performance and durability of the reformulated coatings be maintained or increased. TDA Research, Inc. (TDA) is developing a new class of coating materials that may address the environmental and performance issues for reformulated coatings. TDA has developed chemistries to covalently bond organic functional groups to 50-100 nm boehmite particles. The functional groups can then be used to prepare phenolic, epoxy, acrylic and urethane coatings containing up to 60% solids. We will present our results from the preparation, polymerization and coating studies using the hybrid inorganic-organic materials.

187. FUNCTIONALIZED POLYDIACETYLENES WITH POLAR, HYDROGEN BONDING AND METAL-CONTAINING GROUPS. I. H. Jenkins, C. Murray, W. E. Lindsell, P. N. Preston and T. A. J. Woodman. Heriot-Watt University, Department of Chemistry, Riccarton, Edinburgh, EH14 4AS, UK.

Polydiacetylenes [C(R)C≡CC(R)]$_n$ (PDA's) are of interest for their third-order nonlinear optical behaviour and for their thermo- or solvato- chromic properties. We have previously described synthetic and physical studies on soluble PDA's, including polymers with coordinated metal units. We now report the preparation and characterization of new PDA's incorporating polar, hydrogen bonding and metal-containing groups. In this contribution, the following polydiacetylenes will be described:- (1) ionic ammonium and phosphonium derivatives and ammonium sulfonate inner salts. (2) adenine- and thymine-substituted polymers with the potential for generation of intermolecular hydrogen bonded species in solution. (3) phosphine-and 2, 2'-bipyridyl-containing derivatives and metal complexes derived therefrom. Comparative nonlinear optical studies on selected new materials will be described.

188.

Local motion in NLO main-chain polymers with enhanced orientational stability

Martin Döbler, Christoph Weder, Oscar Ahumada, Peter Neuenschwander, Ulrich W. Suter*
Department of Materials, Institute of Polymers, ETH, CH-8092 Zurich, Switzerland
Stéphane Follonier, Christian Bosshard, Peter Günter
Laboratory of Nonlinear Optics, Institute of Quantum Electronics, ETH, CH-8093 Zurich, Switzerland

ABSTRACT: Polyamides based on 2',5'-diamino-4-(dimethylamino)-4'-nitro-stilbene (DDANS) and aliphatic diacid, and polyamideesters based on DDANS, 4-(bis-(2-hydroxyethyl)amino)-benzaldehyde-1,1-diphenylhydrazone (BBDH) and adipic acid represent a class of polymers for NLO and photorefractive applications, where the nonlinear optical units are fixed in the polymer backbone with their dipole moments oriented transversely to the main chain. The orientational relaxation behaviour of a series of random copolymers and blockcopolymers was investigated at different temperatures below the glass transition by the decay of the nonlinear optical susceptibilities of corona poled thin films. The time dependence of the decay was found to be well represented by the Kohlrausch-Williams-Watts stretched exponential function. The temperature dependence of the decay could be correlated with the glass transition temperature T_g using a normalized relaxation law with $(T_g - T) / T$ as the relevant scaling parameter. The polyamides investigated exhibit an enhanced orientational stability when compared to side-chain or guest-host systems.

189.

NEW CHELATING RESINS CONTAINING AMIDE FUNCTIONAL LINKAGES. Aliecia R. McClain, You-Lo Hsieh, Fiber & Polymer Science, University of California, Davis, California 95616

The preparation and characterization of new chelating resins with selected chelating functional groups containing amide linkages has been investigated. The polymer substrate used in this study is polystyrene crosslinked with 2% divinylbenzene. The synthesis involves the functionalization of the polystyrene substrates containing the reactive intermediates, e.g. acetylated and carboxylated polystyrene. The carboxylated polystyrene was then converted to the acyl-substituted polystyrene in the presence of thionyl chloride. Finally, the chelating ligand (primary amine) is attached to the acylated polystyrene via amide linkages. Characterization of these resins was done by fourier transform infrared spectroscopy (FTIR). Once the conversion process was carefully controlled, the absorption of these chelating resins for metal ions was investigated using atomic absorption spectrophotometry (AAS).

190.

NEW OPTIONS VIA CHEMICAL MODIFICATIONS OF POLYOLEFINS:
PART 1. SYNTHESIS AND PROPERTIES OF NOVEL PHOSPHONIUM IONOMERS FROM POLY(ISOBUTYLENE-CO-BROMOMETHYLSTYRENE). P. Arjunan, H-C. Wang, Exxon Chemical Company, 5200 Bayway Dr, Baytown,
Tx 77522, and J.A. Olkusz, Exxon Chemical Company, 1900 E. Linden Avenue, Linden, NJ 07036.

A new ionomer, - Poly(Isobutylene-co-(4-methylstyrenyl) triphenyl phosphonium tetraphenyl borate) was synthesized by reacting the Poly(Isobutylene-co-(4-bromomethyl styrene) with triphenyl phosphine and sodium tetraphenyl borate in tetrahydrofuran as solvent. This phosphonium ionomer exhibited interesting mechanical properties that were in the range of typical ionomeric thermoplastic elastomers such as Kraton G, Santoprene, etc: tensile strength = 1200 to 1600 psi & % elongation at break = 275 - 652. The thermomechanical behavior of these ionomers indicated the presence of strong ionic interaction which resulted in higher glass transition temperature and retention of modulus upto 70 ^{0}C. A ten fold increase in melt viscosity of these ionomers with reference to their starting material, indicated the strong inter molecular attraction of their ionic clusters. The mechanical, thermal, and rheological properties of the above ionomers were controlled by their molecular structure characteristics such as the molecular weight, ionic content, and type of the counter ion. These novel phosphonium ionomers could be useful in various applications such as impact modifiers, tire-tread components, hoses, adhesives, polymer-bound catalysts, biocides, flame retardants, and membranes.

191.

FREQUENCY DEPENDENCE OF ELECTRICAL BEHAVIOR OF POLYPYRROLE. **Mamoun M. Bader** Department of Chemistry, S. Abdul-Jawad, A. Al-Najjar, Department of Physics; UAE University, P. O. Box 17551, Al-Ain, United Arab Emirates

The effect of frequency at room temperature of the AC resistivity, permitivity and electric modulous for pure polypyrrole and KBr dispersed in polpyrrole powders with different ratios was investigated in the frequency range 1Hz-106 Hz and zero dc bias. The frequency dependence of the AC impedance is similar to that observed in RC network in parallel. The real component of the ac impedance is frequency independent in the low frequency range and becomes proportional to the inverse frequency at high frequencies. The equivalent bulk resistance of the material decreases with increasing KBr content while the capacitive component remains nearly unchanged. Discussion of the effect of electrostatic interactions on the electrical properties and relaxation transitions of polypyrrole will be presented.

192.

MODIFICATION OF PU/PE MEMBRANES USING PHOSPHORYLCHOLINE AND THEIR NON-THROMBOGENIC PROPERTIES Yi-Chang Chung Jui-Hsiang Liu, Department of Chemical Engineering, National Cheng Kung University,Tainan Taiwan 70101, R.O.C.

Abstract

In this investigation, we synthesized a reactive choline dichlorophosphate (CDP), from choline hydroxide. In the other hand, the UV-irradiated graft-polymerization of HEMA onto PE/PU membranes were carried out to generate OH groups on the surfaces. The CDP was then introduced onto the surfaces so that the PE/PU having PC hydrophilic groups were prepared. The synthesis condition of CDP was identified from FTIR and NMR spectra. The surface properties of the grafted or CDP modified PE/PU were all analyzed by ATR, ESCA, SEM, and contact angle measurement for determining the reaction content so as to find out the dependence on the properties and the compositions. Through platelet adhesion tests, the CDP modified PE could restrain the platelet adhesion obviously but the modified PU didn't display such a good compatibility.

193. A NEW CLASS OF POLY(1,6-HEPTADIYNE)-BASED PHOTOREFRACTIVE MATERIALS BY METATHESIS POLYMERIZATION. J.-H. Lee, I. K. Moon[*], H. K. Kim[*], and S.-K. Choi. Department of Chemistry, Korea Advanced Institute of Science and Technology, Taejon, Korea; [*]Department of Macromolecular Science, Hannam University, Taejon, Korea.

A new class of photorefractive polymers based on poly(1,6-heptadiyne) derivatives were synthesized by metathesis cyclopolymerization of 1,6-heptadiyne derivatives containing a carbazole unit and a NLO chromophore. All functional groups are covalently linked to the polymer backbone. These polymers show two maximum values of photocurrent around 350 and 700 nm. The former one might be due to the carbazole groups and the latter might be based on the photodetrapping in shallow electron traps and charge carrier generation in the charge transfer band. Also, the electro-optic coefficient, r_{33}, of the the poled polymer films at the wavelength of 1.3 μm were in the range of 1.7 - 11 pm/V. The nonresonant values of the third-order NLO coefficient, $\chi^{(3)}$, of the resulting polymers were found to be 2.1 x 10^{-11} esu at the incident wavelength of 1.907 μm.

194. SYNTHESIS AND THERMAL PROPERTIES OF PROCESSABLE POLYIMIDES CONTAINING DIACETYLENE GROUPS. Njeri T. Karangu, Mary E. Rezac, Haskell W. Beckham *Polymer Education and Research Center, School of Textile and Fiber Engineering, Georgia Institute of Technology, Atlanta, GA 30332-0295.*

Diacetylene-functionalized processable polyimides have been prepared by the oxidative coupling of a diethynyl diimide monomer based on 6FDA. By controlling polymerization temperature and time, either oligomers or high-molecular-weight (>100,000 g/mol) polyimides can be produced in high yield. High-resolution ^{1}H NMR is used to monitor the polymerization and determine molecular weight via end-group analysis. The 6FDA-based diacetylene-containing polyimide is soluble in many common solvents including acetone. Upon thermal annealing at 200 °C for 24 h, the polyimide may be crosslinked via the diacetylene groups. TGA/DTA curves reveal the disappearance of the diacetylene reaction exotherm for the annealed material. Crosslinking occurs without evolution of volatile byproducts to give a polyimide insoluble in all common organic solvents.

195.

AC ELECTRICAL BEHAVIOR OF A NOVEL ELECTRO-OPTIC POLYIMIDE. **S. Abdul-Jawad** A. Al-Najjar, Department of Physics; M. M. Bader, Department of Chemistry, UAE University, P. O. Box 17551, Al-Ain United Arab Emirates; H. A. Saadeh and L. Yu, Department of Chemistry, University of Chicago, Chicago, IL, 60637, USA

AC electrical behavior of a novel aromatic electro-optic polyimide was investigated in the temperature range 25-300Co and a frequency range 1Hz-1MHz. The measurements were performed on a polymer thin film casted onto ITO glass. The imaginary and real components of resistivity, permitivity, and electric modulous were determined. The results show that the polymer has high thermal stability below 200Co, were the resistivity, dielectric constant and permitivity are nearly temperature-independent indicating highly rigid structure. Above this temperature, however, a well defined broad peak corresponding to a relaxation transition process was observed for which the activation energy was calculated to be ~ 8.5 Kcal/mole. This relaxation is associated with a restricted local rotational motion of the side chain chromophore.

196. **CATALYTIC HYDROLYSIS OF ACTIVE ESTERS USING CROSS-LINKED POLY(BUTYL METHACRYLATE) LATEX PARTICLES**

Paul D. Miller and Warren T. Ford
Department of Chemistry, Oklahoma State University
Stillwater, OK 74078

Polystyrene latexes containing quaternary ammonium sites are highly active as co-catalysts with *o*-iodosobenzoate ion (IBA) for the hydrolysis of *p*-nitrophenyl diphenylphosphate (PNPDPP) (J.-J. Lee and W. T. Ford, *J. Am. Chem. Soc*. **1994**, *116*, 3753) but are inactive for hydrolysis of diisopropyl fluorophosphate (DIFP) (J. Walker and F. Hoskin, Natick RD&E Center). Although PNPDPP is a good simulant for the reactivity of the fluorophosphonate nerve agents, it is a poor simulant on the basis of solubility in polymers. The aromatic PNPDPP is completely absorbed from water by polystyrene latexes, but DIFP is hardly absorbed at all. With the aim of increasing activity for the hydrolysis of aliphatic substrates, monodisperse latexes have been synthesized by copolymerization of 83-63% butyl methacrylate with 15-33% vinylbenzyl chloride, 1% ethylene dimethacrylate as a crosslinker, and 1% (vinylbenzyl)trimethylammonium chloride as a surface stabilizer. Treatment with trimethylamine gave quaternary ammonium groups both on the surface and in the interior of the latex particles. Tests of their activity for hydrolysis of aliphatic *p*-nitrophenyl carboxylic and phosphoric esters in weakly basic buffer solutions will be reported.

197. ANIONIC SYNTHESIS OF MACROMONOMER CARRYING AMINO GROUP USING DIPHENYLETHYLENE DERIVATIVE. J. C. Cho, K. H. Kim, K. U. Kim, J. Kim, Division of Polymer, Korea Institute of Science and Technology, P. O. Box 131 Cheongryang, Seoul, Korea, W. H. Jo, Dept. of Fiber and Polymer Science, Seoul National University, Seoul, Korea, R. P. Quirk, Institute of Polymer Science, The University of Akron, OH 44325-3909.

Macromonomer carrying amine-functional group was synthesized by the crossover reaction of poly(styryl)lithium with 1,3-bis(1-phenylethenyl)benzene (MDDPE). All reactions were carried out in all glass, sealed reactor, using break-seals and standard high vacuum techniques. 1-[4-[N,N-Bis(trimethylsilyl)-amino]phenyl-1-phenylethylene was reacted with *n*-butyllithium in benzene/tetrahydrofuran (THF) for the preparation of the functionalized oligomeric alkyllithium, followed by polymerization of styrene in

198.

NOVEL WELL-DEFINED SIDE CHAIN LIQUID CRYSTALLINE POLYMERS BY ANIONIC POLYMERIZATION. **Wenyue Zheng** Thomas Epps and Paula Hammond, Department of Chemical Engineering, Massachusetts Institute of Technology, Cambridge, MA 02139

A series of chiral smectic C side chain liquid crystalline polymers was synthesized via the anionic polymerization of optically active methacrylate monomers containing alkyl and oligo(oxyethylene) spacers and (S)-2-methyl-1-butyl-4'-(((4-hydroxyphenyl)carbonyl)oxy)-1,1'-biphenyl-4-carboxylate mesogens. Well defined polymers with narrow molecular weight distribution could be approached in this technique. The mesomorphic behaviors in the polymers were characterized by using DSC, optical microscopy and WAXD. All polymers exhibit a smectic C* phase (Sc*) over a wide temperature range, even near room temperature. The Sc* phase can exhibit ferroelectric properties which have great potential in applications such as display devices, light valves and optical switches. Free radically synthesized polymers with analogous molecular weights were also studied to reveal the effects of polydispersity and tacticity on the liquid crystalline phase behaviors.

199. DESIGN OF POLYMERIC SYSTEMS FROM FUNCTIONALIZED NANOSCOPIC GRID-LIKE COORDINATION ARRAYS. Ulrich S. Schubert, Jean-Marie Lehn, Jörg Hassmann, Cathérine Y. Hahn, Paul Müller, Laboratoire de Chimie Supramoléculaire, Institut Le Bel, Université Louis Pasteur, 4 Rue Blaise Pascal, 67000 Strasbourg, France; Physikalisches Institut III, Universität Erlangen-Nürnberg, Erwin-Rommel-Str. 1, 91058 Erlangen, Germany

The design of two or three dimensional coordination arrays -- formed by programmed self-assembly of inorganic architectures from mixtures of organic ligands and metal ions - represents a major current goal in supramolecular chemistry. Such systems may potentially display new magnetic, photochemical or redox properties and as a result be used as building blocks for novel functional supramolecular polymers. A recent investigation into synthesis and properties of tetranuclear [2 x 2]-grid complexes incorporation metal ions such as cobalt(II) or zinc(II) has been extended to include peripherally functionalized analogues. The hydroxy-functionalized coordination arrays were found to form extended molecular assemblies, as visualized by microscopic techniques. First results in this area will be presented.

200.

POLYPROPYLENE REINFORCED WITH LONG GLASS FIBERS. Reinoud J. Gaymans, University of Twente, PO Box 217, 7500 AE Enschede, The Netherlands

The tensile properties of Polypropylene can strongly be enhanced with glass fiber reinforcement. If the fibre length is increased the properties are further enhanced. Particular the impact strength increases strongly with the fibre length. The long fiber reinforced materials were prepared with a melt impregnation set up and subsequently injection moulded. The deformation behavior of fibre reinforced PP were studied as function of fibre length. The main energy absorbing process is fibre pull out. With increasing test speed the fracture energy first decreased but increased again at high test speeds. During the fracture the deformation zone is warmed up, as was observed with an Infra Red Camera. The critical fiber length and the pull out length were found to increase with test speed.

201. EFFECT OF COMONOMER DISTRIBUTION OF LLDPE ON MISCIBILITY BEHAVIOR WITH HDPE. S. Y. Lee and J. Y. Jho, Department of Chemical Technology, Seoul National University, Seoul 151-742 Korea; Y. C. Lee, Chemical Process Engineering Team, KAITECH, Seoul 153-020 Korea

The relation between the comonomer distribution of LLDPEs and the miscibility in the blends with HDPE was investigated. LLDPEs prepared with Ziegler-Natta catalysts (ZNLLD) have heterogeneous distribution of comonomer units along the chain, while LLDPEs from metallocene catalyst (MLLD) possess relatively uniform distribution. The miscibility of ZNLLD/HDPE and MLLD/HDPE blends were examined by investigating morphology, thermal and mechanical behaviors. The results indicated that ZNLLD was miscible in a wider range of compositions with HDPE than MLLD was. Crystallinity was, however, observed to be insensitive to miscibility due to cocrystallization. We concluded that LLDPE with heterogeneous comonomer distribution was more miscible to HDPE than the LLDPE with homogeneous comomer distribution.

202.

SILOXY SUBSTITUTED METALLOCENE CATALYSTS FOR OLEFIN POLYMERIZATION Reko Leino, Hendrik J. G. Luttikhedde, Carl-Eric Wilén and Jan H. Näsman*, Laboratory of Polymer Technology, Åbo Akademi University, Porthansgatan 3-5, Åbo, Finland, FIN-20500

Metallocene catalysts have received considerable academic and industrial interest in olefin polymerization. Recently, we reported the preparation of the first siloxy substituted bis(indenyl) ansa-zirconocene. We describe here the synthesis and characterization of new siloxy substituted metallocenes and their application in polymerization of olefins. Structure/property relationships and different substitution patterns are discussed. The paper summarizes the present status of our research on hetero atom functionalized metallocenes.

203. LIQUID CRYSTAL POLYMER FABRIC REINFORCED POLYMERIC COMPOSITE MATERIALS, Seong Hun Kim, Dong Keun Lim, * Seung Goo Lee, Young Youp Choi, Department of Textile Engineering, College of Engineering, Hanyang University, Seoul, Korea 133-791, *Chungnam National University, Taejeon, Korea 305-764

The liquid crystal polymer (LCP) fabric instead of short fiber was applied to obtain high mechanical strength polymeric composites. The mechanical properties of the LCP/epoxy and LCP/polyimide (PI) composites were investigated. The LCP/PI composite has poor adhesion between LCP fabric and PI matrix, however the LCP/PI composite shows greatly improved interfacial adhesion and mechanical property after $NH_3(95)/H_2(5)$ plasma treatment. The interlaminar shear strength (ILSS) of LCP/epoxy composites were decreased after heat treatment at 220°C for 1hour. The ILSS of LCP/PI composite was increased with heat treatment. The plasma treated LCP/PI composites have higher ILSS compared to that of untreated composites.

204. THERMOPLASTIC VULCANIZATES BASED ON POLYOLEFINES AND RUBBER WASTE. E.V.Prut, O.P.Kuznetsova, D.D.Novikov, Institute of Chemical Physics of RAS, st.Kosygin 4, Moscow 117977, Russia.

Rubber wastes are valuable, not yet fully used resourses for industry. Proper processing of worn-out tires allows to realise the recovering and reciclying of rubber materials. Blends of isotactic polypropylene (iPP) and low density polyethylene (LDPE) with rubber powders and/or regenerated materials have been investigated. Rubber powders and regenerated materials are the industrial products obtained from worn-out tire. Mixture of rubber wastes and original rubbers have been also used. These are materials which have the properties of conventional vulcanized elastomers. The influence of the different compositions of thermoplastic resin and rubber waste, various curavities, the mixing methods and conditions on the ultimate mechanical properties and modulus has been shown.

205. SOME ASPECTS OF STRUCTURE AND PROPERTIES OF THERMOPLASTIC ELASTOMERS AS FUNCTION OF THEIR PREPARATION. E.V.Prut, N.A.Yerina, L.M.Chepel, A.N.Zelenetskii, Institute of Chemical Physics of RAS, st.Kosygin 4, Moscow 117977, Russia.

Thermoplastics elastomers possess two-facedness in mecanical behavior: vulcanized rubbers around room temperature and thermoplastic resin in the higher temperature. This is caused by the material structure of blend in which the thermoplastic component forms the matrix and the vulcanized rubber component consists of many smoll domens dispersing in the resin matrix. The dynamic vulcanization of blends based on EPDM elastomer with isotactic polypropylene (iPP) in the presence of sulfur as curavite has been investigated. The effect of the presence of oil in blend has been also analyzed. It has been shown the influence of the different composition of EPDM/iPP and the mixing methods on the mechanical (ultimate tensial strength, ultimate elongation and tension set), dynamic and reological properties. The mathematical model of the dynamic vulcanization kinetic and models discribing the modulus have been considered.

206. LOW-TEMPERATURE RELAXATION OF POLYMERS PROBED BY PHOTOCHEMICAL HOLE BURNING S. Machida, S. Tanaka, and K. Horie, *Graduate School of Engineering, The University of Tokyo, 7-3-1 Hongo, Bunkyo-ku, Tokyo 113, JAPAN*

Photochemical hole burning (PHB) is one of the effective methods to study structure and properties of glassy polymers at low temperatures. We measured thermally induced irreversible hole broadening to study local segmental motion of mainchain for several polymers. As polymer matrices we used polyethylene, poly(vinyl alcohol), poly(vinyl chloride), poly(vinylidene chloride), poly(methyl methacrylate), poly(methyl acrylate), polystyrene, and poly(ether ketone) and poly(ether sulfone) derivatives. They are doped with a tetraphenylporphine dye molecule by solvent casting. Polymers with chlorine atoms showed smaller extent of the hole broadening or the segmental motion of the mainchain. For polymers with aromatic mainchain larger local free volume should induce the local motion of mainchain. Poly(vinyl alcohol) showed the smallest extent of relaxation, which suggests that the intermolecular hydrogen bonding of hydroxyl groups suppresses the low-temperature relaxation of the mainchain.

207. PHYSICAL AGING OF POLYSTYRENE: VOLUME AND ENTHALPY RECOVERY. Sindee L. Simon, Donald J. Plazek, Brant J. Harper, and Thomas Holden, University of Pittsburgh, Pittsburgh, PA 15261

Volume and enthalpy recovery measurements have been performed to study the physical aging behavior of a polystyrene using capillary dilatometry and differential scanning calorimetry, respectively. Isothermal aging temperatures from 94 to 100°C were studied with aging times ranging up to several days. The volume decreases during physical aging and levels off at equilibrium. Analogously, the difference between the enthalpy of aged and unaged DSC curves increases and levels off at equilibrium. For comparison purposes, both sets of data are normalized to yield the departure from equilibrium which varies from unity at very short aging times to zero when equilibrium is reached. It appears that both volume and enthalpy come into equilibrium at the same time at 97°C. However, the approach to equilibrium differs, with volume recovery being faster than the enthalpy recovery at short times.

PMSE

208. **CONVERSION-TEMPERATURE-PROPERTY RELATIONSHIPS IN THERMOSETTING SYSTEMS: PROPERTY HYSTERESIS DUE TO MICROCRACKING OF AN EPOXY/AMINE THERMOSET-GLASS FIBER COMPOSITE.**
Amy S. Vallely and John K. Gillham, Department of Chemical Engineering, Princeton University, Princeton, NJ 08544

A single specimen of an epoxy/amine thermoset-glass fiber composite was examined for different conversions (as measured by T_g) from $T_{g0} = 0°C$ to $T_{g\infty} = 184°C$ by temperature scans. T_g was increased by heating to higher and higher temperatures. The data was used in two ways: i) vs. temperature for a fixed conversion to obtain transitions, modulus and mechanical loss data, and ii) to obtain, by crossplotting, isothermal values of the mechanical parameters vs. conversion (T_g). Hysteresis between cooling and subsequent heating data was observed in temperature scans of essentially ungelled material ($T_g < 70°C$) and was attributed to microcracking. Hysteresis was analyzed in terms of the following three parameters: T_{crack}, the temperature corresponding to the onset of microcracking on cooling; T_{heal}, the temperature at which the specimen heals on heating; and the difference between isothermal cooling and heating data vs. conversion. Results were incorporated into a conversion-temperature-property diagram which serves as a framework for relating transitions (relaxations) to macroscopic behavior.

209. UNDERSTANDING AND CONTROLLING MELT CRYSTALLIZATION OF GLASSY PHOSPHATE POLYMERS. J. U. Otaigbe, Department of Materials Science and Engineering, Iowa State University, Ames, Iowa 50011; B. C. Sales, Oak Ridge National Laboratory, Oak Ridge, TN 37831; G. H. Beall, Science and Technology Group, Corning Inc., Corning, NY 14831

To discover how glassy phosphate polymers crystallize under melt processing conditions, we used a combination of simultaneous rheology, XRD, and HPLC measurements to selectively observe and to characterize the poorly understood shear flow-induced melt crystallization of glassy phosphate polymers. This method has provided characterization strategy that can be used to observe real-time development of both molecular and microstructural features in glassy phosphate polymers as it is being sheared and heated. The glassy phosphate polymer was found to be amorphous at room temperature and semi-crystalline above its glass-transition temperatures. Higher temperatures and shear distortion rates increased the crystallization rate of the glasses, producing significant levels of crystalline matter after melt processing at 400 °C. The exponential rise in viscosity at short times under oscillatory shear flows confirmed the shear-induced crystallization. This process-induced crystallization of the glasses must be controlled, possibly during processing and/or glass formulation, to produce new materials and devices.

210. **SYNTHESIS AND CHARACTERIZATION OF NOVEL TRIBLOCK COPOLYMERS OF NYLON 6.**
M.V.Pandya* and Mahesh Subramaniyam, Department of Chemistry, I.I.T, Powai, Bombay 400-076 INDIA.

Nylon 6 is one of the most important engineering thermoplastic having many applications. However due to its high water absorption and poor notch impact strength its application is limited. To overcome these limitations and to widen the scope of applications block copolymers of' A-B-A' are synthesized inserting elastomeric soft segments.. The commonly used elastomeric blocks like polyether type or polybutadiene type either increase the water absorption or undergoes gross phase separation. To avoid this problem a novel hybrid soft segment of polyurea urethane with polyether (PEG, Mn =400 or PPG, M_n = 4000) and polybutadiene (HYCAR ATBN) have been inserted.

In this paper two series of block copolymers are synthesized from ε caprolactum, Hexa methylene diisocyanate (HDI) and polyether diol (Series I, PEG and Series II, PPG). The detailed synthetic procedure is discussed. Various polymers are prepared by varying the concentration of soft segment from 10 to 40 weight percent. The polymers are characterized by various spectroscopic techniques. The physical, mechanical, thermal and SEM data of the polymers are reported.

211. INVESTIGATION OF GLASS-IONOMER CEMENTS USING XRPD, SEM AND DSC.
Safaa H. Khalil*, Y. Aboush[+], E.D. Atkins*, J.S. Shah*, H. H. Wills Physics Laboratory, [+]Dept. of Oral and Dental Science, University of Bristol, Bristol, BS8 1TL, UK

Six commercial glass-ionomer cements (GICs) commonly used for various dental applications have been characterised using X-ray power diffraction (XRPD), scanning electron microscopy (SEM) and differential scanning calorimetry (DSC). XRPD patterns were recorded for the original powder component, ashed powder components at 450°C and powdered set cement of each product. SEM micrographs were obtained for the fractured surfaces of matured cements. The heat flow behaviour and heat capacity of the cements were measured during isothermal (at 37°C) setting reactions. XRPD results reveal that two products of GICs are amorphous and four are crystalline. The crystalline structure of these products are unexpected of glass-like structures. SEM micrographs show that fractured surfaces have different sizes and shapes of filler grains, some with jagged edges, and the cracks propagate the intergrain regions. Some materials showed grain failure. DSC results show that all materials undergo an exothermic setting process, but with different enthalpies of reactions and different heat capacities.

212. OPTICAL LOSS IN CURED EPOXY SYSTEM   J.-H. Park and H. H. Song, Department of Macromolecular Science, Hannam University, Taejon 300-791, S. Korea

Polymers are one of the promising materials to meet the requirements of the effective nonlinear optical (NLO) materials. In second order NLO materials, chromophores are dispersed in the amorphous polymer or attached to the polymer backbone as a side unit. The polymers are then electrically poled to give second order NLO signals. To minimize the relaxation of the poled chromophores, the cross-linking epoxies and polyurethane are used as the host matrices and proven to be very effective to reduce the relaxation of the poled chromophores. In the cross-linking system, however, the high transmission loss due to the structural inhomogeneity has limited the use of cross-linking polymers. In this work, we have tried to study the optical loss and the changes along with the structural changes during the cross-linking process of epoxy/EDA. A UV-VIS spectrometer was used to monitor the cross-linking reactions and the optical loss changes simultaneously. The initial homogeneity of the mixture epoxy/EDA turned out to be the key factor to determine the optical transparency of the cured epoxy. The slow heating allowed the mixture uniform, resulting in a cured epoxy having the low transmission loss.

213. MISCIBILITY BEHAVIORS OF ALKYLSULFONYLMETHYL SUBSTITUTED POLY(OXYALKYLENE) BLENDS. Jong-Chan Lee, Morton H. Litt, and Charles E. Rogers, Department of Macromolecular Science, Case Western Reserve University, Cleveland, OH 44106

Miscibility behaviors of alkylsulfonylmethyl substituted poly(oxyalkylene)s were studied. Di-alkylsulfonylmethyl substituted poly(oxytrimethylene)s were found to be miscible with alkylsulfonylmethyl substituted poly(oxyethylene)s over quite large solubility parameter ranges, even though no strong specific interaction is expected because the polymers have similar backbones and the same side chains. This miscibility behavior is probably due to volume contraction upon mixing caused by the free volume differences of the polymer pairs.

214. A STUDY OF THE EVOLUTION OF MECHANICAL PROPERTIES AND STRUCTURAL HETEROGENEITY DURING UV-INITIATED POLYMERIZATIONS OF DIMETHACRYLATE DENTAL RESINS Anandkumar R. Kannurpatti[++], Sheldon M. Newman[+] and Christopher N. Bowman[++,*], [+]Department of Restorative Dentistry, University of Colorado Health Sciences Center, Denver, CO 80262, [++]Department of Chemical Engineering, University of Colorado, Boulder, CO 80309-0424.

Photocured crosslinked dimethacrylate networks are used as dental restorations. To use effectively these glassy materials a fundamental understanding of the evolution of mechanical properties as well as structural heterogeneity is essential.   In this work, a recently introduced technique employing living radical polymerizations to photopolymerize these networks is used. By this technique, the problem of radical trapping encountered while using conventional initiators is avoided. Because there is no radical trapping, mechanical properties can be studied as a function of double bond conversion. From such studies, quantitative conclusions regarding the degree of heterogeneity of the networks are drawn. Specifically, p-xylene bis(N,N-diethyl dithiocarbamate) has been used as the initiator to study the evolution of properties and the structural heterogeneity of the crosslinked polymers formed by photocuring a comonomer mixture containing 2,2-bis(4-(2-hydroxy-3-methacryloxyprop-1-oxy)propane (BisGMA) and triethylene glycol dimethacrylate (TEGDMA) in the ratio 3:1.

215. GAS PERMEATION AND FREE-VOLUME PROPERTIES PROBED BY POSITRON ANNIHILATION. Xiaoyong Hong, Jen-Pwu Yuan, Chia-Ming Huang, Huimin Cao, and Yanching Jean, Department of Chemistry, University of Missouri - Kansas City, Kansas City, Missouri 64110-2499; and Hsinjin Yang, Eastman Chemical Company, Kingsport, TN 37662.

Free-volume hole sizes, fractions, and distributions in gas-exposed polymers have been measured by using positron annihilation lifetime spectroscopy. Hysteresis is observed during CO_2 sorption/desorption, while no such effect is observed in a N_2 system. Microstructural changes are discussed in terms of free-volume hole dynamics.

216. FILM FORMATION OF LATEX POLYMERS: THE ROLE OF BASE POLYMER AND pH. P. Nagarajan, Expanded Incorporation, Pawne, New Mumbai - 400 705, M. K. Trivedi and C. K. Mital, Department of Chemical Engineering, Indian Institute of Technology, Bombay - 400 076, India.

Three different latex interpenetrating polymer networks (LIPN) of varying hydrophilicity were prepared using polyacrylate (methyl methacrylate (MMA), butyl acrylate, methacrylic acid and tetraethylene glycol dimethacrylate - 43:57:2.5:0.7 by wt.) seed (I) and second polymer comprised either of polystyrene (PS) or poly(methyl methacrylate) (PMMA) or a copolymer of MMA and styrene (P(MMA-co-S)), with polymer I and II in the ratio of 65:35. The films were cast from latexes as such at a pH of 3.3 and neutralized with alkali at a pH of 9.0 at 30°C. The non-neutralized latexes of PS and (P(MMA-co-S)) LIPN yielded translucent films, while transparent films were obtained from PMMA based counterpart. However, all the neutralized latexes produced high strength transparent films indicating better coalescence. Latexes were characterized by surface tension measurements, and films by ATR-FTIR, tensile test and optical microscopy. These measurements on non-neutralized LIPN clearly indicate that adsorption of surfactant (sodium dodecyl sulphate) molecules on the particles of styrene based latexes prevents complete coalescence leading to translucent films, whereas, the exudation of surfactant molecules in PMMA based ones results in better coalescence.

217. The Fundamental Aspects of Sample Preparation and Physical Aging Below the Beta Transition of a Poly(amide-imide)

George Dallas*, Thomas Carl Ward**, James Rancourt***
*Beloit Manhattan Research and Development, Ivy Park Clarks Summit, Pa 18411; **NSF Science and Technology Center: Department of Chemistry, Virginia Polytechnic Institute and State University, Blacksburg, VA 24061; ***Polymer Solutions 1872 Pratt Dr. Blacksburg, VA 24060

The intent of the project was to distinguish between the effects of water and low temperature aging on the dynamic mechanical properties of a poly(amide-imide). Proper drying above 190°C was necessary to remove water and NMP. This was determined by thermogravimetric analysis-mass spectroscopy. The effect of water was concentration dependent. At concentrations, below 2 weight percent water appeared as a low temperature transition from -120 to 50°C. Above 2 weight percent the water distorted the beta transition. The enthalpy of activation for the beta transition was dependent upon aging temperature and time, analogous to physical aging of glass transitions. Elevated aging temperatures, closer to the beta transition temperature resulted in higher activation enthalpies at shorter times. A comparison of these results to others lead to generalizations on water ingression and physical aging below the beta transition.

218. PHYSICAL AGING OF MISCIBLE PS/PPO BLENDS AS MEASURED BY VOLUME, ENTHALPY, AND MECHANICAL PROPERTY CHANGES
Christopher G. Robertson and Garth L. Wilkes
Chemical Engineering Department, Polymer Materials and Interfaces Laboratory
Virginia Polytechnic Institute and State University; Blacksburg, VA 24061-0211

The compositional dependence of the physical aging process, as followed by isothermal enthalpy, volume, and mechanical creep compliance changes, was examined for miscible blends formed from atactic polystyrene (PS) and poly(2,6-dimethyl-1,4-phenylene oxide) (PPO). The volume relaxation rate (β_V) exhibited a nearly linear increase with increasing PPO content for aging performed 60°C below the inflection glass transition temperature (T_g) but exhibited a sharp rise at high PPO contents for aging at temperatures closer to T_g. The presence of a secondary dynamic mechanical relaxation in PPO, which appears to manifest itself in the blends at the same temperature/frequency as in PPO, was considered responsible for this dependence of β_V on composition and aging temperature. The variation of the enthalpy relaxation rate (β_H) with blend composition was similar to that found for the corresponding volume relaxation rate for aging performed 30°C below T_g. Mechanical creep testing during aging at T_g-30°C indicated that the mechanical shift rate (μ) reached a compositional minimum near 75 wt.% PPO, a trend clearly in contrast to the β_V and β_H data. It was suggested that the μ data reflected compositional variations in mobility, variations influenced by specific interactions and the associated negative change in volume upon mixing for the PS/PPO blends. These mobility variations were experimentally evident in differences in the damping characteristics at the α transition.

219. DIFFUSION OF AZO-DYES IN GLASSY POLYMERS PLASTICIZED BY SUPERCRITICAL CARBON DIOXIDE. S.G. Kazarian, B.L. West, M. F. Vincent, N. H. Brantley and C. A. Eckert, School of Chemical Engineering and Specialty Separations Center, Georgia Institute of Technology, Atlanta, GA 30332-0100.

Supercritical carbon dioxide induces plasticization of glassy Poly(methyl methacrylate) (PMMA), increases polymer free-volume, and swells the polymer matrix. These phenomena result in increased mobility of the ester side groups of PMMA, thus enhancing diffusion of solutes in the polymer matrix. *In situ* UV/Vis spectroscopy has been used to study diffusion of azo-dyes (such as Disperse Red 1 and 4-(diethylamino)-4'-nitroazobenzene (DENAB) in PMMA films plasticized by supercritical CO_2. The diffusion of azo-dye can be "tuned" simply by changing pressure of CO_2 surrounding the polymer film. Higher pressure of CO_2 enhances diffusion of a dye in PMMA. The diffusion of a dye in CO_2-swollen PMMA can also be influenced by specific intermolecular interactions. We found the diffusivity of Disperse Red 1 in plasticized PMMA to be about an order of magnitude lower than the diffusivity of DENAB under the same conditions. Presumably H-bonding between hydroxyl group of Disperse Red 1 and the carbonyl group of PMMA causes retardation of dye diffusion even in CO_2-swollen PMMA.

220. TIME - TEMPERATURE AND TIME - STRAIN SUPERPOSITION IN POLYCARBONATE BELOW Tg. P.A O'Connell and G.B McKenna, Polymers Division. National Institute of Standards and Technology, Gaithersburg, MD 20899

Data are presented on the stress relaxation response of a polycarbonate under torsional load. Test were performed on samples over a range of strains and temepratures. A stretched exponential form was found to describe the relaxation response over time scales of 3-4 decades. time-temeprature superposition was found to be applicable in the sub-Tg region, though the master curve was found not to be described by the stretched exponential function. For strains up to 0.07 the data at each temeprature could be superimposed to form a master curve following the principle of time-strain superposition. Interestingly, the master curves found from time-strain and time-temperature superposition did not have the same form.

Keywords : glass transition, KWW function, polycarbonate, stress relaxation, time-strain superposition, time-temperature superposition.

221. TOUGHENING OF EPOXY RESINS WITH DESIGNED CORE-SHELL PARTICLES. King-Fu Lin* and Yeow-Der Shieh, Institute of Materials Science and Engineering, National Taiwan University, Taipei, Taiwan 10617, Republic of China

Various types of reactive core-shell particles (CSP) were prepared by soapless emulsion polymerization with butylacrylate as a core and methylmethacrylate copolymerizing with various concentration of glycidyl methacrylate (GMA) as a shell. Ethylene glycol dimethacrylate was used to crosslink either core or shell. The prepared CSP were then employed as a toughener for the DGEBA epoxy resin crosslinked with m-phenylene diamine. Their interfacial bonding to the epoxy matrix and degree of coagulation, both controlled by the content of GMA and the crosslinking status of CSP, were found to be a key role to regulate their toughening effect. After the fracture test, the plastic flow of epoxy matrix surrounding the CSP due to the strong interfacial bonding was observed. Local small-scale coagulation of CSP was capable to fuse into large particles, whose deformation triggered more plastic flow. The observed plastic flow was believed to initiate the local yielding, and thus absorbed more fracture energy.

222. THE DEFORMATION AND FRACTURE BEHAVIOR OF PMMA FILLED WITH HETEROCOAGULATED COMPOSITE PARTICLES Y.J. Park H.J. Ha, and J.H. Kim, Department of Chemical Engineering, Yonsei University, Seoul, Korea, 120-749

The morphology of the rubber toughened composite particles provide strong influence on the mechanical behavior of the base polymer matrix. Also, the properties along the interface, for example, interface thickness, grafting efficiency, adhesion between phases, etc., is essential for improvement in the impact strength. To clarify the toughening behavior, a series of model composite particles having different chain structure and functionality were prepared by heterocoagulation technique, whereby the interfacial properties of the composite can be modified. This synthesis method made it possible to vary the degree of grafting, thickness of interface, molecular interdiffusion and interaction between phases without changing the other physical or chemical properties. The measurement of tensile strength, impact strength, and dynamic mechanical thermal properties were made to evaluate the deformation and fracture behavior of PMMA filled with heterocoagulated composite particles.

223. MOISTURE EFFECTS ON THE MOLECULAR RELAXATIONS IN SEMI-AROMATIC POLYAMIDES. Y. Park, J. Ko*, T.-K. Ahn & S. Choe, Dept. Chem. Engg., Inst. Poly. Sci. & Engg., Inha Univ., Inchon, Korea, *Keum-Ho Petroleum Co., Ltd., Yu-Chun, Chun-La Namdo, Korea

The influence of moisture absorption on the primary (glass) transition (T_α or T_g) and the low temperature relaxations of semi-aromatic amorphous polyamides has been studied using DSC and DMTA. The glass transition of semi-aromatic polyamides was lowered due to the water absorption, and the β and the γ relaxations were as well. The depressed β relaxation observed in the specimen containing a few percent of moisture was splitted into two transitions due to the reduction of water content, of which one was the elevation of the T_β and another was the simultaneous appearance of the T_γ, and then the single T_γ solely was observed for the completely dried specimen. The T_γ seemed to be merged into or not to be observed by the large and broad T_β transition when the sample was governed by a few percent of water, then it was emerged from the T_β due to water desorption. Thus the T_β is believed to be arised from the intermolecular hydrogen bonding between water molecules or between water and amide groups in wet polyamides. In addition, the γ relaxation originated from the peptide groups is attributable to the inter- and intramolecular hydrogen bonding between amide groups.

224. MISCIBILITY AND FREE VOLUME IN LIQUID CRYSTALLINE POLYMER BLENDS G. P. Simon*, T.T. Hsieh [+], C. Tiu [+], *Materials Engineering and [+] Chemical Engineering, Monash University, Australia 3168

Thermotropic main-chain liquid crystalline polymers have in themselves advantageous properties of low viscosity, die swell and thermal expansion and high modulus. Due to their expense and useful properties, they have mainly been blended with random coil thermoplastics. Blends of two liquid crystalline polymers have been much less studied, despite often showing synergistic properties of low viscosity and high modulus, even greater than either component. This poster will look at blends of LCPs with different copolymer constituents and copolymer ratios to determine which seem to result in miscibility in rigid rod mixtures. This will be done using relaxational techniques such as dielectric relaxation, dynamic mechanical analysis and the free volume technique, positron annihilation spectroscopy which will be compared to density measurements.

225.

PREPARATION OF POLYPHOSPHAZENE HYDROGEL MICROSPHERES BY COACERVATION Alexander K. Andrianov, Jianping Chen and Lendon G. Payne, Virus Research Institute, Inc., 61 Moulton St., Cambridge, Massachusetts 02138

A new "coacervation" method of preparing ionically cross-linked polyphosphazene hydrogel microspheres with a controlled microsphere size distribution has been developed. The synthesized microspheres can be used for controlled release applications and as vaccine delivery vehicles. The process is highly reproducible and generates microspheres with a more narrow microsphere size distribution compared to the previous technologies. The effect of microencapsulation parameters on the physico-chemical properties of microspheres is studied and will be discussed.

226.

WATER-SOLUBLE IONIC COPOLYMERS WITH POLYPHOSPHAZENE BACKBONE AS MATERIALS FOR CONTROLLED RELEASE TECHNOLOGIES. Alexander K. Andrianov, Jonathan R. Sargent, Sameer S. Sule, Mark P. LeGolvan and Lendon G. Payne, Virus Research Institute, Inc., 61 Moulton St., Cambridge, Massachusetts 02138

A series of mixed-substituent poly(organophosphazenes) containing alkyl ether and carboxylic acid groups was synthesized by the sequential modification of poly(dichlorophosphazene) with sodium salts of the corresponding alcohols followed by treatment with base. These polymers have potential for use in controlled release and vaccine delivery systems. The effect of polyphosphazene composition on the physico-chemical properties of the polymer such as solubility, polymer conformation, and degradation in aqueous solution was studied.

227.

DEGRADATION OF POLY[DI(CARBOXYLATOPHENOXY)PHOSPHAZENE] IN AQUEOUS SOLUTION. Alexander K. Andrianov, Mark P. LeGolvan, Sameer S. Sule and Lendon G. Payne, Virus Research Institute, Inc., 61 Moulton St., Cambridge, Massachusetts 02138

The degradation of poly[di(carboxylatophenoxy)phosphazene] (PCPP) macromolecules in aqueous solution was studied by monitoring molecular weight changes and the release of hydroxybenzoic acid HBA. PCPP was incubated at pH 7.4 at various temperatures from 4° C to 55° C. Model PCPP polymers, containing hydroxyl groups, were also studied. The results suggest a weak link mechanism for the degradation of PCPP.

228. 17β-ESTRADIOL, 3-KETODESOGESTREL TRANSDERMAL DRUG DELIVERY SYSTEM
J.D. Audett, T.F. Chen, M.L. Park, A. Ramdas, O. Wong, CYGNUS Inc., Redwood City, CA 94063

A hormone replacement therapy (HRT), transdermal drug delivery system has been formulated using an acrylate pressure sensitive adhesive and polyvinylpyrrolidone blend. The two drug substances, 17β-estradiol and 3-ketodesogestrel, are completely solubilized in the dry polymer blend. The solubility of 17β-estradiol in the polymer blend and its impact on the in vitro drug delivery characteristics will be discussed. This formulation provides substantial drug delivery without the use of skin permeation enhancers because the solubility of 17β-estradiol in the adhesive matrix decreases as the matrix is hydrated. It is hypothesized that hydration during wear decreases the solubility of 17β-estradiol in the polyvinylpyrrolidone, temporarily supersaturating the adhesive matrix. Supersaturation leads to higher skin flux until the extent of supersaturation decreases because of estradiol crystallization. Some pharmacokinetic data, consistent with the in vitro skin flux results, is presented.

229.
SYNTHESIS AND CHARACTERIZATION OF LACTIC/GLYCOLIC ACID OLIGOMERS
Nuo Wang[1], Xue Shen Wu[1,2], Hannia Lujan-Upton[3], Edward Donahue[3], Asiful Siddiqui[1]
1. Division of Pharmaceutics and Industrial Pharmacy, Arnold & Marie Schwartz College of Pharmacy and Health Sciences, Long Island University. 2. Herman F. Mark Polymer Research Institute, Polytechnic University. 3. Department of Chemistry, Long Island University, Brooklyn, NY 11201, USA.

A series of lactic/glycolic acid oligomers (PLGA) has been synthesized. These PLGA oligomers have advantage of faster biodegradation and are more suitable for short term drug delivery applications. Molecular weight of the oligomers ranges from 895.8±48.7 to 1368±0 daltons and intrinsic viscosity varies from 0.0513 to 0.0814 dl/g. The oligomer composition correlates to the feed ratios of lactic to glycolic acid. Glass transition temperatures of the PLGA oligomers are lower than physiological temperature (37 °C), which results in a rubbery state of the oligomers at the physiological temperature.

230. SUSTAINED RELEASE OF THERAPEUTIC AGENTS FROM ERODIBLE POLYMERIC HYDROGELS. M.S. Goldenberg, A. Beekman and G. Scott, Pharmaceutics Department, Amgen Inc., Thousand Oaks, CA 91320

Multiple injections of aqueous solutions of therapeutic agents for sustained delivery can be both inconvenient and inefficient. These are among the difficulties that have stimulated research into polymer based delivery systems. The topic to be discussed will be release of active ingredients from erodible hydrogels.

231.
CARRIER MEDIATED ORAL BIOAVAILABILITY OF PROTEINS
John E. Smart, Ph.D., Vice President of Research
Emisphere Technologies, Inc.
15 Skyline Drive, Hawthorne, NY 10532

As a part of ongoing efforts to develop oral drug delivery agents, our laboratory continues to design and synthesize an extensive library of carrier molecules that have the ability to facilitate oral delivery of proteins and other therapeutic compounds, which are currently administered through injection. A positive correlation has been demonstrated between carrier mediated stabilization of intermediate protein conformations and carrier mediated oral bioavailability. The efficacy of these carriers has been demonstrated in multiple *in vivo* and *in vitro* models. Carrier mediated oral delivery of proteins occurs without any evidence of damage to the gastrointestinal tract.

232.
EFFECT OF GRAFT YIELD ON THE TEMPERATURE-RESPONSIVE PERMEABILITY OF PNIPAAM-G-PE POROUS MEMBRANE   Y-L Cheng, T. Peng, Department of Chemical Engineering and Applied Chemistry, University of Toronto, Toronto, Ontario, Canada M5S 3E5

Abstract   Grafting a temperature-responsive polymer, poly(N-isopropylacrylamide) (PNIPAAm), onto porous polyethylene (PE) membranes by UV irradiation was investigated. A wide range of graft yields was achieved by varying irradiation time (20-240 minutes) and monomer concentration (1.2-3.6 wt%). Characterization by XPS and SEM showed that the graft polymers were located both on the external surfaces as well as inside the pores of the membranes. Diffusional permeation experiments showed that (1) membrane permeability decreased with increasing graft yield, (2) permeability varied with temperature, (3) the responsiveness of membrane permeability to temperature increased with graft yield, and (4) two distinct types of temperature responses were observed depending on the graft yield.

233.
CHRONSET® ORAL OSMOTIC SYSTEM CAPABILITIES AND APPLICATIONS,
Joseph.A. Fix, Liang C. Dong, Crystal Pollock, Keru.O. Shafi, Philippe. J.M. Dor and Patrick S.L. Wong Alza Corporation, 950 Page Mill Road, Palo Alto, CA 94303.

The ability to reliably delay drug release, target drug release to a specific region of the g.i. tract or deliver high local drug concentrations would provide pharmaceutical and therapeutic advantages for certain drug classes and disease treatments.
CHRONSET® is an oral osmotically-controlled drug delivery system which reproducibly delivers a bolus drug dose (up to 500 mg) in a time- or site-specific manner to the g.i. tract. CHRONSET® performance and formulation stability are unaffected by the g.i. milieu. An excellent correlation was observed between in vitro opening times (2.46 ± 0.18 hr and 5.97 ± 0.4 hr) and in vivo opening times (2.54 ± 1.1 hr and 6.63 ± 1.4 hr) in a human clinical evaluation.
Incorporation of an active drug expulsion layer affords >80% drug release in less than 15 minutes following CHRONSET® opening. Pharmacokinetic analysis T_{max}, C_{max}, AUC) of 2-hr and 6-hr acetaminophen CHRONSET® systems in human volunteers, compared to standard acetaminophen tablet administration, validated device performance.

234.   A GLUCOSE-SENSING POLYMER.   Guohua Chen, Vidyasankar Sundaresan and Frances H. Arnold, Division of Chemistry and Chemical Engineering, MC 210-41, California Institute of Technology, Pasadena, CA 91125

We have prepared a robust polymer that can be used to measure glucose concentrations in complex biological media. At alkaline pH, this metal-complexing polymer binds glucose and instantly release protons in proportion to the glucose concentration over a clinically relevant range (0-25 mM). The inexpensive polymer is sufficiently selective to provide an easily measurable response to glucose in porcine plasma. The polymer's ability to function at nonphysiological pH's (at which the buffer capacity of biological samples is small) makes it possible to design simple and inexpensive sensing devices based on measurement of changes in proton concentration.

235. **The Wettability of LaRC Colorless Polyimide Resins on Casting Surfaces.**
G. A. Miner, D. M. Stoakley, A. K. St. Clair, NASA Langley Research Center, Hampton, VA 23681; P. A. Gierow and K. Bates, SRS Technologies, 500 Discovery Drive, Huntsville, AL 35806

The effect of inherent viscosity, solids concentration, and solvent choice on the wettability of two colorless polyimides was investigated on various casting surfaces. LARC™-CP1 and LARC™-CP2 are optically transparent, radiation resistant, soluble polyimides developed at NASA Langley Research Center that may be used to make transparent, thin polymer films for large space reflector/collector antennas, solar arrays, and radiometers. Structures such as these require large area, seamless films necessitating spin casting or spray coating on mandrels or large casting substrates. From this study, the wetting of the substrate decreased (average contact angle increased) with increasing polymer inherent viscosity and with increasing imide concentration in DMAc. The trend for decreased wettability for increased imide concentration was not observed for the LARC™-CP1 polyimide in diglyme or NMP which could neatly simplify spin casting or spray coating from these solvents.

236.

BREAKABLE FUNCTIONALIZED EPDM TERPOLYMER NETWORKS. C. K. Jones, Nalco/Exxon Energy Chemicals, L.P., Houston, Texas 77478

Breakable epoxy-alkoxy silane (EAS)-modified ethylene-propylene-diene monomer (EPDM) sulfonate terpolymer networks were prepared in hydrocarbon and solution rheology measured as a function of concentration and time. Results indicate gelation time decreases with increasing EAS concentration and base EPDM terpolymer number average molecular weight. Network formation was confirmed to be reversible with change in apparent pH.
The base EPDMs selected (prior to functionalization) in this study were low-to-high (10k to about 300k) number average molecular weight, substantially non-crystalline commercially available grades.
Finally, results from networks prepared with differing EPDM terpolymer diene monomer ("DM") levels (zero to about 9wt%) within a relatively similar number average molecular weight range are also presented, displaying increased tendency towards gelation with unsaturation as anticipated, while maintaining essentially complete network reversibility.

237. NEW NONIONIC POLYMERIC HYDROGELS VIA GRAFT COPOLYMERIZATION.
Kevin D. Belfield* and Guillermina Garcia, Department of Chemistry, University of Detroit Mercy, P.O. Box 19900, Detroit, MI 48219-0900.

Hydrogels are materials that exhibit the ability to swell in water and retain a significant fraction of water within its structure. Their ability to absorb water is due to the presence of hydrophilic groups. Polymeric hydrogels are important in a number of fields including, e.g., matrices for enzyme immobilization, biosorbents in preparative chromatography, materials for agriculture, development of various biomedical systems ranging from biosensors to artificial muscles, site-specific drug delivery systems, and superabsorbant polymers. We wish report a new class of nonionic superabsorbant hydrogels based on the graft copolymerization of cellulose derivatives or poly (vinyl alcohol) with poly (ether urethane-urea) polymers, along with their thermal analysis and swelling properties under several physiologically and commercially important conditions.

238.

IN-SITU POLYMERIZATION METHOD FOR THE PREPARATION OF BONDED POLYMERIC POROUS LAYER OPEN TUBULAR COLUMNS FOR GC APPLICATION   Cindy Zhou, Thomas C. Shen, SETI Institute, NASA Ames Research Center, Moffett Field, CA 94035-1000

A new method to prepare polymeric PLOT columns by using in situ polymerization technology is described. The method involves a straight forward in situ polymerization of the monomer, which eliminates many of the steps needed in conventional polymeric PLOT column preparation. Several columns have been developed for analyzing hydrocarbons and inert gases in the space. A recent improvement has allowed us to produce bonded polymeric PLOT columns. The chemically bonded PLOT metal columns were made by pre-coating trialkoxysilyl-azo initiator onto the inner surface of stainless steel tubing through a formulation with Sol-Gel A, followed by an in-situ polymerization process. The syntheses of trialkoxysilyl-azo initiators, the conditions of polymerization, as well as the behaviors of the columns for separating hydrocarbon light gas mixture are also described in this presentation.

239. DEVELOPMENT OF AQUEOUS ACETYLENE POLYMERIZATION SYSTEMS: SYNTHESIS OF STEREOREGULAR POLY(PHENYLACETYLENES) BY RHODIUM COMPLEXES IN AQUEOUS MEDIA. Ben Zhong Tang,* Wan Hong Poon, Shuk Mei Leung, Wa Hung Leung, and Han Peng, Department of Chemistry, The Hong Kong University of Science & Technology, Clear Water Bay, Kowloon, Hong Kong

Polymerizations of phenylacetylenes (HC≡CC$_6$H$_4$-R, R = H, p-Me) are effected in water by a number of organorhodium complexes including Rh(diene)(tos)(H$_2$O), [Rh(diene)Cl]$_2$, [Rh(cod)Cl]$_2$(pda), [Rh(cod)(mid)$_2$]$^+$PF$_6^-$, Rh(cod)(NCL)Cl, and Rh(cod)(bbpmt), where diene = 1,5-cyclooctadiene (cod), norbornadiene (nbd); tos = p-toluenesulfonate; pda = o-phenylenediamine; NCL = NH$_3$, t-BuNH$_3$, piperidine, N-methylimidazole (mid); and bbpmt = bis(4-t-butyl)-2-pyridylmethylthiolate. Compared to toluene, water has been found to be a much better solvent. Thus, while the polymerizations in toluene yield trace amounts (<3%) of poly(phenylacetylenes), the polymerizations in water are very fast and produce polymers with high molecular weights ($\overline{M}_w$ up to 220x10^3) and high stereoregularity (up to 100% cis) in high yields (up to 98%).

240. SYNTHESIS OF NITROXY-CONTAINING INITIATORS FOR MEDIATED RADICAL POLYMERIZATION. B.A. Howell, B. Pan, and D.B. Priddy, Center for Applications in Polymer Science, Central Michigan University, Mt. Pleasant, MI 48859.

Benzyloxyamines have proven to be very effective initiators/ mediators for styrene and other radical polymerizations. These have often consisted of adducts of hydrocarbon-derived benzylic radical with the stable 2,2,6,6-tetramethylpiperidinyl-]-oxy (TEMPO) stable free radical. Most often these have been prepared by the thermal decomposition of di-t-butylperoxide at]30° in the presence of an appropriate hydrocarbon and TEMPO. However, at this temperature, the decomposition of the derived benzyloxyamine is appreciable such that yields are always modest. It has now been demonstrated that benzyl-oxyamines derived from toluene, ethylbenzene, and cumene can be prepared in essentially quantitative yield using di-t-butylperoxalate as the source of t-butoxy radicals.

241. PHOTOINDUCED ELECTRON TRANSFER IN PYRENE-LABELED POLYNORBORNENES. Renae D. Fossum and Marye Anne Fox, Department of Chemistry and Biochemistry, University of Texas at Austin, Austin, Texas, 78712

Photoinduced intramolecular electron transfer has been observed in polynorbornenes substituted with blocks of acceptors that are terminated with pyrene as the donor. The polymers have been prepared according to living ring opening metathesis polymerization (ROMP) of substituted norbornene acetals and ketals. Homopolymers of cyanonaphthalene and dicyanobenzene terminated with pyrene (2a and 2b, respectively) show fluorescence quenching of pyrene and exciplex formation in the steady-state emission. Copolymers incorporating dicyanobenzene and cyanonaphthalene in varying block lengths and terminated with pyrene (3) show fluorescence quenching as well as the formation of two different exciplexes. The transient absorption spectra of the pyrene-labeled homopolymers show the pyrene radical cation absorption whose lifetime varies according to the nature of the acceptor present. The lifetime of the pyrene radical cation absorption in the copolymers has a slow component that is the same magnitude as that measured in 2a and a long component similar to that measured in 2b. The synthesis, characterization and results of the flash photolysis will be discussed.

242.

THE EFFECTS OF GREEN DENSITY ON FRONT VELOCITY AND PRODUCT MORPHOLOGY IN THE FRONTAL POLYMERIZATION OF ACRYLAMIDE. **Dionne I. Fortenberry*** and John A. Pojman, Department of Chemistry and Biochemistry, University of Southern Mississippi, Hattiesburg, MS 39406

Frontal polymerization has been observed in solid monomers such as acrylamide. The initiator used is potassium persulfate. The effect of green density on front velocity, product morphology, and front temperature has been studied. While green density had little effect on morphology of the polymer, it had an interesting effect on front velocity. Front temperature, as well as thermal diffusivity, also appeared to be affected by green density. NMR, IR, and TGA gave insightful information about polymer structure and side reactions.

243. **NOVEL NAFION®/ORMOSIL AND TELECHELIC POLYMER/ORMOSIL HYBRIDS VIA IN-SITU SOL GEL CHEMISTRY.** S. K. Young, K. Page, K. A. Mauritz, Department of Polymer Science, University of Southern Mississippi, Hattiesburg, Mississippi 39406-0076

Nafion®/ORMOSIL hybrids were generated via *in-situ* sol-gel reactions for tetraethoxysilane:diethoxydimethylsilane, tetraethoxysilane:triethoxyvinylsilane, tetraethoxysilane:aminopropyltriethoxysilane, tetraethoxysilane:hydroquinone, and triethoxyvinylsilane:diethoxydimethylsilane binary monomer mixtures within the polar clusters of perfluorosulfonic acid membranes. Light microscopy was utilized to ensure that ORMOSIL formation occurred within the membrane as opposed to on the surface. FTIR-ATR and ^{29}Si NMR spectroscopy established structural incorporation of the di- and tri- functional silanes into a copolymer network. Preliminary DSC and TGA thermal analyses were performed. Structure-property relationships for these *in-situ* comonomer inclusions were established through gas diffusion and fluorescence probe studies. In addition, telechelic, hydrolytically-degradable polyesters were generated. Further alteration of these telechelic endgroups will allow for covalent attachment of the inorganic network, producing property enhancements.

244. PREPARATION AND PROPERTIES OF AZO AND AZOXY GROUPS CONTAINING AROMATIC POLYETHERS. D. K. Mohanty, A. Bhatnagar and A. Schroder, Department of Chemistry and Center for Applications in Polymer Science, Central Michigan University, Mt. Pleasant, MI 48859

Low molecular weight azo and azoxy containing compounds are of considerable interest. We have prepared a series of azo and azoxy aromatic polyethers. The azo polymers were synthesized by the reactions of bisphenols with 4,4'-difluorobenzophenone in N-methylpyrrolidinone (NMP) in the presence of anhydrous potassium carbonate. Moderate to high molecular weight azoxy aromatic polyethers were prepared by the reactions of four bisphenoxides with 4,4'-difluoroazoxybenzene in NMP. The polymers are amorphous and exhibit trans to cis isomerization upon irradiation. They undergo crosslinking reaction at elevated temperatures. In addition, the azoxy containing polymers rearrange to azo groups with the concurrent production of a phenolic moiety.

245. **NOVEL WATER-SOLUBLE TAPES**
Matthew J. Armitage and Norman C. Billingham
Department of Chemistry, Physics and Environmental Science
University of Sussex, Brighton BN1 9QJ
England

The micro-electronics industry has long used tape casting as a method for producing thin, flat plate ceramics. The process involves using a doctor blade to cast a suspension of ceramic powder in a liquid system which contains a binder polymer. The resulting 'green' tape can be stored, shaped then fired to the hard ceramic.

Polyvinylpyrrolidone has been used as a binder to develop a water-based tape casting system producing water-soluble tapes. Other constituents of the tapes have included water-soluble salts and insoluble inorganic powders. The effect of varying formulation on tape properties will be reported. These tapes could form the basis of water-based ceramics technology or may be useful for controlled delivery of bioactive components.

246. CHAIN FOLDING IN RIGID ROD SYSTEMS Portia D. Yarborough and Dotsevi Y. Sogah, Department of Chemistry, Cornell University, Ithaca, New York 14853

Copolymerization of a rigid xanthene based moiety with mesogenic units provides architectural control of the system. The xanthene based diacid initiates a 180° reversal of the polymer chain leading to a stacked sheet conformation. Preliminary investigations of U-turn containing polyamides using p-phenyleneterephthalamide as the building block show a system possessing a pleated sheet morphology, solvent induced liquid crystalline behavior dependent upon the length of the spacing between the folds, and a correlation between the degree of order and the amount of U-turn segment in the polymer.

247. CHARACTERIZATION OF ASPHALT BINDERS MIXED WITH EPOXY TERMINATED ETHYLENE TERPOLYMER. Y. J. Lee, L. M. France, and M. C. Hawley, Department of Chemical Engineering, Michigan State University, East Lansing, Michigan 48824

An epoxy terminated ethylene terpolymer was used to modify asphalt binders and its optimum content and curing condition in the asphalt blends were determined based on the rheological properties of the binders and economic concerns. The blends were cured at 180°C for 2 hours. A protocol for fingerprinting polymer modified asphalt binders was established by using high performance gel permeation chromatography (GPC) and Fourier transform infrared spectroscopy (FTIR). Fundamental themo-mechanical and rheological properties of the polymer modified asphalt binders were also investigated. Only 1 or 2% (w/w) epoxy polymer in the blend could significantly improve the binder's properties. The goal of polymer modification is to increase the temperature range of both the elastic and viscous properties of the asphalt binders by a formation of network structure and to enhance its performance at both high and low temperatures. It has been shown that polymer-modified asphalts have potential to improve pavement performance.

248. FLAME RETARDANT PHOSPHONIUM SALTS CONTAINING HIGH LEVELS OF BROMINE. B. A. Howell, F. M. Uhl, and K. J. Johnston, Center for Applications in Polymer Science, Central Michigan University, Mt. Pleasant, MI 48859

Highly effective flame retardants for polymeric materials continues to be sought. Those containing both phosphorus and bromine are often particularly effective. A series of phosphonium salts containing high levels of halogen has been prepared by treating 1,3,5-tri(bromomethyl)benzene with selected phosphines. Both highly brominated phosphites and phosphines have also been prepared for evaluation as flame retardant agents.

249. THERMAL STABILITY OF POLY((4-t-BUTOXYCARBONYLOXY)PHENYL) METHYL ACRYLATE). B.A. Howell and B. Pan, Center for Applications in Polymer Science, Central Michigan University, Mt. Pleasant, MI 48859.

Comonomers capable of consuming a mole of evolved hydrogen chloride to expose a phenolic group which might scavenge chlorine atoms have been utilized for the stabilization of vinylidene chloride barrier resins, i.e., to control degradative dehydrochlorination. However, vinylidene chloride polymers containing (4-(t-butoxycarbonyloxy)phenyl) methyl acrylate as comonomer display thermal stability lower than that of the vinylidene chloride homopolymer. Therefore, the degradation of poly((4-(t-butoxycarbonyloxy)phenyl)methyl acrylate) has been examined. The polymer undergoes major decomposition at 150° and 420° to evolve gaseous products and leave a residue of poly(acrylic)acid)

250.

AN ACRYLATE MONOMER CONTAINING THE ACID-SENSITIVE t-BUTYL PHENYL CARBONATE FUNCTIONALITY. B.A. Howell and N.L. Moore, Center for Applications in Polymer Science, Central Michigan University, Mt. Pleasant, MI 48859.

Vinylidene chloride polymers occupy a place of prominence in the barrier plastics packaging industry. At processing temperatures these materials undergo degradative dehydrochlorination. A means to stabilize these materials would be to incorporate a comonomer containing functionality that would react with evolved hydrogen chloride to expose a phenolic site capable of scavenging chlorine atoms. A suitable comonomer is 2-(4-(t-butoxycarbonyloxy)phenyl)ethyl acrylate which contains an acid-sensitive t-butyl phenyl carbonate moiety.

251. **THE PREPARATION OF PIGMENTED MELAMINE-FORMALDEHYDE POLYMER BEADS (PMFB).** J. Shen and G. Crews, Melamine Chemicals, Inc., P. O. Box 748, Donaldsonville, LA 70346

The spherical colored polymer beads of melamine-formaldehyde resins with uniform diameter of 3-20 microns were prepared by water-in-oil (w/o) emulsion polymerization of colored methylolmelamine aqueous solutions. Dyes, when combined with methylolmelamine, provide excellent hue stability. Moreover, this polymeric pigment with crosslinked melamine-formaldehyde resin, as an organic pigment, not only retains higher tinctorial strength and brightness but also exhibits good weather resistant, heat stability and chemical resistant properties. Because the PMFB contains functional groups such as amino group $-NH_2$, hydroxyl group $-OH$, and oxygen bridge $-CH2OCH2-$, it is very easy to obtain a variety of colors with various dyes and easy to disperse in application medium by hydrogen bond or chemical bond or complex bond.

252. SYNTHESIS AND PROPERTIES OF NEW BIODEGRADABLE POLYESTERS DERIVED FROM DIACIDS AND DIGLYCIDYL ETHERS.
Michael N. Mang, Jerry E. White, Shari L. Kram, David L. Rick, Robert E. Bailey, Paul E. Swanson, The Dow Chemical Company, Central R&D, Midland, MI 48674

Materials that combine biodegradability with practical mechanical characteristics have been a synthetic objective for many years. We describe here the synthesis of a series of new biodegradable polyesters derived from the reaction of diglycidyl ethers with aliphatic diacids, using quaternary ammonium halides as initiators in ether solvents. The nucleophilic addition of a carboxylic acid to an oxirane ring produces polymers with both ester linkages and pendent hydroxyl groups. This procedure yields thermoplastic materials that have a broad range of T_g's (-34 °C to 110 °C), with both crystalline and amorphous morphologies. The materials have generally good mechanical properties, and can be readily processed using conventional fabrication techniques. The biodegradability of the polymers was evaluated using a combination of soil burial testing, bacterial growth measurements, and carbon dioxide respirometry.

253. **BIODEGRADABLE COPOLY(ESTER-AMIDE)S BASED ON SHORT, UNIFORM AND SYMMETRICAL AMIDE BLOCKS,** H.R. Stapert[1], M. van der Zee[2], P.J. Dijkstra[1] and J. Feijen[1]. [1]University of Twente, Department of Chemical Technology, Polymer Chemistry and Biomaterials, P.O. Box 217, NL-7500 AE, Enschede, the Netherlands. [2]ATO-DLO, Agrotechnological Research Institute, Bornsesteeg 59, P.O. Box 17, NL-6700 AA Wageningen, The Netherlands.

Well defined block copoly(ester-amide)s were synthesised from aliphatic diesters, diols and a short, uniform and symmetrical diamide block. During polycondensation the molecular structure of the amide blocks was unchanged, indicating that ester-amide interchange reactions were minimal. The poly(ester-amide)s crystallized fast from the melt. Furthermore, optical microscopy revealed a biphasic melt resembling the known batonnet structure. Temperature dependent wide angle X-ray diffraction showed the existence of long-range correlations in the melt. It is hypothesised that due to strong segregation of the amide-blocks a layered structure is obtained consisting of very thin alternating ester and amide layers, which are stabilized by hydrogen bonds. Biodegradation of the monomers and polymers was shown with modified Sturm, Biological Oxygen Demand and soil burial tests. It was shown that the diamide blocks were completely degraded within 40 days. Biodegradation of the polymers decreased with increasing amide content. Fast biodegradation of the polymers was observed in soil burial tests.

254. **PREPARATION OF STARCH SUCCINATES BY REACTIVE EXTRUSION;** Linfu Wang, Randal L. Shogren, and Juliious L. Willett; Plant Polymer Research, United States Department of Agriculture, Agriculture Research Service, Midwest Area, National Center for Agricultural Utilization Research, 1815 N. University, Peoria, Illinois 61604.

The succinylation of cornstarch by extrusion processing has been studied. Several different factors affecting the succinylation of starch were investigated including water content, succinic anhydride quality, catalyst type and quality, and reaction temperature. The degree of substitution (DS) increases with decreasing water content and increasing succinic anhydride quality. Higher temperature favors higher DS using $NaHCO_3$ as catalyst while the opposite is true using $Mg(OH)_2$ as catalyst. Among all the selected catalysts, $NaHCO_3$ is the best. Aqueous solubility and viscosity of the starch succinates tend to increase with increasing DS. Based on the results of aqueous solution viscosity, some samples appear to be promising as commercial thickening agent.

255. **PACKAGING MATERIALS FROM AGRICULTURAL HARVESTING RESIDUES BY STEAM EXPLOSION.** Sheryl D. Baldwin[1], Joseph M. Genco[2], and Wolfgang G. Glasser[3], [1]Philip Morris USA R & D, Richmond, VA 23261, [2]University of Maine, Orono, ME 04469, and [3]Virginia Tech, Blacksburg, VA 24061.

Rye straw, wheat straw, and sugar cane bagasse are the agricultural harvesting residues that were tested for their utility in liner board-type packaging materials by steam explosion and fractionation. Steam explosion was carried out at 216°C for 2.3 minutes corresponding to a severity factor (log R_o) of 3.9. Washing with hot water and extracting with hot alkali under standard conditions produced unbleached pulp fibers in 35 to 50% yield. Component analysis revealed that the fractionation was cleanest for rye straw (less than 5 to 10% of the non-cellulosic components remained in the pulp fibers after fractionation) and least clean for bagasse which retained one-third of its lignin and three-fourths of its ash in the pulp fibers. Evaluation for packaging materials revealed that the cleanest fibers produced the strongest paper. Paper from rye straw was more than twice as strong as paper from bagasse in terms of burst, tear, and tensile strength. The results suggest that steam explosion pulps from agricultural harvesting residues may potentially meet all liner board requirements except tear strength, but the drainage rate of freeness would be lower.

256. **THE INFLUENCE OF EXTRUSION OPERATIONAL VARIABLES ON INTRINSIC VISCOSITY AND EXPANSION CHARACTERISTICS OF STARCH FOAMS.** P. D. Tatarka, Biomaterials Processing Research, USDA-ARS-NCAUR, 1815 N. University Street, Peoria, Illinois 61604

Extrusion cooking technologies are used to produce environmentally benign starch-based foam products. The impact of changes in a myriad of operational variables on performance must be quantified to enable the manufacture of economically competitive industrial packaging and other products. The molecular weight and expansion characteristics of starch-based foams are dependent on many twin-screw processing variables, including moisture and amylopectin content, temperature, screw speed, fill factor and die orifice diameter and L/D ratio. Foams made with 50% amylose starch had the highest volumetric expansion. Its intrinsic viscosity did not decrease during extrusion as much as corn or waxy maize starch, but exhibited the most shear-thinning, pseudoplastic behavior. Molecular weight degradation can be reduced by increasing moisture content or die orifice diameter and reducing die L/D or residence time. Extrusion conditions necessary for molecular weight preservation are usually unfavorable to those needed for expansion. With low residence time, molecular weight degradation of starch extruded under high shear and temperature conditions are minimized. High screw speeds and fill factors and low moisture content reduced residence time.

257. LIPASE-CATALYZED RING-OPENING POLYMERIZATION OF CYCLIC CARBONATES

Kirpal S. Bisht[†], Yuri Y. Svirkin[†], Richard A. Gross[†]*, David L. Kaplan[+] and Graham Swift[✤]; [†]University of Massachusetts Lowell, Department of Chemistry, One University Avenue, Lowell, Massachusetts 01854. [+]Tufts University, Chemical Engineering, 4 Colby Street, Medford, MA 02155. [✤]Rohm & Haas Company, Norristown Road, Spring House, Pennsylvania 19477.

The use of lipases as catalysts for trimethylene carbonate (TMC) ring-opening polymerization was reported. Of the four lipases screened for bulk polymerization of TMC (70 °C, 120 h), Novozym-435 gave almost quantitative monomer conversion (97 %) and poly(TMC) with an M_n=15000 g/mol (MWD or M_w/M_n=2.2) without decarboxylation during propagation. The lipases from *Pseudomonas* (AK and PS-30) and *Mucor* (MAP-10) species also exhibited high polymerization activity and monomer conversions of >80%. However, lower molecular weight polymers with broad MWD were obtained. Analyses from ^{1}H NMR spectrum suggested that poly(TMC) prepared by lipase-catalyzed polymerization had terminal -CH_2-OH functionalities at both chain ends. A monotonic increase in monomer conversion in the Novozym-435 catalyzed TMC bulk polymerization at 70 °C implied that consumption was not zero order with respect to monomer concentration. The relationship observed between M_n versus time and conversion are consistent with a chain polymerization where propagation was much faster than initiation. The mechanism for chain initiation and propagation for lipase-catalyzed TMC polymerization was proposed.

258.

TRANSPORT PROPERTIES OF STIFF CHAIN GLASSY POLYMERS. B. D. Freeman, Department of Chemical Engineering, North Carolina State University, Raleigh, NC 27695-7906

Ultrahigh permeability polymers are stiff chain, glassy polymers which pack very poorly in the solid state, having high values of fractional free volume. High barrier materials may be prepared from stiff chain glassy polymers which pack efficiently in the solid state, with low fractional free volume values. Manipulation of solid state chain packing can change permeability coefficients over several orders of magnitude. These composite observations suggest that control of primary and higher order structure of stiff chain glassy polymers can be used to rationally tailor permeation properties between those of high barriers to those of extremely permeable membranes.

259. SORPTION IN GLASSY POLYMERS SEEN BY POSITRONIUM, Yasuo ITO, Research Center for Nuclear Science and Technology, The Univ. of Tokyo, Ibaraki 319-11, Japan

Positronium (Ps), with its small mass, is the unique quantum mechanical tool of probing molecular size vacancies in polymers. Briefly it is a "seeker and digger of holes". The lifetime, τ_3, of the long-lived o-Ps component can be correlated with the size of the vacancies in which Ps is tapped, and its intensity, I_3, is a measure of the number of the vacancies in which Ps can be trapped. Owing to this characteristics it is possible to distinguish between the Langmuir- and the Henry-type sorptions in quite a contrasting way, *i. e.* both τ_3 and I_3 decrease for the former and increase for the latter. Thus in typical glassy polymers like polyimides Ps normally tells us about spontaneous occurrence of the former and about delayed and slow occurrence of the latter. In some glassy polymers like polycarbonate, however, I_3 decreases as expected for the Langmuir-type sorption, while τ_3 increases spontaneously with the sorption. This is an indication of local plasticization at the sorption site. Ps is believed to provide incomparable method of the mechanism of sorption in glassy polymers.

260. THE SEGMENTAL MOTION AND GAS PERMEABILITY OF GLASSY POLYMER POLY[1-(TRI-METHYLSILYL)-1-PROPYNE] MEMBRANES. T. Nakagawa, T. Watanabe and K. Nagai, Department of Industrial Chemistry, Meiji University, Kawasaki 214 Japan

Poly [1-(trimethylsilyl-1-propyne) (PMSP) is a typical glassy polymer at room temperature that was first synthesized by Masuda and Higashimura in 1980s. The biggest problem of this glassy plymer is a decrease in its gas permeability with age. Aging is due to the relaxation of the unrelaxed volume. Relaxation can be attributed to physical aging, which is most likely related to a change in molecular motion. In this study, the authors describe the effects of physical aging on the molecular motion of the PMSP membranes synthesized using various catalysts. In order to correlate the physical aging with the molecular motion of PMSP the spin-lattice relaxation, T_1, was investigated using the solid-state NMR. The molecular motions of methyl group in the side chain did not reduced by aging, whereas the molecular motions in the main chain carbons reduced by aging.

261. STRUCTURE, RATE, AND MECHANISM OF METHANE DIFFUSION IN GLASSY ATACTIC POLYPROPYLENE AS DETERMINED BY COARSE-GRAINED MOLECULAR MODELING. Michael L. Greenfield and Doros N. Theodorou, Department of Chemical Engineering, University of California at Berkeley, Berkeley, CA 94720

Hierarchical molecular modeling has been used to simulate methane diffusion in glassy atactic polypropylene over time scales much longer than those accessible by conventional molecular dynamics. Multidimensional diffusion pathways calculated with transition-state theory revealed a wide distribution of jump rates. Methyl groups within 5 Å of the jumping penetrant were displaced by ca. 1 Å. Relatively slow jumps were unlikely when the penetrant was near a chain end. Angles between consecutive jumps were similar to those expected from a random distribution, except for fewer pairs near 0 and 90°. Reverse Monte Carlo was used to create large networks (without periodic boundary conditions) whose instantaneous site coordination, inter-site distance, and three-site angle distributions matched those obtained from the atomistic simulations. Continuous-time discrete-space Monte Carlo simulations on these networks revealed anomalous diffusion regimes both when the rate constants equaled unity and when their distribution matched that obtained from the atomistic simulations, demonstrating a relation between jump network topography and anomalous diffusion.

262. CORRELATIONS BETWEEN SMALL ANGLE X-RAY SCATTERING AND POSITRON ANNIHILATION LIFETIME SPECTROSCOPY MEASUREMENTS OF GLASSY POLYMERS, H. A. Hristov, C. L. Soles, B. A. Bolan, D. W. Gidley[†], and A. F. Yee, Department of Materials Science & Engineering, [†]Department of Physics, The University of Michigan Ann Arbor, MI 48109-2136

Quantifying the structure of glassy polymers in an manner that is useful for describing and predicting both physical and mechanical properties has been a difficult task in the past due to the limited number of experimental techniques available. Recently, positron annihilation lifetime spectroscopy (PALS) has been gaining a lot of attention in field of glassy polymers. PALS has the unique capability that it can quantify both the number and size of holes, or electron deficient regions, that are on the order of 2 to 20 Å in radius. In this work, we focus on describing exactly what PALS is capable of measuring as well attempt to establish an experimental link between the holes that PALS sees and the electron density fluctuations that can be observed from small angle X-ray scattering.

263. XENON-129 NMR AS A PROBE OF SOLID POLYMER SORPTION SITES: A NEW VIEW OF STRUCTURE AND TRANSPORT. Jeff M. Koons, Wen-Yang Wen, Paul T. Inglefield and Alan A. Jones, Carlson School of Chemistry, Clark University, Worcester, MA. 01610

Xenon-129 NMR is used to study sorption sites in poly(4-methyl-1-pentene) and high density polyethylene. At room temperature, only a single resonance is observed for xenon dissolved in these polymers. However, as temperature is lowered several resonance lines are observed. For the poly(4-methyl-1-pentene), two lines are observed at 223 K which are interpreted as xenon dissolved in the amorphous phase and xenon dissolved in the crystalline phase. Exchange by translational diffusion collapses these two lines to a single line at room temperature. At a temperature 223 K, polyethylene remains a single line consistent with xenon being dissolved in only the amorphous phase. At 203 K, the xenon resonance associated with the crystalline phase of poly(4-methyl-1-pentene) further splits into two lines indicating two sorption sites as predicted by computer simulation. At 183 K, xenon in amorphous polyethylene also splits into two lines indicating two sorption sites consistent with the dual mode model for polymeric glasses.

264.
MOLECULAR ANALYSIS OF THE DISSOLUTION OF GLASSY POLYMERS. Balaji Narasimhan and Nikolaos A. Peppas, School of Chemical Engineering, Purdue University, West Lafayette, IN 47907-1283.

The dissolution mechanism of glassy polymers in solvents was analyzed using a combination of molecular theories and continuum mechanics arguments. The effect of molecular parameters like the disentanglement rate of the polymer chain and the diffusion coefficient on the dissolution mechanism were established through the solution of the model equations. Approximate solution of the model equations under pseudo steady state conditions yielded information on the temporal evolution of the gel layer thickness. The model predictions were compared to experimental data for poly(ethylene glycol) dissolution in water and for polystyrene dissolution in methyl ethyl ketone. The predictions agreed well with the data within experimental error.

265. OXYGEN TRANSPORT THROUGH ELECTRONICALLY CONDUCTIVE POLYANILINES, Y. S. KANG, H.J.LEE, J. NAMGOONG, H.LEE AND U.Y. KIM, Div. of Polymer Sci. & Eng., Korea Institute of Science and Technology, P.O.Box 131 Cheongryang, Seoul, Korea

The oxygen and nitrogen permeabilities through the doped polyanilines decreased with their doping level, while their oxygen selectivity over nitrogen increased and reached 12.2 when doped with 0.0150 M HCl solution. When doped with 0.0175 M HCl, its selectivity is expected to be higher but it was unmeasurable because of extremely low permeability of nitrogen. The origin of such high selectivity is explored in terms of the facilitated transport and free volume change upon doping. Because the polaron, generated upon doping, reacts with oxygen specifically and reversibly, it can act as an oxygen carrier and the facilitated oxygen transport may occur. It is, therefore, expected that the oxygen permeability increase with the polaron (carrier) concentration. However, it decreased with increasing the polaron concentration. Instead, it correlated well with the d-spacing or free volume, obtained from an x-ray diffractometer. It is found that, in conclusion, although facilitated oxygen transport may occur, its contribution to the oxygen permeability is insignificant. Instead, the free volume change upon doping seems to play a major role in determining permeability.

266. GAS BARRIER PROPERTIES OF ALKYLSULFONYLMETHYL SUBSTITUTED POLY (OXYALKYLENE)S. Jong-Chan Lee, Morton H. Litt and Charles. E. Rogers, Department of Macromolecular Science, Case Western Reserve University, Cleveland, Ohio 44106

Gas barrier properties of alkylsulfonylmethyl substituted poly(oxyalkylene)s and their miscible blends were discussed. Oxygen permeability coefficients of methylsulfonylmethyl substituted poly(oxyalkylene)s were generally lower than those of poly(oxyalkylene)s with longer side chains. Among the methylsulfonylmethyl substituted poly(oxyalkylene)s, poly[oxy(methyl sulfonylmethyl)ethylene] (MSE) showed lowest permeability. Some of the pairs of methylsulfonylmethyl substituted poly(oxyalkylene)s were found to be miscible. Oxygen permeability coefficient of the 50/50 miscible blend was lower than that of either homopolymer. The oxygen permeabilities of MSE and the miscible blends were comparable to those of EVAL-F and lower than those of any other commercial high barrier polymers.

267. ULTRADEFORMABLE, SELF-OPTIMIZING LIPID AGGREGATES, TRANSFERSOMES^{TM}, FOR INTRA- AND TRANSCUTANEOUS DRUG DELIVERY. Cevc Gregor,* Schätzlein Andreas,[†] Gebauer Dieter.* *Medizinische Biophysik, Klinikum r.d.I., TU München, D-81675 München, and [†]IDEA, Innovative Dermal Applications GmbH, D-80807, München, Germany.

The evolutionary perfectioned skin barrier minimizes trans-cutaneous material transport. Therefore, only few, small drugs can be delivered across the skin for human therapy. We overcame this problem by mimicking the mechanism by which a living cell crosses the intra-corporal barriers: an active search for the resistance minimum in the barrier is followed by the penetrant deformation and passage across the weak spot. Within the framework of our concept of rational membrane design we developed lipid aggregates, Transfersomes^{TM}, that reach such goal. These novel carriers 1) are extremely sensitive to the (transcutaneous) water activity gradient and 2) penetrate through the pores (in the skin) much smaller than aggregate's own diameter, owing to the high, and deformation dependent, aggregate membrane flexibility. A Transfersome therefore crosses the skin spontaneously after the non-occlusive administration and transports associated drugs into the body through the (normally closed) 'virtual pores' in the barrier with a short delay and efficacy >50 %.

268.
MEASUREMENT OF GLUCOSE WITH NONINVASIVE TRANSDERMAL EXTRACTION
I. IDENTITY AND RESPONSE OF INTERFERING SPECIES BY A PT ELECTRODE
Chia-Ming Chiang, Janan Jona, Tung-Fen Chen and Bret Berner
Cygnus Inc., Redwood City, CA 94063

The desire for convenience, precision, and good specificity has brought about the
development of many methods for analyzing glucose for diagnosis and treatment of
diabetics within the past several decades. A quantitative relationship between
serum and transdermally extracted glucose in diabetics was shown by Tamada et al.
The glucose level in the noninvasive transdermal extract was measured by an HPLC
method. Uric acid, tyrosine and tryptophan are the three major interfering species
being found in the in vivo transdermal extract. The concentrations and responses
of uric acid, tyrosine and tryptophan have been determined. With an optimal potential,
it is possible to diminish the interfering species response o f the Pt sensor.

269. pH OSCILLATION OF A DRUG FOR TEMPORAL DELIVERY

Steven A. Giannos and Steven M. Dinh, Transdermal Product Development
CIBA-GEIGY Corp., Suffern, NY 10901

Currently, drug delivery modulation is being accomplished by external means, such as ultrasonic modulation, magnetic modulation and iontophoresis. A novel technology, using chemical oscillators, is being developed to generate a periodic release of a drug or active ingredient without external power sources and/or electronic controllers. One limitation for the formulation of oscillator-diffusion systems has been the need for a pH oscillating reaction that has a long enough periodic time, and a longer low pH state within a cycle, to allow diffusion across a membrane. Recently, Rabai and Hanazaki described the bromate oscillator system, using marble as the acid accepting component. The oscillation period has been lengthened to approximately an hour, as opposed to minutes. The coupling of this system with membrane diffusion offers a unique opportunity to design novel temporally controlled drug delivery systems which are internally regulated.

270. pH-HYSTERESIS OF GLUCOSE PERMEABILITY IN ACID-DOPED LCST HYDROGELS--A BASIS FOR PULSATILE, OSCILLATORY DRUG RELEASE. X. Zou, J.-C. Leroux, J.P. Baker, and R.A. Siegel. Departments of Biopharmaceutical Sciences and Pharmaceutical Chemistry, University of California, San Francisco, CA 94143-0446.

An increasing number of hormones are found to be released in periodic pulses, and hormone replacement therapies that do not mimic endogenous patterns often fail. This observation has motivated the construction of periodic, pulsatile delivery devices. We report progress towards a device that pulsates autonomously, without prompting from an external energy source. The device is conceived by combining understanding of oscillating chemical reactions and of the properties of gels which undergo first-order swelling phase transitions. Chemical oscillators typically include a reaction whose rate law shows hysteresis, coupled to a "feedback" reaction which causes the system to switch back and forth between branches of the former reaction. First order swelling transitions in hydrophobic acidic gels also feature hysteresis with respect to pH. Coupling swelling hysteresis to transport of the reactant (glucose) and product (H^+) of an enzymatically (glucose oxidase) driven reaction, we predict periodic, pulsatile behavior of the gel/enzyme system. Methods to determine pulse widths and frequencies for a simple model system will be presented, along with experimental evidence.

271.
DUROS™ OSMOTIC IMPLANT SYSTEM FOR THE DELIVERY OF HUMAN PEPTIDES AND PROTEINS
James E. Brown
Alza Corporation, 950 Page Mill Road, Palo Alto, CA 94303.

DUROS™ is a small, zero-order, retrievable, drug delivery system specifically designed for the long-term parenteral delivery of potent therapeutic agents to humans. The system consists of an impermeable, cylindrical drug reservoir, piston, a semi-permeable membrane and an osmotic engine. The system is implanted subcutaneously using a trocar in an outpatient setting with local anesthetic. Site specific delivery is achieved by replacing the orifice with a catheter directed to the desired location.

Once the DUROS™ is implanted, water from the surrounding tissue is imbibed into the osmotic engine, generating the pressure which drives the implant. Systems have been designed to achieve a range of delivery rates from 10 days to 12 months. The system allows for zero order drug delivery rates ranging from 10 to 2,000 ug/day over these periods. Stable protein & peptide concentrations in excess of 30% w/v have been achieved for periods of 3 to 12 months.

A placebo system study was completed this past April at Stanford. This was an eight subject study. Both the implant and explant procedures were well tolerated. It took less than 45 minutes to remove the implants from all eight subjects. The subjects had the implants for 8 weeks, no adverse events were reported.

272. CYCLODEXTRINS FOR CONTROLLED DELIVERY. Z. H. Qi, Center of Expertise - Cyclodextrins, Cerestar USA, Inc., Hammond, Indiana 46320

Cyclodextrins are a group of cyclic oligosaccharides produced enzymatically from starch. Because of their unique molecular structures, cyclodextrins can form inclusion complexes with a large number of compounds. Many beneficial effects have been discovered by utilizing cyclodextrins in widespread applications including food and flavors, pharmaceuticals, cosmetics and toiletry, agricultural chemicals, consumer products, process aids, analytical and diagnostics. As a novel vehicle for controlled delivery, cyclodextrin technology compliments other methods with its distinct characteristics of molecular encapsulation. For instance, decrease of volatility, change of solubility or dissolution rate, improvement of stability, enhancement of bioavailability, and selective release of ingredients can be achieved through inclusion complexation with cyclodextrins. In this paper, properties of cyclodextrins will be reviewed along with theories and techniques of forming cyclodextrin complexes. Applications of cyclodextrins for controlled delivery as well as the advantages will be illustrated through examples.

273. **POLYMER-POLYMER COMPOSITES FABRICATED BY THE IN SITU RELEASE AND COALESCENCE OF POLYMER CHAINS FROM THEIR INCLUSION COMPOUNDS WITH UREA INTO A CARRIER POLYMER PHASE.** L. Huang, N. Vasanthan, A. E. Tonelli, Fiber and Polymer Science Program, College of Textiles, North Carolina State University, Raleigh, NC 27695-8301

Several small molecules can be co-crystallized with polymers to form inclusion compounds (ICs). Urea, perhydrotriphenylene, and the cyclodextrins are examples, and serve to form the host crystalline lattice containing the guest polymer chains in their ICs. The guest polymer chains are confined to narrow, cylindrical channels created by the host, small-molecule lattice, where the polymers are highly extended, as a consequence of being squeezed, and are separated from neighboring polymer chains by the IC channel walls composed exclusively of the small-molecule lattice. The net result is an unique solid state environment for polymers residing in IC channels which can be utilized as model systems for ordered, bulk polymer phases. Comparison of the behavior of polymer chains isolated and extended in IC crystals with the behaviors observed for ordered, bulk phases of polymers is beginning to permit an assessment of contributions made by the inherent, single chain and the cooperative, interchain interactions to the properties of ordered, bulk polymers. It is also possible to release and coalesce polymers from their IC crystals in a manner which leads to their consolidation with a chain-extended morphology. Embedding polymer-IC crystals into a carrier polymer, followed by in situ release and coalescence of the included polymers from their IC crystals, offers a means to obtain polymer-polymer composites with unique morphologies. Several such polymer-IC generated composites are described and it is suggested that their unique morphologies might translate into useful, tailorable properties, as well as provide a means for addressing several questions that are fundamental to the behavior of both phase-separated and homogeneous polymer solids.

274.
RELEASE RATE CONTROL OF TABLETS MANUFACTURED BY THE THERIFORM PROCESS. J. Yoo C.W. Rowe, B.M. Wu, R.D. Palazzolo, M.J. Cima, D.C. Monkhouse, Therics Inc. 115 Campus Drive, Princeton NJ 08540, Department of Materials Science and Engineering, Massachusetts Institute of Engineering, Cambridge, MA 02139

The TheriForm™ process has the capability of producing unique drug delivery systems with chronopharmacologic release profiles that are difficult or impossible to manufacture using conventional tablet pressing techniques. Sustained release tablets were constructed using the TheriForm™ process to deposit a drug substance in one or more reservoirs inside the tablets. The process conditions were varied to produce tablets with various reservoir wall thicknesses. The wall thickness and density were shown to have direct effects on the delay time and release rate of the active. Decreasing tablet porosity and increasing tablet density led to longer delay times and improved mechanical properties of the tablets.

275. **CHEMICAL AND BIOCHEMICAL SENSORS BASED ON SOL-GEL DERIVED, LAMINATE PLANAR WAVEGUIDE STRUCTURES.** L. Yang, M. J. Huskey, N. R. Armstrong, S. S. Saavedra, Department of Chemistry, University of Arizona, Tucson, Arizona 85721

An optical sensing platform based on a combination of planar waveguiding and sol-gel processing technologies is described. The sensing element is a planar integrated optical waveguide (IOW) composed of two, submicron thick glass layers coated on glass substrate; both layers are fabricated via the sol-gel method. The lower layer is a densified titania-silica composite. The upper layer is an undensified glass doped with an optical indicator that is physically entrapped yet sterically accessible to dissolved analytes that can diffuse into the pore network. Formation of an analyte-indicator complex is detected via attenuated total reflection (ATR) of light guided in the IOW. The sensor response is both sensitive and rapid, features that are difficult to achieve simultaneously in monolithic sol-gel glass sensors. In the IOW-ATR geometry, these features are realized simultaneously because the primary axes of light propagation and analyte diffusion are orthogonal. The overall approach is technically simple and applicable to a wide variety of indicator chemistries.

276.
BIOENGINEERING OF MATERIAL SURFACES H. Sigrist CSEM, Centre Suisse d'Electronique et de Microtechnique SA, Jaquet-Droz 1, CH-2007 Neuchâtel, Switzerland

Biomolecules performing specific biological functions on material surfaces are progressively employed in the development of miniaturized bio-based assays, biosensors, bioelectronic devices and medical equipment. Guidelines for surface bioengineering are discussed. The unique advantages of using light-controlled reactions to achieve biomolecule immobilization on surfaces are addressed. Immobilization by light is compatible with biological functions, allows surface patterning and molecular coating of materials. Procedures leading to selective functionalization of surfaces are presented, and current strategies and protocols are illustrated with selected examples of biomolecule photoimmobilization.

277. UV-LASER MACHINED POLYMER SUBSTRATES FOR THE DEVELOPMENT OF MICRO-DIAGNOSTIC SYSTEMS Matthew A. Roberts, Paul Bercier, Joel Rossier, Brian J. Seddon, Hubert H. Girault, Laboratoire d'Electrochimie, EPFL, CH-1015 Lausanne, Switzerland

This report describes an UV-laser photoablation method for the production of miniaturized reagent handling systems on polymer substrate chips. The fabrication of such systems is accomplished via pulsed UV-laser energy impinging on substrates moving in pre-defined computer-controlled patterns. This method was used for producing micro-channels (40 - 100 µm wide, 20 - 100 µm deep) and wells (100 - 1000 µm i.d.) in polystyrene, polycarbonate, cellulose acetate, and poly(ethylene terephthalate). Sealing of the resulting microstructures is accomplished using a low-cost readily-available lamination technique. Due to ablatively generated surface characteristics, microstructures are shown to be capable of generating electro-osmotic flow which is used in this system for liquid handling operations, i.e., sample injection, reagent mixing, analyte delivery to a detector, etc. A number of test receptors and enzymes are bound in these channels and the interaction of moving sample plugs with these immobilized zones is studied under a range of flow conditions.

278. **DNA DENDRIMERS: ASSEMBLY AND SIGNAL AMPLIFICATION**

Helen H. Vogelbacker, Robert C. Getts, Nian Tian, Robert Labaczewski, Thor W. Nilsen*
Polyprobe Inc., P.O. Box 2675, Bala Cynwyd, PA 19004

Nucleic acid blots are the method of choice for genetic diagnosis, forensic identification and determination of parentage. DNA dendrimers provide an excellent method for amplifying signal in nucleic acid assays. We have developed and assembled dendritic DNA with hundreds to thousands of single stranded binding sites. Specificity to a particular target sequence is conferred to dendrimers by hybridizing and covalently crosslinking oligonucleotides to the outer surface of the dendrimers. Addition of DNA dendrimers in conventional nucleic acid blot assays results in signal amplification greater than 100-fold compared to oligonucleotides alone.

DNA dendrimers with specificity to seven different DNA sequences have been used to amplify signal(s) in DNA blots. We have constructed DNA dendrimers with specificity to HIV1 LTR, β_2-microglobulin, FSHr (follicle stimulating hormone receptor) and XRCC4 (X-ray repair) sequences and utilized these dendrimers in conventional Southern and dot blot assays. Southern blots probed with DNA dendrimers having 50 bases of sequence specificity to FSHr and ^{32}P-labeled oligonucleotides have identified the same bands in human genomic digests as a traditionally labeled 700bp cDNA probe. Amplification of signal by dendrimers occurs independent of the label used in the assay. ^{32}P, digoxigenin, biotin and alkaline phosphatase labeled probes all yield signal enhancement when used with DNA dendrimers.

279. POLYMERS FOR USE ON BULK ACOUSTIC WAVE DNA HYBRIDIZATION BIOSENSORS. N. C. Fawcett, Jeffrey A. Evans, Richard D. Craven, Ping Zhang, Kirby Harvey and Robert B. Towery, Department of Chemistry and Biochemistry, University of Southern Mississippi, Hattiesburg, MS 39406

Many methods have been reported for attaching nucleic acid probes to polymer modified surfaces of bulk acoustic wave biosensors. None of these are ideal. The ideal polymer would provide: 1) optimum probe surface concentration and activity; 2) resistance to attachment bond hydrolysis at elevated temperature; 3) minimal non-specific binding; 4) simple, reliable, well characterized and reproducible attachment chemistry; 5) reproducible coating characteristics with outstanding adhesion as an ultra thin film; 6) low swelling. We will describe several methods for attaching DNA probes using polyethylene co-acrylic acid (PEAA). This will include a method for increasing the density of surface functional groups and a description and comparison of several different attachment reactions. Results will be compared with those obtained using polystyrene coacrylic acid. and using a photocrosslinkable polyvinylcinnamate. Application of the various probe attachment methods to biosensor detection of *E. coli* O157:H7 will be illustrated.

280. EFFECT OF STERIC STABILIZERS ON THE PERFORMANCE OF COLLOIDAL IMMUNOREAGENTS. Mark R. Pope, Brent Putman, David R. Mees, Mary Joseph, Daniel Subotich, Terry Pry and Peter J. Tarcha, Abbott Laboratories Diagnostics Division, North Chicago, IL 60064-3500

Polypyrrole latexes have been uniquely applied as visual labels in chromatography-based immunoassays. Their main advantages stem from their intrinsically black color and ability to effectively immobilize protein-based ligand binding species, such as antibodies. The colloids are readily synthesized in aqueous media in the presence of a polymeric steric stabilizer. We have shown that the stabilizer can adversely affect the binding of the colloidal immunoreagent to complimentary antigen immobilized on another solid phase. On the contrary, the ability of the immunocolloid to bind soluble antigen appears to be unaffected. By selective chemical degradation of the stabilizer on these antibody-coated particles, the specific binding ability to surface-immobilized antigen can be improved over 100-fold. Additionally, such methods significantly improve the reproducibility of producing such immunoreagnets. A mechanism based on steric repulsion is proposed to explain these results.

281. STRUCTURE-FUNCTION RELATIONSHIPS OF IMMUNOCHROMATOGRAPHIC SUBSTRATES. Abigail S. Fisher and Debra A. Alcox, Millipore Corporation, Bedford, MA 01730

Structural and chemical characteristics of membranes affect their function and performance in lateral flow diagnostic applications. Microporous membranes made of different polymers bind protein by different mechanisms. Consequently, ionic and chaotropic detergents have differential effects on protein binding to various membrane polymers. The protein binding capacity of a membrane is dependent on its total surface area which is determined by pore size, thickness, and porosity. The lateral flow rate of the membrane is often the most significant membrane characteristic affecting the sensitivity of the immunochromatographic reaction. Flow rate is determined by the membrane's pore size, pore size distribution, and porosity (number of pores). Flow rate is not affected by surfactant concentration once membrane wettability is achieved. The structural features of an immunochromatographic substrate membrane must be carefully specified in order to achieve the desired functional result.

282. A DISPOSABLE ELECTROCHEMICAL SENSOR BASED ON A POLYMER-MODIFIED CARBON ELECTRODE AND MULTI-ENZYME AMPLIFICATION

Judith Rishpon and Dmitri Ivnitski, Department of Molecular Microbiology and Biotechnology, Tel-Aviv University, Ramat-Aviv 69978, Israel

We developed a disposable, electrochemical cell that is based on a polymer-modified, carbon electrode and uses a non-toxic mediator: 5-aminosalicylic acid. A vibrating plate that efficiently enhances the mass transfer in the solution and to the electrodes, resulting in an immediate response of the electrode. A disposable antibody graphite electrode, with both immobilized glucose oxidase and the relevant antibody on the polyethylenimine film, was placed in a cell (V = 0.3 ml) containing 5-ASA; the sample of analyte and peroxidase conjugate and glucose were added to the cell simultaneously, and changes in the current were monitored continuously and recorded for 1 min. A competition assay was used to detect the HIV antigen gp120. *Salmonella* was detected in a sandwich assay using CSA-1 antibody. In both assays the response of the sensor was immediate. The measurement is very easy to perform and is completed in less than 3 minutes.

283. BIODEGRADABLE WATER-SOLUBLE POLYMERS, G. Swift, Research Division, Rohm and Haas Company, Spring House, PA 19422

Interest in biodegradable polymers began several years ago with major attention given to the waste-management of plastics, especially wrappers, which were considered to be responsible for a growing shortage of landfill space and potentially harmful litter in both terrestrial and freshwater and marine environments. Since then, media panic has subsided and a rational scientific approach has added a better perspective to the problems resulting in the development of a number of complementary solutions that are applicable in different situations. These include recycling, volume reduction, incineration, and the development of truly biodegradable plastics. With the new awareness of polymers in the environment, industry has also begun to focus on water-soluble polymers. These polymers are usually discarded as water solutions and are, therefore, not visible as they enter the environment. Ultimate fate in the environment is difficult to predict and biodegradable substitutes are being sought for those that are known to be recalcitrant. Research to date has indicated directions for the development of truly biodegradable polymers, and this will be discussed.

284. ACCELERATED AEROBIC BIODEGRADABILITY OF PAPERBOARD. Anthony L. Andrady and Ye Song, Camille Dreyfus Laboratory, Research Triangle Institute, Research Triangle Park, NC 27709

Bleached paperboard substrates are almost totally cellulosic and are easily mineralized under laboratory biodegradation conditions. Using a specially designed respirometer we investigated the relative rates and extents of mineralization of bleached and unbleached paperboard materials. A soil medium inoculated with aerobic sewage sludge effluent was used. The mineralization data analyzed using an empirical equation allows relative rates of mineralization of the samples to be compared. The mineralization data was compared to weight loss data obtained in a second experiment using larger test pieces of the same set of samples. The rates of degradation and the extent of biodegradability of these materials will be discussed.

285. MOLECULAR WEIGHT CHANGES CORRELATED WITH FORMATION OF DEGRADATION PRODUCTS IN OXIDIZABLE AND HYDROLYSABLE POLYMERS. A-C. Albertsson and S. Karlsson, Dept. of Polymer Technology, the Royal Institute of Technology (KTH), S-100 44 STOCKHOLM, Sweden.

The rate of environmental degradation of polymers is dependent on several factors. Oxidizable polymers show a slower degradation than hydrolysable but are generally considered to have better material characteristics. Hydrolysable polymers form few and well-defined degradation products while the oxidizable ones form series of products with varying patterns and amount. Natural polymers may be both oxidizable and hydrolysable, but generally there are enzymes with specific action in every naturally occuring biopolymer. Biological oxidation is done by hydroxylases, oxygenases or (per)oxidases. PLA and LDPE with starch and/or pro-oxidants were degraded and the resulting degradation products identified by GC-MS and the molecular weight changes monitored by SEC. The type of degradation products formed are correlated with the molecular weight changes and the degradation environment.

286.

BIODEGRADATION OF C-14 LABELED POLY (ETHYLENE-CO-VINYL ALCOHOL) B. Erlandsson, S. Karlsson, and A.-C. Albertsson, Dept. of Polymer Technology, the Royal Institute of Technology (KTH), SE-100 44 STOCKHOLM, Sweden

Polymers that are all carbon in their main chain usually degrade slowly, with the exception of Poly (vinyl alcohol) (PVA). The fairly rapid degradation of PVA is due to the OH groups in the polymer backbone. A copolymer between ethylene and vinyl alcohol (EVOH) will give a material suitable for packaging, but with an increased rate of degradation as compared to polyethylene. The degradation of EVOH was studied by exposing EVOH samples to a mixture of two fungi, *Fusarium redolens* and *Pencillium simplicissimum*. The samples contained 44 mol-% ethylene, and the samples were labeled with C-14, so the degradation could be followed very exactly by measuring the evolved $^{14}CO_2$. Some of the samples were exposed to UV radiation for 60 h to initiate the degradation. The degradation was also followed by measuring changes in functional groups with ATR-FTIR, changes in crystallinity was measured with DSC and changes in molecular weight was measured with light scattering connected on-line to a SEC

287. **DECOMPOSITION OF MODEL COMPOUNDS OF PHENOL RESIN WASTE WITH SUPERCRITICAL WATER.**
Y. Suzuki, N. Tohji, H. Tagaya, M. Karasu, J. Kadokawa and K. Chiba, *Department of Materials Science and Engineering, Yamagata University, 4-3-16 Jonan, Yonezawa, Yamagata 992 Japan.*

Chemical recycling process of thermosetting resin such as phenol resin has scarcely reported. In this study, aqueous solutions of model compounds of phenol resin waste such as substituted diphenylmethanes and prepolymers of phenol resin were reacted in a 10ml tubing bomb reactor at 300~430°C. Seven prepolymers and p- and o-bis(hydroxyphenyl)methanes as model compounds of phenol resin were effectively decomposed into their monomers by the reaction with supercritical water. The decomposition reactions were accelerated by the addition of alkali salts such as Na_2CO_3. Furthermore, we added alkali salts in the decomposition reaction of phenol resin molding materials. The addition of alkali salt was also effective even for the decomposition reaction of phenol resin molding materials.

288. **BIOSYNTHETIC STRATEGIES TO TAILOR THE STRUCTURE OF BIOEMULSIFIERS FROM *ACINETOBACTER CALCOACETICUS* RAG-1**
‡Jinwen Zhang, ‡Richard A. Gross, ◊Alfred L. Allen, and ♦David L. Kaplan; *‡University of Massachusetts Lowell, Department of Chemistry, One University Ave., Lowell, MA 01854; ◊Biotechnology Division, U.S. Army Natick RD&E Center, Kansas St., Natick, MA 01760-5020; ♦ Tufts University, Chem. Eng. Dept., 4 Colby Street, Medford, MA 02155*

Acinetobacter calcoaceticus RAG-1 was cultured on the following different chain length saturated 2-hydroxyl fatty acid (2-HOFA) carbon sources: C12:0 (2-OH), C14:0 (2-OH), C16:0 (2-OH) and C18:0 (2-OH). The content of C12:0 (2-OH) EM substituents reached high levels (306 nmol/mg-EM, 64.4 mol% of total FAs) by selectively feeding this FA. Substantial quantities of 2-HOFAs with chain lengths ≥ C14 {up to 96 nmol/mg-EM or 15.2 mol% for C16:0 (2-OH)} were also incorporated in EMs by providing the corresponding 2-HOFA carbon source in the medium. By increasing the medium 2-HOFA concentration large increases in EM total FA contents resulted. The EM FA content was as high as 955 nmol/mg-EM or 23 wt % for a culture containing 0.75 g/100 mL C18:0 (2-OH). Addition of the enzyme alkylating agent iodoacetamide to cultures was used to: i) enhance the incorporation into EMs of both C12:0 (2-OH) and C16:0 (2-OH) substituents and ii) increase by 1.3 to 1.8 fold (by wt.) the total EM FA content. The relationship between bioemulsifier structure and properties for o/w dispersions will be discussed.

289. **BIODEGRADATION OF REACTIVE BLENDS OF ETHYLENE ACRYLIC ACID COPOLYMER AND EPOXIDISED NATURAL RUBBER.** Golok B. Nando and Subhra Mohanty, *Rubber Technology Centre, Indian Institute of Technology, Kharagpur-721302, India.*

Miscibility of the blends of poly(ethylene co-acrylic acid) and epoxidised natural rubber through reactive processing has been studied in details by the authors. Biodegradability of PEA and one of its blends with ENR has been thoroughly investigated by soil burial method. A blend of 30:70 ENR and PEA has been chosen for convenience because maximum chemical reaction takes place between the blend constituents. The results of biodegradation after every two months for a period of six months has been reported. The effect of starch in the proportions varying from 5 to 50 parts by weight and the effect of ferric stearate on the biodegradation process has been critically evaluated by means of IR spectroscopy, mechanical property measurements and SEM study of the surfaces. Presence of starch in the PEA and its blends in the proportion at and beyond 30 pbw enhances the biodegration process rapidly. Degradation of PEA was 75 percent after 4 months of soil burial where as that of N30 blend was total i.e. unmeasurable. The reason for such variation in the mechanical properties due to biodegradation has been assigned to the loss of crystallinity in the polymer matrix. This has been supported by the disappearance of the 720 cm-1 peak in the degraded matrix. Surface topography of the matrix show very interesting results. It shows first thread like structures after either 2 or 4 months of biodegradation which disappear after 6 months of soil burial due to complete degradation.

290. POLYMER GLASSES: THERMODYNAMIC AND RELAXATIONAL ASPECTS. R. Simha, *Department of Macromolecular Science, Case Western Reserve University, Cleveland, Ohio 44106-7202*

A structure function, defined as a hole fraction $h(V,T)$ in the lattice-hole model is uniquely determined in the equilibrium melt, and can be derived from the equation of state (eos). We discuss its application in the glass. The interpretation of h as a free volume quantity yields connections between eos and other physical properties in the steady state glass. Such correlations are extended to time dependent physical processes in terms of a time dependent structure function h. Finally, we discuss a theory of free volume relaxation relevant to physical aging processes. A simulation by this theory of the melt to glass cooling experiment establishes the existence of a glass transition region as a function of cooling rate and the characteristic interaction parameters of the polymer.

291. **INVESTIGATION OF ISOTHERMAL PHYSICAL AGING OF A FULLY CURED EPOXY-AMINE THERMOSETTING SYSTEM.** Shellee L. Maddox and John K. Gillham, Department of Chemical Engineering, Princeton University, Princeton, NJ 08544

The rate and effects of isothermal physical aging of a fully cured epoxy-amine, glass-fiber composite specimen were studied for a wide range of isothermal aging temperatures (-180 to 200°C) using a freely oscillating torsion pendulum (TBA) operating at about 1 Hz. As assigned from the maxima in the mechanical loss vs. temperature, T_g was 182°C (0.9 Hz) and the secondary transition temperature (T_β) was ≈ -30°C (1.9 Hz). Plots of the increase in isothermal modulus and of the decrease in isothermal mechanical loss were linear vs. log aging time: slopes provided the aging rate. Two maxima in the isothermal aging rate vs. the aging temperature (T_a) were displayed. A correlation presumably exists between the two maxima in aging rate and the two transitions. This is not surprising since both mechanical loss maxima (i.e., transitions) and aging rate maxima correspond to restricted specific localized submolecular motions. Effects of isothermal physical aging after aging were investigated in terms of change of modulus of the specimen vs. temperature. The effect of isothermal aging at T_a vs. temperature existed only in a narrow temperature region localized about T_a. The majority of the aging effect can be eliminated by heating to temperatures above T_a but below T_g. Theoretical and practical implications of this observation are discussed.

292. **DYNAMIC MONTE CARLO SIMULATION AND DIELECTRIC MEASUREMENTS OF THE EFFECT OF PORE SIZE ON THE GLASS TRANSITION TEMPERATURE AND THE STATE OF THE GLASS.** N. Haralampus, J.P. Northrop, C. Scordalakes, E. Ashmore, W. Martin, D. Kranbuehl Department of Chemistry, College of William and Mary, Williamsburg, Virginia 23187-8795; P. H. Verdier, Polymers Division, National Institute of Standards and Technology, Gaithersburg, Maryland 20899

With the increasing interest in nanostructures has come an interest in the effect of confinement in pores on the dynamic behavior and properties of these materials. Much of this interest has been generated by the recent studies of Jackson and McKenna which showed that glasses formed in small pores had a lower Tg than in the bulk. The reduction in Tg increased as the pore size decreased. Both experimental and theoretical experiments are reported on to better understand this effect of confinement on Tg. In experimental studies, the dielectric rotational relaxation spectrum has been measured of several glass forming small molecules both in the bulk and in 4 nm porous Vycor glass. Theoretical studies have been conducted using Monte Carlo simulation techniques to study the effect of confinement in 2 sizes of cavities on the formation of the glassy state as the system is cooled from the melt using different cooling rates. The activation energy of the rotational relaxation rate is greater in the bulk then in the pore causing a crossover in the rates. The simulation results show a lower Tg in the smaller pores.

293. **PHYSICAL AGING OF AMORPHOUS POLYESTERS.** James M. O'Reilly, Eastman Kodak Company, Rochester NY, 14650-2116

Physical aging of amorphous films of polyethylene terephthlate (PET), polycyclohexane dimethanol terephthlate (PCHDMT), and polyethylene napthalate (PEN), have been studied by a low heating rate DSC7 protocol. The Tool-Narayanaswamy relaxation parameters were determined by specific heat and fictive temperature calculations. The activation energies range from 225 to 275 Kcal/mol and do not follow the usual trend of increasing with increasing Tg. The fractional exponent of the KWW distribution function, β, is 0.50 ± 0.05 for all the polyesters indicting that the microhetereogeniety in these polymers is similar. However, the average relaxation time at Tg is shortest for PEN and longest for PCHDMT. This is unusual because PEN has the highest Tg. The structure dependence of the relaxation times, x, ranges from 0.27 to 0.42 for these polyesters. These results will be compared with mobilty data obtained from other relaxation methods.

294. AMORPHOUS ORIENTATION IN GLASSY POLYCARBONATE
M. D. Shelby[1], G. L. Wilkes[2], M. R. Tant[1], J. Zawada[1], T. J. Bastow[3] and A. J. Hill[3]
[1]Eastman Chemical Company Research Laboratories, PO Box 1972, Kingsport TN 37662 USA
[2]Polymer Materials and Interfaces Laboratory, Virginia Polytechnic Institute and State University, Blacksburg, VA 24060 USA
[3]CSIRO Division of Materials Science & Technology, Private Bag 33 S. Clayton MDC, VIC 3169 AUSTRALIA

Amorphous polycarbonate films with a varying degree of orientation (draw ratio: 1 to 2.5; Herman's orientation function f_H: 0 to 0.2) are studied. Orientation causes an initial increase in permeability for oxygen ($f_H \leq 0.03$) at room temperature followed by a decrease in oxygen permeability with increasing orientation. Similarly, density displays an initial decrease ($f_H \leq 0.03$) followed by an increase, and Phillips et al. (J. Appl. Polym. Sci., 21, 1859 (1977)) have shown an initial decrease in the velocity of ultrasound measured at room temperature prior to an increase with further orientation. In contrast, diffusivity, free volume measured by positrons, and attenuation of ultrasound decrease monotonically with orientation. Molecular mobility measurements (mechanical loss modulus and solid state ^{13}C nuclear magnetic resonance spin lattice relaxation time) indicate heterogeneous mobility response of molecular moeities to orientation. The present study addresses why (in terms of the effect of orientation on molecular environment), two related properties, such as permeability and diffusivity; density and free volume; or ultrasonic velocity and attenuation, show distinct behaviours with initial orientation in the glassy state.

295. REFRACTIVE INDEX: A PROBE FOR MONITORING VOLUME RELAXATION DURING PHYSICAL AGING OF GLASSY POLYMERS
Christopher G. Robertson and Garth L. Wilkes
Chemical Engineering Department, Polymer Materials and Interfaces Laboratory
Virginia Polytechnic Institute and State University; Blacksburg, VA 24061-0211

A positive correlation between refractive index and density has been experimentally illustrated in the literature for numerous materials, including polymers. This relationship was exploited in an attempt to follow the densification of a glassy polymer during the physical aging process. Atactic polystyrene films were isothermally aged at a temperature 30°C below the glass transition temperature following a quench from the equilibrium liquid state, and the refractive index (n) was evaluated as a function of aging time (t_e) using an Abbe refractometer. The refractive index increased linearly with respect to log(t_e), and this dependence was used in conjunction with experimentally determined values of the thermal expansion coefficient in the glassy state and the sensitivity of refractive index to temperature for unaged polystyrene to determine the volume relaxation rate. This rate was similar to that obtained for the same polystyrene material using a precision dilatometer, indicating the possibility for using refractive index measurements to quantitatively assess volume relaxation during physical aging.

296. ENTHALPY RECOVERY OF POLYETHERIMIDE: EXPERIMENT AND MODEL CALCULATIONS. Sindee L. Simon, Department of Chemical Engineering, University of Pittsburgh, Pittsburgh, PA 15261

Enthalpy recovery of polyetherimide is measured during heating with differential scanning calorimetry (DSC) after cooling at various rates. The resulting annealing peaks are fit using the Moynihan-Tool-Narayanaswamy model of structural recovery. A self-consistent phenomenological equation is used to describe the experimentally observed structure and temperature dependence of the relaxation time in both glass and equilibrium regimes. Temperature gradients in the DSC sample are incorporated into the model calculations. When no thermal gradients are assumed, model parameters are found to vary with thermal history despite the use of the self-consistent equation for the relaxation time. Accounting for the presence of thermal gradients in the DSC sample is found to affect the values of the model parameters needed to fit the data. However, thermal gradients are unable to account for the thermal history dependence of the model parameters or the discrepancy between the observed and calculated shapes of the DSC annealing peaks.

297. DIELECTRIC RELAXATION SPECTROSCOPY OF POLYCARBONATE AND ITS BLENDS. G.J.Pratt, Mech. & Manuf. Engineering, Univ. of Melbourne, Parkville 3052, Australia and M.J.A.Smith, Physics, University of Warwick, UK

Dielectric relaxation spectroscopy identifies eight distinct absorption regions in nitrogen-equilibrated commercial bisphenol-A polycarbonate. Bound water causes a further absorption in unequilibrated material. These overlapping absorptions confer high impact strength even at low temperatures. Two intermediate-temperature losses are differentiated and delineated; their origins and reason for diminution by annealing are examined. For u.v.-irradiated polycarbonate the dielectric data is broadly consistent with a sequence of photo-Fries reactions and subsequent photo- oxidation, and with depletion of the stabilizing additive in the u.v.-resistant grade. For impact-modified PC/PBT and PC/PET blends evidence is provided for partial miscibility of the component polymers and for a two-phase morphology with a polyester-rich dispersed phase in a continuous matrix rich in polycarbonate. Other absorptions are attributed to MWS interfacial polarization, the presence of the impact modifier and of a phosphite processing stabilizer respectively.

298. RUBBER TOUGHENED BLENDS OF POLYCARBONATE AND STYRENE-MALEIC ANHYDRIDE COPOLYMERS. H. A. Stretz, P. E. Cassidy, Dept. of Chemistry, Southwest Texas State University, San Marcos,TX 78666, and D. R. Paul, Dept. of Chemical Engineering, Univ. of Texas, Austin, TX 78712.

Polycarbonate (PC) blends with impact-modified styrene-maleic anhydride (SMA) copolymers combine the toughness of PC with the economics and processibility of SMA materials. This study compares the mechanical properties and morphology of PC/SMA blends with those of PC/SAN (styrene-acrylonitrile) blends previously studied, and similarly compares rubber toughened PC/ SMA blends with PC/ABS blends. PC/SMA blends showed mechanical properties similar to PC/SAN blends. Rubber-toughening of PC/SMA blends improves the standard notch toughness and sharp notch toughness considerably relative to the unmodified blends. Rubber particles remained in the SMA phase in all blends.

299. LIPOSOMES: DELIVERY SYSTEMS FOR NOVEL THERAPEUTIC AGENTS. P.W.Taylor, Ciba Pharmaceuticals, Horsham, W.Sussex, RH12 4AB, United Kingdom.

Liposomes as vesicles consisting of one or more concentric phospholipid bilayers entirely enclosing a water-filled space, have properties that render them attractive candidates for delivery and controlled release of a wide variety of drugs. They are biodegradable, relatively non-toxic and can be used to entrap hydrophilic drugs in the aqueous compartment;

in addition, lipophilic drugs can frequently be incorporated into the lipid bilayers with minimal bilayer perturbation. They are increasingly being considered for the delivery of the new generation of biopharmaceuticals that include proteins, peptides, genes and antisense sequences and recent advances in liposome technology will facilitate more precise delivery of drug payload to the target tissue. Thus, in this symposium we will consider strategies for the effective delivery of novel therapeutics via systemic and transdermal routes.

300. LIPOSOME-BLOOD PROTEIN INTERACTIONS IN RELATION TO LIPOSOME CLEARANCE. Arcadio Chonn, Inex Pharmaceuticals, Vancouver, BC, Canada V6P 6P2.

Our previous studies with anionic liposomes have suggested an apparent inverse relationship between the amount of blood proteins associated with liposomes and their circulation half-life. High levels of protein binding are associated with rapidly cleared liposomes, indicating that blood proteins play an important role in mediating cellular uptake. More recently, we have shown that this relationship holds for a number of variables affecting liposome clearance, including the lipid fatty acyl chain length and saturation, the cholesterol content, as well as the dose of the liposomes. The surface adsorption property of liposomes to blood proteins, therefore, is an important factor to consider when developing circulation-stable liposomes for systemic drug delivery.

301. STRATEGIES FOR INCREASING THE CIRCULATION TIMES OF LIPOSOMES T.M. Allen and S. Zalipsky, Dept. of Pharmacology, U. of Alberta, Edmonton, AB, T6G 2H7 and SEQUUS Pharmaceuticals Inc., Menlo Park, CA, 94025.

A number of surface modifications of liposomes have been described in recent years which lead to increased circulation times. The inclusion in lipid bilayers of such molecules as monosialylganglioside GM_1, and lipid derivatives of flexible polymers like polyethylene glycol, polymethyloxazoline, polyacrylamide and polyvinyl pyrrolidone increase the surface hydrophilicity of liposomes, protecting them from recognition and removal from circulation by scavanger cells in the body. These surface modifications result in significant shifts in the pharmacokinetics and biodistribution of the liposomes and any associated drug molecules, shifts which can be exploited in the development of therapeutic applications for liposome-associated drugs. In particular, polymer-engrafted long-circulating liposomes deliver increased concentrations of entrapped anticancer drugs to solid tumours resulting in increased anticancer efficacy compared to non-entrapped drugs, an observations which has led to the clinical approval of one formualtion of long-circulating liposomes.

302. TARGETED LIPOSOMES WITH OPTIMIZED PAYOUT CHARACTERISTICS FOR CANCER CHEMOTHERAPY. P.R. Cullis, Department of Biochemistry and Molecular Biology, University of British Columbia, 2146 Health Sciences Mall, Vancouver, BC V6T 1Z3 and Inex Pharmaceuticals Corporation, 1779 West 75th Ave, Vancouver, BC, Canada V6P 6P2

Three parameters which can strongly influence the efficacy and toxicity profile of liposomes containing anticancer drugs concern the ability of the liposomes to accumulate at the tumour site (disease site targeting), the payout rates of the encapsulated drug and the ability to target specific cells in the disease site (cell-specific targeting). Disease site targeting to tumours can be achieved by employing small ($\leq$ 100 nm diameter) liposomal carriers with long circulation lifetimes as has been noted by a number of investigators. The importance of payout rates, which has received less attention, will be emphasized for liposomal formulations of vincristine. Finally, cell-specific targeting protocols including antibody conjugation protocols and small molecule targeting procedures will be discussed.

PMSE

303. **INTRACELLUAR DRUG DELIVERY USING pH- AND LIGHT-ACTIVATED DIPLASMENYLCHOLINE LIPOSOMES.** O.V. Gerasimov, M. Qualls, Y. Rui, & D. H. Thompson, Department of Chemistry, Purdue University, West Lafayette, IN 47907.

A plasma-stable liposome, comprised of naturally-occurring diplasmenylcholine lipid (1,2-di-O-(Z-1'-hexadecenyl)-sn-glycero-3-phosphocholine or DPPlsC), that rapidly decomposes upon exposure to oxidative environments or endosomal pH's has been developed. We have used these liposomes to efficiently deliver calcium ions, fluorescent dyes, antimetabolite drugs, and water-soluble sensitizers to the cytoplasm of KB cells. Control experiments indicate that endosomal acidification is required to effect cytoplasmic delivery of the contents encapsulated by DPPlsC liposomes. Experiments using bacteriochlorophyll a as membrane-bound sensitizer indicate that photooxidative triggering is an alternative pathway for cytoplasmic delivery of liposomal cargo. The implications of this triggered release strategy for the delivery of macromolecules to the cytoplasm will be discussed.

304.
CONDITIONALLY STABLE LIPOSOMES FOR DRUG DELIVERY. S. C. Davis and F. C. Szoka Jr., Department of Biopharmaceutical Sciences, School of Pharmacy, University of California, San Francisco, CA 94143

The dispersal of lipids in an aqueous media can give rise to a number of different molecular arrays (or phases) which are dependent on lipid structure, temperature, pressure, and concentration. We have designed and synthesized a series of lipids which can be induced to undergo a phase change under defined, physiologically relevant conditions. These lipids form stable liposomes, but upon receipt of specific stimuli rearrange to the H_{II} hexagonal phase with concomitant release of the liposomes contents. Delivery of the stimulus by the target cell enables the liposome contents to be targeted to a specific cellular location. One example of this protocol is the use of a pH change to destabilise the liposome. Therefore liposomes which are stable at pH 7.4 but unstable below pH 6 will undergo a phase change in the endosomal vesicle following endocytosis releasing the drug into the cell. Other strategies will also be presented. This work was supported by NIH grants GM 26691, DK 46052, DK 47766.

305. **ARTIFICIAL VIRAL ENVELOPES FOR GENE THERAPY.** Hans Schreier. Advanced Therapies, Inc., 371 Bel Marin Keys #210, Novato, CA 94949

Delivery of DNA and oligonucleotides for the treatment of infectious, metabolic and neoplastic disorders has been accomplished using recombinant viruses and cationic liposomes. Both approaches suffer from toxicity and poor gene expression. Enveloped viruses fuse with cell or endosomal membranes upon binding to cell surface receptors. This process requires a specific fusogenic lipid composition, fusogens and protein surface markers. In an effort to exploit the (patho)physiologic process of viral infectivity, we designed an artificial viral envelope in whose payload compartment human therapeutic gene products can be delivered to the nucleus of target cells with high efficiency of gene expression. Selective delivery and expression of marker genes have been demonstrated *in vitro* and *in vivo* in small animals. The artificial viral envelope gene delivery system may be an attractive alternative to recombinant viruses and cationic liposomes.

306. **NOVEL SUPRAMOLECULAR ASSEMBLIES FOR GENE TRANSFER**, Leaf Huang, University of Pittsburgh, Department of Pharmacology, W1351 Biomedical Science Tower, Pittsburgh, PA 15261

Lipofection has become a simple, effective and safe method to introduce DNA, RNA and proteins into cells. One of the cationic liposome formulations developed by us, DC-chol/DOPE liposomes, is relatively non-toxic and efficient in transfection. Recently, this liposome formulation has been used in two separate clinical trials for the immunotherapy of malignancy and for treatment of cystic fibrosis genetic defects. The structure of DNA/liposome complexes and their mode of action in transfection will be reviewed. Based on this information, two types of novel condensed structure containing DNA polycation and lipids have been developed. The LPD (lipid-entrapped, polycation-condensed DNA) particles are small (<100 nm in diameter), monodisperse, and colloidally stable. The transfection activity of LPD is similar to that of adenovirus and is about 10-100 fold higher than that of the first generation cationic liposomes. The LPD I particles are cationic and used primarily in local and regional delivery routes. The LPD II particles are anionic and can be target specific by attaching specific ligand molecules on the surface. The parenteral use of these novel particles for systemic gene transfer is under development. Recently, reconstituted chylomicron remnants have been used to solubilize DNA/cationic lipid complexes. This new non-viral vector induced high level transgene expression in the liver. These formulations will be discussed in terms of their efficiency, toxicity and potential uses in gene therapy.

307.

POLYMER AND MATRIX EFFECTS IN THE CONSTRUCTION AND FUNCTION OF MEMBRANE BIOSENSORS.

John R. Woodward and Jay Johnson. YSI Inc. 1725 Brannum Lane, Yellow Springs, OH. 45387

Membrane biosensors can be constructed in a number of ways. Conventionally the membrane consists of a cellulose acetate base inner membrane which controls access of electroactive molecules to the electrode, and a diffusion limiting outer membrane, with an adhesive enzyme layer immobilized between the two. These membranes can be manipulated to change the characteristics of the sensor. Also the composition of the enzyme layer can be changed thus altering the performance of the membrane. Alternative polymers which can be used for membrane biosensor construction include methyl methacrylate and the controlled release polymers such as ethylene vinyl acetate. Performance of these membranes and the form in which they function best will be discussed.

308. GLUCOSE AND LACTATE BIOSENSORS BASED ON OXIDASES "WIRED" WITHIN HYDROGELS MADE WITH POLYVINYL PYRIDINE AND POLYVINYL IMIDAZOLE COMPLEXED WITH $[(Os-4,4'-DIMETHOXY\ 2,2'-BIPYRIDINE)Cl]^{+/2+}$. Gregg Kenausis, Chris Taylor and Adam Heller, Department of Chemical Engineering, University of Texas at Austin, Austin, Texas 78712-1062.

Glucose and lactate electrodes based on hydrogels made by crosslinking glucose oxidase and lactate oxidase with the redox polymers formed upon complexing $[Os(dmo-bpy)_2Cl]^{+/2+}$ (dmo-bpy = 4,4'-dimethoxy-2,2'-bipyridine) with polyvinyl pyridine (PVP) and polyvinyl imidazole (PVI) on vitreous carbon electrode surfaces were investigated. The redox potentials of the hydrogels were +35mV and -65 mV (SCE), respectively, and their glucose electrooxidation current reached a plateau at +150 mV and +50 mV(SCE), respectivley. At these operating potentials, urate and acetaminophen were not electrooxidized at rates that would interfere with the glucose and lactate assays.

309. Electrochemical Glucose and Lactate Sensors Based on "Wired" Thermostable Soybean Peroxidase Operating Continuously and Stably at 37°C. Qiang Chen, Gregg Kenausis and Adam Heller, Department of Chemical Engineering, The University of Texas at Austin, Austin, TX 78712

Glucose and lactate sensors maintaining stable output under continuous operation at 37°C for 12 and 8 days, respectively, were built. The vitreous carbon base of the sensor was coated with four polymer layers. The first was made by cross-linking thermostable soybean peroxidase and the redox polymer formed through complexing part of the rings of polyvinyl pyridine (PVP) with $[Os(bpy)_2Cl]^{+/2+}$ (bpy = bipyridine) and quaternizing part of the rings with bromoethylamine. The second was an insulating and H_2O_2 transport controlling cellulose acetate layer; the third was an immobilized glucose oxidase or lactate oxidase layer; and the fourth, a substrate transport controlling cellulose acetate layer. In the case of the glucose sensor, the current output was independent of potential between -0.2 V and +0.3 V (vs. SCE), and the response time ($t_{10/90}$) was less than two minutes when the concentration was raised from 0 to 5mM glucose. The current was independent of the O_2 partial pressure above 15 torr. At a glucose concentration of 4 mM, the addition of 0.1 mM ascorbate decreased the current by less than 1%. The operational stability was glucose oxidase loading dependent. Though the current decreased by 85% after 100-hour operation at 37°C when the 3mm diameter electrode was loaded with only 1.3 µg of glucose oxidase, it decreased by less than 1% after such operation when loaded with 52 µg of the enzyme. Similar results were obtained for the lactate sensor.

310.

GALACTOSE BIOSENSORS USING COMPOSITE POLYMERS TO PREVENT INTERFERENCES. **P. W. Stoecker, UALR, Chemistry Department, Little Rock, AR 72204.** Paul Manowitz, UMDNJ, Department of Psychiatry, Piscataway, NJ 08854. Alexander M. Yacynych, Rutgers University, Department of Chemistry, New Brunswick, NJ 08903.

A biosensor using a composite polymer to prevent interferences was used in a flow injection analysis system for the detection of galactose in human plasma. The biosensor consisted of galactose oxidase immobilized on a platinized carbon electrode that had been modified with a composite polymer. The composite polymer showed improved selectivity to hydrogen peroxide compared to either of its individual polymeric components, Nafion and a copolymer of diaminobenzene and resorcinol. The composite polymer minimized the effect of possible interferences from urate, ascorbate, and acetaminophen. This analytical system has a minimum detection limit of 50 µM, linearity to 6 mM, a storage stability of greater than 30 days, and a high sample throughput (approx. 120 samples/h).

311.

BINDER PASTE: NEW COMPOSITE MATERIAL FOR BIOSENSORS AND ELECTROCHEMICAL BIOREACTORS. J. Parellada, E. Domínguez & I. Katakis*. Department of Analytical Chemistry, University of Alcalá, E-28871 Alcalá de Henares (Madrid), *Department of Chemical Engineering, University Rovira i Virgili, E-43006 Tarragona, SPAIN

A new composite material is demonstrated comprised of conducting particles held together with a hydrophobic or hydrophilic polymeric "binder". In this work, this composite, called "binder paste", is used for the construction of mediated amperometric enzymatic biosensors with different kinds of oxidoreductases (oxidases, peroxidases and bi-enzyme systems). Its performance is compared with traditional carbon paste electrodes and is found to be superior in current densities, sensitivity and stability. The increase in current densities reached two orders of magnitude. The superior performance of the material is attributed to its high permeability to the enzymatic substrate (in the case of hydrophilic binders). It is also demonstrated that the material retains superior characteristics for electrochemistry in organic solvents and additionally, a flow-through bioelectrochemical mini-reactor is demonstrated with glucose oxidase immobilized in it.

312. **STRUCTURALLY AND CHEMICALLY MODIFIED, SOL-GEL CARBON THICK FILM GLUCOSE SENSORS.** Prasad V.A. Pamidi, Doeg Su Park and Joseph Wang, Department of Chemistry and Biochemistry, New Mexico State University, Las Cruces, NM 88003

The microfabrication of disposable enzyme electrode strips, such as those used by diabetic patients relies on thick film technology. Such a fabrication route involves printing of different electrode patterns through the pre-designed mask and baking them at elevated temperatures in order to remove the solvent. As enzymes cannot withstand high temperature curing associated with the fabrication process, a separate immobilization is required. We report the development of sol-gel derived disposable biosensors by coupling the sol-gel and thick film technologies. Such a coupling offers a one-step fabrication process, as it obviates the need for high temperature curing. The sol-gel derived carbon electrodes can be cured rapidly at lower temperatures. The resulting strips are amenable for a wide variety of chemical or structural modifications. Sol-gel derived thick film electrodes can serve as hosts for various heat sensitive biomaterials in connection with the microfabrication of biosensors for other analytes of interest.

313. BIOMIMICRY OF SMELL USING ORGANIC ELECTROCONDUCTIVE POLYMER SENSOR ARRAYS: APPLICATIONS IN AN ELECTRONIC NOSE .T. D. Gibson, Department of Biochemistry and Molecular Biology, Astbury Building, University of Leeds, Leeds, LS2 9JT, UK.,J N Hulbert, R W Marshall, J Peace, E A Ruck - Keene, Bloodhound Sensors Ltd, Q Laboratory, Worsley Building, Clarendon Road, Leeds, LS2 9JT, UK., P Corcoran, Sensing and Control Research Group, School of Engineering, University of Derby, Kedleston Road, Derby, DE22 1GB, UK.

The detection and simultaneous identification of 3 different yeasts by measuring the volatile compounds produced from plate cultures has been carried out using 16 different electroconductive polymers configured as a chemoresistive sensor array and incorporated into an electronic nose. Headspace samples were taken from static atmospheres formed from inoculated agar plates after a suitable growth period at 37°C and analysed using the sensor array operating in transient flow mode. The response of the sensor array to water and to the control media in the absence of yeast growth was also determined, allowing greater discrimination of yeast volatiles. The response curves produced were processed using standard back propagation neural network techniques to provide identification. The overall classification rate for the 3 similar yeast cultures were compared and correctly classified at a level of 96.9% with no pre-processing to remove the sample signal generated by the media.

314. POLYANILINE BASED POTENTIOMETRIC BIOSENSORS. A. A. Karyakin, E. E. Karyakina; Chemical Department of Moscow State Univeristy, 119899, Moscow, RUSSIA

Application of semiconductor polyaniline (PAn) for potentiometric biosensor development provides certain advantages comparing with the known systems. At one hand enzyme immobilization inside pH sensitive matrix is expected to improve sensor sensitivity. At the other hand semiconductor PAn possesses an increased potentiometric response towards pH changing. Since the common polyaniline loses its electroactivity in neutral aqueous solutions (pH > 5) we synthesized the self-doped polymer electrochemically active even in basic media up to pH 12. Using self-doped polyaniline instead of common polymer as pH transducer the stable potentiometric response of 70 mV/pH was obtained. Potentiometric biosensors were made either by enzyme entrapment into polyaniline film using simple adsorption or by creating of the enzyme containing Nafion layer on the top of polyaniline modified electrode. Taking as an example glucose biosensor we showed that the response of polyaniline based electrode was three-four fold increased as compared with the glucose-sensitive field-effect transistor. Biosensor for environmental control was made by immobilization of Organophosphate hydrolase (OPH) into self-doped polyaniline film.

315. REDUCTION OF REMAINING MONOMER IN POLY-L-LACTIDE BY SOLID-STATE POLYMERIZATION AND CRYSTAL NUCLEATION. Kazuya Shinno, Masatoshi Miyamoto, and Yoshiharu Kimura, Department of Polymer Science and Engineering, Kyoto Institute of Technology, Matsugasaki, Sakyo-ku, Kyoto 606, Japan

Solid-state polymerization of L-lactide was studied for reducing the remaining monomer in poly-L-lactide (PLLA). The polymerization was examined by two different ways with 0.1 mol% of tin 4-ethylhexanoate as the catalyst. In a two-step method, the ordinary melt polymerization of L-lactide was carried out at temperature higher than the crystallization temperature (Tc) of PLLA and followed by the post-polymerization around Tc of PLLA. Without occurring crystallization of PLLA during the latter step, the remaining monomer ratio exceeded 5 mol% owing to the monomer / polymer equilibrium preserved in the homogeneous super-cooling state. When the reaction temperature was changed in the second step (e.g., from 140 to 120°C), the remaining monomer ratio decreased to zero with PLLA crystallized. In another one-step method, the polymerization was continued around Tc of PLLA. The monomer consumption was found to reach 100 % with promotion of the polymer crystallization, although the molecular weight did not increase. This should be because various oligomers were formed in the later stage of solid-state polymerization by ester interchange reaction. The polymerization kinetics for this system was successfully analyzed assuming that the reactions would proceed in the amorphous region of polymer. The effectiveness of some newly designed nucleators to the crystallization of PLLA and to the solid-state polymerization will also be described.

PMSE

316. BIOACTIVE MOLECULES RELEASE CONTROLLED BY GRAFTED LACTIC ACID OLIGOMERS HYDROLYSIS

Karine VALLEE-REHEL, Valérie LANGLOIS, Alain LE BORGNE, Philippe GUERIN, Su Ming LI *
Laboratoire de Physico-Chimie des Biopolymères, UMR 27 - Université PARIS XII - 2 à 8, rue Henry DUNANT 94 320 THIAIS, FRANCE, * Centre de Recherche sur les Biopolymères artificiels, URA CNRS 1465, Faculté de Pharmacie, 15, avenue Charles FLAHAULT, 34060 MONTPELLIER, FRANCE

The need for polymeric devices in the sustained release of bioactive molecules is increasing in several domains, besides the therapeutic domain. Encapsulation is concerned in agriculture for fungicides and seed protection , in sea application for biocides in surfaces protection as in the case of ship hulls or pipes. The liberation of the bioactive materials is controlled by the diffusion of water in the polymeric matrix which depends of its hydrophilic/ hydrophobic balance. Acrylic or methacrylic acid derivatives copolymers are good candidates for such application, because of the possibility to combine different types of chemical structures in the macromolecular chain. Two approaches will be discussed for the synthesis of the polymeric materials. In the first one, copolymers have been prepared by using a combination of hydrophobic monomers as butyl (meth)acrylate, and hydrophilic monomers as N,N dimethylaminoethylmethacrylate, N vinyl-2- pyrrolidinone. The second way concerned the development of degradable systems and, at first, grafted copolymers containing lactic oligomers as hydrolyzable segments have been retained. After formulation ,their ability for releasing lactic acid has been tested through casted films which have been deposited in water solution.

317. THE INFLUENCE OF AUTOCATALYSIS ON THE HYDROLYSIS OF DEGRADABLE POLY(LACTIC ACID)

Georgette L. Siparsky, Kent J. Voorhees, Department of Chemistry and Geochemistry, Colorado School of Mines Golden, Colorado 80401. Fudu Miao, Chronopol Inc., 4400 McIntyre Avenue, Colorado 80403.

Poly-(lactic acid), PLA, is currently being developed as a commodity plastic because it is a hydrolytically degradable polymer. The rate of polymer hydrolysis increases with time, and that has been attributed to both the high reactivity of the terminal ester, as well as the influence of catalysis resulting from the increasing acid end-group concentration. Previous solution hydrolysis studies of PLA were carried out with the addition of an external acid, and the process of "self catalysis" was not evaluated. The kinetics of autocatalysis were examined in an acetonitrile solution of the polymer in the absence of an external acid, and an equation was derived to describe the process. Hydrolysis carried out at 50°C and 60°C demonstrates the influence of acid end-group catalysis on the reaction rates. The results are compared with previous solution hydrolysis studies conducted in acetone and dioxane in the presence of hydrochloric acid and acetic acid.

318. ALIPHATIC POLYESTER BLENDS OF POLY(LACTIC ACID) AND OLIGOMERIC POLY(HEXAMETHYLENE SUCCINATE). Samuel J. Huang and Michael R. Lostocco, Institute of Materials Science, University of Connecticut, Storrs, CT 06269-3136

In an attempt to gain a degree of control over the mechanical and degradation properties of poly(lactic acid)[PLA], large-scale efforts are underway to alter the phase morphology of PLA through chemical and physical modification. Consistent with this theme, our work aims to adjust the molecular architecture of highly amorphous PLA with an increasing concentration of oligomeric poly (hexamethylene succinate)[PHS]. Gel-permeation chromatography(GPC) verifies the enhanced presence of PHS in the blends with a concomitant reduction in number-average molecular weight as the weight fraction of PHS is raised from 0.10 to 0.40. Differential scanning calorimetry(DSC) indicates amorphous phase compatibility between PHS and PLA at weight compositions of 10/90 and 20/80. However, as the amount of PHS approaches 30 and 40 weight %, the PHS exhibits the ability to crystallize independently from the induced PLA crystalline phase. Hydrolytic degradation results appear to depend upon the specific multiphasic character of the blends.

319. SYNTHESIS AND BIODEGRADATION OF MODIFIED POLYGLYCOLIDE

B. Belenkaya, V. Sakharova, B. Sinevich, S. Belousov, A. Kuptcov, Karpov Institute of Physical Chemistry, Moscow, 103064, Russia

It was found that modified polyglycolide annealed fibers were significantly more stable in vivo than unmodified polyglycolide ones at the same extension degree. At the same time the extension limit for modified polyglycolide fibers was higher than for unmodified, and initial mechanical properties and there retention in vivo were proportional to extension degree. From the other side, from DSC data it follows that the differences of crystallinity degrees are insignificant for both types of fibers and slightly depended on the extension of the fiber and during biodestruction the crystallinity increased. It means that modificator blocks didn't penetrate into the crystal structure of the polymer but are displaced into amorphous areas and destruction occurred in amorphous areas also. From electron micrografs it's followed that predominatingly degradation occurred inside the fiber and erosion of its surface is neglectable. Raman spectroscopy studies of the samples of fibers with different extension degrees from 1 to 5.5 and pellets from the initial polymers showed that not only orientation degrees of polymer molecules increased but addition intermolecular interaction appeared by ~CH O=C~ intermolecule hydrogen bonding. So, the introduction of the modificator containing flexible blocks and hydrophobic groups in the polyglycolide chain facilitate orientation and interchains interactions of polymer molecules and hydrofobized the amorphous areas which results in decrease of water/medium penetration and polymer chains destruction.

320. PREPARATION AND CHARACTERIZATION OF POLY(LACTIC ACID) AND MALEIC ANHYDRIDE GRAFT COPOLYMER (PLA-g-MAH) AND ITS BLENDS WITH PEG
Wenguang Ma[1], Linfu Wang[1], Richard A. Gross[2] and Stephen P. McCarthy[1], Department of Plastics Engineering[1] and Chemistry[2], NSF Center for Biodegradable Polymer Research, University of Massachusetts Lowell, Lowell, MA 01854

Poly(lactic acid) (PLA) and 5-20 wt% maleic anhydride (MAH) were melt reactive blended in the Haake twin screw mixer chamber at 160 °C and 60 rpm. T-butyl peroxide was used as an initiator. Products were purified by using acetone as an solvent and pentane as a precipitant, then were titrated for determining grafting ratio of MAH. The purified products and benzyl amine reacted at 40 °C for 2 hr.

NMR and FTIR analyses of the amidated products after purification and the titration results show that the grafting ratio of MAH is about 0.5-2 %. PLA with 5%wt MAH product (PLA-g-5MAH) was extrusion blended with PEG (Mn=10000). NMR analysis of the extrudate purified showed the PLA and PEG graft copolymer has formed by reaction of PEG and appended MAH on PLA.

From GPC testing results, it is seen that the molecular weight of the PLA-g-PEG is higher than that of PLA/PEG blends. Enzymatic degradation test (protein K) results indicate that the degradation rates of graft copolymer are higher that of pure PLA.

321. A DYNAMIC PROBE OF TRIBOLOGICAL PROCESSES AT METAL / POLYMER INTERFACES: TRANSIENT CURRENT GENERATION, J. T. Dickinson, L. Scudiero, and S. C. Langford, Department of Physics and Materials Science Program, Washington State University, Pullman, WA 99164-2814

An important component of friction during rubbing of two surfaces arises in the rapid, transient making and breaking of adhesive bonds between asperities. When conductors are drawn across polymers, continuous detachment between the two surfaces generates charge separation due to contact electrification. We have devised sensitive electronic circuits for detecting instantaneous transient currents generated by this process while simultaneously measuring the normal and lateral forces as a conducting stylus is moved across insulating polymeric surfaces. These measurements are extensions of experiments on dynamic transient currents generated by propagating cracks at metal-polymer interfaces. The experiments are performed in high vacuum as well as in controlled atmosphere. We present results on conducting metal tips on softer substrates (polymers such as PMMA, polycarbonate, polystrene, and polyethelene). We show how these currents are related to the extent of damage to the substrate, the contribution of adhesion to the frictional force, and the physics of contact charging. Present tip radii used are a few μm in dimensions with efforts to extend down to smaller sizes. Time resolved measurements reflect temporal statistics of make-break interactions - presently, time resolutions of μs have been achieved. The analysis of both magnitude and fluctuations in the current are informative regarding the micromechanics of asperity/substrate interactions.

322.
STRUCTURAL DEVELOPMENT IN HIGHLY ORIENTED GLASSY POLY(ETHYLENE TEREPHTHALATE) WITH TEMPERATURE AND MECHANICAL STRESS. U. Göschel, Department of Polymer Chemistry and Technology, Eindhoven University of Technology, P.O. Box 513, 5600 MB Eindhoven, The Netherlands

Isotropic and X-ray amorphous poly(ethylene terephtalate) (PET) film strips were cold drawn at well-defined parameters to achieve noncrystalline structures with an extraordinarily large chain orientation. These oriented structures have been studied with respect to temperature- and stress-induced changes of the microstructure and thermo-mechanical behavior as well. *In-situ* X-ray diffraction and DSC experiments revealed a large decrease in the crystallization temperature and increase in the crystallization velocity with orientation. Highly oriented and glassy PET crystallizes already at 56 °C compared to 94 °C for isotropic PET. Dynamic mechanical thermal, *in-situ* shrinkage strain and stress investigations confirmed the crystallization results and provided additional information about the relaxation of local molecular orientation.

323. THE EFFECT OF ORIENTATION BY SOLID STATE DRAWING ON THE GLASSY REGIONS IN POLY(VINYL ALCOHOL)

R.M. Hodge[1,2], T.J. Bastow[3] and A.J. Hill[3,4]
[1] Department of Mechanical Engineering, Victoria University, PO Box 14428, MCMC, Melbourne Vic 8001, Australia.
[2] Department of Materials Engineering, Monash University, Clayton Vic 3168, Australia. [3] CSIRO, Division of Materials Science and Technology, Private Bag 33, Sth. Clayton MDC, Clayton Vic 3169, Australia. [4] Faculty of Engineering, Monash University, Clayton Vic 3168 Australia

Cold-drawn films of poly(vinyl alcohol) PVOH have been characterized at room temperature using positron annihilation lifetime spectroscopy, (PALS), ^{13}C solid state nuclear magnetic resonance (NMR) and various x-ray diffraction techniques. The first stages of drawing lead to an increase in molecular orientation and strain induced crystallinity, while also increasing the relative free volume in the amorphous regions of the polymer. As the draw ratio approaches the maximum achievable value, the crystallinity and molecular orientation undergo no further changes. The free volume, however, decreases to a value below that of the unstrained sample, indicating that the amorphous regions are constrained and aligned by the oriented crystals. Throughout the drawing process, the molecular mobility remains essentially constant.

324. MODELING OF PRESSURE EFFECTS ON STRUCTURAL RELAXATION IN GUEST-HOST POLYMERS FOR NONLINEAR OPTICAL APPLICATIONS. Grisha A. Medvedev, James M. Caruthers, Hilary S. Lackritz, School of Chemical Engineering, Purdue University, West Lafayette, IN 1283 - 47907

Recently we have reported on a stochastic model of structural relaxation in polymer glasses which successfully explains experimentally observed properties of the SHG decay in guest-host systems. In the present paper, the effect of pressure on the local fluctuations and relaxation of the specific volume, and its manifestation in the rate of the SHG decay is considered. The Adam and Gibbs theory of cooperative rearrangement expanded to the nonequilibrium states is used to describe the shift factor for the specific volume relaxation. An equation of state, differing from the Simha-Somcynsky result, is derived from the lattice model using an isobaric rather than an isochoric ensemble. Model predictions for the glass transition temperature, the shift factor, the compressibility, and the SHG decay parameters as functions of pressure are compared with experimental data.

325. EFFECT OF TOPOLOGY ON PMMA RELAXATION. Hsing-Yeh Parker, Rohm and Haas Co., 727 Norristown Road, Spring House, PA 19477

The entanglement of high molecular weight polymer chains is a very important aspect of polymer structure because it is the origin of many useful physical properties of polymeric materials. In this work we found that various entanglement structure and/or chain conformation of the polymers is generated by emulsion polymerization where the chains are formed in a space comparable to its own dimension. Several poly(methylmethacrylate) (PMMA) emulsion polymers were compared experimentally. It is found that variation in entanglement structure significantly affect the chain relaxation and inter-mixing process of the high molecular weight PMMA in a high viscosity matrix. We thus conclude that polymer topology, which describes how macromolecules are organized in a collective state, should be recognized as one of the fundamental structural parameters such as molecular weight, composition and chain architecture, of polymers.

326. EFFECTS OF PHYSICAL AGING ON CRAZE INITIATION AND FAILURE IN POLY(ETHYLENE TEREPHTHALATE) BOTH WITH AND WITHOUT EXPOSURE TO CHEMICAL ENVIRONMENTS. M. R. Tant, E. J. Moskala, M. K. Jank, and P. P. Shang, Research Laboratories, Eastman Chemical Company, P.O. Box 1972, Kingsport, TN 37662 and A. J. Hill, CSIRO, Division of Materials Science & Technology, Private Bag 33, South Clayton MDC, Victoria 3169, AUSTRALIA

Applications of glassy or semicrystalline polymers often require that these materials be exposed to corrosive chemical environments. Such exposure may lead to premature brittle fracture of the material by environmental stress cracking. The time required for brittle failure to occur as a result of chemical exposure depends on the chemical nature of the polymer, its morphology, and the chemical environment. Physical aging below the glass transition temperature results in a decrease in molecular free volume and hence a reduction in molecular mobility which, in turn, results in the molecules becoming increasingly unable to respond to imposed stresses and leads directly to a transition from ductile to brittle behavior. We report here the results of studies on glassy poly(ethylene terephthalate) to determine the effects of both chemical exposure and physical aging on the critical strain for crazing. In addition, we report the effects of physical aging on both yield stress and craze stress in the absence of an aggressive chemical environment. The implications of these results for the mechanical behavior of these materials are discussed.

327. CELLULOSE MICROSTRUCTURE STUDIED BY POSITRON ANNIHILATION LIFETIME SPECTROSCOPY. Huimin Cao, Yongming Lou, and Yanching Jean, Department of Chemistry, University of Missouri - Kansas City, Kansas City, Missouri 64110-2499; Alexandra Pekarovicova, Department of Wood, Pulp, and Paper, Slovak Technical University, Bratislava, Slovakia; and Richard A. Venditti, Department of Wood and Paper Science, North Carolina State University, Raleigh, North Carolina 27695-8005.

Positron annihilation lifetime (PAL) measurements have been performed on Avicel and Whatmann celluloses and their ball-milled derivatives as well as on cellulose acetates. Correlations between crystallinity and fractional microstructures are found. Combined studies using PAL and Fourier Transform Infrared (FTIR) on deuterated celluloses and moisture sorptions will be presented.

328. PROPERTIES AND TRANSITIONS DURING CURE OF DENSELY CROSSLINKED ACRYLATE POLYMERS
Jakob Lange, John W. Davison[†] and Anders Hult[*]
Department of Polymer Technology, Royal Institute of Technology, S-100 44 Stockholm
[†]*Davison Chemographics Ltd, 28 Woolmer Way, Bordon, Hants, GU35 9QF, UK*
[*]*To whom correspondence should be addressed*

Gelation and vitrification during isothermal cure at different temperatures of an acrylate material have been investigated using torsional dynamic mechanical analysis. The acrylate was UV-cured, which simplified varying the cure temperature. Vitrification during cure of the chain-wise reacting acrylate system was found to proceed differently from what is generally observed for step-wise reacting systems. In the acrylate vitrification began just after gelation and lasted throughout the reaction, leaving the sample in the transition region, at all temperatures where it occurred. The modulus at the end of isothermal cure at temperatures below the ultimate glass transition temperature was found to be far below the glassy level. Cure of the acrylate at room temperature thus produces a material with significantly lower room temperature modulus than does cure at higher temperatures. This could indicate that chain-wise reacting systems cured far below their ultimate glass transition temperature do not obtain full mechanical properties.

329. SYNTHESIS AND CHARACTERIZATION OF AMORPHOUS NYLON 6. M.V.Pandya[*] and Mahesh Subramaniyam, Department of Chemistry, I.I.T. Powai, Bombay 400-076 INDIA.

Nylon 6 is popular for engineering applications. It has good mechanical strength, oil resistance and moderate oxygen permeability, there by qualifying as suitable packaging material. However it has one drawback i.e it is opaque in nature. If the nylon 6 is made amorphous it can have excellent transparency with good mechanical properties. Amorphous nylon 6 has applications in packaging, automotive, construction and electrical industries.

This paper reports two series of nylon using isophorone diamine and metaxylene diamine along with isopthalic acid and ε caprolactum as the comonomer. The diamine concentration is varied from 5 to 20 weight % in each series. The polymers are characterised by IR., NMR, Xray, thermal and viscometric studies. The mechanical and electrical properties of the polymer are also reported. The high T_g and high amorphous content with good mechanical and electrical properties are the outstanding features of the polymer with 15 weight % of Isophorone diamine.

330. FUNDAMENTAL STUDIES OF DNA ADSORPTION AND HYBRIDIZATION ON SOLID SURFACES   David J. Graves*, Lynn Anne Sanguedolce, Vincent Chan, Steven E. McKenzie[†], Saul Surrey[†], and Paolo Fortina, Departments of Chemical Engineering and Pediatrics, The University of Pennsylvania and Children's Hospital of Philadelphia, and [†]Jefferson Medical College and the duPont Hospital for Children, *311A Towne Building, 220 South 33 St., Philadelphia, PA 19104-6393

Arrays of DNA or oligonucleotide sequences immobilized on a solid surface are becoming important tools for identifying genetic defects, pathogens, forensic samples, gene expression, etc. by hybridization to solution-phase complements. However, little fundamental work has been done on such areas as solution and surface diffusion, steric hindrance due to the solid support, immobilized probe density, spacer arm effects, and other parameters which can affect hybridization. We are studying some of these relevant questions. Our current results include data on adsorption equilibria and desorption rates of oligonucleotide on various types of silanized surfaces, surface diffusion constants, and preliminary data on spacers and the location of a hybridizing sequence within a larger piece of DNA. Surface concentrations which cause fluorescence quenching are surprisingly easy to attain, indicating close packing and revealing a potential problem which may exist in certain situations. These results and instrumentation used to obtain them will be discussed.

331. THERMOCAPILLARY PUMPING IN MICROFABRICATED LIQUID ANALYSIS DEVICES

Timothy S. Sammarco[1], David T. Burke[2], Carlos H. Mastrangelo[3], and Mark A. Burns[1]

[1]Department of Chemical Engineering, [2]Department of Human Genetics
[3]Department of Electrical Engineering and Computer Science
The University of Michigan
Ann Arbor, MI 48109

Thermocapillary pumping (TCP) is a non-mechanical technique that can be used to move nanoliter fluid drops in microfabricated analysis devices. The technique uses the temperature sensitivity of surface tension at the interface of discrete drops to propel them in particular directions. By heating one interface of a discrete droplet, the corresponding decrease in surface tension induces a pressure difference which moves the drop. Experiments using various liquids in microfabricated devices, however, have shown that contact angle hysteresis can prevent TCP in some channel designs. In addition to addressing the microfabrication issues relating to reducing contact angle hysteresis, this talk will discuss the ability to successfully control temperature profiles within the device.

332. MICRO-FLUIDIC **MIXER** D. Liepmann and J. Evans, Department of Mechanical Engineering, University of California, Berkeley, California 94720

Mass produced microelectromechanical (MEMS) devices for mixing tens to hundreds of nanoliters of fluid have applications in fields ranging from medicine, to printing, to chemical analysis. Mixing fluids in MEMS devices is extremely difficult because these devices are generally too large to rely on upon diffusion alone yet too small to allow turbulent flow ($Re \sim O(1)$). These constraints eliminate standard approaches to mixing. We present research on a planar, laminar device that employs chaotic advection to mix fluids. The system was designed using a numerical model based on Hele-Shaw flow and infinite Schmidt number. A fully integrated prototype has been constructed containing five mixing chambers and associated pumps that allow testing of a range of system configurations that span the numerically investigated designs. Testing of the completed device will include Laser Induced Fluorescence for two-dimensional mixing and Digital Particle Image Velocimetry for velocity measurements.

333. MICROCHANNEL ELECTROPHORESIS IN INJECTION MOLDED SUBSTRATES. Herbert Hooper, Randy McCormick, Goretty Alonso, HongYing Wang, Dominic Benvegnu, Soane BioSciences, Inc. Hayward, CA 94545

The use of micromachining techniques in electrophoretic separations is increasing rapidly. Various laboratories are employing photolithography to directly manufacture such devices on glass and other substrates. Work in our labs centers on using microfabrication techniques to produce injection molding tools for high throughput, reproducible, and low-cost production of cassettes for microchannel electrophoresis. Channel dimensions reproduced in plastic substrates are 10 micrometers and up, while fabricating submicron dimensions is also possible. We successfully produce cassettes and obtained high efficency separations of DNA and other biomolecules in under 3 minutes.

334. DETECTING THE ACTIVATION OF RECEPTORS ON LIVING CELLS USING A HIGH-PERFORMANCE MICROPHYSIOMETER. J. C. Owicki, L. J. Bousse, D. L. Modlin, and J. W. Parce, Molecular Devices Corporation, 1311 Orleans Drive, Sunnyvale, CA 94089

Microphysiometry is an analytical method that noninvasively detects changes in cell physiology in vitro by monitoring cellular metabolic activity. It does so by using a semiconductor-based sensor to measure the rate of extracellular acidification, a quantity that is related to the rate of energy metabolism (glycolysis, oxidative phosphorylation) and also to intracellular pH homeostasis [see McConnell et al. (1992) Science 257:1906; Owicki et al. (1994) Ann. Rev. Biophys. Biomol. Struct. 23:87]. The Light Addressable Potentiometric Sensor, or LAPS, that is at the heart of a microphysiometer is compatible with micromachining; this enables it to be simultaneously sensor and fluid pathway. In this talk we describe a high-performance microphysiometer based on a LAPS chip containing eight micromachined fluid channels, each 1 mm wide by 100 μm deep by 1 cm long. Each channel has four sites at which cells can be stationed, so that there are 32 samples of cells (all potentially of different types) per chip. About 15 μL of test fluid is sufficient to obtain responses from the four cell samples in a channel. We will illustrate the performance of the instrument with data drawn from the principal application of microphysiometry, detecting substances that modulate the activation of cellular receptors. Supported in part by ARPA, Contract MDA972-92-C-0005.

335.
MEASUREMENT OF GLUCOSE WITH NONINVASIVE TRANSDERMAL EXTRACTION
II. CORRELATION BETWEEN SENSOR RESPONSE AND BLOOD GLUCOSE

C. M. Chiang, Tung-Fen Chen and Bret Berner
Cygnus, Inc. Redwood City, CA 94063

It has been demonstrated previously that sufficient quantities can be detected from noninvasive transdermal extract while the glucose was measured by HPLC. In this study, the glucose was extracted from 6 healthy subjects by using reverse iontophoresis. The response profiles suggested that with a 15 minute delay, the sensor response tracks closely with the blood glucose level. A linear regression was performed with these data excluding the first data point. This is consistent with the equilibrium period observed by Tamada et al. Error Grid analysis was also used for determining clinical accuracy for the measurement of glucose level. A linear regression was performed with these data noninvasive method can be developed for glucose measurement.

336. APPLICATION OF NEURAL NETWORKS TO OPTIMIZE CUMULATIVE RELEASE PROFILES FOR TRANSDERMAL SYSTEMS.
R. T. Kurnik, J. Jona, Cygnus Inc., Redwood City, CA 94063

Neural networks are being increasingly used in the pharmaceutical field for many different applications due to their ability to model processes that cannot be modeled by classical mathematical physics. These include such diverse areas as structure - function relationships for transdermal enhancer selection, determination of peptides that will undergo iontophoresis through the skin, determination of outcomes from clinical trials, optimization of manufacturing processes, and process control. In this paper, we describe a new application of neural networks which is to determine the chemical composition of a transdermal device such that the resultant cumulative release profile will exactly overlap a pre-determined profile.

337.

ON THE IMPORTANCE OF THE BURST EFFECT DURING DRUG RELEASE FROM POLYMER FILMS. Balaji Narasimhan and Robert Langer, Department of Chemical Engineering, Massachusetts Institute of Technology, Cambridge, MA 02139.

An analysis is presented to describe the role of the burst effect in a controlled release polymeric slab. Asymptotic solutions of the model show that the burst effect is controlled by the drug solubility in the release medium. It was shown that as drug solubility increased, the drug released faster and the velocity of the interface between dissolved and dispersed drug is higher. The ratio of the amount of drug released during the burst to the amount released at steady state was shown to be dependent on the drug solubility in the release medium. The model was used to establish conditions under which the burst effect could be maximized or minimized. The model predictions were compared to experimental studies of sodium salicylate release from polyethylene slabs.

338. ELECTROCHEMICAL ENZYME SENSORS FOR EX VIVO AND IN VIVO ON LINE LACTATE ANALYSIS: Dorothea Pfeiffer, Jan Szeponik, Barbara Möller, BST Bio Sensor Technologie GmbH, D - 13156 Berlin/Germany and Liu Yang, Peter Kissinger, BAS West Lafayette, IN 47906/USA

Monitoring of lactic acid is of significant importance in critical care. During hypoxia the energy production has to switch from glucose degradation to a less effective mechanism. NADH produced during glucose degradation has to be used to form lactate. This may be followed by dangerous metabolic stress. Therefore, during resuscitation as well as monitoring of coronary instability or shock, intensive care would be supported in the realization of impending problems by the lactate state measured as rapid as possible, which is ideally a real-time monitoring.
This talk will discuss the development of a lactate oxidase polymer matrix for application in biosensors. The enzyme matrix will be characterized and the application to different types of sensors will be discussed: 1) Lactate oxidase membrane electrode for the analysis of whole blood, serum and fermentation solutions. 2) Ex vivo on-line monitoring of lactate based on microdialysis sampling and lactate oxidase membrane sensor. 3) Thinfilm lactate oxidase probe for in vivo lactate monitoring.

339. MORE BIOCOMPATIBLE ELECTROCHEMICAL SENSORS THROUGH THE USE OF COMBINED NITRIC OXIDE RELEASE/ION SENSING POLYMERIC FILMS. K. Mowery, C. Espadas-Torre, V. Oklejas, and M.E. Meyerhoff, Department of Chemistry, The University of Michigan, Ann Arbor, MI 48109.

The potential for fabricating more blood compatible *in vivo* electrochemical sensors by preparing devices that release the potent platelet anti-aggregation and vasodilation agent nitric oxide (NO) while simultaneously maintaining useful analytical response properties is examined. The approach is demonstrated using classical ion-selective polymer membrane electrodes for H^+ and K^+, prepared by doping either polyurethane or poly(vinyl chloride) films with the ionophores, tridodecylamine and valinomycin, respectively. When various diazeniumdiolate adducts of NO are further added to the films, the resulting membranes emit NO over periods of days, with rates of release highly dependent the nature of the polymer and plasticizer used to cast the film. The same films maintain excellent potentiometric ion response toward H^+ and K^+, with slopes and selectivities comparable to films that do not contain the NO adducts. Most importantly, the resulting films that release NO display a marked decrease in platelet adhesion (compared to films without NO release), as measured via SEMs taken after exposure of the films to platelet rich plasma. The approach suggested may provide the ideal means to reduce the risk of thrombus formation on intravascular electrochemical sensors while concomitantly enhancing blood flow at the implant site.

340.
DEVELOPMENT OF A BIOSENSOR-BASED, PATIENT-ATTACHED BLOOD GLUCOSE MONITORING SYSTEM. **J.Y. Lucisano,** S.V. Edelman, B.D. Quinto, and D.K. Wong, VIA Medical Corporation, San Diego, CA 92121

A novel biosensor-based system for near-continuous monitoring of blood glucose concentration has been developed and is now commercially available. The system consists of a sterile disposable electrochemical sensor module, an i.v. tubing set, an i.v. bag of infusible calibrant, and a monitor containing a pumping mechanism and electronics. The system is attached to either an intravenous or intra-arterial catheter and functions by withdrawing blood from the patient, performing an analysis and reinfusing the blood sample. The system responds linearly to glucose concentration over the range 30-600mg/dl. Systems tested with aqueous solutions over 84 hours (20 sensors, total n=1465) demonstrate r2 of 0.997, bias of -1.4%, and precision of 5.1%. Results from paired samples (41 patients, total n=1830) during clinical application (glucose clamp study) show r2 of 0.95, bias of -0.23%, and precision of 4.35% compared to off-line analysis.

341.   WETTABILITY AND SURFACE STRUCTURE OF HYALURONIC ACID AND HYALURONIC ACID ESTERS FOULING-RESISTANT COATINGS. M. Morra, C. Cassinelli, Nobil Bio Ricerche, 14018 Villafranca d'Asti, ITALY

Hyaluronic acid (HA) and HA esters covalently linked to materials surfaces produce highly hydrated surface structures, that minimize non-specific binding. While it is well known that an array of intramolecular hydrogen bonding affects properties of hyaluronan molecules in solution, the structure of surface bonded hyaluronans is much less known. We have used wettability techniques to try to gain insights on the structure-properties relationship of surface bonded HA and HA esters. In particular, test were performed on surface-bonded HA, 50% benzyl ester of HA, 75% benzyl ester of HA. Data were compared to those obtained on a 100% benzyl ester film. Surfaces were probed by contact angle measurement and calculation of surface free energy components, ESCA analysis and resistance to cell (fibroblasts) adhesion. Results show that increasing the esterification degree produces an increase of the electron acceptor character of surfaces, probably due to decoupling of the acidic acetamido group from the basic carboxylate anion. Contrary to the 100% ester, HA and partial esters completely inhibit cell adhesion. The role of interfacial mobility of carbohydrate units bridging surface-bonded sites is discussed.

342.   DIFFUSION PROPERTIES OF TISSUE WHICH ENCAPSULATES SUBCUTANEOUS IMPLANTS. A. Sharkawy, B. Klitzman, G. Truskey, and W. Reichert, Center for Emerging Cardiovascular Technologies, Department of Biomedical Engineering, Duke University; Durham NC 27708-0295

Polyvinyl alcohol (PVA) specimens with 60 and 350 micron mean pore size (PVA-60, PVA-350), nonporous PVA (PVA-skin), and stainless steel cage (SS) specimens were implanted in the subcutis of Sprague Dawley rats for four weeks to elicit a range of capsular wound healing tissues. Histological examination showed the capsular tissue formed around PVA-skin and SS specimens to be densely fibrous and avascular, while that forming around PVA-60 and PVA-350 was less densely fibrous and vascular. Using sodium fluorescein (Mw 376 g/mol) as a model analyte, the two most fibrous capsular tissues exhibited diffusion coefficients that were statistically ($p<<.05$) less than that determined for rat subcutis by 50 and 25% for PVA-skin and SS, respectively. The vascularized capsular tissue produced diffusion coefficients that were not statistically different from subcutis. Measured diffusion coefficients of the three most fibrous capsular tissues were closely predicted by a simple two component diffusion model, while diffusion coefficients for the least fibrous tissues were overestimated by the model. The effect of a densely fibrous capsule on sensor response time was predicted to increase sensor response time from 5 min to 20 min.

343.

HYDROGELS FOR IONTOPHORETIC EXTRACTION OF GLUCOSE THROUGH SKIN
W. Abraham, P. Joshi, R. Kurnik, B. Berner, Cygnus Inc., Redwood City, CA 94063

Hydrogels are matrices made from water soluble polymers that are cross-linked to maintain three-dimensional structure with high water content. We report the development of hydrogels that function as a collection medium in an iontophoretic device. The hydrogels were characterized for their ability to accommodate a large protein such as glucose oxidase in stable form. Diffusion coefficient of glucose in these hydrogels was shown to be an order of magnitude less than that in water. The effect of irradiation dose on enzyme activity and cross-link density in hydrogel will be discussed.

344.

IN VITRO TESTING OF A NON-INVASIVE GLUCOSE MONITOR
Kathleen Cogan Farinas, Charles Wei, Janet A. Tamada, Michael J. Tierney, Bret Berner, Russell O. Potts
Cygnus, Inc., 400 Penobscot Drive, Redwood City, CA 94063

Reverse iontophoresis enables extraction of nanomolar quantities of glucose through intact skin using low levels of electrical current. This has been coupled with an amperometric sensor to form a non-invasive glucose monitor in a wristwatch format capable of making frequent glucose readings. A hydrogel provides the electrical contact between the sensor and the skin and serves as the reservoir for collection of glucose. The hydrogel also contains glucose oxidase, an enzyme that makes the sensor specific to glucose by converting it to hydrogen peroxide which is measured by the sensor. In vitro studies have been performed using sensors to measure the glucose extracted across human cadaver skin. The results display a strong correlation between measured signal and donor glucose concentrations for clinically relevant glucose concentrations. In addition, a linear relationship between detected glucose and iontophoretic current density is observed, consistent with theoretical predictions in the literature.

345.

CHEMISTRY, RECOGNITION AND FUNCTION OF A NATURAL SHEAR STRESS BIOSENSOR. G. Siegel, M. Malmsten, D. Klüßendorf, A. Walter and A. Schmidt, Institute of Physiology, Biophysical Research Group, The Free University of Berlin, D-14195 Berlin, Germany

Flow-dependent dilatation may be controlled by a sensor at the endothelium–blood interface. Probably, the flow sensor is identical with the integral membrane protein heparan sulfate proteoglycan (syndecan) of the endothelial cells, which is a viscoelastic anionic polyelectrolyte. With increasing flow, the biopolymer passes through a shear stress-dependent conformational change from a random coil to an unfolded filament-helix structure. By this process, additional Na^+ ions from the blood are bound to previously masked intrachain binding sites. After counterion migration along the polysaccharide chain and transmembrane Na^+ influx, these ions mediate the signal transduction into the endothelial cell for a vasodilatatory vessel reaction. On the other hand, due to the elastic recoil forces of the biosensor, a decrease in flow causes entropic coiling again, the release of Na^+ ions into the blood, and thus an interruption of the signal transduction chain. – The dependence of flow-induced vasodilatation on shear stress and sodium implies that the flow sensor molecule has to meet the requirements of a mechanoelectrical transducer. Indeed, with application of a 1.25 mmol/l $[Ca^{2+}]_o$ solution, measuring of the 39.1 nm long proteoheparan sulfate sensor yielded a shortening of the macromolecule by 8.9 nm, whereas Na^+ ions effected an elongation by 13.7 nm. The Ca^{2+}-induced contraction leads to a decrease in Na^+ binding, and the Na^+-induced extension to a promotion in Na^+ binding. This means that Na^+ ions amplify the sensitivity of the flow sensor for Na^+ in an autocatalytic process. Ca^{2+} ions impair the sensitivity of the sensor to bind Na^+. Since less Na^+ ions are bound after the Ca^{2+} shortening of the heparan sulfate chains, Ca^{2+} ions are obviously capable of adjusting the susceptibility of the sensor via a conformational change and, in this way, of controlling the signal transduction cascade that regulates the blood flow.

346.

THE POTENTIAL CONTRIBUTION OF STEAM-EXPLOSION TECHNOLOGY TO RECYCLING, Wolfgang G. Glasser, Brecc K. Avellar, and Robert S. Wright Department of Wood Science & Forest Products, and Biobased Materials/Recycling Center Blacksburg, VA 24061-0324.

The application of steam explosion technology to a variety of raw materials consisting of agricultural crop residues and hardwood-based waste and recycled materials (paper and board products and shipping pallets) is described as the basis for a variety of useful materials. Primary and secondary post treatment methods are reviewed that include fractionation by water and aqueous alkali (primary post treatment); and enzyme hydrolysis, linear board production, and mixed cellulose/lignin ester production (secondary post treatment). The technology favors the installation of small production units at local recycling centers.

347. RECYCLING OF ENGINEERING THERMOPLASTICS. Rudolph D. Deanin and Balaji T. Srinivasan, Plastics Engineering Department, University of Massachusetts Lowell, Lowell, MA 01854.

Thirteen recent studies at UMass Lowell have advanced the recycling of engineering thermoplastics: polybutylene terephthalate, polyethylene terephthalate, polycarbonate, polyphenylene ether, nylon 66, and polyphenylene sulfide. At the least, they defined the amount of recycled resin that could be added to virgin resin. At best, they identified properties which were actually improved by recycling.

348. CATALYTIC CONVERSION PROCESS FOR RECYCLING NAVY SHIPBOARD PLASTIC WASTES, R.E. Allred, L.D. Busselle, T.J. Doak, B.W. Gordon, L.A. Harrah, and A.E. Hoyt, Adherent Technologies, Inc., 11208 Cochiti SE, Albuquerque, NM 87123

A novel low-temperature catalytic tertiary recycling process has been developed as an economical means for recycling polymeric waste materials including post-industrial scrap, post-consumer waste, scrap composites, scrap electronics and tires. This process converts polymers into low molecular weight hydrocarbons at temperatures below 200°C. The hydrocarbons produced as a result of this process are suitable for reuse as chemicals, fuels or monomers. Metals, glass, ceramics, fibers and fillers can separated from the hydrocarbons for further reclamation. This paper focuses on the tertiary recycling of Navy Shipboard Plastic Waste Processor (SPWP) product, an extremely complex and contaminated waste stream. Parameters affecting reproducibility of the process, analysis of recycling products and process economics will be discussed.

349. INFLUENCE OF GIBBERELLIC ACID AND GIBBERELLIC ACID-CONTAINING POLYMERS IN THE REESTABLISHING OF THE EVERGLADES, C. Carraher, A. Gaonkar, H. Stewart, S. Miao, M. Colbert, S. Casanova-Clark, T. Hale and D. Sterling, Florida Atlantic University, Boca Raton, FL 33431, Florida Center for Environmental Studies, Palm Beach Gardens, FL 33410 and Everglades System Research Division, South Florida Water Management District, West Palm Beach, FL 33406
The Everglades is undergoing rapid change where the "river of grass", sawgrass, is being replaced by cattail. Sawgrass is critical in controlling soil erosion, trapping pollutants, holding rain water, decreasing flooding, recharging groundwater and filtering out toxins and other water pollutants helping to maintain water quality for some 5 million people of south Florida. Cattails germinate within 3-10 days at a rate of about 50 %. Sawgrass germinates in 4 to 8 weeks at a rate of 0 to 5 %. Through using GA3 and GA3-containing polymers the germination time for sawgrass has been lowered to 11 days and the rate increased to 20 %. The organotin-GA3 polymer delays the onset of algae and seedlings treated with it are much larger. It is envisioned that treated sawgrass seeds may be spread using aerial dispersion from aircraft.

PMSE

350. USE OF POLYMER FLOCCULATING AGENTS TO CONTROL AGRICULTURAL SOIL LOSS
William J. Orts, Gregory M. Glenn, U.S.D.A. Western Regional Research Center, 800 Buchanan St., Albany, CA 94710

In recent advance in agricultural technology, charged polyacrylamide polymer (PAM) has been added to irrigation water because it removes sediment from run-off water preventing up to 90% soil loss from fields. Although PAM is generally recognized as safe, the long term effects of this "nondegradable" polymer merit further study, considering that it is applied at rates of ~4-8lbs per acre per year. In this paper, the benefit/risk issues for PAM that went into USDA's guidelines for its use are discussed. Data from light scattering, neutron scattering and sedimentation tests are presented to establish the mechanistic properties of PAM critical for its success in this application. Alternatives to PAM which are potentially more biodegradable, such as starch/polyacrylamide copolymers, chitosan, modified starches, and charged proteins will be evaluated. For example, chitosan is virtually as effective as PAM in flocculating soil, but does not stabilize soil in the same manner.

351. THE DEINKING OF POLYETHYLENE PACKAGING FILMS FOR ENHANCED RECYCLING,
Ellis Lim, Department of Chemical Engineering, and Dale Teeters, The Department of Chemistry, The University of Tulsa, Tulsa, OK 74104-3189

The ability to recycle plastic packaging films is adversely affected by ink labels printed on the films. Thus an understanding of the surface chemistry of inks and plastic film as it pertains to ink adhesion is important in film recycling. We have found that deinking of plastic packaging film can be correlated with basic surface chemistry values such as the surface tensions of inks and the critical surface tension of the plastic substrate. Inks which have higher surface tensions than the critical surface tension of the plastic films are easily removed from the film. A standard deinking technique was developed and the results of deinking are quantitatively examined by using (ATR)FT-IR spectroscopy. Deinking of plastic films which are coated with multiple layers of inks have also been investigated and the results can be correlated with the critical surface tension of the plastic film.

352. MOLECULAR DYNAMICS SIMULATION OF POLYMER FLOW IN NANO-CHANNELS. R. E. Tuzun, D. W. Noid, B. G. Sumpter, Chemical and Analytical Sciences Division, Oak Ridge National Laboratory, Oak Ridge, TN 37831-6197; J. U. Otaigbe, Department of Materials Science and Engineering, Iowa State University, Ames, IA 50011

The molecular dynamics method is used to perform computational experiments on the flow of polymers in nano-channels. The results clearly show a dynamical dependence on the size of the channel: the smaller the radius of a concentric tube the more restricted the dynamics. The mechanism underlying this behavior is relatively straight-forward, involving mechanical restriction of the vibrational modes that lead to coiling. As the radius of the concentric channel is reduced, the nonbonded forces between the polymer chain and the channel increase and become sufficient at some critical radius to effectively constrain the large amplitude longitudinal modes of the chain, confining the chain to undergo structural changes in a restricted volume.

[†]Research sponsored by the Division of Materials Sciences, Office of Basic Energy Sciences, U.S. Department of Energy under contract DE-AC05-96OR22464 with Lockheed Martin Energy Research Corp.

353. MODELING OF MICROSTRUCTURAL DEVELOPMENT IN BIAXIALLY DRAWN POLYMER FILMS Lee M. Nicholson, Alan H. Windle, Department of Materials Science and Metallurgy, University of Cambridge, Cambridge, CB2 3QZ, UK

Modelling and computer simulation are now a main-stream activity in polymer science and engineering. The modelling of the processing of polymeric materials, and the imparted microstructure as a consequence of the process, is also rapidly emerging. It is the aim of this paper to illustrate the use of microstructural modelling simulations to enhance our understanding of the three-dimensional behaviour involved in the phenomenological process of strain-induced crystallization in polymer films. A 3D lattice model has been developed and a novel Monte Carlo technique is employed to describe the kinetic behaviour and characterization of the molecular orientation in a system under the influence of an external mechanical field. The initial development provides an appraisal of classical crystal nucleation and growth theory with the progression into representation of polymer morphology and the incorporation of chain folds and entanglements.

354. CORROSION INHIBITION OF ALUMINUM ALLOYS COATED WITH POLY(2,5-BIS(N-METHYL-N-ALKYLAMINO) PHENYLENE VINYLENES. Peter Zarras, John D. Stenger-Smith and Melvin H. Miles, Research and Technology Group (Code 4B2200D), Naval Air Warfare Center Weapons Division, China Lake, CA 93555.

Several long term constant current experiments were conducted with polymer coated and uncoated aluminum plates using poly(2,5-bis(N-methyl-N-alkylamino)phenylene vinylenes as the coating material. The studies have shown that the corrosion pits were significantly less for the polymer coated plates.

355.
THE FORMATION OF PROTECTIVE COATINGS ON STEEL BY SPONTANEOUS POLYMERIZATION. X. Zhang, and J. P. Bell, U-136, Polymer Program, Institute of Materials Science, University of Connecticut, Storrs, CT 06269

We have developed a new method of forming protective coatings on metal surfaces. This paper describes the spontaneous polymerization process as it occurs specifically on steel. The process is very simple, and environmentally friendly. Polymerization occurs spontaneously on the steel surface after immersion into a dominantly aqueous monomer solution. A uniform layer of coating is formed in-situ, with thickness ranging from 1-50 microns. The monomer system studied is primarily a 4-carboxyphenyl maleimide/styrene system. Kinetic studies showed that the process depends on solution pH, monomer concentration, and monomer feed composition. The coating obtained has excellent thermal and dielectric properties. The proposed initiation mechanism involves the electrochemical reduction of the monomer by the substrate to generate the propagating free radicals.

356. SYNTHESIS, PROCESSING, AND CHARACTERIZATION OF TESLIN® FILMS
R.W. Pekala, R.A. Schwarz, R.G. Swisher, R.C. Wang, R.O. Ondeck, and M.O. Okoroafor, PPG Industries Inc., 440 College Park Drive, Monroeville, PA 15146

Teslin® film is a microporous material having high porosity (> 60%), high surface area, small pore size, and a solid matrix that is composed primarily of precipitated silica and polyethylene [1]. The weight ratio of silica to polyethylene in Teslin® film ranges from 1.8 -2.5 depending upon the desired stiffness and properties of the product. Teslin® film is manufactured in thicknesses of 175-350 micrometers, and it has major applications as a printing medium. Teslin® film offers good ink receptivity and excellent image quality when printed with a wide variety of techniques (e.g., flexography, ink jet, laser, lithography, thermal transfer). Unlike ordinary paper, Teslin® film exhibits good chemical resistance and excellent peel strength when laminated to other polymer films.

While Teslin® film has mainly been used in printing applications, we are now examining new characteristics of this material that have not been fully exploited. In this paper, the physical, dielectric, and thermal properties of Teslin® film are discussed.

357. THE SYNTHESIS AND THERMAL PROPERTIES OF LINEAR FERROCENYLENE-SILYLENE/SILOXYL-DIACETYLENE POLYMERS. Eric J. Houser and Teddy M. Keller Materials Chemistry Branch, CODE 6120, Chemistry Division, Naval Research Laboratory, Washington, D.C. 20375-5320.

New ferrocenylene-silylene/siloxyl-diacetylene linear polymers have been prepared from the reaction of dimethyldichlorosilane or 1,7-bis(chlorotetramethyl)-m-carborane with dilithiobutadiyne and dilithioferrocene tmeda. The structures of these polymers is supported by FTIR, proton, and carbon NMR spectroscopic studies. Thermal analyses showed the new polymers exhibit excellent thermal stabilities with 77-78% weight retention at 1000 °C under inert conditions. The polymers were found to undergo a thermally induced crosslinking reaction of the diacetylene groups near 300 °C. This prepyrolysis crosslinking is determined to be responsible for the high thermal stability of these materials. The oxidative stability of the ceramic chars obtained from pyrolysis of the polymers to 1000 °C was also examined and found to be excellent in the carborane containing system.

358.

CHEMICALLY AMPLIFIED PHOTOLITHOGRAPHY FOR THE FABRICATION OF HIGH DENSITY OLIGONUCLEOTIDE ARRAYS. **Jody E. Beecher,** Glenn H. McGall, and Martin J. Goldberg, Affymetrix, 3380 Central Expressway, Santa Clara, CA 95051

We have developed a chemically amplified photo process for the fabrication of high density oligonucleotide arrays. Using photolithography and chemical amplification to generate acid within a polymer coating, the acid-labile protecting groups on the 5´ hydroxyl of the growing oligonucleotide can be selectively removed, allowing for site specific coupling in the next growth step. The process can be tuned to have high sensitivity (20 mJ/cm2) or high contrast (~3.0) and has been used to synthesize oligonucleotides on a glass substrate in relatively high yields.

359.

SYNTHESIS, CHARACTERIZATION AND INITIAL BIOLOGICAL CHARACTERIZATION OF THE SYNTHETIC NUCLEIC ACID DERIVED FROM 3,9-DICHLORO-2,4,8,10-TETRAOXA-3,9-DIPHOSPHASPIRO[5.5]UNDECANE-3,9-DIOXIDE AND 5-IODO-2'-DEOXYURIDINE, C. Carraher, C. Parkanyi, G. Deng and D. Louda, Florida Atlantic University, Boca Raton, FL 33431 and Center for Environmental Studies, Palm Beach Gardens, FL 33410   The synthesis and characterization of the synthetic nucleic acid from 5-iodo-2'-deoxyuridine and 3,9-dichloro-2,4,8,10-tetraoxa-3,9-diphosphaspiro[5.5]undecane-3,9-dioxide is described. Characterization is by FTIR, light scattering, thermal, and elemental analysis. The polymer has a weight average degree of polymerization of about 870 so it is a high polymer. It inhibits growth of Balb/3T3 cells so it does show cellular activity as hoped for.

360.

NONRESONANT SHG ON A LIQUID-CRYSTAL COPOLYMER. **S.M. Cohen,** F.H. Long, Department of Chemistry, Rutgers University, Piscataway, New Jersey, 08855-0939

We have observed nonresonant second-harmonic generation (SHG) originating from a polymer. Using 100-fs pulses from a Ti:sapphire laser with λ_{input} = 800 nm, unusually strong SHG was observed at 400 nm from a thermotropic liquid-crystal random copolymer (Vectran A910™). The second-harmonic light was polarized primarily along the fiber direction. Other polymers studied did not give strong second-harmonic signals. The origin of such intense SHG from Vectran will be discussed.

361.
TIME RESOLVED STUDY IN PHASE SEPARATION DYNAMICS IN POLYMER DISPERSED LIQUID CRYSTAL. **J.B. Nephew,** T. Nihei, S.A. Carter, Department of Physics, University of California, Santa Cruz, CA 95064

The dynamics of addition polymerization induced phase separation is examined via confocal microscopy in PDLC systems used for reflective display technologies in which the final morphology consists of spherical liquid crystal domains suspended in a polymer matrix. We find the phase separation dynamics and final morphologies inconsistent with a simple binary fluid separation model for this system. We observe instead two distinct mechanisms during the polymerization; nucleation and growth, and a brief hydrodynamically active period in which phase separation accelerates and domains rapidly coalesce. We speculate that both processes are driven by polymer formation and segregation. We present a time resolved study of the phase separation as a function of cure temperatures, polymerization rates, and solvent concentrations and compare to temperature induced phase separation of the same system.

362. SHEAR-INDUCED ORIENTATION AND RELAXATION IN SMECTIC SIDE CHAIN LIQUID CRYSTALLINE POLYMERS. G. Wiberg, M. L. Skytt and U. W. Gedde. Department of Polymer Technology, Royal Institute of Technology, 100 44 Stockholm, Sweden.

The intensive research on side-chain liquid crystalline polymers has mainly been focused on the possibility of using them in applications such as non-linear optics, data storage and displays. The desired properties rely on the samples possessing a high degree of macroscopic orientation. One early approach was to polymerize monomers aligned by a surface treated substrate. This method is only applicable on very thin films. Attempts have also been made to macroscopically orient polymers using electric or magnetic fields. The shortcoming from an industrial point of view is here that it requires very high field strengths in combination with long annealing times. More recently attention has been paid to create the orientation using simple shear flow. We have built an apparatus in which samples are subjected to shear flow while measuring the order parameter by IR-dichroism. Results will be presented from a study on smectic side-chain liquid crystalline polyvinyl ethers. Shear induced orientation of the mesogens perpendicular to the shear direction was seen which partly relaxed after cessation of shear.

363.
REAL-TIME OPTICAL CHARACTERIZATION OF DEGRADATION IN POLYMER LIGHT EMITTING DIODES. **S. A. Carter***, S. Bailard*, J. C. Scott**, *Physics Dept., University of California, Santa Cruz, CA 95064, **IBM Almaden Research Center, San Jose, CA

We present optical and electrical studies of degradation in MEH-PPV OLEDs with the goal of understanding the mechanisms that lead to device failure. Two regimes of aging are observed, a short-term regime which results in order-of-magnitude drops in quantum efficiency within a few hours and long-term regime which results in slow decay over weeks. We concentrate on real-time simultaneous current-voltage (IV) and confocal microscopy images to understand how features in the IV-curves depend on morphological changes in the luminance and electrodes. In addition, we measure how the decay of the radiance and device efficiencies depend on the operating temperature, driving voltage and frequency. We conclude by comparing our results to theories on charge transport and injection in OLEDs, discussing possible mechanisms for device degradation in both regimes and ramifications on the design of long-lasting OLEDs.

364.
SYNTHESIS AND CHARACTERIZATION OF BIS(2,2'-BIPYRIDINE)RUTHENIUM-CONTAINING POLYDYES FOR USE IN SOLAR ENERGY CONVERSION, **C. Carraher** and A. T. Murphy, Florida Atlantic University, Boca Raton, FL 33431 and Palm Beach Gardens, FL 33410 The conversion of solar energy into a more usable form of energy is critical. Ruthenium compounds have been studied as integral agents in this quest. Most studies

include the Ru-containing complexes as side-groups dangling off the backbone chain. The present polymers have the ruthenium-containing moiety as part of the polymer backbone. The characterization of the polymer is by FTIR, UV-VIS, HRMS, thermal, and light scattering photometry. The polymer has a weight average degree of polymerization of about 33,000 so it is a high polymer. The polymer absorbs light over the UV-VIS region so it is a good candidate to act as a "light harvester" in a solar energy conversion scheme.

365. BIOFUNCTIONAL MEMBRANES: ELECTRON PARAMAGNETIC RESONANCE STUDIES OF THE ACTIVE SITE STRUCTURE OF ENZYMES SITE-SPECIFICALLY IMMOBILIZED ONTO POLYMERIC SUPPORTS THROUGH MOLECULAR RECOGNITION. D. Allan Butterfield[1,3]*, Ram Subramaniam[1,3], Dibakar Bhattacharyya[2,3], Shekhar Vishwanath[2,3], Wei Huang[1,3], and Leonidas Bachas[1,3], Depts of [1]Chemistry and [2]Chemical Engineering and [3]Center of Membrane Sciences, University of Kentucky, Lexington., KY 40506

The applications of biofunctional membranes range from catalysis (enzyme-based bioreactors), separations (affinity membranes), analysis (biosensors), and artificial organs. Most often, biomolecules have been immobilized onto porous polymeric membranes via non-specific covalent or physical adsorption methods, but because of the randomly-immobilized biomolecules, decreased activity is often the result. In the current study, molecular recognition chemistry was used to enhance activity of the bound molecule. Site-specific immobilization, in which the active site of the biomolecule is directed away from the polymeric surface, was employed. Site-directed mutagenesis and avidin-biotin coupling were used to ensure a reproducible orientation of the immobilized enzyme, with higher specific activity relative to that of randomly-immobilized biomolecules. Electron paramagnetic resonance (EPR), in conjunction with active site-specific spin labels, was used to investigate the structure of the active site of the proteases, papain and subtilisin, immobilized onto polymeric supports using molecular recognition methods. Site-specific immobilization of enzymes on modified PES membranes or modified silica beads led to significant improvements in all structural and kinetic parameters relative to randomly-immobilized enzymes. This research offers the possibility to significantly enhance existing membrane-based technology for biofunctional membranes. Such studies are in progress. Supported by NSF grant (CTS-9307518).

366. FORMATION OF TWO-DIMENSIONAL CRYSTALS OF HISTIDINE-TAGGED PROTEINS USING Ni^{2+}-CHELATING LIPID MONOLAYERS. Y. Rui[1], J. Mc Dermott[2], E. Barklis[2], S. Wilkens[3], & D. H. Thompson[1], [1]Department of Chemistry, Purdue University, West Lafayette, IN 47907; [2]Department of Microbiology, Oregon Health Sciences University, Portland, OR 97201; [3]Institute of Molecular Biology, University of Oregon, Eugene, OR 97403.

• We have developed a system for immobilizing histidine-tagged (his-tagged) proteins onto lipid monolayers consisting of egg L-α-phosphatidylcholine (EPC) plus the novel synthetic lipid, 1,2-di-O-hexadecyl-sn-glycero-3-N-(5-amino-1-carboxypenyl)iminodiacetic acid (DHGN). DHGN was shown to tightly bind nickel by atomic absorption spectrometry. DHGN-containing monolayers were also observed to specifically bind gold conjugates of his-tagged proteins. Using mixed monolayers of EPC and DHGN, we have examined monolayer-bound, two-dimensional (2D) arrays of several N-terminal his-tagged proteins by electron diffraction methods. These studies have provided 2D structures of protein crystals that are resolved to 9.5Å. The power and simplicity of this technique for determination of 2D protein structures will be summarized.

367. ALKALINE PHOSPHATASE ACTIVATABLE POLYMERIC CROSSLINKERS AND THEIR USE IN THE STABILIZATION OF PROTEINS. C. Bieniarz*, M.J. Cornwell, and D.F. Young, Abbott Laboratories, D-97D, 100 Abbott Park Road Abbott Park, IL 60064-3500

Although many methods of protein stabilization have been reported, there is a continuing need for more efficient means of preventing denaturation of proteins. We report synthesis of polymeric crosslinking agents, poly(glutamic acid) poly(phosphorothioates), which upon enzymatic activation by alkaline phosphatase, react covalently with the electrophilic functional groups on the surface of the protein. The talk will focus on thermal stabilization of two enzymes of different structural and functional characteristics, bovine alkaline phosphatase and glucose oxidase. Incubation at 45 °C for 7 to 14 days resulted in 35% greater loss of activity of native alkaline phosphatase as compared to covalently crosslinked enzyme. In case of alkaline phosphatase the process is self-catalyzed because this enzyme catalyzes the activation of the polymer and concomitant self-crosslinking. Glucose oxidase stabilization was even more notable, ranging from 800% at 37 °C, pH 9.0 to 3,000% at 37 °C, pH 7.4. Data will be presented showing construction of stabilized immunoconjugates based on crosslinked alkaline phosphatase.

368. STABILISATION OF ACETYLCHOLINESTERASE USING CHARGED POLYMERS: TOWARDS SHELF STABLE PESTICIDE BIOSENSORS. J.J.Rippeth, T.D. Gibson, Department of Biochemistry and Molecular Biology, University of Leeds, Leeds LS2 9JT, U.K., J.P. Hart, Faculty of Applied Sciences, University of the West of England, Coldharbour Lane, Frenchay, Bristol BS16 1QY. U.K., G.Nelson British Textiles Technology Group, Shirley House, Wilmslow Road, Didsbury, Manchester M20 2RB. U.K.

The enzyme acetylcholinesterase has been successfully stabilised using polyelectrolyte - polyhydroxyl combinations as unmodified native enzyme and immobilised onto electroactive surfaces. The use of the stabilising combinations enabled enzyme electrodes to be fabricated without loss of activity and to be stable with respect to time in storage. The native enzyme lost 90% of its activity on dehydration whilst in the presence of stabilisers no loss of activity was observed on dehydration. Furthermore the stabilised enzyme retained 100% of its activity after 43 days storage at 37°C. Acetylcholinesterase was immobilised on to screen printed, cobalt phthalocyanine impregnated, carbon electrodes for use in flow injection format for indirect amperometric organophosphorous pesticide detection. Using the flow injection system the electrodes were able to measure the organophosphorous pesticide Dichlorvos down to a concentration of 3×10^{-8} M.

369. ENZYMATIC SYNTHESIS OF MULTIFUNCTIONAL POLYPHENOLS FOR BIOSENSOR APPLICATIONS. Madhu S. Ayyagari and Joseph A. Akkara, US Army Soldier Systems Command, Biotechnology Division, Natick RD&E Center, Natick, MA 01760-5020.

Horseradish peroxidase is used to catalyze the synthesis of a series of multifunctional polyphenols and aromatic amines. Functional groups such as carboxyl, amino, sulfonate and/or hydroxyl groups that are available in the polymer structure are further functionalized for desired applications. The polymers are derivatized with a number of ligands to impart specificity for interactions with proteins and other biomolecules. Because of their backbone conjugation, these polymers serve as conducting matrices under doped conditions. Molecular weight of the polymers and number density of the functional groups in the polymer structure can be controlled. Molecular imprinting methods to build specificity into the polymer structure for recognizing biological components such as amino acids will be presented. Spectroscopic and thermal characteristics of the polymers and their use in biosensor applications will be discussed.

370.
DEVELOPMENT OF AN ELECTROCHEMICAL BIOREACTOR FOR NITRATE REMEDIATION USING NITRATE REDUCTASE/REDOX POLYMER TECHNOLOGY. **Graham Ramsay** and Stephen M. Wolpert, Wolpert Polymers, Inc., Virginia Commonwealth University, Biomedical Engineering Dept., Box # 980694, Richmond, VA 23298-0694

This talk concerns the application of thin-film, redox polymer technology to the development of an electrochemical bioreactor for nitrate remediation. Recent work has demonstrated that: Viologen-based redox polymers can effect electrochemical reduction of nitrate reductase.

Nitrate can be reduced to nitrogen gas via a cascade of reductase enzymes.

The effectiveness of merging these two technologies will be discussed. Nitrate/ nitrite pollution is an increasingly serious problem. In the U.S. 4.5 million people use drinking water that contains more than the EPA standard of 10 ppm. Ingested nitrate is reduced to nitrite in the human gut and can cause neonate death by hampering oxygen transport ("blue baby syndrome"). No effective remediation method is commercially available.

371.
ELECTROCHEMCIALLY DRIVEN ION-SELECTIVE SENSING AND EXTRACTION BASED ON PVC MEMBRANE-COATED ELECTRODES. Stephen G. Weber and Lifang Sun , Department of Chemistry, University of Pittsburgh, Pittsburgh PA 15260

A Poly(vinyl chloride) membrane containing supporting electrolyte, 2-nitrophenyl octyl ether (NPOE) and an additional redox species such as benzoquinone was coated on a glassy carbon electrode. Such a membrane-coated electrode was polarized in an aqueous solution of LiCl in a potential range of +100 mV to -1100 mV. However, when tetraalkylammonium ions were present in the aqueous phase, cyclic voltammetric waves were observed. This indicates cation transfer from water to the membrane. The peak potential E_p of the cathodic wave for a number of tetraalkylammonium ions were highly correlated to the hydrophobicity of those ions. Experimental data also indicates the involvement of a chemical driving force in the anion tranfer. We hypothesize that this driving force is ion-pair formation between the transferred anion, and NPOE radical anion. The amount of charge transferred is dependant on the concentration of cation and potential sweep rate. potential applications include analytical sensing and extraction, industrial separations and environmental clean-up.

American Chemical Society
DIVISION OF PROFESSIONAL RELATIONS

213th ACS National Meeting

San Francisco, CA
April 13-17, 1997   T. J. Kucera, Program Chairperson

SUNDAY AFTERNOON

- **Restructuring, Retooling and Reinventing Careers in Chemistry**
J.D. Burke, Presiding   Papers 1-7

MONDAY MORNING

- **Doing More With Less:**
Educating Future Professional in Chemistry
A. Nalley, Presiding   Papers 8-12

MONDAY AFTERNOON

- **Doing More With Less:**
Educating Future Professionals in Chemistry
A. Nalley, Presiding   Papers 13-17

TUESDAY MORNING

- **Project Seed**
A.C. Nixon, Presiding   Papers 18-22

TUESDAY AFTERNOON

- **The Impact of the Americans with Disabilities Act (ADA) on Chemistry in the Workplace**
T.A. Blumenkopf, Presiding   Papers 23-27

WEDNESDAY AFTERNOON

- **Incidents in the History of Chemistry in California**
A.C. Nixon, Presiding   Papers 28-33

DIVISION OF PROFESSIONAL RELATIONS

001. WORKPLACE TECHNOLOGICAL CHANGE AND EFFECTS ON EMPLOYMENT: AN HISTORICAL OVERVIEW. D. Chamot, National Research Council, 2101 Constitution Ave., NW, Washington, D. C. 20418

At least since the early nineteenth century, invention and application of new technologies have had profound effects on individual job content, employment levels, and the rise and fall of whole industries. Internationalization of the chemical industry in modern times and its many implications for American employment of chemists and chemical engineers can be seen as a being shaped, or at least enabled, by these longer term trends. A brief analysis of these issues will be presented that sets trends in chemical employment in a broader context in an attempt to understand both the scope and permanence of current developments.

002.
INDUSTRIAL ADJUSTMENT IN THE AGE OF GLOBAL COMPETITION
Michael G. Borrus, BRIE, 2234 Piedmont Ave, Berkeley, CA
The emergence of new players (firms and economies) in new locations (especially Asia) with fundamentally new approaches to production and organization, is combining with the rapid, pervasive spread of information technologies to create new forms of international competition. These, in turn, change the way jobs and competencies are located among different places. Though the new forms of competition create dramatic new potential for economic growth and job creation, the interim adjustments can be excruciating for employees and communities, particularly in the advanced economies. This presentation will set a context around the employment and organizational changes in the chemical industry by locating them within the larger economic shifts in Asia, Europe and the US affecting all industries and national economies.

003. **THE FUTURE OF CHEMICAL RD&T**, Charles J. Shearer, Glenn L. Taylor, and William P. Rothwell, Shell Chemical Company, P.O. Box 1380, Houston, TX 77251-1380.

The chemical industry was once virgin territory for product R&D. Substitution of synthetic for natural materials was the big push. Few implementational hurdles existed. Customers had similar needs; balance between function, cost, performance, safety and the environment was easier to achieve.

Today and in the future, customers have fragmented needs, products must be clearly differentiated, customers and competitors are global, and speed and cost are essential. Now the industry is replacing synthetic materials with improved or different synthetics, development costs are very high, and restrictions by regulatory bodies are considerable. These changed conditions call for a changed operating paradigm for individuals and organizations, part of which is the establishment of technology partnerships.

All of these changes, in turn, are causing changes in the way that research, development, and technical service (RD&T) are accomplished. Now the norm is interfunctional teamwork, self-directed teams, and an emphasis on speed and value to the business. Work has been re-engineered and a much higher value is being placed on technologies that improve the process.

004.
A REPORTER'S 30-YEAR PERSPECTIVE ON EMPLOYMENT OF CHEMISTS. Madeleine S. Jacobs, Chemical & Engineering News, Washington, D.C. 20036.

Employment of scientists and engineers, in particular chemists and chemical engineers, is a cyclical phenomenon. In the early 1960s, the outlook for employment of chemists was rosy, and many people receiving their bachelor's degree in chemistry went on to earn a Ph.D., leading to what appeared to be an oversupply of Ph.D.s by the early 1970s. At the same time the Ph.D. "glut" was materializing, the chemical industry-- the major employer of chemists--was undergoing a recession. Since that time, there have been perceived shortages and surpluses of chemists. The employment situation for chemists and chemical engineers will be reviewed from the perspective of a reporter who has covered or followed this issue since 1969.

005.   SURVIVING IN AN INSECURE JOB MARKET: Donald A Upson, Molecular Probes, Inc., Eugene, OR 97402

Job security, over which we have no control, is now a misnomer. Career security means maintaining and enhancing competence within your field and having the ability to sell yourself effectively when laid off or otherwise changing jobs. Today's chemists should expect to have perhaps a half-dozen career transitions. This presentation will provide steps and attitudes that will help you survive a layoff, move on to a better next situation, be more prepared for the next change, and maintain or enhance self-confidence in the process. There are steps that if taken before initiating a new job search will substantially increase your likelihood of success. Similarly, there are ways to optimize the first contact with the potential employer, the interview itself and the transition to the new job that can significantly influence both professional and personal satisfaction.

006.   EXPERIENCES OF A RESTRUCTURER/RESTRUCTUREE.   John T. Lowe, 5700 Greenledge Cove, Austin, TX   78759-6244

Corporate restructuring can lead to very satisfying new career directions for a technically trained person. Finding those new directions, however, requires both good preparation and good luck. I will talk about my own experiences both as a restructurer and as a restructuree, as well as the after life from those experiences.

Observations will be shared on how an individual can improve preparation and promote good luck in a career transition. Specifics of personal experience in moving from management in a large chemical company to leading a university-centered industry consortium completely dependent on volunteer efforts in member companies will be discussed and generalized to the broader persoective of the symposium.

007.   LIFELONG EDUCATION. Ronald Breslow, Department of Chemistry, Columbia University, New York, New York 10027.

The United States cannot afford to be the low wage country, so our future demands that we be the high technology/advanced science country. This means that chemists must get an education that is both broad and deep while they are students, but they must also continue education throughout their professional life. The way to avoid being "downsized", the way to get a succession of even better jobs, is to continue intellectual growth. One CEO said that his technical personnel are the only asset in his company that increases in value with time. Make sure you are one of those assets. Professional chemists need to keep reading and keep going to technical meetings. They will become too valuable to lose, and able to make the contribution to U.S. technical strength that ensures our country's future.

PROF

008. THE FUTURE, OF NSF FUNDING FOR BASIC RESEARCH IN CHEMISTRY. John B. Hunt, National Science Foundation, Arlington, Virginia 22230

Increases in NSF funding for research and education in chemistry over the next few years are highly unlikely. NSF investments in academic chemistry will be aimed at upholding world leadership in chemical research, promoting the employment of research results in service to society, and achieving excellence in chemical education at all levels. Education through research will continue to receive high priority, as will efforts to integrate research and education. NSF's research directorates will cooperate with its Education and Human Resources Directorate in promoting new paradigms for graduate, undergraduate and teacher education. Advanced instrumentation and the development of new instruments will be given high priority. Academic-industrial research collaborations and industrial input to chemical education will be promoted by increased investment in the GOALI Program. The ability of chemists to deal with systems of increasing complexity will result in increased investments in research in materials and bio-related chemistry. Broad-ranging NSF initiatives in computational science are likely, bringing new opportunities for research collaborations for chemists.

009. FUNDED RESEARCH WITH B.S. AND M.S. STUDENTS. Patrick E. Cassidy, Polymer Research Group, Department of Chemistry, Southwest Texas State University, San Marcos, Texas 78666

A meaningful funded research program can, of course, be attained in a department where only B.S. and M.S. candidates are available. Even a reasonable number of publications, patents and grants can develop from such a program. However, this will likely require a significant amount of effort in view of the classroom teaching requirements at the universities in this situation. There are federal and private programs set aside for undergraduate programs, and industrial funding sources, although difficult to find, can be excellent sources of funds and resources for students. A very important issue to facilitate research at this level is university support; e.g., release time, matching funds and an aggressive Office of Research and Sponsored Programs.

010. INDUSTRIAL SUPPORT FOR UNDERGRADUATE RESEARCH, Dale Teeters, Department of Chemistry, The University of Tulsa, Tulsa, OK 74104-3189

Securing funding for research in an undergraduate department of chemistry can be quite challenging. One source of funds is industrial projects which can provide a way to support such research and which are exciting for the students. This presentation will offer some suggestions for stratagies to gain support from private industry. The discussion will include the possible benefits to industry for funding both small and large research projects and will discuss how to sell industry on these benefits. The importance of establishing contacts with industrial researchers in your area will also be covered. Examples of research at The University of Tulsa that have been supported by the private sector will be given.

011. TEACHING THE SCIENTIFIC METHOD: RESEARCH WITH UNDERGRADUATES AND ELEPHANTS. Thomas E. Goodwin, Department of Chemistry, Hendrix College, Conway, Arkansas 72032

Discovery-based" labs are currently in vogue, and are often viewed as useful mechanisms for teaching the scientific method. Many believe that the epitome of discovery-based, undergraduate education in science is the engagement of our students in a collaborative, basic research project with an experienced mentor. This talk will be a recounting of the lessons we've learned in the establishment and maintenance of an undergraduate research program at a liberal arts college. Topics to be addressed include helpful hints for success, funding strategies, pitfalls to avoid, and benefits that result. Specific examples of research projects and program accomplishments will be presented, including an opportunistic research project involving undergraduates and elephants.

012. SUPPORTING RESEARCH THROUGH SMALL GRANTS, Dr. Adriane G. Ludwick and Barbara G. Rackley, Chemistry Department, Tuskegee University, Tuskegee, AL 36088

Small grants can have large impacts on educational programs, both curriculum and research. Continuity is a major factor in the success of a small grant. Examples of grants, small and large, will be given. The effect of these grants on the chemistry program at Tuskegee University will be analyzed. The relationship between size and effectiveness will be discussed.

013. INSTITUTIONAL, STATE AND FEDERAL SUPPORT FOR CHEMICAL EDUCATION AT LIBERAL ARTS COLLEGES. David. J. Oostendorp, Department of Chemistry, Loras College, Dubuque, IA 52001

Securing funding for chemical education and research at small liberal arts colleges is a time consuming task. Institutions at this level have the task of preparing a large number of students for graduate school and industry with limited resources. It is important for us to be able to provide our students with quality instruction and experience with research they will need to succeed at the next level. At the institutional level, we are in competition with other departments and programs for very limited funds. At the state level, we are in competition with state run schools. At the federal level our competition is with the major research institutions. This presentation will explore current sources of funding being used to support research and education at small schools throughout the Midwest.

014. USING LIMITED RESOURCES MORE EFFECTIVELY: INDUSTRY-ACADEMIC PARTNERING IN CHEMICAL EDUCATION. Ned D. Heindel and Natalie Foster, Department of Chemistry, Lehigh University, Bethlehem, PA 18015

Universities and companies are increasingly collaborating on R&D. "Doing More With Less" is a resonant message in the research arena. It is also an increasingly important phenomenon in the education arena. The preparation of young scientists for careers in highly pragmatic industrial laboratories requires rethinking traditional university training. Industrial research employment opportunities for young PhDs are slim. Where historically chemical industry employed 60-70% of doctoral graduates, only about 20% of the current doctoral graduating class obtained such posts. About 60% of recent PhDs took temporary or postdoctoral employment. In many locales, industrial employment is easier to obtain as a BS, thereby raising the question "should I take a job now and do an advanced degree part time?" With electronic media there are distance education options over the Internet, satellite video, or by tape and CD. Today's BS/BA can obtain advanced degrees in chemical engineering, chemistry, and business while holding an entry position in industry. In many cases, industry provides part of the educational experience in a clear partnership with academia.

015.
RESEARCH AT THE UNIVERSITY-INDUSTRY INTERFACE: IT WORKS BUT BOTH PARTIES MUST ADJUST!. Jean M.J. Frechet Department of Chemistry, University of California, Berkeley, Berkeley, CA 94720-1460

The research interface between universities and Industry is becoming more important today as university researchers look to new sources of research support and many companies look to the wealth of university resources to complement their highly focused research staff. This presentation will focus on issues that must be considered when establishign such relationships and explore the working details of several case histories. The importance of direct one-on-one relationships and the need for flexibility from both parties cannot be overemphasized and issues such as the compatibility of industrial research with graduate education, or the sharing of intellectual property will be broached.

016.
Declining Research & Educational Funding: Give it All to the Big Schools, J. Wiesenfeld

R&E funding is decreasing in the U. S. and will continue to do so. The following arguments are given to support the preposition that all academic A&E funds should be given to the larger schools and programs in the 21st century:

- Funding decisions become simple since only a handful of schools get all the money; this will decrease administrative costs.
- Larger schools are more cost effective: they put through larger numbers of students with fewer support staff and faculty.
- Larger schools have a better infrastructure and instrumentation for doing R&E.
- Larger schools have the best faculty and therefore give the best education.
- Larger schools do the best research and provide the best research training.
- Larger schools have the best reputations, therefore providing better job opportunities for their graduates.

Larger schools attract the best students, eliminating less qualified graduates

017.
Declining Research and Education Funding: Give it All to the Smaller Schools, L. Mathias

Small schools play a unique role in R&E graduate training in America. For the following reasons, they should be given all of available funds:

- More students graduate from smaller schools combined than from the smaller number of larger schools.
- Smaller class sizes allow better, more individual instruction.
- Fewer projects per faculty allow more focused research effort and training.
- Faculty and research groups interact more and collaborate more effectively, thus providing better training for real-world industrial positions.
- Faculty at smaller schools are more dedicated, and spend more time with their students in all aspects of R&E.
- The quality of research at smaller schools is just as good as at larger schools as confirmed by number and place of publication.
- Cost per graduate student is one-third to one-half that at larger schools.

018.
THE BIRTH OF SEED   by ALAN C NIXON & GLENN FULLER

During the 60,s civic disturbances occurred in all our major cities, occasioned by the ongoing Vietnam war and the depressed conditions of our inner cities. It was called the "Urban Crisis" and it involved mainly black neighborhoods. As a result many groups, mainly churches, started to provide help to these areas, including my own church, Arlington Community and I took a minor part in this.

But it occurred to me that if every group; religious, social civic and business pitched in and did what it could the problem could be solved. Then it occured to me "Why not the ACS?" So I started talking about it within the CA section, wrote an article in the VOrtex(1/68) and put it on the agenda of the Western Councillors Caucus, of which I was Chair. They passed with only two no votes, Bill Johnson of SoCA helped me shape it for the CPC, who refined it and passed it on to the Council Unanimously as did that body. The whole proceedure took just 39 hrs!

019. SEED-THE EARLY YEARS. S. T. Quigley, Science Policy & Management Consultant, 2908 Upton St., N.W., Washington, D.C. 20008

In the Spring of 1968, the American Chemical Society (ACS) Council approved a resolution recommending that an ACS Program be established to assist economically disadvantaged high school students. The Society's Board-Council Committee on Chemistry and Public Affairs (CCPA) initiated a social action program to allow rising seniors in disadvantaged high schools to spend ten weeks during the summer in an academic, industrial, or governmental research laboratory participating in meaningful research under the personal supervision of a research scientist. The objective of the program was to raise these students' goals and expand their horizons. The ACS Board of Directors approved the policy that all administrative expenses of the program would be under-written by the ACS. Student stipends would come from volunteer contributions from individual ACS members, ACS local sections, corporations, and foundations. The planning, development, and logistics of the early years of SEED will be presented.

020. PROJECT SEED: EXPERIENCES OF THE CALIFORNIA SECTION. E. S. Yamaguchi, Chevron Chemical Company, Richmond, California 94802

As SEED enters its 29th year, it is instructive to examine the history of this program in the California Section. SEED, under the leadership of Dr. Alan Nixon, started modestly. Over the years it has grown to be amongst the largest SEED efforts in the country. This presentation will detail the history of the SEED program in the California Section, with special attention to the following issues: (1) growing the program, (2) finding mentors, (3) finding students, (4) finding money, and (5) keeping the interest level high.

021. PROJECT SEED: TODAY AND TOMORROW. Herbert B. Silber, Chemistry Department, San Jose State University, San Jose California 95192 and Christine B. Brennan, Education Division, American Chemical Society, 1155 16th Street, NW, Washington, DC 20036.

Project SEED has existed for more than twenty-five years to provide summer research experiences in chemistry to economically disadvantaged high school students at colleges, national laboratories and industrial sites. Although admission into the SEED Program is based upon family income, about two-thirds of past participants have also been under-represented minority students. Initially Project SEED provided one summer's experience (SEED I), but the program has recently been broadened to provide a second summer's experience (SEED II). Graduates of SEED are eligible to apply for national scholarships to use in their first year of college. The ACS has conducted an outside evaluation of the SEED Program in which the evaluators surveyed past participants with the goal of obtaining a statistical sampling of students. We will discuss the results obtained from this survey.

022. HOW TO MOVE SEED INTO THE CLASSROOM S. Rosenlund & AC Nixon 511: 2140 SHATTUCK AVE BERKELEY CA 94704

SEED is a very good program within its scope. But it is limited to just 12 weeks during the summer even though it is very intensive. But how do we, in effect, take it into the classroom? This paper discusses some ways we have thought of that should be explored.

023.
ADA IN THE ACADEMIC WORKPLACE. Paul D. Grossman, Chief Regional Attorney US Department of Education, Office for Civil Rights San Francisco, California; Adjunct Professor of Disability Law, Hastings College of Law, University of California.

Profound changes in the education of elementary and secondary school children with disabilities is producing a generation of undergraduate and graduate students who are entitled to accessible chemistry departments, laboratories, and workplaces. Access may require something as simple as the lowering of a laboratory bench or as complex as acquiring adaptive technology that makes the internet independently useable by blind persons or the seminar class effectively available to a deaf students. Not all required changes are physical, they may also be programmatic such as accommodation in the manner in which examinations are given or the length of time permitted to obtain a degree. Not every accommodation sought by an individual with a disability is required by law. For example, academic institutions are not required to "fundamentally alter" the nature of their programs. The sources, breadth, and limits of these requirements will be fully explored.

023A   SIMPLE, AFFORDABLE, AND REASONABLE ACCOMMODATIONS. Bonny Decker, OSHA Salt Lake Technical Center Library, 1781 South 300 West, Salt Lake City, UT 84165.

This paper will focus on a variety of accommodations for the mobility impaired in the analytical laboratory of the OSHA Salt Lake Technical Center. Some of the accommodations were made with the inexpensive materials such as plywood and rope. Many of the accommodations will be discussed in a slide show with a cost analysis of these products. The inaccessible features of a couple of laboratory products purchased with wheelchair accessibility in mind will be discussed. This paper will show that making reasonable accommodations for the physically challenged chemist is both affordable and and an asset to the laboratory as a whole.

024.   THE IMPACT OF THE AMERICANS WITH DISABILITIES ACT (ADA) ON THE RESEARCH AND LEARNING ENVIRONMENT IN UNIVERSITY SCIENCE DEPARTMENTS. O.J. Cooks, Facilities Planning, Purdue University, 1694 Freehafer Hall, West Lafayette, IN 47907

This presentation is a discussion of the ADA, its application in the sciences and its effects on the attitudes of students and employees of Universities. The responsibilities of Administrators are covered and a proactive, empirical philosophy of implementation of the regulations is described. A case study presentation of the compliance efforts at Purdue University is presented including an examination of the relevant regulations for compliance in a University setting. Specific examples of progressive responses to the legislation described, including the VISIONS Lab in Chemistry, in which visually impaired students use Braille and technology to perform calculations will be demonstrated.

025.   WORKING CHEMISTS WITH DISABILITIES: EXPANDING OPPORTUNITIES IN SCIENCE. Todd A. Blumenkopf, Central Research Division, Pfizer Inc., Groton, CT 06340

The American Chemical Society has recently published a book entitled *Working Chemists with Disabilities: Expanding Opportunities in Science*. The idea for this book was triggered by numerous inquiries from employers, corporate recruiters, faculty members, career counselors, students with disabilities, and newly disabled scientists for useful strategies by which people with disabilities can function productively and safely in the chemistry workplace. Since the passage of the ADA, new technology and new equipment has been introduced into the marketplace and is being used successfully by scientists with disabilities in the performance of their jobs. This talk will provide practical examples of assistive technologies, architectural modifications, and other reasonable accommodations used today by scientists with disabilities, who can and do work productively and safely in a variety of chemistry and chemistry-related jobs.

026.   IMPACT OF THE AMERICANS WITH DISABILITIES ACT ON CHEMISTS IN THE WORKPLACE - ACCESS TO INFORMATION AND COMMUNICATION, Deborah Kaplan, Issue Dynamics, Inc. 50 California St., Suite 1500, San Francisco, CA 94111

Access to information and communication is achieved, more and more, through the use of computers and private or public telecommunications networks. For chemists with a variety of disabilities, this access can be frustrated or precluded by inadvertant barriers in the design of hardware, software or network applications. For example, the popular graphic user interface can be nearly impossible for a blind or visually impaired person to navigate around, even with voice synthesis or braille output. The Americans with Disabilities Act requires that employers make "reasonable accomodations", eliminating barriers to the effectiveness of a disabled employee. This includes ensuring that the technology used in the workplace is useable, especially if there are accessible products available in the marketplace.

027. **RESPONSIVE LABORATORY DESIGN**, Bonnie Blake-Drucker, AIA, Blake-Drucker Architects, P.O. Box 11246, Oakland, CA 94611

The design of a laboratory which is responsive to the needs of the disabled student can be looked at as the design of a door to all of the sciences.

The basic sciences such as freshman or organic chemistry are a student's door to the world of science. These courses are the pre-requisites to upper division science courses. To be denied access to the sciences is a tragic loss of academic potential.

The design of a laboratory which is accessible to students with a disability must include working areas, circulation areas and operation of equipment which is required by the class work. The focus of this presentation is the design of an accessible chemistry workstation.

The design of the prototypical disabled accessible workstation will be presented with its elements which contribute to its universal use. Additionally, examples of laboratory planning which go hand-in-hand with accessibility will be presented to advance the thesis that accessible laboratory design is no more difficult than safe laboratory design.

028. DRAKE'S PLATE OF BRASS by ALAN C NIXON and <u>EVALDO KOTHNY</u>

In 1579 Sir Francis Drake paid a visit to the Pacific Coast to see what the Spaniards were up to. After running into an ice storm off the Oregon coast he decided to return south to careen his ship. This he did in what is now Drakes Bay. While this was going on his chaplain got a brass plate and chiseled a message on it regarding Drake's visit and claiming the surrounding real estate for Queen Elizabeth-(I).
   In the course of time the plate disappeared but then almost miraculously it was found and then discarded, then refound and brought to the UCB Dept of History.
   Then the arguments began- was it authentic? The decision rested mainly on chemical analyses for physically it seemed kosher. The first analysis said NO, the second said YES, the most recent comes down on the NO side. Will a more modern try reverse this ?
   These matters will be discussed.

029. THE ASSOCIATION OF INDUSTRIAL SCIENTISTS by ALAN C NIXON & PAUL H WILLIAMS
       2140 Shattuck Ave ,BERKELEY CA 94704, Ste 511

During the late 1930s not many chemists belonged to labor unions, although the idea was spreading. This paper relates how the professional employees at one of the major research organizations in the US (Shell Development Co) came to be represented by a union called the Association of Industrial Scientists. It came about because of the stupidity of the management, which lead to the strengthening of a CIO union(the FAECT) which sought to represent all of the workers in a single unit. Most of the professional(profs) wanted to have a seperate prof unit. This led to an NLRB hearing in which the ACS interviened on behalf of the profs (mostly ACS members). The NLRB ruled in favor of 2 units. In the subsequent election the CIO won the nonprofs but lost the profs.
   The powerfull ventilation of the idea of the benefits of unionization persuaded the profs that they should forn an independent union of their own, which they did so successfully that it withstood a charge of company domination & remained bargaining agent till 1972.

030. TEMPEST IN A TEST TUBE <u>Fred Stross</u> and Aubrey McClellan
          ·Ste 511 2140 Shattuck Ave, Berkeley Ca 94704

During the 1950's, there were very few science programs on television and none focusing on Chemistry. The third author, Chair of the California Section at the time, thought this was deplorable and decided to do something to rectify it. So he appointed the first author to head a Television Program Committee. Arrangements were made with the local TV station KQED to show our stuff, if good enough. We were successful: we wrote the programs, produced them at local labs, and presented them. Our MC's, Harry Sello and Aubrey McClellan caught the fancy of the public, and KQED said that the Tempest was one of their most successful events.

Attempts are being made to transfer the material from film to tapes so they can be reshown -- at meetings and in classrooms.

031. INFLUENCE OF CHEMISTRY AND TECHNOLOGY ON AGRICULTURE IN CALIFORNIA. Glenn Fuller and Thomas A. McKeon, Western Regional Research Center, ARS, USDA, Albany, CA 94710.

Although California may be considered an urban state, it is also the premier agricultural state in the U.S., producing and processing many horticultural and field crops. Agricultural technology came to the state with the earliest European settlers, who produced wine grapes and processed cattle hides for export. Because of strong research programs at the University of California and the USDA, agriculture has maintained its preeminent position here. Some examples of agricultural progress have been the development of frozen foods through research in enzyme chemistry, preservation of foods by new methods such as freeze drying and irradiation, and use of crop plants as raw materials for medicinal drugs. Recently, molecular biology and genetic engineering have produced improved crop varieties. These new plants are the first of many crops which will have increased resistance to disease and pests, improved nutritional quality, and enhanced storage stability. Some specific examples will be discussed.

032.
COMPARISON OF THE 1981 AND 1996 RETIREE SURVEYS
By Harald Drews and Aldo DeBenedictis

In 1981 the Senior Chemists Committee of the CA Section of the ACS conducted a survey of all retired chemists on the Pacific Coast. Results were reported at the '83 Seattle meeting. The questions asked dealt with the conditions of retirement, if they thought they were treated with fairly, what they thought of the ACS, what were their activities in the ACS, how they were making out, etc.

Substantially the same survey was sent to a portion of the same group in 1996. The results have been compared and some conclusions drawn. Since employment conditions for chemists have changed considerably over the 15 years that have elapsed it is not surprising that some differences cropped up. These will be discussed.

033. RETIREES AS CONSULTANTS by Frank M. McMillan & Ewell E. McDole
2140 Shattuck Ave., Suite 511, Berkeley, CA 94704

It is absolutely clear that fewer and fewer chemists are going to stay with one or two employers during their working life. Also, the threat of early retirement looms ever larger. So what to do: they will not be able to live decently on their pension (if any) plus social security (if any).

One thing a retiree can do is become a consultant. This is quite demanding for a sole practitioner unless he/she can hook up with a company that retains him/her. Otherwise, joining an association of retired individuals is an attractive alternative. There are quite a few such around the county.

Calsec Consultants Inc. is appropriate to feature for this meeting. Headquartered in Berkeley, It was organized in 1979 and incorporated in 1980. It has a Board of Directors with the usual officers. Retirees can become members and pay a small annual fee. Membership has varied from about 20 to about 100. Since it is a non-profit corporation, consultants do both pro bono and for fee jobs. Some of these will be discussed. Jobs are assigned on the basis of interest and competence.

American Chemical Society
DIVISION OF SMALL CHEMICAL BUSINESSES

213th ACS National Meeting

San Francisco, CA
April 13-17, 1997   N.H.Giragosian , Program Chairperson

MONDAY MORNING

- **Biosensor Technology**
  N.H.Giragosian, Presiding   Papers 1-5

MONDAY AFTERNOON

- **Biosensor Technology**
  N.H.Giragosian, Presiding   Papers 6-9

MONDAY EVENING

- **Sci-Mix**   Papers 10-11

TUESDAY MORNING

- **Independent Testing Laboratories**
  N.H.Giragosian, Presiding   Papers 12-16

TUESDAY AFTERNOON

- **True Stories of Small Chemical Businesses**
  G. Austin, Presiding   Papers 17-20

DIVISION OF SMALL CHEMICAL BUSINESSES

001.
BIOSENSOR TECHNOLOGY: PRESENT STATE AND FUTURE PERSPECTIVES. H.H. Weetall, Biotechnology Division, National Institute of Standards and Technology, Gaithersburg, MD 20899

Biosensors technology has been available for over 20 years. However, this technology has made, only minor inroads in the areas of human diagnostics, environmental monitoring, bioprocess control and food and beverage contamination monitoring. The reasons are numerous and include: cost, sensitivity, interferences, reproducibility, ease of use and precision. This presentation will discuss the characteristics necessary for this technology to gain a major share of the existing markets where biosensors would compete. Biosensors have unique capabilities that could make this technology highly competitive with existing technologies presently found in the diagnostic and analytical laboratories. Biosensor technologies approaching single molecule detection have the potential for replacing amplification techniques used in genetic analysis. The sensor arrays "genosensors" have the potential for replacing the expensive, time consuming processes presently used for gene sequencing and detection of gene mutation. The ability of biosensors to measure in real-time and under remote situations make this technology potentially valuable to the environmental monitoring, bioprocessing and food and beverage industries. These potential advantages will be discussed.

002. DETECTING AND CHARACTERIZING ANALYTES WITH AN ION CHANNEL
John J. Kasianowicz, NIST, Biotechnology Division, 222/A353, Gaithersburg, MD 20899

This review summarizes our recent efforts to adapt channels formed by wild-type[1-4] and genetically engineered mutants[5] of *Staphylococcus aureus* α-hemolysin for detecting and/or characterizing a wide variety of analytes. These include monovalent[1] and transition metal divalent[5] cations, nonelectrolyte polymers[2], polynucleotides[3] and other species[4]. The interaction between the analyte and the pore causes a change in the channel's ionic current which otherwise flows freely. The analyte-induced perturbations in the current permit the measurement of the association and dissociation rate constants for the reaction[1], the analyte's concentration[1-5] or the physical size of the molecule in question[2,3]. Other more highly specific detection schemes[4] will also be discussed.

References: 1) Bezrukov, S.M. & J.J. Kasianowicz. 1993. *Phys. Rev. Lett.* **70:** 2352; Kasianowicz, J.J. & S.M. Bezrukov. 1995. *Biophys. J.* **69:** 94; 2) Bezrukov, S.M., I. Vodyanoy, R.A. Brutyan & J.J. Kasianowicz. 1996. *Macromolecules* **29:** in press; 3) Kasianowicz, J.J., E. Brandin, D. Branton & D.W. Deamer. 1996. *PNAS, in press;* 4) Kasianowicz, J.J. & H.H. Weetall. *Patent pending;* 5) Walker, B., J.J. Kasianowicz, M.V. Krishnasastry & H. Bayley. 1994. *Protein Eng.* **7:** 655; Kasianowicz, J.J., B. Walker, M.V. Krishnasastry & H. Bayley. 1994. *MRS Symp.* **330:** 217.
Support: NRC (JJK), NSF (D. Branton), NIH (D. Branton & D.W. Deamer), DOE (H. Bayley), & the ONR (HB & V.A. Parsegian).

003. A LOOK TOWARD TOMORROW'S COMMERCIAL BIOSENSORS: WHAT'S COMING FROM THE RESEARCH LABORATORIES. Frank V. Bright, Department of Chemistry, State University of New York at Buffalo, Buffalo, NY 14260-3000

Significant advances in biotechnology, genetic engineering, small scale instrumentation, and combinatorial library development are clearly changing key aspects of the chemical sciences. For example, in the biosensing area these advances are helping to create a wave on which new biosensors are being developed in today's research laboratories. This presentation will highlight several recent "laboratory-based" biosensor schemes, discuss the speakers view of biosensing for the future, and outline the key questions/issues that remain to be addressed/solved.

004.
ENVIRONMENTAL APPLICATIONS OF BIOSENSORS: OPPORTUNITIES AND FUTURE DIRECTIONS. **E.J. Poziomek**, Department of Chemistry and Biochemistry, Old Dominion University, Norfolk, Virginia 23529-0126

This paper addresses opportunities and future directions in biosensor development for environmental applications. The potential of biosensors remains high because of the promise of high selectivity and sensitivity. Immunochemical methods continue to gain acceptance for rapid field screening applications. A number of immunoassay methods have been added recently in EPA's method compendium, SW-846. Sucesses with immunoassays such as in field kit form for environmental pollutants have led to increased interest in the development of biosensors (biological sensing elements combined with signal transducers). Interest by the U.S. EPA and other government agencies in biosensor development is expected to remain high. However, it is not clear how large the market actually is for environmental biosensors.

005. COMMERCIAL APPLICATIONS OF BIOSENSOR TECHNOLOGY James J. Valdes, U.S. Army Edgewood Research, Development and Engineering Center, SCBRD-RT, Aberdeen Proving Ground, MD 21010-5423

Biosensors are analytical devices which combine the exquisite selectivity and sensitivity of biological recognition molecules (BRM) with the amplification and signal processing capability of electronic or optical microsensors. They can be configured for varying degrees of specificity by using different BRM's or even whole cells. Similarly, the particular application dictates whether optical or electronic transduction is the optimal mode. Applications of biosensors include: military, for the detection of biological and chemical weapons; agriculture, to monitor food supplies; environment, for monitoring pollutants or for assessing the efficacy of bioremediation; industry, to monitor fermentation and other processes; fishery industry, to detect marine toxins in deep water catches; medical, to monitor physiological markers for diagnostics. The field of biosensors is rapidly maturing and the advantages which they have over traditional analytical techniques, especially when fully exploited with microelectro-mechanical and nanotechnologies, will make biosensors commercially viable.

006.
WHY I AM VERY SKEPTICAL ABOUT REUSABLE BIOSENSORS. Peter T. Kissinger, Bioanalytical Systems, Inc., 2701 Kent Avenue, West Lafayette, IN 47906

One shot biosensors for glucose are a fabulous commercial success with a total market of some one billion dollars per year. The market is dominated by three very large corporations. Over the last decade or so, substantial publicity has been given to electrochemical and optical biosensors, a great deal of government and commercial funds have been expended, and wildly optimistic projections have been put forth for commercial prospects. Unrealistic expectations are followed by deep disappointments. While there certainly are cases where a (bio)sensor selectively monitoring a single analyte is of real value (thank you glucose and lactate), this is very difficult to achieve. Instability in inventory and use, limited dynamic range, difficulty of calibration, and variations related to "matrix" (ionic strength, pH, . . .) are very difficult to overcome in reusable commercial devices. Some skeptics even see "biosensors" as a euphemism for "not another glucose sensor!" In spite of these concerns, our company is very interested in biosensors and has commercialized several kinds. The commercial world demands realism since customers are unforgiving (granting agencies are not).

007.
APPLICATION OF MEMBRANE BASED BIOSENSORS IN THE BIOPROCESS INDUSTRY. John R. Woodward YSI Inc. Yellow Springs Ohio, 45387

The development of biosensor systems over the last 20 years has allowed the bioprocess industry to expand its control of process variables. Biosensors have provided rapid measurements for glucose, lactate, glutamate and glutamine in the monitoring and control of cell culture and fermentation. The ability to produce on-line measurement within seconds of sampling makes biosensors far more flexible than conventional off-line measurement techniques. Similarly the development of systems which can be used in the food and beverage industry with a minimum of sample treatment has led to more rapid analysis of samples and, in many cases, substantial improvements in accuracy and reduction in the cost of tests. The importance of construction criteria, and the stable properties of the enzyme matrix, will be discussed with reference to providing the customer with a robust and reliable system.

SCHB

008.
ADVANCED FLOW IMMUNOASSAY TECHNIQUES. **Helen K. Powell,** Elric W. Saaski, and David A. McCrae, Research International Inc., Woodinville, WA 98072. Anne W. Kusterbeck, Center for Bio/Molecular Science and Engineering, Naval Research Laboratory, Washington, D.C. 20375.

Although specific and quantitative, most fluorescence-based immunoassay techniques are not suitable for real-time field applications such as chemical agent detection. A portable continuous-flow immunoassay system is described that allows the quantitation of a wide variety of analytes with parts-per-billion detection limits. In this assay, an antibody is bound to a membrane and then saturated with fluorophore-labeled antigen. Samples are carried through the membrane by a flow stream. Displacement of the labeled antigen occurs which is directly proportional to the amount of antigen in the sample. Downstream detection of the label is performed with a proprietary fiber-optic fluorimeter. Data analysis is carried out in real-time with an integrated portable computer to provide typical assay times in the range of one to two minutes.

009. PROBING RECEPTOR PHARMACOLOGY AND CELL METABOLISM WITH A MICROPHYSIOMETER. J. C. Owicki, Molecular Devices Corp., 1311 Orleans Drive, Sunnyvale, CA 94089

This talk is a case study of a novel technology, microphysiometry, from inception to commercialization. Microphysiometry is an analytical method that noninvasively detects changes in cell physiology in vitro by monitoring cellular metabolic activity. It does so by using a semiconductor-based sensor to measure the rate of extracellular acidification, a quantity that is related to the rate of energy metabolism (glycolysis, oxidative phosphorylation) and also to intracellular pH homeostasis [see McConnell et al. (1992) Science 257:1906; Owicki et al. (1994) Ann. Rev. Biophys. Biomol. Struct. 23:87]. Among other applications, this is a good method for detecting the functional activation of receptors on living cells relatively independently of their signal-transduction mechanisms. Accordingly, its main use to date is pharmacological, in drug-discovery assays. Beyond describing the technical aspects of microphysiometry, I will discuss the sociological and business obstacles that had to be overcome before a commercial incarnation of microphysiometry, the Cytosensor™ Microphysiometer System, could be produced and accepted in the marketplace.

010. How To Construct A Successful Business Plan. NH Giragosian. Delphi Marketing Services, Inc. 400 East 89th Street NYC 10128. 212-534-4868.

The elements of a successful business plan are listed and discussed.

011. SUCCESS FACTORS IN STARTING YOUR CHEMICAL BUSINESS. E. HUMPHREY. HUMPHREY CHEMICAL COMPANY. NEW HAVEN, CT.

The same success factors involved in starting your own business are discussed.

012. LABORATORY SERVICES AND TESTING IN THE TWENTY-FIRST CENTURY, Gary E. Clapp, Ph.D., OREAD Laboratories, Inc., Lawrence, Kansas 66047.

An overview of the laboratory testing industry will be presented as it relates to Agricultural and Food Chemistry, Environmental Testing, Pharmaceutical and Life-science Chemistry, Contruction Materials/Geotechnical Testing and all other testing services. Problems and strategies will be discussed as the industry prepares itself for the 21st century. A brief look at the recent history of the laboratory testing industry will be used to elucidate future trends and movements in this constantly changing business arena.

013. ENVIRONMENTAL TESTING MARKET REBIRTH ENTERING THE 21ST CENTURY, Steve Vincent, President, Columbia Analytical Services, Inc., Kelso, WA 98626.

The commercial environmental testing industry emerged in the late 1970's as a result of increased concern about pollution. the 1980's saw rapid growth in the industry as new environmental regulations were promulgated. The zenith of the industry was the early 1990's. Beginning around 1992, due to over capacity the industry entered into a mature state where growth at any one laboratory was the result of taking market share away from a competitor. Columbia Analytical Services, Inc. (CAS) was founded in 1986 and currently operates nine laboratories throughout the United States. Sales exceeding over $20M annually make CAS one of the largest environmental testing organizations in the country. During its history, CAS has shown consistent growth and profitability in the face of change within the industry. As the 21st century approaches, change continues. As US industry continues to improve operations and focuses on waste minimization, providing financial benefit will play an ever increasing role over regulatory drivers in defining the environmental testing business. This newly defined environmental laboratory market is much bigger and broader than the current one. Those laboratory companies able to make the transition

014. PREPARING FOR THE TWENTY-FIRST CENTURY. Jane V. Thomas.
Wyoming Analytical Laboratories, Inc., 1660 Harrison, Laramie, WY 82070.

Wyoming Analytical Laboratories, Inc. will celebrate its 20th year of operation in 1997!
Successful strategies in surviving for 20 years and hopefully for another 20 years will be discussed in terms of facing the exciting and ever-changing opportunities and challenges that are occurring as we approach the 21st century. A number of these opportunities and challenges will be discussed. For example, in order to be successful in the future, the small, independent laboratory will need to be cognizant of changing environmental regulations -- both from the standpoint of attracting business, and also with regard to the impact these regulations (federal, state, and local) will have on the operation of the laboratory. In addition, the laboratory will have to meet the demands of ever-shortening turnaround times and ever-decreasing detection limits. In other words, the laboratory must, somehow, acquire expensive, up-to-date equipment yet keep operating costs down so that it can be competitive in the marketplace. The bottom line is and will continue to be -- the small, independent laboratory must be able to produce timely, quality results at a low price so that it can compete successfully.

015. CONCERNS OF THE CONTRACT TESTING LABORATORY. Kathy Savage, Ph.D. Applied Analytical Industries, Inc., Wilmington, NC 28405.

The cost effectiveness and efficiency of using a contract testing laboratory to assist in the development of a pharmaceutical product has been demonstrated. Contract laboratories can provide multiple advantages to virtual, developing and mature pharmaceutical companies. These advantages include reduced overhead costs, expanded staffing, specialized skills and equipment, focused project management and execution and access to in-depth regulatory and development expertise. To achieve the maximum benefits from these advantages requires a strong commitment by the contract laboratory to the client. The elements which comprise this commitment are the primary concerns of the contract laboratory. These include developing a defined workscope, implementing acceptable materials handling procedures, establishing efficient effective approaches for methods transfer and providing communications channels for problem areas. Effective management of these issues leads to a confident working relationship between the client and contract laboratory

016. V-LABS, INC., A CONSULTING LABORATORY OVERVIEW. Sharon V. Vercellotti, V-LABS, INC., 423 North Theard Street, Covington, Louisiana 70433.

V-LABS, INC., was established in 1979 as an independent, consulting chemical laboratory in Covington, LA. V-LABS, INC., offers consultation, in-house laboratory problem solving, and contract research for industry and academia in food, pharmaceutical, and cosmetic materials. As an active leader in biotechnology, they synthesize and sell carbohydrates, polysaccharides, and glycosidic enzymes used in research. V-LABS is the North American distributor for Dextra Laboratories, Ltd., the world's premier supplier of medically important oligosaccharides and neoglycoproteins. They also offer analytical services for carbohydrates and polysaccharides, including gas and liquid chromatography, molecular weight determination, dietary fiber determination, custom synthesis and polysaccharide modification. They have studied extensively the development of rancidity in oils and oil-containing natural products. Purge and trap chromatography is offered, including food-packaging interactions, food flavor volatiles, residual solvents and off-flavors. They have received three Small Business Innovation Research Grants from the National Science Foundation and the National Institutes of Health. Home page: http://www.wild.net/~v-labs

017. FROM BROAD TECHNOLOGY TO FOCUSED MARKETS: CHALLENGES FACED BY TECHNOLOGY-BASED ENTREPRENEURIAL VENTURES. Ricardo B. Levy, Catalytica, Inc., 430 Ferguson Drive, Mountain View, California 94043

Many companies, like Catalytica, start with technology platforms that have broad potential impact in many areas. The breadth of the science represents the strength of the venture, embodies the excitement and unlimited scope of the opportunity, and provides the basis for lowering the risk profile of the initial investments. However, it is not until the activities can be focused to very specific commercial targets and well defined markets that a real business is born. This presentation will use the story of Catalytica as a case study of how we maneuvered through this complicated path, and highlights the solutions that have worked for us. It embodies interesting lessons that are applicable to many other entrepreneurial situations.

018. THE STORY OF MYCOGEN. Jerry Caulder, Mycogen Corp., 5501 Oberlin Drive, San Diego, California 92121.

Mycogen is a diversified agribusiness and biotechnology company that is using technology to develop biological alternatives to chemical pesticides and add value to major crops. Mycogen Crop Protection markets environmentally compatible biopesticides for high value fruit, vegetable and vine markets, and provides crop protection services to growers of high value crops. Mycogen Seeds is now the 4th largest marketer of hybrid seed corn in North America, and is among the top five in soybean, sunflower, sorghum and alfalfa. In 1996, Mycogen sold seeds for the first corn hybrids with genetically enhanced-resistance to European corn borers, which cause a billion dollars worth of damage to the U.S. corn crop each year. By the year 2000, Mycogen expects to be marketing seeds for insect-resistant varieties of several major crops. Additional development efforts are focused on enhancing the quantity and value of agricultural outputs, such as vegetable oils, starch, fiber and livestock feed.

019. Oxford Asymmetry - The Complete Chemical Solution

Within the pharmaceutical industry, the chemical development function is coming under increasing pressure. In traditional pharmaceutical companies, the increased number of lead compounds in development is placing a strain on resources. In the emerging biopharmaceutical companies, new skill sets are becoming necessary as compounds enter development. Across the industry, chemical development is becoming a rate-determining step at a time when "speed to market" is ever more important. To meet the changing needs of these industries and others, Oxford Asymmetry has positioned itself as the pharmaceutical industries partner offering "The Complete Chemical Solution" of integrated chemistries ranging from combinatorial chemistry for lead candidate generation through complex multi-step organic synthesis. Oxford manages the chemical development and allows the client to focus on other activities.

Stuart Needleman
Vice President, Business Development USA

020. AN UPDATE ON AN EVOLVING CHEMICAL COMPANY, Robert Louis Steed and D. Larry Brotherton, Ph.D., Ortec, Inc., P.O. Box 1469, Easley, SC 29641

In the last four years, Ortec has experienced substantial growth which has resulted in significant new challenges. The greatest challenge for Ortec is to reinvent itself as market needs shift. Issues to be addressed in the paper are: (1) keeping the entrepreneurial spirit alive as the corporation grows (2) Maintaining close customer contact and customer service without building bureaucracy and ensuring that all employees share a sense of urgency and a focus on customer needs (3) selecting business opportunities and building manufacturing capabilities with minimum debt (4) maintaining growth rate while complying with reporting thresholds and environmental regulations (5) ensuring that growth is relevant to the bottom line and (6) planning market direction and developing staff for the next fifteen years.